ENCYCLOPEDIA OF MATERIALS SCIENCE AND ENGINEERING

BVOL3-A

HONORARY EDITORIAL ADVISORY BOARD

ENCYCLOPEDIA OF MATERIALS SCIENCE AND ENGINEERING

Volume 3

F–I

Editor-in-Chief

MICHAEL B. BEVER

Massachusetts Institute of Technology, USA

PERGAMON PRESS

OXFORD · NEW YORK · TORONTO · SYDNEY · FRANKFURT

U.K.	Pergamon Press Ltd., Headington Hill Hall, Oxford, OX3 0BW, England
U.S.A.	Pergamon Press Inc., Maxwell House, Fairview Park, Elmsford, New York 10523, U.S.A.
CANADA	Pergamon Press Canada Ltd., Suite 104, 150 Consumers Road, Willowdale, Ontario M2J 1P9, Canada
AUSTRALIA	Pergamon Press (Aust.) Pty. Ltd., P.O. Box 544, Potts Point, N.S.W. 2011, Australia
FEDERAL REPUBLIC OF GERMANY	Pergamon Press GmbH, Hammerweg 6, D-6242 Kronberg, Federal Republic of Germany
JAPAN	Pergamon Press Ltd., 8th Floor, Matsuoka Central Building, 1–7–1 Nishishinjuku, Shinjuku-ku, Tokyo 160, Japan
BRAZIL	Pergamon Editora Ltda., Rua Eça de Queiros, 346, CEP 04011, São Paulo, Brazil
PEOPLE'S REPUBLIC OF CHINA	Pergamon Press, Qianmen Hotel, Beijing, People's Republic of China

First edition 1986

Library of Congress Cataloging in Publication Data

Main entry under title:
Encyclopedia of materials science and engineering.
Includes bibliographies.
1. Materials—Dictionaries. I. Bever, Michael B. (Michael Berliner). II. Massachusetts Institute of Technology
TA402.E53 1986 620.1′1′0321 85–23192

British Library Cataloguing in Publication Data

Encyclopedia of materials science and engineering
1. Materials science—Dictionaries
I. Bever, Michael B.
620.1′1′0321 TA402

ISBN 0–08–022158–0 (set)

Distributed in North and South America by The MIT Press, Cambridge, Massachusetts

Computer typeset by Page Bros. (Norwich) Ltd.
Printed in Great Britain by A. Wheaton & Co. Ltd., Exeter.

CONTENTS

SUBJECT EDITORS

F

Factor Dominance in Industry

The production and cost characteristics of industries are heavily influenced by their basic technologies and, in particular, by the comparative roles of labor, materials and capital facilities in determining their production capabilities. Actual industrial processes cover a wide range in respect of such patterns of "factor dominance."

This concept is fundamentally different from the traditional idea of "factor intensity," which relates to cost proportions. Thus, classifying an industry as "capital intensive" implies a high ratio of capital costs to total costs, as in the case of electricity generating plants where such ratios may be 40% or more. But "capital intensive" is also often mistakenly applied to operations requiring large amounts of fixed investment, although these need not result in large ratios of capital costs. For example, this term is frequently applied to the steel industry because of its enormous capital requirements. But these investments yield such huge productive capacities that total depreciation and interest charges for major steel companies in the USA usually average less than 10% of total costs. On the other hand, steel may well be classified as a "capital-dominated" industry because the scale and modernity of its capital facilities and equipment are the overwhelming determinants both of the effectiveness with which materials, energy and labor inputs can be utilized and of the production capacity of the mill.

The "capital-dominated" extreme would include processes in which production workers are chiefly engaged in starting, stopping, loading, unloading, setting controls or otherwise simply facilitating the functioning of the equipment and capital facilities. In this category, materials merely play the passive role of providing the relatively standardized inputs being acted upon by the capital goods. Similarly, the "labor-dominated" extreme would encompass processes in which tools and other capital facilities serve essentially as accessories to wage earners, whose manual efforts and related skills represent the fundamental core of the production process—materials once again serving passively as the standardized inputs being acted on by manual operations. Similarly, a "materials-dominated" extreme may be envisaged, covering operations in which productive capacity is dominated by the resources to which labor and capital facilities are applied—both of the latter merely facilitating attainment of the production capacity determined by materials and related inputs, as in the case of changes in the metallic yield of smelters owing to variations in ore tenor, or in the case of farms differing in soil fertility.

What is the significance for management of the distinction between "factor dominance" and "factor intensity"? Unit costs are determined, of course, both by the volume of various inputs per unit of output and by their respective factor prices. Because factor intensity refers to cost proportions, or the relative magnitude of various unit costs, changes are attributable in large measure, especially in the short run, to market-determined adjustments in factor prices. Factor dominance, however, reflects the relationship among the various unit input requirements as determined by management's choice of production technologies. Thus, factor dominance is an active determinant of factor intensity, and tends to exert an increasing influence on it over the longer run as technological changes are introduced, whereas changes in factor intensity are merely the passive result of internal changes in factor dominance and of external changes in factor prices. Thus, despite material cost proportions exceeding 80%, blast furnaces, petroleum refineries and flour mills are basically capital-dominated operations.

1. Factor Dominance and Productivity Relations

1.1 Single Factor Dominance

The powerful influence of factor dominance on productivity relations is readily illustrated. In operations approximating the capital-dominated extreme (e.g., power plants, petroleum refineries, cement mills and blast furnaces), increases in labor productivity are more likely to reduce man-hour requirements than to increase productive capacity or the productivity of fixed capital; increases in the productivity of fixed capital are likely to leave direct labor requirements per unit of fixed capital input unchanged or reduced, and hence induce parallel increases in physical output per man-hour; and major increases in the productivity of the operations as a whole are thus much more likely to come from—and to be better measured by—adjustments in the productivity of fixed capital than in the productivity of labor. In operations approximating the labor-dominated extreme (e.g., bricklaying, custom tailoring and carpentry), increases in the productivity of fixed capital are more likely to reduce requirements for such equipment than to increase productive capacity or the productivity of direct labor; increases in labor productivity are likely to leave fixed-capital requirements unchanged, and hence induce parallel increases in the productivity of fixed capital; and major increases in the apparent productivity of the operation as a whole are thus much more likely to come from—and to be better measured by—adjustments in the productivity of direct labor than in the productivity of fixed capital.

Moreover, both in labor-dominated and in capital-dominated fabrication processes, materials requirements tend to vary with output levels, thus tending to increase the ratio of materials inputs to labor inputs with an increase in the productivity of labor, and to increase the ratio of materials inputs to effectively utilized fixed-capital inputs with an increase in the productivity of fixed capital. Also in such processes, decreases in the input requirements of materials per unit of output are more likely to reduce the volume of the input requirements of materials than to increase the productivity of other factors, or the productive capacity of the operation as a whole.

Materials-dominated processes are encountered most commonly in the extractive industries. Such dominance tends to be greatest in operations involving the direct extraction of natural resources, including agriculture, fisheries, forestry and mining. In each of these, the output resulting from the application of given levels of labor and capital inputs varies with the quality of the natural resources being worked, namely, the fertility of the land, the abundance of fish in the seas being combed, the volume of timber on the lands being lumbered, and the richness of the ore deposits being mined.

The degree of dominance by material inputs is, however, being reduced even in these industries, primarily as a result of growing reliance on capital goods to increase attainable rates of production. More machinery enables farmers to increase the productive capacity of their operations, not only through more intensive cultivation of existing farmland, but also through farming additional acreage. Fishing vessels with improved fish-locating, catching, handling and refrigeration equipment, as well as larger storage capacity and more powerful engines, can increase their average catch per month. Similar means exist for increasing the production capabilities of mines and lumbering operations through greater investments in facilities and equipment.

Increases in production rates from unchanged geographical boundaries are usually generated by greater capital inputs accompanied by smaller increases, or even significant decreases, in labor inputs. Thus, output per unit of material resources being worked increases, output per man-hour tends to increase either more or less rapidly than the preceding, and output relative to capital investment tends to decrease. The effects on total unit costs obviously depend on the extent to which decreases in unit material and unit labor costs are offset by increases in average capital charges per unit of output. Increases in the production of a farm or mine or timber operation due solely to extending the boundaries of such operations, and applying unchanged rates of labor and capital to similar additional areas, would obviously leave all productivity relationships unchanged. But in the more common case where such boundaries are extended because the production capabilities of new machinery, which increases output and reduces labor requirements, are not fully utilized by applications to past working areas, such extensions may be expected to increase output per man-hour further and also to reduce the initial decrease in capacity relative to fixed investment.

1.2 Jointly Dominated Processes

Within the enormous diversity to be found in industry, the extremes of capital-dominated, labor-dominated and materials-dominated processes may well account for only a limited proportion of actual operations. A large number of these might well be characterized as "jointly dominated," representing the determination of production capabilities either jointly by labor and capital, or jointly by materials and capital. The former may be illustrated by a wide array of manufacturing operations involving significant contributions either by skilled labor utilizing essentially general purpose equipment to make a variety of products, or by operators on assembly lines involving substantial reliance on the speed and quality of labor efforts to determine acceptable output. Jointly dominated materials and capital operations are most commonly exemplified in manufacturing by operations involving the initial refining of natural raw materials, such as ore smelting, petroleum refining and the processing of milk into dairy products.

Such essentially extractive operations in manufacturing are jointly dominated because changes in the quality of the materials input may alter the relationship of attainable output to fixed investment within a significant range. This results from the fact that any given capital facilities limit the flow of materials which can be processed within a given period. Hence, even if such facilities are fully utilized, the marketable output will depend on the metal content or chemical composition or butter-fat content of the raw materials being treated. On the other hand, changes in the technology, processing capabilities and speed of the capital facilities can also alter the marketable output of a plant, while leaving the quality of the materials being processed unchanged. Thus, the actual capacity will be determined jointly by the processing capabilities of the capital facilities and by the quality of the materials being processed, but the relative contributions of each are usually readily determinable.

In smelting operations, for example, an increase in the metal content of the ores being processed by a given plant would tend to increase output relative to man-hours and fixed investment as well as materials volume. Investments in improving the extractive capabilities of the facilities as well as the volume of materials which can be processed would tend to have the following effects in the absence of any change in the quality of the ores being treated; a significant increase in output per man-hour, some increase in output relative to material volume, and a probable decrease in capacity relative to fixed investment.

2. *Cost Effects of Factor Dominance*

The cost effects of changes in the proportioning of inputs under different conditions of factor dominance depends not only on accompanying interactions with input factor prices, but also on any resulting changes in outputs (see *Productivity and Technological Improvements: Managerial Evaluations*). If output per unit of materials volume is increased by decreasing scrap and wastage, resulting savings will be paralleled by lower unit material costs. If such economies are also effected by other users through increasing diffusion of the technologies and practices involved, resulting decreases in demand may engender reductions in the price of such materials, thereby tending to enhance the initial reductions in unit material costs.

On the other hand, if increases in output per unit of materials volume are achieved through the use of better or more highly processed materials or purchased components, resulting benefits in unit material costs may be at least partly offset by the higher prices charged for such improved inputs. In such instances, however, two additional ramifications would be of concern to management. First would be the possibility of partially or wholly offsetting reductions in other processing costs per unit of output. Secondly, consideration would also have to be given to the possibility that the changes in the quality of material inputs make possible improvements in the attractiveness of the final products to customers which permit increases in the prices charged for them or in the quantities which can be sold.

Changes in output per man-hour may also produce a variety of cost effects, depending on accompanying changes in wage rates and in other unit costs. But increases in output per man-hour are more likely to engender at least partially offsetting increases in wage rates, although they are less likely to be attributable to increases in the direct contributions of labor to production per man-hour than to the partial replacement of such contributions by inputs of capital and of better or more highly processed materials. Moreover, product quality improvements in most industrial processes are seldom due to enhanced labor contributions.

The cost effects of changes in the productivity of capital goods (i.e., in productive capacity per dollar of net fixed assets) depend heavily on the capacity utilization rate, of course, as well as on the rate of charges to cover interest and capital recovery. But these cost effects may be affected to an even greater degree by resulting changes in other unit input requirements and in product qualities. Major changes in the productivity of fixed assets are usually the result of advances in technological processes, or in the scale or modernity of facilities and equipment. Such innovations frequently involve at least temporary decreases in the productivity of net fixed assets because the facilities being replaced may have cost less initially and because they have been largely depreciated—as a result of which the increase in new investment is likely to exceed the gain in capacity. Moreover, interest and depreciation rates may also be higher because of inflation and recognition of the need to adjust to faster rates of technological obsolescence in many industries. Hence, the key determinants of the economic desirability of such new capital inputs center around the magnitude of resulting decreases in other unit costs and increases in output and average price levels due to improvements in product qualities and mix. Detailed illustrations of the cost effects of various kinds of productivity adjustments in industries representing differing patterns of factor dominance and under alternative assumptions concerning attendant capacity and output adjustments are available in Gold (1955 pp. 202–32).

3. *Competitive Effects of Technological and Productivity Improvements*

The economic effects of improvements in a firm's productivity and technology cannot be determined by comparisons with its past performance. Resulting benefits and burdens can only be appraised within a broader framework which encompasses recent and prospective changes in market demand, in factor prices and in the developmental efforts of competitors. Changes in the product composition of demand or in the relative prices of different input factors often affect different producers unevenly; and although early adopters of major technological advances may gain significant cost and market share benefits, late adopters may merely reduce their hitherto growing competitive disadvantages.

Moreover, prospective means of improving productivity and technology may be rooted not only in managerial efforts to improve current performance, but also in anticipation of unfavorable pressures from factor or product markets. In such cases, as illustrated by major jumps in energy prices and by shortages of certain critical materials, achievements may be measured by slower or smaller increases in unit costs rather than by decreases.

See also: Productivity and Technological Improvements: Managerial Evaluations; Productivity and Materials; Value Added in Processing

Bibliography

Baird R N 1977 Production functions, productivity and technological change. In: Gold 1977, pp. 11–41

Dogramaci R (ed.) 1981 *Productivity Analysis: A Range of Perspectives.* Martinus Nijhoff, Boston, Massachusetts

Erdilek A 1977 Productivity, technological change and input–output analysis. In: Gold 1977, pp. 43–85

Gold B 1955 *Foundations of Productivity Analysis: Guides to Economic Theory and Managerial Control.* University of Pittsburgh Press, Pittsburgh, Pennsylvania

Gold B 1971 *Explorations in Managerial Economics: Productivity, Costs, Technology and Growth.* Macmillan, London

Gold B (ed.) 1977 *Research, Technological and Economic Analysis.* Heath-Lexington, Lexington, Massachusetts

Gold B 1982 *Productivity, Technology and Capital: Economic Analysis, Managerial Strategies and Governmental Policies.* Heath-Lexington, Lexington, Massachusetts

Gold B, Peirce W S, Rosegger G, Perlman M 1984 *Technological Progress and Industrial Leadership.* Heath-Lexington, Lexington, Massachusetts

Gold B, Rosegger G, Boylan M G 1980 *Evaluating Technological Innovations: Methods, Expectations and Findings.* Heath-Lexington, Lexington, Massachusetts

Hogan J D, Craig A M (eds.) 1981 *Dimensions of Productivity Research.* American Productivity Center, Houston, Texas

Nabseth L, Ray G F (eds.) 1974 *The Diffusion of New Industrial Processes: An International Study.* Cambridge University Press, London

B. Gold

Failure Analysis: General Procedures and Applications to Metals

This review of failure-analysis guidelines and procedures is concerned with the failure analysis of metallurgical components. The same guidelines apply in general to components made of ceramic or polymeric materials.

The two main goals of a failure analysis are (a) to find the primary causes of the failure, and (b) to define a corrective action which will prevent repetition of the same failure.

A failure analysis requires the integration of the work of experts in different disciplines of engineering such as mechanics, physics, chemistry, metallurgy, fracture, manufacturing, quality control and statistics. Depending upon the degree of complexity of the failure under investigation, and also depending upon the confidence level required of the analysis, a failure analysis may take one hour, one day, one week or even one month to be completed. The time-scale available for a given failure analysis will control each of the steps to be followed. The guidelines given below are in principle time and cost independent.

1. General Procedures

It is important to make sure that all the information needed to perform a detailed failure analysis is collected very quickly from all possible sources. This information includes the manufacturing, processing and service history of the failed component, as well as the reconstruction of the sequence of events resulting in the failure. The different types of failure which are likely to be encountered are very varied and a listing of the main types is given in Table 1.

Table 1
Types of failure

Ductile failures	Wear-fretting failures
Brittle failures	Liquid metal embrittlement failures
Fatigue failures	Gas embrittlement failures
Stress-corrosion failures	High-temperature failures
Corrosion failures	

1.1 Reconstruction of Failure

In the case of failure of a complex system such as the explosion of a petrochemical plant or the crash of an aircraft, it is important to establish unambiguously the location of the initial failure source. This wreckage analysis is assigned to "reconstruction" experts. The word reconstruction assumes that the experts who have a practical knowledge of the different functions of a component are able to establish the failure sequence. For instance, if the tail section of an aircraft is found a mile and a half away from the crash site, it will be concluded that it is likely that the tail section failed first, prior to impact of the aircraft. The same is true of a helicopter blade: if a single blade is found away from the main crash site, it is a sure indication that the blade attachment or the blade itself led to the crash. The reconstruction work may take a fair amount of time. One of the many findings derived from the reconstruction work should be the clear identification of the primary fracture path and of all the secondary fracture paths. The identification of a primary fracture path leads to the determination of the origin of the fracture and subsequently to the cause of the failure.

This can be done through the reconstruction of the sequence of the accident and also by fractography. A rapid visual evaluation of different but characteristic fracture features will give the local direction of crack propagation. Examples of such features are chevron markings, beach marks, crack branching and shear lips (see Sect. 2.2). The identification of the primary fracture path and of all the secondary fracture paths is critical because it will guide the next steps in the failure analysis, and much valuable time could be wasted if the primary fracture path is not correctly identified.

The reconstruction experts will also analyze in detail all of the available data on the operation of the component prior to the failure.

1.2 Component History

At the same time as the reconstruction experts are doing their work, it is important to review the history of the failed part. This will include the following factors.

(*a*) *Design criteria.* A component is usually designed on the basis of engineering design guidelines, codes and specifications. The original design and the modifications brought to the design of a part over many years of service should be reviewed. It is important to note that a design code serves only as a guideline for minimum or typical performance. A good designer will add a factor of safety to the design parameters recommended by the code. Furthermore, a code is usually based on a body of technical knowledge which may be outdated by the time the part is manufactured or put into service. Thus the fact that a component may have met a code at a given time does not ensure that it is fail-safe. The design codes used in different countries or by different manufacturers may have to be compared to assess the technical soundness of the design criteria used at the time of the design.

(*b*) *Materials selection.* During failure analysis, the alloys used in the fabrication of a component should be carefully reviewed. For a given part, it is not uncommon to see marked modifications in the alloys used for the initial fabrication of the part. For instance, a part made with air-melted steel in the early 1970s would now be made with vacuum arc-melted steel in order to ensure better performance in fatigue and stress-corrosion applications as well as greater toughness. A change of alloy does not necessarily mean that the old alloy was unsuitable for the part at the time of the design, it just implies a higher degree of safety. The selection and purchase of an alloy is usually based on a set of specifications which are agreed upon between the vendor and the buyer of the material. All the specifications should be reviewed to see whether the material specifications were adequate, up-to-date and complete, and whether the materials used in the failed component met the specifications.

(*c*) *Process history.* A detailed review of the processing history of the key component involved in the primary fracture path is a critical part of failure analysis. This review will include an evaluation of all the manufacturing steps as well as all the quality-control tests performed on the failed part. A detailed understanding is required of all the different correlations and interactions between manufacturing procedures and variables and the structure and properties of the part.

(*d*) *Actual service history.* This is one of the most important pieces of information needed for each failure analysis. The detailed service history of the failed component and a reconstruction as far as possible of the loads, displacements, temperature and environment "service envelope" versus time is needed. It is not very common to have a log book giving a recorded history of the component, and thus in general it will always be difficult to know the exact service history of the failed part.

In most cases a component is designed within a narrow service envelope so that a mean set of service conditions can be obtained with a typical standard deviation. It is usually also possible to go back to a similar component to review the service conditions, or even to set up a new record of the current operating conditions.

(*e*) *Statistics of failures of the component.* Up-to-date statistics giving the number of components which have failed in a similar manner should be sought. These can be obtained from trade associations, or for example in the case of an aircraft from the Federal Aviation Authority, and in the case of a ground vehicle, from the Department of Transportation. Other sources of failure statistics are the manufacturers and the users. At the same time as the statistics of failures are collected, it is necessary to review the cause of each previous failure. The modifications implemented following each previous failure need to be reviewed.

1.3 Failure Stages

The failure history of a component can be divided into four stages.

(*a*) *Local degradation of the component.* This stage can take the form of wear, corrosion, fretting corrosion, pitting, oxidation, local microplasticity and other degradation processes.

(*b*) *Crack initiation.* This is the stage at which a microfissure or microcrack is initiated. This microcrack may have the depth or length of one, two or three grains ($<100\ \mu m$) and may be difficult to find optically, but it can be detected with modern nondestructive test methods. The microcracks formed as the result of the material degradation stages may or may not grow into longer cracks. There are many components in which numerous microcracks can be generated without further crack growth. In this case, one of the questions to be answered is whether the component containing microcracks can remain in service for a finite lifetime without any probability of failure.

(*c*) *Slow crack growth.* Once initiated, in general, a microcrack will grow by ductile tearing, fatigue, creep, stress corrosion or liquid metal embrittlement. These processes of slow crack growth usually take place during a fraction of the lifetime of the component. One of the most important problems in each failure analysis is to identify the fraction of component life spent in growing a crack from a size equal to two or three grains to the final stage of fast crack propagation. A knowledge of the crack-length–time history can be used to define a quality control inspection period. It can also be used to assess the risk of leaving the other components in service.

(*d*) *Fast crack propagation.* A growing crack will eventually reach a critical length which, when combined with local stresses, leads to rapid crack propagation by brittle fracture, ductile tearing or fully plastic collapse. This process of fast crack propagation occurs over a very small time scale, usually of the order of less than a second.

1.4 Incipient Failure

The processes of degradation and crack growth may occur at any time during the life of a component. Incipient failure may occur:

(a) During manufacture. Here failure may be caused by forging cracks, forging laps, quench cracks, hydrogen flakes, solidification defects and shrinkage porosity, among others.

(b) During quality control or proof testing. Some quality-control procedures are such that the components are subjected to vibratory stresses or to large peak stresses during proof testing. It is not uncommon to see cracks generated during quality-control handling. For instance, stress corrosion cracks in pressure vessels made of high-strength steels have been initiated during pressurized proof testing with water.

(c) In prototype tests. Very often a prototype component is tested during its manufacture. The prototype testing program is designed to provide a component which leads all the other parts in service time by simulating service loads in an accelerated time frame. If the prototype fails during testing, the failure analysis should be especially painstaking in order to plan the corrective action and to derive a reliable prediction of the probability of failure of all the other components in service.

(d) Failure of the part in service. Most failure analyses deal with parts which fail in service, for which a "decision/action" will be needed very quickly. In this case the reliability of the failure-analysis conclusion is very important to the decision process.

1.5 Decision/Action Following Failure Analysis

Following failure analysis, and on the basis of the understanding obtained by the failure-analysis team, the following decisions/actions are possible.

(a) Do nothing—this is not an uncommon approach when the cause of failure is not clearly understood, or when the failure is blamed on improper use of the component.

(b) Proceed with a detailed and careful quality-control inspection of all the components in service—such a decision requires careful planning of the steps to be taken, i.e., locations to be inspected, samples to be taken, time intervals for inspection. There will always be a conflict between the cost of the quality-control inspections, including the downtime cost of the component, and the degree of reliability of the inspections.

(c) Perform accelerated tests—following the failure analysis, it may be critical to verify the results of the analysis by performing a series of accelerated tests on a component similar to the failed one, or on a smaller scaled-down part, taking into account the size effect. With modern mechanical testing equipment (e.g., servohydraulic machines) it is possible to simulate very rapidly a set of loads, displacements or strains on a given component. Furthermore, it is also easy to reproduce the size of the flaw which may have led to the failure. Using linear-elastic and elastic–plastic fracture mechanics, correlations can readily be established for the cracking rates and failure probabilities between a crack in the test component and a crack in the real component.

(d) Repair the parts—the conclusions of the failure analysis may be such that repairs of all the components in service are feasible and can be performed quickly, without any detrimental effects. However, it should always be remembered that a "quick fix" might be the source of other failures if it is not carried out properly and tested carefully.

(e) Cost versus probability of failure—in each case, the decision/action recommendations are based on an evaluation of the costs of the action to be taken. This is part of the field of risk analysis and risk assessment. Depending upon the risk of further failures, it must be decided either (i) to rework the failed component, (ii) to take all the other components out of service or (iii) to leave them in service. Removing all the components from service is a drastic action which may be necessary if the probability of failure is high. In this case a redesign of the component will be needed. An alternative choice is to change the service conditions to ensure a safe life.

2. Metallography and Fractography

2.1 Chemical Analysis and Metallography

For a metallic component it is crucial to know its overall chemistry, and also the chemical gradients near the origin of the crack. It is also recommended to characterize fully the macrostructure of the part in order to take into account the gradients of microstructure (see *Macrostructure of Metals*). For instance, the orientation of the different processing planes in a forging (e.g., short transverse plane, long transverse plane) or the directions of solidification in a casting should be identified. The "flow" lines of the grains in a forging are very important because they introduce a strong mechanical anisotropy. A detailed understanding of the influence of the processing history on the macro- and microstructure of the material and of the component is required to account for the fracture behavior. The microstructure at the location of the initiation site should be determined without entirely destroying the fracture features which may be needed for further investigation. The microstructure analysis includes grain size, grain shape, inclusions (size, shape and volume fraction), voids and chemical gradients.

2.2 Macrofractography

A macrofractographic analysis should be carried out in parallel with the stress analysis. It should include the following steps.

(*a*) *Fracture path versus load direction.* It is important to know the load directions near to and at the location of the fracture origin and the relative orientation of the fracture plane with respect to the loading direction. In fracture-mechanics terminology the modes of crack-tip opening are referred to as mode I (normal opening), mode II (shear opening) and mode III (antiplane-strain opening). The cracking plane is directly related to the cause of cracking. For instance for shear overload, it is a plane of maximum shear stress. On the other hand, a stress-corrosion or a fatigue crack propagates in a plane normal to the maximum-principal-stress axis. Textbooks in applied mechanics give the principal-stress directions for simple loading conditions (e.g., bending, torsion, internal pressure) for different structural components.

(*b*) *Brittle fracture versus ductile fracture.* A macrofractographic analysis gives an evaluation of the degree of plasticity associated with the fracture path in the component. It is important to differentiate between a fully plastic fracture and a very localized brittle fracture where there is little or no plastic deformation (see Sect. 3).

(*c*) *Fracture markings.* The identification of different fracture markings is critical to the macrofractographic analysis. Fracture origins are identified by tracing back all the directions of crack growth. An origin is often characterized by a difference in texture, roughness and color, indicating a long exposure to oxidizing or corrosive environments. Its location is usually associated with a high-stress or high-strain region such as the edge of a hole, the fretting marks near a bearing surface, machining defects and processing pores.

In many components having the shape of a sheet or a plate, there will be a unique propagation direction of the crack. In this case the fracture surfaces have a series of markings, referred to as chevron markings. These are initiated along the center-line of the plate and run to the edges. The chevrons point in the direction away from the crack propagation direction (see Fig. 1). Chevron markings are extensively used in failure analysis of large or small components to trace the local directions of crack propagation and thus to find the fracture origins. There are some alloys for which the chevron markings may not always be clearly defined; this is the case for clean steels or some alloys which contain only a limited volume fraction of inclusions and which do not show a gradient of inclusion content from the center of the plate to its edges. In these alloys there will always be some "river" fracture lines normal to the crack front, similar to chevron markings. In most cases these lines can also be used to identify the main directions of crack propagation.

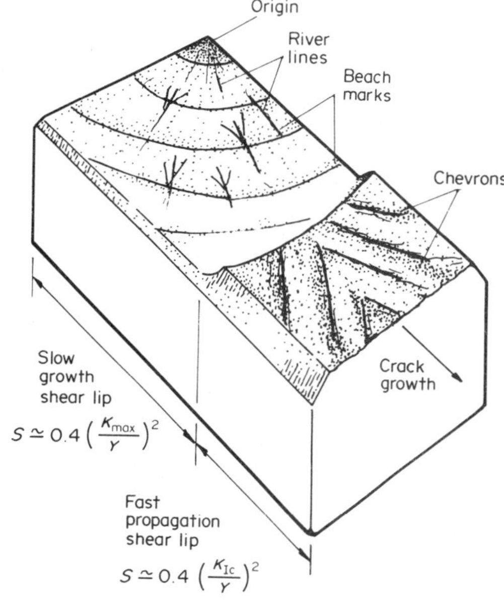

Figure 1
Sketch of typical fracture features including origin, river lines, beach marks, chevrons and shear lips

Beach marks represent the successive positions of the crack front as a function of time. They represent markers in the history of the progression of the crack front and are due to transient and/or cyclic variations in loads, corrosion and temperature. Beach marks are extensively used to reconstruct the life history of the component by correlating specific beach marks with particular events during service. Whenever it is difficult to understand the sequence of beach marks, it is recommended that similar beach marks be produced in an accelerated time frame on the fracture surfaces of a test component or test bar subjected to a history similar to the service history.

Shear lips represent the near-surface zone which is in the pseudo-plane-stress condition and where fracture proceeds along planes at ±45° to the plane of the surface (Fig. 2). The width or thickness of the shear lips is very important because it gives a measure of the toughness of the component during crack growth. The empirical correlation between fracture toughness and shear-lip width is given by

$$w_{\text{shear lip}} = 0.4(K_{\text{Ic}}/Y)^2 \qquad (1)$$

where K_{Ic} is the fracture toughness index and Y is the yield stress. When the failure of a plate is such that the fracture proceeds under a plane-stress condition, the secondary cracks resulting from the change of orientation of the shear lips can also be used to trace the direction of crack propagation. This fracture feature is very useful when chevron markings are not readily visible. In the case of slow crack growth, the shear-lip width gives a measure of the size of the plastic zone at the crack tip which can in turn be related to the local stress-intensity factor or to the J parameter (see Sect. 3.) and then to the crack propagation rates.

Compression spalls occur in components that are broken in bending. The bending fracture of a part leads to fracture from the tensile side to and through the compression side; branching of the crack in the compression region is referred to as a compression spall. This fracture marker is used to find the direction of crack propagation under conditions of bending. The origin is usually 180° away from the compression spall.

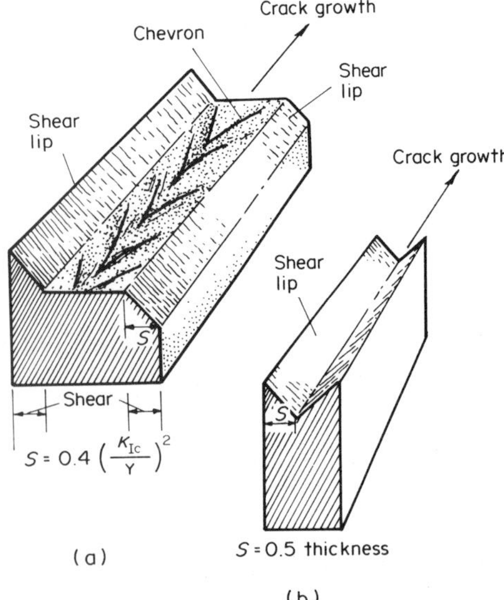

Figure 2
Sketch of typical fracture features for (a) plane-strain and (b) plane-stress fractures

The texture of fracture surfaces, that is, the roughness and the color, gives a good indication of the interactions between the fracture path and the microstructure of the alloy. For instance, at low stress, a fatigue fracture is typically silky and smooth in appearance. Stress-corrosion fractures show extensive corrosion features and corrosion beach marks. A discontinuous ductile fracture shows some stages of crack-tip blunting, crack arrest and "pop-in." All the features related to the texture of the fracture surfaces can, if necessary, be duplicated in a prototype or coupon fracture test.

(d) Reproducibility of macrofractographic features in accelerated tests. It is, in general, quite easy to reproduce the macroscopic fracture features of a component with a test sample of the same material subjected to an accelerated test program. This is very important when the correlation between the microstructure and the fracture features of a material is not well understood. All the fracture features mentioned above can be readily reproduced in an accelerated test program.

2.3 Microfractography

Assuming that there is a clear understanding of all the macrofractographic features, a need often exists for a high-magnification electron-microfractographic analysis to provide further information about the micromechanisms involved in the material failure (see Table 2). The scanning electron microscope, with or without plastic replicas, is an ideal tool for this type of work. The guidelines for microfractographic analysis are given in detail in the American Society for Metals Fractography Handbook. Figure 3 gives a sketch of the fracture features associated with different fracture conditions.

Table 2
Micromechanisms of fracture

Transgranular crystallographic cleavage
Interface decohesion
Intergranular brittle decohesion
Microvoid initiation, growth and coalescence
Rupture by single shear or by alternating shear
Slip plane shear decohesion by stage 1 fatigue
Transgranular or intergranular fatigue cracking with fatigue striation
Transgranular or intergranular decohesion by stress corrosion cracking
Mixed shear and cleavage modes on atomic scale
Creep cavity nucleation and growth

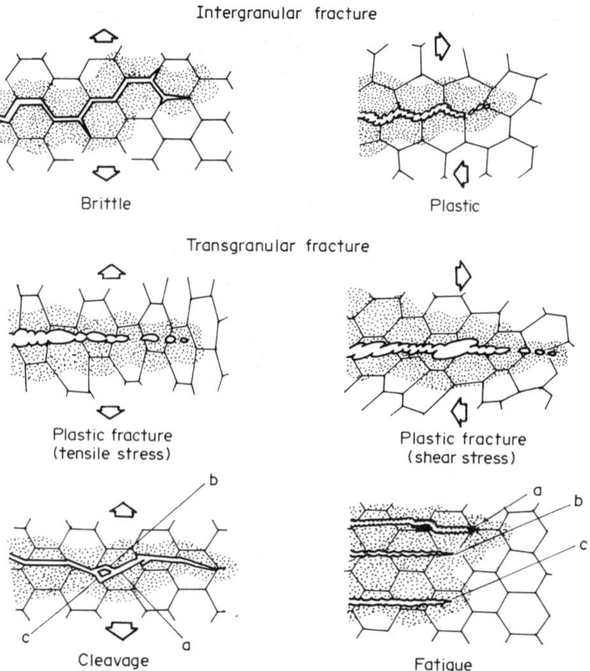

Figure 3
Sketch of intergranular and transgranular fracture paths and associated fracture features. The hexagonal cells represent the grain size

Great care should be exercised not to take the microfracture features out of context, that is, at too high a magnification, or to view artifacts which may be irrelevant and misleading. It is important to avoid conclusions based on a single high-magnification SEM fractograph. In cases where an unknown feature could be a key to the overall fractographic analysis, it is necessary to perform a duplicate test on the same material under similar conditions to provide a clean and undamaged fracture surface for a comparative fractographic analysis. It is only when all the fracture features have been reproduced under controlled conditions that conclusions can be drawn from the microscopic fractographic observations. As an illustration of the problems that may arise from fractographic analyses, a case occurred in which pearlite lamellae etched by seawater on the fracture surfaces in a low-carbon steel were mistaken for fatigue striations. The pearlite spacing was assumed to represent the fatigue crack growth per cycle, and a detailed but worthless fracture mechanics analysis was based on that erroneous interpretation. In the case of fatigue failures, measurements of fatigue striation spacings (when the striations are well-defined) provide a powerful measure of the crack growth rates for a given crack length, and the loading history. These data should be used in the reconstruction of fracture sequences.

3. *Mechanical Analysis*

A failure analysis is not complete without a detailed study of the mechanical conditions which led to failure. There are certain steps which may be taken to perform this analysis. First, however, the mode of fracture must be determined. There are two different types of fracture: fully plastic and elastic–plastic.

Fully plastic fracture is typically the cause of failure of a component made of a low-strength material such as pure aluminum, pure copper or pure brass. A fully plastic fracture can be due to a buckling instability or to an overload. There is always extensive yielding of the component over its entire cross section. A plastic-limit analysis will give an approximate value of the critical load at failure. This estimated load should be compared to the actual service loads.

In elastic–plastic fracture, plastic deformation is limited to the region around the crack tip and crack flanks. An elastic–plastic crack is initiated in a region of localized plastic strain. The crack initiation criterion can be verified by using the following procedures:

(a) Estimate the loads and displacements applied to the part, taking into account the directions as well as the intensities of these parameters.

(b) Calculate the local stress at the sites of stress concentration, taking into account, if needed, the residual stresses. This can be done either by calculating the stress concentration factor at the root of the notch or by using finite element analysis. Note that an elastic analysis will in most cases be too simplistic. K_t, the elastic stress concentration factor, can be obtained from tables or from handbooks, and the K_t can then be converted following Neuber's equation into a stress concentration and a strain concentration factor:

$$(K_t)^2 = K_{\text{stress conc.}} \times K_{\text{strain conc.}} \qquad (2)$$

The Neuber analysis is quite accurate, and the strain concentration factor gives a first estimation of the total local strain existing at the root of the notch. The finite element analysis gives the stress and strain gradients around the notch. This information is used to predict whether a crack will or will not grow. Following the notch analysis, it should be possible to predict whether a crack will be initiated either by low-cycle fatigue, by stress-corrosion cracking or by creep deformation. A fracture criterion will be needed to predict the initiation of a crack at the root of a notch or at the stress concentration site. Several textbooks on fracture provide the needed guidelines (*Metals Handbook* 1974, 1975).

3.1 *Fracture Mechanics Analysis of Crack Growth*

Assuming that it has been clearly demonstrated that the stress and strain conditions at the initiation sites are such that a crack will be initiated, the next step is to predict the rates at which the crack will grow. A crack may grow under conditions of ideally linear elastic fracture mechanics, where the plastic-zone size at the tip of the crack is very small compared with the crack length. On the other hand, if there is extensive plasticity near the crack tip, the crack growth conditions are referred to as elastic–plastic conditions. In that case the analysis of crack growth becomes more difficult. Finally, a crack may grow under a condition of fully plastic yielding.

In order to estimate the rate of crack propagation, one should have access to a handbook of fracture mechanics data which will provide the crack growth rate data for fatigue (da/dn), creep crack growth (da/dt), stress corrosion cracking (da/dt) and fracture toughness (K_{Ic} or J_{Ic}). The data given in the handbooks are never directly applicable to the failure analysis at hand, but they can be interpolated or extrapolated depending upon the need. Correlations between the mechanical properties of the material and the crack growth data are also needed to derive rapidly the magnitude of the crack growth rates.

3.2 Crack Growth Reconstruction

Having considered fractography, fracture mechanics and crack growth, the crack growth history of the failure can be reconstructed. This is best done by estimating the time and/or the number of cycles it takes to grow a crack from a length a_0, of the order of a few grains at the initiation site, to a final fracture length a_f, under the service loads or conditions reported. The analysis should quickly verify within a factor of 2 to 4, whether the failure history is plausible. This calculation can be done by direct integration or by iteration with a computer.

Following this first approximation, it is necessary to perform a second calculation to take into account all the secondary factors such as residual stress, corrosion effects and microstructural variations which may account for marked differences in crack growth rates.

3.3 Accelerated Test Plan

It may be crucial, in order to verify the validity of the crack growth analysis, to perform a series of accelerated mechanical tests to obtain the rates of crack growth and the fracture conditions for the material or component at hand. With the servo-hydraulic equipment available, this can be done within a very short time. Accelerated test results provide valuable information to clarify and complete the failure analysis.

3.4 Estimated Costs versus Risk of Failure

In order to be able to provide some guidelines on the decision or action to be taken with respect to the other components in service, it is necessary to have a complete review of the analysis of the time spent in crack growth in the failed component and in other components in service. Once this analysis is performed it should be possible to define a final set of crack-detection inspection intervals to ensure safe operation and planned retirement from service of cracked components.

The final step in a failure analysis is to review the probability of failure and to estimate the costs of failure versus the cost of quality-control inspections and the cost of a fracture control plan. This is somewhat beyond the scope of the failure analyst, but the reliability of the conclusions arrived at in the failure analysis is very important in the risk-analysis task.

4. Conclusions

There should be a preparation of a report which takes into account all the facts that have been clearly established. This may take the form of a preliminary report, in order to have it reviewed and coordinated by the team of experts who have participated in the analysis. A preliminary report will give everyone an opportunity to verify that a consensus of opinion has been reached and that there is no misunderstanding about the facts on which the analysis is based. The final analysis report should not only include the results of analysis but should also set guidelines for a short-term solution to avoid a repetition of the failure, as well as a long-term solution to the fracture problem. Typically, the final report would include the following sections:

(a) description of the failed component,

(b) service conditions at the time of failure,

(c) wreckage reconstruction,

(d) service history,

(e) design, manufacturing and quality-control history,

(f) mechanical and metallurgical conditions of failure,

(g) summary of causes and sequence of failure, and

(h) recommendation for prevention of similar failures and guidelines for corrective action for similar components in service.

See also: Failure Analysis of Ceramics; Failure Analysis of Plastics; Failure Analysis: Nondestructive Evaluation; Mechanics of Materials: An Overview

Bibliography

Hutchings F R, Unterweiser P M (eds.) 1981 *Failure Analysis: The British Engine Technical Reports.* American Society for Metals, Metals Park, Ohio

Metals Handbook, 8th edn., Vol. 9, *Fractography and Atlas of Fractographs*, 1974. American Society for Metals, Metals Park, Ohio

Metals Handbook, 8th edn., Vol. 10, *Failure Analysis and Prevention*, 1975. American Society for Metals, Metals Park, Ohio

Nauman F K 1983 *Failure Analysis: Case Histories and Methodology.* American Society for Metals, Metals Park, Ohio

Pomey G, Rabbe P 1968 *Ruptures de fatigue de pièce de machines classification et analyse.* Société Francais de Metal, Dunod, Paris

R. Pelloux

Failure Analysis: Nondestructive Evaluation

Engineering failure analysis of structural metals and alloys usually involves an investigation into the causes of unexpected breakage. The broken item may be a component of a large engineering structure (e.g., a girder in a bridge or a blade in a turbine), or a part of a consumer product (e.g., a bicycle fork). Failure analysis can also be used in production to determine why a process suddenly begins turning out faulty products. The results of a failure analysis can help to explain why something broke, and more importantly, how to proceed so that further problems are avoided.

Failures occur due to three basic causes: (a) abuse (including natural catastrophies), (b) design deficiencies, or (c) materials problems. Abuse accounts for most failures in service. One example is the overloading of engineering structures due to a larger than expected earthquake. Dropping a glass container is another. Design deficiencies often result simply because of poor application of known technology; for example, a keyway in a rotating shaft is designed with a sharp right angle instead of with a fillet, setting the stage for fatigue-crack initiation and growth. Materials selection can be another design deficiency (e.g., a less-than-adequate material is selected simply because the more appropriate material cannot be supplied in time). Materials problems are by far the most interesting causes of failure. They often result from unexpected corrosion processes, for example, intergranular cracking of a sensitized stainless steel.

1. Destructive Testing

When a sufficiently important failure occurs, or when a significant failure seems imminent, a failure analysis investigation is usually undertaken to provide the basis for remedial action (American Society for Metals 1975, Colangelo 1974). Failure analysis usually involves destructive testing to measure strength, ductility, toughness and other mechanical properties. Chemical analysis and metallographic analysis are two more forms of destructive testing commonly used in failure analysis; these methods require much less material than mechanical testing. Test results are usually compared with specification requirements.

2. Nondestructive Evaluation

Traditional nondestructive evaluation (NDE) techniques have application in failure analysis. The most commonly applied techniques include visual–optical, dimension measurement, ultrasonic testing and x-radiography; hardness testing is also widely used. Occasionally, hydrostatic testing is used as an NDE method for failure analysis of pressure vessels and piping. Other NDE techniques sometimes used in failure analysis include: magnetic particle inspection of ferrous metal; liquid-penetrant inspection and eddy-current inspection to detect surface discontinuities; γ-radiography to detect internal discontinuities; and x-ray diffraction to measure surface residual stresses. New techniques currently under development include tomography to image internal defects (see *Three-Dimensional Radiography*), and time-of-flight ultrasonic techniques to determine the position and size of such defects.

In some cases, NDE techniques may be applied to a component which is in service. A typical scenario may be as follows. During a routine inspection of a system, it is found that the surface of a component is dented or abraded. Closer examination with the aid of a magnifying glass would follow. Possibly some chemical etching would be done to reveal whether or not the damaged area is in the vicinity of a weld. If sufficient concern develops, the damaged area could be examined by ultrasonic techniques. Should ultrasonic testing suggest intolerable loss of thickness or intolerable internal flaws, it is likely that the damaged area would be either reinforced or cut out and replaced.

If the damaged area is cut out, the opportunity exists for further failure analysis. For example, the pieces may be examined further by x-radiography to supplement ultrasonic results hinting at internal flaws. In general, ultrasonic test results indicate the presence and size of a two-dimensional flaw, but cannot always characterize the defect as "sharp," like lack of fusion in a weld, "blunt," like an inclusion in steel, or "round," like a void in a casting. Radiographic results can help to make these distinctions.

Furthermore, if the damaged area is cut out, metallographic sections may be prepared to look directly at the flaws detected by ultrasonic testing and x-radiography, and to determine the size of any cracks. It is likely that the flaws detected by NDE techniques would be a mix of internal cracks, inclusions and other microstructural features.

3. Examples of NDE Techniques in Failure Analysis

3.1 Microcracks in Fire Extinguishers

X-radiography (see *Radiographic Nondestructive Evaluation*) is a useful technique for detecting internal microcracks. In an investigation of brazed joints in soda–acid fire extinguishers (Christ et al. 1978), radiography was used to detect intergranular cracking (see *Intergranular Corrosion*). In the manufacture of extinguishers, a red-brass collar about 0.25 cm thick is brazed to a stainless steel dome of the same thickness around a circumference of about 30 cm. Prior metallographic analysis had revealed intergranular cracking in a sensitized zone of the stainless steel. The path of attack for corrosive fluid (aqueous sulfuric acid or sodium bicarbonate) was between the broken fillet of the braze metal and the stainless steel. It was presumed that the fillet broke away from the stainless steel dome because of high torque when the threaded cap of the fire extinguisher was tightly twisted onto the collar to make a pressure-tight seal (operating pressure was about 0.7 Pa). Once this finding was established through metallographic techniques on one fire extinguisher, other extinguishers were examined by x-radiography to identify those with intergranular cracking in the region of the circumferential braze.

One limitation of this technique is that it is difficult to detect tight cracks, that is, cracks with openings less than 1–2% of the wall thickness (about 0.005 cm in this instance). Another limitation is that the crack plane must be oriented almost parallel to the x-ray beam for maximum contrast. Finally, material thickness is a possible limitation. Thicknesses of 2.5–5 cm, which are typical of bridge girders or pressure vessels used in the petrochemical industry, require high accelerating voltages on the x-ray source. In the case of the fire extinguisher analysis, thickness and orientation were not limitations, but tight cracks were.

3.2 Oil Pipeline Defects

Another instance where x-radiography proved useful in failure analysis was an investigation of a rupture in the Trans-Alaska Oil Pipeline. The rupture occurred as the line was being filled with petroleum for the first time in July 1977. Nitrogen gas from a liquid-nitrogen source was pumped into the line ahead of the moving oil front, to establish a nonoxidizing atmosphere and thereby minimize the explosion hazard. Localized low temperatures, and accompanying thermal stresses in a region cooled by the nitrogen, caused cracking of parent metal in the vicinity of a girth weld. Visual and optical inspection of the fracture surfaces led to the view that the crack originated internally at some feature of the microstructure in the girth weld. Review of a girth-weld radiograph taken at the time of fabrication revealed a nonmetallic inclusion at the suspected location of the fracture origin. Presumably the inclusion acted as a crack nucleator because the girth weld had been subjected to unusual conditions during filling of the pipeline. The remedy was to cut out the cracked section and weld in a new one. Procedures to ensure that the nitrogen did not excessively cool the line were also instituted.

3.3 Defects in Compressed-Gas Containers and Pressure Safety Devices

Dimensional measurement is an NDE technique sometimes used in failure analysis. Wall thickness of pressure vessels can be measured to a precision of about 1–2% using an ultrasonic thickness gauge. For this technique to work, it is necessary that the inside and outside surfaces are approximately parallel and are reasonably smooth. In an investigation of the wall thickness of compressed-gas containers, measurements with an ultrasonic thickness gauge indicated that wall thickness did not meet requirements. Verification of the measurements was made by cutting out a wall section from one container, and measuring it directly with a hand-held micrometer. One possible outcome of this investigation was a reduction in the maximum allowable operating pressure of containers of low wall thickness.

In an investigation of a ruptured steel compressed-oxygen container (Christ et al. 1979), wall thickness measurements made with a hand-held micrometer were used to determine the ductility in the vicinity of the fracture. A decrease in wall thickness of 0.018 cm led to the result that the fracture exhibited 3% reduction in thickness. The conclusion was that the material exhibited the ductility that was intended according to design requirements.

Subtle changes in thickness resulting from corrosion can sometimes be measured with a hand-held micrometer. For example, the decrease in thickness of a beryllium–copper rupture disk used in a pressure-relief safety device amounted to 15 μm. Metallographic analysis showed that this loss in thickness was attributable to corrosion pitting and intergranular attack, followed by flaking and spalling of metal. Possible remedies for this problem included application of protective coatings and redesign to prevent corrosive liquids from reaching the disk.

See also: Failure Analysis: General Procedures and Applications to Metals; Failure Analysis of Ceramics; Nondestructive Evaluation: An Overview

Bibliography

American Society for Metals 1975 *Metals Handbook*, 8th edn., Vol. 10, *Failure Analysis and Prevention*. American Society for Metals, Metals Park, Ohio

Christ B W, Smith J H, Brady C H 1978 Analyzing fire extinguisher failures. *Met. Prog.* 11(9): 28–32

Christ B W, Smith J H, Hicho G E 1979 Fracture analysis of a pneumatically-burst seamless steel compressed gas container. In: Smith C W (ed.) 1979 *Fracture Mechanics*, ASTM Special Technical Publication 677. American Society for Testing and Materials, Philadelphia, Pennsylvania pp. 734–45

Colangelo V J 1974 *Analysis of Metallurgical Failures*. Wiley, New York

B. W. Christ

Failure Analysis of Ceramics

Failure analysis entails examination of a fractured component or test specimen and correlation of the observed features with design, with analytical and instrumentation data and with prior fabrication information. The aim is to elucidate the cause of the failure. It is critical in component application to determine whether the failure was the result of a design deficiency, a material deficiency (arising from the fabrication process) or an abnormal off-design condition. Examination of the fracture surfaces and surrounding regions of a ceramic component can aid in this determination.

The techniques for failure analysis pertinent to glasses and fine-grained ceramics are relatively well established. These materials exhibit distinctive topographical features in the vicinity of the fracture origin which can be observed by a combination of macroscopic and microscopic methods. The essence of these procedures is briefly described in this article. The corresponding analysis for large-grained ceramics, especially those ceramics with elastic and thermal expansion anisotropy (such as Al_2O_3), is appreciably more difficult. The indications are more subtle and experience with a specific material is often required.

1. *Morphology of Fracture*

The fracture of ceramic components typically arises either from defects introduced during processing or machining, or from service (especially impact) damage. The specific fracture-initiating defect responsible for the failure of a component can usually be located, and its origin identified, by recognizing the various morphological features by the following procedures.

The first step is to reassemble the fragments whenever possible, to determine, macroscopically, distinctive crack patterns that suggest the position or cause of failure initiation. A frequently encountered fracture pattern is illustrated in Fig. 1a. The fracture origin is invariably located near the center of the planar central section, between O and O′ in Fig. 1a. This feature develops because, as the crack accelerates from the origin, it begins to bifurcate (Fig. 1b). This typically occurs when the stress intensity K (quasi-statically defined) exceeds a critical value. The bifurcations occur repeatedly from both crack tips, resulting in substantial fragmentation of material in the plane beyond the fracture origin, outside OO′ (Fig. 1). Consequently the loss of small fragments is not detrimental to the failure analysis, because such fragments do not contain the failure origin. The larger fragments, on both sides of the planar initiation zone, are now separated from the assemblage and can be exposed to closer scrutiny. (Occasionally a highly defective component separates into only two fragments. In such instances, the general region of the failure origin cannot be so readily located. This can also be true of failure caused by thermal shock.)

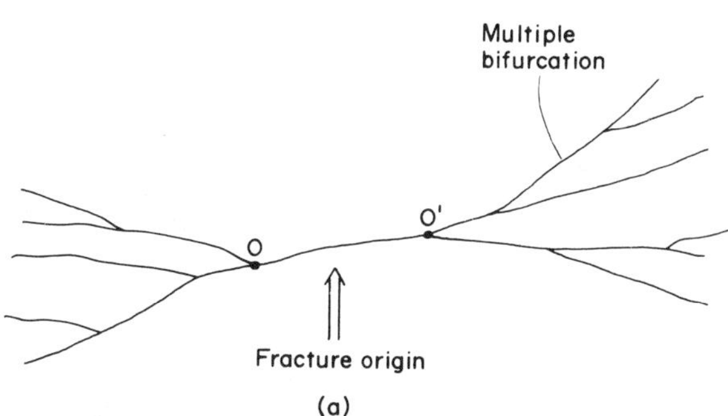

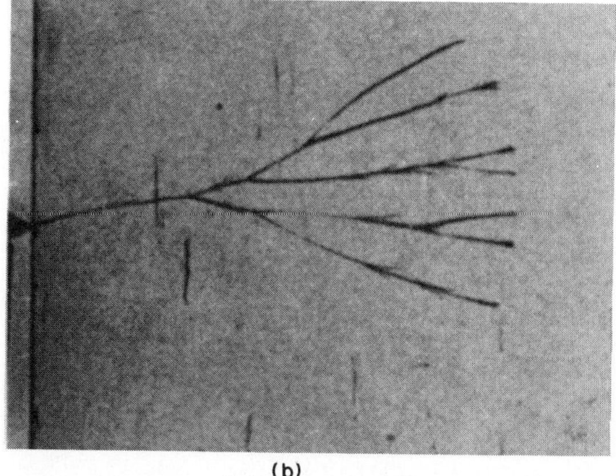

Figure 1
(a) Schematic illustrating a typical crack morphology in the vicinity of the origin, and (b) crack bifurcation in glass from an edge-initiated failure (Schardin 1958)

In some instances, additional or alternative diagnostic surface features are observable at the macroscopic level. These include

(a) fractures intersecting (or radiating from) a single point or small area;

(b) a "witness mark," consisting of a debris-coated or burnished area indicating abnormal localized surface distress during service;

(c) fractures extending along or parallel to machining grooves;

(d) a fracture extending inwards from an edge of a component;

(e) a fracture which appears to be associated with a contour or discontinuity in the component configuration;

(f) a fracture intersecting a hole in the component (e.g., a bolt hole, feed-through hole); and

(g) a fracture intersecting or passing through a point of attachment.

Often the fundamental cause of failure can be deduced from such macroscopic features. For instance, fracture radiating from a point can often result from localized impact or some other form of mechanical interference. A fracture at a hole, edge, contour or joint can suggest an unacceptable design with excessive stress concentration. Alternatively, an abnormal defect may be involved in the fracture process. To distinguish between a stress concentration and an abnormal defect as the primary cause of failure, the fracture surfaces adjacent to the suspected region of fracture must be investigated.

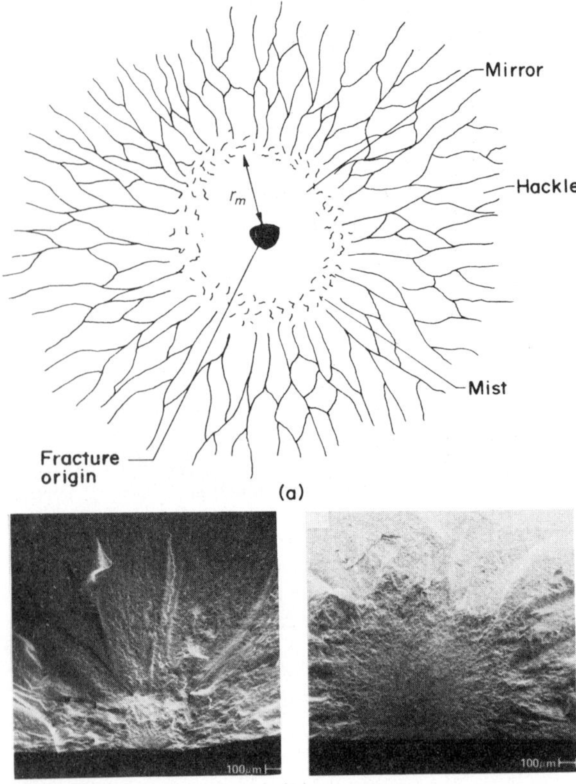

Figure 2
(a) Schematic illustrating river markings on a fracture surface; (b) failure origin in silicon nitride near the surface

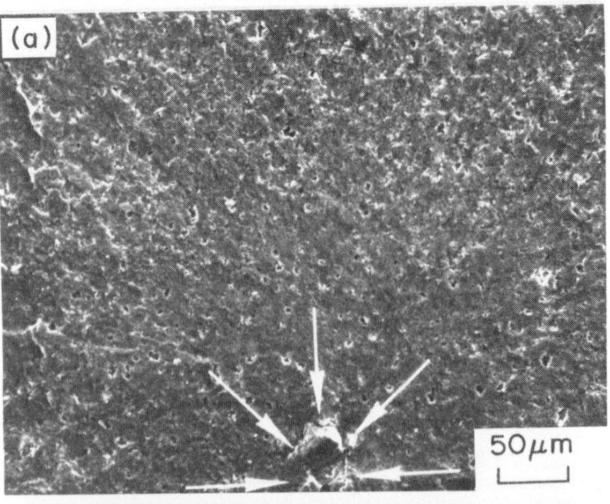

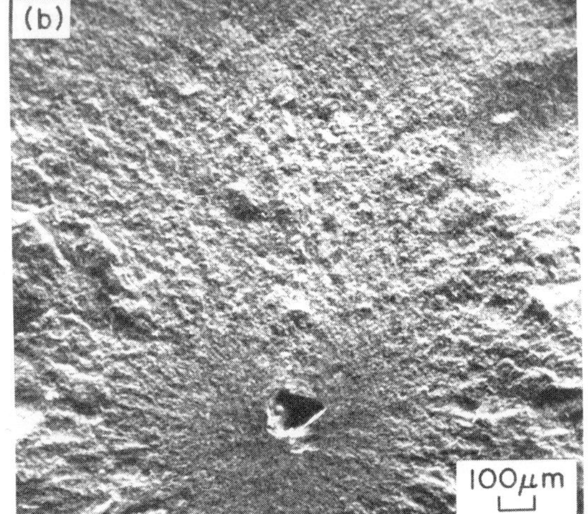

Figure 3
Scanning electron micrographs of (a) an Si inclusion and (b) a large pore as failure origins in Si_3N_4

The examination of the fracture surfaces thus constitutes the second step in failure analysis. This phase is achieved visually, by low-magnification optical microscopy and sometimes by scanning electron microscopy (SEM). The examination is predicated on the recognition that, as a crack initiates and propagates through a glass or ceramic, distinguishing features are created on the fracture surface. Some of these features are illustrated in Fig. 2.

The features of interest are referred to as river markings. These markings radiate toward the fracture origin but terminate in a relatively featureless, essentially circular zone before converging on the actual origin. This featureless central zone has been referred to as the "mirror" region in the glass litera-

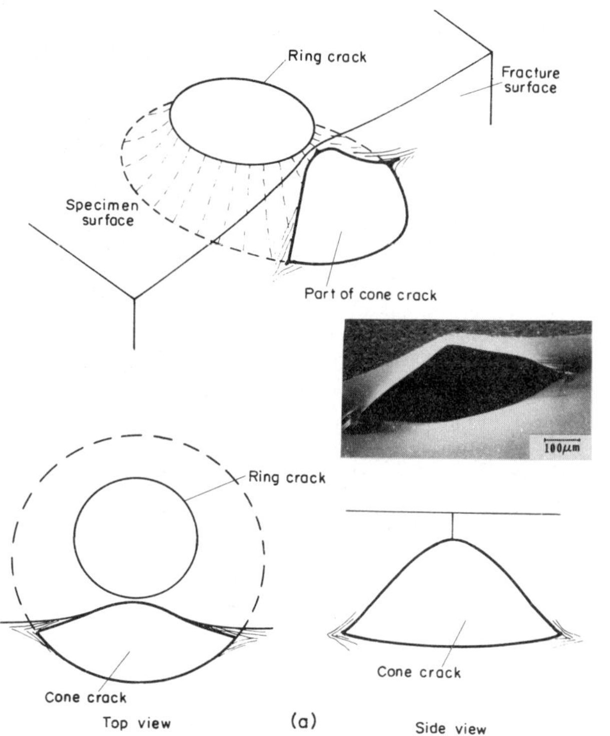

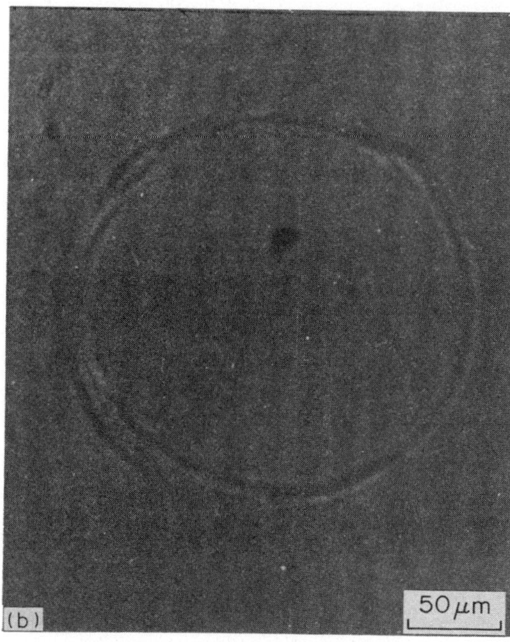

Figure 4
(a) Partial Hertzian cone crack revealed by the fracture process, and (b) ring crack observed on the surface

ture while the zone containing the river markings has been termed the "mist and hackle" region. The river markings are created while the crack is moving dynamically. The dynamic crack tip stress field causes the crack to deflect onto parallel planes, in adjacent regions of the front. When these planes overlap, steps are created. These steps are the source of the river markings. The crack velocity in the mirror region is too low to cause crack bifurcation and deflection. Consequently the crack remains relatively planar and the associated fracture surface is comparatively featureless. The perimeter of the mirror zone coincides, in fact, with the initial bifurcation point OO′ on the surface crack pattern (Fig. 1a).

Once the river markings have been identified and their center traced (from the point of intersection of the radial lines), the actual failure origin can be sought by inspection at higher magnification of the center of convergence. This phase of the investigation is important, because the nature of the fracture origin establishes whether failure was caused by a design deficiency, a material deficiency or an abnormal service condition. The appearance of a variety of fracture origins and the implications regarding the probable cause of failure are as follows.

The facility with which the origin can be identified is contingent on the type of origin involved. If the material is defective or if the component has been subject to impact damage, the origin tends to be relatively large (compared with typical microstructural features) and is easily distinguished. Common, readily detected defects (Fig. 3) include large pores or pore clusters, inclusions, isolated large grains or preexisting cracks (cracks present prior to fracture). Pores derive from deficiencies in the fabrication process. The size, shape or distribution of the pores often provides insight regarding the specific processing step which initiates the pores and thus has implications for material improvements. Inclusions typically result from contamination during the fabrication process. Chemical analyses of the inclusion by an electron microprobe technique can determine its likely origin. Isolated large grains originate from inadequate powder quality, inhomogeneous mixing or insufficient control of the densification parameters. The size, shape and crystal structure of the large grains can indicate the specific source of such defects.

Preexisting cracks may originate in several ways. Each source produces a crack or crack network having distinct characteristics visible on the fracture surface adjacent to the fracture origin. Sources of cracks include laminations formed during pressing, and shrinkage separations during drying or binder removal.

Some failures result from damage produced in the material during handling or during operation. Impact and oxidation/corrosion-induced damage are two examples of defects that can be readily detected by examination of the fracture origin. In the case of impact, some of the characteristics of the impacting body can also be deduced. Two principle types of impact damage occur: that induced by blunt projectiles (e.g., spheres or soft projectiles) and sharp projectile damage. Blunt projectile impacts elicit elastic response of the ceramic and result in the creation of Hertzian cone cracks. Such cracks can be readily identified at the failure origin as a cone partially exposed by the final fracture process (Fig. 4a). A thin circular ring on the surface is also generally associated with this type of origin (Fig. 4b). In contrast, irregular projectiles may cause plastic deformation of the ceramic and result in an array of radial cracks (Fig. 5). Evidently only one of these cracks causes the final fracture. Consequently, when this type of damage is the source of failure, several radial cracks remain on the surface around the fracture origin and can be readily identified by inspection of the surfaces. In this instance, fracture surface observations are of minimal utility, because the profile of the specific radial crack that caused the failure is difficult to distinguish, being morphologically similar to the fracture surface created during final failure.

Oxidation and corrosion-related damage processes include surface recession, pitting and reaction layers usually visible on the fracture surface. An example

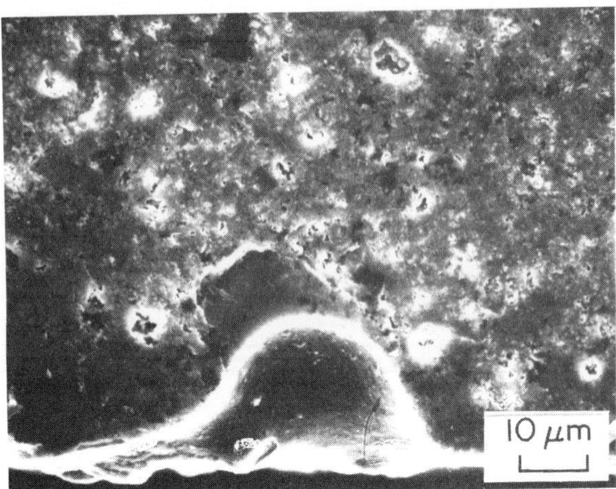

Figure 6
Failure from an oxidation-induced surface pit in Si_3N_4

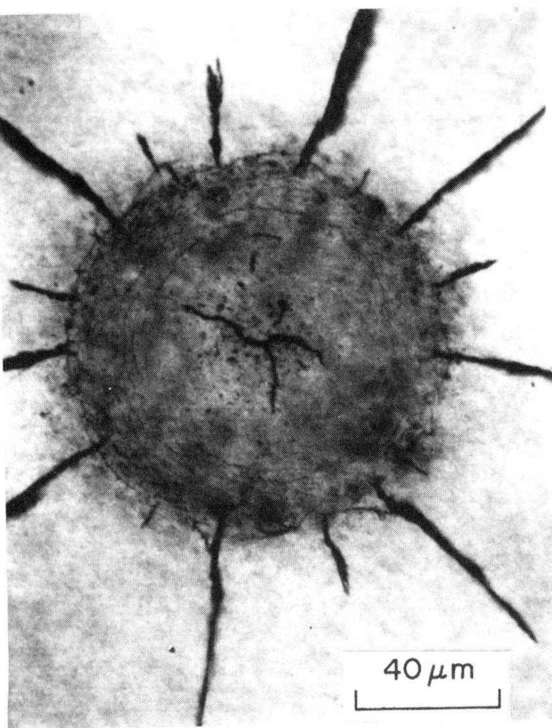

Figure 5
Radial cracks caused by the impact of an irregular projectile

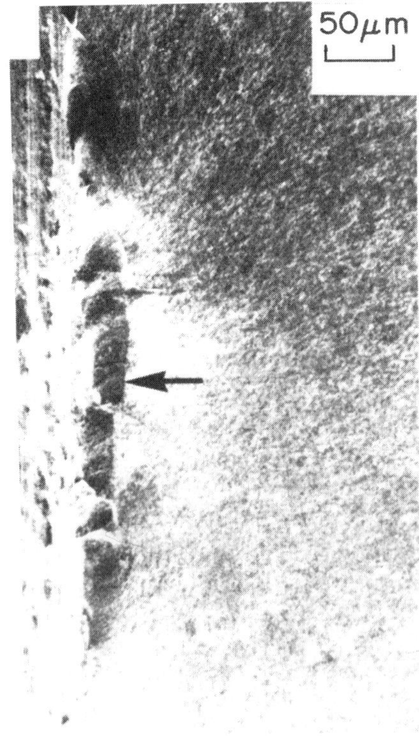

Figure 7
Scanning electron micrograph of machining-induced cracks in MgF_2 (courtesy R W Rice)

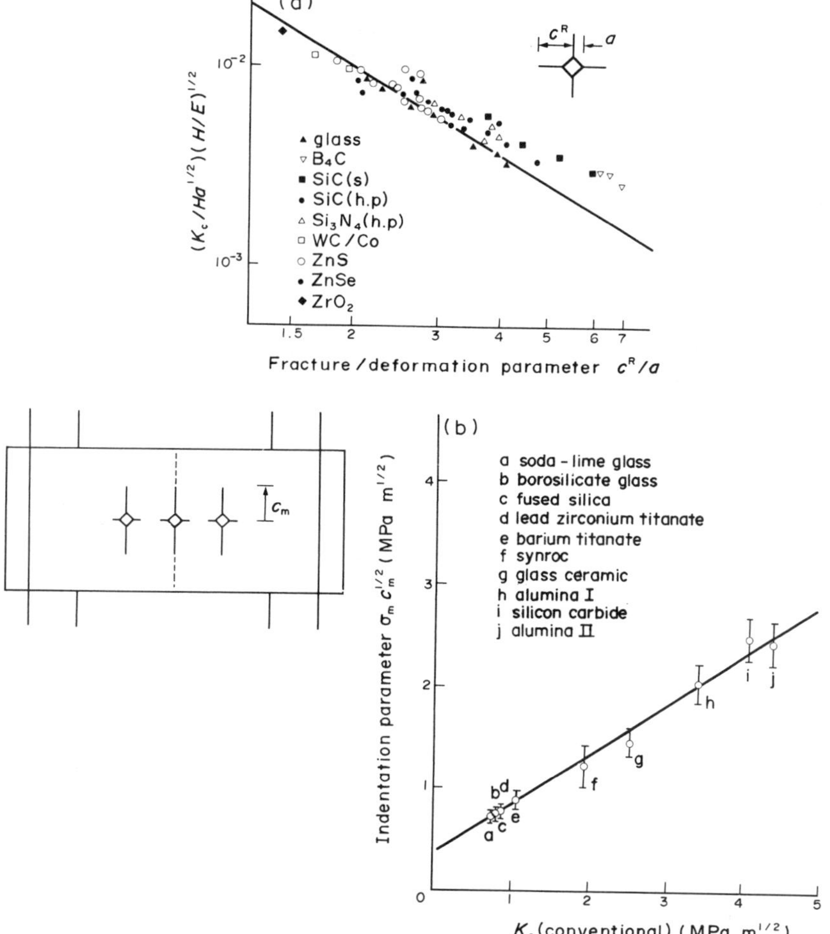

Figure 8
Synopsis of the indentation technique for determining the toughness on small fragments of the failed ceramic component: (a) direct method; (b) indirect method

of a fracture which has initiated at a surface pit caused by oxidation/corrosion is illustrated in Fig. 6.

The fracture surfaces discussed so far are relatively easy to diagnose. Some fracture surfaces are much more difficult to evaluate, because the features are less distinct. This is the case for large-grained ceramics, low-strength ceramics, bimodal microstructure ceramics and fractures initiated by machining damage, contact stress and thermal shock.

Failure from machining damage is probably the most difficult to distinguish. Machining generates small, subsurface cracks (Fig. 7). The cracks develop along failure surfaces similar to those associated with the final, unstable failure. Consequently, fracture surface observations do not readily reveal the profile of the critical machining flaws. Relatively subtle distinguishing features must be sought. For example, in Si_3N_4, machining damage exhibits a larger proportion of transgranular fracture than the subsequent, load-induced stable growth. The machining flaws can thus be most effectively identified using optical microscopy with oblique illumination, by taking advantage of the greater reflectivity of the transgranular segments of the failure surface. Scanning electron microscopy is of minimal utility, because the transgranular regions are not readily discerned at the low magnification needed to locate the critical flaws. In other materials, changes in trajectory between the machining flaws and the failure surface can be used to define the flaw profile, but unambiguous identification is difficult.

2. *Analysis of Failure*

In addition to flaw identification, measurements of various dimensions at the failure origin can provide knowledge of the local stress that caused the failure. In particular, the radius r_m of the mirror (Fig. 2a) is uniquely related to the fracture stress σ_f by

$$\sigma_f = \lambda K_c r_m^{-1/2} \tag{1}$$

where K_c is the fracture toughness and λ is a material-independent constant ($=3.5 \pm 0.3$). Consequently, an independent measurement of the toughness, in conjunction with determination of r_m, allows the fracture stress to be determined a posteriori.

Furthermore, the toughness can be conveniently ascertained on a fragment of the actual failed component, using the indentation technique (Fig. 8). The estimate of the failure stress obtained in this manner is probably the least ambiguous, because it relies on the measurement of a characteristic dimension of the *failure* crack (i.e., the bifurcation length), which in turn is apparently determined by a characteristic stress intensity factor $K \approx 4.5K_c$. However, additional complexity arises when the component is highly defective. The mirror radius then becomes a significant fraction of the component dimension in the fracture plane and λ in Eqn. (1) is modified due to the effect of the compliance on the stress intensity factor K. The modified λ can be computed from the pertinent stress intensity factor solutions, some of which are available in handbook form.

Separate measurements of the dominant flaw dimension can also be informative. Clearly a relationship exists between flaw size and strength. In principle, therefore, the flaw size should provide an independent estimate of the failure stress. However, such measurements are of limited utility, because the strength–flaw size relationship depends on the flaw type (Fig. 9). Furthermore, unique relationships have not been developed for many important flaw types, notably pores, low-strength inclusions and large grains. Nevertheless, flaw size measurements can be advantageous in the analysis of impact damage or failures associated with machining damage. In particular, cracks generated by sharp projectile impact and machining are subject to stable growth on load application, due to plasticity-induced residual stress. Consequently the crack dimension at instability is straightforwardly related to the stress. For example, a circular crack with instability radius a_m relates to the stress by

$$K_c = (2/\pi^{1/2})\,\sigma\, a_m^{1/2} \tag{2}$$

Measurement of the radius a_m when clearly delineated on the fracture surface thus provides the requisite estimate of the failure stress.

3. Diagnosis of Failure

The information obtained in accordance with the above procedures may be assembled and used to specify the cause of failure, thereby permitting recommendations for avoiding a recurrence of the failure. The unambiguous identification of the failure-initiating flaws provides direct evidence of the defective nature of the material. In particular, a large processing flaw at the origin would be a tentative basis for classifying the material as defective. Conversely, a flaw attributed to impact or machining would indicate that the material had been abused after processing. Furthermore, evaluation of the stress at which the fracture occurred establishes whether the material satisfies the strength specification afforded by the supplier and, additionally,

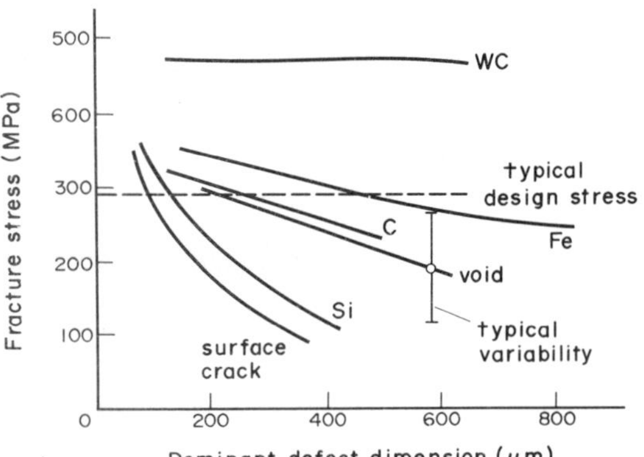

Figure 9
Effects of flaw type on the strength–flaw size relation for Si_3N_4 (curves labelled WC, Fe, C and Si are for inclusions of these substances)

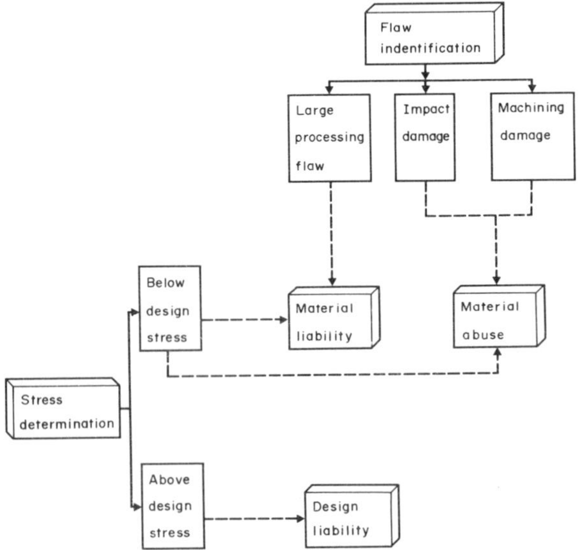

Figure 10
Procedure for assigning liability based on the two principal steps in the failure analysis: flaw identification and failure stress evaluation

whether the component was subject to an overload condition (i.e., by comparison with the design stress). These implications are summarized in Fig. 10.

The recommendations based on the preceding diagnosis are self-evident, with liabilities assigned to the material manufacturer, to the handling or installation engineer or to the design engineer. Corrective action based on these findings can then be instituted.

See also: Ceramics: Relation of Microstructure to Processing History; Failure Analysis of Plastics; Failure Analysis: General Procedures and Applications to Metals

Bibliography

Evans A G, Charles E 1976 Fracture toughness determinations by identification. *J. Am. Ceram. Soc.* 59: 371–72

Kirchner H P, Gruver R M 1976 In: Bradt R C, Hasselman D P H, Lange F F (eds.) 1976 *Fracture Mechanics of Ceramics,* Vol. 1. Plenum, New York, p. 309

Lawn B R, Evans A G, Marshall D B 1980 Elastic/plastic indentation damage in ceramics: The median/radial crack system. *J. Am. Ceram. Soc.* 63: 574–81

Rice R W 1978a In: Herman H (ed.) 1978 *Treatise on Materials Science and Technology*, Vol. 11. Academic Press, New York, p. 199

Rice R W 1978b In: Palmour H, Davis R F (eds.) 1978 *Processing of Crystalline Ceramics.* Plenum, New York, p. 303

Rice R W 1984 *Fractography of Ceramic and Metal Failure,* ASTM STP 827. American Society for Testing and Materials, Philadelphia, Pennsylvania

Richerson D W 1983 *Modern Ceramic Engineering: Properties, Processing and Use in Design.* Dekker, New York

Schardin H 1958 In: Averbach B L, Filbeck D K, Hahn G T, Thomas D A (eds.) 1958 *Fracture.* Wiley, New York, p. 304

A. G. Evans; D. W. Richerson

Failure Analysis of Plastics

Failure analysis of plastics is frequently an investigation of the occurrence of brittle failure in a material which is generally accepted as ductile. Failures do occur in the ductile mode but their appearance can usually be explained satisfactorily on the basis of simple mechanical properties, such as the yield stress. Account must be taken, however, of the effect of variables, including the time of loading and temperature, on such properties, since the load-bearing capability of plastics in general, and thermoplastics in particular, decreases with time and with increasing temperature.

In this article, factors which may induce brittleness in otherwise ductile plastics are identified, and relevant examples are discussed.

1. Ductile Failure

1.1 Ductile Failure in Air

This type of failure is found in pressure vessels and pipes, but is perhaps observed most frequently in polyethylene film: overloaded garbage sacks extend several hundred percent locally until the seriously thinned polymer can no longer support the load and breaks. In more critical engineering applications, the appropriate design calculations must be made, involving stress (or strain) level, time under load and temperature and also environmental effects, if any. A ductile failure is characterized by the occurrence of cold drawing, often accompanied by stress whitening.

1.2 Influence of Liquids on Ductile Failure

The mechanical properties of plastics which absorb appreciable quantities of specific liquids can be affected adversely; for example, the stress at yield of high density polyethylene is reduced by 10% if the polymer is immersed in gasoline. This polymer is suitable, nevertheless, for the manufacture of fuel tanks for road vehicles. The ductile failure stress for polyamides (nylon) is reduced even more seriously by the absorption of water, but the toughness is improved.

1.3 False Yielding of Particulate Structure

Polymers in which the particulate structure is preserved after melt shaping (especially poly(vinyl chloride) and polytetrafluoroethylene), and some polymers containing particulate fillers, may suffer premature ductile failure at an unexpectedly low stress. This is ascribed to shear failure of the links between the particles, and is illustrated in Fig. 1.

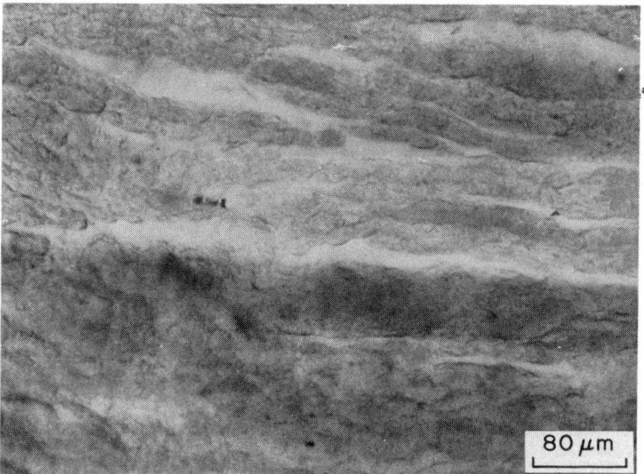

Figure 1
Thin section of polytetrafluoroethylene after tensile failure, showing interparticle separation. Normal light micrograph

2. *Factors Which May Induce Brittle Failure*

2.1 *Materials Factors*

(*a*) *Choice of polymer*. If brittle failure constitutes a hazard which should be minimized, it is reasonable that one should choose a plastic with a low propensity to brittle failure; or, if the choice is between two brittle materials, the one with the higher strength under the operating conditions should be chosen. Brittle failures produce smooth surfaces, with characteristic zones generated by an accelerating crack in the material. These zones are shown in Fig. 2.

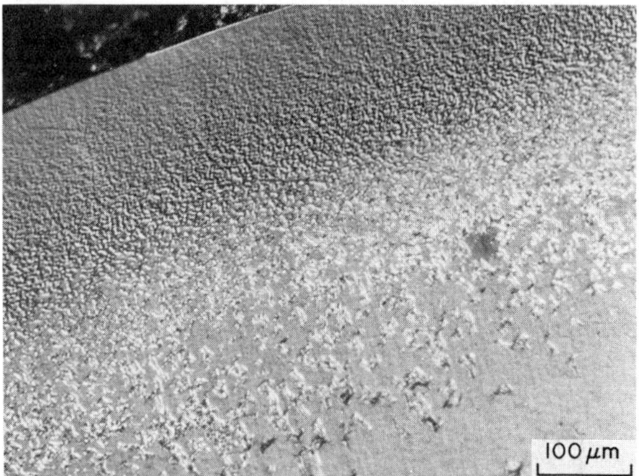

Figure 2
Fracture surface of polystyrene. Reflected light, differential interference contrast

(*b*) *Wrong grade of polymer*. Within a family of plastics, there are frequently many different "grades:" these may be based on polymers of different molecular chain length (often identified as "molecular weight") or polymers with different additives to improve particular properties (e.g., impact modifiers). The incidence of brittleness is increased in polymers of lower molecular weight, as illustrated by the cracking of a strip cut from a pressure vessel which was inadvertently manufactured from an injection-molding grade of propylene copolymer, instead of the higher-molecular-weight blow-molding grade. The failure in Fig. 3 resulted from 100 h loading at 100 °C, while a similar specimen of the blow-molding grade remained intact.

(*c*) *Inclusion of hard particulate fillers*. Embrittlement is an inevitable consequence of adding particulate fillers, which are of higher modulus (hardness), to a ductile plastics material; the elongation at break decreases rapidly with increasing filler content. The effects are particularly disastrous if the particle size exceeds 10–20 μm, or when the polymer matrix–filler particle interface is weak.

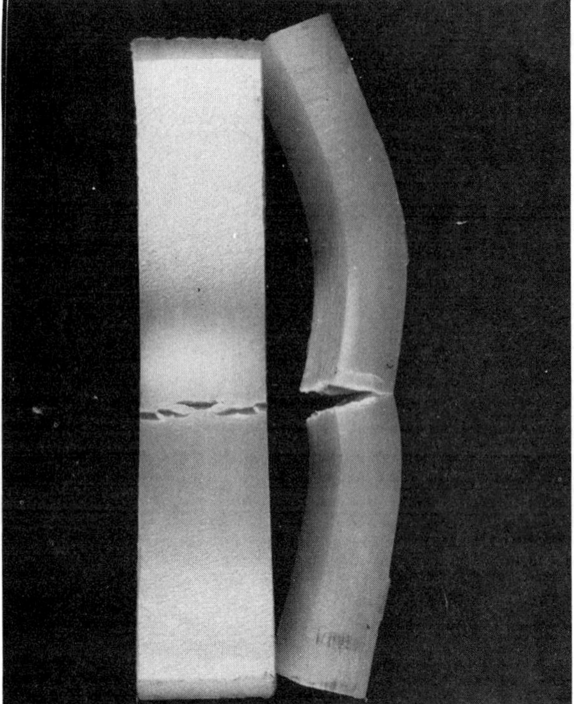

Figure 3
High-temperature stress cracking in low-molecular-weight polypropylene

2.2 *Design Factors*

(*a*) *Poor design and stress concentrations*. Stress concentrations cause engineers many problems, yet designers insist on their inclusion in many products. The example cited here in Fig. 4 is an automobile coolant reservoir tank, which incorporated a concave section to accommodate an electric cable. Proof test-

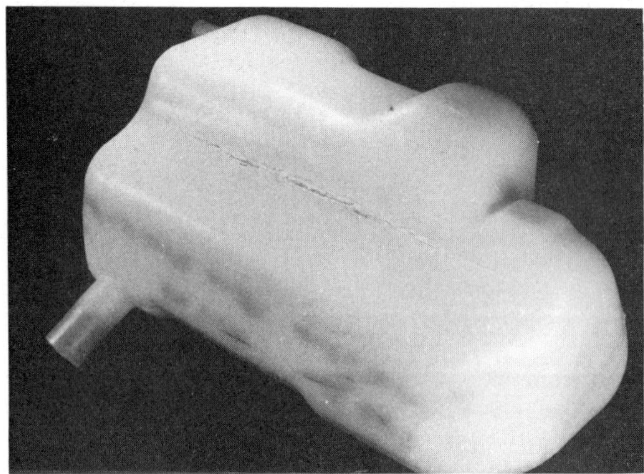

Figure 4
Stress cracking in polypropylene

ing, and stress analysis by a finite elements method, both highlighted this feature as a serious stress concentrator when the vessel was pressurized. Sharp corners are even more widespread examples of stress concentrators which can induce brittle failure.

(*b*) *Poor mold or die design*. Features of mold design which lead most frequently to unexpected failure are associated with a lack of understanding of the behavior of polymer melts, particularly where multiple flow paths are involved. There are numerous examples of such failures; for instance, a washing machine tub which is fed by eight gates located circumferentially around the central hub. The requirement that the multiple flows join together leads to eight lines of weakness along the hub, providing ample explanation of the cracks which develop in service. Another common feature which encourages failure is an abrupt change in thickness.

(*c*) *Unsatisfactory assembly—welds*. Many plastics products are fabricated by joining several subcomponents; the methods employed are many and include solvent welding, adhesives and hot welding. There are many variants of the last, which we illustrated here by hot plate welding. In this process the two parts to be joined are brought into contact with a "hot plate," maintained at a temperature above the softening or melting temperature of the plastics material. After a suitable time, the hot plate is removed and the softened parts brought into contact and kept under a steady pressure until the polymer has cooled sufficiently to harden. Variables such as hot plate temperature, time of heating, pressure of the components on the hot plate, pressure exerted at the welding stage and the time of welding all affect the finished weld, as does the texture of the original components. Even for the optimum weld there are regions of weakness associated with the demarcation between polymer which was melted during the welding and the original texture, and with the presence of welding beads, made up of material which was forced out of the weld. An ideal weld should retain about 50% of the softened polymer in the weld, with the other 50% being forced into the beads, carrying any contaminant or degraded polymer at the junction out of the weld, into the beads. Figure 5 shows a far from perfect weld, made at too high temperature, or too high welding pressure, or both. Noteworthy are the sharp notches formed by the welding beads, leading into the weakened region at the boundary of the weld. This component failed catastrophically at the weld during proof testing.

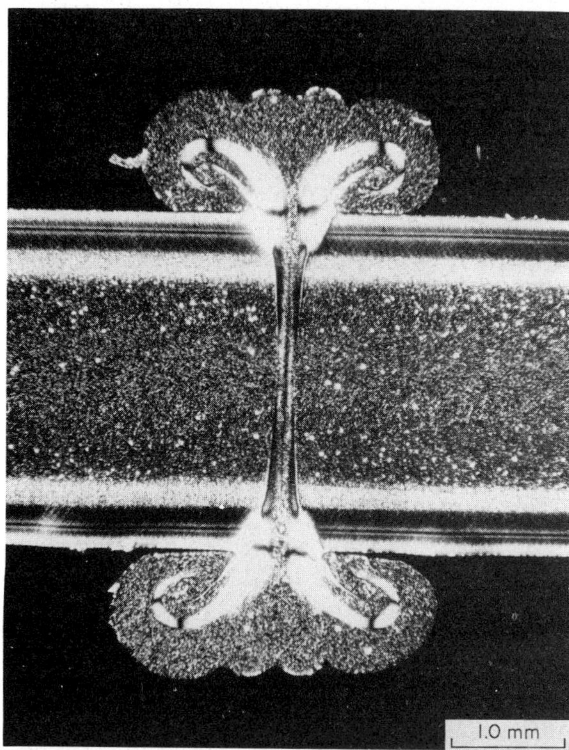

Figure 5
Unsatisfactory polypropylene weld. Crossed polars

2.3 Processing Faults

(*a*) *Too low temperature—inhomogeneous melt*. One of the important considerations in the shaping of thermoplastics is that the melt should be homogeneous with respect to consistency and temperature. If the melt is prepared at too low temperature, with a minimum amount of shear, it is possible that unmelted granules pass through the melting zone into the shaping process and are incorporated in the final product. For crystalline plastics particularly, such inclusions in the finished product frequently initiate brittle failure, as illustrated in Fig. 6, which

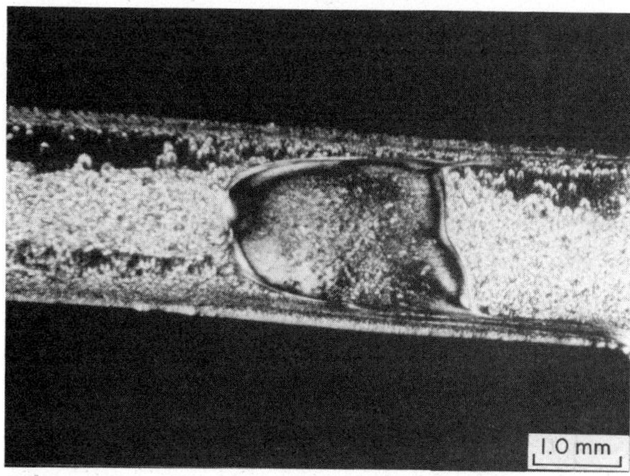

Figure 6
Unmelted granule in acetal injection molding. Crossed polars

shows such a granule of unmelted acetal polymer "contaminating" an injection molding.

(*b*) *Too high temperature—degraded polymer*. Many plastics are unstable at elevated temperatures, degrading by mechanisms which include depolymerization (unzipping) and random chain scission; oxidation during processing is a further powerful route for degradation. Figure 7 shows infrared spectra of two regions of a 250 liter container, shaped by rotational casting of medium-density polyethylene: both spectra are from the base of the container; that labelled "outside" had been in contact with the mold surface, whereas that coded "inside" had free access to air during the molding operation. The latter spectrum shows evidence of heavy oxidation, indicating that an unreasonably high temperature, possibly with a long molding cycle, had been employed. This product split across the base when abused by impact blows. Degraded regions in poly(vinyl chloride) (PVC) can be located using fluorescence, as shown in Fig. 8.

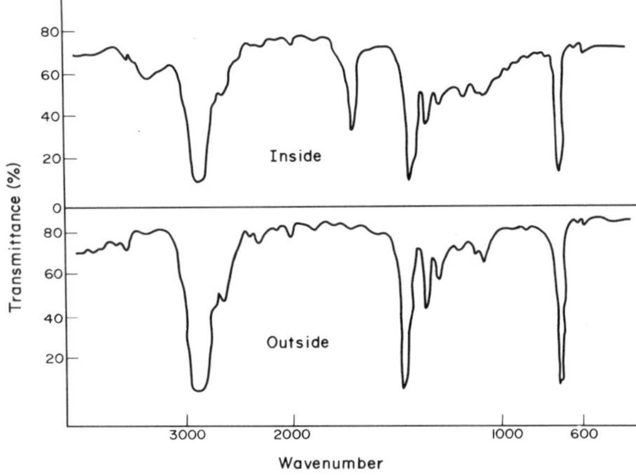

Figure 7
Infrared spectra of oxidized and normal polyethylene

Figure 8
Degradation and splitting at an interface in poly(vinyl chloride). Fluorescence microscopy

(*c*) *Pressure too low—surface defects*. The shaping process of extrusion blow molding utilizes gas pressure—usually air—to blow a hollow extrudate to conform to the shape of a mold. During the cooling stage the molding seeks to shrink and to part from the mold. Should this occur, the surface quality of the molding is impaired, and this can contribute to failure. Figure 9 illustrates the surface texture of blow moldings: micrographs (a) and (b) were direct observations of the surface by scanning electron microscopy, whereas micrographs (c) and (d) were obtained by a surface replication technique, using carboxymethyl cellulose, and photographed in normal light. The latter technique is applicable when a nondestructive examination is required.

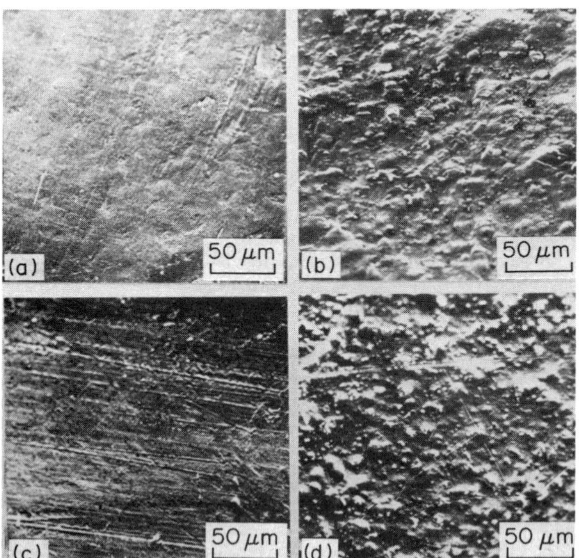

Figure 9
Micrographs (a) and (c) are of the surface of a satisfactory blow molding; (b) and (d) are from a blow molding at too low pressure

(*d*) *Inadequate dispersion of additives*. This is one of the most common causes of the embrittlement of ductile plastics and often involves fillers or pigments, which are frequently hard particulate solids. Sometimes the dispersion is so poor that mere visual examination is sufficient to pinpoint this as a prime cause of brittle failure, but it is more usual to cut thin sections from the faulty product (3–10 μm thickness) and examine these under normal light illumination in a microscope. Figures 10 and 11 are micrographs which illustrate two aspects of poor dispersion, both of which have led to brittle failure in service. Figure 10 shows typical pigment smears which constitute a plane of weakness; the pigment is carbon black, which is a difficult material to disperse satisfactorily. Figure 11 is titanium dioxide in PVC, demonstrating in this case the difficulty of breaking down pigment agglomerates, the large particle acting as a stress concentrator; a crack can be seen originating at the particle. Streaks of high pigment concentration passing through the entire thickness of the product may give rise to high gas and vapor permeability.

(*e*) *Incompatible masterbatch polymer*. The practice of incorporating additives by means of concentrates in a carrier polymer, a masterbatch, is growing. It allows the accurate dispensing of additives which may be present at very low concentration in the final product. In principle, the polymer in the masterbatch should be similar to the bulk polymer to which it is added, but the use of low-molecular-weight polymers facilitates the mixing of the masterbatch. Low viscosity impedes dispersion, however. There is an increasing tendency to regard masterbatches as "universal," with some manufacturers recommending the use of "alien" polymer masterbatches. Thus low-density polyethylene is employed as a masterbatch carrier for incorporating additives in polypropylene. The incompatibility of these two polymers, coupled with poor mixing, results in heterogeneous texture which, in critically stressed regions, leads to premature failure. Such an inclusion is shown in Fig. 12; the micrograph, taken with crossed polars, shows very clearly the undispersed low-density polyethylene. This contributes to failure in two ways; first, by acting as a stress concentrator, and second, by being a region of low-molecular-weight polymer, which

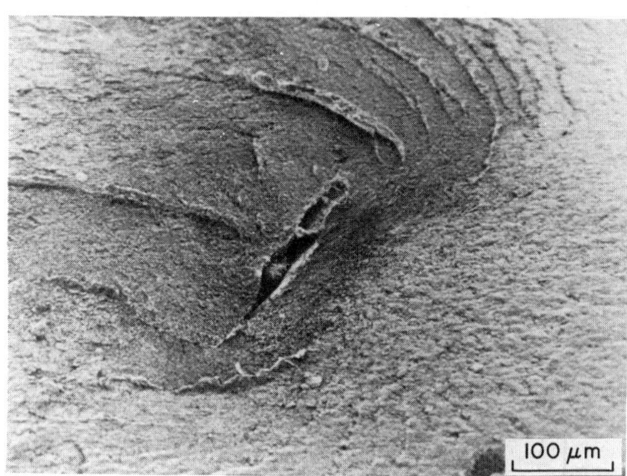

Figure 11
Pigment agglomerate in PVC. Scanning electron micrograph

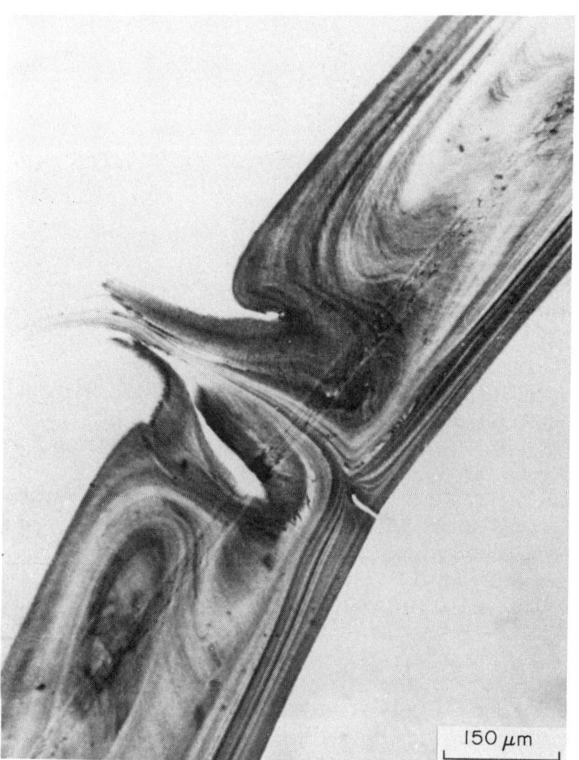

Figure 10
Section of carbon-black-filled polyethylene. Normal light

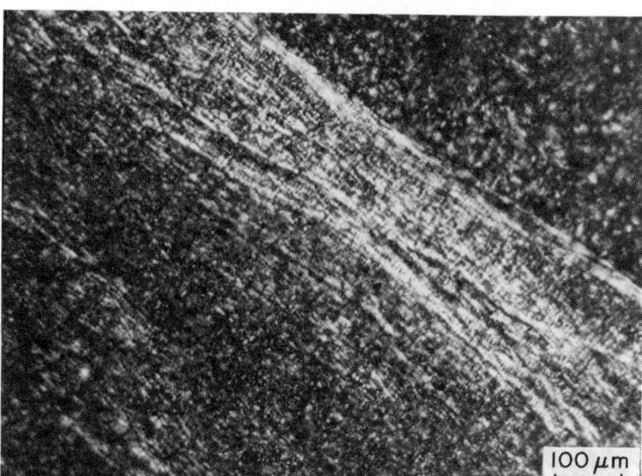

Figure 12
Low-density polyethylene inclusions in polypropylene. Crossed polars

is weaker than the higher-molecular-weight matrix polymer.

(*f*) *Particulate contamination*. This type of failure is similar to that resulting from inadequate dispersion of additives which are incorporated deliberately, except that contaminants are included accidently, often as a consequence of poor working practice. Such inclusions act as stress concentrators, and are effective at sizes of 5–10 μm and upwards. They are thus visible in a light microscope, and this method, together with x-ray dispersive analysis, is used in their identification. Contaminants are frequently inorganic, often minerals, but may be alien polymer, used elsewhere in the plant.

(*g*) *Voids*. Should they occur, voids are obvious stress concentrators, and thereby contribute to premature brittle failure. The origin of voids is frequently associated with poor processing, and their appearance is promoted by parts of thick section. Figure 13 is a micrograph taken in polarized light showing voids in an injection molding of acetal copolymer, giving rise to a crack.

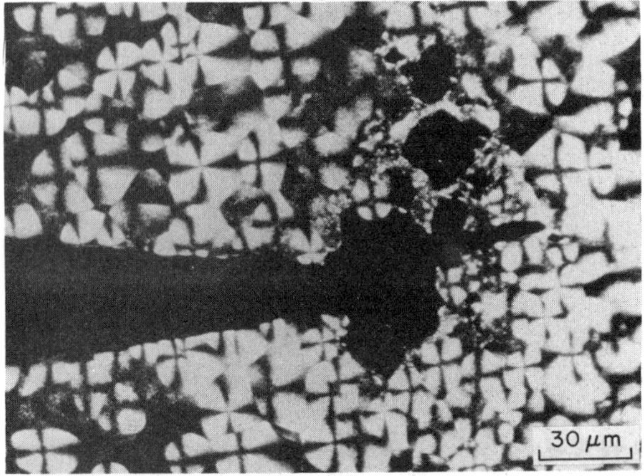

Figure 13
Voiding and crack in injection molded acetal polymer. Crossed polars

(*h*) *Unsuitable cooling conditions*. Blow moldings are often cooled only by their contact with a cold mold which, to reduce cycle time and thereby increase productivity, is chilled in a temperature marginally higher than the dew-point of the surrounding air. For artefacts of considerable thickness, 4 mm and upwards, the cooling time may be a minimum of 2 min, imposed by the low cooling rate of the polymer at the inner surface. Such a slow cooling rate allows the growth of a very coarse texture, exemplified here by an automobile coolant reservoir tank, blow molded in polypropylene. Figure 14 is a scanning electron micrograph of the inner surface of the tank in a region where the thickness was 6 mm, implying a very low cooling rate when the cooling route is by natural convection. Cracks can be seen developing between the spherulites, which have grown to large size, approximately 150 μm.

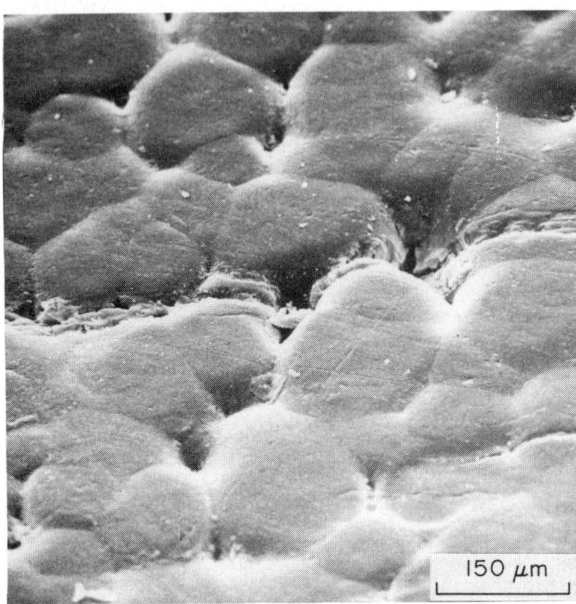

Figure 14
Slow-cooled inner surface of blow-molded polypropylene tank. Scanning electron micrograph

(*i*) *High residual orientation*. High shear rate shaping processes, such as injection molding, inevitably give rise to orientation of the polymer chains; further, the mold is used to cool and solidify the plastic, so that the polymer in contact with the mold surface is solidified rapidly, before there is opportunity for the orientation to relax. This region of high molecular alignment is a fruitful source of cracks, particularly when the injection molding is subsequently remelted in part, as in welding. Figure 15 is a micrograph, taken with crossed polars, of a weld joining two sections of subcomponents injection molded in high-molecular-weight, high-density polyethylene. The very high residual orientation of the original texture can be seen, together with the very confused morphology this leads to in the weld, which itself was near optimum for these subcomponents. To improve the quality of the injection molding, it is necessary to change the processing conditions, increasing cycle times; however, this reduces the profit to the molder, and therefore this change is avoided.

(*j*) *Imperfect internal welds*. When two flow paths meet head on, as they do after passing round an

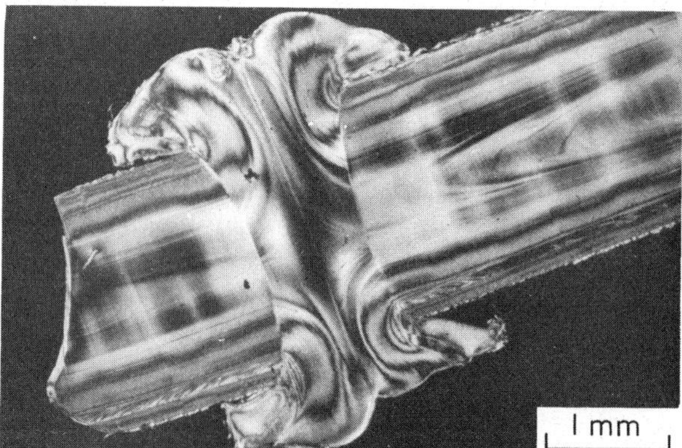

Figure 15
High residual orientation in high-molecular-weight, high-density polyethylene

obstacle, a weak region is generated unless they integrate perfectly, which they do only rarely. Similar problems can arise when flows from multiple gates come together, so that internal welds are common features in injection moldings. Attention must be focused on minimizing their effect, both at the mold design stage and in subsequent processing. Internal welds can be readily identified by microscopic examination of a thin section, usually between crossed polars, and exemplified here by a molding in acetal polymer (Fig. 16). Remedies at the mold design stage include thickening of the section at the expected meeting point, allowing greater opportunity for the mixing of the two flows.

Figure 16
Internal weld in molded acetal polymer

(*k*) *Faults with gates and runners*. The gate is a unique feature in any molding, since it is the region where packing the mold to compensate some of the shrinkage in cooling is likely to lead to a textural hiatus, and consequently a region of stress concentration. This is demonstrated in Fig. 17, which is the gate area of an injection-molded nylon sealing ring. Cracks which initiated brittle failure were generated in this region.

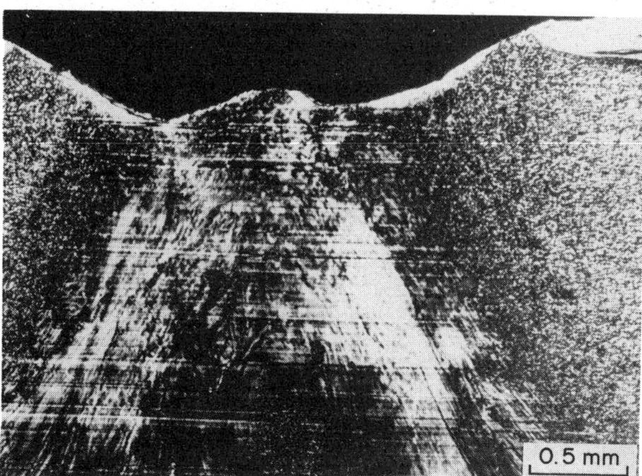

Figure 17
Gate region of nylon injection molding. Crossed polars

(*l*) *Orientation of fillers, especially fibers*. Anisotropic fillers, and especially fibers, react to flow by turning their major axes into a preferential direction, which depends on the nature of the flow (e.g., whether it is converging or diverging). The effect on the properties of the composite is to make them anisotropic, and in extreme cases this can promote failure.

(*m*) *Morphological interfaces*. Most thermoplastics products are of nonuniform texture, a feature which is often caused by the different rates of cooling experienced. Thus, when a cold mold is used in either injection or blow molding, the product has a quenched zone of fine texture adjacent to the mold, with the interior, which is cooled more slowly, exhibiting a coarser texture. The change in texture (and density) may be abrupt and the mismatch appears to give a plane of weakness from which cracks may generate. Figure 18, which is a section taken from blow-molded polypropylene, shows this phenomenon.

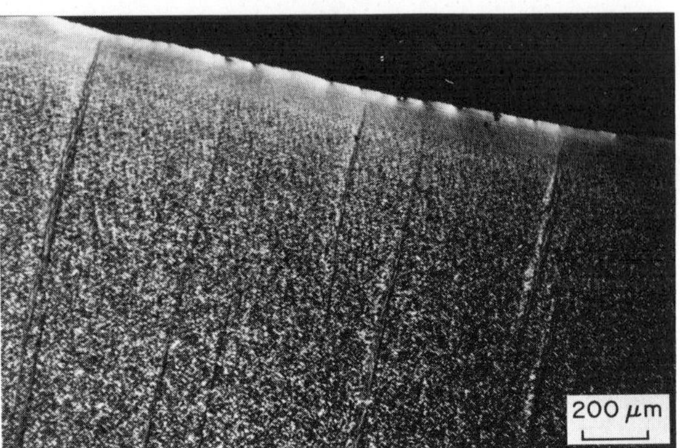

Figure 18
Thin section of blow-molded polypropylene. Crossed polars

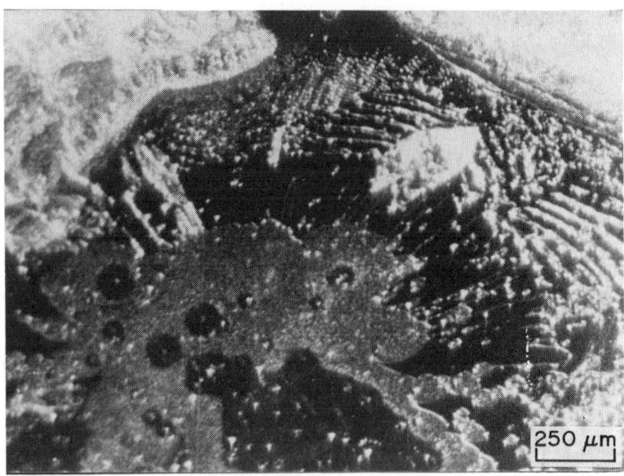

Figure 19
Fatigue crack in polystyrene

2.4 Service Factors

(*a*) *Prolonged loading*. The possibility of brittle failure is increased by increased times under load, the failure occurring apparently at low deformation, although microscopic examination reveals evidence of partial ductility. The classic example of brittle failure brought about by long-term loading can be found in the proof testing of high-density polyethylene pipes; failure of the pipes under internal pressure is characteristically ductile at short times, but at long times, particularly at high temperature, brittle failures at low strain are encountered.

(*b*) *Cyclic loading*. This mode of stressing, often known as "fatigue" or "dynamic fatigue," leads to embrittlement, and is a cause of failure of some materials which are otherwise very durable and can withstand almost unlimited impact abuse. Thus, although acrylonitrile–butadiene–styrene (ABS) is generally regarded as a very tough plastic, finding application where impact stresses are likely, for example, as a cover for an air-cushion grass cutter, it is rapidly embrittled on being subjected to cyclic (alternating) stresses. The illustration in Fig. 19 is a fatigue crack in polystyrene, the micrograph being taken in the dark field, reflected light mode. Examination of the fracture surface reveals the periodic undulations, which record the progress of the crack with each cycle of stress.

(*c*) *Thermal/photochemical degradation*. Most polymers will degrade if heated to sufficiently high temperature, which is usually, however, above the melting or softening point. This behavior, therefore, mainly concerns the processor. Light energy, and especially ultraviolet radiation, is potentially more damaging and, at ambient temperatures, is the more usual cause of degradation. The requirements for photochemical reaction are that the radiation quanta are absorbed by the substrate, and that the quanta contain sufficient energy to cause reaction to occur. Some PVC compounds are susceptible to photochemical attack which reduces the molecular weight of the polymer, embrittling it. Further, volatile products are evolved, decreasing the volume, the two effects together giving severe surface cracking, which is exemplified in Fig. 20, a micrograph obtained by scanning electron microscopy.

(*d*) *Thermal/photochemical oxidation*. As with degradation in service, again the most insidious effects are photochemical in origin; natural weathering causes deterioration in many unprotected plastics, although stabilized grades of most are available and give satisfactory lifetimes. An example is the use of propylene–ethylene sequential copolymer in crates for storing and transporting beers and soft drinks. When not in use, such crates are frequently stored

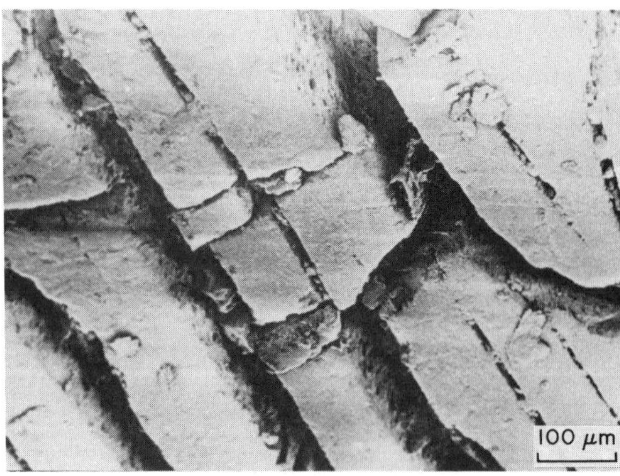

Figure 20
Typical surface cracking of PVC

outside and subjected, therefore, to weathering. Suitably protected grades can withstand such treatment for many years, but unprotected grades degrade and embrittle by photochemical oxidation in a year or two. Oxidized polypropylene residues can be found in the surface layers of such exposed crates; reprocessing them spreads and activates the oxidized products, resulting in products which are far less stable to further thermal or photochemical attack.

(*e*) *Attack by environment.* It is the responsibility of the designer to choose materials for a product which are resistant to the environment in which the product is expected to operate. These "rules" are generally followed, except in the case of water, which is often not considered to be an active environment. Several important plastics react with water, however, particularly at high temperatures, although deterioration at temperatures near the ambient is not unknown. Susceptible polymers include polyamides and polyesters, including polycarbonate, which is taken here as an example.

Polycarbonate (PC) is a tough transparent plastic, with a softening point of 145–150 °C, which implies that it can be sterilized by boiling water or steam techniques. Further, it absorbs only a small amount of water, even after prolonged immersion, and the mechanical properties are affected only marginally by water content. However, water causes degradation of PC by hydrolysis; thus, a baby's feeding bottle, for which PC seems the ideal material, giving lightweight and tough transparent products, embrittles after a few weeks' use, because of degradation by hydrolysis resulting from the repeated sterilization. A similar mechanism degrades nylon when the common practice of boiling completed moldings in water to stabilize the water content, and thereby the dimensions and properties, is followed.

Figure 21
Environmental stress cracking of polypropylene in ethylene glycol–water mixture. Temperature 115 °C

(*f*) *Environmental stress cracking.* Although plastics are generally resistant to chemical attack, and can be loaded over long periods without failure, the combination of certain environments with stress, applied externally or generated internally, or both combined, can induce brittle failure in particular plastics. Polypropylene is a plastic favored for the manufacture of automobile coolant reservoir tanks, although the ethylene glycol added to the water as "antifreeze" is a mild stress cracking agent for the polymer. This can be appreciated by reference to Fig. 21, which shows samples taken from the same regions of three similar tanks. Sample (a) is the control, which was not subjected to any test, while samples (b) and (c) were subjected to the same proof test (identical pressure, time and temperature). For (b) the tank had been filled with a 50:50 mixture of ethylene glycol and water, while for (c) the tank contained water only. The deterioration observed in (b) is typical of this type of failure.

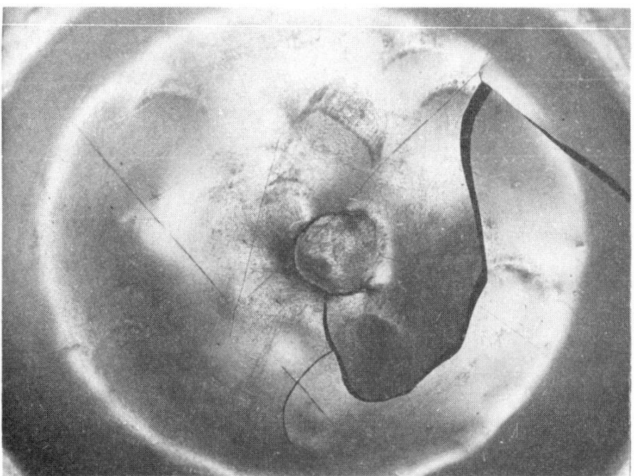

Figure 22
Much abused base of large polyethylene container

(*g*) *Abuse of product.* Most products are designed, and expected, to withstand limited abuse; toys provide many examples where maltreatment is almost inevitable. Hammer blows frequently simulate the main type of abuse; the example in Fig. 22 was hit repeatedly with a heavy hammer, in an attempt to prove that the vessel was brittle! By counting the number of scars which did not cause failure, it can be deduced that it was the twelfth blow which precipitated the failure. Fortunately, there are not so many people about with this destructive urge, so this level of abuse need not be catered for. Nevertheless, the designer of a product must try to foresee the types of abuse most likely to be encountered and to design against failure under such circumstances;

economics will inevitably curtail what can be achieved.

3. Concluding Comment

The foregoing may seem to be a very full and possibly alarming catalogue of the failures that can, and indeed have, affected plastics products, since all those cited have been failures in commercial products. On the pessimistic side, the list is probably incomplete, as further contributors to failure are being identified as more is learned about plastics. On the credit side, many of the known types of failure can now be prevented.

See also: Failure Analysis: General Procedures and Applications to Metals; Mechanics of Materials: An Overview

Bibliography

Birley A W, Scott M J 1982 *Plastics Materials: Properties and Applications*. Blackie, Glasgow

Brown R (ed.) 1984 Polymer Testing—A Special Issue. *Meas. Tech. Polym. Solids* 4 (2–4)

Engel L, Klingele H, Ehrenstein G, Schaper H 1981 *Atlas of Polymer Damage*. Wolfe Science, London

Haslam J, Willis H A, Squirrel D C M 1972 *Identification and Analysis of Plastics*. Butterworth, London

Rodriguez F 1982 *Principles of Polymer Systems*. McGraw-Hill, New York

Tadmor Z, Gogos C G 1979 *Principles of Polymer Processing*. Wiley, New York

D. A. Hemsley; A. W. Birley

Failure of Composites: Stress Concentrations, Cracks and Notches

In designing structural components from brittle fibrous composites (e.g., glass fibers in epoxy), it is assumed that for an acceptable level of survival probability the operating stress does not exceed the strength of the composite. Unfortunately, the situation is more complex in practice. A fibrous composite containing a notch or hole in monotonic loading exhibits premature cracking at stresses significantly lower than the ultimate strength; this may take the form of interfacial shear cracking, delamination or splitting, fiber fracture, matrix cracking and so forth. The formation of a damaged region is due to the concentration of localized tensile and shear stresses close to the notch front. The precise mode of failure depends on the orientation of fibers and stacking geometry of the laminate; it is also sensitive to the properties of the matrix and fiber–matrix interface, and to the stress state and environment.

1. Failure Modes in Monotonic Loading

A typical unidirectional fibrous composite (glass fibers in epoxy) loaded in tension exhibits stable delamination or splitting at the root of a notch. As the split extends parallel to the fibers and direction of applied load, a matrix crack propagates from the notch tip, passing still-intact fibers. A second split nucleates at the crack tip and the crack becomes arrested. Further crack growth in the original notch orientation requires extension of the two splits, followed by fracture of the intact fibers somewhere along their debonded length. The extent of delamination, fiber debonding and fiber fracture in the localized damage zone is related to the applied load. The split that forms can be considered as rendering the sharpest inherent defect or crack effectively equal to a hole or notch (Mandell 1971).

In a monotonic tensile test on a crossply (0°/90°) laminate, the transverse (90°) layers hinder longitudinal splitting at the root of a notch. Those layers perpendicular to the direction of applied load, and adjacent to an interfacial shear crack, may also delaminate slightly during shearing between the two sliding surfaces of a longitudinal split. Premature fracture of the localized 0° load-bearing fibers may lead to subcritical crack growth in the original notch direction. Behind the crack front lies an array of longitudinal, parallel-shear cracks, which mark the position of successive crack arrest points. The length of the delaminations and spacing between them are related in some way to the laminate stacking geometry.

For a quasi-isotropic laminate (0°/ ± 45°/90°), delamination in the 45° layers at the root of a notch reduces the concentration of stress on the 0° plies. An increase in thickness of the 45° layer decreases the constraining effect on the 0° plies, which permits further delamination (Bishop and MacLaughlin 1979). The formation of a damage zone at the notch tip having a lower modulus than the surrounding material can be thought of as decreasing the intensification of localized stress. In effect, the notched strength and fracture toughness of the laminate is increased. For some, perhaps all, laminates containing glass, Kevlar or carbon fibers, the damage zone can be considered as increasing the effective radius ρ of the notch (Potter 1978). The fracture stress of a notched composite σ_f can then be estimated using measurements of ultimate strength σ_u, together with a suitable stress concentration factor K_t:

$$K_t = \frac{\sigma_u}{\sigma_f} = 1 + (c/\rho)^{1/2}\left[2\left(\frac{E_{11}}{E_{22}}\right)\nu_{12} + \left(\frac{E_{11}}{G_{12}}\right)\right]^{1/2} \qquad (1)$$

where c is the length of the notch, E_{11} and E_{22} are the longitudinal and transverse Young's moduli, ν_{12} is the Poisson ratio, and G_{12} is the shear modulus.

At the microscopic level, the breakage of a fiber at the tip of a notch may induce the sequential failure of longitudinal (0°) fibers by the transfer of load from the broken fiber to an adjacent intact fiber. The

Table 1
Typical strength and fracture toughness data for various laminates

Laminate construction[a]	Tensile strength σ_u ($MN\ m^{-2}$)	Measured fracture toughness K_c ($MN\ m^{3/2}$)	Estimated damage zone size C_0 (mm)
High tensile strength carbon fiber–epoxy			
(0°)	1045	90.0	1.86
$(0°/90°)_{2S}$	530	42.1	1.78
$(0°/\pm45°)_{2S}$	460	33.2	1.02
$(0°/\pm45°/90°)_{2S}$	270	32.8	1.49
E-glass fiber–epoxy			
(0°)	1035		
$(0°_2/\pm45°)_{2S}$	600		
$(0°/90°)_{2S}$	540	30.7	
$(0°/\pm45°/90°)_{2S}$	350	24.3	2.54
Boron fiber–epoxy			
(0°)	1325		
$(0°_2/\pm45°)_{2S}$	700	67	2.54
$(0°_2/\pm45°/90°)_{2S}$	420	38.7	2.79
Kevlar 49 fiber–epoxy			
(0°)	1378		
$(0°/90°)_{2S}$	578		
$(0°/\pm45°/90°)_{2S}$	393	~30	
(±45°)	119	~14	
Boron fiber–aluminum			
(0°)	~1000	90	
$(0°/\pm45°)_{2S}$		50	

a The subscript S denotes a stacking sequence symmetrical about the midplane of the laminate; the subscript 2 indicates a laminate constructed from two sets of repeating sequences of laminae

localized concentration of stress on the unbroken fiber is sensitive to the strength of the fiber–matrix interface, and the matrix. For example, a toughened matrix, epoxy dispersed with elastomeric spheres, would be expected to reduce the amount of delamination at the notch front, and this is observed. However, reducing splitting raises the localized stress at the notch tip, and the fracture toughness corresponding to crack propagation perpendicular to the 0° fibers is lowered. The initiation of cracking parallel to the original notch direction may be prevented if, as a result of the stress distribution due to the notch, the initial difference in stress carried by a fiber that has just broken and the adjacent intact fiber is sufficiently large. The interaction between the distribution of stress in the vicinity of a notch tip and the concentration of localized stress in an unbroken fiber adjacent to one that has just failed leads to an effect of notch size on strength and fracture toughness (Potter 1978). The smaller the notch, the more localized the perturbed stress field becomes and the greater is the applied stress required to initiate fiber fracture and transfibrillar crack propagation. It is this subtle balance between the transfer of load to an intact fiber next to a broken fiber, the distribution of flaws in the fiber and the variability of fiber strength, combined with the localized stress field ahead of a notch tip that determines the strength and notch sensitivity of the laminate.

The fracture toughness K_c can be related to the damage zone size C_0 and notch geometry as follows:

$$K_c = \sigma_u[\pi(C + C_0)(1 - \xi^2)]^{1/2} \qquad (2)$$

where $\xi = C/(C + d_0)$ and d_0 is a critical distance ahead of the discontinuity (Nuismer and Whitney 1975).

The concept of a damage zone is invoked so that K_c reaches a maximum value as the notch length C becomes very small, that is, as $K_c \rightarrow \sigma_u(\pi C_0)^{1/2}$. Typical values of K_c, σ_u and C_0 for various laminates and stacking geometries are listed in Table 1.

2. Cyclic Failure and Residual Strength

Load cycling of fibrous composites containing brittle fibers, carbon fibers or glass fibers in epoxy, for example, brings about fiber–matrix decohesion and delamination at notches and inherent flaws. These damage zones are larger than those induced in monotonic loading. The nucleation and growth of damage zones in cyclic failure reduces the localized stress, together with a corresponding improvement in residual strength. The residual strength of a laminate depends on the stress level, and increases with time in cyclic loading, eventually reaching the unnotched or inherent strength in monotonic fracture. Fatigue lifetime and residual strength are therefore affected by interactions between microstructure, distribution of flaws, and the formation of damage zones, shear cracking and so forth. The micromechanisms of failure are sensitive to the chemistry of the resin and nature of the fiber–matrix bond. Extrinsic variables, such as temperature, humidity, and time in an aging experiment, are likely to have considerable influence on the properties of the fiber–matrix interface, and therefore on residual strength and fatigue lifetime of the laminate.

See also: Multiple Fracture; Strength of Composites; Composite Materials: An Overview

Bibliography

Bishop S M, McLaughlin K S 1979 *Thickness Effects and Fracture Mechanisms in Notched Carbon Fibre Composites*, Royal Aircraft Establishment Technical Report 79051. HMSO, London

Mandell J F 1971 Fracture toughness of fiber reinforced plastics. Ph.D. thesis, Massachusetts Institute of Technology

Nuismer R J, Whitney J M 1975 Uniaxial failure of composite laminates containing stress concentrations. *Fracture Mechanics of Composites*, ASTM Special Technical Publication 593. American Society for Testing and Materials, Philadelphia, Pennsylvania, pp. 117–42

Potter R T 1978 On the mechanism of tensile fracture in notched fibre reinforced plastics. *Proc. R. Soc. London, Ser. A* 361: 325–41

P. W. R. Beaumont

Fast Firing of Ceramics

As costs of fuel and raw materials continue to rise, the ceramic industry is faced with the necessity of combating the increase. To mitigate the problem of rising production costs, a method of firing ceramic ware has been developed which requires only 60 min (floor and wall tiles) or 2 h (dinnerware) from heating to cooling and which ensures that the quality of the finished ware is comparable with or in many cases superior to that of ware fired in a conventional manner. The replacement of normal kiln firing schedules with these newer firing programs has already proved economically advantageous in the production of ceramic ware.

The design of the kiln and the kiln furniture are new developments which allow energy conservation. Special consideration must be given to the selection of refractory materials for the insulation of the kiln as well as the transport system and the method of heating the kiln. The components of the charge, or batch, must be carefully chosen and milled to ensure the completion of phase transformations and reactions during fast firing, so that the required properties of the ceramic body are attained. The method of fast firing has the potential to save 20–25% of the energy required by normal firing procedures used in recent decades.

1. Definition of Fast Firing

The terms fast firing or rapid firing need precise definition in relation to conventional firing. Fast firing of sanitary ware (12–17 h) takes more time than that of dinnerware (2 h), that of wall or floor tiles (1 h) or finally the sintering of special alumina under laboratory conditions (several minutes). Fast firing, although frequently taken to refer solely to the firing process itself, should be understood to include the interdependence of batch composition with its grain size distribution and mineral composition, forming and shape of the body, drying, firing, finishing, and the resulting properties of the product. The firing temperature and time, the kiln design, the furniture, the stacking of the ware and the fuel consumption are further factors. Many problems of fast firing are problems mainly because of insufficient scientific and technical understanding of the phenomena involved.

2. General Aspects

2.1 Batch Composition

In general the body formulation depends on the kiln to be used. To ensure complete reaction during firing the particle size distribution of the raw materials in the batch must be very fine and uniform in most cases. This requires expensive milling and/or raw materials of higher quality. On the other hand, the maximum firing temperature required for complete sintering of the body is reduced, lowered by the use of finer raw materials.

2.2 Drying

For fast firing the water content of the batch must be very low. For floor and wall tiles, which need only 45–60 min for the whole cycle, the water content should be <0.5%. For porcelain (dinnerware) it must be <1%. Therefore, dry pressing methods are preferable. In the porcelain industry the isostatic dry pressing of plates and cups instead of any other technique which uses raw material of higher water content has gained new importance as a means of saving energy. Special preheating units upstream of the tunnel kiln in the tile industry prevent the formation of a black core, which had become a serious problem in fast firing of tiles and split tiles. However, high-speed dryers for wall tiles, for instance, need $500\,kJ\,kg^{-1}$ compared with $1700\,kJ\,kg^{-1}$ for the firing—a highly ineffective ratio.

2.3 Firing Schedule

The rapid increase in temperature, producing a greater temperature difference between the kiln and the core of the body than in conventional firing, may result in stresses in the body. To reduce the reject rate, time must be allowed for completion of the reaction in distinct temperature regions during heating (oxidation and burning out of organic impurities, dehydration of clay minerals, quartz inversion from low to high quartz and other reactions) and also during cooling (quartz inversion from high to low quartz). In regions where no reaction takes place and no stresses can occur, rapid heating is possible to conserve time and energy.

These considerations determine a modified firing schedule which depends on the nature of the batch. In addition, no long soaking period is needed, since the higher temperature permits completion of the reactions in a short time. The conventional firing schedule, in contrast, shows a linear temperature increase to a maximum, a constant temperature at the maximum for several hours and linear cooling. Figure 1 shows schematically the fast firing schedule in comparison with conventional firing.

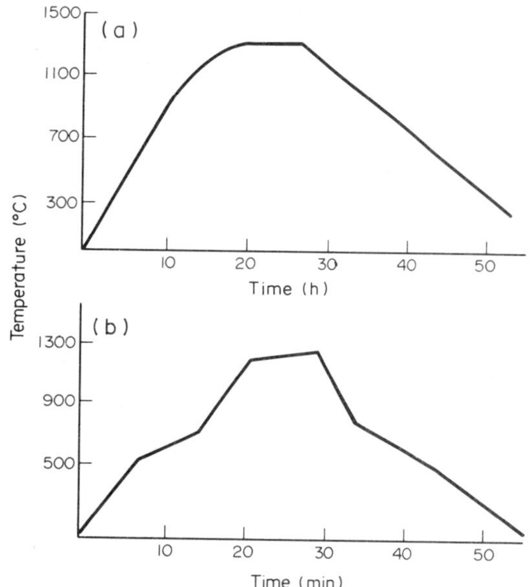

Figure 1
Schematic firing schedules for (a) conventional and (b) fast firing

2.4 Kiln Design and Kiln Furniture

Fast firing technology differs markedly from conventional firing in kiln design and kiln furniture as well as in the burners used and in the stacking of the ware (Harms 1979a,b). In conventional kilns as much ware as possible is packed tightly in capsules of heavy refractories. The burners are situated in the firing zone, with the natural result that there is a temperature gradient to both ends of the kiln. During slow firing the temperature gradient between the kiln and the core of the body is small. The ratio of refractories to ware is ~6:1.

In fast firing plants on the other hand, as few refractories as possible are used, so the heat is transmitted directly to the ware without wasting energy in warming up refractories. The firing of ware with or without small quantities of refractories is made possible by the new kiln design, which permits the firing of single layers or a maximum of three layers. The ratio of ware to refractories can therefore be reduced to 1:1 or less.

The reduction of refractories also benefits the units for transportation through the kiln. Instead of the heavy cars needed to transport heavy stacks of refractories surrounding the ware, light cars are now possible, or roller conveyors made of ceramics or special alloys (Inconel), or pusher slides and skids. Which transport system is used depends on the type of product to be fired and on the firing temperature.

Faster firing became possible only after much experience had been gained in transmitting sufficient heat to the product uniformly and at accelerated rates. In modern kiln design, burners are also mounted in the early stages of the preheating zone. They are supported by additional air inlets to obtain the necessary volume and circulation. High-velocity burners and light insulating materials are used in the kilns. The entire kiln construction is much lighter than that of a conventional kiln. The walls have less than half the heat storage and the thermal conductivity is also appreciably lower. The use of such insulating refractories results in a great improvement in flexibility of kiln operation. This in turn permits better adjustment and better control of the temperature profile. Figure 2 shows schematic cross sections of a fast firing tunnel kiln and a classical tunnel kiln with their cars.

The kiln furniture in fast firing has to fulfill several requirements; in particular, the refractories must have a high thermal shock resistance, high strength, low weight and a long lifetime. For firing wall tiles and floor tiles, refractories are available which fulfill most of these requirements adequately. For the required temperatures of 1140–1300 °C several materials are available, for instance those containing

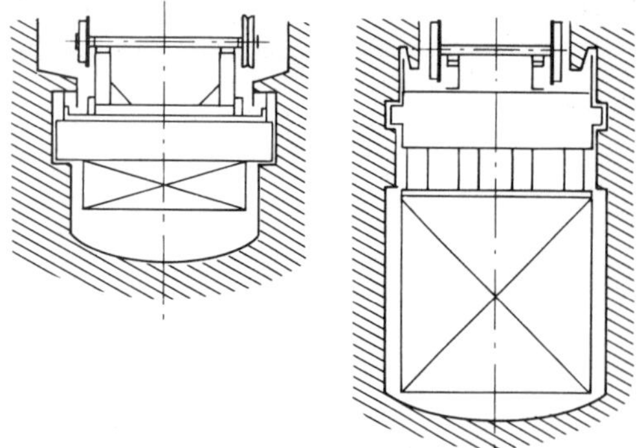

Figure 2
Schematic comparison of (a) a modern tunnel kiln with (b) a fast firing kiln with special cars

mullite and cordierite. At present only recrystallized silicon carbide and silicon nitride refractories are in use above 1300 °C, where they show some promise. Their lifetime should be at least 1000 cycles to be economical, but currently only several hundred cycles are achievable.

The bulk and shape of the kiln furniture also play an important role. The refractories should not shadow parts of the ware to be fired, otherwise warping of plates, for instance, will result from the temperature differences between different parts of the body. Figure 3 shows kiln furniture used in conventional and fast firing.

3. *Fast Firing of Tiles*

Different firing units are available, depending on the type of tile. Roller kilns with heat-resistant steel rollers for firing temperatures up to 1140 °C work without slabs (ceramic support plates). The main advantage of this type is the complete absence of slabs and the number of burners, which permit rapid adjustment of the firing schedule if the product changes. Such kilns are constructed to a length of 70 m with a short predryer. Their output is ~1500 m^2 per day. These kilns are particularly suitable for fast firing of wall and floor tiles (water absorption >3–4%). Tiles whose edges have been glazed cannot be processed in such kilns, as they may touch and stick together.

Other roller kilns use ceramic rollers for firing temperatures up to 1200 °C without slabs. Preheating and firing zones are equipped with upper and lower burners which are combined in groups. Each group is controlled by temperature regulators.

Vitrified tiles, which are fired at higher operating temperatures without slabs, become plastic in the firing zone and tend to adhere slightly to the rollers. This can be avoided by spraying the underside of the tile with kaolin. This, however, creates subsequent problems with the on-site fixing of the tiles, because the kaolin coating prevents a satisfactory bond with the fixing cement.

Figure 3
Kiln furniture for fast firing: (a) for firing temperatures <1300 °C; (b) in comparison with a capsule for faster firing

Ceramic tiles have indentations on their underside in contact with the rollers. The resulting abrasions are fired onto the rollers and cause irregularities on them. When the size of tile is altered, the underside of the new tiles may no longer correspond to the earlier size and they become misaligned by running over the irregularities on the rollers, which therefore have to be cleaned or replaced.

Other roller kilns use slabs. These kilns are used up to 1250 °C. The slabs have webs on which the product sits, which have to be cleaned and reground if they become contaminated by glaze. These slabs must be highly resistant to temperature changes. The extruded type has proved more satisfactory than the cast variety. To obtain greater capacities it is necessary to arrange several tracks above one another and side by side (up to 24 tracks).

Other systems use single and double track kilns for

firing temperatures up to 1300 °C, with special kiln cars. The single track kiln has an upper tier in which the product is fired and a lower return channel in which the fired material is conveyed to the unloading machine. Special raising and lowering platforms are fitted at the kiln entrance and exit which transfer the kiln cars from the upper to the lower channel and vice versa. The furniture on the cars is a mullite–cordierite material which is very durable at these temperatures, lasting up to 10 000 cycles. Such kilns can fire 3500 m^2 of white body earthenware with a water absorption of <1% in a cycle of 70 min at 1250 °C maximum temperature in 24 h. The energy consumption is 4200 kJ kg^{-1} of fired material.

The tiles are loaded and unloaded by special machines. There are two different systems: in one, the tiles are stored in the setter, and in the other there are no separate setters but instead the storage frames themselves are designed to accommodate the individual tiles.

Easily fusible glazes are used for rapid firing; these generate emissions when they contain lead and selenium. If the lead content is >7% it causes the formation of droplets on the roof inside the kiln. These droplets fall not only onto the product and spoil it but also onto the rollers. The support plates then stick to the rollers and are prevented from moving, making it necessary to remove them manually. This means that all rapid firing techniques require modification of normal tunnel kiln glazes if no slabs or cars are used. In kilns with cars and/or slabs, the normal body formulations can be used.

Rapid firing kilns discharge much less fluorine than tunnel kilns. This is a significant advantage in relation to occupational health and environment protection.

In all the types of rapid firing kiln the cycle time and therefore the kiln output are restricted by the formation of black cores in the tiles. The formation of these cores can be avoided by predrying or prefiring and in this way a higher output can be achieved.

The features of the different rapid firing systems are summarized in Table 1. Figure 4 shows a kiln for firing tiles. A good review on this topic is given by Speer (1981).

4. Fast Firing of Porcelain

Kilns used for fast firing of porcelain are similar to those used in the tile industry. Roller kilns with slabs, kilns with special cars or skid-type kilns may be used. Fig. 5 shows a typical example.

Today the fast firing of porcelain is possible for hotel ware, especially for hollow ware and plates up to a diameter of 19 cm. Biscuit firing and also glost firing are common. The firing of inglaze and onglaze decors is possible and is generally used without serious problems. In fast firing, the onglaze technique leads to a remarkable reduction in the leachability of

Table 1
Features of different rapid firing systems

	Roller hearth kiln			
	Ceramic rollers with slabs 1	Ceramic rollers without slabs 2	Steel rollers without slabs 3	Channel kiln with special kiln cars with slabs 4
Maximum firing temperature (°C)	1250	1200	1140	1300
Lowest possible water absorption in the product (%)	<1	<1	3	<1
Size of the product	limited to certain sizes	minimum size 200 × 200 mm		limited to certain sizes
Firing of tiles with glazed edges	possible	not possible	not possible	possible
Lead content of the glaze (%)	<17	<7	<7	optional
Supervision	intensive	more intensive than 1		insignificant
Difficulties specific to kiln	wear of rollers and slabs	wear and contamination of rollers; tiles tend to stick on the rollers and turn off course		none
Consumption of fuel	approx. as 4 but higher than 2 and 3	lower than 1 and 4		approx. as 1 but higher than 2 and 3
Space necessary	1, 2 and 3 need an extra space equal to the width of the kiln			normal

Source: Speer (1981)

Figure 4
Fast firing kiln for firing tiles

lead and cadmium. This is why fast firing is now common for the decoration of porcelain. Special glazes and colors have been developed for this particular application. Decoration with gold is also carried out successfully.

The fast firing of smaller objects of technical porcelain should be free from problems, as shown by Mörtel (1977a,b, 1978). During firing, the reaction sequence differs from that in conventional firing. Depending on the rate of temperature rise, the evaporation of the chemically bonded water, which is liberated between 585 and 695 °C, may sometimes be incomplete, so that the water content in the microstructure rises. This enhances the formation of melt at lower temperatures, possibly <900 °C. The attack by the melt on the quartz is slight or undetectable. In a batch which contains more fluxing agents, the formation of cristobalite surrounding quartz crystals may be detected. The melt also does not affect the primary mullite. Secondary mullite grows only from the melt, which is formed from the original feldspar or similar fluxing agents.

Pores in fast firing are normally not well developed. They show sharp edges and they are badly rounded and much more numerous and finer than in conventionally fired porcelain. The quartz crystals appear to be dispersed due to the rapid heating and cooling. The microstructure is apparently not mature. The aggregates of primary mullite suggest that the clay minerals have exploded during dehydration. The crystals formed are very small and poorly crystalline. The secondary mullite which grows from the melt is also of very small grain size; on the one hand this improves the strength and thermal shock resistance of the material, but on the other hand it reduces the transparency. The strength is normally 30–40% higher than that obtained by conventional firing. Figures 6 and 7 illustrate the differences in microstructures.

Figure 5
Fast firing kiln for decorating dinnerware

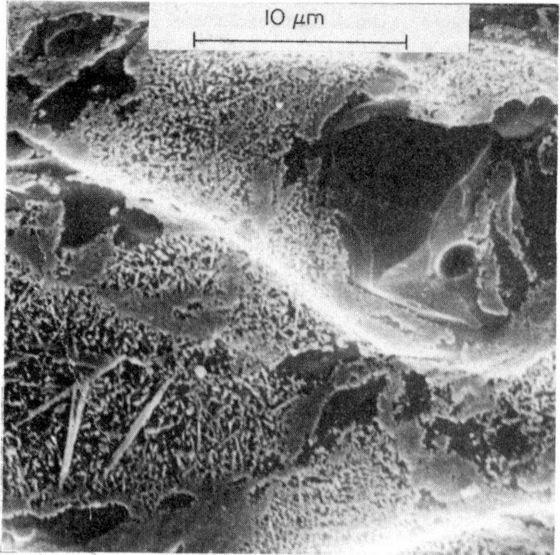

Figure 6
Scanning electron micrograph of a fractured porcelain surface conventionally fired, showing a quartz grain with a rounded pore, primary mullite (fine gray areas) and coarse-grained secondary mullite needles (dark areas are glassy phase)

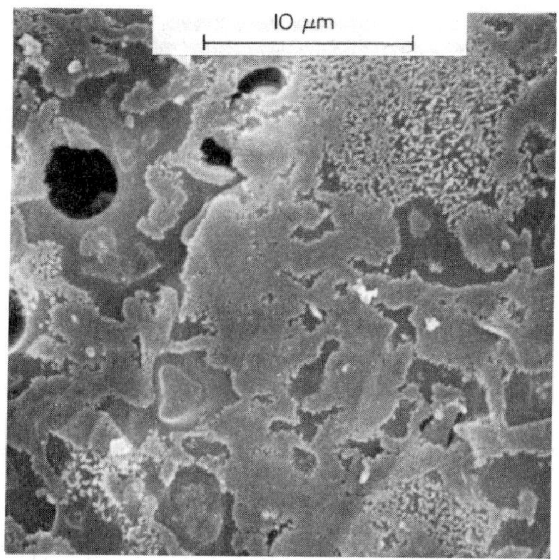

Figure 7
Scanning electron micrograph of fractured porcelain, fast-fired, showing some well rounded and small poorly rounded pores in a light gray matrix of glassy phase, very fine primary mullite (dense gray areas) and secondary mullite (coarser light grains)

Other properties are also affected. The strength during firing is normally reduced as a result of the low viscosity of the melt which is formed, particularly in the early stages of sintering. The addition of grog or alumina alleviates this problem and raises the strength (and thermal shock resistance). The transparency is then reduced. The whiteness is equal to that obtained by conventional firing or can be made to approach that of conventionally fired ware. Large plates, for instance, conventionally fired, show a different whiteness to hollow ware containing the identical glaze, fired in fast firing kilns. Small changes in glaze composition for the fast fired hollow ware result in a color identical to that of conventionally fired ware. The stress of the glaze on the body is reduced, however, making it impossible to raise the strength of a body by glazing to a large degree. The linear thermal expansion coefficient is reduced. The leaching of lead is reduced and leaching of cadmium is undetectable.

The use of clays instead of kaolin permits a reduction in the firing temperature to 1300 or 1200 °C instead of the 1400 °C or more which is normally required for fast firing of porcelain. The dehydration of the clay minerals starts at 200 °C. This leads to the formation of a melt at ~800 °C and explains the reduced emission of fluorine.

As fluxes, glass as well as wollastonite, albite, anorthite, nepheline syenite and other agents can be used instead of potassium feldspar, normally in mixtures of at least two components except for nepheline syenite, which may be used alone. These fluxes also reduce the firing temperature needed; however, the temperature range is not shortened as much in fast firing as in conventional firing (Greines et al. 1984).

Another possibility of reducing the firing temperature lies in reducing the particle size of the raw materials by milling or by hydrocyclone separation. Very finely sized batches of the classical mineral composition, which consists of 50% kaolin, 25% quartz and 25% potassium feldspar, can be fired to a high-quality porcelain at 1300 or even at 1200 °C. These porcelains are characterized by a combination of very high strength (two or three times that of conventionally fired porcelain) and high thermal shock resistance, as well as good electrical properties, which are equal to or better than the standard values. The quartz is completely dissolved, thus increasing the viscosity of the melt during firing (Mörtel 1977a, 1977b, 1982).

The energy saving in firing porcelain is less than in firing tiles, only 10–20% being possible, but the investment costs are reduced, since smaller buildings and shorter kilns can be used. In addition, no storage space for capsules and other kiln furniture is needed, because all the kiln furniture remains on the transporting system. The manpower required is also reduced substantially.

5. *Fast Firing of Sanitary Ware*

Faster firing of sanitary ware necessitates the development of new batch compositions and special kilns with car transport systems. The particle size of the batch must be fine and very uniform. Firing times of 17 h at 1230 °C are reported by Kukusheva et al. (1979). One factory which fires the ware 12 h only, uses naturally extremely fine raw materials.

In sanitary ware, thick walls of the body are common and the temperature gradient between the kiln and the core of the body becomes a serious factor. In regions where reactions take place, time must be allowed for their completion. This applies especially to outgassing reactions. Otherwise the glaze may close and pinholes, at the very least, or more serious flaws may consequently occur. This problem becomes serious when raw materials containing organic matter are used.

6. *Fast Firing of Advanced Ceramics*

For fast firing of ferrites special kilns are needed, which allow careful control of the protective firing atmosphere. These kilns therefore have special seals at the inlet and outlet.

Otherwise in the field of advanced ceramics no firing on an extended industrial scale has been reported in the literature. Some papers deal with the scientific possibility of sintering alumina or zirconia within a few minutes to dense bodies of high strength, combined to some extent with transparency or translucency in the case of alumina (Harmer 1979, 1981). Leaver and Mistler (1972) report the use of a gas-fired roller hearth kiln for fast firing of high-alumina substrates. The firing cycle is 2½–7 h at a peak temperature of 1550 °C. The kiln is used to fire tape-cast 99.5% Al_2O_3 for thin film applications. In the hot zone, rollers of silicon carbide with a thin mullite coating to increase oxidation resistance are used.

A purely scientific publication which deals with the ultrarapid sintering of β-Al_2O_3 within 15 s to 99% of the theoretical density, using a gas plasma, is the monograph by Johnson et al. (1981).

See also: Ceramic Tile Manufacture; Dinnerware; Firing of Ceramics; Kilns for Firing Advanced Ceramics; Ceramics Process Engineering: An Overview

Bibliography

Greines M, Meger A, Mörtel H, Wolfes M 1984 Influence of the fluxes on the properties of fast fired porcelain. *Sci. Ceram.* 12: 125–130

Harmer M, Roberts E W, Brook R J 1979 Rapid sintering of pure and doped Al_2O_3. *Trans. J. Br. Ceram. Soc.* 78: 22–25

Harmer M, Brook R J 1981 Fast firing—Microstructural benefits. *Trans. J. Br. Ceram. Soc.* 80: 147–49

Harms W 1972 Recent experiences in fast firing ceramic products. *Sprechsaal* 105: 765–71

Harms W 1979a New results on the use of kiln furniture for fast firing processes. *Ber. Dtsch. Keram. Ges.* 56: 223–26

Harms W 1979b the infra-structure of ceramic production as seen from kiln design and application. *Ceram. Ind. J.* 88: [1015] 26–27, 29–30

Johnson D L, Kim J S, Lynch D C, Sanderson W B 1981 Ultra rapid sintering of ceramics. In: *Materials Science Monographs*, Vol. 4, *Sintering—New Developments*. Elsevier, Amsterdam

Kukusheva M, Radkova A, Nedev V 1979 Preliminary study of accelerated firing of sanitary porcelain. *Stroit. Mater. Silik, Promst.* 20: 17–19

Leaver N E, Mistler R E 1972 A gas-fired roller hearth kiln for fast firing high alumina substrates. *Ceram. Bull.* 51: 845–46

Mörtel H 1977a Influence of the batch composition on the reaction behaviour and properties of fast-fired (2 h) porcelain. *Sci. Ceram.* 9: 84–91

Mörtel H 1977b Porcelain for fast firing. *Ceramurgia Int.* 3 (2) 65–69

Mörtel H 1978 Influence of the processing parameters on the properties of rapid fired porcelain. *Mater. Sci. Res.* 11: 253–62

Mörtel H 1982 Mechanical and electrical properties of fast fired porcelain (2 hours, 1300 °C). In: *Materials Science Monographs*, Vol. 14, *Sintering—Theory and Practice*. Elsevier, Amsterdam pp. 463–69

Speer M 1981 Rapid firing plants. *Interceram.* 30(2): 106–09, 187–89

H. Mörtel

Fatigue Crack Growth: Macroscopic Aspects

Fracture by the progressive growth of incipient flaws under cyclically varying loads, that is, metal fatigue, must now be considered as the principal cause of in-service failures in engineering structures and components. However, the process of fatigue failure itself consists of several distinct processes including initial cyclic damage (cyclic hardening or softening), formation of an initial "fatal" flaw (crack initiation), macroscopic propagation of this flaw (crack growth) and final catastrophic failure or instability. The purpose of this article and the article *Fatigue Crack Growth: Mechanistic Aspects* is to focus attention primarily on the macroscopic crack propagation stage of fatigue failure, and to attempt to relate the microscopic (metallurgical) and macroscopic (continuum mechanics) descriptions which have been presented for this process.

The physical phenomenon of fatigue was first seriously considered in the mid-nineteenth century when widespread failures of railway axles in Europe prompted Wöhler in Germany and Fairbairn in England to conduct the first systematic investigations into the fracture of materials under cyclic stresses around 1860. In 1917 the first reported observations of corrosion fatigue were made by Haigh concerning the effect of seawater on the failure of steel cables. However, the main impetus for research directed at the crack propagation stage of fatigue failure, as opposed to mere lifetime calculations, did not occur until the mid-1960s when the concepts of linear elastic fracture mechanics and so-called "defect-tolerant design" were first applied to the problem of subcritical flaw growth. This approach recognizes that all structures are flawed, and that cracks may initiate early in service life and propagate subcritically. Lifetime is then assessed on the basis of the time or the number of loading cycles for the largest undetected crack to grow to failure, as might be defined by an allowable strain, or limit load or fracture toughness (K_{Ic}) criterion. Implicit in such analyses is that subcritical crack growth can be characterized in terms of some governing parameter (often thought of as an effective "crack driving force") which describes local conditions at the crack tip yet may be determined in terms of loading parameters, crack size and geometry. Linear elastic and elastic–plastic fracture mechanics have, to date, provided the most appropriate methodology for such analyses to be made, and consequently, considerable effort has been directed towards defining parameters which uniquely describe stress and strain fields at the crack tip over length scales characteristic of the local fracture mechanisms involved.

The concept of directly applying fracture mechanics to subcritical fatigue crack growth was first suggested by Paris et al. (1961) in their famous

"Rational Analytic Theory of Fatigue." Despite difficulties in finding a technical journal which would publish the manuscript, their proposal of correlating fatigue crack propagation rates (da/dN) with the stress intensity factor (K_I) has remained the basis of the defect-tolerant fatigue design approach ever since.

The current widespread acceptance of the fracture mechanics approach to fatigue, as utilized in defect-tolerant design philosophies, has meant that a major emphasis in fatigue research in recent years has been directed towards the study of macroscopic crack propagation, particularly with regard to the collection of engineering materials data on cyclic crack growth rates. The macroscopic aspects of such fatigue crack propagation behavior are reviewed in this article.

1. Fracture Mechanics Characterization

For cracks subjected to cyclically varying loads, maximum and minimum stress intensity factors, K_{max} and K_{min}, defined at the extremes of the cycle, are assumed to prescribe crack growth. According to the original analysis of Paris and his co-workers, the crack growth increment per cycle in fatigue (da/dN) can be described in terms of a power law function of the range of K_I, given by the alternating stress intensity $\Delta K(=K_{max} - K_{min})$, thus:

$$\frac{da}{dN} = C\Delta K^m \tag{1}$$

where C and m are scaling constants for a particular material–heat-treatment condition. It is important to note here that such an asymptotic continuum mechanics characterization does not necessitate detailed quantitative microscopic models to be known for the individual fracture events, and in view of the complexities of these processes on the microstructural scale, this must be regarded as fortunate, at least for a macroscopic description of fatigue crack extension.

One of the principal limitations in this approach (and, in fact, the criterion that K_I or ΔK are valid descriptions of crack tip fields) is that a state of small-scale yielding (ssy) exists, as dictated by the size of the crack tip plastic zone in relation to the overall dimensions of the body. For a cyclically-stressed elastic–perfectly-plastic solid, plastic superposition of loading and unloading stress distributions can be used to compute the extent of plastic zones ahead of a fatigue crack (Fig. 1). On loading to K_{max}, a monotonic or maximum plastic zone is formed at the crack tip of dimension:

$$r_{max} \approx \frac{1}{2\pi}\left(\frac{K_{max}}{\sigma_y}\right)^2 \tag{2}$$

where σ_y is the yield strength. However, on unloading

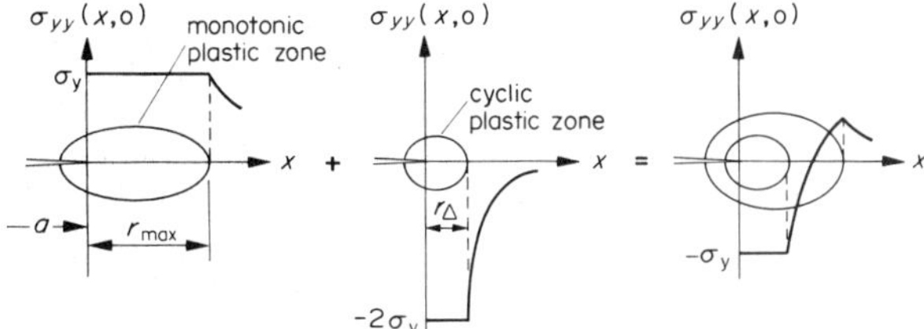

Figure 1
Plastic superposition of loading and unloading stress distributions for an elastic–perfectly-plastic solid of flow strength σ_y during fatigue crack propagation used to estimate the extent of the maximum plastic zone (r_{max}) and cyclic plastic zone (r_Δ). $\sigma_{yy}(x,0)$ is the maximum principal stress directly ahead of the crack (after Rice 1967)

from K_{max} to K_{min}, superposing an elastic unloading distribution of maximum extent $-2\sigma_y$, results in a region ahead of the crack tip where residual compressive stresses of magnitude $-\sigma_y$ exist. This region is known as the cyclic plastic zone size r_Δ and is approximately one fourth the size of the monotonic zone, i.e.:

$$r_\Delta \approx \frac{1}{2\pi}\left(\frac{\Delta K}{2\sigma_y}\right)^2 \tag{3}$$

where, strictly speaking, σ_y is now the cyclic yield strength. The correlation of K_I (or ΔK) to crack extension by fatigue will thus be a valid approach provided that r_{max} and r_Δ are small compared to the in-plane dimensions of crack length and ligament depth.

In situations where the above conditions are not met, such as for fatigue crack growth under large-scale yielding at high stress intensity ranges or for very short cracks, alternative macroscopic descriptions have been proposed based on elastic–plastic (nonlinear) fracture mechanics. One such approach is based on the J integral, where despite difficulties in its precise meaning as applied to cyclically stressed nonstationary cracks, where regions of elastic unloading and nonproportional loading exist, some authors have proposed a power-law correlation of the fatigue crack growth rate under elastic–plastic conditions to the range of J, i.e.,

$$\frac{da}{dN} \propto \Delta J^{m'} \tag{4}$$

as discussed in Sect. 4.

An alternative approach to elastic–plastic fatigue crack growth, which is not necessarily limited by the problems associated with the nonlinear elastic definition of J (i.e., based on deformation theory of plasticity), can be derived using the concept of crack-tip opening displacement (CTOD), δ_t. For proportional loading:

$$\delta_t = Jd(\varepsilon_y, n)/\sigma_y \quad \text{(elastic–plastic)}$$
$$\propto K_I^2/\sigma_y E' \quad \text{(linear elastic)} \quad (5)$$

where E' is the appropriate elastic modulus and d is a proportionality factor dependent upon the yield strain ε_y, the work hardening exponent n, and whether plane strain or plane stress is assumed. Since, like J, δ_t can be taken as a measure of the intensity of the elastic–plastic crack tip fields, it is feasible to correlate the rate of fatigue crack growth to the range of δ_t, the cyclic crack tip opening displacement (ΔCTOD), as

$$\frac{da}{dN} \propto \Delta\text{CTOD} \quad \text{(elastic–plastic)}$$
$$\propto \frac{\Delta K_I^2}{2\sigma_y E'} \quad \text{(linear elastic)} \quad (6)$$

Approaches based on J and δ_t are fundamentally equivalent for proportional loading situations, and can be converted to K_I-based analyses for linear elastic behavior. However, unlike J and K_I, the CTOD approach perhaps offers more physical insight into the mechanisms of crack advance since it can be more readily related to the physical crack tip failure processes involved.

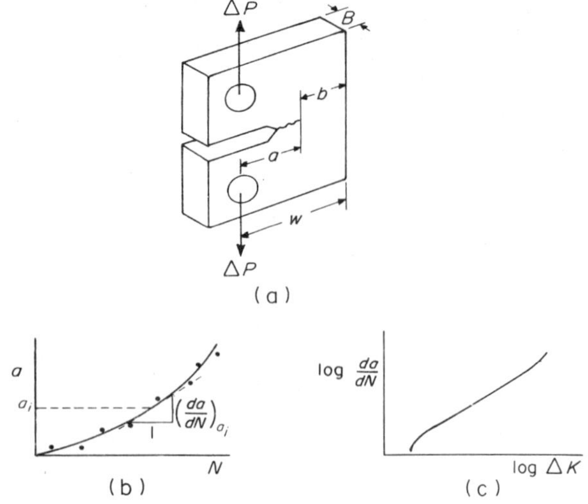

Figure 2
Schematic illustration showing procedures for measuring fatigue-crack propagation rates: (a) compact specimen stressed under cyclic loads ΔP, (b) crack length a versus number of cycles N curve differentiated to give growth rate (da/dN) at particular crack length a_i, and (c) resulting log–log plot of da/dN vs alternating stress intensity ΔK

2. *Experimental Measurement*

Fatigue crack propagation rate data are generally measured using standard fracture mechanics-type specimen geometries, such as the compact tension, edge-notched bend, edge-notched tension, center-cracked sheet test-pieces, and so forth. Starting from a mechanically sharpened crack, specimens are subjected to a cyclically varying load ΔP, generally under constant load control (increasing stress intensity K_I), and the increase in crack length is monitored as a function of time or number of cycles N at constant frequency. Numerous crack monitoring techniques have been employed, such as optical procedures using calibrated travelling microscopes or high-speed photography, electrical procedures using eddy current probes or resistivity measurements (electrical potential techniques), compliance procedures using strain gauges or crack mouth displacement gauges, and ultrasonic or acoustic emission detectors. Data are used to construct crack length a versus number of cycles N plots, which are then differentiated, either graphically or numerically, to determine the rate of crack growth $(da/dN)_{a_i}$ for each crack length a_i (Fig. 2). Corresponding to each crack length a_i, the value of $(\Delta K)_{a_i}$ is computed from the applied loads and the relevant K_I calibration for the particular test geometry, such that the data are finally presented in the form of log–log plots of da/dN versus ΔK (Fig. 2). Empirical expressions, of the form of Eqn. (1), are then numerically fitted to the data to define the crack growth relationship (i.e., $da/dN \propto \text{fn}(\Delta K)$) for the particular combination of material, testing conditions, and environment in question. These procedures have now been standardized by the American Society for Testing and Materials in their ASTM Standard Test Method E647-81 for constant-load-amplitude fatigue crack growth rates above 10^{-8} m per cycle.

To determine crack growth behavior at lower growth rates, i.e., in the so-called near-threshold regime where stress intensities approach a threshold ΔK_{th}, below which cracks remain dormant or propagate at experimentally-undetectable rates (Fig. 3), it is generally more realistic to monitor growth rates under load-shedding (decreasing K_I) conditions, in order to minimize transient residual stress effects. Threshold levels are approached using a procedure involving successive load reduction followed by crack growth. Measurements of da/dN are taken at each load level, over typical increments of 1–1.5 mm increase in crack length, after which the load range is reduced by not more than 10%, and the same procedure followed. Larger reductions in load are liable to give premature crack arrest from retardation effects owing to residual plastic deformation (see Sect. 6). The increments, over which measurements of growth rate are taken, are chosen to represent distances at least four times larger than the maximum plastic zone dimension generated at the previous load level. The threshold ΔK_{th} for no crack growth is then

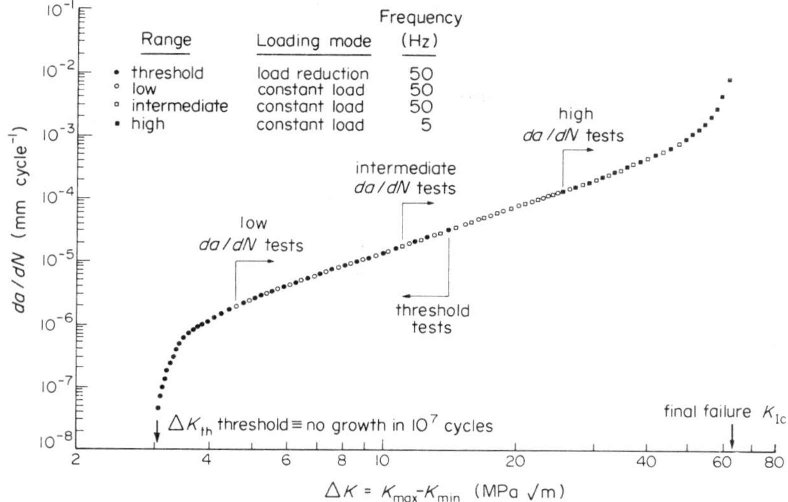

Figure 3
Typical test conditions for obtaining fatigue crack propagation data spanning entire range of growth rates from threshold levels (ΔK_{th}) to final failure (defined by the fracture toughness K_{Ic}). da/dN vs ΔK data measured for an ultrahigh-strength 300-M steel ($\sigma_y = 1700$ MPa)

measured as the value of ΔK where the growth rate is infinitesimal, which is generally operationally defined in terms of minimum growth rate, say 10^{-11} m per cycle, calculated from the accuracy of the crack monitoring technique and the number of cycles elapsed. For example, using electrical potential crack length measurements where an accuracy of 0.1 mm on absolute crack length can be conservatively assumed, detecting no crack extension in say 10^7 cycles could be employed to define the threshold as the stress intensity range to give a growth rate not exceeding 10^{-11} m per cycle. Following ΔK_{th} measurement, the load may be increased in increments and procedures again used to measure da/dN. In this way, low growth rate data are monitored under decreasing and increasing K_I conditions, and provided that care is taken to minimize transient effects caused by too severe load changes, growth-rate behavior should be similar. A typical test plot of da/dN versus ΔK for fatigue crack propagation in an ultrahigh strength steel, representing data spanning the entire range of growth rates from threshold levels to final failure determined by these procedures, is shown in Fig. 3.

Such fatigue crack growth measurement procedures can be easily automated using a suitable crack monitoring technique and a computer-controlled testing machine. This is particularly relevant for near-threshold testing under load-shedding conditions, where a programmed constant decrease in the normalized K_I gradient, i.e., $|\Delta K^{-1}.d\Delta K/da|$, is utilized. Such procedures are currently being standardized by the ASTM.

3. Fatigue Crack Growth Under Linear Elastic Conditions

As described above, the application of linear elastic fracture mechanics and related small-scale crack tip plasticity has provided a basis for describing the phenomenon of fatigue crack propagation, where the crack growth increment per cycle (da/dN) is found to be principally a function of the stress intensity range ΔK through power-law expressions of the form $da/dN = C\Delta K^m$ (Eqn. (1)). This basic relationship

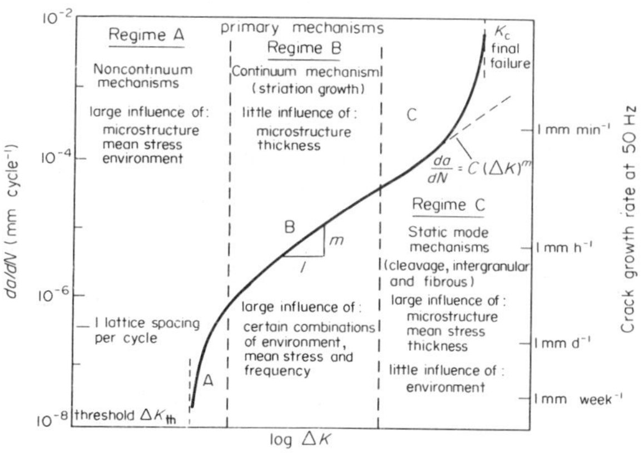

Figure 4
Schematic variation of fatigue-crack propagation rate (da/dN) with stress-intensity range (ΔK), showing regimes of primary growth rate mechanisms, and effects of several major variables on crack-growth behavior

provides an adequate engineering description of behavior at the mid-range of growth rates, typically $\sim 10^{-5}$ to 10^{-3} mm per cycle. At higher growth rates, however, as K_{max} approaches instability conditions, defined by the fracture toughness (K_{Ic}, K_c) or the limit-load, Eqn. (1) often underestimates the propagation rate, whereas at lower (near-threshold) growth rates it is generally conservative, as ΔK approaches ΔK_{th}. In general, the sigmoidal variation of growth rates with ΔK can be characterized in terms of three regimes, governed by different primary mechanisms of failure, as shown schematically for ambient environments in Fig. 4. From this figure, it is apparent that crack growth behavior can be dependent upon a number of different variables (to an extent dependent upon the growth rate regime), such as mean stress (characterized by the load ratio $R = K_{min}/K_{max}$), microstructure, frequency, environment, test-piece thickness and so forth. Mechanisms describing such kinetics of fatigue crack extension, together with the specific influence of such variables, are discussed in detail in the article *Fatigue Crack Growth: Mechanistic Aspects*.

4. Elastic–Plastic Fatigue Crack Growth

The majority of fatigue crack propagation data is obtained in conditions where the scale of local crack tip plasticity is small compared to the in-plane test-piece dimensions of crack length and ligament depth. As described in Sect. 1, under such circumstances the conditions of small-scale yielding exist and characterization of the rate of crack growth in terms of a linear elastic stress intensity factor is appropriate. There are, however, several situations where crack growth is accompanied by more extensive plasticity such that linear elastic analyses become invalid. These situations can arise for crack propagation at high stress intensity ranges (i.e., regime C in Fig. 4), for initial crack growth from notches where the crack is advancing within the strain field of the notch, and for the growth of so-called short cracks where local crack tip plastic zone sizes are comparable with crack lengths (see Sect. 5).

Early approaches to characterizing such elastic–plastic fatigue crack growth were based on the plastic strain range ($\Delta\varepsilon_p$) where the rate of crack extension was empirically approximated by expressions of the form:

$$\frac{da}{dN} \approx (\Delta\varepsilon_p)^2 a \qquad (7)$$

However, the difficulties associated with measurement of a local parameter such as $\Delta\varepsilon_p$ in the vicinity of the crack prompted Dowling (1976) to propose a more global approach relying on elastic–plastic (nonlinear) fracture mechanics. Based on the concept that crack tip stress and deformation fields in an elastic–plastic strain hardening solid can be described by the Hutchinson–Rice–Rosengren (HRR) singularity and thus uniquely characterized by the J-integral, Dowling suggested that fatigue crack propagation in the presence of extensive plasticity could be correlated to the range of J, that is, ΔJ, which under linear-elastic conditions would be equal to $\Delta K^2/E'$. The success of this approach is evident from Fig. 5 where both linear elastic and elastic–plastic crack growth data in nuclear pressure vessel steel are shown to closely correspond, even though plastic zone sizes above $\sim 10^{-3}$ mm per cycle are of the order of 25 mm, that is, comparable with overall test-piece dimensions. However, the fundamental validity of this approach has often been questioned since it appears to violate the basic assumption in the definition of J that stress must be proportional to current plastic strain. This follows because J is defined from the deformation theory of plasticity (i.e., nonlinear elasticity) which does not allow for the elastic unloading and nonproportional loading effects inherent in crack extension. However, despite these limitations the use of ΔJ does provide a good fit to fatigue crack propagation data over a very wide range of growth rates, as shown in Fig. 5, and therefore must still be regarded as a feasible, yet perhaps empirical, characterization parameter for cyclic crack advance.

An alternative approach to elastic–plastic fatigue crack growth, which is not subject to these nonlinear elastic restrictions, is to correlate crack extension to the range of crack tip opening displacement ΔCTOD. This is appealing since it permits a clearer physical interpretation of the macroscopic crack propagation data in terms of the microscopic fracture events involved, although, somewhat surprisingly, it has not been extensively utilized in the literature, possibly because of difficulties in measurement and interpretation.

5. Short Fatigue Cracks

Current defect-tolerant design codes have been developed to estimate the service life of cyclically-loaded structures in terms of the subcritical growth rates of incipient flaws utilizing the existing database of laboratory-determined fatigue crack propagation results. However, the majority of these data has been determined from test-piece geometries containing cracks of the order of 25 mm or so, whereas many defects encountered in service are far smaller than this, for example, in turbine disk and blade applications. In the relatively few cases where the behavior of such short cracks has been studied, it has been found, almost without exception, that at the same nominal driving force, the growth rates of short cracks are greater than (or equal to) the corresponding growth of long cracks. Thus, a potential exists for nonconservative defect-tolerant predictions in components where the growth of short flaws represents a large proportion of the life.

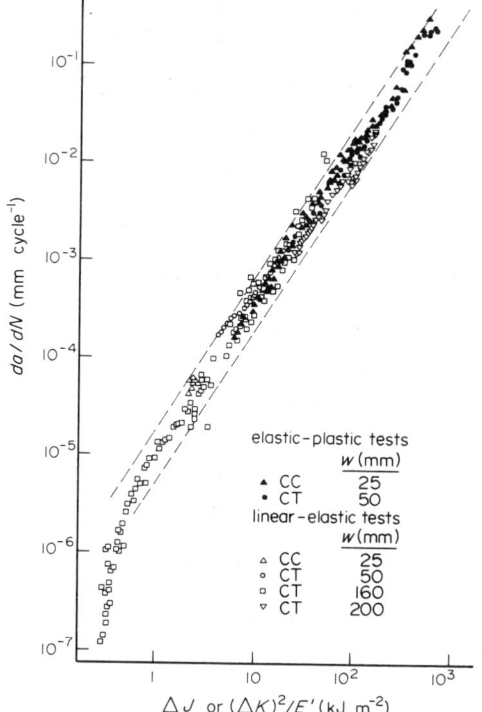

Figure 5
Fatigue crack propagation rates (da/dN) under linear elastic and elastic–plastic conditions in A533B nuclear pressure vessel steel ($\sigma_y = 480$ MPa) as a function of ΔJ for test geometries of compact (CT) and center cracked (CC) specimens of varying width w. Note ΔJ is equal to $\Delta K^2/E'$ under linear-elastic conditions (after Dowling 1976)

Cracks may be considered short if their length is (a) comparable to the microstructural size-scales (e.g., of the order of a grain size), (b) comparable to the scale of local plasticity (e.g., ~1 μm in ultrahigh strength materials ($\sigma_y \sim 2000$ MPa) to ~1 mm in low strength materials ($\sigma_y \sim 300$ MPa)), or (c) simply physically small (<0.5–1 mm).

For the growth of cracks small compared to the scale of plasticity, at intermediate growth rates (i.e., regime B in Fig. 4), the use of elastic–plastic fracture mechanics, specifically involving ΔJ, has shown a one-to-one correspondence between short and conventional long crack data. However, this is only one aspect of a very complex problem, since the kinetics of short flaw crack growth may be influenced by microstructural and environmental factors in ways markedly different from long flaws. In particular, it has been shown that short cracks can propagate at stress intensities below the long crack threshold ΔK_{th}, both from smooth surfaces and when emanating from notches. Such cracks then advance at progressively decreasing growth rates before arresting or joining the long crack propagation curve (Fig. 6). Further details concerning the microscopic aspects of the short crack problem can be found in the bibliography and in the article *Fatigue Crack Growth: Mechanistic Aspects*.

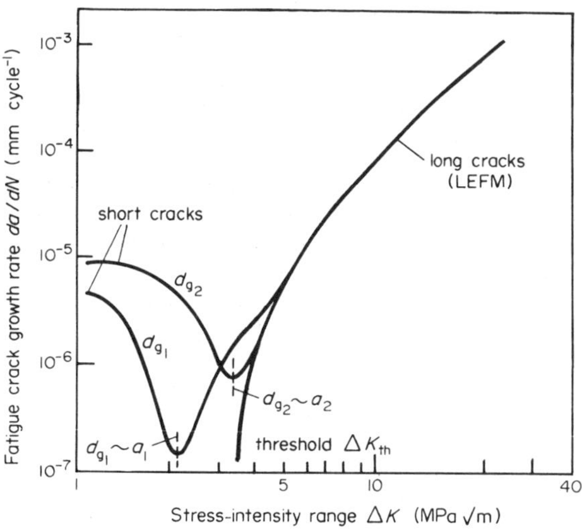

Figure 6
Data showing the growth rate behavior of microscopically-short cracks, as a function of grain size (d_g), in 7075-T6 aluminum alloy ($\sigma_y = 515$ MPa). Note how short cracks propagate below the long crack threshold (ΔK_{th}) at progressively decreasing growth rates, with the minimum velocity occurring when the crack length is of the order of the grain size (i.e., $a \sim d_g$) (after Lankford 1982)

Short crack studies are currently of high priority in fatigue research since the problem represents a breakdown in the similitude concept ingrained in conventional fracture mechanics analyses. Extreme care, therefore, must be taken in the application of existing (long crack) fatigue crack propagation data for defect-tolerant lifetime calculations in components containing small defects, since without making allowances for the possible faster propagation rates and lower thresholds associated with such short flaws, predictions may be dangerously nonconservative.

6. *Variable Amplitude Loading*

Since the majority of components in service are subjected to variable amplitude, spectrum or random loading, rather than constant amplitude loading, an appreciation of the influence of periodic overloads and block loading sequences on fatigue crack propagation is clearly of practical significance. Classically, the approach taken for variable amplitude loading in fatigue design is to use the Palmgren–Miner relationship, where for a body subjected to n_i number of cycles of different stress amplitudes $\Delta\sigma_{a_i}$, with $i = 1$ to k,

$$\sum_{i=1}^{k} n_i/N_i = 1 \tag{8}$$

where N_i is the number of cycles to failure had all the stress amplitudes been of magnitude $\Delta\sigma_{a_i}$ (computed from S/N diagrams). This approach, which is utilized for total life prediction and thus includes both crack initiation and crack propagation stages, predicts that the cycles to failure will be decreased if a large single positive (spike) overload is applied during constant amplitude cycling. However, as shown schematically in Fig. 7, the effect of such an overload on mode I crack propagation rates is to initially retard crack growth, sometimes to the point of arrest, and thus to increase the propagation life. Hence it is apparent that the application of Miner's rule (Eqn. (8)) to crack propagation behavior under variable amplitude loads is often inappropriate.

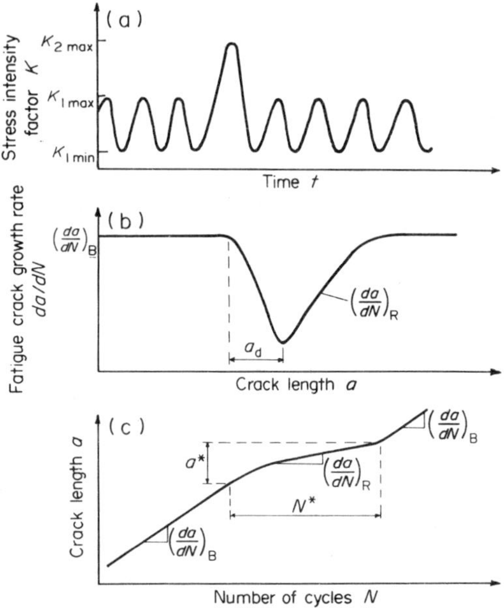

Figure 7
Schematic illustration showing the transient deceleration in growth rates during mode I fatigue crack propagation resulting from the application of a single positive (spike) overload

For crack growth studies, however, the above approach can be utilized for narrow band random loading where each cycle does not differ significantly from its predecessor. Here interaction effects can be considered small and it is reasonable to assume that each cycle causes the same amount of crack extension as if it were applied as part of a sequence of constant amplitude loads. For broad band loading, on the other hand, cyclic crack extension has been estimated on the basis that the load, or stress intensity, ranges are constant at their root mean square value. However, since loading sequence interaction effects are more relevant here, this procedure often yields conservative predictions for mode I crack advance.

The nature of these interaction effects is shown in Fig. 7. The application of a single positive (spike) overload of sufficient magnitude (generally >50% of the baseline ΔK) can result in significantly retarded crack growth over a crack length increment a* which is of the order of, or greater than, the overload plastic zone size, whereupon crack growth rates return to their original baseline level. Sometimes cracks overloaded in this fashion totally arrest, where presumably the effective stress intensity actually experienced at the crack tip is reduced below ΔK_{th}. However, the onset of the retardation or crack arrest often does not coincide with the actual overload cycle but is delayed until the crack has grown some distance into the overload plastic zone (so-called delayed retardation). For fully reversed overload cycles, the application of an underload (compressive overload) prior to a single overload does not appear to have much effect on subsequent retardation, whereas an underload following an overload tends to minimize retardation. Block loading sequences are also known to produce large transient effects on fatigue crack propagation. For example, high-low block loading sequences, where the stress or stress intensity range is reduced from one constant level to a lower one, can result in an initial retardation or crack arrest which is actually much greater than for an equivalent single overload.

The origin of these interaction effects is associated with mechanisms of crack tip blunting and branching, premature crack closure and the fact that the overloaded crack is growing into the residual stress field of the overload plastic zone. In defect-tolerant design, however, the role of variable amplitude cycles is incorporated through the use of more empirical models, such as due to Wheeler and Willenborg. The Wheeler model assumes that the extent of crack growth retardation following the overload cycle is related to the distance the retarded crack has transversed into the overload plastic zone, whereas the Willenborg approach computes an effective stress intensity, based on the nominal K_{max} being reduced due to the presence of residual stresses within this plastic zone. However, since in many materials, the onset of retarded crack growth does not coincide with the point of application of the overload cycle (i.e., delayed retardation), both models are somewhat artificial physically. Nevertheless, they are utilized practically in the development of more complex crack growth relationships (based on Eqn. (1)) which are in widespread use for defect-tolerant design procedures, particularly for airframe components.

7. Design Philosophies

The conventional approach to design against fatigue failure during a finite lifetime has classically involved the use of S/N curves (representing the life N resulting from a stress amplitude S), suitably adjusted to take into account the effects of mean stress (using Goodman diagrams), effective stress concentrations at notches (using fatigue-strength reduction factors), variable-amplitude loading (using the Palmgren–Miner cumulative-damage law), multiaxial stresses, environmental effects, and so forth. In materials that show a well-defined fatigue limit, such as steels in benign atmospheres, infinite life might be expected when applied alternating stresses are below this limit, with due consideration of the factors listed above. This approach essentially represents design against crack initiation, since near the fatigue limit, especially in smooth specimens, this may occupy the major portion of the lifetime (~80%). Crack initiation here may represent either the nucleation of a microcrack that becomes the "fatal" macrocrack by continued propagation or the coalescence of several microcracks to form the macrocrack—whichever is the critical event for the particular microstructure in question. This approach, although simply empirical in nature, is still in widespread use for design.

For safety-critical applications, especially with welded and riveted structures, there has been a growing awareness that the presence of defects in a material below some certain size must be assumed and taken into account at the design stage. Under such circumstances, the integrity of a structure will depend on the lifetime spent in crack propagation since the crack-initiation stage can be assumed to be small, and conventional S/N analyses, where remaining lifetime is based on the number of cycles both to initiate and to propagate the crack, may dangerously overestimate the life. Such considerations have led to the widespread adoption of the defect-tolerant fatigue design approach based on a fracture mechanics description of crack propagation behavior. In this approach, maximum possible initial defect sizes are estimated using nondestructive inspection or proof-testing procedures so that fatigue lifetimes can be computed as the number of cycles for this defect (assumed to be a crack) to propagate to some critical size at failure (defined by the fracture toughness, limit load or allowable strain criteria, for example). This analysis involves integration of the crack-growth law for that material (e.g., of the form of Eqn. (1)) starting from some small initial crack size, suitably modified to take into account mean-stress effects, variable-amplitude loading, environmental data and so on, as required. It must be appreciated, however, that successful application of this approach in design for finite life demands periodic nondestructive-testing evaluation both to define initial defect sizes and to detect cracks before they reach critical dimensions—a procedure that is feasible for aircraft structures, large pressure vessels, power-plant shafts and rotors, large turbines and so forth. However, from economic considerations alone the defect-tolerant approach is unlikely ever to be generally adopted. For many applications, including such everyday commodities as washing machines and automobiles, analyses using S/N data will remain the principal design assessment criteria.

In certain applications, where large numbers of small-amplitude, high-frequency stress reversals are involved for lifetimes in excess of 10^{10} to 10^{12} cycles (beyond the range of available data used to construct S/N curves), the defect-tolerant approach has been utilized in design for infinite fatigue life based on no crack growth. In such cases, initial defect sizes are estimated and a design stress is chosen such that the stress intensity factor range (ΔK) is maintained below the crack propagation threshold (ΔK_{th}). Although overly conservative for many applications, this approach has proved essential for the design of turbine blades and in the operation of gas circuitry in nuclear-reactor systems subject to high-frequency (~1300 Hz) vibrational stresses from the gas circulators ("acoustic fatigue"), where growth rates as minute as 10^{-7} mm per cycle, if allowed, would cause failure in a small fraction of the required life. Accordingly, calculations must err on the side of pessimism, and the size of permissible defects must be such that the overwhelming majority of the stress cycles produce ΔK ranges that are less than ΔK_{th} and presumably do not cause growth. This threshold concept can, however, lead to difficulties if the size of the "threshold defect" at the design stress amplitude is so small that uncertainty in the value of ΔK_{th} arises from the short crack problem and further that nondestructive inspection cannot be depended upon to detect sightly larger cracks with 100% reliability. Here, other means, such as proof testing, realistic inspection and repair schemes, must be sought to achieve the required durability.

8. Conclusion

This article attempts to provide a basic framework for understanding the macroscopic aspects of fatigue crack propagation in metals and alloys from a continuum mechanics viewpoint. Further details may be obtained with reference to the bibliography.

Although much progress has been made in the analysis of fatigue with the continuum mechanics characterization of cyclic growth rates through the application of linear elastic and elastic–plastic fracture mechanics, particularly with the widespread use of defect-tolerant design codes, many uncertainties still exist with mechanistic interpretations. A comprehension of the micromechanisms of fatigue crack growth, aside from the obvious benefits for the design of new and improved fatigue-resistant alloys, is vital to the definition of the parameters that govern the growth of short cracks, near-threshold crack growth, and crack extension under variable amplitude loading. These issues are described from both continuum and metallurgical viewpoints in the article *Fatigue Crack Growth: Mechanistic Aspects*.

See also: Fatigue Cracks: Resistance to Initiation; Fatigue Cracks: Resistance to Propagation; Mechanics of Materials: An Overview

Bibliography

Dowling N E 1976 *Cracks and Fracture*, ASTM STP 601. American Society for Testing and Materials, Philadelphia, Pennsylvania, p. 19

Elber W 1971 *Damage Tolerance in Aircraft Structures*, ASTM STP 486. American Society for Testing and Materials, Philadelphia, Pennsylvania, p. 230

Frost N E, Pook L P, Denton K 1971 A fracture mechanics analysis of fatigue crack growth data for various materials. *Eng. Fract. Mech.* 3: 109–26

Hudak S J 1981 Small crack behavior and the prediction of fatigue life. *J. Eng. Mater. Technol.* 103: 26–35

Johnson H H, Paris P C 1968 Subcritical flaw growth. *Eng. Fract. Mech.* 3: 1–45

Kocańda S 1978 *Fatigue Failure of Metals*. Sijthoff and Noordhoff, The Hague

Paris P C, Erdogan F 1963 A critical analysis of crack propagation laws. *J. Basic Eng.* 85: 528–34

Paris P C, Gomez M P, Anderson W E 1961 Rational analytic theory of fatigue. *Trend Eng.* 13: 9

Rice J R 1967 *Fatigue Crack Propagation*, ASTM STP 415. American Society for Testing and Materials, Philadelphia, Pennsylvania, p. 247

Rice J R 1968 In: Liebowitz H (ed.) 1968 *Fracture: An Advanced Treatise*, Vol. 2. Academic Press, New York, p. 191

Ritchie R O 1979 Near-threshold fatigue-crack propagation in steels. *Int. Met. Rev.* 20: 205–30

Suresh S, Ritchie R O 1984 In: Davidson D L, Suresh S (eds.) 1984 *Fatigue Crack Growth Threshold Concepts*. Metallurgical Society (AIME), Warrendale, Pennsylvania, p. 227

Wheeler O E 1972 Spectrum loading and crack growth. *J. Basic Eng.* 94: 181–6

R. O. Ritchie

Fatigue Crack Growth: Mechanistic Aspects

Micromechanical modelling of fatigue crack propagation behavior to develop predictive growth rate relationships from first principles still is an imprecise science. This is due to the complex nature of the fatigue process, involving striation formation, additional static fracture modes at high growth rates, the three-dimensional and noncontinuum nature of low growth rates, environmental interactions, crack closure effects and so forth. In fact, no single model for fatigue crack propagation currently embodies all these salient factors.

McClintock (1963), however, concluded that fundamentally there were two limiting mechanisms by which a fatigue crack could grow. If the crack advanced by a mechanism of damage accumulation or renucleation over the volume of the crack tip plastic zone, a fourth-power dependence of crack growth rates (da/dN) on the stress intensity range (ΔK) would result, whereas if crack growth occurred by kinematically irreversible crack tip plastic deformation based on the instantaneous crack tip opening displacement range (ΔCTOD), a second-power dependence would result.

Experimentally determined growth rate behavior (typically between 10^{-6} and 10^{-3} mm per cycle) tends to fall between these two extremes. Experiments involving (a) interrupted crack propagation tests, where the fatigue damage was annealed out and the test restarted at the same ΔK, and (b) variable amplitude loading, where the striation spacings following the retardation period after single spike overloads were found to be identical to the pre-overload spacings, have tended to favor the instantaneous CTOD models.

While perhaps valid over the range of data on which they are based, the majority of these theoretical fatigue crack growth models suffer from the fact that they generally provide poor agreement with measured propagation rates and furthermore are unable to account for such factors as near-threshold behavior, crack closure effects and the influence of aggressive environments. This is evident from the fact that the crack growth relationships most often used in engineering defect-tolerant design and lifetime calculations are based on mere empirical curve fits to the relevant experimental crack propagation data. Perhaps the safest approach to this problem has been to develop semiempirical relationships based on physical models for crack growth and then to determine the constants in the model by fitting to relevant experimental data.

1. General Characteristics of Fatigue Crack Growth

The application of linear elastic fracture mechanics and related small-scale crack tip plasticity has provided a basis for describing the phenomenon of fatigue crack propagation. The crack growth increment per cycle (da/dN) is found to be principally a function of the stress intensity range ΔK through power-law expressions of the form:

$$\frac{da}{dN} = C\Delta K^m \tag{1}$$

where ΔK is defined as the difference between the maximum and minimum stress intensities in the cycle (i.e., $\Delta K = K_{max} - K_{min}$). This basic relationship provides an adequate engineering description of behavior in the mid-range of growth rates, typically $\sim 10^{-5}$–10^{-3} mm per cycle.

At higher growth rates, however, as K_{max} approaches instability conditions, defined generally by the fracture toughness (K_{Ic}, K_c) or the limit load, Eqn. (1) often underestimates the propagation rate, whereas at lower (near-threshold) growth rates it generally is conservative, as ΔK approaches a threshold stress intensity range (ΔK_{th}) below which cracks remain dormant or propagate at experimentally undetectable rates. In general, the sigmoidal variation of growth rates with ΔK can be characterized in terms of three regimes, governed by different primary mechanisms of failure, as shown schematically in Fig. 1. It should be noted that the form of the relationship illustrated in Fig. 1 may be changed markedly in the presence of active environments (i.e., in corrosion fatigue) and elevated temperatures (i.e., in creep fatigue).

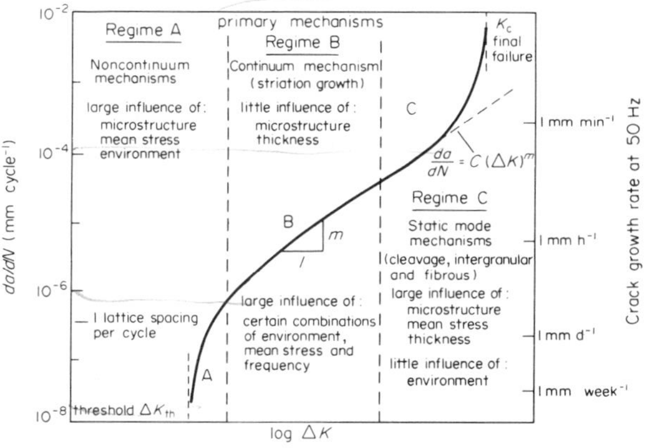

Figure 1
Schematic variation of fatigue crack propagation rate da/dN with stress intensity range ΔK, showing regimes of primary growth rate mechanisms

1.1 Mid-Range of Growth Rates (Regime B)

In the mid-range of growth rates, typically $\sim 10^{-6}$–10^{-3} mm per cycle, where the simple power-law relationship of Eqn. (1) applies (regime B in Fig. 1), fatigue failure is generally characterized by a transgranular ductile striation mechanism, as shown in Fig. 2a. Such striations represent local crack growth increments per cycle and it has been suggested that they occur by a mechanism of opening, advancing and blunting of the crack tip on loading, followed by resharpening of the tip on unloading. An example of this mechanism is shown in Fig. 3, although the precise details of this process are still uncertain.

Several theoretical models for such macroscopic striation crack growth (often termed stage II crack propagation) have been proposed which rely on the fact that, where plastic zones are sufficiently large compared with microstructural dimensions, plastic blunting at the crack tip is accommodated by shear on two slip systems at roughly 45° to the crack plane. As such sliding-off is largely irreversible, new crack surface can be created during cyclic crack advance by simultaneous or alternating slip on these two systems. An example of the latter process is shown in Fig. 4, based on observations of fatigue crack tips in copper single crystals subjected to bending or push–pull loading. The observations that striations are formed less readily, and that cracks propagate far slower, in vacuo often are considered to be consistent with such models, since due to the lack of crack surface oxidation, the irreversibility of crack tip shear should be less efficient.

Regardless of the precise details of the mechanism, such crack tip sliding-off models predict that the increment of crack advance per cycle should be proportional to the cyclic crack tip opening displacement (ΔCTOD). From fracture mechanics considerations, the fatigue crack growth rate for the striation mode of propagation should then be given by

$$\frac{da}{dN} \propto \Delta\text{CTOD} = \beta \frac{\Delta K^2}{2\sigma_y E'} \qquad (2)$$

where σ_y is the cyclic flow stress, E' the appropriate elastic modulus and β a proportionality constant reflecting the efficiency of crack tip blunting (taken theoretically to be in the region 0.5–1). Equation (2) predicts that growth rates in regime B should be dependent on ΔK^2, and this certainly is consistent with most data in this range, which show the power-law exponent m to be in the region 2–4.

Furthermore, the inverse dependence of da/dN on elastic modulus has also been verified, as shown in Fig. 5 for a wide range of metallic systems. However, the inverse dependence on cyclic yield strength is rarely seen experimentally (Fig. 6). Moreover, in absolute terms, experimentally measured fatigue crack growth rates (under small-scale yielding conditions) invariably indicate a rate of incremental crack extension which is a far smaller fraction of ΔCTOD, particularly at lower growth rates (Fig. 7). Numerous explanations have been advanced for the lack of direct correlation between da/dN and ΔCTOD (including the minimal role of strength), although none is entirely satisfactory. The discrepancy appears to be related to a number of factors, including the nonuniformity of crack advance in three dimensions, the occurrence of other fracture modes, the role of crack closure and the micromechanisms of crack tip blunting. The latter factor specifically

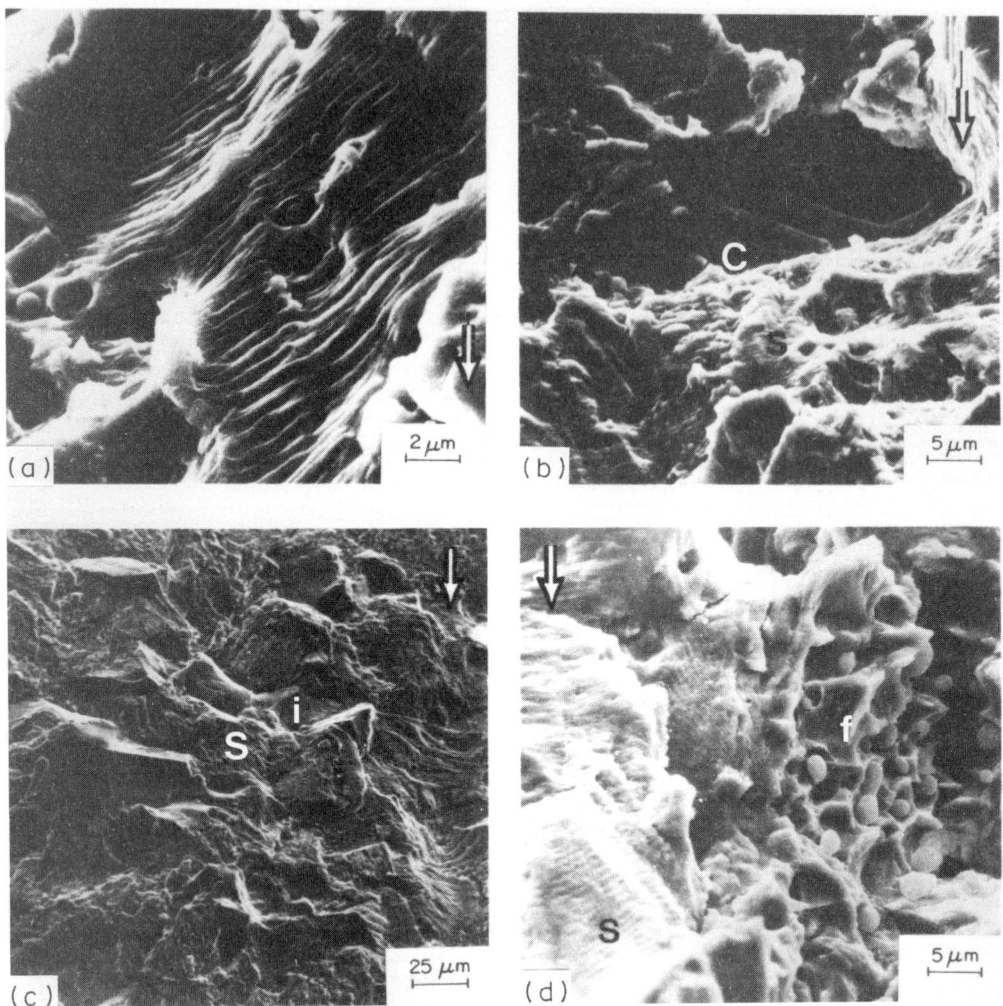

Figure 2
Fractography of fatigue crack propagation at intermediate (regime B) and high (regime C) growth rates in steels tested in moist room air at $R = 0.1$: (a) ductile striations in mild steel at $\Delta K = 30\ \mathrm{MPa\ m^{0.5}}$; (b) additional cleavage cracking (C) during striation growth (S) in mild steel at $\Delta K = 19\ \mathrm{MPa\ m^{0.5}}$; (c) additional intergranular cracking (i) in low-alloy Ni–Cr steel at $\Delta K = 15\ \mathrm{MPa\ m^{0.5}}$; (d) additional microvoid coalescence and fibrous fracture (f) with striations (S) in AISI 316L stainless steel at $\Delta K = 30\ \mathrm{MPa\ m^{0.5}}$ (arrows indicate general direction of crack growth)

refers to the degree of strain localization at the tip as a function of its opening, and has been used to rationalize the absence of a yield strength effect of da/dN by assuming $\beta \propto \sigma_y/E'$.

Aside from the strong dependence on modulus, rates of fatigue crack extension by this striation mechanism are largely insensitive to metallurgical microstructure in the mid-growth rate range, as shown in Figs. 6 and 7. In Fig. 6, data are plotted for a wide range of ferrous alloys representing an order of magnitude change in yield strength from simple ferritic to tempered martensitic structures. It is apparent that where crack growth occurs via a striation mechanism (indicated by the letters), growth rates vary by a factor of <10. Furthermore, propagation rates in this regime appear relatively independent of the mean stress, characterized by the load ratio $R = K_{min}/K_{max}$. Data for a 2.25%Cr–1%Mo and pressure vessel steel in Fig. 8 reveal that growth rates between $\sim 10^{-6}$ and 10^{-3} mm per cycle are almost totally insensitive to changes in load ratio from 0.05 to 0.75, although it should be noted that strong mean stress effects are present at lower and higher (not shown) growth rates. Behavior in the mid-growth regime, in the absence of environmental interactions, is also essentially independent of changes in cyclic

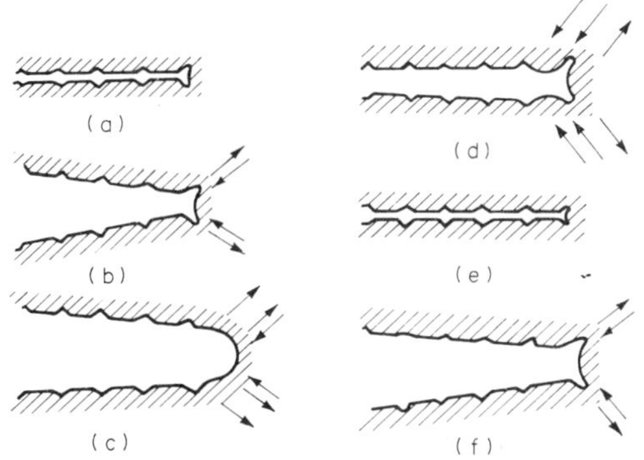

Figure 3
Diagrammatic representation of the sequence of formation of ductile fatigue striations according to the plastic blunting model (Laird 1967), involving sequential blunting, crack extension and resharpening

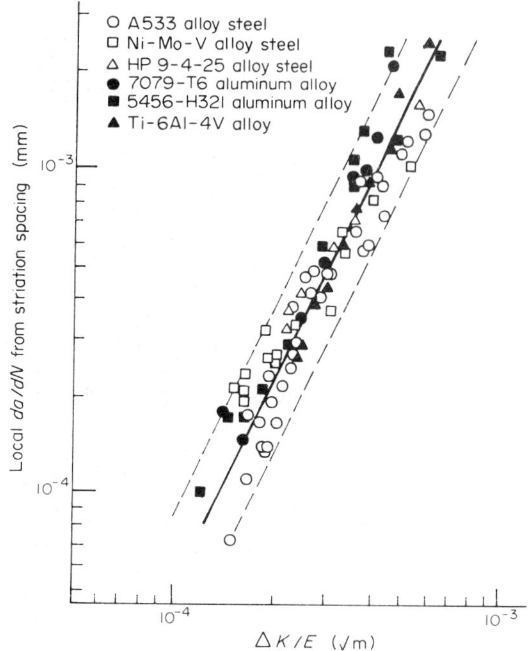

Figure 5
Experimental data showing inverse dependence of local fatigue crack propagation rate da/dN, measured from striation spacings, on elastic modulus E (note how normalizing ΔK with respect to E yields similar mid-growth rate behavior for all materials) (after Bates and Clark 1969)

frequency, wave shape and whether deformation is in plane stress or plane strain.

Such load ratio, frequency and wave shape effects are, however, important in the presence of certain active environments. Under such circumstances, fatigue crack growth rates in this regime become environmentally assisted through additional contributions to crack advance from such mechanisms as active path corrosion (metal dissolution) and hydrogen embrittlement. An example of such corrosion fatigue behavior can be seen in Fig. 9 for a bainitic 2.25%Cr–1%Mo steel tested in air and hydrogen. Below a critical frequency, sufficient to permit the chemisorption of hydrogen at the crack tip, fatigue crack growth rates are seen to be markedly accelerated in hydrogen (above $\sim 10^{-5}$ mm per cycle) compared with air, the onset of the acceleration being a function of the load ratio. In this instance, hydrogen-assisted crack propagation often is coincident with a change in fracture mode from predominantly transgranular to intergranular and can be attributed to hydrogen-induced degradation of material ahead of the crack tip. Analogous behavior has been reported in water, hydrogen sulfide and other corrosive environments, although the resulting corrosion fatigue crack growth behavior observed tends to be very specific to the particular material, environment and test conditions in question.

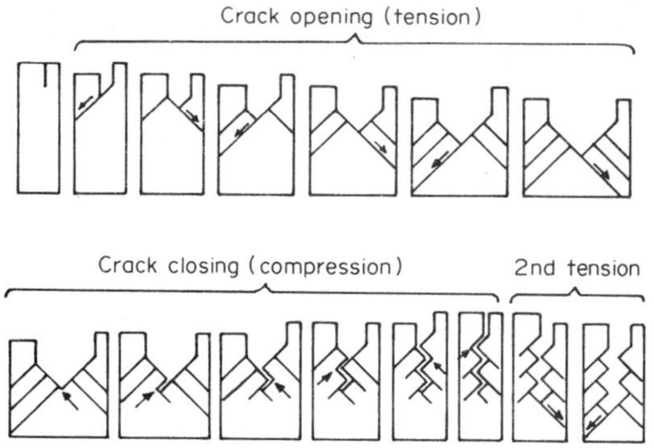

Figure 4
Model for the alternating slip processes involved in fatigue crack extension during a loading cycle (after Neumann 1974)

1.2 High Growth Rate Regime (Regime C)

At high fatigue crack growth rates, typically above $\sim 10^{-3}$ mm per cycle, where K_{max} approaches K_{Ic} or limit load failure (regime C in Fig. 1), Eqn. (1) generally underestimates measured growth rates, due to the occurrence of brittle fracture mechanisms (so-called static modes) which replace or are additional to striation growth. Such static modes include cleavage, intergranular cracking and microvoid coalescence (Fig. 2) and their presence results in growth rate behavior which is markedly sensitive to microstructure, mean stress and testpiece thickness. Since both cleavage and intergranular cracking are stress-controlled fracture mechanisms and microvoid

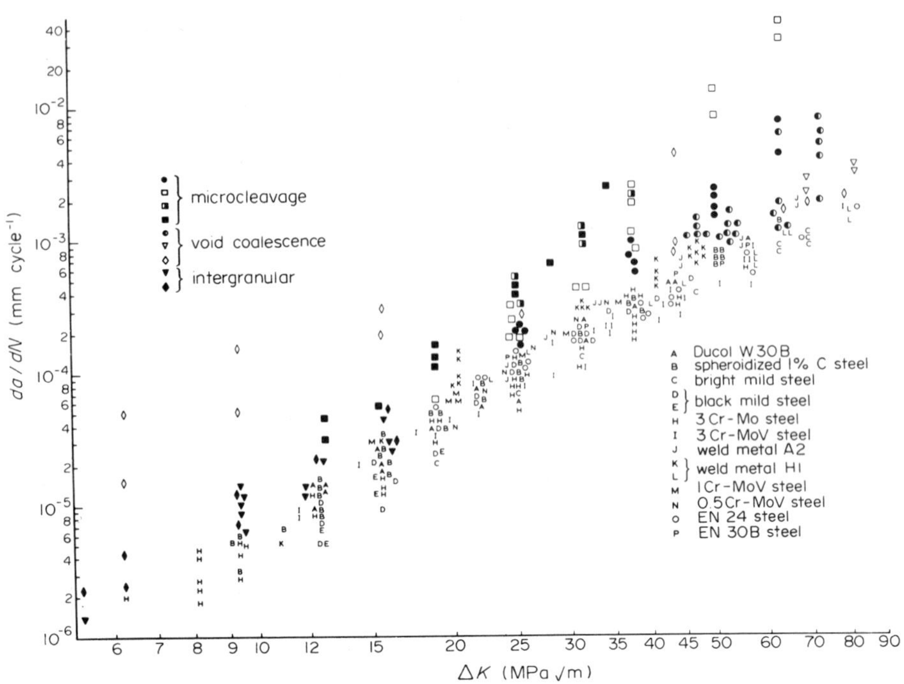

Figure 6
Fatigue crack propagation rate da/dN a function of ΔK for a wide range of ferrous alloys ($\sigma_y \sim 200$–2000 MPa). Note little change in intermediate propagation rates where mechanism of growth involves striations (after Lindley et al. 1976)

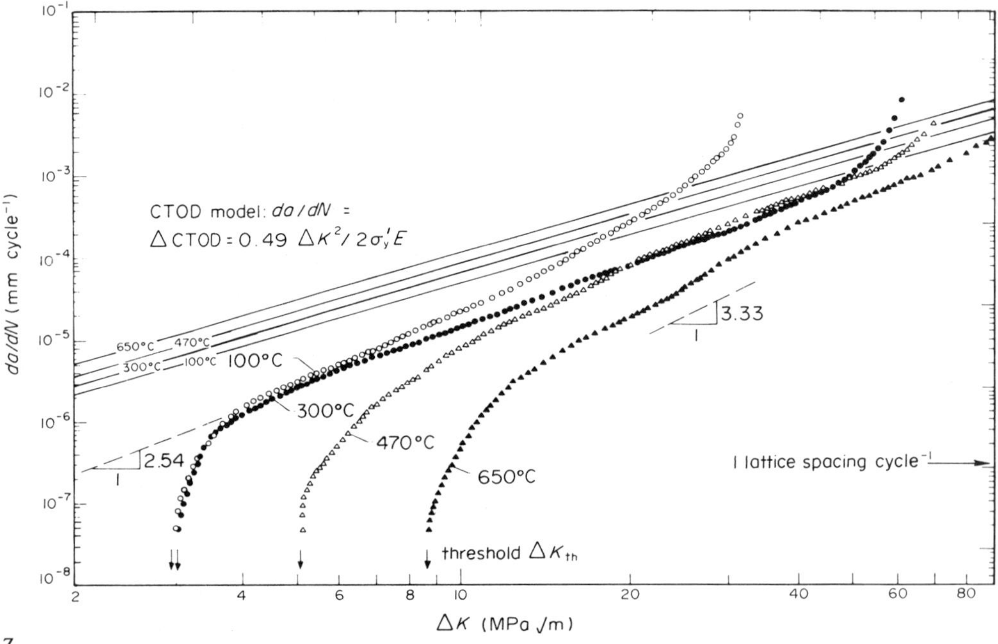

Figure 7
Variation of fatigue crack propagation rate da/dN in moist air at $R = 0.05$ and 23 °C with ΔK for ultrahigh-strength 300-M martensitic steel austenitized at 870 °C, oil-quenched and tempered between 100 and 650 °C to vary tensile strength from 2300 to 1190 MPa, respectively, showing smallest effect of strength level in the mid-growth rate regime. Solid lines indicate predictions based on the CTOD model for crack growth, Eqn. (2)

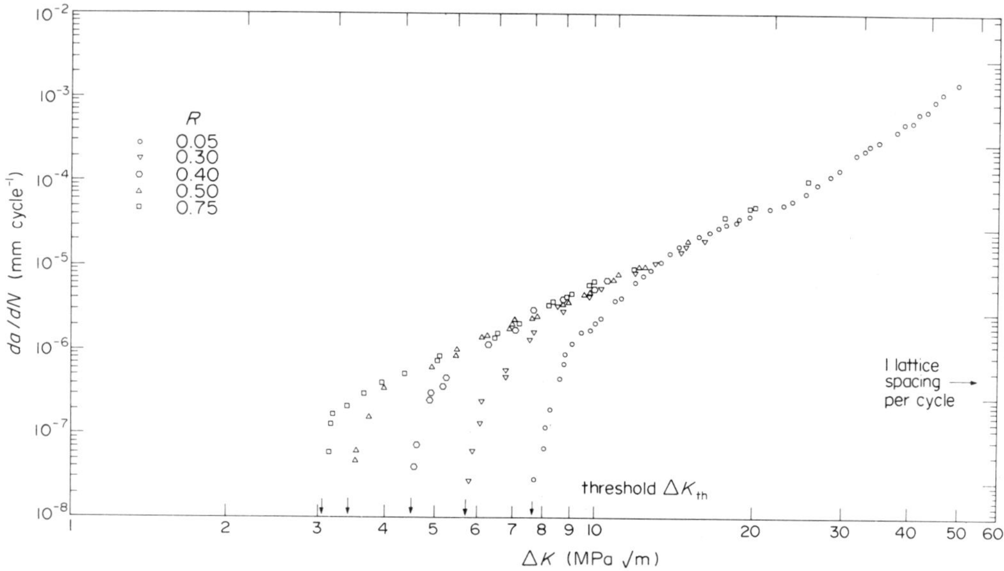

Figure 8
Influence of load ratio R on fatigue crack propagation rate da/dN as a function of ΔK in a bainitic 2.25%Cr–1%Mo pressure vessel steel SA 542 class 3 ($\sigma_y = 500$ MPa) at 50 Hz (after Ritchie and Suresh 1983)

coalescence is dependent upon the nature of the hydrostatic stress, their occurrence is promoted by high mean stresses (high R values), thick specimens (promoting plane strain conditions) and low toughness microstructures (e.g., when $K_{max} \rightarrow K_{Ic}$). This is apparent in Fig. 10, where high growth rate data for a low-toughness (temper-embrittled) low-alloy steel can be seen to be markedly dependent upon load ratio. Here, brittle intergranular cracking was found to be responsible for the marked acceleration in growth rates with increasing load ratio as K_{max} approached K_{Ic}. However, because crack velocities

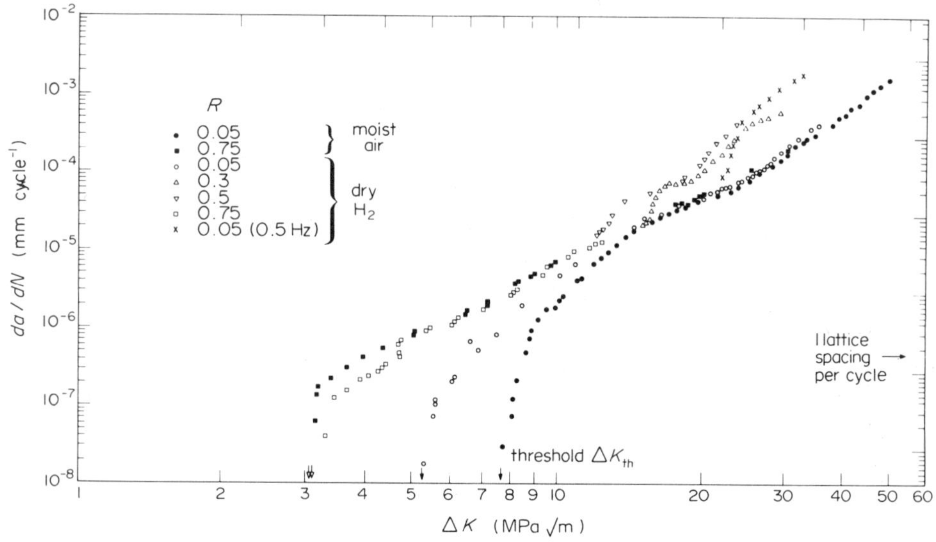

Figure 9
Influence of load ratio R and environment on fatigue crack propagation in bainitic 2.25%Cr–1%Mo steel SA 542 class 3 ($\sigma_y = 500$ MPa, $f = 50$ Hz), tested in moist air and dry hydrogen at ambient temperature gas (after Ritchie and Suresh 1983)

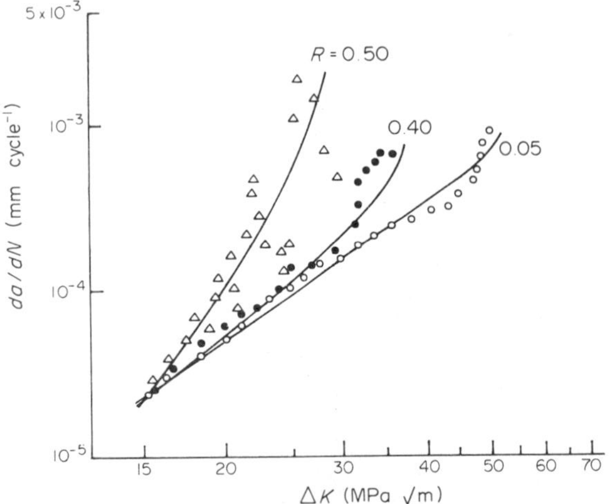

Figure 10
Influence of load ratio R on fatigue crack propagation at high stress intensities (regime C) for a low-toughness Ni–Cr En 30A steel ($\sigma_y = 735$ MPa), tested in ambient temperature moist air (after Ritchie and Knott 1973)

are so fast, there is generally insufficient time per cycle for strong corrosion effects to be active, and behavior in this regime is considered to be largely insensitive to environment.

1.3 Low Growth Rate (Near-Threshold) Regime (Regime A)

At ultralow fatigue crack growth rates, typically below $\sim10^{-6}$ mm per cycle, approaching the threshold stress intensity range ΔK_{th} (regime A in Fig. 1), Eqn. (1) generally overestimates measured growth rates and behavior becomes markedly sensitive to mean stress, microstructure and environment. In addition, at such near-threshold levels, the scale of local plasticity approaches that of the microstructure and measured propagation rates become less than an interatomic spacing per cycle, indicating that crack extension is not occurring uniformly over the entire crack front. This deviation from continuum crack growth mechanisms is evident in Fig. 7 from the large discrepancy, at near-threshold stress intensities, of measured growth rates from predictions

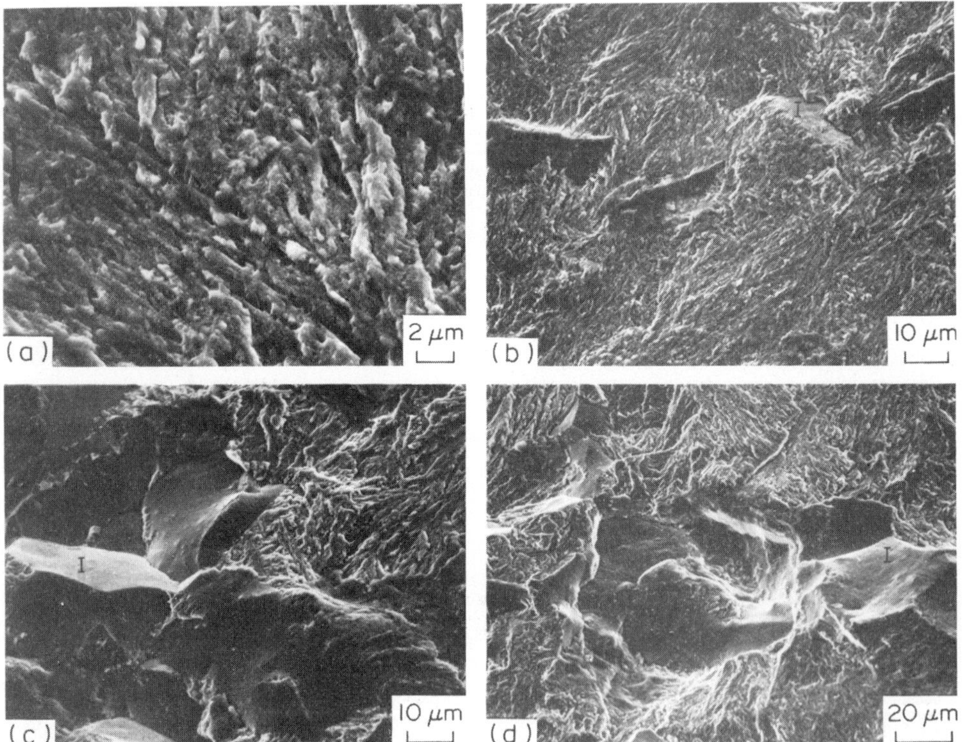

Figure 11
Fractography of near-threshold fatigue crack propagation (regime A) in high-strength martensitic 300-M steel (σ_y = 1500 MPa) tested in moist ambient temperature air: (a) transgranular fracture mode close to threshold of ΔK = 5.1 MPa m$^{0.5}$ segments of additional intergranular facets at ΔK = (b) 5.2, (c) 7.6 and (d) 11 MPa m$^{0.5}$ ($R = 0.05$) (arrows indicate general direction of crack growth)

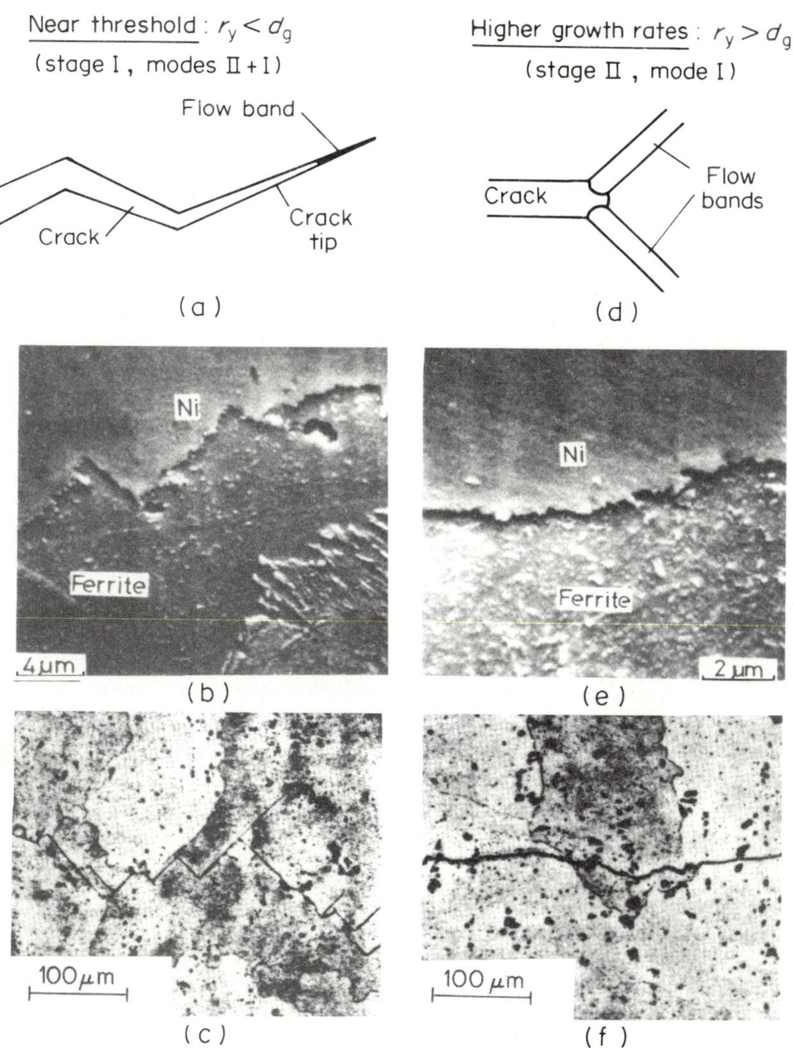

Figure 12
Crack opening profiles and resulting crack path morphologies corresponding to (a–c) near-threshold (Stage I) and (d–f) higher growth rate (Stage II) fatigue crack propagation: (b, e) nickel-plated fracture sections of fatigue crack growth in 1018 steel (after Minakawa and McEvily 1981); (d, f) metallographic sections of crack growth in 7075-T6 aluminum alloy (after Louwaard 1977) (r_y is maximum plastic zone size, compared with average grain diameter d_g)

based on the crack tip opening displacement model, Eqn. (2).

Mechanistically, comparatively little is known about near-threshold crack propagation, although fractographically surfaces tend predominantly to be transgranular with evidence of isolated intergranular or "cleavage-like" facets (Fig. 11). Such faceted-type crack advance, sometimes referred to as "microstructurally sensitive" or "crystallographic" fatigue, is observed at ultralow growth rates, where maximum plastic zone sizes are typically less than the grain size. The low restraint of the elastic surroundings on cyclic slip will promote single slip at the crack tip on the slip system with the highest critical resolved shear stress, resulting in a slip-band cracking mechanism (stage I growth) with associated mode II + mode I displacements. This gives rise to a faceted or zigzag fracture path, unlike ductile striation growth at higher ΔK levels (stage II growth), where fracture morphologies are macroscopically more planar (Fig. 12).

This transition in fracture mode follows from the fact that, at higher stress intensities, the crack tip plastic zone encompasses many grains and the maintenance of slip-band cracking in a single direction becomes incompatible with a coherent crack front.

Thus, the accompanying increase in cyclic plasticity activates further slip systems, leading to the noncrystallographic mode of striation crack advance by alternating or simultaneous shear. Since the single shear type crack advance is accompanied by significant mode II crack tip displacements in addition to mode I, this combination of faceted fracture morphologies and local shear displacements at near-threshold levels has particular importance in the development of fatigue crack closure effects, as discussed in detail below.

In terms of macroscopic growth rate behavior, near-threshold propagation rates are strongly sensitive to mean stress, unlike behavior in regime B. With rising load ratios, growth rates are increased significantly and threshold ΔK_{th} values are reduced, as shown in Fig. 8. Similarly, in contrast to regime B behavior, near-threshold growth rates are markedly higher with increasing strength level in steels (Fig. 7) and generally slower in coarser-grained microstructures (Fig. 13). Furthermore, the nature of the environment plays a prominent role in this regime, as illustrated in Fig. 9: near-threshold growth rates in dry hydrogen are significantly faster (at $R = 0.05$) than in moist air.

In general, however, environmental interactions with cyclic crack extension at ultralow growth rates are particularly complex, since in lower-strength steels, for example, seemingly aggressive environments such as water and wet hydrogen result in marginally slower propagation rates compared with moist air, whereas crack propagation rates in inert gases, such as dry helium, appear to be substantially faster (Fig. 14). It is pertinent here, though, that the marked effects of strength level (Fig. 7), grain size (Fig. 13) and environment (Figs. 9, 14) all predominate at low mean stresses and are largely absent at high load ratios, behavior which is indicative of major contributions from crack closure, as discussed in the following section.

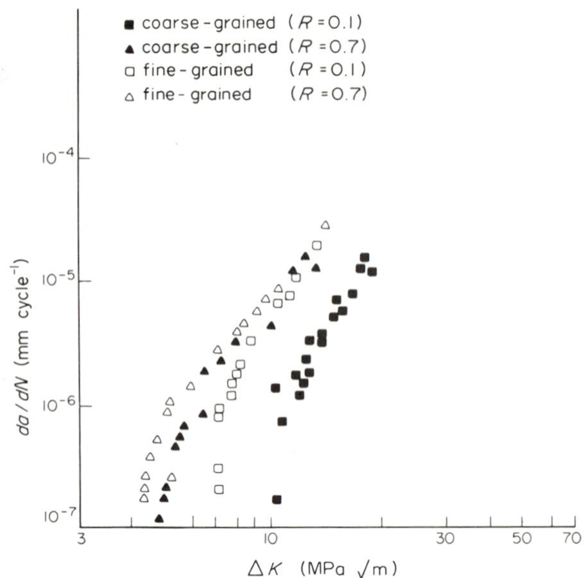

Figure 13
Fatigue crack propagation in a fully pearlitic hot-rolled eutectoid rail steel ($\sigma_y = 410$ MPa), tested in moist ambient temperature air at $R = 0.1$ and 0.7 and a frequency of 30 Hz, showing lower near-threshold growth rates and higher threshold ΔK_{th} values in coarser-grained microstructure ($d_g = 130$ μm) than in finer-grained structure ($d_g = 25$ μm) (note that major influence of grain size is at low load ratios, with little effect at $R = 0.7$) (after Gray et al. 1981)

2. *Fatigue Crack Closure*

Crack closure, or more precisely plasticity-induced crack closure, can be considered as arising from interference between crack surfaces in the wake of the crack tip, resulting from the constraint of surrounding elastic material on the residual stretch in material elements previously plastically strained at the tip. This process is aided by the fact that the crack is growing into a reversed plastic zone in which a state of residual compressive stress exists. Since the crack cannot propagate while it remains closed, the net effect of this closure is to reduce the nominal stress intensity range ΔK to some lower effective value ΔK_{eff} actually experienced at the crack tip:

$$\Delta K = K_{max} - K_{min} \quad \text{(no closure)}$$
$$\Delta K_{eff} = K_{max} - K_{cl} \quad \text{(with closure)} \quad (3)$$

where K_{cl} is the stress intensity to close the crack ($\geq K_{min}$).

This phenomenon was demonstrated effectively by Elber (1971) for crack propagation at high stress intensities in aluminum alloys by mounting surface strain gauges of very small gauge length (so-called Elber gauges) at the tips of fatigue cracks. The resulting elastic compliance curves are illustrated schematically in Fig. 15, where it is apparent that a fatigue crack of length a initially "opens" as if no crack were present before following the expected compliance slope for a physically open flaw of length a. The transition point in this curve thus defines the physical opening or closing load and hence the value of K_{cl}. Subsequently, other experimental measurements of closure loads have been performed using crack mouth opening gauges, back-face strain gauges, electrical potential measurements and ultrasonic techniques, although the data from the different procedures are not always totally consistent.

The closure concept has proved to be extremely useful in explaining, at least qualitatively, many aspects of fatigue crack propagation behavior, including the existence of a threshold, the role of variable amplitude loading and the observations of mean

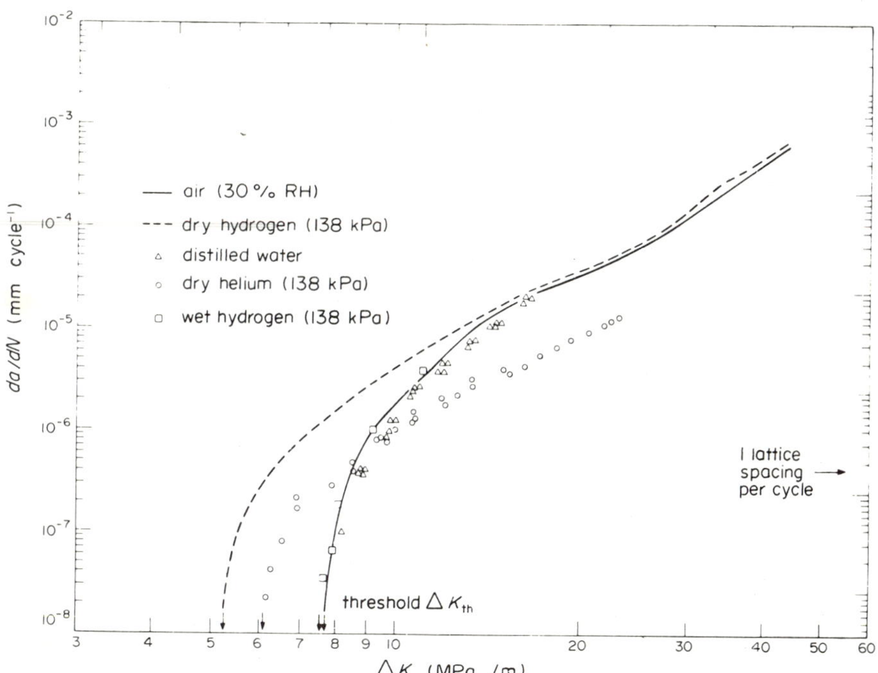

Figure 14
Fatigue crack propagation in bainitic 2.25%Cr–1%Mo, SA 542 class 3 ($\sigma_y = 500$ MPa, frequency = 50 Hz, $R = 0.05$ and ambient temperature), showing influence of environment at near-threshold and mid-growth rate regimes (note that above $\sim 10^{-5}$ mm per cycle, growth rates in hydrogen and air are similar at this high frequency, whereas growth rates in inert helium gas are slower; in contrast, below $\sim 10^{-6}$ mm per cycle, near-threshold growth rates are fastest in dehumidified environments—dry hydrogen and dry helium—and slowest in moist environments—wet hydrogen, moist air and water) (after Suresh et al. 1981)

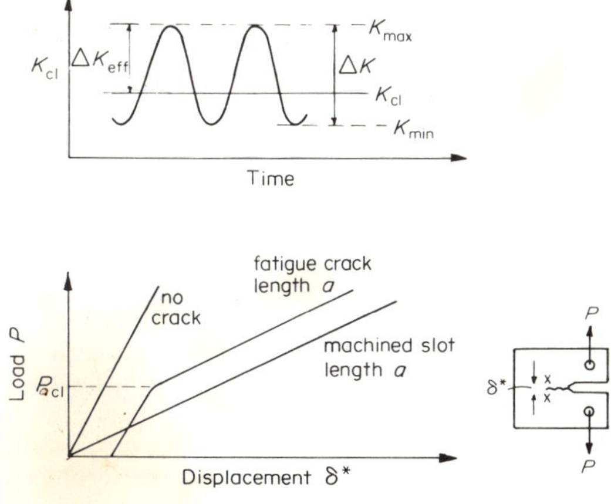

Figure 15
Procedures for experimentally demonstrating plasticity-induced crack closure, showing elastic compliance curves for uncracked test piece, test piece containing a finite width slot of length a, and test-piece containing fatigue crack of length a (note how fatigue cracked specimen does not appear to indicate presence of crack until above the closure load) (after Elber 1971)

stress-sensitive crack growth. For the latter case, since as the mean stress or load ratio is raised, the crack can be considered to remain open for a larger proportion of the cycle, it follows that, at a fixed nominal ΔK, a higher R value results in a higher ΔK_{eff} and hence a faster growth rate. However, once the minimum stress intensity K_{min} exceeds K_{cl}, the crack remains open during the entire loading cycle, so that above the critical load ratio R_{cr}, closure effects become insignificant as $\Delta K_{eff} = \Delta K$, and further increases in the mean stress have little influence.

One of the more serious objections to this concept is that plasticity-induced closure is most prevalent under essentially plane stress conditions, yet in the mid-range of growth rates where such conditions often exist, mean-stress effects are generally minimal (Fig. 8). Furthermore, the major influence of mean stress is usually found at the ultralow growth rates associated with near-threshold fatigue, where invariably plane strain conditions are present. Recent experimental compliance measurements on near-threshold cracks have confirmed that very significant effects of closure are present in plane strain, although it is difficult to rationalize these effects solely in terms of the plasticity arguments of Elber. Accordingly, several "microscopic" mechanisms of crack closure have been identified recently, based on the role of crack surface corrosion deposits and fracture surface roughness or morphology. These additional closure mechanisms are relevant specifically to the near-threshold regime, since it is principally here that the size of the corrosion debris and the roughness of the fracture surface are comparable with the crack tip opening displacements.

2.1 Oxide-Induced Crack Closure

Oxide-induced crack closure arises from the fact that during near-threshold fatigue crack propagation at low load ratios in moist environments, corrosion deposits of a thickness comparable with the size of crack tip opening displacements can build up near the crack tip, thus providing a mechanism for enhanced closure so that the crack becomes "wedged-closed" at stress intensities above K_{min} (Fig. 16). Such oxide films (Fig. 17) are thickened at low load ratios by a mechanism of fretting oxidation—a continual breaking and reforming of the oxide scale behind the crack tip due to a smashing together of the crack surfaces as a result of plasticity-induced closure aided by the strong mode II displacements characteristic of near-threshold crack growth.

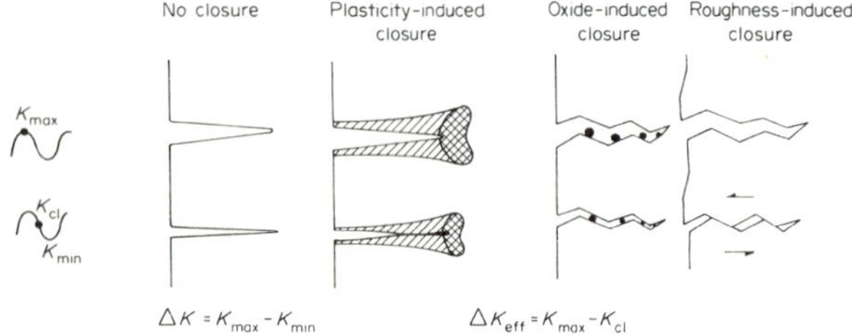

Figure 16
Schematic illustration of mechanisms of crack closure arising from plasticity, oxide debris and fracture surface roughness

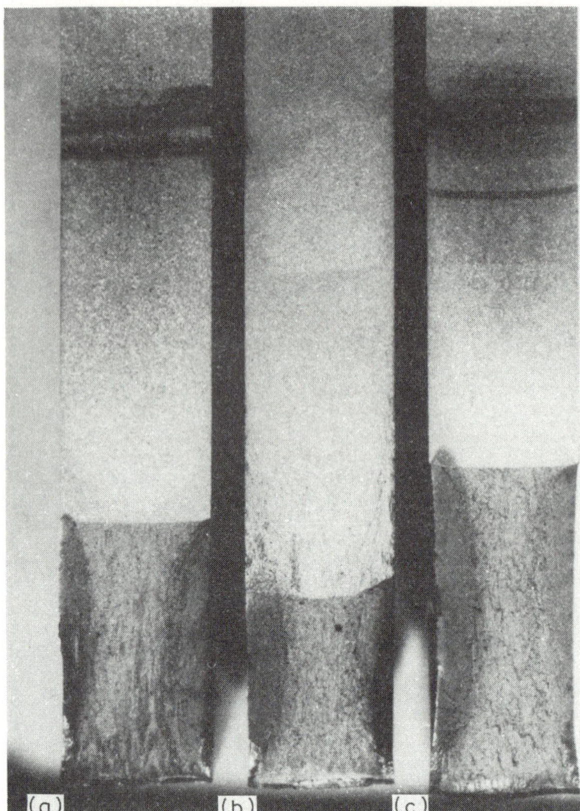

Figure 17
Bands of visual corrosion deposits from close to ΔK_{th} during fatigue crack propagation in a HP 9-4-20 (9% Ni–4%Co) steel tested in moist air: (a, c) near-threshold tests ($da/dN < 10^{-6}$ mm per cycle); (b) mid-growth rate test ($>10^{-5}$ mm per cycle) (all tests at low load ratios of $R = 0.05$–0.10)

For example, in lower-strength steels tested in moist air at $R = 0.05$, oxide films have been observed near ΔK_{th} with a maximum thickness some 20–40 times the limiting thickness of films formed naturally on metallographically polished samples exposed to the same environment for similar periods. The closure effect promoted by such deposits has been substantiated by Auger measurements of oxide thicknesses (Fig. 18), direct experimental closure measurements using ultrasonics and simple modelling studies involving rigid wedges inserted into elastic cracks. Since the formation of enhanced crack surface oxide deposits is far less predominant in dry, oxygen-free environments, at high load ratios (where plasticity-induced closure effects are minimal) and in higher-strength materials (where fretting oxidation is less effective), the resulting concept of oxide-induced crack closure provides an attractive explanation of many of the somewhat surprising effects of microstructure and environment on near-threshold behavior (outlined above).

First, the occurrence of oxide-induced closure can be associated directly with the large load ratio effects observed for near-threshold growth and the value of ΔK_{th} (Fig. 8), since enhanced crack face oxidation cannot occur (except in very oxidizing environments) at high load ratios (above R_{cr}), as the crack is sufficiently open to prevent fretting oxidation. This is consistent with observations of a substantial reduction in load ratio effects in the mid-range of growth rates where oxide-induced closure effects are minimal.

Second, the reported reduction in threshold values and increase in near-threshold crack propagation

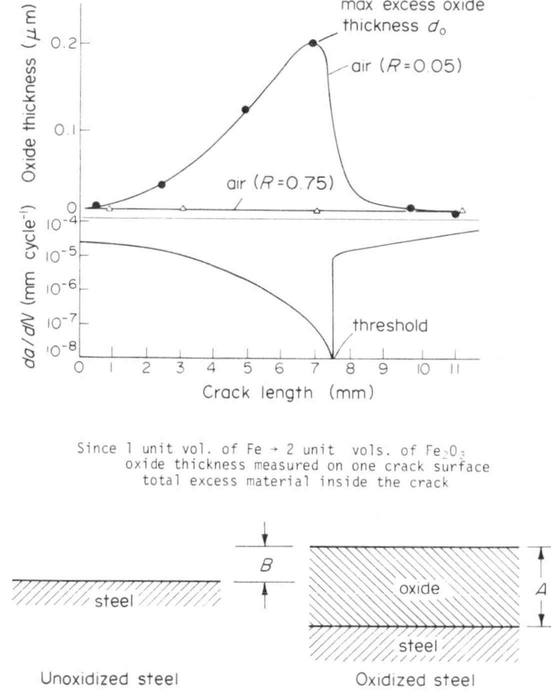

Figure 18
Measurements of fatigue fracture surface oxidation products in lower strength steel (2.25%Cr–1%Mo steel, SA 542 class 3), showing oxide film thickness at ambient temperature and a frequency of 50 Hz as a function of crack length (measured from notch) and crack growth rate da/dN during typical load-shedding near-threshold testing procedure; growth rates measured in moist air and dry hydrogen at $R = 0.05$ and 0.75, and oxide data generated from Ar^+ sputtering analysis using Auger spectroscopy (note enhanced oxidation close to threshold ΔK_{th} at low load ratios compared with minimal excess oxidation at high load ratios) (after Suresh et al. 1981)

rates in steels with increasing strength level (Fig. 7) is also in accordance with this mechanism, since oxide formation inside near-threshold cracks is increasingly restricted in higher-strength materials (presumably because plasticity-induced closure is reduced and fretting oxidation is less predominant). The fact that the influence of strength level on near-threshold behavior is far less apparent at high load ratios is totally consistent with this.

Third, and most important, the concept of oxide-induced crack closure provides a very feasible explanation for the complex role of environment at near-threshold levels. Environmentally influenced near-threshold fatigue crack propagation can be interpreted in terms of competition between two concurrent processes: crack closure due to corrosion debris (which decreases growth rates) and hydrogen embrittlement or active path corrosion arising from the production of hydrogen vis-a-vis the oxidation process (which increases growth rates). In lower-strength steels ($\sigma_y < 1000$ MPa), for example, the susceptibility to hydrogen embrittlement is relatively low, particularly at the high frequencies ($\gtrsim$50 Hz) commonly employed for near-threshold measurements, and thus the retarding influence of oxide-induced closure appears to dominate overall behavior. Hence, faster near-threshold crack growth rates are to be expected in dry hydrogen or helium, compared with the moist environments of air, wet hydrogen and water, because behavior in the dry atmospheres is retarded less by closure (Figs. 9, 14).

Furthermore, these differences should become insignificant at high load ratios, since closure phenomena will be ineffective. This is shown by the similar near-threshold growth rates and ΔK_{th} values for 2.25%Cr–1%Mo steel tested at $R = 0.75$ in dry hydrogen and moist air, compared with the large effect of environment for tests at $R = 0.05$ (Fig. 9). Thus, because of this important influence of closure mechanisms at low stress intensity ranges, where cyclic CTOD values approach the size of the oxide scales, the nature and characteristics of environmentally influenced crack propagation behavior at near-threshold levels is often radically different from the more customary corrosion fatigue behavior observed at higher growth rates.

Finally, the concept of oxide-induced closure may provide one important rationale for the very existence of a fatigue crack propagation threshold. Comparisons between the cyclic crack tip opening displacements and the maximum thickness of the excess oxide scale indicate that there is a one-to-one correspondence at the threshold (Fig. 19). Since the

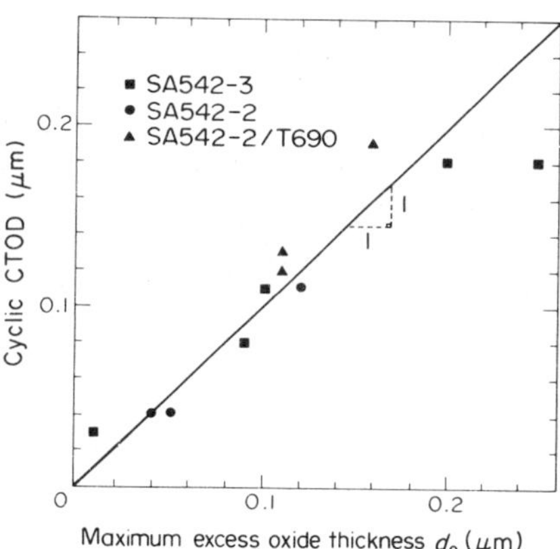

Figure 19
One-to-one correspondence of maximum excess crack surface oxide thickness and cyclic crack tip displacement ΔCTOD at the threshold (i.e., at ΔK_{th}, $d_o \sim$ ΔCTOD), for a range of lower-strength 2.25%Cr–1%Mo steels (σ_y = 290–770 MPa) tested in air, water, hydrogen and inert gas environments (after Ritchie and Suresh 1983)

crack is unable to propagate once it is totally wedged-closed with debris, this result suggests one physically appealing explanation for the threshold—an explanation which is consistent with data which indicate that near-threshold growth rates in vacuo, where presumably little oxide debris are generated, show no well defined threshold and actually can exceed growth rates in air at very small crack tip displacements.

2.2 Roughness-Induced Crack Closure

A further mechanism of crack closure of specific relevance to near-threshold behavior is that induced by fracture surface roughness or morphology (Fig. 16). Such roughness-induced closure can arise if the fracture surface roughness is comparable in size with crack tip opening displacements and if significant mode II displacements exist, such that the crack can become wedged-closed at discrete contact points along the crack faces (Fig. 20). Like oxide-induced closure, this mechanism is most prevalent at low stress intensity ranges, where maximum plastic zone sizes typically are smaller than microstructural dimensions. In this regime, crack advance proceeds by a single shear mechanism, with associated mode I plus mode II displacements, resulting in serrated fracture paths (Fig. 12) which enhance the closure effect.

Such roughness-induced closure provides an important contribution to the role of microstructure in influencing near-threshold crack growth and in fact has wide significance for a large range of engineering materials (e.g., titanium alloys which form little oxide debris). For example, with regard to the well known observations of improved near-threshold crack growth resistance in coarser-grained materials (e.g., Fig. 13), coarse microstructures tend to promote rougher fracture morphologies and further delay the transition from a serrated (single shear) to a more planar (alternating or simultaneous shear) fracture surface to higher ΔK values when the maximum plastic zone size is of the order of the grain size. Thus, coarse microstructures generate greater closure, making effective ΔK values lower, at the same nominal ΔK, than in finer-grained microstructures. This explanation is entirely consistent with the fact that at high load ratios where closure effects are minimal, the effect of grain size on near-threshold behavior is reduced markedly (Fig. 13).

Simple analytical expressions for both oxide-induced and roughness-induced crack closure mechanisms are summarized in Fig. 21. Both mechanisms predominate at near-threshold levels, but it is worth noting that their effect is not necessarily unique to this regime. Given a sufficient degree of faceted fracture morphology, such as in the case of "crystallographic" fatigue crack propagation in nickel-base superalloys, or substantial crack face oxidation, which may occur for elevated temperature creep or fatigue crack growth, such mechanisms may be active at the larger ΔCTOD values associated with cyclic crack growth at higher ΔK levels. Their effect, however, is maximized at near-threshold stress intensities, due to the single shear (mode I + mode II) character of the crack extension mechanisms there and to the small magnitude of the associated crack tip opening displacements.

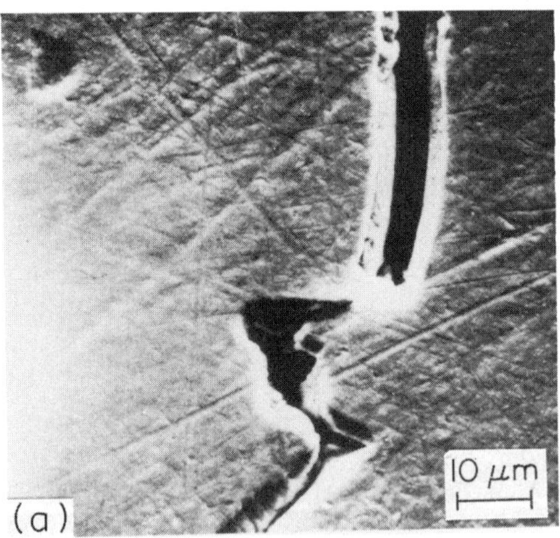

Figure 20
Example of crack closure due to near-threshold fracture morphologies from replicas of a fatigue crack in α-titanium ($\Delta K \approx 9$ MPa $m^{0.5}$, $R = 0.1$), taken at minimum load fatigue cycle at (a) 9.2 mm and (b) 14.5 mm from crack tip (note contact between faceted crack surfaces at discrete points, aided by the mode II crack tip displacements, leading to roughness-induced crack closure) (after Walker and Beevers 1979)

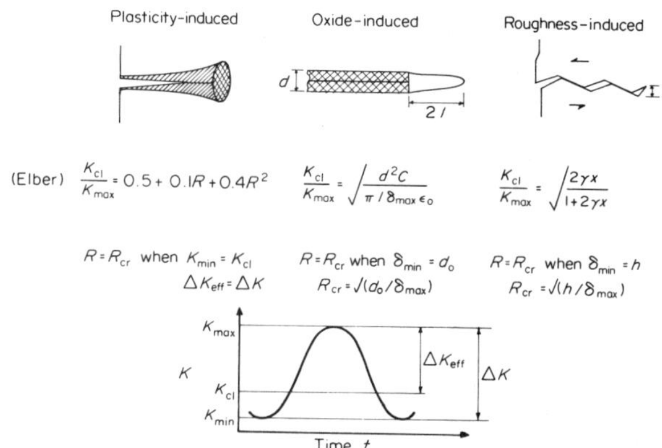

Figure 21
Summary of the various mechanisms of fatigue crack closure and their respective analytical expressions (δ = crack tip displacement; ε_0 = yield strain; d_o = total oxide thickness at ΔK_{th}; C = constant ($\sim 1/32$); x = ratio of mode II to mode I crack tip displacements; γ = nondimensional roughness factor, defined as the ratio of the height h of fracture surface asperity to its width w; R_{cr} = critical load ratio above which closure effects are minimal, at given ΔK) (after Ritchie and Suresh 1983)

3. *Short Fatigue Cracks*

As outlined in the article *Fatigue Crack Growth: Macroscopic Aspects*, short fatigue cracks have been observed experimentally to show crack growth kinetics which can be markedly different from those commonly observed for long cracks (i.e., typically $\sim$25 mm in length in commonly used geometries such as the compact specimen). There are several factors which define a short crack, namely cracks which are (a) of a length comparable with the scale of microstructure, (b) of a length comparable with the scale of local plasticity, and (c) simply physically small (<0.5–1 mm).

The first of these cases, the microstructurally small crack, represents a continuum mechanics limitation to current analysis procedures. Although much metallurgical research has focused on this topic, it is generally not a major problem in the engineering sense, since microcracks of this size are not readily detectable by nondestructive testing and their formation and growth are generally classified in terms of the crack initiation stage of an engineering-sized flaw. It is important to note, however, that metallurgically the microstructures which show optimum resistance to such microcrack growth (or crack initiation) do not necessarily guarantee optimum resistance to the propagation of long cracks. A classic example of this is the well known fact that the smooth bar fatigue limit in steels (which represents an effective threshold for crack initiation) increases with increasing strength level, whereas the crack propagation threshold ΔK_{th} decreases. This poses something of a dilemma in the selection and design of alloys, such that the choice of a material for a given fatigue-sensitive application often depends critically upon whether design is based on crack initiation (i.e., using *S–N* curves) or on crack propagation (i.e., using defect tolerance).

The majority of short crack research has focused on cracks which are comparable in size with the scale of local plasticity. This case arises when the crack length is of the order of the plastic zone size or when cracks emanating from notches are growing in the strain field of the notch, and represents a linear elastic fracture mechanics limitation to current design methodologies. An example of this is given in Fig. 22, where the threshold conditions for no fatigue failure are shown as a function of crack size. For long cracks, the threshold condition is simply given, in fracture mechanics terms, as one of a constant threshold stress intensity range ΔK_{th}, whereas for very small cracks the data approach the smooth bar fatigue limit, or endurance strength $\Delta\sigma_e$, of a constant threshold stress. It is readily apparent that extending the linear elastic fracture mechanics analysis of a constant ΔK_{th} to small crack sizes will result in dangerously nonconservative predictions. The transition between long and short crack behavior

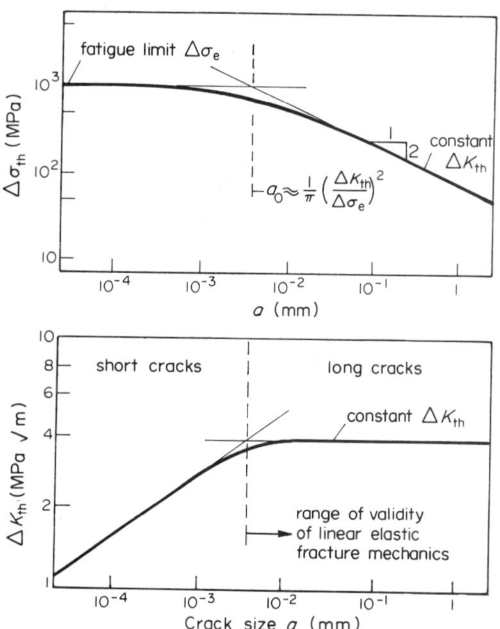

Figure 22
Variation with crack size a of threshold stress range $\Delta\sigma_{th}$ and threshold stress intensity range ΔK_{th} for no crack growth in 300-M steel (σ_y = 1400 MPa) at $R = 0$

here is often known as the intrinsic crack size a_0 and is given by

$$a_0 \approx \frac{1}{\pi}\left(\frac{\Delta K_{th}}{\Delta \sigma_e}\right)^2 \tag{4}$$

where $\Delta\sigma_e$ is the fatigue limit corrected for the appropriate load ratio. Substitution of typical values into Eqn. (4) reveals a_0 to be of the order of 1 μm ultrahigh-strength materials ($\sigma_y \sim 2000$ MPa) and ~1 mm in low-strength materials ($\sigma_y \sim 300$ MPa), indicating that this problem is most predominant in lower-strength alloys.

Allowing for the geometrical factors associated with notches, the discrepancy between long and short crack behavior in this case has been traced largely to inappropriate fracture mechanics characterization. Whereas the use of the stress intensity range ΔK may be valid for long cracks, when crack sizes approach the dimension of the plastic zone, conditions of small-scale yielding no longer apply and analyses based on a linear elastic K_I essentially are meaningless. Accordingly the growth of such short cracks can be treated using elastic–plastic fracture mechanics, so that when compared on the basis of ΔJ or ΔCTOD, the propagation rates of long and short cracks show a significantly closer correspondence (Fig. 23).

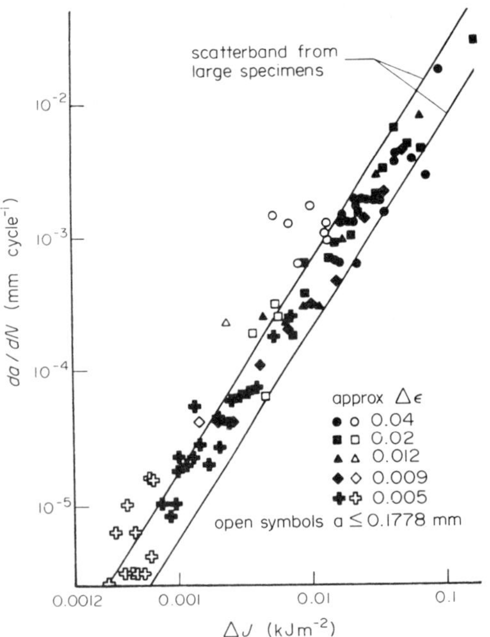

Figure 23
Comparison of fatigue crack propagation rates da/dN in A533B nuclear pressure vessel steel ($\sigma_y = 480$ MPa) obtained from experiments with long cracks ($a \sim 25$ mm) and short cracks ($a < 0.18$ mm) (after Dowling 1978)

The third type of short crack, the physically short flaw, perhaps is the most dangerous, since it may be considered "long" in terms of both continuum mechanics and linear elastic fracture mechanics analyses, yet it may still show faster propagation rates than corresponding long cracks at the same ΔK level. Two reasons have been advanced for such behavior. The first is associated with crack closure. Since the origin of the differences in behavior resulting from a contribution from crack closure arises from the fact that such closure effects predominate in the wake of the crack tip, and since, by definition, short cracks possess only a limited wake, it is to be expected that in general such cracks will be subjected to less closure. Thus at the same nominal ΔK, short cracks may experience a larger effective ΔK compared with the equivalent long crack. This may arise from a restricted envelope of prior plastic zones in the wake of the tip, less corrosion debris or less roughness-induced closure generated on the smaller crack surface. In fact the latter effect has been reported: experimental measurements of the crack opening displacements (at zero loads) of small surface cracks in titanium alloys clearly have shown far less closure associated with the short flaw. Thus at equivalent nominal ΔK levels, the short crack would be expected to propagate faster (and show a lower threshold ΔK_{th}) than the corresponding long crack because the effective stress intensity range experienced at the crack tip would be greater.

The second factor is associated with electrochemical effects and is relevant to the growth of physically short flaws in aqueous corrosive environments. Experiments by Gangloff (1981) on high-strength 4130 steel tested in NaCl solution revealed that corrosion fatigue crack propagation rates of short cracks (<0.8 mm) were more than two orders of magnitude faster than corresponding rates of long cracks (>25 mm) at the same ΔK level, although behavior in inert atmospheres was essentially similar. A complete understanding of this phenomenon is as yet lacking, but preliminary analysis indicated that the effect could be attributed to different local crack tip environments in the long and short flaws, principally resulting from the influence of crack length on crack surface reactions and on the solution renewal rate at the crack tip region.

4. Variable Amplitude Loading

Mode I fatigue cracks subjected to broad band variable amplitude loading spectra show transient crack growth behavior in the form of local accelerations or decelerations, depending upon the magnitude and sequencing of the applied loads (see *Fatigue Crack Growth: Macroscopic Aspects*). For example, the application of a single positive (spike) overload of sufficient size (generally >50% of the baseline ΔK) can result in significant retardations in crack growth over crack lengths comparable with the overload plastic zone size. Furthermore, low–high block loading sequences can yield local accelerations, whereas high–low sequences can produce local decelerations (and sometimes even arrest). The origin of these interaction effects is currently uncertain, but mechanisms involving crack tip blunting, residual compressive stress fields and fatigue crack closure concepts have been suggested which appear, at least qualitatively, to rationalize the data.

Recent analysis of the post-overload retardation effect by Suresh (1982) have suggested that during the overload cycle, the crack tip profile becomes blunted, often with one or two 45° branch cracks subsequently emanating from the tip. The reduction in effective stress intensity range resulting from this blunted and branched crack, coupled with the fact that the crack is now propagating into the compressive residual stress field of the overload plastic zone, acts to retard crack growth to near-threshold levels and in certain cases to arrest it. Furthermore, the retardation can be enhanced during the initial post-overload region because the crack is experiencing effective stress intensities ranges in the near-threshold regime, such that the process of crack advance becomes influenced by additional (near-threshold) crack closure mechanisms (see Sect. 2).

Specifically, the branched morphology of the overloaded crack together with the single shear faceted nature of the subsequent (near-threshold) crack growth can promote significant roughness-induced crack closure, evidenced by severe abrasion between the crack faces in the post-overload zone (region C in Fig. 24). It is also apparent from Fig. 24 that the stretch zone (region B) created at the application of the overload is free from abrasion marks, which suggests that the initial blunting of the crack tip minimizes the effect of plasticity-induced crack closure (due to the residual plastic displacements created during the overload cycle) in contributing to the retardation in crack growth. Similar arguments involving crack tip blunting and branching, residual compressive stress and crack closure mechanisms have been advanced to rationalize transient crack growth behavior following spectrum and block loading sequences.

One final point of interest with regard to the influence of variable amplitude loading is that for fatigue crack growth in antiplane shear (i.e., in mode III rather than in mode I), the response to single overload cycles and high–low block loading sequences is almost exactly opposite to that observed for mode I cracks. This difference in the transient growth rate behavior of mode III and mode I cracks follows from the fact that, in mode III, cracks are not influenced by such mechanisms as crack tip blunting or branching and fatigue crack closure (other than rubbing between sliding fracture surfaces), which so profoundly affect the behavior of mode I cracks. Accordingly, analyses of the role of variable amplitude loading sequences on mode III crack growth can be based more on the Miner's rule approach, in that the transient behavior can be rationalized simply in terms of the damage accumulated within the reversed plastic zones for each individual load reversal.

5. Conclusion

This article has attempted to provide a simplified basis for understanding the physical mechanisms associated with fatigue crack propagation in metals

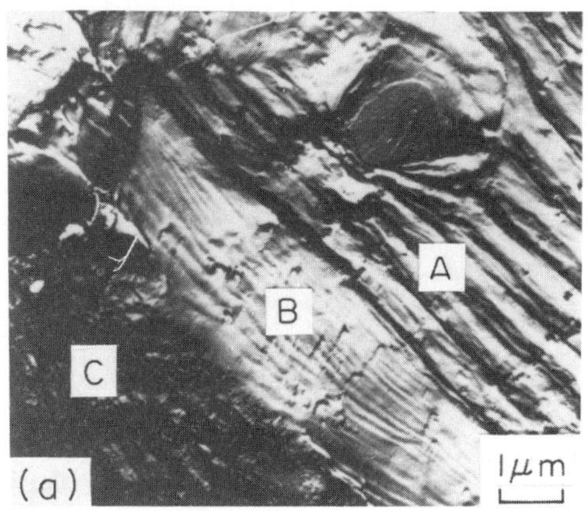

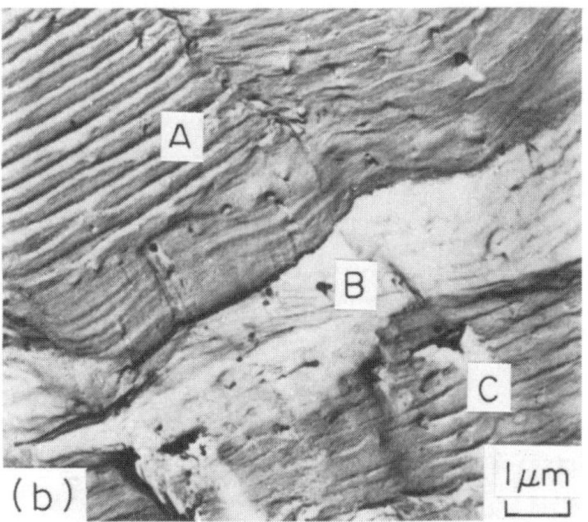

Figure 24
Transmission electron micrographs showing fracture surface appearance of a 2024-T3 aluminum alloy ($\sigma_y \approx 380$ MPa) during a single positive (spike) overload sequence: the crack is propagating from the top right-hand corner to the lower left-hand corner and indicates ductile striations in the pre-overload zone (A), a relatively featureless stretch zone (B) at the overload cycle and a post-overload zone (C) showing severe signs of crack surface abrasion (after Hertzberg 1976)

and alloys from both microscopic (metallurgical) and macroscopic (continuum mechanics) viewpoints. Although fatigue represents one of the major causes of failure in engineering service, a complete understanding of how a fatigue crack actually propagates has not yet been reached. Good progress has been made with the continuum mechanics characterization of cyclic crack growth rates and fracture mechanics analyses are now in widespread use for defect-tolerant design codes. Similarly, an understanding of the role of microstructure in improving the resistance to fatigue crack growth has emerged to the point where alloy design guidelines exist for the production of alloys with optimum resistance to fatigue failure.

However, much work remains in the definition of mechanisms associated with environmentally influenced crack growth, with the effect of variable amplitude loading and particularly with the problem of the short flaw. These problems demand an interdisciplinary approach to fatigue research involving applied mechanics, materials science and surface chemistry studies, and clearly offer substantial opportunities for further investigations, both of a fundamental nature and to provide reliable engineering data needed in the design and maintenance of fatigue-critical structures.

See also: Fatigue Cracks: Resistance to Initiation; Fatigue Cracks: Resistance to Propagation; Mechanics of Materials: An Overview; Fatigue Crack Growth: Macroscopic Aspects

Bibliography

Beevers C J 1977 Fatigue crack growth characteristics at low stress intensities of metals and alloys. *Met. Sci.* 11: 362–67

Elber W 1971 *Damage Tolerance in Aircraft Structures*, ASTM STP 486. American Society for Testing and Materials. Philadelphia, Pennsylvania, p. 230

Fine M E, Ritchie R O 1979 *Fatigue and Microstructure*. American Society for Metals Metals, Park, Ohio, p. 245

Gangloff R P 1981 The criticality of crack size in aqueous corrosion fatigue. *Res. Mech. Lett.* 1: 299–306

Hertzberg R W 1976 *Deformation and Fracture Mechanics of Engineering Materials*. Wiley, New York

Laird C 1967 *Fatigue Crack Propagation*, ASTMP STP 415. American Society for Testing and Materials, Philadelphia, Pennsylvannia, p. 131

Lindley T C, Richards C E, Ritchie R O 1976 Mechanics and mechanisms of fatigue crack growth in metals: A review. *Metall. Met. Form.* 43: 268–80

McClintock F A 1963 In: Drucker D C, Gilman J J (eds.) 1963 *Fracture of Solids*. Interscience, New York, p. 65

McEvily A J 1983 *Quantitative Measurement of Fatigue Damage*. American Society for Testing and Materials, Philadelphia, Pennsylvania, p. 283

Minakawa K, McEvily A J 1981 On crack closure in the near-threshold region. *Scr. Metall.* 15: 633–65

Neumann P 1974 New experiments concerning the slip processes at propagating fatigue cracks—I. *Acta Metall.* 22: 1155–65

Paris P C, Erdogan F 1963 A critical analysis of crack propagation laws. *J. Basic Eng.* 85: 528–34

Pelloux R M N 1969 Mechanisms of formation of ductile fatigue striations. *Trans. Am. Soc. Met.* 62: 281–85

Ritchie R O 1982 In: Bäcklund J, Blom A, Beevers C J (eds.) 1982 *Fatigue Thresholds*, Vol. 1. Warley, UK p. 503

Ritchie R O, Knott J F 1973 Mechanisms of fatigue crack growth in low alloy steel. *Acta Metall.* 21: 639–48

Ritchie R O, Suresh S 1983 Behavior of short cracks in airframe components. *Proc. 55th Specialists Meeting of AGARD Structural and Materials Panel*, AGARD Vol. CP328. North Atlantic Treaty Organization, Advisory Group for Aerospace Research and Development, Paris, p. 1.1

Starke E A, Lütjering G 1979 *Fatigue and Microstructure*. American Society for Metals, Metals Park, Ohio, p. 205

Suresh S, Zamiski G F, Ritchie R O 1981 Oxide-induced crack closure: An explanation for near-threshold corrosion fatigue crack growth behavior. *Metall. Trans. A* 12: 1435–43

Tomkins B 1968 Fatigue crack propagation—An analysis. *Phil. Mag.* 18: 1041–66

Weertman J 1979 *Fatigue and Microstructure*. American Society for Metals, Metals Park, Ohio, pp. 279

Wei R P, Speidel M O 1971 In: Devereux O, McEvily A J, Staehle R W (eds.) 1971 *Corrosion Fatigue*. National Association of Corrosion Engineers, Houston, p. 433

Wei R P, Simmons G W 1982 Fatigue—environment and temperature effects. *Proc. 27th Sagamore Army Materials Research Conf.* AMMRC, Watertown, Massachusetts

R. O. Ritchie

Fatigue Cracks: Resistance to Initiation

The fatigue of metals involves cyclic plastic deformation, and fatigue cracks initiate at locations where the plastic strain is higher than the average. In many cases, cracks result from flaws or poor design features which act to concentrate stress and cause plastic flow at low loads. However, fatigue cracks may also form under cyclic loading in the absence of stress risers when the plastic strain amplitude is high enough to cause irreversible deformation. Under this condition fatigue crack initiation is strongly influenced by both microstructure and environment. Microstructure controls the slip and/or twinning modes and subsequent crack initiation behavior. Conversely, cyclic deformation may alter the microstructure, leading to instabilities that determine the fatigue performance of a material.

Fatigue cracks have been observed to initiate at (a) inclusions, (b) second phase–matrix interfaces, (c) intense slip bands, (d) grain boundaries and grain-boundary triple junctions. These all usually involve some form of strain localization, and consequently microstructures that distribute the plastic strain homogeneously give better fatigue crack initiation resistance. Selected alloy modifications and new primary processing procedures offer promise for developing the desired microstructures.

1. Inclusions

Many commercial alloys have impurity inclusions (constituent particles) that may affect the fatigue crack initiation behavior, especially at low strain amplitudes. In the high strain, low-cycle regime, plastic strains are usually large enough to create intense slip bands which may interact with inclusions and second-phase particles to produce crack nuclei. However, inclusions are most detrimental in the low strain, high-cycle regime, when bulk cyclic plastic strains are small or absent, because they intensify the applied stress and cause local regions of plastic deformation. Stress and strain localization also occurs as a result of differential thermal contraction during cooling and concentration of applied stress, this differential contraction being due to elastic constant differences between the matrix and inclusion. The localized stress σ_T due to thermal contraction is of the form

$$\sigma_T = \phi(\alpha_2 - \alpha_1)\Delta T$$

where ϕ is a function of the elastic moduli of the matrix and inclusion, and also of the inclusion size, shape and location; α_1 and α_2 are the coefficients of thermal expansion of the inclusion and matrix, respectively; and ΔT is the cooling range. For most oxides and intermetallics in a metal matrix, the stress field in the matrix is tensile (i.e., $\alpha_1 < \alpha_2$) and has a symmetry corresponding to that of the inclusion. The applied stress is concentrated in the vicinity of the inclusion by a factor β that also depends on the size, shape and location of the inclusion and the ratio $E_{\text{inclusion}}/E_{\text{matrix}}$. Large tensile stress concentrations may occur at the polar points of inclusions when the Young's modulus ratio is greater than one. Inclusions giving a ratio less than one are usually not detrimental.

The size of the inclusion is also important in determining whether it has a detrimental or beneficial effect. Larger inclusions induce larger tensile stress under cyclic loading and hence are prone to earlier crack nucleation. The probability of an inclusion initiating a fatigue crack decreases rapidly as the size decreases below a few μm, but the critical size depends on its location with respect to the surface. Most current work shows that inclusions larger than 5 μm normally have an adverse effect on the fatigue life, whereas those smaller than 0.1 μm have no effect or are beneficial. The beneficial effect of the very small particles arises from their role in homogenizing deformation in planar-slip-mode materials.

Two other factors concerning inclusions must also be considered as they can be related to the fatigue crack initiation mechanisms. In addition to the slip band–particle interaction that is important in low-cycle fatigue, crack initiation in the high-cycle regime occurs by debonding of the particle–matrix interface or by spreading of cracks from precracked inclusions. Consequently, the matrix–inclusion bonding and the formability of the inclusion are important factors. If inclusions are deformable, matrix–inclusion bonds may remain unbroken, and those inclusions which do not debond do not appear to nucleate fatigue cracks.

Improvements in fatigue crack initiation resistance in the high-cycle regime can be obtained by lowering impurity levels, increasing the formability of the inclusions, or reducing their size below some critical value. The most dramatic improvement in the endurance limit of engineering materials has come through the reduction of the volume fraction of impurity inclusions. Increases in the fatigue strength of 10–20% are commonly obtained by carefully controlling composition and using improved processing techniques. In many materials, however, it is impractical to attempt to elminate all inclusions and other approaches must be taken. A reduction in the size of inclusions is an effective method of reducing their influence. One way a small size can be obtained is by rapid quenching from the melt of fine particulate, and powder metallurgy consolidation. Rapid solidification of prealloyed liquids leads to improved homogeneity, structural refinement, and control of grain size and precipitate distributions in the resulting solid.

2. Second Phases

2.1 Coherent Precipitates

The high strength of many engineering materials is due to the interaction of dislocations with precipitates that may be coherent, partially coherent or incoherent with the matrix. Depending on the size, spacing and degree of coherency, the precipitates are either sheared or looped and bypassed by dislocations during plastic deformation. The strengthening associated with coherent and partially coherent particles is reduced when they are sheared by dislocations. This decrease results successively from a local decrease in resistance to further dislocation motion, planar slip, concentration of slip in narrow bands, and destruction of the strengthening particles. The strain localization leads to slip offsets at surfaces (Fig. 1). Fatigue cracks can then initiate either at the grain boundaries or at the surface irregularities and propagate along the intense slip bands. Such strain localization also enhances environmental effects.

Although overaging changes the coherency of the precipitates and thus changes the deformation mode from coarse planar slip to fine wavy slip, overaging is not an effective method of homogenizing deformation since it is usually accompanied by a significant decrease in strength. However, in many cases the coherency can be modified by minor alloying

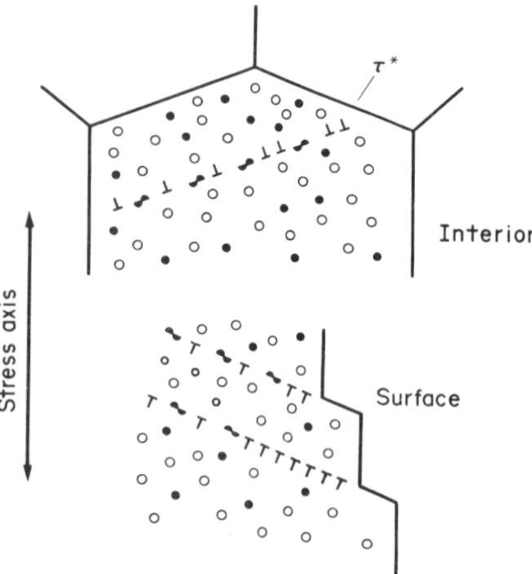

Figure 1
Schematic representation of strain localization associated with shearable precipitates

additions if the additions participate in the precipitation process. For example, the degree of coherency of the strengthening precipitates in Al–Zn–Mg alloys can be modified by additions of copper. Copper participates in the precipitation process during aging, changes the character of the precipitates and alters the type of interaction between precipitate and dislocation. In the maximum strength temper, the shearable precipitates of copper-free Al–Zn–Mg alloys promote strain localization; whereas in the same condition the precipitates in the Al–Zn–Mg–Cu alloys are looped by dislocations, deformation is more homogeneous and fatigue crack initiation is delayed (Fig. 2).

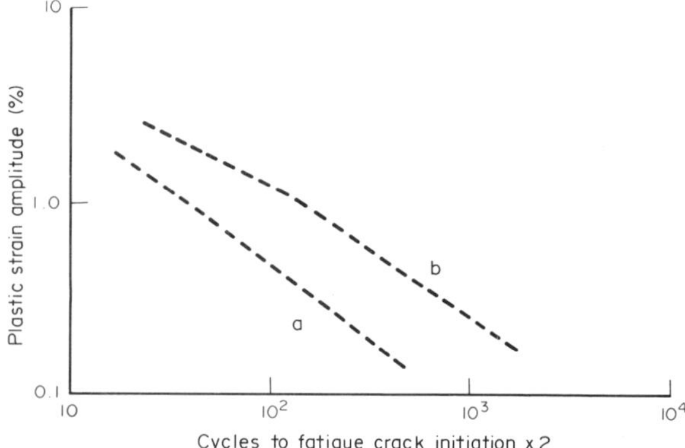

Figure 2
Plastic strain amplitude versus cycles to fatigue crack initiation tested in water for Al–6Zn–2Mg–XCu alloys having (a) shearable (0.0% Cu) and (b) nonshearable (2.1% Cu) precipitates

The degree of softening, the magnitude of the stress concentrations τ^* produced at grain boundaries, and the magnitude of the slip step at the surface all depend on the number of dislocations that can be accommodated on a given slip plane. This is controlled by the slip length and the reduction of particle hardening in the slip plane, both of which may be altered by modifications of the microstructure. The slip length can be minimized by reducing the grain size, by adding grain refiners, by applying thermal mechanical treatments or by using powder metallurgy processes. The effectiveness of grain boundaries in inhibiting slip depends on the degree of preferred orientation of the product, being somewhat ineffective in highly textured materials. The subgrains produced in many materials during hot working, and random dislocation networks developed during stretching and forming operations, may also aid in preventing the formation of intense slip bands. However, the networks must be stabilized by precipitates or else they break down during cyclic loading. In many commercial aluminum alloys, dispersoids are used for controlling recrystallization, grain size and shape. The dispersoids also homogenize slip and inhibit the formation of intense slip bands, which increases the fatigue crack initiation resistance.

2.2 Grain-Boundary Precipitates and Precipitate-Free Zones

Age-hardened alloys obtain their beneficial mechanical properties from homogeneous metastable precipitation below some critical temperature. However, heterogeneous precipitation is always possible below the normal solvus line and occurs concurrently with the homogeneous precipitation. Since all boundaries may lower the surface and strain energies of precipitates, nucleation of the stable incoherent phases occurs preferentially at grain boundaries. In many cases this results in a solute-depleted precipitate-free zone (PFZ) adjacent to the grain boundary. This zone is weaker than the matrix and can be the site of preferential deformation, which leads to high stress concentrations at grain-boundary triple junctions and hence to early crack nucleation. The extent of strain localization will depend on the difference in flow stress between the grain-boundary region and the adjacent matrix, and the magnitude of the stress concentration will depend on the extent of strain localization and the slip length. Soft grain-boundary phases such as the α phase along prior β grain boundaries in Widmanstätten α–β titanium alloys, or α-ferrite along prior austenite boundaries in martensitic steels may also lead to strain localization and low fatigue crack initiation resistance similar to that associated with PFZs.

In the event that formation of soft grain-boundary phases or PFZs cannot be prevented, their effect can be minimized by reducing the grain size and flattening the grains to a pancake shape. A reduction in grain size reduces the slip distance and lowers the stress concentrations at grain-boundary triple junctions. This reduces the probability of early fatigue crack initiation due to strain localization in these regions. Equiaxed grains normally have boundaries favorably oriented for slip. Many commercial alloys, such as high-strength aluminum alloys, have dispersoids that inhibit grain growth during high-temperature processing and subsequent heat treatment. For these alloys, the resulting grain shape is characteristic of the processing treatment; for example, rolled-plate grains are pancake shaped. If the stress axis is parallel or perpendicular to the long-grain dimension, there will be no shear stress parallel to the grain boundary, and preferential deformation within the grain-boundary region will be restricted. β processing of α–β titanium alloys also produces beneficial elongated grains. After the forming operation of Widmanstätten α–β titanium alloys is completed, thermal excursions above the β transus should be avoided since they cause equiaxed β grains with resultant strain-localization problems. Very fine grained equiaxed α–β titanium alloys can be produced by working the alloy in the α–β phase field followed by recrystallization treatments. These fine-grained structures are normally more resistant to fatigue crack initiation than the coarse Widmanstätten structures produced by β processing.

3. *Surface Structure Modification*

Fatigue crack initiation is usually associated with the cyclic deformation behavior of the surface layer. Any surface modification that reduces the magnitude or increases the homogeneity and reversibility of cyclic deformation is expected to improve the fatigue crack initiation resistance. It is well known that manufacturing processes that produce surface-compressive residual stresses (e.g., shot peening) can have a substantial effect on stress-controlled fatigue resistance. However, these beneficial effects may be removed by periodic overloads. Recent studies have shown that ion implantation may be used to modify the composition and microstructure of the surface layer to obtain that desired for fatigue crack initiation resistance. This can be accomplished without modifying the bulk properties (see *Ion Implantation Metallurgy*).

Microstructures that are resistant to strain localization are desirable for fatigue crack initiation resistance as is a small grain size. Soft grain-boundary regions and inhomogeneous microstructures are not desirable. At elevated temperatures, soft grain boundaries become the sites for the initiation of fatigue cracks. If shearable precipitates are necessary to produce the desired strength, they should be accompanied by small nonshearable dispersoids or a small or unrecrystallized grain structure in order to prevent strain localization. It should be remembered, however, that fatigue crack initiation is quite different from fatigue crack propagation, and microstructures that improve the resistance to one may lower the resistance to the other.

See also: Fatigue Cracks: Resistance to Propagation; Fatigue Crack Growth: Macroscopic Aspects; Fatigue Crack Growth: Mechanistic Aspects

Bibliography

Cohen M, Kear B H, Mehrabian R 1980 Rapid solidification processing—An outlook. In: Mehrabian R, Kear B H, Cohen M (eds.) 1980 *Rapid Solidification Processing—Principles and Technologies*, Vol. 2. Claitor, Baton Rouge, Louisiana, pp. 1–23

Fine M E 1980 Fatigue resistance of metals. *Metall. Trans. A* 11: 345–79

Fine M E, Ritchie R O 1979 Fatigue crack initiation and near-threshold crack growth. In: Meshii M (ed.) 1979 *Fatigue and Microstructure.* American Society for Metals, Metals Park, Ohio, pp. 245–78

Kujore A, Chakrabortty S B, Starke E A Jr, Legg K O 1981 The effect of ion implantation on the fatigue properties of polycrystalline copper. *Nucl. Instrum. Methods* 182/183: 948–59

Lankford J 1977 Effect of oxide inclusions on fatigue failure. *Int. Met. Rev.* 22: 221–28

Starke E A Jr, Lütjering G 1979 Cyclic plastic deformation and microstructure. In: Meshii M (ed.) 1979 *Fatigue and Microstructure.* American Society for Metals, Metals Park, Ohio, pp. 205–39

E. A. Starke Jr.

Fatigue Cracks: Resistance to Propagation

The fatigue process can be divided into two stages: crack initiation and subsequent crack propagation. Crack propagation can occur from cracks nucleated by slip activity (see *Fatigue Cracks: Resistance to Initiation*) or from preexisting flaws. In either case, the crack propagation behavior is determined by load–microstructure–environment interactions at the crack tip. The relationship between applied load and propagation rate depends on crack size and geometry, and not simply on a critical value of applied stress such as the fatigue limit associated with fatigue crack initiation. The stress distribution at the crack tip can be calculated using linear elastic and elastic–plastic fracture mechanics. The parameter that combines the information about the applied load and the crack is the stress-intensity factor $K = f(\sigma, a)$, where σ is the applied stress and a is the crack length. The functional relationship depends on the configuration of the cracked component and the manner in which the load is applied. The application of fracture mechanics to fatigue crack propagation has systematized testing and aided in the clarification of load–microstructure–environment interactions and the identification of microstructures resistant to fatigue crack propagation.

1. Crack Growth Equations and Microstructure

For a particular material, the fatigue crack propagation rate is a function of the stress-intensity factor, and experimental results are normally presented as log–log plots of fatigue crack growth rate per cycle (da/dN) against the stress-intensity-factor range (ΔK). Such plots usually have a sigmoidal shape, and using the notation of Fine and Ritchie (1979) can be divided into three regions, characterized by the operating fatigue mechanism. The low-ΔK region (designated A) begins at a threshold ΔK_{th}, below which crack growth cannot be detected experimentally. The threshold should not be confused with the fatigue limit since the former is related to crack growth, whereas the fatigue limit is related to a crack initiation. The da/dN value increases rapidly above ΔK_{th}, and is sensitive to microstructure, environment and the mean stress as determined by the $K_{min}:K_{max}$ ratio (R). As R increases, ΔK_{th} decreases and da/dN increases. Similar changes can occur by microstructural modifications or when the environment is made more aggressive. The transition from the microstructurally sensitive region A to the less microstructurally sensitive region B is thought to occur when the maximum or reversed plastic zone size approaches the size of the controlling microstructure parameter (e.g., the grain size). Rice (1967) relates these zones to the stress intensity by

$$\text{maximum plastic zone size} = K_{max}^2/3\pi\sigma_y^2 \quad (1)$$

$$\text{reversed plastic zone} = \Delta K^2/12\pi\sigma_y^2 \quad (2)$$

where σ_y is the yield strength of the material.

The intermediate range of ΔK (region B) is dominated by the striation growth process and can be described by the Paris equation:

$$da/dN = C\Delta K^m \quad (3)$$

where C and m are scaling constants. This region is less sensitive to modifications in microstructure, environment and mean stress. The Paris equation underestimates the crack growth rates in the incremental tearing region C, but since an engineering material usually spends a negligible fraction of its fatigue life in this stage, it is less important than either region A or B. The fracture toughness determines the limiting stress intensity in region C.

Until recently fatigue crack growth was considered to be independent of microstructure and mainly a function of the elastic modulus E of the material, such that

$$da/dN = C'(\Delta K/E)^{m'} \quad (4)$$

This equation is based on the assumptions that the crack grows simply by kinematically irreversible plastic flow in the neighborhood of the tip and that there is no other fracture mechanism. It predicts the fatigue crack growth rate fairly accurately for the high crack growth range ($\gtrsim$1 μm per cycle) of a clean ductile material. For this case, a material with higher strength and modulus may show a reduction in growth rate when compared with a material with lower values of these parameters. However, for the slower growth range, modifications in microstructure have been observed to change substantially the fatigue crack propagation behavior, and Eqn. (4) does not predict these effects. There are currently no accepted models that accurately predict the effect of microstructural parameters on the propagation of fatigue cracks, although the model of Chakrabortty and Starke (1979) appears promising. Consequently, the design of fatigue-resistant alloys must be done empirically based on experimental observations.

Such observations of microstructural effects have been somewhat conflicting and frequently rationalized in terms of environmental effects. Microstructure controls the deformation behavior of a material and may influence its environmental sensitivity under fatigue conditions. More recently, the influence of some microstructural features, such as precipitate coherency and grain size, on the fatigue crack propagation behavior of materials has been established, and the interaction with environmental effects has been clarified. Although the exact mechanisms associated with the fatigue failure in the microstructurally sensitive region A are under debate, current knowledge suggests methods by which the constitution and microstructure of some alloys can be modified to improve their resistance to fatigue crack propagation. The modifications have their greatest effect in region A but may also affect region B.

2. Deformation Mode, Fracture Path and Fatigue Crack Growth Rate

Microstructure plays a dominant role in the fatigue crack propagation of age-hardenable alloys, including those of aluminum, titanium, nickel and iron, by its influence on deformation mode and fracture path. The effect of deformation mode on the fatigue crack growth rate is illustrated in Fig. 1 using an age-hardenable aluminum alloy as an example of generally observed behavior. The fatigue crack growth rate of the underaged alloy is lower than that of the overaged alloy at an equivalent ΔK. The coherent precipitates of the underaged alloy are sheared by dislocations, thus promoting coarse planar slip and heterogeneous deformation, both of which favor slip plane cracking. The incoherent precipitates of the overaged alloy are looped by dislocations promoting more homogeneous deformation, noncrystalline fracture and a straight-running crack lying normal to the stress axis. Shear fracture along slip bands should be easier than that associated with homogeneous deformation and noncrystallographic growth. However, slip band cracking enhances zigzag crack growth and crack branching, both of which increase the total crack path and lower the effective stress intensity at the main crack tip. In addition, slip band cracks must usually be renucleated at grain boundaries. These factors, along with a higher degree of slip reversibility, result in lower fatigue crack growth rates for the underaged alloy.

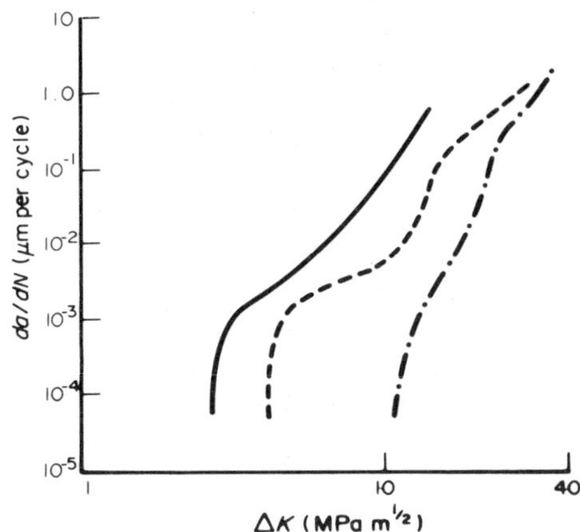

Figure 1
Effect of deformation mode and environment on the fatigue crack growth rate in Al–6Zn–2Mg–0.1Zr: ——, overaged sample in dry air; ---, underaged sample in dry air; —·—, underaged sample in vacuum

Lamellar structures also enhance crack branching and result in improved fatigue crack propagation resistance. Widmanstätten structures in $\alpha+\beta$ titanium alloys have a higher resistance to fatigue crack propagation than either martensitic, equiaxed α, or bimodal equiaxed α-Widmanstätten microstructures. Crack propagation takes place in or parallel to intense slip bands in Widmanstätten colonies. Extensive zigzag crack growth occurs because orientation differences cause the crack to change direction as it proceeds from colony to colony. Fracture along colony boundaries and at boundaries between Widmanstätten α plates and the β matrix results in extensive crack branching, thus lowering the effective stress intensity and reducing the fatigue crack growth rate.

The fatigue crack propagation resistance can be greatly affected by environmental reactions, especially in the microstructurally sensitive region of the crack growth curve. Even dry air, which is not a particularly aggressive environment, accelerates fatigue crack propagation of the underaged aluminum alloy over that in vacuum (Fig. 1). The actual mechanisms of environmentally accelerated fatigue crack growth are not well understood but it appears that the environment may modify the plastic behavior near the crack tip and environmental sensitivity varies with deformation mode. An example of the magnitude of this effect is illustrated in Fig. 2, which shows plots of the ratio of crack propagation rate in water to that in dry air as a function of the stress-intensity factor for both an underaged and an overaged aluminum alloy. The higher sensitivity of the underaged alloy is associated with the coarse planar slip and strain localization described above. The deformation of the overaged alloy is more homogeneous, and it is much less sensitive to environmental conditions. Likewise, the maxima in the curves of Fig. 2 are associated with a transition from single-slip to multiple-slip behavior. Multiple slip increases the homogeneity of deformation and reduces environmental effects. These results are general, that is, strain localization and coarse planar slip enhance environmental effects. Consequently, microstructures which have the highest resistance to fatigue crack propagation in inert environments may not offer this advantage in aggressive environments.

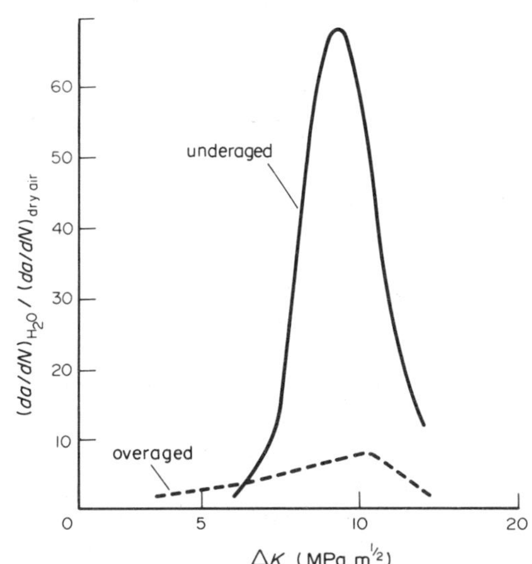

Figure 2
Effect of deformation mode on the environmental sensitivity in Al–6Zn–2Mg–0.1Zr

3. Grain Size, Fracture Path and Fatigue Crack Growth Rate

Early studies on the influence of grain size on the fatigue crack propagation behavior gave contradictory results. However, in most cases no special precautions were taken to separate environmental effects or determine the influence of deformation mode. More recent results have shown that grain size effects are most pronounced in planar slip materials for which grain boundaries act as major barriers to slip and thus determine the slip length. These materials are also most sensitive to aggressive environments and vacuum tests are necessary in order to separate grain size and environment effects.

A strong positive correlation between ΔK_{th} and grain size has been found to exist for titanium and aluminum alloys and low-strength steels. The fatigue crack growth rates are also affected over a significant range of ΔK with large-grain materials having growth rates as much as an order of magnitude slower than fine-grain ones. Examples of the influence of grain size and environment on fatigue crack growth rates of aluminum and titanium alloys are given in Figs. 3 and 4. The crack propagation rates in a vacuum are quite different, with the large grain size reducing the propagation rate at a given ΔK with respect to the fine-grain material for both alloys. However, when the aluminum alloy is tested in a NaCl solution, no grain size effect is observed. This is due to a change in fracture mode from transgranular to intergranular that accompanies the addition of the aggressive environment. Such a change in fracture mode is often observed in aggressive environments and hinders the determination of microstructural effects on fatigue crack growth rates. No such transition occurs for the titanium alloy and the effect of grain size on fatigue crack growth rate is the same in both environments.

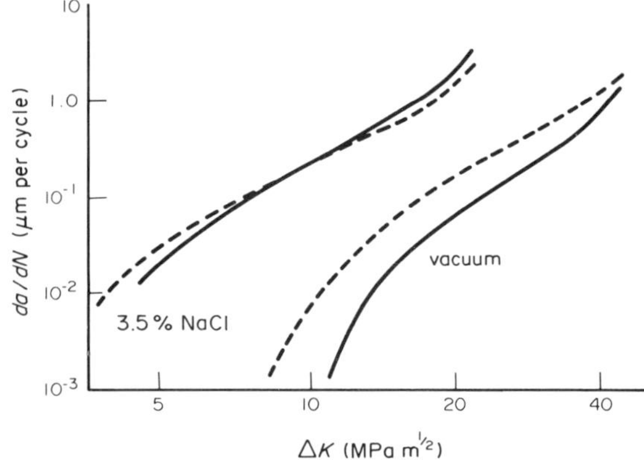

Figure 3
Effect of grain size and environment on the fatigue crack growth rate of an Al–5.7Zn–2.5Mg–1.5Cu alloy: ——, grain size 200 μm; ---, grain size 46 μm (after Lindigkeit et al. 1979)

Mechanistic studies suggest three major effects of grain size on deformation at the crack tip and subsequent crack propagation. At low ΔK, when the plastic zone is smaller than the grain size and only one slip system is operative, slip is fairly reversible and crack propagation is delayed. When the plastic

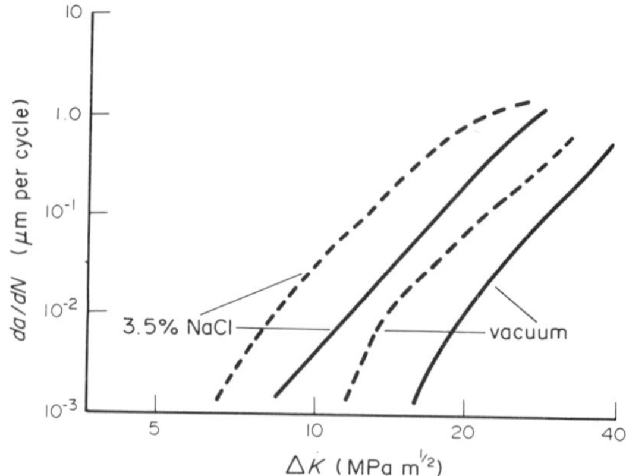

Figure 4
Effect of grain size and environment on the fatigue crack growth rate of a Ti–8.6Al alloy: ——, grain size 100 μm; ---, grain size 20 μm (after Lindigkeit et al. 1979)

zone exceeds the grain size, multiple slip must occur, and slip reversibility is reduced. These events predict a higher ΔK_{th} for the larger-grain-size materials. In addition, the degree of zigzag crack growth and crack branching that accompanies the slip band cracking increases with increasing grain size, thereby increasing the number of cycles to reach the critical plastic strain for fracture. All these factors suggest an improvement in the resistance to fatigue crack propagation with increasing grain size, at least up to the striated growth regime.

4. Microstructures Resistant to Fatigue Crack Propagation

In general, planar slip and slip restricted to one system reduce crack propagation because these types are more reversible than wavy or multiple slip. Alloying additions that lower the stacking-fault energy or restrict the number of slip systems operating improve crack propagation resistance. Heterogeneously deforming microstructures such as those containing coherent precipitates result in lower crack growth velocities than homogeneously deforming microstructures because of enhanced crack branching and zigzag crack growth. Two-phase lamellar structures produce similar results. In some cases dispersoids added to homogenize deformation in age-hardening alloys may also improve fatigue crack propagation resistance provided there is a change in fracture mode from low-energy intergranular fracture to higher-energy transgranular fracture. In most cases, microstructural features that promote crack branching (e.g., inclusions, phase boundaries and grain boundaries) usually improve the resistance to crack propagation. Inclusions cause branching of the main crack, often into nonpropagating secondary cracks, as long as the mean inclusion spacing is greater than the plastic zone size. The effectiveness of phase boundaries and grain boundaries in causing cracks to deflect or branch depends on strength differentials and orientation relationships. Crystallographic texture may also play a large role in crack propagation behavior, especially in anisotropic materials such as titanium. Increasing the grain size, or the microstructural feature that controls the slip length, increases the resistance to fatigue crack propagation and raises ΔK_{th}. This is probably associated with increased slip reversibility, increased probability of crack branching, and reduced plastic strain in the grains at the crack tip. Again it should be remembered that fatigue crack propagation is quite different from fatigue crack initiation, and there may be no optimum microstructure that simultaneously improves the resistance to both fatigue crack initiation and fatigue crack propagation.

See also: Fatigue Crack Growth: Macroscopic Aspects; Fatigue Crack Growth: Mechanistic Aspects; Environmentally Assisted Crack Growth; Mechanics of Materials: An Overview

Bibliography

Chakrabortty S B, Starke E A Jr 1979 Fatigue crack propagation in metastable beta titanium–vanadium alloys. *Metall. Trans. A* 10: 1901–11

Fine M E, Ritchie R O 1979 Fatigue crack initiation and near-threshold crack growth. In: Meshii M (ed.) 1979 *Fatigue and Microstructure.* American Society for Metals, Metals Park, Ohio, pp. 245–78

Irving P E, Beevers C J 1974 Microstructural influences on fatigue crack growth in Ti–6Al–4V. *Mater. Sci. Eng.* 14: 229–38

Lin F S, Starke E A Jr 1980 The effect of copper content and deformation mode on the fatigue crack propagation of Al–6Zn–2Mg–Cu alloys at low stress intensities. *Mater. Sci. Eng.* 45: 153–65

Lindigkeit J, Terlinde G, Gysler A, Lütjering G 1979 The effect of grain size on the fatigue crack propagation behavior of age-hardened alloys in inert and corrosive environment. *Acta Metall.* 27: 1717–26

Rice J R 1967 Mechanics of crack tip deformation and extension by fatigue. *Symposium on Fatigue Crack Propagation*, ASTM Special Technical Publication No. 415. American Society for Testing Materials, Philadelphia, Pennsylvania, pp. 247–309

E. A. Starke Jr.

Fatigue: Engineering Aspects

Wide use has been made of metals and their alloys in engineering components and structures over the past two centuries. A major factor has been the ability of metals to withstand tensile and torsional loads, thus enabling rotating machinery and lightweight structures to be developed. The cyclic and transient tensile loading of such components and structures has revealed a major failure process—fatigue. The physics of this process, which can produce failure after many cycles under a repeated fraction of the yield load, is now well understood. It concerns the initiation and subsequent growth of cracks. Other articles deal in detail with these aspects in terms of material microstructure, mechanisms and mechanics (see *Fatigue Cracks: Resistance to Initiation; Fatigue Cracks: Resistance to Propagation; Fatigue Crack Growth: Macroscopic Aspects; Fatigue Crack Growth: Mechanistic Aspects*): this article describes the important aspects of fatigue as seen by the engineer who has to provide some assurance that structures will have adequate fatigue resistance. Understanding and quantitative description of the process is of great help in underpinning engineering judgement of this process, so some reference is made where appropriate to basic principles.

1. Behavior of Materials Under Cyclic Loading

Tensile or compressive straining of a metal induces dislocation movement and multiplication, with the eventual development of discrete dislocation substructures. Cyclic straining, either by repeated tension or alternate tension/compression modifies such substructures and leads to hysteretic behavior in the stress–strain plane (Fig. 1). For many metals and alloys, a unique relationship exists between the applied cyclic stress range $\Delta\sigma$ and cyclic plastic strain range $\Delta\varepsilon_p$, often of power law form:

$$\Delta\sigma = k\,\Delta\varepsilon_p^{n'} \qquad (1)$$

This is the cyclic analogue of the static tensile stress–strain curve. However, there can be significant differences between monotonic and cyclic stress–strain

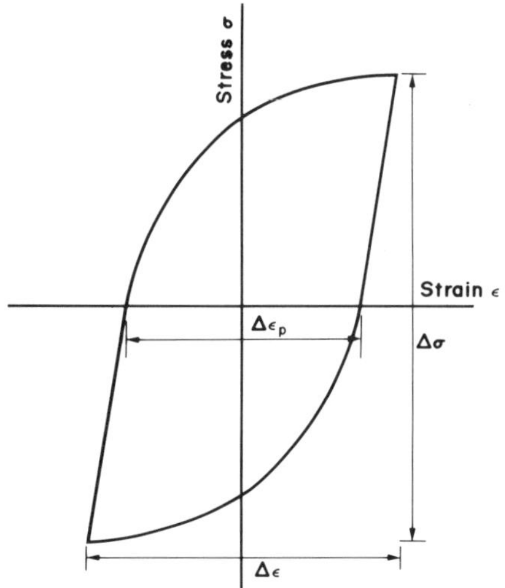

Figure 1
Typical cyclic stress–strain hysteresis loop

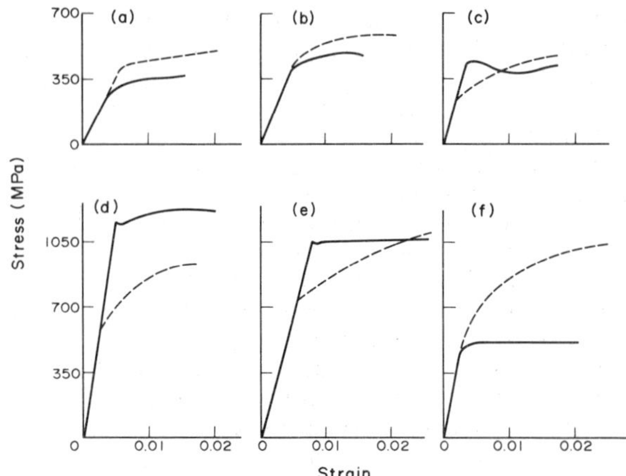

Figure 2
Monotonic and cyclic stress–strain curves for (a) 2024-T4 Al alloy, (b) 7075-T6 Al alloy, (c) Man-ten steel, (d) SAE 4340 (350 BHN) steel, (e) Ti-811 titanium alloy, (f) Waspaloy A nickel alloy (– – – cyclic stress, —— monotonic stress)

curves, as shown in Fig. 2 for a range of engineering alloys at ambient temperature. Both cyclic "hardening" and cyclic "softening" behavior are observed.

The cyclic stress–strain curve is usually described in terms of the steady-state stress and strain ranges, arrived at after a number of cycles. Steady state is often defined in terms of half-life values. Figures 3 and 4 show extreme hardening and softening behavior for two high-temperature alloys. The initial hardening period is associated microstructurally with the generation of a stable dislocation substructure. In this sense it is analogous to primary creep. The rapid softening always observed towards the end of life is caused by the presence of macrocracks and is not solely a material effect. Surface crack initiation is often noted to occur at the end of the initial hardening period. Precipitation-hardened alloys may exhibit dramatic cyclic softening as shown in Fig. 4. This is thought to be due to precipitate shearing by dislocation channelling. Cold-worked alloys also exhibit cyclic softening. The cyclic stress induced generation of dislocation substructures can occur at stress levels below general yield and in some alloys this is thought to be the basis of the material's fatigue strength (fatigue limit), via the link between such substructures and crack initiation. Other alloys (e.g., austenitic stainless steels) can sustain significant cyclic plastic straining beyond general yield without initiating cracks, because the dislocation substructures generated are irregular. The fatigue strength of an alloy is therefore a material structure sensitive property (see *Fatigue Cracks: Resistance to Initiation*).

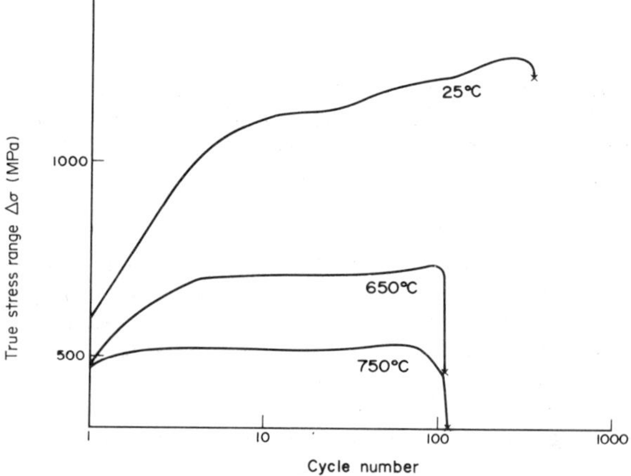

Figure 3
Variation of true stress range $\Delta\sigma$ with cycles, 20%Cr–25%Ni/Nb stainless steel ($\Delta\varepsilon_p = 0.4$) for various test temperatures

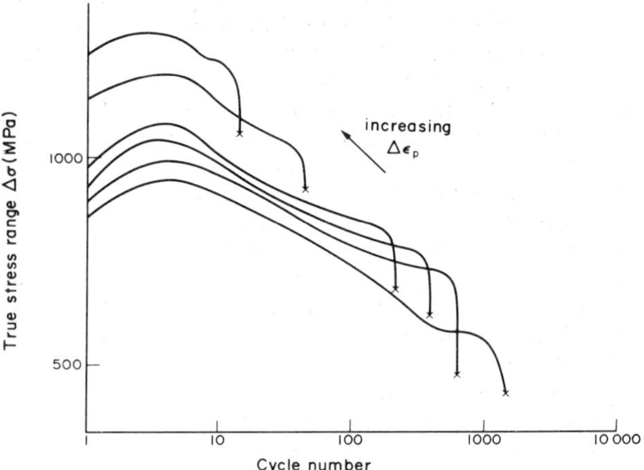

Figure 4
Variation of true stress range $\Delta\sigma$ with cycles, aged PE16, test temperature 750 °C

2. Endurance Relationships

2.1 Material

The basic endurance relationships for a material are expressed in terms of the applied cyclic stress/strain range ($\Delta\sigma$, $\Delta\varepsilon$) or amplitude versus endurance (cycles to failure, N_f). In terms of elastic and plastic components these are often of power law form,

$$\Delta\varepsilon_e N_f^{\alpha'} = A_1 \qquad (2)$$

$$\Delta\varepsilon_p N_f^{\alpha} = A \qquad (3)$$

and the curves are shown schematically in Fig. 5. Equation (2) represents the well known *S–N* or Basquin relationship whereas Eqn. (3) is the Coffin–

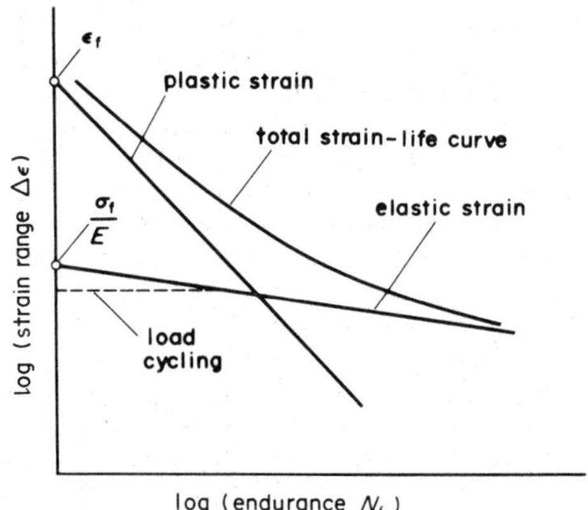

Figure 5
Lifetime as a function of elastic, plastic and total strain amplitude

Manson law. It is worth noting that above general yield, the Basquin relation holds only for strain cycling conditions as load cycling leads to excessive accumulation of plastic deformation (ratchetting) and tensile rather than fatigue failure. For materials where crack initiation is rapid, these relationships are effectively integrated crack-growth laws between an initial crack size a_o (which may be only a few micrometers) and final crack size a_f of several millimeters. Figure 6 shows such an integration compared with endurance data for mild steel and a high-strength low-alloy steel. Increase in material yield strength tends to raise the fatigue strength, which depends on crack initiation, but leads to a decrease in lifetime at high strain levels. The latter is caused by an increase in crack growth rate, and reduction in a_f because of reduced toughness.

The material endurance curve is also influenced by other factors. An increase in tensile mean stress tends to reduce fatigue strength, by promoting crack initiation, whereas a compressive mean stress marginally increases it. A rise in temperature also tends to reduce fatigue strength as well as increasing crack growth rates, thus leading to a dropping of the whole fatigue curve. Multiaxial straining induces different effects in the initiation and propagation dominated regions. Where initiation is by dislocation processes alone, described by a "cyclic" yield point, it may be represented by a typical multiaxial yield criterion such as von Mises, expressed in cyclic terms. The crack growth region, however, is described by a more complex criterion, dependent on both shear and normal stresses or strains (Brown and Miller 1973). If "natural" crack initiation is overcome by some other process, then the fatigue strength can be drastically reduced, to a level determined solely by crack growth considerations. Examples of such reductions are encountered in corrosion fatigue, where surface pitting induces initial surface "cracks," wear-induced fatigue and fatigue of materials with mechanical surface damage (e.g., scratches, machining marks).

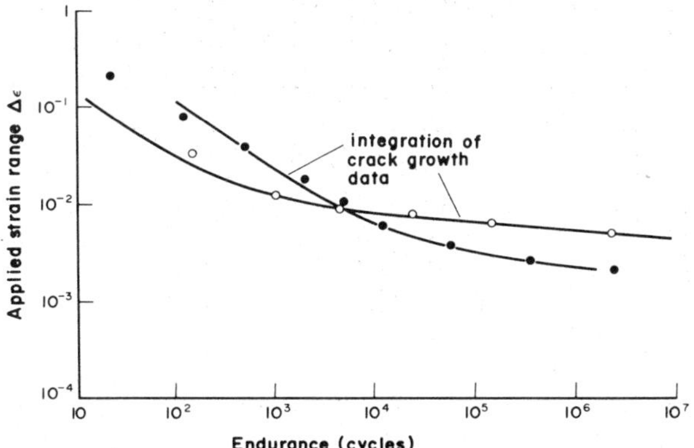

Figure 6
Strain endurance curves for two ferritic steels: ○, En 25 steel, 200 °C temper; ●, mild steel

2.2 Components

Endurance relationships for components often reflect the sensitivity of fatigue life to surface damage. Welded joints, for example, contain microflaws (e.g., slag, porosity, undercut) which act as already initiated cracks. Figure 7 shows the endurance data for fillet and butt welded joints in a structural steel as a function of the nominal stress at the weld location. A comparison is made with plain material behavior. The typical scale of initial flaws in the welded joints producing such dramatic effects on life and strength is of order 0.1–0.5 mm. The fillet weld curve is of interest in that it represents an integration of the Paris linear elastic fracture mechanics (LEFM) crack growth law (see *Fatigue Crack Growth: Macroscopic Aspects*).This is of the form

$$da/dN = C\Delta K^m \quad (4)$$

where ΔK is the cyclic stress intensity factor ($\propto \Delta\sigma a^{1/2}$) and C and m are material constants. For structural steel, m is ~ 3. Integration of Eqn. (4) then gives an S–N curve of the form

$$\Delta\sigma N_f^{-1/3} = \text{constant} \quad (5)$$

consistent with the curve shown in Fig. 7. The fatigue limit would be related to the LEFM threshold for crack growth ΔK_{th}, based on the initial crack size and making due allowance for weld residual stresses. It should be noted that this value is less sensitive to material yield strength than the material fatigue strength. In fact, higher strength materials tend to have lower ΔK_{th} values (Lindley 1981). For butt welds, the lower initial stress condition at the weld toe means that initial flaws do not behave as cracks from the outset and some cycles are spent initiating cracks from them, giving a higher fatigue strength

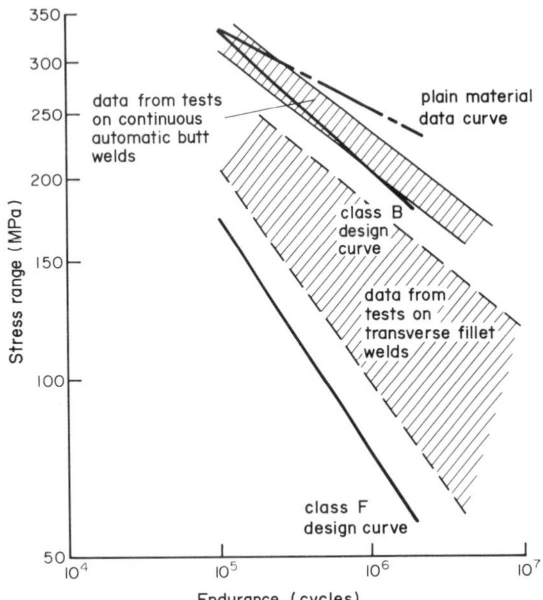

Figure 7
Weldment fatigue endurance data and design curves (after Gurney and Maddox 1973)

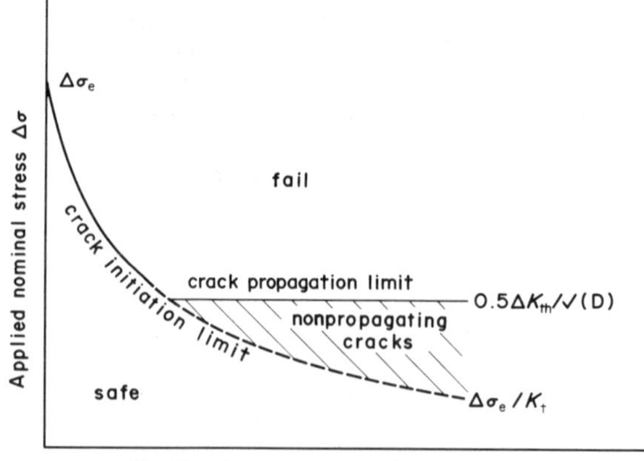

Figure 8
Variation of fatigue strength of notched components with K_t

and shallower *S–N* slope than that observed for fillet welds.

If flaws are not of a sharp crack profile, they behave as blunt notches which have different stress concentrating effects dependent upon the notch root radius ρ. The notch is the strongest feature controlling fatigue life in components, whether it is on a macro- or microscale. This has been recognized for many years and careful determination of the fatigue strength reduction due to notches is a major element in the design of rotating machinery. A notch stress concentration is characterized by two dimensions, ρ and the notch depth D. Under elastic conditions, the stress concentration factor K_t is given by

$$K_t = 1 + 2(D/\rho)^{1/2} \tag{6}$$

and the static stress field at a notch root,

$$\sigma = K_t\,\sigma_a\,[\rho/(\rho + 4x)]^{1/2} \tag{7}$$

where σ_a is the nominal applied stress and x the distance from the root. The fatigue strength of a notched component varies with K_t as shown in Fig. 8. The actual fatigue strength $\Delta\sigma_f$ is defined by two limits. At low values of K_t, the limit is determined by crack initiation, where

$$\Delta\sigma_f = \Delta\sigma_e/K_t \tag{8}$$

where $\Delta\sigma_e$ is the plain material fatigue strength. At higher values of K_t, cracks are readily initiated and the actual limit is defined by the crack growth threshold ΔK_{th}, as

$$\Delta\sigma_f = 0.5\,[\Delta K_{th}/(D + a)^{1/2}] \approx 0.5\,(\Delta K_{th}/D^{1/2}) \tag{9}$$

In the region between the two limits, nonpropagating cracks are observed.

3. Design Assessment of Fatigue

The design of an engineering structure or component is related to its primary function but it involves consideration of issues pertinent to likely service history, one of which may be fatigue. Whether this is a major or minor issue, it is a matter of design assessment rather than a design basis. When the assessment of a detailed design shows fatigue as a potential problem, the designer has several options. These include reduction of cyclic stresses, improvement of component details and modification of service history. However, the simple expedient of using stronger material to raise fatigue strength seldom improves overall fatigue behavior.

In considering fatigue, the designer usually adopts one of two philosophies; "safe life" or "damage tolerance." A third philosophy, "fail safe," is really a subset of damage tolerance, although historically its development, primarily for aircraft structures, preceded the more generalized concept. The safe-life philosophy relies on the provision of a valid endurance curve, which is factored to provide a design curve. The endurance curve may be for the material (e.g., as in Fig. 6) or a critical structural detail (e.g., as in Fig. 7), expressed in terms of a relevant stress range. The design factor is chosen to provide an adequate margin on failure taking account of considerations such as accuracy of stress analysis, relevance of testpiece to structure and inherent scatter of test data. In addition, the factor is increased for safety related structures, for example, nonredundant structures whose catastrophic fatigue-induced failure is unacceptable.

Although the safe-life approach is that most widely

used for fatigue assessment, it has an inherent uncertainty in the applicability of the test data to an actual structure or component, whose initial state, particularly with regard to flaws, is unknown. This has led to increased development and use of the damage-tolerance approach. This philosophy, which has become feasible with the advent of fracture mechanics, relies on knowledge and awareness of acceptable or "tolerable" levels of fatigue damage in the structure. The fail-safe condition is a special case of this approach, where the structure has sufficient redundancy to tolerate partial failure which can be defined as damage. Any initial failure is then to a safe state involving stress redistribution. However, it must be possible to detect this change before further damage leading to a critical unsafe state occurs. Examples of fail safe are in some airframe structures, offshore structural joints and leak-before-break in pressure boundary components (pipes, vessels). Damage tolerance is a wider concept, but it always relies on damage or flaw detection. Nondestructive examination and fracture mechanics methodology are the basis for the development and utilization of the approach. Damage tolerance is used in the assessment of nuclear pressure vessels and aero-engine components.

The main features of the damage-tolerance approach have been outlined elsewhere (see *Fatigue Crack Growth: Macroscopic Aspects*). In practice, limitations are currently found in several areas:

(a) Detection and accurate measurement of initial flaws or cracks induced in service.

(b) Determination of the stress intensity factor range ΔK for a crack in its specific geometric location. This includes consideration of crack shape.

(c) Determination of a relevant material crack growth curve. This should cover the range of interest, which often includes the threshold and tearing regions, noting any dependence on mean stress. It may also include an environmetal factor.

(d) Determination of a_f.

For many structures, limitations of knowledge in these areas mean that unduly conservative assumptions are made.

The safe-life approach also has limitations which include the ability to cope adequately with:

(a) Stress concentrations.

(b) Stress gradients—"size" effects are most often related to stress gradients.

(c) Load history—the variable load history experienced by most components and structures still provides one of the major uncertainties in fatigue assessment. The linear Palmgren–Miner rule is still the most widely used rule for coping with variable fatigue loads. It can be derived from a crack growth analysis, but mechanics/material interactions under variable amplitude conditions lead to nonlinearities (see *Fatigue Crack Growth: Mechanistic Aspects*). Developments in waveform analysis and simulative testing are producing some more realistic rules for specific applications.

(d) Stress state—mean and multiaxial stresses.

4. Fatigue Interactions

Major uncertainties in the estimation of fatigue behavior are introduced if time-dependent effects are present which can interact with the fatigue crack initiation and growth process. Corrosion was mentioned earlier as a factor which can seriously erode fatigue limits based on crack initiation. Crack growth rates can also be strongly influenced by an environment, gaseous or liquid. Figure 9 shows an example of such an influence for a pipeline steel tested in a water environment. The effect is accentuated by a high mean stress (R ratio) and low cyclic frequency.

At elevated temperature, creep-induced cavitation damage can interact with a growing fatigue crack to produce accelerated crack growth by effectively reducing the K_{max} level for tearing. Figure 10 shows this for an austenitic stainless steel where a crack is

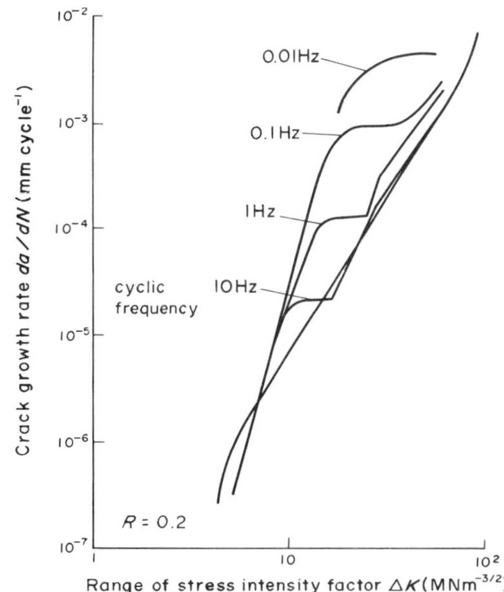

Figure 9
The influence of a seawater environment on fatigue crack growth in a C–Mn steel (after Vosikovsky 1975)

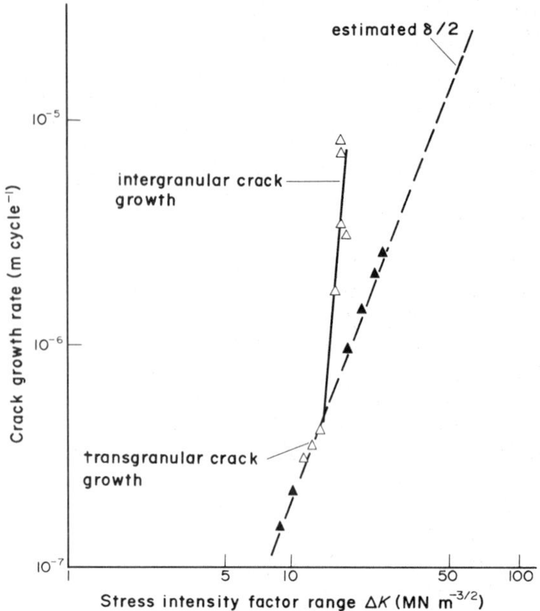

Figure 10
Effect of creep on fatigue crack growth in type 316 stainless steel at 625 °C, $R = 0.6$ (▲, sawtooth cycle, 10^{-12} Hz; △, tensile dwell cycle, 10^{-4} Hz) (after Lloyd and Wareing 1979)

growing under a load cycle involving long dwell periods at maximum tensile load.

See also: Failure Analysis: General Procedures and Applications to Metals; Mechanics of Materials: An Overview

Bibliography

Brown M W, Miller K J 1973 A theory for fatigue failure under multiaxial stress–strain conditions. *Proc. Inst. Mech. Eng.* 187: 65

Gurney T R, Maddox S J 1973 A reanalysis of fatigue data for welded joints in steel. *Weld. Res. Int.* 3(4): 1

Lindley T C 1981 Near threshold fatigue crack growth. *Proc. ASFM3 Ispra.*, Applied Science, London, p. 167

Lloyd G J, Wareing J 1979 Stable and unstable fatigue crack propagation during high-temperature creep—fatigue in austenitic steels. *J. Eng. Mater. Technol.* 101: 274

Vosikovsky O 1975 Fatigue crack growth in an X-65 line pipe steel at low cyclic frequencies in aqueous environments. *J. Eng. Mater. Technol.* 97: 298

B. Tomkins

Fatigue of Composites

Composites are, in general, highly resistant to fatigue. Composite fatigue behavior is also significantly different from that of metals in the types and multiplicity of failure modes, sensitivity to notches and behavior in compression. On the basis of equivalent weight, unidirectional composites have two to three times the fatigue strength of structural metals. Even practical cross-plied laminates are more fatigue resistant than metals and have replaced them in such fatigue-critical applications as helicopter rotor blades and high-speed textile machines.

Whereas metals fail in fatigue by crack initiation and growth, composites exhibit several modes of damage including matrix crazing, fiber failure, delamination, debonding, void growth and multidirectional cracking. As seen in Fig. 1, these damage modes often occur very early in the fatigue life to fracture. This phenomenon, coupled with the greater dispersion of the damage in the volume of material compared with metals, gives rise to a gradual loss of stiffness during fatigue, as depicted in Fig. 2. The gradual softening behavior has significant implications in terms of safe design. It provides early indication of fatigue damage to allow repair or removal. It also has implications for stiffness critical structure such as springs or aircraft wings for which the failure criterion is stiffness change rather than fracture. Fatigue data should be reported as stress or strain against cycles to a given change in stiffness (Fig. 3) rather than cycles to fracture. Stiffness change indicating fatigue damage is usually coupled with a loss of residual strength as seen in Fig. 4. For design, nondestructive inspection data are needed which link damage indications with residual stiffness and strength. Dynamic stiffness measurements, ultrasonics, penetrant x radiography, holographic interferometry and thermography (many composites exhibit extensive self-heating due to fatigue damage) have been used to detect damage.

Two further characteristics which distinguish composite fatigue behavior from that of metals are the effect of stress concentrations and compression.

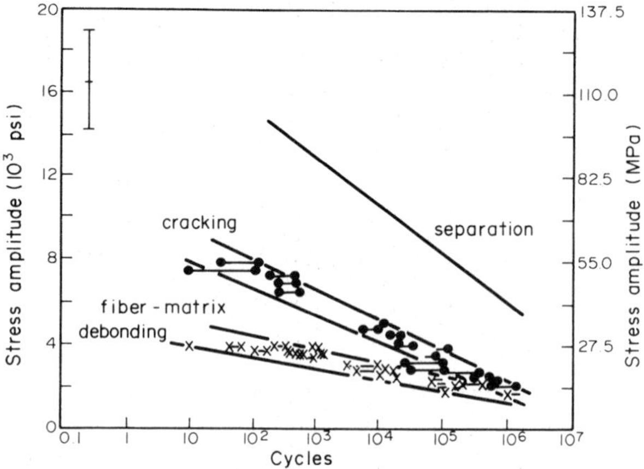

Figure 1
Fatigue behavior of glass-mat-reinforced polyester (after Smith and Owen 1969)

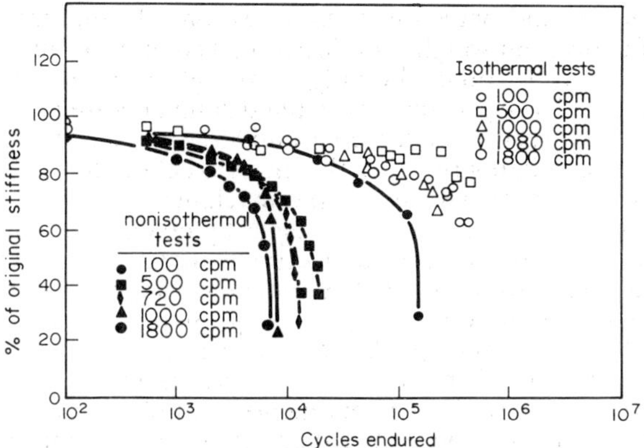

Figure 2
Stiffness decay during fatigue of glass-reinforced polypropylene (after Cessna et al. 1969)

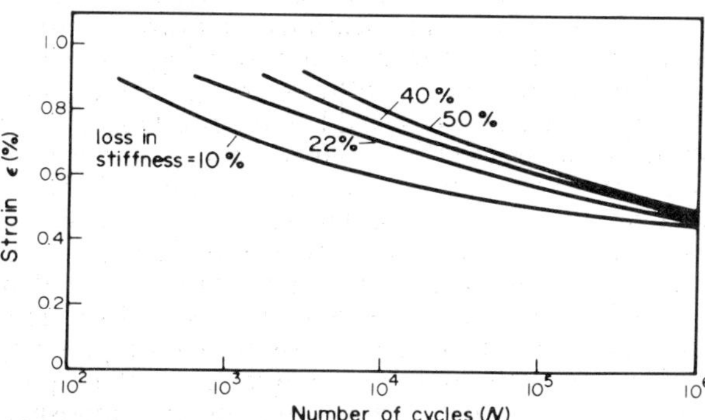

Figure 3
Fatigue behavior of 10° glass epoxy showing different stages of damage (after Agarwal and Joneja 1981)

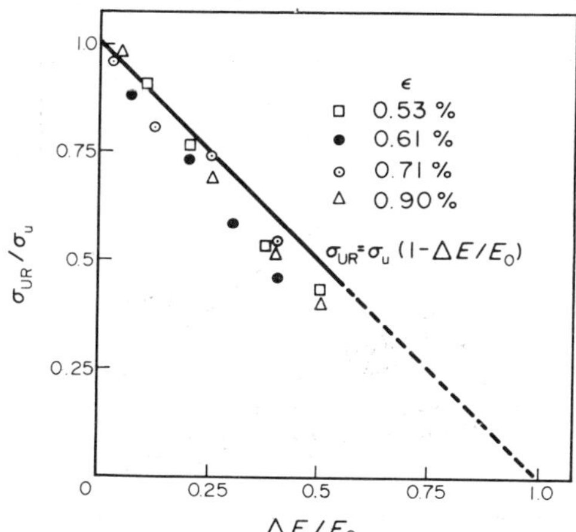

Figure 4
Residual strength related to stiffness decrease of fatigued 10° glass epoxy (after Agarwal and Joneja 1981)

Whereas metals exhibit a greater sensitivity to notches in fatigue than statically (or in low-cycle fatigue), composites can exhibit the opposite behavior as seen in Fig. 5. Composites also exhibit lower resistance to fatigue in compression than in tension, apparently due to cooperative buckling of adjacent fibers with concomitant matrix shear.

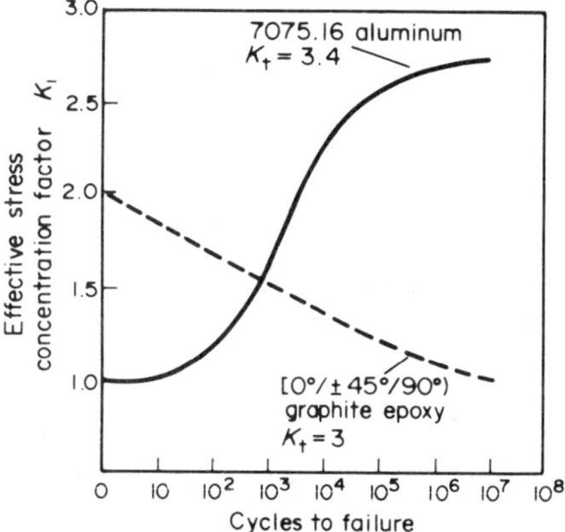

Figure 5
Composites and metals exhibit opposite sensitivity of stress concentration to fatigue (after Salkind 1976)

See also: Failure of Composites: Stress Concentrations, Cracks and Notches; Strength of Composites

Bibliography

Agarwal B D, Joneja S K 1981 Strain controlled off-axis (10°) flexural fatigue of a unidirectional composite. *Fibre Sci. Technol.* 14: 69–80

American Society for Testing and Materials 1975 *Fatigue of Composite Materials*, Special Technical Publication 569. American Society for Testing and Materials, Philadelphia, Pennsylvania

Broutman L J, Krock R H (eds.) 1974 *Composite Materials*, Vol. 5. Academic Press, New York

Cessna L, Levens J, Thomson J 1969 Flexural fatigue of glass-reinforced thermoplastics. *Proc. 24th Tech. Conf. of the SPI Reinforced Plastics Division.* Society of the Plastics Industry, New York, Sect. 1-C, pp. 1–12

Nevadunsky J J, Lucas J J, Salkind M J 1975 Early fatigue damage detection in composite materials. *J. Comp. Mater.* 9: 394–408

Ogin S, Smith P, Beaumont P 1985 Matrix cracking and stiffness reduction during the fatigue of a $(0/90)_S$ GFRP laminate. *Comp. Sci. Technol.* 22: 23–31

Reifsnider L, Lauraitis K (eds.) 1977 *Fatigue of Filamentary Composite Materials*, Special Technical Publication 636. American Society for Testing and Materials, Philadelphia, Pennsylvania

Salkind M J 1972 Fatigue of composites. *Composite Materials: Testing and Design (Second Conference),* Special Technical Publication 497. American Society for Testing and Materials, Philadelphia, Pennsylvania, pp. 143–69

Salkind M J 1973 VTOL aircraft. In: Salkind M, Holister G (eds.) 1973 *Applications of Composite Materials*, Special Technical Publication 524. American Society for Testing and Materials, Philadelphia, Pennsylvania, pp. 76–107
Salkind M J 1976 Fiber composite structures. *Proc. Int. Conf. Composite Materials*, Vol. 2. American Institute of Mechanical Engineers, New York, pp. 5–30
Smith T R, Owen M J 1969 Progressive nature of fatigue damage in RP. *Mod. Plast.* 46: 128–33

M. J. Salkind

Fatigue of Polymers

The term fatigue refers to the reaction of materials to alternating loading. Under such conditions, common in many engineering applications, polymeric components may fracture after many cycles of loading, even though the applied stress amplitude is well below the tensile strength of the material. The fatigue process involves a crack initiation phase and a crack propagation phase. Which phase dominates depends on the type of polymer and the test conditions. The fatigue resistance depends on many material variables, such as composition, molecular weight, extent of orientation and presence of additives, and on many external variables, such as frequency, temperature, stress and environment.

Figure 1
Part of fatigue fracture surface of polystyrene sample tested at 17.2 MPa and 2 Hz

1. General Nature of Fatigue Fracture

Observation of polymer specimens that have fractured in service or in laboratory fatigue tests yields much information about the principal characteristics of fatigue fracture (Hertzberg and Manson 1980). There is little macroscopic plastic deformation and the fracture surface, generally perpendicular to the stress direction, has rather characteristic markings.

Part of a typical fatigue fracture surface for a glassy polymer, polystyrene, is shown in Fig. 1. The smooth region surrounding the source is indicative of crack propagation through a surface craze (see *Crazing and Cracking of Polymers*). The craze, consisting of oriented fibrillar elements interspersed with voids, forms after a certain number of alternating cycles, as a result of a high local stress concentration at a surface flaw. Hence the resistance offered by different polymers to the development of a craze, or localized plastic zone, greatly affects their fatigue performance.

The series of concentric rings, called discontinuous growth bands, is a distinctive feature of many fatigue fracture surfaces. The crack is arrested near the end of each band, usually for hundreds of cycles. Meanwhile, the craze in front of the crack tip grows continuously until some critical length or some critical level of cyclic damage is reached. At this stage the crack jumps through about two-thirds of the existing craze and is again arrested (Döll 1983). The stability of the craze to cyclic breakdown is a determining factor in the resistance of different polymers to fatigue fracture. Molecular rearrangements can occur during fatigue cycling. Changes have been observed in infrared absorption bands, in mechanical loss spectra, in extent of local packing and in defect structure (Kausch 1978).

2. Effects of Material Variables

Crystalline polymers are more resistant to fatigue fracture than are the glassy amorphous polymers. Many amorphous polymers, like polystyrene (PS) and poly(methyl methacrylate) (PMMA), have a fatigue strength that is only about one-fifth of the tensile strength of the material. In crystalline polymers, such as Nylon 66 and polyoxymethylene, this ratio is appreciably higher (0.3–0.5). Figure 2 shows the fatigue crack propagation (FCP) rate da/dn, where a is the crack length and n the number of cycles, as a function of the stress-intensity-factor range ΔK for a number of thermoplastics. The crystalline nylon shows lower FCP rates than the amorphous polymers (Hertzberg et al. 1975). The greater resistance of the crystalline polymers to fatigue is attributed to the presence of the many small crystallites which hinder crack growth.

It is also evident from Fig. 2 that ductile amorphous

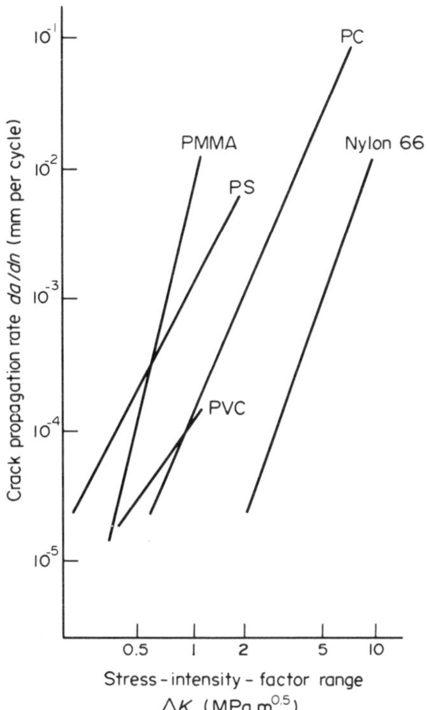

Figure 2
Fatigue crack propagation rate vs stress-intensity-factor range for various thermoplastics

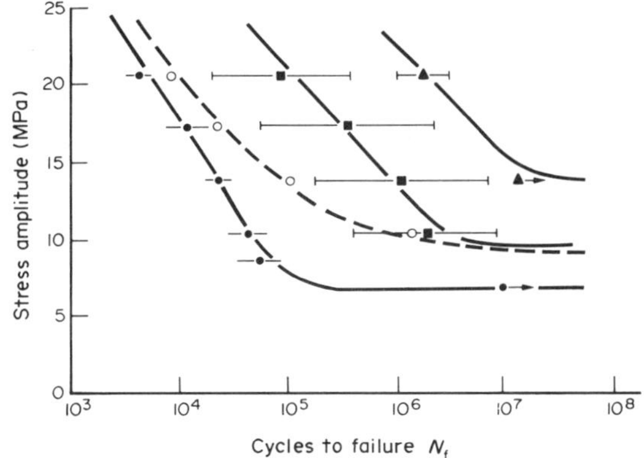

Figure 3
Stress amplitude vs number of cycles to failure for polystyrene samples of different molecular weight (MW): ●, 1.6×10^5; ○, 2.74×10^5; ■, 8.6×10^5; ▲, 2×10^6; —— narrow, --- broad MW distribution

polymers, such as poly(vinyl chloride) (PVC) and polycarbonate (PC), are superior in fatigue performance to the brittle glassy polymers like PS and PMMA, where crazing is the predominant deformation mode.

Molecular weight is an important material variable. This is demonstrated by Fig. 3 for PS specimens. As the molecular weight increases from 1.6×10^5 to 2×10^6, the average lifetime is increased by more than two orders of magnitude and the fatigue strength is doubled. Reductions in FCP rate with increase of molecular weight have been observed in other polymers as well, such as PVC, PMMA and polyethylene (PE). The beneficial effects of molecular weight are attributed to an increased craze stability as a result of an increase in chain entanglements and a reduction in the number of chain ends.

Fatigue resistance of polymers can be increased by glass-fiber reinforcement and by lamination. For many polymers thus modified, the ratio of fatigue strength at 10^7 cycles to tensile strength lies between 0.3 and 0.5, or considerably above that of the amorphous glassy homopolymers. The better fatigue performance arises largely because of the impediments to crack propagation produced by the reinforcement or by the laminations.

Orientation in polymers tends to hinder craze development and thus can have a beneficial influence on fatigue performance. The fatigue response of highly oriented polymers, such as fibers, is complex; different deformation modes may be encountered, depending on the polymer, extent of orientation, stress state, microstructure and frequency (Bunsell and Hearle 1974).

Rubber modification is a standard method for raising impact strength, but it has less effect on fatigue response. Tests on notched specimens, as well as fractographic studies of fractured fatigue samples, show that resistance to FCP is improved. However, unnotched rubber-modified polymers fracture sooner in fatigue than the unmodified polymer (Chen et al. 1981). This is an example of where the rubber modification, though enhancing FCP resistance, alters the crack initiation phase adversely by accelerating craze initiation.

Addition of diluents, such as plasticizers, generally decreases fatigue resistance. Plasticizers induce greater chain mobility; thus, there may be localized heating near the crack tip. This can lead to blunting of the crack tip and hence to a decrease in FCP rate. Both effects have been observed.

3. *Effects of External Variables*

When polymers are subject to alternating stress, the temperature may rise above ambient, owing to hysteresis. The seriousness of the resulting thermal softening effect depends on the magnitude of the applied stress, on the test frequency and on the viscoelastic nature of the polymer. For polymers having low internal friction, such as PS, the temperature rise is negligible. However, when more viscoelastic polymers, like PE, are subject to cyclic stresses and high frequencies, appreciable temperature rises are observed and failure may occur, not by progressive growth of a fatigue crack but by thermal rupture or even melting (Sauer and Richardson 1980).

The effects of frequency on fatigue life and on FCP rate depend on the polymer, on its mechanical relaxation spectra and on the ambient temperature. In some polymers, such as PS, PMMA and PVC, an increase in test frequency tends to increase average fatigue lifetimes and to decrease FCP rates. In other polymers, such as Nylon 66, PC and polysulfone, no effect of frequency on FCP has been noted. It has been suggested that the frequency sensitivity of polymers is a maximum when the frequency of segmental jumps of chain segments associated with the principal secondary transition is in the range of test frequencies.

The effect of increasing the test temperature on fatigue performance is to shift the stress–lifetime curve to shorter fatigue lifetimes and to lower stresses, without altering its general shape.

The stress-intensity-factor range, as Fig. 2 shows, is an important factor controlling FCP rate. However, other factors, such as the mean stress or the mean stress intensity, are also important. In general, as these factors increase, the fatigue resistance decreases. Many different forms of rate equation have been proposed; their possible application to fatigue crack growth in polymers has been reviewed (Radon 1980).

Environment may also play a role in fatigue failure of polymers. Liquid coatings, for example, such as silicone oils for PS, can reduce fatigue lifetime even when the environment in the absence of stress is inert. Low-viscosity liquids, which more readily penetrate surface crazes and cracks, and environments that are more compatible with the polymer and thus readily plasticize can severely reduce fatigue performance. On the other hand, fatigue resistance can be enhanced by protecting the polymer surface against aggressive environments.

Other means of extending the fatigue lifetimes of polymers include improving surface finish, developing an oriented surface layer and applying thin adherent elastomeric films. All of these inhibit development of a localized damage zone such as a surface craze and hence enhance fatigue performance. More detailed information on these matters, together with fatigue data on many types of polymers, including rubbers, fibers, plastics and composites, is given in the references cited in the bibliography.

See also: Mechanical Properties of Polymers; Polymers: An Overview of Structure, Properties and Structure–Property Relations

Bibliography

American Society of Testing and Materials 1975 *Proc. Symp. Fatigue of Composite Materials,* ASTM Special Technical Publication No. 569. American Society for Testing and Materials, Philadelphia, Pennsylvania

Andrews E H 1969 Fatigue in polymers. In: Brown W E (ed.) 1969 *Testing of Polymers*, Vol. 4. Interscience, New York, pp. 237–96

Bunsell A R, Hearle J W S 1974 The fatigue of synthetic polymeric fibers. *J. Appl. Polym. Sci.* 18: 267–91

Chen C C, Chheda N, Sauer J A 1981 Craze and fatigue resistance of glassy polymers. *J. Macromol. Sci., Phys.* B 19: 565–88

Döll W 1983 Optical interference measurements and fracture mechanics analysis of crack tip craze zones. *Adv. Polym. Sci.* 52–53: 105–68

Hertzberg R W, Manson J A 1980 *Fatigue of Engineering Plastics*. Academic Press, New York

Hertzberg R W, Manson J A, Skibo M 1975 Frequency sensitivity of fatigue processes in polymeric solids. *Polym. Eng. Sci.* 15: 252–60

Kausch H H 1978 *Polymer Fracture*. Springer, Berlin

Radon J C 1980 Fatigue crack growth in polymers. *Int. J. Fract.* 16: 533–52

Sauer J A, Richardson G C 1980 Fatigue of polymers. *Int. J. Fract.* 16: 499–532

J. A. Sauer

Fermi Surfaces

The simplest model of a metal is a system in which free electrons are constrained to move in a large box representing a macroscopic piece of metal. In this system the energy of an electron with momentum $\boldsymbol{p}$ is $E(\boldsymbol{p}) = p^2/2m$, where $p = |\boldsymbol{p}|$ and m is the electron mass. The different possible quantum mechanical states for an electron in this system are distinguished by the differing values of the electron momentum; that is, a complete set of electron states may be labelled by momentum (the spin degree of freedom may be ignored in this discussion). When a large number of electrons, say N, are placed in the box, the lowest energy state is found by occupying the N lowest energy electron states. This is analogous to the case of an atom, where the ground state of an N-electron atom is determined by occupying the N lowest energy atomic orbitals; no state may be occupied by more than one electron. Thus the ground state of the idealized metal will have all momentum states with $|\boldsymbol{p}| < p_F$ occupied, and all states with $|\boldsymbol{p}| > p_F$ unoccupied, where p_F is the Fermi momentum which depends on the density of electrons in the box. The surface in momentum space which separates unoccupied electron states from occupied electron states (in this case a sphere of radius p_F) is known as the Fermi surface.

1. Basic Theory

The above discussion is an oversimplification in two important respects. First, the electrons inside a real metal interact with each other via the Coulomb repulsion between like charges. The consequences of this interaction are extremely difficult to take into account with any great precision. Fortunately, it has been shown on very general grounds that the presence of this interaction does not invalidate the concept of a Fermi surface. In fact, the electron–electron interactions are normally included in describing electron states in real crystalline metals as a state-independent contribution to the effective interaction between an electron, and the underlying lattice of nuclei and the surrounding electrons. There is a good deal of evidence that this approach is entirely adequate as far as the Fermi surface is concerned.

The second oversimplification of the introductory discussion, namely the omission of the interaction between the electron and the lattice, may now be considered. The most important modification of the introductory discussion required by the presence of electron–lattice interaction follows from the observation that it is no longer possible to label the electron eigenstates by momentum. Formally, the possibility of labelling states by momentum will exist whenever it is possible to find simultaneous eigenstates of the momentum operator $\boldsymbol{p}(\boldsymbol{r}) = -\mathrm{i}\hbar\boldsymbol{\nabla}$ and the electron energy operator or Hamiltonian $H = -\hbar^2\boldsymbol{\nabla}^2/2m + V(\boldsymbol{r})$, where $V(\boldsymbol{r})$ is the potential in which an electron moves. It is easily seen that simultaneous eigenstates can be found if, and only if, H and $\boldsymbol{p}$ commute, that is, if $[\boldsymbol{p}, H] \equiv \boldsymbol{p}H - H\boldsymbol{p} = -\mathrm{i}\hbar\boldsymbol{\nabla}V(\boldsymbol{r}) = 0$. Thus, the electrons of the "electron-in-a-box" model ($V(\boldsymbol{r}) \equiv 0$) can be labelled by momentum, but those in a real metal cannot. However, as discussed below, for crystalline metals a generalization of the momentum can be introduced which leaves the introductory theory substantially unchanged.

The crystal momentum operator $\boldsymbol{P}_\mathrm{c}$ can be loosely defined as an operator which picks out the part of the spatial variation of a (wave) function which does not follow the periodic variations of the underlying lattice. Consider the Fourier transform of some function $\psi(\boldsymbol{r})$:

$$\psi(\boldsymbol{r}) = (2\pi\hbar)^{-3/2} \int d\boldsymbol{p} \exp(\mathrm{i}\boldsymbol{p} \cdot \boldsymbol{r}/\hbar)\, \psi(\boldsymbol{p}) \qquad (1)$$

Corresponding to the lattice vectors of the crystal $\boldsymbol{L}_\gamma$, there is a discrete set of momentum vectors $\boldsymbol{P}_n$ such that $\exp(\mathrm{i}\boldsymbol{P}_n \cdot (\boldsymbol{r} + \boldsymbol{L}_\gamma)/\hbar) = \exp(\mathrm{i}\boldsymbol{P}_n \cdot \boldsymbol{r}/\hbar)\exp(\mathrm{i}\boldsymbol{P}_n \cdot \boldsymbol{L}_\gamma/\hbar) = \exp(\mathrm{i}(\boldsymbol{P}_n \cdot \boldsymbol{r}))$ for any n or γ. These momentum vectors are said to form the reciprocal lattice of the crystal. Note that the spatial variations in $\psi(\boldsymbol{r})$ corresponding to $\psi(\boldsymbol{p})$ and $\psi(\boldsymbol{p} + \boldsymbol{P}_n)$ are the same, apart from a component which follows the periodicity of the lattice, and that the effect of the crystal momentum operator on these two spatial variations should, therefore, be the same. It is possible to define a volume in momentum space known as the Brillouin zone (see *Brillouin Zones*) which when translated by reciprocal lattice vectors fills all momentum space, but within which no two momenta are separated by any reciprocal lattice vector. Therefore $\boldsymbol{P}_\mathrm{c}\psi(\boldsymbol{r})$ may be expressed in terms of the Fourier transform of $\psi(\boldsymbol{r})$ by

$$\boldsymbol{P}_\mathrm{c}\psi(\boldsymbol{r}) = (2\pi\hbar)^{-3/2} \int d\boldsymbol{p} \exp(\mathrm{i}\boldsymbol{p} \cdot \boldsymbol{r}/\hbar)\boldsymbol{p}|_{\boldsymbol{P}_n}\, \psi(\boldsymbol{p}) \qquad (2)$$

where $\boldsymbol{p}|_{\boldsymbol{P}_n}$ denotes the momentum in the Brillouin zone which is related to $\boldsymbol{p}$ by some reciprocal lattice vector. $\boldsymbol{p}|_{\boldsymbol{P}_n}$ is often referred to as the reduced momentum.

Since the crystal potential possesses the translational symmetry of the lattice (i.e., $V(\boldsymbol{r} + \boldsymbol{L}_\gamma) = V(\boldsymbol{r})$), it can be shown that $[\boldsymbol{p}_\mathrm{c}, H] = 0$, and therefore that the electron eigenstates may be labelled by the available values of crystal momentum in the Brillouin zone. In general there will be more than one eigenstate of a given crystal momentum and so an additional index, the "band" index, is used to label the complete set of electron states: $H\psi_{n,\boldsymbol{p}}(\boldsymbol{r}) = E_n(\boldsymbol{p})\psi_{n,\boldsymbol{p}}(\boldsymbol{r})$. Just as in the free-electron case, however, the ground state for a given density of electrons is obtained by occupying the states of lowest energy. For metals, the maximum occupied energy E_F will lie above the minimum energy and below the maximum energy for at least one band. The surface (or surfaces) in momentum space defined by the equation $E_n(\boldsymbol{p}) = E_\mathrm{F}$ forms the Fermi surface of a metal and, just as in the free-electron case, separates occupied and unoccupied states.

The Fermi surface is not merely an abstraction in the mind of the scientist. Several techniques, notably those based on the de Haas–van Alphen effect and positron annihilation, are available to measure its properties (see *de Haas–van Alphen Effect; Positron Annihilation*). Fermi surfaces have been mapped out for most pure crystalline metals, as well as for many intermetallic compounds and alloys. (In disordered alloys $[H, \boldsymbol{p}_\mathrm{c}] \neq 0$ but the exact electron eigenstates may have only a small width in momentum space, in which case no qualitative alteration of the preceding discussion would be required.) The details of the Fermi surface can have a profound effect on most electronic properties of metallic systems.

Fermi surfaces can be and, in most cases where experimental data are available, have been evaluated theoretically. There is a variety of techniques available for determining the eigenstates and eigenvalues associated with a given value of crystal momentum given a periodic potential $V(\boldsymbol{r})$. The calculation of the Fermi surface then rests on the determination of the energy E_F at which the proper number of states are occupied and the mapping out of the surface or surfaces on which $E_n(\boldsymbol{p}) = E_\mathrm{F}$ is satisfied. Generally, the largest source of uncertainty in such calculations results from the approximations used to treat the electron–electron interaction as a contribution to the effective crystal potential. Although discrepancies between theory and experiment persist on a quantitative level, the wealth of information in experimental mappings of metallic Fermi surfaces is generally reproduced in impressive detail by ab initio electronic structure calculations.

2. Fermi Surfaces in Simple Metals

A simple metal is one in which the ions consist of a nucleus plus closed shells of tightly bound core electrons which do not overlap with those of neighboring ions. These ions are embedded in a sea of conduction electrons (corresponding to the valence electrons of the free atom), which behave largely like free electrons. Generally speaking, in such metals the energy bands show distortion from free-electron behavior only near the Brillouin zone boundary. This is illustrated in Fig. 1 where a typical simple-metal band is displayed. Note that there are no available states in the energy range marked ΔE, referred to as the energy gap. The alkali metals, and magnesium, aluminum and lead are examples of simple metals. The transition metals (e.g., iron) do not fall into this class because of the strong overlap between *d* electrons associated with different ion sites.

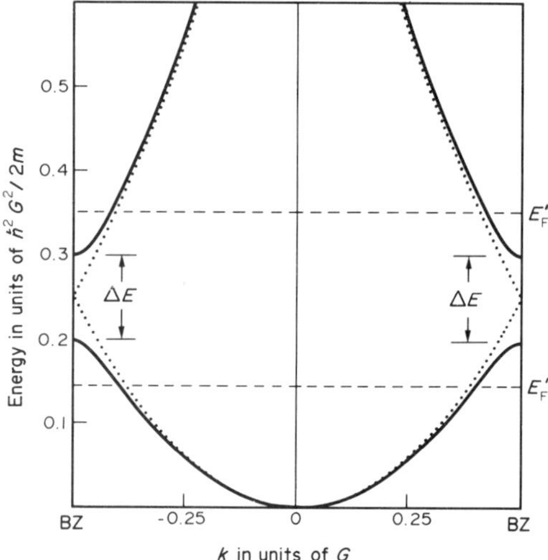

Figure 1
Simplified picture of energy dispersion, in a particular direction in reciprocal space, in a simple metal (solid lines). Dotted lines indicate the parabolic dispersion that occurs in a free-electron system. The wave number k is in units of the nearest reciprocal lattice vector G. Note the presence of the energy gap ΔE at the Brillouin zone boundary (BZ). The energy levels E'_F and E''_F indicate schematically where the Fermi energy might be for monovalent and polyvalent systems, respectively

The alkali metals are monovalent, that is, they have one conduction electron per ion. This means that in momentum space only half the available states in the first Brillouin zone are occupied. Consequently the Fermi energy will be at the level marked E'_F in Fig. 1 where there is very little distortion of the band from the free-electron parabola. (Note that if the energy gap were larger, corresponding to strong scattering of electrons by the ions, the band might well be appreciably distorted at the Fermi energy.) Since the same consideration will hold for all directions in momentum space, the resulting Fermi surface (mapped out by the locus of all points at the intersection of the electron band with E'_F) will be almost a perfect sphere. This is in fact the case for sodium and potassium, where the measured radial distortions of the Fermi surface are less than 0.2% of the average radius. For the other alkali metals, lithium, rubidium and cesium, the distortions are more significant, but are still only a few percent.

If the electron band were distorted the same amount in all directions in momentum space, the Fermi surface would still be spherical, but because the Brillouin zone is not spherical, the boundary would come closer to the Fermi surface in some directions than in others. Hence the Fermi surface tends to bulge out towards nearby surfaces of the Brillouin zone. In the noble metals, copper, silver and gold, the bands are sufficiently distorted in the ⟨111⟩ directions that the Fermi surface touches the zone boundary, forming necks on an otherwise nearly spherical surface.

For polyvalent materials the situation is much more complicated. For a material like aluminum or lead the Fermi energy would be, schematically, at the level indicated by E''_F, meaning that the lower band is fully occupied and levels in the upper band become populated. The Fermi surface now becomes very complex in shape, even within the oversimplified band picture that has been illustrated. This complexity is largely the result of the reciprocal lattice periodicity and is very nicely illustrated by Harrison (1970 pp. 118–19). The periodicity may be used to unfold the Fermi surface into what is referred to as the extended-zone scheme: it can be seen that, using this representation, the surface for simple metals is nearly spherical with distortions only where it intersects zone boundaries. An example is given in the work on lead by Anderson and Gold (1965).

In summary, the Fermi surface is approximately spherical only in the alkali metals; in all other simple metals it is much more complex in shape.

3. Fermi Surfaces in Nonsimple Metals

The term "nonsimple metal" here describes those metals in which the electron–ion interaction is so strong that no discernible remnant of the isotropic parabolic dispersion of the free-electron system remains. The transition metals and their compounds are one important group falling into this category. We have shown in Fig. 2, along one particular direction in the Brillouin zone, a model set of electron bands whose dispersion relations $E_n(\boldsymbol{p})$ are similar to those for the five d bands of a transition metal in the face-centered-cubic structure. (The five atomic d states broaden into bands when the atoms are placed in a metal.) There would, in general, be very little similarity between these curves and those corresponding to another direction in the Brillouin Zone. It may also be seen in Fig. 2 that more than one band may cross the Fermi level. These features lead to Fermi surfaces of very complicated shapes which may, in general, be composed of several sheets. These Fermi surfaces will also be very sensitive to quantitative details of the electron dispersion relations for various bands and may therefore reflect relativistic effects such as the spin–orbit coupling effect illustrated in Fig. 2.

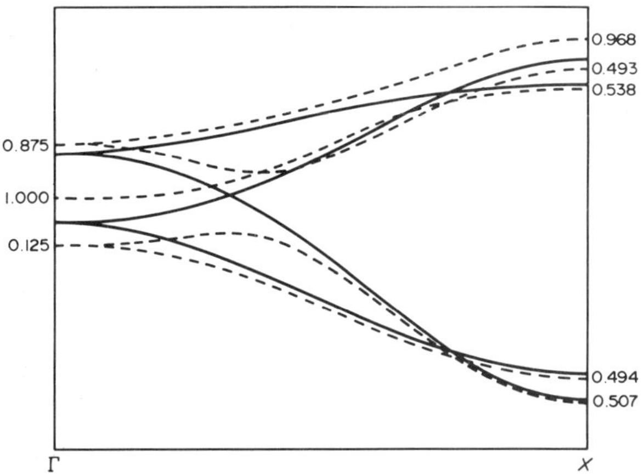

Figure 2
Energy dispersion $E_n(\boldsymbol{p})$ for the five d bands of an fcc transition metal from the center of the Brillouin zone ($\boldsymbol{p} = 0$ at Γ) to the edge of the Brillouin zone along one of the principal crystallographic axes (X). The solid curves show the bands which would be expected if spin–orbit coupling were unimportant; dashed curves show the bands expected when spin–orbit coupling is included

Bibliography

Anderson J R, Gold A V 1965 Fermi surface, pseudopotential coefficients and spin-orbit coupling in lead. *Phys. Rev. A* 139: 1459–81

Azbel M Ya, Kaganov M I, Lifshitz I M 1973 Conduction electrons in metals. *Sci. Am.* 228: 88–98

Harrison W A 1970 *Solid State Theory*. McGraw-Hill, New York, Chap. 2

Harrison W A, Webb M B (eds.) 1960 *The Fermi Surface*. Wiley, New York

Ziman J M 1964 *Principles of the Theory of Solids*. Cambridge University Press, Cambridge, Chap. 3

A. H. MacDonald; R. Taylor

Ferroalloys of Manganese, Silicon, Chromium: Production

A description of the production of ferrochromium, ferromanganese and ferrosilicon is essentially a description of the operation of an electric smelting furnace. Therefore, a general discussion of the electric furnace will precede detailed representations of the production of the individual alloys.

1. Electric Smelting Furnace

The electric smelting furnace has evolved into an efficient unit for performing carbon reduction of oxides at elevated temperatures which yield liquid and gaseous products. A charge composed of ore and reducing agent is fed continuously into a low shaft furnace. Energy for smelting is introduced through carbon electrodes which are immersed in the charge. Gases, resulting from the reaction of ore and carbon deep in the furnace, flow upward and are released at the top of the shaft. Liquid products descend and are drained from the bottom of the shaft.

For efficient transmission of electrical energy, three-phase power is employed. For a symmetrical circuit using three-phase power, three electrodes are disposed at the apices of an equilateral triangle. Three single-phase transformers are each connected across a pair of electrodes obviating the need for a ground return. Exceptions to this configuration are large ferromanganese furnaces employing six electrodes operating in pairs as three single-phase units.

The electrode is the heart of the electric smelting furnace. It must be capable of vertical movement for regulating power to the furnace. Therefore, the portion of the secondary conductor nearest the electrode must be flexible. The electrode is consumed and must be replenished. Elaborate mechanisms have been developed for suspending the electrodes, for providing vertical movement, for supplying electrical contact and for permitting the slipping of the electrode relative to the suspension and contact plates to replace electrodes consumed during the operation of the furnace. Figure 1 shows the suspension system of a large ferrosilicon furnace. Preformed electrodes can be used, in which case the electrode is made continuous by joining lengths with threaded ends or threaded pins.

A more economical technique, particularly for large electrodes, is to make the electrodes in the smelting furnace itself. A thin steel casing with internal fins acts as a form for the electrode, supports the unbaked paste and carries current initially until the carbon is sufficiently conductive. Electrode paste, a mixture of pitch and granular carbon, is fed into the top of the column where it melts, filling the casing. In the vicinity of the contact clamps, the pitch is baked to solid carbon. The temperature continues to rise and baking continues as the baked mass descends into the furnace. The casing is replenished by welding additional sections at the top of the electrode column. Electrode breakage, usually caused by thermal shock associated with furnace shutdowns, can be a severe operational problem.

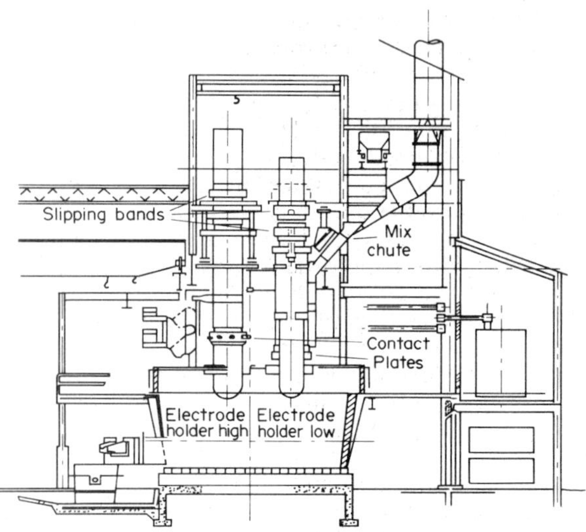

Figure 1
The suspension system of a large ferrosilicon furnace

The distance between centers of the electrodes is between two and three electrode diameters. Closer spacing is impossible because of the bulk of the suspension gear. Wide spacing is desirable because it increases resistance; it may be undesirable because it increases reactance and may destroy contiguity between reaction zones under each electrode. Therefore there is an optimum spacing which depends on the product being smelted. Furnaces have been built in which the electrode suspension rides on a carriage which can be moved radially, resulting in variable spacing.

The furnace shell, which is grounded, must be sufficiently remote from the electrodes so that current does not flow to it. Furnace depth is a compromise between efficient energy utilization and increased risk of electrode breakage and increased reactance. Some operations are conducted with a large reservoir of slag which increases the depth of the furnace. The hearth of the furnace is usually constructed of carbon. Subhearth cooling is required. The walls are made of carbon in the tapping area and carbon or refractory elsewhere. The hearth and walls are contained within a steel shell. The tops of furnaces are frequently closed with a water-cooled cover which has openings for electrodes and mix delivery. The cover permits the collection of CO which can be utilized as a fuel or a chemical. Gas flow is assured by employing fans, and gas cleaning is accomplished by using wet scrubbers or bag filters. In some installations the surface is left open and the furnace gases are burned at the top of the charge. The combusted gases are collected in a hood and cleaned. Energy recovery has been accomplished on some open furnaces. In covered units the furnace charge is fed through chutes or around the electrodes. In open furnaces the mix can be introduced by chutes extending through the hood or by a charging car. The ore and carbon are proportioned stoichiometrically. Accurate weighing is essential and can be made difficult by the presence of variable amounts of water. This is particularly a problem with porous materials such as coke and fine ore or reducing agent. The burden above the reaction zone should be porous to permit the flow of product gases. It must also not bridge but descend freely into the furnace. Therefore fines should be removed from the ore, fluxes and reducing agents. These can be agglomerated before charging to the furnace. The optimum size of mix varies from operation to operation. Large particle size mix may be more difficult to obtain and transport to the furnace through chutes. It also may cause low electrical resistance and chemical reactivity. Small particle size may lead to loss by gaseous entrainment, low bed porosity and mix bridging.

The energy input is controlled by changing voltage or electrode position. Stoichiometric adjustments are frequently necessary due to the difficulty in determining the precise amounts of the constituents fed or due to poor operating conditions. Energy and stoichiometry have been successfully controlled by computers.

The liquid products are removed from the hearth through a taphole. There are both continuous and batch tapping operations. The furnace is opened by drilling, oxygen lancing, or heating with an auxiliary electrode. The liquid products are received in refractory lined ladles. If a slag is present, it is necessary to remove the slag by decantation, skimming, or bottom tapping of the ladle before casting the metal in chills. Metal refining and additions can be accomplished in the ladle. Most ferroalloys are brittle and can be easily reduced to a desired size by crushing.

In principle, one furnace could be used to produce any product. In practice, large units are usually designed with a given product in mind. This is due primarily to the differences in resistivity of the furnace charge for different alloys. For a given size of electrode the power depends on the alloy because the electrode is usually operated near its current carrying limit.

The suitability of the electric smelting furnace as a tool for smelting ferroalloys is attested to by its being relatively unchanged, except in size, for 70 years. It is an example of art being ahead of science.

2. Ferrochromium

Although chromium ore occurs throughout the world, by far the greatest reserves are in Southern Africa. The ore is classified according to its Cr/Fe ratio, high chrome ores having a Cr/Fe ratio greater than 2, low chrome ores having a ratio of 1.5 to 2.0. In addition to iron and chromium oxides, chrome ore contains MgO and Al_2O_3, which are not reduced under the conditions required to convert chromium oxide to metal. Because they are refractory, a flux must be added to lower the melting point of the slag so that it can be discharged from the furnace. Coke and noncoking coal are used as reducing agents and should have low sulfur, phosphorus and ash contents as well as high electrical resistivity.

Chrome ore, reducing agent and flux are reacted in an electric smelting furnace. The products of primary smelting include high carbon ferrochromium, charge chromium and ferrochrome silicon. The high-carbon composition shown in Table 1 is attained by using selected lumpy ores and an MgO/Al_2O_3 ratio of approximately unity. Silica is added to the mix to obtain a product slag containing approximately equal amounts of MgO, Al_2O_3 and SiO_2.

The addition of more silica results in a slag richer in SiO_2, metal of higher silicon content and a lower energy consumption. The two grades of charge chromium listed in Table 1 arise from ores of differing Cr/Fe ratios. If the SiO_2 in the charge is increased so that it is approximately equal in weight to the ore, ferrochrome silicon will be produced. The high silicon content lowers the carbon content of the alloy. The smelting of silicon requires more energy than chromium and therefore the energy requirement is greater for this alloy.

The slag and metal are tapped periodically from the furnace into ladles. A crude slag–metal separation can be obtained by allowing slag to overflow from the primary ladle. For charge chromium and high-carbon ferrochromium, the metal and slag are cast in metal chills. The slag is removed from the metal after solidification.

In the production of ferrochrome silicon, liquid slag and metal are held in the ladle to permit SiC to float out of the metal into the slag. The ladle is then bottom tapped to obtain a low-carbon ferrochrome silicon.

The recovery of chromium in the alloy is variable. In good operations approximately 90% of the chromium in the ore will be recovered in the alloy. Of the chromium in the slag, part will be as metal, which is recoverable by physical beneficiation, and the balance as unreduced oxide.

Fine ores, beside making furnace operation difficult, lead to large chromium losses. Fine ores have been agglomerated for furnace charges by briquetting. In addition, alternative processes have been developed for utilizing fine ores. Agglomerates of ore and reducing agent can be prereduced in a kiln and then fed to an electric furnace. Although the energy requirement in the electric furnace is reduced, the overall energy requirement for the prereducer and furnace are similar to the requirement for the electric furnace.

A low-carbon ferrochromium is produced by using ferrochrome silicon as a reducing agent for Cr_2O_3 dissolved in an ore–lime melt which is prepared in an open-arc tilting furnace. Two stages of contacting can be employed in which an intermediate alloy is reacted with an ore–lime melt and an intermediate slag is stripped to less than 1% Cr_2O_3 with ferrochromium silicon.

Simplex, low-carbon ferrochromium, is made by decarburizing ferrochromium with a mixed iron–chromium oxide in a vacuum solid-state reaction. The stoichiometrically balanced powders are briquetted and heated in a large vacuum furnace to 1400 °C at a pressure of approximately 10^{-4} atm. The cycle time is approximately 90 h.

The advent of the argon–oxygen decarburization (AOD) process for making stainless steel has shifted

Table 1
Ferrochromium alloys: composition and energy consumption in production

Ferrochromium	Composition (%) Cr	C	Si	S (max)	P (max)	Theoretical energy consumption ($kWh\ t^{-1}$ alloy)	Typical energy consumption ($kWh\ t^{-1}$ alloy)
High carbon	62–72	4–9.5	3.0 max	0.06	0.03	3300	4600
Charge chrome							
low Fe	52–64	4–6	8–14	0.04	0.03	3300	4000
high Fe	52–58	6–8	6.0 max	0.04	0.03		
Low carbon	67–75	0.75–0.025	1.0 max	0.025	0.03		
Vacuum low carbon	67–72	0.02–0.01	2.0 max	0.03	0.03		
Ferrochrome-silicon	34–38	0.06 max	38–42	0.03	0.03	5100	7600

Source: American Society for Testing and Materials 1980 *Annual Book of Standards*

the market for ferrochromium to low Cr/Fe charge chrome grades. Therefore, the present emphasis in ferrochromium smelting is on improving the charge chrome practice and not on refining.

3. Ferromanganese

Manganese ores are widely distributed, occurring in Australia, Brazil, Gabon, India, the Republic of South Africa and the USSR. High-grade ores contain more than 35% Mn and varying amounts of SiO_2, Al_2O_3, MgO, CaO, BaO and K_2O. Large deposits of nodules, which contain manganese, have been found on the ocean floor. Several processes have been recommended for manganese recovery but none has reached a commercial scale.

The less easily reduced oxides form a slag during smelting. MnO itself can be used as a flux to form low-melting slags. Coke is the usual reductant. Manganese ore frequently contains oxygen in excess of MnO. This leads to reaction between the ore and CO in the shaft of the furnace so that the product gas contains significant amounts of CO_2. To promote this reaction, bed porosity is essential. Coke and ore fines are screened from the charge. The ore fines are frequently agglomerated by sintering, which also reduces their oxygen content.

Formerly most of the high-carbon ferromanganese was smelted in blast furnaces; however, much of the production has shifted to large electric smelting furnaces which are rated up to 45 MW.

The charge is conductive leading to a low voltage, low power factor, low power density operation. The slags and ores have relatively low melting points resulting in low temperature operation, which necessitates an enlarged reaction zone. Some large manganese furnaces are run with a slag bath, established by using a slag taphole at an elevated position relative to the metal taphole.

High manganese recoveries are difficult to obtain in a single-stage of smelting because of the relatively low melting points of manganese ores and slags, the high electrical conductivity of the manganese-bearing slags, and the relatively high vapor pressure of manganese. Therefore, the usual practice is to smelt manganese ore so that a high-carbon ferromanganese and a high-manganese slag result. The analysis of the metal is shown in Table 2. The charge is composed of manganese ore, coke, and flux if required. The slag contains between 30 and 40% Mn and is equal to 0.4 to 0.7 of the weight of the alloy. Manganese recovery in the alloy ranges between 70 and 80%. Under certain conditions, if there is no use for the slag or the ore contains large amounts of CaO, single-stage smelting is practised and manganese recoveries of 85–90% are realized.

The overall recovery of manganese can be raised to over 95% by smelting the slag from the high-carbon ferromanganese operation in a mixture containing manganese ore, coke and silica to make silicomanganese in an electric furnace. Because of the higher resistance of the charge, higher temperatures are realized and the manganese in the slag is reduced to below 10% which permits it to be discarded. Table 2 shows a typical composition of this alloy. Silicomanganese is frequently smelted in the same furnace as high carbon ferromanganese. Higher power levels are possible because of the higher resistance of the mix.

The carbon content of manganese alloys is inversely proportional to the silicon content. Alloys containing approximately 30% Si have a carbon content of approximately 0.1%, and are usually obtained by smelting silicomanganese with additional carbon and silica. Alloys which are not saturated in carbon can be produced by refining carbon-saturated alloys. Medium-carbon ferromanganese alloys containing 1–1.5% carbon can be made by oxygen refining high-carbon ferromanganese or by desiliconizing silicomanganese in a reaction with a melt of manganese ore and CaO. Low-carbon ferromanganese is made by using ferromanganese-silicon instead of silicomanganese in a reaction with an ore–lime melt. In an integrated manganese plant, producing high-carbon ferromanganese, silicomanganese and medium-carbon ferromanganese, slag generated in high-carbon ferromanganese and medium-carbon ferromanganese production are circulated into the silico-

Table 2
Ferromanganese alloys: composition and energy consumption in production

Ferromanganese	Composition (%)					Theoretical energy consumption ($kWh\ t^{-1}$ alloy)	Typical energy consumption ($kWh\ t^{-1}$ alloy)
	Mn	C (max)	Si (max)	P (max)	S (max)		
High carbon	78–82	7.5	1.2	0.35	0.05	1500	2200
Medium carbon	80–85	1.5	1.0	0.30	0.02		
Low carbon	85–90	0.75–0.1	2.0	0.20	0.02		
Silicomanganese	65–68	2.0	16–18.5	0.20	0.04	3200	4000
Ferromanganese-silicon	63–66	0.08	28–32	0.05			

Source: American Society for Testing and Materials 1980 *Annual Book of Standards*

manganese operation in which the manganese content of the slag is brought low enough that it can be discarded. The slag and metal from a high-carbon ferromanganese furnace can be separated by tapping from separate tapholes. If the slag and metal are withdrawn from a single taphole a skimmer can be employed which separates metal and slag in the tapping launder. The carbon content of silicomanganese can be lowered by holding metal in the ladle, allowing SiC to float to the slag.

In addition to alloys produced in furnaces, refined manganese is also produced by fused salt electrolysis. Prereduced manganese ore is fed to an electrolyte containing CaO and CaF_2. A carbon anode is consumed and liquid manganese is periodically tapped from the cell.

4. *Ferrosilicon*

The raw material for producing silicon alloys is SiO_2. This is preferably a lump quartz or quartzite of high purity containing at least 96% SiO_2 which does not decrepitate on heating. Because the process is slagless, the ash content of the reducing agent should be a minimum. Both coke and coal are employed. The most desirable reducing agents are reactive toward SiO_2 and of high electrical resistivity. Charcoal possesses the best combination of these properties but is too expensive. Gas cokes and low-ash coals are effective reducing agents. In addition, appreciable amounts of wood chips are used as a bulking agent. Much of the weight of the wood chips is volatilized so that they contribute only 10–20% of their weight as fixed carbon. Iron is usually added in the form of fine scrap such as turnings and punchings which can be easily handled in the mix system and can be blended with the other ingredients.

There are several grades of ferrosilicon, ranging from 22 to 90% Si. Alloys in the low-silicon range are less attractive as silicon units because of the greater conductivity of the furnace burden and the cost of transporting iron. Alloys in the range of 30–45% and 50–65% Si are not produced because they are very friable. Ultimately the choice of ferrosilicon is made on economic grounds. If the price of scrap iron is low there is a tendency to produce 45–50% FeSi, otherwise 75% FeSi is favored. The presence of iron greatly facilitates the reduction of SiO_2 with carbon.

The problems of silicon smelting are the recovery of volatile SiO, the formation of hearth accretions of SiC which force the electrodes out of the furnace, and the encrustation of the charge in the shaft. Volatile SiO losses can be minimized by using a reactive reducing agent, employing bulking agents such as wood chips to promote bed porosity, maintaining deep electrodes and producing high-iron alloys. Recovery of silicon ranges from 95% in 50% ferrosilicon to approximately 80% in 90% ferrosilicon. Poor recovery results in higher energy consumption as may be seen by comparing the energy consumed per unit of silicon in 50% ferrosilicon and 90% ferrosilicon in Table 3.

Hearth accretions can be controlled by maintaining the proper stoichiometry in the mix. Rotation of the hearth relative to stationary electrodes is also helpful. The rate is slow: 100–150 h rev^{-1}. The electrodes sweep out a continuous path which prevents the development of buildup between electrodes. Rotation of the hearth necessitates a movable tapping arrangement. Because of the tendency of the mix to fuse, ferrosilicon furnaces are usually operated with an open top so that the charge can be mechanically stoked and top accretions broken up. Recently, differential rotation has been established on a split shell. The upper portion of the shell is rotated relatively more rapidly to aid in the descent of mix.

The usual impurities in ferrosilicon are aluminum and calcium. For certain applications these are undesirable. They can be removed by oxygen refining and slag treatment of the liquid alloy.

Calcium silicon can be produced by co-smelting CaO and SiO_2 with carbon. Calcium contents of approximately 30% can be obtained in this manner. A calcium silicate slag containing CaC_2 is produced along with calcium silicon alloy and the slag is recycled. The energy consumption for this alloy is high because recovery of both calcium and silicon

Table 3
Ferrosilicon alloys: composition and energy consumption in production

Ferrosilicon	Composition (%) Si	C (max)	S (max)	P (max)	Al (max)	Mn (max)	Theoretical energy consumption (kWh t^{-1} Si)	Typical energy consumption (kWh t^{-1} Si)
50% Ferrosilicon	47–51	0.1	0.025	0.04	1.25	0.75	8200	10300
75% Ferrosilicon	74–79	0.1	0.025	0.035	1.50	0.40	8900	11900
90% Ferrosilicon	92–95	0.1	0.025	0.025	2.00	0.25	8900	13200
Calcium-silicon	60–65 28–32(Ca)	1.0	0.07	0.02	1.50		7170[a]	13200[a]

Source: American Society for Testing and Materials 1980 *Annual Book of Standards* a kWh t^{-1} alloy

are low due to vaporization losses. A wide variety of silicon-base alloys are also produced by ladle addition to ferrosilicon. Such alloying is facilitated by the high degree of superheat in tapped ferrosilicon as a result of the smelting temperature being much higher than the melting point of the alloy. Magnesium, barium, aluminum and strontium, for example, can be added to produce alloys for use in the treatment of cast iron.

See also: Manganese Production; Silicon Production; Chromium Production; Metals Production: An Overview

Bibliography

Edneral F P 1979 *Electrometallurgy of Steel and Ferroalloys*, Vol. 2. Mir, Moscow

Robiette A G E 1973 *Electric Smelting Processes.* Wiley, New York

Volkert G, Frank K-D (eds.) 1972 *Metallurgie der Ferrolegierungen.* Springer, Berlin

J. H. Downing

Ferroelastic Materials

The property of ferroelasticity, which has been recognized only since 1969, is most simply viewed as a form of mechanical twinning in which the lattice reorients in response to mechanical stress. Many materials are now known to be ferroelastic. The ferroelastic property may be coupled with other properties, such as ferroelectricity, or it may be independent. The property itself gives rise to a variety of effects as crystal axes are interchanged under the application of stress with resulting changes in the optical propagation and other crystal constants. Coupling allows these effects to be produced either by the application of a suitable stress or by a field appropriate to the coupled property. Devices using ferroelastic–optic effects include modulators, displays, memories and pattern generators.

1. Definition of Ferroelasticity

A crystal containing two or more equally stable orientational states, in the absence of mechanical stress, is ferroelastic if the application of stress along an appropriate direction reproducibly transforms one state into another. The minimum stress required for transformation is the coercive stress E_{ij}, where i, j denote the stress and transformed axis directions. The experimental magnitude of E_{ij} is a function of temperature and pressure, and is also dependent on the defect distribution, with a common range $10^4 < E_{ij} < 10^8\ \mathrm{N\,m^{-2}}$. Ferroelasticity is always associated with a small distortion from a higher-symmetry phase. The spontaneous strain e_s is a measure of this distortion, which is typically parts per thousand or less. A simple example is given by a slightly distorted tetragonal crystal in which the normal equality of the tetragonal a_1 and a_2 axes is replaced by orthorhombic a and b axes such that $e_s = (a - b)/(a + b) \simeq 10^{-3}$.

2. Ferroelastic Subgroup of Prototype Symmetry

The point group of each ferroelastic crystal lacks certain symmetry elements that become transiently present in the course of reorienting e_s. These same symmetry elements are generally present in the stable phase formed at higher temperatures. The higher-symmetry phase is known as the prototype structure, and the ferroelastic crystal symmetry is hence a subgroup of the prototype point group. In the example above, if the prototype point group were $4mm$, then the only permissible orthorhombic subgroup would be $mm2$. By analogy with ferroelectricity (see *Ferroelectric Materials*), the prototype phase may be referred to as the paraelastic phase.

3. Atomic Displacement Relationship

In addition to the requirement that e_s be small in a ferroelastic crystal to allow reorientation of the crystal axes, it must simultaneously be possible to rearrange the atomic distribution. Within each unit cell, there must hence be pairs of pseudosymmetrically related atoms such that for every atom at $x_1y_1z_1$ there is another atom of the same kind at $x_2y_2z_2$ with $x_1y_1z_1 = f(x_2y_2z_2) + \Delta$, where Δ is an atomic displacement of the order of 1 Å (see Table 1) and $f(x_2y_2z_2)$ is a transformation that results in reorientation of the lattice vectors. In the example above, if $x_1y_1z_1 = (\frac{1}{2} - y_2, \frac{1}{2} + x_2, z_2) + \Delta$, then the function $f(x_2y_2z_2)$ indicates that the a and b axes are interchanged, with a change in the sense of the transformed a axis, accompanied by a displacement Δ for each atom.

Table 1
Characteristics of some ferroelastic materials[a]

Formula	$10^6 e_s$	E_{ij} (MN m^{-2})	ij	Δ_{max} (Å)
$CsFeF_4$	320	22	13	0.45
$K_2Cd_2(SO_4)_3$	58	b	23	1.19
KIO_2F_2	3100	2	13	2.40
$LaFeO_3$	240	300	12	0.79
$Mg_3B_7O_{13}Cl$ (boracite)	550	120	12	0.60
PtGeSe	4300	16	12	2.02
$SmAlO_3$	56	50	12	0.64
$Tb_2(MoO_4)_3$ (TMO)	2000	1.5	12	0.67
$Gd_2(MoO_4)_3$ (GMO)	1600	1.0	12	0.71

a All values are for room temperature b Frozen ferroelastic (see Sect. 5)

4. Temperature Dependence of E_{ij} and e_s

The magnitudes of E_{ij} and e_s in the prototype structure are, by definition, zero. For second-order phase transitions from the ferroelastic to the prototype phase, both E_{ij} and e_s decrease monotonically as a function of increasing temperature to zero at the phase transition. In the case of a first-order phase transition, E_{ij} and e_s generally decrease with increasing temperature but may have substantial values to within a very small temperature interval of the transition before becoming zero at the transition.

5. Coupling of Other Properties with Ferroelasticity

The development of ferroelasticity at the phase transition, on cooling from the prototype structure, is often accompanied by the appearance of other tensor properties such as ferroelectricity or ferromagnetism. In the case that the other property is coupled, then ferroelastic reorientation will cause a reorientation in that property. Thus, in $Tb_2(MoO_4)_3$ with $x_1y_1z_1 = (\frac{1}{2} - y_2, x_2, 1 - z_2) + \Delta$, the crystal *a* and *b* axes are interchanged simultaneously with a change in the sense of the polar *c* axis resulting in a reversal of P_s (see *Ferroelectric Materials*). The coupling in this material between ferroelasticity and ferroelectricity is experimentally demonstrated by the identity of the normalized temperature dependence of P_s and e_s.

A crystal may show other cooperative phenomena in addition to ferroelasticity without a coupling between them. Thus, in the transition from a piezoelectric prototype to a lower temperature piezoelectric and ferroelastic phase, no coupling need be present. In the case of $K_2Cd_2(SO_4)_3$, the transition from point group 23 to 222 is first order with ferroelasticity developing below the Curie temperature T_c. The three identical piezoelectric d_{ij} coefficients (see *Piezoelectric Materials*) above T_c transform to become three independent coefficients below, but each remains uncoupled to e_s.

6. Cohesive Strength and Coercive Stress

The maximum load a crystal is capable of sustaining without rupture is a measure of the cohesive strength along the load direction. If the coercive stress exceeds the cohesive strength at some temperature, then e_s cannot be reoriented at that temperature. In such a case it is possible that the coercive stress will become less than the cohesive strength at other temperatures: if not, the crystal is referred to as a frozen ferroelastic. It is always possible to reorient e_s by the application of a mechanical stress along the appropriate direction if the crystal is cooled through T_c under stress.

7. Materials Preparation and Poling

Single-crystal ferroelastic materials are prepared by a variety of techniques (see *Crystal Growth of Ceramics: An Introduction*). These include growth from aqueous solution for KIO_2F_2, flux growth for $SmAlO_3$, vapor-phase growth for the boracite family, Bridgman solidification from the melt for $K_2Cd_2(SO_4)_3$, and Czochralski pulling from the melt for TMO and GMO (see Table 1). Some materials may be crystallized by more than one technique.

It is usually necessary to detwin as-grown crystals in order to obtain single-domain material. Most ferroelastic materials may be detwinned by the application of compressive stress, often at room temperature, although it is necessary to cool frozen ferroelastic crystals through T_c while under stress. If ferroelasticity is coupled with ferroelectricity then electric field poling, as described for ferroelectric materials (see *Ferroelectric Materials*), may be more convenient. Similarly, magnetic poling could be used if the ferroelasticity is coupled with magnetic ordering, as in ferrimagnetic Fe_3O_4.

8. Uses of Ferroelastic Materials

Most uses have been proposed for the specific cases of GMO and TMO; these materials can be prepared readily as large, high-optical-quality crystals. By applying an electric field to produce ferroelectric switching, thereby reversing the direction of the *c* axis, the coupled ferroelasticity results in a simultaneous interchange of the *a* and *b* axes. An electric field can hence be used to obtain precise control over the domain-wall motion and has been utilized to construct ferroelastic–optic switching devices including optical switches, adjustable optical slits and pattern generators. It is even possible to make a color modulator that provides saturated interference colors switchable from green to blue to red. Domain-wall motion has also been employed in variable acoustic delay devices in which an acoustic wave is reflected from the movable domain wall. Electrically–elastically induced changes in crystal dimensions have also been used in micropositioners to obtain precise displacements larger than those possible with piezoelectric devices but smaller than those with mechanical devices. Ferroelastic materials that are also piezoelectric, pyroelectric and/or ferroelectric can, in addition, be used in the various devices described under these headings.

See also: Electronic Materials: An Overview

Bibliography

Abrahams S C 1971 Ferroelasticity. *Mat. Res. Bull.* 6: 881–90

Hahn T (ed.) 1983 *International Tables for Crystallography*, Vol. A. Reidel, Dordrecht

Klassen-Neklyudova M V 1964 *Mechanical Twinning of Crystals*. Consultants Bureau, New York
Lines M E, Glass A M 1977 *Principles and Applications of Ferroelectrics and Related Materials*. Clarendon Press, Oxford
Newnham R E 1975 *Structure–Property Relations*. Springer, New York

S. C. Abrahams; K. Nassau

Ferroelasticity

Ferroelastic crystals are those crystals that undergo small spontaneous atomic displacements that form a basis for a subgroup of the symmetry group of the original crystal as a result of a temperature change or application of external stress (Aizu 1970). The name "ferroelasticity" is derived from "ferromagnetism" which refers to the spontaneous appearance of magnetic polarization in a crystal, and is also analogous to "ferroelectricity," the spontaneous appearance of electrical polarization. More than one of these "ferroic" transitions may occur simultaneously and are then regarded as being coupled. Ferroelastic phenomena of pyroelectricity and pyromagnetism, or the possession by some materials of a temperature-dependent spontaneous electric or magnetic dipole moment, have been known since ancient times because some materials attracted others when heated. In the late nineteenth century, systematic empirical studies of various ferroelastic materials resulted in a series of other discoveries: piezoelectricity, ferromagnetism, thermoelasticity and, in the twentieth century, ferroelectricity and ferrimagnetism. Systematic research on the physics of these materials began only in the early 1940s, when reproducible series of crystals were made. Careful study revealed that important cross-coupling occurs between the thermodynamic constraints and the various spontaneous deformations and polarizations that occur near the critical point.

Theoretical thermodynamic treatments of these systems near the critical point began in the late 1930s by Landau and others, who used a phenomenological truncated Taylor expansion of the free energy density in an order parameter function. The current standard theory of the nonlinear time-dependent ferroelastic material is based on the conservation laws of linear and angular momentum, and of energy together with four basic assumptions:

(a) The material's behavior is fully specified by the elastic strain and stress fields, electric field and displacement, and the magnetic and inductive fields.

(b) The electromagnetic fields are very slowly varying functions of time (quasistatic).

(c) The material possesses some state in which the mechanical stress and the electromagnetic fields all vanish.

(d) The total internal energy density is itself an analytic function of the components of the above six fields.

Further discussion of the polynomial expansion in these functions and the critical properties can be found elsewhere (see *Landau Theory of Second-Order Transformations*). In the 1950s and after, theoretical progress permitted an understanding of the microscopic dynamics of the transformations, in the context of lattice dynamics. This was especially the case in the 1960s, when inelastic neutron scattering became available to explore the entire Brillouin zone. A crystal undergoing a structural phase transition in which there is a displacement of the equilibrium positions of the atoms in the primitive cell is characterized according to the behavior of the collective lattice phonon modes upon changing one or several external thermodynamic parameters. Ferrodistortive crystals have optical phonons which soften at the Brillouin zone center. Antiferrodistortive crystals have optical phonons which soften at the zone boundary. A more complete exploration of the theory can be found elsewhere (see *Soft Mode Phenomena; Central Peak Phenomena*).

Let us imagine a ferroelastic material which is in contact with a heat reservoir at a certain temperature, and can be acted on by any one of three external fields or constraints: mechanical stress, electric field, and magnetic field. Suppose we are close to one of the critical points of the material's structural phase transition. The ferroelastic material is classified according to the effect of a change in the extenal fields on the thermodynamically conjugate fields of elastic strain, of electrical displacement and of induction. The terms used depend on which of the fields are coupled, as summarized by Table 1.

With such a wide variety of phenomena occurring in ferroelasticity, a simple classification scheme is needed to enumerate the various materials, the nature of the transformations, and a characterization of the different phases P in terms of their point groups. Several such schemes are used but perhaps the simplest, which also recognizes the cross coupling between elastic, electromagnetic and thermal properties, is one by Keve and Abrahams (1970). Each phase of a crystal is represented by a triplet of letters followed by the point group of that phase. The triplet corresponds to the first three columns of Table 1. The letters of the scheme follow each adjective entry. The first symbol refers to the dielectric state of the crystal:

F: for actual or potential ferroelectric in which the spontaneous polarization can be reversed or reori-

Table 1
Ferroelastic phenomena. The terms used to describe the phase transformation depend on the material's response to an applied field as indicated in the table. The letters refer to the Keve–Abrahams classification scheme

	Electric displacement	Strain	Magnetic induction	Entropy
Electric field	ferroelectric: F antiferroelectric: A paraelectric: P	electrostrictive: π	electromagnetic	electrocaloric: F
Stress	piezoelectric: π	ferroelastic: F antiferroelastic: A paraelastic: P	piezomagnetic: π	piezocaloric: π
Magnetic field	magnetoelectric	magnetostrictive: π or magnetoelastic	ferromagnetic: F ferrimagnetic: F antiferromagnetic: A paramagnetic: P diamagnetic: P	magnetocaloric: F
Temperature	pyroelectric: F	thermoelastic: F	pyromagnetic: F	

ented, either experimentally or conceptually (this includes pyroelectrics).

A: for antiferroelectric, characterized by the existence of ordered local dipoles in the structure but a zero net polarization (including antipolar).

π: for piezoelectric, encompassing all piezoelectric materials except those of A and of F above, in which mechanical strain is produced by the application of an electric field in the absence of a stress and of a magnetic field. Conversely, these crystals exhibit electrical polarization when subject to mechanical stress.

P: for paraelectric, characterized by zero polarization, containing no ordered dipoles in the structure.

The second symbol refers to the elastic state of the crystal:

F: for ferroelastic, characterized by the existence of a net spontaneous strain which can be reoriented, either experimentally or conceptually (including thermoelastics). The crystal has two or more orientation states in the absence of mechanical stress (and of electric or magnetic fields) and can be shifted from one to another of these states by mechanical stress. The two orientational states are identical in crystal structure but have different components of the mechanical strain tensor at zero stress (mechanical and electromagnetic). In terms of the collective lattice modes, if upon changing an external thermodynamic parameter the frequency of an optical phonon mode becomes zero (softens) at the center of the Brillouin zone, the crystal is called ferrodistortive. This category includes cooperative Jahn–Teller transformations in insulators. In an insulator composed of ions with orbitally degenerate ground states it is possible to lower the electric energy by splitting these degenerate levels apart by means of a lower symmetry distortion. The amplitude of the distortion depends upon the nature of the electronic state and the size of the increase in the elastic energy (see *Jahn–Teller Effects*). Closely related is the formation of incommensurate lattices and of charge density waves as described in the article on that subject (see *Incommensurate Structures and Charge Density Waves*).

A: for antiferroelastic, defined by the existence of a single stable mechanical state in which the unit cell contains, in the absence of stress, an equal number of opposite local spontaneous strains with a zero resultant.

P: for paraelastic, encompassing all other crystals with a single stable mechanical state and zero net strain in the absence of stress. Such a crystal is neither ferro- nor antiferroelastic.

The third symbol refers to the magnetic state:

F: for ferro- or ferrimagnetic characterized as possessing a nonzero moment.

A: for antiferromagnetic.

π: for piezomagnetic.

P: for lack of magnetic order (para- or diamagnetic).

Consider three examples of this notation. First, a material undergoing a series of ferroelastic transitions, barium titanate ($BaTiO_3$):

$$\mathrm{PPP}(3mm)\mathrm{FFP}(4mm)\mathrm{FFP}(2mm)\mathrm{FFP}(3m)$$

Second, a ferroelastic showing ferromagnetism, rubidium iron fluoride ($RbFeF_4$):

$$PPP(4/mmm)PFP(mmm)PFA(mmm)$$

Finally, a ferroelastic showing both electrical and magnetic responses, barium cobalt fluoride ($BaCoF_4$):

$$PPP(mmm)FPP(2mm)[FFA(2) \text{ or } FAA(2mm)]$$

Obviously, there are many relationships between these categories. For example, it is sufficient that a ferrodistortive crystal have a polar soft mode to be a ferroelectric; however, it is not necessary. Crystals may be highly anisotropic and ferroelastic in one direction, yet antiferroelectric in another. All ferroelectrics are piezoelectrics. Focusing on the behavior of these materials at the critical point one distinguishes a handful of universality classes to which this myriad of materials belong. Thermodynamic functions and derived quantities, such as susceptibilities, all undergo certain power law singularities at the critical point. The exponents of these singularities determine to which of the classes the phase transformation belongs. An example of a theoretical "mean field" universality class and its critical exponents is given elsewhere (see *Landau Theory of Second-Order Transformations*).

A common feature of ferroelastic crystals is the formation of domains of uniform polarization, deformation or magnetization, separated by domain walls. This is a consequence of the various competing contributions to the free energy of the crystal. For example, in ferromagnets, the short-range exchange and anisotropy energies dominate the ferromagnetic ordering except at very long distances, where the sum over many dipoles, whose individual interactions are weak but of very long range, yields a dominant energy. Each domain wall raises the total exchange energy but lessens the dipolar energy more. The width (and hence inertia) of the wall can range from several nanometers, in ferromagnetic materials, to an interatomic spacing, in the case of order–disorder boundaries in binary alloys or of dislocations in the thermoelastic martensitic phase transformation. The exact number and distribution of domains is sample preparation dependent because morphology, defects and applied field history are all important. Application of an aligning external field energetically favors certain domains and at sufficiently high fields, the sample's domain structure undergoes an irreversible change with some or all of the unfavored domains disappearing and the sample polarizing. The reverse field necessary to reduce the polarization to zero is the coercive field. The irreversible nonlinear response to the applied field (e.g., magnetic induction versus field) is known as hysteresis. Domain walls and other defects may be observed by x-ray diffraction and by inelastic neutron scattering, among other techniques where they give rise to characteristic central peaks (see *Central Peak Phenomena*).

Recognizing that elastic displacements of charged atoms give rise to local dipolar fields makes it easy to understand how cross coupling between elastic and electrical polarization comes about. Less evident is how magnetic moments couple to the elastic field. Magnetoelasticity arises microscopically from the strain dependence of any of several interactions which influence the magnetic behavior of the material. It is important not only in ferro-, ferri- and antiferromagnetic materials, but also in paramagnetic (electronic and nuclear) materials. Often the dominant contribution of the magnetoelastic interaction is the strain dependence of the crystalline electric field. For the transition metals, this crystal field acts on the electronic orbital motion which, in turn, is coupled to the spin magnetic moments via the spin–orbit coupling. For the rare-earth ions, the spin–orbit interaction is so strong that the spin and orbital moments cannot be considered separately and the crystalline electric field acts on the moments associated with the total angular momentum of the ion. These types of coupling are equally operative on paramagnetic impurities in diamagnetic host lattices. If the magnetic moments are nuclear, the strain can couple to the electronic moments which are coupled (a) to the nuclear ones by the hyperfine interaction, (b) to the internal magnetic field generated by strain-induced itinerant electron currents, or (c) to strain-induced crystal electric-field gradients through the electric quadrupole moment.

Ferro- and ferrimagnetic materials form the cornerstone of the mass-memory and recording industries. Magnetostrictive transducers are used to convert high-frequency electromagnetic radiation into elastic waves in processes of ultrasonic cleaning, homogenization, emulsification, dispersion and agglomeration. Variable delay devices are made with these materials and controlled by the size of the biasing dc magnetic field.

Early ferroelectric applications evolved around the application of electromechanical transducers and of materials with high dielectric constants, for use in specialty capacitors. Later applications exploited the nonlinearity and the size of the ferroelectric polarizability in devices whose index of refraction could be controlled by an electric field, for use as modulators, deflectors and as optical memories. Single crystal devices avoid polarization reversal, high fields and illumination by high-energy photons to prevent elastic fatigue and radiation damage. Modern ferroelectric ceramics seem less prone to these problems.

Pyroelectrics are increasingly used for thermal detectors of infrared radiation and for high frequency response imaging devices.

See also: Brillouin Zones; Displacive and Order–Disorder Transformations

Bibliography

Aizu K 1970 Possible species of ferromagnetic, ferroelectric, and ferroelastic crystals. *Phys. Rev. B* 2: 754–72

Ashcroft N, Mermin D 1976 *Solid State Physics.* Holt, Rinehart and Winston, New York

Christian J W 1975 *The Theory of Transformations in Metals and Alloys*, 2nd edn. Pergamon, New York

Cullity B D 1972 *Introduction to Magnetic Materials.* Addison Wesley, Reading, Massachusetts

Grindlay J 1970 *An Introduction to the Phenomenological Theory of Ferroelectricity.* Pergamon, Oxford

Keve E T, Abrahams S C 1980 Dielectric, elastic and magnetic property symmetry and phase transition representation. *Ferroelectrics* 1: 243–46

Kittel C 1976 *Introduction to Solid State Physics.* Wiley, New York

Lines M E, Glass A M 1977 *Principles and Applications of Ferroelectrics and Related Materials.* Clarendon, Oxford

Mason W P (ed.) 1964 *Physical Acoustics: Principles and Methods.* Academic Press, New York

Rado G T, Suhl H (eds.) 1963–73 *Magnetism: A Treatise on Modern Theory and Materials*, Vols. 1–5. Academic Press, New York

Stanley H E 1971 *Introduction to Phase Transitions and Critical Phenomena.* Oxford University Press, Oxford

White R M, Geballe T H 1979 Long range order in solids. *Solid State Phys., Suppl.* 15

J. F. Currie

Ferroelectric Materials

The permanent electric dipole moment possessed by all pyroelectric materials (see *Pyroelectric Materials*) may, in certain cases, be reoriented by the application of an electric field. Such crystals are called ferroelectric, a term first used by analogy with ferromagnetism. The early designations Seignette-electric and Rochelle-electric are no longer used. All ferroelectric crystals are necessarily both pyroelectric and piezoelectric. Many lose these polar properties at the transition or Curie temperature T_c. A nonpolar phase above T_c is the so-called paraelectric phase.

Ferroelectric single crystals grown in the absence of an electric field are inevitably electrically twinned, with a domain volume containing a given spontaneous polarization direction approximately equal to that containing the antiparallel direction. Ceramics have similar domain structures superimposed on the more general orientational disorder associated with polycrystalline materials. Exposure to an electric field under appropriate conditions can result in complete or partial realignment of the spontaneous polarization. A net dipole moment is not normally detectable in such materials because the surface charges are rapidly neutralized by ambient charged particles. The nonlinear optical properties of noncentrosymmetric materials are generally enhanced if they are also ferroelectric.

Ferroelectricity was discovered in 1921 by J. Valasek during an investigation of the anomalous dielectric properties of Rochelle salt, $NaKC_4H_4O_6 \cdot 4H_2O$. A second ferroelectric material, KH_2PO_4, was not found until 1935 and was followed by some of its isomorphs. The third major substance, $BaTiO_3$, was reported by A. von Hippel in 1944. Since then, this small group has been joined by well over a hundred pure materials and many more mixed crystal systems.

1. Spontaneous Polarization

A crystal is ferroelectric if it has a spontaneous polarization P_s which can be reversed in sense or reoriented by the application of an electric field larger than the coercive field. Reversal is also known as switching. The resulting states for each orientation are energetically and symmetrically equivalent in a zero external electric field, and may be enantiomorphous. Crystalline properties, such as the defect distribution and conductivity, together with temperature, pressure, and electrode conditions, may affect the ferroelectric reversal. Most ferroelectrics have a characteristic value of P_s and T_c. Some materials melt or decompose before T_c has been attained. Reversal or reorientation of P_s is always the result of atomic displacements.

The direction of P_s necessarily conforms to the crystal symmetry and has restrictions identical to those for the pyroelectric p_i coefficients (see *Pyroelectric Materials*).

The spontaneous polarization in most ferroelectric crystals is greatest at temperatures well below T_c, and decreases to zero at T_c. If the high-temperature phase is also polar, P_s may merely pass through a minimum at T_c; similarly, if another phase forms at lower temperatures, P_s may either increase, decrease or become zero below that transition. In a first-order ferroelectric–paraelectric phase transition, P_s may have a substantial value at temperatures very close to T_c whereas in a second-order phase transition, the decrease in P_s as T approaches T_c is more gradual.

2. Dielectric Hysteresis

The application of a dc field higher than the coercive field along a polar direction in a polydomain ferroelectric crystal results in the parallel orientation of all P_s vectors, by moving out of the crystal those domain walls that separate regions with antiparallel P_s. The minimum dc field required to move domain walls is a measure of the coercive field. The initial value of P_s in a polydomain crystal increases with increasing dc field to a maximum that is characteristic of the material. Reversing the field reintroduces domain walls as the sense of P_s in different regions is reversed. At zero applied field, the crystal will have a remanent polarization no larger than the spontaneous polarization. At full reverse field, the final P_s will have magnitude equal to the original full P_s but of opposite sign. The hysteresis thus observed is a function of the work required to displace the domain walls and is closely related both to the defect distribution in the crystal and to the energy barrier separating the different orientational states.

3. *Magnitude of Spontaneous Polarization*

The spontaneous polarization of single-domain materials usually lies within the range 10^{-3} to $1\,\mathrm{C\,m^{-2}}$; numerical values are customarily given in units of $10^{-2}\,\mathrm{C\,m^{-2}}$. The magnitude of P_s in a single crystal is directly related to the atomic displacements that occur in ferroelectric reversal (see Sect. 4) and may be calculated from the atomic positions within the unit cell if known. Designating Δ_i as the component of the atomic displacement vectors joining the ith atom positions in the original and reversed orientations along the direction of P_s, Z_i as the effective charge on the ith ion, and V as the unit cell volume, then $P_s = (1/2V)\,\Sigma_i\, Z_i\, \Delta_i$. The spontaneous polarization may also be calculated directly from the charge density obtainable by careful x-ray diffraction structural measurements, corrected for the charge transferred across unit cell boundaries.

4. *Atomic Arrangement and Ferroelectricity*

The arrangement of the atoms in all ferroelectric crystals is such that small displacements, usually less than 1 Å, result in an equally stable state but with reoriented P_s. The mid-position arrangement corresponds to a higher symmetry structure known as the prototype. The prototype is necessarily paraelectric, since dipoles in this state are either all zero or exactly cancel. A simple example is $BaTiO_3$ for which the prototype is cubic, with Ba atoms at the corners, a Ti atom at the body center and an oxygen atom at each face center of the unit cell. Below a Curie temperature of 393 K, the symmetry transforms to tetragonal as the Ti atom is displaced by about 0.05 Å from its prototype position along the $+c$ direction and the oxygen atoms are displaced in the opposite sense by about 0.08 Å, as referred to the Ba atom position. The resulting displacements give rise to the spontaneous polarization. An electric field applied along the c axis that displaces Ti by about 0.1 Å and O by about 0.16 Å reverses the sense of this axis and also that of P_s.

The paraelectric to ferroelectric transformation at T_c may be viewed in terms of a low-frequency temperature-dependent mode of the crystal lattice, observable by optical or neutron spectroscopy. This so-called soft mode may condense at T_c, causing the formation of long-range polar order below T_c.

5. *Absolute Sense of Spontaneous Polarization*

The relative sense of P_s in a crystal is given by the charge developed on the polar faces as a single-domain crystal is cooled below T_c. This sense can be related to the atomic arrangement by making use of the anomalous scattering in an x-ray diffraction experiment. All known experimental determinations of the absolute sense of P_s are in agreement with the sense as calculated from the effective point charge distribution; thus, in tetragonal $BaTiO_3$, the absolute sense is given by the direction from the oxygen layer toward the nearest Ti ion.

6. *Dimensionality of Ferroelectric Crystals*

Ferroelectric materials may be divided into three classes on the basis of the nature of the displacement vectors Δ_i that produce reversal or reorientation of P_s. The one-dimensional class involves atomic displacements all of which are parallel to the polar axis, as in the case of tetragonal $BaTiO_3$. In this class, $P_s > 25 \times 10^{-2}\,\mathrm{C\,m^{-2}}$. The two-dimensional class involves atomic displacements in a plane containing the polar axis: a typical example is $BaCoF_4$. This class has values of P_s between 10×10^{-2} and $3 \times 10^{-2}\,\mathrm{C\,m^{-2}}$. The three-dimensional class involves atomic displacements of similar magnitude in all three dimensions: a typical example is $Tb_2(MoO_4)_3$. In this class, $P_s < 5 \times 10^{-2}\,\mathrm{C\,m^{-2}}$.

7. *Changes in Properties Near the Curie Temperature*

The properties of many ferroelectric materials show significant changes with temperature near T_c, particularly those represented by the dielectric, piezoelectric and electrooptic coefficients. Thus the dielectric constant of $BaTiO_3$ increases in value with temperature from about 1000 to about 10 000 in the 10 K interval below $T_c = 393$ K; above T_c there is a more gradual decrease. Many of the uses of ferroelectric materials involve such property changes.

8. *Antiferroelectric Materials*

Materials with sublattices containing compensating dipoles may be called antiferroelectric, by analogy with antiferromagnetic materials. Such nonpolar arrangements are found in $NH_4H_2PO_4$ below 148 K and in $PbZrO_3$ and $NaNbO_3$ at room temperature. A ferroelectric state can be induced in these phases by applying a strong electric field.

9. *Representative Ferroelectric Materials*

Some ferroelectric materials are listed in Table 1. Potassium dihydrogen phosphate (KDP) transforms from the orthorhombic ferroelectric phase (space

Table 1
Characteristics of some ferroelectric materials

Material	Formula	T_c (K)	P_s (10^{-2} C m^{-2})[a]
Ammonium dihydrogen phosphate (ADP)	$NH_4H_2PO_4$	148	0[b]
Barium cobalt fluoride	$BaCoF_4$	c	8
Barium titanate	$BaTiO_3$	183, 278, 393	~20
Boracite	$Mg_3B_7O_{13}Cl$	538	0.05
Guanidinium aluminum sulfate hexahydrate (GASH)	$C(NH_2)_3Al(SO_4)_2 . 6H_2O$	d	3.5
Lead titanate	$PbTiO_3$	763	~75
Lead zirconate	$PbZrO_3$	503	0[b]
Lithium niobate	$LiNbO_3$	1473	71
Lithium tantalate	$LiTaO_3$	938	50
Potassium dihydrogen phosphate (KDP)	KH_2PO_4	123	5[e]
Rochelle salt	$NaKC_4H_4O_6 . 4H_2O$	255, 297	0.25[f]
Sodium niobate	$NaNbO_3$	73, 627	0[b]
Terbium molybdate (TMO)	$Tb_2(MoO_4)_3$	436	0.2
Triglycine sulfate (TGS)	$(NH_2CH_2COOH)_3 . H_2SO_4$	322	2.8

a Values of P_s are for single crystals at room temperature unless specified otherwise b Antiferroelectric at room temperature c Melts below T_c d Decomposes at about 273 K e At 100 K f At 280 K

group *Fdd*2) to the nonpolar but piezoelectric tetragonal $I\bar{4}2d$ phase at 123 K. Guanidinium aluminum sulfate, GASH, is trigonal (*P*31*m*) and decomposes at about 473 K without undergoing a transition. Rochelle salt has two Curie temperatures, transforming from nonpolar but piezoelectric orthorhombic $P2_12_12$ at 255 K to ferroelectric monoclinic $P2_1$, returning at 297 K to orthorhombic $P2_12_12$ but with a different structure. Barium titanate has three ferroelectric phases and three Curie temperatures: it is rhombohedral *R*3*m* below 183 K, orthorhombic *Amm*2 between 183 and 278 K, and tetragonal *P*4*mm* between 278 and 393 K; there are also two nonpolar forms at higher temperatures, namely cubic *Pm*3*m* below and hexagonal $P6_3/mmc$ above 1733 K. Sodium niobate transforms from ferroelectric trigonal *R*3*c* to antiferroelectric orthorhombic *Pbma* at 73 K, to nonpolar orthorhombic *Pmnm* at 627 K, and to four additional nonpolar phases at higher temperatures.

10. Material Preparation and Poling

The growth of crystals and preparation of ceramics and thin films is described elsewhere (see *Piezoelectric Materials*). The domain walls invariably present in as-grown ferroelectric materials may generally be removed by the application of an electric field if the temperature is high enough. A poled crystal stable at room temperature may usually be obtained by cooling in the field. Poling is also required for ceramic materials in which single-crystal grains are randomly arranged; here too the polar axis component may be reoriented by similar techniques. The resulting approach to the full value for the single-crystal polar coefficients will depend on many factors, including the density of the ceramic, grain shape and number of possible equivalent polar axis directions permitted by symmetry. The magnitude is also strongly influenced by the presence of additives such as lanthanum in PLZT ceramics, which may also be used to produce optical transparency in the ceramic. Some polarization relaxation may occur at ambient temperature with time.

11. Uses of Ferroelectric Materials

There are uses involving the switching property, i.e., polarization reversal, of ferroelectric materials as well as nonswitching uses, which mostly employ the high values near T_c of the dielectric constant and other properties. Ferroelectric materials are necessarily also pyroelectric and piezoelectric, the coefficients of which are often exceptionally large for $T \sim T_c$, so they have additional uses based specifically on these properties (see *Pyroelectric Materials*; *Piezoelectric Materials*).

Switched ferroelectrics are employed in matrix-addressed memories, shift registers and switches known as transchargers or transpolarizers. Combined devices based simultaneously on switched ferroelectric and piezoelectric properties include memories, oscillators and filters, while the analogous use of electrooptic properties results in light switches, displays and memories. A variety of display devices make use of circuits containing both ferroelectric and electroluminescent components in which light can be stored and emitted in a controlled manner; this is also possible using photoconductor and ferroelectric components, which also permit direct x ray to light conversion displays. Switched ferroelectric materials involving semiconductor properties lead to field effect devices and adaptive resistors and transistors. TANDEL, the temperature autostabilizing nonlinear dielectric element, is used in miniature thermostats, frequency multipliers and a variety of transducer and nonlinear devices.

There are many nonswitching uses for ferroelectric materials. Their high dielectric constant makes them useful as capacitors, and their rapid change of resistivity with temperature as thermistors. The linear or quadratic electrooptic effect in ferroelectrics is employed in light deflectors, modulators and displays, and the nonlinear optical effect in second harmonic generation, frequency mixing and optical parametric oscillation. The photorefractive effect is also used to provide permanent or optically erasable holographic storage; it is even possible to provide selective erasure in multiple holographic storage.

See also: Electronic Materials: An Overview

Bibliography

Burfoot J C, Taylor G W 1979 *Polar Dielectrics and Their Applications.* Macmillan, London

Hellwege K H (ed.) 1981, 1982 *Landolt–Börnstein Numerical Data and Functional Relationships in Science and Technology*, Group III, Vol. 16a,b. Springer, New York

Lines M E, Glass A M 1977 *Principles and Applications of Ferroelectrics and Related Materials.* Clarendon Press, Oxford

Ramaseshan S, Abrahams S C 1975 *Anomalous Scattering.* Munksgaard, Copenhagen

S. C. Abrahams; K. Nassau

Ferrous Physical Metallurgy: An Overview

Ferrous physical metallurgy is the branch of physical metallurgy that is concerned with the design and processing of steels or cast irons with special properties. The design process consists of selecting alloying additions and processing procedures to produce microstructures that result in the desired properties. The alloying elements of importance include major alloying elements (C, Mn, Si, Ni, Cr, Mo, Al, B) and microalloying elements (Nb, Ti, V); in many cases, control of the impurity elements (S, P, O, N, H) is also a critical issue. The processing procedures include melting, casting, hot- and cold-rolling, annealing, thermomechanical treatment, and heat-treatment. The microstructural elements include grain size; dislocation density; crystallographic texture; the size, amount, distribution, and chemical composition of the individual phases; and, possible segregation effects of alloying elements at grain or interface boundaries. The properties that are subject to control may be physical, chemical, mechanical, or manufacturing-related (e.g., weldability, formability).

The design of steels or cast irons with special properties thus involves a complicated interplay between alloying, processing, and microstructural elements. Although scientific principles can be used to investigate the effect of single alloying additions on a single property (e.g., Mn on solid solution hardening), multiple alloying additions and complex microstructures generally require that simplistic assumptions (e.g., additive strengthening factors) or empirically developed regression equations be used in predicting various steel properties. In addition, in many cases, several alternative alloying/processing/microstructure paths may be available to achieve the same properties, and then economic factors may control the final design decision.

In this overview, we shall first discuss the major steel alloying additions and how steels are classified into various types based on their chemical composition, their properties, or their end uses. We shall then discuss various microstructural elements in steel and how processing procedures are used to control them. Next, we shall discuss those steel properties that are subject to control. Finally, we shall discuss the different types of cast irons. For more in-depth information on these subjects, the reader should consult individual articles in the Encyclopedia or the bibliography at the end of this article.

1. Alloying Elements in Steel

1.1 Major Alloying Elements

Major alloying elements in steels consist of C, Mn, Si, Ni, Cr, Cu, Mo, Al, and B.

(*a*) *Carbon.* Carbon is the principal alloying element in steels and is present in amounts between 0.02 and 2%. Its principal functions are to increase hardness and strength in hot-rolled steels and also to increase hardness and strength in quenched and tempered steels. C (and N) segregation to dislocations or precipitates from α-Fe also affects steel properties by strain-aging or quench-aging reactions (see *Quench and Strain Aging*). Ductility and weldability decrease with increasing C content. The importance of C as an alloying element results in the Fe–C phase diagram serving as the basis for the discussion of steel microstructures.

(*b*) *Manganese.* Manganese has several functions. In small amounts (0.2%) it reacts with the S impurity to form MnS inclusions and prevents formation of low melting FeS which causes hot-shortness. In larger amounts (1.0–1.5%) Mn contributes to solid-solution hardening, decreases ferrite grain size in hot-rolled plate steels, and increases the hardenability of quenched and tempered steels.

(*c*) *Silicon.* Silicon in small amounts (0.2%), is used as a deoxidizer, reacting with dissolved O in the liquid steel to form SiO_2. In intermediate amounts (0.5–1.0%) it contributes to solid-solution hardening.

In larger amounts (1–3%) it is used to increase the electrical resistivity and decrease core loss of magnetic steels. Steels containing more than 3% Si may remain ferritic to very high temperatures because of the strong ferrite-stabilizing effect of this element.

(*d*) *Nickel.* Nickel has several functions. In small amounts (0.5–1.0%) it contributes to solid-solution hardening in hot-rolled plate steels and increases the hardenability of quenched and tempered steels. In larger amounts (3–9%), Ni is used to lower the ductile-to-brittle transition temperature of plate steels. Ni is also added in amounts up to 10% in stainless steels to stabilize the austenitic phase at room temperature (see *Cryogenic Steels*).

(*e*) *Chromium.* Chromium is added in small amounts (0.2–2%) to increase the hardenability and to promote secondary hardening in quenched and tempered steels. In large amounts (10–20%), Cr improves corrosion and oxidation resistance, and is the principal alloying element in many high-temperature or stainless steels.

(*f*) *Copper.* Copper additions in small amounts (0.2–0.5%) are used to improve the atmospheric corrosion resistance of plate steels. In larger amounts (1.0–1.5%) Cu results in precipitation hardening after quenching and aging.

(*g*) *Molybdenum.* Molybdenum additions (0.1–0.2%) are used to increase the hardenability and to promote secondary hardening in quenched and tempered steels.

(*h*) *Aluminum.* Aluminum additions (0.01–0.04%) are used primarily as a deoxidizer to react with O impurities and prevent gas evolution during solidification. Al also reacts with N impurities to form a fine dispersion of AlN particles that restrict austenite grain growth during reheating.

(*i*) *Boron.* Boron additions (0.0003–0.003%) are used to improve hardenability in quenched and tempered steels by segregation to austenite grain boundaries which inhibits nucleation of high-temperature decomposition products (see *Boron Steels*).

1.2 Microalloying Elements

Microalloying elements in steels consist of additions of Nb, Ti, or V in amounts generally less than 0.2%. Nb, Ti, V all react with C and N to form a fine dispersion of NbCN, TiCN, or VCN, respectively. These fine dispersions of precipitates have several important functions. First, they contribute to strength by precipitation hardening effects. Secondly, they inhibit austenite grain growth during reheating of steels into the austenite temperature range because of their pinning action. Lastly, they inhibit recrystallization of the austenite during hot rolling and are necessary to achieve optimum properties during controlled rolling. In some tool steels, these elements may be added in larger amounts (1–2%) to form large volume fractions of hard carbides that increase wear resistance.

1.3 Impurity Elements

Impurity elements in many steels must be controlled to achieve optimum properties. The most important impurity elements are S, P, O, N and H. These elements originate from many sources including the iron ore or scrap used in steelmaking, the ferroalloy additions used for major alloying elements, the reaction of the liquid steel with air, the furnace linings, or the slag during melting or casting operations.

(*a*) *Sulfur.* Sulfur is present in all steels in amounts ranging from 0.001 to 0.04%. S is generally deleterious because of the formation of MnS inclusions which act as nucleation sites for initiating ductile fracture. As a result removal of S as an impurity can greatly increase the tensile ductility or toughness of many steels. However, in some free machining steels, S is added as an alloying element (0.1–0.2%) to promote machinability, since ductile fracture is not a critical issue in these steels.

(*b*) *Phosphorus.* Phosphorus is present in all steels in amounts between 0.001 and 0.04%. Its effect is generally considered to be deleterious because of segregation to grain boundaries and the promotion of brittle intergranular failure. However, in some sheet steels P is considered beneficial because of its solid-solution hardening effect. Also, in motor lamination steels, P increases the ease of cold-punching operations.

(*c*) *Oxygen.* Oxygen occurs as an impurity in the form of oxide inclusions since its solubility in solid steel is negligible. Its concentration is sensitive to melting practice but varies from 20 to 200 ppm. Because high inclusion contents lead to lower ductility or toughness, high O contents are always considered detrimental.

(*d*) *Nitrogen.* Nitrogen may occur in the form of precipitates or as dissolved interstitial N. In the form of AlN precipitates, N has a beneficial effect in restricting austenite grain growth. In the form of TiCN, VCN, or NbCN precipitates, N has a beneficial effect in strengthening. In solution, N interacts with dislocations to cause strain aging, and generally leads to a decrease in ductility and toughness. However, N can be added as a deliberate solid solution strengthening or austenite stabilizing element in some renitrogenized sheet steels or in stainless steels. Total amounts range from 0.001 to 0.1%.

(*e*) *Hydrogen.* Hydrogen is a particularly deleterious impurity because of its rapid rate of diffusion to internal defects where it recombines to form molecular hydrogen and generates a high internal pressure that can cause blistering, flaking, or embrittlement. It is present in extremely small amounts of the order of 1–100 ppm. Special vacuum degassing treatments or slow cooling treatments may be used to lower the hydrogen content.

2. *Steel Types*

The major types of steels include carbon, alloy, high-strength low-alloy (HSLA), tool, stainless, heat-resisting, and electrical or magnetic steels (see *Steels: Classification*). These distinctions are partly based on alloying additions and partly on end use or properties. Additional subtypes may exist within these steel types based on special applications, the method of manufacture, or the shape of the product. For instance, there are hot-rolled structural and plate steels; heat-treated alloy steels; cryogenic steels; wire, rod, and bar steels; tubing and pipe steels; ultrahigh-strength steels; or cold-rolled sheet steels. The reader should consult individual articles in the Encyclopedia for discussion of these subtypes (see *Tubing and Pipe Steels*; *Bar, Rod and Wire Steels*; *Low-Carbon Sheet and Strip Steels*; *Cryogenic Steels*; *Ultrahigh-Strength Steels*; *Heat-Treated Alloy Steels*). We shall only briefly describe the major types.

(*a*) *Carbon steels.* Carbon steels are by far the most important type of steel in terms of total tonnage. They are the low-cost choice for many applications including structural beams; wire, rod, and bar products; hot-rolled plates; and hot- and cold-rolled sheet products. Also, the microstructure of carbon steels serves as a basis for understanding the properties of other steel types, since alloying additions only have a minor effect on the amount and distribution of the various phases. C is the only major alloying addition in these steels, although they may contain small amounts of Mn, Si, and Al as deliberate alloy additions. These steels are generally separated into subtypes of low-carbon, medium-carbon, and high-carbon steels. Low-carbon steels contain 0.02–0.2% C; medium-carbon steels contain 0.30–0.50% C; and high-carbon steels contain 0.60–1.2% C.

(*b*) *Alloy steels.* Alloy steels are steels that derive improved properties from the addition of alloying elements or the presence of larger amounts of Mn and Si than are ordinarily present in carbon steels. The most important subtype of these steels are heat-treated alloy steels in which the higher alloy contents promote increased hardenability. Typical alloy additions include Mn, Ni, Mo, Cr, V and B. The importance of these steels has led to the designation of various grades according to SAE or AISI codes. Carbon contents of these grades vary from 0.10 to 0.80% C with the strength increasing and the ductility or toughness decreasing as the carbon content increases. In the heat-treated condition, strength levels can vary from 100 to 250 ksi (690–1723 MPa) (see *Heat-Treated Alloy Steels*; *Boron Steels*).

(*c*) *High-strength low-alloy steels.* These are steels with yield strengths greater than 40 ksi (276 MPa) and with alloy additions designed to provide improved strength, toughness, formability, or corrosion resistance. These steels are designed for applications where higher strengths are required than those that can be obtained in simpler hot-rolled carbon steels. These higher strength levels can be achieved by microalloying additions of Nb, Ti, or V: by additions of Ni, Cr, and Mo; by additions of Cu; or by various combinations of these alloying additions. Also, these steels can be processed by hot-rolling; by controlled rolling; by quenching and tempering; or by quenching and aging treatments. Although higher in cost than carbon steels, this type of steel is finding increasing usage because of weight savings. Total tonnage amounts are about 10% of that of carbon steels. Important applications include structural and bridge, pipeline, and automotive body components. Generally, the carbon content of these steels is low (0.04–0.2% C) to ensure weldability during fabrication of structural components (see *High-Strength Low-Alloy Steels*).

(*d*) *Tool steels.* Tool steels are either carbon or alloy steels capable of being hardened and tempered. Their use is for mechanical fixtures for cutting, shaping, forming, and blanking of materials. AISI designates seven grades: (a) high-speed tool steels; (b) hot-work tool steels; (c) cold-work tool steels; (d) shock-resisting tool steels; (e) mold steels; (f) special purpose tool steels; and (g) water-hardening tool steels. Bearing steels are another important subtype of tool steels. Tool steels contain 0.3–2.5% C, and depending on the grade, may contain large amounts of Cr, W, Mo, V and Co. The heat treatment of these steels is sophisticated because of their high-carbon content and the need to avoid cracking during quenching and subsequent tempering. With the higher C and alloy content steels, hard carbides of Cr, W, Mo, or V may be formed to increase wear resistance. Also, secondary hardening reactions during tempering help to promote hot hardness in high-speed tool steels (see *Tool and Bearing Steels*).

(*e*) *Stainless steels.* Stainless steels are steels that contain substantial amounts of chromium to increase corrosion and oxidation resistance. The actual value of Cr that qualifies a steel as a stainless steel is not fixed. Some AISI standards use 5.5% Cr as a minimum whereas generally in excess of 10% Cr is required to achieve substantial corrosion resistance. Steels containing more than 14% Cr are always classified as stainless steels. Stainless steels are generally separated into subtypes of ferritic, martensitic, austenitic and precipitation-hardening. Ferritic grades contain 17–25% Cr and 0.1–0.2% C and generally are not heat-treatable. Martensitic grades contain 12–17% Cr and 0.15–1.2% C and are generally quenched and tempered to high strengths. Austenitic grades contain 17–25% Cr and 7–20% Ni with 0.15–0.25% C. Additions of Ti or Nb may be added to these grades to tie up C as TiC or NbC and prevent sensitization of these steels by precipitation of Cr_7C_3 or $Cr_{23}C_6$ at grain boundaries. Precipitation-hardening grades contain 16–17% Cr, 4–7% Ni, and additions of Cu or Al to promote precipitation hardening by precipitation of Cu or Ni_3Al intermetallic compounds (see *Stainless Steels*).

(*f*) *Heat-resisting steels*. These are steels that have increased high-temperature strength, and better oxidation resistance compared to carbon steels. They generally contain increased Cr levels for better oxidation resistance. Also, additions of Mo are used to induce precipitation of alloy carbides during high-temperature exposure to improve creep strength. Carbon contents are generally in the range 0.1–0.2% C to ensure weldability (see *Elevated-Temperature Steels*).

(*g*) *Electrical and magnetic steels*. These are steels used in motor and transformer applications. There are three general types: low-carbon steels, non-oriented Si steels, and oriented Si steels. Low-carbon steels are hot-rolled steels that contain less than 0.08% C, 0.25–0.75% Mn, and sometimes additions of P to make cold-punching of the laminations for motors easier. They are the cheapest grade of electrical steels and have the highest core loss. Non-oriented silicon steels contain between 0.8 and 3.5% Si (Si + Al) to increase the electrical resistivity of the steel and decrease core loss. Carbon contents are less than 0.025% or lower to minimize the amount of cementite particles which tend to pin domain walls during the magnetization cycle. No attempt is made to produce a crystallographic texture in these steels, but their core loss is appreciably lower than that of low-carbon grades. Oriented Si steels contain up to 3% Si and undergo a complex cold rolling, annealing, and decarburization cycle to produce a coarse-grained, ultralow-carbon steel with most of the grains having a (110) [100] texture. Such steels have the lowest core loss obtainable because the high Si content increases the electrical resistivity and the texture is such that the easy direction of magnetization ⟨100⟩ lies in the direction of the magnetic field.

3. *Steel Microstructures*

The various microstructural elements that can be controlled in steels to improve properties include grain size; dislocation density; crystallographic texture; the size, amount, distribution and chemical composition of the individual phases; and the possible segregation effects of certain elements at grain or interface boundaries.

(*a*) *Grain size*. Grain size of steels has an important influence on many properties but particularly strength and toughness. As the grain size is decreased, the strength increases and the ductile-to-brittle transition temperature decreases. As a result, refinement of the grain size is one of the most important strengthening mechanisms in steel products because it also increases toughness. Ferrite grain size is indirectly controlled by the austenite grain size since after reheating into the austenite range for normalizing or hot-rolling, the steel subsequently transforms during cooling by nucleation of ferrite at the austenite grain boundaries. Many metallurgical processes such as normalizing, accelerated cooling, or controlled rolling are thus designed to result in a refinement of the austenite grain size since this results in a refinement of the ferrite grain size. Quenching and tempering operations that produce martensitic transformations also result in marked refinement of the microstructure because each austenite grain is subdivided into many small martensite plates or units (see *Steels: Physical Metallurgy Principles*).

(*b*) *Dislocation density*. Dislocation density increases as the steel is cold worked so that the strength of the cold-worked steel products is higher than that of hot-rolled products. The increase in strength is proportional to the square root of the dislocation density. The stored energy in the dislocations also acts as the driving force for the recrystallization process in annealing operations. Transformation of austenite to martensite in quenching and tempering operations produces a high dislocation density (10^{12} cm^{-2}) which is largely responsible for the high strength of heat-treated steels.

(*c*) *Crystallographic texture*. The crystallographic texture of polycrystalline steels may be controlled by proper cold-rolling and annealing treatments. In deep-drawing operations, strong {111}⟨110⟩ textures are beneficial because the resulting plastic properties give an increased resistance to thinning of the steel sheet and delay fracture until higher plastic strains are reached. In electrical steels, the {110}⟨100⟩ texture is beneficial because the easy direction of magnetization of steel is in the ⟨100⟩ direction (see *Texture, r_m Value and Drawability*; *Low-Carbon Sheet and Strip Steels*).

(*d*) *Individual phases*. Individual phases in steels contribute to many of the steel properties. Although the matrix phase is always bcc or fcc Fe, a large number of separate phases can be precipitated from the matrix during heat treatment or solidification including carbides (Fe_3C or alloy carbides), aggregates of carbides and other phases (pearlite, bainite, or tempered martensite), or inclusions (MnS, Al_2O_3, AlN). All of these individual phases influence the properties as their size, distribution, spacing, or chemical composition are changed. For instance, the strength of HSLA steels is increased as the interparticle spacing of microalloy carbides (NbCN, VCN) is decreased. Also, the strength of eutectoid carbon steels is controlled by the spacing between the carbide (Fe_3C) lamellae of the pearlite. On a coarser scale, the strength of hot-rolled carbon steels is determined by the volume fraction of pearlite. The ductility and toughness of steels is also controlled to a large extent by the volume fraction, size and shape of inclusions. In addition, the strength of the Fe matrix depends on the solid-solution hardening effects of the individual alloying elements. In some ways, steels can be considered as a composite material with the overall properties resulting from the combined action of the matrix and precipitated phases (see *Steels: Classification*).

(*e*) *Segregation.* Segregation of alloying or impurity elements to grain boundaries plays an important role in many embrittlement processes in steel. Segregation of Sb, P and Sn impurities in heat-treated alloy steels during tempering in certain temperature regimes leads to severe embrittlement, referred to as temper embrittlement. Certain alloying additions such as Mo help to alleviate this condition by slowing the rate of diffusion of the impurities to the grain-boundary regions. Other alloying elements like Mn increase the susceptibility to such embrittlement. Segregation of hydrogen to grain boundaries may also play a role in hydrogen embrittlement phenomena. The driving force for such segregation is the lowering of strain energy of the impurity species or a lowering of the grain boundary interfacial free energy (see *Temper Embrittlement of Steel*; *Hydrogen Embrittlement of Steels*).

On a coarser scale, segregation of alloying elements occurs during solidification of steel ingots on both a microscale (dendrite arm spacings of about 100 μm) and a macroscale (top to bottom of ingot). Microsegregation can lead to significant differences in transformation behavior. For instance, hot-rolling deforms regions of Mn segregation into bands. During slow cooling of hot-rolled low-carbon steels, pearlite forms in the high Mn regions of these bands. Such "banding" phenomena are responsible for the banded microstructure exhibited by hot-rolled plate steels. Macrosegregation of C, Mn and other alloying elements from top to bottom of an ingot can lead to variation in properties from head to tail of a rolled plate or sheet. The origin of such segregation is the different solubilities of elements in solid and liquid Fe, so that as freezing progresses, many alloying elements are rejected from the solid Fe and the liquid Fe phase becomes enriched in alloying element. Increased casting speeds, as in continuous casting or in rapid solidification processing, can minimize or prevent such segregation, whereas slow casting speeds, as in ingot mold casting, maximize such segregation (see *Ingot and Continuous Casting of Steel*).

4. *Steel Processing*

Steels, like other metals, must be melted, cast and then forged, rolled, extruded, or drawn into useful shapes. Subsequently these shapes or parts formed from them may be heat treated. Many of these steps prior to heat treatment can have a significant effect on properties.

(*a*) *Melting.* Melting practices can be utilized to help remove impurity elements or to make additions that render them less harmful. Also, the type of deoxidation practice has an important influence on the grain-coarsening characteristics of the steel. For instance, the level of S impurity can be drastically reduced (from 0.02 to 0.002%) by treatment of the liquid steel with Ca. Also, additions of rare-earth elements (La, Pr) can react with S to form hard $(RE)_2O_2S$ inclusions that do not deform during rolling and remain spherical in contrast to soft MnS inclusions. Such treatments can result in more nearly isotropic and higher impact energies. Other harmful impurities such as hydrogen can be removed by special vacuum degassing treatments. Additions of Al for deoxidation purposes during melting are beneficial also because of formation of AlN during solidification and subsequent cooling. As discussed above, fine AlN precipitates prevent austenite grain growth during reheating. Al-killed steels are thus finer grained than Si-killed steels (see *Ingot and Continuous Casting of Steel*).

(*b*) *Casting.* Casting processes can also affect steel properties primarily because of segregation during solidification. Faster casting speeds, as in continuous casting, can lead to less segregation during casting. This will generally result in more uniform properties in the final product than ingot casting. Extremely rapid solidification can be achieved by gas jet cooling of liquid droplets. The solidified droplets must then be compacted and sintered to form a larger part, but the final product is uniform and free of segregation (see *Tool and Bearing Steels*).

(*c*) *Hot-rolling.* Hot-rolling of steel ingots is used to produce useful shapes such as structural shapes, plates, bars and rods. Generally, this requires reheating the ingot to high temperatures (1200 °C) followed by a series of roughing or breakdown passes and then a series of finishing operations. The hot-rolling process also causes some rather important metallurgical changes. The austenite phase undergoes repeated recrystallization during hot-rolling and the grain size becomes increasingly smaller as the finishing temperature is lowered. Also, soft MnS inclusions and segregated zones are aligned in the rolling direction into stringered inclusions or bands, respectively. These latter two effects lead to a directionality of properties, such as ductility or toughness, with minimum values occurring in the through-thickness direction, intermediate values in the transverse direction, and maximum values in the longitudinal direction (see *Hot Rolling: A Thermomechanical Process*).

(*d*) *Cold-rolling and annealing*. Cold-rolling operations are used to produce thinner sheet products from hot-rolled slabs. In contrast to hot-rolling, cold-rolling is performed at or near room temperature where the steel has a ferritic structure and where no recrystallization occurs during the working operation. As a result the dislocation density of the ferrite and the strength increase during cold-working operations. Also, the ferrite grains are rotated and deformed to develop a cold-worked texture. In general, the surface finish of cold-rolled products is superior to that of hot-rolled products because of the absence of oxidation. After cold-rolling, steel products are usually annealed to soften them and increase their ductility. In the annealing operation the cold-worked ferrite recrystallizes to produce an equiaxed, dislocation-free ferrite grain structure. A crystallographic texture is still present in the annealed product, although it will be different from that of the cold-worked texture. Control of the annealed texture is important in design of steels for deep-drawing operations (see *Cold Working and Annealing of Steel*).

(*e*) *Thermomechanical treatment*. This term is used to describe steel processing operations that combine a mechanical deformation and heating operation. Hot-rolling can be considered as one form of thermomechanical treatment, but in general the latter term is applied to temperature ranges much lower than those used for hot-rolling. Controlled rolling is a special type of thermomechanical treatment in which the hot-rolling operations occur at lower temperatures in the austenite range where recrystallization does not occur, and the time–temperature deformation steps are carefully controlled to produce flattened austenite grains. Controlled rolling is used to produce ultrafine-grained steels for applications requiring high toughness and high strength. Other thermomechanical treatments are ausforming and isoforming but these have found only limited applications (see *Hot Rolling: A Thermomechanical Process*).

(*f*) *Heat treatment*. Heat treatment of steels is used to improve their strength or toughness. It involves heating steels to elevated temperatures in the austenite range followed by air cooling or quenching and tempering. If air cooling is employed, the operation is called normalizing, and it results in a marked refinement in the grain size and some increase in strength. Normalizing operations are used to lower the ductile-to-brittle transition temperature of many hot-rolled steels. If quenching operations are used, the austenite transforms to a hard martensite phase that results in very high strength products. Normally, quenched steels are subsequently tempered to increase their ductility and toughness with a slight decrease in hardness usually being unavoidable (see *Steels: Physical Metallurgy Principles*).

To produce higher surface hardness for increased wear resistance, the surface of steel parts may be locally heat treated by induction, flame, or laser hardening methods. In this case there is no change in the surface chemistry of the steel. Alternatively, gas nitriding or gas carburizing methods may be used to increase the C or N content of the surface. Carburizing is performed at high temperatures in the austenite range and is generally followed by quenching and tempering. Nitriding is performed at lower temperatures in the ferrite range and is not followed by a heat-treatment operation. (see *Nitriding*; *Steels: Surface Treatment*; *Carburizing and Carbonitriding*).

5. *Steel Properties*

Steel properties that are subject to control are physical, chemical, mechanical and manufacturing-related. Physical properties include magnetic behavior, electrical resistivity and thermal conductivity. Chemical properties include resistance to corrosion and to high-temperature oxidation. Mechanical properties include elastic moduli, strength, ductility and toughness. Manufacturing-related properties include weldability, formability and machinability.

5.1 *Physical Properties*

(*a*) *Magnetic behavior*. Magnetic behavior of steels is sensitive to crystallographic texture because the intensity of magnetization of Fe is a minimum in the ⟨100⟩ direction. Special hot- and cold-rolling and annealing procedures are used to develop (110)[100] textures to produce sheet steels with high permeabilities. High permeabilities are also favored by low C contents ($<0.02\%$) since Fe_3C particles tend to pin domain walls and prevent their movement during the magnetization cycle. Finally, core loss in magnetic steels is decreased by addition of Si (1–3%) to increase the electrical resistivity and decrease the I^2R loss caused by eddy currents.

(*b*) *Electrical resistivity and thermal conductivity*. In certain applications, the electrical resistivity and thermal conductivity of steels are important. For instance, low thermal conductivity is desired in steels used for containers for liquid gases. High electrical resistivity is desired in some transformer steels to decrease I^2R losses.

Several factors determine the electrical resistivity. These include thermal, impurity, and magnetic scattering of electrons. The first two factors are considered to be additive and can be analyzed by use of Mathiessen's rule. The impurity scattering is specific to each element in solid solutions and values for many of the common alloying elements are available.

The impurity scattering is assumed to be temperature independent, with the thermal scattering term accounting for all the temperature dependence of the electrical resistivity. The magnetic scattering term is very large and, as the temperature is increased to the vicinity of the Curie temperature (770 °C), a rapid increase in the electrical resistivity occurs because of the decrease in magnetic ordering. In fcc austenitic steels which are paramagnetic, the electrical resistivity is higher than in bcc ferromagnetic steels because of the absence of magnetic ordering. Also, the temperature dependence of the electrical resistivity of fcc steels is lower because of the absence of a paramagnetic–ferromagnetic transition.

Heat is conducted by both atomic vibrations (phonons) and by conduction electrons. For good electrical conductors like steels, heat transfer by conduction electrons is more important than by phonons. This leads to a parallelism between thermal and electrical conduction as expressed by the Lorentz function. Thus, many of the conclusions reached for electrical conductivity (reciprocal of resistivity) also apply to thermal conductivity. Since electrical resistivity can be measured more easily than thermal conductivity, the Lorentz function can be used to calculate the thermal conductivity from electrical resistivity measurements. The Lorentz constant in steels is not, however, equal to the Lorentz number as predicted by the Wiedemann–Franz law. Rather, the Lorentz constant is equal to 1.16 × Lorentz number (see *Steels: Electrical Resistivity and Thermal Conductivity*).

5.2 Chemical Properties

(*a*) *Corrosion resistance.* The corrosion resistance of hot-rolled carbon structural steels exposed to atmospheric conditions may be increased two to four times by additions of small amounts of copper (0.2–0.5%). The Cu may be added in conjunction with other elements such as P, Mn, or Mo, which not only further increase the corrosion resistance but also increase the strength. For resistance to more severe corrosive environments such as acidic solutions, larger amounts of alloying elements must be added. Cr is the most effective alloying condition for giving adequate corrosion resistance in acidic solutions. Steels containing substantial amounts of chromium (at least 5.5% and generally more than 10%) are designated as stainless steels, and are resistant to corrosion in many severely corrosive environments (see *Stainless Steels*; *Steels: Classification*). Of course, the corrosion resistance of many steel products is improved by simply putting a protective layer of another metal (Zn, Pb, or Sn) on the surface to protect the metal against attack by the corrosive media. This can be accomplished by electroplating (Zn or Sn) or by hot-dipping (Zn or Pb).

(*b*) *Oxidation resistance.* Oxidation resistance is an important consideration in steels designed for high-temperature service as in steam and gas turbines, exhaust valves, and catalytic converter housings. Cr is again the element that is used to improve oxidation resistance by formation of a thin oxide layer that protects the base metal from being exposed to the oxidizing atmosphere. As a result, most steels for high-temperature service contain 5-20% Cr (see *Elevated-Temperature Steels*).

5.3 Mechanical Properties

(*a*) *Elastic moduli.* The elastic moduli of polycrystalline steels are dependent on the degree of crystallographic texture because the elastic constants of single crystals of bcc Fe are anisotropic (C_{11} = 23.7, C_{12} = 14.1, C_{44} = 11.6 × 10^{10} Pa). The maximum elastic modulus, $E\langle 111\rangle = 28.4 \times 10^{10}$ Pa, occurs in the ⟨111⟩ direction whereas the minimum, $E\langle 100\rangle = 11.6 \times 10^{10}$ Pa, occurs in the ⟨100⟩ direction. By suitable hot- and cold-rolling and annealing practices, it is possible to produce sheet steels with a preponderance of (111) planes in the plane of the sheet, and thus a high E value in a direction perpendicular to the plane of the sheet. Such steels are more resistant to thinning and fracture during deep drawing than steels with a random texture.

In general, alloying additions have much less effect on the elastic moduli than texture because the atomic binding energies of the iron lattice cannot be altered significantly by alloying additions. The maximum effect for elements in solid solution is about 1% for each 1 at.% alloying element addition. For example, the isotropic E value for Fe is lowered from 20.82 to 19.25 × 10^{10} Pa by addition of 9 at.% Ni.

The elastic moduli of bcc Fe decrease slowly with temperature up to temperatures near the Curie temperature (770 °C) where a rapid decrease in E occurs as the ferromagnetic bcc Fe undergoes magnetic disordering. This results from the effect of magnetic ordering on the atomic binding energies. At 910 °C, where bcc Fe transforms into fcc Fe, a small increase in the value of E is observed.

(*b*) *Strength.* The strength of steels is dependent on many factors including grain size, solid-solution hardening effects of individual elements, dislocation density, and the size, spacing, and distribution of precipitate particles. In general, several of these factors are operative in any steel alloy so that rigid theoretical analysis or prediction of the strength from first principles is difficult. However, by assuming simple additivity of the various strengthening factors, useful empirical equations based on linear regression analysis of hundreds of steels have been developed. These can be used to analyze or predict effects of various changes in heat treatment or alloying additions.

(*c*) *Ductility.* The ductility of steels generally refers

to the reduction-of-area or total elongation values of a tensile test specimen. It is dependent on both strength level and inclusion content and decreases with increasing tensile strength and with increasing inclusion content. Ductility values are sensitive to testing direction in rolled or extruded products because of the elongation of sulfide inclusions in the primary deformation direction.

(*d*) *Toughness*. Toughness refers to the energy absorbed in breaking notched impact specimens. The impact energy is sensitive to tensile strength, inclusion content, and testing direction in a manner similar to tensile ductility. Toughness is drastically reduced by lowering the temperature of testing because of the ductile-to-brittle transition exhibited by all bcc metals. The high-temperature ductile fracture surface has a fibrous appearance and the energy value absorbed in the 100% fibrous region is called the shelf energy. The low-temperature brittle fracture surface has a shiny appearance with cleavage facets. The temperature at which the steel exhibits a 50% brittle appearance is called the fracture appearance transition temperature (FATT). The FATT is sensitive to many variables including grain size and various embrittlement phenomena such as temper embrittlement.

5.4 Manufacturing-Related Properties

(*a*) *Weldability*. The weldability depends on both carbon and alloy content. High-carbon and high-alloy content promote cracking in the weld bead or weld heat-affected zone. For many steels, a limit of 0.20% C is considered adequate to assure good welding characteristics because the hardenability of low-carbon steels is low enough so that martensite is not formed in the weld heat-affected zone. However, alloying elements also play a role because they too affect hardenability. As a result, the weldability of steels is generally considered in terms of a carbon equivalent equation which includes terms both for C and for other alloying elements. The weldability may also be affected by hydrogen generated from water present in the welding electrodes or in the air because of hydrogen embrittlement mechanisms.

(*b*) *Formability*. Formability is an important consideration in deep drawing operations such as stamping of automobile parts. Steels for these applications are designed on the basis of "r_m" values which are a measure of the relative plastic strains in the plane of the sheet and in the thickness direction in a tension-test specimen. Steels with high r_m values (near 2.0) resist thinning during drawing operations and exhibit good drawability, whereas steels with low r_m values (near 1.0) exhibit poorer drawability. The r_m value is sensitive to crystallographic texture with the $\{111\}\langle 110\rangle$ texture resulting in high r_m values. Careful control of cold-rolling and annealing cycles can be used to produce special deep-drawing steels with this texture (see *Texture, r_m Value and Drawability*; *Low-Carbon Sheet and Strip Steels*).

In other forming operations, control of stretchability is more important. Steels for stretching operations require high work-hardening rates so that necking is delayed. The work-hardening rate is measured in terms of "n" values which is the exponent in the Hollomon work-hardening power law. High n values are promoted by increasing the grain size of the steel. However, grain sizes coarser than ASTM No. 6 (45 μm) are not desirable as they produce an "orange-peel" effect (see *Low-Carbon Sheet and Strip Steels*).

(*c*) *Machinability*. The machinability of steels can be increased by increasing the sulfur content of the steel beyond the normal limits. Such "free-machining" steels may have S contents as high as 0.15%. The high S content makes chip breaking easier because of the low ductility of the steel caused by a high volume fraction of MnS inclusions. The sulfides probably also act as a lubricant between the tool and the workpiece. Other alloy additions for machinability include Pb which occurs as inert Pb particles and also serves as a lubricant. Bi and Te additions are also used to increase machinability. In general, Al_2O_3 particles present in Al-killed steels are detrimental to machinability because of excessive tool wear.

6. Cast Irons

Fe–C–Si alloys, commonly called cast irons, and the liquid forming processes of metal casting, are an ideal combination of material and process that has been used for centuries to provide mankind with metal in desired shapes. Cast iron is of relatively low cost, can be readily melted and cast into intricate shapes, and has very useful properties. It is an excellent engineering material with good strength and hardness, and is readily machined. It has sufficient heat and corrosion resistance for many common applications such as furnaces and the drainage grating in streets. Critical components of the internal combustion engine such as the motor block, head, exhaust manifold, connecting rods, and flywheel are made of iron castings.

All cast irons contain more than 2% C in contrast to steels which contain less than 2% C and usually less than 1% C. Since 2% C is the maximum carbon content at which Fe can solidify as a single phase alloy with all the C in solution, all cast irons solidify as heterogeneous alloys and have more than one constituent in their microstructure. In addition to C, cast irons usually contain from 1 to 3% Si. The high C and Si make cast iron an excellent casting alloy. It is readily melted in a simple furnace using charcoal or coke and ambient temperature air. The molten iron is very fluid and allows the metal to be handled and cast effectively. Also, the precipitation of graphite in the solidifying iron reduces the normal liquid to solid contraction so that very little volume change occurs on solidification. This allows the casting of complex shapes (see *Cast Irons and Casting Processes*).

Cast iron designates a family of five different types: white, malleable (cast white), gray, ductile, and compacted graphite.

(*a*) *White cast iron.* This consists of a mixture of Fe_3C and austenite upon solidification. The austenite transforms to pearlite during cooling. White iron is hard and brittle because of the large amount of Fe_3C it contains. Its name derives from the white crystalline surface which appears when it is fractured. White iron can be produced in selected areas at the surface of a casting by chill casting of these regions. White iron provides very high compressive strength, good wear resistance, and retains its hardness for limited periods up to red heat. Typical compositions are 2.0–3.6% C, 0.5–1.9% Si, 0.25–0.8% Mn, 0.06–0.2% S, and 0.06–0.2% P (see *White and High-Alloy Cast Irons*).

(*b*) *Malleable cast iron.* Malleable cast iron is iron in which carbon occurs in the microstructure as irregularly shaped nodes of graphite. This form of graphite is called temper carbon because it is formed in the solid state during heat treatment. The iron is cast as white iron of a suitable chemical composition. After the castings are removed from the mold, they are given an extended heat treatment starting at a temperature above 885 °C. This causes the iron carbide to decompose and the free carbon precipitates in the solid iron as graphite. Typical chemical compositions are 2.2–2.9% C, 0.9–1.9% Si, 0.15–1.2% Mn, 0.02–0.2% S, and 0.02–0.2% P (see *Malleable Cast Iron*).

(*c*) *Gray iron.* Gray iron is iron in which the carbon separates during solidification to form graphite flakes that are interconnected within each eutectic cell. This graphite grows into the liquid and forms the characteristic flake shape. When this iron is subsequently fractured, most of the fracturing occurs along the planes established by the graphite, thereby accounting for the characteristic gray color of the fracture surface. Because the large majority of the iron castings produced are of gray iron, the generic term, cast iron, is often improperly used specifically to mean gray iron. The properties of gray iron are influenced by the size, shape, and distribution of the graphite flakes, and by the relative amounts of ferrite and pearlite in the matrix. Low cooling rates and high C and Si contents tend to produce more and larger graphite flakes in a matrix with more ferrite and less pearlite. The flake graphite provides gray iron with some unique properties such as excellent machinability at hardness levels that produce superior wear resisting characteristics, the ability to resist galling, and good vibration damping from a nonlinear stress–strain relationship at relatively low stresses. The damping capacity increases as the graphite flakes become coarser. Typical chemical compositions are 2.5–4.0% C, 1.0–3.0% Si, 0.2–1.0% Mn, 0.02–0.25% S, and 0.02–1.0% P (see *Gray Cast Iron*).

(*d*) *Ductile iron.* This material was developed in the 1940s. It is sometimes referred to as nodular iron and is called spherulitic graphite iron in Europe. It had a phenomenal ninefold increase in use as an engineering material during the 1960s. An unusual combination of properties is obtained in ductile iron because the graphite occurs as spheroids or spherulites rather than as individual flakes as in gray iron. This mode of solidification is obtained by adding a very small, but definite amount of Mg to molten iron of a proper composition. The added Mg reacts with the S and O in the melt and changes the way the graphite is formed. Typical chemical compositions are 3.0–4.0% C, 1.8–2.8% Si, 0.1–1.0% Mn, 0.01–0.03% S, and 0.01–0.1% P.

(*e*) *Compacted graphite iron.* Compacted graphite iron is characterized by a graphite structure which is intermediate between that in gray and in ductile irons. The graphite exists as blunt flakes which are interconnected. Compacted graphite iron has been referred to as quasiflake, aggregated flake, semi-nodular iron, and by other names. Compacted graphite is formed by additions of Ce or other rare-earth elements, with the presence of about 0.06 to 0.13% Ti, and a limiting S content of 0.03%. The tensile and yield strengths of compacted graphite iron are comparable to the lower values of ductile iron, while the ductility is limited to a maximum of about 6% elongation in 50 mm. The physical properties of compacted graphite such as thermal conductivity and damping capacity are closer to those of gray iron because the graphite is interconnected. The chemical composition ranges of these irons are 2.5–4.0% C, 1.0–3.0% Si, 0.2–1.0% Mn, 0.01–0.02% S, and 0.01–0.1% P (see *Compacted Graphite Iron*).

See also: Metals Production: An Overview; Metals Processing and Fabrication: An Overview; Physical Metallurgy: An Overview

Bibliography

Angus H T 1976 *Cast Iron: Physical and Engineering Properties.* Butterworths, London

Honeycombe R W K 1981 *Steels: Microstructure and Properties.* Edward Arnold, London

Leslie W C 1981 *The Physical Metallurgy of Steels.* McGraw-Hill, New York

Metals Handbook, 9th edn., Vols 1, 3 and 4, 1978. American Society for Metals, Metals Park, Ohio

Pickering F B 1978 *Physical Metallurgy and the Design of Steels.* Applied Science, Barking, UK

US Steel Corporation 1985 *The Making, Shaping, and Treating of Steel*, 10th edn. Association of Iron and Steel Engineers, Warrenville, Pennsylvania

Walton C F 1981 *Iron Castings Handbook*. Iron Castings Society, Des Plaines, Illinois

G. R. Speich

Fertile Blanket Materials

Breeding in nuclear systems refers to the process by which an isotope with no fuel value is converted by neutron capture to a form useful as reactor fuel. For fission reactions, the principal opportunities for fertile-to-fissile conversion involve $^{238}U \xrightarrow{n} {}^{239}Pu$ and $^{232}Th \xrightarrow{n} {}^{233}U$. Fuel for fusion reactions can be obtained by $^{7}Li \xrightarrow{n} {}^{4}He + {}^{3}T$. The most common applications for materials containing fertile isotopes are for breeder blankets in thermal and fast spectrum fission reactors, and for tokamak or similar fusion reactors.

1. *Breeder Materials for Fission Fuels*

Breeder blankets may be incorporated in a reactor around the core periphery and/or homogeneously mixed with fuel rods or assemblies within the core. Metal alloys, oxides and carbides containing ^{238}U or ^{232}Th are used as blanket materials. $^{238}UO_2$, in essentially the same metal-clad cylindrical form as the reactor fuel elements, is the most common blanket material. In contrast to fuels, blanket materials generate the least amount of heat at the beginning of their reactor life. As conversion progresses, the blanket composition shifts from UO_2 to a UO_2–PuO_2 solid solution. During irradiation physical and chemical properties change with both the actinide metal composition and deposited fission products. Fission rates and blanket temperatures increase throughout irradiation of the blanket, in proportion to the concentration of the bred-in fissile isotope.

ThO_2 and UO_2–ThO_2 solid solutions are utilized for breeding ^{233}U. Their irradiation behavior is analogous to UO_2–PuO_2. Uranium and thorium carbide blankets offer higher metal density and improved breeding efficiency compared with oxides. Breeder materials for high-temperature gas-cooled reactors are prepared as particles of $(U,Th)O_2$ or $(U,Th)C_2$, coated with impervious pyrolitic carbon and silicon carbide, and dispersed in a graphite structure.

2. *Breeder Materials for Tritium*

Lithium irradiated as ceramics or metal alloys is the primary source of tritium for fusion devices and experiments. Lithium–aluminum alloys are used in fission reactors for breeding where low temperatures (~375 K) are maintained. Supplies of tritium to fuel large fusion reactors are based on continuous tritium recovery from a blanket region surrounding the fusion chamber. Liquid lithium metal, Li_7Pb_2 and lithium ceramics are candidate fusion reactor blanket materials. Flowing liquid lithium functions as a coolant while solid materials require external cooling and flowing helium for tritium removal. The efficiency of tritium production is principally a function of lithium atom density. Candidate materials aligned in order of decreasing lithium atom density are: Li_2O (0.93 g cm^{-3}), Li_4SiO_4 (0.54 g cm^{-3}), Li (0.51 g cm^{-3}), Li_7Pb_2 (0.49 g cm^{-3}), Li_2SiO_3 (0.36 g cm^{-3}), and Li_2ZrO_3 and Li_2TiO_3 (both 0.33 g cm^{-3}).

Li_7Pb_2 and Li_2O provide excellent breeding, but are quite reactive with moisture and air. The three-component oxides have refractory properties and are chemically stable and readily fabricated. Tritium diffusion out of these solids is optimized by a microstructure with small grain size and open porosity. Equilibrium of Li_2O–LiOH with water vapor affects the rate of tritium release as T_2O and influences concentration of corrosive LiOH, which in turn affects compatibility with structural materials. Near total release of tritium from LiO_2 and $LiAlO_2$ occurs in the 925–975 K temperature range, with near total retention below 575 K.

See also: Fission Reactor Fuels and Materials; Fusion Reactor Materials; Nuclear Materials: An Overview

Bibliography

Evans T W 1964 *The Technology of Thoria and Thoria–Urania Compositions*, HW-84106. US Atomic Energy Commission–Hanford Works, Richland, Washington

Roberts J T A 1981 *Structural Materials for Nuclear Power Systems*. Plenum, New York

Smith D L, Clemmer R G, Davis J W 1979 Assessment of solid breeding blanket options for commercial Tokamak reactor. In: McGregor C K, Batzer T H (eds.) 1979 *Proc. 8th Symp. Engineering Problems of Fusion Research*, Pt. 1, IEEE Publication No. 79CH1441-5 NPS. Institute of Electrical and Electronics Engineers, New York, pp. 433–38

Waltar A E, Reynolds A B 1981 *Fast Breeder Reactors*. Pergamon, New York

E. T. Weber

Fiber Bundles: Strength Statistics

The tensile failure of a bundle of brittle fibers in a flexible matrix is a complex process involving the failure of fibers at scattered flaw sites, the overloading of neighboring fibers at these sites and the growth of sequences of adjacent fiber breaks to some critical size. Current statistical theory describes a chain-of-microbundles model for this process. An approximate Weibull distribution arises for the strength of the bundle.

1. Description of the Failure Process in a Bundle

On application of a moderate tensile load x (to each fiber in a bundle), fibers fail at random flaws which have strength below x, and these failures are typically quite far apart spatially. At these failure sites, the load in the broken fiber decreases to zero, but the load on each immediate neighbor increases to say K_1x, where $K_1 > 1$ and is called a load concentration factor. Moreover, the length of the region, say δ, over which this overload takes place is typically of the order of a few fiber diameters. K_1 and δ depend on the mechanical properties of the fiber and matrix.

At these failure sites, a few overloaded fibers may fail, and typically some pairs of adjacent breaks will result. At these pairs, the flanking neighbors are then subjected to the more severe overload K_2x, where $K_2 > K_1 > 1$, and a few sequences of three adjacent breaks may occur. Again, the length of the overload region is of the order of δ. This process continues until all failure sequences become stable, or one or more sequences reaches critical size, say $\hat{k}$, and a catastrophic crack propagates through the bundle. In the latter case, the overload $K_{\hat{k}}x$ exceeds a characteristic strength x_δ of a flanking fiber, and failure of this fiber becomes almost certain.

2. The Weibull Distribution and the Strength of Single Fibers

When tension-testing single fibers, one typically finds that the fiber strengths follow a Weibull distribution function of the form:

$$F_l(x) = 1 - \exp[-l(x/x_0)^\rho], \quad x \geqslant 0 \qquad (1)$$

where ρ is the shape parameter, x_0 is the scale parameter for unit length and l is the (dimensionless) gauge length; the Weibull scale parameter being $x_l = x_0 l^{-1/\rho}$. Roughly speaking, this means that if each of the fibers in a very large population were submitted to a load x, then the fraction of fibers that would fail is $F_l(x)$. The mean for the above distribution is $l^{-1/\rho}x_0(1 + 1/\rho)$, the standard deviation is $l^{-1/\rho}x_0 \times [\Gamma(1 + 2/\rho) - \Gamma(1 + 1/\rho)^2]^{1/2}$, and the coefficient of variation (standard deviation divided by the mean) is approximately $1.2/\rho$. Typically, $4 \leqslant \rho \leqslant 15$ is measured experimentally, so that the mean fiber strength diminishes quite rapidly with increasing length l. For example, if $\rho = 6$, the fiber strength is reduced to half when the length is increased 64 times.

In the simple bundle model considered here, we will need the distribution function $F_\delta(x)$ for fiber elements of length δ, where δ is of the order of a few fiber diameters. Note that l in typical tension test is at least a few millimeters, whereas δ is of the order of 100 μm. Because δ is so small, extrapolation is necessary to estimate $F_\delta(x)$. The obvious choice is the Weibull distribution,

$$F_\delta(x) = 1 - \exp[-(x/x_\delta)^\rho], \quad x \geqslant 0 \qquad (2)$$

where by extrapolation $x_\delta = x_l(\delta/l)^{-1/\rho} = x_0\delta^{-1/\rho}$.

Unfortunately, the accuracy of Eqn. (2) in modelling the true $F_\delta(x)$ can only be demonstrated for the lower tail of $F_\delta(x)$, that is, for $x \ll x_\delta$, since strengths in the middle and upper tail are not typically observed at gauge length l. Fortunately, knowledge of the upper tail of $F_\delta(x)$ is not important. To demonstrate this, results can be obtained under the power distribution function

$$F_\delta(x) = (x/x_\delta)^\rho, \quad 0 \leqslant x \leqslant x_\delta \qquad (3)$$

which has a very different upper tail to Eqn. (2).

3. The Model and Basic Assumptions

The bundle is viewed as a planar structure of n parallel fibers, which are partitioned into a series of m short sections called microbundles, each with n fiber elements (Fig. 1); each microbundle has length δ called the ineffective length. Statistically and structurally the microbundles are independent, and the strength of the structure is that of the weakest microbundle (measured on a load-per-fiber basis). The strengths of the mn fiber elements are independent and identically distributed random variables, with a common distribution function $F_\delta(x)$, for $x \geqslant 0$ as discussed earlier.

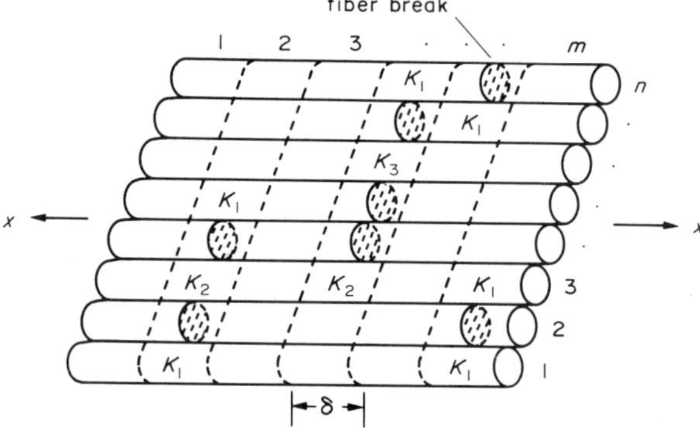

Figure 1
Bundle of fibers in a matrix in the form of a planar tape; failure is localized within microbundles

Each microbundle is a planar arrangement of parallel elements (Fig. 1), which share the load according to the following simple local load-sharing rule: if the bundle load is x (per fiber), a surviving fiber element carries a load K_rx, where K_r is called a load concentration factor and r is the number of consecutive failed fiber elements immediately adjacent to the surviving element. Furthermore, a failed element carries no load. In our numerical calculations, we assume the simplest case $K_r = 1 + r/2$ (where $r = 1,2,3,\ldots$).

In the above description, the (total) load on a microbundle will fall slightly short of nx if a few elements have failed at the edge of a microbundle. However, these boundary effects turn out to be unimportant, and are neglected for simplicity. Also, load increases on fiber elements not adjacent to broken elements are neglected. The consequences of this are also unimportant.

To describe the weakest link, consider a single microbundle, and let $G_n(x), x \geqslant 0$ be the distribution function for its strength. In other words, $G_n(x)$ is the probability of failure of an arbitrarily selected microbundle under the nominal fiber load x (total load nx). Next, let $H_{m,n}(x), x \geqslant 0$ be the distribution function for the strength of the bundle of fibers in a matrix. Since the bundle survives if and only if each of its m microbundles survives, the probability of survival of the bundle is $[1 - G_n(x)]^m$ under fiber stress x. Thus $H_{m,n}(x)$, which is also the probability of failure of the bundle under the fiber load x, is

$$H_{m,n}(x) = 1 - [1 - G_n(x)]^m, \quad x \geqslant 0 \qquad (4)$$

Equation (4) amounts to a statement of the weakest link rule.

4. *Analysis of the Model*

Rosen (1964) was the first to consider a chain-of-microbundles model of this type; however, he assumed equal load sharing among nonfailed elements in each microbundle, an assumption which is not in keeping with those here and which leads to very different results. (Smith and Phoenix (1981) discuss such models.) However, Zweben and Rosen (1970) and Argon (1972) obtain interesting approximate results for versions of the model here.

Harlow and Phoenix (1978a,b) were the first to obtain exact results for the present model, but only for small n. Current computer capacity limits exact calculations to $n \leqslant 14$ with little hope for significant extension to higher n.

Later, Harlow and Phoenix (1981a,b) developed a powerful recursion analysis for the treatment of larger bundles. A major result was that the distribution for bundle strength $H_{m,n}(x)$ is extremely accurately approximated by the weakest link form

$$H_{m,n}(x) = 1 - [1 - W(x)]^{mn}, \quad x \geqslant 0 \qquad (5)$$

where $W(x)$ is a characteristic distribution function, depending on $F_\delta(x)$ and the K_r of the local load-sharing rule. Unfortunately, $W(x)$ cannot be expressed in terms of simple functions, and the procedure for its numerical calculation is cumbersome. However, $W(x)$ can be approximated (see Sect. 6).

Smith (1979, 1980) has developed two distinct asymptotic analyses. His results are quite easy to use, and are accurate for the range of ρ of practical importance. In what follows, we discuss the most useful of Smith's approximations.

5. *Exact Results and Approximations for a Bundle with Three Fibers*

For a bundle with three fibers, the key task is to compute $G_3(x)$ for a microbundle whose three elements have strengths, say X_1, X_2 and X_3. To do this we must enumerate all the distinct ways the bundle can fail under fiber load x, calculate the probabilities for these, and sum the probabilities to obtain $G_3(x)$. One way is by the event $(X_1 \leqslant x, X_2 \leqslant x, X_3 \leqslant x)$, that is, all elements fail under their initial load x, and this has probability $F_\delta(x)^3$. A second way is when exactly two elements fail under load x, and the remaining element fails under the induced overload K_2x; one such event is $(X_1 \leqslant x, x < X_2 \leqslant K_2x, X_3 \leqslant x)$. There are three such events for this way of failure, so that the total probability is $3F_\delta(x)^2[F_\delta(K_2x) - F_\delta(x)]$. A third way for failure to occur is for an end element to fail under load x, the middle element to survive x but fail under K_1x, and the other end element to survive x but fail under K_2x. There are two such events, one of which is $(X_1 \leqslant x, x < X_2 \leqslant K_1x, x < X_3 \leqslant K_2x)$, so that the total probability is $2F_\delta(x)[F(K_1x) - F_\delta(x)][F_\delta(K_2x) - F_\delta(x)]$. A fourth possibility is for the middle element to fail under load x, and each of the end elements to survive x, but fail under their overloads K_1x; this has probability $F_\delta(x)[F_\delta(K_1x) - F_\delta(x)]^2$. Finally, we may have a sequence of failures starting with the middle element, such as the event $(x < X_1 \leqslant K_1x, X_2 \leqslant x, K_1x < X_3 \leqslant K_2x)$. There are two such events, and the resulting probability term is $2F_\delta(x)[F_\delta(K_1x) - F_\delta(x)][F_\delta(K_2x) - F_\delta(K_1x)]$. Summing all the above probabilities yields

$$\begin{aligned} G_3(x) = {} & 4F_\delta(x)F_\delta(K_1x)F_\delta(K_2x) \\ & - F_\delta(x)F_\delta(K_1x)^2 - F_\delta(x)^2F_\delta(K_2x) \\ & - 2F_\delta(x)^2\,F_\delta(K_1x) + F_\delta(x)^3, \quad x \geqslant 0 \end{aligned} \qquad (6)$$

In the analysis of bundle failure, a useful distribution function is $G_n^{[k]}(x)$, $x \geqslant 0$, which we define as the distribution function for the fiber load x, at which k or more adjacent breaks occur in the microbundle of n elements. Note that $G_n^{[n]}(x) = G_n(x)$. To compute $G_3^{[2]}(x)$, we note that $1 - G_3^{[2]}(x)$ is the probability that at most only isolated broken elements occur in the bundle. One possibility is for none of the elements to fail, and this has probability $[1 - F_\delta(x)]^3$. Another way is for exactly one element to fail and this yields the probability $2F_\delta(x)[1 - F_\delta(K_1x)][1 - F_\delta(x)] + F_\delta(x)[1 - F_\delta(K_1x)]^2$, where the first term corresponds to an end failure and the second term to a middle failure. Finally, exactly two elements may fail, but these must be the first and last elements with the middle element as a survivor; this yields the probability $F_\delta(x)^2[1 - F_\delta(K_2x)]$. Summing these probabilities, and subtracting them from unity gives

$$G_3^{[2]}(x) = 4F_\delta(x)F_\delta(K_1x) - 2F_\delta(x)^2 - F_\delta(x)F_\delta(K_1x)^2 - 2F_\delta(x)^2F_\delta(K_1x) + F_\delta(x)^2 F_\delta(K_2x) + F_\delta(x)^3, \quad x \geqslant 0 \quad (7)$$

Lastly, we obtain $G_3^{[1]}(x) = 1 - [1 - F_\delta(x)]^3$, since $1 - G^{[1]}(x)$ is the probability of having no breaks in the microbundle.

Figure 2 shows the various distribution functions above, under both Eqn. (2) and Eqn. (3) with $\rho = 15$. The scaling is that of Weibull probability paper, so that a Weibull distribution plots as a straight line. (Weibull probability paper has coordinates which are linear in $\ln[-\ln(1 - G)]$ versus $\ln(x)$, where G is the cumulative probability and x is the load.) Also shown are the values of the distribution function $H_{m,n}(x)$ for the strength of a bundle of fibers with $m = 1000$ and $n = 3$.

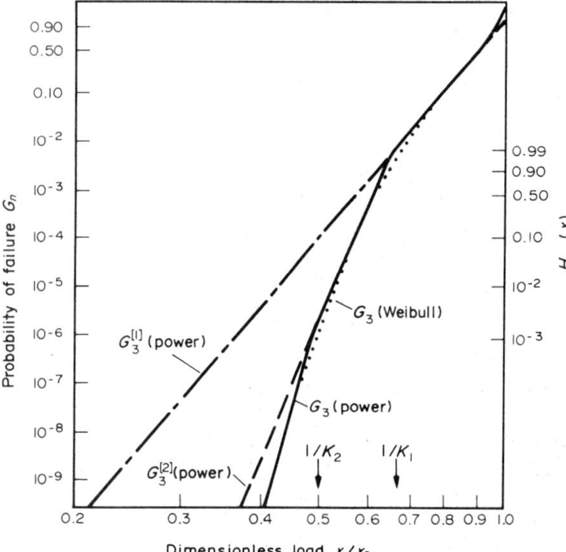

Figure 2
Distribution functions associated with the failure of a bundle of three fibers displayed under the Weibull and power distributions

Notice that it makes little difference whether one uses the Weibull distribution [Eqn. (2)], or the power distribution [Eqn. (3)] for the fiber elements; clearly the lower tail of $F_\delta(x)$ matters most. (When ρ is substantially less than 5, the difference is more marked; however, such low values of ρ occur infrequently for commercial fibers.) Also, the expected branch points in the distributions at the critical loads x_δ/K_1 and x_δ/K_2 should be noted. For loads x such that $K_1x < x_\delta < K_2x$, a pair of adjacent breaks is critical for the bundle, because of the position of its median strength along the load axis.

Figure 2 indicates that the bundle strength approximately follows a Weibull distribution with shape parameter $\hat{k}\rho = 30$. To see why, and to calculate the scale parameter, the lower tails of the $G_3^{[k]}(x)$ are investigated. Since $F_\delta(x) \sim (x/x_\delta)^\rho$ (where $\sim$ indicates that the ratio of the two sides tends to one as $x \to 0$), we can show by direct substitution into Eqn. (6) and Eqn. (7) that for $k = 1,2,3$,

$$G_3^{[k]}(x) \sim (3 - k + 1)d_k(x/x_\delta)^{k\rho} \quad (8)$$

where $G_3^{[3]} \equiv G_3$, $d_2 = 1$, $d_2 = 2K_1^\rho - 1$ and $d_3 = 4K_1^\rho K_2^\rho - K_1^{2\rho} - K_2^\rho - 2K_1^\rho + 1$. Thus we have the Weibull approximation:

$$H_{m,n}(x) \cong 1 - \exp[-m(n - \hat{k} + 1)d_{\hat{k}}(x/x_\delta)^{\hat{k}\rho}], \quad x \geqslant 0 \quad (9)$$

where $\rho = 15$, $m = 1000$, $n = 3$, $\hat{k} = 2$ and $d_{\hat{k}} = 2K_1^\rho - 1$. (Beware that the critical $\hat{k}$ changes if ρ or m are changed significantly.)

The two stumbling blocks to using the above approximations as n increases are as follows: first $\hat{k}$ is determined graphically, and this requires prior knowledge of $G_n^{[k]}(x)$; secondly, $G_n^{[k]}(x)$ and d_k become very difficult to calculate as k increases. Smith (1980) has resolved these issues with his asymptotic analysis as we now see.

6. *Exact and Approximate Results for Large Bundles*

For $k = 1,2,3, \ldots$ we let

$$\gamma(k) = k\ln(K_k) - \sum_{j=0}^{k-1} \ln(K_j) \quad (10)$$

where $K_0 \equiv 1$. We also let $\hat{k}$ be the value of k that solves

$$\gamma(k - 1) < \ln(mn)/\rho < \gamma(k) \quad (11)$$

From Smith (1980) we have the Weibull approximation:

$$H_{m,n}(x) \cong 1 - \exp[-mn\,\tilde{d}_{\hat{k}}(x/x_\delta)^{\hat{k}\rho}], \quad x \geqslant 0 \quad (12)$$

where

$$\tilde{d}_k = 2^{k-1}(K_0K_1 \ldots K_{k-1})^\rho, \quad k = 1,2,\ldots \quad (13)$$

Here, $\hat{k}$ is interpreted as the critical failure sequence size.

This result is consistent with those of the previous section for $n = 3$. Here, we use n in place of $n - \hat{k} + 1$, because n is large. Also, we use $\tilde{d}_k$ as an accurate approximation to d_k; the error is very small for $\rho \geqslant 5$, which is typical.

To evaluate the accuracy of the above Weibull approximation, we recall the Harlow–Phoenix approximation [Eqn. (5)], which for all practical purposes represents $H_{m,n}(x)$ exactly. Figure 3 shows a plot of the characteristic distribution function $W(x)$ for Weibull fiber elements with $\rho = 5$. (This low value of ρ yields a true test of accuracy since the differences are most exaggerated.) Also shown is the distribution function $H_{m,n}(x)$ for a bundle with $mn = 10^6$ elements in total. The coordinates are again Weibull coordinates.

For $k = 1,2, \ldots$ we let

$$\mathscr{F}_j^{[k]}(x) = 1 - \exp[-j\, \tilde{d}_k (x/x_\delta)^{k\rho}], \quad x \geqslant 0 \quad (14)$$

On the scale for $W(x)$, the straight lines on Fig. 3 happen to be the distribution functions $\mathscr{F}_1^{[k]}(x)$, but on the scale for $H_{m,n}(x)$ they are $\mathscr{F}_{mn}^{[k]}(x)$. Clearly one of the $\mathscr{F}_{mn}^{[k]}(x)$ for some k is an accurate approximation of $H_{m,n}(x)$, and it can be seen from Fig. 3 that $k = 5$ is the correct choice. This value of k is also the critical $\hat{k}$ which satisfies Eqn. (11), and thus $\mathscr{F}_{mn}^{[\hat{k}]}(x)$ is simply the Weibull approximation [Eqn. (12)].

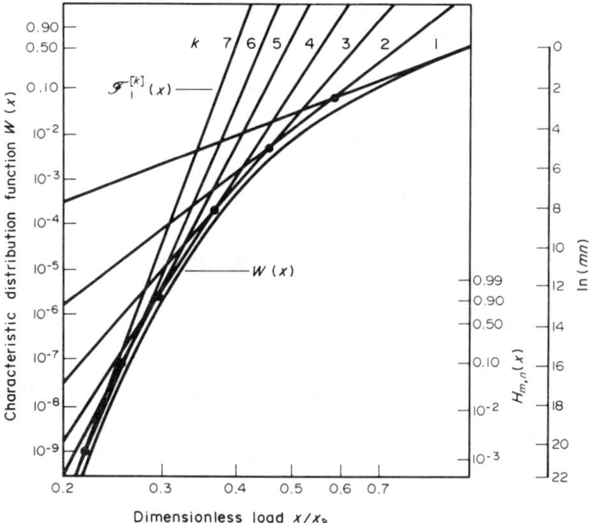

Figure 3
Characteristic distribution function for bundle strength and associated Weibull approximations

An accurate estimate of $W(x)$ is the inner envelope formed by the Weibull distributions $\mathscr{F}_1^{[k]}(x)$, $x \geqslant 0$, that is,

$$W(x) \cong \min[\mathscr{F}_1^{[1]}(x), \mathscr{F}_1^{[2]}(x), \mathscr{F}_1^{[3]}(x), \ldots], \quad x \geqslant 0 \quad (15)$$

The above ideas are discussed in more detail in Phoenix and Smith (1983), and Smith et al. (1983) extend the results to bundles with fibers arranged in a hexagonal array.

7. Conclusions

We have found that the strength of a bundle of fibers in a matrix approximately follows a Weibull distribution. The shape parameter for this Weibull distribution is $\hat{k}\rho$, where $\hat{k}$ is the critical failure sequence size and ρ is the Weibull shape parameter for the fiber. In the example of Fig. 3, we have $\hat{k}\rho = 5 \times 5 = 25$, and the scale parameter for this Weibull distribution is $x_\delta(mn\tilde{d}_{\hat{k}})^{-1/\hat{k}\rho}$, which is clearly much smaller than x_δ. This is typically observed in experiments.

Unlike the model itself, the final results are symmetric in the number of fibers n and the number of microbundles m; that is, the total bundle volume $mn\delta$ is what matters.

See also: Fracture: Statistical Theories; Strength of Composites

Bibliography

Argon A S 1972 Fracture of composites. In: Herman H (ed.) *Treatise on Materials Science and Technology*, Vol. 1. Academic Press, New York, pp. 79–114

Harlow D G, Phoenix S L 1978a The chain-of-bundles probability model for the strength of fibrous materials. I: Analysis and conjectures. *J. Compos. Mater.* 12: 195–214

Harlow D G, Phoenix S L 1978b The chain-of-bundles probability model for the strength of fibrous materials. II: A numerical study of convergence. *J. Compos. Mater.* 12: 314–34

Harlow D G, Phoenix S L 1981a Probability distributions for the strength of composite materials. I: Two-level bounds. *Int. J. Fract.* 17: 347–72

Harlow D G, Phoenix S L 1981b Probability distributions for the strength of composite materials. II: A convergent sequence of tight bounds. *Int. J. Fract.* 17: 601–30

Phoenix S L, Smith R L 1983 A comparison of probabilistic techniques for the strength of fibrous materials under local load-sharing among fibers *Int. J. Solids Struct.* 19: 479–96

Rosen B W 1964 Tensile failure of fibrous composites. *AIAA J.* 2: 1985–91

Smith R L 1979 Limit theorems for the reliability of series-parallel load-sharing systems. Ph.D thesis, Cornell University, Ithaca, New York

Smith R L 1980 A probability model for fibrous composites with local load-sharing. *Proc. R. Soc. London, Ser. A* 372: 539–53

Smith R L, Phoenix S L 1981 Asymptotic distributions for the failure of fibrous materials under series–parallel structure and equal load-sharing. *J. Appl. Mech.* 48: 75–82

Smith R L, Phoenix S L, Greenfield M R, Henstenburg R B, Pitt R E 1983 Lower-tail approximations for the probability of failure of three-dimensional fibrous composites with hexagonal geometry *Proc. R. Soc. London, Ser. A* 388: 353–91

Zweben C, Rosen B W 1970 A statistical theory of material strength with application to composite materials. *J. Mech. Phys. Solids* 18: 189–206

S. L. Phoenix

Fiber Networks: Models for Predicting Mechanical Behavior of Paper

The mechanical behavior of a fiber network, such as a paper sheet composed of pulped wood fibers, depends on the fiber properties and the physical or geometrical structure of the bonded fibrous network. The principal objective of a mathematical model is to predict the mechanical properties of the fiber network in terms of the properties of the fibers and the structure of the fiber network. In paper, the fibers are often collapsed, ribbon-like structures which are bonded together, primarily by hydrogen bonds formed when the sheet is pressed and subsequently dried (see *Hydrogen Bonding in Paper: Theory*).

The first attempt to develop a mathematical model for predicting the mechanical behavior of a fiber network such as paper was made by Cox (1952), who demonstrated how fiber orientation distribution influences the in-plane elastic behavior of the system. The analysis predicted all the in-plane elastic constants of a fiber mat in terms of the fiber orientation distribution expressed as a Fourier series.

Although Cox's analysis for in-plane elastic constants was based on an ideal mat in which the fibers lie in a plane and extend from one edge of the sheet to the other, he also discussed the effect of short fibers in a resin-bonded system. Since Cox's pioneering work, a considerable amount of study has been done. The basic premises of Cox's work are generally accepted as being valid, but it has been subjected to refinements and extensions. The refinements incorporate the effects of fiber network structure, straining during the formation of fiber–fiber bonding and different fiber elastic moduli for tensile and compressive straining. Extension of the analysis allows prediction of the fiber network strength.

Work in this area has been reviewed by Algar (1966), Dodson (1973), Van den Akker (1970, 1972) and Kallmes (1972a, 1972b). The most recent work was reported by Perkins (1980, 1982), Perkins and Mark (1981), and Seth and Page (1981).

In this article, the basic assumptions and results of the current mathematical models for predicting the in-plane mechanical behavior of fibrous networks are presented.

1. *Basic Elements of the Mathematical Model*

A typical fiber shown in Fig. 1 consists of several straight segments λ_{s1}, λ_{s2}, λ_{s3}, ... The sum of the segment lengths equals the total fiber length λ_T. It is assumed that the fibers are so flexible in bending that no appreciable load can be transmitted from one straight segment to the next.

If a fiber has microcompressions or other damaged regions along its length, the locations of these regions determine the ends of the load-bearing segments, even if the fiber is perfectly straight. The segment

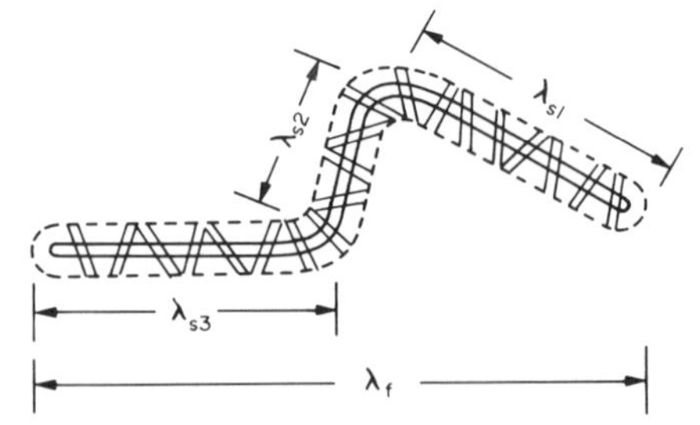

Figure 1
Schematic of fiber having three distinct segments and numerous crossing fibers in contact with it

elements are coupled to the network by means of the crossing fibers. The strains in the sheet are presumed to be transmitted to the segment elements by bending and shearing deformation of the crossing fibers and by shearing deformation of the fiber–fiber bonds. Thus the axial strain in the segment elements is not uniform but varies from the segment ends, where it is zero, to the middle, where it has its maximum value. If the segment is long enough and if the coupling of the crossing fibers is strong, the axial strain in the middle of a segment is the same as the normal component of strain of the sheet for the direction of the segment. On the other hand, if the segment is short and/or the coupling is weak, the segment strain is less than that associated with the sheet.

The model is further illustrated by Fig. 2, which shows a portion of a segment that is coupled by two crossing fibers to the remainder of the network. The boundary between the element and the network is depicted by a dashed line. This boundary is assumed to be located a distance l from the centerline of the segment, where l represents the center–center distance between bonds along a typical fiber; l_b is the bond length along the fiber and w_f and t_f are the width and thickness of the fiber, respectively. If l is small in comparison with w_f, as would be expected in moderately dense paper, the coupling is primarily attributable to the shearing deformations of the fiber–fiber bonds. In a very low-density system such as tissue paper, the bending and shearing deformation of the crossing fibers may be substantial.

On the assumption that the fiber is linearly elastic, the stress σ_f at any point along the fiber is related to the fiber strain ε_f at that point by

$$\sigma_f = E_{af}\varepsilon_f \tag{1}$$

where E_{af} is the effective axial modulus of the fiber. The effective axial modulus depends not only on the inherent properties of the fiber but also on how the fiber is bonded to other fibers in the network. Thus, there is a stiffening effect due to the reinforcement of other bonded fibers. E_{af} can be considered as an inherent property of the paper system, or its value

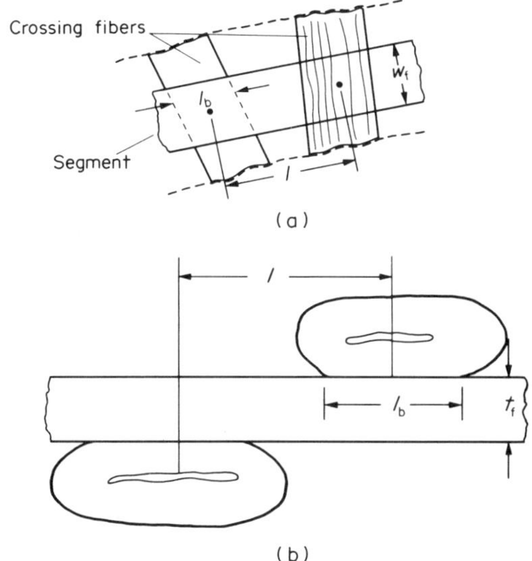

Figure 2
Portion of an element of a fiber segment: (a) plan view; (b) cross-sectional elevation

can be estimated in terms of an idealized paper network. For example, if the fibers are idealized (Figs. 3, 4) as thin rectangular strips (e.g., perfectly collapsed thin-walled springwood fibers), and if the density of the network is low enough so that on average the fiber surface is only half (or less) in contact with other fibers in the network, then Perkins (1980) has shown that

$$E_{af} = \frac{(l/l_b)(E_{fL} + E_{fT})}{C_B + \left(\frac{E_{fL} + E_{fT}}{E_{fL}}\right)\left(\frac{l - l_b}{l_b}\right)\left[1 + \frac{3}{2}\left(\frac{2a_0}{t_f}\right)\right]} \quad (2)$$

where

$$C_B = 1 + \frac{\tanh B}{B} \quad (3)$$

and

$$B = \frac{1}{2}\left[\frac{l_b}{t_f t_b}\frac{G_B}{E_{fT}}\left(1 + \frac{E_{fT}}{E_{fL}}\right)\right]^{0.5} \quad (4)$$

Here E_{fL} and E_{fT} represent the elastic moduli in the axial and transverse directions of the fiber, respectively; G_B is the shear modulus of the bond material; t_f and t_b represent the thicknesses of the collapsed fiber and the bond, respectively; l and l_b are as previously defined; and a_0 represents a deviation from straightness of a fiber (Figs. 2–4).

Equations (1–4) provide a method of estimating the effects of several important structural factors. For example, if the sheet density is low enough that part of the fiber is not bonded on either the top or bottom surface, the tendency of the fiber to deviate from perfect straightness, quantitatively described by the value of a_0, can have a pronounced calculable effect on the effective fiber modulus. This factor can have a significant effect if the fiber has some unbonded length, such as in a sheet with a density in the region of 250 kg m^{-3}.

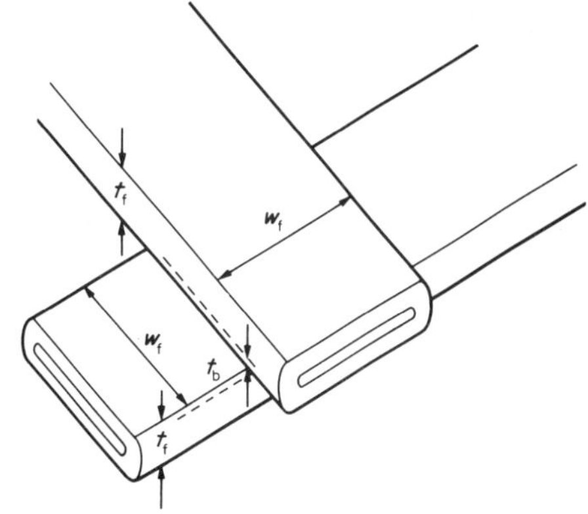

Figure 3
Fiber–fiber bond

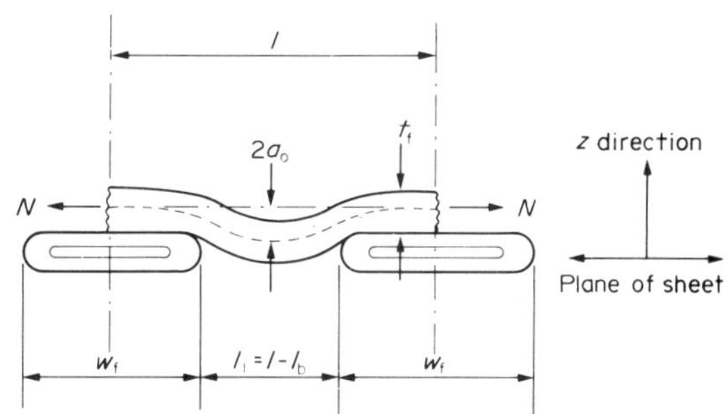

Figure 4
Fiber segment situated between bond centroids

The factor C_B determines the amount of reinforcement associated with fiber bonding. G_B and t_b in the factor B (Eqn. (4)) determine the stiffness of the fiber–fiber bond. It is expected that these factors are related to the degree of fiber refining which roughens the fiber surface and causes damage or separation of the S1 layer of the cell wall (see *Refining of Wood Fibers for Papermaking*). As the bond stiffness becomes very small, C_B tends towards a value of 2 and the reinforcing effect vanishes. At the other extreme, the bond becomes infinitely stiff and the factor C_B tends to unity.

The strain in the fiber ε_f is determined by the sheet strain and the degree of coupling between each fiber and the other fibers in the sheet. As shown in Fig. 5, a fiber of length λ_f is coupled to the rest of the network through the crossing fibers. If the fiber can

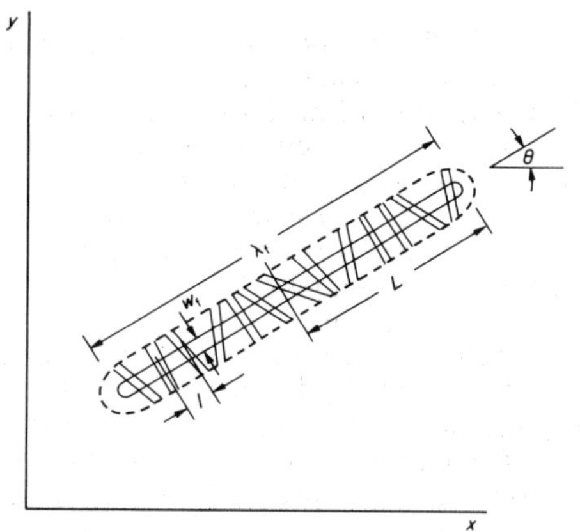

Figure 5
Typical fiber of length λ_f and orientation θ with respect to the machine direction x

be described by several straight segments, λ_f is the length λ_s of the straight segment. Referring to Fig. 6, the equilibrium equation for a fiber segment can be given as

$$\frac{d}{d\xi}\left(E_{af}\frac{d\varepsilon_f}{d\xi}\right) - k_f\varepsilon_f = -k_f\varepsilon_s \tag{5}$$

where

$$\varepsilon_s = \varepsilon_x \cos^2\theta + \varepsilon_y \sin^2\theta + 2\varepsilon_{xy} \sin\theta \cos\theta \tag{6}$$

Equation (6) represents the normal component of strain in the sheet in the direction θ corresponding to the orientation of the fiber. The factor k_f is the effective stiffness of the coupling between the fiber of orientation θ and the other crossing fibers that couple the fiber to the network. The stiffness k_f includes the flexibility of the fibers in bending (which may be discernible for low-density sheets) and the flexibility of the fiber–fiber bond. If the idealized system of rectangular cross section straight fibers is considered, k_f can be estimated as

$$k_f = 2\Big/ l\left[\frac{(l-l_b)^3}{12E_{fL}I_f}\left(1 + \frac{12fE_{fL}I_f}{G_fA_fl^2}\right) + \frac{2t_b}{A_bG_b}\right] \tag{7}$$

Here I_f represents the moment of inertia associated with the fiber cross section, G_f represents the cell wall shear modulus, f is a factor that depends upon the fiber cross-sectional shape (it is 6/5 if the fiber is rectangular), A_b is the area of the fiber–fiber bond, G_b is the bond region shear modulus and t_b is the thickness of the bond region.

Equations (2–4, 7) can be used to estimate values of E_{af} and k_f in an idealized case. These quantities can also be considered as fundamental parameters of the system which are to be determined by mechanical property testing. In the latter case, direct measurement of paper mechanical response is used with the theoretical property relations to calculate "experimental" values of E_{af} and k_f.

For a prescribed sheet strain given by ε_x, ε_y, ε_{xy}, the normal component of sheet strain in the direction of the fiber, ε_s, is known. Equation (5) can be solved by assuming an arbitrary value of ε_s. The factors A_f, E_{af} and k_f should properly be considered as varying in some random fashion. In this case, the solution to Eqn. (5) is very difficult. An alternative procedure would be to select average values for A_f, E_{af} and k_f and to determine an average fiber response. When this procedure is followed, the solution of Eqn. (5) is easily obtained as

$$\varepsilon_f = \varepsilon_s\left(1 - \frac{\cosh a\xi}{\cosh aL}\right) \tag{8}$$

where

$$a = \left(\frac{k_f}{A_fE_{af}}\right)^{0.5} \tag{9}$$

and L is half the fiber length.

The strain energy of a typical segment element is calculated as

$$W_e = \tfrac{1}{2}E_{af}A_f\lambda_s\varepsilon_s^2\eta_L \tag{10}$$

where

$$\eta_L = 1 - \frac{1}{2}\frac{\tanh aL}{aL} - \frac{1}{2\cosh^2 aL} \tag{11}$$

and can be identified as a coupling efficiency. For strong coupling and long length, η_L approaches unity. At the other extreme of very short length or very weak coupling, η_L approaches zero. The total strain energy of the system can be written

$$W = \int_0^\infty \int_0^\pi D_0 f^*_{\theta\lambda_s} W_e \, d\theta d\lambda_s \tag{12}$$

where D_0 represents the number of segment elements per unit area and $f^*_{\theta\lambda_s} d\theta d\lambda_s$ represents the probability

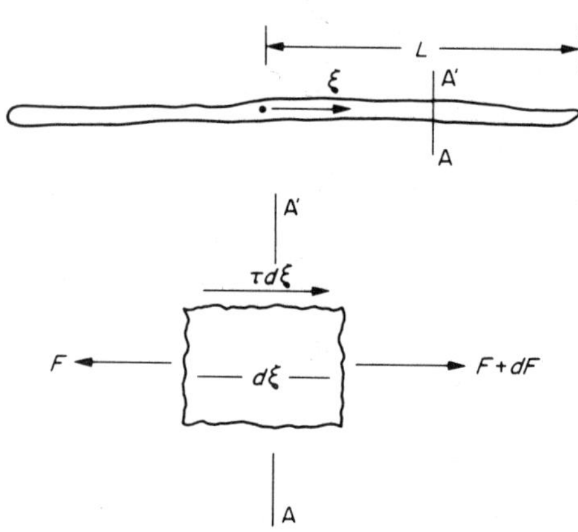

Figure 6
Equilibrium of a typical fiber element

of finding a segment having a length in the interval λ_s, $\lambda_s + d\lambda_s$ and orientation in the interval θ, $\theta + d\theta$. In general, it is assumed that length and orientation effects are coupled (Carroll 1962).

Suppose that $f^*_{\theta\lambda_s}$ can be expressed in the form

$$f^*_{\theta\lambda_s} = [f_\theta(\theta, a_1, a_2, a_3, \ldots, a_m)] \times [f_{\lambda s}(\lambda_s, b_1, b_2, b_3, \ldots, b_n)] \quad (13)$$

Here, f_θ describes the fiber orientation distribution as a function of θ depending on the m parameters, a_1, a_2, . . ., a_m. Likewise, the distribution of fiber lengths λ_s depends on n parameters b_1, b_2, . . ., b_n. In general, the parameters $a_1, a_2, \ldots, a_m$ are assumed to be functions of λ_s, while the parameters b_1, b_2, . . ., b_n are functions of θ.

The evidence available suggests that coupling between the two distributions is weak and therefore as an approximation, valid for at least certain papers, it can be assumed that the length and orientation distributions are independent. For this assumption, the parameters $a_1, a_2, \ldots, a_m$ and $b_1, b_2, \ldots, b_n$ are constants, and the strain energy per unit sheet area can be written in the form

$$W = \frac{1}{2}\frac{\omega_s}{\rho_{af}}\int_0^\pi \phi_{\lambda_s} E_{af}\varepsilon_s^2 f_\theta d\theta \quad (14)$$

where ω_s represents the basis weight, ρ_{af} represents the apparent fiber density and

$$\phi_{\lambda_s} = \int_0^\infty \eta_L f_{\lambda_s} d\lambda_s \quad (15)$$

The length parameter ϕ_{λ_s} and the apparent fiber modulus E_{af} may depend upon the orientation direction θ as a result of the stresses imposed on the fibers during drying of the sheet (see *Drying of Paper and Paperboard*). Analytically, it can be assumed that the apparent fiber modulus E_{af} depends upon the shrinkage and restraint conditions present during drying in the following way:

$$E_{af} = E_{af0}(1 + H\varepsilon_{ND}) \quad (16)$$

where E_{af0}, a constant, represents the apparent fiber modulus in the absence of drying restraint, H is a constant that predicts the magnitude of stiffening due

$$\varepsilon_{ND} = (\alpha_x M + \varepsilon_{xD})\cos^2\theta + (\alpha_y M + \varepsilon_{yD})\sin^2\theta \quad (17)$$

Here $\alpha_x M$ and $\alpha_y M$ represent the sheet shrinkage strains during unrestrained drying in the x and y directions, and ε_{xD} and ε_{yD} represent the sheet strains applied or allowed during drying. Thus, for unrestrained shrinkage conditions, $\varepsilon_{xD} = -\alpha_x M$ and $\varepsilon_{yD} = -\alpha_y M$. It is evident that ε_{ND} is related to the magnitude of restraint that a fiber of orientation θ would be subjected to during the drying of the sheet. In the following analysis, it is further assumed that ε_{ND} is positive, that is, the fibers are loaded in tension as a result of any drying restraint.

With the help of Eqn. (16), integration of Eqn. (14) can be easily carried out, if the length factor ϕ_{λ_s} is independent of θ. In fact, ϕ_{λ_s} does depend on θ, because the coupling parameter a that appears in η_L depends on E_{af}. Fortunately, the influence of variation in E_{af} through a in ϕ_{λ_s} has an insignificant effect on the prediction of the elastic moduli of the sheet. Therefore, there is no serious analytical error committed in assuming that ϕ_{λ_s} is independent of θ.

The stress–strain relations for the sheet can be obtained from the strain energy function W after carrying out the integration through the relations

$$\tau_x = \frac{\partial W}{\partial \varepsilon_x} \qquad \tau_y = \frac{\partial W}{\partial \varepsilon_y} \qquad \tau_{xy} = \frac{1}{2}\frac{\partial W}{\partial \varepsilon_{xy}} \quad (18)$$

where τ_x, τ_y, τ_{xy} represent the force per unit edge length of the paper sheet. Thus, the paper sheet is assumed to be a hyperelastic solid.

To express conveniently the resulting elastic moduli it is desirable to state the orientation distribution $f_\theta(\theta)$ in terms of its Fourier series expansion (Cox 1952). When the x direction is one of the axes of elastic symmetry, for example the machine direction, and $\theta = 0°$ corresponds to this direction,

$$f_\theta(\theta) = \frac{1}{\pi}[1 + a_1\cos 2\theta + a_2\cos 4\theta + a_3\cos 6\theta + \ldots + a_n\cos 2n\theta + \ldots] \quad (19)$$

After substitution of Eqns. (19, 16) in Eqn. (14) and subsequently using Eqn. (18), the Young's moduli E^*_x, E^*_y, the shear modulus G^*_{xy} and the Poisson ratios ν_{xy}, ν_{yx} corresponding to plain stress loading of the sheet are found to be

$$E^*_x = \frac{1}{16}\frac{\omega_s}{\rho_{af}}\phi_{\lambda_s}E_{af_0}[(1 + \langle e\rangle)(6 + 4a_1 + a_2) + \frac{e_\Delta}{2}(8 + 7a_1 + 4a_2 + a_3)][1 - \nu_{xy}\nu_{yx}] \quad (20a)$$

$$E^*_y = \frac{1}{16}\frac{\omega_s}{\rho_{af}}\phi_{\lambda_s}E_{af_0}[(1 + \langle e\rangle)(6 - 4a_1 + a_2) + \frac{e_\Delta}{2}(8 - 7a_1 + 4a_2 - a_3)][1 - \nu_{xy}\nu_{yx}] \quad (20b)$$

$$\nu_{xy} = \frac{(1 + \langle e\rangle)(2 - a_2) + \frac{e_\Delta}{4}(a_1 - a_3)}{(1 + \langle e\rangle)(6 - 4a_1 + a_2) - \frac{e_\Delta}{2}(8 - 7a_1 + 4a_2 - a_3)} \quad (20c)$$

$$\nu_{yx} = \frac{(1 + \langle e\rangle)(2 - a_2) + \frac{e_\Delta}{4}(a_1 - a_3)}{(1 + \langle e\rangle)(6 + 4a_1 + a_2) - \frac{e_\Delta}{2}(8 + 7a_1 + 4a_2 + a_3)} \quad (20d)$$

$$G_{xy}^{*} = \frac{1}{16}\frac{\omega_s}{\rho_{af}}\phi_{\lambda_s}E_{af0} \times \left[(1+\langle e\rangle)(2-a_2)+\frac{e_\Delta}{2}(a_1-a_3)\right] \quad (20e)$$

where

$$\langle e\rangle = \frac{1}{2}H[(\alpha_x M + \varepsilon_{xD}) + (\alpha_y M + \varepsilon_{yD})] \quad (21a)$$

$$e_\Delta = \frac{1}{2}H[(\alpha_x M + \varepsilon_{xD}) + (\alpha_y M + \varepsilon_{yD})] \quad (21b)$$

An inspection of Eqns. (20a–e) indicates that the elastic constants depend only on the first three parameters, a_1, a_2, a_3, that occur in the Fourier series expansion for f_θ. Furthermore, if the influence of drying restraint is not incorporated or if the sheet is dried under conditions of no restraint, the elastic moduli depend only on the parameters a_1 and a_2. For this condition, Eqns. (20a–e) are essentially the same as those given by Cox (1952) when $\theta = 0°$ corresponds to a direction of elastic symmetry in the sheet. (Note, however, that the effects of fiber–fiber bonding and the effect of finite fiber length are taken into account.)

The number of coefficients of the Fourier series expansion for f_θ that are necessary for predicting the elastic properties, therefore, depends on the functional relation between the apparent fiber modulus E_{af} and orientation, which in turn depends on the procedures and conditions of drying. Incorporation of any other phenomena affecting the angular dependence of the strain energy stored in the sheet may change the number of Fourier coefficients appearing in the elastic constants.

It is theoretically possible to describe any fiber orientation distribution in terms of the parameters a_1, a_2, a_3, . . . of the Fourier series expansion. However, a single-parameter function would simplify the expressions for the elastic moduli. The reduction to one parameter can be accomplished very simply by truncating the Fourier series and retaining the single parameter a_1 (Perkins 1980). However, reduction to a single parameter can have some significant effects. For example, for freely dried paper, the shear modulus should decrease with increased anisotropy, but this reduction effect will be predicted only if the coefficient a_2 is retained in the expansion of Eqn. (19). A single-parameter function would predict the shear modulus to be independent of the degree of anisotropy of the paper.

An alternative approach is to employ a distribution function that has only one shape parameter. For a general discussion of fiber orientation distribution refer to Perkins and Mark (1981). The two most suitable single-parameter distributions are the elliptical and the von Mises distributions. The distributions and their corresponding values of a_1, a_2, a_3 are given in Table 1.

2. *Application of Theory to Prediction of Elastic Moduli*

Equations (20a–e) can be used to estimate the effects of changes in structural parameters on the elastic properties of paper, or to estimate from experimental measurement the parameters of the system. As an example, the values of E_x and E_y can be estimated from measurements of the Young's moduli in the machine and cross-machine directions.

To use Eqns. (20a–e) in this example, it is necessary to investigate the influence of drying restraint. The stiffening parameter H of Eqn. (21) has been estimated by experiment (Setterholm and Chilson 1965) to lie in the range 10–30. From a knowledge of the drying restraint conditions and sheet shrinkage behavior, the values of $\langle e\rangle$ and e_Δ can be calculated. The ratio of the experimental Young's moduli for the machine and cross-machine directions can be equated to the ratio E_x^*/E_y^* from Eqns. (20a, b). The elliptical or von Mises distribution can be written in a Fourier series form (Table 1). The first three coefficients of the expansion can then be taken as a_1, a_2 and a_3, respectively. The inversion of this relation would yield a value of the concentration parameters for orientation ζ or κ depending on the choice of distribution to be used. The value of ζ or κ can then be used to obtain a_1, a_2, a_3 and subsequently the

Table 1
Parameters for the elliptical and von Mises fiber orientation distributions

	Parameter		
	a_1	a_2	a_3
Elliptical $f_{(\theta)} = \frac{\zeta}{\pi}\left(\frac{1}{\cos^2\theta + \zeta^2 \sin^2\theta}\right)$	$2\left(\frac{\zeta-1}{\zeta+1}\right)$	$2\left(\frac{\zeta-1}{\zeta+1}\right)^2$	$2\left(\frac{\zeta-1}{\zeta+1}\right)^3$
von Mises[a] $f(\theta) = \frac{1}{\pi I_0(\kappa)} e^{\kappa \cos 2\theta}$	$2\frac{I_1(\kappa)}{I_0(\kappa)}$	$2\frac{I_2(\kappa)}{I_0(\kappa)}$	$2\frac{I_3(\kappa)}{I_0(\kappa)}$

a $I_n(\kappa)$ = modified Bessel function of the first kind and order n

values of ν_{xy}, ν_{yx}, G_{xy}. The easy measurement of two Young's moduli then permits all the in-plane elastic moduli for the sheet to be calculated. The calculation is best performed for several different assumptions regarding the variables in Eqn. (21), to ascertain the reliability of the assumptions.

The choice of fiber orientation distribution function can have significant effects on predictions obtained from Eqns. (20a–e). For a given fiber orientation, predictions of ν_{xy}, ν_{yx} and G_{xy} from experimental measurements of E_x and E_y may be different for the elliptical distribution from those for the von Mises distribution. While the elliptical distribution or the single-parameter cosine distribution is the easiest to apply, the von Mises distribution is considered to provide more accurate estimates (Perkins and Mark 1981).

In the case of fiber segment length distributions, the Erlang distribution has been found to be quite satisfactory. Thus, the function f_{λ_s} can be taken as

$$f_{\lambda_s} = \frac{(\lambda_s/b)^{c-1}\exp(-\lambda_s/b)}{b[(c-1)!]} \tag{22}$$

Here b and c are determined by fitting the distribution to the experimental data. The product bc is equal to the mean value of λ_s. For further information, refer to Perkins and Mark (1981).

In some cases it may be desirable to use the basic approach numerically rather than to utilize Eqns. (20a–e). For example, it may be desired to investigate the consequence of different values of E_{af} for fibers in compression and tension. In the uniaxial tensile case, the fibers with orientations within the angular sector

$$\theta = \pm \arctan\left(\frac{1}{\nu_{xy}}\right)^{0.5} \tag{23}$$

are loaded in tension, whereas those outside this sector are in compression. By dividing the fibers into two classes corresponding to compressive and tensile loading for an arbitrary value of ν_{xy}, the Young's modulus E_x and the Poisson ratio ν_{xy} can be obtained by calculating the total strain energy. The true value of ν_{xy} corresponds to the minimum strain energy value. Because of this assumption, the elastic constants will depend on the manner in which the material is tested. Thus, the elastic constants for tensile, compressive or biaxial loading could all be different. Experimentally, this is not true for papers in the normal density range. Very low-density papers or nonwoven fabrics may experience these phenomena (Dent 1982). Calculations show that the Poisson ratio can be significantly lower when this effect is present.

3. Application of Theory for Prediction of Strength

Theories for predicting the tensile strength of paper have been developed by Page (1969) and Kallmes (1972b) which are based on the relative fractions of fibers that fracture and those that pull out of the network across the tensile fracture zone. El-Hosseiny and Abson (1979) showed that these theories do not adequately predict the variations in paper strength observed when process variables related to the degree of fiber–fiber bonding and sheet density are systematically changed. The inadequacy of the strength theories described above may be due to oversimplification of the fracture process.

The mathematical model in Sect. 1 can be used to analyze possible failure mechanisms in the fiber network. The analysis provides the means for developing a more refined, realistic approach to the study of failure. The failure process can be approached by considering behavior of the system immediately beyond the elastic limit.

As a first approximation, assume that the fibers and fiber–fiber bonds behave as linear elastic elements. The elastic limit of the network is first reached when fibers or fiber–fiber bonds deteriorate or fail completely. It is reasonable to assume that the fiber network does not fail catastrophically even though some fibers or some bonds within the network fail. The assumption is made that the fiber network system remains stable for some straining beyond the elastic limit while the fibers or bonds are progressively fracturing.

A general theory of failure cannot be presented because the process depends on a mechanism of fiber or bond failure events that ultimately lead to network fracture. For each postulated failure mechanism it may be possible to develop a theoretical prediction of the stress–strain curve beyond the elastic limit. The potential of this approach is illustrated by two examples, one involving network breakdown based on fiber–fiber bond failure (Perkins 1980) and the other involving the failure of fibers in edgewise compressive loading (Perkins 1982).

3.1 Biaxial Tensile Test with Fiber–Fiber Bond Failure

The solution of Eqn. (5) enables the stress and strain in both the fiber and the fiber–fiber bonds to be calculated. The maximum fiber stress σ_{fm} occurs at the midpoint of the fiber and has the value

$$\sigma_{fm} = E_{af}\varepsilon_s\left(1 - \frac{1}{\cosh aL}\right) \tag{24}$$

The maximum fiber–fiber bond shearing stress occurs at the fiber endpoint bonds and has the value

$$\tau_{Bm} = \frac{A_f}{A_B} a l_b \varepsilon_s \tanh aL \tag{25}$$

It is assumed that the fiber strength in tension is greater than the fiber–fiber bond strength. Thus, under biaxial tensile loading, the paper behaves elastically until the sheet strain ε_s is large enough to make the bond shear stress τ_{Bm} equal to the bond shear strength (note that the sheet strain is the same in all directions for equal biaxial straining). At this sheet strain, bonds at the ends of fibers commence to break. As the bonds fail, the network structure changes. The center–center distance between bonds increases. As a result, the values of a and L are continually changing as the sheet strain increases beyond the elastic limit. It can be shown that the product Ll remains constant. This makes it possible to calculate all the instantaneous values of l, L and a in terms of sheet strain using the assumption that the shear stress at the fiber end remains constant.

The system is inherently stable because the breaking of bonds changes the network structure and reduces the shear stress at the critical points. Thus, the strain must be continually increased to maintain a constant value of τ_{Bm}. Also, the breaking of bonds leaves the system linearly elastic but with continually changing values of the elastic moduli. This is precisely the assumption made in the stable progressively fracturing model discussed above. This observation makes it possible to generate the entire stress–strain curve. For each value of strain ε_s beyond the elastic limit, the values of l, a and L are determined. These determinations enable the elastic moduli to be calculated and therefore the stress corresponding to each value of strain. For further discussion and a numerical example, refer to Perkins (1980).

3.2 Compressive Uniaxial Test

In edgewise compression loading of paper sheets, fiber failure in the form of the development of slip planes in the cell wall can take place (Perkins 1981). The eventual failure may be explained if the stress–strain curve in compression can be predicted (Perkins and McEvoy 1981). The theory above can predict the compressive stress–strain curve if it is assumed that the failure mechanism is the development of slip planes in fibers of the sheet.

Assume that a fiber develops slip plane failure when the fiber strain reaches a critical value ε_{fsp}. With the use of Eqn. (8) the condition for slip plane development in a fiber is

$$\varepsilon_{fsp} = \varepsilon_s \left(1 - \frac{1}{\cosh aL}\right) \qquad (26)$$

Here

$$\varepsilon_s = \varepsilon_x(\cos^2\theta - \nu_{xy}\sin^2\theta) \qquad (27)$$

Interpreting ε_x as compressive strain of the sheet in the direction of loading, ε_s represents a compressive strain for fibers oriented within the sector

$$\theta = \pm\arctan\left(\frac{1}{\nu_{xy}}\right)^{0.5} \qquad (28)$$

Fibers oriented outside this sector experience a tensile strain. As the loading increases beyond a point satisfying Eqn. (26) when $\theta = 0$, it is assumed that the length L decreases for fibers oriented within a sector of the loading direction. The length decrease is obtained by assuming that Eqn. (26) is satisfied beyond the elastic limit. Physically, this corresponds to reducing the fiber segment length as slip planes develop. The entire procedure assumes that compressive loads cannot be transmitted through a slip plane.

As the sheet strain continues, the failure sector grows and the length L becomes more dependent on the direction angle θ. The system is inherently stable because fiber strains are relieved by the development of the slip plane. The calculation of strain energy must be made numerically for an arbitrary choice of ν_{xy}. The correct ν_{xy} for each sheet strain ε_x imposed is determined by numerically calculating strain energy versus ν_{xy} and selecting the minimum value. Once ν_{xy} is known, the other elastic constants can be calculated and the sheet stress corresponding to the assumed value of sheet strain can be calculated. The process can be continued to generate the entire stress–strain curve. As in the preceding example, it is implicitly assumed that the failures that take place in the sheet merely alter the structure of the sheet and, therefore, change the elastic properties of the sheet. The system is assumed to satisfy the conditions for the model of the stable progressively fracturing solid (Sect. 3.1).

A potential benefit of this analysis is that the effects of changing system parameters such as fiber properties and degree of fiber–fiber bonding on the stress–strain curve can be determined relatively easily.

4. Limitations of Present Models and Areas for Further Development

The basic assumptions of the current models for predicting the elastic behavior of paper are quite similar to those originally made by Cox (1952). The principal differences lie in the assumptions concerning the load transfer mechanism between the fiber network and the individual fiber. The model described above assumes that the energy associated with the uniaxial straining of a fiber and the load transfer from the other fibers in the network composed of shearing and bending deformation are independent processes. Estimates of fiber stiffness along the fiber axis and shear coupling coefficient are based on a model fiber–fiber bond. These estimates appear reasonable when the fiber–fiber bond consists of only two crossing fibers, which is a reasonable assumption for thin-walled paper networks when the sheet density is lower than ~750 kg m^{-3}. For sheets of higher density, the model may be acceptable if each fiber–fiber bond pair makes independent strain energy contributions when more than one fiber is bonded to a given fiber.

The principal differences between models such as those of Perkins and Mark (1981) and Seth and Page

(1981) lie in assumptions concerning the effect of the fiber–fiber bonds. Predictions of elastic moduli from these two models are very similar in form. There are significant differences, however, in their estimates of effective fiber axial stiffness and shear coupling coefficient.

A number of extremely important factors are not incorporated in any current model. For example, the effect of fibers that exhibit a gradual curl in the plane of the sheet has not been investigated. The current models all assume that the fibers are either straight or can be modelled as independent straight segments as depicted by Fig. 1. This limitation is the result of the assumption that fibers transmit only axial loads. Furthermore, current models assume that fibers lie entirely in the plane of the sheet. Out-of-plane orientation and curl of the fibers through the thickness of the sheet is likely to be an extremely important factor, particularly for postelastic and failure phenomena.

There are no known attempts to model the deformation behavior of fibrous networks for cases of loading or deformation in the thickness direction. A number of important processes such as printing, folding and pressing, depend on thickness-direction response. Furthermore, in-plane behavior, such as edgewise compression loading to failure, can be markedly influenced by restraints in the thickness direction.

The inherent inhomogeneity of fibrous systems, especially those of low mass per unit area or basis weight, has been studied extensively. Unfortunately, however, no rational models of practical utility are available to predict the effect of inhomogeneity on elastic and strength characteristics.

See also: Hydrogen Bonding in Paper: Theory; Mechanical Properties of Paper; Paper and Paperboard: An Overview; Paper and Paperboard: Destructive Mechanical Testing

Bibliography

Algar W H 1966 Effect of structure on the mechanical properties of paper. In: Bolam F (ed.) 1966 *Consolidation of the Paper Web,* Vol. 2. Technical Section of the British Paper and Board Makers' Association, Cambridge, UK, pp. 814–49

Carroll C W 1962 Joint probability function relating fiber segmental length and orientation. In: Bolam F (ed.) 1962 *Formation and Structure of Paper,* Vol. 1. Technical Section of the British Paper and Board Makers' Association, Cambridge, UK, pp. 243–45

Cox H L 1952 The elasticity and strength of paper and other fibrous materials. *Br. J. Appl. Phys.* 3: 72–79

Dent R W 1982 *The Initial Modulus of Filament Webs,* ASME Paper No. 80-Tex-6. American Society of Mechanical Engineers, New York

Dodson C T J 1973 A survey of paper mechanics in fundamental terms. In: Bolam F (ed.) 1973 *The Fundamental Properties of Paper Related to its Uses,* Vol. 1. Technical Section of the British Paper and Board Makers' Association, Cambridge, UK, pp. 202–26

El-Hosseiny F, Abson D 1979 A critical examination of theories of paper tensile strength. *Proc. Int. Paper Physics Conf.*, Harrison Hot Springs, British Columbia, pp. 185–90

Kallmes O J 1972a A comprehensive view of the structure of paper. In: Jayne B A (ed.) 1972 *Theory of Design of Wood and Fiber Composite Materials.* Syracuse University Press, Syracuse, New York, pp. 157–76

Kallmes O J 1972b The influence of nonrandom fiber orientation and other fiber web parameters on the tensile strength of nonwoven fibrous webs. In: Jayne B A (ed.) 1972 *Theory of Design of Wood and Fiber Composite Materials.* Syracuse University Press, Syracuse, New York, pp. 177–96

Page D H 1969 A theory for the tensile strength of paper. *Tappi* 52(4): 674

Perkins R W 1980 Mechanical behavior of paper in relation to its structure. *Proc. Conf. Paper Science and Technology: The Cutting Edge.* Institute of Paper Chemistry, Appleton, Wisconsin, pp. 89–111

Perkins R W 1981 Mechanical behavior of wood and fibrous composites. In: Selvadurai A P S (ed.) 1981 *Mechanics of Structured Media.* Elsevier, New York, pp. 449–59

Perkins R W 1982 Models for describing the elastic, viscoelastic and inelastic mechanical behavior of paper and board. In: Mark R E (ed.) 1982 *Handbook of Mechanical and Physical Testing of Paper.* Dekker, New York

Perkins R W, McEvoy R P 1981 The mechanics of the edgewise compressive strength of paper. *Tappi* 64(2): 99–102

Perkins R W, Mark R E 1981 Some new concepts of the relation between fibre orientation, fibre geometry and mechanical properties. *The Role of Fundamental Research in Papermaking, 7th Fundamental Research Symp.* Mechanical Engineering Publications, Bury St. Edmunds, UK

Seth R S, Page D H 1981 The stress–strain curve of paper. *The Role of Fundamental Research in Papermaking, 7th Fundamental Research Symp.* Mechanical Engineering Publications, Bury St. Edmunds, UK

Setterholm V C, Chilson W A 1965 Dry restraint: Its effect on the tensile properties of 15 different pulps. *Tappi* 48(11): 634–40

Van den Akker J A 1970 Structure and tensile characteristics of paper. *Tappi* 53(3): 388–400

Van den Akker J A 1972 Large strains and inelastic behavior of paper. In: Jayne B A (ed.) 1972 *Theory of Design of Wood and Fiber Composite Materials.* Syracuse University Press, Syracuse, New York, pp. 197–218

R. W. Perkins

Fiber-Reinforced Cements

Cement mortar and concrete differ from most other matrices used in fiber composites in that their essential characteristics are those of low tensile strength (generally less than 7 MPa) and very low strain to failure in tension (generally less than 0.05%) which result in materials which are rather brittle when subjected to impact loads.

The binder in the matrix is cement paste produced by mixing cement powder and water which harden after an exothermic reaction to form complex chemical compounds that are dimensionally unstable due to water movement at the molecular level within the crystal structure. Inert mineral fillers known as aggregates are added at up to 70% by volume to increase stability and reduce the cost, and therefore the space available for fiber inclusion may be severely limited.

Fiber volumes in excess of 10% of the composite volume are difficult to include in mortar and, in the case of concrete, 2% by volume of fiber is considered to be a high proportion which is generally insufficient to permit significant reinforcement of the matrix before cracking occurs, particularly where individual fibers may be spaced up to 40 mm apart by the aggregate particles.

The type of fiber composite described in this article is therefore one in which the fiber has little effect on the properties of the relatively stiff matrix until cracking has occurred at the microlevel. The fibers then become effective, resulting in a composite with reduced stiffness and increased strain to failure providing increased toughness compared with the unreinforced matrix.

Notable exceptions to this type of behavior are asbestos-cement and the cellulose fiber reinforced thin sheets introduced to the market in the mid-1980s. In these materials the fibers are used to increase the matrix cracking strain to more than 1000×10^{-6}.

1. Principles of Reinforcement in Tension

Owing to the relatively stiff matrix and its low strain to failure, fibers are not normally included in cement matrices to increase the matrix cracking stress. Therefore the merits of fiber inclusion lie in the load carrying ability of the fibers after microcracking of the matrix has been initiated.

1.1 Critical Fiber Volume

The critical fiber volume, $V_{f(crit)}$, is defined as the volume of fibers which, after matrix cracking, will carry the load which the composite sustained before cracking. For the simplest case of continuous aligned fibers with frictional bond, the critical fiber volume can be expressed in terms of the elastic modulus of the composite, E_c, the ultimate strain of the matrix ϵ_{mu} and the fiber strength σ_{fu}:

$$V_{f(crit)} = \frac{E_c \epsilon_{mu}}{\sigma_{fu}} \qquad (1)$$

Figure 1 shows stress–strain curves for composites with less than the critical volume (OAB) and greater than the critical volume of fiber (OACD).

Efficiency factors for fiber orientation and fiber length increase $V_{f(crit)}$ considerably above that calculated from Eqn. (1). The result is that most concretes are unable to contain the critical volume of short chopped fibers and still maintain adequate workability for full compaction to be achieved. However, many techniques are available which enable the inclusion of more than the critical volume of fiber in fine grained cement mortars.

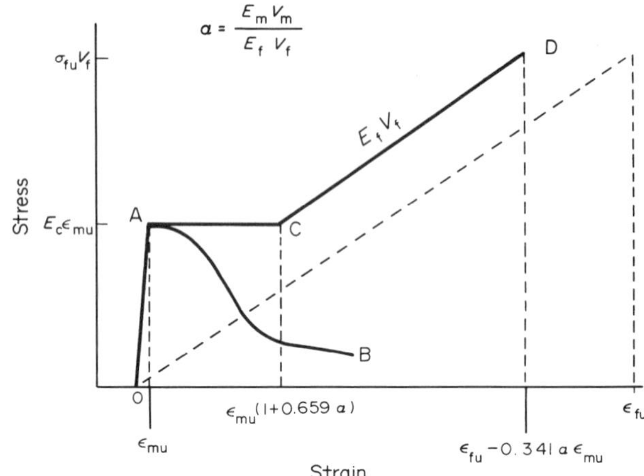

Figure 1
Idealized tensile stress–strain curves for a brittle cement matrix composite

The dashed curve in Fig. 1 results when high volumes of very small diameter, stiff, well bonded fibers are included. The existing flaws are reinforced and may be restrained from propagating at the stress at which they would propagate in an unreinforced matrix, thus leading to an increased cracking strain in the matrix. Typical examples are asbestos-cement and cellulose fibers in autoclaved calcium silicate.

1.2 Stress–Strain Curve, Multiple Cracking and Ultimate Strength

Figure 1 shows the idealized stress–strain curve (OACD) for a fiber-reinforced brittle matrix composite with more than the critical volume of fiber. The assumptions used to calculate the curve are that the fibers are long, aligned with the stress, and that the bond τ is purely frictional. The matrix is assumed to crack at A at its normal failure strain ϵ_{mu} and the material breaks down into cracks in region A–C with a final average spacing S given by

$$S = 1.364 \frac{V_m}{V_f} \frac{\sigma_{mu}}{\tau} \frac{A_f}{P_f} \qquad (2)$$

where V_m, V_f are the volume fraction of the matrix and fiber respectively; A_f, P_f are the cross-sectional area and perimeter of the fiber; and σ_{mu} is the cracking stress of the matrix. It will be noted that crack spacing is not dependent on the elastic modulus of the fiber. Region C–D represents extension of the fibers as they slip through the matrix without further matrix cracking.

The ultimate strength of the composite depends only on the fiber strength σ_{fu} and volume V_f whereas the ultimate composite strain is less than the fiber failure strain by a factor depending on the volume and elastic modulus of the matrix and of the fiber, and the matrix failure strain.

The multiple cracking represented in Fig. 1 is a desirable situation because it changes a basically brittle material with a single fracture surface and low energy requirement to fracture into a pseudoductile material which can absorb transient minor overloads and shocks with little visible damage.

1.3 Efficiency Factors for Fiber Length and Orientation

The efficiency of fibers in reinforcing the matrix in practice may be very much less than that assumed in the calculation of the idealized stress–strain curve. Efficiency depends on many factors including the orientation of the fibers relative to the direction of stress, whether the matrix is cracked or uncracked and the fiber length relative to the critical length (defined as twice the length of fiber embedment which would cause fiber failure in a pullout test).

For short fibers in a three-dimensional orientation, the combined efficiency could be less than one-fifth of the efficiency in the aligned fiber case (Hannant 1978).

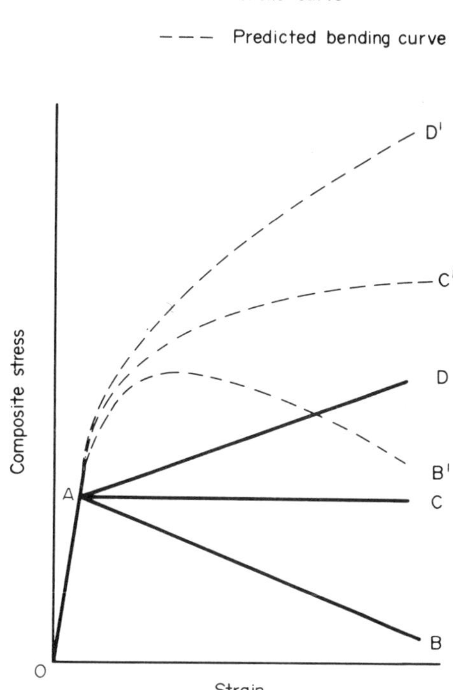

Figure 2
Apparent bending (flexure) curves predicted for assumed direct tensile curves (after Laws and Ali 1977)

2. Principles of Reinforcement in Flexure

The critical fiber volume for flexural strengthening can be less than half that required for tensile strengthening so that the load which a beam of the composite can carry in bending can be usefully increased, even though the tensile strength of the materials has not been altered by the addition of fibers. In fact, the flexural strength is commonly between two and three times that predicted from the tensile strength using an elastic analysis. The main reason for this is that the postcracking stress–strain curve in tension (AB or ACD in Fig. 1) is very different from that in compression (similar slope to OA in Fig. 1). As a result, the neutral axis moves towards the compression surface of the beam and conventional beam theory becomes inadequate because it does not allow for a plastic stress block in tension combined with an elastic stress block in compression (Hannant 1978).

The relationship between idealized uniaxial tensile stress–strain curves and flexural stress–strain curves is shown in Fig. 2.

The importance of achieving adequate post-cracking tensile strain capacity in maintaining a high flexural strength is demonstrated in Fig. 2. For instance, curve OAC is a tensile stress–strain curve and OAC′ is the associated bending curve. If the tensile strain reduces from C to A at a constant stress, then the bending strength will reduce from C′ to A with an increasing rate of reduction as A is approached and the material becomes essentially elastic. Thus, tensile strength on its own cannot be used to predict bending strength and composites which suffer a reduction in strain to failure as a result of natural weathering will also have a reduced bending strength.

3. Asbestos-Cement

Since 1900, the most important example of a fiber cement composite has been asbestos-cement. The proportion by weight of asbestos fiber is normally between 9 and 12% for flat or corrugated sheet, 11–14% for pressure pipes and 20–30% for fire-resistant boards and the binder is normally a Portland cement. Fillers such as finely ground silica at about 40% by weight may also be included in autoclaved processes where the temperature may reach 180 °C.

The most widely used method of manufacture for asbestos-cement was developed from papermaking principles about 1900 and is known as the Hatschek process. A slurry, or suspension of asbestos fiber and cement in water at about 6% by weight of solids, is continuously agitated and allowed to filter out on a fine screen cylinder. The filtration rate is critical and coarser cement than normal (typically 280 $m^2 kg^{-1}$) is used to minimize filtration losses. Also, although chrysotile fibers form the bulk of the fiber, most formulations have included a smaller proportion of the amphibole fiber wherever possible. This is because of the dewatering characteristics of crocidolite and amosite. Mixtures of cement and chrysotile drain very slowly and production rates are consequently slow. By using a blend of chrysotile and amphibole fibers (usually between 20% and 40%), considerable improvements in the drainage rate have been achieved, plus the additional bonus of enhanced reinforcement.

Other types of fiber such as cellulose derived from wood pulp or newsprint have also been added to the slurry to produce different effects in the wet or hardened sheet.

There are several other manufacturing processes such as the Mazza process for pipes and the Magnani process for corrugated sheets. Also the material may be extruded and injection molded.

The tensile strength varies between about 10 MPa and 25 MPa depending on fiber content, density and direction of test whereas the flexural strengths may be between 30 MPa and 60 MPa.

Durability is exceptionally good, a lifetime in excess of 50 years under external weathering being possible. A disadvantage is the brittle nature of the material under impact situations due to its relatively low strain to failure.

Major applications have included corrugated sheeting and cladding for low-cost agricultural and industrial buildings, and flat sheeting for internal and external applications has also proved very successful.

Pressure pipes are another worldwide market and the ease with which the material can be formed into complex shapes has enabled a wide range of special purpose products to be supplied.

The future for asbestos-cement products depends largely on the attitudes taken by Government Health and Safety organizations to the health problems associated with asbestos fibers (see *Asbestos Hazards*).

4. *Glass-Reinforced Cement*

The early development of glass fibers in low alkali cements took place in China and Russia in the 1960s but because of the susceptibility to attack of E-glass fibers by the higher alkali Portland cements used in the West, considerable investment in the UK took place in the 1970s to produce alkali-resistant zirconia glass on a commercial scale.

The fibers are commonly included in the matrix in strands consisting of 204 filaments of 10 μm diameter fibers.

The three basic techniques are premixing which gives a random three-dimensional array unless the mix is subsequently extruded or pressed, the spray method which gives a random two-dimensional arrangement, and filament winding which gives a one-dimensional array or a multi-angled oriented array. The volume of fibers is generally greater than the critical volume for tensile strengthening.

The spray method is the most popular and consists of leading a continuous roving up to a compressed-air-operated gun which chops the roving into lengths of between 10 mm and 50 mm and blows the cut lengths at high speed simultaneously with a cement paste spray onto a forming surface which may contain a filter sheet and vacuum water extraction in certain applications.

The fiber-cement sheet can be built up to the required thickness and demolded immediately, an additional advantage being that it has sufficient wet strength to be bent round radiused corners to give a variety of product shapes such as corrugated or folded sheets, pipes, ducts or tubes. Colored fine aggregate can be trowelled into the surface to produce a decorative finish.

The process can be readily automated for standard sheets. The optimum fiber volume for flexural strength using the spray technique is about 6% because of increase in porosity of the composite at higher fiber volumes.

One of the problems with this type of glass-reinforced cement has been the reduction in strain to failure with time under water storage or natural weathering conditions. This effect results in the material becoming essentially brittle under these conditions but air storage in a dry atmosphere causes little change in the initial properties even after periods of ten years.

Continuing developments in glass-cement systems show promise of reducing the rate of embrittlement under natural weathering conditions.

Typical properties are tensile strengths between 7 MPa and 15 MPa and bending strengths between 13 MPa and 35 MPa.

Applications include cladding panels, permanent formwork for concrete construction, pipes, flat sheeting, sewer linings and surface coatings for dry blockwork.

5. *Steel-Fiber Concrete*

Steel fibers have been used mainly in bulk concrete applications as opposed to cement mortars.

It has been found from experience that, for traditional mixing and compaction techniques to be applicable, the maximum aggregate size should not exceed 10 mm and there should be a high content of fine material. In these circumstances 1–1.5% by volume of fibers typically 50 mm long by 0.5 mm diameter can be mixed with little modification of existing plant.

The fibers are generally deformed to give better "keying" to the matrix but, as the fibers are arranged randomly in two and three dimensions, the critical fiber volume for tensile strengthening is rarely exceeded. However, the flexural strength can be increased up to about 15 MPa and toughness is greatly increased.

Higher volumes of longer fibers can be included by spray techniques known as "gunite" or "shotcrete" and these materials are often used as tunnel linings or for rock slope stabilization.

The increased toughness is utilized in highway and airfield pavement overlays which have reduced thickness and greater impact resistance than traditional construction. The use of stainless steel has proved especially advantageous in refractory applications.

Fiber production processes include spinning from the melt which has reduced the cost of the fiber thus increasing its economic viability.

6. Organic Fibers in Cement

6.1 Man-Made Fibers

Polymer fibers used in cement matrices cover the complete range of fiber elastic moduli. The moduli range from 2 GPa to 12 GPa for polypropylene to 200 GPa to 300 GPa for carbon fibers but the basic principles of reinforcement remain the same for all the fiber types.

Polypropylene, because of its high strength combined with excellent alkali resistance and low price, has had the greatest application in cement and concrete. It is generally used as fibrillated film between 50 μm and 100 μm thick made from isotactic polypropylene with draw ratios between 5 and 20 to produce a high degree of molecular alignment. The stretched fibrillated film with strengths between 300 MPa and 500 MPa may be used in concrete as chopped twine in volumes generally less than 1% or as layers of flat opened networks in fine grained mortar at volumes between 5% and 20%.

The chopped twine is mixed with the concrete in standard mixers but as the volume of fiber is well below the critical volume for tensile strengthening the major benefits are in handling the fresh material or in impact situations such as piling or pontoons where the fibers serve to hold the cracked material together when damaged.

The manufacture of thin cement sheets containing many layers of film requires special production machinery. The sheets, which are intended as alternatives to asbestos-cement, are characterized by high toughness and flexural strength (up to 40 MPa) resulting from the closely spaced multiple cracking.

The higher elastic modulus fibers such as carbon fibers and Kevlar show considerable promise as cement reinforcements but are too expensive for bulk applications. Cheaper, high-modulus polymers, such as high-modulus polyethylene (20 GPa to 100 GPa) will fibrillate in a similar manner to polypropylene and show potential should commercial production become available.

6.2 Natural Fibers

Cellulose fibers produced from wood pulp have been used for many years as additives in asbestos-cement products but their precise physical properties seem to vary widely depending on the authority quoted and this variability may also depend on the pulping techniques used. Fiber lengths between 1 mm and 4 mm have been quoted and the helical structure of the fibers gives a variable diameter between 10 μm and 50 μm. Tensile strengths for the fiber between 50 MPa and 600 MPa have been quoted with moduli between 10 GPa and 30 GPa.

Flat and corrugated sheets containing cellulose fibers in autoclaved calcium silicate have been marketed as alternatives to asbestos-cement. The strength of these products is lower in the wet than the dry state due to the sensitivity of cellulose fibers to water absorption.

Other natural fibers such as sisal, agave, piassava, coconut and bagasse fibers have been used in cement in an attempt to produce cheap cladding sheets for houses in developing countries but although some success has been achieved, the process is labor intensive and there are many problems still to be overcome.

See also: Composite Materials: An Overview; New Cement-Based Materials

Bibliography

Building Research Establishment 1979 *Properties of GRC: 10 Year Results*, Information Paper IP 36/79. Building Research Station, Watford

Cook D J 1980 Concrete and cement composites reinforced with natural fibres. *Fibrous Concrete—Concrete International 1980*. Construction Press, London, pp. 99–114

Hannant D J 1978 *Fibre Cements and Fibre Concretes*. Wiley, Chichester

Laws V, Ali M A 1977 The tensile stress–strain curve of brittle matrices reinforced with glass-fibre. *Conf. Proc. Fibre Reinforced Materials*. Institute of Civil Engineers, London, pp. 101–9

D. J. Hannant

Fiber-Reinforced Ceramics

The development of fiber-reinforced ceramics, including glasses and glass ceramics, has been motivated by the need for materials with the advantages of ceramics combined with increased toughness. The useful properties of ceramics include retention of strength at high temperature coupled with chemical inertness, relatively low density, hardness and high electrical resistance. Their principal disadvantage is their brittleness, failure strains being low and fracture energies (G_{Ic}) lying in the range 30–300 J m^{-2}, compared with values for tough plastics and metals which are commonly in excess of 10^4 J m^{-2}. This renders them susceptible to damage by thermal or mechanical shock, easily weakened by damage in service, and causes them to have large variability in strength. Fiber reinforcement can increase toughness by several orders of magnitude to values as high as 10^4 J m^{-2} (fracture energy, G_{Ic}) and 20 MN m$^{-3/2}$ (fracture toughness, K_{Ic}), through mechanisms similar to those in reinforced plastics.

A large number of fiber and matrix systems have been developed on the laboratory scale but they have so far found little use in practical applications, although this could change in the near future. The results of existing research indicate that a composite containing 50 vol% of unidirectional, continuous, high-strength, ceramic or graphite fibers could have a strength and stiffness of ~1.5 GPa and 200 GPa, respectively, with a density of ~3 × 10^3 kg m^{-3} and a toughness greater than 10 kJ m^{-2}. Chemical compatibility of the fibers and matrix with one another and with the environment is of crucial importance if the materials are to be employed at high temperatures. Tough, high-temperature-resistant fiber-reinforced ceramics could have a role in gas turbines and other high-temperature applications where their lower density compared with metals would be an added advantage. Problems still exist in maintaining properties in the presence of oxidizing atmospheres and this is currently the main obstacle to their wider use, and where further development work is necessary. There are, however, potentially important applications where extreme temperature capability is not required. For example, fiber-reinforced glasses and glass ceramics have mechanical properties comparable with high-performance fiber-reinforced polymers, and could be used at intermediate temperatures higher than those attainable with polymeric systems. They are more stable to ionizing radiations than fiber-reinforced plastics and this could make them more attractive for nuclear and space environments. They are less susceptible to hydrothermal effects and therefore could be attractive for aircraft or marine applications. Their hardness could make them more erosion resistant than fiber-reinforced plastics and therefore attractive as radome materials, while hardness combined with a tailorable toughness could make them useful in armor. Chemical inertness could make them suitable in specialized chemical plants and as biomedical implant materials.

A variety of techniques have been used to incorporate fibers into ceramic matrices and subsequently to consolidate the ceramic. Usually a ceramic powder route has been used. The simplest incorporation method, producing short fiber composites, has been merely to mix the fibers and ceramic powder in a slurry. Some degree of alignment can be achieved by causing the slurry to flow while gelling or drying. Continuous aligned fiber preimpregnated sheet can be produced by drawing continuous fibers through a slurry of powder and organic binder. Unidirectional or angle-plied composites can then be produced from the preimpregnated sheet. Electrophoretic deposition has also been used to impregnate fibers with powder. The best properties have been obtained by hot pressing to consolidate the matrix. Other techniques which have been employed include cold pressing and sintering, reaction bonding, slip casting, plasma spraying and glass casting.

Fibrous composites have also been produced by in situ growth of a precipitated fibrous phase. Increases in toughness by this technique have been only modest because the efficiency of fiber toughening mechanisms decrease with decreasing fiber diameter, and because coherent interfaces are less efficient in deflecting or stopping cracks.

The successful combination of a ceramic and a fiber reinforcement requires that the thermal expansion coefficients of the fiber and matrix are matched, otherwise spontaneous cracking can occur during fabrication or in service. Early work on refractory-metal-wire-reinforced ceramics and glasses was disappointing partly because of this problem and partly because the matrices did not adequately protect the wire from oxidation at high temperatures. More successful composites have been obtained by combining carbon, silicon carbide and alumina fibers with glass and glass-ceramic matrices with better thermal expansion matching. The carbon-fiber composites retain their properties at high temperatures in inert atmospheres but oxidize in air above about 400 °C. Silicon carbide and alumina composites are reported to retain properties above 1000 °C even in air.

The strain to failure of a ceramic matrix is usually less than that of the reinforcing fiber and, on loading, the matrix of the composite can therefore crack at loads lower than the ultimate load. For a unidirectional composite stressed in the fiber direction this manifests itself as an array of regularly spaced cracks. To a first approximation the stress at which matrix cracking occurs is

$$(\sigma_m)_u\left[1 + V_f\left(\frac{E_f}{E_m} - 1\right)\right]$$

where $(\sigma_m)_u$ is the strength of the unreinforced matrix, V_f is the fiber volume fraction, and E_f and E_m are the fiber and matrix Young's moduli. A composite with a multiply-cracked matrix can retain useful properties, even under fatigue conditions, but the cracks lead to easier ingress of aggressive environments. Suppression of matrix cracking to higher stresses can occur and theories due to Aveston et al. (1971) and to Korczynski et al. (1981) appear to provide an explanation of crack inhibition. In practice the use of higher fiber volume fractions and, in the case of laminates, thin plies provides the best method of minimizing multiple matrix cracking.

See also: Composite Materials: An Overview

Bibliography

Aveston J, Cooper G A, Kelly A 1971 Single and multiple fracture. *Proc. Conf. Properties of Fibre Composites.* IPC, Guildford

Donald I W, McMillan P W 1976 Ceramic-matrix composites. *J. Mater. Sci.* 11: 949–72

Korczynski Y, Harris S J, Morley J G 1981 The influence of reinforcing fibers on the growth of cracks in brittle matrix composites. *J. Mater. Sci.* 16: 1533–47

Phillips D C 1981 Fibre reinforced ceramics. In: Kelly A, Mileiko S T (eds.) 1981 *Handbook of Composites*, Vol. 4, *Fabrication of Composites.* North-Holland, Amsterdam

Phillips D C 1985 Fibre reinforced ceramics. In: Davidge R W (ed.) *Survey of the Technological Requirements for High Temperature Materials R and D*, Sect. 3. Commission of the European Communities, Brussels

D. C. Phillips

Fiberboard Containers

Fiberboard containers are generally used as outer shipping containers, although smaller inner containers such as folding cartons are also made from fiberboard. This article concentrates on the two principal types of shipping containers, corrugated boxes and fiber drums, which constitute the major fraction of such containers in common use.

Fiberboard containers, both boxes and drums, began to be accepted during the first World War because of the need for additional containers and the raw material shortages existing at that time. Once their performance had been demonstrated, they continued to supplant more traditional packaging (e.g., wooden boxes) and have had a rapid growth rate over the intervening years. The quality and economics of fourdrinier kraft linerboard, with its high strength-to-weight ratio, have increased the acceptance of these packages, especially in the USA. The different types of cylinder and fourdrinier boxboard drums and cans have been reviewed by Sacharow (1974).

1. Corrugated Boxes

Corrugated boxes are widely used to consolidate and protect smaller containers (e.g., cans, boxes or other unit packages), to protect larger bulky items such as machines or furniture, or as bulk containers. Innovations in construction and treatments have allowed them to penetrate markets formerly held by other forms of packaging.

A variety of box styles may be made from the available types of board (see *Corrugated Paperboard*). The most common style is the regular slotted container formed from a single sheet of corrugated board which is slotted and scored so that all flaps are the same length and the outer flaps meet at the center of the box. Half-slotted containers with flaps on only one end may be strengthened by inserting a smaller box to provide double sidewall thickness. Alternatively, they may be inverted over an object fastened to a pallet or skid, and then attached to the pallet.

Whatever the intended use of the corrugated container, its design is partly governed by transportation regulations on package strength and restrictions on size or weight. In the USA, the National Motor Freight (highway) and Uniform Freight (rail) Classifications give the allowable packaging for listed articles as well as detailed specifications for box manufacture and closure. For shipment of hazardous materials, Title 49 of the Code of Federal Regulations is also applicable. In Europe, the United Nations Recommendations and ADR/RID regulations (i.e., European Regulations for the Safe Transport of Dangerous Goods by Road and by Rail, respectively) give similar requirements. In most cases, the larger manufacturers will select the proper container, though it is the responsibility of the shipper to make sure that all packages meet the applicable regulations in both construction certification, and the approved closing and sealing of the completed package.

2. Fiber Drums

Fiber drums are defined by the rail and motor freight classifications as cylindrical straight-sided shipping containers which traditionally are convolutely, not spirally, wound from multiple plies of high-strength kraft fiberboard. The ends of the drum, which are fastened by adhesive, stitching, or other means, may be of steel, fiberboard or plastic. No single ply of the multilayered sidewall may be less than 0.30 mm thick, and the exterior, if not otherwise protected, must at least be sized to resist the effects of moisture. The sidewalls are generally wound from fourdrinier kraft linerboard which gives excellent strength combined with low tare weight. Drums are available in a number of different diameters in sizes up to 75 US gallons (284 l) and in several different weight constructions to accommodate different packing weights for both nonhazardous and hazardous products.

The thickness and strength of the sidewalls, the materials and thickness of the bottom heads and covers, and the methods used to fasten the heads to the sidewalls are varied in accordance with the type of products and net packing weight, in order to furnish the most economical drum for the required duty. The methods of manufacture also allow for an almost infinite variation in drum height. Thus, capacities may be tailored to net packing weights regardless of product density as compared to packages such as steel drums which are only available in a few capacities (Carpenter 1981).

A variety of protective linings or barriers may be wound into the drum sidewalls and incorporated into fiber headings (Hewitt 1982). Product moisture protection can be obtained by the use of sandwich sheets containing polymeric materials or metallic foil, which are wound into the sidewalls. Impervious linings of fiberboard coated with polymers (e.g., polyethylene) or combinations of polymers and/or foil may be wound as the inner ply of the sidewalls with the vertical seam suitably sealed; lamination of the same material to the inside surface of the bottom heading provides a lined drum for liquids or extremely sensitive dry products. Liquid-holding capability may also be obtained by using a polyethylene insert (thermoformed or blow-molded, open head or tight head), with the drum acting as a protective overpack.

Since exterior protective coatings can be damaged and lose their effectiveness, fiber drums are also produced using a thermoplastic adhesive to combine the multiple plies of linerboard. The adhesive forms interlaminar protective layers which give very good weather resistance and also lower susceptibility to scuffing or scraping.

The most common fiber drum style in worldwide use is made with a convolutely wound single fiber shell having a metal chime or reinforcing band at both top and bottom of the drum. The bottom chime secures the head to the sidewall and provides juncture protection against abuse in handling. The top chime is formed with a groove so that a lever-actuated locking band may be used to secure the fully removable cover conveniently and quickly; a sealing device may also be fitted for security purposes. These drums are normally furnished with (a) steel covers having a coating resistant to both atmospheric corrosion and the packaged product, (b) fiber covers molded to fit over the top chime, or (c) molded plastic covers. The barriers, linings and treatments available, together with the tightness of the band-reinforced juncture of sidewall and heading, allow great versatility for transport and storage of both dry and liquid products.

Fiber drums with no metal parts are also available in either two- or three-piece construction. In these styles a portion of the sidewall is bent inwardly to form a flange which is stitched or glued to fiber disks to form a unit structure. The three-piece construction has an inner straight tube which projects the same distance above the basic body shell as the inner height of the cap. When the cap is applied, it joins directly against the upper edge of the lower shell and provides a smooth, stepless surface for application of sealant taping. The more economical two-piece construction has a lower shell of the full height and a cover larger in diameter so that it may be slipped over the body. This drum is usually used for lighter duty shipments since it has no reinforcing tube.

Light-duty drums primarily for dry products may be constructed with straight fiber sidewalls and formed metal ends having a deep groove to receive the sidewall. The bottoms are pinched against the fiber and are also stitched or caulked for added strength. The covers are normally secured with a clip-type closure which allows easy attachment and removal, and are usually equipped with an interlocking feature for convenient stacking.

Straight-sided drums with recessed heavy fiber ends have been developed for the protection of rolls of material such as films or linoleum. These drums utilize a collar that is stitched to the sidewall to retain the disks which provide edge protection for the rolled product. Covers with a larger diameter to slip over the sidewall tube are constructed in the same way.

A number of modifications or combinations of the above constructions are available from the various drum manufacturers for special applications or products. They can, in addition, be furnished with complete exterior labels or protective and/or decorative coatings, silk-screen printed instructions and customer identification. The light weight and shape of the drum allows facile handling by plant employees and warehousemen. Drums are suitable for automated handling on conveyors, fork-lift trucks and pallets. The natural resilience and strength of the fiberboard provides protection against handling abuse, while the inherent vertical stacking strength of the cylindrical shape allows a large number of drums to be stored empty or filled, thus using warehouse space to good advantage.

As with corrugated boxes, US regulations concerning shipment in fiber drums are to be found in the National Motor Freight Classification and Uniform Freight Classification. Products listed in these publications and shown for shipment "in barrels" or "in bulk in barrels" are also approved for shipment in fiber drums subject to their compatibility with the available linings or interior treatments. The appropriate construction requirements for dry, semiliquid and liquid products are specified, and provide the basis for certification by the drum manufacturers that their drums meet the applicable requirements. The shipment of hazardous materials in fiber drums is governed by the same regulations as for corrugated boxes. In the UK, the major strength criterion for selecting fiber drum walls is the Mullen or burst test (British Standard 1596, 1961). A puncture energy test has been proposed as an alternative (Fava 1968).

Having checked the regulations concerning the use of fiber drums for the product, the shipper and the drum manufacturer determine the optimum drum size for the packing weight, the level of protection necessary, and the special features which may be necessary or desirable. Many dry products such as starches or pigments may require little additional protection or may use a film bag in a plain drum so that the drum remains clean for reuse; more hygroscopic products such as caustic soda or dehydrated onions will need the moisture protection afforded by a foil barrier. Extremely hazardous or noxious materials such as bleaching compounds will require more sophisticated, often laminated, linings.

Semiliquid products, such as grease, filter cake and adhesives, present their own special problems in packaging and require interior linings and heavier constructions. Certain specialized drum constructions have been developed for such products as automotive grease and vegetable shortening.

The ability to wind impervious inner linings and form liquid-tight junctures of sidewalls and headings has enabled drums to be used for the shipment of such liquid products as latex, adhesives, sizing agents, liquid cleaners, and other chemical products, as well as food products (e.g., vinegar, brined fruits and vegetables, and juice concentrates). The technology of thermoforming or blow molding both open-head and tight-head polyethylene inserts has further increased liquid-holding capability to include certain regulated corrosives and similar products.

Fiber drums are applicable to the packaging of a wide range of foods and chemicals of commercial value. This versatility has recently been put to use for the low-cost disposal of waste products, particularly hazardous liquids which are prohibited from landfill disposal and must be incinerated. Fiber drums leave a minimal amount of residue after incineration and their wide range of sizes may be used to decrease thermal loads in the case of highly exothermic products.

See also: Paper and Paperboard: An Overview

Bibliography

American Trucking Associations 1983 *National Motor Freight Classification 100-J.* American Trucking Associations, Washington, DC

Carpenter H L 1981 Containers—Drums and pails, fiber. In: Simms W C (ed.) 1981 *Packaging Encyclopedia.* Cahners Publishing, Denver, Colorado, pp. 166–69

Fava R A 1968 The limitations of the Mullen burst test. *Pap. Maker (London)* 155(3): 52–57

Hewitt N S 1983 Containers—Drums, fiber. In: Simms W C (ed.) 1983 *Packaging Encyclopedia.* Cahners Publishing, Denver, Colorado, p. 172–73

Sacharow S 1974 Rigid and semirigid paperboard packages. *Package Print. Flexography* 19(7): 21–24

Uniform Classification Committee 1983 *Uniform Freight Classification 6000-B.* Uniform Classification Committee, Chicago, Illinois

US Code of Federal Regulations, Title 49: Transportation, Pts. 100–199. US Government Printing Office, Washington, DC

N. S. Hewitt; W. L. Swihart

Fibers and Textiles: An Overview

The first fibers used by man were natural fibers such as cotton, wool, silk, flax, hemp and sisal. The first man-made fiber was probably glass; the Egyptians made vessels out of coarse glass fibers. The first use of glass fibers in textiles took place in the eighteenth century. The idea that man could make an organic fiber is attributed to Hooke, who in 1664 suggested that a "glutinous substance" could be drawn into fibers. However, it was not until the latter part of the nineteenth century that Despassis was awarded a French patent for the production of rayon fibers by the cuprammonium process. Then, in 1893, a British patent was granted to Cross, Bevan and Beadle for making rayon fibers by the viscose process.

It was only at the beginning of the twentieth century, however, that man-made fibers started supplementing and replacing natural fibers. The first important man-made fibers were the cellulose-derived fibers, rayon and cellulose acetate. Then came nylon, the first truly synthetic fiber, followed by polyesters, polyacrylics and polyolefins. Along the way, the man-made elastomeric, glass and carbon fibers became important commercial products.

Man-made fibers are now available, ranging in properties from the high-elongation and low-modulus elastomeric fibers, through the medium-elongation and medium-modulus fibers such as polyamides and polyesters, to the low-elongation, high-modulus glass, carbon and aramid (aromatic polyamide) fibers. With such a wide variety of man-made fibers available, the volume of synthetic fibers consumed in the USA is now greater than that of natural fibers, and it appears that in a few years it will also be greater worldwide.

1. Classification

The major fibers used in textiles can be classified into the groups given in Fig. 1.

Although most man-made fibers are composed of one class of polymers only, some biconstituent or bicomponent fibers are manufactured; these fibers may be spun from different polymers side by side or in other configurations. One purpose of such fibers is to produce a crimp in the fibers by exploiting the differential thermal properties of the two polymers. Also, differential coloring effects may be produced in this manner.

In fabric manufacture, blends of fibers are often used to obtain the desired fabric properties. An example is a blend of polyester and cotton that is treated with a formaldehyde reactant to give a durable press fabric. The tensile strength and abrasion resistance of the polyester fibers counteract the loss of these properties in the treated cotton fibers. Another example is a blend of nylon and wool, which is often used for clothing fabrics where both warmth and wear resistance are important.

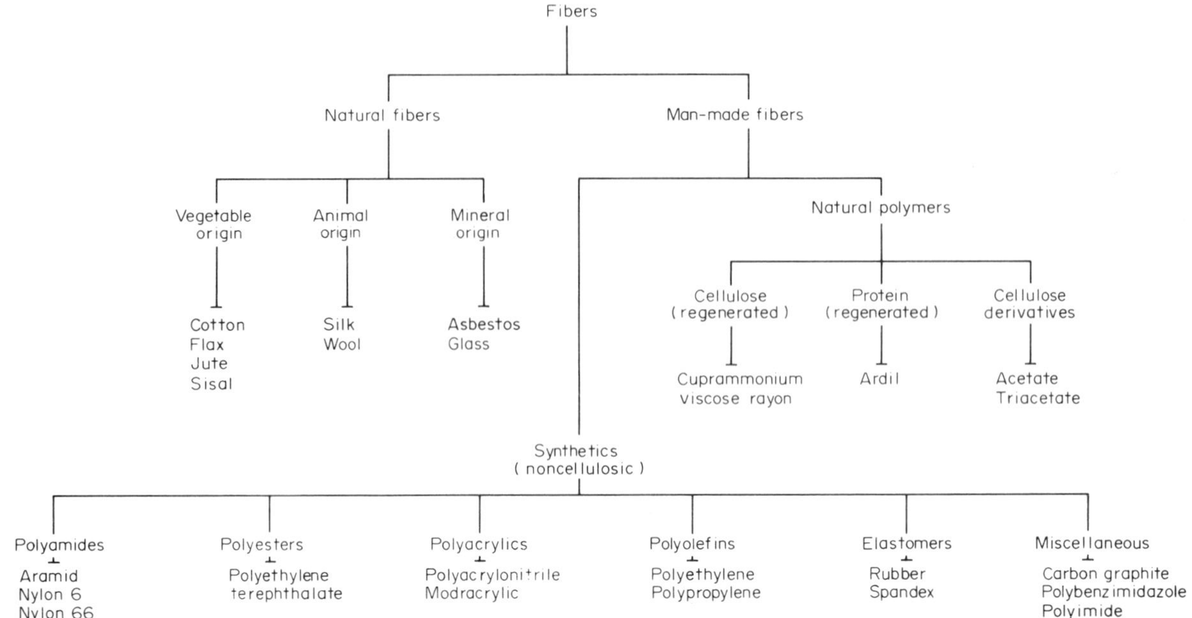

Figure 1
Classification of fibers used in textiles

2. *Physical Properties*

Some of the more important physical properties of the most extensively used fibers are given in Table 1.

Fibers used in textile structures must have sufficient length to be made into twisted yarns, but can be as short as 1.3 cm in staple fibers. The diameter of fibers with the exception of asbestos generally varies from 5 to 45 μm. However, nylon, polyester and polyacrylic fibers are now being manufactured with diameters approaching 1 μm; these fibers give fabrics that are soft and have a good drape. Even smaller diameter fibers have been produced by solvent removal of one component of a bicomponent fiber.

The cross-sectional shape of fibers varies considerably. The cross section of a cotton fiber is a collapsed circle, which is somewhat flat and irregular. Silk has a triangular cross section, which accounts for its sheen and feel. Wool is circular in shape with scales on the surface. Rayon can be circular or serrated-circular, and it often has a center or core which differs from the outer layer. Man-made fibers, particularly melt-spun fibers, can also be made oval, multilobal and even hollow-circular. The shape of the fiber cross section affects the optical properties of the fiber, including luster and translucency. It also affects the feel or hand of fabric made from the fiber.

The surface of fibers also varies considerably. Cotton has a surface that has many irregularities or rugosities. Wool has a scalar structure, which has a tendency to make a wool fabric felt or shrink during washing. Most man-made fibers have relatively smooth surfaces; however, rayon may be serrated.

Some fibers, such as wool, have a natural undulating physical structure or crimp, whereas man-made fibers are frequently subjected to various mechanical and heat-setting processes to provide crimp. These processes include false twist, stuffer box, gear crimping, knife-edge, antitwisting, knit-de-knit and air-jet crimping. Most of these processes are self-descriptive, and all involve heating the fiber or yarn while it is being held in a crimped configuration. Fabrics made from crimped fibers tend to have increased stretch, bulk and warmth.

The density of fibers (see Table 1) varies from about 2.5 $\mathrm{g\,cm^{-3}}$ for glass to 0.9 $\mathrm{g\,cm^{-3}}$ for polypropylene, with polyester being about 1.4 $\mathrm{g\,cm^{-3}}$ and cotton about 1.5 $\mathrm{g\,cm^{-3}}$. The low density of polypropylene makes it desirable for marine ropes, since they will float in water.

Fibers must have sufficient strength to be useful in textile structures. The strength of textile fibers is usually referred to as tenacity, which is defined as the force required to rupture or break the fiber per unit linear density. Tenacity is generally measured in grams-force per denier (where denier is the weight in grams of 9 km of fiber or yarn), or as grams-force or newtons per tex (where tex is the weight in grams per kilometer of fiber or yarn). The tenacity of fibers varies from slightly less than 6–8 $\mathrm{gf\,tex^{-1}}$ for the

Table 1
Physical properties of textile fibers (after Goswami et al. 1977)

Fiber type	Diameter (μm)	Density (g cm^{-3})	Tenacity (gf tex^{-1})	Breaking extension (%)	Initial modulus (gf tex^{-1})	Work of rupture (g tex^{-1})	Moisture regain[a] (%)	Melting point (°C)
Natural								
cotton	11–22	1.52	35	7	500	1.3	7	
flax	5–40	1.52	55	3	1840	0.8	7	
jute	8–30	1.52	50	2	1750	0.5	12	decomposes (cotton to silk)
sisal	8–40	1.52	40	2	2500	0.5	8	
wool	18–44	1.31	12	40	250	3	14	
silk	10–15	1.34	40	23	750	6	10	
glass	⩾5	2.54	76	2–5	3000	1	0	800
asbestos	0.01–0.30	2.5			1300		1	1500
Regenerated								
viscose rayon	⩾12	1.45–1.54	20	20	500	3	13	decomposes
acetate	⩾15	1.32	13	24	350	2	6	230
triacetate	⩾15	1.32	12	30	300	2	4	230
Synthetic								
Nylon 6	⩾14	1.14	32–65	30–55	250	6–7	2.8–5	225
Nylon 66	⩾14	1.14	32–65	16–66	250	6–7	2.8–5	250
Qiana	⩾10	1.03	25	26–36			2.5	274
Nomex	⩾12	1.38	36–50	22–32		7.5	6.5	decomposes above 380 °C
Kevlar	⩾12	1.44	200	2–4	4500–8000		4	decomposes above 500 °C
polyester	⩾12	1.34–1.38	25–54	12–55	1000	2–9	0.4	250
acrylic	⩾12	1.16–1.17	18–30	20–50	650	5	1.5	sticks at 235 °C
polypropylene		0.91	60	20	800	8	0.1	165
polyethylene		0.95	30–60	10–45		3	0	115
spandex		1.21	6–8	444–555		18	1.3	230

a At 65% relative humidity

elastomeric spandex fibers to 76 gf tex^{-1} and even higher for some glass fibers, with nylons and polyesters generally being 25–65 gf tex^{-1}.

In addition to having sufficient strength for its various end uses, a fiber must also have sufficient extensibility to withstand the stresses to which it will be subjected in service. The breaking extension or elongation-at-break varies from 2% for some jute and sisal fibers to over 500% for some spandex fibers, with nylons and polyesters generally 12–66%. For many end uses an elongation of only 2% is not sufficient, so the use of fibers with elongations of this order is limited.

Another fiber property that is important is recovery from deformation. This property of the fiber is one that is also imparted to yarns and to fabric made from them; for example, the ability of a fabric to recover from wrinkling depends to a large extent on the recovery from deformation of the individual fibers. Recovery can be measured by imposing deformation in a cyclic manner and determining the elongation recovered. The initial modulus or slope of the first part of the stress–strain curve can also be used as a measure of this property. Generally, fibers with low elongation (e.g., glass fibers) have high modulus values, and those with high elongation such as spandex fibers have low moduli. Nylons and polyester fibers generally have intermediate elongation and modulus values. The area under a fiber stress–strain curve represents the toughness of the fiber and, if measured to rupture, is called the work of rupture. Values of work of rupture are generally high for spandex, medium to high for nylons and polyesters, and low for glass fibers.

Other important properties of fibers include abrasion resistance, water absorption, dimensional stability, thermal characteristics, flammability, density, and resistance to attack (e.g., from chemicals, sunlight and mildew). Generally, fibers with a low work-of-rupture value, such as glass fibers, have poor abrasive resistance, whereas fibers such as nylon and polyester have good abrasion resistance. Water absorption is an important factor for comfort in clothes. Fibers with high water absorption such as cotton and rayon are usually considered more comfortable than polyamide, polyester, or polyolefin fibers, which have low water absorption. Since it is important that fabrics neither shrink nor stretch excessively in an irreversible manner, man-made fibers are often heat-set. This gives the fibers dimensional stability provided that they are not exposed to a temperature in excess of that of the heat-setting. Some cellulosic fabrics shrink during washing if they have been stretched during manufacture. Finishing of cellulosic fabrics with a formaldehyde reactant may impart dimensional stability as well as recovery from wrinkling.

While some fibers such as the cellulosics do not melt but will burn, others (e.g., nylons, polyesters and polyolefins) melt at relatively low temperatures, but burn with difficulty because the molten polymer falls away from the flame.

3. *Morphology*

The morphology, or spatial arrangement of the individual polymer molecules and groups of molecules, has a bearing on fiber properties. The naturally occurring fibers already have a built-in morphology, which can sometimes be altered. For example, cotton can be mercerized by treatment with caustic to swell the fiber and give it higher luster, and it can also be cross-linked by various formaldehyde reactants to enhance its recovery from wrinkling. Similarly, cellulose can be chemically modified and spun into rayon or acetate fibers, and wool can be chemically treated to decrease its felting properties.

4. *Production*

A large number of organic man-made fibers with a variety of properties have been produced from polymers by various spinning techniques, including melt, dry, wet and emulsion spinning. Polyamide, polyester and polyolefin fibers are formed by forcing the melted polymer through an orifice into air. Polyacrylonitrile fibers are spun from a solvent solution of polyacrylonitrile by either dry spinning (into air), or wet spinning (into a coagulating bath). Spandex fibers can be formed from polyurethanes by either dry or wet spinning.

Generally, after being spun, the fibers are mechanically stretched or drawn. In this process, the molecules are oriented in a direction parallel to the axis of the fiber. The resulting aligned molecules may become closely packed, so that crystallization can occur in some areas. This crystallization stabilizes the fiber and imparts improved properties of strength and recovery from deformation. However, some areas are left relatively unaligned, and in these amorphous areas the molecules are loosely held together by secondary forces including hydrogen bonding, van der Waals forces, and dipole–dipole interactions. These permit more flexibility of movement than in the crystalline regions. The amorphous regions also allow the penetration of dyes and finishes into the fiber.

Fibers are usually processed into yarns, and the yarns are woven or knitted into fabrics. Filament fibers can be made into yarns with little or no twisting, but staple fibers, being short, must be twisted together to form long continuous yarns. Often sizes, which are usually water-soluble polymers, are applied to the yarns to facilitate weaving. Fibers can also be randomly arranged in so-called nonwoven textile structures, without first being made into yarns. Nonwovens can be formed by spun bonding, needle-punching or resin bonding. In spun bonding, bonds between the fibers are formed by heating the fibers to near their melting point.

5. Dyes and Dyeing

Fibers, yarns and fabrics may be colored with dyes (a) containing anionic groups (acid, direct and mordant dyes), (b) containing cationic groups (basic dyes), or (c) requiring chemical reaction during dyeing (azo, reactive, sulfur and vat dyes). Special colorants such as disperse dyes and pigments may also be used. All of the dye molecules have a chromophore group, which is responsible for the color, and usually a functional group which increases the affinity of the color for the fiber.

Acid dyes are water-soluble anionic dyes that are applied to textiles from acid solutions. Fibers which develop a positive charge in acid solutions (e.g., wool, silk and nylon) can be dyed with acid dyes. Direct dyes frequently contain azo groups connecting aromatic chromophores and are mostly used on cellulosic fabrics. Mordant dyes are acid dyes that are chelated with metal ions such as aluminum, chromium, cobalt and copper, and are used chiefly for wool and silk. These dyes are normally applied to a fiber and then mordanted.

Basic dyes are dyes which yield colored cations in aqueous solutions. They can be applied in neutral or mildly acidic dye baths to acrylic, cellulosic, nylon and protein fibers. These dyes may be treated with complexing agents such as tannin to improve their fastness to washing.

Azo dyes are formed in the fiber by interaction between two separate dye components. They are chiefly applied to cotton, but can be used with man-made fibers. Reactive dyes contain groups that can react covalently with a chemical group in the fiber to make the color durable to washing and dry cleaning. They are chiefly applied to cellulosic fibers, with the reaction occurring with the hydroxyl group of the cellulose, but are also used with nylon and protein fibers. Sulfur dyes are products formed by the reaction of aromatic compounds with sodium polysulfide. They are reduced under alkaline conditions before application to a fiber, and then oxidized after the fiber is dyed. They are chiefly applied to cellulosic fibers. Vat dyes are generally water-insoluble. They are reduced under alkaline conditions to form a water-soluble, colorless leuco form, which is applied to the fiber, and then oxidized to an insoluble colored state within the fiber. Vat dyes are chiefly used with cellulosic fibers, but can also be applied to acrylic and nylon fibers.

Disperse dyes are nonpolar, relatively insoluble molecules, usually containing azo or anthraquinone groups, and are applied to the fiber from an aqueous dispersion in water. They were developed originally to dye acetate fibers but are now also used in dyeing other man-made fibers. Pigments are insoluble colors that are normally added to the polymer from which the fiber is spun, or bonded to the fiber surface with a resin binder.

6. Printing

Fabrics may be printed with a design using any of the classes of colorants already described. The colors are dispersed in a thickened printing paste and applied to the fabric in a design. The printed fabric is then treated in some fashion such as by heating or an oxidation process to fix the colors in the fabric. With pigments, a resin binder is used to bind the pigment to the fabric.

7. Finishing

Fabrics are finished in a great variety of ways; the finishing may be to improve either the esthetic or performance properties, or both (see *Finishing of Fabrics*).

7.1 Removal of Impurities

In processing fibers to fabrics, sizes are often applied to the warp yarns. These sizes may be removed in a final washing step. With natural fibers, impurities such as waxes, grease and vegetable matter may remain in the fabric and are removed in a final scouring and bleaching process.

7.2 Changing the Hand or Feel

In the case of woollen fabrics, a felting or compacting operation may be the finishing step. Singeing or passing a fabric over a gas flame or a heated metal plate may be used to burn off the fiber ends of a fabric made from staple yarns. If a fabric is limp, it may be stiffened by adding a nonpermanent starch or other sizing, or a permanent size may be obtained by applying a resin finish.

7.3 Changing the Appearance or Luster

Fabrics may be whitened by bleaching and/or by the application of a fluorescent whitener. A glossy fabric surface can be produced by calendering or passing it through rollers. If a friction calender is used, a glazed finish is obtained. With a Schreiner calender, a soft luster and hand are produced. Embossed designs can be produced with an embossing calender. With the addition of resins to the fabric, a durable calendered finish is produced. Napped fabrics are made by brushing the fibers on the surface of the fabric, and napped or pile fabrics may be sheared to make the nap uniform in thickness.

7.4 Control of Biological Degradation

Cellulosic fabrics are especially susceptible to damage by mildew and bacteria. Various antimildew and antibacterial chemicals including chlorinated phenols and organometallic compounds are applied to these fabrics to retard biological degradation. In addition, treatment of cellulosic fabrics with formaldehyde reactants such as melamine–formaldehyde will also protect the fabrics from biological degradation. Wool is susceptible to damage from moth larvae unless properly treated.

7.5 Durable Press

Since cellulosic fibers do not recover well from deformation, formaldehyde reactants are used to treat these fabrics to improve this property and consequently the recovery of the fabric from wrinkling. If sufficient formaldehyde reactant is applied, the resulting fabric will have both wet and dry recovery from wrinkling, and consequently will have durable press properties.

The formaldehyde reactants used to produce a durable press have at least two functional groups capable of reacting with the hydroxyl groups of cellulose. Such reactants are normally applied in an aqueous solution with a metal salt catalyst by immersion of the fabric in a pad bath, followed by passage through a pad roll. The treated fabric is dried and then cured in an oven. The most frequently used formaldehyde reactant is dimethyloldihydroxyethyleneurea (DMDHEU). This agent can react with the hydroxyl groups of cellulose and cross-link adjacent cellulose chains. Other formaldehyde reactants that are used include the dimethylolcarbamates and dimethylolethyleneurea. The methyl, ethylene glycol or diethylene glycol ethers of DMDHEU are often used because the release of formaldehyde from the etherified products is less than from DMDHEU. Since the formaldehyde reactant tends to lower the tensile strength and abrasion resistance of cellulosic fibers, blends with polyester are commonly used to improve both of these properties. Also, ammonia treatment of the cotton in durable press fabric, either before, or after treatment, is now being practised. With the ammonia treatment, less formaldehyde reactant is required to obtain durable press properties.

In the manufacture of garments from durable press fabrics, the garment is often fabricated before the curing operation is carried out, and then cured with the creases in place. This operation is referred to as postcuring.

7.6 Shrinkage Control

The tendency of some fabrics, especially wool and cellulosic, to shrink during washing can be minimized by finishing treatments. The shrinkage of wool fabrics caused by felting can be reduced by treatment with chlorine, or by resin treatment, or by a combination of the two. The shrinkage of cellulosic fabrics during washing can be reduced by compressive shrinkage of the fabric during finishing; an example of this kind of treatment is Sanforizing. Another method of controlling shrinkage is with formaldehyde reactants such as those used for durable press. In fact, durable press treatments will reduce the shrinkage of cellulosic fabrics during washing.

7.7 Water and/or Oil Repellency

There are a number of chemical compounds that will impart water and/or oil repellency to fabrics. Durable water repellency can be obtained with silicone polymers and with long hydrocarbon chain pyridinium chloride derivatives. Both water and oil repellency can be obtained if a fabric is treated with certain polymeric fluorocarbon containing materials. With these compounds, it is important that the fluorine groups be present at the end of the polymer side chains to impart a low-energy surface to the fabric. The most effective compounds have a terminal trifluoromethyl group in the polymer.

7.8 Soil Release

Fabrics with a certain degree of soil release can be obtained by the application of hydrophilic polymers (e.g., polymers or copolymers of acrylic or methacrylic acid, or hybrid fluorocarbon polymers). These finishes can be applied along with the formaldehyde reactants to give durable press fabrics with soil-release properties.

7.9 Fire Retardancy

Most textile fibers with the exception of glass and asbestos are flammable to some degree. Some of the aramids are thermally stable. Some fibers such as wool and some modacrylics are inherently flame retardant. Others such as nylon, polyester and polyolefins will melt and the molten beads will drop away from the flame. Cellulosic fibers will burn readily unless chemically treated (e.g., with organic compounds containing halogen and/or phosphorus such as tetrakis(hydroxymethyl) phosphonium chloride). The flammability of man-made fibers can be reduced by the addition of flame retardants to the polymer before spinning, although only limited commercial success has been obtained with these fibers. The flame retardant can also be applied to the fabric, as was done in the unfortunate treatment of childrens' sleepwear with tris(2,3-dibromopropyl) phosphate, which was found to be a potential carcinogen.

7.10 Antistatic Properties

The build-up of static electricity on fibers, with resultant clinging and/or shock, is mainly a problem with hydrophobic man-made fibers. This problem can be minimized by the addition of antistatic agents, which discharge the electrostatic charges that tend to build up on fiber surfaces. These agents are surface active materials, and include ethylene oxide derivatives and polyamines.

See also: Acetate and Triacetate Fibers; Elastomeric Fibers; Glass Fibers; Polyacrylic Fibers; Polyamide Fibers; Polyester Fibers; Polyolefin Fibers; Rayon Fibers; Textiles: Structures and Processes

Bibliography

Goswami B C, Martindale J G, Scardino F L 1977 *Textile Yarns: Technology, Structure and Applications*. Wiley-Interscience, New York

Hearle J W S, Grosberg P, Backer S 1969 *Structural Mechanics of Fibers, Yarns and Fabrics*. Wiley-Interscience, New York

Lewin M, Sello S B 1983 *Chemical Processing of Fibers and Fabrics*, Vols. 1 and 2. Marcel Dekker, New York

Mark H F, Atlas S M, Cernia E 1967 *Man-Made Fibers: Science and Technology*. Wiley-Interscience, New York

Mark H F, Wooding N S, Atlas S H 1971 *Chemical Aftertreatment of Textiles*. Wiley-Interscience, New York

Needles H L 1981 *Handbook of Textile Fibers, Dyes and Finishes*. Garland, New York

Tortora P G 1978 *Understanding Textiles*. Macmillan, New York.

T. F. Cooke

Fibrous Composites: Thermomechanical Properties

The development of fibrous composites is one of the great advances in materials. This article is concerned with the thermomechanical properties of one of the most important types, unidirectional composites, in which the reinforcing phase is an array of straight, parallel, continuous fibers.

Although all materials are heterogeneous when examined on a sufficiently small scale, they are invariably treated as homogeneous continua when used as practical engineering materials. This is true of unidirectional composites, which are anisotropic on the macroscopic level. A body of analyses, often called "micromechanics", has been developed relating effective properties of composites to those of their constituents, fibers and matrices. The analyses for elastic, viscoelastic, plastic and thermal properties of unidirectional composites are considered here, along with macroscopic strength theories.

1. Effective Properties

The practice is to treat composites as if they were homogeneous, anisotropic materials. The properties used to represent the materials are commonly referred to as effective properties. The use of effective properties implies that fibers are distributed throughout the material either in a regular array or in a statistically random manner, producing statistical "uniformity." In the latter case, the material is said to be statistically homogeneous.

The concept of effective properties involves an averaging of properties over some region of material. To justify this averaging approach, the dimensions of the region, usually called a representative volume or representative volume element, must be large with respect to the characteristic dimensions of material heterogeneity, such as fiber diameter and fiber spacing. It is further assumed that when effective properties are used the dimensions of the body are large with respect to those of the representative volume element.

2. Elastic Properties

The body of analyses dealing with the elastic properties of composites is far more developed than any other aspect of micromechanics. Numerous approaches have been proposed, and several are considered in this section. Because of the complexity of the subject, discussion focuses on composites with isotropic fibers and matrices. However, several important types of fibers, such as carbon and aramid, are strongly anisotropic and are usually considered to be transversely isotropic. The expressions for elastic properties of composites having isotropic fibers can, with care, easily be converted to include these important reinforcements (Hashin 1979).

This article considers unidirectional composites in which each of the phases in the material is bounded by parallel cylindrical surfaces, so that each cross section perpendicular to the fiber direction is identical. Figure 1 illustrates this geometry. The 1 axis is taken to be parallel to the cylindrical surfaces of the phases that make up the material. This is commonly called the fiber, axial or longitudinal direction of the material.

Figure 2 shows some cross-sectional geometries

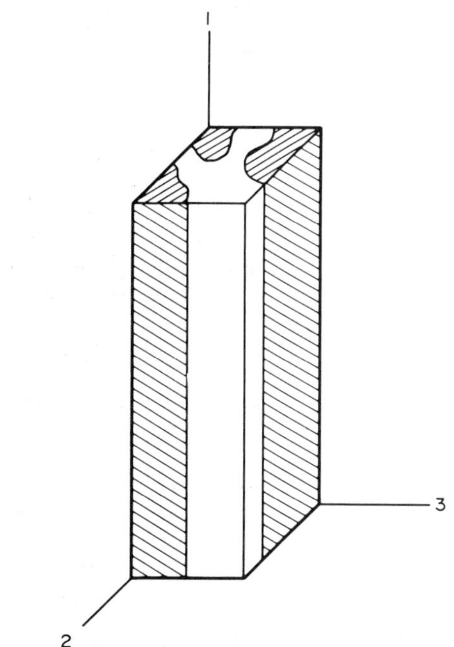

Figure 1
Geometry of a unidirectional composite

which have been examined. In the most general case, the shapes of the boundaries separating the two constituent materials or phases of the composite are completely arbitrary; this is commonly referred to as arbitrary phase geometry (Fig. 2a). Figure 2b represents a special case of arbitrary phase geometry in which fibers of equal diameter are randomly distributed throughout the composite. Figures 2c and 2d show circular fibers arranged in regular and hexagonal arrays, respectively. The square array is a special case of the rectangular array. Although no real process produces truly random fiber dispersion, case 2b is generally considered to be a reasonable representation of many practical materials of interest.

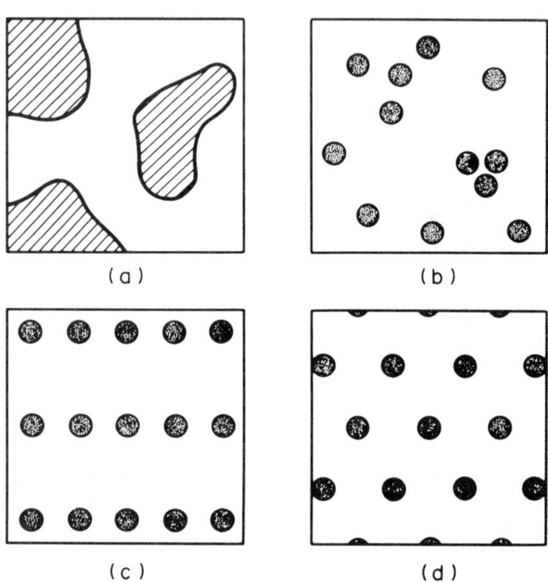

Figure 2
Cross-sectional phase geometries of composites: (a) arbitrary phase geometry; (b) randomly distributed fibers; (c) rectangular fiber array; (d) hexagonal fiber array

The rectangular fiber array produces a macroscopically orthotropic material which has nine independent elastic constants. When the material geometry is random, or when fibers of equal diameter are arranged in a hexagonal array, the composite is transversely isotropic and has five independent elastic constants. The latter class of materials is the focus of this article.

2.1 *Effective Elastic Constants*

The effective elastic constants of a composite material are commonly defined as the quantities that relate the average stress $\bar{\boldsymbol{\sigma}}_{ij}$ to the average strain $\bar{\boldsymbol{\varepsilon}}_{ij}$ in a representative volume element V.

Mathematically, average stress and strain tensor components are defined as

$$\bar{\boldsymbol{\sigma}}_{ij} = (1/V)\int_V \boldsymbol{\sigma}_{ij}\, dV \tag{1}$$

$$\bar{\boldsymbol{\varepsilon}}_{ij} = (1/V)\int_V \boldsymbol{\varepsilon}_{ij}\, dV \tag{2}$$

The effective elastic moduli, or effective stiffnesses, are defined by the relation

$$\bar{\boldsymbol{\sigma}}_{ij} = \hat{\boldsymbol{C}}_{ijkl}\bar{\boldsymbol{\varepsilon}}_{kl} \tag{3}$$

where repeated indices indicate summation.

For simplicity, compact notation is used, in which 11 is replaced by 1, 22 by 2, 33 by 3, 23 by 4, 31 by 5 and 12 by 6.

For a transversely isotropic body, which has five independent elastic constants, Eqn. (3) can be expressed in the following simplified form:

$$\begin{aligned}
\bar{\sigma}_1 &= \hat{C}_{11}\bar{\varepsilon}_1 + \hat{C}_{12}\bar{\varepsilon}_2 + \hat{C}_{12}\bar{\varepsilon}_3\\
\bar{\sigma}_2 &= \hat{C}_{12}\bar{\varepsilon}_1 + \hat{C}_{22}\bar{\varepsilon}_2 + \hat{C}_{23}\bar{\varepsilon}_3\\
\bar{\sigma}_3 &= \hat{C}_{12}\bar{\varepsilon}_1 + \hat{C}_{23}\bar{\varepsilon}_2 + \hat{C}_{22}\bar{\varepsilon}_3\\
\bar{\sigma}_4 &= (\hat{C}_{22} - \hat{C}_{23})\bar{\varepsilon}_4\\
\bar{\sigma}_5 &= 2\hat{C}_{66}\bar{\varepsilon}_5\\
\bar{\sigma}_6 &= 2\hat{C}_{66}\bar{\varepsilon}_6
\end{aligned} \tag{4}$$

The effective elastic constants $\hat{C}_{11}$, $\hat{C}_{12}$, $\hat{C}_{22}$, $\hat{C}_{23}$ and $\hat{C}_{66}$ are related to the more familiar engineering elastic moduli by the following expressions:

$$\begin{aligned}
\hat{E}_a &= \hat{E}_{11} = \hat{C}_{11} - (2\hat{C}_{12}^2/\hat{C}_{22} + \hat{C}_{23})\\
\hat{E}_t &= \hat{E}_{22} = \hat{E}_{33}\\
&= \hat{C}_{22} + \{[\hat{C}_{12}^2(\hat{C}_{23} - \hat{C}_{22})\\
&\quad + \hat{C}_{23}(\hat{C}_{12}^2 - \hat{C}_{11} - \hat{C}_{23})]/(\hat{C}_{11}\hat{C}_{22} - \hat{C}_{12}^2)\}\\
\hat{G}_a &= \hat{G}_{12} = \hat{G}_{13} = \hat{C}_{66}\\
\hat{G}_t &= \hat{G}_{23} = \tfrac{1}{2}(\hat{C}_{22} - \hat{C}_{23})\\
\hat{\nu}_a &= \hat{\boldsymbol{\nu}}_{12} = \hat{\boldsymbol{\nu}}_{13} = \hat{C}_{12}/(\hat{C}_{22} + \hat{C}_{23})\\
\hat{\nu}_t &= \hat{\boldsymbol{\nu}}_{23} = (\hat{C}_{11}\hat{C}_{23} - \hat{C}_{12}^2)/(\hat{C}_{11}\hat{C}_{22} - \hat{C}_{12}^2)
\end{aligned} \tag{5}$$

$\hat{E}_a$ and $\hat{E}_t$ are the effective extensional (Young's) moduli in the axial and transverse directions, respectively; $\hat{G}_a$ is the effective shear modulus in planes parallel to the fiber axis; $\hat{G}_t$ is the effective shear modulus in the plane transverse to the fiber direction; and $\hat{\nu}_a$ and $\hat{\nu}_t$ are the Poisson's ratios in plane parallel to and perpendicular to the fiber direction, respectively. The convention adopted here is that $\hat{\nu}_{12}$ is equal to the magnitude of the transverse strain divided by the axial strain when a stress is applied in the axial direction. $\hat{\nu}_{23}$ is defined in a similar manner.

Another important elastic constant is the plane-strain bulk modulus $\hat{k}$, which is related to the $\hat{\boldsymbol{C}}_{ij}$ by the expression

$$\hat{k} = \tfrac{1}{2}(\hat{C}_{22} + \hat{C}_{23}) \tag{6}$$

The modulus $\hat{k}$ is one of the basic elastic constants evaluated using micromechanical analyses. The following expressions are useful for evaluating other engineering constants:

$$\hat{E}_t = 2(1+\hat{\nu}_t)\hat{G}_t$$
$$\hat{\nu}_t = (\hat{k} - q\hat{G}_t)/(\hat{k} + q\hat{G}_t) \quad (7)$$

where

$$q = 1 + (4\hat{k}\hat{\nu}_a^2/\hat{E}_a) \quad (8)$$

The basic difficulties in developing analytical expressions for elastic constants are the complex and, in practice, undefined composite internal geometry. The problem has been approached in a number of ways, as discussed by Ashton et al. (1969), Hashin (1972), Sendeckyj (1974) and Christensen (1979). These include: finite-difference, finite-element and series solutions for regular arrays; "self-consistent" models consisting of a fiber (or a fiber surrounded by a concentric cylinder of matrix material) in a homogeneous material with the effective properties of the composite; semiempirical equations; rigorous bounding methods; and special models. Three widely recognized approaches are considered here: bounding methods, a special model called the composite cylinder assemblage, and the semiempirical Halpin–Tsai equations.

2.2 *Bounds for Effective Elastic Constants for Arbitrary Phase Geometry*

Upper and lower bounds on the elastic constants of transversely isotropic unidirectional composites with arbitrary internal geometry have been determined (Hill 1964a, Hashin 1965). These expressions involve only the elastic constants of the two phases and the fiber volume fraction c. (The matrix volume fraction is $1-c$.)

The following symbols and conventions are used in expressions for mechanical properties: plus and minus signs denote upper and lower bounds, respectively; subscripts f and m indicate fiber and matrix, respectively.

Upper and lower bounds on composite axial Young's modulus, axial Poisson's ratio, axial shear modulus, transverse plane strain bulk modulus and transverse shear modulus are given by the following expressions:

$$\hat{E}_a(+) = cE_f + (1-c)E_m + \frac{4c(1-c)(\nu_f-\nu_m)^2}{c/k_m + (1-c)/k_f + 1/G_f} \quad (9a)$$

$$\hat{E}_a(-) = cE_f + (1-c)E_m + \frac{4c(1-c)(\nu_f-\nu_m)^2}{c/k_m + (1-c)/k_f + 1/G_m} \quad (9b)$$

$$\hat{\nu}_a(+) = c\nu_f + (1-c)\nu_m + \frac{c(1-c)(\nu_f-\nu_m)(1/k_m - 1/k_f)}{c/k_m + (1-c)/k_f + 1/G_f} \quad (10a)$$

$$\hat{\nu}_a(-) = c\nu_f + (1-c)\nu_m + \frac{c(1-c)(\nu_f-\nu_m)(1/k_m - 1/k_f)}{c/k_m + (1-c)/k_f + 1/G_m} \quad (10b)$$

$$\hat{G}_a(+) = G_f + \frac{1-c}{1/(G_m - G_f) + c/2G_f} \quad (11a)$$

$$\hat{G}_a(-) = G_m + \frac{c}{1/(G_f - G_m) + (1-c)/2G_m} \quad (11b)$$

$$\hat{k}(+) = k_f + \frac{1-c}{1/(k_m - k_f) + c/(k_f + G_f)} \quad (12a)$$

$$\hat{k}(-) = k_m + \frac{c}{1/(k_f - k_m) + (1-c)/(k_m + G_m)} \quad (12b)$$

$$\hat{G}_t(+) = G_f + \frac{1-c}{1/(G_m - G_f) + c(k_f + 2G_f)/2G_f(k_f + G_f)} \quad (13a)$$

$$\hat{G}_t(-) = G_m + \frac{c}{1/(G_f - G_m) + (1-c)(k_m + 2G_m)/2G_m(k_m + G_m)} \quad (13b)$$

These relations hold when $k_f > k_m$ and $G_f > G_m$.

It has been shown that, with the exception of the expressions for transverse shear modulus, these bounds are the best possible when only phase elastic constants and volume fractions are specified (Hashin 1972). To improve on these bounds, more information on internal geometry must be specified. It is not known whether the bounds on $\hat{G}_t$ are the best possible.

For most materials, the bounds on axial Young's modulus tend to be reasonably close together, and the "Rule of Mixtures" predictions are generally accurate enough for practical purposes; that is,

$$\hat{E}_a \cong cE_f + (1-c)E_m \quad (14)$$

This is also true, to a lesser extent, for the axial Poisson's ratio $\hat{\nu}_a$; that is,

$$\hat{\nu}_a \cong c\nu_f + (1-c)\nu_m \quad (15)$$

The Rule of Mixtures generally provides a poor estimate of $\hat{k}_t$, $\hat{G}_a$ and $\hat{G}_t$.

2.3 *Composite Cylinder Assemblage*

If all of the bounds on elastic moduli discussed above were reasonably close for all material combinations, the subject of elastic properties would be closed. However, this is not the case. To improve on the bounds, it is necessary to specify internal geometry.

A variety of models have been proposed (Christensen 1979). One particular model provides closed-form solutions for four of the five elastic constants of a macroscopically transversely isotropic material. This is the composite cylinder assemblage (CCA) (Hashin 1972).

The CCA model consists of an infinite number of concentric circular cylinders of different diameters that are packed together so that they completely fill the volume of the composite (Fig. 3). The inner material of each cylinder has the elastic properties of the fiber, and the outer material those of the matrix. The inner and outer diameters of composite cylinder n are denoted a_n and b_n, respectively. The ratio a_n/b_n is the same for each cylinder and is equal to $c^{1/2}$, where c is the fiber volume fraction; that is, the fiber volume fraction of each cylinder, $(a_n/b_n)^2$, is equal to the overall average fiber volume fraction of the composite.

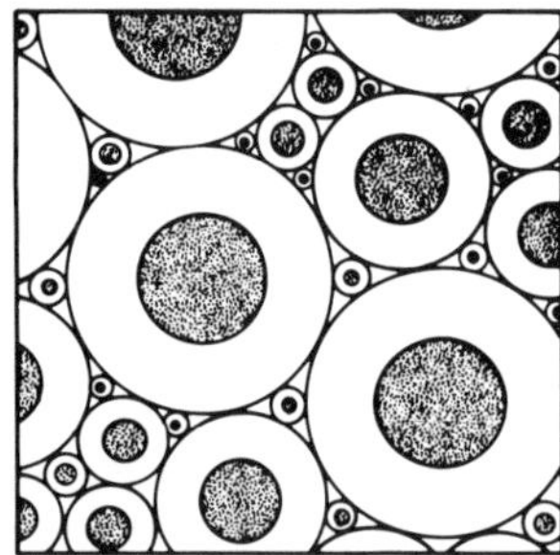

Figure 3
The composite-cylinder-assemblage model

The range of diameters of most real fibers is relatively small, so that the CCA model does not reflect the actual internal geometry found in practice. However, it is argued that it does reflect a certain randomness which is characteristic of composites.

The model provides closed-form solutions for $\hat{E}_a$, $\hat{\nu}_a$, $\hat{G}_a$ and $\hat{k}$, and bounds on the transverse shear modulus $\hat{G}_t$. However, the lower bound on $\hat{G}_t$ obtained for arbitrary phase geometry, Eqn. (1b), is better than that resulting from the CCA model. Since the arbitrary-phase-geometry model is perfectly general, and includes the CCA as a special case, its bounds also apply to the CCA. Therefore, Eqn. (1b) should be used for the lower bound on $\hat{G}_t$ and only the upper CCA bound $\hat{G}_t(+)$ is presented.

The expressions for elastic constants obtained from the composite-cylinder-assemblage model are:

$$\hat{E}_a = cE_f + (1-c)E_m + \frac{4c(1-c)(\nu_f - \nu_m)}{c/k_m + (1-c)/k_f + 1/G_m} \tag{16}$$

$$\hat{\nu}_a = c\nu_f + (1-c)\nu_m + \frac{c(1-c)(\nu_f - \nu_m)(1/k_m - 1/k_f)}{c/k_m + (1-c)/k_f + 1/G_m} \tag{17}$$

$$\hat{G}_a = G_m \left[\frac{G_m(1-c) + G_f(1+c)}{G_m(1+c) + G_f(1-c)}\right] \tag{18}$$

$$\hat{k} = \frac{k_m(k_f + G_m)(1-c) + ck_f(k_m + G_m)}{(1-c)(k_f + G_m) + c(k_m + G_m)} \tag{19}$$

$$\hat{G}_t(+) = G_m \left[\frac{(1 + c^3\alpha_1)(\rho_1 + c\beta_1) - 3c(1-c)^2\beta_1^2}{(1 + c^3\alpha_1)(\rho_1 - c) - 3c(1-c)^2\beta_1^2}\right] \tag{20}$$

where $\alpha_1 = (\beta_1 - \gamma_1\beta_2)/(1 + \gamma_1\beta_2)$, $\rho_1 = (\gamma_1 + \beta_1)/(\gamma_1 - 1)$, $\gamma_1 = G_f/G_m$, $\beta_1 = 1/(3 - 4\nu_m)$, and $\beta_2 = 1/(3 - 4\nu_f)$.

2.4 Halpin–Tsai Equations

The Halpin–Tsai equations represent a semiempirical approach to the problem of the significant separation between upper and lower bounds observed for some properties when the fiber and matrix elastic constants differ significantly (Ashton et al. 1969). The equations employ the Rule of Mixtures approximation for axial extensional modulus and Poisson's ratio [Eqns. (14) and (15)]. The expressions for the three other elastic constants are assumed to be of the form

$$\hat{p} = p_m(1 + c\xi\eta)/(1 - c\eta) \tag{21}$$

where

$$\eta = \left(\frac{p_f - 1}{p_m}\right) \Big/ \left(\frac{p_f}{p_m} + \xi\right) \tag{22}$$

Here, $\hat{p}$ stands for the composite moduli $\hat{E}_t$, $\hat{G}_a$ or $\hat{G}_t$, and p_f and p_m are the corresponding fiber and matrix moduli, respectively. The "reinforcing factors" ξ are arbitrary constants which are, in general, different for the three elastic constants. For example, the Halpin–Tsai expression for the transverse modulus is

$$\hat{E}_t = E_m(1 + c\xi_E\eta_E)/(1 - c\eta_E) \tag{23}$$

where

$$\eta_E = \left(\frac{E_f}{E_m} - 1\right) \Big/ \left(\frac{E_f}{E_m} + \xi_t\right) \tag{24}$$

and ξ_E is the transverse-modulus reinforcing factor.

3. Viscoelastic Properties

Polymers are widely used as matrix materials and as reinforcements. As these materials have viscoelastic characteristics, their influence on the time-dependent properties of composites is an important subject.

The analysis of viscoelastic behavior is more complex than that of elastic behavior. Fortunately, the elastic–viscoelastic correspondence principle permits the use of analytical elasticity solutions in many cases. A thorough discussion of the subject is beyond the scope of this article. Generally speaking, however, the correspondence principle states that the solution of static elastic problems can be converted to Laplace-transformed solutions of viscoelastic solutions by replacing elastic moduli with transformed viscoelastic moduli multiplied by the transform parameter.

For example, expressions for the transformed composite effective relaxation moduli $\tilde{\boldsymbol{C}}_{ij}(s)$ can be obtained from expressions for composite effective elastic moduli $\hat{\boldsymbol{C}}_{ij}$ by replacing the phase elastic moduli by phase relaxation moduli multiplied by the transform parameter s. A similar relationship exists between the transformed composite effective creep compliance $\tilde{\boldsymbol{S}}_{ij}(s)$ and the composite effective elastic compliance $\hat{\boldsymbol{S}}_{ij}$.

Use of the correspondence principle requires that there be an "exact" or approximate analytical solution for the appropriate effective elastic property. The existence of upper and lower bounds on viscoelastic properties has been proven only for a very few special cases (Christensen 1979).

As an example, the Halpin–Tsai approximate expression for the transverse extensional modulus given by Eqns. (23) and (24) can be used to obtain an approximate expression for the transformed transverse modulus $\tilde{E}_t(s)$:

$$\tilde{E}_t(s) = \tilde{E}_m(s)[1 + \xi_E \tilde{\eta}_E(s)c]/[1 - c\tilde{\eta}_E(s)] \quad (25)$$

where

$$\tilde{\eta}_E(s) = \left[\frac{\tilde{E}_f(s)}{\tilde{E}_m(s)} - 1\right] \bigg/ \left[\frac{\tilde{E}_f(s)}{\tilde{E}_m(s)} + \xi_E\right] \quad (26)$$

and $\tilde{E}_f(s)$ and $\tilde{E}_m(s)$ are the Laplace transforms of the fiber and matrix viscoelastic extensional moduli, respectively.

The correspondence principle can also be used to obtain the composite effective complex moduli $\hat{C}^*(\omega)$ for steady-state harmonic oscillation, where ω is the frequency. For example, the Halpin–Tsai relations, Eqns. (23) and (24), can also be used to obtain an approximate expression for the composite effective complex transverse modulus, $\hat{E}_t^*(\omega)$:

$$\hat{E}_t^*(\omega) = E_m^*(\omega)[1 + c\xi_E \eta_E^*(\omega)]/[1 - \eta_E^*(\omega)c] \quad (27)$$

where

$$\eta_E^* = [E_f^*(\omega)/E_m^*(\omega) - 1]/[E_f^*(\omega)/E_m^*(\omega) + \xi_E] \quad (28)$$

4. *Plastic Behavior*

The increasing interest in metal-matrix composites is focusing attention on the subject of plasticity, which has not received as much consideration as other aspects of mechanical behavior. The contribution of metal matrices to composite stiffness, even in the axial direction, can be significant and matrix plasticity effects can have an appreciable influence on composite behavior.

Composite plastic load–deformation behavior is far more difficult to describe than for the elastic or even viscoelastic cases. In the latter two cases, it is possible to use effective composite properties to obtain relations between average stresses and strains and their time derivatives. However, to describe plastic behavior, it is necessary to define the complete internal state of stress for every value of applied load. This requires specification of a geometric model and is extremely complex, even for simple loading conditions. The problem has been approached both analytically and by the use of finite-element models (Hill 1964b, Mulhern et al. 1967, Dvorak and Rao 1976a, Adams 1970, Foye 1973). Although definition of complete load–deformation behavior is difficult, it has been possible to establish analytical bounds on limit loads, and this is discussed in Sect. 4.1.

The complexity of the subject precludes a detailed discussion of plastic load–deformation behavior. However, micromechanical analyses have provided important insights into the general characteristics of composites with plastic constituents and these will be examined in this section. The discussion is based primarily on the work of Drucker (1975) and Dvorak and Rao (1976a,b).

The elastic constants of fiber and matrix usually differ, so the internal stress distribution is not homogeneous, even under hydrostatic loading. Unexpected results follow from this; for example, a composite with elastic fibers in an elastic, perfectly plastic matrix shows an initial stress–strain curve with work hardening caused by the formation of plastic zones which spread as the load is increased. The application of hydrostatic stress produces irreversible volume changes due to yielding of the matrix, although the matrix is plastically incompressible.

Dvorak and Rao (1976b) showed that plasticity effects are often important when composites whose constituents have different coefficients of thermal expansion undergo significant temperature change.

A simple model for the behavior under axial load of a composite with elastic fibers and an elastic, perfectly plastic matrix proposed by Spencer (1972) illustrates some of the effects of plasticity on composite behavior. Figure 4 shows the stress–strain behavior of the composite; the fibers, which have an elastic moduli E_f; and the matrix, which has an elastic modulus E_m, a tensile yield stress Y_m and a compressive yield stress $-Y_m$. Assuming that the composite is originally stress-free, the initial modulus is approximately $cE_f + (1-c)E_m$. The matrix yields when the applied stress is $[cE_f + (1-c)E_m]Y_m/E_m$ (point A), and the effective composite modulus

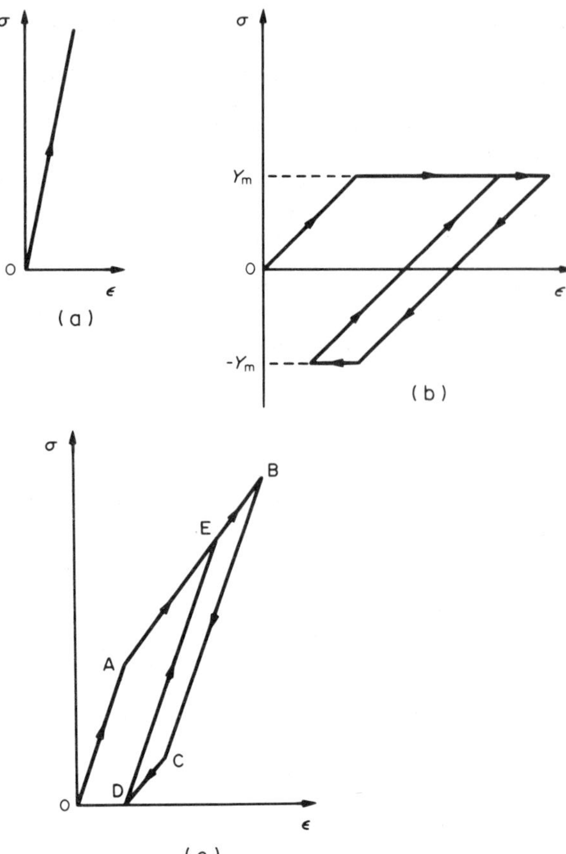

Figure 4
Stress–strain curves for (a) elastic fibers, (b) an elastic, perfectly plastic matrix and (c) the resulting composite

drops to cE_f. The slope of the unloading curve BC equals the initial elastic slope. At point C, the matrix stress reaches $-Y_m$ and it yields in compression, whereupon the effective composite modulus again drops to cE_f. At point D, where the applied stress is zero, the matrix is under a state of residual compressive stress $-Y_m$, the fibers are in tension and there is a macroscopic residual deformation. Subsequent application of tensile stress unloads the matrix and it behaves elastically, so that the slope of the composite stress–strain curve is, again, $cE_f+(1-c)E_m$. At point E, the matrix yields in tension. The cycle EBCD can then be repeated indefinitely.

This simple model does not display the apparent work hardening predicted by more detailed analyses (Dvorak and Rao 1976a,b). Composite behavior under shear and transverse extensional loadings is more complex and is not easily represented by simple models like the one presented here (Adams 1970, Foye 1973).

4.1 Composite Limit Loads

Limit analysis permits the definition of bounds on composite failure loads without consideration of the details of the internal state of stress. In this respect, it is somewhat analogous to the bounding approach to composite elastic properties discussed in Sect. 2.2. Lower bounds on the surface tractions producing unrestricted plastic flow are obtained by defining statically admissible stress fields that do not exceed the constituent yield stresses. Upper bounds are established by assuming kinematically admissible velocity fields and equating work done by surface tractions to the internal plastic dissipation (Hashin 1972).

For example, Shu and Rosen (1967) constructed a composite-cylinder-assemblage model consisting of rigid circular fibers in an elastic, perfectly plastic matrix that obeys the von Mises yield criterion. They found upper and lower bounds on limit loads for various surface tractions. In particular, they found that a lower bound (LB) on the limit load associated with an axial-shear surface traction, $(\sigma_{12}^{L})_{LB}$ is simply the matrix yield stress in shear σ_y. They also obtained an upper bound (UB), subsequently improved by Majumdar and McLaughlin (1973). The upper bound obtained by the latter is

$$(\sigma_{12}^{L})_{UB}=\sigma_y\left[1+c\left(\frac{4}{\pi}-1\right)\right] \qquad (29)$$

This expression predicts that the composite in-plane shear strength can never be more than 27% greater than the matrix shear strength when it is reinforced with circular cylindrical fibers, no matter how strong they are.

In general, surface tractions that produce failure in the unreinforced matrix are lower bounds on composite strength (Hashin 1972). Other upper bounds are more complex than that of Eqn. (29) and are not presented in this article. One other result should be noted: Hashin (1972) showed that the shear strength of a longitudinal plane not passing through any fibers is simply the matrix shear strength.

5. *Coefficients of Thermal Expansion*

The relations between average stress and average strain in a statistically homogeneous composite material undergoing a uniform temperature change θ are

$$\hat{\boldsymbol{\varepsilon}}_{ij}=\hat{\boldsymbol{S}}_{ijkl}\bar{\sigma}_{kl}+\hat{\alpha}_{ij}\theta \qquad (30)$$

and

$$\bar{\boldsymbol{\sigma}}_{ij}=\hat{\boldsymbol{C}}_{ijkl}\bar{\varepsilon}_{kl}-\hat{\boldsymbol{C}}_{ijkl}\hat{\alpha}_{kl}\theta \qquad (31)$$

where the $\hat{\alpha}_{kl}$ are the effective thermal-expansion coefficients. For a transversely isotropic or square symmetric material there are two independent effective thermal-expansion coefficients, the axial $\hat{\alpha}_a$ and the transverse $\hat{\alpha}_t$.

Methods for establishing relations between composite effective expansion coefficients and constituent properties are similar to those used for elastic constants. For example, Schapery (1968) and Rosen and Hashin (1970) used thermoelastic extremum principles to derive upper and lower bounds. A direct method for two-phase composites, proposed by Levin, is reported in Rosen and Hashin (1970).

For a transversely isotropic composite having two phases which are elastically and thermally isotropic, the coefficients of thermal expansion are

$$\bar{\alpha}_{\rm a} = \bar{\alpha} + \left(\frac{\alpha_{\rm f} - \alpha_{\rm m}}{1/K_{\rm f} - 1/K_{\rm m}}\right) \times \left[\frac{3(1-2\hat{\nu}_{\rm a})}{\hat{E}_{\rm a}} - \left(\frac{1}{K}\right)\right] \quad (32)$$

$$\hat{\alpha}_{\rm t} = \bar{\alpha} + \left(\frac{\alpha_{\rm f} - \alpha_{\rm m}}{1/K_{\rm f} - 1/K_{\rm m}}\right) \times \left[\frac{3}{2\hat{K}} - \frac{3\hat{\nu}_{\rm a}(1-2\hat{\nu}_{\rm a})}{E_{\rm a}} - \left(\frac{1}{K}\right)\right] \quad (33)$$

where

$$\bar{\alpha} = c\alpha_{\rm f} + (1-c)\alpha_{\rm m} \quad (34)$$

$$\left(\frac{1}{K}\right) = \frac{c}{K_{\rm f}} + \frac{(1-c)}{K_{\rm m}} \quad (35)$$

and $\alpha_{\rm f}$ is the fiber coefficient of thermal expansion, $\alpha_{\rm m}$ is the matrix coefficient of thermal expansion, $K_{\rm f}$ is the fiber bulk modulus and $K_{\rm m}$ is the matrix bulk modulus.

These expressions for thermal expansion coefficients require knowledge of the composite effective axial modulus $\hat{E}_{\rm a}$, axial Poisson's ratio $\hat{\nu}_{\rm a}$ and transverse bulk modulus $\hat{k}$, as well as constituent thermal expansion constants and bulk moduli. The composite elastic properties can be measured experimentally or evaluated from analytical expressions. Use of bounds for composite effective moduli results in bounds on effective thermal-expansion coefficients.

6. Macroscopic Failure Criteria

Failure of composites results from complex internal processes that depend on the kind of loading applied and numerous constituent material parameters including stress–strain and statistical strength characteristics, interfacial bond strength and internal geometry. Consequently, it is not reasonable to expect that a precise, simple expression should exist for the strength of composites under arbitrary loading. For convenience, a number of empirical failure criteria for unidirectional composites under arbitrary load have been proposed. The most widely used—maximum stress, maximum strain and interaction—are examined here. Stresses are referred to the principal axes of the material.

The primary advantage of the maximum stress and strain failure criteria is their great simplicity. Required strength parameters can be determined from tests involving single-stress components. They also provide information on the mode of failure.

6.1 Maximum-Stress and Maximum-Strain Failure Criteria

The maximum-stress failure criterion assumes that failure occurs when any one of the applied stresses equals the ultimate stress obtained when that stress alone is applied to the composite. That is, it completely ignores any interaction effects between the stresses. For a state-of-plane stress, for which $\bar{\sigma}_{13} = \bar{\sigma}_{23} = \bar{\sigma}_{33} = 0$, the maximum-stress failure criterion is

$$\begin{aligned} \bar{\sigma}^{\rm c}_{11} &\leqslant \bar{\sigma}_{11} \leqslant \bar{\sigma}^{\rm t}_{11} \\ \bar{\sigma}^{\rm c}_{22} &\leqslant \bar{\sigma}_{22} \leqslant \bar{\sigma}^{\rm t}_{22} \\ |\bar{\sigma}_{12}| &\leqslant \bar{\sigma}^{\rm u}_{12} \end{aligned} \quad (36)$$

where the superscripts c and t indicate compression and tension failure strengths, respectively, and $\bar{\sigma}^{\rm u}_{12}$ is the axial shear strength. Extension of this criterion to a more general state-of-stress is obvious.

The maximum-strain criterion assumes that failure occurs when any macroscopic strain value reaches the ultimate strain achieved in a corresponding single-component tension, compression or shear test. For the case of plane stress ($\bar{\sigma}_{13} = \bar{\sigma}_{23} = \bar{\sigma}_{33} = 0$) the criterion is

$$\begin{aligned} \bar{\varepsilon}^{\rm c}_{11} &\leqslant \bar{\varepsilon}_{11} \leqslant \bar{\varepsilon}^{\rm t}_{11} \\ \bar{\varepsilon}^{\rm c}_{22} &\leqslant \bar{\varepsilon}_{22} \leqslant \bar{\varepsilon}^{\rm t}_{22} \\ |\bar{\varepsilon}_{12}| &\leqslant \bar{\varepsilon}^{\rm u}_{12} \end{aligned} \quad (37)$$

where $\bar{\varepsilon}^{\rm c}_{11}$ is the failure strain under a compressive axial stress, $\bar{\varepsilon}^{\rm t}_{11}$ is the failure strain under a tensile axial stress, $\bar{\varepsilon}^{\rm c}_{22}$ and $\bar{\varepsilon}^{\rm t}_{22}$ are similarly defined for transverse extensional loading, and $\bar{\varepsilon}^{\rm u}_{12}$ is the ultimate strain under pure shear loading.

6.2 Interaction Failure Criteria

Numerous interaction failure criteria have been proposed. For example, Sendeckyj (1972) reports eighteen. A description of features of these criteria is beyond the scope of this article. Instead, one of the more general and widely used formulations is considered—that of Tsai and Wu (1971). For convenience, contracted notation will be used.

The most general form of the Tsai–Wu criterion is

$$\boldsymbol{F}_i\bar{\boldsymbol{\sigma}}_j + \boldsymbol{F}_{ij}\bar{\boldsymbol{\sigma}}_i\bar{\boldsymbol{\sigma}}_j = 1 \quad (38)$$

where a repeated index indicates summation. This defines a failure surface in stress space. To ensure that the criterion predicts finite failure stresses, the surface must be closed. (Infinite failure stresses, even for hydrostatic loading, are generally rejected on physical grounds.) This requires that $\boldsymbol{F}_{ii}\boldsymbol{F}_{jj} - \boldsymbol{F}^2_{ij} \geqslant 0$ for $i, j = 1, 2, \ldots 6$, where repeated indices are not summed.

For thin layers in a state-of-plane stress, with $\bar{\sigma}_3 = \bar{\sigma}_4 = \bar{\sigma}_5 = 0$, Eqn. (38) reduces to

$$F_1\bar{\sigma}_1 + F_2\bar{\sigma}_2 + F_{11}\bar{\sigma}_1^2 + F_{22}\bar{\sigma}_2^2 + 2F_{12}\bar{\sigma}_1\bar{\sigma}_2 + F_{66}\bar{\sigma}_6^2 = 1 \tag{39}$$

which has six independent constants. The four parameters F_1, F_{11}, F_2 and F_{22} can be determined from axial and transverse tension and compression tests, and an axial shear test is used to evaluate F_{66}. This leaves F_{12}, which can only be determined from a multiaxial strength test in which $\bar{\sigma}_1$ and $\bar{\sigma}_2$ are both present. To avoid performing multiaxial tests, it is often assumed that $F_{12} = 0$.

Figure 5 compares predictions of the Tsai–Wu and maximum-stress failure criteria for a series of off-axis tensile tests performed on unidirectional boron–epoxy composites.

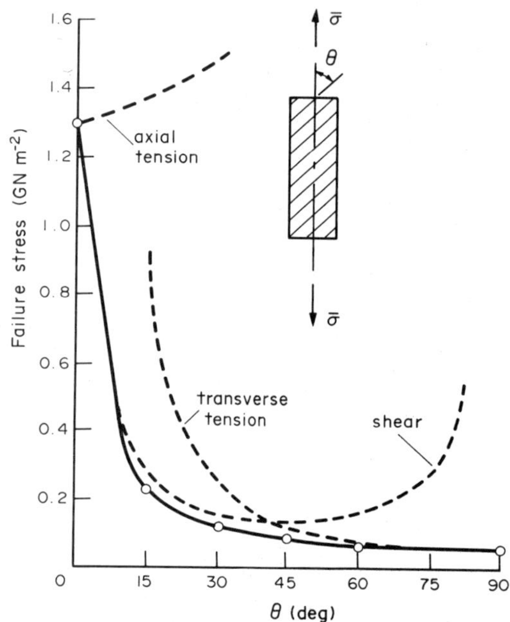

Figure 5
Comparison of the Tsai–Wu and maximum-stress failure criteria with off-axis strength of unidirectional boron–epoxy composites: ——, Tsai–Wu prediction; – – –, maximum stress; ○, experiment (after Pipes and Cole 1973)

See also: Composite Materials: An Overview; Creep of Composites; Fatigue of Composites; Strength of Composites

Bibliography

Adams D F 1970 Inelastic analysis of a unidirectional composite subjected to transverse normal loading. *J. Compos. Mater.* 4: 310–28

Ashton J E, Halpin J C, Petit P H 1969 *Primer on Composite Materials: Analysis.* Technomic, Westport, Connecticut

Christensen R M 1979 *Mechanics of Composite Materials.* Wiley, New York

Drucker D C 1975 Yielding, flow and fracture. In: Herakovich C T (ed.) 1975 *Inelastic Behavior of Composite Materials.* American Society of Mechanical Engineers, New York, pp. 1–15

Dvorak G J, Rao M S M 1976a Axisymmetric plasticity theory of fibrous composites. *Int. J. Eng. Sci.* 14: 361–73

Dvorak G J, Rao M S M 1976b Thermal stresses in heat-treated fibrous composites. *J. Appl. Mech.* 43: 619–24

Foye R L 1973 Theoretical post-yielding behavior of composite laminates: I—Inelastic micromechanics. *J. Compos. Mater.* 7: 178–93

Hashin Z 1965 On elastic behaviour of fibre reinforced materials of arbitrary transverse phase geometry. *J. Mech. Phys. Solids* 13: 119–34

Hashin Z 1972 *Theory of Fiber Reinforced Materials.* National Aeronautics and Space Administration, Hampton, Virginia

Hashin Z 1979 Analysis of properties of fiber composites with anisotropic constituents. *J. Appl. Mech.* 46: 543–50

Hill R 1964a Theory of mechanical properties of fibre-strengthened materials: I. Elastic behaviour. *J. Mech. Phys. Solids* 12: 199–212

Hill R 1964b Theory of mechanical properties of fibre-strengthened materials: II. Inelastic behaviour. *J. Mech. Phys. Solids* 12: 213–18

Majumdar S, McLaughlin P V 1973 Upper bounds to in-plane shear strength of unidirectional fiber-reinforced composites. *J. Appl. Mech.* 40: 824–25

Majumdar S, McLaughlin P V 1975 Effects of phase geometry and volume fraction on the plane stress limit analysis of a unidirectional fiber-reinforced composite. *Int. J. Solids Struct.* 11: 777–91

Mulhern J F, Rogers T G, Spencer A J M 1967 Cyclic extension of an elastic fibre with an elastic–plastic coating. *J. Inst. Maths. Appl.* 3: 21–40

Pipes R B, Cole B W 1973 On the off-axis strength test for anisotropic materials. *J. Compos. Mater.* 7: 246–56

Rosen B W, Hashin Z 1970 Effective thermal expansion coefficients and specific heats of composite materials. *Int. J. Eng. Sci.* 8: 157–73

Schapery R A 1968 Thermal expansion coefficients of composite materials based on energy principles. *J. Compos. Mater.* 2: 380–404

Schapery R A 1974 Viscoelastic behavior and analysis of composite materials. In: Sendeckyj G P (ed.) 1974 *Composite Materials*, Vol. 2, *Mechanics of Composite Materials.* Academic Press, New York, pp. 86–168

Sendeckyj G P 1972 A brief survey of empirical multiaxial strength criteria for composites. *Composite Materials: Testing and Design.* American Society for Testing and Materials, Philadelphia, Pennsylvania, pp. 41–51

Sendeckyj G P 1974 Elastic behavior of composites. In: Sendeckyj G P (ed.) 1974 *Composite Materials*, Vol. 2, *Mechanics of Composite Materials.* Academic Press, New York, pp. 45–83

Shu L S, Rosen B W 1967 Strength of fiber-reinforced composites by limit analysis methods. *J. Compos. Mater.* 4: 366–81

Spencer A J M 1972 *Deformations of Fibre-reinforced Materials.* Clarendon Press, Oxford

Tsai S W, Wu E M 1971 A general theory of strength for anisotropic materials. *J. Compos. Mater.* 5: 58–80

C. Zweben

Field-Ion Microscopy: Atom-Probe Microanalysis

The field-ion microscope (FIM) is, at present, the only microscope which routinely allows the imaging of individual atoms in direct lattice space. The FIM is a point-projection microscope and hence no lenses are required for the formation of an image. The image is a projection of the atoms which reside in the surface of a sharply pointed specimen. When an FIM is combined with a mass spectrometer (to form an "atom probe"), it is possible to determine the mass-to-charge ratio of individual atoms, one at a time. Invented by E. W. Müller, the FIM and atom probe have been used extensively to study, for example, precipitation phenomena (Brenner 1978), surface diffusion of adatoms (atoms adsorbed on a surface) (Ehrlich 1980, Tsong 1980), radiation damage (Seidman 1978, Herschitz and Seidman 1984), phase transformations in steels (Miller et al. 1981), order–disorder alloys (Yamamoto et al. 1972) and segregation (Herschitz and Seidman 1985).

1. *The Field-Ion Microscope*

A basic modern FIM consists of a vacuum system (high to ultrahigh vacuums are employed), a liquid-helium cryostat for cooling the specimen to temperatures in the range 4.2–300 K, a well-stabilized high-voltage supply (0–30 kV), an internal image-intensification system [a 7.5 cm diameter channel electron multiplier array (CEMA) and a phosphor screen] and a gas train to supply the required imaging gas (typically helium or neon). A single CEMA has a gain of 10^3–10^4, thus very little adaptation to a darkened room is required to observe an FIM image. The FIM specimen is mounted pointing towards the CEMA; it is electrically, but not thermally isolated from the cryostat. The distance between the specimen and the CEMA is typically 2.5–8 cm. A wire with a length of ~1 cm and a diameter of 0.1–0.2 mm serves as a specimen. Before it is inserted into the FIM, the wire is electroetched or electropolished to a very fine point with an average radius of only 5–10 nm. The operating procedure consists of first obtaining a background pressure of at least 10^{-8} torr, then cooling the specimen with liquid helium to a temperature between 10 and 70 K, backfilling the FIM with an imaging gas to a pressure between 10^{-5} and 10^{-4} torr and finally applying a voltage to the specimen.

2. *Field Ionization*

With the aid of a high-voltage supply the specimen can be placed at a positive potential with respect to the front surface of the CEMA. Since the average radius of the tip is quite small, a local electric field of 5 V Å^{-1} (5×10^{10} V m^{-1}) can easily be generated with the application of only a few thousand volts to the tip.

In a local electric field of 4.5–5.0 V Å^{-1}, helium atoms are ionized as a result of the tunnelling of an electron from a helium atom into the metal (field ionization). The electron tunnels through the deformed electron potential energy barrier in a time of $<10^{-9}$ s. The positively charged helium ion created by this process is then accelerated along an electric field line to the internal image-intensification system, where its energy is eventually converted into visible light. The actual field-ionization events occur over the sites of individual atoms, where the local electric field—as opposed to the average value of the electric field—is greatest. A simple physical model of the surface of a metal in a high electric field is called the ionic model. In this model, the free-electron gas is pushed back slightly by the field into the bulk of the metal to create a series of positively charged ions on the surface. Thus, each individual metal ion is the source of a small pencil of gas ions. The helium (or neon) ion current from each metal ion is $\sim 10^{-12}$ A and the total current from the tip is $\sim 10^{-8}$ A.

3. *Field Evaporation*

The process of field evaporation is basic to the use of the FIM as a metallographic instrument. Field evaporation is the controlled sublimation of atoms from a specimen under the influence of an electric field. If the voltage on a specimen is increased to a critical value, the atoms on the surface begin to sublime, even though the specimen may be as cold as 4.2 K. The exact value of the electric field at which field evaporation commences is a function of the metal being examined. For example, the evaporation fields of tungsten and gold are ~5.7 V Å^{-1} and ~3.5 V Å^{-1}, respectively. Since the ionization field of the imaging gas is independent of the metal being studied, a gas with an ionization field that is less than or equal to the evaporation field of the metal under examination must be used. The micrograph of gold shown in Fig. 1 was obtained using neon as the imaging gas—a neon atom requires a field of ~3.5 V Å^{-1} for field ionization to occur (Averback and Seidman 1973).

The field-evaporation process can be controlled with great precision by applying the positive potential in the form of short (1–10 ms in width) high-voltage pulses. This technique, called pulse field evaporation, enables dissection of an atomic plane at a rate of one to three atoms per pulse. Thus, the atoms contained

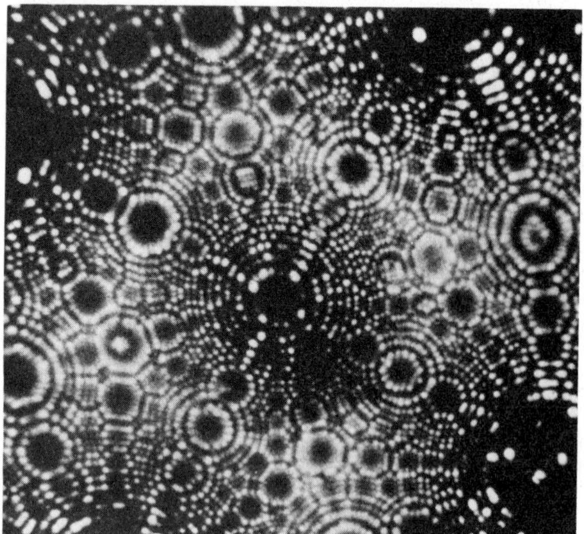

Figure 1
An FIM image of gold: the (001) plane is in the center of the image (note the fourfold symmetry about this plane); specimen temperature 23 K; imaging gas, neon

within the interior of the specimen can be imaged—albeit at the surface—at a determined rate. The area of surface examined at any instant is $\sim 10^{-14}\,m^2$. In practice, $\sim 10^{-23}\,m^3$ of metal can be examined during the course of one day employing the pulse-field-evaporation technique; this corresponds to $\sim 6 \times 10^5$ atoms.

A simple physical picture of the process of field evaporation is obtained with the aid of the ionic model. In this model the applied electric field deforms the potential-energy curve for an ion on the surface of the specimen and introduces a Schottky barrier. The ions then evaporate by either jumping over a small Schottky barrier—as a result of a thermally activated step—or by tunnelling through the barrier (Müller and Tsong 1969, 1973).

4. Resolution

The diameter of the pencil of imaging ions emanating from each atom is a function of the temperature of the tip. The purpose of cooling the specimen to cryogenic temperatures is to accommodate thermally the imaging-gas atoms to the temperature of the tip, prior to the field-ionization event. This temperature accommodation process reduces the transverse velocity of a gas atom—and hence the ion—thereby reducing the diameter of the pencil of imaging-gas ions. In practical terms this means that the effective diameter of the image of an atom is a function of the temperature of the tip. Therefore, the resolution of the FIM is temperature dependent (Chen and Seidman 1971).

The resolution of the FIM is also a function of the radius of the tip; the blunter the tip the poorer is the resolution. The physical origin of this effect is the angular momentum of the gas atom at the instant of its ionization. In the central-field approximation the angular momentum is proportional to the square of the radius of the tip. Hence, the diameter of the pencil of imaging gas ions increases with increasing radius and concomitantly the resolution decreases (Gomer 1961).

From the above it is clear that the best atomic resolution is achieved employing sharp tips (<20 nm) and low temperatures (<20 K). Thus, it is possible in practice to observe individual atoms with interatomic distances approaching 0.2 nm (Müller and Tsong 1969).

5. Atom-Probe Field-Ion Microscope

The time-of-flight (TOF) atom-probe FIM consists of an FIM combined with a special TOF mass spectrometer (Fig. 2). The atom probe permits the identification of the chemistry of any atom that appears in an FIM image. Thus, it is possible to obtain an atomic resolution image of microstructural features and to measure the mass-to-charge ratio (m/q) of single atoms from preselected regions of a specimen, with a lateral spatial resolution (i.e., within the surface) of a few tenths of a nanometer and a depth spatial resolution that is determined by the interplanar spacing of the region; the latter quantity can be hundredths of a nanometer for a high-index plane. An atom probe with a straight TOF tube (Fig. 2) has a mass resolution ($m/\Delta m$) of ~200, whereas an atom probe with a Poschenrieder lens has an $m/\Delta m$ value of >1000. A magnetic-sector atom probe has been constructed to analyze the products of the field-evaporation process—this instrument has an $m/\Delta m$ of ~2000 (Müller and Sakurai 1974). Its main disadvantage at present is that only a small portion of the periodic table can be examined at one time.

The operation of an atom probe can be understood with reference to Fig. 2. When a short (<25 ns) high-

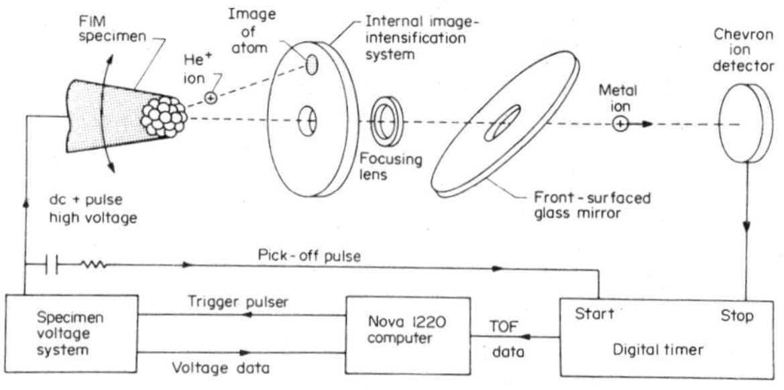

Figure 2
Schematic diagram of a time-of-flight atom-probe field-ion microscope

voltage pulse (V_{pulse}) is superposed on the steady-state imaging voltage (V_{dc}), atoms on the surface of the specimen are field evaporated in the form of ions. The ions that are projected into the probe hole, at the center of the internal image-intensification system, pass down the flight tube to the chevron ion detector. This detector consists of two CEMAs in series and has a gain of $>10^6$. Each ion produces a single voltage pulse, which stops a clock in the digital timer. A portion of the high-voltage evaporation pulse is used to start the clock. A minicomputer is used to control the specimen voltage system and analyze the data. The TOFs of the ions and the voltages applied to the specimen are stored in the minicomputer after each evaporation pulse. The m/q ratios are calculated according to

$$m/q = 2e(V_{dc} + \alpha V_{pulse})(t - t_0)^2/d^2$$

where e is the charge on an electron, α the pulse factor, d the flight distance (>2 m) and $t - t_0$ the actual TOF of the ion. The quantity t is the observed TOF and t_0 is the total delay time for the system. In practice, the FIM specimen is pulsed at a frequency of 60 Hz.

6. Imaging Atom-Probe Mass Spectroscopy

It is also possible to produce an image of the atoms on the surface of a tip by using the field-desorbed ions to create the image. This can now be achieved routinely employing a chevron to detect the individual field-desorbed ions, with a flight distance of 0.12 m between the specimen and the chevron (the short flight distance is necessary to observe the entire surface area of a tip). To form a field-desorption image of an ion with a specific m/q ratio, the chevron detector is time-gated, so only that ratio is detected. To operate in this mode it is necessary to employ a chevron detector with a spherical geometry. This geometry guarantees that each ion will have the same flight distance, independent of its position on the surface of the specimen. An imaging atom-probe mass spectrometer can be used, for example, to determine elemental maps of the distribution of alloying elements in the vicinity of grain boundaries (Waugh and Southon 1979).

7. Pulsed-Laser Atom-Probe Field-Ion Microscope

Kellogg and Tsong (1980) showed that it is possible to field-evaporate a silicon specimen in a highly controlled manner by illuminating it with 6 ns-wide pulses of laser light, while maintaining the specimen at a steady-state dc voltage. This work built on the earlier experiments of Melmed and coworkers who demonstrated that specimens of Si, Ge and GaAs had an enhanced field-evaporation rate when they were irradiated with white light. Kellogg and Tsong measured the mass-to-charge ratios of the silicon isotopes in a conventional time-of-flight atom probe and demonstrated that the pulsed-laser atom probe achieved a mass resolution which is comparable to that for metals employing high-voltage pulses. Thus, the pulsed-laser atom-probe field-ion microscope can be useful for chemically analyzing semiconducting materials.

See also: Field-Ion Microscopy: Observation of Radiation Effects; Investigation and Characterization of Materials: An Overview

Bibliography

Averback R S, Seidman D N 1973 Neon gas imaging of gold in the field-ion microscope. *Surf. Sci.* 40: 249–63

Brenner S S 1978 Application of field-ion microscopy techniques to metallurgical problems. *Surf. Sci.* 70: 427–57

Chen Y C, Seidman D N 1971 On the atomic resolution of a field-ion microscope. *Surf. Sci.* 26: 61–84

Ehrlich G 1980 Quantitative examination of individual atomic events on solids. *J. Vac. Sci. Technol.* 17: 9–14

Gomer R 1961 *Field Emission and Field Ionization.* Harvard University Press, Cambridge, Massachusetts

Herschitz R, Seidman D N 1984 An atomic resolution study of homogeneous radiation-induced precipitation in a neutron irradiated W–10 at.%Re alloy. *Acta Metall.* 32:1141

Herschitz R, Seidman D N 1985 Atomic resolution observations of solute atom segregation effects and two-dimensional phase transitions in stacking faults in dilute cobalt alloys: I. Experimental results. *Acta Metall.* 33 (in press)

Hochman R F, Müller E W, Ralph B (eds.) 1969 *Applications of Field-ion Microscopy in Physical Metallurgy and Corrosion.* Georgia Institute of Technology Foundation, Atlanta, Georgia

Hren J J, Ranganthan S (eds.) 1968 *Field-ion Microscopy.* Plenum, New York

Kellogg G L, Tsong T T 1980 Pulsed-laser atom-probe field ion microscopy. *J. Appl. Phys.* 51:1184

Miller M K, Beavan P A, Smith G D W 1981 A study of the early stages of tempering of iron–carbon martensites by atom probe field ion microscopy. *Metall. Trans., A* 12: 1197–204

Müller E W, Sakurai T 1974 A magnetic sector atom-probe FIM. *J. Vac. Sci. Technol.* 11: 878–82

Müller E W, Tsong T T 1969 *Field Ion Microscopy: Principles and Applications.* Elsevier, New York

Müller E W, Tsong T T 1973 Field ion microscopy, field ionization and field evaporation. *Prog. Surf. Sci.* 4: 1–139

Seidman D N 1978 The study of radiation damage in metals with the field-ion and atom-probe microscopes. *Surf. Sci.* 70: 532–65

Surface Science (Vol. 23, 1970): issue devoted to Field-ion, Field Emission Microscopy and Related Topics

Surface Science (Vol. 70, 1978): issue devoted to Field Emission and Related Topics

Tsong T T 1980 Quantitative investigations of atomic processes on metal surfaces at atomic resolution. *Prog. Surf. Sci.* 10: 165–248

Wagner R 1982 *Field-ion Microscopy.* Springer, Berlin

Waugh A R, Southon M J 1979 Surface analysis and grain-boundary segregation measurements using atom-probe techniques. *Surf. Sci.* 89: 718–24

Yamamoto M, Neno S, Futamoto M, Nakamura S 1972 Field-ion microscope study of early stages of ordering in Ni_4MO alloys. *Jpn. J. Appl. Phys.* 11: 437–44

D. N. Seidman

Field-Ion Microscopy: Observation of Radiation Effects

Defects in metals produce contrast effects in field-ion microscope images which enable useful study of the effects of radiation on metals. As a consequence, the field-ion microscope (FIM) and a variant of this, the atom-probe FIM (see *Field-Ion Microscopy: Atom-Probe Microanalysis*), are widely used in such studies.

1. Capabilities and Advantages of the FIM

The principal capabilities and advantages of the FIM technique are:

(a) the resolution of *individual* vacancies with self-interstitial atoms (SIAs) as well as clusters of these point defects;

(b) the controlled dissection of an irradiated FIM specimen on an atom-by-atom basis employing the pulse-field-evaporation technique;

(c) the ability to irradiate FIM specimens in situ—with atomically clean surfaces—under ultra-high-vacuum conditions employing a magnetically analyzed ion beam;

(d) the observation and irradiation of specimens at cryogenic temperatures—as low as 4.2 K—which are below the onset of the long-range migration of SIAs; and

(e) the in-situ irradiation of FIM specimens at elevated temperatures (up to at least 400 °C seems practical).

2. Capabilities and Advantages of the Atom-Probe FIM

The atom-probe FIM has the following features which make it suitable for the study of the chemical changes associated with the effects of radiation:

(a) the ability to determine the mass-to-charge ratio of a *single* atom that appears in an FIM image;

(b) the possibility of chemical identification of *all* the elements in the periodic table, as the mass-to-charge ratio is determined by a time-of-flight technique;

(c) the easy identification of elements of low atomic number ($Z < 11$)—in particular, the gases hydrogen and helium can be analyzed;

(d) a mass resolution $m/\Delta m$ of ~200 for a conventional atom-probe FIM and ~1500 for one equipped with a Poschenrieder lens;

(e) a spatial resolution in the surface of ~0.3–0.5 nm—that is, the chemical identity of two atoms separated by this distance can be determined; and

(f) a depth resolution, in terms of chemistry, that is determined by the interplanar spacing of the specific (*hkl*) plane being analyzed (this can correspond to <0.1 nm).

The analytical chemistry features of the atom-probe FIM can be combined with the imaging capability of the FIM to obtain a composition profile that is associated with a structural defect, for example, a void in a neutron-irradiated alloy.

3. Some Contrast Effects Caused by Radiation-Induced Defects

The simplest contrast effect is that caused by a monovacancy. A monovacancy is a lattice site that is missing its atom, and its local region is deficient in positive charge. Its image in an FIM micrograph is simply a black spot on a positive print; an atom usually gives rise to a white spot on a positive print. An atomic site can be shown to be vacant by the careful pulse field evaporation of each plane. In practice, this means one to three atoms removed per field-evaporation pulse (Seidman 1973).

An SIA gives rise to a contrast pattern that is a result of the large number of atoms displaced from their equilibrium positions by its presence. Thus, the SIA is sensed at the surface of the FIM specimen when it lies several atomic layers below the surface (Beavan et al. 1971, Seidman and Lie 1972).

A displacement cascade consists of a three-dimensional spatial distribution of vacancies with a local concentration of 1–20 at.%. The individual vacancies within the displacement cascade can be detected by pulse-field-evaporation experiments (Seidman 1973, 1976, 1978, Pramanik and Seidman 1983).

A void detected in fast-neutron-irradiated metals or alloys produces a range of contrast effects that are a function of the crystallographic region of the surface which it intersects (Brenner and Seidman 1975). The simplest contrast effect is a "black hole" that has a diameter equal to the diameter of the void.

4. Examples of Studies

The three-dimensional spatial distribution of vacancies contained within displacement cascades, created by a wide range of different energetic charged particles, has been studied in tungsten and platinum (Beavan et al. 1971, Seidman 1973, 1976, 1978). In all cases, each displacement cascade was produced by a single projectile ion, so the experimental observations were a direct reflection of the manner in which a single ion dissipated its energy elastically. Some of the properties of each displacement cascade determined were:

(a) the absolute number of vacancies;

(b) the average diameter;

(c) the average vacancy concentration;

(d) the distribution of first-nearest-neighbor vacancy clusters;

(e) the radial distribution function out to the ninth nearest neighbor; and

(f) the direction of major elongation.

The spatial distribution of SIAs with respect to the displacement cascades was measured in ion-irradiated tungsten. It was shown that the SIAs were most probably created as a result of focused replacement collision sequences. The isochronal recovery behavior of a number of in-situ ion-irradiated metals and alloys was studied in detail, for example, tungsten, tungsten–rhenium, tungsten–carbon, platinum, platinum–gold, Pt_3Co and molybdenum. The temperature at which uncorrelated long-range migration occurred was determined by directly observing the flux of SIAs to the surface of an FIM specimen; this temperature corresponded to the so-called stage I_E peak. The contrast effect produced by an SIA when it arrived at the surface of an FIM specimen was an extra-bright spot which was found to be metastable over a wide temperature range (Seidman et al. 1975).

Direct evidence was obtained, using the FIM, for the thermally activated release of SIAs from gold atoms in ion-irradiated specimens of dilute platinum–gold alloys. It was shown that there were at least two trapping levels associated with each gold atom. An analysis of the isochronal recovery spectra yielded values of the binding enthalpy for each trapping level. The atom-probe FIM was employed to determine the range profiles of low-energy (0.1–1.5 keV) implanted ^3He and ^4He atoms in tungsten, at 60 K, where the helium atoms had no atomic mobility (Wagner and Seidman 1979a). The diffusive behavior of implanted ^3He and ^4He atoms was measured in the absence of point or line defects and it was shown that the enthalpy change of migration for both isotopes was approximately 24 kJ mol^{-1} (Amano and Seidman 1984).

The atom-probe FIM was also used to obtain direct evidence of the presence of carbon inside a void detected in a molybdenum–titanium alloy irradiated to a fluence of approximately 6×10^{26} m^{-2} at about 713 K (Wagner and Seidman 1979b).

Other applications include studies of sputtering, reflection coefficients of light gas atoms, disordering of order–disorder alloys and displacement cascades produced by ions at several hundred keV.

See also: Lattice Defect Production During Irradiation; Neutron Interactions with Matter; Point Defects in Crystals; Radiation Effects in Metals and Alloys

Bibliography

Amano J, Seidman D N 1984 Diffusivity of ^3He atoms in perfect tungsten crystals. *J. Appl. Phys.* 56: 983

Beavan L A, Scanlan R M, Seidman D N 1971 The defect structure of depleted zones in irradiated tungsten. *Acta Metall.* 19: 1339–50

Brenner S S, Seidman D N 1975 Field ion microscope observations of voids in neutron irradiated molybdenum. *Radiat. Eff.* 24: 73–78

Pramanik D, Seidman D N 1983 Atomic resolution observations of nonlinear depleted zones in tungsten irradiated with metallic diatomic molecular ions. *J. Appl. Phys.* 54: 6352

Seidman D N 1973 The direct observation of point defects in irradiated or quenched metals by quantitative field ion microscopy. *J. Phys. F.* 3: 393–421

Seidman D N 1976 Field-ion microscope studies of the defect structure of the primary state of damage of irradiated metals. In: Peterson N L, Harkness S D (eds.) 1976 *Radiation Damage in Metals*. American Society for Metals, Metals Park, Ohio, pp. 28–57

Seidman D N 1978 The study of radiation damage in metals with the field-ion and atom-probe microscopes. *Surf. Sci.* 70: 532–65

Seidman D N, Lie K H 1972 On contrast patterns produced by self-interstitial atoms in field ion microscope images of a B.C.C. metal. *Acta Metall.* 20: 1045–59

Seidman D N, Wilson K L, Nielsen C H 1975 The study of stages I to IV of irradiated or quenched tungsten and tungsten alloys by field ion microscopy. In: Robinson M T, Young F W (eds.) 1975 *Int. Conf. Fundamental Aspects of Radiation in Metals*, Vol. 1. National Technical Information Service, Springfield, Virginia, pp. 373–96

Wagner A, Seidman D N 1979a Range profiles of 300- and 475-eV $^4He^+$ ions and the diffusivity of ^4He in tungsten. *Phys. Rev. Lett.* 42: 515–18

Wagner A, Seidman D N 1979b Direct observation of solute segregation to voids in a fast-neutron irradiated Mo/1.0 at.% Ti alloy. *J. Nucl. Mater.* 83: 48–58

D. N. Seidman

Fillers and Coatings: Carbonate Minerals

The uses of carbonate minerals in fillers and coatings closely parallel those of clay, siliceous and sulfate minerals. Utilization varies primarily with performance and cost, though it is affected by a myriad of chemical or physical characteristics. Some factors which give the carbonate minerals their popularity include availability, geographically widespread production locations, lack of abrasiveness, whiteness (when required), and low cost.

In the USA, relatively pure (95% or greater $CaCO_3$) limestones are generally the industry standard product, though some dolomites are marketed. Dolomite is more frequently utilized as a filler in Europe.

1. Sources

The carbonate-filler industry should be considered in two parts:

(a) those utilizations requiring a relatively white product such as paints, paper, and in some instances plastics and rubber, and

(b) those utilizations where color may not be a factor such as asphalts and carpet backings.

The white carbonate fillers are almost invariably obtained from deposits of white marble, which are found where extensive metamorphic activity has occurred. In the eastern USA the producers are in or along the Appalachian Mountain chain from Vermont to Alabama. In the far west the deposits are associated with the Rocky Mountain chain.

Limestone with proper characteristics for many filler applications is available throughout the USA. Production tends to come from specialized operations that may also produce limestone for its chemical characteristics. As a result these "filler limestones" can be chemically pure although this purity may not be necessary for the filler application.

Some of the ultrafine-ground limestones may be chemically bleached to improve whiteness, permitting their utilization in applications previously restricted to white marble products.

2. Mineralogy and Physical Properties

This analysis, like the sources, must be divided into the two categories based on color. The marbles, which are most desirable in terms of whiteness, are crystalline with visible crystals varying from about 0.2 to 4 mm. Although calcite is white, most deposits have impurities that include pyrite, graphite, and a large number of silicate minerals. The eastern (Appalachian) deposits are generally bedded, extensively folded, and extend laterally, in some instances, for several kilometers (though the purity may vary substantially from point to point). The western deposits tend to be more recent "pendants" of marble with recrystallization primarily a result of igneous intrusives.

The unmetamorphosed limestones are generally not as white as marble and vary texturally from granular to crystalline, depending on original composition of the limestone and subsequent diagenesis. The chemical purity can vary widely, but most filler-producing facilities operate from relatively pure deposits to satisfy the requirements of other markets. High silica contents (more than 2%) are generally considered detrimental since silica tends to increase machinery wear for both the producer and the user.

Most limestone deposits are distinctly bedded with significant fluctuations in quality of the stone within a formation. In many cases only a single bed will be of acceptable quality. In recent years it has been found that limestones originating in high-energy environments, such as reefal and oolitic deposits, are a basis for relatively pure lighter colored limestones that are desirable for a broad range of filler applications.

3. Mining and Production Technology

Carbonates for filler production are mined by underground and open-pit methods, with systems similar to those used for other minerals. The specific system will depend on the geology, overburden, structure and other characteristics of the individual deposit. The mining method will frequently provide for additional selectivity to produce a whiter, or more chemically uniform, product.

Production technology varies depending on end use. Most marble is crushed, then washed to remove mud or other color-degrading material. In some instances it may then be sorted (by hand or by machine) to remove pieces of darker colored rock. Subsequent crushing steps produce a feed for grinding circuits, and may produce very coarse filler products of size 20–200 mesh (0.84–0.074 mm).

Many grinding circuits incorporate a color beneficiation step, usually selective flotation, which rejects undesirable minerals. The finely pulverized fillers are produced by wet or dry grinding, and in most instances the grinding is performed in closed circuit with a particle-size classifier. The coarser sizes are produced in roller mills, impact mills or hammermills, whereas the finer sizes are generally ground in some form of pebble mill. The most common grinding medium is high-density alumina. If color degradation is to be minimized, silica or alumina mill linings are used. Wet grinding circuits utilize centrifuges for particle classification, whereas dry circuits use an air separator specifically designed for low micrometer-size separations. Ultrafine grinding is accomplished in attrition mills.

Finished products are shipped in bags or bulk, with the utilization of bulk increasing in response to the high costs of bags and bag handling. A new family of products is appearing which are chemically coated to improve their performance, primarily in plastics systems. There has been increased interest in calcium carbonates ground to 1.5 μm or less. The primary market for this product group is paper filling, paper coating and plastics. In addition to meeting a new market demand, it may also substitute for chemically precipitated calcium carbonate.

Ground carbonate testing procedures include sieve analysis, usually as fine as 325 mesh (0.044 mm); sedimentation procedures to measure particle sizes down to 0.3 μm; color, measured as reflectance from a smooth cake of the ground limestone; chemical composition, determined by the usual analytical procedures; and a number of specialized tests that may be required for a particular using industry. Most calcium carbonate filler producers have extensive laboratory facilities for testing their products, assisting customers in solving problems, and developing new product uses.

4. Uses

4.1 Bituminous Products

Carbonate minerals are the primary filler in asphaltic-type products. Typical end uses include roofing shingles, roll roofing, asphaltic coatings and sealing compounds, and asphalt-based patching materials. The filler elevates the softening point, increases viscosity and improves weather resistance. Color is not usually a factor considered. Relatively coarse ground products (50% finer than 200 mesh (0.074 mm)) are generally used for fillers, whereas somewhat finer grinds (50% finer than 325 mesh (0.044 mm)) are used for bitumastic coatings. Since most fillers are mixed at elevated temperatures, the filler must be dry.

4.2 Plastics

The carbonates are the most commonly used filler for plastics compounding. They enjoy this popularity because of their low abrasion, high brightness, ease of dispersion and relatively low cost. Upper particle size limit and average particle size will vary with the specific application but fall into the 0.1–20 μm range. End uses include automotive parts, pipe and plumbing, boats, and floorings. The cultured marble used for bathroom lavatories and table tops is heavily filled with carbonate fillers that are in some instances quite coarse (50 mesh (0.297 mm) and finer).

4.3 Rubber

Carbonate fillers are used in a wide range of rubber products where their use tends to maintain the qualities of softness, resilience, and elongation, even at relatively heavy loadings (see *Fillers in Elastomers*). They can provide whiteness (e.g., in athletic footwear), but in many instances, such as with automotive floor mats, color is not critical and ground limestones that are not white can be used. In some formulations magnesium ions have a chelating effect on the compounds, changing their processing or final characteristics; hence, a high calcium limestone filler rather than a dolomitic one may be specified.

4.4 Protective and Decorative Coatings

Although a number of minerals are used in this group, calcium carbonate is again the leading one. The high brightness and broad range of particle sizes available promote its utilization. The coarser grinds are used in flat (matte) paints, exterior paints and traffic paint, while the finer grinds are used in those products where a glossy finish is desired. They are specified with equal effectiveness in the traditional oil-based coatings as well as the more recently developed emulsion paints.

4.5 Paper

Some calcium carbonate has traditionally been utilized directly in the paper industry, but most was chemically precipitated calcium carbonate. In recent years the calcium carbonate industry has developed ultrafine ground products (finer than 1 μm) that are receiving increased acceptance. Calcium carbonate is included in filling the paper stock when an alkaline binder is employed in paper manufacturing, but is not generally compatible with acidic binders. In coatings, which are applied after the base sheet is formed and filled, the addition of ultrafine calcium carbonate to the traditional kaolin filler improves brightness, opacity and ink receptivity, and may assist in higher production rates on the coating machines.

4.6 Other Applications

Carbonate fillers are common in a host of other applications. Some of these that require substantial quantities include: the latex on the back of tufted carpet and the carpet cushion, whether it be attached to the carpet or a separate underlay; adhesives, mastics and putty that are used in many industries; drywall joint cement that is used to give the smooth final surface to gypsum-board walls; and the sound proofing compounds used in the automotive industry. The versatility of the carbonate fillers is illustrated by their inclusion in chewing gum, concrete curing compounds, children's crayons and oil-well drilling mud.

See also: Fillers and Coatings: Industrial Minerals; Fillers and Coatings: Siliceous Minerals; Fillers and Coatings: Clay Minerals; Fillers and Coatings: Sulfate Minerals

Bibliography

Carr D D, Rooney L F 1983 Limestone and dolomite. In: Lefond S J (ed.) 1983 *Industrial Minerals and Rocks: (Nonmetallics Other Than Fuels),* 5th edn., Vol. 2. American Institute of Mining, Metallurgical, and Petroleum Engineers, New York, pp. 833–68

Galvanek E 1979 *Extender and Filler Pigments*. Fairfield, New Jersey, pp. 85–136

Hall R F 1964 Calcium carbonate pigments. *Off. Dig, Fed. Soc. Paint Technol.* 183–92

Power W R 1983 Dimension and cut stone. In: Lefond S J (ed.) 1983 *Industrial Minerals and Rocks: (Nonmetallics*

Other Than Fuels), 5th edn., Vol. 1. American Institute of Mining, Metallurgical, and Petroleum Engineers, New York, pp. 161–81

Severinghaus N Jr 1983 Fillers, filters, and absorbents. In: Lefond S J (ed.) 1983 *Industrial Minerals and Rocks: (Nonmetallics Other Than Fuels)*, 5th edn., Vol. 1. American Institute of Mining, Metallurgical, and Petroleum Engineers, New York, pp. 243–57

N. Severinghaus Jr.

Fillers and Coatings: Clay Minerals

One of the most widely used minerals for fillers and coatings is kaolinite. Kaolinite is a hydrated aluminum silicate ($Al_2O_3.2SiO_2.2H_2O$), and is the major clay mineral constituent in the rock kaolin. The paper industry is by far the largest user of kaolin, both as a filler and as a coating; it uses over 50% of the total kaolin produced in the world. A second clay that is employed as a filler and coating material is smectite. Smectite is a group name for hydrated sodium, calcium, iron, magnesium and aluminum silicates known as montmorillonites. Industrially the term bentonite is used to describe these clays.

Some physical and chemical properties of a Georgia kaolinite and a Wyoming bentonite are shown in Table 1. Each of these two clay types are discussed separately below as they have very different properties and are used for different purposes.

1. Kaolin

Over 18 Mt of kaolin were mined throughout the world in 1980 and over half came from two areas; 7 Mt from Georgia and South Carolina in the USA and 3.5 Mt from Cornwall, UK. Other areas producing significant quantities of kaolin are Brazil, Czechoslovakia, the German Democratic Republic, France, Russia, Spain and the Federal Republic of Germany. Kaolinite is a unique industrial mineral because, except for catalytic activity in some organic systems, it is chemically inert over a relatively wide pH range; is white in color; has good covering power when used as a pigment or extender in coating and filling applications; is soft and nonabrasive; has low conductivity of heat and electricity; is very fine in particle size and thus disperses readily; and is lower in cost than most materials with which it competes. Many grades of kaolin are especially designed for specific end uses, particularly those used in paper, paint, rubber, ceramics, plastics and ink.

The term kaolin in addition to being a rock name is a group name for the minerals kaolinite, nacrite, dickite and halloysite. Kaolinite is by far the most common mineral of the group. With the exception of the hydrated form of halloysite, which has two more water molecules per unit cell, all these minerals have essentially the same composition. Halloysite has an elongated tubular shape, whereas the other kaolin minerals are pseudohexagonal (Fig. 1).

Kaolins are classed as primary or secondary deposits. Primary kaolins are those that have formed

Table 1
Chemical analysis and physical properties of a Georgia kaolinite and Wyoming bentonite

Property	Georgia kaolinite	Wyoming bentonite
Specific gravity	2.6	2.0
Hardness (Mohs scale)	1.5–2.0	1.5–2.0
Refractive index	1.57	1.64
Bulk density (kg m^{-3})	320–800	640–960
Oil absorption (cm^{-3} g)	0.25–0.60	0.20–0.30
Color	Off white	Brown, tan, cream
Base-exchange capacity (Meq/100 g)	1–2	75–100
Brightness (457 μm%)	78–90	60–75
Chemical composition (%)		
SiO_2	45.30	54.62
Al_2O_3	38.38	20.08
Fe_2O_3	0.30	3.39
TiO_2	1.44	0.11
MgO	0.25	2.15
CaO	0.05	0.40
Na_2O	0.27	2.26
K_2O	0.04	0.35
Loss on ignition (%)	13.97	16.36

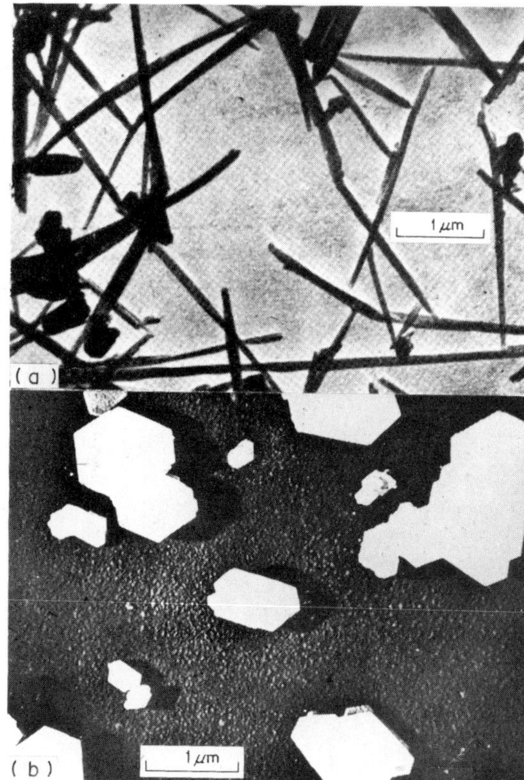

Figure 1
Electron micrographs of kaolin minerals: (a) halloysite, (b) kaolinite

by the alteration of crystalline rocks such as granite, and remain in the place where they were formed. The English china clay deposits are an example of this type. Secondary deposits of kaolin are sedimentary, and have been transported from their place of origin and deposited in beds or lenses associated with other sedimentary rocks, such as sands. The Georgia–South Carolina deposits are examples of secondary or sedimentary deposits. Most kaolins are processed to some extent because none are naturally pure. Beneficiation is accomplished using either a dry or a wet process. All kaolins used for coating paper are wet processed, because the product is more uniform, is relatively free from impurities and has a better color.

The dry process is simple in that the kaolin is crushed to approximately egg size, dried, disintegrated, and air floated to remove the grit particles. This process yields a lower cost and lower quality product than the wet process. In the wet process, the kaolin can be beneficiated to produce a relatively pure product with strict specifications as to particle size, brightness, grit content, and other properties. A general flowsheet for wet processing is shown on Fig. 2. Leaching is to remove iron coloration, delamination is a process to reduce coarse kaolin stacks to thin plates and calcination is thermal treatment of the kaolin to develop certain properties for special uses. With the exception of high-intensity wet magnetic separation, the other processes are widely used in mineral processing. The major impurities that are detrimental to color are iron and titanium oxides and hydroxides, and a major portion of these fine mineral particles can be extracted using high-intensity magnetic separation.

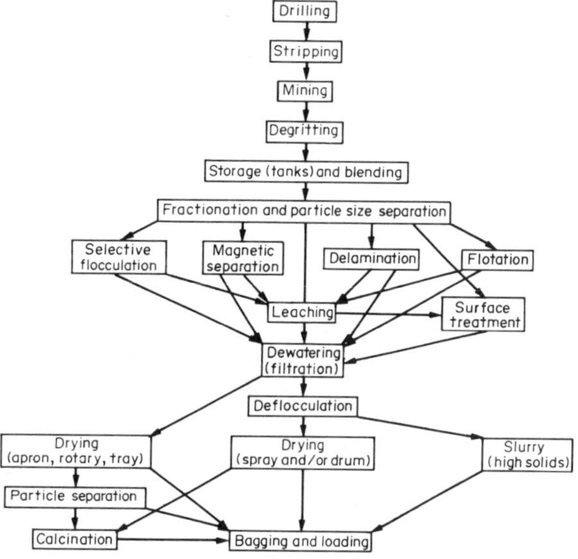

Figure 2
Generalized flow sheet for wet processing of kaolin

1.1 Major Uses

Kaolin used in paper fills the interstices of the sheet, producing better color, higher opacity, better printability and lower costs. As a coating on paper, kaolin gives the sheet better smoothness, gloss, brightness, opacity and printability, in that colored inks have much more fidelity when printed on kaolin-coated sheets. The low viscosity of kaolin at both high and low shear rates is a very important property, because at today's very high-speed production rates, the coating must be applied at a high solids content, and still give the correct coating thickness and smoothness to the paper. High-quality paper-coating kaolin flows readily when applied with very high-speed coating equipment, giving the paper a smooth and even-coated film.

The rubber industry uses large amounts of kaolin as a filler or extender in both natural and synthetic rubber. Kaolin is incorporated in the rubber mix to lower costs and improve properties such as strength, abrasion resistance and rigidity. Through the years, the terms hard and soft kaolins have become commonplace in describing kaolins used by the rubber industry. Hard kaolin has a very fine particle size that improves the tensile strength of the rubber and its resistance to tear and abrasion. Soft kaolin is coarser in particle size, and is used to lower elasticity and improve abrasion resistance.

Kaolin is used as an extender in paints, because it is chemically inert, has a high covering power, gives desirable flow properties, is low in cost, is white and reduces the amount of expensive pigment required. In addition, kaolin has excellent suspension properties and is available in a wide range of particle sizes which can be used in many types of paints. Coarse-particle kaolins are used in paints where a dull or flat finish is required, and fine-particle kaolins are used in high-gloss paints. Large quantities of calcined and water-beneficiated kaolins are used in interior wall paints and metal primers. Calcined kaolin is particularly suited for use in paint as an extender for TiO_2, because of its resistance to abrasion and dry covering properties.

Kaolin is used extensively as a filler in plastics. The desirable characteristics that can be obtained with kaolin include smoother surfaces, more attractive finishes, dimensional stability, and resistance to chemical attacks. Manufacturers of poly(vinyl chloride) (PVC) use kaolin as a reinforcing agent, and it makes the plastic more durable. Calcined and partially calcined kaolins are used as a filler in PVC wire insulation to improve electrical resistivity. In manufacturing glass-reinforced polyesters, kaolin has helped eliminate flow problems which hindered production of large products such as boat hulls. The use of kaolin actually results in a stronger and more uniform body.

Kaolin has many other important filler applications; these include inks, adhesives, insecticides, medicines, catalyst preparations, fertilizers, detergents, pastes, linoleum, textiles and cosmetics.

1.2 Specifications and Grades

For paper filler application, coarse, medium and fine particle sizes are used. The coarse fillers have particle sizes of 20–40% <2 μm, the intermediate fillers have particle sizes of 40–60% <2 μm and the fine filler kaolins have particle sizes in the range of 60–80% <2 μm. In addition to particle size, a brightness of 80–85% is specified by the Technical Association of the Pulp and Paper Industry (TAPPI). Another important specification is screen residue, which is called grit. This is material held on a 325 mesh (44 μm) screen. Because of abrasivity and scratching, only a very small percentage of grit (about 0.01%) can be tolerated. Another property that is important for a paper filler is its abrasion index. For clays this is usually not a problem, although some primary kaolins contain fine quartz and mica that cause abrasion problems. The test used in the paper industry is the Valley Abrasion test which again is described by TAPPI.

The most common coating clay grades are shown in Table 2. The #3 and #2 coating grades are generally used in high-solids formulations, for publication-grade paper and medium-finish lower cost enamels. The #1 grade is used to coat high-quality boxboard and enamel papers. This grade gives excellent gloss and brightness and relatively uniform ink receptivity and good smoothness. The fine #1 grade is used where maximum gloss and very good ink holdout are desired.

Table 2
Particle size and brightness of some coating clay grades

Clay grade	Particle size (% <2 μm)	Brightness (%)
Regular		
#3	73	85–86.5
#2	80–82	85.5–87.0
#1	90–92	87.0–88.0
fine #1	95	86.0–87.5
Delaminated		
regular	80	88.0–90.0
fine	95	87.0–88.0
High-brightness		
#2	80	89–91
#1	92	89–91
Calcined		
dehydroxylated	60	85.5–88
fully calcined	90	90–92

The regular-delaminated grade is used to coat lightweight paper and paper that has a rough surface, where good printability must be maintained. This clay makes up some of the base-sheet deficiencies, and gives coatings that are exceptionally smooth with excellent ink and varnish holdout. The fine-delaminated grade is used to achieve maximum gloss, good runnability and high opacity.

The high-brightness grades are very bright and have a blue-white tint. The coatings using the #1 high-brightness grade are exceptionally smooth, and give excellent gloss, ink holdout and good printability. These high-brightness grades are used to coat premium grade paper.

The dehydroxylated or partially calcined kaolin is used as a paper coating additive to enhance resiliency and opacity in low basis-weight sheets. The addition of dehydroxylated kaolin to a coating formulation provides increased thickness to the coating, increased opacity, a significant reduction in loss of brightness and opacity on calandering, improved ink receptivity and more uniform color printing. The fully calcined kaolin coating clay increases light scattering of the film in proportion to its content. This enhanced light scattering makes it possible to achieve opacity and brightness specifications at lower levels of titanium dioxide, which reduces the pigment coat required.

A very critical specification for coating kaolins is viscosity. The flow properties of kaolin are very important to the paper coater because they affect the functioning of the coating operation, as well as the final quality of the paper product. Two viscosity tests are used, one to measure high-shear viscosity and the other to measure low-shear viscosity. The measurements are normally made on slurry suspensions at 71% solids. A Brookfield four-speed viscometer is used to measure low-shear viscosity, and a Hercules viscometer is used to measure high-shear viscosity.

In addition to the particle size, brightness, pH, moisture and grit specifications, kaolins used in paints and plastics must meet certain electrical resistivity specifications. The resistivity test is designed to given an indication of the amount of residual salts that are contained in the clay; high resistivity values reflect a low soluble salt content. For certain paint applications, the Hegman fineness-of-grind test specification is important. This test is intended to measure the degree of dispersion, and involves the use of a fineness gauge and a scraper. The gauge is calibrated from 0 to 8, and the fineness is reported by number. The higher the number, the finer the particles and the better the dispersion. Some paint manufacturers specify that a pigment or extender must be finer than a particular Hegman number.

2. *Bentonite*

The definition of bentonite proposed by Grim at the International Clay Conference at Madrid, Spain, in 1972 is used here. According to Grim, bentonite is a clay consisting essentially of smectite minerals regardless of origin or occurrence.

One way of classifying bentonite is based on its swelling capacities when wet, or added to water. Bentonite containing sodium as the dominant exchangeable ion, typically has very high swelling capacities and forms gel-like masses when added to water. This type of bentonite is produced in Wyoming, Montana and South Dakota, and is commonly called Wyoming or Western bentonite. Bentonite in which the major exchangeable cation is calcium has a much lower swelling capacity, and is commonly referred to as subbentonite or Southern bentonite, because it occurs in Texas, Louisiana and Mississippi.

Bentonite production in the USA in 1980 was slightly over 4 Mt, and the world production was estimated to be about 8 Mt. The processing of bentonite is rather unusual, in that the mining is openpit using shovels or payloaders. The bentonite is crushed to egg size or smaller, dried in rotary flash or fluidized bed driers, and screened and/or further ground to meet the size specifications required. A very small proportion of the bentonite is wet processed using centrifuges, filter presses or rotary vacuum filters for dewatering, and flash drying to produce very pure special grades. Also some bentonite is surface coated with organic compounds such as amines, for use in paints, greases and oil-based drilling muds.

Bentonite is produced in many other countries of the world including Mexico, Canada, Argentina, Brazil, Peru, Cyprus, Czechoslovakia, France, Greece, Hungary, Italy, Poland, Romania, Spain, the UK, Russia, the Federal Republic of Germany, Yugoslavia, Algeria, Morocco, Mozambique, South Africa, Japan, Pakistan, Turkey and New Zealand.

Although the major uses of bentonites are in drilling muds (see *Well-Drilling Materials: Clay and Nonclay Minerals*), foundry bonding agents and bondants for metal ore pellets, some bentonites are used in cosmetics, pharmaceuticals, paints, detergents, grease and paper. Near Gonzales, Texas, and on the Island of Ponza near Naples, Italy, white bentonites are mined that are particularly good for applications where color is important (e.g., pharmaceuticals, paints, detergents and paper).

A special product, an organic-treated bentonite, is used to make thick nondrip paints, to gel organic liquids and to produce greases having superior adherence to metal, ability to repel water and resistance to high temperatures. This organic-clad bentonite is processed in such a way that the original inorganic exchangeable ion on the smectite is replaced by an alkali amine organic cation. This reaction produces a hydrophobic clay, because the inorganic ions that can be hydrated are removed, and a large part of the mineral surface formerly capable of absorbing water is coated by hydrocarbon chains. These special products are made in the USA, Great Britain, France, the Federal Republic of Germany and Japan. The processing is done in water after a large proportion of the nonsmectite minerals have been removed by centrifuging or sedimentation. The drying is accomplished using special flash dryers.

Another special smectite product is used to coat paper for making multiple copies without using carbon paper. This product is wet-processed and treated with acid to activate the smectite. The activator removes the sodium, calcium and some aluminum and/or iron, making the clay surface highly charged with hydrogen ions. This special smectite clay is used to coat the receiving surface of the copy sheet; it contains about 25% smectite, with starch, dextrin and butyl rubber as the binder. The transfer surface is coated with starch containing minute encapsulated droplets of colorless dyestuffs, such as triethynyl methane derivatives with a lactone structure. Writing or typing breaks the dyestuff globules, so that they penetrate the smectite-coated layer, which catalyzes their conversion into a colored letter or mark. This type of clay is produced in the USA, Mexico, Japan and the Federal Republic of Germany.

See also: Fillers and Coatings: Industrial Minerals; Fillers and Coatings: Siliceous Minerals; Fillers and Coatings: Sulfate Minerals; Fillers in Elastomers

Bibliography

Murray H H 1984 Clay. In: Hagemeyer R W (ed.) 1984 *Pigments for Paper*. Technical Association of The Pulp and Paper Industry, Atlanta, Georgia, pp. 95–193

Patterson S H, Murray H H 1983 Clays. In: Lefond S J (ed.) 1983 *Industrial Minerals and Rocks*, 5th edn., Vol. 1. American Institute of Mining, Metallurgical, and Petroleum Engineers, New York, pp. 585–651

H. H. Murray

Fillers and Coatings: Industrial Minerals

The use of finely powdered minerals to load or fill such materials as paper has been common practice for thousands of years, and for other materials such as rubber and plastics a relatively few tens of years. In addition to minerals being incorporated in the body of the paper itself, minerals are also applied as a thin film or coating on the surface. The choice of mineral fillers and coaters is wide, and the selection is based upon the properties desired in the finished product, and the cost.

Mineral fillers are defined as inert materials that are incorporated in a composition for some useful purpose. Included are a broad group of minerals: asbestos, barite, bentonite, calcite, diatomite, feldspar, kaolin, micas, nepheline syenite, perlite, silica, talc, titanium dioxide, tripoli and wollastonite. Mineral fillers can be included in compounds to attain a variety of resultant properties. Hardness, brittleness, impact strength, softening point, fire resistance, surface smoothness, color, opacity, electrical conductivity and a multitude of other physical characteristics may be modified by the inclusion of fillers.

The characteristics of the product are the result of the physical and chemical properties of the fillers or coaters, including chemical activity, hardness, particle size, shape and distribution, color, density, surface area, structure, chemistry, refractive index, and dispersibility. In addition to the physical and chemical characteristics, a most important consideration is cost. Certainly, the inclusion of a filler that costs 10¢ kg^{-1} will reduce the cost of a plastic whose resin costs \$1 kg^{-1}, or pulp for paper that costs 55¢ kg^{-1}. The decision on the use of a particular filler, however, is not as simple as just reducing the cost. For example, a certain filler may react with a particular ingredient, making the compound exceedingly viscous such that it will not flow into a mold as an injection molding compound, although it may be very useful in certain caulking compounds to prevent sagging. Thus, it is very important to evaluate a filler in any compound with which it is to be used, in order to determine fully the resultant properties.

Fillers can be used to modify cost, flow characteristics and other properties, but seldom is the desired modification achieved without some sacrifice of other qualities. For example, some plastics may become excessively brittle when cost-reducing fillers are incorporated, or a paint may tend to settle out in the can when antichalking fillers are added. Therefore, some properties are usually compromised in order to achieve a degree of modification that is acceptable. In many cases, a combination of fillers is used to attain a number of desirable modifications; for example, a paint may contain kaolin, calcite, talc, bentonite and perhaps other fillers. One filler may be substituted for another in some uses, but as technology advances and specifications become tighter, substitution becomes more difficult. Replacement or partial substitution may require total reformulation, which the consumer is reluctant to undertake unless significant cost and/or performance improvements are likely to be forthcoming.

The physical characteristics of a filler or coater are the result of the inherent characteristics of the mineral and the features imparted by processing. Inherent characteristics include hardness, particle size and shape, color, refractive index, chemical composition and chemical reactivity. The presence of trace quantities of other minerals may preclude the use of a filler or coater in certain applications if that particular trace mineral cannot be removed during processing. Hardness of a mineral affects its utilization as a filler or coater. Silica, for example, has a hardness of 7 on the Mohs scale, and will score and wear blades, knives and rolls, meaning that in most instances a softer mineral such as calcite, kaolin or talc must be used. Particle size and particle size distribution are probably the most important physical characteristics of a filler or a coating material. The properties affected by particle size and size distribution include smoothness, viscosity, binder demand, film integrity, density, gloss, color and, in the case of coated paper, the printing characteristics and ink receptivity. Particle shape affects utilization, in that acicular particles such as asbestos provide reinforcement and strength, while plates of kaolin and mica provide a tough, smooth surface film. Color is very important in many products, so the color of the filler or coating pigment becomes an important characteristic. Most mineral fillers and coaters are more desirable the whiter they are. Index of refraction is closely related to color and is very important in some applications. If the index of refraction of the mineral filler is very close to that of the material it is filling, the color of the filler will be very faint and the product will be translucent, whereas if there is a large difference the product will become opaque. Chemical composition is important and affects the use of many fillers. Calcium carbonate cannot be used in acid systems because it is soluble, so a nonreactive filler such as kaolin must be used. Many other physical characteristics are important and influence the intended use of the filler or coater mineral.

Fillers are used extensively in many applications. They are utilized in bituminous compositions, including asphalt shingles and roll roofing, road-paving asphalt, flooring materials, waterproof coatings and other compositions where up to 50% loadings are not uncommon. Both the thermoplastic and thermosetting branches of the plastics industry use fillers, not just to gain cost savings, but to perform a function in determining the properties of the final product—in some cases up to 80% loadings are used. The rubber industry uses fillers to achieve hardness, abrasion resistance, tensile strength and to modify several other properties. Large quantities of fillers and coaters are used by the paper industry (estimated to be over 3 million tonnes in the USA in 1983) to improve opacity, smoothness, ink receptivity, body, feel and printing fidelity. The paint industry utilizes functional fillers to control flow characteristics, to improve sheen or control gloss or flatness, to enhance film strength, scrubbability and weather resistance, to reduce film tack, and to influence many other properties. Other areas where fillers are employed include fertilizers, adhesives, textiles, cement, caulking, inks, lubricants, food, cosmetics, pharmaceuticals and insecticides.

Coater minerals are used primarily by the paper industry, and to some extent by the wire and cable industry. The coater minerals are suspended in an adhesive–water mixture such as starch or synthetic latex. Normally, the adhesive content is 12–18% based upon the dry weight of the coater minerals. The purpose of coating paper is to improve the printing characteristics, including brightness, opacity, gloss, smoothness and ink receptivity. In wire and cable coatings, minerals are used to improve the electrical characteristics and must have a high dielectric constant and low heat conductivity.

The filler and coater minerals are a necessity in many industries because they are not just inert cost-saving space fillers, but perform significant functions in controlling the physical and chemical properties of the finished product.

See also: Fillers and Coatings: Carbonate Minerals; Fillers and Coatings: Clay Minerals; Fillers and Coatings: Siliceous Minerals; Fillers and Coatings: Sulfate Minerals

Bibliography

Deanin R D, Schott N R (eds.) 1974 *Fillers and Reinforcements for Plastics,* Advances in Chemistry Series No. 134. American Chemical Society, Washington, DC

Hagemeyer R W 1976 *Paper Coating Pigments*, TAPPI Monograph Series No. 38. Technical Association of the Pulp and Paper Industry, Atlanta, Georgia

Patton T C 1973 *Pigment Handbook*, Vol. I, *Properties and Economics*. Wiley, New York

Technical Association of the Pulp and Paper Industry 1958 *Paper Loading Materials*, TAPPI Monograph Series No. 19. Technical Association of the Pulp and Paper Industry, Atlanta, Georgia

H. H. Murray

Fillers and Coatings: Siliceous Minerals

A large number of siliceous minerals are used as fillers and coatings. This article concentrates on the following: asbestos, diatomite, perlite, pumice, pyrophillite, silica and talc.

Industrial minerals in this category have some common characteristics: all have a refractive index in the range 1.4–1.7, a specific gravity of 2.0–2.9 and a pH of 6–10. They all receive processing that enhances their natural beneficial properties and which allows them to be used in a variety of applications. Some receive additional processing: for example, perlite is thermally expanded to a material of low density similar to pumice, and diatomite is flux calcined to improve color. Generally they all contribute economic benefits in their applications. They supply bulk, porosity, selective abrasiveness (hardness), strength and/or reactivity.

1. Sources

Table 1 gives an abbreviated summary of the principal states or countries where there are operating deposits of siliceous minerals. The sources are listed in an approximate order of decreasing annual production. However, since these minerals are used in a broad variety of applications that include nonfiller applications, it is difficult to assign an order of filler production.

2. Mineralogy

Typical mineral compositions of the siliceous fillers discussed in this article are shown in Table 2. Actual chemical compositions of commercial products vary from these values owing to dilution by impurities. Generally, small amounts of impurities do not adversely affect the utility of the product for filler uses, although they may affect color or brightness which is important in some applications, determining a premium price for purer deposits. Impurities may affect expansion characteristics of perlite, and some

Table 1
Main sources of supply of fillers

Asbestos	Diatomite	Perlite	Pumice	Pyrophillite	Silica	Talc
Canada	USA	USA	USA	USA	USA	USA
Quebec	California	New Mexico	Arizona	North Carolina	Illinois	California
Newfoundland	Nevada	Arizona	California	Japan	Missouri	Texas
Ontario	Washington	California	New Mexico	South Korea	Arkansas	Nebraska
British Columbia	Oregon	Idaho	Idaho	Australia	Oklahoma	Vermont
Yukon	Arizona	Colorado	Colorado	Canada	Tennessee	Montana
USSR	France	Nevada	Oregon	India	Georgia	New York
Central Urals	Denmark	Hungary	Nevada	Thailand	Pennsylvania	Belgium
Kazakhstan	USSR	Greece	Iceland	South Africa	New Jersey	Italy
Tuva		USSR	Hungary		West Virginia	Canada
South Africa		Italy	Italy		Texas	France
Zimbabwe		Japan	New Zealand		Canada	Finland
China		China	Greece		Ontario	Japan
Italy		Bulgaria			Quebec	
Swaziland		Turkey				
		Mexico				
		Australia				

applications of silica such as ceramics that require high purity for color.

Asbestos. Asbestos is a name which applies to a group of natural fibrous minerals, of which chrysotile (a type of serpentine) and the amphiboles have the greatest commercial importance, with chrysotile accounting for over 90% of commercial asbestos. Chrysotile probably results from two geologic steps: metamorphism of volcanic ultrabasic rock to serpentine, followed by solution recrystallization of serpentine in cracks and fissures as chrysotile.

Diatomite. Diatomite or kieselguhr is a sedimentary material formed by centuries of life cycles of aquatic diatoms, a simple plant in the algae family with an opaline silica cell wall. Thousands of species of diatoms have flourished and continue to do so in both marine and lacustrine environments. Fossilized skeletal remains of diatoms in commercial quantities are found in many parts of the world.

Table 2
Typical mineral compositions (%) of siliceous fillers

Mineral	MgO	SiO_2	H_2O	FeO
Asbestos				
Chrysotile	40	40	12	
Amphiboles				
Amosite	5	50	4	40
Anthophyllite	30	56	4	10
Crocidolite[a]		50	4	15
Diatomite		86–92	0–4	
Perlite[b]		74	4	
Pyrophillite[c]		67	5	
Silica		97–99+		
Talc				
Theoretical	32	64	5	
Typical	26–33	35–62	5	

a Plus 5% NaO, 18% Fe_2O_3 b Plus 4% Na_2O, 4% K_2O, 12% Al_2O_3 c Plus 28% Al_2O_3

Perlite. Perlite is believed to result from hydration of volcanic glass or obsidian. Generally, hydration is about 2–5%; this water content is important to the expansibility of the perlite, influencing melting point and supplying expansion steam.

Pumice. The rapid expansion of dissolved gases in silica lavas during volcanic eruptions produces the light density pumice or pumicite. The finer pumicite particles are transported by wind away from the source volcano, whereas pumice accumulates closer to the vent.

Pyrophillite. The hydrous aluminum silicate, pyrophillite, is formed by hydrothermal metamorphism of acid tuffs or breccias.

Silica. Silica sand is frequently obtained from the weathering of quartz-containing rock. Decomposition and disintegration of the rock with decomposition of other minerals leaves a primarily quartz sand that has been concentrated by water movement. Induration of sands to sandstone results in another source for silica sand. Amorphous silica, or more properly cryptocrystalline or microcrystalline silica, is formed by the slow leaching of siliceous limestone or calcareous chert.

Talc. Talc is formed by the metamorphic (hydrothermal) alteration of magnesium silicates such as serpentine, pyroxene or dolomite.

3. Physical Properties

These siliceous fillers are generally inert in most applications as shown by pH values in the range

6–10. Asbestos, diatomite and expanded perlite can contribute higher oil absorptions due to their structure or porosity. The other materials have lower oil absorptions of less than 50 g per 100 g. Some grades of talc, silica and diatomite can offer high brightness or white color that enhances certain applications. All of these fillers have refractive indices around 1.5 which is similar to many application liquids, giving them little pigment value. They can, however, contribute by acting as pigment extenders, spacing the individual higher refractive-index pigment particles so that they contribute to the maximum extent. The porosity of diatomite and expanded perlite can give some hiding power when formulated above the critical pigment volume concentration.

Typical filler physical properties are shown in Table 3.

Table 3
Typical physical properties

	pH	Oil absorption (g per 100 g)	Brightness (%)[a]	Specific gravity	Refractive index
Asbestos					
Chrysotile	9–10	50–180	<80	2.56	1.53–1.56
Diatomite	6–9.6	90–300	65–90	2.0–2.3	1.40–1.49
Perlite (expanded)	7.6–9.0	50–275	80–90	2.2–2.6	1.48–1.49
Pumice	7–9	30–40		2.2–2.6	1.49–1.50
Pyrophillite	6–8	25–70	80	2.8–2.9	1.57–1.59
Silica					
Cryptocrystalline	6.8–7.2	15–30	81–90	2.63–2.65	1.54–1.55
Quartz	7.0	13–50	53–89	2.65	1.55
Talc	8.1–9.5	20–52	72–94	2.6–2.9	1.54–1.59

a Using MgO standard

cations. Perlite is always thermally expanded or "popped" like popcorn to a much lighter density material.

Table 4 summarizes and compares the variety of techniques used to mine and produce these fillers. It also shows overall production tonnages. Production includes uses other than fillers and coatings, since it is not possible to develop figures by individual market areas.

Most of the products typically result in over 80% yield on a dry basis. Some loss results from size-reduction processes which grind material into unwanted fines. Loss also occurs due to beneficiation or classification designed to remove some of the impurities included during mining or interlayered in the deposit in a way that selective mining would not avoid. The lowest recovery is asbestos which is found in thin layers or bands in serpentine rock. A great amount of rock must be mined and processed to recover the asbestos with yields of about 10%.

4. *Mining and Production Techniques*

Although a variety of mining and production methods are used with these minerals, they are generally all mined by open pit techniques; asbestos, silica and talc, however, are sometimes obtained by underground mining. The harder minerals like perlite, asbestos, pyrophillite, silica and talc usually require drilling and blasting, although occasionally the minerals can be ripped or bulldozed. All the minerals undergo some drying, crushing, and/or milling and air classification and/or screening. They are marketed in a series of particle size ranges from coarse to fine or ultrafine.

Specialized processing techniques, such as wet classification by froth flotation, hydrocyclones and magnetic separation, are used to achieve color improvements in silica and talc. Fluidized energy milling is employed to achieve extra-fine size reduction in silica and talc. Diatomite often receives flux calcination to improve its color for filler appli-

5. *Safety*

Nearly all these minerals have some hazards in their use if only as "nuisance dusts" that have low levels of hazards, but usually require dust control, ventilation, and use of dust masks. Asbestos, silica (quartz) and flux-calcined diatomite (cristobalite), require more stringent control to ensure that safe dust counts are achieved in dry use areas (see *Asbestos Hazards*).

6. *Industrial Uses*

In coating applications, the properties of these fillers are used to achieve a variety of functions. They can control gloss and sheen, modify consistency and flow,

Table 4
Typical mining and production techniques

Mining and production	Asbestos	Diatomite	Perlite	Pumice	Pyrophillite	Silica	Talc
Mine type							
open pit	×	×	×	×	×	×	×
underground	×					×	×
Ore mining method							
drilling and blasting	×		×		×	×	×
hydraulic jetting						×	
ripping		×	×				
bulldozing		×	×		×		×
dredging						×	
drag line on shovel	×	×		×	×	×	×
trucks	×	×	×	×	×		×
Ore processing methods							
crude blending		×			×		×
primary crushing	×	×	×		×		×
drying	×	×	×	×	×	×	×
crushing/milling	×	×	×	×	×	×	×
classification	×	×	×	×	×		×
benefication		×					×
screening	×		×	×	×	×	×
wet classification						×	×
flash drying							×
air classification	×	×	×	×	×	×	×
fluid energy milling						×	×
calcination		×					
thermal expansion			×				
post calcination/thermal expansion/classification		×	×				
Product size							
coarse	×	×	×	×	×	×	×
medium	×	×	×	×	×	×	×
fine	×	×	×	×	×	×	×
ultra-fine		×				×	×
Typical % yields (dry basis)[a]	10	80–90	80–90	(90)	(90)	(90)	(90)
World production (10^3 t)	5700	1700	1800	16000	1700		5000
US production (10^3 t)	93	650	860	4000	127	30000	1200

a Estimated values are given in parenthesis

influence film permeability, provide "tooth," improve solvent release, and improve sanding resistance. They also can control polishing of flat paints, transparency in clear finishes, corrosion resistance, tint retention, improved touch-up or lapping, suspension, stain resistance, weathering, dispersion and color uniformity. Generally, fillers or pigment extenders are used in combination, in an attempt to achieve the best balance in properties.

In caulking, plastics, asphalt, rubber and adhesives, they can supply viscosity, surface appearance, strength, water resistance, antiblock, bulking, resiliency, and higher softening point.

Other applications of siliceous fillers and coatings include:

(a) Roofing, where they can supply weathering resistance and strength.

(b) Paper, where they can supply "tooth", pitch control, porosity, and color. They can influence printing characteristics, strike-through, and opacity.

(c) Insecticides, where they are used to provide chemical inertness and controlled dilution for even distribution.

(d) Insulation, concrete, plastic, acoustical products, and refractory building blocks, where they supply light density, thermal stability, low heat conductivity, and reaction with cement.

Table 5
Typical uses of mineral filler

Mineral	Uses
Asbestos	Roofing cement, caulking, paints, plastics, cement products
Diatomite	Paints, abrasives, plastics, fertilizers, paper, insecticides, insulation, filter aids
Expanded perlite	Acoustical, insulating and horticultural products, concrete, plaster, abrasives, paints, filter aids
Pumice	Concrete, abrasives, roofing, paints, insecticides, building blocks
Pyrophillite	Refractories, insecticides, ceramics, asphalt, floor tiles, paints, crayons
Silica	Cement, ceramics, paints, plastics, adhesives, abrasives, caulking, plastics, rubber, filter aids
Talc	Paints, paper, cosmetics, plastics, rubber, ceramics, insecticides, roofing

(e) Abrasives, where they supply controlled particle size, friability, and various hardnesses (see *Abrasives: Siliceous Minerals*).

The wide variety of uses for these minerals is outlined in Table 5.

See also: Fillers and Coatings: Carbonate Minerals; Fillers and Coatings: Clay Minerals; Fillers and Coatings: Industrial Minerals; Fillers and Coatings: Sulfate Minerals

Bibliography

Patton T C (ed.) 1973 *Pigment Handbook*, Vol. 1: *Properties and Economics*. Wiley, New York

Severinghaus N Jr 1983 Fillers, filters and absorbents. In: Lefond S J (ed.) 1983 *Industrial Minerals and Rocks: Nonmetallics other than Fuels*, 5th edn. American Institute of Mining, Metallurgical and Petroleum Engineers, New York, pp. 243–57

Speil S, Leineweber J P 1969 Asbestos minerals in modern technology. *Environ. Res.* 2: 166–208

G. Coombs

Fillers and Coatings: Sulfate Minerals

Sulfate minerals, and in particular gypsum and barite, are used on a limited scale as mineral fillers and coatings, chiefly in paints, rubber, plastics, paper and textiles. These sulfate minerals are generally of high purity, are easily ground and classified, are white to near-white in color, and are chemically inert. Barite is especially useful because of its high density. Commercial deposits of gypsum generally contain large reserves and are widespread, whereas barite resources are more limited both in distribution and reserves.

1. Sources

Gypsum reserves are large and commercial deposits are widely distributed in the USA and throughout the world. The mineral has a low unit value, one of the chief attributes of using gypsum as a filler, and transportation costs are kept at a minimum by mining deposits close to industrial and population centers.

Major gypsum-producing districts in the USA and Canada are: the Permian Basin region of New Mexico, Texas, Oklahoma and Kansas; the Michigan Basin and surrounding areas in Michigan, Ohio, New York and Ontario; and the maritime provinces of Nova Scotia and Newfoundland. Other important districts are in southern Indiana, Iowa, southern California, Nevada, Wyoming, Arizona and Colorado. Other North American districts include British Columbia, San Marcos Island and the Caribbean nations of Jamaica and the Dominican Republic. Major gypsum-producing countries on other continents include the UK, Argentina, Brazil, France, Italy, Spain, the USSR, Egypt, South Africa, India and Iran.

US consumption of gypsum in 1980 was about 18.4 Mt. Most of the gypsum was used in the construction industry (wallboard, plasters and cement retarders) or in agriculture, and less than 2% was used as fillers and coatings.

Barite reserves are somewhat limited in the USA and on a worldwide basis. The large demand in recent years for barite in oil-field drilling muds has increased US consumption to about 3.2 Mt, of which about 3% is now used as fillers and coatings.

The major barite deposits in the USA are in Nevada, which yielded about 85% of the total US output in 1980. Other major world producers include the People's Republic of China, the USSR, Ireland, Peru, India, Thailand and Mexico.

2. Mineralogy and Physical Properties

Gypsum is the name given to the mineral that consists of hydrous calcium sulfate ($CaSO_4 . 2H_2O$), and also to the sedimentary rock that consists primarily of this mineral. In its pure state, gypsum contains 32.6% lime (CaO), 46.5% sulfur trioxide (SO_3), and 20.9% water. Single crystals and rock masses that approach this theoretical purity are generally colorless to white, but in practice, the presence of impurities such as clay, dolomite, silica and iron imparts a gray, brown, red or pink color to the rock.

There are three common varieties of gypsum: selenite, which occurs as transparent or translucent crystals or plates; satin spar, which occurs as thin veins (typically white) of fibrous gypsum crystals; and alabaster, which is compact, fine-grained gypsum that has a smooth, even-textured appearance. Most deposits of rock gypsum that are suitable for industrial purposes are aggregates of fine to coarse gypsum crystals that have intergrown to produce a thick, massive sedimentary rock unit that is 90–98% gypsum. Alabaster is highly prized because of its uniformly fine particle size, but the more common deposits of rock gypsum consisting of coarser-grained selenite can generally be crushed and ground to produce a suitable filler and coating material.

Gypsum has a hardness of 2 on the Mohs scale, and can be scratched with the fingernail. Large rock masses are easily crushed and ground to a fine powder. The specific gravity of pure gypsum is about 2.31, and the refractive index is about 1.53. Gypsum is slightly soluble in water, but it is an inert substance that resists chemical change. The oil-absorption capacity of gypsum is fairly low (0.17–0.25 $cm^3\,g^{-1}$).

Raw or crude gypsum is one of the forms used as fillers and coatings, but for some purposes calcined or deadburned gypsum is desired. In calcining, the gypsum is heated to about 120–160 °C to drive off free water and partially remove the water of crystallization. The calcined material, or stucco, has a chemical composition of $CaSO_4 . \frac{1}{2}H_2O$, and it readily takes up water. Calcination at higher temperatures (500–725 °C) results in a product called deadburned gypsum, which has a composition of $CaSO_4$.

Anhydrite, a sulfate mineral and rock that is closely associated with gypsum in nature and has minor uses as a filler, is anhydrous calcium sulfate ($CaSO_4$) containing 41.2% CaO and 58.8% SO_3. It is typically fine grained (like alabaster), and occurs in thick, massive sedimentary rock units. Anhydrite usually is white or bluish gray when pure, but it may be discolored by impurities. Anhydrite has a hardness of 3.5, a specific gravity of 2.98, and a refractive index of 1.57–1.61.

Barite, also called barytes or heavy spar, is a mineral that consists of barium sulfate ($BaSO_4$). Pure barite, which is 65.7% BaO and 34.3% SO_3, is white, but many commercial deposits are discolored by small amounts of impurities.

Commonly, barite is well crystallized and occurs as thin to thick tabular crystals that have intergrown as aggregates. Commercial deposits of barite occur (a) as irregular masses, concretions, or nodules in vein and cavity fillings; (b) as residual masses produced by weathering of preexisting rocks; and (c) as fine-grained, massive rock layers in bedded deposits several centimeters to more than 15 m thick. Any of these forms of barite is readily crushed and ground to produce fillers and coating material, although the vein or residual types are more commonly used for this purpose because of their white or off-white color.

Barite has a hardness of 2.5 to 3.5, and therefore cannot be scratched with the fingernail. The calculated specific gravity of pure barite is 4.5, but inclusions and chemical impurities in commercial deposits cause this value to range from 4.3 to 4.6. Barite is thus the highest density mineral among those commonly used as mineral fillers. The refractive index of barite is about 1.64. The mineral is relatively insoluble in water and acids, and can thus be used as a chemically inert material. The oil-absorption capacity of barite is low, with values of 0.06–0.10 $cm^3\,g^{-1}$.

Celestite is a high-density strontium sulfate ($SrSO_4$) mineral consisting of 56.4% SrO and 43.6% SO_3. With a density of 3.96, celestite may be a suitable substitute for barite in paints and other uses where extra weight is needed. Celestite has a hardness of 3 to 3.5, a refractive index of 1.62, is chemically inert, and typically is white with a faint blue or red tint. Celestite is commonly a secondary mineral associated with some deposits of limestone, gypsum, or rock salt. Principal production of celestite comes from the UK, Mexico, Spain, and Canada; no deposits in the USA are believed to be commercial at this time.

3. *Mining and Production Methods*

Gypsum is produced both from surface pits and from underground mines. Open-pit mining is most desired where the gypsum bed is within 15 m of the surface and/or where the overburden thickness is only one to three times greater than that of the gypsum bed itself. Draglines, scrapers, or bulldozers are used to remove the overburden, and power shovels or front-end loaders are used to load the gypsum. Explosives are used to break up rock overburden and also are used to fragment the gypsum.

In underground mines, gypsum is fragmented with explosives and is loaded on self-loading dump units or on conveyor belts, rail cars or diesel-powered trucks by front-end loaders or gathering-arm loaders. Square or rectangular gypsum pillars 6–9 m wide are left in place to support the mine roof, whereas the mined-out areas surrounding the pillars commonly are 9–12 m across. Such "room-and-pillar" mining methods enable recovery of up to 75% of the gypsum, and are being employed at depths commonly ranging from 15–180 m below ground.

Individual gypsum mines normally produce 0.1–0.6 Mt annually. The mining process is generally highly mechanized, and working conditions are good because of the lack of explosive gases or harmful dust.

Processing of raw or crude gypsum consists merely of crushing and grinding the mine-run rock to specified sizes. Processing of calcined or deadburned gypsum after heating involves only grinding the material and classifying it by particle size.

Barite is also mined by surface or underground methods, depending upon the depth of the ore body and whether the deposit is residual, vein-type or bedded. Residual barite deposits are normally poorly consolidated and just below the topsoil. They are generally mined by open-pit methods. After removal of topsoil, the barite-bearing ore body (consisting of clay, rock and about 10–25% barite) is removed by power shovel, dragline or front-end loader, and is hauled on dump trucks to a washer plant, where the barite is separated from waste material.

Bedded and vein deposits of barite are mined either by open-pit or underground methods. Open-pit mining of shallow deposits is preferred, because of the lower costs and higher recovery, but underground mines can also be operated because of the integrity of adjacent roof and wall rock in most of these types of deposits. Ore from these deposits must normally be crushed, ground and beneficiated by jigging or by flotation of barite with anionic reagents.

Barite is processed either from crude lumps, or from a jigged or flotation concentrate. The barite is crushed and ground either wet or dry. For some filler uses, bleaching with sulfuric acid is necessary to remove objectionable iron stains.

4. Uses

Gypsum is not a major mineral filler or coating material. It is, however, used on a modest scale in paints, paper, cloth, pesticides, pharmaceuticals (such as aspirin) and toothpaste. Properties that make it especially useful as a filler include its widespread distribution, relatively low unit cost, high purity, white to near-white color, softness, freedom from grit, fine particle size, low solubility and chemical inertness. Gypsum may be used raw or in the calcined or deadburned state. Anhydrite also is used sparingly as a higher density filler in paints.

Barite is more widely used than gypsum as a mineral filler or coating material, with much of its demand resulting from its high density. Barite is used in rubber, paints, plastics, paper and textiles. Barite may be used either raw or after being bleached in sulfuric acid to remove iron impurities. The high density of barite is its main value as a filler, but other desirable properties include its relatively low unit cost, high purity, white to off-white color, moderate hardness, freedom from grit, fine particle size, low solubility and chemical inertness. A specialized use of barite, based on its high density and its absorption of γ radiation, is as a shielding material in nuclear reactors to reduce the amount of expensive lead shielding that would otherwise be needed (see *Radiation Shielding*).

See also: Fillers and Coatings: Carbonate Minerals; Fillers and Coatings: Clay Minerals; Fillers and Coatings: Siliceous Minerals; Fillers and Coatings: Industrial Minerals

Bibliography

Brobst D A, Pratt W P (eds.) 1973 *United States Mineral Resources*, US Geological Survey Professional Paper 820. US Government Printing Office, Washington, DC

Severinghaus N Jr 1983 Fillers, filters and absorbents. In: Lefond S J (ed.) 1983 *Industrial Minerals and Rocks*, 5th edn., Vol. 1. American Institute of Mining, Metallurgical and Petroleum Engineers, New York, pp. 243–57

Sibbing E 1976 Barytes: Raw material for fillers and chemicals. *Ind. Miner. (London)* No. 101: 33–43

US Bureau of Mines 1980 *Mineral Facts and Problems*, US Bureau of Mines Bulletin 671. US Government Printing Office, Washington, DC

K. S. Johnson

Fillers in Elastomers

Finely divided solids are used almost universally in formulating elastomeric compositions. Their function may be to reinforce the rubber, to reduce compound cost, or both. The most important fillers used in rubber are carbon blacks, silica and silicates, clays and whiting (calcium carbonate), although many other materials have found limited application.

Reinforcing effects generally increase with decreasing filler particle size or increasing specific surface area up to a limit of about 300 $m^2\ cm^{-3}$. The morphology of the filler particles is another important factor. In the most highly reinforcing fillers (carbon black and silica), the particles consist of more or less random aggregates of fused, roughly spherical subunits or nodules. The spatial arrangement of the nodules within the aggregates determines the filler structure.

1. Filler Characteristics

The most general technique for estimating specific surface areas is the low-temperature nitrogen-adsorption method. However, in some highly reinforcing carbon blacks and silicas, a portion of the surface area is contained in micropores. A more realistic measure of surface area, approximating the surface accessible to macromolecules, may then be obtained with a larger adsorbate molecule. Such measurements are most easily carried out in aqueous solution using a surfactant as the adsorbate; e.g., in ASTM Method D3765, the adsorbate is hexadecyltrimethylammonium bromide, whose molecules are too large to penetrate the micropore space. Iodine number is used widely as a control test for carbon blacks, but it is not always an accurate reflection of the specific surface area, because the adsorption of iodine is influenced also by surface chemical characteristics. Because of the generally broad size and shape distributions of carbon black aggregates, geometric parameters are not easily determined. Tinting strength, used as an empirical measure related to carbon-black aggregate size, is governed by an average aggregate volume. It generally increases with specific surface area and decreases with structure. Filler structure is estimated by "vehicle demand." The vehicle, usually dibutylphthalate (DBP), is added to the dry powder up to the point of incipient formation of a coherent paste. The test (ASTM Method D2414) is performed in a small internal mixer fitted with a device sensing the torque on the rotor blades. The end point is marked by a sharp increase in torque. The DBP number of a carbon black (volume absorbed at end point) contains a contribution from interaggregate packing (Medalia 1970) which is related to the densification history of the material rather than the volume pervaded by the aggregates, which is the relevant descriptor of structure. For this reason, DBP numbers are often measured after a standard precompression treatment (ASTM Method D3493); DBP numbers are also used in the characterization of inorganic fillers.

Electron microscopy was used extensively in the past to measure nodule size distributions of carbon blacks, but the practice has lost favor in recent years. Computerized automatic image analysis methods have been applied with some success to characterize the details of filler morphology.

Common methods for the assessment of surface chemical characteristics of carbon blacks include measuring the slurry pH, diphenylguanidine adsorption or the content of volatiles. These quantities are related to the degree of surface oxidation of the black, which affects the rate of sulfur vulcanization. Toluene extractables, moisture, ash and grit are impurities in carbon black which are monitored regularly in manufacture. Carbon black is sold in the form of pellets (agglomerates) roughly 1–2 mm in diameter. Pellet size distribution, crushing strength, pack point and bulk density are parameters important in bulk handling and the dispersion of the material in rubber. Commonly specified properties of mineral fillers, in addition to specific surface area and vehicle demand, are moisture, sieve residue, chemical composition, pH, specific gravity and optical brightness.

Table 1 lists a few important fillers and some of their characteristic properties. Carbon blacks are available in many combinations of specific surface area (7–150 $m^2\ g^{-1}$) and structure (DBP absorption: 0.30–1.30 $cm^3\ g^{-1}$). The first digit in the ASTM grade designation is intended to follow the arithmetic mean nodule size; hence, in general, the larger the grade number, the smaller the specific surface area. Surface chemical properties in carbon blacks are rarely specified as they are similar throughout the ASTM N category of blacks. There is also an S category comprising oxidized and hence slow-curing blacks. The now largely obsolete channel process carbons belong to this group, as do some oxidized furnace blacks.

2. *Mixing and Dispersion*

Dispersion of fillers in rubbers is usually accomplished in internal mixers of the Banbury type or on two-roll mills. The attainment of a technologically adequate filler dispersion is rendered more difficult the higher the specific surface area of the filler and the lower its structure. Whereas a low-structure carbon black (small DBP number) will be incorporated more rapidly than a high-structure black, the final dispersion of the latter will usually be better. Inhomogeneities of 1 μm or less are not harmful to vulcanizate properties, so that the attainment of a perfect dispersion is not necessary (Boonstra and Medalia 1963). The easiest, and usually sufficient, means of ascertaining the quality of a dispersion is by examining a razor-cut surface of the vulcanized compound under a low-power optical microscope. The method is rendered semiquantitative by establishing a set of standards. The examination of microtome sections at magnifications of about 400 is also used widely. Electron microscopy presents a serious sampling problem and is rarely employed except in research studies.

Carbon black–rubber master batches, prepared by the coagulation of mixtures of a rubber latex and a carbon-black slurry, are available commercially.

Table 1
Characteristics of some important fillers (typical values)

Filler	ASTM grade number	Specific surface area ($m^2\ g^{-1}$)	DBP number ($cm^3\ g^{-1}$)	Principal uses
Carbon black	N339	95	1.00[a]	Tire treads
Carbon black	N330	83	0.88[a]	Tire sidewalls, treads
Carbon black	N326	83	0.70[a]	Tire breaker stocks
Carbon black	N660	35	0.76[a]	Tire carcasses, extruded goods, wire and cable insulation
Carbon black	N990	7	0.33[a]	Mechanical goods
Silica		150	1.93[b]	Light-colored articles, polysiloxane rubbers
Hard clay		22	0.35[b]	Footwear, mechanical goods
Precipitated $CaCO_3$		2–20	0.3–0.6[b]	Mechanical goods

a ASTM D3493 b ASTM D2414

Although the dispersion of the carbon black in a master batch is rarely adequate, the subsequent addition of vulcanizing agents and other compounding ingredients provides sufficient additional mixing to refine the carbon-black dispersion.

Typical filler concentrations are 20% by volume of the mix for highly reinforcing types and greater for mildly or non-reinforcing fillers.

3. *Effects of Fillers in Unvulcanized Rubbers*

In un-cross-linked rubber compositions, fillers increase the batch viscosity; to partly overcome this effect, plasticizing hydrocarbon oils are used routinely in rubber compounding. Common mixture rules for viscosity, such as the modified Einstein equation

$$\eta = \eta_0(1 + 2.5c + 14.1c^2) \tag{1}$$

where η is the viscosity of the mixture, η_0 that of the medium and c the volume concentration of filler, rarely hold for any but low-surface-area fillers, devoid of significant structure, e.g., N990 (medium thermal) black. Highly reinforcing fillers invariably produce viscosity increases greatly in excess of those predicted by Eqn. (1). These result from the occlusion of polymer in the internal void space of filler aggregates (thereby raising the effective filler concentration), adsorption of polymer segments on the filler surface with loss of segmental mobility and filler agglomeration effects. Invariably fillers also accentuate non-Newtonian behavior.

A technologically important aspect of filler structure, especially with carbon blacks, the DBP value of which can be held within narrow specification limits in manufacture, is the control of extrusion shrinkage. The higher the DBP number the lower the shrinkage or degree of die swell. The decrease in recoverable strain results from inclusion in the mix of a rigid phase incapable of elastic recovery. The abundance of this phase increases with the ability of the filler to shield rubber from elastic deformation by occlusion in its internal void space.

4. *Physical Properties of Filler-Reinforced Vulcanized Elastomers*

Figure 1 shows, schematically, the effects of some 20% by volume of various fillers on the uniaxial stress–strain curve of an amorphous elastomer. Curve A represents the unfilled vulcanized rubber, B the same rubber filled with a weakly reinforcing filler of small specific surface area (e.g., $CaCO_3$), C the rubber filled with a highly reinforcing carbon black (surface area ~100 cm^2 g^{-1}) and D the rubber filled with the same black after deactivation by heat treatment at 2700 °C (graphitization). Carbon blacks carry reactive sites capable of grafting rubber molecules onto their surface by several mechanisms; these sites are lost in graphitization. The result of the removal of active sites is a dramatic loss in modulus or stiffness (stress at a given strain), but only a moderate loss in tensile strength. This phenomenon is quite general. Silica, of 100 m^2 g^{-1} surface area, which does not graft to general-purpose rubbers, gives low modulus vulcanizates of good tensile and tear strength, but using an organosilane coupling agent such as 3-mercaptopropyltrimethoxysilane, which provides surface grafts to the vulcanized rubber network, gives results similar to carbon black. The relatively coarse $CaCO_3$ filler (curve B) has neither the surface area for the development of high tensile strength, nor the ability to bond rubber to produce a vulcanizate of high stiffness at the moderate loading of the present example.

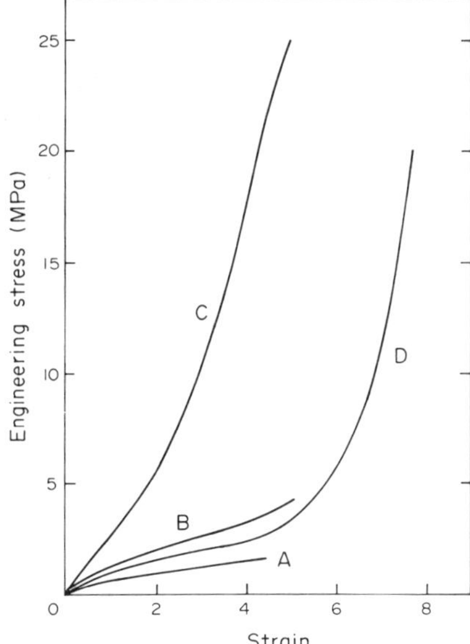

Figure 1
Schematic stress–strain curves in uniaxial tension for A, unfilled rubber; B, rubber filled with coarse filler of low activity; C, rubber filled with highly reinforcing carbon black; D, rubber filled with graphitized carbon black

In structured fillers, the ability to occlude polymer leads to high effective filler concentrations; thus vulcanizate stiffness increases with the DBP number of the carbon black.

Many applications of rubber involve periodic deformations of modest magnitude, so that the role of fillers in the dynamic viscoelastic response of vulcanizates is of great importance. In general, fillers increase both the storage and loss components of the dynamic modulus in a rather complicated way which is strongly dependent on the deformation amplitude and the filler loading. The amplitude dependence, which is absent in unfilled vulcanizates, is illustrated in Fig. 2. It has been explained as the result of filler agglomeration by van der Waals forces, which is increasingly disrupted as the strain amplitude increases, and constitutes a major part of the energy losses occurring in the periodic deformation of filled rubbers. In a given compounding recipe and at fixed strain amplitude, the storage modulus is governed by filler structure, while the loss tangent is essentially a function of specific surface area; consequently, the loss modulus is a function of both morphological characteristics.

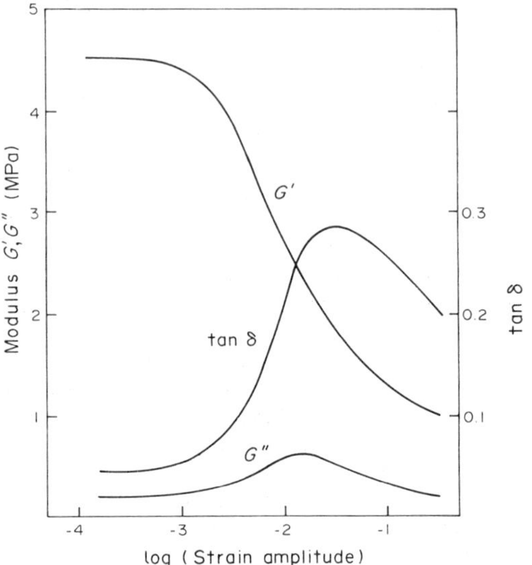

Figure 2
Schematic curves for dynamic storage modulus (G'), loss modulus (G'') and loss tangent for carbon black (20 vol%) reinforced vulcanizate. Scales convey an idea of magnitude only

The agglomeration of carbon black in rubber has another important consequence. When the black concentration reaches a critical value, electrically conducting paths which pervade the entire volume of the rubber are formed, leading to a precipitous drop in electrical resistivity. Certain high-structure blacks of very large specific surface area can form conductive networks at volume concentrations as low as 5%.

The glass-transition temperature of rubbers is raised very little by carbon black and other fillers. However, the transition is broadened on the high-temperature side, this being the result of loss in segmental mobility of the polymer in the vicinity of the filler surface.

The ability of fillers, especially carbon black, to impart abrasion resistance to rubbers is of utmost importance in their use in automotive tire treads. Tread-grade carbon blacks have specific surface areas in excess of $75\,m^2\,g^{-1}$ and DBP values above $0.90\,cm^3\,g^{-1}$. Under all conditions of wear, abrasion resistance increases with specific surface area up to a limit of $\sim 150\,m^2\,g^{-1}$. The carbon-black structure has only a minor effect under mild wear conditions, but becomes a major contributor to abrasion resistance in severe service. The ability of fillers to graft rubber molecules to their surface is an important factor in wear, but it is not a major variable within the ASTM N series of carbon blacks (see Table 1). Graphitization of carbon blacks leads to significant loss in wear resistance. Conversely, the use of coupling agents with silica increases resistance to abrasion, albeit not to the level attainable with the most highly reinforcing carbon blacks.

There are some important distinctions in the response to filler reinforcement of amorphous and strain-crystallizing rubbers (natural rubber, high *cis*-polyisoprene). Strain-induced crystallization provides a reinforcing mechanism which is not simply added to the effects of filler reinforcement. Consequently, the tensile strength of most natural rubber vulcanizates, when measured at room temperature and modest strain rates, is hardly increased by fillers (and may even be lowered). On the other hand, in tire wear, where strain rates are large and temperatures higher so that strain-induced crystallization does not come into play, response to carbon black is similar to that of amorphous butadiene–styrene rubbers. Important differences between crystallizing and noncrystallizing rubbers exist in cut-growth and fatigue behavior.

5. *Mechanism of Reinforcement*

The mechanism of filler reinforcement of rubbers is not completely understood. It is generally agreed that dissipation of strain energy at the tip of growing cracks by viscoelastic processes is important in the mechanical failure of all polymers and that fillers certainly contribute additional loss mechanisms. The stiffening or "modulus reinforcement" of fillers is governed by morphological features of the filler and the formation of strong attachments of polymer molecules to the filler surface.

See also: Elastomers: Additives; Fillers and Coatings: Industrial Minerals

Bibliography

Boonstra B B, Medalia A I 1963 Effect of carbon black dispersion on the mechanical properties of rubber vulcanizates. *Rubber Chem. Technol.* 36: 115–42

Donnet J-B, Voet A 1976 *Carbon Black*. Dekker, New York

Eirich F R (ed.) 1978 *Science and Technology of Rubber*. Academic Press, New York

Kraus G (ed.) 1965 *Reinforcement of Elastomers.* Wiley, New York
Kraus G 1971 Reinforcement of elastomers by carbon black. *Adv. Polym. Sci.* 8: 155–237
Medalia A I 1970 Morphology of aggregates, VI. Effective volume of aggregates of carbon black from electron microscopy; application to vehicle absorption and to die swell of filled rubber. *J. Colloid Interface Sci.* 32: 115–31
Voet A 1980 Reinforcement of elastomers by fillers: Review of period 1967–1976. *Macromol. Rev.* 15: 327–73

G. Kraus

Filters, Sorbents and Ion Exchangers

Mineral filters, sorbents and ion exchangers are employed in a wide variety of special applications. Mineral filters are generally used to remove solid particles from liquid media, and in particular for applications requiring the filtration of high volumes of liquids. Mineral sorbents are employed in the purification by selective absorption of industrial items such as edible oils and fats, and also as carriers to absorb insecticides, herbicides and other agricultural chemicals. These mineral sorbents are also used as sweeping compounds to absorb oil spills and other industrial chemicals. The largest use of mineral absorbents is for pet litter. Ion-exchange minerals are employed mainly in the conditioning of process waters and in the treatment of wastewaters and effluents.

This article serves as a general introduction to filters, sorbents and ion exchangers. Detailed information on siliceous filters such as sand, diatomite, perlite and asbestos, clay sorbents and filters, and minerals with ion-exchange properties can be found elsewhere (see *Filters, Sorbents and Ion Exchangers: Siliceous Minerals*; *Filters, Sorbents and Ion Exchangers: Clay Minerals*; *Zeolite Minerals*).

For industrial purposes, a good filter aid must possess several characteristics. These include the ability to form a very porous cake, a low surface area, correct particle size distribution and non-uniform sizing, and availability in a number of grades. The performance of a mineral filter aid is measured by the clarity of the filtrate, the rate of flow of the liquid through the filter bed and the length of the filter cycle. Mineral filter aids are used in food products, metallurgical processes, petroleum refining and processing, drug and pharmaceutical manufacture, and in many other industrial manufacturing processes. They also find use in the dry-cleaning process, in waste disposal and for water filtration of all types. The most widely used mineral filter aid is diatomaceous earth, and the largest producer is the USA with about 0.7 Mt per year. World production is estimated at about 2 Mt. Diatomaceous earth has a very low bulk density, so freight costs are a very important factor in marketing. Exploration for new deposits has intensified recently because of the high freight cost.

Mineral sorbents are highly absorbent and adsorbent in their natural or modified form. The most important are attapulgite (palygorskite), sepiolite, smectites (bentonites) and diatomite. A high surface area is the leading factor in the absorptive capacity of these minerals, although charged sites on the surface are also a factor for the clays. Absorption is readily reversible by simple low-temperature heating, and most sorbents can be restored to their original absorptive character by such processing. Adsorption, on the other hand, is a selective process and is not readily reversible by low-temperature heating.

The terminology of the clays used as mineral sorbents needs some clarification. The term Fuller's earth normally signifies clay or other fine-grained earthy material suitable for bleaching and absorbent uses. It has no compositional or mineralogical meaning. The term was first applied to material used in cleansing and fulling wool, thereby removing the lanolin and dirt from it. When it was found that some of the earths used for fulling wool would also serve to decolorize and purify mineral, vegetable and animal oils, and as a general sorbent material, the term Fuller's earth was modified to include this usage. The terms bleaching clay and bleaching earth are applied to both naturally active and activated clay, including activated bauxite. Activated clays is a term used for clays which are treated with acid or otherwise altered to improve their desirable properties. Bleaching earths then refers to clays that in their natural state, or after chemical or physical activation, have the capacity to adsorb coloring matter from oils. The term absorbent clay is applied to Fuller's earth used for a wide variety of absorbing purposes different from those of processing oils.

Sorbent clays are produced in the USA, the UK, Germany, Spain, India, Japan, the USSR, Hungary, Rumania, Italy, Morocco, Syria, Argentina, Mexico and Australia. The USA is the largest producer followed by Germany and the UK. The product is either granular or powdered dependent on the use requirements. In many instances, the granules are hardened by thermal treatment so that they can withstand attrition without readily breaking down into fines. The size of the granules range from 30 to 60 mesh (9.8–23.2 μm). In addition to the size of the granules, the amount of oil and water that is absorbed per gram of sample is important.

Two processes are used to decolorize various types of oil, namely percolation and contact. In the percolation process the bleaching clay is placed in a filter bed and the oil to be processed percolates through the bed. In the contact process, the clay is mixed with the oil to be decolorized, and then the clay is separated out in filter presses or some other filtration device.

The future outlook for mineral filters, sorbents and ion exchangers indicates a slow to moderate growth in usage. Since the late 1960s, most companies producing mineral filters expanded their capacity to meet the increased demands. In the late 1970s, the growth rate slowed considerably, but it is anticipated that as developing countries progress, new markets for mineral filter aids will open up. In the case of sorbents much the same market pattern exists. The tremendous growth of absorbent clays as pet litter took place in the 1960s and 1970s, and future growth will be slow but steady. Agricultural uses of absorbents particularly as carriers of pesticides could increase dramatically as the developing countries modernize their agriculture. The use of bleaching clays could increase significantly as more used oils are filtered and decolorized to be reused. The high prices of essential petroleum products will mandate that more used oil will be cleaned, which means that bleaching clays should have a promising future.

See also: Industrial Minerals: An Overview

Bibliography

Lefond S J (ed.) 1983 *Industrial Minerals and Rocks,* 5th edn. American Institute for Mining, Metallurgical, and Petroleum Engineers, New York

Mumpton F A 1983 Commercial utilization of natural zeolites. In: Lefond 1983, pp. 1418–31

Patterson S H, Murray H H 1983 Clays. In: Lefond 1983, pp. 585–651

Severinghaus N Jr 1975 Fillers, filters, and absorbents. In: Lefond 1983, pp. 243–57

H. H. Murray

Filters, Sorbents and Ion Exchangers: Clay Minerals

Clays find widespread use as sorbents, and are also employed in minor quantities as filter aids. The specific clay minerals used as mineral sorbents are members of the hormite and the smectite groups. The major clay mineral used as a filter aid is kaolinite.

Sorbent clays are employed in decolorizing, clarifying and absorbing various types of oil, and in pesticide and chemical carriers, pet litters and carriers for microencapsulated ink in no-carbon copy paper. They are also used in clarification of sugar cane juice, beers and wine. The largest use is as pet litter.

Clay minerals are ideal sorbents for many industrial processes because of their fine particle size, high surface area and surface charge sites. The absorbent markets are thought to tie in closely with the gross national product and sales fluctuate with the economy. The granular pesticide carrier markets have only been fully developed in the USA and Japan, so there is a high potential market in Europe and the rest of the world.

1. Mineralogy

1.1 Hormite Group

The hormite group of clay minerals are hydrated magnesium aluminum silicates. Palygorskite (or attapulgite) has a fibrous, lathlike form, which along with its crystal structure results in unusual colloidal and sorptive properties. Sepiolite is structurally similar to attapulgite, but has a slightly larger unit cell and the individual laths are about one-third wider. Sepiolite in its compacted form is known as meerschaum.

1.2 Smectite group

The best sorbent amongst the smectite minerals is calcium montmorillonite, sometimes called a subbentonite or nonswelling bentonite. Calcium montmorillonite is a hydrated calcium aluminum silicate that normally has considerable iron and magnesium substitution of aluminum and calcium. The mineral consists of thin sheetlike irregular particles that are extremely fine in particle size ($\leq 2\ \mu m$).

2. Sources and Production Techniques

Attapulgite is not as common as many other commercial clays. It is produced in Georgia and Florida in the USA, and in the USSR and Senegal. By far the largest production is from the USA, although Senegal has the most extensive reserves.

Sepiolite is also a relatively rare clay mineral in comparison with other commercial clays such as kaolins and bentonites. It is produced in Spain and Turkey, and in Nevada in the USA. The largest production is from Spain.

Attapulgite and sepiolite are mined by open-pit stripping methods. Typically, the overburden is removed in panels, the clay mined by loading trucks with shovels or endloaders, and the overburden from an adjacent panel shifted to the mined-out area. Most of the hormite group minerals are selectively mined because their properties vary considerably in most deposits. Processing involves simple milling techniques that involve removal of water and grinding to suitable sizes. Granular products are screened, and the granules hardened by heating to 650 °C, which drives out all the absorbed water.

Calcium montmorillonites used as sorbent or bleaching clays are much more widely distributed than the hormite group clays. Production is from the USA, Mexico, Italy, Spain, the UK, the Federal Republic of Germany, South Africa, India, Japan, Turkey and Canada. The largest production is from the USA; the Federal Republic of Germany and the UK are also large producers.

Calcium montmorillonites are mined and processed in a similar fashion to attapulgite and sepiolite. Activated clays for use in decolorizing, for paper coating and other specialized uses require special processing. The calcium montmorillonite is slurried in water at about 10–20% solids and the sand is settled or screened out. The slurry is heated and sulfuric and/or hydrochloric acid is added to make a hydrogen clay; this treatment also removes a small amount of aluminum from the clay mineral structure. The acidified clay slurry is kept hot using steam, and after the appropriate time the slurry is filtered, washed and then dried. Almost all the product is bagged for shipment, although small quantities are now shipped in bulk.

3. Physical and Chemical Properties

Attapulgite and sepiolite absorbent clays are used for many miscellaneous uses, with more than 90 different grades of attapulgite being produced. In the USA, most absorbent granules are prepared to fulfill the requirements outlined in Federal Specification P-A-1056A. A similar set of specifications is outlined in the American Society for Testing and Materials Standard C431-65.

The US Federal specifications require that absorbent granules must be clean, uniform and free of lumps or foreign matter, and that no more than 10% of the granules can pass through an 80 mesh (0.177 mm) sieve in an attrition resistance test. This test is carried out by shaking the granules with steel balls on a screen, according to a specified procedure.

Test methods for evaluating bentonite and other types of Fuller's earth for bleaching soybean and cottonseed oils are outlined by the American Oil Chemists Society. Their specifications contain instructions on bench-type tests, including stirring time, heating rates and temperatures; approved equipment; quantities of raw oil and clay required; and color determination methods. They also require the comparison of the material tested with an official natural bleaching earth. Purchase specifications are based on comparison of the color of oils bleached by the official earth with those bleached by the test clay.

For pet litter the granules must be absorbent and dust free so that the cat does not track the dust, and so that the dust does not fly when the litter box is filled with fresh litter. Also the granules must be hard enough so that they do not break up as they are handled, or when the cat scratches them around in the litter box.

When granules are used as an insecticide carrier, in addition to meeting the absorbent specifications outlined above, specific tests using the particular insecticide chemical must be performed to make sure that the chemical will be released and will not be polymerized or cracked into other compounds. This can only be determined by making absorption tests under controlled conditions.

4. Industrial Applications

The largest market, and perhaps the area of highest competition, is in domestic pet litter, agricultural carrier, and oil spillage applications. In the pet litter markets, the utility of a particular clay is dependent not only on its sorptive characteristics, but also on balling properties and bulk density, whereas in the oil spillage markets the ability of the clay to resist slaking and its wear properties are of importance along with bulk density. Bulk density in both of these applications governs the quantity of material needed in a particular situation and consequently the overall costs. As an agricultural carrier the percentage of chemical that can be absorbed and still flow readily is very important.

Another large market is in the refining of liquids—mainly oils, but also sugar cane juice, beers and wines. Large amounts of naturally active calcium bentonites are used for this purpose, but the acid-activated clays, which have superior absorption properties, are gradually replacing the naturally active clays. The acid-activated clays are more efficient decolorizers because of their larger surface area. However, in the treatment of some very light oils which are particularly sensitive, a natural bleaching clay is usually prescribed. Attapulgite also has a share of the bleaching and decolorizing market, particularly in the USA, and a significant proportion of the sepiolite production from Turkey is used for the extraction of sulfur from light paraffin oils. The refining of glyceride oils of both animal and vegetable origin is a significant market for calcium montmorillonites. Oil recycling has been stimulated by the antidumping laws, particularly in the USA, and also by the high cost of new oils. Many types of industrial oils and lubricants are purified and decolorized using granular attapulgite and activated calcium bentonite. The principal types of oil refined in the mineral-oil field are naphthas, fuel oils, lubricating oils, waxes and greases. The more important vegetable and animal oils refined with bleaching clays are cottonseed, soybean, linseed, coconut, palm, tallow and grease. The principal object of applying bleaching earths is to remove color, but there are other important uses in all fields of refining; for example, in the treatment of naphtha, reduction of the amount of gum and stability are most important factors. In the refining of lubricating oil, improved sludge content, carbon content, oxidation test, acidity, emulsion test and viscosity index are also important. In processing of vegetable oils, stabilization and removal of residual soap, gums, odor and taste are also important improvements after treatment using bleaching earths.

See also: Filters, Sorbents and Ion Exchangers; Filters, Sorbents and Ion Exchangers: Siliceous Minerals

Bibliography

Oultin T D 1965 *Mining, Production, and Uses of Attapulgite Clay Products*, SME Preprint 65H39. American Institute of Mining, Metallurgical, and Petroleum Engineers, New York

Patterson S M, Murray H H 1983 Clays. In: Lefond S J (ed.) 1983 *Industrial Minerals and Rocks*, 5th edn. American Institute of Mining, Metallurgical, and Petroleum Engineers, New York, pp. 585–651

Rich A D 1960 Bleaching clay. In: Gillson J L (ed.) 1960 *Industrial Minerals and Rocks*, 3rd edn. American Institute of Mining, Metallurgical, and Petroleum Engineers, New York, pp. 93–101

H. H. Murray

Filters, Sorbents and Ion Exchangers: Siliceous Minerals

Siliceous filtration materials function in several different ways. Diatomite and perlite are used as filtration aids to remove suspended particles from liquids by straining or trapping the particles on a layer of filter aid supported by a filter septum or screen as the liquid passes through the filter aid. This is accomplished in a variety of units, for example, pressure filters, vacuum filters, plate and frame filters, and rotary precoat filters.

Body addition or controlled addition of filter aid to the "dirty" liquid can help to improve filter cake porosity by spacing the dirt particles with the more porous filter aid particles. The flow rate and filtration ability of the filter aids are normally a function of crude quality and processing.

The sorptive properties of asbestos can be used to improve some filtrations of very fine particles by attracting them to the asbestos surface and allowing filtration of the sorbed dirt with the asbestos.

Silica is used in sand filters to form a porous bed that traps the dirt particles. Filtration in sand beds is often aided by chemical coagulants or flocculents to make dirt more readily trapped by the relatively porous sand.

Garnet is also used in sand filters to form a porous bed that tends to stay at the bottom of the sand bed due to its specific gravity of 4.2.

1. Sources of Supply

Sources of asbestos, diatomite and perlite are discussed elsewhere (see *Fillers and Coatings: Siliceous Minerals*). Sources of quartz filtration sand are extensive. A good filtration sand should be a clean river sand, free of organics, fines, clay and friable particles, as well as low in acid solubles. Rounded sand grains (river sand) work best in filter beds since they require the least backwashing pressure, scour best during backwash, and fluidize more easily during cleaning cycles. The hardness of silica is needed to resist the intense scouring. However, the prime requirement is very tight size gradation, often in the 0.35–2.0 mm size range.

In addition to quartz filtration sands, garnet sands are used due to their much higher specific gravity which allows maintenance of a layered bed after backwashing. Garnet sands are angular since they need crushing to generate the required sizes. Cost and availability considerations restrict the use of garnet in filtration.

2. Mineralogy

2.1 Asbestos

Chrysotile forms by the solution recrystallization of serpentine in cracks and fissures of the host serpentine. The chrysotile fibrils have been shown to be hollow cylinders formed from a magnesium silicate layer. These fibrils occur in fiber bundles that must be opened by careful milling to maximize the usefulness of the asbestos.

2.2 Diatomite

Natural diatomite comprises the opaline silica skeletons of the various species of diatoms that flourished in marine or freshwater environments. Typical commercial processing of diatomite includes either calcination or flux calcination to modify the flow rates of filter aids. Calcination "burns out" some of the finest tertiary silica structure and helps sinter fines together. Flux calcination accentuates these changes and may also improve color. Calcination dehydrates the opaline silica and causes an increase in specific gravity and some conversion to cristobalite. Flux calcination increases the specific gravity to 2.3 and results in as much as 40–60% conversion to cristobalite.

2.3 Perlite

Perlite is typically a sodium, potassium or aluminum silicate glass hydrated with about 2–5% water. It is believed that perlite results from hydration of volcanic glass or obsidian. Size graded perlite, typically in the range 2.00–0.15 mm, expands up to 30 times its original size when heated to 760–1090 °C in either a horizontal or vertical furnace. The expanded perlite has the same composition as the ore, but without the water of hydration.

2.4 Silica

Filtration sand is typically quartz or quartzite sand preferably obtained from river transported sources so that grains are rounded and uniform.

2.5 Garnet

A variety of types of mineral garnets exist, including grossularite, pyrope, almandite, spessartite and andradite. These are all aluminum silicates with calcium, magnesium, iron or manganese (see *Abrasives: Precious and Semiprecious Minerals*).

3. Physical and Chemical Properties

In filtration applications, diatomite, expanded perlite, silica and garnet normally do not react with the liquids being filtered. They can show trace solubilities with acids, and with strong alkalies (generally with pH values greater than 10 or 11), siliceous minerals will show solubility, particularly at increased temperatures and concentrations.

Particle shape and porosity are important properties of diatomite and expanded perlite when used as filter aids. The pore-size distribution of these materials and the particle-size distribution of turbidities can often be related to assist in selecting suitable filter aids.

Sand and garnet filtrations are generally thicker beds than diatomite or perlite. They trap the turbidities in the relatively larger pores between sand grains. The proper use of chemical flocculents can greatly benefit sand filtration.

Chrysotile asbestos contributes to filtration in two ways. Because of its fibrous nature, it can be used to form a strong precoat over the filter septum. It also has a positive zeta potential which aids filtration by attracting generally negatively charged turbidity particles to its surface.

Physical and chemical properties of siliceous minerals used as filter aids are shown in Table 1.

4. Mining and Production Techniques

4.1 Asbestos

Asbestos is recovered from the crushed serpentine host rock through a variety of screening and air clarification techniques. Generally, the shorter fibers are chosen for filtration applications, and preferably, the fiber is well opened to maximize surface area.

4.2 Diatomite

The cyclone fraction of dried, milled and air-classified natural diatomite is used for maximum clarity low-flow-rate filtrations. Straight calcination at 760–1090 °C agglomerates some fines and melts some of the tertiary diatomite structure. After milling and air classification to remove fines, the cyclone fraction has faster flow rate but lower clarity than the natural filter aids. Straight or nonflux calcined diatomite is typically pink due to oxidation of small amounts of iron. Still faster flow rates can be attained by flux calcination with 2–8% soda ash, sodium chloride, or similar fluxes, and calcination at 760–1090 °C. Flux calcination results in substantial agglomeration of fines and almost complete melting of tertiary structure. Generally, white colors are obtained as a result of the flux reacting with the color-causing impurities like iron and aluminum.

4.3 Perlite

Perlite ore is dried of free surface moisture, crushed and size graded to various mesh-sized ores, that may be used as it is or blended together to give a selected distribution of particles. The sized ore is expanded in either horizontal or vertical furnaces. In vertical furnaces, the ore is fed into the flame and is concurrently carried to collection cyclones and/or

Table 1
Some physical and chemical properties of siliceous minerals used as filter aids

Mineral	Specific gravity	pH	Bulk density ($g\,cm^{-3}$)	Surface area ($m^2\,g^{-1}$)	Particle size (mm)	Flow rate ($m^3\,m^{-2}\,h^{-1}$)
Asbestos						
chrysotile	2.56	9–10	0.8–1.2	4–50	fibrous	0.10–2.0
Diatomite						
natural	2.0	6.0–7.5	0.25–0.37	10–30	0.01–0.05	0.16–2.0
calcined	2.1–2.2	5.0–7.5	0.25–0.33	3–7	0.01–0.10	0.04–2.0
flux calcined	2.3	8.5–10.5	0.29–0.37	0.5–5	0.01–0.50	0.04–4.9
Perlite						
milled, expanded	22–23	6.6–8.0	0.13–0.28	1	0.01–0.50	0.05–4.0
Silica sand						
fine	2.6	7.0	1.0–1.3	0.10–0.02	0.26–1.50	5
medium	2.6	7.0	1.0–1.3	0.01–0.02	0.34–1.80	12
coarse	2.6	7.0	1.0–1.3	0.01–0.02	0.41–2.00	50
Garnet	3.9–4.1		1.6–2.0		0.21–2.00	5–50

baghouses. The ore may be preheated in the jacket of the furnace to aid expansion. In horizontal furnaces, the ore may also be preheated in the jacket of the rotating tube. The ore drops across the flame and is carried out of the tube by rotation and gas velocity. The perlite is generally expanded to 0.03–0.10 $g\,cm^{-3}$ for filter aid production. Filter aids are milled to reduce floaters or bubbles, and classified to give various flow rates.

4.4 Silica Sand

Natural bank sand usually needs to be modified to achieve the necessary sizing and uniformity to accuracies as close as economically possible. Generally, oversize is screened out and fines are washed out. This can be done by the sand supplier or by the sand user by running their own sand washers and sand separators or by shovelling off fines washed to the top of the bed after backwashing.

4.5 Garnet Sand

The production of garnet sand is discussed elsewhere (see *Abrasives: Precious and Semiprecious Minerals*).

5. Safety

These minerals are all possible sources of dust that at least have to be treated as a nuisance and require adequate dust control. Asbestos, silica (quartz), and flux-calcined diatomite (cristobalite) require a higher degree of dust control. Generally, the coarser particles used in filtration are not in the respirable size range. However, safe dust levels must be maintained (see *Asbestos Hazards*; *Airborne Hazards from Materials Use: Monitoring and Measurement*).

6. Industrial Uses

Filter-aid filtration is used for a wide variety of applications. Beverages such as beer, wine, fruit and vegetable juices, and spirits are frequently filtered. Cane, beet, and corn sugars, animal and vegetable oils, antibiotics, waxes, gelatin, inorganic chemicals, acids, lube oils, rolling and cutting oils, varnishes, lacquers, jet fuels, and organic chemicals use filter-aid filtration. Other uses are in process, potable and waste waters.

Turbidities removed can vary in characteristics from noncompressible particles to compressible, gelatinous or slimy materials. They also can vary widely in particle size and concentration. In addition to turbidities, filtration is used to remove or recover adsorbents, catalysts, and other solids added to a liquid as part of a chemical processing step. The selection of the proper filter aid and type of filter to use strongly influences filtration rate and clarity of filtrate.

Asbestos is used to give a high degree of clarity or polish to critical filtrations. It is also used to form a precoat layer over the filter septum that has greater stability against loss of filter cake due to pressure or flow-rate fluctuations.

Sand filtration is used primarily for water filtration, for example, in producing potable water, and in swimming pool systems. Generally, low levels of turbidities are involved and filtration is aided by flocculating with, for example, alum, organic polymers and polyelectrolytes.

See also: Filters, Sorbents and Ion Exchangers; Filters, Sorbents and Ion Exchangers: Clay Minerals

Bibliography

Fair G M, Geyer J C, Okun D A 1968 *Water and Wastewater Engineering*, Vol. 2: *Water Purification and Wastewater Treatment and Disposal.* Wiley, New York

Manual of Instruction for Water Treatment Plant Operators. New York State Department of Health, Health Education Service, Albany, New York

Severinghaus N Jr 1983 Fillers, fitters, and absorbents. In: Lefond S J (ed.) 1983 *Industrial Minerals and Rocks*, 5th edn. American Institute of Mining, Metallurgical and Petroleum Engineers, New York, pp. 243–57

Speil S, Leinweber J P 1969 Asbestos minerals in modern technology. *Environ. Res.* 2: 166–208

G. Coombs

Fining of Glass

The fining (or refining) of glass generally refers to the removal of heterogeneities during glass manufacture, but more often specifically describes the elimination of bubbles. Bubbles, known as seed or blisters depending upon their size, are generated initially during reaction of batch materials, and also later in the melting cycle by interactions between glass and the containing refractories, electrolytic reactions and reboil mechanisms. Two mechanisms account for removal of bubbles; flotation out of the melt and dissolution into the glass.

1. Origin of Bubbles

1.1 Gases in Batch Materials

Gases trapped between grains of the batch form the first bubbles. Unless the raw materials themselves generate gases during the batch reactions, these bubbles consist mostly of N_2, the O_2 having reacted with reducing agents in the batch or dissolved rapidly in the melt.

Most glass is melted from batches which contain the modifier oxides as carbonates; the CO_2 from their decomposition amounts to about 200 times the final volume of refined glass. It is not surprising that in these melts the first bubbles are mostly CO_2. Sulfur-containing gases (e.g., H_2S, SO_2, COS) are also present initially if sulfates are used to accelerate melting, and N_2, O_2 or H_2O may also be generated by nitrates, borates and hydrated oxides.

1.2 Reaction of Glass with Refractories

As molten glass wets and slowly dissolves the surface of the oxide refractories which constitute the container, pores which are impregnated spawn mostly N_2 bubbles. Impurities in these refractories may also react with molten glass to generate gases. Certain multivalent transition metal ions sometimes present in glass-contact refractories (e.g., those of iron) generate O_2 bubbles. Bubble-generating characteristics of refractories are compared in the laboratory by equilibrating test pieces against molten glass.

Oxygen bubbles arise from local electrolytic cells which incorporate electronic conductors, and also from variations in temperature or composition of refractory and glass. Thus certain glass-contact refractories containing zirconia or alumina phases cannot be used too near the forming process when a seed-free product is required. Clay bodies used as floating barriers and as submerged draw-bars in sheet glass processes may initially generate O_2 bubbles, until alkali diffusion from glass into the refractory quenches the electrolytic cell. Electrolytic O_2 bubbles also come from concentration or temperature cells involving a refractory metal as the electronic conductor (e.g., platinum of a thermocouple sheath inadvertently grounded to tank steel-work).

1.3 Reboil

Certain mechanisms are somewhat arbitrarily brought together in the category "reboil." Here, reboil is used to describe nucleation and growth of bubbles in previously bubble-free glass, other than at the glass–refractory interface.

Many gases are sufficiently soluble in molten glass to be considered minor components. However, molten glass is usually far from an equilibrium state and may become supersaturated by some of these gases during the melting cycle. Final dissolution of sand from the batch may be considered from this point of view. Sand dissolution unavoidably creates local gradients in silica concentration. The surrounding lower-silica glass contains dissolved CO_2 and SO_3, the equilibrium activities of which are inversely proportional to silica content. Thus, if local silica enrichment near dissolving sand grains causes supersaturation, bubbles may nucleate and grow.

Since most gases have higher solubility with decreasing temperature for any given partial pressure above the melt, supersaturation and bubble nucleation may occur when refined glass is reheated to a temperature higher than the temperature of an earlier conditioning period. Similarly, reboil may occur when the partial pressure of a gas above the melt decreases below the level at which it would be in equilibrium with the dissolved species. The solubility of some gases (e.g., SO_3) depends strongly on the oxidation state of the melt. Thus, for example, reboil may occur when a burner flame impinges upon the glass surface. It has also been shown that water vapor diffusing into molten glass causes dissociation of dissolved sulfate and reboil bubbles.

A difficulty exists in accounting for homogeneous nucleation—that is, bubbles generated in the absence of a boundary between condensed phases; theory predicts that extraordinarily large intermolecular forces must be overcome to initiate bubbles within a homogeneous liquid. Nonetheless, the commonly reported observations of reboil indicate that either small heterogeneities (e.g., undissolved seed) serve as nuclei, or true homogeneous nucleation must occur.

2. Removal of Bubbles

2.1 Flotation

The primary mechanism for removal of bubbles is flotation. Under the usual conditions of bubble size and glass properties, the velocity of the rising bubbles v is described by a relation derived from Stokes' law:

$$v = \tfrac{2}{9}(g\rho r^2)/u \qquad (1)$$

where g is the gravitational acceleration, ρ is the glass density, r is the bubble radius, and u is the glass viscosity.

From Eqn. (1) it can be seen that making bubbles larger greatly accelerates their removal. In the earliest stages of melting, large faster rising bubbles unavoidably coalesce with smaller ones. Deliberate generation of large bubbles from submerged tubes or "bubblers" is reported to be helpful in sweeping out seed early in the melting cycle. Nonetheless, later in the refining stage, growth of individual bubbles becomes a much more important mechanism. Specific additives to the batch dissolve in the initial alkali-rich liquid, but then supersaturate the melt as the solvent composition, temperature and redox state change with time. These supersaturated gases then diffuse into existing bubbles, making them larger.

The well-known use of the batch additive "salt cake" (Na_2SO_4) illustrates the bubble-enlargement mechanism. As pointed out above, dissolving sand grains frequently spawn small bubbles. Thus, in many processes, including those for flat glass and containers, the batch contains about 0.75 wt% salt cake. A portion of this sulfate dissolves rapidly in the initial alkali-rich glassy phase before the sand grains are completely dissolved; however, in general, Na_2SO_4 solubility varies inversely with the silica content of the glass. Thus, as the melt approaches its final composition, it becomes supersaturated in salt cake, causing the formation of an immiscible molten sulfate phase at the sand grain–glass boundary. Reaction of Na_2SO_4 with SiO_2 hastens final sand dissolution, generates SO_2 and enlarges the original small bubbles, accelerating their removal. After initial batch reactions, the molten product in most processes moves into a zone of maximum temperature, the refining section of a conventional tank. Here, more of the dissolved sulfate dissociates thermally, yielding more SO_2 which diffuses into remaining bubbles, enlarging them and speeding their rise to the surface.

Bubbles may also be enlarged by infusion of O_2 from reduction of multivalent oxides in the batch; oxides of arsenic and antimony have been most commonly used. During initial dissolution into the melt at relatively low temperatures, the oxide refining agent is preferentially maintained in its higher oxidation state by the inclusion of an oxidant, usually $NaNO_3$, which dissociates early in the melting process. As the temperature of the glass rises following batch dissolution, the higher oxidation state of the oxide-refining agent becomes more unstable, effectively supersaturating the melt with O_2 as shown by the equilibrium:

$$As_2O_5 \rightleftharpoons As_2O_3 + O_2 \tag{2}$$

Existing bubbles become the sinks for this dissociated O_2 and grow larger, their upward velocity increasing accordingly. This approach yields a second important benefit to refining which will be discussed below.

Alkali halides may also be included in the batch as refining accelerators. At the maximum temperature of the melting cycle, these additives volatilize, thus enlarging existing bubbles. Even helium has been reported as a refining aid: it is introduced into the glass by a dissolution source at atmospheric pressure and diffuses rapidly through the glass into existing seed. In all of these strategies the partial pressure of the new gaseous species increases up to that limit set by its solubility within the molten glass and Henry's law.

Most glass is produced in gas- or oil-fired furnaces, which have carefully defined temperature profiles. At some point within each profile there is a region of maximum temperature that promotes upward flow of glass from the bottom of the tank due to the change in density of glass with temperature. This upward movement of glass corresponds to the so-called spring zone, where the most significant bubble enlargement takes place. Additionally, this upward flow brings the glass into a relatively shallow layer which eventually supplies the forming process. Because of the high temperature (low viscosity) and shallow depth of this layer, flotation becomes more effective here than at any other location in the tank.

2.2 Dissolution

Small bubbles usually disappear from molten glass much more rapidly than can be accounted for by flotation alone. Therefore, dissolution mechanisms must contribute to refining. As a bubble becomes very small, its internal pressure increases due to the energy of the spherical glass–gas interface, according to the following relation:

$$\Delta p = 2\sigma/r \tag{3}$$

where Δp is the pressure increase inside the bubble, σ is the surface tension of the glass, and r is the radius of the bubble.

The internal pressure within a 0.1 mm diameter bubble in an ordinary silicate glass is 10% greater than atmospheric. When the diameter falls to 0.01 mm, the internal pressure rises to twice atmospheric. Clearly, when the internal pressure increases above atmospheric pressure, gases diffuse more readily across the bubble interface into the glass. This limiting mechanism is probably responsible for the characteristic size distribution of bubbles equilibrated within molten glass, as in a large tank. Under this condition, the maximum bubble concentration usually corresponds to a diameter of 0.1–0.2 mm with a decreasing number of bubbles appearing over a range about one order of magnitude larger; however, the number of bubbles smaller than the size at maximum concentration falls sharply, cutting off usually in the range 0.1–0.05 mm. Bubbles smaller than about 0.1 mm evidently dissolve relatively quickly.

The surface-tension mechanism is not the only cause of bubble shrinkage. Bubbles containing mainly CO_2 shrink with decreasing temperature because CO_2 has a negative temperature coefficient of solubility. Bubbles originally inflated by inward diffusion of O_2 or SO_2 during heating of the glass to maximum temperature, but which are not removed by flotation, will shrink as those gases recombine with the glass at lower temperatures. Unfortunately, the real picture is usually made much more complex by the simultaneous diffusion of several gases between bubble and glass. Of particular concern is the back-diffusion of N_2 from glass to bubble while the bubble's original principal species move from bubble into the glass, leaving in the end a small, mostly N_2 seed. In principle of course, a high N_2 partial pressure should then encourage dissolution back into the glass and continued shrinkage. However, in practice, the diffusion coefficient and solubility of N_2 are so much smaller than those of the other gases of interest that this small N_2 bubble only shrinks very slowly at lower temperatures within the melting schedule.

Again, in practice bubble rise and dissolution occur not independently but simultaneously, with significant interaction between the two mechanisms. Thus, we can understand intuitively that a rising bubble will dissolve more rapidly than one at rest. The glass layer around the bubble which has been saturated locally by gas diffusing outward is continually replaced as the bubble rises, thus accelerating transfer of gas from bubble to glass. An important recent contribution to the understanding of glass refining consists of the formulation of theoretical models for gaseous exchange between bubble and glass, and the interaction of flotation and dissolution mechanisms.

See also: Gases in Glass; Glass: An Overview

Bibliography

Kramer F 1983 Mathematische Modelle über das Wachsen und Schrumpfen von Gasblasen in Glasschmelzen. *Thirteenth International Congress on Glass.* Glastechnische Berichte, Frankfurt, pp. 72–75

Mulfinger H O 1980 Gase (Blasen) in der Glasschmelze. In: Jebsen-Marwedel H, Brückner R (eds.) 1980 *Glastechnische Fabrikations-fehler*, 3rd edn. Springer, Berlin pp. 193–268

Scholze H 1969 Gases in glass. *Eighth International Congress on Glass.* Society of Glass Technology, Sheffield, pp. 69–81

Swarts E L 1972 The melting of glass. In: Pye L D, Stevens H J, LaCourse W C (eds.) 1972 *Introduction to Glass Science.* Plenum, New York, pp. 273–327

Union Scientifique Continentale du Verre 1956 *Symposium sur L'Affinage du Verre.* Union Scientifique Continentale du Verre, Charleroi

Union Scientifique Continentale du Verre 1973 *Glass-Making (Melting and Fining): Bibliographic Review.* Union Scientifique Continentale du Verre, Charleroi

E. L. Swarts

Finishing of Fabrics

The purpose of finishing, the last step in the treatment of fabrics, is to improve either the esthetic or performance properties of the fabric or both. Such modification of properties often involves a chemical treatment (e.g., with softeners, whiteners, or fire-retardant compounds), although mechanical finishing processes are also employed, notably to change the appearance of a fabric.

1. Removal of Impurities

In processing fibers and yarns to fabrics, sizes are usually added to the yarn, the removal of which is accomplished in a final water-washing step. In addition, oils added to the surface of the fibers during processing, and also soil accumulated during manufacture, are removed in this washing. The process of washing fabrics after manufacture, called scouring, is accomplished with the use of various washing additives plus agitation.

2. Changing the Hand

There are many methods of changing the hand or handle of a fabric. Wool fabrics may be compacted in a finishing step called fulling. Singeing or passing a fabric over a gas flame or a heated metal plate may be used to burn off the fiber ends of a fabric made from staple yarns. If a fabric is limp, it may be stiffened or given added body by the application of various additives. Nonpermanent additives or sizes that are used include starches, gums, gelatins, poly(vinyl alcohol) and carboxymethylcellulose. Durable stiffening is accomplished by the application of resin finishes. The hand of fabrics may be made softer or more limp by the addition of a softener. Oils, waxes and long chain cationic and nonionic materials are applied as softeners to fabrics, but they are usually not durable to washing or dry cleaning. Increased durability is obtained with polymeric softeners such as polyethylene and silicones. Softeners lubricate the fiber surfaces and reduce friction between fibers, so that in addition to softening the fabrics, improved abrasion resistance may also be provided.

3. Changing the Appearance

The appearance of fabrics may be changed in a variety of ways. Fabrics may be whitened by bleaching or by the application of fluorescent whiteners. Bleaching may be accomplished with the use of a chlorine bleach or other oxidizing agents capable of removing colored material from the fabric. Fluorescent whiteners or optical bleaches whiten fabrics by absorbing ultraviolet light and re-emitting the light in the blue part of the spectrum, counteracting any yellow cast.

The luster of cotton fabrics may be increased by mercerizing, which involves immersion of the fabric in sodium hydroxide solution followed by washing. The mercerization process produces greater luster by swelling the cotton fiber and changing its shape from a collapsed to a cylindrical structure.

A glossy surface can be produced on a fabric by calendering or passing the fabric through rollers to produce a smooth finish by mechanical means. If a friction calender is used, a glazed finish results. With a Schreiner calender, which contains many fine diagonal lines, a soft luster and hand is produced. Moiré finishes are achieved by passing fabric through a calender with a pattern etched into the surface of the roller, which flattens one area more than another. Embossed designs are similarly produced by pressing a pattern onto a fabric. If the fabric is impregnated with a resin prior to calendering, a durable finish can be obtained provided the calendered fabric is subsequently cured.

Napped fabrics are produced by brushing with machines containing small wires, thereby raising the loose ends of fibers on the fabric surface. Napped fabrics are usually made of yarns with a fairly loose twist so that the yarns will brush easily. The brushed fabrics may be sheared to make the nap uniform in thickness. Napping changes both the appearance and the hand of the fabric and also increases the warmth by trapping air which provides an insulating effect.

An effect similar in appearance to a napped fabric can be obtained by flocking, in which short staple fibers are attached to a fabric surface by means of an adhesive. The adhesive may be uniformly coated on the fabric surface or applied in a printed pattern. The fibers are held on that part of the surface coated with the adhesive in a direction perpendicular to the surface of the fabric. The durability of a flocked fabric is dependent on the properties of the adhesive used.

4. Control of Biological Degradation

Synthetic fibers are not normally damaged by mildew and bacteria. However, the natural and regenerated natural fibers are susceptible to microbiological damage, cellulosic fibers being especially prone to damage. Indeed, cellulolytic fungi can destroy cellulosic fabrics that are subjected to moist, warm conditions. To resist biological degradation, cellulosic fabrics may be treated with antimildew agents such as chlorinated phenols, copper naphthenate, and other organometallic compounds. Treatment of cellulosic fabrics with formaldehyde reactants such as melamine–formaldehyde will also protect the fabrics from biological degradation, as will acetylation and cyanoethylation of the cellulose.

5. Durable Press

Since cellulosic fibers do not recover well from deformation, formaldehyde reactants are used to treat these fabrics to improve this property and consequently the recovery of the fabric from wrinkling. If sufficient formaldehyde reactant is applied, the resulting fabric will have both wet and dry recovery from wrinkling and consequently will have durable press properties.

Historically, the treatment of fabrics with formaldehyde-containing reactants to provide recovery from wrinkling started with an invention of Foulds, Marsh, and Wood in 1926, which involved the finishing of rayon fabrics with water-soluble urea–formaldehyde or phenol–formaldehyde resins. The first resin treatment was carried out on rayon which has greater elongation than cotton and is embrittled less by the resin treatment. Subsequently, urea–formaldehyde resins were applied to cotton fabric to give dimensional stability and recovery from wrinkling. Later, blends of polyester and cotton were resin treated to provide durable press fabrics with greater strength and resistance to wear. The resins used included not only urea–formaldehyde but also water-soluble reaction products of formaldehyde and triazine, triazone, uron, ethylene urea, propylene urea and carbamates. The most widely used product at present is dimethyloldihydroxyethyleneurea (DMDHEU (**1**)) or its ethers. Treated fabrics also include all-cotton fabrics that have been finished with liquid ammonia to provide improved strength and wear resistance. Reactions of DMDHEU with the hydroxyl groups of cellulose can occur through the two methylol groups or through the two ring hydroxyl groups by the elimination of water.

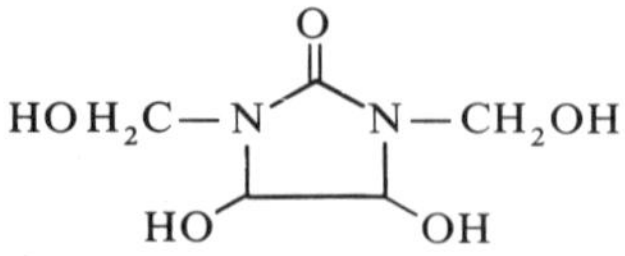

O

HOH_2C-N $N-CH_2OH$

HO OH

(1)

Durable-press treatment involves immersion of the fabric in an aqueous pad bath containing the formaldehyde reactant and a metal-salt catalyst, squeezing the fabric through pad rolls, drying, and curing. The generally accepted mechanism involves cross-linking of adjacent cellulosic chains by reaction of the formaldehyde-containing reactant with cellulose. This cross-linking prevents the slippage of adjacent cellulosic chains that may not be prevented by van der Waals and hydrogen bond forces when stress is applied to the fibers or fabric.

Formaldehyde release at high temperatures and humidities from durable-press fabric treated with formaldehyde reactants depends on many factors. With regard to reactant composition, those reactants having low reaction rates with cellulose such as DMDHEU release less formaldehyde. The rate of hydrolysis of the formaldehyde reactant bears a direct relation to its reaction with cellulose and to formaldehyde release. The leaving group has an important effect; earlier work indicated that of the groups studied, the methyl leaving group resulted in the lowest release of formaldehyde from the treated fabric. Later studies showed that the polyols gave lower formaldehyde release values than the methyl group. Increasingly, therefore, the polyol ethers of DMDHEU, especially the diethylene glycol ether, have been used. Some finishers add the polyol to pad baths containing DMDHEU rather than using the polyol ether of DMDHEU. A strong catalyst such as zinc nitrate is preferable, along with a cure time and temperature which gives optimum cross-linking with a minimum of pendant NCH_2OH groups. Impurities such as urea in the glyoxal–urea reaction product from which DMDHEU is prepared can result in increased formaldehyde release from the treated fabric; the presence of impurities such as alkali in the fabric prior to treatment and the use of additives such as buffers in the pad bath can also have a deleterious effect. On the other hand, formaldehyde scavengers added to the pad bath can reduce the amount of formaldehyde released from the treated fabric. Formaldehyde release can also be decreased by afterwashing the treated fabric or aftertreating with a formaldehyde scavenger.

In the manufacture of garments from durable-press fabrics, the garment is often made before the curing operation is carried out. The garment is then cured with the creases in place. This operation is referred to as postcuring, as differentiated from precuring.

In order to avoid the problem of formaldehyde release, reactants have been developed which either do not contain formaldehyde, or do not decompose and liberate formaldehyde. A product used to a limited extent in Japan, where there are stringent requirements for formaldehyde release, is *N,N'*-dimethyldihydroxyethyleneurea (**2**). This product can react with cellulose through the ring OH groups, but the durable-press properties are not equal to those obtained with a difunctional *N*-methylol compound.

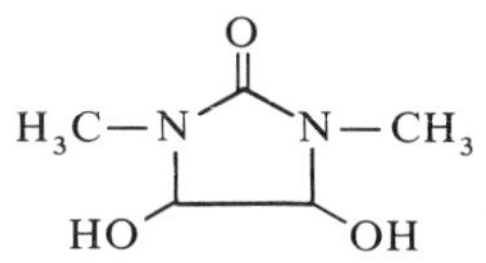

O
H_3C-N $N-CH_3$
HO OH

(2)

Although the main method of increasing the elastic recovery of cellulosic easy-care fabrics is by cross-linking with cellulose reactants inside the fiber, another method, which gives a lesser effect, is surface application. Since the cotton fiber surface has a high rugosity, it has a comparatively high coefficient of friction. Therefore, a topical application to decrease the coefficient of friction of the fiber and yarn will to some extent increase the recovery of the fabric from deformation. Emulsions of silicones and low-molecular-weight polyethylene are used extensively for this purpose as well as to provide a softer hand. If, in addition to the reduction in the coefficient of fiber and yarn friction, a surface application can produce an elastic fiber-to-fiber or yarn-to-yarn bond, a still further increase in the elastic recovery of the fabric may result. Therefore, lattices of elastomers, such as polyurethanes, are sometimes added to the formaldehyde reactant pad bath to impart additional recovery from deformation. Care must be taken, however, not to apply excessive quantities of elastomeric products to the fabric; otherwise an undesirable cold, clammy hand may result.

6. *Shrinkage Control*

Fabric shrinkage may be due to many causes. The shrinkage of cellulosic fabrics during washing is due to the relaxation of stresses imposed during manufacture of the fabric. This type of shrinkage, referred to as relaxation shrinkage, is usually greater in the warp direction of woven fabrics because the warp yarns are subjected to greater tension than the filling yarns. Knit fabrics tend to be stretched more than woven fabrics and therefore tend to shrink more during washing. Mechanical agitation in washing wool fabrics produces a compaction or felting of the fabric. This felting is due to the scalar surface of the wool fiber which causes the fiber to migrate preferentially in the direction of the fiber root since the overlapping scales point toward the fiber tip.

The tendency of fabrics to shrink during washing can be minimized by finishing treatments. The shrinkage of cellulosic fabrics can be reduced by a treatment that causes a compression or compaction in the plane of the fabric. An example of this type of treatment is Sanforizing. In this process the fabric and a backing blanket of wool or rubber held under tension are fed between a feed roller and a curved braking shoe. The tension on the blanket is released after passing the fabric and blanket between the roller and braking shoe. The result of this action is a compaction of the fabric. Another method to reduce the shrinkage is aftertreatment with formaldehyde reactants such as those used for durable-press treatments. In fact, the durable-press treatment itself provides shrinkage control.

The felting of wool fabrics can be reduced by chemical treatment or by application of resins. The most commonly used chemical treatments involve chlorination, which partially dissolves the scalar structure of the wool fiber. As a result, the differential friction effect (the difference between the "antiscale" and "with scale" coefficients of friction) is reduced so that the preferential migration is minimized. Unless carefully controlled, chlorination may reduce the strength of the wool fiber and change the hand of the treated fabric. Various resin treatments have also been developed for controlling the felting of wool fabrics. These treatments have included (a) impregnation with melamine–formaldehyde, (b) interfacial polymerization primarily utilizing polyamide resins, and (c) application of polyurethane resins. The resin treatments also reduce the differential friction effect and further reduce felting shrinkage by spot welding the fibers together.

In the case of fabrics made of thermoplastic man made fibers, shrinkage can usually be controlled by heat setting the fabrics. If the temperatures to which the fabric is exposed during service are below the temperature at which the fabric is heat set, shrinkage of the fabric is not usually a problem.

7. *Water and Oil Repellency*

There are a number of chemicals that can be applied to fabrics to impart water and/or oil repellent properties. One of the earliest and simplest water repellent treatments involves the application of wax to a fabric from an aqueous emulsion to impart a hydrophobic surface. These wax treatments have no durability to dry cleaning and limited durability to washing. If aluminum or zirconium salts are applied in a second treatment, limited durability to dry cleaning as well as washing is obtained. Formaldehyde reactants are also coapplied to improve the durability.

The next type of water-repellent finish involved long-chain pyridinium compounds (**3**), which were applied to cellulosic fabrics via an aqueous padding procedure followed by drying and curing. It is generally believed that during the curing operation, these compounds lose pyridine and hydrogen chloride and react with cellulose. The resulting finishes have fairly good durability to washing, but limited durability to dry cleaning. The coapplication of formaldehyde reactants such as melamine–formaldehyde increases the durability to washing.

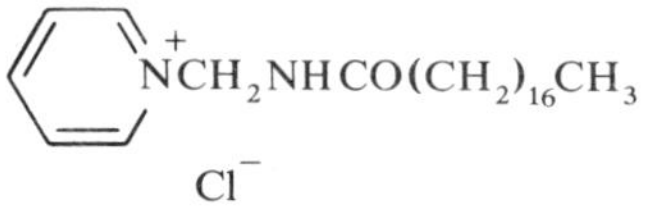

$$C_5H_5\overset{+}{N}CH_2NHCO(CH_2)_{16}CH_3 \quad Cl^-$$

(3)

A number of resin-based water repellents were then developed in which formaldehyde reactants such as melamine–formaldehyde were combined with long-chain hydrocarbon moieties to give durable finishes. Often, quaternized nitrogen moieties were incorporated to give improved emulsification properties.

Historically, the next water repellents developed were based on silicone polymers (**4**), (**5**) which imparted low surface energy properties to fabrics. The Si—H groups are reactive, whereas the Si—CH_3 groups are not. The silicone polymers are normally applied from an aqueous emulsion with a metal soap catalyst. They provide good durability to both washing and dry cleaning.

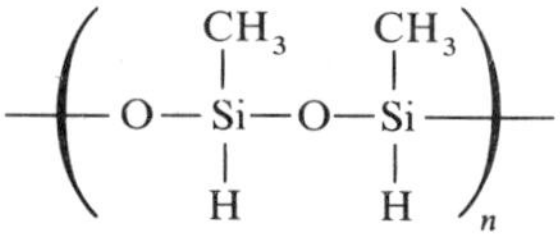

$$\left(O-\underset{H}{\overset{CH_3}{Si}}-O-\underset{H}{\overset{CH_3}{Si}} \right)_n$$

(4)

$$\left(O-\underset{CH_3}{\overset{CH_3}{Si}}-O-\underset{CH_3}{\overset{CH_3}{Si}} \right)_n$$

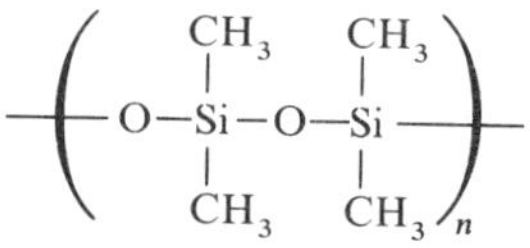

(5)

The most recently developed and most widely used products are based on fluorochemicals. These products are both water and oil repellents because of their very low surface energies. The compounds are usually provided as fluoroacrylate polymers with fluorine constituents present at the end of the polymer side chains as shown in Formula (**6**), where R_F represents a perfluoroalkyl chain. The polymers are usually applied to the fabric in the form of an emulsion, and the treated fabric is then dried and cured.

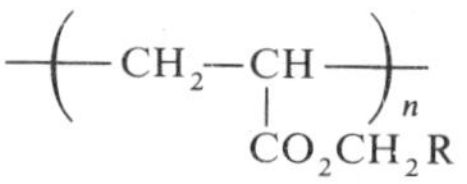

$$\left(CH_2-\underset{CO_2CH_2R}{CH} \right)_n$$

(6)

8. *Soil Release*

Even fabrics treated with oil-repellent fluorochemicals show a propensity to retain a stain when exposed to oils for a long period of time. For soil release, polymers with hydrophilic groups are needed. Fluorocarbon polymers with terminal fluorocarbon groups and modified with hydrophilic segment (containing carboxyl or other hydrophilic groups) when applied to fabrics as aqueous emulsions will provide both oil repellency and soil release. Many of the compounds developed as soil-releasing agents are polymers or copolymers of acrylic or methacrylic acid. They are usually provided as aqueous emulsions and can be applied alone or with formaldehyde reactants used to obtain durable-press properties. It is believed that the soil is held by the hydrophilic groups, perhaps by the acid moiety, and released during washing. The major disadvantages of these finishes are the change in hand of the treated fabric (stiffening effect) and their limited effectiveness as soil releasers.

9. *Fire Retardancy*

Although most organic fibers support combustion, their flammability can be reduced if they are treated with fire retardants. Some organic fibers such as the aramid fibers are thermally stable and withstand the high temperatures at which most organic fibers support combustion. In contrast, polyamide, polyolefin and polyester fibers usually melt before they burn, and the molten polymer may be removed from the source of flame before combustion occurs. Inorganic fibers such as glass and asbestos do not burn.

The flammability of fibers can be reduced by (a) modification of the polymer used for the fiber, (b) addition of a fire retardant to the dope prior to extrusion of the fiber, and (c) treatment of the fiber (usually in fabric form) with a fire retardant. An example of the first type of modified fiber is a fire-retardant acrylonitrile fiber made by the copolymerization of vinyl chloride with acrylonitrile. An example of the second type of fiber is a polyester fiber made from a dope containing brominated biphenyl as the fire-retardant additive. An example of the third treatment is the finishing of cellulosic fabric with tetrakis-(hydroxymethyl)phosphonium chloride (THPC). The fire retardants that have been used are many and varied in composition. However, some of the common ones are halogenated compounds and compounds containing phosphorus and nitrogen.

Several theories have been advanced to explain the modes of action of fire retardants. These include (a) noncombustible gases are generated which dilute the combustible volatiles and the oxygen supply, (b) the fire retardant decomposes endothermally, (c) the fire retardant degrades into free radical acceptors which interfere with flame chain reactions, (d) a nonvolatile char or liquid coating is formed which protects the fibers, and (e) a physical interaction occurs which alters the course of gas-phase reactions.

A great deal of research has been conducted on fire-retardant finishes for cellulosic fabrics. Some of the earlier nondurable finishes for cellulosic fabrics included inorganic salts such as borates and phosphates. Ammonium phosphate and diammonium phosphate are the most widely used nondurable flame retardants today. A flame retardant with some durability that is commonly used for flameproofing of industrial and military fabrics is a mixture of antimony oxide and chlorinated paraffin. Durable flame retardants developed for cellulosic fabrics used in clothing include many phosphorus compounds capable of reacting (a) directly with cellulose, or (b) with a formaldehyde reactant that reacts with cellulose. The most widely used of these fire retardants are THPC, its reaction product with sodium hydroxide (THPOH), and THPS where sulfate replaces chloride in THPC. These flame-retardant finishes are usually applied with formaldehyde reactants to give durable-press properties and greater durability to washing, but they may produce an undesirable hand and reduced wear characteristics.

Although wool is less combustible than cotton, its flammability can be reduced by many of the fire retardants that are used on cotton, and by the application of zirconium or titanium salts. The THPC treatments for cellulose fabrics are also effective on both wool and silk fabrics. Acetate fibers can be made fire retardant by the addition of retardants such as haloalkylphosphonates to the dope prior to spinning. For polyester and polyamide fibers, a fire retardant can be added to the spinning bath or tris(2,3-dibromopropyl) phosphate (Tris) can be applied. However, the use of Tris has essentially ceased since it was found to be mutagenic and a potential carcinogen. Acrylic, polyamide and polyethylene fibers can be made fire retardant by adding an organic phosphorus or halogenated compound to the dope prior to spinning.

10. Antistatic Properties

Cellulosic fibers and most of the natural fibers are hydrophilic, having polar groups on the fiber surface, and are therefore capable of dissipating static charge to the atmosphere so that an antistatic finish is not needed. However, with most synthetic fibers static charge can be built up on the fiber surface, especially under low humidity conditions, causing garments to cling and electric shocks when the wearer touches a conductor such as a metal door knob.

An antistatic finish or the incorporation of a conducting carbon or metallic fiber into the fabric can solve these problems. Antistatic materials can be applied to the fiber as a temporary finish or added to the spinning bath prior to fiber formation to give a more durable finish. These finishes can either coat the surface of the fiber with a conductive material or attract enough moisture to increase the conductivity of the fiber surface.

Antistatic materials applied to polyamides include hydrophilic polymers, polyethylene glycols and salts of amphoteric and cationic surface active agents. Antistats used on polyester include lauryl phosphate, morpholine, polyethylene glycols, organosilicones and polyamines. Antistatic agents for polyacrylic fibers include polyglycol esters, fatty acids and hydroxyl and amino compounds that are cross-linked with aldehyde or epoxy compounds.

See also: Fibers and Textiles: An Overview; Textiles: Structures and Processes

Bibliography

Lewin M, Atlas S M, Pearce E M 1975 *Flame-Retardant Polymeric Materials*. Plenum, New York

Lewin M, Sello S B 1983 *Chemical Processing of Fibers and Fabrics*, Vols. 1 and 2. Dekker, New York

Mark H F, Atlas S, Cernia E 1968 *Man-Made Fibers: Science and Technology*. Wiley-Interscience, New York

Mark H F, Wooding N S, Atlas S 1971 *Chemical Aftertreatment of Textiles*. Wiley-Interscience, New York

Needles H L 1981 *Handbook of Textile Fibers, Dyes and Finishes*. Garland STPM Press, New York

Tortora P G 1978 *Understanding Textiles*. Macmillan, New York

T. F. Cooke

Firing of Ceramics

The firing process involves the application of a heat treatment to formed ceramic products to bring about physical and chemical changes so that the products acquire the desired characteristics. This heat treatment (the firing process) takes place in kilns. For formed products of structural clay ceramics, refractories and advanced ceramics, a single firing process is usually sufficient to obtain a satisfactory product, but for the manufacture of dinnerware two or three heat treatments are often necessary (biscuit firing, glazing, decoration firing).

1. General

In principle the firing process comprises a heating-up stage in which the formed product is heated to a certain temperature (the firing temperature), a soaking period at this temperature and a cooling stage. The supply and dissipation of heat to or from a kiln charge, or "batch," takes place by radiation, convection and conduction.

The heat treatment is tied to certain conditions with regard to the heating-up rate, firing temperature, soaking period and cooling rate (in short, temperature changes, or the firing curve) and also the composition of the gaseous medium surrounding the product (the kiln atmosphere). Failure to satisfy these conditions is detrimental to the quality of the product.

Variations in the heat treatment of ceramic articles fired simultaneously or successively, whether or not accompanied by fluctuations in the composition of the raw material, may affect the consistency of quality so much that the products have to be sorted or adjusted after the firing process. This explains the importance of controlling the firing process as closely as possible. To limit the costs of building and running a kiln for a given production capacity, the firing process must be completed in the minimum time, with the minimum energy consumption and in the simplest installation consistent with maintaining the quality of the fired product.

A distinction is made between intermittent and continuous kilns. Intermittent kilns produce a fixed amount of a product equal to the total kiln capacity periodically. The product remains in a fixed position throughout the heat treatment. As soon as the kiln has been loaded, it is heated up to the firing temperature. The soaking period is followed by cooling, after which the kiln can be emptied. Continuous kilns either produce a continuous flow of fired product or at regular intervals a quantity of the product equal to part of the kiln's capacity. There are continuous kilns in which the product remains in a fixed position during the firing process (continuous chamber kilns) and others in which the product is kept moving (tunnel kilns).

There are numerous versions of kilns. The great variety stems from differences in

(a) type, shape and dimensions of the product to be fired,

(b) heat treatment required,

(c) production capacity of the kiln,

(d) arrangement of the products inside the kiln,

(e) loading and unloading methods,

(f) type of energy (solid, liquid or gaseous fuel, or electricity),

(g) heating and cooling methods (direct or indirect),

(h) position of the heat sources in relation to the batch, and

(i) direction of flow of the kiln gases.

The choice is further determined by requirements relating to flexibility, safety, environmental considerations and the site of the installation.

2. *Heat Transfer in Kilns*

2.1 *Radiation*

Heat exchange by radiation between solid bodies occurs in kilns between the walls and parts of the batch and between parts of the batch themselves. A condition for radiative heat transfer between bodies is that the radiation is not intercepted. If a body with a surface area A_1 and a temperature T_1 is entirely surrounded by another body with a larger surface area A_2 and a temperature T_2 the radiative heat flow is

$$\phi_s = A_1 C_r (T_1^4 - T_2^4) \tag{1}$$

where

$$C_r = \frac{C_z}{\dfrac{1}{\varepsilon_1} + \dfrac{A_1}{A_2}\left(\dfrac{1}{\varepsilon_2} - 1\right)} \tag{2}$$

ϕ_s is the heat flow (W), A_1 is the effective radiating surface (m^2) of the smaller body regardless of the direction of the heat flow, A_2 is the effective radiating surface (m^2) of the larger body, C_r is the resulting coefficient of radiation (W m^{-2} K^{-4}), T_1 is the absolute temperature (K) of A_1, T_2 is the absolute temperature (K) of A_2, C_z is the radiation constant of a black body (5.67×10^{-8} W m^{-2} K^{-4}), ε_1 is the emissivity of A_1, and ε_2 is the emissivity of A_2.

It may be inferred from Eqn. (1) that radiative heat transfer increases markedly with temperature. The emissivities depend on the type of material, the nature of the surface and the temperature of the bodies. For most ceramic surfaces with a temperature of θ °C, ε is given approximately by

$$\varepsilon = 0.95 - f\theta \tag{3}$$

where $f = 3 \times 10^{-4}$ within $\pm 15\%$.

If the radiation can penetrate into a batch largely by way of the open spaces between the products, the batch may be regarded as black body ($\varepsilon = 1$). Equations (1,2) can also be used for surfaces of equal area lying parallel to and directly opposite one another. In this case $A_1/A_2 = 1$. If A_2 is very large in relation to A_1, A_1/A_2 can be made zero and the equations still remain applicable. Slightly different equations apply to other geometries.

Gases such as hydrogen, oxygen, nitrogen and air are practically transparent to radiation: they are diathermanous. A number of other gases, however, including carbon monoxide, hydrocarbons, water vapor, carbon dioxide and sulfur dioxide, absorb radiation in certain limited wavelength ranges and on being heated emit radiation in these characteristic ranges: they are selective radiators. In kilns heated by fuels the selective radiation of water vapor and carbon dioxide in the combustion gases is particularly important. As the thickness of gas, the partial pressures of CO_2 and H_2O and temperature increase, the radiation becomes stronger.

2.2 Convection

Heat exchange by convection occurs between the kiln batch and kiln walls on the one hand and the gaseous medium on the other. The more the kiln gases are in motion relative to the surface of the batch and walls, the greater the amount of heat transferred. Forced convection occurs if the movement is artificially maintained, and natural convection when it is due to internal causes (for example density differences). Both kinds of convective heat transfer occur in kilns.

The heat flow ϕ_c (W) transferred by convection through a gas with a temperature θ_g (°C) to a surface A(m^2) with a temperature θ_0(°C) is

$$\phi_c = a_c A(\theta_g - \theta_0) \qquad (4)$$

where a_c is the heat transfer coefficient (W m^{-2} K^{-1}). The functional connection between a_c and the relevant variables, such as the gas velocity, the geometric relationships and the density, viscosity and heat conduction coefficient of the gas, can be established experimentally by means of dimensionless parameters. Both the numerical value of a_c and that of A are usually difficult to determine because of the great variety of shapes, dimensions and arrangements of ceramic products; an estimate usually has to be made.

The contribution of selective radiation from H_2O and CO_2 in the heat transfer is, for the sake of simplicity, expressed by an analogous formula:

$$\phi_s = a_s A(\theta_g - \theta_0) \qquad (5)$$

where a_s is the coefficient of heat transfer (W m^{-2} K^{-1}) for this gas radiation, which is very much temperature-dependent.

The following formula expresses the heat transfer resulting from convection and gas radiation together:

$$\phi_{c+s} = (a_c + a_s)A(\theta_g - \theta_0) \qquad (6)$$

The value of a_c in kilns for the ceramic industry generally appears to lie between 12.5 and 20 W m^{-2} K^{-1} at 200 °C and between 20 and 30 W m^{-2} K^{-1} at 1400 °C. The value of a_s usually varies from 0.5–1.0 W m^{-2} K^{-1} at 200 °C to 16–25 W m^{-2} K^{-1} at 1400 °C.

2.3 Conduction

In the kiln walls and in the products the transfer of heat takes place by conduction. In the walls of tunnel kilns the heat flow is stationary, because the temperature at every point inside a tunnel kiln does not change with time. For the simple one-dimensional stationary heat transfer in the walls of this type of kiln, which are generally constructed of layers of different refractories and heat-insulating capacity,

$$\phi_g = A\frac{(\theta_1 - \theta_2)}{\Sigma_{i=1}^{n} d_i/\lambda_i} \qquad (7)$$

where ϕ_g is the heat conduction (W), A is the surface under consideration (m^2), θ_1 is the internal wall temperature (°C), θ_2 is the external wall temperature (°C), i is the layer under consideration, d_i is the thickness of layer i (m), λ_i is the thermal conductivity of layer i (W m^{-1} K^{-1}), and n is the number of layers.

In contrast to the walls of tunnel kilns, those of intermittent kilns and continuous chamber kilns are heated and cooled periodically, so the heat flow is not stationary. In the course of heating, some of the heat supplied to the inner surface of the wall accumulates in the wall and the remainder is lost to the exterior. During cooling, heat continues to flow to the exterior but some of the heat accumulated during the heating process flows back inside the kiln (regenerated heat). The extent of the heat flow very much depends on the firing curve and on the thickness and composition of the walls. Nor is the heat flow in the products stationary during heating and cooling; often it is even multidimensional.

2.4 Temperature Lags

An important aspect of heat transfer in the batch is the lags in temperature at different depths. An analysis of one-dimensional heat transfer in a plane-parallel plate heated or cooled on both sides at a constant heating or cooling rate shows that after an initial period a parabolic temperature profile develops over the cross section. The lag of the core behind the surface is directly proportional to the square of half the plate thickness and independent of the heating or cooling rate. The lags arising from the conduction of heat are particularly important for the specification of firing processes for relatively large products and compact product stackings.

In a group of products (the batch in an intermittent kiln for example), temperature differences also occur in the direction in which heat transfer by convection and radiation takes place and manifest themselves as lags in the temperature changes of the different parts of the batch. Lags of this nature are also important factors in the specification for heat treatment. They can be limited by making the batch permeable, by directing a sufficiently large flow of kiln gases through the batch and by selecting the smallest possible dimension for the main direction of heat transfer. With regard to radiation, possible shadow effects should be taken into account.

2.5 Endothermic and Exothermic Reactions

The net effect of the heats of the endothermic and exothermic reactions that take place in the product during the heat treatment appears generally to be relatively small in relation to the total amount of heat supplied. It is therefore usually sufficient to make an overall calculation of these heat effects.

The main endothermic processes that can occur in classical ceramics are:

(a) the evaporation of the remaining free water at 2.40 MJ kg^{-1} of water vapor between 15 and 100 °C,

(b) the dispersal of the chemically bound water at 4.10 MJ kg^{-1} of water vapor between 450 and 650 °C, and

(c) the decomposition of calcium carbonate at 3.77 MJ kg^{-1} of carbon dioxide between 650 and 900 °C.

The most important exothermic reaction is usually the combustion of any humus present, at 18.45 MJ kg^{-1} of humus between 200 and 500 °C.

2.6 Heat Balance and Heat Consumption

Heat balances may have a bearing on the entire firing process or on one part of it (cooling for example). It is therefore necessary to determine precisely the limits of the system under consideration. The heat balance is an aid in the design of kilns. For kilns already in use, the balance gives an idea of possible energy savings or improvements to the firing process.

3. Heat Treatment

3.1 Firing Curve

The firing processes of the various ceramic products can be distinguished from one another in the first place by a wide variation in the temperature–time regime. The heating and cooling rates vary between a few K per hour and several hundred K per hour, and the firing temperatures are also extremely varied (see *Kilns for Firing Advanced Ceramics*).

Figure 1 is a firing curve for a hypothetical ceramic product. The curve has been divided into ranges in which the temperature varies linearly with time, but in reality the curve would be smooth. The number of ranges and their limits are determined fundamentally by requirements relating to the quality of the product. The specification is made on the basis of theoretical considerations and experimental research, whereby account is taken of lags in the temperature changes in the separate products and in the kiln batch as a whole in selecting the temperature limits. The firing curve which in a given case not only ensures the quality of the products but which is also

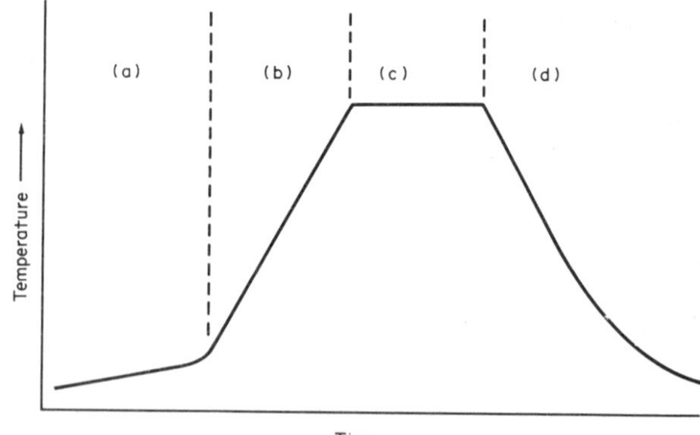

Figure 1
Schematic firing curve for ceramics: (a) burning out of binders, (b) heating up, (c) soaking, (d) cooling

technically and economically the best is the optimum firing curve.

3.2 The Firing Process in Detail

(a) Heating-up period. The variations in the heating-up rate are chosen so that the changes of state to which the product is subject and the stresses that arise from the thermal expansion of the product and the combustion of humus or binders do not cause damage (for example, cracks, discoloration, unwanted porous structure). Shortly before the start of sintering, the heating-up rate can usually be reduced so that the temperature differences arising from lags can be brought back to values that are acceptable at that time (equalization period).

(b) Soaking period. During the soaking period the temperature is generally raised in stages to a certain value which is then maintained for some time. The degree of sintering to which a product is subject is a function of the temperature and time, and also depends very much on the kiln atmosphere and the composition of the raw material. The relation can be established dilatometrically, for example. On this basis and also from what is known about the behavior of the time-dependent temperature differences in the batch, the temperature variations and the total duration of the soaking period can be chosen so that the extent of sintering satisfies the requirements in all parts of the batch.

During sintering, ceramic products may become plastically deformed under their own weight. This deformation is usually greater, the more strongly the product is sintered, and threatens the correctness of shape of the product. Where products are stacked, this also means that there is a danger to the stability of the stack. Although the pressure can be reduced by giving the products local support and by selecting a lower stacking height, the deformation limits the amount of sintering that can be applied and affects the specification of the heat treatment in the soaking period. Depending entirely on the type of product, there are sometimes other quality requirements which are factors in the specification of the soaking process.

(c) *Cooling period.* The cooling rates to be applied for safe and rapid cooling depend on crystallization processes and crystal conversions in the product and on its thermal contraction. At the beginning of the cooling process the material is generally still rather plastic. It can usually be cooled fairly quickly at this stage, because the thermal contraction does not cause the stresses to increase very much.

Below a certain temperature the plastic characteristics disappear. For classical ceramics this is between 800 and 600 °C. In this range a higher cooling rate accompanied by a steep temperature gradient can cause stresses to arise in the product.

In temperature ranges in which crystal conversions accompanied by volume changes take place, too high a cooling rate can produce cracks. Especially when using raw materials rich in quartz, with a quartz conversion at 575 °C and also sometimes a crystobalite conversion between 275 and 200 °C, the specification of the cooling rate requires special care.

The temperature at which the cooling process in the kiln can be ended varies from case to case. It is often possible to interrupt the cooling process in the kiln at ~300 °C and continue it outside.

3.3 Setting

The semifinished products are generally arranged in the desired pattern in the manufacturing area outside the kiln, where possible mechanically. Then they are loaded into the kiln. In many cases, refractory accessories are used to support and protect the product. This kiln furniture is repeatedly subject to the same heat treatment as the product.

To limit the consumption of heat it is desirable to keep the mass of the kiln furniture as low as possible in relation to the mass of the fired product. The ratio of the mass of kiln furniture to the mass of fired product is called the tare number s. The tare number varies very much according to the type of product to be fired and the type of kiln used. In kilns for bricks, the tare number is usually between 0 and 0.1. In kilns in which advanced ceramics are fired, in refractory covers, it may be as much as 5, for example.

Generally the space inside a kiln is never completely full of products. The degree of fill μ is defined as the ratio of the volume within the limiting surfaces of the charge in the unfired state (the setting space) to the total internal volume of the kiln under consideration. The space that is not reversed for the charge serves primarily for the generation and distribution of the heat. Depending on the type of kiln, the heating areas are located under, next to, between or above the charge. The dimensions of the heating areas depend in part on the type and capacity of the heat source and on the dimensions of the charge. The degree of fill μ usually lies between 0.5 and 0.9.

In the setting space the semifinished products are placed in one layer next to one another, stacked on or in each other or accommodated at different levels. Usually the fixed or moving floor of the kiln serves as a base for the stacking of the charge; occasionally, however, the products are fired in a hanging position (rolls of reinforced brick netting, large mural tablets, thin extrusion products).

The lags in temperature within the batch as a whole can be limited by making use of an open setting and limiting the dimensions of the kiln batch in the main direction of heat transfer. An open setting also means that the surface of the separate products is greater for heat transfer access. Temperature lags within the product are consequently also limited, which makes shorter firing times possible.

The setting, however, is not determined exclusively by the need for sufficiently small lags. Other factors that have to be taken into consideration include:

(a) the shape and the dimensions of the product,

(b) the use of kiln furniture,

(c) the aim for a high setting density to limit the extent of the installation and consequently of investment, and

(d) all kinds of technical considerations relating to setting and handling.

Factors that influence the choice of setting height are:

(a) the floor space taken up by the kiln,

(b) the investment, which is generally lower as the batch is made higher,

(c) the stability of the kiln furniture,

(d) deformation of stacked products during sintering,

(e) fracture of the product due to mechanical causes, and

(f) temperature lags within the batch, if the main direction of the heat transport through the batch coincides with the direction in which the products are stacked.

See also: Fast Firing of Ceramics; Kilns for Firing Advanced Ceramics; Sintering of Ceramics; Ceramics Process Engineering: An Overview

Bibliography

Chapman A J 1974 *Heat Transfer*, 2nd edn. Macmillan, New York

Kingery W D 1978 Firing—The proof test for ceramic processing. In: Onoda G Y, Hench L L (eds.) 1978 *Ceramic Processing Before Firing*. Wiley, New York, pp. 291–305
Perry R H, Chilton C H 1973 *Chemical Engineers Handbook*, 5th edn. McGraw-Hill, New York
Reynolds Th G 1976 Firing. In: Wang R Y (ed.) 1976 *Treatise on Materials Science and Technology*, Vol. 9. Academic Press, New York, pp. 199–215
Van der Velden J H 1974 Some aspects of the specification of the sintering process of ceramic products. *Trans. XIIIth Int. Ceramic Congr.*, Amsterdam
Van der Velden J H 1977 Firing behaviour of ceramic products. *ZI Int.* H3

D. J. Perduijn; J. H. Van der Velden

Fission Reactor Fuels and Materials

The growth of nuclear energy has been greatly influenced by the successful development and utilization of fuels and materials for use in power reactors. The requirements of low fuel-cycle costs coupled with high performance and reliability in the fuel elements have been essential factors in the competition with fossil-fuelled power plants. Nuclear fuel elements and materials are subjected to the unique operating conditions set by the designs of the reactor cores. In addition to accommodating the stresses and temperatures generated as a result of reactor operation, the materials in the reactor core must withstand prolonged exposure to nuclear irradiation and to the coolant.

The choice of nuclear fuel materials and designs has been dictated by the characteristics of the reactor cores, namely, the fuel enrichment, the types of coolant and moderator used, the operating temperature and pressure, the average neutron energy (whether thermal, epithermal or fast), the fuel burnups and the operating time in the core. The production of a new, advanced type of fuel element is a costly and time-consuming endeavor and may require the expenditure of millions of dollars and more than a decade of effort. However, tremendous savings in power costs in large reactors can result from improvements that lower the fuel-cycle costs by only a fractional amount per kWh. The engineering designs of nuclear reactors are limited by materials properties. Improvements in materials properties can result in significant increases in the efficiency and reliability of the reactors, as well as reductions in the costs.

1. Power Reactors

The most important nuclear reactor concepts at present include the following.

1.1 Light-Water Reactor (LWR)

Light water moderates neutrons and removes heat from slightly enriched UO_2 fuel rods. In the pressurized-water reactor (PWR), circulating light water at about 14 MPa and 300 °C transfers heat from fuel elements to steam generators via intermediate heat exchangers. In the boiling-water reactor (BWR), circulating light water boils at the top of the reactor core and furnishes steam at about 7 MPa and 300 °C directly to turbines (Fig. 1).

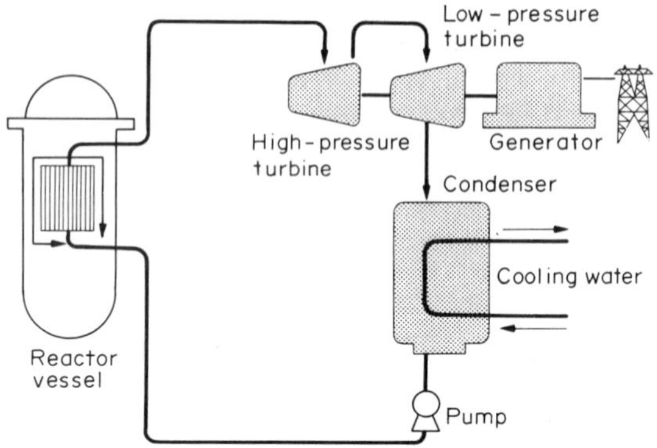

Figure 1
Diagram of a boiling-water reactor system

1.2 Pressurized Heavy-Water Reactor (CANDU)

Heavy water in a calandria moderates neutrons; light water at 10–15 MPa, circulating in pressure tubes, transfers heat from natural-UO_2 fuel rods to steam generators.

1.3 Carbon Dioxide Gas-Cooled Reactor

Magnox or Calder Hall type reactors are graphite moderated and are cooled by circulating CO_2 gas which transfers heat from natural-uranium metallic fuel rods to steam generators at about 400 °C. The name "Magnox" derives from the Mg–0.8%Al–0.05%Be–0.008%Ca alloy cladding developed for the fuel rods.

Advanced gas-cooled reactors (AGRs) are also graphite moderated and CO_2 cooled. Stainless steel cladding is used for slightly enriched UO_2 fuel rods, which permits steam generation at over 500 °C.

1.4 High-Temperature Helium Gas-Cooled Reactor (HTGR)

Circulating helium gas at 700–1000 °C and 5–8 MPa transfers heat to steam generators (550 °C), work to gas turbines or heat to high-temperature chemical processes. Graphite serves as moderator, reflector and core structure material. Coated particle fuel (UO_2, UCO, UC_2, ThC_2 and ThO_2) is used.

1.5 Liquid-Metal Fast Breeder Reactor (LMFBR)

Liquid sodium transfers heat from sealed $(U,Pu)O_2$ fuel rods to an intermediate heat exchanger, from which sodium transfers heat to the steam generator.

2. Classification of Reactor Materials

Nuclear reactor materials may be grouped as follows.

(a) Fuel and fertile materials, fuel elements and their cladding materials (see *Cladding and Assembly Materials*; *Fertile Blanket Materials*).

(b) Fuel systems and fuel cycles, which include the preparation, fabrication, reprocessing and refabrication of the fuel and fuel elements and the management of nuclear waste (see *Nuclear Fuel Element Fabrication*; *Reprocessing of Nuclear Fuel Elements*; *Nuclear Waste Materials*; *Radioactive Waste Management: A Systems Approach*).

(c) Moderator and reflector materials, which may be liquid or solid materials. In power reactors these include liquid H_2O and D_2O and solids such as graphite, beryllium, beryllium oxide and zirconium hydride (see *Solid Moderator and Reflector Materials*).

(d) Coolants are classified as gas (CO_2 or He) or liquid (H_2O, D_2O, sodium, organics or molten salts).

(e) Control materials and components use materials with high neutron absorption cross sections, such as boron, cadmium, indium, silver, hafnium, samarium and europium. They are employed in solid shapes, rods or blades, and in the case of boron compounds as fluids, to adjust neutron flux and to act as reactor shutdown devices (see *Control Materials for Nuclear Reactors*).

(f) Shielding materials attenuate the absorption and scattering of neutrons and γ rays. Moderating materials and steel often provide the internal shielding and concrete is usually the outer shielding material.

(g) Thermal insulation for gas-cooled reactors limits the transport of heat from the reactor core to surrounding structural materials and the containment vessel. Metal foils or metal-supported siliceous and related refractory materials serve as insulators.

(h) Structural materials include fuel element cladding and duct or coolant channels, solid moderator blocks, control element matrices and cladding material, grid structures, shielding, reactor vessels, coolant piping, heat exchange materials and many others (see *Breeder Reactor Core and Coolant Circuit Structural Materials*; *Pressure Vessel and Piping Materials for Light-Water Reactors*).

3. The Nuclear Fuel Cycle

The fissionable isotopes are ^{233}U, ^{235}U, ^{239}Pu and ^{241}Pu. The fertile isotopes are ^{238}U and ^{232}Th, which are converted into fissionable isotopes by neutron absorption (^{238}U into plutonium isotopes and ^{232}Th into ^{233}U). Natural uranium contains 0.71% ^{235}U, 99.28% ^{238}U and 0.006% ^{234}U. Fuel enriched in ^{235}U, ^{233}U or plutonium is used to provide greater latitude in selecting materials for use in the reactor system and to achieve higher burnup. Since ^{233}U and plutonium must be produced from thorium and ^{238}U, respectively, by neutron capture, the neutrons are provided initially by the fission of ^{235}U. The nuclear fuel cycle includes: (a) production of nuclear fuel (mining, milling, enrichment); (b) fabrication of the fuel elements; (c) reprocessing and recycling of the spent fuel to recover and reuse uranium and plutonium; and (d) storage of the radioactive waste. Nuclear fuel in its natural state has a low level of radioactivity and does not present a health hazard when correct procedures and adequate ventilation are maintained. Irradiated fuel is highly radioactive and must be handled and treated in shielded facilities. Special precautions have to be taken to minimize dust formation and contamination in handling and fabricating ceramic fuels in powder or particle form (see *Radioactive Materials: Safe Handling*). Enrichment of the fuel in the fissile isotope ^{235}U is required for power reactors in the USA. Hence, the uranium is extracted from the oxide in the form of the hexafluoride (UF_6), which is processed through an isotope separation plant.

Ceramic fuels can be fabricated into precise shapes (usually cylindrical pellets) that are clad in tubular thin-walled metal sheathing ("cladding"), which is backfilled with helium and end-capped (see *Oxide Nuclear Fuels*). The cladding in water-cooled reactors is Zircaloy or stainless steel. It protects the fuel from the reactor coolant, retains the volatile fission products and serves structurally to provide geometrical integrity. The clad fuel pins are assembled into fuel elements or subassemblies. The fuel elements are held in position by grid plates in the reactor core (Fig. 2). The fuel burnup to which a reactor may be operated is expressed as the thermal output per unit mass of total uranium ($^{235}U + {}^{238}U$) in megawatt-days per kilogram (MWd kg^{-1}). In light-water power reactors the core may be operated to about 35 MWd kg^{-1} (about 3.5% burnup of the heavy-metal atoms) before fuel elements have to be replaced. In fast breeder reactors (LMFBRs) and high-temperature helium gas-cooled reactors (HTGRs), the burnup may exceed 100 MWd kg^{-1} (about 10% burnup of heavy atoms).

The fuel elements in power reactor cores are distributed in zones of different uranium enrichment. The highest enrichment is at the periphery of the core, to compensate for the lower neutron flux toward the periphery and thereby achieve a flatter neutron flux profile and higher power output. At each refuelling period (about once a year), the fuel elements are discharged from the central zone of the core and the elements in the outer zones are moved

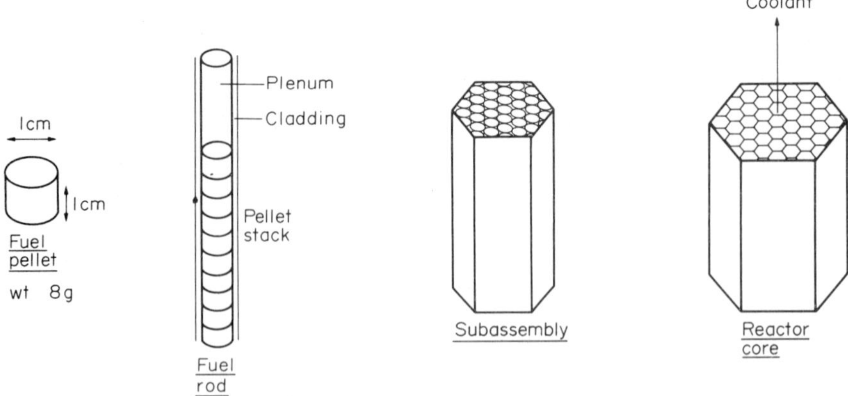

Figure 2
Building blocks of a nuclear reactor fuel assembly

inward. The fresh fuel elements are loaded into the vacated outer zone. The control rods are another core component that is periodically replaced. A 1000 MW light-water reactor (LWR) has an initial fuel loading of low-enriched (2–3%) uranium of approximately 80 t and an annual replacement fuel requirement of about 25 t per year to make up for the fuel that is consumed. Supplying the fuel for a 1000 MW LWR requires the mining of 218 000 t of uranium ore, the processing of about 450 t of UO_2 into feed for the enrichment plants and the fabrication of 83 t of the enriched UO_2 into 72 t of reactor fuel elements for the initial core loading. The spent fuel elements contain several million dollars' worth of unconsumed uranium and plutonium, as well as the fission products. The fuel reprocessing plants are designed to separate the fission products from the remaining fuel and to solidify the liquid radioactive waste for permanent disposal. The recovered fuel is recycled (see *Reprocessing of Nuclear Fuel Elements*).

Operational experience with nuclear fuels in power, test and research reactors is continually being evaluated. The quality-assurance system incorporates the various stages of design, fabrication techniques and performance of the fuel elements. The quality-assurance circuit for fuel elements includes the major tasks of fuel technology, namely, to determine the properties of the materials, to develop fabrication techniques and to establish testing and inspection methods for quality control. The primary factors that limit the performance of fuel rods are pellet–clad interaction, fission gas release and rod bowing. Much effort is being expended on the development of computer codes for fuel rod design and evaluation of the operational limits under steady-state and transient emergency conditions. The design codes have been "bench-marked" against operating experience and can be applied to a wide range of irradiation conditions and fuel parameters.

The important economic effects of fuel design, fuel fabrication methods and quality control on nuclear power generating costs have been assessed in some detail. The major factors that reduce costs include decreasing fuel failure rates, increasing margin to thermal operating limits in the fuel elements and improving fuel utilization. Thus, the performance and reliability of the fuel have a marked effect upon power generating costs. The fuel-cycle costs are approximately 25% of the nuclear operating costs. However, improved fuel performance does influence the costs of plant capital, operation and maintenance and total electrical system generation costs. For example, an increased core output and fuel reliability will increase the plant capacity factor and thereby reduce the total system reserve requirements and costs, particularly by decreasing the need for costly replacement power.

4. *Fuel Performance*

The performance requirements for nuclear fuel elements include the following items:

(a) dimensional stability to high fuel burnups
(b) fission product retention
(c) corrosion resistance
(d) high thermal performance
(e) fabricability
(f) economic advantage
(g) inspectability
(h) chemical reprocessing and recycling

There are numerous variables that influence the complex relationships which govern the operating characteristics of oxide fuel elements. These include the configuration and dimensions of the fuel pellets, the composition of the cladding, fabrication methods, fuel center temperature and heat fluxes. There has to be a compromise between the conflicting requirements of the materials scientist, the thermal designer and the nuclear physicist. Long-term in-pile tests under simulated reactor operating environments are the principal means for evaluating the performance of fuel elements.

The most extensively used ceramic fuels are the oxides, namely, UO_2, $(U,Pu)O_2$ and ThO_2, all of which have the face-centered-cubic fluorite structure and are completely miscible in solid solution (see *Oxide Nuclear Fuels*). A number of reactors have also operated with the carbide fuels UC, UC_2, (U,Pu)C and ThC_2. Nitride fuels have been prepared and irradiated in test reactors (see *Carbide and Nitride Nuclear Fuels*).

The oxide ceramic fuels have a number of advantages and disadvantages compared with other forms of nuclear fuels. The advantages include high neutron utilization, excellent irradiation stability, exceptional corrosion resistance in conventional coolants, high melting point, compatibility with cladding, ease of manufacture and high specific power and power per unit length of fuel pin. The disadvantages include low thermal conductivity, poor thermal shock resistance and relatively low fissionable atom density compared with metallic and carbide fuels. The relatively high melting points of the oxide fuels compensate partially for the low thermal conductivity.

Deviations from stoichiometry have a profound influence on the properties of UO_2: they reduce the already low thermal conductivity, lower the melting point and strength, increase creep and fission-product migration and release, and alter the complex irradiation behavior. The increase in oxygen activity with burnup can be very significant in leading rods in LWRs (5% burnup) and in fast breeder reactor fuels (over 10% burnup).

The allowable values of the thermal conductivity integral and the temperatures within pellets have been estimated from observation of microstructure. For example, the melting-point boundary corresponds to 2865 °C, columnar grain growth to 1700 °C and equiaxed growth to 1500 °C (Fig. 3). The reported integral conductivity values from 500 °C to melting range from 63 to 73 $W\,cm^{-1}$. The thermal conductivity of UO_2 decreases as the O/U ratio is increased.

The melting point of stoichiometric UO_2 is 2865 ± 15 °C. It drops to 2425 °C at an O/U ratio of 1.68 and to 2500 °C at an O/U ratio of 2.25. The lowering of the melting point to 2620 °C at a burnup of 1.5×10^{21} fissions per cm^3 has been reported.

Particularly striking among the behavioral features of UO_2 is the large increase, as the O/U ratio exceeds two, in the rate of creep, sintering, diffusion and

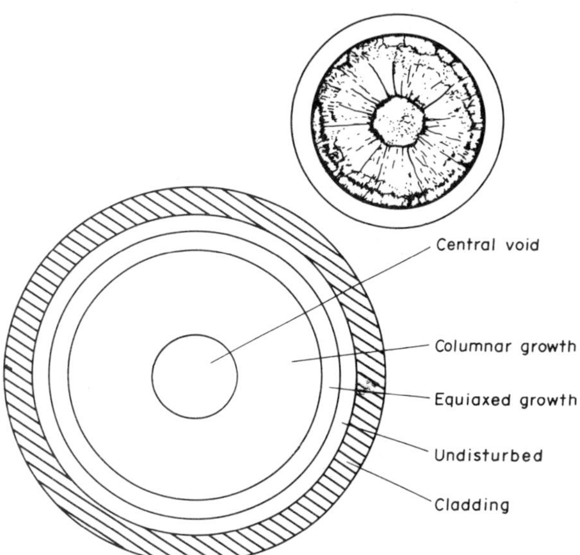

Figure 3
Axial cross section of a fuel element

other processes depending on mobile defects. Creep data on UO_2 demonstrate this effect.

Bulk oxide fuels have been fabricated by a number of different manufacturing processes, primarily by cold compaction followed by sintering at an elevated temperature. Vibratory compaction of particulate fuel (loose particles or optimum mixtures of several sizes of spherical (sol–gel) particles) packed directly into the cladding has attracted increased interest in recent years.

One of the most important quality requirements for fuel pellets is the need to minimize the moisture and fluorine contents (to <10 ppm each) in order to prevent internal corrosion failure of the cladding. The pellets should be stored in a dry environment and preferably heated in a vacuum after loading into the cladding. A major fuel rod failure mechanism is mechanical interaction between the fuel pellets and the cladding in the presence of fission products (e.g., iodine, cesium, tellurium), which results in stress-corrosion or intergranular cracking of the cladding.

The serious problem of UO_2 pellet densification under irradiation was experienced in pressurized-water reactors in the early 1970s. This behavior causes the fuel materials to contract and leads to loss of integrity of the fuel rods by collapse of the cladding in axial gaps in sections of the fuel columns. The solution to fuel densification has been to control the manufacturing process so as to produce fuel pellets with higher density and stabilized pore structure (pore size and grain size). Prepressurizing the fuel rod with helium also avoids clad flattening.

Most practical fuel elements are of cylindrical geometry, as are reactor cores, and for similar reasons. The exception is the spherical element for the experimental AVR helium gas-cooled reactor in Jülich, FRG. The overall active length of an element derives from the neutronic calculations that determine the size and fuel loading of the reactor. The diameter and number of the cylinders, usually called rods or pins, must satisfy two requirements: (a) the total volume of the pins must contain the required mass of fuel and fertile material; and (b) the surface area multiplied by the local permissible heat flux and integrated over the reactor must equal the required power.

A third requirement balances minimizing fabrication costs against maximizing specific power and forces the diameter into the upper range.

Compared with UO_2, UC has a higher uranium density, has at least five times the thermal conductivity and is almost as refractory. The compatibility of UC with stainless steel depends on the stoichiometry and on whether the gap between pellet and cladding is filled with gas or sodium. The cladding acts as a sink for carbon, and sodium enhances the transport of carbon from the fuel to the cladding. The decarburized fuel tends to crack, and the carburized cladding loses ductility quickly, even at 600 °C. With a gas gap, there is no significant interaction with stainless steel cladding below about 800 °C.

5. *Fuel Elements for Individual Reactor Types*

5.1 *Light-Water-Reactor Fuel Elements*

In LWRs the types of fuel failure that have been reported include the following:

(a) cladding failure due to excessive strain from fuel swelling and to fuel–cladding interactions;

(b) internal corrosion of cladding resulting from the presence of moisture, fluoride or hydrogen in the fuel;

(c) wear and fretting of cladding by extraneous metallic pieces;

(d) defects in the cladding or in the welds;

(e) hot spots in the cladding due to deposits of scale or poor heat transfer on corner rods.

It has been estimated that about 0.1–0.3% of the fuel rods have failed in LWRs.

The costs of fuel may be reduced primarily by higher burnup, lower fabrication costs and increase in the maximum specific power output of the fuel rods.

Zirconium alloys were developed for water reactors for cladding and pressure tubes because of their low neutron cross section, adequate strength in the operating temperature range and satisfactory resistance to corrosion by water at high temperatures (see *Pressure Vessel and Piping Materials for Light-Water Reactors*). In the USA over one million Zircaloy-clad UO_2 fuel rods have operated in LWRs.

5.2 *Heavy-Water-Reactor (CANDU) Fuel Elements*

In the CANDU reactor the natural UO_2 pellet fuel is clad with Zircaloy-4 and the fuel rods are separated by Zircaloy-4 spacers brazed to the cladding. The fuel elements are made up of bundles of 28 rods. The maximum fuel rod rating is about 690 W cm^{-1}. The fuel temperature is 400 °C at the surface and 2000 °C at the center. Operational experience with the fuel elements in the CANDU reactor has been highly satisfactory. A recent modification has significantly improved the fuel's performance. The new fuel rods include a thin graphite layer between the fuel and cladding and are designated CANLUB fuel. This has decreased friction and the strain concentrations in cladding expanded over fuel fragments resulting from power cycling. The mean burnup in heavy-water reactors is about 10.5 MWd kg^{-1} U, and the maximum specific fuel rod rating is about 660 W cm^{-1}.

5.3 *CO_2-Cooled-Reactor Fuel Elements*

The first-generation commercial nuclear power reactors in the UK and France were cooled by CO_2 gas. These reactors were graphite moderated and fuelled with natural-uranium metal rods clad in magnesium alloys. The first of these reactors (Calder Hall) started generating electricity in 1956. The second-generation CO_2-cooled graphite-moderated reactors in the UK (the AGRs) use slightly enriched UO_2 clad in stainless steel. The UO_2 fuel is in the form of sintered pellets which are loaded into stainless steel tubes. A cluster of 21 fuelled tubes is supported by stainless steel grids within a graphite sleeve to form a fuel element. In each channel several fuel assemblies are joined together by a central tie bar to form a fuel stringer. The fuel pellets are of stoichiometric composition, $UO_{2.00}$.

5.4 *Helium-Cooled-Reactor Fuel Elements*

The high-temperature gas-cooled reactors (HTGRs) use helium gas at about 800 °C and 5 MPa as the primary coolant, graphite as the neutron moderator and fuel-element structural material, and pyrocarbon and SiC-coated (Th,U) carbide or oxide fuel particles dispersed in a graphite matrix as the fuel.

The choice of graphite as the moderator and core structural material is based on its unique high-temperature chemical, physical and mechanical properties and on its very low neutron cross section, satisfactory radiation stability, ease of fabrication and low cost. The use of the graphite moderator as a diluent of the fuel permits much greater fuel dilution than would otherwise be possible and thereby minimizes radiation damage, increases specific power and greatly extends the heat-transfer surface.

The Th–^{233}U standard fuel cycle (with ^{235}U as the

initial fissionable fuel) is used because of its potential for achieving a higher fuel utilization and lower power cost than any other thermal-spectrum reactor system. The neutronic characteristics of ^{233}U are far superior to those of either plutonium or ^{235}U in thermal systems. A substantial portion of the power comes from fission of the ^{233}U converted from the fertile ^{232}Th. The overall C–Th ratio is optimum at a value of 240.

The use of coated particle fuel allows the high-temperature operation of the core to very high burnup (80%) of the fissile fuel, with extremely high retention of the fission products (see *Coated Particle Fuels for High-Temperature Gas-Cooled Reactors*). Also, the ^{235}U fissile particles are segregated from the ^{233}U-bearing fertile particles and thus the neutron poisons ^{236}U and ^{237}Np can be separated during the fuel reprocessing operation. The average fuel burnup of 100 MWd kg^{-1} obtainable is by far the highest of all existing systems. However, the high radioactivity of by-product ^{232}U present in the ^{233}U requires that the reprocessing and refabrication of the irradiated fuel be carried out remotely in shielded facilities. The ThO_2 fertile material is generally mixed with oxides of the fissile isotopes ^{235}U, ^{233}U or ^{239}Pu (usually <10% UO_2), or it may be used as pure ThO_2 in blanket elements with a seed-and-blanket arrangement. $(Th,U)O_2$ fuels have also been used in the cores of several water-cooled power reactors.

5.5 Liquid-Metal Fast Breeder Fuel Elements

The advantages of sodium as a coolant are its high boiling point, excellent thermal conductivity, large heat capacity and low-pressure operation. Sodium can remove decay heat by convective cooling after shutdown if power is lost. The disadvantages of sodium are its chemical reactivity with air and water, corrosion of structural and fuel materials, opacity, prolonged radioactivity after reactor shutdown, void coefficient if boiling occurs or gas is entrained, and the requirement for heating on shutdown in order to retain fluidity.

The development of sodium-cooled fast breeder reactors has been in progress for about 30 years. The first such reactor was the EBR-I, which operated from 1951 to 1963 in Idaho, USA. It was replaced in 1964 by EBR-II. Prototype demonstration-size LMFBRs of about 300 MW are in operation in the USSR (1972), France (1974) and the UK (1976). Fast test reactors are in operation in Japan (1981) and the USA (1982).

Plutonium is obtained by neutron capture from ^{238}U, since only an insignificant amount occurs in nature. Plutonium serves as a fissile fuel in both fast and thermal reactors. The fissile isotopes ^{239}Pu and ^{241}Pu produced from ^{238}U can replace some of the ^{235}U in thermal reactors. However, the most efficient and economical use of plutonium is in fast breeder reactors, where more ^{239}Pu and ^{241}Pu are produced than are fissioned in situ. Plutonium is separated by chemical reprocessing from irradiated fuel containing ^{238}U.

The fast breeder reactors that utilize plutonium will increase fuel use to over 70% of the uranium employed, compared with about 1% in present thermal reactors. A fast breeder reactor of 1000 MW capacity would require about 23 t of uranium in its lifetime. The fuel-cycle cost is also expected to be significantly lower in fast breeder power reactors than in thermal reactors or fossil-fuelled power stations. The cost of uranium ore will not strongly influence the cost of power from fast breeder power reactors, whereas it strongly affects the cost of power from thermal burner reactors. The doubling time (i.e., the time necessary for the reactor to produce a surplus amount of fissile material equal to that required for its initial fuel loading) can be relatively short (about 10 years) for fast breeder reactors, depending on breeding ratio, power rating and recycle rate.

The composition and properties of the mixed fuel in a fast breeder reactor will change significantly with burnup. These changes include the alteration in the ratio of uranium to plutonium, increase in the chemical potential of oxygen in oxide fuel, and the production of as much as 20 at.% of fission products for a 10% burnup (100 MWd kg^{-1}) of the heavy-metal atoms. The volume increase resulting from the formation of solid fission products may be approximately 5% for a 10% burnup of the heavy-metal atoms in a mixed oxide fuel. The steep temperature gradient in mixed oxide fuels will also cause marked redistributions of the fission products, actinides and oxygen.

The fuel burnup required in fast breeders (over 100 MWd kg^{-1}) is more than three or four times that in thermal reactors. The fuel element has to accommodate the volume changes from approximately 10% of fission products and survive the irradiation damage. At present the types of fuel being studied for fast breeder reactors include alloys such as U–Pu–Zr and the ceramic materials $(U,Pu)O_2$, (U,Pu)C and (U,Pu)N.

The mixed oxides $(U,Pu)O_2$ are the choice for most of the prototype fast breeder fuel elements. The advantages of the oxide fuel are high melting temperature, compatibility with cladding and coolants, and relatively good irradiation stability and fission product retention. The disadvantages are the relatively low metal density, the moderating effect of oxygen, low thermal conductivity, excessive swelling over a critical temperature range at high burnup, and the possibility of fission product and plutonium redistribution during reactor operation.

The mixed carbides (U,Pu)C offer a significant improvement in breeding and a shorter doubling time through their higher metal atom density and thermal conductivity. The disadvantages of the carbide fuels are control of composition (stoichiometry), in order to be compatible with the cladding and the sodium coolant and to minimize swelling, and the lack of adequate irradiation experience at high burnup and elevated temperature.

The rate of swelling of stainless steels and other alloys by fast neutrons is governed by temperature (maximum at about 575 °C for 20% cold-worked type 316 stainless steel), fluence, stress and structure. Hence, inhomogeneous swelling in the duct, cladding and structural parts will result in bowing and distortion (see *Swelling in Irradiated Metals and Alloys*).

Irradiation creep causes the maximum cladding strain to occur in the region of maximum neutron flux rather than maximum temperature. Irradiation creep varies linearly with stress level and fluence and depends on temperature. It is enhanced by swelling.

The main result of the most advanced studies is the indication that the commercial austenitic stainless steels will exhibit excessive swelling at the high fast fluences required for the commercial fast breeder reactor, but that the alloys containing about 40% nickel and minor amounts of Ti and other elements will probably retain their dimensional integrity to the highest fluences. However, the irradiated alloys have very low uniform ductility and this factor will have to be considered in the core design. The maximum fast-neutron fluences achieved in fast test reactors are still about a factor of two below the fluence required for the demonstration plant core structural materials and cladding, and a factor of four below that required for the commercial FBRs.

Although the technical feasibility and advantages of the fast breeder reactor have been demonstrated, the goal of the fast-reactor programs at present is to develop the technology and data to build economically viable fast breeder power reactors during the next decade. Among the major problem areas are the higher-performance fuels and materials, fuel reprocessing and recycling, coolant technology (particularly steam generators), and the physics and engineering for the large fast breeder power reactors.

By the year 2000, when the commercial fast breeder reactors could come on line, about 600 t of plutonium should be available from LWRs. This amount of plutonium is sufficient fuel for as many as 200 fast breeder reactors. The value of plutonium as a fuel is much greater in a fast breeder reactor than in an LWR. The high fuel inventory in the fast breeder reactor cores calls for close control of the cost factor in the design of the core. There is also a strong incentive to shorten the cooling and reprocessing times.

See also: Dispersion Fuels for Nuclear Reactors; Metallic Nuclear Fuels; Radiation Effects on Nuclear Fuels; Nuclear Materials: An Overview

Bibliography

American Nuclear Society 1981 *Proc. American Nuclear Society Topical Meeting on Reactor Safety Aspects of Fuel Behavior*. American Nuclear Society, La Grange Park, Illinois

Cameron I R 1982 *Nuclear Fission Reactors*. Plenum, London

Frost B R T 1982 *Nuclear Fuel Elements*. Pergamon, New York

Glasstone S, Sesonske A 1981 *Nuclear Reactor Engineering*, 3rd edn. Van Nostrand Reinhold, New York

Journal of Nuclear Materials. North-Holland, Amsterdam

Kessler G 1983 *Nuclear Fission Reactors*. Springer, Wien

Olander D R 1976 *Fundamental Aspects of Nuclear Reactor Fuel Elements*, TID-26711-P1. National Technical Information Service, Springfield, Virginia

Roberts J T A 1981 *Structural Materials in Nuclear Power Systems*. Plenum, New York

Simnad M T 1971 *Fuel Element Experience in Nuclear Power Reactors*. Gordon and Breach, New York

Simnad M T, Howe J P 1979 Materials for nuclear fission power reactor technology. In: Libowitz G G, Whittingham M S (eds.) 1979 *Materials Science in Energy Technology*. Academic Press, New York, pp. 31–179

M. T. Simnad

Flammability Hazards of Structural Materials and Furnishings

The flammability hazards of structural materials and furnishings are of great importance: they have been significant factors in many catastrophic fires such as the Cocoanut Grove nightclub fire in 1942, the Beverly Hills Supper Club fire in 1977 and the MGM Grand Hotel fire in 1980, and they continue to be significant factors in less newsworthy fires which occur in residences every week if not every day.

1. Fire Response Characteristics

Flammability hazards of materials can be viewed in terms of nine fire response characteristics: (a) smolder susceptibility, (b) ignitability, (c) flash-fire propensity, (d) flame spread, (e) heat release, (f) fire endurance, (g) ease of extinguishment, (h) smoke evolution and (i) toxic gas evolution. Smolder susceptibility is the tendency to undergo progressive combustion in the solid state without flame. Ignitability is the ease of initiation of burning by a small flame or spark. Flash-fire propensity is the tendency of a heated material to emit flammable gases and vapors which can propagate flame with extreme rapidity. Flame spread measures the progress of flame over a combustible surface. Heat release relates to both the amount of heat released from a unit quantity of burning material and the rate at which that heat is released. Fire endurance is a measure of the elapsed time during which a material or product maintains its design integrity under exposure to heat and fire. Ease of extinguishment is the facility with which a fire can be extinguished in a burning material. Smoke evolution is the tendency of a burning or smoldering material to produce visible mixtures of gases, aerosols and particulates which obstruct light transmission and vision. Toxic gas evolution is the production of gases during burning or smoldering which are poisonous or destructive to body tissues.

2. Structural Materials

The primary functions of structural materials are to support and maintain the integrity of a structure. Fire endurance is the most important fire response characteristic for load bearing and fire barrier materials. Building codes recognize this importance in classifying types of construction. Fireproof, fire-resistant and noncombustible types of construction generally involve steel, iron, concrete and masonry; the concrete and masonry are reinforced by the steel and iron, and the steel and iron are protected from direct fire contact by the concrete and masonry. Unprotected steel performs well in terms of ignitability, flame spread, heat release, smoke evolution and toxic-gas evolution, but is limited in fire endurance. Special recognition is given to heavy timber construction, because the characteristic charring of wood at the surface exposed to fire provides a measure of protection to the underlying material in large wood cross sections. Ordinary wood-frame construction involves a material which has general acceptance because its flammability hazards are widely understood and considered in the design and occupancy of a structure. Unreinforced plastics are generally not used as structural materials because of their tendency to creep under load and heat; substantial reinforcement in the form of glass fibers and carbon fibers is required for even relatively small structural members. The use of combustible materials such as wood and plastics in construction increases the importance of smolder susceptibility, ignitability, flash-fire propensity, flame spread, heat release, ease of extinguishment, smoke evolution, and toxic gas evolution in assessing flammability hazards. Wood and wood-based composites (plywood, laminated beams) are preferred to plastics for structural applications because both performance characteristics and flammability hazards are better understood as a result of long experience and can be predicted with greater confidence in both design of the structure and control of its occupancy.

Thermal-insulation materials are incorporated in structures to reduce the energy requirements for heating and air conditioning. Inorganic insulation materials such as glass fiber and mineral wool batts and blankets are essentially noncombustible, but foam plastics can represent a significant fire hazard if not properly applied. Rigid polyurethane and polyisocyanurate foams present the least hazard when protected by noncombustible materials, as when they are used as the core of sandwich panels or covered by plaster. Rigid polyurethane foam has had an unfavorable history of fires when spray-applied on interior surfaces and left unprotected.

3. Furnishings

Furnishings are designed with a view to utility, comfort and esthetic considerations. The majority of materials used in furnishings are combustible materials, and all nine fire response characteristics are important, each to a degree that varies with the material. Smolder susceptibility is more important in cellulosic fabrics and cushioning materials and in polymeric and elastomeric foam materials than in most other materials. Flash-fire propensity appeared to be more relevant to some early polyurethane foams than to most other materials. Smoke evolution may be more of a hazard with poly(vinyl chloride), polystyrene, acrylonitrile–butadiene–styrene and foam rubber than with many other materials. Toxic gas evolution is a potential hazard from all combustible materials because of their ability to produce potential oxidation products, including carbon monoxide, in a fire. Certain plastics have aroused additional concern because of the possible production of other toxic gases such as hydrogen chloride from poly(vinyl chloride) and hydrogen cyanide and nitrogen oxides from certain nitrogen-containing plastics, wool and leather. The evolution of heat, smoke and toxic gases from furnishings was a significant factor in the Beverly Hills Supper Club fire and in the MGM Grand Hotel fire.

The particular furnishings in which materials are used determine to a large extent their potential exposure to fire and their potential for harm. Approximately 39% of residential fires originate in the bedroom and about 36% in the living room. Upholstered furniture and bedding are the furnishings most likely to be ignited in a fire, and their smolder susceptibility, ignitability and flash-fire propensity are important; they are furnishings which represent a major source of heat and other products of combustion, and so their flame spread, heat release, smoke evolution, and toxic-gas evolution are also important. The susceptibility of certain combinations of upholstery fabrics and cushioning materials to smoldering is a well-known fire hazard. Smoldering can continue for relatively long periods of time, during which it may or may not kindle into flame, but during which it produces smoke and toxic gases. Smolder susceptibility is not a function of the fabric or cushioning material alone, but of the specific fabric–cushion combination. Cellulosic fabrics have been observed to be more smolder-susceptible than thermoplastic fabrics in furniture upholstery, but this susceptibility is greatly reduced when the underlying cushioning material is resinated polyester batting or neoprene foam interliner. Thermoplastic fabrics exhibit no smolder susceptibility except when untreated cotton batting is the cushioning material. Smoldering combustion generally requires a porous fuel, and charring appears to be a prerequisite for smoldering. Cellulosic fabrics such as cotton and rayon are, like wood, prone to char and smolder. Thermoplastic fabrics such as nylon and polyolefin inhibit smoldering by not forming a porous char and by melting down to help seal off areas against smoldering.

In ignitability, as in smolder susceptibility, the physical characteristics of the material are important. Thermal diffusivity and absorptance characteristics are more decisive factors in the rate of increase of surface temperature than differences in chemical composition. The effect of density, which influences thermal conductivity, volume specific heat and thermal diffusivity, appears to override any differences in chemical composition. A dark-colored material would probably ignite more rapidly than a light-colored or clear material in a fire.

Flash-fire propensity is of the greatest concern in occupied compartments. Some early polyurethane flexible foams were noted for their flash-fire propensity. Polychloroprene and poly(vinyl chloride) appear to be less prone to flash fires, since they do not readily yield flammable volatiles on heating.

Flame spread is more relevant to floor coverings, wall coverings and ceiling linings. These materials, which can be classified as interior finish, are discussed in Sect. 4. Flame spread becomes more relevant in furniture when the objects have at least one long dimension, as for sofas and beds.

Heat release is of greatest concern with materials which burn rapidly. In furnishings, untreated polyurethane flexible foam and latex foam rubber have been noted for high heat release. Polystyrene structural foam used in kitchen cabinet doors and in baby cribs, and bathtub–shower enclosures of reinforced thermosetting polyester, have been notable cases of high heat release.

Smoke evolution from furnishings has been of greatest concern with materials such as poly(vinyl chloride), polystyrene, acrylonitrile–butadiene–styrene and foam rubber.

4. *Interior Finish*

Floor coverings, wall coverings and ceiling linings are furnishings in that they are primarily selected for utility, comfort and esthetic reasons, but they may be considered as structural materials to the extent that they are attached to or part of the structure. Because of the large surface area exposed by these materials, flame spread is their most important fire response characteristic. Floor coverings are more likely to be exposed to small ignition sources such as smoldering cigarettes and lighted matches than are wall coverings and ceiling linings. In the event of a substantial fire, however, the ceiling linings and wall coverings are more likely to become involved because of the tendency of the hot combustion gases to rise toward, and rapidly heat materials in, the upper portions of an enclosure. Wainscoting, for example, is not as likely to contribute to flame spread as the upper half of a wall.

Carpet may be the most widely used floor covering material in residential and commercial occupancies. Many carpets are installed with an integral cushion backing or on a separate cushion pad. The addition of a carpet cushion pad under the carpet in many, if not most, cases may have a greater effect of flame spread than differences between carpets. Since the rate of flame spread is dependent on the rate at which succeeding surface areas are heated to their ignition temperature, the thermal characteristics of the carpet–underlay system may prove to be more important than the chemical composition of the carpet.

Various materials are used in wall coverings and ceiling linings. The paints and coatings used on walls and ceilings are important factors in response to fire. The flammability hazards of these materials are generally judged on the basis of their flame spread classification according to the ASTM E84 tunnel test. A flame spread classification of 0–25 is considered the least hazardous, followed by a flame spread classification of 26–75. The smoke-developed classification by this same test is gaining importance.

5. *Legal and Regulatory Aspects*

Product-liability litigation is one of the major economic factors in flammability hazard evaluations. The increasing number of claims, the increasing size of settlements and awards, and the increasing cost of litigation have all multiplied the cost of flammability hazards to manufacturers and their insurance companies. Opponents of such litigation contend that the substantial resources expended in litigation could be better spent in making the product safer. Proponents of such litigation believe that only the economic penalty of prohibitive jury awards is effective in forcing manufacturers to improve the safety of their products.

Some materials which have been the subject of a series of lawsuits, substantial lawsuits or both are rigid polyurethane foam used as thermal insulation, flexible polyurethane foam used as cushioning and poly(vinyl chloride) used as electrical insulation.

Regulatory action has generally paralleled litigation. Misuse and misleading advertising of rigid polyurethane foam have led to action by the Federal Trade Commission, which resulted in a consent decree involving payment of five million US dollars by polyurethane manufacturers. Hazardous flexible polyurethane foam has been banned from upholstered furniture sold in the state of California, and more fire-safe polymeric foams have been developed to meet California state requirements. There has been considerable opposition to permitting the use of poly(vinyl chloride) in electrical conduits because of its potential for producing smoke and toxic gases.

See also: Plastics and Composites as Building Materials; Wood as a Building Material; Hazards and Materials: An Overview

Bibliography

Berl W G, Halpin B M 1979 *Human Fatalities from Unwanted Fires*, NBS GRC 79-168. National Bureau of Standards, Washington, DC

Best R L 1977 *Reconstruction of a Tragedy: The Beverly Hills Supper Club Fire*. National Fire Protection Association, Quincy, Massachusetts

Damant G H, Young M A 1977 Smoldering characteristics of fabrics used as upholstered furniture coverings. *J. Consum. Prod. Flammability* 4: 60–113

Hilado C J 1981 *Flammability Handbook for Plastics*, 3rd edn. Technomic Publishing, Westport, Connecticut

Hilado C J, Huttlinger P A 1980 Toxicity of off-gases from furnishing materials: A review. *J. Consum. Prod. Flammability* 7: 189–99

Hilado C J, Murphy R M 1979 *Fire Response of Organic Polymeric Materials (Organic Materials in Fire: Combustibility)*, ASTM STP 685. American Society for Testing and Materials, Philadelphia, Pennsylvania, pp. 76–105

C. J. Hilado; P. A. Huttlinger

Flammability of Elastomers

The increasing use of polymers has caused considerable concern regarding the hazard these materials may present when brought into contact with an ignition source. Much of the interest originally was in plastics. However, sizable quantities of elastomers also are used in situations which require low flammability. Consequently, these materials, too, have recently received a great deal of attention.

Cellular elastomer compositions are used in seat cushions, rug backing, carpet underlay and as thermal insulation, while dense elastic products find use as conveyor belts in mine shafts, electrical wire and cable insulation, hose and ducting material, and seals and gaskets. All of these applications call for elastic materials exhibiting some degree of flame-retardant behavior.

The flammability of a material is characterized by its response to a specific ignition stimulus. Typically, the flammability is expressed in terms of ease of ignition, surface flame spread and the amount and rate of generation of heat, combustible material, smoke and toxic gases. A wealth of laboratory tests has been designed to obtain relative numerical values for these characteristics under strictly specified test conditions ranging from simple exposures to ignition sources of varying energy (e.g., a burning cigarette or an electrically heated silicon carbide rod) to more elaborate measurements using instruments which permit the determination of more than one of the above parameters at the same time.

Among the more common procedures are the Motor Vehicle Safety Standard 302 (for automotive interiors), a vertical flame test (DMS 1510A) for aircraft interior materials, the Steiner tunnel test (ASTM E84) for carpet and rug underlay, as well as the radiant panel (ASTM E162) and heat-release-rate calorimeter tests (Smith 1972) which attempt a more realistic simulation of actual fire conditions by exposing the material sample to radiant heat fluxes in addition to ignition stimuli.

The formation of combustible volatile matter necessarily has to precede any burning process. The chemical structure of an elastomer will, therefore, greatly determine its flammability characteristics (Fabris and Sommer 1977, Rogers and Fruzzetti 1975). Hydrocarbon rubbers (i.e., natural rubber, ethylene–propylene rubbers, polybutadienes and styrene–butadiene copolymers) produce large amounts of combustible gases at relatively low temperatures (350–400 °C) and ignite and burn with relative ease. Elastomers containing halogen, phosphorus or silicon have a much greater resistance to ignition.

Generally, flame-retardant elastomers can be obtained by two fundamental methods: (a) the addition of flame retardants or (b) the preparation of elastomers in which the flame retardant is an inherent part of their structure. Chamberlain (1978) has reviewed the various mechanisms by which flame retardants operate.

The additive approach to flame retardance is widely used with the low-cost, commodity hydrocarbon rubbers. These rubbers account for approximately 70% of the total elastomer consumption (excluding polyurethanes—see below) in applications requiring some control of flammability. Antimony trioxide in combination with chlorine-containing compatible compounds (e.g., chlorinated paraffins) and specific fillers such as alumina trihydrate are frequently the chosen additives, although in certain cases phosphorus-containing plasticizers are used also (Fabris and Sommer 1977). Extensive research continues to develop improved additives and techniques for their incorporation into the final composite.

Polyurethanes (flexible foams and dense elastomers) represent the largest volume of elastomers used in applications where flame retardancy is a consideration. Flame-retardancy in these polymer classes is achieved both by compounding with additives or by polymer design (Papa 1975, Fabris and Sommer 1977).

Polymers with built-in flame-retardant characteristics have been prepared by polymerization of, or copolymerization with, special monomers or by carrying out suitable post reactions on otherwise flammable polymers (Fabris and Sommer 1973). The polymerization of special monomers yields elastomers with various degrees of flame retardance. These include the well-known polychloroprene (Neoprene) and a group of specialty polymers which are used in a diversity of highly critical applications. Fluoroelastomers, nitroso rubbers, triazine elastomers and polyphosphazenes burn with difficulty even in oxygen-rich atmospheres and are used in spacecraft and in high-velocity aircraft to meet extreme requirements of flame retardance. Silicone elastomers, although they do ignite, burn only slowly and produce no flaming drip and relatively little smoke. Because of their outstanding flexibility at low temperatures, they have found use as seals and gaskets in commercial aircraft and as components in space suits. Polychloroprene has a chlorine content of 40% and, therefore, inherently burns less readily than the hydrocarbon rubbers. The choice of fillers is most critical for the retention of flame-retardant behavior in this polymer. While inert fillers show the normal dilution effect in combustible matter giving a slight improvement, calcium carbonate, a very common filler, has been found to reduce the flame retardance of polychloroprene (Fabris and Sommer 1977).

Appropriate modification of preformed polymers is an alternative way to arrive at elastomers with built-in flame retardance. In particular, the chlorination and chlorosulfonation of polyethylene have led to materials of commercial value (e.g., Dow Chemical Company's "CM" series and Du Pont's "Hypalons"). The chlorine content of these modified materials is generally high enough to pass all but the most severe flammability test requirements.

The copious amount of smoke produced by burning rubbers (particularly those derived from dienes and styrene) obscures vision and contributes to the creation of panic in a fire and, therefore, is a serious deterrent to their acceptability in many applications. Flame retardants added to a polymer generally prevent complete combustion and further tend to increase smoke evolution. Inert inorganic fillers lower the amount of smoke only to the value predicted from the dilution of organic matter in the total composition. Certain hydrated fillers (alumina trihydrate and magnesium hydroxide) appear to be exceptions to this rule. These two fillers are found to lower both the total light obscuration as well as the initial rate of smoke evolution caused by burning styrene–butadiene rubber (SBR) and natural rubber (Hecker et al. 1973, Lawson et al. 1975).

In addition to these flammability factors, increasing attention is being given to the toxicity of airborne matter formed by burning or thermally degrading polymers. Toxic gas (e.g., carbon monoxide, hydrogen cyanide and hydrogen chloride) evolved during burning of specific polymers or polymer compositions have to be considered not only by themselves but also in possibly synergistic combinations. Temperature and oxygen depletion and their effect on the human organism have to be taken into account. No adequate tests have been developed yet to assess fully the hazards related to this important characteristic of burning or pyrolyzing elastomer systems and, although research is being vigorously pursued, sufficient information on this subject will probably not be available in the immediate future.

The field of elastomer flammability is a very dynamic one and much research is directed toward the development of compositions which ignite with greater difficulty and also (at least in the early stages of a fire, when the heat flux is still relatively small) burn at slower rates. Much needs to be done to reduce the smoke and the toxic gases produced during burning. Test procedures will have to be developed which make it possible to predict the burning behavior of materials with sufficient accuracy to allow choosing the best elastomer for any specific application.

See also: Flammability Hazards of Structural Materials and Furnishings; Flammability of Polymers; Flammability of Polymers: Test Methods

Bibliography

Chamberlain D L 1978 Mechanism of fire retardance in polymers. In: Kuryla W C, Papa A J (eds.) 1978 *Flame Retardancy of Polymeric Materials*, Vol. 4. Dekker, New York, pp. 109–68

Fabris H J, Sommer J G 1973 Flame retardation of natural and synthetic rubbers. In: Kuryla W C, Papa A J (eds.) 1973 *Flame Retardancy of Polymeric Materials*, Vol. 2. Dekker, New York, pp. 135–99

Fabris, H J, Sommer J G 1977 Flammability of elastomeric materials. *Rubber Chem. Technol.* 50: 523–69

Hecker K C, Fruzzetti D G, Sinclair E A 1973 Flammability and smoke properties. *Rubber Age* 105(4): 25–32

Lawson D F, Kay E L, Roberts D T Jr 1975 Mechanism of smoke inhibition by hydrated fillers. *Rubber Chem. Technol.* 48: 124–31

Papa A J 1975 Flame-retarding polyurethanes. In: Kuryla W C, Papa A J (eds.) 1975 *Flame Retardancy of Polymeric Materials*, Vol. 3. Dekker, New York, pp. 1–133

Rogers T H Jr, Fruzzetti R E 1975 Flame retardance in rubbers. In: Levin M, Atlas S M, Pearce E M (eds.) 1976 *Flame Retardant Polymeric Materials*, Vol. 1. Plenum, New York, pp. 223–37

Smith E E 1972 Measuring rate of heat, smoke and toxic gas release. *Fire Technol.* 8(3): 237–45

H. J. Fabris

Flammability of Polymers

Polymeric materials are being increasingly used as construction materials in a number of areas such as home furnishings, domestic and industrial buildings, fabrics and vehicles. Selection or design of a flame retardant system for a particular application is often difficult. It is usually desirable that a flame-retardant polymer has high resistance to ignition, low rate of combustion and smoke generation, low toxicity of product gases, retention of low flammability during use, acceptability in appearance and properties, no environmental or health impact, and little or no economic penalty. In order to select or design a polymeric material with desirable flammability properties, a familiarity with the principles of combustion, the relationship of flammability to polymer structure, and the modes of flame inhibition is essential.

1. *Basics of Polymer Flammability*

The presence of fuel, heat and oxygen is necessary to initiate and sustain a fire. A schematic flammability cycle is shown in Fig. 1. When heated sufficiently by the external ignition source, the polymeric material reaches a characteristic temperature at which it begins to degrade. The extent to which the molecular oxygen plays a role in this surface decomposition depends on the type of polymer. Gaseous combustible products may then be formed, with a rate dependent on factors such as the intensity of external heat, temperature and rate of polymer decomposition.

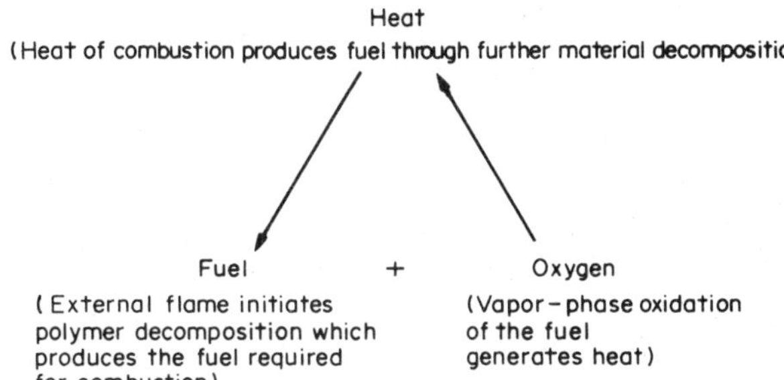

Figure 1
Schematic representation of the flammability cycle

Flammable gases (the fuel) thus produced diffuse to the flame front, where a series of heat-generating complex free-radical chain reactions take place in the presence of surrounding oxygen. Since the flaming of organic polymers simply represents the oxidation of hydrocarbons, the well-understood methane–oxygen combustion system can be used as a model for the more complex polymer flames. The mechanism of methane combustion is a complicated free-radical chain reaction consisting of propagation, chain-branching, and termination steps:

Propagation

$$CH_4 + HO\cdot \rightarrow CH_3\cdot + H_2O \quad (1)$$
$$CH_4 + H\cdot \rightarrow CH_3\cdot + H_2 \quad (2)$$
$$CH_3\cdot + O\cdot \rightarrow CH_2O + H\cdot \quad (3)$$
$$CH_2O + CH_3\cdot \rightarrow CHO\cdot + CH_4 \quad (4)$$
$$CH_2O + HO\cdot \rightarrow CHO\cdot + H_2O \quad (5)$$
$$CH_2O + H\cdot \rightarrow CHO\cdot + H_2 \quad (6)$$
$$CH_2O + O\cdot \rightarrow CHO\cdot + HO\cdot \quad (7)$$
$$CHO\cdot \rightarrow CO + H\cdot \quad (8)$$
$$CO + HO\cdot \rightarrow CO_2 + H\cdot \quad (9)$$

Chain-branching

$$H\cdot + O_2 \rightarrow HO\cdot + O\cdot \quad (10)$$

Termination

$$H\cdot + R\cdot + M \rightarrow RH + M^* \quad (11)$$

It is interesting that oxygen is only found in one step, the major branching step. In the branching step, two chain-carrying radicals are produced for every one consumed, and this accounts for the explosive nature of burning. The main chain-carrying radicals in the propagation steps are the hydrogen atom and hydroxyl radicals. This mechanism clearly indicates that hydrogen, and not carbon, is mainly responsible for the flammability of hydrocarbons.

After ignition and removal of the ignition source, combustion becomes self-propagating if sufficient heat is generated and radiated back to the material to continue the decomposition process. The combustion process is governed by such variables as the rate of heat generation, the rate of heat transfer to the surface, surface area, and the rate of decomposition.

2. *Approaches to Flame Retardation*

Polymer combustion, a highly complex process, is believed to involve a vapor phase in which the reactions responsible for the formation and propagation of the flame take place, and a condensed phase in which fuel for the gas reactions is produced. Flame retardancy, therefore, can be improved by appropriately modifying either one or both of these phases. The approaches to reducing the flammability of polymer systems can be grouped into the following three categories:

(*a*) *Vapor phase.* In the vapor-phase approach, a flame retardant or the modified polymer unit releases a chemical agent on heat exposure that inhibits free radical reactions involved in the flame formation and propagation. For example, the presence of HX (X = halogen), produced by the pyrolysis of a halogen-containing organic material in the polymer, acts as a radical scavenger and replaces less reactive halogen atoms for active chain carriers (Factor 1974):

$$H\cdot + HX \rightarrow H_2 + X\cdot \quad (12)$$
$$HO\cdot + HX \rightarrow H_2O + X\cdot \quad (13)$$

Halogenated flame retardants such as chlorinated paraffins and alicyclic compounds, and chloro- and bromo-aromatic additives, are postulated to function primarily by a vapor-phase flame inhibition mechanism. Flame retardation is implemented by incorporating fire-retardant additives, impregnating the material with a flame retardant substance, or by using flame retardant comonomers in the polymerization or grafting.

(*b*) *Condensed phase.* In condensed-phase modification, the flame retardant alters the decomposition chemistry such that the transformation of the polymer to a char residue is favored. This is achieved by the addition of additives which catalyze charring rather than flammable product formation, or by designing polymer structures which favor char formation. Carbonization which occurs at the cost of flammable product formation also shields the residual substrate by interfering with the access of heat and oxygen. Phosphorus-based additives are typical examples of flame retardants which act by a condensed-phase mechanism.

(*c*) *Miscellaneous.* These approaches include dilution of the polymer with nonflammable materials (e.g., inorganic fillers), incorporation of materials which decompose to give nonflammable gases such as carbon dioxide, and formulation of products which decompose endothermically. A typical example of such a flame retardant is alumina trihydrate ($Al_2O_3 \cdot 3H_2O$). This kind of material acts as a thermal sink to increase the heat capacity of the combustion system, lower the polymer surface temperature via endothermic events, and dilute the oxygen supply to the flame. Thermoanalytical techniques can prove to be helpful in determining the mechanism of flame retardation (Cullis and Hirschler 1983, Pearce et al. 1981).

Of the several test methods for evaluating the burning behavior of different polymers (see *Flammability of Polymers: Test Methods*, and Kishore and Mohandas 1982), Fenimore and Martin's (1966) limiting oxygen index will be used here to illustrate the relative flammability of materials. This test measures the minimum concentration of oxygen in an oxygen–nitrogen atmosphere that is necessary to initiate and support a flame.

3. *Polymer Structure and Flammability Properties*

The flammability of a particular polymer mainly depends upon its structure. The amounts of char and incombustible gases formed on thermal decomposition determine to a great extent the flame resistance of a polymeric material. A material with a limiting oxygen index of $\leqslant 26$ should be considered flammable (van Krevelen 1975). Based on the polymer structure and limiting oxygen index, it appears that the polymeric materials can be broadly classified into three categories. Table 1 presents the structures, thermal decomposition products and limiting oxygen index data for some representative polymers belonging to each of the three classes.

Thermal decomposition of class I polymers gives rise to little or no char residue. This absence of a pyrolysis residue is due to the high hydrogen content of these polymers, which allows the entire decomposition product to be disproportionated into volatiles. High yields of monomers or other hydrocarbons from the thermal decomposition of these systems constitute the fuel needed for the flammability cycle shown in Fig. 1. The CN containing polyacrylonitrile and the OH containing polymers (cellulose and poly(vinyl alcohol)) must be considered as exceptions in that they cyclize, and therefore produce some char. However, at the same time, they give off a large amount of flammable products. The polymers based on structures similar to those included in class I, have limiting oxygen indices lower than 26, and are considered to be flammable.

Class II covers those materials which are known to be high-temperature polymers. Structurally, they are characterized by the presence of either wholly aromatic or heterocyclic–aromatic functional units. An aromatic ring in the polymer backbone is not only relatively unreactive thermally, but also increases the chain rigidity. Stiffer chains resist thermally induced vibrations, and therefore, a higher temperature is required for their decomposition. These materials not only degrade at high temperatures, but also undergo extensive cross-linking and cyclization, as a result of which their pyrolysis generates high char residues; typically 20–70% at 700 °C. The formation of char residue is conveniently measured using thermogravimetric analysis (TGA) in an inert atmosphere (Pearce et al. 1981).

Since the formation of char takes place at the expense of flammable products (i.e., the elements are converted into carbonaceous involatile residues rather than volatile species), it should be expected to result in the reduction of flammability. In fact, it has been demonstrated that there is a significant correlation between the char residue (e.g., at 850 °C) and the oxygen index of polymers (van Krevelen 1975). As shown in Fig. 2, the higher amount of char, the higher the oxygen index, and hence, the lower the flammability. The high temperature polymers of class II, such as those in Table 1, are considered flame retardant (limiting oxygen index >26). The lower flammability is attributed to factors such as reduction in the amount of combustible volatiles, high energy requirements for continuous fuel generation, and insulating effects of the resultant char.

Class III consists of halogen-containing polymeric materials. Some of them form a small char residue, others none at all. These polymers are inherently flame retardant. This is a direct consequence of the fact that the halogen radicals act as radical scavengers in the vapor phase, and therefore inhibit combustion as described earlier. Also, the formation of non-

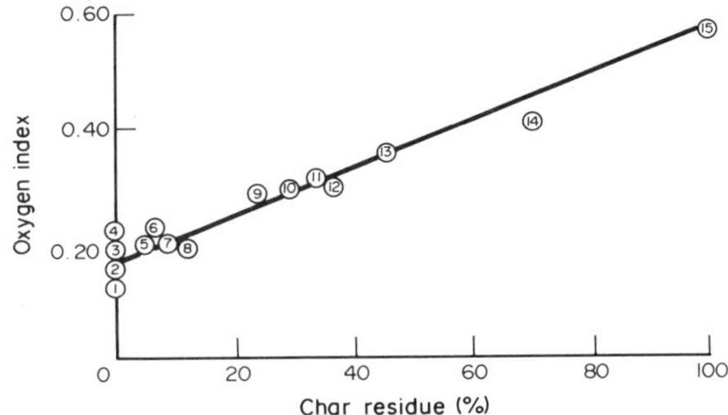

Figure 2
Correlation between oxygen index and char residue: 1, polyformaldehyde; 2, polyethylene, polypropylene; 3, polystyrene, polyisoprene; 4, nylon; 5, cellulose; 6, poly(vinyl alcohol); 7, poly(ethylene terephthalate); 8, polyacrylonitrile; 9, poly(phenylene oxide); 10, polycarbonate; 11, Nomex; 12, polysulfone; 13, Kynol; 14, polyimide; 15, carbon (after van Krevelen 1975 © IPC Business Press, London. Reproduced with permission)

combustible gases such as HCl, HF and C_2F_4, which seal the polymer surface from the air, is partly responsible for the flame retardancy. All these factors thus influence the interaction between pyrolysis and ignition.

4. *Flame Retardation of Polymeric Materials*

The techniques of reducing the flammability of polymers are based on one or more of the three fundamental approaches described earlier. Whether a particular flame-retarding method is suitable or not, is often determined by the nature of polymer material (e.g., thermoplastic or thermoset).

4.1 *Thermoplastics*

Many flame-retardant chemicals have been developed for use in thermoplastics. The majority of these flame retardants are of the additive type, usually halogen- and/or phosphorus-based compounds. Some examples of fire-retardant additives for use in polyolefins (Kuryla and Papa 1973) are: hexabromocyclododecane, octabromobiphenyl, hexabromobiphenyl, chlorowax, chlorinated triphenyl, chlorendic acid, tris(tribromophenyl) phosphite and tris(dibromopropyl) phosphate. The choice of flame retardants is dependent on the nature of the polymer, the method of processing, the proposed service conditions and economic considerations. Although the processing, service and economic factors are important, the flame-retardancy potential of an additive is of primary importance, and this can be readily evaluated by thermal analysis. Einhorn (1971) has described the use of TGA in selecting the appropriate fire retardants. Figure 3 represents the TGA thermogram for a hypothetical polymer which exhibits a simple unimolecular degradation process. Point A represents the start of the decomposition process, and point B represents the temperature of maximum rate of degradation. Fire retardants are screened so as to select a material having thermal characteristics similar to those in Fig. 3. The efficiency of matching the degradation curve of the polymer with the volatilization or degradation characteristics of the fire retardant has been cited as the key to effective flame retardancy. If the flame-retardant additive possesses low thermal stability compared with that of the polymer, it will be lost before its function is needed, and if the additive has greater stability it may be intact at the time its function is needed. Depending on the thermal degradation characteristics of the flammable substrate, more than one fire-retardant additive could also be employed.

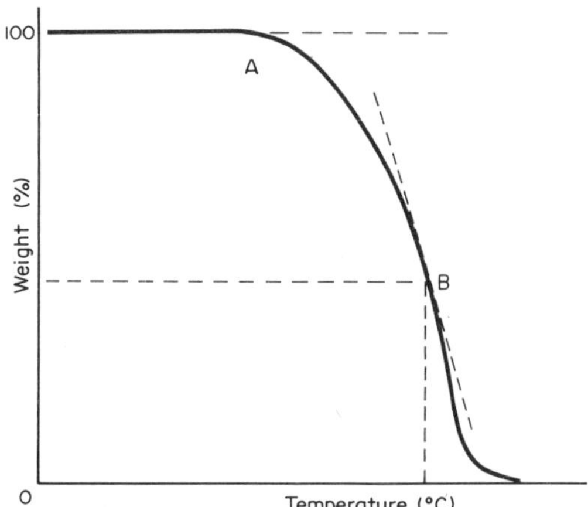

Figure 3
TGA thermogram for a hypothetical polymer (after Einhorn 1971 © National Academy of Sciences, Washington, DC. Reproduced with permission)

The concept of synergism is very important in fire retardation, since it can lead to efficient flame retardants with less expensive polymer systems and minimal effects on other desirable properties. One of the classical illustrations of synergism observed in flame retardation is the addition of Sb_2O_3 to halogen-containing polymers. For example, it is believed that the reaction of a chlorine source with Sb_2O_3 produces SbOCl as an intermediate with a lower energy barrier, which on thermal decomposition evolves $SbCl_3$, the actual flame retardant (Khanna and Pearce 1978). The existence of phosphorus–halogen and nitrogen–phosphorus synergisms has also been suggested. Compounds based on antimony–halogen, nitrogen–phosphorus and phosphorus–halogen combinations have been utilized as flame retardant additives for thermoplastics.

Table 1
Effect of structure on polymer flammability

Class	Name	Structure	Degradation products	Limiting oxygen index
I	Polyethylene	$-\!\!(CH_2\text{–}CH_2)_n\!\!-$	continuous spectrum of saturated and unsaturated hydrocarbons from C_2–C_{90}	17.4
	Polybutadiene	$-\!\!(CH_2\text{–}CH{=}CH\text{–}CH_2)_n\!\!-$	about 2% monomer in addition to saturated and unsaturated hydrocarbons	18.3
	Polystyrene	$-\!\!(CH_2-CH(C_6H_5))\!\!-$	about 41% monomer, 2% toluene; remainder dimer, trimer and tetramer	18.3
	Cellulose		H_2O and small amounts of CO_2, CO and a tar containing principally levoglucosan	19.9
	Poly(ethylene terephthalate)	$-\!\!(OC\text{–}C_6H_4\text{–}COO\text{–}CH_2\text{–}CH_2\text{–}O)_n\!\!-$	acetaldehyde major product with CO_2, CO, C_2H_4, H_2O, CH_4, C_6H_6, terephthalic acid and more complex chain fragments	20.6
	Nylon 66	$-\!\!(H_2N\text{–}(CH_2)_6\text{–}NHCO\text{–}(CH_2)_4\text{–}CO)_n\!\!-$	H_2O, CO_2, cyclopentanone, traces of saturated and unsaturated hydrocarbons	21.5

II	Polycarbonate	$\left[O-C_6H_4-C(CH_3)_2-C_6H_4-O_2C \right]_n$	major products CO_2, bisphenol A; minor products CO, CH_4, 4-alkylphenols; much char	29.4
	Nomex	$-\!\!\left(HN-C_6H_4-NH-CO-C_6H_4-CO \right)\!\!-_n$	high char yield; volatiles contain CO_2, CO, H_2O, 1,3-dicyanobenzene, 3-cyanobenzoic acid	29.8
	Kynol	$\sim CH_2-C_6H_2(OH)(CH_2\sim)-CH_2 \sim$	high char yield; volatiles comprise xylene (76%), traces of phenol, cresol, benzene	35.5
	Polybenzimidazole	$\left[\text{benzobisimidazole (N–H, N)}-C_6H_4 \right]_n$	very high char yield, small volatiles may include NH_3 and H_2	41.5
III	Poly(vinylidene fluoride)	$-\!\!\left(CH_2-CF_2 \right)\!\!-_n$	35% HF and high yields of products involatile at 25 °C; some carbonization	43.7
	Poly(vinyl chloride)	$-\!\!\left(CH_2-CHCl \right)\!\!-_n$	quantitative yields of HCl, aliphatic and aromatic hydrocarbons	47.0
	Poly(vinylidene chloride)	$-\!\!\left(CH_2-CCl_2 \right)\!\!-_n$	high yields of HCl	60.0
	Polytetrafluoroethylene	$-\!\!\left(CF_2-CF_2 \right)\!\!-_n$	>95% monomer, 2–3% C_3F_6, no larger fragments	95.0

4.2 Thermosets

Fire retardancy in thermosetting polymers is achieved largely by the use of reactive fire-retardants, since the common fire-retarding additives lack permanence. The flammability of thermosetting materials can be reduced by the addition of inorganic fillers and/or reactive flame-retardant components. Flame-retardant vinyl monomers, or other cross-linking agents are also frequently employed.

A large number of halogen-containing reactive diols, polyols, anhydrides and other functional group-containing intermediates have been used to produce flame-resistant unsaturated polyester resins. Flame-retardant polyester resins have been made by using bromostyrene as partial replacement of styrene for cross-linking reactions.

As in polyester resins, reactive halogen-containing fire-retardant chemicals are most often used in epoxy materials. Tetrabromobisphenol A is perhaps the most widely used component for flame-retarding epoxy resins.

Polyurethane foams, however, are generally flame-retarded by the incorporation of halogens, phosphorus and/or nitrogenous compounds. Conley and Quinn have reviewed the fire retardation of various resin systems (Lewin et al. 1975).

4.3 Elastomers

Many applications of rubbers (e.g., tires, gaskets and washers) normally do not require flame resistance. Where improvement in flammability is required, it can be achieved by the addition of halogen-containing materials, phosphorus compounds, oxides of antimony, and combinations of these materials. Rubbers containing chlorine and silicon atoms (e.g., neoprene and silicone) have self-extinguishing properties. Rogers and Fruzzetti have described the flame retardance of elastomer systems (Lewin et al. 1975).

4.4 Fibers

The flame retardation of fiber-forming polymers is generally achieved by the incorporation of additives based on halogens, phosphorus and/or nitrogen. The use of antimony oxide as a synergist with halogen is also common.

The most common route for making cellulosic fibers flame-retardant is through the use of catalysts (e.g., antimony trichloride, phosphoric acid) which enhance the formation of char (Kuryla and Papa 1979).

Retardation of combustion of polyamides (e.g., poly(ethylene terephthalate)) can be decreased by using bromo compounds in conjunction with an Sb_2O_3 synergist, red phosphorus and triphenylphosphine oxide (Lewin et al. 1975 pp. 193–221).

The flammability of high-temperature fiber-forming polymers can also be improved. Durette made by chlorination or oxychlorination of Nomex, a polymer of *m*-phenylene isophthalamide, exhibits much improvement in flammability properties over Nomex. Retardation of flammability by the use of halogen-substituted diamines has also been described for aromatic polyamides (Khanna and Pearce 1981). In these high-temperature materials, the resistance to burning is also achieved by thermal treatment (Bingham and Hill 1975). Hirsch and Holsten (1968) have reported that the fabric made from poly(*m*-phenylenebis(*m*-benzamido)terephthalamide) can be stabilized thermally such that the product will not burn even if heated to red hot in a gas flame. Such treatments are believed to involve cyclodehydration of the original polymer.

See also: Flammability of Elastomers; Flammability Hazards of Structural Materials and Furnishings; Thermal Analysis of Polymers

Bibliography

Bingham M A, Hill B J 1975 A study of the thermal behaviour of flame-resistant fibres and fabrics. *J. Therm. Anal.* 7: 347–58

Cullis C F, Hirschler M M 1983 The significance of thermoanalytical measurements in the assessment of polymer flammability. *Polymer* 24: 834–40

Einhorn I N 1971 The chemistry of fire resistant materials and suppression. *Fire Res. Abstr. Rev.* 13(3): 236–67

Factor A 1974 The chemistry of polymer burning and flame retardance. *J. Chem. Educ.* 51: 453–56

Fenimore C P, Martin F J 1966 Candle-type test for flammability of polymers. *Mod. Plast.* 44(3): 141–92

Fristom R M, Westenberg A A 1965 *Flame Structure.* McGraw-Hill, New York

Hirsch S S, Holsten J R 1968 Enhancement of aromatic polyamide thermostability by heat treatment in air. *Polym. Prepr., Am. Chem. Soc., Div. Polym. Chem.* 9: 1240–41

Khanna Y P, Pearce E M 1978 Synergism and flame retardancy. In: Lewin M, Atlas S M, Pearce E M 1978, pp. 43–61

Khanna Y P, Pearce E M 1981 Aromatic polyamides III. Mechanistic studies on the role of substituent chlorine in flame retarding aromatic polyamides. *J. Polym. Sci., Polym. Chem. Ed.* 19: 2835–40

Kishore K, Mohandas K 1982 Laboratory test methods for the evaluation of flammability of polymers. *J. Macromol. Sci., Chem.* 18(3): 379–93

Kuryla W C, Papa A J (eds.) 1973, 1975, 1978, 1979 *Flame Retardancy of Polymeric Materials*, Vols. 1–5. Dekker, New York

Lewin M, Atlas S M, Pearce E M (eds.) 1975, 1978 *Flame-Retardant Polymeric Materials*, Vols. 1 and 2. Plenum, New York

Pearce E M, Khanna Y P, Raucher D 1981 Thermal analysis in polymer flammability. In: Turi E A (ed.) 1981 *Thermal Characterization of Polymeric Materials.* Academic Press, New York, Chap. 8, pp. 793–841

Van Krevelen D W 1975 Some basic aspects of flame resistance of polymeric materials. *Polymer* 16: 615–20

Y. P. Khanna

Flammability of Polymers: Test Methods

All polymers will burn, given sufficient heat and oxygen. Some, such as polyethylene and poly(methyl methacrylate), burn readily under normal ambient conditions, whereas others (e.g., polytetrafluoroethylene) burn only under extreme conditions of applied heat or oxygen content.

Typical characteristics measured by flammability tests are ignitability, surface flame spread, heat-release rate and smoke evolution. All of these flammability characteristics are extrinsic properties of a polymer. Unlike intrinsic properties, they are dependent on external factors such as geometry and orientation, as well as the amount of oxygen available for combustion. Because of this dependency on external conditions, it is difficult to predict "real fire" performance by extrapolation from fire test results.

1. Fire Development Regimes

The stages of a fire from ignition to inferno are characterized in Fig. 1 where fire intensity is plotted versus time. The intensity scale conceptually represents the fire growth parameters of mass burning rate, rate of heat release, temperature, and degree of involvement of combustibles. Three regimes of fire intensity are shown: incipient, developing and fully developed.

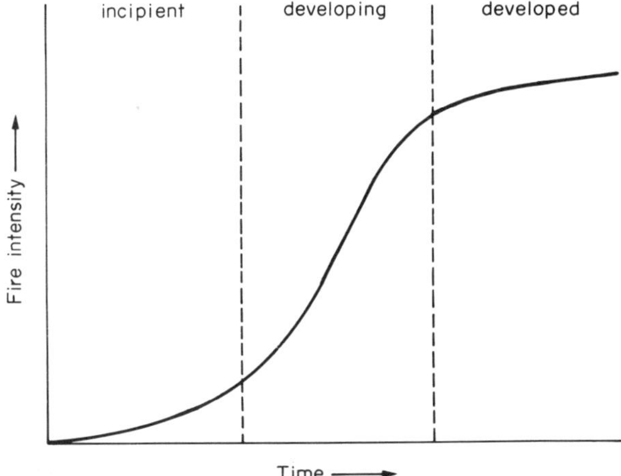

Figure 1
Idealized fire development curve showing the three regimes of fire growth (©Underwriters Laboratories, Melville, New York. Reproduced with permission)

In the incipient regime the fire is small with the materials involved in the fire being subjected to, and producing, relatively low levels of thermal energy. When the rate of heat evolution from the burning materials exceeds the rate of heat loss to the environment, the temperature rises, causing the polymers to burn faster. In the developing regime, the temperature and the heat-release rate increase. Energy feedback (convective and radiative) accelerates the rate of burning until the fire becomes limited by the availability of either fuel or oxygen. The fire is then in the fully developed regime.

Flammability tests for polymers operate in one or more of the three regimes. Most small-scale laboratory tests which utilize small ignition sources such as a Bunsen burner operate in the incipient regime. The performance of a polymer in small-scale tests may relate to its performance in an incipient fire, but such tests are not reliable predictors of performance in developing or fully developed fires because they do not simulate the thermal environment of these larger fires.

Flammability tests which operate in the developing regime typically have higher energy ignition sources than incipient tests, are usually larger in scale, and often provide containment so that heat loss is restricted to provide a severe thermal environment. Perhaps the most widely used test in this regime is the ASTM E84 surface flame spread test which utilizes a 1.4 m long ignition flame and a test sample 7.6 m long.

Flammability tests which operate in the fully developed regime typically have high-flux ignition sources, and provide for containment of energy and full involvement of test samples. Generally they are used to measure the fire endurance of structural elements such as walls.

In the section that follows, some of the widely used flammability tests applicable to polymers have been selected and grouped according to the regime of fire intensity where they are most applicable.

2. Incipient Regime Tests

In ASTM D1929 (Test for Ignition Properties of Plastics) a 3 g sample is exposed to flowing air (Fig. 2), and the air temperature is increased until the plastic ignites. When tests are conducted with a pilot flame, the flash-ignition temperature is measured. Without a pilot light, the self-ignition temperature is determined.

The ASTM D2863 oxygen index test (Fig. 3) determines the relative flammability of plastics by measuring the minimum amount of oxygen, in a mixture of oxygen and nitrogen, which is required to support combustion. Materials can then be rank ordered with the flammable materials having low oxygen indices (below the 21% O_2 in the air) and fire-resistant ones having indices considerably above 21%. It is common practice to specify a minimum oxygen index of 28% for fire-resistant materials.

The oxygen index test is one of the most widely used flammability tests. It is easy to run, reproducible and most importantly, yields a quantitative result,

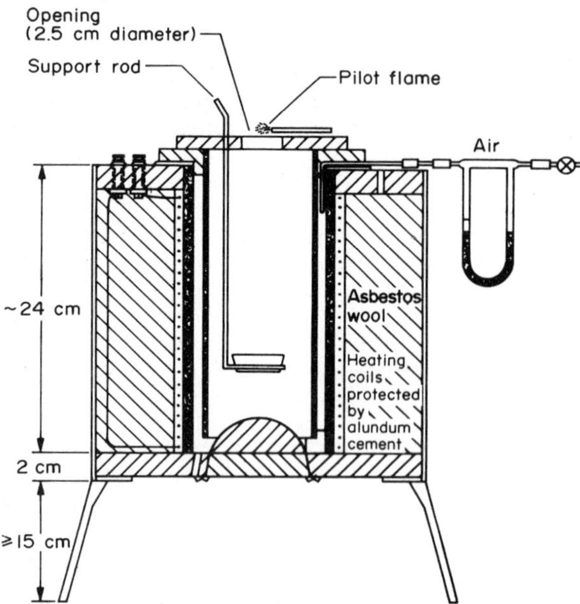

Figure 2
Apparatus for measuring ignition temperatures by ASTM D1929

rather than a pass/fail result for assessing the flammability of materials. It is an excellent test for quality control and research, but provides insufficient information for predicting the performance of a material in a developing fire.

Several factors affecting the oxygen index test help explain the poor correlation between oxygen index test results and developing fire performance. Unlike a developing fire, there is minimal energy feedback from the flame to the polymer substrate—the oxygen level is adjusted so the flame is on the verge of extinguishing. This situation is in contrast to a large fire where radiative feedback is the dominant mode of energy transfer.

The oxygen index decreases with increasing sample temperature. For example, the oxygen index of a polychloroprene compound containing clay and carbon black as the principal additives is 33% at room temperature and 19% at 300 °C (Johnson 1974). The oxygen index also decreases with decreasing sample thickness. Another factor is that thermoplastic materials often drip and flow away from the flame. Since the melted thermoplastic carries heat away, a higher oxygen concentration is needed to maintain combustion and a higher oxygen index results. In a real fire, thermoplastics will often drip and burn in a molten pool.

In UL 94, Tests for Flammability of Plastic Materials for Parts in Devices and Appliances (Underwriters Laboratories 1980), methods are provided for both vertical and horizontal testing. In the vertical test (Fig. 4) a Bunsen burner with a 19 mm flame is applied to test specimens 13 cm long, 13 mm in width and up to 13 mm in thickness. The burner is placed under the sample for 10 s, then withdrawn until flaming ceases, and reapplied for an additional 10 s. The duration of flaming combustion after each flame application is used to rate materials as 94V-0 if flaming combustion averages 5 s or less without dripping flaming particles and 94V-1 if flaming combustion averages 25 s or less without dripping flaming particles. Materials that produce dripping flaming particles that ignite surgical cotton are rated 94V-2.

In the UL 94 horizontal test (Fig. 5) a Bunsen burner with a 25 mm flame is applied to test specimens 13 cm long and 13 mm in width and the rate of flame spread over the horizontal surface is measured. Materials that have flame spread rates less than

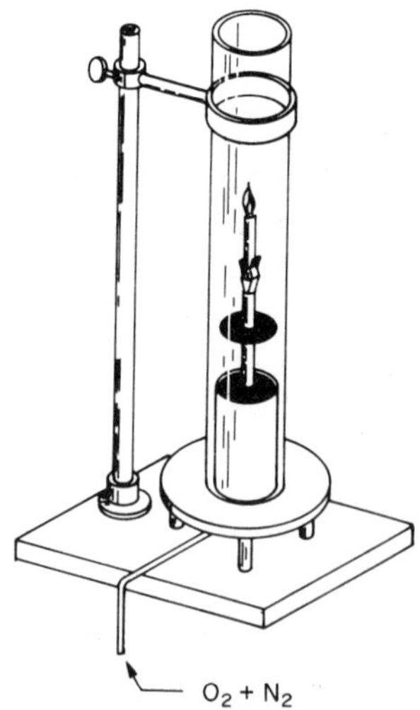

Figure 3
Apparatus for measuring oxygen index by ASTM D2863

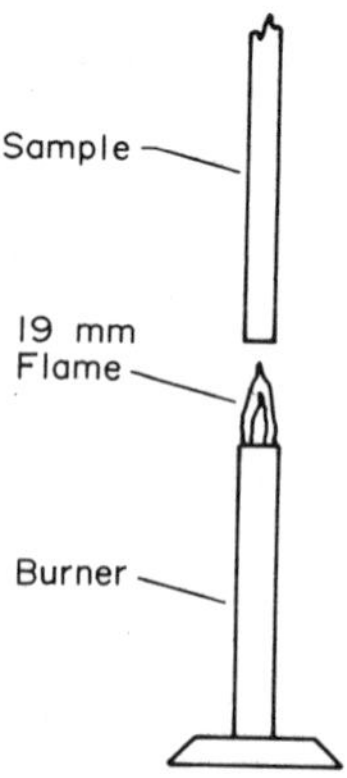

Figure 4
The UL 94 vertical test set-up

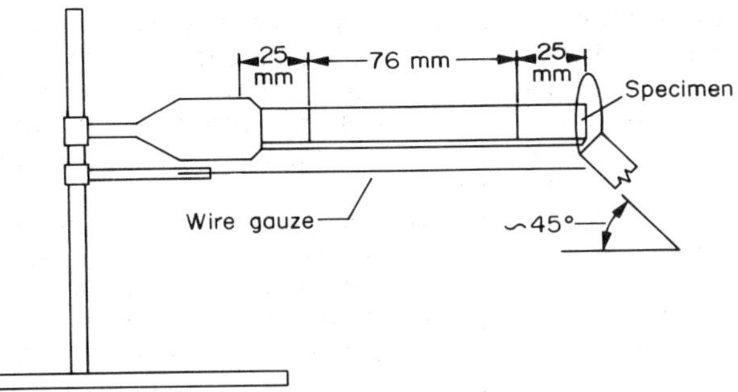

Figure 5
The UL 94 horizontal test set-up

specified maxima are rated 94HB. This test is nearly identical to ASTM D635 (Test Method for Rate of Burning and/or Extent of Time of Burning of Self-Supporting Plastics in a Horizontal Position).

3. *Developing Regime Tests*

ASTM E84 (Test for Surface Burning Characteristics of Building Materials) is the standard test to certify building materials for compliance with building codes. It utilizes the Steiner tunnel furnace (Fig. 6) where a 7.3 m long by 51 cm wide test specimen forms the roof of the furnace. The lower surface of the specimen is ignited by a large (1.4 m) methane flame. A 1.22 m s^{-1} draft and an enclosure, which provides for conservation of heat energy, aid flame spread. The fire is observed through windows in the tunnel wall.

The tunnel is calibrated with red oak, which spreads flame 5.9 m from the end of the ignition flame to the tunnel end in 15.5 min, and with asbestos cement board, which does not burn. Asbestos cement board is assigned a flame spread classification of 0 and red oak 100.

A smoke value is determined by measuring the percent attenuation of a beam of light across the exhaust duct of the tunnel. The area under the percent attentuation versus time curve for red oak in a 10 min test is assigned a value of 100. The smoke values of other materials are reported relative to the red oak value of 100.

The combination of a large ignition flame, with a heat release rate of 88 kW, and an insulated furnace which reradiates heat once it is heated up, provides a severe test environment. Measurements made by Parker (1977) indicate that the sample in the flame impingement area is subjected to 40–60 kW m^{-2} incident heat flux.

Materials with a flame spread classification of 25 or less (less than 1.5 m flame spread) are generally regarded by building codes as providing a high degree of fire safety. The correlation of a low flame-spread classification with good fire performance was established for cellulosic (wood) materials, but recent experiences with thermoplastic foams show that ASTM E84 should not be employed as the sole source of fire hazard rating for polymeric materials.

The rate at which a material burns (i.e., the rate

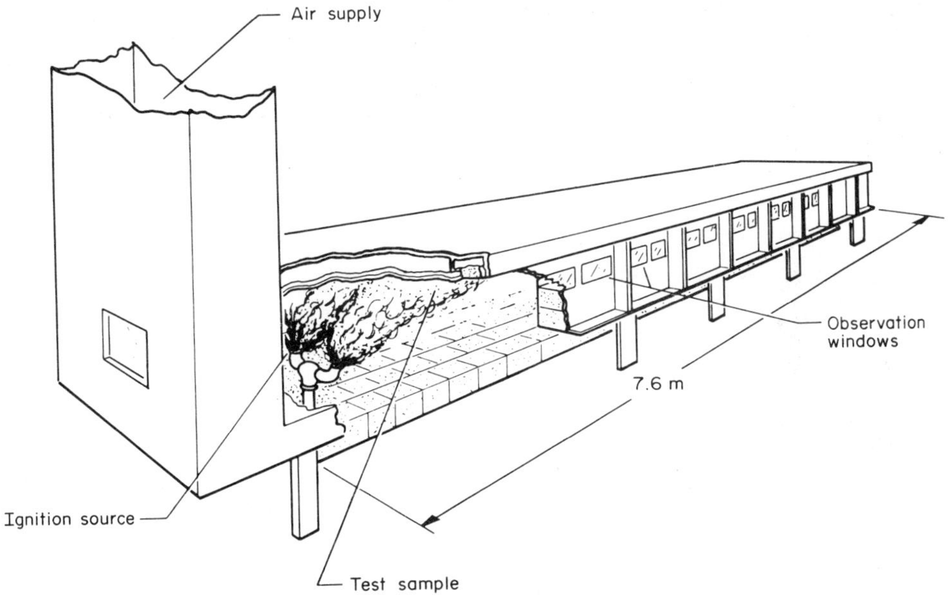

Figure 6
The Steiner tunnel furnace for measuring flame spread and smoke developed by ASTM E84

at which it releases heat) is an important flammability property since it is a measure of fire intensity. Rate of heat release calorimeters measure rates of heat and smoke release of a material subjected to an externally applied heat flux. Unlike most other fire tests which utilize only one exposure condition, heat release rate calorimeters allow the incident heat flux to be varied from test to test, and hence the influence of flux on release rates can be determined.

In the Ohio State University (OSU) calorimeter (Fig. 7), which is used in ASTM E906 (Test Method for Heat and Visible Smoke Release Rates of Materials and Products), incident flux can be varied from a level characteristic of an incipient fire (0–10 $kW\,m^{-2}$) on through those characteristics of developing fires, up to values associated with fully developed fires (80–100 $kW\,m^{-2}$). Rowan and Lyons (1978) have published some experimental results which demonstrate the importance of externally imposed heat flux on the burning behavior of materials. Figure 8 shows peak heat release rates versus incident applied flux for a series of fire retardant polyurethane and polyisocyanurate foams. The rate of burning generally increases with increasing incident flux, and for flame retardant materials there is a minimum flux required to sustain combustion. The rank ordering of materials can change when flux is increased. The material represented by the dashed line is best (i.e., it has the lowest heat release rate) below 40 $kW\,m^{-2}$ incident flux, but it is the worst or next to worst material at incident fluxes above 45 $kW\,m^{-2}$.

Smoke evolution is also dependent on incident heat flux. Figure 9 shows the variation in total smoke output versus incident flux for several polymers tested in the OSU release rate calorimeter.

The most widely used instrument for measuring smoke is the National Bureau of Standards smoke chamber (Fig. 10). Test conditions are specified in ASTM E622 (Standard Test Method for Specific Optical Density of Smoke Generated by Solid Materials). The specimen is subjected to a 25 $kW\,m^{-2}$ incident flux. Tests are run either in the flaming mode where the sample is ignited by a pilot flame or in the smoldering mode where the pilot flame is not used. Smoke is accumulated within the smoke chamber box and the specific optical density of the smoke is reported. Specific optical density D_s is defined by

$$D_s \equiv (V/LA) \log (100/T)$$

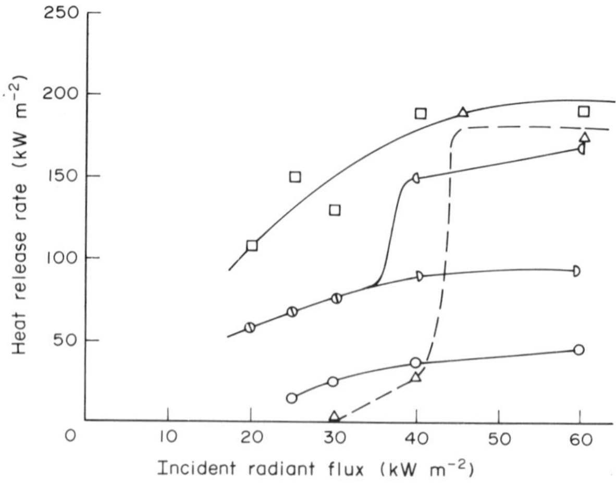

Figure 8
The influence of externally applied flux on peak heat-release rates for a series of polyurethane and polyisocyanurate foams (after Rowan and Lyons 1978. © Technomic, Westport, Connecticut. Reproduced with permission)

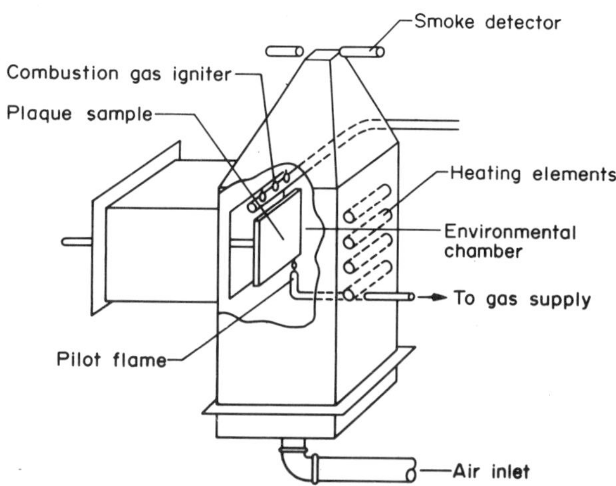

Figure 7
The Ohio State University heat release rate calorimeter (ASTM E906)

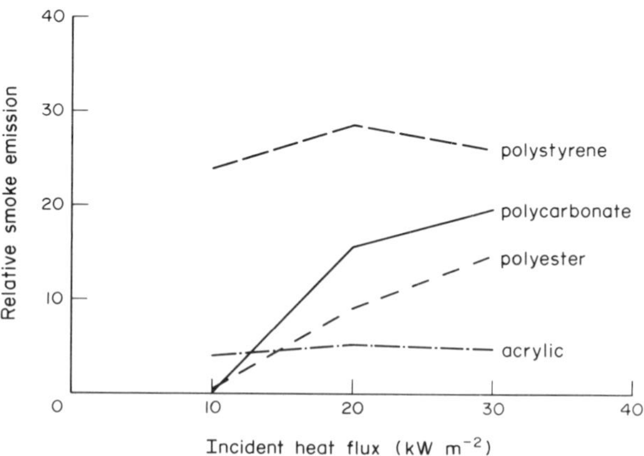

Figure 9
The influence of externally applied flux on total smoke emissions for various polymers (courtesy Professor E E Smith, Ohio State University)

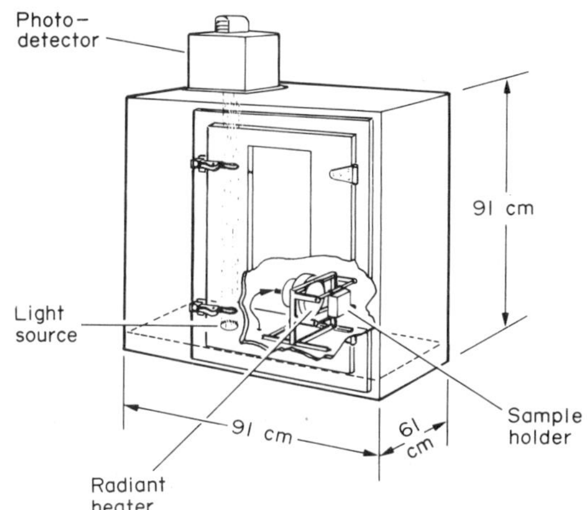

Figure 10
National Bureau of Standards smoke chamber (ASTM E662)

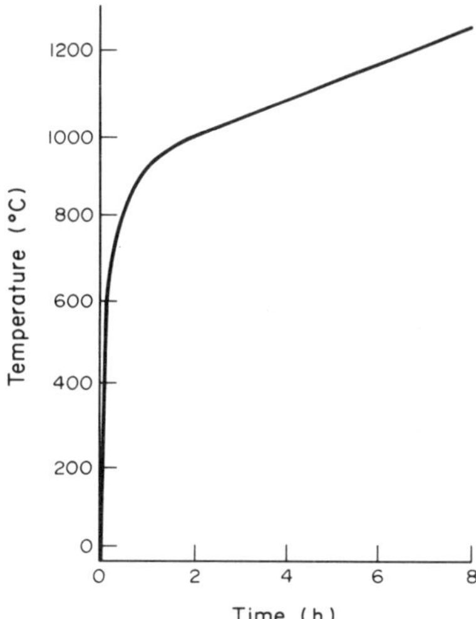

Figure 11
The ASTM E119 time–temperature curve for testing the fire endurance of structures

where V is the chamber volume, L is the optical path length, A is the exposed specimen area and T is the percent transmittance of the light beam.

There are several differences between smoke measurements in the NBS chamber and in the OSU release rate calorimeter. The NBS chamber is normally run at only one standardized flux. Thus it cannot be related to a real fire situation where materials are exposed to a spectrum of fluxes. The NBS chamber measures total smoke by accumulating it within the box whereas the OSU calorimeter operates with an air flow through the combustion chamber and total smoke is determined by integration of the instantaneous smoke values. Both test methods report smoke in optical density, which is linearly related to human visibility through smoke. Both methods differ greatly from the ASTM E84 smoke measurement, where the amount of smoke generated is dependent on the flame spread value of the material tested.

4. Fully Developed Regime Tests

ASTM E119 (Fire Tests of Building Construction and Materials) is used for rating the fire endurance of materials and structures. Large specimens, minimum exposed area 9.3 m^2, are exposed to a standard fire whose time–temperature curve (Fig. 11) is characteristic of a post flashover fire. Assemblies are rated by the elapsed time to failure. Several failure criteria are provided, such as ignition of cotton waste on the unexposed side of the specimen or temperatures exceeding 139 °C on the unexposed side.

ASTM E814 (Fire Test of Through Penetration Fire Stops) is applicable to polymeric materials used in penetrations (e.g., cables, plastic ducts and firestop materials). It utilizes the ASTM E119 time–temperature exposures with a firestop and penetration in place to test the endurance of a firestop assembly.

See also: Flammability of Polymers; Flammability Hazards of Structural Materials and Furnishings

Bibliography

American Society for Testing and Materials 1966 *Symposium on Fire Test Methods—Restraint and Smoke 1966*, ASTM Special Technical Publication 422. American Society for Testing and Materials, Philadelphia, Pennsylvania

American Society for Testing and Materials 1972 *Ignition, Heat Release and Noncombustibility of Materials*, ASTM Special Technical Publication 502. Amercian Society for Testing and Materials, Philadelphia, Pennsylvania

Hilado C J 1973 *Flammability Test Method Handbook.* Technomic, Westport, Connecticut

Hilado C J (ed.) 1974 *Flammability Handbook for Plastics*, 2nd edn. Technomic, Westport, Connecticut

Johnson P R 1974 A general correlation of the flammability of natural and synthetic polymers. *J. Appl. Polym. Sci.* 18: 491–504

Kanury A M, Alvares N J, Martin S B 1977 *Flammability Testing of Polymers.* Manufacturing Chemists Association, Washington, DC

National Materials Advisory Board 1979 *Fire Safety Aspects of Polymeric Materials—II: Test Methods, Specifications and Standards*, Publication NMAB 318-2. National Academy of Sciences, Washington, DC

Parker W J 1977 *An Investigation of the Fire Environment in the ASTM E84 Tunnel Test*, National Bureau of Standards Technical Note 945. US Government Printing Office, Washington, DC

Rowen J W, Lyons J W 1978 The importance of externally imposed heat flux on the burning behavior of materials. *J. Cell. Plast.* 14: 25–31

Smith E E, Harmathy T Z (eds.) 1978 *Design of Buildings for Fire Safety,* ASTM Special Technical Publication 685. American Society for Testing and Materials, Philadelphia, Pennsylvania

Underwriters Laboratories 1980 *Tests for Flammability of Plastic Materials for Parts in Devices and Appliances*, UL94. Underwriters Laboratories, Melville, New York

S. Kaufman

Flax

Flax fiber from the plant *Linum usitatissimum* is one of the most valuable and useful of the fibers, following cotton and jute in importance. Since flax has been cultivated from prehistoric times, its point of origin is unknown. In 1982, 590 kt of flax fibers were produced worldwide. The USSR was the largest producer at 271 kt; China was next at an estimated 121 kt; France produced an estimated 60 kt; and Egypt 26 kt (see Table 1).

Table 1
Production of flax, 1980–1982 (in kt)

Country	1980	1981	1982
Egypt	23	25	26[a]
China	111[a]	116[a]	121[a]
Belgium–Luxembourg	11	9	10[a]
Bulgaria	1	3	1
Czechoslovakia	23	16	20
France	68	39	60
Hungary	3	3	3[a]
Netherlands	7	5	6
Poland	49	34	40[a]
Rumania	33	16	24[a]
USSR	291	268	271[a]
Others[b]	8	8	8
Total	628	542	590

Source: *FAO Production Yearbook*, Vol. 36, 1982. Food and Agriculture Organization of the United Nations, Rome
a FAO estimate b Argentina, Chile, Turkey, Yugoslavia, Australia

1. *The Plant*

The flax plant is an annual herb with a slender, erect, smooth, grayish-green stem which grows to a height of 70–120 cm. The plant produces blue or white flowers. Flax grown for fiber branches only near the top of the stem with small, alternate, grayish-green leaves. It grows best in the temperate regions between latitudes 49° and 53°N, where there is an even distribution of rainfall of 15–20 cm annually. It also grows best on rich, loamy soil. The seed is usually sown between early March and the middle of April in France, and harvested the second week of July. One hectare of flax yields an average of 720–960 kg of fiber.

2. *Harvesting and Processing*

The plants are harvested when the stems change color from green to yellow, usually one month after the appearance of the first flowers. Timing of harvest is an important factor in producing the best fiber. The plants are pulled from the ground by hand or by machine. The stems are gathered together in round bundles known as beets, and left in the field for 10–14 days for drying. The straight, long, golden-yellow stems are then broken by a process known as rippling, and the seeds removed by combing.

A process called retting is used to loosen or remove the noncellulosic portions of the flax straw which holds the fibers together in the stem. Retting is accomplished by three principal methods: dew retting, water retting and chemical retting. The water softens and removes the soluble carbohydrates, glucosides and tannins in the stem. Natural fungi, bacteria and enzymes break down the pectins and other materials holding the fibers together. The length of retting time depends on the temperature, and must be carefully controlled so that the cementing material of the fiber bundles is not affected. In dew retting, the stems are spread in thin layers on the field for 3–7 weeks, while rain, dew or snow allows the retting process to take place. This process is cheaper than the others; however, there is little control over the process, and the fibers may be damaged. In water retting, bundles of stems are placed in slow-moving rivers, pools or ditches. This process is quicker, taking 2–3 weeks. However, this form of retting causes pollution problems in the rivers. More flax is retted by dew than by water. The stems can also be retted in warm-water tanks (30 °C), giving better control to the process and reducing the time to 4–5 days. Chemical retting is quickest and is easily controlled. After the retting process, the stems are piled in the fields to dry.

The dried stems are put through a process of breaking, where the woody parts of the stems are broken into small pieces called shives by mallets or revolving fluted wooden rollers. The fibers, being tough and elastic, remain unbroken. The shives are removed by the process of scutching, by machine buffing with knife blades. The final process, called hackling, separates the long fibers from the short or broken tow fibers by means of sharp steel pins or combs. An average of 6 t of stems are harvested per hectare. Twelve to sixteen percent of the weight of the dried flax stem is fiber. In France, average fiber yield is 1080 kg ha^{-1}; in Belgium, 975 kg ha^{-1}; and in Poland, 622 kg ha^{-1}.

3. *Fiber Characteristics and Uses*

The fiber contains waxes and oils which produce a sheen, and is pure white to steel-blue in color. The fine and durable fibers are usually 70–90 cm in length, flexible, stronger than cotton, rayon or wool, but weaker than ramie. Flax fiber contains 82% cellulose and 2% lignin.

Flax fiber is graded according to fineness, softness, strength, density, color uniformity, luster, oiliness, length, handle and cleanliness. The fineness of the fiber is judged by the diameter of the strands. Good flax fiber has relatively small bulk for its weight, feels cool and oily to the touch, and should have a sheen. The fibers are broken to judge the strength. Water-retted flax varies from white to steel-blue in color; dew-retted flax varies from light brown to gray-black. Uniformity of color is important in grading. The fibers should be about 83 cm in length; however, they should not be shorter than 70 cm and not longer than 90 cm.

Flax is used in the manufacture of linen fabrics (damasks, cambrics, sheeting and lace), which are used for dresses, evening wear, gloves, handbags, handkerchiefs, sheets, sewing threads, twines, canvas bags, fish and seine lines, the finest writing paper and insulating material. The coarse, strong fibers are used for warp yarn, while the fine, soft fibers are used for weft yarn. The short fibers, scutched tow, are used for rope, coarse fabrics, and high-grade papers such as cigarette, bond and currency. Another product, derived from the plant seeds, is linseed oil, which is used in printing inks, varnishes, linoleum and soft soaps.

See also: Cotton; Jute; Natural Fibers in the World Economy; Materials of Biological Origin: An Overview

Bibliography

Bradbury F 1920 *Flax Culture and Preparation.* Pitman, London

Cartin W 1953 *Linen, The Story of an Irish Industry.* Cartin Publishers, Belfast

Dempsey J M 1975 *Fiber Crops.* University of Florida, Gainesville, Florida

Dujardin A 1948 *Retting of Flax.* Flax Development Committee, Northern Ireland

P. Ross

Fluorescence Properties of Materials

Fluorescence light is emitted by a material in response to absorbed light. The emission occurs by an electronic transition localized within a small spatial region which can contain an atom or a chemical group, or in the case of dyes a somewhat larger region occupied by a molecule consisting of a few carbon rings. Absorption of the fluorescence-excitation light can be the result of a transition of the electronic center itself or of a different center, the interaction of different centers, or the fundamental absorption across the material bandgap. Fluorescence is related to luminescence, a more general light-emission phenomenon in which light output may also be excited chemically, electrically, or thermally. When emission occurs at times long after the excitation occurs the response is called phosphorescence.

The localized fluorescence center interacts with the extended states of the bulk material and the interaction can affect the fluorescence in a variety of ways. Conversely, fluorescence can reveal intrinsic material properties such as the atomic-vibration dynamics, the type of chemical bonding and other details of the center surroundings in the material. Generally, the center-material interaction affects the frequency distribution of the fluorescence electronic transitions and the emission rate.

Fluorescence studies are relevant to the development of new laser materials, light-emitting diodes, and phosphors used in lamps and television tubes. Impurities and defects within a material have also been identified using fluorometric instruments. High-irradiance monochromatic lasers operating with pulse widths in the range of nanoseconds provide effective new tools for investigation of fluorescent materials. In biochemical applications fluorescent chemical groups are used to tag compounds in organic processes.

1. Fluorescence-Center Electronic Levels

Sharp line spectra (linewidths of 10^{11} Hz or less) can be observed in the fluorescence of atomic centers which are coordinated as ions or chemical groups in an ionic-bonded crystalline material. The fluorescence transitions occur in the unfilled electron shells of cations, the anion shells being filled in the formation of the compound. It should also be mentioned that lattice defects and the material chemical bond can alter these effects. The center electronic orbital states interact with the crystalline electric field resulting from the bonding anion electrons. The symmetry and strength of the crystalline field are seen as the electronic Stark effect in the absorption and fluorescence spectra. Crystal-field theory analytically describes the influence of the ion surroundings on the electronic spin–orbit states.

Crystal-field theory is also applicable to noncrystalline materials such as glass. Because glass is amorphous there is a site-to-site nonuniformity in electronic-center coordination which causes fluorescence broadening, the detected output being the superposition of the transitions occurring in differently perturbed fluorescence centers. For example, glass doped with neodymium oxide (Nd_2O_3) operates as a laser material on the electronic transitions of the Nd^{3+} ion which is coordinated by the O^{2-} anions in the glass. However, fluorescence linewidth in glass is about fifty times larger than in a single crystal Nd-doped oxide such as $Nd:La_2O_3$ where Nd^{3+} substitutes for La^{3+}. Using narrow-bandwidth laser output lines as excitation source, absorption and fluorescence transitions of particular Nd^{3+} ion coordinations in glass can be detected. With a tunable dye laser, it is then possible to scan the glass coordination nonuniformity and show its effect on the different center-fluorescence-transition frequencies and linewidths.

A fluorescence-center spin–orbit electronic state also interacts dynamically with the surrounding material. The electronic-fluorescence-center position at equilibrium within the coordination sphere of the material corresponds to a free-energy minimum for the system. An increased electronic-orbital radius caused by optical excitation perturbs the system configuration coordinates Q and a new equilibrium position of the center results. In a harmonic approximation the system energy increases quadratically with variation of Q, as illustrated in Fig. 1. A value Q_1 that minimizes ground-level free energy will then differ from the value Q_2 that minimizes the excited-level-system free energy, depending on the radial range, or localization, of the electronic-center spin–orbit state. The shift in Q values is shown in Fig. 1 for the equilibrium ground and excited states of the center.

Figure 1 also shows electronic transitions within an adiabatic approximation. Because the electronic transitions can occur faster than the more massive ion positions can change, there is essentially no variation

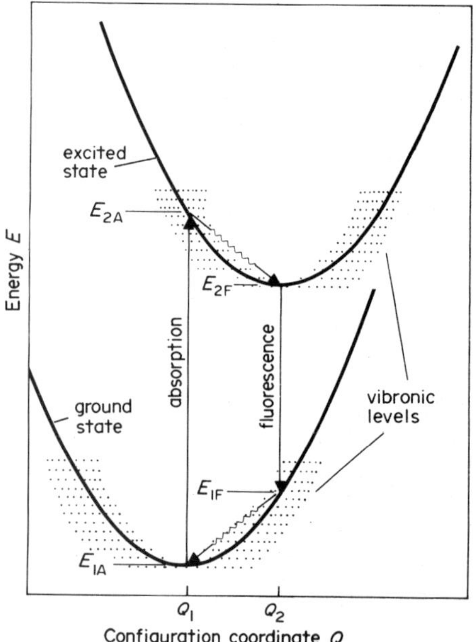

Figure 1
Energy-level diagram for ground and excited electronic states illustrating dependence on configuration coordinate Q, and showing absorption transistions, fluorescence and nonradiative relaxation, and vibronic levels

required in the configuration coordinate for the ground and excited states in a transition; only a corresponding change in electronic energy occurs. The absorbed energy corresponds to a light frequency ν_A given by $h\nu_A = E_{2A} - E_{1A}$, where $h = 6.626 \times 10^{-34}$ J s is Planck's constant. Similarly, the fluorescence transition from excited-state equilibrium Q_2 value occurs at frequency $\nu = h^{-1}(E_{2F} - E_{1F})$, where the optical frequencies are in the range 10^{14}–10^{15} Hz.

2. *Center Interaction with Bulk Material*

In the case of a crystal fluorescence center, the extended energy states that characterize the bulk material are well defined. Complex oscillation modes of a periodic crystal lattice correspond to a system of point masses and the oscillation can be considered as unperturbed by the short-range disorder caused by the localized state of a fluorescing substitutional-impurity electronic center, lattice vacancy or electron–hole pair. However, these lattice-oscillation modes interact with the centers via the crystal field. The lattice normal modes, or phonons, are quantized oscillations about equilibium points with frequencies ranging from zero to 10^{14} Hz.

A few phonon modes may be thermally excited in a material, in which case the fluorescence-center electronic levels are perturbed through the center–lattice coupling. Additional levels, called vibronics, result as the electronic-center energy levels are influenced by the lattice-motion energy states. Again in the harmonic approximation, Fig. 1 shows the vibronic levels as dotted lines at their intersection with the electronic energy state parabolas. Absorption transitions can then occur from thermally populated vibronic levels of the ground state, and fluorescence likewise can originate from thermally populated vibronic levels of the excited states.

In the vibronic picture, the adiabatic approximation can still hold for the fluorescence-center electronic transitions because of the much larger masses involved in lattice vibration. When the system temperature is near absolute zero and the lattice modes are unexcited, the absorption results from the ground state with coordinate value Q_1. Relaxation of the excited state, whereby $Q_1 \rightarrow Q_2$, can occur by transfer of excited-state electronic energy to lattice-phonon modes via the electronic–lattice coupling. Similarly, a fluorescence transition from an excited level to ground can result with value $Q = Q_1$ and ground-state relaxation $Q_2 \rightarrow Q_1$ to equilibrium follows again from electronic–lattice coupling. These nonradiative relaxation transitions are shown in Fig. 1 by the staggered arrows.

A transition between two spin–orbit states where emission occurs at lesser energy (lower frequency) than absorption is called a Stokes shift. It is also possible to see anti-Stokes emission which occurs at higher energy than absorption because populated phonon modes can also transfer energy to electronic states.

The absorption and fluorescence transition linewidths can depend further on lattice motion. When the electronic spin–orbit state is highly localized at a center, the low-temperature linewidths can be sharper than the resolution of grating spectrographs, i.e., about 10^8 Hz. At higher temperatures the linewidths can increase as different ions experience varying lattice vibrations. When the lattice is strained or otherwise perturbed by imperfections, a randomness in electronic-center coordination, or inhomogeneity, occurs which increases the low-temperature linewidth and also influences the frequency distribution over the transition line. This effect is similar to the line broadening observed in glass fluorescence, but for narrowly localized centers in single crystals the linewidths may be two orders of magnitude smaller than in glass and still be inhomogeneously broadened.

3. *Fluorescence Centers*

In covalent-bonded materials the bonding electrons tend to be shared between the constituent atoms. Consequently, the electric field potential seen by the fluorescence center is less precisely defined than for ionic materials. Fluorescence of ionic impurities in diamond is observed, nonetheless, and an extensive theory of electronic–lattice interaction in this covalent material is now under development. Dye fluorescence in polymer materials is yet another example of fluorescence in a covalent-bonded material. In the case of rhodamine B in poly(methyl methacrylate), for example, there is very little interaction between the dye-molecule electronic states and the host polymer material. The red fluorescence of the dye occurs in much the same way as it does in ethanol. Light emission from semiconductors further involves the interaction of the electronic center with electrons of the conduction band.

Returning to ionic fluorescent materials, three different types of electronic centers, which show a wide range of effects caused by interaction with surrounding material, will be considered.

3.1 *Rare-Earth and Actinide Ion Fluorescence Materials*

The ions of rare-earth and actinide elements have unfilled *f*-electron configurations within a noble-gas electron shell. The electronic states are therefore highly localized and all the energy-level configuration coordinates *Q* in Fig. 1 have essentially the same equilibrium values. Nonradiative relaxation occurs by phonon coupling and by coupling to neighboring centers which may show no fluorescence because of a strong lattice coupling. The fluorescence linewidths are generally sharp and fluorescence rates slow because electronic transitions are forbidden except through nonparity interaction with the crystal field.

3.2 *Transition-Metal-Ion Fluorescence Materials*

This ion series is found in the periodic table beginning with Ti^{2+} and ending with Cu^{3+}. The electronic configurations are $3d^n$ and spatially extend out beyond the inner filled shells of the ions into the region of the coordinating anions. Consequently, the transition-metal ions can show more pronounced electronic–lattice interaction than the rare earths or actinides. Broad fluorescence lines and an extensive vibronic spectrum are usually seen. On the other hand, the opposite of both these features occurs in the case of ruby. The strong ruby fluorescence in the red spectral region results from Cr^{3+} with a $3d^3$ electron configuration in sapphire (Al_2O_3) material. The unusual fluorescence features occur because Cr^{3+} substituting for Al^{3+} is coordinated by O^{2-} ions in a symmetrical arrangement that fits the orbital angular dependence of the Cr^{3+} *d*-electron wave function in a particular way.

3.3 *Color-Center Fluorescence Materials*

Color centers in ionic crystals are usually the result of electrons trapped in the material at the site of a defect. To be specific, we may consider the fluorescence center in an alkali halide, such as NaF, where an electron is bound to an anion vacancy. However, the binding is weak and the electron can range out from the vacancy over about ten lattice constants. The electron therefore moves in the potential of the surrounding anions and cations, and the configuration-coordinate model of Fig. 1 aptly describes the observed shift in frequency between absorption and fluorescence, the shift being about half the absorption energy. The very broad transition lines include hundreds of vibronic levels, none being resolvable in the detected spectra.

4. *Laser-Excited Fluorescence*

Laser excitation of fluorescence has revealed fluorescence properties of materials that are not otherwise detectable. By using high-irradiance picosecond-pulse lasers, studies can be made of the nonradiative process whereby excited levels relax and populate fluorescent levels. This technique, and the selective-site fluorescence-excitation method previously described for glass are used to further distinguish differently coordinated fluorescence centers in the material. Lasers are also utilized to observe the electronic Raman effect that causes Stokes and anti-Stokes shifts of scattered light as a result of phonon coupling to an electronic center.

See also: Optical Bistability; Laser Materials

Bibliography

DiBartolo B (ed.) 1980, *Radiationless Processes*. Plenum, New York

Fowler W B (ed.) 1968 *Physics of Color Centers*. Academic Press, New York

Henderson B 1972 *Defects in Crystalline Solids*. Edward Arnold, London

Imbusch G F 1978 Inorganic Luminescence. In: Lumb M D (ed.) 1978 *Luminescence Spectroscopy*. Academic Press, New York

Kaminskii A 1981 *Laser Crystals: Physics and Properties*, 2nd edn. Springer, Berlin

Yen W M, Selzer P M (eds.) 1981 *Laser Spectroscopy of Solids*. Springer, Berlin

D. P. Devor

Fluoride Glasses

Fluoride and other halide glasses were uncommon and treated with little more than scientific interest in the past. Glass formation in BeF_2 had long been known, and in the 1940s extensive work extended the range to include a number of metal fluoroberyllates. However, despite their attractive optical properties (a combination of low refractive index and low dispersion), practical application of these materials was precluded by their extreme hygroscopicity and, as later discovered, toxicity. In the 1970s, particularly in response to the needs of high-power fusion lasers, interest in fluoroberyllate glasses was revived and facilities were established for their safe preparation and handling. Concurrently, in France, a new family of relatively stable glasses based on ThF_4 and ZrF_4 was accidentally discovered. The good infrared transparency and potential low optical attenuation of these glasses stimulated worldwide interest. As a result it is now known that glass formation is possible in a wide range of mixtures of metal fluorides. This article summarizes the current state of knowledge of their formation, properties and potential uses.

1. Glass Formation

Among the fluorides only one, BeF_2, is known to form a single-component glass. Its ready vitrification is a consequence of the high viscosity at the melting point ($10^5\,N\,m^{-2}$), which causes the rate of crystallization to be extremely sluggish. Additions of alkali and alkaline-earth metal fluorides reduce this viscosity; nevertheless, glasses containing up to 50 mol% of some of these fluorides can be prepared by pouring the melt onto a cold metal plate. A survey of glass formation in metal fluoroberyllates was undertaken in the 1940s by Sun, who patented over 1000 compositions, some of which contained as many as ten components.

Fluorozirconate glasses were discovered by accident during attempts to prepare single crystals of $2ZrF_4$–BaF_2–NaF—NdF_3. Subsequently it was discovered that the limited glass-forming range in ZrF_4–BaF_2 binaries can be extended by the addition of a third fluoride from the lanthanide or actinide series.

Figure 1 shows the extensive glass-forming ranges in ZrF_4–BaF_2–LaF_3 and ZrF_4–BaF_2–GdF_3, which are the most thoroughly documented systems. Similar glass-forming ranges exist for HfF_4-containing systems. Additions of small amounts of AlF_3 further broaden the glass-forming range and allow formation of samples up to 10 mm thick. Multicomponent glasses based on these compositions are being studied as potential fiber-optic materials, as described below.

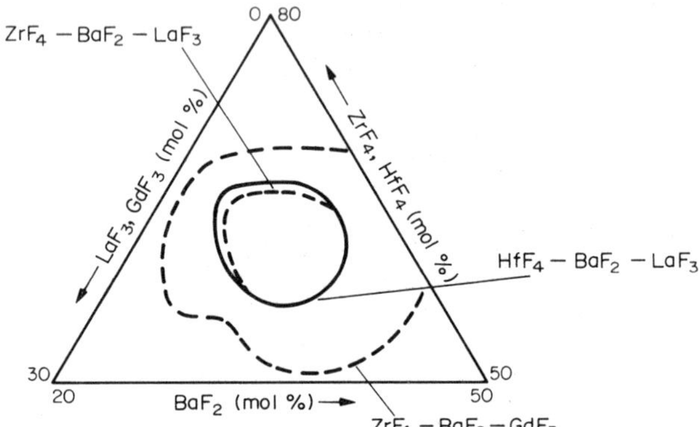

Figure 1
Glass-forming regions in the systems ZrF_4–BaF_2–LaF_3, ZrF_4–BaF_2–GdF_3 and HfF_4–BaF_2–LaF_3

Table 1 lists some representative fluoride glass compositions. With the exception of the transition metal glasses all of these contain varying amounts of either BaF_2 or ThF_4, which appear to play a role analogous to the modifier of oxide glasses. Many of the glasses can be prepared only in sections a few millimeters thick and often only by rapid quenching. Among the exceptions are BaF_2–ThF_4–ZrF_2–LnF_3 glasses and some multicomponent transition metal fluoride mixtures that approach the fluorozirconates in their glass-forming ability.

2. Structure

The crystalline and vitreous phases of BeF_2 are isomorphous with those of SiO_2. The basic structural unit is the tetrahedron $[BeF_4]^{2-}$, which can be corner-linked to produce a three-dimensional crystalline or glassy network, such that each anion is bonded to two cations and each cation to four anions. However, since the valence of the Be^{2+} cation is only half that of Si^{4+}, the Be–F bond is probably more ionic than the Si–O bond. The concomitant weakening of the

Table 1
Some representative fluoride glasses (compositions are given in mol%)

	T_g (°C)	T_x (°C)	Density (10^3 kg m^{-3})	Refractive index
BeF_2	250		1.99	1.275
$80BeF_2$–20NaF	115		2.16	
$64ZrF_4$–$36BaF_2$	300	352	4.66	1.522
$57ZrF_4$–$36BaF_2$–$3LaF_3$–$4AlF_3$	310	390	4.61	1.516
$19BaF_2$–$27ZnF_2$–$27LuF_3$–$27ThF_4$	353	453	6.45	1.534
$40BaF_2$–$30ZnF_2$–$30CdF_2$	290	330	5.53	
$50PbF_2$–$25MnF_2$–$25CrF_3$	261	310	5.90	1.663

structure manifests itself in lower glass transition temperature, hardness, modulus and higher expansivity and chemical reactivity than that of SiO_2. For cations of similar size, polarizability and half the charge, the alkali fluoroberyllates are analogous to the alkaline-earth silicates; NaF—BeF_2 can be considered a weakened version of CaO—SiO_2.

There is little good experimental evidence for the structure of the heavy metal fluoride glasses, so any structure models must be regarded as speculative. In crystalline ZrF_4, each Zr^{4+} ion lies at the center of a square Archimedean antiprism and is surrounded by eight F^- ions. However in crystalline compounds of zirconium, transition metal and lanthanide fluorides, the Zr^{4+} ion can be 6, 7 or 8 coordinated. This fact has led to speculation that in fluorozirconate glasses the network consists of a variable number of corner-sharing ZrF_x polyhedra ($6 < x < 8$). Large cations such as Ba^{2+} are incorporated in the chain, whereas smaller ones such as Al^{3+} occupy octahedral interstices. Alternatively, the network can consist of ZrF_7 and ZrF_8 chains that are cross-linked through Ba^{2+} ions. Such network structures are contrary to the predictions of the classical Zachariasen rules, which require that "the coordination number of anions about the central cation must be small (four or less)." It may therefore be more appropriate to treat these glasses as a random packing of F^- ions with the cations filling the interstices.

3. Thermal Properties

The weakness of the metal–fluoride bond relative to that of Si–O is reflected in the significantly lower glass transition temperatures (T_g) of fluoride glasses (see Table 1). The value of 250 °C for pure BeF_2 is extremely sensitive to small impurity levels, in particular OH^-, and is reduced significantly by alkali fluoride additions (100 °C for 10 mol% LiF). By contrast the T_g of heavy metal fluoride glasses is relatively insensitive to changes in composition. For most fluorozirconates the value is in the range 300–330 °C, with transition metal glasses somewhat lower and the fluorothorates nearer to 350 °C.

The high viscosity of liquid BeF_2 (10^5 N m^{-2} s at the liquidus) renders the liquid highly stable against crystallization. Modifying cations such as Li^+ and K^+ substantially reduce the viscosity but intermediates such as Al^{3+} have little effect. The heavy metal fluorides, however, are extremely fluid at the liquidus (typically 10^{-2} N m^{-2} s) and the viscosity increases rapidly towards T_g. As a consequence the working range of these glasses is limited and they are prone to crystallization. It is conventional to quote a crystallization temperature, T_x, as that at which rapid crystallization occurs during heating, usually during thermal analysis. Typically T_x for these glasses is 50 to 100 °C above T_g; however, it is very sensitive to composition and by careful choice of multicomponent glasses, for example ZrF_4–BaF_2–LaF_3–AlF_3–LiF–PbF_2, values of $T_x - T_g$ as high as 130 °C have been achieved. Even so this value may inspire false confidence in the stability of these glasses since, though crystal growth rates may be reduced, the homogeneous nucleation temperature is often above T_g. Consequently, considerable nucleation can occur during short anneals at temperatures just above T_g such as experienced during fiber drawing. At present there are insufficient data to form any general conclusions about the detailed crystallization mechanisms of fluoride glasses.

4. Physical and Mechanical Properties

Table 2 lists some important physical and mechanical properties of BeF_2, a representative fluorozirconate glass and, for reference, vitreous SiO_2. The densities of fluoride glasses reflect the atomic weights of their constituents and range from 1.9 for BeF_2 to close to 6.0 for some fluorohafnates and fluorothorates. The expansion coefficients, which are greater than those of silicate glasses, are consistent with the looser bonding and lower T_g's of fluoride glasses. As expected, the Young's moduli are lower than that of SiO_2 (typically 25% lower for fluorozirconates) though in some ZrF_4/ThF_4 glasses values as high as that of vitreous SiO_2 have been reported. Unlike both BeF_2 and SiO_2, the shear and bulk moduli of fluorozirconate glasses show a positive pressure and negative temperature dependence.

Table 2
Comparative properties of BeF_2, fluorozirconate and silica glasses

Property	BeF_2	ZBLA[a]	SiO_2
Density (10^3 kg m^{-3})	1.99	4.61	2.21
Glass transition temp. T_g (°C)	250	310	1200
Crystallization temp. T_x (°C)		390	
Refractive index n_D	1.275	1.516	1.46
Abbe number ν_D	107	75	65
Nonlinear refractive index (10^{-13} esu)	0.23	0.8	0.95
Expansion coefficient (10^6 K^{-1})	6.8	15	0.5
Vickers hardness	200	230	900
Fracture toughness (MPa m$^{1/2}$)		0.25	0.75
Youngs modulus (10^4 MPa)	1.3	2.1	3.1
Electrical conductivity at 200 °C (Ω^{-1} cm^{-1})	10^{-16}	10^{-6}	10^{-12}

a $57ZrF_4$–$36BaF_2$–3LaF–$4AlF_3$

Strength measurements on glasses require large numbers of samples for reliable data because of a distribution of flaws. Consequently, data currently available for fluorides must be treated as preliminary. Apparently fluoride glasses have strengths inferior to those of silicates but better than those of some other infrared-transmitting materials. Hardnesses are in the range 200–300 kg mm^{-2} and fracture toughnesses are typically 0.3 MPa m$^{1/2}$, though there is some optimism that with carefully chosen compositions this value can be improved. There is some evidence that fluoride glasses are susceptible to static fatigue.

Fluoride glasses are unusual because they are anionic electrical conductors, unlike silicates in which the conducting species are the alkali cations. High-purity BeF_2 has a very low conductivity (10^{-10} Ω^{-1} cm^{-1} at 200 °C) that is increased significantly by small levels of OH^-. The alkali fluoroberyllates behave anomalously in that at small concentrations the conductivity is lower than BeF_2 but at larger concentrations it is increased. The conductivities of the fluorozirconates are somewhat higher, typically 10^{-6} Ω^{-1} cm^{-1} at the same temperature.

5. *Chemical Properties*

The high hygroscopicity of BeF_2 and the alkali fluoroberyllate glasses severely limits their usage. In ambient air the surface is rapidly attacked and in aqueous solution dissolution is rapid regardless of pH. Atmospheric attack of the heavy metal fluorides is relatively slow but in liquid water the surface is attacked rapidly. However, unlike BeF_2, these glasses are inert to fluorinating agents, a property which suggests usage of the heavy metal fluoride glasses as labware for HF and similar reagents.

6. *Optical Properties*

Much of the impetus for the development of fluoride glasses has derived from their novel and potentially useful optical properties, especially low refractive index, low dispersion, low attenuation and high infrared transparency.

Table 1 lists the refractive indices of selected glasses. Vitreous BeF_2 possesses the unique combination of the lowest known refractive index of any inorganic material (1.275) with the lowest material dispersion (variation of refractive index with wavelength). Associated with this low dispersion is a nonlinear component of refractive index (that part which is proportional to incident intensity) which is only one-quarter that of silica. These properties make BeF_2-based glasses prime candidates for windows in high-power lasers. The refractive indices of the heavy metal fluoride glasses fall in the range 1.48 to 1.54, which is similar to those of most silicates. However, these glasses benefit from significantly lower wavelength dispersion than silica, and a negative temperature coefficient of refractive index.

The biggest attraction of many fluoride glasses is their good infrared transparency and associated low optical attenuation. The wavelength dependence of the intrinsic attenuation α of any glass can be described by:

$$\alpha = A/\lambda^4 + B_1 \exp(B_2/\lambda) + C_1 \exp(-C_2/\lambda)$$

where λ is the wavelength, and A, B_1, B_2, C_1 and C_2 are material constants. The first term describes the Rayleigh scattering loss, which arises from periodic fluctuations of electron density and therefore refractive index on a scale comparable to the wavelength of the light, and has inverse dependence on the fourth power of wavelength. The second term, which predominates in the ultraviolet, describes absorption due to excitation of electrons from the valence to conduction band of the material. The third represents absorption due to coupling of the photon energy with multiphonon modes in the solid and occurs in the infrared. The effect of the two absorption terms is to define an optical window for the material, within which the attenuation is determined primarily by Rayleigh scattering (Fig. 2). For BeF_2 this window stretches from about 0.15 μm in the uv to 4–5 μm in the ir, which is roughly comparable to that of vitreous SiO_2. By contrast, as is shown in Fig. 3, the transmission window of the heavy metal fluoride glasses extends from the uv (typically 0.2 μm) to well into the mid ir (8–9 μm). This good infrared transparency is a consequence of the high atomic weights and relatively weak bonding of these materials.

From Fig. 2 the minimum intrinsic optical attenuation of a glass is determined by the intersection of the Rayleigh scattering curve and the multiphonon absorption edge. For silica the predicted values of about 0.15 dB km^{-1} at 1.6 μm, are routinely

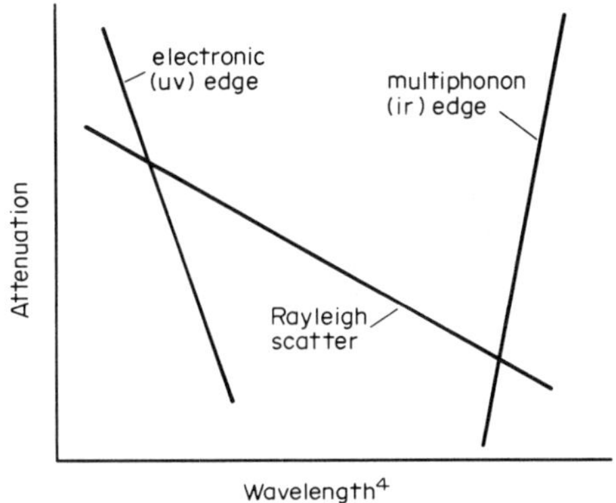

Figure 2
Schematic representation of the wavelength dependence of the optical attenuation mechanisms in a glass

approached in high-purity waveguides. In theory, significantly lower attenuations should be possible in materials such as heavy metal fluoride glasses which have an infrared cut-off at longer wavelengths and therefore reduced scattering at the absorption minimum. Extrapolations of measured infrared absorption edges and scattering losses of ZrF_4-based glasses give 10^{-2}–10^{-3} dB km^{-1} losses at 3–4 μm. However, attainment of these values represents a formidable task in purification and processing of fluoride glasses. Impurities such as Fe^{2+}, Ni^{2+}, Cr^{3+} and OH^- must be reduced to the ppb level, well below those in commercially available materials. The low thermal stability of fluoride glasses renders them prone to crystallization during processing above T_g (such as is required for fiber drawing); even the smallest amount of crystallinity will dramatically increase the scattering loss and reduce its wavelength dependence towards λ^{-2}. Nevertheless, fibers have been produced

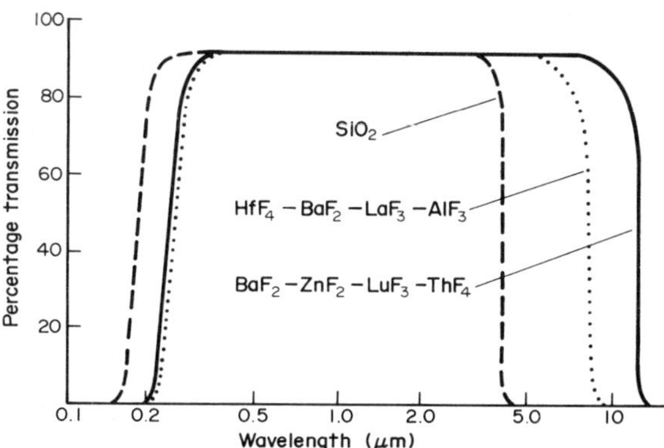

Figure 3
Optical transmission spectra of fluoride glasses and fused silica (after Drexhage 1984)

in which the scattering losses in the visible wavelength are comparable to those of silica, and although the attenuations are high (5–10 dB km^{-1}), the high level of research activity should lead to fluoride glass fibers with attenuations lower than those of silica. Such fibers could increase significantly the spacings required between repeaters in fiber-optic systems and open the possibility of repeaterless transoceanic links.

See also: Nonoxide Glasses; Glass: An Overview

Bibliography

Baldwin C M, Almeida R M, Mackenzie J D 1981 Halide glasses. *J. Non-Cryst. Solids* 43: 309–44

Drexhage M G 1984 Heavy metal fluoride glasses. In: Tomozawa M, Doremus R (eds.) 1984 *Treatise on Materials Science and Technology: Glass IV*. Academic Press, New York

M. G. Scott

Fluoroelastomers

The importance of fluorine in polymer chemistry has been known since the discovery of polytetrafluoroethylene in 1938. Highly fluorinated polymers are very stable and have a remarkable resistance to oxidative attack, flame, chemicals and solvents. This stability has been attributed to the strength of the carbon–fluorine bonds present in the polymers and to the steric hindrance around the carbon atoms.

The synthesis of elastomeric polymers containing enough fluorine to impart a significant degree of stability was not achieved until the mid-1950s. Since then, a multitude of fluoroelastomers has been reported. Three types have become commercially important: copolymers of vinylidene fluoride, fluorosilicone polymers and the copolymer of perfluoro(methyl vinyl ether) (PMVE) and tetrafluoroethylene. Representative properties are given in Table 1.

1. Copolymers of Vinylidene Fluoride

Elastomeric high-molecular-weight copolymers (**1**) of vinylidene fluoride (50–70 mol%) and hexafluoropropylene with and without tetrafluoroethylene as well as copolymers (**2**) of vinylidene fluoride, PMVE and tetrafluoroethylene, are prepared in free-radical

$$(CH_2CF_2)_x(CF_2\underset{\displaystyle R}{\underset{|}{C}}F)_y(CF_2CF_2)_z \quad (\mathbf{1}, R = CF_3; \mathbf{2}, R = OCF_3)$$

emulsion polymerization systems using organic or inorganic peroxy compounds as initiators. Fluorinated soaps may be used and several megapascals of monomer pressure at temperatures of 80–125 °C are required for optimum polymerizations. Mol-

Table 1
Properties of fluoroelastomers

Property	Vinylidene fluoride copolymers	Fluorosilicones	Perfluoroelastomers
Suitable for continuous service at	200 °C	180 °C	260 °C
Fluid resistance			
hydrocarbon oils	very good	good	excellent
esters, ketones, amines, ethers	poor	fair	excellent
Low-temperature properties			
T_g (°C)	−15 to −25	−65	−5 to −10
brittle point (°C)	−35 to −45	<−65	−35 to −45
Ozone resistance	excellent	excellent	excellent
Compression set resistance (%)			
200 °C, 70 h	15–40	60–70	25–50
288 °C, 70 h	100	100	75
Tensile properties at 25 °C			
M_{100} (MPa)[a]	6–8	2–4	8–10
T_B (MPa)[b]	>14	6–8	>14
E_B (%)[c]	~200	~200	~150

a Modulus at 100% elongation b Tensile strength at break c Elongation at break

ecular weight is controlled by the monomer/initiator ratio or by the use of chain transfer agents such as carbon tetrachloride, chloroform, aliphatic alcohols, alkyl esters or selected halogen salts.

These elastomers can be cured either by nucleophiles, such as aliphatic diamines and bisphenols, or by peroxides. In the diamine or bisphenol cure, the first step involves the elimination of hydrogen fluoride from the polymeric backbone to give olefinic unsaturation, and the second step involves the addition of the nucleophile across the double bonds. For peroxide curing, special cure-site monomers containing a labile halogen, such as bromide or iodide, must be incorporated into the polymer. Good cures require both a peroxide (to initiate the radical formation on the polymeric backbone) and a polyfunctional molecule like triallylisocyanurate (to trap these radicals). The addition of metal oxides to both types of cures neutralizes any hydrogen fluoride evolved during curing reactions and provides thermal stability in high-temperature uses. Typical curing conditions are 10–30 min in a press at 160–180 °C followed by 24 h at 230–250 °C in an air oven.

The main applications of this type of fluoroelastomer are in elastomeric seals for industrial, automotive and aerospace uses as well as oil exploration. The seals are mostly O-rings of all sizes, flat or lathe-cut gaskets and lip-type rotating or reciprocating shaft seals.

2. Fluorosilicone Elastomers

These polymers are made by ionic polymerization. The preparation of the basic monomer and elastomer is shown in Eqn. (1). The polymer is in equilibrium

$$CF_3CH{=}CH_2 + CH_3SiHCl_2 \xrightarrow{\text{catalyst}} CF_3(CH_2)_2SiCH_3Cl_2$$

$$\xrightarrow[\text{2. base}]{\text{1. } H_2O} \text{cyclic oligomers} \rightleftarrows \left[\begin{array}{c} (CH_2)_2CF_3 \\ | \\ SiO \\ | \\ CH_3 \end{array} \right]_n \tag{1}$$

with the cyclic trimer and tetramer. The oligomers are thermodynamically favored, hence the polymerization must be stopped before equilibrium is reached. In commercial practice, polymers of a degree of polymerization of about 6000 are obtained.

A small number of vinyl substituents are introduced into the polymer by means of a comonomer in order to facilitate peroxide curing. The amount of fluorine that can be incorporated is somewhat limited by the tendency of α or β fluoroalkyl substituents to undergo thermal cleavage. Thus the γ-fluorinated substituent generated by addition to 3,3,3-trifluoropropene (as shown above) is preferred. Nevertheless, these polymers have very good resistance to aliphatic and aromatic hydrocarbon fluids. Their outstanding property is a low glass-transition temperature (T_g = −65 °C) which makes them particularly suitable for low-temperature applications. They are mainly compounded with reinforcing silicas and cured with peroxides for use in molding, extruding or calendering.

3. Perfluoroelastomers

Copolymers (**3**) of tetrafluoroethylene and PMVE containing a small amount of a cure-site monomer are made by emulsion polymerization with an inorganic peroxy initiator at 40–100 °C and monomer pressures

$$\underset{(3)}{(CF_2CF_2)_x(\underset{\displaystyle OCF_3}{CF_2CF})_y(\underset{\text{cure site}}{CF_2CF})_z}$$

of a few megapascals. The composition range of 20–50 mol% PMVE gives elastomers. Below 20 mol% PMVE, the polymers behave like thermoplastic materials.

The elastomeric polymers have glass-transition temperatures of about −8 °C, which limits their application to areas where good low-temperature properties are not important. The raw polymer is essentially unchanged in appearance after being heated for one month in air at 260 °C, losing only a few percent of its weight. It is soluble in a few highly fluorinated liquids, but it is insoluble in, unreactive toward, and essentially unswollen by, almost all organic liquids, inorganic acids, bases, oxidizing agents and reducing agents. These materials can be cured in a normal press cycle followed by a long oven postcure.

Perfluoroelastomers are used mainly in sealing applications in hostile environments, such as corrosive chemicals or in oil exploration, where hydrocarbons and "sour" gas occur at high temperature and pressure.

4. Other Fluoroelastomers

High-molecular-weight elastomeric copolymers of tetrafluoroethylene with propylene (**4**) as well as hexafluoropropylene and ethylene (**5**) with nearly alternating structures have also been prepared in emulsion polymerizations. Although these materials,

$$\underset{(4)}{\left[(CF_2)_2CH_2\overset{\displaystyle CH_3}{CH}\right]_n} \qquad \underset{(5)}{\left[CF_2\overset{\displaystyle CF_3}{CF}(CH_2)_2\right]_n}$$

when properly compounded and cured by free radicals give comparable physical properties to those of the other fluoroelastomers, they have not as yet become important commercially.

See also: Fluoropolymers; Elastomers: An Overview

Bibliography

Arnold R G, Barney A L, Thompson D C 1973 Fluoroelastomers. *Rubber Chem. Technol.* 46: 619–52

Kalb G H, Quarles R W, Graff R S 1973 A new engineering material for advanced design concepts. *Appl. Polym. Symp.* 22: 127–42

Meals R 1969 Silicones. *Kirk–Othmer Encyclopedia of Chemical Technology*, 2nd edn. Vol. 18. Interscience, New York, pp. 221–60

Pierce O R, Kim Y K 1971 Fluorosilicones as high-temperature elastomers. *Rubber Chem. Technol.* 44: 1350–62

Schmiegel W W, Logothelis D L 1984 Curing of vinylidine fluoride based fluoroelastomers. *Polymers for Fibers and Elastomers*, ACS symposium series 260. American Chemical Society, Washington, DC, pp. 161–82

Thompson D C, Barney A L 1971 Elastomers. *Encyclopedia of Polymer Science and Technology*, Vol. 14. Interscience, New York, pp. 611–17

A. L. Logothetis

Fluoropolymers

Since the discovery of polytetrafluoroethylene (PTFE) in 1938, fluoropolymers have grown in importance from both technological and economic viewpoints. The outstanding combination of the chemical- and heat-resistant properties of these materials has made them indispensable in applications where ordinary plastics or elastomers may perish. In general, PTFE still provides outstanding chemical inertness and temperature resistance. Nevertheless, a number of other fluoropolymers have been developed for special uses. Some of these exhibit advantages over PTFE, such as ease of fabrication, good clarity, and higher stiffness or creep resistance. However, some sacrifice in thermal or chemical resistance is usually encountered. Some commercially important fluoropolymers are listed in Table 1.

1. Fluoropolymer Synthesis

Most commercial fluoropolymers are manufactured by processes employing free-radical polymerization. These processes include emulsion, suspension and solution polymerization techniques. In the laboratory, polymerizations have also been initiated by irradiation from ultraviolet or high-energy sources and by electrolytic techniques.

Emulsion and suspension polymerizations are characterized by an aqueous phase in the reactor. Both methods employ emulsifiers to disperse the monomer, but differ in the amount of emulsifier and type of initiation used. While emulsion polymerizations use high levels of emulsifier and water-soluble initiators (these conditions producing submicron polymer particles), suspension polymerizations use low levels of emulsifier and water-insoluble initiators (producing polymer suspensions with particle sizes very much larger than emulsion polymerizations). The polymer dispersion so produced may be used as it is or the polymer may be isolated and sold in powder or pellet form.

Solution polymerizations require a single liquid phase consisting of the monomer, the initiator, the polymer and the solvent. As the polymerization proceeds the solution viscosity increases and, depending on the polymer, the product may precipitate to yield a slurry. The choice of solvent is critical in these cases

Table 1
Some commercially important fluoropolymers

Chemical name	Structure of monomer(s)	Tradename(s)	Continuous service temp. (°C)	Typical uses
Polytetrafluoroethylene	$CF_2{=}CF_2$	Teflon Hostaflon TF Fluon Aflon TFE	260	electrical insulation, chemically inert gaskets and parts, parts for low friction application
Polychlorotrifluoroethylene	$CF_2{=}CFCl$	Kel-F Hostaflon Fluorothene	150	gaskets, chemically inert parts, coatings
Poly(tetrafluoroethylene–hexafluoropropylene)	$CF_2{=}CF_2$, $CF_2{=}CF{-}CF_3$	FEP resins	200	electrical insulation, chemically inert parts, film
Poly[tetrafluoroethylene–perfluoro(alkylvinyl ether)]	$CF_2{=}CF_2$ $CF_2{=}CF{-}OR_f$	Teflon PFA	260	injection molded parts, valve and pump linings
Poly(tetrafluoroethylene–ethylene)	$CF_2{=}CF_2$ $CH_2{=}CH_2$	Aflon COP Hostaflon ET Tefzel	150	electrical insulation, creep resistant parts, film
Poly(chlorotrifluoroethylene–ethylene)	$ClFC{=}CF_2$ $CH_2{=}CH_2$	Halar	150	molded low-creep parts, sealants
Poly(vinylidene fluoride)	$CH_2{=}CF_2$	Kynar	150	molded parts, tank linings, electrical insulation
Poly[tetrafluoroethylene–perfluoro(methylvinyl ether)] (contains an unspecified curesite)	$CF_2{=}CF_2$ $CF_2{=}CFOCF_3$	Kalrez	320	O rings, gaskets for extremes of temperature or solvent resistance
Poly(vinylidene fluoride–hexafluoropropylene) (may also contain TFE in some grades)	$CH_2{=}CF_2$ $CF_2{=}CF{-}CF_3$	Viton Fluorel	200	fluid-resistant rubber parts, gaskets, O rings, tubing
Poly(vinylidene fluoride–1-*H*-pentafluoropropylene)	$CH_2{=}CF_2$ $HFC{=}CF{-}CF_3$	Technoflon SL	180	gaskets, seals
Poly(tetrafluoroethylene–propylene)	$CF_2{=}CF_2$ $CH_2{=}CH{-}CH_3$	AFLAS	230	gaskets, O rings, seals
Poly(vinyl fluoride)	$CH_2{=}CHF$	Tedlar	110	weather-resistant films, coatings
Poly(tetrafluoroethylene–perfluoroethylene sulfonic acid)	$CF_2{=}CF_2$ $CF_2{=}CF(OCF_2{-}CF(CF_3))_nOCF_2CF_2SO_3H$	Nafion	—	membranes for chloroalkali cells, solid catalysts

and requires the balancing of a number of factors to achieve optimum results. Consideration of the monomer, polymer and initiator solubility are also important, as is the chain-transfer activity, toxicity, and recovery or disposal of the solvent. For many purposes, chlorofluorocarbons are useful solvents and provide reasonable compromises for factors important to the polymerization.

2. *Perhalofluorocarbons*

The oldest members of this family are PTFE (**1**) and polychlorotrifluoroethylene [PCTFE (**2**)]. Other members are copolymers containing a small percentage of hexafluoropropylene [FEP (**3**)] or perfluoroalkylvinyl ether [PFA (**4**)]. The parent member of this group, PTFE, is a rigid linear molecule of extremely high molecular weight; hence, the polymer is extremely viscous in the molten state and special fabrication techniques are required to obtain the best physical properties. Substitution of some of the fluorine atoms on the chain by a much larger chlorine atom, by a trifluoromethyl group or by a perfluoroalkylether group introduces branches which "kink" the polymer chain, thereby lowering its melting point. These products are less viscous and easier to fabricate by conventional processing techniques. The substitution may introduce a weak link into the polymer chain, resulting in lower thermal stability. Perfluoroalkylvinyl ether groups are exceptional in that thermal stability is maintained.

The chemical resistance of the perfluorinated

$$—(CF_2CF_2)_n— \quad (1)$$

$$—(CF_2\underset{\substack{|\\Cl}}{CF})_n— \quad (2)$$

$$—(CF_2CF_2)_n—(CF_2\underset{\substack{|\\CF_3}}{CF})_m— \quad (3)$$

$$—(CF_2CF_2)_n—(CF_2\underset{\substack{|\\O(CF_2)_nF}}{CF})_m— \quad (4)$$

resins is similar to PTFE. Significant attack is only observed with molten alkali metals, fluorine, chlorine trifluoride or oxygen difluoride. The presence of chlorine in PCTFE renders this material subject to attack by some organic solvents, especially highly halogenated and aromatic materials. The radiation resistance of PCTFE is, however, superior to the other fluorocarbon resins.

3. *Fluorinated Ethylene Polymers*

Poly(vinylidene fluoride) [PVDF (**5**)] and poly(tetrafluoroethylene–ethylene) [ETFE (**6**)] provide

$$—(CF_2CH_2)_n— \quad (5) \qquad —(CF_2CF_2CH_2CH_2)_n— \quad (6)$$

an interesting comparison in that these materials are essentially isomeric polymers. Although ETFE is a copolymer, the commercial products contain almost equal amounts of each monomer. In spite of this similarity, ETFE has a melting point approximately 100 °C higher than PVDF. Both polymers are highly crystalline but PVDF is reported to be polymorphic. The structures show different electrical properties: for example, the dielectric constant of PVDF ranges from 8.4 to 6.4 as the field frequency ranges from 60 Hz to 1 million Hz. The dielectric constant for ETFE remains at 2.6 over that frequency range. PVDF also shows a piezoelectric effect. There are no solvents for ETFE below 150 °C. PVDF dissolves at room temperature in polar solvents including dimethylformamide, dimethylsulfoxide and dimethylacetamide. As the polymers contain hydrogen, both materials are sensitive to strong oxidizing agents and to organic bases.

4. *Fluoroelastomers*

Elastomeric fluoropolymers may be prepared from many combinations of randomly copolymerized monomers. Commercially available products include those based on vinylidene fluoride (Viton, Fluorel, Technoflon) [Formula (**7**)] and tetrafluoroethylene (Kalrez) [Formula (**8**)]. In contrast to FEP and PFA, these copolymers contain 30–40 mol% hexafluoropropylene or perfluoromethylvinyl ether. The chemical and heat resistance increases as the fluorine content increases.

$$—(CF_2CH_2)_n—(CF_2\underset{\substack{|\\CF_3}}{CF})_m— \quad (7)$$

$$—(CF_2CF_2)_n—(CF_2\underset{\substack{|\\OCF_3}}{CF})_m— \quad (8)$$

New fluoroelastomers with improved low temperature properties and base resistance are under development. The first examples to be commercially tested are copolymers of ethylene and complex perfluorovinyl ethers (Aftex) [Formula (**9**)].

$$—(CH_2CH_2)_n—(CF_2\underset{\substack{|\\OCF_2\underset{\substack{|\\CF_3}}{CF}O(CF_2)_xF}}{CF})_m— \quad (9)$$

The fluorophosphazenes (**10**) are a new type of elastomer in which the backbone is composed of alternating phosphorus and nitrogen atoms, rather than carbon atoms. This structure yields elastomers with better low-temperature properties but poorer high-temperature properties than the fluorocarbons.

$$—(N{=}\overset{\substack{OCH_2(CF_2)_nF\\|}}{\underset{\substack{|\\OCH_2(CF_2)_nF}}{P}})_m— \qquad (n = 1, 2, \ldots, 10) \quad (10)$$

5. *Specialty Fluoropolymers*

Poly(vinyl fluoride) [PVF (**11**)], sold in film-form, has excellent surface and optical properties. This, together with weather and stain resistance, leads to its use as a surfacing material. Recently, PVF films have been used as glazing on solar panels. Films from poly(tetrafluoroethylene–perfluoroethylene sulfonic

$$—(CH_2CHF)_n— \quad (11)$$

$$-(CF_2CF_2)_n-(CF_2CF)- \\ \quad | \\ (OCF_2CF)_nOCF_2CF_2SO_3H \\ | \\ CF_3$$

(12)

acid) [PSA (**12**)] are strongly acidic materials and function as ion-exchange resins. These properties have permitted the use of these films as membranes in electrolytic cells and as solid catalysts.

See also: Fluoroelastomers

Bibliography

Elias H-G 1977 *New Commercial Polymers, 1969–1975.* Gordon and Breach, New York

Hudlický M 1976 *Chemistry of Organic Fluorine Compounds*, 2nd edn. Halsted Press, New York, pp. 601–10

Mark H F, Othmer D F, Overberger C G, Seaborg G T (eds.) 1980 Organic fluorine compounds. *Kirk–Othmer: Encyclopedia of Chemical Technology*, 3rd edn., Vol. 11. Wiley-Interscience, New York, pp. 1–81

Roff W J, Scott J R 1971 *Handbook of Common Polymers: Fibres, Films, Plastics and Rubbers.* CRC Press, Cleveland, Ohio

Wall L A (ed.) 1972 *High Polymers*, Vol. XXV, *Fluoropolymers.* Wiley-Interscience, New York

B. C. Anderson; R. E. Uschold

Force Measurement in Metal Forming

For optimum utilization under production conditions, a metal-forming press should be used at or near its maximum load and energy capacities, without being overloaded. Continuous overloading of a forming machine may result in blocking or locking in a mechanical press, breakage of tooling, decreased life of moving machine members, and in high maintenance costs and downtime. Overloading is avoided, and improved utilization of press capacity achieved by the measurement and monitoring of the forces in metal-forming processes. In setting up a press for a production run, force monitoring assists in avoiding damage to tooling and reducing setup time. During production, force monitoring supplies valuable information regarding changes in process conditions such as lubrication, tool wear and variations in properties of formed billet or sheet.

1. Force–Displacement Curves

For a given material, a specific forming operation, such as impression die forging, extrusion or bending, requires a certain variation of the forming force over the slide displacement stroke. This is illustrated in Fig. 1, where the area under the curve represents the energy necessary to perform the forming operation. Knowledge of these curves for various operations, part geometries and materials allows the improvement of die and process design.

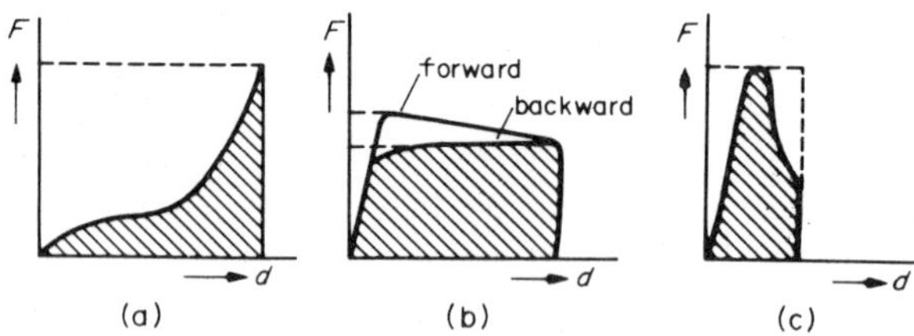

Figure 1
Force versus displacement curves for (a) impression die forging, (b) extrusion and (c) blanking

2. Load-Cell Measurement

Forming forces can be measured with transducers based on piezoelectric, capacitive and inductive effects. However, the large majority of these transducers use electrical strain gauges which transform elastic strain into a change in electrical resistance (see *Pressure-Sensing Materials and Methods*). An electrical load cell has strain gauges, attached to a load-carrying member, which give a voltage output when excited and strained (Lebow 1963). The load cell is placed under the tooling of a forming press and calibrated, that is, the voltage output of the load cell is measured under a set of known load values. Load cells are usually used in research and development, where high accuracies are desirable.

3. Press-Frame Deflection Measurement

A very practical method for measuring forces in forming presses is to measure the straining of the press frame. For this purpose strain gauges can be attached to the weakest section of the press columns. However, since the strain in the press columns is usually small and difficult to measure accurately, it is advantageous to use strain bars, which mechanically amplify the strain by concentrating a relatively large strain over a short length (Fig. 2). In order to compensate for the off-center loading of a press, or to obtain the loading of each column separately, strain bars are attached on each press column and wired appropriately. The accuracy of force measurement via strain bars is quite adequate in a production environment. These sensors have proven to be reliable and effective for force measurement and monitoring in presses.

4. Other Methods

It is also possible to attach strain gauges directly to appropriate locations on the forming tools. This is

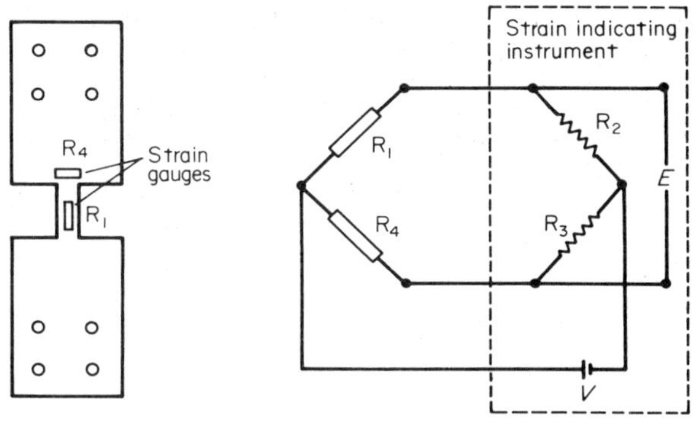

Figure 2
Half wheatstone bridge as used for a strain bar: V is the excitation voltage; E is the output voltage

particularly useful where symmetrical tooling is used and where uniform loading can be assured. In hydraulic presses, for practical purposes it is usually sufficient to measure the hydraulic pressure in the press drive. However, in this case errors are introduced because pressure losses in the hydraulic piping and friction losses in the ram drive are not measured.

The output of a load-measuring device (load cell, strain bar or strain gauge) is essentially a voltage which is proportional to the applied load. The magnitude of this output voltage, or signal, is calibrated such that the relationship between the unit load and unit voltage is linear and known. The output signal (or the load) is then measured, displayed or recorded using appropriate voltage measuring instrumentation. The most important criterion in selecting this instrumentation is adequate frequency response. Most forming operations, conducted in mechanical presses, take place within only 5–100 ms. Therefore, instrumentation, used in recording or displaying the forming load, must respond quickly to represent the correct voltage output from the load sensors.

Chart recording and digital readout are the most practical methods for monitoring the forming load. For recording, it is necessary to use a galvanometer-type light-beam oscillograph, ensuring rapid response time. Under production conditions, it is often convenient to have a digital display of the measured load so that it can be easily read by the press operator. In the mid-1970s, several companies started using digital display units for force monitoring with great reliability and advantages. The digital readout can be handled in a number of ways: most commonly, the reading from a forming stroke is retained and displayed until the subsequent stroke takes place and the new load reading replaces the previous one. Thus, the display is automatic and indicates to the press operator the variation of the forming load from one stroke to the next.

See also: Strain Sensor Materials; Presses in Metal Forming; Metals Processing and Fabrication: An Overview

Bibliography

Altan T, Boulger F W, Becker J R, Akgerman N, Henning H J 1973 *Forging Equipment, Materials and Practices.* Metals and Ceramics Information Center, Columbus, Ohio

Douglas J R, Altan T A 1973 Characteristics of forging presses: Determination and comparison. In: Tobias S A, Koenigsberger F (eds.) 1973 *Proc. 13th Int. Machine Tool Design and Research Conf.* Macmillan, London, pp. 535–45

Lebow M J 1963 Some principles of transducer design. *ISA Trans.* 2: 85–92

T. Altan

Forging of Optical Materials

Forging of optical materials is the process of slow compressive deformation to shape and strengthen optical materials. Forging became widely used in the early 1970s when high-energy laser windows were required. This technique offered a simple alternative to growth of large flat crystal disks of correct dimension for use in windows, since the candidate materials were not readily available in the required shapes and sizes. In addition, deformation raised the strength of cleavage-prone crystalline materials, both by blocking crystal cleavage through formation of differently oriented grains and by strengthening effects of dislocations.

The materials for high-energy laser windows are those with high transparency in the infrared, and include the alkali halides and alkaline-earth halides. For CO_2-laser windows at 10.6 μm, KCl, KBr and NaCl crystals offer low absorption; whereas for the CO laser at 5.3 μm, the neodymium : yttrium–aluminum–garnet (Nd : YAG) laser at 1.06 μm and the HF and DF lasers at 2.8 and 3.8 μm, respectively, CaF_2, SrF_2 and BaF_2 crystals are the candidate window materials. All these materials have been press-forged into large flat blanks for polishing into windows. Starting materials are single crystal boules grown by Bridgman techniques, and oriented crystals are forged in a specific crystal direction. Forging temperatures for the alkali halides are 150–350 °C, and for the alkaline-earth halides 750–1000 °C. Reductions in height due to forging range from 50 to 90%.

Response of these ionic single crystals to forging is different from that of metals where structure and strength properties become highly directional, and vary greatly with the amount of deformation. Ionic crystals polygonize into equiaxed fine grains which are not elongated in the forging flow direction and whose increased strength is only partly dependent upon the amount of reduction.

Table 1
Typical forging conditions and resultant properties for some alkali and alkaline-earth halides

Material	Forging			Flexure strength (MN m^{-2})
	Temperature (°C)	Pressure (MN m^{-2})	Reduction (%)	
KCl crystal				3.4 (yield)
KCl	200	18.6	80	19.3 (yield)
KCl	300	8.2	80	11 (yield)
NaCl crystal				19.3 (yield)
NaCl	300	24.8	70	37.2 (yield)
CaF_2 crystal				75.8
CaF_2	700	34.4	80	
CaF_2	900	12.4	80	75.8
SrF_2 crystal				41.3
SrF_2	900	12.4	85	55.1

Most forging development has been done as flat die forging: compression between two platens without lateral constraints. Lubrication at platen interfaces (teflon, graphite foil, silicone oil) is used to moderate the constraining effects of friction at platen faces. At low forging temperatures edge cracks can form, so modifications of the flat die forging process have been used. These include restraining rings of copper or aluminum, hydrostatic oil pressure and high-pressure gas. The latter two also improve platen interface lubrication, and this minimizes edge cracks (Koepke et al. 1973). Rolling, a type of continuous forging, has also been investigated (Koepke et al. 1974).

A typical forging run involves the following steps:

(a) Preparation of billet. This is cut from a boule with proper orientation, then the ends and sidewalls are polished, and dimensions recorded.

(b) Assemble to forge. A pressing column is built between platens consisting of the billet, interface lubricants, heated plates and heat isolation members. The heaters and thermocouples are then added, the drybox closed and the system flushed with inert atmosphere.

(c) Heat and forge. Reduction adjustment for the desired rate at final thickness is made (for constant rate, the platen speed is readjusted during the run as the piece is reduced in height).

(d) The pressure is reduced and, after cooling, the forged blank is removed.

Typical forging conditions for KCl, NaCl, CaF_2 and SrF_2 are shown in Table 1, along with final forging pressure for 80% reduction.

The temperature of deformation determines the yield strength and grain size. For KCl forged at <280 °C, the yield strength is 19.3 MN m^{-2} and the grain size is 10 μm; forging at >280 °C gives 11 MN m^{-2} and 35 μm grains. Forged KCl, in the presence of humidity over 40%, will undergo grain growth to a size approaching that of the original single crystal. This growth does not occur in KCl crystals prepared by reactive atmosphere processing (see *Reactive Atmosphere Processing*). Grain growth can also be repressed by the addition of alloy elements such as Sr^{2+}, Eu^{2+} or Rb^{+}. These additives increase the forged strength and the temperature required for forging.

Infrared transparency is unaffected by forging. For KCl, where surface damage from polishing causes 90% of total absorption, the harder forged polycrystalline matrix tends to damage less and give lower absorption values. Forged CaF_2 crystals, while fine grained (2 mm) after forging 80% at 900 °C, have a strength of 82.7 MN m^{-2}, which is not appreciably greater than that of single crystal CaF_2. This strength can vary widely due to sensitivity to both internal bubbles and surface flaws, and the unfortunate combination of high thermal expansion (19×10^{-6} K^{-1}) and high elastic modulus (120 GN m^{-2}).

See also: Alkali Halide Crystals; Infrared Laser Window Materials

Bibliography

Dickinson S K 1975 *Infrared Laser Window Materials Property Data for ZnSe, KCl, NaCl, CaF_2, SrF_2, BaF_2*, AFCRL-TR-75-0138. US Air Force Cambridge Research Laboratories, Hanscom Field, Bedford, Massachusetts

Koepke B G, Anderson R H, Bernal G 1973 The structure and properties of hot rolled KCl crystals. *3rd Conf. High Power Infrared Laser Window Materials.* US Air Force Systems Command, Washington, DC

Koepke B G, Anderson R H, Eberwal G, Stokes R J 1974 Hydrostatic press forging of alkali halide crystals for laser window applications. *4th Annual Conf. Infrared Laser Window Materials.* US Air Force Systems Command, Washington, DC

Weertman J, Weertman J R 1965 Mechanical properties, mildly temperature dependent. In: Cahn R W (ed.) 1965 *Physical Metallurgy*. Interscience, New York, Chap. 15, pp. 754–75

R. R. Turk

Forming and Bending of Wood

Giving shape to wood by forming and bending is a technique thousands of years old. Today, forming is used to transform economical straight wood into simply-curved or complex three-dimensional forms. Applications of the technique include the manufacture of furniture, motor vehicles, ships, airplanes and sports equipment, as well as musical instruments and works of art.

Normal forming techniques involve selection and preparation of stock, softening, forming and setting. Softening is omitted when thin laminations are bent and glued together to produce curved shapes (see *Glued Laminated Timber*). Since bending by laminating is a relatively simple process, and since the most common method of softening in forming is by steaming, this article is primarily concerned with steam bending.

Forming has many advantages: it is simple and fast, requires little energy and reduces material waste. Forming and bending do, however, require high levels of expertise and appropriate equipment and facilities. Undesirable side effects can include reduced stability, an increase in the swelling capacity and weakening of certain strength properties.

Permanent form alterations of single pieces are accomplished by the introduction of minute compression failures into the structure. Species showing certain discontinuities in their anatomical structure are prized for their good bendability. Examples of useful discontinuities are ring-porous vessel patterns (black locust, chestnut), broad rays (beech, sycamore), banded parenchyma (elm, sapele), radial series of vessels (jelutong), or combinations of these (oak, ash, afzelia, mansonia). Species without these features or those with relatively high lignin content, which impedes plasticization, have only limited formability. This is the case with most softwoods (particularly compression wood) and many exotic species.

1. Material Selection

For industrial forming, species with good bending qualities are generally chosen; these include hickory, birch, maple, walnut, mahogany and sweetgum, in addition to those named above.

Only straight-grained clear material can be used in production since growth characteristics (e.g., pith, knots and cross grain) and defects (e.g., surface checks and decay) lead to stress concentrations, and thus production defects. However, bending by laminating can be done with any species and with ordinary lumber.

2. Preparation of Stock

The blanks are first dried to the proper moisture content, that is, 8–12% for bending by laminating and 15–20% for steam bending. They are then cut to the desired dimensions with some allowance for the volume lost during forming. The surface is planed to reduce the tendency to crack. For the same reason, mortises and holes are cut only after forming.

Green wood is often flexible enough for bending. Industrially this is rarely done, since it involves long drying times and equipment tie-up, and results in high wastage caused by shrinkage and checking. Wood is therefore plasticized by steaming and boiling in water. This sharply increases compressibility, so that lengthwise-oriented cells tend to buckle and the cell cavities tend to collapse. Plasticization also reduces the tensile strength across the grain, as seen by the buildup of checks and delamination.

In practice, softening is usually performed with saturated steam at about 100 °C and vapor pressures of 1.1–1.3 bar. Since completely softened material easily checks during bending, wood is only partially plasticized. For each millimeter of material thickness, dry wood requires about two minutes to plasticize, and moist wood about one minute.

Wood can also be softened by other means (e.g., with urea which makes it thermoplastic, or with ammonia which gives better results than steaming). However, owing to high investment costs, these methods have not found much acceptance.

3. Forming

In the production of laminated wood, cold bending is used to make glued laminated timber or curved furniture parts from veneer. Cold-formed objects must be glued or mechanically held.

Solid wood is normally steam bent. Severe curvatures require forming with a supporting strap, a steel band placed on the convex side of the piece. This limits the tensile strain and causes most of the strain to shift to the compression side, where it is taken up by minute compression failures. Temperature, applied force and forming time are varied according to species, blank dimensions and the desired radius of curvature. Under expert control, wastage is less than 1%.

In addition, wood can be formed by densification, and partly by molding. In these processes, the cells are crushed and wedged together under pressure and heat to densities of up to 1400 kg m^{-3}. The mechanical properties can be significantly improved while the swelling capacity diminishes. These types of products are mainly used in specialized applications.

4. Setting

Forming gives rise to mechanical stresses in the blank. Therefore, the blanks must be fixed in the new form until the stresses have dissipated. Drying at 60–90 °C, cooling in a fixed form and storage in a climatically controlled chamber for 2–4 weeks, finally lead to form stability.

The curvatures of steam-bent wood are not completely stable, and tend to straighten out with temperature and humidity changes. In practice, a water-repellent surface coating is applied, or some mechanical restraint is introduced to assist in maintaining the shape.

See also: Wood: Macroscopic Anatomy; Wood: An Overview

Bibliography

Bariska M, Schuerch C 1977 Wood softening and forming with ammonia. In: Goldstein I S (ed.) 1977 *Wood Technology: Chemical Aspects*, ACS Symposium Series No. 43. American Chemical Society, Washington, DC, pp. 326–47

Darzinsh T A 1976 *Modification of Wood.* Indian National Scientific Documentation Centre, New Delhi

Stevens W C, Turner N 1970 *Wood Bending Handbook.* HMSO, London

M. Bariska

Forming of Glass

The forming of glass involves the simultaneous extraction of heat from the glass and its shaping to a desired configuration. Forming of soda-lime glass containers is emphasized in this article; however, the basic principles are also applicable to other glass systems such as borosilicates and other end products such as flat glass or tubing.

Certain factors are critical in forming glass. At the start of forming the glass must have the desired composition and, therefore, the viscosity–temperature characteristics that are consistent with process timing parameters. The glass must be at the proper temperature, homogeneous and free of gaseous inclusions (seeds or blisters). At the end of the forming operation the glass article, tube or sheet is heated for a time at a particular temperature and then cooled at a controlled rate, to assure relief of stresses associated with the forming operation.

Thus, forming begins in the batch house, where the raw materials must be correctly proportioned and adequately mixed before introduction to the melting furnace, and includes the melting, fining and temperature conditioning in the melting furnace, as well as the postforming heat treatment and cooling operations.

1. Viscosity of Glass

The viscosity–temperature relationship for a typical soda-lime glass (Fig. 1) shows a rapid rise in viscosity η as the temperature decreases below 1200 °C. This rise is characteristic of all liquids that easily form stable glasses; it prevents the structural reorganization necessary for crystallization as the glass cools through this temperature range.

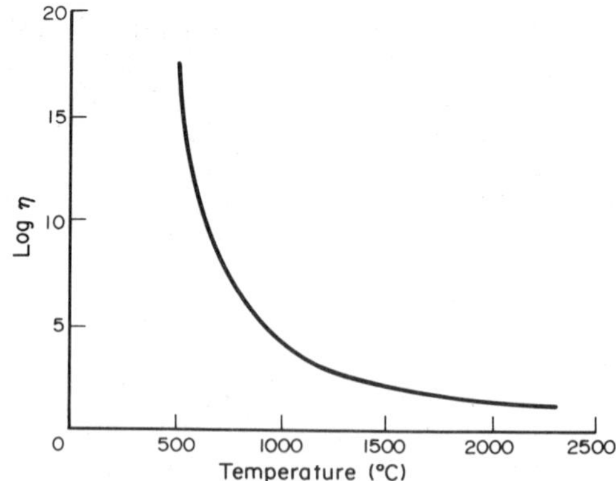

Figure 1
Plot of log η against temperature for a typical soda-lime glass

Viscosity–temperature process parameters for a soda-lime glass are shown in Table 1. The glass is relatively fluid at melting temperatures. Before forming the glass must be "conditioned" by cooling to the temperature where the viscosity is that required for subsequent processing; the glass is then held at this temperature to ensure uniform heat distribution and hence consistent and predictable flow characteristics. The forming process itself takes place in a 300 °C temperature range during which the viscosity increases from about 400 to 10^6 Pa s. After forming, the glass article, sheet or tube is annealed to prevent the development of thermal stresses.

In practically all forming operations, glass behaves as a Newtonian fluid. That is, the viscosity is a function only of the temperature of the glass and not of the rate at which the glass flows in the forming process.

2. Forming of Containers

Glass containers are produced in a variety of machines, the one most commonly used being the

Table 1
Viscosity–temperature process parameters for soda-lime glasses

Process	Temperature (°C)	Log η
Melting	1480–1540	1
Refining	1315–1370	2.6–2.5
Delivery to forming process	1090	3.6
End of forming process	740–760	7.0
Annealing	590–500	11.5–16.0

Emhart IS (Individual Section) machine. Individual units of glass called gobs are delivered to a mold where a preliminary or embyro form of the final shape is made by a pressing or blowing operation. This preliminary shape is called a parison. After the parison is formed, the mold is opened and the chilled glass surface that was in contact with the mold is reheated briefly, after which the parison is transferred to a second mold where the final shape is blown. The resultant bottle is transferred to a conveyor which carries the completed article to an annealing oven (lehr). An IS machine can have as few as four or as many as ten sections, and multiple gobs can run through each section on each cycle.

Table 2 provides more specific viscosity–temperature data for forming of glass containers; these data represent average temperatures for bottles weighing 184 g and having a capacity of 355 ml, made on an IS machine. The gobs, parisons or bottles were removed from the process at the desired step and quickly put into a calorimeter to determine their total heat content, from which the average temperatures at the different steps were calculated. In both molding steps, the glass is in intimate contact with a metal mold which determines the shape of the article and extracts heat from the glass at a controlled rate. When the first- or second-stage mold opens, the surface layer of the glass is considerably colder (typically by 100 °C) than the average temperature of the article. Thus reheating the glass surface by conduction from the bulk glass is an essential step between the first and second molding operations to ensure uniform temperature in the blow mold. At the end of the final blowing operation, this chilled surface layer of higher viscosity, together with the surface tension of the glass, holds the container to the desired shape and dimensions until its bulk has cooled sufficiently to be dimensionally stable.

Table 2
Viscosity–temperature forming parameters for the container process

Process	Temperature (°C)	Log η
Gob shearing	1140	3.3
Delivery to blank mold	1110	3.5
Transfer from blank mold to blow mold	970	4.4
Delivery from blow mold to machine conveyor	690	8.3

The cycle time for a section of an IS machine is typically 10–15 s, during which the glass viscosity increases from 200 to 2×10^7 Pa s. Since the viscosity–temperature relationship depends strongly on glass composition, the glass must be homogeneous in space and time if the highly complex, high-speed forming operation is to give acceptable products.

3. *Heat Transfer During Forming of Containers*

For the high rates of production characteristic of modern forming operations, heat transfer must be consistent and controllable. A modern container machine can form more than 100 t of glass in 24 h, and requires heat extraction rates in the molding operation alone of 2×10^9 J h^{-1}.

Mold material is typically gray cast iron although aluminum bronzes and stainless steel are sometimes used. The metal molds are cooled by blowing air over their external surfaces. In Fig. 2 temperatures in the glass and the mold are shown, including a temperature discontinuity at the glass–metal interface. Heat transfer from the glass to the mold is almost all by conduction, with only a small contribution by radiation.

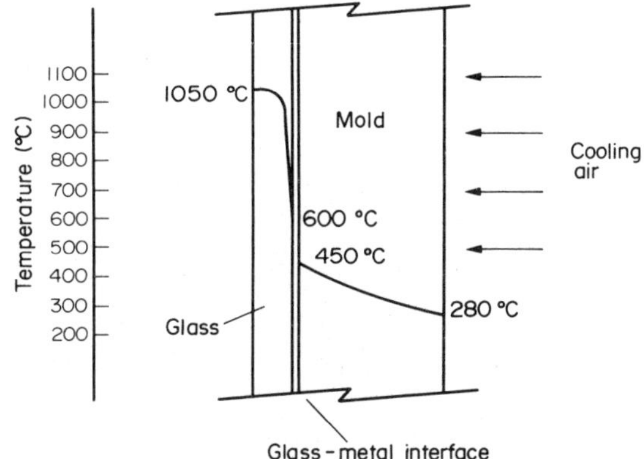

Figure 2
Heat transfer in glass forming

The temperature discontinuity at the glass–metal interface results because the glass contacts the metal only at isolated points. Probably less than 10% of the area between the glass and the metal is in contact, and the average thickness of the air gap is 0.02–0.03 mm. The presence of an air gap means that the contact pressure can strongly influence the rate of heat transfer.

4. *Float Glass*

Table 3 provides viscosity–temperature data for the float process for producing flat glass. In this process the glass is transferred from the melting furnace onto a bath of molten tin in a continuous stream. Because the molten glass spreads onto the tin, a continuous sheet of controlled width and thickness can be formed. Heat is extracted by conduction to the tin

Table 3
Viscosity–temperature forming parameters for the float process (courtesy of E L Swarts, PPG Industries)

Process	Temperature (°C)	Log η
Delivery of glass to float bath	1065–1150	3.2–3.7
End of attenuation zone	760	6.9
Transition from float bath to lehr	580–610	10.8–11.9

bath on one side and by radiation to the atmosphere on the other. The advantage of this process is that the precisely parallel, highly polished surfaces required for plate glass are obtained without costly grinding and polishing. Tables 2 and 3 show that despite the great differences in the container and float processes, the viscosities at critical steps of forming are similar for both operations.

In the float process the glass cools from 1100 to 700 °C shortly after delivery to the tin bath and during the initial spreading phase. Later the glass on the tin is reheated to 850 °C by heaters above the bath, after which it is cooled to 600 °C and delivered to the annealing lehr. The reheating lowers the viscosity so that flow can eliminate any surface imperfections that remain from the delivery and spreading processes.

5. Effects of Microstructure

High rates of shear, as in gob shearing, and structural discontinuities such as phase separation or devitrification can produce non-Newtonian behavior in glass flow. In glass forming, "workability" problems can be caused by microstructure in glass, detectable only in the electron microscope. These problems involve difficulty in controlling the distribution of glass in a container and a tendency for cracks, described as checks or splits, to develop in the container during or immediately after forming. Submicrometer structures resulting from phase separation, devitrification or incompletely melted batch constituents can cause these problems, which are related to abnormal flow and heat removal. The checks or splits can be produced by non-Newtonian flow and associated stress increases.

See also: Fining of Glass; Glass Melting; Viscosity and Annealing of Glass

Bibliography

Giegerich W, Trier W (eds.) 1969 *Glass Machines*. Springer, New York
Pilkington A 1971 Float: An application of science, analysis and judgement. *Glass Technol.* 12: 76–83
McCauley R A 1980 Float glass production; Pilkington vs. PPG. *Glass Ind.* 61: 18–22
McGraw D A 1961 Transfer of heat in glass during forming. *J. Am. Ceram. Soc.* 44: 353–63
Trier W J 1961 Temperature distribution and heat flow in glass in blank molds of container machines. *J. Am. Ceram. Soc.* 44: 339–45

R. J. Ryder

Forming of Paper and Paperboard

Forming of a wet paper web, either on a single continuous fabric/wire (fourdrinier) or between two fabrics/wires (twin wire), is the first step on the papermachine. Process variables in the forming of paper that affect the sheet properties fall into three general categories. The first category consists of those variables related to the headbox. The second comprises those related to the fourdrinier and the third those associated with twin wire forming. The headbox and the associated forming device, be it a fourdrinier or a twin wire, each has its own variables, the one set independent of the other, but in respect of some other variables the two devices are interactive.

This article discusses paperforming by fourdrinier and twin wire methods; recent developments in forming technology are dealt with elsewhere (see *Papermaking: Alternative Wet End Processes*).

1. Headbox

The headbox, through which the dilute paper slurry is dispensed, is a critical part of the forming process determining the ability to produce a paper sheet of uniform basis weight (uniform mass distribution of fibers). Being the first step in the process of forming paper, it must produce a uniform mass distribution of fibers on both the large scale, called basis weight variations, and the small scale, called formation. The subsequent forming components can reduce only the very small-scale variations to improve the formation and cannot correct basis weight variations due to poor delivery from the headbox.

Uniform fiber distribution is a very important factor, because in strength properties the final sheet of paper is only as good as its weakest part. Thus, if there is a large variation of basis weight, the low-weight areas will determine the final strength properties. The fiber–water ratio in the flow from the headbox has a very large effect on the formation and hence the strength properties. The lower the fiber–water ratio, the better the paper is formed.

The turbulence levels in the stock jet from the headbox can have a considerable effect on the fiber orientation and the formation. A highly turbulent jet tends to produce a more random fiber orientation, that is, more uniform strength properties in both the machine direction and the cross direction.

Another important variable is the ability to control the discharge angle of the jet from the headbox. The jet discharge angle determines the angle of impact and the point of impact on the forming wire, both of which affect formation and porosity.

2. Fourdrinier

The jet from the headbox lands on the continuous forming wire or fabric of the fourdrinier, the function of which is to remove as much water as possible from the fiber suspension before it enters the press section. This is accomplished with several drainage components: table rolls, foils, vacuum-augmented foils and suction boxes. All these components are used to control the drainage profile in the machine direction. They are also used to introduce turbulence into the fiber suspension to improve the formation of the sheet.

Which components are used is highly dependent on the grade of paper being produced and the machine speed. At high speeds, table rolls introduce too much disturbance of the slurry and foils are used to introduce a lower turbulence level while still maintaining drainage. Vacuum-augmented foils are used to increase drainage while minimizing the introduction of turbulence. Finally, suction boxes are used at the end of the forming process to drain as much water as possible from the sheet before it enters the press section. The drainage profile greatly influences the formation of the sheet, the strength properties and the removal of filler and pulp fines which affect the porosity of the sheet. Different grades require different drainage profiles to optimize both sheet strength and porosity of the raw stock.

3. Twin Wire Formers

The major differences between twin wire formers and fourdriniers are that (a) two-sided drainage occurs on twin wire formers, and (b) there is no free surface in a twin wire former for large disturbances to develop as there is on a fourdrinier. On twin wire formers there are two wire sides instead of the single wire side on the fourdrinier. Thus twin wire formed sheets have very similar printing characteristics on both sides, whereas there is a difference between the printing characteristics on the wire side and the top side of a sheet made on a fourdrinier machine.

There are two basic types of twin wire former. One is the roll former, which uses a large suction roll as the primary drainage element in the forming zone. The other is the static element former, which uses a forming zone defined by deflectors or foil blades. Some process variables are common to both types and some are unique to each.

Elements common to both formers are wire tension, headbox jet–wire ratio, wire (fabric) mesh and weave, and suction devices, such as suction rolls or normal suction boxes. By changing wire tension on either type of former, the drainage pressure changes according to the equation $p = \sigma/r$, where p is the drainage pressure between the wires, σ is the wire tension and r is the radius of curvature of the wire. For roll formers, r is very closely approximated by the diameter of the suction forming roll. For static element formers, r is the local radius of curvature of the wire near the deflectors. By increasing wire tension, the drainage pressure increases and hence the capacity of the forming zone increases. Drainage pressure can influence sheet properties such as porosity and surface smoothness.

Both formers have a vacuum device of some type. In the case of the roll former it is part of the suction forming roll and in some of the static element formers there are suction boxes as well as a normal suction couch roll. The vacuum areas, depending on whether they are high vacuum or low vacuum, affect the removal of fines in the sheet, thereby influencing the porosity and surface characteristics.

The primary differences between the two types of twin wire formers lie in the magnitude and the manner in which drainage pressure is introduced into the suspension. In a roll former, because the radius is constant, the drainage pressures throughout the forming zone remain constant. In the static element formers, however, there is a pressure pulse created at each supporting deflector blade which can vary in magnitude at different positions along the forming zone. Because of this pressure pulse, static element formers tend to produce a more uniform and better formed sheet than roll formers. The pressure pulse causes the stock between the wires to accelerate and decelerate, producing relative motion between fibers and thereby improving the formation.

On all formers, the fabric mesh and weave can have an important effect on the surface properties of the sheet as well as the drainability of the suspension in the forming zone. The wire characteristics must be matched to the desired quality of the end product; compromises are necessary among surface finish, porosity and drainage. The headbox jet velocity is carefully controlled on all formers because of the need to maintain a jet–wire speed ratio at a fixed value. This is the main control feature for establishing the desirable ratios of strengths in the machine and cross directions.

See also: Paper and Paperboard: An Overview

Bibliography

Bolam F (ed.) 1962 *Trans. Symp. Formation and Structure of Paper*. British Paper and Board Makers' Association, London

Norman B 1979 Principles of twin wire forming. *Sven. Papperstidn.* 82: 330–36

Parker J D 1972 *The Sheet Forming Process*, TAPPI Special Technical Association Publication No. 9. Technical Association of the Pulp and Paper Industry, Atlanta, Georgia

R. E. Hergert

Fossil Energy: Materials for Mining and Preparation of Coal

The systems used for mining coal by surface and underground methods are shown in Figs. 1 and 2. The equipment for surface mining is essentially the same as that for earth moving in construction. Underground mining equipment is specialized. In either case, conventional structural machinery and tool steels and other materials are used.

Operations in coal preparation consist of crushing, screening, various methods of separation by density, froth flotation of fines and magnetic separation. The materials used in these systems are conventional and similar to those used in ore concentration, except that coal is softer than most ores.

1. Equipment

Main frames of coal mining equipment are generally of weldable low-carbon steel. Cast steel used in equipment contains <0.4% C. Shafts are case- or through-hardened steel. Wear pads on crawlers are weldable higher-alloy steels. Pads are either forged or cast steels. Screen materials are high-carbon drawn steel or heat-tempered steel, but stainless steel is occasionally used for high wear resistance, and rubber and plastics have also found increasing use in both coarse and fine screening.

Drill bits are constructed of tungsten carbide and bit holders of through- or case-hardened steel. Where temperature extremes are expected, chromium-bearing steel is specified.

2. Roof Supports

Roof supports can be props or cribs constructed of locally available wood, or roof bolts. Roof bolts are mainly of the anchor or resin types, but other types of roof bolts, such as the split set or pins, are used on a smaller scale. The anchor type of bolt is composed of anchor, bolt, plate and hardened steel washer. The washer is optional, but it does allow greater accuracy in torquing the bolts. The anchors are either cast malleable iron or cast or pressed steel. The bolts are steel of 380–550 MPa yield strength. The plates are medium-carbon steel and the washer is a hardened heat-treated steel.

The resin bolt system comprises a two-part resin, a rebar and a plate. The system can be fully or partly grouted with the resin. The partly grouted bolt allows a torque of 400 Nm to be applied to the bolt by means of a threaded end and a nut. The rebar is steel with a yield strength of 275 MPa or more depending on

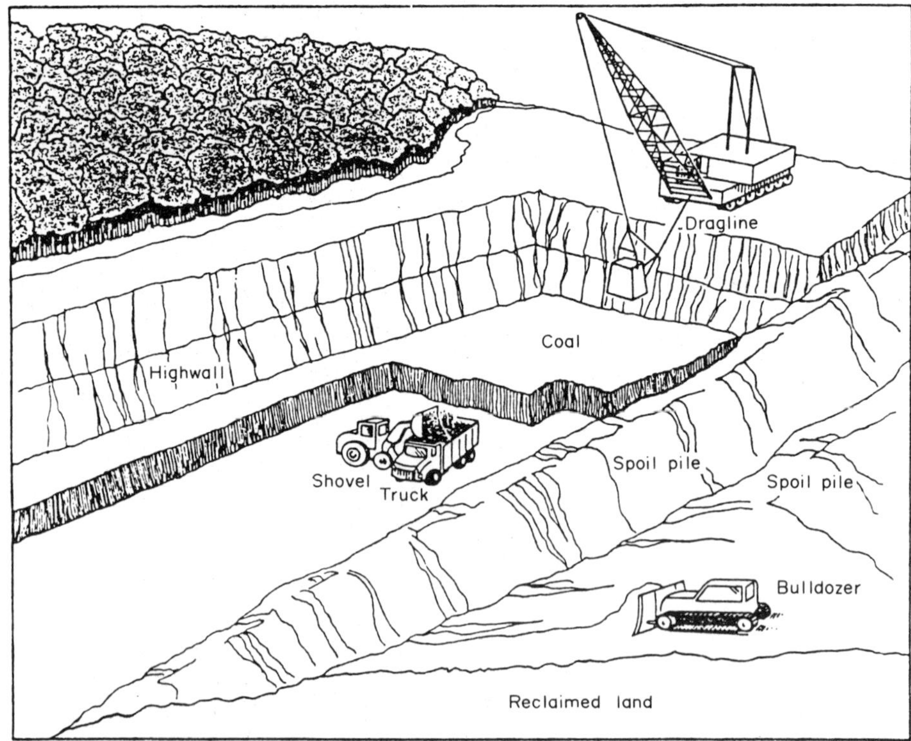

Figure 1
Surface mining system (strip mining)

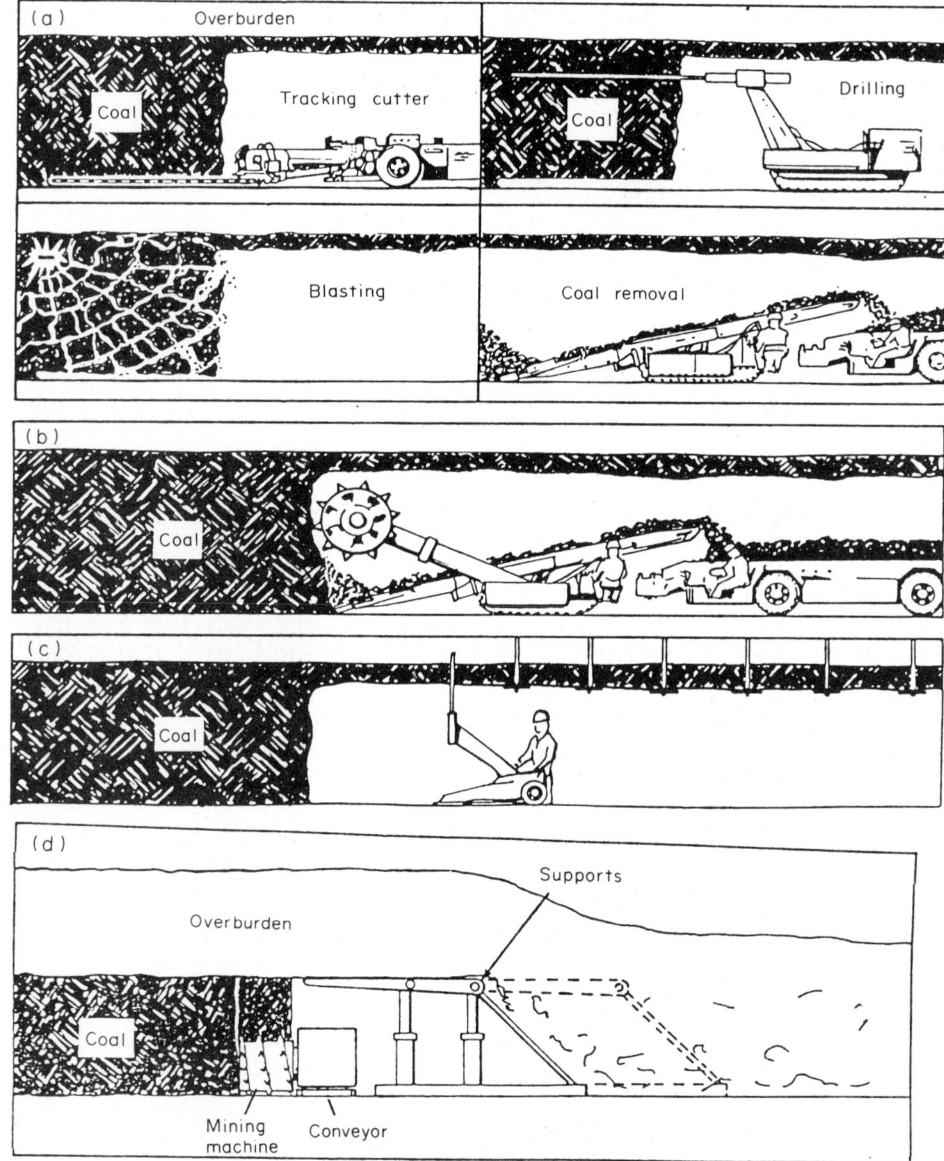

Figure 2
Underground mining systems: (a) conventional, (b) continuous, (c) roof bolting, (d) longwall (continuous)

mine requirements, and the plate is steel. The resin has a fast setting time of <2 min to allow rapid installation of the bolts.

3. Rails

Rail materials are variable, depending on the availability of scrap and new steel to the steel mill. Light rail (13.5–18 kg) is usually used in underground coal mines.

4. Construction

Construction materials used in surface installations, underground coal mines and coal preparation plants are steel, concrete, plastics and ceramics. A wide variety of steel shapes (i.e., plates, I-beams, channels) are in stock and easily available. The availability of steel is of extreme importance to the contractor and mine operator, due to the need for short construction times and rapid repair of existing structures and machinery. The AISI 1020 and 1040 grades are by far the most frequently used. The other steels used are specialty types of limited application. Structural steel is normally hot-rolled, but steel for machine usage is usually cold-rolled because of its better uniformity and surface quality.

Nonferrous metals commonly used in machinery are aluminum and copper; lead and zinc are used for special applications. Most equipment is electrically powered and thus uses these materials in motors, conductors and batteries.

Nonmetallic materials in the US coal industry are concrete, plastics, rubber, brick, ceramics, glass and wood. Concrete is extensively used in mining for foundations, shafts, shotcrete, block, ties, bins, silos, floors and pipes. For general use, a Portland cement of 20 MPa (3000 psi) strength is the choice, for static thickeners a 27.5 MPa (4000 psi) cement is employed and for storage silos a 35 MPa (5000 psi) cement is used. Standard reinforcing steel is widely employed.

Plastics and rubber are employed in coal mining and preparation plants for brattice cloth, tubing, belt conveyors, piping, screens, windows and siding (preparation plants). Neoprene with reinforcing nylon finds use for conveyor belts. Resin or epoxy fiber glass is used for tubing, piping and siding. Abrasion-resistant rubber (or neoprene) is used for slurry pipelines and also in chutes for wear and noise control.

Fireclay brick and insulating brick are used in fireboxes of dryers, and cast and fully fired ceramics for long lifetimes in chutes and at transfer points; abrasion-resistant pipes are also fabricated from ceramics.

5. *Summary*

Materials used for both underground coal mines and preparation plants have changed as new, improved materials have been made available to the coal industry. Improved steels, nonferrous metals, ceramics and plastics have found gradual, increased use. However, changes have been slow, due to an overriding concern of the industry for fast construction, ease of repair and ready availability of supplies to maintain production and minimize downtime. The industry is price-competitive, the two most important considerations for material use being price and availability. However, the coal industry will specify special materials for severe and unusual needs (i.e., under stress and corrosion conditions).

See also: Open-Cut Mining; Underground Mining

Bibliography

Cassidy S M 1973 *Elements of Practical Coal Mining*. Society of Mining Engineers, New York

Leonard J W, Mitchell D R 1968 *Coal Preparation*. Society of Mining Engineers, New York

W. E. Foreman

Fossil Fuel Utilization

This article discusses the materials problems and the selection of materials for fossil fuel utilization systems. Two such systems are considered: heat engines (e.g., gas turbines) and combustion systems (e.g., boilers and fluidized bed combustors).

1. *Materials for Heat Engines*

Heat engines convert thermal energy to mechanical energy. In general, their efficiency is determined by the difference between the maximum and minimum temperatures in the cycle. Usually, the minimum temperature is determined by the available cooling and is thus close to ambient. Improved efficiency consequently means higher maximum cycle temperatures. This can be achieved either by using materials with adequate properties at high temperatures or by cooling the structural components in the high-temperature regions. Cooling itself carries some efficiency penalty, so in practice both approaches may be used simultaneously, employing the best material to minimize the amount of cooling required.

The thermal energy is extracted in a heat engine by having a working fluid do work on the mechanical system, usually by making a hot compressed gas expand and cool. The hot gas can be the product of combustion of the fuel, as in the internal combustion engine or the directly fired gas turbine, or a secondary fluid, as in the steam turbine, the Stirling engine or the indirectly fired gas turbine. In the latter case, the heat is transferred from the combustion gas to the working fluid in a heat exchanger.

It is convenient to discuss each heat engine separately, since the problems and materials selections are system-specific.

1.1 *Gas Turbines*

In a gas turbine, air is ingested and compressed by an axial or radial rotary compressor. Part of the compressed air is mixed with fuel and burned in a combustion chamber. The remainder of the air is mixed into the combustion products as they pass down the combustion chamber, the object being to reduce the overall gas temperature to that which can be accepted by the turbine. The turbine inlet temperature has risen progressively over the years. The present generation of large industrial engines has turbine inlet temperatures of ~1150 °C; aircraft engines have turbine inlet temperatures of 1300 °C and perhaps higher. The hot gas then passes through a row of stationary blades which redirect it. These are called the nozzle guide vanes. Since they are stationary, they are subjected only to stresses from aerodynamic forces, which are in the region of 40 MPa. The gas stream, which cools some 100 °C in passing through the vanes (the exact degree of cooling depends on the details of the design), then strikes the first row of rotor blades, which are fixed to the periphery of a disk which itself is attached to a shaft. The stresses on the rotor blades are significantly higher, in the region of 160 MPa. The combination of the stator and rotor rows is called a stage, and the gas turbine may have several stages.

There are several materials problems in the gas turbine, and in the early days the pace of advance was largely determined by the development of new high-temperature materials for the first stage of the turbine. The major criterion at this early stage was an adequate stress rupture life or, for more ductile materials, the time to a critical creep strain at which the blades would rub excessively on the seals. Early developments were based on two classes of alloy: Ni–20%Cr (Nichrome), an alloy developed for heaters, and Co–35%Cr with some carbon (Stellite), an alloy used originally for surgical implants and later for turbosuperchargers. The cobalt-base alloys were investment-cast, but in the early stages of development the nickel-base alloys were forged.

It was discovered quite early that the nickel-base alloys could be strengthened by the addition of Al and Ti to precipitate the ordered fcc phase $Ni_3(Al, Ti)$, called gamma prime (γ'), which is more or less coherent with the Ni matrix. The degree of coherency depends on the alloy content of the matrix, but more particularly on the Al–Ti ratio in the γ' phase. Increasing strength then depended on increasing the volume fraction of γ' by increasing the total amount of Al+Ti in the alloy. However, the maximum solubility of γ' could be increased by decreasing the chromium content of the alloy.

The chromium serves two purposes. The first is that it leads to the formation of a protective Cr_2O_3 layer on the metal surface in oxidizing environments at elevated temperatures; the second is that it is an effective solid-solution strengthener. The effect of the decrease in chromium on the oxidation resistance can be compensated for if the Al content is high enough, because an even more protective oxide, Al_2O_3, forms; generally, a minimum of 5 wt% Al is required for this. The solid-solution strengthening can be replaced by the addition of one of the refractory metals; since for the blades there are clear benefits in reducing the density, molybdenum was preferred over tungsten. Finally, the addition of cobalt raises the solvus temperature without reducing the maximum solubility. Table 1 illustrates the development of the nickel-base superalloys; only the major constituents are shown.

Table 1
Composition of some early nickel-base superalloys (in wt%)[a]

Alloy	Cr	Al	Ti	Mo	Co	Others
Nimonic 75	20	0	0.4	0	0	
Nimonic 80A	20	1	2	0	0	
Nimonic 100	11	5	1.5	5	20	
Nimonic 115	15	4.5	1.2	5	20	
Udimet 500	19	2.9	2.9	4	18	
Udimet 700	15	4.2	3.5	5.2	18.5	
713C	12.5	6	1	4	0	
IN 100	10	5.5	4.7	3	15	
B 1900	8	6	1	6	10	4 Ta
Mar-M 200	9	5	2	0	10	12.5 W

a Balance nickel

Nimonic 80A and Udimet 500 represent the early, ductile, oxidation-resistant superalloys. The earliest high-strength alloys were Nimonic 100, IN 100 and Alloy 713C. All three ran into a severe problem. In gas turbines used in aircraft flying missions near the sea, they suffered an accelerated form of corrosion induced by sodium sulfate, called hot corrosion (see *Hot Corrosion*). It speedily became clear that low Cr and high Mo were responsible for the attack. As a result, in Nimonic 115, which replaced Nimonic 100, the chromium content was restored to 15%; its hot corrosion resistance was not the best, but still a great deal better than that of the earlier alloys. However, the approach taken by the US engine manufacturers was rather different. Unwilling to give up the high strengths of the newer superalloys, they developed protective coating procedures.

The two last alloys in the list, B1900 and Mar-M 200, are widely used, although their intrinsic hot corrosion resistance is extremely poor. For engines in severe environments, however, such as shipboard engines or industrial engines, a new class of relatively high-strength alloys with good hot corrosion resistance was developed. The first and most widely used of these is IN 738, whose approximate composition is Ni–16%Cr–3.4%Al–3.4%Ti–1.75%Mo–2.6%W–1.75%Ta–0.9%Nb–8.5%Co–0.17%C. More recent alloys in this group are IN 792 (Ni–13%Cr–3.2%Al–4.2%Ti–2.0%Mo–3.9%W–3.9%Ta–9%Co–0.21% C) and Nimonic 939 (Ni–22.5%Cr–1.9%Al–3.7%Ti–2.0%W–1.0%Nb–1.4%Ta–19.0%Co–0.15%C). The later higher strength nickel-base superalloys cannot be forged, and all are now investment-cast.

The earliest coatings depended on the diffusion of aluminum into the surface of the alloy from a pack containing Al, Al_2O_3 as a diluent and an activator such as ammonium chloride. The treatment produced an outer layer of NiAl, typically 75–125 μm thick. By adjusting the thermodynamic activity of the Al in the pack, the composition and structure of the aluminide layer could be optimized. The introduction of platinum into these layers appeared to produce beneficial results, for reasons which are not altogether understood. Newer coatings are applied by physical vapor deposition (PVD) in a vacuum chamber; the coating material is volatilized from a pool by an electron beam (EB). This allows for a wide variation in coating composition.

The most widely used coatings are the M–Cr–Al–Y coatings, where M is usually Co but may be Co + Ni, Ni or Fe. For aircraft engines, the coatings are high in Al (~12 wt%) and relatively low in Cr (8–10%), but for hot corrosive environments the chromium content is increased to 20% or more and the Al content reduced to ~8%. In all cases, the yttrium content is normally 0.5–1.0%. The yttrium serves to improve the adhesion of the protective Al_2O_3 oxide layer to the substrate. More recently, hafnium has been added to the coatings to replace or supplement the yttrium.

An alternative method of coating metals is to spray the coating on. The most effective method is the plasma spray, in which a flame is established and ionized to the plasma. Metal powder is injected into the flame and projected against the surface to be coated. However, the resulting coating is porous and contains much oxide; the adhesion to the substrate is not as good as that of the EB–PVD coatings. Recently, it has been shown that if the plasma spraying is performed in a vacuum, the velocity of the molten metal droplets is greatly increased, and a very dense adherent coating is produced, comparable with the EB–PVD coating.

In comparison, the development of cobalt-base superalloys has been very simple. There is no strengthening phase equivalent to γ', and these alloys are strengthened with solid-solution strengtheners and carbide dispersions. Generally, the cobalt-base alloys have poorer mechanical properties than the nickel-base alloys at moderate temperatures but maintain their strength to higher temperatures; they are mainly used for the stationary vanes, so density is a less important factor. As a consequence, tungsten is commonly used as the solid-solution strengthener. The two most widely used cobalt-base superalloys are X-40 (also called Haynes Stellite 31), whose composition is Co–10%Ni–25%Cr–7.5%W–0.5%C; and Mar-M 509, Co–10%Ni–24%Cr–7%W–3.5%Ta–0.5%Zr–0.6%C. The function of the nickel is to stabilize the face-centered-cubic modification of cobalt.

In practice, the metal temperature cannot exceed ~900 °C, as can be seen from Fig. 1, which illustrates the creep rupture properties of some nickel- and cobalt-base alloys. The higher turbine inlet temperatures are achieved by cooling the vanes and blades; air is bled from the compressor and ducted through passages within the airfoils. In the case of the vanes, these are usually fairly simple passages, with the air exiting through gill holes in the concave (pressure) side near the trailing edge; it is entrained with the gas and thus the associated losses are fairly small. Because of the higher stress levels, the rotating blades have to be cooled more effectively, and since the air used for cooling is effectively not used by that stage at all, there are efficiency benefits in keeping the cooling air flow as low as possible. The cooling passages are very complex, and present considerable fabrication problems.

The castability of the superalloy is an important criterion. New advanced techniques for cooling are being investigated: transpiration cooling, in which the airfoil surface is essentially a metal gauze through which air is blown, the stress being carried by a relatively cool spar up the center of the blade; shell and spar constructions, in which a thin foil airfoil is cooled by impinging jets from the inside, again with a central cool spar carrying the load; and water-cooled blades, in which the blade (or vane) is made of copper with stainless steel water tubes cast in. All these present their own special materials problems, largely connected with the fabrication processes.

There are other materials problems in gas turbines—erosion of the compressor by ingested dust and stress-corrosion cracking of the compressor disks, for example—but most of them are not as critical to the performance as the hot section materials problems. The combustor itself is discussed in Sect. 2.1.

1.2 Internal Combustion Engines

In comparison with gas turbines, internal combustion engines are relatively simple. The air is ingested into the cylinder, fuel is injected, the mixture is compressed and ignited and the resulting expansion drives the piston down; on the next upward stroke the combustion gases are exhausted through an exhaust valve. There are three major identifiable problem areas: the sealing and wear problems between the piston and the cylinder, the thermal shock experienced by the piston head, and the hot gas attack of the exhaust valve.

The major problems occur with diesel engines. These have very high thermal efficiencies, and increasing the thermal efficiency still higher requires reduced cooling of the combustion chamber and preheating of the intake air. In the extreme, the adiabatic diesel has been proposed, which attempts to eliminate all heat loss from the cylinder and extract the maximum heat energy from the exhaust gas by using a turbosupercharger and a regenerative air preheater.

Exhaust valve corrosion resembles hot corrosion. The exhaust gas temperature is in the region of 800 °C or perhaps higher, and there are some ash components, which may include impurities such as S, Pb, V, Na and P together with NaCl arriving with the intake air as a result of road deicing treatments. The exhaust valves are usually made of nitrogen-stabilized stainless steels, such as 21–4N (Fe–21%Cr–4%Ni–9%Mn–0.5%C–0.45%N), but nickel-base superalloys have been used in advanced engines. In addition to the corrosion problem, there may be erosion problems at the valve and seat junction induced by the high-velocity, high-temperature gas stream containing ash particles; the mating surfaces

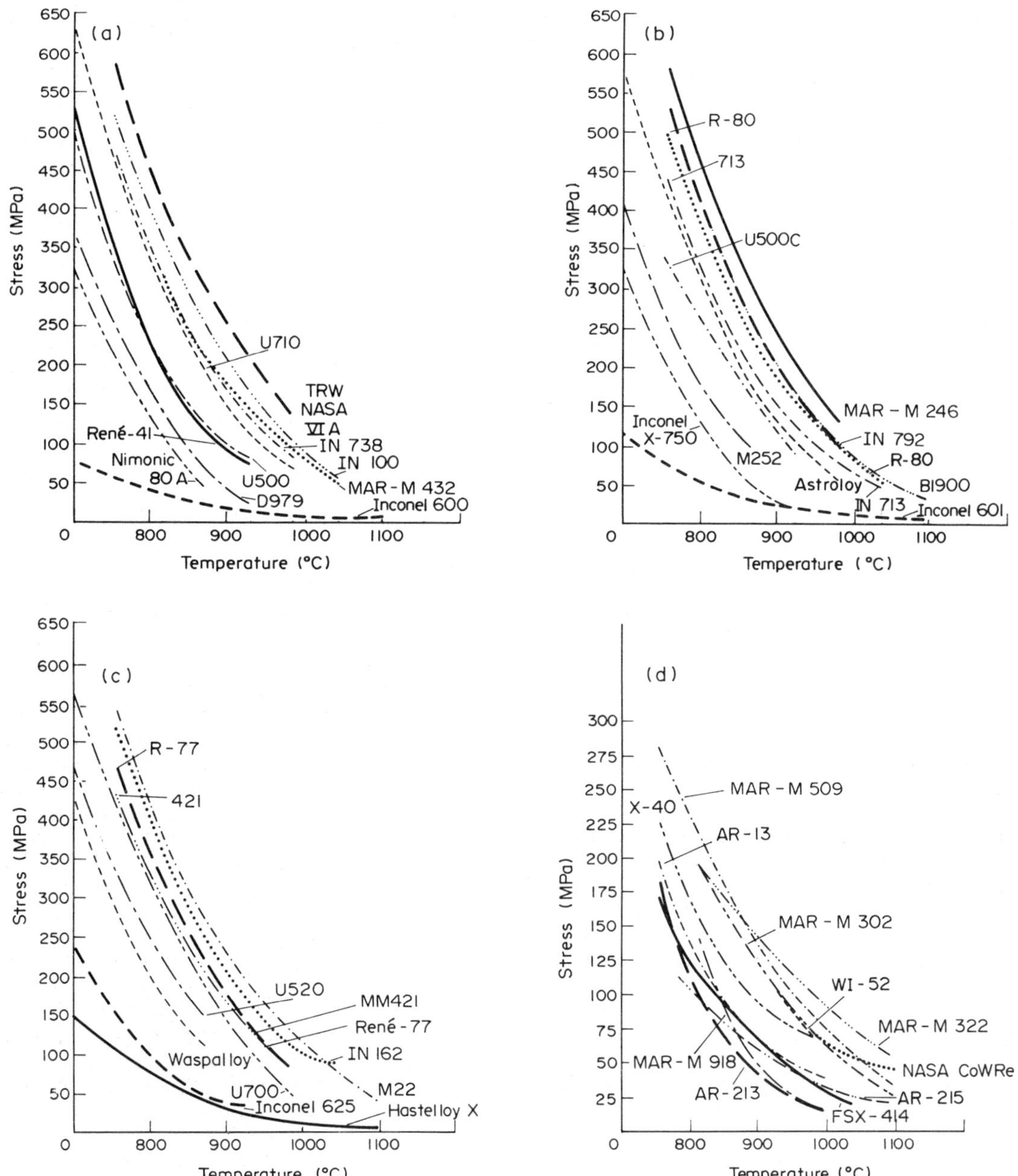

Figure 1
Stress–temperature curves for rupture in 1000 h for selected alloys: (a), (b), (c) nickel-base; (d) cobalt-base (after Sims and Hagel 1972)

may be hardfaced. Finally, the crown of the valve may suffer thermal fatigue craze cracking. Large valves may contain liquid metal (NaK, for example) in an internal cavity to transfer heat from the crown of the valve to the upper stem.

The piston crown also suffers from craze cracking. The newest development in this area is to use a thermal barrier coating. This consists essentially of a ceramic layer of low thermal conductivity; it thus damps out the thermal spikes associated with the cyclic combustion process and so limits the thermal fatigue of the crown. The problem, of course, is in making such a layer which does not itself spall from the substrate. The current thermal barrier coatings are made from zirconia, ZrO_2, and contain a material to stabilize the cubic or tetragonal phases. The most usual stabilizer is yttria, Y_2O_3, but others are possible. The mixed oxides are applied by plasma spraying. Usually, a metal coating such as a Co–Cr–Al–Y is sprayed on first, to act as a transition layer between the substrate and the thermal barrier coating.

An alternative approach is to use ceramic inserts in the engine. The developments leading to the adiabatic diesel will certainly involve extensive use of ceramic inserts and perhaps a ceramic cylinder liner, so long as reasonable sealing ring lives can be attained.

1.3 Steam Turbines

Large utility fossil-fuel-fired generating systems virtually all use steam turbines to drive the generator. Typically, steam enters the high-pressure turbine at 538 °C, 15.5 MPa (although for supercritical units the pressure may be as high as 24.1 MPa), and finally exits at perhaps 20 °C, 2.5 kPa. A single steam turbine may represent in excess of 1000 MW(e) and is thus a very large structure indeed. In the USA, the common practice is to attach the rotating blades to disks which are shrunk-fit on the shaft; in Europe, some manufacturers use a welded construction. The principal materials of construction are low-alloy steels.

The principal materials problems are as follows.

(a) Erosion of the inlet valves and the inlet nozzles on the high-pressure and intermediate-pressure turbines caused by entrained particles of oxide which spall off the inner surface of the boiler superheater and reheater circuits.

(b) Cracking of the rotor shaft. This does not happen very frequently, but a failure has such major consequences that stringent inspection procedures are used and rotors are retired if there is any significant indication of cracking.

(c) Cracking of the disks. This is more common in nuclear turbines than fossil fuel turbines (because of the wet steam in the former case) but is nevertheless a significant problem.

(d) Creep of the high-pressure turbine casing bolts. The casing for the high-pressure turbine is made in two sections which are bolted together. The very high pressure within the casing leads to high bolt stresses, and they relax by creeping. Retightening of the bolts eventually leads to failure.

(e) Corrosion fatigue failure of the next-to-last stage (L-1) blades.

(f) Water droplet erosion of the last stage blades.

Solid particle erosion (SPE) has emerged as a major problem only in the last few years. This is because the damage seldom causes an unscheduled outage; instead, the damaged components are routinely replaced at maintenance outages. There is, however, an efficiency loss associated with the damage, and as fuel prices rise this aspect becomes more important. There are really two approaches to solving the problem: stopping the oxide particles from leaving the boiler (this aspect is reviewed below) or making the turbine more resistant. Some utilities have used hard facings, usually weld overlays of materials such as Stellite 6B, on the components at risk, but the manufacturers are not enthusiastic about this approach for the blades, because of possible deleterious effects on the mechanical properties.

Water droplet erosion is somewhat different in character. The water droplets condensing out of the steam near the low-pressure end of the turbine are very small and do not themselves cause erosion. However, the water collects on the last stage stator vanes, and large droplets may then detach from the trailing edge, travelling relatively slowly; these are then struck by the last stage rotating blades. There are again two methods of approach: the first is to try to prevent the large droplets from detaching from the vanes by encouraging the water to flow toward the turbine casing, where it can be removed; the second is to armor the blades. The method of armoring is to braze hard inserts (cemented tungsten carbide or a Stellite) into the blade. Care must be taken to avoid initiation of fatigue cracks at the corner of the inserts.

The corrosion fatigue of the L-1 blades arises because the first droplets to condense from the steam have a very high concentration of dissolved salts, even though the water used in boilers is exceptionally pure, with impurity levels in the parts per billion range. These droplets can cause pitting corrosion on the Fe–12%Cr blades and the pits then act as stress raisers, initiating fatigue cracking. There are at least three methods to combat this problem. The simplest is to replace the blades with a more corrosion-resistant material; a high-strength titanium alloy, Ti–6% Al–4%V, has been suggested, although there may be problems with its relatively low damping capacity. Improving the water chemistry still further, or modifying it so that the first-formed liquids are not corrosive, is another possibility. Finally, some form of protective coating might be able to resist the attack. Work is currently in progress on all three approaches, but no generally accepted solution is yet available.

Disk cracking appears to be a stress-corrosion cracking (SCC) problem, although the corrosive medium is essentially high-purity water. The cracking initiates at the keyway where the disk is attached to the rotor. The solution to the problem is to design so as to lower the stresses at this location and to maintain as good water chemistry as possible. The disk material in the USA is basically similar to that of the rotor; an example is A469, Class 8: Fe–4%Ni–2%Cr–0.5%Mo–0.1%V–0.28%C (max).

Cracking of the rotor itself appears to be related to the presence of inclusions, principally sulfides, in the ingot from which the rotor is forged. The best solution would appear to be to improve the metallurgical practice in melting, by using such techniques as electroslag remelting (ESR) or vacuum carbon-deoxidized (VCD) low-sulfur or silicon-deoxidized low-sulfur steelmaking. In addition, careful inspection coupled with fracture mechanics studies of the effect of cracks and residual lifetime analyses can extend the permissible life of rotors without unacceptable risk of catastrophic failure.

2. *Materials for Combustion Systems*

The major combustion systems of interest in fossil fuel energy systems are the gas turbine combustor, referred to briefly in Sect. 1.1, the conventional boiler, the fluidized bed combustor and the air heater combustor used for indirectly fired heat engines. Again, the materials problems are sufficiently system-specific that it is sensible to treat each combustor system separately.

2.1 *Gas Turbine Combustors*

Directly fired gas turbines in the USA generally use a combustion chamber which is in line with the engine, situated between the compressor and the turbine itself. All aircraft gas turbines are of this type. The earliest form of combustor was a can, cylindrical in cross section; a typical engine would have eight cans spaced symmetrically round the engine. From the can, the gases are ducted to the nozzle guide vanes through a duct whose cross section changes from circular to sectoral along its length; this is called the transition piece. The can combustor may be direct flow, in which case the direction of gas flow through the combustor is the same as the overall direction of flow through the turbine, or reverse flow.

In recent years, the combustors used in some aircraft engines have changed to an annular design, in which a single annular chamber surrounds the engine. However, there are still discrete fuel nozzles—typically again eight—and the flame pattern may be stabilized by small cans within the annulus. No industrial gas turbines have used annular combustors to date. Some gas turbines, notably some of the European industrial units, use external combustors. The compressed air is ducted to a large combustion chamber standing to the side of the engine, and the hot combustion gases are ducted back. Typically, there are two such combustors to an engine and they are usually arranged vertically, firing downwards, although this is not essential.

The fuels used by gas turbines include kerosene and other light petroleum-derived distillates, no. 2 light fuel oil, no. 6 heavy fuel oil, residual oil, crude oil, natural gas and blast furnace gas. Some limited experience has been gained with coal gas, coal-derived liquids and methanol. There have been several efforts to use coal and peat in directly fired gas turbines, but with no real success. The fuels vary in their combustion characteristics and in their luminosity, and these factors have an effect on the combustor design. In addition, the exhaust from a gas turbine normally contains a relatively high concentration of nitrogen oxides (NO_x); attempts to limit these emissions may also have an effect on combustor design.

As explained in Sect. 1.1, the temperature of the exhaust gas is determined by the permissible turbine inlet temperature; as a consequence there is usually a very great excess of air over that required for stoichiometric combustion. In fact, the excess is usually so great that the fuel will not burn, so only part of the air is injected into the primary combustion zone. The burner nozzle for liquid fuels contains an atomizer to break up the liquid stream into fine droplets and swirl vanes to promote good mixing of the fuel and air. This component is usually investment-cast.

The combustor can is designed to stabilize the flame at the center of the can and to promote good mixing of the hot combustion gas and the cooler secondary air. Ideally, there should be no temperature variation across the exit from the transition piece; the actual temperature variations are quantified by a pattern factor. As a result, the inner combustor can is quite complex. Three different kinds of hole are present: relatively large holes through which dilution air enters; smaller holes whose primary function is to stabilize the flame; and skin cooling holes. The inner can is made in rings, each ring fitting over the one next to it. The inner mating ring may have its end corrugated, so that air is encouraged to flow parallel to the can walls. This is called a wiggle strip. Alternately, there may be an investment-cast ring joining the sections, with holes through it parallel to the can axis performing the same function. The inflowing air is the cooling flow over the inner walls. The incoming air flows over the outside of the inner can and inside the outer can, and cools the outside. The annulus spacing and the size of the different holes have to be carefully calculated so that the proper flows pass through each of the holes.

As a result, combustor cans are usually made of alloys which can be rolled to sheet. The early Nimonics, such as Nimonic 80A, and other relatively early nickel-base superalloys such as Hastelloy X (Ni–22%Cr–9%Mo–18%Fe–1.5%Co) and Waspalloy (Ni–20%Cr–1.4%Al–3%Ti–4%Mo–13%Co) are possibilities, and one or two cobalt-base alloys, notably L-605 (Co–20%Cr–15%W–10%Ni) and Haynes 188 (Co–22%Cr–22%Ni–14%W–1.8%Fe–0.04%La). In the early days, it was sometimes also required that the combustor can should be field weld-repairable, since a common mode of failure was a flame instability which would burn a hole in the wall. However, this is now seldom a requirement, the practice being to replace the can and shop-repair the damaged unit.

The combustor experiences a very large thermal shock on startup and shutdown. In addition, there is a high acoustic loading in operation. High-temperature high-cycle fatigue and low-cycle thermal fatigue should thus be the major failure modes; the presence of all the holes in the can also favors these modes. With higher turbine inlet temperatures, the thermal loads increase. There is relatively less air available for cooling. Higher luminosity flames from burning lower grade fuels or coal-derived liquids also increase the thermal loading on the combustor walls. There are several different approaches to this problem.

First, the thermal barrier coating referred to in Sect. 1.1 can be applied to the combustor walls. This has been done with a considerable degree of success. Next, advanced cooling techniques can be used for the walls of the combustor. The Lamilloy system, developed by Detroit Diesel Allison, is a possibility. Sheet is made from three laminations, which have grooves etched into them forming convoluted passages when they are bonded together; in addition, small holes through the outer layer of the sheet lead out cooling air to form a cool gaseous boundary layer. Inconel 617 (Ni–22%Cr–1.2%Al–0.3%Ti–9%Mo–1.5%Fe–12.5%Co–0.2%Cu) is the current alloy used to make these structures, but there are other possibilities. It would be possible to make a water-cooled combustor using techniques similar to those developed for the water-cooled nozzle guide vanes.

Finally, it may be possible to make a combustor wholly or partly from ceramics. While ceramic rotating blades present many problems, and even ceramic stator blades are difficult, the ceramic combustor problem might be simpler. Two designs, one composed of stacked rings and the other of "barrel-stave" panels, have been proposed. However, obtaining proper air flows and pattern factors in these combustors is a problem. It might be possible to use a ceramic flame tube; this could stabilize the flame, minimizing the risk of flame impingement, and also limit the radiation from a high-luminosity flame to the combustor can wall.

All these problems are significantly reduced in the large external combustor. It would not be impossible to line the whole unit with refractory bricks or tiles, for example, and the flame stability problem is nowhere near as great. However, the hot gas ducting presents a set of problems of its own, and arranging a proper distribution of gas through the nozzle guide vanes is difficult, involving large hot gas scroll pieces. The hot gas path would be extremely difficult to cool and thus it is difficult to see how further increases in the turbine inlet temperature would be possible.

Low-NO_x combustors are based either on staging the combustion or on the use of catalysts. Staged combustion can present more severe corrosion conditions in the primary zone. With catalytic combustors, the catalyst (usually platinum) is supported by a ceramic honeycomb. Several commercial systems exist, but long-term durability at full size has still to be demonstrated.

2.2 Conventional Boilers

A liquid or gaseous fuel boiler injects the fuel through a burner nozzle and mixes it with air, usually preheated to 200–225 °C. Combustion takes place, producing gas at a temperature in the region of 1500–1800 °C. Coal-fired boilers have a variety of firing systems. Large utility boilers pulverize the coal to <74 μm and blow it in through burners in a manner essentially similar to that employed for oil. Smaller units may fire lump coal on grates, using either a fixed grate with a spreader stoker or a moving chain grate stoked from one end with the ash removed at the other. Various other schemes have been used.

Heat is extracted from the hot combustion gas by different stages of heat exchangers. Commonly, the walls of the combustion chamber are formed of tubes linked by membranes; water flows through the tubes and heat is transferred from the hot combustion gases by radiation. From the combustion chamber the cooler combustion gases may pass through tubes round the outside of which passes water (fire tube boiler), or pass over banks of tubes through which pass water or steam (water tube boiler). In both cases, the heat transfer is by convection. For smaller industrial boilers the fire tube design is preferred; for large utility boilers the water tube design is universally used.

If the steam temperature and pressure are such that it is in equilibrium with the liquid phase, the steam is referred to as saturated steam; if the temperature is above this, the steam is superheated. At a pressure of 22.1 MPa, water changes to the vapor phase at a temperature of 374 °C without any distinguishable boundary; this is called a critical point. Boilers producing steam at lower pressures than this (the large majority) are called subcritical boilers; above this pressure the boiler is referred to as a supercritical boiler. In subcritical boilers the water boils somewhere in the heat exchangers, and beyond this point there is a two-phase mixture of water and steam.

As the temperature rises, the proportion of water diminishes, but often it is convenient to pass the two-phase mixture to a steam drum, whose purpose is to separate the two phases. The water-free saturated steam then goes on to the superheater heat exchangers; the water is returned to the boiler. Commonly, the steam drum is at the top of the boiler and the recirculating water passes down through pipes called downcomers to re-enter the water walls; there may be another smaller drum at the bottom called the mud drum. In a supercritical unit there is no need for a steam drum, and the water passes once through the boiler only. It is possible to dispense with the steam drum in a subcritical unit as well, but there are relatively few once-through subcritical boilers. Often, the many tubes of the water wall or of the tubular heat exchangers are manifolded from longer horizontal tubes; these are called headers.

Finally, the steam is taken from the boiler to the point where it is to be used; this is commonly a steam turbine, and the large diameter steam lines are commonly referred to as pipes, to distinguish them from the tubes in the heat exchangers. In a utility boiler, the superheated steam is expanded in the high-pressure turbine; it is then returned to the boiler and reheated in heat exchangers, albeit at a much lower pressure. The reheated steam is expanded through the intermediate-pressure, or reheat, turbine. It is possible to reheat several times; there are a few systems with two stages of reheat, but beyond this point there is little advantage. In the USA, most utility boilers are subcritical, producing steam at 16.5 MPa and 538 °C and reheating once to 538 °C at ~6.9 MPa. The small number of supercritical units produce steam at 24.1 MPa and 538 °C and reheat twice to the same temperature.

The materials for tubing, piping and steam drums are regulated in the USA by the ASME Boiler Code, which is related to mechanical properties requirements. The following criteria are listed in paragraph A-150 of Section 1 for consideration by the Code Committee in establishing allowable stresses. The *lowest* criterion establishes the allowable stress.

(a) 25% of the specified minimum tensile strength at room temperature,

(b) 25% of the tensile strength at the use temperature,

(c) 67% of the specified minimum yield strength at room temperature,

(d) 67% of the yield strength at the use temperature,

(e) 100% of the stress required to produce a creep of 0.01% in 1000 h (a creep rate of approximately $3 \times 10^{-11}\, s^{-1}$),

(f) 67% of the average stress to produce rupture at the end of 100 000 h, or 80% of the minimum stress for rupture as determined from extrapolated data, whichever is lower.

Figure 2 shows the use of these criteria to establish the allowable stresses for a 2.25%Cr–1%Mo steel. Up to ~400 °C, the first criterion determines the allowable stress. Above 510 °C, the creep and rupture strengths are limiting, and these fall very rapidly with increasing temperature. Figure 3 illustrates the allowable stresses for a number of steels used for tubing in US boilers. For a new material to be used, it is necessary for experimental data to be provided to the Code Committee to allow them to assess it according to the above criteria; in practice, an alloy cannot be used in the USA until approved. Other countries have generally similar procedures for estab-

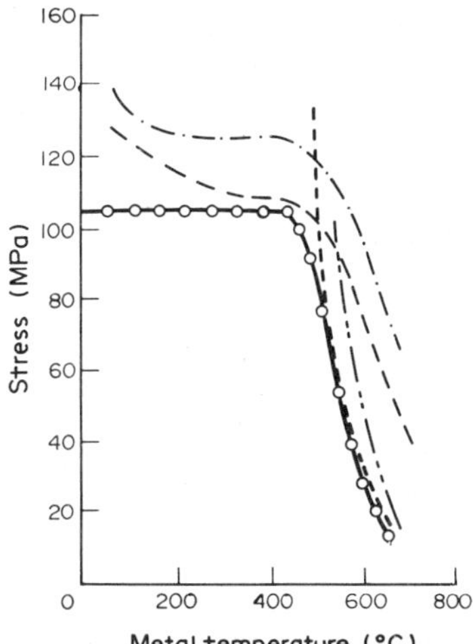

Figure 2
Use of ASME Boiler Code criteria to establish allowable stress for a 2.25%Cr–1%Mo steel (SA-213 Grade T22): ——— maximum allowable stress; —·— 0.666 yield strength; — — — 0.25 tensile strength; — - - — stress for rupture in 100 000 h; - - - - stress for creep rate of 0.01% in 1000 h (after Singer 1981)

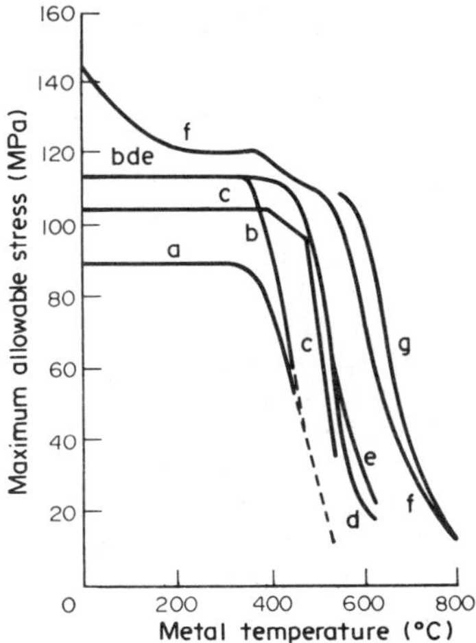

Figure 3
Effect of temperature on ASME Boiler Code allowable stresses for steel tubing: (a) carbon steel SA-192; (b) carbon steel SA-210 A-1; (c) C–0.5%Mo SA-209 T-1; (d) 1.25%Cr–0.5%Mo SA-213 T-11; (e) 2.25%Cr–1% Mo SA-213 T-22; (f) 18%Cr–8%Ni SA-213 TP304H; (g) 18%Cr–10%Ni–Nb SA-213 TP347H (after Singer 1981)

Table 2
Materials used in the construction of US boilers

	ASME specification	Grade	Approximate composition	Max. allowable metal temperature (°C)
Tubes				
Seamless carbon steel	SA-192			
	SA-210	A-1		
	SA-210	C		
Seamless alloy steel	SA-209	Tla	0.5%Mo	
	SA-213	T2	0.5%Cr–0.5%Mo	
		T5	5%Cr–0.5%Mo	
		T11	1.25%Cr–0.5%Mo	
		T3b	2%Cr–0.5%Mo	
		T22	2.25%Cr–1%Mo	
		TP304H	18%Cr–8%Ni	
		TP321H	18%Cr–10%Ni–Ti	
		TP347H	18%Cr–10%Ni–Nb	
Pipes				
Seamless carbon steel	SA-106	B or C		425
	SA-266	11		425
Seamless ferritic alloy steel	SA-335	P2	0.5%Cr–0.5%Mo	510
		P12	1%Cr–0.5%Mo	540
		P11	1.5%Cr–0.5%Mo	560
		P22	2.25%Cr–1%Mo	595
		P5	5%Cr–0.5%Mo	620
		P9	9%Cr–1%Mo	650
Seamless austenitic steel	SA-335	TP304H	18%Cr–8%Ni	815
		TP316H	16%Cr–12%Ni–2%Mo	815
Forged and bored austenitic steel	SA-430	FP304H	18%Cr–8%Ni	815
		FP316H	16%Cr–12%Ni–2%Mo	815
Plate steels (for drums and headers)				
	SA-285		0.22%C	
	SA-515	55	0.2%C (varies with section) 0.2%Si	
		60	0.3%C 0.2%Si	
	SA-302	B	0.2%C 1.2%Si 1.3%Mn 0.5%Mo	

lishing acceptable materials, but there are differences in detail. Table 2 lists the materials commonly used in the construction of US boilers.

In some cases, other criteria determine the use of alloys. Thus, plain carbon and low-alloy steels tend to graphitize and lose strength after long exposures at elevated temperatures; on these grounds, 400 °C is the maximum permissible use temperature for plain carbon steels and 470 °C for 0.5%Mo steel. Current practice is to use chromium–molybdenum steels (for example, 0.5%Cr, 0.5%Mo) because of their improved resistance to graphitization. In some cases, the oxidation resistance of steels determines their maximum use temperature, although in the USA this is a matter of boilermakers' practice rather than a code requirement. Table 3 lists the oxidation limit temperatures for a number of steels. Note that this is a metal surface temperature, rather than a mean metal temperature as is the case for the strength criteria; the former should be reduced by approximately 20 °C to make it comparable with the latter.

The additional problems in boiler materials are:

(a) waterside corrosion

(b) hydrogen embrittlement

(c) fireside corrosion of the water walls

(d) fireside corrosion of the superheater

(e) steam-side oxide exfoliation

(f) erosion by soot blowers

Table 3
Maximum permissible temperature of the surface in contact with flue gas for several tube steels

ASME specification	Grade	Maximum surface temperature (°C)
SA-192		510
SA-210	Al	510
SA-210	C	450
SA-178	A	480
SA-178	C	510
SA-209	Tla	525
SA-213	T2	550
SA-213	T11	565
SA-213	T22	580
SA-213	T9	650
SA-213	TP304H	760
SA-213	TP321H	760

(g) erosion of the economizer

(h) dissimilar weld failures

(i) cold and corrosion

(j) fan erosion

(k) scrubber corrosion

Detailed analysis of all these problems is beyond the scope of this article. In the section of the tubes where boiling takes place, the intention (in subcritical units) is for the boiling to occur by the nucleation of streams of bubbles, the two-phase mixture remaining fairly intimate. However, it is possible for continuous films of steam to form over the inside of the tubes in regions of high heat flux and poor circulation. This is called departure from nucleate boiling (DNB), and there are two consequences. Because the heat transfer of the steam film is much poorer than that of the water, the metal temperature can rise considerably and rupture of the tubes may result. Although boiler water is extremely pure, it may still be possible for salts to precipitate out at the steam–water interface and for accelerated corrosion to occur beneath the deposit. Corrosion processes may also inject hydrogen into the steel, embrittling it and resulting in characteristic tube failures with no thinning associated with the fracture.

In pulverized-coal-fired boilers, the incombustible mineral matter is usually melted in the combustion zone. Some falls down and is removed through an opening in the bottom of the furnace, but some—the fly ash—is entrained with the combustion gases. Deposits form on the superheater tubes at the top of the radiant section, and these have two effects. The heat transfer is reduced, with an adverse effect on the overall boiler performance. The deposit has to be removed periodically by jets of air or steam through lances, and if this is done carelessly, dust particles entrained in the soot-blower jets can result in severe erosion of the superheater and reheater tubes. If the coal is high in the alkali metals sodium and potassium, rapid corrosion can take place under the deposits. This can be combatted by wrapping the tubes with bandages of (for example) type 310 stainless steel (Fe–25%Cr–20%Ni) or, in extreme cases, using a coextruded tube with an inner layer of a strong material and an outer layer of a corrosion-resistant material such as type 310 or Inconel 671 (Ni–50% Cr).

Fireside corrosion of the water walls arises because of a displacement of the combustion zone, with the result that regions of low oxygen activity impact the water walls. Slag deposits containing unburnt carbon and unoxidized iron sulfide form and rapid wastage of the relatively cool (350–400 °C) steel tubes may result. The principal reaction appears to be sulfidation. Some investigators believe that chlorine plays an important part in this reaction. Generally, the best method of control is to correct the combustion pattern; air bled into the furnace along the wall helps. Coextruded tubes could be used and in some cases plasma-sprayed coatings have been applied.

The fly ash can cause erosion of the various heat exchanger tubes. Boilers are usually designed so that the mean free cross-sectional area gas velocity does not exceed ~20 m s^{-1} for an ash which is not particularly erosive; for some very erosive ashes this value may be as low as 10 m s^{-1}. The most severe erosion is of the cold end components: the economizer heat exchanger, which preheats the water before it enters the water walls, and the induced draft fan which draws the gas through the boiler. In severe circumstances, hardfacing protection is used, for example, by weld-overlaying a material such as Stellite 6B. More sophisticated protection systems are currently being developed.

Steam-side oxidation of the superheaters and reheaters is not particularly significant from the point of view of metal wastage, but the magnetite (Fe_3O_4) scale can spall and be entrained in the steam, resulting in erosion of the steam turbine, as mentioned in Sect. 1.3, or sometimes blockage of the pendant superheater tubes, leading to overheating and rupture. This can be combatted by chromizing or applying a chromate conversion coating to the inner surface of ferritic tubes, or by cold-working the inner surface of austenitic tubes.

Because of the restriction on the operating temperature of material for the various reasons listed above, it is often necessary towards the end of the superheater (or less frequently the reheater) exchanger to change over from the ferritic T22 steel to an austenitic, such as type 304 stainless steel. This weld presents a number of problems. The coefficients of thermal expansion differ and this can induce stresses in the region of the weld. The austenitic steel has a much higher solubility for carbon than the ferritic, so during welding the heat-affected zone of the T22 may be decarburized, resulting in a loss of strength. The combination of the local stresses and the local weakening can result in failures. Early practice was to use an austenitic weld metal, type 309 stainless steel (Fe–22%Cr–13%Ni–2.5%Mn–1% Nb+Ta), but to prevent the carbon loss, nickel-base weld metals, which have relatively low carbon solubilities and intermediate coefficients of thermal expansion, have been developed. The most frequently used nowadays is Inconel 82 (Ni–20%Cr–0.75%Ti–3%Fe–3%Mn–2.5%Nb+Ta). An alternative is to use a special transition piece, called a "dutchman"; this is more common in Europe than in the USA.

2.3 Fluidized Bed Combustors

If air is blown through a bed of particles with gradually increasing velocity, at first it simply permeates the pores in the fixed bed. As the velocity increases, the particles start to move a little and then they levitate and begin to move quite substantially and essentially randomly. The bed is then said to be fluidized, and in many ways it behaves like a liquid: the surface will remain horizontal when the bed is tilted; dense objects sink, less dense objects float, and so forth. At slightly higher velocities the excess air over that required for fluidization forms voids in the bed, resembling bubbles. The bed resembles a boiling liquid. At still higher velocities, the distinction between the bubbles and the dense "emulsion" phase vanishes and the particles are in rapid turbulent motion. Finally, as the velocity exceeds the terminal velocity of the particles, they become entrained in the air stream.

If a fluidized bed is heated and a combustible material is introduced, the latter will burn in the air stream. The bed has excellent heat transfer characteristics, with an apparent thermal conductivity many times that of copper, so the heat of combustion is rapidly transported away from the burning material. As a result, the combustion temperature can be kept low; this limits the production of nitrogen oxides, prevents slagging of the ash and may limit the liberation of the alkali metals from the clay minerals. The very efficient solids–gas mixing in the bed also encourages chemical reactions to occur, so that if limestone is added to the bed, sulfur is efficiently converted to calcium sulfate. In addition to these environmental benefits, the efficiency of the fluidized bed as a chemical reactor means that fuels of very low heating value can be burnt.

To keep the bed temperature low, the heat of combustion must be removed. This can be done, as in the gas turbine combustor, by increasing the flow of air above that required for stoichiometric combustion; this system is termed an adiabatic fluidized bed. More commonly, heat is removed by a heat exchanger immersed in the bed; the heat transfer in the bed is very efficient, so in spite of the fact that the bed temperature is relatively low, the total heat exchange surface required in the bed for a given heat removal is usually much less than in the gas passes of a conventional boiler.

There are two modes of operation. In the first, the overall system operates at atmospheric pressure (the AFBC system). The heat of combustion is extracted by heat exchangers in the walls of the combustor, in the bed and in the gas passes. Usually the combustor is a boiler and the steam is superheated in an in-bed heat exchanger. However, it is also possible to use the heat exchanger to heat air, which may be used for space heating or to drive a gas turbine. In the second mode, the system is pressurized (the PFBC system). It is still possible for it to be essentially a boiler; the pressurization has the effect of reducing the size of the unit. More commonly, the energy in the combustion gas is recovered by expanding it through a turbine which may simply drive the compressor or may also produce power. This can be done with an adiabatic bed, but more generally there are heat exchangers within the bed. These may raise steam for a combined cycle system, or heat air which can be mixed with the combustion gases to drive the turbine.

The materials problems, which are different from those described in the preceding sections, are concerned with the in-bed heat exchangers and the gas turbine expander. The environment within the bed is quite complex. It appears that the "emulsion" phase contains burning coal particles and the transport of air to them through the emulsion is relatively slow. As a result, the oxygen activity within the emulsion phase is low: at a mean bed temperature of ~850 °C, oxygen activities as low as 10^{-10} Pa have been measured using solid electrolyte stabilized zirconia oxygen probes. The bubbles on the other hand, have oxygen activities which are higher: at the bottom of the bed they are close to the composition of air. In the presence of calcium sulfate in the bed, the equilibrium sulfur partial pressure at the lower oxygen activities can be quite high, perhaps as high as 1 Pa. At any given point the oxygen activity (and thus, presumably, the sulfur activity) fluctuates quite rapidly.

Under these circumstances, sulfidation–oxidation of alloys may occur. Nickel-base alloys may undergo catastrophic corrosion within the bed, producing liquid nickel sulfides. The effect on conventional stainless steels, such as types 304 and 347, is much less marked. Subsurface sulfides do form; initially these are manganese-rich, but later chromium-rich sulfides appear. In the early stages they are often present as small discrete particles, and these do not appear to affect the oxidation. However, they may begin to develop along grain boundaries and oxidation penetration may develop in their wake. Under these circumstances, the matrix of the alloy may be depleted in chromium and rapid attack may result at longer times.

In addition, the low-alloy ferritic steels such as Fe–2.25%Cr–1%Mo (T22) appear to oxidize rather more rapidly in the bed than they do in the gas pass of a conventional boiler, implying that the oxidation limit temperature may be lower. The weld metal used for austenitic–ferritic welds is nickel based and should not be used for in-bed welds. Uncooled supports present a problem. The candidates appear to be: type 347 stainless steel, although this may be close to its limit; type 310, although this suffers from sigma phase embrittlement (typical bed temperatures of 850–900 °C are as bad as they could be for sigma formation); and perhaps Incoloy 800, although this alloy is at best marginal. If these materials prove to be inadequate, it may be necessary to use a corrosion-resistant cladding: an alloy such as GE 2541 (Fe–25%Cr–4%Al–1%Y) would be suitable.

In addition to the sulfidation–oxidation corrosion, there may be alkali-metal-sulfate-induced hot corrosion of the gas turbine expander in PFBC systems. Although the alkali release is generally less than that in a pulverized coal boiler, the concentration in the combustion gas may still be two orders of magnitude higher than is recommended by the gas turbine manufacturers. Generally, the deposits are higher in potassium than those usually encountered in gas turbine hot corrosion. The actual situation in practice is not well defined, because no large gas turbine has as yet been operated with a PFBC. Some cascade experiments have revealed no hot corrosion in the efflux from small PFBCs; however, in one or two cases quite severe hot corrosion has occurred. The remedy for this problem, should it prove to be a real one, will be to use coatings or claddings, as discussed in Sect. 1.1.

A further problem in FBC systems is erosion. The gas pass erosion should not be qualitatively different from that in pulverized coal boilers, although (depending on the location of the heat exchangers in the system) the particle loading may be as much as an order of magnitude higher and the character of the dust is somewhat different. In a PFBC system, the erosion of the gas turbine by the entrained dust may be a problem. It appears that if the dust loading can be reduced to levels in which <0.1 ppm by weight in the gas is particles >20 μm diameter, <1 ppm in the range 10–20 μm and <10 ppm in the range 4–10 μm, the erosion of the turbine should be tolerable. This is quite a stringent requirement. The gas is cleaned by inertial separators—cyclones—but the required removal is on the borderline of what can be achieved by three stages of cyclones.

Work is currently in progress to develop other hot-gas cleanup techniques, including electrostatically enhanced granular bed filters and woven ceramic filter bags, but these have yet to be demonstrated on the full scale. Some design changes would improve the ability of the turbine to tolerate particles: lowering the maximum relative gas velocities, increasing the leading edge radius and the trailing edge thickness of the airfoils and designing the inlet volute to minimize inertial separation of the particles. Finally, early work directed at the construction of a coal-burning gas turbine showed that some benefits could be obtained by the use of erosion-resistant coatings and by carbide inserts at critical locations. Research is currently in progress to identify improved protection systems.

An interaction between erosion and hot corrosion has been demonstrated. In principle, two situations could be envisaged:

(a) the overall degradation is diminished, either by the erosive component preventing the buildup of a corrosive deposit or the corrosion product resisting the erosion better than the alloy; or

(b) the overall degradation is enhanced, either by the erodant removing the protective oxide, allowing access of the molten salt to the metal, or the corrosion producing a more erodable surface layer.

In practice, the second situation has been demonstrated clearly, although it has not been possible to distinguish between the two suggested mechanisms.

The data available at present suggest that the best material will be a relatively thick coating or cladding of an alloy such as GE 2541 (Fe–25%Cr–4%Al–1% Y), which forms a strong, adherent, erosion-resistant Al_2O_3 scale, has high chromium to resist hot corrosion and is relatively soft and ductile at elevated temperatures, making it more tolerant of the particle impacts.

See also: Materials for Energy Applications: An Overview

Bibliography

Fairbanks J W, Stringer J (eds.) 1979 *Proc. 1st Conf. Materials for Alternative-Fuel-Capable Heat Engines*. US Department of Energy, Washington, DC

Fairbanks J W, Stringer J (eds.) 1982 *Proc. 2nd Conf. Materials for Alternative-Fuel-Capable Heat Engines*. US Department of Energy, Washington, DC

Institution of Metallurgists 1975 *Materials in Power Plant*, Spring Residential Course, Series 3, No. 3. Chameleon Press, London

Kear B H, Muzyka D R, Tien J K, Wilodek S T (eds.) 1976 *Superalloys: Metallurgy and Manufacture*. Clantor's Publishing, Baton Rouge, Louisiana

Meetham G W (ed.) 1981 *The Development of Gas Turbine Materials*. Applied Science, London

Sims C T, Hagel W C (eds.) 1972 *The Superalloys*. Wiley, New York

Singer J G (ed.) 1981 *Combustion: Fossil Power Systems*. Combustion Engineering, Windsor

Wyatt L M 1976 *Materials of Construction for Steam Power Plant*. Applied Science, London

J. Stringer

Foundry Sands: Classification and Sources

Sands are defined as granular particles of diameter 0.05–2 mm that result from the disintegration or crushing of rocks. The most widely used foundry sand, silica sand, is composed mainly of quartz (SiO_2) grains; however, sands based on zircon, chromite, olivine and ground ceramic minerals are also employed in the production of castings (see *Foundry Sands: Silica Sand and Nonsilica Minerals*). In foundry practice, a wide range of sands is used which vary in purity, structure, refractoriness and also in grain size, shape and distribution.

1. Classification

Foundry molding sands are generally classified into two broad categories: naturally bonded sands and unbonded sands. The naturally bonded molding sands are water- or wind-deposited sands which are naturally combined with clay minerals. They occur with a wide range of grain fineness and clay contents depending on their source and origin. The processing of this type of sand consists primarily of screening to remove oversize particles, roots and other deleterious material. For some sands, this treatment is all that is required to meet specifications. Other sands may be dried to approximately 6% moisture and then passed through mullers, cage mills or dry pans to mill the sand and distribute the natural bond. Sands from different pits or different strata in the same pit, having different compositions, are often blended with or without clay additions to produce a range of sands having consistent uniform properties. Sand with a wide range of properties can be produced; specifications are negotiated between producer and consumer.

Many unbonded sands are referred to as "washed-and-dried" silica sands. These sands are prepared for use in casting production by being bonded with the required amount of binders, additives and "temper water." The total sand mixture is prepared in a sand muller in order to gain uniform distribution and dispersion of the clay and additives around the individual sand grains.

2. Mode of Occurrence

Foundry-sand deposits occur naturally as a result of the geological sedimentary processes of weathering, erosion, transportation by wind, water and ice, and deposition. The composition of the deposits depends on the nature of the materials eroded and the manner in which they were deposited. For example, some silica sands were deposited by rivers, whereas others were deposited along coasts of ancient seas.

2.1 Consolidated Deposits

In many places, individual grains of sand have been cemented into a consolidated rock known as sandstone, and these deposits have been covered by other layers of rock. The normal processes of uplift, faulting and erosion eventually cause these sandstones to be exposed at the surface. Where erosion of the material covering the sandstone has been extensive, only a thin layer of overburden will remain, making the sandstone much easier to mine.

2.2 Unconsolidated Deposits

Unconsolidated deposits with a low clay content are generally referred to as "bank sands," although they are usually subjected to considerable processing and blending before shipment. Bank sands are a product of the disintegration of sandstone or other rocks by weathering and erosion. These sands were transported by streams or blown by wind over vast areas and heaped into small banks. Bank sands vary in purity, depending on the foreign materials and minerals with which they are mixed. In many areas they are of high purity and suitable for foundry use. In the processing of bank sands, after stripping of overburden, the sand is hauled to the plant. A front-end loader dumps the sand into a hopper, from where it is conveyed to a rotary screen with 13 mm openings to remove roots. Sand for use as a system sand addition and as a synthetic sand base is dried and shipped without further processing. Where a cleaner and more consistent product is required, the screened sand is washed. The sand to be washed is dropped into a surge bin to assure a uniform feed and then conveyed to a tank where it is mixed with water before being pumped to a scalping screen where fine roots and organic materials are removed. The sand is then pumped to a deslimer to remove the clay and silt. Fresh water is added, and the sand slurry is either pumped to a second deslimer and then to ground storage, or it may be fed to banks of screens to separate it into fine and coarse particle sizes. Each sand grade is collected in a sump, mixed with water, pumped to a deslimer and dewatering cyclone, and piled for drainage and storage. The sand is then dried, cooled and passed over a final scalping screen before being conveyed to storage bins ready for shipment.

Unconsolidated deposits composed of sands produced by the erosion of rock along the shores of lakes are known as "lake sands." (This term also includes dune sands formed from surface sand by wind action.) In a typical lake-sand processing operation, the overburden is first removed and then a front-end loader is used to mine the sand and feed a hopper which discharges to a portable screen in order to remove deleterious material. Conveyor belts transport the sand to the plant site for further processing, and the hopper and screen can be moved along these as required.

During the natural deposition process, classification by size, composition and specific gravity can take place. In general, the degree of purity of a deposit is dependent on the purity of the parent material and the history of its deposition. Based on their origin, sands vary in grain size, shape, composition, surface texture and distribution of sizes. These physical properties, in addition to chemical composition, sintering point and expansion characteristics, are important considerations when using sands as the base molding or core aggregate in metal casting.

See also: Casting of Metals

Bibliography

Ammen C W 1954 Preparation of synthetic sand. *Foundry* 82: 202–08

Goettman F P 1947 Preparation of foundry sands for market. *Trans. Am. Foundrymen's Soc.* 55: 490–92

Goettman F P 1975 Production of sands for the foundry industry. *Trans. Am. Foundrymen's Soc.* 83: 15–24

Hardy T H 1954 Natural sands of mid-America. *Foundry* 82: 114–15

Wilborg H E, Henderson G V 1983 Foundry sand. In: Lefond S J (ed.) 1983 *Industrial Minerals and Rocks*, 5th edn., Vol. 1. American Institute of Mining, Metallurgical, and Petroleum Engineers, New York, pp. 271–78

Winter W P 1960 Molding sand. *Foundry* 8: 103–05

E. L. Kotzin

Foundry Sands: Silica Sand and Nonsilica Minerals

Refractory materials to make molds and cores have been very important to foundrymen since metal casting began hundreds of years ago. Early requirements were simple, but as core and mold sand technology has become more sophisticated and binder technology continues to be developed, the physical properties of foundry sands have become more critical.

1. Silica Sand

Over the years, the most commonly used foundry sand has been silica sand. This is not surprising because it is the most abundant and one of the most easily mined minerals at the earth's surface. Silica sand and other foundry materials are produced from rocks which formed under a variety of geologic conditions and environments. Silica is the common term applied to SiO_2 in the mineral form of quartz. In addition to foundry sands, quartz is used in a number of industrial applications, including abrasives, aggregates, sand blasting abrasives, fillers, filters, well-drilling materials, electronic and optical minerals and refractories (see *Abrasives: Siliceous Minerals*; *Sand and Gravel*; *Fillers and Coatings: Siliceous Minerals*; *Filters, Sorbents and Ion Exchangers: Siliceous Minerals*; *Well-Drilling Materials: Industrial Minerals*; *Electronic and Optical Minerals*).

Although silica sand has the advantages of abundance, ease of bonding with organic or inorganic binders, low cost and ability to be reclaimed for reuse by wet, dry or thermal methods, it also possesses certain disadvantages when used in the production of metal castings. Undoubtedly the major disadvantage of silica sand is its high thermal expansion. It is this expansion behavior which causes casting quality problems (e.g., rat-tails, buckles and scabs), as well as contributing to other expansion-type defects. Figure 1 illustrates the typical thermal expansion of various silica materials, and depicts α-quartz—a polymorph of quartz—expanding at a constant rate until a temperature of nearly 600 °C is reached. As the temperature increases above this point, a sudden expansion takes place due to the change in crystal form from α-quartz to β-quartz. This high thermal expansion requires carefully controlled additions of cushioning materials (e.g., cellulose additives) to minimize the deformation and rupture of mold surfaces in contact with molten metal. In addition, silica sand is unable to resist metal penetration and reaction when in contact with casting surfaces where there are re-entrant angles in hot-spot areas of large iron and steel castings, or when the steel contains high amounts of manganese (e.g., Hadfield's austenitic manganese steel castings) which wet and attack the silica sand mold surfaces. In addition, considerations of occupational health place increased stress on silica sand's potential as a cause of silicosis (see *Silicon Dioxide Hazards*).

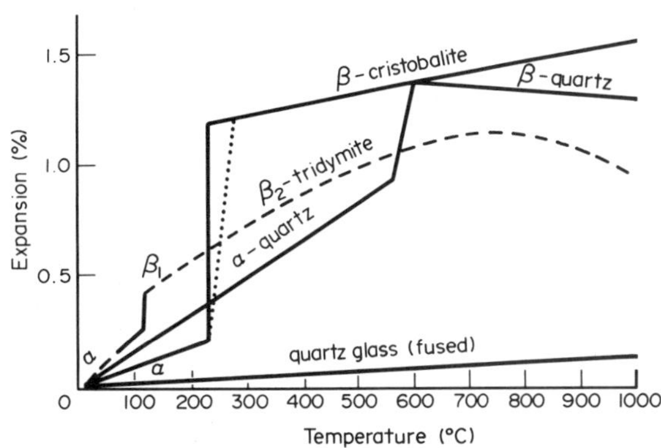

Figure 1
Thermal expansion properties of silica materials

The use of silica sands for metal casting has been the subject of studies by many investigators. However, the continued demands of casting purchasers for improved cast surfaces and closer dimensional tolerances have also emphasized some of the physical and chemical limitations of silica sands and their inability to provide all of the properties required to meet these quality levels.

2. Nonsilica Minerals

A number of speciality sands of material other than silica possess properties which make them superior to silica sand for certain applications. The major nonsilica sands, zircon, chromite and olivine, have gained wide recognition and use throughout the metal-casting world for their individual properties. Although other nonsilica molding media such as aluminum silicate sand and staurolite sand also possess unique properties, their use in casting production has been on a much smaller scale. Figure 2 shows the typical thermal expansion data for the nonsilica materials as compared with silica sand. It is readily apparent that the nonsilica aggregates have a much lower rate of thermal expansion than silica sand, and none of the nonsilica minerals undergoes the phase change which is characteristic for silica sand. These data and experience in the use of the major nonsilica minerals indicate that expansion-type defects are greatly reduced and, in many cases, eliminated. Table 1 shows some of the typical properties of the nonsilica sands in comparison with those of silica sand.

2.1 Zircon Sand

The major sources of zircon sand are found in Australia (east and west coasts), the USA (north central Florida), Sri Lanka, Malaysia and South Africa. Zircon and other heavy minerals, such as tourmaline, spinel, kyanite, sillimanite, corundum, topaz and staurolite, are recovered as by-products in mining of the titanium-bearing minerals, ilmenite, leucoxene and rutile.

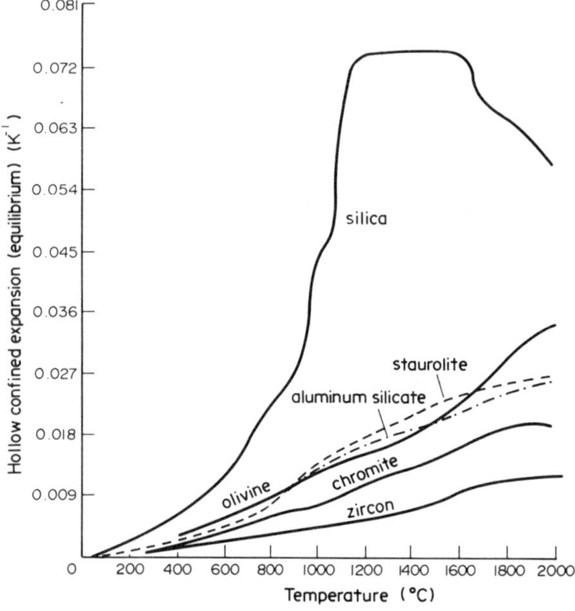

Figure 2
Comparison of thermal expansion properties of silica and nonsilica sands

After stripping of overburden, a typical sand mining operation uses a suction-type dredge capable of producing up to 1100 t of sand per hour. The sand, containing about 4% heavy minerals, is pumped to a wet mill on a floating barge. The wet mill produces a concentrate averaging 85% heavy minerals with a recovery of 80%. Three stages of gravity concentration using spiral classifiers are carried out. The concentrate is then pumped to a land-based dry mill. The scrubbed ore is dried and subjected to various magnetic and electrostatic separations to remove titanium-bearing minerals. The tailings from the titanium separation are fed to high-intensity magnets, which produce sand suitable for foundry use consisting of zircon and staurolite with small percentages of tourmaline, spinel, kyanite and sillimanite.

Zircon sand is chemically and thermally stable and possesses most of the desirable properties for a foundry sand. Its major advantages are as follows:

(a) it possesses the lowest thermal expansion of any foundry sand;

(b) it has a high thermal conductivity and bulk density, giving approximately four times the cooling rate of quartz;

(c) it is completely unwetted by molten metal;

(d) it is chemically unreactive with molten metal;

(e) it requires less binder than any of the other sands when using chemical binders;

(f) it is compatible with all known binder systems (organic or inorganic);

(g) it possesses excellent dimensional and thermal stability characteristics at elevated temperature; and

(h) it has a pH that is neutral or slightly acid.

2.2 Chromite Sand

As indicated in Table 1, the unusual properties of chromite sand lend themselves ideally to fulfilling some of the extreme thermal requirements of heavy-sectioned ferrous castings. In addition, the basic nature of chromite sand also makes its use highly desirable when producing castings of Hadfield's austenitic manganese steel.

Most of the world production of chromite comes from South Africa (Steelport and Rustenburg areas of the Transvaal), the USSR and Zimbabwe, but Albania, Turkey, Finland and the Philippines also

Table 1
Comparison of silica and nonsilica sand properties

Property	Silica	Olivine	Chromite	Zircon	Zircon/aluminum silicate	Staurolite
Origin	USA	USA (Washington, N.Carolina), Norway	Republic of South Africa	USA, Australia	USA (Florida)	USA (Florida)
Color	white–light brown	greenish gray	black	white–brown	black and white	dark brown
Hardness	6.0–7.0	6.5–7.0	5.5–7.0	7.0–7.5	6.5–7.0	6.5–7.0
Dry bulk density ($g\ cm^{-3}$)	1.36–1.60	1.60–2.00	2.48–2.64	2.56–2.96	2.48–2.69	2.29–2.34
Specific gravity	2.2–2.6	3.2–3.6	4.3–4.5	4.4–4.7	3.2–4.0	3.1–3.8
Grain shape	angular/rounded	angular	angular	rounded/angular	rounded	rounded
Thermal expansion	0.018	0.0083	0.005	0.003	0.005[b]	0.007[b]
Apparent heat transfer	average	low	very high	high	high	high
Fusion point (°C)	1427–1760	1538–1760	1760–1982	2038–2204	1815–1982	1371–1538
High-temperature reaction	acid	basic	basic	acid	slightly acid	slightly acid
Wettability with molten metal	easily	not generally	resistant	resistant	resistant	resistant
Chemical reaction	acid-neutral	basic	neutral-basic	acid-neutral	neutral	neutral
Grain distribution[a]	2–5 screens	3–4 screens	4–5 screens	2–3 screens	3 screens	3–4 screens
AFS grain fineness number ranges	25–180	40–160	50–90	95–160	≈80	≈70

a ≥10% retained on USA standard sieve b Clay-bonded sand mixture

produce small amounts (see *Chromite Resources*). Although the USA has a small amount of chromite reserves, over 92% of the material has a high iron content, which is undesirable for foundry purposes. About 60% of the world's production of chromite is consumed in metal production and processing, 20% in refractories, 12% in chemicals and 8% in foundry sand.

Chromite deposits are associated with the basic igneous peridotites and pyroxenes, or with serpentine resulting from the metamorphic alteration of these rocks. Most chromite is mined by underground methods. The ore is drilled, blasted, loaded onto conveyors or mine cars, and conveyed to the processing plant. Hand sorting and/or mechanical screening is used to separate lumps and chips from the sand, followed in some cases by bar or rod milling to reduce the sand to American Foundrymen's Society grain fineness number (AFS GFN) 80 and release the gangue minerals (primarily serpentine and talc). A natural grain size of approximately AFS GNF 55 is generally preferred for metal-casting purposes. The chromite sand is further beneficiated by the use of auger-type washers to remove the slimes, and then gravity concentration using riffling tables or spiral concentrators. The sand is then stored on concrete pads to drain. At this point, the sand is assayed to ensure that it contains a minimum of 45% Cr_2O_3. A turbidity test has been developed to control the working process and assure the production of clean sand. The sand is then transported to a finishing plant where it is dried, cooled and screened to produce grades ranging from AFS GFN 55 to 80. The sand may also be further washed to below 150 ppm turbidity for certain core operations. Quality control tests include the turbidity test, acid-demand values and screen and chemical analyses.

Chromite sand was introduced to the US foundry industry in the early 1960s, when zircon sand was in short supply. Later, in the face of rising zircon sand prices, foundrymen adopted chromite sand as a substitute for zircon sand. Chromite has been successful as a nonsilica foundry sand because it possesses excellent thermal stability and good heat diffusivity characteristics, and is not easily wetted by molten metals, is highly refractory and chemically unreactive with molten metals. One of the major disadvantages in using chromite sands is the presence of hydrous impurities which can contribute to pinhole formation and blows.

2.3 Olivine Sand

Olivine is a constituent of many basic igneous rocks, such as basalt, dunite, gabbro and peridotite, but most commercial production comes from dunite. Large deposits of olivine-bearing dunite are found in Norway, Sweden, the USSR, Austria, Zimbabwe, South Africa and the USA. Norway supplies much of the olivine sand consumed in Europe, while dunite deposits in the states of Washington and North Carolina furnish the major amounts of olivine sand for foundry use in the USA.

In a typical mining operation, rock is broken by blasting and a drop ball to produce fragments less than 1 m in diameter. Ore is passed through a jaw crusher, reduced to <15 cm and stored for feed for the processing plant. At the mill, rock is further reduced to particles <1.5 cm by jaw and gyratory crushers. It is then passed through a screw classifier to wash out clays and slimes and then to a rod mill, from which it is pumped to scalping screens. Oversized material is returned to the rod mill, and undersized material is fed to riffling tables to make a gravity separation of the olivine from the gangue, which is predominantly serpentine and talc. Riffling tables are fed a water slurry containing 25% solids. The amount of water and the tilt and reciprocating speed of the tables control the quality of the product. The olivine concentrate, representing 60% of the table feed, is pumped to a dewatering screen and then to drainage tanks. The sand is dried before screening to produce four basic grades.

Olivine sand is less stable under thermal shock than zircon or chromite sand, but since the thermal expansion of olivine is much less than silica, it has been very successful in nonferrous as well as ferrous casting production where high thermal requirements are not necessary. The basic character of olivine also makes it an ideal aggregate for the production of castings of Hadfield's austenitic manganese steel. The presence of quantities of hydrous magnesium silicates, such as serpentine, may contribute to a pinholing or pockmarking condition when the uncalcined sand is used in the production of low-carbon-steel castings. Because of mineralogical characteristics, olivine sand is less durable than other nonsilica sands.

2.4 Aluminum Silicate Sand

Aluminum silicate (Al_2OSiO_4) is trimorphous and exists in nature as the minerals andalusite, kyanite and sillimanite. Along with zircon, aluminum silicate minerals are produced as by-products of mining of titanium-rich minerals. The three minerals are thermally stable up to a temperature at which they break down to yield 88% mullite and 12% silica (cristobalite). Mullite is stable to 1810 °C and imparts the properties of high refractoriness, low thermal expansion, resistance to thermal shock, intermediate thermal conductivity and resistance to chemical erosion.

The aluminum silicate minerals are processed by means of first scrubbing and washing the sand to remove contaminants and fines to yield a rounded grain sand much like zircon sand that is associated with it. Aluminum silicate sands, being neutral, have been successfully used in the production of Hadfield's austenitic manganese steel castings as well as production of other ferrous metals. These nonsilica sands are also compatible with all known binder systems.

2.5 Staurolite Sand

Staurolite sand is a naturally occurring mineral with the general formula $(Fe, Mg)_2Al_2SiO_4O_{23}(OH)$, and it is produced as a by-product of the same heavy mineral separation which yields zircon and aluminum silicate sands. Staurolite sand consists primarily of the mineral staurolite with small amounts of tourmaline, kyanite, zircon and other silicate minerals. In a typical processing operation, staurolite and the other heavy minerals are removed from the ore body by use of a floating dredge, then concentrated in three stages of spiral separators on a floating wet mill. The heavy mineral concentrates are pumped from a wet mill to a conditioning unit where they are scrubbed with caustic to remove organic and clay coatings from the grain surfaces. After scrubbing, the minerals are rinsed and dried, and the staurolite is removed by means of electric and magnetic separation.

Like other nonsilica minerals, staurolite sand is used in casting production because of its low thermal expansion, refractoriness, durability and compatibility with the various binder systems. Although the thermal expansion of staurolite sand is slightly greater than that exhibited by zircon or chromite sands, it is approximately equal to the thermal expansion of aluminum silicate and olivine sands. Staurolite sands have been used singly as a base aggregate or blended with silica sands in the production of nonferrous castings and ferrous castings of light- to medium-section thickness, although, because of its lower melting point (≈1538 °C), it is not recommended for use in steel-casting production.

See also: Casting of Metals; Foundry Sands: Classification and Sources

Bibliography

American Foundrymen's Society 1978 *Mold and Core Test Handbook*, 1st edn. American Foundrymen's Society, Des Plaines, Illinois

Asanti P 1963 The wetting of quartz and olivine sand by iron at high temperature. *International Foundry Congress, Congress Papers*, Vol. 30. International Committee of Foundry Technical Associations, Zurich

Asanti P 1966 On the interface reactions of chromite, olivine and quartz sands with molten steel. *International Foundry Congress, Congress Papers*, Vol. 33. International Committee of Foundry Technical Associations, Zurich

Beckius K 1967 Chromite sand in the steel foundry. *Foundry Trade J.* 10: 562–66

Beckius K 1973 Olivine sand: Recent research and experience. *International Foundry Congress, Congress Papers*, Vol. 40. International Committee of Foundry Technical Associations, Zurich

Garnar T E Jr 1977 Mineralogy of foundry sands and its effect on performance and properties. *Trans. Am. Foundrymen's Soc.* 85: 399–416

Locke C, Briggs C W, Ashbrook R L 1954 Heat transfer of various molding materials for steel castings. *Trans. Am. Foundrymen's Soc.* 62: 589–99

Marais J J 1966 Foundry techniques using chromite sands. *Mod. Cast.* 50(4): 45–47

Middleton J M, Bownes F F 1970 Properties of chromite and its application in the foundry. *Br. Foundryman* 63: 248–60

Petro A, Flinn R A 1978 Mold–metal interactions between chromite sand and cast steel. *Trans. Am. Foundrymen's Soc.* 86: 357–64

Rassenfoss J A 1977 Mold materials for ferrous castings. *Trans. Am. Foundrymen's Soc.* 85: 583–96

Schaller G S, Snyder W S 1954 Olivine–silica molding sands. *Trans. Am. Foundrymen's Soc.* 62: 413–19

E. L. Kotzin

Fracture and Fatigue of Glass

The design strength of glass is low because it is brittle, and becomes weaker with time. The prediction of usable lifetime from knowledge of glass fatigue is quite uncertain, and necessitates large safety factors in design. Better knowledge of fracture criteria and fatigue life could lead to higher design strengths and much wider application of glass. Special applications of glass, such as in fiber optics, require improved techniques for the prediction of fracture limits and fatigue life.

Until the 1950s, there was little concern given to detailed analytical treatments of fracture and fatigue in brittle solids. The reduction of strength by surface flaws was accepted, but the emphasis in strength studies was on experimental measurements of the effects of sample shape, size and composition on strength. More recently, as a result of the recognition of the importance of the strength of brittle solids, a vast amount of analytical and interpretive work has been done on fracture.

1. Fracture Testing Methods

In brittle materials, failure usually takes place at much lower tensile stresses than compressive stresses. In principle, the simplest test is to subject a rod to uniform tensile stress. In practice, there are difficulties in gripping the sample, and except for fine fibers, one must make the central length of the rod smaller in diameter to make certain that failure does not occur at the grips. Making such samples is difficult and expensive, so other tests are used in which the stress distribution is not as simple as in a tensile test. For long fibers, the direct tensile test is the simplest and most used, since gripping problems can be overcome in a variety of ways.

The most common tests involve bending (flexure), either with three or four points of stress application. Rod samples are better than bars, because chips at the corners of the bars can initiate fracture. Bend tests favor failure starting at the sample surface, where the tensile stress is high. It is always wise to examine the specimen after testing, to make certain that failure did not take place at the contact points, where the stress is uncertain. The four-point bend test is preferable to the three-point test because the former samples more of the specimen surface and is less prone to failure at the contact points.

The diametral compression test provides an attractive alternative to bend tests for some materials. In this test, a right circular cylinder is compressed across its diameter by two flat plates, giving a maximum tensile stress normal to the loading direction across the loading diameter. This stress is present across the diameter of both flat ends of the cylinder, and also across a plane from one end of the cylinder to another, so that both surface and volume failures are possible. A pad of material softer than the compressing plates must be inserted between the sample and plates, to prevent failure at the plates.

Compressive tests are sometimes used for glass, but failure usually takes place at a corner contact of the sample with the compressing plates.

2. Strength Distributions

When the failure stress of a number of glass samples is measured under identical conditions of sample preparation and loading, the failure stresses vary over a wide range, as shown in Fig. 1. It is desirable to express this distribution of strengths in statistical terms. The strengths are often symmetrically distributed, so they can be described by a normal (Gaussian) distribution:

$$P = [1/d(2\pi)^{1/2}]\exp[-(S - S_m)^2/2d^2] \qquad (1)$$

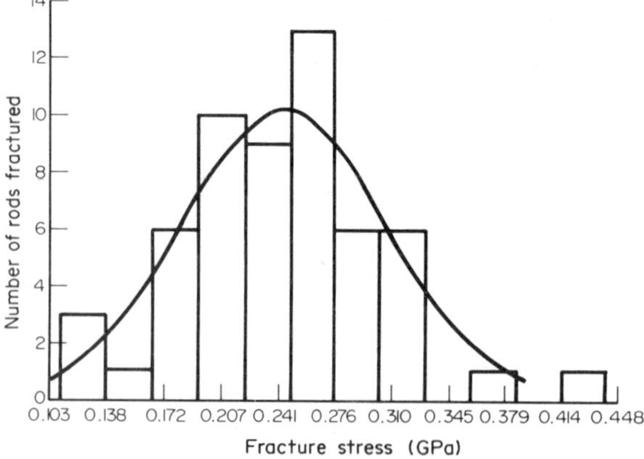

Figure 1
Distribution of fracture strengths of pyrex borosilicate glass rods in four-point bending at −196 °C: line from Eqn. (1) with $S_m = 0.245$ GPa and $d = 0.061$ GPa; 3 mm diameter rods

where P is the probability of finding a sample of strength (failure stress) S, S_m is the strength of greatest probability $[P_m = 1/d(2\pi)^{1/2}]$ and d is the standard deviation. The integral of Eqn. (1) gives the fraction F of samples that break below the stress S.

Recently another distribution function called the Weibull distribution has become quite popular in describing strength distributions:

$$F = 1 - \exp[-(S/S_0)^m] \qquad (2)$$

where S_0 is a scaling factor and m is a measure of the spread of the distribution—a broader distribution has a smaller m. For $m \geqslant 3$, the Weibull distribution is nearly symmetrical and closely resembles the normal distribution.

Strengths of carefully drawn glass fibers are much higher than those of ordinary glass; E-glass can have a strength of 1–4 GN m^{-2} at room temperature, and silica fibers have even higher strengths (Proctor et al. 1967). (Compositions of glasses are given in the introductory article on glass (see *Glass: An Overview*).) The strength distributions of short E-glass fibers can be quite narrow, with a coefficient of variation of 1–2%. On the other hand, high-strength flame-polished or etched glass rods can have very broad and nonuniform distributions, as shown in Fig. 2.

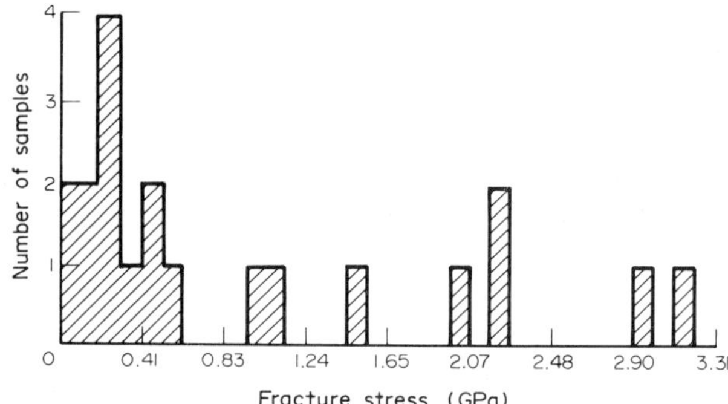

Figure 2
Distribution of fracture strengths of rods of soda-lime glass in four-point bending at −196 °C: 3 mm diameter rods, after removing 9.4 μm by etching with dilute HF

3. Effect of Surface Area and Treatment on Strength

Many investigators have confirmed that as the stressed area of a glass surface increases, the strength decreases. When the data of Griffith (1920) are plotted as the log of breaking strength S against the log of diameter D, the relation is linear. Fused-silica fibers also show a linear relation between log S and log D below 25 μm diameter, with higher than expected strengths at larger diameters. This size dependence of strength can be understood in terms of strength and flaw distributions.

If glass samples are carefully prepared and kept pristine, they will have very high strengths ($\leq 10\ GN\ m^{-2}$). Hillig (1962) measured strengths of up to 15 $GN\ m^{-2}$ at $-196\ °C$ for 1 mm diameter fused-silica rods. These samples were flame-drawn from larger rods, and carefully protected from contact with solids.

The strength of ordinary or abraded glass can be substantially increased by etching with hydrofluoric acid (HF). The distribution of strength of soda-lime glass rods after etching is shown in Fig. 2. Etching gives some samples with very high strengths, but because samples etched in HF are highly sensitive to damage by handling, many samples have strengths lower than before etching.

The strength of an abraded soda-lime glass can be increased by a factor of up to 1.5 by standing in water for a few hours after abrasion.

The strengths of strong glasses increase at lower temperatures. However, abraded glass of low strength often shows little temperature dependence of strength at short loading times. It seems likely that these temperature effects are caused by the same mechanisms as delayed failure. Hillig (1962) found no difference in glass strength between -196 and $-268\ °C$, at which temperatures delayed failure is absent.

Glass composition can influence the strength indirectly through surface chemical reactions, through the introduction of surface stresses and through other secondary effects. However, there is no evidence for direct effects of glass composition on strength (e.g., as a result of changes in bonding in the silicon–oxygen network).

4. *Fracture Surfaces and Origins*

A glass fracture surface shows several characteristic regions—an origin (see Fig. 3), a smooth region called the mirror, a misty region outside the mirror and finally a region of surface roughness called hackle. The distance r from the fracture origin to the mirror–mist boundary is related to the fracture strength S by the empirical equation $Sr^{1/2} = A$, where A is a constant for a particular material and temperature. This constant has a value of about $2\ MN\ m^{-3/2}$ for glass at room temperature. Apparently, A is almost independent of temperature or glass composition. This relation is of considerable practical value, because it allows accurate estimates of fracture stress to be made without direct measurement.

A variety of chemical treatments, such as sodium vapor or a molten lithium salt, give visible crack patterns on a glass surface. However, these cracks result from the treatment, and do not represent prior surface damage. Etching a glass surface with HF reveals elongated pits resulting from surface cracks.

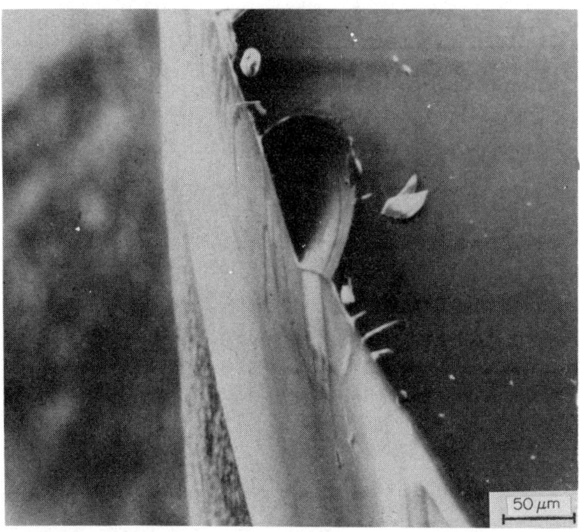

Figure 3
Scanning electron micrograph showing a fracture origin in a soda-lime glass

If a hard sphere is pressed against a glass surface, a crack develops around the contact and then grows into the glass in a conical shape. If the stress is applied with a sharp rather than spherical object, a median vent crack propagates into the glass, directly under the contact area. When the load is released, winglike cracks (lateral vents) propagate radially from the contact region. The impact of a hard spherical projectile causes damage with features of both spherical and sharp indenters. At low impact velocities cone cracks are formed, but at higher velocities the cone cracks resemble radial winglike cracks; at high velocities median cracks also propagate down into the glass.

When a hard object such as a diamond indenter is dragged across a glass surface, median and lateral cracks form along its track in much the same way as described above. Abrasion of a hard surface against the glass also gives this kind of linear damage.

5. *Velocity of Fracture Propagation*

The velocity of fracture propagation in glass can be measured from line markings on the mirror surface known as Wallner lines. Fracture velocities can also be measured from ripple marks on the fracture surface, which result from ultrasonic waves introduced perpendicular to the fracture face. Various techniques of high-speed photography, such as rotating mirror and multiple-spark cameras, can be used to measure crack velocity. Another method is to measure the electrical conductivity of metal strips painted on the glass as they are progressively broken by the fracture front.

6. Fracture Criteria and Theoretical Strength

The experimental observations described above provide strong evidence that the surface condition of glass determines its strength, and that flaws in the surface reduce the strength. This leads to a model of surface cracks as the origin of brittle fracture. Unfortunately it is difficult to observe these cracks directly, and there still is uncertainty about their size, depth and shape that requires more experimental investigation.

The stress σ at the tip of an elliptical crack of depth c and tip radius ρ subjected to a tensile stress S is (Inglis 1913):

$$\sigma/S = 1 + 2(c/\rho)^{1/2} \approx 2(c/\rho)^{1/2} \quad (3)$$

Thus, if the crack is deep and has a sharp tip, the stress at the tip can be much higher than the applied stress. It seems reasonable that various types of surface damage on glass could give rise to cracks with different depths and tip radii, resulting in the influence of surface treatment on strength found experimentally. Experiments show that glass samples with cracks of the same length but with different treatment (e.g., freshly abraded or abraded and aged in water) have different strengths, showing the importance of the crack tip radius.

The theoretical strength σ_t of glass is approximately given by:

$$\sigma_t^2 = E\gamma/4a \quad (4)$$

where E is Young's modulus, γ is the surface energy and a the average interatomic distance in the material. This equation gives a strength for vitreous silica of about 24 GN m^{-2}, or about 60% greater than the highest measured value.

7. Plastic Deformation in Glass

Tensile and bend tests on glasses to stresses close to the theoretical strength show no evidence of any nonrecoverable deformation. However, indentation of glass with a sharp point leads to some regions of deformation that do not recover when the load is removed. In fused silica, this deformation is apparently a densification that recovers completely on heating. In glasses with modifiers (alkali and alkaline earth oxides), this deformation is partly a densification, but only a small part of the deformation recovers on heating. More detailed examination shows that the deformed region is made up of flow or fault lines between which there are only elastic strains. These deformations are very different from plastic deformation in metals and other ductile materials, since they arise only from compressive loads, and do not occur with large-scale stressing in tensile and bending modes. Since crack propagation under normal conditions occurs as a result of applied tensile stress, and high tensile stress at the crack tip, there is no reason to expect densification or faulting under these conditions.

8. Cracks and Energy

From the strain energy change in a stressed material containing a crack, Griffith (1920) derived the equation:

$$S_f^2 = 2E\gamma/\pi c \quad (5)$$

for the fracture stress S_f in a material with a crack of depth c, surface energy γ and Young's modulus E. In a treatment of crack energies and fracture criteria in terms of thermodynamic laws, the Griffith equation provides a necessary but not sufficient condition for fracture, the correct criterion for fracture being provided by Eqn. (3). The second law of thermodynamics shows that the fracture stress cannot be less than that given by Eqn. (5), but can be greater. It also predicts that fracture will only occur when the tip radius is larger than some critical minimum value ρ_m. A measure of ρ_m can be gained by comparing Eqn. (3) and Eqn. (5) for the fracture stress. Then,

$$\rho_m = 8E\gamma/\pi\sigma_t^2 \quad (6)$$

is the minimum tip radius for fracture. From Eqn. (4) and Eqn. (6), $\rho_m = 32a/\pi$. Thus the minimum tip radius for fracture is about ten times the interatomic spacing, and is independent of other parameters.

9. Nucleation of Cracks

How do cracks get into a material? As described above, pressing a hard sphere against a glass surface causes a ring crack and then a cone crack. In some cases, the ring cracks appear to start at preexisting flaws, and then propagate. However, indenting a flaw-free surface of fused silica can nucleate a ring crack if the load is high enough. In soda-lime glass, sharp indenters and projectiles give median and radial cracks, and scribing gives median and lateral cracks. These types of damage are probably common in a glass surface as a result of ordinary handling and contact with other solids. Thus, nucleation of flaws in a pristine glass surface can be understood qualitatively as resulting from the stresses and deformation caused by colliding projectiles, indentations and scribing that are involved in handling.

10. Strengthening of Glass

Etching is not a practical way of strengthening glass because the etched surface is very sensitive to new damage, even if it is protected. Glass can be strengthened by introducing a compressive stress at its surface; then the total tensile stress needed for failure is increased by the amount of the compressive stress. Compressive stress is introduced by either quenching or ion-exchange (chemical strengthening). Quenching is simpler and cheaper, but harder to control, and results in a region of high tensile stress inside the glass. Exchange of large ions for small ones already in the glass gives a thin region of high compressive stress in the glass surface, and only a low tensile stress inside the glass.

11. Fatigue

In principle, the simplest method of investigating fatigue in glass is a static test in which a constant load is imposed on a set of samples and the time of failure of each sample is recorded. Such tests can be done in bending, simple tension or diametral loading; bending is easiest for rods and tubes, and requires simple test equipment. Although static tests are the closest to practical situations and are the least complicated in principle, they have some disadvantages. They require a large number of samples for each of several stresses to define the stress–failure time relationship, because of the large spread in failure times at a particular stress. It is often difficult or expensive to obtain the required number of samples. Static tests must also be carried out over a large spread of times, leading to long delays in obtaining results.

These disadvantages have led to a search for alternative methods to test fatigue in glass and other brittle oxides. Two methods have been widely used. In the first, samples are tested at different loading rates; the dependence of fracture strength on loading rate is then related to the stress–failure time relationship in static loading. This method requires fewer samples and less time, but a testing machine with a linear load application is needed. The main disadvantage of this technique is that it can be used for only a narrow range of failure stresses, because of limits in the loading rates available. Thus, this method is good for surveys of materials, and rapid or preliminary tests, but static fatigue tests are needed for complete definition of strength–failure time relations.

A second method is to measure directly the velocity at which large (~1 cm) cracks propagate under different applied loads. This method has been widely used, but experimental and theoretical treatments suggest that the results it provides are not always directly comparable with the fatigue behavior of materials with much smaller flaws (a few micrometers long).

In Fig. 4, static fatigue results for soda-lime glass in four-point bending are shown. Each point represents the mean of at least twenty samples. The results are plotted as reduced stress S/S_N against the log of fracture time, where S is the maximum stress in four-point bending and S_N is the mean fracture stress in

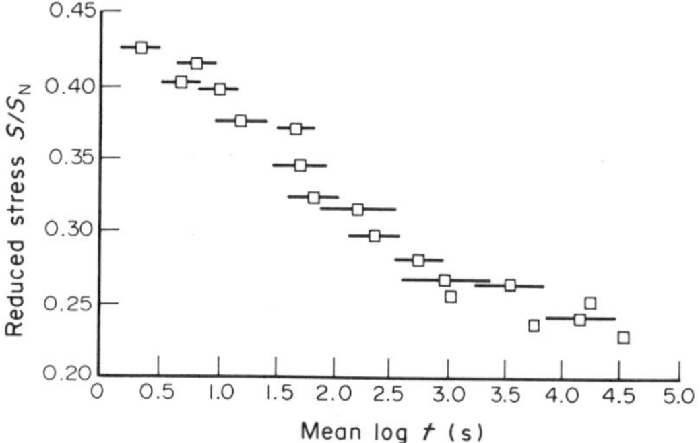

Figure 4
Mean log failure time at 100% relative humidity against relative stress for soda-lime glass rods abraded and then aged 24 h in water before testing; lines show 95% limits for the spread of log t values

the same test at liquid nitrogen temperature, where there is no fatigue.

The influence of pretreatment and humidity on the time for failure $t_{1/2}$ at $S/S_N = 0.5$ for soda-lime glass is summarized in Table 1. The results are taken from different studies which used many samples and included measurements at −196 °C for each condition; each $t_{1/2}$ value is based on measurements at many (five to twenty) stresses, and at each stress twelve to twenty samples were broken.

The results in Table 1 show that fatigue times increase strongly as the ambient concentration of water is reduced. When the samples are not aged after abrasion, their fatigue times are up to one hundred times longer than after soaking in water for 24 h with other conditions the same. Flaw type and severity and direction of abrasion also have a strong effect on both inert fracture strength and fatigue times.

Charles (1958) measured the static fatigue of soda-lime glass at different temperatures from −50 °C to 242 °C in saturated water vapor; he found that failure times decreased with temperature increase. At temperatures above about 50 °C, the failure times no longer fit a simple Arrhenius line, and become longer than expected. At temperatures below about −100 °C, the fatigue becomes very slow compared to experimental time.

12. Life Prediction

An important practical problem is to predict the lifetime of a piece of glass under stress. Several studies have shown that there is no separate effect of cyclical stress on the fatigue of glass, so the fatigue results from a cumulative process depending on the time at different applied stresses. Thus the problem of predicting the lifetime is to find a reliable function that can be extrapolated to longer times and lower stresses than are convenient experimentally.

Table 1
Fatigue times for abraded soda-lime silicate glass

Test	Relative humidity (%)	Average fracture stress at −196 °C ($MN\ m^{-2}$)	log $t_{1/2}$ (s)	Remarks[a]
1	100	118	1.67	not aged
2	50	118	2.46	not aged
3	100	160	−0.66	
4	50	160	0.48	
5	water	75	0.94	aged 20–60 s in water
6	water	86	0.94	
7	43	86	2.30	
8	0.5	86	3.54	
9	water	90	2.30	heated at 470 °C for 3 h in a vacuum of 10^{-5} mm Hg, stored in dry nitrogen, then tested in water
10	50	187	1.47	heated to 400 °C for one h after abrasion, then held 24 h in water

a All samples were aged 24 h in water after abrading and before testing, unless otherwise noted. Tests 1–4 and 10 from Mould and Southwick (1959); tests 5–9 from Pavelchek and Doremus (1976)

Since the inert (liquid nitrogen) strengths are distributed, failure times at a constant stress are also distributed. Several different functional dependencies between failure time and stress have been suggested:

$$\log t = a - b\, S/S_N \tag{7}$$

$$\log t = c - n \log S/S_N \tag{8}$$

$$\log t = d + g\, S_N/S \tag{9}$$

where the values of b, n and g are positive.

The simplest way to examine the equations is to calculate a regression equation by the least-squares method in the form of Eqns. (7–9), and compare the correlation coefficient R^2 from the regression. For pyrex glass, R^2 values for Eqns. (7–9) have been calculated as 0.920, 0.947 and 0.976, respectively. Although the correlation coefficients are all close to one, there is a clear increase in the closeness of fit from Eqn. (7) to Eqn. (9), especially when $1 - R^2$ values are compared.

Often data are not extensive or reliable enough to find any differences between correlation coefficients; a more sensitive method is needed. Pavelchek and Doremus (1976) suggested that the variation of the spread (standard deviation) of log t values with stress is a sensitive measure of the functional dependence of failure time on stress. If the differential of log t is considered to equal the standard deviation of log t, then the following relations are found. For Eqn. (7) the spread in log t values should decrease as the stress decreases, for the power law of Eqn. (8) the spread in log t should remain constant with changes in stress, and for the inverse exponential of Eqn. (9) the spread in log t should increase as the stress decreases.

For every set of static fatigue data on glass that are reliable, the spread in log t values increases as the stress decreases (Pavelchek and Doremus 1976). Figure 4 also shows this increase. Thus Eqn. (9) again fits best.

13. *Theories and Interpretations of Static Fatigue in Glass*

The most widely accepted theory of static fatigue (Hillig and Charles 1965) is that a stress-accelerated chemical reaction of water with the glass leads to a sharpening of the crack tip. Thus, as stress at the tip increases, the reaction rate becomes faster, and finally the stress at the crack tip becomes high enough to cause rapid propagation of the crack. A comparison of the Hillig–Charles equations with experimental fatigue data shows that one expects tip sharpening rather than crack lengthening to take place during fatigue in glass.

Further evidence to support the conclusion that cracks sharpen rather than lengthen during fatigue comes from observations of crack initiation sites (Fig. 3). These sites have the same appearance after fracture at −196 °C and at 25 °C. A comparison of sizes of these sites shows no statistically significant difference between sites formed at −196 °C and at 25 °C. If fatigue at 25 °C caused crack lengthening, one would expect larger sites at this temperature.

There are striking differences between failure times of samples not aged after abrasion and those aged in water (see Table 1). From the Hillig–Charles theory, one would expect only small differences in $t_{1/2}$ for these different samples. Thus this theory, and also other theories of fatigue of glass, give no explanation for this large difference in fatigue time.

Heating the glass to above 300 °C after abrasion can anneal out residual stresses resulting from abrasion, and also decompose hydronium ions. These ions are introduced into the glass as a result of their exchange and interdiffusion with alkali ions in the glass. The decomposition of the hydronium ions leaves a proton in place of the larger sodium ion, so a tensile stress develops in the glass and it becomes weaker. The net result of these two processes can give a $t_{1/2}$ between that of samples not heated and aged in water and that of those not aged, as shown in Table 1.

14. Proof Testing

Brittle parts can be stressed to a higher level than the maximum stress to which they will be subjected in practice to eliminate weak samples. This procedure is called proof testing. If the proof test is made under conditions where fatigue occurs, the surviving samples will be weaker. To reduce this effect, the proof test can be made in a dry atmosphere and/or at a temperature below ambient.

See also: Fracture: Reliability Criteria for Brittle Materials; Fracture Toughness Testing of Brittle Materials

Bibliography

Charles R J 1958 Static fatigue of glass. *J. Appl. Phys.* 29: 1549–60

Doremus R H 1976 Cracks and energy—Criteria for brittle fracture. *J. Appl. Phys.* 47: 1833–36

Doremus R H 1982 Fracture and fatigue of glass. In: Doremus R H, Tomozawa M (eds.) 1982 *Glass III.* Academic, New York, pp. 169–239

Ernsberger F M 1960 Detection of strength-impairing surface flaws in glass. *Proc. R. Soc. London, Ser. A* 257: 213–23

Griffith A A 1920 The phenomena of rupture and flow in solids. *Philos. Trans. R. Soc. London, Ser. A* 221: 163–98

Hagan J T 1980 Shear deformation under pyramidal indentations in soda-lime glass. *J. Mater. Sci.* 15: 1417–24

Hillig W B 1962 Sources of weakness and the ultimate strength of brittle amorphous solids. In: Mackenzie J D (ed.) 1962 *Modern Aspects of the Vitreous State*, Vol. 2. Butterworths, Washington, pp. 152–94

Hillig W B, Charles R J 1965 Surfaces, stress-independent reactions and strength. In: Zackey V F (ed.) 1965 *High Strength Materials.* Wiley, New York, pp. 682–93

Inglis C E 1913 Stresses in a plate due to the presence of cracks and sharp corners. *Trans. Inst. Naval Arch.* 55: 219–30

La Course W C 1972 The strength of glass. In: Pye L D, Stevens H J, La Course W C (eds.) 1972 *Introduction to Glass Science.* Plenum, New York, pp. 451–512

Mould R E, Southwick R D 1959 Strength and static fatigue of abraded glass under controlled ambient conditions: II. Effect of various abrasions and the universal fatigue curve. *J. Am. Ceram. Soc.* 42: 582–92

Pavelchek E K, Doremus R H 1976 Static fatigue in glass—A reappraisal. *J. Non-cryst. Solids* 20: 305–21

Proctor B A, Whitney I, Johnson J W 1967 The strength of fused silica. *Proc. R. Soc. London, Ser. A.* 297: 534–57

R. H. Doremus

Fracture Characteristics of Weldments

Fractures of weldments may be classified according to the mechanisms by which the fractures occur. These include ductile fracture, brittle fracture, fatigue fracture, and stress corrosion cracking. The basic mechanisms of these fractures are discussed elsewhere in the Encyclopedia. The emphasis of this article is on the unique characteristics of weldments and welded structures which may be summarized as follows:

(a) *Metallurgical heterogeneity*. Compared with the base plate, which is often treated as a homogeneous material, a weldment is composed of the weld metal, the heat-affected zone, and the unaffected base metal. The heat-affected zone, which is a narrow zone of the base metal just outside the weld fusion line, is composed of materials with different metallurgical structures and properties.

(b) *Residual stresses and distortions*. A weldment always contains residual stresses, unless it has been stress relieved. Residual stresses in regions near the weld are normally tensile, and their maximum values can be as high as the yield stress of the material. Welding also causes various types of distortions including transverse shrinkage, longitudinal shrinkage, and angular distortion (see *Residual Stresses and Distortion in Weldments*). These residual stresses and distortions often promote fractures in weldments.

(c) *Weld defects*. A weldment often contains various types of defects including cracks, porosity, and slag inclusions, to name a few. Today there are many nondestructive inspection techniques, such as radiography, ultrasonic inspection, and magnetic particle inspection (see *Welds: Nondestructive Evaluation*). However, it is not yet possible to determine the exact shapes and distributions of these defects in weldments, especially in complex welded structures such as ships, bridges, and pressure vessels. A probabilistic approach is frequently taken to assess the reliability of a welded structure, considering the effects of weld defects.

As a result of the combined effects of these factors, mechanisms of fractures in weldments are far more complex than those in a base plate. Therefore, studies on fractures of weldments have been mainly empirical; however, there has been increasing interest in using analytical techniques in these studies.

1. Brittle Fractures of Welded Structures

A brittle fracture is a sudden fracture that can result in the catastrophic failure of a welded structure.

Brittle fractures have occurred during construction, proof testing, and service. The most extensive and best known failures of welded structures are those of the welded merchant ships built in the USA during World War II. Among approximately 5000 ships built, about 1000 experienced structural failures in various degrees. About 20 ships broke in two or were abandoned due to structural failures. Catastrophic brittle fractures have also occurred in bridges, storage tanks, pressure vessels, pipelines, offshore oil-drilling rigs, and aerospace structures. However, incidents of brittle fractures have decreased over the years as a result of combined efforts, including the use of materials with adequate fracture toughness, careful design in reducing stress concentrations, and improved workmanship in reducing weld defects and other stress raisers.

Characteristics of brittle failures in welded steel structures are as follows:

(a) *Fracture appearance*. The surface of a brittle fracture is a plane that is approximately perpendicular to the plate surface and is granular in appearance. The reduction in thickness near the fracture surface is very slight, usually less than 3%. A brittle fracture surface normally contains chevron markings which point back to the origin of the flaw responsible for the fracture initiation (see *Failure Analysis: General Procedures and Applications to Metals*). In the investigation of an actual failure, chevron markings provide important information about fracture initiation and propagation.

(b) *Temperature*. Brittle fractures have most frequently occurred in winter.

(c) *Stress at failure*. According to the statistics about brittle failures of welded ships during World War II, more failures occurred in heavy seas than in moderate and calm seas. However, a number of failures occurred when the average stress in the structure was well below the yield stress of the material used.

(d) *Origins of failures*. According to the statistics, brittle failures always originated from notches that caused severe stress concentrations, and many failures initiated from welds. In fact, a number of serious failures started from unimportant, incidental welds, such as tack welds and arc strikes, made on important, major structural members.

(e) *Fracture propagation*. A brittle failure usually propagates at a very high speed, of the order $1000\ m\ s^{-1}$.

2. *Fracture Toughness of Weldments*

In order to avoid brittle fracture in a welded structure, the materials used, including the base metal, the heat-affected (base metal) zone, and the weld metal, must have adequate fracture toughness.

2.1 Fracture Toughness

Low-carbon structural steel exhibits good ductility in terms of elongation and reduction of the area when tensile tests are performed on ordinary specimens (round bars and flat plates with no notches). However, when the specimen contains a sharp notch and the test temperature is low, a crack may initiate from the notch, causing brittle fracture. This phenomenon is called notch brittleness. The temperature at which the mode of fracture changes from ductile to brittle is called the ductile-to-brittle transition temperature or simply the transition temperature.

A complication here is that the transition temperature is not a definite and fixed temperature for a certain material, as is the melting temperature. In many cases, the ductile-to-brittle transition takes place over a range of temperatures. Furthermore, the transition temperature of a material depends upon a number of factors, including the size and shape of the specimen used, the type of loading (e.g., tensile, bending), the loading speed (static or dynamic), and the criteria used (e.g., energy absorbed, fracture appearance, deformation at the notch root).

Figure 1 shows bands of fracture toughness data on ship steels used during World War II, and weld metals deposited by ordinary electrodes such as AWS E6010 electrodes. The data were obtained using the Charpy V-notch impact test. (The geometry of the specimen used is shown in Fig. 1.) As the temperature decreases, the amount of absorbed energy decreases and the fracture appearance changes from ductile to brittle.

2.2 Metallurgical Factors Affecting Fracture Toughness

Fracture toughness is affected by various metallurgical factors including the crystal structure, the chemical composition, and the microstructure.

Metals and alloys which have a face-centered cubic (fcc) lattice are ductile even at extremely low temperatures. Aluminum alloys and austenitic stainless steels are widely used for cryogenic applications.

A wide variety of steels with different levels of fracture toughness are currently in use for various applications. They are made using different chemical compositions, deoxidation practices (rimmed, semi-killed, and killed), and heat treatments (including as-rolled, normalized, and quenched-and-tempered). As examples of the fracture toughness requirements for low-carbon steels, points D and E in Fig. 1 show the requirements for Charpy V-notch impact values for Grade D and Grade E steels, respectively, as specified by the American Bureau of Shipping (ABS). These steels, which are made by using fully

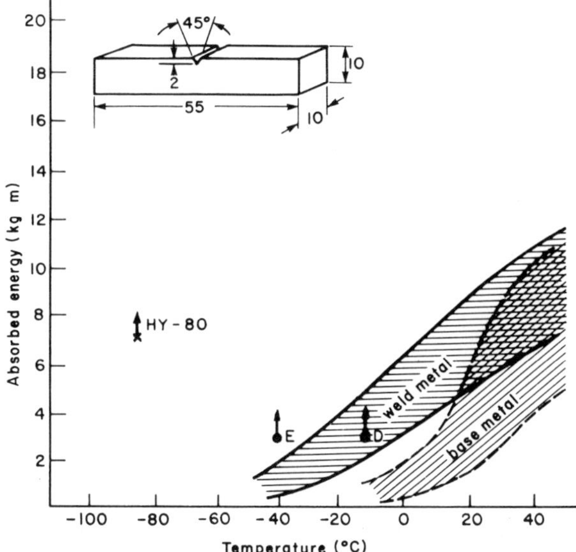

Figure 1
Fracture toughness bands for several steels. Fracture toughness values shown at points D, E and HY-80 represent specified minimum values for samples of grade D and E steels and HY-80 steel plate, respectively. Consequently, fracture toughness values of actual steels are higher than the values at these points, indicated by upward arrows. Charpy specimen dimensions are in millimeters

killed, fine-grain practices, may be regarded as low-carbon construction steels with good fracture toughness.

For a given type of steel, fracture toughness generally decreases with increases in the strength level (in terms of the yield strength and the ultimate tensile strength). A tremendously large amount of effort has gone into developing techniques for increasing strength and retaining fracture toughness. Some low-alloy quenched-and-tempered steels are known to meet this requirement. For example, the point HY-80 in Fig. 1 shows the fracture toughness requirement for an HY-80 steel plate with a thickness of 50 mm or less. The HY-80 steel, which has a specified minimum yield strength of 56.2 kg mm^{-2} (552 MPa), is widely used by the US Navy for submarine hulls. This steel may be regarded as an example of a high-strength steel with excellent fracture toughness.

2.3 Evaluation of Fracture Toughness of a Weldment

The evaluation of the fracture toughness of a weldment is a very complex subject. A few of the major reasons for this complexity are discussed below.

First, there are many tests that have been proposed and used for evaluating the fracture toughness of a material (a base metal and a weldment). No test has yet been selected by a majority of investigators as the most appropriate. For example, the Charpy V-notch impact test has been widely used for practical applications. The crack opening displacement (COD) test also has been fairly widely used, especially in the UK. The US Navy makes extensive use of the drop-weight test and the dynamic tear test, both of which have been developed at the Naval Research Laboratory.

Second, there are several ways of evaluating the fracture toughness of a weldment. If there is interest in determining fracture toughness values at various locations of a weldment, the investigator may decide to prepare test specimens from different locations (the weld metal, the heat-affected zone, and the base metal) of the weldment. However, in other cases one may be interested in evaluating the fracture toughness of a weldment as a whole.

Third, there are considerable differences in the cost of performing tests, as well as in the willingness of investigators to pay the necessary cost. In some cases, one may be interested in performing a simple test even though the test may not yield an accurate assessment of the fracture toughness of the weldment being investigated. In other cases, one may be looking for a test which can simulate as accurately as possible what might actually happen in a specific structure, although the test may be extremely costly. For example, a welded-and-notched wide-plate test, as shown schematically in Fig. 2, has been used as a test which will simulate with reasonable accuracy the brittle fracture of a butt-welded joint. A transverse notch is placed in regions near the weld where high tensile residual stresses exist in the direction parallel to the weld line. The specimen is then subjected to tensile loading. In order to simulate brittle fracture in an actual welded structure, considering the effects of residual stress, a specimen as wide as 1 m is needed, requiring a large-capacity testing machine. Testing machines of the required size and load

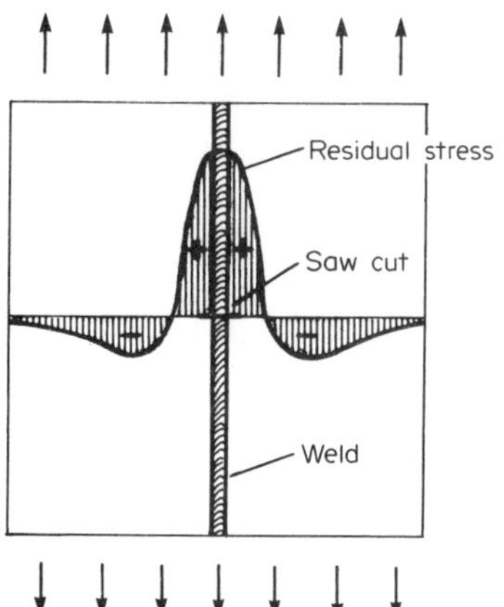

Figure 2
A welded-and-notched wide-plate test specimen

capacity are not commonly available and a test of this kind is very expensive.

2.4 Fracture Toughness of the Weld Metal

An important task given to a welding engineer is to obtain a weld metal that can match the base metal in both strength and fracture toughness. In welding ordinary carbon steel, it is not difficult to accomplish this task, as can be seen from the data shown in Fig. 1. However, it becomes increasingly difficult to accomplish this task when welding steels with higher strength levels. This is especially true for quenched-and-tempered steels for which high strength and excellent fracture toughness are obtained through heat treatments.

3. Theoretical Studies of Brittle Fractures of Weldments

During the last 40 years, there have been many theoretical studies of brittle fractures in weldments. The most widely used are the fracture mechanics theories, especially the linear elastic fracture mechanics (LEFM) theory (see *Brittle Fracture: Linear Elastic Fracture Mechanics*). When an infinitely large plate containing a straight crack of length $2a$ is subjected to a uniform tensile stress σ the stress intensity factor K is expressed as:

$$K = \sigma\,(\pi a)^{1/2}$$

The stress intensity factor expresses the extent of stress concentrations in regions near the crack tip. For example, Fig. 3 shows how σ_{y0}, the normal stress in the y-direction along the x-axis, can be expressed by K and the distance from the crack tip r.

On the basis of LEFM, unstable brittle fracture occurs when the K-value exceeds the critical stress intensity factor K_c:

$$K = \sigma\,(\pi a)^{1/2} \geqq K_c \quad \text{or} \quad \sigma \geqq \frac{K_c}{(\pi a)^{1/2}}$$

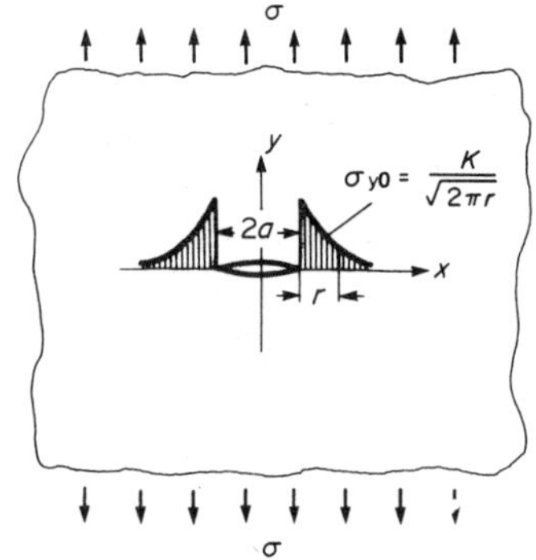

Figure 3
Infinite plate containing a sharp crack of length $2a$ under uniform tensile stress σ

For a certain material with a certain value of K_c, the fracture stress decreases as the length of an initial crack, $2a$, increases. When the initial crack is short, on the other hand, the value of K is small and the crack is stable or would not grow.

When an initial crack is located in regions where high residual stresses exist, as shown in Fig. 2, the K-value at the crack tip is increased significantly while the crack is still short. Consequently, an unstable brittle fracture can occur from a very short crack even when the level of applied stress σ is low. This is why brittle fractures of welded structures have occurred from very short initial cracks, even when service stresses were rather low.

4. Fatigue Fractures of Weldments

Fatigue fractures have plagued engineers associated with the construction and operation of welded structures subjected to repeated loading.

4.1 Fatigue Fracture

When a material is subjected to repeated loading, fracture takes place after a certain number of cycles. The lower the applied stress, the larger the number of cycles at fracture. When the material is exposed to water and other corrosive environments, corrosion fatigue often becomes a serious problem. Compared with the fatigue fracture in air, corrosion fatigue can occur after much fewer cycles.

Figure 4 is a schematic representation of a typical fatigue fracture surface. A fatigue fracture occurs in three stages: (a) the initiation of the crack, (b) the slow growth of the crack, and (c) the onset of unstable fracture. A fatigue fracture originates at the surface. The region surrounding this origin has a smooth silky appearance. Careful examination of this smooth area frequently reveals the existence of concentric rings or beach markings around the fracture nucleus, and radial lines emanating from it. The beach markings or striations are more evident under an electron microscope. As the crack grows, the stresses on the remaining section increase so that there is a corresponding increase in the rate of crack growth. Ultimately, a stage is reached when the remaining

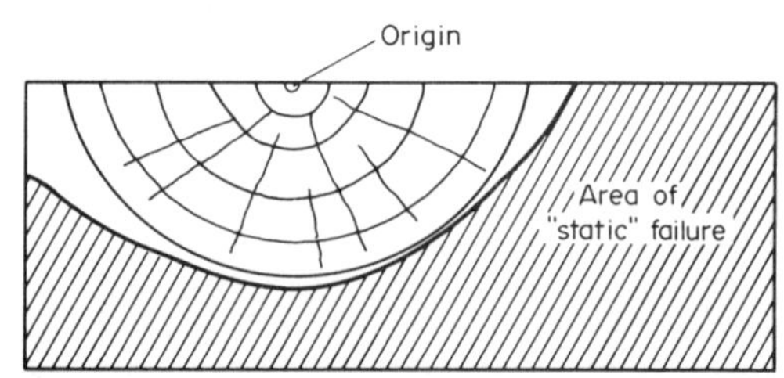

Figure 4
Schematic representation of a typical fatigue fracture surface

area is unable to support the load, and final fracture occurs. The fracture surface of the final rupture may be characteristic of ductile or brittle fracture.

4.2 Fatigue Strengths of Weldments

Numerous research programs have been performed to determine the fatigue strengths of various types of welded joints and weldments. As an example, Fig. 5 shows fatigue-test data for the following four cases:

Case 1: A base plate of HT60 steel, a high-strength steel with a minimum ultimate tensile strength of 60 kg mm^{-2}

Case 2: A transverse butt weld

Case 3: A transverse butt weld in which two vertical plates are attached to a continuous flat plate

Case 4: A transverse fillet weld in which two horizontal plates are attached to a continuous horizontal plate

All welds were made with covered electrodes.

Shown here are relationships between the stress range S and the number of cycles at fracture N_f, both of which are shown in log scales. The results presented here have been obtained after performing tests on a large number of specimens, and the lines indicate 95% confidence levels. For example, when a transverse butt weld is subjected to pulsating stresses of 20 kg mm^{-2}, the probability of the number of cycles at fracture being greater than 2.5×10^5 is 95%.

Among the four cases shown in Fig. 5, the base plate has the highest fatigue strength. Among the three types of welded joints tested, the butt weld has the highest fatigue strength. The fatigue strength of a fillet weld is lower than that of a butt weld, since a fillet weld always contains some stress concentrations, especially in regions near the toe of the fillet. It is in these regions that fatigue cracking frequently initiates. Case 3 has a higher fatigue strength than Case 4, in which the horizontal plate which is subjected to loading is not continuous. In Case 4, stresses must be transmitted only through fillet welds; this causes high stresses in the fillet.

4.3 Methods of Improving Fatigue Strengths of Weldments

The key to improving the fatigue strength of a weldment is to reduce (or eliminate if possible) sources of stress concentrations on the surface. A few examples of the methods for improving the fatigue strength are as follows:

(a) Use a butt weld instead of a fillet weld.

(b) When a butt weld is used, minimize the amount of weld reinforcement, or make the weld contour as flush as possible with the plate surface.

(c) When a fillet weld is used, the contour of the fillet should blend smoothly to the plate, especially at the toe, to avoid stress concentrations at the toe. Grind areas near the toe if necessary.

(d) Use a continuous fillet weld and avoid an intermittent fillet weld.

(e) Avoid a lap weld and a plug weld.

Fatigue fractures frequently initiate from unimportant, incidental welds such as fillet welds used to attach a small piece to a plate. If possible, efforts should be made to avoid making these welds on a major structural member. The fatigue strength can be improved by producing compressive residual stresses in regions near the surface. For example, peening is known to produce compressive residual stresses, thus improving the fatigue strength, if it is performed properly. However, the peening operation must be performed carefully so as not to produce notches while hitting the surface.

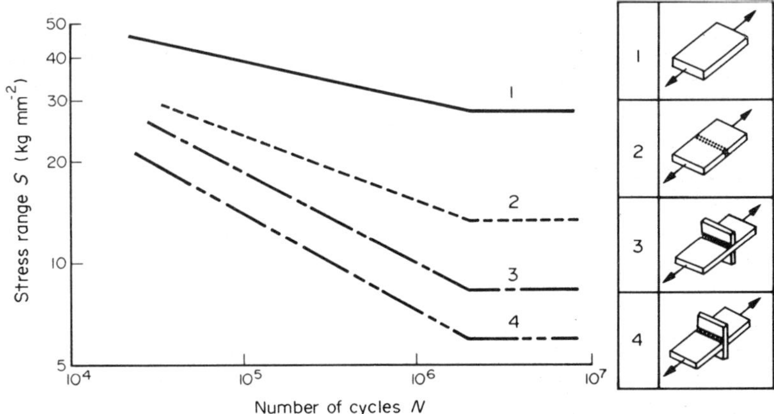

Figure 5
Fatigue strengths of various welded joints (described in the text) made with HT60 steel (95% confidence data)

4.4 Analysis of Fatigue Crack Growth

Recently, efforts have been made to analyze the rate of fatigue crack growth by applying linear elastic fracture mechanics. The rate of fatigue crack growth da/dN may be expressed as follows:

$$da/dN = C\,(\Delta K)^m$$

where $\Delta K = K_{max} - K_{min}$, K_{max} and K_{min} being the stress intensity factors corresponding to the maximum and minimum stress, respectively, during the cyclic loading. C and m are parameters determined by the material and the stress condition.

Figure 6 shows experimental relationships between ΔK and da/dN for the following cases:

Case 1: low-carbon steel base plate 25.4 mm thick

Case 2: heat-affected zone of a butt weld made with electrogas process

Case 3: heat-affected zone of a butt weld made with electroslag process

Case 4: heat-affected zone of a butt weld made with submerged arc welding process from one side only, using three electrodes

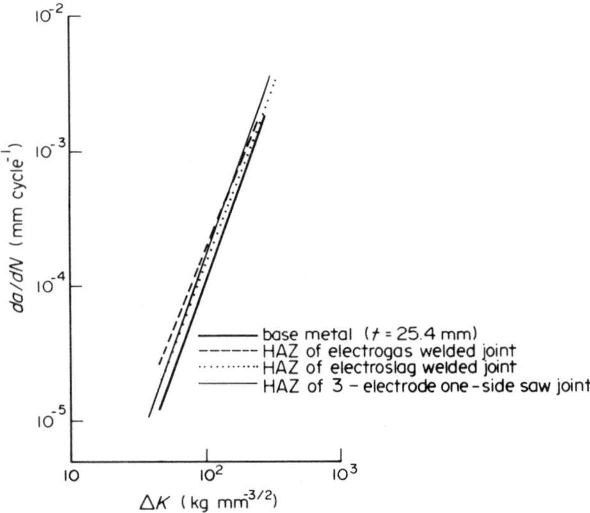

Figure 6
Fatigue crack growth rates in various low-carbon steel weldments as described in the text

The results show that the rate of fatigue crack growth da/dN increases as the crack extends and the value of ΔK increases. The results also show that the rate of fatigue crack growth is about the same for the four test conditions. This indicates that the rate of fatigue crack growth is about the same in various locations in a butt weld.

5. Stress Corrosion Cracking in Weldments

Stress corrosion cracking (SCC) is a sudden fracture, in a brittle manner, of a material exposed to a certain environment. Much research has been performed to determine combinations of materials and environments that cause SCC (see *Stress-Corrosion Cracking*). For example, steel is known to suffer SCC when it is exposed to various environments including alkalis, some nitrates, and hydrogen sulfide.

Since a weldment contains high tensile residual stresses, stress corrosion cracks have occurred in many weldments, even under no external loading. For example, stress corrosion cracks occurred in many welded high-strength steel storage tanks containing petroleum products. The cracks were caused by hydrogen sulfide which was present as an impurity in the petroleum product. Stress corrosion cracks have also occurred in welded stainless steel pipes used for nuclear reactors.

See also: Mechanics of Materials: An Overview; Welding: An Overview

Bibliography

Boeck D 1974 *Elementary Engineering Fracture Mechanics*. Nordhoff, Leyden, The Netherlands

Hall W J, Kihara H, Soete W, Wells A A 1967 *Brittle Fracture of Welded Plates*. Prentice-Hall, Englewood Cliffs, New Jersey

Liebowitz H (ed.) 1968–71 *Fracture*, Vols. I–VI. Academic Press, New York

Masubuchi K 1980 *Analysis of Welded Structures*. Pergamon, Oxford

Pellini W S 1976 *Principles of Structural Integrity Technology*. Office of Naval Research, Arlington, Virginia

Rolfe S T, Barsom J M 1977 *Fracture and Fracture Control in Structures*. Prentice-Hall, Englewood Cliffs, New Jersey

Tetelman A S, McEvilly A J Jr 1967 *Fracture of Structural Materials*. Wiley, New York

K. Masubuchi; H. Yajima

Fracture: Reliability Criteria for Brittle Materials

Design problems arise for two reasons when brittle materials are used as structural components: (a) the strength of brittle materials is not a well-defined quantity, but can vary widely depending on the material; and (b) the strength of brittle materials is time dependent, so that these materials often exhibit a time delay to failure. This time dependence and scatter of strength so typical of most ceramic materials occurs because of the presence of defects such as cracks or cracklike flaws. When subjected to an applied tensile stress, these defects act as stress concentrators, and fracture occurs when the applied stress-intensity factor reaches a critical value. Scatter in the strength of ceramic materials is a consequence of the scatter in the size of the most critical defect. The time dependence of strength results from subcritical crack growth, which gradually lengthens the crack until it reaches critical dimensions, at which time failure occurs. The time delay to failure is the time required for the crack to go from a subcritical to a critical size.

Methods of dealing with design problems involving time-dependent failure were developed during the 1970s through the application of the techniques and concepts of fracture mechanics. These techniques promise to revolutionize the way in which structural components made of ceramic materials are designed, and a complete understanding of these techniques, their application and their limitations is required. This article reviews the application of fracture mechanics theory to the prevention of delayed failure. Three successful applications of this theory in assuring the mechanical reliability of ceramics are also discussed, to demonstrate the viability of the theory for purposes of engineering design.

1. Theory

As a crack grows to its critical size, the crack velocity V is assumed to be dependent on the applied stress-intensity factor K_I as (see *Subcritical Crack Growth in Ceramics*)

$$V = AK_I^N \tag{1}$$

where A and N are crack growth constants that depend on the environment and material composition. From Eqn. (1) and the definition $K_I = \sigma_a Y a^{-1/2}$, it can be shown that the time to failure t_f under constant applied tensile stress σ_a is

$$t_f = BS^{N-2}\sigma_a^{-N} \tag{2}$$

where $B = 1/(AY^2(N-2)K_{Ic}^{N-2})$, K_{Ic} is the critical stress-intensity factor, S is the fracture strength in an inert environment, a is the crack size and Y is a geometric crack constant. From its own definition, B is seen to be a crack propagation constant that depends on the environment and material composition.

In Eqn. (2), t_f represents the time required for a flaw to grow from an initial, subcritical size to dimensions critical for catastrophic propagation; B and N are the material and environment constants that characterize this subcritical crack growth. The initial flaw size is characterized in Eqn. (2) by the fracture strength in an inert environment. (The initial crack size a_i can be calculated from the well-known fracture mechanics relationship $a_i = (K_{Ic}/YS)^2$. In subsequent discussion, a_i is assumed to be small relative to a component dimension, so that Y is a single-valued constant.) From Eqn. (2) it is seen that the time to failure decreases with increasing stress.

From Eqn. (1) a relationship can be derived between the fracture strength σ_f and stressing rate $\dot{\sigma}$. In this case, flaws grow from subcritical to critical size under constantly increasing stress. The result of this analysis is

$$\sigma_f^{N+1} = B(N+1)\,S^{N-2}\dot{\sigma} \tag{3}$$

From Eqn. (3) it is seen that fracture strength decreases with decreasing stressing rate, since the flaws have more time to grow before final fracture.

In addition to constant stress and stressing-rate loading conditions, cyclic loading must also be considered. For most ceramics it is generally thought there is no enhanced effect of cycling on the crack growth rate. With this assumption, it can be derived that the failure time under cyclic stress t_c is directly proportional to the failure time for static stress t_f:

$$t_c/t_f = g^{-1}(\sigma_a/\sigma_c)^N \tag{4}$$

where σ_c is the average cyclic stress and g is a proportionality factor that depends on the type of stress cycle, the amplitude of the cycle and N. The g factor can be evaluated by numerical integration for any periodic load cycle, and has been analytically determined for a sinusoidal and general trapezoidal stress cycle.

The probability of failure F for a given lifetime and applied stress can be obtained from Eqn. (2) by expressing the inert strength in terms of its failure probability distribution. Generally, the inert strength distribution of ceramics can be approximated by the Weibull relationship:

$$\ln\ln(1/1-F) = m\ln(S/S_0) \tag{5}$$

where m and S_0 are empirical constants evaluated by a fit of the inert strength data. Likewise, the strength distribution at a fixed stressing rate can be obtained by substituting Eqn. (5) into Eqn. (3), and the probability of failure under cyclic stressing can be obtained by substituting Eqn. (5) into Eqn. (4).

Since ceramics exhibit a wide spread in strength values (m is typically in the range 4–8), the allowable stress in service can be quite low if low failure probabilities are required. Proof testing offers one means of increasing the design stress for ceramics. The value of proof testing is that it characterizes the largest effective flaw possible in a tested component, since any larger flaws would have caused failure during the proof test.

Assuming that flaw growth during the proof-test cycle is given by Eqn. (1), the inert strength after proof testing S_a is related to the initial inert strength S_i by

$$S_a^{N_p-2} = S_i^{N_p-2} - (D_p/B_p) \tag{6}$$

where

$$D_p = \int_0^t \sigma_p t^{N_p}\,dt \tag{7}$$

Here, D_p represents the amount of strength degradation during the proof stress cycle, and N_p and B_p are crack-propagation constants appropriate for the proof-test environment. For a typical proof test, the component is loaded at a constant rate $\dot{\sigma}_1$, held at the maximum proof stress σ_p for a time t_p, and then unloaded at a constant rate, $\dot{\sigma}_u$. Integrating the D_p equation for this typical proof-stress cycle gives

$$D_p = \sigma_p^{N_p} t_p + [\sigma_p^{N_p+1}/(N_p+1)](1/\dot{\sigma}_u + 1/\dot{\sigma}_1) \quad (8)$$

By substituting Eqn. (5) into Eqn. (6), the inert strength after proof testing can be related to failure probability by

$$(S_a/S_0)^{N_p-2} = (Q_a + Q_p)^{(N_p-2)/m} - Q_p^{(N_p-2)/m} + (S_{min}/S_0)^{N_p-2} \quad (9)$$

where $Q_a = -\ln(1-F_a)$, $Q_p = -\ln(1-F_p)$, F_a is the failure probability after proof testing, F_p is the failure probability of the proof test and S_{min} is the minimum inert strength of a sample that just passes the proof test cycle. It is significant to note that the inert strength after proof testing will be greater than the initial inert strength at all levels of failure probability if $m < N_p - 2$ and is truncated at S_{min}.

The failure probability in the proof test F_p can be determined directly from the number of specimens that break during the proof test, or can be predicted from Eqn. (6), since, for the specimen that just breaks during a proof test, S_a is negligibly small. From Eqns. (5) and (6) it can then be derived that

$$Q_p = (S_0^m)^{-1}(D_p/B_p)^{m/(N_p-2)} \quad (10)$$

For a given material and proof-test environment, S_{min} is determined only by the unloading rate ($\dot{\sigma}_u$) from the proof stress:

$$S_{min} = [\dot{\sigma}_u B_p(N_p-2)]^{1/3}[3/(N_p+1)]^{1/(N_p-2)} \quad (11)$$

If $\dot{\sigma}_u > 0$, Eqn. (10) shows that proof testing always truncates the strength distribution. The higher the value of $\dot{\sigma}_u$, the greater is the strength level S_{min} at which truncation occurs. However, since S_{min} cannot be greater than the maximum stress σ_p during the proof test, σ_p is an upper bound for S_{min}. Substitution of σ_p for S_{min} in Eqn. (10) gives the maximum unloading rate for which σ_p is equal to the truncation strength. Equation (10) also shows that good proof-test conditions (high N_p) result in a high S_{min}.

By coupling Eqn. (9) with Eqns. (2) and (3), the failure time and fracture-strength distributions after proof testing can be predicted for any service conditions. Of particular interest is the minimum lifetime t_{min} of a component in service after proof testing. This is obtained by substituting S_{min} for S in Eqn. (2) to give

$$t_{min} = B S_{min}^{N-2} \sigma_a^{-N} \quad (12)$$

The above equations summarize failure predictions for ceramic materials before and after proof testing. The parameters B and N must be determined in a test environment that simulates the appropriate proof test and service conditions, and can be obtained from one of three types of experiments: crack-velocity experiments, stress-rupture experiments and stressing-rate experiments (see *Subcritical Crack Growth in Ceramics*).

2. *Applications*

2.1 *Stress-Corrosion Cracking of Alumina Substrates*

For alumina substrates fabricated by cofired metallurgy, stress-corrosion cracking can be a serious problem. The metals most commonly cofired with alumina are either molybdenum or tungsten. Because these metals possess lower thermal-expansion coefficients than alumina, their cosintering can cause the ceramic to retain residual tensile stresses of considerable magnitude. Subsequent cleaning and plating baths, necessary to render the metal solderable, expose the substrate to conditions that promote stress-corrosion cracking in the alumina. To statistically assess the susceptibility of polycrystalline alumina to stress-corrosion cracking under processing conditions, the stressing-rate technique can be used to characterize the crack-propagation parameters for alumina in a KOH cleaning solution at 100 °C and a gold-plating solution at 78 °C.

Figure 1 shows the experimental results of strength as a function of stressing rate of alumina in the KOH and gold solutions. For comparison, the fatigue-

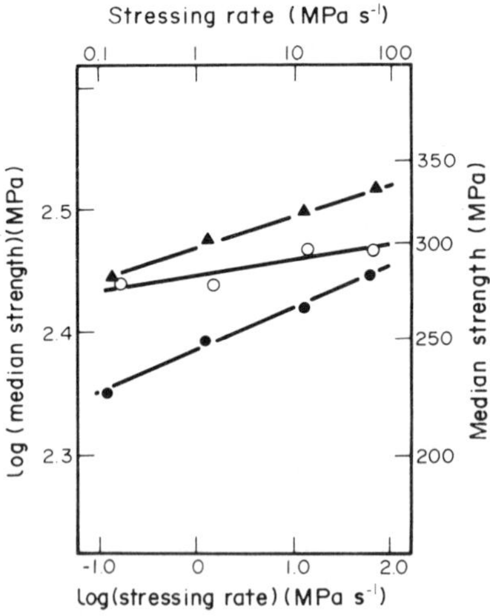

Figure 1
Fracture strength as a function of stressing rate for polycrystalline alumina in a KOH cleaning solution at 100 °C, in a gold-plating bath at 78 °C, and in moist air at 30 °C and 80% relative humidity: ▲, air; ○, KOH solution; ●, gold-plating bath

strength results determined for the same alumina in a moist-air environment are included. From these data, and a knowledge of the inert strength of the samples, the crack propagation parameters B and N were determined from Eqn. (3) for the various environments.

To validate the applicability of fracture mechanics principles in predicting failure probability in the processing solutions, the actual fatigue-strength distributions were compared with predictions based on Eqn. (3), using the appropriate values for B and N and the inert-strength distribution of the samples (Fig. 2). Agreement between the theoretical predictions and the actual fatigue-strength distributions is good. However, it is apparent from these strength distributions that 25 samples are not enough to define clearly the low-strength component of the bimodal population.

The severity of the processing conditions on the alumina substrates can be most easily assessed by means of a lifetime-prediction diagram (Fig. 3), based on Eqn. (2) and the inert-strength distribution of the substrates. Although the strengths of the substrates cannot be measured directly, they can be estimated from the strength distribution of the samples and the Weibull size relation

$$S_1 = S_2(A_2/A_1)^{1/m} \tag{13}$$

where S_1 and S_2 are the strengths of the substrates and samples respectively, and A_1 and A_2 are the surface area of the substrates and samples respectively. It should be noted that it is here assumed that the inert-strength distribution of the substrates is not altered through any interaction between the metal and alumina during sintering. Although the validity of this assumption cannot be proven directly, its fairness can be verified by comparing the predicted failure probabilities based on Eqn. (13) with actual experience.

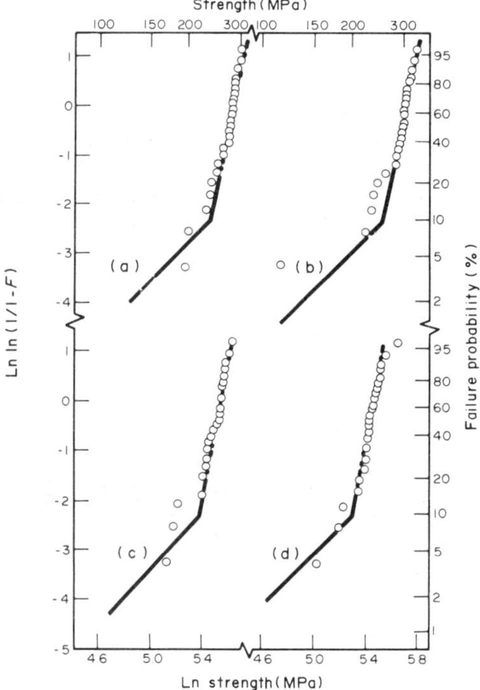

Figure 2
Comparison of theoretical predictions versus measurements of the strength distributions of polycrystalline alumina in a KOH cleaning solution at 100 °C and in a gold-plating bath at 78 °C: (a) KOH solution; ○, $\dot{\sigma}$ = 0.15 MPa s^{-1}; —, predicted. (b) KOH solution; ○, $\dot{\sigma}$ = 66.9 MPa s^{-1}; —, predicted. (c) Gold-plating bath; ○, $\dot{\sigma}$ = 1.31 MPa s^{-1}; —, predicted. (d) Gold-plating bath; ○, $\dot{\sigma}$ = 0.12 MPa s^{-1}; —, predicted

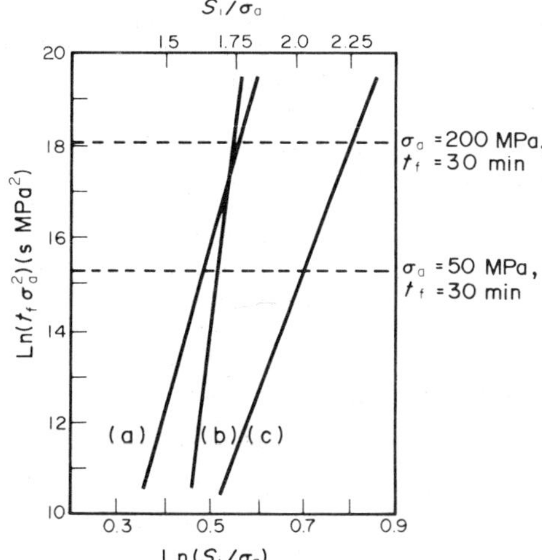

Figure 3
Lifetime-prediction diagram for polycrystalline alumina in (a) moist air, (b) KOH cleaning solution and (c) gold-plating bath

The residual tensile stress in a substrate is dependent on the thermal-expansion mismatch between the metal and the ceramic; in molybdenum–alumina substrates, residual stresses of 200 MPa can be generated. If substrates having such stresses are subjected to the KOH cleaning solution or the gold-plating solution for 30 min, respective S/σ_a ratios of 1.75 and 2.27 are predicted (Fig. 3). Since σ_a is known, S can be calculated and a failure probability determined. With a 200 MPa residual stress, an 8% failure rate would be expected for the substrates in the 10% KOH solution, and rate of 40% for the substrates in the gold-plating bath. Also, substrates having a residual stress of this magnitude would have a failure rate of 8% in moist air. Clearly, failure rates of this magnitude are unacceptable.

The failure probability of the substrates can be lowered by (a) using a stronger alumina, (b) reducing the processing times, or (c) reducing the residual stress in the substrates. The obvious choice is to reduce the residual stress. This can be accomplished by reducing the thermal-expansion mismatch between the metal and the ceramic, by altering the geometry of the system or by a combination of both. Since the thermal-expansion differential can be reduced only by altering the metal configuration and/or the ceramic, this choice alone is not particularly attractive. Similarly, if the stress is lowered only through geometric redesign, the potential geometric density of the metal in the system becomes limited. Thus, a combination of both approaches is generally considered most effective in reducing stress.

Regardless of the technique used, if the stress in the ceramic is lowered to, for example, 50 MPa, it can be determined from Fig. 3 that the failure probabilities become 0.1 and 0.2% respectively. While the possibility of stress-corrosion cracking of substrates during processing is not completely eliminated by reducing the residual stress, the resultant failure rates are now low enough to offer a minimal impact on product reliability.

2.2 Thermal Protection System for the US Space Shuttle

Low-density ceramics have a great potential for use in the thermal-protection system (TPS) for space vehicles. The current TPS in the US Space Shuttle is based on a rigid ceramic tile that has a network structure of silica glass fibers. These materials were developed to withstand severe environments, particularly with respect to re-entry and reusability. In addition to the requirements of low density and low thermal diffusivity, the titles must also withstand a variety of thermal and structural stresses during flight.

The major portion of the TPS for the Space Shuttle is based on glazed silica tiles that are adhesively bonded to a strain-isolation pad (SIP), which in turn is bonded to the aluminum surface of the shuttle. The predominant structural stresses act perpendicular to the aluminum skin, which is unfortunately the weakest direction within the ceramic tile. With the initial TPS (called the undensified tile/SIP system), failure occurred in the ceramic, but very close to the ceramic tile–SIP interface due to interfacial stress concentrations set up by the SIP. To improve the reliability of this TPS, each tile was subjected to a tensile overload proof test. For those tiles that failed the proof test, the strength of the tile in the region near the tile–SIP interface was increased by densifying the tile surface to eliminate the interfacial stress concentration. The densified tile/SIP system possessed a strength approximately double that of the undensified system, and failure occurred in the tile but away from the interface. After the densified tiles were reattached to the Shuttle, they were again subjected to a proof test.

To evaluate proof testing as a means of assuring mission success for the TPS of the Space Shuttle, time-dependent strength data were obtained on the ceramic tile itself, the undensified tile/SIP system and the densified tile/SIP system. These strengths were measured in a uniaxial tensile test as a function of stressing rate. This data, coupled with fracture-mechanics theory, could then define the strength allowables and probability for mission success of the TPS.

Figure 4 gives the strength–stressing rate data for the undensified tile/SIP system, and the data for the densified system are shown in Fig. 5. From these figures, it can be seen that the values of the fatigue parameter N is approximately the same for the undensified and densified systems, although the densified system has a much higher strength. Indeed, as shown in Fig. 5, the strength behavior of the densified system is very similar to that of the ceramic tile alone. Thus, it is believed that the strength behavior of both tile systems is controlled by the ceramic. In support of this, fractographic analysis indicates that failure of the ceramic tile occurs by fracture and fragmentation of the individual silica fibers.

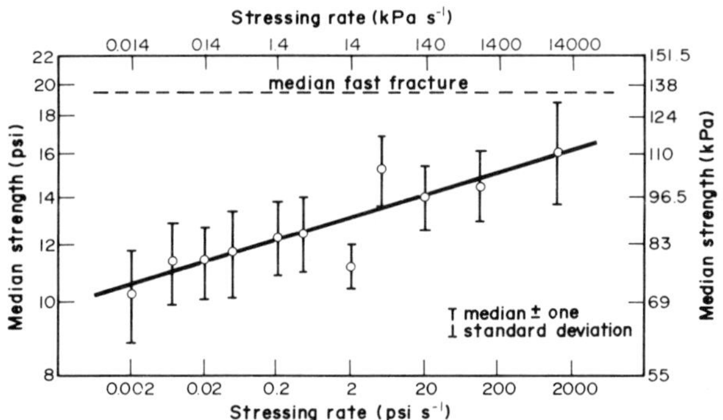

Figure 4
Fracture strength in ambient air as a function of stressing rate for the undensified tile/SIP system: $N \approx 34$, $\ln B \approx 8.9$

To substantiate the applicability of fracture-mechanics theory in predicting the effectiveness of proof testing, the initial strength distribution of the undensified tile/SIP system was compared with the after-proof strength distribution of a set of samples proof tested up to 70 kPa (Fig. 6). With this proof test, approximately 30% of the samples failed, and Eqn. (9) coupled with Eqn. (3) was used to predict the after-proof strength distribution. It is quite evident that proof testing eliminated the weak samples and improved the strength distribution, and that there is good agreement between theory and experiment.

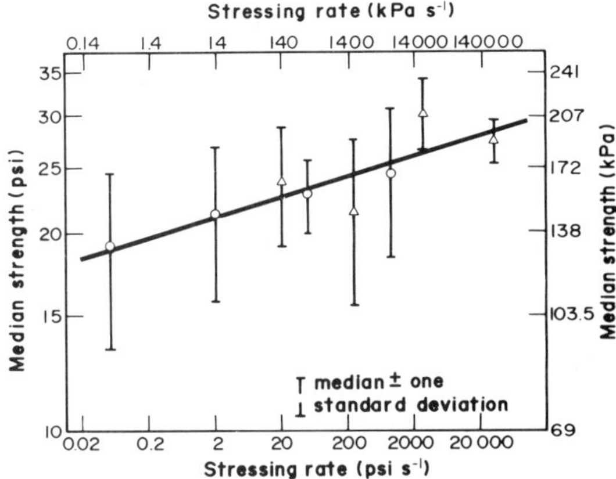

Figure 5
Fracture strength in ambient air as a function of stressing rate for the densified tile/SIP system and the ceramic tile alone: $N \approx 32$; ○, ceramic tile material; △, densified tile/SIP system

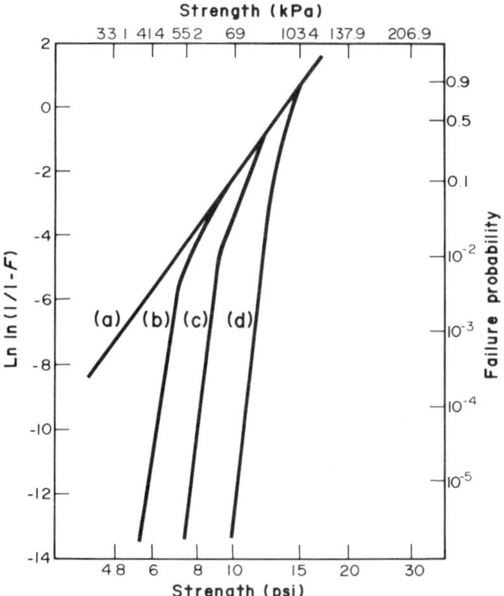

Figure 7
Proof-test predictions for undensified tile/SIP system in ambient air: (a) initial distribution; (b) predicted, 6 psi proof; (c) predicted, 8 psi proof; (d) predicted, 10 psi proof

Figure 7 summarizes proof-test predictions for the undensified tile/SIP system, where the maximum proof stresses are 6 psi (41 kPa), 8 psi (55 kPa) and 10 psi (70 kPa). It is clear from the initial strength distribution why proof testing is necessary for the undensified system. For example, it is expected that many of the tiles will be subjected to 33 kPa stress during flight. For such stresses, failure probabilities of about 10^{-3} would be predicted from the initial strength distribution. By proof testing the tile at 41 kPa, the failure probability for a required strength of 33 kPa is decreased to 10^{-8}. Thus, by eliminating about 1% of the undensified tiles, the failure probability at a strength level of 33 kPa is reduced five orders of magnitude.

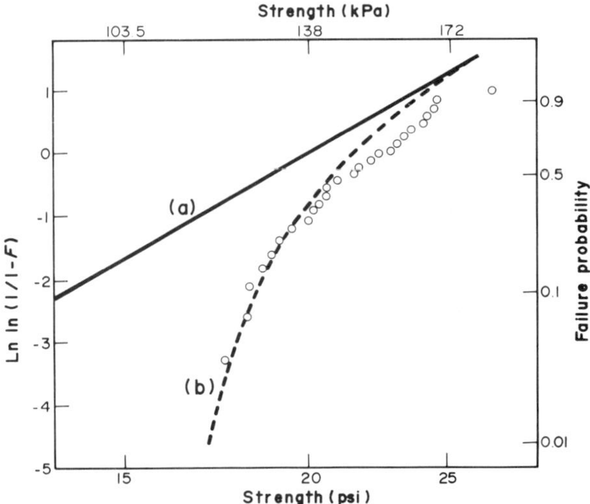

Figure 6
Comparison of the strength in ambient air of the undensified tile/SIP system before and after proof testing: (a) initial distribution; (b) predicted after-proof distribution; ○, after-proof data

2.3 Proof Testing of Grinding Wheels

Overspeed proof testing is used in the grinding-wheel industry to assure individual wheel safety by providing protection against excessive variability in

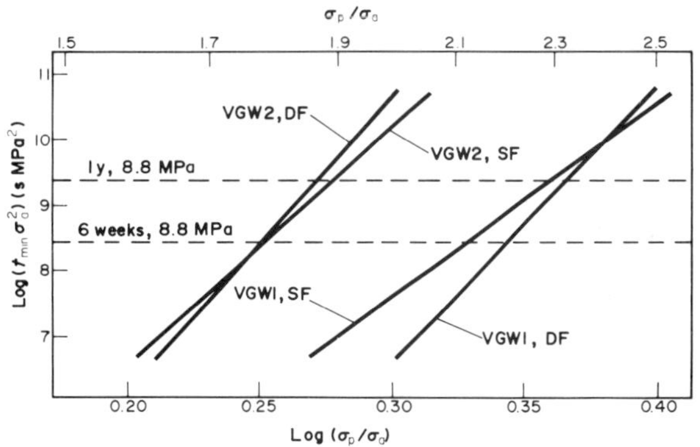

Figure 8
Minimum-lifetime predictions comparing static fatigue and dynamic fatigue techniques for VGW1 and VGW2 in aqueous coolant at 35 °C

wheel strength. Empirical standards have been established for overspeeding on the basis of measurements of relative strengths and experience from wheel usage. To place overspeed proof testing on a more sound, fundamental basis, fracture-mechanics principles were applied to the verification of the use of overspeed proof testing in assuring against failure of vitrified grinding wheels. Two vitrified-wheel specifications were used in this study—17A90-L5-VX2 and 2A601-K4-V9676 (for purposes of identification, these two specifications are referred to as VGW1 and VGW2 respectively).

To measure the crack-growth parameters appropriate to VGW1 and VGW2 in a commercial coolant, stress-rupture and stressing-rate experiments were used. Figure 8 compares minimum-lifetime predictions for these two vitrified grinding wheels based on the two test techniques (labelled static fatigue (SF) and dynamic fatigue (DF)). This figure is based on Eqn. (12), where S_{min} is taken to be equal to σ_p. Since the maximum applied stress in service for these two vitrified grinding wheels was calculated to be 8.8 MPa and the desired minimum lifetime in service was 6 weeks, it is evident from Fig. 5 that for a given vitrified-grinding-wheel composition, the σ_p/σ_a ratios based on the two test techniques are not significantly different. From the σ_p/σ_a ratio, the overspeed proof test can then be calculated.

Table 1 summarizes these proof-test predictions, where the spin-proof-ratio (SPR) is defined to be the overspeed proof divided by the operating speed. The predicted SPRs for the vitrified grinding wheels evaluated in this study can be compared with those actually used—1.80 for VGW1 and 1.50 for VGW2.

Table 1
Prediction of the spin-proof-ratio (SPR) necessary to ensure against delayed failure in service of vitrified grinding wheels for a 6-week lifetime at 8.8 MPa

Material	Test technique	σ_p (MPa)	SPR
VGW1	static fatigue	130	1.57
VGW1	dynamic fatigue	134	1.60
VGW2	static fatigue	108	1.45
VGW2	dynamic fatigue	108	1.45

It is evident from these results that the SPR actually used for these wheel specifications is greater than that predicted on the basis of fracture-mechanics theory. Therefore, it is believed that the overspeed proof test currently used for these vitrified-grinding-wheel compositions is adequate for preventing fatigue failure in service. In support of this conclusion, it is significant to note that there has never been a known fatigue failure in service for these wheel specifications.

See also: Fracture: Statistical Theories; Mechanics of Materials: An Overview

Bibliography

Davidge R W 1979 *Mechanical Behavior of Ceramics*. Cambridge University Press, Cambridge, UK

Davies D G S 1973 The statistical approach to engineering design in ceramics. *Proc. Br. Ceram. Soc.* 22: 429–52

Evans A G 1974 Fracture mechanics determination. In: Bradt R C, Hasselman D P H, Lange F F (eds.) 1974 *Fracture Mechanics of Ceramics*, Vol. 1. Plenum, New York, pp. 17–48

Evans A G, Fuller E R 1974 Crack propagation in ceramic materials under cyclic loading conditions. *Metall. Trans.* 5: 27–33

Evans A G, Langdon T S 1976 Structural ceramics. *Prog. Mater. Sci.* 21: 174–285

Evans A G, Wiederhorn S M 1974 Proof testing of ceramic materials—An analytical basis for failure prediction. *Int. J. Fract.* 10: 379–92

Freiman S W 1980 Fracture mechanics of glass. In: Uhlmann D R, Kreidl N J (eds.) 1980 *Glass Science and Technology*, Vol. 5, *Elasticity and Strength in Glasses*. Academic Press, New York, pp. 21–78

Green D J, Ritter J E, Lange F F 1982 Fracture behavior of low density fibrous ceramics. *J. Am. Ceram. Soc.* 65: 141

Hartsock D L, McLean A F 1984 What the designer with ceramics needs. *Am. Ceram. Soc. Bull.* 63(2): 266–70

Humenik J N, Ritter J E 1980 Susceptibility of alumina substrates to stress corrosion cracking during wet processing. *Bull. Am. Ceram. Soc.* 59: 1205–11

Ritter J E 1978 Engineering design and fatigue failure of brittle materials. In: Bradt R C, Hasselman D P H, Lange F F (eds.) 1978 *Fracture Mechanics of Ceramics*, Vol. 4. Plenum, New York, pp. 667–86

Ritter J E, Wulf S A 1978 Evaluation of proof testing to assure against delayed failure. *Bull. Am. Ceram. Soc.* 57: 186–89

Ritter J E, Wiederhorn S M, Tighe N J, Fuller E R 1980 Application of fracture mechanics in assuring against fatigue failure of ceramic components. In: Burke J J, Gorum A E, Katz R V (eds.) 1980 *Ceramics for High Performance Applications*, Vol. 3, *Reliability*. Plenum, New York, pp. 503–33

Wiederhorn S M, Ritter J E 1979 Application of fracture mechanics concepts to structural ceramics. In: Freiman S W (ed.) 1979 *Fracture Mechanics Applied to Brittle Materials*, ASTM Special Technical Publication 678. American Society for Testing and Materials, Philadelphia, Pennsylvania, pp. 202–14

J. E. Ritter

Fracture: Statistical Theories

The practical importance of statistical fracture theory rests primarily on the fact that brittle materials characteristically exhibit a large dispersion in strength. The strongest and also the most refractory materials tend to be brittle. In addition, semiconductors, insulators and materials transparent to electromagnetic radiation are generally brittle. Consequently, the use of brittle materials in structural components exposed to severe environments and in load-carrying members of electronic and optical devices is often unavoidable. When brittle materials are employed in practical structures, the designer must ensure an acceptably low probability of failure during service. Typically, on the basis of laboratory data on a limited number of specimens loaded in simple tension or bending, the probability of failure must be calculated for structural members of different sizes and shapes and under completely different loading conditions. The tool for accomplishing this is fracture statistics.

The basic assumption of statistical fracture theory is that the reason for the variations in strength of nominally identical specimens is their varying content of randomly distributed flaws. The strength of a specimen thus becomes the strength of its weakest flaw, just as the strength of a chain is that of its weakest link. The first to apply the weakest-link concept quantitatively appears to have been Chaplin (Lieblein 1954), who used it in 1880 to explain the observed length–strength relation for copper wire. It was Weibull (1939a, 1939b), however, who made the subject a standard engineering design tool, and his theories are still the basis for most calculations in this field (Weibull 1977).

Before discussing weakest-link theory, it is appropriate to consider its range of applicability. Clearly, the fracture of a single fiber is similar to the breaking of a chain. The fiber may be regarded as a linear array of very short elements, and the fracture of any element causes failure of the fiber as a whole. Similarly, when an isolated crack in an elastic body is sufficiently stressed normal to its plane, it becomes unstable and grows catastrophically, causing fracture. The fracture stress of the entire body is determined by the weakest crack. On the other hand, there are many situations to which weakest-link theory does not apply. For instance, in a bundle of fibers the first fiber failure does not ordinarily cause failure in the bundle as a whole; rather, failure is the result of damage accumulation (Daniels 1945, Rosen 1964) (see *Fiber Bundles: Strength Statistics*). Similarly, in some stress states (e.g., pure compression), a crack in an elastic body may grow in a direction which results in crack arrest. Here also, failure is a result of progressive damage and weakest-link theory does not apply.

1. Weakest-Link Theory

To derive the fundamental equation of weakest-link theory, let it be assumed that a stressed solid can fail due to any of a number of independent and mutually exclusive mechanisms or causes, each involving an infinitesimal probability of failure $(\Delta P_f)_i$. The probability that the ith mechanism will not cause failure is $(P_s)_i = 1 - (\Delta P_f)_i$. The overall probability of survival is the product of the individual probabilities of survival:

$$P_s = \prod_i (P_s) = \prod_i [1 - (\Delta P_f)_i]$$
$$\cong \prod_i \exp[-(\Delta P_f)_i] = \exp[-\sum_i (\Delta P_f)_i] \quad (1)$$

The sum of the individual probabilities of failure appearing in the final equality above was called by Weibull the "risk of rupture" and was given the symbol B.

In evaluating B, Weibull added together the probabilities of failure of all the elements of volume ΔV in the entire body. The probability of failure of the ith element ΔV_i in simple tension σ, for example, is

$$(\Delta P_f)_i = n(\sigma)\Delta V_i \quad (2)$$

where $n(\sigma)$ is the number of flaws per unit volume with a strength less than σ. If $n(\sigma)$ is less than unity, it can be regarded as the probability that such a flaw will occur in a unit volume. Thus, the probability of failure is given by

$$P_f = 1 - P_s = 1 - \exp[-\int dV n(\sigma)] \quad (3)$$

For a uniformly stressed body

$$P_f(\sigma) = 1 - \exp[-V n(\sigma)] \quad (4)$$

or

$$n(\sigma) = V^{-1} \ln P_s^{-1}(\sigma) \quad (5)$$

If $P_f(\sigma)$ is determined by testing a number of specimens in simple tension, Eqn. (5) can be used to obtain $n(\sigma)$ and Eqn. (3) can then be used to evaluate the probability of failure of a body of arbitrary size and shape and nonuniformly distributed tensile stress. By assuming, as Weibull did, that compressive stresses do not contribute to fracture, the above procedure can also be used to calculate the probability of failure in pure bending or indeed any combination of uniaxial tension and compression. A limitation to this procedure is that in ignoring fracture due to compression, failure in some stress states is not adequately accounted for.

To apply weakest-link theory it is convenient to express $n(\sigma)$ in parametric form. Weibull offered two alternative expressions for $n(\sigma)$:

$$n(\sigma) = (\sigma/\sigma_0)^m = k\sigma^m \quad (6a)$$

and

$$n(\sigma) = \left(\frac{\sigma - \sigma_u}{\sigma_0}\right)^m$$
$$= k(\sigma - \sigma_u)^m \quad (\text{for } \sigma > \sigma_u)$$
$$= 0 \quad (\text{for } \sigma < \sigma_u) \quad (6b)$$

If Eqn. (6a) is inserted in Eqn. (4) and P_f is plotted against σ on probability paper, a slightly curved line results. The curvature depends on m and reverses sign around $m = 4$. Since a straight line would represent the Gaussian error function, it is evident that the relation

$$P_f = 1 - \exp(-Vk\sigma^m) \quad (7)$$

can be regarded as a somewhat skewed Gaussian curve. The two parameters k and m, or σ_0 and m, can be chosen to yield the observed average failure stress and coefficient of variation, respectively. It can be shown that

$$m \cong \frac{1.2}{\text{coefficient of variation}} \quad (8)$$

Thus, low values of m correspond to large dispersions in fracture stress and vice versa.

If the main concerns in analyzing fracture data are the mean fracture stress and standard deviation, it matters little whether the normal distribution or the Weibull distribution is used. However, it can be shown that the normal distribution is incompatible with weakest-link theory. Moreover, the normal distribution is philosophically objectionable in that it implies that a body has a nonvanishing probability of tensile failure when it is stress-free or even in compression. Equation (6a) implies that at zero stress or when in compression, tensile failure cannot occur. Use of Eqn. (6b) is indicated when there is a lower limit $\sigma \neq 0$ for tensile failure. When $\sigma_u = 0$, Eqn. (6b) reduces to Eqn. (6a).

Use of Eqn. (7) leads directly to a size effect. If a body of volume V_1 has a certain probability of failing at stress σ_1 and a second body of volume V_2 has the same probability of failing at σ_2, then

$$V_1\sigma_1^m = V_2\sigma_2^m \quad (9)$$

or

$$\sigma_2 = \sigma_1(V_1/V_2)^{1/m} \quad (10)$$

If the flaws are surface-distributed rather than volume-distributed, the same relation applies with the surface area A replacing volume V. If they are length-distributed, V is replaced by L.

Problems involving nonuniform uniaxial stress distributions can be solved by inserting Eqn. (6a) in Eqn. (3) and carrying out the indicated operation. In this way it is readily shown that for a rectangular beam in pure bending

$$P_f = 1 - \exp\left[-\frac{Vk\sigma^m}{2(m+1)}\right] \quad (11)$$

where σ is the maximum tensile stress in the beam. Similar formulae have been worked out for beams of circular or rectangular cross section in three-point bending, four-point bending and so on. If Eqn. (6b) is used, the same formulae apply with $(\sigma - \sigma_u)$ substituted for σ.

The Weibull parameters are conventionally determined by conducting N tensile tests and designating the observed fracture stresses in ascending order $\sigma_1, \ldots \sigma_N$. It is then usually assumed that

$$P_f(\sigma_j) = j/(N+1) \quad (12)$$

A more sophisticated statistical treatment leads to the relation

$$P_f(\sigma_j) = \frac{j - 0.3}{N + 0.4} \quad (13)$$

A simple technique for determining the parameters m, σ_u and k is to write Eqn. (7) in the form

$$\ln\ln(1 - P_f)^{-1} = \ln Vk + m\ln(\sigma - \sigma_u) \quad (14)$$

Next, $y = \ln\ln(1 - P_f)^{-1}$ is plotted against $x = \ln(\sigma - \sigma_u)$ for various assumed values of σ_u. The value of σ_u adopted is that for which the N data points most nearly lie on a straight line. The slope of the line chosen is m, and Vk is the value of $\ln(1 - P_f)^{-1}$ for $\sigma = \sigma_u + 1$. When Weibull's two-parameter form is used, σ_u is arbitrarily taken to be zero.

It may be noted that a least-squares fit in $\ln\ln(1 - P_f)^{-1}$ versus $\ln(\sigma - \sigma_u)$ space is in general not a least-squares fit in P_f versus $(\sigma - \sigma_u)$ space. Consequently, the Weibull parameters found in the manner described do not yield the best possible fit to the data. However, if the scatter in the data is not large, the fit will be adequate for most practical purposes.

2. Weakest-Link Theory for Polyaxial Stress Conditions

Weibull gave a procedure for calculating P_f for polyaxial stress states when the failure statistics for simple tension are known. Basically, his procedure is to calculate B by averaging the tensile stress in all directions (Weibull 1939a, Shur 1971). This is intuitively plausible but not rigorous. Presumably, in part at least because of doubts concerning Weibull's procedure, some investigators (Freudenthal 1968, Stanley et al. 1977) have chosen to use an approximation in which it is assumed that with respect to fracture the principal stresses act independently:

$$P_s(\sigma_1, \sigma_2, \sigma_3) = P_s(\sigma_1)P_s(\sigma_2)P_s(\sigma_3) \quad (15)$$

where σ_1, σ_2 and σ_3 are the principal stresses.

In this section the explicit assumption is made that the flaws responsible for fracture are microcracks in the material. It is further assumed that the cracks do not interact and that each crack has a critical stress σ_c, defined as the remote tensile stress which will cause fracture when applied normal to the crack plane. Fracture under combined stresses occurs when the effective stress σ_e acting on the crack is equal to or greater than σ_c. The effective stress is some function of the applied stresses at the location of the crack, its precise form depending on the fracture criterion employed. Before particular fracture criteria are discussed, however, the statistical theory of fracture is reformulated for an arbitrary fracture criterion by leaving σ_e unspecified.

For purposes of analysis, it is convenient to group the cracks according to their location, the applied stress state and the crack critical stress. It is assumed that the stress state varies slowly, so that within a volume element ΔV all cracks are subject to the same macroscopic stress. It is also assumed that the cracks are randomly distributed and oriented, so that a function $N(\sigma_c)$ can be defined as the number of cracks per unit volume having a critical stress equal to or less than σ_c. The probability that a crack having a critical stress in the range σ_c to $(\sigma_c + d\sigma_c)$ exists in volume element ΔV is then $\Delta V[dN(\sigma_c)/d\sigma_c]d\sigma_c$. If such a crack is actually present, the probability that it will fracture depends on its orientation, the stress state and the fracture criterion. It is assumed there is a solid angle Ω such that if the normal lies within Ω, then $\sigma_e > \sigma_c$. Since the cracks are randomly oriented, the probability that a crack will fracture under an applied stress Σ is thus $\Omega(\Sigma,\sigma_c)/4\pi$.

Now the probability of failure due to a crack in the critical stress range $d\sigma_c$ located in volume element ΔV is the product of the above probabilities:

$$(\Delta P_f)_i = \left(\Delta V \frac{dN(\sigma_c)}{d\sigma_c} d\sigma_c\right)\left(\frac{\Omega(\Sigma, \sigma_c)}{4\pi}\right) \quad (16)$$

Substituting Eqn. (16) into Eqn. (1) and changing sums into integrals:

$$P_s = \exp\left(-\int dV \int d\sigma_c \frac{dN}{d\sigma_c}\frac{\Omega}{4\pi}\right) \quad (17)$$

Since $\Omega(\Sigma, \sigma_c)$ can be calculated for any stress state Σ when the fracture criterion has been chosen and $P_s(\Sigma)$ can be obtained experimentally for a conveniently chosen stress state, Eqn. (17) can be regarded as an integral equation from which $N(\sigma_c)$ can be obtained. Equation (17) can then be used again to evaluate P_s for any other stress state. Integration by parts leads to the alternative expression

$$P_s = \exp(-\int dV \int d\Omega\, N(\sigma_e)) \quad (18)$$

where σ_e is the effective stress acting on the crack when its normal lies in the element of solid angle $d\Omega$.

3. Relation to Griffith Theory

Griffith (1924) published a theory for the fracture of solids under biaxial stress conditions. He assumed the presence of a crack in the form of a nearly flat elliptical cylinder (the Griffith crack) with its axis normal to the plane of the stresses. He then calculated the maximum peripheral tensile stress on the free surface of such a crack as a function of crack orientation and assumed that fracture would occur whenever this maximum value exceeded the intrinsic strength of the material.

This fracture criterion leads to the failure envelope shown by the solid line in Fig. 1. In the tension–tension quadrant and most of the tension–compression quadrant, fracture is initiated in a crack whose plane is normal to the largest principal tensile stress. This results in a straight-line maximum applied tensile stress failure envelope. The curved part of the envelope near the compression axis corresponds to failure initiation in an obliquely oriented crack.

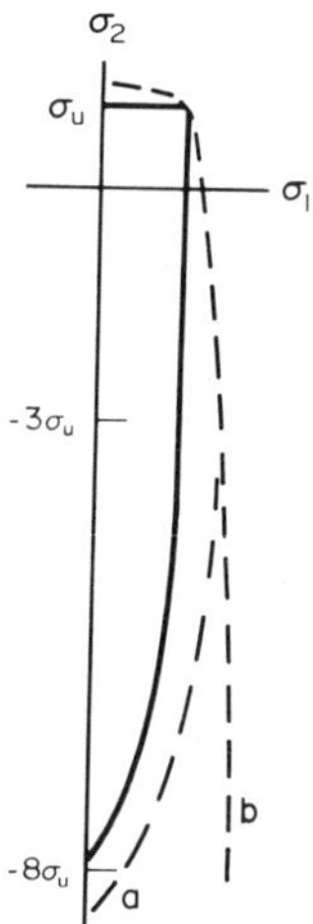

Figure 1
Comparison of fracture theories for biaxial stresses: a,b, curves for shear-sensitive and shear-insensitive cracks, respectively, in weakest-link theory

In any crack-based statistical theory it is most unlikely that the weakest crack will be oriented at exactly the angle of maximum vulnerability, so Griffith's curves must be regarded as the lower limit for fracture; in other words, they represent the contour for $P_f(\sigma_1,\sigma_2) \ll 1$. A typical $P_f(\sigma_1,\sigma_2) = 0.5$ contour, for the case of shear-insensitive cracks, is shown as a dashed curve in Fig. 1. This curve covers both quadrants almost completely but does not provide for failure due to pure compression. In the case of shear-sensitive cracks, the entire $P_f(\sigma_1,\sigma_2) = 0.5$ contour predicted by weakest-link theory falls close to Griffith's failure envelope as shown schematically in Fig. 1. For high compressive stresses, the results differ significantly from those for shear-insensitive cracks. However, as noted earlier, in this region weakest-link theory does not apply. Thus, Weibull theory and the more refined theories to be discussed later are limited to cases in which no principal compressive stress exceeds the maximum principal tensile stress by more than a factor of about three in absolute value.

4. Effective Stress

If σ_e is assumed to be simply σ_n, the component of tensile stress normal to the crack plane, then Eqn. (18) becomes the rule for polyaxial stress states proposed by Weibull. However, this is a rather crude approximation.

The direct stress applied normal to a crack plane results in a very high local stress at the root of the crack, whereas direct stresses in the plane of the crack do not. The shear stress τ applied parallel to the crack plane also results in very high local stresses. Consequently, the effective stress σ_e causing fracture is a function of both σ_n and τ and also depends on the size and shape of the crack, and sometimes on Poisson's ratio ν as well.

Unfortunately, there is as yet no consensus among fracture mechanics experts regarding the proper criterion for mixed-mode fracture. Among the best-known theories are that fracture occurs when:

(a) the normal stress σ_n reaches some critical value;

(b) the tensile stress on some part of the surface of the crack cavity exceeds the intrinsic strength of the material;

(c) the strain energy released during coplanar crack extension is equal to the energy of the newly created surfaces;

(d) the strain energy released during noncoplanar crack extension equals the energy of the newly created surfaces; or

(e) the minimum strain energy density on a circle about the crack tip reaches a critical value.

The way in which statistical fracture theory is affected by the choice of fracture criterion must now be considered. A material having uniaxial fracture properties adequately defined by Weibull's two-parameter representation is assumed:

$$P_s(\sigma,0) = \exp(-Vk_1\sigma^m) \tag{19}$$

The fracture statistics for combined stresses take the same form except for a change in the value of k. Thus, for equibiaxial tension

$$P_s(\sigma,\sigma) = \exp(-Vk_2\sigma^m) \tag{20}$$

It is found that the ratio k_2/k_1 is a function of the crack shape, the fracture criterion selected and also the parameter m. This dependence is illustrated in Fig. 2.

For $P_f \ll 1$,

$$k_2/k_1 = P_f(\sigma,\sigma)/P_f(\sigma,0) \tag{21}$$

Consequently, when $P_f(\sigma,0)$ is known and the assumption $\sigma_e = \sigma_n$ (leading to the Weibull theory) is used to determine $P_f(\sigma,\sigma)$, it is evident from Fig. 2 that the result is very conservative. The best agreement was obtained using energy density theory and assuming penny-shaped cracks. Surprisingly poor agreement was obtained using an energy release rate criterion for noncoplanar crack extension (Palaniswamy and Knauss 1978, Wu 1978). The reasons for this are not clear at present.

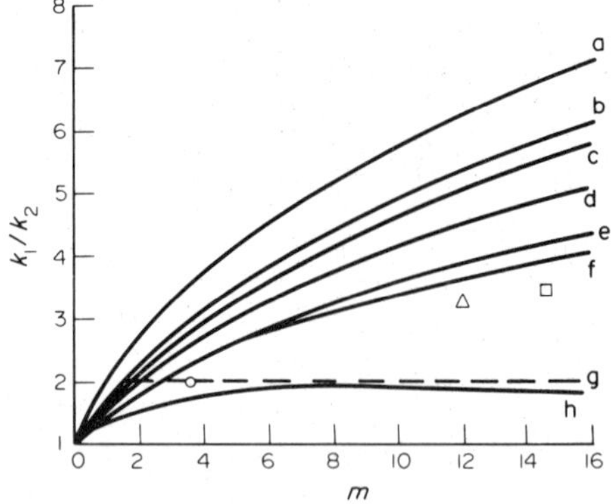

Figure 2
Comparison of fracture statistics for uniaxial and equibiaxial stress states (after Batdorf 1980, Batdorf and Heinisch 1978): a, $\sigma_e = \sigma_n$; b,c, maximum tensile stress for Griffith (G) and penny-shaped (PS) cracks, respectively; d,e, coplanar energy release rate for G and PS cracks, respectively; f, energy density for PS cracks; g, Eqn. (15); h, noncoplanar energy release rate for G cracks

5. Extension to Very Low Probability of Failure

A frequently encountered problem is that of determining the stress level at which a structure has a very small probability of failure, say $P_f \leq 10^{-4}$. It is usually not feasible to test 10^4 full-scale prototypes to establish this stress level experimentally. What can theory do to help?

First of all, it follows from Eqn. (14) that the Weibull plot for a volume V_2 can be obtained from that for V_1 by a vertical translation equal to $\ln(V_2/V_1)$. This is illustrated in Fig. 3, where the ratio V_2/V_1 is assumed to be 100. The important point to note is that if 100 specimens were tested, the Weibull failure curve would be established only over the region shown as a solid line. The vertical translation for a volume ratio of 100 is depicted as another solid line, which reveals the fact that at the lowest stress for which data are available, the probability of failure of the full-scale structure is 0.64.

Under these circumstances designers often succumb to the temptation to assume that the Weibull distribution function continues to apply outside the stress range in which it has been tested. Sometimes the validity of this assumption has been argued by citing the fact that the Weibull distribution is the only one of the three asymptotic forms of extreme-value

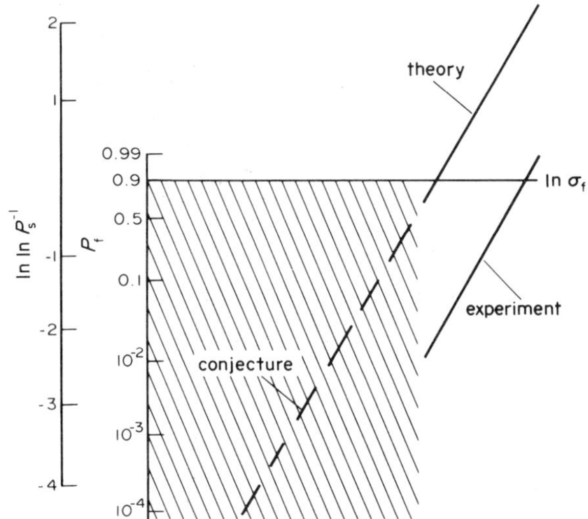

Figure 3
The problem of determining σ_f for $P_f \ll 1$ (schematic)

theory that is physically applicable to fracture phenomena. However, no law of physics or mathematics states that the strength distribution must asymptotically approach a limiting form. It has been shown that physically plausible crack size distributions may lead to fracture statistics that are not of the Weibull form (McClintock 1973, Batdorf 1975). Moreover, the experimental data frequently cannot be represented by a straight line on a Weibull plot. The projection of the failure line could be any monotonic extension into the shaded area of Fig. 3. A reliable extension will be possible only when the crack size distribution can be related to the material microstructure.

6. Concluding Remarks

The preceding discussion brings out the fact that if the flaws causing failure are cracks, the relation between the fracture statistics for different stress states differs somewhat from that postulated by Weibull. There are also other consequences of this assumption. One is that a crack of given geometric form is weaker near a free surface than in the interior. Thus, if cracks are uniformly distributed, there is a surface distribution of flaw strengths differing somewhat from the volume distribution, and both need to be taken into account. The relative importance of surface and interior flaws depends on crack size. There are also other effects associated with crack size. One is that when stress gradients are so high that the applied stress varies significantly over the crack dimension, the crack is weaker than it would be if subjected everywhere to the mean stress. This affects the relative strength in tension and bending. Also, when the thickness of a specimen is less than a few crack radii, strength may decrease as size decreases, thus reversing the usual size effect. Such a phenomenon has been noted in nuclear graphites (Brocklehurst 1977, Price 1975).

The above treatment of fracture statistics assumes the material to be an ideal elastic solid containing randomly distributed and oriented cracks. This type of analysis is the most fully developed and has been found to be useful in determining the short-term strength of many brittle materials at room temperature in a noncorrosive environment. When long-term strength or even short-term strength in a high-temperature or corrosive environment is to be predicted, subcritical crack growth must be taken into account (Wiederhorn et al. 1976). In some materials the flaws causing failure are not cracks but pores, collections of pores or abnormally large grains. For such flaws statistical failure theories are not yet available.

See also: Fracture: Reliability Criteria for Brittle Materials; Mechanics of Materials: An Overview

Bibliography

Batdorf S B 1975 Fracture statistics of brittle materials with intergranular cracks. *Nucl. Eng. Des.* 35: 349–60

Batdorf S B 1980 Comparison of the best known fracture criteria with data on naturally occurring cracks, UCLA-ENG-8040. University of California, Los Angeles

Batdorf S B, Heinisch H L 1978 Weakest link theory reformulated for arbitrary fracture criterion. *J. Am. Ceram. Soc.* 61: 355–58

Brocklehurst J E 1977 Fracture in polycrystalline graphite. In: Walker P L, Thrower P A (eds.) 1977 *Chemistry and Physics of Carbon: A Series of Advances*, Vol. 13. Dekker, New York, pp. 145–278

Daniels H E 1945 The statistical theory of the strength of bundles of threads I. *Proc. R. Soc. London, Ser. A* 183: 405–35

Freudenthal A M 1968 Statistical approach to brittle fracture. In: Liebowitz H (ed.) 1968 *Fracture*, Vol. 2. Academic Press, New York, pp. 591–619

Griffith A A 1924 The theory of rupture. In: Biezeno C B, Burgers J M (eds.) 1924 *Proc. 1st Int. Congress for Applied Mechanics.* Waltman, Delft, pp. 55–63

Lieblein J 1954 Two early papers on the relation between extreme values and tensile strength. *Biometrika* 41: 559–60

McClintock F A 1973 Statistics of brittle fracture. In: Bradt R C, Hasselman D P H, Lange F F (eds.) 1973 *Fracture Mechanics of Ceramics*, Vol. 1. Plenum, New York, pp. 93–114

Palaniswamy K, Knauss W G 1978 On the problem of crack extension in brittle solids under general loading. In: Nemat-Nasser S (ed.) 1978 *Mechanics Today*, Vol. 4. Pergamon, Oxford, pp. 87–148

Price R J 1975 Statistical study of the strength of near-isotropic graphite. *Extended Abstracts, 12th Biennial Conf. Carbon.* American Carbon Society, University Park, Pennsylvania, pp. 145–46

Rosen B W 1964 Tensile failure of fibrous composites. *AIAA J.* 12: 1985–91

Shur D M 1971 Statistical criteria of the danger of material fracture in a complex stress state. *Mashinovedenie* 1971(1): 51–58

Stanley P, Sivill A D, Fessler H 1977 Applications of the four-function Weibull equation to design of brittle components. In: Bradt R C, Hasselman D P H, Lange F F (eds.) 1977 *Fracture Mechanics of Ceramics*, Vol. 3: *Flaws and Testing*. Plenum, New York, pp. 51–66

Weibull W 1939a A statistical theory of strength of materials. *Ingenioersvetenskapsakad. Handl.* 151: 1–29

Weibull W 1939b The phenomenon of rupture in solids. *Ingenioersvetenskapsakad. Handl.* 153: 1–55

Weibull W 1977 *References on the Weibull Distribution*, Report A20: 23. Forvarets Teletekniska Laboratorium, Stockholm, pp. 1–141

Wiederhorn S M, Fuller E R, Mandel J, Evans A G 1976 An error analysis of failure prediction technique derived from fracture mechanics. *J. Am. Ceram. Soc.* 59: 403–11

Wu C H 1978 Maximum-energy-release-rate criterion applied to a tension–compression specimen with crack. *J. Elast.* 8: 235–57

S. B. Batdorf

Fracture Toughness Improvement

The means of improvement of fracture toughness are generally discussed with reference to fracture models. With this method, it is necessary to identify the operative fracture modes, the specific fracture mechanisms and associated failure criteria, the loading conditions that satisfy the failure criteria and the effects of metallurgical variables on the achievement of the failure criteria. Here, this process is illustrated through consideration of cleavage (brittle) fracture and microvoid coalescence (ductile fracture).

1. Stress and Strain Distributions in the Plastic Zone

Elastic stresses at the tip of a sharp crack exceed the yield stress even for small applied loads. As a load is applied to a cracked body, the crack blunts and the material surrounding the crack tip yields, so that a plastic zone develops in the region of the crack. Within the plastic zone, redistribution of stresses accompanies the large stretching associated with the crack-tip blunting process. The fracture process occurs within the plastic zone or at the elastic–plastic zone boundary; consequently, the stress and strain distributions within the plastic zone are of utmost interest.

The dimensions of the plastic zone and the distributions of stresses and strains are related to the crack-tip opening displacement δ_T measured across the crack opening near the blunted tip. δ_T is also related to the fracture-mechanics field parameters, critical values of which are measures of fracture toughness, so δ_T provides a useful link between the fracture process and fracture toughness.

Several different relationships between δ_T, stress intensity K and the J integral have been advanced, the differences probably being due largely to differences of the position at which δ_T is measured. Generally, smooth rounded blunting is a reasonable representation of the crack-tip blunting process, and δ_T (measured at the intersections of the flanks and rounded portions of the blunted crack) can be expressed as $\delta_T = 0.6K^2/\sigma_0 E$ where σ_0 is the yield strength.

At the first instance of plastic zone formation, the maximum value of the longitudinal stress ahead of the crack occurs at the plastic zone boundary and slightly exceeds the yield strength of the material. As δ_T increases, the position of the peak longitudinal stress is located within the plastic zone and the magnitude gradually increases (Griffiths and Owen 1971) and approaches a limiting value of about $3\sigma_0$ to $5\sigma_0$ depending on the strain-hardening characteristics of the material (McMeeking 1977) and flank angle (Santhanam and Bates 1979). The longitudinal strain also increases with increasing δ_T. It reaches very high values at the crack tip but rapidly drops off ahead of the crack tip, so that the intense strain field is virtually confined to the region between the crack tip and the position of maximum longitudinal stress, that is, about $2\delta_T$ ahead of the crack tip (Rice and Johnson 1970). The stress and strain distributions have been determined for blunted notches for a δ_T of five times the original notch opening and are shown in Fig. 1. These distributions are applicable for all values of δ_T greater than about three times the original crack

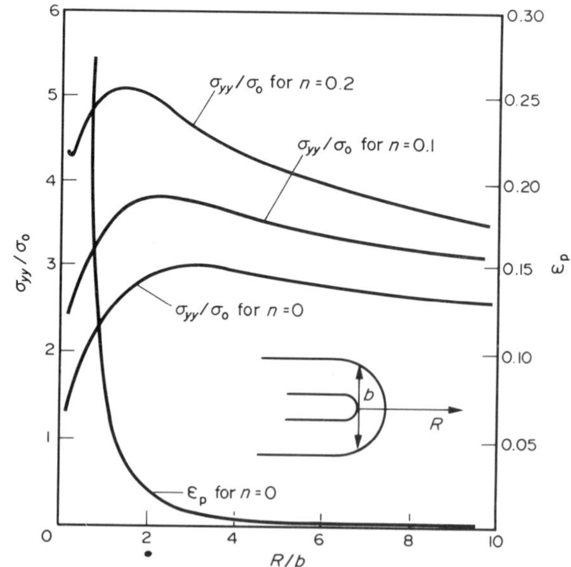

Figure 1
Stress and strain distribution directly ahead of notch for blunting to five times original notch width (after McMeeking 1977)

or notch opening (Bates 1981), so they represent virtually all cases of interest in the fracture process.

2. Failure Modes

The macroscopic fracture behavior and associated fracture appearance of structural metals and alloys are often divided into the three categories of ductile, brittle and mixed-mode fracture. Ductile failures generally exhibit a dull-gray fibrous fracture surface and are usually accompanied by extensive plastic yielding and shear lip formation. Brittle fractures generally appear smooth and flat with little evidence of plastic yielding. Often, brittle fracture surfaces appear bright and exhibit a fine crystalline texture. Mixed-mode fractures exhibit a mixture of ductile and brittle characteristics.

On a microscopic scale, several fracture modes commonly occur which affect, but do not exclusively control, the macroscopic fracture behavior. The most common of these are cleavage, quasi-cleavage, dimpled rupture, ductile cutting or alternate slip, intergranular separation and fatigue. Only cleavage and dimpled rupture are considered here.

Cleavage fracture (Fig. 2a) represents brittle fracture along crystallographic planes and occurs in body-centered-cubic and hexagonal-close-packed metals and alloys. The characteristic feature of cleavage fracture consists of flat facets which are generally about the size of ferrite grains in steel and which usually exhibit "river" markings. These markings are caused by the cleavage crack traversing the crystal along a number of parallel planes which form a series of plateaus and connecting ledges. The direction of the "river pattern" is the direction of local crack propagation.

Dimpled rupture (Fig. 2b) is common to a wide variety of metals or alloys of all strength levels. The characteristic feature of this type of fracture is cup-like depressions which may be equiaxed, parabolic or elliptical, depending on the stress state. In practice, several modes of dimple formation occur, but they are all related in that they occur by nucleation (usually from second-phase particles) and growth of microvoids. Fracture occurs when the microvoids are joined by coalescence (or otherwise), fracturing the ligaments between microvoids. This process is generally a highly ductile one (on a microscopic scale), since the growth of microvoids is strain controlled and the final coalescence occurs by internal necking of the ligaments between voids. Macroscopically, ductility may be high or low depending on the size, number and distribution of voids.

3. Cleavage Fracture Mechanism and Failure Criteria

The development of cleavage fracture involves sequential stages of nucleation (or initiation) and

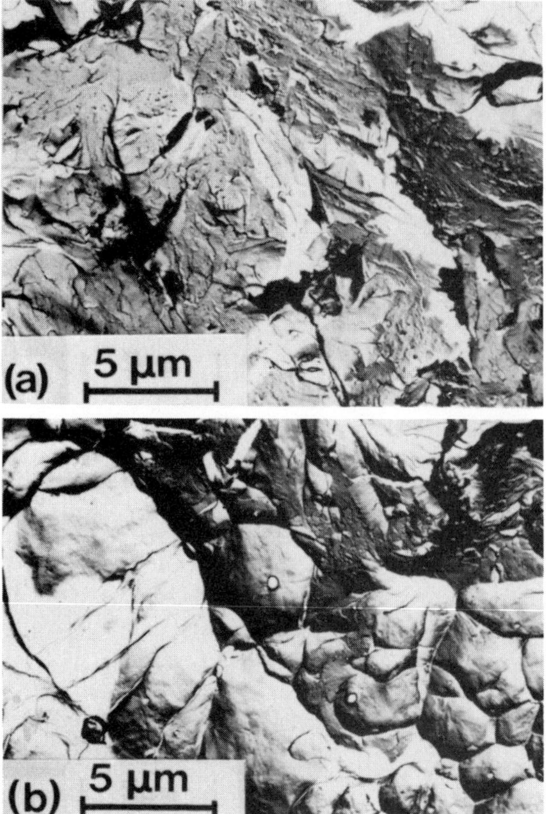

Figure 2
Fractographic features of fractures in a type 403 stainless steel: (a) cleavage fracture; (b) dimpled rupture microvoid coalescence

crack growth (or propagation). Since most materials of concern are highly inhomogeneous on a microscopic scale, the growth stage may be further divided into initial growth into the material immediately surrounding the "microcrack" nucleus and subsequent growth as obstacles are encountered. There is a general recognition that cleavage fractures can be described in terms of a critical stress criterion for failure; that is, failure occurs when the stress exceeds some critical value σ_F over a distance (termed the critical distance) which spans all potential obstacles to crack propagation. Since the stages of crack development are sequential, σ_F is determined by the most difficult obstacles to overcome (Tetelman and McEviley 1967).

The achievement of the critical stress for cleavage-crack nucleation is rarely fulfilled by the applied stress alone. Rather, localized stress enhancement caused by inhomogeneous plastic deformation is required. If twinning occurs, intersections of twins with other twins or grain boundaries are preferred nucleation sites (Hull 1962); however, twinning is frequently absent, so the majority of cleavage cracks are slip induced (Smith 1968). In such cases, nucleation is achieved by cracking of carbides or other hard particles (usually in grain boundaries) at the tips of long dislocation pile-ups. In any event (whether slip or twinning induced), local plastic deformation (yielding) is required for cleavage-crack nucleation.

After a cleavage crack is nucleated, crack growth first requires penetration into the surrounding matrix where the fracture energy may be higher. Subsequently, the first grain boundary may present obstacles to crack growth, and traversing the next grain may require even higher applied stress to achieve the critical resolved stress on available cleavage planes. Thus, the critical distance over which the required stress must be achieved is related to microstructural obstacles to crack propagation. It is usually about one or two grain diameters (Ritchie et al. 1973).

4. Ductile Fracture Mechanism and Failure Criteria

The most important ductile mode for fracture toughness considerations is microvoid coalescence (or dimpled rupture). Dimpled rupture occurs by distinct steps of void nucleation, growth and coalescence.

Virtually all important instances of microvoid nucleation occur at second-phase particles, mostly by interface separation, although sometimes by particle cracking similar to that previously described in the discussion of cleavage cracks. Both critical interfacial stress and energy criteria are applicable to interface separation and must be simultaneously satisfied. Heterogeneous strain aids in the attainment of the critical stress just as it does for cleavage-crack initiation. Heterogeneous strain is important enough that localized yielding may be considered a third requirement for void nucleation.

Typically, the lowest critical stress is exhibited by large particles (on a microstructural scale) or fragile particles that are weakly bonded to the matrix; examples are manganese sulfides, large oxides and silicates. These types of particles are preferred nucleation sites for microvoids. Next in order of preference are densely distributed noncoherent precipitates. In general, the critical stress for interface separation increases as particle size decreases and as spacing between particles increases.

For particles greater than about 10–20 nm in diameter, the energy criterion is satisfied soon after yielding. This criterion (that the release of elastic strain energy equals or exceeds the energy of the generated surfaces) is difficult to satisfy for small coherent particles (less than about 10–20 nm in diameter), even if the critical stress is exceeded. When interface separation is initiated at such particles, continually increasing strain is required to complete the nucleation step, that is, to complete the interface separation.

Void growth is primarily strain controlled, although it is influenced by the stress state and is particularly enhanced by triaxial stresses. In tensile tests, significant void growth occurs only after the onset of necking and primarily in the necked region. In fracture toughness tests, significant void growth occurs only after the void site is enveloped by the intense plastic strain portion of the plastic zone. As crack blunting continues, void growth occurs as a direct function of the initial void size and the ratio $\delta_T : \delta_0$ (where δ_0 is the value of δ_T at which the void growth commences), and is not significantly influenced by microstructural features or material properties (Bates 1981).

The void-growth process is terminated by crack–void or void–void coalescence. This process is controlled by microstructure (through its influence on initial void size and spacing) and δ_T. Once void growth and crack-tip blunting have sufficiently reduced the ligament spacing between a void and the crack tip, the separating ligament becomes unstable. A small additional increase in δ_T completes the void–crack-tip coalescence, causing an incremental change in the crack length. The ligament separation completing this event may occur by necking or by the secondary formation of a sheet of tiny voids. This process is repeated until δ_T becomes sufficiently large to achieve simultaneous instability of ligaments between several successive voids ahead of the blunted crack (Bates 1981). The fracture then becomes unstable, and the value of δ_T at that time determines the fracture toughness. Thus, toughness is determined by initial void size and void spacing, both of which are determined by void nucleation.

The metallurgical means to improve fracture toughness with respect to dimpled rupture can be analyzed in terms of minimizing void nucleation. First, however, it is necessary to examine the effect of stress–strain distributions on the nucleation event. This will be done using an example of a steel containing large sulfide particles and smaller, but much more numerous, carbide particles.

In plain-bar tensile tests, the stress–strain distributions are relatively uniform across the cross section, at least until necking has occurred. The large sulfide particles are preferred nucleation sites, and voids nucleate at these sites first, possibly soon after initial yielding. The critical stress for typical carbides (or other hard noncoherent particles) is much higher than the yield strength, typically 1.5–2.5 times the yield strength (Argon and Im 1975). Appreciable strain hardening is required before these stress levels are achieved; consequently, voids nucleate at the sulfides and exhibit significant growth long before voids nucleate at carbides. Thus, the voids nucleated at carbides contribute only in the later stages of fracture, and the sulfide particle size and distribution largely determine the fracture properties.

On the other hand, the dimpled rupture fracture process is quite different for precracked or sharply notched specimens. The peak stress in the plastic zone in these specimens ($3\text{–}5\delta_0$) easily exceeds the critical stress for void nucleation at many carbides. The carbides, being much more densely distributed, have a much higher probability of being enveloped by the intense strain field than the less densely distributed sulfides. Thus, the carbides are the primary nucleation sites in this case. The key point is that particle spacing relative to the dimensions of the crack-tip intense-strain region must be considered in addition to the critical stress and nucleating particle size.

5. *Methods of Toughness Improvement*

Some methods by which materials can be improved to resist fracture with concomitant improvements of fracture toughness can now be formulated in terms of the failure criteria. Enhanced resistance to cleavage fracture can be achieved by one or more of the following: decrease of the stress in the near-crack-tip region, increase of the critical stress for microcrack nucleation, increase of the critical stress for crack propagation and increase of the critical distance over which the critical stress must be achieved.

The most obvious and direct way to decrease the stress in the plastic zone is to decrease the yield strength of the material, since the stress in the plastic zone is directly proportional to the yield strength. Lower strain-hardening rates also result in lower stresses, but higher strain-hardening rates often accompany decreases in yield strength. The yield strength effect is usually the dominant one, but this depends on the means of achieving the changes of strength level.

The critical stress for slip-induced microcrack nucleation can be increased by reducing the tendency to develop long dislocation pile-ups or eliminating large hard particles in which microcracks can occur. Thus, coarse-grain boundary carbides, coarse lamellar structures (such as pearlite) or similar particles should be avoided. Refinement of the microstructure to reduce grain size or the introduction of substructures or finely dispersed coherent particles reduces the length of potential dislocation pile-ups. Appropriate alloying can also reduce the tendency to form dislocation pile-ups, by reducing planar slip and increasing cross slip, dislocation tangles and cell formation. Typically, austenite stabilizers in steel are beneficial in this respect, whereas ferritic stabilizers have the opposite effect.

The critical stress for cleavage propagation can be increased by some of the same factors that improve the resistance to microcrack nucleation. In particular, alloying with austenite stabilizers to promote cross slip and dislocation tangles and to reduce planar slip are effective. Also, the introduction of finely dispersed particles is beneficial.

The critical distance parameter can be increased by coarser grain size or by reducing the number of available slip planes to extend the propagation difficulty through more than one grain. Grain-size-coarsening benefits, however, are generally more than offset by an accompanying decrease in the critical stress for nucleation or propagation. The number of operative slip planes can be influenced in the same manner as improvements of the critical stress for propagation.

The most important factor in fracture toughness when microvoid coalescence is the fracture mode is to control λ/ρ, where λ is the spacing between nucleated voids and ρ is the initial void size. The obvious and most direct means of achieving low values of λ/ρ are: (a) reducing stress levels in the plastic zones as discussed previously for cleavage fracture; (b) increasing λ by decreasing the number of potential void sites; (c) decreasing ρ by decreasing the size of nucleating particles.

Strengthening, where possible, should be achieved by solid-solution hardening (especially with elements that promote cross slip and dislocation tangles), grain refinement and coherent precipitates. Both λ and ρ can be favorably affected by achieving microstructures containing only small coherent particles for which the energy criteria for interface separation is difficult to achieve and which exhibit high critical stress values.

If noncoherent precipitates are unavoidable, then they should be as small and as evenly distributed as possible, and the volume fraction should be minimized within the constraints of achieving the desired tensile and yield strengths. Particle clustering and banding should be avoided because the localized high-particle areas increase the number of voids, decrease λ/ρ and lower toughness. Clearly, high-volume fractions of manganese sulfides, large oxides, silicates or other low-critical-stress particles should be avoided.

It should be remembered, however, that different failure mechanisms are often competing mechanisms. A modification that improves resistance to only one fracture mechanism may merely shift failure to another mode with little benefit to toughness, so all potential fracture modes should be considered simultaneously. Although only cleavage and microvoid coalescence are treated here, the principles outlined here can be applied to other mechanisms as well.

See also: Fatigue Cracks: Resistance to Initiation; Fatigue Cracks: Resistance to Propagation; Grain Boundary Strengthening; Precipitation Strengthening: General Considerations; Solid-Solution Strengthening

Bibliography

Argon A S, Im J 1975 Separation of second phase particles in spheroidized 1045 steel, Cu–0.6 pct. Cr alloy, and maraging steel in plastic straining. *Metall. Trans., A* 6: 839–51

Bates R C 1981 Mechanics and mechanisms of fracture. In: Tien J K, Elliott J F (eds.) 1981 *Metallurgical Treatises*. The Metallurgical Society (AIME), Warrendale, Pennsylvania, pp. 551–70

Griffiths J R, Owen D R J 1971 An elastic–plastic stress analysis for a notched bar in plane strain bending. *J. Mech. Phys. Solids* 19: 419–31

Hull D 1962 Twinning and nucleation of cracks in body-centered cubic metals. In: Drucker D C, Gilman J J (eds.) 1963 *Fracture of Solids*. Wiley, New York, pp. 417–60

McMeeking R M 1977 Finite deformation analysis of crack-tip opening in elastic–plastic materials and implications for fracture. *J. Mech. Phys. Solids* 25: 357–81

Rice J R, Johnson M A 1970 The role of large tip geometry changes in plane strain fracture. In: Kannien M F, Adler W F, Rosenfield A R, Jaffee R I (eds.) 1970 *Inelastic Behavior of Solids*. McGraw-Hill, New York, pp. 641–72

Ritchie R O, Knott J F, Rice J R 1973 On the relationship between critical tensile stress and fracture toughness in mild steel. *J. Mech. Phys. Solids* 21: 395–410

Santhanam A T, Bates R C 1979 The influence of notch-tip geometry on the distribution of stress and strain. *Mater. Sci. Eng.* 41: 243–50

Smith E 1968 Cleavage fracture in mild steel. *Int. J. Fract. Mech.* 4: 131–45

Tetelman A S, McEviley A J Jr 1967 *Fracture of Structural Materials*. Wiley, New York

R C. Bates

Fracture Toughness Testing of Brittle Materials

Brittle materials fail through the growth of microscopic flaws (usually 20–50 μm in size), which form during processing (e.g., inclusions and pores) or finishing (e.g., machining-induced cracks), or result from surface impacts during service. Under uniform tension, failure takes place from the most severe flaw. The strength is then determined by two parameters: the resistance of the material to crack extension (measured by the stress-intensity factor K_{I} at a critical crack) and the crack severity, that is, its size, shape and state of stress. These parameters are related through the expression

$$\sigma = K_{\mathrm{Ic}}/Ya_{\mathrm{c}}^{1/2} \tag{1}$$

or in terms of strain-energy release rate G_{I},

$$\sigma = (1/Y)(G_{\mathrm{Ic}}E/a_{\mathrm{c}})^{1/2} \tag{2}$$

where the subscript c denotes the point of criticality; a_{c} is the critical flaw size; Y is a factor that depends on flaw shape and location with respect to a free surface; and E is Young's modulus. By measuring K_{Ic} or G_{Ic} directly it is possible, in principle, to separate the resistance of the material to crack growth from variations in flaw size (which, presently, can only be measured by a microscopic examination of the fracture surfaces). The fracture-mechanics techniques commonly employed are discussed here from two points of view: the determination of fracture toughness K_{Ic}, and the measurement of subcritical crack growth.

1. Critical Fracture Toughness

K_{Ic} for brittle materials is defined as the stress-intensity factor at the onset of rapid fracture (i.e., crack velocities between 0.01 and 1.0 m s^{-1}). To a great extent, this is an operational definition, and it raises the question as to whether K_{Ic} is truly a material parameter. Almost all measurement procedures for K_{Ic} involve the continuous loading of a specimen until rapid fracture occurs, accompanied by a drop in load. This maximum load, combined with the geometry of the specimen, is used to calculate K_{Ic}. The occurrence of subcritical crack growth introduces some uncertainty into the determination of K_{Ic} because uncertainties in the measurement of crack velocity at the maximum load for different specimen geometries lead to specimen-to-specimen variations in K_{Ic} values. Fortunately, crack growth curves in the high-velocity range (Fig. 1) are usually steep, and thus the K_{Ic} variations should be relatively small.

There are a number of specimen configurations used for both K_{Ic} determinations and measurements

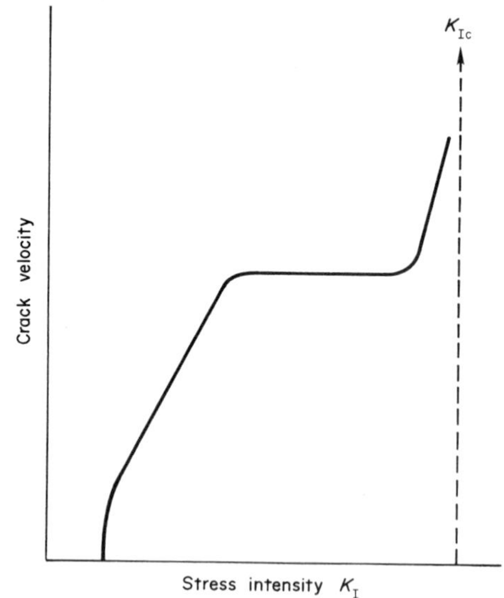

Figure 1
Typical crack velocity–stress intensity plot for brittle materials in a moist environment. Each portion of the curve corresponds to a different mechanism for crack growth

of crack growth rate. The ones most widely used for brittle materials are discussed below.

1.1 Double-Cantilever Beam

The double-cantilever beam (DCB) specimen (Fig. 2) has been used extensively to measure K_{Ic} as well as subcritical crack growth in both glasses and ceramics. The advantages of this technique are that it has been well characterized and requires a relatively small amount of material. By applying a moment rather than a force to the specimen, K_I is made independent of crack length c (Fig. 2c). The specimen can also be wedge loaded (Fig. 2a), in which case K_I decreases with increasing crack length, a feature which enables a number of K_{Ic} determinations to be made on the same specimen. A general disadvantage of the DCB configuration is the prerequisite for a groove down the center of a specimen in order to guide the crack. Also, except for the wedge-loaded configuration (where frictional effects can be troublesome), the attachment of loading fixtures poses problems at elevated temperatures.

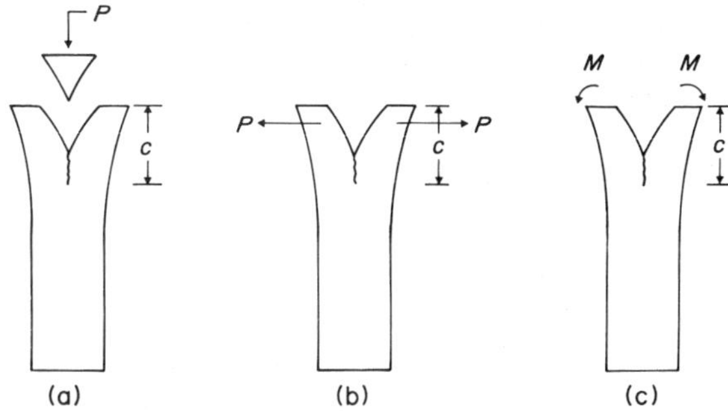

Figure 2
Double-cantilever beam specimen: (a) wedge loaded—K_I decreases as c increases; (b) applied load—K_I increases as c increases; (c) applied moment—K_I independent of c

1.2 Double Torsion

The double-torsion specimen (Fig. 3) has found considerable usage in obtaining crack growth data, especially for materials and environments in which a direct measurement of crack length is very difficult (e.g., opaque ceramics at elevated temperatures). The three features of this specimen which make it particularly useful under these conditions are (a) loads are applied in compression; (b) K_I is independent of crack length; and (c) the change in compliance of the specimen with a change in crack length is large. The latter two features make it possible to obtain crack velocity data by measuring the load relaxation of a specimen under a constant cross-head displacement. The pertinent expression for K_I is

$$K_I = PW_m[3/Wt^3t_1(1-\nu)\xi]^{1/2} \quad (3)$$

where P is the applied load; W, W_m, t and t_1 are specimen dimensions (see Fig. 3); ν is the Poisson ratio; and ξ is a correction factor for thick specimens:

$$\xi = 1 - 0.6302(2t/W) + 1.20(2t/W)\,\mathrm{e} - \pi(2t/W) \quad (4)$$

K_I has been shown to be independent of crack length over the approximate range $L/4$ to $3L/4$, where L is the specimen length. Because of the torsional loading, the crack front in the double-torsion specimen is curved and K_I is constant over only ~75% of the thickness, which can lead to interpretive difficulties. A further disadvantage of the double-

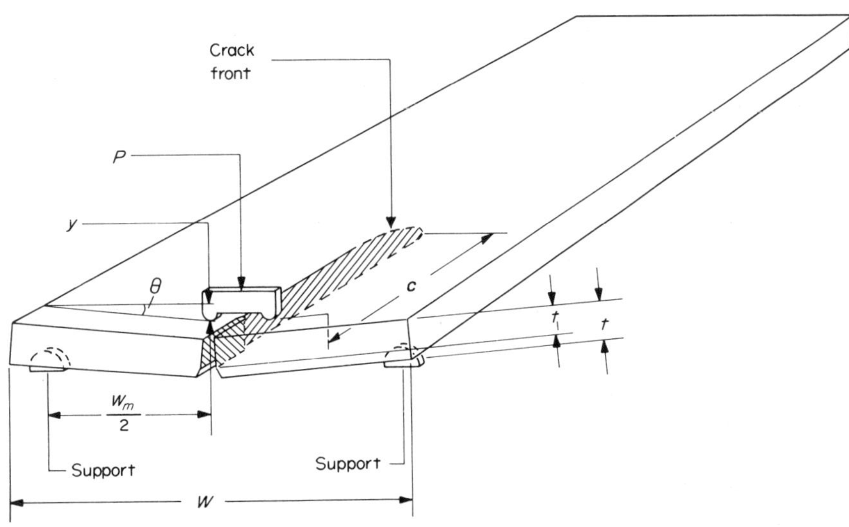

Figure 3
Double-torsion specimen; note that, because of the loading, K_I will vary along the curved crack front

torsion specimen is the relatively large quantity of material needed to conduct a satisfactory test.

1.3 Notch Bend

Notch-bend tests involve the use of a bar of material in which a relatively deep notch has been cut (Fig. 4). The notch-bend technique requires a minimal amount of material and is particularly useful for obtaining data at elevated temperatures. The specimen is loaded in bending, and K_{Ic} is calculated from the load at which failure occurs, the specimen geometry and the notch depth. However, because inserting a sharp crack into the base of the notch can be quite difficult, determinations of K_{Ic} are often made assuming that a crack will "pop in" during loading. This assumption can lead to serious errors. A recently developed concept, based on the use of a chevron-shaped notch, appears to improve the accuracy and precision of this technique since it allows for the smooth development of a crack.

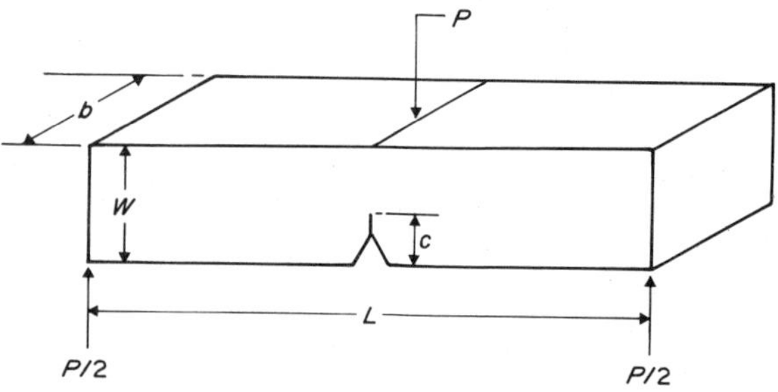

Figure 4
Notch-bend specimen; a crack must be initiated at the base of the notch. Here, K_I is given by $K_I = 3YPLc^{1/2}/ZbW^2$ where $Y = f(L/W, c/W)$

1.4 Short Rod or Bar

A recently developed variation on the double-cantilever beam configuration for which accurate K_I calibrations are available is the short-rod or short-bar specimen (Fig. 5). A crack is initiated from the point of a chevron notch in the specimen, and grows along the midplane. The specimen geometry causes the crack to remain perpendicular to the applied stress, reducing errors caused by twisting. The load can be applied either by expanding a bladder in the opening at the front of the specimen or by applying a tensile load.

1.5 Indentation Techniques

Although the previously described procedures have involved the use of specimens containing macroscopic cracks, it is sometimes advantageous to determine crack growth parameters using cracks which are approximately the same size as those expected in an actual structure. This is especially true for polycrystalline ceramics, in which the ratio of flaw size to grain size can have a significant effect on crack growth. In one such technique, a crack is introduced into the surface of a bend specimen via a Vickers or Knoop hardness indenter. The specimen is then broken rapidly in flexure in an inert environment. The critical fracture toughness is calculated from the failure stress σ_m and the indentation load P at which the crack was introduced. Residual stresses introduced by the indentation are accounted for in the determination of K_{Ic}:

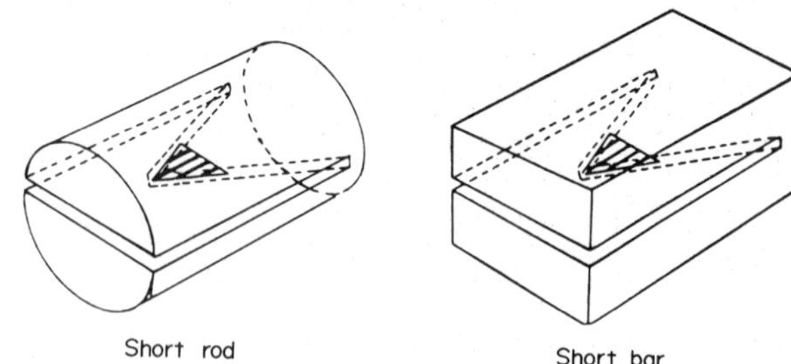

Figure 5
Short-rod and short-bar specimens; a crack initiates smoothly from the tip of the chevron

$$K_{Ic} = \eta_V^R (E/H)^{1/8} (\sigma_m P^{1/3})^{3/4} \quad (5)$$

where η_V^R is a constant depending only on the indenter geometry, E is Young's modulus, and H is hardness. As can be seen from Eqn. (5), this method does not require a crack size measurement.

Another, but less accurate, procedure involves measuring the crack length, c_0, on the surface indented under an inert environment. K_{Ic} is calculated from the expression

$$K_{Ic} = \S_V^R (E/H)^{1/2} (P/c_0^{3/2}) \quad (6)$$

where $\S_V^R$ is an empirical constant. One significant advantage of this indentation technique is the very small quantity of material required.

2. Subcritical Crack Growth

As noted earlier, K_{Ic} is but one point on a crack growth curve. In order to understand the effects of environment on crack growth, as well as to make lifetime predictions for ceramic structures under load, it is necessary to measure crack growth rates to the lowest feasible velocities (see *Subcritical Crack Growth in Ceramics*).

The need to accurately determine low rates of crack growth for use in lifetime prediction can be seen by considering the fact that

$$\text{time to failure} = 2a_i/(n-2)v_i \quad (7)$$

where a_i is the initial flaw size, v_i is the initial crack velocity and n is the slope of a ln v vs ln K_I curve. Taking typical values of $a_i = 20$ μm and $n = 25$, for a structure to survive for 1 year, v_i must be $<5 \times 10^{-14}$ m s^{-1}. However, there are problems related to laboratory measurements, and these are now outlined.

Crack velocities for transparent materials in double-cantilever beam specimens are generally measured optically, using a travelling microscope. Since the minimum resolution of a change in crack tip position is about 10 μm, it would take 2×10^8 s (about 6 years) to obtain a velocity of 5×10^{-14} m s^{-1}. The uncertainty in the crack growth data for polycrystalline ceramics, where smooth motion of the crack does not occur, is much greater.

The double-torsion specimen is commonly used to obtain crack growth data for opaque materials or under conditions in which the crack cannot be viewed directly. One procedure involves rapidly raising the specimen load to a level of K_I at which crack growth is known to occur, then arresting the cross-head and allowing the load to relax as the crack grows. Under these conditions, K_I can be calculated from the load at any time through Eqn. (3), and the crack velocity is given by

$$v = -(c_i P_i / P^2)(dP/dt) \tag{8}$$

where c_i is the initial crack length, P_i is the highest load and dP/dt is the slope of the load–time trace at each point. In these tests, care must be taken to eliminate (or account for) any load relaxation other than that due to crack growth. Because of limitations in the ability to measure very small changes in dP/dt (plus drift in the test machine), accurate values of crack velocity below $\sim 10^{-8}$ m s^{-1} cannot be obtained.

Double-torsion specimens can also be loaded at a constant displacement rate $\dot{y}$. In this case, the load–time trace reaches a plateau corresponding to a constant crack velocity given by

$$v = \dot{y} W t^3 G / 3 P W_m^2 \tag{9}$$

where G is the shear modulus of the material. This procedure is limited by the lowest rate at which the cross-head can be driven, usually equivalent to a velocity $\geqslant 10^{-5}$ m s^{-1}.

Other techniques, such as the short rod/bar or notch bend, in which the crack velocity is calculated from load relaxation are subject to the same crack velocity limitations as the double-torsion method. Indirect measurements of crack growth can also be obtained from static fatigue (delayed failure) data. For small flaws (<75–100 μm) this can be a more time-efficient technique for collecting low-velocity data but may lead to significant inaccuracies due to the statistical scatter in flaw sizes.

See also: Brittle Fracture: Linear Elastic Fracture Mechanics; Brittle Fracture: Micromechanics

Bibliography

Anstis G R, Chantikul P, Lawn B R, Marshall D B 1981 A critical evaluation of indentation techniques for measuring fracture toughness: I, Direct crack measurements. *J. Am. Ceram. Soc.* 64: 533–38

Chantikul P, Anstis G R, Lawn B R, Marshall D B 1981 A critical evaluation of indentation techniques for measuring fracture toughness: II, Strength method. *J. Am. Ceram. Soc.* 64: 539–43

Freiman S W (ed.) 1979 *Fracture Mechanics Applied to Brittle Materials*, ASTM Special Technical Publication 678. American Society for Testing and Materials, Philadelphia, Pennsylvania

Freiman S W 1982 A critical evaluation of fracture mechanics techniques for brittle materials. In: Bradt R C, Hasselman D P H, Lange F F, Evans A G (eds.) 1982 *Fracture Mechanics of Ceramics.* Plenum, New York

Fuller E R Jr 1979 An evaluation of double-torsion testing—analysis. In: Freiman 1979, pp. 3–18

Lawn B R 1982 The indentation crack as a model surface flaw. In: Bradt R C, Hasselman D P H, Lange F F, Evans A G (eds.) 1982 *Fracture Mechanics of Ceramics.* Plenum, New York

Pletka B J, Fuller E R Jr, Koepke B G 1979 An evaluation of double-torsion testing—experimental. In: Freiman 1979, pp. 19–37

Underwood J H, Freiman S W, Baratta F I (eds.) 1984 *Chevron-Notched Specimens: Testing and Stress Analysis*, ASTM Special Technical Publications 855. American Society for Testing and Materials, Philadelphia, Pennsylvania

S. W. Freiman

Fretting Corrosion

Fretting corrosion is the surface damage arising when two surfaces in contact undergo a small oscillatory tangential movement relative to each other termed slip. The amplitude of motion is usually 25 μm or less. When the amplitude exceeds 75 μm, the motion is more properly termed reciprocating sliding, and the consequences resemble normal sliding wear. In machinery the source of the movement may be vibration, whereas in cooling systems it may arise from turbulent flow. In many cases the movement is the result of cyclic stressing of one of the members of the contact (e.g., a press-fitted hub on a rotating loaded axle, or a riveted joint in an aircraft wing).

In oxidizing atmospheres the damage to metal components takes the form of oxide debris (e.g., α-Fe_2O_3 on mild steel). The small slip amplitude ensures that the debris remains in situ, and hence leads to a buildup of pressure resulting from the greater volume of the oxide compared with the original metal volume. This may result in seizure if the joint is one which is required to move periodically as, for example, in the sliding contact of a machine governor. If the debris can escape from the contact, it leads to a spread of the damage and possible loss of fit of the mating components. Also, the debris, being abrasive, may cause damage to other parts of the machine.

The most serious effect of fretting is the initiation of fatigue cracking where the origin of movement is cyclic stressing (see *Corrosion Fatigue*). Strength reduction factors of 2–5 are quite common, with the high-strength materials suffering the most damage. This places a severe limitation on the fatigue behavior of metal structures and steel ropes.

The environment is a significant factor in fretting corrosion. In aqueous electrolytes, fretting continually disrupts passive surface films, which would otherwise give the material adequate protection against corrosion. In high-temperature oxidizing atmospheres, disruption of protective oxide layers may expose underlying material deficient in the element forming the protective layer (e.g., chromium in stainless steels). However, certain materials are capable of developing a "glaze" oxide when fretted at high temperature: this results in low friction and wear, and is self-repairing if local breakdown occurs.

Fretting damage can be avoided or minimized by ensuring at the design stage that contact surfaces are not located at regions of stress concentration. Breakdown of large contact areas into smaller discrete areas by machining grooves in one of the surfaces can be effective. Other methods for reducing fretting damage include surface treatment by shot peening or surface rolling to give a work-hardened layer with a residual compressive stress, as well as the application of coatings (e.g., by diffusion, ion implantation or electroplating).

See also: Oxidation of Metals and Alloys; High-Temperature Oxidation of Metals and Alloys: Engineering Aspects; Design for Corrosion Control; Corrosion Protection Methods; Corrosion of Metals: An Overview

Bibliography

Gordelier S C, Chivers T C 1979 A literature review of palliatives for fretting fatigue. *Wear* 56: 177–90

National Materials Advisory Board 1977 *Control of Fretting Fatigue*, National Materials Advisory Board Report No. 333. Department of Defense, Washington, DC

Waterhouse R B 1972 *Fretting Corrosion*. Pergamon, Oxford

Waterhouse R B (ed.) 1981 *Fretting Fatigue*. Applied Science, London

Waterhouse R B, Taylor D E 1980 High temperature fretting and wear of like metallic contacts. *Rev. High Temp. Mater.* 4: 259–98

R. B. Waterhouse

Friction and Lubrication in Metalworking

In metalworking processes the shape of the workpiece is changed by a tool or die. In general, the die and workpiece move relative to each other under pressure, and this movement is resisted by frictional forces at the die–workpiece interface. Friction makes a significant contribution to die pressures, deforming forces and energy requirements, and often sets a limit to the attainable deformation. An important feature of metalworking is that in the course of deformation the surface area of the workpiece increases, oxide and other protective films present on the old surface break up, and new, highly reactive virgin surfaces are exposed. These can then adhere to the die surface to form cold-welded junctions. To control friction and prevent metal-to-metal contact, a lubricant must be interposed. The success or failure of lubrication has important consequences in terms of the quality of the issuing product, and it also controls the progressive loss (wear) of the workpiece and, more importantly, of the die material.

1. Macroscopic View of Interface

In its simplest description, the die can be regarded as a rigid (or elastic) body, of a well-defined geometry, which contacts the rigid–plastic (or elastic–plastic) workpiece of flow stress σ_f, and thus defines the process geometry (Fig. 1). The interface is regarded as a continuous substance of an interface shear strength τ_i. For convenience of calculation, it is preferable to describe the interface by a nondimensional factor, and one possibility is to adopt the definition of the coefficient of friction: $\mu = F/P = \tau_i/p$ where F is the force required to move the body and P is the normal force; τ_i is the frictional shear stress and p the normal pressure, defined as the force acting on the apparent area of contact between the two bodies. The magnitude of τ_i cannot rise beyond the shear flow strength of the material τ_f, because it then takes less energy for the workpiece to deform by internal shear, and sliding at the interface ceases. This is often referred to as sticking friction, even though no actual adhesion needs to occur. Since $\tau_f = 0.5\sigma_f$, it is often said that for plastic deformation $\mu_{max} = 0.5$. This

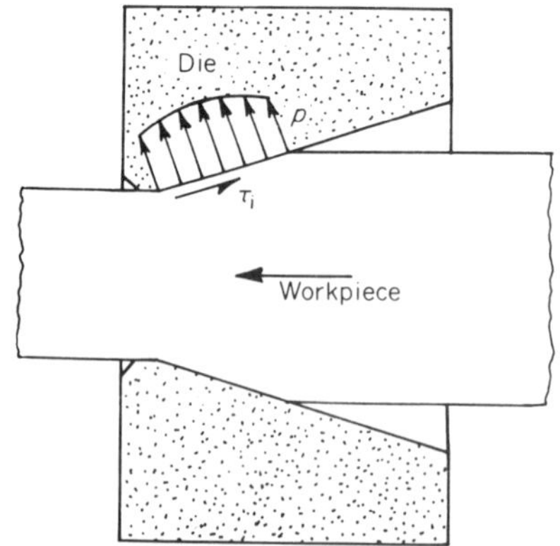

Figure 1
Macroscopic view of an interface

statement is true only at $p = \sigma_f$; at higher interface pressures the maximum value of the coefficient of friction drops, and it is more meaningful to say that it is indeterminate when sliding is arrested.

Because of the difficulties encountered with the coefficient μ, an alternative description of the interface is often favored: $\tau_i = m\tau_f$, where m is the interface shear factor, with a value ranging from 0 (in the absence of friction) to 1 (at sticking). This approach often simplifies analysis, but the concept itself has little physical significance because it assumes that the properties of the interface are in some fixed relation to the properties of the workpiece material, a condition seldom fulfilled.

2. Microscopic View of Interface

While the above treatment is sufficient for purposes of mechanical analysis (see *Plasticity Analysis in Metalworking*), it is quite inadequate to describe the complexities of the real interface. On the microscopic scale (Fig. 2), the surfaces of die and workpiece show peaks (asperities) and valleys (see *Surface Roughness*). The materials can no longer be described by average compositions; the compositions and distribution of individual phases determine the nature of the reaction (oxide) films present on the surface, superimposed on which are adsorbed films from the atmosphere and from the lubricant.

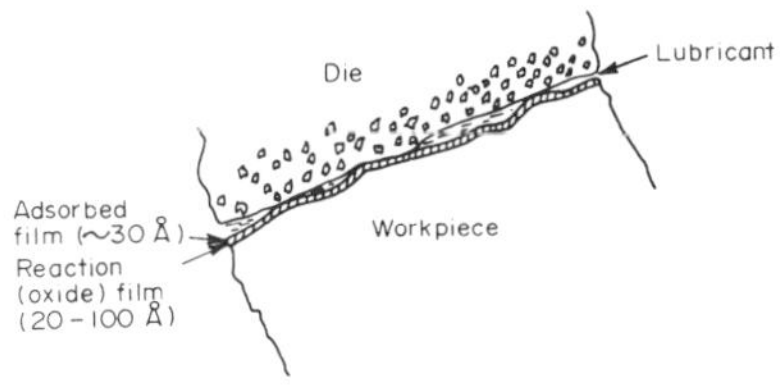

Figure 2
Microscopic view of an interface

The lubricant is exposed to extremely severe conditions. Interface pressures p range from a fraction of σ_f to multiples of σ_f (up to 4 GPa). Sliding velocities range from 0 at sticking friction to 50 m s^{-1} in high-speed rolling or wiredrawing, sometimes combined with approach velocities of up to 20 m s^{-1}. Since much of the deformation and frictional energy is converted into heat (see *Heat Generation in Metal Cutting and Deformation*), temperatures up to several hundred °C are encountered, even in cold working, and up to 2000 °C in hot working.

High pressures and temperatures, intense rubbing, and the high reactivity of virgin surfaces combine to create an environment in which various tribochemical reactions can take place. Therefore, the nature and reactivity of die and workpiece materials become vital variables in the success of lubrication. Useful improvements can sometimes be obtained by surface or diffusion coatings that reduce adhesion; these, however, may have the undesirable side effect of interfering with chemical reactions essential for the function of some lubricants.

3. Lubricating Mechanisms

The lubricant interposed between die and workpiece has to fulfill a number of functions:

(a) separation of the die and workpiece surfaces to prevent adhesion and wear;

(b) control of friction—this usually means as low a value of τ_i as possible, but some minimum value is required in processes that rely on traction to assure metal movement (as in rolling) or support of the workpiece material (as on the punch in deep drawing or ironing);

(c) control of the surface finish of the issuing product, ranging from bright to relatively dull rough finishes;

(d) reaction with the workpiece and/or die surface in a controlled manner to assure the formation of protective films without corroding the die or product; and

(e) control of heat generation and extraction; generally the desirable function is that of cooling, although in some hot plastic deformation processes insulation is required.

These functions can be fulfilled in a number of ways.

3.1 Films of Low Shear Strength

In a few instances, such as the hot rolling of steel, oxides grown on the surface break up, or even deform, to give a reasonably effective parting agent. In general, however, a continuous film of a material with a shear strength $\tau_i < \tau_f$ is relied on.

Soft metals, such as tin, lead or copper, are chosen for low adhesion to the die; they also act as lubricants and make additional lubrication easier by virtue of their composition and surface characteristics. Since their τ_i is largely independent of pressure (Fig. 3), they become particularly useful at high working pressures.

Plastics (polymers) deposited as thin films have a pressure-dependent shear strength, often expressed as a constant μ that seldom drops below 0.05 (Fig. 3). They lack spreadability and cannot prevent pickup when the film is discontinuous or damaged.

Layer-lattice compounds such as graphite and molybdenum disulfide are effective if deposited in a continuous film. Film formation and durability is aided by reaction with the metal substrate, and therefore molybdenum disulfide is more effective on

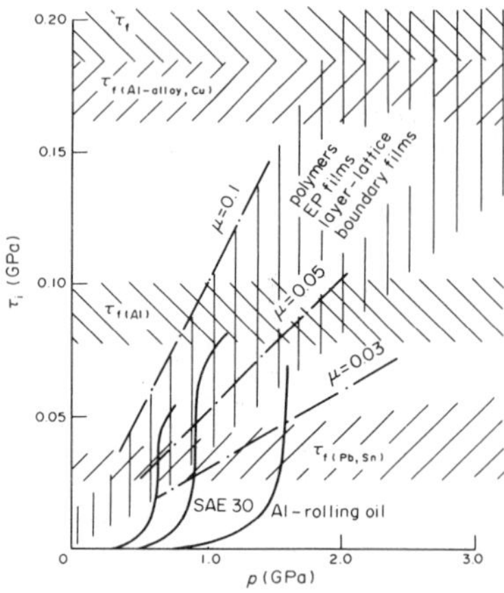

Figure 3
Response of the shear strength of lubricants to interface pressure (after Schey 1979. © Springer, Berlin. Reproduced with permission)

sulfide-forming materials. These films also can be modelled by a constant μ (Fig. 3).

3.2 Extreme-Pressure (EP) Lubrication

Organic compounds of phosphorus and sulfur can react with metals, especially steel, to form surface films which prevent adhesion and, to some extent, reduce friction, essentially by wearing away at a controlled rate. The film re-forms if temperatures are high enough and if sufficient time is available (as in machining, particularly drilling), or if contact is repeated (as in cold heading, wiredrawing, sheet metalworking and, sometimes, rolling).

3.3 Boundary Lubrication

Polar substances, particularly fatty acids, alcohols, and other derivatives of fatty oils and fats, attach themselves to metal surfaces in an oriented fashion, and react with a suitable substrate to form metal soaps. These films have a low shear strength which is pressure and temperature dependent (Fig. 3); they form rapidly under the conditions prevailing at the interface, but break down at the melting point of the polar compound or its soap. Therefore, lubricants are often compounded to include EP additives which are activated when boundary additives break down.

3.4 Full-Fluid Film Lubrication

Full-fluid film lubrication can be performed with viscous fluids if the process geometry creates a converging gap (Fig. 1), and sliding speeds or approach speeds are large enough. Hydrodynamic and elastohydrodynamic theory can be modified to take account of plastic deformation; such plastohydrodynamic theory can describe the effect of major process variables, provided that the rheology of the lubricant is known. Viscosity increases usually exponentially with pressure and drops with temperature; many lubricants also solidify at some critical pressure and then behave like a polymer (Fig. 3). Full-fluid film lubrication is attained or sufficiently approached in hydrostatic extrusion, high-speed wiredrawing and rolling, to make modelling by plastohydrodynamic theory useful. Since the surface deforms through a liquid cushion, individual grains deform as dictated by crystallographic constraints, and the surface roughens.

3.5 Mixed Film Lubrication

In the majority of liquid-lubricated operations, some asperity contact is unavoidable, and only a portion of the total contact area is separated by a thick lubricant film. The remainder of the contact area is in boundary contact, making the addition of boundary or EP additives an absolute necessity in almost all metalworking fluids. Modelling of mixed film lubrication is made more difficult by the change of surface configuration resulting from the presence of entrapped liquid which creates hydrodynamic or hydrostatic pockets in the deforming workpiece surface (Fig. 4).

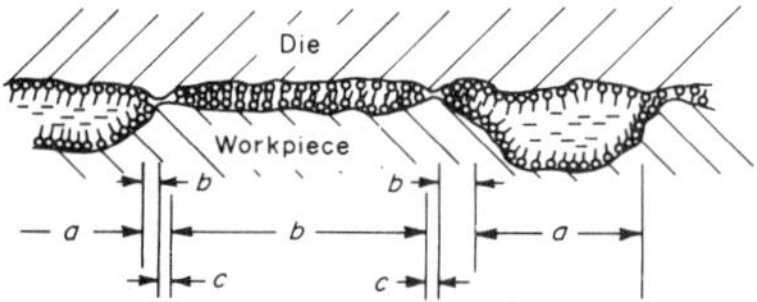

Figure 4
Interface in mixed-film lubrication: a, area in hydrodynamic or hydrostatic contact; b, area in boundary contact; and c, area in metallic contact (after Schey 1980)

4. Lubricating Regimes

The range of operating conditions for various lubricating mechanisms can be conveniently summarized in a diagram based on the Stribeck curve, in which the coefficient of friction μ is plotted as a function of lubricant viscosity and interface sliding velocity v (Fig. 5). For $p = \sigma_f$ the maximum value of μ is limited, by sticking, to 0.5. Solid films of a pressure-dependent shear strength (Fig. 3) reduce μ to about 0.05–0.1, and this value changes little with sliding velocity unless temperatures are also affected. In the presence of a liquid lubricant, mixed-film conditions are attained once velocity and viscosity combine to sustain the pressures required for plastic deformation. With increasing η or v the proportion of surface area lubricated by the fluid film increases and μ drops to typically 0.03–0.05. With increasing interface pressures, μ decreases under sticking conditions (Fig. 5), whereas it remains constant for solid films. There-

Table 1
Typical lubricants and friction coefficients in plastic deformation[a]

Workpiece material	Working	Forging: Lubricant	Forging: μ	Extrusion: Lubricant	Wiredrawing: Lubricant	Wiredrawing: μ	Rolling: Lubricant	Rolling: μ	Sheet metalworking: Lubricant	Sheet metalworking: μ
Sn, Pb and Zn alloys		FO–MO	0.05	FO or soap	FO	0.05	FA–MO or MO–EM	0.05 0.1	FO–MO	0.05
Mg alloys	hot or warm	Gr and/or MoS_2	0.1–0.2	none			MO–FA–EM	0.2	GR in MO or dry soap	0.1–0.2
Al alloys	hot	GR or MoS_2	0.1–0.2	none			MO–FA–EM	0.2		
	cold	FA–MO or dry soap	0.1 0.1	lanolin or soap on PH	FA–MO–EM FA–MO	0.1 0.03	1–5% FA in MO (1–3)	0.03	FO, lanolin or FA–MO–EM	0.05–0.1
Cu alloys	hot	GR	0.1–0.2	none (or GR)			MO–EM	0.2		
	cold	dry soap, wax or tallow	0.1	dry soap or wax or tallow	FO–soap–EM MO	0.1 0.03	MO–EM	0.1	FO–soap–EM or FO–soap	0.05–0.1
Steels	hot	GR	0.1–0.2	GL (100–300) GR			none or GR–EM	ST[b] 0.2	GR	0.2
	cold	EP–MO or soap on PH	0.1 0.05	soap on PH	dry soap or soap on PH	0.05 0.03	10% FO–EM	0.05	EP–MO, EM or soap on polymer	0.05–0.1
Stainless steel, Ni and alloys	hot	GR	0.1–0.2	GL (100–300)			none	ST	GR	0.2
	cold	CL–MO or soap on PH	0.1 0.05	CL–MO or soap on PH	soap on PH or CL–MO	0.03 0.05	FO–CL–EM or CL–MO	0.1 0.05	CL–MO, soap or polymer	0.1
Ti alloys	hot	GL or GR	0.2	GL (100–300)					GR, GL	0.2
	cold	soap or MO	0.1	soap on PH	polymer	0.1	MO	0.1	soap or polymer	0.1

a After Schey 1977 © McGraw-Hill, New York. Reproduced with permission b ST, sticking friction

fore, it is quite possible to attain the condition of sticking even on a well-lubricated surface. In the presence of fluids, increasing pressures increase the proportion of boundary-lubricated areas in the regime of mixed lubrication and μ rises, as shown by the A–A section in Fig. 5.

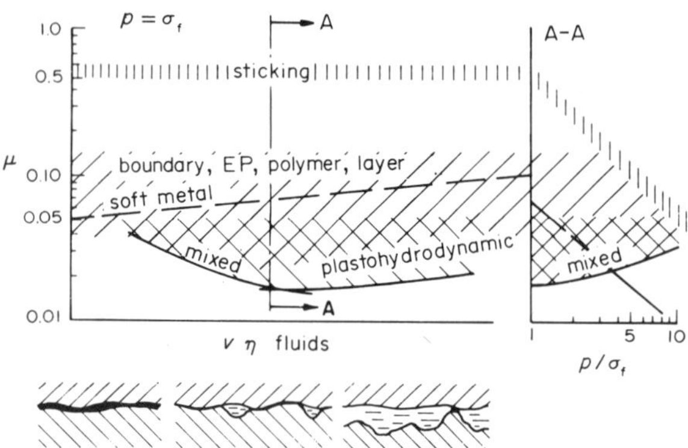

Figure 5
Regimes of lubricating mechanisms (after Schey 1979. © Springer, Berlin. Reproduced with permission)

Any of these mechanisms may become active in plastic deformation processes (see *Deformation Processing*). In metal cutting, lubrication of the cutting process is most effective at low speeds; at high cutting speeds the prime function of the lubricant is that of cooling, combined with lubrication at the rubbing face (flank) of the tool (see *Cutting Fluids*).

5. Wear

The predominance of boundary and mixed lubrication in metalworking processes suggests that wear can become a serious problem.

Adhesive wear is predominant when there is great affinity between die and workpiece material. Cold welding (pickup) to the die surface is followed by shearing or fatiguing of the transferred particles, and the resulting debris enters the lubricant. The cold-welded junction may be stronger than the die, and then separation may take place by wear of the die. Hard constituents of workpieces, oxide films and foreign bodies (including wear particles) present at the interface lead to abrasive wear. High temperatures at the interface promote diffusion of alloying elements into the workpiece, especially in metal cutting, and the die may then lose its strength and suffer severe wear through both adhesion and abrasion. Lubricants of excessive reactivity can cause wear by corrosion. In hot working, contact with the hot workpiece subjects the colder dies to severe thermal shock, and thermal fatigue then contributes to wear and die failure.

6. Lubricants

The choice of lubricant is influenced primarily by the temperature prevailing at the interface, the die and workpiece material, and process geometry. The most widely used lubricants are shown in Table 1. At hot-working temperatures, organic liquids are destroyed and only glasses (GL) can be used for fluid-film lubrication. They behave like Newtonian fluids and can act as true hydrodynamic lubricants or, if the process geometry is favorable (as in extrusion), a thick glass mat may be melted off gradually to provide continuous coating on the deformed product. Their main disadvantage is the difficulty of removal. Much higher friction has to be tolerated if layer-lattice compounds, such as graphite (GR) or molybdenum disulfide are used, usually in an aqueous carrier. Aqueous dispersions or emulsions (EM) of mineral oils (MO), fatty oils (FO) and EP compounds ensure the best cooling, leave sufficient residues on the surface to prevent uncontrolled adhesion, and also reduce friction.

In cold working, the temperatures are low enough to permit a wide variety of organic and inorganic lubricants to be used. The viscosity of natural or synthetic oils is chosen to give mixed film or sometimes almost full-fluid film lubrication, consistent with the required surface finish of the product and the minimum friction necessary to control the process. Almost invariably the oil is compounded with fatty acids (FA), fatty oils and/or EP additives, including chlorinated paraffin (CL). More effective cooling is assured, but at the expense of somewhat increased friction, when the oily lubricant with its additives is dispersed in water to form an emulsion of controlled stability. For the most severe operations, the lubricant is keyed to the surface by means of conversion coating, such as phosphating (PH), which creates a surface layer of intricate geometric configuration favorable for lubricant entrapment, and may also be reacted with soap-type lubricants to give maximum film attachment. Layer-lattice compounds are used mostly as additives to provide a last defense in case of lubricant breakdown.

See also: Tribology; Lubricants; Wear; Metals Processing and Fabrication: An Overview

Bibliography

Schey J A (ed.) 1970 *Metal Deformation Processes: Friction and Lubrication.* Dekker, New York

Schey J A 1977 *Introduction to Manufacturing Processes.* McGraw-Hill, New York

Schey J A 1979 Modelling of the tool–workpiece interface. In: Lippmann H (ed.) 1979 *Metal Forming Plasticity.* Springer, Berlin, pp. 336–48

Schey J A 1980 Tribology in metal-forming processes. *Proc. 4th Int. Conf. Production Engineering.* Japan Society of Precision Engineering, Tokyo, pp. 102–15

Schey J A 1983. Tribology. In: American Society for Metals 1983 *Metalworking Friction, Lubrication, and Wear.* American Society for Metals, Metals Park, Ohio

Wilson W R D 1979 Friction and lubrication in bulk metal-forming processes. *J. Appl. Metalwork.* 1: 7–19

J. A. Schey

Frictional Properties of Polymers

Polymer friction is influenced by both the surface and the bulk properties of the polymer and its counterface. Thus, the friction force F consists of two components: the surface component F_s and the bulk component F_b, as shown in the following equation:

$$F = F_s + F_b \tag{1}$$

Depending upon the type of friction (sliding or rolling, lubricated or unlubricated) the relative importance of these two components varies (Lee 1975).

Polymer friction has also been considered to consist of an adhesion component F_a and a deformation component F_d (Bowden and Tabor 1964). F_a is generally synonymous with F_s; however, F_a is somewhat dependent on the bulk rheological (deformation) properties of the polymer. F_d for a thermoplastic (especially if ductile) represents the effects of plastic deformation and viscoelastic dissipation. In elastomer friction, the viscoelastic properties influence both the adhesion component F_a and the hysteretic component F_h (Kummer 1966). These components are discussed in detail in the following sections.

1. Unlubricated Sliding Friction of Thermoplastics

The friction behavior of a polymeric material sliding on a rigid surface, or vice versa, is different from that of a metal sliding on metal (Bowden and Tabor 1964). Firstly, the contact area depends on the geometry of the surface as well as on the load. Secondly, the sliding speed and temperature can both influence friction (Bartenev and Lavrentev 1981). Thirdly, the deformation component, in general, has a much larger effect than the adhesion component on the friction force (Kuritsyna and Meisner 1961).

However, as with metal friction, the sliding friction force F of a polymer is the force required to shear the interfacial junctions and is thus proportional to the true contact area A and the interfacial shear strength τ_i:

$$F = A\tau_i \tag{2}$$

If the asperities deform plastically, the real contact area will be the quotient of the normal load W and the mean yield (or contact) pressure p of the softer material:

$$A = W/p \tag{3}$$

Then, the friction coefficient μ will equal the ratio of the interfacial shear strength to the yield pressure:

$$\mu = F/W = A\tau_i/Ap = \tau_i/p \tag{4}$$

In some cases, the friction force of clean metals sliding on polymers is about the same as that of polymers sliding on polymers (Shooter and Tabor 1952). Over a moderate load range, the friction is roughly constant and appears to be due primarily to the shearing of interfacial junctions. In most cases, the shear strength calculated from the friction test was found to be approximately equal to the bulk shear strength τ_b. However, in those cases, the shearing could have taken place in the bulk.

At the interface, the shearing rate of the thin film could reach $\sim 10^4\,s^{-1}$, while that in the bulk is 10^{-3}–$10\,s^{-1}$. Furthermore, the shearing is cyclic at the interface, and the direction of yield is constrained to within a few degrees of the displacement vector because of the contact geometry (Briscoe and Smith 1981). Thus, the shear strength of the bulk τ_b is different from that (τ_i) of the thin film.

For polymers, the interfacial shear strength is a function of the yield (or contact) pressure, temperature and sliding velocity. The relationship between τ_i and the yield pressure p is given as follows (Briscoe 1980):

$$\tau_i = (\tau_i)_0 + \alpha p \quad \text{(constant } T, V\text{)} \tag{5}$$

where $(\tau_i)_0$ is the threshold interfacial shear strength at zero yield pressure and α is the pressure coefficient. Under high loads, $\tau_i \gg (\tau_i)_0$ and $\alpha = \mu$.

Polymer rheological properties are generally temperature dependent. The effect of temperature on τ_i is expressed by

$$\tau_i = (\tau_i)_0'' \exp(-E/RT) \quad \text{(constant } p, V\text{)} \tag{6}$$

where $(\tau_i)_0''$ is a constant, E is the energy of activation, R is the gas constant and T is the temperature.

The effect of sliding velocity on τ_i is more complicated. It can be described to a first approximation by

$$\tau_i = (\tau_i)_0''' - \beta \ln(V/V_0) \quad \text{(constant } p, T\text{)} \tag{7}$$

where $(\tau_i)_0'''$ is the interfacial shear strength at the sliding velocity V_0 and β is a constant.

The sliding friction force of a thermoplastic generally contains both adhesion and deformation components. However, it is difficult to determine these two components separately. Through a lubricated sliding friction experiment or through a rolling friction experiment, one can obtain the deformation component F_d from the friction force. The deformation component always involves the loss angle of a polymer and is also affected by the external load W:

$$F_d = K_d(W)^n \tan\delta \tag{8}$$

where $\tan\delta = E''/E'$, E'' is the loss modulus, E' is the storage modulus, K_d is the deformation constant, and n is a constant.

Deformation loss is a collective term, including special cases such as hysteresis and grooving losses. Hysteresis losses occur, for example, in viscoelastic materials, where deformation is not permanent and a part of the deformation energy is recovered (partial energy feedback). Grooving losses are observed with plastics and elastic materials stressed beyond their yield point; they imply permanent deformation (no energy feedback). Deformation in thermoplastics generally involves grooving losses.

2. *Lubricated Sliding Friction of Thermoplastics*

In the presence of a lubricant, the sliding friction force is reduced as a result of the prevention of shearing of interfacial junctions. Thus, the adhesion component diminishes and the total friction force is represented by the deformation component (Greenwood and Tabor 1958). Under lubricated conditions, surface energetics and shear strength of the lubricant can influence the sliding friction.

The mechanism of boundary lubrication in polymers is generally similar to that in metals. Lubricants are known to be ineffective for polymeric surfaces, partly because of plasticization, which can weaken the interfacial bonds and thus reduce both the shear strength and the yield pressure of a polymer. This weakening effect is illustrated by the following equations:

$$F = A[f\tau + (1 - f)\tau_l] \tag{9}$$

and

$$\mu = f\tau/p + [(1 - f)\tau_l]/p \tag{10}$$

where f is the fraction of unlubricated area, τ is the shear strength of the soft polymer (unlubricated) and τ_l is the shear strength of the lubricated film. If plasticization takes place, the first term in Eqn. (10) remains unchanged but the second term will increase due to the lowering of yield pressure.

3. *Unlubricated Sliding Friction of Elastomers*

In principle, the sliding friction of an elastomer should be similar to that of a thermoplastic. However, the viscoelastic properties of elastomers affect both the adhesion component F_a and the hysteretic component F_h. Thus, unlike thermoplastics, the total friction force of elastomers can be expressed in terms of tan δ (Bulgin et al. 1962). An expression for the unlubricated elastomer sliding friction coefficient has been proposed (Kummer 1966):

$$\mu = \mu_a + \mu_h$$

$$= [k_a(E'/p^r) + k_h(p/E')^n] \tan \delta \tag{11}$$

where μ_a and μ_h are adhesion and hysteretic constants for the friction pair, E' is the elastic modulus, p is the nominal pressure, and r and n are constants.

4. *Unlubricated Rolling Friction of Polymers*

Similar to lubricated sliding friction, the rolling friction of thermoplastics (Flom 1960a) and elastomers (Flom 1960b) generally involves only the bulk or the deformation component. However, the coefficient of rolling friction λ can generally be expressed in two components, λ_s (surface) and λ_b (bulk):

$$\lambda = \lambda_s + \lambda_b \tag{12}$$

$$= C_s(a^2/WG^2)^{2/3} + K_b \tan \delta (W/Ga^2)^{1/3} \tag{13}$$

where $C_s = K_s\pi\{\frac{3}{8}[(1 - \nu^2)/(1 + \nu)]^{2/3}\}$, ν is Poisson's ratio, K_s is a surface constant, a is the radius of the ball, G is the shear modulus and K_b is the bulk constant.

For hydrogen-bonding polymers (e.g., nylon) a small surface effect can be detected. In other polymers, the surface component is generally negligible, and λ can be simplified to be a function of tan δ as shown in the second term of Eqn. (13).

See also: Polymers: An Overview of Structure, Properties and Structure–Property Relations; Tribology

Bibliography

Bartenev G M, Lavrentev V V 1981 *Friction and Wear of Polymers.* Elsevier, Amsterdam

Bowden F P, Tabor D 1964 *The Friction and Lubrication of Solids*, Vol. 2. Clarendon Press, Oxford

Briscoe B J 1980 The role of adhesion in the friction, wear and lubrication of polymers. In: Allen K W (ed.) 1980 *Adhesion*, Vol. 5. Applied Science, Barking, UK, pp. 49–80

Briscoe B J, Smith A C 1981 Polymer friction and polymer yield: A comparison. *Polymer* 22: 1587–89

Bulgin D, Hubbard G D, Walters M H 1962 Road and laboratory studies of friction of elastomers. *Proc. Rubber Technol. Conf., 4th.* Institution of the Rubber Industry, London, p. 173

Flom D G 1960a Rolling friction of polymeric materials, I. Elastomers. *J. Appl. Phys.* 31: 302–14

Flom D G 1960b Rolling friction of polymeric materials, II. Thermoplastics. *J. Appl. Phys.* 32: 1426–36

Greenwood J A, Tabor D 1958 The friction of hard sliders on lubricated rubber: The importance of deformation losses. *Proc. Phys. Soc. London* 71: 989–1001

Kummer H K 1966 *Unified Theory of Rubber and Tire Friction*, Engineering Research Bulletin B-94. Pennsylvania State University, University Park, Pennsylvania

Kuritsyna A D, Meister P G 1965 Plastics as anti-friction materials. *Izd. Plastmassy Podshipnikakh Skol'zheniya. Dokl. Soveshch.* Institut Mashinovedeniia, Moscow, pp. 57–64

Lee L H (ed.) 1975 *Advances in Polymer Friction and Wear.* Plenum, New York

Shooter K V, Tabor D 1952 The frictional properties of plastics. *Proc. Phys. Soc. London B* 65: 661–71

L. H. Lee

Fur

Both wild and domesticated animals provide the furs used in garments, rugs, wall hangings and covers. The terms hides, skins, pelts, and furs are often used interchangeably; however, in the fur trade they have special meanings. Hides are the outer covering of raw or tanned pelts of a fully grown animal of the larger species such as cattle, horses, camels and elephants. Skins refer to the outer covering of smaller animals or an immature species of the larger animals with or without hair; the term also refers to the covering of a fur-bearing animal dressed and finished with the hair left on. Pelts are the undressed skins or hides with the underfibers or ground hair and guard hairs. Fur is a piece of the dressed pelt of an animal used as a material to make garments.

From earliest history, man has used animals to provide food, covering, shelter, ornaments and tools. Pelts were probably originally worn in their natural shapes for cover and warmth: Hebrews in the Old Testament are recorded as wearing a garment known as the gaunaki, made from a shaggy lamb skin that covered the body from the waist down. Dwellers in the Nile region wore a long section of sheepskin crossed by a shorter section; a hole in the center being cut for the head. The pelisse, a long rectangle of fur with a hole in the middle for the head, paralleled or preceded the development of the Roman toga; Roman senators were called pellite because they wore pelts. Furs (e.g., sable, ermine, chinchilla and Alaska seal) have often acquired a special status value, being restricted to royalty at various times and in various countries.

1. Sources of Fur Pelts

Primitive man hunted animals for food and clothing. Later, he learnt to domesticate animals, herding them for food and clothing; sheep were among the first animals to be herded in this way. Ranching of wild animals for their pelts is a quite recent development, starting only about 100 years ago: pelts of such species as mink, fox, chinchilla, coypu and rabbit are now produced commercially by ranching. Mink ranching originated in the USA, and has now spread to over 21 countries, with the USA, the USSR and Denmark being the leading producers. During the past 50 years of mink ranching, over 29 major fur colors have been developed by selective breeding. In 1979 the USA produced over 3.4 million mink pelts, while the world production of both wild and ranched mink totalled 21 million pelts.

The chief fur-bearing animals in Canada and the USA (including Alaska) are bear, beaver, calf, ermine, fox, lynx, marten, mink, muskrat, opossum, otter, seal, skunk, ground squirrel, rabbit, weasel, wolf and wolverine. Persian lamb fur from the Karakul sheep comes from the USSR, Afghanistan and southwest Africa. The USSR also produces ermine, kolinsky, sable, squirrel, marten and fox. South America produces jaguar, coypu and chinchilla. Australia and northern Europe produce rabbit skins.

Many of the world's fur-bearing animals are considered threatened and endangered species. Species such as the Giant River Otter of the Amazon Basin, the tiger from Asia, the cheetah from Asia and Africa, and the vicuna from the Andes Mountains are but a few of the endangered species. However, most governments now regulate the hunting and trapping of fur-bearing animals.

2. Pelt Processing

The pelts are prepared by skinning the carcass of the animals. The skin on smaller animals—e.g., mink, marten, otter, fox—is skinned or "cased" by making a cut down the legs and by pulling the skin off the carcass intact, while in larger animals—e.g., wolves and bears—the skin is split down the belly and paws and removed flat. Fur from the tail is usually removed separately. The flesh is scraped off the pelt down to the leather, and the skin is stretched on a form or laid out flat and allowed to cure or dry. The cured pelt is salted to preserve it, and then shipped as soon as possible to auction houses.

Auction houses are located in such world fur centers as New York, Winnipeg and London. Secret bids secured by the auctioneer are the usual sale method. However, some mink ranchers sell their furs directly to the manufacturer, thus by-passing the auction houses.

The stiff, heavy pelts must be tanned (see *Leather*) or dressed by chemicals to produce a light, thin and pliant product. This chemical treatment also fixes the hairs and fibers in their follicles, and permanently protects all parts of the skin against natural bacterial action. The original natural oils of the leather must be replaced to preserve the life, softness and durability of the leather.

The fur pelts may then be modified by processes such as dyeing, printing, blending, plucking, shearing, grooving and tipping. (Plucking of the guard hairs of aquatic furs such as beaver, seal, otter, coypu and muskrat exposes the dense, soft, wavy fiber underneath; sharp, parallel grooves cut into the fiber may help to imitate other more expensive furs; tipping adds selective color to the furs.) After processing, the pelts are ready for manufacturing into garments. Fur garments from the manufacturer are usually presented at special showings in the main manufacturing centers in New York, Chicago, Montreal, Toronto, Frankfurt, London, Paris, Milan, Tel Aviv and Tokyo.

3. Fur Characteristics

The fur of most animals is thickest and glossiest during the winter months, when most of the wild animals are trapped. Next to the skin or hide is a dense (up to 48 000 fibers per cm^2), dull, curly mass of fine fibers much like wool fiber, called ground hair or underfibers, which is usually more dense along the center of the back of the animals. Growing through this dense fiber are longer, straighter, lustrous guard hairs. The oiliness of these fibers imparts the luster to the hair.

Wild mink is light brown or tan to dark chocolate color, while ranched mink colors range from white and pale silver to dark brown. Mink pelts are about 65 cm long, with underfibers usually 1.3 cm and the guard hairs 2.1–2.5 cm in length. Chinchilla pelts are around 18 × 31 cm, with a 2.5 cm deep fur of lustrous blue-gray. Coypu produces a short fur with a blue-brown color. Beaver pelts average 163 cm in length and have a dark brown or black, dense fur with shiny, coarse guard hairs. Seal pelts average 111 cm length and have a short, dense, glossy brown fur. The pelt of the ermine is approximately 38 cm long, and white to creamy white. Lamb and sheep fur is usually flat and glossy, with a curl to the fiber. Karakul from the Karakul sheep of lower central Asia is black, brown or white, with a short, wavy, shiny hair. Mouton lamb fur from sheep reared in the USA, Australia and South America is off-white. Persian lamb fur is black, gray, brown or white in color, and lustrous with a flat, tight curl to the fibers.

4. Uses of Fur Pelts

Because of the warmth of furs and the fashion styles, the major use of fur pelts in the garment industry is for coats, jackets, capes, stoles, neckpieces and trimmings for cloth coats and dresses. Fur is also used for hats, vests, boots, slippers and gloves. Furriers usually use many pelts or a mixture of pelts to make their garments. From 121 to 151 pelts of the chinchilla are used to make a coat; 51–81 pelts are used to make a mink coat. The fibers from fur pelts may be removed for weaving in the case of the chinchilla or beaver for making the finest of felt hats. The hairs may also be used in artists' paintbrushes. Large fur pelts from the bear, mountain lion, wolf and zebra are used as rugs, wall hangings and bed covers. Furs are also used for pillows, cushions and upholstering material.

See also: Wool; Materials of Biological Origin: An Overview

Bibliography

Ashbrook F G 1954 *Furs, Glamorous and Practical.* Van Nostrand, New York

Bachrach M 1953 *Furs.* Prentice-Hall, Englewood Cliffs, New Jersey

Campbell M E W 1968 *The Fur Trade.* Clarke, Irwin, Toronto

Kaplan D G 1951 *The Fur Book.* Donnelly Corporation, New York

Kaplan D G 1974 *World of Furs.* Fairchild Publishing, New York

Tuttle T 1968 *How to Grade Furs*, CDA Report No. 1362. Canadian Department of Agriculture, Ottawa

P. Ross

Fuses and Circuit Breakers

Normally the current flowing in electrical wiring is limited by the load in the circuit. Whatever the load (lights, motors or other appliances), its impedance limits the current. If a fault in a load device develops, this resistance to current flow is lost. Such a short circuit results in a great rush of current which rapidly heats the conductor, possibly destroying the conductor and starting a fire. To avoid such effects, circuit protective devices (fuses and circuit breakers) are placed in the circuit. They sense the rush of current and open the circuit before damage can be done.

A second danger in electrical circuits is the possibility of overload. When too many appliances are put in a single circuit, the multiple current paths permit excess current to flow in the permanent wiring. This overload current—for example, 30 A in a 15 A circuit—does not cause immediate damage, but with time, heat (proportional to the square of the current) builds up and permanent damage can be done to the wiring or a fire can be started. Fuses and circuit breakers protect the circuit by sensing the overload current and opening before damage can be done.

1. Fuses

Fuses provide an element in the current path which melts to open the circuit. The fuse element is made of a low-melting metal and is designed to carry the normal current without overheating and melting. A short circuit causes very rapid melting and vaporization of the fuse element. An overload current causes melting in a longer period of time. This period varies as the square of the current and ranges from ~1–3 min at twice the rated current to ~0.1–0.5 s at 10 times the rated current.

1.1 Types of Fuse

Miniature, plug, cartridge and expulsion-type fuses are in general use today. Miniature fuses are used for supplementary overcurrent protection for individual pieces of equipment and in automatic and similar electrical systems. They are used in circuits of ≤250 V with current ratings of fractions of an ampere up to 30 A. Cylindrical fuses with brass end terminals and a glass body (Fig. 1) have been used for many years.

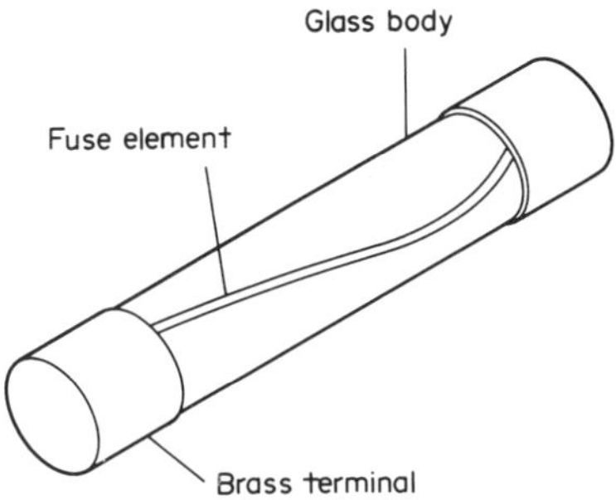

Figure 1
Cylindrical miniature fuse

These fuses normally have a soldered connection between the terminals and the fuse element.

A more recent development is the plug-in miniature fuse (Fig. 2), which consists of a single piece of zinc with a plastic cover. The stab portion of zinc is tin plated for protection from the environment and the fuse element is simply a portion of reduced cross section connecting the two stabs. The cover is transparent, provides insulation and may be color coded to designate the rating.

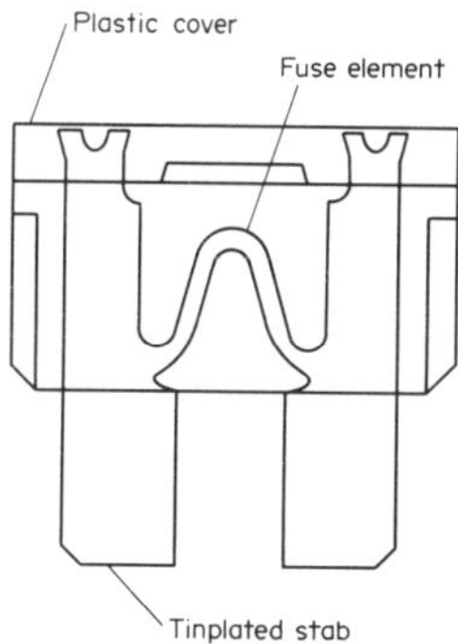

Figure 2
Plug-in miniature fuse

Plug fuses (Fig. 3) are used in residential systems with circuits of ≤125 V. They have a screw base of brass similar to an incandescent lamp, and a glass or mica window in the top which permits observation of the fuse element. The fuse elements are ribbons of zinc or copper which are soldered to the brass terminals. Insulation between the terminals is provided by a plastic base, molded from phenolic resin.

Cartridge fuses (Fig. 4) form a large family, but are limited to use at ≤600 V. They include ferrule and blade types for conventional, current-limiting or time-delay service. The terminals, whether ferrule or blade type, are brass which may be tin or silver plated. The fuse body is a fiberboard or fiberglass

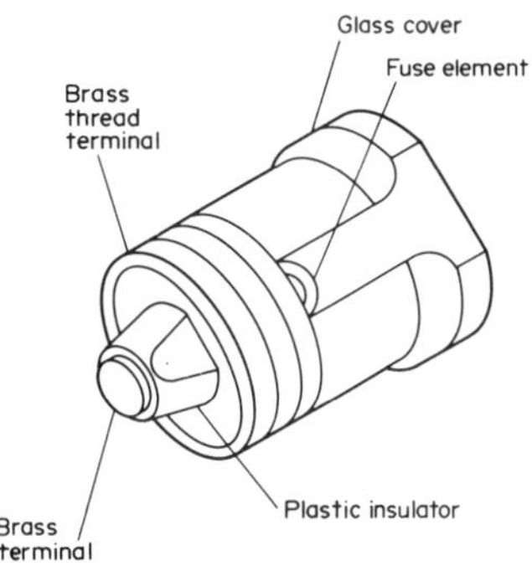

Figure 3
Plug fuse

tube which constitutes an insulated connection between the terminals and which confines the molten globules of fuse metal if the fuse blows. While some cartridge fuses can be used only once, others can be disassembled to permit replacing blown fuse elements.

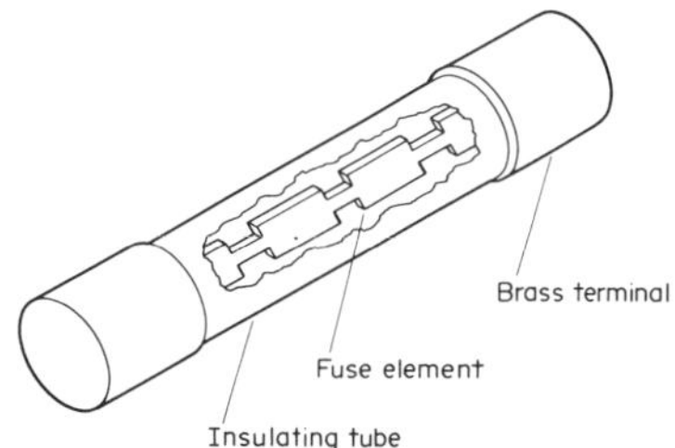

Figure 4
Cartridge fuse (ferrule terminals)

Current-limiting fuses are cartridge fuses which are designed to interrupt the current in less than half a cycle (Fig. 5). This limits the current flowing and the energy that must be dissipated. These fuses use multiple silver elements (Fig. 6). The individual elements have several deep notches along the length which result in high-resistance hot spots that open first under short-circuit conditions. The fuse elements are attached to the end pieces with solder which melts to open the fuse under overload. An alternative design uses an overload trigger of solder or tin in the element itself. The fuse is tightly packed with quartz sand around the fuse elements. When a short circuit occurs, the heat of the arcs created by the melting of the notches is absorbed by the sand and the arcs are extinguished. In this process the sand melts, producing small glasslike nodules called fulgurites at each arc location. Since several notches in many fuse elements fuse simultaneously, the energy is spread more or less uniformly throughout the fuse.

Under overload, the links would have to heat up

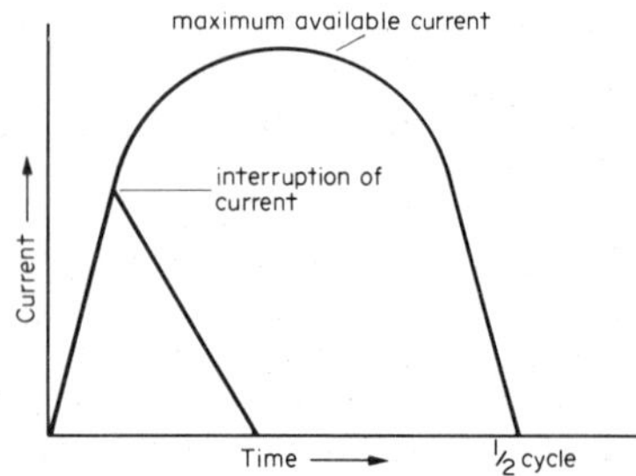

Figure 5
Effect of current-limiting fuse

to the melting point of silver (960 °C) if the low-melting solder or overload trigger designs were not used. Obviously this would result in the entire fuse being too hot for safety. With existing designs, the solder melts at low temperature, avoiding this problem. As the solder in each link melts, the current increases in the remaining links. Thus overload failure is rapid once a few links reach the melting temperature of the tin or solder.

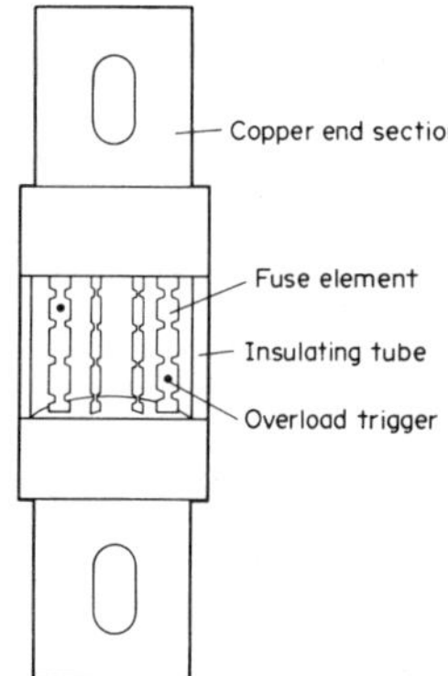

Figure 6
Current-limiting fuse

Time-delay fuses (Fig. 7) are cartridge fuses which are used to avoid open circuits due to momentary current surges such as occur when a motor is started. These fuses have two fuse links in series. There is an overcurrent element of copper or silver for heavy overloads and short circuits and a slow-melting element of solder or tin which opens on small overloads after a time delay. Either link can blow, opening the circuit. An alternative design uses a fuse element shielded with tin. When subjected to an extended overload, the tin shield melts, increasing the resistance of the element, which results in the fuse opening.

Expulsion-type fuses (Fig. 8) are used for high-voltage protection (600 V to 161 kV). (Some current-limited fuses are also used at voltages up to 34.5 kV.) Expulsion-type fuses generate and expel gases to dissipate the energy arising from the fault. The fuse consists of a fusible element, an arcing rod (or rods) and an arc-extinguishing, gas-generating material such as boric acid, bone fiber or liquid solutions. When the fusible element melts, the arcing rod is pulled away by a spring, lengthening the arc, which vaporizes the arc-extinguishing material, producing water vapor and other gases which cool and deionize the arc. The hot gases are expelled from the end of the fuse with tremendous energy. Mufflers are often used to silence, slow down and dissipate the energy of the expelled gases. Expulsion-type fuses are slower than current-limiting fuses, about one cycle being needed to clear the fault.

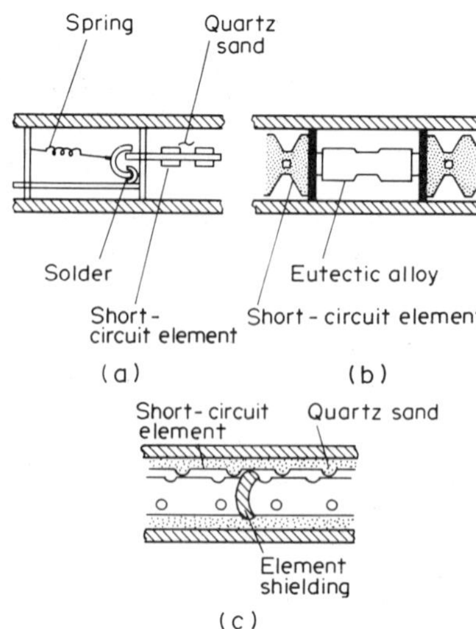

Figure 7
Three types of time-delay fuse: (a) spring-loaded; (b) eutectic alloy; (c) shielded element

1.2 Fuse Element Materials

Whatever the type of fuse, the fuse element or link is the key to the performance of the fuse. It is the fuse element that melts and opens the circuit under overload. Zinc, tin, copper and silver are often used

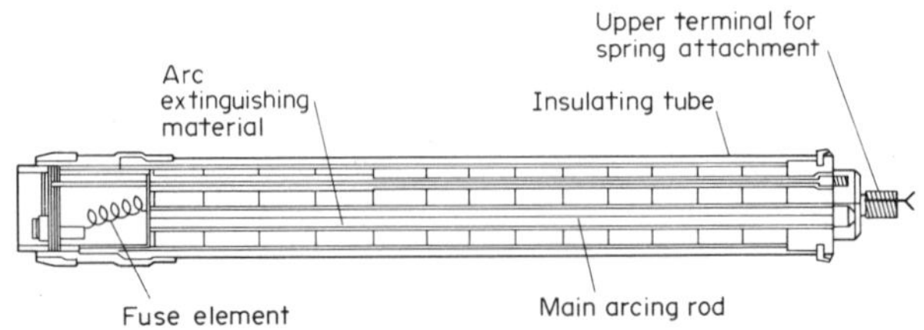

Figure 8
Expulsion fuse

for fuse elements. Solder, usually 60%Sn–40%Pb, is used in time-delay fuses and low-melting alloys, such as 50%Sn–25%Pb–25%Cd and 53%Pb–44%Bi–3%Sn, are used in transformer fuses. Their melting points are sufficiently low to avoid vaporization of the transformer oil.

The ideal material for fuse elements has low melting and boiling points and high thermal and electrical conductivities; it should have a high ionization potential, not sustain an arc and not reignite. Low-melting-point materials melt before the overall fuse gets hot. A low-boiling-point metal volatilizes rather than forming molten globs. High thermal and electrical conductivities contribute to low generation and rapid dissipation of heat. When the element melts, an arc is generated which continues to carry the current until it is extinguished. The fuse element material must minimize the stability of the arc and its tendency to reignite. Table 1 compares the key properties of several common fuse element materials.

No material is the ideal element material. Zinc has only fair conductivity but it has low melting and boiling points and is low in cost. Its use is limited to fuse applications where interrupting faults do not exceed 10 000 A. Silver has excellent conductivity and good arc characteristics. Unfortunately, its melting point is high and it is expensive. Nevertheless, elements of silver are used for high-fault fuses. Copper has similar properties to silver at much lower cost, but it has the disadvantages of tarnishing and being subject to corrosive attack. Many time-delay fuses use copper short-circuit elements in series with the special time-delay element. Tin and solders are used as fuse element materials in time-delay fuses.

2. Circuit Breakers

Circuit breakers sense overloads and short circuits and mechanically open the circuit. Unlike fuses, circuit breakers can be reset and reused repeatedly. Thermal circuit breakers sense overloads by the use of a thermostatic bimetal strip. Short-circuit inrush is sensed by a coil in electromagnetic circuit breakers.

The current path through a typical molded-case circuit breaker (Fig. 9) is as follows: line strap, stationary contact, movable contact, movable contact arm, braid, electromagnetic coil, thermostatic bimetal strip, heater strip and load strap. Currents up to and including the rated current flow through this chain of parts indefinitely. An overload current causes gradual heating of the bimetal strip. Heat also builds up in the adjacent heater strip and is conducted to the bimetal strip. As its temperature increases, the thermostatic strip deflects. Eventually the deflection is sufficient to trip the breaker and the movable contact swings away from the stationary contact, creating an arc which is swept into, broken up and finally extinguished by a series of plates called arc

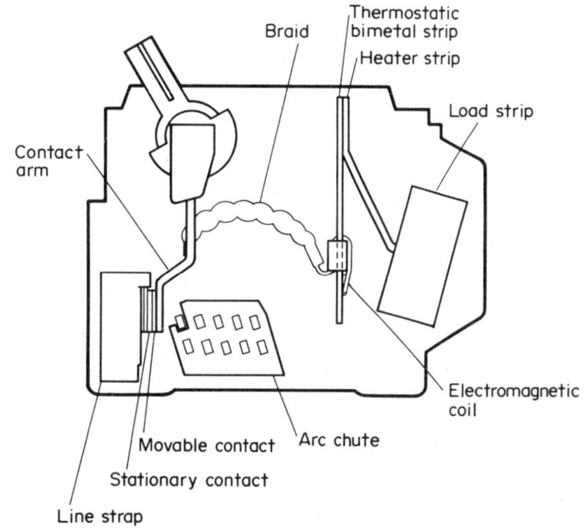

Figure 9
Circuit breaker

Table 1
Significant properties of fuse element materials

Material	Melting point (°C)	Boiling point (°C)	Resistivity (μΩ cm)	Thermal conductivity ($W\ m^{-1}\ K^{-1}$)	Minimum ionization potential (V)
Ag	960.5	2210	1.59	420	7.54
Cu	1083.0	2600	1.67	395	7.68
Sn	231.9	2270	11.5	65	7.30
Zn	419.5	906	5.92	115	9.36
Solder:					
60%Sn–40%Pb	181–191	2270–1740	15–20		7.30–7.38
Tin alloy:					
50%Sn–25%Pb–25%Cd	141–151	2270–1740–765	10–12		7.30–7.38–8.96
Lead alloy:					
53%Pb–44%Bi–3%Sn	105–157	1740–1420–2270	65–75		7.38–8.0–7.30

chutes. After the fault has been corrected, the circuit breaker can be closed and will perform normally. As a short circuit or heavy overload current flows through the electromagnetic coil, it creates adequate magnetic flux to move the armature and trip the breaker.

Important components in the circuit breaker from the materials standpoint include the contacts, the braid, the thermostatic bimetal strip, the heater strip and the arc chutes. The contacts must have high electrical and thermal conductivity and low surface resistance and must resist welding and arc erosion. Copper and silver have the required conductivities. Only silver retains low surface resistance when heated in air. Tungsten carbide, graphite and cadmium oxide are most commonly used to avoid welding problems. Nickel, tungsten, molybdenum and tungsten carbide are used to avoid arc erosion. Most contacts are multiphase materials with copper (oil breakers) or silver (air breakers) dispersed in or dispersing one or more of the other materials mentioned above. Powder metallurgy techniques are used in the production of contacts.

The braid in a circuit breaker must have high conductivity so that it can conduct the rated current without overheating. It must also have excellent flexibility to permit free motion for the movable contact arm. These requirements are met by very fine copper wires braided together. Occasionally the individual copper wires are tin plated.

The heater strip is also in the current path of the circuit breaker. It is designed to have sufficient resistance to contribute thermal energy to the thermostatic bimetal strip (see *Thermostatic Bimetals*). The requirements of the particular device determine the resistance needed. This in turn determines the material required. Brass, bronze and cupronickel have been used. The resistivities of candidate heater strip materials are shown in Table 2.

Table 2
Materials for circuit breaker heater strip

CDA alloy	Common name	Resistivity (Ω cmf^{-1})	Resistivity ($\mu\Omega$ cm)
210	Gilding metal	19.7	3.08
220	Commercial bronze	25.3	3.92
230	Red brass	28.8	4.66
240	Low brass	33.8	5.39
260	Cartridge brass	36.9	6.16
280	Muntz metal	36.4	6.16
464	Naval brass	42	6.63
510	Phosphor bronze	66	11.5
521	Phosphor bronze	84	13.3
651	Low-Si bronze	96	15.2
655	High-Si bronze	169	26.8
706	Cupronickel 10%	120	19.1
715	Cupronickel 30%	232	37.5

The arc chutes are used to extinguish the arc generated when the contacts separate. They consist of steel plates, often nickel or zinc plated, which are held in position by insulating side walls of fiberboard or ceramic. When the contacts part, the arc generated lengthens. The magnetic flux concentration in the chute plates draws the arc into the chute array, where it is broken up into small arclets. The steel plates cool the arc plasma and the gaps between them provide a path for the escape of the hot gases.

See also: Low-Voltage Contact Materials; Electrical Materials: An Overview

Bibliography

Electrical Protection Handbook 1978 Bussmann Manufacturing Division, McGraw-Edison Company, St. Louis, Missouri
Fahnoe H H 1980 Cut plant-protection costs, increase reliability, with the right fuse. *Power* 124(1): 23–26
Freund A 1980 *Overcurrent Protection.* McGraw-Hill, New York
Gould Inc. 1979 *Fault-Trap Fuse.* Electric Fuse Division, Gould Inc., Rolling Meadows, Illinois
Holm R 1967 *Electric Contacts, Theory and Application.* Springer, New York
Lauben R 1979 New trends in industrial molded case circuit breakers. *Ind. Power Syst.* 14–18
Machine Design (May 17 1979): issue devoted to Electrical and Electronics Reference, pp. 137–45
Tuniewicz R M 1979 Cut circuit damage by sizing fuses carefully—1. Low-voltage fuses. *Power* 123(12): 21–27

D. T. O'Brien

Fusion Reactor Materials

The development of fusion power as a viable energy source depends to a large extent on the proper selection of materials for the individual reactor components. These materials will be exposed to a wide variety of conditions, such as severe ion and neutron radiation, high thermal fluxes, large thermal and mechanical stresses, and various chemical environments. In recent years, conceptual designs of fusion reactors have been developed, the materials requirements for the various reactors components have been defined and candidate materials have been identified. Many different types of materials are required in a fusion reactor. As examples, iron-base alloys and refractory metals have been proposed as candidate structural materials, beryllium and graphite for the surface of the plasma chamber, liquid lithium and Li_2O as tritium-breeder materials, beryllium and lead for neutron multiplication, and NbTi and Nb_3Sn as superconducting materials. Other types of materials that are required include electrical insulators, radiation shielding materials, coolants and conventional electrical conductors. The appropriate selection of an integrated combination of materials for the individual reactor system involves numerous trade-offs between the critical materials properties and requirements.

This article presents a brief review of the fusion reactions, proposed fusion reactor concepts, the critical materials requirements for the major reactor components, and candidate materials that are currently believed to offer the greatest potential for use in fusion reactors. The materials discussed are still in an early stage of development. During the next few years, as additional information becomes available, the candidate materials will go through a sorting process with a number of materials being eliminated from further consideration, but with other materials possibly being added. A large effort will then be required to qualify the primary candidate materials for service in fusion devices. A major concern in qualifying materials is the inability to exactly simulate the fusion environment, particularly the high-energy-neutron spectrum. Ultimately, operating experience in fusion reactors is necessary to determine the capabilities and limitations of the materials.

1. The Fusion Process

A variety of elements can undergo fusion reactions, including deuterium–tritium (D–T), deuterium–deuterium (D–D) and deuterium–helium-3 (D–^{3}He). The D–T fuel cycle has the highest reaction rate together with a high energy release per reaction (17.6 MeV), and therefore the largest power density for a given plasma density. It also has the lowest plasma-temperature requirement of any of the fusion fuels. The ideal ignition temperature for the D–T reaction is ~4.5 keV, compared with ~35 keV for the D–D and D–^{3}He reactions. It is generally believed that the first-generation commercial fusion reactors will operate on the D–T fuel cycle and so attention here is focused on this cycle. The fusion reaction of interest is

$$\mathrm{D} + \mathrm{T} \rightarrow {}^4\mathrm{He}\ (3.5\ \mathrm{MeV}) + \mathrm{n}\ (14.1\ \mathrm{MeV}) \qquad (1)$$

To derive useful energy production from fusion, it is necessary to heat a D–T plasma to temperatures of the order of 10^8 K (equivalent to a kinetic energy of ~8 keV) and contain the fuel at a particle number density n for a time τ such that $n\tau \geqslant 10^{20}$ particles m^{-3}. The product $n\tau$ may be attained by relatively long confinement times (of the order of a second) with modest plasma densities (~10^{20} particles m^{-3}) as in the case of magnetic-confinement reactors (Baker et al. 1981), or very short confinement times (~10^{-10} s) at very high plasma densities (of the order of 10^{30} particles m^{-3}) as in the case of inertial-confinement reactors (Monsler et al. 1981). In magnetic-confinement fusion, very intense magnetic fields (typically 5–12 T) are used to contain the hot plasma ions (D^+, T^+). In inertial-confinement fusion, a large amount of energy from either lasers or ion accelerators is deposited on small fuel pellets; rapid compression of the pellets to very high densities provides the desired condition for the fusion reaction. Since a much larger international effort is focused on magnetic-confinement fusion, the materials evaluation presented here is limited to this concept.

Large resources of deuterium are available from sea water. However, since tritium does not occur naturally, it is necessary to produce tritium by neutron reactions in the blanket region surrounding the plasma. Lithium, in the form of liquid lithium or a compound such as Li_2O, is the only element from which adequate tritium production appears feasible. Tritium is produced by two reactions with lithium:

$${}^6\mathrm{Li} + \mathrm{n} \rightarrow \mathrm{T} + {}^4\mathrm{He} + 4.8\ \mathrm{MeV} \qquad (2)$$

and

$${}^7\mathrm{Li} + \mathrm{n} \rightarrow \mathrm{T} + {}^4\mathrm{He} + \mathrm{n} - 2.5\ \mathrm{MeV} \qquad (3)$$

Natural lithium is composed of 92.6% ^{7}Li and 7.4% ^{6}Li. The fact that the breeding reaction in ^{7}Li releases a neutron that is available to induce a tritium-producing reaction in ^{6}Li is the basic reason why a breeding ratio substantially greater than unity can be achieved with natural lithium. The fuel cycle for this system may be more correctly referred to as a D–T–Li fuel cycle.

The non-D–T fusion fuels, often called advanced or alternate fuels, are of interest because they offer the potential advantages over D–T fuels of no tritium breeding, reduced tritium-handling requirements, reduced neutron activation of structural materials and the availability of a larger fraction of the energy released by charged particles, which may permit the use of direct energy conversion into electricity. However, the smaller cross sections and higher temperature requirements imply more difficult plasma-confinement problems and the need for larger magnetic fields.

It is important to note that there are several potential end uses of fusion energy. The focus here is on electricity production in the form of a base-load central station power plant. However, the application of fusion energy to synthetic fuel production (e.g., hydrogen) and the production of fissile fuel and electricity in fusion–fission hybrid reactors may also be feasible.

2. Reactor Concepts

Magnetic-confinement concepts can be divided according to the shape of the plasma, which determines the geometry of the reactor (Baker et al. 1981). The tokamak concept currently represents the mainline magnetic-confinement approach throughout the world. The tokamak topology (Fig. 1) is an axisymmetric toroidal configuration with a relatively low aspect ratio (ratio of the major radius to the minor

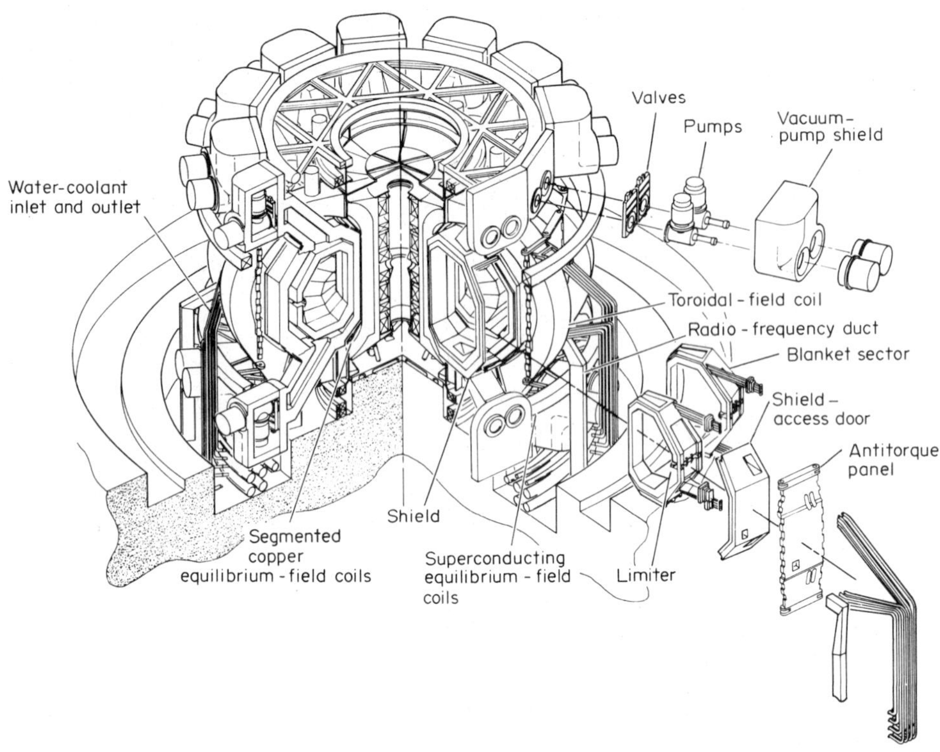

Figure 1
Schematic diagram of a tokamak reactor (STARFIRE)

radius) of 3–4. An alternative magnetic-confinement concept also being considered is the tandem-mirror concept, which is a linear device with various end configurations to minimize plasma leakage out the end of the cylindrical plasma chamber. Several other confinement concepts, such as the Elmo Bumpy Torus reactor, are also being evaluated. Since many of the materials considerations are generic to the different reactor concepts and since the technology of the tokamak is more advanced than that of the other concepts, the materials evaluation presented here is based primarily on the materials requirements for the tokamak reactor concept.

A cross-sectional view of a typical tokamak reactor design illustrates the main systems of the reactor (Fig. 2). The major components and their primary functions are summarized in Table 1. The plasma is surrounded by the first-wall/impurity-control system, which is directly exposed to radiation and energetic particles escaping from the plasma. The blanket system, behind the first wall, captures most of the neutrons produced by the fusion reaction and contains the tritium-breeding material. The shield, surrounding the blanket, serves to reduce the radiation level to an acceptable level for the magnet system.

Table 1
Primary functions of major reactor components

First-wall system (including limiter/divertor)
provide appropriate geometry for plasma
prevent atmospheric contamination
acceptable plasma–wall interaction
maintain structural integrity
provide for removal of impurities from plasma
Blanket
convert neutron energy into sensible heat
incorporate heat-transport system (coolant)
breed tritium and provide for tritium recovery
provide some shielding for magnets
Shield
provide adequate shielding for magnet system
Magnets
contain plasma
control plasma shape and position
heat plasma
Auxiliary components
vacuum systems, plasma heating, plasma fuelling, diagnostics

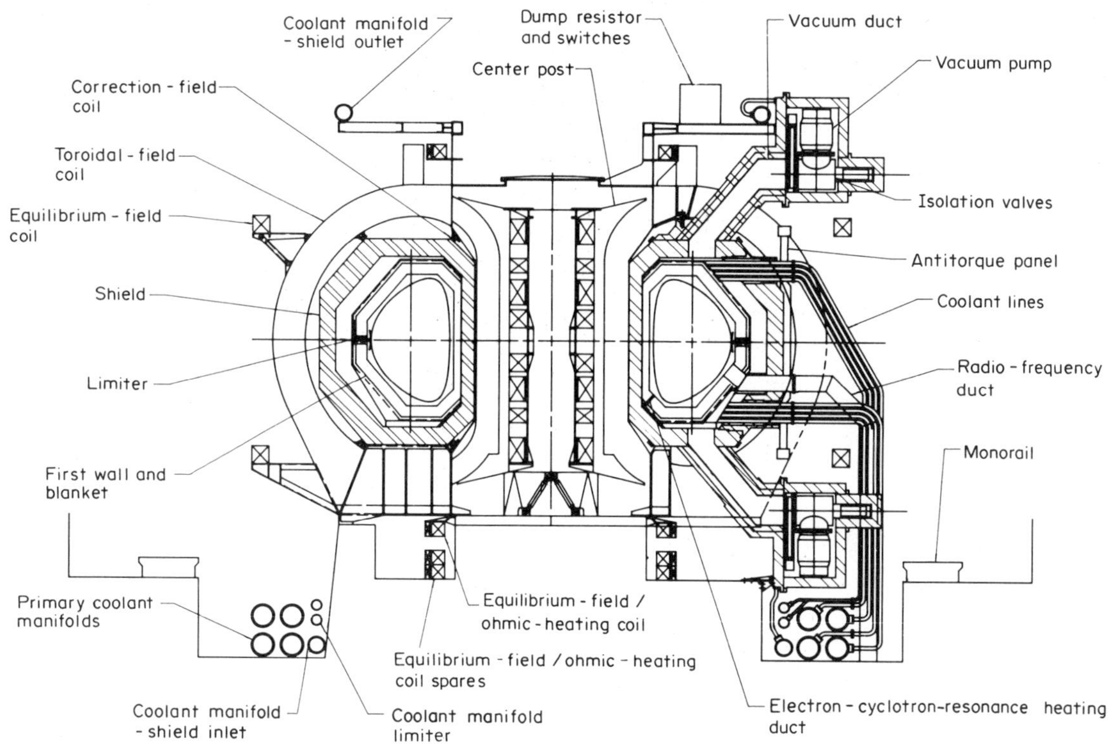

Figure 2
Cross-sectional view of a tokamak reactor (STARFIRE)

The magnets serve to raise the temperature and current in the plasma (ohmic heating coils) and to control the shape and position of the plasma (toroidal and equilibrium field coils). Most of the materials discussed below are considered for use as part of the first-wall system or the blanket system. These two systems experience the harshest operating environment and therefore present the greatest challenges to engineering.

The first-wall system consists of those components that form the plasma chamber and the impurity-control system, and other components that are directly exposed to the plasma. Specific components that make up the first-wall/impurity-control system include:

(a) First wall—forms the plasma chamber, portions of which are exposed to high particle and heat fluxes.

(b) Limiter—defines the plasma edge and is exposed to high particle and heat fluxes. It may be part of the impurity-control system.

(c) Divertor—part of an impurity-control system that is generally exposed to high plasma-particle and heat fluxes. (A limiter may be used instead of a divertor for impurity control.)

(d) Beam dumps—regions that receive neutral beam shine-through from injectors.

(e) Armor—used in special high-heat-flux and high-particle-flux regions of the plasma chamber.

The first-wall/impurity-control system must maintain its integrity for acceptable lifetimes under severe radiation, thermal, stress and chemical environments, and it must not be a source of excessive plasma contamination. Table 2 summarizes the critical requirements of the system components. Three types of materials generally considered for the first-wall components include a structural material, a coolant and a special material for the plasma interface. In early conceptual fusion reactor designs (Badger et al. 1973), the first-wall structural material also served as the plasma interface. Subsequent designs provided for radiatively cooled liners (Badger et al. 1976) or special low-atomic-number (low-Z) coatings (Baker et al. 1980) for the plasma interface. The coating concept provides greater flexibility in materials selection since the coating material can be selected for its

Table 2
Requirements for first-wall impurity-control system

Withstand erosion by energetic plasma particles
physical sputtering
chemical sputtering
blistering
Not be a source of excessive plasma contamination
low erosion rates
low-*Z* materials
Hydrogen effects
tritium inventory
tritium permeation
Withstand plasma disruptions
vaporization of surface
melt layer erosion
electromagnetic forces
Withstand high surface-heat flux
particle flux
electromagnetic radiation
thermal stress (cyclic)
Neutron damage (14 MeV)
swelling
embrittlement
transmutation effects
effect on mechanical properties
effect on physical properties
Coolant compatibility
radioactive mass transport
stress corrosion
Reliability
fabrication difficulty
design complexity
repair and maintenance
Neutronic and environmental considerations
effect on tritium breeding
safety
activation

superior plasma-interaction characteristics with less importance placed on its structural characteristics. Table 3 lists candidate materials most often proposed for the first-wall components. The materials shown in *italic* type are currently believed to offer the most potential.

All blanket concepts that have been proposed require lithium or a lithium-bearing material for tritium breeding, a circulating coolant for heat removal and a structural material to contain and support the blanket system. In some design concepts, the same material (e.g., liquid lithium) may be used as both breeder and coolant. Some concepts require a neutron multiplier to provide for adequate tritium production. In several concepts, tritium is recovered from the recirculating coolant or liquid-metal breeder material. For solid breeder concepts, a separate tritium-processing fluid, usually helium, is utilized for tritium recovery. Reflectors or moderators have been proposed in some blanket designs to improve breeding performance or to reduce the lithium inventory. Certain concepts, particularly those involving water cooling, may require tritium barriers to minimize tritium permeation into the coolant since tritium recovery from water is difficult and expensive.

The primary function of the shield is to reduce the nuclear heating and radiation damage in the magnet system to acceptable levels. In a tokamak, toroidal field coils provide the primary containment for the plasma. Superconducting coils are generally considered necessary for an economically attractive commercial reactor. Poloidal field coils, which may or may not be superconducting, are required to control the plasma shape and position. Additional poloidal field coils can be used to heat the plasma during start-up. Several auxiliary components such as a vacuum system, plasma heating system (neutral beams or radio-frequency heating), plasma fuelling system and diagnostics are required. In addition to the types of materials used in the other major components, special purpose materials such as electrical insulators, vacuum seals and electrical contacts may be required in these systems. In general, the materials requirements for these components are less well defined at present.

Figure 1 is an isometric view of the recent STARFIRE reactor design. This design represents the best present understanding of a commercial tokamak reactor. The STARFIRE design utilizes a pumped-limiter concept for removal of the fusion reaction product (i.e., helium). A poloidal divertor is also being considered for impurity control in tokamak reactors. A significant aspect of the STARFIRE design is the proposed use of radio-frequency (rf) power for the heating of the plasma. Neutral-beam heating, which is used for plasma heating in several physics devices in present use, is the primary alternative. A large current (several MA) must be induced in a tokamak plasma to maintain a stable plasma burn. Conceptual reactor designs have most often used induction coils to obtain the required plasma current. The burn cycle attainable by this method is limited to tens of minutes. An rf current drive, as proposed in the STARFIRE design, would provide for a continuous plasma burn. The potential for steady-state operation has a major impact on the materials and technology considerations since fatigue problems, exacerbated by cyclic operation, would be greatly reduced.

3. *First-Surface Materials*

The materials that interface the plasma edge are exposed to a high-energy-neutron spectrum, electromagnetic radiation and energetic particles (deuterium, tritium, helium and impurities) emitted by

Table 3
Candidate materials for fusion reactor applications[a]

Plasma-side material[b]	First-wall/blanket structure	Limiter/divertor structure	First-wall/blanket coolant	Tritium-breeder material	Neutron multiplier
Beryllium	*Austenitic steel*	*Vanadium alloy*	*H_2O*	*Lithium*	*Beryllium*
Boron	*Ferritic steel*	*Niobium alloy*	D_2O	*Li–Pb alloy*	*Lead*
Graphite	*Vanadium alloy*	Molybdenum alloy	*Lithium*	Li–Pb–Bi alloy	Bismuth
B_4C	Titanium alloy	Tantalum alloy	Li–Pb alloy	Li_7Pb_2	Zirconium
BeO	Nickel alloy	*Copper alloy*	*Helium*[c]	*Li_2O*	Zr_5Pb_3
SiC	Niobium alloy	Aluminum alloy	Steam	$LiAlO_2$	PbO
TiC	Molybdenum alloy			Li_2SiO_3	Pb–Bi alloy
TiB_2				Li_4SiO_4	
Tungsten				Li_2ZrO_3	
Tantalum					

a Leading candidate materials are given in *italic* type b First-wall structural materials are also considered candidates for the plasma-side material c Helium is generally regarded as unacceptable for the limiter/divertor coolant

the plasma. The 14 MeV neutron currents to the first wall are fairly uniform and are projected to be 2–5 MW m^{-2} for commercial tokamak reactors. A 1 MW m^{-2} neutron wall loading corresponds to a 14 MeV neutron flux of 4.4×10^{17} neutrons $m^{-2} s^{-1}$ and a total neutron flux of $\sim 1.4 \times 10^{18}$ neutrons $m^{-2} s^{-1}$ (>0.1 MeV). Since these high-energy neutrons have a mean free path of 0.3 m or greater in typical reactor materials, their energy is deposited throughout the blanket region. The remainder of the plasma energy, which initiates as the 3.5 MeV He (equivalent to ~20% of the total fusion energy), is deposited in the form of photons or energetic particles incident on the surfaces of the first-wall components. The photon energy, which may be 20–60% of the total surface heat load, is deposited fairly uniformly on the first-wall components. However, the transport power from the energetic particles escaping from the plasma will be deposited preferentially in specified regions such as limiters or divertor plates.

As summarized in Table 2, the surface materials of the first-wall system must (a) withstand erosion by the energetic plasma particles, (b) not be a source of excessive plasma contamination, (c) withstand relatively high surface heat fluxes from photons and particles, (d) withstand thermal transients caused by rapid (probably ms) deposition of the plasma energy on the wall during a plasma disruption, (e) not be degraded by hydrogen (D, T) interactions nor create an excessive sink for tritium, (f) withstand neutron damage and (g) be environmentally acceptable. Plasma–wall interactions can potentially lead to significant erosion of the first-wall components by energetic plasma particles. The mechanisms currently believed to be the most serious in a tokamak power reactor are physical sputtering, chemical sputtering and blistering.

For most reactor designs, physical sputtering (see *Sputtering*) is believed to be the most important plasma–wall interaction during reactor operation. Plasma ions (D^+, T^+, He^+) and charge-exchange neutrals (D^0, T^0) from the edge region strike the first-wall components, typically with energies in the range 100–2000 eV. Physical sputtering yields, which are strongly energy dependent, are typically of the order of 10^{-3}–10^{-2} atoms per incident particle for these low-mass fuel (D, T) particles. Effects on the limiter or divertor are generally much more severe because of the much higher particle fluxes projected for these regions. An important aspect of the physical sputtering process relates to the fact that material sputtered from the wall surfaces must redeposit somewhere. Much of this material is expected to redeposit as energetic ions on the limiter or divertor surfaces. If, as can be the case for high-Z wall materials, the self-sputtering yield is greater than unity, a propagating effect can result. This, plus the fact that low-Z wall atoms have a much smaller impact on plasma performance in the event that they penetrate the plasma, provides a basis for proposing the use of low-Z materials such as beryllium and graphite for the first surface. If low particle energies at the plasma edge can be attained, high-Z materials such as tungsten or tantalum may be acceptable because of their very low sputtering yields at low energies (<100 eV).

Chemical interactions between the hydrogenous plasma and certain candidate wall materials may occur. This chemical sputtering may lead to excessive wall erosion or plasma contamination. The classic example is the reaction with graphite to form hydrocarbons such as CH_4. For most other candidate materials, except possibly some carbides such as SiC or TiC, chemical sputtering is not expected to be a major problem.

The blistering ("chunk emission") mechanism (see *Sputtering*) can lead to large erosion rates under certain conditions, but is not considered to be a major

Table 4
Defect production rates in alloys typical of the alloy systems proposed for fusion first-wall applications[a]

Metal or alloy	Produced per MW year m^{-2}		
	Displacements per atom	H (atomic ppm)	He (atomic ppm)
Aluminum	14	296	316
Ti–6%Al–4%V	16	175	142
Ferritic steel (Sandvik HT-9)	11	450	110
Austenitic stainless steel (type 316)	11	532	147
Nickel-base alloy (Nimonic PE-16)	12	780	240
V–15%Cr–5%Ti	11	245	47
Niobium	7	105	29

a A first-wall neutron spectrum was used in the calculation

problem under the conditions expected for most first-wall components.

The first-wall components must also withstand relatively high surface heat fluxes produced by the incident particle fluxes and additional electromagnetic radiation. Typical surface heat fluxes for the first wall of a tokamak reactor are predicted to be 0.2–1.5 MW m^{-2}. Surface heat fluxes on the divertor or limiter may be an order of magnitude higher. Thermal stresses induced in the wall become important if relatively thick walls (~1–2 cm) are required to allow for surface erosion. If the reactor operates in a cyclic burn mode, fatigue of the wall becomes an important factor. Materials with a high thermal conductivity and a low thermal expansion coefficient generally exhibit low thermal stresses.

Potential effects caused by energetic hydrogen ions injected into the first wall include increased tritium inventory, enhanced tritium permeation through the wall and possible hydrogen embrittlement. Certain structural alloys (e.g., titanium-base alloys) may be eliminated from consideration primarily because of this effect. Tritium permeation through the wall is important for the case of water coolant, since it appears to be economically prohibitive to extract large amounts of tritium from water.

In the event of a plasma disruption, the energy in the plasma is expected to deposit in a short time (<50 ms) on a limited region of the first wall, limiter or divertor. Certain regions may receive energy densities of the order of 100–500 J cm^{-2}. Vaporization and melting of surface regions will occur when most materials are exposed to these transient heat fluxes. Since the extent of wall erosion is an important consideration, refractory-type materials such as tungsten or graphite provide an advantage.

The high-energy (14 MeV) neutrons can affect the properties and integrity of the first-surface materials. The physical properties (e.g., thermal conductivity) of certain materials, such as ceramics, are strongly degraded by neutron irradiation. The mechanical properties can also be significantly affected by neutron irradiation. Of particular significance is the fact that all low-*Z* materials have relatively high helium generation rates when exposed to 14 MeV neutrons (see Table 4). Special microstructural tailoring may be required to accommodate this helium or to enhance the release. If the first-surface material is used in the form of a coating or cladding, the mechanical requirements should be less critical since the substrate can provide much of the required structural support.

4. *Structural Materials*

The first-wall structural material provides the mechanical support for the first wall, serves as part of the vacuum chamber, contains the first-wall coolant, and in most designs, is part of the blanket and serves as the blanket containment. The structural material must maintain its integrity at elevated temperatures for acceptable lifetimes while exposed to high neutron fluxes, high heat fluxes, a high-temperature coolant and moderate hydrogen (D, T) pressures. The materials that have most often been considered for first-wall/blanket and limiter/divertor structures are listed in Table 3. The austenitic stainless steels, ferritic steels and selected vanadium-base alloys are currently believed to offer the most potential for the first-wall/blanket structure. Either the refractory metal alloys (V, Nb or Ta) or copper alloys are currently regarded as the primary candidate structural materials for the limiter and/or divertor. The favorable thermal–physical properties of these materials compared with those of the steels appear necessary to accommodate the high thermal fluxes, particularly for cyclic operation. In addition to the relatively high thermal stresses, the structure must withstand gravity loads, magnetic forces, coolant pressure and possibly vacuum/atmospheric pressure.

The structure must provide adequate tensile, creep and fatigue properties at temperatures of interest.

A potentially serious materials problem in a fusion reactor is that of radiation damage to the structural material. As previously mentioned, the total flux at the first wall of a commercial reactor will be 3–8×10^{18} neutrons $m^{-2} s^{-1}$ with about 20% having an energy of 14 MeV. Since the cross sections for nuclear reactions such as (n,α) and (n,p) are high at 14 MeV, the production of helium and hydrogen is much greater than in fission reactors. The production rates of helium and hydrogen in various materials in a fusion reactor first wall with a neutron wall loading of 1 MW m^{-2} are shown in Table 4. In stainless steel, the gas production rate will be 20 to 30 times higher in a fusion reactor than in a fast-fission-spectrum breeder reactor. The high gas production rates coupled with the displacement damage make the fusion reactor environment potentially more severe than that in the more familiar fission reactors. The presence of helium in the lattice can modify the properties of the material. Interstitial helium, like an interstitial atom of the material itself, is highly mobile but it is also virtually insoluble. Therefore, helium is quickly trapped by vacancies, dislocations and other sinks. The helium–defect interactions may affect swelling and the mechanical properties of the structural alloy (see *Swelling in Irradiated Metals and Alloys*; *Helium-Induced Blister Formation of Fusion Reactor First-Wall Materials*). The primary concerns involve loss of ductility and the potential for brittle failure. Since the responses of structural materials to radiation damage are sensitive to temperature, the projected materials performance will depend on the operating temperature, and hence on the type of coolant.

The structure must be compatible at appropriate temperatures with the coolant, breeder material and other materials that it must contact. Selection of a structural material that is compatible with both coolant and breeder at high temperatures is a major problem in many cases. Important considerations include both mass-transfer and stress-corrosion effects.

5. Tritium-Breeding Materials

Lithium is considered to be the only viable tritium-breeding element, with both liquid-metal and solid lithium compounds being considered for the tritium-breeding blanket (see *Fertile Blanket Materials*). Liquid lithium is possibly the most attractive candidate for the tritium-breeding material. In addition to its good breeding capability, lithium has excellent heat-transfer characteristics and can potentially be used as both breeder and coolant. This dual role simplifies the design and eliminates several concerns such as coolant–breeder compatibility and selection of a structure that is compatible with both the breeder and the coolant. Also, radiation damage is not a problem with liquid metals and tritium recovery appears to be acceptable. Major concerns regarding the use of lithium include compatibility with the structure, possible magnetohydrodynamic effects produced by a moving electrical conductor (lithium) in a magnetic field, and safety issues related to its reactivity with air and water.

Liquid lithium–lead alloys (17 at.% Li) offer similar advantages to those of lithium. Major differences relate to better breeding performance because of neutron multiplication with lead, lower chemical reactivity with air and water, lower tritium solubility that leads to higher tritium pressures, higher corrosion rates with most structural materials and higher density. Designs with Li–Pb breeder and a separate coolant have been proposed.

Of the solid-breeder options, the oxide ceramics are believed to offer the most potential. Li_2O is unique because of its high lithium-atom density and hence good breeding performance. The major feasibility issue associated with the solid breeders involves tritium recovery. The difficulties relate primarily to the low tritium diffusivities and low thermal conductivities characteristic of the oxides. The effect of irradiation on the tritium release is a major uncertainty. Some of the ternary oxides (e.g., $LiAlO_2$) are more stable than Li_2O; however, the breeding performance of these ceramics is less favorable. A neutron multiplier is required for acceptable breeding performance with the ternary oxides.

6. Coolants

Lithium, helium or water have most often been proposed for the coolant in fusion reactors. Lithium has the advantage of combining cooling and breeding functions, but magnetohydrodynamic pumping losses are experienced in pumping the conductive lithium across the magnetic fields. There is also a compatibility problem with many structural materials. Stainless steels undergo significant thermal-gradient mass transfer in flowing lithium above ~700 K. Aside from impurity effects, refractory metals such as niobium and vanadium alloys are compatible with lithium to much higher temperatures, but these materials must be protected from air.

Helium of commercial purity is not compatible with niobium or vanadium alloys, but is compatible with stainless steels and nickel-base alloys up to ~1075 K. The major drawback to helium is that relatively large coolant ducts and high pressures are required. Stresses from the high-pressure gas plus high thermal stress could lead to an early failure of coolant ducts in the blanket. Since relatively high helium temperatures (>773 K) are probably required for acceptable energy conversion systems, the mechanical property limitations and irradiation effects of most structural materials at temperatures above

773 K pose more difficult problems. Helium is not generally considered an acceptable coolant for the high-heat-flux (limiter/divertor) components.

Pressurized water has been proposed as the coolant in solid breeder systems. Water can be used with conventional austenitic and ferritic steels and is widely accepted as a heat-transfer fluid. Major concerns regarding the use of water relate to the high-pressure requirement, the modest thermodynamic efficiencies of the energy conversion systems and the difficulty of tritium removal from water. The reaction between pressurized water and the breeding material in the event of a pipe failure is also of concern. Use of water with the liquid-metal breeders may not be acceptable because of the high chemical reactivity.

7. *Neutron-Multiplier Materials*

A neutron multiplier is required to increase the tritium-breeding performance of certain breeder materials, in particular the ternary ceramics such as $LiAlO_2$. Materials with high (n,2n) or (n,3n) cross sections, such as beryllium, lead and bismuth, exhibit the highest neutron multiplication. Compounds or alloys of these materials are potentially acceptable. For example, PbO provides adequate neutron multiplication, but it has very poor thermal conductivity, which causes design difficulty. Beryllium technically provides the best performance of any material because of its excellent neutronics properties (high neutron multiplication and low activation), high thermal conductivity and heat capacity, and low density. Major concerns relate to its known toxicity, limited resources and radiation swelling, but these are not considered prohibitive. Lead is attractive in most respects; primary concerns regarding the use of lead relate to its low melting temperature (600 K), high density and poor compatibility with candidate structural materials.

8. *Moderator/Reflector Materials*

A neutron moderator or reflector has been proposed in some designs to enhance the tritium-breeding performance. The purpose of the moderator is to rapidly reduce the neutron energy in order to minimize neutron capture in the structural materials. The reflector, located behind the blanket, enhances the breeding performance by reflecting some of the neutrons coming out of the blanket back into the breeding region. Candidate materials include water, graphite and beryllium (see *Solid Moderator and Reflector Materials*). Water and graphite are most often considered. Primary concerns related to the use of water include the pressure–temperature limits and the difficulty of tritium recovery. Graphite has many favorable characteristics for this application; chief reservations are related to radiation damage problems and questionable compatibility with the structure at elevated temperatures.

9. *The Shield*

The purpose of the shield is to prevent excessive radiation damage and energy deposition in the superconducting magnets, which operate at a temperature near 4 K, and to allow personnel access soon after reactor shutdown. About 1 kW of electrical energy is required to remove 1 W of thermal energy deposited in the low-temperature magnets; hence, the economic incentive for efficient shielding is obvious. Radiation damage to the magnet insulator currently appears to be the life-limiting factor, and it therefore sets the shielding requirements. Thus, the shield must be effective if the fusion device is to be a reliable producer of energy. Many proven materials such as borated steel, B_4C, or tungsten are available for use. Material cost and activation are important considerations in the selection of the shield since the shield typically contains the largest material inventory of all the reactor components.

10. *Magnet Materials*

Superconducting magnets are required because only they have sufficiently low energy requirements for net power production in magnetically confined reactors. The two superconducting materials presently being considered for fusion reactor applications are the Nb–Ti alloy for use up to 11 T and Nb_3Sn for use up to 16 T (see *Superconductivity: Large-Scale Applications*). In the event that the superconducting magnets go normal, the superconductors are protected by a stabilizer such as copper or aluminum that would then carry the current, avoiding permanent damage to the superconductor. To reduce eddy-current losses, the superconductor must be divided into fine strands.

The magnet structural material should not be ferromagnetic and must be relatively inexpensive in view of the large quantities of structure required. Therefore, at present, most designers consider either type 304LN or a higher strength stainless steel for this application. Very large coils cannot be shop-fabricated and then shipped to the operating site; they must be erected in the field. This implies that operations such as postweld heat treatment, machining to final dimensions and coil winding will be very difficult.

The irradiation dose to the electrical insulation within the coil must be carefully controlled (see *Radiation Effects on Electrical Insulators*). Organic insulators such as Mylar, Kapton and epoxy–fiberglass composites are currently used by commercial manufacturers. The lifetime fluence (radiation damage level) of these insulators is of the order of 10^7–10^8 Gy;

Table 5
Typical insulator components, operating conditions and candidate materials

Component	Peak yearly fluences[a] (neutrons m^{-2})	Peak temperature[b] (K)	Materials
Electrical breaks	2.5×10^{26}	1100	Si_3N_4, Sialon, Si_2ON_2, Al_2O_3, $YAlO_4$, $MgAl_2O_4$, ZrO_2, Y_2O_3
Implosion and compression coil insulation	1.8×10^{26}	375	to be established
Field-shaping-coil insulation	2.5×10^{24}	375	MgO and first-wall insulation materials
Neutral-beam-injector insulators	3×10^{23}	475	Al_2O_3, Y_2O_3, MgO, SiO_2, BeO
Microwave insulators	3×10^{23}	475	
Fuel device insulators	3×10^{23}	70	
Direct-converter insulators	3×10^{22}	375	
Superconducting-magnet insulators	1.8×10^{21}	4	epoxy, polyimides

a 100% duty factor at 1 MW m^{-2} b During machine operation at full power

this corresponds to a neutron fluence of 1.8×10^{21} neutrons m^{-2} ($E > 0.1$ MeV) for the STARFIRE design. Because of their cost and difficult replacement, the superconducting coils must operate for the life of the plant. Thus, very effective radiation shielding is required, especially around penetrations (e.g., vacuum, heating and diagnostic ports) through the blanket.

11. Insulators

A number of insulator requirements for fusion devices have been identified. Components, operating conditions and candidate materials are listed in Table 5. Relative to metals, less is known about the radiation effects on the mechanical, physical and chemical properties of insulators.

See also: Neutron Interactions with Matter; Nuclear Materials: An Overview

Bibliography

Badger B et al. 1973 *UWMAK-I*, UWFDM-68. University of Wisconsin, Madison, Wisconsin

Badger B et al. 1976 *UWMAK-III*, UWFDM-150. University of Wisconsin, Madison, Wisconsin

Baker C C et al. 1980 *STARFIRE—Commercial Tokamak Power Plant Study*, ANL/FPP-80-1. Argonne National Laboratory, Argonne, Illinois

Baker C C, Carlson G A, Krakowski R A 1981 Trends and developments in magnetic confinement reactor concepts. *Nucl. Technol./Fusion* 1: 5–78

Monsler M J, Hovingh J, Cook D L, Frank T G, Moses G A 1981 An overview of inertial fusion reactor design. *Nucl. Technol./Fusion* 1: 302–58

R. F. Mattas; D. L. Smith

G

Gallium, Germanium and Indium Resources

Gallium, germanium and indium are dispersed in trace amounts in many rocks and minerals. Commercial production of these important elements, used in solid-state semiconductor electronics, is limited to minor by-product recovery in refining zinc and aluminum. Approximate world annual production ranges are: gallium, 14–20 t; germanium, 80–90 t; and indium, 45–60 t.

Bauxite, the major aluminum ore mineral, is the principal source of gallium; all three metals are recovered in smelting sphalerite (ZnS_2), the major zinc ore mineral. Caustic leaching of bauxite concentrates gallium in a sodium aluminate solution, with recovery accomplished by fractional carbonation and electrolysis (see *Gallium Production*). In roasting zinc sulfide, gallium and germanium are reduced with the zinc metal, which is distilled off, and the germanium is then separated from the gallium by oxidation, chloridization and fractional distillation (see *Germanium Production*). Indium is recovered from various smelter dusts, anode slimes, and bullion and cadmium recovery wastes (see *Indium Production*).

Crustal abundance estimates ($g\,t^{-1}$) are: gallium, 18; germanium, 1.5; and indium, 0.14. None occur as native elements, and they are essential elements in only a few rare minerals. Initial discoveries in the latter part of the nineteenth century were all made in spectrographic studies of sphalerites.

Gallium is coordinated with aluminum, and is thus present in most aluminosilicate minerals, especially bauxite and clay minerals. Germanium is coordinated with silicon and tin, and is present in many silicate minerals, especially late magmatic and hydrothermal minerals including fluorine-rich pegmatites, greisens and veins. Indium is slightly coordinated with tin. All three enter the crystal lattice of sulfide minerals, especially sphalerite. Gallium and germanium occur in plant ash, and are also present in coal and petroleum ashes. No specific biological function has been reported.

Gallium is more abundant in bauxites derived from granitic and alkalic rocks, and the Ga:Al ratio increases with weathering intensity. In zinc ores, the gallium and germanium contents are higher in replacement deposits formed at lower temperatures, and the indium content is higher in deposits formed at intermediate temperatures. The Tri-State and Upper Mississippi Valley districts have been the source of US production of gallium and germanium, and the western mining districts have provided most of the US and Canadian indium.

The Tsumeb deposit in Namibia, a pipelike replacement body of copper–lead–zinc sulfides, is unique in its production of a germanite–renierite concentrate, which has been the dominant source for European refining of gallium and germanium. This deposit has been the source for most of the other reported germanium and gallium minerals from oxidized ores.

World production potential from zinc and aluminum resources appears to be adequate for projected demand. Additional resources of all three elements may be available in other smelter by-products, notably in the refining of copper and tin. Germanium and gallium may also be recoverable from combustion wastes at major power installations.

See also: Aluminum Resources; Zinc Resources; Mineral Resources: An Overview

Bibliography

Shaw D M 1957 The geochemistry of gallium, indium, thallium—A review. In: Ahrens L H, Press F, Rankama K, Runcorn S K (eds.) 1957 *Physics and Chemistry of the Earth*, Vol. 2. McGraw-Hill, New York, pp. 164–211

Weber J N (ed.) 1973 *Geochemistry of Germanium*, Benchmark Papers in Geology Vol. 11. Dowden, Hutchinson and Ross, Stroudsburg, Pennsylvania

Weeks R A 1973 Gallium, germanium, and indium. In: Brobst D A, Pratt W P (eds.) 1973 *United States Mineral Resources*, US Geological Survey Professional Paper 820. US Government Printing Office, Washington, DC, pp. 237–46

R. A. Weeks

Gallium Production

Gallium is a silver-white metal with a melting point of 29.8 °C, employed in electronic devices with many new uses predicted for the future. It is widely distributed in the earth's crust, but only occurs in low concentrations: bauxite, sphalerite, kaolin, coal and phosphate ores often contain a few parts per million of gallium. It is often feasible to produce gallium as a by-product in the recovery of alumina from bauxite and in the recovery of zinc from sphalerite ores.

The major source of gallium is now the Bayer process for the production of alumina (see *Aluminum Production*). The gallium collects in the sodium aluminate liquor when bauxite is digested with hot sodium hydroxide. One recovery method used involves a two-stage treatment of the sodium aluminate solution with carbon dioxide to remove most of the alumina and sodium carbonate. The gallium concentrate obtained is dissolved in sodium hydroxide, and the resulting sodium gallate solution is then electrolyzed at 40–60 °C using stainless steel electrodes. The gallium collects in liquid form below the cathode and, after washing, yields a 99.9–99.99% pure product. Higher purity can be obtained by heating under vacuum, by electrochemical purification or by fractional crystallization.

Gallium can also be recovered from the sodium aluminate liquor by electrolyzing it with a continuously agitated mercury or sodium amalgam cathode. After electrolysis, the sodium aluminate liquor is returned to the Bayer circuit, and the gallium is then leached from the cathode with hot sodium hydroxide solution, electrodeposited and purified as described above.

Gallium can also be a by-product in the processing of sphalerite ore (see *Zinc Production*). The zinc sulfate solution obtained from leaching the ore with sulfuric acid is purified by neutralizing the excess acid to precipitate "iron mud." This may contain as much as 0.07% gallium plus large amounts of aluminum and iron. The iron mud is leached with sodium hydroxide to dissolve the gallium and aluminum, and the hydroxides are precipitated by neutralizing the leach liquor with acid. The precipitate formed is leached with hydrochloric acid to obtain a solution containing gallium and some aluminum. The gallium trichloride is then extracted with ether, leaving the aluminum and other impurities behind. The ether is distilled off and the gallium trichloride dissolved in concentrated sodium hydroxide. Gallium metal is electrodeposited from the sodium gallate solution and purified.

Gallium has also been recovered from the furnace dust resulting from the electric-furnace smelting of phosphate rock to obtain phosphorus, and also from coal ash, but neither of these sources is currently utilized.

See also: Gallium, Germanium and Indium Resources; Metals Production: An Overview

Bibliography

de la Bretèque P 1980 Gallium and gallium compounds. In: Grayson M (ed.) 1980 *Kirk–Othmer Encyclopedia of Chemical Technology*, 3rd edn., Vol. 11. Wiley, New York, pp. 604–20

Hudson L K 1965 Gallium as a by-product of alumina manufacture. *J. Met.* 17: 948–51

Katrak F, Agarwal J C 1981 Gallium: Long-run supply. *J. Met.* 33(9): 33–36

Petkof B 1980 Gallium. *Mineral Facts and Problems*. US Bureau of Mines, Washington, DC

Thompson A P 1961 Gallium. In: Hampel C A (ed.) 1961 *Rare Metals Handbook*. Reinhold, New York, pp. 178–87

D. Schlain

Galvanic Corrosion

Galvanic corrosion occurs when two or more metals or alloys are electrically coupled in the same electrolyte. The rate of attack on one metal or alloy is usually accelerated while the corrosion rate of the other is decreased.

The extent of accelerated corrosion resulting from galvanic coupling is affected by the following factors:

(a) the potential differences between the metals;

(b) the nature of the environment;

(c) the polarization behavior of the metals; and

(d) the geometric relationship of the component metals.

The potential difference existing between dissimilar metals when they are electrically coupled in a conducting solution causes an electron flow between them. The direction of flow and therefore the galvanic corrosion behavior depends on which metal is the more active. Thus, the more active metal becomes anodic and the more noble metal becomes cathodic. The driving force for galvanic corrosion is the potential difference between the component metals.

The galvanic series is an arrangement of metals according to their relative potentials measured in a particular electrolyte such as seawater (Table 1). It allows one to determine which metal in a couple is the more active. In some cases, the separation between the two metals in the galvanic series gives an indication of the magnitude of corrosive effects.

Table 1

Galvanic series of metals and alloys in seawater

Noble end	Platinum, Gold, Silver, Titanium, 316 stainless steel (passive), 304 stainless steel (passive), 410 stainless steel (passive), Nickel (passive), Cupronickel, Bronzes, Copper, Brasses, Nickel (active), Tin, Lead, 316 stainless steel (active), 304 stainless steel (active), Lead–tin solder, 410 stainless steel (active), Cast iron, Steel, Cadmium, Aluminum, Zinc, Magnesium	*Active end*

The most common experimental method of predicting galvanic corrosion is immersion testing of the couple in a particular environment (see *Corrosion Testing*). However, although this is the most desirable method, it is time consuming. Therefore, it is supplemented by screening tests consisting of certain other electrochemical techniques, namely, potential measurements, current measurements and polarization measurements.

Potential measurements are made to construct a galvanic series. As a first approximation, the series is a useful tool. However, serious drawbacks exist in its use. Metals which form passive films will exhibit potentials which vary with time and it is therefore difficult to position them in the series. Also, the series does not provide a quantitative prediction of galvanic effects.

Measurements of galvanic currents between coupled metals are based on the use of a zero resistance ammeter.

Polarization measurements (see *Corrosion Kinetics*) on the components of a couple can provide precise information on their behavior. The polarization curves and the mixed potential for the galvanically coupled metals in a particular environment can be used to determine the magnitude of corrosion effects as well as the type of corrosion.

Methods of controlling galvanic corrosion include proper materials selection and design (see *Design for Corrosion Control*), electrical isolation of the metals comprising the couple, isolation from the electrolyte using coatings, cathodic protection and the use of transition materials and chemical inhibitors.

See also: Corrosion Protection Methods; Corrosion and Oxidation Study Techniques; Electrochemical Principles of Corrosion; Passivity and Breakdown

Bibliography

Baboian R, France W D, Rowe L C, Rynewicz J F (eds.) 1974 *Galvanic and Pitting Corrosion*, ASTM STP 576. American Society for Testing and Materials, Philadelphia, Pennsylvania

Uhlig H H 1971 *Corrosion and Corrosion Control*, 2nd edn. Wiley, New York

R. Baboian

Galvanized Steel

New galvanized steel surfaces are bright and shiny and generally exhibit a characteristic spangle caused by the crystallization of the coating. This appearance is often desired for esthetic reasons and it demonstrates that the coating is indeed zinc. The spangled appearance need not be present for the coating to provide excellent barrier protection and sacrificial protection to steel. In general, the duration of the protection is directly proportional to the coating thickness.

There are three principal types of galvanized steel. The term hot-dip galvanized is generally applied to steel components which are treated in batches, briefly immersed in molten zinc, withdrawn from the bath, drained and allowed to cool. There is substantial interaction between the zinc and the steel. This provides an excellent bond for the coating which includes a series of iron–zinc alloy phases, iron rich at the steel interface with zinc at the surface. This progression of alloy phases allows a considerable buildup of coating thickness, which is under the control of the manufacturer and is dependent on such variables as the composition of the base steel, time of immersion in the zinc, rate of withdrawal and cooling rate.

Continuous galvanized steel refers to steel sheet and strip which has passed through a molten zinc bath at speeds of 75 m min^{-1} or more. Additions of aluminum to the zinc bath inhibit the reaction between molten zinc and steel. Thus the thickness of the continuous galvanized coating, while substantial, is less than that of hot-dip or batch galvanized steel.

Electrogalvanized steel usually provides a much thinner zinc coating, which is closely bonded to the steel but lacks the iron–zinc alloy layer of the hot-dip galvanized steel. The thickness of the zinc is governed by the economics of electroplating. Developments in the high-speed plating of zinc have improved the relative cost of this process. Electrogalvanizing can be readily applied to only one side of a continuous steel strip and, where this is desirable, electrogalvanized steel can be competitive with continuous galvanized steel.

The coating is often specified in terms of weight per unit area. For continuous galvanized steel, the quoted figures refer to the total amount of zinc in ounces per square foot on both sides of the designated area of sheet or strip; for hot-dip galvanizing they refer only to a single surface. Thus batch galvanized plate has twice the thickness of coating for a stated number of ounces per square foot.

Economic considerations may dictate that electrogalvanized steel is given a relatively thin zinc coating. Therefore, hot-dip or continuous galvanizing is usually specified where greater coating weight and long-term corrosion resistance is important. Corrosion resistance is also obtained when modern high-speed plating techniques are employed for special applications such as the steel, galvanized on one side, used in the automotive and some other industries.

1. Hot-Dip Galvanized Steel

Hot-dip galvanizing is specified for its ability to provide low-cost maintenance-free protection to a wide variety of parts from nuts and bolts to bridge girders. Very large structures can be accommodated in a modern galvanizing plant. Kettles over 15 m long are available and greater length can be handled by the technique of double dipping, where each half of a long girder is galvanized in turn.

The galvanized coating provides excellent barrier protection for the underlying steel but also the zinc can confer galvanic protection across gaps in the coating and prevent rusting of cut edges. The service life of all galvanized coatings is in general proportional to coating weight or coating thickness (see Fig. 1). Good surface preparation of the base steel is essential if the hot-dipping operation is to produce a closely adherent coating of uniformly high quality.

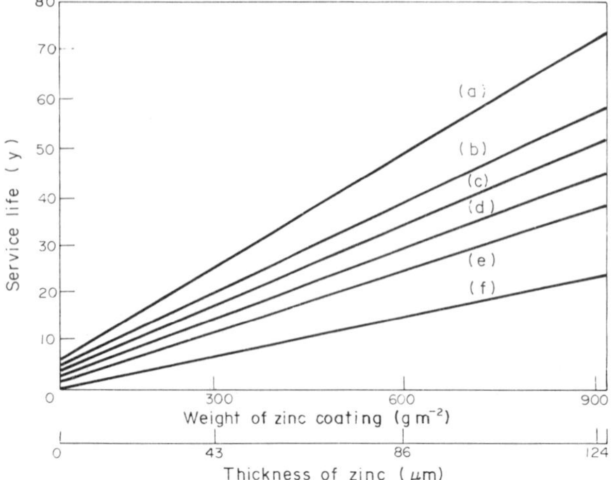

Figure 1
Service life of galvanized steel in different environments: (a) rural; (b) tropical marine; (c) temperate marine; (d) suburban; (e) moderate industrial; (f) heavy industrial

The first operation is usually pickling but any oil present may be removed by immersion in 0.5–1.0% detergent at 60 °C. Grease or paint can be removed in concentrated caustic soda (4.5–7.0 kg in 45 l). Alkali baths maintained at about 85 °C are effective after 1–20 min immersion.

Shot or grit blasting is occasionally employed to remove mill scale or welding slag and blasting may precede, or in some cases replace, the pickling operation if the work is to be galvanized immediately.

In the USA, 10–14% sulfuric acid is generally used for pickling. The bath must be heated to between 60 °C and 65 °C. In the UK it is more common to use 14 wt% hydrochloric acid at ambient temperature. In both cases, an inhibitor should be used to prevent excessive metal removal. Thorough rinsing is essential after degreasing and pickling.

Pickling is followed by either a wet or dry fluxing operation in order to complete the cleaning of the steel surface. Wet fluxing involves passing the work through a flux blanket on top of the molten zinc. In addition to its cleaning action, the flux blanket minimizes the oxidation of the zinc bath. Dry fluxing may rely upon the fluxing action of the dry iron salts which are present after pickling. A new dry-fluxing process involves removing these iron salts by rinsing, then prefluxing the part in a zinc ammonium chloride solution. The work is dried and immersed in a clear zinc bath.

Hot dipping is carried out at a bath temperature of about 450 °C. The work is immersed as rapidly as possible, left until "boiling off" ceases and then withdrawn at a speed of about 1.5 m min^{-1}, according to the thickness of coating required.

When hot-dip galvanized materials are specified, the standards of the American Society For Testing and Materials (ASTM) are useful (see Table 1). There are over 30 ASTM specifications relating to galvanized products. Close liaison is recommended between the design engineer, the fabricator and the galvanizer.

Table 1
Selected standards of the American Society for Testing and Materials

A 123	Specification for zinc coatings (hot galvanized) on products fabricated from rolled, pressed and forged steel shapes, plates, bars and strip
A 153	Specification for zinc coating (hot dip) on iron and steel hardware
A 386	Specification for zinc coating (hot dip) on assembled steel products
A 90	Tests for weight of coating on zinc-coated (galvanized) iron or steel articles
E 376	Recommended practice for measuring coating thickness by magnetic-field or eddy-current (electromagnetic) test methods
A 143	Recommended practice for safeguarding against embrittlement of hot-dip galvanized structural steel products and procedure for detecting embrittlement
A 384	Recommended practice for safeguarding against warpage and distortion during hot-dip galvanizing of steel assemblies
D 2092	Recommended practice for preparation of zinc-coated steel surfaces for painting
A 385	Recommended practice for providing high-quality zinc coatings (hot dip)

The quality of the coating is affected by (a) the composition of the zinc bath, (b) the bath temperature, (c) the time of immersion and (d) the rate of withdrawal. The quality and composition of the base steel are additional factors which will affect the batch galvanized coating. Where the silicon content of the steel is above 0.06%, as in silicon-killed steel, the surface appearance of the coating may vary from gray or matt to bright and shiny often in the same piece. The coating thickness may be greater than usual and, although it is somewhat more brittle, the corrosion resistance is increased. The appearance may, however, be unattractive. Galvanizing at a temperature of 100 °C higher than normal, i.e., at about 550 °C, overcomes this problem.

The designer should provide for venting and draining during the hot-dipping process, and should seek to minimize distortion by following the guidelines in ASTM A 384.

Cold forming may induce strain-age embrittlement in susceptible steels. The high temperature of the galvanizing process can increase the rate of embrittlement and cause early failure in these susceptible steels. Where cold working must be done before galvanizing, a steel with a carbon content below 0.25% should be selected. A reduction of cold working may be necessary where the carbon content of the steel is in the range 0.1–0.25%.

The practice recommended in ASTM A 143 safeguards against strain-age embrittlement. The severity of cold working is the most important factor in the embrittlement of susceptible steels and the following rules should be observed:

(a) For steels having a carbon content in the range 0.1–0.25%, a bending radius of three times the section thickness (3*t*) should be maintained if the material is fabricated cold.

(b) Holes should be drilled rather than punched in material thicker than 12 mm or, if holes are punched, they should be punched undersize and reamed an additional 3 mm.

(c) The cold shearing of plates that will be subject to heavy loads should be avoided. Flame cutting or sawing is preferred.

(d) Where 3*t* cold bending cannot be followed, the work should be fabricated hot (600 °C) or stress relieved (25 min per cm of section thickness) after cold fabricating.

(e) Steels having a carbon content below 0.25% should be used if cold forming is to occur prior to galvanizing.

(f) Steels having a low transition temperature should be selected.

(g) Where a steel may be judged susceptible to strain-age embrittlement, the galvanizing organization should be advised and a sample of the steel should be galvanized and tested for impact resistance.

2. *Continuous Galvanizing*

In contrast to hot-dip galvanizing (which is suitable for such items as nuts, bolts, bridge girders, reinforcing bars and heat exchangers), continuous galvanizing is a high-speed process which produces zinc-coated steel sheet and strip of exceptional formability.

The iron–zinc alloy formed at the steel surface is less ductile than zinc. To reduce the activity of the steel, and thus the thickness of the coating and of the iron–zinc alloy layer, aluminum is added to the bath. This results in improved formability.

As the aluminum content is increased from 0.04% to 0.18%, the adhesion of the coating is maintained by increasing the bath temperature from 410 °C to 500 °C. The batch galvanizing process relies upon a flux to ensure a clean steel surface. Aluminum in the bath tends to react with the chlorides of the flux. Therefore in continuous galvanizing a flux is not used, except in the Cook–Norteman process.

The first successful process combining annealing, surface conditioning and heat treatment was first used commercially in 1936. An outline of this process (the Sendzimir process) is shown in Fig. 2. Steel strip is first oxidized and then reduced to form a thin layer of sponge iron. The furnace can be operated in an annealing mode at 73 °C or in a normalizing mode at 900 °C or higher.

After the Sendzimir development came the gas

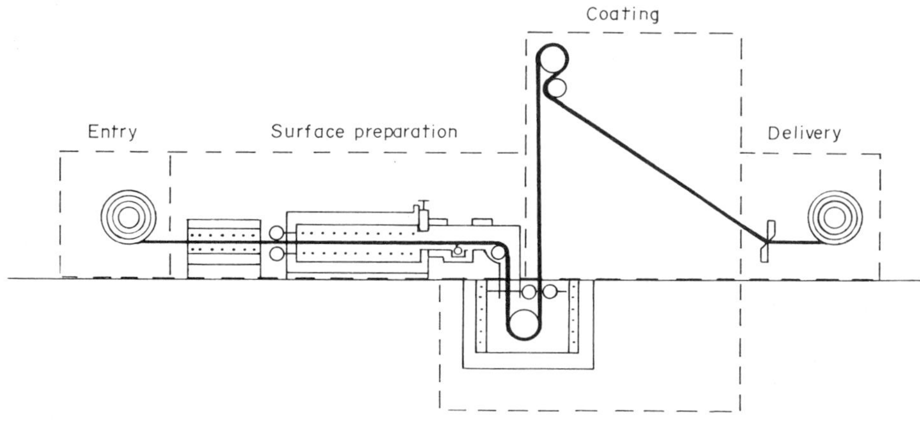

Figure 2
Division of Sendzimir's metal-coating process into functional sections

Table 2
Continuous strip galvanizing[a]

A[b] Sendzimir process	B[b] Gas Pickling process	C[b] US Steel process	D[c] Cook–Norteman process
Burn-off	Anneal in hydrogen–nitrogen gas mixture	Electrolytic cleaning	Solution cleaning Solution pickling
Anneal in hydrogen–nitrogen gas mixture	React with hydrogen chloride gas	Anneal in hydrogen–nitrogen gas mixture	Flux bath
Cool	Cool	Cool	Dry and preheat
Coat	Coat	Coat	Coat

a All processes use an aluminum addition to the zinc bath b Processes A, B and C: the strip is annealed in the process and enters the coating bath at or above pot temperature c Process D: the strip is annealed before processing and enters the coating bath below pot temperature

pickling process, the US Steel process and the Cook–Norteman process (Table 2). Such processes allow the application of a tightly adherent uniform coating of zinc. The surface may have the characteristic spangle or the coating may be produced with a modified spangle or no spangle, and in a range of thicknesses.

The galvanized sheet is produced in four principal qualities: galvanized commercial quality (ASTM A 526); galvanized drawing quality (ASTM A 528); galvanized physical quality (ASTM A 446); and galvanized lock-forming quality (ASTM A 527).

The coating weight, which ranges up to a maximum of $720\,g\,m^{-2}$, should be chosen to provide the required service life and to accommodate any forming or welding requirements. Galvanized commercial quality will provide moderate formability. Galvanized drawing quality is suitable for applications where severe drawing is specified.

For continuous galvanized steel sheet and strip, it is usual to specify, for example, a G90 or G60 coating, where G is an abbreviation for "grade." The number relates to the coating thickness and hence corrosion resistance (ASTM A 525). As an example, G60 refers to a coating of $0.60\,oz\,ft^{-2}$ ($180\,g\,m^{-2}$) of sheet. The G60 coating will last about two thirds as long as the G90 coating in a corrosive environment. It costs less and is easier to resistance weld.

Table 3
Recommended coating thickness for zinc electroplate

Degree of exposure	Minimum thickness (μm)[a]	Chromate finish	Salt spray (hours to white corrosion)	Typical applications
Mild—exposure to indoor atmospheres with rare condensation and subjected to minimum wear or abrasion	5	None		Screws, nuts and bolts, buttons, wire goods, fasteners
		Clear	12–24	
		Iridescent yellow	24–72	
		Olive drab	72–100	
Moderate—exposure mostly to dry indoor atmospheres but subjected to occasional condensation, wear or abrasion	8	None		Tools, zipper pulls, shelves, machine parts
		Clear	12–24	
		Iridescent yellow	24–72	
		Olive drab	72–100	
Severe—exposure to condensation perspiration, infrequent wetting by rain, cleaners	13	None		Tubular furniture, insect screens, window fittings, builders' hardware, military hardware, washing machine parts, bicycle parts
		Clear	12–24	
		Iridescent yellow	24–72	
		Olive drab	72–100	
Very severe[b]—exposure to bold atmospheric conditions, and subject to frequent exposure to moisture, cleaners and saline solutions plus likely damage by denting, scratching or abrasive wear	25	None		Plumbing fixtures, pole line hardware

a After chromate treatment if employed b Although there exist some applications for heavy electrodeposited coatings for very severe service they are most usually satisfied by hot dipped or sprayed coatings

3. *Electrogalvanizing*

Electroplated zinc usually provides a very attractive bright finish with very fine grain. Some zinc plating techniques produce a more functional finish but in all cases corrosion protection is proportional to coating thickness.

Electrogalvanizing provides a relatively pure zinc coating in contrast to hot-dip galvanizing, where an iron–zinc alloy is present at the steel surface; nevertheless, bonding of zinc electroplate to the steel surface is excellent and the plated coating will withstand severe deformation.

Economic considerations often dictate that electroplated zinc coatings are thinner than those normally associated with hot-dip galvanizing. Care must therefore be taken to ensure that an electrogalvanized coating is sufficiently thick to provide the required corrosion protection. Table 3 provides suggested standards for quality zinc coatings on iron and steel products.

While corrosion protection is usually regarded as the main reason for electrogalvanizing steel, it also may be specified for ease of soldering, low electrical contact resistance, good surface conductivity, nonstaining of fabrics or reduced seizing of moving parts.

The approximate corrosion rates for electrodeposited coatings of zinc on steel are given in Table 4. These data are derived from worldwide sources and are taken from ASTM B 633-78. Chromate conversion coatings are commonly used to improve the corrosion resistance of electroplated zinc (see ASTM B 201).

Table 4
Approximate mean corrosion rates of zinc in various environments (from ASTM B 633-78)

Atmosphere	Mean corrosion rate ($\mu m\ y^{-1}$)
Industrial	5.6
Urban nonindustrial or marine	1.5
Suburban	1.3
Rural	0.8
Indoors	$\leqslant 0.5$

Steel stampings, nuts and bolts and other small parts can be coated economically by the barrel plating technique. Here it is not necessary to attach the parts to a rack. A typical hexagonal oscillating barrel 46 × 76 cm can process 136 kg of material. The operation is carried out using a current of 1200–1500 A at 18 V. Larger parts must be processed while attached to racks, and whereas barrel plating uses current densities of 108–160 A m^{-2}, rack plating employs current densities of 860 A m^{-2}.

Plating solutions which are used may be based on the chloride, cyanide, sulfate or fluoroborate salts of zinc, or a zincate electrolyte, depending on economic factors and the required quality and type of work. The American Electroplaters Society has completed a project on the optimization of barrel plating solutions.

Cyanide solutions have long been popular because of their excellent throwing power and smooth deposits. However, environmental considerations have caused much greater attention to be focused on acid plating solutions. Improvements have been made and the acid bath seems likely to displace the cyanide process.

The acid chloride bath can give better-looking deposits than those from a cyanide bath. A typical bath contains 31.5 g l^{-1} zinc chloride and 200 g l^{-1} ammonium chloride, and operates at a temperature of 18–30 °C with a cathodic current density of 1160 A m^{-2}.

Sulfate baths are not used for their decorative appeal but mostly for industrial use, for wire and strip plating. They are operated at high temperatures and offer limited throwing power.

Ammonium and zinc chlorides are used in an almost neutral solution and provide good appearance and performance. Zincate baths produce a bright decorative finish but they require careful control and are most suitable for operations where conditions are standardized.

Fluoroborate, perchlorate and sulfamate baths provide good results but are comparatively costly.

Techniques have been developed for high-speed plating of zinc. Increased temperatures up to 60 °C and rapid movement of the electrolyte over the plating surface have resulted in greatly increased plating rates. Such high-speed plating techniques can reduce considerably the cost of electrogalvanizing.

See also: Corrosion Protective Coatings for Metals; Electrodeposited Metallic Coatings; Metallic Coatings for Corrosion Protection: Selection

Bibliography

General Galvanizing Practice (undated). The Hot Dip Galvanizers Association, London

Proskurkin E V, Gorbunov N S 1972 *Galvanizing, Sheradizing and Other Zinc Diffusion Coatings.* Technicopy Ltd., Stonehouse, UK (translated by Hayler D E)

Satterfield G (ed.) 1981 *The Design of Products to be Hot Dipped Galvanized after Fabrication.* American Hot Dip Galvanizers Association, Washington, DC

Selected Specification for Hot Dip Galvanizing, 1981. American Hot Dip Galvanizers Association, Washington, DC

Welding Zinc Coated Steel, 1972. American Welding Society, Miami, Florida

A. R. Cook

Gamma Radiography

Gamma radiography is a procedure for the nondestructive examination of industrial materials, utilizing the penetrating radiation emitted by radioactive sources. Gamma rays and x rays of any given energy, both being electromagnetic radiation, are indistinguishable except for their origin; gamma rays represent the release of energy from the atomic nucleus, while x rays result from the rearrangement of the electron structure surrounding the nucleus. Most gamma radiography is conducted with gamma rays having energies in the megavolt range, primarily because most radioactive sources emit copious radiation at these energies. By contrast, most x radiography is performed at x-ray tube potentials up to hundreds of kilovolts.

Gamma radiography is used principally for the inspection of welds and castings in structures such as steel tanks and pipelines where sections are thick enough to require megavolt radiation to penetrate them, or where it is advantageous to make use of the portability offered by gamma radiography equipment. A number of applications take advantage of the small size of radioactive sources to radiograph objects from the inside, such as the walls of small-diameter pipes. This technique permits a single exposure to produce a radiograph of an entire circumferential weld, and often considerably simplifies the interpretation of radiographs of complicated structures such as aircraft jet engines where film wrapped around the engine is exposed by a source inserted down the engine shaft.

Hundreds of artificially radioactive isotopes are available, offering a wide choice of characteristics. Naturally occurring radium is no longer used for radiography, having been entirely supplanted by the cheaper, higher-intensity isotope cobalt-60. With the growth of nuclear reactors and the development of handling techniques, practical gamma-ray source intensities have increased by a factor of a hundred over the last 40 years, resulting in correspondingly shorter exposure times necessary to produce satisfactory radiographs.

1. Radiographic Sources

The decay of any radioactive nucleus is characterized by the ejection of an alpha particle or a beta particle (positive or negative electron) or the capture of an external electron, any of which usually leave the nucleus in an excited state. This is followed almost immediately by emission of one or more gamma rays (photons) until the nucleus arrives at a stable ground state. Because the excited nucleus can exist only at particular energy levels, the energies of these gamma rays and the rate of decay are characteristic of the particular isotope. Thus, cobalt-60 always decays by emission of a negative electron followed by two equally abundant gamma rays of 1.17 MeV and 1.33 MeV energy, at a rate described by a half-life of 5.26 years.

Not all radioactive isotopes decay by gamma-ray emission, and not all that do so are useful for gamma radiography. Important factors are the efficiency of production of the isotope, its rate of decay and the amount of gamma radiation that it releases. Only five isotopes (excluding radium) have found much utility in gamma radiography (see Table 1), and of these, only cobalt, iridium and ytterbium are in appreciable use today. Once, cesium was employed extensively, particularly in Europe, but its low

Table 1
Gamma radiography isotopes

Isotope	Half-life	Effective gamma energies (keV)	Gamma intensities (rhm per Ci)[a]	Typical maximum source strength (Ci)	Minimum thickness of steel (mm)[b]
Cobalt-60	5.26 y	1250	1.30	1000	25
Cesium-137	30.2 y	662	0.32	10	18
Iridium-192[c]	74.0 d	310	0.249 }		
		467	0.144 } = 0.46	200	12
		603	0.063 }		
Thulium-170[d]	125 d	53 (x ray)	0.015 } = 0.03	50	1
		84	0.012 }		
Ytterbium-169[e]	30.7 d	52 (x ray)	0.090 }		
		117	0.018 } = 0.19		
		191	0.056 }	50	1
		302	0.022 }		

a rhm = roentgens per hour at 1 meter b For 2% radiography c The 30 gamma rays of iridium-192 can be represented by 3 monoenergetic radiations. For historical reasons, US industry uses 0.55 rhm per Ci d Because of the low yield of x rays and gamma rays from thulium-170, the continuous x radiation generated by the 1 MeV beta rays is an appreciable contaminant that hardens the emitted spectrum e The 24 gamma rays of ytterbium-169 can be represented by 3 monoenergetic radiations

specific activity (25 Ci g^{-1} maximum) severely limits the achievable resolution. With improved methods for replacement of spent sources, ytterbium has effectively taken over the role once played by thulium as a low-energy emitter, since ytterbium can be produced in considerably greater strengths.

Except for cesium-137, a fission product, all radiographic isotopes are made by irradiating the element with thermal neutrons in high-flux nuclear reactors. Pellets of metallic cobalt or iridium or sintered thulium or ytterbium oxide are installed in capsules specially designed to avoid excessive self-absorption in the target material and subjected to fluxes of the order of 10^{15} n cm^{-2} s^{-1} for times that will maximize the yield. Because some of the desired radioactive atoms are lost by decay and by capture of further neutrons during irradiation, the optimum times in the reactor are generally less than a half-life of the isotope.

Radiographic sources are manufactured by loading the radioactive material into stainless steel capsules that are hermetically sealed by welding. For metallic isotopes, the use of a number of pellets, each containing a small fraction of the total activity, not only avoids neutron self-absorption and consequent dilution of the activity that would take place if the source were irradiated as a whole, but also permits the manufacture of sources of predetermined strengths regardless of the age of the raw material. Radiographic resolution considerations dictate that the physical length of the active material approximate its diameter.

2. Exposure Devices

The commonest form of exposure device in the USA today consists of a football-shaped lead or depleted-uranium shield containing an S-shaped tube. As shown in Fig. 1, the sealed source, attached to a short "pigtail," normally resides at the center of the shield; after attaching a cable drive system to the pigtail at the rear of the shield, the remotely located operator can drive the source out of the shield and along the guide tube to a prepositioned radiography location by means of a handcrank. The source is later withdrawn to its storage position at the end of the radiographic exposure. Radiation from the stored source is confined to the shield by the S curve of the internal tube.

The high atomic number, high density (18.7 g cm^{-3}) and ready availability of depleted uranium has led to its almost exclusive use as a primary shielding material in exposure devices and source changers. An iridium-192 exposure device shielded with depleted uranium is half the weight of an equivalent lead-shielded unit. Tungsten–copper alloy is also used for shielding, particularly where the design calls for machined parts.

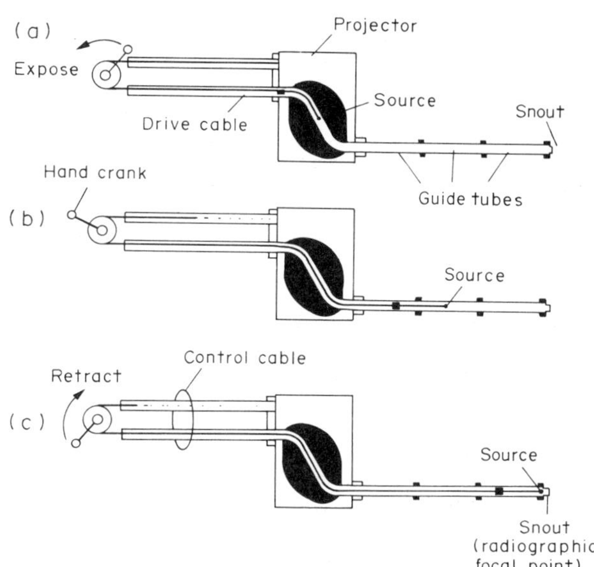

Figure 1
Principle of operation of a typical radiographic exposure device: (a) the source in its stored position, (b) the source in transit and (c) the source at the radiographic site

3. Exposure Techniques

Exposure techniques for gamma radiography are similar to those for x radiography (see *Radiographic Nondestructive Evaluation*), except that exposure times are generally measured in minutes rather than seconds and the energy and hence penetrability of the radiation can be varied only by using different isotopic sources. The variables affecting exposure include, besides the exposure time, the source strength and distance from the film, the thickness of the section to be penetrated and the sensitivity of the detector system. The radiographer usually works with a minimum source–film distance consistent with acceptable geometric distortion and resolution in order to minimize exposure time.

Minimum thicknesses of steel for which flaw detection can be carried out successfully are given in Table 1. Gamma radiography is nevertheless useful in thinner sections where the purpose is to display the presence of relatively large objects, such as reinforcing bars in concrete slabs or hidden internal parts in complicated assemblies.

In general, exposure techniques for gamma radiography are similar to those for high-energy x rays generated at peak voltages approximately double the photon energies. Thus, absorption curves for cobalt-60 radiation in steel are similar to those for 2.5 MVp x-radiation.

See also: Real-Time Radiography; Radiographic Nondestructive Evaluation

Bibliography

Clarke E T 1959 Isotope radiation sources. In: McMaster R C (ed.) 1959 *Nondestructive Testing Handbook.* Ronald Press, New York, Sect. 15

Clarke E T 1976 Radiographing the Liberty Bell. *Foundry Trade J.* 141 (Aug): 223–28

Clarke E T 1980 Radiography of the Cape Hatteras lighthouse. *Technol. Conserv.* 5(1): 20–24

Clarke E T, Coenraads C N 1967 *Radiographic Characteristics of Ytterbium 169,* Report ORNL-IIC-10. Oak Ridge National Laboratory, Oak Ridge, Tennessee

Evans R D 1955 *The Atomic Nucleus.* McGraw-Hill, New York

Glasgow G P, Dillman L T 1979 Specific gamma ray constant and exposure rate constant of iridium-192. *Med. Phys.* 6(1): 49–52

Graham R L, Wolfson J L, Bell R E 1952 The disintegration of thulium-170. *Can. J. Phys.* 30: 459–75

Hatziandreou L, Ladopoulos G 1980 Use of radiography for solution of diagnostic problems existing in marble monuments. *Materialprüfung* 22: 298–300

Pullen D, Hayward P 1979 Gamma radiography of welds in small diameter steel pipes using enriched ytterbium-169 sources. *Br. J. Non-Destr. Test.* 21: 179–84

E. T. Clarke

Gas and Liquid Chromatography

Gas chromatography (GC) has been a revolutionary force in the growth of organic chemistry because of the great simplification in the analysis of volatile samples that it offers. Liquid chromatography (LC) is complementary to GC in that it is effective for nonvolatile materials. LC has had a similar impact on organic chemistry and both techniques have had wide influence, with applications in biochemistry, inorganic chemistry and polymer science. Both are relatively recent techniques and instrumentation and methods have been developed at a rapid rate in the recent decades. In this article the two areas of chromatography are introduced and their current states of development described.

1. Gas Chromatography

The gas chromatograph offers a method of separating the volatile components of a mixture. The primary result is the chromatogram, which is a plot of the intensity of detector response against time. The time from injecting the sample to when a separated component of the mixture appears at the detector is called the retention time. This time is characteristic of a particular compound (under fixed instrumental conditions) and can be used to infer the presence or absence of a particular compound in a mixture. Retention times generally increase with increasing boiling point of the component. A typical gas chromatograph (Fig. 1) consists of several parts:

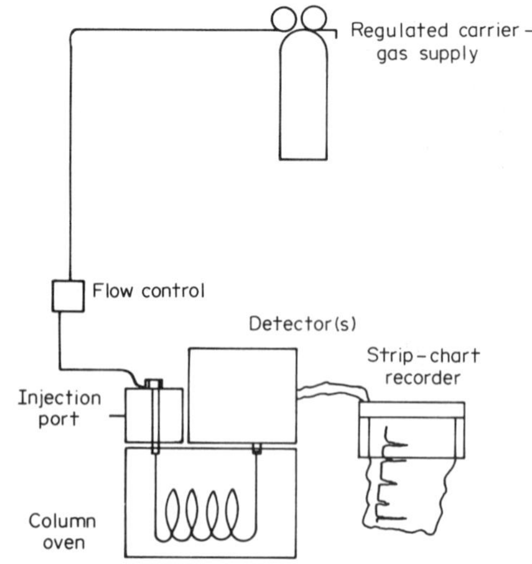

Figure 1
Block diagram of a typical gas chromatograph

(a) A source of clean carrier gas (helium, hydrogen, nitrogen or others) that has a precisely regulated flow rate.

(b) A heated zone containing a rubber-septum-sealed injection port through which the sample is introduced into the flowing carrier-gas stream with a syringe.

(c) A separating column (generally $\frac{1}{8}$ or $\frac{1}{4}$ in. (0.32 or 0.64 mm) diameter stainless steel or glass tube) containing the packing material and mounted in a temperature-controlled column oven.

(d) One or more detectors to sense the appearance of the separated components at the outlet of the column.

(e) An electronic integrator, or strip-chart recorder, to display the chromatogram as an intensity–time plot.

The instruments available today range from fairly simple devices that are frequently set up for one specific analysis to complex microprocessor-controlled instruments with several types of inlet sytems and multiple detectors. Instruments of a moderate level of sophistication have the capability of ramping the column oven temperature in a controlled manner (temperature programming) to allow for elution of mixtures of a wide range of volatilities. A low column temperature at the start of the analysis resolves low-boiling components whereas increasing the temperature elutes the higher-boiling materials with satisfactory peak widths (and thus good sensitivity). The ability to cool the column oven below ambient temperatures with the aid of liquid carbon dioxide or liquid nitrogen is occasionally provided for the analysis of gases or low-boiling species.

Gas chromatography can involve two different separation mechanisms. In gas–liquid chromatography (GLC) the sample is partitioned between a liquid phase held immobile by a solid support and the moving carrier-gas stream. This technique is used for the majority of organic separations. In the second technique, gas–solid chromatography (GSC), the sample is partitioned between a solid adsorbant (which is often a specially prepared porous polymer) and the carrier-gas stream. This is usually used for the separation of inorganic volatiles, gases and light hydrocarbons.

1.1 Columns

There is an enormous variety of liquid phases that have been prepared for GLC. These range in temperature stability to above 350 °C and cover an extremely broad polarity range. However, most separations can be carried out using fewer than ten different liquid phases. A range of polarities for the stationary phase is needed in order to optimize the separation as well as to accommodate special samples such as amines.

The choice of which liquid phase to use to effect a separation is the most difficult aspect of gas chromatography. Frequently, however, a suitable liquid phase can be chosen by consulting examples in the literature published by suppliers of columns or support materials. New developments include bonded-phase columns in which the liquid phase is chemically bonded to the support, giving enhanced stability at higher temperatures. This technology has been extended to fused-silica capillary columns in which a limited number of liquid phases are available which are bonded directly to the capillary surface. Table 1 lists common liquid phases and their characteristics as well as a few column packings for GSC. These cover a wide range of polarities and applications.

A typical nonpolar column packing for GLC consists of diatomaceous earth that has been acid washed, then treated with a silylating reagent such as hexamethyldisilazane to deactivate the active surface sites and finally coated with 1–10 wt% of a silicone fluid such as OV-101. This is then sieved to give a narrow particle-size distribution. Typically, the particle-size range chosen is 100–120 mesh (125–149 μm). This packing material is loaded into carefully cleaned chromatographic tubing. Columns are usually $\frac{1}{8}$ or $\frac{1}{4}$ in. diameter but larger diameters have been used for preparative GC, albeit with an attendant loss in resolution. Aluminum and copper have been used in the past but stainless steel is the most common column material. Even with stainless steel the metal walls can cause decomposition of some materials or adsorption can cause peak tailing. Glass is a much more inert material but should be treated with silicone reagents to remove polar sites. Glass columns are quite fragile and require great care during installation and removal from the chromatograph. Some workers have suggested that nickel is an appropriate substitute for glass in certain applications. A typical chromatogram using a nonpolar column is shown in Fig. 2. Within a homologous series the increased retention times will be in order of increased boiling points.

As might be expected, narrower column bores, finer and more uniform particle sizes of packing, and connections free of volumes where mixing can occur (dead volume) will improve resolution. The highest resolution (up to 3000 plates per meter) can be obtained using capillary columns of stainless steel, glass or fused silica. Capillary columns are of two general types: (a) wall-coated open tubular (WCOT) columns with an internal diameter range of 0.2–0.3 mm with the inside wall coated with a 0.1–1.0 μm layer of a liquid phase, and (b) support-coated open tubular (SCOT) columns where a thin layer of a solid support is coated on the walls of the capillary and followed with the liquid-phase loading. The SCOT columns have a higher level of liquid phase and consequently a higher sample capacity (0.01–1 μg) whereas the WCOT columns can handle efficiently

Table 1
Gas-chromatography packing materials

Stationary phase	Composition	Application	Temperature limit (°C)
Apiezon L	hydrocarbon	general nonpolar	275
OV-17	50% phenyl, methyl silicone	general medium polar	350
Dexsil-300	carborane–methylsilicone	high-temperature phase	450
Carbowax 20M	polyethylene glycol	general-purpose polar	225
FFAP	carbowax + nitroterephthalic	free fatty acids, polar	250
DEGS	diethylene glycol succinate	fatty-acid esters	200
OV-275	dicyanoallyl silicone	highly polar, pesticides	250
Carbowax 20M + KOH		amines	225
Tenax-GC	2,6-diphenyl phenylene oxide	polar, water tolerant	375
Molecular Sieve 5A	zeolite	gas analysis, oxygen/hydrogen	350
Porapak Q	vinyl benzene polymer	medium polar, water tolerant	250

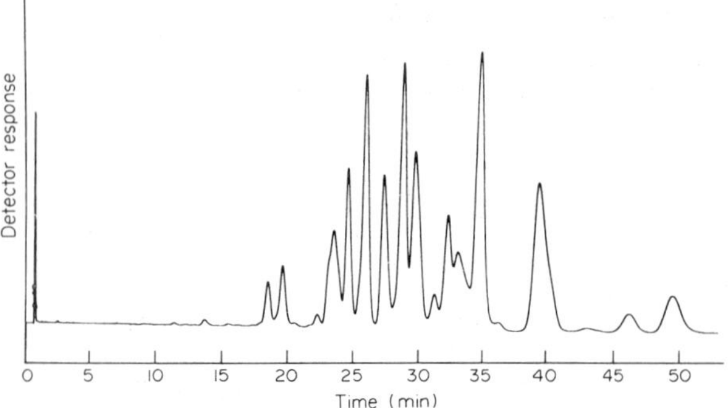

Figure 2
Capillary gas chromatogram of a polychlorinated biphenyl mixture. Column: 30 m fused silica with bonded liquid phase. Temperature program: 40–80 °C at 10 °C min^{-1}, 80–225 °C at 1 °C min^{-1}. Detection: electron capture. Carrier gas: helium (make-up with nitrogen). Sample: Aroclor at 1 μg ml^{-1}, splitless injection

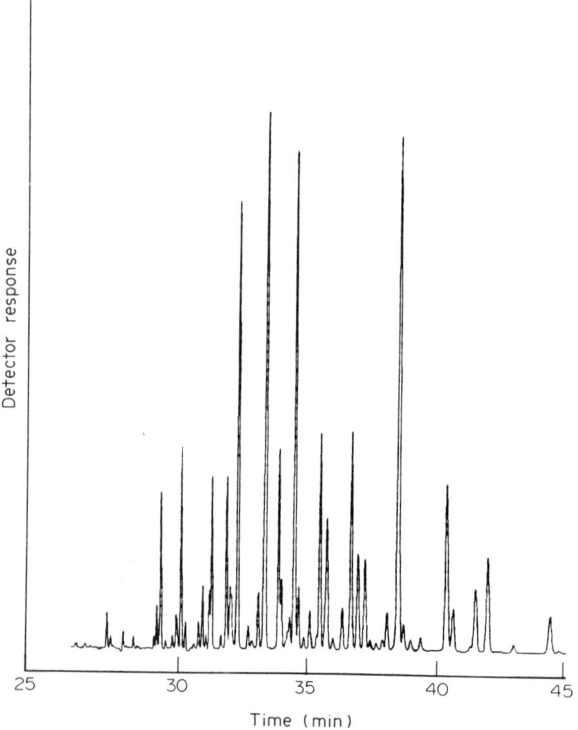

Figure 3
Packed column gas chromatogram of a polychlorinated biphenyl mixture. Column: 2.0 m × 4 mm internal diameter glass packed with 1.5% OV-17 and 1.95% QF-1 on 60/80 mesh Gas-Chrom Q. Temperature program: 150–250 °C at 2 °C min^{-1}. Detector: electron capture. Carrier gas: 5% methane in argon at 60 ml min^{-1}. Sample: Aroclor 1260 at 10 μg ml^{-1}, 1 μl injection

only very small amounts of sample (10–100 ng) and thus require the use of more sensitive detectors. Both types of column require special injection techniques or equipment such as on-column microsyringe injectors or splitters that divert the greater part of the injected sample to waste. Capillary columns of fused silica have found increasing use because of their chemical inertness and inherent ruggedness and ease of handling. However, they are limited in that they are not currently available with some of the more polar liquid-phase materials. A typical capillary separation using a fused-silica column is shown in Fig. 2. This is the same mixture used for the example in Fig. 3. Note the increased resolution possible with capillary columns with no increase in the analysis time.

1.2 Detectors

Gas chromatography utilizes a wide array of general and specific detectors. Specific detectors tend to be very sensitive and their selectivity for particular types of molecules (e.g., those containing halogens, sulfur, nitrogen or phosphorus) makes them especially useful in situations where the sample molecules are in a complex mixture of hydrocarbons. The most common detectors are the thermal-conductivity detector (TCD) and the flame-ionization detector (FID or HFID). The TCD senses the sample by comparing the thermal loss from a hot filament in the effluent from the column with the loss from an identical filament in the effluent from a balancing column carrying only the carrier gas. This detector is almost universal (detecting fixed gases and inorganics as well as organic molecules) but has only moderate sensitivity. The FID burns the sample in a hydrogen flame between two electrodes, one of which is the flame support. A high potential difference is applied to the electrodes and the sensor electronics detect current flow when the burning of carbon-containing species increases the conductivity of the flame. This detector is an excellent choice for organic compounds, for which it has high sensitivity. The FID has no response to fixed gases and is useful for carbon monoxide only when that gas is catalytically converted to methane in a methanizer. The electron-capture detector (used for the examples in Figs. 2 and 3) is particularly sensitive to halogenated organics and has found extensive use in pesticide analysis.

Less commonly used detectors include the flame photometric detector (FPD), the Hall electrolytic-conductivity detector (HECD), infrared detectors and the mass spectrometer. In the FPD, the sample is burned in a hydrogen flame and emission from excited sulfur or phosphorus species is detected by a photomultiplier after being appropriately filtered. This is a very specific detector with a response ratio for carbon : sulfur or carbon : phosphorus of about 10^6.

The HECD can be configured so as to be halogen,

nitrogen or sulfur specific. In this detector, the column effluent is mixed with oxygen (for sulfur) or hydrogen (for halogens and nitrogen) and passed through a reactor tube at 700–900 °C. After passing through scrubbers to remove interfering compounds, the products are detected as they change the conductivity of a circulating electrolyte. The carbon:(halogen, nitrogen, sulfur) ratios are also around 10^6. This quality is particularly useful when analyzing for heteroatom-containing compounds in complex hydrocarbon mixtures.

Infrared detectors can be either simple single-beam single-wavelength instruments set to monitor a particular functional group as the column effluent passes through a heated cell, or sophisticated Fourier-transform instruments that can record the entire infrared spectrum as a peak elutes.

The mass spectrometer is a very specific and sensitive detector. It is usually configured as a stand-alone instrument (GC/MS) and operates by analyzing the mass spectrum of each component as it elutes from the GC column. The mass spectrum is a plot of the mass: charge ratio (usually the charge is unity) for the sample molecular ion and the characteristic fragments it produces in the ion source. The information is stored in computer files for subsequent analysis and graphic display. The mass spectrum defines the molecular weight and probable formulas of the unknown peaks or enables comparison with the mass spectra of known standards stored in computer files. The GC/MS has found extensive use in the analysis of environmental samples where its sensitivity and specificity are a great advantage.

1.3 Sample Introduction

The most common type of injector, the heated injection port, allows for the injection of sample solutions via a syringe through a rubber septum. Rapid vaporization of the solvent and sample at temperatures up to 300 °C provide for efficient transport to the separating column. This procedure can be automated and autosamplers are available that are capable of handling as many as 100 samples and injecting them with greater precision than is possible with manual injection. The coupling of autosamplers with microprocessor-instrument controls and data-collection devices provides for essentially complete automation of the GC analysis. Other types of specialized devices include headspace samplers to analyze the vapor contents in containers, purge-and-trap samplers which collect and concentrate organic volatiles from water for pollutant analysis, and pyrolytic devices that flash-pyrolyze solid samples from a platinum coil or ribbon. This latter device allows analysis of nonvolatile materials such as polymers and geological materials by the study of their thermal-fragmentation products. For gas samples, the gas-sampling valve with fixed-volume sampling loops allows for accurate, precise and contamination-free analysis.

2. Liquid Chromatography

Liquid chromatography is applicable to most types of sample that can be separated by GC but, in addition, has special advantages in the handling of nonvolatile, thermally labile or ionic materials. A block diagram of a typical liquid chromatograph is shown in Fig. 4. The single pump is fed by a microprocessor-controlled mixing valve to produce solvent mixtures or continuously varying solvent compositions (e.g., changing from 50% methanol in water to 100% methanol over a 30 min period). On some instruments, two or more pumps each with its own solvent supply may be controlled to produce solvent gradients. Solvent programming has similar effects in LC to temperature programming in GC in that it shortens analysis time and increases detection limits by sharpening peaks. The pumps are commonly piston pumps and are capable of producing a constant flow rate of 0.1–10 ml min^{-1} at back pressures of up to 400 atm. Because of the pressures encountered, the manual injectors and autosamplers are based on high-pressure valve systems that isolate the sample from the system pressure until the moment of injection.

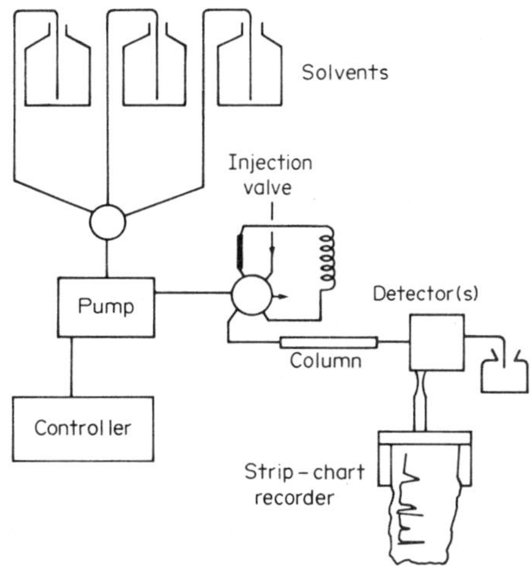

Figure 4
Block diagram of a typical liquid chromatograph

2.1 Detectors

The detectors used for LC are very dependent on the characteristics of the sample to be analyzed. Two of the most common are the ultraviolet-absorbance detector and the differential-refractive-index detector (DRI). The uv detector is usually a miniature double-beam filter spectrometer with a flow cell of 4–10 μl volume. This detector is used only for uv-absorbing samples and, of course, with uv-transparent solvents.

The DRI detector measures the difference between the refractive index of the column effluent and a reference volume of the eluting solvent. This detector is a more general detector but has rather low sensitivity, is sensitive to temperature variations and, of course, cannot be used in gradient-elution systems.

Other LC detectors include fluorescence, infrared, electrochemical and derivitization detectors. Fluorescence detectors irradiate the column effluent with monochromatic light and observe the emission at longer wavelength. The detector is extremely sensitive and specific but requires either a naturally fluorescent molecule or one that can be tagged with a fluorescent label.

Infrared detectors for LC are the same as for GC (Sect. 1.2) but are more limited because of solvent absorbance.

Electrochemical detectors are available in several forms but the most popular is the amperometric type. There is also a polarographic detector available that uses a hanging mercury-drop electrode. The electrochemical detectors require electroactive sample components and are limited to conductive-solvent systems such as buffers. These detectors are as a class extremely sensitive. Common functional groups that are electroactive are phenols, aromatic amines and sulfur- and halogen-containing molecules.

Derivitization detectors are low dead-volume chemical reactors that are used to modify the column effluent to confer detectability or enhance the sensitivity of any of the detectors described above. They usually have the facility for adding reagents and mixing and heating the column effluent without reducing the chromatographic resolution.

2.2 Columns

There are several basic macroscopic classes of separation in LC: liquid–liquid (LLC), liquid–solid (LSC), ion-exchange (IEC), and gel-permeation (GPC) or size-exclusion (SEC) chromatography. Each of these uses a different type of column packing. LLC uses a porous support coated with a liquid phase that is immiscible with the mobile phase (solvent). Column packings that consist of physically adsorbed coatings require care in the choice of solvent and tend to be more difficult to work with. For routine LC, the physically adsorbed phases have been almost completely replaced by chemically bonded phases in which silica substrates are reacted with, for example, a siloxane-terminated 18-carbon hydrocarbon chain. Other functionalized groups are similarly attached to give liquid coatings with a range of hydrophobic character. LSC, also called adsorption chromatography, commonly uses silica gel but may also use alumina, molecular sieves, porous glass or carbon as a stationary phase.

In LSC, sample polarity is of major importance in determining the order of elution. Sample elution takes place more rapidly as the polarity of the mobile phase is increased. LSC is related to thin-layer chromatography (TLC) in which the adsorbant is coated as a thin layer on a rigid substrate such as glass, aluminum or polyester.

IEC takes place on columns packed with resin particles that have ionic functional groups such as quaternary ammonium salts or sulfonic-acid groups attached. The mobile phase can be acids, bases or buffer solutions. The separation mechanism involves variations in affinity of mobile ions in solution to the bound ions on the stationary phase. IEC has been extensively used in amino-acid analysis and for the separation of metal ions, sugars and many biological molecules.

A variant of IEC is ion chromatography (IC) in which an ion-exchange separation of anions or cations is carried out on a low-capacity ion-exchange separating column, followed by a suppressor column that neutralizes the acidic (for cations) or basic (for anions) mobile phase. This allows the sampe ions of interest to be detected by a conductivity detector against a background of deionized water. Detection levels in the ppb range have been demonstrated for chloride, fluoride and sulfate as well as for alkali and alkaline-earth cations. This technique is of particular importance for anion analysis since classical methods for these tend to be tedious and limited in sensitivity. In addition, IC can distinguish between different oxidation states of some anions such as sulfate–sulfite and nitrate–nitrite. Cation analysis by ion chromatography is not as frequently used as it competes with atomic absorption (AA) which is a relatively sensitive technique.

GPC can be carried out on the same instruments as other LC techniques but differs from them in that the stationary phase consists of a controlled-porosity polymer network, a solvent-swelled gel or a controlled-porosity glass. The solvent carries the sample molecules along and they distribute according to their size between the stationary phase and the mobile phase. Larger molecules such as high-molecular-weight polymers tend not to penetrate the pores or gel and elute first. Proportionately smaller molecules elute later. The columns can be calibrated in terms of a certain retention volume (retention time × flow rate) by using known-molecular-weight standards. Standardizations by this method give molecular weights referenced to the standardizing polymer (usually narrow-distribution polystyrene). Other structurally different polymers may deviate significantly in retention volume from polystyrenes of equivalent molecular weights. Figure 5 shows a composite of two GPC traces. The trace in the background is from an experimental polymer and shows a fairly broad molecular-weight distribution as well as the presence of residual monomer and some oligomeric material. The superimposed trace is of a mixture of polystyrene standards labelled with their

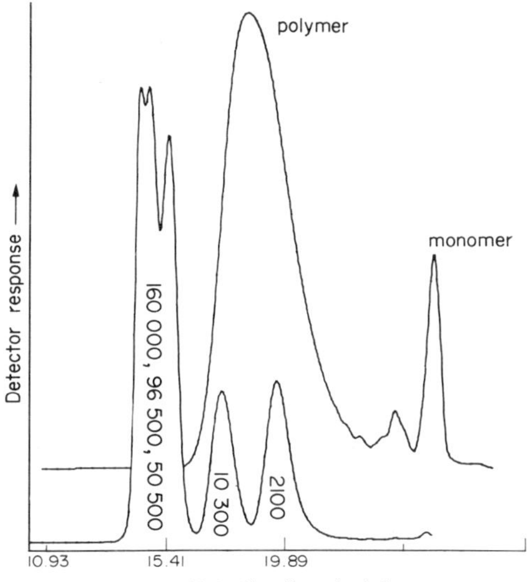

Figure 5
Gel-permeation (size-exclusion) chromatogram. The upper trace is from an experimental polymer, the lower (superimposed) trace is from an analysis of a mixture of narrow-molecular-weight-distribution polystyrene standards (the molecular weights are labelled). The chromatograph utilized four micro-Styragel columns. The solvent was chloroform at a rate of 2.0 ml min^{-1}, uv detection was at 254 nm

molecular weights. Most of the detectors useful for LC are also applicable to GPC. The low-angle laser-light-scattering (LALLS) detector is specific to GPC. This detector measures the instantaneous weight-average molecular weight of the column eluent and calculates the distribution based on a combination of the LALLS and a mass-sensitive detector such as DRI. The use of this detector eliminates the need for calibrations for each specific polymer analyzed.

See also: Investigation and Characterization of Materials: An Overview

Bibliography

Cazes J (ed.) 1977, 1980, 1981 *Liquid Chromatography of Polymers and Related Materials*, Vols. 1–3. Dekker, New York

Grob R L (ed.) 1977 *Modern Practice of Gas Chromatography*. Wiley-Interscience, New York

Hagnauer G L 1982 Size exclusion chromatography. *Anal. Chem.* 53: 265R-76R

Risby T H, Field L R, Yang F Y, Cram S P 1982 Gas chromatography. *Anal. Chem.* 53: 410R-28R

Simpson C F (ed.) 1976 *Practical High Performance Liquid Chromatography*. Heyden, London

Snyder L R, Kirkland J J 1979 *Introduction to Modern Liquid Chromatography*, 2nd edn. Wiley-Interscience, New York

Yau W W, Kirkland J J, Bly D D 1979 *Modern Size-Exclusion Liquid Chromatography: Practice of Gel Permeation and Gel Filtration Chromatography*. Wiley-Interscience, New York

C. Carnahan Jr.

Gas Metal Arc Welding

Gas metal arc (GMA) welding, which was developed in the 1950s and became extremely popular in the 1960s, is an arc welding process which fuses the parts to be welded by heating them with an arc between a continuous, consumable solid wire electrode and the work, with shielding by means of an externally supplied gas or gas mixture. The electrode is melted in the arc and becomes the deposited weld metal. The process is normally applied semiautomatically, but it can be applied by machine or by automatic equipment. It can be used to weld thin and fairly thick metals, both steel and nonferrous. The arc is visible to the welder and it can be used in all positions. A lesser degree of welding skill is required than in manual metal arc welding, but the equipment is more complex.

A variation known as flux-cored arc welding utilizes an arc between a continuous flux-filled electrode wire and the work. Shielding is obtained from gas generated by the decomposition of the flux within the tubular wire, but additional shielding may be obtained from an externally supplied gas or gas mixture.

The GMA process is also known as MIG welding and is sometimes called microwire welding, short arc welding, dip transfer welding, CO_2 welding or other names.

1. Process Features

GMA welding is illustrated in Fig. 1, which shows schematically the electrode wire, the nozzle of the

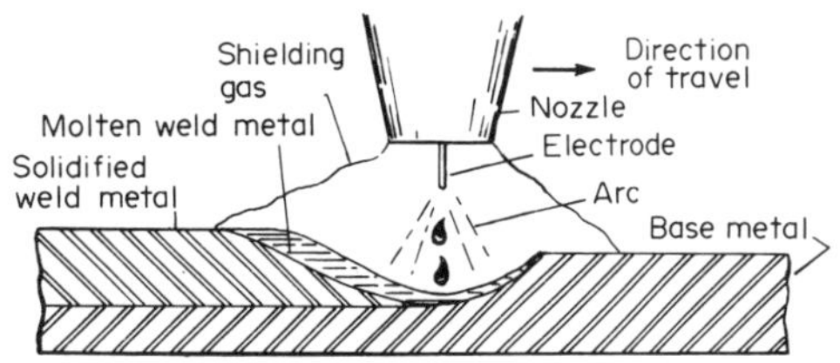

Figure 1
Schematic of gas metal arc welding

welding gun or torch, the shielding gas envelope and the arc between the end of the electrode and the base metal. The flux-cored system is shown in Fig. 2.

There are a number of variations of GMA welding, depending on the type of shielding gas, which relates to the type of metal transfer across the arc:

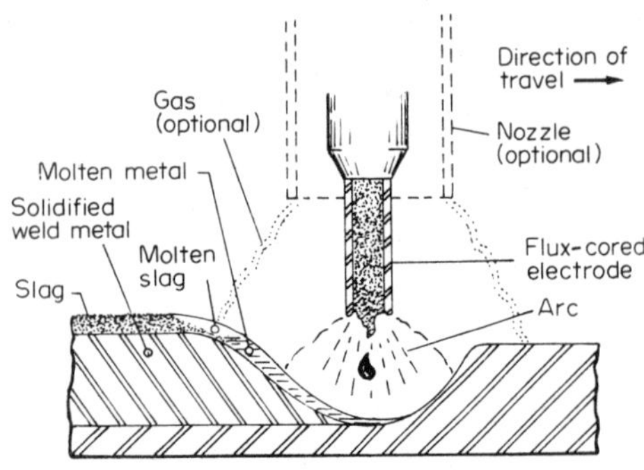

Figure 2
Schematic of flux-cored arc welding

(a) GMA welding using inert gas shielding for welding nonferrous metals;

(b) short-circuiting transfer (microwire) normally using CO_2 gas or CO_2 gas mixtures and small diameter electrode wire for welding in all positions and on thin steel;

(c) CO_2 welding using CO_2 shielding gas and larger electrode wires for welding steels;

(d) spray arc welding, which uses the argon–oxygen shielding gas normally used on steels; and

(e) pulsed arc welding, which provides pulsed metal transfer and uses a special power source.

There are two variations of flux-cored arc welding: the self-shielding and the externally gas-shielded variation, shown dotted in Fig. 2. The flux-cored electrode wire and the arc between it and the base metal are shown. The process normally produces a slag covering which must be removed after welding. The externally gas-shielded variation was the original one and employed CO_2 for external shielding. The other, or self-shielding variation, generates sufficient shielding gas from the decomposition of the ingredients in the core of the electrode wire. (In either case, the gas shield prevents the atmospheric oxygen and nitrogen from reaching the arc area.) This variation was developed in the mid-1950s.

The outstanding features of gas metal arc welding are:

(a) high-quality welds;

(b) minimum postweld cleaning;

(c) the arc and weld pool are visible;

(d) welding is possible in all positions, depending on electrode size;

(e) relatively high speed welding;

(f) little or no slag is produced, depending on the variation; and

(g) it is a low-hydrogen welding process.

Variations of the process offer special advantages. The short-circuiting arc will weld most steels in the thinner gauges. CO_2 welding allows for high-speed travel on steel. The spray arc variation produces high-speed welds with minimum spatter and cleanup, and the GMA process with inert gas shielding can weld most nonferrous metals.

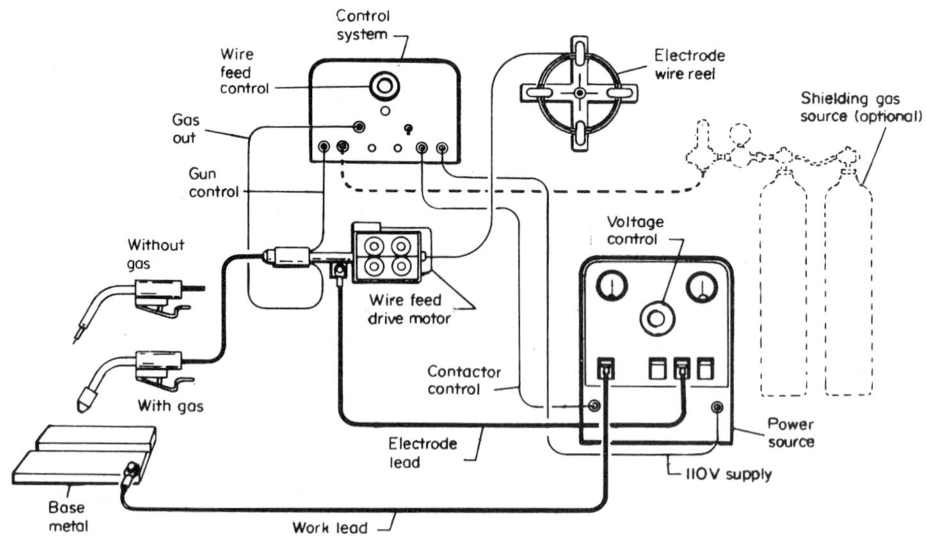

Figure 3
Equipment used for gas metal arc welding

The two variations of flux-cored arc welding provide slightly different welding features. With external shielding gas, the features of the process are:

(a) extremely smooth, sound, high-quality welds;

(b) deep penetration; and

(c) good properties for x-ray-quality welds.

The gasless or self-shielding variation offers the following features:

(a) elimination of gas supply, controls and gas nozzle;

(b) moderate penetration; and

(c) ability to weld in drafts or breezes.

Both variations have the following features:

(a) high deposition rates;

(b) the arc is visible to the welder;

(c) all-position welding, based on the size of the electrode; and

(d) the weld joint design is similar to that used for the other arc welding processes.

Both variations are normally restricted to the welding of carbon and stainless steels and for overlaying. The externally gas-shielded version can be used for welding many low-alloy steels.

2. *Equipment*

Major components required for all variations are shown in Fig. 3. These are the welding machine or power source, the electrode wire-feed system and control, the welding gun and cable assembly (for semiautomatic welding, or welding torch for automatic welding), the shielding gas supply and controls (when used), and the consumable electrode wire (either solid or flux cored).

2.1 Welding Machine

The power source for gas metal arc welding is normally a constant voltage (CV) or constant potential (CP) type. Its characteristic output volt–ampere curve is essentially flat with a small droop. Thus the output voltage is approximately the same even though the welding current changes. The output voltage is adjusted at the power source, which can be a transformer–rectifier, a motor generator or an engine-driven generator. A constant voltage or constant potential power source does not have a welding current control and cannot be used for manual metal arc welding. The welding current output is determined by the electrical load on the machine, which depends on the electrode wire feed rate. Direct current electrode positive is normally used. Machines for this process are available from 150 up to as high as 1000 A and should be rated at 80–100% duty cycle. They should include a contactor and meters (see *Arc Welding Power Sources*).

2.2 Wire Feeder

This mechanism feeds the electrode wire automatically from a coil or spool to the welding gun and into the arc. The wire feed system must be matched to the power supply. The CV system of welding relies on the relation between the electrode wire burnoff rate and the welding current. This relation is fairly constant for a given electrode wire size, composition and shielding atmosphere. At a given wire feed rate, the welding machine supplies the proper amount of current to maintain a steady arc. Thus the wire feed rate control adjusts the welding current. The CV welding system is a self-regulating system and is recommended when using small diameter electrode wires. A miniaturized wire feeder built into the welding gun is popular for welding with small diameter aluminum wire. The wire feed system and controls are essentially the same for semiautomatic, machine or automatic welding. The drive rolls must be selected to feed the type and size of electrode wire being used.

2.3 Welding Gun

The welding gun and cable assembly are used to carry the electrode wire, the welding current and the shielding gas (when used) to the welding arc. For higher-current applications, water-cooled guns are used and the water is also carried through the cable assembly. There are two general types of welding guns: the pistol grip and the curved head (gooseneck) type. The gooseneck type is more popular for small diameter electrode wire. The pistol grip type is usually used for welding with larger electrode wires and for welding with nonferrous electrodes. Guns used for heavy-duty work at high currents and those using inert gas for shielding at medium to high currents are water-cooled. Water cooling is not used for the gasless flux-cored welding guns. Sometimes with the gasless variation, a special insulated extension which adds to the electrical stickout is added to the gun to provide higher deposition rates. Smoke exhaust nozzles are usually included.

For machine or automatic welding, a welding torch is used. The automatic torches are either air- or water-cooled, depending on the welding application. For CO_2 welding, a side delivery gas nozzle is often used with automatic torches. The wire guides in all guns and torches must match the size of the electrode wire being used.

2.4 Shielding Gas

The shielding gas displaces the air around the arc to prevent contamination by the oxygen and nitrogen of the atmosphere. The gas shielding envelope must efficiently shield the arc area in order to obtain a high-quality weld deposit. Various shielding gases can be used, depending on the process variation and the base metal being welded. Carbon dioxide is the least expensive and is very popular. Mixtures of CO_2 and argon and mixtures of argon and oxygen are also used.

The shielding gas must be specified "welding grade." This means that the gas has a high purity and low moisture content, indicated by its dewpoint. The type of gas for shielding must be related to the electrode wire and base metal and is specified in welding procedure tables for welding various metals with the different process variations.

The gas flow rate depends on the type of gas, the metal being welded, the welding position and other factors. When welding outdoors or when air currents disturb the gas shield, higher gas flow rates are necessary. For high flow of CO_2, two or more cylinders are manifolded to avoid freezing of the pressure regulators. Heaters may also be required. External shielding gas is not required for the self-shielded type of flux and wires.

2.5 Electrode Wires

The composition of the electrode for gas metal arc welding must match the metal being welded, the particular process variation and the shielding atmosphere. The diameter or size of the electrode depends on the process variation and the welding position. All GMA steel electrode wires are normally solid and bare, except for a thin, protective coating—usually copper on carbon steel wires. Nonferrous metal electrode wires and the flux-cored wires do not have a protective coating.

The welding procedure tables indicate the proper electrode wire, type and size for welding different metals. Electrode wires are available in a wide variety of diameters, spools, coils and reels and are specified by national standards. The flux-cored wires are classified by strength level, properties, whether or not external gas is used and the deposited weld metal composition.

See also: Gas Tungsten Arc Welding; Shielded Metal Arc Welding; Welding: An Overview

Bibliography

Cary H B 1979 *Modern Welding Technology*. Prentice-Hall, Englewood Cliffs, New Jersey

Metals Handbook, 9th edn., Vol. 6, 1983. American Society for Metals, Metals Park, Ohio

Welding Handbook, 7th edn., Vol. 2, 1978. American Welding Society, Miami, Florida

H. B. Cary

Gas Tungsten Arc Welding

Gas tungsten arc welding (GTAW), also known as tungsten–inert gas (TIG), Heliarc or Argonarc welding, is one of the newer welding processes. It was developed by the aircraft industry in the early 1940s to join "hard-to-weld" materials, particularly magnesium, aluminum and stainless steels. In this process, parts are fused together by forming an electric arc between them and a nonconsumable tungsten electrode. Filler metal may or may not be used. Shielding is obtained from an inert gas or gas mixture. Figure 1 shows the process. The torch holds a tungsten electrode and has a nozzle for directing the shielding gas to the arc area.

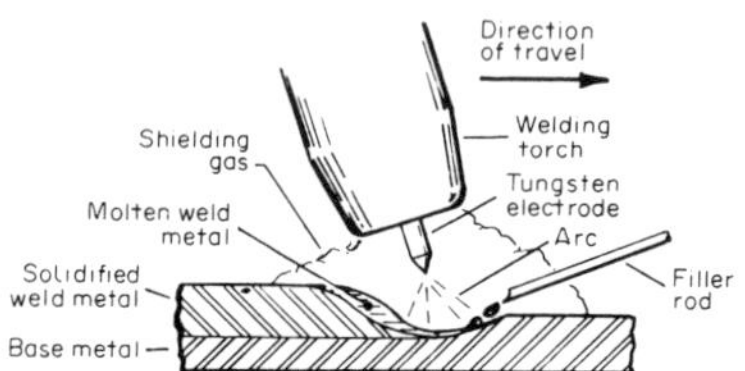

Figure 1
Schematic of the GTAW process

GTAW is normally applied manually and is capable of welding steels and nonferrous metals in all positions. It is commonly used to weld thin metals and for the root pass welding on tubing and pipe. It requires a high degree of welder skill but can produce excellent quality welds. GTAW is also used for automatic and automated welding applications.

1. Application

The features of the gas tungsten arc welding process are:

(a) the ability to produce high-quality welds on almost all metals and alloys;

(b) little or no postweld cleaning is required;

(c) the arc and weld pool are clearly visible to the welder;

(d) there is little or no weld spatter since filler metal does not cross the arc;

(e) welding is done in all positions; and

(f) there is no slag which might be trapped in the weld.

Gas tungsten arc welding can be used for welding aluminum, magnesium, stainless steel, bronze, silver, copper and copper alloys, nickel and nickel alloys, cast iron and steel. It is capable of welding a wide range of metal thicknesses but is most popular on thinner gauges.

Figure 2
GTAW in practice

2. *Equipment*

The major components required for gas tungsten arc welding are shown in Fig. 3. These include the welding machine or power source, the torch (including the tungsten electrode), the shielding gas and controls, and (when required) the filler rod. There are several optional accessories available, in particular a remote-controlled rheostat which permits the welder to control the current while welding. Other accessories include arc timers and controllers, high-frequency units and water circulating systems.

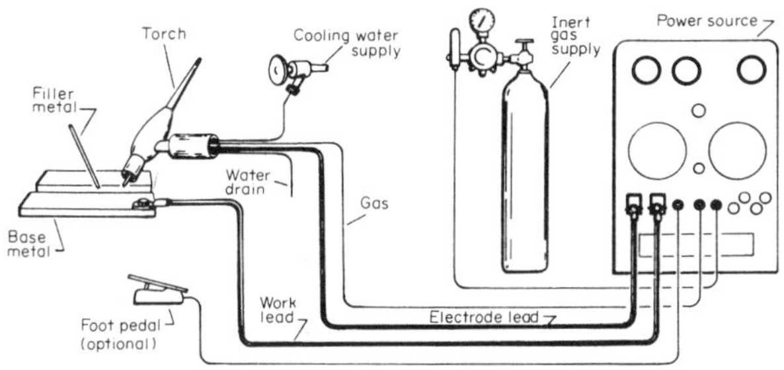

Figure 3
Components of a typical GTAW system

2.1 *Welding Machine*

A specially designed welding machine or power source with a drooping output characteristic is employed. Alternating current and direct current are used. The power source is of the transformer, transformer/rectifier or generator type. The power source usually includes a high-frequency spark gap oscillator which is used to aid arc starting when welding with direct current, and is used continuously when welding with alternating current. The choice of alternating or direct current depends on the material being welded. Alternating current is recommended for welding aluminum and magnesium. Direct current is recommended for welding stainless steels, carbon steels, copper and its alloys, nickel and its alloys and precious metals.

Most GTAW machines include solenoid valves for controlling the shielding gas and cooling water (when used). The welding machine may also include meters and programmers. Some machines provide pulsed current capability.

Welding machines designed for GTAW can also be used for shielded metal arc welding, plasma welding and several other processes.

2.2 *Welding Torch*

The gas tungsten arc welding torch holds the tungsten electrode and directs the shielding gas and welding power to the arc. Torches come in different sizes, the larger sizes usually being water-cooled. The torches normally come equipped with a cable assembly which transmits the gases, welding current and cooling water (when used) from the machine to the torch.

2.3 *Tungsten Electrodes*

Tungsten has the highest melting point of any metal (3410 °C) and is considered nonconsumable. When properly used, the electrode does not touch the molten weld puddle. If the electrode accidentally touches the weld puddle it becomes contaminated and must be cleaned immediately; if it is not cleaned, an erratic arc will result.

Electrodes of pure tungsten and of tungsten alloys are available. Pure tungsten is the least expensive; however, alloys containing 1% or 2% thoriated tungsten are quite popular, despite being somewhat more expensive, and are recommended for welding specific metals. The different tungstens are identified by national codes or specifications, and are color-coded for ease of recognition. Tungsten electrodes come in diameters ranging from 0.020 in. (0.5 mm) up to 0.25 in. (6.4 mm), and in either a ground or cleaned finish. The lengths are normally 3–6 in. (75–150 mm).

2.4 *Shielding Gas*

An inert shielding gas must be used. Argon is the most popular, but helium is used in certain applications and in some cases a mixture of argon and helium is used. Argon is heavier than helium and provides better shielding at lower flow rates. The arc produced with helium is considered hotter and is used for obtaining deeper penetration. Helium is also used for welding in the overhead position. It is usually used at a higher flow rate.

2.5 *Filler Metal*

Filler metal may or may not be used. It is normally used except when welding very thin material. The composition of the filler metal should match that of the base metal. Filler metals may not be available in every alloy, and therefore filler metal charts are employed to select the recommended type for use. The size of the filler metal rod depends on the thickness of the base metal and the welding current. Filler metal is usually added manually to the weld puddle, but automatic feed is sometimes used. National specifications provide information about filler wires available.

See also: Arc Welding Power Sources; Gas Metal Arc Welding; Welding: An Overview

Bibliography

Cary H B 1979 *Modern Welding Technology*. Prentice-Hall, Englewood Cliffs, New Jersey

Metals Handbook, 9th edn., Vol. 6, *Welding, Brazing and Soldering*, 1983. American Society for Metals, Metals Park, Ohio

Welding Handbook, 7th edn., Vol. 2, 1978. American Welding Society, Miami, Florida

H. B. Cary

Gases in Glass

The migration of gases in glass and their reaction with glass have a large effect on many of its optical and physical properties. The choice of a glass for vacuum applications is strongly influenced by its permeability and outgassing behavior. Diffusion-controlled reactions can affect optical properties via the oxidation or reduction of transition metal ions and alter physical properties (e.g., viscosity) by affecting the concentration of water dissolved in the glass. This article deals with the solution, migration and reaction of gases with glasses below the transformation temperature range, with emphasis on the effect of glass composition and morphology on gas mobility.

1. Experimental Techniques

The permeability and diffusivity of gases in glass are usually determined by a flow technique in which the rate of gas migration through a membrane of known area and thickness is measured with a mass spectrometer while the sample is maintained at a known temperature under a fixed pressure gradient. The diffusivity and solubility of gases can also be obtained from measurements of the rate and amount of gas absorbed or desorbed from glass powders, fibers or rods. In general, the flow techniques are preferable for diffusivities greater than 10^{-10} cm^2 s^{-1}, while the absorption/desorption techniques are most suitable for lower diffusitivies or for materials which cannot readily be formed into a membrane.

Occasionally diffusion-controlled reactions also are studied using flow techniques. However, in most cases, the concentrations of the reactants and the reaction products are measured as a function of time or temperature by chemical analysis, optical spectroscopy or electron spin resonance.

2. Gas Solubility

Gas solubilities are usually expressed as (a) a volume of gas at standard temperature and pressure per unit volume of glass or as (b) the number of molecules per unit volume of glass per unit applied pressure. This introduces an artificial temperature dependence into the solubility as a result of the temperature dependence of pressure. The use of the Ostwald solubility, which is the concentration of dissolved gas divided by the concentration of gas in the gas phase, eliminates this problem. The Ostwald solubility can be obtained from the more usual values by multiplying the solubility by $T/273$, where T is the absolute temperature.

The solubility of a variety of gases in the same glass has been measured for only a few glasses. The Ostwald solubilities of He, Ne, Ar, H_2 and D_2 at 200 °C in vitreous silica all lie between 0.023 and 0.043. Similar values have been measured for these gases in other glasses. The heat of solution of the inert gases becomes more negative in the order He, Ne, Ar.

The pressure dependence of gas solubility in vitreous silica is adequately described by the Langmuir adsorption model. This model gives the concentration of dissolved gas C_i as

$$C_i = AFN/(1 + AF) \qquad (1)$$

where A is an equilibrium constant, F is the gas fugacity and N is the concentration of potential solubility sites in the glass. The values of N lie in the range (1–2) $\times 10^{21}$ sites cm^{-3} for He, Ne and D_2 in vitreous silica. Since values for the solubility of these gases in most glasses at atmospheric pressure lie between 10^{16} and 10^{18} atoms cm^{-3}, the value of A is usually between 10^{-5} and 10^{-3}. As a result, the solubility of gases in most glasses has usually been found to vary linearly with gas pressure for pressures below one atmosphere.

The effect of glass composition on gas solubility has been studied extensively only for helium. In general, the solubility of helium decreases as modifiers are added to the glass structure. In alkali silicate and aluminosilicate glasses, the solubility of helium decreases in the order K, Na, Li, whereas the reverse order holds for alkali germanate glasses. Helium solubility in the alkali borate glasses is almost independent of the identity of the alkali oxide present. These studies indicate that gas solubility is a strong function of glass structure and that no simple rules for the compositional dependence of solubility exist at this time. Gas solubility in phase-separated glasses is further complicated by morphology effects.

Although only limited data exist, it appears that gas solubility in glass melts is two to four times that in the corresponding solid glass.

3. Permeation and Diffusion

The correct description of the temperature dependence of gas permeation and diffusion in glass is quite controversial. Results of studies of helium migration in vitreous silica from −30 to over 1000 °C have shown conclusively that the simple Arrhenius equation cannot adequately describe the temperature dependence of permeation and diffusion over the entire temperature range. It has been proposed that a modified form of the Arrhenius equation containing a temperature-dependent preexponential term, i.e.,

$$D = D_0 T^n \exp(-Q/RT) \qquad (2)$$

where D is the diffusivity, D_0 is a preexponential factor, T is the absolute temperature, n is a constant equal to either 0.5 or 1.0, Q is the activation energy and R is the gas constant, should be used to describe the temperature dependence of gas diffusion in glass. Experimental results indicate that a value of $n = 1.06 \pm 0.11$ provides an excellent fit to the data over the entire temperature range studied. Recently, another model, which assumes a distribution of activation energies for diffusion, has been offered to explain the unusual temperature dependence of gas diffusion in glass. This model also provides an excellent fit to the experimental data.

At low pressures, gas permeation varies linearly with pressure and gas diffusion is independent of applied pressure for all gases studied in a wide variety of glasses. This behavior suggests that the diffusion of molecular species such as H_2 occurs without the dissociation of the molecule at the glass surface. At elevated pressures (above 100 atmospheres), Eqn. (1) indicates that a large fraction of the solubility sites should be occupied. As a result, the effect of interactions between diffusing molecules can no longer be ignored. Furthermore, the applied gas can no longer be treated as an ideal gas. It has been found that the permeation of helium in vitreous silica is given by

$$K = \phi K^*(1 - 0.0002P) \qquad (3)$$

where K is the measured permeability, ϕ is the fugacity coefficient, K^* is the permeability at one atmosphere and P is the gas pressure in atmospheres. While this expression adequately describes the experimental data, it is not believed to describe the complex phenomena which lead to this result. It has been found also that stresses near those required to fracture the specimen can lead to unusual effects in gas permeability in some, but not all, glasses even at relatively low gas pressures.

The diffusivity of different gases in the same glass is strongly dependent upon the identity of the diffusing species. While the preexponential term in Eqn. (2) is almost independent of the identity of the gas, the activation energy can vary as much as an order of magnitude, which results in corresponding changes in diffusivity of many orders of magnitude. In general, the activation energy for diffusion increases with increasing diameter of the diffusing species.

The permeability and diffusivity of gases in glasses are strong functions of the atomistic structure of the glass and of the morphology of the sample. As a result, the mobility of gas in a given sample is a function of both the glass composition and its thermal history. In single-phase samples, the diffusivity of helium can decrease by as much as a factor of two when a highly quenched sample is annealed. However, if a change in thermal history affects the morphology of the sample, the mobility of helium has been found to vary by as much as two orders of magnitude in the same sample. Studies of a wide variety of systems have shown that the diffusing gas will tend to migrate along paths through the higher-diffusivity, lower-activation-energy phase. If this phase is continuous, its properties will dominate the overall diffusion process. If a change in thermal history results in a loss in continuity of a high-diffusivity phase, or results in the formation of a continuous phase from a previously discontinuous high-diffusivity phase, the effective diffusivity of gas in the bulk sample will change sharply. The degree of change will depend upon the relative diffusivities of the gas in the two phases, with an increase in the effect as the difference between the diffusivities in the phases increases. It follows that changes in overall glass composition which lead to changes in morphology can also strongly influence the mobility of gases, so that the compositional limits for continuity of the high-diffusivity phase can be determined in systems which exhibit phase separation (e.g., alkali borosilicates) by means of permeation and diffusion measurements. Also it follows that, since the diffusivity of gases in most oxide crystals is much less than that in glasses, the diffusivity of gases in glass ceramics will be determined by the properties of the glassy phase.

Early studies of gas permeation in glasses dealt exclusively with commercial glasses. These studies revealed that the mobility of gases tended to decrease with increasing modifier content. This conclusion was based primarily on the behavior of commercial borosilicate and aluminosilicate glasses of complex compositions. Recent studies have demonstrated that this hypothesis is reasonably valid for silicate glasses, but may be completely incorrect for glasses based on other glass formers. The mobility of helium in alkali germanate glasses actually passes through a minimum as the alkali oxide content of the sample is increased. Even more complex relationships between glass composition and helium mobility have been observed in the alkali borate systems. Definite relationships between gas mobility and glass structure have also been observed in the alkali aluminosilicate and galliosilicate systems. The identity of the modifier present in the glass has also been found to play a role in determining the diffusivity of helium. In the systems studied to date, helium mobility decreases in the order K, Na, Li. Substitution of an alkaline earth oxide for an alkali oxide usually decreases the helium mobility. These general guidelines apply only so long as the change in composition has no effect on the morphology of the material.

4. Gas–Glass Reactions

Although molten glasses can react with a large number of gases, only a few gases diffuse rapidly enough in solid glasses to result in a significant reaction in normal experimental times. As a result, most of the studies of reactions between gases and bulk glasses have involved either hydrogen isotopes or water. Considerable effort has been devoted to the study of the reduction of various ions, e.g., silver, iron, lead and the rare earths, by hydrogen isotopes. Many of these processes apparently result from a diffusion-controlled reaction between the migrating hydrogen molecule and a relatively fixed ion. At elevated temperatures, hydrogen isotopes can react with the silicate network to form hydroxyl and hydride groups, and can exchange at hydroxyl sites, i.e., the hydroxyl groups in a glass can be replaced by deuteroxyl groups by the appropriate heat treatment in D_2. It is also possible to induce reactions involving dissolved hydrogen isotopes at room temperature by exposing the sample to ionizing radiation. These reactions result in the formation of both hydroxyl and hydride groups and in hydrogen isotope exchange at previously existing hydroxyl sites.

Glasses normally contain a few hundred ppm of water in the form of hydroxyl. This water can be removed at temperatures below the transformation range by an appropriate heat treatment. Although there is considerable controversy concerning the mechanism for water diffusion in glasses, it appears that the dehydroxylation of glasses probably is controlled by the diffusion of water molecules through the bulk of the material. Dehydroxylation is controlled also by the local environment of the hydroxyl groups, as is evidenced in the dramatic differences in the dehydroxylation rates for various types of vitreous silica.

See also: Glass: An Overview

Bibliography

Doremus R H 1967 Diffusion in non-crystalline silicates. In: Mackenzie J D (ed.) 1967 *Modern Aspects of the Vitreous State*, Vol. 2. Butterworths, London, pp. 1–71

Doremus R H 1973 *Glass Science.* Wiley, New York

Shelby J E 1979 Molecular solubility and diffusion. In: Tomozawa M, Doremus R H (eds.) 1979 *Treatise on Materials Science and Technology*, Vol. 17. Academic Press, New York, pp. 1–40

J. E. Shelby

Gating and Risering of Castings

Most liquid metals and alloys readily react with oxygen in the air to form oxides which may contribute to the formation of casting discontinuities if allowed to enter the mold cavity. In addition, most metals are prone to absorb hydrogen from moisture present in the air, resulting in defects in the casting if excessive quantities of such gases are absorbed. Most castings are therefore produced with a gating system that minimizes the contact of liquid metal with air. There are additional reasons why a gating system is required: slag and dross from the ladle as well as eroded mold refractories are prone to be carried along by the liquid metal and will enter the mold cavity unless restrained by a suitable gating system. Thus, a gating system should introduce metal into the mold cavity in a manner that avoids the formation of casting defects such as laps and folds, sand erosion and spalling of the mold, slag or dross inclusions, as well as gas cavities. Finally, because the gating system affects the heat flow during solidification of the casting, the design of the gating system has to include the choice of the riser system.

Figure 1 illustrates a gating and riser system for a simple casting shape. Liquid metal enters the system via a pouring cup designed to facilitate pouring liquid metal without splashing. It descends the downsprue from which it enters the runner, and in turn passes through in-gates into the mold cavity. Risers are frequently placed such that the liquid metal enters through the riser before passing into the mold cavity. Gating systems are designed to minimize turbulence and to promote the entry of clean metal. Special consideration is given to the dimensions of the gating system to regulate the flow rate of the metal and to

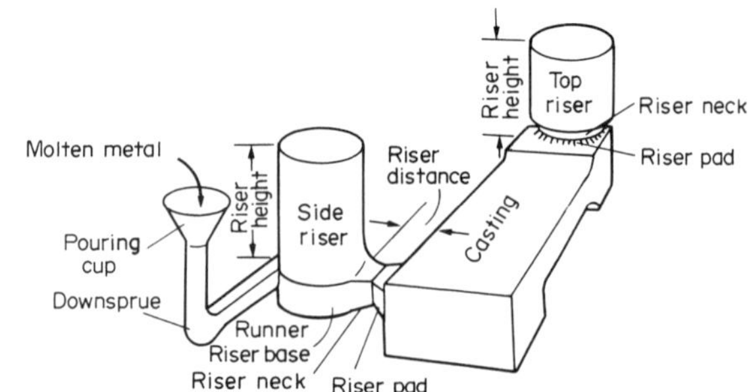

Figure 1
Riser and gated casting with side and top risers (courtesy AFS Gating and Risering Committee)

facilitate the removal of slag, or to prevent erosion of the sand from which the gating system is molded.

1. In-Gate and Runner Systems

A great variety of in-gate and runner systems exist. These specialized systems have evolved for the various liquid metals being poured. Some of these may include special ceramic components that act as skimmers to retain nonmetallics. Screens produced from metals or ceramic filters have been used successfully, particularly for aluminum and several other nonferrous alloys. Ceramic filters are now being introduced for ferrous alloys as well as cobalt- and nickel-base superalloys. In addition, traps have been incorporated successfully which operate on the principle of centrifuging light nonmetallic particles to the center of the so-called "whirl gate," thus preventing their entry into the mold cavity.

After taking into account the required pouring rate for the casting and the fluidity of the metal being poured, the laws of fluid flow are considered in determining the dimensions of the gating system. Figure 2 illustrates the forces relevant to the application of Bernoulli's theorem to metal casting. This theorem states that the sum of the potential head wz, pressure head wp/ρ, velocity head $wv^2/2g$ and frictional loss wF is constant for all locations in a flowing system; that is,

$$wz + wp/\rho + wv^2/2g + wF = K$$

where w is the total weight of fluid flowing, ρ is the density of the fluid, g is the acceleration due to gravity, z is the height of the head above a reference plane, p is pressure, v is the average velocity of molten metal and F is the friction loss per unit weight.

The principal conclusions from the application of these hydraulic principles to the design of gating systems are as follows:

(a) A pouring cup of sufficient depth and size should be selected to facilitate flow control without vortex formation in the sprue while maintaining a low pouring height. Skimming devices should be used for alloys that are particularly prone to drossing.

(b) Tapered sprues are the most suitable design as they choke the flow, enhance filling of the sprue and minimize the aspiration of air. A large sprue basin is desirable to cushion the impact of the metal and so reduce turbulence and erosion as the metal changes direction and enters the runners.

(c) Reduced velocity, lower turbulence, and increased opportunity for dross to rise and become retained at the top surface of the runner are facilitated by increased runner cross sections. To maintain uniform rates of metal entry in the mold cavity from multiple in-gates the runner cross section is reduced proportionally after each in-gate.

(d) Multiple in-gates are desirable for large and/or complex castings, and to assure equal and rapid metal distribution in the mold cavity. The total in-gate area should equal or exceed that of the runner. For metals which dross readily (high-chromium steels, and certain aluminum and magnesium alloys) the in-gates should be placed at the bottom of the casting in order to prevent cascading of the metal into the mold cavity (which exposes a large metal surface to oxygen in the mold atmosphere, enhancing undesirable oxide formation).

(e) Single in-gate systems are preferred where successful mold filling requires high metal velocities, where only one riser is employed, or where favorable thermal gradients can be established to promote directional solidification.

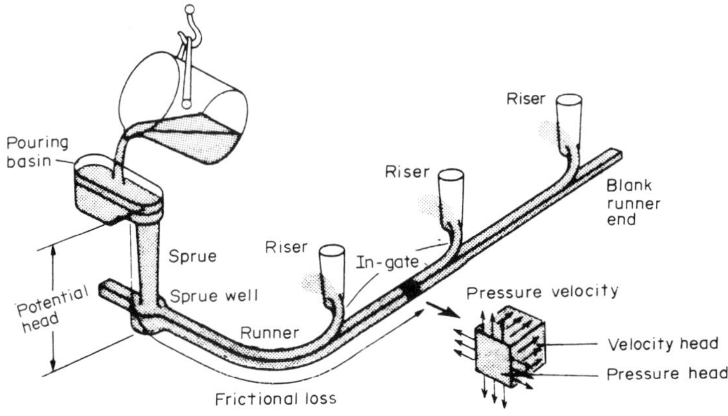

Figure 2
Forces and pressures in a gating system (after Wallace and Evans 1957)

2. *Risering*

Upon pouring liquid metal into a mold, shrinkage occurs in three consecutive steps. During cooling from the pouring temperature to the start of solidification the metal undergoes liquid contraction. Solidification contraction is then experienced during freezing. Solid contraction follows during cooling of the metal to room temperature. Figure 3 illustrates these consecutive stages of shrinkage for 0.35% carbon steel. To compensate for shrinkage a supply of liquid feed metal from risers is required (Fig. 1). In this fashion a casting can solidify without shrinkage porosity.

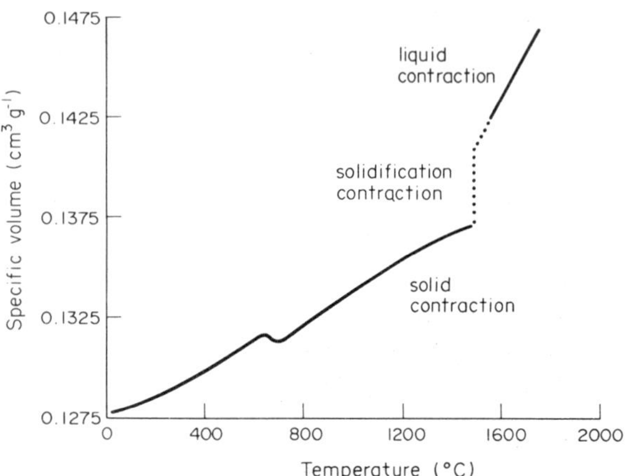

Figure 3
Contraction of 0.35% carbon steel on cooling

There are two more essential requirements that must be met for the riser(s) to supply liquid metal effectively. First, the riser–casting connection must be designed to ensure that it does not complete solidification prior to that of the casting itself. Second, the liquid feed metal must have free access to all positions along the solidification front in the casting itself (Fig. 4). Two factors have evolved as important in risering of castings: the solidification time of the casting and the feeding distance.

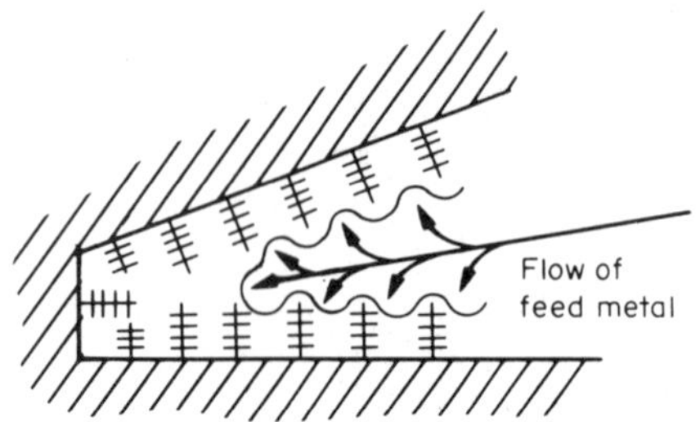

Figure 4
Schematic of unrestricted flow of feed metal to the solidification front

The computation of solidification time for dimensioning risers can be avoided because it is only the solidification time of the riser in relation to that of the casting which is important. Based upon heat-transfer considerations formulae have been developed, such as the following for steel:

$$R = 1.2\,(V_c/A_c)\,[(1 + 2x)/x]$$

where R is the riser diameter, V_c is the casting volume, A_c is the casting surface in contact with the mold and x is the ratio of riser height to riser diameter. Relations such as these are useful in many instances because solidification time is the primary criterion for determining riser dimensions. However, modifications are needed if high pouring temperatures are employed; that is, when consideration must be given to the liquid contraction, and when exothermic or insulating riser sleeves are used to delay the onset of solidification of metal in the riser. The amount of solidification shrinkage differs for different metals. For instance, it is approximately 6.6% for pure aluminum, 4.9% for pure copper, 2.5–4% for carbon steel (depending on the actual carbon content), 0 to −1.9% for gray cast iron (depending on graphite formation) (Heine et al. 1967).

Most castings require the use of several risers because the distance over which liquid metal can be supplied from an individual riser is limited. The feeding distance is limited for shapes with parallel walls, or those with a subcritical taper, because liquid metal flow is restricted when the solidification fronts, which advance from the casting walls, meet near the thermal centers. Empirical relationships have been developed for the feeding distance of a riser. More recently, computations of the feeding distance have become possible based upon the heat flow in the casting and the mold. Either the thermal gradient or the solidification gradient is used in these calculations as a criterion for feed metal flow from the riser to all areas of the solidification front.

See also: Casting of Metals; Metals Processing and Fabrication: An Overview

Bibliography

Briggs C W, Gezelius R A 1934 Studies on solidification and contraction in steel castings. *Trans. Am. Foundrymen's Assoc.* 42: 449–70

Heine R W, Loper C R Jr, Rosenthal P C 1967 *Principles of Metal Casting*, 2nd edn. McGraw-Hill, New York

Wallace J F, Evans E B 1957 Gating of gray iron castings. *Trans. Am. Foundrymen's Soc.* 65: 267–75

Wieser P F 1984 *Filtration of Liquid Steel*, TMS Paper F84-4. American Institute of Mining, Metallurgical and Petroleum Engineers, New York

P. F. Wieser

Geochemical Distribution of the Elements

The manner of occurrence, distribution and abundance of almost all the naturally occurring chemical elements in the accessible parts of the Earth are well known. This has come about largely as a result of the researches initiated by Victor Moritz Goldschmidt, summarized in his posthumous text (Goldschmidt 1954) and further updated in the *Handbook of Geochemistry* (Wedepohl 1969-79). The latter allocates a chapter to each element or to coherent groups of elements (e.g., lanthanides) and should be consulted for details beyond the scope of this article.

Goldschmidt's work, and that of his colleagues and students, was influenced profoundly by earlier and contemporary research by F W Clarke in the USA and V I Vernadsky in the USSR. More recent work has established with some confidence the abundances of many elements in the solar photosphere, other stellar bodies, meteorites, the Moon, other planetary bodies and interstellar gas and dust, but only terrestrial occurrences are considered here.

Following geochemical practice, abundances of the elements are given in weight percent or more generally in weight fractions. It is convenient to divide the periodic table into:

(a) major elements, in the general range 1–100%;

(b) minor elements, in the range 0.1–1%; and

(c) trace elements, at lower concentrations, often expressed as parts per million or grams per tonne, but here (Table 1) expressed for convenience in floating point fractional notation.

Questions of analytical accuracy and precision are not considered here, but the number of digits quoted in the tables gives an indication of the confidence to be attributed to the values; in some cases (Table 2) average values are taken from the original references without rounding. In general the least confidence is to be attributed to abundance values for trace elements, but this is not universally the case: some rare elements (e.g., U) are well characterized and some commoner ones (S, Cl) are not.

Familiarity of an element is of course no guide to its abundance. Some of the rarest elements are familiar to all (Au, Hg, Pt). Some of the commonest are unknown except to scientists and engineers (Rb, V, La); such elements largely owe their unfamiliarity to their manner of occurrence as dispersed elements, forming few natural chemical compounds in which they are enriched. Many natural compounds are familiar to the public without any reference to the elements composing them (emerald, asbestos, fuller's earth).

1. *Distribution of the Elements*

Certain elements are abundant in the atmosphere, in natural waters and in biological materials. The principal constituents of the atmosphere and of normal salinity ocean water are reported in Table 1, omitting reference to complexes and organic compounds. Biologically major elements are few in number and are not further considered here, although many trace elements have important roles, both essential and pathological, in biological systems.

Only a minuscule fraction of the solid Earth is accessible for sampling and direct study. Geophysical measurements nevertheless indicate the Earth to be made up of roughly concentric units (Table 3).

The nature of the core has been controversial, although its high density suggests a similar composition to the nickel-bearing iron phase of meteorites. Although an inner solid core exists (radius 1217 km), the outer core is in a liquid state. Other proposals for the possible composition of the core include condensed solar matter (mostly hydrogen) and mantle silicates transformed to the metallic state at high pressure (the pressure is ~1.3 Mbar at the mantle–core interface).

Recent measurements of bulk sound velocities in a variety of materials at ultrahigh pressures, compared with seismic information from the core, indicate the average atomic number of the outer core material to be in the range 24–26 (see Ringwood 1979 Chap. 3). Adjacent elements (atomic numbers from 23 to 29) are V, Cr, Mn, Fe, Co, Ni and Cu, of which Fe has by far the greatest abundance in stellar photospheres and meteorites and is therefore the most likely major element in the outer core. The density of the outer core is nevertheless about 11.5% lower than that of pure Fe. This and other measurements indicate the core also contains some 5–15% of elements of lower atomic weight than Fe. The most likely candidates are S, Si and O.

The mantle of the Earth comprises material above the liquid outer core and below the seismic discontinuity at the base of the crust. Its density and seismic velocities match rather well with those of compressed iron–magnesium silicates having a ratio MgO/(MgO + FeO) of ~0.9 and containing minor amounts of other components. In detail, the seismic structure of the mantle is complex and additional discontinuities occur at various depths: the lower mantle is taken as lying between depths of 2900 and 1000 km, overlain by a transition zone of some 600 km and the upper mantle, about 400 km thick. The composition of the mantle may also be expected to vary and there is much indirect evidence to suggest that the upper mantle may be rather heterogeneous in composition, though probably not so much so as the crust.

Geophysical evidence as to the nature of the mantle is supplemented by information from samples of it found within surface rocks, including:

(a) small xenoliths or inclusions brought to the surface

Table 1
Element abundances in different Earth units (weight fractions)[a]

Atomic number	Symbol	Atmosphere[b]	Sea[c]	Crust[d]	Mantle[e]
1	H		0.1082	15E-4	
2	He	73E-8	72E-13		
3	Li		18E-8	22E-6	21E-7
4	Be		60E-13	28E-7	1E-7
5	B		46E-7	1E-5	33E-8
6	C	46E-5	29E-6	52E-4	
7	N	0.755	15E-6	2E-5	
8	O	0.232	0.859	0.4688	0.439
9	F		13E-7	62E-5	163E-7
10	Ne	127E-7	12E-11		
11	Na		1076E-5	213E-4	25E-4
12	Mg		1294E-6	233E-4	0.246
13	Al		2E-9	808E-4	18E-3
14	Si		29E-7	0.2686	0.208
15	P		6E-8	66E-5	105E-6
16	S		904E-6	5E-4	319E-6
17	Cl		1935E-5	18E-5	23E-6
18	Ar	13E-3	4E-9		
19	K		399E-6	255E-4	252E-6
20	Ca		412E-6	499E-4	184E-4
21	Sc		6E-13	7E-6	11E-6
22	Ti		1E-9	50E-4	15E-4
23	V		25E-10	53E-6	84E-6
24	Cr		3E-10	35E-6	31E-4
25	Mn		2E-10	53E-5	10E-4
26	Fe		2E-9	509E-4	62E-3
27	Co		5E-11	12E-6	1E-4
28	Ni		6E-10	19E-6	24E-4
29	Cu		3E-10	14E-6	1E-5
30	Zn		3E-9	71E-6	62E-6
31	Ga		3E-11	15E-6	28E-7
32	Ge		6E-11	15E-7	85E-8
33	As		3E-9	18E-7	1E-7
34	Se		9E-11	5E-8	41E-9
35	Br		67E-6	25E-7	11E-9
36	Kr	327E-8	21E-11		
37	Rb		12E-8	11E-5	86E-8
38	Sr		79E-7	316E-6	21E-6
39	Y		1E-12	21E-6	29E-7
40	Zr		3E-11	24E-5	78E-7
41	Nb		15E-12	26E-6	6E-7
42	Mo		1E-8	15E-7	
44	Ru				6E-10
45	Rh				2E-9
46	Pd				43E-10
47	Ag		8E-12	7E-8	19E-9
48	Cd		1E-10	98E-9	4E-8
49	In		1E-13	8E-9	18E-9
50	Sn		1E-11	2E-6	5E-7
51	Sb		33E-11	2E-7	25E-9
52	Te				22E-9
53	I		6E-8	5E-7	1E-8
54	Xe	403E-9	47E-12		
55	Cs		4E-10	12E-7	19E-9
56	Ba		2E-8	107E-5	49E-7
57	La		34E-13	32E-6	89E-8
58	Ce		1E-12	66E-6	225E-8
59	Pr		6E-13		19E-8
60	Nd		28E-13	26E-6	97E-8
62	Sm		4E-13	45E-7	32E-8
63	Eu		1E-13	94E-8	17E-8
64	Gd		7E-13	28E-7	42E-8
65	Tb		1E-13	48E-8	79E-9
66	Dy		9E-13	3E-6	52E-8
67	Ho		2E-13	62E-8	13E-8
68	Er		9E-13	28E-7	34E-8
69	Tm		2E-13		48E-9
70	Yb		8E-13	15E-7	33E-8
71	Lu		1E-13	23E-8	52E-9
72	Hf			58E-7	23E-8
73	Ta			2E-6	4E-8
74	W		4E-12	15E-7	31E-8
75	Re				4E-10
76	Os				51E-10
77	Ir			24E-12	36E-10
78	Pt				7E-8
79	Au		5E-12	18E-10	27E-10
80	Hg		2E-11	9E-8	8E-8
81	Tl		1E-11	75E-8	11E-9
82	Pb		3E-11	21E-6	2E-7
83	Bi		2E-11	13E-8	1E-8
90	Th		15E-13	1E-5	8E-8
92	U		33E-10	26E-7	2E-8

a Concentrations exceeding 10 wt% are given as weight fractions; remainder are in floating-point notation (e.g., 13E-7 denotes 13×10^{-7} weight fraction) b Holland (1978, Chap. 6), Smith (1977, Table 2) c Holland (1978, Chap. 5) except H, O (Smith 1977, Table 2) d Heinrichs et al. (1980), Shaw (1980, Table 1), Smith (1977, Table 2) e Heinrichs et al. (1980), Morgan et al. (1980a, 1980b), Jagoutz et al. (1979), O'Nions et al. (1979), Shaw (1980, Table V), Smith (1977, Table 2), Taylor (1980, Table 1), Wood (1979).

(i) by erupting basaltic lavas, or
(ii) in diamondiferous kimberlite pipes; and

(b) larger masses outcropping over many km^2 and believed to represent wedges or slices of upper mantle, intruded in mountain belts (Alpine peridotites).

The principal mineral components of these rocks are olivine, orthopyroxene, clinopyroxene and in some cases garnet, spinel and other minerals. When fresh, such rocks are called peridotite; hydrous alteration converts them to serpentine. Their major elements are O, Si, Mg, Fe, Al, Ca and their minor elements are Cr, Ni, Na, Ti, Mn, K, both groups in order of decreasing abundance.

Since the mantle constitutes >67% of the mass of the Earth, it is the most important reservoir of most of the metallic elements and many of the nonmetallic ones. It is probably the source from which the crust

Table 2
Major element composition of important rocks, expressed mostly as oxides (wt%)

	GR	GD	AND	MORB	SH	SST	LST	UCC	UM
SiO_2	72.08	66.88	59.5	49.14	58.9	78.7	6.9	64.93	44.66
TiO_2	0.37	0.57	0.70	1.17	0.78	0.25	0.05	0.52	0.19
Al_2O_3	13.86	15.66	17.2	15.64	16.7	4.8	1.7	14.63	2.80
Fe_2O_3	0.86	1.33		2.64	2.8	1.1	0.98	1.36	
FeO	1.67	2.59	6.10	6.66	3.7	0.3	1.3	2.75	7.92
ΣFeO				9.14					
MnO	0.06	0.07	0.12	0.16	0.09	0.03	0.08	0.068	0.12
MgO	0.52	1.57	3.42	8.22	2.6	1.2	0.97	2.24	40.97
CaO	1.33	3.56	7.03	11.84	2.2	5.5	47.6	4.12	2.59
Na_2O	3.08	3.84	3.68	2.40	1.6	0.45	0.08	3.46	0.30
K_2O	5.46	3.07	1.60	0.20	3.6	1.3	0.57	3.10	0.12
H_2O+	0.53	0.65		0.75	5.0	1.3	0.84	0.79	
H_2O-				0.34				0.13	
P_2O_5	0.18	0.21		0.12	0.16	0.08	0.16	0.15	
S				0.04	0.24		0.11	0.06	
Cr_2O_3				0.04					0.45
CO_2					1.3	5.0	38.3	1.28	
NiO									0.31
F								0.05	
Total	100.00	100.00	99.35	99.36	99.67	100.01	99.64	99.588	100.43

GR average of 72 granites (Nockolds 1954); GD average of 137 granodiorites (Nockolds 1954); AND average of andesite (Taylor 1968); MORB average of 387 mid-ocean-ridge basalts (Wedepohl 1981); SH average of 277 shales (Wedepohl 1969); SST average of 253 sandstones (Clarke in Wedepohl 1969); LST average of 93 limestones (Wedepohl 1969); UCC upper continental crust (Shaw et al. 1967); UM upper mantle, as average of 20 garnet peridotites (Carswell 1968)

of the Earth formed early in its history and from which additions to the crust are still provided, in the form of volcanic magmas which generate lava and ash deposits. These magmas form by partial melting, in part of mantle rocks and in part of recycled crustal materials, carried back down into the mantle by plate tectonic processes. For these reasons it is important to attempt estimates of elemental abundances in the mantle. Such estimates are obviously of limited value, being made on sparse sampling of those surface rocks which are believed to be mantle-derived, but they are better than nothing. The values provided in Table 1 should be considered appropriate to the upper mantle, since even less is known of the lower mantle. The number of digits in these values relates to precision of analysis more than to the range in composition between mantle samples of different mineralogy.

Table 3
Structural make-up of the Earth, according to geophysical evidence

	Radius/ thickness (km)	Mass (10^{25} g)	Mean density (g cm^{-3})
Core	3471	192	11.0
Mantle	2880	403	4.5
Crust	20	2.5	2.8
Oceans	3.8	0.14	1.03
Whole earth	6371	597.6	5.52

As regards the crust of the Earth, it is evident that there is a great diversity in the rocks exposed at the surface, probably of much greater degree than within the upper mantle. This arises from the fact that crustal rocks include a variety of low-temperature products, formed by interaction of igneous rocks with the atmosphere and hydrosphere. Such sedimentary rocks may persist for long periods or, in certain cases, are subjected to alteration produced by high temperature and pressure. This alteration, or metamorphism, may also affect igneous rocks produced from magma, giving rise to metamorphic rocks. In extreme cases metamorphism may lead to partial melting, giving rise to regenerated magma of various kinds.

Among the great diversity of crustal rocks there are a number of commoner varieties. These include granite, granodiorite, andesite, basalt, shale, sandstone and limestone. The average abundances of major and minor elements in these rocks are given in Table 2, mostly expressed as oxides. In addition there is an estimate of the upper mantle and an estimate of the average surface composition of a large area of the Canadian Precambrian shield, which has been shown to be remarkably close to similar measurements for several other Precambrian shields. Such data are commonly taken as representing the composition of the continental crust, though it is acknowledged that they are of course surface estimates and do not take account of possible composition changes with depth. It is inferred that composition does change with depth for some elements, both from studies of rocks believed to represent lower crustal varieties and by comparing surface heat flow with surface abundances of U, Th and K, which generate heat by radioactive decay.

The crustal abundances of trace elements in Table 1 are representative only of the continental crust. The oceanic crust is both thinner and less variable in composition than the continental crust. Apart from a thin veneer of sediment at the surface, the bulk of the ocean crust is constituted by basaltic rocks (column headed MORB, Table 2) of fresh or altered character.

By far the major parts of both continental and oceanic crust are composed of the rocks whose composition is shown in Table 2 and by related varieties of intermediate composition. Economic mineral and fuel deposits are of negligible dimensions and have been produced by special conditions working at locally favorable situations, which resulted in abnormally high enrichment processes.

2. *Principles of Element Distribution in the Solid Earth*

The major achievement of Goldschmidt (1954) and his successors was to demonstrate that element distributions exhibit regularities based for the most part on simple chemical and crystallographic properties.

The principal rock types composing the crust and mantle are aggregates of silicate minerals, with smaller amounts of oxides, carbonates, phosphates and other compounds. The major rock-forming silicate minerals are listed in Table 4, with their general formulae. X-ray diffraction and crystal chemistry have demonstrated that minerals of this kind form extensive solid-solution series and are largely indifferent to the chemical nature of the cation components, so long as these ions are of the right size for a particular crystallographic site and provided that charge differences can be compensated so that the structure is electrically neutral. The right size is determined by the number of coordinating oxygen ions, but a purely ionic model is too simple and has to be modified to some degree for bonding polarization effects.

This behavior explains the main principles of cationic element distribution. The bulk composition of a particular chemical system which crystallizes to form an igneous, sedimentary or metamorphic rock determines the distribution of the elements within the rock. The particular mineral phases to crystallize are determined by the major elements present and by the phase relations appropriate to the system, but external factors largely control the crystallization process (equilibrium or fractionation, partial or complete, for example). The trace elements (or non-structure-determining elements) are picked up and held in appropriate crystal sites under the thermodynamic and kinetic control of solution chemistry, according to ionic size (Table 5). Since fractional crystallization is common in natural processes, minerals formed at different stages (early, middle, residual) may differ in kind or, for a given species, in chemical composition.

An example of this process is the typical crystallization of basaltic magma, widely documented and illustrated by the typical trace element abundances in Table 6. The early prôducts of fractional crystallization are peridotitic, consisting largely of

Table 4
Structural formulae of the common silicates (generalized and simplified)[a]

Olivine	Y_2ZO_4	Z mostly Si
Clinopyroxene	XYZ_2O_6	
Rhombic pyroxene	$Y_2Z_2O_6$	Z mostly Si
Clinoamphibole	$(OH,F)_2X_{2\text{-}3}Y_5(Z_4O_{11})_2$	
Mica	$(OH,F)_2Y_{2\text{-}3}Z_4O_{10}\cdot W$	W mostly K
Chlorite	$(OH)_8Y_6Z_4O_{10}$	
Chloritoid	$(OH)_4Y_2Z_4O_{10}$	Y mostly Fe, Al
Staurolite	$(OH)Y_{11}Z_4O_{23}$	Y approximately Fe_2Al_9
Garnet	$X_3Y_2Z_3O_{12}$	garnet is unusual in that Fe, Mg and Mn occur in octahedral (X) sites
Cordierite	$Y_2Z'_3Z''_6O_{18}$	Z′ is Al; Z″ is mainly Si
Alkali feldspar	$(X,W)Z_4O_8$	Z_4 usually $AlSi_3$
Plagioclase feldspar	XZ_4O_8	mostly a mixture of $NaAlSi_3O_8$ and $CaAl_2Si_2O_8$
Epidote	$(OH,F)X_2Y_3Z_3O_{12}$	Y is normally restricted to Al and Fe
Nepheline	$WX_3Z_8O_{16}$	W mostly K; X mostly Na
Scapolite	$W_4Z_{12}O_{24}(Cl_2,CO_3,SO_4)$	
Sillimanite, kyanite, andalusite	Y_2ZO_5	Y is all Al except in andalusite, which may be ferrian

a W,X,Y,Z denote cations in coordination >8,8,6,4, respectively

Table 5
Some ionic radii in groups of similar size

Symbol	Coordination	Size (Å)	Ion[a]
Z	4	0.55–0.64	Si^{4+}, (Al^{3+}), (Ti^{4+}), Ge^{4+}
Y	6	0.65–0.90	Mg^{2+}, Fe^{2+}, Fe^{3+}, Al^{3+}, Li^{+}, Ti^{4+}, Mn^{3+}, Ni^{2+}, Co^{2+}, Cr^{3+}, V^{3+}, Zn^{2+}, Zr^{4+}, Sc^{3+}, Sn^{4+}, Ga^{3+}, Mo^{4+}, Nb^{5+}, Ta^{5+}, (Mn^{2+})
X	8	0.90–1.22	Ca^{2+}, Na^{+}, Mn^{2+}, (K^{+}), (Sr^{2+}), (Pb^{2+}), La^{3+}, Y^{3+}, $Ce^{3+} \rightarrow Lu^{3+}$
W	8–12	1.20–1.65	K^{+}, (Na^{+}), Rb^{+}, Cs^{+}, Ba^{2+}, Sr^{2+}, Pb^{2+}

a Elements in brackets are transitional and may be found in two coordinations

Table 6
Typical trace element[a] abundances (ppm) in the products of igneous fractional crystallization

	Peridotite	Basalt	Granite
Cr	3100	120	10
Ni	2400	50	5
Co	150	50	5
Zr	8	100	300
U	0.1	1	5
La	3	15	100
Ba	0	680	1300
Rb	1	30	250

a The elements are arranged in sequence of increasing size from top to bottom

olivine, pyroxenes and perhaps some plagioclase, removing preferentially the major elements Mg and Ca. Crystal sites in olivine and pyroxenes favor sixfold coordinated cations such as Cr, Ni and Co, which therefore accumulate in such rocks. The main phase of crystallization typically shows pyroxenes and plagioclase precipitating, depleting the magma in Mg, Ca and Fe but enriching it in Na, K and Al. Trace components such as the ferrides continue to be accepted in the minerals, but at lower concentrations.

The later stages of crystallization are characterized by sodic plagioclase, alkali feldspar, amphibole, micas and quartz. These minerals, except the last, have larger cation sites and so accommodate the larger cations such as La, Ba, Rb and Cs which have been excluded (by their size) from the earlier mineral phases. In addition, small amounts of accessory minerals rich in some minor or trace element (e.g., zircon, $ZrSiO_4$) also now begin to form. Throughout the crystallization process Fe–Ti oxides and phosphate may also crystallize; the former picks up other ferrides, while the latter takes in the rare earth elements and Sr.

The remark made earlier, that the chemical nature of a cation plays no role in the process just described, is of course an oversimplification. When other anions are present in the system, such as S, certain metals preferentially bond with them rather than with O. Some transition metals, such as Fe, Ni, Co, Cu, Zn and Pb, characteristically form sulfides, and such processes dominate their occurrence as economic deposits (except for Fe). Both in these cases and in ordinary magmatic crystallization the redox potential of the system greatly affects the behavior of polyvalent metals.

The distribution of elements may also follow different patterns in aqueous environments, in minerals formed from the sea, from aqueous alteration of preexisting rocks, from hot springs and from hydrothermal waters percolating through rock pores and fissures. In addition to redox potential, pH is of great importance here, and in sedimentation processes so also is adsorption of metals (or their complexes) on fine-grained particles. In products of chemical sedimentation, such as carbonates, clays, phosphorites, bauxites and Fe oxides, these minerals also take up trace cations of appropriate size and charge to substitute for major cations, just as in silicates.

In general, therefore, the distribution of elements in rocks is controlled by the principles:

(a) that major elements determine the mineral structures that can form; and

(b) that every natural mineral is impure, but the impurities are determined by thermodynamic properties, kinetic limitations and availability of the trace impurities.

3. *Economic Deposits*

It is appropriate to consider the relationship between the crustal abundance of an element and the concentration in a host rock at which the rock would be commercially exploitable as an ore deposit. For certain metals (Au, Hg, Cd, Bi, Ag, Sb, Mo, As, Pb, Cu, Zn, P, K, Fe), it has been shown by McKelvey (1974) that the domestic reserves in the USA in tonnes are in fact in the range 10^9–10^{10} times the percentage abundance in the crust.

This apparent relationship suggests, however, a more direct expression of economic value than in fact exists, and suggests also that the supply of every element is virtually unlimited, so that if demand increases it can always be met by exploiting a lower-grade source. Such a cornucopian view of resources has been subjected to close analysis by Cloud (1975), who has identified the flaws inherent in this view of mineral wealth and has shown how energy expenditures, population pressure and social attitudes affect the expression of "need" for raw materials, and also that sources are not in fact unlimited in practical terms.

It is consequently not easy to indicate how much of a mineral resource is necessary to make extraction worthwhile. First of all the word "worthwhile" relates to the need of a particular society at a particular time. For example, uranium was long a valueless by-product from the extraction of radium. Secondly, the costs of extraction are closely related to geography, energy availability, transportation costs, political stability and so forth, as well as to technical questions of metallurgical treatment. In Arctic Canada, for example, a lead–zinc mine must be of higher grade, or larger in tonnage, than one close to energy and markets. Thirdly, the chemical and geological nature of a metallic deposit influence its value. Some metals occur in minerals of which they are major components (e.g., Pb or Zn in simple sulfides), whereas others occur in problematical compounds at low concentrations even though in vast tonnages (e.g., U in phosphorites, Mo in black shales). Some elements form few or no minerals of their own (e.g., Rb, Hf, Br) and can be obtained only as by-products from other extractions.

But beyond these economic and technical considerations, the prime influence is the needs of society for a particular resource, and these needs continually wax and wane as a function of the current technology; for example, Li, Zr and B are currently in high demand, whereas Tl and In are not. Similarly, biogenic fuels such as hydrocarbons and coal are now of immense importance as energy sources but will eventually be largely replaced by fission and fusion power, if mankind survives.

It appears, therefore, inappropriate to try to relate elemental natural abundances to commercially exploitable ore grades.

See also: Mineral Resources: An Overview; Mineral Resources: Definitions, Uses, Classification and Future Availability; Ore Minerals

Bibliography

Carswell D A 1968 Possible primary upper mantle peridotite in Norwegian basal gneiss. *Lithos.* 1: 322–55

Cloud P E 1975 Mineral resources today and tomorrow. In: Murdoch W W (ed.) 1975 *Environment*, 2nd edn. Sinauer Publishers, Sunderland, Massachussetts, pp. 97–120

Goldschmidt V W 1954 *Geochemistry*. Oxford University Press, Oxford

Heinrichs H, Schulz-Dobrick B, Wedepohl K H 1980 Terrestrial geochemistry of Cd, Bi, Tl, Pb, Zn and Rb. *Geochim. Cosmochim. Acta.* 44: 1519–34

Holland H D 1978 *The Chemistry of the Atmosphere and Oceans*. Wiley, New York

Jagoutz E, Palme H, Baddenhausen H, Blum K, Cendales M, Dreibus G, Spettel B, Lorenz V, Wanke H 1979 The abundances of major, minor and trace elements in the earth's mantle as derived from primitive ultramafic nodules. *Proc. 10th Lunar Planet. Sci. Conf.* Pergamon, Oxford, pp. 2031–50

McKelvey V E 1974 Mineral resource estimates and public policy. In: Utgard R O, McKenzie G D (eds) 1974 *Man's Finite Earth*, 1st edn. Burgess Publishing Co., USA pp. 246–65

Morgan J W, Wandless G A, Petrie R K, Irving A J 1980a Earth's upper mantle: Volatile element distribution and origin of siderophile element content. *Proc. 11th Lunar Planet. Sci. Conf.* Pergamon, Oxford, pp. 740–42

Morgan J W, Wandless G A, Petrie R K, Irving A J 1980b Composition of the earth's upper mantle: II Volatile trace elements in ultramafic xenoliths. *Proc. 11th Lunar Planet. Sci. Conf.* Pergamon, Oxford, pp. 213–34

O'Nions R K, Evensen N M, Hamilton P J 1979 Geochemical modeling of mantle differentiation and crustal growth. *J. Geophys. Res.* 84: 6091–6101

Nockolds S R 1954 Average chemical compositions of some igneous rocks. *Bull. Geol. Soc. Am.* 65: 1007–32

Ringwood A E 1979 *Origin of the Earth and Moon*. Springer, Heidelberg

Shaw D M 1980 Development of the early continental crust. Part III. Depletion of incompatible elements in the mantle. *Precambrian Res.* 10: 281–99

Shaw D M, Reilly G A, Muysson J R, Pattenden G E, Campbell F E 1967 The chemical composition of the Canadian precambrian shield. *Can. J. Earth Sci.* 4: 829–54

Smith J V 1977 Possible controls on the bulk composition of the earth: Implications for the origin of the earth and moon. *Proc. 8th Lunar Planet. Sci. Conf.* Pergamon, Oxford, pp. 333–69

Taylor S R 1968 Geochemistry of andesites. In: Ahrens L E (ed.) 1968 *Origin and Distribution of the Elements*. Pergamon, Oxford, pp. 559–83

Taylor S R 1980 Refractory and moderately volatile element abundances in the earth, moon and meteorites. *Proc. 11th Lunar Planet. Sci. Conf.* Pergamon, Oxford, pp. 333–48

Wedepohl K H (ed.) 1969–1979 *Handbook of Geochemistry*. Springer, Heidelberg

Wedepohl K H 1969 Composition and abundance of common sedimentary rocks. *Handbook of Geochemistry*, Vol. 1. Springer, Heidelberg, pp. 250–71

Wood D A 1979 A variably veined suboceanic upper mantle—Genetic significance for mid-ocean ridge basalts from geochemical evidence. *Geology* 7: 499–503

D. M. Shaw

Geochemical Exploration for Mineral Deposits

Geochemical exploration is the systematic sampling and analysis of natural materials to detect anomalous concentrations of chemical elements derived from mineral deposits. The materials commonly sampled are rocks, soils, stream sediments, vegetation and water. The chemical elements analyzed may be the ore metals or other elements that are associated with a mineral deposit. Element concentrations are plotted on maps and contoured at various intervals to delineate anomalous areas that are favorable for ore deposits. Geochemical prospecting can thus be thought of as a means of extending our ability to detect mineral deposits by using modern analytical techniques. As geochemical exploration methods are nondestructive to the environment, they are ideally suited for use in modern mineral exploration programs.

1. *Geochemical Anomalies*

1.1 Primary Processes and Anomalies

Primary geochemical anomalies are those formed when an ore deposit is emplaced. The ore metals or their compounds are transported in hydrothermal solutions, until changing physicochemical conditions result in the precipitation of new minerals that form the deposit. These may be deposited in veins, stockworks, open fillings or replacements, or may be disseminated through large volumes of rock. Although there are many other types of ore deposits (e.g., laterites, sedimentary deposits, pegmatites and magmatic segregations), hydrothermal deposits are the most common, most widely distributed and most amenable to discovery by geochemical exploration methods.

Although an ore deposit is limited to a volume of rock from which metals can be economically extracted, lesser concentrations of the metals or associated elements are transported for considerable distances into the surrounding rocks, forming halos around the deposit. These halos or geochemical anomalies are commonly much larger than the deposit, and extend both laterally and vertically. This is illustrated by a geochemical study done around a porphyry copper deposit at Ely, Nevada, USA. The main ore metal, copper, is concentrated in and near the quartz–monzonite intrusive rock, and forms the core of the deposit. Molybdenum also occurs in the core rocks in much smaller amounts. Gold, in small amounts, is found just outside the central core, while zinc and lead extend even further out. More volatile (or mobile) elements, such as As, Ag, Sb, Hg, Tl, Te and Mn, are enriched in halos further out from the core, resulting in zonal patterns. These zones are not sharply defined, but overlap.

1.2 Secondary Processes and Anomalies

After erosional processes have removed most of the overlying rock from an ore deposit, weathering processes begin to alter its physical and chemical makeup. Ore deposits exposed at the surface are dispersed by weathering, and the constituent minerals and trace elements are transported downstream to produce anomalies that can be detected in the sediments for many kilometers. Transportation can occur either through suspension of fine grains, or through dissolution of the minerals. Thus, sampling stream sediments can effectively indicate mineralized areas upstream.

In some cases, nonore elements associated with the ore are more mobile than the ore elements, and produce more extensive, broader anomalies, which are more useful in exploration. Geochemical anomalies in soils result from weathering and disintegration of the bedrock. Where the bedrock is mineralized, anomalous concentrations of the ore elements are commonly found in the overlying soil.

Stream or lake sediment, water and vegetation anomalies are also secondary. They are usually weaker in amplitude than bedrock anomalies, owing to dispersion and dilution. Most geochemical surveys conducted today are based on the detection of secondary anomalies.

2. *Geochemical Surveys*

Geochemical exploration surveys are used for both regional and local prospecting. Regional or reconnaissance prospecting is conducted to detect mineral provinces or deposits, whereas local geochemical surveys are conducted when a single deposit or mineralized area is sought and usually cover only a few square kilometers.

Stream sediments are effective for regional exploration, because the entire area of a drainage basin can be assessed with carefully selected samples. In the USSR, geochemical sampling of soil is done routinely in conjunction with other geological studies to prospect regions covering hundreds of thousands of square kilometers.

In local geochemical surveys, rock or soil samples are preferred to other media, and sample spacing may be only a few meters. The disseminated gold deposit at Cortez, Nevada, USA was discovered by sampling surface rocks, and soil sampling has been successfully used for exploration in all parts of the world.

Large disseminated deposits such as molybdenum or porphyry copper deposits may be detected by sample spacings as great as 1 km, but more commonly samples are taken at intervals of ten to a few hundred meters. When vein deposits are suspected, smaller sample intervals may be required. Traverses are usually made normal to the strike of the vein or tabular bodies, or on a grid. The sample interval can be determined by doing an orientation study using close sample spacing, and plotting the analytical data to show the extent of the geochemical halo. In general, the smaller the body of exposed ore, the shorter the sample interval required.

Rock samples have the distinct advantage of providing geochemical data on primary dispersion halos.

Sampling quartz or other material in veins or fractures will show the extent of the hydrothermal "plumbing" system.

Geochemical anomalies in soils are often broader than the underlying mineralized area. Where the overburden is transported (e.g., glacial till or loess), surface soil sampling may be of limited usefulness. In Scandinavia, mineralized boulders in glacial till have been successfully used to find ore in bedrock.

Not uncommonly, the principal ore metal will have a less extensive or more subdued geochemical anomaly than other, associated trace elements. This pattern is exemplified by a geochemical survey for cobalt deposits in northern Idaho, USA. The cobalt occurred as cobaltite (CoAsS), but the ore also contained abundant pyrite. Soil samples were collected and analyzed for cobalt and arsenic. The cobalt anomalies were less useful in outlining the deposits than were the arsenic anomalies, because the acid solutions generated by the dissolution of pyrite during weathering leached the cobalt out of the surface soils, leaving the arsenic behind. In addition to cobalt, copper and zinc tend to be depleted in surface material during weathering.

Where soils are on flat terrain, geochemical anomalies will be found directly over the ore in the bedrock. Where the terrain is hilly, geochemical anomalies will be transported downslope by soil creep. Soils developed on transported overburden, such as alluvium or glacial till, may give false anomalies that reflect mineralized rock in the overburden, but not in the bedrock.

2.1 Coordination with Other Methods of Exploration

Other exploration methods are commonly used in conjunction with geochemical methods. Airborne magnetic surveys are often used in the early stages of exploration. Large tracts of land can be evaluated rapidly, and favorable target areas can be identified for later detailed ground exploration. Widely spaced geochemical sampling accomplishes the same purpose. Sophisticated techniques for analyzing satellite imagery reveal areas of subtle alteration, which may be mineralized.

2.2 Climate in Study Area

Climate influences the dispersion of trace elements and determines the type of sampling and the trace elements that are likely to be most useful in tracing mineralization. In arid climates, dispersion of trace elements will be largely by physical transport, whereas in areas of considerable precipitation these elements will be dissolved and transported in water. If a stream sediment survey is done in an arid environment, the trace elements will be largely confined to discrete mineral grains in the sediment. Orientation studies of stream sediments in arid environments have shown that a coarse fraction (e.g., 600–180 μm) is often most useful in showing anomalies.

In wet environments where chemical weathering and dispersion predominate, the fine fraction of the stream sediment (e.g., $<180\ \mu m$) usually produces the best anomalies, because fine particles and clays have a greater surface area and readily adsorb metal ions or compounds from water. The adsorption of iron and manganese on fine particles results in enrichment of their oxides with concomitant adsorption of many trace elements, and frequently results in enhanced anomalies. Although these rules-of-thumb are helpful, an orientation study should be conducted to determine the best size fraction for the survey. Water sampling may also be done in these areas.

2.3 Surface Exposures

Rocks or minerals in mineralized districts or provinces are often enriched in trace elements and can be used to delineate these areas. Analyses for trace metals in specific minerals (e.g., biotite) in intrusive rocks have been used to indicate the favorability for porphyry copper provinces in regional exploration.

2.4 Planning a Survey

All geochemical surveys involve sampling and chemical analysis. Sample locations are plotted on topographic maps or on aerial photographs. In reconnaissance (regional) prospecting, stream sediments are the most widely used sample medium, because the mineral potential of large areas can be assessed with few samples. Although less used, lake-bottom sediments have the same advantage. In reconnaissance surveys with widely spaced sample sites (one sample per 5–20 km^2), about 25 samples per day may be collected by one sampler or sampling team. In local surveys as many as 100 samples per day may be collected.

Perhaps the most important aspect in planning geochemical sampling is to be consistent, as some elements are more concentrated in one rock type or soil horizon than in another. Therefore, it is necessary to sample consistently to obtain meaningful anomalies.

2.5 Rock and Minerals

Although nonconsolidated media (e.g., residual soil) may reflect primary halos, weathering removes the more soluble elements and may obscure primary zoning patterns. Selective altered or mineralized material reflects the hydrothermal plumbing system. Vein material, fracture coating and jasperoid are examples of this type of material. Individual minerals in rocks are sometimes found to contain anomalous concentrations of trace elements in the vicinity of ore deposits. In the southwestern USA porphyry copper province, biotite in ore-related intrusives contains several hundred ppm Cu or more, whereas whole-rock analysis reveals no such anomaly. Mineral analysis may be most useful in identifying large regions favorable for mineralization. Manganese and iron oxides tend to coprecipitate or adsorb many trace metals. As sample media, these take advantage of a natural process of concentration and may reveal ore-related anomalies that whole-rock analysis fails to detect.

2.6 Soil

Soil samples are used more for local than for regional exploration. Residual soils reflect the underlying bedrock, and the mineral deposits it contains. Some ore elements are readily leached from soils by the action of surface water, and therefore analyzing the soils for indicator elements is often more appropriate for locating mineral deposits. Chemical analysis shows the best horizon to sample, namely the one where the anomaly-to-background ratio is greatest. The element or elements best correlating with mineral deposits are then selected for the soil survey. Transported soils, such as those in glaciated areas, may be of little use.

2.7 Water

Water sampling is used for both regional and local exploration programs. Analysis of stream waters, like that of stream sediments, can indicate the mineral deposits of an entire drainage basin. However, trace-element concentration in water is usually very low—in the ppb range—and dilution limits the distance an anomaly may be detected downstream. Natural springs issuing from favorable stratigraphic horizons have been used to detect uranium and other metals. Well samples have been successfully used to locate molybdenum deposits in arid environments.

2.8 Vegetation

Plants are used in exploration in several ways. Essential nutrients for plants include trace elements derived from the soil and rock on which they grow. Hence, the presence of certain plant species indicates that specific trace elements occur in the soil. This characteristic has been used in the Colorado Plateaus province in the USA to prospect for selenium-rich uranium ores, by mapping the distribution of the plant *Astragalus*.

More commonly, plants are sampled and chemically analyzed for trace elements like other sample media. Widespread root systems may extend to considerable depth, and elements dissolved in ground water are taken up and deposited in the plants. Thus, geochemical anomalies in plants reflect mineralized areas in the same manner as soils, rocks or water.

Another manner in which plants are used for geochemical exploration is to examine changes in plants that are caused by abnormal concentrations of metals and other elements. Changes in color, such as yellowing of leaves or foliage deformities, can sometimes be seen in aerial photographs or satellite imagery, and can indicate mineralized areas.

2.9 Other Media

Airborne dust from minerals exposed at the surface has been collected and analyzed in South Africa, and geochemical anomalies in these particulates are found to correlate with base metal deposits. Other particulates are organic, and are given off from the surface of vegetation. These also contain anomalous concentrations of trace metals, which correlate with exposed and buried mineral deposits.

Anomalous concentrations of He, CO_2, SO_2, H_2S, COS, CH_4 and other gases are found in soil over buried deposits. Gas chromatographic or mass spectrometric techniques are used to analyze constituent gases. A soil gas survey over a massive sulfide deposit in glaciated terrain of the northern USA showed anomalous concentrations of CO_2 and hydrocarbon gases over the deposit; these anomalies were detected through as much as 60 m of transported glacial overburden.

3. Sample Analysis

Analytical methods useful in geochemical exploration must (a) have sufficient sensitivity to obtain positive values on most samples, (b) be rapid, and (c) be sufficiently reproducible to distinguish relatively small differences in trace element concentration. The methods in common use are semiquantitative, and give results that are $\pm 30\%$ of the true value. The absolute concentration of an element is usually of less importance than the differences between samples. The element or elements selected for analysis will depend on the type of mineral deposit sought, the sample media and the local weathering environment. Analyzing the samples for a single element may define a deposit, but in many cases, analyses for several elements more fully define the type and extent of the deposit. An example is the disseminated gold deposits in the Basin and Range province of the western USA, where the elements Au, As, Sb, Hg and W are characteristic.

3.1 Colorimetric Methods

Colorimetric methods were among those first used in geochemical exploration. The trace element concentration of an element is determined by observing the change in color of an organic reagent; the change being proportional to the trace element content of the sample. Colorimetric methods have been used for the following individual elements: Cu, Pb, Zn, Mo, W, V, Mn, Fe, Co, Ni, Ge, As, Se, Nb, Ag, Sn, Sb, Te, Au, Bi and U.

Some of these methods have been adapted to the analysis of stream sediment or soil without a pretreatment or sample digestion step; these are called cold-extraction methods. They require little equipment and few reagents; they are convenient and find extensive use in onsite soil or stream sediment analyses.

3.2 Emission Spectrographic Methods

These methods can be used for the simultaneous determination of 30 or more elements, and are often used when multielement analysis is sought. The limit of sensitivity is in the low ppm range for many elements, including Cu, Pb, Ag, Mn, B, Ba, Be, Bi, Co, Cr, Mo, Ni, Sc, V, Y and Zr. The advantage of emission spectrographic methods is that many elements can be determined simultaneously; however, the instruments are costly and require skilled operators. As many as 50 samples can be analyzed per day by one spectrographer. It is necessary to have a finely divided, homogeneous sample, since a very small sample (10 mg) is used for the analysis.

3.3 Atomic Absorption Methods

These methods have largely superseded colorimetric methods of analysis because of their greater sensitivity and specificity. Most atomic absorption instruments determine one element at a time, although a single sample solution can be used to determine several elements. The sample is usually digested in acid, and is introduced into a flame that produces a fine spray with large numbers of atoms in the atomic state. A second method used to atomize samples is direct evaporation of the element at elevated temperatures. These methods are termed flame and flameless atomic absorption, respectively.

The advantages of these methods are their great sensitivity, freedom from interferences and relative simplicity of operation. Instrument costs are moderate to high, but highly trained personnel are not required. They are currently the preferred methods of analysis in most geochemical laboratories.

3.4 Other Methods

In addition to the commonly applied methods given above, others used include x-ray fluorescence spectrometry, partial extractions, and several methods with limited applications. Radiation detection instruments, such as Geiger counters, are used in exploration for U and Th. Portable instruments containing radioactive sources have been used to detect Be and Ag. Specialized instruments are also used to detect volatile elements (e.g., Hg) or compounds in air or soil gas. These include gas chromatographs, mass spectrometers and mercury detectors.

4. Data Handling and Interpretation

Geochemical data are best shown by plotting the analytical data for each element on a map of the area, and drawing contours to show the geochemical anomalies. Contours are drawn on the map to enclose different ranges of element concentration, thus delineating the geochemical anomalies.

The values or ranges to be contoured are selected by examination of all of the analytical data. Statistical and graphical techniques are used to determine the range and distribution of each element. The background or threshold value for an element is the concentration that marks the dividing line between normal and anomalous geochemical values in a group of samples. This value varies from area to area and largely depends upon lithology.

Geochemical exploration programs, either local or regional, frequently involve the collection of hundreds or even thousands of samples, and analysis for several elements results in a large amount of data. Computers are being increasingly used to store and analyze the data, and even to plot the geochemical maps. The use of computers not only saves a great deal of time, but provides the geologist with powerful programs which greatly facilitate interpretation.

5. Costs of Geochemical Surveys

The costs of geochemical surveys are lower than those of many other exploration methods. Commonly the cost of sample collection is about half of the total, the other half being for analysis and interpretation of data. In countries where the cost of trained personnel is high, sophisticated machines and instruments are employed to minimize the number of skilled people utilized. In countries where the cost of manpower is low, more people are employed to conduct geochemical programs. Thus the local availability and cost of manpower and equipment should be considered in planning exploration programs.

Sample preparation and analysis can also be done with emphasis on greater use of manpower or instrumentation. The decision to build and equip a geochemical laboratory will depend on the size and extent of the exploration program and on the availability of skilled manpower and equipment. In areas where little or no prospecting has been done and mineralized outcrops exist, simple methods of sampling and analysis may be used to discover ore deposits.

6. Environmental Impact

There is an ever-increasing concern about the effect of exploration on the environment. In the past, exploration programs often involved building new roads and clearing brush from lines with a bulldozer. These practices are now being restricted in many countries. One example is new guidelines for exploration enacted by Botswana in 1977. At least 85% of the land surface in Botswana is dune sand, a surface that is particularly vulnerable to surface disturbances which accelerate erosion. The new act specifies exploration methods that may be used, and geochemical exploration is one of the accepted (specified) methods.

7. Other Uses

Although the primary use of trace element geochemical data is for mineral exploration, these data have also found application in other fields. A correlation has been noted between incidence of certain diseases in man and animals and the trace element content in foodstuffs and water. It is also becoming increasingly evident that many trace elements are essential to human health. Thus, in addition to exploration for minerals, geochemical data can be used to evaluate the effects of geology on human health.

See also: Geochemical Distribution of the Elements; Geological Exploration for Mineral Deposits; Geophysical Exploration for Mineral Deposits; Remote Sensing in the Search for Mineral Deposits

Bibliography

Beus A A, Grigorian S V 1977 *Geochemical Exploration Methods for Mineral Deposits.* Applied Publishing, Wilmette, Illinois

Brooks R R 1972 *Geobotany and Biogeochemistry in Mineral Exploration.* Harper and Row, New York

Levinson A A 1974 *Introduction to Exploration Geochemistry.* Applied Publishing, Calgary, Canada

Rose A W, Hawkes H E, Webb J S 1979 *Geochemistry in Mineral Exploration*, 2nd edn. Academic Press, New York

Siegel F R 1974 *Applied Geochemistry.* Wiley-Interscience, New York

J. H. McCarthy Jr.

Geological Exploration for Mineral Deposits

Geological exploration is the fundamental method of searching for ore deposits. Based on geological concepts and field observations, it provides the context for supporting work in remote sensing, geophysics and geochemistry (see *Remote Sensing in the Search for Mineral Deposits; Geochemical Exploration for Mineral Deposits; Geophysical Exploration for Mineral Deposits*).

Exploration may begin with the reconnaissance of a metallogenic province, or favorable region for a particular kind of orebody, and may involve examination of a single prospect or of a few known occurrences of ore minerals. Exploration may also be initiated within a mining district where the characteristics of recently discovered or nearly depleted orebodies provide incentives and geological guidelines for finding additional ore. Regardless of the initial step, an exploration effort is predicated on a conceptual model or image of the orebody that is being sought.

Based on genetic and environmental characteristics of well-studied orebodies in geologically similar terrain, an exploration model comprises an expected association of minerals in a particular setting of rock type, stratigraphy and geological structure. Relevant geophysical and geochemical patterns or signatures are expected; these are part of the exploration model, and they furnish means for recognizing the presence of ore mineralization.

In field exploration work, an array of signals or observations is obtained that can be related either to ordinary geological terrain or to an anomaly—a condition that may possibly be associated with an orebody. An anomaly or a group of anomalies that seem to fit a particular exploration model may derive from an orebody or from a halo of related geological features surrounding an orebody. In a succession of field investigations, anomalies are tested in relation to the exploration model, and because no two orebodies are exactly alike, the model itself is tested against new field data. The model and its working hypotheses are revised if need be, and the anomalies are evaluated by further testing until they are verified by an ore discovery, or until the search in that locality is abandoned.

1. Conceptual Models in the Subsurface Environment

Pure conceptual models of ore deposits are images of undisturbed mineral associations at a certain depth below the influence of the zone of weathering. The geological evidence for an orebody obtained from exploration at the earth's surface is more likely to have been affected by surface processes, such that all that is exposed is a weathered derivative of the deeper ore. A complete exploration model pertains to the entire system of an orebody, with its leached surface expression, with a zone of secondary or supergene mineralization resulting from near-surface processes and with the deeper primary or hypogene zone that represents the basic conceptual model.

Figure 1 shows a representative conceptual model in cross section. In this instance, the expected orebody is a hydrothermal vein-type copper deposit that has occupied a fault zone in a limestone host rock; ore has replaced some of the walls of the fault zone and also a thin zone of host rock beneath a relatively impermeable shale cover. If normal faulting is envisioned, steeper portions of the fault zone should be associated with greater widths of ore. Satellitic orebodies and ore-filled gash fractures are expected to occur in the vicinity of the main orebody.

In this model, zonation of metals and minerals is anticipated, with manganese minerals dominating the outer limits, lead and zinc minerals characterizing an intermediate zone and copper minerals predominant in the central part of the orebody. Wall rock alteration is expected to occur in a halo around the orebody, possibly in a zonal pattern with an inner zone of silica–jasperoid rock and an outer zone of recrystallized dolomite or dolomitic marble. Leakage of hydrothermal ore-forming solutions through the controlling fault zone is expected to have extended into less favorable overlying rock for some distance from the orebody, possibly to the level of the present-day surface, where it may have formed thin veinlets and weak alteration halos.

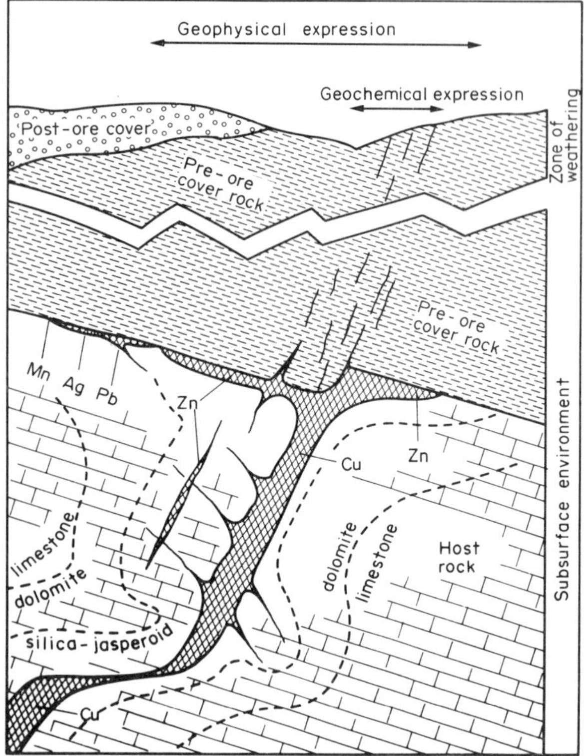

Figure 1
Representative conceptual model of a copper deposit, as used in the design of an exploration program (cross section, not to scale)

In an exploration model based on the particular conceptual model shown, a geochemical expression would be expected with the thin leakage veinlets. If a geophysical expression were to be detected at the surface, it would represent as much of the orebody, the satellitic features and the extension of the ore to depth as would come within the range of the particular geophysical method.

In a regional exploration program where several kinds of geological terrain are considered, there may be several conceptual models of ore deposits. In addition, conceptual models need not be based solely on the most thoroughly established and accepted principles of ore genesis; new theories of ore deposition can provide models for innovative exploration programs in new terrain.

2. Zone of Weathering

As a result of continual erosion, orebodies that were in equilibrium with a deeper geological environment are exposed to the surface zone of weathering and to an environment characterized by meteoric water, atmospheric pressure and atmospheric temperature. Some ore minerals such as gold and cassiterite (SnO_2), and some of the associated gangue minerals such as quartz and tourmaline, are relatively stable in the zone of weathering; most, however, are unstable and are destroyed or changed into minerals that can remain in equilibrium with the new environment.

Oxidation, the breakdown of subsurface ore minerals, takes place from the surface down to the water table, as shown in Fig. 2, which represents the conceptual model of Fig. 1 as it would appear in the zone of weathering. The depth of oxidation may extend to 30 m in temperate humid climates, and to 200 m or more in deserts. In the upper or leached part of the zone of oxidation, a cellular accumulation of iron oxide and quartz is left behind by the weathering of a sulfide ore-bearing vein. The remaining gossan does not normally contain ore minerals of the base metals, but it may contain a residual accumulation of native gold and also silver halide minerals.

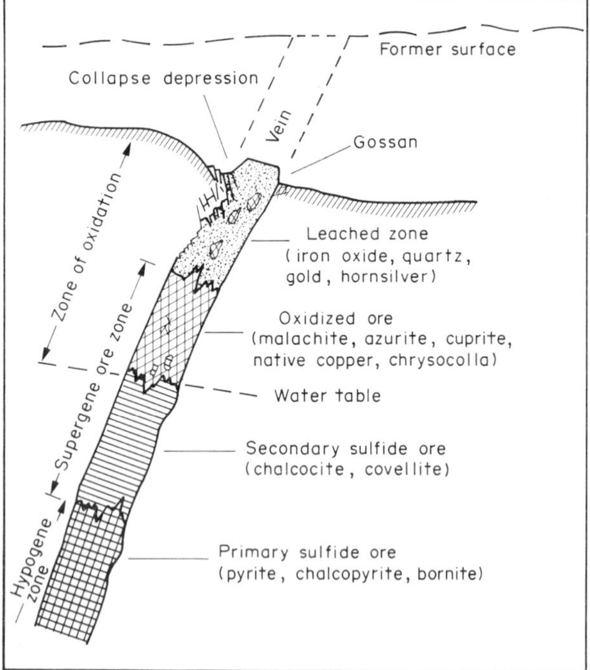

Figure 2
Diagrammatic representation of the oxidation and supergene enrichment of a copper–quartz vein with gold–silver values

In the oxidized ore zone of Fig. 2, gold, silver and base-metal mineralization occurs in the form of oxides, silicates, carbonates, elements and other minerals that have formed in situ or have been formed by supergene processes from solutions descending through the entire zone of leaching. Beginning at the water table and extending downward from a few to several hundred meters is an interval of supergene or secondary sulfide ore. Supergene ore, both in oxidized and sulfide form, commonly provides an enriched zone; in early mines this was often the only zone that was mined, because of a sharp decrease in ore grade at the top of the hypogene or primary zone.

3. Nature of Exploration

The time-honored approach to geological exploration has been to examine ore occurrences or suspected gossan reported in the literature, or prospects brought to the attention of a mining firm by prospectors or promoters. A geologist or mining engineer would investigate previous findings, project the geological conditions into a potentially larger or deeper orebody, acquire the mineral rights for his or her firm, and test the projections in new mine workings or drill-holes. The examination of prospects still has a fundamental place in mineral exploration, but it is being supplanted by, and combined with, systematic programs of stepwise investigation in broader regions and districts.

Emphasis on concept-oriented regional exploration has increased because of the rarity of ore outcrops. A regional search for hidden and blind orebodies relies heavily on geophysical and geochemical techniques, and on drill-hole investigation, with the programming of the techniques and the interpretation of the data done in the context of geology and a currently accepted exploration model.

Although artisanal mining and individual prospecting are still important elements of mineral exploration in many of the developing countries, exploration in the industrialized countries has become an effort that is more commonly carried out by large organizations. The acquisition of mineral rights from prospectors and small-scale miners continues to be a factor in all exploration, but the role of the individual prospector as a discoverer has shrunk with the depletion of near-surface high-grade ore.

4. Objectives of Mineral Exploration

An exploration program is related to objectives which are in turn related to the specific needs of an organization and to the financial risk the organization is willing to accept in searching for an orebody. The need may be for a mining opportunity in a specific region, or it may be for a particular commodity at any location in the world. A depleting orebody or an inadequate supply of mineral material for an established industrial complex may define an objective in terms of both commodity and region.

The financial risk or uncertainty inherent in exploration has several facets. It may be largely a geological risk, the 1000:1 or 10:1 probability that the anticipated orebody will not be found. It may also involve the risk that the orebody will be diminished in value or made uneconomic by low mineral prices, unfavorable political climate or overwhelming technological problems.

The decision to explore in a particular place for a particular mineral commodity is sensitive to trends in mineral markets, access to geographical regions and new concepts in science and technology.

5. Exploration Programs and Patterns

Figure 3 shows the stages involved in a regional exploration program. In its preliminary or design phase, an exploration program is organized and budgeted on the basis of an objective, a region and an exploration model. During this phase, metallogenic maps provide indices for the selection of a favorable region, and orientation surveys provide information for use in the planning of appropriate geochemical and geophysical techniques (see *Metallogenic Provinces*).

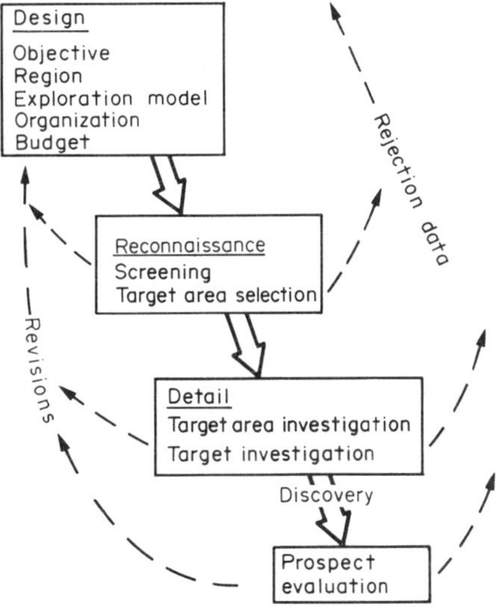

Figure 3
Stages involved in a regional exploration program

The initial field phase in a regional exploration program is reconnaissance, a screening of the entire region to define smaller areas of specific interest. Geological mapping is the main method used to gather reconnaissance information, with a strong dependence on airborne or satellite imagery, geochemical sampling in major drainages, and airborne geophysics.

Geological field work in the reconnaissance phase begins with the preparation of a small-scale synoptic map of the region, so that major elements in geomorphology, structure and lithology can be evaluated in relation to the exploration model. A mapping scale of about 1:250 000 is common in the initial stage of reconnaissance. At this scale, field geological observation sites within a spacing of 1 km can be plotted on topographic maps and on satellite imagery, and can be related to principal topographic and cultural features. Patterns of aeromagnetic, airborne radiometric and reconnaissance geochemical data serve as guides to the outlining and mapping of major geological domains in the field.

Using the information obtained from a synoptic map, the region is screened to provide areas of specific interest for further reconnaissance mapping at a scale in the range of 1:50 000–1:25 000. At this scale, geological observations at intervals of 50–100 m can be plotted on maps, and aerial photographs, geochemical stream sediment sampling sites and airborne geophysical flight lines can be shown. Other information such as individual mine workings, stratigraphic boundaries, local structural conditions and major areas of rock alteration can also be discriminated.

A further appraisal of areas and further reconnaissance geological mapping at 1:10 000 scale is sometimes needed in order to provide a sufficient resolution of geological features for designating target areas.

Geological mapping for target area investigation is commonly done at a scale in the range of 1:5000–1:1000, as shown in Fig. 4. This permits the boundaries of a fault zone or a dike a few meters in width to be shown without exaggeration, and also permits the plotting of pits or trenches and of individual stations in detailed geophysical and geochemical work. Geological observations are most commonly recorded in the field on aerial photographs, and then transferred to large-scale topographic maps. Considerable detail is recorded, with emphasis given to significant evidence of mineralization, zones of rock alteration, and the attitudes of bedding, jointing, rock cleavage and schistosity. As a result of detailed investigations within target areas, one or more specific targets are expected to be located.

Geological information from drill holes, though sometimes used in reconnaissance investigations where regional changes in stratigraphy are significant, is most commonly obtained during target investigations. Detailed geological mapping and geophysical and geochemical work provide the outline of a target that can be related to the exploration model and ultimately to an expected orebody at a particular depth. Drill-holes are then used to evaluate the target. If ore mineralization is discovered, drilling continues until a significant body of ore can be outlined. With the recognition of significant ore mineralization, the target becomes a prospect.

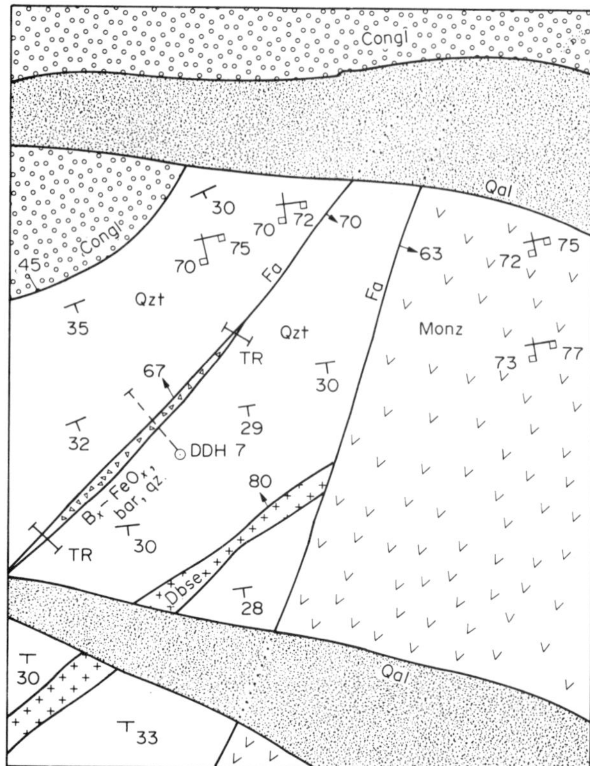

Figure 4
Part of a geological map of an exploration target area

In the evaluation of a prospect, a close pattern of drill-holes is used to outline and sample the ore, and to determine whether the ore mineralization can be designated as an orebody. In addition to the close pattern of drilling, shafts or adits and underground mine workings are often required for an adequate evaluation.

During evaluation, geological mapping of the immediate area may be done at a larger scale than during target investigation. Surface geological maps with scales of about 1:500 are commonly used to show the information in drill holes and trenches. Geological maps and sections of underground mine workings are commonly made at a scale of 1:250 or 1:500.

Prospect evaluation (see *Mineral Deposits: Examination and Evaluation*) involves economic and engineering considerations as well as geology. The final step in a successful exploration program and in a favorable prospect evaluation is marked by a decision to enter into a preproduction program to develop a mine.

Each stage in an exploration program, from reconnaissance to preproduction, is related to a series of decisions. On this basis the program is continued, revised or halted. Revisions, based on the accumulation of new information, may make it necessary to re-examine a target area that was once rejected. With a decision to halt or abandon an exploration program, data are accumulated for future use in the formulation of exploration models and the design of new exploration programs.

See also: Geochemical Distribution of the Elements; Ore Minerals

Bibliography

Kužvart M, Böhmer M 1978 *Prospecting and Exploration of Mineral Deposits*. Academia, Prague

Lacy W C 1983 *Mineral Exploration*. Van Nostrand Reinhold, New York

Pearl R M 1973 *Handbook for Prospectors*, 5th edn. McGraw-Hill, New York

Peters W C 1978 *Exploration and Mining Geology*. Wiley, New York

Reedman J H 1979 *Techniques in Mineral Exploration*. Applied Science, London

W. C. Peters

Geophysical Exploration for Mineral Deposits

Ground, airborne, well-logging and marine methods are commonly used in geophysical prospecting. All of the methods and their survey applications are complex, but a simple approach to a prospecting problem usually consists of the following stages:

(a) flying over the area and utilizing available airborne techniques;

(b) carefully evaluating the data before sending in ground exploration teams and applying ground techniques; and

(c) collecting subsurface information by monitoring geophysical instruments or probes placed in wells (well logging).

General textbooks on the subject are included in the bibliography.

1. Electrical Techniques

Electrical techniques can be divided into two basic categories: (a) those which depend on naturally occurring influence fields, such as self-potential and magnetotelluric methods; and (b) those based on introducing fields into the earth artificially, such as electromagnetic and induced-polarization methods. Electrical techniques are used in metals and minerals exploration, in shallow exploration (<500 m) and sometimes in oil exploration. They also aid in geothermal exploration, engineering geology and hydrology.

The three basic electrical properties of rocks on which these techniques are based are resistivity, electrochemical activity (including the chemical composition and electrolyte concentration) and the dielectric constant.

Electrical methods are summarized below, starting with methods based on introducing fields into the earth artificially and concluding with methods based on naturally occurring influence fields.

1.1 Electromagnetic Methods

Electromagnetic methods are based on the induction of electric currents in conductors buried in the earth. These produce a magnetic field which induces a current in a pick-up coil. This induction, a function of the resistivity of the buried conductor, results from magnetic components of electromagnetic waves generated at the earth's surface or in the air, and which originate from alternating field currents of varying frequency. An optimum depth versus frequency relationship exists, given by $h(f/\rho)^{1/2} = 10$, where h is the depth, f is the frequency (Hz) and ρ is the resistivity (Ω m).

A number of specific electromagnetic techniques have been developed for specialized applications. Examples of these are summarized below.

(a) *Audio-frequency magnetic field (AFMAG) method.* Variations (inclination and phase changes) in the natural horizontal electromagnetic field due to buried conductors are measured. This natural electromagnetic field is the result of electrical storms in the atmosphere and has a frequency of 30–20 000 Hz, in the audio range.

(b) *Very low frequency (VLF) method.* A strong electromagnetic field is produced by employing vertical antennae. The operation range is 15–30 kHz, and variations in the field due to near-surface buried conductors are measured.

(c) *Sequential electromagnetic signals (SES) method.* Induced currents produced at two or three frequencies are used to give interfering electromagnetic fields. The resultant field induces a current in a pick-up coil, which is then measured.

(d) *Induced-pulse transient (INPUT) method.* Short ac wave-crest pulses, separated by 2 ms pauses, induce a field in the earth. This field produces charged conductors which act like capacitors. Interrupting the current allows these capacitors to discharge, thus producing secondary electromagnetic fields, which are then measured. A good conductor has a long current-decay curve.

(e) *Turam method.* A straight conductor produces electromagnetic waves using two or three ac frequencies. Measurements consist of determining the phase differences and amplitude ratios of the resultant polarized electromagnetic waves.

(f) *Tilt-angle method.* A stationary transmitter produces a primary electromagnetic field, which in turn induces current flow in a buried conductor. A mobile pick-up coil is used to measure the angle between the ground surface and the resultant vector (tilt angle) arising from interaction between the primary and secondary fields. The minimum tilt angle is observed immediately above the buried conductor.

1.2 Resistivity Methods

Resistivity methods are based on applying a current to the ground through electrodes and measuring the change in electrical potential at the surface. Basically, the method yields a measure of the resistance to current flow through the ground. Plots of apparent resistivity ρ_a are developed from the fundamental resistivity equation:

$$\rho_a = (2\pi V/I)D \tag{1}$$

where V is the potential difference between two surfaces of constant potential, I is the current in the conducting body and D is some function of the electrode spacing which varies depending on the exploration technique employed.

Field procedures involve profiling or sounding techniques. Electrical profiling, in which the electrode spacing is fixed, provides a rapid ground surveying technique. It is generally used in locating sand and gravel, ore bodies, faults, and steeply dipping contacts between different rock bodies. The fixed spacing of electrodes yields a reading which represents a weighted average of all resistivities present. In electrical sounding, the center of a spread is fixed and the electrode spacing is increased. This technique is generally used to monitor ρ_a variability with depth, to differentiate a sequence of zones with high and low ρ_a, to determine the depth of a layer, and to estimate the thickness of a layer. This method assumes that all ρ_a variability is vertical.

1.3 Induced-Polarization Methods

Induced-polarization methods are the most widely used ground-based geophysical techniques in the mining industry, and are generally used in searching for disseminated sulfide ores and in ground water exploration (controlled by clay dispersal). Principally, the technique involves passing a variable or constant frequency, ac or dc current into the ground, which gives rise to exchange-ions at surface contacts between metallic substances and electrolytes in pore-space fluids. This electrochemical exchange produces a voltage which opposes the current flow through the metallic substance. A voltage, termed the overvoltage, is used to overcome this opposition. After stability is reached, the applied current is removed, and the electrochemical voltages at metallic grain surfaces allowed to decay. The decaying voltages are measured, and the ratio of the overvoltage amplitudes after and before shut-off gives an indication of the concentration of the subsurface metallic substance. Electrode configuration in the field is similar to that used for resistivity measurements.

1.4 Self-Potential Method

The self-potential or spontaneous polarization method is based on the existence of electric potentials which naturally develop within the earth as a result of electrochemical action. Ore bodies are good conductors which carry currents from oxidizing electrolytes above a water table to reducing electrolytes below a water table surface. Potential anomalies thus produced over graphite, magnetite and sulfide (pyrrhotite and pyrite) ore bodies are negative. Field detection is based on profiling and sounding techniques using two electrodes and a potentiometer. Applications involve shallow-depth prospecting, and require terrain corrections which may indicate tilting of the water table.

1.5 Telluric-Current Method

The telluric-current method is based on the presence of natural, large, sheet-like currents which flow within the earth in a fixed pattern with respect to the location of the sun. These currents are induced below the earth's surface by ionospheric currents which correlate with diurnal changes in the earth's magnetic field. Potential differences, controlled by variations in subsurface resistivity, are measured employing self-potential methods at two differing sites. At each site, the amplitudes of simultaneous oscillations of the sheet-like currents are compared, and the potential difference is measured in two orthogonal directions.

1.6 Magnetotelluric Method

The magnetotelluric method is based on the simultaneous measurement at the same location of an east–west and north–south component of the oscillating electric (sheet-like) field within the earth, and an east–west, north–south and vertical component of the magnetic field at the site. Measurements are made with nonpolarizing electrodes and buried induction coils. The method is generally used for large area reconnaissance (rapid basin delineation), and provides greater depth information than any other electrical technique.

2. Magnetic Techniques

Magnetic techniques are based on the fact that the earth has a magnetic field which is variable and easily measured at or above the surface. Contributing to this field are the main field (approximating to a dipole and originating within the outer (liquid) core of the earth) and the nondipole field, which includes nondipole components of the total field. These components generally result from (a) secular variations in the field due to long-term interaction of the outer core fluid with the lower mantle interface, (b) micropulsations and diurnal variations in the field due to short-term magnetic field characteristics arising from currents induced within the earth by ionospheric currents, and (c) field components produced by variations in the natural remanent magnetic properties of the earth's crust.

The natural magnetization of rocks generally resides in a suite of important iron-bearing minerals, including iron oxides (magnetite, hematite), iron–titanium oxides (ilmenite), iron oxyhydroxides (goethite), iron carbonates (siderite) and iron sulfides (pyrrhotite). Magnetic methods are particularly well suited to detection of concentrations of, and variations in, these minerals.

Two types of magnetic methods are employed in exploration for deposits. By far the most important of these is measurement of local variations in the magnetic field, at or above the ground surface, which result from buried magnetic bodies. The magnetic properties of rocks can also be directly measured, but are not often used in exploration for deposits. The three types of magnetometers in general use in exploration geophysics are discussed below.

2.1 Flux-Gate Magnetometers

These magnetometers are composed of two balanced alternating current (primary) coils wound in opposite directions around bars of high-permeability magnetic material, all surrounded by a single direct current (secondary) coil. A net inbalance between the primary coils induced by the earth's ambient field is measured using the secondary coil. The component of the field which is measured in this way is that which is parallel with the axis of the cores. In order to measure the total field, three orthogonal elements must be measured. This can be accomplished remotely using servo motors. The method is not absolute and has a sensitivity of about 1 nT.

2.2 Proton Magnetometers

A bottle of fluid is surrounded by a coil through which a current is passed. Protons in the fluid are aligned with the magnetic field thus produced, and begin to precess when the current is removed. This precession is proportional to the total magnetic field, and can be measured using the energizing coil. The method yields an absolute estimate of the total field, requires periodic current flow and decay cycles and can have a sensitivity of 50 pT.

2.3 Optical Pumping Magnetometers

These instruments yield an absolute, continuous reading of the total magnetic field and have a sensitivity of 1–10 pT. They are, however, extremely expensive and cumbersome. Measurement is based on the resonance frequency of rubidium, cesium or helium vapors being proportional to the ambient magnetic field.

3. Gravimetric Techniques

Variations in the near-surface gravity field are a function of horizontal and vertical variations in density. Therefore, gravimetric techniques have application where such variations are geologically important. Generally, only few applications exist in ore exploration, but gravimetric techniques are important in oil exploration, where variations in density can help to delineate salt domes and other structures.

Gravimeters can be airborne, carried at sea or positioned on the ground, and gravity interpretation usually entails producing contour maps of density variability. The data are often modelled in cross section, but some knowledge of the geology must exist to make meaningful interpretations, since many structures that lie at different depths and have different densities can have similar gravity anomaly patterns. By utilizing filtering and smoothing techniques, a regional (long wavelength: deep) anomaly pattern is defined from which local (short wavelength: shallow) anomalies can be subtracted. The resulting residual maps are useful in exploration.

Great care must be taken while conducting gravity surveys to determine station location, elevation, local variations in terrain, and estimates of density, based on rock type. For exploration purposes, two types of gravity anomaly are calculated: (a) a free air anomaly, where the observed gravity is corrected for the decrease in gravity due to elevation above sea level, with $C_F = g_o - (g_\phi - f_g)$, where g_o is the observed gravity value, g_ϕ is the theoretical gravity value at the site latitude, and f_g is the free air correction; and (b) a Bouguer anomaly, where the observed gravity is corrected for the attraction of crustal material (a slab) between a base level and the station elevation h above the base level. It is given as $C_B = g_o - (g_\phi - f_g + S - T)$, where S is the slab correction and T is a terrain correction.

Generally, negative anomalies are observed over sedimentary basins, salt domes, granites and grabens. Positive anomalies can be correlated with uplifts (horsts, reverse faults) and mafic rocks.

4. Seismic Techniques

Seismic methods are sea or ground-based exploration techniques which can give an accurate indication of subsurface depth of rock units, as well as giving a reasonably comprehensive picture of subsurface structure. These exploration techniques are based on the premise that contrasts in the seismic impedence of rocks reflect and refract elastic waves within the earth. These waves are termed body (P and S) and surface (Rayleigh and Love) waves. The velocity of propagation of these waves is a function of the elastic properties and densities of the subsurface rocks. These are controlled by such factors as compaction, cementation and metamorphic grade.

Principal applications in industry are associated with exploration for oil. The method involves generating seismic waves at the surface of the earth by using a source (e.g., explosives), recording the arrival times of these waves at receivers (geophones or by hydrophones usually set out in a straight line around the source) and then processing the data. The time it takes a wave to travel the distance to the receiver is the basic information available for analysis. Since the earth is not homogeneous and propagating waves tend to move with varying speeds in different directions, and since different types of elastic waves move at different velocities, such analysis can become complex.

5. Radiometric Methods

Certain radioactive elements in near-surface rocks give off α and β particles and γ rays during radioactive disintegration. Important examples are ^{238}U, ^{235}U, ^{232}Th and ^{40}K. While the α and β particles are readily absorbed by very thin layers of sediment, γ rays can be more readily detected. This is done by using a device such as a scintillation or a Geiger counter. Certain rocks (granite) or sediments (beach sand) intrinsically contain or accumulate greater concentrations of uranium, thorium and radioactive potassium than do other rocks, and γ-ray emission from these units is often readily detectable. These data can then be compiled into maps which provide geological information on near-surface rock types and structure. Such maps are also useful in uranium-ore exploration.

6. Thermal Methods

Thermal methods are based on the generation of heat within the earth as a result of radioactive decay. The flow of heat Q out of the earth can be measured from the relation:

$$Q = -K \operatorname{grad} T \tag{2}$$

where K is the thermal conductivity of the rock and grad $T = dT/dZ$, the rate of temperature increase with depth ($\mathrm{W\,m^{-2}\,s^{-1}}$).

Local variations in Q may be the result of (a) high concentrations of radioactive heat sources, (b) variations in thermal conductivity of the rocks, (c) subsurface geothermal sources, or (d) subsurface chemical reactions. Thermal exploration is currently being employed for shallow-level prospecting, and is useful in delineating subsurface aquifers, salt domes, fault contacts, thermal reservoirs, granites, sulfide ore bodies and other geological features.

7. Spectral Techniques

Spectral methods involve detection of electromagnetic waves, including visible light, ultraviolet and infrared rays and high-frequency radio waves. Methods involve simple photography (black and white, color and ordinary infrared), optical–mechanical scanning (thermal infrared), side-scan and direct-scan radar (radio detection and ranging), satellite imagery, and spectral reflectance measurements.

Air and satellite images and radar techniques provide extremely important geological data, which can give rapid estimates of structure and lithologies. These are based on information such as linear trends and affinities of some vegetation types for soil chemicals present in certain lithologic units. Combined with thermal infrared data, areas of minor heat emissions can be identified, which arise from subsurface chemical reactions such as those known to occur in association with porphyry copper deposits. Other uses exist, such as remote identification of hydrothermal sources and of variations in water table level.

Spectral reflectance measurements can be used to detect changes in rocks and soils caused by mineralization, metal stresses in plants and groundwater-related phenomena. Spectral reflectance information is possible because certain minerals when excited by sunlight emit a characteristic wavelength of visible light. This data has proven useful in directly delineating zones of hydrothermal alteration where distinct iron and clay spectral bands are emitted. It has been demonstrated that specific mineral identification is possible using such methods, especially when dealing with clay minerals such as kaolinite and montmorillonite. Alunite and iron minerals are also easily identified.

8. Well Logging

Well logging is a prospecting tool generally used in oil exploration and prospect evaluation. It is not extensively applied to metallic minerals exploration, probably due to the greater structural complexity in locations where ore deposits are found. Well-logging techniques are designed to provide direct information from the monitoring of instruments lowered into drill holes, and in conjunction with instruments distributed around the drill holes, to evaluate the subsurface geology. Generally, the rock properties evaluated using electrical and acoustic well-logging methods are porosity, permeability and the nature of fluids within the rock. Using magnetic susceptibility, radiometric and heat-flow methods, mineral constituents of the rock can be determined, and using gravimetric methods, density is estimated. Instruments designed to be inserted into drill holes generally provide continuous output graphs as they are lowered into the hole. These graphs, which show the variations of physical properties with depth, are usually interpreted as stratigraphic logs. Such interpretations are qualitative and rely on independent checks on the data, such as rock cuttings from cores and data from nearby wells.

See also: Geochemical Distribution of the Elements; Geochemical Exploration for Mineral Deposits; Geological Exploration for Mineral Deposits; Remote Sensing in the Search for Mineral Deposits

Bibliography

Dobrin M B 1976 *Introduction to Geophysical Prospecting*. McGraw-Hill, New York

Kazmitcheff A 1974 *Modern Mineral Prospecting*. Casterman, Tournai, Belgium

Kearey P, Brooks M 1984 *An Introduction to Geophysical Exploration*. Blackwell Scientific, London

Keller G V, Frischknecht F C 1977 *Electrical Methods in Geophysical Prospecting*, International Series of Monographs on Electromagnetic Waves, Vol. 10. Pergamon, Oxford

Nettleton L L 1976 *Gravity and Magnetics in Oil Prospecting*. McGraw-Hill, New York

Parasnis D S 1979 *Principles of Applied Geophysics*, 3rd edn. Chapman and Hall, London

Sharma P V 1976 *Geophysical Methods in Geology*, Methods in Geochemistry and Geophysics, 12. Elsevier, Amsterdam

Telford W M, Geldart L P, Sheriff R E, Keys D A 1976 *Applied Geophysics*. Cambridge University Press, Cambridge

B. B. Ellwood

Geothermal Energy Systems: Materials for Conversion

In general, the process of materials selection for geothermal energy conversion equipment is somewhat easier than for downhole equipment (see *Geothermal Energy Systems: Materials for Downhole Environments*). Fluid temperatures continually decrease as energy losses and extraction proceed. The steam separated in flash tanks is purified by the expansion process, leaving the majority of the impurities with the condensate. Noncondensible gases such as H_2S and CO_2 remain with the steam. They should essentially be removed before the steam enters the turbine, as they significantly affect corrosion and plant operating characteristics. At higher concentrations the H_2S is a health hazard; even at low concentrations it remains an environmental and corrosion problem. After scrubbing, the separated steam proceeds to the turbine generator where energy is extracted in a manner similar to fuel-fired steam turbine-generator units. Such steam, because of its complex origins, is always more corrosive than that obtained in well-maintained coal or oil fired systems.

Heat is extracted in a different manner from lower-temperature, nonflashing resources. Generally the hot fluid flows through corrosion-resistant heat exchangers.

It is much easier to sample fluid and perform field corrosion tests in surface equipment. Although materials selection problems are not as severe as for downhole environments, care must be taken. High costs ensue when excessive corrosion causes lengthy downtime and replacement of major surface components. Machine efficiency can also suffer because of wear and erosion of turbine blading.

1. Piping

Piping from the wellhead is subject to erosion from sand particles entrained in the fluid. Low spots in piping runs must be avoided because liquid accumulations in such locations become saturated with salts. Special traps are generally designed into the system for drainage. Insulation on pipes to decrease heat losses must be periodically inspected to avoid local salt solution buildups near leaking flanges and joints.

Low-carbon steel steam transport piping has shown good service for most applications. In some instances wellhead separators or mildly alkaline water scrubbing is necessary to reduce corrosion to tolerable limits. Lines should be designed with adequate corrosion allowances (2.5 mm was specified for a 30 year design lifetime in El Salvador).

2. Separators, Flash Tanks and Silencers

Steam separators, flash tanks and silencers are all subjected to a wide variety of erosive, scaling environments. In some instances erosion from sand can be significantly abated by letting a new well flow freely for 5–10 days. After most of the sand has been eliminated the well can be put online to a steam separator. In other instances erosion can be eliminated through good design practice. For applications where elbows might be specified, blind tee fittings are preferable. The blind end of the tee fills with liquid which serves as a hydraulic damper while still diverting the main flow.

Because of severe corrosion of low-carbon steel, silencers at Cerro Prieto (Mexico) are now constructed of wood or fiberglass on concrete slab foundations. Polymer concrete liners are being tested for steam separator applications in Iceland. Here also low-carbon steel has not been satisfactory for longer-term reliable service.

It is especially important to keep piping tight. The

fluid often has low but deleterious concentrations of hydrogen sulfide and perhaps low pH with significant levels of dissolved salts. Oxygen in such an environment can promote stress-corrosion cracking and/or accelerated pitting of the steam separators and flash tanks.

3. Heat Exchangers

Heat exchangers are typically employed for binary cycles where cyclic expansion to a gas of an organic liquid (isobutane or pentane) drives a turbine-generator unit. Such systems, also used as bottoming cycles for industrial waste heat recovery, have potential for the exploitation of moderate- to low-temperature geothermal electric resources. The three types being evaluated for geothermal systems—traditional shell-and-tube, flat plate and direct contact—have specific advantages and disadvantages. Shell-and-tube exchangers can be difficult to clean in scaling geothermal fluids, may be more expensive to fabricate on site and often must be treated chemically to restore their heat transfer capabilities. Flat plate heat exchangers are more easily opened up, have a larger heat transfer area per mass of material, are more easily fabricated from large sheets of material and may be mechanically scrubbed, with consequent longer lifetime.

Direct contact heat exchangers provide a mixing chamber where the geothermal fluid intimately contacts the organic fluid. Thermodynamic phase separation—the two liquids are immiscible—is then employed to keep brine out of the binary turbine and organics out of the discharge water. The advantages of this system are that scale does not materially affect the heat transfer process and the chamber can be lined with a relatively thin layer of corrosion-resistant composite cement or other more expensive corrosion-resistant material. Unfortunately, carryover of small amounts—parts per million—of binary fluid into the discharge waters gradually causes environmental pollution and excessive costs due to organic fluid loss. Such small relative amounts from flows of 1.3–2.6 m^3s^{-1} for several years can have serious impact on the environment. Continued engineering research with scrubbing and secondary recovery systems is required to solve this problem.

Before heat exchanger materials are finally selected, the resource developer should establish resource chemistry through initial and extended production well flow tests. These also provide a field corrosion testing opportunity for evaluation of the most appropriate plant materials. Although each geothermal fluid has its own unique characteristics, it will generally belong to one of five or six broad classes, on which worldwide information has been compiled. Corrosion tests followed by careful selection should prevent unexpected leaks into the organic side which can cause deterioration of seals and turbine equipment. Long-term operating data for binary geothermal systems are limited. Inexpensive low-carbon steels have proved inadequate in very limited service. Lower-temperature systems have successfully employed titanium plate heat exchangers. This metal and its alloys offer promise for moderate-temperature applications as well.

4. Geothermal Turbines

Relatively few, if any, insurmountable materials problems have been reported in geothermal steam turbine performance. Severe erosion which has occurred in some situations has been attributed to either intermittent operations (where temperature cycling breaks off scale particulates) or water droplet erosion of blade edges. The latter problem has been solved to a large extent by welding either 316 stainless steel or Stellite strips on the eroded surface. Crevice corrosion of 19%Cr–2%Ni stainless steels occurred in early Italian turbines between the fixed blade material and the base material. Local hardening of 13%Cr steel, with subsequent cracking susceptibility resulting from bronzing, has been avoided at Wairakei by mechanically attaching shrouds and lacings to the turbine blades.

At Cerro Prieto an important part of the normal maintenance schedule involves inspecting for and sand blasting scale deposition from the turbine blades. Such nonuniformly distributed additional weight, if not removed, causes turbine imbalances, reduces efficiency and can lead to blade failure. Scale cleaning of discharge drains, valves and production wells, the recalibration and repair of electronic and other monitoring and control equipment as well as the painting and repair of normally inaccessible structures are also fitted into a shutdown period of up to 2 weeks.

In general, geothermal steam turbines are fabricated from the same materials as are used for fossil fuel applications. These include carbon or low-alloy steels for rotors, AISI 403 or 410 for rotating buckets and AISI 405 for diaphragm blades. AISI 405 has improved weldability, so it is more easily fabricated and field weld-repaired. To reduce sulfide stress cracking, turbine components are tempered to optimize their performance. Fatigue tests in Geysers (California) steam of typical AISI 403 turbine steels with various heat treatments have confirmed the beneficial effects of such special processing.

Binary turbines rely to some extent on the state of the art of refrigeration and bottoming cycle technology. Experience of their long-term reliability when operated as part of a geothermal system is minimal. Existing experience indicates that small heat exchanger leaks can allow significant amounts of brine to enter into the secondary system. Condensation is then possible into relatively small quantities of concentrated and corrosive brine. Methods for removal and detection of such fluids should be included in binary system design. Neglect of this could allow pitting or crevice corrosion to proceed and endanger the whole system with catastrophic failure.

Other turbine designs in the prototype or pilot trial development stage may provide significant benefits for some applications. For instance, the helical screw expander relies on a thin scaling layer to provide a seal during steam expansion. This unit has performed well at a number of sites worldwide in a variety of corrosive and highly scaling fluids. The fine scale layer has also served as a protective coating against corrosion.

5. *Cooling Towers*

Cooling towers provide a difficult environment for materials because they oxygenate the cooling fluid. Since water is at a premium in geothermal locations, the hot mineral-rich condensate leaving the flash tanks after steam separation is usually recycled, generally five to seven times, through the cooling tower before discharge. In addition, in some locations pollution control requirements mean chemical treatments to prevent hydrogen sulfide release into the atmosphere. Addition of ferric iron partially accomplishes this, but at the same time significantly increases corrosion of cooling tower materials. Other treatments, including acidification to a pH of 5–6, have not been established long enough to be fully evaluated. Materials in current use at the Geysers include AISI 316 or fiber-reinforced plastic pipe, epoxy-coated concrete channels and redwood for cooling towers. Poly(vinyl chloride) covers are now employed to protect closed concrete channel surfaces. Coal tar epoxy proved unsatisfactory in practice, failing after 2 years of service.

Dry cooling towers have not had much application because they need to be so much larger than wet systems. It is expected their use will be restricted to smaller geothermal systems by capital costs and will also to some extent depend on the development of more durable and corrosion-resistant plastics.

6. *Injection Pumps and Well Casings*

Experience of injection pump production capacity is relatively limited but is expected to increase significantly in the future. At many sites, injection will be required to maintain reservoir pressure and working fluid volumes. In general, injected fluid will have a wide pH range and high corrosivity. Flashed geothermal condensate in many cases has relatively high chloride concentrations, opportunity for oxygen contamination and varying residual amounts of hydrogen sulfide. Typically, a dual flash system can be expected to handle fluid prior to injection at temperatures from slightly above the boiling point to 140 °C. Such an environment is conducive to stress-corrosion cracking of susceptible materials such as austenitic stainless steels. Fluid that has passed through a cooling tower or injection pretreatment mineral removal process will have characteristics that must be evaluated through corrosion testing for each process. It is expected that such treatments will be required to make fluids compatible with the reservoir, thus permitting long-term injection. High-cost but available materials will most probably initially be required to resist stress-corrosion cracking and abrasive erosion from some fine particulates for injection pump applications. Acid-resistant polymer concrete liners and plastic pipe have promise as economical alternatives for lower-temperature injection well casings.

7. *Drains and Holding Ponds*

The oxygenated hot condensates of varying pH provide corrosive environments for drains and holding ponds. In addition, condensate temperatures promote the growth of sulfate-forming microorganisms which give rise to sulfuric acid. Although the microbiological growth may be controlled by additions of hypochlorite, this also lowers the pH and increases corrosivity. Problems have been severe at Larderello (Italy) with Portland cements and calcareous aggregates. These are rapidly destroyed by cooling water and steam condensate. Pozzolanic cements and slag cements with siliceous aggregates, although still affected, have significantly better service life. They must be slowly cured for up to 60 days for optimum performance. Newly developed polymer concrete shows significant promise as a replacement material or liner in this application. A wide range of laboratory and field corrosion tests has been successfully performed with these materials. Their durability in low-pH oxygenated fluids relies on reaction between vinyl groups in the monomer with CaO from the Portland cement–silica aggregate.

8. *Pollution Control and Mineral Recovery*

Pollution control and mineral recovery equipment can either be upstream of the turbine or put in place after energy extraction has occurred. Hydrogen sulfide gas is the primary pollutant for which control technology is required. Although preferable to remove it upstream of the turbine, current technology relies on a Stretford Unit to treat noncondensible gas with an iron chelate/peroxide process for the condensate. Such treatments generally complicate an already complex corrosive medium. A prototype upstream process utilizes copper sulfate to precipitate the sulfide. Titanium liners and titanium trim have performed satisfactorily in this environment. A stainless steel, 20 3Cb with high chromium and nickel content, although not attacked in laboratory and initial pilot field tests, suffered severe crevice corrosion at welds of field-welded sections. It was thought that better welding technique might solve this problem. Corrosion tests in the scrubbed steam versus unscrubbed indicated significant reductions in corrosion rates downstream for turbine alloys and piping compositions.

Mineral recovery from condensate streams would utilize materials suitable for drains and tanks (Sect. 7). Lower temperatures permit the use of fiberglass and polymer composite cements of lower service temperature.

9. Instrumentation

Recommendations for electrical instrumentation are dependent on the presence or absence of hydrogen sulfide. Platinum, gold, rhodium and gold plating have been satisfactory for relays. Other materials should be sealed off in inert atmospheres. Terminal boxes in Italy have been protected by thick cadmium or nickel coatings. The use of silver is discouraged for all except medium-voltage interrupters. These have a self-cleaning effect from the high pressures and currents present during operation.

10. External Structures, Corrosion and Scale Control

External structures at Larderello have had satisfactory service when coated with a tar and tar-oil-based paint on top of a galvanically protective, flame-sprayed metallic zinc coating next to sand-blasted metal. Atmospheric corrosion tests at the Geysers (California) indicated severe corrosion of mild steel, copper-based alloys and galvanized coatings. Ferritic and martensitic stainless steels had light general attack. Aluminum alloys with and without coatings all proved satisfactory for structural applications.

Corrosion control is the objective of a large part of maintenance procedures. Localized corrosion can occur where stagnant concentrations of geothermal brine accumulate. Careful system design including an adequate number of drain valves minimizes such effects. Hydrazine, other oxygen scavengers and corrosion and scale inhibitors should be added during standby periods. Such control procedures are even more important when parts of the system are opened or drained during maintenance. An appropriate number of isolation valves facilitates chemical additions and at the same time minimizes the amounts required. Other corrosion control procedures relate directly to the specific geochemistry of the produced fluid. As mentioned earlier, in some Italian fields the fluids produced are so acidic and corrosive that it has been found cost-effective to raise the pH by scrubbing with clean water. Reasonable piping lifetimes enable the field to be operated, although somewhat less efficiently with the reduced-enthalpy steam.

A serious problem in any scale and corrosion control program is the lack of online chemical information. Prototype instruments and reliable prototype sensors for pH, corrosion, specific ion chemistry and gas–liquid ratios in the fluid are in various stages of development, testing and application. Field tests of such equipment have already shown their worth by reporting upset conditions of major parts of the first prototype binary plant at East Mesa (California). Corrective actions turned what could have been expensive corrosion failures into routine shutdowns for remedial action. It is expected that such instrumentation with improved sensors will significantly increase the reliability of geothermal and fossil-fuelled plants.

11. Repair

Repair is required for components subjected to particulate and water droplet erosion. Valve faces and turbine blades may be built up to original specifications with plasma-deposited hard metals. Drainage channels and piping can be internally recoated with coaltar epoxies. Some recent successes have been achieved on a trial basis with polymer concrete relinings for silencers and steam separators. Polymer impregnation also appears feasible for cement and concrete systems which have not excessively deteriorated.

12. Direct Use

Materials for direct use of geothermal fluids as a heat source represent a larger proportion of the total plant costs than they do in electric generation systems (where well drilling is deeper and consequently more expensive). Materials selection is also generally more conservative because of the expense of digging up large underground piping systems. Low-carbon steels have been used extensively in Iceland for transporting relatively benign geothermal fluids at 85 °C. It is preferable in the case of more complex chemistries to use titanium or other corrosion-resistant material for plate heat exchangers at or near the wellhead. Water circulating through the closed-loop extended pipelines can then be chemically treated to minimize scaling and corrosion. Low-cost fiberglass and polymer concrete piping, tanks and pressure vessels now under development show significant promise for these applications.

See also: Corrosion in the Power Industry; Corrosion of Metals: An Overview; Materials for Energy Applications: An Overview

Bibliography

Anderson D A, Lund J (eds.) 1979 *Direct Utilization of Geothermal Energy: A Technical Handbook*, Geothermal Resources Council Special Report No. 7. Geothermal Resources Council, Davis, California, pp. (4)42–(4)75

Cabibbo S V, Costello R M, Ammerlaan T 1979 *Economic Assessment of Using Nonmetallic Materials in the Direct Utilization of Geothermal Energy*, BNL Report No. 51112. US Department of Energy, Washington, DC
DiPippo R 1980 *Geothermal Energy as a Source of Electricity*. US Government Printing Office, Washington, DC
Ellis P F II, 1983 *Review of Shell and Tube Heat Exchanger Fouling and Corrosion in Geothermal Power Plant Service*, DOE/SF/11503-2 (DE84004539). US Department of Energy, Washington, DC
Ellis P F II, Conover M F 1981 *Materials Selection Guidelines for Geothermal Energy Utilization Systems*, DOE/RA/27026-1. US Department of Energy, Washington, DC
National Materials Advisory Board 1981 *Materials Needs for the Utilization of Geothermal Energy*, NMAB-375. National Materials Advisory Board, Washington, DC
Reeber R R 1980 Coatings in geothermal energy production. *Thin Solid Films* 72: 33–47
Smith C S, Ellis P F II 1983 *Addendum to Materials Selection Guidelines for Geothermal Energy Utilization Systems*, DOE/ET/27026-2 (DE83017656). US Department of Energy, Washington, DC

R. R. Reeber

Geothermal Energy Systems: Materials for Downhole Environments

Borehole materials for geothermal wells may be defined as those employed in drilling and completing wells, and in production itself. Information concerning the well or reservoir may be required during any stage of development and is generally obtained by lowering logging tools and instruments into the well. Since these components are exposed to the borehole environment, materials selection can significantly affect their usability and lifetime. The materials concerned include steels, cements, elastomers, ceramics and alloys, which are exposed to the highest temperature and untreated fluids as produced from the formation or as modified by stimulation chemicals.

1. Geothermal Environments

Materials required for geothermal equipment are subjected to a wide variety of corrosive environments related to the geological situations in which the geofluids are produced. Geothermal fluids range from those of high pH and low corrosivity to those of low pH (approaching 1.5) containing various concentrations of chloride ions, carbon dioxide, hydrogen sulfide, methane, ammonia, scale-forming carbonates, silicates and sulfates. Copper, lead, mercury and other heavy metal ions may also occur in trace amounts. Geothermal fluids that become aerated generally accelerate corrosion of most engineering alloys—the few exceptions include titanium, molybdenum-containing austenitic stainless steels, high-chromium alloys, certain nickel-base alloys and chromium-plated steels.

Useful fluid temperatures range from near 50 to 350 °C, though new discoveries in Italy have temperatures greater than 400 °C. Flow rates of some wells are considerably over 65 kg s^{-1} of steam and water at wellhead pressures greater than 2070 kPa. A few geothermal resources bear a superficial resemblance to seawater, but the majority of these do not contain equivalent amounts of calcium and magnesium. These cations provide a buffering action to stabilize seawater pH to a near-neutral 8 and also form tenacious scale deposits that at least partly protect iron and steel from corrosion. Another difference important for materials selection is the almost total absence of oxygen in uncontaminated high-temperature geothermal fluids.

In some systems the geofluid chemistry has been modified through additions of very low levels of calcium carbonate inhibitors for scaling control and steam washes with weakly alkaline water for corrosion control. Because of the large volumes required to operate geothermal plants, chemically effective treatments must be carefully considered before specifying expensive chemicals as corrosion and scale inhibitors. Fortunately, the majority of geothermal resources, with a few important exceptions (Salton Sea, Cesano in Italy), can be economically extracted by making use of a framework of careful materials selection and systems design.

2. Drilling Fluids

Drilling fluids carry chips and rock cuttings out of the well. In addition, they cool and lubricate the rock bit and drill pipe. Such fluids should keep cuttings in suspension if circulation is discontinued temporarily. The weight of the drilling fluid controls the pressure of a well. Adequate weight for safety is important if a high-pressure-producing formation is drilled. On the other hand, if the mud is too dense it can fracture the formation and flow into and eventually plug permeable zones.

Mud-based drilling fluids (see *Well-Drilling Materials: Clay and Nonclay Minerals*) are generally used below 120 °C. They contain varying proportions of material such as bentonite, sepiolite and brown coal and are stabilized to increase their life at temperature and to reduce corrosion of drill bits, drill pipe and ancillary equipment. In the higher-temperature range, which may occasionally reach 140 °C, mud-based fluids are subject to gelling, increased viscosity and chemical deterioration. These changes can be accelerated by chemical interactions with the geological formation, especially gypsum contamination. Additives such as tannins, sodium carboxymethylcellulose, ferrochrome–lignosulfonates and chrome–lignins to some extent stabilize high-temperature properties. Sodium hydroxide is a common additive for pH control, with preferred pH maintained at about 8. The solids content in the fluid is maintained at a low level by surface filtration and settling tanks.

At higher temperatures, water and air drilling fluids are preferred because of the rapid dehydration and chemical deterioration of mud. Water is not always available at the well site and, in addition, does not have the best lubricating properties for reducing drill bit wear. Air, although frequently employed in high-temperature wells, has its own limitations. Severe drill pipe erosion can occur from air-driven rock cuttings. This can be abated to some extent by foam additives. Such additives cure at well temperatures to form an elastomeric coating on the cuttings. Experimental drilling with nitrogen gas has also reduced erosion.

Low-temperature wells (40–120 °C) are generally drilled with existing water well technology and state-of-the-art materials. Either air rotary or cable rigs can drill from 50 to 550 m in such resources. Cable rigs are preferred for the deeper sections because they allow more control over surface release of hot well flows.

3. Drill Pipe

Drill pipe, typically ranging from 88.9 to 127 mm o.d., is required to perform in an environment that is conducive to corrosion, mechanical abuse, fatigue and wear. For that reason, mechanical properties, microstructure and hardness are subjected to quality control checks before pipe is put in service. Pitting corrosion, hydrogen embrittlement, sulfide stress cracking and erosion (especially in air drilling situations) are additional common failure modes. Pipe is initially ultrasonically inspected for manufacturing defects and then reinspected after each drilling operation. Standard procedures are used to downgrade pipe into restricted use categories. During service, drill pipe is subjected to frequent stress reversals which occur at several locations in the drill string. Surface imperfections, corrosion pitting, dents and notches are stress concentrators at such locations and elsewhere and can lead to premature fatigue cracking. At predictable transition zones such as near the bottom of the hole, heavier drill pipe provides a gradual change in stiffness than can significantly improve fatigue life. Special attention is given to stress relief of drill collars (generally chromium–molybdenum alloy steels) with drill collar threads cold rolled to extend their fatigue life. Such treatment places the thread root metal in compression and thereby improves bending stress resistance. Areas of the pipe that are subject to the severest wear and erosion are also often banded with hard material (tungsten carbide). Epoxy and other organic coatings, when properly applied, significantly reduce pitting corrosion.

4. Geothermal Well Cements

Geothermal well cements provide vertical and horizontal support for the production casing. When impermeable they protect the steel from corrosion and act as a seal to isolate porous formations from one another. The mechanical integrity of the cement bond between casing and formation rock provides well control assurance during further drilling and eventually long-term production. Chemical, thermal and pressure stability are also important. Circulating temperatures during completion (the casing and cementing operation) are significantly lower than production temperatures. Cements must therefore be able to withstand the severe thermal shock of initial production and later occasional shutdown.

Silica flour additives to Portland cement result in the formation of higher-strength cement phases such as xonotlite ($6CaO \cdot 6SiO_2 \cdot H_2O$). Such cements have been successfully placed in high-temperature, highly saline geothermal fluid environments. Polymer cements, including styrene–acrylonitrile–acrylamide with divinyl benzene as a crosslinking agent and semi-inorganic polymers such as organosiloxane, have also shown promise for some specialized applications. Attainment of good properties in place requires that the cement be capable of 3–4 h retardation at placement temperatures. The cement must also displace drilling mud as well as undergo minimal detrimental chemical reactions with the mud. Since some formations, especially those that will later produce fluid, are highly permeable and may be easily fractured, the cement must seal these without excessively flowing into them. Final cement properties are the result of a compromise between these various requirements.

Commercial additives to change setting times, slurry densities and/or viscosities and to control lost circulation are commonly used. Others cause the hardened cement to expand slightly, thereby significantly lowering the risk of interzonal communication and aquifer contamination. Many of these chemicals exhibit erratic behavior in higher-temperature geothermal wells (>200 °C) and may also reduce the chemical and temperature resistance of the cement.

5. Well Casing

The casing string for a geothermal well is usually composed of several sections, including surface, intermediate and production casing. Normal practice is to start at the surface with 335 mm o.d. and reduce to 300 mm and/or 240 mm further down. Often the well is left open at the bottom (production zone). Higher temperatures and a different chemical environment require caution with some steels that would normally be considered satisfactory for many oilfield applications. Where hydrogen sulfide is present in the geothermal fluid, brittle delayed fracture has occurred in well casings at the 552 MPa yield stress range (N-80 grade casing). J-55 grade casing (379 MPa yield stress range) is currently most commonly used in US geothermal wells. Laboratory experiments indicate that improved performance is obtainable from steels having lower inclusion contents. AISI 4130 and 4135 modified (high purity for both grades) have been shown in laboratory tests to have significant advantages in break strength over presently manufactured commercial varieties of these steels. A 4135 steel modified with columbium (niobium) also had improved properties, indicating that some detrimental effects may be controlled without the necessity and costs of ultrapurity.

Casing failures are often the result of improper cement jobs. If fluid is left in a pocket of the annulus between the casing and wall rock, the casing is subject to buckling when the well is producing. The trapped fluid is transformed to high-pressure steam which often stresses the casing beyond its yield strength. For very saline, low pH (1.0–2.5) fluids, severe corrosion can occur at the higher bottomhole temperatures.

A variety of nonmetallic casings, including the thermosetting plastics (epoxy and polyester) and thermoplastics (polypropylene and acrylonitrile–butadiene–styrene polymer), are adequate for low-temperature geothermal fluid service. Such materials have been sparingly used but will have increasing applications because of their improved corrosion resistance and fluid flow characteristics. Present drilling practices will have to be modified somewhat to permit the use of significantly thicker, fiberglass-reinforced casings. Also, bentonite additions to the grout can prevent overheating of the polymer by lengthening cement curing time. Since the casing cannot be driven into rock, the driller, if rotary drilling, must also modify installation practices somewhat. Polymer concrete is another promising material that has been extensively laboratory corrosion-tested. It has had limited field tests in resources of low to moderate temperature.

6. *Stimulation*

Stimulation is accomplished by isolating and fracturing a specific formation zone in the well so as to increase permeability and subsequent well productivity. Packers serve as seals to isolate the fracture region from other zones. Immediately after a successful stimulation job, proppants are pushed or forced into the newly created fractures to keep them open during later production or injection. Oil and gas packer technology is inadequate for geothermal conditions. Geothermal packer materials are required to withstand bottomhole temperatures of 250–300 °C and differential pressures from 20 to at least 35 MPa. Elastomeric systems have been developed based upon EPDM (Nordel) and EPDM–FKM (Nordel–Viton) blends which have survived under these conditions in steam–brine–H_2S–CO_2 environments. Such seals are generally an integral part of a resettable metal frame unit which has either spring- or pressure-actuated controls. Although some problems still exist because of poor high-temperature metal–elastomer adhesion, these can generally be accommodated through mechanical design and reduced clearances between packer and casing. Adequate formation packers are still under development.

Proppants encounter an especially severe chemical environment. To be effective they must have a combination of good chemical resistance and compressive strength. Alumina and other relatively insoluble high-temperature minerals are most commonly employed.

The fluid used for stimulation is generally water (hydrofracture) but may also contain acids (for example, hydrofluoric for silica, hydrochloric for limestone-bearing rocks). Much work must be done to understand and interpret the complex high-temperature geochemical–chemical interactions required for optimizing stimulation fractures.

7. *Logging*

Geothermal well logging is an essential step for determining reservoir properties at various stages of field development. These stages include activities such as exploration, reservoir testing, production testing, well condition assessment and injection evaluation. During a logging operation the instrument package is lowered into the well, generally by means of a logging cable. The well may be static or flowing, with the tools, sensors, instruments and cable exposed in lesser or greater degree to the downhole environment.

Initial attempts to employ oil and gas logging equipment failed because of seal, cablehead and cable degradation. In general, the deterioration of metal parts is not a significant problem because the limited quantities of materials involved permit the use of Hastelloy C-276, Ti–1.5%Ni, Inconel 625 and Allegheny–Ludlum 29-4 or equivalents. The 300-series stainless steels, including 316, in contrast have suffered severely from pitting and stress-corrosion cracking in certain geothermal fluids of moderate to high salinity and are not recommended. Elastomeric materials described earlier for packer applications have also been successfully employed as seal materials. Specially compounded elastomers of EPDM (Nordel), Viton and the Japanese elastomer AFLAS have shown good properties in laboratory and field logging tests. Logging service has now been accomplished at temperatures of 275–300 °C.

Single-conductor logging cable under development has received a limited amount of field testing. The wire construction consists of a monoconductor surrounded by ceramic insulation which is itself sheathed with a corrosion-resistant flexible metal tube. Multiconductor cables of more traditional construction with Teflon insulation have been successfully employed at slightly lower temperatures. Above 250 °C their long-term life is expected to be considerably less than that of the metallic sheath experimental cable.

The actual sensors which measure temperature (Pt resistance, 1.0 °C accuracy), pressure (quartz transducer), flow and pH, specific ion electrodes, conductivity probes and downhole fluid samplers have additional problems relating to high-temperature stability and the availability of appropriate electronics. Newer equipment incorporates high-temperature diodes, resistor inks with low thermal coefficients of resistivity and higher-temperature connection and joining technology.

8. *Downhole Pumps*

Pumps are essential for increasing the productivity of moderate-temperature geothermal wells. Such pumps, either line shaft or downhole, must be capable of continuous operation over long periods of time at temperatures of 125–160 °C. By maintaining fluid pressure through the wellbore and heat exchangers they reduce or eliminate scaling problems. Such problems occur when high-pressure fluid flashes to relatively pure steam and more concentrated brine at a reduced pressure. Prior knowledge of the well chemistry is important in specifying pumps. High downhole carbon dioxide levels, for instance, can create extremely high vapor pressures in the fluid which may destroy pump impellers by cavitation.

Each pump type has advantages and disadvantages. The line shaft motor, being out of the brine stream and outside the well, does not have the size constraints of the downhole pump, which must fit into a 175–330 mm diameter casing. The line shaft pump has either a closed or open shaft design with shaft bearings spaced at 1.5–3 m intervals along the length of the shaft (up to 300 m). In the open shaft design the bearings are lubricated by the pumped fluid. This also contains suspended particles which erode the bearing surfaces. The enclosed shaft version has an inner tube separating the bearings from the pumped fluid. This design also allows for injection of a lubricant, either water or oil, through the tube to the bearings. Because of the large number of bearings required by line shaft pumps, bearing reliability is essential. Specially hardened steel, Ti alloy, ceramic and ceramic-coated bearings are under development for this application. An additional possibility being considered relies on fabricating elastomeric cutlass bearings from brine-resistant materials.

When setting depths deeper than 300 m are necessary, the downhole pump is required. Materials specified for existing oil and gas pumps have not proved adequate for the geothermal environment. Pumps operating downhole need specially designed motors, seals, corrosion-resistant containers and support pipes or cable. Stress-corrosion cracking and seal degradation have limited these pumps to significantly shorter than optimum operating times. Longer-life seals of EPDM and Nitanol (titanium–nickel memory alloys) are under development.

Cable for the downhole pump (rated presently at 205 °C and 10 MPa) is generally armored with interlocked galvanized steel and a lead metal wrap and has EPDM insulation. No special problem exists with main pump housings, especially if materials selection procedures screen for sulfide cracking.

In some instances special sensors have been employed to detect worn pump vibrations prior to failure. Automatic shutdown procedures prevent a major catastrophic breakup of the pump, which could permanently block the well.

9. *Wellhead and Other Valves*

Valves at the wellhead must withstand erosion, sulfide cracking and scale buildup. Valve reliability is essential for safe operation of the well. At the Ahucahapan site it is customary to protect the master wellhead valve with two secondary valves. These are cycled from fully open to fully closed and back once a month. After 5 or 6 months, the well is shut down and the eroded gates are repaired by welding.

Scaling can render many valve designs inoperative. Gate, butterfly and spring-type safety valves must be routinely cycled to minimize sticking. At Wairakei, low-strength steels with stainless valve trim have given satisfactory service. High-strength API grade N-80 steels have serious stress-corrosion problems and are avoided for this application. In more erosive environments containing entrained sand and rock, hard metal and Stellite trim increase erosion resistance sufficiently for service.

Care must be taken to avoid materials susceptible to stress-corrosion cracking where chloride and hydrogen sulfide are present. Wells are usually kept flowing to prevent accumulations of cold chloride-containing brines saturated with hydrogen sulfide. Such environments have been shown to promote stress-corrosion cracking. Asbestos packing and asbestos gaskets have been successfully used in Italian hot water and steam control valves.

Where possible, field operation practices should be such that oxygen contamination of the fluid is totally prevented. The oxygen content should be monitored where possible. Precautions taken to regulate line pressures can prevent oxygen contamination through Venturi effect leakage. This has been observed in low-pressure steam systems when expansion occurs through butterfly valves. In this situation, slight increases in line pressure eliminate the problem by reversing the effect. Smaller bleed valves on wellheads and design precautions can ensure that pressure buildups do not lead to catastrophic failure when wells are restarted after maintenance shutdowns. In some instances, valves were originally specified to protect water–steam separators from overpressure. Because of vapor leakage in the seat, they quickly became inoperative, and so they have been replaced by rupture discs. Paints, other coatings (including epoxies) and/or galvanic methods are feasible for external valve surface protection against steam and salt sprays.

10. Inspection

Several areas relating to the wellbore require inspection. During drilling it is essential to test drill pipe periodically by standard nondestructive methods (see *Nondestructive Evaluation: An Overview*). These include ultrasonics, magnetic flux, acoustical damping, wall thickness gauges and any other NDE system that will detect a defective pipe before it breaks in the hole. The cyclic loading and corrosive–erosive environment present during the drilling operation often cause such failure by fatigue, erosion, sulfide stress cracking and/or localized corrosion. After the well is completed, it is also important to inspect the contiguity of the cement job. As mentioned in Sect. 5, trapped water behind a casing can cause it to collapse from the excessive steam pressure generated during initial production. A new generation of temperature-hardened ultrasonic devices is being developed to solve this problem.

After casing failures occur, it is necessary to determine the exact type of failure and its specific location. Such failure analyses have been carried out with lead seal impressions, depth caliper electric logs and boroscope cameras. Caliper electric logs are especially useful for finding corrosion-thinned casing. Periodic inspections of wellhead flanges are important for detecting leaks from hot chloride solutions. When aerated, these can lead to classic cases of stress-corrosion cracking. The situation is particularly dangerous if salt concentrates accumulate under insulation. Valves should be periodically inspected for scale buildups, which often make a valve inoperable.

New plants generally have relatively frequent scheduled inspections (9–12 month intervals). The complex geochemistry of a new field can then be evaluated in terms of where and how much scale accumulates.

11. Maintenance and Repair

A normal maintenance schedule generally involves shutting down the turbine generator system for up to 2 weeks. When possible, wells are kept flowing, with the flow either diverted or blown off. This prevents the severe thermal shock to cement and casing that occurs during a complete cooldown and startup.

Well repairs, when needed, are relatively complex operations requiring a workover rig capable of maintaining well control during the shutdown operation. If the well has just been cemented and logging has found fluid pockets indicative of a poor cement job, it is possible to shoot steel balls explosively through the casing wall and subsequently squeeze pressurized cement ("squeeze job") into the cavities. Pressure is maintained in the localized area of interest by isolating that portion of the well with retrievable sealing elements (packers). When casing has been damaged and protrudes into the wellbore, it may be mechanically milled out with a bit attached to a drilling string of tubulars. Often after such an operation the simplest and most reliable procedure is to cement in a smaller diameter production casing.

In some situations large-scale deposits partly or completely block the well. These are generally drilled through. A more difficult problem arises when scaling occurs at the wellbore rock face in the production zone. Redrilling and acidizing operations can sometimes recover well production. In these cases attention should be given to future production flow rates and pressures. Reduced flows and increased wellbore pressures may move the scaling out of the wellbore and into surface equipment where it can be more conveniently handled.

See also: Geothermal Energy Systems: Materials for Conversion; Materials for Energy Applications: An Overview; Underground Corrosion

Bibliography

Cabibbo S V, Costello R M, Ammerlaan T 1979 *Economic Assessment of Using Nonmetallic Materials in the Direct Utilization of Goethermal Energy,* BNL Report No. 51112. US Department of Energy, Washington, DC

Ellis A J, Mahon W A J 1977 *Chemistry and Geothermal Systems*. Academic Press, New York

Ellis P F II, Conover M F 1981 *Materials Selection Guidelines for Geothermal Energy Utilization Systems,* DOE/RA/27026-1. US Department of Energy, Washington, DC

Ellis P F II, Smith C C, Keeney R C, Kirk D K, Conover M F 1983 *Downhole Materials Selection*, DOE/SF/11503-1 (DE83010306). US Department of Energy, Washington, DC

National Materials Advisory Board 1981 *Materials Needs for the Utilization of Geothermal Energy,* NMAB-375. National Materials Advisory Board, Washington, DC

Reeber R R 1980 Coatings in geothermal energy production. *Thin Solid Films* 72: 33–47

Veneruso A F 1981 *Sourcebook on High Temperature Electronics and Instrumentation*, NTIS Report No. SAND-81-2112. US Department of Energy, Washington, DC

R. R. Reeber

Germanium

Although germanium formed the basis of the early semiconductor device industry, it has been essentially superseded by silicon in all the common devices, but it still finds some specialized applications. Some properties of germanium are given in Table 1.

Table 1
Properties of germanium

Dielectric constant:	16.3
Effective mass, 300 °C	
electrons:	0.19 m
holes:	0.3 m
Intrinsic resistivity, 300 °C:	0.47 Ω m
Specific heat, 300 °C:	310 J kg^{-1} K^{-1}
Band gap:	0.67 eV
Lifetime, 300 °C	
pure electrons:	600–1000 μs
pure holes:	200–400 μs
Mobility	
electrons:	0.39 m^2 V^{-1} s^{-1}
holes:	0.19 m^2 V^{-1} s^{-1}
Intrinsic carrier concentration:	2.4×10^{19} m^{-3}
Density of states	
conduction band:	1.04×10^{25} m^{-3}
valence band:	6×10^{24} m^{-3}
Ionization potential:	8.13 eV
Melting point:	937.4 °C

1. Technology

Germanium does not appear in commercial quantities as a native ore but is produced principally as a by-product of zinc processing, with a smaller amount from the processing of copper. The germanium-containing metallic zinc is first distilled under nonoxidizing conditions. The residue, which contains the germanium, is leached with chlorine water to convert the germanium into germanium tetrachloride. The tetrachloride is hydrolyzed into germanium oxide, which may then be reduced to metallic germanium. The impure polycrystalline germanium is next purified by zone refining, in which a succession of molten zones are caused to move along the ingot, carrying the impurities along with them, to accumulate at one end (see *Zone Refining, Zone Levelling and Temperature-Gradient Zone Melting*). The resistivity of the purified ingot may be controlled by doping it with any of several impurities such as indium or antimony. In this technique a dopant is introduced into one end of an ingot and is swept along by a single molten zone, which leaves a uniform deposit in its wake.

2. Physics

The germanium structure is tetrahedral, in which each atom has four nearest neighbors, with which it shares its valence or bonding electrons. There are thus two electrons per bond. At room temperature there is enough thermal excitation to break a small number of bonds, thus releasing a few electrons, which will be free to act as conduction electrons. In their place, in the so-called electron sea surrounding the nuclei, will be regions of net positive charge, called holes, which are also able to conduct electricity, because they too can move from atom to atom under the influence of an electric field. The conductivity of germanium is thus made up of two terms:

$$\sigma = ne\mu_n + pe\mu_p \tag{1}$$

where n and p are the densities of electrons and holes, respectively; e, the electronic charge; and μ_n and μ_p, the mobilities of electrons and holes.

In order to make useful semiconductor devices, the resistivities of the semiconductor must be capable of being controlled at will. This is done (Adler et al. 1964) by doping the material with elements having five valence electrons for electron conduction, or n-type material, or three valence electrons for hole conduction, or p-type material. Recall that four electrons are needed for bonding. The excess electron of arsenic, for example, is only loosely bonded at room temperature, so is free to conduct. The electron deficit of indium likewise easily forms a hole, so it also is free to conduct.

According to modern theories of atomic structure, the electrons are grouped in shells about the nucleus, with the maximum number of electrons or electron energy levels per shell, each of which is unique, given by $2n^2$, where n is the serial number of the electron shell, starting outwards from the nucleus. When a crystal is formed, the electron energy levels group into discrete energy bands (Shockley 1951) separated by empty, or forbidden, bands. The outermost filled band is called the valence band. The next band, which is higher in energy, is the conduction band, which in germanium is nominally empty. The impurity energy levels, or dopant energy levels, lie in the forbidden band above the valence band, normally at about a few hundredths of an electron volt from the band edges (Neuberger 1965). Donors of electrons lie just below the conduction band, and acceptors of electrons just above the valence band. The density of electrons in the conduction band is given by (Grove 1967)

$$n = \frac{N_c}{\exp[(E_c - E_f)/kT] + 1} \tag{2}$$

The density of holes in the valence band is given by

$$p = \frac{N_v}{\exp[(E_f - E_v)/kT] + 1} \tag{3}$$

where N_c and N_v are called the densities of states of the conduction and valence bands (and represent effective densities of states or electron energy levels at the band edges) and E_c and E_v are the energies of the bottom of the conduction band and top of the valence band. The quantity E_f is called the Fermi level. For normal doping levels, say below $10^{24}\,m^{-3}$, the material is said to be nondegenerate. The Fermi level lies in the forbidden gap and is called the chemical or thermodynamic potential. Since it is a measure of the average electron energy, it will move towards the conduction band as the electron concentration increases, and towards the valence band for holes. From Eqns. (2, 3) it can be seen that the product np is a constant at a given temperature, usually denoted as n_i^2.

If the doping becomes high enough, the impurity levels form a band which can overlap the top of the valence or bottom of the conduction band for acceptors (holes) and donors (electrons), respectively. At liquid helium temperatures, conduction is in the impurity band, where the effective mass rises to around 1000 m (where m is the free electron mass) depending on the doping.

3. Devices

The first single-crystal semiconductor devices which became available commercially, in particular the original transistors, were made from germanium (Bridgers et al. 1958). Except for a few devices used for special purposes such as lithium drift detectors and Esaki diodes, all of the germanium devices are essentially only of historical interest.

3.1 Diodes

The early diodes, rectifiers, power diodes and high-frequency detectors were based on simple *pn* junction rectification. One device which was not, and which still uses germanium, is the Esaki or tunnel diode, which has a negative resistance to its forward characteristic before it assumes normal conduction. In this case, both the *n* and *p* materials are so heavily doped that the valence and conduction bands overlap in energy, and there is direct band-to-band tunnelling.

3.2 Transistors

(*a*) *Point contact*. The transistor of the original invention was formed by pressing two fine, sharply pointed doped wires onto a block of *n*-type polycrystalline germanium, a few mils apart, and bonding the wires into the germanium. This device demonstrated a current gain greater than unity, but was generally not suited to mass production (Bridgers et al. 1958).

(*b*) *Grown junction*. The next type of transistor was based on Shockley's comprehensive theory of junction transistors and was known as a grown-junction transistor. In this technique, *p*- then *n*-type dopants (or *n* and *p*) are successively introduced into an appropriately doped melt of germanium as a crystal is drawn from it in such a way as to produce three narrowly separated *npn* (or *pnp*) layers, from which the transistor is made. These performed well but were only suitable for low-frequency use below the megahertz range.

(*c*) *Alloyed junction*. In this form of transistor two buttons of indium, for a *pnp*, are alloyed into opposite sides of a thin slice of *n*-type germanium to form a transistor. Proper choice of crystal orientation keeps the front surfaces of the alloyed buttons parallel. This was the first low-cost, commercially successful transistor. It could be used for both switching and amplifying, although the frequency response was still low. These devices are still being made, but only to a small extent. Antimony-doped lead buttons could be used to produce *npn* transistors.

(*d*) *Diffused base*. With the advent of diffusion techniques, reliable high-frequency transistors became available. Arsenic was diffused into the germanium (with very tight tolerances) to form the base or central layer of the transistor. Aluminum was alloyed into the top to form the emitter or input region. The bottom collector layer or output was, of course, the starting material. Gigahertz performance became available, and reliability was of such a level that these devices were used in submarine cables, and in the first American satellites, Vanguard and Telstar (Sze 1981).

3.3 Radiation Detectors (Lithium Drift)

Some of the most interesting uses of germanium today lie in the field of detectors and gauges. A particularly successful device has been the lithium drift detector. This is a *pin* structure in which the wide *i*, or intrinsic, region is compensated. Lithium is deposited onto, and diffused slightly into, ultrapure *p*-type germanium. Then at a temperature below 60 °C the lithium will drift under the influence of an applied field and will tend to compensate exactly any acceptor impurities which are present. Thus a very large intrinsic region can be formed. In fact, it can be a few centimeters wide. To preserve the charac teristics, storage must be at liquid nitrogen temperatures.

In operation, a hole–electron pair is produced in the *i* region for an expenditure of about 3 eV energy. The detector has a useful range of from 10 keV to several MeV. It has extremely good resolution. Peaks separated, for example, by 14 keV can be clearly resolved (Metz 1974).

3.4 Temperature Gauges

Since the Seebeck coefficient (Metz 1974) for germanium is very high (about 400 $\mu V\,K^{-1}$), it should make a very good thermal detector; in fact, its sensitivity is at least five times that of commonly used materials. Thin films of germanium form one element of a thermocouple in ultrasensitive heat transfer gauges.

3.5 Strain Gauges

Germanium shows a very considerable piezoelectric effect and consequently makes a very sensitive strain gauge. Thin filament crystals show a sensitivity at least two orders of magnitude better than conventional metal gauges (Metz 1974).

3.6 Solar Cells

Since the band width between the valence and conduction bands of germanium is only 0.67 eV as compared with the 1.2 eV of silicon, germanium has been proposed as a companion solar cell to be used with silicon to collect the long-wavelength solar radiation and convert it to useful electrical energy.

See also: Electronic Materials: An Overview; Gallium, Germanium and Indium Resources; Germanium Production; II–VI and IV–VI Semiconductors

Bibliography

Adler R B, Smith A C, Longini R L 1964 *Introduction to Semiconductor Physics*, Vol. 1. Wiley, New York

Bridgers H E, Scaff J H, Shive J N 1958 *Transistor Technology*, Vol. 1. Van Nostrand, Princeton, New Jersey

Grove A S 1967 *Physics and Technology of Semiconductor Devices*. Wiley, New York

Metz F I 1974 *New Uses for Germanium*. Midwest Research Institute, Kansas City, Missouri

Neuberger M 1965 *Germanium Data Sheets AD610828*. Hughes Aircraft Company, Culver City, California

Shockley W 1951 *Electrons and Holes in Semiconductors*. Van Nostrand, New York

Sze S M 1981 *Physics of Semiconductor Devices*, 2nd edn. Wiley, New York

F. A. D'Altroy

Germanium Production

Germanium is produced principally as a by-product from the processing of zinc and copper–zinc ores. In the USA germanium is recovered from cadmium-rich dusts or retort-furnace residues produced during the treatment of zinc concentrates, whereas in Africa recovery is from copper concentrates obtained during conventional selective flotation of copper–zinc ores. The African ores yield a magnetic concentrate containing about 1% germanium which is treated in an electric furnace, giving off fumes containing germanium dioxide. In another process, the copper concentrates are pelletized and then treated in a vertical retort furnace where germanium sulfide is volatilized in a reducing environment and the fume is collected for further treatment. In both processes the fume contains germanium suitable for treatment in hydrochloric acid and subsequent purification by distillation.

Germanium is separated chemically during the purification of electrolytes prior to electrolysis in electrolytic zinc plants. Economic recovery of germanium is possible if the germanium concentration is high enough in the separated solids. Because germanium oxide concentrates react readily with concentrated hydrochloric acid and the resultant $GeCl_4$ product has a convenient boiling point of 83.1 °C, chlorination is the standard refining step, chlorine usually being added to the primary distillation or to the subsequent fractionation to suppress the volatilization of arsenic. Since the germanium product is used in electronics and standards for the purified $GeCl_4$ product require a maximum of about 1 ppm impurities, the fractionation is usually carried out in a glass or quartz vessel. Hydrolysis of the purified $GeCl_4$ with deionized water produces GeO_2 which is removed by filtration and then dried.

Production of germanium metal involves the reduction of the dioxide to metal powder in graphite boats at 650 °C in an atmosphere of hydrogen or cracked ammonia. The temperature is increased to 1100 °C following reduction in order to melt the powder. The molten metal is cooled to a solid ingot and cleaned in an etching solution. For the production of intrinsic or electronic grade germanium, the ingot is purified by zone refining in a graphite boat. The boat is passed through a series of induction coil heaters, each of which produces a molten zone in the ingot. Impurities remain in the molten zone and become concentrated at the rear end of the ingot which is subsequently cut off and returned to the circuit for additional purification. Polycrystalline germanium resulting from the zone-refining operation usually contains less than 100 ng total impurities and less than 0.5 ng of electrically active impurities per gram of germanium. Boron and silicon must be removed prior to the zone-refining step.

Refined germanium is grown into single crystals for use as a semiconductor. The crystals are sliced and fabricated into devices. Germanium scrap from the process is recycled through the refinery into intrinsic grade germanium. Zone-refined germanium used in infrared optics is recast and grown in forms suitable for lenses and windows. The germanium in this case is annealed, cut and ground into lens or window blanks, polished, coated and assembled into an infrared system.

See also: Germanium; Metals Production: An Overview

Bibliography

Adams J H 1980 Germanium and germanium compounds. *Kirk-Othmer Encyclopedia of Chemical Technology*, 3rd edn., Vol. 2. Wiley, New York, pp. 791–802

Battelle-Columbus Laboratories 1980 *Germanium Supply and Demand*, Final Report. US Bureau of Mines, Washington, DC, pp. 7–15

Germanium Information Center 1964 *Brochure on Germanium*. Germanium Information Center, Kansas City, pp. 8–9

H. W. Sheffer

Glass: An Overview

Glass is an essential material in modern technological society. In addition to traditional uses such as windows and containers, glass is used in a host of specialized ways in lamps, optics, composites and electronics. Recent developments of fiber-optic waveguides, laser optics for initiating fusion reactions, and repositories for radioactive waste—all of glass—demonstrate the versatility, importance and promise of glass for future applications.

This article serves as an introduction to the articles on glass in the Encyclopedia. It discusses the history, definition, uses and main types of glass. The other articles cover the following fields: (a) the structure and formation of glass; (b) special types of glasses; (c) the manufacturing of glass; (d) transport in glass; and (e) the mechanical, thermal, chemical, electrical, magnetic and optical properties of glass. The reader is referred to the index volume of this work for a complete listing of these articles.

1. History

Archaeological evidence shows that natural glasses such as obsidian have been used by man from the earliest times. Synthetic glass and glazes were made far back in human history. The earliest known glaze (coating of glass) dated from about 12 000 BC, and the earliest solid glass from about 7000 BC; both were found in Egypt, which continued to be a center of glass and glaze manufacture throughout ancient history. At first, glass was used only for decorative purposes, but later it was molded or pressed into vessels. The invention of glass blowing in about the first century BC greatly increased the use of glass for practical purposes in Roman times, mainly for vessels but later for windows also. In Europe, glassmaking dispersed to isolated sites after the fall of the Roman Empire, but was continued in the Byzantine Empire (Syria especially), and then by the Arabs in Syria, Egypt and Persia. Venice became the center of a resurgent glass industry in Europe after about 1300. Artistic development paralleled that of the Renaissance in the fifteenth century, the most significant advance in Venice being the manufacture of a clear, colorless glass called cristallo. With this glass Venice became the premier center of glass manufacture and export until the seventeenth century, when centers sprang up throughout Europe, especially in Bohemia, Nuremberg and the UK.

Of special interest in this period was the publication of *L'Arte Vetraria* by Neri in Florence in 1612; this work released the art and technology of glassmaking from the secret hold of the Venetian artisans and made it available to the world.

After 1600 the development of glassmaking was rapid. Cut glass was first made about 1600, and the use of coal instead of wood for fuel gave higher temperatures and a more reliable fire. Late in the seventeenth century, lead oxide was added to glass in England, resulting in lower melting temperatures, a wider temperature range for working the glass, and "sparkle" from a higher refractive index. Better selection and some purification of raw materials in the eighteenth century gave clearer glasses and better control of melting and working conditions and of the color of the glass (see *Color and Optical Absorption in Glasses*).

In the nineteenth century, the quality of glass produced in Europe increased tremendously in response to the increased affluence of industrial society. At the end of the nineteenth century, automatic methods of making glass were developed, such as the bottle-blowing machine.

Traditionally, window glass was made by hand by either the crown or cylinder process. In the crown process a glob of glass was blown out and one side of the resulting globe flattened. A solid iron rod was attached to the flat part and the blowing pipe detached. Then the globe was reheated and rotated until it formed a flat disk about three feet in diameter. Panes of glass were cut from the disk after it was slowly cooled. The part attached to the rod was the "bull's-eye"—still seen in some older windows. In the cylinder process the blower made a large cylinder, which was then split open and flattened. Cylinder glass was also made by machine.

In the early part of the twentieth century, processes for drawing sheet glass directly from the glass melt were developed (the Fourcault and Colburn processes). When linked with a continuous-glass-melting furnace, this process is capable of producing large quantities of flat glass of reasonable quality. Plate glass of the highest quality was, until recently, made by flowing the glass from the furnace through rollers, which gives a rough surface, and then grinding and polishing it by large automatic machines. This process requires a large capital investment, but is fairly economical because of the large quantity of glass that is drawn directly from the melt. The sheet-glass surface has a fine fire-polished finish, but shows some surface distortion because of variations in processing conditions as the glass is drawn from the melt.

In the 1950s, an ingenious new method of making relatively inexpensive flat glass of high quality was developed by Alistair Pilkington of the Pilkington Glass Company of England. In this float process, a continuous strip of glass from the melting furnace floats onto the surface of a molten metal, usually tin, at a carefully controlled temperature. The flat surface of the molten metal gives a smooth, undistorted surface to the glass as it cools. After sufficient cooling, the glass is rigid and can be handled on rollers without damaging the surface finish. High forming speeds are possible, and the cost is much less than for similar quality glass made by grinding and polishing. As a result, one glass manufacturer after another has converted to the float process and today most flat glass is made by this process.

Glass containers such as bottles and jars are made by blowing hot glass into a mold on a continuous machine (see *Forming of Glass*). Lamp bulbs are made in a similar way by blowing hot glass into a mold; however, in this case the glass is fed from the melting furnace as a ribbon, rather than blowing out an individual glob of glass as in the container machines. A special nozzle blows glass in the ribbon into a mold. The high-speed "ribbon machine" for lamp bulbs is a spectacular sight as lamp bulbs pop out from it at a rate of 2000 per minute.

Certain glass objects such as plates, tumblers and vases can be made inexpensively by pressing the hot glass in a mold. This pressed glass was especially popular in the US in the nineteenth century because it was much cheaper than cut-crystal glass imported from Europe. In this method, a glob of hot glass is placed into a metallic mold, and a metallic plunger is forced into the mold to shape the glass in the desired way. Patterns in the mold surface are imprinted in the glass and make it look like an expensive article. Pressed glass can also be made continuously using automatic feeders for molds on a rotating bed.

The continuous manufacture of glass in a large furnace, called a "tank," was developed in response to the demand for greater, cheaper and more uniform production of glass. This furnace requires improved refractory materials to line it, and is fueled with natural gas or fuel oil.

A variety of new methods for making special types and forms of glass have been developed. Glass fibers as fine as 10 μm in diameter are drawn and coated at high speeds. Fused-silica tubing is drawn at temperatures above 1800 °C. Highly pure and clear glass is made for a number of optical purposes. A large variety of glass compositions have been developed for diverse applications. The infinite variability of glass compositions leads to a great assortment of possible properties which are only beginning to be exploited. Many of these glasses, their properties, and methods of manufacture are described in more detailed articles (see *Glass Ceramics; Glass Fibers; Laser Glass; Glass Optical Fibers*).

2. Definition

Glass is an amorphous solid (see *Glasses: Structure*). A material is amorphous when it has no long-range order, that is, when there is no regularity in the arrangement of its molecular constituents on a scale larger than a few times the size of these groups. For example, the average distance between silicon atoms in vitreous silica (SiO_2) is about 3.6 Å, and there is no order between these atoms at distances above about 10 Å. A solid is a rigid material: it does not flow when it is subjected to moderate forces. More quantitatively, a solid can be defined as a material with a viscosity of more than about 10^{15} P (poise).

Many glass technologists object to the above definition of a glass. These workers prepare a glass by cooling a liquid in such a way that it does not crystallize, and feel that this process is an essential characteristic of a glass. Many earlier writers insist on this criterion: "glass is an inorganic product of fusion which has been cooled to a rigid condition without crystallization" as taken from the ASTM standards for glass. The difficulty with this view is that glasses can be prepared without cooling from the liquid state. Glass coatings are deposited from the vapor or liquid solution, sometimes with chemical reactions. Thus sodium-silicate glass can be made by evaporating an aqueous solution of sodium silicate (water glass) and baking the deposit to remove water. The product of this process is indistinguishable from sodium-silicate glass of the same composition made by cooling from the liquid. It seems wise to use the same name for materials with the same molecular structure and properties no matter how they are made; consequently, the broader definition given here is preferred.

3. Uses

The original use of glass for decorative and artistic objects continues to this day. One need only visit the Stueben exhibit in New York City, the Corning museum in Corning, New York, or glass exhibits in Czechoslovakia to sense the originality and excitement of present-day artistic work in glass.

The second use of glass was for containers, for food, perfume, oils and many other substances; this application still uses the largest quantity of glass today. The production of flat glass, mainly for windows in buildings and vehicles, is now the second largest item of glass manufacture. Lamp envelopes and seals are another major area of use (see *Incandescent Lamp Materials*). There are many special applications of glass, some in small quantity but of high value, most of which have been developed in the last few decades. Some of these special applications are glass ceramics and surface-strengthened glass for higher strength and chemical durability; lightweight composites of fiberglass in polymer matrixes; glass for laser hosts and optic waveguides for long-distance communication; fused silica for melting semiconductors, telescope mirrors and arc lamps; amorphous silica layers on silicon in electronic devices; encapsulation of these devices; and as a medium for consolidating radioactive wastes. New

Table 1
Some types of silicate glasses and their uses

Glass	Use	Approximate softening temperature (°C)	Coefficient of thermal expansion (per °C × 10^6)
Soda lime	Containers, windows, lamp bulbs	700	9.0
Pyrex borosilicate	Headlamps, cookware, laboratory ware	830	3.2
Vitreous silica	Semiconductor crucibles, lamps, optical components, fiber optics	1600	0.5
Alkali lead	Lamp tubing, sealing	620	9.3
"E" lime aluminosilicate	Fibers	830	6.0
Lime–magnesia aluminosilicate	High temperatures, cookware	900	4.6

compositions of silicate glasses, and even nonoxide glasses, are being rapidly developed for a variety of new applications.

These uses of glass are based on a variety of desirable properties, such as ease of forming into many different shapes, cheap and widely available raw materials, chemical durability, transparency, high-temperature durability, wide "solubility" of constituent oxides and low electrical conductivity.

4. *Main Types of Glass*

The possibility of incorporating a large number of different oxides in a silicate glass has led to a wide variety of commercial glasses. Nevertheless, there are a relatively small number of glasses that make up the large majority of glass production. Some of these glasses are listed in Table 1, and their approximate compositions given in Table 2. The softening temperature in Table 1 is either the temperature at which a rod 25 cm long and 0.76 mm in diameter elongates by 1 mm min^{-1}, or the temperature above which the glass can be easily deformed and worked.

By far the most common glass is based on the soda-lime–silicate (sodium calcium silicate) system. All ancient glasses contained the oxides of sodium, calcium and silicon. These glasses are cheap, chemically durable and relatively easily melted and formed. Many minor additions to the basic composition (not far from 70% silica, 15% soda, 10% calcium oxide and magnesium oxide, and 5% other oxides) are made to improve properties of melting and forming; alumina for improved chemical durability and reduced devitrification (crystallization), borates for easier working and lower thermal expansion, zinc oxide for lower melting temperatures, and arsenic and antimony oxides for fining (removal of bubbles). Soda-lime glass is often termed "soft" glass because of its relatively low softening temperatures.

Pyrex-borosilicate glass was developed by the Corning Glass Works to be more resistant to thermal shock and more chemically durable than soda-lime glass, yet still melt at a similar temperature. The borate in this glass reduces its viscosity and the coefficient of thermal expansion, and allows low sodium content, which increases chemical durability. Pyrex-borosilicate glass is often called "hard" glass because of its higher softening temperature compared to soda-lime glass, and is somewhat more expensive than soda-lime glass because of its higher melting temperature and more expensive borate raw material. The mirror for the Mount Palomar telescope was made of Pyrex-borosilicate glass because of its low thermal expansion; nevertheless, it is necessary to correct minute distortions of the mirror surface caused by temperature differences.

Table 2
Approximate compositions (wt%) of the glasses in Table 1

Glass	SiO_2	B_2O_3	Al_2O_3	CaO	MgO	PbO	Na_2O	K_2O
Soda lime	72	1	2	5	4		15	
Pyrex borosilicate	81	13	2				4	
Vitreous silica	100							
Alkali lead	77			1		8	9	5
"E" lime aluminosilicate	55	7	15	21			1	
Lime–magnesia aluminosilicate	64	5	10	9	10		1	1

Vitreous silica is made of pure silica, SiO_2, giving it excellent high-temperature stability, optical properties and thermal-shock resistance. A separate article in the Encyclopedia is devoted to this glass (see *Vitreous Silica*).

A variety of lead glasses are important as low-melting reading and solder glasses with a wide working-temperature range. Lead glass for fine "crystal" contains much more lead than these glasses.

See also: Fluoride Glasses; Nonoxide Glasses; Glass as a Building Material; Glasses in Medical Applications

Bibliography

History of glass

Charleston R J 1980 *Masterpieces of Glass: A World History from the Corning Museum of Glass.* Abrams, New York

Douglas R W, Frank S 1972 *A History of Glass Making.* Foulis, Henley-on-Thames, UK

Morey G W 1954 *The Properties of Glass,* 2nd edn. Reinhold, New York, Chap.1

Neri A 1612 *L'Arte Vetraria.* Florence, Italy, and many subsequent editions and translations

Polak A B 1975 *Glass: Its Tradition and Its Makers.* Putnam, New York

Zerwick C 1980 *A Short History of Glass.* Corning Museum of Glass, Corning, New York

Monographs on glass

Babcock C L 1977 *Silicate Glass Technology Methods.* Wiley, New York

Doremus R H 1973 *Glass Science.* Wiley, New York

Dunken H H 1981 *Physikalische Chemie der Glasoberfläche.* VEB Deutscher Verlag für Grundstoffindustrie, Leipzig

Holloway D G 1973 *The Physical Properties of Glass.* Wykeham, London

McMillan P W 1979 *Glass–Ceramics,* 2nd edn. Academic Press, New York

Morey G W 1954 *The Properties of Glass,* 2nd edn. Reinhold, New York

Scholze H 1965 *Glas: Natur, Struktur, und Eigenschaften.* Vieweg, Braunschweig

Vogel W 1979 *Glaschemie.* VEB Deutscher Verlag für Grundstoffindustrie, Leipzig

Volf W B 1961 *Technical Glasses.* Pitman, London

Wong J, Angell C A 1976 *Glass Structure by Spectroscopy.* Dekker, New York

Zarzycki J 1982 *Les Verres et L'État Vitreux.* Masson, Paris

Collections of review articles

Mackenzie J D (ed.) 1960, 1962, 1964 *Modern Aspects of The Vitreous State,* Vols. 1–3. Butterworths, London

Pye L D, Stevens H J, LaCourse W C (eds.) 1972 *Introduction to Glass Science.* Plenum, New York

Tomozawa M, Doremus R H (eds.) 1977, 1979, 1982 *Treatise on Materials Science and Technology,* Vols. 12, 17, 23, *Glass I, II, III.* Academic Press, New York

Uhlmann D R, Kreidl N J (eds.) 1980 *Glass: Science and Technology,* Vol. 5: *Elasticity and Strength in Glasses.* Academic Press, New York

Reference works

Eitel W H J 1965, 1976 *Silicate Science,* Vols. II, VII, VIII. Academic Press, New York

Mazurin O V 1971–1980 *Properties of Glass and Glassy Melts,* Vols. 1–4 (in Russian). Science, Leningrad

Journals on glass

Glass Technology and Physics and Chemistry of Glasses. Society of Glass Technology, Sheffield

Glastechnische Berichte (in German). Deutsche Glastechniche Gesellschaft, Frankfurt Main

Journal of the American Ceramic Society. American Ceramic Society, Columbus, Ohio

Journal of Noncrystalline Solids. North Holland, Amsterdam

Soviet Journal of Glass Physics and Chemistry (in Russian, also translation into English). Consultants Bureau, New York

Yogyo Kyokaishi, Journal of the Ceramic Association of Japan (mostly in Japanese, with English abstracts). Ceramic Association of Japan, Tokyo

R. H. Doremus

Glass as a Building Material

The window, a word believed to be derived from the Norse meaning "eye of the wind," originally was just an open hole, but over the centuries it has become a barrier between the internal, man-created environment and the weather. However, it is only in the twentieth century that architecture and glass manufacturing technology have developed glasses which can control and enhance many aspects of the environment. Large and transparent windows can now connect living spaces to the outside and also insulate those spaces. The worst effects of solar heat gain can be mitigated by glasses that either absorb or reflect the heat. Silent internal environments can be created.

Windows can be made fire-resistant as well as bullet-proof. In recent years, they have proved able to conserve energy and reduce the need for some internal energy sources. The twentieth century has also seen developments that are not related to the basic manufacturing method for glass. The future will bring greater development of processed glass, coatings and combinations of glasses with different properties, and greater use of the natural properties of glass; all of these will improve the standard of living.

1. Types of Glass

The major glasses available today for architectural use are as follows.

(a) Clear transparent glass. In the first part of the twentieth century, manufacturers concentrated on more economical ways of producing flat, transparent glass. First came the crown process, then the cylindrical process, various methods of drawing glass vertically (the sheet process), grinding and polishing rolled glass (the polished plate process) and most recently floating glass on molten metal (the float process). Each of these processes has successively improved the quality of glass. Clear transparent glass can be manufactured up to 25 mm thick, and thicker polished plate glass is still available. The size of glass is determined by the building design, energy conservation considerations and loading requirements.

(b) Clear translucent glass. This glass bears an impressed pattern on one or both surfaces. Such textured glass is available in a variety of designs and provides various degrees of privacy and internal natural lighting.

(c) Colored transparent glass. This glass, colored during manufacture by very small modifications to the basic composition, is usually made as float glass. The most popular colors are gray, bronze, and green; the precise color depends on the composition of the glass.

(d) Colored translucent glass.

(e) Transparent polished, wired glass. In this glass, wire, in a mesh or in another configuration, is incorporated during manufacture. Such wire is centered in the thickness of the glass; this makes the product fire and impact resistant. It is made as cast glass and the resulting pattern is polished away to give transparency.

(f) Translucent wired glass. Patterned, wired glass offers privacy as well as the attributes described above.

(g) Colored translucent wired glass. Such glass is intended for internal use, because it is particularly weak when subjected to thermal stress.

(h) Colored surface-modified transparent glass. Metallic ions are injected near one surface to produce color.

(i) Colored, coated transparent glass. While types (a–h) are manufactured "on-line," this type can also be produced when glass already cut from the basic manufacturing line is subsequently coated. In the future, most technological advances will affect surface treatments. Initially the enhanced reflection properties of the coatings were merely an additional method of controlling solar heat gain. However, recent developments have led to new coatings which modify the surface coefficient of convective and radiative heat transfer. Such coated glasses can influence the insulation of the facade of the building.

In addition to the basic products, additional types of glass have been developed using three secondary manufacturing processes:

(a) Toughening (tempering). Glass is heated; under controlled cooling, residual stress is frozen in. The result is a stronger, safer product which, when broken, fractures into small, relatively harmless pieces.

(b) Laminating. Glasses are bonded together with plastic interlayers (or more recently, special inorganic interlayers) to provide glass that is difficult to penetrate.

(c) Insulation glazing. Glass is incorporated into a hermetically sealed, factory-made multiple glazing unit which provides thermal insulation. Special low-emissivity coatings or particular gases enhance the insulation ability. Gases can also give increased acoustic insulation.

The range of functions which can be provided by these secondary processes is immense. Almost any requirement can be met.

A third group of glasses offers particular attributes not found in ordinary flat glass. Included in this group are opal glasses, antique glasses, borosilicate glasses, bullions, channel glass, copper lights, glass blocks, glass domes, leaded lights, and lenses (roof and pavement).

2. *Performance of Glass*

Glass must meet a range of design specifications appropriate to the particular building. Often an analysis of the needs leads to contradictory requirements, and compromises are required. No one requirement is usually sufficient to specify the glass to be used.

Research in many diverse fields is continuing, so that relatively simple design rules can be formulated. Each subsection below has its own field of research. Many publications, including national standards and regulations, assist in understanding local requirements for glass performance.

2.1 *Natural Lighting*

This is one of the most primitive reasons for using glass in windows. Although artificial lighting has been relied upon within buildings, concern for energy conservation is encouraging a return to natural lighting and a better integration of this with artificial lighting.

There are well-established methods of predicting the natural light that will enter a building, and recommendations for relating lighting levels to the building type and the task to be performed.

The design procedures for natural light often are separated into considerations of sunlight and of daylight. Glare also needs to be accounted for and separate consideration is usually given to (a) direct glare from the sun, (b) sky glare, and (c) reflected glare from adjacent water or other highly reflecting surfaces. Although sky glare and reflected glare can be controlled by glass with special transmittance properties, direct glare can be reduced only by permanent shading devices.

2.2 Thermal Performance

Heat passes through glass by conduction and by radiation, in both the visible region and the near infrared. However, glass is opaque to far infrared radiation produced by low-temperature objects. Some of the incident solar radiation is absorbed and reradiated both inwards and outwards; the proportions depend on the wind speed over the outer surface.

2.3 Thermal Insulation

Glass is a relatively poor conductor of heat and it is usually used in buildings as a thin membrane. Nowadays it is more often integrated into insulating glass units which provide greater insulation. Figure 1 indicates how the insulation changes with the width of the air gap; it also shows the benefits of low-emissivity metallic coatings and insulating gases.

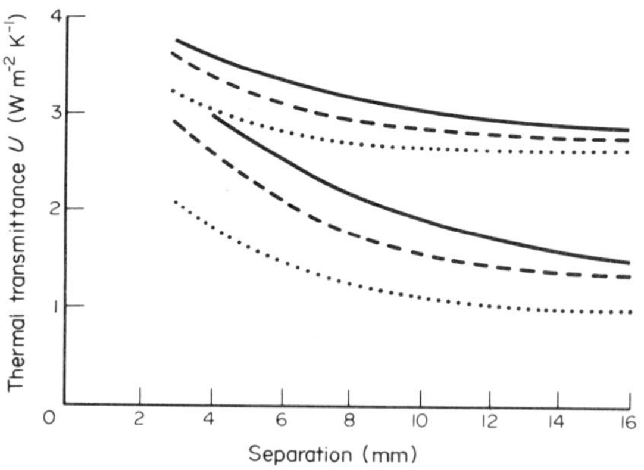

Figure 1
Effect of distance between glass sheets, low-conductivity gases and low-emissivity metallic coating on thermal transmittance of double glazing: —— air; – – – argon; · · · krypton; lower set of curves is for low-emissivity coated glass

The degree of exposure to wind affects the actual thermal insulation obtained, and design guidelines usually divide exposure into three categories indicating sheltered, normal and exposed building sites.

2.4 Solar Heat Gain

The transmission properties, including reradiation from the glass itself, can create a significant amount of heat within buildings. This solar heat gain is the reason for moves to replace heat absorbing glasses with heat reflecting glasses. Methods of designing the glass and determining its effects on air conditioning loads in terms of glass size, location and thermal performance are available.

2.5 Energy Balance of Windows

Until recently, the thermal insulation aspects and solar heat gain aspects were rarely considered complementary benefits of windows. However, with the growing interest in energy conservation, computer techniques have been developed to assess the energy consumption of buildings based on the type of glass, window orientation and building size for any specific form of construction.

2.6 Assessment of Thermal Safety of Glass

In certain special, and infrequent, climatic conditions, all glass types, except toughened glass, are vulnerable to thermal breakage. Such breakage occurs because of the tensile stress from the cooling effect of the window frame on the edge of the glass. Prediction methods are available from glass manufacturers, so that building designers can avoid risks by selecting a suitable glazing system (see Fig. 2).

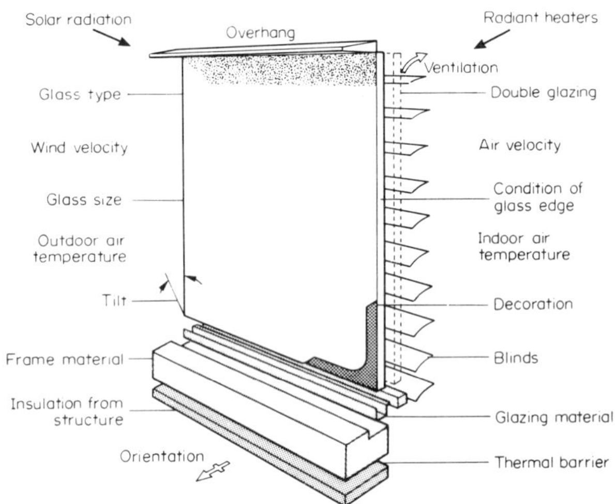

Figure 2
Factors considered in assessing the thermal safety of a glazing system

2.7 Thermal Expansion

Soda–lime glass of all colors, toughened or not, has a thermal expansion of about $8.0 \times 10^{-6}\,K^{-1}$. The effects of expansion are allowed for by selecting an appropriate gap between the glass and the frame.

2.8 Condensation

Condensation on windows is a problem of considerable importance because it can damage building components and thus increase maintenance costs. Methods have been evolved for predicting the risk of condensation as a function of the temperature of the glass, the surrounding air temperatures and the relative humidity.

Because their inner pane is at a higher temperature, hermetically sealed insulating glass units reduce the risk of condensation. Only recently has the added benefit of energy conservation been emphasized.

2.9 Sound Insulation

The reduction of transmitted noise requires an assessment of:

(a) the noise source,

(b) the human response, and

(c) the glass and its surround.

Each of these elements is frequency dependent; some sounds are more objectionable than others.

Glass can be used in several ways to provide acoustic insulation. Some of the factors for assessing the characteristics of different glasses are:

(a) for single glazing:

- (i) mass per unit area, in other words, glass thickness;
- (ii) coincidence effect;
- (iii) closure or sealing (one of the most important);
- (iv) edge support conditions;
- (v) lamination (an acoustic absorbing interlayer); and
- (vi) glass dimensions;

(b) for multiple glazing, in addition:

- (vii) the pane separation;
- (viii) the acoustic insulation of the surroundings between the glass panes;
- (ix) differences in glass thickness;
- (x) nonparallel panes (not usually practical); and
- (xi) mechanical separation.

Some of these factors operate over a narrow band of frequencies, while others can affect the whole frequency range (see Fig. 3).

2.10 Mechanical Strength of Glass

Simply, glass can be considered as a pure brittle material whose surface is covered with small fissures—Griffith flaws. There has been much research on the strength of glass as a material, but relatively little has been published in relation to its use in windows. The behavior of glass in glazing has only recently begun to be understood. Because window glass is flimsy, it is subjected, even under moderate wind loads, to membrane stressing. Although much is now understood of how glass behaves under the short duration pressure loading experienced in a wind storm, most published wind loading design charts are based on an approximation. These charts avoid the need for designers to resort to complex computer techniques.

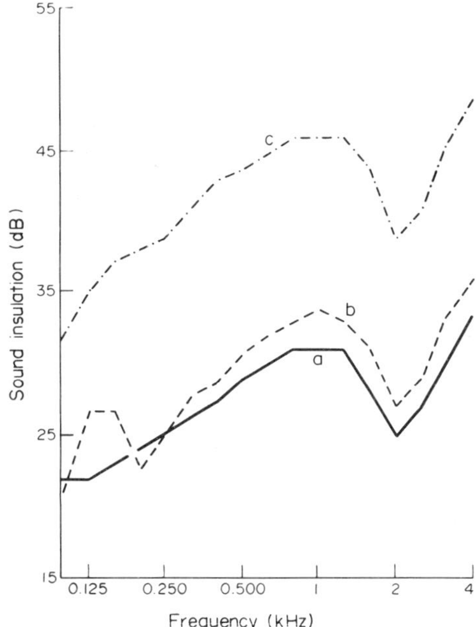

Figure 3
Variation in sound insulation of glazing with frequency, showing effect of double glazing and width of air space: (a) 6 mm single glazing; (b) 6–12–6 mm double glazing; (c) 6–200–6 mm double glazing

The mechanical behavior of glass in a window depends on:

(a) the magnitude of the loading;

(b) the size and thickness of the glass;

(c) the duration of the load and the time for which any part of the pane is subjected to a particular stress; and

(d) the support conditions (although most windows are considered to contain glass which is simply supported, some resistance to rotation exists at the edges and some edge deflection occurs).

With glasses combined in insulating glass units, or laminated glass, a more complex partition of loading occurs. Toughened glass behaves in a similar way to other forms of single glass, but it is up to five times as strong as normal float glass.

For building design, the thickness of glass depends on the predicted maximum wind load. However, the precise relationship between load and glass area varies from country to country, and depends on the basis of the national wind loading code—in particular, the choice of load duration and the statistical return period of such loads.

In many building uses, the maximum deflection of the glass needs to be controlled, not for structural integrity but to allow the glass to appear safe to onlookers. Glass can be used safely in large thin panes, particularly where the wind loading is low, but a certain thickness is needed to cope with any accidental impacts.

2.11 Safety Glass

In some specifically defined risk areas, glass in buildings is subject to accidental human impact. In such situations, glass which either will not break under the expected impacts or breaks safely when hit is recommended for use in many national building regulations or standards.

Throughout the world, the major risk area is in the home (about 2.5% of all home accidents can be attributed to flat glass). The major hazardous situations are:

(a) glass in sliding or hinged doors;

(b) glass in adjacent panels which can be mistaken for doors;

(c) glass fitted near ground level, particularly where many people use the adjacent areas;

(d) glass in balustrade panels; and

(e) glass near swimming pools, where people can slip on wet surfaces.

The appropriate glass for such areas is usually defined by its ability to perform satisfactorily under a simulated human impact. The US test method known as "Z97" has been accepted in slightly modified forms throughout almost the whole world; this test defines glasses that break safely under prescribed impact conditions. The prime examples of safety glasses are toughened glass and laminated glass, but other glass types have also met the particular national standards, for example wired glass, the prime function of which is to provide fire resistance.

2.12 Fire Resistance

In almost all buildings, glass is required to be used in areas where resistance to the passage of flames is required. Although too much glass can increase the fire risk, in some situations the use of glass is advantageous. For instance, a transparent area in a fire door can give advance warning of the presence of a fire. In such areas, glass must also be able to contain the fire for a specified period (30 minutes being a typical requirement).

Traditionally, wired glass has been used as fire-resistant glass, as a consequence of its success in meeting the appropriate specification over many decades. However, like all transparent glass, it transmits radiation created by fires, and this in itself can spread the fire. To overcome this drawback, special laminated glasses have recently been developed. They are normally transparent but become opaque to radiation when subjected to the high temperatures which occur near fires.

All soda–lime glasses are incombustible and will not themselves contribute to a fire. Hence, fires cannot spread by burning of the glass surface; this is particularly important on the outside of buildings and on roofs.

2.13 Security

Security glass is required more today than ever before. Such glass, which is almost always a complex laminated glass, is classified in terms of its ability to withstand impacts from bullets, malicious attack or explosions.

2.14 Glass Durability

Glass made from the standard soda–lime composition is highly resistant to deterioration. On site, glass surfaces can be attacked if allowed to remain wet, and the effects are very difficult to remove except by polishing the surface. Glass should therefore be stored in dry, warm conditions. Glass is attacked by alkaline solutions, such as the effluent from rain water draining over concrete, and by welding "splatter."

2.15 Control of Reflections

Reflections can reduce a clear, uninterrupted view through a shop window. All glass reflects light, and the degree of reflection depends on the angle at which the glass is viewed. By positioning the glass at an angle to the general direction of viewing, or by using horizontally curved glass, such reflections can be avoided. The extent of the problem also depends on the relative lighting levels on the two sides of the glass. Wherever possible, the lighting level should be higher in the area to be viewed.

In other buildings, highly reflective glass creates privacy, as long as the lighting level on the outside of the building is greater than the level inside. Similarly, privacy can be obtained by lowering the lighting level in a private area. Security viewing windows work in this way.

2.16 Antifade Properties of Solar Control Glasses

Although fading of colors is attributed to the effects of ultraviolet radiation, modern organic pigments can also be affected by wavelengths in the visible spectrum and by heat. Only 3% of solar radiation reaching the earth's surface is contained within the uv range (10–380 nm) and normal clear float glass rejects about 50% of this; a polyvinyl butyral interlayer in laminated glass rejects about 95%.

Some research has been carried out on how tinted glass can reduce the risks of fading, but for irreplaceable works of art, windows need to be located so that direct solar radiation does not fall on the object.

2.17 Cleaning and Maintenance of Glass in Buildings

To preserve its surface, glass should be cleaned regularly. Dust can reduce the thermal and visual performance of glass, increase its thermal absorption and reduce its life. In extreme cases, the glass may become discolored. Glass should be cleaned with nonalkaline solutions.

3. Installation of Glass

Glass is installed in numerous ways, some being suitable only for certain climates. In areas of severe weather, more sophisticated methods of glazing are required. In some parts of the world, different factors, such as temperature rather than driving rain, are more important.

The current methods have evolved from tradition, and only in the last 20 years have internationally accepted methods of fixing glass become prevalent. Changes in glazing systems have also resulted from the changes in the materials used for window frames. First wood, then aluminum and steel were common. Now that plastic frames are gaining in use, coping with expansion has become a major problem.

With the introduction of adhesive glazing compounds which have considerable strength as well as an ability to survive on buildings for more than 15 years, interest is growing in locating glass in preformed openings without any mechanical fixings. Structural integrity can be maintained solely with these compounds.

Most countries have published guidelines to ensure that glazing is carried out to a sufficiently high quality.

3.1 Glazing of Single Glass

The major requirements of the frames are:

(a) to cope with the quality of glass cutting and the resulting variations in size and shape;

(b) to prevent glass–hard metal contact, as this could lead to glass fracture;

(c) to allow for the differences between the thermal expansions of the glass and the frame material;

(d) to allow the glazing compound to survive (many compounds require a considerable thickness of fillets to withstand temperature and humidity); and

(e) to withstand the weight of the glass and the structural effects on the glass of the many environmental factors (e.g., wind loading).

These requirements have been translated into simple glazing rules which have been almost universally adopted:

(a) The glass size should be about 3 mm less than the dimensions of the space within the window frame. This difference is increased as the glass size and thickness increase.

(b) The glass should rest centrally on two blocks set at about the quarter points. The width of these blocks corresponds to the glass thickness; their height is adjustable. As frames have become lighter in construction, the position of the "setting blocks" has been adjusted and now they can be positioned anywhere away from the corner of the glass, but they should preferably be close to the quarter points.

(c) In the past, no location pieces were used between the vertical edges of the glass and the frame. Now that glass is transported already glazed, side location pieces are necessary to prevent glass–frame contact.

(d) Clearance is ensured between both the front and back faces of the glass and the frame, by using distance pieces. The gap allowed depends on the glazing method but is rarely less than 3 mm.

Nowadays there must be additional considerations. Traditionally, a very hard glazing compound, linseed oil putty, has been used to hold the glass in position; over many years, its suitability has been proved in countless buildings. However, putty sets very hard and is not able to cope with glass expansion. With the increasing use of, for example, heat-absorbing solar control glasses, glazing compounds have been introduced. The same requirements apply, but the importance of distance pieces is increased.

3.2 Glazing of Sealed Insulating Glass Units

For hermetically sealed, insulating glass units, a new, additional set of glazing parameters has arisen. First, a further source of glass movement was introduced. The sealed airspace in the unit changes volume as the external pressure and temperature change. Suitable glazing compounds have been developed to cope with the resulting edge rotation.

An even more important design requirement soon became evident. In almost every case, the life of sealed insulating glass units depends critically on diversion of water from the edge of the sealed unit. At first, full bedding of the unit was carried out, but the creation of a satisfactory long-life seal depends on choosing suitable materials for such thick volumes of glazing compound and achieving the necessary glazing quality and workmanship. However, at about the same time it was found that the accuracy with which aluminum frames could be extruded allowed for the incorporation of channels to hold preformed rubber-based gaskets. The use of such dry methods opened up a new era of glazing design. On-site work could be reduced to a minimum and consistent glazing quality was easier to achieve.

3.3 Drained Glazing Systems

Such systems are now almost universally accepted as the best means of glazing. Although the principles are easy to describe, achievement in practice is difficult to perfect. The system is designed so that any water entering behind the outside glass–frame seal is allowed to drain out of the frame rather than forced into the building. Because the chamber beneath the glass must always remain at external air pressure, the seal between the inside face of the glass and the frame must be pressure-proof. Thus, no water enters the building and any water getting past the outer seal drains away. The glass edge is not subjected to continuous immersion in water, a situation which would considerably reduce the life of insulating glass units.

The two most popular methods for drainage are (a) beads which directly connect to the outside, and (b) internal channels. However, experience shows that the problems associated with drainage should not be underestimated. Not only is good initial design necessary, which takes into account erection tolerances, but also a high quality of erection skill is required, particularly in terms of achieving the essential inner waterproof seal.

See also: Glazing; Glass: An Overview; Structural Sealant Glazing

Bibliography

American Society of Heating, Refrigeration and Air Conditioning Engineers *ASHRAE Guide*. American Society of Heating, Refrigeration and Air Conditioning Engineers, Atlanta, Georgia

Charles R J 1959 The strength of silicate glasses and some crystalline oxides. In: Averbach B L, Felbeck D K, Hahn T T, Thomas D A L (eds.) 1959 *Fracture*. Wiley, New York

Chartered Institution of Building Services *CIBS Guide*. Chartered Institution of Building Services, London

Collins B L 1975 *Windows and People: A Literature Survey—Psychological Reaction to Environments with and without Windows*, Building Science Series 70. National Bureau of Standards, Washington, DC

Glass and Glazing Federation 1978 *Glazing Manual*. Glass and Glazing Federation, London

Hastings S R, Crenshaw R W 1977 *Window Design Strategies to Conserve Energy*, Building Science Series 104. National Bureau of Standards, Washington, DC

Hopkinson R G, Petherbridge P, Longmore J 1966 *Daylighting*. Heinemann, London

Lynes J A 1968 *Principles of Natural Lighting*, Elsevier Architectural Science Series. Elsevier, New York

McGrath R, Frost A C 1961 *Glass in Architecture and Decoration*. Architectural Press, London

Pilkington Environmental Advisory Service 1976 *Glass and Noise Control*. Pilkington Flat Glass, St Helen's, UK

Pilkington Environmental Advisory Service 1981 *Calculating Thermal Transmittance*. Pilkington Flat Glass, St Helen's, UK

Pilkington Technical Advisory Service 1980 *Glass and Thermal Safety*. Pilkington Flat Glass, St Helen's, UK

Turner D P (ed.) 1977 *Window Glass Design Guide*. Architectural Press, London

Walsh J W T 1961 *The Science of Daylight*. MacDonald, London

D. W. Armstrong

Glass Bonding in Advanced Ceramics

Oxide mixtures, like glass, are widely used as a bonding agent for joining ceramics and forming ceramic-to-metal seals in the fabrication of electron tubes, magnetic recorder heads, high-pressure alkali vapor lamps and high-voltage feedthroughs. The oxide mixture can be composed in such a way that it melts either like a metal alloy or like a glass, which means that the mixture can have a melting point or a melting range that can be very short or very long.

Melting of a glass in contact with a ceramic or a metal causes interaction at the interface that can finally lead to a reaction. The reactions of glass with a ceramic or with an oxidized metal are dissolution reactions, which decisively influence the quality of the bond.

In conventional glass-to-metal joining, the metal is preoxidized so as to be easily wetted by the molten glass. The dissolution of metal oxide in the glass has to be controlled in such a way that the glass at the interface becomes saturated with the low-valent oxide of the metal before the oxide layer is completely dissolved (Pask and Fulrath 1962). Processes involving a liquid phase with the bond formation between ceramic parts or between a metal and a ceramic part may be associated either with capillary action or with mechanical deformation between the surfaces to be bonded.

Since nonequilibrium reactions are involved in the bond formation, the reactivity of the partners in the joint should be known. In the final bond the properties of the reaction products should also be known, because they determine to a great extent the strength and other properties of the joint. In the following descriptions emphasis is therefore placed on the engineering and the processing conditions for glass bonding of advanced ceramics.

1. Use of Glass as a Bonding Agent

In the following description, the term glass is used for every oxide mixture that is used as a bonding agent between ceramic parts and between ceramic–metal parts. Although joining of materials by a liquid phase at elevated temperatures is a nonequilibrium process from the thermodynamic point of view, the equilibrium diagrams of the components involved are very useful to an understanding of the joining process.

This may be illustrated by the reactions at the interface between alumina and two different melts in the Al_2O_3–CaO–SiO_2 system (Fig. 1). First consider a melt with a noneutectic composition indicated by A in Fig. 1, in contact with Al_2O_3 ceramic at 1600 °C. This melt is not in equilibrium with Al_2O_3, so the melt starts to dissolve Al_2O_3 from the ceramic. The composition of the melt changes towards the Al_2O_3 corner, indicated by the dashed line. As a result, the melting temperature drops until the composition coincides with the composition on the eutectic line, and then increases with further dissolution of Al_2O_3 until equilibrium is established at 1600 °C. This is indicated by B in Fig. 1.

On cooling, the crystalline phases 2CaO . Al_2O_3. SiO_2, CaO . $2Al_2O_3$ and CaO . $6Al_2O_3$ segregate from the melt (Fig. 2), and so also does CaO . Al_2O_3. $2SiO_2$, due to the selective reaction of CaO with the ceramic. The alumina-rich compounds CaO . $6Al_2O_3$ and CaO . $2Al_2O_3$ close to the interface are formed at the Al_2O_3–melt interface by diffusion of CaO from the melt, and they are in the solid state during the reaction process. Thus, the melt reacts not with the Al_2O_3 from the ceramic surface, but with the CaO . $2Al_2O_3$ compound.

Since the formation of the solid Al_2O_3-rich compounds is diffusion-controlled (Oishi et al. 1965), the reaction rate at the interface is controlled by the formation rate of the compounds. The reaction rate has to be determined experimentally, because it depends not only on the composition of the ceramic but also on the microstructure. This can be done by measuring the displacement d of the melt–ceramic interface at different temperatures. For a parabolic reaction rate the reaction constant $K = d^2/t$ is determined, and with $K = D\exp(-Q/RT)$, the diffusion constant is found (Kingery 1960). Q is the activation energy for diffusion, R the gas constant, T the absolute temperature and t the reaction time.

There can be wide differences in the physical and chemical properties of the new phases in the solidified melt, for example in the thermal expansion and melting temperature or in resistance to aggressive media (Chu 1979, Jonker et al. 1965). Therefore it is pointless to start with a joining agent in the glassy state that matches the thermal expansion of the materials to be joined, if crystallization occurs during the joining operation. The thermal expansion values ($10^{-7}\,K^{-1}$) of the crystalline phases in the above example are 70 for CaO . $6Al_2O_3$, 46 for CaO . $2Al_2O_3$ and 72 for 2CaO . Al_2O_3 . SiO_2, with the respective melting temperatures 1830, 1762 and 1600 °C. The latter show that the physical stability of a bond made with this glass is much higher than that of the original glass.

Now consider a melt with the eutectic composition

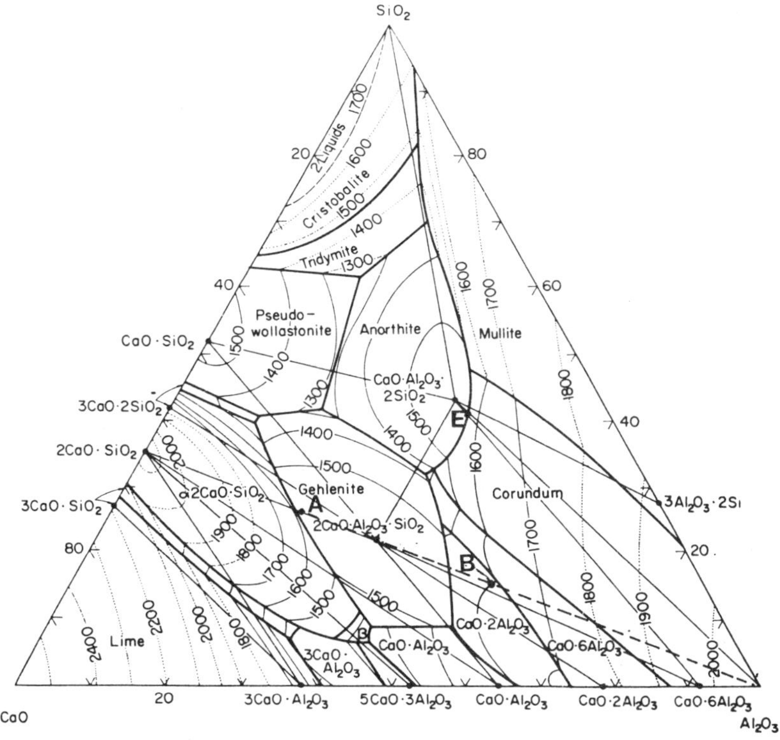

Figure 1
Phase diagram of Al_2O_3–CaO–SiO_2 system (Levin et al. 1979)

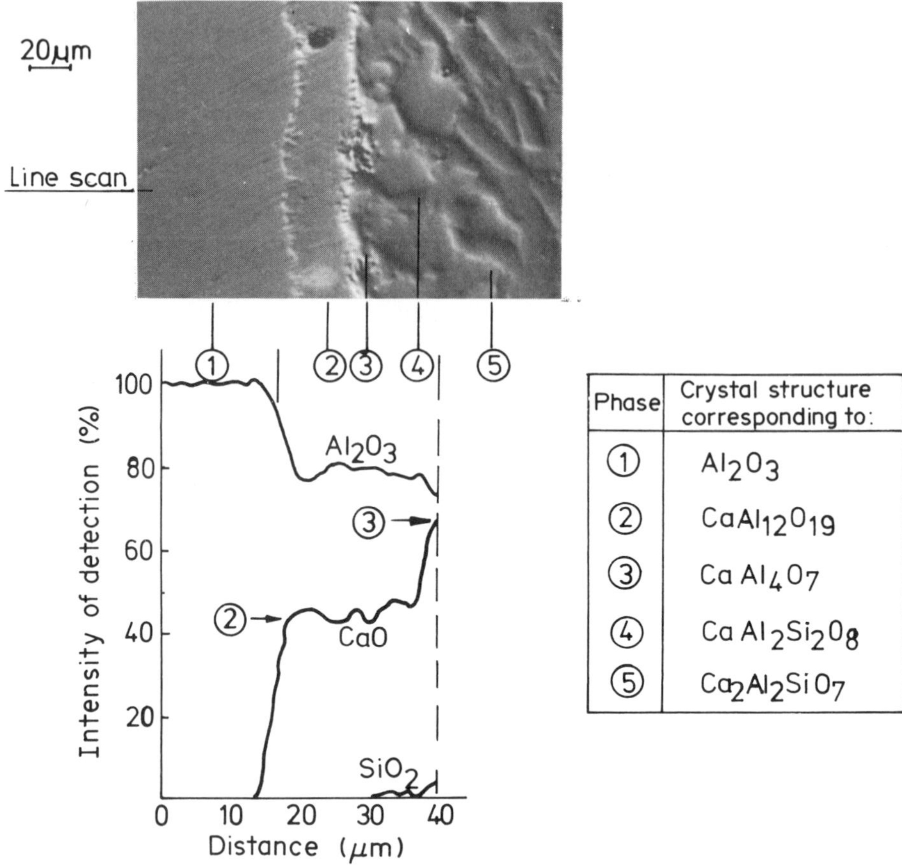

Phase	Crystal structure corresponding to:
1	Al_2O_3
2	$CaAl_{12}O_{19}$
3	$CaAl_4O_7$
4	$CaAl_2Si_2O_8$
5	$Ca_2Al_2SiO_7$

Figure 2
Interface between Al_2O_3 ceramic and Al_2O_3–CaO–SiO_2 (Klomp 1980)

indicated by E in Fig. 1, in contact with Al_2O_3 ceramic at 1550 °C, that is, at the eutectic temperature. At this temperature the melt is in equilibrium with alumina and no dissolution of Al_2O_3 takes place. On cooling, Al_2O_3 and $CaO.Al_2O_3.2SiO_2$ crystals will segregate from the melt and the physical stability of the bond will be identical with that of the original glass. If the melt is heated to a temperature above the eutectic temperature, Al_2O_3 starts to dissolve into the melt. On cooling there is a solidification range, and the solidified melt resists higher temperatures than the eutectic melt.

Thus, it is possible for the temperature that a joint can resist to be higher than the original melting temperature of the glass if the glass composition and/or the joining temperature are properly chosen. The formation of crystalline phases, which increases the thermal stability of the bond, can also be realized without any reaction with the ceramic to be bonded if glass is used which easily crystallizes or if crystallization nuclei are introduced into the glass (McMillan 1964, Monneraye 1970). This has led to the development of glass ceramics. Such materials can be used to make hermetic seals between ceramics and between ceramic and metal (Monneraye 1970, Partridge et al. 1984, Lay et al. 1984).

In joining ceramic to metal, it is necessary to consider reactions not only between the glass and the ceramic but also between the glass and the metal. In common glass-to-metal joining practice, the metal is preferably preoxidized in such a way that a well-adhered oxide layer is formed on the metal surface. Typical materials treated in this way are FeCr, FeNi and FeNiCo. These materials are commonly joined in air. Preoxidation of the metal not only improves wetting by the glass but also provides favorable conditions for achieving a gradual transition from metal to glass, since oxide from the layer and components from the glass dissolve mutually. In this way a chemical bond is formed between the glass and the metal (Pask and Fulrath 1962). If the glass contains components that can be easily reduced by the metal, the latter will continue to oxidize and the reduced component can dissolve in the metal, stay in the glass, where it may finally be present in crystalline form (Borom and Pask 1966), or evaporate as, for example, sodium does (Pask and Tomsia 1981).

If the metal cannot be preoxidized and the joining process has to take place in an atmosphere with low partial pressures of oxygen and hydrogen, there can be a reaction of the metal with the glass at the interface, depending on the chemical nature of the metal and the glass. This is illustrated in Fig. 3a, which shows a photomicrograph of an interface between niobium and glass, composed of 38% Al_2O_3–28% B_2O_3–22% La_2O_3–12% SiO_2 (mol%). The interface is formed at 1500 °C in argon containing 3 ppm oxygen. The reaction at the interface has resulted in the formation of a boundary layer in the niobium, which contains the reaction products. Electron probe microanalyses (EPMA) reveals the presence of aluminum and silicon in the reaction layer (Fig. 3b). The oxygen released from the glass components dissolves in the niobium, as also does the oxygen present in the argon. The reactions occurring are a combination of redox and dissolution reactions that can be represented by the reaction equations:

$$Al_2O_3 \rightarrow 2Al + \tfrac{3}{2}O_2 \quad (1)$$

$$SiO_2 \rightarrow Si + O_2 \quad (2)$$

$$2Nb + O_2 \rightarrow 2NbO \quad (3)$$

$$Nb + Al \rightarrow Nb.Al \quad (4)$$

$$Nb + Si \rightarrow Nb.Si \quad (5)$$

The energy required for (1) and (2) to take place is less than the energy supplied by reactions (3–5), as can be calculated from thermodynamic data for reactions (1–3) (Barin and Knacke 1973) and from the heat of alloy formation for reactions (4) and (5) (Miedema et al. 1977). The above consideration holds for metals with high oxygen affinity, in particular for metals of the IVb and Vb groups of the periodic table. The dissolution of oxygen, aluminum and silicon makes these metals brittle and consequently less resistant to mechanical stresses.

To illustrate the absence of an interaction layer when a metal with low oxygen affinity is used, consider the interface between molybdenum and the above mentioned glass formed under the same conditions as the niobium–glass interface (Fig. 4). Figure 4a shows a photomicrograph of the interface and Fig. 4b the results of EPMA across the interface. The oxygen partial pressure of 0.3 Pa is higher than the equilibrium partial pressure of MoO_2, so that MoO_2 is formed. This in turn dissolves in the glass (Klomp and Vrugt 1981). From this it may be speculated that MoO_2 is also present at the molybdenum–glass interface, although in a very thin layer.

The phenomena described above occur in any ceramic–glass–metal system and they are basic to the development of joining techniques. However, a vast

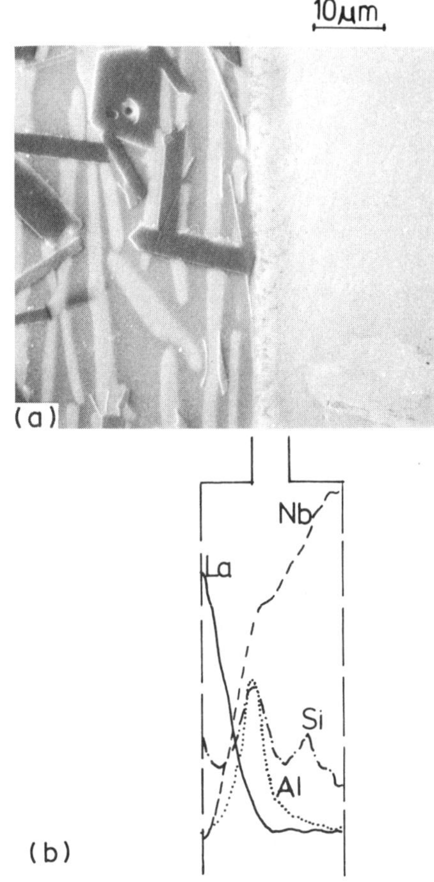

Figure 3
(a) Interface between niobium and Al_2O_3–B_2O_3–La_2O_3–SiO_2; (b) EPMA of aluminum and silicon concentration profiles over the same interface

number of advanced ceramics are multicomponent materials, such as ferrites, titanates and nitrides, so the glass can be more reactive with one component than with another. Therefore instrumental analyses are indispensable in the development of joining techniques. The choice of the glass will thus be based on properties that reflect the operating conditions of the final structure.

If chemical properties are the important factor that determines the life of a structure, for example the metal–ceramic joints in metal-vapor discharge lamps, the glass composition is based on the chemical resistance of the glass and that of the reaction products of the joint formation. This requires knowledge of the reactions that can take place, and in particular of the kinetics of the reactions, because the latter determine the life of a joint.

Useful glasses for making seals in high-pressure sodium vapor lamps may be synthesized from the system Al_2O_3–CaO–MgO–BaO with minor additions

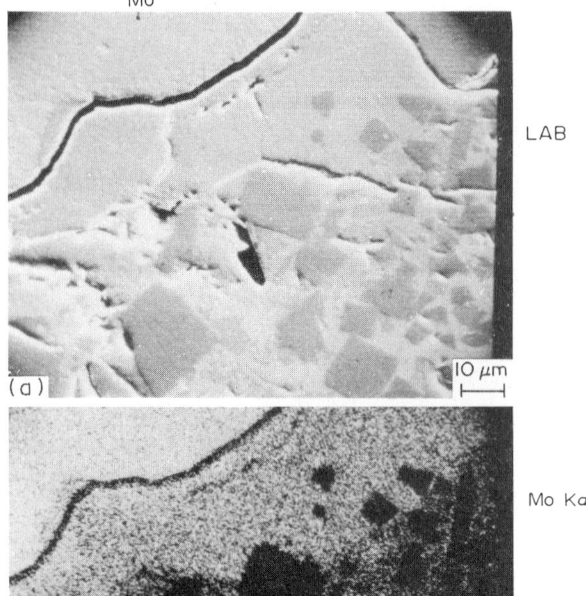

Figure 4
(a) Melt interface between molybdenum and Al_2O_3–B_2O_3–La_2O_3–SiO_2 (LAB); (b) EPMA of molybdenum over the same interface

of SiO_2, B_2O_3 and Y_2O_3 to modify the properties (van Vliet and de Groot 1981). The melting temperature is 1350 °C or higher, depending on the composition.

2. Joint Design

A glass joint can be made between ceramic parts or between a metal and a ceramic part either by filling a capillary space between them with a free flow of liquid glass or by applying the glass between the parts and squeezing it out by application of an external load. The latter can be done at lower temperatures, that is, at a higher glass viscosity, which has the advantage that the glass then has a lower reactivity and a lower reaction rate than the more fluid glass. This method is widely applied in the fabrication of magnetic recorder heads (see Sect. 3). For cylindrically shaped joints this method cannot be applied, and the gap then has to be filled by capillary action. This requires the glass to be in a low-viscosity state.

Filling of the joint gap by capillary action occurs if the gap is small. This follows from wetting and spreading principles (Milner 1958). In the following it is assumed that the glass does not react with the ceramic. The capillary is considered, in a first approach, as a gap with spacing d between horizontal parallel plates of ceramic that has a surface energy γ_c, to be filled with glass with surface energy γ_g, filling the capillary and forming an interface with energy γ_{cg}. Then the pressure drop p across the liquid surface is given by

$$p = \frac{2(\gamma_c - \gamma_{cg})}{d} \tag{6}$$

The pressure p causes the glass flowing through the capillary to oppose the viscous friction in the glass. The filling rate of the capillary with length l depends on the viscosity η of the glass, and is given by the relation

$$l^2 = \frac{pd^2t}{6\eta} \tag{7}$$

where t is time. With $\gamma_c - \gamma_{cg} = \gamma_g \cos\theta$, where θ is the equilibrium wetting angle of the glass on the ceramic, Eqn. (6) becomes

$$p = \frac{2\gamma_g \cos\theta}{d} \tag{8}$$

and with this, Eqn. (7) becomes

$$l^2 = \frac{2\gamma_g \cos\theta dt}{6\eta} \tag{9}$$

The time for filling then follows from Eqn. (9).

In the filling of a vertical joint by upward flow of the glass, the driving force is opposed by the hydrostatic pressure of the glass column and the viscous friction. Then

$$p = \frac{2\gamma_g \cos\theta}{d} - \rho g l \tag{10}$$

where ρ is the density of the glass and g the gravitational constant, and the time for filling, t, is

$$t = \frac{12}{\rho^2 g^2 d^3}\left[-\rho g l d - 2\gamma_g \times \cos\theta \ln\left(\frac{2\gamma_g \cos\theta - \rho g l d}{2\gamma_g \cos\theta}\right)\right] \tag{11}$$

Identical calculations can be carried out for a joint between glass and metal and for ceramic–metal systems to be joined by glass.

In systems where chemical reactions occur, the surface and interface energies change as well as the viscosity of the glass. If new phases form at the interface (see Sect. 1) that are solid at the process temperature, the capillary can even be blocked. Preference will therefore be given to systems that are less reactive. In systems where the gap is functional for the product, as for example in magnetic recorder heads, even the smallest change in the gap dimension is not tolerated and the glass has to remain a non-magnetic material. In such cases the capillary filling technique can be applied only using glasses with a low liquidus temperature.

Methods to prevent the glass reacting with the ceramic have been developed, in which thin layers of metal or oxides, strongly adhering to the ceramic, are used in order to protect the ceramic from dissolution in the glass. If glasses are used that have a viscosity of <10 Pa s at ~ 1000 °C, a Cr or SiO_2 film fully protects the ceramic from reaction with the glass for at least 15 minutes. The glass can also be applied by sputtering on top of the Cr or SiO_2 film and the joint can be made by exerting a pressure on the parts to be joined. The force F required depends on the viscosity of the glass η, the surface area to be bonded, the gap width h and the deformation rate v. For circular surfaces of radius r, the force is given by the relation (Cottrell 1963)

$$F = \frac{3\pi\eta r^4}{2h^3} v \tag{12}$$

The parts to be bonded have to be designed in such a way that the bonding pressure does not introduce cracks in the ceramic or such high flexural stresses that failures occur in the joints after removal of the load (Klomp and Van der Ven 1980).

3. Joining Procedures

The glass to be used in the joining of ceramic parts or ceramic and metal parts can be applied as a powder with a suitable binder, as a sintered or premelted piece or as a sputtered layer, depending on whether the free capillary flow technique or the pressure bonding technique is applied. A few examples will illustrate this. In the capillary flow technique, the glass is placed at the entrance of the capillary, as indicated in Fig. 5, which shows an alumina–glass–niobium joint assembly with a capillary space averaging 80 μm width and 5 mm length. The assembly is heated by radiation in a neutral gas up to 1450 °C, where the Al_2O_3–CaO–MgO glass is a liquid that flows into the capillary spaces between the metal and the ceramic, in which process the ceramic ring moves simultaneously towards the ceramic tube.

In the example of Fig. 6, several joints were simultaneously made between a debased alumina disk and molybdenum wires using an Al_2O_3–MnO–SiO_2 glass, applied as small rings made by pressing the glass powder with nitrocellulose as a binder. The gap width was 35 μm, the length 2 mm. The joints were made by heating in a forming gas-shielded furnace at 1200 °C. A ceramic–glass–Kovar joint, as indicated in Fig. 7, was made by applying a devitrifying glass powder near the gap between the alumina and Kovar. The glass was composed of Al_2O_3–B_2O_3–CaO–SiO_2–ZnO, which liquefies with a viscosity of ~ 1 Pa s at 1000 °C. On heating in a forming gas furnace at 1000 °C the glass filled the gaps, and during cooling, Zn_2SiO_4 and $ZnAl_2O_4$ crystallites segregated from the melt, while at the same time the viscosity of the glass increased.

Figure 8 shows a joint between two MnZn ferrite parts to be used in a magnetic recorder head. The ferrite surfaces are protected from dissolution in the

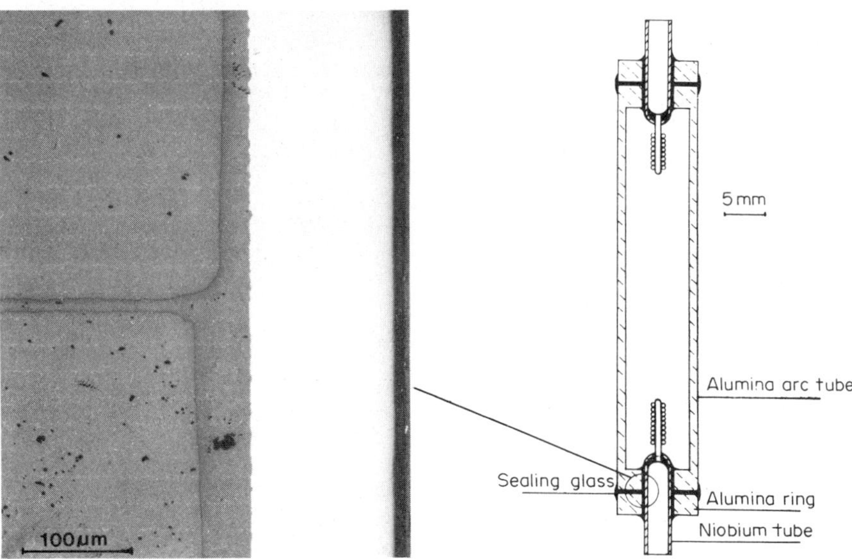

Figure 5
Al_2O_3–glass–niobium joint, gap length 5 mm, average width 80 μm, with Al_2O_3–CaO–MgO melt (Van Vliet and De Groot 1981)

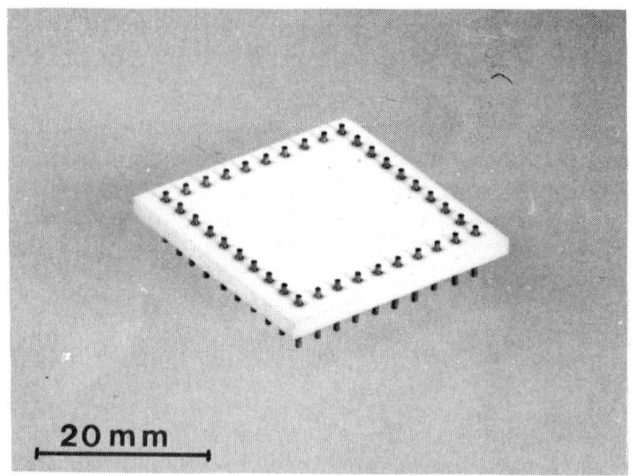

Figure 6
Molybdenum feedthroughs in debased alumina, with Al_2O_3–MnO–SiO_2 melt (Klomp 1980)

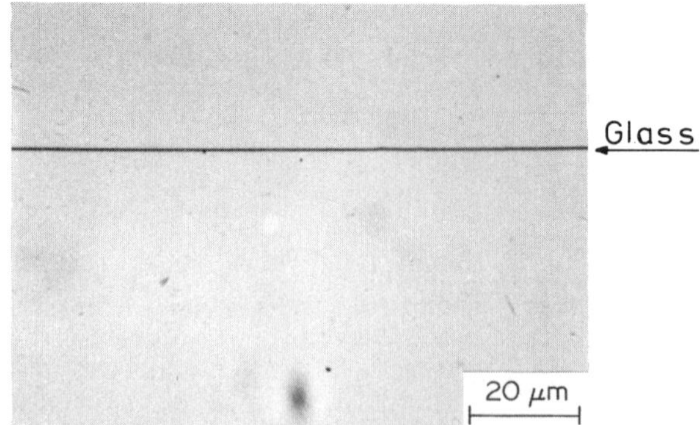

Figure 8
MnZn ferrite parts joined by pressure bonding with Na_2O–B_2O_3–SiO_2 glass; gap width 0.5 μm

glass by an SiO_2 sputtered layer. The glass is also applied by sputtering on top of the SiO_2 layer, forming a final layer thickness of ~250 nm. The coated surfaces are joined by heating in air and simultaneous application of a pressure of 0.4 MPa, when the viscosity of the glass is 10–100 Pa s. The final gap width is controlled by the pressure applied.

Figure 9 depicts a joint made by capillary filling with glass of a 1.8 μm gap between NiZn ferrite parts. The gap width is maintained by spacers. The solid glass is placed in the wedge-shaped entrance of the capillary. The glass flows into the gap at a temperature of 950 °C, its viscosity being ~2 Pa s.

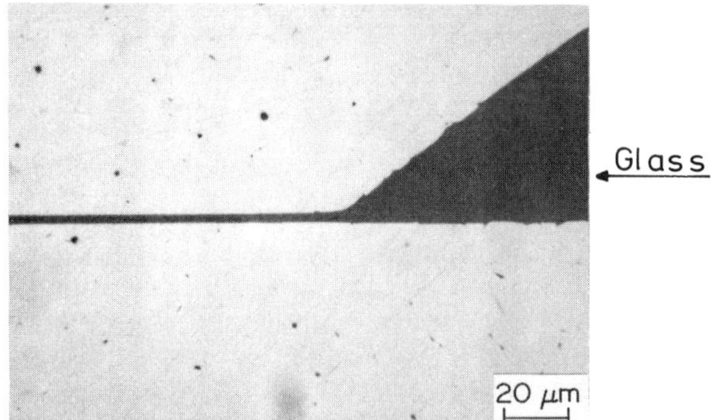

Figure 9
NiZn ferrite parts joined by capillary flow of Na_2O–B_2O–SiO_2 glass in the 1.8 μm wide gap

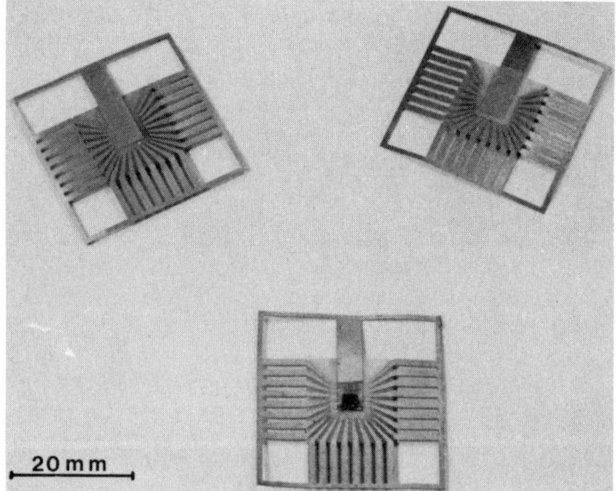

Figure 7
Alumina–glass–Kovar joint with Kovar frame on top of an alumina substrate and Al_2O_3–B_2O_3–CaO–SiO_2–ZnO devitrified glass (Monneraye 1970)

4. Properties of Joints

The properties of glass-bonded joints of ceramic and ceramic–metal systems depend on the type of glass used in the bonding process. Since glass cannot relieve stresses at room temperature, the residual stresses from the joining process and those caused by differences in thermal expansion between the glass and the joined parts determine to a great extent the mechanical loadability of the joint and its resistance to thermal shock. Furthermore, the quality of the joint (that is, its homogeneity) is important, whether or not cracks develop in the joint. Pores in the outside area of the joint act as stress raisers when the joint is loaded.

Thus, the strength figures given in the literature are generally related to the loadability of the particular test structure rather than to the bond strength of the joint. The strength of glass joints depends on the morphology of the glass, the grain size of the ceramic and the thermal stresses in the joint. The latter can

Table 1
Strength of glass joints between alumina and metals

Materials joined	Glass[a]	Joining temperature (°C)	Strength (MPa)
Al_2O_3–Al_2O_3	ACM	1500	2.5[b]
	ASM	1200	2.6[b]
	ABCSZ	1000	2.6[b]
Al_2O_3–molybdenum	ACM	1500	1.5
	ASM	1200	1.2
Al_2O_3–Kovar	ASM	1200	1.7
	ABCSZ	1000	1.6
Al_2O_3–niobium	ACM	1500	2.5
	ACS	1450	2.4

a ACM, Al_2O_3–CaO–MgO (Klomp 1980); ASM, Al_2O_3–SiO_2–MnO (Klomp 1980); ABCSZ, Al_2O_3–B_2O_3–CaO–SiO_2–ZnO (Monneraye 1970); ACS, Al_2O_3–CaO–SiO_2 b Failure in the ceramic

be generated not only by differences in thermal expansion between the materials but also by the presence of new crystalline phases in the joint, formed during the joining process. If the glass is crystallized in the joint, the strength is higher than in the amorphous state. Thus, without a structural and physical characterization of the material in the joint, it is difficult to interpret strength figures. Nevertheless, Table 1 indicates what values can be achieved. The glass in these joints was crystallized and the alumina was $\geqslant 99.5\%$ pure with an average grain size of 30 μm. The strength was measured by a three-point bending test.

The use of a recently developed fracture mechanics concept to characterize the mechanical properties of a joint makes it possible to obtain data of the fracture resistance of the joint which do not depend on the geometry of the test sample (Suga et al. 1984).

The thermal shock resistance depends predominantly on the thermal expansion differences between the materials and between the crystalline phases in the ceramic. The latter can give rise to microcracks in the joint, weakening it. This is a well-known effect in glass ceramics (McMillan 1964).

The behavior of a joint when exposed to chemical aggressive media, whether or not at high temperature, depends on the composition of the glass in the joint. If new crystalline phases are formed during the joining process, the chemical properties of these phases in relation to the application have to be considered.

See also: Glass Sealing; Joining of Ceramic–Metal Systems: General Survey; Joining of Ceramic–Metal Systems: Procedures and Microstructures

Bibliography

Barin I, Knacke O 1973 *Thermomechanical Properties of Inorganic Substances.* Springer, New York

Borom M, Pask J A 1966 Role of adherence oxides in the development of chemical bonding at glass-metal interfaces. *J. Am. Ceram. Soc.* 49: 1

Chu G P K 1979 Properties and evaluation of HPS lamp seals. *J. Illum. Eng. Soc.* 8(4): 250–56

Cottrell A H 1963 *The Mechanical Properties of Matter.* Wiley, New York

Jonker G H, Klomp J T, Botden T P J 1965 High temperature seals on pure and dense alumina. In: Stewart E (ed.) 1965 *Science of Ceramics*, Vol. 2. Academic Press, New York, pp. 295–302

Kingery W D 1960 *Introduction to Ceramics.* Wiley, New York

Klomp J T 1980 Interfacial reactions between metals and oxides during sealing. *Am. Ceram. Soc. Bull.* 59: 794–800

Klomp J T, Van der Ven A J C 1980 Parameters in solid-state bonding of metals to oxide materials and the adherence of bonds. *J. Mater. Sci.* 15: 2483–88

Klomp J T, Vrugt P J 1981 Interfaces between metals and ceramics. *Mater. Sci. Res.* 14: 97–105

Lay L A, Morell R, Wallis R 1984 Joining "REFEL" silicon carbide to porcelain. *Proc. Br. Ceram. Soc.* 34: 232–39

Levin E M, Robbin C R, McMurdie H F 1979 *Phase Diagrams for Ceramists.* American Ceramic Society, Columbus, Ohio

McMillan P W 1964 *Glass-Ceramics.* Academic Press, New York

Miedema A R, Boer F R de, Boom R 1977 Model predictions for the enthalpy of formation of transition metal alloys. *CALPHAD: Comput. Coupling Phase Diagrams Thermochem.* 1: 341–59

Milner D R 1958 A survey of the scientific principles related to wetting and spreading. *Br. Weld. J.* 5: 90–105

Monneraye M 1970 Le scellement métal céramique. Un renouveau des scellements non métalliques. *Tech. Philips* 5: 23–35

Oishi Y, Cooper A R, Kingery W D 1965 Dissolution in ceramic systems, III. Boundary layer concentration gradients. *J. Am. Ceram. Soc.* 48: 88–95

Partridge G, Elyard C A 1984 Glass–ceramic bonds to metals. *Proc. Br. Ceram. Soc.* 34: 219–29

Pask J A, Fulrath R M 1962 Fundamentals of glass to metal bonding, VIII. Nature of wetting and adherence. *J. Am. Ceram. Soc.* 45: 592–96

Pask J A, Tomsia A P 1981 Reactions at liquid/solid interfaces. *Mater. Sci. Res.* 14: 411–19

Suga T 1984 Bruch mechanische charakterisierung und bestimmung der Haftestigkeit von materialübergängen. Thesis, Max Planck Institut für Metallforschung, Stuttgart

van Vliet J A J M, de Groot J J 1981 High pressure sodium discharge lamps. *IEE Rev.* 128: 415–41

J. T. Klomp

Glass Ceramics

Glass ceramics are glasses in which fine, uniform crystals are grown by controlled heat treatment. They have higher strength, chemical durability and electrical resistivity than the parent glass, and can be made with very low thermal expansions, giving excellent resistance to thermal shock. Glass ceramics are one of the few really new engineering materials developed in the last few decades, and important applications include their use as cookware, stove tops, radar domes, electrical insulators and telescope mirrors.

To make a glass ceramic, the starting material is formed into a desired shape by conventional methods of fabricating glass (see *Forming of Glass*), and is subsequently heated first to nucleate and then to grow the crystals throughout the glass. The resultant composite structure of fine crystals held together in a glassy matrix has improved properties over conventional glasses. The manufacture of glass ceramic products takes advantage of the high-speed, high-volume economics of glass melting and forming, and provides materials with new and often unique properties.

In this article the microstructures and properties of several different types of glass ceramics, selected on the basis of their commercial importance, are discussed. Some compositions are given in Table 1.

1. *β-Spodumene Glass Ceramics*

Glass ceramics containing the crystal phase β-spodumene (solid solution) ($Li_2O.Al_2O_3.nSiO_2$) are particularly useful for cookware and counter-top cooking surfaces because of their low thermal expansion, as shown in Table 2. Commercial compositions utilize TiO_2 as a nucleating agent and partial substitution of MgO and ZnO for Li_2O to improve the working properties of the parent glass while also lowering the materials cost. The alumina–modifier ratio is quite critical and is held close to 1.1–1.2, ensuring that the residual glass phase is stable and sufficiently low in thermal expansion to match the crystalline phase.

The heat treatment of these glass ceramics allows enough time for glass-in-glass phase separation followed by the precipitation of the first crystalline phase of a pyrochlore structure ($Al_2Ti_2O_7$). Formed vessels are held at 780 °C for 1.5 h to complete nucleation. This treatment leads to precipitation upon the titanate nuclei of hexagonal β-quartz solid solution (sometimes called β-eucryptite) crystals of 40–100 nm diameter. Local stresses created by density and thermal expansion differences between the β-quartz solid solution crystals and the residual glass must be relaxed by viscous flow of the glass phase. As the temperature is raised to about 1100 °C, the viscosity of this glass is lowered to about 10^{10} Pa s, allowing relaxation.

A temperature of 950 °C promotes the conversion of the metastable quartz to a β-spodumene solid solution, accompanied by an increase in grain size of about 1 μm. The completion of the heat treatment at

Table 1
Chemical composition of glass ceramics (wt%)

Component	β-Spodumene	β-Quartz	Cordierite	Mica	Lithium silicate	Slagsitall
SiO_2	69.7	68.5	56.0	46.2	79.3	58.4
Al_2O_3	17.9	19.2	19.7	16.6	4.2	7.0
Li_2O	2.6	2.8			9.4	
Na_2O	0.3	0.13			1.5	4.6
K_2O	0.1	0.1		9.5	4.1	0.5
B_2O_3				8.5		2.6
MgO	2.8	1.75	14.7	14.8		2.6
BaO		0.75				0.5
CaO			0.11			23.4
ZnO	1.0	1.0			1.0	
Fe_2O_3		0.08				0.14
TiO_2	4.8	2.6	9.0			0.2
ZrO_2		1.7				
As_2O_3	0.9	0.75	0.5			
Others		$0.5Nd_2O_3$, $0.0005V_2O_5$		7.6F	$0.4Sb_2O_3$	1.1MnO, 0.42S

Table 2
Thermal properties of glass ceramics

Property	β-Spodumene	β-Quartz	Cordierite	Mica	Lithium silicate	Slagsitall
Thermal expansion coefficient α (10^{-7} K^{-1})						
25–200 °C	10.2	5.7	64.6	88.4	10.75	68.5
25–400 °C	10.9	6.5	48.6	88.5	96.2	68.4
25–600 °C	11.0	6.8	44.5	101.5	105.8	72.8
25–800 °C	12.5	7.7	43.5	128.5	127.3	93.1
Thermal conductivity (W m^{-1} K^{-1})						
25 °C	2.1		4.0	2.1	2.6	
200 °C	2.2		3.4	2.1	2.0	
600 °C	2.3		3.0	3.3		
Specific heat c_p (J kg^{-1})						
25 °C	79.9	78.2	80.3	76.1	87.0	76.9
200 °C	100	100	98.3	93.7	97.5	97.0
600 °C	122	125	116	~109	~118	123
Thermal diffusivity (10^{-6} m^2 s^{-1})						
25 °C	1.11	0.84	1.7	0.68	1.46	0.75
200 °C	0.92	0.68	1.4	0.55	0.99	0.55
600 °C	0.78	0.59	1.1	0.55	0.73	0.48
Softening temperature (°C)	1248	—	>1300	1110	877	—

1125 °C in 0.5 h leads to only minor grain growth (Chyung 1969). A fine-grained structure is essential to prevent microcracks from forming due to the strong crystal anisotropy in β-spodumene solid solution. During the conversion of β-quartz to β-spodumene, formation of the minor phases TiO_2, $MgAl_2O_4$ and $ZnAl_2O_4$ occurs. These high-index crystals along with the larger β-spodumene grains promote opacity.

The microstructure of a β-spodumene glass ceramic (Corning Code 9608) is shown in Fig. 1. The uniformity of grain size and nearly complete crystallization is evident. Small, needle-like rutile crystals and siliceous residual glass are concentrated at the grain boundaries and help to inhibit grain growth, thus contributing to the high-temperature dimensional stability of these materials.

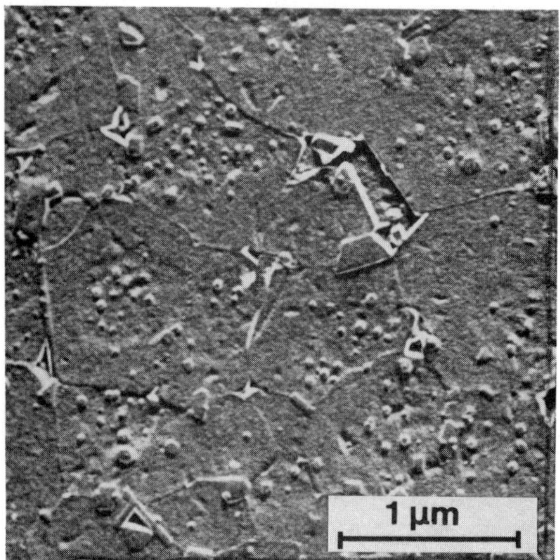

Figure 1
Electron micrograph of a β-spondumene glass ceramic

Production of large counter-top cooking surfaces requires not only a low expansion and thermally stable material but also high strength (Table 3). Inclusion of up to 0.5% fluorine in the composition leads to volatilization at the surface during the crystallization process. This composition gradient in fluorine promotes a slightly smaller unit cell volume in the interior, resulting in surface compression and strengths greater than 140 MPa.

Other methods for strengthening β-spodumene glass ceramics involve producing compression in the outer surface by postcrystallization ion-exchange treatments such as Na^+ or K^+ for Li^+, or in the case of β-quartz glass ceramics, $2Li^+$ for Mg^{2+}. Another process for compression strengthening is lamination and differential densification of the skin and core compositions.

2. *β-Quartz Glass Ceramics*

Although chemically very similar to the β-spodumene glass ceramics, the materials comprised mainly of β-quartz solid solution crystals are ultrafine grained and very low in thermal expansion, as shown in Table 2. Their fine grain size of 40 nm and characteristically low birefringence permits visible light transmission with little scattering.

In commercial compositions, such as Schott's Zerodur or Corning's Code 9607, a mixture of nucleation catalysts TiO_2 and ZrO_2 is used. Mixtures of nucleating agents are frequently more efficient in producing larger numbers of nuclei than either agent alone. The enhanced nucleation precipitates the β-quartz phase at lower temperatures and a correspondingly higher viscosity, resulting in a finer-grained and more transparent body. Minimization of light scattering is also achieved by matching the index of refraction of the residual glass to that of the β-quartz phase through incorporation of modifiers such as BaO.

Heat-treatment schedules for production of β-quartz glass ceramics typically employ a hold at

Table 3
Mechanical properties of glass ceramics at room temperature

Property	β-Spodumene	β-Quartz	Cordierite	Mica	Lithium silicate	Slagsitall
Density (g cm^{-3})	2.73	2.58	2.6	2.52	2.41	2.66
Modulus of rupture, abraded (MPa)	83	70	240	100	150	100
Modulus of elasticity (GPa)	80.7	93.4	119	65	74.5	95.0
Young's modulus (GPa)	80.7	93.4	119	65	74.5	95.0
Shear modulus (GPa)	33.3	39.4	47.8	26	31.9	38.6
Poisson's ratio ν	0.21	0.19	0.24	0.26	0.17	0.23
Microhardness (KNH_{100})	660	685	698	250	488	577
Critical stress intensity (MPa $m^{1/2}$)	0.8	0.7	2.1	1.1	1.5	0.7
Fracture energy (J m^{-2})	3.7	2.4	17.2	8.2	14.0	2.3

750 °C for 0.5–1 h to ensure proper nucleation. Crystal growth temperatures are only 900 °C, and higher temperatures (≥950 °C) result in a conversion of the β-quartz phase to β-spodumene. The ultrafine microstructure is shown in Fig. 2.

Figure 2
Electron micrograph of a β-quartz glass ceramic

The ultralow thermal expansion coupled with very fine grain size has made β-quartz glass ceramics useful for telescope mirror applications. More recently they have been used for transparent cookware and for dark-colored counter-top cooking surfaces.

3. Cordierite Glass Ceramics

Glass ceramics in the $MgO–Al_2O_3–SiO_2$ system found one of the earliest applications for a glass ceramic as a radome or missile nose cone. This nonalkali glass ceramic combines high electrical resistivity (Table 4) with high mechanical strength (Table 3) and moderately low thermal expansion (Table 2).

While most studies have involved compositions close to stoichiometric cordierite ($2MgO.2Al_2O_3.5SiO_2$) nucleated with either TiO_2 or mixtures of TiO_2 and ZrO_2, commercial compositions such as Corning Code 9606 contain an excess of SiO_2 and utilize approximately 9 wt% TiO_2 for nucleation.

The crystallization sequence in these materials can be somewhat complicated. However, the phases produced by heat treatments in excess of 1200 °C are primarily cordierite, rutile, magnesium aluminum titanate and cristobalite. The microstructure of a cordierite glass ceramic (Corning Code 9606) is depicted in Fig. 3. The predominant phase with somewhat rounded grains is cordierite. The sharp-angled, needle-shaped crystals have been identified as magnesium aluminum titanate or a pseudobrookite solid solution. Cristobalite, although present in the x-ray diffraction traces, is not discernible by replica electron microscopy.

The inclusion of approximately 11 wt% cristobalite

Table 4
Electrical and chemical properties of glass ceramics

Property	β-Spodumene	β-Quartz	Cordierite	Mica	Lithium silicate	Slagsitall
Log volume resistivity ρ (Ω cm, dc)						
25 °C	13.9	13.7			14.7	11.1
200 °C	8.8	8.2	10.9	11.0	10.0	7.0
400 °C	6.3	5.2	8.5	7.9	6.7	4.7
Dielectric constant (25 °C, 100 kHz)	6.93	7.44	5.71	6.15	5.94	7.92
Loss tangent tan δ (25 °C, 100 kHz)	0.0058	0.0102	0.0021	0.0026	0.0057	0.0171
Dielectric strength ($kV\ mm^{-1}$, dc, 25 °C, 0.25 mm sample)	150	142	160	173	169	79
Weight loss per unit area ($mg\ cm^{-2}$)						
5% HCl (95 °C, 24 h)	0.01	<0.01	10	110	0.08	0.21
H_2O (95 °C, 24 h)	0.01	0.01	0.02	0.01	0.02	0.02
N/50 Na_2Co_3 (95 °C, 24 h)	0.01	0.18	0.02	0.5	0.6	0.04
5% NaOH (95 °C, 24 h)	4	3.1	10	40	4	2.2

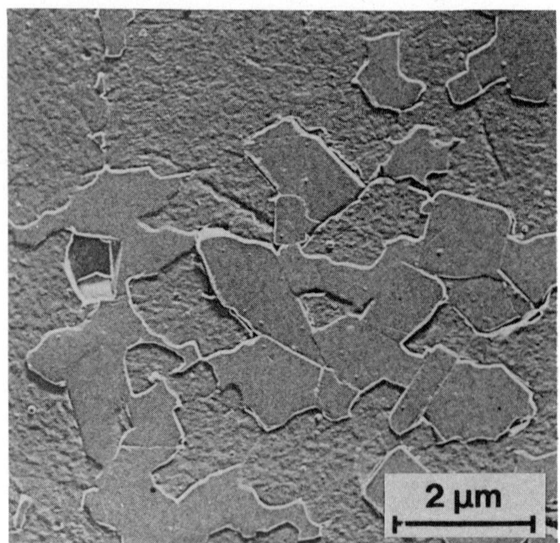

Figure 3
Electron micrograph of a cordierite glass ceramic

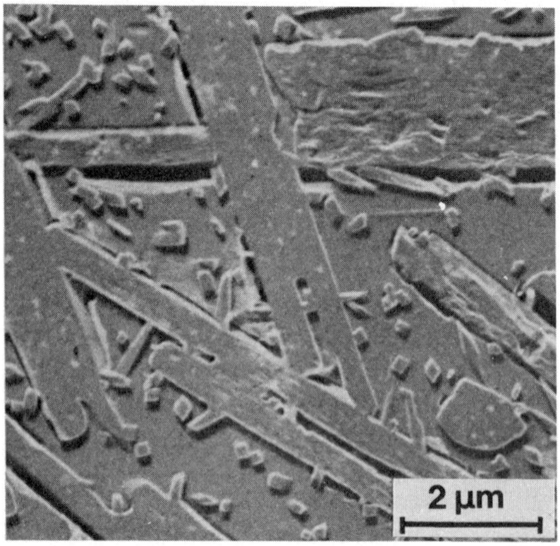

Figure 4
Electron micrograph of a mica glass ceramic

in the structure of the radome glass ceramic provides for a unique strengthening mechanism known as fortification. The crystallized material is exposed to an alkali solution which leaches the cristobalite phase from the surface of the material leaving a porous layer. This process removes the mechanical surface damage due to grinding and increases the strength from 120 to 240 MPa. In addition, the porous layer acts as an abrasion buffer, providing protection to the underlying glass ceramic material.

4. Mica Glass Ceramics

A family of nonporous machinable materials is based on fluorine-containing mica crystals. Glass ceramics with unique microstructure result from the growth of plate-like crystals within a glass matrix. These micas include fluorphlogopite, $KMg_3AlSi_3O_{10}F_2$; potassium tetrasilicic mica, $KMg_{2.5}Si_4O_{10}F_2$; and nonalkali micas, $(Ca,Sr,Ba)_{0.5}Mg_3AlSi_3O_{10}F_2$.

The highly interlocked microstructure of mica glass ceramics is revealed in Fig. 4 for Corning Code 9658 MACOR. Since mica can be readily delaminated due to its low cleavage energy, fractures propagate mainly parallel to the thin, flat crystals. The random intersections of the crystals cause crack deflection, branching and blunting which help to prevent microscopic fractures from propagating beyond the local cutting area. This fracture containment accounts for the machinability, strength and thermal-shock resistance of the material. These glass ceramics can be machined to very tight tolerances (± 0.01 mm), using regular high-speed tools.

The parent glass of the fluorphlogopite glass ceramic is phase-separated into fluorine–silica-rich droplets dispersed in a continuous boroaluminate glass. On heating to 750 °C, the dispersed droplets crystallize to a body-centered-cubic form of chondrodite ($Mg_5Si_2O_8F_2$) which at 825 °C recrystallizes to small, plate-like crystals of norbergite ($Mg_3SiO_4F_2$). The mica crystals are found to grow epitaxially on these earlier crystals. Further heating to 950 °C results in highly anisotropic growth of the mica phase to plate diameters averaging 10–15 μm. The small residual needle-shaped crystals are mullite ($3Al_2O_3 . 2SiO_2$).

Compositional variations, particularly a reduction in the K_2O content, can result in a wide variety of mica plate diameters. The lower alkali levels delay nucleation of the phlogopite crystals to higher temperatures where the growth rates are more rapid. Also, the lower concentration of the cross-bonding species K^+ favors lateral plate-like growth as opposed to thickening. The larger crystal size, up to 250 μm, results in improved thermal-shock performance.

5. Lithium Disilicate and Photosensitive Glass Ceramics

Glass ceramics in the lithium silicate family hold a special place in the history of crystallized glass. By overheating a photosensitively opacified glass in this system, Stookey (1959) found an opaque polycrystalline ceramic with improved mechanical, electrical and chemical properties. This discovery led to the search for other nucleating agents and other glass-forming systems and eventually produced a wide variety of glass ceramic materials.

These materials are marketed today as complex shapes with holes as small as 0.05 mm in a variety of patterns and designs. Up to 8000 holes cm^{-2} can be etched to tolerances within 0.03 mm. The process involves the use of ultraviolet radiation through a photographic mask to form a pattern or latent image through the entire thickness of the glass. The glass contains CeO_2 as a sensitizer and silver metal to serve as a nucleation catalyst.

On heat treatment below 540 °C, silver crystals grow in irradiated areas by the photoreduction of silver ions to silver atoms. These atoms act as nuclei and provide sites for the epitaxial growth of a hexagonal lithium metasilicate ($Li_2O.SiO_2$) phase which precipitates on the silver crystals and takes on a dendritic growth habit as the temperature is increased. The concentration of nuclei, and therefore the size of the metasilicate crystals, depends on the ultraviolet exposure time and intensity. The appearance of the metasilicate phase is unusual for a high-silica composition and is not obtained with other nucleating agents such as platinum or P_2O_5.

Complete development of the image is attained when the crystal boundaries are in direct contact with one another. The enhanced solubility of lithium metasilicate over the unexposed glass regions provides a method for chemically machining the material. Additional exposure and heat treatment to higher temperatures can convert the remaining glass to a glass ceramic with improved physical properties. During this conversion, the lithium metasilicate phase transforms to lithium disilicate ($Li_2O.2SiO_2$) with quartz and a feldspathic residual glass. The conversion temperature depends on the initial metasilicate grain size and can vary from 800 to 850 °C. The pattern of crystallization retains a relic structure of the earlier metasilicate dendritic form (Fig. 5).

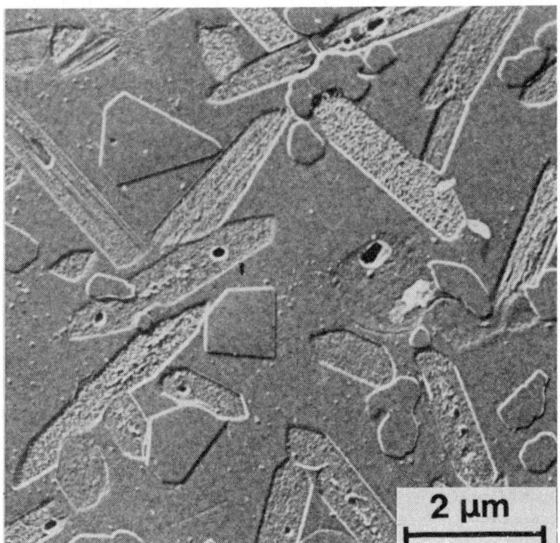

Figure 5
Electron micrograph of lithium disilicate glass ceramic

6. *Slagsitall or Slagceram*

Glass ceramics derived from steelmaking slags have been produced mainly in the form of rolled sheet for architectural or building materials, such as flooring, and both interior and exterior cladding. In 1971, over 2×10^6 m^2 of slagsitall material were reported to be in service in the USSR, where these materials were developed.

The slags are made more siliceous by addition of 20–40% sand and up to 10% clay. Other additives are 4–6% sodium sulfate and 0.2–1.5% nucleating agents. The composition of the slags varies depending on the source but is primarily 40–50% CaO, 2–8% MgO and MnO, 10–20% Al_2O_3, 0.5–2% Fe_2O_3, 0.5–2% S, and 35–45% SiO_2 (Table 1).

Nucleation is achieved primarily by the separation of heavy-metal sulfides. The sulfates are reduced through the use of powdered coke or other reducing agents added to the batch. X-ray diffraction as well as electrical and hardness measurements indicates an optimum metal sulfide level of 0.3 wt%.

The slag-based glasses are relatively fluid, typically with a viscosity of <100 Pa s at the liquidus. New machinery was developed for production of a continuous sheet as wide as 3 m and 5–25 mm thick.

Heat treatments generally involve heating at 720 °C for 2 h followed by crystallization at 850–900 °C for 3–4 h. More rapid crystallization of slagceram materials results by the addition of 1.5% fluoride as fluorspar, and cuspidine ($3CaO.CaF_2.2SiO_2$) forms early in the heat treatment. Further heating to 880 °C causes the formation of gehlenite ($2CaO.Al_2O_3.SiO_2$) followed by pseudowollastonite (α-$CaSiO_3$). In some higher alumina compositions, the phase anorthite ($CaO.Al_2O_3.2SiO_2$) is also present.

Slag-based glass ceramics have been produced primarily as opaque white or dark gray materials. A dark gray variety from Konstantinovka contains wollastonite crystals in the form of short prismatic rods approximately 0.5 μm long by 0.1 μm diameter (Fig. 6). The crystallinity is fairly low at about 70%. The slagceram materials have demonstrated excellent wear resistance coupled with good durability and have proved useful in industrial applications in addition to their decorative utilization.

7. *Properties*

The coefficient of thermal expansion of glass ceramics can vary over a wide range (Table 2). Thermal conductivities and specific heats are similar to those of other oxides. The softening temperatures are much higher than those of soda-lime (700 °C) and Pyrex borosilicate (820 °C) glasses.

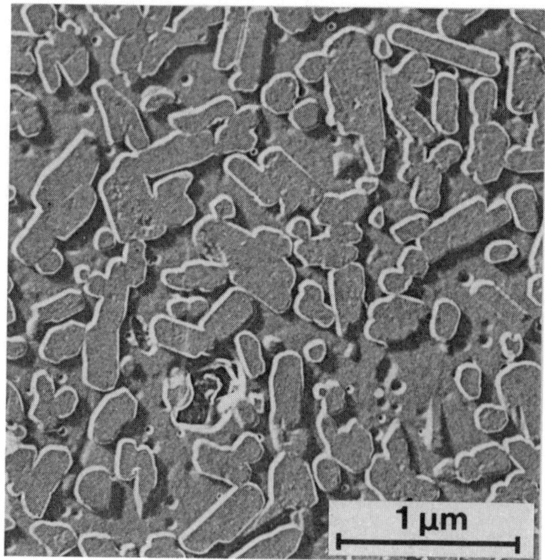

Figure 6
Electron micrograph of slagsitall from Konstantinovka, USSR

Strengths (moduli of rupture) vary widely depending upon composition and microstructure (Table 3), but are mostly higher than for common glasses (30–100 MPa). Hardness is also quite variable, but can be higher than for most glasses (KNH_{100} = 400–600). Moduli are somewhat higher than for glasses (~70 GPa).

Electrical resistivity is quite variable, depending upon alkali content and microstructure (Table 4); comparable values for Pyrex borosilicate glass are 15 Ω cm at 25 °C and 8.8 Ω cm at 200 °C. Dielectric constants are higher and loss tangents lower than in most commonly used glasses. The chemical durabilities are better than for most glass, except for the mica glass ceramic.

See also: Glasses: Structure; Glass: An Overview

Bibliography

Barry T I, Cox J M, Morrell R 1978 Cordierite glass-ceramics: Effect of TiO_2 and ZrO_2 content on phase sequence during heat treatment. *J. Mater. Sci.* 13: 594–610

Beall G H, Karstetter B R, Rittler H L 1967 Crystallization and chemical strengthening of stuffed β-quartz glass-ceramics. *J. Am. Ceram. Soc.* 50: 181–90

Berezhnoi A I 1970 *Glass-Ceramics and Photo-Sitalls*. Plenum, New York

Chyung C K 1969 Secondary grain growth of Li_2O–Al_2O_3–SiO_2–TiO_2 glass-ceramics. *J. Am. Ceram. Soc.* 52: 242–45

Doherty P E, Lee D W, Davis R S 1967 Direct observation of the crystallization of Li_2O–Al_2O_3–SiO_2 glasses containing TiO_2. *J. Am. Ceram. Soc.* 50: 77–81

Grossman D G 1972 Machinable glass-ceramics based on tetrasilicic mica. *J. Am. Ceram. Soc.* 55: 446–49

McMillan P W 1979 *Glass-Ceramics*, 2nd edn. Academic Press, London

Stookey S D 1959 Catalyzed crystallization of glass in theory and practice. In: von Bülow N (ed.) 1959 *Int. Glaskong., 5th, Glastechnische Berichte, Sonderband, 32K*. Deutsche Glastechnische Gesellschaft, Frankfurt, pp. 1–8

Vogel W 1971 *Structure and Crystallization of Glasses*. Pergamon, Oxford, pp. 158–202

D. G. Grossman

Glass Electrodes

Electrodes of glass are widely used to measure the hydrogen ion concentration (pH) in aqueous solutions. These electrodes give accurate, reproducible pH measurements over wide ranges of solution composition and temperature. Recently glass electrodes for the measurement of concentrations of other ions such as the alkalis and silver have been developed.

1. Equipment for Ion Concentration Measurements

The glass electrode is usually a thin (about 0.1 mm thick) bulb attached to a thicker glass tube. The inside electrical connection can be made in various ways; a common one is to use a silver–silver chloride electrode on the end of a platinum wire. This wire is inserted into an ionic solution inside the tube, which is then sealed with an epoxy cement to reduce water loss and electrical leakage. An external reference electrode such as silver–silver chloride or calomel completes the circuit in the solution. The potential difference between the glass and reference electrodes is read with a potentiometer.

The glass must have low enough resistance so that an accurate potential difference can be measured with an inexpensive electronic potentiometer. The glass must "condition" readily by soaking in water for a few days, after which it gives steady and reproducible potentials. This property of "hygroscopicity" must be balanced by enough resistance to attack by aqueous solutions to prevent damage, dissolution or irregular readings. The glass must also have as little alkaline error as possible.

One of the first glass compositions to satisfy these conditions was Corning 015 (72.2 mol% SiO_2, 21.4% Na_2O and 6.4% CaO) glass, which has a low softening point for easy working, low resistivity and is easily conditioned to give stable and reproducible potentials. Above about pH 9 with sodium ions present, this glass shows an increasing nonlinearity between the measured potential and the hydrogen ion concentration. Other glass compositions have been found that have much less alkaline error; these glasses contain lithium oxide rather than sodium oxide. Addition of other oxides gives further reduction in resistivity, chemical reactivity and alkaline errors, even up to pH 12. A typical composition is 65 mol% SiO_2, 28% Li_2O, 3% Cs_2O and 4% La_2O_3 (Bates 1973).

2. *Origin of the Glass Electrode Potential*

When a silicate glass containing alkali is placed in water the alkali ions in the glass exchange with hydronium ions from the solution (see *Chemical Durability of Glass*). It is this ion exchange that leads to a potential difference between the glass and the solution which depends on the hydronium ion concentration in solution.

The potential difference V between two different solutions in contact with two different faces of a glass bulb or membrane is made up of two contributions, a surface or Donnan potential V_s and a diffusion potential V_D. Consider two ions A and H, A being initially in the glass and H initially in the solution. The activities a and h of these ions in the two solutions 1 and 2 are related to the potential V_s by

$$V_s = \frac{R_0 T}{F} \ln \frac{a_1 + Kh_1}{a_2 + Kh_2} \tag{1}$$

with certain conditions on the ionic activites in the glass. Here R_0 is the molar gas constant, T is the temperature, F is the Faraday constant and the distribution coefficient K is given by $K = ah_g/a_g h$, where subscript g refers to activities in the glass in equilibrium with a solution with activities a and h (Eisenman 1967, Doremus 1973).

The diffusion potential for interdiffusion of two ions A and H in a membrane between two solutions 1 and 2 can be derived from the Nernst–Planck equations (Doremus 1973):

$$V_D = \frac{R_0 T}{F} \int_{C_A(1)}^{C_A(2)} \frac{U_H/U_A - 1}{C_H(U_H/U_A) - C_H} \frac{\partial \ln a}{\partial \ln C_A} dC_A \tag{2}$$

where U is the mobility and C the concentration. If the mobility ratio U_H/U_A is not a function of relative ionic concentration, and the activity factor $\partial \ln a/\partial \ln C_A = 1$, Eqn. (2) can be integrated to

$$V_D = \frac{R_0 T}{F} \ln \frac{C_A(2) + C_H(2)(U_H/U_A)}{C_A(1) + C_H(1)(U_H/U_A)} \tag{3}$$

The diffusion potential depends only on the relative ionic concentration in the glass at the two membrane surfaces [$C_A(1)$ and $C_A(2)$] and the mobility ratio. The total membrane potential including both surface and diffusion contributions is the sum of Eqn. (1) and Eqn. (3):

$$V = \frac{R_0 T}{F} \ln \frac{a_1 + K(U_H/U_A)h_1}{a_2 + K(U_H/U_A)h_2} \tag{4}$$

in which it is assumed that $h_g/a_g = C_H/C_A$ in the glass.

If the mobility ratio and thermodynamic term are functions of concentration, Eqn. (2) cannot be integrated without knowledge of this functionality. However, even in this case, the potential is not a function of time, but only of the surface concentrations, some function of the mobility ratio and the distribution coefficient K. Therefore, as soon as equilibrium between the solutions and glass surfaces is established, the potential is fixed with time, no matter what the concentration profile of ions in the glass. The potential should change with time only if there is some change in the structure of the glass with time, such as progressive hydration of the glass surface or increasing internal stress.

3. *Comparison of Theory and Experiment*

Equation (4) is consistent with the experimental result that the alkaline error of a glass electrode is related to the reduction of hydrogen ion concentration and competition with another ion such as sodium. In Corning 015 glass in alkaline solution with ammonium ions and no sodium, there is no alkaline error up to pH 13; with sodium ions (0.1 M) the error is appreciable at about pH 10 (Hubbard et al. 1939).

The functional dependence of voltage on relative ionic concentrations given by Eqn. (4) has been found by different investigators; however, a rigorous test of various theories for the electrode potential is possible only if there are independent measurements of K and U_A/U_H. Two tests of this sort have been reported, one on vitreous silica in fused salts (Doremus 1968) and the other on potassium-sensitive aluminosilicate glass in aqueous solution (Eisenman 1967). Both gave satisfactory agreement with Eqn. (4), supporting the model leading to this equation.

4. *New Electrode Types*

When the value of $K(U_H/U_A)$ in Eqn. (4) is low enough so that the electrode is sensitive to the ion A over a wide pH range, the glass can be used to measure activities of ion A in solution. Certain aluminosilicate glasses have been found to be suitable as electrodes for measuring sodium and potassium ions. Eisenman found that optimum glass compositions were 11% Na_2O, 18% Al_2O_3, 71% SiO_2 for sodium and 27% Na_2O, 4% Al_2O_3, 69% SiO_2 for potassium. Glass electrodes have also been suggested for divalent ions but these electrodes have not found widespread use. Certain glass compositions, such as soda-lime silicates, are sensitive to other monovalent ions such as silver and thallium and are commercially available as electrodes. A wide variety of solid-state and liquid membrane electrodes have been developed for measuring concentrations of particular ions; glass is the best material for measuring hydrogen, sodium, potassium and silver ions in aqueous solution.

Glass electrodes have been made in a variety of shapes for special applications. Very fine electrodes are useful for analysis of biological fluids and even within a living cell.

See also: pH Sensor Materials; Glass: An Overview

Bibliography

Bates R G 1973 *Determination of pH*. Wiley, New York, Chap. 11

Doremus R H 1968 Electrical potentials of glass electrodes in molten salts. *J. Phys. Chem.* 72: 2877–82

Doremus R H 1973 *Glass Science*. Wiley, New York, Chap. 14, pp. 253–77

Eisenman G (ed.) 1967 *Glass Electrodes for Hydrogen and Other Cations*. Dekker, New York

Hubbard D, Hamilton E H, Finn A N 1939 Effect of solubility of glass on the behavior of the glass electrode. *J. Res. Natl. Bur. Stand.* 22: 339–49

R. H. Doremus

Glass Fibers

In the space of about 50 years, glass fibers have grown from a novelty product with little commercial application to a material whose production is now measured in billions of pounds annually and that is used extensively in our modern lives. There are two basic types of glass fibers, with distinctly different uses and some important differences in methods of manufacture. The first consists of short, discontinuous fibers, which are formed into wool to make boards and mats for insulation and allied applications. The second consists of continuous fibers, which are used to reinforce plastics and composites, to weave various fabrics, or chopped for nonwoven mats.

1. Definition of a Glass Fiber

In basic terms, glass fiber is glass that has been drawn into fine filaments. The important element in glass formation is its rapid cooling, which prevents crystallization.

The substances of greatest interest in fiber production are the primary glass formers or glass network formers. These are oxides, such as SiO_2, B_2O_3, GeO_2, P_2O_5 and As_2O_3, that will readily form into glass when rapidly cooled from the molten liquid state. They do not require the presence of additional oxides to achieve glass formation. Of these, SiO_2 is the most commonly used. To understand its excellent ability to form glass and the failure of some other oxides to do so, even with rapid cooling, we can turn to Zachariasen's empirical rules for glass formation based on molecular structure:

(a) Each oxygen atom should be coordinated with not more than two metal atoms.

(b) The metal atoms should have a small coordination number.

(c) The oxygen polyhedra must share corners only, not edges or faces.

(d) Three or more corners of each polyhedral coordination unit must be shared.

Relatively few oxides meet all of these requirements. However, several, besides the known network formers mentioned above, can enter a network to a limited extent. Alumina is the most common example. Aluminum atoms substitute for some of the silicon atoms in the tetrahedral positions. Zirconium, titanium, vanadium, antimony and lead atoms behave similarly.

In addition to the primary glass formers, other oxides are widely used as network modifiers in glass formation. These modifiers, including Na_2O, K_2O, CaO, MgO, BaO and PbO, open up the three-dimensional framework structure. They are important in glass fiber technology because they make the glass more workable by increasing its fluidity, as well as contributing to optical and thermal properties. Because these modifiers increase the ionic character of glass, resulting in more rapid crystallization, there are definite limits to the amounts that can be added.

A qualitative index of the glass-forming tendencies of an oxide is provided by its bond strength. The bond strengths of the network formers are greater than 335 kJ per bond, compared with a strength of less than 250 kJ per bond for the network modifiers.

Using these basic formers and modifiers, there are a great number of commercial glass compositions, each made to provide specific desired properties. The choice, for fibers as well as other applications, depends on striking a balance between the requirements for processing and the properties of the material.

In considering glass fibers, it is important to understand that glass in its modern forms is not necessarily a brittle material, unable to withstand shock. It can be pliable, light, soft and strong; it can withstand shock and vibration and can be formulated to resist moisture and the action of acids and alkalis. It is just such properties that have made certain glasses useful in various fiber applications. Key properties of glass fibers include high thermal endurance, excellent dimensional stability and excellent electrical characteristics. These properties have resulted in the proliferation of uses for glass fibers.

2. Glass-Fiber Applications

Table 1 lists some of the primary areas of application for glass fibers in three major categories: fiber-glass-reinforced plastics and composites, fiber-glass fabrics and discontinuous fibers.

To form reinforced plastics and composites, glass

fibers are mixed with a liquid polymer. Polyesters, epoxies, phenolics, vinyl esters and furans are among the polymers used. Catalytic chemicals in the mixture act to cure it to a rigid solid. This catalytic action may take place either at room temperature or at elevated temperature and pressure. Once cured, the impregnated fibers add high strength to the bonding power, processability and chemical durability of the polymer. Numerous molding processes are used to form the glass-reinforced material into a useful product. These products offer excellent molded surface finish and light weight, as well as superior thermal and electrical insulation.

As seen in Table 1, there is a wide variety of applications for these materials. In aircraft, glass-reinforced materials are most frequently used for molded interior parts, for example panels, but also for exterior parts such as nose cones and farings. On smaller aircraft, they have been used for the entire external fuselage.

Table 1
Typical glass-fiber applications

Fiber-reinforced plastics/composites
Aircraft components
Appliance housings
Construction materials
Consumer goods
Marine products
Automobile body parts
Fiber-glass fabrics
Protective apparel
Printed circuit boards
Orthopedics
Cable and wire insulation
Industrial fabrics
Screening
Filters
Structural materials
High-temperature materials
Yarns and threads
Tire cord
Discontinuous fibers
Building insulation
Appliance insulation
Pipe insulation
Acoustical insulation

Because of their high strength-to-weight ratio, dimensional stability and high dielectric strength, glass-reinforced materials are often used for housings and bases of appliances, such as computers, television sets, air conditioning units, typewriters and calculators.

Fiber-glass-reinforced materials are also used extensively for interior and exterior residential, commercial, industrial and farm structures. Modular building panels, tub and shower units, patio covers and garage doors are examples.

Increasing use in consumer goods is tied largely to leisure and recreational activities—furniture, swimming pools, playground equipment, sporting goods, pleasure boats and snowmobiles are typical consumer products. The controlled flexibility of glass-fiber-reinforced materials is important in other products, such as vaulting poles and golf-club shafts. Other favorable properties that encourage use in consumer products are high strength, light weight, ease of forming, durability, surface finish, molded-in color, and corrosion and wear resistance.

Fiber-glass-reinforced materials are finding increasing use in boat construction thanks to their ease of molding, seamless construction, high strength, durability, minimum maintenance and resistance to weathering. Nearly three-quarters of all outboard pleasure boats longer than 4.5 m are constructed of these materials, along with many other marine products.

Because of the weight savings, glass-reinforced materials are widely used for automotive parts, including fiber-glass bodies. In the 1975–77 period, these materials accounted for only about 3.7% of a typical car's curb weight. This figure is expected to increase to 10%. In addition to appearance items (hood covers, fender extensions and instrument panels, for example), functional parts such as leaf springs, valves, gear trains and fuel pumps are now made from glass-reinforced materials.

The techniques used to fabricate textiles and fabrics from continuous glass fibers are discussed later. Once fabricated, these materials are used in a wide variety of ways (Table 1). Printed circuit boards, laminated to the desired thickness primarily with epoxy resins, make up the largest single use. Fiber-glass textiles have proved superior to other materials for these boards because of their uniformity, ease of handling and punchability. These fabrics are also used in many other electrical applications, where their excellent dimensional stability and superior electrical insulating properties offer important advantages.

In another primary application, fiber-glass fabrics are widely used as construction materials. For example, fiber-glass fabric impregnated with Teflon fluorocarbon resin is a flexible material that can be employed as roofing for buildings, stadiums or swimming pools, or as an entire inflated structure. In these structures, the flexible membrane is elevated either by forced air pressure or by supporting or tension members. Important advantages are noncombustibility, weather resistance, translucency, ease of fabrication and low operating and maintenance costs.

A further important application is in automobile tire construction. For tire fabrication, fiber-glass fabric, impregnated with a rubber bonding material that is compatible with the tire, is formed as the tire-cord reinforcement. The advantages of tires reinforced with fiber-glass plies and belts include a softer ride and lower reinforcement cost.

As already noted, discontinuous fibers are formed into wool, board and mat products. These materials are used in many configurations for the thermal insulation of buildings, appliances and pipes, and for acoustical insulation in buildings.

3. Origin and Growth of Glass Fiber

Interest in glass fiber dates back thousands of years. The Egyptians made vessels by winding coarse glass fibers onto a suitably shaped mandrel of clay. They allowed the fibers to flow together under heat and then removed the clay core after cooling, leaving the glass-fiber vessel. A similar technique was employed by Venetian glass makers in the sixteenth and seventeenth centuries.

The first uncertain steps in using glass fibers in textiles were taken during the eighteenth century. By the late 1800s and early 1900s, there were some very limited commercial applications. The real advances came during the 1930s when major research and development efforts resulted in important advances in fiber forming technology.

Growth of the glass-fiber industry was stimulated by the discovery that glass-wool products have excellent thermal and acoustical insulating properties, thanks to the large volume of air that they contain.

Another important advance in glass-fiber technology was the development of continuous filaments for electrical insulation of fine wires in the 1930s. Previously, all fiber manufactured on an industrial scale was discontinuous, i.e., glass wool. The advent of continuous filaments allowed expansion into the electrical industry and eventually a wide variety of other applications through the development of woven textiles and chopped fiber mats.

4. Manufacture of Glass Fibers

As noted, there are two basic types of fibers and it is necessary to consider them separately in describing manufacturing processes. The manufacture of continuous glass-fiber products involves three stages, that is, glass melting, drawing of the fibers and conversion of the fibers into a usable form. The first two stages were at one time separate, but are now capable of being integrated into a continuous process.

In the manufacture of glass, raw materials (such as sand, limestone, boric acid and various fluxes) are mixed and blended in different formulations and fed into a furnace for melting. The exact formulation determines the characteristics and properties of the glass, and hence the glass fiber, produced. Other steps in the process also affect final properties.

After melting, the liquid glass from the furnace is fed directly to fiber drawing channels and drawn through multiple hole "bushings" (Fig. 1). Each bushing will normally have approximately 200 nozzles or some multiple thereof. Because the molten glass is fed directly to the bushings, this is called the direct-melt process. An alternative is to manufacture glass "marbles" that are remelted in a separate bushing furnace.

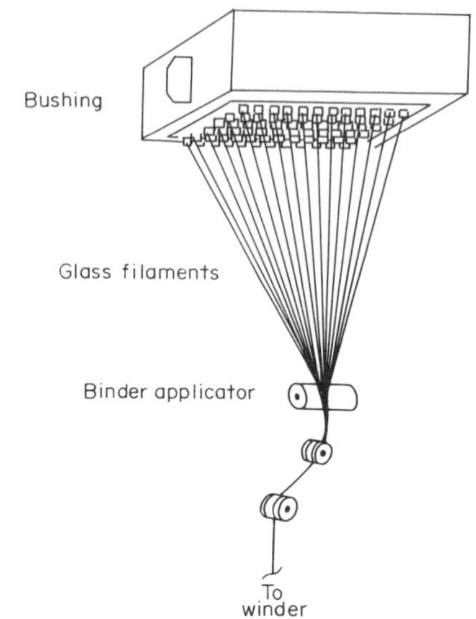

Figure 1
Formation and collection of glass filaments

Schematic representations of these two basic processes are shown in Figs. 2 and 3. The direct-melt process is more energy efficient since reheating is not required.

In either process, the molten glass is drawn through the nozzles into filaments. The filament diameter is controlled by the viscosity of the glass melt and the attenuation rate, which is a function of the winders discussed below. After drawing, the filaments pass through a cooling water spray and a chemical sizing is applied. The sizing protects and lubricates the filaments during processing. Without it, their resistance to abrasion is low.

After sizing, the filaments pass over a gathering shoe into a strand. These strands are packaged on a collet by high-speed winders. The resulting product is called a forming package or "cake." The winders are actually the attenuating or drawing devices, determining the rate of extrusion, and ultimately the fineness of the filament. A traverse mechanism on the winder locates the fiber on the collet to facilitate further processing that will require unwinding of the filament. Winding proceeds at speeds around 2500 m min^{-1}.

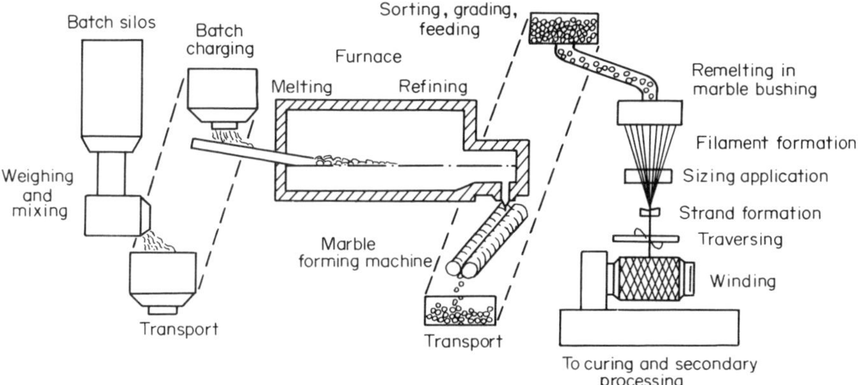

Figure 2
Older, marble-melt process used to produce continuous filaments

The glass-fiber cakes may be further processed like conventional textile yarns to produce continuous glass-fiber yarn in a form suitable for use by the customer. These yarns are used for woven fabrics. In this process, the fiber-glass strands are twisted and plied. Twisting into yarn produces a uniform product that can be processed more smoothly. The strands are twisted a predetermined number of turns per unit length in a single direction to obtain the desired yarn characteristics. Plying involves twisting two or three of the yarns together by turning in the opposite direction. This increases yarn strength and diameter, but decreases flexibility. Several types of yarns are described in Table 2.

As with continuous fibers, the discontinuous fibers used for wool and mat products may be formed by either direct-melt processes or from marbles. Whichever melting approach is used, there are several basic processes for fiber formation.

Table 2
Definitions of strands and yarns

Basic strand	Strands composed of filaments which have not been twisted
Single yarns	Yarns made from one or more basic strands subjected to one twisting operation. The number of basic strands used constitutes the type of construction—single construction is two strands twisted together to produce a single yarn
Plied yarns	Yarns made by twisting two or more single yarns together in a second twisting operation
Cabled yarns	Yarns made from two or more plied yarns which have been twisted together in a third twisting operation
Multiple wound yarns	Two or more single or plied yarns gathered and wound on a single package without further twisting

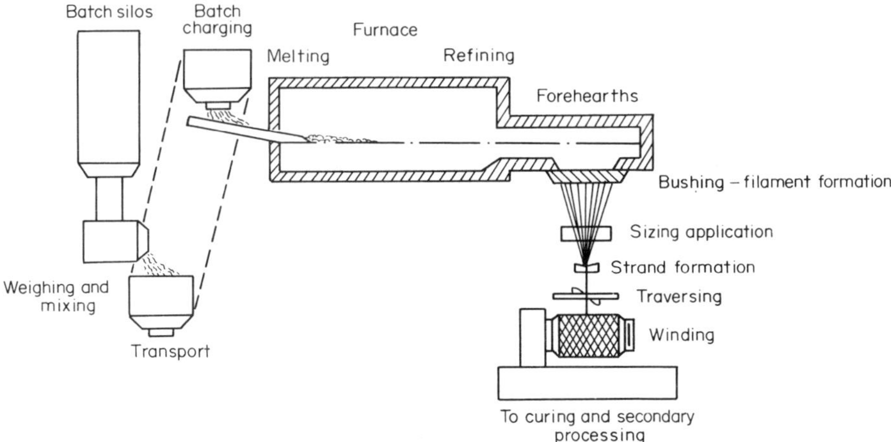

Figure 3
Direct-melt process for collection of continuous filament glass

4.1 Rotary Wool Forming

In the rotary process (Fig. 4), the molten glass from the melting tank drops into a rotating cylindrical unit, or centrifuge, that has a large number of holes in its sidewalls. Centrifugal force breaks up the molten glass and it streams laterally from the holes into the path of a high-velocity gas stream (air, steam or combustion gases) that breaks them into discontinuous fibers. The fibers are sprayed with binder as they fall to a collecting belt that moves them into a curing oven. As the cured wool leaves the oven, it is chopped to the desired size.

insulating fine wires at high temperatures. Its use has spread much wider and over 99% of the continuous glass fiber produced today is comprised of E glass. The glass, as intended, provides superior electrical characteristics and high heat resistance.

C glass is used in applications where high resistance to the corrosive action of acids is required. Examples include industrial batteries and FRP (fiber-glass-reinforced plastic) tanks.

S glass is used to produce lamination fabrics for high-performance structures such as aircraft components. S glass can offer strength-to-weight ratios higher than those of most metals.

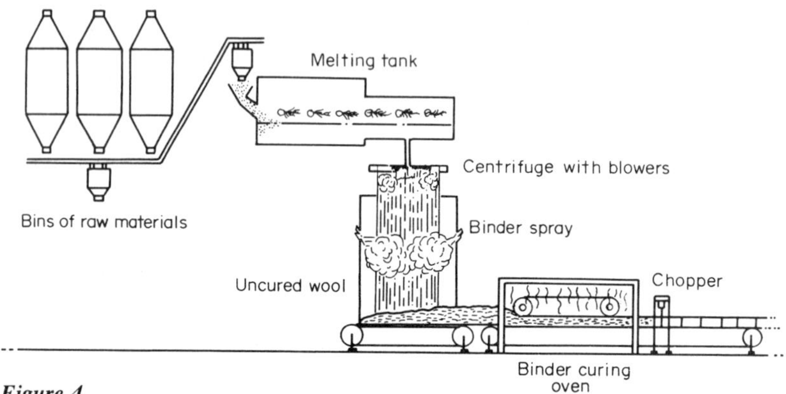

Figure 4
Rotary wool forming process

4.2 Steam Blowing

In steam blowing, molten glass leaves the melting tank in small streams through metal alloy bushings in the bottom of the tank. High-pressure steam jets adjacent to the bushing break the glass into discontinuous fibers, which again are sprayed with binder as they fall to a conveyor belt for transport to the curing oven.

4.3 Flame Attenuation

For flame attenuation, primary fibers, approximately 1 mm in diameter, are drawn through bushings in the bottom of the melting tank. These filaments are then aligned in an exact, uniformly juxtaposed array and exposed to a jet flame from an internal combustion burner. The flame breaks the filaments into very fine, long fibers that pass into the binder spray.

5. Types of Fibrous Glass

Three common types of glass now provide the predominant materials for production of textile-grade continuous glass fiber. They are E glass (electrical), C glass (chemical) and S glass (high tensile strength).

E glass was first formulated in the 1930s when the requirement arose for a continuous fiber for

Table 3 lists the general chemical compositions and some of the properties of glass types E and S.

Key properties for E glass are the relatively high tensile strength, zero creep and low elongation at break with 100% recovery. The major advantages of E glass compared to other synthetic textile fibers are:

(a) High strength (15.3 gpd).

(b) Retention of 75% of strength at 340 °C while other synthetics either melt or decompose at 200–260 °C.

(c) Low water absorption, high dimensional stability and excellent resistance to bleaches, solvents, mildew and sunlight.

(d) Good electrical properties.

(e) Resistance to many acids and bases under moderate conditions; severe conditions require the use of specially designed glass.

E-glass fiber does have certain shortcomings for some applications, however:

(a) Dyes will only lightly tint the fibers unless the glass is specially prepared.

Table 3
Composition of glasses used in fiber manufacture and their basic properties

	E-glass	S-glass
Composition (wt%)		
SiO_2	52–56	65
Al_2O_3	12–16	25
B_2O_3	5–10	
MgO	0–5	10
CaO	16–25	
Na_2O	0–2	
K_2O	0–2	
Fe_2O_3	0–0.8	
F_2	0–1	
Tensile strength of single fiber (MPa)		
at 22 °C	3500	4600
at 370 °C	2600	3700
at 540 °C	1800	2800
Tensile strength of strand (MPa)	2400	3100
Young's modulus of fiber at 25 °C (MPa)	75×10^3	85×10^3
Density ($g\,cm^{-3}$)	2.52–2.62	2.47–2.49
Refractive index n_D	1.546–1.560	1.524–1.528
Coefficient of linear thermal expansion per °C	5×10^{-6}	3×10^{-6}

(b) Fiber extensibility is low compared to most synthetics.

(c) The fibers are susceptible to abrasion.

With such strengths and weaknesses, it is important when designing a product to consider not just the fibers themselves but also the properties of the glass-fiber composite. For example, the high tensile strength and low extension at break, coupled with the excellent recovery and high stiffness of E glass yield a material with excellent dimensional stability, which simplifies fabrication. Although glass fibers are much stiffer than other synthetics of the same diameter, they are also available in much finer diameters. Because of this, glass-fiber textiles can be produced that are soft and flexible with good dimensional stability and high heat resistance, but low extensibility.

6. *Forms of Textile Fibers*

Blown, or discontinuous, fibers are manufactured in the form in which they will be used: wool, mats and boards. Continuous fibers, on the other hand, can be used as reinforcement for plastics or composites, or in two basic forms to fabricate textiles: continuous yarns, either cakes or rovings, for woven fabrics and reinforcing scrim, and chopped strands used to make nonwoven mats or fabric.

The weaving of twisted and plied continuous glass-fiber yarns uses basically the same processes used in other textile weaving operations. The woven glass fabrics are manufactured by interlacing two systems of yarns at right angles. Vertical, or lengthwise, threads comprise the warp of the textile and the horizontal, or crosswise, threads the filling. The warp ends may be fed directly to the loom from a creel, which holds the forming package, or may be pulled from warp beams. The fill yarns may be supplied in several ways, depending on the type of loom.

Chopped-strand mats and fabrics may be fabricated by dry laid and wet processes. In the dry process, multiple strands from the forming packages or cakes are drawn together into a chopper and cut into 2.5–7.5 cm lengths. A plastic binder is applied to the chopped strands as they pass into a low-temperature oven. The binder softens in the oven. After leaving the oven, the strands and binder pass through pressure rollers and are bonded into a continuous mat. In other dry processes, the mat may be formed directly from the bushing or fine glass fiber may be carded into mats and bonded conventionally.

In the wet processes, a nonwoven web is formed from a water slurry using what is essentially paper-making equipment. Chopped fibers, which can vary in diameter and length, are suspended in a water slurry and then collected on a forming screen. The requirements of the finished mat determine the filament and chop size to be used. This wet-formed mat is saturated with a binder, dried and cured into a mat or "paper" product. The wet-process glass fibers may be treated with a selected chemical size to facilitate processing and improve physical characteristics. The process is simple and economical in comparison with other mat manufacturing processes. The fiber can also be blended quite simply with other glass or nonglass fibers to combine the advantages of different materials.

The most important fiber characteristic to be considered in the wet-laid process is dispersibility, i.e., the extent to which filaments can be suspended in water and the extent to which the suspension can be maintained without fiber reagglomeration. A bare glass-fiber bundle will wet-out very rapidly in water (zero contact angle) and completely disperse into a suspension of individual filaments. However, these unprotected fibers will rapidly reagglomerate into clumps, within which fiber damage due to abrasion occurs quite rapidly. As the fibers break up into short lengths, a good fiber dispersion depends upon the fiber consistency and severity of agitation. Thus, diluted, gently agitated suspensions will be maintained longer than more concentrated, vigorously agitated ones. The concentration of fiber (or "consistency") normally used when dispersing glass fiber is considerably lower (0.3–0.5%) than that used in paper-making with wood pulp, and somewhat lower than is common with other synthetic fibers. When either poorly protected glass fiber or fibers with a large length-to-diameter (L/D) ratio are used, these fiber consistencies may be reduced even further.

The application of a chemical size to the fiber during fiber forming modifies the glass surface. While this chemical size generally slows down the fiber wet-out, it protects the fiber and prevents reagglomeration and fiber damage. It also allows the use of higher fiber consistencies. The rate of fiber wet-out can be increased by the use of surfactants in the dispersing medium. Viscosity modifiers are also used in some systems to reduce fiber reagglomeration, as well as to improve mat formation.

The maximum fiber L/D ratio that can be handled by a given machine is primarily related to the amount of water flow available. Thus, machines designed to handle synthetic fibers can generally accommodate glass fibers with L/D ratios of 800–1600. Apparently, glass fibers have not yet been formed on conventional paper-making equipment. However, there is evidence that, in blends, glass fibers with L/D ratios on the order of about 500 can be handled on this type of equipment.

See also: Continuous-Filament Glass Fibers; Glass-Reinforced Plastics: Thermoplastic Resins; Glass-Reinforced Plastics: Thermosetting Resins

Bibliography

Budd S M, Franckiewicz J 1961 The mechanisms of chemical reaction between silicate glass and attacking agents: Pt. 2. Chemical equilibria at glass-solution interfaces. *Phys. Chem. Glasses* 2: 115–18

Burgman J A 1970 Liquid glass jets in the forming of continuous glass fibers. *Glass Technol.* 11: 110–16

Clark D E, Pantano C G, Hench L L 1979 *Corrosion of Glass*. Books for Industry, New York

Doremus R H 1973 *Glass Science*. Wiley, New York

Hartlein R C 1971 New coupling concepts for glass reinforced thermoplastics. *Ind. Eng. Chem., Prod. Res. Develop*. 10: 92–99

Loewenstein K L 1961 Studies in the composition and structure of glass possessing high Young's modulus: Pt. 1. *Phys. Chem. Glasses* 2: 69–82; Pt. 2, *Phys. Chem. Glasses* 2: 119–25

Loewenstein K L 1973 *The Manufacturing Technology of Continuous Glass Fibers*. Elsevier, New York

Loewenstein K L, Dowd J 1968 Glass fiber tensile strength. *Glass Technol.* 9: 164–71

Mohr J G, Rowe W P 1978 *Fiber Glass*. Van Nostrand Reinhold, New York

Morey G W 1954 *The Properties of Glass*, 2nd edn. Reinhold, New York

Schrader M E 1977 Surface properties of glass fibers. In: Schick M J (ed.) *Surface Characteristics of Fibers and Textiles*, Part II. Dekker, New York

Thomas W F 1960 An investigation of factors likely to affect the strength and properties of glass fibers. *Phys. Chem. Glasses* 1: 4–18

Tooley F V 1974 *The Handbook of Glass Manufacture*, Vols. 1, 2. Books for Industry, New York

Van Vlack L H 1964 *Physical Ceramics for Engineers*. Addision-Wesley, Reading, Massachusetts

Zachariasen W H 1932 The atomic arrangement in glass. *J. Am. Chem. Soc.* 54: 3841–51

S. L. Mikesell; W. W. Wolf

Glass Formation

Glasses can be formed using a variety of techniques, including cooling from the liquid state, condensation from the vapor, pressure quenching, solution hydrolysis, anodization, gel formation, and bombardment of crystals by high-energy particles or by shock waves. Of these, the technique of cooling from the liquid state is the most important and most widely used.

1. Glass Formation Models

The best-known model of glass structure and glass formation is that due to Zachariasen (1932) and Warren (1937). Speaking of oxides, Zachariasen expressed the "ultimate condition" for the formation of glasses as: "the substance can form extended three-dimensional networks lacking periodicity with an energy content comparable with that of the corresponding crystal network." This led to the formulation of his celebrated rules for glass formation. According to these, a glass will be formed "(1) if the sample contains a high percentage of cations which are surrounded by oxygen tetrahedra or by oxygen triangles; (2) if these tetrahedra or triangles share only corners with each other; and (3) if some oxygen atoms are linked to only two such cations and do not form further bonds with any other cations." The requirement of a three-dimensional network led to a fourth rule: "(4) at least three corners in each oxygen polyhedron must be shared."

The structural implications of this random network model led to the classification of cations as network-formers, network-modifiers and intermediates. The physical basis for such models of glass formation was provided by Zachariasen (1932): "Glasses which do not devitrify very rapidly will have an energy only slightly greater than that of the (corresponding) crystal." Materials with small differences in energy between liquid and crystal will, at a given undercooling, have smaller driving forces for crystallization than those with large differences in energy, and hence should crystallize more slowly.

As an approach to glass formation, the criterion of a small energy difference between liquid and crystal is inadequate. Examples of good glass-formers with large energy differences include B_2O_3 and $K_2O\,.\,3SiO_2$. Further, materials with large entropies of fusion crystallize more slowly at a given undercooling and viscosity than those with small entropies of fusion (Uhlmann 1972a).

An alternative approach recognizes that glasses have been formed of materials with all types of bonding. These include covalent (SiO_2), ionic ($0.4Ca(NO_3)_2$–$0.6KNO_3$), metallic (0.4Fe–0.4Ni–0.14P–0.06B), van der Waals (toluene) and hydrogen (H_2O) bonding. In all cases, the formation of glasses requires cooling to a sufficiently low temperature—below the glass transition—without the occurrence of detectable crystallization. From this perspective, the critical question in discussing glass formation is not whether a liquid will form an amorphous solid when cooled from the liquid state, but rather how fast must it be cooled to avoid detectable crystallization. The critical concerns in discussing glass formation are then with the rates at which materials are cooled, and with the kinetic processes involved in crystallization.

2. Kinetic Treatments

For liquids of low viscosity in which a single nucleation event can lead to rapid and complete crystallization of a sample, Turnbull (1969) suggested the condition for glass formation as the absence of even a single nucleation event. This was expressed by

$$N = V \int_0^t I_v \, dt \quad (N < 1) \tag{1}$$

where N is the number of nuclei which form in a sample of volume V cooled over a time t, and I_v is the nucleation frequency per unit volume.

Application of this treatment to metals indicates that copious nucleation of crystals is expected. Hence a glass would not be formed even at very fast cooling rates, save for very small samples or materials with exceptionally high glass-transition temperatures (as T_g/T_m is ~0.8, where T_g is the glass-transition temperature and T_m is the melting point). For typical nonmetals, glass formation should be much easier, albeit not certain, and will be favored by high values of T_g/T_m.

A more predictive approach to glass formation (Uhlmann 1972b) treats crystallization by considering both nucleation and crystal growth, and estimates the cooling rates required to form glasses of various materials. This approach identifies a just-detectable degree of crystallinity and considers a body to be glassy if it contains smaller degrees of crystallinity. A volume fraction crystallized, V_c/V, of 10^{-6} has often been taken as the just-detectable degree of crystallinity, but estimated cooling rates are not very sensitive to the assumed V_c/V.

In treating the crystallization of an amorphous body, the formal theory of transformation kinetics (Avrami 1941) can be employed. The theory relates the fraction crystallized to the nucleation frequency I_v, the crystal growth rate u and time t as

$$V_c/V = 1 - \exp\left[-\int_0^t I_v \left(\int_{t'}^t u \, d\tau\right)^3 dt'\right] \tag{2}$$

Knowing the nucleation frequency and the growth rate at a given temperature, the degree of crystallinity can be calculated as a function of time. The results are represented graphically as time–temperature–transformation (TTT) curves. Such curves have extrema which represent the least time required to form a given degree of crystallinity. The extrema result from competition between the driving force for crystallization, which increases with increasing undercooling, and the molecular mobility, which decreases with increasing undercooling.

Realistic, albeit approximate, estimates of the critical cooling rate $\dot{T}_c$ required to form a glass (to avoid the just-detectable degree of crystallinity) can be obtained by an averaging method which employs the isothermal information contained in the TTT curves to construct continuous cooling (CT) curves. Onorato and Uhlman (1976) have applied this method to the problem of glass formation.

An alternative and more precise description of crystallization on cooling a body was advanced by Hopper et al. (1974), who introduced a crystal distribution function $\psi(r,t,R)$ defined such that the number of crystals dn in a volume dV at r having radii between R and $R + dR$ at time t is given by

$$dn = \psi(r,t,R) \, dV dR \tag{3}$$

This function contains detailed information about the crystal sizes, numbers of crystals and volume fractions crystallized in any portion of a body. Using this approach with a computer, the number and size distributions of small crystallites in a supercooled liquid can be calculated as a function of temperature; the analysis is known as the technique of crystallization statistics. From the nucleation frequency and the crystal growth rate, the calculations yield a time and temperature for each cooling rate at which the degree of crystallinity reaches the just-detectable value of 10^{-6}.

In addition to describing crystallization on cooling a liquid and providing estimates of the cooling rates required to form glasses, the analysis of crystallization statistics has also been used to describe crystallization on reheating a glass (Onorato et al. 1980). It is found that glasses at temperatures below the glass transition contain a finite number of crystal nuclei which depends on the material characteristics and thermal history. On reheating, three processes take place: (a) the nuclei already present grow; (b) new nuclei are formed; and (c) since the size of the critical nucleus (the smallest crystalline embryo stable at a given temperature) increases with increasing temperature, some nuclei which were stable at low temperature become unstable at higher temperatures and melt out. Such nuclei are no longer included in the calculations.

When such calculations are carried out, a maximum is seen in the calculated overall crystallization rate which can be compared with the maximum of a differential thermal analysis (DTA) crystallization peak. Knowing the crystal growth rate and viscosity as functions of temperature, the barrier to crystal nucleation B can be evaluated from the variation with heating rate of the temperature of maximum crystallization rate, T_{cr}. This is done by calculating T_{cr} for the experimental heating rates at various values of B. The value which provides the best fit to the measured dependence of T_{cr} on heating rate is taken as the nucleation barrier.

The analysis has been applied to a variety of materials, both organic and inorganic, and is found to predict closely both the values of T_{cr} and the dependence of T_{cr} on heating rate. To illustrate the use of this technique for estimating nucleation barriers, Cranmer et al. (1981) estimated a barrier of $80 \pm 2\,kT^*$ for anorthite ($CaO \, . \, Al_2O_3 \, . \, 2SiO_2$), where $T^* = 0.8\,T_m$. This is in excellent agreement with the value of $82kT^*$ obtained from measurements of the nucleation frequency as a function of temperature.

Applications of the analysis of crystallization statistics to a range of materials have indicated that T_{cr} depends on the cooling rate used to form the glass, as well as on the heating rate. The dependence on cooling rate is less pronounced than that on heating rate, and is most apparent when poor glass-formers are heated relatively rapidly.

Consideration of only homogeneous nucleation in kinetic treatments of glass formation directs attention to the minimum cooling rates required to form glasses. If a material will not form a glass at a given cooling rate when nucleating heterogeneities are absent, it will certainly not form a glass when such heterogeneities are present.

The effects of nucleating heterogeneities on glass formation have been investigated by Onorato and Uhlmann (1976). Heterogeneities characterized by contact angles higher than about 100° have no effect on the critical cooling rates for glass formation, at least for concentrations of heterogeneities as large as $10^7\,cm^{-3}$.

In systems containing distributions of heterogeneities, the nucleation process on cooling is dominated by different types of heterogeneities in different ranges of temperature. As the heterogeneities with small contact angle are depleted, those with larger contact angle become effective. For materials which are reasonably fluid over a range of temperature below the melting point, the crystallization process can be dominated by a very small fraction of heterogeneities having the smallest contact angles.

The kinetic treatment of glass formation has been applied to a range of liquids, including oxides, organics and metals. In general, the agreement between predicted and measured critical cooling rates has been found to be quite good (Davies 1976).

Application of the kinetic treatment to various materials has indicated that glass formation is favored by: (a) a large viscosity at the melting point and a viscosity which increases strongly with falling temperature below the melting point; (b) the absence of nucleating heterogeneities, or at least the absence of potent nucleating heterogeneities with small contact angles; (c) a large barrier to crystal nucleation, which implies a large crystal–liquid interfacial energy; and (d) the necessity of substantial redistribution of solute in crystallization, as for liquid compositions which differ appreciably from those of the crystallizing phase(s).

Consideration of these factors suggests that glass formation is also favored by a large value of T_g/T_m or T_B/T_m (where T_B is the boiling point); by compositions which are at or near low-lying eutectics, which contain a diversity of components, or which are effective solvents for second-phase material; and by superheating melts well above the melting point before cooling.

Although kinetic treatments of glass formation have proven quite successful in elucidating the processes involved in forming glasses, the ultimate descriptions of glass formation will require considerable expansion on the present kinetic views. Further insight into transport and crystallization behavior seems essential; and the perspective provided by structure and intermolecular interactions should prove most valuable in this regard.

See also: Glasses: Structure; Phase Separation in Glass; Glass: An Overview

Bibliography

Avrami M 1941 Granulation, phase change and microstructure. Kinetics of phase change. III. *J. Chem. Phys.* 9: 177–84

Cranmer D, Salomaa R, Yinnon H, Uhlmann D R 1981 Barrier to crystal nucleation in anorthite. *J. Non-Cryst. Solids* 45: 127–36

Davies H A 1976 The formation of metallic glasses. *Phys. Chem. Glasses* 17: 159–73

Hopper R W, Scherer G W, Uhlmann D R 1974 Crystallization statistics, thermal history and glass formation. *J. Non-Cryst. Solids* 15: 45–62

Onorato P I K, Uhlmann D R 1976 Nucleating heterogeneities and glass formation. *J. Non-Cryst. Solids* 22: 367–78

Onorato P I K, Uhlmann D R, Hopper R W 1980 A kinetic treatment of glass formation: IV. Crystallization on reheating a glass. *J. Non-Cryst. Solids* 41: 189–200

Turnbull D 1969 Under what conditions can a glass be formed? *Contemp. Phys.* 10: 473–88

Uhlmann D R 1972a Crystal growth in glass-forming systems—a review. In: Hench L L, Freiman S W (eds.) 1972 *Advances in Nucleation and Crystallization in Glasses*, American Ceramic Society Special Publication No. 5.

American Ceramic Society, Columbus, Ohio, pp. 91–115
Uhlmann D R 1972b A kinetic treatment of glass formation. *J. Non-Cryst. Solids* 7: 337–48
Warren B E 1937 X ray determination of the structure of liquids and glass. *J. Appl. Phys.* 8: 645–54
Zachariasen W H 1932 The atomic arrangement in glass. *J. Am. Chem. Soc.* 54: 3841–51

D. R. Uhlmann

Glass: Hardness, Wear and Polishing

The hardness of glass is usually defined by indentation hardness. Wear is caused by fracturing of glass in the fabrication process. Glass is polished by hydration of the glass surface and mechanical scratching of the hydrated layer. In this article, the relationships between hardness and wear and that between polishing rate and physical and chemical properties of glass are explained.

1. Hardness

The hardness of glass is usually expressed in terms of indentation hardness such as Vickers or Knoop hardness. There are two theories of indentation hardness: the densification theory and the microplastic theory.

1.1 Indentation Hardness of Optical Glass and Densification

Neely and Mackenzie (1968) observed that Vickers indentation in a silica glass decreased with heat treatment below the transition temperature and concluded that indentation is primarily due to densification of a volume of glass in the vicinity of the indenter.

Similar experiments have been made on optical glasses but recovery in optical glass is very small and virtually no change is observed in dense flint glass containing a large amount of PbO.

1.2 Indentation Hardness of Optical Glass and Microplasticity

The Knoop indentation on dense flint glass produced under a load of 1 kg for 15 s has been observed with a Leitz reflection interference microscope (500×), which indicated a hump around this indentation. The indentation depth was 3.3 μm and the hump was 85 μm wide and rose 0.3 μm above the glass surface. Similar humps have been observed in SK and BK glasses. This phenomenon shows that Knoop indentation is due to plastic flow.

Thus, it is concluded that densification and plastic flow occur simultaneously beneath the indentation on the glass surface. The indentation for glass containing a high silica content (e.g., silica glass) is mainly caused by densification, while the indentation for glasses containing a high content of modifiers (such as optical glasses) is mainly induced by plastic flow.

2. Wear

Wear is defined by the volume loss during a lapping or grinding process. Lapping or grinding is a phenomenon leading to mechanical rupture of glass. There are two kinds of wear processes. One comes from lapping with loose abrasives and the other comes from grinding with bonded diamond grains.

2.1 Lapping Hardness

Lapping hardness is determined from the volume of glass lost during the lapping process. Lapping grains form indentations on the glass surface and microfractures then extend from each of the indentations. Therefore glass lapping is regarded as an accumulation of microfractures generated by lapping grains.

Lapping hardness is affected by water. When the effect of water is removed by using kerosene as a lapping liquid in a dry nitrogen atmosphere, the lapping volume loss decreases with increasing mechanical strength in the case of a given size of indentation. The effect of water on lapping hardness occurs because the mechanical strength is affected by stress corrosion.

Although the lapping hardness is greatly affected by water, the Knoop hardness is not affected by ambient atmosphere. The lapping volume loss increases with increasing Knoop indentation when the effect of water is removed using kerosene in a dry nitrogen atmosphere. However, there are still some glasses with the same indentation size but different lapping volume losses. This is due to the difference in the intrinsic mechanical strength. The relationship between indentation and microfracture is discussed quantitatively by Swain and Hagen (1976).

Lapping hardness is a measure of resistance to fracture by lapping grains. It is concluded that, removing the effect of water, lapping hardness is dependent upon the indentation hardness and intrinsic mechanical strength of glass.

2.2 Diamond Grinding

A diamond pellet is used as a diamond grinding tool. A mixture of metallic powder (bronze) and diamond grains is pressed, sintered and pressed again to form the pellet. Pellets are affixed to a cast iron plate with epoxy resin.

In the diamond grinding process, there are three different glass removal mechanisms. Glass is removed by fracturing, scratching and both fracturing and scratching. Generally, the glass removal rate in diamond grinding is proportional to that in lapping. However, some glasses, for example dense flint glasses containing a large amount of lead oxide, have low grinding rates in spite of high lapping rates.

This is because glass is removed by scratching as well as fracturing in the grinding process, while it is removed only by fracturing in the lapping process.

3. Polishing

3.1 Polishing Mechanism

There are several proposed polishing mechanisms of glass: the wear mechanism, the flow mechanism and the chemical mechanism. According to the wear mechanism, molecular layers of the glass surface are mechanically removed. According to the flow mechanism, a glass surface is heated by friction between the surface and the polishing compounds under high pressure, resulting in viscous flow of the glass. According to the chemical mechanism, a silica gel layer is formed in the glass surface layer by chemical reaction with water, and the hydrated layer is removed with polishing compounds.

There is no correlation between the polishing rate and hardness. This observation invalidates the wear mechanism. Also, there is no correlation between the polishing rate and softening point, so the flow mechanism is not applicable. Figure 1 shows the linear relationship between polishing rate and chemical durability (weight loss of glass samples immersed in a 0.01 N nitric acid solution). A linear relationship shows that a chemical reaction between the glass and the polishing liquid plays an important role in the polishing process, because the chemical durability is a measure of the rate of reaction of water with the glass.

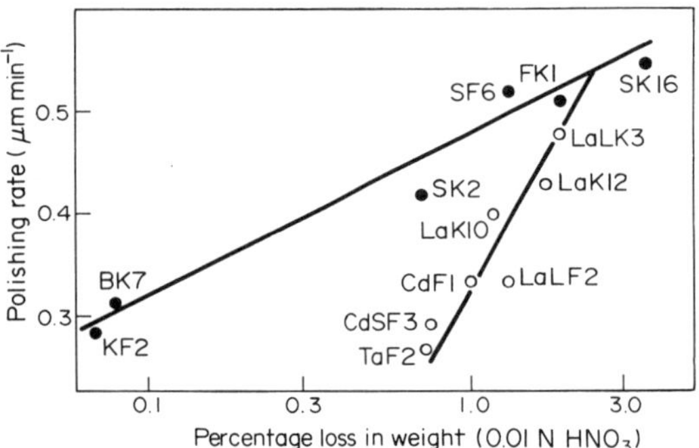

Figure 1
Relationship between polishing rate and weight loss of glasses that have been immersed in 0.01 N nitric acid: •, silicate glasses; ○, borate glasses (after Izumitani and Harada 1971)

When a glass surface is leached in 0.1 N hydrochloric acid, the surface hardness decreases with leaching time. This indicates that a soft layer is formed by chemical reaction. The polishing rate is inversely proportional to the hardness of the hydrated layer. The softer the surface layer, the higher the polishing rate.

The polishing process of an optical glass surface has been observed using the differential interference microscope (Izumitani 1979). To make the micrographs (Fig. 2), the glass surface was leached by 0.1 N hydrofluoric acid for 1 min and then chromium sputtered. As shown in Fig. 2a, the lapped glass surface consists of fractured concoidal surfaces. As the polishing proceeds, fine scratches appear and the area of fractured concoidal surface is gradually reduced. Finally a smoothed surface covered with scratches is formed (see Fig. 2d). The role of polishing grains is to scratch the rough glass surface and then to form the smooth surface. Thus polishing is the process of forming a hydrated layer by chemical reaction and then cutting away the hydrated layer with polishing grains.

3.2 Polishing Compounds

Cerium oxide is commonly used as a polishing compound for glass. Zirconium oxide and iron oxide are also sometimes used. Duncan (1960) showed that the polishing rate increases with the hardness of the polishing grains, and is greatest at a Mohs hardness of 5–7. Because MnO, Fe_2O_3, TiO_2 and ZrO_2 are good polishing compounds, Kaller (1956) suggested that polyvalent metallic compounds are effective in polishing.

A lapped surface becomes a mirror surface only when polishing grains are used. The roughness of the glass surface is unchanged when it is polished only with water without cerium oxide. The polishing grains are therefore necessary for effective polishing. The function of the polishing grains is to scratch and remove the soft hydrated layer. The removal mechanism is mechanical cutting rather than chemical adsorption at the interface between the polishing grains and hydrated glass surface.

Cerium oxide (CeO_2) compounds that are fired to get the optimum hardness, soft enough to be crushed during polishing but harder than the glass surface, give the maximum polishing rate. The optimum firing temperature is about 900 °C.

3.3 Polisher

Glass is usually polished with polishing grains such as cerium oxide and a polisher such as pitch. Pitch, rosin, beeswax and felt have been used as polishing material. Recently, synthetic polymeric materials such as polyurethane and polyesters have been used. The polisher must (a) coat and hold the polishing grains together, (b) transmit the applied load to the polishing grains and (c) deform by itself or fit to the lapped glass surface geometry. In order to satisfy the conditions (a) and (c), the polisher must be a viscous or elastic material. In order to satisfy condition (b), it must have high rigidity. Therefore, the polisher must consist of a viscoelastic material or elastic ma-

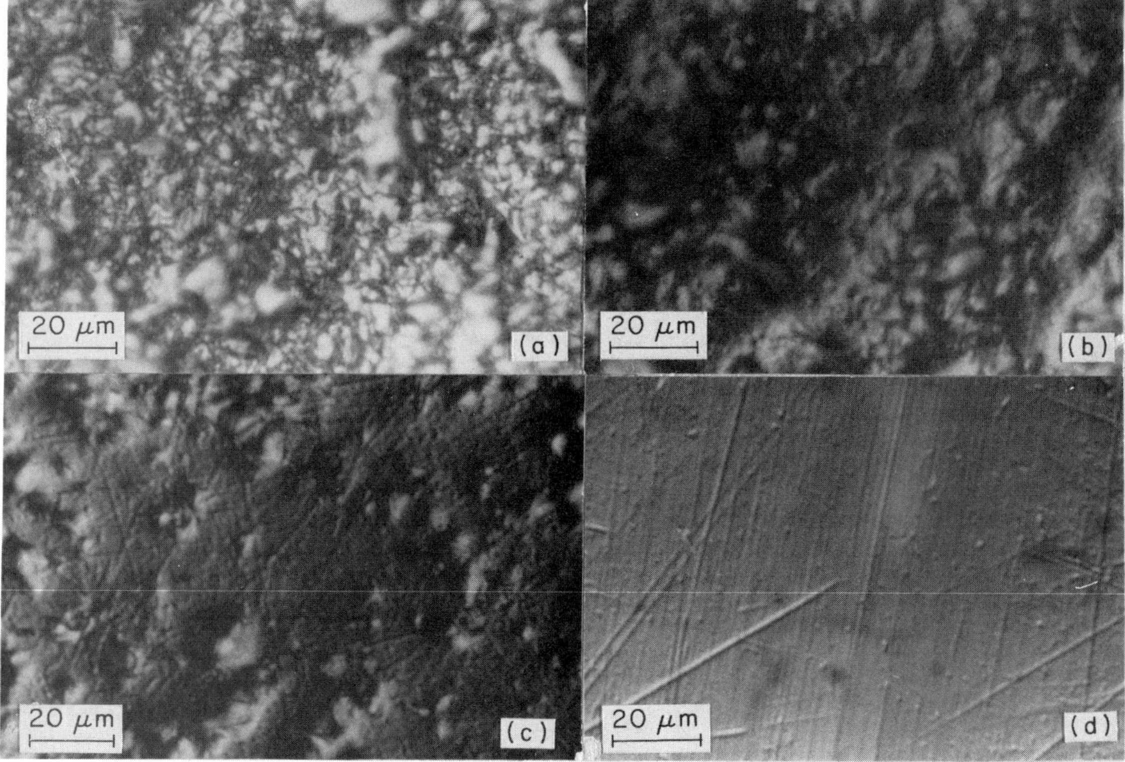

Figure 2
(a) Lapped surface, (b) 2 min polished surface, (c) 8 min polished surface, (d) 14 min polished surface

terial. In the viscoelastic polishers, there is an optimum creep compliance that gives the maximum polishing rate for each type of glass (Izumitani 1979). The optimum value is found at a higher creep compliance for soft glass and at a lower creep compliance for hard glass. A viscoelastic polisher such as pitch gives a ripple-free surface and a lower creep compliance polisher gives an edge-roll-off-free surface.

See also: Glass Surfaces; Glass: An Overview

Bibliography

Charles R J, Hilling W B 1962 Kinetics of glass failure by stress corrosion. *Symp. Resistance Mecanique du Verre et les Moyens de l'Améliorer, CR*. Union Scientifique Continentale du Verre, Charleroi, Belgium, pp. 511–27

Dick E 1973 Plastische Verformung, Anrissbildung and statische Ermüdung bei Ritzversuchen an Tafelglas. *Glastech. Ber.* 46: 120

Duncan L K 1960 Cerium oxide for glass polishing. *Glass Ind.* 41: 387

Gilman J J 1973 Flow via dislocations in ideal glasses. *J. Appl. Phys.* 44: 675–79

Holland L 1964 *The Properties of Glass Surfaces*. Chapman and Hall, London

Imanaka O 1961 Studies on lapping of glass. *Bull. Electro-Tech. Lab. (Japan)* 25: 1

Izumitani T 1979 *Glass II*, Treatise on Materials Science and Technology Vol. 17. Academic Press, New York, p. 126

Izumitani T, Harada S 1971 Polishing mechanism of optical glass. *Glass Technol.* 12: 131–35

Izumitani T, Suzuki I 1973 Indentation hardness and lapping hardness of optical glass. *Glass Technol.* 14(2): 35–41

Izumitani T, Suzuki I 1974 Diamond grinding of optical glass. *Proc. 10th Int. Cong. Glass*. Ceramic Society of Japan, Tokyo, pp. 35–41

Kaller A 1956 Zur Poliertheorie des Glases. *Silikattechnik* 7: 380–90

Kaller A 1983 Einfluss der chemischen, kristallographischen und physikalischen Eigenschaften der Poliermittel berim Polieren des Glases. *Silikattechnik* 34(1): 15–17

Koucky J et al. 1984 A kinetic interpretation of the glass polishing process. *Czech. Glass Technol.* 25(5): 240–43

Neely J E, Mackenzie J D 1968 Hardness and low-temperature deformation of silica glass. *J. Mater. Sci.* 3: 603–09

Peter K W 1970 Densification and flow phenomena of glass in indentation experiments. *J. Non-Cryst. Solids* 5: 103–15

Swain M V, Hagen J T 1976 Indentation plasticity and ensuing fracture of glass. *J. Phys. D* 9: 2201–14

T. Izumitani

Glass Melting

Glass melting is the conversion of a batch of discrete mineral particles of several different compositions into a melt sufficiently homogeneous and bubble-free to be useful when it is cooled into a solid glass. The more refractory components of the batch, for example quartz, rarely melt in the strict sense; rather, they dissolve in the eutectic liquids formed on heating. Although various glasses of different composition are of commercial importance, the great majority of current production, including flat glass, container glass, tableware and a considerable share of tubing and fiberglass, is from a fairly narrow range of compositions termed "soda-lime glass." This has the approximate chemical composition (wt%) 72 SiO_2, 15 ($Na_2O + K_2O$), 12 ($CaO + MgO$), 1 Al_2O_3. This composition is similar to that of the earth's crust in the sense that there is no class of species ($K_2O + Na_2O$, $CaO + MgO$, Al_2O_3, SiO_2) whose concentration in soda-lime glass is more than about twice its concentration in the earth's crust.

Soda-lime glass requires less energy for its production than do most other construction materials produced by high-temperature melting. For example, per unit volume it requires about one-tenth as much energy to melt glass as it does to produce iron. Thus, in terms of raw material and energy requirements, glass occupies a particularly advantageous position.

Modern glassmaking is continuous. The batch (the mixture of raw materials) is introduced into one end of the "glass tank" while molten glass is removed from the other end. From a large tank of melting area ~350 m^2 and a glass depth of ~1.5 m, the daily production may be as high as 300 m^3, equivalent to 700 t. Usually, heat for melting is supplied by the combustion of natural gas with preheated air. It is not surprising, therefore, that glass melting is the largest single industrial use of natural gas in the USA. Ohmic heating can also be used, either to supplement fuel ("electric boosting"), or to provide all the heat.

As several texts exist which give good descriptions of glass melting and the equipment currently used, this article is mainly devoted to a somewhat more analytical discussion of glassmaking.

1. History

Uncertainty exists as to whether glassmaking originated from curiosity about slags from copper or lead smelting processes, from natural evolution of highly siliceous ceramics (faience) coated with alkali glazes (Brill 1963) or, as the Roman naturalist Pliny suggested, from the observation of glass beads in the ashes of a wood fire enclosed in blocks of niter on a sandy desert (Morey 1954). There is no doubt, however, that the credit for continuous glassmaking furnaces is due to the Siemens brothers. In the middle of the nineteenth century they fostered three developments: (a) the production of gaseous fuel from coal, (b) the utilization of regenerators to preheat this producer gas and combustion air, and (c) the separation of the melting chamber from the refining and working chambers in a glass melting pot in order to permit continuous charging and withdrawal of molten glass. Taken together, these three developments allowed them to design, construct and operate successfully the first continuous glass tank in 1867. Glass tank designs have been altered in the past century, but not enough to destroy the resemblance to that original.

Although improvements in glass melting technology have been evolutionary, they have been impressive. For example, between 1920 and 1970, the energy required to melt a unit mass of glass decreased by a factor of three to four, and both furnace lifetime and the amount of glass melted daily per unit area increased by a factor of four. These achievements can be attributed to the higher melting temperature permitted by the development of improved refractories.

2. Enthalpy Balance

Enthalpy accounting provides a means of assessing the performance of a glass melter. Figure 1 shows a schematic (Cooper 1980). Enthalpy input at a rate $\dot{H}_0$ is supplied to the melter directly by combustion of fuel and/or dissipation of electric energy. Additional enthalpy can be recycled to the melter from heat recovered by a regenerator or recuperator, and occasionally from heat recovered by a batch

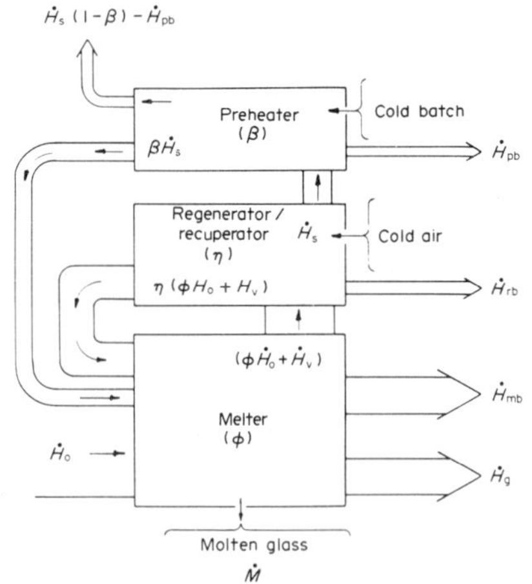

Figure 1
Schematic showing enthalpy balance for a glass melting system: ϕ, β and η are the parameters determining the performance of the melter, regenerator/recuperator, and preheater, respectively

preheater. The respective performance of each chamber is determined by the following parameters: ϕ, the fraction of $\dot{H}_0$ carried out of the melter by the flue gases; η, the efficiency of the regenerator; and β, the efficiency of the batch preheater. There are two types of losses from the system: a convective flux of heat carried out of the stack by flue and batch gases, $\dot{H}_s$, and conductive fluxes through the boundaries of the melter, the regenerator and the preheater. Even if perfect insulation were possible, the conductive flux through the melter boundaries, $\dot{H}_{mb}$, could not be reduced to zero, because melting reactions require a temperature higher than that at which molten glass can be removed for forming. The associated enthalpy difference must be lost through the boundaries. All other boundary fluxes are almost independent of the rate at which glass is being melted. Thus, the improved thermal efficiency noted in Sect. 1 was no doubt associated with the higher rates of production.

The enthalpy flux H_g required to convert the batch into molten glass at the exit temperature of the furnace is the useful part of $\dot{H}_0$. Expressions for $\dot{H}_g$ based on the specific heat of glass and melt and on the heat of the batch reactions are available (Kroger 1953, Bacon 1977).

Part of the batch is converted to gas, largely CO_2. The enthalpy flux $\dot{H}_v$ required to heat the batch vapor to the flue gas temperature can be determined from the heat of vaporization of any water added to the batch and the specific heats of the batch gases.

An overall enthalpy balance equation is achieved by summing algebraically all the enthalpy fluxes into the system. Local balances can be taken for each unit.

3. *Effects of Higher Temperatures*

The rate of glass melting reactions increases markedly with temperature. Is it reasonable then to expect that future improvements will be achieved by the use of still higher temperatures? The answer is complex. Increased melting rates $\dot{M}$ enhance the thermal efficiency of glass melting by decreasing the boundary losses per unit mass of glass melted, $\dot{H}_{mb}/\dot{M}$. An equivalent reduction can be achieved by reducing $\dot{H}_{mb}$ by better insulation. When $\dot{H}_{mb}$ is a small fraction of $\dot{H}_0$, as is the case in modern fuel-fired furnaces, the benefits from decreasing $\dot{H}_{mb}/\dot{M}$ are likewise small.

Also, increased temperatures require more-corrosion-resistant refractories, and it is not obvious that such refractories are available at costs which justify the benefits. In fuel-fired tanks there are other disadvantages associated with higher melting temperatures: evaporation from glass surfaces increases markedly with increasing temperature; the volatilizing species may be costly ingredients and they may contribute to air pollution; moreover, incongruent volatilization of species causes demixing of the melt. Higher temperatures also result in a higher content of NO_x in the flue gas. These deleterious factors suggest that, at least for fuel-fired tanks, increased melter temperatures will not provide the basis for more efficient glassmaking in the future. If historical precedents are viewed broadly, it is reasonable to expect that better understanding of the engineering fundamentals of the glass melting process will provide the basis for future improvement. For example, Fig. 1 reveals the possible savings from batch preheat. The following brief discussion of current practice may suggest other alternatives.

4. *Mixing of Batch*

After weighing, the batch components (sand, limestone, soda ash and feldspar) are introduced into a rotary mixer which produces (a) random motions of the particles, by causing particles either to roll down irregular inclines of other particles or to suffer airborne collisions, and (b) shear flows which minimize the distances required for transport by random motions. At best, batch mixers produce a random mixture. Evaluation of the homogeneity of a mixture can be based on the value of the variance V. For a random mixture with all particles of each component of the same size, the variance is given by $V_j = C_j(1 - C_j)/N$, where C_j is the expected number fraction of component j, and N is the number of particles in the sample. A real mixture has a higher value of variance than a random mixture because particles of different components are not identical and tend to segregate. Mixing for too long a time may cause an increase in V.

There are experimental techniques for assessing the value of V for each component in the batch. In evaluating the results, it is necessary to account for the variance caused by sampling and by measurement errors.

It is evident that in a random mixture the variance decreases with sample size, and for a given volume of sample, the smaller the particle size, the lower is the variance and hence the more homogeneous is the batch. To assess a real mixture reasonably, values of V at several different sample sizes are required.

5. *Melting and Homogenization*

After mixing, the batch enters the furnace and floats on molten glass. In conventional furnaces, the batch is heated from above primarily by radiation from the flame and superstructure and from below by radiation from the molten glass. The rate of conversion of the batch into a semimolten mixture sufficiently fluid to join the melt is given by the heat flux to the batch divided by the amount of heat necessary to transform the batch at the entering temperature to this semimolten condition. Kinetics of the batch reactions influence the rate because they determine the temperature at which the mixture is sufficiently fluid to leave the batch pile.

Once the batch is "converted," it is necessary that remaining particles be dissolved, bubbles be removed and the melt be homogenized. In the melt, random motions caused by species interdiffusion and heterogeneity attenuation from shear determine the rate of homogenization. In a multicomponent system, like soda-lime glass, interdiffusion is described by a matrix. The eigenvalues of the matrix have an Arrhenius temperature dependence and the smallest eigenvalue limits the rate of the homogenization process. Shear orients striations and streamlines. Once alignment is attained, simple shear is no longer effective in producing further attenuation. Stirrers are often introduced to produce additional attenuation of heterogeneities (Cable 1969).

Techniques for measurement and evaluation of the homogeneity of glass are diverse and interesting (Cable 1969).

6. Heat and Fluid Flow in Glass Tanks

Performance of glass tanks depends on the velocity and temperature distribution. The thermal conductivity k of molten glass is determined almost entirely by photons (Gardon 1961), and is thus given by $k \cong (16/3)n^2\sigma\lambda T^3$, where σ is the Stefan–Boltzmann constant, n is the refractive index (~1.5) and λ is the appropriately weighted average mean free path for photons (~70 mm for "clear" glass). At melting temperature, $k \cong 250\,\mathrm{J\,m^{-1}\,K^{-1}}$. This expression needs modification for regions within a distance λ or so of a surface. However, all dimensions of a glass tank are many times greater than λ.

Momentum conductivity, the dynamic viscosity η, shows behavior given approximately by the Fulcher–Vogel–Tamman equation, $\eta = \eta_0 \exp A/(T - T_0)$, with values of $\eta_0 = 0.0023$ Pa s, $A = 10000$ and $T_0 = 510$ K being typical for soda-lime glass. Therefore, at melting temperatures ~1800 K, η is ~5 Pa s, and the viscosity approximately doubles for each ~70 K decrease.

The temperature and velocity distributions are determined respectively by the heat balance equation and by the continuity and the momentum balance equations. Although the velocity is typically small enough to allow the "creeping flow" approximation to the Navier–Stokes equation to apply, the considerable interaction between heat and momentum transport makes simultaneous analytical solution essentially impossible. Three-dimensional numerical solutions have been obtained. They still require considerable computer time and are not easy to understand intuitively. Thus, numerical solutions in one and two dimensions and physical modelling are frequently used to gain information about the flow patterns.

From these treatments and from measurements in operating furnaces, a qualitative understanding of the flow behavior in glass tanks has emerged which is similar to that envisaged half a century ago by Gehlhoff and Schild. There are two superimposed flow systems: throughput flow caused by introduction of batch and removal of melt, and circulation caused by horizontal temperature gradients. Their interaction is indicated schematically in Fig. 2.

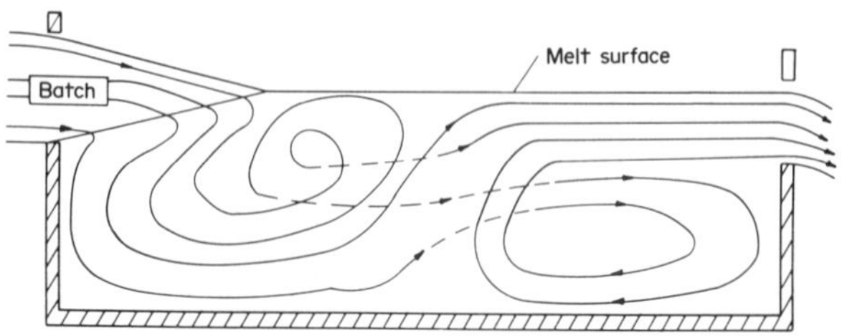

Figure 2
Schematic showing streamlines in a longitudinal cross section of a glass melting tank

Introduction of the cold batch into the entrance of the furnace and cooling of glass at the exit end causes a "hot spot" to occur near the center, resulting in two opposite circulating flows as indicated in Fig. 2. On the entrance side of the hot spot there is an upstream flow toward the batch. It convects heat to and transports partly molten material away from the batch pile. This incompletely melted glass moves forward beneath its previous path until it approaches the region of maximum temperature, where it flows upward in the so-called "spring zone." Here the throughput flow gives rise to interaction between the two circulating systems, causing some of the now nearly homogeneous and bubble-free melt to be transferred to the downstream cell and carried along the upper portion of the glass tank (in some designs following Siemens's scheme, this stream is forced to move downward to enter a throat which separates a melting zone from a refining zone). Some of the forward-moving stream is removed at the exit, but due to circulation a portion of the stream reverses direction and flows back again to the hot spot, where it is reheated and once again flows forward. Two dimensions are inadequate to represent the streamlines fully. Thus, in Fig. 2, broken lines have been used to allow streamlines to cross in order to portray accurately the fact that there are no endless circulating regions when a glass tank is in production.

7. Electric Melting

Cornelius demonstrated in 1925 that, as long as an alternative enthalpy source is provided to create the melt initially, it is practicable to dissipate sufficient heat by electrical conduction through the melt itself to melt the incoming batch. The rate of energy dissipation per unit volume in an electric melter is the square of the current density divided by the conductivity. The positioning, shape and size of the electrodes in the melt and the phase relationship between them determine the distribution of dissipated energy and the resulting circulating flows (Staněk 1977).

All electric melters have two persuasive advantages: (a) they are not linked to petroleum-based fuels, and (b) because they can operate with cold batch covering the surface, species vaporized from the melt are often condensed on the batch, reducing the raw material costs and the amount of volatile species entering the environment.

See also: Fining of Glass; Forming of Glass; Glass: An Overview

Acknowledgement

The author acknowledges financial support from the US Department of Energy under Grant No. EY-76-S-02-4075.

Bibliography

Bacon C H 1977 High temperature heat content and heat capacity of silicate glasses. *Am. J. Sci.* 277: 109
Brill R H 1963 Ancient glass. *Sci. Am.* 209(5): 120–30
Cable M 1969 The physical chemistry of glassmaking. *8th Int. Congress Glass*. Society of Glass Technology, Sheffield, pp. 163–78
Cooper A R 1980 An overall view of continuous glassmaking. *Glass Technol.* 21: 87–94
Doyle P J (ed.) 1979 *Glass Making Today*. Portcullis, Redhill
Frischat G H 1975 *Ionic Diffusion in Oxide Glasses*. Trans Tech, Bay Village, Ohio
Gardon R 1961 A review of radiant heat transfer in glass. *J. Am. Ceram. Soc.* 44: 305–12
Gunther R 1958 *Glass-Melting Tank Furnaces*. Society of Glass Technology, Sheffield
Kroger C 1953 Theoretischer warmebedarf der glasschmelz-prozesse. *Glastechn. Ber.* 26: 202
Morey G W 1954 *The Properties of Glass*, 2nd edn. Reinhold, New York
Scholes S, Greene C 1975 *Modern Glass Practice*, 7th edn. Cahners, Boston
Staněk J 1977 *Electric Melting of Glass*. Elsevier, Amsterdam
Tooley F V (ed.) 1974 *Handbook of Glass Manufacture*. Books for Industry, New York

A. R. Cooper Jr.

Glass Optical Fibers

The development of lasers in the early 1960s stimulated much research into how light, with its vast potential bandwidth, could be used for communications. Of the many approaches considered, glass optical fibers as dielectric waveguides have progressed through the various stages of research and development into practical application in the early 1980s. In addition to high information-carrying capacity, optical fibers offer other advantages over conventional metallic cables, such as longer distances between signal amplifiers, smaller and lighter cables, and immunity from electromagnetic interference.

In telecommunications the use of optical fiber technology is largely determined by the cost compared with conventional technology. For this reason, present-day applications are chiefly in long-distance and high-capacity intercity routes. If the predicted information technology "revolution" takes place, it is possible that optical fibers will be commercially competitive at all levels of the telecommunications systems hierarchy, even down to links to individual subscribers, if these are required to have high bandwidths. Optical fibers are already finding applications in other areas such as communications within buildings, aircraft, power plants and computers where light weight or immunity to electrical interference is of paramount importance. Another rapidly growing application of optical fibers is as sensors, where the polarization properties of specially designed fibers can be used to measure properties such as electromagnetic fields and currents, pressure and temperature.

1. Properties of Optical Fibers

Three properties are critical to the performance of optical fibers: optical loss, dispersion, and strength.

1.1 Optical Loss

Optical fibers should have low attenuation from absorption and scattering in the glass. The initial goal was $\sim$20 dB km^{-1}. (A loss of this magnitude means that 1% of the optical power is transmitted through 1 km; 10 dB km^{-1} corresponds to 10% transmission, and so on.) Today, because of reductions in extrinsic attenuation from impurities, fibers with losses of $<$0.2 dB km^{-1} can be fabricated from silicate glasses. However, the scope for further reduction in this glass system is limited because the intrinsic loss level of the glass constituents has probably been reached.

The various intrinsic loss components are shown in Fig. 1, compared with a typical transmission spectrum of present-day fiber. At short wavelengths the intrinsic losses from Rayleigh scattering from the glass inhomogeneities and absorption by electronic excitations dominate. At long wavelengths the infrared absorption caused by vibrations of atomic bonds is the major factor. These intrinsic properties provide some guidance as to which glass systems are best for a particular application. For the lowest loss the decrease in Rayleigh scattering as wavelength increases indicates that it is advantageous to move the infrared edge as far as possible to longer wavelengths.

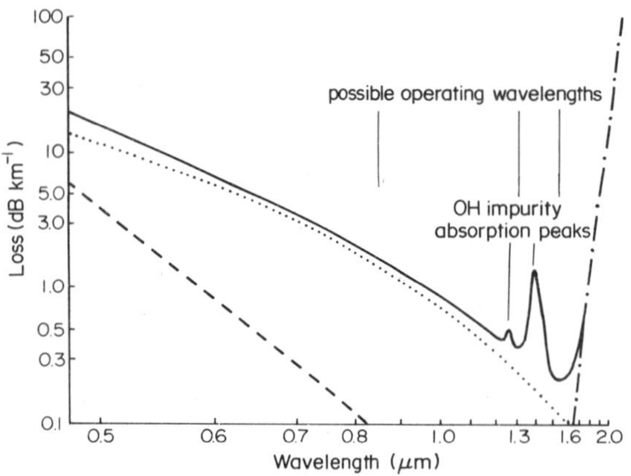

Figure 1
Intrinsic optical attenuation in glass and typical fiber transmission spectrum: – – –, uv absorption; · · ·, Rayleigh scattering; – · –, ir absorption; ——, spectrum of GeO_2–SiO_2 fiber

This factor has led to the choice of fibers based on the GeO_2–SiO_2 glasses. Where loss is not of prime importance, as in short-distance data links, the position of the infrared edge is of less consequence. Thus other glass systems with potential advantages in cost and ease of fabrication have been investigated, such as Na_2O–B_2O_3–SiO_2 glasses.

Other major concerns are extrinsic absorptions from transition metal and OH impurities. These are a significant problem for direct melt techniques (e.g., with the Na_2O–B_2O_3 glasses) and limit losses to >3 dB km^{-1} and the wavelength of operation to <1.3 μm. In the vapor-phase technologies (e.g., with the GeO_2–SiO_2 glasses) the transition metal impurities can be reduced to less than 1 ppb by distillation of volatile starting materials such as $SiCl_4$, $GeCl_4$ and $POCl_3$, and hence the loss from transition metal absorption is negligible. In some cases, where extreme care has been exercised to prepare dry glasses, fibers with negligible OH absorption have been fabricated. However, it is difficult to exclude all OH, and absorption from OH vibrations is usually seen at 1.39 and 1.24 μm. The typical transmission spectrum in Fig. 1 shows OH absorption peaks. At first, most fiber optic systems operated at 0.85 μm, but as longer-wavelength lasers have become available and the OH content of fibers has been reduced, the advantage of operating at 1.3 or 1.55 μm is well illustrated in Fig. 1.

1.2 Dispersion

As the majority of applications for optical fibers involve digital transmission, the information capacity is determined by the spreading of the light pulses as they progress along the fiber. The pulse spreading is caused by dispersion—either chromatic dispersion from the glass constituents or dispersion from the waveguide structure. The minimum dispersion (and hence maximum information capacity) should be designed to occur at the same wavelength as a minimum in the transmission loss.

The transmission of light along an optical waveguide is schematically represented in Fig. 2. The waveguide is composed of two regions: the outer cladding region composed of a material with refractive index n_2, and an inner region with index n_1. If $n_1 > n_2$, the light propagates by total internal reflection at the core–cladding interface; the angle of the light to the optical axis, θ, must be such that $\sin\theta < (n_1^2 - n_2^2)^{0.5} = \mathrm{NA}$, where NA is the numerical aperature of the fiber. For multimode fibers, with typically n_1–$n_2 \leqslant 0.02$ and a core diameter d of 50–100 μm, the light has many different possible propagation paths along the fiber, corresponding to many "modes" if the more correct wave analysis is applied. The propagation times for these different paths or modes is different, and hence there is another pulse spreading mechanism—"modal dispersion." Typically pulse broadening is 50 ns km^{-1}, limiting the information rate to 10–20 Mbits s^{-1} over 1 km, equivalent to ~200 telephone channels. A considerable improvement may be achieved by grading the refractive index profile of the core as a function of the radial distance from the fiber axis. Using this graded index, multimode fiber 140 Mbits s^{-1} systems can operate over 10–30 km without repeaters (pulse detectors and retransmitters).

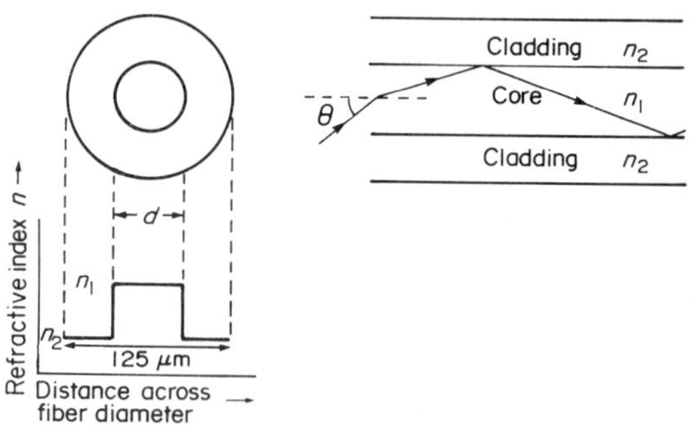

Figure 2
Light transmission in optical waveguides

If the refractive index difference and core diameter are chosen such that $2.405 > 2\pi d\mathrm{NA}/\lambda$, where λ is the wavelength of operation, there is only one path or mode for light propagation, that is, the fiber is monomode. Typically n_1–n_2 ~0.005 and d ~8–10 μm. As there is no modal dispersion in this case, the only significant dispersion is caused by materials dispersion, which passes through zero at 1.3 μm. Thus monomode fibers offer the best performance in terms of simultaneous achievement of high information rate and low loss. In laboratory trials, systems have already been demonstrated which could carry about 400 Mbits s^{-1} (~6000 telephone channels) on one fiber over distances in excess of 100 km without repeaters.

1.3 Strength

The predominant failure mechanism in fibers is due to surface contamination and microcracks; hence at all stages of fabrication such damage must be minimized in order to achieve high fiber yields. Polymer coatings are applied during fiber drawing to minimize surface damage. It is impossible to guarantee that the fiber surface will be dry during its use, so surface defects will suffer from static and dynamic fatigue if the fiber is strained during cabling, installation or service. To ensure that fibers do not fail by fatigue, they are proof-tested at a sufficiently high level so that the subsequent strains will not cause delayed failure of the weakest remaining flaw over the lifetime of the system. For land telecommunication systems the proof test strain is usually in the range 0.5–1.0% (1% represents ~700 MPa), and for submarine applications the requirements are even more stringent. Alternative coatings which reduce or prevent attack on the fiber by water (e.g., metals and ceramics) are at present under investigation. These would allow proof test levels to be reduced and fiber yields to be increased.

2. Fiber Fabrication

2.1 Vapor-Phase Technology

This technology is used to make all present-day high-performance optical fibers from the GeO_2–SiO_2 glass system, and in the simplest case consists of a GeO_2–SiO_2 core and a pure SiO_2 cladding. Often a small amount of P_2O_5 is added to the cladding for ease of fabrication; because P_2O_5 raises the refractive index, fluoride is usually also added to reduce the index to that of SiO_2. There are many variations of vapor-phase technology in use today, but they have several common features:

(a) use of distillation of volatile halides (e.g., $SiCl_4$, $GeCl_4$ and $POCl_3$) and high-purity gases (e.g., Cl_2, O_2, $C_2Cl_2F_2$) to provide a vapor stream containing the reactants in the required ratios;

(b) after an oxidation or hydrolysis reaction, glass is deposited onto a suitable substrate;

(c) a solid preform consisting of both core and cladding glasses is produced and subsequently drawn into fiber.

The chief differences between the various techniques lie in the method by which the energy is supplied for the oxidation. The most common technique is modified chemical vapor deposition (MCVD). Figure 3 shows schematically the various stages of preform preparation by MCVD. A pure-silica tube is rotated on a lathe and heated, usually by traversing an oxyhydrogen torch. The input vapor stream reacts in the hot zone and deposition occurs on the tube wall downstream where the gas is hotter than the tube wall. At this stage the deposit is in the form of a white "soot" of small particles. As the torch progresses down the tube, this soot is sintered into a glassy layer as the wall temperature increases. Initially the cladding material is deposited—some of the optical power travels in the cladding, and as the starting tube material is not of sufficiently low loss, pure material must be deposited. When sufficient cladding has been deposited, the input reactants are changed so as to produce the desired core glass composition. After deposition is completed, the temperature of the flame is increased and the tube is collapsed by surface tension forces to form a solid preform rod.

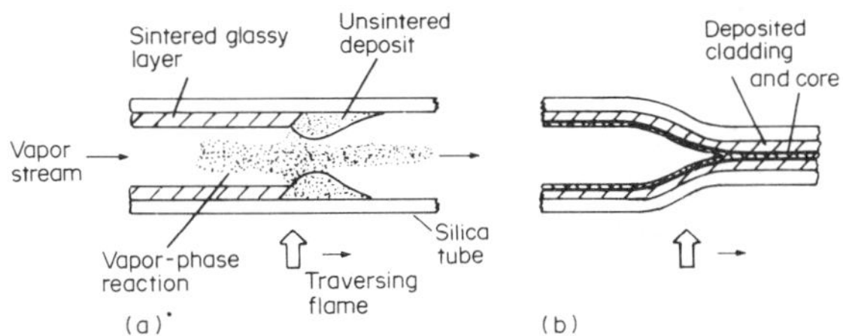

Figure 3
Stages in modified chemical vapor deposition (MCVD): (a) deposition; (b) collapse

In two other techniques, outside vapor-phase oxidation (OVPO) and vapor-phase axial deposition (VAD), the reactants are hydrolyzed on passing through a methane–oxygen or oxyhydrogen flame. The white soot is deposited on the side of a rotating starting rod which traverses the flame (OVPO) or onto the end of a rotating rod which is raised away from the flame as the deposit grows (VAD). In both cases a porous preform is obtained which is sintered into a solid glassy preform before fiber pulling. The OH content of the glass made by these techniques can be greatly reduced, in some cases to insignificant levels, by protracted heat treatment in a chlorine environment while in the unsintered form.

Another two variations involve the use of a plasma generated inside a starting tube by either radio-frequency or microwave radiation. In general, higher reaction efficiencies may be obtained with these plasma oxidation techniques than with standard MCVD.

After deposition, the solid preform is pulled into fiber using either a carbon resistance furnace or a zirconia induction furnace, as shown schematically in Fig. 4. The diameter of the fiber is monitored at the furnace exit and used to control, by a feedback circuit, the draw speed to ensure that the fiber diameter is maintained at the desired value. The bare glass fiber is protected by one or more layers of polymer as soon as possible after it emerges from the furnace. To minimize surface damage, the preform is usually etched and flame-polished before drawing. Also, the environment during fiber drawing, both inside and outside the furnace, is kept as clean as possible so that particles do not become embedded in the fiber surface or trapped between the fiber and the polymer coating. By careful attention to such details it is possible to pull lengths of fiber 5–10 km long with minimum breaking strains in the range 2–3%.

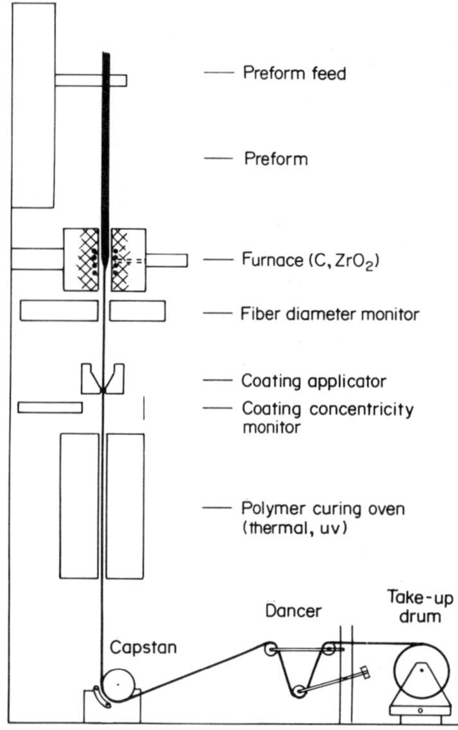

Figure 4
Preform fiber drawing

Although these vapor-phase technologies are well suited to the preparation of high-performance mono- and multimode fibers, it is difficult to produce large-core fibers of very high numerical aperture (suited to short-distance links) because of expansivity mismatches between the core and cladding glasses. In principle the potential market for such relatively low-performance, but necessarily cheap, fiber is larger than that for the high-performance fiber. It is in this area of the market that alternative fiber making techniques, such as direct melt methods, may offer commercial advantages.

2.2 Direct Melt Techniques

Several glass systems have been studied with this technology; typical are the Na_2O–B_2O_3–SiO_2 glasses, which produce some of the lowest losses ($\sim$3 dB km^{-1}), and Na_2O–BaO–GeO_2–B_2O_3–SiO_2 glasses, from which fibers with numerical aperture >0.5 can be made. This numerical aperture is considerably larger than that achievable by vapor-phase technologies ($\sim$0.3), and as a consequence it is easier to connect such fibers and launch power into them efficiently.

The starting oxides and carbonates must be ultrapure, for example, <10 ppb total transition metal impurity. The powdered raw materials are melted in a silica crucible. The two glass compositions (cladding and core) are transferred to a double crucible which contains two concentric chambers with nozzles at the bottom, and fiber can be pulled directly from the melt.

Other approaches include plastics-clad silica fibers, in which a pure-silica rod is used as the core and the protective polymer also acts as the optical cladding. Also, losses of $\sim$20 dB km^{-1} can be obtained with fibers fabricated entirely from plastics.

3. Recent Research and Future Technologies

For low-cost fibers, preparation of glass by the sol–gel technique is being evaluated. Gels of the appropriate starting materials, such as tetraethoxysilane, are first dried and then sintered. Reduction of the OH content is critical, but the use of a high-temperature Cl_2 treatment has led to the achievement of losses of $\sim$6 dB km^{-1}.

With high-performance fibers the main research and development topics are concerned with optimizing production processes and designing new monomode fibers with superior performance.

There is much research interest in glass systems for which the infrared edge is at longer wavelengths than for the GeO_2–SiO_2 system. Figure 1 shows that for these glasses the minimum intrinsic losses can be much lower, because the contribution of Rayleigh scattering decreases rapidly as wavelength increases. Theoretical predictions are for 10^{-2}–10^{-3} dB km^{-1} at 2–4 μm for fluoride glasses based on ZrF_4, such as ZrF_4–BaF_2–GdF_3–AlF_3. Important practical considerations under investigation are the minimization of impurities, the optimization of properties relevant to fiber pulling (e.g., viscosity and stability against crystallization), and the demonstration of adequate strength and fatigue properties. Although this work is at an early stage, losses of <10 dB km^{-1} and fiber lengths of >1 km have been separately achieved.

See also: Scattering of Light in Glass

Acknowledgements

The author thanks the Director of Research of British Telecom for permission to publish this article.

Bibliography

Ainslie B J, Beales K J, Day C R, Rush J D 1982 The design and fabrication of monomode optical fiber. *IEEE J. Quantum Electron.* 18: 514–23
Beales K J, Day C R 1980 A review of glass fibres for optical communications. *Phys. Chem. Glasses* 21: 5–21
Giallorenzi T G, Bucaro J A, Dandridge A, Sigel G H Jr, Cole J H, Rashleigh S C, Priest R G 1982 Optical fiber sensor technology. *IEEE J. Quantum Electron.* 18: 626–66
Midwinter J E 1979 *Optical Fibers for Transmission*. Wiley, New York
Miller S E, Chynoweth A G (eds.) 1979 *Optical Fiber Telecommunications*. Academic Press, New York
Miyashita T, Manabe T 1982 Infrared optical fibers. *IEEE J. Quantum Electron.* 18: 1432–50

J. D. Rush

Glass-Reinforced Plastics: Thermoplastic Resins

Glass-fiber-reinforced thermoplastics (GFRTP) fall into two distinct categories: injection-moldable compounds in which the reinforcement is in the form of short fibers, and thermoformable sheet materials containing either continuous fibers or relatively long chopped strands. The former are well-established materials which may be molded into complex load-bearing components, such as gears and small machine parts, while the latter have been only recently made available and are intended to be shaped, by hot stamping, into parts of similar form to metal pressings.

1. GFRTP Injection Molding Compounds

The molding compound is prepared by melt blending the polymer with glass fibers. The most common process is to charge a dry blend of chopped glass strand and polymer granules into a screw extruder, which feeds a "spaghetti" die. The extrudate is chopped, either hot or cold, into pellets, which may then be injection molded. The compounding process disperses the glass strands into individual filaments and breaks them into short lengths, typically <0.5 mm. In the injection molding operation the pellets are remelted and injected at high pressure (~50–200 MPa) into a cold mold. This process may cause further fiber breakage. Alternative compounding processes are available which retain longer fibers (~1–6 mm) in the molding granules; these compounds offer enhanced properties but are not widely exploited.

The molded material consists of an array of short fully dispersed fibers of different lengths, oriented in a pattern determined by the flow of material into the mold. The fiber orientation is complex, generally parallel to the flow direction in convergent passages but normal in divergent regions. Frictional drag at the mold walls also induces a through-thickness variation, especially in the thin platelike sections which are very common in injection moldings.

Glass-fiber reinforcement enhances both stiffness and strength at the expense of ductility and, to some extent, toughness. The in-mold shrinkage and thermal expansion is greatly reduced, and dimensional stability and elevated temperature performance of the material is very greatly improved. The material is anisotropic, however, on account of the flow-induced fiber orientation.

The short fibers can be stressed only by a shear transfer mechanism from the strained matrix. The tensile stress in the fiber builds up, from zero at the fiber ends, over a transfer length l_t, which depends on the average matrix strain ϵ_c and the effective shear strength of the fiber–matrix interface τ, and is given by $l_t = \epsilon_c E_f r_f/2\tau$, where E_f is the tensile modulus and r_f the radius of the fiber. If the fiber length l_f exceeds $2l_t$, then the central portion is strained to the same extent as the matrix; where $l_f \gg l_t$, reinforcing efficiency approaches that of a continuous fiber, but where $l_f < 2l_t$, it is less than 50%. Ideally, l_f should approach $10l_t$, when the efficiency is about 90%. Reinforcing efficiency is also reduced if the fibers are misaligned relative to the principal stress direction. A planar random array is only 30% as effective as an aligned array but would, of course, give planar isotropy. In practical terms τ is limited by the shear yield strength of the matrix (typically ~35 MPa for a polyamide), so that, for fibers of 12 μm diameter, l_t would be of the order of 60 μm at a strain of 1% and 120 μm at 2%. This indicates an ideal fiber length of 1.2 mm for a 90% reinforcement efficiency, assuming perfect alignment. In commercial materials the average fiber length is often only 150–200 μm, so that efficiency is comparatively low. It is desirable therefore that the bond between matrix and fiber should be as effective as possible. This is accomplished by the use of silane-based coupling agents, which are incorporated into the glass-fiber finish, and sometimes by incorporation of comonomers into the thermoplastic (especially in the case of polypropylene) to enhance fiber–matrix adhesion. The properties of some typical GFRTP materials are given in Table 1.

Table 1
Typical properties of GFRTP materials (molded test bars)

Material	ρ^a	E^b	σ^c	HDT^d	SK^e
Unfilled polymers					
Polypropylene	0.91	1.35	34	60	1.8
Polyamide 6.6	1.14	3.2	80	75	2.0
Polycarbonate	1.20	2.3	60	140	0.7
Filled with 20 vol% fiber					
Polypropylene	1.14	5.7	45	130	0.35
Polypropylene (chemically coupled)	1.14	5.7	103	155	0.35
Polyamide 6.6	1.46	10.0	210	260	0.50
Polycarbonate	1.45	9.0	135	160	0.20

a Density ($10^3\,kg\,m^{-3}$) b Flexural modulus (GPa) c Tensile strength (MPa) d Heat distortion temperature (°C) e In-mold shrinkage (%)

2. *Thermoforming Sheet Materials*

These materials consist of glass-fiber reinforcement impregnated with a thermoplastic resin. The reinforcement may consist of unidirectional continuous fiber strands (UD), woven roving (WR), continuous random mat (CRM), chopped strand mat (CSM) or a combination of these forms. Resins currently being exploited include polypropylene, polyamides, polysulfones, polyimides, and more recently polyether ether ketone (PEEK), the latter three for higher-temperature applications.

These materials are intended to compete with wet lay-up systems and with the sheet molding compounds (SMC) and "prepreg" based on thermosetting resins. The principle of fabrication is to preheat the sheet, and then to shape it by pressing, or stamping, between cold dies. This offers the possibility of a faster production rate, since there is no chemical curing operation as in the competing processes. This process is comparable with metal pressing and uses similar equipment. The replacement of metal components by reinforced plastics offers the possibility of weight reductions, improved corrosion resistance and lower total energy consumption, especially in automobile applications.

The constitution of the material is determined by the mechanical property and surface finish requirements, and by the amount of flow necessary during the shaping operation. Continuous aligned strands and woven cloth give the highest mechanical properties, but limited flow possibilities, so that materials containing this form of reinforcement would be suitable only for panels of shallow curvature. For more complex shapes, especially where stiffening ribs and flanges are to be formed, CSM and/or CRM should be used. The CSM allows the maximum flow, but there is a danger of nonuniform glass distribution, as only the polymer may flow into the extremities of the mold. The CRM type of reinforcement gives much greater uniformity of fiber distribution.

The sheet material may be manufactured by melt-compounding chopped strand and polymer and extruding or calendering into sheet, or by passing a sandwich of reinforcement and thermoplastics film through a hot calender-type mill. The surface finish may be enhanced by placing unreinforced layers of thermoplastic at one or both surfaces of the pack. Currently available products are based on polyamide 6 and polypropylene matrices with 40–50% of chopped or continuous random glass strand. They are less stiff (flexural modulus 5–6 GPa) than SMC due to the lower glass content and matrix stiffness but are generally tougher. A material based on continuous carbon fibers in PEEK resin has recently become commercially available but no glass fiber version using this matrix system has yet been offered.

See also: Glass-Reinforced Plastics: Thermosetting Resins; Glass Fibers; Thermoplastic Elastomers

Bibliography

Broutman L J, Krock R H (eds.) 1967 *Modern Composite Materials.* Addison-Wesley, Reading, Massachusetts
Hull D 1981 *An Introduction to Composite Materials.* Cambridge University Press, Cambridge
Ogorkiewicz R M (ed.) 1974 *Thermoplastics: Properties and Design.* Wiley, New York
Parratt N J 1972 *Fiber Reinforced Materials Technology.* Van Nostrand Reinhold, London
Titow W V, Lanham B L 1975 *Reinforced Thermoplastics.* Applied Science Publishers, London

M. G. Bader

Glass-Reinforced Plastics: Thermosetting Resins

Glass-reinforced plastics (GRP) consist of fibers of low-alkali E glass embedded in a thermosetting plastics matrix, such as polyester or epoxide resin. During fabrication, glass fibers are incorporated into the liquid thermosetting resin to which catalyst and hardener have been added. These cause the resin to polymerize and after curing give a solid plastic matrix reinforced with glass fibers. The fibers enhance the low stiffness and strength of the resin, whose main purpose is to transmit the load into the stiffer, brittle fibers and to protect them from damage.

Single E glass fibers, typically 8–15 μm in diameter, are easily damaged and difficult to handle (see *Glass Fibers*; *Continuous-Filament Glass Fibers*). For protection, they are coated with a silane or polyvinyl acetate size, which binds the fibers into strands containing about 200 filaments and acts as a coupling agent to provide a good fiber resin bond. To facilitate fabrication of GRP components, glass strands are incorporated into rovings, mats and fabrics. Glass fiber rovings consist of up to 120 untwisted strands, usually supplied wound together on a spool and suitable for the unidirectional (UD) fiber reinforcement of resins. Woven rovings (WR) are glass fiber rovings woven into a coarse fabric, usually with a balanced square weave. Glass fabrics are woven on textile machinery from twisted glass fibers (glass yarn) and are available in several weaves such as plain, square, twill and satin. Chopped strand mat (CSM) consists of chopped glass strands about 30 mm long held randomly orientated in a mat by a small amount of resin binder.

1. *Glass-Fiber-Reinforced Resins*

The principal resins used in GRP materials are unsaturated polyester resins, with epoxides being used in more limited quantities for high-technology applications, and phenolics, furanes and vinyl esters finding use for specialist applications (see *Resin Matrices Used in Composites*). Polyesters are less dangerous to handle, easier to catalyze and much cheaper than the other resins. They may be cold cured and are suitable for low-pressure, ambient-temperature fabrication techniques. Epoxide resins have better mechanical properties than polyesters, with higher stiffness and strength, particularly at high temperatures, and lower shrinkage on curing. Phenolics and vinyl esters are used in high-temperature applications, with continuous temperatures up to 260 °C, and furanes are noted for their chemical resistance at high temperatures. These specialist resins must be cured at high temperature and GRP components require hot-press molding. The higher performance resins are generally used with UD glass fibers and glass fabric reinforcement, while CSM and WR reinforcements are used almost exclusively with polyester resins. Recent surveys show that polyesters accounted for 92% of resin consumption in the USA and 99% in the UK, with the remainder being mainly epoxide. Thus the important GRP materials in high-volume applications are CSM/polyester, WR/polyester and, to a lesser extent, UD glass/polyester. Glass fabric/epoxide and UD glass/epoxide composites are used in low-volume, high-technology, applications.

Glass fiber content is related to reinforcement type and is shown, by weight, in Table 1 for the main GRP materials. A high fraction of aligned fibers (60–80% by weight) can be packed into a UD composite, while more resin is needed to impregnate all the fibers in CSM and WR laminates, giving glass contents as low as 25–35% by weight in CSM/polyester materials. In addition to the fibers, resin, catalyst and hardener, glass/resin composites may contain fire-retardant additives to reduce surface flame spread, uv absorbers to improve outdoor durability, pigments, dyestuffs and thixotropic additives which thicken resins to aid fabrication on inclined surfaces.

2. *Glass-Reinforced-Plastic Molding Compounds*

Several GRP molding compounds are now available in which the resin and glass fibers are premixed before molding. Component fabrication is achieved by placing the uncured compound in a closed mold and hot-press molding. The three main types of molding compound are sheet molding compound (SMC),

Table 1
Typical short-term mechanical property data for GRP materials

		DMC	SMC	CSM/ polyester	WR/ polyester	UD glass/epoxide longitudinal	UD glass/epoxide transverse
Glass content (wt %)							
	average	20	30	30	50	70	
	range	15–25	20–40	25–35	45–60	60–80	
Modulus (GPa)							
tensile	average	9	13	7.7	16	42	12
	range	8–11	9–16	6–9	12–22	30–55	8–20
flexural	average	8	11	6.3	13		
	range	7–9	7–14	5–8	10–18		
shear		3	3	3	4	5	
Strength (MPa)							
tensile	average	45	85	95	250	750	50
	range	35–60	50–120	60–150	195–350	600–1000	25–75
flexural	average	100	180	170	290	1200	
	range	85–120	90–240	100–300	160–500	1000–1500	
in-plane shear	average			80	95	65	
	range			60–100	80–120	50–80	
interlaminar shear	average		15	25	20	40	
	range		12–20	20–30	10–30	30–50	
Impact strength[a] (kJ m^{-2})		20–40	50–75	40–80	100–200		

a Measured by Izod-type test, unnotched specimen, according to British Standard BS2782, Sect. 306A

dough molding compound (DMC), sometimes referred to as bulk molding compound (BMC), and preimpregnated glass fiber sheet (prepreg).

Sheet molding compound consists of E glass fibers reinforcing a mixture of catalyzed polyester resin and a mineral filler, such as particulate, chalk, limestone or clay. The filler thickens the resin sufficiently for uncured sheets of SMC to be stored, handled and cut to the correct size for molding. Sheet molding compound usually contains chopped glass strands 20–50 mm in length randomly orientated in the plane of the sheet and is supplied in several nominal glass weights, containing 20, 25, 30 or 35% glass fiber, with filler contents of 30–50% by weight. Fillers reduce material costs and modify the handling characteristics of the SMC sheet and the flow properties in the mold. High-performance SMC materials may also contain aligned chopped glass fibers or continuous fibers, at glass contents up to 70% by weight. These high-performance materials, sometimes termed HMC or XMC, are being used in automotive and other demanding high-volume applications. Their further development and a rapid growth in their usage are expected.

Dough molding compound contains catalyzed polyester resin, up to 50% mineral filler and 15–25% short glass fibers, 3–12 mm in length. The usual filler is chalk though some resins are chemically thickened with alkali metal oxides. Shorter fibers and less glass than SMC means that DMC flows more easily and is thus used to make small articles of intricate shape by hot-press or injection molding.

Prepregs consist of sheets of glass fiber reinforcement about 1 mm thick preimpregnated with catalyzed resin and ready for hot-press molding. The reinforcement is usually UD fibers or glass fabric, with epoxide rather than polyester resin and component thickness is achieved by laminating together several sheets in the mold. The glass fiber content is high, typically 50–80% by weight. Accurate alignment of fibers is possible with prepregs, because there is minimal flow during molding and hence they are used mainly in high technology applications. Prepreg tapes are also available for selective stiffening of complex moldings and for the fabrication of GRP tubes and vessels by a dry filament winding process.

3. *Sandwich Materials and Laminates*

Different types of GRP material are frequently combined together in laminates or with metals, wood or foamed plastics as sandwich materials. Glass fibers may also be combined with carbon or polyamide fibers in a hybrid composite (see *Hybrid Fiber–Resin Composites*). A composite structure may thus be built up with material properties tailored to design loads. A widely used GRP laminate consists of CSM combined with WR or UD fibers. The CSM/polyester plies improve the interlaminar shear strength and provide a corrosion barrier to the WR or UD fibers which carry the main loads. To simplify fabrication, glass reinforcement may be supplied in a combination mat, consisting of CSM and WR or UD fibers stitched together. With UD reinforcement such as prepreg, or in filament wound structures, fibers may be orientated in different directions through the thickness. A common construction is a cross-ply laminate, with alternate plies orientated at 0° and 90° to a fixed axis, or an angle-ply laminate with plies at $\pm\alpha°$ for some fixed angle α.

Glass-reinforced-plastic sandwich materials are used widely in panel applications where the main loading is flexural. An ideal sandwich panel should consist of thin stiff skins with a thick low-density or low-cost core. For low and medium technology applications, GRP skins of CSM/polyester or WR/polyester are combined with cores of end-grain balsa wood or foamed plastic such as PVC or polyurethane. A typical construction for a meter square cladding panel might be CSM/polyester skins 3 mm thick with a balsa core 25 mm thick. For high-technology applications where weight saving is a priority, special low-density honeycomb core materials have been developed. These consist of a hexagonal cell structure arranged in a honeycomb pattern as shown in Fig. 1.

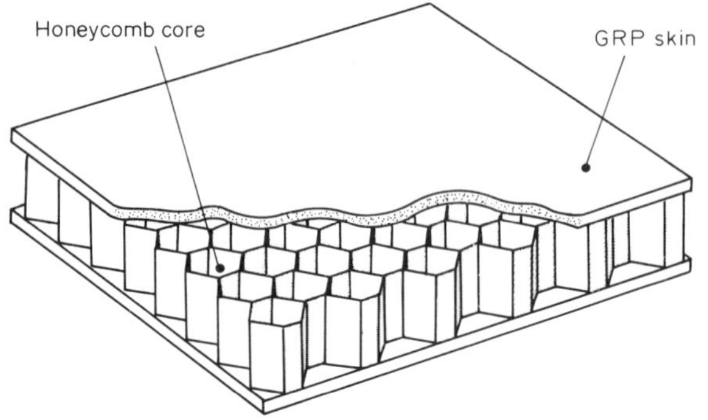

Figure 1
Schematic diagram of GRP/honeycomb sandwich material

The main honeycomb core materials are aluminum foil and a thin resin-impregnated paper. High-performance GRP skin materials are used with honeycomb cores, such as glass fabric/epoxide and cross-ply or angle-ply UD glass/epoxide laminates. A typical construction for a meter-square aircraft floor panel might be cross-plied skins 1 mm thick, with an aluminum honeycomb core 10 mm thick.

4. *Applications of Glass-Reinforced Plastics*

Glass-reinforced-plastic materials are used in low-, medium- and high-stress applications in industries as diverse as boat building, chemical plant, motor vehicles and aerospace. Important properties of these materials which have led to their wide usage are ease of fabrication into complex shapes, high strength to weight ratio, excellent corrosion resistance, good weathering properties, low taint and toxicity for foodstuffs and good thermal and electrical insulation. The selection of a particular material is determined mainly by material properties, costs and the suitability of fabrication and quality-control procedures for use in a particular industry.

In low-stress applications such as cladding panels, shower cabinets, small boats and meter boxes, GRP materials act as space-filling panels, supporting their own weight but not subjected to any significant external loads. Stress analysis is not usually carried out and the main design requirements are a good surface finish, good weathering properties and the ability to withstand accidental damage. The most widely used material is CSM/polyester with fabrication by hand lay-up, spray-up or vacuum-bag molding. Sheet molding compound materials are used for longer production runs of larger moldings such as business-machine cases and cladding panels, and DMC for smaller components such as electrical switchgear, lampholders and meter boxes. Glass-reinforced-plastic sandwich materials containing balsa or foamed-plastic core with CSM/polyester skins are used for boats and for some of the larger panel applications.

In medium-stress applications, GRP materials may be subjected to significant loads of short duration or low sustained loads. Some stress analysis is necessary based on the short-term behavior of the material and product design may be governed by codes of practice or standards. Large design factors are imposed so that stresses in the material are often less than 10% of the short-term strength. Applications in this category include: pressure vessels and storage tanks, with associated pipework, for use in chemical plant and the food processing industry; larger marine applications such as workboats, small hovercraft, minesweepers and small submersibles; and land-transport applications such as rail coachwork and larger automotive moldings for car bumpers, body panels and truck cabs. Resins are usually polyester, reinforced by WR or UD fibers with CSM/polyester surfaces for improved weather and chemical resistance. Most of the vessels, tanks and the marine applications are fabricated by hand lay-up, with filament winding becoming increasingly important for chemical plant pipework, sewage pipes and cylindrical tanks and pressure vessels. Sheet molding compound is used for the volume production of such items as car bumpers, lorry cab panels and rail seat shells. Where low weight is required, as in storage tank covers, hovercraft body panels and rail carriage doors, GRP/foam sandwich materials are used with fabrication by hand lay-up or vacuum-bag methods.

The high-stress applications of GRP materials are in the aerospace and defense industries. Here, low weight is important and hence materials are highly stressed and used efficiently. Design factors are kept as low as 1.5 in some applications and detailed stress analysis is thus required (see *Design of Composite Structures*). Typical applications include power boats, gliders, which are fabricated by hand lay-up using WR/polyester or UD glass/polyester, and helicopter rotor blades and rocket motor casings, which are filament wound usually with epoxide resin. Sandwich materials consisting of angle-ply prepreg GRP skins with foam or honeycomb cores find use in lightweight structures such as aircraft flooring, radomes and radar antennae.

5. *Fabrication by Hand Laminating*

Hand laminating or contact molding is the traditional fabrication technique for GRP materials. It is the method used for about 50% of the UK and 30% of the US consumption of GRP materials. Glass fiber reinforcement, in the form of CSM, WR, fabric or combination mat, is placed on a mold, impregnated with catalyzed resin and consolidated by hand rolling or brushing. The required thickness is built up by adding further layers of reinforcement and resin. Alternatively, chopped glass rovings are sprayed with a controlled quantity of resin from a special gun onto the mold and then consolidated by hand. The resulting "spray-up" material has a random distribution of fibers in the plane of the molding, giving similar mechanical properties to CSM/polyester. After an initial gelation stage, the GRP molding may be removed from the mold and left to cure, during which time the material attains its full mechanical properties. Curing takes about a week at 20 °C, but is accelerated by postcuring at higher temperatures. Typical postcure times vary from 3 h at 80 °C to 30 h at 40 °C, and depend on the resin used.

Contact-molded laminates have a single molded surface usually protected by a 0.5 mm thick resin layer termed the gel coat, which is brushed or sprayed onto the mold before the main laminate. For additional protection, the gel coat may be backed by a thin glass surfacing tissue. The main advantage of hand laminating is its versatility. There is no size limitation and metal inserts or additional glass reinforcement can be added where they are needed. Because it is a low-pressure, ambient-temperature process, inexpensive molds of wood, plaster or GRP may be used, which makes it suitable for small production runs. The main disadvantages are that moldings usually only have one good surface and the process is labor intensive with quality dependent on the laminator. Control of glass content, thickness and the cure schedule is poor, which can lead to variability in mechanical properties in a single molding or between moldings in a batch.

6. *Automated Fabrication Techniques*

To eliminate the variability in hand laminating and to permit longer production runs, several automated and semiautomated fabrication techniques have been developed. The semiautomated methods are based on matched die molding which permits critical control of thickness and glass content, and produces a molding with two good surfaces. Cold-press molding makes use of relatively inexpensive tools in a hydraulic press with low pressures up to about 200 kPa. Glass fiber reinforcement such as CSM is cut or preformed to shape and placed by hand in the mold with the required quantity of catalyzed polyester resin. The mold is closed and the resin allowed to cure for about 5 min. Two variations on cold-press molding are vacuum-bag molding and resin injection. In the vacuum-bag method, the matched molds, or an open mold, are sealed inside a rubber bag, which is evacuated so that atmospheric pressure is exerted over the surface of the molding. Resin injection or transfer molding involves the injection of catalyzed resin into the closed mold containing a preformed glass fiber mat. Injection pressures are up to 200 kPa and may be combined with vacuum assistance to improve resin flow.

In hot-press molding, matched metal dies are used and a hot curing polyester resin. Pressures are 1 MPa and temperatures in the range 100–130 °C, with a mold cycle time of 2–5 min. Because of the higher pressures, entrapped air is forced out of the mold giving lower void content than with cold-press molding. In conventional hot-press molding, the operator must still handle the glass fibers and the liquid resin, but the process can be made cleaner by using molding compounds. These have a high viscosity and thus require higher molding pressures, typically 1–7 MPa, with temperatures in the range 120–170 °C. Prepregs are cut accurately and each ply is positioned in the mold to have the required orientation. Because of the short fibers SMC, and more particularly DMC, are able to flow into the mold. With these materials, a weighed charge is positioned by hand centrally in the mold and on closing flows into the mold causing significant fiber reorientation. Sheet molding compound has the better mechanical properties and is used in larger panel applications, with DMC retained for smaller, bulkier components. Because of the good flow characteristics, DMC components may also be fabricated by injection molding.

Fully automated methods have been developed for the large volume production of GRP pipe, sheet and rod or tube stock. In the filament winding process, glass fiber rovings are fed under tension through a resin bath onto a revolving mandrel. Several layers of reinforcement, with the same or different winding patterns, may be placed on the mandrel in a controlled way. The mandrel is removed after curing, which usually takes place at elevated temperatures. A variation, which eliminates the resin bath, is dry winding using UD prepreg tape. Filament winding is used for GRP pipes and cylindrical tanks and enables high glass contents to be achieved with close control of fiber orientation. In order to improve chemical resistance, the glass fiber reinforcement may be wound over a chemically resistant liner of CSM/polyester or a thermoplastic such as PVC or polypropylene.

Continuous laminating processes have been developed for the manufacture of GRP sheet for applications such as electrical insulating panels, corrugated panels and translucent roof sheets. A moving sheet of CSM or WR reinforcement is impregnated with catalyzed polyester resin and sandwiched between layers of plastic film. This is then passed continuously through a die, to impart the required shape, and a curing oven before being cut into lengths. Pultrusion is a similar process suitable for the production of long lengths of GRP in such forms as rods, tubes or I beams. In pultrusion, glass rovings are impregnated with resin from a bath or by injection and pulled through a shaped, heated die. Other forms of glass reinforcement such as WR or continuous-filament mat may also be pulled through the die with the UD rovings to improve the transverse mechanical properties of the pultruded section. Another recently developed automated process is rotational molding, which is used for hollow lighting and telegraph poles. Here the laminating pressure is derived from centrifugal force in the rotating mold. Methods are also being developed to automate hand lamination, for example, by machine-controlled spray-up.

Automated fabrication techniques are suitable for volume production of good quality components through careful control of glass and resin contents, fiber orientation and the cure cycle. Many of these fabrication methods are suitable for control by microprocessor, which can lead to further cost savings and help overcome the styrene emission problem by eliminating manual operations. The main disadvantage is the high capital cost of the processing equipment. Some of the versatility of GRP materials is also lost, since these methods produce GRP stock in the form of pipes or sheets which require bonding or jointing to make into a component, whereas contact or press moldings are often integral GRP structures.

7. *Glass-Reinforced-Plastic Material Properties*

On loading a GRP material such as CSM/polyester in a short-term tensile test, it is found that the behavior is approximately linearly elastic to failure, which occurs by a brittle fracture at a strain of 1.5–2.5%. The short-term mechanical properties may thus be characterized by an elastic modulus and a failure stress or strength. A more detailed study of failure (see *Strength of Composites*) shows that there is progressive damage to the material at stresses below the ultimate. The first sign of damage under tensile loading is transverse fiber debonding, that is, separation takes place between the resin and those fibers perpendicular to the load direction. In CSM/polyester and WR/polyester, this occurs typically at 0.3% strain and at a stress level of about 30% of the tensile strength. Under increased loading, the debonds initiate resin cracks and as these cracks spread, more of the load is transferred to the fibers which eventually fracture or pull out at ultimate failure. Resin cracking occurs at approximately 50–70% of the tensile strength and is observed as crazing and by a change of slope in the stress–strain curve. In this article, ultimate failure stress data are given although it should be noted that, for many applications, failure data based on alternative criteria such as resin cracking or fiber debonding may be necessary. Limited data of this type are given, for example by Johnson (1978), or alternatively a suitable design factor may be applied to the ultimate strength data.

With a knowledge of the appropriate moduli and strengths, elastic design formulae may be used for stress and deformation analysis of GRP materials under short-term loads (see *Design of Composite Structures*). For CSM/polyester and SMC materials, conventional isotropic design formulae may be used, while for WR, fabric or UD reinforcement, account must be taken of the anisotropy in stiffness and strength properties arising from fiber orientation. For GRP materials, it is not usually possible to quote a single value of modulus and strength because of the inhomogeneous structure of the material. Thus, in addition to the effects of anisotropy, mechanical properties also depend on the quantity and type of glass reinforcement and on the fabrication method used.

Comparative short-term mechanical property data are summarized in Table 1 for the principal GRP materials and typical physical properties are given in Table 2. The tables are abstracted from Johnson, who has collated material property data on several types of GRP material from UK resin and glass fiber suppliers, GRP fabricators and research laboratories. Table 1 gives mean values of modulus and strength for DMC, SMC, CSM/polyester, WR/polyester and UD glass/epoxide at a typical glass content for each type of material. Below each mean value approximate bounds are given for the quantities. These bounds are usually wide because they refer to a full range of glass weight fractions, to different resins and glass fibers, fabrication techniques and test methods.

Reference to Table 1 shows that DMC, SMC and CSM/polyester, which all contain chopped fibers at low glass contents, have corresponding low moduli and strength values. The higher moduli and lower strength values for the molding compounds, compared with CSM/polyester at similar glass contents, are explained by the mineral filler which enhances the modulus but lowers the strength of the molding compound. Higher glass contents are found in WR/polyester laminates with average tensile modulus and strength about twice as high as the corresponding values for CSM/polyester. The data given in Table 1 refer to a fiber direction in a plain weave WR, with an equal number of rovings in the warp and weft directions. The highest glass contents are found in UD glass/epoxide composites and these materials show significant anisotropy in properties. Average tensile moduli range from 42 GPa along the fibers to 12 GPa transverse to them, with corresponding tensile strengths of 750 MPa and 50 MPa. Data on other GRP materials lie within the spectrum of properties shown in the table. Glass fabric reinforced resins and UD glass/polyester, fabricated by pultrusion or filament winding, usually have glass contents and mechanical properties intermediate between the WR/polyester and UD glass/epoxide values shown. Other points to note from Table 1 are the low interlaminar shear strength of all GRP materials and the differences observed between tensile and flexural properties, with flexural modulus usually below, and flexural strength above, the tensile

Table 2
Typical physical property data for GRP materials

	SMC and CSM/ polyester	Fabric and WR/ polyester	UD glass/ epoxide
Density (kg m^{-3})	1300–1600 (SMC, 1600–1900)	1500–1900	1800–2000
Linear thermal expansion coefficient (10^{-6} K^{-1})	18–35	10–16	5–15
Thermal conductivity (W m^{-1} K^{-1})	0.16–0.26	0.2–0.3	0.28–0.35
Specific heat (J kg^{-1} K^{-1})	1200–1400		950
Maximum heat distortion temperature (BS2782, 102 G) (°C)	175	250	300
Dielectric strength (BS2782, 201 A, C) (kV mm^{-1})	9–12	13–16	
Permittivity at 1 MHz (BS2782, 207 B)	4.3–4.7	4.1–5.2	
Power factor at 1 MHz (BS2782, 205 B)	0.015	0.016	
Dry insulation resistance (MΩ) (BS2782, 204 A)	10^6	10^6	

values, as a result of the inhomogeneous structure of the material.

Under long-term loads, fatigue loads and when subjected to high temperature or a wet environment, GRP stiffness and strength properties are below the short-term values given in Table 1 (see *Fatigue of Composites*; *Composite Materials: Nonmechanical Properties*).

8. Calculation of GRP and Sandwich Properties

A wide range of theoretical formulae are available for predicting the mechanical properties of composite materials in terms of microstructure and fiber and matrix properties (see, for example, Ashton et al. 1969, Jones 1975). Although these methods give satisfactory predictions of GRP properties, particularly moduli, they are usually only applicable to UD fiber reinforcement. The more rigorous results are also difficult to apply and often require correlation coefficients to be determined. It is therefore of interest to consider a simpler "rule-of-mixtures" approach which accounts for the main trends in the GRP properties reported here. Let E denote the Young's modulus of the composite, and E_g and E_r the Young's moduli of the glass fibers and resin matrix. If v_g, v_r are the volume fractions of glass fiber and resin in the composite then the Halpin–Tsai equation may be written in the modified form

$$E = \alpha E_g v_g + E_r v_r \qquad (1)$$

where α is a parameter which depends on the efficiency of the reinforcement. Suitable values are $\alpha = 1$ for UD reinforcement, $\alpha = 0.5$ for WR, with E then referring to the modulus along the fiber directions, and $\alpha = 0.3$ for CSM, which has a random fiber distribution. Taking $E_g = 75$ GPa, $E_r = 3$ GPa as typical modulus values for glass fiber and polyester resin and assuming fiber volume fractions v_g are 0.2 for CSM laminates, 0.35 for WR laminates and 0.5 for UD fibers, the following estimates for the GRP modulus are obtained from Eqn. (1): $E = 6.9$ GPa for CSM/polyester, $E = 15.1$ GPa for WR/polyester and $E = 39$ GPa along the fibers in UD glass/polyester. Reference to Table 1 shows that these predicted values agree well with typical GRP data.

An analogous expression to Eqn. (1) may be used for estimating GRP strength properties, in the form

$$\sigma = \alpha \sigma_g v_g + \sigma_r v_r \qquad (2)$$

Here σ is the tensile strength of the composite, σ_g that of the glass fiber and σ_r the tensile stress in the resin at the fiber failure strain. Because of damage, the strength of a glass fiber in a composite may be only half the fresh drawn fiber strength (Parkyn 1970, Chap. 10). Taking as a nominal fiber strength $\sigma_g = 1500$ MPa, with $\sigma_r = 30$ MPa and the v_g, v_r values used above, Eqn. (2) gives the following estimates of GRP tensile strength properties: $\sigma = 114$ MPa for CSM/polyester, $\sigma = 278$ MPa for WR/polyester and $\sigma = 765$ MPa along the fiber direction in UD glass/polyester. Again, these values are seen to be in reasonable agreement with measured data.

The mechanical property data and the prediction methods described above refer to GRP materials containing a single type of reinforcement. In practice, GRP structures often contain several types of reinforcement laminated together or consist of angle-ply or sandwich materials and it is not feasible to provide detailed mechanical property data on each material combination. However, theoretical formulae are available which enable the designer to calculate the stiffness and strength properties of combined materials and these have been found to be in good agreement with measured properties. Laminated plate theory permits the calculation of the full anisotropic stiffness and strength properties of laminated composite materials, consisting of plies of different materials or the same material oriented in different directions (see *Elastic Properties of Laminates*). The analysis method is rigorous and requires a complete knowledge of the anisotropic stiffness and strength properties of each lamina. The calculations are lengthy and best carried out by a small computer program or with prepared tapes on a desk calculator. A simpler, strength of materials, method for estimating combined laminate tensile and flexural properties is described by Johnson (1978) and by Allen (1969) for calculating the properties of sandwich structures. When the flexural rigidity of the core material is negligible compared with that of the stiffer skin materials, a simple design formula may be derived for the flexural rigidity per unit width, D, of a sandwich panel, which is an important quantity for the design of sandwich structures. For a symmetric sandwich with thin skins

$$D = \tfrac{1}{2}Eh(h + c)^2 \qquad (3)$$

where E is Young's modulus of the skin material in the load direction, h is the skin thickness and c the core thickness. Similar design formulae are available for calculating strength properties and the minimum weight design of sandwich panels.

See also: Composite Materials: An Overview; Glass-Reinforced Plastics: Thermoplastic Resins

Bibliography

Allen H G 1969 *Analysis and Design of Structural Sandwich Panels*. Pergamon, Oxford

Ashton J E, Halpin J C, Petit P H 1969 *Primer on Composite Materials Analysis*. Technomic, Westport, Connecticut

Holmes M, Just D J 1983 *GRP in Structural Engineering*. Applied Science, London

Hull D 1981 *An Introduction to Composite Materials*. Cambridge University Press, Cambridge

Johnson A F 1978 *Engineering Design Properties of GRP*. British Plastics Federation, London

Jones R M 1975 *Mechanics of Composite Materials*. McGraw-Hill, New York

Knoll K T 1980 TI-59 programs for analysis of advanced composite materials. In: Bunsell A R, Bathias C, Martrenchar A, Menkes D, Verchery G (eds.) 1980 *Advances in Composite Materials*, Vol. 1. Pergamon, Oxford, pp. 826–39

Modern Plastics International 11(1), 1981. Materials Survey, pp. 34–35

Parkyn B 1970 (ed.) *Glass Reinforced Plastics*. Iliffe, London

Rubber and Plastics Research Association 1977 *Survey of Fibre-reinforced Plastics Markets and Needs*. Rubber and Plastics Research Association, Shawbury, UK

Tsai S W, Hahn H T 1980 *Introduction to Composite Materials*. Technomic, Westport, Connecticut

A. F. Johnson

Glass Sealing

Products of modern technology often demand material properties which cannot be met by a single material alone. Rigid structures formed by a suitable combination of materials, such as a glass and a metal, are invariably the economic answer. Glass-to-metal seals are the primary focus of this article; many of the principles, however, apply to glass-to-glass, glass-to-ceramic and ceramic-to-metal sealing.

1. Interfacial Adherence and Failure

The weakest link in any material combination is the interface. Glass-to-metal seals fail because of lack of adherence between the glass and the metal, glass fracture, and metal failure. Adherence requires good wetting characteristics of glass on the given metal, followed by the development of a chemical bond or mechanical interlocking. The contact angle θ is a measure of wetting tendency; good wetting occurs when $\theta \ll 90°$.

The theory of interfacial oxide saturation for a good chemical bond was developed by Pask and coworkers (Pask 1964). The glass near the interface becomes saturated with the low-valence oxide of the substrate metal (the oxide should bond to the metal as well as to the glass). Controlled preoxidation or borating by fusing a thin layer of borax on the surface of the metal is a common technique to improve wetting of the glass.

An obstacle to wetting and adherence is outgassing of the metal, which generally results from the oxidation of carbon on the surface. Carbon tends to increase the contact angle. Decarburization of the metal by heating in wet hydrogen before sealing can help greatly.

When good wetting and adherence have been achieved at a high temperature, cooling to ambient temperature generates stresses in the sealing components as a result of the differential contraction between the materials. Fracture in glass initiates when the magnitude of the tensile stresses, interacting with Griffith flaws (usually on the glass–metal interface), exceeds the strength of glass. Fracture in glass propagates roughly perpendicular to the direction of the maximum principal tension. Observed fracture modes can be used to pinpoint the cause and to rectify the problem. Fracture probability of glass in the seals is difficult to estimate; hence, the usual practice is to maintain the level of tension in the glass below certain values established by experience.

Metal failure is quite rare. Most metals can withstand high stresses, tensile or compressive, and resist buildup of stresses by plastic deformation. A common reason for metal failure is oxidation (e.g., after hotspot formation in electric conductor wires) or a gradual chemical reaction with glass. Occasionally the viscous drag of glass during sealing can be transferred to the metal and cause metal tearing if the transferred force exceeds the hot tensile strength of the metal.

2. Classification of Seals

Seals are classified according to the level of stress in the glass.

(a) Matched seals. The thermal expansions of the sealing components match closely over the entire temperature range of sealing. Some matched systems in common commercial use are summarized in Fig. 1.

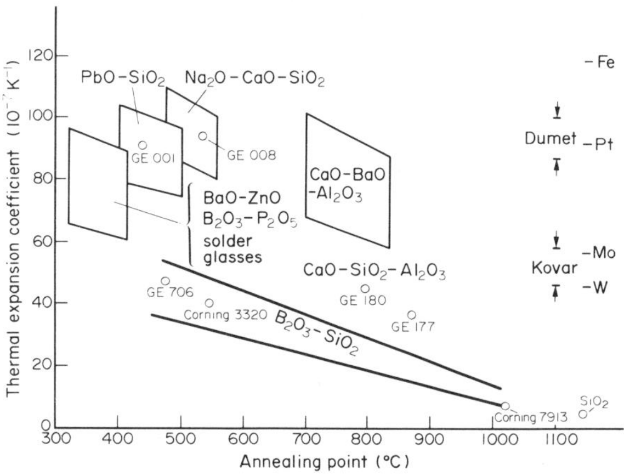

Figure 1
Thermal expansion coefficients of seal glasses and metals, and annealing temperatures of the glasses

(b) Unmatched seals. The thermal expansions of the sealing components differ considerably. However, fracture in the glass is avoided either by confining the high stresses to a limited region (thus reducing the probability of the occurrence of a critical flaw), for example by the use of small diameter metal wire

or a ductile metal (see Housekeeper seals, Sect. 6) or by distributing the stresses over a large region, for example by the use of intermediate expansion glasses ("graded seals").

(c) *Soldered seals*. The metal is soldered to a thin metallic film previously sealed to the glass.

(d) *Compression seals*. Molten solder is poured on the outside of the glass, or glass is poured on the inside of the metal with a higher expansion coefficient. On cooling, the metal shrinks around the glass.

3. Set Point

The set point T_Q of glass is defined as shown in Fig. 2. The thermal contraction ($\Delta L/L_0$) curves of the metal and the glass are matched at T_Q; the difference of the ordinates at the ambient temperature yields δ, often called the differential contraction or the mismatch. Thus

$$\delta = (\alpha_g - \alpha_m)(T_R - T_Q) \tag{1}$$

Manufacturers of seal components often use 5 °C above the strain point of glass, 15 °C below the annealing point or the annealing point itself as the set point; the strain point T_s is the temperature at which the glass has a viscosity of $10^{14.5}$ poise and the annealing point the temperature where the viscosity is $10^{13.0}$ poise (see *Viscosity and Annealing of Glass*). Varshneya (1980) has shown that these definitions of the set point are inadequate and that it is a strong function of the cooling rate given approximately by the empirical relation

$$\log \eta_{T=T_Q} = 13.0 - 0.94B \tag{2}$$

where $\eta_{T=T_Q}$ is the viscosity (poise) at T_Q and B the cooling rate (K s^{-1}).

Slow cooling of seals does not necessarily anneal out the stresses; it can increase the stress. The effect of the cooling rate on T_Q should be evaluated to estimate δ. Occasionally, glass fracture can occur from mismatch around point D (Fig. 2c). The fracture mode in this case is different from that at room temperature because the sign of the local mismatch is changed. Cooling or heating the metal with respect to glass can help reduce the mismatch during transient stages.

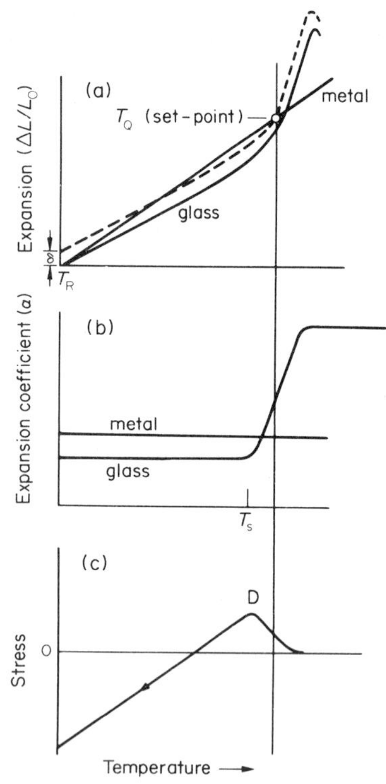

Figure 2
Expansion of glass (—— slow cooling; – – – faster cooling) and metal as a function of temperature (b) Expansion coefficients of glass and metal as a function of temperature. (c) Stress between glass and metal as a function of temperature

4. Development of Stresses in Seals

If the sealing components have equilibrated their dimensions at a high temperature T_δ, then by cooling through a temperature ΔT an incompatible strain equal to $\Delta\alpha\Delta T$ develops, where $\Delta\alpha = \alpha_g - \alpha_m$ is the difference in the true thermal expansion coefficients α $(= d\ln l/dT)$ between the glass (g) and the metal (m). At the same time, some of the stresses developed in glass at high temperatures can undergo viscous relaxation. A convenient expression for isothermal relaxation of stress σ is

$$\sigma/\sigma_o = \exp[-(t/\tau)^b] \tag{3}$$

where t is the elapsed time, τ the relaxation time and b is a constant. When $b = 1.0$, Eqn. (3) represents Maxwellian relaxation. Generally b is 0.5 for shear stress relaxation and 0.68 for volume relaxation.

Generation of stresses in a glass-to-metal seal undergoing nonisothermal viscoelastic relaxation of stress in the glass can be computed (Rekhson 1979, Varshneya 1980).

A simple equation for calculating the maximum stress σ in a glass-to-metal seal with mismatch δ (Eqn. (1), Fig. 2) is:

$$\sigma \simeq -E_g\delta/2 \tag{4}$$

where E_g is Young's modulus of the glass. Since E_g is ~70 GPa for most glasses and σ should be no more than ~7 MPa, δ is <200 ppm. Exact analytical calculations for the stresses are available for ideal geometries such as sandwich and bead seals.

For an infinitely long and wide sandwich seal (length and breadth $\gg$ thickness) the in-plane stresses σ_x and σ_z (denoted by σ_∞^c for the core and σ_∞^s for the skin) are equal and given by

$$\sigma_\infty^c = -\delta \frac{E_s}{1-\bar{\nu}} \left(\frac{a}{2c} - \frac{E_s}{E_c} \right)^{-1}$$

$$\sigma_\infty^s = -\delta \frac{E_c}{1-\bar{\nu}} \left(\frac{2c}{a} + \frac{E_c}{E_s} \right)^{-1} \quad (5)$$

where a and c are the thickness of the core and each skin layer respectively, E_c and E_s are their elastic moduli and $\bar{\nu}$ is a composite Poisson ratio given by

$$\bar{\nu} = \frac{\left(\frac{a}{2c}\right)\nu_c + \left(\frac{E_s}{E_c}\right)\left(\frac{1+\nu_c}{1+\nu_s}\right)\nu_s}{\frac{a}{2c} + \left(\frac{E_s}{E_c}\right)\left(\frac{1+\nu_c}{1+\nu_s}\right)} \quad (6)$$

Glass sealed over a metal wire in concentric cylinder geometry is commonly referred to as a bead seal. Analysis of stresses in an infinitely long bead seal was first carried out by Poritsky (1934) and later corrected by several others (see Varshneya 1982). At a distance r from the center, the stresses are given by

$$\sigma_{rr} = A + B/r^2$$

$$\sigma_{\theta\theta} = A - B/r^2$$

$$\sigma_{zz} = C \quad (7)$$

where the subscripts rr, $\theta\theta$ and zz refer to the radial, hoop and axial stresses, respectively, and the constants A, B and C are given by the relations (subscript m for metal and g for glass)

$$A_m = A_g R, \quad B_m = 0, \quad C_m = C_g R, \quad R = 1 - (R_g/R_m)^2$$

$$A_g = -\delta \left(\frac{1+\nu_g}{E_g} - R\frac{1+\nu_m}{E_m} \right) \Big/ D, \quad B_g = A_g R_g^2$$

$$C_g = -\delta \left(\frac{1+\nu_g}{E_g}(2-R) - \frac{1+\nu_m}{E_m} R \right) \Big/ D$$

$$D = \begin{vmatrix} \frac{1+\nu_g}{E_g}(2-R) - \frac{1+\nu_m}{E_m} R & R\frac{1+\nu_m}{E_m} - \frac{1+\nu_g}{E_g} \\ 2\left(\frac{\nu_m}{E_m} R - \frac{\nu_g}{E_g}\right) & \frac{1}{E_g} - \frac{R}{E_m} \end{vmatrix} \quad (8)$$

These formulae must be corrected further if the seals are "narrow" (length or breadth $\leqslant 10 \times$ thickness for the sandwich seal, and length $\leqslant 10 \times$ radius for the bead seal). Approximate correction formulae have been developed by Gulati and Hagy (1978).

For complex geometries, numerical methods such as the finite element techniques are often helpful (Varshneya and Petti 1978, Gulati and Hagy 1978). The calculations are matched to the gross features of the photoelastic pattern and later expanded to reveal the individual stress components at a fine spatial resolution.

Analytical methods are good for "infinite" geometries, which generally do not fit practical seals. For instance, in a seal composed of two short, high-expansion epoxy beads sandwiching an infinite glass plate, the in-plane stress in glass is compressive except at the ends of the contact with the beads, where it rapidly becomes tensile and can cause the glass to spall.

5. Measurement of Stresses in Seals

To estimate the seal reliability, the tensile components of stresses in the glass in the vicinity of the seal interface need to be measured (or calculated). Two methods are generally considered: stress birefringence and strain gauges. Measurement of stress birefringence is the most popular method because it is easy, but it suffers from the severe limitation that only the difference in principal stresses can be measured and not the individual stress components. Often mathematical principles are of help (e.g., stresses normal to a free surface must vanish). At the very least, photoelastic stress patterns can be recognized and correlated with subsequently observed fracture modes for quality control.

Strain gauges find limited use in glass-to-metal seals; because they can be used only to measure strains on external surfaces. In seals, the interest lies in stresses near the glass-to-metal interface, which is generally a subsurface. The low-temperature capability is yet another serious disadvantage. Most gauges have an upper operating limit of ~200–400 °C. This is far below the usual strain-free temperature of seals. In such cases, the strains can be measured only destructively: the gauge is applied at room temperature and the strain relieved is measured as one of the seal components is cut or ground away.

Strain gauges do have the inherent advantage over photoelastic techniques that they can be used to measure individual stress components in a complex biaxial system. In seals such as porcelain enamels, strain gauge techniques can be quite valuable.

6. Applications

Sealing between a ductile metal and a glass with widely differing expansion coefficients can be achieved by using the metal in the form of a thin foil (Housekeeper 1923). The shear stresses which arise in the metal on cooling reach the yield strength at some point, causing plastic flow on further cooling. Further buildup of tensile stresses in the glass is therefore greatly reduced. The fracture probability is reduced further by feathering the foil edges (achieved conveniently by dipping the entire foil in a suitable electro-etchant).

Fusion and bonding of a thin layer of glass frit (generally alkali borosilicate-based) onto a metal substrate (generally steel or aluminum) is termed porcelain enamelling. When the substrate is ceramic, the process is more commonly called glazing. Water vehicle frits can be applied by dipping, spraying, or screen-printing; dry frits can be applied electrostatically. The object is heated, thus oxidizing the metal and causing the glass to bond to it. Enamelling is often carried out in two steps: first a ground or base coat is applied which ensures good adhesion to the metal and then the finishing coat of the desired appearance effect is applied. The enamel should have an expansion slightly less than that of the substrate, so that the enamel is under compression, providing increased impact resistance at the surface. Enamels and glazes with an expansion coefficient higher than that of the substrate tend to crack. Often, use of a thinner coat alleviates cracking.

Sealing of contour-matched and polished surfaces of glass and metal can be achieved at temperatures well below the glass softening point by the application of a dc field (200–2000 V) to the components with the metal as anode (Wallis and Pomerantz 1969). The thermal expansion coefficients of the components should be well matched. The technique has an obvious advantage in the manufacture of electronic devices where low sealing temperatures are critical to the proper functioning of the device.

High fringing electric fields exist where *p–n* junctions intersect the surface. Containment of these fringing fields, or "passivation", can be achieved by the application of a glass coating which essentially encapsulates the device. Encapsulation also reduces the risk of weathering. The chief problems are that the glass should have a near-zero Na content, a very low softening point such that the device is not damaged during the encapsulation and a matching expansion coefficient.

Glass frits are often incorporated in conductor, resistor and capacitor thick-film pastes which can be screen-printed and fired-on much like the enamelling principles described above. Sealing of TV tubes, magnetic recording heads and many other products of today can be achieved by using suitable low temperature softening glass frits. Several glass frit compositions have been discussed by Takamori (1979). Some of these compositions can be devitrified during firing which usually increases the strength of the bond and the maximum service temperature of the product in use.

See also: Glass Bonding in Advanced Ceramics; Joining of Ceramic–Metal Systems: Procedures and Microstructures; Joining of Ceramic–Metal Systems: General Survey; Porcelain Enamelling Technology

Bibliography

Borom M P, Pask J A 1966 Role of "adherence oxides" in the development of chemical bonding at glass–metal interfaces. *J. Am. Ceram. Soc.* 49: 1–6

Gulati S T, Hagy H E 1978 Theory of the narrow sandwich seal. *J. Am. Ceram. Soc.* 61: 260–63

Housekeeper W G 1923 Art of sealing base metals through glass. *J. Am. Inst. Elec. Eng.* 42: 954–60

Pask J A 1964 In: Mackenzie J D (ed.) 1964 *Modern Aspects of the Vitreous State*. Butterworths, Washington, DC, pp. 1–28

Poritsky H 1934 Analysis of thermal stresses in sealed cylinders and the effect of viscous flow during anneal. *Physics* 5: 406–11

Rekhson 1979 Annealing of glass-to-metal and glass-to-ceramic seals: Part 1. *Glass Technol.* 20: 27–35

Takamori T 1979 Solder glasses. In: Tomozawa M, Doremus R H (eds.) 1979 *Glass II*. Academic Press, New York

Varshneya A K 1980 The set point of glass in glass-to-metal sealing. *J. Am. Ceram. Soc.* 63: 311–15

Varshneya A K 1982 Stresses in glass-to-metal seals. In: Tomozawa M, Doremus R H (eds.) 1982 *Glass III*. Academic Press, New York

Varshneya A K, Petti R J 1978 Finite element analysis of stress in glass-to-metal foil seals. *J. Am. Ceram. Soc.* 61: 498–503

Wallis G, Pomerantz D I 1969 Field assisted glass–metal sealing. *J. Appl. Phys.* 40: 3946–49

A. K. Varshneya; S. T. Gulati

Glass Surfaces

Many of the valuable properties of glass products depend on the character of their surfaces, and not on their bulk composition. A knowledge of the physical and chemical properties of glass surfaces is thus essential. Methods of making products with specified surface properties are based on the removal of an imperfect glass layer, or the reaction of this outer surface with various substances to attain unique characteristics.

1. Surface Structure

The surface structure of oxide glasses depends on the reactions of the oxide bonds. For example, a pristine silicate-glass surface contains unsatisfied Si and SiO bonds, which react rapidly with atmospheric water to form SiOH groups. Thus, the surface of an oxide glass is normally composed of metal hydroxyl groups. The thickness and structural arrangement of this hydrated surface layer depends on the glass composition, its thermal history, humidity and surface treatment after melting and cooling (Doremus 1973).

The exact structure of the surface layer on glass is not known. Recent ion-exchange studies on silicate glasses have shown that replacement of Na^+ in the dry glass by H_3O^+ occurs at room temperature. The outer surface layer of a silicate glass is thus a hydrated gel which forms on top of a second layer, in which there is a variable Na^+/H_3O^+ content (Fig. 1). These layers control the rate of dealkalization and dissolution of the bulk glass. Depending on compo-

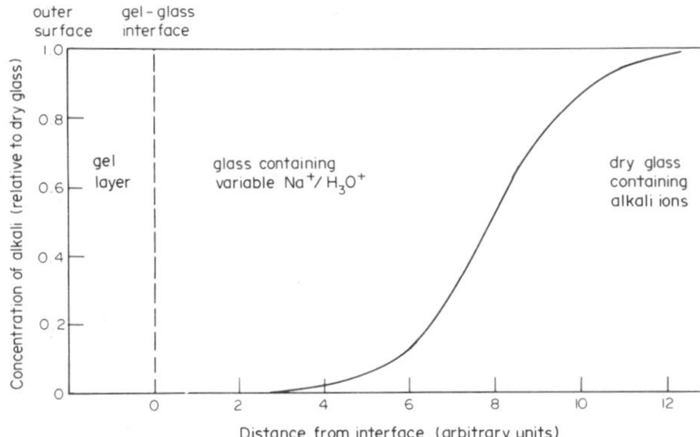

Figure 1
Structure of silicate glass surface produced by corrosion

sition, they may either protect the glass from further corrosion or be dissolved themselves.

2. Molecular Considerations

Most commercial glasses contain at least 60 wt% silica and, therefore, some idea of the molecular chemistry of glass can be obtained by first obtaining data for pure silica surfaces, and then treating glasses as an impure form of silica.

The effect of increasing temperature on the removal of water from a pure silica surface has been observed using infrared spectroscopy (Hair 1967) and is shown in Fig. 2. At room temperature, a silica surface consists of surface hydroxyl groups and physically adsorbed molecular water. As the silica is heated, the physically adsorbed water and some of the hydroxyl groups are removed, leaving behind two types of surface hydroxyl groups: those that are hydrogen bonded to each other, and those in a single configuration. This dehydration of the silica surface is reversible to 400 °C, but at higher temperatures an alteration in the silica surface occurs which makes water adsorption increasingly difficult. Above 800 °C, a silica surface readsorbs water very slowly.

The surfaces of mixed-oxide silicate glasses have similar SiOH groups on their surface, but if other glass-formers exist in the glass network, they will also provide sites for hydroxyl groups. Thus, AlOH, BOH and POH may also exist on the surfaces of glasses that contain these additional components. Moreover, the presence of these atoms can increase the reactivity of the original SiOH group. Monovalent cations such as Na^+ and K^+ will exchange with the H^+ of the surface OH group. This exchange may also change the reactivity of the surface. Ion-exchange reactions with more highly charged cations are much weaker, although adsorption of multivalent ions is reported.

3. Adsorption onto Glass Surfaces

Silica surfaces are weakly acidic. Polyvalent metal ions are adsorbed from solution if the solution pH is not far below the point where the metal hydroxide precipitates. If silica gel adsorbs sodium hydroxide, then the adsorbed sodium ions may be replaced by silver, copper or iron ions, by treating the surface with an appropriate solution of the corresponding metal salt. Polyvalent ions on silica surfaces can be replaced by sodium and ammonium ions from dissolved salts. The adsorption of long-chain anions such as fatty-acid anions and long-chain sulfonate ions on a silica surface is greatly enhanced if polyvalent cations are first adsorbed. Metal ions such as

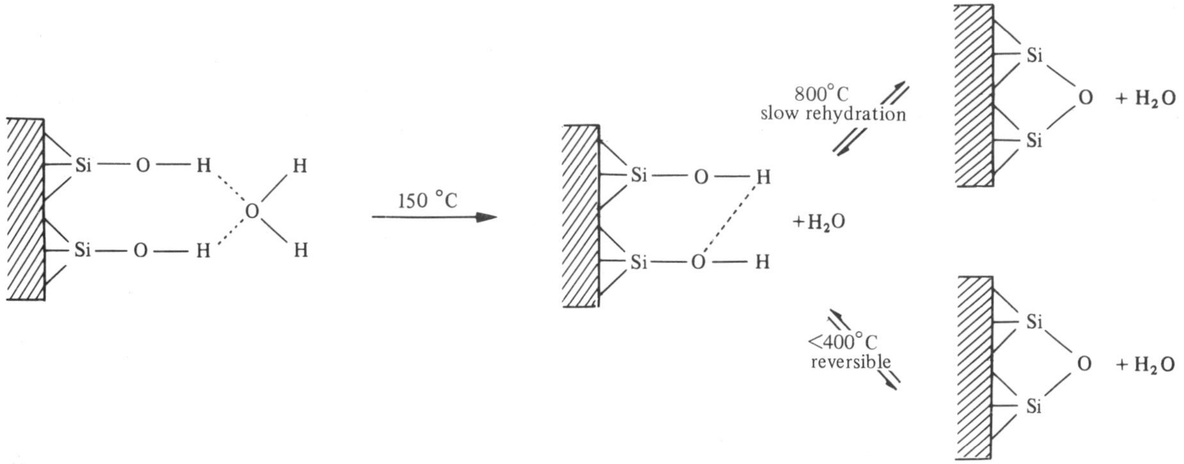

Si—O—H H
O
Si—O—H H
150 °C
Si—O—H
Si—O—H
$+H_2O$
800°C slow rehydration
Si
O $+ H_2O$
Si
<400°C reversible
Si
O $+ H_2O$
Si

Figure 2
Dehydration–rehydration of a silica surface

Al^{3+} and Fe^{3+} are adsorbed on silica and activate the surface by the formation of acidic groups.

The adsorption of organic compounds on the surface of silica depends on an interaction between polar groups in the molecule and the polar SiOH surface. The association between organic compounds and silica is much stronger if there are several polar groups on a given molecule. In aqueous systems, water also competes for the SiOH surface, and the amount of organic compound adsorbed is a function of parameters such as temperature and concentration. Proteins bind strongly to the surface of silica in weakiy acid solution, but such molecules do not attach strongly if Na^+ ions, with attendant water molecules of hydration, are present on the surface.

Silica gel and porous glass differ in one important respect: porous glass contains boron and, therefore, a significant number of BOH sites occur. When dehydrated, these boron sites become Lewis acid sites, and also increase the acidity of the SiOH groups. Although the ratio of boron to silicon in the bulk porous glass is 1:18, the ratio on the surface of the glass is 1:3. The adsorption of ammonia on hydrated porous glass at room temperature apparently occurs through an interaction with surface SiOH groups. However, if a dehydrated porous glass is exposed to ammonia, the adsorption is due to ammonia molecules coordinately bonded to the Lewis acid sites. At elevated temperatures (>400 °C), ammonia condenses with surface SiOH groups to form direct Si–N bonds, a structure that remains even after the porous glass has been consolidated. There is also evidence for direct B–N binding.

Helium, argon, and nitrogen adsorb onto porous glass and other porous materials. Studies of nitrogen, oxygen, argon, krypton, methane and ethane adsorption onto porous glass and a silica–alumina catalyst have also been reported (Barrer and Rees 1961). All adsorptions were of a physical nature and, therefore, no structural considerations were involved.

The adsorption of gaseous hydrogen fluoride on porous glass causes an elimination of all infrared hydroxyl bands, but the bands can be restored with water vapor. This demonstrates that the water content of porous glass resides on the surface (Low and Ramasubramanian 1966).

4. Silane–Glass Reactions

Silanes, molecules with both organic and inorganic characteristics, react readily with siliceous surfaces. The mechanism of reaction involves condensation of ester, halide or SiOH functional groups of the silane with SiOH groups on the silica surface. In the absence of water on the silica surface, the halide or ester groups on the silane condense with the freely vibrating groups on the silica. In the presence of water, the alkoxy groups on the silane hydrolyze to form hydroxyls (Eqn. (1)), which react with the SiOH groups on the glass surface (Eqn. (2)).

$$(RO)_3Si(CH_2)_3NH_2 \xrightarrow{H_2O} (HO)_3Si(CH_2)_3NH_2 \quad (1)$$

$$\begin{array}{l} \diagdown \\ \text{—Si–OH} + (HO)_3Si(CH_2)_3NH_2 \rightarrow \\ \diagup \end{array}$$

$$\begin{array}{l} \diagdown \\ \text{—Si–OSi(OH)}_2(CH_2)_3NH_2 + H_2O \\ \diagup \end{array} \quad (2)$$

Ideally, it would be desirable for all of the hydroxyl groups on the silane to react with the surface SiOH groups. This does not occur, however, because of steric factors. Instead, a partial condensation between silanes occurs with the formation of siloxane polymers.

5. Modified Surfaces

The availability of a large number of silanes allows a wide capability in modifying silica-type surfaces for further reaction. This is most useful if adhesion to another material is involved as, for example, in preparing glass–polymer composite materials. Covalent coupling of biological compounds to inorganic carriers involves the chemical treatment of the support with an organosilane, thus placing an organic group on the glass surface. The method is illustrated in Fig. 3. Using the proper organosilane and appropriate covalent coupling procedure, almost any biological compound can be attached to the inorganic material. Two derivatives have been used for the majority of the coupling studies, alkylamine glass and arylamine glass.

The quantity of biological compound that can be covalently attached to the inorganic carrier is a function of pore diameter and total available surface area; it appears to be relatively independent of particle size. With porous glasses, as much as 50 mg protein per gram of carrier has been coupled, but not all of the biological compound is active. Some portion is sterically hindered or inactivated by the coupling procedure.

6. Wetting

The wetting characteristics of glass surfaces are important in many applications. The advancing contact angle between distilled water and crystalline quartz is as low as 5°; hydrated glass is about 23°. This can be substantially increased by treating the glass surface to make it hydrophobic. The contact angle of water on glass depends on the treatment history of the surface and on the method of cleaning it.

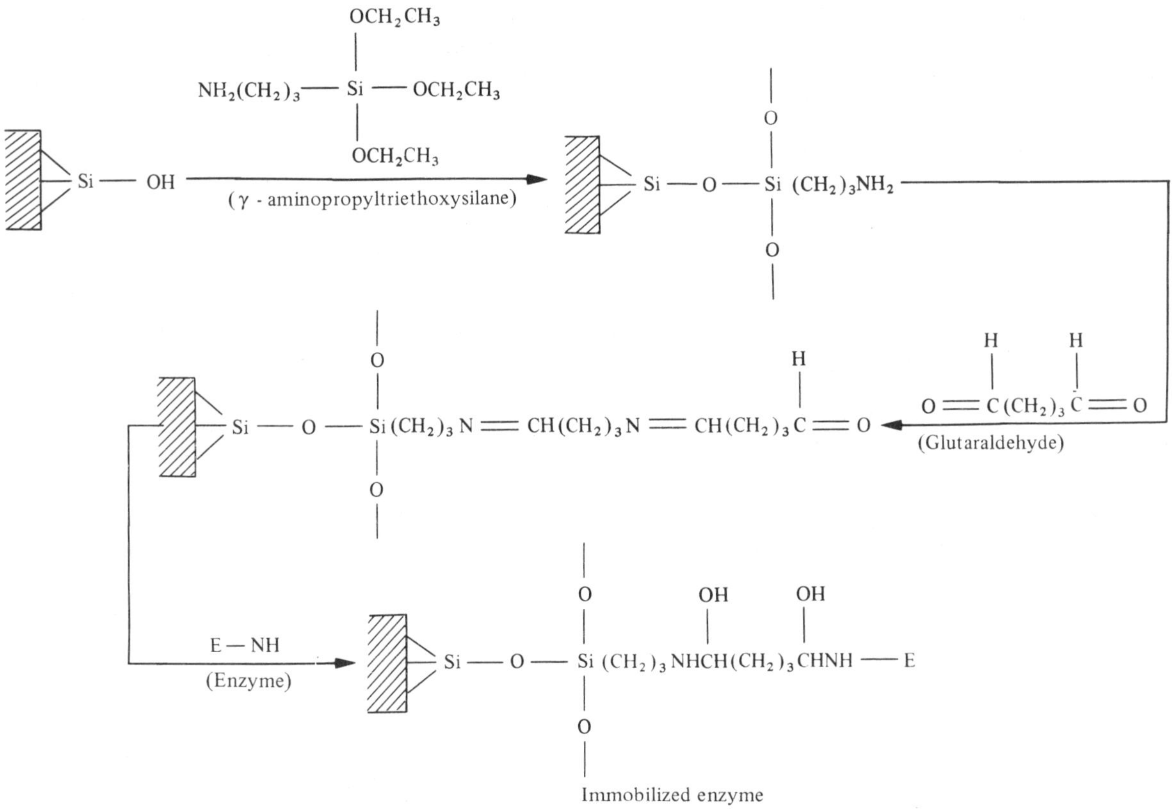

Figure 3
Immobilization of an enzyme on a glass surface

7. Surface Characterization

The methods used in analyzing glass surfaces are many and varied (see *Chemical Analysis of Solid Surfaces*). Chemical and structural information is needed for the outermost atomic layers and to depths of approximately 1000 Å beneath the surface. Each analytical method is sensitive to a particular characteristic of the surface. Thus, ellipsometry has provided a means for observing changes caused by polishing or chemical modification. Auger electron spectroscopy provides identification of chemical species present at the glass surface, and electron spectroscopy for chemical analysis is very valuable for the analysis of chemical bonding characteristics of glass surfaces with interacting molecules. Infrared spectroscopy is particularly suited to the study of hydrated layers. Scanning and transmission electron microscopy are valuable tools for the study of glass surfaces, and resolution down to 50 Å is obtained. The electron microprobe enables microanalysis of a selected area. In this method, a 1 μm electron beam scans the surface; penetration depths range from 200 to 20 000 Å.

8. Optical Properties

The optical properties of glass surfaces are important as glass is used in optical instruments for both reflecting and refracting elements.

Glass surfaces are usually polished so that they reflect rather than scatter light. The reflecting surface must only be smooth enough to satisfy optical requirements. If a glass surface is covered with prominences of a height h above the general surface plane, then a number of deductions may be drawn:

(a) As h is increased, light scattering first occurs at the shortest wavelength. Thus, a partly polished surface may diffusely reflect blue light.

(b) A surface which diffusely reflects light in one wavelength region will specularly reflect in a longer-wavelength region. A ground glass surface reflects infrared light almost as well as a polished surface.

(c) A diffusely reflecting surface can become specularly reflecting as the angle of incidence is increased, with regular reflection initially occuring for red light.

Detailed discussions on reflection and transmission of light from smooth and rough glass surfaces have been presented (Holland 1964). This reference also gives details of the effects of the thickness of surface films on refractive index of glasses, and the optical assessment of surface finish on the performance of glass optical surfaces.

See also: Glass: An Overview

Bibliography

Barrer R M, Rees L V C 1961 Henry's law adsorption constant. *Trans. Faraday Soc.* 57: 999–1007

Doremus R H 1973 *Glass Science*. Wiley-Interscience, New York, Chap. 12, pp. 213–28

Doremus R H 1979 Chemical durability of glass. In: Tomozawa M, Doremus R H (eds.) 1979 *Treatise on Materials Science and Technology*, Vol. 17, *Glass II*. Academic Press, New York, pp. 41–69

Dunken H H 1981 *Physikalische Chemie der Glasoberfläche*. VEB Duetscher Verlag für Grundstoffindustrie, Leipzig

Dunken H H 1982 Glass surface. In: Tomozawa M, Doremus R H (eds.) 1982 *Treatise on Materials Science and Technology*, Vol. 22, *Glass III*. Academic Press, New York, pp. 1–74

Ernsberger F M 1972 Properties of glass surfaces. *Annu. Rev. Mater. Sci.* 2: 529–72

Hair M L 1967 *Infrared Spectroscopy in Surface Chemistry*. Dekker, New York

Holland L 1964 *The Properties of Glass Surfaces*. Chapman and Hall, London

Iler R K 1979 *The Chemistry of Silica*. Wiley, New York.

Low M J D, Ramasubramanian N 1966 Infrared study of the nature of the hydroxyl groups on the surface of porous glass. *J. Phys. Chem.* 70: 2740–46

M. L. Hair; A. M. Filbert

Glass Transition

The term glass transition denotes a transition from a liquid to glassy state. In the liquid state a substance flows under any shear stress and its structure changes with changing temperature. In the glassy state, the substance, a glass, is solid, i.e., does not flow irreversibly, and the structure is frozen. A glass has a molecular structure with order in a short range (a few molecular dimensions), but no long-range order at many molecular distances. In this article the thermodynamics and kinetics of the transition from a liquid to a glassy state are discussed.

1. Glassy State

Figure 1 shows schematically the change in volume V, enthalpy H, or entropy S of a liquid during cooling.

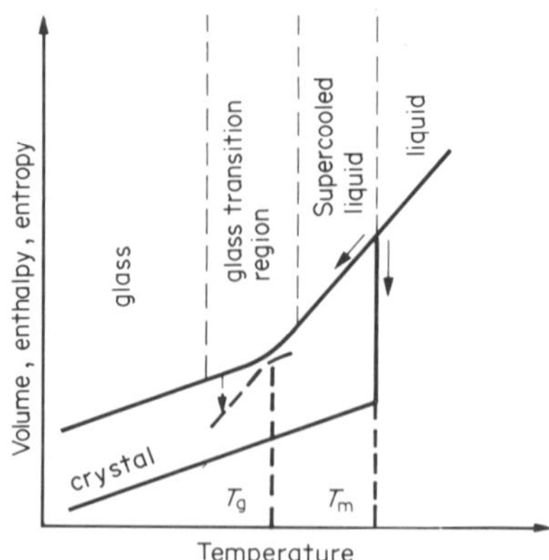

Figure 1
Temperature dependence of properties of liquid, glass and crystal

For substances which can exist in both glassy and crystalline states, there are two possible paths below the melting temperature T_m. One of these paths is the usual one for a crystalline solid below T_m; the other is a supercooled liquid below T_m. From the relation $\partial G/\partial T = -S$, where G is the Gibbs free energy, we can derive from Fig. 1 the plot shown in Fig. 2. Similarly, a plot of G versus pressure P is derived from the relation $\partial G/\partial P = V$. It follows from Fig. 2 that to form glass the liquid must make a thermodynamically unfavorable choice and follow the path with higher free energy, for kinetic reasons.

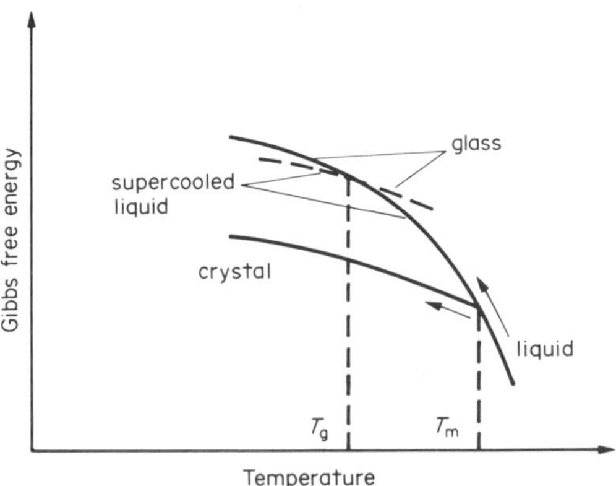

Figure 2
Temperature dependence of the Gibbs free energy for liquid, glass and crystal

At the temperature T_m crystallization is impeded by a finite viscosity of the liquid and therefore takes some characteristic time t^*. If the dwelling time at T_m is made shorter than the time t^* by rapid cooling, crystallization is kinetically arrested. Inorganic glass-forming melts have a high viscosity at T_m and can be cooled slowly. In contrast, molten metals require cooling rates of the order of $10^5\,°C\,s^{-1}$ to form glass.

A supercooled liquid is in metastable equilibrium with respect to a crystalline solid. The free energy profile can be pictured as one with a minimum for a crystalline state and a "trough," above the minimum, for the supercooled liquid state. Below the glass transition temperature T_g the structure of a glass is not in equilibrium; it is frozen in by a viscosity 10^{15}–10^{20} times greater than that of ordinary liquids. The liquid is never completely frozen; even at temperatures below T_g its volume decreases slowly in the direction of its equilibrium value. This process, called stabilization, evolves on a geological time scale and can be observed only with the help of very sensitive instruments. At the temperatures closer to T_g, stabilization can be monitored more easily.

2. *Universality and Nature of Glass Transition*

The glass transition is characteristic of all glass formers: molecular liquids, organic polymers, molten salts, and metallic alloys. The glass transition temperature varies from $-180\,°C$ for alcohol, $30\,°C$ for selenium and $550\,°C$ for sodalime–silicate to $1200\,°C$ for fused silica. However, both the qualitative and many quantitative features of the glass transition of different materials are strikingly similar.

Many other properties of glasses show transitions similar to that shown for thermodynamic properties in Fig. 1; for example, the index of refraction, electrical conductivity, and diffusion coefficient. If a liquid near its glass transition temperature is subjected to a sudden change in temperature, all of these properties exhibit an instantaneous change followed by a slower evolution to a new equilibrium value. The common reason is an evolution of structure. An instantaneous change in properties corresponds to a change in interatomic distances concurrent with the change in temperature. The time evolution of a property indicates that the structure relaxes to a new configuration. During this time the atoms change their relative positions, which can affect coordination numbers, the number of broken bonds and bond angles—everything that taken together constitutes a "structure." This process is similar to viscous flow or diffusion.

A characteristic time, called the structural relaxation time τ, can be somewhat loosely defined as the time needed for a given property p to reach two-thirds of its equilibrium value following a stepwise change in temperature. After a time much longer than τ, the structure will be in equilibrium. If the time is much shorter than τ, the structure remains frozen. A key to understanding the glass transition is that the relaxation time τ increases exponentially with decreasing temperature, whereas the dwelling time for a constant cooling rate is the same at each temperature. For example, cooling at a rate of 3° per minute is equivalent to 3° temperature jumps and 1 minute isothermal holds. At a temperature $100\,°C$ above T_g, the relaxation time τ for a silicate glass with $T_g = 550\,°C$ is about 0.055. Therefore at this cooling rate stabilization will be complete, since both glassy and liquid relaxation contributions to the change of the property will take place rapidly and the liquid will follow the equilibrium curve. At a temperature $100\,°C$ below T_g, τ is on the order of 2 months, and the structure is frozen; the only contribution to the property change is the glassy one.

According to classical thermodynamics, the slopes of the V–T, V–P, and S–T diagrams are

$$\left.\frac{\partial V}{V\partial T}\right|_P = \frac{\partial^2 G}{\partial P\partial T} = \alpha, \qquad \left.\frac{\partial V}{V\partial P}\right|_T = \frac{\partial^2 G}{\partial P^2} = -\kappa$$

and

$$\left.\frac{\partial S}{\partial T}\right|_P = -\frac{\partial^2 G}{\partial T^2} = \frac{C_P}{T} \qquad (1)$$

the thermal expansion coefficient, compressibility and heat capacity divided by temperature, respectively.

The glass transition is controlled by thermodynamics and kinetics in close association. Thermodynamics determines the direction of stabilization (toward a minimum of free energy) and the relationship between the properties of glass and liquid; kinetic processes determine rates.

3. *Thermodynamics of Glass Transition*

Glass can be obtained by a rapid increase in pressure as well as by rapid cooling, as shown in Fig. 3. The thermodynamics and kinetics of these processes are similar. The properties of a liquid during these

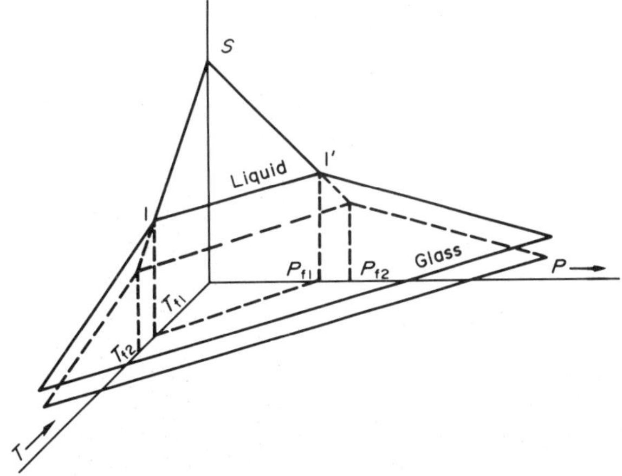

Figure 3
Entropy–temperature–pressure relationship for a liquid and glass formed by instantaneous freezing

changes are defined by two external parameters, P and T. This is not true for a glass. We can rapidly cool the liquid from the temperature T_{f1} or T_{f2} and obtain glasses with different properties, because different structures were frozen in. The same result occurs with rapid pressurizing at P_{f1} and P_{f2}. Therefore, to specify the glass, one additional parameter called an order parameter is needed. The temperature T_f or the pressure P_f can be used as this parameter. T_f and P_f are called the fictive temperature and pressure and are defined as intersections of the glassy and liquid portions of the entropy–temperature and entropy–pressure curves. The concept of fictive temperature was first introduced by Tool and Eichlin (1931). During stabilization, as a property approaches its equilibrium value, T_f approaches T. Davies and Jones (1953) showed that this result is a consequence of an entropy principle. They obtained the expression for the irreversible production of entropy:

$$\Delta S_{\text{irr}} = \Delta C_P \left(\frac{1}{T_f} - \frac{1}{T}\right) dT_f \geqslant 0 \qquad (2)$$

where ΔC_P is the difference between the heat capacity of liquid C_{Pl} and that of glass C_{Pg}. Since $\Delta C_P > 0$, T_f will increase or diminish according to whether $T > T_f$ or $T < T_f$ (i.e., during stabilization, T_f always approaches T). The changes in thermodynamic properties of the liquid in the glass transition region are given by:

$$\begin{aligned} TdS_{\text{irr}} &= -V\Delta\alpha\,(T - T_f)\,dP + (\Delta C_P/T_f)\,(T - T_f)\,dT_f \\ dH &= C_{Pg}dT - VT\alpha_l dP + \Delta C_P dT_f \\ dV/V &= \alpha_g dT - K_l dP + \Delta\alpha dT_f \end{aligned} \qquad (3)$$

where Δ denotes, as in Eqn. (2), the difference between the values of the property for the liquid and for the glass. Other notation is the same as in Eqn. (1). Similar formulae are available in terms of the fictive pressure P_f.

Freezing in of the parameters T_f or P_f by a fast experiment is described by glass entropy surfaces (Fig. 3). The liquid and glass surfaces intersect along the line l,l′, the projection of which on the plane P, T has the slope given by

$$\left(\frac{dP}{dT}\right) = -\frac{\partial\Delta S}{\partial T}\bigg/\frac{\partial\Delta S}{\partial P} = \frac{\Delta C_P}{TV\Delta\alpha} \qquad (4)$$

The result shows that a sudden isobaric change in temperature is equivalent thermodynamically to a pressure increment. Substituting some typical values for a sodalime–silicate glass: $\Delta C_P = 0.33\ \text{J g}^{-1}\ \text{K}^{-1}$, $\Delta\alpha = 2 \times 10^{-5}\ \text{K}^{-1}$, $V = 0.4\ \text{cm}^3\ \text{g}^{-1}$, $T_f = 800\ \text{K}$, we obtain $(P - P_f)/(T - T_f) = 50\ \text{MPa K}^{-1}$. This means that an instantaneous change in temperature by 10 K is equivalent to creating 500 MPa of internal pressure—a large driving force for stabilization.

A diagram for volume is similar to the one shown in Fig. 3 for entropy. The projection of the intersection line on the plane P, T has the slope

$$\left(\frac{dP}{dT}\right) = -\frac{\partial\Delta V}{\partial T}\bigg/\frac{\partial\Delta V}{\partial P} = \frac{\Delta\alpha}{\Delta\kappa} \qquad (5)$$

Experimentally, this slope is 2–5 times lower than that for entropy (Eqn. (4)). The quantity Π:

$$\Pi \equiv \frac{\Delta C_P \Delta\kappa}{TV(\Delta\alpha)^2} \qquad (6)$$

is equal to this ratio of slopes, and was defined by Prigogine and Defay (1954), who showed that $\Pi = 1$ if a single order parameter, along with T and P, is sufficient to specify the system. Davies and Jones (1953) and Gupta and Moynihan (1976) proved that $\Pi > 1$ when more than one order parameter is required. Therefore, the experimental evidence that $\Pi > 1$ suggests that more than one order parameter is needed to specify the state of the glass.

4. Kinetics of the Glass Transition

Figure 4 depicts in greater detail the changes of properties and their derivatives in the glass transition region. Stabilization of the liquid rapidly cooled to a given temperature is faster than stabilization of the liquid rapidly heated to the same temperature (Fig. 4a). This is because the cooled sample arrives with a more open structure and higher mobility than the heated one. The property curve for reheating (Fig. 4b) does not follow the cooling curve and forms a hysteresis loop, because reheating provides additional time for stabilization. Once the reheating curve crosses the cooling curve, it must reverse the direction of stabilization and, after a short delay, converges with the cooling curve. An increase in the cooling rate q is equivalent to shortening the isothermal holds in the representation of continuous cooling as a series of temperature jumps ΔT followed by isothermal holds $\Delta t = T/q$. The whole glass transition is shifted to higher temperatures, that is, to shorter relaxation times (Fig. 4b,c).

The behavior shown in Fig. 4 is predicted by equations suggested by Tool (1946) which include the second and the third of Eqns. (3), and its derivatives for isobaric conditions, for example,

$$\frac{\alpha(T) - \alpha_g|_{T'}}{\alpha_l - \alpha_g|_{T_f}} = \frac{dT_f}{dT} \qquad (7)$$

The equations for the fictive temperature are:

$$\frac{dT_f}{dT} = \frac{T - T_f}{q\tau}$$

or

$$\frac{dT_f}{dt} = \frac{T - T_f}{\tau} \qquad (8)$$

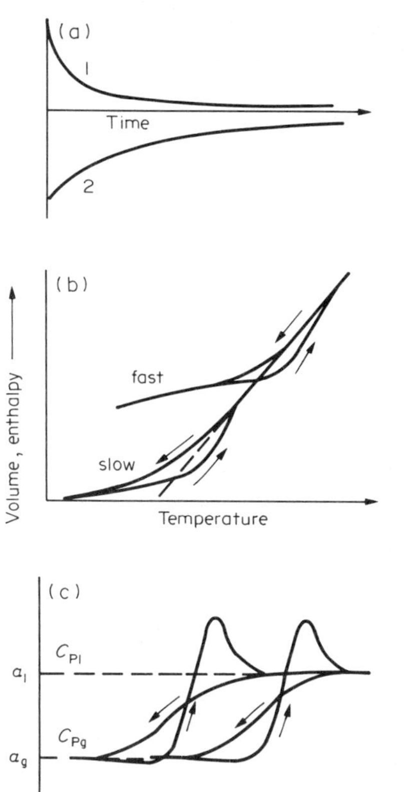

Figure 4
Property evolution: (a) in the isothermal condition following a sudden cooling (1) and heating (2); (b) and (c) during fast and slow cooling and reheating

and for the relaxation time τ as modified by Narayanaswamy (1971):

$$\tau = \tau_0 \left(\frac{x\Delta h}{RT} + \frac{(1-x)\Delta h}{RT_f}\right) \qquad (9)$$

where τ_0 is a constant, Δh is the activation energy, R is the ideal gas constant, and x is a constant ($0 \leqslant x \leqslant 1$).

Ritland (1956) scrutinized the notion of T_f. Literally following the definition, he quickly heated the glass to the temperature where it is supposed to be in equilibrium; the fictive temperature T_f should be equal to T and (dT_f/dT) equal to zero. Contrary to this prediction, he observed changes in T_f as a function of time, showing that the T_f given by Eqn. (8) is the average of a number of T_{fi} (see Macedo and Napolitano 1967, Narayanaswamy 1971, Kovacs et al. 1979). Thus both kinetics and thermodynamics lead to the conclusion that more than one order parameter is needed to characterize the glassy state.

See also: Glass-Transition Phenomena in Polymers

Bibliography

Davies R O, Jones G O 1953 Irreversible approach to equilibrium in glasses. *Proc. R. Soc. London, Ser. A* 217: 26–42

Gupta P K, Moynihan C T 1976 Prigogine–Defay ratio for systems with more than one order parameter. *J. Chem. Phys.* 65: 4136–40

Kovacs A J, Aklonis J J, Hutchinson J M Ramos A R 1979 Isobaric volume and enthalphy recovery of glasses. II. A transparent multiparameter theory. *J. Polym. Sci. Polym. Phys. Ed.* 17: 1097–162

Macedo P B, Napolitano A 1967 Effects of distribution of volume relaxation times in the annealing of BSC glass. *J. Res. Natl. Bureau of Stds. Sect. A* 71: 231–38

Narayanaswamy O S 1971 A model of structural relaxation in glass. *J. Am. Ceram. Soc.* 54: 491–98

Prigogine L, Defay R 1954 *Chemical Thermodynamics*. Longmans, New York

Ritland H N 1956 Limitations of the fictive temperature concept. *J. Am. Ceram. Soc.* 39: 403–6

Tool A Q 1946 Relation between inelastic deformability and thermal expansion of glass in its annealing range. *J. Am. Ceram. Soc.* 29: 240–53

Tool A Q, Eichlin C G 1931 Variations caused in heating curves of glass by heat-treatment. *J. Am. Ceram. Soc.* 14: 276–308

S. M. Rekhson

Glass-Transition Phenomena in Polymers

The glass-transition temperature T_g is the single most important property of an amorphous polymer to consider in practical applications of amorphous plastics. At T_g, the polymer changes from a rigid disordered solid into a viscous or rubbery liquid. For a rubber whose T_g is below room temperature, cooling in liquid nitrogen will lead to a brittle glass that is readily shattered. For a thermoplastic, such as polystyrene or poly(methyl methacrylate), heating in boiling water will soften the plastic sufficiently to allow viscous flow near T_g (100 °C). The lower and upper service temperatures of a plastic are dictated by T_g as well as by other chemical and physical properties.

1. Effects on Physical Properties

The glass transition in polymers is accompanied by large changes in physical properties such as specific heat, thermal expansivity, elastic modulus, viscosity, and dielectric constant and loss. These changes can be used to define the glass-transition temperature. Perhaps the easiest method of measuring T_g is by differential thermal analysis (DTA or differential scanning calorimetry) techniques, which detect the change in C_p at T_g (see *Thermal Analysis of Polymers*). Methods sensitive to changes in expansivity and/or modulus are also widely used. Practical methods, such as those involving heat-distortion temperature (ASTM D1525-82) or Vicat softening temperature, correlate reasonably well with T_g for amorphous polymers.

In the glass state below T_g, the monomer repeat units in the polymer chain are frozen into fixed positions. X-ray scattering from glassy polymers shows amorphous haloes indicative of short-range order (i.e., C—C bonds along the chain and second-nearest-neighbor distances both along the chain and between the chains), but no long-range order. The local arrangements of the chain backbone (conformations) are frozen in the glass, and the chain dimensions of the random coil frozen in at T_g should approximate those of a random coil in a solution at T_g. Although the positions and shapes of polymer chains are frozen at T_g, significant motion and relaxation occur in the glass. At temperatures $\leqslant 50\,°C$ below T_g, both the density and the enthalpy as well as other properties change by small but significant amounts with time. Some polymers become brittle upon annealing below T_g. Various aging processes occur at temperatures below T_g, affecting many phenomena such as diffusion, creep and fatigue. Side groups or even small main-chain units (e.g., —CH_2—) undergo rotation, vibration or libration in the glassy state. These motions can be observed by mechanical, electrical and nuclear magnetic resonance relaxation measurements at different frequencies (1–10^6 Hz). Molecular motions in the glassy state are important in many end-use mechanical and electrical properties (Bailey et al. 1981).

In the liquid state a few degrees above T_g, the thermodynamic properties are in equilibrium with respect to temperature, pressure and other intensive variables. When the temperature changes, the free volume (the free space between assumed hard-sphere molecules) changes, allowing diffusive motion to occur, and affecting the overall response time of the polymer chain. The conformation or shape of the polymer chain changes with temperature, the population of high-energy states (usually gauche) increasing with increasing temperature. Although the conformation changes are important in certain properties, such as C_p, the changes in free volume with temperature dominate the relaxation behavior in the polymer liquid. Furthermore, the strong coupling between polymer chains larger than a critical size, referred to as entanglements, is a primary factor in the viscous and elastic behavior of polymer melts above T_g (Graessley 1974).

The viscosity of a polymer melt is strongly dependent upon the number of degrees the melt temperature is above T_g. A rule of thumb suggests that the viscosity at T_g is 10^{12} Pa s, and at 100 °C above T_g the Williams–Landel–Ferry (WLF) equation, Eqn. (1), predicts a viscosity of 10 Pa s. The large change in viscosity and viscoelasticity of polymers is important practically in the processing of plastics by, for example, injection molding, extrusion and thermoforming.

The WLF equation or an equivalent form, the Vogel or Tammann equation, is given as

$$\log\left[\eta(T)/\eta(T_g)\right] = \frac{-C_1(T-T_g)}{C_2+(T-T_g)} = A + \frac{B}{T-T_\infty} \tag{1}$$

where

$$C_1 = 2.3/f_g$$
$$C_2 = f_g/\alpha_f = T_g - T_\infty$$
$$B = 2.3\,C_1/C_2 = 1/\alpha_f$$

where η is the viscosity, A is a constant, f_g (= 0.025) is the fractional free volume at T_g, α_f is the expansivity of free volume ($\simeq$ expansivity of liquid − expansivity of glass = $4.8\times10^{-4}\,K^{-1}$), and T_∞ is the temperature where the free volume goes to zero. This equation is widely used to describe the temperature dependence of viscosity and of mechanical and dielectric relaxation times.

2. *Effects of Chemical Structure*

Chemical structure affects the glass transition in a direct but as yet unquantified manner. Two important structural features are the intermolecular forces (exemplified by the cohesive energy density) and intramolecular forces (exemplified by chain stiffness, which is related to energy differences and energy barriers between chain conformations). Increasing polarity or cohesive energy density leads to an increase in T_g. Flexible chains with low barriers to rotation about chain bonds have low T_g, whereas stiff chains with high energies between conformations have high T_g. Although qualitative descriptions of T_g are possible, quantitative treatments of the effect of structure on it are lacking.

Other chain structure factors affect T_g more predictably. The number-average molecular weight M_N affects T_g according to the equation $T_g(M_N) = T_g(\infty) - K/M_N$. For a typical value of $K = 2\times10^5$, $T_g(M_N)$ approaches within 1 ° of $T_g(\infty)$ when $M_N = 2\times10^5$. T_g changes by 9 °C as M_N changes from 200 000 to 20 000, and by 90 °C as M_N changes from 20 000 to 2000. An analysis of the constant K in terms of the excess free volume associated with chain ends leads to reasonable values of 50–100 Å^3 mol^{-1} which approximate the molar volume of the repeat unit. Copolymerization leads to polymers with T_g values intermediate to the T_g values of the homopolymers. A variety of equations such as

$$T_g = \phi_1 T_{g_1} + k\phi_2\,(T_{g_2} - T_{g_1})$$

where ϕ_1, ϕ_2 are the volume fraction of components 1 and 2, and k is a constant, describe the dependence of T_g on copolymer composition. The coefficients can be related to expansion coefficients of free volume of the components or to the change in C_p of the components. An equivalent equation describes the T_g of mixtures with plasticizers or blends of other polymers. The blending of compatible polymers provides a means of controlling T_g and other physical properties. A compatible polymer functions better than a low-molecular-weight plasticizer because migration of plasticizers is eliminated and the physical properties of the polymer diluent are superior to those of the plasticizer.

The glass-transition temperature can also be increased by pressure, cross-linking and incorporation of ionic groups. Other polymer structure variables, such as cis–trans isomerization and stereoregularity, can also lead to changes in T_g (Table 1).

Table 1
Glass-transition temperatures

Material	T_g (°C)
Polymeric	
poly(dimethyl siloxane)	−125
cis-poly(1,4-butadiene)	−100
trans-poly(1,4-butadiene)	−40
poly(methyl methacrylate)	
isotactic	45
atactic	105
syndiotactic	115
polycarbonate	140
poly(phenylene oxide)	220
Organic	
methyl cyclohexane	−186
toluene	−168
phenyl ether (5P4E)	−35
o-terphenyl	−27
Inorganic	
Se	30
$ZnCl_2$	103
B_2O_3	250
BeF_2	320
SiO_2	1200
Metallic	
$Te_{15}Ge_2As_3$	110
$Te_{80}Ge_{15}Sn_5$	140
$Fe_{82}B_{18}$	140
$Pd_{48}Ni_{32}P_{20}$	312
$Pd_{82}Si_{18}$	375

3. Mechanical Properties of Glasses

Historically, glasses are considered as brittle materials because of the behavior of inorganic glasses. However, recent work on polymeric and metallic glasses has shown that not all glasses are brittle. The elastic modulus of a glass is high (10^2–10^4 MPa) because the forces between neighboring atoms are high, but is not as high as in crystals because the deformation of primary bonds in crystals involves greater stresses than the deformation of secondary bonds in a glass. The tensile strength of amorphous polymers is 30–60 MPa, and amorphous polymers such as polycarbonate show ductile yielding and cold flow because the chain segments can orient and flow at these stresses in the glassy state. In addition to brittle and ductile behavior, intermediate behavior involving local failure and flow, known as crazing, can appear (see *Crazing and Cracking of Polymers*).

For many practical applications, the impact strength of polymers is important. A weakness of most amorphous polymers is their low impact strength. Furthermore, impact strength is sensitive to aging and to fabrication conditions of the polymer (see *Impact Strength of Polymers*).

4. Nonequilibrium Properties

One of the most challenging and troublesome characteristics of glasses is that they exist in a nonequilibrium state that changes with time. In some cases the changes are small and can be ignored, but there are many cases where properties change as a function of time and temperature, preventing other physical phenomena from occurring. For instance, if a polymer is quenched from $T_g + 20$ °C to $T_g - 20$ °C, the volume of the sample will continue to change when held at $T_g - 20$ °C. If a mechanical property measurement (e.g., stress relaxation or creep) is made on the sample while volume relaxation is occurring, there will be a contribution to the modulus (or compliance) from the volume relaxation. In fact, all measurements below T_g are nonequilibrium measurements. Although glasses prepared by lengthy annealing below T_g may not show property changes with time, measurements on these glasses can be compared reliably with measurements on glasses with the same thermal history.

A major effect of these phenomena on the properties of glasses is the large temperature coefficient of viscosity or relaxation time. There are a number of equations (e.g., WLF, Vogel–Tammann, Gibbs–Adams) which describe the observed behavior. A free-volume model describes the behavior and assumes that the viscosity or relaxation time depends exponentially on the reciprocal of the fractional free volume. At T_g the fractional free volume is ~2.5% and increases to 7.5% at $T_g + 100$ °C.

The relaxation time for volume recovery depends on free volume. During contraction, the free volume decreases and the volume contraction slows down because of this decrease. During volume expansion, the opposite occurs—volume expansion accelerates because the free volume is increasing. This effect is described as the nonlinear behavior near T_g because the behavior differs depending upon direction and distance from equilibrium. In addition, the relaxation behavior of glasses requires that there is a distribution of times required for the properties (e.g., volume, enthalpy and modulus) to approach equilibrium. The distribution of relaxation times required to describe real glasses extends over many decades (five or more) (Kovacs et al. 1979). Even when a glass is specified by a temperature dependence of relaxation times, a distribution of relaxation times and a nonlinearity parameter, it is not clear that these parameters uniquely specify the state of the glass.

5. Theories of Glass Formation

Many treatments of glass formation have been proposed since Kauzmann (1947) pointed out that the entropy of the supercooled liquid would become less than the entropy of the crystal if glass formation did not intervene. In fact, the statistical mechanical theory of Gibbs and DiMarzio (1958) concludes that the entropy of polymer molecules and holes (free volume) becomes vanishingly small at a temperature T_2. T_2 resembles a true second-order transition temperature and is estimated to be 50 °C lower than the experimentally observed T_g. Nevertheless, the observation of a true second-order transition has not been possible because the kinetic slowdown at T_g causes the system to depart from equilibrium. Computer-simulated molecular-dynamics calculations have shown the existence of a glasslike transition for hard spheres. The simulation corresponds to very high cooling rates and short time scales, but many features of the glass transition are observed, including hysteresis (Fox and Anderson 1981).

Free-volume treatments of glass formation have been the most widely used and accepted. Originally, the molecular-weight dependence of the specific volume and T_g led to the concept that the glass transition occurs at a constant fractional free volume. An advantage of the free-volume approach is the quantitative description of the temperature dependence of viscosity and relaxation times in terms of the WLF equation; however, there is not a unique definition of free volume, and the fractional free volume at T_g ranges from 2.5 to 12%. Free-volume and entropy theories will continue to be useful and will be refined. However, a more comprehensive and general theory which treats the interdependence of the (hole) energy and flex (orientational) energy upon packing volume is needed to describe real glasses.

See also: Amorphous Polymers; Secondary Transitions in Polymers; Transition Temperatures in Elastomers; Polymers: An Overview of Structure, Properties and Structure–Property Relations

Bibliography

Bailey R H, Petrich R A, North A M 1981 *Molecular Motion in High Polymers*. Oxford University Press, New York

Ferry J D 1980 *Viscoelastic Properties of Polymers*, 3rd edn. Wiley, New York

Fox J R, Andersen H C 1981 Molecular dynamics simulation of the glass transition. *Ann. N.Y. Acad. Sci.* 371: 123–35

Gibbs J H, DiMarzio E A 1958 Nature of the glass transition and the glassy state. *J. Chem. Phys.* 28: 373–83

Goldstein M, Simha R (eds.) 1976 The glass transition and the nature of the glassy state. *Ann. N.Y. Acad. Sci.* 279: 1–246

Graessley W W 1974 The entanglement concept in polymer rheology. *Adv. Polym. Sci.* 16: 1–179

Haward R N (ed.) 1973 *The Physics of Glassy Polymers*. Applied Science, London

Kauzmann W 1947 The nature of the glassy state and the behavior of liquids at low temperatures. *Chem. Rev.* 43: 219–56

Kovacs A J, Aklonis J J, Hutchinson J M, Ramos A R 1979 Isobaric volume and enthalpy recovery of glasses. *J. Polym. Sci., Polym. Phys. Ed.* 17: 1097–116

Mackenzie J D (ed.) 1960–1964 *Modern Aspects of the Vitreous State*, Vols. 1–3. Butterworth, London

O'Reilly J M, Goldstein M 1981 Structure and mobility in molecular and atomic glasses. *Ann. N.Y. Acad. Sci.* 371: ix–xi

Shen M C, Eisenberg A 1966 Glass transition in polymers. *Prog. Solid State Chem.* 3: 407–81

Shen M C, Eisenberg A 1970 Recent advances in glass transitions in polymers. *Rubber Chem. Technol.* 43: 95–155

Van Krevelen D W 1976 *Properties of Polymers*, 2nd edn. Elsevier, Amsterdam

J. M. O'Reilly

Glasses in Medical Applications

There are two major categories of medical applications of glasses: (a) *in vitro*, where glass is used as a passive substrate for growth of cell cultures or immobilization of enzymes and antigen–antibodies; and (b) *in vivo*, where surface-active glasses or glass-ceramics are used as implants to replace living tissues. Glasses used *in vitro* are usually alkali borosilicates or porous silica, which are very resistant to surface chemical attack. In contrast, glasses or glass-ceramics used as implants are designed to have controlled rates of surface reaction with body fluids, resulting in a chemical bond to living tissues.

1. In Vitro Applications

1.1 Tissue Culture Substrates

Glasses have been used in the development and production of vaccines and therapeutic hormones since the 1930s, when it was discovered that cell cultures could be made to grow on glass surfaces if the proper nutrients were supplied. Alkali borosilicate glasses, such as Pyrex, are used because the glass surface can be cleaned and sterilized repeatedly with little chemical damage. A thin, protective, dense

SiO_2-rich film forms on the glass during cleaning, which allows cells of various types to attach, spread rapidly, grow and multiply. Limitations, however, of tissues grown *in vitro* are: (a) inability to maintain differentiated cells, (b) failure to establish cell aggregation and physiological three-dimensional structuring, and (c) alteration of cell morphology and function with time. Use of glasses with bioactive surfaces may solve some of these problems.

1.2 Enzyme Carriers

By chemically adhering enzymes to porous glass, such as Corning 96% SiO_2 with 900 Å pores, it is possible to make the enzymes water-insoluble. Such insoluble enzymes retain activity for many days. Either direct enzyme–glass binding, or covalent coupling of the enzyme to the glass by amino-functional silane coupling agents is possible, but the coupling agents offer more versatility. Effective enzyme activity has been retained in this manner for trypsin, papain, ficin, urease, glucose oxidase, peroxidase, L-amino acid oxidase and alkaline phosphatase.

1.3 Antigen–Antibody Carriers

Similar techniques are used to bond antigens and antibodies irreversibly to insoluble porous-glass carriers without loss of specific combining capacity. Insoluble antigens or antibodies are termed immunoadsorbents: they are used for antibody and antigen isolation, purification, and study; enzyme removal or isolation; virus and microorganism detection; and radioimmunoassay. Human γ-globulin, covalently coupled to porous-glass particles, has yielded recovered antibodies that were 90–100% precipitable.

2. In Vivo Applications

2.1 Compositions

Surface-active glasses, also termed Bioglasses, based on a certain range of Na_2O–CaO–P_2O_5–SiO_2 compositions, have been developed for use as medical and dental implants. Partially or fully crystallized glasses are termed either Bioglass-ceramics, surface-reactive glass-ceramics, or bioactive glass-ceramics. These materials are unique in that they form a bond with living tissues by means of controlled chemical reaction at the surface of the implant. Thus, the implant becomes quickly and permanently fixed at the interface with the host, before fibrous encapsulation can occur. Loosening of the implant is prevented because of the interfacial bond.

For Bioglasses, controlled biological reactions occur at the interface only within specific limits of composition and surface reactivity. Within these limits the material bonds to tissues (region A in Fig. 1). When these limits are exceeded, the material behaves in the physiological environment either as a nearly inert material (region B) eliciting fibrous capsule formation, or resorbs in the tissue (region C). Compositional modifications, such as addition or substitution of K_2O, MgO, CaF_2, B_2O_3 or Al_2O_3, generally contract region B.

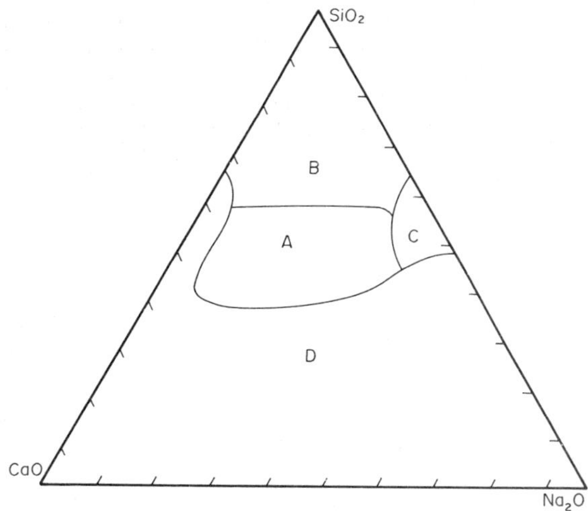

Figure 1
Bioglass compositional range for bonding to rat bone: A, bonding at ≤30 days; B, nonbonding, reactivity is too low; C, nonbonding, reactivity is too high; D, nonbonding, non-glass-forming

2.2 Mechanism of Bioglass–Tissue Bonding

A comparison of the surface chemistry of glasses within region A with those of region B in Fig. 1, show that all bone-bonding Bioglasses develop a large increase in surface area when exposed to physiological solutions. The ultrapores of 3–30 nm size that result from formation of an active silica-rich gel on the implant surface, nucleate hydroxylapatite crystals on that surface. The hydroxylapatite agglomerates provide a means of incorporating collagen, mucopolysaccharides and glycoproteins within the surface-active layer of the implant. The resulting composite organic–inorganic interface is responsible for the mechanically strong bond at the interface exhibited by Bioglasses within region A, the bone-bonding boundary. Glass compositions that fall outside the boundary (region B), have too little reactivity to form a stable hydroxylapatite surface-layer and active bond-sites for the organic molecules. Glasses in region C, are too reactive, surface layers breakdown and total dissolution occurs.

Examination by transmission electron microscopy of well-mineralized rat-bone bonded to 45S5 Bioglass shows a gradual change in electron contrast across the interface (Fig. 2). This is due to a gradation of hydroxylapatite concentration across the glass–bone bond. The inserts of electron diffraction patterns of bone (Fig. 2a) and glass reaction zone (Fig. 2b) show the continuity of the bonded interface for this type of biomaterial. The high strength of the bond is

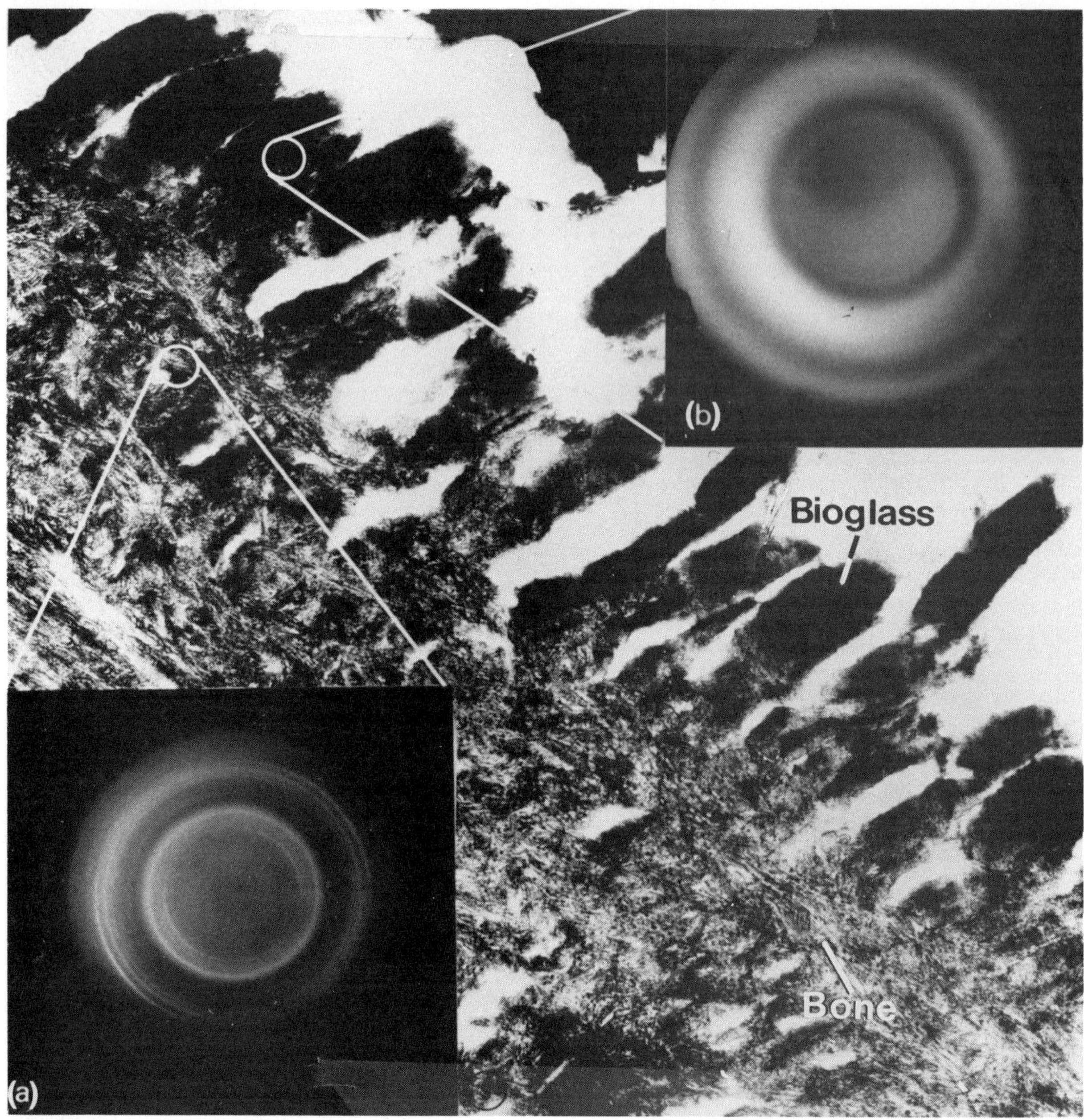

Figure 2
Transmission electron micrograph of a bonded bone–Bioglass interface: (a) electron diffraction pattern of the bone; (b) electron diffraction pattern of the glass reaction layer (Courtesy of T K Greenlee)

indicated by the fracture artifact (due to microtoming) in the glass reaction layer, which has failed to propagate into the bone.

2.3 Toxicology and Biocompatibility

Evidence for a lack of toxicity of various Bioglass and surface-active glass-ceramic formulations has been established from results of a variety of *in vivo* and *in vitro* test models (Table 1). The surface activity critical in bone adhesion is also seen in soft connective tissues, and without accompanying toxic effects in either type of tissue. Adherence to muscle and other soft tissues occurs if the implant is immobilized. Even if motion and wear should occur and produce

Table 1
Toxicology and biocompatibility tests for Bioglass

Type of test	Species	Material form	Results
In vitro tests Bone cells	rat	solid	cells grew and divided normally (30 days)
Fibroblasts	rat	solid	cells grew and divided normally (30 days)
Fibroblasts	chick	solid	cells grew and divided normally (21 days)
Fibroblasts	human	solid	cells grew and divided normally (21 days)
Fibroblasts	hamster (CHO)	solid	grew slowly for 7 days, resumed normal rate when removed to more usual substrates
Fibroblasts	hamster (NIL)	solid	grew slowly for 7 days, resumed normal rate when removed to more usual substrates
Lymphocytes	human	solid	RNA systhesis remained normal
Lymphocytes	human	extract	RNA synthesis remained normal
Macrophages	mouse	solid	phagocytic activity remained normal
Macrophages	mouse	extract	phagocytic activity remained normal
Macrophages	rat	powder	cells with ingested material continued to behave normally in culture
In vivo tests Tibia	rat	solid	bonded to bone with no toxic effects (up to 2.5 years)
Femur	rat	solid	bonded to bone with no toxic effects (up to 2.5 years)
Cervical spine	dog	solid	bonded to bone with no toxic effects (3 months)
Femur	dog	solid	bonded to bone with no toxic effects (6 months)
Femur	primates	solid	bonded to bone with no toxic effects (up to 1.5 years)
Mandible	primates	solid	bonded to bone with no toxic effects (up to 6.5 years)
Ossicles	mouse	solid	bonded to periosteum with no toxic effects (up to 1 year)
Subcutis	rat	solid, powder	adhesion to collagen, no toxic effect
Muscle	rat	solid	no toxic effect
Muscle	rabbit	fibers	no toxic effect, some adhesion
Peritoneal cavity	rat	powder	eliminated without effect, no fibrosis
Peritoneal cavity	rat	solid	thin capsule, no fibrosis, normal cell growth on surface (up to 4 weeks)
Pulmonary embolus	mouse	powder	no granuloma or thrombosis

particulate Bioglass, the material is quickly eliminated without adverse effect.

2.4 Orthopedic Applications

Because of low flexural strength, surface-active glasses must be used as coatings on high-strength ceramics, such as Al_2O_3, or metals, such as 316L stainless steel or Co–Cr alloys, for load-bearing devices such as segmental bone replacements and joint prostheses. The main motivation for the use of surface-active glass-coatings is the replacement of poly(methyl methacrylate) (PMMA), commonly used as a self-polymerizing grouting agent between bone and the orthopedic device. Although excellent short-term stabilization of devices is achieved with PMMA, there are complications: low blood pressure due to release of monomer into the blood stream, necrosis of bone due to the heat generated in the exothermic polymerization reaction, and eventual loosening of devices due to lack of interfacial bonds. These problems are eliminated with use of surface-active glass-coatings. However, improved mechanical designs and surgical techniques are needed to achieve the glass–bone bond within the first few weeks after surgery. If motion at the interface between glass and bone occurs, mineralization of the bond is prevented, and a nonadherent fibrous capsule results. Figure 3 is an x-radiograph of a successful 45S5 Bioglass-coated hip prosthesis in a monkey. Mechanical tests of the femoral stem–bone bond showed that the distal condyles were torn off the bone with no loosening of the implant. Three-point bend-tests causing fracture of the bone also did not loosen the interface. Use of a square cross section

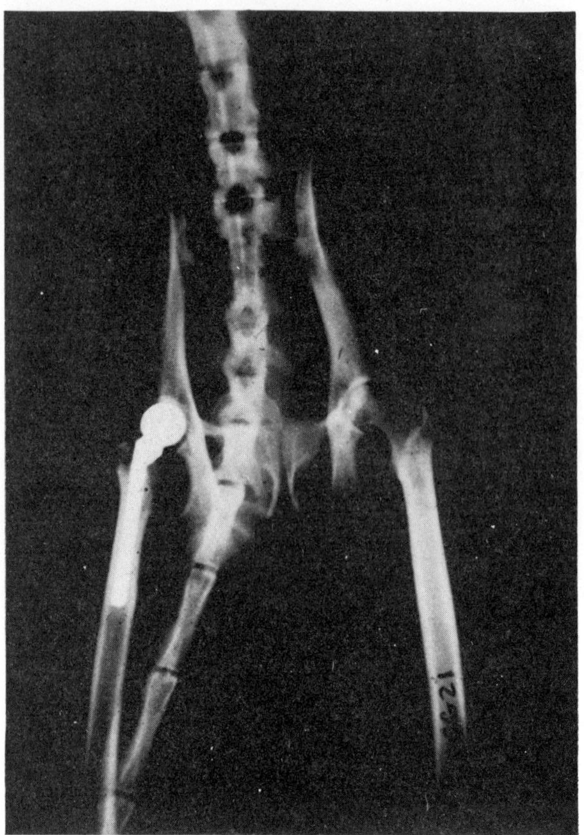

Figure 3
Radiograph of a Bioglass-coated hip-prosthesis in a monkey

for the stem, and a tight fit at the time of surgery, gave a functional, load-bearing prosthesis within days after surgery. Bonding of the coating to bone occurs as early as eight weeks. Bulk 45S5 Bioglass implants in rats develop a strong bond as early as two weeks. Less reactive Ceravital surface-active glass-ceramics achieve a strong bone bond by the end of four weeks. Maturation of the bone bond continues for several months for most compositions.

Surface-active glass and glass-ceramic coatings can be applied to surgical metal alloys by several techniques. Flame-spray coatings are easily applied to simple or complex shapes, but the process is difficult to control. Rapid immersion of a preoxidized metal device in molten glass is the easiest and most acceptable procedure for simple geometries, but is difficult to apply to complex devices. Modified enamelling methods are also used with a top coat of either bioglass powder or Ceravital glass-ceramic granules fused to an undercoat.

An intermediate approach to eliminate PMMA, but still retain the surgical ease of the grouting technique, is the use of Ceravital granules in PMMA, to form a bioactive cement. Use of $\leqslant 70$ vol% of the bone-bonding active glass-ceramic phase significantly reduces the temperature rise during *in situ* polymerization, and apparently reduces free monomer transport in the circulatory system as well.

Total joint components and segmental bone replacements made of Nucerite glass-ceramic coated stainless steel are also used with PMMA cementing procedures. Although the glass-ceramic coating can decrease metallic corrosion, and perhaps wear, elimination of the cementing media is not possible with this composition because it lies within region B of Fig. 1, and therefore behaves as a nearly inert biomaterial.

2.5 Dental Applications

Several formulations of surface-active glasses and glass-ceramics have performed successfully as implants for tooth replacements. Although breakage of bulk glass implants is a problem, it has been dealt with by using Bioglass coatings on dense Al_2O_3 or metal substrates. Even when gingival level breakage was observed in animal studies, the root portion of the implant remained solid. The mandibular ridge was restored distal to the fracture, and healed mucosa and healthy gingival tissues were present after two-year trials. A mandibular ridge augmentation in a baboon using 45S5 Bioglass remained solidly bonded and fully functional up to six years, when the experiment was terminated.

Orthodontic anchors in monkeys, composed of 45S5 Bioglass coatings on dense Al_2O_3-blade implants, show no movement of the implants in the bone and no resorption of bone adjacent to the implant even with forces of 425–950 g applied by both lingual and maxillary arch supports. The direct bond of bone to the surface-active glass-coating apparently allows force to be applied to the bone without initiating osteoclast activity. Finite-element and histological analysis of bonded Bioglass tooth implants, show evidence of a 100–200 μm thick, elastically compliant inorganic-interface which dissipates the applied load. Electron microprobe and microhardness measurements across the bonded tooth implants confirm this conclusion. These favorable dental responses to surface-active glasses require firm fixation in the implant site and appropriate steps to prevent mobility as the bond develops. When motion occurs, the bond is disrupted, fibrous encapsulation follows, and complications leading to loosening and implant removal are almost certain. However, if the initial placing of the implant in its host tissue is correct, and immobilization during the 2–4 week bonding period is complete, then the surface-active glass-coated dental implant will function as an ankylosed tooth and may be treated as such.

Bioglass implants are being used in patients to maintain the alveolar ridge after multiple extractions. These are placed in the socket after extraction, bond to the surrounding bone and help support the dentures. In this way the resorption of alveolar bone under dentures is prevented.

2.6 Other Applications

Polymer and metal implants used in repair of the ossicular chain of the middle ear are often unsatisfactory due to development of excessive scar tissue, break-up or migration of the implant, chronic inflammation or infection, or extrusion of the prosthesis from the ear. Middle ear implants made of 45S5 Bioglass and Ceravital show excellent adherence, absence of mobility, restoration of function, and little or no tendency for extrusion or exacerbation of infection in extensive animal tests and human trials. The bonding interface of the Bioglass implants consists of a very thin, tightly adherent, collagen layer similar to the bond between normal articulating ossicles. Ceravital bonds are osseous in character and these implants exhibit osteogenesis in this application. Surface-active glasses and glass-ceramics appear especially suited to repair in the head and neck region, because of excellent interfacial adherence, absence of color to show through skin tissues, and low strength requirements. Successful animal results have been obtained in cranial repair, augmentation of the zygomatic arch, and mandibular and maxillary reconstruction.

Cervical spinal fusion is another area where standard metallic and polymeric biomaterials have not been successful, and autogenous bone grafts from the iliac crest of the pelvis (which do succeed) have painful side effects. Animal experiments using 45S5 Bioglass plugs produce fusion of the cervical vertebrae and therefore could provide an alternative to autogenous grafts.

One of the most intriguing potential applications of surface-active glasses is in the field of cell and tissue culture, and genetic engineering. Cell-culture studies show that cell types are sensitive to the active surfaces. Cells with very few membrane-attachment complexes are very slow to attach, spread and divide. Thus they can be maintained for long times in a resting state without modification of cellular morphology and without contact or confluence with other cells. Subsequent removal of the cells from the bioactive surface restores their normal behavior. Therefore, manipulation of primary cell lines in this stabilized resting state may result in unique modification of the cells.

See also: Glass Ceramics; Biocompatibility of Biomedical Materials; Biomedical Materials: An Overview

Bibliography

Blencke B A, Bromer H, Deutscher K K 1978 Compatibility and long-term stability of glass-ceramic implants. *J. Biomed. Mater. Res.* 12: 307–16

Boretos J, Eden M (eds.) 1984 *Contemporary Biomaterials*. Noyes Data, Park Ridge, New Jersey

Gross U, Brandes J, Strunz V, Bab I, Sela J 1981 The ultrastructure of the interface between a glass ceramic and bone. *J. Biomed. Mater. Res.* 15: 291–305

Grote J J (ed.) 1984 *Biomaterials in Otology* 1984 Martinus Nijhoff, Boston, Massachussetts

Hench L L 1980 Biomaterials. *Science* 208: 826–31

Hench L L, Ethridge E 1982 *Biomaterials: An Interfacial Approach*. Academic Press, New York

Hench L L, Paschall H A 1973 Direct chemical bond of bioactive glass–ceramic materials to bone and muscle. *J. Biomed. Mater. Res. Symp.* 4: 25–42

Hench L L, Paschall H A, Allen W C, Piotrowski G 1975 Interfacial behavior of ceramic implants. In: Horowitz E, Torgesen J L (eds.) 1975 *Biomaterials*, NBS Special Publication 415. National Bureau of Standards, Washington, DC, pp. 19–35

Hench L L, Splinter R H, Greenlee T K, Allen W C 1971 Bonding mechanisms at the interface of ceramic prosthetic materials. *J. Biomed. Mater. Res. Symp.* 2: 117–41

Hench L L, Wilson J 1984 Surface active biomaterials. *Science* 226: 630–36

Merwin G, Atkins J, Wilson J, Hench L L 1981 Comparison of ossicular replacement material in a mouse ear model. *J. Am. Acad. Otol. (Head and Neck Surgery)* 90: 461–69

Pigott G H, Ishmael J 1970 A comparison between *in vitro* toxicity of PVC powders and their tissue reactions *in vivo*. *Ann. Occ. Hyg.* 22: 111–26

Piotrowski G, Hench L L, Allen W C, Miller G J 1975 Mechanical studies of the bone Bioglass interfacial bond. *J. Biomed. Mater. Res. Symp.* 6: 47–61

Reck R 1981 Tissue reactions to glass ceramics in the middle ear. *Clin. Otolarynogol.* 6: 63–5

Smith J R 1977 Bone dynamics associated with the controlled loading of Bioglass coated aluminum oxide endosteal implants. Masters Thesis, University of Washington

Stanley H, Hench L L, Bennett Jr C G, Chellemi S J, King III C J, Going R E, Ingersoll N J, Ethridge E C, Kreutziger K L, Loeb L, Clark A E 1982 The implantation of natural tooth form Bioglass in baboons—long term results. *Int. J. Oral Implantol.* 2 (2): 26–36

Stanley H, Hench L L, Going R, Bennett C, Chellemi S J, King C, Ingersoll N, Ethridge E, Kreutzinger K 1976 The implantation of natural tooth form Bioglasses in baboons. A preliminary report. *Oral Surg., Oral Med., Oral Path.* 42: 339–56

Weetall H H 1972 Insolubilized antigens and antibodies. In: Hair M L (ed.) 1972 *The Chemistry of Biosurfaces,* Vol. 2. Dekker, New York, pp. 597–631

Weetall H H, Messing R A 1972 Insolubilized enzymes on inorganic materials. In: Hair M L (ed.) 1972 *The Chemistry of Biosurfaces,* Vol. 2. Dekker, New York, pp. 563–95

Wilson J, Pigott G H, Schoen F J, Hench L L 1981 Toxicology and biocompatibility of Bioglasses. *J. Biomed. Mater. Res.* 15: 805–17

L. L. Hench; J. Wilson

Glasses: Structure

With the increasing uses and importance of glasses it is surprising that so little is known about the structure of the glassy state. Molecular structures of crystalline materials can be precisely defined by diffraction or scattering of radiation, as a result of the long-range ordering of atoms characteristic of crystalline materials. Glasses, on the other hand, are characterized by an absence of atomic periodicity, and it is this inherent disorder which complicates structural studies and interpretations.

While glasses do not contain long-range ordering of atoms they do possess short-range ordering on a scale of 2–10 Å. This localized ordering can be investigated by specialized techniques involving x-ray, neutron and electron diffraction. These methods are often coupled with other analytical techniques and computer analyses which match proposed models of the glass with experimental data. Hence, structural characterization of specific glass-forming systems can be obtained, although models that emerge are not unique. Reviews and discussions of glass structure can be found in the following works: Doremus (1973), Kingery et al. (1976), Pye et al. (1972), Porai-Koshitz (1966), Pye et al. (1978), and Wright and Leadbetter (1976).

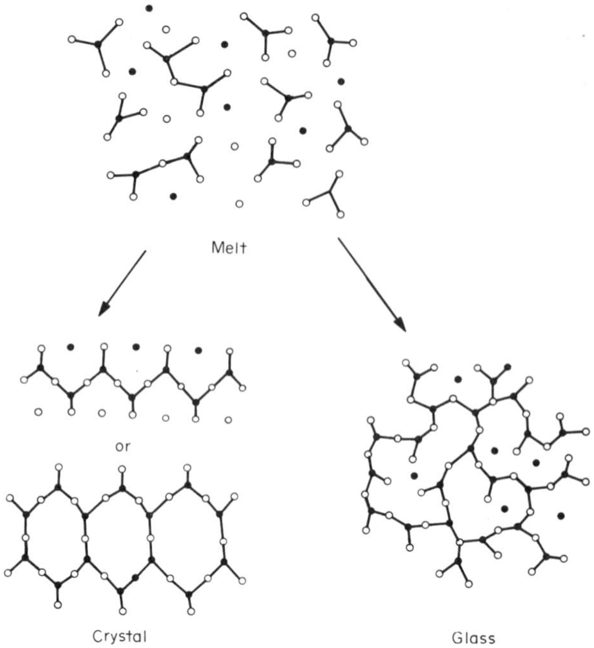

Melt

or

Crystal

Glass

Figure 1
Structural arrangements of atoms cooled from the melt for the crystalline and glassy states

1. Structural Arrangements of Atoms

The structure of glasses can best be understood on an atomic scale, and the structure for a specific glass will depend on how it was produced.

Being liquids, the melts from which glasses are produced are highly disordered: however, small areas of localized order do exist on the scale of 2–10 Å. These areas of local ordering may consist of individual atoms, groups of atoms or oxygen polyhedra (as in the case of most inorganic oxide glasses). The manner in which these units are hooked together determines whether a crystal or a glass will form during solidification.

During cooling, atoms will tend to rearrange themselves into the lowest energy configuration; that is, into a crystalline lattice. Solidification occurs at the melting temperature, and causes a discontinuous change in properties such as the specific volume. Upon crystallization, the atoms rearrange into rings, segments or long chains, all configurations of which have long-range order.

If the liquid is cooled quickly enough to prevent the rearrangement of atoms into periodic crystalline arrays, a glass is formed. In this case some of the disorder characterizing the melt is frozen into the solidified material. The liquid is cooled to a supercooled state and below the glass-transition temperature it is like a solid. At this temperature, there is a continuous change in properties such as the specific volume. The structural arrangements of atoms cooled from the melt for the crystalline and glassy states are depicted in Fig. 1.

2. Radial Distribution Function

The basis for structural studies of amorphous materials is the radial distribution function. Consider an aggregate of atoms and allow an average atom to be the center of a coordinate system, where $\boldsymbol{R}$ represents a radial vector emanating from the origin (Fig. 2a). Spherical shells at the end of the radial vector have a thickness $d\boldsymbol{R}$, and the density of atoms as a function of $\boldsymbol{R}$ is obtained by counting atom centers in each shell defined by $\boldsymbol{R}$ and $\boldsymbol{R} + d\boldsymbol{R}$. The number of such atoms is $4\pi R^2\rho(R)dR$, where $\rho(R)$ is the number of atoms per unit volume and is called the radial distribution function (RDF).

X-ray diffraction is the tool most commonly used to study the structure of glasses. Since diffraction is most conveniently described in terms of reciprocal space, a Fourier transformation is required to obtain the density distribution of atoms in real space. The RDF found from x-ray analysis is shown in Fig. 2b: ρ_0 is the average density, and after the first few maxima $\rho(R)$ approaches ρ_0 due to the lack of long-range order. However, the maxima at low R values indicate high-density differences or short-range ordering. From the positions of these peaks, interatomic distances can be obtained, and from the area under the peaks, the number of atoms responsible for the observed scattering can be calculated. Hence, from crystal data and atomic radii the scattering pairs responsible for the low R maxima can usually be determined, and consequently the short-range structure of the vitreous material defined.

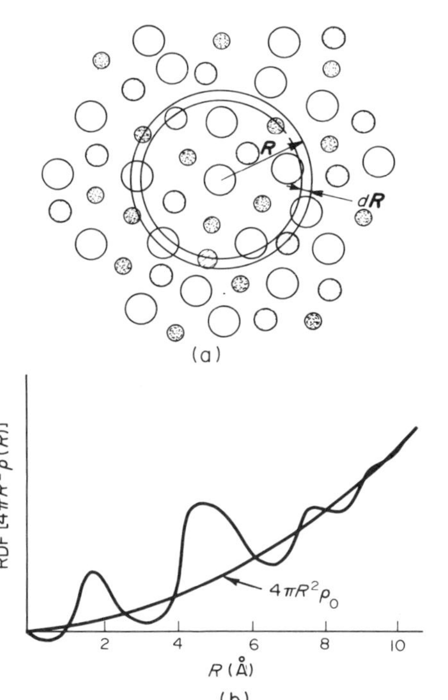

Figure 2
Radial distribution function (RDF): (a) origin of the RDF and (b) typical radial distribution for a simple inorganic glass

A more detailed and exact method for analyzing scattering for amorphous materials composed of different atom types was developed by Warren (1969), using the concept of pair functions as initiated by Finbak:

$$\sum_{uc}\sum\frac{N_{ij}}{r_{ij}}P_{ij}(r)=2\pi^2 r\rho_e\sum_{uc}Z_j+\int_0^{k_m}ki(k)e^{-\alpha^2k^2}\sin r\,k\,dk \qquad (1)$$

where

$$P_{ij}(r)=\int_0^{k_m}\frac{f_jf_i}{g^2(k)}e^{-\alpha^2k^2}\sin k\,r_{ij}\sin k\,r\,dk$$

and

$$i(k)=\frac{I_{eu}/N-\sum_{uc}f_j^2}{g^2(k)}$$

Here, $k=4\pi\sin\theta/\lambda$, r_{ij} is the distance from atom j to an atom in shell i of neighboring atoms, k_m is the maximum value of k from which measurements can be made, N_{ij} is the number of atoms in the ith shell from a central j atom, Z is the number of electrons in the atom or ion, ρ_e is the average electron density, $g(k)$ is a sharpening factor chosen to be the average scattering factor per electron [i.e., $g(k)=f_e=\Sigma_{uc}f_j/\Sigma_{uc}Z_j$], I_{eu}/N is the unmodified intensity in electron units per unit of composition, P_{ij} is the pair function for atoms i and j, $e^{-\alpha^2k^2}$ is a convergence factor, uc denotes the unit of composition which for vitreous silica would be one silicon and two oxygen atoms, and f_i and f_j are atomic scattering factors of atoms i and j, respectively.

The right-hand side of Eqn. (1) is an experimentally determined quantity. The first term is obtained from a measure of the sample density in the absence of appreciable porosity. The function $i(k)$ is related to measured coherent intensity, which is normalized after correcting for background, noise, absorption and polarization effects. With appropriate sharpening and convergence factors, the right-hand side of Eqn. (1) can be plotted as a function of r.

From the positions of the experimental peaks, r_{ij} can be obtained, and from known atom sizes, scattering efficiencies and crystalline interatomic distances, the atom–atom combination responsible for each peak can be assigned. It is usually easy to assign the atom combinations responsible for peaks in simple inorganic glasses at short interatomic distances, but this task becomes more difficult at higher values of r due to peak overlapping, more possible interatomic combinations, and the diffuse and weak nature of the modulations. The same sharpening and convergence factors are used to evaluate the pair function P_{ij} for each r_{ij} and, by trial and error, N_{ij} values for each r_{ij} are used to evaluate the left-hand side of Eqn. (1). From this treatment, a "best" model (although not unique) can be derived by matching the experimental curve. By proper assessment of thermal effects, r_{ij} distance distributions can be obtained by a deconvolution of peak shapes.

3. Models of Glass Structure

During the past 50 years there have been many attempts to describe the structure of the glassy state. Since glasses are characterized by short-range ordering of atoms and an absence of long-range order, structural models have been derived mainly from the liquid and crystalline states. The two most prevalent models are the microcrystalline and the random network.

3.1 The Microcrystalline Model

The microcrystalline model considers glasses to be composed of very small crystallites. The support for this approach is derived from x-ray diffraction data which produce very broad and diffuse peaks, indicative of glassy materials. Within the region of these wide maxima are locations of the sharp Bragg peaks, characterizing the crystalline structure of the same glass composition. Since it is known that the breadths of x-ray maxima increase and become more diffuse as the crystallite sizes decrease below the 1000 Å range, glasses were thought to be composed of very small crystallites.

Studies attempting to quantify the size and extent of the microcrystallites present have led to serious questions concerning the validity of this model as originally proposed. The first challenge was made by B. E. Warren, when he calculated the size of crystals in glass from the width of the Bragg peaks. He found that the size of proposed microcrystallites for vitreous silica (SiO_2) would be about 8 Å. Since this is about the same size as the corresponding unit cell of the crystalline phase crystobalite, microcrystallites in glass would be about one unit cell in size. This result contradicts a characteristic feature of crystals involving long-range periodicity of atoms.

In addition to the concept of microcrystallites of only unit cell sizes, it was also found that small-angle scattering did not occur for vitreous silica. Since maxima about 0° 2θ are normally observed for systems composed of particles, Warren concluded that the structure of glass is better described as a continuous medium.

3.2 The Random-Network Model

The random-network structure was first proposed by W. H. Zachariasen in 1932, and later supported by experimental and analytical procedures of x-ray diffraction developed by Warren and coworkers. The model is based on the concept of building blocks such as atoms, molecules or oxygen polyhedra, and on energy similarities between the crystalline and liquid states.

For inorganic oxide glasses, the building blocks or units of composition are oxygen polyhedra in the form of triangles or tetrahedra. Cations can play three basic roles in the glass structure, depending on considerations such as the cation–oxygen bond strength. For example, cations such as silicon which help form the backbone structure of the glass and about which oxygen polyhedra are coordinated, are called network formers. Others, such as the alkali and the alkali-earth cations, are referred to as modifiers and are usually added to glasses as oxides. These additions increase the oxygen to network-former ratio within the glass and result in a modification or breaking of the three-dimensional network. The modifier cations fit into holes within the network structure and associate with nonlinking or singly bonded oxygens. Finally, cations can also act as intermediates and either join into the framework or occupy holes within the network.

Based on energy arguments, W. H. Zachariasen postulated four rules governing the formation of oxide glasses: (a) an oxygen ion is linked to not more than two cations, (b) the coordination number of oxygen ions about the glass-forming cations is small (four or less), (c) oxygen polyhedra share corners (not edges or faces), and (d) the polyhedra are linked in a minimum of three corners, thus generating a three-dimensional random-network structure. While each of these conditions is desirable, they are not all always necessary to produce a glass structure.

Oxides MO and M_2O do not satisfy these conditions. Oxide glasses MO_2 and M_2O_5 (SiO_2, GeO_2, P_2O_5, As_2O_5) satisfy these rules by forming tetrahedra with oxygen on each of the corners surrounding the central M cation. Oxide glasses M_2O_3 (e.g., B_2O_3) yield triangles as units of composition. These units are found not only in the glassy state but also in their crystalline structures.

Other structural models have been developed over the years in attempts to describe the glassy state. L. W. Tilton described the structure of silica glass as being composed of five-membered rings and pentagonal dodecahedral units of composition. These units consist of twenty SiO_4 tetrahedrons. This concept was probably derived from suggestions of the structure of the liquid state as originated by J. D. Bernal. Bernal's concept was that the structures of liquids are based on random close-packing of atoms yielding a liquid composed of pentagonal icosahedrons. Each of these approaches involves structural units containing fivefold symmetry so that they cannot totally fill space, and inherent disorder is built into the models.

Another concept visualizes glasses as composed of paracrystals containing ordering somewhere between the crystalline and liquid states. These units are also combined with varying degrees of ordering.

Mathematical models describing the structures of glasses have also been proposed. In the Rutgers model, suggested by Pye et al. (1972), simple glass structures are represented mathematically, and important physical properties such as density, viscosity, mechanical and electrical characteristics of the glass are derived from this model as a function of temperature and composition.

4. Vitreous Silica

The structure of vitreous silica was studied by x-ray diffraction using improved experimental and analytical techniques. The Warren–Mavel method of fluorescence excitation was used to eliminate experimentally Compton or inelastic scattering, while pair functions were used to computer model-match experimental data with proposed models. Based on existing data, the best structural picture of vitreous silica is that of a continuous random network composed of SiO_4 tetrahedra as defined by Mozzi and Warren (1969). These tetrahedral units are linked to other units by sharing corners in a manner producing a relatively wide distribution of Si–O–Si bond angles. These tetrahedra can then be twisted with respect to each other, so that a continuous network can easily be generated.

From the first two peaks of the pair function, the basic unit of composition (the Si–O tetrahedron) is defined as consisting of silicon in the center and oxygen on each of the four corners. The Si–O distance is 1.625 Å and the edges of the tetrahedron or O–O distances are 2.66 Å. By deconvoluting the third peak, it can be determined that the tetrahedra are attached at corners with a distribution of intertetrahedral angles (Si–O–Si angles) of 130–180° and a mean angle of 143°. With this same technique, the next peak ($Si–O^2$) can be analyzed and shown to represent a random twisting of one tetrahedron with respect to an adjacent unit.

Network formation in simple oxide glasses has been described by Uhlmann and Wicks (1979) as resulting from two major modes. Mode 1 is the generation of a random network as a direct result of a distribution of intertetrahedral angles, while Mode 2 signifies the relative twisting distributions of neighboring tetrahedra. Each of these modes builds-in inherent disorder, characterizing the glassy structure. Both modes appear possible for vitreous silica. The modes of formation have a direct influence on the ring morphology, ease of glass formation and subsequent properties of the material.

5. *Vitreous Germania*

The structure of vitreous germania has been studied by neutron and x-ray diffraction. The basic unit of composition is again the tetrahedron. For this structure, four oxygen ions tetrahedrally surround a central germanium cation with a Ge–O distance of 1.74 Å and an O–O distance of 2.48 Å. This unit is larger in size than its SiO_4 counterpart due to the larger size of the germanium ion.

The half-widths of the Ge–Ge peaks are very similar for the experimental and model-matching curves. Hence, the measured mean intertetrahedral angle of 134° is characteristic of the structure and only a small variance about this angle is observed.

If one considers the modes of forming the random network, a distinct difference arises between vitreous silica and germania. Germania does not show the large distribution of intertetrahedral angles exhibited for SiO_2; hence, to generate the GeO_2 random network, the major mode of formation apparently involves the twisting of the tetrahedra. In addition, the Mode 2 type of network formation should place larger constraints on the network, including ring statistics as well as cage formation. It would be interesting to generate a physical model with the above criteria and observe the resulting ring formation, as performed by R. J. Bell and P. Dean and by D. L.

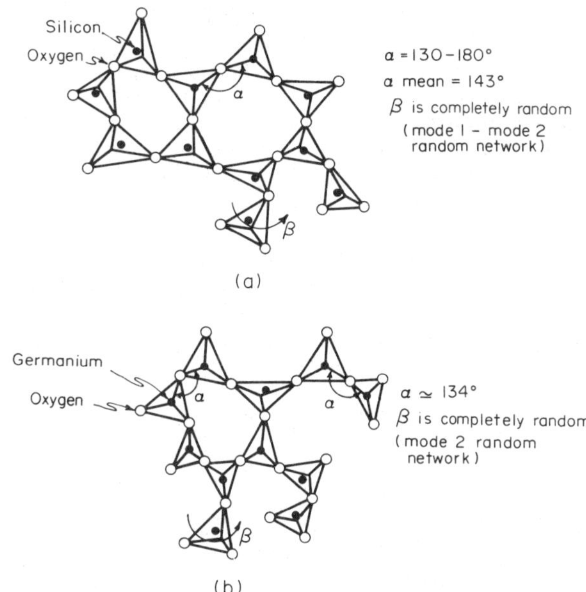

Figure 3
Structures of (a) vitreous silica and (b) vitreous germania

Evans and S. V. King for SiO_2 glass. The random-network structures of vitreous silica and germania are illustrated in Fig. 3.

6. *Vitreous Boron Oxide*

X-ray diffraction and nuclear magnetic resonance studies have defined the basic building blocks of vitreous boron oxide as triangles. These units of composition have been determined by Mozzi and Warren (1970) to consist of oxygen atoms triangularly coordinated about a central boron atom with a B–O distance of 1.37 Å and an O–O distance of 2.40 Å. This structure differs from vitreous silica and germania in that the building blocks are not linked together to form a continuous, three-dimensional, interconnected random-network structure (Fig. 4).

There are currently two leading structural interpretations for vitreous B_2O_3. The first structure involves linking triangles together to form B_3O_6 boroxol groups. These groups consist of hexagonal rings of three boron and three oxygen atoms, with three corner oxygen atoms outside the ring. A Mode 2 type of random-network structure can be generated by hooking together both boroxol groups and triangles. These rings can be attached together in a random orientation and with a B–O–B inter-ring angle of 130° and an intra-ring angle of 120°. Each of these groups can have neighboring groups not directly connected through B–O and O–B bonds.

An alternative structure for vitreous B_2O_3 has been

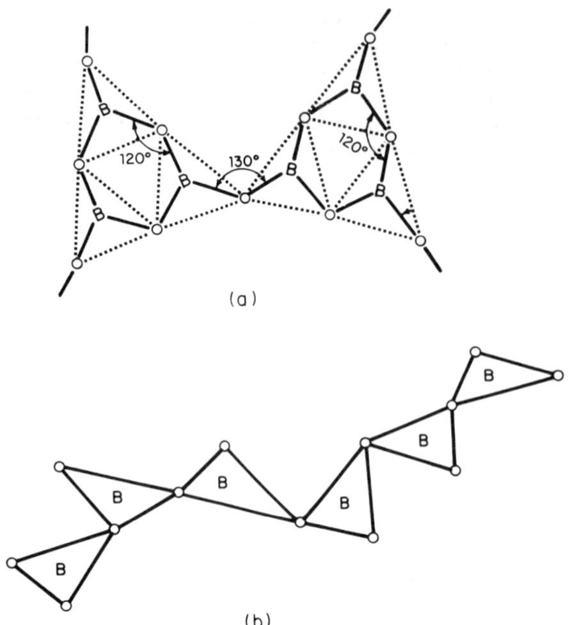

Figure 4
Proposed structures of vitreous B_2O_3: (a) boroxyl structure and (b) twisted-ribbon structure

proposed, in which triangles are hooked together to form long chains. These chains would be twisted and randomly distributed to avoid long-range periodicity of atoms.

For either the boroxol or twisted ribbon models for vitreous B_2O_3, an important contributing factor to the random glass structure is the significant number of incomplete or broken B–O and O–B bonds.

7. Multicomponent Glasses

The addition of alkali and alkali-earth oxides to silica glass results in a breaking of the random-network structure. This is due to the increase in the ratio of oxygen ions to the ratio of network cations, resulting in the creation of charged singly bonded oxygen ions. The alkali and alkali-earth cations are incorporated into cages of the random network and are attracted towards the singly bonded oxygen ions for charge neutrality. In potash silica glasses, a K–O peak and a K–K peak appeared in the pair-function curve and increased as a function of increasing potash content, suggesting that the alkali ions exist in pairs about two nonlinking oxygen ions (Wicks 1975). In other glass systems, such as Tl_2O–SiO_2, data suggest that clustering of cations may exist (Blair and Millberg 1974). An alkali-silicate-glass structure is depicted in Fig. 5.

The addition of alkali and alkali-earth oxides to borate glass results in a change of BO_3 triangles to BO_4 tetrahedra (Bray 1967). From nuclear magnetic resonance studies, these changes occur progressively up to about 30 mol% of alkali oxide. At this concentration about half of the boron atoms are in tetrahedral coordination and half are in triangular coordination. For higher modifier oxide content, the fraction of boron atoms in tetrahedral coordination decreases with increasing modifier content, indicating that the number of singly bonded oxygens increases.

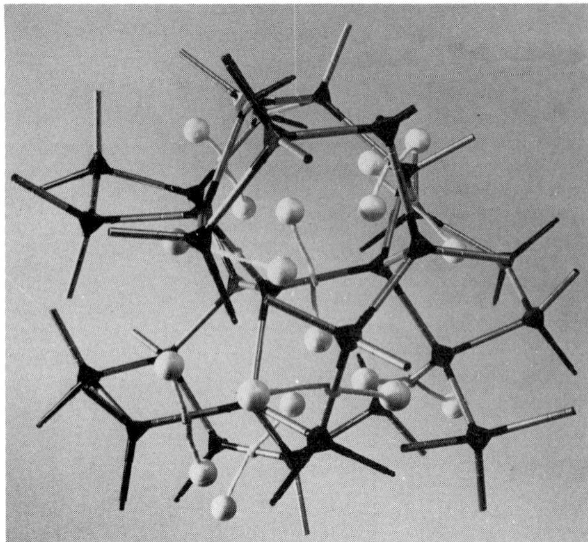

Figure 5
Alkali-silicate-glass structure model. The white spheres represent alkali atoms, whereas the black centers represent silicon with oxygen atoms tetrahedrally coordinated about silicon

Unlike silicate glasses, many important properties of borate glass systems change unexpectedly as a function of increasing alkali content. This is called the boron oxide anomaly, and is most likely related to changes in the triangular to tetrahedral structures described above. X-ray diffraction studies of borate and silicate glasses containing heavy atoms (e.g., Cs, Ba, Pb, Tl) show pairing or clustering effects.

Although most structural studies have been concerned with inorganic oxide glasses, many other glass-forming systems have also been investigated. However, the structure of many of these systems are less well defined.

Glasses made from ionic salts such as beryllium fluoride and zinc chloride are composed of tetrahedra as units of composition, which generate a random network believed to be similar to vitreous silica. The chalcogenide glasses are more complex: several of these, containing, for example, sulfur and selenium, produce structures composed of long chains as well as eight-membered rings. Long chains are also the characteristic structural feature of many organic glasses. These chains are believed to get tangled up in the glassy state into a random coil configuration. In this configuration, some cross-linking could occur, as well as mutual alignment of chain segments producing some small crystalline regions.

See also: Glass: An Overview; Metallic Glasses: Structure

Bibliography

Blair H D, Millberg M E 1974 Structure of thallium silicate glass. *J. Am. Ceram. Soc.* 57: 257–60

Bray P J 1967 Magnetic resonance studies of bonding, structure, and diffusion in crystalline vitreous solids. In: Bishay A (ed.) 1967 *Interaction of Radiation with Solids*. Plenum, New York, pp. 25–54

Doremus R H 1973 *Glass Science*. Wiley-Interscience, New York

Kingery W D, Bowen H K, Uhlmann D R 1976 *Introduction to Ceramics,* 2nd edn. Wiley-Interscience, New York

Mozzi R L, Warren B E 1969 The structure of vitreous silica. *J. Appl. Crystallogr.* 2: 164–72

Mozzi R L, Warren B E 1970 Structure of vitreous boron oxide. *J. Appl. Crystallogr.* 3: 251–57

Porai-Koshitz E A (ed.) 1966 *The Structure of Glass,* Vol. 6, *Properties, Structure and Physical–Chemical Effects*. Plenum, New York

Pye L D, Frechette V D, Kreidl N J (eds.) 1978 *Borate Glasses*, Materials Science Research Vol. 12. Plenum, New York

Pye L D, Stevens H J, LaCourse W C (eds.) 1972 *Introduction to Glass Science*. Plenum, New York

Uhlmann D R, Wicks G G 1979 Structural analysis using fluorescence excitation. *Wiss. Z. Friedrich-Schiller-Univ., Jena, Math.-Naturwiss. Reihe* 28: 231–42

Warren B E 1969 *X-Ray Diffraction*. Addison-Wesley, Reading, Massachusetts

Wicks G G 1975 Structural studies of amorphous materials. Sc.D. Thesis, Massachusetts Institute of Technology, Cambridge, Massachusetts

Wright A C, Leadbetter A J 1976 Diffraction studies of glass structure. *Phys. Chem. Glasses* 17: 122–45

G. G. Wicks

Glazing

Glazing is the trade practised by glaziers. A glazier is a craftsman who prepares and places glass sealants or gaskets, positioning shims and setting blocks in sash or curtailwall rabbets to ensure desired long-term appearance, weather-tightness and structural integrity. When practices of the Construction Specifications Institute are followed, glass and glazing specifications are included in contract documents.

Glazing systems must satisfy a remarkable combination of owner, occupant and public expectations. Glazing is subject to critical, close-up inspection, to hands-on contact with windows and doors, to esthetic and comfort responses to appearance, and to glare, drafts, noise, condensation, heat gain, heat loss and dirt accumulation. Reliable long-term performance requires design materials, installation and maintenance that address conservatively and cope practically with a multitude of significant changes in details and relationships. Changes occur from minute to minute (wind, solar radiation, rain) as well as from year to year (live loads, concrete curing, building settlement).

Knowledgeable and skilled craftsmanship by each glazier and careful planning by the owner are essential if design and performance goals are to be achieved. The most important considerations for proper glazing are listed in Table 1.

Table 1
Glazing design and planning considerations for proper glazing

Barometric pressure differentials	Mockup tests
Buffeting	Post-installation inspection and tests
Cascading of broken glass	Roof gravel abrasion
Concrete curing	Settling
Earthquake resistance	Stains and etching alkali
Explosions	Storage
Falling objects	Thermal expansion/contraction
Human safety	Tolerances
Impact	Ultraviolet radiation disintegration at adhesive interface
Maintenance pressures and chemicals	Vandalism
	Wind and stack pressure

1. Weather Tightness

Corners and other rabbet fabrication joints must be sealed in the fabrication plant or by other trades, since this operation is not normally included in glazing contracts.

Weather tightness requires that moisture penetration be limited, by a long-lasting, watertight seal between glass perimeter and sash rabbet, to an amount that will flow by gravity from the building through weep holes. On a tightly sealed rabbet, usually three 1 cm diameter weep holes in each sill rabbet which lead outdoors from under the glass provide adequate relief. Adequate seals can be achieved and maintained with gaskets, glazing tape and wet sealants. Effective glazing gaskets provide weather tightness by establishing continuous "window-wiper" contact between gasket tip and smooth glass surface and by maintaining compression of 7–18 N per linear cm of tip contact. Lock strip gaskets that comply with ASTM standard C542 provide this performance (Fig. 1).

Wet sealants (Fig. 2) and tape sealants require that

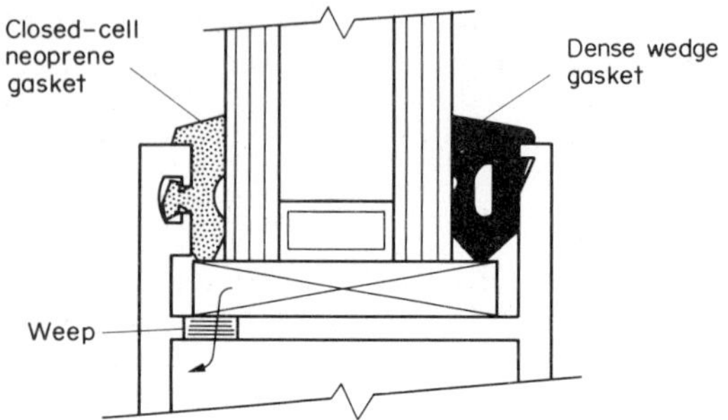

Figure 1
Typical dry glazing system

glass and sash sealing surfaces be absolutely clean, dry, free of oily deposits and not colder than 4.5 °C during application. Some sealants require that glass and rabbet surfaces be treated with special primers after careful cleaning before application.

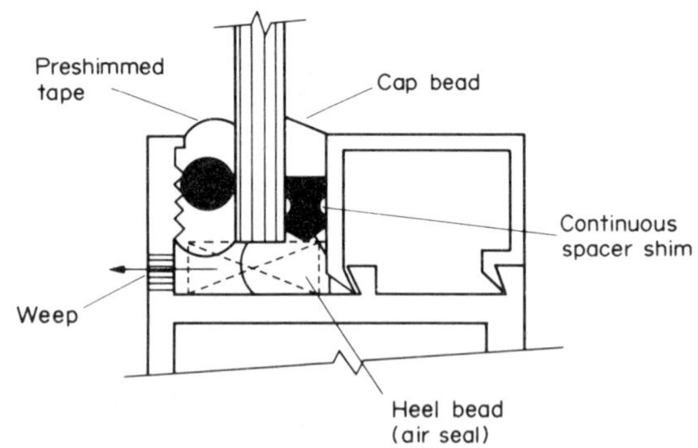

Figure 2
Typical wet glazing system

Usually, sealant sections must be controlled in the 3–6 × 6–13 mm range by rabbet design, proper placement of glass and proper shimming. Resilient shims positioned at regular intervals within the rabbet at indoor and outdoor glass surfaces near glass edges limit movement between glass and rabbet (Fig. 3). This restricts sealant squeezeout and gives the sealant an opportunity to perform.

Resilient shims may be continuous or on 45–65 cm centers. Structural glazing systems require extra care, including temporary structural stabilization during a 2–3 week curing period.

2. Structural Reliability

Annealed glass manufactured by the continuous float process is a brittle material. In practical use, wind resistance may be halved by even mild surface abrasion. Slight damage to a skillfully clean-cut edge,

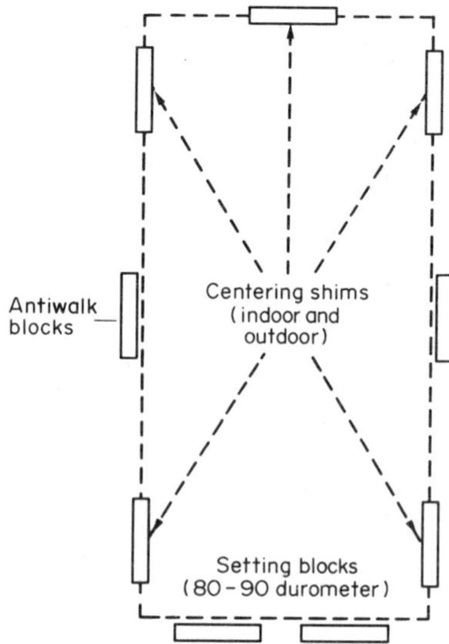

Figure 3
Positioning of centering shims and setting blocks

caused by contact with hard materials during installations, may reduce thermal break resistance by 80% or more.

As shown in Fig. 4, breaking strength can be doubled by heat strengthening or quadrupled by tempering. However, where significant safety or economic losses may result from spontaneous failure, tempered glass should be utilized in laminates with annealed or heat-strengthened lites. A laminated lite (or piece of flat glass) has about 60% of the load-carrying ability of a monolithic lite of the same total thickness. A sealed insulating glass unit (Fig. 5) will

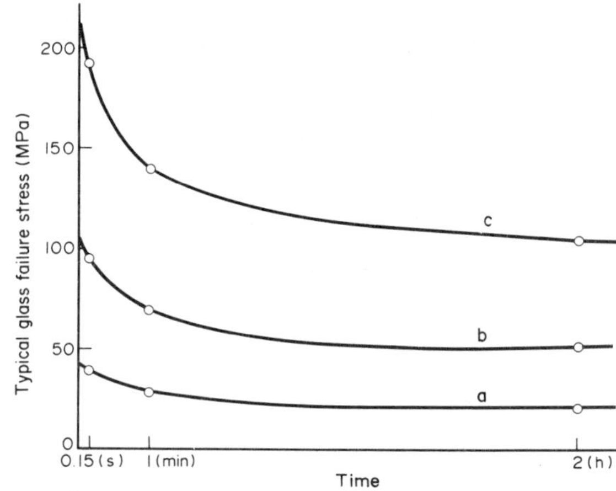

Figure 4
Typical failure stresses for glasses: (a) annealed; (b) heat-strengthened; (c) tempered

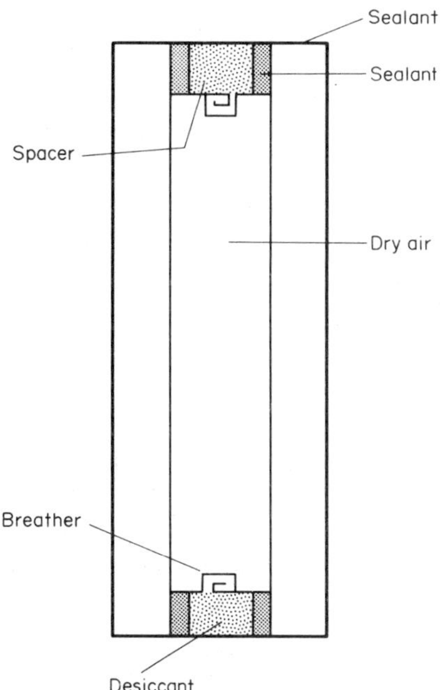

Figure 5
Sealed insulating glass unit

carry ~80% more wind load than the thinner lite of which it is fabricated.

Usually, architects rely on manufacturers' charts for selecting glass thickness. To select proper glass type and thickness required to meet load and safety requirements, the architect must specify job goals and regulatory requirements. If a lite is to perform up to its nominal structural capacity, edges must be resiliently and uniformly supported in a stable glazing rabbet. Minimum clearance between lite and rabbet must be 3 mm at every point. For good performance where earthquakes may occur, continuous clearance of at least 6 mm at the perimeter, as required by the Uniform Building Code, is recommended.

Only two setting blocks of 80–90 durometer neoprene or ethylene–propylene terpolymer meeting ASTM C542 should be placed at quarter points of the sill (not closer than 15 cm to the nearest corner). Each setting block must be at least 10 cm long, equal to full rabbet width, and thick enough to prevent glass contact with the sill under all circumstances. 20–40 durometer centering shims placed on 45–65 cm centers against indoor and outdoor glass surfaces and soft, antiwalk blocks placed in the perimeter space between glass and rabbet are essential.

Shims, spacers and sealants must remain resilient over a long period of time, especially at corners. Under positive and negative wind loads, lites must be free to "curl" in or out at corners. If flexing of corners is not permitted, stress reversals may lead to breaks originating near corners.

To reduce thermal stress, especially in tinted and reflective lites and in insulating units, edges should be insulated to prevent heat loss to adjacent heat sinks such as outdoor concrete or aluminum fins (Fig. 6).

3. *Satisfactory Appearance*

When lites are exposed to sunlight at grazing angles, glass surfaces are viewed critically and close up by building occupants. Even minute surface dirt, scratches, abrasions, or etchings may become visible and unattractive. For this reason, glass surfaces must

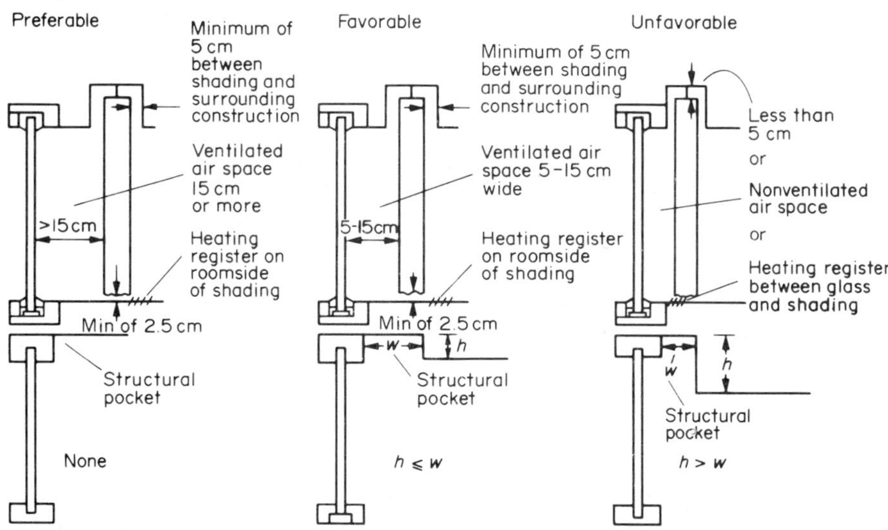

Figure 6
Preferable, favorable and unfavorable conditions of indoor shading

be protected carefully during storage, handling, installation and service. Glass surfaces may be pitted by welding splatter or etched by alkaline or fluoride deposits from adjacent masonry or concrete surfaces after rainstorms or maintenance washing. Strength may be reduced significantly. In situ surface repairs are not economically feasible.

Like all manufactured products, glass has subtle appearance characteristics related to its production line and fabrication orientation. To minimize "checkerboarding" in a facade, all lites should be cut and installed to achieve the same up–down, in–out, right–left relationships. All blocks, shims and so on must be uniform in dimension and installed in uniform positions and with uniform tightness of fit.

Facade appearance may vary significantly in color and flatness depending upon lighting and viewing location as well as upon color and geometry of images reflected. With barometric and temperature changes, sealed insulating units bow in or out and may cause significant differences in reflected images from day to day.

All heat-strengthened and tempered glass is distorted to some degree by the heating process that imparts increased strength. Also under some north-sky (partly polarized) viewing conditions, quench patterns related to processing may become visible. These characteristics vary between processes but cannot be eliminated.

4. Lessons from Experience

The following considerations should always be borne in mind by glaziers:

(a) explain risks to the owner so that he understands the design and agrees with cost–performance compromises,

(b) show glazing details in shop drawings two to four times full scale,

(c) label each item and material,

(d) explain the function of each,

(e) dimension each item and show the maximum permissible tolerance,

(f) protect glass from impact and abrasion, against alkalis in fluoride washes,

(g) replace damaged glass promptly, and

(h) clean glass regularly.

See also: Glass as a Building Material; Structural Sealant Glazing

Bibliography

Adams F W 1942 *Behavior of Glazing Materials Subject to Explosion,* ASTM Bulletin No. 122. American Society for Testing and Materials, Philadelphia, Pennsylvania

American Society for Testing and Materials. *Recommended Practices and Standards on Glass and Glazing*, ASTM Committees C24 and C26. American Society for Testing and Materials, Philadelphia, Pennsylvania

Architectural Aluminum Manufacturers' Association 1972 *Joint Sealants in Glass and Glazing,* Curtainwall Monograph No. 6. Architectural Aluminum Manufacturers' Association, Chicago

Architectural Aluminum Manufacturers' Association 1971 *Lock Strip Gaskets,* Curtainwall Monograph No. 4 Architectural Aluminum Manufacturers' Association, Chicago

Beason W L 1980 *A Failure Prediction Model for Window Glass.* Institute for Disaster Research, Lubbock, Texas

Cook J P 1970 *Construction Sealants and Adhesives.* Wiley, New York, pp. 121–25

Ernsberger F N 1960 Detection of strength impairing surface flaws in glass. *Proc. R. Soc., London* 257: 213–23

Flat Glass for Glazing, Federal Specification DDG 451D. US General Services Administration, Washington DC

Heat Strength and Fully Tempered Flat Glass, Federal Specification DDG 001403. US General Services Administration, Washington, DC

Flat Glass Marketing Association 1978 *Glazing Manual.* Glass and Glazing Federation, London

Frownfelter 1959 *Structural Testing of Large Glass,* ASTM Special Technical Publication No. 251. American Society for Testing and Materials, Philadelphia, Pennsylvania

Orr L 1972 Practical analysis of fractures in glass windows. *Mater. Res. Stand.* 12: 21–47

Panek J 1976 *Building Seals and Sealants.* American Society for Testing and Materials, Philadelphia, Pennsylvania, pp. 81, 134–150

Purchase S 1981 You can become an expert glazer. *Boston Sunday Globe,* Oct. 18th Issue

Further information on glazing can be found in the following publications issued by PPG Industries, Pittsburg, Pennsylvania:

Butt Joint Glazing System, PPG Brochure

Care and Storage of Glass, PPG Technical Service Memo, Feb. 11, 1980

Design Windloads, ANSI A58.1, PPG Technical Service Memo, June 21, 1974

Framing and Glazing—Does it Float? PPG Technical Service Memo, Sept. 30, 1977

Installation Recommendations for Tinted and Reflective Glass, PPG Technical Service Report No. 130

Installation Recommendations for Windows, PPG Technical Service Report No. 230

Mock-Up Test Requirements versus Glass Design Loads, PPG Technical Service Memo, May 8, 1981

Recommended Glazing Practices for Insulating Units over 20 square feet in Area, PPG Recommended Practice

Recommended Glazing Practices for LHR Spandrelite and Hestron Glasses, PPG Technical Service Memo, Aug. 18, 1980

R. W. McKinley

Glued Joints in Wood

Glued joints are used in wood construction to increase the size of available materials and also for product assembly. Natural adhesives have been used for furniture and other nonstructural products throughout most of recorded history. Development of durable synthetic resin-based adhesives prior to World War II allowed the use of adhesives in demanding structural products and applications.

Three types of joints are readily defined with respect to the grain directions of the members joined together: edge-to-edge grain, end-to-end grain and end-to-side grain (where edge grain also called side grain, refers to radial or tangential surfaces, and end grain to the cross section). Edge-to-edge- and end-to-end-grain joints are widely used in glued, laminated beams and other structural products. The use of adhesives in other types of building construction is expanding. The greatest variety of joints occurs in furniture construction, where not only edge-to-edge- and end-to-end-grain joints are used, but also end-to-side-grain joints. Most of the following discussion will deal with furniture joints; for applications of glued joints in building construction and laminated beams, see *Design with Wood* and *Glued Laminated Timber*.

1. Edge-Grain Joints

Edge-to-edge-grain joints are used for furniture parts such as solid wood seats and tops, and lumber cores for panels. The material is laid up in large, revolving glue-clamp carriers called reels, which hold the glued stock under pressure while the adhesive dries. High-frequency electronic glue presses are also used. Pressure is used to secure proper alignment of the parts and to ensure intimate contact between the wood and the glue. Animal glues, cold- and hot-setting urea formaldehyde, and poly(vinyl acetate) adhesives are commonly used to bond the wood together (see *Adhesives for Wood*). Boards are carefully machined to mate smoothly along their entire lengths without gaps, to prevent uneven glue lines and nonuniform distribution of pressure (see *Machining of Wood*). Machined edges are preferred to sanded. Machining is carried out just prior to gluing, to remove surface resins and surface oxidation products. Pressures of 0.35–2 kPa are used. Wide boards are placed towards the outsides of the panels to help distribute pressures uniformly throughout the entire assembly. Temperature of the glue, the wood and the surroundings are controlled to produce joints with thin but adequate glue lines which retain sufficient adhesive to prevent starved joints. Assembly times are held as short as possible to reduce subsequent cure times. Hold-downs are used to prevent panels from bowing while being pressed. Pneumatic torque wrenches lessen the time required to tighten clamps, and ensure that clamping forces are applied uniformly. The wood is conditioned to a uniform moisture content prior to gluing to prevent subsequent problems with unequal shrinking and swelling (see *Shrinking and Swelling of Wood*). Glued panels are allowed to set until the swollen glue lines have dried sufficiently to allow machining.

2. End-Grain Joints

Furniture joints which require the connection of end-to-end-grain or end-to-side-grain surfaces are termed assembly joints. Included are dowel, mortise-and-tenon, finger, corner-block and gusset-plate joints (see Fig 1). An illustration of finger joints is given in the article *Glued Laminated Timber*. Several of these assembly joints are also used in light frame building construction.

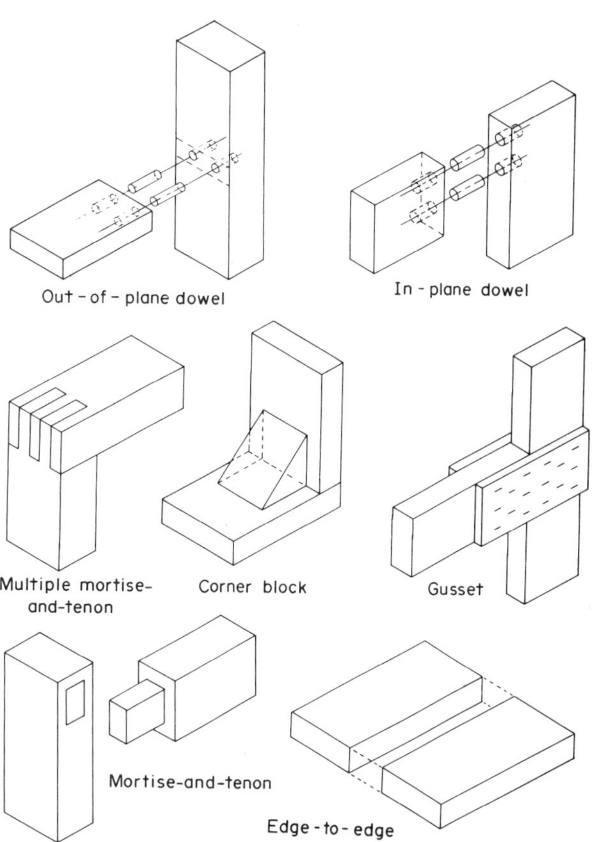

Figure 1
Common types of assembly joints

2.1 Dowel Joints

Owing to their simplicity, dowel joints are the most common connection employed in furniture construction. Withdrawal strength of a dowel from the side grain of a wooden part is directly proportional to the diameter of the dowel and to the shear strengths parallel to the grain of the dowel and of the materials joined together; withdrawal strength is slightly less than proportional to depth of embedment of the dowel. For highest strengths, both the dowel and the walls of the hole are coated with glue; a close fit between dowel and hole is essential. The in-plane bending strength of joints constructed with two symmetrically spaced dowels is directly related to the withdrawal strength of the dowels and to the width of the spacing between them. The out-of-plane bending strength of dowel joints is much less than in-plane strength; thickness of the end-grain member is the dominant strength parameter (Eckelman 1979). Dowel joints also carry shear loads in furniture. Strength is limited largely by the splitting strength of the wood. Typical joints constructed with two dowels 1 cm in diameter carry loads in excess of 2.5 kN. Methods of calculating the strengths of most of the dowel joints described have been developed (Eckelman 1978); therefore, these joints can be designed to meet specified in-service requirements.

2.2 Mortise-and-Tenon Joints

Mortise-and-tenon joints have largely been displaced by dowel joints in furniture because of the relative ease of construction of the latter. Adhesives are used in nearly all mortise-and-tenon furniture joints. Well-made joints develop somewhat more strength than comparable dowel joints (Sparkes 1968). Bending strength is related to the thickness, width and length of the tenon, and the closeness of fit of the tenon into the mortise. Bending strength of the tenon itself provides the ultimate constraint on strength.

2.3 Corner Blocks

Corner blocks are used in furniture construction to reinforce points of high shear and bending stress. Ordinarily, corner blocks are cut with each of the two glue faces at angles of 45° to the grain; as a result, both faces form bonds intermediate to side grain and end grain. Blocks may also be cut with one face parallel to the grain to allow a side-grain-to-side-grain bond on one face. Properly fitted glue blocks can effectively resist bending loads which tend to close a joint, but may be less effective in resisting bending loads which tend to open it (Eckelman 1978).

2.4 Finger Joints

End-grain finger joints are used both to increase the longest length of lumber available and to salvage material by creating long members from short scraps. Strongest joints occur when the fingers have a small slope and are sharply pointed. Corner finger joints are more difficult to cut and clamp than end joints, although joints with slight angles are readily fabricated since they are essentially end-grain joints. Such joints are regularly used in items such as the back posts of chairs. Techniques for cutting geometrically complex high-strength finger joints have been developed (Richards 1962), but are little used commercially.

2.5 Multiple Mortise-and-Tenon Joints

Multiple mortise-and-tenon joints consist of a number of tenons with parallel sides, which are cut into the ends of two mating members. A common example of this construction is the box joint. They are also used to form visible decorative joints in fine furniture. Well-made multiple mortise-and-tenon corner joints may develop up to 50% of the strength of the wood (Richards 1962). Dovetail joints are basically similar in design, although they are not usually thought of as multiple mortise-and-tenon joints. Primary use of dovetail joints is in fine-furniture construction, where they are used for drawer construction. Close fit of the parts and use of adequate glue are necessary for high strength.

2.6 Staple Glued Plywood Gusset Joints

Staple glued plywood gussets are used in hidden areas in furniture to reinforce heavily stressed joints (Eckelman 1978). Such joints were once widely used in the construction of roof trusses, but have largely been supplanted by metal-tooth connector plates (see *Design with Wood*; *Wood Joints with Mechanical Fastenings*). Strength of such joints is dependent on the size of gusset used and the rolling shear strength of the plywood.

See also: Wood: An Overview

Bibliography

Eckelman C A 1978 *Strength Design of Furniture*. Tim Tech, West Lafayette, Indiana

Eckelman C A 1979 Out of plane strength and stiffness of dowel joints. *For. Prod. J.* 29: 32–38

Feirer J 1967 *Cabinetmaking and Millwork*, 2nd edn. C A Bennett, Peoria, Illinois

Gillespie R H, Countryman D, Blomquist R F 1978 *Adhesives in Building Construction*, US Department of Agriculture Handbook 516. US Government Printing Office, Washington, DC

Pound J 1973 *Radio Frequency Heating in the Timber Industry*, 2nd edn. Clowes, London

Richards D B 1962 High strength corner joints for wood. *For. Prod. J.* 12: 413–18

Selbo M L 1975 *Adhesive Bonding of Wood*, US Department of Agriculture Bulletin No. 1512. US Government Printing Office, Washington, DC

Slaats M A 1979 Glue bond quality in wood. *Int. J. Furniture Res.* 1: 24–26

Snider R 1975 *Gluing and Furniture Design*. Franklin Chemical Industries, Columbus, Ohio

Sparkes A J 1968 *The Strength of Mortise and Tenon Joints*, Report 33. Furniture Industry Research Association, Stevenage, UK

US Forest Products Laboratory 1974 *Wood Handbook*, US Department of Agriculture Handbook 72. US Government Printing Office, Washington, DC

C. A. Eckelman

Glued Laminated Timber

Glued laminated timber used for structural purposes is an engineered stress-rated product comprising assemblies of suitably selected and prepared wood laminations securely bonded together with adhesives. The grain of all laminations is approximately parallel longitudinally. The individual laminations are of lumber thicknesses. Laminations may be composed of pieces end-joined to form any length, of pieces placed or glued edge-to-edge to make wider components, or of pieces bent to curved form during gluing.

Glued laminated timber (often referred to by the generic term glulam) was first used as early as 1893 in Europe, where laminated arches (probably glued with casein adhesives) were erected for an auditorium in Switzerland. Improvements in casein adhesives during World War I created further interest in the use of glued laminated timber structural members for aircraft and, later, as framing members for buildings. The development of durable synthetic-resin adhesives during World War II permitted the use of glued laminated timber members in bridges, truck beds and marine construction, where a high degree of resistance to severe service conditions is required. Today, synthetic-resin adhesives—primarily of the phenol–resorcinol and melamine types—are the principal adhesives used for structural laminating (see *Adhesives for Wood*).

Typically, glued laminated structural timbers are rectangular in cross section. They may be either straight or curved between supports. Straight beams can be designed and manufactured with horizontal laminations (load applied perpendicular to wide faces of laminations) or vertical laminations (load applied parallel to wide faces of laminations), as shown in Fig. 1. Horizontally laminated beams are the most widely used. Curved members are horizontally laminated to permit bending of the laminations to a curved form during gluing.

The advantages of glued laminated timber construction include the provision of the ability to manufacture larger structural elements from smaller commercial sizes of lumber, giving better utilization of available lumber resources; the ability to achieve architectural effects through the use of curved shapes; and the ability to design structural elements varying in cross section between supports, and in accordance with strength requirements. Advantages also include minimization of checking or other seasoning defects associated with large sawn-wood members, because the laminations can be dried before being glued, thus permitting designs to be based on the strength of seasoned wood. In addition, lower grade lumber can be used for less highly stressed laminations without adversely affecting the structural integrity of the member.

Factors determining the assigned strength values of structural lumber will also affect glued laminated timber, since glulam is made up of individual pieces of structurally graded lumber (see *Lumber: Behavior Under Load*; *Lumber: Types and Grades*). The mere gluing together of pieces of wood does not, of itself, improve strength properties in laminating material of lumber thicknesses over those of comparable sawn timbers. Three of the benefits inherent in the laminating process do, however, form the bases for assigning higher glued laminated timber design values: (a) drying or seasoning of the laminations; (b) positioning or placement of the laminations in the member; and (c) distribution of growth characteristics.

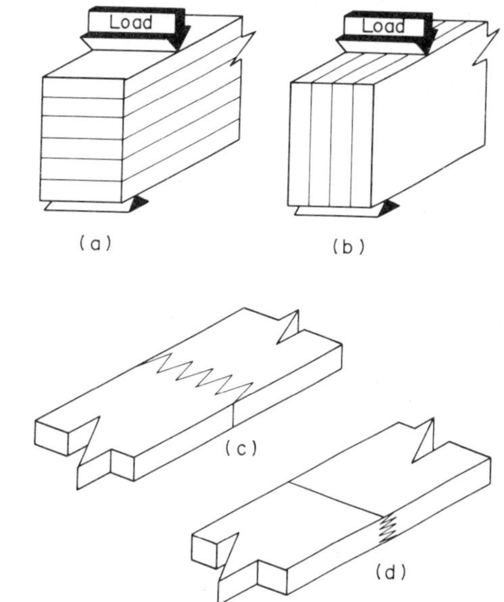

Figure 1
Glued laminated timber forms: (a) beam, horizontally laminated; (b) beam, vertically laminated; (c) end joint, vertical finger joint; (d) end joint, horizontal finger joint

As wood seasons or dries under controlled conditions, it increases in strength. It is difficult to thoroughly dry pieces of wood with least dimensions greater than about 10 cm within a reasonable time. But laminated timbers are fabricated of individual laminations of lumber thickness, dried to a moisture content of 16% or less; therefore, regardless of the size of the finished laminated timber, it is dry, and design values can be based upon the strength of dry wood.

The bending strength of horizontally laminated timbers depends on the positioning of the various grades of lumber used as laminations. High grade laminations are placed in the outer portions of the member where high strength is effectively used, and lower grade laminations are placed in the inner portion, where low strength will not greatly affect the overall strength of the member. By selective placement of the laminations, knots are scattered and improved strength can be obtained. Studies have indicated that knots are unlikely to occur one above another in several adjacent laminations.

A comparison of the allowable design values for a typical laminated-timber grade for bending members, and a No. 1 grade of sawn timber of the size classification "beams and stringers," both made from Douglas fir, is shown in Table 1.

Principal steps in the manufacture of glulam members are:

(a) Selection and preparation of laminations. Lumber used for structural laminating is commonly of the same softwood species used for other construction purposes. In the USA, for example, the softwood species most commonly used are Douglas-fir, Southern pine and hem-fir (a marketing combination of western hemlock and true firs). The lumber is graded for knot size and location, rate of growth, density, slope of grain and other characteristics that affect its use as a structural material. The laminations are surfaced prior to gluing, to assure that the wide faces can be brought into close contact for adhesive bonding.

(b) End joining of laminations. Lumber in the long lengths necessary for large laminated timbers is not normally available; therefore, the fabrication of structural end joints is an important part of the laminating process. The most common type of end joint used by the laminating industry consists of a series of intermeshed tapered "fingers" cut into the ends of laminations. These end joints, known as finger joints, can be well bonded and develop the high strengths needed for structural laminating (Fig. 1).

(c) Spreading of adhesives. Adhesives are spread on the laminations by means of rollers through which the laminations pass, or, more commonly, by an extruder which spreads beads of adhesive along the length of the lamination as it passes beneath.

(d) Clamping. After the laminations have been spread with adhesives, pressure is applied in forms that have been preset to the shape required in the finished glulam product. Pressure is applied, usually by means of a series of evenly spaced clamps, to bring the laminations into close contact, to pull the member to its final shape, to force out excess adhesive and to hold the laminations together during the adhesive curing period. After this curing period, which may range from 8 to 24 h, the glulam members are removed from the forms and are surfaced by large planing equipment to the specified width dimension.

(e) Finishing. In the finishing area of a laminating plant the completed glulam members are cut to the proper length, appropriate taper cuts are made, holes and daps for connectors are formed, end sealers are applied and wrapping material is applied as specified.

(f) Quality control. Throughout the laminating process, samples of production are regularly checked to be sure that the products are in conformance with established laminating standards. Quality control is an essential part of the laminating process, since the strength of the glulam members is dependent upon the quality of the glue joints. Special equipment, plant facilities and manufacturing skills are needed to assure that this high quality is maintained.

The advent of techniques for glue laminating of wood provided a practical means for manufacturing wood structural members not limited by the size and shape of a tree. The use of glued laminated timber permits the construction of timber buildings with long clear spans and a variety of shapes.

See also: Glued Joints in Wood; Wood: An Overview

Bibliography

American Institute of Timber Construction 1983 *Structural Glued Laminated Timber*, ANSI/AITC 190.1. American Institute of Timber Construction, Englewood, Colorado

American Institute of Timber Construction 1984 *Standard Specifications for Structural Glued Laminated Timber of Softwood Species*, AITC 117-84. American Institute of Timber Construction, Englewood, Colorado

American Society for Testing and Materials 1983 Standard method for establishing stresses for structural glued laminated timber (glulam) manufactured from visually graded

Table 1
Comparison of design values, Douglas-fir

Design values	Bending (MPa)	Compression perpendicular to grain (MPa)	Shear (MPa)	Modulus of elasticity (MPa)
Glulam, typical grade	16.55	3.10	1.14	12410
Sawn timber, No. 1 "beams and stringers"	9.31	2.65	0.59	11030

lumber, ASTM D3737-83. *1983 Annual Book of ASTM Standards*, Vol. 4:09, *Wood*. American Society for Testing and Materials, Philadelphia, Pennsylvania, ASTM D3737-83, pp. 645–63
Chugg W A 1964 *Glulam: The Theory and Practice of the Manufacture of Glued Laminated Timber Structures*. Benn, London
National Forest Products Association 1982 *National Design Specification for Wood Construction*. National Forest Products Association, Washington, DC

R. P. Wibbens

Gold and Gold Alloys

Gold, being found in various parts of the earth's surface in native form, was undoubtedly one of the first metals used by man. However, its first use was purely ornamental and this use based on its esthetic appeal has become deeply imbedded into man's culture. Its singular ease in being reduced to a pure form and in being fabricated into conveniently handled quantities (bullion and coins), its constancy in quality, its remarkable imperviousness to atmospheric or chemical attack, plus its relative scarcity make it the monetary base par excellence. In addition it has sufficiently unique physical properties to assume a role in a large variety of industrial and technological applications.

Gold was probably instrumental in stimulating the growth of early science. Burke (1978) attributes to gold the initiation of practical physical science in the touchstone technique for determining the karat content of a gold alloy. It was also this need to determine gold alloy quality that induced Archimedes to discover the buoyancy law for determining the density of solids. Alchemy, the forerunner of chemistry, had its origin in the incentive to produce gold from base metals.

Gold is found, although not generously, in all parts of the earth's crust, including seawater (Wise 1964). Its concentration in the earth as a whole is about 0.25 ppm (estimated from gold found in meteorites), it is much less so in the earth's crust: 1–9 ppb, and still less so in sea water: 0.004 ppb. Worthwhile ores contain 5–30 ppm, although, as a by-product in silver, copper, nickel and lead mining, extraction can come from ores of lower concentration. Other precious metals such as the platinum group metals are also frequently found in association with gold. Placer mining seeks stream-deposited gold in nugget, flakes and specks form. Lode mining seeks gold via hard rock mining, frequently in quartz veins.

Various methods of concentration and smelting render the gold from mining into some form containing silver, copper and other base metals that requires further refining. Most base metals can be slagged off by melting under oxidizing conditions. An electrolytic process is usually resorted to for the last stage of refining, although the Miller process which involves melting and bubbling chlorine through the melt can be used to remove the residual base metals and silver, giving a gold of about 99.6% purity.

The 1980 world production (Lucas 1980) approached 40 million troy ounces, the major producing countries (in millions of troy ounces) being: the Republic of South Africa, 22; the USSR, 8; Canada, 1.5; Brazil, 1.3; and the USA, slightly less than 1 (see *Gold Resources*)

1. Properties

General, mechanical, thermal, electrical and magnetic properties of gold are listed in Table 1. The reflectance and emittance are shown in Figs. 1 and 2, respectively.

Table 1
Properties of gold

General	
Atomic number	79
Atomic weight	196.9665
Radioactive isotopes (none in nature)	^{197}Au, 1.83 days half-life ^{198}Au, 2.7 days half-life ^{199}Au, 3.15 days half-life
Crystal structure	fcc; a = 0.40786
Density at 25 °C	19.302 Mg m^{-3}
Mechanical (99.99% purity)	
Young's modulus	80 GPa
Tensile strength	131 MPa
Proportional limit	3.4 MPa
Elongation	50%
Solid surface tension	1400 mN m^{-1}
Liquid surface tension at 1200 °C	1070 mN m^{-1}
Thermal	
Melting point	1064.43 °C
Boiling point	2857 °C
Linear coefficient of thermal expansion at 20 °C	14.2 μm m^{-1}K^{-1}
at 700 °C	15.42 μm m^{-1}K^{-1}
Specific heat	
at 25 °C	128 J kg^{-1}
at 1227 °C	159 J kg^{-1}
Latent heat of fusion	62.762 kJ kg^{-1}
Latent heat of vaporization	1.6987 kJ kg^{-1}
Thermal conductivity at 0 °C	317.9 W m^{-1} Kg^{-1}
Vapor pressure	
at 1770 °C	0.1013 kPa
at 2857 °C	101.3 kPa
Electrical and magnetic	
Electrical conductivity at 20 °C	73.4% IACS
Electrical resistivity at 0 °C	20.1 nΩ m
Magnetic susceptibility volume	1.79 × 10^{-6} mks

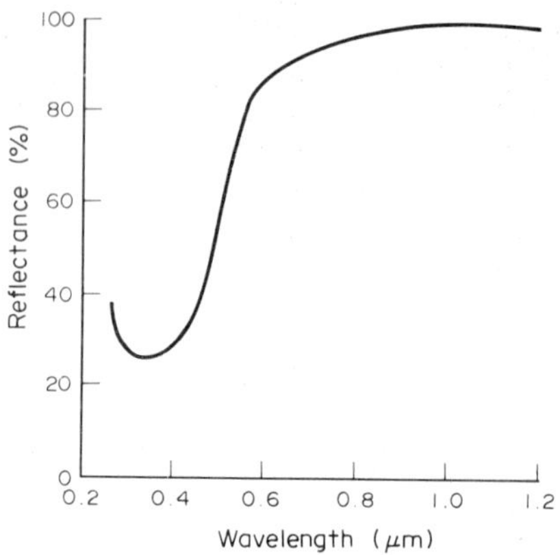

Figure 1
Reflectance of gold as a function of wavelength

2. Usage

Gold has been a monetary commodity since the Egyptian era and used from the very beginning of coinage by the Lydians. To this day, governments, banks and individuals can buy gold for investment in the form of coins or bullion, or by trading at various types of commodity exchanges. Gold in this form is at least 99.5% pure and is quoted today in US dollars per troy ounce.

Semifinished products are manufactured as sheet, strip, foil, leaf, wire, tubing, ribbon or powder. Unalloyed gold is applied in electroplating, sputtering and vapor deposition. However, most gold used in manufacturing is in alloy form.

2.1 Jewelry Alloys

These are karat golds, the consumer being protected by a law that assures the gold content is as claimed by the seller. Generally this is done by karat designation, the karat being 1/24th part by weight of gold in the alloy. In the US, the lowest legal karat designation for jewelry gold alloys is 10 kt. In some other countries the lower limit may be as low as 8 kt, while in others as high as 18 kt. An alternative gold alloy designation for jewelry (and bullion) is by fineness; a 925 fine gold being of 92.5% gold content.

The balance of the composition of the karat golds is made up of copper, usually some silver, zinc and occasionally nickel. The copper retains the yellow color and has a marked strengthening effect by virtue of a second phase (an ordered structure) tending to form a precipitate on cooling. Strength is important since pure gold is too soft for some jewelry items. Silver is also a strengthener to some extent and it is important for color modification to obtain interesting hues for various jewellery designs. Zinc in small quantities is a deoxidizer and in larger concentrations also a color modifier, especially to offset the red color of high copper alloying. Nickel is primarily a whitener for producing the white gold alloys. All these additions are welcome as a diluent for gold, because they reduce the intrinsic value for marketing purposes. Silicon is sometimes added to give a "clean cast" effect to castings, that is, the castings come out of the molds with a clean metallic color. Cadmium, tin, indium and gallium may be added to the so-called jewelry solders. Jewelry solders must be color matched, and there is also a legal restriction in their use if karat content is lowered in the alloying to produce the solder.

As the karat content is lowered, there is a tendency to tarnish, the copper in the alloying tending to oxidize, while the silver tends to form sulfides. The problem exists in alloys of about 10 kt.

Some "poisons" in the jewelry golds are lead, phosphorus and sulfur, all of which embrittle gold. Oxygen as a contaminant is also a problem.

2.2 Dental Alloys

The dental gold alloys followed the jewelry golds and were developed for making cast dental restorations (1910–1940) using copper, silver and zinc for strengthening and color control. The higher content gold alloys were relatively soft and were used for inlays and single crowns, while the lower content golds which were stronger were used for multiple crowns and bridges in which clasps for abutment teeth were incorporated in the prostheses. Karat content played little or no role in the decision by the dentist in the choice of alloy. In due course the strength demands for the alloys increased over those of the jewelry alloys and alloys appeared with platinum additions for age hardening. Platinum–iron additions yield a particularly good age-hardening effect. After some time, palladium began to be used as a substitute for the expensive platinum. Palladium was discovered to have a particularly good effect in

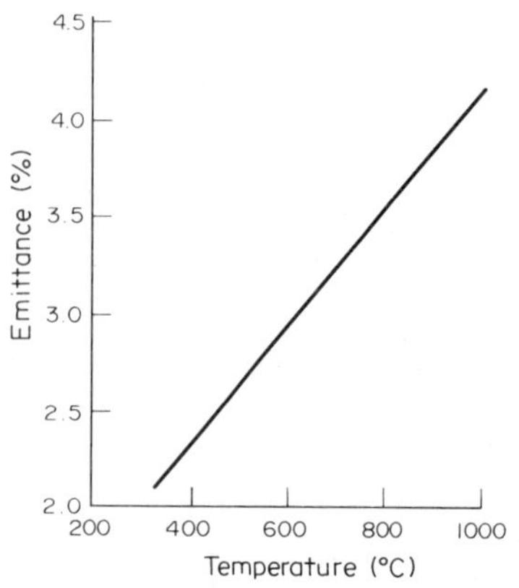

Figure 2
Total hemispherical emittance of gold as a function of temperature

that it reduced tarnish in the silver-bearing gold alloys, leading to dental yellow gold alloys with gold content as low as 42% (see *Dental Noble-Metal Casting Alloys: Composition and Properties*).

In recent years, a new requirement has emerged for dental alloys: compatibility with porcelain veneering. Since the metallic portion of the restoration could be hidden with the porcelain veneering, the yellow color was no longer important and the palladium content of the gold alloys could be increased, making them generally white in color. This trend is so strong that in due course gold will no longer be found in dental alloys. Currently, more than half of the dental casting alloys are both gold and silver free.

2.3 Other Applications

Jewelry and decorative applications for gold are holding their own in spite of the price rise of gold since 1968 when the US government put gold in the US on the free market, while dental applications of gold alloys are on the wane. Industrial uses, however, seem to be on the increase, but with considerable pressure on technology to make gold go as far as it can. Thus lower gold content alloys are continually emerging to substitute for the high-content gold alloys, thinner coatings are used more and more, and bulk deposits are becoming increasingly miniaturized.

In instrument applications (resistance and potentiometer wires) Au–Pd and Au–Fe–Pd alloys are used with iron as high as 9%. Some Au–Pd, Au–Co, and Au–Fe alloys are used for thermocouple leads. Gold alloys are still used to a considerable extent for electrical contacts, where cladding with gold alloys is important for conserving gold. The field of microelectronics is growing and so is the use of gold deposited primarily by sputtering for this application. Gold is used to some extent in semiconductor devices for doping of silicon and in MOSFET devices for use as infrared detectors.

In engineering applications, such as for diffusion seals, for joining tungsten, and for electrical components such as waveguides and electron tubes, gold brazing alloys sometimes compete successfully with base metal brazing alloys. Au–Pt–Rh alloys are used for spinnerets. The optical properties of gold are sufficiently unique (low reflectivity in the ultraviolet range, high reflectivity of the infrareds), to render them useful in some reflector applications. Solar-energy collectors utilize the property of thin gold films to be selectively transparent to solar wavelengths, while retaining the low emissivity at longer wavelengths. An interesting space application is based on gold's capability, when applied as a thin coating on windows, to protect space capsule interiors from short-wave solar radiation. Other special uses of various gold alloys are gold-coated glasses for buildings and transport vehicles, electrically heated transparent gold coatings on glass, gold blacks as coatings for thermocouples and radiometers, spark plug electrodes, gold-plated condenser surfaces for dropwise steam condensation, thermal fuses, gold coatings for diffusion barriers and for fuel cell construction.

See also: Gold Resources; Precious Metals Production

Bibliography

Burke J 1978 *Connections.* Little, Brown and Co., Boston, Massachusetts, pp. 15–16

Lucas J M 1980 *Bureau of Mines Minerals Yearbook.* US Bureau of Mines, Washington, DC, p. 21

Metals Handbook, 9th edn., Vol. 2, 1979. American Society for Metals, Metals Park, Ohio, pp. 737–38

Rapson W S 1978 *Gold Usage.* Academic Press, London, pp. 111–76

Wise E M 1964 *Gold.* Van Nostrand, Princeton, New Jersey, pp. 1–23

J. P. Nielsen

Gold Resources

Gold is a soft yellow metal with a melting point of 1063.0 °C. Its discovery and first use by people are shrouded in the mists of antiquity, but by 1000 BC it had attained wide usage as a metal for currency. Since that time gold has played a significant role in the history of many individuals and nations. Today the importance of gold continues undiminished, for it not only provides a valuable standard against which various currencies can be measured, and a metal of superb beauty for decorative purposes, but its unique properties of malleability, conductivity and resistance to corrosion are finding increasing utilization in modern technology. These factors, combined with worldwide inflationary pressures, have led to sharp increases in the price of gold in recent years.

1. Crustal Distribution and Geochemistry

The average gold content of the earth's upper lithosphere is approximately 5 ppb. Rocks of basaltic composition contain an average 7 ppb, those of andesitic composition 5 ppb, and those of rhyolitic composition 3 ppb (Boyle 1979).

Sedimentary rocks understandably exhibit a broader range of gold content, but on average conglomerates and sandstones contain 30 ppb, shales 4 ppb, and limestones 3 ppb. However, much higher values, up to 2 ppm, are found in certain graphitic and sulfidic schists, iron-rich chemical sediments, and some sandstones and conglomerates. Native gold is extremely insoluble in virtually all surficial environments, and hence occurs in only very small amounts in natural fresh waters (~0.03 ppb) and seawater (~0.012 ppb) (Boyle 1979).

The concentration of gold to levels at which it can be economically mined takes place by either mechanical or chemical means, but even in the former case a preconcentration of gold into particulate form by chemical means is required. For this reason considerable attention has been focused on gold solubility in aqueous solutions. Gold occurs mainly in two ionic forms, Au^+ (aurous) and Au^{3+} (auric), and can form a series of complex ions. The most important of these in terms of transport of gold in natural environments appears to be $AuCl_2^-$ (Helgeson and Garrels 1968), but $AuCl_4^-$, $Au(OH)_4^-$, AuS and $Au(HS)_2^-$ may also be significant. These complex ions dominate the aqueous chemistry of gold because Au^+ and Au^{3+} are unstable in water due to their high oxidation potentials. On the basis of natural geochemical associations it seems probable that thioarsenito and thioantimonito complexes, such as $[Au(AsS_3)]^{2-}$ and $[Au(SbS_3)]^{2-}$ are also locally important in gold transport.

The atomic radius of gold (1.44 Å) is identical to that of silver, and as a consequence native gold inevitably contains at least a few percent silver, and gold–silver alloys (electrum) are not uncommon. Gold also forms alloys with copper and platinoid metals. The only other important group of gold minerals are the gold tellurides (e.g., calaverite, $AuTe_2$) and a series of gold–silver tellurides (e.g., petzite, Ag_3AuTe_2). Finally, the mineral aurostibite ($AuSb_2$) has been reported from a number of gold deposits that contain antimony.

2. Principal Types of Gold Deposits

It is estimated that to the present time approximately 3 billion ounces of gold have been extracted from the earth's crust. A small percentage of that amount has come as a by-product of the mining of other metalliferous ores, especially of base metals and nickel. The bulk of the remainder has come from placer deposits, of both modern and fossil type. The word placer means pleasure in Spanish, an allusion to the ease of mining the unconsolidated auriferous materials that constitute modern placer deposits.

The development of gold placers requires the presence of particulate gold in veins or disseminations in preexisting rocks, followed by a period of deep weathering and subsequent concentration of the free gold by some agency, generally running water. Modern placer deposits occur in many parts of the world, but most are now mined out. Those in California yielded about 42 million ounces, those in the Lena–Amur area, USSR 40 million Troy ounces, and those in Colombia about 32 million ounces of gold. The rich gold placers along the Klondike in the Yukon yielded 9 million ounces of gold. As noted by Henley and Adams (1979), these giant placers and many others appear to be related to young uplifts around the Pacific rim.

2.1 Fossil Placer Gold Deposits

Quartz pebble conglomerates of lower Proterozoic age are the most important sources of gold in the world. Auriferous ancient conglomerates of this age are known in South Africa, Ghana and Brazil. Although of similar type, the deposits of Jacobina in Brazil and Tarkwa in Ghana are dwarfed by the gold-rich conglomerates of the Witwatersrand Basin in South Africa. The Witwatersrand Supergroup actually was deposited 2.8–2.5 billion years ago, but is of typical lower Proterozoic style in terms of its clastic sediments. Altogether 1400 m of clastic sediments and volcanics are represented, but virtually all the gold mineralization is associated with six large fluvial fan systems, the sites of which were controlled by structural domes to the north and west of the basin (Pretorius 1981a). Maximum gold concentrations are in medium–small pebble conglomerates in the mid-fan facies, but gold values also occur in fanhead and fanbase facies. Such conditions prevailed a number of times during basin-filling as gold-bearing "reefs" or blankets occur at a number of stratigraphic levels. Recent work has shown that algal mats along the fanbase facies were important traps for the finest gold (and uranium) particles, and possibly some gold in solution.

2.2 Other Types of Gold Deposits

The large majority of the remaining deposits are either of Archean or Tertiary age. Those of Archean age (≥2.6 billion years) can be divided into two broad types: Homestake-type deposits associated with iron-rich chemical sediments (Sawkins and Rye 1974), and hydrothermal vein-type deposits.

Homestake-type gold deposits exhibit a close association with sulfide bearing, cherty, carbonate iron formation and exhibit a strong stratigraphic control. Prime examples, in addition to the Homestake Mine in South Dakota, include Morro Velho in Brazil, the Chesterville mine in Canada, the Kolar goldfields in India, and the Copperhead mine in west Australia.

An important characteristic of these deposits is the continuity of the auriferous horizons. Morro Velho and the Kolar goldfields, for example, have been exploited to depths of over 3000 m. It appears that the gold was deposited by hotspring activity at the time of chemical sedimentation, although in most examples folding and metamorphism may have caused some redistribution of the gold values, especially into hinge zones of fold structures.

The Archean hydrothermal vein-type deposits are widespread in greenstone belts, and are the major source of Archean gold (Colvine et al. 1984). Some appear to have been formed prior to metamorphism, whereas many others formed during the metamorphism of these belts. Typical examples include the Campbell Red Lake, Giant Yellowknife, East

Malartic and Pamour mines in Canada, and those at Kalgoorlie in western Australia.

Hydrothermal gold deposits of Tertiary age can also be divided into two distinct types: the Bonanza gold–silver vein deposits, and disseminated gold deposits of Carlin-type. Bonanza-type gold–silver vein deposits are widespread in the western USA (e.g., Comstock Lode) and notable examples occur in Mexico (Tayoltita), the Philippines (Baguio) and Fiji (Emperor Mine). Recent research on these deposits has demonstrated that the ore fluids contained a major meteoric water component, but it is not yet clear whether the precious metals are of magmatic origin or leached from the surrounding, typically volcanic, rocks.

Carlin-type disseminated gold deposits are mainly known in Nevada, USA, and appear to have formed in the deeper portions of meteoric water-dominated, hotspring systems (Radtke et al. 1980). Mineralization is associated with replacement of carbonate units and the ores are characterized by anomalous amounts of As, Sb, Hg and Ti in addition to gold. A number of disseminated deposits of this type have been discovered in Nevada and exploration for additional examples is active in other areas of similar geologic setting.

Finally, mention should be made of a series of significant vein deposits of Paleozoic or Mesozoic age that occur largely within siltstones and shales or their metamorphosed equivalents. These include the Mother Lode, California, the Muruntau Mine, USSR, and the Ballarat and Bendigo goldfields of Victoria, Australia. Satisfactory genetic models for deposits such as these remain to be worked out, but in some instances hydrothermal derivation of the gold from the pelitic host-rocks either during metamorphism, or at the time of original sedimentation seems most likely.

3. Production and Reserves

Total world gold production in 1977 was 47 million ounces. As can be seen from Table 1, South Africa produced slightly over one half of this amount. Close to 60% of the remainder comes from the USSR.

South Africa has dominated world production of gold for many decades, and has produced about 1 billion ounces of the yellow metal. Given maintenance of the current gold price, there should still be tonnages of gold ore that will allow South Africa to produce an additional 500 million ounces or more in the future.

After South Africa and the USSR, Canada and the USA are currently the most important gold producers, but their combined output only totals ~6% of total world supply. Other significant producers include Brazil (2%), Papua New Guinea (1.5%), Philippines (1.5%), and Australia (1.4%). It is noteworthy that virtually all production from Papua New Guinea and most of that from the Philippines comes from gold-rich porphyry copper deposits.

Table 1
World gold production in 1979 (after Pretorius 1981b)

Country	Production (10^6 oz)	(%)
South Africa	24.83	52.57
USSR	13.29	28.16
Canada	1.73	3.71
USA	1.00	2.15
Brazil	0.92	1.99
Papua New Guinea	0.70	1.51
Philippines	0.67	1.47
Australia	0.65	1.42
Zimbabwe	0.42	1.09
Ghana	0.41	0.90
Dominican Republic	0.39	0.86
Colombia	0.35	0.79
Japan	0.25	0.56
Mexico	0.19	0.45
Peru	0.17	0.39
India	0.09	0.24
Nicaragua	0.07	0.18
Zaire	0.05	0.15

The great bulk of Canada's 1979 production (1.7 million ounces) came from mines in the Archean greenstone belts of the Superior Province. It is estimated that Canadian lode gold reserves are slightly in excess of 30 million tons with an average grade ot close to 0.2 oz/ton (Kavanagh 1980). In 1979 US gold production was 1 million ounces. Of that about 25% came from the Homestake Mine in South Dakota and almost as much from the Bingham porphyry copper deposit in Utah. A significant fraction of the remainder came from disseminated deposits of Carlin-type in Nevada.

4. Future Production and Global Resources

Although it appears that future production of gold from Canada, the USA, Brazil, Australia and the southwest Pacific will be enhanced, a downward trend in total gold production in the western world is predicted by Pretorius (1981b). South African production is expected to decline slightly through the 1980s and then decline to about 50% of its present value by the year 2000. By that time USSR production is expected to exceed that of South Africa, and Pretorius predicts that within the first quarter of the twenty-first century, output from communist countries will exceed that of the western world.

These predictions do not, of course, allow for major new discoveries in the western world, but as Pretorius points out, the balance cannot be redressed by enhanced production from deposits of the type mined in Canada and the USA. Definitely planned new mines in Canada and the USA will add over 100 million tons to reserves, but the average grade of these in the USA is only 0.08 oz/ton. Kavanagh (1980) estimated that by 1985 Canadian gold production should reach 2.5 million ounces, won from ore with an average grade of 0.2 oz/ton.

The main problem with future gold supply, especially in the western world is the indicated fall-off of South African gold production during the 1990s. If this situation is to be averted it is important that another major early Proterozoic fossil placer basin similar to the Witwatersrand be discovered. It is unlikely in the extreme that increased gold production from porphyry or Carlin-type deposits and from conventional vein deposits could make up for a serious decrease in South African gold production.

Global resources of gold, assuming an average price of $600 per ounce over the next two decades, are estimated at 2118 million ounces (Pretorius 1981b). This will represent a distinct shortfall with respect to anticipated demand. If energy and labor costs continue to rise at a rate of about 10% per year and world demand for gold remains strong, a major increase in the price of gold within the next decade appears likely.

See also: Gold and Gold Alloys; Precious Metals Production; Mineral Resources: An Overview

Bibliography

Boyle R W 1979 *The Geochemistry of Gold and its Deposits*, Bulletin 280. Geological Survey of Canada, Ottawa

Colvine A C et al. 1984 *An Integrated Model for the Origin of Archean Lode Gold Deposits*, Open File Report 5524. Ontario Geological Survey, Ontario

Helgeson H C, Garrels R M 1968 Hydrothermal transport and deposition of gold. *Econ. Geol.* 63: 622–35

Henley R W, Adams J 1979 On the evolution of giant gold placers. *Inst. Mining and Metall. Trans. B* 88: B41–B50

Kavanagh P M 1980 The world's gold reserves and production capacity. *Geoscience Canada* 7: 73–81

Pretorius D A 1981a *Gold and Uranium in Quartz-Pebble Conglomerates*, Information Circular 151. University of the Witwatersrand, Johannesburg

Pretorius D A 1981b *Gold, Geld, Gilt: Future Supply and Demand*, Information Circular 152. University of the Witwatersrand, Johannesburg

Radtke A S, Rye R O, Dickson F W 1980 Geology and stable isotope studies of the Carlin gold deposit, Nevada. *Econ. Geol.* 75: 641–72

Sawkins F J, Rye D M 1974 Relationship of Homestake-type gold deposits to iron-rich Precambrian sedimentary rocks. *Inst. Mining and Metall. Trans. B* 83(810): B56–B59

F. J. Sawkins

Grain Boundary Strengthening

The understanding and control of grain boundary properties offer one of the most complex and challenging problems facing materials science today, and although significant progress is being made in studying and modelling them, many of the engineering practices and advances still lie beyond adequate theoretical interpretations.

Grain boundaries have advantages and drawbacks in terms of material strength. On the one hand, they offer one of the simplest and cheapest ways to strengthen a metal or alloy that is known. This is true at low homologous temperatures. On the other hand, they can offer paths of easy failure in many materials at elevated temperatures, due to creep or impurity effects. Much time and effort has been expended on mitigating these effects, but even so, the exceptionally costly expedient of single crystal production has been found advantageous in certain instances.

1. Intrinsic Grain Boundary Strengthening

In 1942, L Bragg proposed an interdependence between strength and grain size based on a d^{-1} relationship (where d is the grain size). From subsequent work it has been realized that the exponent can vary between -1 and $-\frac{1}{3}$ but is usually around $-\frac{1}{2}$. This is found to apply to yield, flow and ultimate tensile strengths in both metals and alloys, assuming that all other forms of strengthening (e.g., solid solution, precipitation) are kept constant, and is expressed in the well-known Hall–Petch relationship:

$$\sigma = \sigma_0 + kd^{-1/2} \qquad (1)$$

where σ is the yield or flow stress and σ_0 and k are material constants. Obviously, the effects of other forms of strengthening are reflected in the magnitude of σ_0. Figure 1 replots the data from Hall's original paper (Hall 1951) for mild steel specimens. The lowest datum is that obtained from a single-crystal sample. Not only does this plot show the relationship to be well obeyed, but it illustrates the considerable strengthening that can be achieved by grain refinement.

Various theories have been advanced to explain this relationship; most are based on dislocation theory. One, for example, envisages that yield occurs by the operation of grain boundary dislocation sources due to stress concentrations at the head of pileups in neighboring grains. This stress is proportional to the half power of the pileup length, and this introduces the correct dependence. For a review of such theories, see the articles referenced by Hirth (1972).

In practical terms, grain size strengthening provides a major source of the advances achieved in the strength of many mild steels since World War II; microalloying and controlled rolling have successfully minimized grain growth and led to a doubling of yield strengths in materials with grain sizes ~10 μm. Similar grain sizes are also found in the strongest

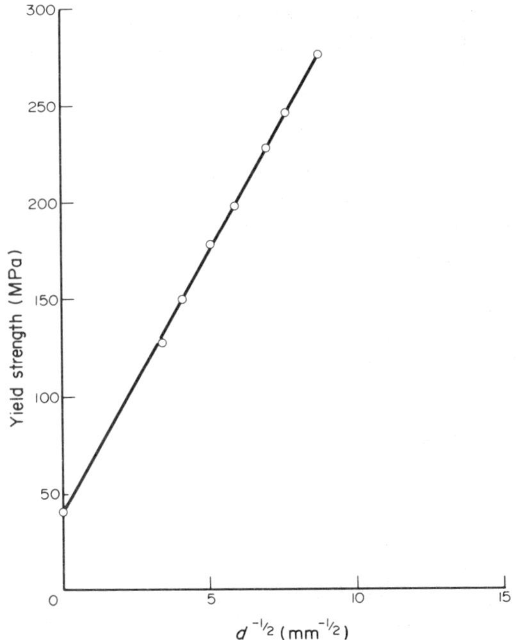

Figure 1
Variation in yield strength of mild steel with grain size, showing the Hall–Petch $d^{-1/2}$ relationship (based on Hall 1951)

nickel-based alloys currently produced by a powder metallurgy route for aircraft engine disk applications.

2. Extrinsic Grain Boundary Strengthening

At high temperatures ($>0.4\ T_m$), metals and alloys can fail intergranularly in creep either by the production and interlinkage of "round" grain boundary cavities (r-type cracking) or by the opening of wedge cracks at grain boundary triple points (w-type cracking). The former is favored by low stresses and high temperatures, the latter by high stresses and lower temperatures. A good example of w-type cracking is shown in Fig. 3a. However, these processes are not independent and may occur together, and indeed many theories have linked the occurrence of grain boundary sliding, which is certainly responsible for w cracking, to the production of r cracks. The deposition of certain particles can markedly slow or inhibit grain boundary sliding and suppress w-type cracking. In some cases these precipitates also reduce r-type cavities, but in other cases they can serve as nucleation sites for enhanced cavitation. This has led to the proposal to use different heat treatments for some engineering alloys, depending on their service conditions.

One area in which great success has been achieved in structural materials for high-temperature applications is in limiting grain boundary sliding by producing serrated boundaries by controlled precipitation. Such boundaries resist sliding by the large local accommodations needed and also, perhaps, by the far greater diffusive paths needed for the inward penetration of harmful species. This approach has been adopted with great success in austenitic stainless steel and a wide range of nickel-base superalloys. In the steels, $M_{23}C_6$ precipitates on the boundaries produce the serrations, while in the nickel-base superalloys, slow cooling through the γ' solvus is required to produce grain boundary γ' precipitates which occasion the serrated boundaries. Figures 2a and 2b show examples of the smooth and serrated boundaries formed in Nimonic 115 by quenching and slow cooling (0.5 K min^{-1}), respectively. In this manner, striking increases in stress rupture lifetimes have been attained, sometimes in the region of two orders of magnitude under identical conditions, and

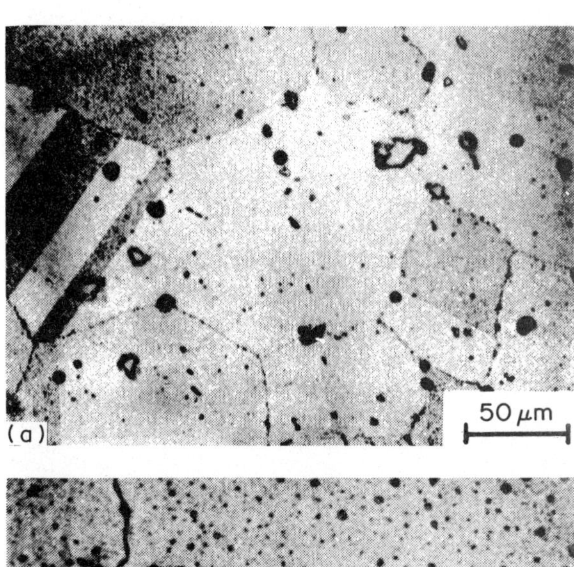

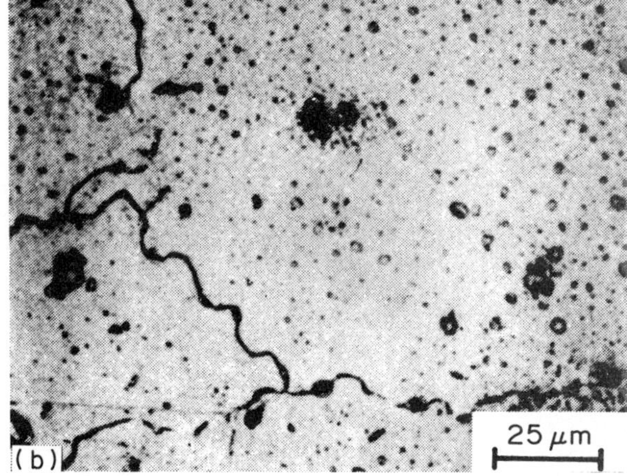

Figure 2
Example of the production of serrated grain boundaries in a nickel-base superalloy, Nimonic 115: (a) after quenching; (b) after slow cooling (0.5 K min^{-1}) (after Koul and Gessinger 1983)

slow cooling through the γ' solvus is the specified heat treatment for many nickel-base superalloys.

Exposure of many metals and alloys in a critical temperature range can lead to severe embrittlement due to segregation of certain detrimental elements to grain boundaries. Principal among these is sulfur, but phosphorus, lead, antimony and many others can also be deleterious. This segregation tends to occur only in a temperature range where the matrix solubility is exceeded or where some precipitate is dissolved, and also where the kinetics allow diffusion to grain boundaries.

Such segregation may be prevented by small, typically 0.01–1.0 wt%, chemical additions. These elements may be effective in preventing grain boundary segregation either by competing for grain boundary sites with the embrittling species or by combining with this species to produce intragranular precipitates. An excellent example of this strengthening is afforded by pure nickel, which was regarded as a brittle metal due to sulfur segregation to grain boundaries on cooling from the melt. This problem was solved about a century ago by small additions of either magnesium or manganese, which combine with the sulfur to form stable sulfides. This same practice is still followed today for all but the purest grades of nickel.

Sulfur is also thought to be one of the elements responsible for the temper embrittlement of steels which occurs above 600 °C. Here the situation is very complex, involving many segregating species, and no unambiguous mechanism has evolved; phosphorus appears beneficial in some situations and harmful in others. However, the reduction of S, P, As, Sb, and Sn certainly tends to improve resistance to temper embrittlement.

In nickel-base superalloys, small additions of B, Zr and/or Hf are routinely made, to improve grain boundary strength (this practice is so standard that these elements are referred to as grain boundary strengtheners). It is thought that they remove sulfur from the boundaries either by sulfide formation or by site competition, but again the situation is very complex and the effects of each element appear maximized in a very narrow composition range.

Oxygen embrittlement is apparently one of the most pervasive mechanisms of grain boundary weakening and failure in metals and alloys. It is characterized by a marked reduction in ductility coupled with intergranular failures in materials which have high grain boundary oxygen contents due either to processing route or to the diffusion of oxygen down grain boundaries on high-temperature exposure. Figure 3 shows the drastic effect of such oxygen grain boundary penetration on the tensile properties of an iron-base superalloy, IN903A. Similar effects are seen in some pure metals as well as in nickel, copper and cobalt-base alloys.

Considerable progress has been made in strengthening grain boundaries against this form of degradation in high-temperature service. Such strengthening may take one of several forms:

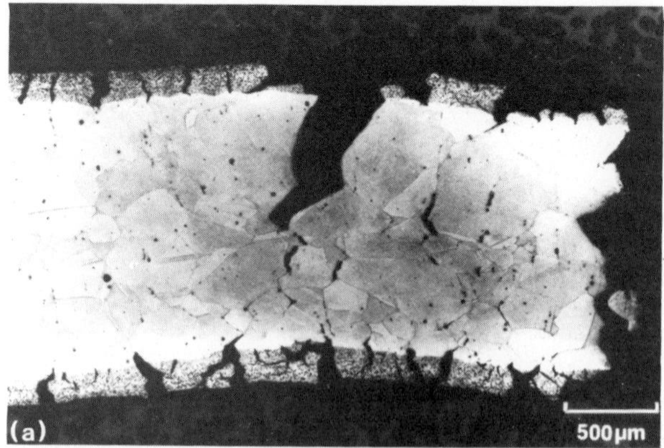

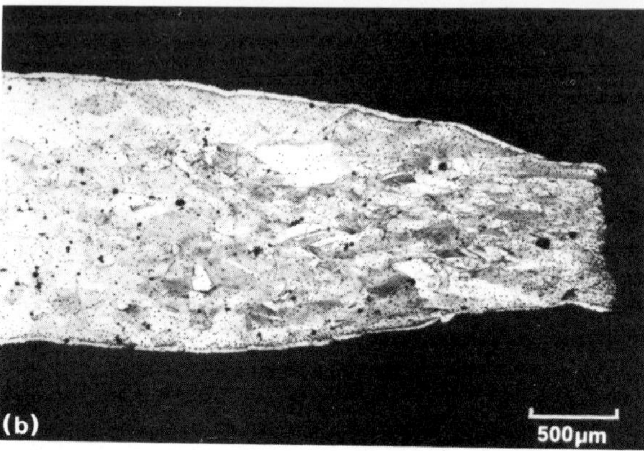

Figure 3
Example of grain boundary oxygen embrittlement in the iron-base alloy IN903A, revealed in tensile tests at 800 °C: (a) intergranular failure with w-cracking in oxygen-embrittled material; (b) ductile, transgranular failure in undamaged material (Bricknell and Woodford 1981)

(a) application of an external coating of a stable oxide former to reduce the oxygen activity to very low levels in the alloy beneath, for example, aluminide or NiCoCrAlY coatings;

(b) enhancement of the alloy's own Al or Cr content to produce its own stable oxide; or

(c) addition of known grain boundary segregants such as B, Zr and Hf. Such additions have prevented intergranular failures with stress rupture lifetimes raised by several orders of magnitude. Studies on model systems indicate that these additions are effective by occupying the sites used by oxygen for rapid grain boundary diffusion.

The most complete form of high-temperature grain boundary strengthening can be achieved by either aligning all the grain boundaries parallel to the applied stress (as in directionally solidified alloys) or, in the extreme case, by removing all grain boundaries from the material (as in single-crystal alloys). This final approach has been found cost-effective, despite the high processing charges, in the case of certain blade alloys for aircraft engine applications. Whether grain boundaries can be strengthened by means based on theory or experiment to match the benefits obtained by their removal in this very extreme application remains a challenge for the future.

See also: Precipitation Strengthening: General Considerations; Solid-Solution Strengthening; Dispersion Strengthening; Directionally Solidified Eutectics: Strengthening

Bibliography

Briant C L et al. (eds.) 1983 *Embrittlement of Engineering Alloys*. Academic Press, New York

Bricknell R H, Woodford D A 1981 Grain boundary embrittlement of the iron-base superalloy IN903A. *Metall. Trans.* 12A: 1673–80

Burke J J, Weiss V (eds.) *Ultrafine Grain Metals*. Syracuse University Press, New York

Hall E O 1951 The deformation and ageing of mild steel III. Discussion of results. *Proc. Phys. Soc., London, Sec. B* 64: 747–53

Hirth J P 1972 The influence of grain boundaries on mechanical properties. *Metall. Trans.* 3: 3047–67

Institution of Metallurgists 1976 Grain boundaries. *Proc. Spring Residential Conference*. Institution of Metallurgists, London

Koul A K, Gessinger G H 1983 On the mechanism of serrated grain boundary formation in Ni-based superalloys. *Acta Metall.* 31: 1061–69

Perry A J 1974 Cavitation in creep. *J. Mat. Sci.* 9: 1016

Stevens R N 1966 Grain-boundary sliding in metals. *Met. Rev.* 11: 129–42

Walter J L et al. (eds.) 1974 *Grain Boundaries in Engineering Materials*. Claitor's, Baton Rouge, Louisiana

Woodford D A, Bricknell R H 1983 *Environmental Embrittlement of High Temperature Alloys by Oxygen*, Treatise on Materials Science and Technology Vol. 25. Academic Press, New York, p. 157

R. H. Bricknell

Grain Boundary Structure

Grain boundaries are the interfaces between neighboring crystals; they play an important role in controlling many physical, mechanical and electrical properties of polycrystalline solids. At present the structure of grain boundaries is not as well understood as those of crystals and their surfaces. It may well be that not only the coarse features of the boundary (e.g., dislocations and ledges) but also some aspects of the detailed atomic structure of the boundary influence the strength of a solid. Only recently have experimental tools become available that can provide information on the fine-scale structure of a grain boundary. This article is concerned with structural models for grain boundaries.

1. Models of Grain Boundary Structure

The earliest model of the structure of grain boundaries involved neighboring grains in a polycrystalline solid being joined together by an amorphous cement layer. This model was then modified by the suggestion that boundaries consist of a transition lattice containing a regular arrangement of atoms. Grain boundaries were subsequently described in terms of an array of dislocation lines, so that, for example, a tilt boundary is described in terms of a periodic array of edge dislocations. This model was strongly supported by, for example, etch-pit studies using optical microscopy, which confirmed that small-angle tilt boundaries consist of periodic arrays of dislocations. The concept of the dislocation array was then combined with the coincidence-site-lattice model to describe the structure of large-angle grain boundaries.

1.1 Small-Angle Boundaries

For a tilt boundary, the axis describing the misorientation lies in the plane of the boundary. Figure 1a shows two simple cubic crystals which have been misoriented with respect to each other by a rotation of θ about a $\langle 001 \rangle$ axis which is normal to the diagram.

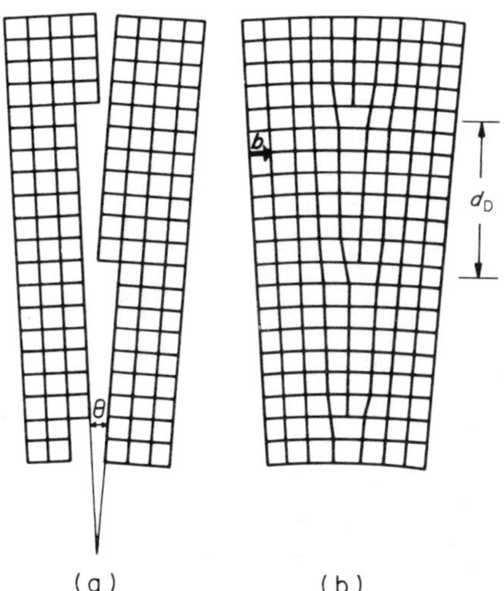

Figure 1
The formation of a small-angle tilt grain boundary: (a) unrelaxed configuration, (b) relaxed configuration (after Read 1953 © McGraw Hill, New York. Reproduced with permission)

The two crystals have been cut along a vertical plane, the boundary plane. It is seen that on an atomic scale the surface of the boundary-plane-to-be in each crystal is stepped in a periodic fashion. Figure 1b shows the result of bringing the two misoriented single crystals together and allowing the atoms in the vicinity of the interface to relax and assume low-energy positions. It is then seen that the boundary consists of a periodic array of edge dislocations with Burgers vector $\boldsymbol{b}$ and spacing $d_D = |\boldsymbol{b}|/2 \sin \theta/2$. For small angles, this expression becomes $d_D \cong |\boldsymbol{b}|/\theta$. Between the dislocations are regions of normal cubic material. Because of the periodic nature of the boundary structure, the strain field associated with the boundary falls off exponentially along the direction normal to the interface. It is expected that at a distance d_D from the boundary plane, the atomic structure is that of the undistorted perfect crystal (Cottrell 1953). In other words, the boundary region is a thin layer with thickness of $2d_D$ or less, inside of which is distorted material.

In a similar fashion, the structure of a small-angle twist boundary can be obtained. In this case, the axis describing the misorientation lies normal to the plane of the boundary. Figure 2a shows two simple cubic crystals which have been misoriented with respect to each other by a rotation of θ about a ⟨001⟩ axis normal to the diagram. (This boundary is termed a ⟨001⟩ twist boundary.) In the unrelaxed configuration it is possible to pick out regions of good cubic stacking (or good matching) and regions of bad matching. If the atoms at the interface are allowed to relax, then the configuration in Fig. 2b is obtained. The regions of good match expand out, and all the mismatch is squeezed into narrow bands, which consist of a square array of screw dislocations with Burgers vector $\boldsymbol{b}$ and with a spacing d_D given by the same expression obtained for tilt boundaries. Figure 3 shows a transmission electron micrograph of a small-angle ⟨001⟩ twist boundary. A square array of screw dislocations is present, in agreement with the prediction of Fig. 2b.

1.2 Large-Angle Boundaries

When the models discussed in the previous section on small-angle boundaries are extended to the large-angle regime, the expected dislocation spacing (for $\theta = 30°$) is $\sim 2|\boldsymbol{b}|$, which for most metals is less than 0.6 nm. It seems unrealistic to consider discrete dislocations to be present at such small spacings, since their cores must be overlapping. A physically appealing approach to the structure of large-angle boundaries is obtained through the use of the concept of the coincidence site lattice (CSL) (Bollmann 1970).

The CSL and its relation to grain boundary structure can be understood using the small-angle ⟨001⟩ twist boundary just considered as an example; extension to large-angle boundaries is relatively simple. The CSL for the ⟨001⟩ twist boundary is obtained by allowing the two misoriented crystal lattices adjoining the boundary to interpenetrate and translate so that lattice points of each crystal coincide. The space lattice made up of the coincident lattice points defines the CSL, and the number of such lattice points

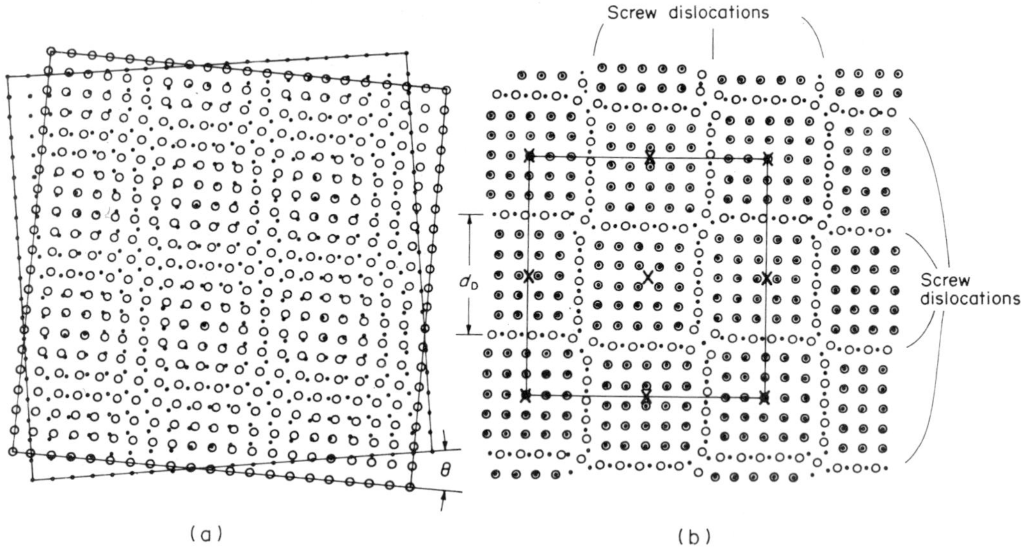

Figure 2
(a) Unrelaxed and (b) relaxed atomic configurations in the planes just above and below the boundary plane in a small-angle ⟨001⟩ twist grain boundary, illustrating formation of the screw-dislocation network. The periodicity is that of the coincidence-site-lattice model, the unit cell of which is indicated by the square in (b)

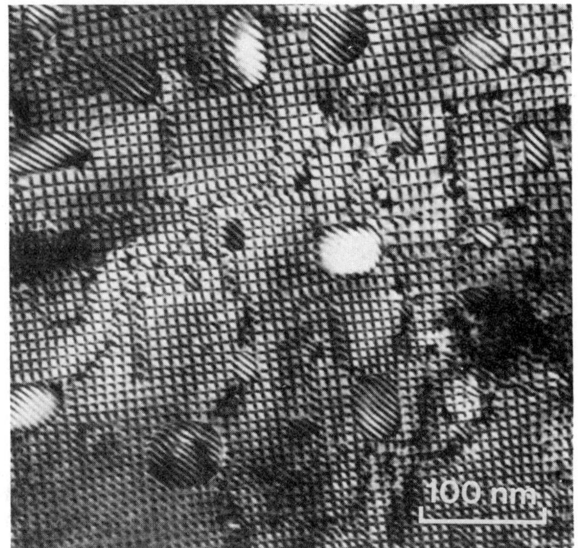

Figure 3
Transmission electron micrograph showing a small-angle ($\theta = 2°$) ⟨001⟩ twist grain boundary in gold

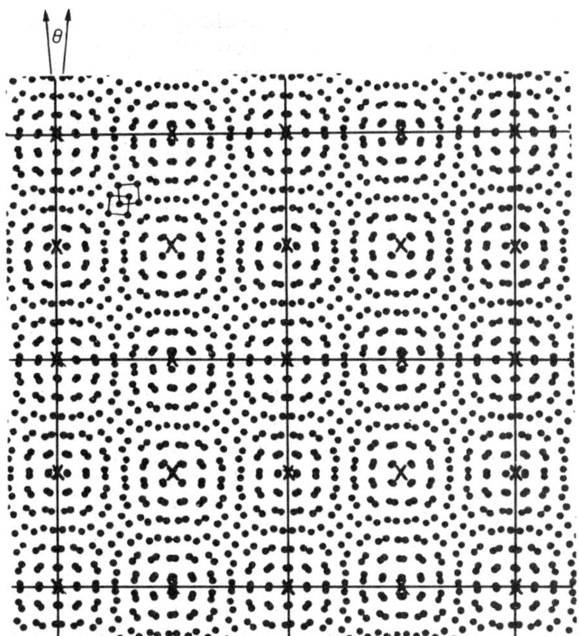

Figure 4
The unrelaxed atomic configuration in the planes just above and below the boundary plane in a small-angle ⟨001⟩ twist grain boundary with $\theta = 9.5°$. The periodicity is that of the coincidence site lattice which is indicated by the square net. The regions of good matching are marked by crosses, which are O-lattice points

expressed as a fraction of the lattice points in one crystal is defined as Σ^{-1}. A grain boundary can then be constructed by passing a plane through the interpenetrating crystals and removing all the lattice sites from one crystal on one side of the boundary plane and from the second crystal on the other side of the boundary plane. The two-dimensional CSL in the boundary plane shown in Fig. 4 describes the actual periodicity of the grain boundary structure, even when the atoms move from their perfect crystal sites to take on a lower-energy configuration as in Fig. 2b.

There are additional locations, besides the CSL points, where the two crystals are in good coincidence (or matching). These locations lie on a sublattice of the CSL which is called the O lattice, and it is easily seen that the two crystals can rotate with respect to each other at the O-lattice points (marked by crosses in Figs. 2b and 4) to achieve good matching on a local scale. In the small-angle case, the O-lattice points are separated by bands of mismatch (with spacing d_D), which are screw dislocations possessing Burgers vectors of the perfect crystal (and hence are called primary dislocations). Figure 2b shows that the periodicities of the screw dislocations and of the O lattice are identical.

For large misorientations and special angles θ_c, the dimensions of the CSL unit cell can be relatively small. It is believed that boundaries with small unit cells and low Σ values (see Table 1) have a low energy per unit area. Figure 5 shows the CSL unit cell for the $\theta_c = 22.62°$ ($\Sigma = 13$) boundary in a face-centered-cubic (fcc) material. In this figure it is seen that one atom in thirteen is in good fcc stacking, and hence $\Sigma^{-1} = 1/13$, or $\Sigma = 13$.

What now can be said about the atomic structure of large-angle boundaries? There have been numerous studies involving computer modelling of large-angle grain boundaries which have tried to simulate their structure (see, for example, Vitek et al. 1980). Only recently has it been possible to combine the computer modelling approach with experimental observations on boundaries in order to determine the detailed atomic structure with some confidence. For example, in an x-ray-diffraction and computer modelling study, the atomic structure of the large-angle $\Sigma = 13$ twist boundary in gold was obtained (Bristowe and Sass 1980). Figure 6 shows the unrelaxed and relaxed

Table 1
Examples of special grain boundaries with a ⟨001⟩ misorientation axis in the cubic system

θ_c	Σ
22.62°	13
28.07°	17
36.87°	5

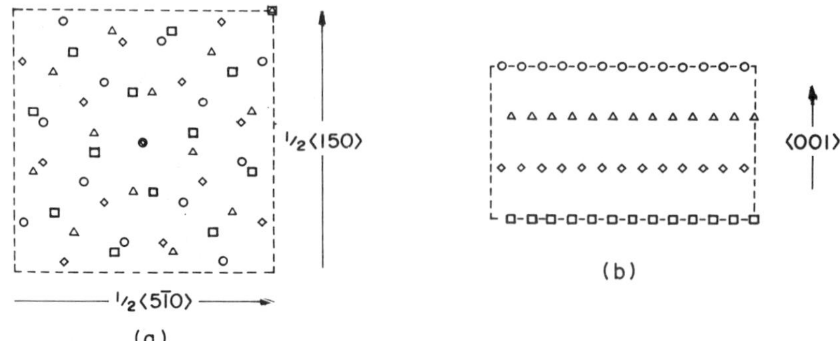

Figure 5
(a) Plan and (b) side views of the $\theta_c = 22.62°$ ($\Sigma = 13$) $\langle 001\rangle$ twist grain boundary in the unrelaxed state. Four $\langle 001\rangle$ planes are shown, each represented by a different symbol. The large dashed square outlines the unit cell of the coincidence site lattice

structures of this boundary in gold with the plane spacing normal to the interface enlarged by a factor of four. It is clear that the originally flat (001) planes in Fig. 6a have become puckered in the relaxed structure (as indicated by shaded pyramids) in Fig. 6b. The relaxed structure is also shown in Fig. 7 with the correct plane spacing normal to the boundary. It is seen that certain polyhedral configurations of atoms are present in the boundary. A severely distorted archimedean antiprism and an octahedron are indicated in Fig. 7. Polyhedral arrangements of atoms are frequently found in both tilt and twist

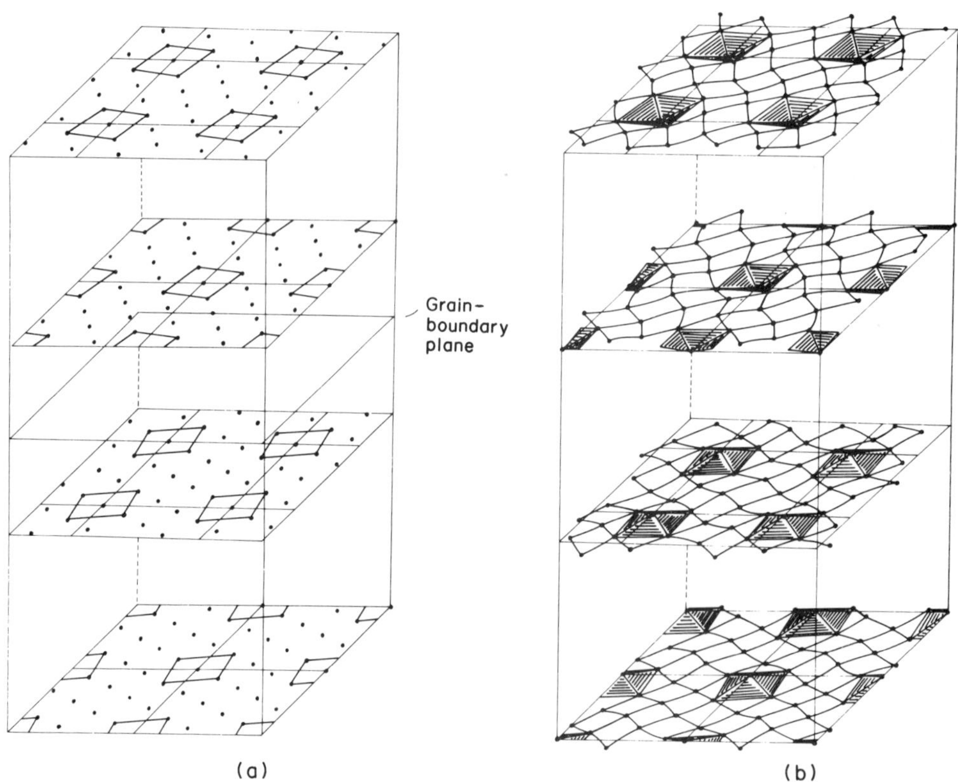

Figure 6
Perspective representation of the $\theta_c = 22.62°$ ($\Sigma = 13$) twist-grain-boundary structure in gold. Two $\langle 001\rangle$ planes above and below the boundary plane are shown. (a) and (b) show the relaxed and unrelaxed structures, respectively, with the interplanar spacing increased by a factor of four

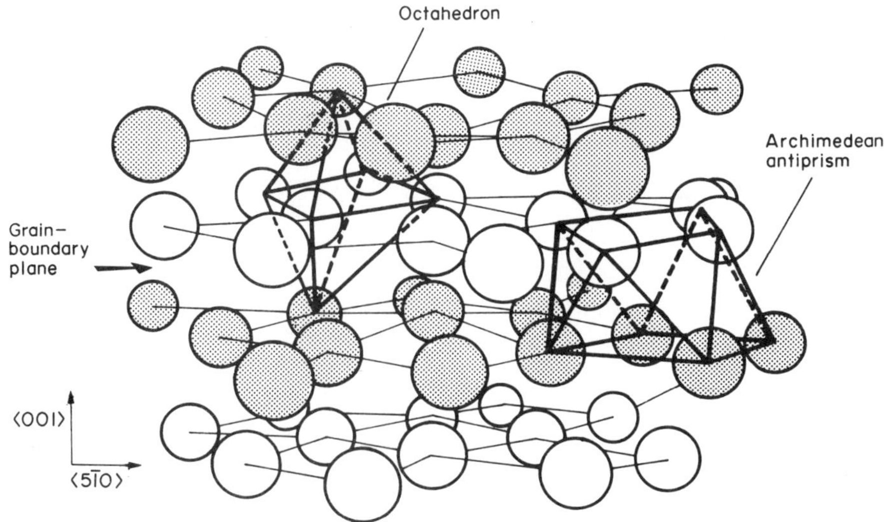

Figure 7
The $\theta_c = 22.62°$ ($\Sigma = 13$) twist-grain-boundary in gold as in Fig. 6b, but with the correct interplanar spacing

boundaries; it is appealing to consider that these polyhedra are an important component of the boundary structure.

In the discussion of the structure of small-angle boundaries, it has been shown that local regions of good matching occur which are separated by dislocations possessing Burgers vectors of the perfect crystal (see Fig. 2b). In a similar manner, it is expected that in boundaries with misorientations that deviate from the special values θ_c there will be local rotations to achieve local regions of low Σ structure, separated by grain boundary dislocations (Schober and Balluffi 1970). These dislocations have Burgers vectors which are not present in the perfect crystal and, therefore, are called "secondary" dislocations, to distinguish them from the "primary" dislocations in Figs. 1b and 2b. The spacing d_s of the secondary dislocations with Burgers vector $\boldsymbol{b}_s$ is given by the expression $d_s \cong |\boldsymbol{b}_s|/|\theta - \theta_c|$. Figure 2b can, therefore, also represent the structure of a large-angle twist boundary with the regions of good matching replaced by regions of low Σ, and the primary dislocations replaced by secondary dislocations.

The structures of tilt and twist boundaries are the easiest to visualize and have been the subject of the most research. For the more general type of boundary, where the axis describing the misorientation between the two crystals is not lying either in or normal to the boundary plane, but inclined to it, the boundary structure can be visualized as having both a tilt and a twist component.

See also: Grain Boundary Strengthening; Grain Growth and Secondary Recrystallization; Grain Size: Nondestructive Evaluation; Surfaces and Interfaces: An Overview

Bibliography

Balluffi R W (ed.) 1980 *Grain Boundary Structure and Kinetics.* American Society for Metals, Metals Park, Ohio

Bollmann W 1970 *Crystal Defects and Crystalline Interfaces.* Springer, New York

Bristowe P D, Sass S L 1980 The atomic structure of a large angle (001) twist boundary in gold determined by a joint computer modelling and x-ray diffraction study. *Acta Metall.* 28: 575–88

Cottrell A H 1953 *Dislocations and Plastic Flow in Crystals.* Oxford University Press, London, pp. 93–94

Hageye S, Nouet G, Sainfort G, Sass S L (eds.) 1982 Structure and properties of intergranular boundaries. *J. Phys. (Orsay, Fr.)* 43 (Colloque C-6)

Read W T Jr 1953 *Dislocations in Crystals.* McGraw-Hill, New York

Schober T, Balluffi R W 1970 Quantitative observation of misfit dislocation arrays in low and high angle twist grain boundaries. *Philos. Mag.* 21: 109–23

Vitek V, Sutton A P, Smith D A, Pond R C 1980 Atomistic studies of grain boundaries and grain boundary dislocations. In: Balluffi 1980, pp. 115–48

S. L. Sass

Grain Growth and Secondary Recrystallization

The control of grain size and shape in metals and alloys is important in determining their mechanical and physical properties. Although cold work is used principally to effect changes in shape of the material, and annealing is used to promote recovery of mechanical or physical properties, these processes produce changes in the grain size during primary recrystallization of deformed metals. Grain size changes can also be induced by phase transformations in polymorphous metals and alloys. However, the grain size can be further changed following primary recrystallization or allotropic transformation by the process of grain growth. The energy changes accompanying grain growth are discussed here, and attention is drawn to the important effects of grain size heterogeneity on this growth. The general features of normal grain growth are included. The grain-growth-inhibiting effects of second-phase particles are described in some detail as these have an important bearing in differentiating between normal grain growth and secondary recrystallization.

1. Grain Growth as a Discontinuous Process

The driving force for grain growth is generally recognized to derive from the reduction of grain-boundary area, and hence grain-boundary energy, that accompanies the grain growth process. The grain-boundary surface area per unit volume in an infinite solid comprising regular cubo-octahedra is approximately $1.5/R_0$, where R_0 is one-half the distance between opposite hexagonal faces of the polyhedron. Thus arrays of larger grains have lower interfacial areas and hence lower grain-boundary energies. This provides the basic driving force for grain growth processes, but it is important to realize that in a regular array of polyhedral grains not all grains can grow. The growth of certain grains implies the shrinkage or complete disappearance of other grains.

1.1 Grain-Boundary Energy

The energy of clean grain boundaries is known to vary depending upon the relative orientations of the adjoining grains and upon the orientation of the boundary with respect to the two lattice structures (see *Grain Boundary Structure*). In general, however, the majority of grain boundaries have similar energies, as the occurrence of low-energy boundaries requires special orientation relationships. Here, it is assumed that grain-boundary energies are equal. This assumption does not invalidate the conclusions drawn, but may modify quantitative aspects of the treatment.

1.2 Shape of Metal Grains

The shape and size of metal grains in a real material are not uniform or regular. Arrays of metal grains contain a variety of shapes and a distribution of sizes. The majority of grains have a shape that permits most planar sections to have 5–8 sides and that gives angles of about 120° between boundaries meeting at a grain edge. The cubo-octahedron is such a shape, but there are many polyhedra which also possess these properties. Although many of these polyhedra are not space-filling, their existence is made possible by the mixing of grain shapes, and the fact that the grain shapes do not have to show complete symmetry in a real system. Figure 1 shows a soap froth which is analogous to a spatial array of metallic grains. The shapes before and after grain growth result from the spatial distribution of grain nuclei and collision events between growing grains, whether these arise from primary recrystallization, phase transformation or the grain growth process itself. The presence of small, infilling tetrahedral grains could arise from the collision events of four uniformly spaced polyhedral grains. The presence of different grain shapes with a distribution of sizes in a real array has important consequences for the grain growth process.

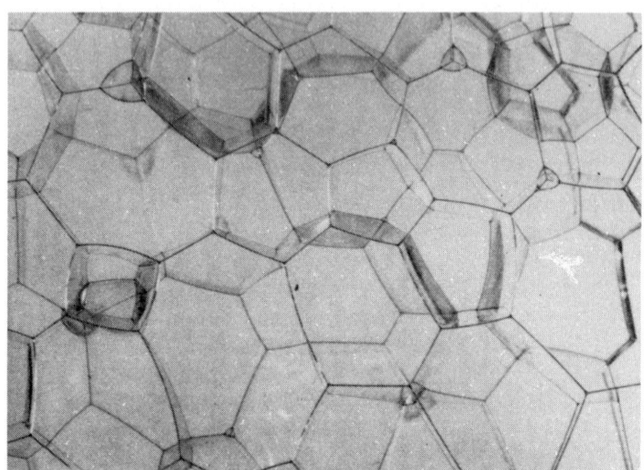

Figure 1
Soap froth: a translucent analogue of real metal grain shapes and sizes

1.3 Energy Changes Accompanying Grain Growth

When one grain grows in a matrix of grains, there are two important energy changes involved. First, there is an increase in the grain-boundary area (and hence energy) of the grain which is growing, and secondly, there is a decrease in the grain-boundary area (and energy) resulting from the disappearance of neighboring grains. The energy change ΔE per unit boundary area of the growing grain, accompanying the growth of a single grain in a matrix of cubo-octahedral grains, is given by (Gladman 1966):

$$\Delta E = \gamma S(2R^{-1} - \tfrac{3}{2}R_0^{-1}) \tag{1}$$

where γ is the grain-boundary energy per unit area, S is the boundary displacement, R is the radius of the growing grain, and R_0 is the mean radius of the neighboring grains being consumed. It is clear that the growing grains must have a distinct size advantage to produce a decrease in energy. In other words, growth of large grains produces a decrease in energy, whereas growth of small grains increases the energy of the system.

The energy changes accompanying the shrinkage

of a small tetrahedral grain with the concomitant growth of the four neighboring grains have been considered (Gladman 1980). The important conclusion from this work is that the total grain-boundary energy decreases as the small grain disappears, and the rate of decrease in energy increases as the size of the small grain decreases.

The energy changes given above were evaluated for specific grain geometries, but the findings can be generalized as follows:

(a) If a large grain begins to grow in a matrix of smaller grains there is a decrease in grain-boundary energy which enables growth to continue.

(b) If a small grain begins to grow in a matrix of larger grains there is an increase in energy; that is, the growth of small grains is energetically unfavorable.

(c) The shrinkage of small grains is energetically favorable, and the rate of decrease in energy becomes progressively greater as the small grain shrinks, so that small grains disappear.

1.4 General Features of Grain Growth

It is clear from the energetic approach considered above that grain growth in a polycrystalline specimen is very dependent upon the grain size heterogeneity. Large grains with a suitable size advantage grow, consuming their neighbors, whereas sufficiently small grains shrink. The grain growth process is in fact a discontinuous process, in the sense that the disappearance of a small grain might not produce an assemblage of grains in which further growth is favored. Also, the growth of any grain with a sufficient size advantage continues until it collides with other large growing grains; the driving force for continued growth then depends upon the conditions of heterogeneity established in the matrix of large grains. The increase in the average grain size accompanying grain growth decreases the driving force for further grain growth. The nature of grain growth can be demonstrated by using a soap froth and also by direct observation using thermionic emission microscopy.

The significance of grain size heterogeneity becomes even greater when considering grain growth inhibition and secondary recrystallization.

2. Grain Growth as a Continuous Process

2.1 Kinetics of Grain Growth

Metallurgical significance is commonly attached to the average grain size in a metallic body, and has led to continuous treatments of the change of "average" grain size. The general treatment of the kinetics of grain growth is based on the reduction in driving energy when the average grain size increases; that is, the driving force is inversely related to the grain size. The process of grain growth involves the diffusion of atoms from the shrinking grain across the grain boundary to a lattice site belonging to the growing grain. This latter process is temperature dependent, the functional relationship being of the Arrhenius type. These ideas lead to the following relationship:

$$D^2 - D_0^2 = kt \exp(-Q/RT) \qquad (2)$$

where D is the grain size after time t at temperature T, D_0 is the initial grain size, and Q is the energy differential across the boundary.

Although some data show excellent agreement with Eqn. (2), there are many instances in which the index relating grain size and time shows a marked variation with temperature. It is possible that other factors such as impurities may influence the kinetics of grain growth.

2.2 Effect on Impurities in Solid Solution

Most impurities show a tendency to segregate to grain boundaries, the extent of segregation being inversely related to the solubility of the particular impurity (see *Surfaces and Interfaces: Composition*). The impurity atoms can exert a drag effect at certain critical temperatures, depending upon whether the solutes can diffuse with the moving boundary or whether breakaway of the boundary from the impurity atoms occurs. The rate of grain-boundary movement, however, can be dramatically reduced by the presence of extremely small impurity contents (Lücke and Stüwe 1971).

3. Grain Growth Inhibition by Second-Phase Particles

It has long been established that grain growth can be inhibited by certain grain-refining additives, such as aluminum additions to steel. A feature of such additives is that they show limited solid solubility in the matrix grains, and are available in the form of relatively fine particles over the temperature ranges of interest. The mechanism of grain growth inhibition is due to the reduction in grain-boundary area that occurs when a grain boundary intersects a second-phase particle (Fig. 2). Grain-boundary migration results in a bowing of the boundary and an increase in the grain-boundary area as the boundary is pulled away from the particle, as shown in Fig. 2. According to Zener, the maximum force required to release the boundary from the particle is of the order of $\pi r \gamma$, where r is the particle radius and γ is the grain-boundary energy (Smith 1948). The pinning force is directly proportional to the radius of the particle, but since the volume of each particle increases as the cube of the radius the overall pinning force of small particles is greater than that of large particles for a given volume fraction of particles. The exact relationship between the pinning effects and particle size is

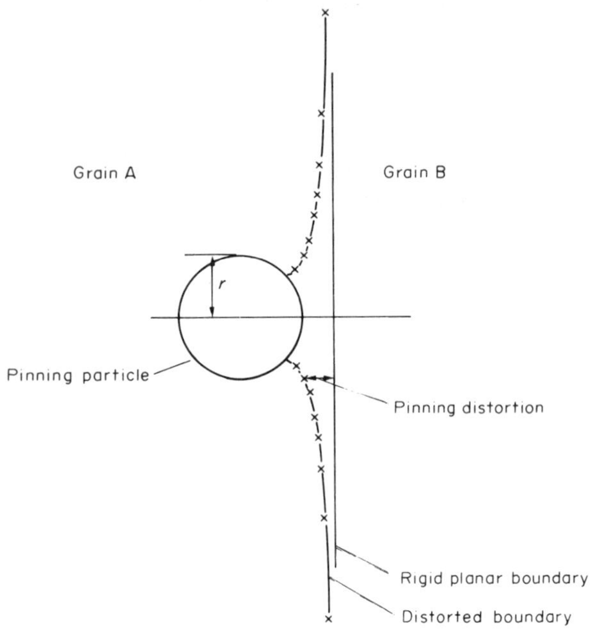

Figure 2
Particle unpinning: separation of the grain boundary from the particle increases the grain-boundary area

considered in more detail in the following section. The main point is that the pinning effects of a sufficient volume fraction of second-phase particles can inhibit the grain growth process. Ashby (1980) has drawn attention to the possibility that a moving grain boundary may drag particles along with it, but this process is extremely slow in relation to normal grain-boundary migration.

3.1 Unpinning of Grain Boundaries

When second-phase particles are present, the pinning force exerted by the particles opposes the driving force for the grain growth process. In order to determine whether the grain boundary is effectively pinned, giving rise to grain growth inhibition, or whether unpinning occurs giving rise to a grain-growth process, it is necessary to consider the combined energy changes accompanying the unpinning process (Gladman 1966, 1980).

Effectively, the driving force per unit area of grain boundary must be compared with the effective pinning force of the number of second-phase particles associated with the unit area of the growing grain boundary. It has been found that there is a critical second-phase particle size r_g above which grain growth can occur, and below which the grain growth process is totally inhibited. The critical particle size is given by

$$r_g = 6(R_0f/\pi)(\tfrac{3}{2} - 2Z^{-1})^{-1} \qquad (3)$$

where Z is a heterogeneity factor ($Z = R/R_0$), and f is the volume fraction of particles.

It is clear from Eqn. (3) that the critical second-phase particle size is dependent upon the matrix grain size, the volume fraction of particles and the grain size heterogeneity. The initial particle size increases as the matrix grain size and volume fraction of particles increases, but decreases as the grain size heterogeneity (R/R_0) increases.

Similarly, there is a critical particle radius r_s connected with the shrinkage of small grains. This is not numerically equivalent to the critical size which inhibits the growth of large grains, as the driving force for grain growth and the shrinkage of small infilling grains is different in magnitude. The critical particle size for small-grain shrinkage is given by

$$r_s = \sqrt{3}hf/(\sqrt{2} - 1) \qquad (4)$$

where h is the edge length of the shrinking tetrahedron. The critical particle size for shrinkage increases as the size of the tetrahedron increases and as the volume fraction of second-phase particles increases.

A summary of the critical particle size for these processes is given in Fig. 3. The major contribution of grain size heterogeneity (either h/R_0 or R/R_0) on the critical particle size can be clearly seen. The model for shrinking tetrahedra is valid up to $h = \sqrt{2}R_0$, whereas the model for the growth of tetrakaidecahedral grains is valid down to $R = 4R_0/3$. The shrinkage of very small grains can occur at smaller particle sizes than will permit the growth of large grains. Thus, in certain two-phase structures with $r = 1.0R_0f$, the shrinkage and disappearance of small grains may occur without any growth of the larger grains. Although the shrinkage of the small infilling grains implies growth of the adjacent grains, the growth of these is only small in proportion to

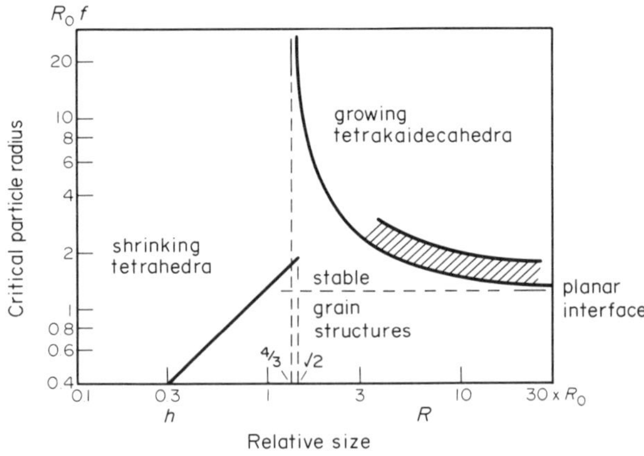

Figure 3
Grain growth inhibition by second-phase particles: the critical particle sizes separate the domains of grain growth, grain shrinkage and grain growth inhibition

their own size, and would not continue after the total annihilation of the small grains.

At a somewhat larger particle size (e.g., $r = 2R_0f$), the growth of very large grains could occur as well as the shrinkage of small grains.

3.2 Practical Implications of Grain-Boundary Unpinning

When materials containing second-phase particles are heat treated, two important effects occur which relate to grain growth inhibition and the breakdown of the inhibiting effect: second-phase particles may undergo partial or total dissolution, and secondly, such particles as remain undissolved undergo particle coalescence and growth. The solubility products of various grain-refining particles, such as aluminum nitride, vanadium nitride, niobium carbonitride and titanium carbide in austenite are well established (Gladman and Dulieu 1974), but it is equally well established that the commencement of austenite grain growth in steels can occur without substantial dissolution of the second-phase particles. Under such conditions, it is the Ostwald ripening (see *Ostwald Ripening*) which controls the onset of grain growth. The particles when freshly precipitated and small are effective in inhibiting the grain growth processes. However, with prolonged holding, the particles coalesce and grow until they exceed the critical particle size appropriate to the volume fraction of particles, the matrix grain size and the matrix heterogeneity. When the particles exceed the critical size, grain growth commences. As already pointed out, grain shrinkage of small grains may precede the commencement of grain growth but would in general merely cause some slight increase in the overall size of the fine-grained structure.

In order to produce alloys that are resistant to grain growth, the objective is to increase the temperature range or the heat-treatment time over which the particles are effective. This involves increasing the critical particle size by increasing the volume fraction of particles, or selecting particular types of particle which are resistant to coarsening. It is clear from Eqn. (3) that the volume fraction of particles required to impart a given resistance to grain coarsening increases as the matrix grain size decreases.

4. Secondary Recrystallization

Secondary recrystallization, sometimes termed abnormal or exaggerated grain growth, refers to a phenomenon in which a few grains commence to grow and achieve an extremely large size (perhaps some centimeters), while the matrix grains remain virtually unchanged until they are consumed by the secondary grains. The grain size attained by the secondary recrystallization process is much larger than that produced by the normal grain growth process at comparable times and temperatures.

4.1 Mechanism of Secondary Recrystallization

Secondary recrystallization is frequently observed as the method of grain growth when grain-controlling particles become ineffective through particle growth. The mechanism of the effect can be understood from Fig. 3. An agglomerate of grains shows a distribution of shapes and sizes. The heterogeneity factors for a particular array of grains may range from less than unity up to a value of 2 or 3. Initially all of the grains are effectively pinned by fine particles if present in sufficient quantity. As particle coarsening progresses, some shrinkage and disappearance of small infilling grains may occur, but a particular particle size will be reached at which only the grain with the largest size advantage (i.e., the largest heterogeneity factor) will begin to grow. At this stage, the remaining grains in the array are effectively pinned by the particles and thus the growing grain effectively consumes a stable fine-grained matrix. As the growing grain increases in size, the size advantage increases and the driving force for continued growth also increases. The growth of this single grain continues until the particles grow sufficiently to allow the grain with the second largest size advantage to become unpinned and begin to grow. The final secondary recrystallized grain size depends upon the time interval between the release of grains, which in turn depends upon the volume fraction of second-phase particles and the size of the matrix grains (according to Fig. 3), and also upon the kinetics of particle coarsening, which is a function of the type of particle used.

Secondary recrystallization therefore occurs when the particle size just exceeds the critical value as indicated by the shaded area in Fig. 3.

4.2 Comparison of Normal Grain Growth and Secondary Recrystallization

In the normal grain growth process, when there are no effective pinning particles, all grains with a sufficient size advantage can grow with an accompanying reduction in surface energy, as described by Eqn. (1). As these grains grow and collide, the general matrix grain size is increased and the driving force for grain growth is reduced (see Sect. 2.1). Thus, general grain growth is accompanied by an overall decrease in the driving force for grain growth.

In the case of secondary recrystallization, the growing grain consumes the unchanged fine-grained matrix and the driving force, being related to the size advantage, increases progressively as the grain grows.

The consequences of these differences in general grain growth are illustrated in Fig. 4. Here, the grain-coarsening characteristics of a grain-refined low carbon steel are shown. Secondary recrystallization, occurring at temperatures around 1050 °C, gives rise to mixed grain structures containing extremely coarse grains and fine grains. This structure is a consequence of the partial dissolution of aluminum nitride and the particle coarsening occurring at this temperature. At higher temperatures, up to 1250 °C, the dissolution process is more advanced and normal grain growth occurs. This gives rise to a more uniform grain size which is considerably smaller than that produced by secondary recrystallization.

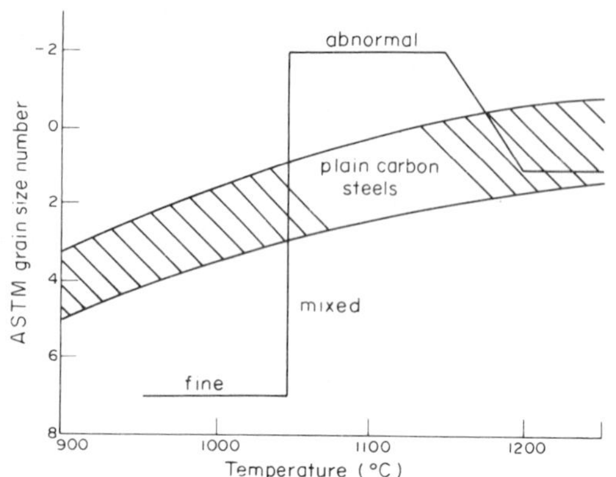

Figure 4
Austenite grain growth characteristics of an aluminum-treated low-carbon steel

5. Limiting Grain Size

The normal grain growth process can be inhibited by a number of factors. Impurities in the form of nonmetallic inclusions are generally too large to provide effective pinning of fine-grained structures (Gladman 1979). However, the normal grain growth process increases the matrix grain size, and consequently the critical particle radius (see Eqn. (4)) until this exceeds the inclusion size. At this stage, grain growth ceases as the inclusions tend to have a high thermal stability. The limiting grain size in high-purity materials is frequently observed to be much larger than in low-grade materials.

A limiting grain size is also believed to exist when the grain size approaches the dimensions of the specimen (Beck 1954). This effect may be attributable to thermal grooving (Mullins 1958). When a grain boundary intersects a free surface, the topography of the surface changes and produces a groove where the boundary intersects the surface in order to comply with the equilibrium energy balance that satisfies the relation

$$\sigma_b = 2\sigma_{fs} \cos(\theta/2)$$

where σ_b is the surface tension of the grain boundary, σ_{fs} is the surface tension of the free surface, and θ is the angle subtended between the two free surfaces at either side of the boundary.

If the boundary is moved away from the groove, grain-boundary area must be created to replace that eliminated in the grooving process. This requires an energy supply which must derive from the grain growth process. When the average grain size is small compared with the metal specimen thickness, the thermal groove effect is insignificant and has little effect on the overall growth rates. However, as the grain size approaches a size above about 10% of the specimen thickness, growth slows down and may subsequently cease.

6. Free-Surface Effects

Many of the products subjected to grain growth treatments (e.g., electrical steels) are in the form of thin strip. The surface area-to-volume ratio of these products is very high and there is evidence to show that the differences in surface energies of differently oriented grains can contribute to grain growth (Hondros and Stuart 1968). The surface energy of a given grain is dependent on the atomic planes exposed (i.e., is dependent upon the crystallographic orientation). The preferential adsorption of impurity atoms such as oxygen on certain planes can change the relative surface energies of (100) and (110) planes. These phenomena result in a change of texture which is dependent upon the oxygen partial pressure of the annealing atmosphere. It would appear that, in coarse-grained specimens, free-surface energy effects and their dependence upon orientation and adsorption effects can play a part in further growth, though these may be irrelevant in materials of low surface-to-volume ratio and in fine-grained steels.

See also: Grain Size and Shape Control in Metals and Alloys; Recrystallization of Deformed Metals (Primary); Thermodynamics of Surfaces and Interfaces

Bibliography

Ashby M F 1980 The influence of particles on boundary mobility. In: Hansen N, Jones A R, Leffers T (eds.) 1980 *Proc. 1st Risø Symp.* Risø National Laboratory, Roskilde, Denmark, pp. 325–36

Beck P A 1954 Annealing of cold-worked metals. *Adv. Phys.* 3: 245–324

Cahn R W 1983 Recovery and recrystallization. In: Cahn R W, Haasen P (eds.) 1983 *Physical Metallurgy*, 3rd edn. North-Holland, Amsterdam, pp. 1595–1672

Feltham P, Copley G J 1958 Grain growth in α-brasses. *Acta Metall.* 6: 539–42

Gladman T 1966 On the theory of the effect of precipitate particles on grain growth in metals. *Proc. R. Soc. London, Ser. A* 294: 298–309

Gladman T 1979 The effect of inclusions on recrystallization and grain growth. In: Pickering F B (ed.) 1979 *Inclusions,* Institution of Metallurgists Monograph No. 3. Sheffield City Polytechnic, Sheffield, UK, pp. 172–81

Gladman T 1980 The effect of second phase particles on grain growth. In: Hansen N, Jones A R, Leffers T (eds.) 1980 *Proc. 1st Risø Symp.* Risø National Laboratory, Roskilde, Denmark, pp. 183–92
Gladman T, Dulieu D 1974 Grain-size control in steels. *Met. Sci.* 8: 167–76
Hondros E D, Stuart L E H 1968 Interfacial energies of textured silicon iron in the presence of oxygen. *Philos. Mag.* 17: 711–27
Lücke K, Stüwe H P 1971 On the theory of impurity controlled grain boundary motion. *Acta Metall.* 19: 1087–99
Mullins W W 1958 The effect of thermal grooving on grain boundary motion. *Acta Metall.* 6: 414–27
Novokov V Yu 1981 On conditions for secondary recrystallization in a material with random orientation. *Acta Metall.* 29: 883–87
Smith C S 1948 Grains, phases and interfaces: An interpretation of microstructure. *Trans. Am. Inst. Min. Metall. Eng.* 175: 15–51

T. Gladman

Grain Size and Shape Control in Metals and Alloys

The properties of a crystalline material are strongly influenced by the presence of grain boundaries (i.e., by their character, density and position), and hence close control of the grain structure is extremely important. A fine grain size increases the strength and ductility of an alloy and is a prerequisite for superplasticity. A fine grain structure is often required in commercial alloys to facilitate forming operations or to ensure good service properties (e.g., fracture toughness and corrosion resistance). Large grains with a specific orientation are sometimes also required (e.g., for the optimization of magnetic properties).

1. Grain-Structure Formation

Solid metallic substances are usually built up from crystals with different orientations separated by high-angle grain boundaries. Such a grain structure is formed during solidification either from the liquid or vapor phase. Exceptions, or rather limiting situations, are the growth of a single crystal, where the whole body consists of a single grain, and the formation of an amorphous solid phase. In the solid state, variation of the grain structure is possible by any reaction which involves the movement of an incoherent interface (grain boundary). This includes, in principle, solid-state transformations such as the decomposition of a super saturated solid solution, eutectoid reactions and order–disorder transformations. However, the most efficient mechanism for the control of grain size and shape in the solid state is the recrystallization of a deformed microstructure. The main driving force for recrystallization (including grain growth) is the decrease in free energy due to the elimination of lattice defects (high- and low-angle boundaries, dislocations and vacancies). To alter the grain structure in the solid state, the material is subjected to plastic deformation (usually cold working) and subsequently annealed. If the degree of cold working is sufficiently large, the alloy shows a very high density of homogeneously distributed dislocations, but no recognizable high-angle grain boundaries.

Grain size may also be controlled by the crystallization of an amorphous solid phase. As for solidification from the melt, spontaneous (homogeneous) and heterogeneous nucleation of the grains have been observed. Although this mechanism has not yet been used for grain size control in commercial applications, the crystallization of an amorphous phase offers the possibility of producing extremely fine-grained materials with potentially new and interesting properties (see *Metallic Glasses: Crystallization*). Deliberate control of grain size and shape involves varying the fabrication procedure to favor either one of the nucleation reactions, and to control the conditions for the nucleation and growth of the new grains.

2. Grain Size Control During Solidification

2.1 Nucleation Kinetics

The liquid–solid transformation is a discontinuous process; a nucleus of the solid phase has to exist which grows into the liquid phase. If the temperature of the melt drops progressively below the freezing point, the solid phase nucleates spontaneously at a certain undercooling. However, in foundry practice such homogeneous nucleation is never observed. Instead, rapid heterogeneous nucleation generally occurs at low undercoolings. Appropriate nucleation sites are the surfaces of, for example, foreign solid particles (inclusions), the container walls, gas bubbles and fragments of the already solidified material swept into the bulk liquid from the advancing solidification front (see *Nucleation from the Melt*).

As the melt cools below the freezing point, two processes take place concurrently. The solid phase nucleates on the available substrate surfaces at a rate which increases exponentially with the undercooling and, when the temperature has fallen below a certain limit, nucleated particles start to grow and evolve latent heat. The cooling rate decreases as nucleation and growth accelerate (Fig. 1), until the temperature reaches a minimum and recalescence occurs. Negligible nucleation and growth occur during the first part of the cooling curve. However, the nucleation rate rises rapidly to a maximum just before the minimum temperature is reached. The number of nucleation events then decreases quickly as the available particles are exhausted and particularly as recalescence begins. Consequently, nucleation is almost complete just beyond the minimum temperature, after which only a process of growth is observed. The grain density varies according to the nucleation rate and reaches the final level by the time recalescence begins. However, the growing particles are still relatively far apart and the volume fraction of solid metal is still small after the nucleation events have finished.

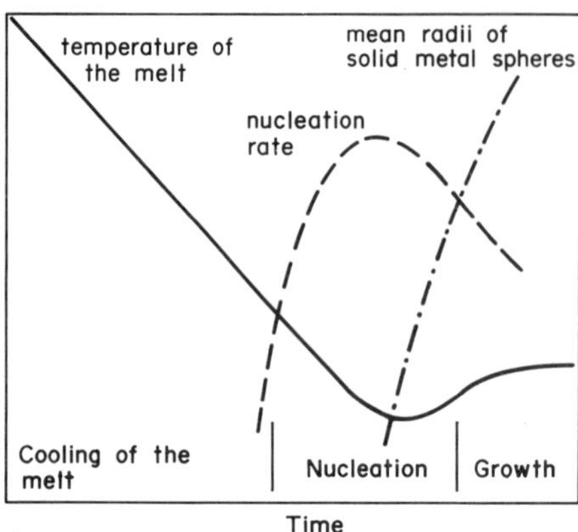

Figure 1
Schematic diagram showing nucleation and growth of the solid phase during cooling from the liquid state

For each set of freezing conditions, a critical particle density exists above which negligible further grain refinement occurs. If this limit is high (and the necessary particles are present), the as-cast grain size will be as fine as possible for the particular casting conditions. Maxwell and Hellawell (1975) have presented a quantitative treatment of grain refinement in aluminum alloys. The ability of a substrate to act as a surface for heterogeneous nucleation (measured by the wetting angle θ) strongly influences the final grain size. Efficient heterogeneous nucleation occurs only for $\theta < 10°$; a decrease of about 3° increases the final grain density by one order of magnitude.

A similar increase in the final grain density is obtained by the variation of the local cooling rate by a factor of five within the range typical for semicontinuous casting (0.1–2.5 K s^{-1}). The influence of solute undercooling as it varies from system to system is also very important. The grain-refining activity of a good potential catalyst is precluded if the growth temperature is not significantly depressed below the freezing point of the alloy. Strong undercooling associated with the presence of a solute is, for example, the reason for the pronounced effect of titanium additions on the grain size in cast aluminum alloys. On the other hand, the influence of the substrate size is comparatively small.

2.2 Casting Conditions

Apart from nucleation kinetics, a very wide degree of control can also be exercised over grain size and shape through the casting conditions during the growth of the nuclei. The rate of heat removal and the temperature of the liquid determine the solidification rate and the prevailing temperature gradients; hence, combinations of these factors govern the growth behavior. Generally, the solid phase grows into the melt in the form of dendrites. Large-scale commercial ingots usually have three characteristic regions: (a) a chill zone at the surface containing approximately equiaxed, fine crystals formed at large undercoolings when the molten metal strikes the cold mold walls; (b) a columnar zone of crystals originating from the chill zone, whose continued growth is due to their favorable orientation for fast growth; (c) a central equiaxed zone formed when the liquid at the center of the ingot becomes supercooled, and hence nucleates. Thus, it is possible to control the extension of the columnar structure by the correct choice of casting conditions (see *Ingot and Continuous Casting of Steel*; *Rheocasting*; *Casting of Metals*; *Cast Structure of Alloys*; *Macrostructure of Metals*).

For most commercial applications, grain refinement by the introduction of suitable nuclei is desirable during casting. A fine grain structure in the as-cast state facilitates and ensures reproducible properties for subsequent working, reduces internal stresses by dispersing the brittle constituents, allows rapid solidification, and promotes a uniform solidification pattern and sound product.

3. Grain Size Control by Recrystallization

Recrystallization processes are either continuous (recovery, in-situ recrystallization, normal grain growth) or discontinuous (primary recrystallization, secondary recrystallization). The essential microstructural processes involved in primary recrystallization, which is the most important process in controlling grain size, are described elsewhere (see *Recrystallization of Deformed Metals* (*Primary*)).

Severe impediment of nucleation leads to continuous recrystallization (i.e., a continued process of subgrain coarsening and subgrain coalescence without excessive growth of selected grains). Continuous recrystallization is found when the existing grain boundaries are held by preferential precipitation and/or when no substructural rearrangement to form such boundaries can occur owing to the impurity drag on the subgrain structure (dislocation pinning by segregated solute atoms). The resulting microstructure is extremely fine-grained and relatively stable with respect to grain coarsening.

Inhomogeneities in the deformed (or recovered) microstructure usually lead to ordinary discontinuous recrystallization. The final grain size is given by the nucleation rate (i.e., the number of active nucleation sites per unit volume and time) and the growth rate of the nuclei. For small degrees of cold working and particularly in single-phase alloys, only former grain boundaries are potential nucleation sites. Thus, the initial grain size prior to deformation also determines the final grain size. At intermediate and large amounts of deformation, deformation bands with large angles of misorientation between the subgrains are much more frequent nucleation sites. The result is a decrease of the mean grain size after recrystallization with increasing cold work. However, in single-phase alloys, relatively large strains (cold deformation $\gtrsim 60\%$) are necessary to produce a fine and homogeneous grain structure; without this the fine grains become preferentially located at former grain boundaries and deformation bands.

Preexisting second-phase particles have a distinct influence on the recrystallization behavior. The microstructure in the deformed state depends strongly on the size, form and dispersion of the precipitates. Adjacent to large ($\gtrsim 1$ μm) particles, the deformed matrix is locally rotated and characterized by the presence of small cells or subgrains and large lattice misorientations. Thus, the formation of recrystallization nuclei around large particles or clusters of smaller particles is strongly favored. In heavily deformed alloys with a uniform dispersion of coarse particles, rapid recrystallization leads to a final grain size of the order of the interparticle spacing, provided that the growth process is not hindered by a retarding force. In this manner, the final grain size may be varied considerably by close control of the distribution of large second-phase particles in a relatively pure aluminum matrix.

Finer (0.1–1 μm) particles have little influence on the nucleation and growth kinetics of the new grains. If the precipitates are homogeneously dispersed, the main structural difference after cold working relative to a single-phase alloy is a greater homogeneity of the dislocation structure, with a higher dislocation density for a given strain. Also, the retarding force of these particles on a moving recrystallization front is relatively small although the rearrangement of subgrain boundaries may be hindered. In principle, a fine grain structure could be produced by restricting the growth rate using a dense distribution of fine particles. However, microstructural inhomogeneities usually result in the preferential growth of just a few grains. The effect of a fine and dense particle dispersion is generally an increase in the recrystallization temperature, and also a coarse final grain structure.

Even smaller ($\ll 0.1$ μm) particles do not appreciably affect the deformed microstructure except for a preferred formation of deformation bands in alloys with coherent precipitates or very small incoherent particles. However, such precipitates may be formed by the decomposition of a supersaturated solid solution during the annealing treatment. The presence of solute elements leads to processes which may precede or accompany and interact with recrystallization. These include:

(a) solute segregation to the deformation and/or recovered structure,

(b) solute segregation to the recrystallization front,

(c) precipitation at the defects constituting the deformed and/or recovered structure, and

(d) precipitation at the recrystallization front.

Segregation and precipitation processes influence both the nucleation and the growth stage of recrystallization. When these reactions occur in the deformed and/or recovered material, the substructure rearrangement necessary for the production of viable nuclei is strongly hindered owing to the pinning action of segregated atoms and preferential precipitation at the dislocations. Thus, the recrystallization temperature is considerably increased. However, as soon as the driving force locally exceeds the retarding forces, anomalous growth of a few grains leads again to a coarse final grain structure.

The influence of segregation on the moving recrystallization front is relatively small. The retarding action of the solute drag effect is strongly concentration and temperature dependent in the low-velocity range and disappears instantaneously at a critical concentration or temperature. Therefore, the effect of solute atoms on primary recrystallization is evident only in special cases, but is considerably more important for grain growth and secondary recrystallization. The impeding effects of segregation and precipitation are observed for instance in Al–Mn alloys, where they lead to intermittent grain-boundary motion if the cold-worked alloy is annealed in a certain temperature range. This process is characterized by the motion of a boundary which is partially loaded with solute atoms. Its velocity gradually decreases owing to the impurity drag effect as an increasing number of solute atoms segregate from the supersaturated solid solution until the boundary is completely pinned by the formation of precipitates. Subsequent particle coarsening decreases the pinning force, allowing the boundary to break loose and the process to start again. In the same temperature range, the simultaneous occurrence of discontinuous (cellular) precipitation and recrystallization is observed. The deformed, supersaturated solid solution decomposes into a lamellar array of a recrystallized, saturated solid solution and an intermetallic phase. The result of these partially impeded recrystallization processes is a highly inhomogeneous grain structure.

A fine grain structure in a deformed supersaturated solid solution is best produced by an annealing treatment at high temperatures where recrystallization precedes precipitation and the two processes do not interact (Fig. 2). This is generally the case for commercial aluminum alloys. Apart from the aluminum matrix with solute atoms, wrought aluminum alloys normally contain a relatively high density of large particles (formed during solidification) and a dispersion of finer particles precipitated from the supersaturated solid solution during a subsequent annealing treatment. In principle, the large particles, the former grain boundaries and the deformation bands provide a very high density of potent nucleation sites in cold-worked commercial aluminum alloys. However, many nuclei will never grow into grains of the final size. Nuclei are developed consecutively during the recrystallization process and individual nuclei show very different growth velocities. Therefore, a number of slowly developing nuclei will be isolated and consumed by faster-growing neighbors.

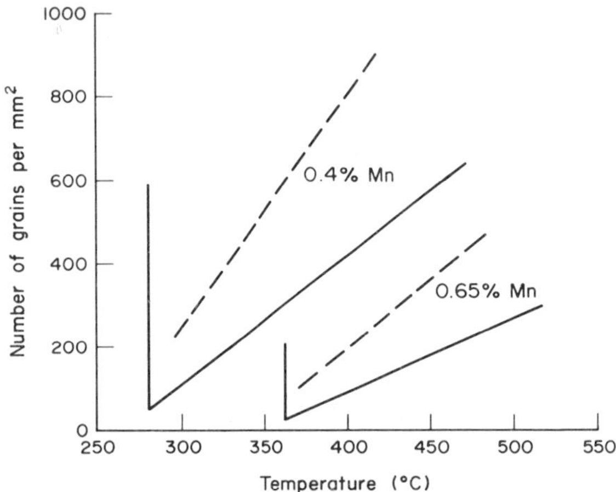

Figure 2
Grain density of a commercial Al–1%Mn alloy when recrystallization and precipitation processes interact (0.65% Mn in solid solution) and when precipitation occurs only after completion of recrystallization (0.4% Mn in solid solution): ——, lower limit; – – –, mean value

Many factors limit the number of nuclei which succeed in growing into fully formed grains. Such factors strongly decrease the final grain density in comparison with the value which could be expected from the density of the available potent nucleation sites.

Additional factors influencing the recrystallized-grain size are the annealing temperature and the heating rate used in the annealing treatment. A high annealing temperature results in a high density of viable nuclei and a fine grain size provided that grain growth and secondary recrystallization are excluded. A high heating rate results in a higher temperature for the start of recrystallization. Thus, the density of the active nucleation sites is increased and the probability of a strongly preferred nucleus starting to grow at relatively low temperatures is reduced. Excessive growth of a few nuclei during slow heating results in an extremely coarse and inhomogeneous grain structure.

See also: Dynamic Recovery and Recrystallization of Metals; Subgrain Coalescence in Metals; Grain Growth and Secondary Recrystallization

Bibliography

Cotterill P, Mould P R 1976 *Recrystallization and Grain Growth in Metals.* Surrey University Press, London

Furrer P, Warlimont H 1978 The effects of segregation and precipitation on the annealing behaviour and grain size of aluminium alloys. *Aluminium (Duesseldorf)* 54: 135–42

Haessner F (ed.) 1978 *Recrystallization of Metallic Materials,* 2nd edn. Riederer, Stuttgart

Hansen N, Jones A R, Leffers T (eds.) 1980 *Recrystallization and Grain Growth of Multi-Phase and Particle Containing Materials.* Risø National Laboratory, Roskilde, Denmark

Maxwell I, Hellawell A 1975 Heterogeneous nucleation and grain refinement in aluminium castings. *Acta Metall.* 23: 229–37

Metals Society 1979 *Solidification and Casting of Metals.* Metals Society, London

Metals Society 1983 *Solidification Technology in the Foundry and Casthouse.* Metals Society, London

P. Furrer

Grain Size: Nondestructive Evaluation

Ultrasonic and magnetic methods have been used to evaluate grain size in single-phase polycrystalline solids. The ultrasonic method is based on the scattering of acoustic energy from anisotropic grains. The magnetic method, restricted to ferromagnetic materials, is based on the interaction of magnetic-domain-wall motion with grain boundaries. The use of these methods can lead to the determination of grain size to within approximately 1–2 ASTM grain sizes.

1. Principle of Ultrasonic Techniques

In polycrystalline material, the propagation of ultrasonic waves is affected by discontinuities in the characteristic acoustic impedance ρc (ρ is the density and c is the phase velocity) at the grain boundaries due to the crystallographic anisotropy of c. At each grain boundary, the wave is partly reflected and partly mode converted between longitudinal (L) and transverse (T) waves. The contribution to attenuation due to grain-boundary scattering is the sum of these scattering encounters over the many grains traversed during propagation. This subject has been reviewed by Papadakis (1968) and by Goebbels (1980).

The ultrasonic attenuation α is the amplitude decrement coefficient in waves of the form $e^{-\alpha x}$, where x is the space coordinate. The L and T contributions to attenuation due to grain scattering (α_L^s and α_T^s) have been calculated for the case when anisotropy is small, multiple scattering can be ignored, grains are equiaxed and of single phase, and texture can be ignored. For Rayleigh scattering of bulk waves in cubic crystals, that is, when the wavelength $\lambda \gg 2\pi\bar{D}$ where D is the average grain diameter,

$$\alpha_L^s = \frac{8\pi^3\mu^2 V f^4}{375\rho^2 c_L^3}\left(\frac{2}{c_L^5}+\frac{3}{c_T^5}\right) \tag{1}$$

$$\alpha_T^s = \frac{2\pi^3\mu^2 V f^4}{125\rho^2 c_T^3}\left(\frac{2}{c_L^5}+\frac{3}{c_T^5}\right) \tag{2}$$

Here, f is the frequency, c_L and c_T are the ultrasonic phase velocities for L and T waves respectively and the elastic anisotropy factor $\mu = c_{11} - c_{12} - 2c_{44}$, where c_{nm} are crystallographic elasticity constants. The quantity $V = \frac{4}{3}\pi(\langle R^6\rangle/\langle R^3\rangle)$ is a characteristic grain volume calculated from the grain size distribution $N(R)$ using moments

$$\langle R^n\rangle = \int_0^\infty R^n N(R)dR \Big/ \int_0^\infty N(R)dR$$

and $\langle R\rangle$ is $\frac{1}{2}\bar{D}$. Similar formulae exist for hexagonal and orthorhombic crystallites. Expressions for α^s in the intermediate wavelength (stochastic) region where $\lambda \approx 2\pi\bar{D}$ and the short wavelength (diffuse) region where $\lambda \ll 2\pi\bar{D}$ are also known.

Papadakis's analysis indicates that Eqns. (1) and (2) are qualitatively correct in so far as the attenuation follows a Vf^4 law, and that they yield quantitatively accurate results for the ratio α_L^s/α_T^s. However, quantitative agreement between actual values of α_L^s or α_T^s predicted by Eqns. (1) and (2) and experimental results is at best within a factor of 2. Owing to this lack of quantitative agreement, generalized equations ($\alpha_L^s = k_L d^3 f^4$ and $\alpha_T^s = k_T d^3 f^4$, where k_L and k_T are empirically determined grain scattering coefficients) are used to analyze experimental results. In most studies the statistical nature of the grain size is suppressed with V or $\bar{D}$ simply being replaced by the cube or value, respectively, of any one of the metallographically determined average grain diameters d. Goebbels determined that measurements of grain size should be made in the region $0.03 < d/\lambda < 0.3$. Frequencies below 20 MHz have been used on materials with grain diameters over 10 μm. Accuracies of no better than ±15%, corresponding to 2 ASTM grain sizes, are claimed. Measurement error arises from factors such as diffraction effects, sample geometry, the presence of large isolated scatterers (inclusions), the effects of the actual grain size distribution and, especially, the presence of the small number of relatively large grains.

For either L or T waves, experimental measurements determine the total attenuation α, of which α^s is only a part. The remainder, referred to as the absorption α^a, is attributed to the excitation of lattice defects and to thermal and magnetic dissipation effects among other mechanisms. Absorption is modelled by $\alpha^a = \Sigma a_n f^n$, where usually $n = 1, 2$ and the a_n are phenomenological constants.

There are at least three distinct techniques for determining grain size from measurements of total attenuation, $\alpha_R = \alpha^a(f) + \alpha^s(d, f)$, in the Rayleigh region.

2. *Single-Frequency Ultrasonic Technique*

The attenuation α_R is determined at a single frequency for samples identical except for grain size. A least-squares analysis of the results of measurements on samples with known grain size allows the determination of α^a and k_L or k_T. With these quantities known d can be deduced from measurements of α_R in similar samples. No assumption on the form of α^a is required, but the technique is sensitive to differences in α^a among the samples.

3. *Two-Frequency Ultrasonic Technique*

The attenuation α_R is measured at two frequencies, α_{R1} at f_1 near the long and α_{R2} at f_2 near the short wavelength portions of the Rayleigh region, and the results are analyzed according to Eqns. (1) and (2). Most commonly the assumption $\alpha^a = a_1 f$ is made leading to $k_{L,T}d^3 = [\alpha_{R1} - \alpha_{R2}f_2/f_1]/[f_1^4 - f_1, f_2^3]$, where $k_{L,T}$ represents either k_L or k_T. Regression analysis of measurements from samples of known grain size is used to determine $k_{L,T}$ and thereafter d for similar samples. This technique requires careful attention to diffraction effects and, owing to the high orders in f, accurate frequency determinations. It is also sensitive to deviation from the (f^1, f^4) empirical model.

4. *Backscatter Ultrasonic Technique*

Since a polycrystalline grain structure scatters ultrasonic waves, the conventional measurement of attenuation, or forward scattering, is complemented by the ultrasonic scattering in other directions. For the backscatter case in homogeneous material, the received amplitude $A_s(X)$ as a function of propagation distance x is

$$\ln[A_s(x)] = b + \tfrac{1}{2}\ln[\alpha^s] - (\alpha^a + \alpha^s)X$$

Analysis of backscatter signals in combination with techniques described in Sects. 2 and 3 permits determining α^s independent of sample geometry. This combined technique requires spatial averaging over large numbers of backscatter waveforms (Goebbels 1980).

While the preceding discussion deals with bulk waves, it is also applicable to measurements with surface waves in the case of both direct attenuation measurements (Kettler et al. 1974) and backscatter measurements (Hecht et al. 1981). The abovementioned techniques have also been used with measurements in the diffuse region by using appropriate forms of α^s and α^a (Kettler et al. 1974). In addition to grain-size measurement in both bulk and sheet metals, applications have been made to the determination of grain-size-dependent properties such as the degree of heat treatment or state of anneal (Papadakis 1968) and mechanical properties, e.g., yield strength (Klinman et al. 1980).

5. *Magnetic Methods*

Experimental evidence points to the influence of grain size on the magnetic properties of steels (Littman 1971). However, the fundamental phenomenological relationships have not been developed. Empirical approaches based on linear regression analysis with coercive field, remanent magnetization, demagnetizing field or maximum permeability as independent variables and either grain size or a grain-size-related property as dependent variable, have achieved 15% accuracies in special cases. Factors such as chemical composition, microstress state and the presence and morphology of secondary phases can severely influence the results obtained by this approach.

6. *Barkhausen Technique*

The discontinuous movement of Bloch walls during domain growth generates Barkhausen noise (see *Barkhausen Effect*). Since domain growth is influenced by grain boundary pinning sites, statistical analysis of Barkhausen noise must include grain size as a parameter (Otalo and Saynajakangas 1972). Because of the shielding of the skin effect, interrogation depth is limited and application is primarily to sheet geometries for which grain size has been determined to better than ±1 ASTM unit to a sample depth of approximately 0.4 mm. Accuracy is strongly affected by crystallographic texture, strain state, chemistry and sheet thickness.

See also: Ultrasonic Nondestructive Evaluation; Magnetic Nondestructive Evaluation

Bibliography

American Society for Testing and Materials 1980 Standard method for estimating the average grain size of metals, ANSI/ASTM Standard E112-80. *Annual Book of ASTM Standards*, Pt. II, *Metallography, Nondestructive Testing*. American Society for Testing and Materials, Philadelphia, Pennsylvania, pp. 186–220

Goebbels K 1980 Structure analysis by scattered ultrasonic radiation. In: Sharpe R S (ed.) 1980 *Research Techniques in Nondestructive Testing*, Vol. 4. Academic Press, London, pp. 88–157

Hecht A, Thiel R, Neumann E, Mundry E 1981 Nondestructive determination of grain size in austenitic sheet by ultrasonic backscatter. *Mater. Eval.* 39(10): 934–38

Kettler R, Latiff R, Fiore N F 1974 Surface ultrasonic measurement of grain size of silicon iron sheet. *Metall. Trans.* 5: 952–53

Klinman R, Webster G R, Marsh F J, Stephenson E T 1980 Ultrasonic prediction of grain size, strength and toughness in plain carbon steel. *Mater. Eval.* 38(10): 26–32

Littman M F 1971 Iron and silicon–iron alloys. *IEEE Trans. Magn.* 7: 48–60

Otalo M, Saynajakangas S 1972 A new electronic grain-size analysis for technical steel. *J. Phys. E* 5: 669–72

Papadakis E P 1968 Ultrasonic attenuation caused by scattering in polycrystalline media. In: Mason W P (ed.) 1968 *Physical Acoustics*, Vol. IV, Pt. B. Academic Press, New York, pp. 269–328

R. Klinman

Granulation of Ceramic Powders

In the ceramics industry, granulation techniques are used in the compounding of press mixes. Powders, which are generally very fine (particle size 1–4 μm or slightly larger), are converted to a granular form; the granule size required is 0.1–1 mm or larger.

Granules here signify a granular material in which the grain size is as nearly uniform as possible. Granulation can be achieved either by agglomeration or by comminution. For some decades, the term granulation has been frequently used for specific processes for increasing the grain size. A distinction is made between layering granulation and press granulation. The granules produced by these processes are referred to as layer agglomerates or pellets, and tablets or briquettes, respectively. The size of the agglomerates is ~0.3–50 mm, depending on the production process. The most advantageous shape of grain or particle is in most cases a sphere, obtained, for example, when the granules are produced in a fluidized bed, spray tower or pan.

1. *Bonding Mechanisms*

The structure and properties of solid agglomerates in a powder are often the dominant factors in powder processing parameters and bulk properties. The most common types of agglomerates in a ceramic powder are formed during calcination. Such agglomerates are strong enough to retain their identity during green forming. The strength of the agglomerates is due to adhesion between the particles. The principal bonding mechanisms in agglomerates are as follows.

1.1 Solid Bridges

Solid bridges are formed by crystallizing salts or by sintering. In the case of crystallizing salts, the strength of agglomerates is very sensitive to the drying conditions employed.

1.2 Van der Waals Forces

Van der Waals forces are well known as bonding forces between molecules, depending on the seventh power of the distance. The same forces—the London–van der Waals forces—cause adhesion of solids to solid surfaces. From a model of stochastic electromagnetic fluctuations in solids, a theory of van der Waals forces between two adherends separated by a vacuum has been developed. This theory can be extended to three-phase systems in which the two adherends are separated by a third medium. Such theory must be adapted to the case of true adhesion by introducing an adjustable parameter z_0.

1.3 Electrostatic Forces

Particles with charges of opposite polarity attract one another. Opposite charges can arise during particle contact, as a result of transfer of electrons from one particle to another (contact potential); however, the particles may already carry an excess charge, for example owing to previous friction, comminution or electron adsorption.

When two different materials come into contact, electrons tend to flow from one to the other because of differences in the electronic work functions at both surfaces. This gives rise at equilibrium to a contact potential difference U in the range 0–0.5 V, depending on the two materials. The work functions of surfaces depend on local impurities and are often unknown.

The adhesion forces between nonconductors are smaller than between conductors. In nonconductors the accumulated charges may extend up to a depth of ~1 μm, while conductors may have charges concentrated in a layer a few angstroms thick on the surface.

1.4 Liquid Bridge Forces

Most information is available for the liquid bridge bonding mechanism. The shape of a liquid bridge is such that the capillary pressure in the bridge is the same throughout the volume (if gravity can be neglected). This implies that the mean radius of curvature is the same at all points of the surface. The exact solution of the corresponding differential equation for all radially symmetrical geometries of the contacting bodies with variable distances, bridge volumes and contact angles can be calculated.

Isolated liquid bridges form only if there are relatively small amounts of liquid in the agglomerate. As the fraction of liquid increases, a transition state is reached at which liquid bridges and liquid-filled pores can exist side by side, until finally a liquid-filled condition is reached. The usual measure of the amount of liquid is the degree of liquid saturation S, which indicates the ratio of liquid volume to void space of the agglomerate. The change from the bridge state to the transition state occurs when S is between 0.2 and 0.4. Both in the transition state and in the capillary state, the adhesion force between individual particles can no longer be defined.

If more than ~80% of the void space between the particles is filled with liquid (capillary state, liquid saturation $S \geqslant 0.8$), a capillary suction (pressure p_k) is generated in the liquid space and the agglomerate is held together by the external pressure. The tensile strength σ_z equals the product of p_k and S.

1.5 Comparison of Adhesion Forces

In Fig. 1 the calculated adhesion force is plotted against sphere diameter for a sphere adjacent to a plane surface. The Lifshitz–van der Waals constant $\hbar\bar{\omega} = 5$ eV is relatively high; the contact potential difference $U = 0.5$ V and the surface charge density $\sigma = 10^2$ e μm^{-2} are both maximum values; the liquid bridge angle $\beta = 20°$ gives a moderate value of the liquid bridge force. The liquid bridge forces are about four times the van der Waals forces, which are an order of magnitude greater than electrostatic adhesion forces between conductors due to contact potential, and these again are ~10–100 times as great as adhesion forces due to maximum surplus charges of opposite sign.

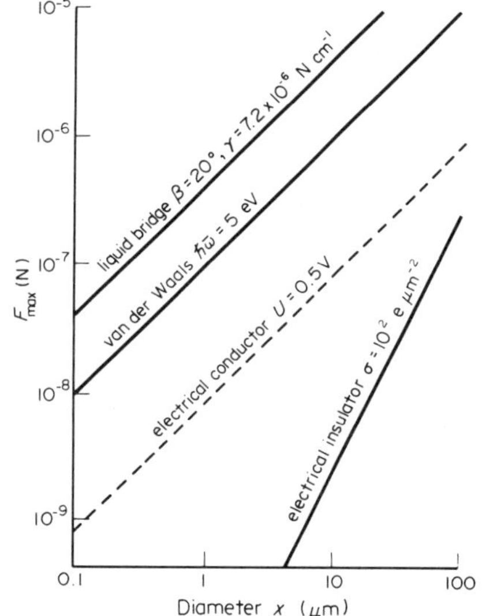

Figure 1
Adhesion force F_{max} of liquid bridge, van der Waals and electrostatic interaction between a smooth sphere and a plane surface as a function of the sphere diameter x. The surfaces of the two interacting bodies are in contact; $z_0 = 4 \times 10^{-10}$ m (after Rumpf and Schubert 1978)

2. Granulation Techniques

2.1 Layering Granulation

Widely used systems include granulating drums, granulating mixers and granulating pans. The finely dispersed material is introduced, with the addition of liquid, into an inclined rotating pan or into a rotating drum and subjected to a rolling motion. The random motion results in spherical agglomerates. The liquid provides the requisite high adhesion forces, through liquid bridges and capillarity.

In a granulation pan, the grain nuclei and small granules move about near the bottom of the pan and are transported upwards because the initially irregular particles exhibit greater friction. The larger granules roll more easily over the smaller granules and are finally discharged with a well-defined size. The particle size of the material which is discharged through overflow is so uniform that a subsequent classification by screening is often unnecessary.

In a granulating drum, the axial inclination is too low to result in classification. Rather, it serves to transport the material through the cylinder. Accordingly, the size distribution of the agglomerates which leave the drum is substantially broader than in the case of a granulating pan.

Granulating pans are now constructed in sizes of up to 7.5 m diameter, and not only in cylindrical but also in conical shapes. Granulating drums have been operated successfully in sizes up to a diameter of 3.6 m and a length of 3.7 m. If the drum is used at the same time for drying or cooling, diameters of up to 4.2 m and lengths of up to 15 m are feasible.

In mixers, the product is agitated either by moving paddles or by rotation of the wall, as in the case of trough mixers. If powder and liquid are introduced, agglomerates can be built up by the same bonding mechanism as that which operates in a granulating pan or drum. High-speed cutter heads ensure that agglomerates which have become too large are comminuted. Trough mixers and vertical mixers operate batchwise, while paddle mixers are either batchwise or continuous.

Provided that the process is carried out suitably, agglomerates can also be produced by drying a moist loose mass of material, such as a suspension or a solution. Equipment extensively used for this method includes spray dryers, fluidized bed dryers and drum dryers. In virtually every case, crystallization processes are important. The crystals formed are in general bonded together by solid bridges. However, solid bridges between particles are also formed merely by drying of a moist loose mass of material or of a suspension.

Towards the end of the drying process, dissolved substances are principally concentrated in liquid bridges between the particles and finally crystallize out in these bridges. The agglomerate strength can be varied by varying the drying conditions. In general, the more rapid the drying, the greater the strength. However, excessively fast drying can cause crazing of the agglomerates and thereby reduce the strength. Other properties, such as the porosity, can also be widely varied by varying the processing conditions.

2.2 Press Granulation

Moist masses such as filter cakes and pastes are granulated in die presses. The material is forced through holes and is thereby compacted to form granules. In the case of a roll press, material is drawn into the nip between a perforated and a solid roll and is forced through cylindrical holes into the interior of the perforated roll. The granules which protrude into the latter roll are cut off by a knife and flow axially out of the drum. A variety of designs of roll press exists.

In press granulation of dry powders, the material is so highly compacted that agglomerates of adequate strength result. The compaction process increases both the number of contact points between the particles and the adhesion of the particles by deforming them in their contact zone. Compaction of dry powders, which is sometimes carried out with binders, uses roll presses with smooth and briquetting rolls. Screws are employed to feed the product through a hopper between two rotating rolls, which compact it to form a ribbon or shaped piece.

3. Applications

Granulation is used in the traditional ceramics industry for dry pressing of floor and wall tiles. It is also frequently applied to advanced ceramics. Most of the manganese–zinc ferrite cores and steatite and carbide bodies are dry-pressed using granules so as to achieve homogeneous die filling.

See also: Ceramic Powders: Packing Characterization; Ceramics Process Engineering: An Overview

Bibliography

Moser S, Sommer K 1977 Calculation of van der Waals forces in adhering systems. *Powder Technol.* 17: 191–95

Onada G Y, Hench L L 1977 *Ceramic Processing Before Firing*. Wiley, New York

Pietsch W 1976 *Roll Pressing*. Heyden, London

Rumpf H 1962 In: Knepper W (ed.) 1962 *Agglomeration*. Wiley, New York

Rumpf H, Schubert H 1978 *Ceramic Processing Before Firing*. Wiley, New York, pp. 357–76

Schubert H 1975 Tensile strength of agglomerates. *Powder Technol.* 11: 107–19

Sherrington P J, Oliver R 1981 *Granulation*. Heyden, London

Stanley-Wood N G 1983 *Enlargement and Compaction of Particulate Solids*. Butterworths, London

K. Sommer

Gray Cast Iron

By far the oldest, in respect of discovery and utilization, of the cast irons is gray cast iron, in which the graphite occurs in a flake morphology. These flakes become interconnected within each eutectic cell as the gray iron solidifies and exert a marked effect on the properties of the iron. Because of the presence of these interconnected flakes, gray iron has often been called a composite material and does show many of the properties characteristic of composites.

1. Solidification Behavior

The solidification and cooling conditions which produce gray iron exert a marked influence on the structure of the graphite and some effect on the matrix. The variations in graphite structures have been classified, together with the length of the flakes, by standards in use for a long time. The five types of flake graphite are shown in Fig. 1. The solidification and growth conditions prevailing during the solidification of gray iron are the primary factors determining which type of eutectic graphite is obtained. Type C is composed of longer flakes that are straight and relatively thick. These flakes grow in the liquid iron and occur as the hypoeutectic graphite phase when the carbon equivalent exceeds 4.3%, the eutectic composition. Their growth is not impeded by the simultaneous, side-by-side growth of austenite that occurs with the flake graphite during the eutectic reaction.

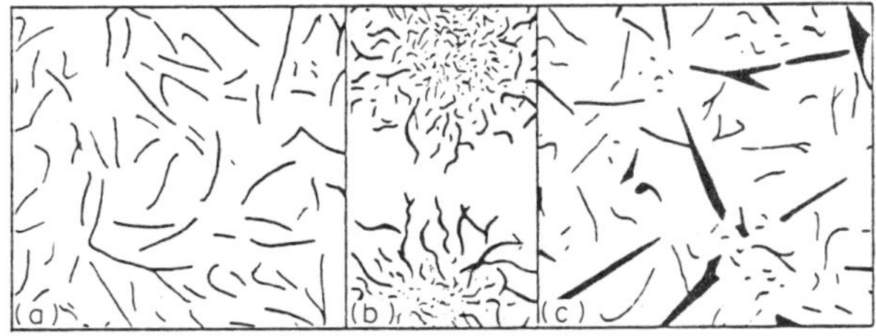

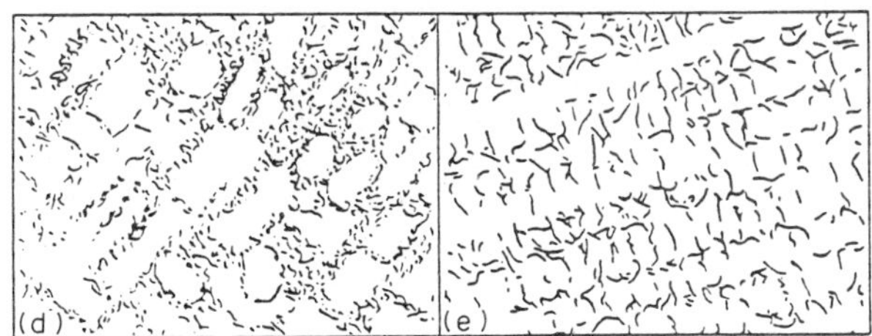

Figure 1
Graphite types in iron castings (American Society for Testing and Materials 1967): (a) uniform distribution, random orientation; (b) rosette groupings; (c) superimposed flake sizes, random orientation; (d) interdendritic segregation, random orientation; (e) interdendritic segregation, preferred orientation

The formation of the eutectic flake graphite (Types A, B, C, and D) is greatly influenced by the amount by which the iron melt cools below the equilibrium temperature for the austenite–graphite eutectic before appreciable solidification occurs. Gray iron with Type A graphite undergoes only small amounts of undercooling, whereas that with Type D graphite undercools significantly below this equilibrium temperature. The undercooling that occurs with Type B graphite is intermediate between the two, producing fine graphite flakes, like Type D, in the center of the eutectic cells or rosette and a coarser type like Type A at the outer cell boundaries. Type E graphite occurs in strongly hypoeutectic gray irons with carbon equivalents well below 4.3%.

Undercooling of the liquid iron before solidification can result from rapid cooling of the melt, such as occurs in thin cast sections, or from poor nucleating conditions for graphite solidification. If the undercooling becomes severe, the melt can cool below the metastable eutectic temperature for austenite–iron carbide or Fe_3C, as shown in Fig. 2. This metastable eutectic temperature is always below the stable austenite–graphite equilibrium eutectic temperature, but the temperature span between the two is significantly affected by the chemical composition.

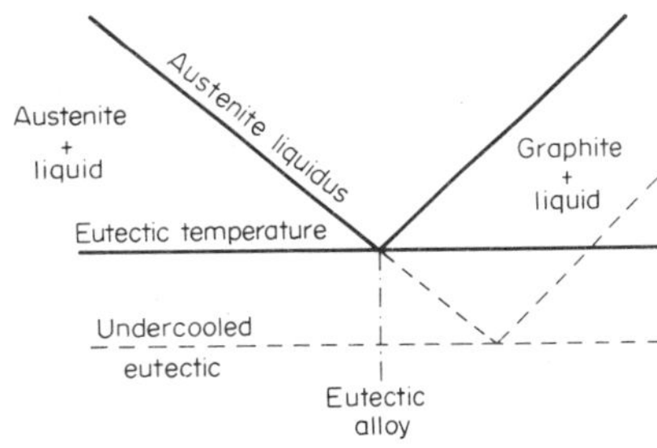

Figure 2
Diagram showing effect of undercooling of the stable austenite–graphite eutectic during solidification

The nucleating conditions depend on the presence in the melt of solid phases which have a similar crystal structure to or epitaxy with the graphite phase. Undercooling can be minimized by the deliberate addition of small amounts of materials which form such phases. This procedure is termed inoculation and is widely used to avoid undercooling and possible formation of the metastable eutectic with its objectionable hard, brittle properties. The effective substrates for graphite have been shown to be sulfides, and the more effective inoculants contain elements which form sulfides, such as strontium, rare earths, barium and calcium. Manganese sulfides also act as substrates but are less potent.

2. Effect of Minor Elements

Numerous elements present in small amounts in the molten iron exert marked effects on the structure and resulting properties of the gray iron by affecting its solidification behavior. Sulfur forms sulfides that are effective substrates for graphite nucleation, so the presence of some sulfur is needed to form these substrates. However, excess sulfur, beyond that needed to form substrates, restricts the growth rate of the graphite flakes, thereby affecting the graphite morphology and in extreme cases causing sufficient undercooling for some carbides to form. The sulfur content of gray irons is generally about 0.03–0.12%. Small amounts of other elements can produce restrictive effects on the growth of the flake graphite, resulting either in degenerate flake structures or carbide solidification. Rare earths are very potent in this respect. They form effective substrates for graphite nucleation when combined with sulfur, but in excess amounts can lead to severe undercooling and carbide formation. A similar effect on undercooling, because of eutectic growth restraint, occurs with tellurium. Lead and bismuth also produce undercooling by similar mechanisms and can result in degenerate graphite structures. Titanium additions to molten gray iron gradually change the graphite form Type A to Type D. This behavior is related to the interaction of titanium with sulfur and nitrogen to produce the undercooling needed for Type D graphite to form.

The effects that these elements exert on the nucleation and growth conditions of the eutectic have marked influences on the resulting mechanical properties. In general, the tensile strength of gray iron increases with decreasing carbon equivalent [%C + (%Si + %P)/3] and with lighter section sizes or faster solidification and cooling times, as illustrated in Figs. 3 and 4. A decrease in carbon equivalent reduces the amount of graphite present and increases the number of dendrites of austenite that form. Smaller section sizes reduce the size of the eutectic cell and generally increase the amounts of both austenitic dendrites and pearlite. When undercooling of the solidifying iron occurs at any given carbon equivalent, such as when Type D graphite forms, larger amounts of austenitic dendrites form, as demonstrated in Fig. 2. Since the dendrites do not contain any graphite, they are stronger than the graphite-containing portion of the matrix and result in higher strength levels. The importance of solidification behavior in determining the properties of the iron has led to the use of thermal analysis based on cooling curves during the solidification process to determine composition and structure.

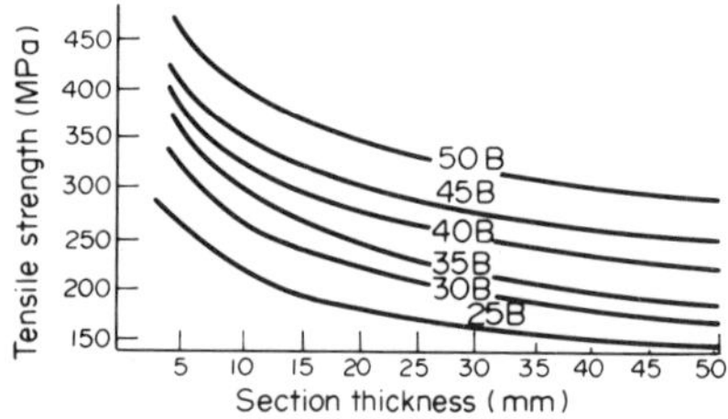

Figure 3
Effect of section thickness of casting on tensile strength of gray irons; symbols on curves denote the ASTM A48 classes (American Society for Testing and Materials 1976) determined by as-cast strength of a 30 mm diameter bar

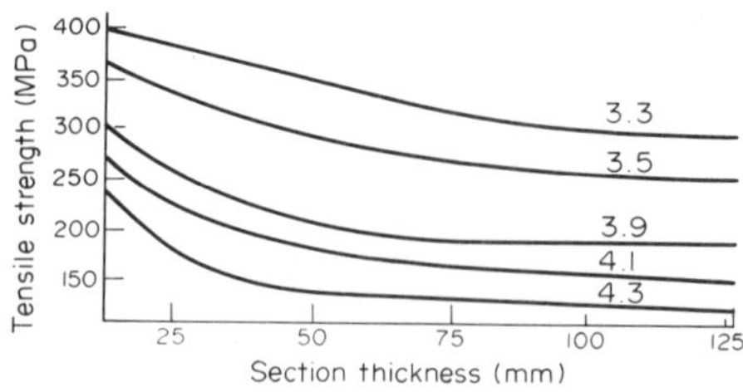

Figure 4
Effect of section thickness of casting on tensile strength of gray irons of different carbon equivalents as shown by figures on curves

3. Composition and Alloying Matrix Effects

Gray irons are produced in plain and low and higher alloy grades to serve a number of different applications. The carbon content of plain gray cast iron grades varies from 2 to 4%, and the silicon content from 1 to 3%. Appreciable amounts of manganese (0.65–1.0%), sulfur (0.02–0.25%) and phosphorus (0.05–1.0%) are also present. Low and higher alloy grades may contain additional amounts of chromium, molybdenum, nickel, copper or other alloying elements, depending on the application.

The tensile strength of the plain and lower alloy grades ranges from 69 to 414 MPa. The carbon equivalent and solidification behavior exert considerable effects on this strength level. The matrix structures may also be varied over a considerable range of hardness and this can exert a significant effect on the properties. Matrix structures are affected by the carbon and silicon contents as well as the presence of alloying elements and can be changed by heat treatment. The matrix structures vary from ferrite, with a low hardness and strength, to mixtures of ferrite and pearlite, completely pearlitic, bainitic and martensitic with steadily increasing hardness and strength. High carbon and silicon contents, slow cooling, the absence of alloying elements and annealing favor a ferritic matrix. Lower carbon and silicon contents, higher manganese contents and faster cooling produce a pearlitic matrix. Small additions of tin and antimony result in a stable pearlitic structure. Bainitic irons are obtained in the as-cast condition by significant alloying with nickel, copper and molybdenum. Martensitic irons are generally obtained by quenching and tempering treatments.

When alloying elements are added to irons, their effect is generally analyzed by their influence on the solidification behavior of the iron and their effect on cooling through the critical or eutectoid transformation temperature range. Elements which favor the formation of the stable and austenite–graphite eutectic are termed graphitizers and those that tend to form the metastable austenite–carbide eutectic are called carbide formers. These elements generally exert their effect by altering the temperature range between the stable and metastable eutectics. Graphitizers such as silicon and nickel increase this temperature range, while carbide formers such as chromium and vanadium reduce it. Nearly all alloying elements except silicon, aluminum and small amounts of titanium tend to increase the amount of pearlite formed in preference to ferrite in the matrix.

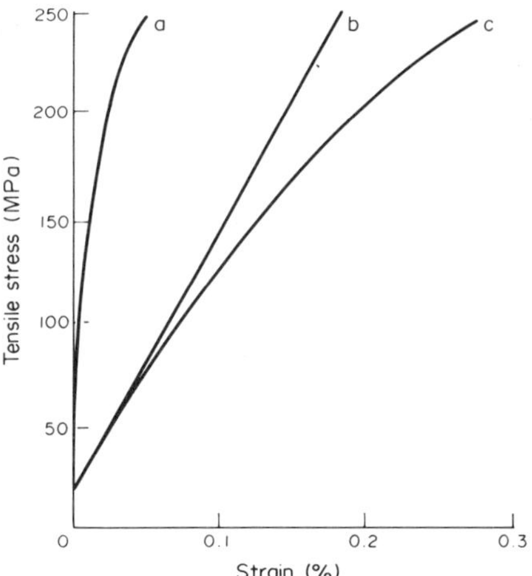

Figure 5
Tensile stress–strain behavior of gray iron: (a) plastic extension; (b) elastic extension; (c) total extension

4. Mechanical Properties

The presence of the flakes of graphite in an iron–silicon matrix causes the iron to exhibit some unusual properties for a metal. The size of the graphite flakes, the matrix structure and the amount of prior austenite dendrites present in the iron structure influence these properties. As mentioned above, the tensile strength of gray iron varies from 69 to 414 MPa, higher strength being attained with less graphite, finer flakes, stronger matrices and more dendrites. However, gray irons do not exhibit a straight stress–strain curve. Because of the stress concentrations developed at the edges of the graphite flakes (notch effect), some localized plastic deformation occurs there during loading and produces the curved stress–strain plot shown in Fig. 5. The amount of graphite, the graphite structure and the matrix also affect the modulus of elasticity, as measured by the tangent to the stress–strain curve at its origin. No true yield strength is measured or reported for gray irons. The modulus of elasticity varies from about 69 to 138 GPa. The value of Poisson's ratio is also affected by the graphite structure and tends to be about 0.24 for the soft irons and 0.27 for the higher-strength irons, at the start of loading. The values decrease at higher stresses.

The graphite flakes exert considerably less influence on the compression properties than on the tensile properties. For this reason, the compression strength at fracture of gray irons is three to four times the fracture strength under tension. For the same reason the relation between the tensile strength and hardness is affected by the amount of graphite present. The ratio of tensile strength to hardness is higher for the higher-strength and lower-carbon-equivalent gray irons. Combined tension and compression stresses encountered in bending and other types of loading are affected by the different response of gray iron to tensile and compressive stresses. Transverse tests provide a considerably higher breaking load for gray iron than calculated values based on simple beam formulae and the tensile strength of the iron. The tension or shear strength of gray iron is also higher than determined from the tensile properties of homogeneous metals. The shear strength of gray iron is reported to be 1.1–1.5 times the tensile strength.

The fatigue properties of gray iron are affected by the graphite and matrix structures, tensile strength and mode of stressing. Because gray irons are stronger in bending than in tension, the endurance ratios under bending stresses exceed those under tension by a considerable amount. Irons with more dendrites in the structure or with a Type D graphite have higher endurance ratios than those with less dendrites or Type A graphite. The endurance limit generally increases with tensile strength but is affected by the stressing mode and graphite structure, as discussed above.

The presence of the graphite flakes increases the damping capacity of gray iron and this makes it useful for preventing objectionable vibrations and noise in machine bases, brake components and engine cylinder blocks. The damping capacity of gray iron is due to the energy absorbed at the edges of the flakes. For this reason, the damping capacity is much greater for the lower-strength irons, which are very significantly better than steel in this respect.

See also: Cast Irons and Casting Processes; Malleable Cast Iron; White and High-Alloy Cast Irons; Spheroidal Graphite Cast Irons

Bibliography

American Society for Testing and Materials 1967 *Standard Test Method for Evaluating the Microstructure of Graphite in Iron Castings*, ASTM Designation A247-67. American Society for Testing and Materials, Philadelphia, Pennsylvania

American Society for Testing and Materials 1976 *Gray Iron Castings*, ASTM Specification A48-76. American Society for Testing and Materials, Philadelphia, Pennsylvania

Ruff G F, Wallace J F 1976 Control of graphite structure and its effect on the mechanical properties of gray iron. *Trans. Am. Foundrymen's Soc.* 84: 705–28

Ruff G F, Wallace J F 1977 Effects of solidification structures on the properties of gray iron. *Trans. Am. Foundrymen's Soc.* 85: 179–202

Sachar H, Wallace J F 1982 Effect of microstructure and testing mode on the fatigue properties of gray iron. *Trans. Am. Foundrymen's Soc.* 90: 777–93

Wallace J F 1974 Cast iron solidification. *Electr. Furn. Conf. Proc. AIME* 32: 201–9

Wallace J F 1975 Effects of minor elements on the structure of cast irons. *Trans. Am. Foundrymen's Soc.* 83: 363–78

Walton C F (ed.) 1971 *Gray and Ductile Iron Castings Handbook.* Gray and Ductile Iron Founders' Society, Cleveland, Ohio

Walton C F (ed.) 1981 *Iron Castings Handbook.* Iron Castings Society, Des Plaines, Illinois

Weiser P F, Bates C E, Wallace J F 1967 *Mechanism of Graphite Formation in Iron–Carbon–Silicon Alloys.* Malleable Founders Society, Cleveland, Ohio

J. F. Wallace

Grinding

Grinding is one of the most versatile methods of removing material from machine parts to provide precise geometry. However, the process is very complex and difficult to study because of the small size of the individual chips produced by hard abrasive particles having a wide range of shape, spacing and relative elevation. Grinding operations may be classified in a variety of ways. One convenient classification is whether the wheel is dressed or whether the wear of the wheel is sufficiently high that it is self-dressing. In form-and-finish grinding (FFG), individual chips are relatively small and wear-flats develop on the active grains in the wheel surface. Periodically, the dull grains are removed or "sharpened" by dressing the wheel with a diamond tool. Typical FFG operations are: horizontal-spindle surface grinding, internal grinding, cylindrical grinding and centerless grinding. In stock-removal grinding (SRG), wheel wear is relatively high and the wheel is self-dressing. Examples of SRG include abrasive cut-off processes, conditioning of slabs and billets in steel mills, and vertical-spindle surface grinding.

In grinding, unwanted material is removed by small, extremely hard (about 2000 kg mm^{-2}), brittle refractory particles. In general, abrasive tools consist of silicon carbide, aluminum oxide, or of aluminum oxide with additions of oxides such as titanium oxide, chromium oxide or zirconium oxide. The individual particles are blocky crystals that roughly resemble a sphere with many sharp edges.

The size of an abrasive grain is expressed in terms of a screen number S which corresponds to the number of openings in a mesh per square inch. The mean diameter g of an abrasive grain is related to the screen number approximately as follows:

$$Sg = 0.7 \qquad (1)$$

Thus, a number 8 grain size has a diameter of about 0.09 in. (2.3 mm) and a number 46 grain size is about 0.015 in. (0.39 mm) diameter.

The depth of penetration of an abrasive grain into the workpiece is a small percentage of its diameter. For example, in the conditioning of steel, which is one of the coarsest grinding operations, the grain size is about 8 and the maximum depth of grain penetration t is about 0.004 in. (0.10 mm). The maximum value of t/g is thus about $0.004/0.09 = 4.5\%$. In horizontal-spindle surface grinding, a 46 grain size is common and the maximum grain penetration is about 0.0003 in. (0.01 mm). The corresponding value of t/g in this case is $0.0003/0.015 = 2\%$. It is thus evident that only an extremely small percentage of the surface of a grain is operative at any one time.

Abrasive particles may be (a) free, (b) mounted in resin on a belt (coated product), or (c) close packed into a relatively dense body held together by a resin or vitrified bonding material (bonded product). The grinding action is basically the same in all three cases but with important differences due to the spacing of active grains and their rigidity and degree of fixation. The grains have the closest spacing and are held with the greatest degree of rigidity in the bonded form. Consequently, bonded products are capable of producing surfaces of the greatest precision.

The structure of a grinding wheel is characterized by 3 parameters:

(a) The mean force required to dislodge a grain from the surface of the wheel. This is designated by the so-called grade of the wheel, which is a letter of the alphabet (A–soft, Z–hard).

(b) The mean void size and distribution. This is designated by the structure number, where 1 represents a very dense structure and 12 a very open one.

(c) The mean spacing of active grains in the wheel surface. This depends on the grain size g, the amount of bonding material, the dressing technique and the grain type (friability).

Some of these items are indicated in the wheel designation which is a collection of numbers and letters as indicated in Fig. 1.

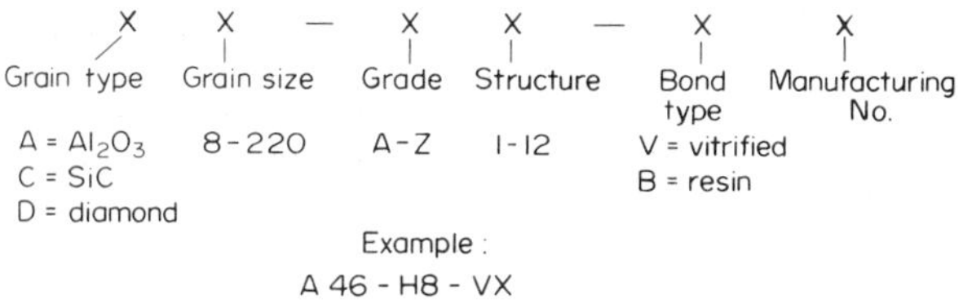

X	X	—	X	X	—	X	X
Grain type	Grain size		Grade	Structure		Bond type	Manufacturing No.
A = Al_2O_3 C = SiC D = diamond	8-220		A-Z	1-12		V = vitrified B = resin	

Example:
A 46 - H8 - VX

Figure 1
Grinding-wheel designation

Grinding wheels are dressed by machining the wheel surface with a diamond tool. This produces a round wheel and removes flats from worn grains by grain fracture or removal. The number of active cutting edges on the wheel face is altered by the depth of cut and the feed in the dressing operation.

The rate of wear of a grinding wheel is usually important to the economics and performance of the process. In the case of SRG, wear is usually expressed in terms of a grinding ratio

$$G = \frac{\text{volume of work removed}}{\text{volume of wheel consumed}}$$

In such cases, the ratio of change of wheel diameter is relatively large and G may be easily measured. In FFG, however, there is usually a negligible change in wheel volume during grinding and essentially all of the wheel wear is associated with dressing; in this case, the grinding ratio is not a convenient way of measuring wheel life. Instead, some means of accurately measuring the change in wheel radius, or of measuring the development of wear-flats on the active abrasive grains, must be devised.

1. *Undeformed-Chip Thickness*

In grinding, metal is removed as individual chips and the size (i.e., the undeformed depth t) of these chips plays a very important role. Some of the implications of chip size are as follows:

(a) Chip size determines the area of contact between the chip and the grain, and hence the force on the entire grain. This in turn determines the bond strength necessary to hold the grains in place.

(b) The specific grinding energy u (the work required to remove a unit volume of material) increases as t decreases. One reason for this is that the mean rake angle becomes more negative as t decreases.

(c) The specific energy has a strong influence on grain-tip temperature, which in turn influences the rate of wear and the workpiece temperature (of importance relative to possible surface damage).

(d) The finished surface is a composite of many individual scratches; surface roughness increases with the depth of the individual scratches t.

(e) The radial force on a grain increases with the contact area and hence with t. The resulting elastic deflection of both the wheel and the workpiece alters the local geometry of the wheel.

Because of the importance of undeformed chip thickness to the grinding process, it is important to be able to estimate this value for different grinding conditions.

The four basic kinematic arrangements used in grinding are shown in Fig. 2. In each case, the wheel speed is designated V, and the work speed v. The quantity d is the infeed per revolution or pass over the work, while $\dot{d}$ is the infeed rate in the case of the abrasive cut-off process.

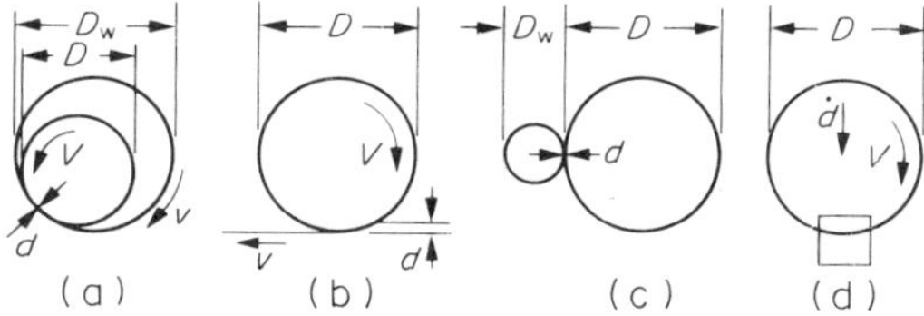

Figure 2
Representative grinding operations: (a) internal, (b) surface, (c) external, (d) abrasive cut-off

Figure 3 shows the shape of an individual surface grinding chip which has a maximum thickness t and a mean thickness $\bar{t}$. The chip is actually much longer relative to its thickness than shown and may be considered to be a slender triangle. The mean volume

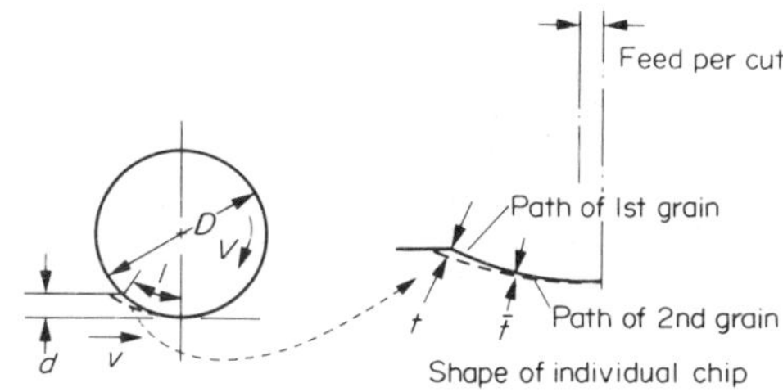

Figure 3
Shape of individual chips in surface grinding

B of the chip will be the product of its length l, mean thickness $\bar{t}$ and mean width $\bar{b}'$. The mean chip volume may also be expressed as follows:

$$B = \frac{\text{total volume removed/time}}{\text{number of cuts/time}} = \frac{vbd}{VbC} \quad (2)$$

where v = work speed, b = wheel width of cut, d = wheel depth of cut, V = wheel speed, and C = mean number of active cutting edges per unit wheel face area. Equating these two values for chip volume

$$B = \frac{vd}{VC} = l\bar{t}\bar{b}' = l\frac{tb'}{4} \tag{3}$$

or

$$t = \frac{4vd}{CVb'l} \tag{4}$$

Let r be the ratio of scratch width b' to scratch depth t, then substituting into Eqn. (4) gives

$$t = \left(\frac{4vd}{CVrl}\right)^{1/2} \tag{5}$$

If we ignore local wheel and work deflections, the length of cut will be as shown in Fig. 3, and to a good approximation will be

$$l = \sqrt{Dd} \tag{6}$$

where D is the wheel diameter. Substituting Eqn. (6) into Eqn. (5), we obtain

$$t = \left(\frac{4v}{VCr}\sqrt{\frac{d}{D}}\right)^{1/2} \tag{7}$$

It may be readily shown that Eqns. (6) and (7) hold for external and internal cylindrical grinding, providing that an equivalent diameter D_e is used in place of the actual wheel diameter D, where

$$D_e = D_w D/(D_w \pm D) \tag{8}$$

and D_w is the work diameter. The minus sign is used for internal grinding and the plus sign for external grinding. For surface grinding $D_w = \infty$; substituting this into Eqn. (8), we find that $D_e = D$ in this case.

2. *Fine-Grinding Mechanics*

The value of t in FFG is generally about 2.5 μm. For such small cuts, a very small part of the abrasive grain is active and the removal mechanism differs substantially from the concentrated shear mechanism that operates in cutting. In fine grinding, the removal mechanism is more akin to that of an indentation hardness test. Figure 4a shows the plastic zone that develops beneath a spherical indenter. The very small radius at the active point of the grain is responsible for the effective rake angle as shown in Fig. 4b. Here, the action is similar to that of a spherical indenter subjected to an inclined load. The effective radius of this indenter will of course be much smaller than that of the dotted circle shown in Fig. 4c. Material at the front of the "indenter" of Fig. 4b will be plastic but unsupported and will flow upward to generate the chip.

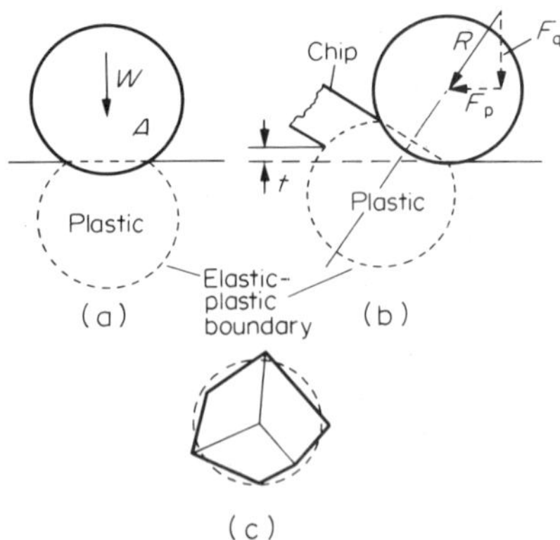

Figure 4
Schematic representation of fine grinding mechanism: (a) plastic zone in indentation hardness test, (b) plastic zone in fine grinding, (c) blocky abrasive particle showing equivalent diameter of entire particle (dotted circle)

The consequences of the chip forming mechanism shown in Fig. 4b are:

(a) Deformation of a much larger volume of material than escapes as a chip.

(b) Considerably greater subsurface flow than with the cutting mechanism of metal cutting, leading to a greater tendency for subsurface cracking in grinding than cutting, and to a greater tendency for residual stresses to develop.

(c) The proportion of the total energy that ends up in the workpiece surface in fine grinding is greater than that in metal cutting. In high-speed cutting, 90% or more of the energy consumed is removed in the chip, whereas in fine grinding most of the energy remains in the workpiece.

The fact that the specific energy in fine grinding is at least 50 times greater than that for cutting the same material is in agreement with the extrusion-like mechanism shown in Fig. 4b, as are other experimental observations. Also, adding to the specific grinding energy, is the fact that metal is pushed from side to side several times before leaving as a chip.

As the undeformed chip thickness increases, a higher percentage of the abrasive particle is active and the effective rake angle increases. For SRG where the chips are quite large, the removal mechanism approaches that of metal cutting and there is much less danger of overheating the work since more of the energy consumed ends up in the chip rather than in the workpiece. In the abrasive cut-off operation, chips are relatively large and the specific grinding energy will be only 6900 MPa instead of about 205 GPa for horizontal surface grinding and as much as 690 GPa for internal grinding.

3. Grinding Forces and Energy

The specific energy u for surface grinding is

$$u = \frac{\text{work/time}}{\text{volume removed/time}} = \frac{F_p V}{vbd} = \frac{F'_p V}{vd}$$

where F_p is the tangential force on wheel, V the wheel speed, v the work speed, b the grinding width, d the wheel depth of cut, and F'_p the tangential force per unit grinding width. The quantity d corresponds to the amount removed per revolution of the work and not to the infeed per revolution of the work since there may be a difference due to wheel wear or shaft deflection.

When u is plotted against t using log–log coordinates, a curve such as that shown in Fig. 5 is obtained. When plotting this curve, values of t were computed using Eqn. (7) in which C was assumed to have a constant value depending on the grain size.

It is significant that the specific energy in finish internal grinding is 100 times as great as for billet conditioning, and the value of u in fine surface grinding is about 30 times that in billet conditioning. The main reason for this lies in the difference in undeformed chip thickness for these three types of grinding and the consequent change in the chip-forming mechanism.

The specific grinding energy is a useful quantity for estimating grinding forces and power. For example, the chips in a billet conditioning operation will be about 0.1 mm thick and u will be about 6900 MPa. A metal removal rate (M) of 450 kg h^{-1} or 1082 cm^3 min^{-1} (a value sometimes achieved by modern conditioning machines) will require considerable power. Just how much power may be estimated as follows:

$$\text{Power} = uM = \frac{6900 \times 1082}{60 \times 1000} = 124\,\text{kW} \qquad (9)$$

The tangential force on the wheel (F_p) will depend

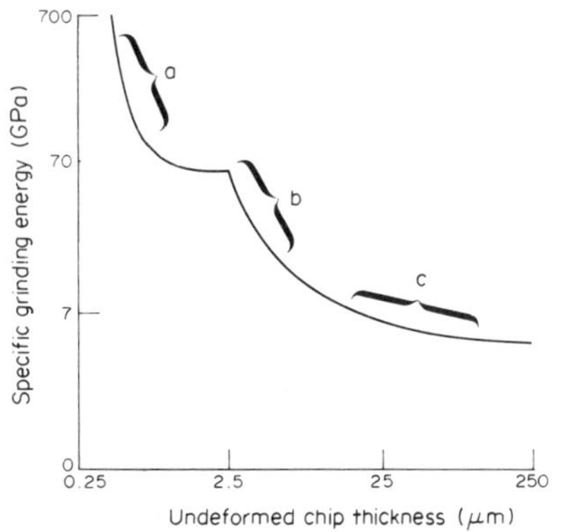

Figure 5
Variation of specific grinding energy u with undeformed chip thickness t for: (a) internal, (b) fine-surface and (c) external grinding operations

on the wheel speed V. If V is 81 m s^{-1}, then F_p is found to be

$$F_p = \frac{\text{power}}{V} = \frac{124 \times 10^3}{81} = 1531\,\text{N} \qquad (10)$$

The radial force on the wheel (F_Q) is normally about twice the magnitude of the tangential force and hence for this case F_Q will be about 3060 N.

At a wheel speed of 61 m s^{-1}, the wheel forces would be

$$F_p = (81/61) \times 1531 = 2033\,\text{N}$$

$$F_Q = 2 \times 2033 = 4066\,\text{N}$$

As another example, consider the power required in horizontal spindle surface grinding under conditions where t is about 5 μm. The value of u in this case will be about 40 GPa (Fig. 5). When grinding under the following conditions

wheel speed $V = 30\,\text{ms}^{-1}$

work speed $v = 0.3\,\text{ms}^{-1}$

cross feed per stroke $b = 1.2$ mm

wheel depth of cut $d = 0.05$ mm

removal rate $M = vbd = 0.018 \times 10^{-6}\,\text{m}^3\,\text{s}^{-1}$

the required power will be

$$uM = 40 \times 10^9 \times 0.018 \times 10^{-6} = 0.72\,\text{kW}$$

The forces on the wheel may be estimated for this case as follows:

$$F_p = \text{Power}/V = 720/30 = 26\,\text{N}$$

$$F_Q = 2\,F_p = 52\,\text{N}$$

The tangential force per grain F'_p for a horizontal spindle surface grinding operation will equal the product of the specific energy u and the cross sectional area of the cut ($rt^2/4$). Substituting into Eqn. (7) gives

$$F'_p = ur\frac{t^2}{4} = \frac{uv}{VC}\left[\frac{d}{D}\right]^{1/2} \qquad (11)$$

This equation is useful in determining which variables to change to make a given wheel change its apparent hardness. To make the wheel behave harder (i.e., to lose grain less readily) conditions should be changed to decrease the force per grain F'_p. From Eqn. (11) it is evident this may be done by decreasing the work speed v, increasing the wheel speed V using a finer dressing procedure to make C larger, or by decreasing the wheel depth of cut d.

4. Abrasive Cut-Off

In the case of abrasive cut-off, the chips are of constant depth and $\bar{t} = t$. Equating the volume per chip computed in two ways as previously, we obtain

$$lt\,(rt) = vd/VbC \qquad (12)$$

or

$$t = (\dot{d}/VCrl)^{1/2} \tag{13}$$

where l is the length of cut (i.e., the work width), rt is the chip width, $\dot{d}$ is the downfeed per unit time, V is the wheel speed, and C is the number of active grains per unit area of wheel face.

Unlike the types of grinding considered previously (i.e., surface, external and internal), many whole grains and large fractions of grains are lost from the wheel in abrasive cut-off. The rate of wheel loss is so great that dressing is unnecessary. The grinding ratio G (ratio of volume of work removed to volume of wheel consumed) is a convenient measure of wheel wear. When a given wheel cuts a given work material at constant wheel speed V, but with different values of d and l, G is found to vary with $\dot{d}$ as shown in Fig. 6. Each of these curves is found to have a maximum value and these maximum values are found to decrease as l increases.

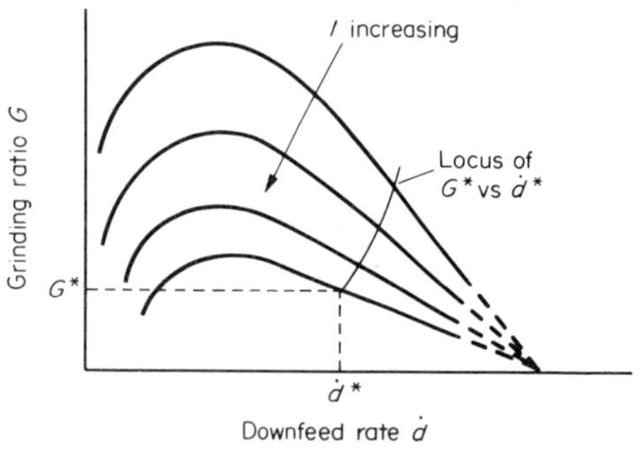

Figure 6
Variation of grinding ratio g with downfeed rate $\dot{d}$ for an abrasive cut-off operation: l is the length of cut in the circumferential direction and * values are for cost optimum conditions

The maximum arises as a result of two actions, one of which is dominant at low values of $\dot{d}$ whereas the other is dominant at high values of $\dot{d}$. At low values of $\dot{d}$, a thermal front moves into the work at a rate greater than $\dot{d}$ and the temperature rises as the cut proceeds. At low values of $\dot{d}$, grains are lost due to softening of the first row of resin bond-posts. As $\dot{d}$ increases, this tendency to lose grains due to bond-post softening decreases and hence G rises. Above a certain value of $\dot{d}$, the volume per chip exceeds the volume available for chip storage between the grains. The force on the grain then becomes excessive and G tends to decrease with increasing $\dot{d}$. The combination of these two actions produces the observed maximum.

The fact that the maximum occurs at lower values of $\dot{d}$ as l increases is due to the fact that the volume per chip is $l\dot{d}/VC$, and thus as l increases, the corresponding value of $\dot{d}$ for constant chip volume must decrease. The reason for the decrease in G_{max} with increased l is that the temperature at the first row of bond posts increases with increased l.

The tangential force per grain for an abrasive cut-off operation may be found as follows (referring to Eqn. (13)).

$$F'_p = urt^2 = u\dot{d}/VC \tag{14}$$

To make an abrasive cut-off wheel behave softer, to avoid the development of excessive wear flats which cause overheating, F'_p should be increased. This may be achieved by increasing the downfeed rate $\dot{d}$, or decreasing the wheel speed V. It is also evident that a given wheel will behave as though harder when wheel speed is increased but other variables are kept the same.

5. *Grinding Fluids*

The application of grinding fluids is a very complex business. In many operations, use of a grinding fluid of some description is advantageous. However, there are situations when it is best to grind in dry air: the abrasive cut-off operation is frequently such a case. In this instance, cooling is unnecessary since a large percentage ($\geqslant$50%) of heat generated goes off with the chips when the downfeed rate $\dot{d}$ is adjusted for cost optimum operation. Furthermore, the lubricating action which decreases the tendency for chip–grain welding is offset by a decreased tendency towards workpiece softening when a fluid is used. If a fluid is used in abrasive cut-off, it should be directed so as to cool the wheel after it leaves the cut, as in friction sawing. An important consideration with resin-bonded abrasive cut-off wheels, is that the temperature of the first row of bond posts must not be permitted to reach the resin softening temperature which will allow the grains to be released.

In surface, internal and external cylindrical grinding, a fluid is generally useful. Here, two types of fluid are used: oil-based and water-based. The choice depends upon the relative importance of lubrication and cooling action. When grinding an easy to grind material at normal speeds (30–40 m s^{-1}), or where surface finish is not critical, a water-based material will often suffice and is to be preferred over an oil-based material since it is less messy. However, when the grinding stress increases (e.g., when grinding stainless steel or a high-temperature alloy), the tendency for chips to weld to the wheel increases and a better lubricant is required. Under such conditions, an oil-based material containing fatty adsorbable substances or chemically active chlorine, sulfur or phosphorus additives is preferred. Better lubrication is also required when a better surface finish is desired.

Oil-based fluids usually give a lower specific grinding energy and hence lower grinding forces and power. Water-based fluids on the other hand are generally better coolants. The combined effect of better cooling action but higher specific grinding energy for water-based fluids tends to give grinding temperatures, at conventional wheel speeds, that are about the same as those obtained with an effective grinding oil.

In general, it is preferable to shift to a grinding fluid of decreased water content as the wheel speed is increased. This is due to the greater need for lubrication rather than cooling, at high wheel speeds. While this statement may at first appear paradoxical, the explanation lies in the fact that there is insufficient time for effective cooling of the grain tips at high wheel speeds, whereas there is a noticeable improvement in performance with increased lubrication at such speeds.

When grinding metals which do not soften at elevated temperatures, such as stainless steel and high-temperature alloys, it is frequently best to use a reduced wheel speed and an active grinding oil that provides some protection against welding. This is particularly true when grinding titanium alloys and materials such as molybdenum which have a strong tendency to form strong primary bonds with the abrasive materials that are commonly used.

In high-speed grinding, it is difficult to make the fluid penetrate the air boundary-layer on the wheel face so as to allow the surface grains to become wetted. It is customary to use a scraper that "peels off" the air film before injecting the fluid onto the wheel face at high velocity. Unless the fluid penetrates to the wheel face, the conditions may resemble those in dry grinding.

General cooling of the workpiece is important for dimensional accuracy, otherwise dimensional errors due to thermal expansion may become excessive. When grinding large-diameter workpieces to a high degree of accuracy, it is sometimes necessary to use a second, low-velocity, high-flow-rate stream of fluid so as to completely immerse the work in coolant.

Grinding temperatures are particularly important when grinding heat-treated parts at high wheel speeds. A grinding oil which provides a less drastic cooling action generally leads to fewer grinding cracks and other surface damage than if a water-based fluid is used.

The success of a fluid in laboratory tests does not mean it will be satisfactory under workshop conditions. A practical grinding fluid must be stable with respect to structural breakdown and bacteriological growth and must also be nontoxic to the operator. There must be no adverse performance characteristics such as a tendency to foam, form gummy or salt-like deposits, remove paint or cause other damage to the grinding machine.

6. *Surface Integrity*

In many cases, the quality of the finished surface is an item of major concern. This involves such factors as surface finish, residual surface stresses, thermally induced damage (e.g., oxidation and burn, overtempering, surface cracks), and mechanically induced surface cracks and highly strained areas.

From the standpoint of brittle fracture and fatigue, residual surface stresses, if present, should be compressive. When surface integrity is a problem, the use of sharp wheels, and grinding conditions to avoid built-up edge formation or overheating of the surface are principal items for consideration. Since the cutting temperature is proportional to $V^{1/2}$, high-temperature problems are apt to occur in grinding since V will normally be about 30 m s^{-1} or greater. When excess grinding temperatures are a problem, a shift to a lower wheel speed (~7.5 m s^{-1}), use of a sharp wheel (frequent dressing), lower removal rates and an active oil-based lubricant will usually be helpful.

See also: Chip Formation Micromechanics and Metal Cutting; Machinability of Metals; Metals Processing and Fabrication: An Overview

Bibliography

Klautsch D N, Zorev N N 1968 Eigenheit des Schleifvorganges beim Intensiven Ermüdungsadhasiven Verschleiss der Schleifkörner. *CIRP Ann.* 16: 267–75

Optiz H, Gürhing K 1968 High speed grinding. *CIRP Ann.* 16: 61–73

Reichenbach G S, Mayer J E, Kalpakcioglu S, Shaw M C 1956 The role of chip thickness in grinding. *Trans. Am. Soc. Mech. Eng.* 78: 847–59

Shaw M C 1971 A new theory of grinding. *Harold Armstrong Conference on Production Science in Industry: Papers.* The Institution of Engineers, Sydney, Australia

Shaw M C (ed.) 1972 *New Developments in Grinding.* Carnegie Press, Pittsburgh, Pennsylvania

M. C. Shaw

Group Technology

Group technology is a manufacturing philosophy which identifies and exploits the similarities of parts and manufacturing processes. In conventional batch-type manufacturing, each part is treated as being unique from initial design through to the final product. However, by grouping similar parts into part families based on either their geometries or manufacturing processes, it is possible to reduce costs through (a) more effective design rationalization and design data retrieval, (b) fewer stocks and purchases, (c) simplified and improved production planning and control, (d) reduction of tooling and setup times, (e) semi-flow-line production by machine groups/cells, (f) less in-process inventory, (g) reduction of total throughput time, (h) reduction of numerical control (NC) programming and more efficient NC machine and machine center utilization, and (i) effective data retrieval systems for successful implementation of computer-aided process planning.

The basic concepts of group technology have been practised for many years. Applications of the concepts are usually identified under different names and in various forms of engineering and manufacturing functions. Traditionally, group technology practices were limited to batch-type manufacturing for productivity improvement, with different degrees of success either in the design or manufacturing areas. For many years, group technology did not receive formal recognition and was not rigorously practised as a systematic scientific technology. In recent years, however, an intensified effort in computer-integrated manufacturing (CIM) has stimulated a stronger interest in the subject, since it provides the essential means for higher manufacturing productivity.

1. Part Families

A part family or group is defined as a collection of related parts which are nearly identical or similar in shape and/or processing requirements. Alternatively, they may be dissimilar in shape, but related by sharing all or some common production operations. The grouping of related parts into families is the key to the group technology concept. There are three basic methods used to form part families: (a) manual visual search or functional names, (b) production flow analysis, and (c) classification and coding systems.

2. Classification and Coding

Classification arranges items into groups according to some principle or system whereby the items are brought together by virtue of their similarities, and then separated by a specific difference. A code can be a system of symbols used in information processing in which numbers and/or letters are given a certain meaning.

Many varieties of classification and coding systems are in use. An ideal coding system is one which is properly adapted to meet the specific needs of the company where it is being used. A company can devise its own unique system based upon publicly available systems or adapt a commercially available system to meet its needs. It is essential that an adapted system be applicable and usable by management and all concerned departments in the company, including design/engineering, planning/control and manufacturing/tooling.

An important application of a good classification and coding system is in connection with design data retrieval and design rationalization. Not only design information, but material specifications and production planning information can be readily available. All relevant production information can be retrieved for functions such as scheduling, group machining, group tooling and setup. A classification and coding system also facilitates a part reduction and standardization program valuable to the company and its customers.

3. Plant Layout and Tooling Setups

There are three basic types of plant layout: (a) line layout, (b) functional layout and (c) group layout. In the practice of group technology, a group of machines for one or more part families may be formed such that it can perform all the operations required for the family or families of parts. The machines themselves are arranged in a semi-flow line to minimize transportation and waiting. The result is very similar to a modern machining center, and if conditions warrant, may be used to replace a group of single-purpose machines. These machine–tool groups may be defined as machine cells or machining centers.

For maximum utilization of tooling setups, tooling for the operations within a part family should be arranged so that all or as many as possible of the parts in a family can be processed with one group-type fixture and/or one setup. Group jigs and fixtures are designed to accept every member of the family, they are supplemented by adapters which can accommodate some differences between the parts in the family. It is evident that tooling and setup costs can be reduced using group technology.

4. Numerical Control Part-Family Programming

One of the important applications of group technology is software development for NC machining. This is referred to as part-family programming. Part-family programming is an NC program system that groups common or similar program elements into a single, master computer program. Therefore, part-family programming increases the productivity of costly NC operations by saving programming time, manpower and tape proveout time. It also reduces lead time, simplifies maintenance and requires fewer computer reruns.

5. Computer-Aided Process Planning

An essential requirement for the implementation of CIM is computer-aided process planning. The use of an automated process planning technique is a basis for a rational and logical approach to improve manufacturing productivity in a CIM system. The part-family concept of group technology is an essential element in the successful execution of computer automated process planning. The development of this concept, using suitable classification and coding systems, will make it possible to systematically form part families, and thus rationalize and develop standards of shape and process within the part families, enabling the automatic generation of process plans.

6. *Production Control and Scheduling*

Production scheduling is greatly simplified by group technology. The scope of the problem is reduced to a small group of machines instead of involving a large portion of the shop. Within the group of machines, the scheduling problem is again reduced to simply scheduling the given jobs through the machines in the cell. A computer program can be developed to schedule jobs of a part family to a corresponding machine group/cell. Jobs can be properly sequenced in the part family and the families properly sequenced through the machine groups/cells for the optimum conditions by minimizing total throughput time.

Proper scheduling may be considered an integral part of group technology. When combined with setup time and transportation, two significant cost-reduction benefits can be effected. Most obvious is shorter total production time, which means that inventories can be reduced substantially and production work can proceed on schedule. At the same time, scheduling becomes more reliable, guaranteeing the delivery date with greater assurance. Production scheduling associated with group technology application is called group scheduling. Since group technology applications directly relate to and influence various planning and inventory activities, it is important to consider the interrelationships between group scheduling and material requirements planning (see *Materials Requirements Planning*).

7. *Management Problems*

When a job is assigned to a machine group/cell, it is processed by that group/cell. The flow of materials or parts into or out of the shop becomes smoother and more accurate with the group production method. Since all the jobs in a group are confined to a small group of operators in a relatively small area of the shop, the group supervisor can have more effective control.

An additional factor of a social nature which group production methods provide is job satisfaction derived from factors such as worker involvement in decision making, personalized work relationships, variety in tasks and freedom to determine group production methods.

Implementation of group technology calls for a high degree of cooperation between a number of groups or departments in the company. Personnel in design engineering, planning controls, tooling, and the production shop itself must realize how dependent each group is on the others.

8. *Future Trends*

It is evident that new technological innovations, such as direct numerical control (DNC), computer numerical control (CNC), machining centers, industrial robots and microprocessors, will be continuously introduced toward a more automated computer-integrated manufacturing systems involving computer-aided design/computer-aided manufacturing (CAD/CAM). This will lead to more integrated applications of group technology for higher manufacturing productivity.

As evolution of CAD/CAM leads to generative design and to generative process planning, certain part classification and coding systems will become an integral part of the total generative system evolving with CAD/CAM. The optimum design and effective operation of a fully automated manufacturing cell system and/or factory systems requires a careful analysis of all the requirements which the systems must satisfy in a group technology environment.

See also: Machine and Process Control in Metals Processing; Computer Applications in Metal Forming; Design and Materials Selection and Processing

Bibliography

Burbidge J L 1970 *Group Technology: Proc. Int. Seminar.* International Centre for Advanced Technical and Vocational Training, Turin

Burbidge J L 1975 *The Introduction of Group Technology.* Wiley, New York

Computer-Aided Manufacturing International 1977 *CAM-I Training Material for CAPP CAM-I Automated Process Planning,* Vol. 1. Computer-Aided Manufacturing International, Arlington, Texas

Edwards G A B 1971 *Readings in Group Technology: Cellular Systems.* Machinery Publishing, Brighton, UK

Gallagher C C, Knight W A 1973 *Group Technology.* Butterworth, London

Ham I, Hitomi K, Yoshida T 1984 *Group Technology: Applications to Production Management.* Kluwer-Nijhoff, Boston, Massachusetts

Mitrofanov S P 1966 *Scientific Principles of Group Technology.* National Lending Library for Science and Technology, Boston Spa, UK

Mitrofanov S P 1970 The group machining method for increasing productivity. *Sov. J. Opt. Technol.* 37: 217–19

Opitz H 1970 *A Classification System to Describe Workpieces.* Pergamon, Oxford

I. Ham

Gun Drilling

Gun drilling is a hole-cutting process employing single-lip, self-guided tools and high-pressure cutting fluid with either external or internal chip removal. It is used where other hole cutting processes would require special handling (e.g., interruption for chip removal) or secondary qualifying operations (i.e., reaming, honing, straightening).

A gun drill with external chip removal consists of a carbide tip with a single V-shaped cutting edge and coolant hole attached to a single-flute tubular shank and driver. The V-flute runs the length of the tool up to the driver and acts as the chip removal channel. The coolant hole is approximately one-quarter of the tip diameter and runs the full length of the tool. The tip cuts past center from one side, and is either brazed to the shank (typical for drills for holes less than 3 cm) or to an attached insert for large holes.

The tips are manufactured to within +0.000/−0.005 mm of nominal diameter with a back taper of approximately 0.0006 mm to produce a hole tolerance of 0.025 mm over the life of the tool (i.e., after multiple resharpenings).

The internal chip removal drill contains a tip connected to a tubular shank for chip removal. The tip consists of a hollow steel body with a brazed carbide cutting edge and bearing surfaces which are ground to the hole diameter. The bearing surfaces are strips on the tip periphery located at 180° to each other. The remainder of the drill has a smaller diameter than the bearing surfaces to allow passage of coolant. The tool nose is designed with a point offset from the tool axis. Negative rake and tool-lip steps are employed for strength near the point and chip-size control, respectively.

See also: Drilling; Metal Removal Processes; Metals Processing and Fabrication: An Overview

Bibliography

Dallas D B (ed.) 1976 *Tool and Manufacturing Engineers Handbook*, 3rd edn. Society of Manufacturing Engineers, Dearborn, Michigan, Chap. 3, pp. 30–40

A. L. Hoffmanner

Hafnium Production

Hafnium belongs to the same subgroup of the periodic table as titanium and zirconium. It always occurs in association with zirconium minerals because it has the same charge and practically the same ionic radius as zirconium. The atomic weight of zirconium was long in doubt because variable amounts of hafnium attended all the early preparations, and it was not until 1923 that hafnium was identified as a separate chemical species by means of x-ray spectroscopy. Its extractive metallurgy is thus intertwined with that of zirconium. Hafnium has found its principal use in nuclear reactor control rods because of its high neutron-absorption cross section (nearly 600 times that of zirconium). Hafnium carbide is the most refractory binary composition known and has found use as a substitute for tantalum carbide. Hafnium metal is also used as an alloy addition in superalloys. Annual world production of hafnium is estimated to be about 50 t.

1. Occurrence

Zircon is the only hafnium-containing mineral of commercial significance (see *Zirconium Resources*). Zircon, nominally $ZrSiO_4$, contains 1–5% hafnium as a replacement for zirconium. Zircon most often occurs as an accessory mineral in granites and syenites, but as a result of weathering and erosion processes zircon concentrations are also found in sands.

Hafnium is in very limited supply since it is a by-product of zirconium metal production and only about 2–5% of all zirconium mined is processed for metal production.

2. Concentration

Zircon is usually a coproduct of upgrading of rutile (titanium). Beach sand is mined by dredges and treated by wet-gravity methods to effect heavy-mineral separation. Zircon is then extracted from the other heavy minerals by combinations of screening, wet-gravity, electrostatic and electromagnetic processes.

3. Metal Extraction

Hafnium is always a by-product of zirconium extraction, and is separated after the initial crude chlorination of zircon in the presence of petroleum coke. In the separation step, crude $ZrCl_4$ is mixed with water containing ammonium thiocyanate (NH_4CNS), and the solution is brought into contact with methyl isobutyl ketone (MIBK). The hafnium, along with impurities, enters the MIBK phase, and the impurities are then removed by bringing the MIBK phase back into contact with an aqueous phase that is adjusted to remove the impurities selectively. The hafnium is subsequently stripped back into an aqueous phase, precipitated as a hydroxide, and calcined to HfO_2.

The HfO_2 is reacted with petroleum coke and chlorine in a fluid-bed chlorinator. The $HfCl_4$ product is then injected into a molten bath containing KCl, NaCl and a small amount of sodium. High-purity $HfCl_4$ is volatilized out of the bath, and reacted with high-purity magnesium in a Kroll-type reaction at 800 °C. The hafnium sponge so formed is vacuum distilled to remove residual salts and unreacted magnesium.

Further refining of hafnium is usually necessary. Refining may be accomplished in one of three different ways:

(a) Electron-beam melting—this process is carried out under a hard vacuum. All residual salts and magnesium are removed by the vacuum system, as are all tramp materials having low vapor pressures. All impurities except carbon, nitrogen and refractory metals can be removed by this method.

(b) Electrorefining—in this process, unrefined hafnium forms the anode in an electrolytic cell containing a fused-salt electrolyte. Impurities drop out as the hafnium passes over to the cathode.

(c) Iodide process—this is the well-known de Boer–van Arkel process. Hafnium metal is first reacted with iodine vapor to form HfI_4, which is then transported to a chamber containing a hot tungsten filament where the HfI_4 decomposes, forming crystal-bar hafnium, and iodine which is recycled.

Commercially, both electron-beam melting and iodide processing are used to effect refining.

The complexity and cost of the MIBK–thiocyanate process has stimulated research into new processing methods, but to date none has been proved to be commercially viable. Separation by distillation is a strong theoretical possibility. The triple points of $ZrCl_4$ and $HfCl_4$ are only 3 °C apart, but the vapor pressures are 14 200 mm Hg and 24 200 mm Hg, respectively. The separation coefficient for single-stage sublimation at 437 °C is thus 1.70. Recent Japanese work indicates the possibility that the formidable engineering materials problems can be resolved. The key is to use a double-shelled reactor with the space between the shells pressurized to levels necessary to prevent stress corrosion of the inner chamber where the distillation is carried out.

See also: Zirconium Production; Metals Production: An Overview

Bibliography

Lynd L E 1980 *Zirconium and Hafnium*, US Bureau of Mines Bulletin 671. US Bureau of Mines, Washington, DC

Megy J 1978 Method of separating hafnium from zirconium. US Patent No. 4,072,506 (7 February 1978)

Spink D R 1977 Extractive metallurgy of zirconium and hafnium. *Can. Min. Metall. Bull.* November: 145

Thomas D E, Hayes E T (eds.) 1960 *The Metallurgy of Hafnium*. Naval Reactors, Division of Reactor Development, US Atomic Energy Commission, Springfield, Virginia

H. W. Rosenberg

Halide Metallurgy

Halogens are used both in the primary extraction of metals from their ores and in the refining of metals to high purity. There are several reasons for this: (a) many metals have oxides that are very difficult to reduce, but halides that are not so refractory; (b) oxides of carbide-forming metals cannot be reduced by carbon or carbon monoxide; (c) halogens are less soluble in metals than oxygen and are more easily removed; and (d) the relatively high vapor pressures of many metal halides permit the use of distillation in the elimination of metallic impurities.

Some metals form several halides. Generally, the higher the valence, the more covalent is the compound, i.e., more volatile and easier to reduce to metal. For a given metal, the stability of its halides is proportional to the electronegativity of the halogens and increases in the order: iodide, bromide, chloride, fluoride. Care must be taken to avoid contact of the metal halides with water—the alkali and alkaline-earth halides are hygroscopic, and the transition-metal chlorides hydrolyze to produce oxides or oxychlorides and liberate hydrogen chloride gas. Thus, for the latter, subsequent dehydration can be achieved only by carbochlorination or by the use of powerful chlorinating agents such as thionyl chloride.

1. Preparation of Metal Halides

Anhydrous metal halides are prepared by several routes. The most common method, especially for chlorides, is halogenation of metal oxide in the presence of a reducing agent such as carbon, hydrogen or sulfur. This is typically performed in a shaft furnace. Alternative methods include halogenation of ferroalloys or carbides, and crystallization from aqueous solution. Crystallization works only for alkali chlorides and some alkaline-earth chlorides as the fluorides are typically insoluble in water and reactive metal halides hydrolyze.

Metal halides can be purified by several methods. In the case of volatile chlorides either fractional distillation or fractional reduction with hydrogen is used. In the case of water-soluble salts, dissolution and reprecipitation or recrystallization are feasible. Solvent extraction and ion exchange are also used.

2. Reduction of Metal Halides

There are four major methods of reducing metal halides to metal: metallothermic reduction, hydrogen reduction, thermal decomposition and electrolysis. Metallothermic reduction is the most common reduction method and is used primarily for producing reactive metals. The metal halide, either in solid or vapor form, is made to react with a metallic reducing agent, usually magnesium, calcium or sodium. In the Kroll process for titanium, titanium tetrachloride vapor reacts with magnesium in a closed reactor at 900 °C to produce solid titanium sponge and molten magnesium chloride. In the Hunter process, titanium tetrachloride vapor reacts with sodium at 900 °C to produce solid titanium sponge and molten sodium chloride. The metal sponge from both processes is subsequently melted to remove occluded salt. In the production of zirconium and hafnium there are similar processes in which the tetrachloride vapors are reacted with metallic magnesium, sodium or aluminum. Metallothermic reduction can also work when the metal halide is involatile. Uranium is produced by the reaction of uranium tetrafluoride powder with calcium metal chips in a calcium fluoride crucible at 1800 °C. There is no need for an inert atmosphere, as the molten calcium fluoride product makes this reaction self-fluxing. Beryllium is produced by the reaction of beryllium fluoride with metallic magnesium in a graphite crucible at 1300 °C. Calcium chloride is added as a fluxing agent.

Reduction of metal halides by hydrogen is another route to metal. Tungsten hexachloride borne by an argon carrier gas can be reacted with hydrogen in a two-zone furnace to produce tungsten powder of specified size. Process variables include the gas composition, hydrogen flow rate, and reaction temperature. Solid ferrous chloride has similarly been reduced with hydrogen to yield iron powder.

Both metallothermic reduction and hydrogen reduction of metal halides occur in primary metal extraction operations. In contrast, thermal decomposition of metal halides is used in the refining of metals. The prime example is the van Arkel–de Boer process for purifying titanium or zirconium. Impure metal is charged into a glass reactor where it is kept next to the reactor wall. The reactor, which contains a small quantity of iodine, is evacuated, sealed and heated to approximately 200 °C, at which temperature volatile titanium tetraiodide forms. A titanium filament which can be electrically heated is positioned in the center of the reactor. When molecules of titanium tetraiodide collide with the incandescent filament, they decompose to titanium metal which deposits on the filament, and to iodine gas which can again react with the titanium feedstock on the wall of the reactor.

A second example of refining by thermal decomposition of a metal halide is the subhalide process for aluminum. Aluminum trichloride vapor reacts with aluminum metal at 1200 °C to produce volatile aluminum monochloride. At temperatures below 700 °C the monochloride decomposes to give aluminum metal and trichloride vapor. In this way aluminum can be purified by transport out of an impure charge.

The final principal reduction method for metal halides is fused-salt electrolysis (see *Electrometallurgy*).

See also: High-Purity Metals; Metals Production: An Overview

Bibliography

Kroll W J 1956 The pyrometallurgy of halides. *Metall. Rev.* 1: 291–337

Minkler W W, Baroch E F 1981 The production of titanium, zirconium, and hafnium. In: Tien T K, Elliott J F (eds.) 1981 *Metallurgical Treatises*. The Metallurgical Society (AIME), Warrendale, Pennsylvania, pp. 171–89

D. Sadoway

Halides

Halides (from the Greek *halos*, salt) are highly ionic salts of metals (M) and halogen gases (X_2). For the purpose of this article, halides are a class of nonmetallic, usually binary, inorganic solids incorporating F^-, Cl^-, Br^- or I^- ions in combination mainly with ions of group IA, IB or IIA metals. (Astatine is rare and highly radioactive, and astatides are therefore not covered in this article.)

The three most prominent members of this class are the alkali halides, the silver halides and the alkaline earth fluorides, all of which have been intensively studied since the early nineteenth century because of the relative purity of their naturally occurring mineral forms and the ease of preparation of synthetic analogues by crystallization from solution or the melt. These solids have in common a strong ionic bonding (binding energies ~1 MJ mol^{-1}), moderately high melting points (T_m ~ 1000 K), a wide electronic band gap (~8 eV), ionic conduction and easy cleavage on prominent crystallographic planes. They frequently serve as model systems for the study of relationships between bonding and structure, surface energy, epitaxy, ionic diffusion, point defects and color centers, solidification, plastic deformation and radiation damage. Commercial applications include use as optical windows, scintillator hosts, photographic and electron lithographic media, dosimeters, lasers, thin film substrates, fast ion conductors and geologically stable rock formations for storage of high-level solidified nuclear wastes.

Mixed-valence halides (e.g., $KMnF_3$, $KMgF_3$, $RbMgF_3$) have been investigated for their spectroscopic and in some cases magnetic properties. Other halides have been much less studied in the solid state and are not reviewed extensively here. Selected physical properties are listed in Table 1.

1. Structure

At normal temperature and pressure, seventeen of the twenty alkali halides, together with the silver halides AgCl and AgBr, crystallize with 6:6 coordination in the face-centered-cubic rocksalt structure. CsCl, CsBr and CsI crystallize with 8:8 coordination in the simple cubic CsCl structure. Cuprous halides and AgI are found in the 4:4-coordinated cubic zincblende and (at higher temperatures) hexagonal wurtzite structures. The alkaline earth fluorides CaF_2, SrF_2 and BaF_2, as well as $SrCl_2$, $BaCl_2$ and PbF_2, crystallize with 8:4 coordination in the face-centered-cubic fluorite structure, while MgF_2 is found with the (approximately) 6:3-coordinated rutile structure.

Most other MX_2 divalent metal halides crystallize in distorted versions of the rutile structure (e.g., $CaCl_2$), chain structures ($CuCl_2$) or layer structures ($CdCl_2$, CdI_2). BeF_2 and $BeCl_2$ have the 4:2-coordinated network structure of β-cristobalite at high temperature and can be produced in other SiO_2 polymorphs. Many trifluorides, like AlF_3, have a 6:2-coordinated ReO_3 structure in which MF_6 octahedra are centered at the corners of a cubic unit cell. Other trivalent metal halides have higher-coordination three-dimensional structures or layer structures. Mixed-valence halides like $KMgF_3$ have the cubic perovskite structure with 12:6:6 coordination.

Extensive and variously successful attempts have been made to account for choice of structure, based on the lattice energy of a point ion model with various choices for the form of ion core repulsion and corrections for polarizability. The best approach to date appears to be the use of the shell model to accommodate polarization and a set of empirical potentials matched to elastic and dielectric properties (Catlow et al. 1977). Computer codes embodying this approach (Catlow 1980) have been utilized to generate the best model information on point and

Table 1
Selected physical properties of some halides

Halide	Structure	Melting point (K)	Dielectric constants ε_0	ε_∞	Exciton energy (eV)	Infrared cutoff (μm)	Thermal disorder	Shear modulus (GPa)	Principal glide systems	Cleavage
LiF	rocksalt	1121	9.01	1.96	12.9	6.2	Schottky	58.8	$1/2\langle 1\bar{1}0\rangle\{100\}, \{001\}$	{100}, {110}
LiCl	rocksalt	887	11.95	2.78	8.7		Schottky		$1/2\langle 1\bar{1}0\rangle\{100\}, \{001\}$	{100}
LiBr	rocksalt	825	13.25	3.17	7.2		Schottky		$1/2\langle 1\bar{1}0\rangle\{100\}, \{001\}$	{100}
LiI	rocksalt	710	16.85	3.80	5.9		Schottky		$1/2\langle 1\bar{1}0\rangle\{100\}, \{001\}$	{100}
NaF	rocksalt	1285	5.05	1.74	10.7	10.5	Schottky		$1/2\langle 1\bar{1}0\rangle\{100\}, \{001\}$	{100}
NaCl	rocksalt	1074	5.90	2.34	8.0	16	Schottky	23.7	$1/2\langle 1\bar{1}0\rangle\{100\}, \{001\}$	{100}
NaBr	rocksalt	1028	6.28	2.59	6.7		Schottky	11.6	$1/2\langle 1\bar{1}0\rangle\{100\}, \{001\}$	{100}
NaI	rocksalt	935	7.28	2.93	5.6		Schottky		$1/2\langle 1\bar{1}0\rangle\{100\}, \{001\}$	{100}
KF	rocksalt	1119	9.33	1.85	9.9		Schottky		$1/2\langle 1\bar{1}0\rangle\{100\}, \{001\}$	{100}
KCl	rocksalt	1049	4.84	2.19	7.8	20	Schottky	10.5	$1/2\langle 1\bar{1}0\rangle\{100\}, \{001\}$	{100}
KBr	rocksalt	1003	4.90	2.34	6.7	25	Schottky	8.8	$1/2\langle 1\bar{1}0\rangle\{100\}, \{001\}$	{100}
KI	rocksalt	959	5.10	2.62	5.9		Schottky	7.0	$1/2\langle 1\bar{1}0\rangle\{100\}, \{001\}$	{100}
RbF	rocksalt	1048	6.48	1.96	9.5		Schottky		$1/2\langle 1\bar{1}0\rangle\{100\}, \{001\}$	{100}
RbCl	rocksalt	988	4.92	2.19	7.5		Schottky		$1/2\langle 1\bar{1}0\rangle\{100\}, \{001\}$	{100}
RbBr	rocksalt	955	4.86	2.34	6.6		Schottky		$1/2\langle 1\bar{1}0\rangle\{100\}, \{001\}$	{100}
RbI	rocksalt	911	4.91	2.59	5.7		Schottky		$1/2\langle 1\bar{1}0\rangle\{100\}, \{001\}$	{100}
CsF	rocksalt	955		2.16	9.3		Schottky		$1/2\langle 1\bar{1}0\rangle\{100\}, \{001\}$	{100}
AgCl	rocksalt	728	9.5	3.97	3.2	>10	cation Frenkel	9.8	$1/2\langle 1\bar{1}0\rangle\{100\}, \{001\}$	none
AgBr	rocksalt	705	10.6	4.25	2.7	>10	cation Frenkel	9.0	$1/2\langle 1\bar{1}0\rangle\{100\}, \{001\}$	none
CsCl	CsCl	915	7.20	2.62	7.9		Schottky		$\langle 001\rangle\{110\}, \{112\}$	none
CsBr	CsCl	905	6.67	2.42	6.8		Schottky		$\langle 001\rangle\{110\}, \{112\}$	none

CsI	CsCl	894	6.59	2.62	5.3	50	Schottky		$\langle 001\rangle\{110\}$, $\{112\}$	none
TlCl	CsCl	700					Schottky		$\langle 001\rangle\{110\}$, $\langle 1\bar{1}0\rangle\{110\}$	none
TlBr	CsCl	733					Schottky		$\langle 001\rangle\{110\}$	none
TlI	CsCl	713					Schottky		$\langle 001\rangle\{110\}$	none
TlBr–TlI	CsCl	414		~5.6		35	Schottky			none
CaF_2	fluorite	1668	6.81	2.04	11.2	9.0	anion Frenkel	44.4	$1/2\langle 1\bar{1}0\rangle\{001\}$, $\{110\}$	$\{111\}$
SrF_2	fluorite	1723		2.07	10.6	10.3	anion Frenkel	35.0	$1/2\langle 1\bar{1}0\rangle\{001\}$, $\{110\}$	$\{111\}$
BaF_2	fluorite	1628		2.16	10.0	11.5	anion Frenkel	25.1	$1/2\langle 1\bar{1}0\rangle\{001\}$, $\{110\}$	$\{111\}$
CdF_2	fluorite	1373		2.40	7.6	9.8	anion Frenkel	36.3	$1/2\langle 1\bar{1}0\rangle\{001\}$, $\{110\}$	$\{111\}$
PbF_2	fluorite	1095		~2.9	5.7	10.9	anion Frenkel		$1/2\langle 1\bar{1}0\rangle\{001\}$, $\{110\}$	$\{111\}$
$SrCl_2$	fluorite	1145		2.72	<5.8		anion Frenkel	16.1	$1/2\langle 1\bar{1}0\rangle\{001\}$, $\{110\}$	$\{111\}$
$BaCl_2$	fluorite	1233					anion Frenkel			
AgI	zinc blende	830	7.0	4.72	2.1	>10	cation Frenkel			
MgF_2	rutile	1536	4.83[a], 5.50[b]	1.90	11.8	7.0	Schottky			
MnF_2	rutile	1129		~2.0		8.4	Schottky			
BeF_2	cristobalite	1070								
$BeCl_2$	cristobalite	678								
$CaCl_2$	distorted rutile	1055								
$CuCl_2$	chain	810[c]								
$CdCl_2$	layer	841								
CdI_2	layer	660								
AlF_3	ReO_3	>1545								
LaF_3	trigonal	1766				9.0				
$KMgF_3$	perovskite	1343			~12	~7			$\{110\}$	$\{100\}$
$KMnF_3$	perovskite								$\{110\}$	$\{100\}$
$RbMgF_3$	$BaTiO_3$	1143				~7				none

a parallel to *c* b perpendicular to *c* c decomposition temperature

extended defects in halides and many other ionic crystals. These static calculations cannot, however, simulate dynamic crystal lattice effects which have been investigated experimentally, for example by inelastic neutron scattering.

Glass formation is unknown, except in 4:2-coordinated halides like BeF_2, because of the combination of high ionicity and preference for higher coordination, which impose a topological specificity incompatible with aperiodicity.

2. *Optical Properties*

The countermotion of ions of opposite charge in halides leads to strong infrared absorption associated with optical phonons (Table 1). Nevertheless, NaCl and KCl are still transparent in the 10 μm region (125 meV) and are therefore used as windows for CO_2 lasers; CaF_2 is used as an attenuator. Infrared transmission is maximized with heavy ion masses, and for this reason mixed TlBr/TlI crystals are useful in window applications farther into the infrared. The large band gap renders most halides usefully transparent over a wide spectral region in the visible, as shown in Fig. 1, where E_x corresponds to anionic exciton creation in alkali and alkaline earth halides and to cationic transitions in silver halides, CdF_2 and PbF_2, and E_G to a direct gap transition responsible for the onset of photoconductivity.

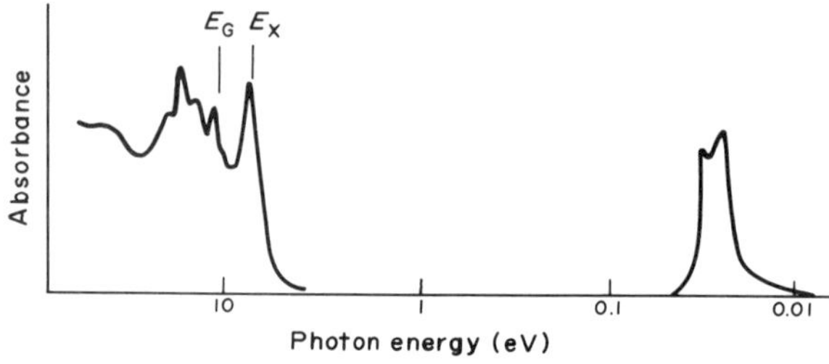

Figure 1
Optical absorption spectrum of a typical halide, showing transparency between the infrared and ultraviolet (see text)

The onset of absorption at the ultraviolet edge is associated with a more or less symmetric excitation of electrons from a valence halogen *p* orbital into bound electron–hole (exciton) states with energies near to or degenerate with conduction energies. Exciton energies for a few halides are listed in Table 1 and are generally large (5–11 eV). In alkali and alkaline earth halides, screening of the electron–hole interaction by lattice polarization is weak, and the optical dielectric constant is the appropriate one to use (Table 1). Excitons in these halides are correspondingly self-trapped and behave as small polarons.

Silver halides are anomalous in that the Madelung potential within the crystal raises the Ag^+ 4*d* and lowers the halogen *p* ion levels to near degeneracy. Mixing of these states bends the valence band upward at the k_{111} zone boundary, creating a smaller indirect gap of 3.2 eV for AgCl and 2.7 eV for AgBr, most nearly corresponding to the transition of an Ag^+ 4*d* electron. A similar situation is encountered in CdF_2 and PbF_2. The indirect transition is rendered probable by coupling with plentiful acoustic and optical phonons and by doping with I^- ions and has obvious implications for the photographic process (see Sect. 4).

3. *Defects and Mass Transport*

The large electronic band gap in halides, which renders them optically transparent, makes them ideal hosts for various impurity ions whose optically excited electronic transitions are easily studied by optical spectroscopy in the visible region and impart coloration to the host solid. Various intrinsic point defects also affect their host halide crystals in similar ways. Study of these color centers has proved a fruitful line of research in solid-state defect physics (Fowler 1968, Townsend and Kelly 1973) and has been exploited commercially for the development of scintillators (such as NaI–Tl), tunable lasers (Mollenauer 1981) and various information storage devices and solid-state dosimeters (Pooley 1975).

Many isovalent cation impurities have appreciable solid solubility and many mixed halide systems (e.g., NaCl–KCl, KBr–KI) exhibit complete miscibility. Incorporation of aliovalent (other-valent) cations requires formation of additional, thermodynamically costly charge-compensating defects. A well-studied example is the substitutional accommodation of divalent cation dopants in alkali halides with consequent formation of cation sublattice vacancies. Aliovalent dopants and their compensating defects are oppositely charged and readily associate; at high enough density and low temperature, ordering of associated defects may occur, precipitating in some cases long-range ordered superlattice phases, known as Suzuki phases (e.g., $MnCl_2.6NaCl$), which are coherent or semicoherent with the matrix structure. Such impurity phases can be important to the light sensitization of silver halides in the photographic process. Prominent anion impurities are H, O and S, but—with the exception of hydrogen—have been much less studied because the same level of spectroscopic activity is not present.

Of more fundamental interest are intrinsic defects, many of which can be rendered optically or paramagnetically active by ionizing radiation. The equilibrium thermal defects (Table 1) in halides, as in all strongly ionic solids, must maintain both stoichiometry and charge neutrality. In alkali halides, the major thermal defects are Schottky pairs (anion and cation vacancies), while in silver halides cation Frenkel defects (Ag interstitials, Ag vacancies) predominate. In fluorite-structure fluorides, anion sublattice Frenkel defects predominate, while in MgF_2 the Schottky trio appears to be favored. Because the equilibrium thermal defect fraction is usually small even near the melting point (~50 ppm), defects created to charge-compensate impurities present even at the 1–100 ppm level usually dominate at lower temperatures.

Mass transport occurs by diffusion of those defects with the combination of highest concentration and highest mobility. In halides, charge transport is predominantly by ionic diffusion, so electrical conductivity is similarly constrained, which means that conductivity is controlled by cation vacancies in alkali halides, cation interstitials in silver halides and anion interstitials in fluorite-structure halides (Corish and Jacobs 1973). Fluorite halides exhibit a vacancy–interstitial order–disorder transition on the anion sublattice considerably below the melting point which gives rise to a specific heat anomaly and the onset of a large increase in electrical conductivity known as superionic conduction. The fraction of F^- ions leaving fluorine sites in PbF_2, for example, exceeds 40% even 300 K below the melting point, though the disorder fraction is smaller in other fluorite halides (Hayes 1980).

A second well-studied class of native defects is that resulting from imposition of an excess of anion or cation species over nominal stoichiometry by equilibration with metal or halogen gas environments (additive coloration). These defects may also be generated by solid-state electrolysis or by irradiation (see Sect. 4). An excess metal content up to ~0.1% can be sustained in halide crystals by equilibration with metal vapor at elevated temperatures. The excess is accommodated substitutionally with the simultaneous formation of halogen vacancies and electrons which associate to form the well known F (for *Farbe*, German for color) centers. Optical transitions of the associated electron(s) are responsible for the characteristic absorption colors of such crystals. The center is also paramagnetic, and its structure has been confirmed by electron paramagnetic resonance.

Multiple aggregates (F_2, F_3, F_4 centers) of F centers form at intermediate temperatures where halogen vacancies are mobile, but further association leads to precipitation of the cation species into metal particles of colloidal size. Light scattering from colloidal sodium particles, formed by aggregation of F centers created by natural radiation sources, is responsible for the blue color of natural rocksalt. Incorporation of excess halogen at similar concentration levels, by equilibration with high pressures of halogen gas at elevated temperature, generates molecular halogen defects, called V centers because they were historically identified with cation vacancy defects. The current model for such defects (Catlow et al. 1980) favors halogen molecules substitutionally occupying adjacent sites on both anion and cation sublattices.

4. Radiation Sensitivity

A property which distinguishes halides from most other inorganic solids is extreme sensitivity to ionizing or near-ionizing radiation. Halide crystals so exposed undergo a complex sequence of structural changes, beginning with point defect creation and ultimately resulting in chemical decomposition. The degradation sequence parallels that of solid-state electrolysis and is thus termed photolysis when photons in the visible range are involved and radiolysis for uv, x-ray or γ-ray photon, or fast charged particle, irradiations. Photolysis of silver halides forms the basis of the photographic process (see *Photographic Materials*), while radiolysis of alkali and alkaline earth halides is conveniently used to generate F center and F-center-aggregate defects and underlies the electron beam lithographic process for halide resists. Electron-irradiation-induced F center production was utilized in dark trace radar display tubes (skiatrons) in World War II. The mechanisms in silver halides and alkali halides appear to be different, but both rely on the creation of electron–hole pairs by irradiation and the subsequent separation of these charges.

In alkali halides, the detailed radiolysis mechanism (Fig. 2) has been well characterized by time-resolved picosecond laser pulse spectroscopy (Williams 1978).

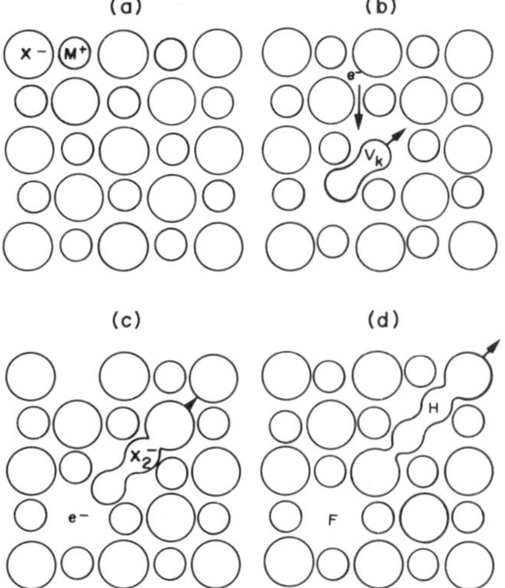

Figure 2
Radiolysis mechanism in halides of rocksalt structure: (a–d) show successive steps in the creation of an F center vacancy and an H center interstitial defect through mechanical dissociation of a V_k center exciton

The initial excitation decays to one or more self-trapped excitons, with energies of the order of the band gap, in which the hole is localized in a homonuclear bond between adjacent halogen ions (V_k center) and the electron in diffuse *s*-like orbitals associated with surrounding alkali ions. The V_k center (which resembles then an X_2^- halogen molecular ion residing in an F_2^+ center) is configurationally unstable in the Coulomb field of the electron, and migrates away by a combination of hole tunnelling and crowdion or interstitial diffusion to a distant site, creating an X_2^- molecular ion occupying a single halogen site (H center interstitial) and leaving behind an electron in a single halogen vacancy (F center). The F center and H center together constitute a Frenkel defect pair on the halogen sublattice.

H centers are exceptionally mobile and aggregate to form dislocation loops, substitutional halogen molecules (V centers) and eventually halogen inclusions, while at temperatures above which F centers are mobile, F center aggregation generates alkali metal colloidal precipitates which complete the decomposition. Similar mechanisms are thought to operate in halides with CsCl, fluorite and rutile structures. Radiolytic efficiency approaches 50% for some halides, and decomposition is most pronounced at homologous temperatures of about 0.2–0.3T_m. Details of this degradation sequence have been followed by transmission electron microscopy (TEM) in alkali and alkaline earth halides; the extreme radiation sensitivity of these materials necessitates cryogenic TEM techniques to retard the degradation kinetics during observation of radiolytic and other extended defect microstructures (Hobbs 1975, 1976).

The photolytic decomposition of silver halides (AgCl, AgBr, AgI) involves absorption of visible light photons to reduce silver ions to metallic silver. Successful operation of the mechanism requires simultaneous absorption of multiple photons, anomalies in the band structure of silver halides, large surface effects provided by the small dimensions of silver halide grains (0.1–1 μm) and at least one impurity effect (Hamilton 1973).

As outlined in Sect. 2, the indirect band-to-band transition in silver halides occurs at ~3 eV, providing spectral sensitivity down to ~500 nm wavelength. The hole is self-trapped, as in alkali halides, but on a silver ion site instead. The photoelectron is not easily trapped in the interior of the halide grain but can become trapped at the surface, converting a surface silver ion to an atom and completing the charge separation. An aggregation of about four of these surface silver atoms constitutes a latent image, rendering the grain developable, that is, completely convertible to metallic silver under subsequent action of organic reducing agents, resulting in an effective gain exceeding 10^8.

Such a concentration of surface silver atoms requires deep surface traps, and there is evidence that surface segregation of sulfur impurity serves this function. The availability of excess mobile interstitial Ag^+ ions is also a requirement, met by the predominance of Frenkel thermal defects in silver halides and the enhancement of positively charged defects near negatively charged grain surfaces. The requirement of about four photoexcitation events in close time and space proximity leads to the nonlinear response which characterizes the density–exposure relationship of photographic emulsions.

The response of most halides to intense high-energy radiation provides a basis for submicrometer-scale electron beam lithography using halide resists. It is possible with intense focused electron beam probes to etch troughs less than 1 μm wide in halide films 50 nm thick, using electron doses between 100 and 10^6 C m^{-2} depending on the halide. The mechanism is not well understood but probably involves the synergism of radiolysis, halogen gas generation and beam-induced desorption (Isaacson and Murray 1981). Halides also undergo damage from high-intensity short-duration laser pulses, even for photon energies far below the exciton edge, for example in the infrared. The mechanism in bulk is an induced avalanche breakdown with an energy density threshold of ~15 TJ m^{-3} for nanosecond pulse widths in NaCl; near-surface breakdown occurs at a lower threshold (~5 TJ m^{-3}) in NaCl and has been attributed to rapid water desorption (Newnam et al. 1980).

5. *Mechanical Properties*

The elastic properties of halides are determined by the long wavelength acoustic phonon mode properties. Neither elastic isotropy ($c_{11} - c_{12} = c_{44}$) nor the Cauchy relation ($c_{12} = c_{44}$) hold well for the second-order elastic constants c_{ij}. (In the fluorite structure, and many more complex halide structures, one or more ion sites lack inversion symmetry, so the Cauchy relation cannot be expected to hold). Voigt-average values for the shear moduli G are given in Table 1 and reflect the greater elastic compliance of ionic halides compared with, for example, most metals.

Crystalline halides are plastic at room temperature and were among the first crystalline solids whose plasticity was systematically investigated. Early rheological investigations gave way in the 1950s to extensive characterization of dislocation propagation and multiplication in single crystals by classic etch pit methods (Gilman and Johnston 1957). Dislocations in halides are also observed by decoration techniques, photoelastic birefringence, x-ray topography and TEM. The relevant literature is reviewed by Sprackling (1976). Prominent glide systems and dislocation Burgers vectors for a number of halides are listed in Table 1. The requirement of four or five independent glide systems for generalized deformation of polycrystals is satisfied by most halides, resulting in polycrystalline ductility at temperatures exceeding $0.2–0.4T_m$.

Macroscopic yield occurs when the applied shear stress τ is sufficient to move large numbers of dislocations. Dislocation velocities in halides are observed to be sensitive functions of applied stress ($v \approx \tau^m$ with $1 < m < 100$), so that dislocations accelerate rapidly with increasing stress and the onset of yield is sudden. Dislocation multiplication occurs readily by multiple cross glide at low applied stress. Subsequent dislocation interactions induce work hardening. Considerable effort has in fact been expended in explaining the form of stress–strain curves, for example the classic three-stage work hardening curves observed for deformation of rocksalt single crystals, in terms of glide system activation and dislocation intersections (Strunk 1977). Stress–strain curves for polycrystals are generally parabolic and depend on grain size. The flow stress throughout the work-hardening regime is often observed to be roughly proportional to the square root of dislocation density.

Plastic deformation generates copious point defect debris, largely through nonconservative motion of jogs on intersected screw dislocations, which has been studied in halides by spectroscopic techniques. Additionally, dislocations in ionic crystals can acquire charge, and this phenomenon has been much studied in alkali halides (Whitworth 1975). The sign of the charge depends on the dislocation velocity, core structure and jog content, the impurity level (through charge-compensating point defects which, attracted to or swept up by the dislocation, can neutralize the dislocation charge) and the temperature of deformation. Dislocations moving fast enough to break away from their compensatory defect clouds can transport charge. At a sufficiently high temperature, point defect fluxes permit creep deformation to be sustained by dislocation climb in single crystals and by dislocation climb, Coble or Nabarro–Herring creep mechanisms or grain-boundary sliding in polycrystals; corresponding deformation maps have been constructed for several halides (Ashby and Frost 1982). As a consequence of plentiful deformation data, their lower melting temperatures and their higher attainable purity, halides have frequently served as models for high-temperature deformation behavior of more refractory isostructural oxides.

Halides, in common with most ceramic materials, are brittle as well as plastic, and many exhibit well-defined cleavage on prominent crystallographic planes (Table 1) as the principal mode of crack propagation under an applied tensile stress (mode I failure). Extensive plastic deformation at low temperature can generate so-called Stroh cracks at the intersections of slip bands which serve as critical flaws. At higher temperatures, polycrystalline ductility may be further reduced by initiation of grain boundary sliding, which may lead to nucleation of intergranular cracking at four-grain junctions.

See also: Alkali Halide Crystals; Point Defects in Ceramics

Bibliography

Ashby M F, Frost H J 1982 *Deformation-Mechanism Maps: The Plasticity and Creep of Metals and Ceramics*. Pergamon, Oxford

Catlow C R A 1980 Computer modelling of ionic crystals. *J. Phys.* 41(Colloque C6): 53–60

Catlow C R A, Diller K M, Hobbs L W 1980 Irradiation-induced defects in alkali halide crystals. *Philos. Mag.* A42: 123–50

Catlow C R A, Diller K M, Norgett M J 1977 Interionic potentials for alkali halides. *J. Phys. C.* 10: 1395–1411

Corish J, Jacobs P W M 1973 Point defects in ionic crystals. In: Roberts M W, Thomas J M (eds.) 1973 *Surface and Defect Properties of Solids*, Vol. 2. Chemical Society, London, pp. 162–228

Fowler W B 1968 *Physics of Color Centers*. Academic Press, New York

Gilman J J, Johnston W G 1957 The origin and growth of glide bands in lithium fluoride. In: Fisher J C et al. (eds.) 1957 *Dislocations and Mechanical Properties of Crystals*. Wiley, New York, pp. 116–61

Hamilton J F 1973 The photographic process. *Prog. Solid State Chem.* 8: 167–88

Hayes W 1974 *Crystals with the Fluorite Structure: Electronic, Vibrational and Defect Properties*. Clarendon, Oxford

Hayes W 1980 Superionics. *J. Phys.* 41(Colloque 6): 7–12

Hobbs L W 1975 Transmission electron microscopy of extended defects in alkali halides. In: Roberts M W, Thomas J M (eds.) 1975 *Surface and Defect Properties of Solids*, Vol. 4. Chemical Society, London, pp. 152–250

Hobbs L W 1976 Point defect stabilization in ionic crystals at high defect concentrations. *J. Phys.* 37(Colloque C7): 3–26

Mollenauer L F 1981 Progress in color center lasers. In: Shvarts K K, Tuchkevich V M (eds.) 1981 *Defects in Insulating Crystals*. Springer, Berlin, pp. 524–41

Newnam B E, Nowak A V, Gill D H 1980 Short pulse CO_2 laser damage of NaCl and KCl. In: Bennett H E, Glass A J, Guenther A H, Newnam B E (eds.) 1980 *Laser Damage in Optical Materials 1979*, NBS Special Publication 568. National Bureau of Standards, Washington, DC, pp. 209–28

Pooley D 1975 Application of radiation damage in dosimetry and information storage. In: Dupuy C H S (ed.) 1975 *Radiation Damage Processes in Materials*. Noordhof, Leiden, The Netherlands, pp. 503–34

Sprackling M T 1976 *Plastic Deformation of Simple Ionic Crystals*. Academic Press, London

Strunk H 1977 Transmission electron microscopy of plastically deformed [001]-orientated NaCl single crystals. *Mater. Sci. Eng.* 27: 225–38

Townsend P D, Kelly J 1973 *Colour Centres and Imperfections in Insulators and Semiconductors*. Sussex University Press, Brighton, UK

Whitworth R W 1975 Charged dislocations in ionic crystals. *Adv. Phys.* 24: 203–304
Williams R T 1978 Photochemistry of F-center formation in halide crystals. *Semicond. Insul.* 3: 251–83

L. W. Hobbs

Hall Effect

When a magnetic field is applied perpendicular to the current in a conductor, a transverse voltage appears across the conductor, perpendicular to both the current and field. This is the Hall effect, named after Edwin Hall who discovered it in 1879. It is one of the many galvanomagnetic effects that occur when a conductor is subjected to a magnetic field. The effects result from the Lorentz force that the current carriers experience in the applied magnetic field.

In the standard configuration (a flat plate or film as used originally by Hall, Fig. 1) the current density $J_x = I_x/wt$, where I_x is the current and w and t are the width and thickness of the sample, respectively. Thus, $J_x = nqv_x$, where n is the number density of charge carriers (charge q), which have a drift velocity v_x (see *Electrical Conductivity of Metals and Alloys*). The magnetic flux density B_z gives rise to the transverse force qv_xB_z, which for electrons is in the $-y$ direction. Since there can be no current in the y direction, an electric field, the Hall field E_H, is set up such that the net force in the y direction is zero. Thus, $E_H = v_xB_z = V_H/w$, where V_H is the Hall voltage. In analogy to the normal electrical resistivity $\rho = E_x/J_x$, the Hall resistivity is defined as $\rho_H = E_H/J_x = v_xB_z/nqv_x = B_z/nq$. (In terms of measured quantities $\rho_H = (V_H/w)/(I_x/wt) = V_Ht/I_x$.) This leads to the Hall constant $R_H = \rho_H/B_z = 1/nq$, which is a direct measure of the carrier density in the conductor. Since n is normally independent of temperature, so too are R_H and ρ_H. In Fig. 1 the direction of deflection and sign of the Hall voltage for electrons is shown. If the current is carried by positively charged carriers the deflection is the same, but the Hall voltage is of opposite sign. Thus, the Hall effect can be used to measure not only the charge density in the conductor but also the sign of the majority charge carriers.

The Hall voltage is small in metals because of the inverse dependence on electron density. For a fixed current the Hall voltage increases as the thickness is decreased. For a typical metal such as copper a flat plate (1 mm thick), carrying a current of 1 A in a field of 1 T, produces a Hall voltage of only about 50 nV.

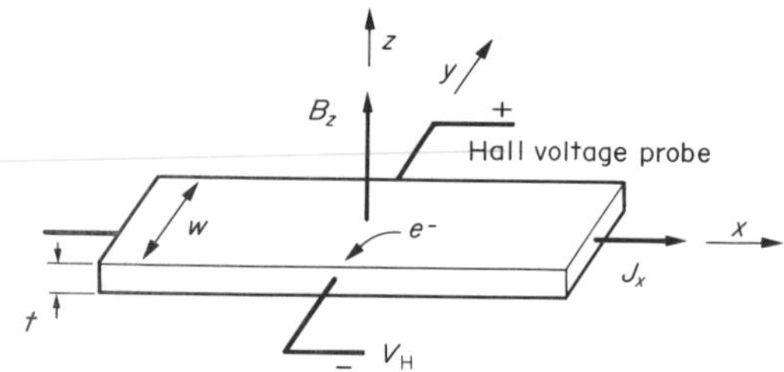

Figure 1
Hall effect

In high magnetic fields at low temperature the Hall resistivity of a pure metal in the form of a single crystal reflects the anisotropy of the Fermi surface (see *Fermi Surfaces*). It may be either positive or negative depending on the major orbits of the Fermi surface appropriate to the orientation of the field in the crystal. Thus, the Hall effect can help in the study of the Fermi surface. At low temperatures the sample may be in the high-field regime where a large section of the Fermi surface is traversed between scattering events, whereas at high temperatures the scattering should dominate (low-field regime). In aluminum, for example, the Hall voltage changes from positive at low temperature to negative at room temperature for these reasons; $R = 1/nq$ is clearly an inadequate description in this case.

In ferromagnetic materials the Hall effect is much larger than in nonmagnets because of an extra component that has a field dependence which follows the internal magnetization. This extra component, known as the extraordinary or anomalous Hall effect, was a puzzle for many years and is still the subject of research. It arises from spin–orbit coupling of the itinerant electrons with the magnetic system, and not from the Lorentz force. In most cases the anomalous component varies as ρ^2 which helps in disentangling the two contributions from the measured effect.

Hall measurements are indispensable in characterizing semiconductors. The effect is much larger in semiconductors than metals because of the large difference in n. If the carriers are predominantly electrons the semiconductor is designated n-type, but if the effective charge has a positive sign the carriers are holes and the semiconductor is p-type (see *Electrical Conductivity of Semiconductors*). Charge carriers of both signs with differing dynamic properties are generally present in semiconductors, making a two-band approach appropriate. Hall-effect studies enable deduction of valuable information about the conduction mechanisms, especially when the conduction is unusual such as in polaron or hopping conduction. R_H in semiconductors varies approximately exponentially with temperature in the intrinsic regime, reflecting the excitation of carriers across the gap. When the conduction is extrinsic, the variation with temperature is minimal. In very low conductivity materials the Hall effect may be measured by photoinjecting either holes or electrons to determine exactly which carriers are responsible for the observed effect.

Unusual phenomena occur when the Hall effect is studied in semiconductors in the quantum-mechanical limit, that is, when the energy levels of the carriers break up into well-separated Landau levels. This occurs in the two-dimensional electron gas in the inversion layer of a MOSFET (metal-oxide semiconductor field-effect transistor) in high magnetic fields and at low temperatures. By varying the gate voltage of the transistor, the Fermi level can be moved through the Landau levels to produce steps in the Hall voltage. The separation of these steps is directly related to h/e^2, enabling the fine structure constant α to be determined to better than 1 in 10^6.

Although the Hall effect is a product of solid state physics it is valuable in other fields. As long as a medium conducts it should produce a measurable Hall effect. The most obvious application of the Hall effect is in gaussmeters. These devices use semiconductors such as InSb to produce a large Hall voltage, which is a direct measure of the field. With the advent of the integrated circuit, Hall devices are now being produced which consist of a Hall-effect sensor and electronics in the same integrated circuit. Such products are widespread, mostly in the form of contactless switches for computer terminal keyboards.

See also: de Haas–van Alphen Effect

Bibliography

Beer A C 1963 *Galvanomagnetic Effects in Semiconductors*, Solid State Physics Supplement No. 4. Academic Press, New York

Chien C L, Westgate C R (eds.) 1980 *The Hall Effect and Its Applications*. Plenum, New York

Hurd C M 1972 *The Hall Effect in Metals and Alloys*. Plenum, New York

Putley E H 1960 *The Hall Effect and Related Phenomena*. Butterworth, London

Wieder H H 1979 *Laboratory Notes on Electrical and Galvanomagnetic Measurements*, Materials Science Monographs Vol. 2. Elsevier, New York

S. McAlister

Halpin–Tsai Equations

The Halpin–Tsai equations constitute a method of estimating the elastic moduli of an aligned linearly elastic orthotropic composite where it is assumed (a) that the fibers are parallel to axis 1 in a matrix; (b) that the Young's modulus in this direction is given by

$$E_{11} = E_f V_f + E_m V_m$$

(where E_f, E_m are the respective Young's moduli of the fibers and matrix, and V_f, V_m are the respective volume fractions); and (c) that the principal Poisson's ratio is

$$\nu_{12} = \nu_f V_f + \nu_m V_m$$

(where ν_f, ν_m are the respective Poisson's ratios of the fibers and matrix). Then the composite moduli E_{22}, G_{12} or G_{23} can be estimated from the Halpin–Tsai equations,

$$\frac{\bar{p}}{p_m} = \frac{1 + \xi\eta V_f}{1 - \eta V_f}$$

where

$$\eta = \left(\frac{p_f}{p_m} - 1\right) \Big/ \left(\frac{p_f}{p_m} + \xi\right)$$

Here $\bar{p}$ is the composite modulus E_{22}, G_{12} or G_{13}; p_f, p_m are the corresponding fiber and matrix moduli E or G; and ξ depends on the geometry of the fiber arrangement.

See also: Composite Materials: An Overview

Bibliography

Ashton J E, Halpin J C, Petit P H 1969 *Primer on Composite Materials*. Technomic Publishing, Westport, Connecticut

A. Kelly

Hard-Fiber Elastomers

Traditionally, elastic polymers have been characterized as rubbers and their mechanical properties have been explained by the entropic mechanisms of rubber elasticity theory. Since 1965, growing interest has developed in a new class of crystalline polymer structures that manifest elastic behavior. These hard-fiber elastomers possess the unusual combination of high-elastic modulus, high extensibility and high elastic recovery from large extensions. The elasticity of these hard elastic polymers, however, seems to be primarily controlled by energetic rather than entropic mechanisms. Polyoxymethylene, isotactic polypropylene, polyethylene, polybut-1-ene, poly-4-methylpentene, acetal copolymer and polypivalolactone all produce hard elastic structures.

The conversion of a normally nonelastic crystalline polymer to a hard-fiber elastomer is extremely process sensitive. A row-nucleated crystal morphology, which forms during stress crystallization, is typically observed with these materials. Since stress crystallization is particularly sensitive to such processing parameters as spinning temperature, shear rate, draw-down ratio and annealing temperature, the elasticity or lack of elasticity of the crystalline polymer will depend on the proper choice of these processing parameters.

During room-temperature extension of a hard-fiber elastomer, microcavitation of the row, nucleated structure predominates in the high-extension elastic region, while plastic deformation predominates at still higher extensions. At least five basic structural models have been proposed to explain the elastic properties of these polymers.

These are (a) a leaf-spring model involving the elastic bending of lamellae, (b) reversible shear of lamellae between fixed tie-points, (c) a general model based on a change in entropy in the intermolecular layer and an increase in surface energy during extension, (d) a combined structural model that attributes the stress for extension to the pulling of fibrils from lamellae, and the retractive force resulting in the particular elastic properties to surface energy and entropy effects in the fibrils, and (e) a microparacrystalline model that associates the driving force for elasticity to the surface free energy of destroyed microparacrystals. These models are not consistent and predict different mechanical properties and structures in the strained state. Also, noncrystalline molecular orientation changes have been observed during the crystalline polymer deformation which have not been included in the above models. Certainly, both energetic and entropic mechanisms contribute to the observed elasticity, and a truly quantitative model has yet to be developed.

The sensitivity of hard-fiber elastomers to morphology also affects the relation between their elasticity and strength. For example, the observed elastic recovery at 50% cyclic extension of isotactic polypropylene hard-fiber elastomer was found to decrease linearly with increasing noncrystalline orientation (and hence, increasing tenacity) in the fiber. Thus, isotactic polypropylene fibers with tenacities of 1–2 grams per denier had elastic recoveries of 100% at 50% room-temperature cyclic extension, while a high-strength fiber having a tenacity of 5.5 grams per denier had an elastic recovery of only 70% at 50% room-temperature cyclic extension. This elastic recovery was, however, still far superior to the approximate 40% elastic recovery at 50% room-temperature extension normally observed for nonelastic-drawn fibers of the same polymer.

See also: Elastomers: An Overview

Bibliography

Cannon S L, McKenna G B, Statton W O 1976 Hard-elastic fibers (a review of a novel state for crystalline polymers). *J. Polym. Sci., Macromol. Rev.* 11: 209–75

Celanese Corporation of America 1965 Filamentous material and a process for its manufacture. Belgian Patent 650,890 (January 1965)

Hashimoto T, Nagatoshi K, Todo A, Kawai H 1976 Deformation mechanism of "hard elastic" polyethylene films. *Polymer* 17: 1063–74

Ishikawa H, Numa H, Nagura M 1979 Deformation in hard elastic polypropylene fiber. *Polymer* 20: 516–17

Loboda-Čačković J, Čačković H, Hosemann R 1979 Hard elastic fibers. 1. Four-component analysis of the NMR spectra of isotactic polypropylene. *J. Macromol. Sci., Phys. B* 16: 127–44

Miles M, Petermann J, Gleiter H 1976 Structure and deformation of polyethylene hard elastic fibers. *J. Macromol. Sci., Phys.* 12: 523–34

Noether H D, Brody H 1976 Low modulus "hard" elastic materials. *Text. Res. J.* 46: 467–78

Samuels R J 1979 High strength elastic polypropylene. *J. Polym. Sci., Polym. Phys. Ed.* 17: 535–68

Wool R P, Lohse M I, Rowland T J 1979 Broad-line NMR studies of deformation and recovery in hard elastic polypropylene. *J. Polym. Sci., Polym. Lett. Ed.* 17: 385–89

R. J. Samuels

Hard Magnetic Ferrites

Ferrites with high coercivity are called hard ferrites. The most important hard ferrites are the hexaferrites or M-type ferrites, characterized by the compound $MeFe_{12}O_{19}$ (Me = Ba, Sr, Pb) with the hexagonal magnetoplumbite or M structure. Their hard magnetic nature is due to the large magnetocrystalline anisotropy, with the hexagonal *c* axis as the preferred direction. M-ferrites are widely used as permanent-magnet materials, well known under various trade names, such as Ferroxdure, Koerox, Roboxid and Magnadure (Stäblein 1982).

This article concentrates on M-ferrites for permanent-magnet applications, but other applications of M-ferrites and other hard ferrites are discussed briefly. Unless specifically noted, the term "ferrite" is to be understood to mean M-ferrite.

1. General Characteristics of Ferrite Magnets

Ferrite magnets (Van den Broek and Stuijts 1977) were first introduced commercially by Philips under the trade name "Ferroxdure," the isotropic form in 1952 and the anisotropic (crystal-oriented) form in 1954. Since then, ferrites have acquired an ever-increasing share of world production of permanent magnets and have now largely surpassed the metallic magnets in tonnage (about 150 and 7 kt, respectively, in 1981). Compared with Al–Ni–Co magnets, ferrites have high coercivity H_c and a moderate remanence B_r, giving rise to moderate $(BH)_{max}$ values and low $B/\mu_0 H$ values at $(BH)_{max}$. The latter has led to applications where large demagnetizing fields are present, such as flat loudspeakers and compact dc motors. Additional advantages of ferrite magnets are high chemical stability and high electrical resistivity ($>10^4\ \Omega$ m), which has allowed new applications at high frequencies (where eddy currents are a major problem for metallic magnets). There are also disadvantages, in particular brittleness and a high temperature coefficient of coercivity and remanence.

The most characteristic feature that underlies the great economic success of ferrite magnets is the low price per unit of available magnetic energy. This is particularly due to the relatively low cost and wide availability of the raw materials.

Table 1
Crystallographic data and primary magnetic properties (at room temperature) of M-ferrites

Property	BaM	SrM	PbM
Crystallographic data			
Cell constants[a]			
a (nm)	0.588	0.586	0.588
c (nm)	2.317	2.303	2.302
Molecular weight	1111	1062	1181
Theoretical density *d* ($kg\ m^{-3}$)	5330	5140	5700
Magnetic properties[b]			
Saturation polarization J_s (mT)	478	478	402
Specific saturation polarization σ_s ($\mu T\ m^3\ kg^{-1}$)	90.4	93.4	70.8
Anisotropy constant K_1 ($kJ\ m^{-3}$)	325	357	220
Curie temperature T_C (K)	740	750	725

a Adelsköld (1938) b Shirk and Buessem (1969) for BaM and SrM, Pauthenet and Rimet (1959) for PbM

2. Crystal Structure

The magnetoplumbite crystal structure (Kojima 1982) is characterized by the space group $P6_3/mmc$. There is a close packing of oxygen and barium ions with iron ions at interstitial sites. The structure is built up from cubic blocks with the spinel structure and hexagonal blocks containing the barium ions. The iron ions are located at five different crystallographic positions. The dimensions of the unit cell, which contains two formula units, and the theoretical densities are given in Table 1 for the three main ferrites (BaM, SrM and PbM).

The M structure allows various substitutions. The most important ones are listed in Table 2. If the substituting ion does not have the nominal charge, it must be compensated. Some examples are $BaFe_{12-2x}Ti^{4+}_xCo^{2+}_xO_{19}$ $BaFe_{12-x}Fe^{2+}_xO_{19-x}F^-_x$, $Ba_{1-x}La^{3+}_x Fe_{12-x}Fe^{2+}_x O_{19}$ and $Ba_{1-2x}Na^+_xLa^{3+}_xFe_{12}O_{19}$.

3. Intrinsic Magnetic Properties

The magnetism of $BaFe_{12}O_{19}$ is due to the Fe^{3+} ions, each carrying a spin moment of five Bohr magnetons (μ_B) (Kojima 1982). These are aligned by exchange interaction, either parallel or antiparallel. Ions of the same crystallographic position are aligned parallel, thus constituting a ferromagnetic sublattice. The interaction between neighboring ions of different sublattices is due to superexchange by way of oxygen. The theory predicts that the ionic moments are inclined to be antiparallel, the more so the nearer the angle Fe–O–Fe approaches 180° and the distance Fe–O–Fe becomes smaller. The resulting most-probable spin configuration leads to a net magnetic moment per cell of eight Fe^{3+} ionic moments, that is, 40 μ_B. This corresponds to a saturation polarization of 668 mT at 0 K for BaM, which agrees with experimental values.

Table 2
Reported substitutions in $BaFe_{12}O_{19}$

$Ba^{2+}_{1-x}A^{k+}_xFe^{3+}_{12-y}B^{l+}_yO^{2-}_{19-z}C^{m-}_z$ $[(k-2)x+(l-3)y=(m-2)z]$							
A			B				C
$k=1$	2	3	$l=2$	3	4	5	$m=1$
Na	Sr	La	Mg	Cr	Ti	Sb	F
K	Pb	Pr	Mn	Mn	Ir	As	
Rb	Ca	Nd	Fe	Co	Ge	V	
Ag		Sm	Co	Al	Sn	Ta	
Tl		Eu	Ni	Ga	Zr	Nb	
		Bi	Zn	In			
			Cu	Sb			
				Sc			
				Ru			

The energy associated with the (anti)parallel coupling of the ionic moments is characterized by the exchange constant A. The latter is difficult to calculate exactly for superexchange in complex structures. Experimentally, it has been found that $A \simeq 6.6 \times 10^{-12}\,\mathrm{J\,m^{-1}}$, as derived from domain-width observations (on PbM). The temperature dependence of A, as shown in Fig. 1, is very small at room temperature.

The temperature dependence of the saturation polarization J_s for BaM is shown in Fig. 1. The J_s–T curve is much less curved than the usual Brillouin function, resulting in a relatively low J_s value at room temperature (478 mT) and a relatively high temperature coefficient ($-0.2\%\,\mathrm{K^{-1}}$).

The magnetization is strongly bound to the hexagonal c axis, owing to a high magnetocrystalline anisotropy. The latter originates from spin–orbit coupling of the iron ions and is characterized by the anisotropy constant K_1. Higher-order constants (K_2, K_3) are negligibly small. The temperature dependence of K_1 is shown in Fig. 1 for BaM. Room temperature values of K_1 for BaM, SrM and PbM are given in Table 1, together with other primary magnetic properties.

Based on the values for A, J_s and K_1 (Fig. 1), a number of secondary properties may be derived which govern the magnetic state (domain structure) or the way of magnetization reversal. For instance, for BaM at room temperature, the anisotropy field $H_A = 2K_1J_s^{-1} = 1375\,\mathrm{kA\,m^{-1}}$, the (Kittel) critical particle diameter for single domain behavior of spherical particles $D_c = 72\mu_0\,(AK_1)^{0.5}\,J_s^{-2} = 0.6\,\mu\mathrm{M}$, and $H_{c(max)} = H_A - NJ_s\,\mu_0^{-1} = 1175\,\mathrm{kA\,m^{-1}}$, where $\mu_0 = 4\pi \times 10^{-7}\,\mathrm{T\,m\,A^{-1}}$ is the permeability of free space and N is the self-demagnetizing factor parallel to the c axis, being about 0.6 for the usual platelet-shaped ferrite particles. $H_{c(max)}$ is the field required for magnetization reversal by coherent rotation of a single domain particle or an assembly of aligned, noninteracting particles. Thus, $H_{c(max)}$ represents an upper limit to the coercivity. However, magnetization reversal in ferrites generally occurs by nucleation and propagation of domain walls and the values of coercivity fall considerably below $H_{c(max)}$ (see also Sect. 4).

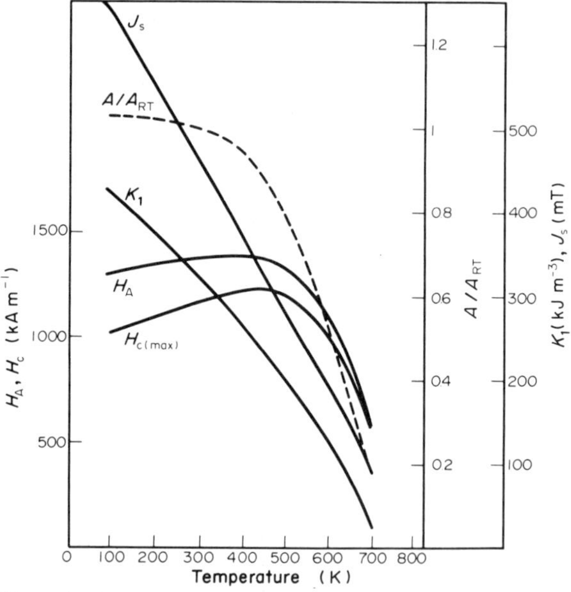

Figure 1
Temperature dependence of J_s, K_1, H_A and $H_{c(max)}$ for BaM (Smith and Wijn 1959) and of A/A_{RT} for PbM (Gemperle et al. 1963)

The temperature dependence of H_A and $H_{c(max)}$ is also shown in Fig. 1. It appears that $H_{c(max)}$ increases with increasing temperature up to about 300 °C. This agrees with the temperature dependence of experimental H_c values.

The intrinsic properties may be modified by substitution (Table 1). Barium can be fully replaced by strontium or lead, and partly by calcium (<40 mol%). CaM, stabilized with 0.03 mol% La_2O_3, is also possible. The intrinsic properties of these M-ferrites vary somewhat (Table 1) and other factors, such as sintering behavior and price of raw materials, often dictate their commercial viability. Although magnets based on CaM or PbM are feasible, large-scale production is concentrated on BaM and SrM. High-quality magnets are generally based on SrM, and somewhat lower priced magnets on BaM.

Substitution for Fe^{3+} has a drastic effect on the intrinsic magnetic properties. Partial substitution by aluminum or chromium decreases J_s without affecting K_1 seriously, resulting in larger H_A and H_c values. Substitution by Ti^{4+} and Co^{2+} causes a considerable decrease in K_1; the uniaxial anisotropy ($K_1 > 0$) may even change into planar anisotropy ($K_1 < 0$). Intermediate magnetic structures are also possible; for example, a preferred direction which spirals along the c axis is observed for In^{3+} substitution. There are a few substitutions by which the K_1 value is increased while the J_s value is hardly affected, such as substitution of Fe^{3+} by Ru^{3+} or by Fe^{2+} compensated by La^{3+} at barium sites. Finally, there are some substitutions which may reduce the temperature coefficient of the remanence: $As^{3+/5+}$ compensated by Fe^{2+}; and Cu^{2+} compensated by Me^{4+} or Me^{5+}.

4. Microstructural Aspects

Apart from the intrinsic properties, microstructural factors are also important for the realization of the desired magnetic properties. The most important microstructural factors are particle size and shape, volume fraction of ferrite phase and degree of alignment of the c axes (Stäblein 1982).

Table 3
Commercial unsubstituted barium and strontium ferrite materials[†]

		Me‡	B_r (mT)	$_BH_c$[§] (kA m^{-1})	$_JH_c$[§] (kA m^{-1})	$(BH)_{max}$ (kJ m^{-3})	μ_{rec}[¶]	H_s (kA m^{-1})	$\alpha(B_r)$[‖] (% K^{-1})	$\alpha(H_c)$[‖] (% K^{-1})	ρ (Ω m)	d (kg m^{-3})	Typical application	Market share (%)
Sintered														
Isotropic	F_1	Ba	220	135	220	7.6	1.2	800	−0.2	0.4	10^4	4900	rotors for cycle dynamos	~10
Anisotropic low $_JH_c$	F_2	Ba	400	160	165	29.5	1.1	560	−0.2	0.48	10^4	4900	flat rings for loudspeakers	~70
medium $_JH_c$	F_3	Sr	370	240	245	25.5	1.1	875	−0.2	0.39	10^4	4650	segments for windscreen wiper motors	
			410	265	275	31.5		955		0.35	10^4	4800		
high $_JH_c$	F_4	Sr	340	265	335	21.5	1.1	1115	−0.2	0.28	10^4	4600	segments for starter motors	
			380	280	320	27.0		1115		0.30	10^4	4700		
Plastic-bonded														
Flexible Isotropic	F_5	Ba	125	88	190	2.8	1.1	800	−0.2	0.4	10^7	3100	doorcatches (e.g. for refrigerators)	~10
			145	96	190	3.6	1.15	800			10^6	3700		
Anisotropic	F_6	—	200	140	200	8.0	—	—	−0.2	0.4	—	—	small dc motors (e.g., for fuel pumps)	
			250	168		12.0								
Rigid Isotropic	F_7	Ba	65	50	190	0.7	1.05	800	−0.2	0.4	10^8	2800	correction magnets for TV	~10
			155	104	190	4.4	1.15	800	−0.2	0.4	10^4	3900		
Anisotropic	F_8	Ba	240	175	240	11.0	1.05	800	−0.2	0.4	10^5	3500	small dc motors	
			270	190	220	13.0		800			10^5	3900		

† Philips Data Handbook (1982); F6 from McCaig (1977) ‡ Normally applied M-ferrite § See Fig. 2 ¶ Recoil permeability ‖ α = Temperature coefficient (0.8 kA $m^{-1}K^{-1}$ for BaM and 0.95 kA m^{-1} for SrM)

The remanence B_r is a product of saturation polarization J_s, fraction of ferrite phase s and degree of alignment f. Typical values of s are 0.9 for sintered materials and 0.6 for plastic-bonded materials, the low value for the latter being due to the presence of the binder. The alignment factor f ranges from a value of 0.5 for isotropic magnets to about 0.9 for anisotropic magnets.

The coercivity H_c strongly depends on the microstructure. In most cases detailed relationships are not known. We concentrate here on fired or annealed materials where crystal imperfections do not play a significant role.

Typical ball milled powder particles are somewhat larger than D_c and magnetization reversal proceeds by nucleation and growth of reversed domains. Nucleation is the limiting factor when the powder is sufficiently fine, say 0.5–1.5 μm (Schippan and Hempel 1985). In that case, the expression for the coercivity of a single crystallite becomes $H_c = H_n - NJ_s\mu_0^{-1}$, where H_n is the internal field needed for nucleation. There are theoretical and experimental indications that H_n increases with decreasing crystallite thickness (Holz 1970).

For an assembly of crystallites f and s are also involved (Kojima 1982, Stäblein 1982). In the case of noninteracting single domain particles, random orientation causes a decrease of the polarization coercivity (${}_JH_c$), in theory by a factor of 0.48. The ${}_JH_c$ of specially prepared powders may indeed be close to this theoretical value. On the other hand, owing to the different reversal mechanism, the ${}_JH_c$ of typical ball milled powders increases when f decreases. Mutual particle interactions give rise to an increase of the term $NJ_s\mu_0^{-1}$. In theory, this increase is proportional to s, but in practice this is only found after special dispersing techniques (Schippan and Hempel 1985).

With respect to typical polycrystalline materials (grain size ~1 μm) some complications arise: (a) the H_n values of the individual crystallites may be lowered by exchange interaction across the grain-boundary; and (b) magnetization reversal on macroscale may be non-uniform, proceeding by the initiation and growth of large multicrystal reversed regions. Only a few studies are available; they refer to grain oriented materials. The influence of exchange is usually neglected, i.e., the H_n of the individual crystallites remains important. Starting from a distribution in H_n and a uniform distribution of reversed regions, a quite realistic demagnetization curve can be calculated (Heinecke 1966). On the other hand, the existence of non-uniform reversal has been demonstrated and on this basis, an approximate expression for ${}_JH_c$ has been derived:

$${}_JH_c = \bar{H}_n - N(B_r + J_s)\mu_0^{-1}$$

where $\bar{H}_n$ refers to the average H_n (Kools 1985).

This expression shows the influence of various microstructural factors: grain size (on $\bar{H}_n$), grain shape (on N), s and f (on B_r). The different implications, except the effect of grain shape, have been verified to a first approximation.

5. *Commercial Ferrite Magnet Materials*

The ferrite crystallites are bonded to solid bodies either by sintering or by plastic bonding and may be randomly oriented (isotropic) or aligned with the c axis in one direction (anisotropic). Depending on the binder used, plastic-bonded ferrites may be flexible or rigid (Stäblein 1982). The anisotropic sintered form is by far the most important one. It is produced in three main groups: high B_r, high B_r in combination with medium polarization coercivity ${}_JH_c$ and high ${}_JH_c$. These groups represent different application fields, as shown in Table 3.

Table 3 summarizes the most important characteristics of different material groups (F1–F8) and Fig.

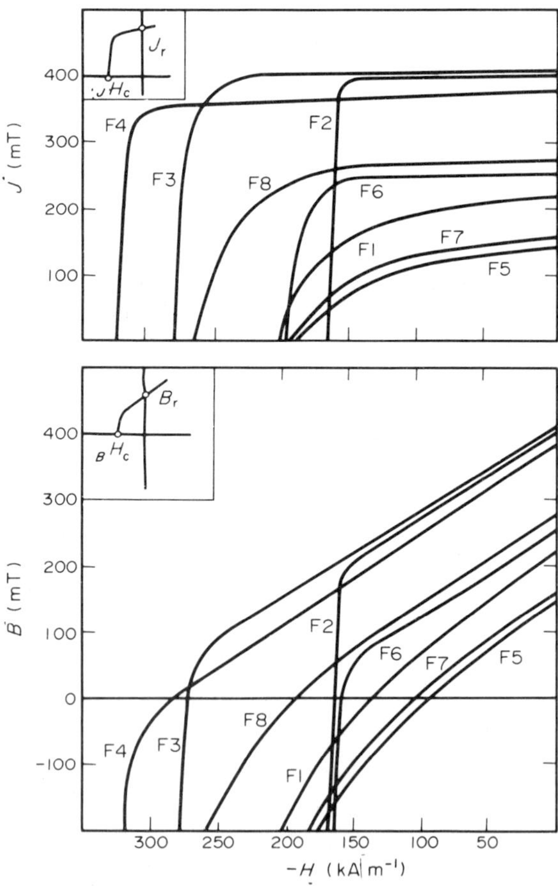

Figure 2
Typical demagnetization curves for the material groups of Table 3 (highest-quality values) (Philips Data Handbook 1982, McCaig 1977 for F6)

2 shows typical demagnetization curves. Where more than one grade exists within one group, the extreme values are given in Table 3. Compared with plastic-bonded ferrites, sintered materials have higher magnetic quality and closer dimensional control (when machined). Plastic-bonded ferrites may be preferred because of their lower price or their special mechanical properties and shaping capabilities. Anisotropic materials are used when high magnetic quality is essential, whereas the cheaper isotropic material is preferred when low magnetic quality is acceptable or when the properties have to be isotropic.

Figure 3 provides a survey of typical processing routes employed in the production of ferrite. The manufacture of high-grade anisotropic SrM (Van den Broek and Stuijts 1977) is taken as an example. The raw materials are dry powders of $SrCO_3$, Fe_2O_3 and additives such as SiO_2 and B_2O_3. The desired quantities (Fe_2O_3–SrO molar ratio $\simeq 5.5$) are weighed and dry-mixed. The mixture is granulated in a disk pelletizer to granules of about 5 mm and subsequently prefired at about 1250 °C in air. During prefiring, the raw materials react to form the desired compound ($SrFe_{12}O_{19}$). The hard prefired granules are wet-milled in steel ball mills to a fine powder. The resulting thick suspension (slurry) is processed to a crystallographically aligned pressed product by pressure filtration (wet pressing) in a magnetic field (see *Pressure Filtration for Forming Advanced Ceramics*). After drying, the compacts are sintered at about 1250 °C in air. During sintering, anisotropic shrinkage occurs, ~15% perpendicular and ~25% parallel to the preferred direction. Since accurate dimensional control is not possible, the pole faces are ground afterwards.

The microstructural requirements impose demands on the sintering process and preceding operations. During sintering, considerable densification has to occur without significant grain growth. These more or less contradictory demands can be achieved satisfactorily by having small particles in the slurry and by using sinter additives such as SiO_2. For a high degree of alignment in the pressed product, the particles in the slurry should be single crystals and mobile. This in turn requires a small grain size of the prefired granule. Control of grain growth during prefiring can be attained by using excess SrO in combination with additives such as B_2O_3 and SiO_2. In addition, conversion to SrM must be sufficient, otherwise a high B_r cannot be obtained.

6. *Applications*

Owing to its low price, ferrite has replaced other magnet materials in existing systems, either with or without modifications to the system (Van den Broek and Stuijts 1977). This is particularly the case with static applications where small demagnetizing fields are involved. A typical example is the application of flat ferrite rings instead of high-metallic centre core magnets in loudspeaker systems. The high H_c has stimulated the development of new systems, especially in dynamic applications where periodically high demagnetizing fields are present. A typical example is the electric motor with its strong armature reaction fields (Van Heffen 1980). New electric motors are being developed which require very high $_JH_c$ values lying far beyond the range of Al–Ni–Co materials, for example, the starter motor, requiring $_JH_c > 320$ kA m^{-1}.

The main ferrite products are anisotropic segments for motors in cars (F3, F4), anisotropic rings for

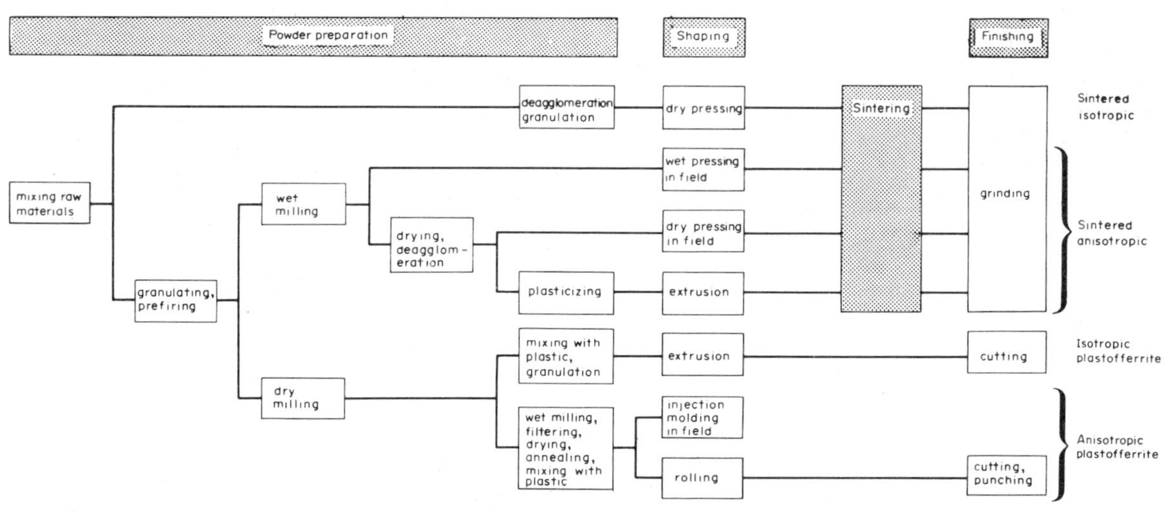

Figure 3
Typical processing routes for M-ferrite magnets (after Van den Broek and Stuijts 1977)

loudspeakers (F2) and large anisotropic blocks for ore separators (F3). These products together represent about 70% of the total production. It appears that large-scale production is concentrated on systems requiring relatively large magnets. In this case the greatest advantage is taken of the low material price. The remainder of production is distributed over a wide range of applications, involving mostly small magnets, such as shavers, mixers and coffee grinders.

Applications of ferrites other than permanent magnets are in the field of microwaves, magnetic bubble memories, magnetic tape recording and magneto-optics. M-ferrites are of interest for resonance-type microwave devices (e.g., isolators, filters, circulators) (Winkler and Dötsch 1979). Below 20 GHz such devices normally employ garnet or spinel ferrites in combination with a bias magnet. At high frequencies the required bias field becomes impracticably high ($>570\ \mathrm{kA\ m^{-1}}$). M-ferrites are then preferred because of their large anisotropy field, which acts as built-in bias field and provides a resonance frequency of about 50 GHz in a small tuning field. For broader frequency coverage, various substitutions can be applied by which H_A is altered. For instance, ZnTi substitution is suitable for the 10–50 GHz range and aluminum substitution for the 50–100 GHz range. The efficiency of the resonance device is improved when the width of the resonance absorption line is reduced. For this reason sintered M-ferrites must at least be in the anisotropic form. Much better results, however, can be obtained with single-crystal materials.

Magnetic bubbles are cylindrical reversed domains in a thin crystal with its dominant preferred axis normal to its wide surface (see *Bubble Domain Materials*). They were first observed in a thin BaM crystal. The mobility of M-ferrite bubbles, however, has proved to be rather slow. The dominant bubble materials today are the garnets.

7. Other Hard Ferrites

The definition of hard ferrites as iron-base oxide compounds possessing significant intrinsic magnetic anisotropy means that most other hexagonal ferrites as well as cobalt ferrite are to be included among the hard ferrites. The first commercial ceramic magnets, such as "vectolite," were based on Fe_3O_4–$CoFe_2O_4$ mixed crystals. Minimum magnet specifications were: $B_r = 0.16\ \mathrm{T}$ and ${}_JH_c = 80\ \mathrm{kA\ m^{-1}}$. The magnetic hardness stems from an induced anisotropy, introduced by slowly cooling from 300 °C in a magnetic field, thus effecting ordering of the highly anisotropic cobalt ions. With the rise of M-ferrite magnets, those based on cobalt ferrite have gradually been ousted. Today, cobalt ferrite, in the form of γ-Fe_2O_3–$CoFe_2O_4$ mixed crystals, is sometimes used in tape recording.

In addition to M-ferrites, other hexagonal ferrites, such as $BaZn_2Fe_{12}O_{27}$ (Zn_2W), $BaZnFe_{14}O_{23}$ (ZnX) and $Ba_2ZnFe_{18}O_{30}$ (ZnU), also possess uniaxial anisotropies comparable to the M-ferrites, while values of J_s can be even higher. Thus, these compounds should in principle make comparable or even superior magnets. Commercial development, however, has been lacking, due to more difficult processing technology.

See also: Soft Magnetic Ferrites; Magnetic Materials: An Overview

Bibliography

Adelsköld V 1938 X-ray studies on magneto-plumbite PbO. $6Fe_2O_3$, and other substances resembling "beta-alumina", $Na_2O.Al_2O_3$. *Ark. Kemi Mineral. Geol.* 12A (29): 1–9

Gemperle R, Shtolts E V, Zelený M 1963 The temperature dependence of the domain structure of magnetoplumbite. *Phys. Status Solidi* 3: 2015–28

Heinecke U 1966 Über ein Modell zur Berechnung der Entmagnetisierungskurven von Hartferriten. *Phys. Status Solidi* 18: 569–79

Holz A 1970 Formation of reversed domains in plate-shaped ferrite particles. *J. Appl. Phys.* 41: 1095–96

Kojima H 1982 Fundamental properties of hexagonal ferrites. In: Wohlfarth 1982, Chap. 5, pp. 305–91

Kools F 1985 Factors governing the coercivity of sintered anisotropic M type ferrites. *Proc. M.M.A.*, Grenoble, 1985

McCaig M 1977 *Permanent Magnets in Theory and Practice.* Pentech, London

Pauthenet R, Rimet G 1959 Variations thermiques des constantes d'anisotropie et de l'aimantation spontanée de la magnétoplombite. *C. R. Hebd. Seances Acad. Sci.* 249: 1875–77

Philips Data Handbook 1982: Components and Materials, Pt. 16B, *Permanent Magnet Materials.* Philips, Eindhoven, The Netherlands

Schippan R, Hempel K A 1985 The influence of microstructural properties on the coercivity of presintered strontium hexaferrite. *Proc. ICF$_4$*, San Francisco, 1984

Shirk B T, Buessem W R 1969 Temperature dependence of M_s and K_1 of $BaFe_{12}O_{19}$ and $SrFe_{12}O_{19}$ single crystals. *J. Appl. Phys.* 40: 1294–96

Smit J, Wijn H P 1959 *Ferrites.* Philips Technical Library, Eindhoven, The Netherlands

Stäblein H 1982 Hard ferrites and plastoferrites. In: Wohlfarth 1982, Chap. 7, pp. 441–602

Van den Broek C A M, Stuijts A L 1977 Ferroxdure. *Philips Tech. Rev.* 37: 157–75

Van Heffen H J H 1980 Ceramic permanent magnets for d.c. motors. *Electron. Compon. Appl.* 3: 22–30, 120–25

Winkler G, Dötsch H 1979 Hexagonal ferrites at millimeter wavelengths. *Proc. 9th European Microwave Conf.* Microwave Exhibitions and Publishers, Sevenoaks, UK, pp. 13–24

Wohlfarth E P (ed.) 1982 *Ferromagnetic Materials. A Handbook on the Properties of Magnetically Ordered Substances,* Vol. 3. North-Holland, Amsterdam

F. Kools

Hardboard and Insulation Board

Hardboard and insulation board are rigid or semirigid sheet materials with thicknesses of less than 1 in. (~25 mm). Summarily known as fiberboards they are a category of the large class of wood composition-boards, made by various methods of manufacture from wood elements ranging from veneers to fibers (Fig. 1). The term fiber refers to any wood or other lignocellulosic element possessing the size and shape of the wood cell, the biological building block of all woody material.

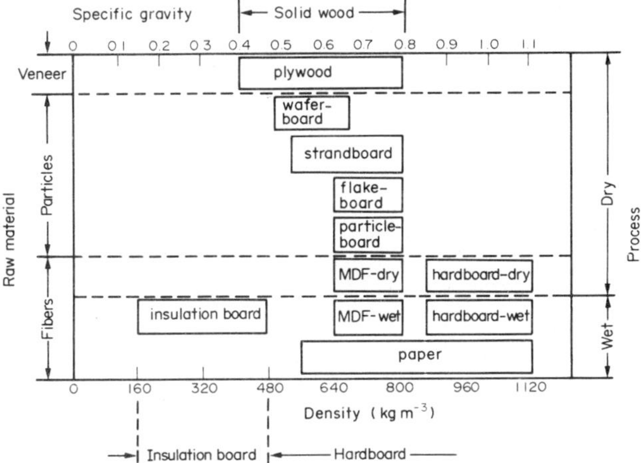

Figure 1
Classification of wood composition-boards

In the terminology of the American Hardboard Association the boundary between insulation board and hardboard is a board density of 480 kg m^{-3} (30 lb ft^{-3}). In the trade, however, a distinction is made between medium-density fiberboard, often referred to as MDF, and higher-density fiberboard, called hardboard in the narrower sense.

In the manufacturing process solid wood is first reduced to fiberlike elements (pulping) which are subsequently recombined in sheet form with or without the addition of binders.

Fiberboard processes utilize the lower-quality end of the available range of raw wood material. Owing to the small size of the elements, these processes achieve a higher degree of homogenization and randomization of the nonisotropic raw material than any other wood utilization process, with the exception of paper.

The most important fiberboard markets are for interior and exterior cladding, and industrial applications in the furniture and automotive industries.

1. Bonds in Fiberboard

Although the tensile strength of the wood fiber is very high, it is generally not utilized to its fullest potential in the structural configuration of a paper sheet or a fiberboard. This is due to the total bonding area between fibers being incapable of transmitting the shear stresses required to stress the fiber to its limit, either because of the limited size of the total shear area or because of the quality of the bond. In low- and medium-density fiberboard, stress failures occur predominantly in the bond. At high board densities failures do occur in the fiber, reflecting both more intimate contact between fibers and possible modification of fiber characteristics under the severe conditions in the hot press.

The establishment of maximum quality fiber bonds is, therefore, one of the predominant goals of process design. Three different types of bonding occur in fiberboard, either individually or simultaneously, depending on the process.

1.1 Hydrogen Bonding

Hydrogen bonds depend on extremely short-range attractive forces associated with surface molecules, and develop between positively charged hydrogen atoms and any negatively charged atoms such as oxygen. This type of bonding occurs primarily in wet fiberboard processes, where the surface tension of the evaporating water pulls the fibers together within the range of molecular forces.

1.2 Lignin Bonding

Lignin bonds are attributed to the thermal softening of the lignin, a major wood component concentrated in the outer region of the cell wall, and a fusing together of lignin-rich surfaces under the conditions existing in the hot press (Spalt 1977). It has also been proposed that such bonds are formed by the condensation of phenolic substances derived from lignin. Lignin bonds are limited to wet-process fiberboards densified in hot presses, although they may occur in dry-process hardboard as a result of severe pulping conditions (Suchsland et al. 1983). Like the hydrogen bond, the lignin bond is unavailable to any of the other wood composition-board products.

1.3 Adhesive Bonding

Adhesive bonds formed by added binders are essential in the manufacture of dry-process fiberboard and all other wood composite products, and are used to enhance board quality in wet processes. The most common binders are phenol–formaldehyde (dry-process hardboard) and urea–formaldehyde resins (dry-process medium-density fiberboard). Other adhesives include drying oils and thermoplastic resins (Table 1).

Table 1
Binder systems used in fiberboard manufacture

Fiberboard	Bonding (listed in increasing order of usage) Primary	Secondary
Wet process		
insulation board	hydrogen	starch, asphalt
S1S Masonite	lignin	
		phenol, thermoplastic
S1S	lignin	resins
S2S	lignin	thermoplastic resins,
MDF	lignin	drying oils
Dry process		
S2S	phenol	
MDF	urea	

2. *Fiberboard Processing*

Figure 2 gives a simplified outline of the entire range of fiberboard processes. Primary defiberizing of the raw material occurs by one of three mechanical pulping processes: (a) the Masonite process, subjecting the wood chips to high steam pressure (7 MPa), which upon sudden release explodes them into pulp; (b) the pressurized refiner, separating fibers between rotating grinding disks while under elevated steam pressure; and (c) the atmospheric refiner, grinding pretreated (steamed or water cooked) or untreated chips at atmospheric pressure. Chemical pulping is not used in fiberboard manufacture.

The fibers are either diluted with water and the mat formed on fourdrinier machines (wet process), or dried and formed by air-felting machines (dry process). Binders and sizing (chemicals to reduce water absorption) are added to liquid furnish and precipitated on fibers by pH reduction (wet process), or are distributed on fibers by spraying or by mechanical action (dry process).

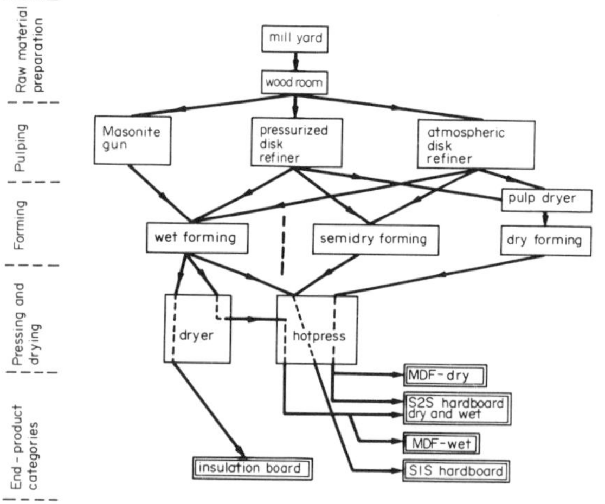

Figure 2
Simplified flowchart of fiberboard processes

Insulation board is made by drying the wet-formed mat. It relies almost entirely on hydrogen bonding. All other fiberboards are hot pressed. The hot press densifies the mat, provides the condition for bond development (plasticizing of lignin, curing of adhesives) and reduces the water content of the wet mat. Water and steam removal in the hot press from the wet and moist (semidry) mats is facilitated by inserting a screen between the bottom mat surface and press platen. The screen pattern is embossed on the back of the finished board (S1S—smooth one side). Dry-formed mats are pressed without screens (S2S—smooth two sides); S2S boards can also be made from dried wet-formed mats. The design of press cycles (pressure–, temperature–time relationships) influences density distribution over the board cross section and thereby affects strength properties. Thicker boards (dry medium-density boards) are particularly sensitive to such effects (Suchsland and Woodson 1976).

Heat treatment (baking at 260 °C) of pressed boards with or without prior application of drying oils (tempering) increases bending strength and reduces water absorption and swelling. Bending strength improvement is believed to be due to fiber bond enhancement at high temperatures. Reduction of water absorption follows conversion of hemicelluloses into less hygroscopic furfural polymers (Stamm 1964). The surface application of tempering oils and subsequent heat treatment increases surface hardness substantially, a property of great importance in the manufacture of premium quality factory-finished wall panelling. Excessive heat treatment causes pyrolytic breakdown of the cell wall. Insulation board and thick medium-density fiberboard are neither heat treated nor tempered.

3. *Fiberboard Properties*

Most strength properties of fiberboard are strongly affected by raw material and manufacturing variables such as fiber characteristics, board density, binder content and heat treatment, and are, therefore, controllable within considerable margins. Real fiberboard properties are compromises between market demands and economic limitations, and are often adjusted to the requirements of particular applications. Table 2 lists minimum and maximum hardboard properties prescribed by the commercial standard.

Properties related to the hygroscopicity of the wood cell wall are much more difficult to control. All fiberboards, by absorbing and giving off water, maintain a moisture content in equilibrium with the surrounding air (Fig. 3). Important consequences of moisture content fluctuations are dimensional

Table 2
Classification of hardboard by thickness and physical properties[a]

Class	Nominal thickness (in.)	Water resistance (maximum average per panel): Water absorption based on weight (%)	Water resistance (maximum average per panel): Thickness swelling (%)	Modulus of rupture (minimum average per panel) (MPa)	Tensile strength (minimum average per panel): Parallel to surface (MPa)	Tensile strength (minimum average per panel): Perpendicular to surface (MPa)
Tempered	1/10					
	1/8	25	20			
	3/16			41.3	20.7	0.9
	1/4	20	15			
	5/16	15	10			
	3/8	10	9			
Standard	1/12	40	30			
	1/10					
	1/8	35	25			
	3/16			31.0	15.2	0.6
	1/4	25	20			
	5/16	20	15			
	3/8	15	10			
Service-tempered	1/8	35	30			
	3/16	30	30	35	13.8	0.5
	1/4	30	25			
	3/8	20	15			
Service	1/8	45	35			
	3/16	40	35			
	1/4	40	30			
	3/8	35	25			
	7/16	35	25			
	1/2	30	20			
	5/8	25	20	20.7	10.3	0.3
	11/16	25	20			
	3/4					
	13/16					
	7/8	20	15			
	1					
	1–1/8					
Industrialite	1/4	50	30			
	3/8	40	25			
	7/16	40	25			
	1/2	35	25			
	5/8	30	20			
	11/16	30	20	13.8	6.9	0.2
	3/4					
	13/16					
	7/18	25	20			
	1					
	1–1/8					

a Source: US Department of Commerce (1980a)

Table 3
Property requirements for medium-density fiberboard for interior use[a]

Nominal thickness (in.)	Modulus of rupture (MPa)	Modulus of elasticity (MPa)	Internal bond (tensile strength perpendicular to surface) (MPa)	Linear expansion (%)	Screwholding Face (kg)	Screwholding Edge (kg)
13/16 and below	20	2000	0.62	0.30[b]	148	125
7/8 and above	19	1723	0.55	0.30	136	102

a Source: US Department of Commerce (1980) b For boards having nominal thicknesses of $\frac{3}{8}$ in. or less, the linear expansion value is 0.35%

changes (thickness swelling and linear expansion), of which the linear expansion is critical in fiberboard exterior siding. Also affected by moisture content changes are most strength properties, but most severely the interlaminar shear modulus. The shear modulus of $\frac{1}{4}$ in. (~6.25 mm) tempered hardboard at 80% relative humidity (RH) averaged only 70% of the modulus at 65% RH, and only slightly more than half the value at 30% RH. The modulus of elasticity in bending varied less than 20% of the value at 65% RH over the range 0–90% RH. Strength properties are similarly affected by increased temperature (Haygreen and Sauer 1969). Fiberboard, like solid wood, is viscoelastic, and the time-dependent responses to stress of fiberboard require special consideration in structural applications (McNatt 1970, Moslemi 1964).

Thick medium-density fiberboard properties as recommended by the commercial standard are listed in Table 3. Mechanical properties are less critical for insulation board, with the exception of insulation sheathing, which requires tensile strength values up to 2.07 MPa and racking loads of at least 2360 kg. Minimum thermal conductivity values for insulation board vary from 0.0548 to 0.0692 W m^{-1} K^{-1}.

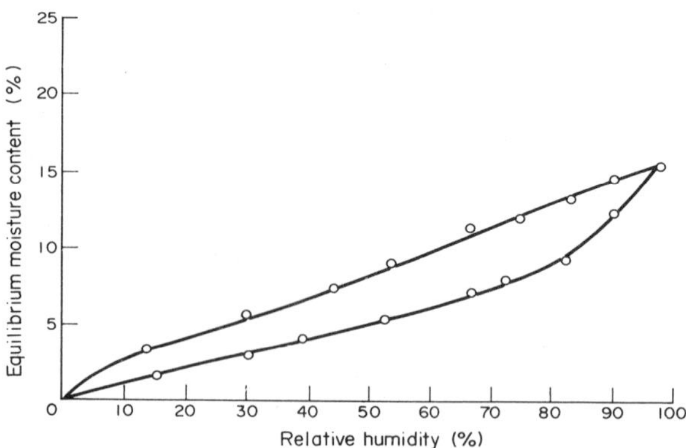

Figure 3
Adsorption and desorption isotherms between 0 and 98% RH at 20 °C for heat-treated hardboard (after McNatt 1974)

See also: Particleboard and Dry-Process Fiberboard; Paper and Paperboard: An Overview

Bibliography

Haygreen J, Sauer D 1969 Prediction of flexural creep and stress rupture in hardboard by use of time–temperature relationship. *Wood Sci.* 1: 241–49

McNatt J D 1970 Design stresses for hardboard—Effect of rate, duration and repeated loading. *For. Prod. J.* 20: 53–60

McNatt J D 1974 Effects of equilibrium moisture content changes on hardboard properties. *For. Prod. J.* 24: 29–35

Moslemi A A 1964 Some aspects of visco-elastic behavior of hardboard. *For. Prod. J.* 14: 337–42

Spalt H A 1977 Chemical changes in wood associated with wood fiberboard manufacture. In: Goldstein I S (ed.) 1977 *Wood Technology: Chemical Aspects,* ACS Symposium Series 43. American Chemical Society, Washington, DC, pp. 193–219

Stamm A J 1964 *Wood and Cellulose Science.* Ronald, New York

Suchsland O, Woodson G E 1976 *Properties of Medium Density Fiberboard Produced in an Oil Heated Laboratory Press,* US Department of Agriculture Forest Service Research Paper SO-116. US Government Printing Office, Washington, DC

Suchsland O, Woodson G E, McMillin C W 1983 Effects of hardboard process variables on fiberbonding. *For. Prod. J.* 33(4): 58–64

US Department of Commerce 1973 *Cellulosic Fiber Insulating Board,* Voluntary Product Standard PS57-73. US Government Printing Office, Washington, DC

US Department of Commerce 1980a *Basic Hardboard,* American National Standard ANSI/AHA A135.4. US Government Printing Office, Washington, DC

US Department of Commerce 1980b *Medium Density Fiberboard for Interior Use,* US Government Printing Office, Washington, DC

US Department of Commerce 1982 *Prefinished Hardboard Paneling,* American National Standard ANSI/AHA A135.5. US Government Printing Office, Washington, DC

US Department of Commerce 1983 *Hardboard Siding,* American National Standard ANSI/AHA A135.6. US Government Printing Office, Washington, DC

O. Suchsland

Hardness Characterization

Hardness determinations are among the most frequently used tests for the assessment of mechanical properties. The value of the test depends on the fact that it affords an inexpensive, nondestructive means of evaluating a number of useful engineering properties, such as tensile strength, heat treatment or the state of cold working in structural metals. The relationship between these properties and hardness, while not an exact one, is close enough for a wide range of industrial controls.

Hardness has been loosely defined as resistance to permanent indentation, but it is well known that the resistance is significantly affected by the shape of the tool used to produce the indentation (Tuckerman 1925). This dependence strongly suggests that hardness is not a distinct property but a combination of a number of properties, such as yield strength in shear and bulk modulus. The penetration of an indenting tool into the surface of a material is influenced by these properties according to the shape of the tool. A hardness number, a ratio of load to impression area, is not a simple compression stress, but a highly complex value in which a number of basic properties are summed in proportions dependent on the shape of the indenting tool.

Although there is probably no such property as "hardness," it is a convenient generic term. Properly, the term hardness has meaning only in terms of a specific test. One may speak of Brinell hardness or Rockwell hardness, but they should not be considered common measures of the same thing.

A number of different tests have been developed. They range from the simple Mohs scratch test, proposed in 1822 and still used as a basic, practical test for the field mineralogist, to the highly sophisticated microhardness tests that can evaluate individual crystal constituents of a metal matrix. Nearly all of these tests involve some kind of impression made in the surface of a specimen. The impression may be in the form of a scratch, an indentation made by a steel or diamond tool under static forces or a dynamic rebound test. Among substances tested, metals have probably received the most attention, but rubber and manufactured plastics are also important areas of application. Methods are available to test minerals, glass, hard carbides and ceramic materials. Tests on such diverse materials as paint and varnish coatings have proved valuable.

The following sections describe specific tests that have gained general acceptance. For most, standard methods and procedures have been adopted by various national standardizing organizations and by organizations such as the International Organization for Standardization (ISO) and the International Organization for Legal Metrology (OIML).

1. The Brinell Hardness Test

The Brinell test, introduced in 1900 in Sweden by J A Brinell, is the oldest of the modern hardness tests. It consists of forcing a steel or hard-metal ball into the surface of the specimen under load. After removal of the load, the diameter of the circular impression left in the specimen is measured with a low-powered microscope and the surface area of the impression calculated on the assumption that it has the shape of the undeformed ball. A hardness number is calculated as the indenting load in kilograms-force (kgf) divided by the surface area of the impression in square millimeters. The Brinell hardness number, like other hardness numbers, is always quoted without units of measurement.

A ball diameter of 10 mm and an indenting load of 3000 kgf are accepted as standard for tests of materials such as steels and similarly hard alloys. For softer copper and aluminum alloys, a load of 500 kgf is generally used with the 10 mm ball. Brinell hardness numbers range from values as low as 1 or 2 for very soft metals and some plastics up to about 600 for hardened steels. Steel balls are not recommended for materials above about Brinell 450, but hard-metal balls such as tungsten carbide can be used up to about Brinell 600. Tests above Brinell 630 are not recommended because of the difficulty of measuring the very small and shallow impression. This value is less than the maximum hardenability of high-carbon steels.

The Brinell hardness number is calculated from the formula

$$\mathrm{HB} = \frac{2F}{\pi D[D - (D^2 - d^2)^{1/2}]}$$

where HB is the Brinell hardness number; F the indenting load in kgf; D the ball diameter in mm; and d the impression diameter in mm. Convenient tables converting impression diameters directly to Brinell numbers are available from many sources, e.g., the *Standard Test Method for Brinell Hardness of Metallic Materials* (ASTM E10), issued by the American Society for Testing and Materials.

The requirement for the unique use of the millimeter and the kilogram-force introduces a problem when a change is made to SI units. Here, the unit of force becomes the newton and in some jurisdictions its use in preference to the kilogram-force is mandatory. The tendency in international standards has been to express F in newtons along with a multiplying factor in the formula converting the force into kilogram-force units.

The sensitivity of the Brinell test to load is a factor of importance in test procedures. Since the Brinell number is essentially a stress or pressure value, i.e., load per unit area, one might expect to obtain the same result in tests made at different loads on a homogeneous piece of material. In practice the results are not the same. Tests made at loads of 3000 and 500 kgf with a 10 mm ball do not agree and, furthermore, the amount of disagreement, even at the same hardness level, is different for different materials. The discrepancy arises because impressions made at two different loads are not geometrically similar. A vertical section through impressions at 500 and 3000 kgf would show that the heavier load penetrated more deeply in relation to the diameter of the impression than did the 500 kgf load. In effect, the two impressions were made with different shaped indenters even though the same ball was used for both. As pointed out earlier, the sampling of basic properties will be different and the composite value that forms the hardness number will, consequently, be different. Because of this effect, it is important that in reporting the results of a Brinell test the load and size of the indenting ball be clearly stated. The usual practice is to annotate the load and ball diameter if they differ from the standard values of 3000 kgf and 10 mm.

A solution to the load sensitivity problem is to change both load and ball diameter in such proportions that impressions will be geometrically similar. If the ratio of load to the square of the ball diameter is kept unchanged, this condition will be achieved. For example, the 3000 kgf load and 10 mm ball have a F/D^2 ratio of 30, as does the combination of a 750 kgf load and 5 mm ball. Thus, if a homogeneous material is tested, the results will be comparable for both combinations. This approach permits tests on material too thin for the standard test load.

The Brinell hardness test is widely used (although it is somewhat slow and leaves an impression readily apparent to the unaided eye) particularly in testing large forgings and materials having a coarse grain structure. The test equipment is rugged and well suited to use in the open mill area. Ball indenters are readily available from ball bearing manufacturers and are produced in dimensional tolerances far closer than needed for the test.

The accuracy of a Brinell tester can be verified by checking its readings on a standardized test block. Such blocks can be obtained from the machine manufacturer or from special standardizing laboratories. Another method is separate verification of the indenting load by means of elastic proving devices and measurement of impression diameters by a high precision measuring microscope.

2. *The Rockwell Hardness Test*

The Rockwell test was developed about 1922 by Stanley P Rockwell, a metallurgist, who saw the need for an instrument capable of testing the very hard steels used in ball bearing races. The test is performed by a machine that initially forces a penetrator into the surface of the specimen under a light load called the *minor load*. A depth-measuring indicator coupled to the penetrator is set to a null reference and the load on the penetrator is increased smoothly to a higher value, called the *major load*, and then reduced again to the minor load value. The depth measuring device then shows the increased depth of indentation caused by the application of the major load. The scale is graduated to read the hardness number directly.

The minor load is 10 kgf and the major load may be either 60, 100 or 150 kgf. The penetrator is a diamond cone having an included angle of 120° with a spherical point of 0.2 mm radius. For testing softer metals and plastics a series of ball penetrators of $\frac{1}{16}$, $\frac{1}{8}$, $\frac{1}{4}$ and $\frac{1}{2}$ in. diameter is available. Each combination of penetrator and major load constitutes a separate scale and a Rockwell hardness number is always quoted with a letter symbol to indicate the scale. One Rockwell number, the smallest graduation on the indicator dial, represents an increase in depth of 0.002 mm.

The Rockwell test is extremely useful in the control of sheet material such as aluminum, brass or steel. Because of the economic importance of very thin sheet, a special tester known as the Rockwell Superficial Tester was developed. In this machine, the minor load is reduced from 10 to 3 kgf. The major loads are 15, 30 and 45 kgf. The sensitivity of the depth-measuring device is increased to 0.001 mm per division.

The Superficial Tester uses the same series of indenters as the standard machine and can successfully test hard sheet materials as thin as 0.125 mm. The deeper impressions in soft sheet require a greater thickness. Most procedural standards provide tables showing minimum thickness limits for various scales and hardnesses. The general rule is that no bulge or other marking showing the effect of load should be visible on the side of the piece opposite the impression.

The Rockwell test is rapid and does not require a skilled operator. The equipment can be adapted readily to automation and use with data acquisition systems. Daily verification of the accuracy of the tester is accomplished by calibrated test blocks available from the manufacturers of the machines. In the UK and in some European countries, test blocks certified by official government bodies are available.

3. *The Vickers Hardness Test*

The Vickers test, named after the original manufacturer but sometimes called the diamond pyramid hardness test, was designed in England in 1922 by R Smith and G Sandlund. The use of a square-based pyramid as the indenter insured that impressions made at different depths would all be geometrically similar. To simulate the Brinell test, an included face angle of 136° was selected. This angle is the angle of tangency at the edge of a Brinell impression having a diameter of 3.75 mm made with the standard 10 mm ball, the midpoint of the recommended range of impression diameters. As a result of this design, Vickers tests made in a homogeneous material over a range of loads from less than 1 to 120 kgf all give essentially the same hardness number.

In the test, the indenter is forced into the surface of the specimen, the resulting square impression, after removal of the indenter, is viewed in a measuring microscope and the two diagonals measured. A hardness number, defined as the indenting load in kilograms-force divided by the surface area of the impression, is calculated from the formula

$$HV = 1.854\, F/D^2$$

where HV is the Vickers hardness number; F the indenting load in kgf; and D the average diagonal length in mm.

The Vickers test is well adapted to testing very hard steels and covers the range down to very soft metals in a single scale of numbers. The impressions are small, typically less than 0.5 mm on the diagonal, and not readily apparent to the unaided eye. Because of this, a more elaborate reading microscope is required than for the Brinell test. Usually the microscope is built into the testing machine and arranged so that the impression is automatically centered in the viewing field for measurement. The microscope has a magnification of the order of 100 and is usually equipped with a micrometer eyepiece graduated to read directly to the nearest 0.001 mm. Because of the higher magnification, a more careful preparation of the specimen surface is required than for the Brinell or Rockwell tests.

The test is capable of resolving small differences in hardness and its accuracy is easily controlled either by primary standardization of load and indenter shape or by standard test blocks. The test is widely used in the UK and on the Continent but has had less popularity in the USA.

4. *The Scleroscope Test*

The Scleroscope test differs from the usual indentation hardness tests in that it is a rebound test. A diamond tipped rod, called the *hammer*, is dropped from a fixed elevation to the surface of the specimen and the height of the rebound noted. The difference in elevation between the release point and the top of the rebound is a measure of the energy lost in deformation of the specimen. The hammer is quite light and leaves only a barely perceptible mark.

The Scleroscope offers a considerable advantage in testing large objects such as the rolls used in steel and paper mills because it need not be rigidly clamped to the test piece as in the case of the conventional indentation testers. Various forms of support are offered so that the tester can be supported in a vertical position during a test. The mark left is so slight that there is little likelihood that the indentation will affect an opposite roll while the two are running together in service.

Scleroscopes are manufactured in two models. In both, the hammer falls within a closed tube, but in one the rebound is read by eye against a scale in the tube. In the other, the height of rebound is indicated by a maximum pointer.

A special Scleroscope tester, known as the Forged Roll tester, has been developed to compensate for the anomalously high readings obtained on a chilled cast iron roll when compared to a forged steel roll.

5. *Microhardness Tests*

Microhardness tests are indentation hardness tests made under unusually low indenting forces, usually in the range from one gram-force to one kilogram-force. The resulting indentation is of microscopic size, in the range of 0.01 to 0.10 mm, and hardly visible to the unaided eye. It is possible to test such objects as small watch pinions, needle points and even individual crystals in a metal matrix.

Because of the problems of making, locating and measuring these small impressions, special testing machines have been developed that permit the operator to locate under the microscope the particular area where the indentation is to be made, transfer it automatically under the indenter, make the impression and return it to the center of the microscope field for measurement. The measurements are usually made by a microscope equipped with a vertical illuminator and a filar measuring eyepiece. Very careful preparation of the specimen is required.

The indenter is a diamond pyramid, either the 136° Vickers indenter or the Knoop indenter. The Vickers indenter is preferred for testing metals since the test is almost independent of load and the results are comparable to Vickers tests made under heavier loads. The Knoop indenter permits tests to be made on frangible materials such as glass, gem stones and hard carbides, which could not be tested by other forms of indenter.

The Knoop indenter was developed at the National Bureau of Standards in the USA (Knoop et al. 1939). It is a four-sided diamond pyramid of elongated shape in which the angle between two opposite edges is 172° 30′ and between the other two edges 130° 0′. The impression left by the indenter is a rhombus with one diagonal about seven times the length of the other. The depth of the impression is about $\frac{1}{30}$ of the long diagonal. In practice, only the long diagonal is measured. The Knoop hardness number is defined as the indenting load in kilograms-force divided by the projected area of the impression in square millimeters. The Knoop number can be calculated from the formula

$$HK = 14.229\, F/D^2$$

where HK is the Knoop hardness number; F the indenting load in kgf; and D the length of the long diagonal in mm.

Knoop hardness values are unfortunately not independent of the value of the indenting load as might be expected from the example of the Vickers indenter. If a homogeneous specimen is tested at successively lower loads, the Knoop number increases markedly. This reaction has been attributed to elastic recovery at the ends of the long and shallow impression and to the inability of the microscope to resolve the fine tip of the impression. Observers have disagreed on whether or not the Vickers test suffers from a similar effect at very low loads. Whatever the result, a Vickers or Knoop test report should always cite the value of the indenting load used.

6. Portable Hardness Testers

A number of small, lightweight testers have been developed for soft metals and nonmetallic materials. A simple mechanical principle used in many of these testers is a spring-loaded indenter, usually a truncated cone or a spherical shape with a flattened tip, mounted within a compact housing with a dial indicator arranged to show the travel of the indenter. The housing, with its slightly protruding indenter, is pressed down by hand against the surface of a specimen until the base of the housing is firmly in contact with the specimen. The dial indicator then shows a value dependent upon the distance the indenter has penetrated into the specimen surface. Usually the dial is graduated into a scale that shows higher numbers for harder materials.

An excellent example of such a tester is the Shore Durometer, an instrument widely used for testing rubber. It comes in several models to cover the wide range of rubber hardnesses. The Durometers have their own scale of hardness numbers. Other testers operating on the principle of the spring-loaded indenter are used for the softer metal alloys and afford a rapid, convenient means for identifying material in stock bins and similar situations where the use of a larger tester would be impractical. Some of these testers have unique scales of their own but others are graduated to read Brinell, Rockwell or Vickers numbers. Some of the latter actually apply the standard loads and use the standard indenters but others do not. The user is warned that even though a nonstandard machine may read correctly on a test block, it may give an incorrect result when used with another material, particularly if there is a hardness gradient from the surface to the interior.

Other types of portable testers operate on principles that seem at first to be far removed from the conventional indentation test. The eddy current test, which responds to a combination of the magnetic and electrical properties of the specimen, the Sonadur tester in which a rod tipped by a diamond indenter is loaded against the specimen and driven into a natural resonant frequency responsive to the area of contact between indenter and specimen, and the Eseway tester in which an indenter impacts the specimen surface at a fixed speed and the deceleration of the indenter is measured by a piezoelectric sensor are all examples of new techniques in the field. Whether or not they will be accepted in time as formal hardness testing devices remains to be seen, but they do fulfill the basic premise behind all hardness testing—they separate materials by a means that correlates with a variety of useful engineering properties.

7. Hardness Conversion Tables

Because hardness is extensively used in procurement specifications, there is inevitably a demand for an authorized conversion relationship among the major types of test. The concept persists that there must be a fixed relationship among them. The factual situation, outlined earlier, that each test measures a different sampling of a number of different fundamental properties, is often ignored or rejected under pressure for a solution to a problem.

Obviously, one can select a specific material, such as a tool steel, heat treat specimens to a series of graduated hardnesses and test each specimen with Brinell, Rockwell and Vickers tests to derive a well-defined conversion relationship among the three tests. If the experiment is then repeated with a steel of different alloying type, such as a high-strength aircraft steel, the results, while close, will differ from the first experiment by significant amounts. The fact that the agreement is so close gives the impression that a universal conversion table is just a matter of a little more care in the laboratory measurements.

Conversion tables have been worked out for some specific materials. The American Society for Testing and Materials provides tables for cartridge brass, copper, austenitic stainless steel, nickel and high-nickel alloys, and a general table for nonaustenitic steel. Other national standardizing organizations have issued similar tables, all with the warning that they are at best only approximations and may introduce substantial error if used outside of the stated scope. The myth of the accurate hardness conversion table is not easily dispelled, however, despite these warnings.

Bibliography

American Society for Testing and Materials 1984 *Annual Book of ASTM Standards,* Pt. 3. American Society for Testing and Materials, Philadelphia, Pennsylvania, pp. 164–74

Knoop F, Peters C G, Emerson W B 1939 A sensitive pyramidal diamond tool for indentation measurements. *J. Res. Natl. Bur. Stand.* 23: 39–61

Lysaght V E 1949 *Indentation Hardness Testing*. Reinhold, New York
Lysaght V E, DeBellis A 1969 *Hardness Testing Handbook*. American Chain and Cable Company, Bridgeport, Connecticut
Mott B W 1956 *Micro-Indentation Hardness Testing*. Butterworths, London
O'Neill H 1967 *Hardness Measurement of Metals and Alloys,* 2nd edn. Chapman and Hall, London
Tabor D 1951 *The Hardness of Metals*. Clarendon Press, Oxford
Tuckerman L B 1925 Hardness and hardness testing. *Mech. Eng*. 47: 53–55
Williams S R 1942 *Hardness and Hardness Measurements*. American Society for Metals, Cleveland, Ohio

D. R. Tate

Hazards and Materials: An Overview

The use of materials to prepare the goods and structures of the modern world carries with it potential hazards at all steps of processing, manufacturing, transport, use and disposal. The goods and structures so produced are themselves subject to failure from faults in materials, design, construction or application—any of which could produce hazards. Control of such hazards and failures is essential for the continued functioning of an industrial society.

Proper practices must be established to allow the use of materials and the products made from them, while avoiding as much as possible the harm that can arise from misuse, carelessness, error and mismanagement. The ideal is safety, health and environmental protection at all stages in technology and the use of the products. Care and the application of scientific understanding are necessary from the bottom to the top of the production process—from the iron ore deposit to the steel bridge, from the oil well to the molded plastic part or from the Douglas fir seedling to the wood frame building. Properly prepared materials, correctly selected, processed and assembled are sought to obtain the goods and structures for the functioning of the economy and society, while at the same time protecting the workers, the public and the environment.

1. Definitions

Some definitions needed for this article are grouped here.

Hazard. A condition in which the possibility of injury or harm exists. The degree of hazard may also involve severity and frequency factors.

Critical components. A structure or system which is totally relied on to function as planned and whose unexpected failure could pose serious or deadly hazard. Failure of a noncritical component would be expected to pose only inconvenience and economic loss.

Hazardous material or chemical. A substance that can be injurious or harmful to life, health, property or the environment. The potential hazards of a substance depend on its inherent chemical and physical properties and its interactions with biology. The actual hazards of a substance depend not only on its inherent properties and interactions with living things but also on the circumstances under which it is used or confronted. See Table 1.

Risk. The probability that a hazard will be realized. The degree of risk depends on how severe the resulting harm, how many exposed and how available the circumstances for harm. An unreasonable risk is a personal/social/political/economic value judgement of a degree of risk that is not acceptable. This term has frequent legal use.

Table 1
Some circumstances of use and surroundings that determine actual hazards with chemicals

Amount	Site conditions:	Labelling and placarding
Concentration	Facilities design	Airborne levels monitoring
Temperature	Housekeeping and maintenance	Medical surveillance
Pressure	Ventilation used	Emergency measures available:
Subdivision and dispersion	Ignition source present or absent	First-aid-trained staff
Presence of other substances	Electrical codes followed	Sprinkler system
Barriers used:	Odor warning properties	Fire extinguishers
Containment	Who is confronting chemical:	Eyewash fountain
Shielding	Training and experience	Alarm system
Hoods	Personal hygiene	Storage practices
Firewalls	Habits	Diked storage tanks
Clothing	Hypersensitivity	Waste handling methods
Goggles	Illness past and present	Disposal methods
Respirator	Pregnant female	Incinerator with scrubber
Safety shoes	Worker with anosmia	Secure landfill

Risk assessment. A qualitative or, preferably, a quantitative determination of what adverse effects would be expected for individuals or populations or the environment from specific hazardous circumstances.
Risk management. The process of the matching of control options and priorities to assessed risks which involves making the value judgment of what is, or is not, an unreasonable risk.
Safe. A condition of low risk; a value judgment that a specific status of substances/circumstances is an acceptable risk. Safe is a relative term. It cannot be expected to be "zero risk."

2. *Codes, Standards and Regulations*

> The engineer must "hold paramount the safety, health and welfare of the public in the execution of his duties" (National Society of Professional Engineers 1981).
>
> "The public must be convinced that when we have a reasonable belief in an unreasonable risk we will move to reduce it, swiftly and decisively . . .." (William D. Ruckelshaus, EPA Administrator, in an address to the Chemical Manufacturers' Association, November 1983).

Codes, standards and regulations are used to help ensure safe, reliable conditions. To proceed without excessive fear or anxiety, society must establish rules and be able to trust the integrity of its professionals, inspectors and regulators who deal with its materials, goods and structures. Bridges, buildings, elevators and passenger aircraft are examples of structures used in faith by the lay public, trusting that the associated professionals know what they have been about and have followed established good practice and that the inspectors and regulators have checked their performance.

Codes and standards of good practice such as those of the American Society for Testing and Materials (ASTM), American National Standards Institute (ANSI), National Fire Protection Association (NFPA), Underwriters' Laboratories (UL), National Bureau of Standards (NBS) and many others provide guidance, specifications and test methods for the professional and the public to help ensure proper performance in the preparation and use of materials, goods and structures. Even a voluntary standard becomes legally binding as part of a purchase order, contract or certification or when included in a local building code. (There are about 400 standards-making organizations in the USA, as well as such important international ones as the British Standards Institution (BSI), the Deutsches Institut für Normung (DIN) and the International Organization for Standardization (ISO).)

For example, consider a critical component such as an eyebolt, which is screwed into heavy equipment for overhead lifting by cranes. Certain eyebolts are covered under ASTM D-489, "Standard specification for carbon steel eyebolts," which references ANSI B18.15, "Forged eyebolts," for its dimensional requirements. When specified for purchasing or use, this standard requires specific composition and processing, followed by standard material tests, such as tensile strength and grain size, and by breaking strength, bending and proof load testing to establish acceptability of a manufacturer's processing and product (see *Eyebolts: Integrity*). Routine inspection is needed before use in a lift. Any eyebolt showing nicks, distortion, wear or cracks should be removed from service and destroyed (by melting, crushing or cutting across the eye to prevent use).

Some further examples of critical components or systems which require nondestructive proof testing for safety assurance can include threaded fasteners (see *Threaded Fasteners: Application and Integrity*), steel cylinders for high-pressure gases (see *High-Pressure Gas Cylinders: Integrity*), and pipelines (see *Pipelines: Safety in Design, Construction, Operation and Maintenance*). (Note that the carrying out of pressurized testing can itself be hazardous.) The comprehensive *Boiler and Pressure Vessel Code* of the American Society of Mechanical Engineers and the NFPA *National Fire Code* are examples of detailed, broad, extensively applied standards and guidelines that promote safe practice. Standard-setting and regulation help to impose the same safety-related business costs on all producers and help to avoid "short-cutting" of safety to obtain improved cost competitiveness.

Legislatures and code developers are currently being pressured to protect fire fighters and the public by restricting the use of materials in buildings on the basis of their relative emission of smoke and toxic gases in qualification burning tests. The motivation is to reduce the level of fatalities in fires, since ~80% of fire deaths are attributed to inhalation of the airborne combustion and thermal degradation products of fire and not to the direct heat effects of the flames. Older test methods are being considered and new ones are being developed to assess combustion toxicological hazards of building materials and furnishings (see *Flammability Hazards of Structural Materials and Furnishings*; *Flammability of Polymers: Test Methods*; *Wood and Fire*).

In the USA when the public becomes concerned (rightly or wrongly) that its anxieties and interests are not being satisfactorily addressed by those using materials to prepare goods and structures, legislation has been used to initiate regulation to carry out what the public wills. As a legacy of the industrial revolution and the political process, such law and regulation on health, safety and the environment began to be developed slowly in the USA at about the turn of the century, continuing to the present. The explosion of technology during and after World War II led to much regulatory response. And especially in the early 1970s the floodgates opened to establish the law, regulatory powers and agencies for the US government to be involved in all stages of the use of materials in the production process, for the worthy purpose of ensuring worker health and safety, the public welfare and the restoration and protection of the environment:

> An employer must "furnish to each of his employees employment and a place of employment which are free from recognized hazards that are causing or likely to cause death or serious physical harm to employees" (*Occupational Safety and Health Act of 1970*, Sect. 5(a)(1))

> "The Administrator shall establish national programs for the prevention, reduction, and elimination of pollution . . . in cooperation with other Federal, State and local agencies . . . ." (*The Federal Water Pollution Control Act, 1972,* Sect. 104(a))

> "The term 'hazardous air pollutant' means an air-pollutant . . . which in the judgement of the Administrator may cause, or contribute to, an increase in mortality or an increase in serious irreversible, or incapacitating reversible illness . . .. The Administrator shall prescribe an emission standard for such pollutant . . . (and) establish any such standards at the level which in his judgement provides an ample margin of safety to protect the public health from such hazardous air pollutants." (*Clean Air Act, as Amended 1972*, Sect. 112)

> "It is declared to be the policy of Congress . . . to improve the regulatory and enforcement authority of the Secretary of Transportation to protect the Nation adequately against the risks to life and property which are inherent in the transportation of hazardous materials in commerce." (*Transportation Safety Act of 1974*, Sect. 102)

> "It is the policy of the United States that adequate authority should exist to regulate chemical substances and mixtures which present an unreasonable risk of injury to health or the environment . . . ." (*Toxic Substances Control Act, 1976*, Sect. 2(b)(2))

> "Subtitle C of the Solid Waste Disposal Act, as amended by the Resource Conservation and Recovery Act of 1976, as amended (RCRA), directs the Environmental Protection Agency (EPA) to promulgate regulations to protect human health and the environment from the improper management of hazardous waste." (Initial Regulations under the Resource Conservation and Recovery Act (RCRA), *Federal Register*, 1980)

> "An Act to provide for liability, compensation, cleanup, and emergency response for hazardous substances released into the environment and the cleanup of inactive hazardous waste disposal sites." (*Comprehensive Environmental Response, Compensation and Liability Act of* 1980 (*CERCLA*) or "The Superfund")

The regulatory agencies have proceeded to carry out their legislative mandates, providing an increasing background of regulations for those using materials to prepare goods and structures. The various administrators of the agencies have proceeded according to their philosophies and frequently their politics, being restrained or prodded at various times by the Executive Department and by adversaries through the court system.

Along with the regulatory powers comes the necessity under the law to determine just what hazards need controlling and which risk is an "unreasonable risk." Necessary regulation may be delayed or a regulation may be quickly instituted which is unnecessary but emotion-driven. The "carcinogen of the week" has been a phenomenon pushed by the news media. Regulation may not at times be wise, fair, consistent or efficient, but it will inexorably pass its costs on to society. The public must eventually pay for regulatory controls and for the degree of risk reduction they require. Through its representatives, the public is the ultimate authority for risk management in the USA.

3. Risk Assessment and Risk Management

Identification of potential hazards and the level of hazard in the use of materials to produce goods and structures is the first step in avoiding unacceptable risks in production or the environment (see Table 2). The scope of a hazard evaluation may be for a specific chemical, the airborne sampling and measurement of a workplace (see *Airborne Hazards from Materials Use: Monitoring and Measurement*), an industry-wide assessment of beryllium exposures (see *Beryllium Hazards*), a continental study of acid rain effects, a global study of the biological hazards arising from the persistence of DDT and polychlorinated biphenyls (PCBs) in the environment (see *Polychlorinated Biphenyls: Hazards*), or the global effects arising from halocarbon-influenced depletion of ozone in the earth's upper atmosphere. A study may be brief and of immediate value to those concerned or, at the other extreme, it may be an extended study with results of value only to future generations.

Risk assessment requires a full review of what is known about a substance (its potential hazards) plus an indication of what may require further study to fill gaps in knowledge. The literature, which provides a heritage of past experience on the interaction of substances and biology, may provide information such as epidemiology and clinical studies, animal tests, dose–response data on various types of exposure (e.g., inhalation, ingestion, skin contact, eye contact) and microorganism effects. Information is also needed on conditions of production and use, environmental conditions (waste disposal procedures, decomposition rate, sinks and fate in the environment, food chain involvement) and population exposures.

It has been estimated that only ~2% of 67 000 chemicals in use have been adequately tested for toxic effects; for 70%, there are no human effects data available. Animals are tested as surrogates and models for man in such cases to indicate possible toxicity hazard. This is done even though it is recog-

Table 2
Some hazard sources to be considered in production

Hazard sources	Examples—properties—controls
Substances which become part of goods and structures	Rubber, wood, metal, ceramics, plastics, composites, fibers, fillers, adhesives, coatings, plating[a], lubricants. Identify hazards of each raw material; handling and processing characteristics. Material safety data sheets (MSDS) are needed
Substances needed for processing	Acids, bases, cyanides, solvents, cleaning agents, sterilants[b], machining coolants and lubricants. MSDS are needed
Processing conditions	Heat and cold, vibration, noise[c], silica dust[d], molten metal, welding, eye injury situations[e], airborne hazards (dust, fume vapors, CO)[f], laser light, skin contact. Provide protective clothing/equipment; train workers
Peripheral substances	Fuels and lubricants for processing equipment, janitorial supplies, water treatment chemicals, marking inks, packaging. MSDS are needed
Spills and upsets	Flammability, reactivity, corrosiveness, volatility, radioactivity[g], containment and cleanup procedures. Regulatory reporting
Waste, waste reduction, waste treatment, waste disposal	"Solid" wastes, emissions, effluents[h], heavy metal precipitation, stack gas scrubbing, flammability, reactivity, corrosiveness, volatility, radioactivity[h], landfills, incineration
Handling, storage and transport of raw materials, in-process and finished goods and wastes	Flammables and combustibles, incompatibles, reactive materials, corrosives, pressurized gases, cryogenic liquids, radioactive materials[i]. MSDS are needed
Products	Design failures, material failures, critical part failures, combustibility and toxic smoke emissions[j], product liability
Eventual scrapping, junking or demolition of goods or structures	What hazards have been built in? Radioactivity[k]? Deteriorated asbestos insulation and gasketing[l]?

Relevant encyclopedia articles are: a *Cadmium Hazards*; b *Ethylene Oxide as a Product Sterilant: Hazards*; c *Noise: Hazards and Control in Materials Processing*; d *Silicon Dioxide Hazards*; e *Eye Injury Hazards in Materials Processing and Use*; f *Airborne Hazards from Materials Use: Monitoring and Measurement*; g *Radioactive Spillage, Responses to*; h *Electroplating Plants: Prevention of Pollution in Effluents*; i *Radioactive Waste Disposal*; j *Flammability Hazards of Structural Materials and Furnishings*; k *Flammability of Polymers: Test Methods*; *Wood and Fire*; *Ionizing Radiation: Acute Effects*; *Ionizing Radiation: Late Somatic Effects in Humans*; *Ionizing Radiation: Genetic Effects in Humans*; l *Asbestos Hazards*

nized that wide differences can exist between species and between animals and man in response to a particular toxic insult. Broad assumptions are often made in choosing statistical models to relate very high-exposure results in animals to very low exposures in man for risk assessment purposes.

Hazards arising from clear cause and effect are the easiest to identify and act upon. Risk assessment and management is straightforward for these. It is the more subtle, borderline hazards, the multisubstance exposures, the low-level threshold and latent effects and the variations in response of individuals to the same exposures that make risk assessment and management difficult problems. It is usually the inadequacy of information and the sparseness and uncertainty of the current scientific information that form the basic problem.

In most cases there is little scientific evidence available on the biological interaction of substances at the part per billion (ppb) level. Determination of ppm levels is routine now (and ppb is becoming common), but the toxic effects of ppb exposures are not known, even though the substance may produce known toxic effects at higher levels. When it is not firmly tied to scientific evidence, fear can often be extrapolated to excessively low "toxic levels" by rhetoric. Scientific concern should certainly be directed at even ppb levels of a toxic substance, for example, in groundwater, but this should be done while avoiding the instigation of unjustified public fear that can generate unnecessary, wasteful, regulatory action when there are so many more immediate hazards to be addressed. In most cases there is no need to regulate as low as the limit of detection.

It is not possible to evaluate all hazards at once. Priorities are needed to establish where limited resources can best be applied to achieve the greatest hazard reduction potential. In many cases in the past, the news media, special interest groups and politicians (rightly or wrongly) have set risk assessment priorities by excessively mobilizing public anxiety. Chemical hypochondria is a problem we do not need.

An example of rational and deliberate priority-setting for materials testing is that of the Chemical Industry Institute of Toxicology (CIIT), which prioritizes by the following criteria the chemicals on which it carries out in-depth studies. Candidate chemicals for in-depth animal study must be commodity chemicals, used in substantial quantities with substantial human exposure associated with production and/or use. Further, the chemical structure of a candidate for study must be similar to that of known toxic materials. Each priority candidate thus selected undergoes an extensive, thorough literature search and the writing of a current status report. A testing advisory council advises on the test methods and protocols that need to be done with selected candidates from the list, as follows:

(a) a dose ranging study (acute toxicity properties with test animals),

(b) 90 day chronic testing study (sublethal statistical testing of toxicity with animals),

(c) a full 2 year bioassay (lifetime statistical study involving 700–900 rats), and

(d) a public report and data compilation (report made public *before* reporting to sponsors).

The overall process can take 5 years and costs millions of dollars, but the results of such a definitive study are useful in more clearly pointing the way for risk management effort.

One of CIIT's recent studies established that formaldehyde can cause increased development of nasal cancer in rats from prolonged exposures (2 years) as low as 6 ppm atmospheric concentration. Present epidemiological study has identified no significant cancer hazard for man from inhalation of formaldehyde in industrial and professional (pathologists, undertakers) activity, but the fact that nasal cancer has developed in the rat labels formaldehyde as a "suspected carcinogen" for man. Even though CIIT has shown that the rat is a poor model for man in this case, increased regulation of formaldehyde may result from this study.

Risk management takes into consideration the pertinent risk assessment(s) and then must perform a balancing act to provide the right amount of control to protect safety, health and the environment (see Table 3). The control options selected must provide a margin of safety and caution for the unknown; benefits and costs must be considered. Social/political/economic factors also strongly influence risk management and sometimes dominate the process, especially when scientific evidence is inadequate.

4. *Waste Disposal*

Most waste materials (including many hazardous ones) can be eliminated from the biosphere by natural processes of decay and chemical and/or biological transformation. Sewage systems and rubbish landfills function in this manner when not overloaded. When the return to nature occurs in a controlled manner,

Table 3
Some chemical hazard management considerations

Risk management determinations		Some risk management control options
A. Using risk assessment data,		Ban
determine "acceptable" versus		Control production
"unacceptable" risk, including		Control use
social/political/economic		Regulate severely
considerations		Guidelines
		Regulate selectively
B. If a risk is "unacceptable,"		Warnings
what are the areas of risk?		Intelligence bulletins
Health effects:	Chemical and physical	Labeling, placarding
Route of entry	effects:	Reports on production, use and exposure
Acute or chronic	Explosibility	Worker protection:
Accumulative	Flammability	Measuring, monitoring
Mutagenic	Reactivity	Medical surveillance
Carcinogenic	Corrosivity	Protective clothing, equipment
Teratogenic	Environmental effects:	Airborne exposure limits
Reproductive	Distribution	Prohibit contact
hazards	Persistence	Respirator use
Multiple substances	Fate and sinks	Engineering controls
exposure effects	Population exposed	Ventilation
Heavy metals		Hoods and closed systems
Radioactivity		Training and education
(MSDS available)		Eyewash fountain
What is the severity and frequency		Safety shower
of risk?		Handling, shipping and storage
Who is at risk?		requirements based on
		properties
C. What control measures need to be		Waste disposal requirements,
applied? When?		based on properties

at a reasonable rate without producing hazardous side effects, it becomes part of the life cycle and the "harmony of nature." However, some of the wastes from the use of materials in the preparation of goods and structures require special treatment and control for disposal to avoid creating health, safety and environmental hazards.

Materials containing only carbon, oxygen and hydrogen can usually be incinerated, either neat or in a mixture. Proper combustion can return such waste to nature as water and carbon dioxide. If the waste contains chlorine, sulfur or phosphorus, for example, burning liberates acidic substances which must be removed from emissions by scrubbers to protect the environment. Heavily halogenated materials do not burn readily and could spread toxic materials into the environment if the combustion is incomplete. (High-temperature incineration equipment, suitable dwell time at high temperature and an excess of oxygen are needed to satisfactorily complete the incineration of these materials.) A scrubber must be used to remove the acidic gases from the emissions (except that special incinerator ships have successfully burned this type of waste far out at sea without scrubbing, using the sea itself as the acid absorber). Cement kiln incineration of PCB's has been a satisfactory disposal method, the chloride content of the PCBs being transferred to the cement as part of its formulation.

Heavy metal pollutants, including elements such as mercury, lead and cadmium (see *Mercury Hazards*; *Lead Hazards*; *Cadmium Hazards*), fluorides and radioactive species cannot be oxidized to harmless products. They can be concentrated, precipitated into water-insoluble form and immobilized. Encapsulation techniques are available to confine toxic elements and prevent any hazardous leaching to cause water or other pollution. Fixation or solidification processes must be used on such hazardous wastes when they are deposited in landfills which do not conform to the US Environmental Protection Agency (EPA) secure landfill standards. Radioactive species require special isolation procedures (see *Radioactive Waste Disposal*; *Radioactive Spillage, Responses to*).

Failure to isolate or destroy properly these and other pollutants such as PCBs allows them to change location freely in the environment—in the soil, water and air and into and out of plant and animal life—in accordance with their chemical and physical properties, producing health, safety and environmental hazards. Environmental concerns are not only with present disposal practices but also with the shortcomings of past waste disposal. Even when conscientiously done, using the "good practice" of their day, the inadequate disposal of hazardous wastes (and/or unsatisfactory control of sites) has made inactive waste disposal sites pollution sources that must now be cleaned up. Society sensitive to hazardous waste has set about recovering from past environmental mistakes in waste disposal and is trying to avoid new ones.

EPA assessment and management of hazardous "solid" waste are of two basic types in the USA. The first, under RCRA regulations, involves studies of waste and waste-generating situations to determine those which could pose a hazard to the environment if improperly handled. Such potential hazardous wastes are managed by licensing those who generate waste, those who operate disposal facilities and those who transport waste (using a manifest system at each step in the process to document a legal waste disposal). RCRA's before-the-fact stance also emphasizes the reduction of wastes by improved processing and by recovery and recycle of values from the wastes before disposal. The second type, under CERCLA regulations, deals with hazardous wastes which have already been released into the environment from accidental spills, from irresponsible dumping or from prior placement in inadequate landfills (or other land disposals) that have failed to isolate the waste from the environment. CERCLA hazard assessment researchers seek to define any waste problems associated with a specific disposal site (and also those comparable with other sites) to work out possible solutions for cleanup. The CERCLA waste management function chooses from the proposals a solution for engineering design and implementation. (When possible, EPA seeks out, and bills the site cleanup costs to, those responsible for the wastes.) Both the RCRA and CERCLA approaches emphasize the protection of groundwater from contamination by hazardous wastes.

Many small emissions can collectively be hazardous. NO_x, CO and Pb emissions from automobile engines can produce urban air pollution. Halocarbon releases from aerosol spray cans, from leaking or scrapped refrigeration systems or from urethane foam production can be damaging to the ozone layer of the upper atmosphere. Volatile organic carbon (VOC) releases from cleaning operations with solvents, painting, adhesives production and use, for example, can result in air pollution or smog. The burning of fossil fuels for power production emits SO_2 and other contaminants to the atmosphere and can contribute to acid rain problems. Emissions to the air and liquid effluents from industrial plants are forms of waste disposal which are EPA-regulated. Hazardous components removed from emissions and from effluents still require recycle or proper disposal somewhere.

When a product fails to give reliable, safe, economical usefulness, all the risks in its production, its resources usage and its waste disposal problems may have to be repeated (hopefully for an improved design). Doing a better job the first time has significance for safety, health and the environment as well as for economics.

5. Substitution?

Traditionally the main driving force for substitution of processes or materials in industry has been technology or economics (see *Substitution: Technology*; *Substitution: Economics*). Now the requirements of safety, health and the environment are becoming of increasing importance in forcing substitution, as hazards become evident and/or regulation occurs. The carcinogenic or cocarcinogenic properties of asbestos fibers (see *Asbestos Hazards*) have led to their replacement in many applications, such as plastics reinforcement and thermal insulation, by glass, ceramic or other fibers. The environmental persistence of, and waste disposal regulations for, PCBs have led to their replacement as dielectric fluids by silicones or special hydrocarbons, for example, in new transformer or capacitor applications in the USA, even though the replacement fluids may be technologically inferior or have higher purchase cost. PCBs have been legally banned in most other applications in the USA. The replacement of solvent-based coatings and paints with water-based, or with high-solids or powder-type coatings may allow better compliance with the VOC emissions requirement of the Clean Air regulations for industrial finishing operations.

Such substitution requires development of a whole new materials processing technology, which might be regarded as a regulation-pushed engineering opportunity to provide improved products, rather than devoting the available capital to emission combustion or control. Substitution of a robot for a worker at a hazardous or undesirable work station is one method of meeting otherwise difficult protection requirements. Robotics is increasingly available and this type of substitution usually receives the blessing of labor unions. The use of inorganic zinc-based primer systems on cleaned steel can eliminate the need to sandblast to bare metal surfaces for later maintenance.

In making such substitutions it is necessary to examine thoroughly the new process or the property profile of the substitute material to avoid "jumping from the frying pan into the fire." Substitutions could initiate unexpected problems even greater than those that an attempt is being made to solve. It may sometimes be better to stick with a well known hazardous material or process and strive to correct its shortcomings and improve controls to meet regulations than to replace with a promising but somewhat unknown situation.

A well documented hazardous material whose problems are known and controllable may provide a safer situation than a new, seemingly safer material without comparable use or hazard experience. Engineering materials with years or decades of experience are abandoned only with reluctance, since their functional properties are so well known and can be relied upon. But if such a forced change is looked upon as an opportunity, good engineering may be able to produce improved economics and improved technology, as well as meeting safety, health and environmental requirements. In any case, these latter requirements will be increasingly important factors in the engineering equation in future.

See also: Aluminum Refining Hazards; Copper Smelting and Refining Hazards; Silicon Dioxide Hazards; Noise: Hazards and Control in Materials Processing

Bibliography

ASTM Standardization News 12(3), 1984. Issue devoted to Hazardous Waste Management

ASTM Standardization News 12(6), 1984. Issue devoted to Environmental Standards

Clayton G D, Clayton F E (eds.) 1978–82 *Patty's Industrial Hygiene and Toxicology,* 3rd edn., Vol. 1, *General Principles*; Vol. 2A, B, C, *Toxicology.* Wiley, New York

Conservation Foundation 1984 *Report on the State of the Environment: as Assessment at Mid-Decade.* The Conservation Foundation, Washington, DC

Department of Transportation 1975–84 *Regulations for Transportation of Natural and Other Gas by Pipeline: Minimum Federal Safety Standards,* 49 CFR Pt 192. US Department of Transportation, Washington, DC

Fawcett H H, Wood W S (eds.) 1982 *Safety and Accident Prevention in Chemical Operations,* 2nd edn. Wiley, New York

Hearfield F 1980 Hazards of Pressure Testing. *European Federation of Chemical Engineering 3rd Int. Symp. on Loss Prevention and Safety Promotion in the Process Industries.* Swiss Society of Chemical Industries, Basel

Henschel A F, Butler J, Logo L N, Tabershaw I R, Ede L (eds.) 1977 *Occupational Diseases, a Guide to their Recognition,* Rev. edn. National Institute of Occupational Safety and Health, Washington, DC

Kurajian G M (ed.) 1983 *Failure Prevention and Reliability—1983, Design and Production Engineering Tech. Conf.* American Society of Mechanical Engineers, New York

Lawrence W W 1976 *Of Acceptable Risk.* Haufmann, Los Altos, California

McElroy F E (ed.) 1981 *Accident Prevention Manual for Industrial Operations,* 8th edn. National Safety Council, Chicago, Illinois

Minter S G 1984 Are fires becoming more deadly? *Occup. Hazards* 46(6): 47–50

National Fire Protection Association 1980 *A Compilation of NFPA Codes, Standards, Recommended Practices and Manuals* (16 vols). National Fire Protection Association, Boston, Massachusetts

National Research Council 1983 *Report of Committee on the Institutional Means for Assessment of Risks to Public Health: Risk Assessment in the Federal Government—Managing the Process.* National Academy of Science, Washington, DC

National Safety Council *Industrial Safety Data Sheets.* National Safety Council, Chicago, Illinois

Powers P W 1976 *How to Dispose of Toxic Substances and Industrial Wastes.* Noyes, Park Ridge, New Jersey

Proctor N H, Hughes J P 1978 *Chemical Hazards of the Workplace.* Lippincott, Philadelphia, Pennsylvania

J. M. Nielsen

Health and Safety in the Workplace: An International Perspective

Technological advances in industrial hygiene and improved safety practices have to a large extent brought under control many physical and chemical hazards in the working environment that were highly prevalent earlier during the industrial revolution. Several occupational hazards and diseases are now considered past history in many industrialized countries, although there are still instances in modern industries of serious episodes of occupational diseases, some of which could have been prevented. In the developing countries, modern measures to protect the workers' health have been introduced in some new industries, but in most cases risks continue to exist; in the hurry to increase industrial production, many plants are built without due concern for occupational health and safety, mainly because of lack of experience in this field.

1. Occurrence of Occupational Diseases and Injuries

Occupational disease reporting is generally inadequate. However, investigations carried out in countries in the process of industrialization often show a high incidence of occupational diseases. Fibrotic pneumoconioses resulting from inhalation of dusts in mining, quarrying, sandblasting, foundries and the pottery industry, for instance, cause considerable disability and mortality in many countries, particularly among workers (Cho and Lee 1978).

Poisoning by toxic gases is frequent. Carbon monoxide poisoning occurs frequently in coal burning, in the iron and steel industry and in gas plants. Respiratory irritants such as sulfur dioxide, nitrogen dioxide, chlorine, ammonia, acroleine and alkaline mists result in a high incidence of acute respiratory disease and in chronic obstructive pulmonary disease from prolonged exposure.

Industrial solvents are encountered in many working processes. Toxic exposure to solvents may cause dermatitis, anemia (e.g., benzene), liver and kidney damage (e.g., carbon tetrachloride), neurological disorders, atherosclerosis and heart disease (e.g., carbon disulfide).

Intoxication by heavy metals also occurs. In Latin America, reports often show a high prevalence of lead absorption among exposed workers and advanced poisoning in a considerable proportion, particularly in workers in the smelting and battery industries and in foundries (El Batawi 1974). Exposure to mercury, arsenic, beryllium, vanadium, chromium, nickel, uranium and zinc is encountered in many industrial operations. Reports from some North African countries show the occurrence of severe damage to the central nervous system by manganese poisoning.

Inhalation of organic and other vegetable dusts throughout the developing world causes occupational asthma, bronchitis and infections, not only in the workers but sometimes also among the members of their families, including children. Large numbers of people are exposed to dusts from flax, hemp, jute, cotton, coconut fibers, rice germs, bagasse, tobacco, tea, cocoa, paprika and wood in industrial and agricultural processes. Although information is lacking on the pathological effects of many of these dusts, investigations carried out in some producing countries show a high prevalence of respiratory diseases and byssinosis in the cotton and flax industries and pulmonary irritation and allergy in certain other agricultural processes.

Occupational dermatoses remain among the most common occupational diseases, affecting workers exposed to mineral oils, solvents, cement and other substances.

Cancer-producing chemical and physical agents remain uncontrolled in many workplaces in developing and industrial countries. Despite the possibilities of prevention, the reporting and control of occupational cancer do not receive sufficient attention.

Statistical information from different countries (International Labour Office 1983) shows clearly that the rates of occupational accidents are high despite efforts to reduce them. For example, in the Philippines in 1982, there were 5200 work injuries with 30 deaths; of these injuries 810 occurred in agriculture, 476 in mining (14 deaths) and 289 in building (1 death). In Chile, 101 872 occupational accidents were reported in the year 1981.

Some other examples of the toll taken by occupational accidents may be mentioned. France in 1980 recorded a total of 2 008 410 accidents (1423 deaths); the Federal Republic of Germany in 1981 recorded a total of 1 960 780 accidents (3638 deaths); and Malaysia in 1981 reported 64 537 accidents (546 deaths).

Examples of the occurrence of occupational diseases follow: in the UK in 1979, there were 14 499 cases of compensable occupational diseases. There were, in the same year, 16 000 cases in Sweden, 5032 in Switzerland, 8080 in France and 30 000 in the USA (US Department of Labour 1980).

Occupational mortality studies are another source of information. In an increasing number of developed countries, population censuses are used as points of departure for detailed investigations into occupational differentials in mortality; relevant information for less developed countries is still scarce. In developed countries the risk of premature death is consistently high among miners, transport workers, laborers and unskilled workers (World Heath Organization 1980b). It must be emphasized, however, that hazards in the working environment are not the only factors involved.

2. Health and Safety Problems in Special Occupational Groups

Certain occupational sectors in the developing countries are much less privileged than their counterparts in industrial nations. Workers in agriculture, small industries and mines are examples. Although exposed to a wide variety of health and safety risks, they usually have no regular occupational health services and they may not be covered by the administration of occupational health and safety acts.

2.1 Agriculture

In Sri Lanka in 1977 alone, the number of cases of intoxication by pesticides used in agricultural work, reported from hospital admissions, exceeded 113 000 cases. It would appear that 511 000 agricultural workers are hospitalized annually for pesticide poisoning with an average case fatality rate of 9.2% (Jeyaratnam et al. 1982).

2.2 Small Industries

Small industries and mines constitute the second most important sector after agriculture in developing countries in terms of their number and significance to the national economy. However, the general working conditions involved have tended to be worse than those in larger concerns. This is attributed to several factors, including a lower level of technology (often with primitive and overcrowded workplaces), limited financial resources, a lower level of education among the workers and owners and the absence of regular health and safety care (El Batawi 1981).

2.3 Miners

The most recent miners' health problem with significant magnitude, in addition to fibrotic pneumoconioses from silica and coal dusts, relates to exposure to radon and radon daughters in uranium, iron ore and other mines. This exposure causes lung cancer among the miners. Smoking of tobacco may shorten the induction-latency time for lung cancer (Edling and Axelson 1983).

2.4 Migrant Workers

Jobs taken by migrant workers are generally the least attractive. Many work in mines, quarries, agriculture, forestry, building and various industries, most of which carry specific disease and accident risks against which no particular preventive measures may have been taken. The problem exists in Western Europe, which has more than 12 million migrant workers, in the Middle Eastern oil-producing countries and in many other parts of the world, including Singapore, Australia and North America.

2.5 Other Vulnerable Groups

The extent of the occupational health problems of other vulnerable groups of workers such as the young, the old and the handicapped has yet to be fully investigated. Statistics on them are on the whole lacking, particularly in developing countries. These groups are mostly employed in agriculture and small-scale industries, or are self-employed. Indiscriminate employment without a preliminary medical examination often makes them particularly vulnerable to toxic hazards, industrial accidents and other occupational risks. The health of working women is also an area of special concern.

3. Main International Concerns at Present

With respect to materials usage and technology, the main international concerns in the field of occupational health and safety that require attention by international and national authorities are the following:

(a) The transfer of technology, industry and industrial chemicals to developing countries without adequate control over the health risks involved requires an international approach among the United Nations agencies concerned with health, industry, agriculture, labor and finance, aiming at a healthy development of industry.

(b) One of the elements in occupational hygiene protection is to substitute a severely toxic substance with a less toxic material; for example, parathion (an organophosphorous insecticide) has been replaced by malathion which is much less toxic. Soluble lead oxides have been replaced in glazes of ceramics with lead silicates, and asbestos in many instances with man-made vitreous fibers.

(c) The field of occupational safety has, to a substantial extent, been given limited attention; safety guides are needed and adaptation of protective devices produced in industrial countries is necessary to allow their use in developing countries where the anthropological characterists may differ.

(d) While a good deal of knowledge is available on prevention and control of occupational diseases and injuries, there is a limited effort in intercountry coordination and in the adaptation, transfer and application of this knowledge in developing and industrialized countries.

(e) Synergism and potentiation of harmful effects in combined exposure in the work environment have hardly been investigated in real life despite the frequent occurrence of such exposure (World Health Organization 1981).

(f) Occupational health and hygiene standards, commonly known as occupational exposure limits, vary widely from one part of the world to another. Most of these limits are promulgated in the USA and the USSR. There is a need for harmonization of these standards, based at least on health indicators (World Health Organization 1977).

(g) Many new chemicals are produced annually, adding to the long list of hazardous chemicals now in use by industry and various trades. Limited or no information is available on the toxicity of these chemicals to the workers exposed. These chemicals are produced and exported sometimes with inadequate labelling and precautions for use. Many of the cases of poisoning in industry are caused by the lack of information to the workers exposed on the nature of chemicals, their toxicity and means of protection, first aid and treatment. Labelling has now become a legal requirement of industrial chemicals in many countries, but much still remains to be done.

4. International and National Action

All the foregoing concerns have become the preoccupation of international organizations including the World Health Organization and the International Labour Organization (ILO). The WHO's Programme of Action on Workers' Health (World Health Organization 1980a) initiated in 1979–80 several collaborative efforts aiming at the development of appropriate occupational health technology and occupational health care programmes.

At the national level, there is a good deal of pioneer work by individuals and governments in various parts of the world to improve working conditions, prevent occupational safety and health hazards and protect workers' health. Occupational health centers and units have recently been established in Algeria, Bahrain, Egypt, Indonesia, Singapore, Sudan, Thailand, Malaysia, the Republic of Korea and other countries. University programmes have been started or expanded in the Philippines, Singapore, Sri Lanka, Nigeria and several countries in Latin America. New legislation for labor protection against occupational health risks has been enacted in Burma, Sri Lanka, Malaysia, Indonesia, Sudan and the Republic of Korea. These are merely examples that are by no means comprehensive. They demonstrate that the international situation is promising and that there will be more efforts to protect and promote workers' health and safety.

See also: Hazards and Materials: An Overview

Bibliography

Cho K S, Lee S H 1978 Occupational health hazards of mine workers. *Bull. W.H.O.* 56: 205–18

Edling C, Axelson O 1983 Quantitative aspects on radon daughter exposure and lung cancer in underground miners. *Br. J. Ind. Med.* 40: 182–87

El Batawi M A 1974 *Study of Occupational Health in the Andean Countries*, OCH/74.3. World Health Organization, Geneva

El Batawi M A 1981 Special problems of occupational health in developing countries. In: Schilling R S F (ed.) 1981 *Occupational Health Practice*, 2nd edn. Butterworth, London, pp. 27–46

International Labour Office 1983 *Year Book of Labour Statistics*. International Labour Office, Geneva, pp. 709–52

IRPTC 1983 *International Register of Potentially Toxic Chemicals*, IRPTC Legal File 1983. United Nations Environment Programme, Geneva

Jeyaratnam J, Alwis Seneviratine de R S, Copplestone J F 1982 Survey of pesticide poisoning in Sri Lanka. *Bull W.H.O.* 60(4) 615–19

US Department of Labour 1980 *An Interim Report to Congress on Occupational Diseases*. Office of Secretary of Labor, Washington, DC

World Health Organization 1977 *Methods used in Establishing Permissible Levels in Occupational Exposure to Harmful Agents*, WHO Technical Report Series No. 601. World Health Oganization, Geneva

World Health Organization 1980a *Programme of Action on Workers' Health for 1979–84*, OCH/80.2. World Health Organization, Geneva

World Health Organization 1980b *Sixth Report on the World Health Situation*, Pt. 1, *Global Analysis*. World Health Organization, Geneva, p. 146

World Health Organization 1981 *Health Effects of Combined Exposures in the Work Environment*, WHO Technical Report Series No. 662. World Health Organization, Geneva

World Health Organization 1984 *Legislative and Administrative Procedures for the Control of Chemicals*, Rev. edn. 5. World Health Organization, Geneva

M. A. El Batawi

Heat Capacity and Thermal Conductivity of Glass

The heat capacity and thermal conductivity of glass increase as temperature increases and depend on the composition of the glass. These properties exert controlling influences on heat-transfer processes throughout the range of interest from 0 °C up to melting temperatures. Two arbitrary temperature zones of applications are distinguished: one from 0 °C to about 500 °C; the other from about 500 °C up to melting temperatures. Particular attention is drawn to the use of thermal properties in heat-transfer processes in glass manufacturing.

1. Specific Heat Capacity

The specific heat capacity of a substance is its thermal capacity per unit mass and is usually expressed in $\mathrm{W\,kg^{-1}\,K^{-1}}$ or $\mathrm{cal\,g^{-1}\,{}^{\circ}C^{-1}}$. Mean specific heat capacity c_m is generally determined using the method of mixtures, which involves the measurement of the

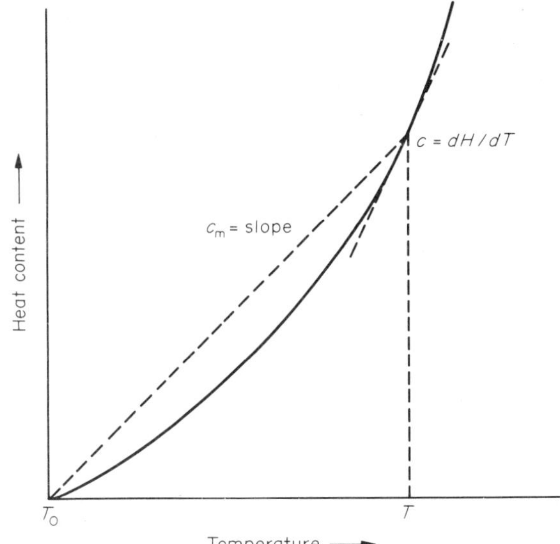

Figure 1
True and mean specific heats (after Babcock 1977. © Wiley, New York. Reproduced with permission)

difference in heat contents, at constant pressure, of the substance at some elevated temperature and of the calorimeter. Figure 1 defines c_m and true specific heat capacity c. Sharp and Ginther (1951) reviewed existing specific heat data and developed equations for their representation in terms of composition and temperature. Table 1 shows their oxide factors used in calculating specific heat capacities. They state that accuracy is generally better than 1% in the range from 0 to 1300 °C. Glass compositions are expressed in weight fractions. The following equations are used in the calculations:

$$c_m = (aT + c_0)/(0.00146T + 1) \quad (1)$$

$$c = (aT + c_m)/(0.00146T + 1) \quad (2)$$

where T is the temperature (°C) and a and c_0 are constants. Following this method the mean specific heat c_m of a glass with weight-fraction composition 0.75 SiO_2, 0.10 CaO and 0.15 Na_2O is represented by

$$c_m = (0.0005164T + 0.1748)/(0.00146T + 1) \quad (3)$$

Similarly, c_m for vitreous silica is given by

$$c_m = (0.000468T + 0.1657)/(0.00146T + 1) \quad (4)$$

2. Thermal Conductivity

Heat conduction in glasses is effected by two main processes. From low temperatures up to about 400 °C heat is transferred principally by so-called regular conduction. Heat is transferred through a continuous material by means of thermoelastic waves. Conductivity coefficients of glasses are low as a result of scattering by the random glass structure. At higher temperatures, this conduction process is augmented by so-called radiation conduction. As Planck (1932) noted, "The equalization of temperature in bodies is effected not by heat conduction alone but also by heat radiation, which is a totally different mode of propagation of heat." Generally speaking, at about 800 °C and above, radiation conduction overshadows regular conduction and the total conduction is designated as effective conductivity.

Table 1
Oxide factors for the calculation of specific heat capacity in cal g^{-1} K^{-1} (after Sharp and Ginther 1951. © The American Ceramic Society, Columbus, Ohio. Reproduced with permission)

Oxide	a	c_0
SiO_2	0.000468	0.1657
B_2O_3	0.000598	0.1935
Al_2O_3	0.000453	0.1765
CaO	0.000410	0.1709
MgO	0.000514	0.2142
PbO	0.000113	0.049
BaO	0.0003863	0.0728
ZnO	0.00045	0.1063[a]
Na_2O	0.000829	0.2229
K_2O	0.000445	0.1756

a Valid only up to 3% ZnO

2.1 Regular Conduction

The thermal conductivity of a material is the rate of heat flow through unit cross-sectional area under unit temperature gradient. The conductivity of soda-lime–silica glasses is about 1.0 W m^{-1} K^{-1} (0.0024 cal s^{-1} cm^{-1} $°C^{-1}$) and that of vitreous silica can be taken as near 1.3 W m^{-1} K^{-1} (0.0032 cal s^{-1} cm^{-1} $°C^{-1}$). There is a lack of data on thermal conductivity of glasses and published values are inconsistent. This is a result of difficulties in making accurate measurements. Evaluating thermal boundary conditions in both the measurement and the application of the data is no easy matter. Data on some glasses show that conductivity increases slowly with temperature (Shand 1958). This property is also dependent on composition.

2.2 Radiation Conduction

In accordance with the above statement of Planck, radiant heat transfer occurs between regions in a continuous material that are at different temperatures. The effects of wavelength, temperature and glass composition on absorption, emission, transmission and reflection have been studied by several investigators. McMahon (1951) published research on thermal radiation characteristics of nine commercial silicate glasses in the spectral range 2.2–13 μm and at temperatures from 260 to 816 °C. Compositions covered a range of major oxide contents and, in some, minor amounts of colorants. Figure 2 shows information obtained on a 0.32 mm thick flint (almost colorless) glass at 538 °C. Property curves for all glasses were similar in that transmissivity (dependent on thickness) was reduced to zero in the range 4–5 μm and showed sharp reflection peaks near 9.5 μm. Research on silicate structures associates this peak with primary Si–O linkages. The strong water-vapor absorption near 2.8 μm was present in all glasses. Presence of colorants had minor effects on property curves in the very near infrared spectral region. Temperature dependence of emissivity and transmissivity was small and almost independent of wavelength. These data show that thermal radiation properties of glasses closely approach those of a black body. In addition to heat transfer by vibrational modes, as in regular conduction, heat may be transferred by absorption and reradiation of electromagnetic energy (Gardon 1961).

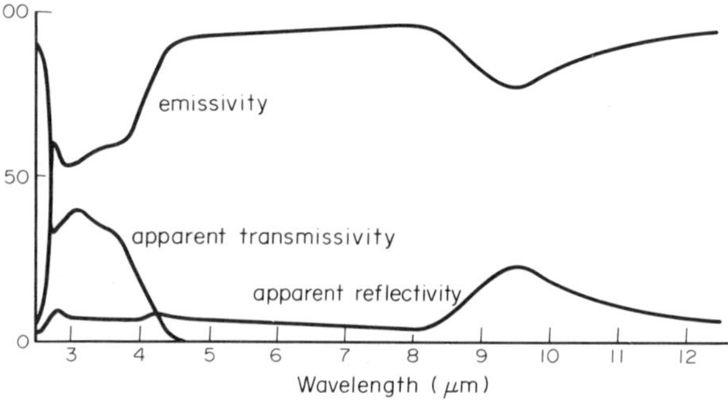

Figure 2
Radiation characteristics of Na_2O–CaO–SiO_2 glass (after Babcock 1977. © Wiley, New York. Reproduced with permission)

2.3 Thermal Diffusivity and Effective Conductivity

The thermal diffusivity coefficient D determines the rate at which a temperature wave is propagated within a conducting substance. It is related to effective thermal conductivity k, specific heat c and density ρ by

$$k = D\rho c \tag{5}$$

Van Zee and Babcock (1951) obtained thermal diffusivity data for two typical commercial Na_2O–CaO–SiO_2 glasses (labelled 2 and 5) in the range 700–1400 °C; chemical analyses are given in Table 2. Their paper gives detailed descriptions of furnace construction, the method of controlling sinusoidal heat input, methods of measurement and related mathematical theory. The glass was contained in a long crucible with cylindrical symmetry. A sinusoidal temperature variation was superimposed on the mean temperature of the sample, and the time for the heat wave to travel from a point on the circumference of the sample to the center was measured. Figure 3 shows that diffusivity values were not affected by the cycle time in the range 1–4 h. Diffusivity values ($cm^2\,s^{-1}$) for glasses 2 and 5 can be obtained from

$$\log_{10}(1000D) = 0.12683 + 0.001586T \quad \text{glass 2} \tag{6}$$

$$\log_{10}(1000D) = 0.04752 + 0.001607T \quad \text{glass 5} \tag{7}$$

Table 2
Chemical analyses of glasses (after Van Zee and Babcock 1951. © The American Ceramic Society, Columbus, Ohio. Reproduced with permission)

Component	Glass 2 (wt%)	Glass 5 (wt%)
SiO_2	67.98	72.30
Al_2O_3 + TiO_2	4.01	0.35
Total Fe as Fe_2O_3	0.038	0.033
CaO	10.82	14.40
CuO	0.0009	0.0005
Na_2O	16.63	12.50
SO_2	0.12	0.14
Total	99.60	99.72
FeO	0.0059	0.0075

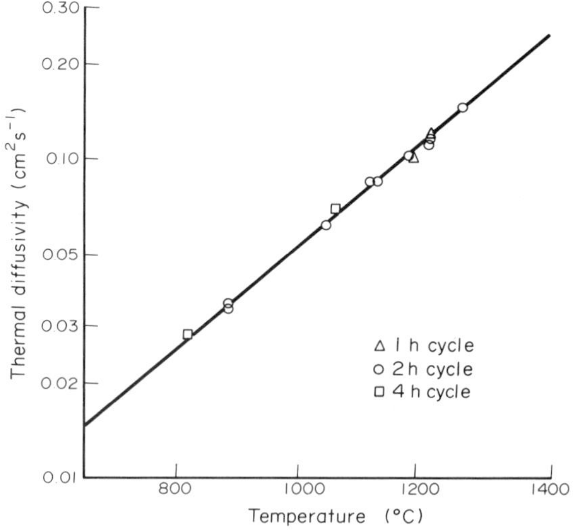

Figure 3
Thermal diffusivity of Na_2O–CaO–SiO_2 glass (after Babcock 1977. © Wiley, New York. Reproduced with permission)

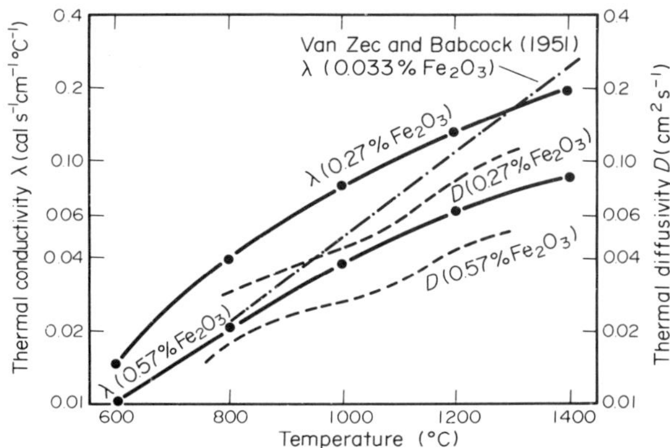

Figure 4
Measured thermal diffusivity and calculated effective thermal conductivity of glasses as functions of temperature (after Babcock 1977. © Wiley, New York. Reproduced with permission)

This research was carried out before any theoretical analysis was available and clearly demonstrated the very rapid increase of thermal diffusivity with temperature. Derived values of effective conductivity using Eqn. (5) increased with temperature in the same manner. Values of both diffusivity and effective conductivity depend on sample thickness.

Charnock (1961) measured thermal diffusivity of plate-glass compositions using a method similar to that outlined above. The data are shown in Fig. 4 and compared with those of glass 5 of Van Zee and Babcock. The data indicate that diffusivity and effective conductivity decrease as the Fe_2O_3 content increases. A review of data by Gardon (1961) clearly shows that effective conductivity increases rapidly with sample thickness.

Table 3 shows heat-transfer data for a typical commercial Na_2O–CaO–SiO_2 glass in the high-temperature range of glass manufacturing. The density data are from Sawai and Inoue (1940), specific heat from Babcock and McGraw (1957) and thermal diffusivity data on their glass 5 from Van Zee and Babcock (1951). Effective conductivity data are calculated from Eqn. (5). These data may be safely extrapolated to 1500 °C. It is noted that thermal diffusivity and effective conductivity data are for a glass sample thickness of 7.62 cm (the thickness used in the Van Zee and Babcock measurements). It is well to remember that values of both properties vary with glass composition, temperature, and sample thickness.

See also: Thermal Expansion of Glass; Glass: An Overview

Bibliography

Babcock C L 1961 Symposium on heat-transfer phenomena in glass. *J. Am. Ceram. Soc.* 44: 301

Babcock C L 1977 *Silicate Glass Technology Methods.* Wiley, New York

Babcock C L, McGraw D A 1957 Application of glass properties data to forming operations. *Glass Ind.* 38: 137–42, 144–46, 148–51, 161

Charnock H 1961 Experimental and theoretical comparison of radiation conductivity predicted by steady-state theory with that effective under periodic temperature conditions. *J. Am. Ceram. Soc.* 44: 313–16

Gardon R 1961 A review of radiant heat transfer in glass. *J. Am. Ceram. Soc.* 44: 305–12

Hirao K, Soga N, Kunugi M 1979 Low-temperature heat capacity and structure of alkali silicate glasses. *J. Am. Ceram. Soc.* 62: 570–73

McMahon H O 1951 Thermal radiation characteristics of some glasses. *J. Am. Ceram. Soc.* 34: 91–96

Muratov A V 1978 Specific heat of silicate glasses at low temperatures. *Fiz. Khim. Stekla.* 4: 741–43

Muratov A V, Chernyshov A V 1979 Thermal conductivity of silicate glass in the low-temperature region. *Fiz. Khim. Stekla.* 5: 119–22

Planck M K E L 1932 *Introduction to Theoretical Physics*, Vol. 5: *Theory of Heat.* Macmillan, New York

Primenko V I, Gudovich D D 1978 Relation between temperature and heat capacity of glasses containing lead oxide. *Fiz. Khim. Stekla.* 5: 84–87

Primenko V I, Gudovich D D 1979 Heat capacity of silicate glasses containing Zn and Ti oxides. *Fiz. Khim. Stekla* 5: 249–51

Table 3
Heat-transfer data for a typical commercial Na_2O–CaO–SiO_2 glass (after Babcock 1977. © Wiley, New York. Reproduced with permission)

Temperature (°C)	Density (g cm^{-3})	Specific heat (cal g^{-1} K^{-1})		Thermal diffusivity (cm^2 s^{-1})	Effective conductivity (cal s^{-1} cm^{-1} K^{-1})
		c_m	c		
500	2.474	0.2445	0.3136	0.007	0.005
700	2.438	0.2695	0.3397	0.015	0.012
900	2.398	0.2882	0.3594	0.030	0.026
1100	2.367	0.3031	0.3751	0.065	0.058
1300	2.344	0.3167	0.3893	0.130	0.119
1400	2.335	0.3264	0.3995	0.200	0.187

Sawai I, Inoue S 1940 Specific gravity of ternary glasses $CaO–Na_2O–SiO_2$ at high temperatures. *J. Soc. Chem. Ind. Jpn.* 43 (Suppl.): 47–49

Shand E B 1958 *Glass Engineering Handbook*, 2nd edn. McGraw-Hill, New York

Sharp D E, Ginther L B 1951 Effect of composition and temperature on the specific heat of glass. *J. Am. Ceram. Soc.* 34: 260–71

Van Zee A F, Babcock C L 1951 A method for the measurement of thermal diffusivity of molten glass. *J. Am. Ceram. Soc.* 34: 244–50

C. L. Babcock

Heat Generation in Metal Cutting and Deformation

In almost all cases, the heat generated during cutting, shaping and forming of metals is an unavoidable consequence of the process rather than something sought after. There are generally two sources of heat during such operations. Heat may arise from purely frictional effects when the workpiece slides relative to the tooling during a metalworking operation. The second and principal source of heat is the internally generated heat of plastic deformation that accompanies the permanent change in shape of metals during deformation processing. In many operations, of which machining and grinding are good examples, it is difficult to separate the frictional heating effects from the effects of high-rate plastic deformation. This depends on a complete understanding of the local metal removal process. Nevertheless, no matter how the heat is produced, it is the rate at which this heat is generated compared with the rate at which it is removed that is the key to whether or not it is a problem.

1. Heating in Machining and Grinding

Metal removal operations such as machining and grinding epitomize high local deformation rate processes in which heating can be a problem. Fortunately, rather than in the part, the major heating is in the discard—the sparks produced by the uncooled grinding of steels or the blue smoking chips from a high-rate turning operation. Nevertheless, there is sufficient heat generated in the system to cause some problems. In machining, lubricants are used to reduce frictional effects but the high rate of plastic deformation during chip formation causes the chip to be heated intensely. The continuous sliding of the hot chip across the tool face under high local pressure produces substantial tool erosion and wear. The effects of the heat generated by machining are generally negligible in the machined part, but the effect on chip formation is very strong. The nature of the chip is controlled by factors such as the material and machining rate. If there is a strong tendency for the deformation and heat to localize, the chips segment; otherwise, a continuous chip forms.

In grinding, the deformation rate, as well as the magnitude of the local surface deformation, is so high that without adequate cooling the surface layers of the part being ground may be severely damaged. Because of their transformation characteristics, steels are particularly susceptible to the formation of hard brittle surface layers that are also frequently cracked. Without adequate cooling, enough surface-generated heat can also be conducted into the bulk of the tool to temper or soften its hardened steel structure.

2. Heating in Deformation Processes

In deformation processes such as forging, rolling and extrusion, material is not removed. Thus, nearly all the heat generated by plastic deformation is retained in the product. However, the tooling may become heated through continuous or repetitive contact with the deforming metal. If this is a problem the tooling may be cooled. In multipass operations, the deformed metal may be cooled between passes to reduce thermal effects. Major problems may arise from deformation-produced heating, despite attempts at cooling, because the heat generation is occurring throughout the metal wherever plastic deformation is taking place, while heat can be removed only from the surface or internally by conduction to cooler regions (i.e., those that are either undeformed or subjected to considerably smaller amounts of deformation). It is clear, therefore, that local temperatures at any point in a metal being deformed depend on the rate at which heat is being generated locally compared with the rate at which it is conducted away. In thermodynamic terminology, if none of the heat generated in a system is allowed to escape through the boundaries of that system, conditions are "adiabatic." This term is also loosely applied to conditions where deformation is so rapid that almost all of the deformation-generated heat is retained locally, sometimes resulting in dramatic increases in temperature (see *Adiabatic Heating During Deformation of Metals*). If the plastic strain throughout the zone of deformation were the same, the general temperature rise would not be large enough to cause significant problems. The characteristics of the mechanical behavior of metals, however, usually lead to strain localization in certain areas, even if the deformation is not initially inhomogeneous as in metal cutting, for example. The strength of a metal when deformed is not a fixed value but is strongly dependent on a number of deformation variables, the most important of which are temperature, rate of deformation and amount of prior deformation. The strength goes up as the latter

Figure 1
Isothermally side-pressed cylinder of titanium alloy having a β microstructure (deformed at 913 °C at a strain rate of $10\,s^{-1}$) (after Semiatin and Lahoti 1981. © The Metallurgical Society (AIME), New York. Reproduced with permission)

two variables increase and goes down with increasing temperature, according to

$$\frac{d\tau}{d\gamma}=\left(\frac{\partial\tau}{\partial\gamma}\right)_{T,\dot{\gamma}}+\left(\frac{\partial\tau}{\partial\dot{\gamma}}\right)_{T,\gamma}\frac{d\dot{\gamma}}{d\gamma}+\left(\frac{\partial\tau}{\partial T}\right)_{\gamma,\dot{\gamma}}\frac{dT}{d\gamma}$$

where τ is the shear stress, γ is the shear strain, $\dot{\gamma}$ is the shear strain rate and T is temperature.

3. Adiabatic Strain Localization

During deformation, whenever local conditions reach the point where the thermal-softening term $(\partial\tau/\partial T)_{\gamma,\dot{\gamma}}dT/d\gamma$ is greater than the hardening terms, the deformation localizes, with strain, strain rate and temperature increasing rapidly. Where this occurs, it becomes impossible to produce general deformation in a metal; moreover, catastrophic local failure may ensue in the high deformation region.

Such adiabatic strain localization may occur not only at normal temperatures but also at elevated temperatures whenever the material and processing conditions permit. This can be a particular problem in hot forging or upsetting. At all temperatures, friction between the tooling and the workpiece effectively prevents lateral metal flow at the contact surface. This produces regions known as "dead metal zones" under the tooling where the metal is essentially prevented from deforming. When a portion of the workpiece is thus prevented from deforming, the strain in the remainder becomes very inhomogeneous with much of the deformation occurring in narrow zones adjacent to the dead metal regions. If the deformation occurs at high rates, the localization is further enhanced by adiabatic heating (Fig. 1). This tendency to localization and inhomogeneous deformation in forged billets is greater in conventional hot forging where cold tools are used. The chill effect creates temperature gradients that are superimposed on the frictional effects to further promote deformation localization (Semiatin et al. 1981). These authors have derived a flow localization parameter α, based on the mechanical and thermal properties of the material to be forged, by which the tendency to flow localization in hot forging can be predicted. In this analysis, strain-rate sensitivity plays an important role, although it is usually assumed to be a minor factor in flow localization at room temperature.

See also: Chip Formation Micromechanics and Metal Cutting; Deformation Processing; Friction and Lubrication in Metalworking; Metals Processing and Fabrication: An Overview

Bibliography

Bedford A J, Wingrove A L, Thompson K R L 1974 The phenomenon of adiabatic shear deformation. *J. Aust. Inst. Met.* 19: 61–73

Bever M B, Marshall E R, Ticknor L B 1953 The energy stored in metal chips during orthogonal cutting. *J. Appl. Phys.* 24: 1176-79

Recht R F 1964 Catastrophic thermoplastic shear. *J. Appl. Mech.* 86: 189–93

Rogers H C 1979 Adiabatic plastic deformation. *Annu. Rev. Mater. Sci.* 9: 283–311

Semiatin S L, Lahoti G D 1981 Deformation and unstable flow in hot forging of Ti–6Al–2Sn–4Zr–2Mo–0.1Si. *Metall. Trans. A* 12: 1705–17

H. C. Rogers

Heat-Treated Alloy Steels

Alloy steels are defined as steels which derive improved properties from the addition of alloying elements or the presence of larger amounts of manganese and silicon than are ordinarily found in carbon steels. Heat-treated alloy steels refer to those alloys with properties further improved as a result of heat treatment and which contain about 1–4 wt% alloying elements. The principal uses for alloying elements in these steels are to develop improved mechanical properties, resistance to tempering, low notch sensitivity, and better machinability in the hardened and tempered condition than is obtainable in a carbon steel of equivalent carbon content similarly heat-treated. Alloy steels are generally not specified for use without an appropriate heat treatment. This article discusses the classification of alloy steels and the effect of alloying elements on both hardenability and tempering.

1. Classification of Alloy Steels

A steel is considered to be an alloy (a) when the maximum of the range given for the content of alloying elements exceeds 1.65% manganese, 0.60% silicon or 0.60% copper; or (b) when there is a definite range or minimum quantity of chromium (up to 3.99%), nickel, molybdenum, aluminum, cobalt, niobium, titanium, tungsten, vanadium or zirconium. The alloy steels most commonly used for heat-treated parts have been classified by several standards organizations. For the purpose of this discussion, the US Society of Automotive Engineers (SAE) and the American Iron and Steel Institute (AISI) classifications are used. In the SAE–AISI four-digit designation system, the first two digits indicate the alloying-element family and the last two digits indicate the nominal carbon content in hundredths of a percent; also, B denotes boron steel, as in 51B60, and BV denotes boron–vanadium steel, as in 43BV12. In the Unified Numbering System (UNS), a five-digit system is used preceded by the letter G standing for steel. The basic numbers for the various types of SAE–AISI steels are listed in Table 1.

2. Alloy Effects on Hardenability

Hardenability is usually the single most important criterion for the selection of an alloy steel, and the response to heat treatment is the most important function of the alloying elements in these steels. Hardenability is that property of steel which determines the depth and distribution of hardness induced by quenching from temperatures above the transformation range to form martensite or lower bainite (see *Steels: Physical Metallurgy Principles*). The general effect of alloying elements that are soluble in austenite is to decrease the rates of transformation of the austenite at subcritical temperatures. Given this effect, pieces can be cooled more slowly or large

Table 1
SAE–AISI four-digit designation system for alloy steels

Series designation	Alloy composition
10XX	nonresulfurized carbon steel
11XX	resulfurized carbon steel
12XX	rephosphorized and resulfurized carbon steel
13XX	1.75%Mn
15XX	1.00–1.65%Mn
23XX	3.50%Ni
25XX	5.00%Ni
31XX	1.25%Ni–0.65%Cr
33XX	3.50%Ni–1.55%Cr
40XX	0.25%Mo
41XX	0.50%Cr or 0.95%Cr–0.25%Mo
43XX	1.55%Ni or 1.80%Ni–0.25%Mo–0.5%Cr or 0.8%Cr
44XX	0.40%Mo or 0.53%Mo
46XX	1.55%Ni or 1.80%Ni–0.25%Mo
47XX	1.05%Ni–0.45%Cr–0.20%Mo
48XX	3.50%Ni–0.25%Mo
50XX	0.28%Cr or 0.48%Cr
51XX	0.80%Cr, 0.90%Cr, 0.95%Cr, 1.00%Cr, 1.05%Cr
5XXXX	1.00%C–(0.50%Cr, 1.00%Cr, 1.45%Cr)
61XX	0.80%Cr or 0.95%Cr–0.10%V or 0.15%V
81XX	0.30%Ni–0.40%Cr–0.12%Cr
86XX	0.55%Ni–0.50%Cr or 0.65%Cr–0.20%Mo
87XX	0.55%Ni–0.50%Cr–0.25%Mo
88XX	0.55%Ni–0.50%Cr–0.35%Mo
92XX	0.85%Mn–2.00%Si
93XX	3.25%Ni–1.20%Cr–0.12%Mo

Source: US Steel, 1971, *The Making, Shaping and Testing of Steel*

pieces can be quenched in a given medium without subsequent transformation of austenite to the undesirable high-temperature transformation products, ferrite and pearlite. To ensure adequate hardenability, the alloying elements must be in solution in the austenite, especially the carbide-forming elements, chromium, molybdenum and vanadium. In annealed steels, these elements occur predominantly as carbides which usually dissolve only at higher temperatures and more slowly than iron carbide. Heating schedules, therefore, should take this factor into account.

The ideal diameter is a true measure of the hardenability of a steel and serves to compare the hardening response of different steels to the same quenching medium. The ideal diameter D_I is affected by austenite grain size, carbon content and alloy content, and an increase in any of these factors reduces the rate at which the diffusion-controlled transformations occur, thereby encouraging the formation of martensite. Calculation of D_I for a given steel requires two types of factors: (a) a base diameter D_C that depends on carbon content and austenite grain size, and (b) multiplying factors for each alloy addition. The ideal critical diameter D_I is the product of D_C and all the multiplying factors for the alloying elements in the given steel (Krauss 1980). The multiplying factors are a good indication of the relative contribution of each alloying element to hardenability, and the reader is referred to Siebert et al. (1977) for a compilation of multiplying factors for the common alloying elements.

In presenting an overview of the effect of alloying elements on hardenability, it is important to show systematic effects. Alloying elements can be separated according to whether they are austenite stabilizers, such as manganese, nickel or copper, or ferrite stabilizers (i.e., γ-loop formers), such as molybdenum, silicon, titanium, vanadium, zirconium, tungsten and niobium. Because of its overriding effect on hardenability, carbon is accepted as being the most potent alloying element. In general, ferrite stabilizers require much less alloying addition than the austenite stabilizers for an equivalent increase in hardenability. However, in many of these ferrite stabilizers the competing process of carbide precipitation in the austenite depletes the austenite of both carbon and alloy addition, thus lowering hardenability. Figure 1 is a representation of the average alloying effects on the multiplying factors. The general effects of alloying elements on hardenability are as follows:

(a) Manganese contributes markedly to hardenability, especially in amounts greater than 0.8%. The effect of manganese up to 1.0% is stronger in low- and high-carbon steels than in medium-carbon steels. The use of manganese is an inexpensive way to increase hardenability.

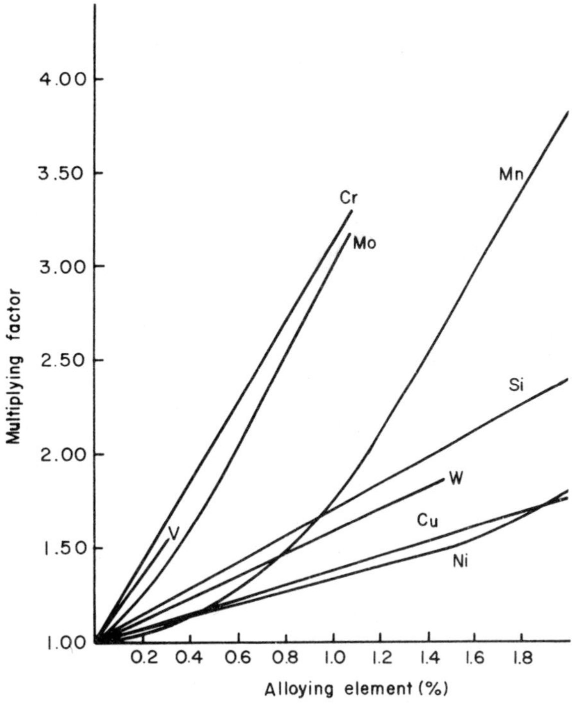

Figure 1
Effect of alloy concentration on the multiplying factors of several common alloying elements (after Siebert et al. 1977)

(b) Nickel is similar to manganese at low alloy additions, but is less potent at the high alloy levels. Nickel is also affected by carbon content, the medium-carbon steels having the greatest effect. There is an alloy interaction between manganese and nickel that must be taken into account at lower austenitizing temperatures.

(c) Copper is usually added to alloy steels for its contribution to atmospheric-corrosion resistance and at higher levels for precipitation hardening. The effect of copper on hardenability is similar to that of nickel, and in hardenability calculations it has been suggested that the sum of copper plus nickel be used with the appropriate multiplying factor of nickel.

(d) Silicon is more effective than manganese at low alloy levels; however, at levels greater than 1% this element is much less effective than manganese. The effect of silicon also varies considerably with carbon content and other alloys present. Silicon is relatively ineffective in low-carbon steel but is very effective in high-carbon steels.

(e) Molybdenum is most effective in improving hardenability. Molybdenum has a much greater effect in high-carbon steels than in medium-carbon steels. The presence of chromium decreases the multiplying factor, whereas the presence of nickel enhances the hardenability effect of molybdenum.

(f) Chromium behaves similarly to molybdenum and has its greatest effect in medium-carbon steels. In low-carbon steel and carburized steel the effect is less than in medium-carbon steels but is still significant. As a result of the stability of chromium carbide at lower austenitizing temperatures, chromium becomes less effective.

(g) Vanadium can increase hardenability if the steel is austenitized at a time and temperature that ensures that the vanadium is in solution. However, solution is possible only if small amounts of vanadium are added. In constructional alloy steels, vanadium is not added for hardenability but to provide secondary hardening during tempering (see Sect. 3).

(h) Tungsten has been found to be more effective in high-carbon steels than in steels of low carbon content (less than 0.5%). Alloy interaction is important in tungsten-containing steels, with Mn–Mo–Cr having a greater effect on the multiplying factors than silicon or nickel additions.

(i) Titanium, niobium and zirconium are all strong carbide formers and therefore should behave similarly to vanadium. In addition, titanium and zirconium are strong nitride formers, a characteristic that affects their solubility in austenite and hence their contribution to hardenability.

Although the hardenability of alloy steels has received a great deal of attention, it has become increasingly evident that quantitative predictions of hardenability require a more detailed analysis of alloy interactions. As an initial estimate, the average effect of alloys on hardenability can best be determined using the data outlined by Siebert et al. (1977) or the slide rules developed by a number of steel companies. Recently, several workers have developed regression analyses that take into account the alloy interactions and result in more accurate predictions of hardenability.

3. Alloying Elements in Tempering

Alloy steels are generally given a subcritical heat treatment (i.e., tempering) after quenching from the austenite. The purpose of this subsequent heat treatment is to improve the toughness of the as-quenched martensite microstructure. As a result of the tempering process the steel is softened, the strength lowered and the ductility increased. Alloying elements generally retard the softening rate.

Individual alloying elements show significant differences in the magnitude of their retarding effect on softening. The alloying elements that are not carbide formers (e.g., nickel, silicon, aluminum and manganese) have only a minor effect on the hardness of tempered steel. As solid-solution strengtheners, such elements remain dissolved in the martensite and do not significantly retard the softening effect. The most effective elements in retarding the rate of softening during tempering are the carbide formers such as chromium, molybdenum and vanadium. These elements, particularly at the higher temperatures, migrate to form a carbide phase when diffusion is possible.

The process of softening in plain-carbon steels with increasing temperature is due to the rapid coarsening of cementite by the diffusion of carbon and iron. Increasing the tempering temperature at constant time results in increasing the softening of tempered steel. Steels containing more than 1% carbon show an initial increase in hardness at low tempering temperatures (Bain and Paxton 1962) that finally gives way to softening at higher temperatures. As the tempering temperature is raised to 250 °C, stage I tempering occurs: the as-quenched martensite transforms to a complex carbon-rich phase called ε-carbide and to martensite containing 0.25% carbon in solid solution. In stage II tempering, which occurs between 200 and 300 °C, retained austenite is transformed to ferrite and cementite. Above 250 °C, the beginning of stage III tempering, the transitional ε-carbide and low-carbon martensite decompose into ferrite and cementite. In alloy steels containing sufficient quantities of strong carbide-forming elements, Fe_3C may finally be replaced by stable complex-alloy carbides. A more detailed analysis of the tempering stages has been reported (Krauss 1984).

The carbide-forming elements not only raise the tempering temperature to obtain an equivalent hardness level, but when present in sufficient quantities also affect the rate of softening in such a way that it is no longer a continuous function of the tempering temperature. The softening curves of these steels show a range of tempering temperatures in which (a) the softening is retarded, or (b) for relatively high alloy contents, the hardness actually increases with increasing tempering temperature. This latter process is known as secondary hardening and results from the formation of fine-alloy carbides that produce the hardness increase.

Figure 2 shows the secondary-hardening effect in a series of molybdenum-containing steels at a carbon level of 0.35% carbon. As the molybdenum content increases, the secondary-hardening peak becomes more discernible, with a maximum in secondary hardening occurring at the highest molybdenum content. In the lowest molybdenum alloy (0.5%), no peak is seen but retardation of the softening effect is still evident. Since secondary hardening depends on the diffusion of carbide-forming elements to produce

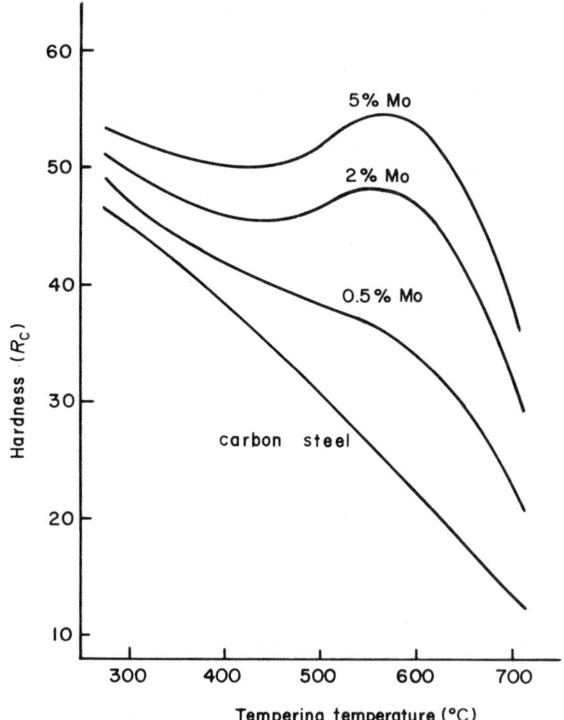

Figure 2
Retardation of softening and secondary hardening during tempering of a 0.35%C steel with various additions of molybdenum (Bain and Paxton 1962)

alloy carbides (a more sluggish reaction than that of carbon diffusion), the secondary-hardening peak occurs only at high tempering temperatures. As a result, the alloy carbides formed have a fine dispersion and are very resistant to coarsening.

For results such as those found in Fig. 2, a constant time of one hour is usually assumed. However, since tempering is a thermally activated process, it depends on both temperature and time. The interchangeability of time and temperature is often accomplished by the use of a tempering parameter: $T(C + \log t)$, where T is the absolute temperature and t is the time (h). The term C is a constant which has values between 15 and 20. From a single set of hardness–temperature data for a constant time, a master curve can be constructed for the tempering response of a given alloy. This type of curve may be used to find the required time–temperature specifications to produce a given hardness. Other curves can be used to determine the amount of additional tempering necessary to bring an undertempered piece to a specified hardness as well as to calculate the hardness of hardened-steel parts subjected to high temperatures for long periods of time.

The effect of alloying elements on tempering is shown in Fig. 3. The alloys are contrasted at two different temperatures: 260 °C and 540 °C. The carbide-forming elements vanadium, molybdenum and chromim are found in the lower curve, whereas phosphorus, manganese, silicon and nitrogen are compared in the upper curve. The effects of various elements on the retardation of tempering are as follows:

(a) Manganese increases the hardness of tempered martensite by retarding the coalescence of carbides that prevent grain growth in the ferrite matrix. These effects cause a substantial increase in the hardness of tempered martensite as the percentage of manganese in the steel increase.

(b) Nickel has a relatively small effect on the hardness of tempered martensite, which is essentially the same at all tempering temperatures. Since nickel is not a carbide former, its influence is considered to be due to a weak solid-solution strengthening.

(c) Silicon increases the hardness of tempered martensite at all tempering temperatures. Grange et

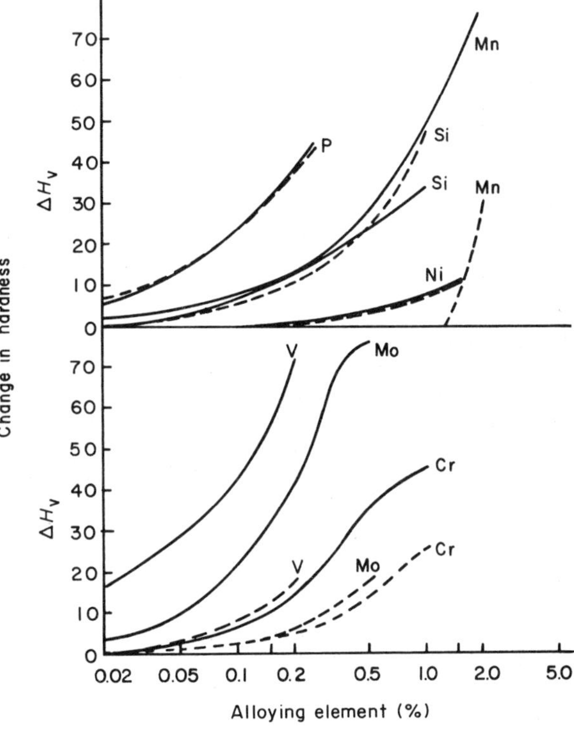

Figure 3
Effect of alloy concentration on the retardation of softening during tempering at 260 °C (– – –) and 540 °C (———) relative to iron–carbon alloys (after Grange et al. 1977)

al. (1977) found that silicon has a substantial retarding effect on softening at 316 °C. They attributed this effect to silicon inhibiting the conversion of ε-carbide to cementite.

(d) Phosphorus increases the hardness of tempered martensite at all tempering temperatures above 204 °C. The increase in hardness has been attributed (Grange et al 1977) to solid-solution hardening of the ferrite matrix.

(e) Molybdenum retards the tempering of martensite at all tempering temperatures. Above 540 °C, molybdenum partitions to the carbide phase and thus keeps the carbide particles small and numerous.

(f) Chromium, like molybdenum, is a strong carbide-forming element that can be expected to retard the tempering of martensite at all temperatures. By substituting for some of the iron in cementite, chromium retards the coalescence of carbides.

(g) Vanadium is a stronger carbide former than molybdenum and chromium and can, therefore, be expected to have a much more potent effect at equivalent alloy levels. The strong effect of vanadium is probably due to the formation of an alloy carbide that replaces cementite-type carbides at high tempering temperatures and persists as a fine dispersion up to the A_1 temperature.

See also: Steels: Classification; Steels: Structure–Property Relations; Temper Embrittlement of Steel

Bibliography

American Society for Metals 1979 *Worldwide Guide to Equivalent Irons and Steels.* American Society for Metals, Metals Park, Ohio

Bain E C, Paxton H W 1962 *Alloying Elements in Steel,* 2nd edn. American Society for Metals, Metals Park, Ohio

Doane D V 1979 Application of hardenability concepts in heat treatment of steel. *J. Heat Treat.* 1: 5–30

Doane D V, Kirkaldy J S 1978 *Hardenability Concepts with Applications to Steel.* The Metallurgical Society, Warrendale, Pennsylvania

Grange R A, Hribal C R, Porter L F 1977 Hardness of tempered martensite in carbon and low-alloy steels. *Metall. Trans.* 8A: 1775–85

Krauss G 1980 *Principles of Heat Treatment of Steel.* American Society for Metals, Metals Park, Ohio

Krauss G 1984 Tempering and structural change in ferrous martensitic structures. In: Marder A R, Goldstein J I (eds.) 1984 *Phase Transformations in Ferrous Alloys.* American Institute of Mining, Metallurgical and Petroleum Engineers, Warrendale, Pennsylvania, pp. 101–24

McGannon H E (ed.) 1971 *The Making, Shaping and Treating of Steel*, 9th edn. US Steel Corporation, Pittsburgh, Pennsylvania

Siebert C A, Doane D V, Breen D H 1977 *The Hardenability of Steels: Concepts, Metallurgical Influences, and Industrial Applications.* American Society for Metals, Metals Park, Ohio

Society of Automotive Engineers 1979 *SAE Handbook 1979.* Society of Automotive Engineers, Warrendale, Pennsylvania

A. R. Marder

Helicopter Materials

The requirements imposed on structural materials by the helicopter are considerably different from those needed for fixed-wing aircraft or ground transport vehicles. The requirements for minimum structural weight, high fatigue resistance, slow crack growth rates, notch or defect insensitivity, and immunity to environmental degradation are paramount.

Figure 1 is a typical modern helicopter depicting the major structural groups: airframe, rotor system, drive system, steel-forged landing gear and controls. The airframe requires major load beams for concentrated load distribution in areas such as the transmission and landing gear, but is otherwise lightly loaded and constructed of minimum gauge materials. The rotor blades are designed for fatigue resistance, damage tolerance, and ease of forming into complex aerodynamic shapes. The rotor hub carries highly concentrated loads and must be fatigue resistant. Gearboxes and driveshafts must be stiff and lightweight, the landing gear is designed for energy absorption and strength, and controls for stiffness and damage tolerance. The cockpit enclosure, radome and floor panels are constructed of lightweight materials. These design requirements lead to specific materials usage.

1. Airframe

The major portions of most helicopter airframes currently in production are made from aluminum. 2024 alloy is commonly used for airframe skins and several of the 7000 series alloys used where loads are concentrated such as the landing gear and transmission mounting structures, which frequently contain machined aluminum forgings. Most basic airframe structures are of conventional frame and stringer (with minimum gauge sheet skin) construction containing thousands of aluminum rivets. Airframe secondary structures such as fairings, covers, engine inlets, cooling ducts and interior trim are made from fiber-reinforced organic resin bonded materials of the thermosetting, and more recently the thermoplastic type. Typically, epoxy thermoset resins are used, though thermoplastic materials such as P1700 polysulfone are under development. This technology has grown rapidly in recent years. The Sikorsky UH-60 Blackhawk, and the Hughes AH-64 helicopters now have 20–25% of their air passage surfaces fabricated from composite materials. The

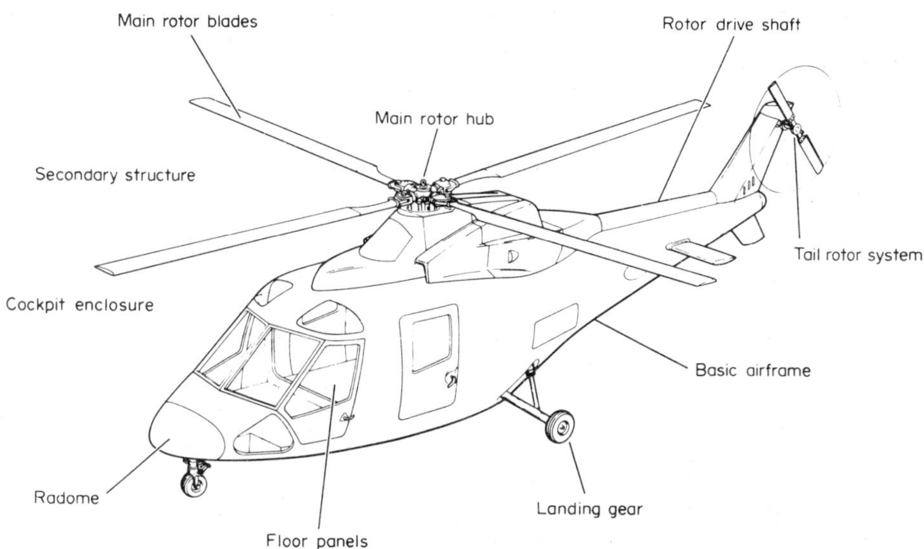

Figure 1
A typical helicopter

Sikorsky S-76 has over 30%, and similar trends are being followed by Messerschmitt–Boelkow–Blohm (MBB), Aerospatiale and Westland in Europe. Boeing Vertol has incorporated composite materials extensively in its commercial derivative of the US Army CH-47 helicopter, the Model 234 (Fig. 2). This large, 22 700 kg gross weight machine contains over 1800 kg of glass, graphite and Kevlar composites which is 23% of the airframe structural weight. Aerospatiale is experiencing a similar trend with current airframe designs containing 18–22% composite structures.

Major programs are underway in the USA to extend the application of composite materials to helicopter fuselage heavily loaded primary structures. Bell and Sikorsky are participating in a US Army funded research and development program to develop an all composite helicopter fuselage. The purpose of the Advanced Composite Airframe Program (ACAP) is to demonstrate major reductions in fuselage weight and fabrication costs through the extensive application of composite structures. Preliminary design and fabrication cost estimates have shown that reductions of 24% in weight and 27% in fabrication costs may be achievable.

A similar major fuselage component program is being completed by Sikorsky for the UH-60 Blackhawk, which will result in the construction and evaluation of the rear fuselage portion of the aircraft shown in Fig. 3. The currently produced conventional metallic section, which is 1.83 m high, 2.74 m long, and 2.44 m wide, contains over 1200 detail parts and approximately 17 000 fasteners. The composite structure, which must be fitted and joined to the forward and aft portions of the currently produced metal fuselage will consist of approximately 460 components and require some 6000 metallic fasteners. It appears that the program will result in a 10% reduction in structural weight and an approximate 30% reduction in fabrication cost. The reduction in parts is a major factor in the anticipated cost reduction. The predominant material used was graphite-fiber/epoxy, with glass and Kevlar reinforced material used where structural requirements were less stringent.

2. *Rotor Systems*

The 1940s saw the development of what may be considered to be the first practical helicopters. The main rotor blades produced were, in general, constructed of steel, wood, and fabric. A hollow steel tube was used for the spar, or main load carrying member. The area forward of the steel spar, comprising the blade leading edge, was frequently made from a solid wood laminate, while the aft portion or trailing edge fairing consisted of built-up wood ribs covered with fabric. A troublesome problem with blades constructed in this manner was moisture absorption with subsequent changes in blade mass, and consequent balance, vibration, and control problems. However, the materials and fabrication techniques did allow the construction of highly efficient rotors with variable and sometimes complex geometry. Blades with nonlinear twist, chordwise taper, and varying airfoil shapes were made in order to achieve the optimum lift to drag ratio required for successful flight with piston engine power. By the

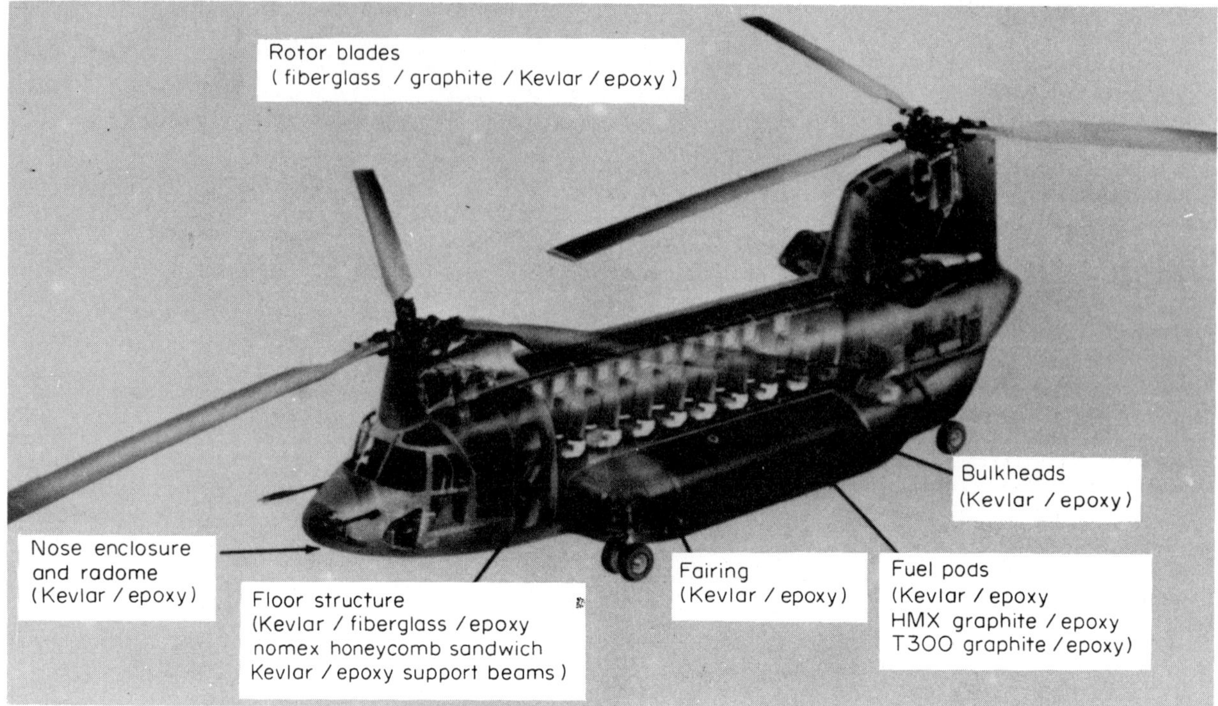

Figure 2
Composite structure detail in a Boeing 234 helicopter

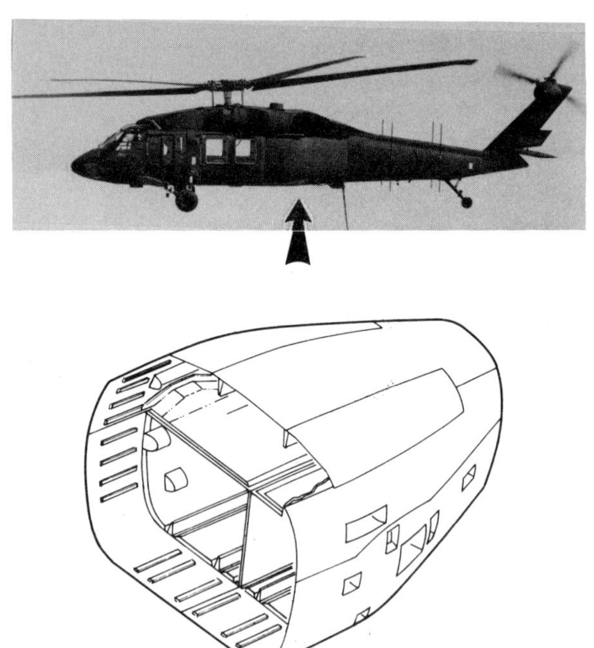

Figure 3
Composite rear fuselage, Sikorsky UH-60

end of the 1940s hybrid steel and wood rotor blades were accepted practice. However, some all-metal blades were being developed. Hiller had developed an all-metal, mechanically fastened rotor which was flown on the XH-44, and Sikorsky had developed an all-metal adhesively bonded blade for its S-51 helicopter. In the late forties, epoxy resins became available and wood spar blades with fiberglass–epoxy components were developed and flight demonstrated.

The advent of the turboshaft engine in the 1950s removed the severe power limitation which had been a major design constraint. Constant chord, constant airfoil, all-metal blades were generally used. Extruded aluminum and roll formed steel tube spars were the central load carrying members. Adhesives became the most popular assembly technique with the blade trailing edge fairings composed of aluminum ribs and skins. Corrosion problems, and subsequent fatigue damage became of concern and aluminum anodizing, sacrificial zinc plating, and chromate conversion coatings came into use. Sophisticated inspection techniques for finding microcracks or inclusions were developed. Methods of sealing the hollow spar tubular members were developed, and differential pressure change used to detect the onset of fatigue cracking. These blades were damage-tolerant designs based on measured crack propagation rates. Blade trailing edge fairings were segmented to allow individual segment or "pocket" replacement in the field. Increased fatigue resistance of the trailing edge and fairing skins was made on the CH-47 Chinook blades by incorporating fiberglass skins in place of the aluminum skins. Continued fatigue problems with the underlying aluminum ribs, and requirements to improve blade aerodynamic surface smoothness resulted in the use of full depth aluminum honeycomb in place of the sheet metal ribs. During the 1960s, many companies built thousands of aluminum spar, honeycomb and single-piece aluminum skin blades. Adhesive bonding using 450 K cure epoxy-based adhesive films became virtually an industry standard and is still in use today.

Later, more advanced metal working technologies allowed the introduction of titanium to metal spar blade designs. The Sikorsky UH-60 Blackhawk has main rotor blade spars fabricated from 6Al–4V titanium alloy using electron-beam welding of tapered sheet followed by hot forming to shape. The Hughes AH-64 uses stainless steel in fabrication of main rotors for the Advanced Attack Helicopter. These, however, may well be the last metal based blades, as substantial development of fiber-reinforced composite materials within the last 25 years and their unique applicability to helicopter rotor design has virtually revolutionized the industry.

The 1960s saw significant development of composite structures in main- and tail-rotor applications. Kaman had transformed an all-wood blade used on the H-43 to wood plus fiberglass and subsequently to all fiberglass which was flight demonstrated. Boeing Vertol completed a 150 h endurance whirl test of an all-fiberglass rotor for its new CH-47 helicopter, and MBB developed fiberglass main blades for its BO-105.

In 1964, Boeing began a research and development program to explore the application and potential benefit of continuous fiber-reinforced composite materials to main rotor blades. Unidirectional glass-fiber tape and woven fabrics made from Owens Corning S glass were used for the entire structure. The blades were fabricated using a precisely machined aluminum core and incorporated a complex airfoil geometry, and were successfully fatigue and whirl tested, and flown on a CH-47 helicopter.

The flight evaluation significantly extended the performance envelope of the helicopter in terms of both gross weight and speed. Later, under sponsorship of the US Air Force, the same advanced geometry main rotor blades (AGB) were fabricated from all-boron fibers. These blades were flown on the more heavily loaded rear rotor of the CH-47, with the Boeing-developed glass blades on the forward rotor (Fig. 4). Both of these programs demonstrated the superior fatigue resistance of composite materials, the potential performance improvement available (Fig. 5), the relative ease with which a rather complex airfoil could be constructed, and that practical production techniques could be developed.

Figure 4
Chinook helicopter equipped with boron fiber rear rotor blades and "S" glass fiber forward blades (after Hoffstedt 1971)

Later, Boeing (under NASA sponsorship) combined the two materials technologies in the fabrication of a rotor set for a tilt-wing program. These blades used S glass for most of the spanwise reinforcement, with the outer blade shell comprising boron fibers oriented at ±45° for high torsional stiffness. During this same period, Sikorsky fabricated and flew a set of tail rotor blades containing boron-fiber spar elements for its S-61 helicopter.

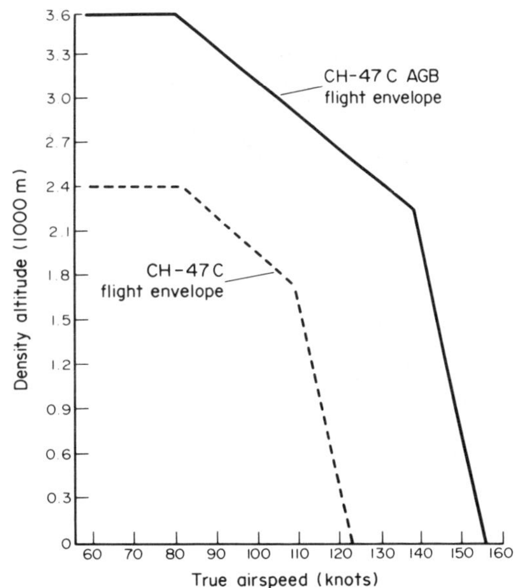

Figure 5
Velocity–altitude envelopes CH-47C and CH-47C with AGB (advanced geometry blade) (AGD 46 000 lb flight envelope evaluation, 245 rpm rotor speed) (after Hoffstedt 1971)

In the second half of the 1970s, MBB and Aerospatiale in Europe were in high-rate production of all-composite main rotor blades using R glass fibers supplied by Saint Gobain and work was well underway in the USA towards the development of rotors using lower cost fibers such as the common E glass, S2 glass, graphite, and Kevlar fibers. Hughes and the Fiber Science Company both fabricated Kevlar fiber rotor blades using filament winding techniques under US Army development contracts. Sikorsky completed a composite blade set for the US Navy. Later, as part of an Army product improvement program, Hercules and Kaman developed main rotors for the Bell AH-1Q helicopter.

Currently, Aerospatiale, MBB, Boeing, Bell and Kaman–Hercules are in full-scale production of all-composite main rotor blades, and Sikorsky and Hughes have active contracts for the development of all-composite rotors for their military helicopters, the UH-60 and the AH-64.

3. Drive System Components

The major drive system components of most helicopters in service are made from metallic materials. Main and tail rotor hubs are fabricated from titanium or steel forgings. Fully articulated hub systems use conventional roller and ball bearing assemblies to accommodate blade flapping, lead–lag and pitch motions. Several hundred separate parts may be included in such a fully articulated hub system, as shown in Fig. 6.

Main and tail rotor drive shafts are made from steel, aluminum or titanium. Others are combinations of steel and aluminum, with aluminum used in the center section of the shaft for weight reduction. Transmission cases or gearboxes are almost universally fabricated from magnesium alloy castings, such as AZ-91 and ZE41A. Internal transmission components are almost entirely fabricated from steel (with great emphasis placed on metal cleanliness, freedom from inclusions and surface defects) such as AISI 9310, AISI 52100, M-50 and Vasco X-2.

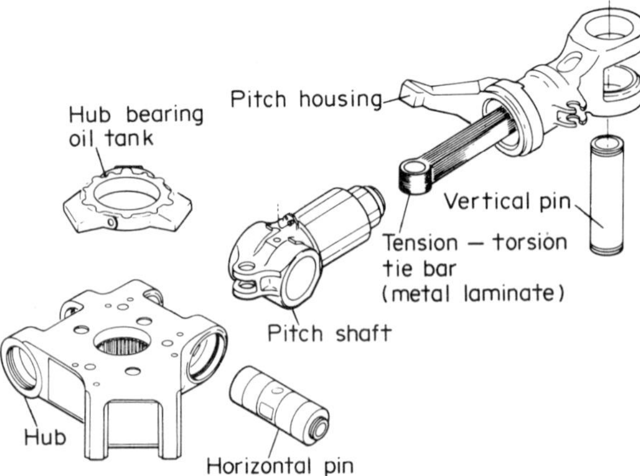

Figure 6
Major components of the CH-47 hub

4. Bearingless Rotor Systems

The expanding utilization of composite materials in flight-critical structural-fatigue applications is further illustrated by the development of rotor systems in which one or more of the principal dynamic motions (blade flapping, lead–lag and torsion, or angle of attack changes) are accommodated and controlled by the bending and twisting of the composite hub structure. Tail rotor blade and hub systems have been developed in which all motions of the blades are taken through the composite structures and all bearings of any type are completely eliminated. The Boeing Vertol UTTAS and US Army/Sikorsky Blackhawk helicopters use such a system.

These new rotor system designs are made possible by the fact that composites can have significant anisotropy, with Young's modulus to shear modulus ratios of more than 30 : 1 being possible. In addition, because the type, amount, and orientation of the fiber can be chosen by the designer, the ratio of bending to torsional stiffness can be selected. This allows designs in which elastic deformation can accommodate the necessary blade motions, rather than using bearings. Such designs are not possible with metals because the ratio of Young's to shear modulus is relatively low and not adjustable over a wide range.

The same factors which have prompted the development of bearingless tail rotor systems (i.e., reliability, maintainability, cost and weight) have prompted development of similar main rotor systems. Here, total bearingless systems have as yet not been committed to production, although several systems in which one or more of the bearing elements have been eliminated are in use. In addition, the development of highly reliable elastomeric laminated bearing elements and their incorporation into rotor systems along with the use of composite materials has virtually eliminated the use of conventional oil lubricated rolling element bearings in new rotor designs. The MBB BO-105, for example, uses the inboard portion of the main rotor blade glass fiber spar as the flexible hinge to accommodate blade flap and lead–lag motions. Rolling element bearings are retained only for blade pitch change.

The Aerospatiale Starflex rotor head uses a combination of these techniques to eliminate all lubricated rolling element bearings and, as a result, over 97 bearings, seals and lubricators are no longer required (Cassier and Sieffer 1978). A fiberglass three-axis star plate is the principal load carrying element in the Aerospatiale design. The plate carries the blade centrifugal loads through an elastomeric bearing which acts as the pitch-change torsion bearing. Flapping and lead–lag motions are also accommodated by this bearing, and are controlled by an elastomeric block located at the end of each of the star arms (Fig. 7).

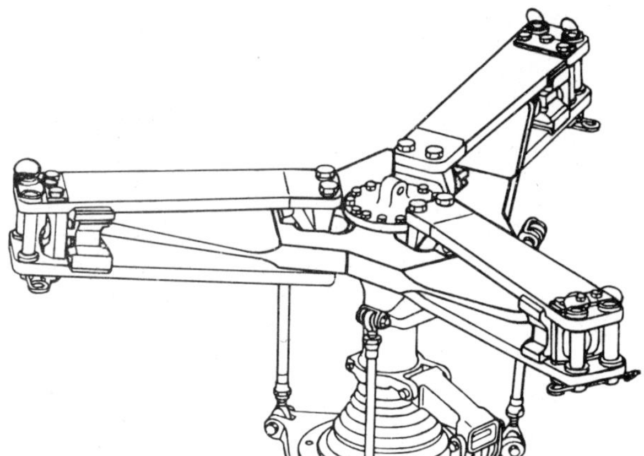

Figure 7
Aerospatiale Starflex rotor head (after Cassier and Sieffer 1978)

Other development programs are underway aimed at eliminating all bearings, rolling element and elastomeric, from the main rotor hub system. Both Aerospatiale and Boeing Vertol have developed and flight demonstrated true bearingless rotor systems.

The Aerospatiale Triflex hub consists of three flexible arms made of glass fibers, fabric, and elastomeric resin materials molded and bonded into a one piece hub component. The central hub area and the blade attachment areas at the outboard end of each arm are reinforced and stiffened with glass fabric and epoxy resin. In the arms themselves, the glass fiber rovings are stabilized and separated by the elastomeric matrix which reduces the torsional stiffness of the hub arms and yet provides sufficient stability for the fibers to withstand the compression loads introduced by arm bending.

The true hingeless triflex rotor hub was successfully flight demonstrated on a Gazelle helicopter, and the aircraft had good flying qualities with excellent control power and response, though the rotor lag damping was found to be below optimum. Further development of elastomeric matrix materials incorporating higher viscoelastic damping is expected to improve this parameter.

The Boeing Vertol/US Army BMR illustrated in Fig. 8 is a four bladed soft-in-plane rotor with no flap, lag or pitch bearings. These were replaced by twin, parallel beam fiberglass flexures having low torsional stiffness. Blade pitch motion was introduced at the flexure to blade junction through a filament-wound graphite torque tube, rigidly attached to the blade attachment point and supported at its inboard end by a pivot which reacted the control loads input by a conventional swashplate/pitch link control system. The BMR was sized to fly on the BO-105 helicopter and was designed to replace that rotor system directly without any modifications to the basic aircraft or control system.

Figure 8
Bearingless main rotor hub on a BO-105 helicopter

Future development of the BMR should include efforts to reduce aerodynamic drag and improve performance by integrating the flexure with an advanced airfoil rotor blade. The use of a single element beam of higher modulus material would result in a smaller flexure profile and permit replacement of the torque tube by a sleeve which could serve as an aerodynamic fairing.

Bearingless main and tail rotor systems have been proved to be feasible and their excellent potential for providing a simpler, lower cost, long life, and reduced maintenance rotor system has been demonstrated.

5. *Drive Shafting*

Rotating drive shafts appear to be one of the more attractive structural applications of high-modulus composite materials. Because the loading is primarily torsion, the fibers can be oriented at $\pm 45°$ to the tube axis to react the principal stresses. Such structures are common in helicopters in the form of tail rotor drive shafts, engine input shafts, and main rotor shafts, and in tilt-wing and tilt-rotor vertical take-off and landing (VTOL) aircraft in the form of cross-shafting. The higher structural efficiency of high-modulus composites, compared with present metal shafts, promises weight savings of more than 50%, and the design flexibility of composites allows the design of longer shafts between bearings, thus reducing the number of bearings, and allows the possibility of having supercritical shafting. In addition, the composite material shafts have shown superior fatigue resistance and ability to tolerate ballistic damage.

The helicopter industry has fabricated prototype shafts from a great variety of composite fibers and fiber combinations including boron, graphite, Kevlar and glass. All-boron fiber shafts have been fabricated and evaluated; however, the relatively high cost of boron fibers compared with other materials has eliminated them from the more recent development programs. A principal design advantage of composite drive shafts derives from the fact that larger unsupported shaft lengths may be used. The incorporation of high-modulus fibers parallel to the shaft axis allows the design of long components which do not encounter high rotational, critical bending modes. The resulting reduction in required end fittings, support bearings, flexure couplings, and airframe support structure and flexible mounts can effect considerable cost, maintenance and weight savings. For example, changing a constant diameter interconnecting shaft system from aluminum tubing to a composite tube design allowed the shaft segment length to increase from 1.74 m in aluminum to 2.60 m in composites. This change resulted in a reduction in total shaft system weight from 305 to 227 kg (a weight reduction of 25%).

6. Transmission Housings

The application of composite materials to helicopter transmissions or gear cases is being explored. The Bell Helicopter Company and the Boeing Vertol Company have both evaluated experimental housings as part of their research and development programs. Both programs used graphite materials as the reinforcing fiber. The Bell program evaluated the main transmission gear case on the UH-1 helicopter (Fig. 9). The Boeing program used one of the engine gearboxes on the CH-47 helicopter. The Bell UH-1 program utilized epoxy resins for the composite matrix and subsequent adhesive bonding, while the Boeing Vertol program used polyimide resins to produce a single piece unitized case. Both cases were essentially duplicates of their production cast magnesium counterparts but demonstrated the ability to increase stiffness and improve gear wear.

In summary, composite materials are well on the way to dominant utilization in the helicopter industry. In addition to the performance improvements, freedom from corrosion and fatigue resistance have made composites very popular among helicopter operators. The turbine engine made the helicopter a practical reality. Composite materials will make it a reliable, efficient means of transportation. The challenge now is to make the most efficient use of these technologies in their implementation in future vertical takeoff and landing aircraft.

Figure 9
Graphite–epoxy housing for Bell UH-1 helicopter (after Chase 1973)

See also: Aerospace Materials; Airframe Materials

Bibliography

Battles R A 1975 *Dynamic Testing of a Composite Material Helicopter Transmission Housing*, USAAMRDL-TR-75-47. Bell Helicopter Company, Washington, DC

Brunsch K 1972 Winding techniques for rotor blade applications. *Filament Winding II, PI-RPG Conf.* Plastics Institute, London

Cassier A, Sieffer J C 1978 Trends in helicopter head design. *12th International Helicopter Forum.* SNI Aerospatiale, Paris, France

Chase V A 1973 *Investigation of the Use of Carbon Composite Materials for Helicopter Transmission Housing Applications*, USAAMRDL-TR-73-7. US Army Air Mobility Research and Development Laboratory, Fort Eustis, Virginia

D'Ambra F D 1981 The impact of new materials in the development of rotary wing aircraft. *J. Am. Helicopter Soc.* 26(4): 16

Dixon P G 1980 *Design, Development and Flight Demonstration of Loads and Stability Characteristics of a Bearingless Main Rotor,* USAMMRDL-TR-80-D3. US Army Air Mobility Research and Development Laboratory, Fort Eustis, Virginia

Hoffstedt D J 1971 *Research and Development of Helicopter Rotor Blades Utilizing Advanced Composite Materials,* USAFML-TR-71-125. National Technical Information Service, Springfield, Virginia

Kiraly R L, Lund H T, Yao S S 1981 Fabrication methodology for a composite main rotor blade for the YAH-64 advanced attack helicopter. *J. Am. Helicopter Soc.* 26(4): 31

Pinckney R L, Freeman R B 1971 *Determination of Physical and Structural Properties of Mixed-Modulus Composite Materials,* USAAMRDL-TR-71-7. Boeing Company, Seattle, Washington

Sciarra J J, Howells R W, Lenski J W Jr, Drago R J 1978 *Helicopter Transmission Vibration and Noise Reduction Program,* USARTL-TR-2B. US Army Air Mobility Research and Development Laboratory, Fort Eustis, Virginia
Weiland E F 1968 Development and test of the BO-105 rigid rotor helicopter. *Proc. 24th Annual National Forum of the American Helicopter Society*. American Helicopter Society, Washington, DC
White R W 1981 Composite technology in the U.K. helicopter industry. *J. Am. Helicopter Soc.* 26(4): 34
Whiteside W, Ray J D 1977 Airframe design of the S-76 helicopter, Sikorsky Aircraft Division, United Technologies Corporation, Preprint No. 77.33-85. Paper presented at the 33rd Annual National Forum of the American Helicopter Society
Zinberg H 1973 An advanced composite tail boom for the AH1G helicopter. Paper presented at the 29th Annual National Forum of the American Helicopter Society
Zinberg H, Symonds M F 1970 The development of an advanced composite tail rotor drive shaft, Preprint No. 451. Paper presented at the 26th Annual National Forum of the American Helicopter Society

R. L. Pinckney; M. J. Salkind

Helium-Induced Blister Formation of Fusion Reactor First-Wall Materials

All metals, including those with an amorphous structure, appear to react in a similar way when bombarded with monoenergetic helium ions; after a critical helium dose—a function of ion energy—very fine-scale blistering or flaking of the surface takes place. Since helium would be a reaction product of the deuterium–tritium (D–T) reaction in any successful fusion device, blister formation has to be considered as a potential wall-erosion mechanism. The phenomenon is not limited to metal surfaces but also occurs on other materials of interest with respect to fusion (e.g., low-atomic-number ceramics), although the patterns of behavior are less well defined for these materials. In addition, blister formation can be induced by protons, deuterium ions and the heavier inert gases. However, inspired to a large extent by the fusion-technology implications, helium has remained the major focus for fundamental work in this area, leading to a large volume of literature: early studies are covered in the review by Das and Kaminsky (1976); this is complemented by the excellent up-to-date review by Scherzer (1983).

1. Blister Morphology and Relevant Parameters

An example of blister formation on a metal sample is seen in the scanning electron micrograph of the helium-irradiated molybdenum surface in Fig. 1. Although this is reasonably representative, the morphology of blisters can vary widely; even in this example variations are apparent between blisters with complete lids and those with lids partially removed or completely flaked away. In some materials, as illustrated in Fig. 1b, surface flaking without any blister formation is dominant, but all these different morphologies are considered to reflect the same underlying mechanisms. Among the parameters of importance for blister formation must be included the helium energy, dose, specimen temperature, state of surface layer (e.g., hardness) and surface-grain orientation. However, the influence of several of these parameters can be rationalized by the observation that blister formation only occurs (a) when a peak of helium concentration is set up

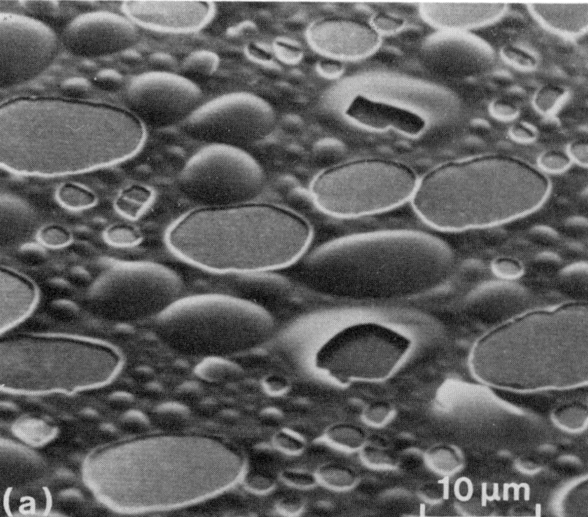

Figure 1
Scanning electron micrographs showing examples of (a) blister formation (molybdenum) and (b) flaking (Inconel 625) after irradiation with 100 keV helium ions

beneath a specimen surface (and not at the surface), and (b) when the concentration at the peak reaches about 30 at.% helium. The helium-peak requirement usually demands a monoenergetic ion, giving blisters over a wide helium energy range from ≃1 keV up to at least 3 MeV, while the concentration requirement defines the helium dose at which blister formation is initiated. This dose, typically 5×10^{21} helium ions per m^2 at 30 keV, increases up to 10^{22} helium ions per m^2 at 200 keV, reflecting the increase of the standard deviation in the helium range distribution as the energy increases (and hence a higher dose to reach a given peak concentration). For the same reason the critical dose increases when the surface orientation favors channelling of the helium ions. The specimen temperature is relatively unimportant up to about 0.4–0.5 of the melting point T_m, but above this temperature the blisters are replaced either by pinholes or a sponge-like surface.

The parameters which do not affect helium blister formation are of equal interest. One is the helium dose rate which is not normally considered important. This is in contrast to the situation in proton-induced blister formation, where a high dose rate is mandatory at normal temperatures (which is one reason why blister formation from the hydrogen isotopes is not usually considered to be a potential erosion mechanism in fusion devices). A second parameter is simply that of metal identity, which only appears to affect the morphology of blisters rather than the underlying behavior. Even here, except in one or two cases (e.g., see Bauer 1978) there are insufficient data to map out, or predict, the effect of parameters such as alloying, cold working or temperature. However, the morphology of blistering is rather important when the erosion potential is considered, particularly at high dose levels of helium. For example, it is possible that surfaces can blister and release helium with negligible surface loss—even when further gas is injected. On the other hand, a completely flaked surface has what might be termed a 100% erosion potential; at the critical dose the whole surface can be lost, revealing a new surface for the cycle to be repeated. A similar cyclic process, on a very localized scale (but equally important), can occur when blister lids are successively lost from an individual blister site.

In reality, the erosion due to blistering or flaking cannot be considered in isolation; surface sputtering can severely moderate, and often totally suppress, its occurrence (see *Sputtering*). The buildup of the distinct high-concentration helium peak required for blistering can always be prevented by a sufficiently fast velocity of surface erosion from sputtering.

2. *Physics of Blister Formation*

The association of blister formation with a high concentration of inert insoluble gas, and the release of gas when blister covers crack, made it natural at an early stage to connect the phenomenon with the internal release of helium. The increase of blister size with helium energy generally fitted a gas-pressure model but considerable discussion ensued when it was suggested that lateral-stress considerations might lead to a similar relationship. Although there is no doubt that the injection of helium does lead to a lateral stress, the stress has not been proved to be essential for blister formation; however, this does not preclude some small effect. For the gas models of blister formation the absence of detailed information on the state of the helium in the metal inhibited progress in understanding how the helium accumulated to form an internal sub-blister cavity. The application of transmission electron microscopy to this area has helped this situation considerably by showing that for temperatures $<0.4\,T_m$, high concentrations of helium are always present in metals (and even metallic glasses) as a high concentration ($\simeq 10^{25}\,m^{-3}$) of small bubbles some 2 or 3 nm in diameter. It is worth noting that all metals, apart from those with an amorphous structure, appear to show alignment into bubble lattices, an unexpected feature analogous to void lattices in metals, and now attributed to the two-dimensional diffusion of interstitial atoms.

The fact that helium bubbles form well below vacancy-migration temperatures, and have concentrations and radius parameters independent of temperature over a wide range of vacancy mobility suggests strongly that bubble growth processes are independent of temperature and hence, rather suprisingly, independent of the vacancies created in the displacement damage associated with the energetic helium ions. Several proposals have been made as to how bubbles gain vacancies, including gas-driven Frenkel-pair formation, interstitial-loop punching, and a mechanism involving transformation from helium platelets (van Veen et al. 1981), but exact details are still unproved. However, it has certainly been shown experimentally that the helium exists in the bubbles at extremely high pressures and packing densities (Rife et al. 1981), while recent work with the heavier inert gases has demonstrated that bubble pressures are high enough to form solid bubbles, even above ambient temperatures. This supports the older proposal that the internal release of helium is initiated by the helium pressure attaining values which allow adjacent bubbles to cooperate to cause interbubble fracture and hence bubble coalescence (Evans 1978). It is easy to show that >10% of the injected helium over the area of a blister is required to satisfy the pressure requirement for surface deformation. Therefore, a large number of bubble "layers" need to percolate their gas via interbubble cracks into the potential blister cavity. The usual peak in the helium distribution with depth, together with lateral stress, must induce the cavity to form a penny-shaped crack parallel to the surface; when sufficient pressure is available, the overlying material then rapidly deforms to create a blister. However, a uniform helium distribution up to the surface can prevent blister formation by allowing interbubble fracture to set up a direct pathway for gas release. The flaking variant is thought to occur when the original interbubble crack parallel to the specimen surface propagates faster than the local flow of gas into the cavity.

At high temperatures two effects become pronounced: bubble coalescence can take place by Brownian motion, while thermal vacancies can allow bubbles to reduce their internal pressure and attain equilibrium. Blister formation or flaking is thus prevented and, instead, bubbles grow until they either bisect the surface or migrate to the surface to give pinhole or sponge-like effects, respectively.

3. *Relevance to Fusion Devices*

In any magnetically confined high-temperature plasma device some leakage of ions from the plasma out to the containment wall (the first wall) is inevitable. This situation already occurs in present-day experimental fusion machines which normally use plasmas of hydrogen or deuterium, and much effort is being devoted to the understanding of plasma constituent behavior after trapping on the first wall and subsequent release back to the plasma. It has been stated already that blister formation from the hydrogen isotopes is unlikely for the ion dose rates and wall temperatures found in practice, but the helium-ion case in future devices requires close examination. It is worth distinguishing between the thermalized helium ions escaping the plasma edge, and the nascent 3.5 MeV helium ions from the D–T reaction which have some probability of directly escaping the magnetic field. The thermalized ions will have a range of energies and angles of incidence at the first-wall surface and are thus very unlikely to meet the requirement of a distinct helium peak as discussed in Sect. 2.

The situation with the direct 3.5 MeV ions is less clear; these ions in uncontained orbits hit the wall at rather low angles and can give a deposited-helium profile peaking within a few micrometers of the surface (Bauer et al. 1979). Thus, the potential for blistering or flaking is present, although its extent must depend initially on the strength of the magnetic confinement and subsequently on the sputtering parameters. Bauer et al. have discussed particular cases of this "competition" between sputtering and blister formation in detail. It would seem that either process can impose a minimum erosion rate on the first wall; in particular, a high blister-erosion rate may cause the advantage of a low-sputtering-rate material to be lost.

Future work in the area of helium-induced blister formation must inevitably concentrate on the synergistic effects of competing erosion mechanisms. However, of equal importance in a D–T fusion device may be the first-wall effects that could result from the simultaneous deposition of helium alongside deuterium and tritium. That these effects would increase the probability of blister erosion is perhaps unlikely, but additional effects from neutron irradiation, temperature cycling and the pulsed operation of early devices (overall, a formidable mix of conditions) may prove significant and will also need detailed consideration.

See also: Fusion Reactor Materials; Ion Bombardment of Materials; Swelling in Irradiated Metals and Alloys; Nuclear Materials: An Overview

Bibliography

Bauer W 1978 Surface processes in plasma-wall interactions. *J. Nucl. Mater.* 76/77: 3–15

Bauer W, Wilson K L, Bisson C L, Haggmark L G, Goldston R J 1979 Alpha transport and blistering in tokamaks. *Nucl. Fusion* 19: 93–103

Das S K, Kaminsky M 1976 Radiation blistering in metals and alloys. In: Kaminsky M (ed.) 1976 *Radiation Effects on Solid Surfaces*, Advances in Chemistry Series 158. American Chemical Society, Washington, DC, pp. 112–70

Evans J H 1978 The role of implanted gas and lateral stress in blister formation mechanisms. *J. Nucl. Mater.* 76/77: 228–34

Rife J C, Donnelly S E, Lucas A A, Gilles J M, Ritsko J J 1981 Optical absorption and electron energy loss spectra of helium microbubbles. *Phys. Rev. Lett.* 46: 1220–23

Scherzer B M U 1983 Development of surface topography due to gas ion implantation. In: Behrisch R (ed.) 1983 *Sputtering by Particle Bombardment*, Topics in Applied Physics Vol. 52. Springer, Berlin, pp. 271–355

van Veen A, Caspers L M, Evans J H 1981 Room temperature precipitation of helium in molybdenum observed by TEM and THDS; helium platelet formation. *J. Nucl. Mater.* 103/104: 1181–86

J. H. Evans

Hemp

True hemp (*Cannabis sativa*) is one of the oldest cultivated plants known to mankind for the production of fiber. This plant is a native of Central and Western Asia, but it is extensively cultivated in China, India and Turkey in Asia; the USSR, Bulgaria, Poland, Romania, Hungary and Yugoslavia in Europe; and Chile in the Western Hemisphere. The world production of hemp fiber in 1982 is estimated to have been 284 kt (see Table 1). The peak year of hemp production was 1940, when 832 kt were produced.

Table 1
Production of hemp, 1980–1982 (kt)

Country	1980	1981	1982
Chile	4[a]	4[a]	4[a]
China	88[a]	88[a]	92[a]
India	54	61	61[a]
Turkey	14[a]	12	15[a]
Bulgaria	3	4	3
Hungary	11	9	9[a]
Poland	4	5	5[a]
Romania	26	23	20[a]
Yugoslavia	6[a]	6[a]	6[a]
USSR	33	40	50[a]
Others[b]	15	15	19
Total	258	267	284

Source: *FAO Production Yearbook*, Vol. 36, 1982. Food and Agriculture Organization of the United Nations, Rome. a FAO estimate b Japan, Korea, Czechoslovakia, Bangladesh, Spain, Pakistan

1. The Plant

The hemp plant is a stout, branching annual with slender, erect, woody stems varying from 1 to 5 m in height and 4 to 20 mm in diameter. When cultivated for fiber in thick plantings, the stems do not branch. Hemp is a dioecious species (separate male and female plants) with palmate leaves and 5–11 rough, dark-green leaflets. Hemp grows best in a mild climate where temperatures range between 13 and 22 °C, and in a rich, loamy soil with an abundance of humus. The plants are raised from seed planted early in the spring.

2. Harvesting and Processing

The hemp plant is harvested by hand or machine 80–150 days after planting, or when the male plants are in flower. The cut stems are spread in layers on the ground for drying. The average yield of dried stems is 2.75–7.5 t ha^{-1}, 25% of which is fiber.

The fibers are separated from the rest of the plant by the process of retting (see *Flax*) which rots the gums that hold the fibers together. The plant stems are soaked in water or spread in thin layers on the ground to expose them to dew. Dew retting, the predominant practice in the USSR, takes 3–5 weeks depending on the temperature and humidity. Water retting is practiced in Italy, Spain, Hungary and Yugoslavia. Bundles of the stems placed in streams or rivers will ret in 10–15 days. Light snow is also favorable for retting.

After retting, the stalks are dried and the fibers are easily removed from the stem either by hand-breaking and hackling, or by mechanized breaking and scutching. In both cases, the woody interior is broken into short pieces called hurds, which are removed by hand-hackling (pulling the stems through a set of sharp pins), or the stems are scutched by passing them over drums with projecting bars that beat and scrape the fibers from the hurds. The yield of retted fiber is variable from 370 kg ha^{-1} in the USSR to 1.9 t ha^{-1} in the Federal Republic of Germany.

3. Fiber Characteristics and Uses

Hemp fibers which are fine and lustrous consist of many strings of long cells twisted together. Dew-retted fiber is gray, while water-retted fiber is creamy white. Fibers found primarily in the inner bark along the stem are 77% cellulose and 16% hemicellulose. Hemp is more nearly like flax than any other commercial fiber. The fiber is valuable because of its length (which varies from 1 to 3 m), its strength and its great durability. Hemp has a tensile strength of 90 kg m^{-2}, twice that of abaca (see *Abaca*); however, it lacks flexibility and elasticity. The fiber has a specific gravity of 1.48 g cm^{-3}. Hemp endures heat, moisture and friction with less injury than most fibers. Hemp fiber is graded with positive characteristics being high density and strength, color in pale to silvery shades, soft "handle," freedom from strain, minimum length of 2.5 m, and high degree of fineness.

Because of the great strength and durability of the fibers, hemp is used for ropes and marine cordage. It is also used for twine, carpets, sailcloth, binder twine, sacks, bags and webbing. Italy, Hungary and adjacent countries produce a fine fiber which is spun into yarn for fabrics. In South Korea, a yellow, coarse, linenlike fabric is produced for farm clothing which may last more than 3 years. Hemp is also used as a fine textile fiber resembling flax for bedsheets and tablecloths. Hemp waste, the woody fibers of the stem, and the hemp fibers themselves are used in making high-quality strong paper. The short fibers, or tow, and ravellings are used extensively in oakum for caulking the seams between planks in shipbuilding, in cooperage and as a packing for pumps and engines. The seeds yield 30–35% of a drying oil that is useful in the soap and paint industries as a substitute for linseed oil. Dried leaves and flowering tops of the male and female plants yield a resinous narcotic alkaloid known as hashish or charas. The dried leaves and tops of the wild plants are known as marijuana, and a sedative and hypnotic, ganja, is produced from dried unfertilized female flowers grown in India.

See also: Sunn Hemp; Natural Fibers in the World Economy; Materials of Biological Origin: An Overview

Bibliography

Ash A L 1948 Hemp—production and utilization. *Econ. Botany* 2: 158–69

Dempsey J M 1975 *Fiber Crops*. University of Florida, Gainesville, Florida
Dewey L H 1943 *Fiber Production in the Western Hemisphere*, Miscellaneous Publication No. 518. United States Department of Agriculture, Washington, DC
Haarer A E 1953 Hemp. *World Crops* 5: 445–48
Kirby R H 1963 *Vegetable Fiber*. Leonard Hill, London
Ochse J J, Soule J, Dijkman M J, Wehlburg C 1961 *Tropical and Subtropical Agriculture*. Macmillan, New York

P. Ross

High-Damping Alloys

Damping capacity in construction materials is an important property to be taken into account together with other physicomechanical characteristics when designing machines, mechanisms or components for work under vibration. The application of high-damping alloys greatly depends on three problems of modern technology. The first is the increased speed in the moving parts of modern machines and power units, which cause increases in undesirable vibrations and a growing danger of failure. If two alloys have similar strength characteristics, that with higher damping capacity will be the more reliable. The second problem is the influence of vibration on the quality of the parts manufactured on fast machines. The third problem is harmful noise, the intensity of which is increasing with the growing number of high-speed industrial installations.

1. Damping Characteristics: Classification

Damping is the ability of solids to dissipate the energy of mechanical vibrations: an equivalent term is internal friction. The dissipation of mechanical energy indicates that a body under external stresses undergoes some changes resulting in a loss of applied energy. In ideal bodies under reversed loads, the relation between stress σ and strain ε in the region of elastic deformation is described as a single-valued function, while in real bodies it is a many-valued function in the form of a hysteresis loop. The loop area in elastic hysteresis is equal to the loss of mechanical energy for a loading cycle (ΔW).

Relative characteristics are usually used to evaluate the damping capacity of a body quantitatively. One of these is specific damping capacity ψ, which is expressed in percentage terms as $100\Delta W/W$, where W is the specific elastic energy stored at a given stress amplitude σ_0. The intensity of damping is also characterized by the logarithmic decrement $\delta = \ln(\varepsilon_m/\varepsilon_{m+1})$, where ε_m and ε_{m+1} are the two subsequent strain amplitudes.

Apart from these characteristics, another property is often used—the inverse mechanical Q factor $Q^{-1} = \Delta W/2\pi W$, which is called internal friction. There is a simple correlation between the damping characteristics:

$$\psi = 1 - \exp(2\delta) \approx 2\delta \qquad (1)$$

Methods of measuring damping capacity are based on defining the damping of torsional and lateral vibrations. The disadvantage of these methods is that the strain along the cross section of the specimen is not uniform. When damping depends on the amplitude of vibration, the internal friction is expressed as the average value, which is, however, related to the amplitude value of the strain ε on the surface of the specimen. To define real internal friction Q_{g}^{-1}, samples of thin-walled tubes are used, or recalculation formulae are applied. For torsional vibrations:

$$Q_{\mathrm{g}}^{-1} = Q^{-1} + K\varepsilon \frac{dQ^{-1}}{d\varepsilon} \qquad (2)$$

where K is a numerical coefficient equal to one quarter for specimens of round section and one third for specimens of parallelepiped form.

Internal friction is of three origins: hysteretic, relaxational and resonant. The relaxational and resonant types do not depend on the amplitude of vibrations, but they are very dependent on the frequency of deformation. Maximum internal friction is at $\omega\tau = 1$ in the former case (where ω is the circular frequency of vibration and τ the relaxation time) and at resonant frequency in the latter.

In high-damping alloys, the internal friction is determined by losses of hysteresis type. The internal friction of this type does not depend on the frequency, but essentially depends on the amplitude of vibration.

Metallic materials can be divided into high-damping materials with $\psi \geqslant 10\%$, moderate-damping materials ($\psi = 1-10\%$) and low-damping materials ($\psi < 1\%$). At least one of the following hysteretic mechanisms of internal friction occurs in high-damping alloys: displacement of irreversible ferromagnetic domain walls, elastic twinning, displacement of interphase coherent boundaries, and microplastic strain.

The dependence of damping on the amplitude of vibration is generally shown by a curve having two specific sections. The initial amplitude-independent internal friction is followed by the sudden rise in internal friction where some hysteretic mechanisms operate. At higher strain amplitudes, the rate of damping change decreases, as a rule, but sometimes it reaches a maximum when irreversible structural changes in the material reach saturation. Characteristic examples of the dependence of damping on amplitude for a number of high-damping alloys are given in Fig. 1.

When using high-damping alloys, it is necessary to take into account the value of alternating stresses and also their distribution over the volume of the

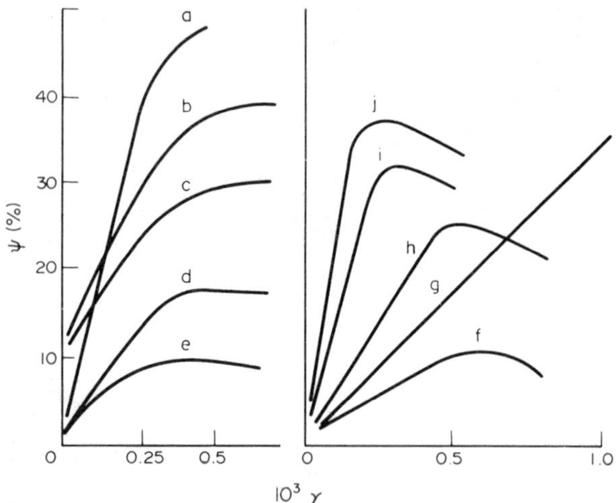

Figure 1
Damping capacity of some high-damping alloys upon optimal heat treatment: (a) cast Mg + 0.4% Zr; (b) Mn–25%Cu; (c) Nitinol; (d) modified austenitic cast iron with flake graphite; (e) gray cast iron with lamellar graphite; (f) steel 403; (g) Nivco-10 alloy; (h) Silentalloy; (i, j) Gentalloys ((i) Fe–6%Mo; (j) Fe–30% Co)

body. For instance, for turbine blades the damping capacity under low and moderate stresses is of highest importance, whereas in the case of crankshafts, damping must be low under relatively low stresses, as required by regular working conditions, and it must rapidly increase at resonance.

2. Ferromagnetic Alloys

Losses due to magnetomechanical hysteresis are the main source of high damping in ferromagnetic alloys. The role of other sources of magnetic losses (eddy currents) is small.

Magnetostrictive strain λ appears in a ferromagnetic sample under loading, together with a Hookean strain. Magnetostrictive strain results from the redistribution of vectors $\boldsymbol{I}_s$ of the domains, following the displacement of 90° domain walls. If external stresses caused irreversible shifts of the domain walls, then on unloading, the remanent magnetostrictive strain λ_r would be observed, and under reversed loading there will be magnetomechanical hysteresis. Under low stresses, when the magnetoelastic analogue of Rayleigh's law applies, ΔW is proportional to σ^3 and Q_h^{-1} to σ.

The correlation between Q_h^{-1} and the material characteristics is clearly seen from the following equation:

$$Q_h^{-1} = A\,\frac{\nu\lambda_l^3 E}{\boldsymbol{I}_s^3}\,\sigma \qquad (3)$$

where ν is Rayleigh's constant, characterizing the mobility of the 90° boundaries of the domains, λ_l is the magnetostriction constant along the axis of easy magnetization, E is Young's modulus, A is a numerical factor, dependent on the nature of the material and on the initial distribution of the domains with various directions of vectors $\boldsymbol{I}_s$. Specifically, if the specimen has a magnetic texture, where there are no 90° domain walls, then the factor $A = 0$ and, consequently $Q_h^{-1} = 0$ (curve 1, Fig. 2). As follows from Eqn. (3), only alloys possessing a rather high mobility of the domain walls and high λ_l value can be high-damping ferromagnetic alloys. Alloys with $\lambda_l = 0$ lack magnetoelastic damping at room temperature (e.g., Fe–6.5%Si).

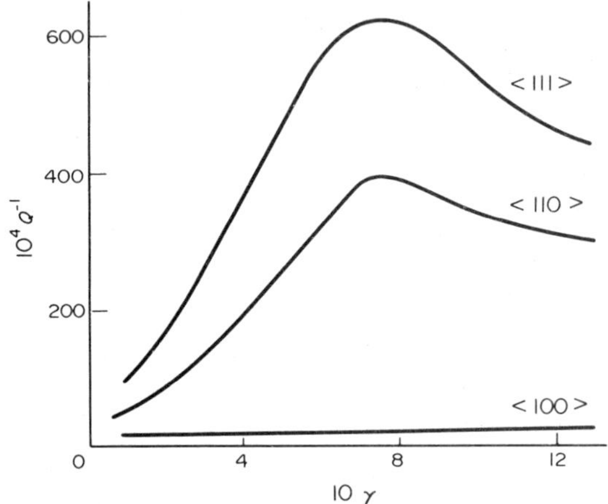

Figure 2
Internal friction of single-crystal specimens of Fe–3%Si alloy cut as narrow platelets along various crystallographic directions

Under relatively large amplitudes of vibration, the increase in internal friction of ferromagnetics is retarded as a result of magnetomechanical saturation, and after reaching its maximum, the friction decreases (curves 2 and 3, Fig. 2).

The important characteristic of magnetomechanical damping is its ability to be suppressed by a magnetic field and static stresses. The effect of static stresses is of particular importance when using high-damping ferromagnetic alloys for turbine blades, as these components undergo large centrifugal forces.

Stainless chromium alloys of the martensitic iron-based type have wide application in turbine engineering. A great advantage of these materials is their high damping capacity compared with steels of austenitic type. Chromium steels of high damping capacity usually contain ~12% Cr. To increase their strength, certain other metals are added, such as V, Ni, Mo and W; the additions usually amount only to ~10%. A high-damping chromium steel has the following typical composition: 12.3% Cr, 0.6% Ni, 0.5% Mo, 0.5% Mn, 0.3% Si, 0.2% C (alloy type AISI 403).

The damping capacity of chromium steels is greatly dependent on heat treatment. These steels acquire the maximum damping capacity upon annealing, when the structure is devoid of martensite, and the minimum upon hardening by quenching from the γ region. With increasing temperature of tempering, the damping capacity increases and the strength decreases. In chromium steels of high strength containing alloying additions of Ni, W, Mo and V, the maximum damping capacity is observed upon tempering at temperatures 710–730 °C, when intensive relaxation of internal strain takes place. At still higher tempering temperatures, the damping decreases because of the transition through point A_{C1} and the formation of austenite which transforms to martensite in the process of cooling.

The Fe–13%Cr alloys with 2–5% Al (Silentalloys) possess high and stable damping over a wide range of temperatures (from −50 to +350 °C). The aluminum additions considerably increase λ_1. In the annealed state under a shear strain $\gamma = 0.25 \times 10^{-3}$ and ψ is ~25%. A favorable effect on all properties of the alloys of this type is produced by microadditions of rare earth elements.

Precipitation-hardened stainless steels of martensitic type based on Fe–12%Cr–10%Ni and alloyed with Al, Ti, W, Mo and V possess not only high strength properties but a satisfactory damping capacity as well. Steel of the following composition may serve as an example: 12.3% Cr, 7% Ni, 0.74% Al, 0.5% Mo, 0.02% C. This steel, even upon cooling in air, has martensite in its structure. Its hardening by tempering occurs as a result of the precipitation of intermetallic phases of the type NiAl or Ni_3Al. Tempering at 500–550 °C causes a loss of strength due to the loss of the coherency, coalescence of phases and relaxation of stresses. At these tempering temperatures, the magnetic as well as the nonmagnetic components of internal friction increase suddenly, but the strength remains adequate.

For some branches of technology, malleable graphitized steels are potentially useful materials, as their damping capacity can be similar to that of a number of cast irons and of steel 403. These steels with no more than 1.5–1.7% C have high damping capacity mainly because of the high level of losses by magnetomechanical hysteresis and the more compact form of graphite than in cast iron. For example, graphitized steel of the composition 6.3% Co, 2.1% Si, 1.27% C, 0.04% Al has good properties.

High damping properties are typical for steels of the Fe–Co system when their composition is close to equiatomic (these alloys have high λ_1). Rapid cooling, while suppressing the processes of ordering, decreases the damping capacity. Other examples, of iron-based high-damping alloys are alloys containing ~6% Mo (Gentalloys), and also Fe–12%Al–2% Co and 60%Fe–Cr(<40%)–Al(<10%)–Co(<1%)–Si(<5%) alloys.

Precipitation-hardened alloys developed on an Ni–Co basis possess greater high-temperature strength and damping properties than the alloys based on iron. It is quite essential that high damping capacity is retained in Ni–Co alloys up to the high temperatures at which turbine blades are employed. Nivco-10 alloy, suggested by Cochardt (1959) has become commercial. It possesses a record damping capacity and has the composition 73.56% Co, 22.5% Ni, 1.1% Zr, 1.8% Ti, 0.22% Al, 0.3% Fe, 0.15% Si, 0.02% C. High mechanical and damping properties are also characteristic of alloys of the following composition: 63.4% Co, 30.5% Ni, 1.5% Al, 3.4% Ti, 0.19% Fe; plus a number of admixtures containing 9.65–10% Mo, 0.1–0.3% Ni, 0.5–0.9% W and 0.5–0.7% Si, in which the balance is cobalt.

3. *Manganese–Copper Alloys*

The alloys containing 50–70% Mn have the best combination of damping and mechanical properties after the optimal heat-treatment regime (quenching and then tempering at 400–500 °C). After quenching, the alloys of this composition consist of a high-temperature γ solid solution with a face-centered-cubic (fcc) lattice (γ_c phase). The tempering results in a sudden increase of damping ($\psi \approx 40\%$ at a stress $\sigma = 0.1\sigma_{0.2}$) and also in a strength increase ($\sigma_f = 450$–500 MPa), while preserving a satisfactory level of plasticity ($\varepsilon \sim 20\%$). The increase of damping upon tempering in the alloys under consideration is connected with the process of thermoelastic transformation when the fcc lattice is converted into the tetragonal phase ($\gamma_c \rightarrow \gamma_t$ transformation). The martensite γ_t phase formed provides high damping in the alloys under low as well as high amplitudes of vibration.

The structure of the γ_t phase consists of alternating macrotwins inside which there are easy-moving microtwins. Under the influence of external stresses, nuclei of new microtwins appear and they begin to grow. The loss of energy of mechanical vibrations on nucleation and redistribution of microtwin volume determines the high damping properties of these alloys.

The major disadvantage of high-damping Mn–Cu alloys is their temperature and time instability. Temperature instability is accounted for by the process of reverse martensite change upon heating ($\gamma_t \rightarrow \gamma_c$). The highest temperature at which high damping is provided is weakly dependent on the alloy composition and usually does not exceed 100 °C. The negative aging effect can be weakened by thorough homogenization of the alloys and reducing the tempering temperature.

To increase the mechanical and anticorrosive

properties of high-damping Mn–Cu alloys, additional alloying is carried out, with elements such as Al (2.5–6%), Ni (<3.5%), Pb (≤1%), Fe (<5%) and Sn (<2%).

4. Magnesium-Based Alloys

Under cyclic loading in Mg and Mg-based alloys, the energy of mechanical vibrations dissipates rapidly, due to the process of direct and reverse elastic twinning. At low amplitudes of vibration, the background internal friction in Mg and Mg-based alloys, as contrasted with Mn–Cu alloys, is small, because they possess an equilibrium structure and elastic twinning cannot take place. The main contribution to damping in this case is provided by microplastic shear deformation and the relaxation of thermally activated twin boundary displacement. Above some critical stress ($\sigma < 1.0$ MPa), the mechanism of elastic twinning emerges and the material attains a higher-damping state. However, under high stresses due to the breakdown of coherence at the twin–matrix boundary, the contribution of elastic twinning to damping decreases.

Pure Mg, especially when cast, possesses very high damping ($\psi = 60\%$ at $\sigma = 0.1\sigma_{0.2}$), but its low mechanical qualities make it unsuitable as an engineering material. However, Mg-based alloys with the addition of 0.4–0.6% Zr have wide application as high-damping materials. The tensile strength of the Mg–0.6%Zr alloy is twice that of pure Mg, and its specific strength calculated per unit weight approaches the strength of steel. These alloys are widely applied in aircraft and spacecraft engineering. The positive Zr effect on the strength of Mg is said to be connected with reduction in grain size. This effect is especially evident in deformed alloys. There are also high-damping Mg alloys, containing Mn and Si. For example the Mg–0.75%Si alloy has $\psi \approx 50\%$, and the Mg–0.9%Mn alloy has $\psi \approx 25\%$.

Mg-based alloys possess high temporal stability, and the thermal stability of damping is also much higher than that of Mn–Cu alloys.

A case alloy of high mechanical, casting and damping properties has the following composition: 0.4–0.7% Zr, 0.1–0.3% Zn, 0.4–0.75% Cd, 0.01% Ni, 0.10% Cu, 0.08% Fe, 0.20% Si, 0.10% Ca.

5. Alloys with a Thermoelastic Martensite Transformation

Thermoelastic martensite transformation is observed in many alloys, but the alloys in which martensite transformation occurs at normal temperatures and for which the mechanical and damping properties make application in technology possible are particularly interesting. The alloys that meet these requirements are mainly from the Ni–Ti and Cu–Al–Ni systems, with the following compositions: Ni–(45–50%)Ti and Cu–(10–14%)Al–(2–4%)Ni. The former have the name Nitinol.

Damping in the alloys under discussion has similar characteristics and at room temperature ψ is ~30%. At temperatures above that of martensite transformation, damping is very small. At the martensite transformation the damping suddenly increases and reaches its maximum. The maximum at the transition point can be accounted for by energy dissipation caused by the motion of coherent interphase boundaries effected by the elastic deformation in the transformation process. For the Cu–Al–Ni alloy, maximum damping is observed with 50% martensite when both the transformation rate and the extent of easy-moving interphase boundaries are large.

The sudden increase of damping at the level of thermoelastic martensite transformation is also connected with the "lattice softening" effect of the high-temperature phase. This effect in the pre-martensite temperature range of the initial phase causes an instability. The transition into the final martensite structure involves some metastable states, so that a pre-martensite phase transition of the second type may be said to take place.

On completion of the martensite transformation, the high damping is provided by the peculiarities of the morphology of the martensite phases. In some alloys (Ni–Ti) the source of losses can be the easy-moving interphase boundary surfaces or twin boundaries (packing defects) inside a martensite plate, whereas in others (Cu–Al–Ni, Cu–Zn–Al) it is the easy-moving boundaries which divide the regions of various martensite modifications, differentiated by the sequences of atom packing.

Apart from high damping capacity, it is also characteristic of the alloys with thermoelastic martensite to exhibit superelasticity, ferroelasticity and shape-memory effect.

6. Alloys of Very Heterogeneous Structure

Primarily, gray cast irons form this group of alloys, in which various mechanisms lead to high damping capacity. First, there is the plastic deformation of graphite inclusions as the "soft" component of structure. Second, there is the microplastic deformation of the cast iron metallic base in the regions where high stresses are generated by graphite inclusions. Third, magnetomechanic hysteresis also contributes to the damping of gray cast irons. All these mechanisms bring about a great dependence of damping on the form, distribution and quantity of graphite inclusions. Thus, in the cast irons with spheroidal graphite, damping is always less than in those with flake, lamellar or crab forms of graphite. The second and third mechanisms cause the damping of cast iron to depend on the type of metallic base: in cast irons with a hard metallic base (pearlite) the damping, other conditions being equal, is lower than in cast irons with a ferrite or austenite base. To illustrate this, the damping properties of three types of cast iron for the same level of amplitude deformation ($\sigma = 34.5$ MPa) are:

	ψ
modified austenitic cast iron with flake graphite	19.3
ferritic cast iron with spheroidal graphite	2.8
pearlitic cast iron	1.4

In leaded bronzes (up to 20% Pb) an almost linear correlation between the Pb content and the damping is observed. Lead grains function as a "soft" component, influenced by microplastic deformation in the region of macroelastic deformation.

In Al–Zn alloys with ~40% Zn, high damping (up to 10%) is connected with microplastic deformation of the β solid solution, rich in Zn; the α solid solution, rich in Al, functions as an elastic frame.

7. Other High-Damping Materials

Ni-based alloys with small (a few per cent) additions of Th have interesting damping qualities. From room temperature to 500 °C their damping capacity increases more than three times. Composite materials can also be long-term high-damping materials. Their damping capacity is greatly dependent on energy dissipation at the matrix–filler boundary. Multilayer materials produced by cold welding may also be noted. The damping capacity of some materials is appreciably influenced by electrolytic coating (Ni, Fe, Cr, Cu and others). Polymers (viscoelastic materials) have a damping capacity considerably higher than that of metallic materials.

Bibliography

Cochardt A W 1959 *Magnetic Properties of Metals and Alloys*. American Society for Metals, Cleveland, Ohio, pp. 251–79

Favstov Ya K, Shulga Ya N, Rakhshtadt A G 1980 *Physical Metallurgy of High Damping Alloys*. Metallurgiya, Moscow

Kekalo I B 1973 Magnetoelastic phenomena. *Physical Metallurgy and Heat Treatment*, Vol. 7. VINITI Information Publications, Moscow, pp. 5–88

I. B. Kekalo

High-Energy-Rate Forming

High-energy-rate forming (HERF) is concerned with deformation processes carried out with a rapid release of energy. Thus, high-energy-rate forming processes involve higher power than more conventional deformation processes, and this is achieved with a higher than normal workpiece velocity. As a general rule, high-energy-rate forming processes are concerned with high-velocity deformation processing at velocities in excess of $10\,\mathrm{m\,s^{-1}}$.

The high deformation velocity may be produced either with chemical or electrical energy. A common class of high-energy-rate forming equipment (Dynapak) utilizes the expansion of a highly compressed gas, whereas another version (Petroforge) is powered by the chemical energy from the ignition of a fuel–air mixture. There is a well-developed technology utilizing the energy from the detonation of explosives (see *Explosive Forming*). Electrical energy is utilized when large capacitor banks are discharged rapidly or when a sudden electrical pulse generates a high magnetic field in a coil which in turn produces a force on the workpiece (see *Electromagnetic Forming*). Typical values produced by these processes are compared with normal deformation processes in Table 1.

1. Characteristics

The major advantage of high-energy-rate forming processes over conventional deformation processes is its economy. By using high velocities rather than large masses to achieve the required power levels, the machines tend to be smaller and more self-contained. Thus, the cost of a unit of energy is lower compared with that of conventional press equipment. The high deformation velocity inherent in high-energy-rate forming processes influences the material response and process conditions in several ways. The high deformation velocity increases the intrinsic flow resistance of the metal, especially under hot-working conditions. Because there is less time for heat loss, the high velocity produces a larger temperature rise due to adiabatic heating. The high deformation velocity also affects lubrication between the workpiece and the tooling. In general, the coefficient of friction decreases with increasing relative velocity between the mating surfaces, but whether better formability results depends on such factors as whether a lubricant film can be maintained and how the properties of the lubricant change at high pressure, high velocity and with increasing temperature.

Pneumatic–mechanical and Petroforge machines have been used chiefly in closed-die forging operations. Under carefully controlled conditions, materials that are normally difficult to work can be forged successfully, often with decreased draft allowances and with thinner attainable sections. High-energy-rate forming equipment is also used for powder compaction. Extrusion at high velocity is difficult owing to inertial separation. Explosive energy is used most typically for forming plates and thick sheet. Similar applications are achieved in thinner gauge sheet with electrical energy sources.

2. Material Response

The deformation imparted by high-energy-rate forming processes is due to stress-wave propagation and complex wave interactions. Deformation tends to

Table 1
A comparison of various deformation processes

Process	Details of equipment	Typical impact velocity ($m\ s^{-1}$)	Approximate process duration (ms)	Typical power (W)
Press forging	0.02 m stroke; 700 t	0.05	300	2×10^5
Hammer forging	0.02 m stroke	4	10	3×10^7
High-energy-rate forming machine (pneumatic–mechanical)	150 000 J; 0.02 m stroke	20	3	2×10^8
Electrospark	150 000 J	50	1	4×10^8
Explosive forming	100 g of explosive	200	1	6×10^9

be localized, with the areas of localization varying rapidly with time.

Variation of metal ductility with velocity is a subject of some controversy. However, there is considerable indication that ductility increases when velocities exceed about $100\ m\ s^{-1}$, depending on the specified metal. At still higher velocities the material ruptures. This critical impact velocity ranges from 30 to $250\ m\ s^{-1}$, depending on the metal.

3. *Pneumatic–Mechanical Machines*

The Dynapak, Hermes, CEFF and similar machines use the adiabatic expansion of highly compressed gas to supply energy. Subsequent to the blow, the gas is recompressed. Maximum available energy is about 500 kJ at a maximum velocity of $20\ m\ s^{-1}$. Once the controls have been adjusted, reproducibility is very good. However, because of the need to compress the gas the cycle time is relatively long (3–10 s).

This disadvantage has been overcome in the Petroforge machine. The ram is driven by the controlled chemical explosion of a fuel–air mixture. Since the piston–ram assembly is a scaled-up version of an internal combustion engine, the return of the ram takes place by the pressure of air compressed in the cylinder. The cycle time is therefore reduced to about 1 s and a separate hydraulic system is not needed to recompress the gas. Maximum energy capacity of the Petroforge is about 100 kJ, with an impact velocity of $20\ m\ s^{-1}$.

Although forging with pneumatic–mechanical machines can result in precision-sized parts from a single blow, forging at high velocity requires considerable skill and experience. Stress-wave propagation must be considered. Thus, the best results are obtained with symmetrical parts. The high impact velocities and the possibility of undesirable rebounding can lead to higher die wear than is encountered with more conventional processes. There are many examples in which high-velocity forging enabled the working of difficult-to-form metals or the manufacture of components impossible to produce conventionally. Unfortunately, there is little fundamental information about the high-energy-rate forming processes which can lead to rational design and understanding.

4. *Electrohydraulic Deformation*

In electrohydraulic metalworking processes the energy stored in large capacitor banks is rapidly released either by discharging across a gap between electrodes or exploding and vaporizing a wire across the electrode gap. This spark discharge takes place in a water transfer medium, producing a shock wave and a pressurized gas bubble. This is similar to the events in explosive forming, but the electrohydraulic process offers opportunity for better control with reflectors to direct and control the pressure pulse.

Energy stored in capacitors does not exceed 150 kJ in commercial units. Thus, the energy available is not as great as in explosive forming. Consequently, electrohydraulic forming is used for forming small parts from flat sheets and tube.

See also: Deformation Processing; Metals Processing and Fabrication: An Overview

Bibliography

Blazynski T Z (ed.) 1983 *Explosive Welding, Forming and Compaction*. Applied Science, London

Brunno E J (ed.) 1968 *High Velocity Forming of Metals*. American Society of Tool and Manufacturing Engineers, Dearborn, Michigan

Davies R, Austin E R 1970 *Developments in High Speed Metal Forming*. Industrial Press, New York

Orava R N, Whittman R H 1978 High-energy-rate deformation processing and its TMP applications. In: Burke J P, Weiss V (eds.) 1978 *Advances in Deformation Processing*. Plenum, New York, pp. 485–533

G. E. Dieter

High-Field Superconductors

Superconductivity is displayed by many materials in bulk form in the presence of magnetic fields up to a present maximum "upper critical field" of approximately 60 T. Superconductivity in high magnetic fields is of both fundamental and technical interest. Important basic questions concern the nature of high-field electronic interactions in the complex alloys and compounds which form the outstanding high-field superconductors. Technical interest is associated primarily with the utilization of superconductors for power transmission and in the winding of lossless (zero-resistance) solenoids to produce intense magnetic fields for research, electrical power apparatus (generators and motors), energy storage, magnetic levitation and plasma confinement.

1. Type II Theory of Upper Critical Fields

In 1950 Ginzburg extended the Landau theory of second-order phase transitions (see *Landau Theory of Second-Order Transformations*) to the case of superconductors in the presence of applied magnetic fields near the superconducting transition temperature T_c. The theory treats the superconducting transition as an order–disorder transition characterized by a complex order parameter ψ which has many of the characteristics of a quantum-mechanical wave function. The Ginzburg–Landau equations form the cornerstone of the current theory of high-field superconductivity and provide the basis for the modern distinction between type I and type II superconductors (Saint-James et al. 1969) (see *Superconducting Materials: Theory*).

In type I superconductors the Ginzburg–Landau (GL) order parameter ψ is relatively "inflexible": only relatively slow spatial variations of ψ are possible. As shown in Fig. 1a, the GL "coherence distance" ξ_{GL} over which the order parameter ψ may vary from 0 (nonsuperconducting or "normal") to 1 (completely superconducting) is greater than $\sqrt{2}\,\lambda$ where $\lambda \approx 10^{-7}$ m is the magnetic field penetration depth. Thus the GL parameter $\kappa = \lambda/\xi_{GL} < 1/\sqrt{2}$. This relative inflexibility of ψ results in an energetically unfavorable sacrifice of the superconductive condensation energy at a super–normal interface because, as shown in Fig. 1a, ψ is low over a wide boundary region. The resulting "positive surface energy" dictates, as shown in Fig. 1b, that a needle-shaped type I specimen in a longitudinal applied magnetic field (such that the demagnetizing coefficient $N = 0$) exhibit a "perfectly" diamagnetic "Meissner-state" magnetization M (M is here defined such that $B \equiv \mu_0(H + M)$) up to a "critical field" H_c where the kinetic energy of the screening currents overbalances the superconductive condensation energy within the specimen. Thus at H_c a transition to the normal state occurs with a first-order discontinuous collapse of M to the nearly zero value of the normal state. Figure 1c shows the corresponding critical-field curve $H_c(T)$ for a zero-N type I superconductor.

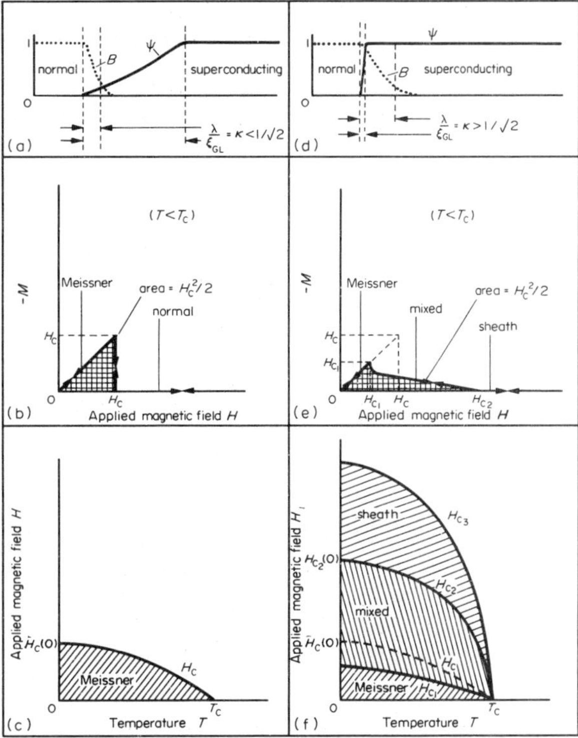

Figure 1
Characteristics of (a–c) type I and (d–f) type II superconductors

For a type II superconductor, on the other hand, the GL order parameter ψ is relatively flexible, as shown in Fig. 1d. The coherence distance ξ_{GL} is less than $\sqrt{2}\lambda$, the GL parameter $\kappa > 1/\sqrt{2}$, and the surface energy is negative. As indicated in the magnetization curve of Fig. 1e, magnetic flux penetration is now energetically favored in a zero-N specimen above the "lower critical field" H_{c1}. Flux penetration is in the form of a regular lattice of flux-enclosing supercurrent vortices (the "mixed state"), each vortex containing one flux quantum $\phi_0 = h/2e = 2.07 \times 10^{-15}$ T m^2. As the applied field H is increased, the vortices move closer together until at the "upper critical field" H_{c2} the near-normal vortex cores overlap, flux penetration is complete, and the specimen undergoes a second-order transition (no discontinuity in M) to a state that is entirely normal except for a very thin ($\approx\xi_{GL}$) surface layer which usually has negligible effects on the magnetization above H_{c2}. The surface layer is finally quenched at the "sheath critical field" H_{c3}. Figure 1f shows the corresponding critical-field curves for type II superconductors. It is important for the thermodynamic treatment and for considerations of the paramagnetic limitation (see below) to emphasize that the conventionally presented type II magnetization curve of Fig. 1e is not a realistic representation for superconductors with $\kappa > 5$–10, where high-field, mixed-state magnetization is usually paramagnetic, merging smoothly with the paramagnetic magnetization of the normal state at H_{c_2}.

From thermodynamics, it can be shown that the zero-H Gibbs superconductive condensation energy is equal to the area between the normal- and superconducting-state magnetization curves up to the transition field, H_c for type I and H_{c_2} for type II. These areas are shown as crosshatched in Fig. 1b and 1e where, in the low-H cases depicted, $M_n(H) \approx 0$. For both type I and type II superconductors the areas are just $H_c^2/2$, but for the type II superconductor the so-called "thermodynamic critical field" H_c (shown as the dashed line in Fig. 1f) is not directly observable as a transition field. It is, however, measurable in the sense that it may be derived from the area between reversible normal- and superconducting-state magnetization curves, or by means of double integration of specific heat vs temperature curves, provided that these are independent of the H–T history of the specimen.

According to Abrikosov's classic analysis of the GL equations (see *Superconducting Materials: Theory*),

$$H_{c_2}(T) = \phi_0[2\pi\mu_0\xi_{GL}^2(T)]^{-1} = \sqrt{(2)}\kappa H_c(T)$$

The first equality is consistent with the simple picture of near-normal vortex cores of radius $\approx\xi_{GL}$ overlapping at H_{c_2} so that the average macroscopic flux density at the upper critical field is just

$$\mu_0 H_{c_2} \approx n\phi_0/n\{\pi[\xi_{GL}(T)]^2\}$$

where n is the number of flux lines within the specimen. The second equality indicates that if $\kappa > 1/\sqrt{2}$ (Fig. 1d) then $H_{c_2} > H_c$ (Fig. 1f) and the specimen is type II. If $\kappa < 1/\sqrt{2}$ (Fig. 1a) then $H_{c_2} < H_c$, the specimen is type I, and H_{c_2} becomes the maximum field at which supercooling can persist in decreasing H.

In the original GL theory, the vital parameter κ was given only in terms of the penetration depth λ and the thermodynamic critical field H_c. An important advance was made by Gor'kov, who translated the microscopic Bardeen–Cooper–Schrieffer (BCS) theory of "weak-coupling" isotropic superconductors into a Green's function language suitable for the description of the spatial variation of the order parameter in the temperature region just below T_c. Subject to these weak-coupling, isotropic and close-to-T_c limitations, Gor'kov derived an expression for κ in terms of normal-state electronic parameters, which can be presented as composed of two parts: $\kappa = \kappa_0 + \kappa_l$, where κ_0 is a "clean" or "intrinsic" contribution which exists even in the absence of electronic scattering, and κ_l is a "dirty" or "extrinsic" contribution which is inversely proportional to the electron mean free path $l = v_f(\tau_{tr})$ (v_f is the Fermi velocity and τ_{tr} is the mean transport scattering time). Some pure elements, such as niobium and vanadium, and some high-purity, well-ordered compounds are "clean" type II superconductors with $\kappa \approx \kappa_0 > 1/\sqrt{2}$. Very high-field superconductors are usually high-electrical-resistivity ($\approx 10^{-6}\,\Omega$ m) "dirty" materials with $\kappa \approx \kappa_l \approx 20$–120. The Ginzburg–Landau–Abrikosov–Gor'kov (GLAG) theory serves to predictH_{c_2} in terms of normal-state electronic parameters, but is limited to isotropic, "weak-coupling" superconductors in the temperature region just below T_c. (For extensive lists of formulae connecting superconducting- and normal-state parameters, see Hake (1967) and Orlando et al. (1979).)

Werthamer, Helfand and Hohenberg and independently Maki (WHHM) extended the GLAG theory of H_{c_2} to all temperatures by solving the linearized Gor'kov equations in the "dirty limit" where the electron mean free path l is much shorter than the zero-temperature BCS coherence length ξ_0, a coherence distance which would exist in the absence of electronic scattering. They also included the important electron–spin and spin–orbit coupling effects discussed below (Fetter and Hohenberg 1969).

According to the BCS theory as extended to "dirty" ($\xi_0/l \gg 1$) materials by Anderson, superconductivity is associated with a pairing of electrons of opposite spin in time-reversed states. Clogston and Chandrasekhar recognized that as H is increased to values beyond a few tesla in high-κ superconductors, such spin-paired electrons must eventually desert the superconducting state to take energetic advantage of spin alignment along the applied field. Their ideas implied the possibility of a first-order transition to the normal state when the field depression of the normal-state free energy was of the order of the superconductive condensation energy, and an associated "paramagnetic limitation" on H_{c_2}. The latter was indeed indicated by early H_{c_2} measurements on high-κ type II superconductors, but first-order transitions were not observed and the paramagnetic limitation was often less than implied by the early theory. Such mitigation of the paramagnetic limitation was evidently due at least in part to spin–orbit coupling induced electronic spin-flip scattering.

2. *Comparison of Experiment and Theory*

The rather stringent conditions for the applicability of the WHHM theory are perhaps best met in high-electrical-resistivity, random-solid-solution, transition metal alloys. Measured upper-critical-field curves for such polycrystalline and amorphous alloys are usually in fair accord with the theory for reasonable (but, unfortunately, not directly measured) values of the spin–orbit parameter λ_{so} (proportional to the spin-flip scattering frequency τ_{so}^{-1}). However, the effective λ_{so} required to obtain a fit of the data to the theoretical $H_{c2}(T)$ often decreases somewhat with decrease of T (higher H_{c2}). This may imply that various real-metal effects must be included for proper treatment of the paramagnetic limitation. Near T_c, where the latter is unimportant, the WHHM theory yields rather good agreement with experiment, not only for random transition metal alloys, but also for binary and even ternary compounds.

Apparent difficulties in the paramagnetic-limitation aspect of the WHHM theory have recently come to light in $H_{c2}(T)$ studies of the A15 compound Nb_3Sn. Measured upper-critical-field curves suggest the absence of paramagnetic limitation, and the consequent λ_{so} values required to fit the data to the WHHM theory seem to imply an unreasonably high spin-flip scattering frequency τ_{so}^{-1}. Similar discrepancies have been noted in measurements of the spectacularly high ($< \approx 60\,T$) upper critical fields of ternary molybdenum sulfides. Orlando et al. (1979) have suggested that neglect of proper renormalization of normal-state parameters in the WHHM theory may result in underestimation of $H_{c2}(T)$ for any finite λ_{so}.

Electron localization and interaction effects, omitted from WHHM theory, have apparently been detected as very weak but striking anomalies in the low temperature resistivity and magnetoresistivity of highly disordered bulk alloys (Naugle 1984). Such effects might also influence H_{c2} (Fukuyama 1984) but definitive experimental evidence on bulk materials is currently lacking (Karkut and Hake 1983).

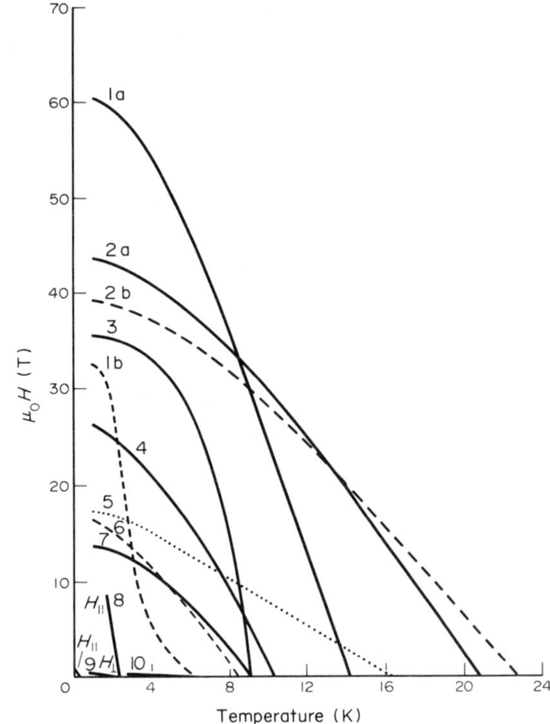

Figure 2
Upper critical fields of ten prominent classes of high-field superconductors. (1) Ternary molybdenum sulfides: (a) $PdGd_{0.2}Mo_6S_8$; (b) $Sn_{1.2(1-0.8)}Eu_{0.8}Mo_{6.35}S_8$. (2) A15 compounds: (a) $Nb_{79}(Al_{73}Ge_{27})_{21}$; (b) Nb_3Ge. (3) "Granular": Nb_3Ge (sputtered thick film). (4) C15 compound: $V_2Hf_{0.4}Zr_{0.6}$. (5) B1 compound: NbN. (6) Amorphous alloy: $Mo_{0.6}Ru_{0.4}$. (7) Bcc alloy: $Ti_{56}Nb_{44}$. (8) Layer: $TaS_{1.6}Se_{0.4}$ (pyridine)$_{0.5}$. (9) Chain polymer: $(SN)_x$. (10) Element: pure vanadium

3. *Upper Critical Fields of Some Prominent High-Field Superconductors*

Figure 2 shows upper-critical-field curves of ten prominent classes of high-field superconductors. The extrapolated $H_{c2}(T=0)$ values range from $\approx 0.3\,T$ for pure vanadium to the current record high of $\approx 60\,T$ for the pseudo-ternary molybdenum sulfide $PbGd_{0.2}Mo_6S_8$. Listings of H_{c2} values for a wide variety of high-field superconductors have been published (Roberts 1976, 1978).

The influence of ionic magnetic moments is apparently responsible for the high-T positive curvature of $H_{c2}(T)$ for $Sn_{1.2(1-0.8)}Eu_{0.8}Mo_{6.35}S_8$. The behavior has been ascribed (Decroux and Fischer 1982) to the Jaccarino–Peter effect in which a negative exchange field due to field-aligned Eu moments acts on electron spins to alleviate in part the paramagnetic limitation. The related interaction of high-field superconductivity with ionic ferro- and antiferromagnetic ordering in ternary superconductors is of intense current interest (Maple 1983).

The technologically important high-field behavior of the A15 compounds and bcc alloy representatives is especially noteworthy. It would seem that further enhancement of $H_{c2}(T)$ in such materials might be possible by increasing the atomic disorder and the spin–orbit interaction. The ultimate in metallic disorder may be achieved in amorphous alloys, and prospects for the practical use of Ti-, Zr- and Hf-based amorphous materials are encouraging.

The markedly anisotropic and very high initial critical-field slopes in the chain polymer $(SN)_x$ and in the layered compound $TaS_{1.6}Se_{0.4}$ (intercalated with pyridine) are intriguing, and the possibility of observing high-field triplet (electron-spin-parallel) superconductivity in materials such as the latter has been suggested by Klemm and Scharnberg (1981). Even higher initial critical-field slopes (e.g., $\approx 40\,TK^{-1}$ in UBe_{13}) have been observed in certain low T_c ($T_c < 1K$) "heavy fermion" superconductors (Stewart 1984). Here again the possibility of triplet pairing has been raised.

In general it would appear that high-field superconductivity, such as that portrayed in Fig. 2, can be roughly understood on the basis of type II theory. However, satisfactory understanding and control of the details of H_{c2} magnitude and temperature dependence has yet to be achieved. Technical utilization of present and future high-field superconductors depends, of course, not only on high H_{c2} but also on artful metallurgical processing so as to enhance their stability, critical currents and tensile strengths.

See also: Superconducting Materials: Metallurgy; Bulk Superconductors: Physical Properties; Superconductivity in Alloys, Compounds, Mixtures and Organic materials; Amorphous Superconductors; A15 Compound Superconductors; Superconductivity: Large-Scale Applications; Superconducting Materials: An Overview

Acknowledgement

The author acknowledges financial support of his work under National Science Foundation Grant DMR 80–24365.

Bibliography

Decroux M, Fisher Ø 1982 Critical fields of ternary molybdenum chalcogenides. In: Maple M B, Fischer Ø (eds.) 1982 *Superconductivity in Ternary Compounds II*. Springer, Berlin, pp. 57–98

Fetter A L, Hohenberg P C 1969 Theory of type-II superconductors. In: Parks R D (ed.) 1969 *Superconductivity*, Vol. 2. Dekker, New York, Chap. 14, pp. 817–923

Fukuyama H 1984 Localization and superconductivity. *Physica* 126B: 306–13

Hake R R 1967 Paramagnetic superconductivity in extreme type-II superconductors. *Phys. Rev.* 158: 356–76

Karkut H G, Hake R R 1983 Upper critical fields and superconducting transition temperatures of some zirconium-base amorphous transition-metal alloys. *Phys. Rev.* B28: 1396–1418

Klemm R A, Scharnberg K 1981 Possible triplet superconductivity in thin films and layered compounds in a strong parallel magnetic field. *Phys. Rev. B* 24: 6361–82

Maple M B 1983 Experiments on magnetically-ordered superconductors. *J. Magn. Magn. Mater.* 31-34: 479–83

Naugle D G 1984 Electron transport in amorphous metals. *J. Phys. Chem. Solids* 45: 367–88

Orlando T P, McNiff E J, Foner S, Beasley M R 1979 Critical fields, Pauli paramagnetic limiting, and material parameters of Nb_3Sn and V_3Si. *Phys. Rev. B* 19: 4545–61

Roberts B W 1976 Survey of superconductive materials and critical evaluation of selected properties. *J. Phys. Chem. Ref. Data* 5: 581–821

Roberts B W 1978 *Properties of Selected Superconductive Materials*, NBS Technical Note 983. National Bureau of Standards, Washington, DC

Saint-James D, Sarma G, Thomas E J 1969 *Type II Superconductivity*. Pergamon, Oxford

Stewart G R 1984 Heavy-fermion systems. *Rev. Mod. Phys.* 56: 755–87

R. R. Hake

High-Frequency Welding

High-frequency welding is a high-speed resistance welding technique (see *Resistance Welding*) used for both continuous and finite-length welding. Frequencies normally used are approximately 400 kHz, with some special welding operations being performed at 10 kHz. Typical welding operations include the welding of longitudinal-seam pipes and tubes of all kinds of metals, structural shapes, fins helically welded to tubes, longitudinal fins welded to tubes, and helically welded pipes. Commercial welding speeds run from 25 to 500 ft min^{-1} (7.5–150 m min^{-1}). In all cases, the weld is produced without adding metal, the faying surfaces being heated and then forged together.

The electric current is produced in the workpiece by means of contacts or, in the case of small tubes, sometimes by induction. High-frequency current is used because it has three unusual characteristics. The first is the skin effect, by which currents at a frequency of about 400 kHz only flow on the surface to a depth of $\frac{1}{32}$ in. (0.8 mm) in hot steel. The second is the proximity effect, which makes possible the accurate control and positioning of the flow of current on a surface by the position of the adjacent return current. The third is the fact that very high currents may be applied to a workpiece via very small, low-pressure contacts as compared with those required for low frequencies (~60 Hz). For steel, no scale removal is required.

High-frequency welding is most widely used for the contact and induction welding of pipes and tubes, in which the longitudinal seams of tubes with outer diameters from 0.5–52 in. (1.3–132 cm) and wall thicknesses from 0.005 to 1.0 in. (0.01–2.5 cm) are continuously welded. The basic system is shown in Fig. 1, which depicts a tube being formed from continuous strip. High-frequency current is applied to the edges of the tube just upstream from the weld point by means of small, low-pressure contacts or by induction. The high-frequency current flows on the faces of the faying edges, creating sufficient heat to permit a forge weld to be produced by the squeeze rolls. Most metals can be welded at high speed by this technique in the open air. A few reactive metals require an argon atmosphere.

High-frequency welding is also widely used in the fabrication of I, T and H structural beams from various steels or, in some cases, exotic metals. Here, too, the high-frequency current is established by means of small contacts on the web and flange as

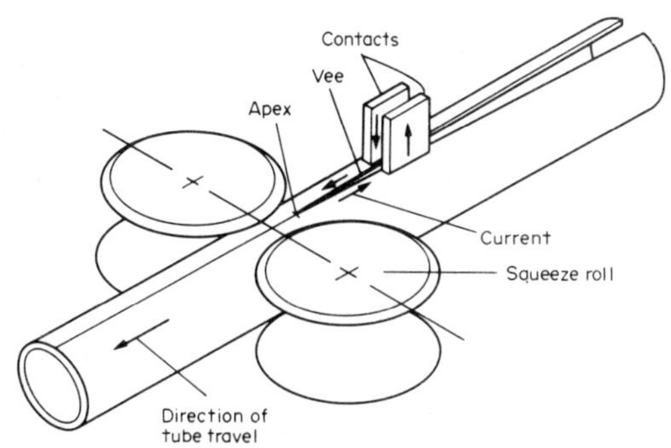

Figure 1
Longitudinal tube welding with contacts

shown in Fig. 2. The current flows in a narrow band on the face of the flange and on the edge of the web.

Another common use of high-frequency welding is continuous helical fin-to-tube welding at approximately 150 ft min^{-1} (45 m min^{-1}). The fin welded to the tube in this way (Fig. 3) acts as an extended surface for heat-exchange purposes.

Very large diameter pipe can be made helically with diameters up to 110 in. (280 cm) using high-frequency welding with the contact system as shown

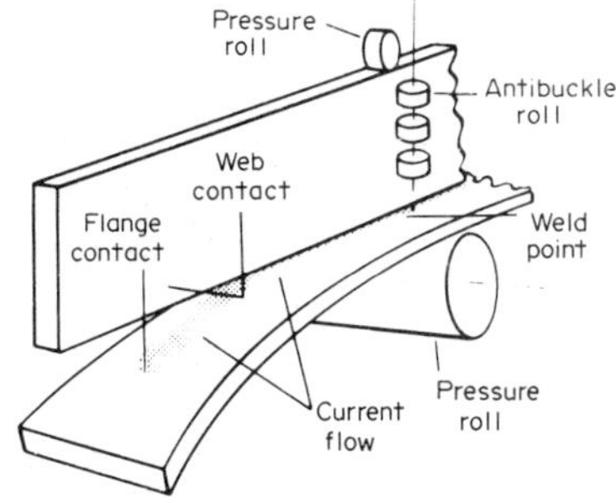

Figure 2
High-frequency current flow in longitudinal structural T welding

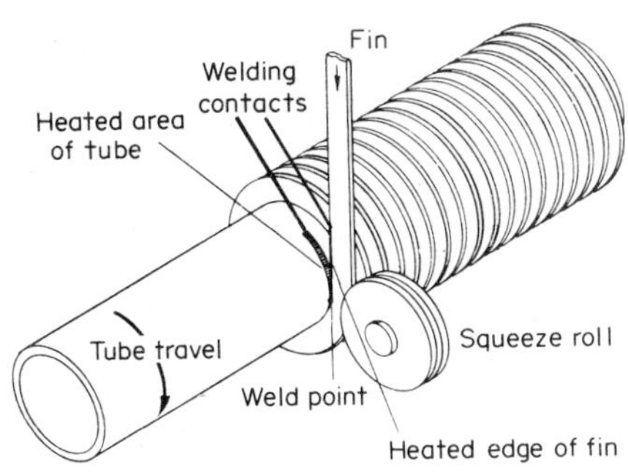

Figure 3
Spiral fin-to-tube welding

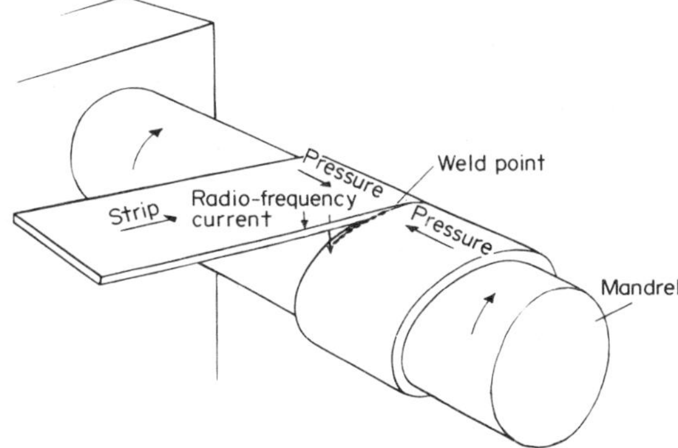

Figure 4
Helical tube welding

in Fig. 4. This has the advantage that one strip width can make several diameters.

Longitudinal fins are welded to boiler tubes for the fabrication of water-boiler walls as shown in Fig. 5. Both side fins are welded simultaneously.

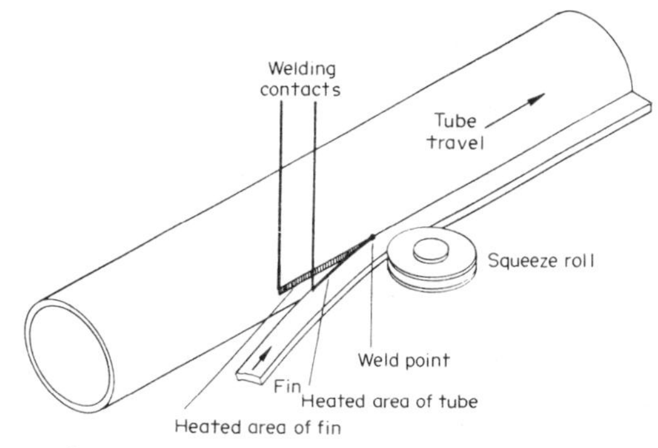

Figure 5
Longitudinal fin-to-tube welding

Roller-bearing cages are welded by use of a 10 kHz current penetration variation of high-frequency welding. In this case, the whole length of the weld is heated all at once and then forged together. Welding time is about 1 s.

High-frequency welding is the highest-speed commercial welding process in the world, and with some 1000 installations in industry produces about 20 million feet (6 million meters) of weld per day in various metals.

See also: Welding: An Overview

Bibliography

Welding Handbook, 7th edn., Vol. 3, 1980. High-frequency welding. American Welding Society, Miami, Florida, pp. 145–67

W. C. Rudd

High-Modulus Fibers

Polymeric synthetic fibers with moduli equivalent to those of glass, asbestos or carbon fibers are a comparatively recent development. The first such fibers to be prepared were based on rigid-chain polyamides. More recently it has been shown that flexible chain polymers, in particular linear polyethylene, can also give high-modulus products including fibers when highly oriented.

Until the late 1960s, the largest moduli to be found among commercial fibers consisting of oriented organic high polymers were those for highly oriented regenerated cellulose and poly(ethylene terephthalate). These products, made for use in tire cord and some related industrial composites, had Young's moduli up to about 20 GPa. Outside the class of oriented organic high polymers, fibers of much higher modulus were commercially available. These included graphitized cellulose and polyacrylonitrile fibers, generally known as carbon fibers (which attained moduli up to about 400 GPa), steel fibers (moduli up to about 200 GPa) and glass fibers (moduli up to about 70 GPa).

The commercial situation changed with the introduction by Du Pont in 1970 of Fiber B, a polyamide containing only *p*-phenylene rings between amide groups. At that stage the product may well have been based upon poly(*p*-benzamide) (**1**) but Du Pont then

$$\left[-NH-C_6H_4-\overset{O}{\overset{\|}{C}}-\right]_n$$

(**1**)

introduced PRD-49, and later replaced both products with Kevlar, Kevlar 29 and Kevlar 49, all of which are based upon the isomeric polyamide, poly(*p*-phenylene terephthalamide) (**2**).

$$\left[-NH-C_6H_4-NH-\overset{O}{\overset{\|}{C}}-C_6H_4-\overset{O}{\overset{\|}{C}}-\right]_n$$

(**2**)

These products, which fall into a generic class of fiber known as aramids (see *Aramids*), are characterized by unusually high tensile strength and modulus (Table 1). Although of lower Young's modulus than glass or steel, Kevlar 49 has a higher specific modulus than either owing to its substantially lower density of 1.45 g cm^{-3}.

The technology used for production of Kevlar 49 is unusual. Important points are:

(a) Use of a polymerization solvent system that permits attainment of a high molecular weight. A mixture of *N*-methylpyrrolidone and hexamethylphosphoramide was the solvent used initially (Kwolek 1972).

(b) Formation of a liquid-crystalline (nematic) spinning solution. The solvent used is concentrated sulfuric acid (approximately 100%), which yields solutions of remarkably low viscosity at high concentrations of polymer, and allows orientation of the nematic phase at quite low shear rates.

(c) Formation of filaments by a wet-spinning process (i.e., using a liquid precipitant, in this case water) with a short air gap between spinneret face and coagulating bath. The air gap permits attainment of a relatively high spin–stretch ratio and wind-up speed, and so promotes high shear-induced orientation in the yarn (Blades 1973).

(d) Subjection of the yarn to a brief annealing process at a temperature of about 500 °C. This process leads to morphological changes such that the molecular chains become still more highly oriented, giving rise to higher tensile strength and modulus of the product.

Table 1
Mechanical properties of Kevlar fibers

Fiber	Tensile strength (GPa)	Young's modulus (GPa)	Extensibility (%)
Kevlar 29	2.7	60	4
Kevlar 49	2.7	130	2.8

Yarns of high strength and modulus can be made with the omission of some of these conditions (e.g., using an isotropic solution or an immersed jet, or omitting the annealing process), but to attain the highest values it is important to use all of the conditions.

Morphologically, the fibers with the most extreme properties consist essentially of sheets of extended parallel polymer chains, which are associated within the sheets by hydrogen bonds (Dobb et al. 1977). The sheets are arranged radially within the fibers, with the polymer chain axes nearly parallel to the fiber axis. The sheets are not quite planar, being slightly pleated so that pleats recur at regular intervals along the direction parallel to the fiber axis.

The stress–strain relationship in the filaments on extension is almost Hookean, and failure under tension is accompanied by fibrillation of filaments. The strength in shear is superior to that of carbon fibers (see *Carbon Fibers*). Quite low compressive strains, however, lead to irrecoverable distortion due to formation of kink bands in the sheet structure (Dobb et al. 1981). A consequence of this behavior is that knots can sometimes be tied in yarns or filaments without exceeding the extension-to-break on the outside of the knot, because compressive distortion occurs on the inside.

Major uses include tire cords and ropes for the yarns of lower modulus, and reinforcement of plastics for the yarns of higher modulus. Problems in use include poor adhesion, overcome by using epoxy or isocyanate surface treatments; interfilament abrasion, overcome by applying a silicone or other lubricant; and poor photostability, overcome by avoiding exposure to light. The thermal stability of the physical properties is high.

The 1970s also saw the preparation of ultrahigh-modulus oriented fibers from flexible polymers. Here, there were two major approaches: the tensile drawing of melt-spun fibers, and the seeded crystallization of fine fibers from dilute solution under conditions of extensional flow. These two approaches have come together in the gel-spinning and high-temperature-drawing route. All these methods will now be discussed in turn.

The first successful preparation of high-modulus polyethylene fibers was reported by Capaccio and Ward (1973). It was shown that very high degrees of molecular alignment, and hence high modulus (~70 GPa), could be achieved by tensile drawing of melt-spun fibers provided that a very high draw ratio (~30) was imposed. Guidelines for the drawing process were established, including the choice of temperature and rate of draw, molecular weight and molecular-weight distribution and, for low-molecular-weight polymers, the initial thermal treatment (i.e., the initial morphology). It was proposed that drawing involves the extension of a molecular network in which molecular entanglements and crystalline regions can form the semi-permanent junction points in the structure. High draw ratios have also led to high moduli in two other polymers, polypropylene and polyoxymethylene, although the helical chain structure of these polymers gives a much lower intrinsic chain modulus than polyethylene, so that the results are less spectactular in absolute terms.

In a totally different approach, Zwijnenburg and Pennings (1976) described the preparation of high-modulus polyethylene fibers by withdrawing a seed crystal from a very dilute solution (~0.5%) in xylene at high temperatures. Here, the seed crystal was located between the inner and outer cylinder of a Couette apparatus, and the growing fiber was brought into contact with the surface of the inner rotating cylinder. At the highest solution temperatures (~120 °C), fibers with room-temperature moduli up to 100 GPa and tensile strengths of 4 GPa were obtained.

More recent work by Smith and Lemstra (1980) describes the preparation of high-modulus polyethylene fibers by first spinning a 2 wt% solution in decalin, and quenching the as-spun filament in cold water. Structural studies show that this gives a gel fiber—a network of polymer molecules with crystallites acting as cross-links. These gel fibers are subsequently drawn in a hot air oven at 120 °C to a very high draw ratio (~30) to give a fiber with a Young's modulus of 90 GPa and a tensile strength of 3 GPa. This approach is especially interesting because it suggests that there is a common theme in the production of high-modulus fibers from flexible polymers. This is the deformation of a molecular network that is of optimum structure to achieve a high degree of extension. The drawing condition is also critical if flow drawing, where only macroscopic deformation occurs without the concomitant molecular alignment, is to be avoided. In the dilute solution method, a high-molecular-weight polymer is mandatory and the molecular network is stretched at appropriate strain rates in an elongational flow field produced by a Couette or Poiseuille viscometer. In solid-phase deformation, lower molecular weights are desirable so that the number of molecular entanglements is not so great that the extensibility is limited.

There is no general agreement regarding the structure of these fibers. Studies of the solution-grown fibers have suggested these might possess a core of extended-chain crystals surrounded by a lamellar overgrowth of chain-folded crystals. For the solid-phase-produced material two very different models have been proposed. Barham and Arridge (1977) have suggested that these materials are akin to an aligned fiber-reinforced composite, where the fiber phase consists of needlelike extended-chain crystals and the matrix is the remaining chain-folded and amorphous material. Gibson et al. (1978) have invoked a variety of structural evidence to support an alternative model in which the crystalline blocks typical of low-draw-ratio material become increasingly linked with intercrystalline bridges as the draw ratio is increased. This model is a natural successor to the Peterlin model for a drawn fiber, with the crystalline bridges replacing taut tie molecules.

The high modulus of these polyethylene fibers implies a high degree of crystal continuity, and this is shown by the very low thermal coefficient of expansion, which is similar to that of the crystalline regions, and by an axial thermal conductivity which at 100 K is close to that of stainless steel. In the solid-phase-deformed fibers, the extension of the original molecular network is demonstrated by the complete recovery on heating above the melting point, and by very high shrinkage forces observed at temperatures below the melting point. Other properties which have received some attention are tensile strength and creep behavior. In solid-phase-deformed fibers, low-molecular-weight materials show permanent flow except at low stress levels, but fibers from homopolymers with an average molecular weight $\overline{M}_w >$ 200 000 and some copolymers can sustain stresses of ~0.1 GPa without any signs of nonrecoverable deformation.

See also: High-Modulus High-Strength Organic Fibers

Bibliography

Barham P J, Arridge R G C 1977 A fiber composite model of highly oriented polyethylene. *J. Polym. Sci., Polym. Phys. Ed.* 15: 1177–88

Black W B 1979 Stiff-chain aromatic polymer solutions, melts, and fibers. *Midl. Macromol. Monogr.* 6: 245–306

Blades H 1973 Dry-jet wet spinning process. US Patent No. 3,767,756 (30 June 1972)

Capaccio G, Ward I M 1973 Properties of ultrahigh-modulus linear polyethylene. *Nat., Phys. Sci.* 243: 143

Ciferri A, Ward I M (eds.) 1979 *Ultra High Modulus Polymers*. Applied Science, London

Dobb M G, Johnson D J, Saville B P 1977 Supramolecular structure of a high-modulus polyaromatic fiber (Kevlar 49). *J. Polym. Sci., Polym. Phys. Ed.* 15: 2201–11

Dobb M G, Johnson D J, Saville B P 1981 Compression behavior of Kevlar fiber. *Polymer* 22: 960–65

Gibson A G, Davies G R, Ward I M 1978 Dynamic mechanical behavior and longitudinal crystal thickness measurements on ultra-high modulus linear polyethylene: A quantitative model for the elastic modulus. *Polymer* 19: 683–93

Kowlek S L 1972 Optically anisotropic aromatic polyamide dopes. US Patent No. 3,671,542 (23 May 1969)

Smith P, Lemstra P J 1980 Ultra-high strength polyethylene filaments by solution spinning/drawing. *J. Mater. Sci.* 15: 505–14

Ward I M (ed.) 1975 *Structure and Properties of Oriented Polymers*. Applied Science, London

White J L, Fellers J F 1978 Macromolecular liquid crystals and their applications to high-modulus and tensile-strength fibers. *J. Appl. Polym. Sci., Appl. Polym. Symp.* 33: 137–73

Zwijnenburg A, Pennings A J 1976 Longitudinal growth of polymer crystals from flowing solutions—IV. The mechanical properties of fibrillar polyethylene crystals. *J. Polym. Sci., Polym. Lett. Ed.* 14: 339–46

I. M. Ward; J. E. McIntyre

High-Modulus High-Strength Organic Fibers

High-modulus high-strength organic fibers are members of a relatively new class of synthetic fibers characterized by their unusually high specific tensile modulus and high specific tensile strength. They are used in a wide variety of both flexible and rigid composites as well as in the more traditional end uses for synthetic fibers. All have one serious deficiency, that is, poor compressive strength, which precludes their use in composites for most primary structural applications and requires well-designed composite assemblies for many others. Nevertheless, where weight is a primary concern, these fibers are often chosen.

1. Historical Background

In the 1960s, fiber scientists at the Du Pont and Monsanto Companies independently discovered that fibers could be made with tensile moduli more than five times that of the stiffer conventional fibers such as poly(ethylene terephthalate). Moreover, these high-modulus fibers were much stronger than the strongest of the synthetic fibers, nylon and polyester tire yarns. Within several years both research groups reported specific tensile strengths four to five times higher than that of steel wire and tensile stiffness about three times that of E-glass fibers. Scientifically, the discoveries rival in importance the discoveries of Ziegler and Natta concerning the regulated polymerization of ethylene and propylene.

Each of the Du Pont and Monsanto discoveries were to some extent an outgrowth of long-term studies of making more thermally stable fibers by incorporating increasingly higher percentages of axially oriented aromatic ring structures into the backbones of wholly aromatic polymers.

More recently it was discovered in several industrial laboratories that high-modulus fibers could also be produced from the melt of certain polymers having a high content of axially oriented aromatic rings in the polymer backbone. For the most part such polymers have been aromatic polyesters. Seemingly, all the polymers that result in the high-modulus fibers from melts do so by way of liquid crystallinity in the melt. It is believed that commercialization of one or more such polymers is imminent.

Polyethylene fibers have been made which have the same level of stiffness and strength as the better aromatic high-modulus fibers. These developments, however, followed the discovery of the aromatic fibers by several years, and polyethylene fibers have not yet found commercial application.

Almost as exciting as the discovery of the greatly increased tensile stiffness and tensile strength in the fibers was the finding by Du Pont workers that many wholly aromatic polyamides consisting almost totally of axially oriented aromatic rings gave liquid crystalline solutions once the concentration reached a critical level.

The potential for both high modulus and liquid crystallinity in linear synthetic polymers had been predicted well ahead of discovery, the former by Mark in 1936 and the latter by Flory in 1956. Mark calculated an upper modulus limit of 2.44 GN m^{-2} (~2800 g per denier) for a straight-chain hydrocarbon, a value about 300 times higher than that actually attained more than 25 years later: Mark's theoretical value is now considered low. Polyethylene has now been made with moduli almost half the theoretical value.

Flory, using lattice-model phase diagrams for monodisperse rigid rods in solution, accurately showed the type of phase relationships observed for wholly para-oriented aromatic polyamides. He also showed that liquid-crystalline behavior was possible for the melt of a stiff-chain polymer.

2. *Chemical Structure and Solution Properties*

The unifying structural characteristic of the aromatic polymers yielding high-modulus fibers is that the valence bonds of the aromatic rings emanate oppositely from the ring structure (Fig. 1). Generally the aromatic structures are joined by some bond conventionally used for condensation polymerization (e.g., amide, hydrazide and ester). There are a number of examples, however, of polymers in which the condensation is accomplished by the formation of a heterocyclic ring along the backbone as is done in making polyimides, polybisbenzimidazoles, polybisbenzoxazoles, polybisbenzothiazoles and certain ladder polymers. Such heterocyclic polymers suffer in general from intractability: they are nonfusible and are exceedingly difficult to dissolve, making fabrication into useful shapes an arduous task, if possible at all. Only a minor portion of the rings can have other than axial alignment with respect to the backbone if high modulus is to be obtained.

The rodlike nature of para-oriented aromatic polymers, in addition to making the polymer difficult to dissolve, dramatically increases the viscosity of the solution for a given concentration over that of nonrigid structurally isomeric polymers such as poly(*m*-phenylene isophthalamide), the meta analog of poly(*p*-phenylene terephthalamide), the polymer used for Kevlar aramid (the lone commercial fiber).

Two very practical benefits are derived from polymer liquid-crystalline behavior, involving the process and the product. First, as the concentration is increased above the critical value, the viscosity falls rapidly until a minimum is reached at a substantially higher concentration than that at the critical value. Thus, it is possible to obtain a much higher concentration of polymer in the spinning solution, thereby decreasing the quantity of spinning solvent to be recycled or disposed of. For polymers of useful molecular weight, the concentration in isotropic solution is limited to 4–6 wt%, whereas concentrations of 15–20 wt% are common for liquid crystalline solutions.

The second consequence of liquid crystallinity of the rodlike polymer solutions is equally important. It permits the attainment of very high tensile strength with intermediately high tensile modulus simply by extrusion into a coagulant bath and winding up the solvent-free yarn—that is, no hot-draw step is required. Upon hot drawing, the modulus is doubled,

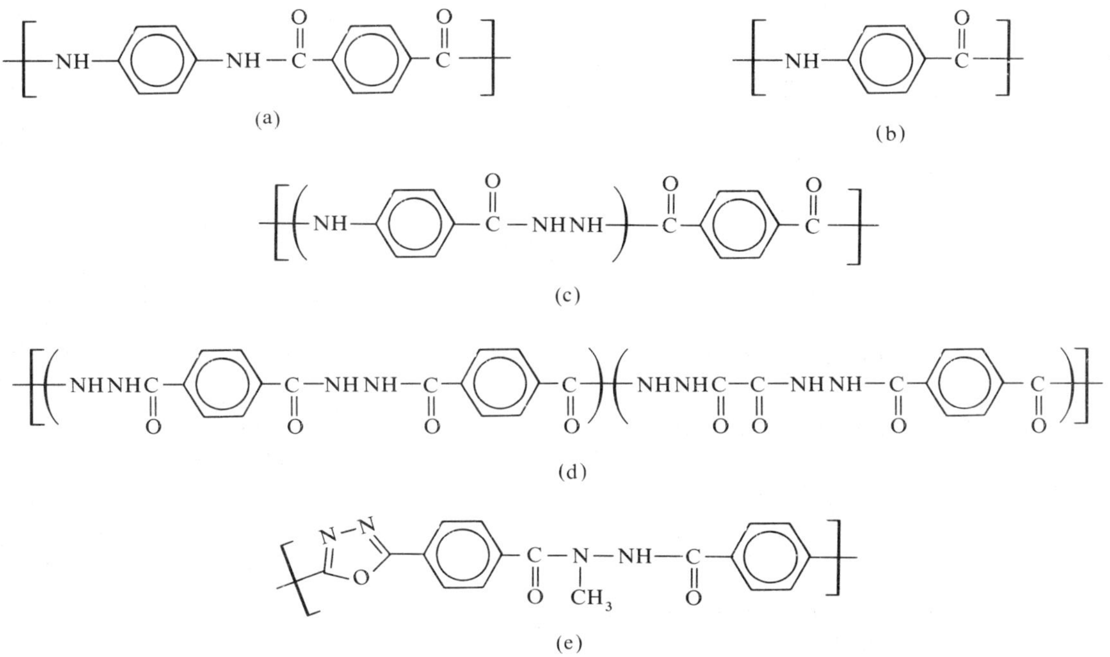

(a)

(b)

(c)

(d)

(e)

Figure 1
Examples of polymers used for high-modulus high strength fibers: (a) poly(*p*-phenylene terephthalamide) and (b) poly(*p*-benzamide), both used for aramid fibers; (c) X-500 polyamide hydrazide; (d) H-202 copolyhydrazide; (e) polyoxadiazole hydrazide

Table 1
Yarn tensile properties of several significant high-modulus high-strength aromatic fibers[a]

Tensile property	Poly(*p*-phenylene terephthalamide)[c]				Poly(*p*-benzamide)	X-500[g] Polyamide hydrazide	H-202[g] Copoly-hydrazide	Poly-oxadiazole hydrazide
	Kevlar 49[c]	Kevlar 29[d]	Fiber B[e]					
Cross-section basis								
tensile strength (GPa)[b]	2.8	2.8	2.8[f]	2.6[g]	1.8	2.4	2.7	2.5
modulus (GPa)	124	61	61[f]		114	106	57	54
Weight basis (specific)								
tensile strength (g den^{-1})[b]	22	22	22[f]	21[g]	14	19	21	21
modulus (g den^{-1})	980	480	480[f]		890	820	440	450
Elongation at break (%)	2.5	3–4	3.6[f]	4.0[g]	1.6	2.9	5.7	5.5
Denier per filament	1.5	1.5		1.5		2.7	1.9	0.3
Total yarn denier	400	1500		1000			1600	
Filament diameter (mm)	0.013	0.013		0.013				
Density (g cm^{-3})	1.44	1.44				1.47	1.47	1.36

a Fibers known to have been developed to the extent that tests in tires were made b Tensile strength values would have been about 20% higher if determined on single filaments c Kevlar 49 was tested as PRD-49 d Du Pont Bulletin A-95059, circa 1973 (undated) e Kevlar 29 and Fiber B are possibly identical except for the total denier of the yarns f Du Pont data g Monsanto data

but the strength remains the same. This increased modulus is obtained, however, at the expense of elongation-to-break and energy-to-break. Where only an intermediately high modulus is needed (e.g., in tire-yarn reinforcement), the higher energy-to-break variant is preferred.

Although aromatic fibers with very high modulus can readily be made from isotropic solution, it has not been possible to obtain both the high-strength high-modulus and the high-strength intermediately high-modulus variants from the same composition. Both types of yarn have been prepared, however, from isotropic solutions of different polymers (e.g., X-500 polyamide hydrazide and H-202 polyhydrazide, respectively, see Fig. 1 and Table 1).

3. *Major Aromatic Fibers*

The structural formulae in Fig. 1 are for the compositions of several aromatic fibers that have been evaluated extensively in either rigid or flexible composites, or both. The fiber properties for these compositions are given in Table 1.

Characteristically, the manufacturing cost of synthetic fibers is highly dependent on volume. Other than in the rigid reinforcement field dominated by glass fibers, a large market did not exist. Thus, in order to obtain the necessary volume base, the area of tire yarn had to be considered sine qua non by companies serious about producing a high-modulus fiber commercially.

For belt plies in tires only intermediate tensile stiffness is required (in the order of that of glass fibers). Even that level of modulus was not available in organic fibers, however, before the appearance of the high-modulus aromatic fibers.

Since only intermediately high modulus is required for tire yarn, the option of choosing from a wider range of compositions existed. This was the impetus for the development of the polyhydrazide and polyoxadiazole hydrazide fibers.

3.1 Polyamide (Aramid) Fibers

The aramid poly(*p*-phenylene terephthalamide) is the polymer of Du Pont's commercially produced Kevlar class of aramid fibers—Kevlar 49, Kevlar 29 and Kevlar tire yarn. From a mechanical property point of view, however, there are only two basic fiber variants. Kevlar tire yarn and Kevlar 29 are members of an intermediately high-modulus variant. Kevlar 29 is used primarily for such products as tensile members (ropes, cables and webbings) and ballistic cloth.

Kevlar 49 is the hot-stretched, very high-modulus variant designed primarily for rigid reinforcement application. Its specific (weight basis) modulus is several times higher than that of E-glass and is almost double even on a cross-sectional basis. All Kevlar yarns have essentially the same strength irrespective of the tensile modulus. Du Pont scientists have estimated that Kevlar poly(*p*-phenylene terephthalamide) fibers will sustain over half their one-second breaking load for longer than 100 years based on the absolute rate theory of creep failure for a fixed load. In contrast, high-modulus polyethylene fiber has a very high level of creep, precluding it from applications requiring sustained load-bearing capability and thus greatly limiting its utility.

3.2 Aromatic Polyamide Hydrazide, Polyhydrazide and Polyoxadiazole Hydrazide Fibers

None of these fibers were made from liquid crystalline solution and yet all have high modulus and high strength. Although making fibers from liquid crystalline solution has two significant advantages over the use of isotropic solution, there is one substantial disadvantage—that of usually requiring a strong acid (e.g., fuming sulfuric or hydrofluoric acid) as the spinning solvent, as is used for Kevlar fibers. These are not ideal solvents either to recycle or dispose of in commercial practice; they also present substantial corrosion problems and the spinning solution must be made in a separate step from the polymerization one. The polyamide hydrazide X-500 and the polyhydrazide X-702, in contrast, are synthesized in an organic solvent and the fibers spun directly from the polymer solution made in polymerization.

4. Stress–Strain Characteristics of High-Modulus Fibers

The stress–strain relations of the fibers of Table 1 and Fig. 1 are shown on a specific property or weight basis (Fig. 2) and on a cross-sectional basis (Fig. 3). Included for comparison are the stress–strain curves of E-glass, S-glass and steel, and in Fig. 2 also several conventional high-strength organic fibers. In Fig. 3 an expanded strain scale has been used which makes it possible to show fine differences in the stress–strain characteristics of the various fibers.

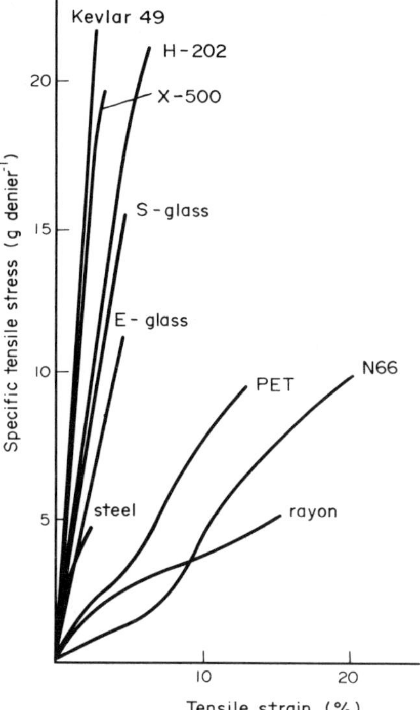

Figure 2
Specific (weight) basis stress–strain curves for several high-modulus high-strength fibers in comparison with those for S-glass, E-glass, steel and several conventional synthetic high-strength fibers. Tensile properties determined on yarns. PET, poly(ethylene terephthalate) polyester; N66, Nylon 66 polyamide (after Black W B 1980 © Annual Reviews Inc., Palo Alto, California. Reproduced with permission)

5. General Physical and Environmental Properties

In general the density is in the order of 1.44–1.47 g cm^{-3}, a range about the same as that for cellulosic fibers. As a class, the hydrogen-bond containing aromatic high-modulus fibers tend to be among the most thermally stable organic fibers known. Hydrolytic stability and stability to common organic solvents (e.g., oils and fuels) is not a problem. None, however, have good light stability. The dielectric properties of Kevlar 49 and X-500 epoxy composites were found to be slightly better than those of comparable E-glass composites, particularly at radar frequencies.

Part of the exceptional resistance of these fibers to the various environments is attributable to their very high degree of crystallinity and both crystalline and amorphous orientation.

6. Fiber Morphology

A distinctive feature of high-modulus organic fibers is extreme anisotropy of mechanical properties. In the transverse direction the strength is only about 20% of the strength along the fiber axis and the modulus is less than 10% of the fiber-axis tensile modulus. In essence, the filaments are composed of a large number of fibrils or microfibrils highly aligned in the draw direction with relatively little covalent-bond (i.e., polymer backbone) interconnections between the fibrillar entities in the filament.

In compression the situation is much the same as in the transverse direction. The compressive yield as determined in 60 vol% Kevlar epoxy composites was less than 20% of the tensile stress for failure. In compression the analogy to pushing on a steel cable is not a gross over-simplification.

The high-modulus fibers as a class have one other negative characteristic that is strongly related to compressive weakness, namely, relatively poor flex fatigue. Also, the fibers tend to abrade moderately easily.

The high-modulus polyethylene fibers are (morphologically) essentially identical in a gross sense to those from rigid aromatic polymers in that they suffer the same transverse and compressional weaknesses. The poor transverse strength and modulus can be

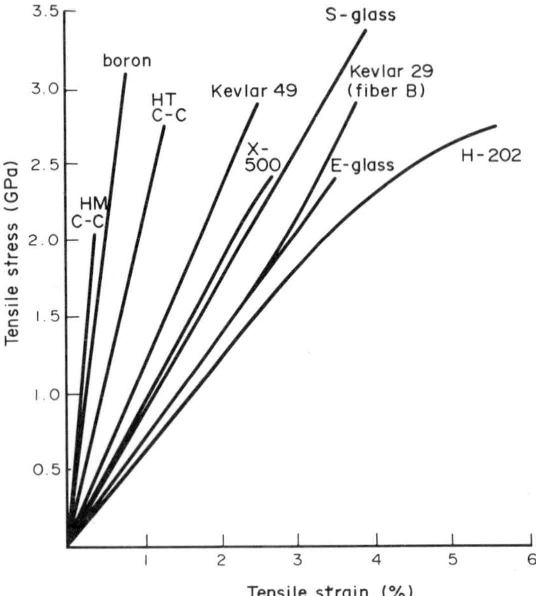

Figure 3
Cross-section based stress–strain curves for several high-modulus high-strength fibers in comparison with those for each high-modulus and high-strength graphite (HM C–C and HS C–C, respectively), boron, S-glass and E-glass. Tensile properties determined on yarns (after Black W B 1980 © Annual Reviews Inc., Palo Alto, California. Reproduced with permission)

ameliorated greatly, however, by crossplying in lamination and similar processes. Fabrication by filament winding of high-modulus organic fibers is an excellent way to obtain an approximation of a quasi-isotropic structure. Filament-wound casings for rocket engines and pressure vessels for space ships provide the pre-eminent examples in that the products combine a superbly appropriate fabrication technique with a product which utilizes the best properties of the yarns, i.e., high specific strength and high specific tensile stiffness. Weaving fabrics for layups is a frequently used approach to crossplying.

7. Applications in Rigid Composites

The applications for which greatest hope was originally held, namely, rigid composites for primary structural components in aircraft, proved to be their Achilles' heel as a consequence of their compressional weakness. In one regard, however, the organic fibers are superior to graphite and glass fibers, their primary competitors: they exhibit noncatastrophic failure in bending. The organic-fiber composites behave more like aluminum in this regard. The noncatastrophic failure characteristic combined with a Charpy impact resistance of epoxy–Kevlar composites approximately equal to that of epoxy–glass and aluminum, has left a window for application of such fibers in certain structurally significant areas. For example, the aromatic fiber is the one of choice for helicopter blades.

Fortunately, there is one way in which the compressive weakness of the aromatic high-modulus fibers can be compensated for in a practical manner—the use of a strong compressive material such as boron, graphite or glass on the compressive side of sandwich composite structures.

Unquestionably, with regard to rigid composites, it is in competition with glass rather than graphite or boron that the high-modulus organic fibers have made their greatest inroad. The key factor has been performance for a given weight of yarn rather than cost per unit stiffness; there is a factor approaching 10 between the cost of Kevlar and E-glass. Thus in areas such as sporting goods (e.g., sailboat and planing craft hulls, canoes and kayaks), aircraft nonstructural and semistructural parts (e.g., seats, flooring, doors, trim, control surfaces, exterior farings, radomes, and antennae), and personnel protection (e.g., ballistics armor), where performance and not cost is the primary consideration, the true value of these fibers is clearly seen. Several years ago Du Pont workers reported that every major type of interior and exterior part (in aircraft) fabricated of glass fiber had been successfully made from PRD-49 (i.e., Kevlar 49). Over 1100 kg of Kevlar are used in each Lockheed L-1011 aircraft.

8. Conclusion

While it is obvious that the high-modulus aromatic fibers are not ideal for all applications in which reinforcement fibers are needed, the overall distinctive value of these materials can be adequately appreciated only within the context of the wide range of applications for which they are not only competing successfully against other fibers, but also in the many instances where they are filling previously unmet needs such as a sorely needed substitute for asbestos (providing in many such applications a far superior product). Similarly, the combination of dimensional stability and very high specific strength has solved a wide variety of problems in the area of deep-sea mooring lines and underwater electromechanical and fiber optic cables.

Most future progress in composites of the high-modulus organic fibers can be expected to be the result of more sophisticated composite design rather than from gross improvements in the mechanical properties of the fibers.

See also: Aramids; High-Modulus Fibers

Bibliography

Bair T I, Morgan P W, Killian F L 1977 Poly(1,4-phenyleneterephthalamides): Polymerization and novel

liquid-crystalline solutions. *Macromolecules* 10: 1396–400
Black W B 1975 High modulus organic fibers. *Int. Rev. Sci: Phys. Chem., Ser. 2* 8: 33–122
Black W B 1979 Stiff-chain aromatic polymer solutions, melts, and fibers. In: Miller R L (ed.) 1979 *Flow-Induced Crystallization.* Gordon and Breach, New York, pp. 245–306
Black W B 1980 High modulus/high strength organic fibers. *Annu. Rev. Mater. Sci.* 10: 311–62
Ciferri A, Ward I M (eds.) 1979 *Ultrahigh Modulus Polymers.* Applied Science, London
E I Du Pont de Nemours 1978 *Kevlar 49 Aramid*, Du Pont Technical Bulletin K-2. E I Du Pont de Nemours, Wilmington, Delaware
Kavesh S, Prevorsek D C 1983 High tenacity, high modulus polyethylene and polypropylene fibers and intermediates therefor. US Patent No. 4,413,110 (1 November 1983)
Kwolek S L, Morgan P W, Schaefgen J R, Gulrich L W 1977 Synthesis, anisotropic solutions, and fibers of poly(1,4-benzamide). *Macromolecules* 10: 1390–96
Morgan P W 1977 Synthesis and properties of aromatic and extended chain polyamides. *Macromolecules* 10: 1381–90
Panar M, Beste L F 1977 Structure of poly(1,4-benzamide) solutions. *Macromolecules* 10: 1401–6
Smith P, Lemstra P J, Kalb B, Pennings A J 1979 Ultrahigh-strength polyethylene filaments by solution spinning and hot drawing. *Polym. Bull.* 1: 733–36

W. B. Black

High-Pressure Gas Cylinders: Integrity

The permanent gases, such as oxygen, are stored and transported at high pressure to provide an economically feasible distribution system. At high pressure, the kinetic energy of the stored gas is large and therefore the design and maintenance of the container are critical, to ensure structural integrity. Distribution efficiency, on the other hand, requires that the weight of the container per unit of contained gas volume be as low as practicable. To be commercially acceptable, the container must meet these contrary requirements. The design, testing and maintenance requirements of a cylinder manufactured for the containment of high-pressure permanent gases and complying with regulations of the US Department of Transportation are reviewed here. The discussion is limited to the principal factors which ensure the integrity of a steel cylinder for pressures above 14 MPa. There are more than 50 million such cylinders in the USA and at least an equal number in use elsewhere in the world. The principles discussed apply equally to cylinders used for transportation and storage of a multitude of other gases, including liquefied gases such as carbon dioxide and nitrous oxide.

1. Regulatory Controls

The energy required to compress a permanent gas into a transportable gas cylinder of the common "large" size ($\approx$40 dm^3) is $\approx$1.4 MJ. If such a cylinder suddenly ruptured longitudinally, the rate of energy release would approach that of an explosive; the stored gas can therefore be compared to an equivalent explosive charge, $\approx$0.35 kg of TNT in this example. Because of this high stored energy, compressed-gas cylinders are controlled as a "hazardous material" by regulations of the US Government and virtually all governments in the world. Within the USA, governmental control has virtually existed since the distribution of industrial gases began at about the turn of the century. Cylinders of the type under discussion were manufactured in substantial quantities in the USA as early as 1907.

2. Basic Design Principles

The essence of the primary design of a high-pressure steel gas cylinder can be stated as: a defined steel composition and heat treatment selected to provide maximum ductility (toughness), with adequate tensile strength for the allowed service stress. While this brief statement contains the key factors which provide design integrity, their application to design and manufacture is far more exacting.

The first requirement to ensure integrity of the mass produced cylinder which is repeatedly pressurized over many years and used under diverse environmental conditions is the selection and control of the material. Virtually all the high-pressure gas cylinders discussed here are produced from only two specific steels, and the large majority are of one steel. This also applies to production outside the USA. The steel chemistry and steelmaking controls have been precisely developed to guarantee the required material quality. The critical attributes of heat treatment resonse and freedom from flaws in both steelmaking and cylinder production are demonstrated by the hundreds of millions of use cycles virtually without cylinder failure.

The principal mechanical property of the steel which ensures integrity of the DOT 3AA cylinder is ductility. The term "ductility" has been regarded as synonymous with "toughness" and in current technical terms more precisely means fracture toughness. The principle of maximizing ductility ensures that the steel possesses adequate toughness so that the cylinder will tolerate flaws induced by the manufacturing process or by service abuse and withstand thousands of pressure cycles. In essence, the steel is not brittle, even at very low ambient temperature (below −45 °C). Ductility (toughness) is ensured in the finished cylinder through the steel chemistry and heat treatment and must be verified by sample testing. The specifications require proof of high ductility, as measured by both the tensile test and a flattening (bend) test. The elongation, as measured on the tensile test bar, must be ⩾20% in a 50 mm gauge length (width about 38 mm). In the flattening test, a finished cylinder is crushed in a plane perpendicular to the cylinder axis between rounded-knife edges at about the midlength. To pass the test, the steel must not crack when flattened to six times the wall thickness. The pattern of bending at the exterior cylinder surface, under the knife edges, subjects the exterior metal fibers to severe triaxial stress loading which is further compounded by the reinforcement effects of crushing the entire container. Charpy impact tests, at both room temperature and −45 °C, are used as additional verification of toughness. Currently, fracture toughness measurements by the *J*-integral method are being introduced as the appropriate measure of toughness.

The toughness of steel decreases with increasing tensile strength. The maximum tensile strength allowed in the design of a DOT 3AA cylinder is 725 MPa (N mm^{-2}), at which level the toughness is not compromised. This strength level is far below the capacity of the steel and therefore provides considerable tolerance for variation in heat treatment and permits a high tempering temperature to maximize ductility. The manufacturing limits for tensile strength are ≈725–860 MPa, the upper value being limited by ability to meet the elongation and flattening test requirements. It should be further noted that the strength levels are such that stress corrosion or hydrogen embrittlement is not of concern, so these cylinders are used for gases known to induce stress corrosion such as hydrogen, compressed natural gas and hydrogen sulfide.

The service stress is limited to either 290 or 308 MPa with certain controls. Thus, the service stress is less than 50% of the yield strength (≈620 MPa), which ensures normal behavior in cyclic fatigue. Furthermore, at this comparatively low service stress, the stress intensity developed at a theoretical crack tip on the inside surface is sufficiently low that many thousands of pressure cycles would be required for such a crack to propagate through the wall. Service experience verifies this statement; there is no known case of fatigue failure in service. As another verification of the integrity of the DOT 3AA cylinder, recent fracture mechanics analysis indicates that such crack growth would result in a leak-before-break performance; that is, the crack would break through the wall and leak gas without rupture of the cylinder. It should also be noted that the ratio of minimum tensile strength to service stress provides the classical safety factor of about 2.5:1. This is a comparatively low safety factor for industrial equipment but is essential to provide minimum cylinder weight, necessary for reasonable distribution efficiency.

3. *Structure*

The integrity of the high-pressure steel gas cylinder is significantly enhanced by its shape. Distribution and use requirements restrict the diameter–volume ratio while maintaining a stand-alone configuration. These constraints have led to a cylindrical container with a l/d ratio of about 6.0. The vessel is symmetrical around the longitudinal axis, without abrupt changes in thickness or shape, and has only one opening, at the top center.

The top head is a near-ideal shape to minimize stresses from pressure, normally being hemispherical and concave to pressure. There is a gradual thickening from the sidewall to the center top opening which further reduces stresses in the head and accommodates the permanent markings indented into the exterior head which are essential for identification in use and maintenance control. The integral stand-alone base configuration of the cylinder, which is essential to use efficiency, is formed while the metal is at forging temperatures and is provided with a gradual transition in thickness and shape for proper stress transition and without stress concentrations.

Another key attribute is a seamless structure, formed from one piece of metal with integral top and bottom and without welding. Such cylinders are manufactured by two processes. One process, billet piercing, starts with a cube of steel which is heated to forging temperature and then successively formed into a closed-bottom cup and elongated through dies into a cylinder with a forged stand-alone base. The open top is then closed by hot forming. The other process starts from a seamless steel pipe and proceeds with bottom closure by hot spinning and forge shaping and with top closure as in the billet piercing process.

This seamless symmetrical geometric shape without notches or abrupt changes of thickness or shape provides a near ideal structure for pressure containment. The manufacturing processes, both of which require piercing and forging of the steel, severely work the steel, requiring control of integrity of the starting stock.

4. *Metallurgy*

As mentioned in Sect. 2, only two steels are used in appreciable quantities to produce cylinders of the DOT 3AA type throughout the world. Their compositions are close to those of standard AISI 4130 (chromium–molybdenum) or 1045 (carbon–manganese) steels. The cylinder steels which have definitely prescribed limits for each element are referred to as "similar to 4130X or carbon manganese". The precise limits of some elements such as carbon, chromium or manganese have been altered from those of the commercial steels to provide the attributes required by gas cylinders.

The steel must be produced by processes known to ensure the required quality, such as basic oxygen furnace or electric furnace, and the pouring and blooming practices must be suitable for the steel to withstand the forging processes and yield a cylinder free of flaws. The steel is aluminum-killed and produced to a fine grain, deoxidized condition. The grain size is ASTM-8 or finer in the finished heat treated cylindrical wall.

The chromium–molybdenum steel is most commonly used and is heat treated by quenching and tempering. The heating and quenching practices are designed to produce a full martensitic structure, and since quenching takes place from only the outside surface, both the heat treatment and steel composition demand careful control procedures.

After quenching, the cylinders are tempered, usually at 620–675 °C. This high tempering temperature requires a complete quench to martensite and the tempered martensite structure ensures excellent ductility (toughness) and other specified mechanical properties.

Charpy V-notch impact test data show that the transition temperature for the 4130X steel is well below −45 °C. Also, the absorbed energy and expansion under the notch are consistently high at both room temperature and −45 °C. The half-size Charpy V-notch specimen normally shows >380 μm lateral expansion and 40 J absorbed energy at either temperature. This is true for all cylinder manufacturers, which verifies compliance with the steel specification.

The carbon–manganese steel may be heat treated by the quench and temper or by the normalizing process. In either case, the essential physical attribute of ductility (toughness) is achieved and the principles discussed for chrome–molybdenum steel are applied.

5. *Testing*

The integrity of steel gas cylinders is controlled by tests conducted during manufacture and is monitored by periodic tests during service.

5.1 *Tests During Manufacture*

During manufacture, tests (inspections) are carried out to monitor steel composition, heat treatment, physical properties, quality of inside and outside surface and structural integrity. These tests are performed or witnessed by an independent inspection agency which must prepare a report and certify compliance.

The steel composition is controlled through certified and check analyses, and traceability is maintained from heat to finished cylinder. Thus, a cylinder can be traced back to the manufacturer, inspector, heat of steel and production lot. Consistency of manufacture and heat treatment are controlled by sample tests of production lots. The lot is defined as a group of cylinders not to exceed 200 produced of the same material, of the same size and of the same design, consecutively produced on the same equipment and heat treated by the same procedure. The flattening test and tensile test are conducted on a finished cylinder from each lot. The tensile test involves determination of yield strength, ultimate strength and elongation.

Each cylinder must be inspected for wall thickness and for surface faults inside and outside. This may be accomplished by visual inspection or by mechanical inspections such as shear wave ultrasonic inspection or magnetic particle testing (see *Nondestructive Evaluation: An Overview*). The wall thickness is measured over the full length, and virtually the entire surface is examined by x-ray or ultrasonic equipment.

Finally each cylinder is subjected to a volumetric expansion test under hydrostatic pressure (Compressed Gas Association, Pamphlet C-1). In this proof test, the cylinder is placed in a sealed water jacket and the expansion during and after pressurization is measured. The cylinder must be pressurized to $1\frac{2}{3}$ times the marked service pressure of the cylinder. Data from this test verify several critical factors including:

(a) elastic expansion, which determines the average effective wall thickness and also confirms consistency of stand-alone base;

(b) permanent expansion, which proves that no part of the structure is yielding and that the cylinder will operate in the elastic range as designed;

(c) a safety factor of at least $1\frac{2}{3}$;

(d) leak tightness.

A number of other tests are used during manufacture but are not obligatory. These include for example the Jominy hardenability test to establish heat treatment parameters, ASTM grain size and inclusion ratings and Charpy impact tests.

5.2 *Tests During Service*

The continued integrity of the high-pressure steel gas cylinder is ensured by tests during service. Each time a cylinder is filled, the exterior is visually inspected. The original markings verify original compliance and possible alterations. Service-induced flaws, which include arc strikes, torch or fire burns, cuts, corrosion or dents can be detected by visual inspection. As discussed above, the low service stresses mean that a substantial exterior flaw and many pressure cycles are needed to initiate crack growth. Visual inspection can detect flaws of such size. Internal corrosion, which rarely occurs today because of noncorrosive (dry) gas conditions, is detected by a sound test. Internal contamination is revealed by an odor test.

Each cylinder is also periodically subjected to the

volumetric expansion test (proof test) discussed above. By measurement of the expansion—total, elastic and permanent—the integrity for another period of use is established. Specifically, the test proves that the average effective wall thickness has not decreased below the specified limit (Compressed Gas Association, Pamphlet C-5), that the cylinder is not yielding significantly (totally or in local areas) and that a safety factor of at least $1\frac{2}{3}$ remains.

In the USA, cylinders not used for toxic gases are equipped with a pressure relief device on the valve, which is designed to release 90–100% of the hydrostatic test pressure; thus, the pressure in the cylinder can never exceed the pressure reached during the hydrostatic test ($1\frac{2}{3}$ times the service pressure).

During service, cylinders may be subjected to rough handling and impacts at low temperatures. The low transition temperature and Charpy impact values at −45 °C ensure that such cylinders withstand such abuse. Furthermore, a test of rupture under gas pressure and a "gunfire" test specified by the Military for "nonshatter" qualifications have verified this performance.

6. Safety

The integrity of the high-pressure steel gas cylinder is proved by its safety record. Since the 1940s there has been virtually no case of such a cylinder failing in service under gas pressure, and there have been hundreds of millions of pressurization cycles, because each cylinder is charged several times each year.

A valve, appropriate for the cylinder contents and containing the pressure relief device, is attached by a threaded connection to the top of the cylinder (Compressed Gas Association, Pamphlet V-1). This valve is protected from accidental mechanical damage by a screw-on cap which should always be in place except when the cylinder is supported and secured, ready for gas delivery.

Proper care and caution by those using high-pressure gas cylinders will ensure than the safety designed into this gas system is realized in practice.

7. Future Trends

Technological advance in steelmaking and development of the science of fracture mechanics will permit a significant increase in the efficiency of steel gas cylinders in the near future. Steel production methods now provide low sulfur content and inclusion shape control, which in combination markedly increase toughness in the transverse direction. In a cylinder, the direction of major stress is transverse (hoop), so an increase in toughness in this direction will permit an increase in tensile strength. With a higher tensile strength, the service stress can be increased and thus the weight of the steel required per unit of gas shipped can be reduced. The production of a more efficient gas cylinder is made necessary by the rise in energy costs, which increase the cost of steel manufacture and the cost of transporting a cylinder during its useful lifetime. Fracture mechanics calculations, checked by physical tests on cylinders, indicate that the cylinder of the future, while more efficient, will be at least as safe as the current DOT 3AA cylinder.

See also: Hazards and Materials: An Overview

Bibliography

American Society for Testing and Materials 1981 *Standard Test for J_{Ic}, A Measure of Fracture Toughness*, ASTM Designation E813-81, Pt. 10. American Society for Testing and Materials, Philadelphia, Pennsylvania

Compressed Gas Association. *Methods for Hydrostatic Testing of Compressed Gas Cylinders*, Pamphlet C-1. Compressed Gas Association, Arlington, Virginia

Compressed Gas Association. *Cylinder Service Life Seamless, High Pressure Cylinders*, Pamphlet C-5. Compressed Gas Association, Arlington, Virginia

Compressed Gas Association. *Safe Handling of Compressed Gases*, Pamphlet P-1. Compressed Gas Association, Arlington, Virginia

Compressed Gas Association. *American National, Canadian, and CGA Standard Compressed Gas Cylinder Valve Outlet and Inlet Connections: ANSI B57.1; CSA B96*, Pamphlet V-1. Compressed Gas Association, Arlington, Virginia

US Department of Transportation. *Transportation*, Code of Federal Regulations, Title 49, Pts. 100–199. Office of Federal Register, Washington, DC

R. O. Tribolet

High-Purity Metals

The quest for specimens of progressively purer substances is generated by a need for determining their true properties and by hopes that new properties and phenomena obscured by impurities will be revealed. It is known that an extremely small concentration of impurities—about one impurity atom per million or even billion atoms of the matrix—significantly affects physical properties of elements and their compounds. From the standpoint of materials science, a substance is thought to be pure if individual properties or a group of properties are determined by its atomic-crystalline structure and the whole complex of intrinsic defects of the crystal lattice (e.g., vacancies, interstitials, dislocations, stacking faults, grain boundaries), and the relative role of impurity defects is negligible owing to their low concentration. The sensitivity of a property to the concentration level of individual chemical impurities and to their state as well as to the concentration level of individual intrinsic defects depends on that property. That is why different kinds of purity, such as electrical, optical and nuclear, are indicated.

1. Methods of Producing High-Purity Metals

The greatest success in improving methods for producing high-purity materials has been achieved in the field of elemental semiconductors and various metals. For example, silicon is now produced on a large scale with a very low content of electrically active impurities (Fe, Co, Mn, 0.01 ppm; Cu, 0.2 ppm), and indium in single-crystal form is one of the purest metals presently produced.

All the modern methods of high purification of materials can be divided, according to the nature of the effect at their basis, into two groups, which utilize chemical and physicochemical techniques, respectively. In chemical methods the separation is based on the difference in chemical interaction of the parent material and impurity with other components of the system. Chemical methods include crystallization from solutions using coprecipitants or complexing agents, sorption, ion exchange, extraction, selective oxidation–reduction of the impurities, electrolytic refining, and chemical transport reactions with the use of carrier gas.

In the physicochemical methods the separation is based on the difference in physicochemical properties of the matrix and of the impurity. They involve processes such as crystallization from the melt, distillation, redistillation (rectification), crystallization from the gaseous phase, thermal diffusion and annealing in a high vacuum. A high total purity of the substance is attained, as a rule, by employing a complex multistaged purification process.

The purity levels of various metallic elements realized at present are listed in Table 1. The γ parameter, the ratio of the electrical resistivity of the sample at room temperature to that at liquid helium temperature, is often a useful index of the purity. This parameter depends mainly on the residual resistivity at low temperatures, but is easier to measure; both the residual resistivity and the γ parameter characterize the material with respect to electrically active impurities, and thus define an electrical purity. However, if used in conjunction with well-developed methods of analysis of impurities at low concentrations (e.g., mass spectroscopy, radioactivation, atomic absorption), the parameter is quite reliable as an overall measure of chemical purity.

In Table 1, metals are classified by their melting points T_m into three groups. Such classification is convenient, since for the metals of each group similar methods and equipment for deep purification are used.

Table 1
The purity of metal samples according to the γ parameter

Metal groups	Resistivity ratio $\gamma_{4.2K} = \rho_{298K}/\rho_{4.2K}$			
$T_m \leq 1000\,°C$	Hg	75000	Tl	12000
	Cs	4000	Pb	30000
	Ga	110000	Cd	100000
	Rb	3500	Zn	100000
	K	7000	Sb	7000
	Na	10000	Mg	100000
	In	30000	Al	55000
	Sn	110000	Ag	30000
	Bi	1000		
$T_m = 1000–1500\,°C$	Au	1000	Fe	8700
	Cu	62000	Pd	10000
	Mn	24	Th	4200
	Ni	4000	Y	1100
	Co	410	Sm	260
$T_m > 1500\,°C$	W	100000	Ti	2000
	Mo	100000	Cr	5000
	Ta	15000	V	2800
	Nb	100000	Re	55000
	Hf	5000	Zr	500
	Os	2500	Ru	2500
			Pt	15000

1.1 First Group

This low-melting group includes gallium, indium, antimony, thallium, lead, tin, bismuth, zinc, cadmium and alkali metals. These metals are commonly purified by refining the starting compounds (volatile halides, hydrides, oxides) using recrystallization, extraction, redistillation (rectification) and zone refining; releasing the elementary form from purified compounds by electrochemical or hydrogen reduction (or more rarely by disproportionation); and finally zone refining.

The impurity concentrations in high-purity indium are given in Table 2 as an example of the low levels achievable for these metals.

1.2 Second Group

This group includes iron, cobalt, nickel, copper, gold and rare earth metals. Preliminary ion-exchange or extraction refining of metal compounds—as a rule, chlorides (except where anhydrous chlorides undergo sublimation)—is used. Prerefined chlorides are converted to metals by both electrolysis of aqueous solutions or hydrogen reduction of dehydrated chlorides. Purification is completed by vacuum zone refining and metal annealing (iron, for instance) in high-purity hydrogen, resulting in removal of carbon, nitrogen, sulfur and oxygen by high-temperature hydrogenation.

As an example of success in refining the metals of this group, the impurity content of high-purity iron is presented in Table 2.

1.3 Third Group

In this group similar deep purification methods are utilized for tungsten, molybdenum, tantalum, niobium, hafnium, rhenium, zirconium, titanium, chromium and vanadium. Preliminary purification of

Table 2
Impurity concentrations (ppm) in three high-purity metals

Metal matrix	Purification method	Impurity content
Indium	synthesis of monochloride, redistillation (rectification), precipitation of indium by the reaction of disproportionation, remelting and zone refining in a high vacuum	Mg, Al, Tl, <0.1; Cu, 0.05; Mn, Sn, Pb, <0.01; Fe, Ga, 0.005; Ag, Cr, 0.001
Iron	zone refining	Mo, Cr, Bi, Ga, Ci, ≤0.01; Tl, Sn, Sb, Cu, <0.03; Mg, Ni, ≤0.10; Al, 0.49; Mn, 0.007
Tungsten	production and sublimation of tungsten dioxydiiodide, thermal decomposition, reduction in high-purity hydrogen, electron-beam zone refining in a deep vacuum	Re, 10; Na, K, Ca, Mg, Al, P, Ta, F, S, Mn, Fe, Co, Mo, Sn, Tl, <0.3; Cu, Ti, As, Be, Se, Sr, Pb, Bi, rare earth metals, <0.05

these metals is achieved most effectively by redistillation (rectification) of extremely high chlorides and fluorides (WF_6, $NbCl_5$, $TaCl_5$, etc.). Sometimes a zone-refined oxide serves as a starting material and metal production is accomplished by reduction in high-purity hydrogen (MoO_3, Re_2O_7, etc.). The most important final stage is electron-beam zone refining in a high vacuum.

The analysis shown in Table 2 for tungsten is an example of the high purities attainable in the refractory metals.

2. *Properties of High-Purity Metals*

The most impurity-sensitive properties of metals are electrical (electrical resistivity) and mechanical (strength and resistance to plastic deformation).

2.1 *Electrical Properties*

An increase in metal purity causes a decrease in the number of scattering centers responsible for the residual (low-temperature) resistivity and a corresponding increase in electron mean free path, transport relaxation time and electron lifetime in a given state. The most important electronic property of a metal, manifesting itself only in studies of high-purity samples, is the functional dependence relating electron energy to its momentum, the graphical description of which is called the Fermi surface (see *Fermi Surfaces*). This isoenergetic surface separates occupied electron states from free ones. The Fermi surface in the simple case of alkali metals can be represented graphically by a sphere in momentum space. For other metals the relation between the Fermi energy and the momentum is more involved, and the Fermi surface differs drastically from a sphere. Each metal has its own Fermi surface with all the symmetry elements of a point group of an appropriate metallic crystal as well as with the translation symmetry.

All electronic properties of metal are determined by the response of electrons on the Fermi surface to external forces. Electron scattering at point defects (at impurity atoms of host elements and at intrinsic defects—vacancies and interstitials) is due to electrical, elastic and magnetic fields produced by the defects. The maximum contribution to scattering is made by magnetic fields typical of atom impurities of transition metals. Scattering at point defects is isotropic, permitting the total cross section of electron–defect scattering to be determined immediately from measurements of electrical resistivities. The magnitude ρ/C (where ρ is the electrical resistivity and C is the concentration of point scattering centres) is proportional to the total cross section, and for metals lies within wide limits. For example, for zinc as an impurity in cadmium, ρ/C is 0.08 μΩ cm at.%$^{-1}$, and for bismuth as an impurity in silver, 7.3 μΩ cm at.%$^{-1}$.

Scattering by dislocations is, in principle, anisotropic, varying with the direction of motion of the electron relative to that of the dislocation line and with the nature of the dislocation. The scattering may be divided into core effects (thought to make the main contribution) and to effects from the elastic field of the dislocation. Scattering by dislocations is principally low-angled.

Differences in scattering by dislocations and point defects are demonstrated in Fig. 1, which shows the results of simultaneous measurements of the de

Haas–van Alphen (dHvA) effect (see *de Haas–van Alphen Effect*) and the helicon resonance line (see Chambers 1969) in annealed and deformed aluminum samples (the method most sensitive to low-angle scattering). The magnitude of the transport cross section of electron scattering on dislocations $\sigma_{tr} \approx \rho_d/N$ (where ρ_d is the electrical resistivity increment due to dislocations and N is the density of dislocations), is dependent on the behavior of the dislocation structure. For copper single crystals, it is found to be in the range 10^{-17}–$10^{-19}\ \Omega\ cm^3$. It has been shown that electron scattering occurs for the most part in the vicinity of the dislocation core.

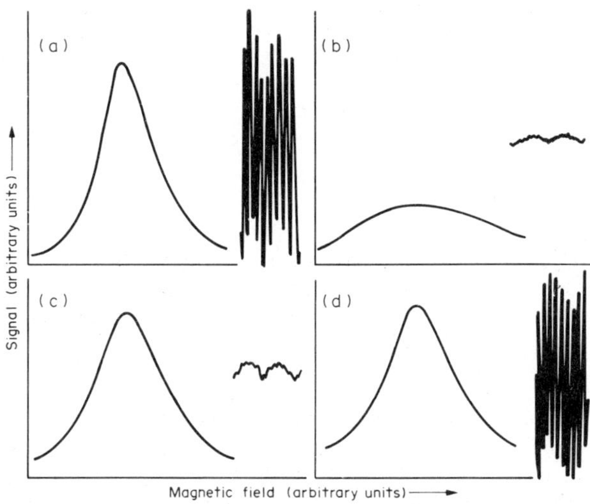

Figure 1
Damping of helicons and the de Haas–van Alphen effect during low-temperature plastic deformation in superpure aluminum: (a) before deformation; (b) after 2% elongation; (c) after annealing at 140 °C; (d) after annealing at 500 °C. The helicon resonance was recorded at a magnetic field of ~0.7 T: the oscillations at ~1.35 T (Volsky et al. 1974)

The probability of electron scattering in a phonon gas is proportional to T^3 and is dictated by the temperature dependence of the concentration of phonons and their energy. At sufficiently low temperatures, the phonon contribution to the scattering proves to be insignificant, as reflected by the absence of temperature dependence of electrical conductivity. The purer the material, the lower is the temperature at which temperature dependence of the electrical conductivity can be detected.

2.2 Mechanical Properties

Impurities in metals occurring in solid solution and as precipitates of other phases greatly affect mechanical properties. This influence is most notable in metals with a body-centered-cubic (bcc) lattice, such as tungsten, molybdenum, chromium, niobium, tantalum and iron. Interstitial impurities (carbon, oxygen, nitrogen and hydrogen) occupying octahedral and tetrahedral vacancies of bcc metals cause tetragonal distortions which considerably impede plastic deformation at low temperatures. This effect can be seen in Fig. 2, where it is shown that metals with an appreciable content of interstitial impurities exhibit a marked additional temperature dependence of the yield point at temperatures below $0.1T_m$. This dependence is closely associated with the phenomenon of low-temperature brittleness in bcc metals. Decreasing the temperature can lead in such metals to a situation where the yield point exceeds the brittle fracture stress so that brittle failure then occurs. Removal of interstitial impurities decreases the yield point temperature dependence and promotes plastic deformation up to failure, leading to a decrease in the brittle fracture temperature (Fig. 2c). It should be noted, however, that even in highly purified bcc metals the intrinsic temperature dependence of the yield and flow stresses remains relatively large at low temperatures; this is usually attributed to the high lattice resistance offered to the motion of screw dislocations. Moreover, the effects of impurities, both interstitial and substitutional, are sometimes complex; as the concentration of a given impurity is increased, there may be an initial softening, followed by a hardening of the solvent metal.

The effect of interstitials on the ductile–brittle transition temperature of bcc metals differs for different metals and is due to different values of solubility of interstitials in the matrix. For example, for the relatively ductile group V metals (vanadium, niobium and tantalum) hydrogen is the most deleterious impurity, followed by nitrogen, carbon and oxygen in that order. In the group VI metals, the solubilities of interstitial solutes are much smaller, and the tendency to brittle behavior at low temperatures is much more marked; the effect of hydrogen is now small, but large effects may be caused by oxygen, nitrogen and carbon impurities, with oxygen being the most potent.

Brittle failure in polycrystalline bcc metals frequently occurs along the grain boundaries and may be due to impurities segregated to grain boundaries, or to precipitates formed on the boundaries. The segregation to the boundaries is a complex phenomenon and often results from heat treatment in a critical temperature range, in which case it is called "temper embrittlement."

2.3 Properties of Defects

Impurities present in a metallic matrix have a pronounced effect on the properties of dislocations and grain boundaries. The effects of dislocation pinning due to the formation of impurity atmospheres (the Cottrell atmosphere and the Suzuki atmosphere) are

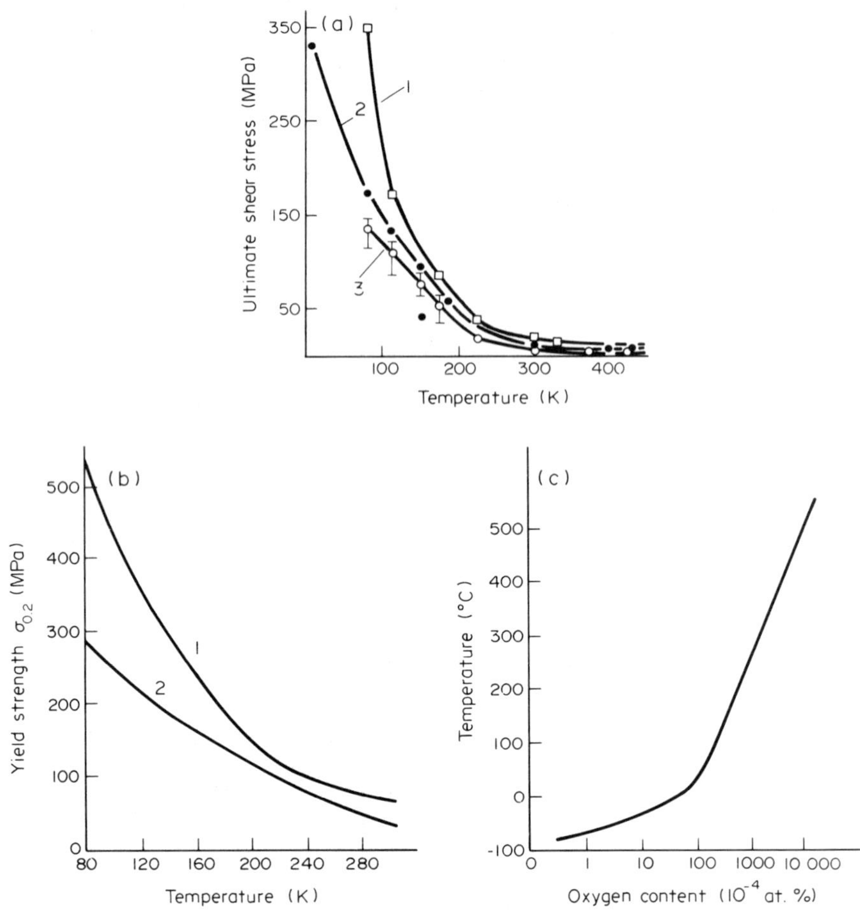

Figure 2
(a) Temperature dependence of ultimate shear stress in niobium single crystals of different purity: 1, zone refined, with an interstitial impurity content of 0.03 at.%; 2, electron-beam zone refined and degassed in vacuo near the melting temperature; 3, degassed in vacuo, with an interstitial impurity content of 0.0005 at.%. (b) Temperature dependence of yield strength $\sigma_{0.2}$ of iron single crystals containing: 1, 0.0044 wt% C and 0.0002 wt% N; 2, $<5\times10^{-7}$ wt% C and 0.0002 wt% N. (c) Ductile–brittle transition temperature of iron as a function of oxygen content

well known. As already stated, tetragonal distortions in bcc metals brought about by interstitials lead to a decrease in the mobility of individual dislocations. Motion of individual dislocations in high-purity molybdenum (γ~50000) at liquid helium temperature takes place at stresses of ~1 kg mm^{-2}, and in samples with $\gamma \sim 300$ at ~10 kg mm^{-2}.

The orientation dependence of mobility of high-angle grain boundaries in metals is mainly determined by their impurity content. This is due to differential absorption of dissolved impurities at special and random boundaries. This can be seen, for example, in aluminum samples of different total purity, such as 99.99995 at.%, 99.9992 at.% and 99.98 at.% (Fig. 3). The minima in the curve of activation energy for migrations versus misorientation for the 99.9992 at.% sample agrees well with the coordinates of special boundaries in the theory of the coincident site lattice.

In experiments at a low impurity content in metals a moving boundary is observed to break away from the adsorbed impurity (Fig. 4). This phenomenon was predicted by Lücke and Detert (1957). Jumps in mobility followed by alternations in the activation energy of the 46.5° ⟨111⟩ tilt boundary migration in aluminum bicrystals correspond to a successive breaking away of the boundary from adsorbed atoms of iron (samples I–III were alloyed with iron in amounts of 0.8, 0.6 and 0.5 ppm, respectively; sample IV was undoped with an iron content <0.1 ppm). The tilt boundary ⟨11$\bar{2}$0⟩ in high-purity zinc, containing a total amount of impurities of

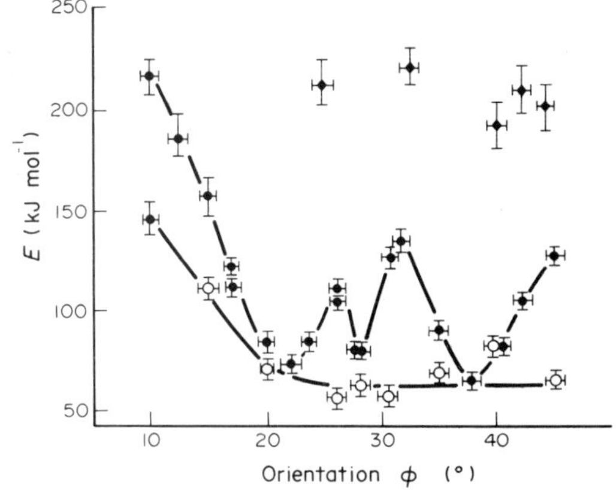

Figure 3
Orientation dependence of activation energy for ⟨100⟩ tilt boundary motion in aluminum of different purity: ○, 99.99995 at.%; ●, 99.9992 at.%; ◆, 99.98 at.%

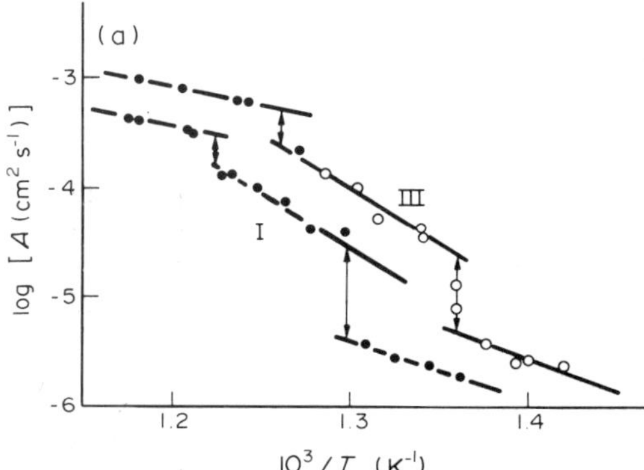

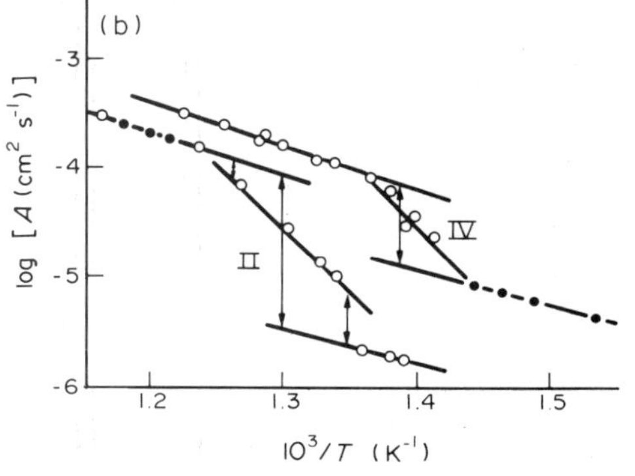

Figure 4
Temperature dependence of the ⟨111⟩ boundary motion in aluminum bicrystals for a misorientation angle of 46.5° for different iron contents (ppm): (a) I, 0.8; III, 0.5; (b) II, 0.6; IV, <0.1. Here the boundary motion is expressed by the parameter A, the mobility multiplied by the surface energy per unit area (Molodov et al. 1981, Demianczuk and Aust 1975)

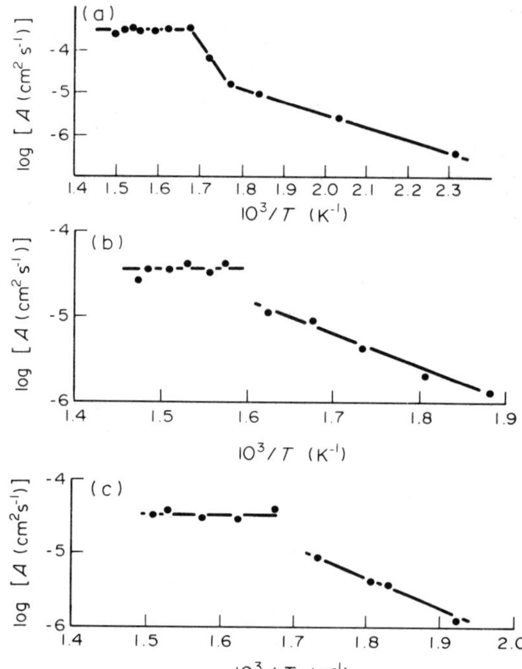

Figure 5
Temperature dependence of the ⟨11$\bar{2}$0⟩ boundary motion in zinc bicrystals for different misorientation angles: (a) 34 ± 1°; (b) 54 ± 1°; (c) 56 ± 1°. The parameter A is defined as for Fig. 4 (Kopezky et al. 1979)

$<5 \times 10^{-4}$ at.% at misorientation angles of 34, 54 and 56°, demonstrates the phenomenon of athermal motion in the temperature range from 320 °C to the melting point (Fig. 5). The results of research on boundary mobility in high-purity metals have been interpreted in terms of both single atom and group transitions across the boundary.

See also: Metals Production: An Overview; Silicon (Semiconductor): Preparation; Zone Refining, Zone Levelling and Temperature-Gradient Zone Melting

Bibliography

Chambers R G 1969 In: Ziman J M (ed.) 1969 *Physics of Metals,* Vol. 1. Cambridge University Press, Cambridge, pp. 175–249

Demianczuk D W, Aust K T 1975 Effect of solute and orientation on the mobility of near-coincidence tilt boundaries in high-purity aluminum. *Acta Metall.* 23: 1149–62

Gantmakher V F, Petrashov V T 1976 Conduction electron scattering in pure metals. *High Purity Metals.* Nauka, Moscow, pp. 31–59

Kamenezkaya D S, Pylezkaya I B, Shirjajev V I 1978 *High Purity Iron.* Metallurgiya, Moscow

Kopezky Ch V 1976 Some problems of metal science of pure metals. *High Purity Metals*. Nauka, Moscow, pp. 5–31

Kopezky Ch V, Plusheva S V, Satunkina L F, Zepkova ZA, Klimenko G L 1981 The preparation of high purity tungsten. *Izv. Acad. Nauk SSSR Met.* 2: 59–61

Kopezky Ch V, Sursayeva V G, Shvindlerman L S 1979 *Mobility of High-Angle Grain Boundaries in Zinc*. Academy of Sciences of the USSR, Chernogolovka, Moscow

Lücke K, Detert K 1957 A quantitative theory of grain-boundary motion and recrystallization in metals in the presence of impurities. *Acta Metall.* 5: 628–37

Molodov D A, Kopezky Ch V, Shvindlerman L S 1981 Double successive breaking-away of moving grain boundaries from adsorbed impurity. *Dokl. Acad. Nauk. SSSR* 260: 1111–14

Volsky E P, Levchenkova L G, Petrashov V T 1974 The de Haas–van Alphen effect and damping of helicons at plastic deformation in aluminium. *Sov. Phys. JETP (Engl. Transl.)* 38: 106

Ch. V. Kopezky

High-Speed Machining

Productivity in metal removal operations is governed by a complex interaction of many diverse factors. One of the more fundamental considerations is the direct relationship between productivity and the cutting rate that can be economically sustained. Over the years, evolutionary advances in machines and tooling have led to steady escalation of the rate at which a given material and shape can be produced. Major pacing elements have been improved cutting tools, stiffer and more powerful machines, and increased control system sophistication. However, the technology of high-speed machining (HSM) has produced increases in cutting speed well beyond those associated with evolutionary developments. Productivity improvements related to HSM are considered here, emphasis being placed on known process limitations and requirements, and on practical aspects.

1. Basis

There are two basic effects associated with the concentrated shear zone produced during chip generation (see *Chip Formation Micromechanics and Metallurgy*). One is that heat is generated as a result of the deformation, and the second is that work hardening of the emerging chip is observed. The degree to which these effects occur depends on the properties of the work material, tool design and cutting parameters.

Most of the heat generated when cutting at very low speeds is dissipated in the workpiece and the tool. Cutting forces are moderate, and tool wear is low. The net effect is that the emerging chip is somewhat work hardened by passage through the shear zone. Assuming that the chip cross-sectional area and the tool geometry are held constant, and that a reasonably effective cutting fluid is applied to the cutting zone, then as cutting speed is increased, the higher strain rate in the shear zone produces greater levels of work hardening and heating. The cutting fluid removes heat from the free surface of the chip and preserves the work-hardening effects. The rest of the heat flows into the workpiece and across the contact zone into the tool. The contact area on the tool face also tends to increase with the degree of work hardening, thus providing a higher heat flux into the tool. Heat flowing into the workpiece has less time to dissipate from the cutting zone, so the tool is exposed to a second heat source on the flank face as it moves over the cut surface. In this situation a stronger chip is being deflected away from the work zone, and the tool is heated from two sources which effectively surround the tool nose. This process continues as the speed is increased through the conventional cutting range. Feed- and cutting-force components become larger, and thermally activated tool wear mechanisms operate at escalating rates.

Heat flow is material dependent, however, and can only occur at finite rates. Continued increase of cutting speeds eventually produces a condition where the heat generated in the shear zone cannot fully escape and is carried away in the chip. The shear zone represents a planar heat source moving at high velocity, so that the heat flux into a given element of the uncut workpiece surface is sharply curtailed by the short time this source is over the element.

A somewhat different situation exists in the case of heat flow across the contact area between the chip and the tool. The chip stream can be regarded as a heat source of infinite extent moving over the contact zone. This is not quite the same as the static case in which no relative movement exists. Under static conditions, a gradient would develop in the chip adjacent to the contact zone, and heat would flow according to the laws of thermal conduction. In the dynamic case, however, chip material at the maximum temperature is being constantly presented at the contact interface. The heat flux into the tool is actually greater than for the static case. Before drawing any conclusions regarding the impact this heat flux might have on tool life, there is an additional heat-flow-related factor to be considered. Bulk chip temperatures will remain high because the cutting fluid applied to the free chip surface will not begin to remove any significant amount of heat until elemental chip volumes are well clear of the cutting zone. This situation, when combined with limited heat flow towards the workpiece, results in local chip temperatures which are essentially in the hot-working range; the effective flow stress is low, and work-hardening effects are negligible. The low-strength chip is easily deflected by the tool, which has the critically important effect of reducing the chip contact area on the tool face. Thus the net heat flow across the tool–chip interface is reduced despite the higher unit heat flux produced by the moving source. The overall effect is to retain most of the shear-induced heat in the chip stream and reduce heat input into the tool. The significance of these changes in heat flow patterns with increasing speed is that HSM becomes feasible once these conditions are established.

The HSM phenomenon is therefore essentially based on an adiabatic system in which most of the heat leaves the cutting zone harmlessly with the chip stream, sharply reducing the flow stress of the work material. The chip temperature never reaches the melting point because of the self-limiting nature of the shear-induced heating process. As the temperature rises shear resistance decreases, which in turn produces less deformation-induced heating. This simplistic concept provides a qualitative explanation for the observed drop and subsequent consistency of the specific energy for chip generation. Representative specific energy levels are presented in Table 1 for common workpiece materials.

Table 1
Representative specific energies for some common materials

Material	Specific energy ($W\ m^{-3}\ s^{-1}$)	Threshold speed[a] ($m\ s^{-1}$)
Aluminum 6061	305–410	20
Brass	715	25
AISI 4340	3580	30.5
Titanium 6Al–4V	1535	4.0

a Tool-geometry dependent

The threshold speed beyond which the specific energy becomes constant is dependent on the effective rake angle of the cutter. The threshold rises with increasingly greater positive rake angles and is related to the development of shear-induced heating within the chip. Frictionally generated heat still exists, as does some finite heat flow from the chip. This heat must still be considered when selecting a compatible tool material for HSM. Some investigators have postulated that a very thin film of molten work material forms at the tool–chip contact zone as a result of frictional heating. The heat flow into the tool would be thereby reduced by the heat of fusion necessary to melt this film. Evidence supporting or disproving this concept has not yet been conclusively demonstrated. The theory is attractive, however, in helping to explain the semi-infinite tool life experienced for aluminum HSM. This effect is difficult to observe for other materials owing to interferences by chemically related tool wear reactions.

2. *Utilization*

It must be recognized that utilization of high-speed machining is not feasible for all materials and all part geometries. Feasibility may be limited by metallurgical, mechanical or economic factors, or a combination of these. For example, there is no question but that HSM can be effectively applied to most aluminum alloys, but some part geometries sharply limit aluminum HSM applications. Metal removal rates may be limited by available spindle power for heavy cuts, or part geometries may require table motions that cannot be accurately maintained at the high traverse rates that represent the major advantage of HSM. Materials other than aluminum are subject to restrictions of varying severity owing to factors such as near catastrophic reactivity with tool materials or threshold cutting speeds beyond practicable limits.

Considerable incentive remains, however, for adopting HSM in production. The product throughput increases that are possible are well worth the efforts necessary for their achievement. Successful HSM implementation into production requires a serial analysis first to determine process feasibility and then to define specific process and equipment requirements, followed by an economic analysis. The economic analysis has a dual purpose—to formally justify the effort, and to give strong incentive to those involved, including management, to complete effectively the implementation.

Metals having face-centered-cubic (fcc) structures tend to be more readily machined by HSM techniques, followed by hexagonal-close-packed (hcp) and then body-centered-cubic (bcc) materials. The only simple relationship that can be offered is that metals with extensive slip systems and a more extensive plastic-strain range seem to be better suited to HSM, rather than bcc alloys which generally have more restricted slip systems.

Part geometry is a critical factor in assessing candidates for HSM; complex parts having details of very short extent prove difficult. Table traverse rates increase in direct proportion to cutting speed, as the chip loads per tooth must remain essentially the same as for conventional machining. Hence, a conventional process involving a 0.005 in. (0.127 mm) chip load, a cutting speed of 500 ft min^{-1} (2.6 m s^{-1}) and a traverse rate of 1 ft min^{-1} (5 mm s^{-1}) translates to a traverse rate of 480 ft min^{-1} (2.5 m s^{-1}) when an HSM cutting speed of 20 000 ft min^{-1} (100 m s^{-1}) is used. Only a few machine tools provide such a traverse speed or achieve accurate positioning when profiling small details involving multiple direction changes over short distances. Larger, simple parts or straight, two-dimensional cuts on smaller parts are far more desirable. An analysis of permissible data transfer rates for computer-numerical-control axis drives is necessary to define realistic traverse rates for reproducible tolerance control. However, the entire exercise should be performed, even on complex parts, to be certain that an opportunity to use HSM is not lost. Although the largest benefits are achieved when cutting rates can be increased over 30-fold, doubling or tripling cutting rates is not undesirable either.

3. *Vibration and Tool Design Considerations*

Another HSM-related concern involves chatter and resonance conditions that can develop during milling operations. Tooth impact rates are more likely to be in the excitation range of part or machine resonance frequencies during HSM than during conventional machining. Machine-related resonance conditions may be identified by measuring vibration levels with an accelerometer over the spindle speed range. The part program may then be written to avoid operation at or near the particular speeds where resonance is observed. Management of part-related resonance is not as straightforward in that natural frequencies vary as metal is removed from the part. On higher-production runs, resonance conditions that may be encountered may be avoided through one or more adjustments to the program. It should be emphasized that increases in cutting rate should be tried as well as decreases to avoid the common practice of slowing down when a problem occurs. Single or batch runs present a similar situation, but changes must be made on line.

Tool design for HSM must allow for ejecting a greater volume of chips from the work zone per unit time. Ideal cutters have fewer but larger flutes and steeper flute angles; large positive axial and radial rake angles are also essential to eject chips and prevent recutting and chip wraparound problems. The high positive rakes will raise the critical specific-energy speed threshold over that of negative-rake tools, but the chip management problem is a more important concern. Chips are generated at ten or more times the conventional rate with HSM, and improved chip handling is essential to avoid "burying" the machine during operation. Streams of cutting fluid have been highly effective in washing chips from the cutting zone into chip conveyors. The cutting fluid streams also perform a second function—cooling of the chips. Chips emerging from HSM operations are considerably hotter than those from conventional operations. In the case of aluminum HSM, the primary functions of a cutting fluid are chip flushing and cooling. A cutting fluid for aluminum HSM is unnecessary for tool wear control as almost all the heat remains in the chip. Tools should be as stiff as possible to avoid deflection and chatter. Carbide should almost always be selected over high-speed steel for HSM of aluminum, even though tool life with high-speed steels is extremely long. Balancing and tool retention are also critical factors. Vibration due to unbalanced cutter assemblies will be devastating to spindle bearing life. Tool designers having experience with HSM should be consulted when insert tooling is required as the shedding of a carbide insert travelling at 10 000–30 000 ft min^{-1} (50–150 m s^{-1}) can be devastating to both machine and personnel alike. However, intelligent attention to the design of an HSM application will reduce such risks to well within those presented by ordinary machining operations.

In addition to chip management considerations, when materials other than aluminum are used, tool wear, or more specifically chemical reactivity, must be carefully analyzed. Equilibrium chip temperatures necessary to produce low flow stresses in other materials are considerably higher in comparison with those of aluminum alloys. The large variety of workpiece materials and possible cutting conditions dictate application-specific analyses to identify specific tool-material recommendations. It cannot be overemphasized that tool-materials selection is one of the most critical aspects for HSM implementation (see Table 2); this question must be examined as early as possible in the process of evaluating a potential HSM application.

Table 2
Guidelines for HSM tooling

Workpiece	Tool material	Rake angle
Aluminum	uncoated carbide/diamond	15–20°
Brasses	uncoated carbide/diamond	10–15°
Steels	coated carbide/cubic boron nitride	5–10°
Titanium	uncoated carbide	5–10°

Cutting edges should be unhoned or at least as sharp as is practicable. Flutes and gullets should be large and smooth to facilitate chip ejection away from the surface being machined.

One factor that may not be obvious relates to spindle power requirements. Table 3 compares conventional and HSM parameters for milling a one-inch-wide (25.4 mm) steel bar. Although it is obvious that the process proceeds at slightly more than thirteen times as fast, the metal removal rate under HSM conditions dictates that at least 24.5 kW will be needed to sustain the process, a significant increase in machine capacity requirements. Metal removal rates for well-engineered aluminum HSM can easily reach 200–300 $in.^3$ min^{-1} (55–80 cm^3 s^{-1}), with spindle power requirements necessary to sustain such cuts of at least 50–75 kW. Power and speed capabilities of present or future machine spindle systems must be carefully evaluated to be certain that

Table 3
Comparison of conventional and HSM parameters for milling a 25.4 mm steel bar

Parameter	Conventional	HSM
Cutter diameter (m)	0.15	0.15
Cutter rpm	318	6370
Cutting speed ($m\ s^{-1}$)	2.5	50
Number of teeth	6	4
Feed (mm per tooth)	0.0762	0.0762
Table feed ($mm\ s^{-1}$)	2.4	32.3
Depth of cut (mm)	3.2	3.2
Metal removal rate ($mm^3\ s^{-1}$)	136	2608
Power required (kW)	2.2	24.5

adequate capacity will be available for the expected application.

The integrity and finish of surfaces obtained from HSM is an additional advantage beyond the potential machine productivity improvements. The distinction between rough and finish machining rates common for conventional practice disappears in HSM operations. Surface finishes are usually in the 10–20 μin. (250–500 μm) AA range for all materials, provided chip recutting and interference effects are controlled. Metallurgical damage occurs since virtually all cutting heat remains in the chip. Residual stresses induced by machining are negligible. Evidence of this can be demonstrated by the fact that a full-width cut can be taken with a 1 in. diameter end mill, 1–1.5 in. in depth, to leave a 0.015 in. wall in aluminum without support and without distortion. Further, if the grain size is large enough, grain boundaries can easily be seen on the cut surface.

The qualitative model advanced for the HSM phenomenon, based on adiabatic-like heat retention in the chip stream, offers an explanation for the feasibility of cutting at rates well above conventional speeds. Classic tool wear curves indicate that lifetimes would drop to essentially zero with substantial speed increases. Changes in heat-transfer mechanisms and reduction of work-hardening effects above the HSM threshold alter this trend.

HSM offers a technology to substantially increase machine throughput rates. It is a new but proven technology which can provide major productivity improvements for those who are dedicated to its implementation. When HSM is integrated into other productivity improvement efforts, such as automation of noncutting manufacturing operations, major advances in manufacturing efficiency are possible.

See also: Metal Removal Processes; Heat Generation in Metal Cutting and Deformation; Cutting-Tool Materials; Cutting Fluids; Machinability of Metals; Metals Processing and Fabrication: An Overview

Bibliography

Arndt G 1972 Ultra-high-speed machining: Notes on metal cutting at speeds up to 7300 feet per second. *Proc. 12th Int. Machine Tool Design and Research Conf.* Macmillan, London, pp. 203–208

Arndt G 1973 Ultra-high-speed machining: A review and analysis of cutting forces. *Proc. Inst. Mech. Eng.* 187: 625–34

Arndt G, Brown R H 1973 Design and preliminary results from an experimental machine tool cutting metals at up to 8000 feet per second. In: Tobias SA, Koenigsberger F (eds.) 1973 *Proc. 13th Int. Machine Tool Design and Research Conf.* Macmillan, London, pp. 217–23

Flom D G 1981 *Advanced Metal Removal in Program*, US Air Force Contract No. 33615-79-C-5119. General Electric Company, Schenectady, New York

King R I 1976 High-speed production milling of nonferrous materials. *Proc. 4th North American Metalworking Research Conf.* Society of Manufacturing Engineers, Dearborn, Michigan, p. 334

King R I, McDonald J G 1976 Project design implications of new high-speed milling techniques. *J. Eng. Ind.* 98: 1170–75

Kuznetsov V D 1945 Super high-speed cutting of metals. *Iron Age* 155(19): 66–69; 142

McGee F J 1978 *Manufacturing Methods for High-speed Machining of Aluminum*, Final Technical Report for DRDMI-EAT Requirement No. 6089. US Army Missile R & D Command, Washington DC

Recht R F 1964 Catastrophic thermoplastic shear. *J. Appl. Mech.* 31: 189–93

Salomon C 1931 Critical machining velocities. German Patent No. 523594

C. F. Barth

High-Strength Low-Alloy Steels

High-strength low-alloy (HSLA) steels are steels having yield strengths >275 MPa with alloying additions designed to provide specific desirable combinations of properties such as strength, toughness, formability, weldability and atmospheric corrosion resistance. HSLA steels are superior alternatives to traditional low-carbon mild steel used for structural steel in the form of shapes, plates and bars for use in riveted, bolted or welded construction of bridges and buildings. A low-carbon mild steel, for example ASTM A36, can contain up to 0.29% C and up to 1.20% Mn and has a minimum yield strength of 250 MPa. Similarly, HSLA steels are superior alternatives to low-carbon sheet steels traditionally used in automotive applications. A typical low-carbon sheet steel, hot-rolled SAE 1010 steel, contains 0.08–0.13% C and 0.30–0.60% Mn and has a minimum yield strength of ~205 MPa.

The designation HSLA is accordingly given to low-alloy steels of higher strength than the traditional mild steels produced as hot-rolled or cold-rolled sheet and strip. However, it is also applied to low-alloy normalized steels and to low-alloy steels strengthened by cold work, aging, accelerated cooling or intercritical annealing. Over 600 proprietary HSLA steels were listed in a complete review of this subject by Fletcher (1979).

Several special terms are used to describe certain types of HSLA steel, for example:

(a) weathering steels, designed to exhibit superior atmospheric corrosion resistance;

(b) control-rolled steels, hot rolled according to a predetermined rolling schedule designed to develop a highly deformed austenite structure that will transform to a very fine equiaxed ferrite structure on cooling;

(c) pearlite-reduced steels, strengthened by very fine-grained ferrite and precipitation hardening but with low carbon content and therefore little or no pearlite in the microstructure;

(d) microalloyed steels, with very small purposeful additions (generally <0.10% each) of such elements as niobium, vanadium, titanium, boron, zirconium, rare earths and calcium, added for refinement of grain size, precipitation hardening, hardenability and inclusion shape control;

(e) acicular ferrite steels, very low-carbon steels with sufficient hardenability to transform on cooling to a very fine high-strength acicular ferrite (low-carbon bainite) structure rather than the usual polygonal ferrite structure; and

(f) dual-phase steels, processed to a microstructure of ferrite containing small uniformly distributed regions of high-carbon martensite, resulting in a product with low yield strength and a high rate of work hardening, thus providing a high-strength steel of superior formability.

1. Weathering Steels

The first HSLA steels developed were the weathering steels. The origin of weathering steels is generally attributed to Williams (1900), who called attention to the increased corrosion resistance of copper-containing mild steel when alternately wet and dried. Such a steel containing 0.15–0.30% copper was first marketed by the American Sheet and Tin Plate Company in 1911. Although this steel exhibited about twice the corrosion resistance of copper-free structural carbon steels, it did not in fact contain sufficient copper to attain a yield strength >205 MPa.

It was not until 1929, when attempts to improve the atmospheric corrosion resistance of copper-containing steels were initiated, that the added bonus of high strength in combination with markedly improved atmospheric corrosion resistance materialized and the first HSLA weathering steels were developed. The increased strength largely resulted from the addition of elements that solid-solution strengthen and refine the ferrite microstructure while providing corrosion resistance. The solid-solution strengthening effect of several alloying elements is shown in Fig. 1. Of these, copper, phosphorus and silicon provide corrosion resistance in addition to solid-solution strengthening.

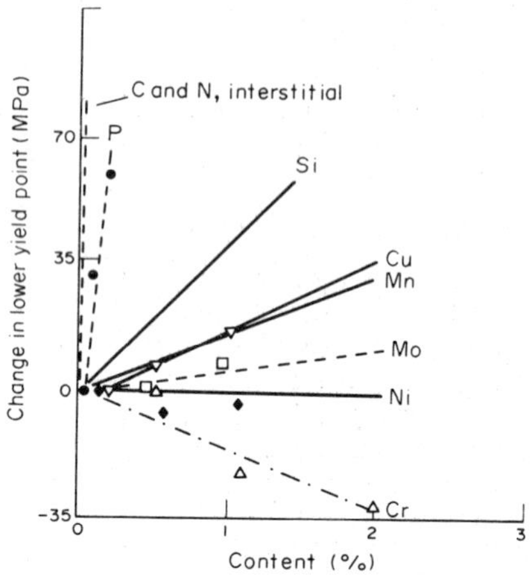

Figure 1
Effect of content of substitutional solid-solution elements on the lower yield point of hot-rolled mild steel: ▽, Cu; ●, P; △, Cr; ◆, Ni; □, Mo (after Pickering 1976)

The optimum balance of corrosion resistance and strengthening elements provides these HSLA weathering steels with not twice but four to eight times the corrosion resistance of carbon steels and with a minimum yield strength of 345 MPa in plates 12.5 mm thick at a carbon content of ~0.10%. A minimum yield strength of 345 MPa is possible in plates up to 200 mm in thickness by increasing the carbon and manganese contents, and adding vanadium. The first commercial steel of this type (US Steel Corporation's COR-TEN A) (Larrabee and Coburn 1962) was introduced to the market in 1933.

In 1979, proprietary grades of HSLA weathering steels manufactured by six different companies in the USA and Canada were based on a high-phosphorus multiple-alloy solid-solution-strengthening approach. The range of chemical compositions generally encompassed by these grades is shown in Table 1. Also, over 35 weathering steels manufactured by fifteen producers in the USA and Canada (Fletcher 1979) provided atmospheric corrosion resistance four to six times that of carbon steel by combining the multiple-alloy corrosion resistance approaches in steels of normal low phosphorus content with strengthening primarily from manganese, vanadium and niobium. Typical applications for weathering

Table 1
Composition range for multiple-alloy weathering steels containing added phosphorus (wt%)

C (max)	Mn	P	S (max)
0.12–0.16	0.50–1.00	0.05–0.15	0.035–0.05
Si	Cu	Cr	Ni
0.15–0.90	0.20–1.3	0.30–1.25	0.00–1.0

steels are in railroad cars, bridges and unpainted buildings.

2. Vanadium-Containing Steels

Vanadium and niobium are two of the most frequently used microalloying elements. The development of vanadium-containing steels occurred shortly after the development of weathering steels. Vanadium contributes to strengthening by forming fine precipitate particles (5–100 nm diameter) of V(CN) in ferrite during cooling after hot rolling. The strengthening from vanadium averages between 5 and 15 MPa per 0.01wt% V depending on plate thickness and therefore on rate of cooling from hot rolling. Figure 2 shows how strengthening by vanadium additions varies with cooling rate from hot rolling (plate thickness). In a 0.15% V steel, optimum strengthening occurred at a cooling rate of 170 K min^{-1}. At lower cooling rates the V(CN) precipitates coarsen and are less effective in strengthening. At higher cooling rates, more V(CN) remains in solution; thus a smaller fraction of V(CN) precipitate forms and strengthening is reduced.

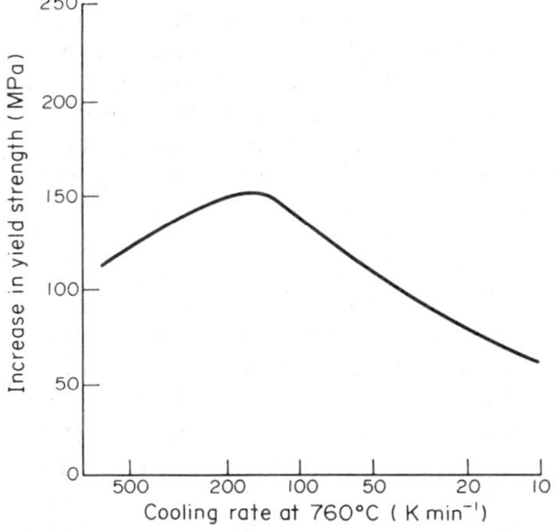

Figure 2
Effect of cooling rate on the increase in yield strength due to precipitation strengthening in a 0.15% V steel (after Bucher and Grozier 1965)

Hot-rolled Mn–Cu–V steels contribute a substantial part of HSLA tonnage; they provide an increase in strength of up to 40% and, as a result, an economical way to reduce weight, particularly in railroad cars and vehicles.

3. Niobium-Containing Steels

Vanadium was the preferred microalloying element in HSLA steels until the late 1950s, when niobium became readily available and lower in cost for the amount required for strengthening than vanadium. The first niobium steels were introduced commercially in 1958 by Great Lakes Steel Corporation (*Mechanical Engineering* 1959). Like vanadium, niobium increases strength through precipitation hardening, and also by refining ferrite grain size. Niobium precipitation in austenite prevents recrystallization during hot rolling and results in the formation of very fine ferrite from the flattened austenite grains. Moreover, niobium remaining in solution in austenite increases hardenability and this also aids in the formation of fine-grained ferrite. These two grain-refining mechanisms, combined with Nb(CN) precipitation strengthening of ferrite, account for the marked strengthening effects of niobium. The usual niobium addition is 0.02–0.04% which is about one-third of the optimum vanadium addition. Strengthening by niobium is 35–40 MPa per 0.01% addition.

In plates thicker than 125 mm, niobium-containing steels processed by conventional hot rolling exhibit poor notch toughness. This embrittling tendency can be eliminated if very fine-grained ferrite is formed by use of low-temperature rolling. As a result, niobium-treated steels are produced as sheets and plates up to 127 mm thick, although advanced control rolling procedures have now raised the thickness range to 250 mm.

Vanadium or niobium used either separately or in combination provide a full range of proprietary economical high-strength steels with yield strengths from 310 to 550 MPa and with good weldability and notch toughness, thus allowing cost savings as high as 30% compared with plain carbon steels.

4. Control Rolling

The object of control rolling is to achieve a very fine ferrite grain size. Figure 3 shows that as ferrite grain size is refined a double benefit accrues. Not only is the strength of the steel increased but also the ductile–brittle transition temperature of the steel is lowered. Moreover, ductility and toughness at a given strength level are improved.

On transformation of austenite, ferrite grains are nucleated primarily at the austenite grain boundaries. (Other nucleation sites are nonmetallic inclusions and deformation bands in the austenite.) Accord-

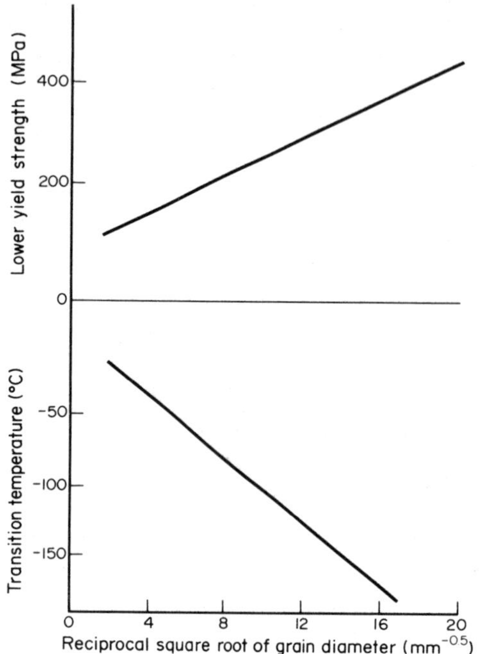

Figure 3
Effect of grain diameter on yield strength (Morrison 1966) and fracture appearance transition temperature (Heslop and Petch 1957) of plain carbon steel

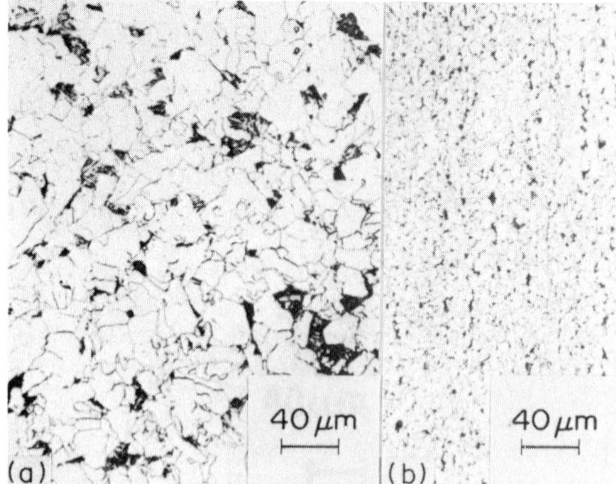

Figure 4
Grain structure resulting on transformation of austenite: (a) undeformed (30 μm grain size); (b) deformed by 90% reduction in the region (880–820 °C) where austenite does not recrystallize (after Cuddy 1981)

ingly, the finest ferrite grain size can be achieved for a given nucleation rate when the austenite grains are flattened so that the ratio of the grain boundary area to grain volume is maximized and the thickness of the flattened grains is minimized. (The resulting ferrite grain size is approximately half the austenite grain thickness.) To obtain minimum ferrite grain size, the objective is to provide a fine-grained austenite at the temperature where recrystallization of austenite between rolling passes ceases, and then to provide the maximum total reduction possible in the temperature interval where the austenite deforms but does not recrystallize and before ferrite transformation begins (that is, maximum flattening of a fine-grained equiaxed austenite). The very fine-grained structure resulting on transformation of highly deformed austenite is shown in Fig. 4.

The precipitation of Nb(CN) in the austenite during hot rolling raises the temperature at which recrystallization of austenite ceases (recrystallization stop temperature) and thus provides a broader temperature range in which the steel can be hot worked to produce highly deformed austenite. The optimum amount of niobium to suppress recrystallization between passes can be as little as 0.02%. Titanium, zirconium and vanadium are not as effective as niobium in raising the recrystallization stop temperature. Titanium and zirconium nitride formed during solidification and on cooling of the slab do not readily dissolve on reheating to hot-rolling temperatures. Although these nitrides may prevent grain coarsening on reheating, they are not effective in preventing recrystallization, because insufficient titanium or zirconium remains in solution at the rolling temperature to precipitate on deformed austenite grain boundaries during hot rolling and thus suppress austenite recrystallization. Vanadium, on the other hand, is so soluble that precipitation does not readily occur in the austenite during hot rolling. The concentrations of Nb, Ti, V, C and N, the degree of strain, the time between passes, the strain rate and the temperature of deformation all influence recrystallization during hot working.

Small amounts of nickel, chromium, copper and especially molybdenum are also present in many control-rolled HSLA steels. These elements increase hardenability and in this way promote high strength at low carbon levels in plate steels such as line pipe steels.

The superior toughness, ductility and weldability of low-carbon steels (<0.10% carbon) containing very little fine pearlite have made control-rolled HSLA microalloyed steels very attractive. These steels are often referred to as pearlite-reduced steels to distinguish them from the older HSLA vanadium and niobium-containing microalloyed steels of higher carbon content.

5. *Acicular Ferrite Steels*

Steels that transform to an acicular structure on cooling generally have very low carbon contents (typically 0.03–0.06%), so they exhibit good toughness and excellent weldability even though they have high strength. Strength and a low ductile–brittle transition temperature result from the very fine grain structure. Additional strengthening (with some loss in toughness) results from dislocation strengthening developed during the transformation. These steels are called acicular ferrite steels, although some prefer to call this structure low-carbon bainite.

The alloying elements particularly useful for providing the hardenability required to produce the acicular structure are manganese, molybdenum, niobium and boron. A typical composition and the mechanical properties for an acicular ferrite plate steel proposed for arctic line pipe are shown in Table 2.

Table 2
Typical composition and mechanical properties of an acicular ferrite arctic line pipe steel[a]

Chemical composition (wt%):						
C	Mn	Si	Mo	Nb	Al	N
0.05	2.0	0.10	0.35	0.07	0.030	0.005
Mechanical properties (9.5 mm thick plate):						
yield strength (MPa) 480–550						
FATT[b] (°C) −25 to −60						

a Coldren et al. (1973) b 50% shear fracture appearance transition temperature (⅔-size Charpy V-notch specimens, transverse orientation)

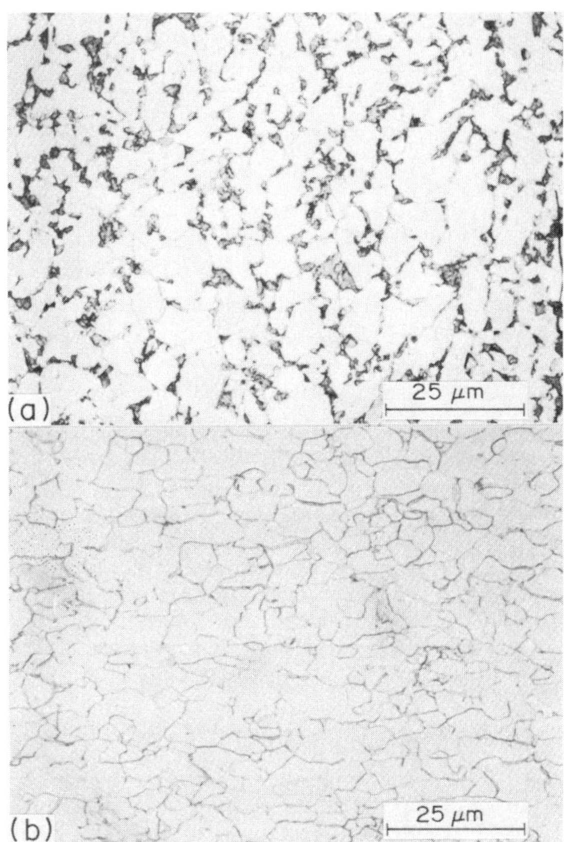

Figure 5
Comparison of microstructures of (a) dual-phase steel (ferrite–martensite) and (b) conventional HSLA steel (precipitation-hardened ferrite)

6. *Dual-Phase Steels*

A US patent on dual-phase steels was taken out in 1968, but the full potential of these steels was not widely recognized until Hayami and Furukawa published a description of their composition, microstructure and properties (Hayami and Furukawa 1976). In 1976, General Motors Research Laboratory (Rashid 1976) found that a simple annealing treatment could produce a sheet steel with 550 MPa minimum yield strength after 2–3% deformation. The steel has the same tensile strength as conventional HSLA steel having a yield strength of 550 MPa, yet it exhibits ductility and formability approaching those of plain carbon steel. This unique combination of properties results from the production of a "dual-phase" microstructure consisting of a ferrite matrix containing dispersed regions of martensite. The microstructure is produced by annealing a low-carbon HSLA steel in the intercritical region where pools of high-carbon austenite form in equilibrium with essentially carbon-free ferrite. On cooling, the high-carbon austenite transforms to about 20 vol% martensite. The resulting microstructure compared with that of a conventional HSLA steel is shown in Fig. 5.

Intercritically annealed dual-phase steels can be produced on a number of different facilities normally used for box annealing, continuous annealing and hot-dip galvanizing. In addition, dual-phase steel can be produced as hot rolled. Because the formation of the martensite pools is highly dependent on the cooling rate and composition, the HSLA steel composition necessary to develop dual-phase steel varies markedly, depending on the cooling rate characteristics of the facilities used to produce the steel. Typical compositions for intercritically annealed dual-phase sheet produced on a stainless continuous annealing line and typical compositions suitable for producing as-hot-rolled dual-phase steel are shown in Table 3.

7. *Optimization of Ductility and Toughness*

Since ductility and toughness usually decrease as strength increases, the optimization of ductility and toughness in HSLA steel is an important consideration. In addition to those factors already discussed

Table 3
Typical chemical compositions for intercritically annealed and as-hot-rolled dual-phase sheet steels (wt%)

C	Mn	Si	Mo	Cr	V
		Intercritically annealed			
0.09	1.40	0.50			
0.11	1.55	0.60			0.04
0.07	1.20	0.25		0.50	
0.11	1.20	0.60	0.12		
		As hot rolled			
0.05	0.90	1.30	0.40	0.60	
0.06	1.70	1.00		1.00	
0.06	1.70	1.00	0.50		

(i.e., the use of very low carbon content to minimize the presence of carbides in excess of those contributing to strengthening and the use of grain refinement as the only strengthening mechanism that lowers ductile–brittle transition temperature while increasing strength), high-cleanliness steelmaking practices such as ladle degassing and stream protection during casting and the use of very low-sulfur steels and/or sulfide shape control to minimize anisotropy are necessary for the production of high-quality HSLA steel.

Microalloying elements such as rare-earth metals, calcium, zirconium and titanium have the ability to change the composition of the sulfides or the manner in which they form so that the sulfides resist deformation during rolling. For instance, additions of rare earth metals (R's) change the precipitated sulfide from the low-melting-point, soft MnS to the high-melting-point, hard R_2O_2S. As a result the flattened MnS inclusions formed during hot rolling of untreated steels are changed to spheroidal R_2O_2S. These spheroidal R_2O_2S sulfides give a higher ductility in the transverse testing direction of sheet product and eliminate the normal anisotropy in ductility or toughness observed in hot-rolled sheet. Accordingly these elements may be added to HSLA steels for sulfide shape control.

See also: Ferrous Physical Metallurgy: An Overview

Bibliography

Bucher J H, Grozier J D 1965 Toughness of high-strength hot rolled steels. *Met. Eng. Q.* 5: 1–6

Buck D M 1965 A review of the development of copper steel. *AISI Year Book*, Vol. 10. American Iron and Steel Institute, Washington, DC, pp. 373–94

Coldren A P, Cryderman R L, Smith Y E 1973 Microstructure and properties of low-carbon Mn–Mo–Cb steels. *Processing and Properties of Low-Carbon Steels.* The Metallurgical Society (AIME), New York, pp. 163–89

Cuddy L J 1981 Microstructures developed during thermomechanical treatment of HSLA steels. *Metall. Trans., A* 12: 1313

Fletcher E E 1979 *A Review of the Status, Selection and Physical Metallurgy of High-Strength Low-Alloy Steels.* Metals and Ceramics Information Center, Battelle, Columbus, Ohio, pp. 78–89

Hayami S, Furukawa T 1976 A family of high-strength, cold-rolled steels. *Microalloying 75, Proc. Int. Symp. High-Strength Low-Alloy Steels.* Union Carbide Corporation, New York, p. 78

Heslop J, Petch N J 1957 Dislocation locking and fracture of α-iron. *Philos. Mag.* 2: 649–58

Larrabee C P, Coburn S K 1962 The atmospheric corrosion of steels as influenced by changes in chemical composition. *1st Int. Congress Metallic Corrosion.* Butterworths, London, pp. 276–85

Mechanical Engineering 1959. Columbium-treated mild-carbon steel. 81: 53

Morrison W B 1966 Effect of grain size on stress strain relationships in low carbon steel. *ASM Trans. Q.* 59: 824–46

Pickering F B 1976 High-strength low-alloy steels—A decade of progress. *Microalloying 75, Proc. Int. Symp. High-Strength Low-Alloy Steels.* Union Carbide Corporation, New York, pp. 3–24

Rashid M S 1976 *A Unique High-Strength Sheet Steel with Special Formability*, SAE Preprint 760206. Society of Automotive Engineers, New York

Speich G R, Dabkowski D S 1979 Effect of deformation in austenite and austenite–ferrite regions on the strength and fracture behavior of C, C–Mn, C–Mn–Cb and C–Mn–Mo–Cb steels. *The Hot Deformation of Austenite.* The Metallurgical Society (AIME), New York, p. 557

Williams F H 1900 Influence of copper in retarding corrosion of soft steel and wrought iron. *Proc. Eng. Soc. Western Pa* 16: 231–33

L. F. Porter

High-Temperature Oxidation of Metals and Alloys: Engineering Aspects

A number of engineering systems require some of their components to be exposed to high-temperature atmospheres which are oxidizing and may also contain other substances that accelerate the oxidation or react directly with the material of the components. Such systems include fossil-fuel-fired boilers, gas turbines, internal-combustion engines, incinerators, coal gasifiers, coal liquefiers, various types of chemical plant, and more advanced systems such as high-temperature gas-cooled nuclear reactors, nuclear fusion reactors, magnetohydrodynamic (MHD) generators and solar thermal systems. In many cases, some of the components exposed to the potentially corrosive environment may be required to have other specific properties, such as strength, toughness, rigidity, and perhaps lightness, thermal-shock resistance and thermal conductivity (high or low). The selected material must be fabricable into the shape required for the component, and it must be possible to attach it to the rest of the system. In most cases there are also the requirements that the system as a whole should be reliable and economic. Faced with these problems, the designer may adopt one or more of several approaches:

(a) Attempt to relax the environmental conditions.

(b) Cool the component below the temperature at which attack is unacceptably rapid.

(c) Design in corrosion allowances, or easy replaceability to accommodate limited component lifetime.

(d) Use inert materials such as structural ceramics.

(e) Select the metal or alloy with the best combination of corrosion resistance and the other required properties, and design to its limitations.

(f) Use a composite structure with the bulk material selected for the other required properties and a second covering material selected for its corrosion resistance.

Which of these approaches a designer uses ultimately depends on the criteria detailed above, particularly cost, reliability, fabricability, and perhaps also repairability. However, in many instances all these criteria interact, as will be shown by considering some specific examples.

1. Gas-Turbine Hot-Section Components

In a gas turbine, air is ingested and compressed. Fuel is then injected and burnt in a combustion chamber to raise the gas temperature; the hot compressed gas then passes from the combustion chamber through a transition piece which directs it against a row of stationary airfoil blades (nozzle guide vanes), which turn the gas and accelerate it. The gas then strikes a row of airfoil blades fastened to the rim of a wheel which drives a shaft. The blades in this row are called the rotor blades or buckets. There may be several stages of vanes and blades in the turbine, and the gas cools and expands as it passes through the stages. As with most heat engines, the efficiency is largely determined by the maximum temperature in the cycle, so that the aim in developing gas turbines is to raise the turbine inlet temperature as high as possible. For twenty or so years after the introduction of the gas turbine the major factor limiting this temperature was the creep-rupture strength, and to a lesser extent the corrosion resistance, of the alloys used for the first-stage rotor blades. It might appear that the combustion chamber is the most critical component, but because not all the air is involved in the primary combustion, it is possible to use a sheet-metal construction in which some of the air from the compressor is made to flow along the outside and inside walls of the combustor, keeping it relatively cool; in addition, the stress levels are relatively low. A major change was the introduction of cooling of both the nozzle guide vanes and the rotor blades, ducting cool air from the compressor through passages within the airfoils. Increasing complexity of the cooling channels has resulted in a progressive increase in the inlet gas temperature, while the metal remains at a fairly constant temperature.

However, the early gas-turbine designs were based on alloys such as Ni–20%Cr which depend for their oxidation resistance on the formation of a Cr_2O_3 scale. Two problems developed. First, increasing strength requirements led to the development of Ni–Cr–Al–Ti superalloys, and a reduction in the chromium content; second, as the temperature exceeded ~800 °C it became apparent that in high-velocity gas streams the Cr_2O_3 was further oxidized to the volatile CrO_3 and protection was lost. Some of the stronger alloys contained sufficient aluminum to develop a protective Al_2O_3 scale, or a mixed Cr_2O_3–Al_2O_3 scale (see *Oxidation of Metals and Alloys*). It appears that approximately 5 wt% Al is required for this. However, a new problem appeared. If the fuel or the intake air contains certain impurities, notably sodium (or potassium), sulfur and vanadium, an accelerated form of attack is experienced called hot corrosion (see *Hot Corrosion*). The reaction to the appearance of this attack involved several approaches:

(a) The fuel specifications for sodium and potassium were tightened; where necessary, fuels not meeting the specification are washed. The specifications for vanadium (and some other impurities) were also tightened; the use of additives based on MgO reduces the effect of the vanadium.

(b) The air filtration for land-based turbines was improved to reduce the ingestion of airborne impurities.

(c) The accumulation of salts on the vanes and blades is limited by regularly washing them with water.

(d) Engines exposed to severe conditions (e.g., marine hovercraft) are usually derated, that is, operated at lower than normal inlet temperatures.

(e) A new class of alloys has been developed, with strength equivalent to the low-chromium alloys, but containing enough chromium to confer hot corrosion resistance—IN 738 has been the most important product of this approach.

(f) It is now common practice to use protective coatings.

In practice, all of these things are done: each has disadvantages, limitations or economic penalties.

Protective coatings for these situations are almost invariably Al_2O_3 formers. For simple oxidation resistance in aviation engines burning very clean fuels, aluminide coatings, formed by diffusion from a pack containing aluminum, Al_2O_3 as a diluent and a halide activator, are widely employed. Recently, these have been improved by the incorporation of platinum or rhodium. For more severe conditions, a range of M–Cr–Al–Y coatings have been developed, where M is cobalt, or cobalt–nickel, or less commonly nickel or iron. The coatings are deposited by evaporating material from a molten pool using an electron beam; the vapor then condenses on the component surface. For simple oxidation resistance, the coatings are relatively high in aluminum and low in chromium; for hot-corrosion resistance the chromium is increased and the aluminum lowered, but there is always enough aluminum to form an Al_2O_3 scale. The problem with Al_2O_3 scales is that they tend to spall (break off) on cooling; the function of the small amount of yttrium (typically 0.5 wt%) is to improve the scale adhesion; the platinum in the platinum aluminides appears to perform the same function.

It is possible to use these very sophisticated approaches to the materials problems in the gas turbine because the vanes and blades together are a very small (albeit critical) part of the engine: coating the airfoils is a trivial fraction of the cost of a large turbine.

2. *Coal-Fired Boilers*

In a coal-fired boiler the economic situation is quite different from that for a gas turbine. A very large part of the system is exposed to severe oxidizing conditions, and only minor cost escalations could easily be accepted.

A utility boiler is very big, perhaps 40 m high, with the combustion chamber having a 15 m^2 cross section. It may burn 10 000 t of coal a day, producing 2000 t of ash. The combustion chamber is a box whose walls are formed by vertical tubes ~60 mm in diameter, linked by 20 mm steel webs. Water enters the bottom of these tubes at approximately 300 °C. The pulverized coal and the preheated combustion air are blown in through burners in the walls, and a combustion zone is established in the center of the furnace enclosure. About 20% of the ash falls to the bottom of the furnace; the rest is entrained in the combustion gases as fly ash and passes through the system. The combustion gases are initially at ~2200 °C and transfer heat by radiation to the water in the wall tubes. The water boils, and at the top of the walls the steam is separated from the water. The water returns to the bottom of the walls, and the dry steam is then further heated (superheated) in tubular heat exchangers which cross the gas stream. The finishing superheater is typically in the form of pendant loops in the top of the furnace section, whereas the reheating heat exchangers are in the later gas passes where the gas has cooled and the heat transfer is mainly by convection. The pressure inside the tubes may be as high as 25 MPa, although 17 MPa is more usual.

The maximum temperature of the steam is defined by materials properties. For the superheater, the limitation is creep rupture of the available materials as was the case for gas-turbine rotors (Sect. 1), but the candidate materials are different: low-alloy ferritic steels such as Fe–2.25%Cr–1%Mo, austenitic stainless steels such as Fe–18%Cr–10%Ni, and less commonly more highly alloyed materials such as IN800H and IN617. A boiler may well contain 50 km of tubes, and although only a small fraction of this length is exposed to the most severe conditions, the cost implications of using a more expensive material are very great.

The high-temperature corrosion problems are:

(a) Molten salt corrosion of the superheaters and reheaters—this is similar in principle to hot corrosion in gas turbines, and involves alkali sulfates derived from the coal.

(b) Steam-side oxidation of the superheaters and reheaters—the problem here is not the metal wastage, but the fact that the oxide formed spalls off periodically and can pass through with the steam to the turbine, where it causes erosion.

(c) Water-wall corrosion—although the water walls are relatively cool, if the combustion zone is not properly located in the furnace, conditions at the water wall may become very reducing. Under these circumstances, accelerated wastage can occur by a sulfidation–oxidation mechanism.

Obviously, it is desirable to limit the alkali content of the coal, but it may not be possible for an operator to do this. It turns out that the maximum attack of the superheaters is at metal-surface temperatures in the range 550°–725 °C, with a maximum at 650 °C. The steam temperature is typically approximately 40 °C less than the outer-surface temperature of the tube. It has been the standard practice in the USA for the last twenty years to limit maximum steam temperature to 538 °C, corresponding to a metal-surface temperature of the order of 580 °C, which certainly reduces the problem. The attack increases with increasing gas temperature, so that if it is known that a corrosive coal will be burned the boiler will be designed with the hotter tube banks further back along the cooling gas stream, at the expense of requiring a greater surface area in the heat exchanger. In severe cases, coextruded tubes may be used with an inner layer of a strong material and an outer thinner layer of a weaker, corrosion-resistant material. This is an expensive solution, but justified in some cases if a cheaper coal can be used. Additives to the fuel have demonstrated some success, but control is not easy. Coatings other than claddings have not so far been very useful, but this approach is currently being studied.

Water-wall corrosion is clearly due to poor combustion. Improving the burner design can often cure the problem, but where a different coal is being burned than the one for which the boiler was specifically designed, this approach may be inadequate. Bleeding air in along the walls to keep the local environment oxidizing has had some success, but is not simple. Plasma-sprayed coatings have been used; they do not last long, but they are relatively easy to replace in corrosion-prone regions on periodic boiler shutdowns. Again, coextruded tubes have been very successful where the problem is severe. Where corrosion is less severe, careful inspection and periodic retubing may be the least expensive approach.

Steam-side oxidation and exfoliation is the most recent problem to be identified. The approach here has been surface treatments of the inner surface of the tube. Two different methods have been developed recently: chromate-conversion coating and chromizing. The first of these involves depositing a chromium-rich oxide phase by decomposition of a chromate solution, the Cr_2O_3 being incorporated in the growing oxide, and has been successful in boiler trials. Chromizing is the high-temperature diffusion of chromium into the surface of the tubes, and in practice this also appears to be a good solution. These two methods are intended for low-alloy ferritic steels. For austenitic steels, the best approach is to use a fine-grained steel (at the expense of creep strength) or to cold work the inner surface by shot peening: both methods promote protective Cr_2O_3 scale formation as opposed to magnetite formation. However, alternative approaches again are possible; for example, the exfoliation may be accepted, and the turbine modified to be more erosion resistant.

3. *Coal Gasification*

The last example to be considered here is coal gasification. Coal is gasified by reacting it with steam and enough oxygen to provide the heat: the gas produced has an oxygen activity as low as 10^{-23} atm at 1000 °C, with a sulfur activity as high as 10^{-7} atm; the carbon activity is, of course, close to unity. Such an environment can cause severe corrosion of many metals by a sulfidation or sulfidation–oxidation process, and a great deal of work has been done in recent years to understand the mechanism of the corrosion process and to try to identify alloys or coating systems capable of resisting attack. These have largely failed.

The solution has been to design the systems such that as far as possible no metallic material is in contact with the gas at elevated temperatures. The gasifier vessels are lined with insulating refractory layers and any components such as stirrers or cyclones which must have metal bodies are water-cooled and (where possible) refractory lined. The gas is cooled before the sulfur is removed, and a cool clean gas goes to the next stages of the process. Indeed, any proposed gasification system that involved a component which would have to be of metal and could not be cooled would probably be rejected as impracticable.

However, there is still a possible problem in the cooling of the gas. In some gasification systems a significant part of the total energy is present as sensible heat in the hot gas, and it is undesirable from an efficiency point of view to throw this away; recovery of heat from relatively low temperature quench water is questionable, although not impossible. There is considerable interest, therefore, in the construction of metallic heat exchangers to boil water, and thus the question is how high a temperature may be achieved with alloys which are compatible with boiling water and of a cost compatible with the very large structures required? It seems probable that the water-compatibility criterion will require the inner surface of the tube to be a low-alloy ferritic steel: if the external surface was also a low alloy steel, the limiting temperature would be very low, perhaps corresponding to a surface temperature of 350–400 °C maximum. Chemically, the conditions are very similar to those experienced during water-wall corrosion in boilers. The solutions available are much the same—coating the metal surfaces, perhaps with an aluminum or chromium-rich diffusion coating, or using coextruded tubes. In any event, the problem of whether to use such a heat exchanger and select the materials combination and maximum surface temperature consistent with an acceptable lifetime, or to abandon this method and quench the hot gas, recovering some of the heat from the quench liquor, requires careful analysis.

See also: Corrosion in the Power Industry; Corrosion of Stainless Steels; Design for Corrosion Control

Bibliography

Beltran A M, Shores D A 1972 Hot corrosion. In: Sims C T, Hagel W C (eds.) 1972 *The Superalloys*. Wiley, New York, pp. 317–39

Grisaffe S J 1972 Coating and protection. In: Sims C T, Hagel W C (eds.) 1972 *The Superalloys*. Wiley, New York, pp. 341–70

Reid W T 1971 *External Corrosion and Deposits: Boilers and Gas Turbines*. Elsevier, New York

Stringer J 1977 Hot corrosion of high-temperature alloys. *Annu. Rev. Mater. Sci.* 7: 477–509

Wasielewski G E, Rapp R A 1972 High-temperature oxidation. In: Sims C T, Hagel W C (eds.) 1972 *The Superalloys*. Wiley, New York, pp. 287–316

J. Stringer

High-Voltage Ceramic Insulators

A high-voltage insulator is a nonconducting mechanical support for an electrical conductor. A ceramic whiteware body is mechanically strong and has superior electrical insulating properties. It is inert and can withstand wide ranges of temperature. It does not corrode, absorb moisture, polymerize when exposed to ultraviolet radiation or char as a result of electrical arcing. Since its adoption in the 1890s, new designs, improved processing equipment and better materials technology have met the increasing demands for electrical energy, safety and reliability.

1. High-Voltage Insulator Standards, Tests and Specifications

Voltage requirements for transmission and distribution insulators are nominal figures established by operating personnel of the electric utility companies. Service conditions determine design and operating voltages. The National Electrical Manufacturers Association (NEMA), the American National Standards Institute (ANSI) and the International Electrotechnical Commission (IEC) are concerned with establishing standards, specifications and tests for the electrical industry. Figure 1 shows various standard types of high-voltage insulator.

2. Basic Design Considerations

Insulators normally operate under conditions where they are immersed in some type of insulating fluid. Usually it is air, but it may be an oil or another demanding environment. When an insulator fails electrically, it is due to the failure of one or more of the following design factors:

(a) the dielectric strength of the insulator material;

(b) the leakage distance path over the actual surface of the insulator;

(c) the arcing distance through the fluid surrounding the insulator.

Certain relationships between these three insulating paths must be maintained if the insulator is to perform satisfactorily. An insulator of good design is self-protecting at the applied voltage. By making the fluid path weaker than the dielectric path, the insulator will flashover instead of puncture. Additional design factors which should be considered are corona noise, wet flashover, impulse flashover and current surges due to lightning strikes. Mechanical design considerations must include the dead weight of the conductor, effects of ice and wind, conductor vibration and stresses due to torsion and cantilever loading. There are manufacturing limitations arising from forming techniques, handling, shrinkage and firing.

3. Composition of Wet Process Ceramic Body

The mechanical and electrical strengths of high-tension insulators can be controlled by the use of certain ceramic raw materials. The choice of these materials is a compromise between their cost, availability, composition and physical properties. Two standard compositions are shown in Table 1. Composition A-100 can fulfill most insulator requirements. The replacement of flint by calcined alumina in composition B-100 permits the insulator body strength to be doubled. Table 2 gives typical physical properties of insulators obtained from composition A-100.

Table 1
Wet process body and glaze compositions for high-voltage insulators

Ceramic	A-100 (%)	B-100 (%)	C-100 (%)
Ball clay	30	20	14
China clay	15	10	9
Flint	20	–	33
Calcined alumina	–	35	–
Feldspar	35	35	18
Calcium carbonate	–	–	16
Color oxides	–	–	10

Table 2
Typical physical properties of wet process high-voltage insulators (refer to American Society for Testing and Materials, D-116-63)

Modulus of rupture, 1.3 cm diam. bar (MPa)	
unglazed	70
glazed	110
Flexural strength (MPa)	80
Tensile strength (MPa)	50
Compressive strength (MPa)	390
Modulus of elasticity (GPa)	80
Dielectric constant	5.75–6.00
Dielectric strength, 5 mm (kV mm^{-1})	28
Coefficient of linear expansion, 20–800 °C (K^{-1})	6×10^{-6}
Resistivity (Ω cm)	
20 °C	10^{14}
200 °C	10^{8}
Power factor at 1 MHz	0.008
Hardness (Mohs scale)	7
Porosity, fuchsine dye penetration	0
Moisture absorption	0

4. Glaze Composition

High-voltage insulators are installed in various types of environment. The surface of the insulator must remain clean; should it become contaminated, conditions resulting in flashover or continuous arcing could develop. To help prevent this, the insulator body is covered with a smooth glassy surface or glaze. Composition C-100 shown in Table 1 is a typical glaze

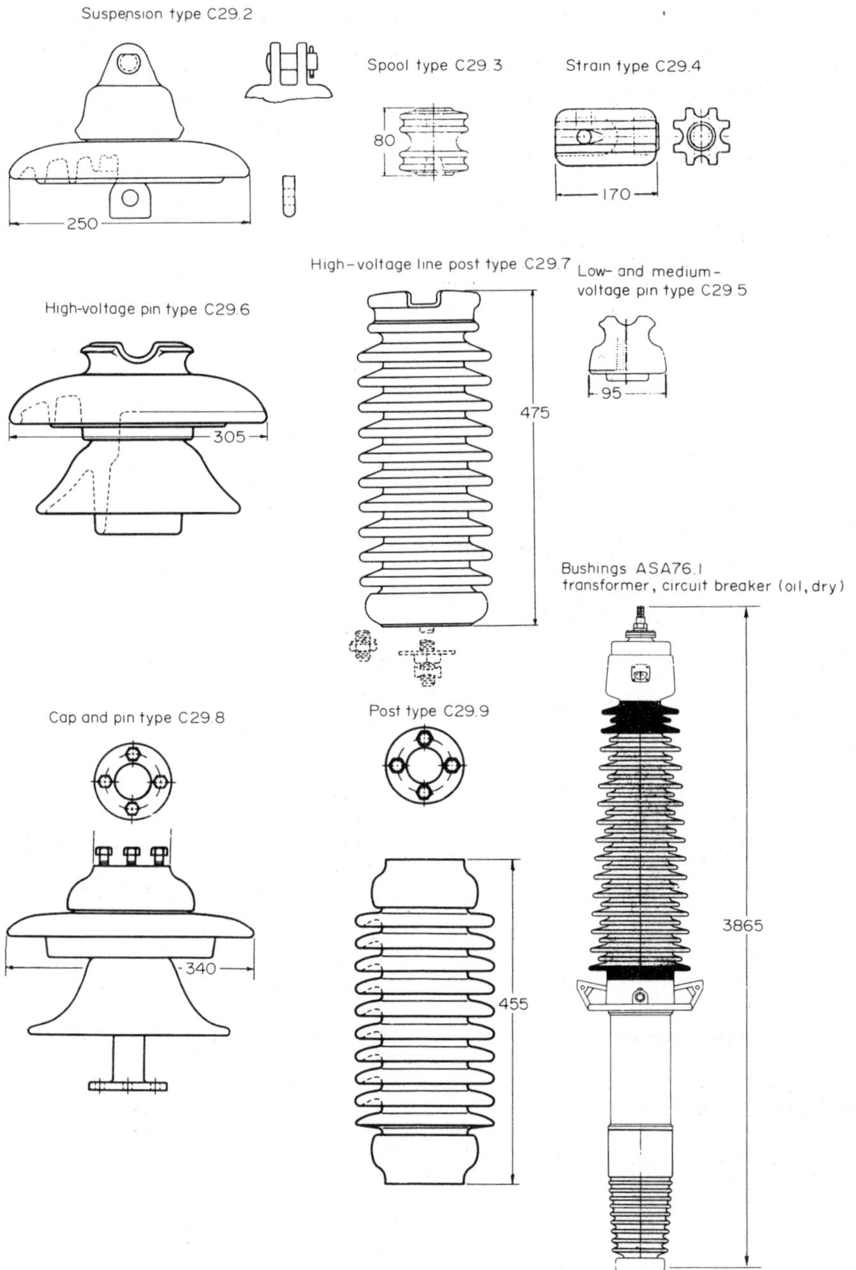

Figure 1
Types of wet process porcelain high-voltage insulator, with dimensions in mm (after ANSI C29)

formula. It requires the same basic raw materials as the porcelain body, but fluxing materials like calcium carbonate, zinc oxide or barium carbonate must be added to increase the glass-forming properties. Parmelee (1948) gives glaze formulae calculations and descriptions of raw materials.

Glazes can be made in various colors, but dark colors are preferred. Cracked or broken insulators can be detected by the color contrast between the white insulator body and the dark glaze. The different coefficient of thermal expansion of a glaze can place the surface of the insulator in either compression or tension. A glaze in compression can increase the body and impact strength by as much as 50%.

5. *Manufacturing Materials*

Inorganic, organic and metallic materials are selected on the basis of their physical properties which facilitate the manufacture of a uniform and reliable product.

Ball clays and china clays have three main functions: they give the insulator body its plastic forming properties; they act as a binder which controls the dry strength of the unfired body; and they prevent the segregation of coarse particles during the manufacturing process, owing to their large surface area.

Feldspar is the basic glass-forming material. During the firing process it melts, binding the refractory clays, flint and calcined alumina particles together.

Flint is a refractory material with a melting point of 1750 °C. It gives the insulator body a structure that remains rigid during the firing process.

Calcined alumina is a refractory material with a melting point of 2000 °C. When used to replace flint in the insulator body, it increases the fired strength. It is more stable during the firing process and it has better thermal shock properties. Economic reasons limit its use, as it is about fifteen times as expensive as flint.

The basic raw materials used to make the insulator body are used to make the glazes. Percentages differ, and auxiliary fluxes have to be used to promote glass formation, control surface texture and provide corrosion and abrasion resistance. The addition of zirconium silicate or tin oxide, called opacifiers, makes the glaze opaque.

A bonding agent, type I Portland cement, is used to hold metal hardware to various types of high-tension insulator. When Portland cement is mixed with water and cured properly, it develops a very hard and strong bond. When certain types of insulator require aluminum hardware, sulfur cement can replace the Portland cement as the bonding agent.

Coatings on metal surfaces for protection from corrosion and for stress relief may be an asphalt compound or an organic polymer. For corrosion protection, they are applied to the metal surfaces which make contact with the Portland cement. The coating absorbs the differences in thermal expansion between the metal hardware and the insulator body. The coating acts as a high-pressure lubricant, allowing the hardware and insulator body to make minor adjustments to equalize loading stresses.

Gripping sand is a sized material made from crushed insulator body. It is applied during the glazing operation to certain areas of the unfired insulator body. During the firing process, the sand is fused to the body. This provides a gripping surface for Portland cement when bonding metal parts to the insulator body.

Metal hardware parts for high-voltage insulators are made from malleable iron castings, steel forgings, drawn steel parts or aluminum alloys. These metals lend themselves to designs where strength, complicated shapes, varying thickness, dimensional control and light weight are required. All ferrous parts are zinc-galvanized for corrosion resistance.

Plaster molds are used to support the plastic body while it is being formed and dried. The outside contour of the insulator is determined by the mold. Whiteside (1966) describes the manufacture and quality control of plaster molds.

6. *Manufacturing Operations*

The raw materials are weighed batchwise and mixed with water in a 50:50 blend called "slip." The wet process method ensures maximum dispersion and blending of all ingredients. The slip is aged and blended in storage wells before being screened to remove coarse material. Filter pressing removes 80% of the water. The resulting plastic filter cake is shredded and extruded through a wad mill to form large solid cubes of plastic insulator body. The cubes are then passed through a vacuum pug mill, the output from which is either sent to a dryer to remove all the water or placed in a plaster mold to be shaped, while still plastic, into an insulator form. If a dryer is used, the dried blank of insulator body is turned to a specific shape on a lathe. Both the plastic-formed units and the dry-turned units are transferred to the glaze department, where glazes and gripping sand are applied. The glazed insulators are moved to the kiln for the firing process. Depending on the shape and mass, the insulator is fired in either a tunnel kiln or a periodic kiln. Firing cycles range from 30 to >100 h.

Insulators removed from the kiln are inspected visually, electrically and mechanically. Those requiring cement-bonded hardware are sent to the assembly and steam curing department. After 3–7 days of steam curing, the insulators are cleaned and are given a final electrical test. Quality control samples are taken for evaluation, and on the basis of the results the insulators are packed for shipment. The flowsheet for the manufacture of wet process high tension insulators is shown in Fig. 2.

7. *Future Requirements*

Present materials and designs, while adequate, are being modified to meet new demands. New concepts are needed to solve contamination and corrosion problems resulting from harsh environments, direct-current transmission and extra-high-voltage energy fields. There is a need for increased dielectric strength, mechanical strength and impact strength. Insulators need to be lighter in weight and immune to the heat of flashovers and arcing. Underground transmission and distribution pose special corrosion,

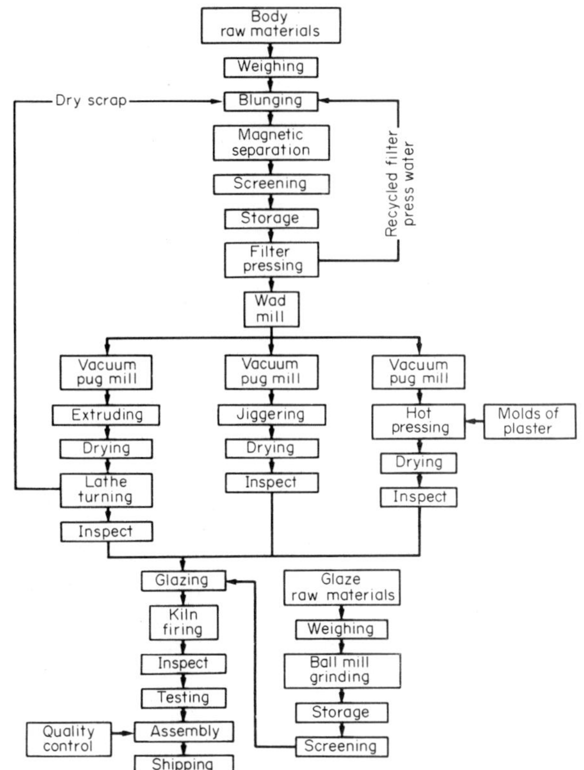

Figure 2
Flowsheet for manufacture of wet process high-voltage insulator

repair and detection problems. High-voltage insulators made from elastomeric and fiberglass materials are lightweight and strong but cannot endure the heat of arcing or deterioration from ultraviolet exposure.

See also: Low-Voltage Ceramic Insulators; Spark-Plug Insulators

Bibliography

Andrews A I 1948 *Ceramic Tests and Calculations*. Wiley, New York

Mills B 1970 *Porcelain Insulators and How They Grew*. Canfield and Tack, Rochester, New York

Norton F H 1970 *Fine Ceramics: Technology and Applications*. McGraw-Hill, New York

Parmelee C W 1948 *Ceramic Glazes*. Industrial Publications, Chicago, Illinois

Robson J T 1954 *Operating the Tunnel Kiln*. Industrial Publications, Chicago, Illinois

Von Hippel A R (ed.) 1954 *Dielectric Materials and Applications*. Wiley, New York

Whiteside E L 1966 Quality control in the plaster mold shop. *Am. Ceram. Soc. Bull.* 45: 1022–26

D. R. Brown

Holographic Materials

Holography is a wavefront reconstruction imaging process involving photographic film-like recording materials. When a suitable coherent reference light wave is present simultaneously with the light diffracted by an object, information involving both the amplitude and phase of the diffracted light waves can be recorded on film in spite of the fact that the recording film responds only to intensity. Such a recorded interference pattern is called a hologram, meaning "total recording," and an image of the original object can be obtained from it by several reconstruction procedures (Smith 1975). Dennis Gabor in 1948 first introduced the hologram concept of true three-dimensional recording into a flat photograph (Born and Wolf 1959). Since the advent of lasers, which are used as coherent light sources in holographic recording and reading, holography has become an important imaging technique. The three most commonly used materials for holographic recording are silver halides, dichromated gelatins and photoresists. The current technology is such that high diffraction efficiency holographic recordings can be routinely achieved.

1. Materials Properties Necessary for Holography

The criteria for holographic materials can be simply understood from the recording scheme of Lippmann (Wood 1967) shown in Fig. 1. Light is focused onto photographic film and then reflected back on itself to produce an interference pattern in the film layer. Such a film would have to be optically transparent in order for the light to be able to interact with the reflector; thus, Rayleigh scatter from grains must be small. Also, the pattern of interference in the film layer (see inset, Fig. 1) must have a definite fidelity and not be smeared. The antinodal zones in the pattern are separated by $\lambda/2$ where λ is the wavelength of light in the film. Thus the recorded patterns must exhibit a precision of the order of $\lambda/10$. The ability of photographic materials to resolve such detail is constrained by grain structure, or by dif-

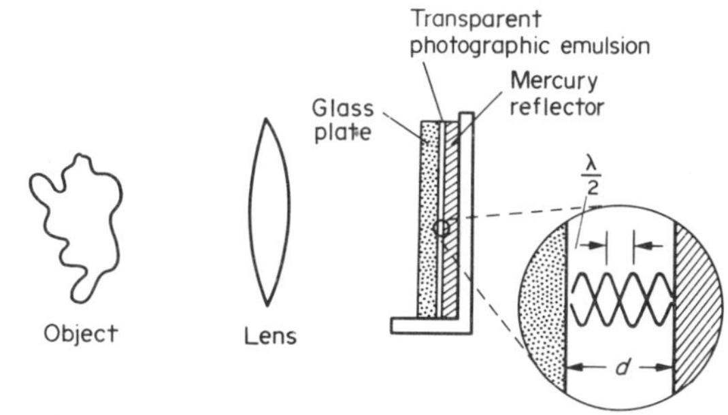

Figure 1
Lippmann's arrangement for holography

fusion and molecular size. For example, when using a recording wavelength of 514.5 nm (in air), the wavelength is reduced to 336 nm in the photographic gelatin (refractive index of 1.53) and detail must be approximately $\lambda/10$ or 34 nm. Crawford (1960) found theoretically that a true Lippmann emulsion should have a grain size of no larger than 10 nm.

Holographic materials are conventionally classified in terms of spatial frequency response for a pattern recorded by two plane waves of coherent light at angle θ incident on the same side of the recording layer. The spatial period of the recording is given by

$$\Delta = \lambda/(2 \sin \theta)$$

for two waves of equal intensity. The ability of the material to record fine spatial patterns is used as a criterion for the resolving ability of the material. This property is critically dependent on processing chemistry as well as the preexposed characteristics of the recording material.

Control of the physical size of photographic grains in film is very important for holography. Rayleigh scattering of light from small halide grains of radius a ($a \ll \lambda$) in gelatin is

$$I_s \propto \frac{a^6}{\lambda_0^4}\left[\frac{(n_H^2 - n_G^2)}{(n_H/n_G)^2 + 2}\right]^2$$

where n_H is the refractive index of the halide grain, n_G is the refractive index of the gelatin and λ_0 is the vacuum wavelength (van de Hulst 1957). Thus, although the wavelength of light shortens as the light enters the gelatin layer, the index-matching effect of the gelatin host helps by reducing the scatter cross section. Fine-grain holographic emulsions have typical grain diameters as listed in Table 1.

Table 1
Grain sizes of holographic silver halide materials

Material	Grain diameter (μm)
Agfa-Gavaert 8E75HD, 8E56HD	0.035
10E75, 10E56	0.09
Soviet/Bulgarian Lippmann	0.01

Light scattered by the emulsion does not contribute to the interference structure of the hologram and fogs the recording material. The mean free path for Rayleigh scatter is comparable with the emulsion layer thickness (~6 μm) for some commercial materials (i.e., Agfa 8E75HD) whereas for other materials (i.e., Agfa 10E75), there is significant scatter of the recording light wave in its transit across the layer due to larger grain sizes. Large grains are used to increase photographic speed, but there is no need to increase the grain size beyond the value where the scatter mean free path in the film layer becomes comparable to the optical path. The blackening of the emulsion during development may otherwise not be consistent with the revelation of image information. Extra photographic speed, as found in coarser grained material, is not useful because the silver produced by development in the faster material is in part the result of scrambled exposure due to light scatter. It is preferable to use the finest grained material and then enhance its speed by either hypersensitization or special chemical treatment during development.

2. *Silver Halide Developable Materials*

The most common holographic materials are those using silver halide grains. The reason for this popularity is that once light has been received by the halide material, the effect is amplified by development so that tiny latent specks of the pattern image are converted into observable grains of silver often of filamentary or worm-like construction. This built-in gain through exposure and processing is the basis of modern photography (see *Photographic Materials*).

The dynamic range of silver halide recording materials is an important consideration for holography. The dynamic range of photographic film is given by the plot of optical density versus log of the exposure. This is called the Hurter–Driffield curve and is shown in Fig. 2. For example, the datum exposure for a typical holographic emulsion (i.e., Agfa 8E75HD) is about 300 μJ cm^{-2}. This response curve leads to the inability of holograms to cope with several (approximately four) incoherent superposed exposures. Fundamentally, the reason for the lack of dynamic capability is the reliance on the amplification of the image by development where the filamentary silver structures are dispersed from the original sites of sen-

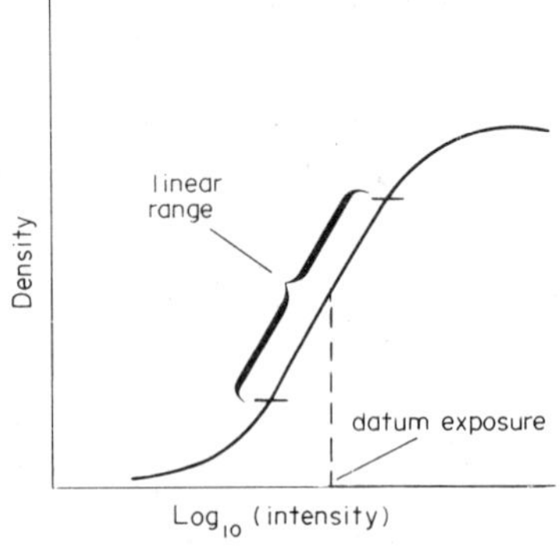

Figure 2
Hurter–Driffield curve showing how density of the developed image changes with light intensity

sitization and become intertwined. Loss of halide in an uncontrolled way during processing is the primary reason for low diffraction efficiency when silver halide materials are used for the recording of holograms.

Recent studies have indicated that by using very high exposure levels, as compared to the datum level, followed by very brief development (~10 s), compact silver structures can be produced. The implication is that if very high elemental exposure is given, then the compact silver can coexist with that produced by other exposures and the dynamic range need not be overloaded as by the conventional techniques (moderate exposure and long development time). Thus the capabilities of silver halide materials to record incoherently superposed images is now improving as better techniques and chemistry allow more exposure and rapid development. The trend is to reduce the amplification during development and increase the exposure accordingly.

Bleaching and tanning are two significant chemical processes used to improve halide materials for holographic recording. Bleaching methods can minimize image solution, with the vapor phase bleach method being of great importance. Graube (1974) nearly eliminated spurious solution effects by this vapor phase technique, obtaining holographic grating efficiencies up to 80% where 60% was the best efficiency obtained by other methods. Solymar and Cooke (1981) indicate that 99.9% efficiency in silver halide materials is obtainable for simple two-plane beam recording by optimizing the development process and using the vapor bleach method. Also tanning methods can be used to improve halide materials where all halides are removed from the hologram during processing so as to leave a tanned and voided gelatin material (Chang and Winick 1980, Graver et al. 1980). The tanned images produced are devoid of grain scatter in the final product, which can be thought of as lying between the silver halide and dichromated gelatin technologies.

Recent progress in silver halide technology for holography has made it competitive with the other materials. For some applications, however, grain scatter in the hologram is unacceptable and attempts to reduce grain size to overcome this scatter have met with difficulties. Soviet research on ultrafine grain materials (Solymar and Cook 1981) demonstrated that there are severe problems in dealing with such tiny grain sizes (~10 nm). The solubility of the ultrafine grains is so high that conventional development and bleaching techniques cannot be used, and new colloidal development techniques must be used. Also the photographic speed is relatively low, one or two orders of magnitude below those of Agfa and Kodak commercial films, which limits their application. In addition, once ultrafine grain holograms are processed, they have to be bulked up on drying so as to thicken the layer and increase the diffraction efficiency. At the present time the use of silver halide materials in holography requires a virgin grain size near 30 nm, and it is clear that on either side of this size there is a rapid degeneration by either scatter or excessive solubility.

3. Dichromated Gelatin

Dichromated gelatin (DCG) is a gelatin layer soaked in either potassium or ammonium dichromate and then, after exposure, rapidly dried in isopropanol and hermetically sealed to prevent attack by moisture (Smith 1977). Very high intensities (~300 mJ cm^{-2}) of monochromatic light are needed for holographic recording using DCG. The physical mechanism that modulates the refractive index in DCG holograms is not precisely known, but it is generally believed that the index variation necessary to cause the required phase shifts in the hologram is associated with the voids and cracks in DCG. Large phase shifts can be induced in DCG without excessive scatter so that the modulation can be taken beyond the limits set by the grain scatter of silver halide materials. In general, DCG is an insensitive medium being about three orders of magnitude less sensitive than silver halide but is clearly the material of choice for holographic optical elements where high volume diffraction efficiency is required.

The chief application of DCG holograms is for heads-up displays for aircraft. Lippmann-type reflective DCG layers are fabricated on both flat and curved surfaces with a high reflection efficiency over a narrow band of wavelengths. This holographic optical element can be used as a beam splitter allowing a pilot to look both at and through the screen. The DCG material can be processed to yield recordings of either narrow- or broadband reflectance.

4. Photoresist

Photoresist is a medium which is developable by the preferential removal of nonhardened molecular structures from those which have been either hardened or not softened by light. Resist is normally coated by spinning or dipping onto a substrate such as glass and then exposed to blue or violet light, where the sensitivity is optimal. When a line grating is recorded, the preferential solution of part of the coating leads to the formation of relief. This effect is due to a difference of removal rates and is thus subtle and relies on experience and some art for its repeatable optimization. This procedure is inherently nonlinear and transmission image holograms can have no more than a few percent efficiency (Smith 1977). The nonlinearity is due to both the structure and the chemical processing.

The surface of the resist can be copied after processing by plating or using resins. The aim is to produce a relief structure (as shown in Fig. 3) with optical path differences generating a modulation effect. The maximum optical path difference $(n-1)(h_1-h_2)$ and the incident wave on the left is diffracted into various orders by the relief structure. Maximum first-order diffraction occurs when the Bessel function $J_1\{2\pi n(h_1-h_2)/\lambda_0\}$ is maximized. This sets a modulation depth requirement given by

$$2\pi n(h_1-h_2)/\lambda_0 = 1.84$$

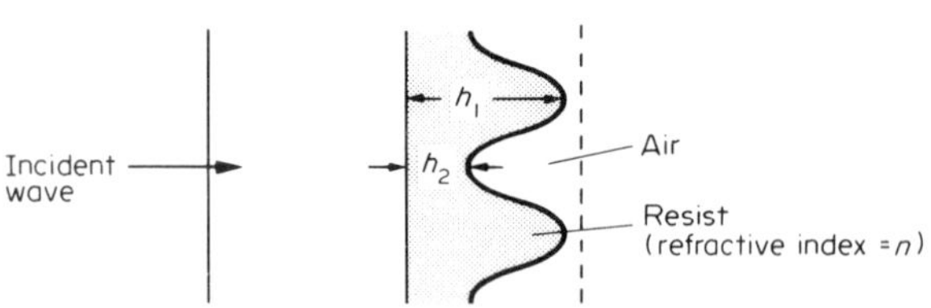

Figure 3
Relief surface of photoresist used in transmission

Other applications have been proposed by Petit (1980) in which a transmissive hologram is made in the form of a grating which loses all of its zero-order light into the higher diffraction orders, thus requiring

$$J_0\{2\pi n(h_1-h_2)/\lambda_0\} = 0$$

or that

$$2\pi n(h_1-h_2)/\lambda_0 = 2.4$$

Petit's proposition is then the subsequent superposition of a further relief structure which permits light to return into the zero-order direction to generate a transmissive two-dimensional color slide. The importance of this work lies in the possibilities of the fabrication of archivally stable color slides in which color encoding is performed by superimposed grating effects. An obvious constraint is imposed by the high zero-order modulation depth requirement.

The photoresist material has a major application in the replication of optical elements, and mass replicated optical elements using surface relief. Transmissive elements require relatively high modulation depths so photoresists play an important role for these elements. The use of this technology in the manufacture of digital disks for sound and vision is a spinoff from this holographic work. Of popular interest is the use of reflective gratings for display holograms. These can be made by coating the undulatory surface of the resist with a reflective layer; thus, the maximal phase difference is greater than the transmissive case by the factor $2n/(n-1)$.

5. Photopolymers

Materials which polymerize under intense light are extremely interesting for holography because they offer the possibility of development in situ. In some materials anaerobic conditions are required, whereas in others the coating can be applied to a substrate, baked and then exposed in air. The modern ultraviolet curing cements are by and large anaerobic and thus can only be exposed in an air-tight sandwich. In most respects, photoresists are just photopolymers and are therefore in the same category, but the role of resists is hardly if ever in applications other than the formation of surface relief elements.

There is a distinct difference between the use of photopolymers as surface relief generators and as volume holograms. Relief may be generated in many photopolymers, but volume refractive index modulation is much more subtle and usually requires high exposure light levels, sometimes near $1\,J\,cm^{-2}$. Jenney (1970) demonstrated a large volume index change for a 25 μm thick photopolymer film. Soviet scientists have used polymers activated by benzoquinone which exhibit volume refractive index changes. It is found, however, that photopolymer holograms need either heating or postexposure irradiation to stabilize the recording.

6. Other Holographic Materials

Thermoplastic materials make use of surface relief which is initiated by the exposure of a plastic to light followed by induction of the relief by electrostatic attractive effects (Smith 1977). The convenience of this method lies in the thermal erasability and the reusable nature of such layers. Thermoplastic holograms tend to have a band-limited character to their range of spatial frequencies, although the band does not necessarily begin at zero frequency. The thermoplastic system has found application in holographic interferometry for small-aperture quick-access work often with pulsed-laser recording light. This medium is "writable and erasable" in a slow dynamic fashion, but in no way lends itself to sufficient speed for flicker-free picture presentation in real time.

Bismuth silicate is a crystal capable of real-time recording without the need for a process procedure (Huignard and Herriau 1980). Grating effects can be induced in the material just by exposure to light permitting the replay in real time of a holographic image recorded in the medium by the phase-conjugated reference wave. Such replay allows images to be formed through phase-distorting media and also permits easily accessed real image interferometric recording. The diffraction efficiency of holograms produced in bismuth silicate is relatively low but the exposure level required to induce modulation is also low, at about $300\,\mu J\,cm^{-2}$ from an argon laser at 514.5 nm.

Holograms can be prepared by ablation of a thin film of copper or bismuth deposited on a flat substrate. The ablation can produce a spatial modulation consistent with the interference pattern presented to the thin-film layer. This method is useful for recording holograms at 10.6 μm with carbon dioxide lasers.

See also: Photographic Materials

Bibliography

Born M, Wolf E 1959 *Principles of Optics*. Pergamon, Oxford

Chang B J, Winick K 1980 Silver-halide gelatin holograms. *Proc. Soc. Photo-Opt. Instrum. Eng.* 215: 172–77

Crawford B H 1960 *Small-Scale Preparation of Fine-Grain, Colloidal, Photographic Emulsions,* National Physical Laboratory Notes on Applied Science No. 20. HMSO, London

Graube A 1974 Advances in bleaching methods for photographically recorded holograms. *Appl. Opt.* 13: 2942–46

Graver W R, Gladden J W, Easter J W 1980 Phase holograms formed by silver-halide (sensitized) gelatin processing. *Appl. Opt.* 19: 1529–36

Huignard J P, Herriau J P 1980 Dynamic holography and coherent four wave mixing in $Bi_{12}SiO_{20}$ crystal. *Proc. Soc. Photo-Opt. Instrum. Eng.* 215: 178–82

Jenney J A 1970 Holographic recording with photopolymers. *J. Opt. Sci. Am.* 50: 1155–61

Petit R 1980 *Electromagnetic Theory of Gratings*. Springer, Berlin

Smith H M 1975 *Principles of Holography*. Wiley, New York

Smith H M 1977 *Holographic Recording Materials*. Springer, Berlin

Solymar L, Cooke D J 1981 *Volume Holography and Volume Gratings*. Academic Press, New York

Van de Hulst H C 1957 *Scattering of Light by Small Particles*. Wiley, New York

Wood R W 1967 *Physical Optics*. Dover, New York

N. J. Phillips

Holographic Nondestructive Evaluation

Optical holographic nondestructive evaluation (NDE) enables material flaws or inhomogeneities to be detected through their effect on surface deformation when a test object responds to some mechanical or thermal load. Optical holography is a method by which complex light waves can be recorded and later reconstructed with such fidelity that they can be used to form interference fringe patterns. Using two holographic recordings—one of the light scattered by an object in some initial state and one of the light scattered by the same object while it is subjected to some additional load—an image of the object with superimposed interference fringes is formed. These fringes are contour curves of constant surface displacement and therefore constitute a whole-field display of the deformation of the test object in response to the applied load. This deformation may be caused by the presence of surface cracks or subsurface defects such as voids, delaminations or material nonhomogeneities. If one compares the fringe pattern for a flawed test object with that for a defect-free master object, it is possible to detect the presence of a subsurface defect, and, possibly, to estimate its size and depth. The key to applying holography to a given NDE problem is the development of a loading method such that the defect will cause a detectable anomaly in the deformation of the test object. Because this is an interferometric technique, its sensitivity is high. Deformation measurements are made on a scale comparable to the wavelength of light, 0.5 μm. Specimens or components of relatively complicated shape and nearly arbitrary surface finish and cleanliness can be inspected.

1. Holography and Holographic Interferometry

The NDE methods discussed in this article require that complex light waves be recorded and later reconstructed. In order to reconstruct a light wave, it is necessary to record the spatial distribution of both its amplitude and phase. All information about phase is lost in ordinary photography. In 1949, Gabor demonstrated that distributions of both phase and amplitude can be recorded on a photographic plate if a coherent reference wave of light is added to the wave to be recorded. The amplitude and phase distributions are then encoded as an interference pattern. When the plate is developed and reilluminated by a light wave similar to the reference wave, the plate acts as a diffraction grating from which three waves emerge. One of these waves is a precise replica of the original object wave. Gabor first referred to this process as wavefront reconstruction, but soon the term *holography* was coined based on his reference to the developed photographic plate as a hologram, indicating that it contains the total optical description of the object. The Nobel Prize in physics was awarded to Gabor in 1971 for this invention.

Virtually all holographic NDE systems utilize off-axis holography which was developed by Leith and Upatnieks (1962). A system for recording an off-axis hologram suitable for NDE applications is shown in Fig. 1. Light from a laser is divided by a beamsplitter into two parts. One part, the object wave, illuminates the test specimen and is then reflected and scattered toward the photographic plate. The other, the reference wave, is a simple plane or spherical wave which impinges on the plate from a different direction. For virtually all NDE applications, it is necessary to use laser light because of its high coherence, that is, its facility for forming interference patterns. Holograms are most often recorded on high-resolution (~1000 lines per mm) silver halide photographic emulsions coated on glass plates; however, other suitable recording media are described in Sect. 2. The hologram is recorded by simultaneously exposing the plate to the reference and object waves.

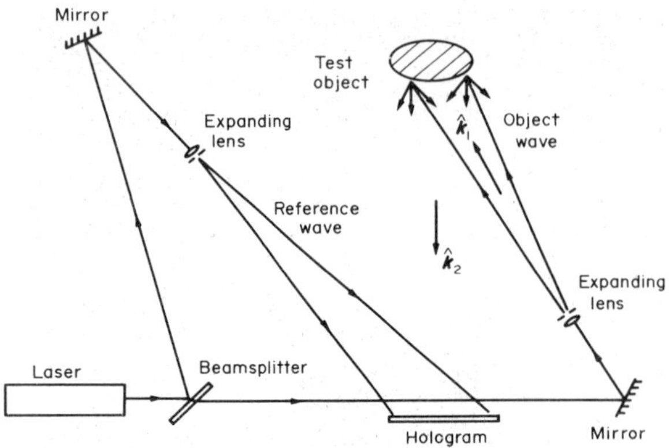

Figure 1
System for recording an off-axis hologram of a test object for NDE

To reconstruct the object wave, the developed photographic plate, referred to as a hologram, is illuminated by a wave identical to the reference wave, as shown in Fig. 2. This is referred to as the reconstruction wave. The hologram transmits an attenuated portion of the reference wave and, more importantly, diffracts part of it into the form and propagation direction of the original object wave. Analysis of the process predicts a third wave, but it is usually trapped by multiple reflections in the photographic emulsion. If one looks through the hologram while it is illuminated in this manner, a high-fidelity image of the object is seen. This image appears to occupy the original object space, is fully three dimensional and exhibits parallax if the observer's viewing position is shifted. It can be viewed from all directions compatible with the aperture of the hologram.

Nondestructive evaluation uses *holographic interferometry*, which is the interferometric comparison of two or more light waves at least one of which is a holographic reconstruction. A holographic interferogram can be formed by using the system depicted in Fig. 1. A holographic exposure is made by simultaneously exposing the plate to the reference wave and the light scattered by the test object in its initial configuration. Some mechanical load is then applied to the object and a second holographic recording is made on the same plate. When this two-exposure hologram is illuminated as in Fig. 2, both object waves are reconstructed simultaneously and interfere with each other to form a three-dimensional image which is modulated by a pattern of interference fringes. The interference is caused by changes, between exposures, of the optical path length from the source to the object and then to the photographic plate. Hence, the fringes are directly related to the displacement and deformation of the object surface. If the object is illuminated, and viewed, along the direction normal to its surface, the fringes are contours of constant normal displacement. In general, it is possible to view the object from three different directions and calculate the vector displacement of any point on its surface. However, detailed quantitative analysis of fringe patterns is not required in most NDE applications.

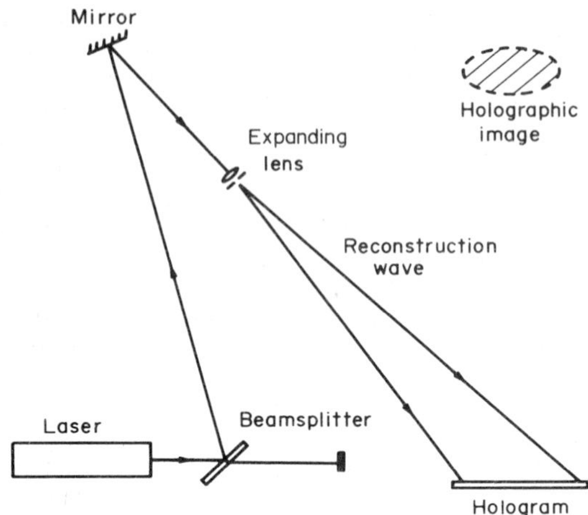

Figure 2
System for reconstructing a holographic image of the test object

In addition to two-exposure holographic interferometry discussed above, various NDE schemes utilize real-time and time-average holographic interferometry. In the real-time technique, a single-exposure hologram of the object in its initial state is recorded. After development, the hologram is returned to its precise original position in the system and is illuminated simultaneously by the reference and object waves. The operator then observes interference between the reconstructed wave from the hologram and the light instantaneously scattered by the object. In this way, changes of surface deformation can be observed as loads are varied.

Time-average holographic interferometry is used to display modal patterns of vibrating structures. A single holographic exposure is recorded with an exposure time which is much longer than the object's vibrational period. When this hologram is developed and viewed, the image appears to be modulated by a fringe pattern such that a very bright fringe outlines the nodal region and fringes of lower brightness form contours of constant vibrational amplitude.

The principles and techniques of holography are presented in detail by Collier et al. (1971), and the field of holographic interferometry is reviewed by Vest (1979), Ostrovsky et al. (1980) and Jones and Wykes (1983).

2. Principles of Holographic Nondestructive Evaluation

Many investigators have used holographic NDE to detect the presence of poor bonds between the core and skin of honeycomb structures. Consideration of this application will serve as an introduction to the principles and procedures of the technique. The front surface of a honeycomb panel bows outward in response to external suction or internal pressurization. If a large, uniformly bonded, region of the skin is subjected to a reduced external pressure it will deform into a smoothly curved surface. If the bond between the skin and core is weak or nonexistent in some region there will be a perturbation of this curved surface in the form of a local bulge of the unrestrained region. Such a bulge can be detected and conveniently displayed by holographic interferometry. If the panel is flat, a thick glass plate can be placed in front of the panel, separated from it by an O-ring. The desired deformation can be induced by drawing a slight vacuum in the space between the glass plate and the panel surface.

The surface deformation of the honeycomb panel can be examined by recording a two-exposure holographic interferogram. Some initial vacuum is applied at the time of the first exposure. The level of the vacuum is then changed slightly and a second exposure is made. The hologram is recorded using an off-axis system like that depicted in Fig. 1. The holographic image of the panel can be viewed or photographed when the developed hologram is illuminated by a reconstruction wave of laser light, perhaps by replacing it in the system and blocking the object wave. Alternatively, the viewing can be done elsewhere as long as the reconstruction wave is identical in wavelength and geometrical form to the reference wave. The holographic image will include a pattern of fringes which, if the panel was illuminated from a nearly normal direction, can be interpreted as contours of constant normal displacement. If the bond integrity is uniform over the test region, interference fringes will be smooth, closed, concentric curves, and the spacing between fringes will vary continuously and systematically across the surface. The presence of a detectable debond will be indicated by a perturbation of this pattern, usually in the form of a "bull's eye" pattern of closed, concentric, closely spaced fringes in the neighborhood of the defect.

Figure 3 is a photograph of a holographic interferogram of a honeycomb panel tested by this procedure. The panel was formed by bonding a light aluminum skin to a resin-impregnated paper core. An 1800 cm^2 area of the panel was illuminated by light from a 50 mW, continuous wave, helium–neon laser in a system similar to that of Fig. 1. These fringes, which display normal displacement with a contour interval of approximately 0.3 μm, roughly outline a square region behind which there is no bond.

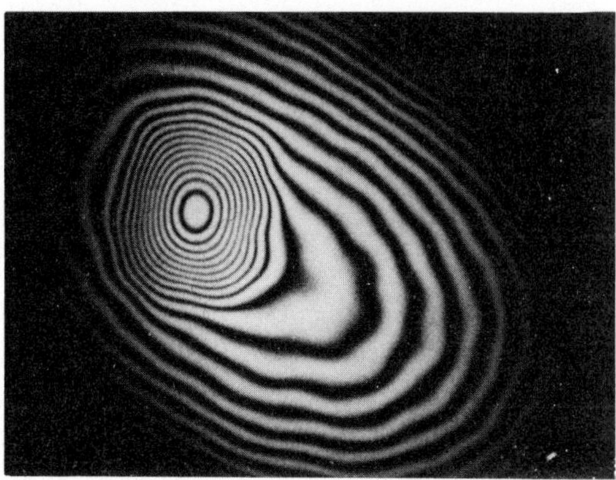

Figure 3
Holographic interferogram indicating the presence of a debonded region between the skin and core of a honeycomb panel

3. Procedures of Holographic Nondestructive Evaluation

The preceding example indicates that the development and utilization of a holographic NDE scheme or device for a particular application involves three interrelated tasks: selecting a loading method, designing the optical system and interpreting the results.

The loading method must be such that if a defect of the type expected is present, it will cause a detectable anomaly in the fringe pattern, which is a map of surface deformation. Each interference fringe is a contour along which the quantity $(\hat{\boldsymbol{k}}_2 - \hat{\boldsymbol{k}}_1) \cdot \boldsymbol{L}/\lambda$ is constant, where $\boldsymbol{L}$ is the vector displacement of the surface, λ is the wavelength of laser light and $\hat{\boldsymbol{k}}_1$ and $\hat{\boldsymbol{k}}_2$, which are indicated in Fig. 1, are unit vectors in the direction of illumination and viewing, respectively. Hence fringes are contours of constant components of $\boldsymbol{L}$ along the bisector of the illumination and viewing directions. In most setups, these two directions are nearly parallel to each other and are normal to the object surface, so the fringes are approximately contours of constant normal displacement with a contour interval of $\lambda/2$. The loading scheme must therefore be selected so that the defect causes an anomaly in normal deformation of the surface.

Five basic types of loading or excitation are used for holographic NDE: direct mechanical stressing, pressure or vacuum stressing, thermal stressing, vibrational excitation and impact loading. Direct mechanical stressing refers to the application of simple bending, tension or point loads. This type of loading has been used to detect cracks in metals, ceramics, concrete and various composite materials.

Tests have been conducted under extreme environmental conditions such as elevated temperatures and corrosive atmospheres. Components such as aircraft wing planks and turbine blades have been inspected for cracks and fabric cuts, and delaminations have been detected in fiber-reinforced materials.

Direct mechanical stressing usually yields a test whose sensitivity strongly depends on flaw orientation, and, in the case of heterogeneous materials, on structural orientation. Defects such as cracks, which might be inspected for by direct mechanical stressing, may best be detected by examining in-plane, rather than normal, deformation. Sensitivity to in-plane deformation is improved by illuminating and viewing the test object obliquely, or by using a nonholographic optical inspection method like speckle photography or speckle interferometry (Erf 1978).

Inspection for flaws which respond to internal pressurization or external pressure reduction is a task for which holographic interferometry is well suited because of its high sensitivity to normal deformation. Honeycomb panels and laminate structures of many types have been inspected by this method, as have been rocket nozzle liners, solid propellant fuel grains and cylindrical and spherical pressure vessels manufactured from materials ranging from Kevlar/epoxy composites to high-energy-rate-formed (HERF) stainless steel. Flaws have been detected in welds of pipe joints and of hermetically sealed cans for implantable microelectronic medical devices. A variety of pipes and artillery shells has been inspected for internal voids or large cracks. Tubes and shells of several composite materials have been studied.

A tire can be inspected for flaws such as ply separations, broken belts, debonds and voids by observing the response of its inner surface when it is placed under a differential vacuum between holographic exposures. Commercially available holographic tire testing machines can be used to inspect approximately 20 tires per hour. These devices are used routinely to inspect aircraft tires when they are retreaded, to test large truck and earthmoving equipment tires and to aid in the adjustment of equipment for production of automobile tires.

Some flaws can be detected when a structure responds to thermal stressing which may be induced by radiant sources, jets of hot air or evaporation of a volatile fluid sprayed on the component. This technique has been most commonly used to locate poor bonds in laminate structures or test specimens including rocket nozzle liners, honeycomb panels, turbofan blades, pressure vessels, tires and entire aircraft wing assemblies. Sensitivity to delaminations is due to a combination of locally reduced stiffness and decreased thermal conductivity through the void. Cracks have been detected by thermal stressing of ceramic tubes, glass plates and weld joints. Poor interference fits of metal fasteners have been detected. Delaminations between a clutch plate and its facing have been detected by induction heating of the surface of the facing. Multilayer electronic circuit boards and encapsulated electronic modules have been inspected for faulty resistors or transistors and bad connections by examining thermally induced distortions when they are energized. Minute distortions due to changes of ambient temperature or humidity have even been used to detect delaminations in fifteenth-century Florentine panel paintings.

Vibrational excitation can be used in two ways to detect debonds, delaminations or certain nonhomogeneities. Bending stresses can be induced by causing an entire structure to resonate. Flaw-induced perturbations of the surface deformation caused by these stresses can be observed by using time-average holographic interferometry. Alternatively, debonds and delaminations can be detected by inducing resonance of the locally flexible region created by the flaw. It is often advantageous to use real-time holographic interferometry for these evaluation techniques so that the surface response can be observed as the transducer or sonic source which drives the vibration is tuned through a range of frequencies. If it is necessary to observe an exceedingly small vibrational amplitude, sensitivity can be increased by using an electrooptical modulator to phase shift the holographic reference wave (Vest 1979 pp.210–17, 224–25). Vibrational techniques have been used to detect flaws in a variety of laminates and honeycomb materials as well as in structures and components including airframe panels, turbine blades, helicopter rotors, rocket propellant fuel grains, cathode ray tube faceplates and brake disks. The observation of resonant modes can also disclose manufacturing errors, for example, improper castings or incorrect or uneven wall thicknesses in hollow structures such as turbine blades.

Some work has been done on holographic NDE techniques based on the observation of the response of test objects to impact loading, for example, striking the object with a pendulum. In this case, the response is a travelling stress wave, rather than a steady vibration, so it is necessary to use a pulsed laser with which a hologram can be recorded in approximately 10 ns. Two-exposure holographic interferograms can be used to detect cracks because rather abrupt perturbations in the fringe pattern occur as the stress wave crosses the cracked region. Cracks have been detected in turbine blades, wing plank splices and steel and aluminum test plates. If the test object is flat, the sensitivity of the method can be increased by covering the object with a thin liquid layer and measuring the instantaneous deformation of the liquid surface, which tends to amplify the stress wave of the object surface. Impact loads have also been used to locate bad bonds between foam insulation and large structures such as spacecraft fuel tanks.

4. *Optical Equipment for Holographic Nondestructive Evaluation*

Optical systems for NDE are generally fairly simple, but care must be taken when designing them for use in industrial environments where reliability, ease of operation, short cycle times, and insensitivity to extraneous motions are important. A particular problem is the formation of extraneous interference fringes because building vibrations induce unwanted translations and rotations of the test object. These effects can be minimized by appropriate choices of illumination and observation directions, by reflecting the reference beam from a mirror which is attached to the object, or by several fringe control techniques which involve adjustments of object or reference waves during recording and viewing. Another approach is to record the two holographic exposures on different photographic plates which, after development, are bonded together to form a sandwich hologram. By changing the orientation of a sandwich hologram relative to the reconstruction wave, the fringes can be manipulated dynamically to compensate for extraneous motions, leaving only the fringes caused by deformation.

Several types of laser are used for holographic NDE. The most common are continuous-wave helium–neon, argon-ion or krypton lasers, with optical powers ranging from 15 mW to 10 W. Powers above 50 mW are generally necessary only when areas larger than about 1000 cm^2 are to be inspected. When impact loading is used, or two-exposure interferograms of rapidly vibrating objects are to be recorded, a pulsed laser must be used. Q-switched ruby lasers and neodymium–YAG lasers (frequency doubled to generate visible light) with energy outputs of about 1 J and pulse widths of 10 to 30 ns are appropriate for NDE work.

Most laboratory and commercial systems use high-resolution films or plates based on traditional silver halide photographic chemistry. These are generally satisfactory, but require time-consuming and relatively cumbersome development. This may introduce a time lag of several minutes between recording and viewing. Other recording systems, notably photothermoplastic cameras suitable for holography, have been developed and commercially marketed. These systems can be used with either continuous-wave or pulsed lasers to produce holograms which can be viewed within seconds after recording. When 35 mm film or photothermoplastic films are used, the small format may be inconvenient for viewing, so it is common to use a video system for viewing holographic interferograms on a large monitor screen. Some work is being done on digital processing and pattern recognition which may be used to more fully automate holographic NDE.

5. *Interpretation of Holographic Fringe Patterns*

The interpretation of fringe patterns for NDE is usually qualitative, consisting of searching for abrupt changes of curvature near cracks or for closely spaced "bull's eye" fringes near debonds. In other cases, one may visually compare the fringe signature of the component being inspected with that of an unflawed master or standard component. Some inspection techniques, however, do require quantitative analysis of interferograms based on well-established equations. Quantitative analysis is required when the inspection is designed to detect changes of material properties rather than flaws. Examples are the determination of strain levels and detection of plastic deformation or residual stresses in pressure vessels.

Some simplified quantitative analysis is useful for estimating flaw size, or, conversely to estimate load levels or excitation frequencies necessary to detect a flaw of a given type and size. For example, debonded regions in the skin of a honeycomb panel or laminate material will deform approximately like a plate with clamped edges, so the maximum deflection could be estimated to be

$$W_{\max} = 0.7\,\frac{\Delta p r^4}{E t^3} \tag{1}$$

where Δp is the differential pressure across the region, r is the radius of the defect, E is the modulus of elasticity and t is the thickness of the skin. The fundamental resonant frequency of such a defect in a horizontal surface can similarly be estimated by

$$f = \frac{10.2}{2\pi r^2}\left(\frac{gD}{\rho t}\right)^{1/2} \tag{2}$$

where g is gravitational acceleration, the stiffness $D = Et^3[12(1-\nu^2)]$ and ρ is the weight per unit area. Such formulae should be expected to be only approximate because geometry, edge clamping, partial bonding and intermittent contact with adjacent parts of the structure will cause deviations from the ideal cases they describe.

6. *Summary*

Holographic NDE is based on the ability of holographic interferometry to generate a whole-field fringe pattern which displays minute deformations of the visible surface of a test object. Flaws or nonhomogeneities of properties lead to perturbations of this pattern when the object is subjected to appropriate mechanical, thermal or vibrational excitation. The basic principles and techniques are well developed. Most interpretation of results is qualitative and requires some operator judgement. Future developments should include more convenient rapid holographic recording media, lasers of increased reliability and ease of repair, increased quantitative analysis for both interpretation of interferograms and design of testing procedures, setting of standards and automated processing systems to remove extraneous fringes and reduce the degree of operator judgement in evaluation of test results.

See also: Holographic Materials; Nondestructive Evaluation: An Overview

Bibliography

Collier R J, Burckhardt C B, Lin L H 1971 *Optical Holography*. Academic Press, New York

Erf R K 1974 *Holographic Nondestructive Testing*. Academic Press, New York

Erf R K 1978 *Speckle Metrology*. Academic Press, New York

Gabor D 1949 Microscopy by reconstructed wavefronts. *Proc. R. Soc. (London) Ser. A* 197: 454–87

Jones D, Wykes C 1983 *Holographic and Speckle Interferometry*. Cambridge University Press, Cambridge

Leith E N, Upatnieks J 1962 Reconstructed wavefronts and communication theory. *J. Opt. Soc. Am.* 52: 1123–30

Ostrovsky Yu I, Butusov M M, Ostrovskaya G V 1980 *Interferometry by Holography*. Springer, Berlin

Vest C M 1979 *Holographic Interferometry*. Wiley, New York

C. M. Vest

Homogenization of Metal Powders

Powder metallurgical processing of alloys uses either prealloyed powders or blends of powders as starting materials. In the former, all powder particles have essentially the same composition, as typically produced by the atomization of a molten alloy by impingement of a fluid stream (gas or liquid). The fluid disperses the liquid alloy into small droplets and causes the droplets to solidify into powder particles. In the latter, two or more types of particles are blended together in the proportions necessary to achieve the desired alloy composition. The constituent powders of the blend may be composed either of pure elements (elemental powders) or alloys containing two or more elements (master alloy powders). Since a homogeneous alloy is usually the goal of the powder fabrication process, alloys produced from blends of powders must undergo one or more elevated temperature processing steps following the blending and compaction operations in order to achieve the required degree of homogenization. The homogenization process takes place by interdiffusion of the chemical elements between the constituent powders of differing composition. Although homogenization is typically a solid-state diffusion process, some liquid may be present in the powder blend at the onset due to either low-melting constituent powders or constitutional effects.

1. Advantages and Disadvantages

Homogenization processing has several advantages over prealloyed powder processing. The hardness and strength of prealloyed powder particles can result in low densities and low green strengths in powder compacts. Blends of powders, with the major component of the alloy present as an elemental powder (typically soft and ductile), often exhibit high densities and high green strengths in the as-compacted condition. Furthermore, adjustments to alloy composition can be made readily by varying the proportions of constituent powders in the blend, eliminating the need for procurement of additional powders of different composition. Powder blends composed entirely of elemental powders can be processed into alloys having very low concentrations of undesirable elements, since elemental powders can normally be obtained with high purity. Prealloyed powders, on the other hand, may contain elements other than the main components of the alloy due to contamination with impurities and trace elements during the melting operation prior to atomization.

The disadvantages of homogenization processing center about the need to remove compositional heterogeneities by thermal processing (often in concert with mechanical working) subsequent to compaction. The homogenization processing usually involves longer exposure to elevated temperatures than is required for normal sintering; this added processing expense is minimized by using processing conditions no more extensive than needed to provide the required homogeneity.

Homogenization processing has two important concerns: designing and carrying out the homogenization processing step(s) to achieve a state of negligible (or acceptable) compositional heterogeneity, and developing appropriate quality control procedures and experimental tests to assure that the necessary homogeneity is obtained. Processing must, therefore, allow for enhanced homogenization kinetics through understanding, control, and/or specification of (a) powder variables (particle size and composition), (b) constitution variables for the alloy system of interest (phase diagram considerations) and (c) processing variables (i.e., efficiency of blending, thermal conditions (time and temperature), and mechanical working (type of working, reduction and sequencing with thermal processing)). The progress of homogenization or the evaluation of the state of homogeneity in the alloy produced may be assessed by direct or indirect analysis. Direct experimental techniques include microscopy (qualitative), composition analysis of the microstructure via electron microprobe analysis or scanning electron microscope instrumentation, and x-ray compositional line broadening. The degree of homogeneity may also be assessed qualitatively from indirect techniques, such as mechanical behavior, corrosion resistance or response to heat treatment, but interpretation of the results can be ambiguous.

2. Homogenization Mechanism

Typically, as-compacted blends of powder exist as solute-rich particles (either elemental or master alloy) surrounded by a matrix of solvent particles (normally elemental particles of the base element of the alloy). The density to which the powder is pressed determines the extent of residual interparticle porosity. A schematic representation of the microstructure of such a compact is shown in Fig. 1. The dispersion of the solute-rich particles is important for rapid homogenization and can be optimized through use of proper blending procedures and determination of the best particle-size ratio (the finer the matrix particle size the greater the dispersion of the solute-rich particles). Overall alloy composition and composition of the solute-rich particles can also affect the dispersion, but are sometimes specified for a given situation and cannot be considered to be independent variables.

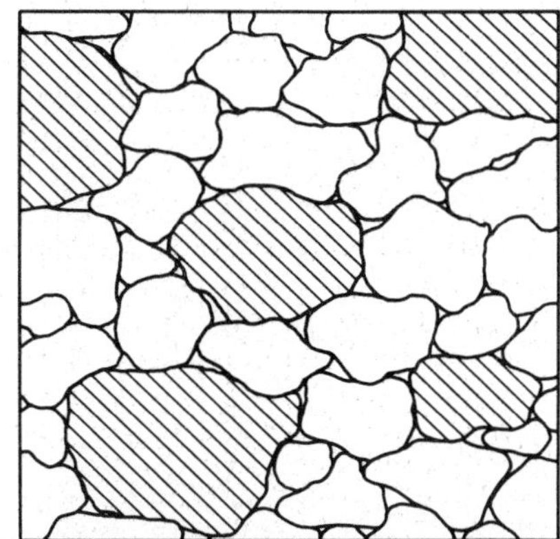

Figure 1
Schematic representation of a compacted blend of solute-rich particles (shaded) dispersed in a matrix of solvent-rich particles

Thermal processing causes the solid-state diffusion (lattice diffusion) of the solute atoms into the solvent matrix and solvent atoms into the solute-rich particles. The progress of homogenization as a function of time and temperature may be described in terms of concentration–distance profiles (Fig. 2). In general, the compact will behave as a collection of many small diffusion couples that are allowed to undergo interdiffusion to the extent that the atomic

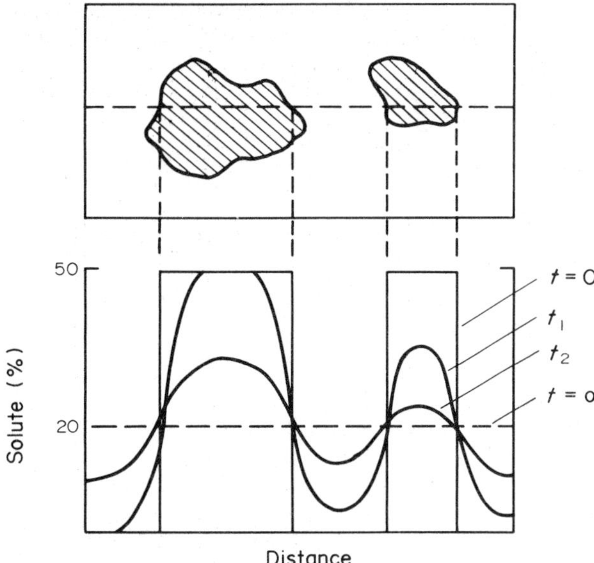

Figure 2
Schematic representation of the changing solute concentration–distance profiles during solute homogenization

fluxes become negligible and homogenization is achieved.

A reduction in interparticle porosity will normally occur through sintering during the thermal processing. In addition, Kirkendall porosity may form due to a net flux of vacancies into or away from solute-rich particles. For example, if the flux of solute atoms away from the dispersed particles is greater than the flux of solvent atoms into the dispersed particles, the net flux of vacancies to the dispersed particles will cause the development of a fine Kirkendall porosity within the compact regions which originally were solute-rich particles.

3. Powder, Alloy System and Processing Variables

The characteristics of the powder particles which compose the blend exert a significant effect on the homogenization kinetics. The sizes of the particles in the blend primarily establish the distance over which interdiffusion must occur to achieve homogeneity. In the usual instance where the elemental solvent particles form a continuous matrix around the dispersed solute-rich particles, it is the particle size of the dispersed particles which becomes the important parameter in establishing the interdiffusion distance (assuming that the proportion of the powders in the blend is already established by the alloy composition desired). Also, it should be noted that very fine, intimate mixtures of powders, such as those produced by reduction of submicron mixtures of oxides and mechanical alloying, result in diffusion distances small enough to render homogenization processing unnecessary.

The composition of the dispersed particles can also influence the homogenization kinetics; solvent alloyed into the solute-rich particles reduces the extent of inhomogeneity in the blend and can shorten the interdiffusion distance. For example, if the dispersed master alloy particles shown in Fig. 2 contained 35% solute (rather than 50% solute), the amount of solvent (matrix) needed in the blend to achieve the same overall alloy composition (20% solute) would be less, thus shortening the interdiffusion distance. The relative advantages of solvent alloyed into the solute-rich particles and elemental solvent as the matrix can be established on the basis of compact green strength (adequate amount of matrix) versus kinetics of homogenization.

Alloy systems with more than two components and/or which contain more than a single phase give rise to complexities in the analysis of homogenization kinetics. Interdiffusion in single-phase ternary systems must be described by four diffusion coefficients in order to consider the diffusion of both solutes, each as influenced by its own concentration gradient and that of the other solute. Multiphase, binary systems must be described by the solute solubilities and interdiffusion coefficients of each of the phases present. In most instances of alloy complexities, the parameters necessary to analyze the homogenization process are not available and reasonable simplifications (i.e., assumption of binary and/or single-phase behavior) are made. Furthermore, the presence of a transient liquid phase at the outset of the homogenization process is also a complexity which is difficult to analyze and reasonable approximations are again necessary.

The duration of thermal processing necessary to achieve a desired degree of homogeneity is critically dependent upon temperature, since the interdiffusion coefficients D are exponentially dependent upon temperature:

$$\tilde{D} = D_0 \exp(-Q/RT) \qquad (1)$$

where D_0 is the preexponential factor, Q is the activation energy, R is the universal gas constant and T is the absolute temperature. (For example, the interdiffusion coefficient for most solid-solution elements in nickel doubles for every 50 K increase in temperature for temperatures near the melting temperature.) The total duration t of elevated temperature treatment to achieve essentially complete homogenization is inversely proportional to $\tilde{D}$:

$$t = Kl^2/\tilde{D} \qquad (2)$$

where l is the average diameter of the dispersed particles and K is a constant dependent upon alloy system parameters, powder particle compositions, and overall alloy composition. (For example, K is approximately 0.2 for a 4:1 blend of elemental powders which exhibit complete solid solubility.) Thus, it can be seen that both the homogenization temperature and the size of the dispersed particles are the most significant variables in controlling homogenization kinetics.

The blending and mechanical working steps in the powder fabrication of an alloy can also influence the homogenization kinetics. Ideal mixing is assumed in the use of Eqn. (2) to predict homogenization time. An inadequate blending technique or incompatible powder particles (significantly different sizes and/or densities) result in poor mixing and greatly enlarged interdiffusion distances which, therefore, dictate longer homogenization treatments. The problem of poor mixing is of great importance as homogenization time is proportional to the square of the interdiffusion distance.

The effective interdiffusion distance may be reduced, and homogenization kinetics increased, by mechanical working. Such working can be considered in much the same manner as a reduction in the particle size of the dispersed particles. However, the effect of a given working reduction will be less than that indicated by the same reduction in l in Eqn. (2) since the spherical-diffusion flux geometry in a compacted blend of powders will be changed to the less efficient unidirectional flux (by rolling) or cylindrical flux (by extrusion). It is important to note that any enhancement of homogenization kinetics will result only if the working operation reduces the thicknesses of the solute-rich regions. If the working operation merely causes the matrix to flow past the solute-rich regions, little, if any, increase in homogenization kinetics will be exhibited during subsequent thermal processing.

See also: Diffusion in Crystalline Metals: Phenomenology; Powder Characterization; Powder Metallurgy

Bibliography

Heckel R W 1978 Diffusional homogenization of compacted blends of powders. In: Kuhn H A, Lawley A (eds.) 1978 *Powder Metallurgy Processing: New Techniques and Analyses*. Academic Press, New York, pp. 51–97

Lenel F V 1980 *Powder Metallurgy: Principles and Applications*. Metal Powders Industries Federation, Princeton, New Jersey

Rudman P S 1960 An x-ray diffraction method for the determination of composition distribution in inhomogeneous binary solid solutions. *Acta Crystallogr.* 13: 905–9

R. W. Heckel

Honing

Honing is a finishing operation which normally follows grinding and improves surface finish, dimensional accuracy, surface texture and oil-retention capability. Either flat or cylindrical surfaces may be honed but the most common application is in finishing the bores of internal combustion engines. In this case, ceramically bonded abrasive sticks that conform to the surface are mounted in a floating head and loaded radially outward. The head is given a complex combined axial and circumferential motion relative to the workpiece and this produces a crosshatched pattern in the honed surface. Honing speeds are low ($0.25–1.5\ m\ s^{-1}$) relative to grinding speeds ($20–50\ m\ s^{-1}$). Abrasive grain sizes employed in honing are relatively small (80–600 mesh) and relatively little material is removed (0.1–0.7 mm). Surfaces are normally honed to a desired size within a few ten-thousandths of an inch ($\sim 2\ \mu m$).

Superfinishing is a special honing operation in which the bonded abrasive stone is subjected to very light pressure and a short, high frequency stroke.

All honing is carried out using a fluid to wash away the debris and to cool and lubricate the work. In addition, solid lubricants known as grinding aids are frequently incorporated into the pores of honing tools to reduce noise and improve surface finish. The abrasives used are chemically the same as those used in grinding—Al_2O_3 for ferrous metals and SiC for ceramics and nonferrous metals.

See also: Grinding; Metals Processing and Fabrication: An Overview

Bibliography

Beaudet E C 1952 Honing used for greater, faster stock removal. *Iron Age* 170: 170–71

Henry A F 1944 High production honing without fixtures. *Tool Eng.* 13: 93–94

Polatsek R 1955 Fundamentals of the honing process. *Grind. Finish.* 1: 57–60

Swigert A M et al. 1953 *Honing, Lapping and Superfinish Collected Papers.* American Society of Tool and Manufacturing Engineers, Dearborn, Michigan

Wilford J S 1943 Honing: A precision production process. *Tool Eng.* 12: 77–79

M. C. Shaw

Hot Corrosion

At elevated temperatures (usually 600–900 °C), the reactions between gases and metals are not the only significant corrosion reactions. Frequently, molten (i.e., fused) salts come into contact with a metal or its protective oxide film. The fused salts may be unintentionally condensed byproducts of a high-temperature system such as a gas turbine, an incinerator, a petroleum processing reactor or a heat exchanger. Otherwise, many energy systems and extraction systems intentionally use molten salts as electrolytes, or as heat treatment, heat storage or heat transfer media.

These fused salts are generally composed of alkali or alkaline earth cations, since these have the greatest thermodynamic stabilities, and cations of the alloy, which are introduced as corrosion products. Three important classes of fused salts are: simple anions such as halides; simple oxyanionic species such as sulfates, carbonates, vanadates and nitrates; and complex polymeric oxyanions such as phosphates, borates and silicates. These particular molten salts are highly dissociated, with high electrical (i.e., ionic) conductivities, high thermal conductivities and reasonably low vapor pressures and viscosities.

The accelerated hot corrosion of metals by fused salt films in contact with an oxidizing gas is generally attributed to an attack by the salt on the protective oxide film of the metal, either by dissolution of the oxide or else by its penetration through a preferential attack at the grain boundaries.

A thin fused salt electrolyte film on a metal permits the existence of local galvanic cells (see *Galvanic Corrosion*) between phases or heterogeneities in the metal which exhibit a difference in electrode potential. Thus, hot corrosion shares common geometric and electrochemical features with the atmospheric corrosion of metals by a condensed water film. But, in fact, the solubilities of oxides and gases in the salt and certain diffusion parameters are greatly different. In addition, fused salts exhibit an acid–base property; a salt is considered basic when the concentration of oxide ions, or the activity of an alkali or alkaline earth oxide, is relatively high. Further, an oxyanion salt, such as an alkali sulfate, may form a nonprotective sulfide product in addition to the oxides.

Research has shown that rapid hot corrosion attack requires the presence of a liquid, not a solid, salt film. The corrosion rate is determined by the chemical (acid–base) compatibility of the protective oxide film with the fused salt film, by the thickness of the salt and the transport parameters of the gaseous oxidants in the salt film, and by the chemical shifts in salt composition as corrosion products are dissolved from the alloy. The most severe corrosion may not occur in the hottest section of a system where the temperature can exceed the boiling temperature of the salt, but at lower temperatures, where condensed salt films, including dissolved corrosion products, can be very aggressive.

Engineering work is emphasizing the evolution of alloy and coating compositions to minimize attack, the purification of fuels or filtering of combustion air (especially for naval vessels) and the introduction of buffering salts into the combustion system. Current research efforts are involved with the quantifying of the acid–base properties of salts and oxides, and the adaptation of electrochemical methods to study thin-film electrolyte systems at high temperature.

See also: Corrosion Kinetics; Oxidation of Metals and Alloys; High-Temperature Oxidation of Metals and Alloys: Engineering Aspects; Sulfidation

Bibliography

Luthra K L, Shores D A 1980 Mechanism of Na_2SO_4 induced corrosion at 600–900 °C. *J. Electrochem. Soc.* 127: 2202–10

Stringer J 1977 Hot corrosion of high-temperature alloys. *Annu. Rev. Mater. Sci.* 7: 477–509

R. A. Rapp

Hot Extrusion

Hot extrusion is the process of forcing a heated billet to flow through a die opening of desired shape. This process is used to produce long, straight semifinished metal products of constant cross section such as bars, solid and hollow sections, tubes, wires and strips. There are three basic types of hot extrusion: (a) nonlubricated, (b) lubricated and (c) hydrostatic. These are illustrated in Fig. 1. This article is concerned with nonlubricated and lubricated hot extrusion; the hydrostatic process is detailed elsewhere (see *Hydrostatic Extrusion*).

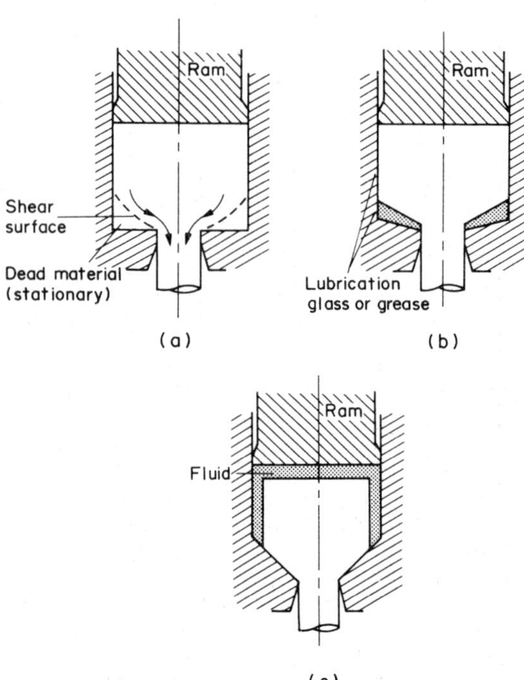

Figure 1
Various extrusion processes: (a) nonlubricated, (b) lubricated and (c) hydrostatic

1. Nonlubricated Hot Extrusion

This method uses no lubrication between the billet, container and die, and is able to produce very complex sections with mirror surface finish and close dimensional tolerances. A flat- or shear-faced die is used to extrude the metal by direct or indirect extrusion (Fig. 2). In direct extrusion, the ram travels in the same direction as the extruded section and there is relative movement between the billet and the container. In indirect extrusion, the billet does not move relative to the container, and the die is pushed against the billet using a hollow stem.

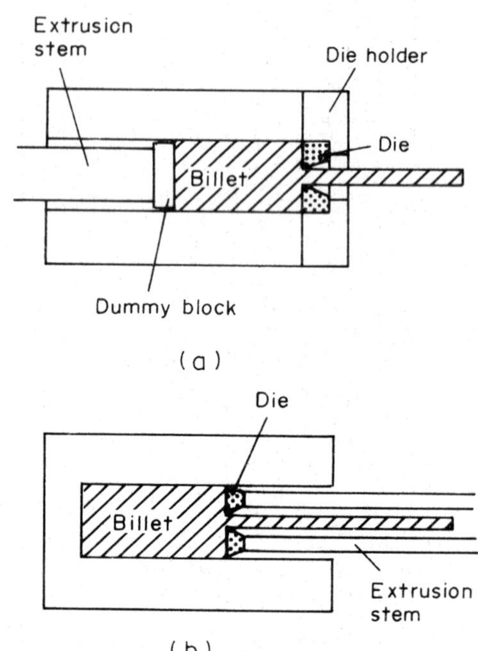

Figure 2
Methods of nonlubricated hot extrusion: (a) direct extrusion and (b) indirect extrusion

Typical load versus ram-displacement curves for direct and indirect extrusion processes are shown in Fig. 3. In direct extrusion the load initially increases very rapidly as the billet upsets to fill the container. With further increase in pressure, extrusion commences and a cone-shaped deformation zone develops in front of the die aperture. The extrusion pressure then falls as the billet length decreases, until a minimum is reached. Finally, a rapid pressure increase occurs when only a disk of the billet remains, since the metal then has to flow radially towards the die aperture. The deformation resistance increases considerably with decreasing thickness.

In indirect extrusion (Fig. 2b), there is no friction at the billet–container interface. Therefore, the extrusion load and heat generated by deformation and friction are reduced (Fig. 3).

Indirect extrusion offers a number of advantages which can be summarized as follows:

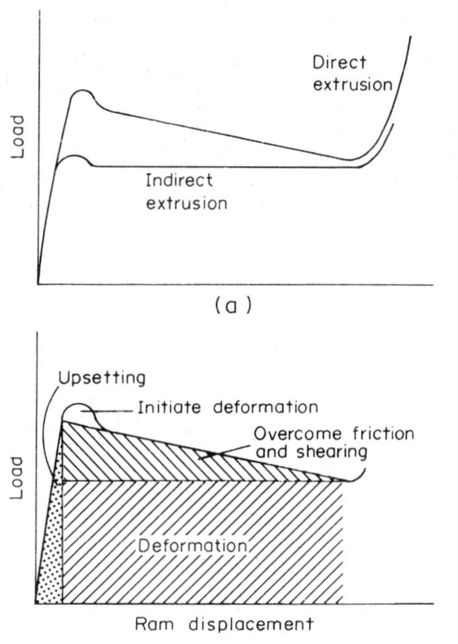

Figure 3
Typical load versus ram-displacement curves for nonlubricated extrusion processes: (a) dependence of the extrusion load on ram displacement and (b) components of extrusion energy

(a) A 25–30% reduction in maximum load, compared with direct extrusion.

(b) The extrusion pressure is not a function of the billet length because there is no relative displacement between the billet and the container. The billet length is therefore not limited by the load required for this displacement, but only by the length and stability of the hollow stem needed for a given container length.

(c) No heat is produced by friction between the billet and the container, and consequently no temperature increase occurs at the billet surface towards the end of extrusion, as is typical in the direct extrusion of aluminum alloys. Therefore, there is a lower tendency for the surfaces and edges to crack in the indirect process, and significantly higher extrusion rates can be used.

(d) The service lifetime of the tooling is increased, especially that of the inner liner, because of reduced friction and temperatures.

The disadvantage of indirect extrusion is that impurities or defects on the billet surface affect the surface of the extrusion and are not automatically retained as a shell or discard in the container. Therefore, machined billets have to be used in many cases. In addition, the cross-sectional area of the extrusion is limited by the size of the hollow stem.

1.1 Extrusion of Tubes and Hollow Sections

To produce tubes or hollow sections by extrusion, the metal must flow through an annular gap formed by the die and a mandrel. There are two basic methods for extruding tubular products. The first uses a billet with an internal bore, machined or pierced in the press. In this case, the mandrel is guided to provide concentricity and stays stationary or moves with the ram during extrusion. In the second method, special multihole dies are used. The multihole die designs have openings in the top face of the die from which material is extruded into two or more segments and then, beneath the surface of the die, welded and forced through the final shape configuration to form a part (Laue and Stenger 1981). The tubular portion of an extruded shape is formed by a mandrel attached to the lower side of the top die segment. This provides a fixed support for the mandrel and a continuous hole in the extrusion.

Higher extrusion pressures are required with multihole dies compared with flat-faced dies for extruding the same section. The material has to shear to flow through various segments, and form a sound weld before extruding out. Therefore, this process is limited to materials such as aluminum alloys, which have low shear strengths at extrusion temperatures. Copper alloys, for example, are not generally extruded using multihole dies.

1.2 Materials and Metal Flow

The flow stress and the workability at extrusion temperature are the main characteristics which determine the extrudability of a given material without lubrication. Materials most commonly processed by the nonlubricated extrusion method are given in Table 1. Most copper alloys are first extruded and then cold-drawn to obtain sections with sharper corners. A variety of aluminum alloys (1000 to 7000 series) are extruded for a large number of commercial and military applications.

The metal flow during extrusion varies considerably, depending on the material, the material–tool interface friction and the shape of the section. The flow patterns observed have been classified into four different types (Fig. 4). Flow pattern S is found in extrusion of homogeneous materials when there is no friction at the container walls or the die surface, whereas flow pattern A results from friction on the die surface, but not on the container wall. Pattern A is characterized by a dead-metal zone between the die face and the container wall; the material in this zone does not move during extrusion. Flow pattern B is obtained in homogeneous materials when friction is present on both container wall and die surface. A dead-metal zone of larger size, compared with that

Table 1
Materials commonly extruded by nonlubricated hot extrusion

Materials	Minimum section thickness (in.)	(cm)	Extrusion temperature (°C)	Extrusion pressure (MPa)	Extrusion rate (m s^{-1})	Extrusion ratio $\ln(A_0/A_1)$[a]
Aluminum alloys	0.04–0.30	0.10–0.76	290–570	280–900	0.02–1.52	4–7
Copper alloys	0.05–0.30	0.13–0.76	650–900	280–900	0.41–5.08	5–6.5
Magnesium alloys	0.04	0.10	300–430	690–900	0.02–0.51	4–5.3
Zinc alloys	0.06	0.15	200–350	620–760	0.38–0.51	4–5

a A_0 is the cross-sectional area of the billet and A_1 is the cross-sectional area of the extruded shape

in flow pattern A, is formed. Flow pattern C is obtained for billets with inhomogeneous material properties, or with an uneven temperature distribution over the billet cross section. The flow stress of the material near the surface of the container is much greater than near the center. The angle of the dead-metal zone is relatively large.

1.3 Extrusion Rate and Temperatures

Heat produced during extrusion greatly influences the rate at which the process can be carried out. This is especially true for extrusion of hard aluminum alloys (2000 and 7000 series). Thus, as soon as the heated billet is loaded into the preheated container and extrusion started, a complex thermal situation exists, with the temperature influenced by:

(a) heat generation due to plastic deformation;

(b) heat generation due to internal shear and friction between the deforming material and the tooling;

(c) heat transfer within the billet;

(d) heat transfer between the billet and the tooling; and

(e) heat transported with the extruded product.

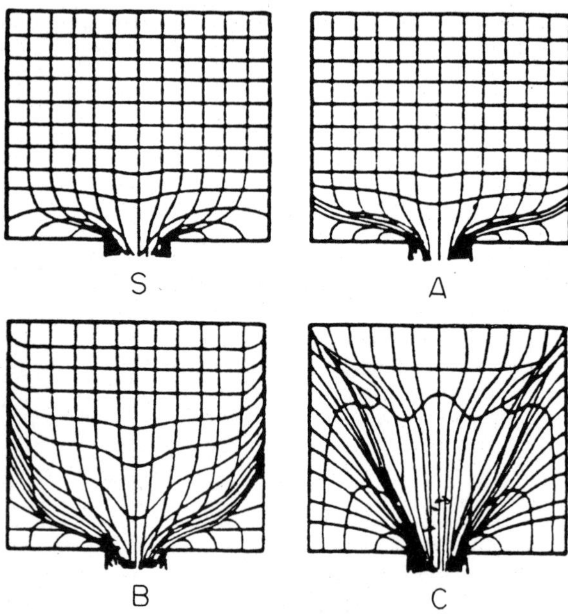

Figure 4
Flow patterns observed in nonlubricated hot extrusion

In order to increase the production rate, it is desirable to increase the extrusion ratio (cross-sectional area of billet: cross-sectional area of product) and the extrusion rate, while maintaining the extrusion pressure at acceptable levels. For this purpose, the flow stress of the extruded material must be kept relatively low; for example, the billet preheat temperature must be increased. The combination of high billet temperature, a large reduction in area and high extrusion rate causes considerable temperature rise in the extruded material, especially near the section surface, because most of the plastic deformation and friction energy are transformed into heat. This can cause surface defects or hot shortness, especially with materials that are difficult to extrude (e.g., hard aluminum alloys of the 2000 and 7000 series). Therefore, exit rates for soft alloys are relatively high and those for harder alloys are quite low (~0.01–0.02 m s^{-1}).

Temperature rise and distribution in extrusion have been widely investigated. For practical purposes, it has been estimated that in extruding high-strength aluminum alloys the maximum temperature rise likely to be encountered will not exceed 100 °C (Akaret 1967). For the soft alloys where lower specific pressures are required, the temperature rise under normal production conditions is not likely to exceed 50 °C.

In extrusion of aluminum alloys, temperature variations in the emerging product can be reduced by imposing a temperature gradient in the billet (Chadwick 1969). The hot end of the billet is fed into the container such that it is extruded first, while the temperature of the cooler end increases during the extrusion. This practice is not entirely satisfactory because of the relatively high thermal conductivity of aluminum alloys, so that if any delays occur in the extrusion sequence, the temperatures in the billet tend to become uniform throughout the billet length. A better method is to water quench the container feed-table. Another approach that has been used to increase the extrusion rate is to use water or nitrogen to cool the die.

2. Lubricated Hot Extrusion

Aluminum alloys are usually extruded without lubrication. However, copper alloys, titanium alloys, alloy steels, stainless steels and tool steels are extruded on a commercial basis, using a variety of graphite and glass-based lubricants. Commercial grease mixtures containing solid-film lubricants, such as graphite, often provide little or no thermal protection to the die; therefore, die wear in conventional extrusion of steels and titanium alloys is very significant.

The process most commonly used for extruding steels and titanium alloys is the Sejournet process (Sejournet and Delcroix 1955). In this, the heated billet is commonly rolled over a bed of ground glass or sprinkled with glass powder to provide a layer of low-melting glass on the billet surface. Prior to insertion of the billet into the hot-extrusion container, a suitable lubricating system is positioned immediately ahead of the die. This may consist of a compacted glass pad, glass wool, or both. The prelubricated billet is quickly inserted into the container, together with appropriate followers or a dummy block, and the extrusion cycle started (Fig. 5).

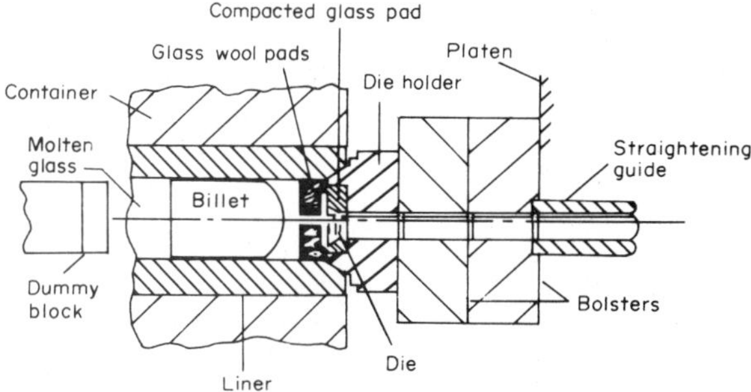

Figure 5
Hot-extrusion setup using glass lubrication

The unique features of glass as a lubricant are its ability to soften selectively during contact with the hot billet and, at the same time, to insulate the hot billet material from the tooling, which is usually maintained at a temperature considerably lower than that of the billet. In extruding titanium and steel, billet temperature is usually 1000–1250 °C, whereas the maximum temperature which the tooling can withstand is 500–550 °C. Thus, the only way to obtain compatibility is to use appropriate lubricants, insulative die coating and ceramic die inserts, and to design dies to minimize tool wear. To date, glass lubricants have worked successfully on a production basis in extruding long lengths.

3. Characterization of Extruded Shapes

Extruded shapes from aluminum and other alloys are generally characterized according to their geometric complexity. The size of an extruded shape is measured by the diameter of the circle circumscribing the cross section of the shape. This is commonly referred to as the circumscribing-circle diameter (CCD). In extrusion, metal tends to flow more slowly at die locations far away from the axis of the billet. Therefore, the larger the CCD of a shape is, the greater control is required to maintain the dimensions of the extruded shape. Special care is needed in extruding large and thin shapes, and especially with thin portions of a shape near the periphery of the die. Thus, size is one of the factors describing the complexity of a shape.

The complexity of an extruded shape can be described in terms of a shape factor, defined as follows:

$$\text{Shape factor} = \frac{\text{perimeter}}{\text{cross-sectional area}}$$

This factor is a measure of the amount of surface generated relative to the amount of metal extruded. The shape factor affects the production rate, and the cost of manufacturing and maintaining the dies. It is often used as a basis for pricing, and gives the designer a means of comparing the relative complexity of alternate designs.

See also: Cold Extrusion and Forging; Deformation Processing; Friction and Lubrication in Metalworking; Heat Generation in Metal Cutting and Deformation

Bibliography

Akaret R 1967 A numerical analysis of temperature distribution in extrusion. *J. Inst. Met.* 95: 204–11

Akaret R, Stratman P M 1973 Unconventional extrusion processes for the harder aluminum alloys, Pts. I, II. *Light Met. Age* 31(3): 6–10; 31(4): 15–18

Byrer T G (ed.) 1973 *Design Guide for Use of Structural Shapes in Aircraft Applications,* Technical Report AFML-TR-73-211. Battelle Columbus Laboratories, Columbus, Ohio

Chadwick R 1969 Developments and problems in package extrusion press design. *Met. Mater.* 3: 162–70

Lahoti G D, Altan T 1974 Prediction of temperature distributions in tube extrusion using a velocity field without discontinuities. *Proc. 2nd North American Metalworking Research Conference.* Society of Manufacturing Engineers, Dearborn, Michigan, pp. 209–24

Lahoti G D, Altan T 1978 Prediction of metal flow and temperatures in axisymmetric deformation processes.

In: Burke J J, Weiss V (eds.) 1978 *Advances in Deformation Processing*. Plenum, New York, Chap. 3, pp. 99–125
Laue K 1963 Possibilities of material related design of light alloy sections (in German). *Z. Metallkd.* 54: 667–71
Laue K, Stenger H 1981 *Extrusion: Processes, Machinery, Tooling*. American Society for Metals, Metals Park, Ohio
Sejournet J, Delcroix J 1955 Glass lubricant in the extrusion of steel. *Lubr. Eng.* 11: 389–96
Singer A R E, Coakham J W 1960–61 Temperature changes occurring during the extrusion of aluminum, tin and lead. *J. Inst. Met.* 89: 177–82
Singer A R E, Al-Samarrai S H K 1960–61 Temperature changes associated with speed variations during extrusion *J. Inst. Met.* 89: 225–31

T. Altan

Hot Isostatic Pressing

Hot isostatic pressing (HIP), also known as gas pressure bonding, was conceived as a specialized hot-pressing technique for diffusion bonding. Early equipment consisted of a hot-wall pressure vessel which, together with its workload, was heated in a conventional furnace. Obviously such units had severe temperature and pressure limitations, but major advances in both equipment and processing technology have enabled more widespread use of hot isostatic pressing techniques.

1. Process

The concept of hot isostatic pressing is very simple; high-temperature, high-pressure gas applies a uniform compressive force on any gas-tight assembly, regardless of geometry. There is no die-wall friction, hence no length–diameter limitations. Moreover, the gas functions as a flexible platen, that is, when all surfaces come into contact, forces balance and no further straining occurs. In conventional platen pressing, loading must cease upon initial asperity contact, leaving unconsolidated regions in the structure, or continue until all areas are in intimate contact, thereby "overstraining" initial contact areas. This is of particular concern in multiphase systems in which one or more constituents are brittle.

The part to be pressed must have a gas-tight skin or must be "canned," for if gas can penetrate through the part, forces are equal and no consolidation takes place. Generally, the part is evacuated and then sealed. This can be accomplished via a stem which is subsequently crimped or forge-welded shut, or by simultaneous evacuation and sealing in an electron-beam welder.

Subject to certain equipment limitations (e.g., heating and cooling rates, and pump capacity) a wide variety of processing cycles have been used. Usually, temperature and pressure are increased simultaneously. However, it is also possible to pressurize before heating, or vice versa. Typical hot isostatic pressing cycles last for many hours. Production capacity can be expanded by loading and unloading hot workpieces, thus using less costly furnaces for preheating, and if required, controlled cooling.

2. Equipment

A basic hot isostatic pressing system comprises a cylindrical steel pressure vessel, a pressurization system, a heating system and a control/monitoring system. Cold-wall pressure vessels have become the industry standard, so a cooling jacket, either internal or external, is incorporated to maintain the vessel-wall temperature at a safe level. The working cover or plug can be threaded, with either a continuous or interrupted (breach) thread for rapid turn-around, or can be pinned in place. Commercial hot isostatic pressing units are generally rated at 200 MN m^{-2}, although research and development units with capacities greater than 690 MN m^{-2} are in existence.

The gas used as the working fluid must be non-reactive with the pressing equipment as well as the workpiece; argon is most common, but both helium and nitrogen have also been used. Pressurization is accomplished with piston or diaphragm compressors although thermal expansion during the heating cycle also increases the gas pressure. Depending upon the volume of gas consumed, recycling systems may be required; a few very large systems employ liquid argon with vaporizers and cryogenic reclaim units. Depending upon the type of heater used and the nature of the product being processed, purity of the gas may be of considerable importance. Thus, the use of gas analysis equipment is becoming increasingly popular.

Every equipment manufacturer has developed a somewhat different type of heating system. The elements could be an iron- or nickel-base alloy, molybdenum, or graphite; all are resistively heated. Iron and nickel alloys are restricted to temperatures below 1100 °C, but offer the advantage of being tolerant of exposure to air, thus facilitating hot loading and hot unloading when rapid cycling is important. Molybdenum and graphite offer a higher temperature capability, but air must be excluded to prevent furnace damage.

Control systems cover the spectrum from completely manual to total computer control. The same is true for data monitoring and recording. A skilled operator is needed to operate in a manual mode, whereas computer-controlled systems with integrated data acquisition can operate virtually unattended, alerting the operator only to deviations from a preprogrammed time–temperature–pressure cycle or an anomalous operating condition.

3. *Applications*

3.1 *Powder Processing*

A major application of hot isostatic pressing is the consolidation of powders, both metal and ceramic. Initial use was for simple shapes encapsulated in cylindrical tubes or formed sheet metal cans. As canning technology became more sophisticated, shapes made by hot isostatic pressing grew increasingly more complex. In addition to sheet metal cans, a variety of glass and ceramic cans have been developed. Some of the cans contain a nondeforming member or mandrel (e.g., to define a hole in the finished part). Depending upon the mandrel material used, the details so defined can be very precise. Other surfaces, those generated by the can, are free surfaces; that is, they all move during consolidation of the powder with the extent of movement a function of the powder's packing density. Localized nonuniformities in powder packing plus deviations from true isostatic pressure due to corners and welds, as well as wall thickness variations in the cans make prediction and control of isostatically hot-pressed shapes quite challenging, and the common practice is to process near-net shapes which are sufficiently oversize to provide a minimum part envelope.

A "can-free" approach to hot isostatic pressing is also used in powder processing. If the powder is first precompacted (e.g., by cold pressing and sintering to a density such that pores are no longer interconnected), a subsequent hot isostatic pressing treatment of the bare part will close all but the surface pores, provided that they are free of any contamination.

Materials processed commercially by hot isostatic pressing methods include various specialty steels (tool, die and stainless grades), superalloys, hard metals (carbides), refractory alloys, beryllium and ceramics.

3.2 *Healing Porosity*

A growing use of hot isostatic pressing is upgrading properties by the elimination of residual porosity in castings and weldments; fatigue properties are particularly improved. Canning is not needed, as the volume of pores is low and they do not connect to the surface. It is imperative, however, that the surfaces of such pores be free of any contamination which would preclude bonding, otherwise a gently radiused pore could be compressed to form a sharp unbonded notch.

A related application being investigated is the use of hot isostatic pressing to heal cracks (e.g., fatigue cracks), thereby salvaging an otherwise useless part. The criteria for internal cracks are the same as those for pores. Surface cracks invariably contain foreign matter which must be removed and then the crack bridged with a gas-tight seal before salvaging by hot isostatic pressing can be effective. Reliable procedures plus suitable post-hot-isostatic-pressing inspection procedures will be needed to make this process a commercial success.

3.3 *Diffusion Bonding*

Although hot isostatic pressing was developed as a diffusion bonding process, this has become its least important commercial application (see *Diffusion Welding*). Solid assemblies such as laminates or clad parts are readily bonded, but hollow parts must be supported to preclude collapse.

See also: Hot Isostatic Pressing of Advanced Ceramics; Hot Pressing of Metals; Metals Processing and Fabrication: An Overview

Bibliography

Hanes H D, Seifert D A, Watts C R 1979 *Hot Isostatic Processing*. Battelle Press, Columbus, Ohio

James P J (ed.) 1983 *Isostatic Pressing Technology*. Applied Science, London

Lenel F V 1980 *Powder Metallurgy: Principles and Applications*. Metal Powder Industries Federation, Princeton, New Jersey

Metal Powder Report 1978 *1st Int. Conf. Isostatic Pressing*. Metal Powder Report, Shrewsbury, UK

J. N. Fleck

Hot Isostatic Pressing of Advanced Ceramics

Hot isostatic pressing (HIP) is a materials fabrication process whereby the starting powder or a premolded shape is simultaneously subjected to both high temperatures and high isostatic pressures using a gas transfer medium. This technique developed from an invention conceived more than two decades ago (Sadler et al. 1964), for the primary purpose of fabricating clad nuclear fuel elements, into a process which is being employed for a broad range of metallic and nonmetallic materials.

The principal factor which differentiates HIP from other processing techniques is the use of a gas as the pressure transmitting medium to effect equivalent changes in three dimensions in the materials which it surrounds. Until recently, argon was used almost exclusively; however, presses are now available in which powders and bulk pieces can be processed in oxygen-rich or other reactive atmospheres. Provisions for fiber–matrix (e.g., carbon–carbon) processing also now exist (see *Carbon–Carbon Composites*). The pressure range commonly used in HIP is 70–300 MPa, which is higher than the 30–50 MPa employed in uniaxial hot pressing.

As a result of this elevated and isostatic application of pressure, a virtually pore-free product can be produced at temperatures which are commonly 25–50% lower than those used in standard isothermal sintering procedures. This is true even for high-purity ceramic powders, some of which cannot be fully densified by normal sintering or even hot pressing techniques. As such, the amounts of sintering aids and grain growth inhibitors which are frequently added to ceramic powders can often be reduced to a minimum when HIP is used.

HIP also has the ability to produce simple and complex parts to net or near-net shapes, which reduces the amount of expensive postfiring machining that must often accompany other process techniques. This allows the overall cost of fabricating bulk ceramic shapes to be substantially reduced and the process to be competitive with other fabrication methods when used in an on-line batch process. Finally, the combination of microprocessor-controlled batch processing and isostatic application of the same pressure to all the parts of the batch produces improved property uniformity within each part, from piece to piece and from batch to batch.

It is important to note that HIP is a final stage in the fabrication of a densified piece. If the prior materials preparation and compaction are poorly done, HIP will not usually eliminate the defects thus introduced (see *Ceramics: Relation of Microstructure to Processing History*).

1. HIP Equipment

For successful hot isostatic pressing, three variables—pressure, temperature and time—must be sufficiently controlled. To this end, HIP systems, as shown in Fig. 1, are composed of four basic units: pressure vessel, internal heater (furnace), pressurization unit and control unit. Gas is pumped from storage through compressors into the pressure vessel. The pressure vessel is also penetrated, normally through the end closures, to provide power, instrumentation and auxiliary services to the furnace. Present day HIP systems are of the cold wall design, in which the internal resistance heater (furnace) is insulated from the wall of the vessel in some manner. Temperature is measured by thermocouples placed throughout the volume of the furnace.

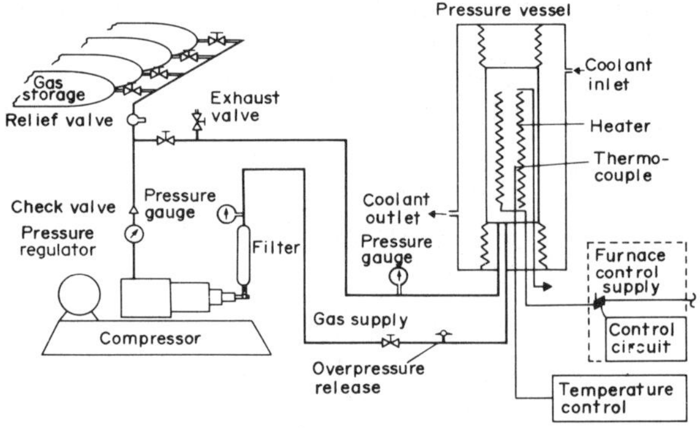

Figure 1
Schematic diagram for a hot isostatic pressure system (after Wills et al. 1984)

Cold wall autoclaves for HIP applications must be designed to contain high-temperature, high-pressure gases with maximum safety, yet possess the capability of fast loading and unloading. Configurations in general use include the following types: monolithic or multiwall forced (Fig. 2), wire-wound (Fig. 3) and multiple ring liner (Fig. 4). All these HIP vessels are fluid-cooled by the addition of external or internal cooling jackets (Smith and Zimmerman 1980).

Figure 2
Example of a monolithic forged HIP vessel with external cooling (courtesy National Forge Company)

The end closures, which must withstand an axial force proportional to the square of the vessel diameter, can be supported in two ways. The threaded or threadless pin closure types add to the axial stresses in the vessel wall, while the yoke frame supports the closure independently of the wall.

A number of pressurization units are now available to supply the required gas pressures to the HIP vessel. The utilization of rapid pressurization and of pressurization at different points in the heating cycle has become important in many HIP cycle applications.

Multiple stage piston, oil-lubricated compressors have been the industry's standard machines since the initial development of HIP. These machines are

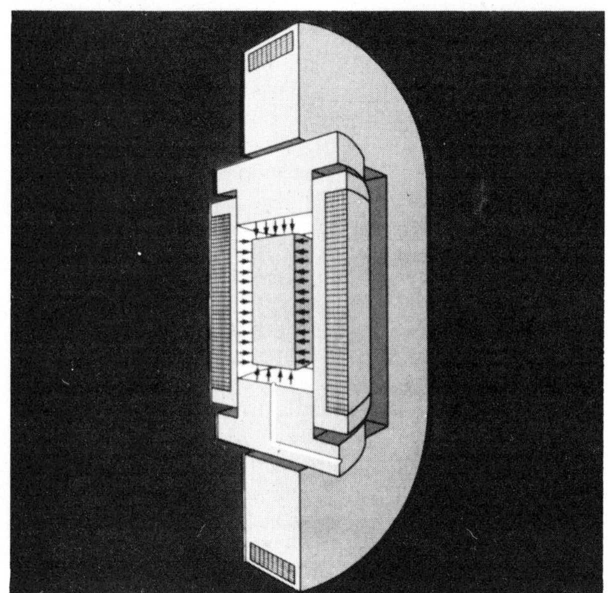

Figure 3
Sectional view of a wire-wound pressure vessel and supporting frame (courtesy ASEA Pressure Systems, Inc.)

Figure 4
HIP unit with multiple ring liner design (courtesy Autoclave Engineers, Inc.)

highly reliable for pumping helium. They have a high pumping capacity and can operate with lower inlet gas pressures, allowing the use of lower gas storage pressures. However, the industry has largely switched to the use of argon as a pressurizing medium for HIP. Because the density of argon in the 100–200 MPa regime approaches that of the lubricating oil, there is a major problem with oil carryover into the autoclave. Significant amounts of oil transport can result in contamination of the workpiece as well as damage to the electrical components in the HIP system.

Currently, many of the HIP production systems use diaphragm compressors because they are capable of pumping the gas without contamination. This type of compressor uses a hydraulically driven diaphragm to compress the gas; the diaphragm separates the gas from the oil. A short diaphragm life has been a major concern throughout the industry, since rupture can cause severe oil contamination. To minimize this problem, most machines now use double diaphragms along with leak detection devices.

A more recent compressor design, which better meets the requirements of the HIP process for large volumes of oil-free gas at high pressure, is now being used in production systems. This compressor, Fig. 5, has a double-acting hydraulic cylinder with a high-pressure gas cylinder attached to each end. The hydraulic fluid is separated from the gas being pumped by an isolation chamber and redundant seals. The inlet pressure can vary over a wide range, but to obtain the maximum outlet pressure, inlet pressures of >7.0 MPa are required. This compressor can be started under load because the hydraulic drive has low inertia and is unloaded automatically. It is also adaptable to different gases with little or no modification.

An additional pressurization approach which has been successfully employed is the use of a liquid argon pumping unit. The stored liquid argon is precharged into the cryogenic pump by a small centrifugal pump. The cryogenic pump delivers the

Figure 5
Double-acting pressure pump of the type commonly employed in HIP

pressurized liquid argon to an electrically heated vaporizer, which in turn delivers ambient heated argon gas directly into the system. This type of pressurization unit provides contamination-free argon gas at rates up to 20 $m^3 min^{-1}$ to a pressure of 100 MPa. This is nominally five times as fast as the capability of the largest piston compressor and ten times as fast as the largest diaphragm compressor. The major drawback of the cryopump is that there is no proven economical means for recycling the gas recovered from the autoclave and then converting it back into a liquid state for subsequent use with the cryopump.

The heater is the most complex unit of the HIP system. The resistance heating elements of modern HIP furnaces are usually located so as to provide uniform heating throughout the work zone. They are constructed of Fe–Cr–Al alloy, molybdenum or graphite. The Fe–Cr–Al alloys are widely applied throughout the industry and are preferred for operation up to a maximum of 1250 °C because of their oxidation resistance, long lifetime and ease of repair. However, they tend to grow with time, either by creep under their own weight or by thermal cycling.

Molybdenum heating elements are used at temperatures from 1200 to 1500 °C. These elements are extremely stable dimensionally, but major disadvantages include the embrittlement that occurs upon recrystallization and the lack of oxidation resistance. However, with proper design, long lifetime can be achieved and the poor oxidation resistance of molybdenum can be tolerated even with hot loading. Hot loading with molybdenum elements requires a bottom loading system to keep the element protected by the hot gas trapped in the inverted furnace chamber or inverted furnace chamber pressure vessel.

Fiber-reinforced graphite elements are required for satisfactory life when processing temperatures are in the 1500–2000 °C range. These are of single or multiple heating zone design. One design utilizes a fan unit at the bottom of the heater to facilitate heating and cooling.

2. *HIP Methods for Ceramic Materials*

Selection of the appropriate method of HIP is dependent primarily upon the desired material properties, the component geometry and acceptable manufacturing costs. There are two primary methods of HIP for ceramics: (a) HIP powder consolidation and (b) sintering plus HIP.

The classical method of producing parts from metal (and ceramic) powders by HIP has been to pour the milled powder into a deformable metal container (usually called a can), evacuate and seal the container and subject it to pressure and heat in the HIP vessel. The can prevents the surrounding gas from penetrating into the sample via the open porosity and equilibrating the pressure inside the sample with the externally applied pressure. The resulting part duplicates the shape of the can. In the process, parts and materials are simultaneously compacted and sintered.

Because the average particle diameter of ceramic powders employed in the HIP process has steadily decreased so that it is now in the submicrometer range, and because these powders have a low bulk density when loose, shape distortion during HIP presents a problem. To minimize this possibility, ceramic powders are now initially preformed to shape by cold pressing (uniaxial or isostatic) or injection molding. The resulting compacts may also be lightly sintered at low temperature to provide sufficient strength for handling. However, the open porosity is maintained. These precompacted shapes are then sealed by conventional or electron-beam welding in one of the various types of container and placed in the HIP system. After HIP, the container must be removed by mechanical or chemical means without damaging the component. The elements of this overall process are shown in Fig. 6.

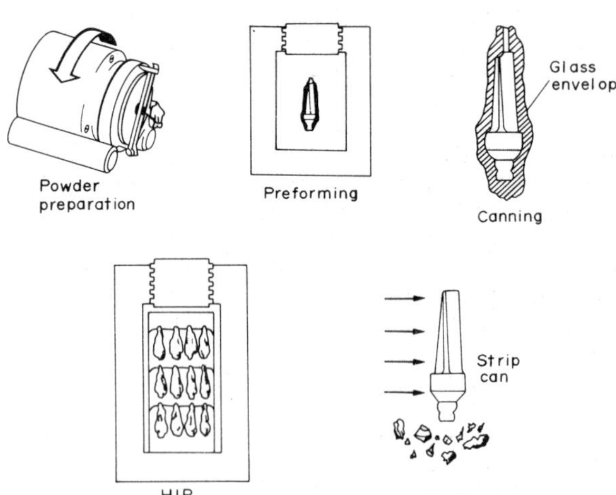

Figure 6
Individual steps of milling, cold preforming, canning, HIP and stripping in the standard HIP process

The container materials for the precompacted powders must be selected on the basis of mechanical properties and reactivity. Under reactivity, both lack of reaction with the compacted powder and easy removal (usually by solution) after compaction must be considered. The mechanical requirements are simply a low flow stress, compared with the hydrostatic stress at the temperature of compaction. Ferrous alloys, either mild steel or stainless steel, have traditionally been used up to ~1100–1200 °C and refractory metals such as tantalum and molybdenum at temperatures above this. However, the latter two metals are relatively expensive and require electron-beam welding to seal them. The thermal expansion of these containers is not usually considered in this application because of the high pressures which are applied.

In recent years, considerable attention has been given to glass containers, primarily because of cost. Much of the effort derives from the interest in HIP of structural ceramics such as silicon carbide (SiC) and silicon nitride (Si_3N_4). Glass ampoules and powder (often called a frit) have been successfully employed, especially for Si_3N_4 (Hüther et al. 1980). The ampoules are particularly useful for technique development and for making a limited number of simple shapes. However, the normally high viscosity of the glass places considerable stress on weak protruding parts when it folds over an unfired ceramic body of a complex shape. To circumvent this problem, glass powder has been placed over the entire surface of the workpiece. The powder is subsequently sintered to an impermeable state before the high external pressure is applied. However, this does impose certain constraints on the HIP equipment, since during the initial part of the cycle the equipment must operate at low pressure or under vacuum. Not all industrial equipment is capable of this. Another potential problem is the flow of the glass into the open porosity of the workpiece. To circumvent this, the viscosity of the glass must be high at the working temperature and the average diameter of the pores on the surface must be small. Furthermore, nonuniform packing of the powder particles at the surface must be minimized. Chemical reaction between the glass and the ceramic workpiece can also be a problem, especially if the ceramic is an oxide system.

Another advantage of the gastight container in the case of Si_3N_4 is the prevention of dissociation and consequent weight loss upon sintering (Wills et al. 1980, Wills and Brockway 1981a). Silicon nitride has a nitrogen partial pressure of 0.1 MPa at ~1870 °C. As the nitrogen gas cannot escape through the surface barrier during HIP, automatic adjustment of the nitrogen partial pressure inside the powder body to the equilibrium pressure takes place and no further dissociation occurs.

It is, however, important to realize that processing methods which require that gaseous species leave the body after the surface barrier has been sealed will not be successful. For example, processes like deoxidation of SiC by excess C, forming volatile CO, must be carried out before the body is isolated from the environment by the gastight surface barrier.

A very clear demonstration of the isolation of the powder body from its environment during HIP was experienced during work on large canisters of α-Al_2O_3 (Tegman 1982). The water content of the alumina powder in the mild steel container had to be reduced to <200 ppm to avoid exaggerated grain growth during the HIP process at 1350 °C. With a content of ~1000 ppm of confined water, an abnormal grain growth up to a rate of 100–200 μm h^{-1} was found. A bend strength reduction from 500 to 75 MPa was experienced. This example demonstrated that a thorough knowledge of the process is needed for the best results from HIP. However, generally the good control of the process offered by the presence of a gas-impermeable barrier is of great advantage.

The classical method of HIP for powders in a metal can described above is limited to fairly simple shapes. It is uneconomical to fabricate cans in complex shapes, and the use of glass containers has its own attendant problems, as noted above. Thus, other pre-HIP processing steps must be used. As such, a combination of processes called sinter plus HIP is gaining popularity. In this process, powders are compacted into net or near-net shape preforms by uniaxial or isostatic cold pressing (best for large preforms with complex shapes), hot forming (suitable for cutting-tool insert preforms) or injection molding (best for small- and medium-sized complex preforms required in production quantities), as noted above for the powder consolidation process. (Injection molding of the preforms is made possible by using a wax or plastic binder.) After forming, sintering at relatively high temperatures fully closes the preform's surface pores, thus preventing entry of gas during HIP. The steps in this process are illustrated in Fig. 7.

The prime objective of this mode of HIP is to remove the residual porosity in sintered ceramics. The microstructure of the ceramics must contain only closed porosity, to ensure that densification occurs during HIP. The presence of interconnected micro-

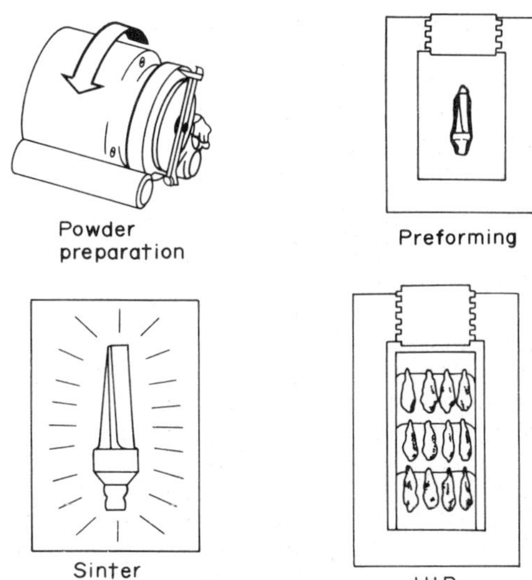

Figure 7
Individual steps of milling, cold forming, sintering and HIP in the sinter plus HIP process

crack networks in parts of the samples prevents complete densification. Otherwise healing of individual microcracks or flaws is expected.

Sintered parts are easily handled and do not require canning. In contrast to the HIP powder consolidation process, the packing density of components in the autoclave is significantly higher, because the can or encapsulant used in the former process occupies part of the available space and the green pressed component occupies a larger volume than the equivalent sintered part. There is, however, less control over the final microstructure and properties in this mode of HIP.

The relation between the processing procedures and the porosity size and concentrations is shown in Fig. 8. There is a significant size effect for normally sintered pieces: larger pieces reach slightly higher densities than smaller pieces. Vacuum hot pressing does offer an advantage over normal sintering, but the best process is HIP. Improvements in the strength of sintered silicon nitride and silicon carbide and of alumina–zirconia (Lange 1982) composites have also been reported. Details of some of these studies are given in Sect. 3.

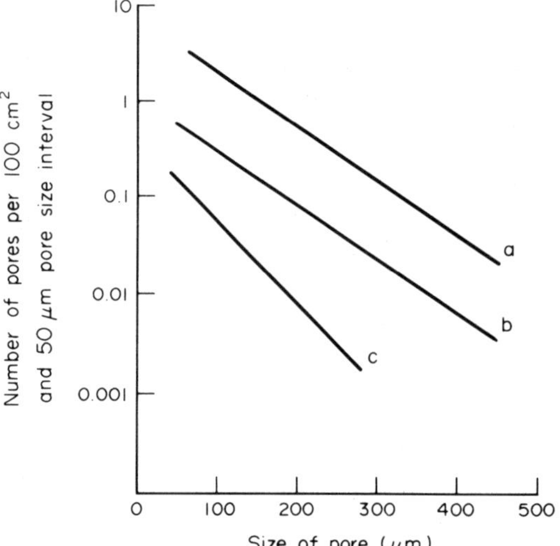

Figure 8
Effect of high pressure on the porosity of tungsten carbide at high temperatures: (a) large workpieces, pressed and normally sintered; (b) large workpieces, pressure-sintered in graphite die, vacuum hot-pressed; (c) large workpieces, HIP at 101 MPa and 1320–1350 °C (after Nillsson 1974)

3. *Examples of HIP of Ceramics*

A partial list of the applications for HIP of ceramics is given in Table 1. The number of uses is growing. These applications encompass those for a wide array of ceramic materials including piezoelectrics and ferrites, as shown in Table 2. Magnetic tape heads and cutting tool bits (Traf and Skotte 1976) are made commercially by the sinter plus HIP method and a similar approach has been adopted in the development of fast-ion-conducting β-alumina tubes (May et al. 1980) which are candidate electrode materials for new storage batteries. More complex shapes also densified on a laboratory scale using HIP include Si_3N_4 ball bearings and turbine blades (Wills and Brockway 1981a) and turbine rotors (Larker 1980) and large Al_2O_3 containers for housing nuclear waste (Larker 1981).

Table 1
Applications for hot isostatically pressed ceramics

Letdown valves for coal liquefaction	Drill bit inserts
Turbine disks	Nuclear waste consolidation
Turbine blades	Nuclear waste containers
Turbine vanes	Bearings
Stirling heater heads	Radomes and infrared domes
Cutting tools	β-alumina tubes
Orthopedic implants and dental ceramics	Nuclear reactor core supports
Magnetic tape heads	Fusion reactor insulators
Transducers	Special dies
Laser windows	Multilayer capacitors
Sputtering targets	

Early work on HIP of alumina (Hodge 1964) described the fabrication of transparent, pore-free material while avoiding grain growth. More recent work has explored HIP of β-alumina tubes (May et al. 1980) and alumina cutting tools, the latter of which are more promising for high-speed machining. Most of the HIP alumina is presintered to >95% of theoretical density (closed porosity). However, a process has been developed to encapsulate presintered alumina of only ~80% density (open porosity) in a titanium film before HIP. This has the advantage of the ability to presinter at much lower temperatures and still reach high HIP final densities, as shown in Fig. 9. Even for material presintered to ⩾95% density, titanium encapsulation develops a slightly higher final density and hardness in the final product.

Alumina ceramic tool materials encompass a broad range of compositions. The compositions include additions of up to 50 vol% of a number of nitrides and borides. Compositions patented by Hitachi Metals KK are shown in Table 3. These materials were all made by nominally the same procedure. The starting powders were sintered in vacuum, hydrogen or inert gas at 1400–1700 °C to >95% density. Materials to be titanium-encapsulated are presintered at 1350–1500 °C to a lower density. HIP conditions necessary to achieve nearly full density are in the ranges 1400–1700 °C and 70–200 MPa for 1 h.

Silicon nitride is a ceramic of significant interest

Table 2
Ceramics investigated for HIP research and development

Al_2O_3	Al_2O_3 and ZrO_2	Al_2O_3 and TiC	$MnZnFe_2O_4$
Al_2O_3 and VN	Al_2O_3 and NbN	Al_2O_3–Nb	Al_2O_3 and TiN
ThO_2	VC	TiC	UO_2
SiC and MgO	Si_3N_4	MgF_2	SiC
Sialons	Si_3N_4–W	Si_3N_4 and Y_2O_3	Si_3N_4 and MgO
HfB_2	Si_3N_4–SiC composites	ZrB_2	ZrO_2
BN	MgO	PLZT	ZnSe
Calcined nuclear wastes	TiB_2	Niobates	$BaFe_2O_4$
Quartz composites	$NiZnFe_2O_4$	WC–Co	$BaTiO_3$
BeO	ZnO	Graphite	Y_2O_3
UC	Aluminosilicates	Carbon–carbon composites	
Fe_3O_4–Al	Glasses		

for applications such as hot-section components of high-temperature engines, such as gas turbines and uncooled diesel engines. Examples of a Si_3N_4 turbine wheel and turbocharger impeller are shown in Figs. 10 and 11, respectively. However, this material does not readily sinter unless the powder contains a critical oxygen content and a sintering aid. The oxygen content of the powder and the additive type and concentration are major factors controlling the sintering kinetics. The reason for this lies in the liquid phase sintering mechanism, in which silicon nitride grains first dissolve in the silicate liquid formed at the sintering temperature by reaction between the oxide on the surface of each silicon nitride grain and the additive. The sintering kinetics depend on the solubility of silicon nitride in the melt, the rate of reaction at the silicon nitride–liquid interface, the diffusion rate through the melt, the volume of liquid present during sintering and the temperature and pressure.

The strength characteristics of two different types of hot isostatically pressed powders with 5wt% Y_2O_3 addition are shown in Table 4. The values given in this table do not represent maximum values obtained to date for Si_3N_4 (up to ~600 MPa for the bend strength) but are used here to compare two processing procedures. Procedure A was a common sequence for materials preparation: ball milling using

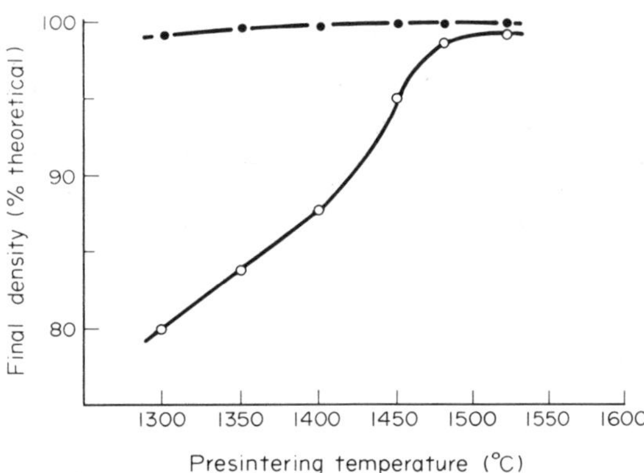

Figure 9
Comparison of final density after HIP of titanium-coated (●) and uncoated (○) alumina after different presintering temperatures

Si_3N_4 media, cold isostatic pressing at ~200 MPa, canning in tantalum and HIP at 2000 K for 1 h under a pressure of 200 MPa. In procedure B the approach was similar, except that the material was encapsulated in high-silica glass. The glass container with its contents was outgassed and sealed by melting. It

Table 3
Alumina-base ceramic cutting tool compositions patented by Hitachi Metals

Al_2O_3 + ~20 vol% TaN + ~15 vol% TiN + ~5 vol% Si_3N_4, AlN or BN
Al_2O_3 + ~20 vol% TaN + ~15 vol% VN
Al_2O_3 + ~20 vol% TaN + ~15 vol% NbN
Al_2O_3 + ~20 vol% TaN + ~15 vol% NbB_2
Al_2O_3 + ~20 vol% TaN + ~15 vol% TiB_2
Al_2O_3 + ~20 vol% TaN + ~15 vol% TaB_2
Al_2O_3 with the following additives:

5–10 vol% ZrN	5–10 vol% VB_2	5–7 vol% HfN
5–15 vol% VN	5–15 vol% NbB_2	5–15 vol% TiB_2
5–15 vol% NbN	5–15 vol% TaB_2	
3–7 vol% HfB_2	3–7 vol% AlN	
3–7 vol% ZrB_2	3–7 vol% BN	5–10 vol% Si_3N_4

Figure 10
Integrated turbine wheel with twisted blades made by the ASEA method for United Turbine, Sweden: (a) injection-molded silicon nitride preform; (b) hot isostatically pressed to full density and cleaned by sand blasting (after Larker 1984)

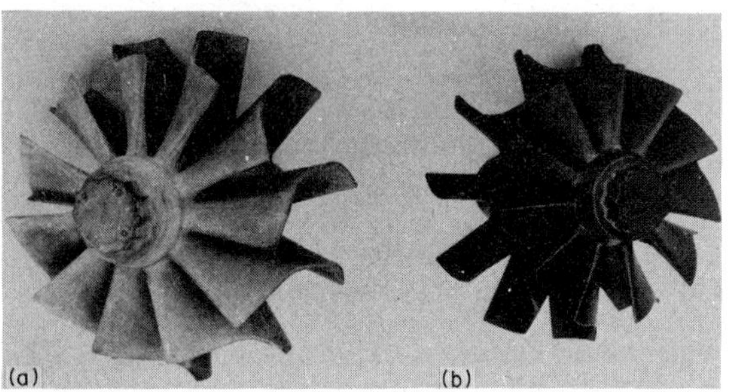

Figure 11
Turbocharger impeller: (a) injection molded with binder; (b) after HIP to full density (after Larker 1984)

was then heated to soften the glass. Pressure was gradually applied, with the result that the glass container folded over the Si_3N_4 green body and assumed its shape. The outside gas pressure and the temperature were then increased to the processing parameters, typically 1750 °C and 200 MPa. After a desired dwell time, usually 0.5–2 h, the system was cooled and depressurized. The glass remaining after cooling was removed by sand blasting.

It should be noted that, for comparative purposes, the three-point bend strength data of Larker et al. have been converted to four-point values. The high-temperature strength values are not strictly comparable, however, because the stressing rate used by Larker et al. is not known and there is a slight difference in test temperature. The high Weibull modulus values found by the Swedish workers (Larker et al. 1977, Larker 1980, 1984) on some batches of hot isostatically pressed silicon nitride are particularly noteworthy; they indicate that with good powder and well-controlled processing, HIP gives a reliable ceramic for structural applications.

Silicon carbide requires higher HIP temperatures than silicon nitride, because of its greater strength and the absence of a liquid phase at high temperatures. This has slowed development to some extent, because furnaces capable of reaching temperatures of >1700 °C, preferably ≥2000 °C are required. These furnaces, usually of graphite, are not widely available.

The influence of HIP temperature on the density and flexural strength of silicon carbide is shown in Table 5. Bare specimens were processed at 138 MPa for 2 h in argon. The material processed at >1850 °C had an average strength and density higher than those of the as-received unprocessed material. As the HIP temperature was increased to 2050 °C, the density and uniformity of the material increased. The flexural strength increased also but not markedly. At 1950 °C there appeared to be a drop in strength, perhaps indicating a wide scatter band in the property data.

Other workers have observed that bare silicon

Table 4
Mechanical properties of hot isostatically pressed silicon nitrides

	Procedure[a]	
	A	B
Creep at 1400 °C and 70 MPa (h^{-1})	nil	10^{-5}
Room temperature fracture toughness (MPa $m^{1/2}$)	4.1	
Mean four-point bend strength (MPa)	636	72–115[b]
Weibull modulus		11–40
Mean high-temperature strength (4-point bend) in air (MPa)	449 (1400 °C)	360–470 (1370 °C)
High-temperature Weibull modulus		26
High-temperature fracture toughness at 1400 °C (MPa $m^{1/2}$)	4.7	

a A, Wills and Brockway (1981a,b); B, Larker (1980), Larker et al. (1977) b Values converted from results of three-point bend test

Table 5
Effect of HIP temperature on flexural strength of α-SiC

Density before HIP ($g\ cm^{-3}$)	HIP temperature (°C)	Density after HIP ($g\ cm^{-3}$)	Room temperature flexural strength[a] (MPa)	1200 °C flexural strength[b] (MPa)
3.10	1750	3.11		
3.10	1750	3.12	400	425
3.11	1850	3.12		
3.08	1850	3.09	500	425
3.11	1950	3.14		
3.09	1950	3.14	450	375
3.09	2050	3.15		
	2050	3.16	540	510

a 475 MPa as-received (before HIP) b 410 MPa as-received (before HIP)

carbide specimens hot isostatically pressed in nitrogen rather than argon have improved strength properties. It was found (Table 6) that after HIP in nitrogen at 1700 and 1850 °C the flexural strength was much higher than after HIP in argon. At 1850 °C the nitrogen HIP material was stronger than the argon HIP material, even though their densities after HIP were the same.

Table 6
Strength of silicon carbide after HIP in argon and nitrogen atmospheres

Conditions of HIP treatment	Density before HIP ($g\ cm^{-3}$)	Density after HIP ($g\ cm^{-3}$)	Flexural strength (MPa)
Prior to HIP	3.14		470
1700 °C, Ar	3.15	3.15	430
1700 °C, N_2	3.15	3.15	590
1850 °C, Ar	3.15	3.15	505
1850 °C, N_2	3.15	3.15	570

A reduced creep rate in hot isostatically pressed SiC compared with sintered material has also been noted. This may have been caused by a reduction in porosity in the former material which allowed a reduction in the effective applied stress.

Exploratory HIP studies have been conducted (Larker 1984) on structural ceramics other than silicon nitride and silicon carbide, for example hexagonal boron nitride, boron carbide and partly stabilized zirconia. Larker has shown that the method can, with advantage, be used and parts processed to shape with modifications of the glass powder encapsulation process. The density obtained is in many cases higher or the temperature needed for full densification lower than is reported to be needed in uniaxial hot pressing of similar materials. Hexagonal boron nitride of high purity was pressed to a density of 2.22 $g\ cm^{-3}$ at 1900 °C and 200 MPa for 1 h. Boron carbide (B_4C) with ~0.1% impurities and 0.25% N was pressed to a density of 2.48 $g\ cm^{-3}$ (99% of theoretical density) at 2000 °C and 200 MPa for 1 h. The hardness H (0.5 kg) was 42 GPa and the three-point bending strength 590 MPa. Partly stabilized zirconia with 2.5% Y_2O_3 was densified at 1300 °C and 200 MPa for 1 h to 6.1 $g\ cm^{-3}$.

In addition, it has been reported (Hodge 1965) that both tantalum carbide and tungsten carbide have been fabricated to theoretical density by HIP at temperatures of 1600–1760 °C and pressures of 70–103 MPa.

Finally, three compositions of NbC_x ($x = 0.865$, 0.785 and 0.724) have been hot isostatically pressed within the temperature range 1750–1790 °C at a pressure of 207 MPa (Chevacharoenkul et al. 1982, Davis 1984). Although the average initial particle size, the unfired density and the temperature played interrelated roles in influencing the final density, the unfired density was the most important factor, particularly at the lower temperatures. However, at 1790 °C, the HIP density was ⩾99.4% of the theoretical density, regardless of the average initial particle size or unfired density.

There is also a continued growth in HIP of piezoelectric materials such as $BaTiO_3$, $SrTiO_3$, $Pb(Zr,Ti)O_3$ and YFe_5O_{12} for applications such as acoustic wave filters and oscillators (see *Piezoelectric Materials*). HIP significantly improves the desirable physical and strength properties of these materials compared with those obtained by sintering in air. However, care must be taken to avoid deoxidation of the piezoelectric material if it is pressed bare in argon. One method is glass encapsulation. This has been successfully used to obtain a 0.2 mm thick, pore-free translucent body of $Pb(Zr_{0.8}Ti_{0.2})O_3$. Another method is to place the specimens in an alumina crucible with ZrO_2 powder packed around them.

Nickel–zinc and manganese–zinc ferrites of high quality are needed for magnetic recording heads (see *Magnetic Head Materials*). A fully dense, fine-grained material is required for the most desirable high-frequency properties, machinability and abrasion resistance. The full densification of these ferrites under relatively low-pressure HIP conditions (1200 °C and 20 MPa for 4–4.5 h) was initially reported some years ago (Hardtl 1975). A more extensive study of densification and grain growth verified this work, showing that full density was obtained for ferrites with presintered densities >95% under HIP conditions of 30 MPa at 1200 °C and 103 MPa at 1100 °C (Anon 1976). In addition, the grain size was only about twice as large after HIP than after sintering. Sumitomo Specialty Metals of Japan now use HIP for essentially all their production of ferrites, because it provides greater uniformity from piece to piece within a batch and from batch to batch.

4. Future Developments

There is little doubt that interest in HIP of ceramics has increased sharply in the last few years. Much of this has been due to the strong interest in structural ceramics such as Si_3N_4 and SiC. This will probably accelerate as more heaters capable of ≥2000 °C are installed. With the success of HIP of ferrites, piezoelectrics and ceramic tool materials in Japan, an increasing level of research and HIP of these materials will be conducted in the USA and Europe.

It is important to stress that HIP of ceramic powders or sintered pieces is the final step of a multistage process in which the starting material is produced, usually altered in size, pressed and often sintered. The chemistry used in creating the material and the subsequent steps employed to bring it to the HIP stage can affect the final properties as much as can HIP. Thus improvements in pre-HIP powder processing, especially in materials such as structural ceramics, must also accompany advances in HIP technology, for the latter process will not usually erase problems introduced in the earlier steps.

See also: Cold Isostatic Pressing for Forming Advanced Ceramics; Hot Isostatic Pressing; Ceramics Process Engineering: An Overview

Bibliography

Adlerborn J, Larker H 1978 US Patent No. 4,112,143

Anon 1976 Presintering and HIP of ferrites. *Nippon Tungsten Rev.* 9: 9–16

Chevacharoenkul S, Davis R F, Peterson J 1982 Hot isostatic pressing of niobium carbide. *Mater. Sci. Eng.* 55: 289–92

Davis R F 1984 Hot isostatic pressing of ceramics. *J. Mater. Ed.* 6: 157–206

Hardtl K H 1975 Gas isostatic hot pressing without molds. *Bull. Am. Ceram. Soc.* 54: 201–7

Hodge E S 1964 Elevated temperature compaction of metals and ceramics by gas pressures. *Powder Metall. Int.* 14: 168–201

Hodge E S 1965 Hot isostatic pressing of refractory compounds. *Mater. Des. Eng.* 61: 92–96

Hüther W, Schweitzer K, Rossman A 1980 Method for encapsulating a molded ceramic member. US Patent No. 4,242,294

Lange F F 1982 Processing related fracture origins in sintered and HIP treated Al_2O_3/ZrO_2 composites. *Proc. 84th Annual Meeting American Ceramic Society*. American Ceramic Society, Columbus, Ohio

Larker H 1980 HIP silicon nitride. *Proc. 1980 AGARD Meeting*. NATO, Neuilly-sur-Seine

Larker H 1981 HIP beta-alumina tubes. *ASEA J.* 54: 85–90

Larker H 1984 Dense ceramic parts hot pressed to shape by HIP. In: Davis R F, Palmour H, Porter R L (eds.) 1984 *Emergent Process Methods in High Technology Ceramics*. Plenum, New York, pp. 548–59

Larker H, Alderborn J, Bohman H 1977 Fabrication of dense silicon nitride parts by hot isostatic pressing. *Proc. Society of Automotive Engineers Meeting*. Society of Automotive Engineers, Warrendale, Pennsylvania

May G J, Tan S R, Jones W 1980 Hot isostatic pressing of beta-alumina. *J. Mater. Sci.* 15: 2311–16

Nillsson M 1974 Cold and hot isostatic pressing of cemented carbides. *Interceram* 23: 55–62

Sadler H A, Paprocki S J, Dayton R W, Hodge E S 1964 Method of elevated-temperature compaction of metals and ceramics by gas pressures. Canadian Patent No. 680,160

Smith C W, Zimmerman F X 1980 *Design of Yoke Frame Multiple Ring Pressure Vessel*, Metal Powder Report No. 35. Metal Powder, Princeton, New Jersey

Tegman R 1982 HIPing large α-Al_2O_3 containers. In: Backman C M, Johannison T, Tegner L (eds.) 1982 *High Pressure in Research and Industry*. Plenum, New York, pp. 803–6

Traf A, Skotte P 1976 Isostatic pressing of ceramic materials—Methods and trends. *Powder Metall. Int.* 8: 65–68

Wills R R, Brockway M C 1981a Hot isostatic pressing of ceramics. *Proc. Brit. Ceram. Soc., Spec. Ceram.* 7: 233–47

Wills R R, Brockway M C 1981b *Hot Isostatic Pressing of Silicon-Based Ceramics*, AFWAL-TR-80-4193. Air Force Wright Aero Laboratories, Dayton, Ohio

Wills R R, Brockway M C, McCoy L G, Niesz D E 1980 Preliminary observations on the hot isostatic pressing of silicon nitride. *Ceram. Eng. Sci. Proc.* 1: 534–39

Wills R R, Brockway M C, McCoy L G 1984 Hot isostatic pressing of ceramics. In: Davis R F, Palmour H, Porter R L (eds.) 1984 *Emergent Process Methods in High Technology Ceramics*. Plenum, New York, pp. 537–50

R. F. Davis

Hot Machining

Hot machining is a metal-cutting process in which the workpiece is heated prior to machining to lower resistance to chip formation. The mechanism involves thermal softening or inducing a transformation of the material to be removed to a softer condition or one less inclined to strain harden during chip formation. In ordinary high-speed machining, considerable energy is converted to thermal energy (about 95% of the total energy ends up as thermal energy). About 75% of this thermal energy is dissipated on the shear plane in the chip in front of the tool, about 25% on the tool face and a small amount on the clearance face if the tool is sharp. This results in about 90% of the energy passing off with the chip, about 5% being conducted into the tool and about 5% to the workpiece. Hot machining involves the application of additional thermal energy to the material as it approaches the shear plane.

The additional thermal energy involved in hot machining may be added by preheating the entire workpiece before machining, but this is inconvenient and wasteful since only the energy in the outer skin of the workpiece is involved in hot machining. Usually some means of in situ surface heating is employed such as the application of a gas flame or plasma torch, induction heating or resistance heating, or even a laser beam.

The tools used in hot machining must be able to withstand the additional temperature without softening, wearing at an excessive rate, or failing due to thermal shock. Tungsten carbide tools best meet these requirements. However, care must be taken to ensure that heating is carried out at the right rate relative to the rate of material removal, so that the advantage of thermal softening is not offset by a decrease in tool life. As might be expected, hot machining is not always successful in improving tool life. Because of this and the fact that its application is troublesome and represents an added manufacturing cost, hot machining is rarely used, and then only if production runs are sufficiently long to warrant the required optimization study that must precede its use.

An attractive method of applying the additional hot-machining energy is by resistance-heating the outer surface of the workpiece. This is conveniently done by allowing current to flow through an insulated tool, along the surface of the work and back to the power supply through a rotating contact on the surface of the work. This calls for a very high current ($\sim$400 A) at relatively low voltage ($\sim$12 V). A resistance-welding power supply will usually serve this purpose.

See also: Chip Formation Micromechanics and Metallurgy; Heat Generation in Metal Cutting and Deformation; Machinability of Metals

Bibliography

Caminada A A 1952 Hot machining methods for difficult-to-machine metals. *Mater. Methods* 36: 98–100

Friedman L T 1956 Hot spot machining. *Iron Age* 165: 71–76

Krabacher E J, Merchant M E 1951 Basic factors in hot machining of metals. *Trans. ASME* 73: 761–69

Rajogpal S 1982 Laser assisted hot machining. *Laser Focus. Fiberoptic Commun.* 18: 29

Tour S 1951 Hot metal machining. *Met. Prog.* 59: 793–94

M. C. Shaw

Hot-Melt Adhesives

A hot-melt adhesive is a material used to bond surfaces together which functions by changing from a fluid at elevated temperatures during bonding to a strong solid at the use temperature. Hot-melts have been used since prehistory when early man first heated a resinous substance obtained from vegetable matter and found that after cooling, it formed a tough material. Sealing wax is an example of a hot-melt adhesive of more recent times. Early hot-melt adhesives were barely functional because of impurities and poor aging resistance of natural resins. Modern hot-melts are based on more stable synthetic polymers.

The primary advantage of a hot-melt compound compared with traditional fasteners and other adhesive systems is its rapid setting time. Depending upon the adhesive formulation and operating conditions, hot-melts can form a strong bond in a fraction of a second. This allows product assembly operations to be run at speeds which are not limited by the adhesive.

Hot-melts have a number of other advantages. In the molten state, they have low viscosities and flow readily to fill gaps between the substrates, thus sealing the assembly from vapor transmission. Like other adhesives, a continuous hot-melt film is superior to mechanical fasteners for distributing stresses and damping vibrations. Hot-melts can also be designed and used to provide thermal and electrical insulation.

In some respects, hot-melt adhesives are more difficult to use than other assembly methods. They require specialized equipment to transport and deliver the molten mass to the substrate. Heaters, pumps and coating or extrusion equipment are needed. The entire system must be insulated to prevent heat losses. An interruption in the process can result in plugged lines if cooling occurs, or degradation of the molten adhesive if it is exposed to the atmosphere for an extended time.

In the past few years, pressure-sensitive adhesives have been developed which are applied as 100% solids from a hot-melt. These products are not traditional hot-melt adhesives and are thus dealt with elsewhere (see *Pressure-Sensitive Adhesives*).

1. Applications

Hot-melt adhesives are used in industries which can benefit from the advantages of fast operation or where bonding of large surfaces is desirable to seal an assembly or to distribute stresses. They are used widely in packaging for bonding paper or cardboard surfaces, and for lamination of foil and plastic film. Hot-melts are used in the sealing of cartons and cases and for construction of paper bags. Disposable nonwoven products such as diapers, sanitary napkins and incontinence pads are assembled with hot-melts, as are books (binding), shoes (counters), furniture (edge veneering and general construction), textile products (seaming, hemming, cuffing) and miscellaneous wood and plastic products. The total hot-melt sales in the USA in 1981 were estimated at about 205 000 t, distributed as follows: 132 000 t (64%) in packaging, 18 000 t (9%) in general product assembly, 14 000 t (7%) in bookbinding, 14 000 t (7%) in disposable nonwoven products, 9000 t (4%) in textiles, 7000 t (3%) in furniture and 11 000 t (6%) miscellaneous. The entire market is estimated to be growing at about 8% per year.

2. *Principles of Use*

2.1 *Variables in Product Assembly*

Hot-metal adhesives function simply by cooling from the molten state, during application to the substrate, to a solid which provides strength at the use temperature. The viscosity of the adhesive must be low enough during application to spread and wet the surface of each substrate. The required viscosity depends upon the operating temperature and speed, and the shear applied during the coating or wetting step. If the adhesive is too fluid, it may spread excessively on the surface or penetrate and even bleed through porous substrates, leaving too little to form an adequate bond. If the adhesive is too viscous, it may not conform to the microscopic surface roughness of the substrates, resulting in poor contact.

The relationship between the viscosity of the adhesive and its temperature is important. If the viscosity does not increase fast enough after the two substrates are joined (the set time is too long), the bond may deform or be destroyed before setting by the stresses in a continuous manufacturing line. This problem can be recognized in a bonding failure when both substrates have been wet by the adhesive and the hot-melt has failed cohesively. If the viscosity increases too fast (the set time is too short), the adhesive may harden before there is contact with the second substrate, preventing wetting and bonding. This condition exhibits itself as adhesive failure between the hot melt and the second substrate.

The rate of increase in viscosity is controlled by both the viscosity–temperature relationship for the adhesive and the cooling rate after bonding. The rate of cooling is governed by the amount of adhesive used, the heat capacity of the substrates and heat losses due to conduction to the processing equipment and the surrounding atmosphere. The time between application of the adhesive to the first substrate and contact with the second substrate (the open time) for a particular piece of equipment must be shorter than the set time. Ideally the set time should be marginally longer than the open time, so that the substrates bond immediately after contact.

The set time of the adhesive in a specific operation varies somewhat because of uncontrolled variables such as the ambient temperature and variations in the thickness of the substrate and quantity of adhesive applied. These variables can be accommodated to eliminate poor bonding by adjusting the operating speed and the volume of adhesive used. This may not be cost effective with respect to productivity. It may be preferable to accept some level of rejected parts to achieve lower costs through high production rates and reduced consumption of adhesive.

2.2 *Adhesive Properties and Performance*

An approach can be made to understanding the performance of hot-melt adhesives by considering the bulk properties and rheology of the adhesive. Surface characteristics of the adhesive and substrates are also important for proper wetting and spreading. However, these considerations of surface tension and work of adhesion are the same for all adhesive systems and are therefore not discussed here.

The important performance characteristics of a hot-melt adhesive are its strength and flexibility over its useful temperature range and its viscosity at the temperature of application to the substrates. The useful temperature range is defined by the appearance of brittleness at low temperature and the loss of strength at elevated temperature.

The low-temperature limit of the adhesive is related to the glass transition temperature (T_g) of the composition. Below T_g, where molecular loss processes become minimal, even small deformations result in brittle failure (fracture). The exact temperature at which brittleness appears depends upon the rate of deformation used in the test. If the deformation rate is increased, the brittleness temperature is higher.

The T_g of the adhesive is a function of the values for the individual compatible components. Modifying resins used in hot melts are generally low molecular weight (≈500–2000) rigid molecules which often have T_g values above room temperature, higher than those of the base polymers. Consequently, addition of resins increases T_g of the adhesive. Addition of compatible oils and waxes of lower T_g, lowers T_g of the adhesive. The T_g of the adhesive can be estimated from the T_g values of the individual compatible components by the following equation:

$$T_{gA} = (W_1 T_{g1}^{-1} + W_2 T_{g2}^{-1} + \ldots + W_n T_{gn}^{-1})^{-1} \quad (1)$$

where T_{gA} is T_g of the adhesive, T_{g1}, T_{g2}, . . . are the values for the components and W_1, W_2, . . . are the weight fractions of the components; the temperatures are in K.

The stiffness or flexibility of the adhesive within its useful temperature range corresponds to its modulus at the use temperature. Use temperatures are generally in the rubbery plateau region of the system between the transition zone and the melt or terminal zone. The modulus in the plateau is related to the average molecular chain length of the amorphous polymeric system between points of restricted molecular motion. In thermoplastic compositions, polymer mobility can be restricted by chain entanglements or segmental crystallinity. The shorter the amorphous chains, the higher is the modulus of the system in the rubbery plateau. Also, the crystallinity itself increases the modulus by acting as a reinforcing filler. A limit is reached when the crystallinity level is so high and the amorphous chains are so short that the temperature at which brittleness appears is too high for the intended application. Both compatible resins and oils will lower the modulus and increase the flexibility of the adhesive in its normal use range. These additives act as solvents for the amorphous polymeric chains, and the amorphous fraction can be considered to exist as a concentrated solution of polymer. The resins and oils reduce the number of chain entanglements per volume of amorphous polymer and increase the molecular weight between entanglements. The plateau modulus is reduced, according to the classical theory of elasticity, because the modulus is proportional to the concentration of network strands. The number of strands between entanglements and crystalline segments is reduced as the number of entanglements is reduced.

The melt temperature is that temperature at which restrictions to molecular flow disappear. This means that the composition must be above the melting point of the crystalline segments and that entanglements do not impede flow. Each entanglement has a characteristic disentanglement time, depending upon molecular motion, which is related to the temperature of the system. The rate of deformation must be slow enough to allow disentanglement and flow to occur. Therefore, the flow characteristics of a hot-melt adhesive depend upon the shear rate as well as the temperature of the system. As indicated earlier, compatible resins and oils reduce the temperature at which flow is observed, because they reduce the number of chain entanglements.

The attainment of the ultimate bond strength of a hot-melt adhesive involves large deformation processes. Cohesive failure depends on the yield or fracture strength of the polymer mixture. Adhesive failure occurs if there is poor contact between adhesive and substrate or if the interfacial work of adhesion is low.

3. Components of Hot-Melt Adhesives

Hot-melt adhesives are composed of a polymer generally combined with modifying ingredients such as low molecular weight resins, waxes, oils and fillers. Stabilizers are usually added to combat degradation which can occur both during the molten application stage and as a result of long-term aging. The components and their functions are summarized in Table 1.

Table 1
Components of hot-melt adhesives

Polymer	
Examples:	ethylene–vinyl acetate, ethylene–ethyl acrylate, amorphous polypropylene, polyethylene, polyamides, polyesters
Functions:	provides cohesive strength and stiffness, determines melt temperature
Resin	
Examples:	petroleum resins, rosin-based resins
Functions:	improves specific adhesion and surface wetting, reduces melt viscosity
Wax	
Examples:	paraffin and microcrystalline waxes derived from petroleum; synthetic waxes, primarily low molecular weight polyethylene
Functions:	reduces melt viscosity, eliminates surface tack, reduces cost

3.1 Polymers

The polymer imparts strength to the composition as a result of its molecular weight, crystallinity or hydrogen bonding. Higher molecular weight polymers have more chain entanglements per molecule and so have higher melt flow temperatures. If crystallinity is present, the crystalline regions maintain the integrity of the system until the crystalline melting point is reached. Chain entanglements may continue to provide strength for an additional temperature span above the crystalline melting point. Hydrogen bonding in polar polymers produces high melting points, which maintains bond strength and performance at higher temperatures.

(*a*) *Olefin copolymers.* Ethylene–vinyl acetate copolymers (EVA) are the most universally used polymers in hot-melt adhesives. Selection of the proper EVA is based on the vinyl acetate (VA) content and the viscosity at elevated temperatures. A broad spectrum of ethylene–vinyl acetate ratios is available, $\approx$10–40 wt% VA, the most commonly used concentrations being 25–30 wt%. Vinyl acetate increases the polarity of the polymer and improves wetting and specific adhesion to substrates such as cellulosics. Although the polymers are manufactured in a random process, the polyethylene forms crystalline blocks whose size decreases with increasing vinyl acetate content. The crystalline polyethylene blocks have a melting point of about 75 °C in the polymer, providing strength up to that temperature.

The higher-VA polymers (≥28 wt% VA) have less-crystalline polyethylene blocks and are therefore softer and more flexible. They have improved adhesion to nonporous substrates. On the other hand, the lower-VA polymers (<28 wt% VA) have higher polyethylene crystallinity and shorter amorphous segments. This gives a harder, stiffer polymer which has improved compatibility with hydrocarbon additives such as waxes and low molecular weight hydrocarbon resins.

The other important factor in choosing the proper EVA is its flow characteristics at elevated temperatures. These are related to the molecular weight and molecular weight distribution of the polymer. Flow is measured by the melt index (ASTM D1238) which is the weight of polymer extruded in a specified time through a designated orifice at a specified temperature and load. The melt index is inversely related to molecular weight. Products having melt indices of ≈1–500 are available, but those of ≈5–30 are the most used. Generally, the higher molecular weight polymers (lower melt index) provide the strongest composition and have more "hot tack" and shorter open time than the lower molecular weight polymers. The polymer having the lowest melt index is generally selected, consistent with the handling requirements of the application equipment.

Ethylene–ethyl acrylate copolymers (EEA) are alternatives to the EVA polymers. Poly(ethyl acrylate) has a lower glass transition temperature than poly(vinyl acetate), so EEA copolymers should have a lower T_g than EVA copolymers at the same weight concentration. On the other hand, there is a higher concentration of crystalline polyethylene chains in the EEA copolymers, which results in higher performance temperatures.

(*b*) *Olefin homopolymers*. Amorphous polypropylene (APP), a by-product of the polypropylene process, is used in large volume in hot-melt adhesives. Refined APP is estimated to contain ≈15–30% crystallinity by differential scanning calorimetry, much lower than that of polypropylene, but it is still not strictly an amorphous product. The molecular weight is estimated to be in the range 2000–15 000 in the commercially available products used in adhesives.

(*c*) *Polyamides*. These are used in specialty, high-performance hot-melt adhesives. The basic products are made from "dimer acid," which is the dimer of unsaturated C_{18} fatty acids and ethylene diamine. Other modifying acids and amines can be included. The polyamides have high service temperatures because of the polar bonding of the amide links.

(*d*) *Polyesters*. These, like the polyamides, are used in low-volume, specialty adhesives because of their high strength and high service temperatures. They are made from organic dicarboxylic acids and dialcohols.

3.2 *Resins*

A low molecular weight resin (MW ≈500–2000) is usually added to the olefinic polymers to reduce the melt viscosity and improve adhesion. The resin must be compatible with the polymer to provide the expected property changes. Resin polarity, molecular weight and molecular weight distribution are important considerations in the functioning of the resin. The softening point of the resin depends upon its molecular weight and molecular weight distribution. The viscosity at the softening point is in the range 0.5–1.0 kPa s. The glass transition temperature of the resin is about 40–50 °C below its softening point. For the so-called "hard resins," T_g is well above room temperature, so T_g of the polymer–resin blend is higher than that of the polymer.

There are two major types of resins used in hot-melt-adhesives: petroleum-derived and resin-derived. Polyterpenes and alkylated phenol resins are also employed.

(*a*) *Petroleum resins*. The most useful of these are manufactured from petroleum cracking streams, mostly C_5 hydrocarbons, or pure aromatic monomers such as styrene, α-methylstyrene and vinyltoluene. Mixtures of the pure monomers are often used. The streams are polymerized to low molecular weight oligomers (MW 500–2000) by acid catalysts such as acid clays, aluminum chloride or boron trifluoride. The lower molecular weight resins from the same hydrocarbon feed are more readily compatible than the higher molecular weight products. The broad distribution products are more compatible than products having a narrow molecular weight distribution.

(*b*) *Rosin-based resins*. Rosin, a mixture of C_{20} diterpenoid acids which is produced by pine trees, is the raw material for an important group of resins used in hot-melt adhesives. The most widely used resins are the esters prepared by reaction with polyhydric alcohols such as glycerine and pentaerythritol. The rosins may be stabilized by hydrogenation or disproportionation prior to esterification, to improve retention of viscosity and color of the adhesive during high-temperature application or long-term aging. The residual carboxylic acid and hydroxyl contents are important considerations in compatibility with polymers. The rosin derivatives in general have a broader range of compatibility than the hydrocarbon resins.

The ester resins have a narrow molecular weight distribution because they are mixtures of simple esters. The rosin acids are bulky molecules having molecular weights of ≈300. The triester of completely esterified glycerine would have a molecular weight of ≈940, while that of completely esterified pentaerythritol would be ≈1260. Despite these relatively low molecular weights, rosin-based resins have comparatively high T_g values, well above room temperature for the glycerine and the pentaerythritol esters.

3.3 Wax

A wax may be added to the olefinic polymers to reduce the melt viscosity further, to influence the melting temperature and to eliminate residual tack at room temperature. Wax also reduces the cost of the composition. The wax can be isolated from petroleum or manufactured synthetically, principally from polyethylene by high pressure polymerization.

3.4 Typical Formulations

Typical examples of hot-melt adhesive formulations are found in the trade literature of suppliers to the adhesives industry. However, finished compositions are considered proprietary by the adhesive manufacturer. A starting point for general EVA-based hot-melt adhesives is equal parts of EVA, resin and wax. Appropriate adjustments to this composition for a specific EVA would be to increase the proportion of polymer for greater strength, increase that of resin for improved adhesion and increase that of wax for reduced viscosity. Some typical formulations are given in Table 2.

Table 2
Typical hot-melt formulations[a]

	Composition (wt%)
General packaging	
EVA (28% VA, MI 6)	30
Aromatic copolymer resin (SP 120 °C)	25
Paraffin wax (MP 65 °C)	45
Carton sealing	
EVA (28% VA, MI 25)	45
Glycerine esters of hydrogenated rosin	25
Microcrystalline wax (MP 90 °C)	30
Bookbinding	
EVA (28% VA, MI 6)	35
Glycerine ester of hydrogenated rosin	40
Microcrystalline wax (MP 70 °C)	25
Carpet seaming tape	
EVA (28% VA, MI 6)	50
Pentaerythritol ester of hydrogenated rosin	15
Paraffin wax (MP 70 °C)	35

a MI melt index; SP softening point by ring and ball method; MP melting point

4. Testing

Adhesives are tested for quality assurance and to guide the formulator toward performance improvements during development of new compositions. In general, they are tested both for performance as an adhesive and to determine the bulk properties of the composition. The performance tests commonly used are those for shear strength, peel strength and shear adhesion failure temperature. Another important consideration is the temperature at which brittleness occurs. In addition, hot-melt adhesives are often examined for melt viscosity, open time, tensile strength, modulus and flexibility at use temperature. Stability tests are also generally required, stability at application temperatures being particularly important for hot melts. Long-term stability of the assembled bond may or may not be important, depending upon the final use.

4.1 Shear and Peel Tests

The most common test for shear strength of a hot-melt adhesive is the lap shear test, in which the bond is tested to failure. Adhesive peel strength is measured at 180° from a rigid substrate or as a T-peel if both substrates are flexible. These two peel tests do not yield the same results, because the peeling angle affects the force required for separation. The rate of peel must be controlled because it affects the measured strength. In the shear and peel tests, the thickness of the adhesive layer must also be carefully controlled.

4.2 Shear Adhesion Failure Temperature

This test measures the temperature at which the viscous resistance of the adhesive in shear can no longer support an applied stress. The test specimen is an overlap shear configuration which is hung from one tab with a weight attached to the other. The assembly is placed in an oven and the temperature is raised either in increments or continuously until at least three of five replicate specimens fail. The temperature at which this occurs is reported as the shear adhesion failure temperature.

4.3 Brittleness

Brittleness or low-temperature flexibility tests measure the temperature at which the adhesive first exhibits brittle characteristics as it is cooled. Various tests have been developed by adhesive manufacturers, but the mandrel bend test is the most common. In this test, a strip of the adhesive is bent around a rod of selected diameter at a set rate at different temperatures. The highest temperature at which the adhesive cracks is recorded as the brittleness temperature. The brittleness temperature is influenced by the thickness of the adhesive, the diameter of the rod and the bending rate. This test is usually done by hand and is therefore carried out by experienced testers, to achieve consistent results.

4.4 Melt Viscosity

The viscosity of hot-melt adhesives at application temperatures is generally measured in continuous shear using a Brookfield Thermosel viscometer. Viscosity–temperature curves are often constructed to cover the temperature range of interest.

4.5 Open Time

Measurement of open time allows the user of hot melts to judge how long an adhesive will remain fluid enough to bond surfaces together. In evaluating adhesives for paper lamination, a molten film of adhesive of controlled thickness is cast onto a paper substrate and strips of a second substrate, such as paper, plastic film or foil, are brought into contact with the molten adhesive after successive 5 s intervals. A non-heat-conductive weight is placed on each bond for 5 s to ensure intimate contact. After the test pieces have cooled, they are lifted to determine the longest time before bond formation that still gives paper tear—in other words, the adhesive is still fluid enough to form a bond when the second substrate is applied to the adhesive surface. This test allows adhesives to be ranked according to open time, but the results do not necessarily relate to specific manufacturing experiences, because of different machine speeds and cooling rates.

4.6 Tensile Strength, Modulus and Flexibility

Hot-melt compositions may be cast into plaques and tested for bulk physical properties by standard test methods. The properties most frequently measured are tensile strength, modulus and elongation at break. Torsional rigidity is sometimes determined at various temperatures to indicate the temperature at which compositions become hard or brittle. There is also a mandrel bend test for flexibility (ASTM D3111-76).

4.7 Dynamic Mechanical Properties

As with any polymeric system, hot-melt adhesives can be examined with a mechanical spectrometer to obtain storage and loss moduli and the loss tangent (tan δ) over a temperature range, or the two moduli and complex viscosity in the melt region. Data also may be obtained at a specified temperature as a function of frequency. This information indicates T_g, the modulus or the stiffness at the use temperature, crystalline melting point and viscosity at application temperature. Compatibility of additives can be assessed by their effects on the dynamic mechanical properties.

4.8 Stability

Hot-melt adhesives are especially subject to changes if they are held molten for extended periods before application. The viscosity of the adhesive may change because of chemical reaction, degradation or volatilization of a low-boiling component. If the viscosity changes, the flow characteristics of the adhesive change. In some applications, color and odor are important, and these may change if the adhesive is held in the molten state. Testing is straightforward: a sample is held molten at the selected temperature in a suitable container and the viscosity, color and odor as well as the development of a skin on the surface of the melt are monitored.

See also: Adhesion: Fundamental Principles; Adhesives: Standard Mechanical Test Methods; Joining by Adhesives: An Overview

Bibliography

Skeist I (ed.) 1977 *Handbook of Adhesives*, 2nd edn. Van Nostrand Reinhold, New York

TAPPI Hot Melt Short Course Notes (1978 onwards). Technical Association of the Pulp and Paper Industry, Atlanta, Georgia

Adhesives Age (1972 onwards): issues devoted to Hot Melt Adhesives

J. B. Class

Hot Pressing of Advanced Ceramics: A Survey

Sintering of ceramic powders to acceptable densities can often be facilitated by application of pressure at high temperatures. In general, application of pressure allows the sintering temperature to be lowered by up to 500 °C. For many materials, hot pressing is the only practicable route to obtaining a dense compact. This article deals mainly with the practical aspects of hot pressing; theoretical considerations are discussed in the article *Hot Pressing of Advanced Ceramics: Technology and Theory*, where emphasis is placed on the various densification mechanisms.

In its simplest form, hot pressing consists of the uniaxial movement of a plunger in a die cavity which contains the powder. Variations of this, whereby two opposing plungers are used or additional pressure is applied to the die body, are also known.

Hot pressing is a very valuable research tool for ceramic materials in that it allows dense solid ceramics to be produced rapidly and reproducibly with minimum contamination. It also enables high-temperature deformation and densification processes to be studied in detail. As a production tool, however, it should be emphasized that the process is restricted to a few firms. Government-sponsored development programs on ceramics for gas turbines, batteries and defense applications often use hot pressing for fabricating the high-performance components. However, the economics of hot pressing are such that there is strong commercial pressure on any firm mass-producing ceramics to find a less expensive alternative.

Nevertheless, some hot-pressed ceramics are commercially available, such as silicon nitride, silicon carbide, bearings and turbine pieces, boron carbide abrasion-resistant parts, plates of various materials for ceramic armor and materials with special electronic and/or optical properties.

The situation is in a continual state of flux,

however, as pressureless sintering methods and compositions are developed for ceramics which up to now have been obtained only by hot pressing.

1. Advantages of Hot Pressing

The most obvious advantage of hot pressing lies in the improvement in the densification kinetics of the ceramic powder. This usually means that the densification cycle is much shorter than in the case of pressureless sintering and that the temperature of operation is significantly lower (by several hundred °C).

One corollary of this improvement is the minimization of grain growth and pore growth which can occur in many ceramics during prolonged sintering treatments. Submicrometer powders, for example, often show strong grain growth, producing a final average grain size of several μm. In addition, the disadvantages associated with inhomogeneous grain growth are avoided by hot pressing.

Ceramic materials containing volatile species such as lead oxide, zinc oxide and some nitrides are more readily densified in the relatively rapid hot pressing process than in pressureless sintering.

A further advantage is that direct pressing to shapes with good dimensional accuracy is possible. The final grinding and polishing of the finished part are thereby minimized.

2. Disadvantages of Hot Pressing

The major disadvantage of hot pressing as a shaping technique for ceramics lies in the high cost and low productivity of the equipment. If research and development materials are excluded, the main industrial use of hot pressing is the production of very specialized electronic, optical, nuclear and engineering ceramics of very high unit cost.

Die wear is a significant problem in the hot pressing of harder ceramic powders such as alumina, silicon nitride and silicon carbide. When nonoxide dies are used, degradation of the die can occur by oxidation. Some combinations of pressing powder and die material can react at high temperatures.

Special powders are normally required, usually of very fine homogeneous grain size to take advantage of the rapid sintering kinetics. All low-temperature volatile constituents such as adsorbed water must be carefully removed from the powders before processing. These factors add further to the overall cost of parts made by this route.

Dimension and shape limitations of the parts produced are additional important constraints on the widespread use of the process. Large plates up to 300 × 300 mm have been produced routinely by some US companies using very specialized and expensive equipment. Quite large cylindrical ceramic pieces including tubes can be made by use of inserts, provided that the length–wall thickness ratio does not exceed ~5. Subsequent machining is necessary to produce thin-walled tubes. The objects produced are almost without exception limited to relatively simple shapes with cylindrical or orthogonal symmetry.

3. Equipment and Die Materials

In normal uniaxial hot pressing (Fig. 1), the die and plunger(s) are heated by a furnace which surrounds the whole die assembly. If the die body is electrically conducting, as for example with graphite, inductive heating of the die is often used. Silicon carbide resistive heating elements provide a particularly suitable system for most pressing temperatures (i.e., up to 1700 °C). Metallic heating elements are often more cost effective at lower temperatures. The thermal expansion of the die material should preferably be lower than that of the compact, to facilitate the extraction of the latter after cooling. The temperature of the die and compact is measured continuously with either a thermocouple or pyrometer.

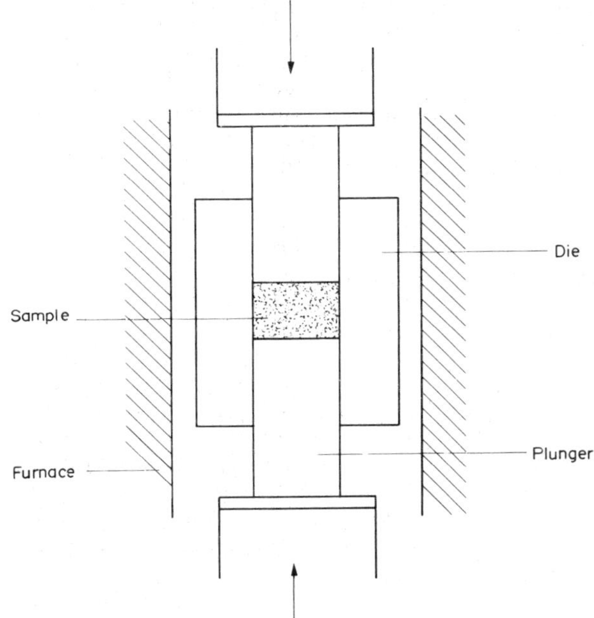

Figure 1
Diagram of a typical laboratory hot press

Table 1 gives information on die materials which have been used.

The clearances between die and punch are chosen with due regard to the precision of the final component, the amount of motion corresponding to compaction and the difficulty of compact extraction. For graphite, tolerances of the order of ±50 μm are used. For really difficult materials, the compact can

Table 1
Die materials for hot pressing

Die material	Maximum use (°C)	Maximum pressure (MPa)	Comments
Nimonic alloys	1100	High	Creep and erosion problems
Mo or Mo alloys	1100	21	Oxidation unless protected
W	1500	24	Oxidation and galling
Al_2O_3	1400	207	Expensive, brittle
BeO	1000	104	Expensive, brittle
SiC	1500	276	Expensive, brittle
TaC	1700	55	Expensive, brittle
WC, TiC	1400	69	Expensive, brittle
TiB_2	1200	104	Expensive, brittle
Graphite (standard)	2500	34.5	Severe oxidation above 1200 °C
Graphite (special)	2500	138	Severe oxidation above 1200 °C

be made inside an inert powder bed which is in turn pressed by the plungers.

4. Materials

Powders for hot pressing are usually fine and uniform. Care must be taken that the materials are homogeneous, since there will be little time for diffusion reactions to take place in the normal process. In some cases powders sinter rapidly without additives. However, for many of the more refractory materials, densification aids such as those listed in Table 2 are necessary.

Table 2
Densification aids for hot pressing

Matrix	Aid
MgO	1–3% LiF, NaF
CaO	1–2% LiF
ThO_2	10–25% fluoride
BeO	0.5% Li_2O
UC, TaC	1–20% U, Co
TaC, TiC, WC, ZrC	1–10% Fe, Ni, Co, Mn
Si_3N_4	1–3% MgO, Y_2O_3
ZrB_2TiB_2	1–10% Ni, Cr
SiC	B_4C, Al_2O_3

The fluoride additives have a tendency to evaporate at high temperatures, so the final composition of the compact is often quite different from that of the starting material. Sometimes the volatile constituent can be eliminated by postfiring. Most of the additives work by forming a liquid phase at high temperatures which speeds up densification, often by a process of solution and reprecipitation when large quantities of liquid phase are involved. Care must be taken to avoid segregation effects whereby the liquid is not uniformly distributed in the compact.

Reaction hot pressing (see *Reaction Sintering of Ceramics*) is analogous to reaction sintering in that two or more different powder phases are mixed together and subsequently undergo a reaction at or near the pressing temperature. Clearly, gas-phase reactions are not compatible with hot pressing.

In the standard piston–die hot pressing techniques, differences in density at various points in the pressed body occur, similar to those encountered in cold pressing.

Anisotropic microstructural effects occur in some hot-pressed compacts. These can be due to preferential crystallization in one or more directions or the plastic deformation of the individual particles.

The selection of pressing parameters (temperature, pressure and time) can be either entirely empirical or based on the various theories of hot pressing densification mechanisms. The latter have now been formalized to the extent that deformation maps exist which can be useful in suggesting approximate processing parameters, with the proviso that practical trials can be used to specify the optimum.

5. Special Hot Pressing Processes

Several processes have been developed to overcome some of the various disadvantages of standard hot pressing. For example, in order to raise the productivity of the method, Oudemans (1969) developed a unit for continuous hot pressing of ceramics such as ferrites. The die is made of alumina because of its high wear resistance. It is subjected to lateral compression to avoid breakage. Simple rod shapes several hundred mm long have been produced routinely. Disks can also be produced automatically by introducing spacers.

Hot isostatic pressing of ceramics can also be more productive than the standard method, in that many pieces can be treated as a batch. The pressure-transmitting medium is usually an inert gas. The materials to be densified can either be enclosed in a flexible container of some kind which is placed in the pressure chamber or can be presintered in such a way that only unconnected isolated pores at grain boundaries remain. In the latter case, theoretical densities can be attained. Density variations within the material are avoided. The latter technique is quite compatible with the fabrication of delicate and complex shapes; for example, turbine blades with edges 0.3 mm thick have been successfully made (see *Hot Isostatic Pressing of Advanced Ceramics*).

Electronics firms have been particularly concerned with the application of the hot isostatic pressing techniques to the production of high-performance electronic ceramics such as $BaTiO_3$, $SrTiO_3$, $Pb(Zr,Ti)O_3$, $Y_3Fe_5O_{12}$(YIG) and various ferrites.

See also: Ceramics Process Engineering: An Overview

Bibliography

Leipold M H 1976 Hot pressing. In: Wang F F Y (ed.) 1976 *Treatise on Materials Science and Technology*, Vol. 9: *Ceramic Fabrication Processes*. Academic Press, New York, pp. 95–134

Oudemans G J 1969 A continuous hot-pressing technique. *Proc. Br. Ceram. Soc.* 12: 83–98

Stuijts A L 1970 New fabrication methods for advanced electrical materials. *Sci. Ceram.* 5: 335–62

J. Briggs

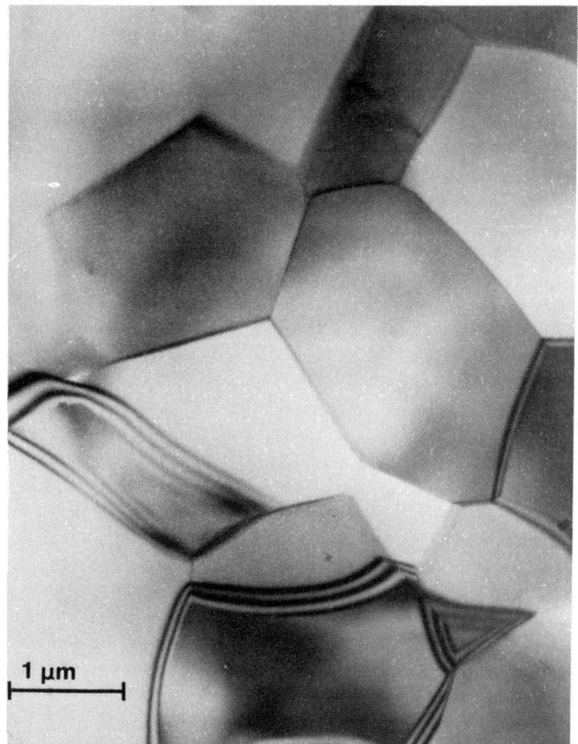

Figure 1
Microstructure of hot-pressed α-Al_2O_3

Hot Pressing of Advanced Ceramics: Technology and Theory

Hot pressing is the densification of a loosely packed powder or a compacted preform by the simultaneous application of heat and pressure. The specific benefits of hot pressing are seen in terms of the achievement of high final density and microstructural improvement such as fine grain size (see Fig. 1). Hot-pressing procedures include simple linear (uniaxial) hot pressing, continuous pressing and hot isostatic pressing. The main disadvantages of the process are its high cost and its limited capacity to fabricate a variety of shapes at mass-production rates.

The most successful approach to the modelling of hot pressing has been to use creep models suitably modified to account for the effects of porosity and external pressure. On this basis, a clear, qualitative understanding of the effects of the principal process variables involved (time, temperature, applied pressure and particle size) has been reached. Quantitative agreement between model predictions and experiment seems very encouraging; particularly noteworthy is the potential being shown by hot pressing for providing the experimental conditions under which a single densification mechanism can be made to predominate for separate study.

1. Technological Aspects

The majority of hot pressing uses a simple uniaxial (linear) arrangement consisting of a punch-and-die assembly and furnace (induction or resistance) mounted in a press frame. Typical operating pressures range from about 10 to 70 MPa. For moderate pressures ($\leq$40 MPa) graphite dies can be used, and for higher pressures ($\leq$120 MPa) more expensive refractory metal (Mo(TZM),W) and ceramic (Al_2O_3, SiC) dies can be used. Graphite is usually preferred because of its low cost, ease of machining and excellent creep resistance at high temperatures. Furthermore, the low expansion coefficient of graphite helps to prevent thermal expansion differences from producing stresses on cooling, thus simplifying the removal of the sample from the die; if thermal mismatch is a problem then hot extraction methods can be used. The main disadvantage of graphite is its high reactivity with other ceramic systems; to limit this problem, the die can be coated with a thin protective layer of boron nitride.

The ideal powder for hot pressing should contain no agglomerates, have a narrow particle size distribution and a fine particle size (<1 μm) for promoting rapid densification. Additives such as MgO (1–5%) with Si_3N_4, and LiF (1–3%) with MgO are also used for enhancing densification. The occurrence of chemical reactions, for example, ZrC + TaC→ (ZrTa)C (solid solution), and thermal decompositions such as $Mg(OH)_2 \rightarrow MgO + H_2O$ have also been seen to enhance densification, in this case through the formation of intermediary compounds existing in an activated state.

Impurities are not always beneficial to hot pressing, however, and adsorbed gases in particular (e.g., H_2O, CO_2 and S) can resist pore closure during the late stages of densification and can be the cause of bloating during subsequent annealing. Substantial bloating can also result from the presence of small amounts of trapped carbon impurity; in this case, the bloating during annealing is caused by the inward diffusion and subsequent reaction of oxygen with carbon (usually at grain boundaries) to form cavities as a result of CO_2 and CO gas evolution. Outgassing, vacuum hot pressing and prior annealing in H_2 to remove carbon helps, but not always sufficiently.

The actual pressing conditions are determined largely by the powder sinterability and the desire to form a particular microstructure. In general, the use of higher temperatures and longer times promote both densification and grain growth while higher pressures promote only densification without affecting grain growth. High final density and small grain size is thus favored by using very fine powders, very high pressures, and temperatures that are low but sufficient to provide the necessary densification.

Some examples of materials to which uniaxial hot pressing has been successfully applied are: optically transparent materials, MgO, Y_2O_3 and Sc_2O_3, ferrites, large (125 mm) transparent piezoelectrics for electrooptical devices, β-Al_2O_3 fast-ion conductors, and many other ceramics including ZrO_2, $BaTiO_3$, Cr_2O_3, Sm_2O_3, CoO, MnO, NiO, ZnO, Fe_2O_3, TiO_2, Nb_2O_5, CaO, Si_3N_4, AlN and SiC.

Attempts have been made to automate hot pressing through the development of continuous hot-pressing procedures in which the lower punch is continuously lowered and the upper punch is removed periodically to introduce additional powder charge. Specimens in the form of long solid rods or tubes can be fabricated in this way with, in principle, infinite length. The process has been successfully applied to the manufacture of dense rods of $Ba_2Ti_9O_{20}$, transparent Al_2O_3 tubes, fully dense potassium sodium niobate as well as a number of ferroelectric, dielectric and piezoelectric materials.

Another rapidly developing form of hot pressing is hot isostatic pressing (HIP). This process is being used to manufacture special ceramics of high cost with complicated shapes; the process itself is also very expensive. During HIP treatment a powder compact or preform is densified by the simultaneous application of heat and high gas (Ar) pressure. To prevent gas penetration through the pores of the compact, it must either be hermetically sealed in a deformable sheet (such as Al, glass or mild steel) or prefired to the closed porosity stage. Typical pressures range from 100 to 350 MPa with temperatures up to 1800 °C. Materials that have been fabricated by HIP to near theoretical density include $BaTiO_3$, $SrTiO_3$, $Pb(ZrTi)O_3$, Al_2O_3, MgO, UO_2, $Y_3Fe_5O_{12}$ (YIG), ferrites and carbon–carbon composites.

The article *Hot Isostatic Pressing of Advanced Ceramics* discusses HIP in detail. A more general and comprehensive review of the practical aspects of hot pressing is given in the article *Hot Pressing of Advanced Ceramics: A Survey*.

2. Mechanistic Aspects

The possible processes that can lead directly to densification during hot pressing are particle rearrangement, lattice diffusion, grain-boundary diffusion and plastic flow including viscous deformation. It is now well recognized that in most cases more than one of these processes occurs during hot pressing. As a consequence, mechanistic interpretations of hot pressing are usually given in terms of the dominant mechanism occurring under a given set of microstructural and processing conditions such as grain size, applied pressure and temperature; such data are frequently displayed in the form of hot-pressing deformation maps such as the one shown in Fig. 2 for the case of α-Al_2O_3.

Hot-pressing mechanisms can be identified by studying the kinetics of densification and the associated microstructural changes involved. Hot-pressing

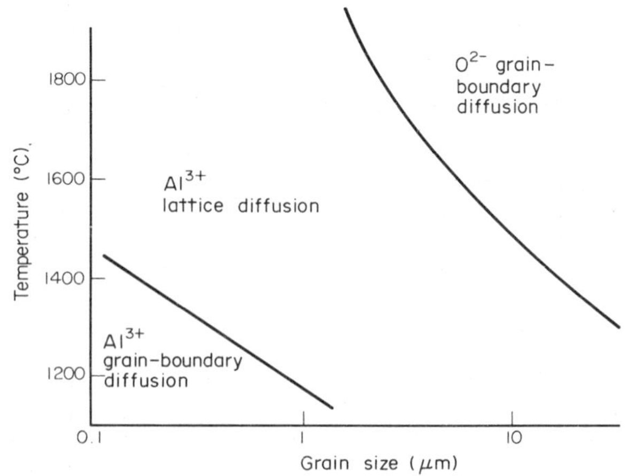

Figure 2
Hot-pressing deformation map for pure α-Al_2O_3 (after Harmer and Brook 1980)

kinetic data are best interpreted using the preexisting theories developed for describing the creep deformation of fully dense solids. The only process that cannot be treated in this way is that of particle rearrangement, for which no adequate mathematical treatment presently exists.

The creep equations when suitably modified for the case of hot pressing can be represented for the later stages of the process by the general equation

$$\frac{1}{\delta}\frac{d\delta}{dt} = \frac{CD}{kT(G)^m}\left(\sigma_{\text{eff}}^n + \frac{2\gamma}{r}\right)$$

where C is a constant, δ is the density, t is time, D is the controlling process diffusion coefficient, γ is the surface energy, r is the pore radius, G is the grain size, n and m are constants determined by the mechanism involved (see Table 1) and σ_{eff} is the effective stress acting on the grain boundaries between the particles. The effective stress can be related to the applied stress σ_A through the Coble relationship $\sigma_{\text{eff}} = \sigma_A/(1-P)$ where P is the volume fraction of porosity.

Table 1
Summary of hot-pressing mechanisms

Mechanism	Stress exponent n	Grain-size exponent m	Appropriate[a] diffusion coefficient D
Lattice diffusion (Nabarro–Herring creep)	1	2	D_L
Grain-boundary diffusion (Coble creep)	1	3	D_b
Grain-boundary sliding	1 or 2	1	D_b, D_L
Interface reactions	2	1	D_b, D_L
Plastic flow	$\geqslant 3$	0	D_L
Particle rearrangement			

a D_L = lattice diffusion coefficient; D_b = grain-boundary diffusion coefficient

Concerning the often-argued issue of diffusion versus plastic flow, there is now considerable evidence to show that the hot-pressing behavior of most ceramic materials at moderate stresses is diffusion controlled. This has been attributed to the use of fine powders (hence rapid diffusion rates) and to the high flow stress for plastic flow in ceramics. The diffusion pathways will depend on many factors, with grain-boundary diffusion (Coble creep) being expected to be more important in comparison with lattice diffusion (Nabarro–Herring creep) at small grain sizes and low temperatures owing to the larger grain size exponent and generally lower activation enthalpy of the former process.

Grain-boundary sliding is needed to accommodate shape changes that occur during diffusional creep. This means that grain-boundary sliding and diffusional creep processes operate sequentially (as must interface reactions with diffusion) such that the slower of the two processes determines the overall rate. This is in contrast to the operation of independent mechanisms (such as plastic flow in conjunction with Nabarro–Herring creep), for which the fastest process determines the overall rate. As a consequence of the differences between the grain size and stress exponents for the different processes, grain-boundary sliding and interface-controlled reactions are expected to increase in relative importance with decreasing stress and grain sizes.

Attempts to correlate the experimental data of real systems with the model predictions have met with a fair degree of success. Perhaps the best support for the validity of the models has come from measurements that show a change in mechanism due to some change in conditions. Examples of such anticipated changes in mechanism that have been observed include (a) changes from lattice to boundary diffusion with changing grain size as evidenced by a corresponding change in the measured grain size exponent from 2 to 3, and (b) changes from diffusion to plastic flow with increasing stress as evidenced by a measured change in the stress exponent from 1 to 3.

The handling of more complex systems (such as those involving chemical reactions and liquid phases) has also been tried with the aid of models. For liquid-phase hot pressing the Coble creep model has been found to be appropriate for some systems and it can be used by simply replacing the boundary in the Coble representation by the intergranular liquid phase present in liquid-phase hot pressing. In systems undergoing a chemical reaction, it has been found possible (under certain conditions) to relate the densification rate to the chemical reaction rate provided that the same atomic species can be held responsible for both processes. Under such circumstances, the balance between the rate of densification and chemical reaction can be used to indicate the extent to which particle rearrangement plays a role in densification (Di Rupo et al. 1978).

Finally, mention should be made of the value of hot pressing as an aid for the study of pressureless sintering mechanisms. Two features of pressureless sintering make it difficult to interpret data; these are (a) the occurrence of simultaneous changes in the microstructure by processes which compete with densification such as grain growth and surface diffusion, and (b) the coexistence of simultaneous densification mechanisms. Hot pressing has the ability to overcome the first and most significant problem by eliminating grain growth and greatly suppressing surface diffusion relative to densification. The second problem can also be alleviated by designing the experimental conditions to isolate the process of interest. In this respect, the additional variable of pressure can be both helpful and informative.

See also: Advanced Ceramics: An Overview; Solid-State Sintering

Bibliography

Di Rupo E, Carruthers T G, Brook R J 1978 Identification of stages in reactive hot pressing. *J. Am. Ceram. Soc.* 61: 468–69

Harmer M P, Brook R J 1980 The effect of magnesia additions on the kinetics of hot pressing in alumina. *J. Mater. Sci.* 15: 3017–24

Notis M R, Spriggs R M 1978 Emerging areas in hot pressing. *Sci. Sintering* 10: 175–90

Vasilos T, Spriggs R M 1966 Pressure sintering in ceramics. *Prog. Ceram. Sci.* 4: 95–132

Vieira J M, Brook R J 1984 Hot pressing of high purity magnesium oxide. *J. Am. Ceram. Soc.* 67: 450–55

M. P. Harmer

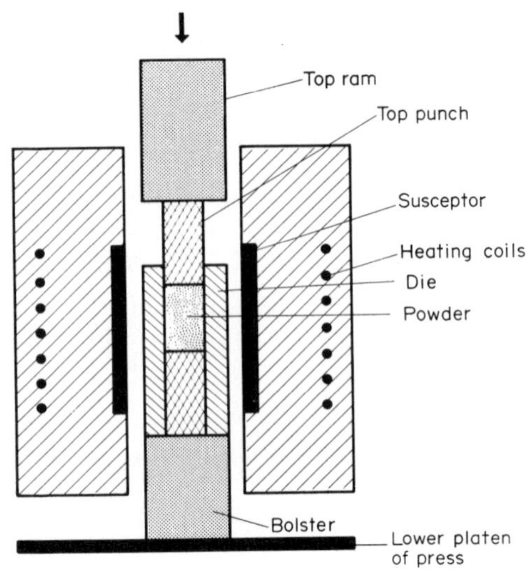

Figure 1
Schematic diagram of a vertical, single-acting press

Hot Pressing of Metals

Hot pressing consists of axially compressing a material at an elevated temperature to produce consolidation and/or bonding. The process is widely used for the consolidation of ceramic powders and selectively used to consolidate metal powders (vacuum hot pressing) and to bond metallic structures (press diffusion bonding). Isostatic hot pressing is discussed separately (see *Hot Isostatic Pressing*).

1. Equipment

A vertical single-acting hot press is shown in Fig. 1. Double-acting and horizontal versions are also available although less common. The presses consist of either flat platens or configured dies, a uniaxial loading system, a heating system and usually an atmosphere chamber enclosing the working area (for vacuum hot pressing the entire assembly is placed in a vacuum chamber). In a double-acting press, the lower bolster becomes a movable ram. For low to moderate temperatures, resistance heating can be used, but induction is more common for higher temperatures. Die materials are a major limitation in hot pressing. At very modest temperatures (e.g., up to 500 °C), conventional hot-work die steels can be used. Of course, working pressures must not exceed the compressive yield strength of the die material. For intermediate temperatures (e.g., up to 1100 °C) superalloys and molybdenum alloys can be used to moderate stress levels. Above these temperatures, nonmetals such as graphite must be used, and unit pressures are restricted to typically 35 MN m^{-2}.

2. Powder Consolidation

Two methods may be used for powder consolidation: encapsulation and loose-powder hot pressing. The former requires that the powder be sealed, usually under vacuum, in a container which will readily deform at the consolidation temperature but not react with the powder being processed. To achieve full densification, lateral constraint is needed during hot pressing. A common way to accomplish this is to consolidate in a blanked-off extrusion press. Encapsulation, which obviates the need for atmosphere control in the press, is frequently used for very reactive metals (e.g., aluminum) or toxic materials (e.g., beryllium). Pseudo-isostatic hot pressing of encapsulated powders using conventional metalworking presses is a recent development. The encapsulated powder is surrounded by a medium such as a salt or alloy which, at the consolidation temperature, behaves as a fluid, thereby transmitting the uniaxial force to the powder in a near-isostatic manner. Tooling must be designed so as to preclude extrusion of the pressure-transmitting medium around the ram, as this would reduce the applied pressure.

An alternative method is to charge loose powder directly into a die cavity, evacuate, bring to temperature, and press. Loose-powder hot pressing is satisfactory for billets or simple shapes, but hot pressing of complex shapes has met with limited success. Major problems have been uniformity of powder fill in the die, die distortion, and cycle time and cost.

Spark sintering is a version of hot pressing in which an electric current is passed through the powder charge via the top and bottom punches. This causes localized arcing (sparking) and heating at contact points. The heated powder is then pressed.

3. Press Diffusion Bonding

Hot pressing can be used to diffusion-bond metallic structures. Workpieces need not be flat, but the configuration must be such that the resolved forces from uniaxial compression will be sufficient to effect the bond. As with powder consolidation, the operation can take place in a vacuum, in a protective atmosphere, or in air. In the latter case, the assembly is usually encapsulated in a retort. Since the extent of deformation in bonding is considerably less than that in consolidation, it becomes practical to work with reusable retorts for atmosphere control.

Technologies in which press diffusion bonding plays a major role include diffusion-bonded structures for aerospace applications and metal-matrix composites. Limitations are similar to those for powder consolidation, namely, die materials, press capacity (and therefore part size), cycle time and cost.

See also: Hot Pressing of Advanced Ceramics: A Survey; Hot Pressing of Advanced Ceramics: Technology and Theory; Powder Metallurgy; Metals Processing and Fabrication: An Overview

Bibliography

Lenel F V 1980 *Powder Metallurgy: Principles and Applications.* Metal Powder, Princeton, New Jersey

Rahaman M N, Riley F L, Brook R J 1980 The preparation of Si_3N by hot pressing. *J. Am. Ceram. Soc.* 63: 648–53

Weissler G A 1981 Resistance sintering with alumina dies. *Int. J. Powder Metall. Powder Technol.* 17: 107–18

J. N. Fleck

Hot Rolling

Hot rolling can be defined as the plastic deformation of workpieces at elevated temperatures by means of power-driven rolls. It is undertaken primarily to change the geometrical shape of the workpiece, but it can also be used to create metallurgical changes in the rolled product, such as converting a casting into a wrought structure or refining the grain size of the workpiece in finishing operations, a process known as controlled rolling or thermomechanical processing (see *Hot Rolling: A Thermomechanical Process*). In addition, hot rolling is used to bond dissimilar metals, as in the cladding of stainless-steel sheets onto carbon-steel sheets, a process often described as sandwich rolling. Furthermore, some mills are used partly for breaking off the scale that forms on a workpiece during heating in air, in which case the stands are often described as scale breakers.

1. Categorization of Hot Mills

Primary hot mills are used to hot-roll metals cast either as conventional ingots or as continuous strands into semifinished products such as slabs, blooms, and billets (Fig. 1). Slabs are further hot rolled into plates or strip, whereas blooms can be rolled directly into structural and other shapes, or processed into billets for later conversion into bars, rods or tubular products.

In the rolling of finished products, hot mills can be categorized as flat and nonflat, the former using basically cylindrical rolls and the latter employing grooved rolls, with the grooves mated in each pair of work rolls to form roll passes, the cross sections of which closely match the cross sections of the workpieces as they emerge from the mill stands (see *Shape Rolling*).

2. Types of Mill Stands

The basic roll configurations used in hot rolling are such that the mills are classified as two-, three- and four-high, and universal, although six-high mills are currently being considered for the rolling of strip. The two-high type (Fig. 2) consists of two horizontal work rolls, both of which are usually driven by spindles, sometimes individually driven and at other times coupled to a pinion stand. In the case of primary mills used for the rolling of slabs, blooms and billets, the drive is generally reversible and the mill is equipped with a motor-driven roller table on each side to facilitate the back-and-forth movement of the workpiece through the mill. The final dimensions of the product are obtained by a number (usually an odd number) of drafts or reductions. To control the thickness of the workpiece, the horizontal rolls are positioned by means of either screws or hydraulic cylinders as well as by roll-balancing systems. To obtain the desired width of the workpiece, box-type passes may be used or the piece may be turned on its side as in high-lift mills. Alternatively, vertical rolls may be used in conjunction with the horizontal rolls to constitute what is generally described as a universal mill.

In the case of three-high mills (Fig. 3), only the center roll is usually driven, and the rotation of this roll is not usually reversible. The workpiece is passed between the lowest and the center rolls for rolling in one direction, and between the topmost and center rolls for rolling in the reverse direction. To achieve this mode of operation, the roller tables can be raised or lowered to correspond in level to the roll bite used or, as in the case of more-recently engineered mills, the mill stand can be correspondingly raised or lowered (the so-called "jumping" mills).

For the flat rolling of relatively thin products, the four-high mill is generally employed, and the two work rolls in the center are prevented from excessive

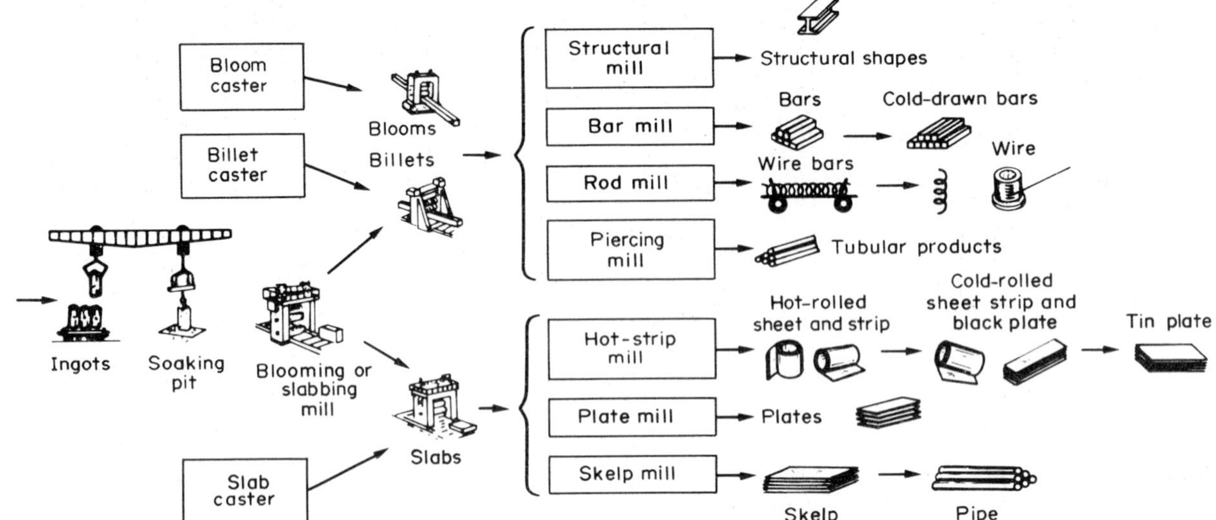

Figure 1
Utilization of various types of hot mill

bending under the action of the rolling forces by the use of larger-diameter backup rolls (Fig. 4). Again, screws or hydraulic cylinders are used to position the rolls so that the height of the roll gap corresponds to the desired thickness of the workpiece.

In many instances the workpiece, instead of undergoing a number of passes in the same mill stand, is subjected to single-pass reductions in a number of stands arranged in tandem, as in the case of modern rod and bar mills. Such an arrangement of stands is designated a continuous mill, and frequently such mills can roll a number of workpieces side by side (multistrand mills). In the area of flat rolling, a tandem arrangement of stands is used in the finishing trains of hot-strip mills and light-gauge plate mills.

Flat-rolled products, such as plates and strip, must exhibit the desired cross-sectional profile as well as the requisite flatness. To attain these qualities in the

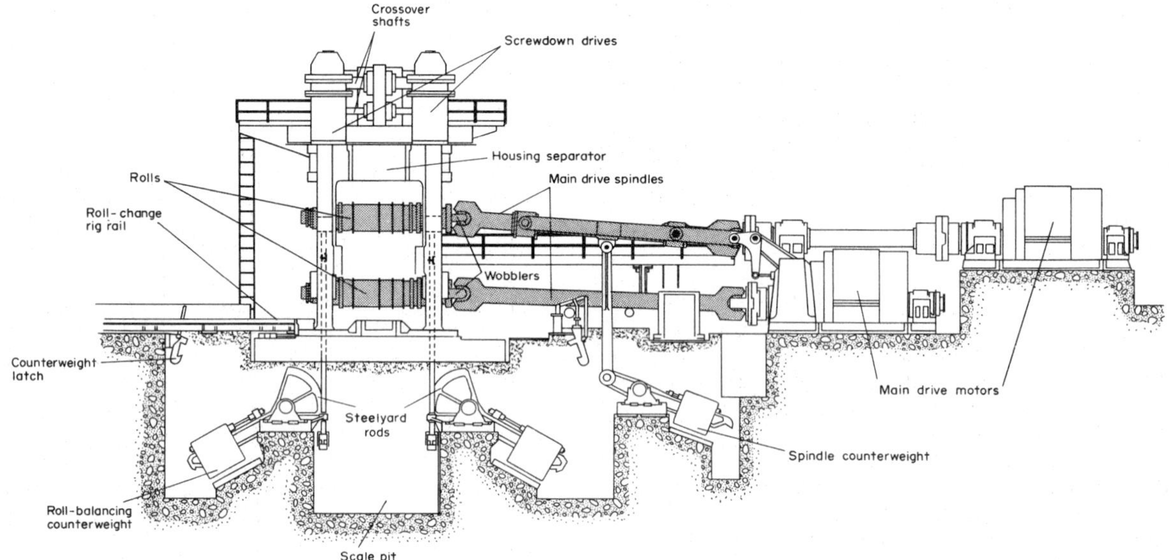

Figure 2
High-lift, two-high, reversing blooming mill

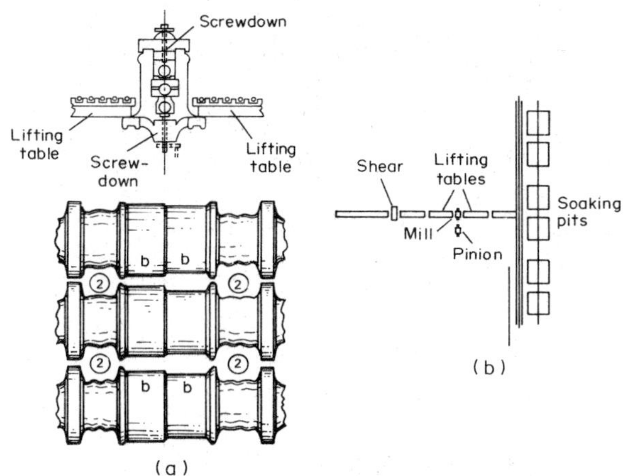

Figure 3
Three-high blooming mill, showing (a) elevation of mill and rolls and (b) plan view. Circled numbers indicate pass sequence

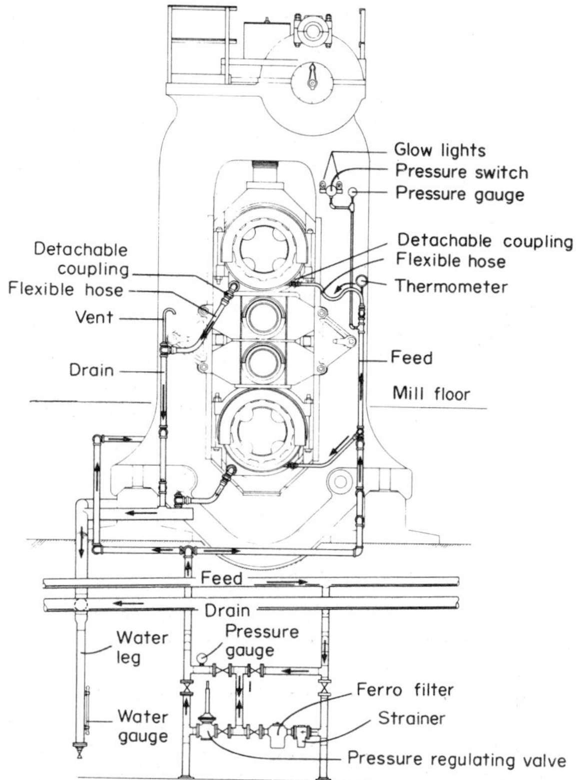

Figure 4
Four-high mill stand showing backup-roll bearing lubrication system

rolled product, the geometry of the roll gap may be changed by crowning the rolls (as discussed later), by using thermal crowns (attributable to nonuniform temperature distributions within the rolls), by work- and backup-roll bending, or by using variable-crown backup rolls (the sleeves of which may be made slightly barrel-shaped by hydraulic pressurization).

In tubular hot-rolling operations, the workpiece is often given a rotational as well as a translational movement, and the process is designed to change the cross-sectional dimensions of the workpiece symmetrically with respect to its centerline. Thus, for example, round billets can be reduced in diameter on a planetary mill or the wall thickness of a hollow shell can be reduced by processing on a multistand pipe mill in which a mandrel is inserted into the shell to act as an anvil. In sinking processes, the outer and inner diameters of the shell are reduced but the wall thickness is increased. In other cases, hot rolling is combined with a piercing operation, as in the use of Mannesmann and three-roll piercing mills (see *Piercing and Tube Production Processes*).

3. Rolls and Bearings

The rolls (particularly the work rolls) of hot-rolling mills are generally made from either cast iron, cast steel or forged steel. However, in the finishing stands of high-speed rod and bar mills, carbide (usually tungsten carbide) rolls are finding increasing use. All these materials provide the requisite hardness and wear resistance and, in the case of iron and steel rolls, the necessary toughness or ductility. Rolls that are cast are generally made by either the bottom-swirl or the centrifugal-casting method and, in the case of iron rolls, often by the double-pour method which permits the production of tough, shock-resistant roll necks and a core of similar properties combined with a hard, wear-resistant shell. In carbide rolls, the carbide generally constitutes only the outer rim of the roll.

If rolls are cast in a vertical position, the segregation of inclusions that occurs at the top of the casting weakens the neck. Accordingly, this end of the neck should not be driven and is identified as the cope neck, in contradistinction to the other end known as the drag neck.

Rolls are maintained in their appropriate positions in the mill housings by means of bearings accommodated in blocks of steel, known as chocks, which fit within the mill windows and, in the case of the work rolls of four-high mills, within the "wings" of the backup roll chocks. The backup roll bearings, which must withstand the entire rolling force, are generally of the oil-film type which, during operation at low speed, are hydrostatically lubricated in many cases to prevent "wiping" of the babbitt bearing. Work-roll bearings of four-high mills are subjected only to limited radial and thrust loading and these are therefore mainly of the tapered roller-bearing type. Such bearings also find extensive use in other hot-rolling mills where the loading is not excessive and where accurate positioning of the rolls is mandatory.

Before use in a hot-strip mill, the rolls are processed on grinding machines to remove surface damage or wear caused by prior use so that diameters of pairs of rolls can be closely matched. To compensate for the bending and thermal expansion that occurs during their use in rolling operations, the work rolls and often the backup rolls are crowned or slightly barrel-shaped. Where the thermal effects predominate, the ground-in crowns may be negative, making the roll surface slightly concave in an axially parallel direction.

In the hot rolling of aluminum without proper lubrication, work rolls become coated with aluminum and aluminum oxide. Because this accumulation, if allowed to proceed uninterrupted, would impart a rough profile to the roll surfaces, rotating wire brushes are used to smooth out this coating.

To dissipate the heat absorbed in mill rolls because of their contact with the workpiece and because of frictional energy dissipated at the roll-workpiece interfaces, hot-mill rolls are usually cooled by the application of water sprays, which are generally located on the exit side of the roll bite.

4. *Frictional Effects*

Without the existence of friction between the roll and the workpiece surfaces, rolling would not be possible because the workpiece could not enter the roll bite. Coefficients of friction at this interface under dry conditions usually lie in the range 0.2–0.5 and, for well-descaled workpieces, the coefficient increases with increasing temperature. Oxides or scale on the workpiece surface decrease the value of the coefficient to an extent dependent on the thickness and type of the scale. At the same time, the rougher the work-roll surfaces, the greater the friction.

In primary mills, where relatively large drafts must be taken, the coefficient of friction must be high to drag the workpiece into the bite. Accordingly, work rolls are often ragged or roughened to increase the coefficient. On the other hand, where relatively small drafts are to be taken, as in the finishing trains of hot-strip mills, the normal frictional level is excessive and rolling forces and drive-power requirements may be decreased if the frictional levels are reduced. Although some decrease in friction could be achieved by not removing the scale from the workpiece, this would mean that the scale would be rolled into the surface creating an undesirable finish. Accordingly, the workpiece surface is usually descaled by high-pressure water sprays to remove the oxide layers as thoroughly as possible.

To minimize the frictional effects in the rolling of hot, well-descaled workpieces, hot-rolling lubricants are being used in structural mills and the finishing trains of hot-strip mills. Those being used for the hot rolling of steel are principally of a fatty-type material and are applied neat (as in the case of structural mills) or as oil–water emulsions (as is generally the case for hot-strip mills). The application of such lubricants reduces rolling forces and spindle torques by 10–30%, increases roll lifetime by decreasing the wear rate and the number of regrinds necessary in the rolling of a given quantity of metal, and in the case of strip, provides a better surface that can be more efficiently pickled.

5. *Metallurgical Aspects*

Carbon steels are usually hot rolled at temperatures at which the workpiece exhibits a face-centered-cubic or austenitic structure. The deformation experienced by the workpiece causes the grains to be broken up and elongated in the rolling direction, resulting in the creation of intense dislocation networks. Immediately subsequent to the deformation, recovery involving the annihilation of point defects in the crystal lattices begins. This is followed by recrystallization, a process activated by the stored energy resulting from the deformation (see *Hot Working*). If the workpiece is at a sufficiently high temperature, this recrystallization is followed by grain growth. Thus, in multiple-pass hot rolling at elevated temperatures, complex metallurgical changes occur, and the flow stress required to deform the steel is dependent on the temperature, chemistry and prior strain history of the workpiece and the rate at which it is deformed (strain rate). The final austenitic grain size is also dependent upon these same parameters, becoming smaller when the finishing temperature is reduced, as in the case of controlled rolling. Alternatively, the rate of grain growth after deformation can be retarded by the use of certain microalloying elements, such as niobium, vanadium and titanium.

The final microstructure in carbon steels at ambient temperatures is dependent upon the rate at which they are cooled and the nature of the continuous-cooling curves that indicate their behavior upon the decomposition of the austenite. Very rapid cooling results in a transition from ferrite to martensite, whereas slower cooling can result in softer structures such as pearlite or ferrite and carbide.

On some mills, such as rod mills, the rate of cooling is controlled by conveying the product in overlapping loops from the finishing stand to the coiler, sometimes under heat shields to further retard the cooling. The rate at which small bars lose heat on a cooling bed is controlled by varying their stacking arrangement on the bed. In hot-strip mills the rate of cooling of the strip is controlled by the proper use of sprays or laminar-flow jets on the run-out table.

To obtain desired metallurgical properties in the

hot-rolled products, they are often subjected to a heat treatment, such as quenching and tempering, annealing, normalizing, grain refinement, carburizing or nitriding. Steel products that are often heat-treated include structural shapes made of special grades, rails, bars (especially those of alloy steels), plates, oil-country goods, and line pipe.

6. Mathematical Models

Models have been developed to describe hot-rolling operations in mathematical terms for use in mill design, establishment of production rates, and control of rolling operations by computer. Generally, these models are intended to predict such parameters as rolling forces and spindle torques (energy requirements). However, a number of associated models have been developed, particularly in the case of hot flat rolling, to predict temperature changes in the workpiece and its metallurgical properties, both as it undergoes deformation and at room temperature. Similarly, other models have been used to determine the thermal crowning, bending and wear of mill rolls so that the cross-sectional profile of the workpiece and its flatness can be computed.

7. Further Processing of Hot-Rolled Products

The majority of hot-rolled products are further processed by heat treatment and/or by pickling or grit blasting to remove oxide or scale, and by further deformation and forming operations. Pickling involves the removal of the oxide layers on the surfaces of the products by acids, such as sulfuric or hydrochloric in the case of carbon steels and nitric and hydrofluoric in the case of stainless steels. These operations may be of the batch type where workpieces such as bars or plates are involved, or of the continuous type, as in the case of strip products. In the grit-blasting process, the impact of small, hard abrasive particles on the workpiece removes the scale from the surfaces. Subsequent to descaling, oil is generally applied to the clean surfaces, not only to retard rusting but to facilitate later operations such as cold rolling or pressing.

See also: Cold Rolling; Deformation Processing; Hot Working; Structural Shape and Plate Steels; Thermomechanical Processing

Bibliography

McGannon H E (ed.) 1971 *The Making, Shaping and Treating of Steel,* 9th edn. US Steel Corporation, Pittsburgh, Pennsylvania

Parkins R N 1968 *Mechanical Treatment of Metals.* Elsevier, New York

Roberts W L 1982 *Hot Rolling of Steel.* Dekker, New York

W. L. Roberts

Hot Rolling: A Thermomechanical Process

Traditionally, hot rolling of steels has been concerned almost entirely with the production of shapes within specified tolerances and has therefore been an exercise in mechanics rather than in physical metallurgy (see *Hot Rolling*). More recently, because of stringent property specifications placed on many steel products, coupled with facility limitations and awareness of a need for energy conservation, the importance of the mechanical properties of as-rolled products has been recognized. Consequently, the hot-rolling process is now considered to be a thermomechanical treatment that must be understood and controlled in every aspect if properties as well as shape are to be optimized. This has led to examination of the interaction among the several metallurgical phenomena that occur during hot working, and of their effects on the structure and properties of the final product.

This article outlines what is known about the effects of processing and composition variables on the metallurgical changes (recovery, recrystallization and grain growth, particle–boundary interaction, transformation, precipitation and dislocation strengthening) that occur in slabs, blooms or billets during reheating, hot working and subsequent cooling. Emphasis is placed on low-carbon, low-alloy C–Mn steels that comprise over 80% of current steel tonnage. Prominent among these steels are the microalloyed steels for which thermomechanical treatments are designed to refine the microstructure to meet strength, toughness and formability specifications in hot-rolled plate and strip.

1. Grain Coarsening During Reheating

Control of final structure and properties begins with control of the austenite microstructure established in the workpiece by the reheating program, especially the reheating temperature. Grain coarsening in C–Mn steels between 900 and 1200 °C follows a well-behaved continuous grain-growth law of the type $D = kt^{\alpha} \exp(-\beta/T)$, where D is the average grain diameter, t is time, T is temperature, and k, α and β are constants (Miller 1951). Grain coarsening in microalloyed steels in the range 900–1300 °C is, however, characterized by temperature regions of continuous growth at high and low temperatures, between which is a rather narrow temperature range where abnormal growth occurs producing a duplex distribution of austenite grain sizes. In extreme cases the duplex structure may be difficult to remove during subsequent hot rolling. Accordingly, common practice is to restrict reheat temperatures to ranges that result in uniform structures while avoiding temperatures so low as to create excessive mill forces during subsequent rolling.

The reheat temperature necessary to achieve a

uniformly fine initial grain structure in the slab varies with the type and amount of microalloy carbide and/or nitride forming elements, since the abnormal grain growth is controlled by grain-boundary–particle interactions (Gladman 1966). Boundary pinning occurs when a critical fraction of the grain-boundary area is occupied by particles. Thus, pinning is favored by high volume fractions of particles and by fine particles. As particles ripen (grow) during elevated-temperature exposure, the area covered by a fixed volume fraction decreases until some boundaries are released and abnormal growth occurs. Consequently, for a given volume fraction of fine particles, less soluble particles that ripen more slowly are effective in pinning boundaries to higher temperatures.

These principles offer reasonable explanations for observations related to reheating of microalloyed steels that contain a volume fraction of about 0.0005 of alloy carbide or nitride (Fig. 1). Vanadium-bearing steels generally resist coarsening only up to 1000–1050 °C, above which VCN is sufficiently soluble in austenite to begin ripening. Niobium-bearing steels resist coarsening to temperatures up to 1100–1150 °C, since NbCN is less soluble in austenite than VCN. Although these grain-coarsening temperatures are below the temperature necessary for complete solution of these precipitates, sufficient solution occurs to allow some ripening of the particles. On the other hand, steels containing a fine dispersion of the very stable TiN can resist grain coarsening to temperatures above 1200 °C since TiN has such an extremely low solubility in austenite that ripening is inhibited at lower temperatures.

2. Grain Refinement During Hot Rolling

If the reheated grain structure has been coarsened, refinement can be effected by repeated recrystallization during hot rolling if the process is adequately controlled. Three parameters are particularly important: the rolling temperature and the reduction per pass (Fig. 2) and the number of recrystallization passes. At temperatures above 1000 °C, rapid recrystallization occurs in most low-carbon, low-alloy steels. In tandem hot-strip rolling, where reductions per pass of 25–30% are common, recrystallization may be dynamic (coincident with deformation); under these conditions the recrystallized grain size after several passes is quite fine (~20 μm). At the opposite extreme, heavy-slab rolling on a reversing mill frequently begins with reductions of only 5% per pass. Such low reductions can actually produce coarsening of the grains by strain-assisted growth during the time interval between the early passes. This trend can be reversed if there are sufficient subsequent passes with higher reductions, as is common in constant-draft schedules. Thus roughing schedules of 8–10 passes with reductions in the last passes approaching 20% will produce, by multiple

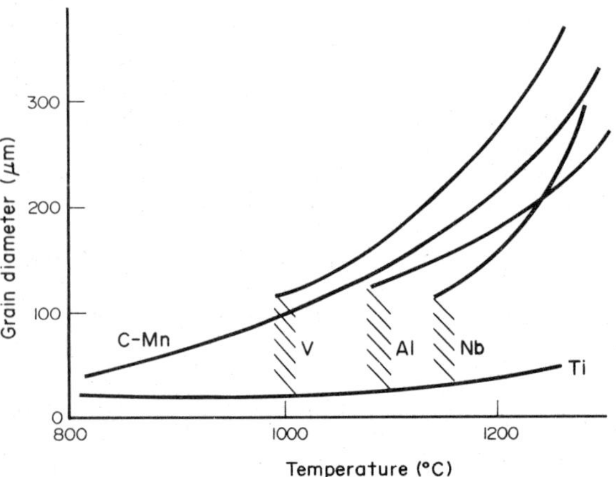

Figure 1
Grain coarsening in C–Mn steels with microalloy additions (30 min holds)

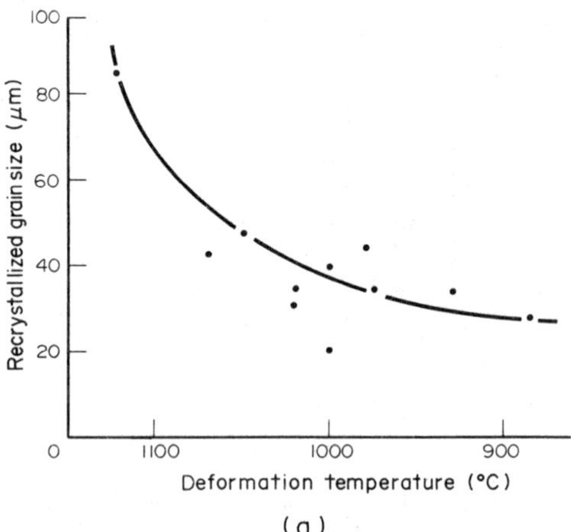

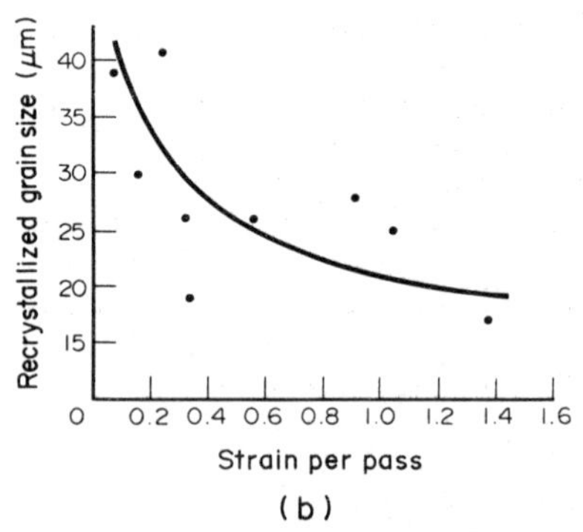

Figure 2
Effect on recrystallized grain size in 0.05 Nb steel of (a) deformation temperature and (b) strain per pass. Total reduction is about 60% in multipass deformation (after Cuddy 1981)

recrystallization, austenite grains of 30–40 μm diameter. However, final light-reduction shaping passes must be avoided since they can again cause strain-assisted grain coarsening, thereby destroying the grain refinement produced by the preceding passes.

The general rule seems to be that any adjustment of the rolling parameters that increases the stress to which the workpiece is subjected will result in finer recrystallized grains, presumably by increasing the density of recrystallization nuclei. Thus large reductions per pass are desirable, as are high rolling speeds and the lowest rolling temperatures in the recrystallization range (Cuddy 1981).

These generalizations are, of course, limited by several factors, notably the capacity of the mill and the recrystallization temperature range of the steel. Plain C–Mn steels will continue to recrystallize down to temperatures near 800 °C and thus will produce quite fine recrystallized austenite of 20–30 μm diameter on most reversing mills. However, austenite of such size may not transform, under typical mill cooling rates, to ferrite that is sufficiently fine to meet modern strength/toughness requirements. Consequently, it is desirable to flatten the refined austenite grains by rolling below the recrystallization range in order to increase the density and proximity of the austenite grain boundaries that act as ferrite nucleation sites. Altering the recrystallization temperature range of a steel is one of the roles of microalloy additions.

The temperature below which recrystallization of C–Mn steels ceases may be so near Ar_3 (the temperature at which the transformation from austenite to ferrite begins) that no flattening of austenite grains can be effected before transformation. However, the recrystallization stop temperature can be increased by the formation of microalloy carbides and/or nitrides that pin sub-boundaries and grain boundaries. Thus, by selection of the proper type and concentration of microalloy, recrystallization and refinement of austenite grains can be achieved in the first stage of hot rolling (roughing) between about 1100 and 1000 °C, and flattening of these refined grains can then be effected in a second stage of rolling (finishing) between about 900 and 700 °C. Generally a 3:1 or 4:1 reduction is performed in the finish passes to assure flattening of the recrystallized 30 μm austenite grains to heights (the austenite grain-boundary separation in the through-thickness direction) of 8–10 μm. To avoid the mixed grain structures that result from partial recrystallization, a delay is generally introduced after the roughing passes to allow the plate to cool before finishing to a temperature below which only grain flattening occurs. Figure 3a shows the relation between the finishing reduction and the refinement of ferrite grain size in the product as a result of the flattening of the austenite grains. The importance of this treatment to final toughness is shown in Fig. 3b.

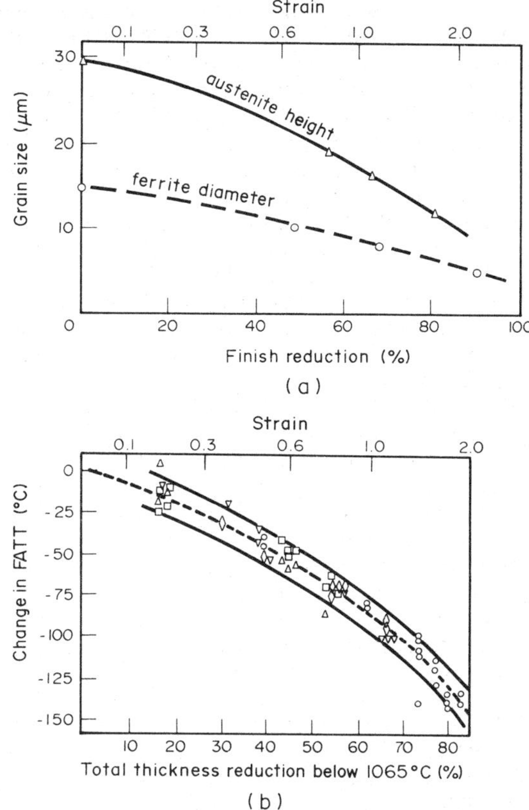

Figure 3
Effect of the amount of finishing reduction on (a) the height of flattened austenite grains and the diameter of the ferrite grains to which they transform (after Cuddy 1981) and (b) the change in the fracture appearance transition temperature (FATT) for a family of microalloyed steels (after Repas 1975)

Several variations in scheduling are used to avoid a decrease in production caused by the hold or delay between roughing and finishing. Since little or no grain growth occurs in microalloyed steels during this hold, the roughed plate may be held in line to air cool while a second slab is being roughed and a previously held plate is being finished. Water cooling is also employed to decrease the hold time. In a novel variation used by Sumitomo (Tanaka et al. 1978) to produce finer, more uniform austenite grains for flattening, roughed plates are taken out of line, rapidly cooled below Ar_1 (the temperature at which the transformation from austenite to ferrite ends), rapidly reheated to just above the Ac_3 (the temperature at which the transition from ferrite to austenite ends) to produce very uniform 10–20 μm austenite grains, and then finished.

To meet the requirements of rolling with lower reheat temperatures, large roughing drafts, and large total reduction at low finishing temperatures, the new generation of rolling mills must be larger and stronger. Four-high mills (i.e., rolling mills comprising a set of two work rolls reinforced by a set of two backup rolls) are an absolute minimum requirement, and considerable investigation of Sendzimir, planetary and other cluster mills is underway. In addition to the rolling itself there is considerable effort in developing new technology for control of cooling through and after the transformation.

3. *Controlled Transformation*

Further strengthening of hot-rolled products may be achieved by intercritical rolling to produce work-hardened ferrite. There is, however, disagreement on how far this process can be carried before the dislocation density begins to deteriorate ductility and toughness properties. Some data indicate that after an initial rise in impact transition temperature (ITT) with increase in intercritical rolling, there is a decrease (improvement) in ITT (Nagumo 1980). These changes may be related to the development of well-defined subgrains with misorientations of several degrees that act as fine grains, and to the development of a strong (100) texture.

As a result of these same changes, however, with increasing intercritical rolling there is also an increase in the tendency for splits, or separations, to occur in the plane of the plate when it is loaded in tension in the through-thickness direction. The loss in through-thickness ductility (which has been variously attributed to stringered inclusions, impurity segregation to elongated austenite boundaries, textured ferrite, and weak interfaces between coarse and fine ferrite) may, depending on product specifications, place a limit on the amount of intercritical rolling that can be performed.

Severe low-temperature rolling of unrecrystallized austenite (control rolling) markedly accelerates transformation, thus raising the ferrite start temperature (Ar_3) during the continuous cooling that accompanies hot rolling. This offsets to some extent the grain refinement achieved by control rolling. To counter this effect small additions of hardenability agents (molybdenum, boron) are being investigated.

Alternatively, high-strength control-rolled products with less severe low-temperature rolling can be obtained by accelerated cooling through the transformation range. The faster cooling rate suppresses the Ar_3, thereby refining the transformation products and changing their morphology and distribution. Faster cooling also retains solute for subsequent precipitation strengthening of ferrite. If the austenite entering the transformation is uniformly fine, the transformation products will be fine polygonal ferrite–pearlite or acicular ferrite–fine bainite with good strength and toughness. Any regions of coarse austenite can, however, transform to large patches of upper bainite that will produce a deterioration in toughness. Consequently, accelerated cooling can be used only in combination with control rolling that is effective in refining the austenite (Ouchi et al. 1981).

Interruption of accelerated cooling below Ar_1 followed by slow cooling in the ferrite region, as in coil cooling of strip, will allow full precipitation strengthening of the ferrite by those microalloy constituents that have been retained in solution by the rapid cooling through the transformation. Since accelerated cooling makes more efficient use of both hardenability agents and microalloys, leaner, more economical steels with lower carbon equivalent can be used to meet the requirements for strength, weldability and ductility.

At the other extreme, deliberate slowing of the cooling rate through the transformation, or an intercritical hold, is being investigated for the production of as-hot-rolled dual-phase steels. The longer times in the intercritical region are necessary to promote sufficient partitioning of hardenability elements from the newly formed ferrite to the remaining austenite pools so that the austenite transforms to martensite or bainite on subsequent cooling.

4. *Reactions in Ferrite*

Control of subcritical cooling affects the mechanical properties of the ferrite. For example, reduction of

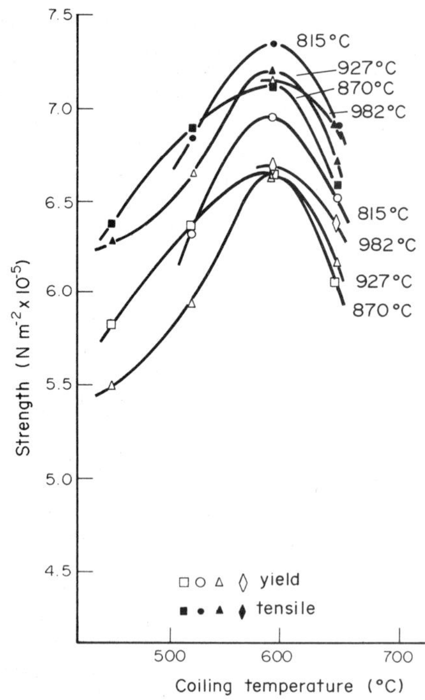

Figure 4
Effect of coiling temperature on the strength of hot-rolled sheets of 1.0%Mn–0.11%Nb–0.048%C–0.002%B with different finish temperatures

coiling temperature, down to about 600 °C, increases the strength of sheet products by limiting ferrite grain growth, by increasing the volume fraction of acicular ferrite, and, in microalloyed steels, by retaining solute to lower temperatures, thereby refining precipitate size and increasing its number density. Rapid cooling to too low a temperature, however, will suppress part or all of the precipitation reaction, thereby causing some decrease from maximum yield strength (Fig. 4).

High coiling temperatures, or slow cooling of heavy sections, particularly in material that has been intercritically rolled, will in addition enhance degeneration of pearlite colonies. Should these simply spheroidize, there will be some loss in pearlite strengthening. If, however, the pearlite degenerates to grain-boundary cementite a loss in ductility and toughness results. The morphology of the degenerate pearlite is apparently affected by steel composition but the phenomenon is poorly understood.

See also: High-Strength Low-Alloy Steels; Structural Shape and Plate Steels; Steels: Physical Metallurgy Principles; Steels: Structure–Property Relations

Bibliography

Cuddy L J 1981 Microstructures developed during thermomechanical treatment of HSLA steels. *Metall. Trans., A* 12: 1313–20

Gladman T 1966 On the theory of the effect of precipitate particles on grain growth in metals. *Proc. R. Soc. London, Ser. A* 294: 298–309

Miller O O 1951 Influence of austenitizing time and temperature on austenite grain size of steel. *Trans. Am. Soc. Met.* 43: 260–89

Nagumo M, Morikawa H, Okumura M, Kawashima Y, Sogo Y, Mantani O 1980 Rolling parameters and their significance on the combination of strength and toughness in HSLA steels rolled at austenite–ferrite two phase regions. *Proc. Int. Conf. Steel Rolling*, Vol. 2. Iron and Steel Institute of Japan, Tokyo, pp. 1075–84

Ouchi C, Okita T, Yamamoto S 1981 The effect of interrupted accelerated cooling after controlled rolling on the mechanical properties of steels. *Trans. Iron Steel Inst. Jpn.* 22: 608–16

Repas P E 1975 Control of strength and toughness in hot-rolled low-carbon manganese–molybdenum–columbium–vanadium steels. In: Korchynsky M (ed.) 1975 *Micro Alloying 75*. Union Carbide Corporation, New York, pp. 387–96

Tanaka T, Nozaki N, Bessyo K, Fukuda M, Hoshimoto T 1978 SHT plates for low temperature service. *Sumitomo Search* 19: 47–61

L. J. Cuddy

Hot Shortness

Hot shortness is the fracture of hot metal without appreciable deformation. This term is usually applied to the cracking experienced during mechanical working of cast sections well above half the melting point. The cracking may occur throughout the worked section or may be manifest as localized surface, edge or corner cracking. Hot shortness may also occur over a range of temperatures at about half the melting point. This intermediate temperature embrittlement is of more concern in finishing operations than during hot breakdown.

Hot short cracking at both high and intermediate temperatures is normally grain-boundary related. Coarse grain size, high flow stress, increased proportion of grain-boundary sliding and lower interfacial energy reduce hot ductility as these factors decrease the strain at which an intercrystalline crack grows in an unstable manner. Interfacial energy is the most important factor, as it alone varies by more than one order of magnitude as a consequence of impurities and segregation. The primary cause of hot shortness is segregation or impurity-related boundary liquation or decohesion.

The problem of liquation associated with segregation is more often seen in highly alloyed materials, particularly when there is a broad freezing range or when the segregated species has a substantially lower melting point. This is characteristic of copper-containing steels, tin-containing copper alloys, and the magnesium- and zinc-containing aluminum alloys. Under these conditions adequate homogenization during heating for hot working is difficult to achieve. Furthermore, highly alloyed metals often require hot working at the highest possible temperatures to provide the necessary flow-stress reduction and temperature range to complete hot working. Higher working temperatures increase the risk of liquation in areas where segregation persists after the homogenization treatment. The maximum hot-working temperature is set below the equilibrium solidus of the alloy to account for this segregation.

Even if segregation is limited, incipient melting and subsequent cracking at grain boundaries can result if furnace controls are inadequate and normal hot-working temperatures are exceeded. A related cracking problem in overheating metal results from oxygen penetration down grain boundaries when the furnace atmosphere is not properly maintained. Liquation can also develop during hot working as a consequence of the adiabatic heat of deformation. This heat rise causes the problem of chatter cracking at the corners of extruded sections where redundant work and friction are high. This aspect of hot short behavior limits extrusion speed to maintain an acceptable working temperature in the die.

Sulfur, even at impurity levels, forms low-melting grain-boundary eutectics in both steels and nickel-base alloys which cause hot short cracking during mechanical working. Manganese additions combine with sulfur as high-melting-point sulfides present as globules in the matrix and, thereby, eliminate the problem. As a rule, manganese is added at about ten times the expected sulfur content. Cerium, calcium and magnesium also are used as sulfur "getters." These metals form more stable sulfides and can be added in amounts closer to stoichiometry.

Certain impurities, even if present in minute quantities, can cause hot short cracking. The action of bismuth in copper is a classic example, and the reason for the potent effect can be understood in part from the relevant phase diagram. There is practically no solubility of the impurity in the host, so the impurity becomes concentrated at the grain boundary. The impurity can be present as a liquid in the grain boundary or an absorbed film on the grain boundary at normal hot-working temperature. In either case there is a reduction in surface energy which promotes fracture as the impurity spreads over the grain boundary. The grain boundary is covered by the liquid in accordance with its dihedral angle as determined from the balance of interfacial free energies for the specific constituents involved.

Complete wetting of the boundary, 0° dihedral angle, causes the most complete embrittlement. With increasing dihedral angle, bridges of solid–solid contact increase and ductility increases accordingly. The relative embrittlement effects of bismuth (0° dihedral angle) and lead (70° dihedral angle) in copper can be understood in this context. This impurity-related aspect of hot short behavior can only be avoided by proper control of the charge makeup and the materials used for melting and casting.

Reduced ductility at intermediate temperatures results from active grain-boundary sliding without sufficient accommodation by boundary migration or matrix flow to relax the localized stresses developed. Severe embrittlement (hot shortness) occurs when reduced boundary cohesion from impurity segregation is superimposed on this condition. For example, the room-temperature tensile elongation of a complex cupronickel alloy is about 40% when produced from either commercial (90 ppm Pb and 10 ppm Bi) or high-purity (6 ppm Pb and 1 ppm Bi) charge materials. At about 900 K a ductility trough exists; the tensile elongation is 15% for the high-purity alloy, whereas the commercial alloy is hot short (1% elongation). The embrittlement of the commercial alloy extends to about 1150 K, above which restoration processes become more active and adequate ductility returns. Consequently, hot-breakdown practice must be scheduled to maintain a satisfactorily high finishing temperature to avoid intermediate-temperature hot short cracking.

See also: Grain Boundary Structure; Metals Processing and Fabrication: An Overview

Bibliography

Reynolds K A 1968 The use of a simple upsetting test in metal working studies. In: Tegart W J M (ed.) 1968 *Proc. Conf. Deformation Under Hot Working Conditions*. Iron and Steel Institute, London, pp. 107–16

The Royal Society 1980 *Residuals, Additives and Materials Properties*. The Royal Society, London

E. Shapiro

Hot Working

The hot working of metals is generally carried out with two primary objectives in mind. The first is to reduce the cross section of the starting material or ingot to form a product (or intermediate product) with a desired shape. This operation is usually performed on a rolling mill, or on a forging or extrusion press. The second objective is to break down the relatively coarse grained ingot structure, which generally contains large intermetallic particles in the vicinity of the grain boundaries. Such a structure, if unaltered, is relatively brittle during subsequent forming or service. In addition, the working process can be carefully controlled so that particular sequences of cooling and deformation are followed. This leads to an improvement in the final mechanical properties and is frequently referred to as thermomechanical treatment (TMT) (see *Thermomechanical Processing*).

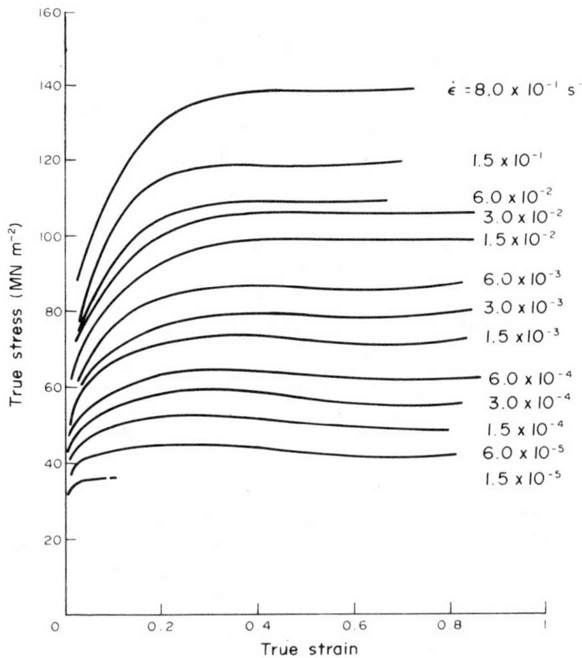

Figure 1
Typical true stress–true strain curves for Armco iron, illustrating the strain-rate ($\dot{\varepsilon}$) dependence of the flow stress at 700 °C (0.54T_m) (after McQueen and Jonas 1975)

These operations are carried out at elevated temperatures because the flow stress of most metals is much lower (e.g., up to one order of magnitude lower) than at room temperature, and also because the ductility is greatly increased. Furthermore, the use of high temperatures promotes rapid microstructural change by such processes as recrystallization in its various forms.

1. Flow Stress at Elevated Temperatures

At elevated temperatures, the flow stress depends primarily on the strain, strain rate and temperature (Figs. 1, 2). It is also affected by the composition of the material and the occurrence of recrystallization or precipitation during or after deformation (Sellars and Davies 1980). For simplicity, consider first the dependence on strain ε in the absence of microstructural changes other than work hardening, that is, the increase in dislocation density that accompanies straining. Under these conditions, the flow stress σ can approximately double, as given by

$$\sigma = \sigma_s\{1 - \exp[-(\varepsilon + \varepsilon_0)/\varepsilon_c]\} \qquad (1)$$

where σ_s is the saturation or asymptotic stress; ε_0 is a "prestrain", which is included in the above expression so as to give a nonzero yield stress at the initiation of straining; and ε_c is the characteristic strain, a measure of the curvature of the flow curve.

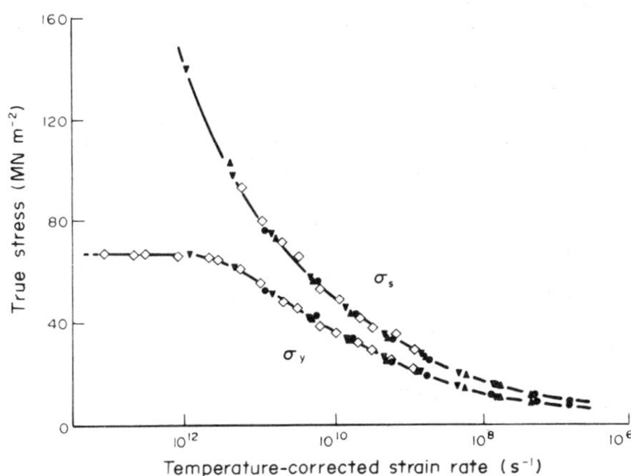

Figure 2
Collected data for yield (σ_y) and steady-state (σ_s) flow stress for α-zirconium determined over the homologous temperature range 0.44–$0.63T_m$ and plotted in terms of temperature-corrected strain rate $Z = \dot{\varepsilon}\exp(Q/RT)$. As the temperature is reduced (leftward motion on the diagram), the yield stress σ_y stops increasing and reaches a limiting (athermal) value. By contrast, the steady-state flow stress σ_s increases rapidly as the athermal flow regime is approached (after McQueen and Jonas 1975)

The saturation stress σ_s depends on the strain rate $\dot{\varepsilon}$ and absolute temperature T as follows:

$$\sigma_s \propto \dot{\varepsilon}^m \qquad (2)$$

$$\ln \sigma_s = [A + B(1/T)] \qquad (3)$$

where m is the strain-rate sensitivity, which falls in the range 0.10–0.20 under hot-working conditions (typically 0.13–0.15 in steel rolling); A and B are empirical constants which permit the rapid decrease in σ_s with increasing T, which has a complex analytical description, to be given an approximate algebraic form.

The dependence of the characteristic strain ε_c on $\dot{\varepsilon}$ and T is as follows:

$$\varepsilon_c \propto \dot{\varepsilon}^m \qquad (4)$$

$$\varepsilon_c \propto \exp(C/T) \qquad (5)$$

where C is a constant.

At elevated temperatures, the yield stress σ_y has dependences on $\dot{\varepsilon}$ and T similar to those of σ_s, as given by Eqns. (2) and (3) (Fig. 2). For example, an increase in $\dot{\varepsilon}$ of an order of magnitude generally leads to increases in σ_y of 30–50%.

1.1 Effect of Solute Content

Pure metals are soft, at elevated as well as ambient temperatures. When alloying elements are added, they generally increase the strength, particularly when present as solutes, but also when they form precipitates or inclusions. Typical additions can increase the strength by factors of as much as two to five. The effect of elements in solution depends primarily on the size misfit parameter $(1/a)(da/dc)$, where a is the lattice parameter and c is the atom fraction of addition (see *Solid-Solution Strengthening*).

1.2 Effect of Precipitates and Inclusions

Most inclusions are fairly coarse and do not contribute directly to high-temperature strength, except by inhibiting grain growth. The presence of precipitates, however, can lead to strength increases, as long as these are fine enough (e.g., <100 Å in diameter). Particles in this size range generally nucleate on dislocations and are thus only found in deformed metals. This type of strengthening is only infrequently of practical importance because of the tendency of precipitates to coarsen at elevated temperatures (see *Precipitation Strengthening: General Considerations*).

1.3 Effect of Dynamic Recrystallization

The exponential strain dependence of the flow stress described by Eqn. (1) only applies to materials which do not undergo recrystallization during straining. This includes all body-centered-cubic (bcc) metals and most hexagonal-close-packed (hcp) metals and alloys, but only aluminum from the face-centered-cubic (fcc) group. In the rest of the fcc metals, recrystallization is initiated during deformation after a critical strain is reached (Fig. 3) (see *Dynamic Recovery and Recrystallization of Metals*). This microstructural process leads to a decrease in the flow stress of the material, and to two types of flow curve: (a) the single-peak curves associated with grain-refinement processes; and (b) the multiple-peak curves associated with grain coarsening (Fig. 4). Under industrial conditions, only the single-peak or grain-refinement type of mechanism is observed; it is of particular importance in processes such as extrusion or planetary hot rolling, in which large strains (in excess of the critical strain for dynamic recrystallization) are imparted in a single operation.

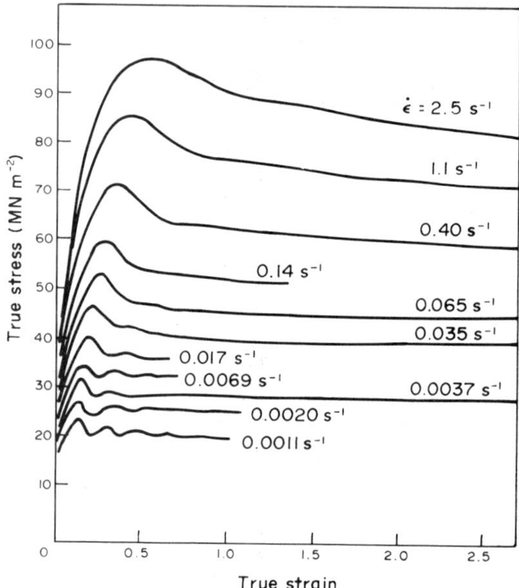

Figure 3
Flow curves for a plain 0.25% C steel in the austenite state (fcc) at 1100 °C ($0.76T_m$), illustrating the strong influence of strain rate. At the higher strain rates, they exhibit the typical shape for dynamic recrystallization: strain hardening to a peak stress followed by work softening to a steady-state level. The work softening is caused by the start of dynamic recrystallization, and the stable flow results from continuous dynamic recrystallization. At low strain rates, the flow curve is periodic owing to recurrent cycles of recrystallization. The single- and multiple-peak curves correspond to the occurrence of grain refinement and grain coarsening, respectively (after Sakai and Jonas 1984)

1.4 Flow Softening and Strain Localization

When flow softening takes place—brought about, for example, by the occurrence of dynamic recrystallization—it tends to induce strain localization, a process which can become serious enough to produce defects within a workpiece. In addition to recrystallization, several other microstructural mechanisms can lead to flow softening. These include: precipitate coarsening, spheroidization of lamellar structures, breakup of Widmanstatten and martensitic structures and substructures, dislocation unpinning phenomena, and strain path change effects. Flow softening can also be caused by nonmicrostructural processes such as texture softening and deformation heating. The latter, particularly, in combination with one of the other mechanisms acting as a trigger, is a frequent cause of localization (Semiatin and Jonas, 1984).

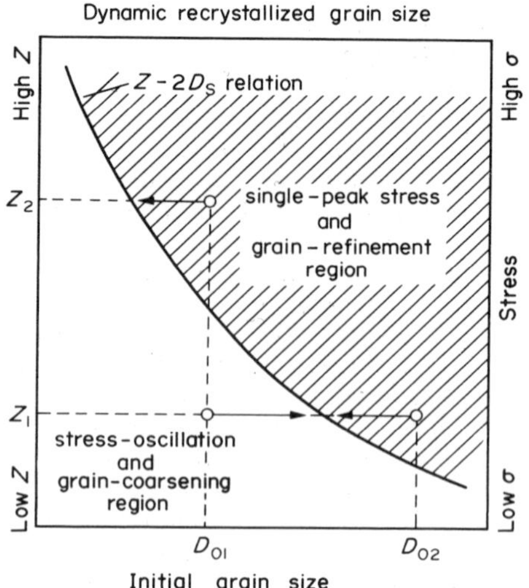

Figure 4
A microstructural mechanism map for distinguishing between the occurrence of the two types of dynamic recrystallization. The curve separates the single-peak (grain-refinement) from the multiple-peak (grain-coarsening) region. Here $Z = \dot{\varepsilon}\exp(Q/RT)$ is the temperature-corrected strain rate and D_s is the stable dynamic grain size. When straining at a given Z, the grain size evolves horizontally leftwards (grain refinement) or rightward (grain coarsening) from the initial D_0 to the stable D_s value (after Sakai and Jonas 1984)

2. Microstructural Processes Associated with Hot Working

The first distinction to be made concerning the various processes of microstructural change that operate under hot-working conditions is to separate those occurring during deformation (and therefore called the dynamic processes) from those taking place between intervals of straining or after metal flow has ceased (termed the static processes). This distinction applies not only to recovery and recrystallization, but also to processes involving the formation of new phases, such as precipitation.

2.1 Dynamic Processes

(*a*) *Dynamic Recovery.* This is the basic mechanism that leads to the annihilation of pairs of dislocations during straining, and is therefore responsible for the flow stress under hot-working conditions being up to one order of magnitude lower than during cold working (see *Dynamic Recovery and Recrystallization of Metals*). When this is the sole restoring mechanism operating (i.e., in the absence of dynamic recrystallization), the flow curve assumes the exponential shape described by Eqn. (1). The low dislocation densities associated with high-temperature deformation can in turn be attributed to the ease of cross-slip, climb and node unpinning in this temperature–flow stress domain. The ease of operation of these mechanisms leads to the formation of well-developed substructures, that is, clearly delineated subgrains, containing relatively low densities of dislocations in their interiors. One of the interesting features of such high-temperature substructures is their equiaxed nature, even at large strains. By contrast, the grains of which they form a part become considerably elongated and flattened. The reason for this difference is that the subboundaries (unlike the grain boundaries) are continuously being broken up and reformed during rolling, forging and extrusion.

(*b*) *Dynamic recrystallization.* This mechanism of dislocation annihilation only intervenes when the density of dislocations approaches very high levels. It is therefore observed in the fcc metals and alloys of moderate to low stacking-fault energy (i.e., all fcc metals except aluminum), and is associated with relatively low rates of dynamic recovery and therefore higher than average rates of work hardening. Under industrial conditions (relatively high temperature-corrected strain rate Z), dynamic recrystallization is nucleated at the grain boundaries of the existing microstructure and, because of the high nucleus density, generally leads to grain refinement (Fig. 4). Two consequences of dynamic recrystallization are of practical importance:

(i) The massive dislocation annihilation produced by the moving grain boundaries leads to local softening, which is reflected in the overall drop in the flow stress (flow softening). The local softening, in turn, promotes flow localization and unstable flow in the regions which are the first to recrystallize. Such localization does not generally become catastrophic because the rate sensitivity is high at elevated temperatures and acts to distribute the strain more uniformly. However, catastrophic localization can still occur if dynamic recrystallization is accompanied by another destabilizing process such as localized deformation heating (Semiatin and Jonas 1984).

(ii) The materials in which dynamic recrystallization can occur have higher than average dislocation densities and flow stresses. The latter, in turn, make the workpiece more susceptible to processes of internal damage, such as crack and cavity formation, and therefore lead to a reduction in the ductility and workability. However, once dynamic recrystallization is initiated, the process of grain-boundary migration tends to isolate the cavities and cracks (which form on the previously static boundaries), thus preventing them from joining up and producing a catastrophic failure.

(*c*) *Dynamic precipitation.* When commercial alloys are heated before working, many of the finely distributed second-phase particles are dissolved because of the exponential increase in solid solubility with temperature. Depending on the alloy system, such particles may consist of carbides, hydrides, nitrides, oxides and, to a lesser extent, silicides and sulfides. During processing, as the temperature slowly falls, the solubility limits of the various species are exceeded in turn and a driving force for their reprecipitation is produced. In unstrained materials, such precipitation is nucleated at grain boundaries as well as within the grains, but is relatively slow and leads to the formation of somewhat coarse particles. The dislocations introduced during straining, on the other hand, provide a very high density of nucleation sites, and increase the rate of precipitation by one to two orders of magnitude. The rate of precipitation is enhanced, not only because of the high density of sites, but also because pipe diffusion along the newly introduced dislocations is much more rapid than the bulk diffusion which is the only atom transport process available in the undeformed material. The high site density also reduces the particle size from several hundred angstroms or more in unstrained materials to well under a hundred angstroms in deformed materials. The practical importance of dynamic precipitation, and of strain-induced precipitation more generally, lies in the role it plays in thermomechanical processing. These precipitates, forming as they do on dislocations and subboundaries, inhibit their motion and thus impede recovery as well as the nucleation of recrystallization. It is in this way that recrystallization is prevented and that the various strengthening consequences of retaining the deformed structure are preserved (Greday 1983).

(*d*) *Dynamic transformation.* Although there are no phase changes involved in the commercial processing of the fcc metals, nor of the simpler hcp (Cd, Mg, Zn) or bcc (Cr, Mo, V, W) metals, the γ to α transformation occurring during the processing of steel and the equivalent transformations involved in the hot forming of Ti, U and Zr alloys are of considerable importance. Normally these take place under static conditions, that is, during simple cooling in the absence of deformation. However, when the temperature range of forming operations such as rolling and forging is extended downwards into the transformation interval, the phase change takes place at least partly under dynamic conditions. The principal effect of superimposing deformation on the transformation process is to accelerate it, partly because of the acceleration of diffusion occasioned by the introduction of dislocations and vacancies, as described previously, and also because deformation increases the nucleus density. Rapid diffusion increases the rate at which the solute partitioning, frequently involved in the formation of new phases, can take place. The practical result of accelerating the transformation is to reduce the difference between the equilibrium and actual transformation temperatures, that is, to reduce the undercooling and increase the temperature at which the change occurs. This in turn generally means that the dynamic transformation product is coarser than the static one that forms at a lower temperature. When the reaction product is softer than the reactant (as in the γ to α transformation in steel), flow softening occurs; this is a factor in promoting flow localization.

2.2 Static Processes

(a) Static recovery. The occurrence of static recovery between intervals of deformation leads to a loss of dislocations, and therefore to a decrease in yield strength or flow stress. This decrease is small, however, falling in the range 15–25%, a proportion which is generally reduced by increasing the prestrain. Moreover, its effect on the load in the subsequent rolling or forging pass is even smaller, because the dislocations that are lost through recovery are rapidly replaced during the first few percent of restraining. Static recovery does not lead to a change in the grain structure and hence is not detectable by means of optical metallography. Nevertheless, it has two consequences of practical significance: (i) it is responsible for the formation of recrystallization nuclei and so is a necessary precursor of this second, more important, softening mechanism; (ii) it leads to the relaxation of internal stresses, by means of the microscopic mechanisms that are also responsible for stress-relief annealing after cold working (see *Recovery of Mechanical Properties*).

(b) Static recrystallization. This mechanism, together with dynamic recovery, is principally responsible for the low loads required to carry out hot-working operations. At the highest processing temperatures, it can take place in tens or hundreds of milliseconds; as the temperature of the deformation is reduced, the recrystallization time gradually increases to seconds, and then to decades and hundreds of seconds (Roberts 1984). Once several minutes are required, recrystallization can be readily prevented and the generally beneficial effects of the deformed structure retained. Although temperature has the greatest effect on the kinetics of this process, as it does on the rate of most metallurgical phenomena, other parameters also play an important role: the magnitude of the prestrain, to a lesser extent the prestraining strain rate, and finally the alloy composition. A certain minimum prestrain, termed the critical strain, is required to initiate static recrystallization; it generally falls in the range of 5–10% and leads to the slowest kinetics. As the amount of prestrain is increased, the rate of recrystallization also increases, by up to two orders of magnitude. This is because the recrystallization rate increases with dislocation density or stored energy. The prestrain strain rate also influences the rate of softening through its effect on the dislocation density introduced per unit strain (see *Recrystallization of Deformed Metals (Primary)*).

(c) Solute and precipitate effects. Recrystallization occurs very quickly in pure metals because of the absence of the impurity atoms which would otherwise lead to "solute drag" and hence slow down the migration rate of the boundaries (Stüwe 1978). In a similar manner, small particles when present lead to boundary or subboundary pinning. The effectiveness of various solutes in retarding recrystallization is apparently related to the size misfit parameter $(1/a)(da/dc)$. This arises from the inverse dependence of the recrystallization time on the ease of static recovery (required for the nucleation of recrystallization).

When a fixed atomic proportion is added to steel in the high-temperature (fcc) form, the size misfit parameter and the solute retardation of recrystallization decrease in the following order: Zr, Nb, Cu, Mo, W, Ti, Al, V, Ni, Mn, Cr, Si. The addition of conventional amounts of the above elements to low-alloy steels (e.g., less than 0.15 at.% Al, Mo, Nb, Ti or V) can retard recrystallization by a factor of up to two orders of magnitude in time with respect to a plain carbon steel in the temperature range 900–1100 °C (Fig. 5). In general, the elements with large misfit parameters (and appreciable retardation and strengthening effects) have low solid solubilities, and vice versa.

As the temperature is decreased during processing and the solid solubility is reduced, precipitates gradually form. If these are nucleated on dislocations prior to the removal of the latter by recrystallization, they are very fine (less than 100 Å in diameter) and can prevent recrystallization entirely within the time frame of the forming operation. Under these conditions, the retardation of recrystallization attributed to precipitation is even greater than that associated with the solutes and can exceed a further two orders of magnitude in time (Fig. 5). In steel, the most common precipitates playing this role are NbCN, followed by TiC, AlN and VN. Other compounds

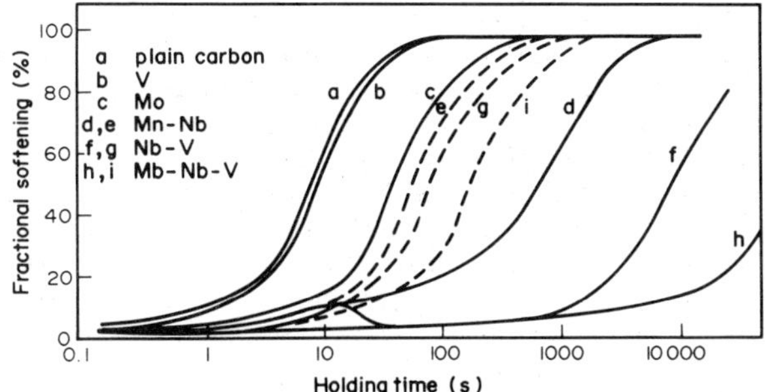

Figure 5
Static recovery and recrystallization in a series of microalloyed steels deformed 25% at 900 °C. Comparison of the plain-carbon steel behavior (curve (a)) with that of the V and Mo steels (curves (b) and (c), respectively) reveals the extent of the solute retardation of recrystallization. The estimated solute retardations for the Nb, Nb–V and Mo–Nb–V steels are presented as broken lines (curves (e), (g) and (i), respectively). The effect of NbCN precipitation in the latter three alloys can be seen by comparing curves (d), (f) and (h) with (e), (g) and (i), respectively (after Akben and Jonas 1984).

form in the nonferrous alloys, with the common feature that at least one component is an interstitial or other rapidly diffusing species.

(*d*) *Postdynamic* (*metadynamic*) *recrystallization.* When the workpiece is unloaded after the critical strain for static recrystallization has been exceeded, recrystallization is not instantaneous but requires an incubation time. During this interval, the static recovery processes participate in the formation of recrystallization nuclei in the lattice regions of highest curvature and dislocation density. This incubation period is no longer required when dynamic recrystallization is taking place during straining because the necessary nuclei are formed during as opposed to after deformation, and so are instantly available. Postdynamic recrystallization is consequently about an order of magnitude more rapid than classical static recrystallization. It also produces finer grain structures than the classical mechanism, essentially because of the higher nucleus density produced by dynamic nucleation.

See also: Dynamic Recovery and Recrystallization of Metals; Warm Working; Workability

Bibliography

Akben M G, Jonas J J 1984 Influence of multiple alloy additions on the flow stress and recrystallization behavior of HSLA steels. *Proc. 1983 Int. Conf. High Strength Low Alloy Steels.* American Society for Metals, Metals Park, Ohio pp. 149–61

Costa P 1981 *Les traitements thermomécaniques: aspects théoriques et applications.* Institut National des Sciences et Techniques Nucléaires, Saclay, France

Greday T 1983 Routes to higher strength and ductility of steels. In: Gifkins R C (ed.) 1983 *Strength of Metals and Alloys*, ICSMA 6. Pergamon, Sydney, pp. 1075–88

MacQueen H J, Jonas J J 1975 Recovery and recrystallization during high temperature deformation. In: Arsenault R J (ed.) 1975 *Plastic Deformation of Materials*, Treatise on Materials Science and Technology Vol. 6. Academic Press, New York, pp. 393–493

Roberts W 1984 Dynamic changes during hot working and their significance as regards microstructure development and hot workability. In: Krauss G (ed.) 1984 *Deformation, Processing and Structure*, 1982 ASM Materials Science Seminar. American Society for Metals, Metals Park, Ohio, pp. 109–84

Sakai T, Jonas J J 1984 Dynamic recrystallization: Mechanical and microstructural considerations. *Acta Metall.* 32: 189–209

Sellars C M, Davies G J (eds.) 1980 *Hot Working and Forming Processes.* Metals Society, London

Semiatin S L, Jonas J J 1984 *Formability and Workability of Metals: Plastic Instability and Flow Localization.* American Society for Metals, Metals Park, Ohio

Stüwe H P 1978 Driving forces for recrystallization. In: Haessner F (ed.) 1978 *Recrystallization of Metallic Materials.* Riederer, Stuttgart, Chap. 2, pp. 11–21

J. J. Jonas

Hume-Rothery Phases

When the noble metals copper, silver and gold are alloyed with the B-subgroup elements of the periodic table (Fig. 1), sequences of alloy phases frequently occur in the corresponding phase diagrams which exhibit remarkable similarities. Figure 2 shows such a typical phase diagram, in which a face-centered-cubic (fcc) primary solid solution is followed by a sequence of intermediate phases with characteristic crystal structures. The early foundations for the understanding of the occurrence of these structures were laid by Hume-Rothery (1936), when he pointed out striking connections between the observed structures and electron concentration (i.e., the ratio n_e/n_a

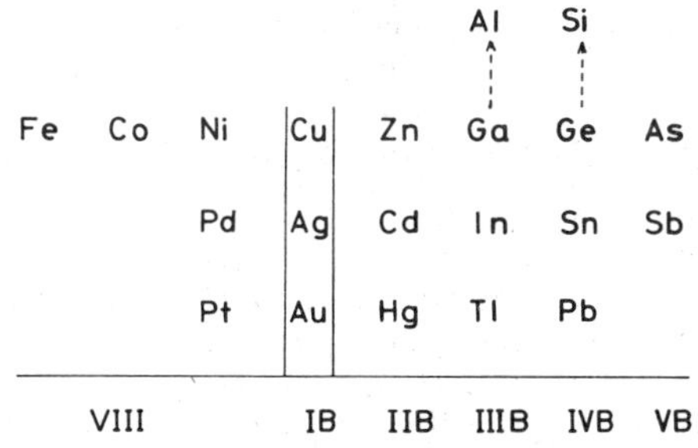

Figure 1
Portion of the periodic table representing alloy phases of the noble metals copper, silver and gold. Roman numerals indicate group valency (after Massalski and Mizutani 1978)

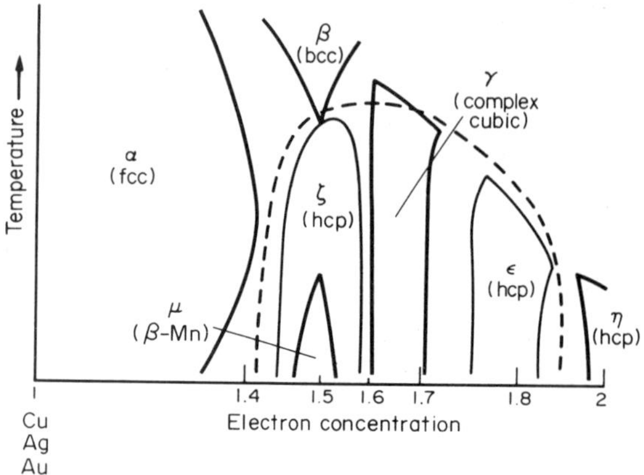

Figure 2
Typical alloy phase diagram characteristic of the noble metals. Thick-line boundaries represent the phases with cubic symmetry α, β, μ and γ. Hcp structures are represented by ζ, ε and η. The broken line shows a possible range of hcp phases (after Massalski and Mizutani 1978)

of valence electrons to the number of atoms in the structure). As a result the phases are often referred to as electron phases or Hume-Rothery phases. Considerable progress has now been made in various systematic studies of a number of physical properties of Hume-Rothery phases (Massalski and King 1961, Massalski and Mizutani 1978). These include elastic constants, specific heats and other thermodynamic data, lattice spacings, electronic transport properties, optical constants, magnetic susceptibilities, and properties directly related to the Fermi surface (see *Fermi Surfaces*).

Since the B-subgroup elements follow the noble metals in the respective rows of the periodic table, the *d* bands in the free atoms of these elements are fully occupied by electrons. It has long been thought that a similar situation is also likely to occur in the alloys of the noble metals. Hence, it may be considered to a first approximation that changes in the electronic structure due to alloying involve only the *s* and *p* electrons; that is, electrons which occupy the conduction band and control the n_e/n_a ratio. The nearly free electron behavior frequently observed in Hume-Rothery phases partially confirms this supposition.

Hume-Rothery phases with cubic symmetry include the body-centered-cubic (bcc) β phase (and also the ordered β' phase), the μ phase with the β-Mn structure, the complex cubic γ phase with 52 atoms per structure cell, and the δ phase (cubic with many vacant lattice sites). The disordered β phases are stable only at high temperatures. This behavior has been coupled with an unusually high value of the (110) $[\bar{1}10]$ shear compliance modulus in the bcc structure, which allows for a large amplitude of lattice vibrations and a correspondingly high vibrational entropy. This in turn influences the entropy term $(-T\Delta S)$ in the free-energy equation, particularly at high temperatures. Conversely, a decrease in temperature causes a rapid decrease in phase stability, resulting in the characteristic V-shaped phase fields observed in all β phases (Fig. 3). This clearly shows that other factors, in addition to the n_e/n_a ratio, can play a role in the stability of Hume-Rothery phases.

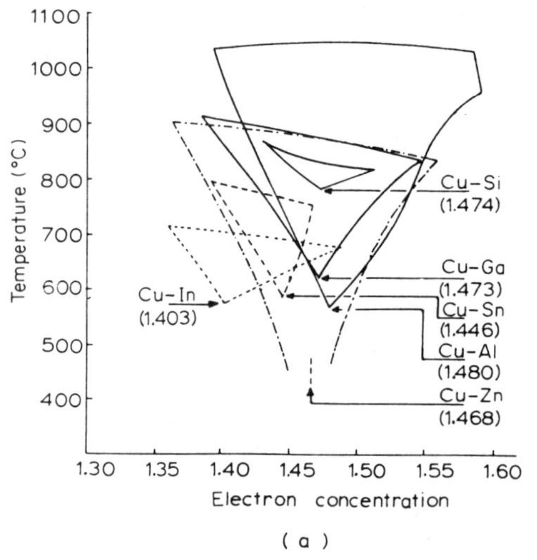

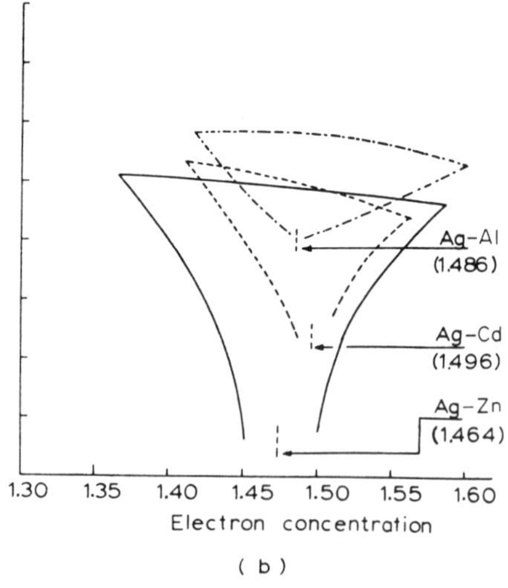

Figure 3
Shape of the β-phase field plotted as a function of electron concentration: (a) in copper alloys, (b) in silver alloys (after Massalski and King 1961)

The hexagonal-close-packed (hcp) phases (ζ, ε and η) are the most numerous Hume-Rothery phases. Each occurs within a characteristic electron-concentration range (Fig. 2), but, taken together, they may occur anywhere within the n_e/n_a range between about 1.32 and 2. Variations in lattice spacings and axial ratio (c/a) with n_e/n_a in hcp phases provide particularly interesting information about the electronic structure of such alloys. The actual value of the axial ratio in numerous hcp phases that have been studied is strikingly dependent on the n_e/n_a ratio, irrespective of the solvents and solutes involved (Fig. 4). This observed behavior has been interpreted in terms of interactions between the Fermi surface of the conduction electrons and the Brillouin zone (see *Brillouin Zones*). As the number of electrons increases on alloying, the Fermi surface contacts and overlaps related sets of zone faces. The energy of the system can be lowered if the zone suffers a distortion which brings certain zone faces closer to the origin in reciprocal lattice space (see *Reciprocal Lattice*). Changes of the zone shape produce corresponding changes in the axial ratio in real space. The proposed details of the Fermi surface contours are consistent with the measured values of the electronic specific heat coefficient, and have also been "directly" observed with the positron annihilation technique (Massalski and Mizutani 1978, Koike et al. 1982).

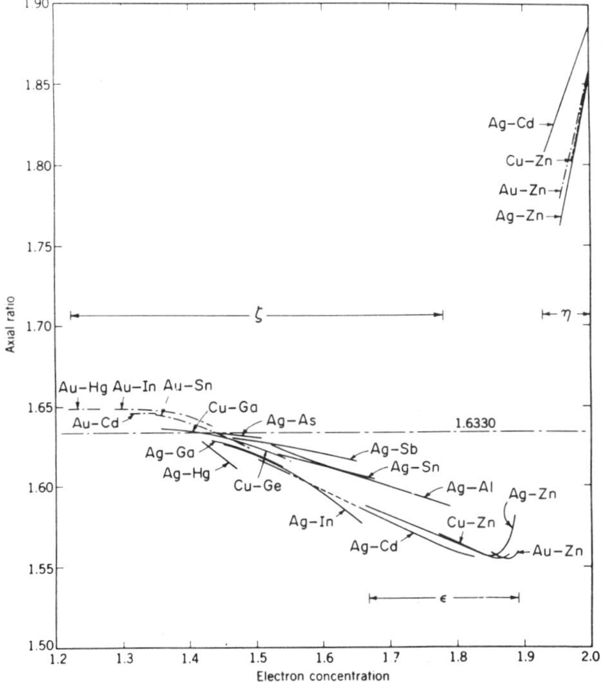

Figure 4
The trend of axial ratio as a function of electron concentration in various hcp alloy phases (after Massalski and Mizutani 1978)

On the whole, the available results on the Fermi surface topography in numerous Hume-Rothery phases with different crystal structures, and its relation to the corresponding Brillouin zones and density-of-states curves, suggest that such features change in a relatively simple manner on alloying, conforming to a general rigid-band description of alloying behavior. Positron annihilation experiments, de Haas–van Alphen effect experiments, electronic specific heats, band calculations and other data support this general conclusion. Nevertheless, the exact relationships between this simple electronic picture, the observed phase-stability ranges, and the factors which contribute to total energy are not yet fully understood and several questions remain unanswered (Massalski and Mizutani 1978).

See also: Alloy Crystal Structures and Their Stability; Phase Diagrams

Bibliography

Hume-Rothery W 1936 *The Structure of Metals and Alloys*, Institute of Metals Monograph and Report Series No. 1. Institute of Metals, London

Massalski T B, King H W 1961 Alloy phases of the noble metals. *Prog. Mater. Sci.* 10: 1–78

Massalski T B, Mizutani U 1978 Electronic structure of Hume-Rothery phases. *Prog. Mater. Sci.* 22: 151–262

Koike S, Hirabayashi M, Suzuki T, Hasegawa M 1982 The Fermi surface of Hume-Rothery phases in Ag–Al and Cu–Ge alloys studied by a positron annihilation method. *Philos. Mag., B* 45: 261–72

T. B. Massalski

Humidity Sensor Materials

This article is concerned chiefly with solid materials used for the electrical sensing of humidity. Other means of sensing the presence of water vapor, such as those relying on physical changes (i.e., of dimension or phase) or thermometry, are not discussed. The various materials that may be used are considered, as well as their mechanisms of sensing and some measurement principles that permit the desired sensory information to be extracted as electrical signals. The electrical properties that change with increasing water absorption are surface and volume resistivity, which decrease, and dielectric constant and loss, which increase.

1. Metal Oxides

This group includes mixed oxides and ceramics as well as anodically grown films, the distinction being made because of the radically different properties and modes of preparation.

1.1 Mixed Oxides and Ceramics

These have been used as humidity-sensing materials and include spinel structures like $Fe(II)_xFe(III)_yO_z$ and $Co(II)_xCo(III)_yO_z$ as well as the related $MgCr_2O_4$–TiO_2 single-phase system, and perovskite structures as exemplified by the aluminates (e.g., Li_5AlO_4), niobates (e.g., $LiNbO_3$), titanates (e.g., $BaTiO_3$, $SrTiO_3$) and tungstates (e.g., $MnWO_4$). Simple oxides have also been used as the basis of a humidity-sensitive ceramic composition that often includes an alkali metal oxide dopant to enhance conduction. Among these materials are TiO_2, α- and γ-Fe_2O_3, V_2O_5, ZnO and SnO_2. Advantage may be taken of a slightly defective stoichiometry to introduce semiconduction. These materials are generally poor electronic conductors and poor ionic conductors in bulk form, so that their use as resistive sensing materials usually relies on ionic surface conduction and moderately high surface area and porosity. However, capacitive responses to humidity have been studied and exploited in the case of the iron oxides, the cobalt oxides and the $MgCr_2O_4$–TiO_2 ceramic.

All of the foregoing oxide and ceramic materials are commonly prepared as comparatively massive compacts, but radio-frequency sputtering has been used to prepare thin films of lower resistance.

1.2 Anodically Grown Films

These are metal oxides grown on a pure metal or an alloy substrate that serves as the anode in an electrochemical cell. Anodic oxides that have been used for electrical humidity sensing include those formed on Al, Hf, In, Nb, Sb, Sn, Ta, Ti and Zr. All of these metal/metal oxide systems share the following features: the extent of the nonporous oxide film growth is limited to a few thousand angstroms; the metal is usually in its highest oxidation state in the film; the films are usually adherent, of less than bulk density, poorly crystallized (if at all), and range from good to poor insulators. Those systems which have been well studied—Al/Al_2O_3, Ta/Ta_2O_5 and Ti/TiO_2—show further subtleties in their structure, growth kinetics and chemical properties (Diggle 1973).

The comparative thinness and insulating properties of the anodic oxide films suggest their use in capacitive structures. Commercial aluminum oxide sensors are anodized to obtain a layered dielectric structure comprising a porous outer region and a nonporous inner layer. An equivalent circuit for this structure has been calculated (Booker and Wood 1960). This kind of structured dielectric appears to be unique among the anodic oxides. The porous layer facilitates condensation of water, leading to large but very nonlinear impedance changes.

Very little has been reported specifically on the humidity response of the other anodic oxides, although they have appeared extensively in Japanese patent literature.

2. Semiconductors

There appears to have been very little exploitation of traditional materials (i.e., silicon and germanium) as humidity-sensing materials per se. Instead, hydrophilic polymers coating the active region of a semiconductor device have been employed.

Other semiconductor materials find use as humidity sensors in certain temperature regimes. For example, tin oxide shows a resistance response at room temperature that is selective to water vapor, whereas at elevated temperatures it responds more to hydrocarbon gases.

3. Polymers

3.1 Natural Polymers

The electrical response of natural polymers has not been greatly exploited for humidity sensing, probably because their physical form does not lend itself to electrical sensing arrangements. However, derivatives of natural products have found significant use as electrical materials. The most important sensing materials are the acetates, butyrates and mixed esters of cellulose. A reduction in affinity for water, and hence in humidity sensitivity, generally accompanies increasing esterification. The same is true for adding an increasing fraction of nonpolar side chains.

The dielectric isotherms of cellulose ester films generally parallel the gravimetrically determined sorption isotherm when they are measured at audio frequencies. The capacitance response to humidity is linear in the range 0–65% RH. Above 65% RH the slope of the sorption curve increases continuously, as does that of the capacitance response at audio frequencies. However, at megahertz frequencies, a near-linear relationship exists between capacitance and relative humidity because the frequency dispersion also increases in this regime of humidity.

3.2 Synthetic Organic Polymers

Synthetic organic polymers of interest as sensing materials are those polymers capable of sorbing several percent water by weight. This usually implies the presence of polar groups such as CO, COOH, OH and NH. Linear polymers with pendent polar groups also tend to absorb more water than heavily cross-linked polymers, and amorphous (disordered) structures are more conducive to water uptake than are well-crystallized (ordered) polymers.

(*a*) *Poly*(*methyl methacrylate*). This material absorbs a few percent water. The variation of conduction current i with water sorption follows the equation log $i/i_0 = Bm$, where i_0 is the current in the dry polymer, B is a proportionality constant and m is the moisture content. This equation is valid up to saturation.

(*b*) *Poly*(*vinyl alcohol*). Poly(vinyl alcohol) is a very hydrophilic polymer; it absorbs more than 50 wt%

water at saturation. Its resistance response is exponential for a water content up to 20–25 wt%. At higher water contents, the current flow becomes linear with m and follows the expression: $i/i_0 = 1 + \alpha m$, where α is a constant.

(*c*) *Poly*(*ethylene oxide*). This is also very hydrophilic and miscible with water in all proportions. It has been used as a film on SiO_2, the sheet resistance of which is measured by a depletion-mode *n*-channel MOSFET on the same chip. The sheet resistance of such films varies almost exponentially from about 30–70% RH and deviates from this relation at lower humidities.

(*d*) *Poly*(*p-aminophenylacetylene*). Poly(*p*-aminophenylacetylene) has been incorporated into the charge-flow transistor, which senses variations in the sheet resistance of surface films. The turn-on time of the device (and presumably the sheet resistance) varies over five orders of magnitude in going from 10 to 50% RH, and is nearly exponential in form over that range.

(*e*) *Poly*(*ethylene glycol*) (*PEG*). Low-molecular-weight (~200) PEG has been studied as a humidity-sensing material by means of impedance measurements over the range 10^{-2}–10^5 Hz, in conjunction with resistivity measurements of bulk PEG–H_2O solutions. The volume resistance of PEG sensing materials decreases with increasing humidity. The change of resistance with humidity is largest at the highest frequency (10^5 Hz) where a total change of about four orders of magnitude is observed. The resistivity of 5–10 μm films is approximately the same as that found for bulk PEG–H_2O solutions, which indicates that the same ionic mobility effects are operative in both cases. The ionic conduction and mobility are governed by viscosity, which, for the films, varies with relative humidity.

(*f*) *Delrin*. This polyacetal resin absorbs a small percentage of water (<1%) at saturation and maintains a high volume resistance. Thus as a material for humidity sensing it has been used as a capacitor dielectric. At 25 °C the effective dielectric constant ε is a near-linear function of relative humidity, rising from 3.6 at 10% RH to 4.0 at 90% RH.

(*g*) *Nylons*. The water absorption properties of nylon are well known: the moisture content of nylon strongly affects its mechanical as well as its electrical properties. There is a considerable range of rate and amount of saturation depending on the nylon type (Table 1). Dry Nylon 6, for example, has a volume resistivity of 10^{15} Ω cm, but this drops seven orders of magnitude at 100% RH; Nylon 11 shows a change of only two orders of magnitude in resistivity over the same range of humidity. The dielectric properties of most dry nylons are similar ($\varepsilon \sim 4$) and have only a small frequency dependence. The frequency dependence of the dielectric constant becomes more pronounced as the moisture level increases because the sorbed water responds poorly to higher frequencies. The dissipation factor of nylons is not monotonic with frequency, due to the existence of several dielectric relaxations (see Sect. 5.1). The electrical response of some common nylons to humidity is summarized in Table 2. These data indicate the potential utility of the longer chain nylons as capacitive sensing materials, particularly if relatively high detection frequencies are used.

Table 1
Absorption of water by nylons (wt% dry basis)

Nylon type	After 24 h[a]	Equilibrium at 50% RH	Saturation
6	1.6	2.7	9.5
7	1.65 (1h boil)		
66	1.5	2.5	8.0
610	0.4	1.5	3.5
612	0.4	1.3	3.0
11	0.25	0.8	1.9–2.9
12	0.25	1.7	1.4–2.5
13	1.04		
1313	0.75		
TMDT		3 (65% RH)	
HPXD8			3.5 (24 h boil)

a American Society for Testing and Materials standard test D-570

(*h*) *Polyimides*. Polyimides are noted for their very high thermal and hydrolytic stabilities, but nevertheless absorb approximately 0.7 wt% water after 24 h immersion. The capacitance response to humidity is quite similar to that of Delrin, but the stability of polyimide films is superior in all respects, and the dielectric loss is lower.

(*i*) *Polyelectrolytes*. Typified by polystyrene sulfonate with its pendent acidic groups, polyelectrolytes have recently been used as the basis for resistive humidity sensors. Analogous materials with pendent basic groups have been prepared as random copolymers of quaternary ammonium cations and vinyl monomers. A volume conductivity of 10^{-6} $(\Omega\ \mathrm{cm})^{-1}$ is typical for the dry polyelectrolyte, rising five or six orders of magnitude as humidity increases toward saturation. Through cross linking, these materials can be made insoluble and thermally and hydrolytically stable.

3.3 Silanes, Siloxanes and Organosilane-Based Polymers

These have been used as resistive materials, sometimes with carbon added to increase conductivity. Very little information is available on these materials, as the emphasis is usually on the use of a semiconductor device coated with the polymer.

Table 2
Electrical properties of nylons

Nylon type	Condition (% RH)	Volume resistivity[a] (Ω cm)	Dielectric constant[a]			Dissipation factor		
			60 Hz	10^3 Hz	10^6 Hz	60 Hz	10^3 Hz	10^6 Hz
66	dry	10^{15}	4.0	3.9	3.6	0.02	0.03	0.02
	50	10^{13}	6.0	5.1	4.1	0.04–0.18	0.04–0.18	0.05–0.09
	100	10^{9}	31	29	18	0.50	0.23	0.28
6	dry	10^{15}	3.8	3.7	3.4	0.01	0.02	0.03
	50	10^{13}	13	8.3	4.2	0.18	0.20	0.12
	100	10^{8}			25			
610	dry	10^{15}	4.0	3.9	3.6	0.04	0.04	0.03
	50	10^{13}	7.4	6.1	3.7	0.04	0.04	
	100	10^{11}	15	12	4.0	0.03		0.1
612	dry	10^{15}	4.0	4.0	3.5	0.02	0.02	0.02
	50	10^{13}	6.0	5.3	4.0	0.04	0.04	0.03
	100	10^{11}	14					

a American Society for Testing and Materials standard test D150

4. Other Materials

Materials that do not fit into the above categories have also been used in electrical humidity sensing, but little information is available. Inorganic/organic polymer composites have been used. Sintered silicon carbide has been reported to exhibit a dielectric response to humidity. Cadmium selenide doped with alkali-metal or alkaline-earth ions has been used for resistive sensing. Thin-film barium fluoride resistive sensors have been characterized in terms of their conductance isotherms and sorption equations.

5. Mechanisms of Sensing

There is no complete treatment available of the mechanisms underlying the electrical response of sensing materials to humidity, and such a treatment would be far beyond the scope of the article. However, a brief overview is presented here.

5.1 Dielectric Materials (Capacitive Sensors)

Solid nonconductors of interest for capacitive sensing have dielectric constants ranging from 2 to about 100, depending on the electronic polarizability and the relative number of polar groups present. The sorption of water, with its large permanent dipole, should increase the effective dielectric constant of the water–dielectric system. The extent of the change depends not only on the relative humidity, but also on the water affinity of the solid, the state of binding, the degree of electronic interaction with the solid and the extent of water–water interactions.

For α-Fe_2O_3, for example, which has been studied in detail, five stages of interaction with water vapor are discernible, with varying electrical effects in the dielectric isotherm.

Stage 1: Irreversible chemisorption of water. Upon initial contact of the solid α-Fe_2O_3 with water vapor, a hydroxylated surface is formed through dissociative chemisorption at cationic sites. This process is irreversible in the sense that the chemisorbed layer is so tightly bound as to be unaffected by subsequent changes in relative humidity. The hydroxyl groups can be removed by thermal desorption at a high enough temperature and are restored first following thermal drying. The chemisorbed layer is dielectrically inactive, that is, it does not contribute to the capacitance at ac frequencies.

Stage 2: Formation of the first physisorbed monolayer. In the first portion of the isotherm a physically adsorbed monolayer of water is formed by strong hydrogen bonding to the underlying hydroxylated surface. This first physisorbed layer is also dielectrically inactive, because the dipoles are immobile. Thus, the capacitance is virtually constant in this region. The dielectric loss (measured by $\tan\delta$ or parallel conductance) rises slightly with the humidity, possibly due to dissociation of a small fraction of the water molecules and the resultant protonic conduction (see Sect. 5.2). At 25 °C this monolayer is complete at about 15–20% RH.

Stage 3: Formation of the second monolayer. As more water is adsorbed, the outer layer becomes less strongly bound to the previous one. On average, a single hydrogen bond is formed per molecule. Since these dipoles can orientate to follow an applied field, they are dielectrically active, and a sharp rise in audio frequency capacitance occurs throughout this region of the isotherm. Characteristic frequencies (f_{char}) for dielectric relaxation of this water on α-Fe_2O_3 are so low (10–300 Hz) that mobility of the dipoles over a considerable area is inferred. On more hydrophobic materials, water clusters form at hydrophilic sites, and have much higher f_{char}, that is 10^3 Hz if structured like ice or 10^{10} Hz as a liquid. The dielectric relaxation at a given relative humidity and temperature is describable by a Cole–Cole dispersion relation for the complex dielectric constant:

$$\varepsilon^* = \varepsilon_\infty + (\varepsilon_s - \varepsilon_\infty)/[1 + (i\omega\tau_0)^{1-\alpha}] \qquad (1)$$

where ε_s and ε_∞ are the static and infinite frequency dielectric constants, respectively, ω is the angular frequency, $\tau_0 = 1/2\pi f_{char}$ is the most probable relaxation time and α is the distribution coefficient for the dispersion. The imaginary component ε'' goes through a maximum at f_{char}. As the humidity rises and the number of active dipoles increases, both f_{char} and the maximum value of ε'' rise correspondingly. The second physisorbed monolayer is completed at about 50% RH and 25 °C.

Stage 4: Multilayers of water. Beyond the second layer, the adsorbate should remain sufficiently mobile to be dielectrically active, and tend gradually toward bulk properties. A continued rise in capacitance is in fact observed as additional layers form, indicating continued mobility, and f_{char} rises smoothly toward a limiting value of about 10^3 Hz with three to six adsorbed layers in place. This is interpreted to mean that fairly strong hydrogen bonding—stronger than in liquid water—has developed within the built-up multilayers. The multilayer adsorbate molecules are viewed as ordered, somewhat as in ice, but still able to rotate in response to an ac field.

Stage 5: Condensation to liquid water. As the relative humidity approaches 100%, additional water condenses, overcomes the ordering influence of the solid, and forms a liquid phase of finite thickness. The contribution to the dielectric loss from surface conduction gradually becomes important, and the electrical properties of the system can be described by

$$\varepsilon^* = \varepsilon_\infty + (\varepsilon_s - \varepsilon_\infty)/[1 + (i\omega\tau_0)^{1-\alpha}] + \sigma_0/i\omega \qquad (2)$$

where σ_0 is a dc conductivity term representing surface conduction. When σ_0 is appreciable the frequency dispersion of the capacitance becomes large.

Dielectric materials other than α-Fe_2O_3 do not share all details of the dielectric response outlined above. Those with a less ionic character, for instance, are less subject to surface conduction effects. Also, while physisorption of water is reversible on α-Fe_2O_3, γ-Al_2O_3, ThO_2 and Ta_2O_5, it is not reversible on SiO_2 or some porous materials. Also, several polymers exhibit near-linear capacitance response at very low humidities, evincing a mobile adsorbate throughout the isotherm. Glassy polymers often exhibit very low frequency relaxation behavior which complicates the interpretation of dielectric measurements.

5.2 Resistive Materials (Conductive Sensors)

Materials of interest for resistive sensing are characterized by a conductance G which is much larger than their capacitive susceptance ωC, this being true at most humidities for frequencies in the audio range or lower if the sensor geometry is chosen to minimize the geometric capacitance. When these conditions are met, the dominant electrical response to water vapor is a moisture-dependent decrease in surface resistance that is primarily a function of relative humidity. Frequency effects are minor in comparison, and are not considered here, but this is an area where further development of understanding is required. A unified mechanistic interpretation of the conductance isotherm at 1 kHz has been advanced by Fleming (1981) and applied by him to the specific case of doped ionic metal oxides. This interpretation, with suitable modification, may be applicable to most resistive materials.

Stage 1: Irreversible chemisorption of water. Depending on the reactivity of the virgin surface, initial contact with water vapor results in a hydroxylated surface layer of either water molecules or hydroxyls, or both, tightly bound at cationic sites. This layer is practically incapable of charge transport and is immobile because of the strong bonding.

Stage 2: Protonic conduction. In the first portion of the isotherm, electrolytic conduction is impossible, even when soluble salts are present on the surface. It is suggested that a small amount (~1%) of the physisorbed water dissociates under the influence of strong electrostatic fields near surface cations. Doping with Li^+, noted for its small ionic radius and high charge density, provides especially strong electrostatic fields which can promote both physisorption and dissociation of water molecules. The hydronium ions so formed can participate in a chain mechanism wherein protons are transferred between them within the physisorbed layers. The basic equation expressing the dependence of this protonic conduction on humidity, derived from equilibrium considerations, is $G = G_0(p/p_{sat})^n$, where G_0 contains a collection of constants at constant temperature, p/p_{sat} is the relative humidity ratio, and n is the average number of physisorbed water molecules per surface cation site. The validity of this expression is restricted to humidities where this conduction mechanism dominates, which can be checked by examining a log–log plot of G against p/p_{sat} for a given sensor. The slope of such a plot gives n directly. For several metal oxides, n ranges from less than unity to about four, depending on doping, and a linear plot is obtained up to about 40% RH.

Stage 3: Electrolytic conduction. Once several monolayers of physisorbed water have been completed, a second conduction mechanism becomes important. The remainder of the log–log conductance isotherm is also linear, but with a higher slope. It has been suggested that electrolytic conduction is facilitated by sufficient liquid water in, for example, the capillary pores in the solid. Since both mechanisms are contributing to the conductance, a larger slope results. The significance of n in this region of the isotherm is not entirely clear. No further change in slope is observed even up to saturation.

As with dielectric materials, resistive materials will vary in the details of the isotherm. Polyelectrolytes, for instance, which have covalently bonded cationic sites, may show a smaller region of dominant protonic conduction.

A consideration of the mechanisms makes it clear that the presence of trace ionic impurities can produce large effects. Thus it is imperative to shield conductive sensors from sources of ionic contamination.

6. *Detection*

A very brief treatment of considerations involved in detecting the electrical response of a humidity-sensing material is given here. Firstly, it is recommended that either ac or pulsed dc be used as excitation sources. The problems associated with continuous dc levels (electrolytic polarization, space–charge effects, drift) are best avoided. Secondly, even with ac, attention should be paid to the amplitude of driving signals, since in the presence of mobile water a variety of electrochemical reactions can take place at almost any electrode subjected to several volts applied potential.

If an ac mode of operation is selected, the flexibility of operating at a chosen frequency gives several advantages:

(a) A frequency can be selected for compatibility with other circuit components, if these are restrictive.

(b) Either the capacitive or resistive component can be emphasized or selected.

(c) Occasionally, discrimination against undesirable effects is possible. For example, linearity of a capacitive response can be improved by choice of a frequency high enough to suppress contributions from dc surface conduction.

See also: Electrical Materials: An Overview

Bibliography

Booker C J L, Wood J L 1960 Further electrical effects of the adsorption of water vapor by anodized aluminum. *Proc. R. Soc. London, Ser. A* 76: 721–31

Daniel V 1967 *Dielectric Relaxation*. Academic Press, London

Diggle J W (ed.) 1973 *Oxides and Oxide Films*, Vol. 2. Dekker, New York

Fleming W J 1981 A physical understanding of solid state humidity sensors. *Soc. Automot. Eng. Spec. Publ.* SP-486: 51–62

Garverick S L, Senturia S D 1982 A MOS device for AC measurement of surface impedance with application to moisture monitoring. *IEEE Trans. Electron Devices* 29: 90–94

Hijikigawa M, Miyoshi S, Sugihara T, Jinda A 1983 A thin-film resistance humidity sensor. *Sens. Actuat.* 4: 307–15

Kawasaki K 1961 On the variation of electrical resistance of a polymer as a function of the extent and nature of sorbed water. *J. Colloid Sci.* 16: 405–10

Licari J J 1970 *Plastic Coatings for Electronics*. McGraw-Hill, New York

Nicholas M E, Pilkanen D E, Lavine C F, Zook J D, Hagen G P 1976 Electrical properties of iron oxide–polyethylene glycol humidity-sensitive elements. *J. Appl. Phys.* 47: 2191–99

Nitta T, Terada Z, Hayakawa S 1980 Humidity-sensitive electrical conduction of $MgCr_2O_4$–TiO_2 porous ceramics. *J. Am. Ceram. Soc.* 63: 295–300

Seiyama T, Yamazoe N, Arai H 1983 Ceramic humidity sensor. *Sens. Actuat.* 4: 85–96

Texter J, Klier K, Zettlemoyer A C 1978 Water at surfaces. *Prog. Surf. Membr. Sci.* 12: 327–403

Uchikawa F, Miyao K, Shimamoto K 1984 Surface OH concentration and electrical resistance of humidity-sensitive silicone composite films. *Am. Ceram. Soc. Bull.* 63: 1043–46

J. Brace

Hybrid Fiber–Resin Composites

The concept of mixed fiber composites is an extension of the composites principle of combining materials to optimize their value to the engineer, exploiting their better qualities while mitigating the effects of their less desirable properties. Mixing two or more types of fiber in one matrix allows a closer matching of composite properties to specific requirements than can be achieved with a single type of fiber.

Many hybrids of current interest represent attempts to reduce the cost of expensive composites, which contain reinforcements like carbon or boron, by incorporating a proportion of cheaper, lower quality fibers, such as glass, without too seriously reducing the mechanical properties of the original composite. At a lower level, the glass–jute–resin composites reflect the same purpose. Of equal importance is the reverse principle, that of stiffening a glass-reinforced plastic (GRP) structure, like a car body or canoe shell, with a small quantity of judiciously placed carbon or Kevlar 49 fiber, without inflicting too great a cost penalty.

In high-technology fields, the question of cost may be insignificant in comparison with the advantages of optimizing properties, as in the use of carbon–glass–resin hybrids for helicopter rotor blades. In demanding aerospace applications, one of the most important reasons for using hybrids is that the natural toughness of GRP offsets the brittleness of typical carbon or boron composites.

A widely debated issue is whether or not it is possible to obtain greater improvements in some properties than might be predicted from calculations based upon properties of the separate components. This so-called hybrid effect has obscured the interpretation of much experimental work. Hybrid properties higher than a mixture-rule prediction have often been hailed as evidence of synergism, regardless of the validity of parallel-connected element models for the composite properties in question.

The elastic properties of hybrid composites can be calculated to within limits acceptable for most design purposes by the application of the principles of composite mechanics. For unidirectional composites, mixture-rule calculations are satisfactory, while, for complex laminates, classical laminate theory provides adequate elastic and thermal strain predictions. Elasticity calculations are independent of the nature of fiber distributions. In contrast, predictions of properties such as strength, toughness and failure strain, are more difficult because they depend on micromechanisms of damage accumulation which, in turn, are determined by the construction of the laminate and the scale of dispersion of the mixed fibers.

At the grossest level of hybridization, strips of GRP incorporated into a carbon-fiber-reinforced plastic (CFRP) laminate act as efficient crack arresters, provided the width of the strips is sufficient to dissipate the energy of a crack moving rapidly in the CFRP by localized debonding and splitting (Bunsell and Harris 1976). This concept is referred to as the introduction of "softening strips" (Sun and Luo 1985). The commonest type of hybrid, however, and that which is most easily achieved in practice, is made by laminating some balanced sequence of plies within each of which there is only a single species of fiber. In such laminates, the failure strain of the higher elongation (HE) component is usually reduced by the presence of the other while that of the lower elongation (LE) component is often increased. This behavior results from the state of elastic constraint in the fabricated laminate (the effect being greater the thinner the plies) and it has been explained by Aveston and Kelly (1980) in terms of energy. An approximate statistical analysis by Zweben (1977) shows that the failure process is affected by the statistical spread of failure strains of the two fiber species and predicts that the addition of HE fibers to a low failure strain composite raises the strain level needed to propagate fiber breaks because the HE fibers behave like crack arresters at a micromechanical level. The strength of such composites does not follow a mixture rule. Addition of small quantities of HE glass fiber to an LE CFRP will reduce the strength because there is insufficient glass to carry the load when the carbon breaks. Carbon failure leads to a single fracture failure of the composite. There is a critical level of mixing, however, at which there is sufficient glass to bear the load when the carbon breaks and beyond this level the hybrid strength rises again with further increases in glass content. Multiple fracture of the carbon then occurs and the fracture energy of the hybrid is substantially raised. The question of synergism is thus confused because any measurement of hybrid strength which appears to agree with a mixture-rule prediction is already showing a greater value than expected.

When the fibers are intimately mixed, strength calculations are more difficult because micromechanisms of failure change as the pattern of nearest neighbors around a given fiber alters, and strength is governed either by failure of single fibers or of small bundles of fibers (Parratt and Potter 1980).

In general, toughness or impact strength of brittle composites like CFRP can indeed be substantially improved by additions of glass or Kevlar 49 fiber but there is evidence to suggest that the fracture energy is satisfactorily predicted by the mixture rule, as energy arguments would suggest. The fatigue resistance of GRP is also improved by carbon-fiber additions, roughly in proportion to the amount of carbon added (Bunsell and Harris 1974, Harris and Bunsell 1975). Results relating to interlaminar shear strength (ILSS) are not clear cut, but it has been shown that the addition of high-modulus carbon fiber to a high strength carbon–epoxy composite does not significantly alter the ILSS (Arrington and Harris 1978).

Recent work on compression of hybrids containing various mixtures of fibers indicates no simple pattern that would permit prediction of compression strengths and moduli, although strengths of most of the hybrids studied fell reasonably close to mixture-rule values (Piggott and Harris 1981). This work emphasizes that changes in local failure modes resulting from hybridization are probably responsible for this uncertainty.

See also: Carbon-Fiber-Reinforced Plastics; Glass-Reinforced Plastics: Thermoplastic Resins; Glass-Reinforced Plastics: Thermosetting Resins

Bibliography

Arrington M, Harris B 1978 Some properties of mixed fibre CFRP. *Composites* 9: 149–52

Aveston J, Kelly A 1980 Tensile first cracking strain and strength of hybrid composites and laminates. In: Watt W, Harris B, Ham A (eds.) 1980 *New Fibres and Their Composites*. The Royal Society, London, pp. 111–26

Bunsell A R, Harris B 1974 Hybrid carbon and glass fibre composites. *Composites* 5: 157–64

Bunsell A R, Harris B 1976 Hybrid carbon/glass fibre composites. In: Scala E, Anderson E, Toth I, Noton B R (eds.) 1976 *Proc. 1st Int. Conf. Composite Materials*, Vol. 2. American Institute of Mining, Metallurgical and Petroleum Engineering, New York, pp. 174–90

Harris B, Bunsell A R 1975 Impact properties of glass fibre/carbon fibre hybrid composites. *Composites* 6: 197–201
Parratt N J, Potter K D 1980 Mechanical behaviour of intimately mixed hybrid composites. In: Bunsell A R, Bathias C, Martrenchar A, Menkes D, Verchery G (eds.) 1980 *Advances in Composite Materials,* Vol. 1. Pergamon, Oxford, pp. 313–26
Piggott M R, Harris B 1981 Compression strength of hybrid fibre-reinforced plastics. *J. Mater. Sci.* 16: 687–93
Sun C T, Luo J 1985 Failure loads for notched graphite epoxy laminates with a softening strip. *Compos. Sci. Technol.* 22: 121–34
Zweben C 1977 Tensile strength of hybrid composites. *J. Mater. Sci.* 12: 1325–37

B. Harris

Hydride Metallurgy

Hydride metallurgy involves the synthesis of a volatile hydride of a metal, or in some cases a nonmetal, purification of the hydride and subsequent decomposition into the pure element and hydrogen. The method is suitable for the production of elemental boron, silicon, germanium, tin, phosphorus, arsenic, antimony, sulfur, selenium, tellurium and iodine. The advantages of the method are that very pure elements can be obtained and decomposition of the hydride can be controlled to form monocrystalline and polycrystalline films, bulk polycrystals and whiskers. Thus, hydride metallurgy has found some application in the production of elemental semiconductors, especially silicon.

1. General Background

Reactive elements with high melting and boiling points are difficult to purify because of contamination from container materials at high temperatures. High-purity materials are best obtained by initial purification of a chemical compound of the element which has a low melting point or high volatility. Purification of the compound is then carried out by chemical methods or fractional distillation at relatively low temperatures. The element may then be obtained from the pure compound by thermal decomposition, photolysis or hydrogen reduction. Compounds may be chlorides, iodides, hydrides and, more recently, volatile organometallics. Preference would be given to compounds from which the element may be obtained at relatively low temperatures. Simple compounds are preferred: organometallics may give a product contaminated by carbon. Hydrides show covalent bonding and have much lower melting and boiling points than halides or organometallic compounds.

2. Hydride Properties

The properties of hydrides which are important in hydride metallurgy are volatility, thermodynamic stability, interaction with water and oxygen, and solvent action for other substances. Table 1 gives some properties of important volatile inorganic hydrides. All the hydrides in Table 1, except hydrogen sulfide, are thermodynamically unstable and decompose at relatively low temperatures into the element and hydrogen. The least stable are hydrides of such heavy elements as tin, antimony and tellurium. The most stable are those of light elements such as silicon, phosphorus and sulfur. The most unstable hydrides, stibine, stannane, hydrogen telluride and germane, may decompose explosively, for example, when ignited by a spark. The hydrides of sulfur, selenium, tellurium and iodine can be decomposed by ultraviolet light.

Table 1
Properties of volatile inorganic hydrides

Name	Formula	Enthalpy of formation $\Delta H^\circ_{f(298.15\,K)}$ (kJ mol^{-1})	Melting point (K)	Boiling point (K)
Diborane	B_2H_6	38.4	108.3	180.6
Silane	SiH_4	34.7	88.48	161.2
Germane	GeH_4	90.8	107.3	184.7
Stannane	SnH_4	162.7	123	221.0
Phosphine	PH_3	5.4	139.4	185.7
Arsine	AsH_3	66.4	156.2	210.7
Stibine	SbH_3	145.1	178.9	254.8
Hydrogen sulfide	H_2S	−20.6	187.6	212.8
Hydrogen selenide	H_2Se	29.70	207.4	231.8
Hydrogen telluride	H_2Te	99.7	222.9	270.9
Hydrogen iodide	HI	26.3	222.5	237.9

The hydrides, especially silane, react readily with oxygen in air and thus present fire hazards. Higher hydrides, existing as impurities in diborane, silane and phosphine, cause ignition. Hydrides are soluble in water and some also hydrolyze. All the volatile hydrides are extremely toxic.

3. Synthesis and Purification of Hydrides

There are many methods for the preparation of hydrides. Those of sulfur, selenium and iodine may be obtained by direct synthesis from the elements at a relatively high temperature; for example:

$$Se + H_2 \xrightarrow{550\,°C} H_2Se \tag{1}$$

Phosphine, arsine and hydrogen telluride are obtained by acid decomposition of the phosphide, arsenide and telluride of magnesium; for example:

$$Mg_3P_2 + 6HCl \rightarrow 2PH_3 + 3MgCl_2 \tag{2}$$

Silane and germane may be obtained by interaction of magnesium silicide and magnesium germanide with ammonium chloride or bromide in liquid, anhydrous ammonia or hydrazine.

There is also a general method based on the reduction of chlorides and some other compounds of hydride-forming elements by mixed, salt-forming hydrides. An example is:

$$GeCl_4 + 4NaBH_4 + 12H_2O \rightarrow GeH_4 + 4H_3BO_3 + 4NaCl + 12H_2 \quad (3)$$

Electrochemical methods are known for the majority of the hydrides, but have found practical application only for hydrogen telluride. There are many other specific methods for synthesis of the volatile hydrides.

Depending on the method of synthesis, hydrides contain many impurities, in concentrations from 10^{-5} to 10^{-1} mol%. Impurities, which may be present in solution or as finely dispersed solids, include hydrocarbons and chloride derivatives, other hydrides, permanent gases, carbon dioxide, water and nonvolatile metallic compounds. Purificiation methods include chemical methods, distillation, crystallization and adsorption.

Chemical methods usually convert impurities into nonvolatile compounds, followed by distillation of the hydride of interest. Thus, phosphine and diborane may be removed from silane by passing the gas through a solution of metallic sodium in ammonia, as a result of which phosphine is converted to sodium phosphide and diborane to compounds of diborane with ammonia.

Distillation, using a rectifying column, is the most common purification method and has been used successfully both on a laboratory and on an industrial scale for purification of most of the volatile hydrides.

Fractional crystallization from the liquid is also a possible purification method. Adsorption, using activated charcoal or zeolites, can reduce the water content of germane, arsine and phosphine by two or three orders of magnitude.

Each of the purification methods is efficient with respect to one, often rather limited, group of impurities. To obtain the purity (10^{-8}–10^{-10} mol%) required in hydrides used for the preparation of high-purity metals and semiconductors, a multistage purification process is necessary.

4. *Decomposition of Volatile Hydrides*

Most methods of recovering the element from the purified hydride are based on the thermal decomposition reaction, such as

$$GeH_4 \rightarrow Ge + 2H_2 \quad (4)$$

The particular equipment and technique used depend on the thermal stability of the hydride and the melting point of the element. If the hydride-forming element has a low melting point, a tubular reactor, usually of silica, can be used. The length of the reactor and the process temperature are selected to provide complete decomposition of the hydride, the element being deposited on the inner wall at the cooler end. Decomposition temperatures of 300–500 °C are used for stibine, stannane and hydrogen telluride and temperatures of 700–900 °C for arsine, phosphine and hydrogen selenide. With high-melting elements, such as silicon and boron, decomposition takes place on a heated seed rod of the pure substance. The polycrystalline material obtained may be converted into a single-crystal specimen by zone refining or directional solidification. Floating-zone refining may be used to prevent contamination from crucible materials.

5. *Applications of Hydride Metallurgy*

All the elements which form volatile inorganic hydrides have been obtained in high-purity form by the hydride method, purer than can be obtained by other methods. As an example, germanium hydride, obtained by the reduction of germanium tetrachloride by sodium tetrahydroborate in an aqueous medium, was purified by impurity condensation, filtration and rectification in succession. The only impurity detected in the hydride was water, at a level of 5×10^{-4} mol%. After decomposition of the hydride at 660 °C in a silica reactor, the germanium was fused into a compact, polycrystalline ingot. The content of each of 30 impurity elements was $<3 \times 10^{-6}$ at.% and of each of 10 others, $<10^{-5}$ at.%, much lower than the levels obtainable by hydrogen reduction of germanium oxide.

In industry, the hydride method has found widest application in the preparation of high-purity silicon. By the decomposition of silane purified by rectification, compact, polycrystalline silicon is obtained. After single-pass floating-zone refining, single-crystal silicon with an electrical resistivity of $>10^4\ \Omega$ cm is obtained.

See also: Germanium Production; High-Purity Metals; Silicon Production; Metals Production: An Overview

Bibliography

Devyatykh G G 1966 Preparation of high purity elements by hydride method (in Russian). *Poluchenie i analiz veshchestv osoboi chistoty*. Nauka, Moskva

Devyatykh G G, Zorin A D 1974 *Volatile Inorganic Hydrides of High Purity* (in Russian). Nauka, Moskva

Hartford A J, Huber E J 1980 Laser purification of silane: Impurity reduction to the sub-part-per million level. *J. Appl. Phys.* 51: 4471–74

Stock A 1957 *Hydrides of Boron and Silicon.* Cornell University Press, Ithaca, New York

Stone F G A 1962 *Hydrogen Compounds of the Group IV Elements.* Prentice-Hall, London

G. G. Devyatykh

Hydroform Process

In the hydroform process, shallow or deep parts are formed from relatively thick blanks using fluid pressure (Fig. 1). The blank is placed between the male punch and a rubber diaphragm, above which is an oil-filled cavity. As the punch is moved upward the pressure builds up in the cavity, pressing the part being formed against the punch walls. This pressure is controlled throughout the forming cycle, reaching a maximum of about 100 MPa.

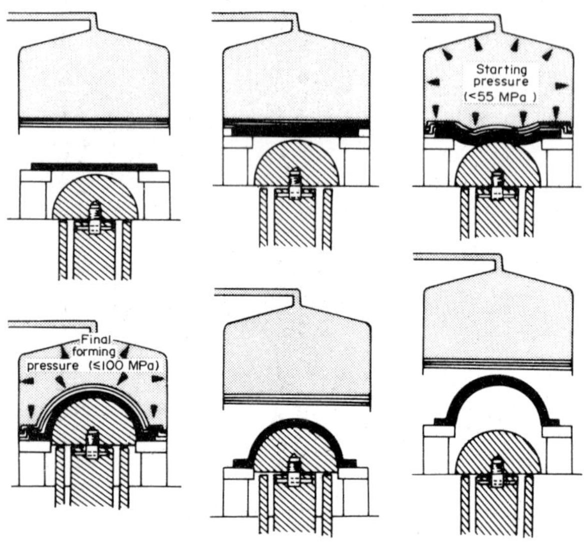

Figure 1
The hydroform process

Deeper draws are obtained in the hydroform process than in conventional deep drawing. This is due to the fact that the pressure on the exterior of the part being formed presses the cup wall against the punch, thus increasing the friction at the workpiece–punch interface and allowing the part to travel up with the punch. This decreases the longitudinal tensile stresses in the cup wall and eliminates or delays fracture of the part. The hydroform process is particularly suitable for deep drawing of critical parts.

See also: Deep Drawing; Sheet Metal Forming; Metals Processing and Fabrication: An Overview

Bibliography

Metals Handbook, 8th edn., Vol. 4, 1969. Rubber-diaphragm forming. American Society for Metals, Metals Park, Ohio, pp. 214–15

Tool and Manufacturing Engineers' Handbook. 4th edn., Vol. 2, 1984. Forming. Society of Manufacturing Engineers, Dearborn, Michigan, Chap. 5, pp. 80–82

S. Kalpakjian

Hydrogen Bonding in Paper: Theory

To understand the significance of hydrogen bonding in paper and paperboard, the nature of pulp fibers, the importance of the fact that paper is made from a water suspension and the characteristics of hydrogen bonds must be reviewed briefly.

Paper typically has many constituents: pulp fibers, clay, titanium dioxide and other minerals, starch, size, alum and other chemicals depending on the functions and properties demanded in making and using it. However, attention is here concentrated on the chief constituent, the fiber.

The paper fiber has many organic and inorganic constituents: cellulose, hemicellulose, lignin, extractives and ashforming substances. The cellulosic and hemicellulosic components are the primary contributors to the integrity and mechanical properties of paper and paperboard. Thus, for present purposes paper is considered as bleached, lignin-free fibers, composed of cellulose and hemicellulose molecules.

1. Significance of Hydrogen Bonding in Paper

An important characteristic of paper is its interaction with water. Paper is made by laying down a web of cellulose fibers from a water slurry, removing water mechanically by suction and pressing and then drying. During drying, air enters pores of decreasing diameter, forcing the menisci to travel towards the intersections of fibers. Surface tension forces pulp fibers together until their outermost hydroxyl groups are less than 0.3 nm apart. At this point, the hydrogen atom of one hydroxyl group can join in a weak binding force with the oxygen of another such group or the ether oxygen of the chain. This is the hydrogen bond that gives paper its integrity and most of its mechanical properties.

How much of the overall properties of paper may be attributed to hydrogen bonding? Figure 1 supplies one answer. It shows the load–elongation curves for paper samples immersed in liquids of different H-bond breaking capabilities. Curve (g), for a sample immersed in water, shows a maximum force <5% of the corresponding force for air-dry paper.

Another answer is given in Fig. 2, which is a composite plot of three pairs of variables:

(a) w versus $(E_w/E_0)^3$, where E_w is the modulus of elasticity of paper at a water–cellulose mass ratio w and E_0 is the modulus at $w = 0$;

(b) DS versus $(E_{DS}/E_0)^3$, where E_{DS} is the modulus of paper made from acetylated pulp with degree of substitution DS and E_0 is the modulus at $DS = 0$; and

(c) DS versus RE_{DS}/RE_0, where RE is the rupture energy of paper made from acetylated pulp and DS has the same significance as in (b).

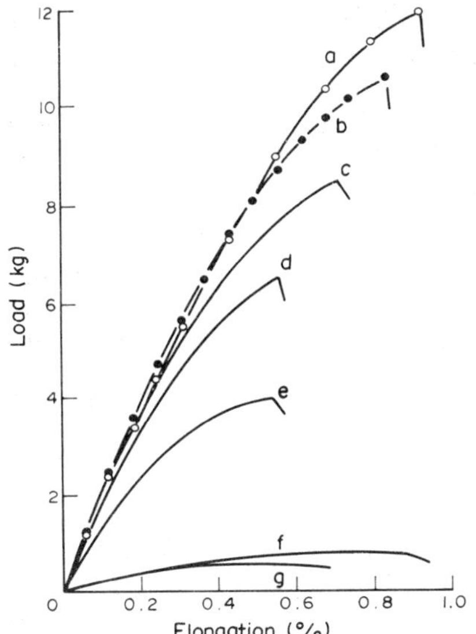

Figure 1
Load–elongation curves for paper samples immersed in different fluids: (a) control in air; (b) benzene; (c) vinyl acetate; (d) nitromethane; (e) ethanol; (f) formamide; (g) water (after Robertson 1964)

The plots indicate that H-bonds are broken by water as w increases or are eliminated by acetylation as *DS* increases. It is seen that both modulus and rupture energy are reduced to nearly zero when enough H-bonds are broken by water or eliminated by acetylation.

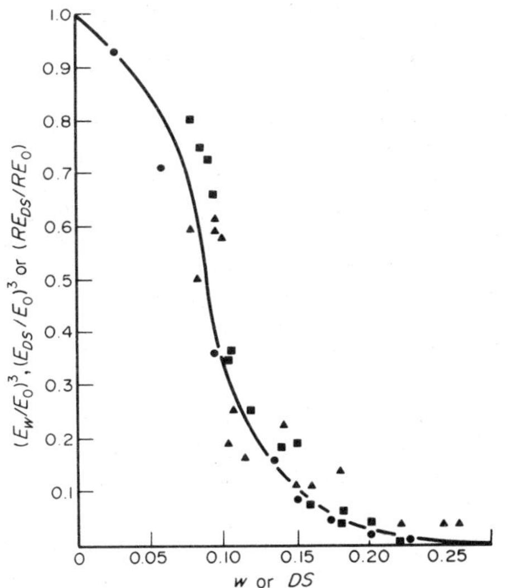

Figure 2
Dependence of modulus E on water–cellulose ratio w and on degree of substitution of hydroxyl groups by acetylation DS, and of rupture energy RE on DS: ● E_w vs w; ▲ E_{DS} vs DS; ■ RE_{DS} vs DS (after Nissan 1977)

The two graphs provide conclusive evidence that the hydroxyl group interactions—the H-bond—account for most of the mechanical properties of paper and paperboard. Thus, paper is a member of the class of solids which may be called "hydrogen bond dominated solids," defined as solids in which, to a first approximation, the contribution of the H-bond to the mechanical properties may be taken as the sole contribution; that is, the contributions of covalent, ionic, van der Waals or other types of bonds as well as the entropic contributions can be ignored without serious error.

In this article, paper is therefore considered as a network of H-bonds joining cellulose and hemicellulose chains to each other, ignoring other materials, components and bonds. For simplicity, paper is assumed, for a sufficient unit volume, to be isotropic, that is, with H-bonds distributed uniformly in all directions.

2. Nature of the Hydrogen Bond

The Hydrogen bond was conceptualized by Latimer and Rodebush (1920) as follows:

> Water occupies an intermediate position and shows tendencies both to add and give up hydrogen, which are nearly balanced. Then, in terms of the Lewis Theory, a free pair of electrons on one water molecule might be able to exert sufficient force on a hydrogen held by a pair of electrons on another water molecule to bind the two molecules together. Structurally, this may be represented as
>
> $$\mathrm{H{:}\ddot{\underset{\cdot\cdot}{O}}{:}H{:}\underset{H}{\overset{H}{\ddot{\underset{\cdot\cdot}{O}}}}{:}}$$
>
> Such combinations need not be limited to the formation of double or triple molecules. Indeed, the liquid may be made up of large aggregates of molecules, continually breaking up and reforming under the influence of thermal agitation.

Such an explanation amounts to saying that the hydrogen nucleus held between two octets constitutes a weak "bond." Later they conclude:

> If our picture of the association of water is correct, a hydrogen nucleus may be held between two oxygen octets by forces which, for quite a distance, obey Hooke's law.

Since then, many studies have confirmed Latimer and Rodebush's original idea. For the purpose of the present discussion, the hydrogen bond is defined as follows.

(a) A hydrogen bond is formed between A–H and B as a new combination A–H---B, where A = F, N or O and B = O, NH or OH

(b) The strength of the bond A–H---B depends on other atoms attached to A.

(c) Ideally, bond A–H---B is linear.

(d) In general, the distance A–H is shorter than the distance H---B; rarely, H is midway as in –F–H–F–.

For the case of hydrogen bonds formed among cellulosic and hemicellulosic molecules, that is, bonds of the nature O–H---O, there can be a range of bond properties, depending on what other atoms are attached to the O atoms, on the distance between the two oxygen atoms and on the angle formed by the bond. Pimentel and McClellan (1960) illustrate the variations in the various parameters characterizing O–H---O bonds, as reproduced in Figs. 3–5. As Fig. 5 shows, these parameters are closely related to each other. For instance, ΔH is the same function of R for all O–H---O bonds (Lippincott et al. 1959).

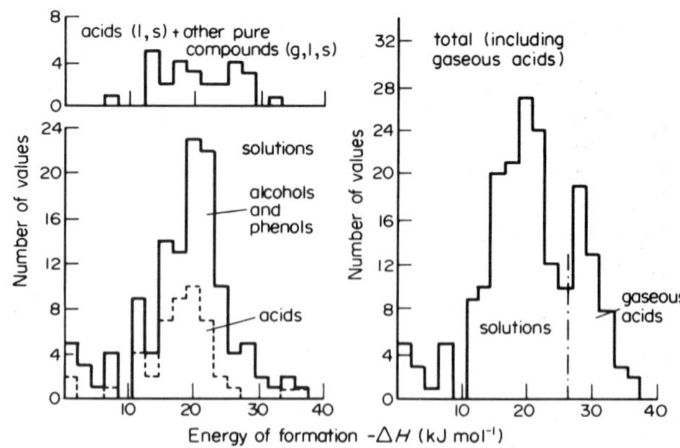

Figure 4
Distribution of energy of formation for the OH---O bond (after Pimentel and McClellan 1960)

3. The Hydrogen Bond in Paper

Infrared spectroscopic studies show that no free hydroxyls exist in either dry paper or paper wetted to different degrees up to saturation. Studies on deuteration of cellulose by liquid or vapor D_2O and rehydrogenation with H_2O show the existence of several types of H-bonds in cellulosic networks. Thus, it is not possible to speak of "the hydrogen bond" between cellulose molecules, having a precise set of values for its parameters. Instead, there are several types of hydrogen bond with a statistical mean and a variance for each parameter. As discussed below, the hydrogen bond network in paper is likely to have the following parameter ranges:

$\Delta H = -19$ to -21 MJ kg^{-1} mol^{-1}

$= -3.13 \times 10^{-20}$ to -3.48×10^{-20} J per bond

O–H distance $r_1 = 0.099$–0.101 nm

O---O distance $R = 0.26$–0.29 nm

O–H distance in unbonded, or "free," state = 0.097 nm

The cellulose network is composed of cellulosic molecules joined by H-bonds. Cellulose itself may be

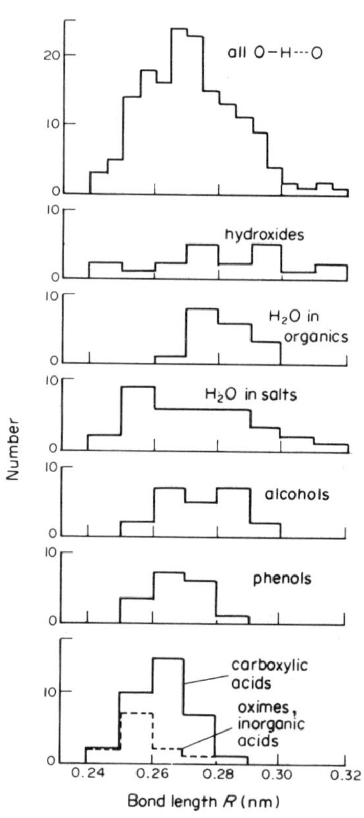

Figure 3
Distribution of O---O bond lengths (after Pimentel and McClellan 1960)

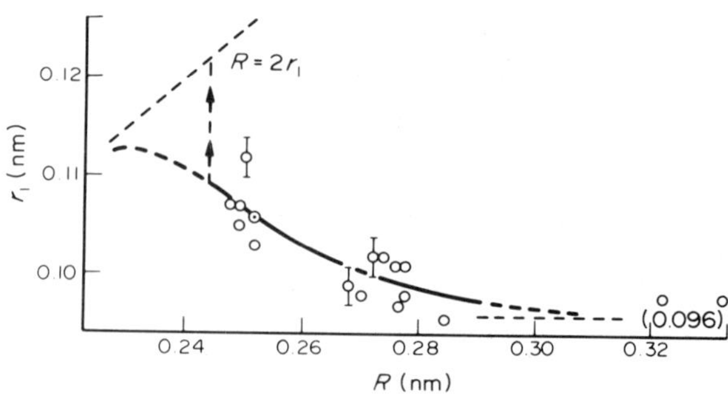

Figure 5
O–H bond length r_1 as a function of O---O distance R in O–H---O bonds (after Pimentel and McClellan 1960)

considered as a polyanhydrocellobiose, as depicted in Fig. 6. This shows that some of the OH groups form intramolecular bonds with an O atom on the molecule. Such intramolecular bonds increase chain stiffness but do not contribute to intermolecular bonds in formation of networks. This is further illustrated in Figs. 7 and 8, which depict the crystalline structure of cellulose microcrystals.

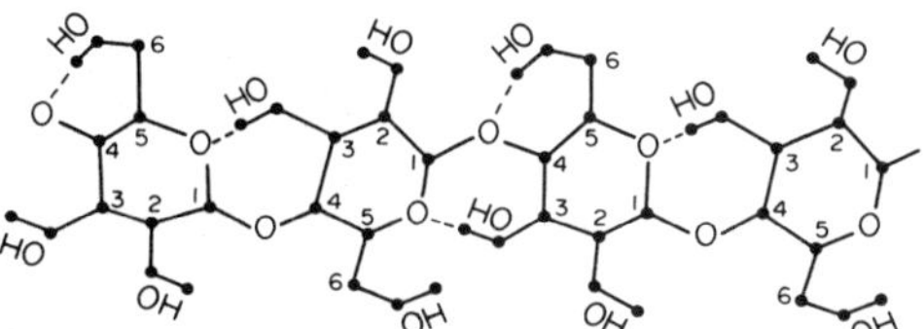

Figure 6
Cellulose molecule, showing possibilities for intrachain H-bonds (after Hermans 1940)

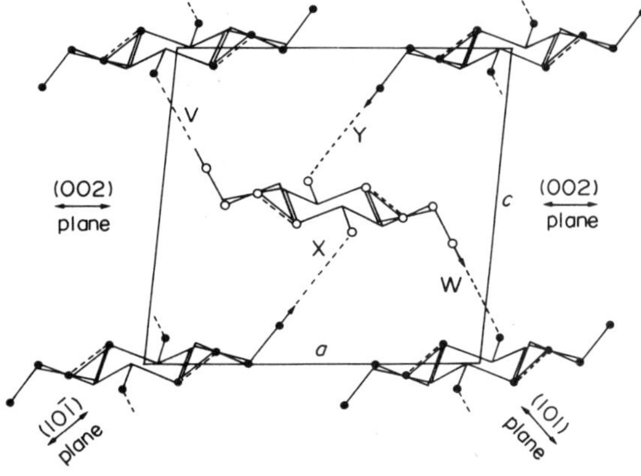

Figure 8
End view of cellulose chains in a unit cell: V, W, H-bonds in (101) plane; X, Y, H-bonds in (10$\bar{1}$) plane (after Liang and Marchessault 1959)

Figures 7 and 8 reveal that of the six OH groups on an anhydrocellobiose, only two are engaged in intermolecular bonding. In more disordered regions, there may be three OH groups participating. This important fact was neglected in an earlier theory of mechanical properties of paper based on the hydrogen bond (Nissan 1957). The necessary corrections are introduced in this article by taking the ratio of intermolecular bonds to total H-bonds, β, to be between $\frac{1}{2}$ and $\frac{1}{3}$.

Because paper is essentially microfibrils and elementary fibrils joined randomly by H-bonds, the network is visualized as depicted in Fig. 9. Here, it is shown that the O–H---O bond itself has a length $R = r_1 + r_2$. But the distance between one O–H---O intermolecular bond and the next one is, on average, $l = r_1 + r_2 + r_3$, where r_3 is the length of the "infinitely rigid" segment of the molecule (i.e., compared with the easily deformable lengths r_2 and hence with R). The reciprocal of l is the number of intermolecular H-bonds per unit length effectively taking part when the material is deformed, $N_E^{1/3}$, where N_E is the number of effective H-bonds per unit volume of material. As shown below N_E is only a fraction of the total number per unit volume in the network.

Consider a randomly distributed hydrogen-bonded system subjected to an infinitesimal normal stress (σ_x) on one rectangular face of the macroscopic body, taken to be a 1 cm cube for convenience. At infinitesimal strains, the hydrogen bonds are Hookean in behavior. If the sheet is isotropic, the hydrogen bond distribution is a spherical surface of uniform bond distribution as indicated in Fig. 10.

Using this isotropic model, the number of oriented bonds can be calculated. The number of bonds (dN')

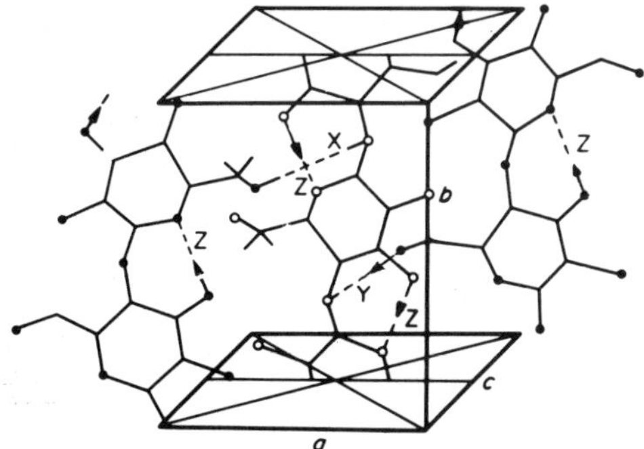

Figure 7
Side view of chain segments in a cellulose-1 crystal: X, Y, H-bonds in (10$\bar{1}$) plane; Z, intrachain H-bonds (after Liang and Marchessault 1959)

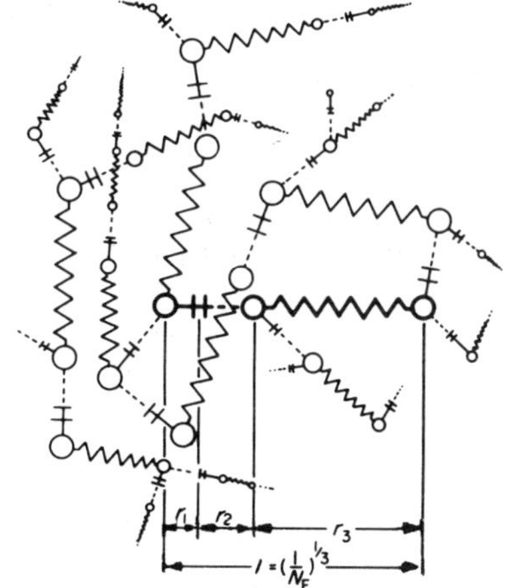

Figure 9
Typical hydrogen bond in network or randomly oriented bonds of an isotropic H-bond dominated solid (after Nissan 1977)

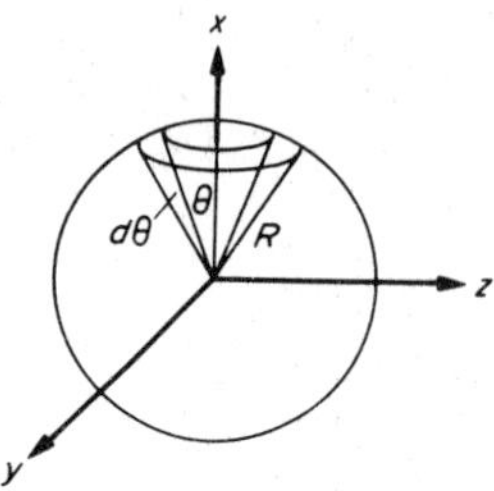

Figure 10
Directional distribution of hydrogen bond orientation in an isotropic hydrogen-bonded solid at an angle θ to the x axis (after Nissan 1977)

at the angle θ contained within a circular hoop on the surface of a sphere of radius R, with the x axis being the central axis normal to the plane of the hoop, is given by

$$\frac{dN'}{N'} = \frac{(2\pi R \sin\theta)R d\theta}{4\pi R^2} = \frac{\sin\theta\, d\theta}{2} \tag{1}$$

where N' is the total number of effective hydrogen bonds in the system. The term (dN'/N') is a probability function with respect to θ of finding bonds at that angle, and it integrates to unity for the limits 0 and N':

$$\frac{1}{N'}\int_0^{N'} dN' = \frac{N'}{N'} = 1 = 2\int_0^{\pi/2} \frac{\sin\theta\, d\theta}{2} = 1 \tag{2}$$

The stress on the bond is the projection of σ_x on the bond line of action, $\sigma_x \cos\theta$. (The bond is assumed irrotational.) The energy stored in the bond is

$$U = \frac{1}{2}\frac{(\sigma_x \cos\theta)^2}{k'} \tag{3}$$

where k' is the Hookean force constant.

The total energy stored in all the bonds in the circular hoop

$$U_\theta = U dN' \tag{4}$$

$$= \frac{N'}{2}\frac{(\sigma_x \cos\theta)^2}{k'}\frac{\sin\theta\, d\theta}{2} \tag{5}$$

The total energy stored in all the bonds is obtained by integrating from 0 to $\pi/2$ and multiplying by 2:

$$U' = \frac{N'}{2}\frac{(\sigma_x)^2}{k'}\int_0^{\pi/2} \cos^2\theta \sin\theta\, d\theta \tag{6}$$

$$U' = \frac{N'}{2}\frac{(\sigma_x)^2}{k'}\frac{1}{3} \tag{7}$$

Now consider a system with N_E bonds all oriented along the x axis. If such a system is subjected to stress σ_x, the energy stored in the system is given by

$$U_E = \frac{N_E(\sigma_x)^2}{2k'} \tag{8}$$

If the models are equivalent,

$$U' = U_E \tag{9}$$

$$N_E = N'/3 \tag{10}$$

An isotropic, randomly oriented network of hydrogen bonds is thus mathematically equivalent to one with a third of the bonds oriented along each of three orthogonal, principal axes. This holds true only for infinitesimal strains, when the energy functions are quadratic with respect to strain. This type of network is shown in Fig. 11, where the suffix E is dropped from N_E. N is then $\frac{1}{3}$ the total number of bonds randomly oriented, but in the model all the N bonds are parallel and in the direction of the stress vector. There are $N^{2/3}$ parallel chains of bonds, each chain consisting of $N^{1/3}$ bonds in series and oriented in the direction of the unidirectional stress imposed on the 1 cm cube.

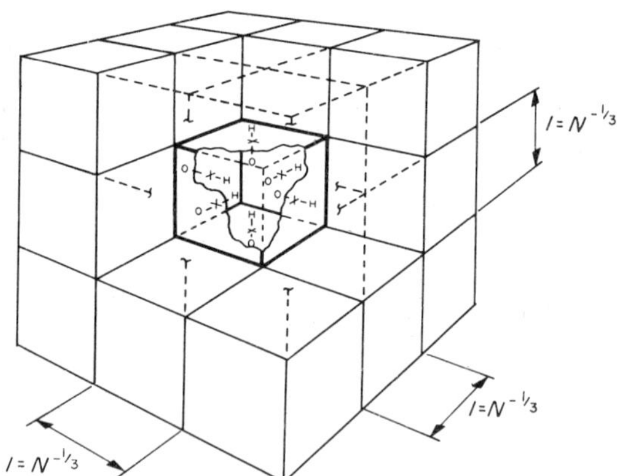

Figure 11
Idealized hydrogen bond orientation, with each unit cube held to adjacent cubes by one H-bond through each face (after Nissan 1977)

If this cube is subjected to a stress σ in the direction of the bonds, the cube side will stretch from 1 to $(1 + \Delta l)$ cm, where $\Delta l/1$ is the strain $\varepsilon = \sigma/E$, E being Young's modulus. The stretch Δl is due to $N^{1/3}$ bonds, each of original length $r_1 + r_2 + r_3 = R + r_3$ cm, stretching by ΔR (Fig. 9). Thus

$$\Delta l = N^{1/3}\Delta R \tag{11}$$

Each bond in the chain is under a force f due to this stretch of ΔR. This force equals $k\Delta R$, where k is the "force constant" of the H-bond, that is the force required to stretch the bond unit length. But there are $N^{2/3}$ such chains in parallel. Hence, the total force

F on the face of the cube and the stress σ are given by

$$\sigma = F/1 = N^{2/3}k\Delta R \quad (12)$$

$$E = \sigma/\varepsilon = N^{2/3}k\Delta R/N^{1/3}(\Delta R/1) = N^{2/3}k\Delta R/N^{1/3}\Delta R = kN^{1/3} \quad (13)$$

Thus, the Young's modulus of paper, paperboard and other H-bond dominated solids consists of the product of the force constant of the H-bond, k, multiplied by $N^{1/3}$, the cube root of the effective number of bonds per unit volume, or of the cube root of one-third of the total randomly oriented intermolecular H-bonds per unit volume of the isotropic material.

The force constant of the hydrogen bond may be determined in a variety of ways. It can be calculated from infrared or Raman absorption spectra or determined, for simple systems, from ab initio molecular orbital calculations. When the present theory was first published (Nissan 1957), a particular potential function for the H-bond was assumed which gave $k = 8\ \mathrm{N\,m^{-1}}$ for the H-bond in cellulose. Later (Nissan 1977), on the basis of other speculative calculations, k was found to be of the order of $10\ \mathrm{N\,m^{-1}}$. In the meantime, other studies have suggested that k is still higher. Thus, Eisenberg and Kauzmann (1969) report values for k ranging from 17–19 $\mathrm{N\,m^{-1}}$ (at −173 °C) to 15.5 $\mathrm{N\,m^{-1}}$ (extrapolated to 0 °C). Nissan, Byrd, Batten and Ogden have collected spectroscopic and molecular orbital calculations of k for H-bonds of the OH---O type. They found k to be in the range of 15.3–19.9 $\mathrm{Nm^{-1}}$ for bonds of energies U in the range of −19 to −21 MJ $\mathrm{kg^{-1}\,mol^{-1}}$ with $k = 18.4\ \mathrm{N\,m^{-1}}$ being the probable value for the average cellulosic H-bond. Thus,

$$E = (18.4 \pm 3)\,N^{1/3}\ \mathrm{Pa} \quad (14)$$

In fact, the bond for cellulose does appear to be the same as that joining water molecules, as a first approximation. The reported values for the energy of formation, the distances involved and for absorption spectra are very nearly the same for both systems (Lippincott et al. 1959). Additional evidence may be adduced from the following.

Luner et al. (1961) measured the apparent density and Young's modulus for a large number of papers. Their correlation was

$$E = 7.812\rho^2 + 0.211\ \mathrm{GPa} \quad (15)$$

where ρ = apparent density ($\mathrm{g\,cm^{-3}}$). Using $\rho = 1.6$, the density of cellulose crystallites, gives $E = 20.2$ GPa as the modulus of a "paper" mosaic of crystallites without voids. This is for papers conditioned at 50% relative humidity where they absorb about 8% water. The modulus for dry paper with $w = 0$ is given (Nissan 1977) by

$$\ln E_0 = \ln E_w + 6.7\,w - 5.71\,w_c \quad (16)$$

where $w_c = 0.045$ and $w = 0.08$. Thus, E_0 is 26.73 GPa.

A crystallite of cellulose with six bonds per anhydrocellobiose segment, of which a fraction β is engaged in intermolecular bonding, will have a total number of $6(6.02 \times 10^{23})\,\rho\beta/324 = 1.11 \times 10^{22}\,\rho\beta$ intermolecular bonds per $\mathrm{cm^3}$, where 6.02×10^{23} is Avogardro's number and 324 is the molecular weight of anhydrocellobiose. For $\rho = 1.6$, the total number of such bonds is $1.78 \times 10^{22}\,\beta$. The effective number is $\frac{1}{3}$ of this, $5.95 \times 10^{21}\,\beta$, or $5.95 \times 10^{27}\,\beta$ per $\mathrm{m^3}$. Upon substitution of this latter value in $E = kN^{1/3}$, $k = 26.73 \times 10^9/(5.95 \times 10^{27}\,\beta)^{1/3}$.

Thus, if $\beta = 1.00$, $k = 14.8\ \mathrm{N\,m^{-1}}$
$\beta = 0.67$, $k = 16.9\ \mathrm{N\,m^{-1}}$
$\beta = 0.50$, $k = 18.6\ \mathrm{N\,m^{-1}}$
$\beta = 0.33$, $k = 21.3\ \mathrm{N\,m^{-1}}$

In the original formulation of this theory (Nissan 1957, 1977), the number of bonds was not divided by 3 and it was assumed that $\beta = 1$. More recent and detailed studies (Liang and Marchessault 1959), have shown β to be 0.33 for cellulose crystals. For disordered cellulose, β may be higher. Hence, k is most probably between 17 and 21 $\mathrm{N\,m^{-1}}$, in close agreement with results from more fundamental calculations using quantum mechanical and spectroscopic methods. As yet unpublished personal work, relating the area of hysteresis loops to the reduction in the values of E from those before to those after the loops, also suggests that k is of this magnitude.

Thus, there is evidence that Young's modulus of paper can be fully accounted for by the hydrogen bonds forming a random network.

See also: Paper and Paperboard: Effects of Moisture and Temperature; Paper and Paperboard Substrates: Composition; Paper and Paperboard: An Overview

Bibliography

Eisenberg D S, Kauzmann W 1969 *The Structure and Properties of Water.* Oxford University Press, Oxford

Hermans P H 1940, *Physics and Chemistry of Cellulose Fibers.* Elsevier, New York

Latimer W M, Rodebush W H 1920 Polarity and ionization from the standpoint of the Lewis theory of valence. *J. Am. Chem. Soc.* 42: 1419–33

Liang C Y, Marchessault R H 1959 Infrared spectra of crystalline polysaccharides Part II, Native celluloses in the region from 640–1700 $\mathrm{cm^{-1}}$. *J. Polym. Sci. Sci.* 39: 269–78

Lippincott E R, Finch J N, Schroeder R 1959 Potential function model of hydrogen bond systems. In: Hadži D, Thompson H W (eds.) 1959 *Hydrogen Bonding.* Pergamon, Oxford, pp. 361–74

Luner P, Karna W E U, Donofrio C P 1961 Studies in interfiber bonding of paper, the use of optically bonded areas with high yield pulps. *TAPPI* 44: 409–14

Marrinan H J 1959 The fine structure of cellulose. In: Honeyman J (ed.) 1959 *Recent Advances in the*

Chemistry of Cellulose and Starch. Interscience, New York, pp. 147–87
Nissan A H 1957 Rheological behavior of hydrogen-bonded solids. I. Primary considerations. *Trans. Faraday Soc.* 53: 700–09
Nissan A H 1977 *Lectures on Fiber Science in Paper*. Joint Textbook Committee of the Paper Industry. Atlanta, Georgia
Nissan A H, Higgins H C 1959 Molecular Approach to the Problem of Viscoelasticity. *Nature* 184: Suppl. No. 19, 1477–78
Pimentel G C, McClellan A L 1960 *The Hydrogen Bond*. Freeman, San Francisco, p. 219, 259, 282
Robertson A A 1964 Cellulose-Liquid Interactions, *Pulp Pap. Mag. Can.* 65: T171–8

A. H. Nissan

Hydrogen Embrittlement

Hydrogen embrittlement is a generic term used to describe a wide variety of fracture phenomena having a common relationship to the presence of hydrogen in the alloy as a solute element or in the atmosphere around the material as a gas (H_2) or a constituent of a gas (e.g., H_2S). Although a single mechanism cannot account for all of the failures within this class, they generally have a common phenomenology. These failures exhibit significantly decreased macroscopic ductility (strain to failure), although in many cases the ductility remains sufficiently large that the alloy would not usually be considered brittle. In many other cases the alloy is converted from a ductile to a truly brittle material by the presence of hydrogen. The decrease in ductility is enhanced by slow strain rates and is most significant in a limited temperature range. Hydrogen effects are often minimal at low temperatures, where the hydrogen diffusivity is low, and at high temperatures, where the strength of the alloy is reduced. Exceptions to this are known; for example, at low temperatures nickel-base alloys are embrittled owing to segregation of hydrogen at grain boundaries, and at high temperatures steels are embrittled by hydrogen attack of carbides to form high-pressure methane bubbles. The effect of hydrogen usually increases as its concentration in solid solution or its fugacity in the atmosphere increases, and it is generally believed that a critical concentration or fugacity must be exceeded to cause embrittlement. This critical concentration decreases as the strength of the alloy or stress intensity at the crack tip increases. Crack propagation due to hydrogen occurs at stress intensities below the critical stress intensity at which catastrophic failure occurs, giving rise to the phenomenon of static fatigue. These kinetics are rather complicated and depend on the metal system and the source of hydrogen. Embrittlement by solute hydrogen is controlled by the kinetics of hydrogen transport by diffusion or by dislocation-enhanced diffusion. Where hydrogen enters from a gaseous or aqueous environment the additional processes of absorption must be considered as possible rate-controlling steps.

Fracture may proceed by an intergranular or transgranular mode (or in some systems by a combination of the two), and the fracture surfaces often appear to have resulted from a rather brittle fracture process. However, there is often significant plasticity in the vicinity of the crack tip during the fracture. The energies of fracture clearly indicate that plastic work accompanies the fracture. In a number of systems, particularly when the fracture proceeds in a more rapid fashion, the fracture surface appears rather ductile (microvoid coalescence) and the effect of hydrogen is to reduce the ductility rather than change the fracture mode. The presence of other solute elements often plays an important role in hydrogen embrittlement; for example, segregation of elements such as sulfur, arsenic and antimony at boundaries in the nickel alloys and steels can alter the fracture mode and the stress intensity at which hydrogen embrittlement is observed.

There does not appear to be a single mechanism of hydrogen embrittlement. Even in one system the mechanism may change depending on the source of hydrogen and the nature of the stress applied. Alloys of metals such as niobium, zirconium or titanium, which can form stable hydrides, appear to fracture by a stress-induced hydride-formation and cleavage mechanism. It has been suggested that non-hydride-forming systems such as iron and nickel alloys (which do not form hydrides under the conditions in which they are embrittled) fail because hydrogen decreases the atomic bonding (decohesion), although this mechanism is in some dispute. In many of these systems the fracture seems to be associated with hydrogen-induced plasticity in the vicinity of the crack tip. Other mechanisms, such as adsorption-decreased surface energy and high-pressure H_2 gas-bubble formation, have also been suggested and may play a role in specific systems. Not only is the detailed mechanism of fracture in some systems in dispute, but in some cases the fact of hydrogen embrittlement is itself disputed. Thus the role of hydrogen in stress-corrosion cracking (see *Stress-Corrosion Cracking*) is not established and is a source of great controversy.

See also: Hydrogen Embrittlement of Steels; Corrosion: Metallurgical Aspects; Corrosion of Metals: An Overview

Bibliography

Birnbaum H K 1979 Hydrogen related failure mechanisms in metals. In: Foroulis Z A (ed.) 1979 *Environment-Sensitive Fracture of Engineering Materials*. The Metallurgical Society (AIME), Warrendale, Pennsylvania, pp. 326–60

H. K. Birnbaum

Hydrogen Embrittlement of Steels

Although there are now known to be many alloy systems, the stress (or strain) behavior of which can be altered by the presence of hydrogen, observed effects are still most pronounced in steels, particularly high-strength steels. These effects can take the form of reduced ductility, ease of crack initiation and/or propagation, the development of hydrogen-induced damage, such as surface blisters and cracks or internal voids, and in certain cases changes in the yield behavior. This article reviews the phenomenology of each of these areas for iron-base alloys in different strength ranges: low, medium and high.

Although the response of a given steel to hydrogen can be quite variable, there are common elements of entry, transport and subsequent interactions with microstructural features and other heterogeneities that can be quite general and are in fact found in most other alloy systems. A summary of the generalized processes which occur during hydrogen embrittlement is presented in Fig. 1.

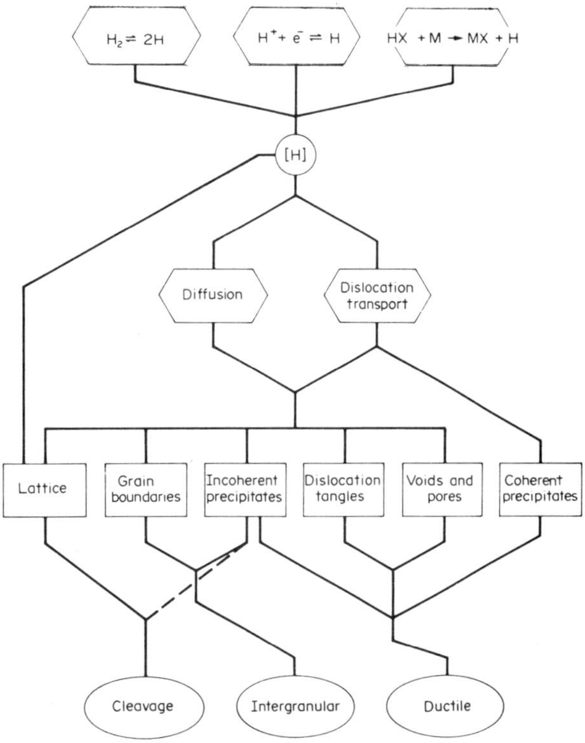

1. Hydrogen Sources, Transport, Traps and Resultant Fracture

Hydrogen may be derived from hydrogen gas molecules which adsorb onto the metal surface and dissociate, by recombination of hydrated protons, H_3O^+, with electrons at the metal surface, or by reaction between hydrogen-containing molecules such as alcohols and the metal (or oxide) surface to release hydrogen. Since all of these sources give rise to hydrogen in solution, it is usually irrelevant which one is operative. There can, however, be kinetic effects on, for example, crack propagation, if the hydrogen-entry step happens to be rate-controlling.

Hydrogen diffuses rapidly in most metals and extremely rapidly in iron and steel. Another transport process can also occur which can be even faster: motion of hydrogen in dislocation cores or as associated Cottrell atmospheres. Calculations show that dislocation transport of this type can be several orders of magnitude more rapid than lattice diffusion. Although transport is not often the rate-controlling step in steels, there are certain phenomena, such as the development of nonequilibrium internal gas pressures, where it could be important.

Hydrogen tends to accumulate at a wide variety of locations in a microstructure—grain boundaries, inclusions, voids, dislocations, dislocation arrays and solute atoms—as well as in solid solution in the iron lattice. Whichever of these locations is the most sensitive to fracture is likely to control the magnitude of hydrogen effects, although in general all of them accumulate hydrogen. However, in many cases this accumulation can be beneficial, if it reduces local concentrations at potential crack nuclei.

Figure 1
Summary of hydrogen sources, transport and microstructural locations with corresponding end processes. The dashed line at the bottom refers to cleavage of hydrides (after Thompson and Bernstein 1980)

Hydrogen-induced fracture is usually thought of as brittle, and intergranular fractures are in fact common; but many materials exhibit ductile fracture (with reduced ductility) in the presence of hydrogen. Cleavage can also occur, either in the matrix or through a precipitated hydride phase. In fact, all the classical fracture modes are observed in the presence of hydrogen.

This brief description indicates how hydrogen effects can occur in relation to the microstructure. Of particular interest is the (potential) control of fracture behavior by varying the microstructural locations of hydrogen. The locations most critically affected by hydrogen will form the fracture nuclei and path; this may accentuate the fracture mode which occurs in the absence of hydrogen, or it may produce a different mode.

After entry and subsequent distribution of the hydrogen within the steel, a number of nonexclusive effects can occur. Each of these will be described as a subset of the general phenomenon of hydrogen embrittlement.

2. *Surface and Internal Cracking; Void Formation and Blistering*

The introduction of large quantities of hydrogen through plating, acid pickling or other cathodic processes, or perhaps as a result of a transient buildup of a large supersaturation of molecular hydrogen, can lead to the development of internal pressure because of the formation of molecular hydrogen at internal heterogeneities. Developed pressures may exceed the local fracture stress, perhaps assisted by a more direct effect of hydrogen on the cohesive strength of neighboring iron–iron bonds. In this case, internal inter- or intragranular cracks or voids can form, reducing the subsequent toughness of the material, owing to one or both of the following: (a) the effective cross section of the steel is reduced because of the defects, thereby raising the operative stress; (b) subsequent cracking is made easier either by the hydrogen-induced defects acting as critical crack nuclei, or because of local stress concentrations at the crack tip.

If these damage processes occur near the surface, the crack may intersect the surface or cause local upsetting leading to blistering and flaking. Such irreversible damage was a common problem a number of years ago in such applications as enamelling steels, rails and plated parts. Proper outgassing procedures and compositional control have reduced the problem in these applications, although such defects are often a common and sometimes degrading occurrence in line-pipe steel exposed to gas conditions where, for example, the presence of H_2S and moisture can promote high hydrogen fugacities.

3. *Hydrogen Attack*

A problem associated with the previous case is the formation of internal voids in steels, at modest operating temperatures, when exposed to a hydrogen-producing environment. Specifically, prolonged exposures to high-pressure hydrogen at temperatures in excess of 200 °C can result not only in surface decarburization but also in the formation internally of methane gas by the reaction between hydrogen and iron carbide. Under certain conditions these voids can link up, usually at grain boundaries, leading to fissures and swelling. These voids not only affect dimensional stability, but can promote premature failure at temperature, and hydrogen attack can be a limiting design problem in such applications as oil processing plants and coal gasifiers and liquefiers. A set of empirical operating curves, the Nelson diagram, has been devised (from many years' experience of plant failures attributable to hydrogen attack) in order to predict the hydrogen partial pressure and temperature limits for safe plant operation with selected carbon and alloy steels. Although this is a useful guideline, a safer approach is to use alloys containing solutes that stabilize the carbides, thereby making methane formation thermodynamically and kinetically more difficult.

4. *Hydrogen-Induced Ductility Losses*

The presence of dissolved hydrogen can often lead to a loss in ductility as measured, for example, by a decrease in the reduction-of-area (RA) value in a smooth tensile test. Such ductility losses can occur with or without a change in the fracture mode with respect to a hydrogen-free test. The embrittlement index is given by the RA loss which is defined by

$$\text{RA loss} = (\text{RA} - \text{RA}_\text{H})/\text{RA}$$

where RA and RA_H are the values without and with hydrogen, respectively. These kinds of tests have long been useful for predicting the potential sensitivity of a given alloy to embrittlement. Detailed studies have shown that at least in mild steels the greatest sensitivity is found near room temperature, for low strain rates, with the RA loss increasing monotonically with hydrogen content.

In recent years the slow strain-rate test performed on tensile samples has become an important diagnostic tool in assessing susceptibility to hydrogen or to other possible stress-corrosive media. In these tests the tensile ductility is measured in the specific environment of interest under a series of continuously decreasing strain rates. Alloys which appear quite resistant at normal strain rates, such as $10^{-4}\,s^{-1}$, can be quite susceptible at slower rates of the order of $10^{-7}\,s^{-1}$ and below, and this may signal potential service problems under, for example, loading conditions which are static or quasi-creep in nature.

Detailed studies of how hydrogen can cause loss of ductility in tensile specimens without a change in fracture mode from microvoid coalescence has led to somewhat conflicting results for different investigators. This is explainable only in part by alloy chemistry and local stress-state variations. However, there is a strong indication that the major role of hydrogen in accelerating ductile fracture, as yet unidentified, occurs late in the fracture process, well after necking, when a large triaxial stress state exists. By contrast, hydrogen appears not nearly as important in promoting early void nucleation.

Although it has long been believed that hydrogen has at best a very nominal effect on the plastic behavior of steel, that is, yield strength, tensile strength and work-hardening rate, recent data suggest that there can be significant changes in these properties under particular conditions. For example, work on high-purity iron has shown that hydrogen can soften iron at temperatures between 200 and 300 K but can harden it at lower temperatures. More general hardening is observed over a broad temperature range in both impure iron and mild steel.

These effects have been interpreted as resulting from the detailed interaction between edge and screw dislocation segments in high-purity material, while in the impure material the hardening is attributed to an interaction between hydrogen, dislocations and impurity atoms like carbon and nitrogen.

Work-hardening effects have also been observed but under different test conditions, namely notched-bar bend tests, where the plastic constraint existing below the notch root is believed to amplify plastic response. Although a function of temperature, the work-hardening rate has been found to be considerably greater in the presence of hydrogen, with a maximum effect near room temperature. It seems clear that an understanding of how hydrogen modifies the plastic response of steel will provide critical insight into how ductile-type fractures can be accelerated in the presence of hydrogen. Further, as described in the next section, hydrogen-induced modification of the plastic response of steel, particularly near notches and other stress raisers, and in regions of plastic constraint, can have an important role in subsequent crack initiation and propagation response.

5. *Hydrogen-Induced Cracking*

In terms of technological importance, the role of hydrogen in accelerating crack nucleation and growth is of greater interest than its effect on ductile fracture, since it can be the single most important design constraint for such structural applications as pressure vessels, pipelines and power equipment. Further, hydrogen-induced cracking has been found to be almost ubiquitous in many types of welding practice, and subsequent weld cracking has been observed under service conditions where the base metal is effectively immune. Studies suggest that the highly sensitive nature of the weld region results from a combination of very sensitive microstructural types (e.g., twinned martensite) and solute partitioning to create local galvanic cells or to provide sensitive regions due to solute depletion.

Hydrogen has been shown to affect the threshold level for the detection of cracking (crack initiation). A particularly insidious manifestation of this effect is delayed failure where a structure under load shows no evidence of cracking until fracture suddenly ensues. This is clearly related to slow crack growth to reach the critical size for catastrophic propagation, and hydrogen probably controls this subcritical crack growth. A typical representation of delayed failure in a high-strength steel is shown in Fig. 2. The applied stress intensity K is usually of the form, for mode I (tensile loading), $K_I = \text{constant} \times (\sigma\sqrt{a})$, where a is the effective crack length for a given applied stress σ (see *Fracture Toughness Testing of Brittle Materials*). Although this test is relatively easy to perform and can illustrate threshold behavior (denoted usually by K_{th}), below which the material is effectively immune to hydrogen, it does have a number of drawbacks. In particular, it is difficult to separate out crack initiation, crack propagation and unstable or fast final fracture, as all occur in the same test.

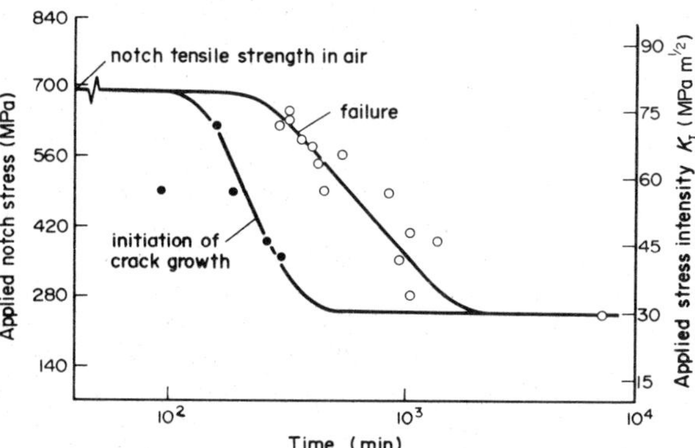

Figure 2
Example of delayed failure in a high-strength steel. Times and scatter in data are typical of such tests. "Initiation" refers to start of crack motion in precracked specimens; K_{th} is the threshold stress intensity below which failure is not observed

Much more information is obtained by measuring not only K_{th} but also the crack velocity in the environment of interest. The results of a typical test of this type are illustrated in Fig. 3. The three-stage behavior shown is often observed in steels and other alloy systems and has been interpreted as follows. Region I is controlled by a combination of mechanical parameters and the interaction of the environment with the controlling and varying stress or strain tensor, usually at the crack tip. In region II the crack rate is primarily controlled by the ability of the embrittling species to remain in contact with the region undergoing fracture; thus, mechanisms like hydrogen diffusion can be quite important. In region III, the failure criterion is satisfied by the mechanical environment alone and unstable fracture follows.

While designers of structural components potentially susceptible to hydrogen embrittlement would like to ensure that K_{th} is never attained under any service condition, this is often an impractical requirement, as defects are invariably preexisting in real alloys. The development of metallurgical strategies to reduce crack velocities is therefore of both practical and scientific interest.

6. *Control of Hydrogen Embrittlement*

There are many approaches to controlling hydrogen embrittlement. In principle the most effective would be to ensure that melting practice and subsequent

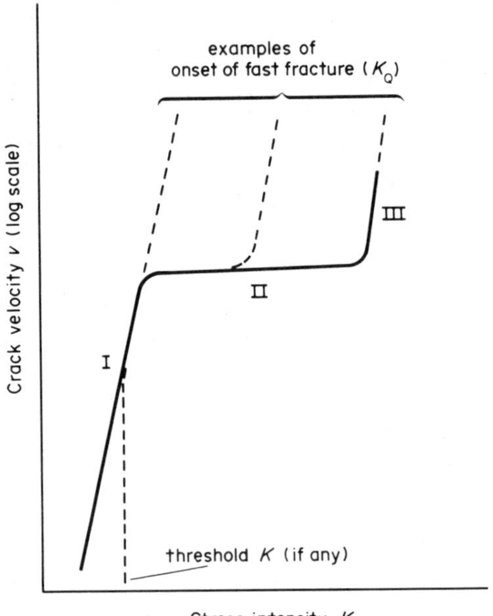

Figure 3
Typical stress-intensity crack-velocity dependence for materials like steel in an embrittling environment. The dashed lines indicate that the extent of the environmentally controlled stage II can vary widely for different alloys and environments

finishing operations like pickling, plating and welding are so well understood and controlled that the steel is effectively free of residual internally dissolved hydrogen. In addition, the electrochemical control and the nature of the surface must be such that any hydrogen generated on the surface, either advertently or inadvertently, will never enter the material. These constraints are in reality as impractical as they appear, but advances in vacuum melting and remelting, controlled cooling after high-temperature finishing operations and surface control ranging from anodic protection to painting and to ion implantation of species noble to hydrogen have reduced the severity of the problem for many applications. Nevertheless, the problems continue to arise and as operating conditions become more severe, as for example in deeper, more-sour (higher H_2S concentration) gas and oil wells, the best approach towards obtaining reliable protection seems to lie with improved alloy design, rather than relying solely on external controls.

This is an area of active, ongoing research and while much has recently been learned, the findings are still too preliminary to warrant their formal inclusion in alloy design methodology. Nevertheless, the following appears to be the most likely framework to provide effective, hydrogen-resistant alloys. Compositional control should include more careful use of manganese, which, although an effective ferrite strengthener and hardenability agent, has been found to lower resistance to hydrogen in many cases. Unnecessary increases in carbon content may also be dangerous, as this appears to lower K_{th}. On the other hand, silicon and titanium appear capable of improving performance, with the latter effective both as a substitutional solute and as finely distributed titanium carbide.

Microstructural control also appears to be a very promising route. In general, it is now possible to describe the structural types which are to be preferred for hydrogen-related applications. Here, strength level can also be important, but the overall ranking of steels extends from quenched-and-tempered martensite (or bainite), being most resistant to hydrogen, through spheroidized structures having intermediate resistance, to pearlitic or ferritic (normalized) structures, usually with somewhat less resistance, and finally to untempered martensite, with the least resistance. The spheroidized and pearlitic structures do not always have the same ranking but it is important to note that most line-pipe steels, and some pressure-vessel steels, are used in the normalized condition. In addition, such microstructures are often "banded" as a result of chemical inhomogeneities originating during solidification which persist through processing. These bands are frequently rich in manganese; as the composition effect discussed above would predict, cracking in service tends to follow these bands.

It should be emphasized that most high-strength and medium-strength steels, with alloying element content sufficient to give good hardening response, are quenched and tempered (sometimes normalized and tempered). When well tempered, such microstructures are relatively resistant to hydrogen, although increasing the strength level can reduce this resistance. At medium strength levels, the normalized and tempered structures containing pearlite can be fairly susceptible to hydrogen. Such microstructures therefore require caution in use. Another important microstructural factor is grain size. Both strength and toughness are increased by decreases in grain size; fortunately, resistance to hydrogen embrittlement is generally increased also.

It should be emphasized that the interrelation of composition, microstructure and strength, or any two of these, can be very complex. Any specific situation of interest would require careful assessment. For example, there are many known cases where a steel is heat treated to the same strength level but with different thermal histories giving rise to different microstructures. The resistance to hydrogen embrittlement can vary widely and, in fact, with different tempering treatments the lower-strength condition can still be more susceptible than the higher-strength condition. For this reason alone, the two well-known "rules of thumb" regarding strength-level effects in hydrogen embrittlement must be used with caution. One is that embrittlement becomes more severe as strength level increases; the other is that steels with strengths below about 700 MPa show no significant embrittlement. Like most rules of thumb, these are true in many cases, but there are many exceptions, as described above.

7. Hydrogen-Embrittlement Models and Mechanisms

Although many models of hydrogen embrittlement have been proposed, confusion and uncertainty still exist, owing to continued inability to understand analytically bonding between iron atoms, even in the absence of hydrogen, and perhaps also because of nonexclusivity, in that more than one or different combinations of mechanisms may dominate for different mechanical or environmental conditions.

Proposed models can be broadly classifed as follows.

(a) pressure assistance in which the internal stresses resulting from recombined molecular hydrogen promote premature stress-assisted fracture;

(b) reduction of the cohesive strength or bond strength between iron–iron bonds at a crack tip, matrix–particle interfaces or grain-boundary atoms;

(c) a change in the plastic response of the material, ranging from reduced plasticity promoting brittleness to enhanced localized plasticity leading to early shear-controlled fractures;

(d) formation and subsequent cracking of a brittle metal hydride.

Except for the last of these, persuasive arguments have been put forward for the validity of each in steels. In fact, it is by no means certain that a hydride mechanism, modified to consider its formation under transient nonequilibrium conditions, may not have some applicability to ferrous alloys. Whether one of the above models, some combination of these, or an entirely new model, will be shown to control the fracture response of steel in a hydrogen environment remains to be answered. What is clear is that understanding of how hydrogen promotes the embrittlement of steels has progressed considerably in recent years, and such environmental effects can now be addressed in a predictive manner.

See also: Corrosion of Iron and Steel; Hydrogen Embrittlement; Steels: Structure–Property Relations

Bibliography

Beachem C D (ed.) 1977 *Hydrogen Damage: A Discriminative Selection of Outstanding Articles and Papers from the Scientific Literature.* American Society for Metals, Metals Park, Ohio

Bernstein I M, Thompson A W (eds.) 1981 *Hydrogen Effects in Metals.* The Metallurgical Society (AIME), Warrendale, Pennsylvania

Birnbaum H K 1979 Hydrogen related failure mechanisms in metals. In: Foroulis Z A (ed.) 1979 *Environment-Sensitive Fracture of Engineer Materials.* The Metallurgical Society (AIME), Warrendale, Pennsylvania, pp. 326–60

Hirth J P, Johnson H H 1976 Hydrogen problems in energy related technology. *Corrosion* 32: 3–26

Louthan M R, McNitt R P, Sisson R D Jr (eds.) 1981 *Environmental Degradation of Engineering Materials in Hydrogen.* Virginia Polytechnic Institute Printing Department, Blacksburg, Virginia

Oriani R A 1978 Hydrogen embrittlement of steels. *Ann. Rev. Mater. Sci.* 8: 327–57

Thompson A W, Bernstein I M 1980 The role of metallurgical variables in hydrogen-assisted environmental fracture. In: Fontana M G, Staehle R W (eds.) 1980 *Advances in Corrosion Science and Technology*, Vol. 7. Plenum, New York, pp. 53–175

Thompson A W, Bernstein I M 1981 Stress corrosion cracking and hydrogen embrittlement. In: Tien, J K, Elliot J F (eds.) 1981 *Metallurgical Treatises.* The Metallurgical Society (AIME), Warrendale, Pennsylvania, pp. 589–603

I. M. Bernstein; A. W. Thompson

Hydrometallurgy

Hydrometallurgy is a method of producing metals in which the key process steps take place in liquid phases, predominantly aqueous solutions. Two basic unit processes are involved: (a) the leaching or dissolution of the value(s) contained in solids and (b) the subsequent isolation and recovery of these values. As shown in Fig. 1, actual hydrometallurgical flowsheets often incorporate additional unit operations that must be effected if the overall process is to be a commercial success. These steps may include preleach treatment of the feed, separation of liquids and

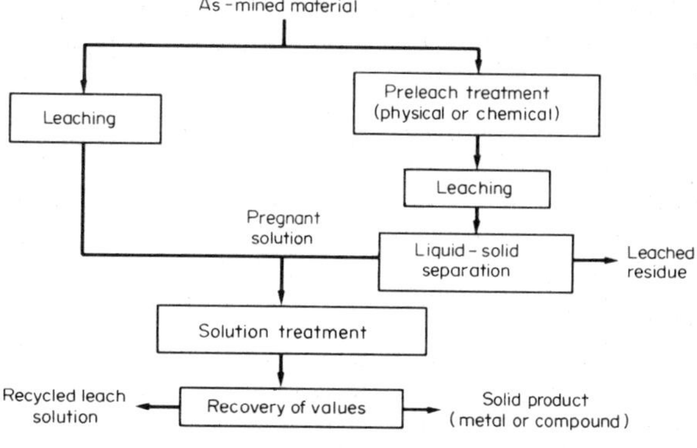

Figure 1
Flowsheet for a hydrometallurgical process

solids, conditioning of the pregnant leach solution, disposal of leach residues, and treatment of barren solution before discharge or recycle.

1. Hydrometallurgical Feeds and Products

Hydrometallurgical systems have been designed to handle a wide range of feed materials and to produce many different metals. These include copper, zinc, cadmium, gold, silver, uranium, vanadium, nickel, cobalt, aluminum, magnesium and beryllium. The hydrometallurgical product may be of high purity, like zinc produced in a leach-electrowinning circuit. It may also be impure, requiring upgrading, as in the case of Merrill–Crowe precipitate derived by using zinc dust to cement silver and gold from cyanide leach liquors. Finally, the product may be a high-grade intermediate such as Bayer process aluminum hydroxide, $Al(OH)_3$, or yellow cake, U_3O_8, derived from uranium leaching.

The material being leached can vary widely in grade. However, process chemistry is sensitive to the characteristics of the feed. Thus, each process must be designed for a specific input and no single process can be used to treat a wide variety of feeds. These can be lean overburden stripped simply to expose higher-grade ore. Richer materials may be mined expressly for leaching. Even high-grade materials such as concentrates are amenable to hydrometallurgical processing. This may involve a fairly sophisticated and expensive operation, but is justified by the high unit value of the feed.

The wide range of feeds that can be treated by hydrometallurgy contrasts sharply with pyrometallurgy (see *Pyrometallurgy*). Smelting is generally limited to a high-grade feed because of the cost of melting and slagging off the gangue.

2. Preleach Treatment

Preleach treatment often starts with size reduction via crushing and grinding so that the values are liberated or at least exposed. This can accelerate leach extraction or make the values amenable to concentration. Flotation is a preferred method of concentration (see *Mineral Processing*), particularly where values exist as sulfides. Wide differences in the specific gravity of the values and gangue allow for gravity concentration. Thus, mechanical classification systems have been developed for recovering free gold from placers or cassiterite (SnO_2) from alluvial deposits. Although recovery may not approach 100%, concentration is frequently justified because it produces a small amount of upgraded material. This is less expensive to process than a large amount of low-grade material.

Chemical treatment may also be undertaken before leaching. This may restrict the solubility of gangue constituents or enhance the leaching of values. Such steps often involve high temperatures, as in the oxidation roasting of chalcopyrite ($CuFeS_2$) concentrate or reduction roasting of nickel laterite ores. In the former, roasting in air converts the copper to a water soluble sulfate, and the iron is transformed to an insoluble oxide. In the latter, nickel and cobalt are reduced to elemental form while the iron remains as a silicate. In the subsequent ammonia leach, nickel and cobalt are solubilized as amine complexes. Instead of pyrometallurgical steps, aqueous oxidation or reduction may also precede leaching. By treating manganiferous silver ores with an aqueous solution of sulfur dioxide (SO_2) the refractory mineral structure is broken by producing water soluble $MnSO_4$. This renders silver soluble in a subsequent cyanide leach.

3. Leaching

Following preliminary treatment, the material is subjected to leaching. Water alone solubilizes copper or zinc sulfate formed during oxidation roasting of the sulfides. Mineral acids, however, are the most common lixiviants. Sulfuric acid (H_2SO_4) can be used on oxide or silicate ores to extract copper, zinc and uranium. Where effective, sulfuric acid is preferred because of its low cost and availability as a by-product of sulfide smelting. Other favorable properties of sulfuric acid include very low volatility and high chemical stability in aqueous solutions. The use of other acids such as nitric or hydrochloric acid is limited to specific systems owing to undesirable properties such as high cost, volatility, chemical instability and corrosiveness.

Alkaline lixiviants are also used, the most common probably being sodium hydroxide which is used in the Bayer process to convert calcined bauxite to purified $Al(OH)_3$. Ammonia is also used on account of its selectivity for nickel, copper and cobalt, and because it does not attack steel vessels. Certain salts can also act as solvents; for example, alkaline cyanide has long been employed in leaching gold and silver.

Regardless of the specific chemistry involved, the solvents should be characterized by the following properties.

(a) Selectivity for the values involved: in general, if the gangue is attacked, reagent consumption will be too high to sustain an economic process. Gangue dissolution can also add impurities to the leach solution or cause decrepitation of the host rock. The latter generates slimes that can lead to serious practical problems in subsequent liquid–solid separations or in washing of leach residues to completely recover the values.

(b) Capacity for rapid extraction: if leaching is too slow, in-process inventories will be so high that the size of the required containment system is not practical.

(c) Low cost: either reagent costs must be low relative to the commodity being recovered or the lixiviant must be regenerated at low cost as part of the process.

Other important characteristics of the solvents include such factors as low corrosiveness, stability during handling and lack of toxicity.

From a chemical standpoint, lixiviation may involve either simple dissolution or a redox reaction. In the latter, a supplemental oxidant or reductant may be required or the lixiviant itself may perform these functions. For example, when leaching tetravalent uranium with sodium carbonate, hydrogen peroxide or air (oxygen) must be used to first convert the uranium to the hexavalent state.

Even bacteria may perform important redox reactions. In leaching copper from low-grade sulfidic mine waste, the active lixiviant is the oxidant ferric sulfate. During leaching this is reduced in reactions such as

$$CuS + Fe_2(SO_4)_3 \rightarrow CuSO_4 + 2FeSO_4 + S^0 \quad (1)$$

However, to account for the observed copper extraction the ferrous iron must be repeatedly converted back to the ferric state for more leaching within the waste dump. Microorganisms such as *Thiobacillus ferrooxidans* catalyze this conversion, which would be too slow otherwise.

Leaching kinetics are of fundamental importance in hydrometallurgy. This is a significant difference between hydro- and pyrometallurgical systems since the latter are often limited by thermodynamic considerations. Kinetically, most leach systems are characterized by rapid initial dissolution of values, followed by a continuously declining leach rate. Extraction asymptotically approaches some ultimate level which is usually less than 100%. Such behavior derives from the mechanism which controls each stage of the leaching process. Initially, the values are exposed directly to the lixiviant and the rate-controlling step may be the reaction kinetics. However, the unleached mineral grains rapidly become insulated from direct contact with the bulk lixiviant, perhaps as a result of a buildup of a protective film on the mineral surface, such as the formation of a coherent sulfur layer on chalcopyrite leached in acidic ferric sulfate. Alternatively, a leached rim may develop on the rock fragment with an unleached core at the center. In either case, diffusion of lixiviant or dissolved values through such a layer becomes rate-controlling. This causes a substantial decrease in the leaching rate. Finally, the inability to solubilize 100% of the values can be due to several factors, such as the presence of refractory minerals which do not leach in the lixiviant being used, or complete encapsulation of the values by an impervious coating.

Leach systems vary widely depending on the nature of the feed. Several types are shown in Fig. 2. Low-grade mine waste is often leached in the run-of-mine condition in dumps or heaps. Leach times as long as several years may be employed to extract all available values from the rock, although such dumps are built to minimize mine costs rather than optimize extraction. At moderate ore grades, the values allow for crushing and use of heaps designed to promote leaching. Here extraction extends from a few weeks to a few months. With rich ores, fine grinding is justified to promote rapid extraction and high recovery; the leaching system generally involves agitation in mechanically agitated tanks or air agitated (Pachuca) columns.

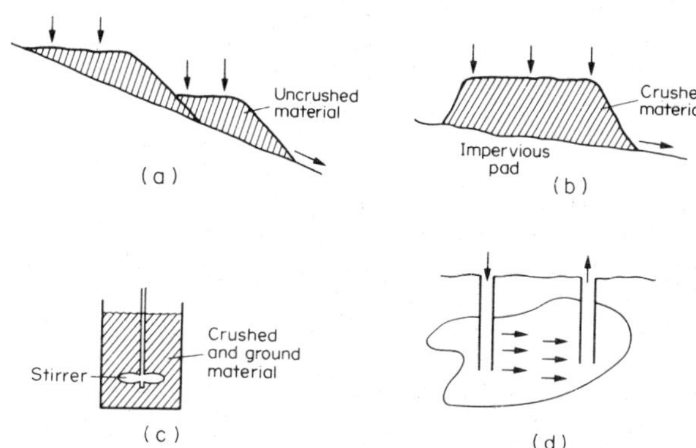

Figure 2
Types of leaching: (a) dump leaching, (b) heap leaching, (c) agitation leaching, (d) in-situ leaching

When concentrates are to be treated, batch or continuous reaction vessels can be employed economically. Such systems offer two principal advantages: enhanced solubility of gaseous reactants, and the use of elevated temperatures. This combination often increases leaching rates by several orders of magnitude over ambient conditions. Extraction is also maximized when the feed particle size is small. Such rapid recovery limits the hidden cost associated with in-process inventory and permits use of small, relatively inexpensive pieces of equipment with high specific throughputs. On the negative side, the combination of high pressure and temperature, together with the use of concentrated lixiviants, can make selection of construction materials a critical factor.

Another type of leach system gaining in importance involves in-situ treatment or solution mining, which avoids the costs of large-scale materials handling and concentration. Since these two steps are the most expensive in many extractive processes, profitability can be quite high if the ore is amenable to in-situ treatment. As in other leach systems, the key is effective contact between the lixiviant and metal value. Thus the host rock must be broken in place to allow access to the minerals, or the matrix must be inherently permeable. To date, most successful in-situ operations have been of the latter type and involve the leaching of uranium from roll front deposits contained in a porous sandstone matrix. These are often too small or low grade to permit easy conventional mining.

The technology necessary for solution mining has been developed by borrowing heavily from the petroleum industry in such areas as drilling methodology, well-construction techniques and reservoir engineering. The result is a system of injection and recovery wells that permits close control over the flow of lixiviant within the formation. Owing to the permeability of the sandstone, relatively small pressure gradients are required to move the solution so that the uranium is effectively solubilized and flushed to the recovery wells.

4. Separation of Liquids and Solids

The leaching process is not economical if pregnant leach solution cannot be completely recovered from the leached solids. Particularly where agitated leach systems are employed, liquid–solid separations are often difficult because the leach residue contains decrepitated feed or fine chemical precipitates. These are not easily filtered or settled without entraining a significant amount of pregnant solution. When large tonnages are involved, liquid–solid separations can be more expensive than any operation except mining. Such separations generally require multiple-stage countercurrent washing in large thickeners. For small tonnages various filtration/wash systems can be employed to fully recover soluble values.

Where vat, heap or dump leaching of as-mined material is practised, the analogous step is to flush the soluble constituents from the rock. This is commonly achieved by progressively advancing solution from leached to fresh material. Recycling of barren solution is another way of maximizing recovery of dissolved values.

If the target metal is of high value, soluble losses due to poor wash efficiency are intolerable. As a result, systems have been developed in which a sorbent for the dissolved values is added directly to the agitation leach circuit. Owing to intimate contact with the slurry and the high affinity of the sorbent for the dissolved species, the leach liquor is continuously stripped as values are dissolved. Differences in particle size or density are used to separate the loaded sorbent from the slurry. Typical commercial systems employing this principle include the carbon-in-pulp process for gold cyanide leaching and the resin-in-pulp process for uranium leaching. Both sorbents are recovered by screening from the finely ground ore.

5. Solution Treatment

Treatment of pregnant solution is often required before final recovery of the solubilized values. This may be as simple as a final filtration step, deaeration, or adjustment of solution Eh or pH. However, the treatment is often complex, involving concentration and/or isolation of the values or removal of impurities. Generally, chemical precipitation is used for impurity removal. Considerations are the same as those associated with chemical recovery of values (see Sect. 6).

An innovative approach to solution treatment is to combine concentration and isolation of the values in a single step. Two chemically similar methods of achieving this are liquid–liquid (solvent) extraction and ion exchange using a solid resin. The first involves contacting the leach solution with an immiscible organic liquid consisting of the active extractant, appropriate modifiers and an inert diluent. The chemical reaction involves an exchange between the extractant molecule and the ion or ionic complex containing the metal value. The extractant must have several specific characteristics in addition to being virtually insoluble in the leach liquor. A key is that loading be specific for the recovered metal and be reversible to permit stripping and regeneration of the organic. The extractant should also have a high loading capacity and should not present special problems due to toxicity, flammability or degradation. Effective dispersion of the two liquid phases in the mixer will maximize mass transfer rates. The emulsion must then break quickly in the settler to allow for rapid coalescence and separation of the two phases. Entrainment of organic material in the barren leach liquor (raffinate) is important since such a loss can be a major operating expense.

One widely used commercial solvent extraction system involves the recovery of copper from acidic sulfate leach liquors (pH 1.5). The loading reaction is

$$2RH(org) + CuSO_4(aq) \rightleftharpoons R_2Cu(org) + H_2SO_4(aq) \quad (2)$$

Here, R represents the extractant, often an oxime or hydroxyoxime. Loading is achieved in two or more countercurrent extraction stages operated in series. Following loading, lean electrolyte (0.15–0.17 $kg\,l^{-1}$ H_2SO_4) is used to strip the copper from the loaded organic by reversing the above reaction. This generates a purified rich electrolyte in which the copper has been concentrated from as little as 0.0005 $kg\,l^{-1}$ in the leach solution to 0.04–0.05 $kg\,l^{-1}$ in the electrolyte. The only impurity transferred with the copper is a small amount of ferric iron.

The term ion exchange is generally applied to a similar chemical system in which the extractant is contained in a solid resin. Pregnant solution is passed through a bed of resin beads until loading is complete.

Then the resin is stripped to produce a small volume of concentrated and purified solution. A major advantage of ion exchange is recovery of metals from very dilute solutions. As with solvent extraction, a potential problem is poisoning of the extractant with impurities that are not easily stripped.

6. *Recovery of Values*

As a result of the metallic nature of the values, the recovery step in most hydrometallurgical processes involves reduction, either chemical or electrochemical. The commonest routes are electrowinning, cementation and hydrogen reduction. In a few cases where compounds rather than pure metals are the final products, other techniques may be employed, such as crystallization by evaporation (e.g., nickel sulfate recovery from decopperized electrolyte), precipitation by pH adjustment (e.g., addition of lime to precipitate calcium molybdate), disproportionation (e.g., steam stripping of copper amine solution to recover ammonia and precipitate copper oxide) and reagent addition (e.g., precipitation of copper chloride by the addition of chlorine).

In electrowinning the imposed voltage drop between the inert anode and the cathode provides the reducing potential. The cathodic reaction is

$$M^{x+} + xe^- \rightleftharpoons M^0 \qquad (3)$$

where M is a metallic ion with a valency of x and e^- denotes an electron supplied by the external power source. In order to maintain electrical neutrality, an electron-generating reaction must occur at the anode. This involves decomposition of some component of the electrolyte, generally with discharge of a gas.

Copper provides a good example of an electrowinning system. Here, the electrolyte is a sulfuric acid–copper sulfate solution, the anode is an inert lead alloy and the cathode blank is usually a thin copper starter sheet. The half-cell reactions and voltages at 25 °C (relative to the standard hydrogen electrode) are as follows:

$$\text{cathode: } Cu^{2+} + 2e^- \rightleftharpoons Cu^0 \qquad (+0.34\ V) \quad (4)$$

$$\text{anode: } H_2O \rightleftharpoons \tfrac{1}{2}O_2(g) + 2H^+ + 2e^- \qquad (-1.23\ V) \quad (5)$$

$$\text{overall: } Cu^{2+} + H_2O \rightleftharpoons Cu^0 + \tfrac{1}{2}O_2(g) + 2H^+ \qquad (-0.89\ V) \quad (6)$$

Operating parameters such as acid and copper concentration and electrolyte temperature must be controlled to obtain good cathode quality. Costs are minimized by maintaining a high current efficiency (i.e., using virtually all of the applied current to deposit copper). One of the chief detriments to this is the presence of impurities in the electrolyte which exhibit a multiple valency. These will consume current by establishing a redox couple, being oxidized at the anode and reduced at the cathode.

A shortcoming of hydrometallurgical processes is that electrowinning is more energy intensive than electrorefining of the same metal. In the latter, the half-cell reactions are dissolution of the impure metal from the anode and cathodic deposition of that same metal in refined form. Since the voltage required for dissolution is less than the overvoltage needed for electrolyte decomposition, the power required to produce a given weight of electrorefined metal is less than that required to produce the same weight of metal by electrowinning. Thus, when combined with the autogenous or semiautogenous nature of most smelting processes, hydrometallurgy often has a higher energy requirement than pyrometallurgical treatment of the same feed. In addition, in the case of base metal sulfides, the smelting and electrorefining processes collect and concentrate any precious metals. In contrast, precious metals generally remain in the solid residue when leaching base metals. Recovery would entail a second round of steps involving feed preparation, leaching, liquid–solid separation.

Cementation is probably the oldest hydrometallurgical process, being used by the ancient Chinese to produce copper at least as early as 150 BC. Chemically, cementation involves a galvanic reaction to reduce ions in solution. The standard electromotive series can generally be used to predict which metal-ion pairs are candidates for cementation. The process involves contacting the solution with a metal that is electrochemically less noble than that to be recovered. The more noble metal is therefore displaced from solution and no external reducing potential is required.

Cementation can be used either as a purification or a recovery step. When used for solution purification the precipitant must be the metal that is recovered in the final step. This avoids adding one impurity while removing others, as illustrated by the use of powdered zinc to remove cadmium from zinc electrolyte. The half-cell reactions are:

$$\text{anode: } Zn^0 \rightarrow Zn^{2+} + 2e^- \qquad (+0.76\ V) \quad (7)$$

$$\text{cathode: } Cd^{2+} + 2e^- \rightarrow Cd^0 \qquad (-0.40\ V) \quad (8)$$

$$\text{overall: } Zn^0 + Cd^{2+} \rightarrow Cd^0 + Zn^{2+} \qquad (+0.36\ V) \quad (9)$$

The overall process is spontaneous since the voltage change is equivalent to a negative free energy of reaction. Kinetically, the rate-controlling step in most cementation reactions is diffusion of ions through a stagnant liquid film adjacent to the metal surface. In an actual system the overall rate will depend on the temperature, available metal surface area, and solution parameters such as pH and the degree of agitation.

Commercially, the precipitant must be less expensive than the metal to be recovered: zinc and iron are commonly used. Aluminum would also appear to be an excellent choice, but the thin oxide layer found on most aluminum renders it passive unless chloride or fluoride ions are present.

The third widely used recovery process involves reduction with gaseous hydrogen. Hydrogen will reduce any metallic ion below it in the electromotive series. Nickel and cobalt are the metals most frequently precipitated this way. An autoclave is required to provide the high temperature and hydrogen partial pressure needed for rapid reduction. Returning seed crystals of the metal to the autoclave will also accelerate precipitation by providing nucleation sites and avoiding supersaturation. Since hydrogen is the reductant, acid will be produced along with the metal according to the following reaction:

$$M^{x+} + (x/2)\,H_2(g) \rightarrow M^0 + xH^+ \qquad (10)$$

This acid may require removal or neutralization to maintain effective metal reduction.

7. *Environmental Considerations*

Hydrometallurgical processes present various environmental problems. Principal concerns include the disposal of leach residues (often acidic or basic), the rejection of sulfur and/or iron, the loss of leach solution to the environment, the presence of toxic impurities in solid or liquid effluents and the need to operate water treatment plants in order to recycle or discharge solutions. These concerns are covered elsewhere (see *Environmental Control in Metals Production*) but are obviously a costly and important part of any commercial hydrometallurgical process.

See also: Electrometallurgy; Mineral Processing; Metals Production: An Overview

Bibliography

Aplan F F, McKinney W A, Pernichele A D (eds.) 1974 *Solution Mining Symposium.* American Institute of Mining, Metallurgical, and Petroleum Engineers, New York

Burkin A R 1966 *The Chemistry of Hydrometallurgical Processes.* Van Nostrand, Princeton, New Jersey

Evans D J I, Shoemaker R S (eds.) 1973 *2nd Int. Symp. on Hydrometallurgy.* American Institute of Mining, Metallurgical, and Petroleum Engineers, New York

Gill C B 1980 *Nonferrous Extractive Metallurgy.* Wiley, New York

Habashi F 1969 *Principles of Extractive Metallurgy.* Gordon and Breach, New York

Hepworth M T (ed.) 1972 *Hydrometallurgy: Theory and Practice.* University of Denver, Denver, Colorado

Sohn H Y, Wadsworth M E (eds.) 1979 *Rate Processes of Extractive Metallurgy.* Plenum, New York

W. J. Schlitt; H. Y. Sohn

Hydrostatic Extrusion

The process of pure hydrostatic extrusion differs from conventional extrusion in that the billet is completely surrounded by a fluid which is sealed off and pressurized sufficiently to extrude the billet through the die (Fig. 1). The presence of a pressurized fluid completely enveloping the billet above the deformation zone prevents the billet from buckling or upsetting against the container wall. This single distinguishing feature results in a number of important advantages over conventional methods.

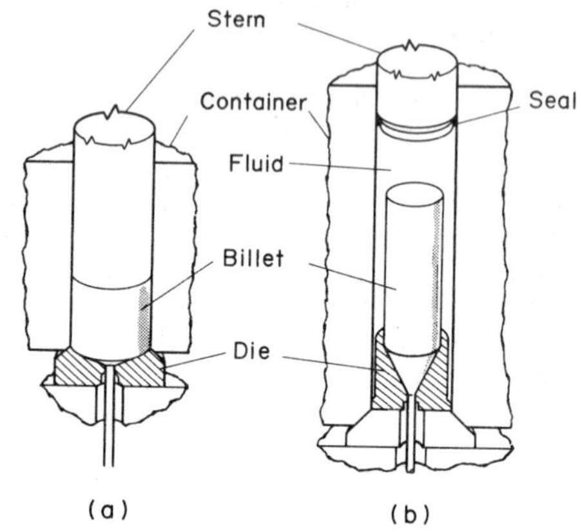

Figure 1
Extrusion processes: (a) conventional and (b) pure hydrostatic

1. *Advantages over Conventional Extrusion*

The prevention of the billet upsetting against the container wall essentially eliminates container friction. Die friction is probably also reduced because of the pressurized lubrication resulting from the hydrostatic fluid medium. Extrusion pressures are therefore minimized. This provides the options of extruding at lower pressure, increasing the extrusion ratio, or lowering the extrusion temperature. Also, because much longer billets can be extruded without increasing pressure requirements, higher extrusion productivity (i.e., more billet weight per extrusion cycle) can be achieved. In addition, the product yield is likely to be increased because, with longer billets, the unextruded butt portion constitutes a smaller percentage of the overall billet. The economic significance of this latter point increases with costly billet materials such as superalloys, some clad products and composites, and multifilament superconductors.

The use of hydrostatic fluid also allows the billets to be almost any shape, even a coil of wire. The process can even be used to extrude very long lengths of relatively small diameter rod (e.g., with a length-to-diameter ratio of over 100:1) that is too stiff to be coiled and too strong to be extruded by the various continuous methods. This cannot be done by conventional extrusion because the length of the ram required would far exceed its critical buckling ratio. In hydrostatic extrusion, however, this problem can be solved by resorting to a stepped-bore container. The larger-bore portion of the container is sized to hold enough fluid to displace the billet volume with a relatively short stroke of the ram. For example, by making the cross-sectional area of the large-bore portion, say, twenty times that of the billet rod, 1 cm of ram travel would extrude 20 cm of billet. This approach has been used for hydrostatic extrusion of stainless steel rod into precision profiles.

The necessity of a pressurized fluid also presents some disadvantages of the process, but most of them can be accommodated without too much difficulty. Containment of the fluid under high pressure requires seals to be used on the ram and die. However, the technology to achieve dependable sealing at these points is considered routine. In fact, under some conditions a seal on the ram is not required. Sealing between the billet nose and the die is achieved by chamfering or tapering the billet nose to closely match the entry angle of the die. Other disadvantages of the process arise when a relatively large volume of fluid is used compared with the billet volume to be extruded. These include:

(a) increased handling for injecting and removing the fluid for each extrusion cycle;

(b) reduced control of billet speed and stopping due to potential stick-slip and excessive stored energy in the compressed fluid;

(c) reduced process efficiency in proportion to the ratio of billet-to-container volume;

(d) increased complications when extruding at elevated temperatures.

The problems of billet speed and stopping control can be reduced by special damping devices as well as improvements in lubrication at the billet–die interface. Another approach to minimizing this and the other problems cited above is to keep the fluid to an absolute minimum as practised in the "thick-film" or "Hydrafilm" process (see Sect. 2).

2. Process Variations and Applications

The process is described as "pure" hydrostatic extrusion when enough fluid is present to prevent the ram from contacting the billet during any part of the extrusion operation. An outstanding application of this method is in the production of copper tubing. Long lengths of near-net-size copper tubing, approximately 13 mm diameter × 1 mm wall thickness, are extruded at a ratio of 500:1 at about 500 °C using castor oil as the fluid. Aside from producing highly concentric tubing, the process offers an advantage over conventional methods in that the need for several bull-block reductions of the extruded tube shell is eliminated. A fixed-mandrel arrangement is used in this case, the mandrel being supported in a fixed position relative to the die by an internal support tube (Fig. 2). Enough fluid must obviously be used between the ram and the mandrel collar to displace the billet from within the support tube. This tooling arrangement functions satisfactorily, but it necessarily keeps the billet length quite short in relation to the container length, thus reducing the efficiency of the process as compared with longer-billet processes.

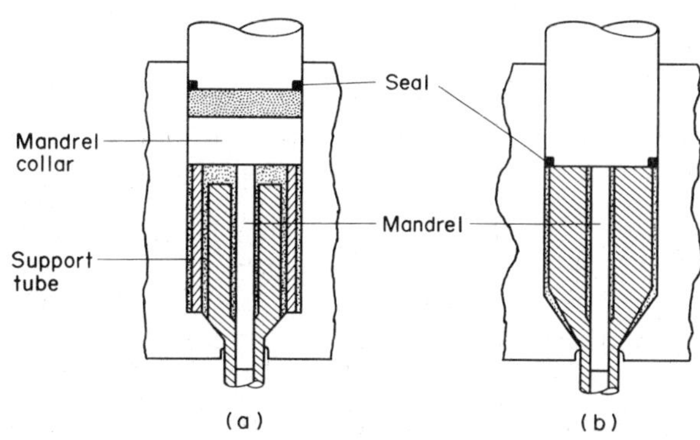

Figure 2
(a) Fixed-mandrel arrangement in pure hydrostatic extrusion, and (b) floating-mandrel arrangement in Hydrafilm extrusion

In spite of this, the overall elevated-temperature operation for making copper tubing functions quite well. It should work equally well for billets of about the same or lower strength. However, it is doubtful that it would work as well for making tubing from materials more difficult to extrude such as steels, superalloys and refractory metals. This is because billet temperatures much greater than 500 °C would be required. Such temperatures would present serious problems in fluid handling, fluid stability and billet chilling. The problem of die and mandrel lifetime could also arise if the billet–tooling contact time were significantly greater than that typically experienced in conventional hot extrusion.

Another variation of the hydrostatic extrusion process is called "billet-augmented" hydrostatic extrusion (Fig. 3). In this case, the ram contacts the end of the billet throughout the extrusion stroke. Because the billet axial stress is always maintained greater than the fluid pressure exerted radially on the billet, the billet stops immediately when the ram stops. Thus, unintentional complete extrusion of the billet through the die is prevented. This approach also minimizes the problem of billet stick-slip.

In the early stages of process development, rather complex tooling was designed to carefully maintain a constant ratio of billet axial stress-to-fluid pressure during the extrusion stroke, so as to prevent billets, typically much smaller than the container diameter, from upsetting to the container bore (Fig. 3).

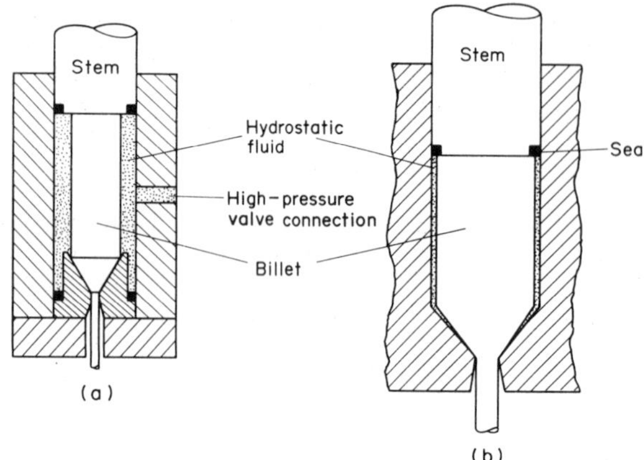

Figure 3
(a) Billet-augmented extrusion with valving, and (b) Hydrafilm extrusion during the billet-augmented stage

However, this method was later greatly simplified by the Hydrafilm process (Fig. 3) in which only billets with a very small annular clearance to the container are used along with a minimal amount of fluid, and the ram is free to contact the billet after some initial fluid pressurization. This initial pressurization of fluid is enough to prevent or minimize billet upsetting. Further ram travel soon increases the fluid pressure between the billet and container to equal the ram–billet contacting pressure, at which time the process changes to pure hydrostatic extrusion. Any billet upsetting that may occur is minor and presents no special problems since container friction is still essentially nonexistent. Thus, the Hydrafilm method (and other variations of this approach) retains all the important advantages of the basic hydrostatic extrusion process. In addition, however, it overcomes many of the problems associated with the other process variants described. Among its advantages are:

(a) the ability to use relatively high billet temperatures (e.g., up to 1200 °C) with much less severe problems of fluid handling and billet chilling;

(b) potential for achieving extrusion cycle rates close or equal to that presently obtained in conventional extrusion, thus increasing productivity;

(c) the possibility of improved mandrel and die lifetimes when extruding at billet temperatures well above 500 °C because of reduced billet–tooling contact time with potentially faster cycle rates;

(d) the ability to minimize problems of stick-slip and billet-stopping control without the use of special damping devices, particularly since the amount of fluid is small and fully compressed when the process finally changes from the billet-augmented to pure stage;

(e) the ability to use a floating or travelling mandrel (Fig. 2) without serious problems of billet upsetting if the ratio of mandrel-to-billet cross-sectional area is too high for pure hydrostatic extrusion;

(f) the use of the floating mandrel also allows a very high billet–container volume ratio, thus increasing process efficiency over that with fixed-mandrel arrangement.

In addition, because fluid volume is minimized, the Hydrafilm process makes it possible for hydrostatic extrusion to be conducted on mechanical presses and potentially at the fast rates typical for such equipment. The Hydrafilm process has been used in the laboratory for extrusion of many difficult materials and complex products, including superalloys, refractory metals, bimetallic tubes, metal-matrix composites and multifilament superconductors.

Selection of hydrostatic over conventional extrusion will depend upon the relative technical and economic advantages offered for a given material and product. Also, the process economics can be affected greatly by hydrostatic extrusion equipment cost, which can vary widely, depending on whether new equipment is essential or if existing equipment can be converted at a low cost. Such factors should be considered carefully in order to select the best process for a particular application.

See also: Cold Extrusion and Forging; Hot Extrusion; Metals Processing and Fabrication: An Overview

Bibliography

Asari A 1977 Development and improvement in extrusion press of nonferrous metal. *Proc. 2nd Int. Aluminum Extrusion Technology Seminar*, Vol. 1. Aluminum Association, Washington, DC, pp. 149–67

Chadwick R 1980 Developments in design and application of extrusion presses for metal processing. *Int. Met. Rev.* 25(3): 94–136

Decours J, Gavinet J, Segurens L, Weisz M 1974 Adaptation of a conventional extrusion press for hydrostatic operation and some applications at high and low temperatures. In: Pugh 1974, pp. 19–23

Fiorentino R J, Meyer G E, Byrer T G 1974 Some practical considerations for hydrostatic extrusion. In: Pugh 1974, pp. 85–93

Fiorentino R J, Smith E G Jr 1980 Hydrostatic extrusion of precision pinions. *US Army ManTech J.* 5: 3–10
Larker H T, Nilsson J O H 1974 Hydrostatic extrusion in production and laboratory—Some examples of the influence on process parameters and equipment design. In: Pugh 1974, pp. 19–23
Nishihara M 1974 Hydrostatic extrusion for production applications. In: Pugh 1974, pp. 24–42
Pugh H Ll D 1970 Hydrostatic extrusion. In: Pugh H Ll D (ed.) 1970 *Mechanical Behavior of Materials Under Pressure*. Elsevier, London, pp. 391–521
Pugh H Ll D (ed.) 1974 *NEL/AIRAPT Int. Conf. Hydrostatic Extrusion*. Institution of Mechanical Engineers, London
Sauve C 1964–65 Lubrication problems in the extrusion process. *J. Inst. Met.* 93: 553–59

R. J. Fiorentino

Hygroscopicity and Water Sorption of Wood

In common with other lignocellulosic materials, wood is hygroscopic; that is, it can adsorb or desorb water in response to changes in the relative vapor pressure of the atmosphere surrounding it. This affinity of wood for water is due primarily to hydroxyl groups which are accessible within the lignocellulosic cell wall of wood.

1. *Definitions*

Hygroscopicity of wood is expressed quantitatively in terms of the wood moisture content m, which is usually defined on a dry weight basis (kg of water per kg of dry wood) at equilibrium with a given relative vapor pressure h or percentage relative humidity H, where $H = 100h$. The curve relating m to h is defined as the sorption isotherm, and the equilibrium value of m for a given h is designated as the equilibrium moisture content.

Moisture content based on wet weight is related to m by

$$m_w = m/(1 + m) \tag{1}$$

Moisture content is frequently expressed on a percentage basis, designated as M, where $M = 100m$. It may also be expressed in terms of concentration c, defined as $c = \rho_0 m$, where ρ_0 is the wood density based on dry mass and volume at m.

Water in green wood (i.e., fresh from the living tree) occurs in three forms: "free" or capillary water contained in the cell cavities, "bound" or hygroscopic water contained in the cell walls, and water vapor contained in the air spaces in those cell cavities not completely filled with liquid water. When green wood is subjected to normal atmospheric conditions, it loses all of the water in the cell cavities as it dries to a moisture content at equilibrium with the water vapor in the atmosphere.

The moisture content at which all of the cell cavity or free water has been lost but the cell walls are still saturated with water has been designated by Tiemann as the fiber-saturation point m_f or M_f. This is a critical moisture content because below M_f changes occur in most of the important physical properties of wood which vary with moisture content. These include changes in dimensions, mechanical properties, electrical properties and treatability with preservatives. This article is concerned primarily with wood containing hygroscopic or bound water only (except for the small amount of water vapor in the cell cavities), since this is the moisture condition in which wood is generally used.

The fiber saturation point M_f is of the order of 25–30% of the dry wood weight for most woods of the USA, but may be considerably lower for woods with high extractive content such as many tropical woods. M_f is not a precise point since capillary condensation may occur in fine interstices in the cell wall at high relative vapor pressure close to unity. Furthermore, M_f is lower for adsorption than for desorption and tends to decrease with increasing temperature at a rate of approximately 1% per 10 °C temperature rise above 0 °C. Below 0 °C the reverse phenomenon occurs, that is, M_f decreases with decreasing temperature.

2. *Measurement of Wood Moisture Content*

Wood moisture content has been measured by more than a dozen different methods. Only those more commonly used are mentioned here; these include the oven-drying, distillation and electrical measurement methods.

In the conventional oven-drying method, the weight loss from a moist sample is determined gravimetrically after it has been dried in an oven at 103 ± 2 °C to constant weight. The weight loss, presumed to be equal to the moisture loss, is divided by the oven-dry weight to obtain m, or M on a percentage basis. There are several errors which may be associated with this method, such as the loss of volatile extractives which cannot be distinguished from moisture loss and the presence of residual water vapor in the drying oven.

The distillation method is a modification of the oven-drying method wherein the moist wood is heated in a water-immiscible liquid which is a solvent for the volatile extractives. The evaporated water is condensed in a calibrated trap and measured volumetrically, thus eliminating errors due to extractives.

Electrical moisture meters are often used to measure the moisture content of wood because they give immediate results and are nondestructive to the wood. Two principal types are resistance meters, which are generally dc operated, and dielectric or power-loss meters which use ac, often at radio or microwave frequencies. The resistance type is most effective from $M = 6$ to M_f. The dielectric meters are effective at all moisture contents but are sensitive to wood density variations. Both types require correction for wood temperature and are sensitive to moisture gradients. They also require calibration for wood species, although this is less important for the resistance meters than for the dielectric or power-absorption types.

3. *Factors Affecting Moisture-Sorption Isotherms*

If green wood is allowed to equilibrate at each of several successively decreasing relative vapor pressures, to zero, all at constant temperature, the sorption isotherm obtained is defined as the initial desorption isotherm. Subsequent equilibration at successively higher values of h, to unity, gives the adsorption isotherm, which is always lower than the desorption isotherm. Repeating the desorption process produces the secondary desorption isotherm which lies below the initial desorption at higher values of h, but tends to converge toward it as h decreases (see Fig. 1).

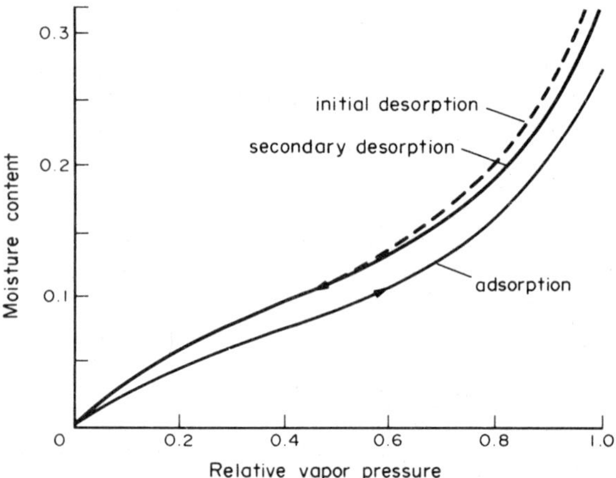

Figure 1
Typical initial desorption, adsorption and secondary desorption moisture isotherms for wood at 25 °C

Further repetition of the sorption cycle results in replication of the adsorption and secondary desorption isotherms. The difference between the reproducible adsorption and desorption isotherms is termed sorption hysteresis, usually expressed in terms of the ratio M_a/M_d, where M_a is the adsorption moisture-content value and M_d the repetitive desorption value for a given value of h. The M_a/M_d ratio usually ranges between 0.8 and 0.85.

In general, moisture-sorption isotherms are measured at room temperature. An increase in the temperature affects the hygroscopicity of wood in two ways. The first or immediate effect is to reduce the equilibrium moisture content at any constant relative vapor pressure as is shown in Fig. 2. This is a more or less reversible effect. The second or long-term effect of exposing wood to high temperatures, particularly above 100 °C, is a permanent reduction in hygroscopicity when the wood is brought back to room temperature. Unfortunately, the mechanical properties of wood are also reduced by long-term exposure to high temperatures (Stamm 1964).

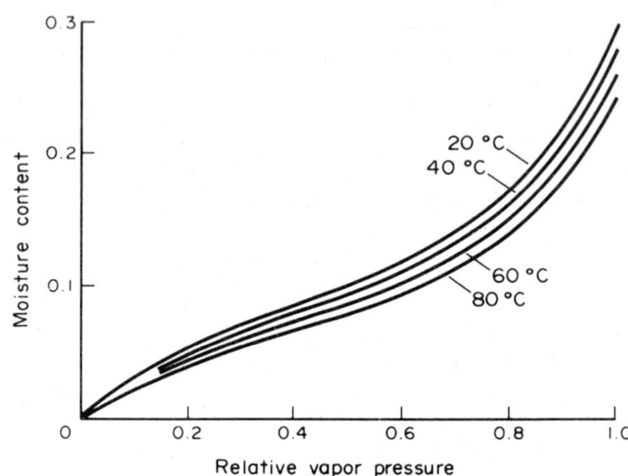

Figure 2
Effect of temperature on typical moisture isotherms for wood

4. *Thermodynamics of Moisture Sorption*

The three forms of water found in wood differ in their energy levels, as is shown schematically in Fig. 3. Water vapor has the highest level, followed by liquid water in the cell cavities with essentially the same energy as ordinary liquid water. Bound water in the cell walls of wood is in the lowest energy state, the level decreasing with decreasing moisture content below m_f. At moisture contents above m_f the energy level is essentially equal to that of liquid water.

The energy level difference Q_0 between water vapor and liquid water is the heat of vaporization of liquid water. The difference Q_v is the heat of vaporization of bound water, equivalent to $Q_0 + Q_L$, where Q_L is the differential heat of sorption of liquid water by wood. This is defined as the energy (kJ) released when 1 kg of water is taken up by wood of sufficiently large mass that m remains essentially constant. Two different methods are used to measure Q_L for wood, the isosteric or thermodynamic method and the calorimetric method.

The isosteric method for measuring Q_L depends on the fact that a linear relationship results when the sorption isotherms shown in Fig. 2 are replotted as $\ln h$ vs the reciprocal of the absolute temperature

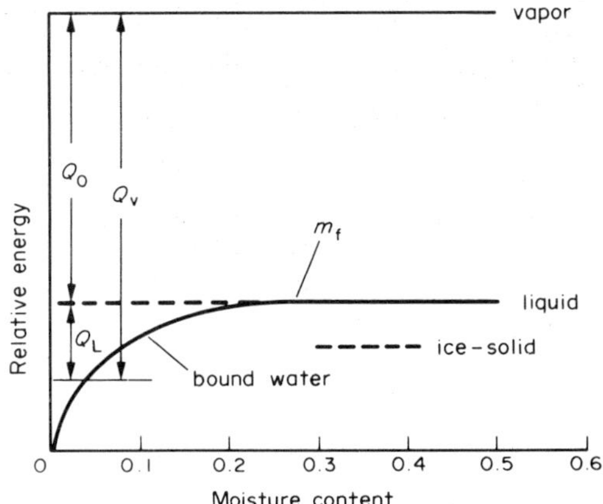

Figure 3
Relative energy levels of water vapor, liquid water and bound water in wood as functions of wood moisture content

T, the resulting curves being isosteres of constant moisture content. The slope of one of these linear isosteres is related to Q_L by

$$Q_L = -(R/0.018)(\partial \ln h/\partial T^{-1})_m \quad (2)$$

where R is the molar gas constant (kJ mol^{-1} K^{-1}) and 0.018 is the mass (kg) of water per mole. The slopes decrease with increasing m and the derived value of Q_L is a maximum for dry wood, decreasing exponentially as m increases (Fig. 4). An approximate empirical equation which applies up to m_f is

$$Q_L = \exp(a - bm) \quad (3)$$

where a and b are empirical constants, approximately equal to 7.07 and 15, respectively, when Q_L is in kJ kg^{-1}.

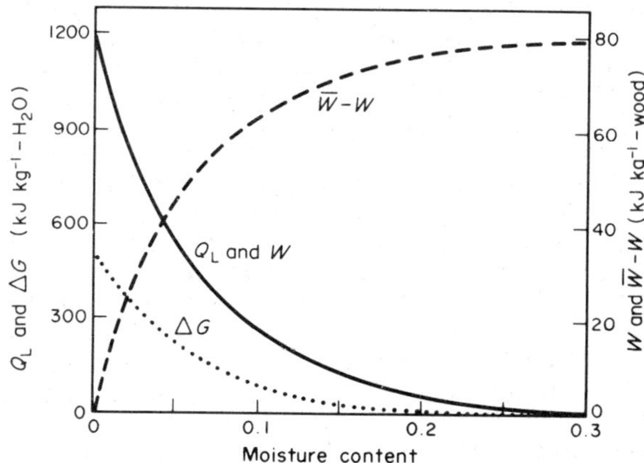

Figure 4
Variation of Q_L, ΔG, W and $\bar{W} - W$ with wood moisture content

In the calorimetric method, Q_L is not measured directly because of the experimental difficulty involved in distributing a small increment of water uniformly throughout a large quantity of wood. Rather, it is derived from measurement of the heat of wetting W, which is defined as the heat (kJ) released when 1 kg (dry weight) of finely divided wood particles, conditioned to a uniform moisture content, is wetted in an excess of water to a moisture content much greater than m_f. The relationship of Q_L and W is given by

$$Q_L = -dW/dm \quad (4)$$

If Q_L is known, W can be evaluated as

$$W = \int_m^{\infty} Q_L dm \quad (5)$$

or, using Eqn. (3),

$$W = Q_L/b = (1/b)\exp(a - bm) \quad (6)$$

It is less difficult to measure W experimentally by calorimetric methods than to measure Q_L by the isosteric method. This is because of the difficulty in measuring sorption isotherms over a range of temperatures as is required to measure Q_L.

The total heat of wetting $\bar{W}$ is the value of W when $m = 0$. Therefore, $\bar{W} = (1/b)\exp(a)$. It gives a measure of the overall hygroscopicity of the wood and appears to be proportional to the fiber saturation point m_f.

The difference $(\bar{W} - W)$ between the total heat of wetting and that measured at m greater than zero is known as the integral heat of sorption. It is the complement of W with respect to variation with m (Fig. 4).

The free-energy change ΔG (kJ kg^{-1}) associated with moisture sorption can be calculated from

$$\Delta G = (RT/0.018)(\ln h) \quad (7)$$

Since h is less than unity, ΔG is negative, the absolute magnitude increasing with decreasing h (Fig. 4). The corresponding decrease in entropy ΔS associated with sorption is given by

$$\Delta S = (Q_L - \Delta G)/T \quad (8)$$

5. *Theories of Moisture Sorption*

The several theories which have been proposed to explain the mechanism of water sorption by wood and hygroscopic polymers fall into two general categories. In one of these, the water molecules are assumed to be condensed in one or more layers on sorption sites or internal surfaces within the wood cell wall. In the second category, the polymer–water system is treated as a solution in which some of the water molecules form hydrates with sorption sites within the cell wall and the remaining water molecules form a solution.

The surface-sorption theory most often applied to wood is the Brunauer–Emmett–Teller (1938) or BET theory. This has recently been modified by Dent (1977) into a theory which better fits the sorption isotherms of wood as well as those of other natural hygroscopic materials such as cotton and wool, for which it was originally developed. The Dent theory will, therefore, be discussed here rather than the better-known BET model.

The Dent theory, in common with the BET theory, postulates that water is sorbed in two forms. The first is in the form of "primary" water molecules strongly attached to specific or primary sorption sites such as hydroxyl groups in accessible portions of the cell wall. The second form is as "secondary" water molecules attached to sites already occupied by primary water molecules or other secondary molecules. These are held by much weaker forces than those attached directly to primary sorption sites. Three fundamental constants determine the sorption isotherm according to Dent. Two of these are equilibrium constants: k_1 for the equilibrium between the primary water and liquid water, and k_2 for the equilibrium between the secondary water and liquid water. The third constant, designated here as m_0, is the moisture content corresponding to complete occupation of all the primary sorption sites with primary water (i.e., one molecule on each available site).

The Dent model predicts a sigmoid sorption isotherm of the form

$$m = h/(A + Bh - Ch^2) \quad (9)$$

where A, B and C are constants determined empirically from the experimental sorption isotherm, plotted as h/m vs h. Figure 5 shows a typical sorption isotherm. The constants are related to the three fundamental constants k_1, k_2 and m_0 as follows:

$$A = 1/(k_1 m_0)$$

$$B = (k_1 - 2k_2)/(k_1 m_0)$$

$$C = (k_1 - k_2)/(k_1 m_0) \quad (10)$$

The moisture content m_1 corresponding to the primary water content at any value of h can be calculated from

$$m_1 = m_0 k_1 h/(1 - k_2 h + k_1 h) \quad (11)$$

Likewise, m_2 is calculated from

$$m_2 = m_1 k_2 h/(1 - k_2 h) \quad (12)$$

The sum of m_1 and m_2 gives the total moisture content m. Figure 5 shows m_1 and m_2 plotted against h, as well as the total isotherm of m vs h.

The free-energy changes (kJ kg^{-1}) associated with the sorption of primary and secondary water are given by

$$\Delta G_1 = -(RT/0.018) \ln k_1$$

$$\Delta G_2 = -(RT/0.018) \ln k_2 \quad (13)$$

The values of k_1 and k_2 for wood at room temperature are approximately 10 and 0.75, respectively, and the corresponding values of ΔG_1 and ΔG_2 are -319 and $+40$ kJ kg^{-1}, respectively. The total free-energy change associated with sorption at any moisture content m can be calculated from

$$\Delta G = \Delta G_1(\partial m_1/\partial m) + \Delta G_2(\partial m_2/\partial m) \quad (14)$$

where

$$\partial m_1/\partial m = (1 - k_2 h)^2/[1 + k_2 h^2(k_1 - k_2)] \quad (15)$$

and

$$\partial m_2/\partial m = 1 - (\partial m_1/m) \quad (16)$$

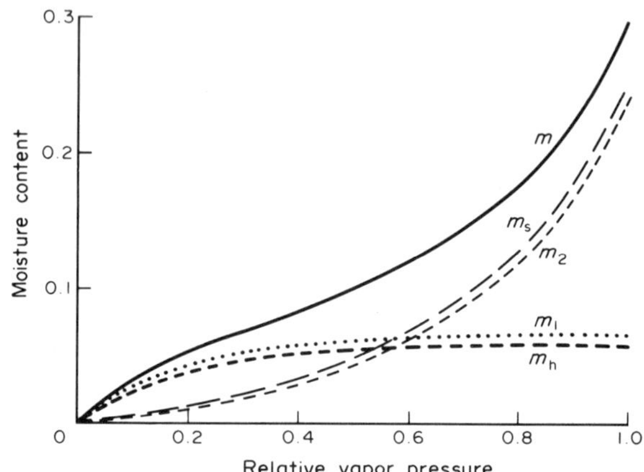

Figure 5
Typical sorption isotherms for wood (at 25 °C) and for the components m_1, m_2, m_h and m_s

The solution theory most often applied to the moisture-sorption isotherm for wood is the Hailwood–Horrobin (1946) theory. Although this is a solution theory in contrast to the Dent theory, it has certain common features with the latter. It predicts the identical sorption isotherm [Eqn. (9)] as the Dent theory with identical values for the constants A, B and C. Furthermore, two of the three fundamental constants are identical with k_2 and m_0 of the Dent theory although the third constant is somewhat larger than k_1. It also predicts two forms of water: water of hydration (m_h), analogous to the primary water (m_1) of the Dent model, and water of solution (m_s), similar to the secondary water (m_2) of the Dent model. However, the magnitude of m_h and m_s are slightly different from m_1 and m_2 at all values of m, although the general shapes of the corresponding curves are similar. Figure 5 includes curves of m_h and m_s, together with m_1 and m_2 predicted by the Dent model for the same isotherm.

See also: Drying of Wood; Shrinking and Swelling of Wood; Wood: Fluid Transport; Wood: An Overview

Bibliography

Brunauer S, Emmett P H, Teller E 1938 Adsorption of gases in multimolecular layers. *J. Am. Chem. Soc.* 60: 309–19

Dent R W 1977 A multilayer theory for gas sorption. Part I: Sorption of a single gas. *Textile Res. J.* 47: 145–52

Hailwood A J, Horrobin S 1946 Absorption of water by polymers: Analysis in terms of a single model. *Trans. Faraday Soc.* 42B: 84–92

Skaar C 1972 *Water in Wood.* Syracuse University Press, Syracuse, New York

Stamm A J 1964 *Wood and Cellulose Science.* Ronald Press, New York

C. Skaar

I

Igneous Ore Deposits: Formation Processes

Today, mineral resources have become almost synonymous with industrial power, and industrial power is in turn dependent upon ownership of, or access to, large quantities of mineral resources. It is startling to realize that the rise of the industrial age has so accelerated the demand for minerals that the world has dug up and consumed more of its mineral resources since the beginning of World War II than in all preceding periods. This insatiable demand for minerals to feed industry has made some sources of supply that we used to think were adequate now look rather small, and sources capable of meeting large demands are being depleted rapidly.

More than a century and a half ago, the rate of increase in the consumption of copper and iron was directly proportional to the increase in population. If the population doubled, the consumption of these metals doubled. In contrast, during the first hundred years of the Industrial Revolution, roughly from 1812 to 1912, the population increased fourfold whereas the consumption of copper increased 80 times and that of iron 100 times.

The quest for the wealth of minerals wrested from the earth for man's vanity, necessities, or comforts has always been a powerful incentive to discover, explore and trade. Geologists study the occurrences of mineral deposits, their mode of formation, the types of rocks in which they occur, and use geophysical tools and geochemical processes as aids in attempts to locate deposits.

Geologists have been aided in the search for mineral deposits by devising several means of classifying the variety of mineral and rock concentrations. Some classifications refer to the geometry of a deposit, e.g., tabular or veined, flat or stratigraphic, irregular or "bunched," and numerous fine veins or fissures. Another classification is by time of formation. Deposits formed at the same time as the host rocks are called syngenetic, and if the mineralization invaded pre-existing host rocks, epigenetic.

Classifications based on temperatures, pressures, reduction–oxidation conditions and partial pressures of fluids provide much information on conditions of formation of mineral deposits. Of course, these factors are not easily determined but relating the results of laboratory experiments to detailed observations of field relationships of minerals and rocks has provided much understanding of the conditions of formation of specific types of deposits.

Lastly, as many mineral deposits form only in specific kinds of rocks, a classification based on the mode of formation and associated rock types can aid the exploration geologist in the search for new deposits. Rocks are separated into three groups based on their mode of origin. Igneous rocks are formed from molten magma, plus its dissolved fluids, either intrusive (formed below the surface of the earth) or extrusive (formed on the surface of the earth); sedimentary rocks are formed by deposition in seas or lakes, or are deposited as thin layers on the surface of the earth; and metamorphic rocks are formed from pre-existing rocks in environments of high temperatures and pressures existing below the surface of the earth.

Igneous rocks are aggregates of minerals that form by precipitation or crystallization from a melt or magma, i.e., molten rock plus its dissolved fluids (gas and liquid). Evidence from plate tectonics suggests that the provenance of most magmas is at plate margins, a result of sea floor spreading or subducting plates. At depths from some tens to several hundred kilometers, magmas are formed by partial melting of specific constituents which then rise from the mantle or lower crust as somewhat plastic masses and may adsorb water when reaching upper crustal, water bearing sediments. Igneous rocks are classified by: (a) texture, or the size, shape and arrangement of the mineral grains in the rocks, and (b) the composition of the minerals in the rocks.

1. *Intrusive Rocks*

The felsic rocks generally contain about 70–80 wt% SiO_2 with a high content of quartz and alkali feldspars ($KAlSi_3O_8$ and $NaAlSi_3O_8$). Since Si, Al and alkalis are important determinants of viscosity, the magma that forms with these minerals has a high viscosity. In contrast, mafic magmas containing higher charged ions such as Ca, Fe or Mg, and Si-poor or late stage differentiates of felsic magmas which are OH-rich, are less viscous or moderately fluid. Rhyolite magma, high in SiO_2, is highly viscous and even when cooled slowly forms a glass which leaves the silica tetrahedrons $(SiO_4)^{4-}$ in random positions. In contrast mafic magma may be cooled at the same rate or even more rapidly but because of its high fluidity, mineral crystals have sufficient time to form. Because of this high fluidity, mafic glasses are rare.

2. *Magmatic Deposits and Processes*

Magmatic deposits result from simple crystallization or from concentration by differentiation of mafic or ultramafic magmas. Magmatic deposits are characterized by their close relationship with ultramafic deep-seated igneous rocks commonly derived from the upper mantle and include diamonds, nickel, platinum group minerals, titanium–iron, chromite and other valuable ores.

The valuable metals deposited are intermixed or juxtaposed with the igneous body. They may form only during each crystallization in the igneous rock, or they may form in a continuous manner as the magma slowly crystallizes, or they may remain in the liquidus until the body has almost completely crystallized. The times of formation will result in different concentrations of the metals because of varying solubilities of the minerals in the solution melt. A diagram showing this process and the effects of gravity settling and forming a liquid deposit is shown in Fig. 1.

Some deposits consist of disseminated ore minerals. If such crystals are valuable and concentrated enough, the result is a magmatic mineral deposit. Diamonds disseminated in the kimberlite pipes of South Africa are examples. Other bodies form as either early or late segregations. The Bushveld Igneous Complex (BIC) with chromite and platinum minerals, and Allard Lake, Quebec, with titaniferous hematite are examples. The Muskox intrusion, Northwest Territories, Canada, the Duluth Complex, Minnesota, the Stillwater Complex, Montana, USA, and the Skaergaard intrusion, Greenland, are additional examples with specific economic materials. The Skaergaard and the Muskox intrusions have occurrences of sulfides but not in economic concentrations, and the Stillwater complex is being explored for economic platinum–palladium deposits. There are many other ultramafic igneous bodies throughout the world which are of interest because of their size and, therefore, their potential for economic mineral production.

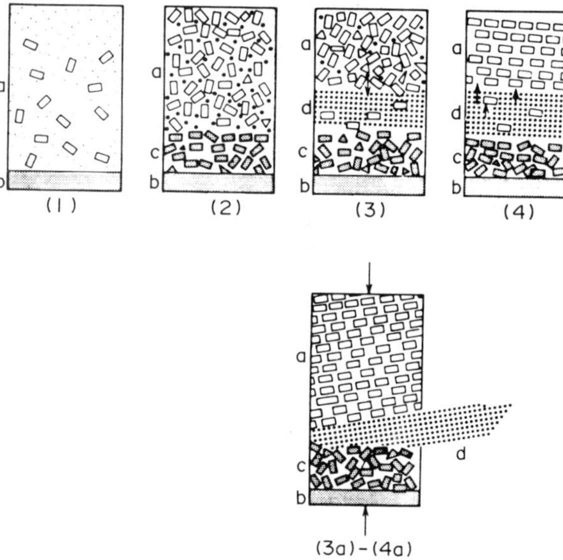

Figure 1
Idealized diagrammatic representation of late gravitational liquid accumulation. (1) Early stage of crystallization of basic magma a, after formation of chill zone b; (2) layer of sunken early-formed ferromagnesian crystals c, resting on chill zone b, with mesh of later silicate crystals above, whose interstices are occupied by residual magma enriched in ore oxides; (3) mobile, oxide-rich, residual liquid draining down to layer d, and, floating up, later silicate crystals; (4) formation of concordant oxide ore body in which a few late silicate crystals are trapped as mobile, enriched gravitational accumulations are squeezed out or decanted to form the magmatic injections (after Jensen and Bateman 1980)

3. *Hydrothermal Deposits and Processes*

The most common intrusive rocks are granite, granodiorite, diorite and monzonite. They generally form nearer the surface than ultramafic rocks and are associated with subducting plate boundaries. They generally form as a magma at some depth and rise until they begin to crystallize in the overlying crustal rocks. Those portions of the magma that reach the earth's surface are extrusive igneous rocks. They cool rapidly and develop either a glassy or fine-grained texture. Felsic magmas become enriched in water as the magma moves upward. This water is gained either from the formations that are intruded or partially melted or it is magmatic water, residually concentrated by the crystallization of other components. Water content may increase from less than 1% in the upper mantle to 10% in the upper crust. This water acts as a fluidizer and mineralizer and allows the magma to crystallize at relatively low temperatures, 625 °C and less. In the sequence of magmatic crystallization, the nonvolatile minerals crystallize first with anorthite-rich plagioclase, pyroxenes and amphiboles. Na-rich plagioclase (albite), orthoclase, micas and quartz form late with volatile-rich aqueous solutions the last to crystallize.

As the volatile content increases in amount, the metals begin to form complexes such as $CuCl_4^{2-}$, $PbCl^+$ and $PbCl_3^-$. Laboratory research has indicated that complexes of metal chlorides, especially lead, silver and many other metals, are relatively soluble at temperatures of a few hundred °C. In contrast, these metals alone or as sulfides are essentially insoluble in aqueous solutions at many hundreds of °C. The phenomenon of metal transport into veins, fractures, or pores of rocks has been a major problem to geologists. Certainly strong acids could dissolve some of the metals and metal sulfides but these veinlet paths are commonly seen crossing limestone beds with no reaction.

In the last few decades, deep wells in the Salton Sea, brines of the Red Sea and fluid inclusions have shown that the concentrations of the metals in these solutions are surprisingly high. For more than a century, fluid inclusions (Fig. 2) have been studied in thin slices of minerals with a polarizing microscope. These fluid filled cavities are microsamples of the solution in which the mineral was growing. As the fluid was trapped by crystal growth, it was retained and is available now for analysis. The inclusions contain predominantly water and/or liquid carbon dioxide but brine solutions in the inclusion sometimes make up from a few percent to 30–40% NaCl.

Figure 2
Large primary inclusions just under striated surface of clean, zoned sphalerite crystal from Cripple Creek, Colorado (after Roedder, US Geological Survey Professional Paper 440JJ)

Recently, laboratory and computer studies have indicated that hydrochloric acid (HCl) does not dissociate into H^+ and Cl^- ions at temperatures of a few hundred °C. Metals can be transported as relatively soluble chloride complexes in only slightly acid solutions because the HCl with its low dissociation constant is not an acid at all at the mineralizing temperatures. As the temperature decreases, the HCl begins to dissociate into H^+ and Cl^- ions, with a resulting decrease in pH; the metal chlorides react with H_2S resulting in the deposition of metal sulfides.

4. Hydrothermal Alteration

Released H^+ ions of the hydrothermal solution react with existing minerals which results in chemical changes and the formation of new minerals. As the mineralizing fluids are much more abundant and more widespread than the smaller hidden ore deposit, recognition of hydrothermal alteration of surface bed rock is a prime tool that may lead to a buried economically viable ore deposit.

Feldspar minerals, such as orthoclase, are prime examples of minerals that undergo hydrothermal alteration. The supposed reaction is:

$$3KAlSi_3O_8(\text{orthoclase}) + 2H^+ \rightarrow KAl_2AlSi_2O_{10}(OH)_2(\text{sericite}) + 2K^+ + 6SiO_2$$

Note that only the H^+ ion is needed for the reaction. Because of this, some types of hydrothermal alteration are referred to as hydrogen ion metasomatism. Sericite is a fine grained, white mica. It is soft and is easily distinguished from unaltered orthoclase which is hard and cannot be scratched with a knife.

Silicification is often associated with hydrothermal alteration or mineralization. Note in the reaction above that silica (SiO_2) is released during the alteration of orthoclase to sericite. Thus silicification commonly occurs with mineralization.

Argillic alteration results from hydrothermal alteration which leaches lime and forms clay minerals. Commonly, kaolinite, montmorillonite, dickite, halloysite, and illite develop in altered rocks with the argillic alteration varying from weak to advanced. This type of hydrothermal alteration is common in porphyry copper and molybdenum deposits.

Stable isotopic studies are now applied to problems of sources of fluids, especially water (D/H and $^{18}O/^{16}O$ ratios) and are used to determine temperatures of mineral formation and hydrothermal alteration. Because of isotopic exchange equilibrium reactions, H_2O may exchange oxygen isotopes with minerals forming the wall rocks. Figure 3 illustrates the isopleths or equal temperature zones, determined by isotopic measurements with the highest temperature near the ore where the lowest $^{18}O/^{16}O$ ratios were found.

Other types of wall-rock or hydrothermal alteration exist, for example, pyritization, aluminization jasperoid and many others.

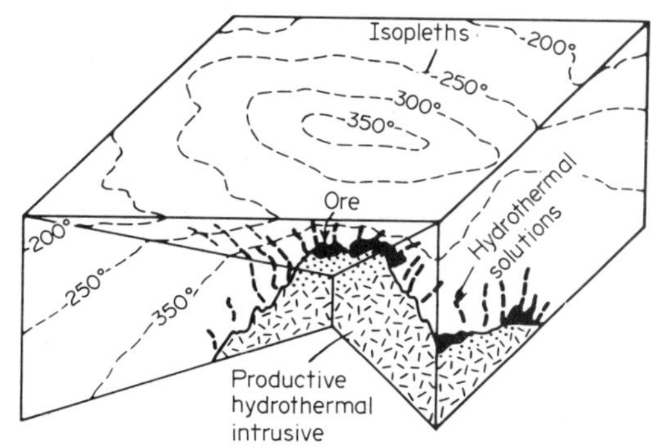

Figure 3
Idealized potential application of ^{18}O studies to hydrothermally altered contact zones near a mineralizing fluid source of water permeating carbonate rocks (after Jensen and Bateman 1980)

5. Replacement and Vein Deposits

The large replacement lead–zinc deposits of Leadville and Gilman, Colorado, were some of the first deposits to be recognized as replacement of limestone by galena (PbS) and sphalerite (ZnS). The process is not fully understood, but the evidence is conclusive that pre-existing limestone and dolomite formations had been selectively removed and replaced by sulfide minerals. Some original fine textures are retained indicating the process must have been a removal of minute portions of the carbonate rocks occupied almost simultaneously by the sulfide minerals. If the process continued for long periods, for example, hundreds of thousands of years, large deposits resulted.

In the Tintic mining district, Juab County, Utah, silver–lead–mineralization exists in fissure veins and selective replacement deposits. The replacement deposits formed predominantly in the Devonian Bluebell dolomite that is mineralized over a horizontal distance in excess of 3 km. Other formations are also replaced but of the total $425 000 000 gross value of ore mined in the district, $110 000 000 was from the Bluebird dolomite.

Hydrothermal replacement and vein deposits are not limited to sulfide ore bodies. Gold exists as hydrothermal replacement occurrences in the Homestake, South Dakota mine and in the Carlin Mine, Nevada. The Mother lode of California is a hydrothermal lode (closely spaced veins) system, as is the famous Virginia City silver vein, Nevada. Tin occurs in hydrothermal vein systems; the Llallagna deposit in Bolivia is a good example. Iron ore replacement deposits are abundant, as in the Iron Springs magnetite deposits in southern Utah where magnetite has formed contact metasomatic, hydrothermal replacement deposits in the host Paleozoic limestone formation.

6. Porphyry Deposits

These deposits, whether porphyry copper, molybdenum, or porphyry tin, are closely associated with intrusions of felsic magma, usually quartz monzonite or quartz diorite, generally porphyritic and almost all Mesozoic or Tertiary in age. Porphyritic igneous rocks have large phenocrysts (crystals) surrounded by a finer grained groundmass. The New Guinea highland deposits are comparably much younger being mid-to-late Miocene in age with some ore deposits believed to be Pleistocene.

Mineralization occurs as disseminations of sulfide minerals within or near the igneous body, the parent source of the ore minerals, but formed in the last phase of the crystallization of the intrusion. In some cases, the volatile pressure is so high and with a thin over-burden, that shattering of the cupola of the intrusion results. This provides innumerable (planar) fractures that provide the "plumbing system" for the alteration fluids and ore bearing fluids to permeate the cupola and form disseminated economic ore deposits. The grade rarely exceeds a few percent sulfides, predominantly pyrite, and ore sulfides of less to slightly more than 1%. A large porphyry copper deposit should contain no less than 0.4 to 0.6% copper to be profitable to mine.

Many of the deposits have undergone leaching from meteoric waters that react with pyrite to produce ferric sulfate $Fe_2(SO_4)_3$ and sulfuric acid (H_2SO_4) that dissolve copper sulfides (Cu_2S, $CuFeS_2$ and CuS) and form soluble sulfates. These percolate downward until the oxidation–reduction boundary is reached; there, the metal-carrying solutions replace pyrite and commonly form chalcocite pseudomorphs after pyrite. These are supergene deposits containing several percent copper and are zones of secondarily enriched, high-grade material.

About 80% of US copper is derived from porphyry copper deposits. Their origin, especially the origin of the felsic intrusions that gave rise to the deposits, is intimately related to plate tectonics and subduction zones that were active at the edges of continental plates. The deposits are numerous in the Andes along the western portion of South America; from there, they extend into North America in New Mexico, Arizona, Utah and Nevada, and into the cordillera of British Columbia.

Subducted oceanic plates in the east Pacific have resulted in numerous deposits in the Philippine Islands, Bougainville and Indonesia. Deposits also occur along the Himalaya–Alpine chain, in Afghanistan, Iran, Turkey and the Balkans.

The search continues for these large volume, multimillion tonne, low-grade porphyry deposits. Some contain copper ore that may have a gross value of about $10 per tonne but where 100 000 tonnes per day may be mined by comparatively low-cost, open-pit methods, they may provide a gross production of about $1 000 000 per day.

7. Extrusive Igneous Rocks

Many mineral deposits are related to extrusive igneous rocks, specifically volcanic rocks. Many epithermal, that is, low temperature and near surface deposits, are often enclosed in altered volcanic rocks of felsic composition, generally latites, dacites, and andesites. Many of the silver and gold deposits of Nevada are replacement fissure vein or disseminated deposits in volcanic rocks. Examples in Nevada are Carlin, Goldfield, Tonopah, Virginia City, Rhyolite, Golden Arrow, Reveille and the Divide district, as well as many other smaller deposits. Portions of the Ruth porphyry copper deposit, Nevada, involve pre-ore volcanic rocks that are hydrothermally altered and mineralized.

The Frisco mine in Millard County, Utah, formed

along a vertical fault separating limestone from felsic volcanic rocks. The limestone was a much less favorable host replacement rock than the extensively replaced and mineralized volcanic rock.

8. Exhalative Submarine Volcanogenic Ores

Ore deposits of this genesis are being recognized in large numbers. This is not only because of new discoveries but also as a result of reclassifying some deposits formerly referred to as massive sulfide deposits.

The fairly recently described process of volcanogenic ore formation applies to deposits formed in shallow seas by submarine volcanism accompanied by mineralizing fluids that result in stratabound, massive sulfide deposits. Generally, the ore zones are sandwiched between overlying felsic volcanic rocks and underlying mafic volcanic rocks. As this process may extend laterally, associated with volcanic ores during a volcanogenic epoch, the search and recognition in continents for such deposits can be concentrated in former volcanic arcs along the contacts of andesitic and rhyolitic rocks.

In the Mattagami district, Quebec, Canada, such contacts can be traced along a westward plunging anticline for more than 35 km. Several ore bodies, such as Bell Channel, Radiore "A," Bell Allard, Orchan, New Hosco, and Mallard Lake occur between the two volcanic groups of the younger Wabasse group (andesite, basalt, dacite and rhyodacite) that form the outer limbs of the anticline with the older Watson Lake group, predominantly rhyolite. The ore bodies of massive Fe–Zn–Cu sulfides are stratabound, of Archean age, and highly metamorphosed.

Other exhalative submarine volcanogenic deposits include many formerly classified as massive sulfide deposits, such as the deposits of the Noranda area, the Bathurst–Newcastle area, New Brunswick, and Japanese Koroko deposits. The Flin Flon district, Saskatchewan and Manitoba, Kidd Creek, Timmons district, Ontario, and Mount Isa, Queensland, Australia are also considered to be volcanogenic.

See also: Metamorphic Ore Deposits: Formation Processes; Sedimentary Ore Deposits: Formation Processes; Mineral Resources: An Overview

Bibliography

Barnes H L (ed.) 1967 *Geochemistry of Hydrothermal Ore Deposits*. Holt, Rinehart and Winston, New York

Jensen M L, Bateman A M 1980 *Economic Mineral Deposits*, 3rd edn. Wiley, New York

Lindgren W 1933 *Mineral Deposits*, 4th edn. McGraw-Hill, New York

Park C F, McDiarmid R A 1975 *Ore Deposits*, 3rd edn. McGraw-Hill, New York

Peters W C 1978 *Exploration Mining and Geology*. Wiley, New York

Ridge J D 1968 *Ore Deposits of the United States 1933–1967*. American Institute of Mining and Metallurgical Engineers, New York

Stanton R L 1972 *Ore Petrology*, McGraw-Hill, New York

M. L. Jensen

Image Quality Indicators

The International Standards Organization (ISO) defines an image quality indicator (IQI) as a device which, appearing on a radiograph, is used to judge the overall quality of that radiograph. Other definitions are similar, but the design of the IQI varies throughout the world. The IQI is normally placed in the source side of the object and radiographed with it. Thus, the image of the IQI appears on the radiograph.

1. Image Quality Indicators Recommended by the International Standards Organization

The ISO recognizes two types of IQI, the wire type (Fig. 1a), and the step-and-hole type (Fig. 1b). The step-and-hole type may be as shown or it may be arranged in a circular or simple stair-step geometry. The indicators are made of the same material as the part being inspected or of a material with an almost equivalent coefficient of absorption.

The wires are mounted parallel and arranged in order of increasing diameter, each being larger than the preceding one by a factor of $10^{0.1}$ or, for large material thicknesses, $10^{0.05}$.

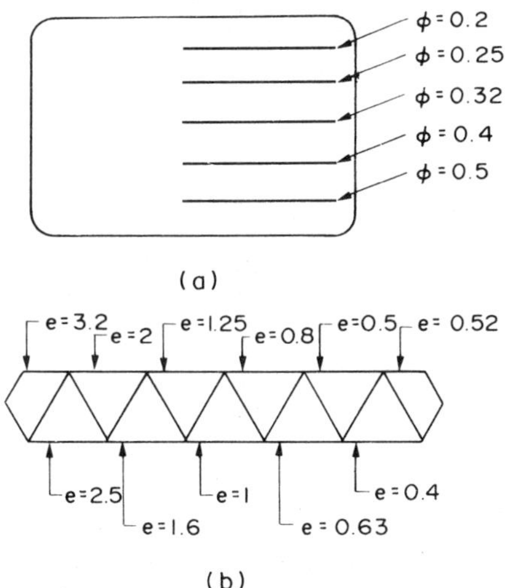

Figure 1
Typical ISO designs for IQIs: (a) wire type; (b) step-and-hole type (in each step, holes are drilled as required). ϕ is the wire diameter and e is step thickness (both in mm)

In the step-and-hole indicator, each step has one or more circular holes of a diameter equal to the step thickness T. The thickness of successive steps again increases by the factor $10^{0.1}$. Steps with a thickness greater than 0.8 mm have a single hole while smaller steps may have two or more holes. The distance from the center of the hole to a step edge or between the edges of two holes is not less than the hole diameter plus 1 mm. For the inspection of thick materials, the thickness of successive IQI steps increases by a factor of $10^{0.05}$.

Each IQI carries symbols or numbers that identify the material from which it is made, the specification or standard with which it complies, the largest diameter or step thickness and the factor chosen to differentiate between steps.

2. *Image Quality Indicators Recommended by the American Society for Testing and Materials*

The IQI recommended by the American Society for Testing and Materials (ASTM), shown in Fig. 2, is commonly called a plaque-and-hole penetrameter. (The word penetrameter is synonymous with image quality indicator.) The indicators are made of the same or equivalent materials as the part being inspected. The identification numbers indicate the IQI thickness in thousandths of an inch, e.g., 5 represents an indicator thickness of 0.005 in. (0.127 mm). For all indicators from 5 through 10, the hole diameters are constant and equal to 0.010 in. (0.254 mm), 0.020 in. (0.508 mm) and 0.040 in. (1.016 mm), respectively. For indicators above 10, the holes have diameters of $1T$, $2T$ and $4T$ where T is the indicator thickness.

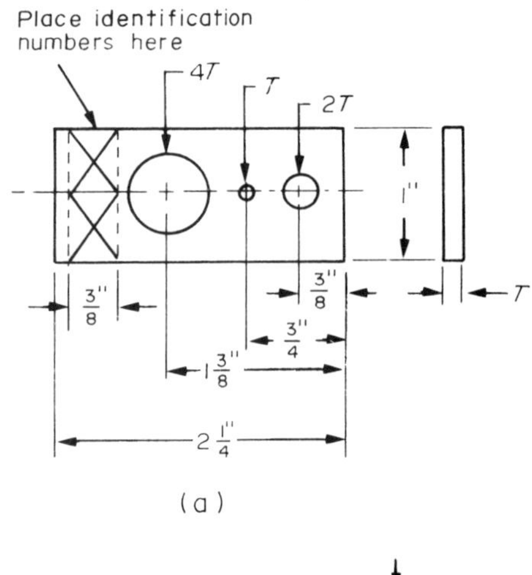

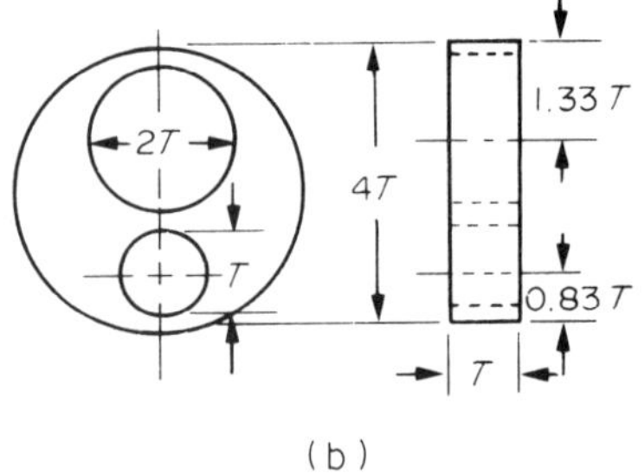

Figure 2
Typical ASTM penetrameter designs: (a) for thicknesses of 0.06–0.16 in.; (b) for thicknesses in excess of 0.18 in.

3. *Equivalence of Image Quality Indicator Systems*

The radiographic sensitivity that is obtained can have different numerical values depending upon the type of IQI used. With the wire type, the radiographic sensitivity is equal to the diameter of the smallest wire visible expressed as a percentage of the total thickness of material radiographed. With the step-and-hole type, the sensitivity is equal to the diameter of the smallest hole visible expressed as a percentage of the total thickness of material radiographed. With plaque-and-hole type penetrameters, the measure of sensitivity is based on the following definition. For 2% radiographic sensitivity a hole whose diameter is twice the plaque thickness shall be detected in a plaque whose thickness is 2% of the material being inspected.

The relation between the ASTM indicator and the ISO indicator can be calculated from the following formula which gives an equivalent indicator sensitivity in terms of the ASTM definition:

$$\text{Sensitivity}\ (\%) = (AB/2)^{0.5}$$

where A is the smallest plaque thickness detectable as a percentage of the total material thickness, and B is the smallest hole diameter detectable as a percentage of the total material thickness. Setting A and B equal to 2 (2% ISO sensitivity), we find that the equivalent penetrameter sensitivity in ASTM terminology would be 1.4%. The relation between the wire and the plaque-and-hole IQI is given by the formula

$$(Fd)^3 l = \frac{\pi}{4} M^2 h^4$$

where F is 0.79, a form factor for the wire; d is the wire diameter; l is the field of appreciation of the eye (7.62 mm); M is the ratio of hole diameter to depth; and h is the hole depth. Thus for a given technique where, for example, a 0.50 mm hole in a 0.50 mm plaque is detectable through 25.0 mm of material (ISO 2% sensitivity) then the detectable wire diameter is 0.235 mm (ISO 0.94% sensitivity). Therefore, in order to compare radiographic sensitivity data from different countries and facilities one must determine which type of image quality indicator was used.

See also: Nondestructive Evaluation: An Overview; Radiographic Nondestructive Evaluation

Bibliography

American Society for Testing and Materials 1979 *Annual Book of ASTM Standards,* Part 11. American Society for Testing and Materials, Philadelphia, Pennsylvania

Association Française de Normalisation. *Radiographic Examination of Steel of Thickness up to and Including 180 mm*, AFNOR A04-304. Association Française de Normalisation, Paris

British Standards Institute. *Image Quality Indicators for Radiography and Recommendation for Their Use*, BS3971. British Standards Institute, London

Criscuolo E L 1963 Correlation of radiographic penetrameters. *Mater. Res. Stand.* 3: 465–71

Deutsches Institut für Normung. *Nondestructive Testing: Image Quality of X-Ray and Gamma-Ray Radiographs of Metallic Materials, Definitions, Wire Penetrameters*, DIN54109. Deutsches Institut für Normung, Berlin

International Standards Organization 1969 *Radiographic Image Quality Indicators: Principles and Identification*, ISO 1027. International Standards Organization, Geneva

Standards Association of Australia 1979 *Image Quality Indicators (IQI) and Recommendations for Their Use*, AS2177 Pt. 2. Standards Association of Australia, Canberra

D. Polansky

Impact Strength of Polymers

The impact strength of a polymer is its ability to withstand shock loading (or impact). Since most polymeric components will be subject to impacts of one kind or another during their service life, it is essential that the chosen polymer possesses at least a minimum resistance to impact fracture. Impact fractures generally originate at points of high stress concentration, such as at changes of section, surface flaws, or notches. When such polymers are fractured by impact blows, fracture is frequently brittle, even though the same polymers may show considerable ductility before fracture in a standard tension test.

1. Impact Testing

The determination of impact energy (or impact strength) of a solid polymer is somewhat arbitrary, in view of the unknown nature of the impact blows likely to be encountered in handling and in service. Two of the standard test methods are the Izod and Charpy impact tests (Fig. 1), which are described in BS 2782 and ASTM D256-73. In the Izod apparatus a notched specimen is struck by a pendulum striker at a point 22 mm above the notch. In the Charpy apparatus, the notched specimen is placed between supports 95.3 mm apart and struck at the center directly behind the notch. Neglecting other losses, such as those due to friction, the impact strength of the specimen is determined from the energy given up by the pendulum in striking the test sample per unit of specimen width. Alternatively, the impact strength may be taken as the fracture energy per unit area of fracture surface. With appropriate modification, the Izod and Charpy apparatus can be used for investigating impact behavior under tension rather than in bending. Another type of impact test, used mostly for films, determines impact fracture energy by dropping weights, such as a ball or a dart, onto a clamped specimen (ASTM D1709-75).

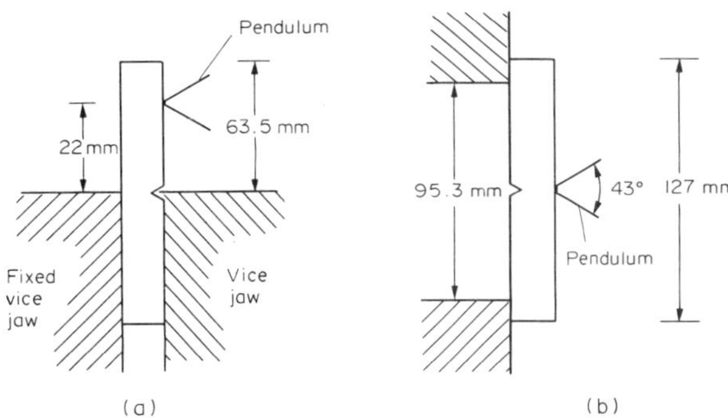

Figure 1
Schematic representation of (a) Izod and (b) Charpy impact test arrangement

2. Effect of Composition and Material Variables

The impact strengths of polymers cover a wide range; various impact-strength values can be obtained for a particular polymer, depending on the fabrication conditions, the presence of additives and such external conditions as temperature, the strain rate and the geometry of the notch. Nevertheless, some general conclusions as to the impact strength of different polymers at ambient temperature can be made. One classification, with examples, is:

(a) Brittle, even unnotched—polystyrene, poly(methyl methacrylate) and cast phenolic.

(b) Brittle, if bluntly notched—polypropylene and polyoxymethylene.

(c) Brittle, if sharply notched—poly(vinyl chloride), high-density polyethylene, nylon (dry) and poly(2,6-dimethyl-1,4-phenylene ether).

(d) Tough, even sharply notched—low-density polyethylene, polycarbonate and nylon (wet).

The impact strength of many polymers, even brittle ones, can be increased by appropriate modification. Two examples of this are evident from the above classification; low-density polyethylene has superior impact resistance to the more crystalline high-density polyethylene, and nylon saturated with water is superior in impact strength to dry nylon. Impact strength may also be increased by copolymerization (e.g., polypropylene shows increased impact strength on copolymerization with ethylene, owing to reduced crystallinity and increased ductility), and by rubber modification, particularly if the rubber particles are well bonded and well dispersed. The second-phase particles cause increased matrix crazing and plastic deformation, and these energy-absorbing processes lead to better impact resistance. Thus acrylonitrile–butadiene–styrene (ABS) has much higher impact strength than unfilled styrene–acrylonitrile copolymer (SAN). Rubber modification is used to raise the impact strength of many polymers, including amorphous ones like poly(methyl methacrylate) and poly(vinyl chloride) (PVC), crystalline ones such as polypropylene and nylon, and thermosets such as epoxy resins.

Other types of fillers also affect impact strength. Reinforcement of polystyrene with 15% polyester fibers is reported to increase impact strength by a factor of about four, while addition of 30% glass fibers can raise the impact strength of brittle materials, such as epoxy and phenolic resins, by ten times or more. However, for ductile resins, like ABS or polycarbonate, addition of glass fibers tends to lower impact strength. Plasticizers, like water in nylon, raise ductility and increase impact resistance.

Molecular weight and molecular orientation also play a role in impact performance. With higher molecular weight, there are more molecular entanglements and impact strength rises. Ultrahigh-molecular-weight polyethylene, for example, has exceptionally good impact resistance. Orientation raises impact strength if flow lines are normal to the plane of the notch but lowers impact resistance when flow lines are parallel to the plane of the notch.

Many investigators have reported a correlation between impact strength and damping behavior for a variety of polymers. Figure 2 shows the temperature dependence of impact strength of five different thermoplastics. Of these, polycarbonate has the largest low-temperature mechanical loss, while poly(2,6-dimethyl-1,4-phenylene ether) (PPO) has several overlapping low-temperature damping peaks. PVC and many other polymers show a rapid rise in impact strength over a relatively narrow temperature range; at this temperature (T_b), they essentially undergo a transition from a brittle fracture mode to a ductile fracture mode. For polystyrene and poly(methyl methacrylate), T_b is close to the glass-transition temperature, for PVC it is about 40 °C, for PPO about 20 °C, and for polycarbonate about −20 °C. The ductile–brittle transition temperature for polycarbonate can be further lowered to −110 °C by forming a block copolymer with 25% silicone.

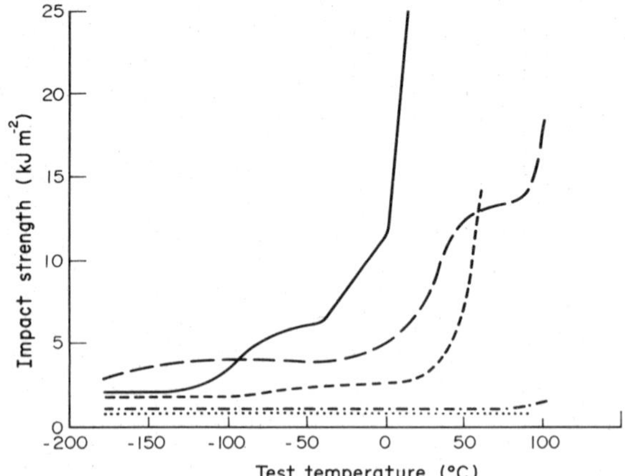

Figure 2
Impact strength against test temperature for five thermoplastics tested at a notch radius of 0.25 mm: ——, polycarbonate; — —, poly(2,6-dimethyl-1,4-phenylene ether); - - -, poly(vinyl chloride); –·–, poly(methyl methacrylate); ····, polystyrene (after Vincent 1974. © John Wiley and Sons, Chichester. Reproduced with permission)

3. *Effect of Other Variables*

The geometry of the test piece can greatly alter the recorded impact strength of a given polymer. For example, impact strength will be increased by decreasing the specimen width, as this accentuates plane stress conditions rather than plane strain; decreasing the notch depth, as this lowers the stress concentration factor; and increasing the notch radius, as this blunts the notch and makes it more difficult to initiate a crack.

The influence of notch tip radius on impact strength for a variety of polymers is shown in Fig. 3. For a notch radius of 0.125 mm, the impact strength of the homopolymers is relatively low, and ABS is superior in impact strength to the other polymers. As the notch radius is increased, however, the increase in impact strength is more rapid for PVC, nylon and acetal than ABS. Hence these polymers become superior to ABS as the notches become blunter. This behavior results from the fact that ABS, with its dispersed rubber particles, offers greater resistance to crack propagation than the other polymers, which offer greater resistance to crack initiation.

The impact strength of polymers also depends on fabrication conditions. Rapid cooling frequently leads to a polymer which is more ductile and has higher impact resistance. Annealing near the glass-transition temperature tends to raise the Young's modulus but to reduce ductility and impact toughness. In rubber-modified polymers, control of rubber particle size is important for optimum impact performance. Impact strength of thermoplastics, such as

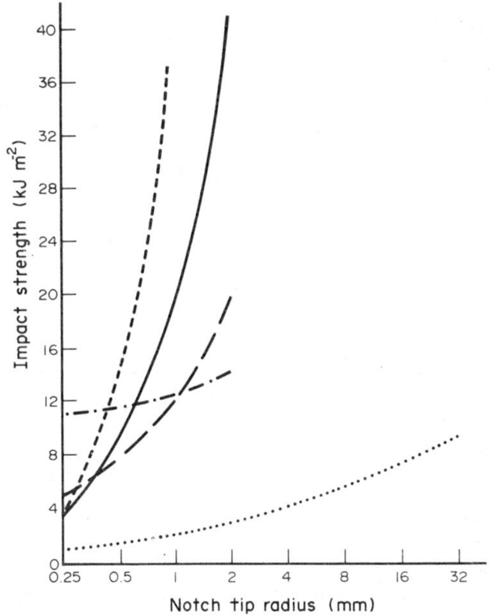

Figure 3
Impact strength against notch tip radius for various polymers: ——, nylon; — —, acetal; - - -, poly(vinyl chloride); -·-, acrylonitrile butadiene–styrene; ····, acrylic (after Vincent 1971)

PVC, polycarbonate, PPO and polystyrene, can be raised by subjecting specimens to a thickness reduction by cold rolling. The enhancement that results is attributed to the introduction of beneficial residual compressive stresses at the surface.

See also: Mechanical Properties of Polymers; Mechanical Properties of Polymers: Test Methods; Tensile Properties of Polymers

Bibliography

American Society for Testing and Materials 1970 Standard method of test for tensile-impact energy to break plastics and electrical insulating materials, ASTM D1822-68. *Annual Book of ASTM Standards*, Pt. 27. American Society for Testing and Materials, Philadelphia, Pennsylvania, pp. 526–35

American Society for Testing and Materials 1975 Standard methods of test for impact resistance of plastics and electrical insulating materials, ASTM D256-73. *Annual Book of ASTM Standards*, Pt. 35. American Society for Testing and Materials, Philadelphia, Pennsylvania, pp. 73–91

American Society for Testing and Materials 1976 Standard methods of test for impact resistance of polyethylene film by the free-falling dart method, ASTM D1709-75. *Annual Book of ASTM Standards*, Pt. 36. American Society for Testing and Materials, Philadelphia, Pennsylvania, pp. 185–93

British Standards Institution 1970 *Methods of Testing Plastics*, Pt. 3, *Mechanical Properties*, BS 2782. British Standards Institution, London

Bucknall C B 1977 *Toughened Plastics*. Applied Science, London

Reed P E 1979 Impact performance of polymers. In: Andrews E H (ed.) 1979 *Developments in Polymer Fracture*, Vol. 1. Applied Science, London, pp. 121–53

Sauer J A 1971 The influence of structure and other factors on molecular motions in solid polymers from 4 °K to 300 °K. *J. Polym. Sci., Polym. Symp.* 32: 69–122

Vincent P I 1971 *Impact Strength and Service Performance of Thermoplastics*. Plastics Institute, London

Vincent P I 1974 Short-term strength and impact behavior. In: Ogorkiewicz R M (ed.) 1974 *Thermoplastics—Properties and Design*. Wiley, London, pp. 68–83

J. A. Sauer

In-Situ Composites

The term in-situ composites is applied to aligned two-phase materials, particularly directionally solidified eutectics, in which the reinforcing phase is produced simultaneously with the matrix (i.e., in situ). The phases take on regular duplex microstructures consisting of parallel fibers or lamellae of one phase in a matrix of another phase over extended distances. The use of the term in situ was intended to contrast these melt-grown materials with the more conventional composite materials in which fibers or lamellae are formed by one process and the matrix is then added in another.

There are hundreds of in-situ eutectic composites already characterized consisting of metals, intermetallics, ceramics, semiconductors and organic compounds. There are potentially thousands more awaiting characterization as phase equilibria of ternary, quaternary and multicomponent systems are researched using both computer predictive and experimental techniques. Furthermore, other compositions of matter, for example, off-eutectics, monotectics, eutectoids and spinodals, can form analogous structures when subjected to controlled melt growth, transformation or deformation. It is the controlled duplex microstructure which endows these materials with interesting new properties—magnetic, electrical, optical, thermal, chemical and mechanical. By combining the properties of several materials in one body, the path has been opened for the development of a new class of materials with a wide range of usefulness.

1. Solidification and Structures

The schematic phase diagram in Fig. 1 shows the eutectic invariant point at temperature T_e and composition C_e. At this point, solid phases α and β simultaneously solidify from the liquid L. For most eutectic systems, if this solidification is done direc-

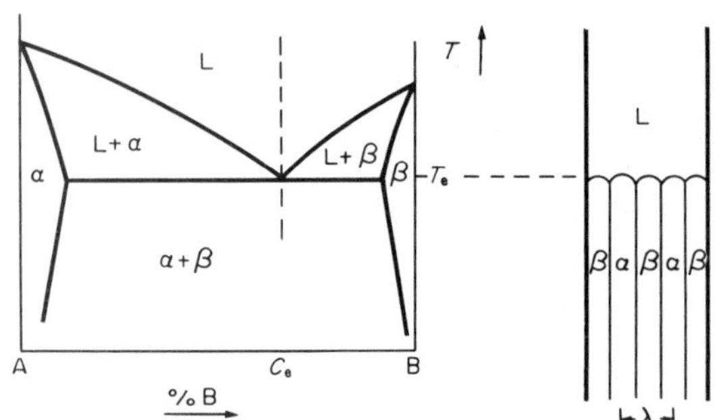

Figure 1
Schematic phase diagram characteristic of a eutectic system

tionally, an aligned two-phase solid can be produced. At least three conditions must be simultaneously satisfied, however, to produce the desired parallel duplex microstructure. The first is that the heat is removed from the melt in a unidirectional manner, the second that a sufficiently positive temperature gradient be maintained ahead of the solidifying interface to prevent nucleation and maintain interface planarity, and the third that cooperative nucleation and growth processes occur between the phases. The α and β phases usually grow normal to this interface, with certain crystallographic directions in one or both of the phases tending to line up parallel to the solidification direction. On examining this process, it becomes clear that the major solidification parameters—the thermal gradient G at the liquid–solid interface and the growth rate R or the velocity at which the liquid–solid interface advances—are keys to the control of structure and hence the chemical, mechanical and physical properties.

With the normal temperature gradients ($G \simeq 10$–$100\ \mathrm{K\,cm^{-1}}$) used in unidirectional solidification, aligned composites with inter-rod or lamellar spacings λ typically in the range 0.1–10 μm are produced. This comes about because the most stable value of λ is determined during the solidification process by the balance between that component of the total driving force needed to provide the excess free energy for forming interphase surface area and that required to drive the mass-transport process in the liquid by diffusion. This free-energy balance leads to the relationship

$$\lambda^2 R = \text{constant} \qquad (1)$$

where λ is the spacing of the microstructure (Fig. 1), R is the solidification velocity and the constant is dependent on the material system and on such parameters as the diffusion coefficient and the interphase surface energy.

If the melt composition is far removed from the eutectic point, solidification of one phase before the other occurs in the form of coarse primary dendrites (see *Dendritic Solidification*). In most cases these dendrites will produce undesirable cast properties; however, under appropriate growth conditions, compositions deviating in both directions from the binary eutectic composition C_e can be directionally solidified to produce aligned eutectic-like composites with varying volume fractions of the phases.

Another approach to vary the volume fraction of the phases as well as their composition lies in the biphase solidification of monovariant eutectic alloys. These compositions lie on or near the eutectic troughs in the liquidus surface of ternary alloys. The solidification considerations to promote planar coupled growth are identical to those for off-eutectic composite growth and are given by the simplified constitutional supercooling equation

$$G/R \geqslant \Delta T/D \qquad (2)$$

where G is the temperature gradient in the liquid, D is the diffusion coefficient and ΔT is the temperature range between start and completion of solidification. (Note that in the simple binary system of Fig. 1 ΔT is zero and increases monotonically as the composition departs from C_e.)

The ability to depart from the precise eutectic composition is extremely important from the technological point of view, as it permits the addition of various alloying elements to modify both the chemistry of the phases and their volume fractions. As will be discussed further, this "alloyability" in multicomponent systems played an important role in the development of high-temperature in-situ composites with complex nickel-base superalloy matrices.

2. Physical Properties and Nonstructural Applications

Directionally solidified in-situ composites generally consist of coarse columnar grains. X-ray and electron diffraction studies often show preferred crystallographic growth directions and preferred-orientation relationships between the phases, together with preferred habit planes, usually low-index, for lamellae or faceted fibers. The lamellar and fibrous microstructures shown in Fig. 2 contain numerous faults (10^5–$10^6\ \mathrm{cm^{-2}}$ terminations or branches) which reflect small random variations between adjacent lamellae or fibers of 0.2–2° about the growth direction and larger variations across the faults of 1–5°. These factors may influence the anisotropic properties which are characteristic of specific in-situ composites. Albers (1973) has distinguished between properties dependent specifically on microstructure and morphology and those which depend on special combinations of properties of the component phases (see *Composite Materials: Nonmechanical Properties*).

In the former category are properties dependent on the periodicity of the microstructure, such as

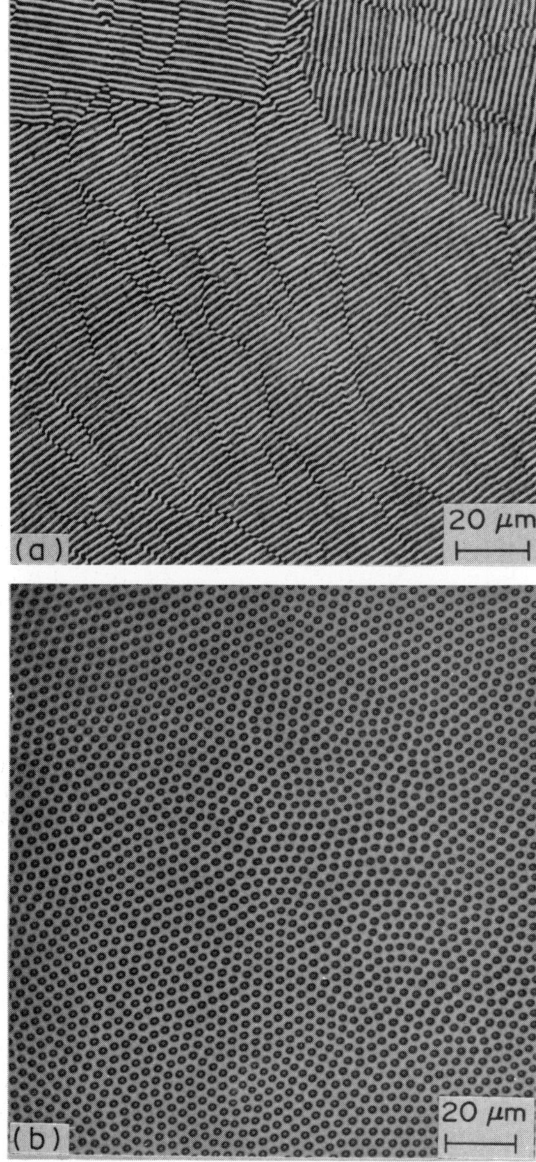

Figure 2
Eutectics: (a) lamellar Al–Cu; (b) fibrous Al–Ni

diffraction of electromagnetic waves. Other properties depend on the anisotropy of the microstructure, such as anisotropic electrical or thermal conductivity. Others again depend on the shape and size of the phases, such as coercive force of a composite containing a ferromagnetic dispersed fibrous phase.

In the second category, that of the combination-type properties, differences between "sum" and "product" properties are distinguished. Any physical property of a material can be considered as the action of a physical input quantity X producing a physical output Y. Sum properties are those in which the X–Y effects of the phases combine to produce an X–Y effect on the composite. Examples are electrical resistivity and elastic modulus. Product properties are those in which an X–Y effect in one phase and a Y–Z effect in the other are combined through a coupling of the Y quantity to produce an X–Z effect of the composite. A somewhat less inclusive classification indicating the diversity of nonstructural applications of in-situ composites is indicated by the set of matrix–dispersed-phase possibilities presented in Table 1.

2.1 Electrical Materials

The most successful applications of nonstructural in-situ composites were found for the system InSb–NiSb reviewed by Weiss (1976). In this eutectic, conducting fibers of NiSb lie within a semiconducting matrix of InSb. When these fibers and applied electric and magnetic fields are all mutually perpendicular, the NiSb fibers short-circuit the InSb Hall voltage, thereby allowing the electrons in the InSb to respond to the Lorentz force and flow at an angle to the applied field. This Hall angle increases as magnetic field increases, and the increasing length of the current path produces a large increase in resistance. For example, a field of 80 000 $A\ m^{-1}$ may produce nearly a twentyfold increase in resistance. This resistance change, or magnetoresistance, is larger than for any other material at room temperature and has been used in a wide variety of commercial applications including field sensors, current transducers and contactless potentiometers and switches.

2.2 Electronic Materials

For another physical reason InSb–NiSb has found application as an infrared detector and an infrared polarizer. The transparent semiconductor InSb is the embedding medium which keeps the small NiSb dipoles aligned by mechanical support. In a similar way ZrO_2 and UO_2 matrices may act as a support for $\sim 10^7$ single-crystal filaments of tungsten per cm^2 of matrix. The insulating matrix may be partially etched away to expose and point the multineedle field emitters, as shown in Fig. 3. Cochran et al. (1980) have reported more complex hemispherically tipped low-voltage field-emitter arrays produced from ZrO_2–W running for over 3000 h at 5 $A\ cm^{-2}$ with others operating at current densities exceeding 20 $A\ cm^{-2}$ and emitting 100 mA for over 700 h before being terminated. The emitter arrays have also been recently applied to atomizing liquid fuels via electric charge transfer to improve combustion efficiencies. It has been found that these emission cathodes suffer from instabilities, especially those derived from inhomogeneities of length and spacing distribution as well as shape of the emitting needles.

As the minimum dimension of electronic circuits is

Table 1
Applications of nonstructural eutectics

Dispersed phase	Matrix: Conductor	Matrix: Semiconductor	Matrix: Insulator
Conductor	Superconductor magnets Field emitters Filters Strengthened electrical conductors	Magnetic probes Galvanomagnetic devices Infrared detectors Optical polarizers	Anisotropic heat conductors Electrostatic atomizers Capacitors
Semiconductor		Variable bandgap semiconductors Optical polarizers	Optical devices
Insulator	Strengthened electrical conductors Capacitors	Optical polarizers Acoustic devices	Photovoltaic devices Magnets

reduced below 1 μm, perfect periodic submicrometer structures are of great interest. Eutectic thin films have been solidified at rates of 0.0015–0.15 cm s^{-1} with both lamp and laser heat sources to fabricate fault-free structures (Lemkey et al. 1982).

2.3 Superconducting Materials

Optical and superconducting properties also suffer as a result of faults, terminations and irregularities of eutectic structures which to date have prevented their application. Superconducting cables with continuous <10 μm niobium–tin filaments within a copper matrix have become commercially available and represent another class of in-situ composites which are produced by cooperative deformation rather than growth of dendritic two-phase alloys. Cu–Nb alloys are converted to Nb_3Sn–Cu superconducting wires possessing excellent critical current values.

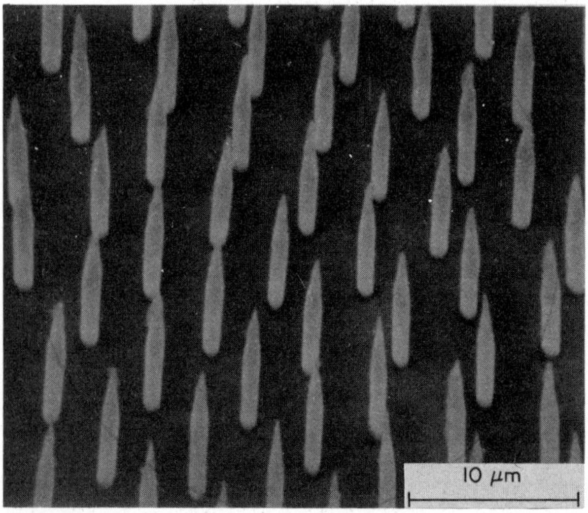

Figure 3
Field emitter fabricated from ZrO_2–W eutectic

2.4 Magnetic Materials

In-situ composites containing dispersed ferromagnetic fibers whose size approaches that of individual magnetic domains have been investigated. Anisotropic ferromagnetic properties can be achieved in the systems Mn–InSb, Fe–FeS, Fe–Fe_xSb, Bi–MnBi, Sb–MnSb and Au–Co. However, none of these systems is yet of commercial interest because of limitations in the size, volume fraction and/or ferromagnetic properties of the dispersed fibers. On the strength of the results obtained, ferromagnetic eutectics based on crystal anisotropy are considered more promising for development than ferromagnetic eutectics based on shape anisotropy.

3. Mechanical Properties and Eutectic Superalloys

Interest in two-phase composite structural materials began with the discovery of whiskers in the 1950s and of high-performance continuous filaments in the 1960s and 1970s. Whiskers are microscopic, needle-like crystals which, because of their high crystalline perfection, often exhibit very high strengths. Their great strength can be exploited by using them as reinforcement in metals, resins and ceramics. But to do this, the whiskers, which generally grow on a substrate by condensation from the vapor, must be harvested, sorted, inserted and bonded into the matrix, and for maximum reinforcement, aligned in one direction. These problems, although not insurmountable, have proved difficult to solve.

With the rodlike or lamellar dispersed phase and the matrix growing simultaneously from the melt, the solidification of eutectic alloys provides a way to obtain a composite whose reinforcement is aligned, evenly distributed and naturally well bonded. The process thus avoids the problem of handling whiskers or filaments separately and gives a composite structure which exhibits true whisker reinforcement. The first eutectic system to behave mechanically as a composite, with significant reinforcement of the matrix by 11 vol% aligned whiskers, was the Al–Al_3Ni system examined by Lemkey et al. (1965). When aligned, the strength is approximately four times as great as that measured in the conventionally cast eutectic material. Other eutectic systems with greater volume fraction of carbide whiskers as reinforcements in a niobium or tantalum matrix resulted in stronger and more-creep-resistant alloys than those commercially available. These observations, coupled with the long-term stability of eutectic microstructures at elevated temperatures, justified the prediction that these materials might be useful in high-temperature structural applications such as in gas-turbine engines. During the past decade eutectic-alloy programs have been undertaken all over the world to develop superior gas-turbine materials (Albers 1973, Weiss 1976, Walter et al. 1979, Lemkey et al. 1982, McLean 1983). Further information concerning the mechanical properties of in situ composites can be found elsewhere (see *Directionally Solidified Eutectics: Strengthening*).

The strongest eutectics developed so far contain the γ' precipitate Ni_3Al within the γ-Ni matrix phase and are reinforced by the phases Ni_3Nb and Mo and the carbide phases TaC, NbC and Cr_3C_2. Their potential suitability for use at temperatures 40–110 °C higher than currently available alloys offers the engine designer opportunities to improve engine performance, increase component lifetime and reduce fuel consumption. Although numerous eutectic systems have been investigated during the past decade, only two are currently receiving extensive evaluation in the USA, France and UK for near-term application, that is, γ–γ' + TaC, MC (M = Ta or Nb) and γ–γ' + Cr_3C_2.

The TaC and NbC fiber-reinforced eutectics have important advantages for turbine application. They are ductile (in the range 10–20% reduction in area at room temperature) and resistant to impact damage (V-notch Charpy above 13.5 J). They can also be alloyed with chromium, aluminum, tungsten and rhenium, for example, to produce reasonable oxidation resistance and impressive creep rupture strength and thermal fatigue resistance. They consist of approximately 3 vol% aligned carbides in a ductile γ–γ' nickel alloy matrix but must be directionally solidified at ~0.6 cm h^{-1}. Because of this slow freezing rate necessary to achieve coupled growth combined with excessive liquid-phase reactivity to foundry molds and cores, the economics of turbine-blade production presents the largest technological challenges.

The Cr_3C_2 fiber-reinforced nickel-base eutectics on the other hand can be produced at freezing rates equivalent to current superalloys in complex shapes, such as that of the turbine blade shown in Fig. 4. They can be alloyed with rhenium, tungsten, molybdenum, yttrium and aluminum to produce excellent oxidation resistance and good creep rupture strength and thermal fatigue resistance. They consist of approximately 11 vol% aligned carbides in a ductile γ–γ' nickel alloy matrix and can be directionally solidified at rates of ~10–25 cm h^{-1}. Further research and development programs are examining the numerous steady-state and vibratory properties required to qualify this new class of materials for engine application. It is expected that the first commercial use of γ–γ' + Cr_3C_2 eutectic airfoils will be as solid blades or vanes in advanced gas-turbine engine applications.

Figure 4
Directionally solidified eutectic superalloy turbine blade

A final attractive in-situ composite candidate for gas-turbine components is the M_7C_3 (M = Cr, Mn, Fe) carbide-reinforced austenitic iron system which can be produced under the same processing conditions as directionally solidified superalloys. These alloys exhibit exceptional surface stability when combined with minor active elements such as yttrium, cerium and lanthanum and possess lower densities than existing superalloys. Much current development centers on achieving improved elevated-temperature mechanical properties together with retaining the outstanding oxidation resistance of the austenite matrix. Iron-base in-situ composites promise considerable cost savings if substituted for nickel and cobalt superalloys. Because of the higher carbon levels required for eutectic growth, chromium additions can be made with high-carbon ferrochromium derived from chromite deposits and thus manufactured with low-cost nonstrategic elements which may in turn accelerate their entry into the later stages of gas-turbine engines during the 1980s.

See also: Eutectic Solidification; Composite Materials: An Overview

Bibliography

Albers W 1973 Physical properties of composite materials. In: Lemkey F D, Thompson E R (eds.) 1973 *Conf. In Situ Composites*, Vol. 3. National Academy of Sciences, Washington, DC

Chadwick G A 1963 Eutectic alloy solidification. *Prog. Mater. Sci.* 12: 99–182

Cochran J K, Chapman A T, Feeney R K, Hill D N 1980 Review of field emitter array cathodes. *Proc. Technical Digest International Electron Devices Meeting*. Institute of Electrical and Electronics Engineers, New York, pp. 462–66

Hogan L M, Kraft R W, Lemkey F D 1971 Eutectic grains. In: Herman H (ed.) 1971 *Advances in Materials Research,* Vol. 5. Wiley Interscience, New York, pp. 83–216

Kurz W, Sahm P R 1975 *Gerichtet Erstarrte Eutektische Werkstoffe: Hersfellung, Eigenschaften und Anwendungen von In-Situ Verbundwerkstoffen.* Springer, Berlin

Lemkey F D 1984 Advanced in situ composites. In: Murr L E (ed.) 1984 *Industrial Materials Science and Engineering*. Dekker, New York, pp. 441–69

Lemkey F D, Cline H E, McLean M (eds.) 1982 *In Situ Composites IV*. Elsevier, New York

Lemkey F D, Hertzberg R W, Ford J A 1965 The microstructure, crystallography, and mechanical behavior of unidirectionally solidified Al–Al_3Ni eutectic. *Trans. Metall. Soc. AIME* 233: 334–41

McLean M 1983 *Directionally Solidified Materials for High Temperature Service*. Metals Society, London

Thompson E R, Lemkey F D 1974 Directionally solidified eutectic superalloys. In: Kreider K G (ed.) 1974 *Composite Materials,* Vol. 4. Academic Press, New York, pp. 101–58

Walter J L, Gigliotti M F, Oliver B F, Bibring H (eds.) 1979 *Conf. In Situ Composites III*. Ginn Custom, Lexington, Massachusetts

Weiss H 1976 Physical properties of in situ composites. In: Jackson M R, Walter J L, Lemkey F D, Hertzberg R W (eds.) 1976 *Conf. In Situ Composites II*. Xerox, Lexington, Massachusetts

F. D. Lemkey

Incandescent Lamp Materials

The incandescent lamp has reached its present form through more than 100 years of major advances in materials and manufacturing practices. The fascinating history of these material developments can be found in the texts listed at the end of this article.

The lamp itself is rather simple, consisting essentially of a conducting filament enclosed within a glass bulb and heated to incandescence (white hot) by an electric current so that it emits light. The original lamp pioneered by Edison and announced in 1879, used a carbon filament in a bulb which was evacuated to prevent filament oxidation. From that day forward, the search for better materials, including the filament, leads, bulb and enclosed atmosphere, has never ended.

The carbon filament was first replaced by osmium, then tantalum and finally tungsten, which is still used today. The tungsten filament is operated at 2850 K, but does not burn up because it is surrounded by inert gases (or vacuum in a low-wattage lamp) that do not react with the metal; in a standard 100 W lamp, the bulb is normally filled with argon and 5% nitrogen to 600 torr. Tungsten wire required extensive development to be useful as a lamp filament, a development that has led to advances, such as the tungsten–halogen lamps.

Halogen lamps use tungsten filaments and employ a regenerative chemical cycle inside the glass or quartz (fused SiO_2) envelope to prevent bulb blackening. In this lamp, the filament can be operated several hundred degrees higher (usually at 3000–3200 K) than the standard lamp, greatly improving the efficacy.

More than 10 000 different incandescent-lamp types are available commercially. Besides the tungsten filament, these lamps use other special materials, chosen and processed not only for mechanical and electrical properties, but also for high-temperature properties in the various lamp atmospheres. Transparency in the envelope, proper expansion coefficient and high purity are needed in all lamp materials; often these factors outweigh the ease of working and cost.

1. Tungsten Filament

Improvements in the filament of the incandescent lamp led methodically to the use of tungsten wire. In modern designs, the wire is formed into a closely wound helical coil to reduce the cooling effect and to compact the lamp (see Fig. 1). Tungsten has stood the test of time because it has four attributes that make it the most suitable filament material. Firstly, it has a high melting point. The light output of a solid body depends primarily on how high its temperature can be raised while maintaining a long life. Edison chose the highest melting element, carbon, and

Figure 1
Enlarged view of a coiled-coil 100 W tungsten filament

others subsequently investigated refractory oxides, which also have high melting temperatures. Tungsten melts at 3650 K, somewhat lower than carbon and some of the other materials, but still a very high melting temperature. Secondly, the filament needs to have a low rate of evaporation, even when a regenerative chemical cycle is employed. Carbon, particularly, had to be operated well below its melting point because of the high rate of evaporation. This resulted in low efficacy. Tungsten has a low rate of evaporation at normal operating lamp temperature and, when combined with inert-gas filling, the evaporation rate can be retarded further. Thirdly, high melting temperature and low evaporation rate are not enough if the material does not also exhibit high mechanical strength. With proper manufacturing procedures, tungsten can be made strong and ductile and can maintain these characteristics at high temperature. Finally, the filament must radiate efficiently in the visible part of a spectrum. All materials at high temperatures produce a visible spectrum but tungsten is particularly favorable. Tungsten selectively radiates with a slightly higher percentage of its energy in the visible; that is, it emits more visible light than an ideal, perfect hot incandescent radiator at the same temperature. At its melting point, tungsten has an efficacy of 53 lm W^{-1}. In practice, however, a 100 W incandescent lamp has an efficacy of only 17.5 lm W^{-1} due, primarily, to its lower operating temperature but also to gas connection losses and heat losses out of the connecting leads.

Tungsten wire is produced from sintered ingots containing very small quantities of aluminum, potassium and silicon compounds. The ingots of tungsten are rolled, swaged and drawn into wire of the desired diameter. The diameter used depends on the lamp voltage and wattage. For a 100 W lamp, the diameter is about 5×10^{-3} cm.

The wire is then coiled double on a molybdenum mandrel and sintered at 1800 K. The "coiled-coil" filaments obtained retain their shape and the mandrel can be dissolved. Inspection of the coils is performed using a projection microscope before automatic mounting of the coils on the lamp leads. A coiled coil filament is shown in Fig. 1.

2. *Glass and Fused Quartz Envelopes*

Incandescent lamps utilize an envelope of a soft glass (lime glass), a hard glass (borosilicate), or fused quartz. A typical soft glass composition is 72.6% SiO_2, 16.6% Na_2O, 3.6% MgO, 5.0% CaO and 2.2% miscellaneous oxides, mainly Al_2O_3. The flare and vacuum exhaust tubes in standard soft-glass lamps usually contain 20–30% lead oxides. The envelope materials need to have high transmission of visible radiation and therefore the bulb glasses must be low in iron oxide.

Fused quartz is commonly used as an envelope material for the compact and highly loaded (W cm^{-2}) halogen lamps, as it can withstand high-operating temperatures. Hard glasses are used for high-wattage lamps and for outdoor lamps. Most common incandescent lamps are manufactured of lime glass because of cost factors.

The electrical leads which supply current to the filament must be hermetically sealed through the glass envelope. The mass production of reliable glass–metal seals has been the subject of considerable engineering and material investigation. One of the results of this work has been the development of Dumet (a copper clad, nickel–iron) lead seal used for soft-glass lamps. The coefficient of expansion of the glass and the metal are not well matched, but stresses are minimized by using small-diameter leads and proper annealing techniques.

The borosilicate glasses use a nickel–iron alloy for the electrical lead. The quartz–metal seal usually uses a thin, feathered section of molybdenum foil.

Despite tailoring of both glass and metal compositions, it is inevitable that somewhere over the forming range of the glass–metal seal a mismatch in expansion occurs, producing a strain in the seal at room temperature. The thermal treatment in the glass–metal seal area is of prime concern to lamp makers. It requires that the surface oxide bonding the glass and metal adhere firmly to both materials, and that the seal area be annealed to reduce the stresses in the area. If the stresses are too great, cracks will develop destroying the integrity of the lamp atmosphere.

3. Getters

Getters are employed in almost all incandescent lamps to remove traces of impurities. Getters absorb the residual contamination in the lamp after sealing and contaminants released during the operation of the filament. The gas impurities in lamps are mainly trace amounts of H_2O, CO_2, CO, hydrocarbons and CN radicals. Water vapor is the most prevalent and most harmful impurity, causing early blackening of the envelope. Water vapor can be left in from the exhaust and inert-gas filling process or can come from the glass structure and envelope itself. Water vapor is released easily from soft glass above 300 °C, and these high temperatures either have to be avoided or the water vapor must be continuously gettered.

Getters are of the bulk or flashed type. Bulk getters, such as aluminum–zirconium mixtures, are applied to the lamp structure, usually the leads. Flash getters, such as red phosphorus or P_3N_5, are usually sprayed directly onto the filament. After the lamp is exhausted, the filament is heated momentarily and the phosphorus (usually about 0.5 g) is flashed into the atmosphere. These getters are effective in reducing common lamp impurities, particularly water vapor.

Some household incandescent lamps employ a white silica powder (finely divided SiO_2) to diffuse the light. This powder, on the inside surface of the envelope, can be troublesome since it can retain large amounts of water. Getters need to be especially active chemically to keep water vapor at a low level in these lamps.

Impurities in the lamp atmosphere reduce performance by causing bulb blackening, filament sag and short life. Amounts of water vapor over 10 ppm in the argon–nitrogen atmosphere are particularly harmful to incandescent-lamp life. The water vapor can result in a degenerate transport cycle, continuously dissociating into oxygen and hydrogen and reacting with the hot tungsten to form volatile tungsten oxides which condense on the cool bulb wall. The hydrogen then reduces these tungsten oxides on the wall, producing a tungsten deposit and reforming the water molecule which can diffuse to the filament and repeat the cycle.

4. Lamp Metals

Copper, nickel, iron and aluminum, and their alloys, find extensive use in lamp designs. The materials used inside the lamp need to be of vacuum quality, that is, made in such a manner as to minimize the amount of sorbed gas. Even so, outgassing at elevated temperatures is usually required on the lamp-making machine. The final tensile strength of these metals (partly obtained by cold working) is critical to lamp operation.

Lead wires, in particular, need a narrow range of tensile strength values. It is desirable to use a small-diameter lead wire to reduce heat conduction from the ends of the hot filament, and at the same time reduce cost. However, the lead must be strong enough to support the filament and to withstand the mechanical stresses encountered in shipping and handling. Copper and its alloys are used as the lead wires because of their good electrical conductivity and freedom from porosity.

External bases are made of either brass or aluminum. The preferred choice for corrosion-resistant, outdoor application is brass. Aluminum is quite satisfactory for indoor use and is less expensive.

A large number of special techniques in drawing, forming, purification and plating are used to improve the vacuum quality and reduce the cost of these metals. For example, it is common to seal the base to the bulb by using a cement cured by heat at the time of lamp manufacture. These cements are outside the bulb itself and are in a paste form during manufacture. They consist of marble flour together with alcohol, resin and shellac. For higher temperature operation, mechanical methods are used to fasten the base to the bulb.

5. Atmosphere Fill Gases

All early lamps employed a vacuum to prevent oxidation of the filament. The degree of this vacuum was important for the life of the lamps, particularly tungsten lamps.

In 1913, Langmuir found that filling the bulb with an inert gas such as argon can reduce the rate of evaporation of the tungsten from the filament, and that this effect was due to thin viscous layers of the gas surrounding the filament. This led to the discovery that coiling the filament would improve the overall efficacy of the lamp. Filaments coiled in a double helix are commonly used today.

At first, nitrogen was used as the inert fill gas since it was the only gas available in large quantities. Argon later came into extensive use. The small amount of nitrogen used in 120 V lamps is necessary to prevent arcing across the filaments. Low-voltage lamps can use almost 100% argon filling. The amount of nitrogen employed is minimized because it has a higher thermal conductivity than that of argon.

The lower the heat conductivity of the gas, the better the performance. Krypton would be a better gas than argon, and xenon even better. Krypton is, in fact, used in certain battery-operated lamps where power costs are very high. For ordinary lamps, xenon is too expensive. Krypton has been used in some household lamps, although the higher cost of krypton does not justify its widespread use. Hydrogen (high heat conductivity) is used in flashing lamps where rapid decay of the light emission is desired.

The fill gas of the tungsten–halogen lamps is particularly interesting. These lamps utilize a regenerative chemical cycle to prevent blackening of the bulb wall. This permits very compact lamps which can be safely pressurized with an inert gas. The results are higher efficacies, better maintenance and longer life. The filaments of these lamps are usually mounted in a linear, fused quartz tubular envelope containing an inert gas with bromine or a gaseous iodine compound. As the tungsten evaporates from the hot filament and diffuses through the fill gas toward the wall, it reacts with the halogen or oxyhalides in the cooler regions of the gas on the fused quartz wall. These reaction products, which are primarily oxyhalides of iodine or bromine, are volatile compounds and diffuse to the vicinity of the filament. At the filament, they dissociate and the tungsten is redeposited on the coil. The net result is that the bulb wall remains clean and the lumen maintenance is near 100% through the lamp life. The lamp life is still limited, however, because the tungsten does not redeposit preferentially on the hottest sections of the filament where the evaporation rate is the greatest.

See also: Photolamp Materials; Discharge Lamp Materials; Electrical Materials: An Overview

Bibliography

Bright A A 1949 *The Electric-Lamp Industry: Technological Change and Economic Development from 1800 to 1947.* Macmillan, New York

Howell J W, Schroeder H 1927 *The History of the Incandescent Lamp.* Maqua, Schenectady, New York

Kaufman J E (ed.) 1981 *IES Lighting Handbook*, 5th edn. Illuminating Engineering Society of North America, New York, Chap. 8

Knoll M, Kazan B 1959 *Materials and Processes of Electron Devices.* Springer, Berlin

Morey G W 1954 *Properties of Glass.* Reinhold, New York

Partridge J H, Turner W E S 1949 *Glass-to-Metal Seals.* Society of Glass Technology, Sheffield

Smithells C J 1953 *Tungsten: Its Metallurgy, Properties, and Applications.* Chemical Publishing, New York

G. Reiling

Inclusion Control in Steel

Nonmetallic inclusions in steels have been the subject of much study during the last half century. This interest originated from observations that the performance of steels can be strongly influenced by inclusions, especially in electrical, free-machining and structural steels.

1. Some Effects of Inclusions

The electrical and free-machining grades are good examples of steels in which the presence of appropriate inclusions can actually improve the performance. Structural steels, on the other hand, can exhibit a marked degradation in properties when they contain certain types of inclusions.

Inclusions, principally MnS or AlN, play an important role in the production of the so-called grain-oriented silicon steels. In these steels, the inclusions inhibit normal grain growth and hence promote secondary recrystallization. This causes the formation of a sharp (110)[001] texture which is highly desirable in sheet steels to be used in transformers. This texture would be difficult to achieve without the proper size, shape and distribution of inclusions (Dunn and Walter 1966).

With regard to the free-machining steels, it is well known and widely accepted that the machinability of steel can be greatly enhanced by the presence of an appreciable volume fraction of globular manganese sulfide inclusions. In this case, the presence of inclusions of suitable morphology can actually improve the performance of the steel during machining operations. Interestingly, it has also been shown that the morphology of the MnS present in the steels is a significant factor affecting machinability; good machinability is related to a globular MnS morphology whereas poor machinability is related to an elongated morphology (Marston and Murray 1970).

Beyond question, however, the effect of inclusions on the mechanical properties of structural steels has received the most attention over the last two decades. The flat-rolled structural steels, in particular, have been the subject of numerous studies. This research has shown that mechanical properties such as Charpy shelf energy, fracture toughness, ductility, formability, fatigue resistance and hot ductility can be remarkably reduced by inclusions. Generally, two approaches have been adopted to mitigate the effect of inclusions. In one approach, the shape of the inclusions is altered from highly damaging, flattened, elongated sulfides to less damaging spheroidal sulfides by additions of special agents such as rare earths or calcium. In the other approach, the total volume fraction of inclusions is reduced by desulfurization of the steel.

2. Morphology of Inclusions

In general, it has been shown that the extent of property degradation of plate products is related to the amount, size, shape and distribution of inclusions. Among the most damaging inclusions are those which have a planar morphology; hence, the highly elongated ribbon-like inclusion morphology ("stringers") found in hot-rolled steels can lead to very low levels of ductility and toughness. This morphology can be responsible for highly directional ductile fracture properties, where the extent of property degradation is dependent upon the test direction and the lowest values are those measured normal to the rolling plane (Luyckx et al. 1970, Baker and Charles 1971, DeArdo and Hamburg 1974, Kozasu and Tanaka 1974). A somewhat analogous situation exists in cast steels, although in this case the inclusion-related problem is connected with clusters of several small particles. These clusters form as a result of interdendritic and/or eutectic solidification patterns. The influence of these clusters is associated more with the size of the clusters than with the size of the individual inclusions.

It is well known that steels with inclusion arrays of similar volume fraction can have very different effects on the short- and long-transverse properties of hot-rolled plates, depending upon the size and shape of the inclusions. The relation between inclusion shape and mechanical properties derives from the level of stress and strain intensification which can exist in the matrix in the vicinity of the inclusion (Gladman et al. 1971). The intensification is inversely related to that radius of curvature of the inclusion which is perpendicular to the maximum principal stress or strain, and it is directly related to the length of the inclusion. Hence, highly elongated inclusion ribbons have extremely small radii of curvature at their ends, and this together with their large length can lead to very high stress or strain intensification in tests run normal to the rolling plane. This leads to low levels of ductility and toughness in this test direction. Longitudinal specimens do not experience this high intensification because the maximum principal stress and strain are parallel to the length of the ribbon and, therefore, interact with a rather large radius of curvature. Long-transverse specimens experience an intermediate level of intensification.

Spheroidal inclusions, on the other hand, have rather large radii of curvature in any test direction. This leads to a low level of stress and strain intensification in the short-transverse direction and, therefore, a high level of ductility in that direction. A dramatic increase in the short-transverse ductility would be expected if the inclusion shape were changed from highly elongated to globular. It was this potential for improvement in short-transverse ductility and its positive effect on formability, toughness and weldment integrity that provided much of the impetus for studies on controlling the shape of inclusions in steel.

3. *Formation and Nature of Inclusions*

The formation of inclusions in steel has been exhaustively studied and excellent reviews are available (e.g., Kiessling 1978, Wilson and McLean 1980). There are two general classes of inclusions which are found in steels: exogenous and indigenous inclusions. Exogenous inclusions enter the liquid steel as a result of either the erosion or fracture of the surrounding refractories, or the entrapment of slag particles. Exogenous inclusions are generally characterized by a larger size, sporadic occurrence, complex structures and irregular shapes.

Indigenous inclusions, on the other hand, result from the physical chemistry of steelmaking and are a product of the chemical reactions which take place. These inclusions can form in either the liquid or solid steel, and do so as a direct result of the lowering of the relevant solubility product with decreasing temperature. Only indigenous inclusions are considered here.

3.1 Inclusions in Cast Steel

The inclusions which are the most troublesome in steel castings are those which form as clusters of rather small particles. There are two principal sources of these clusters. They are the so-called "galaxies" of Al_2O_3 particles (Hilty and Farrell 1975) and the "picket fence" arrangement of type II MnS (Kiessling 1978). Since these clusters have no particular preferred orientation, their effect on mechanical properties can be expected to be nondirectional.

Galaxies of Al_2O_3 form as a result of the solidification of Al-treated steel. The first phase to solidify is Al_2O_3, which is precipitated from the liquid metal. As cooling progresses, the metal and the oxide freeze as a binary eutectic. The Al_2O_3 particles in the galaxies are in fact dendrites. Type II MnS particles are inclusions, found in fully deoxidized steels, which form in a eutectic pattern from liquid which was located in the boundary regions separating the primary dendrites.

3.2 Inclusions in Wrought Steel

Both Al_2O_3 galaxies and type II MnS can, of course, also form in hot-rolled steel. In this case their effect is even more damaging than in cast steel, since the hot rolling causes the clusters to change to essentially two-dimensional planar defects. Thus, these rolled-out clusters approach the shape of the ribbon-like inclusions normally found in hot-rolled steels and lead to similar directionality in mechanical properties.

A more prevalent and damaging problem is caused by large individual inclusions which are globular in the as-cast steel but deform with the matrix during hot rolling and, hence, exhibit a ribbon-like morphology in the as-rolled condition. The inclusions that deform with the matrix are those that are relatively soft at hot-rolling temperatures. Theory and experiment have shown that there is a direct relation between the relative plasticity of an inclusion ($\varepsilon_{inc}/\varepsilon_{mat}$) and its relative hardness (H_{inc}/H_{mat}); the higher the relative hardness of the inclusion, the lower is its relative plasticity. In addition, there is no inclusion deformation when the inclusion is more than twice as hard as the matrix (Gove and Charles 1974).

The inclusions which deform at hot-rolling temperatures are the manganese silicates and manganese sulfides, which are soft at high temperatures. Silicate inclusions, however, are not a problem in Al-deoxidized steels because the oxygen level is too low for their formation.

Deoxidation with Al may eliminate the silicate problem, but it tends to aggravate the MnS problem. There are several types of MnS which can form in a steel. Type I is a hard inclusion found in rimmed steels, type II is a soft inclusion normally found in a eutectic-like pattern in killed steels and type III is a soft inclusion found in steels which contain an excess of deoxidizing agent. Types II and III are often found in aluminum-killed steels, and these inclusions can lead to serious problems because they are very soft at hot-rolling temperatures and, hence, become greatly elongated during deformation.

4. Inclusion Shape Control

The purpose of inclusion shape control is to eliminate large, planar inclusions in hot-rolled steels through control of steelmaking practice. There are essentially four types of inclusions which can be troublesome in hot-rolled steels: Al_2O_3 galaxies, silicates, and manganese sulfides of types II and III. Since most high-strength hot-rolled steels are fully deoxidized with Al and since silicate inclusions do not form in these steels, silicate inclusions will not be further discussed. Inclusion shape control, therefore, means in effect the elimination of Al_2O_3 galaxies and manganese sulfide of types II and III. Fortunately, there are several elements which, when added to liquid, deoxidized steel, can perform one or both of these tasks.

Within the last 15 years it has been found that appropriate combinations of aluminum with elements such as calcium, the rare-earth metals, titanium or zirconium can eliminate planar inclusions and clusters when added to liquid steel. Elements such as Ca and the rare-earth metals (REM) are believed to alter the solidification pattern of Al_2O_3 by shifting the metal–oxide eutectic to the oxide corner of the metal–oxide–sulfide ternary phase diagram. This shift means that the initial precipitation of oxide is in the form of separate, uniformly distributed liquid pools. Upon subsequent freezing, these pools lead to uniformly dispersed globular oxide particles in Al-treated steels containing Ca or rare-earth metals. This morphology and distribution is much less detrimental to mechanical properties than is the angular, dendritic oxide of the galaxy type found in Al-treated steels (Hilty and Popp 1970, Hilty and Farrell 1975).

Stringers or ribbons of MnS are not normally found in Al-deoxidized, hot-rolled steels which have been treated with Ca, Zr or rare-earth metals. Although the mechanism varies with the precise treatment, each of the additions acts to increase dramatically the high-temperature strength of the sulfur-bearing inclusions, thereby reducing their relative plasticity and their elongation during hot rolling.

The addition of Ca can strengthen the sulfide inclusions in two principal ways. The first is by alloying with the MnS to form (Mn.Ca)S. Since MnS and CaS are completely soluble, the sulfides found in Ca-treated steels could range from MnS to (Mn.Ca)S to CaS (Kiessling 1978, Wilson and McLean 1980). CaS is much more stable than MnS and the hardness of the sulfide can be expected to increase with its Ca content. The second strengthening mode is a form of composite strengthening in which the sulfide (Mn.Ca)S is contiguous with the very hard oxides CaO and Al_2O_3. In this case, even a relatively soft Mn-rich (Mn.Ca)S is not able to undergo extensive deformation because of the presence of the neighboring nondeformable oxide inclusions. Ca can also form complex oxysulfides of the general form (Ca.Al)(O.S). There is some question as to whether these complex inclusions are actually a mixture of CaS, CaO and Al_2O_3. The important point, however, is that the sulfur is contained in an inclusion that is very hard and nondeformable at high temperatures.

The rare-earth metals act in a manner similar to Ca, that is, they also form a series of alloy sulfides (Mn.Re)S and $(Mn.Re)_2S_3$ (Kiessling 1978, Wilson and McLean 1980). Thus, a steel treated with these metals might contain any of the following sulfides: MnS, (Mn.Re)S, ReS, $(Mn.Re)_2S_3$ and Re_2S_3. The rare-earth-metal-bearing sulfides appear to be essentially nondeformable during hot rolling. In addition, oxysulfides of the type Re_2O_3S can be found in rare-earth-metal-treated steels. These inclusions are also very hard and can lead to composite strengthening of softer sulfides as described above.

Unlike Ca or the rare-earth metals, there does not seem to be any appreciable solid solubility between MnS and either TiS or ZrS. The inclusions expected in steels treated with either Ti or Zr are MnS, TiS or ZrS, TiO_2 or ZrO_2, TiN or ZrN and $Ti_4C_2S_2$ or $Zr_4C_2S_2$. The elimination of elongated sulfides in Ti or Zr-treated steels appears to be the result of at least two factors. First, the sulfides and carbosulfides are themselves nondeformable. Second, the MnS which does form in steels treated with Ti or Zr appears to nucleate on the oxide and nitride particles which are present when the MnS precipitates. This seeding effect leads to very small MnS particles which are associated with very hard oxide and nitride particles. The deformation of the soft MnS is again limited by the hard particles and this is another example of the composite strengthening effect described earlier (DeArdo and Hamburg 1975).

In summary, inclusion shape control practice comprises a set of steelmaking procedures which result in the absence of highly elongated inclusions in hot-rolled steels. These procedures include efforts both to reduce the amount of troublesome inclusions and to alter the remaining ones so that they are nondeformable at hot-rolling temperatures. Through various mechanisms, the use of elements such as Al, Ca, the rare-earth metals, Ti and Zr can in fact lead to the elimination of the troublesome elongated inclusions in hot-rolled steels.

5. *Improvement in Mechanical Properties by Inclusion Shape Control*

Inclusion shape control can lead to significant improvements in mechanical properties. When the morphology of inclusions changes from ribbon-like to spheroidal or globular, the ductile fracture properties are moderately improved in the long-transverse direction and greatly improved in the short-transverse direction. One of the mechanical properties strongly influenced by inclusion size and shape is the upper shelf energy in the Charpy V-notch impact test.

Typical property improvements resulting from various inclusion shape control practices are shown in Table 1. Note that in most cases the long-transverse shelf energy is approximately doubled by the change in inclusion size and shape. The improvement in the short-transverse direction would be expected to be even more pronounced. The steels which have received an inclusion shape control treatment, therefore, can be expected to have much more isotropic ductile fracture properties. This represents an improvement over cross rolling, where the properties are made nearly equal in the rolling plane but are still very poor in the short-transverse direction.

Table 1
Improvement in notch toughness by inclusion shape control

Treatment	Charpy fracture energy (J)	Reference
As-cast steel		
0.113 wt% Al	37.6[a]	b
0.113 wt% Al from Ca–Si–Ba–Al	63.3[a]	
Hot-rolled steel		
None	78.5	c
Ca	196.2	
None	20.3	d
REM	43.4	
None	(trans/long = 0.40	c
Ti	= 0.82)	
None	34	c
Zr	44	
None	50	e
Zr	88	
None	34	c
REM	68	

a Notch toughness at −18°C; all others represent upper shelf energy data; long-transverse testing direction b Hilty and Popp (1969) c Wilson and McLean (1980) d Luyckx et al. (1970) e DeArdo and Hamburg 1974

6. *Improvement in Mechanical Properties by Desulfurization*

Recent advances in steelmaking (Turkdogan 1981) also make it possible to improve mechanical properties of plate steels by desulfurizing the steel to very low levels (0.003%S, compared with normal levels of 0.010–0.020%). This is an alternative to inclusion shape control. An example of the effect of desulfurization on the impact properties of plate steels for the short- and long-transverse testing directions is given in Table 2. As can be seen, decreasing the sulfur content of the steel from 0.013 to 0.004% doubled the impact energy in the long-transverse testing direction. The effect in the short-transverse (through-thickness) testing direction is even more marked, the impact energy being increased by a factor of five.

Table 2
Improvement in notch toughness of hot-rolled steel by desulfurization[a]

Sulfur (%)	Testing direction[b]	Charpy fracture energy[c] (J)
0.013	L	292
	T	124–152
	TT	50–54
0.004	L	298
	T	296
	TT	192–256

a 0.10%C–1.5%Mn steels of tensile strength 400 MPa (Spitzig and Sober 1981) b L longitudinal; T long transverse; TT short transverse (through-thickness) c Upper shelf energy

See also: Steels: Electrical Resistivity and Thermal Conductivity; Steels: Surface Treatment; Ferrous Physical Metallurgy: An Overview

Bibliography

Baker T J, Charles J A 1971 Influence of deformed inclusions on the short transverse ductility of hot-rolled steel. *Proc. Conf. Effect of Second-Phase Particles on the Mechanical Properties of Steel.* Iron and Steel Institution, London, pp. 79–87

DeArdo A J, Hamburg E G 1975 Influence of elongated inclusions on the mechanical properties of high-strength steel plate. In: deBarbadillo and Snape 1975, pp. 309–37

deBarbadillo J J, Snape E (eds.) 1975 *Sulfide Inclusions in Steel.* American Society for Metals, Metals Park, Ohio

Dunn C G, Walter J L 1966 Secondary recrystallization. In: Margolin H (ed.) 1966 *Recrystallization, Grain Growth and Textures*. American Society for Metals, Metals Park, Ohio, pp. 461–521
Gladman T, Holmes B, McIvor I D 1971 Effects of second-phase particles on strength, toughness and ductility. *Proc. Conf. Effect of Second-Phase Particles on the Mechanical Properties of Steel*. Iron and Steel Institution, London, pp. 68–78
Gove K B, Charles J A 1974 Further aspects of inclusion deformation. *Met. Tech.* 1: 425–31
Hilty D C, Farrell J W 1975 Modification of inclusions by calcium. *Ironmaking Steelmaking* 2(5): 17–12; 2(6): 20–27
Hilty D C, Popp V T 1970 Improving the influence of calcium on inclusion control. *Proc. Electric Furnace Conf.* The Metallurgical Society (AIME), New York, pp. 52–66
Kiessling R, Lange N 1978 *Non-Metallic Inclusions in Steel*, 2nd edn. The Metals Society, London
Kozasu I, Tanaka J 1975 Effects of sulfide inclusions on notch toughness and ductility of structural steels. In: deBarbadillo and Snape 1975, pp. 286–308
Luyckx L, Bell J R, McLean A, Korchynsky M 1970 Sulfide shape control in high strength low alloy steels. *Metall. Trans.* 1: 3341–50
Marston G J, Murray J D 1970 Machinability of low-carbon free-cutting steel. *J. Iron Steel Inst.* 208: 568–75
Spitzig W A, Sober R J 1981 Influence of sulfide inclusions and pearlite content on the mechanical properties of hot-rolled carbon steels. *Metall. Trans. A* 12: 281–91
Turkdogan E T 1983 Ladle deoxidation, desulphurization, and inclusions in steel. *Proc. 2nd Int. Conf. Clean Steels*. Metals Society, London, pp. 75–121
Wilson W G, McLean A 1980 *Desulfurization of Iron and Steel and Sulfide Shape Control*. The Iron and Steel Society (AIME), Warrendale, Pennsylvania

A. J. DeArdo

Incommensurate Structures and Charge Density Waves

Incommensurate structures are peculiar solids which are not strictly crystalline in that they lack periodic translational symmetry. Neither are they amorphous or glassy, since the translational symmetry is broken not in a haphazard way, but rather because two (or perhaps more) elements of translational symmetry are present which are mutually incompatible. In other words, there spatially coexist two or more lattices with spacings which cannot be brought into registry. More formally, suppose $A(\boldsymbol{r})$ and $B(\boldsymbol{r})$ represent the respective spatial distributions of two characteristic properties of a material and that

$$A(\boldsymbol{r}) = \sum_{\{G\}} A_G e^{i\boldsymbol{G}\cdot\boldsymbol{r}}, \quad B(\boldsymbol{r}) = \sum_{\{G'\}} B_{G'} e^{i\boldsymbol{G}'\cdot\boldsymbol{r}} \qquad (1)$$

The structure is said to be incommensurate if along at least one crystallographic direction the sets of reciprocal lattice vectors $\{\boldsymbol{G}\}$ and $\{\boldsymbol{G}'\}$ have no elements in common (except for the trivial $\boldsymbol{G} = \boldsymbol{G}' = 0$). Various physical situations are possible depending upon which properties A and B represent, as shown in Table 1.

1. Modulated Structures

Modulated structure is a somewhat imprecise term used in metallurgical and mineralogical literature to describe a basic crystal structure having any periodic perturbation with a repetition distance appreciably greater than the basic cell dimensions. Modulated structures are thus either commensurate or incommensurate depending upon whether the imposed repetition distance is or is not a rational multiple of a basic cell dimension. Both commensurate and incommensurate modulated structures give rise to (often weak) satellite Bragg reflections which can be conveniently studied by electron, x-ray or neutron diffraction techniques.

Incommensurate compositionally modulated structures can be viewed as being built up from smaller commensurate subunits of varying length arranged in a random or chaotic pattern. It is only the average superperiod that is incommensurate. For example, AuCu II consists of alternating domains of varying lengths of AuCu I, an ordered alloy, with anti-AuCu I, a structure in which Au and Cu atoms exchange positions.

2. Charge Density Waves

An incommensurate structure does not exist in all cases only as an average over short sequences of commensurate units. For example, there are displacively modulated incommensurate structures in which the displacement $\boldsymbol{u}_l$ of the lth atom in the structure away from its nominal crystalline lattice position $\boldsymbol{l}$ is to a good approximation sinusoidal; that is,

$$\boldsymbol{u}_l = \boldsymbol{A} \cos(\boldsymbol{q}_0 \cdot \boldsymbol{l} + \phi) \qquad (2)$$

Often, but not always, the incommensurate wave vector $\boldsymbol{q}_0$ lies along a high-symmetry direction of the underlying lattice. Also, it is not uncommon for the modulation amplitude $\boldsymbol{A}$ to disappear above a certain temperature, the material thereby transforming into a normal commensurate material with the average structure.

The form of Eqn. (2) suggests that displacive incommensurate phase transformations should be viewed as a condensation or "freezing in" of an unstable phonon mode of wave vector $\boldsymbol{q}_0$. In particular, in metals such a transformation may result from a charge density wave (CDW) instability (Overhauser 1968) in the conduction electrons near the Fermi surface. The occurrence of such an electronic instability requires a large electronic susceptibility $\chi(\boldsymbol{q}_0)$,

Table 1
Classification of incommensurate structure types

Type	$A(\boldsymbol{r})$	$B(\boldsymbol{r})$	Example
Magnetic	nuclear density ρ	magnetic density M	Cr, rare earth metals
Compositional	average density $\langle\rho_1 + \rho_2\rangle$	differential density $\langle\rho_1 + \rho_2\rangle$	CuAu II
Surface overgrowth	bulk atom density ρ_b	surface atom density ρ_s	Ar on graphite
Intergrowth	lattice density ρ_1	lattice density ρ_2	*n*-alkanes in thiourea
Displacive	average density $\langle\rho\rangle$	displacement field $u(\boldsymbol{r})$	K_2SeO_4
Charge density wave	average density $\langle\rho\rangle$	charge density $\langle\sigma\rangle$	2H–$TaSe_2$

which is favored when two large portions of the Fermi surface are separated by the wave vector $\boldsymbol{q}_0$. This "nesting" of the Fermi surface is most probable for quasi one- or two-dimensional metals since the Fermi surface is then independent of some components of the electron momentum. The potential resulting from the sinusoidally modulated charge density drives the displacement of the atoms $\boldsymbol{u}(\boldsymbol{r})$ through electrostatic coupling to the charged ion cores.

Displacive incommensurate structures associated with CDWs may be viewed alternatively as intergrowth compounds in which the second component is the CDW itself, the displacement field being induced on the first lattice by intermodulation effects between the two lattices. One of the most fascinating properties of such systems is that, at least for sufficiently weak and short-ranged interactions between lattices, it is possible to slide or rigidly displace the CDW lattice relative to the atomic lattice along the incommensurate direction without cost of energy. This gives rise to a new, low-frequency, branch of excitations, the so-called "sliding modes," which are, in effect, extra acoustic phonon modes. In the case of CDWs, the sliding modes transport electronic charge and form the basis for a novel and as yet unrealized type of superconductivity (Fröhlich 1954). In practice, sliding CDW modes appear to be readily pinned by impurities, and only recently have experiments begun to probe their influence on conductivity.

3. Incommensurate Insulators and "Lock-In" Transformations

Incommensurate modulated structures are not driven exclusively by Fermi-surface effects, nor are they restricted to metals. Incommensurate spiral spin structures, such as occur in several of the rare earth metals, result from a competition between first- and second-nearest-neighbor exchange interactions. Similarly, displacive modulated structures occur in insulators owing to competition between elastic forces of varying ranges. Often a material changes from incommensurate to commensurate with variation in pressure or temperature. An example of such a *p–T* phase diagram is shown in Fig. 1. Although

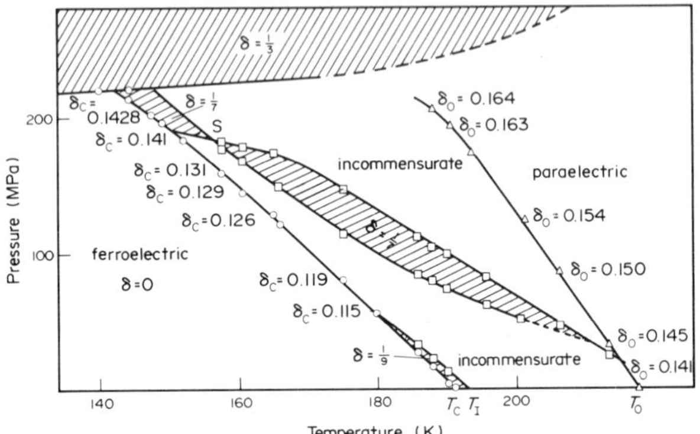

Figure 1
Pressure–temperature phase diagram of thiourea showing the wave vector δ of the displacive modulation. There are three commensurate phases, $\delta = \frac{1}{9}$, $\frac{1}{7}$ and $\frac{1}{3}$. The variations of δ at the paraelectric, δ_0, and ferroelectric, δ_C, phase boundaries are shown. (Courtesy of A H Moudden)

purely sinusoidal modulation (Eqn. (2) with ϕ constant) cannot take advantage of the periodic potential of the average lattice, with the proper spatial variation of $\phi(\boldsymbol{r})$ a tendency for local "locking-in" of the displacements can be accomplished. A careful analysis of this effect shows that this produces a gradual pulling away of the average wave vector from $\boldsymbol{q}_0$ toward one commensurate with that of the average lattice.

See also: Displacive and Order–Disorder Transformations

Acknowledgement

This article has been authored under contract DE-AC02-76CH00016 with the Division of Basic Energy Sciences, US Department of Energy.

Bibliography

Axe J D 1978 Charge density wave instabilities and incommensurate structural phase transformations. *Neutron*

Inelastic Scattering, Vol. II. International Atomic Energy Agency, Vienna, pp. 101–21
Fröhlich H 1954 On the theory of superconductivity: The one-dimensional case. *Proc. R. Soc. London, Ser. A* 223: 296–305
Overhauser A W 1968 Exchange and correlation instabilities of simple metals. *Phys. Rev.* 167: 691–98
Wilson J A, Di Salvo F J, Mahajan S 1975 Charge-density waves and superlattices in the metallic layered transition metal dichalcogenides. *Adv. Phys.* 24: 117–201

J. D. Axe

Indigenous Building Materials

Indigenous building materials have two inherent advantages over other materials: low cost and local availability. As such, they have long been employed throughout the world. Centuries of use have resulted in the attainment of high levels of skill in the use of these materials, as attested by the great pyramids of Egypt, the Greek Parthenon and the Taj Mahal of India.

Developing countries today continue to rely on indigenous materials. In hot and humid lands, which support the growth of plants and forests, a variety of materials such as grasses, leaves, straws, reeds, bamboo, timber and allied materials are found. Hot and dry areas yield inorganic materials like clays, laterite, sand and stone for building purposes.

An example from India illustrates the widespread reliance on mud as a building material. According to the 1981 census, there was a total housing stock of 90 million dwelling units in India. Of these, 57% or 51.2 million employed mud as a principal building material, particularly for walling.

The wide range in type and use of indigenous materials and their corresponding importance in sheltering much of the world's population is evident from the following discussion.

1. Inorganic Materials

1.1 Soil—An Overview

The type of soil available in a given area depends on the climatic region and topography. Silty soils, composed of very fine sand and clay, are found in river deltas. Such soils are well suited for construction of walls and production of mud bricks, burnt clay bricks and roofing tiles, for instance. Muram types of soil, containing gritty material, are common in mountainous plateaus. These soils are suitable for construction of road surfaces, embankments and dams. Laterite soils which contain gravel in compressed formation are also used for building. When laterite soils are stabilized with lime, good quality blocks can be produced.

1.2 Adobe

Adobe bricks are one of humankind's oldest building materials. These unburnt clay bricks are formed from highly plastic soils containing >70% clay and dried in the sun for 2–3 weeks.

Mud bricks are made in the following manner. Soil is collected in heaps and water is poured into it. The soil is then puddled into plastic consistency by trampling with bare feet. Straw or chopped hay is spread over the top in order to allow the bricks to "breathe" while curing or drying. The entire mass is then kneaded to distribute the binder.

1.3 Pisse-de-Terra

The conventional method of construction of mud walls is to convert the soil available at the site into mud with the addition of water. Lumps of mud are then piled one over another to form a wall. For protection and finishing, the mud wall is plastered with a paste of the same soil mixed with small quantities of cow dung and straw.

To make mud walls strong and durable, techniques of soil stabilization have been evolved using materials such as lime, cement and emulsified bitumen. Nearly any soil can be made into a better building material with the addition of the correct stabilizer. Briefly, stabilizers perform the following actions:

(a) they cement the particles of soil together so that they are stronger and more durable;

(b) they increase resistance to weathering from sun and wind;

(c) they render the mud wall watertight, thus protecting it from rain and erosion; and

(d) they protect soil from shrinkage and swelling.

Stabilization thus eliminates the need for frequent repair or complete rebuilding of walls or roofs after rain. Stabilizers have been developed to lengthen the life of bricks also.

1.4 Water-Repellent Mud Plaster

The application of water-repellent mud plaster can also increase the life of soil-based walls and roofs. To prepare this plaster, bitumen emulsion composed of 80/100 grade bitumen, kerosene and creosote is mixed with ordinary mud plaster prepared with cow dung and straw. This treatment lasts for 4–5 years.

1.5 Laterite Blocks

A final type of sun-dried brick is the laterite block. Laterite is an abundant subtropical soil or rock formed by the weathering of igneous rocks. When combined with other locally available materials such as lime, it can be processed to obtain good-quality building bricks with a wet compressive strength of ~5–6 MPa.

1.6 Burnt Clay Bricks

Not all soil-based materials rely on the sun to dry them to their finished state. Many soil products such as clay roofing tiles require high-temperature burning.

The alluvial soils, such as those of the Indo-Gangetic plain, are ideally suited for the manufacture of very good quality burnt bricks and tiles. To produce such products, the soil is dug out and mixed with water and kneaded well to make the soil mass homogeneous. After the soil has been kept wet for several days, bricks are molded in 230 × 115 × 75 mm wooden molds, although other brick sizes can be produced. The wet bricks are left in the open to dry in the sun for about a week. After drying, they are fired in a clamp or kiln, depending upon the scale of production and types of bricks required, using firewood and coal as fuel. For regular production of bricks large kilns called Bulls' Trench kilns are adopted, using coal or oil firing.

Many improvements have been made in the preparation of soil, molding of bricks and firing in kilns to improve the quality of bricks and to reduce the production cost. For production of bricks of high compressive strength, mechanical methods are now used, such as table molding using simple machines, and extrusion. A variety of structural clay products have also been produced for use in construction, such as hollow clay blocks, perforated bricks and flooring tiles.

1.7 Clay Roofing Tiles

Plastic clays are used for producing half-round country tiles with the help of a potter's wheel. After drying, the tiles are burnt in a clamp using firewood, cow dung, straw or other agricultural waste as fuel. These tiles are used extensively for covering sloping roofs, mostly in rural areas. However, country tiles are very fragile and are easily displaced or broken. For this reason, roofing is often made of bamboo or local timber rather than tile.

Improved types of clay roofing tiles have been evolved. One example is pantiles, which are formed in a wooden mold, dried and then burnt in clamps or kilns. Even more durable clay tiles are produced by an extrusion process. They are dried and then burnt in kilns. These tiles interlock into place.

In addition to roofing tiles, burnt clay bricks and tiles are also manufactured with dimensions 230 × 115 × 38 mm. They are used for masonry wall construction as well as for paving floors. Square clay tiles of 60 mm size are also produced and used for flooring in low-cost houses. Finally, vitrified clay roofing tiles having high compressive and abrasive strength are used for paving of footpaths and streets.

Research has been undertaken to upgrade the quality of clay roofing tiles by adopting better methods of molding and improving the design.

1.8 Clay Pozzolana (Surkhi)

To make clay pozzolana, burnt brickbats and overburnt bricks are ground to powder and mixed with lime for preparation of mortars, plasters and foundation concrete. Because the pozzolana prepared in this manner has a low lime reactivity, suitable clay deposits for producing reactive pozzolana have been sought. When calcined at optimum temperature, these specially selected clays produce high pozzolanicity. If used in conjunction with lime for mortars, the reactive pozzolana gives strength similar to that of cement mortars. Development of economical processes for reactive clay pozzolana has hastened the spread of this material within the developing world.

The lime–clay pozzolana mixture can replace cement in mortars, plasters and foundation concrete and is better than the cement–sand mortars. In rural areas, lime pozzolana mixtures could completely replace cement in single and double storey constructions.

1.9 Building Lime

In addition to soil bricks and blocks, rock and stone of various types make a valuable contribution to the building industry. Limestone is often available in the form of rock in the hills or as small pieces of limestone in the soil called Kankar lime. When burnt, limestone provides quicklime, the source of building lime, widely used as a binding material for preparation of mortars, plasters and concrete, with additives such as sand, pozzolanic material obtained from clay, fly ash and other materials.

Lime is produced in small kilns of 3–5 tonnes capacity with wood or coal generally used as fuel. For regular production, continuous kilns of higher capacity are adopted. However, nowadays dry hydrated lime is available in ready-to-use form.

1.10 Gypsum

Large deposits of gypsum of various grades are indigenously available in some developing countries. This material is commonly used in preparation of plasters and the production of cement. It can also be calcined to obtain gypsum plaster, which is used for internal decorative work. In addition, gypsum blocks can be manufactured and used for nonloadbearing internal partitions with good thermal insulation. However, because gypsum absorbs water, it cannot be used for external applications.

Fibrous gypsum plasterboard reinforced with jute, reeds, glass, wood, galvanized wire mesh or other materials is used for floors, ceilings and light partitions. Manufacture of this product could be undertaken on a cottage industry basis.

Acoustic tiles of gypsum are also produced and used in auditoria and other types of room where reduction in reflected sound is required.

1.11 Sand

Sand is made of crushed particles of rock of all sizes. Although it is usually obtained from river beds, rock can be crushed to form sand of required sizes.

Sand is a siliceous material with widely varying characteristics which depend on its place of origin. It is used as an admixture for preparation of mortars and plasters for masonry work and for production of concrete in conjunction with lime or cement in specified proportions. The quality of the local sand determines the amount needed in a particular mortar or cement.

Sand is also used for production of building components such as solid and hollow concrete blocks and other types of structural products. Cement–sand blocks as well as sand–lime bricks produced by autoclaving are often used for loadbearing masonry construction.

Lightweight concrete blocks and other structural building components are produced from sand, cement or lime with the aid of air-entraining agents. These prefabricated products are used for construction of walls, roofs and floors.

1.12 Building Stone

A variety of stone is used in construction, including igneous rocks such as granite, basalt and trap; sedimentary rocks like sandstone, limestone and laterite; and metamorphic rocks such as slate and marble.

Stone is used for construction of thick and massive walls, of which random rubble masonry walls are an example. For fine work, ashlar masonry walls are built. In olden days, stone arches formed the support system for the roof. Also, flat stone slabs from materials such as slate were adopted for roofing.

For stone masonry work, lime-based mortars and plasters are generally adopted. Lime concrete with stone aggregates, for example, is used for foundation work. For flooring and paving, stone is also used to provide a hard durable surface. One such example is marble, a costly stone which finds use in buildings of architectural character.

Nowadays stone is used in the production of cement concrete, a popular building material. Stone spalls can be used with cement concrete to produce stone blocks used for masonry work. Using molds, these blocks can easily be produced at the construction site or factory by skilled workers. Stone masonry blocks are made of loose cement concrete and stone pieces of 50–250 mm, which should be sound, durable and free from defects. The concrete mix used is generally 1:5:8 (cement, sand and concrete aggregate, respectively) to obtain blocks with a compressive strength of >6–7 MPa. The use of stone masonry blocks results in a saving of 15–20% in the cost of walls, compared with random rubble masonry.

2. Organic Materials

2.1 Grass Thatch

In rural areas, grasses from forests and fields are extensively used for thatch roofing, formed by binding the grasses to a bamboo or timber roof frame. Yet, despite its widespread usage, thatch is highly vulnerable to fungi, insects and fire.

A number of preservatives have been developed in response to these threats. To protect thatch from fungi and insects, a solution of copper sulfate, sodium dichromate and acetic acid is often coated on the thatch. For protection from fire as well as decay, an application of a solution of boric acid, copper sulfate, zinc chloride and sodium dichromate is used. However, although the service life of thatch can be extended by 3–5 years with these treatments, the required chemicals are costly and often not available locally, so these treatments have not found wide application. A cheap alternative treatment can be made from mud mixed with 5% bitumen emulsion.

2.2 Palm Leaves

Palm leaves and midribs of dead palms are often used for roofing. Palm leaves are used for roof thatch, whereas palm midribs are used for reinforcing concrete in floors, roof coverings, walls and columns of low-cost housing. However, water-repellent and fire-retardant treatments should be applied to extend the life of palm thatch.

2.3 Husks and Straw

Husks from rice or wheat are available in large quantities after the threshing season. These husks are commonly used as an admixture in mud plaster to provide strength and prevent cracking. They can also be used with cement to form roofing sheets for low-cost housing.

Rice-husk ash provides siliceous material of pozzolanic properties. It is therefore used in conjunction with lime or cement to provide a low-cost binding material used in mortars and plasters.

The straw from rice and wheat is also useful, both as a component of building materials and as a fuel. Often, chopped straw is an admixture for mud plaster. Straw mixed with mud is also used to produce small roofing tiles. At the same time, straw and other agricultural wastes are used as fuel for burning bricks, tiles and lime.

2.4 Jute Stalk

Numerous building materials come from the jute stalk: building boards for low-cost roofing as well as strings and ropes for tying bamboo and timber frames. In addition, jute bags are commonly used for packaging of cement.

2.5 Coir

Like jute, coconut fiber, called coir, is used for making strings and ropes used for tying bamboo and timber frames. In addition, cement-bonded coir can be employed to manufacture building boards used for internal application, such as for acoustic treatment. Cement-bonded coir and roofing sheets can also be produced to provide low-cost roofing material.

2.6 Cow Dung

Cow dung is commonly used as an admixture with soils that are clayey and silty, to reduce shrinkage in preparation of mortars and plasters used to repair cracks in mud walls.

2.7 Reeds

The main types of reed used for building—cane (*Phragmites karka trin*) and papyrus (*Typha latifolia*)—are found in abundance on the edge of marshes. Both of these types have been used as building materials since time immemorial, in part because few if any tools are needed to fashion them into a shelter or building component.

Reeds are most commonly used in the form of sheets and boards. Reed sheets, made of reeds bound with string, are commonly used for roofing over timber or bamboo rafters, despite their high susceptibility to rain and high winds. Reed sheets are also used to enclose spaces in the form of party walls, while reed panels are used to enclose spaces between timber frames. Both sheets and panels are normally finished with mud or cement–sand plaster.

A recent innovation in the use of reeds is the reed board, formed by compressing a small bunch of reeds together and wiring them tightly. These boards, varying from 25 to 50 mm in thickness, are used between timber framework for walling or for roofing. They are coated with mud, lime or cement plaster.

Reed sheets as well as reed boards are vulnerable to termites and thus require antitermite treatment.

2.8 Bamboo

Bamboo is one of nature's most versatile products. By virtue of its ready availability, comparative cheapness, high strength–weight ratio and easy workability, it has continued to play a vital role in rural housing. Bamboo is an excellent material for the framework of houses, particularly in seismic areas. It is also used for foundations, flooring, walls, partitions, ceilings, roofing, doors, windows and drainage systems. Because of its high tensile strength, bamboo has also been used for manufacturing a variety of boards and as a reinforcement for concrete, although its bond strength is low.

Despite its many attributes as a building material, bamboo also has some shortcomings which limit its wider application in the field of building construction. One of the most serious disadvantages of bamboo is its low durability; another is its extreme fissility. Except for certain thick-walled types, most bamboos have a tendency to split easily. This prevents the use of normal joints or even the use of nails. However, cutting a notch or mortice in a bamboo drastically reduces its ultimate strength. The only remedy is the use of nodes as places of support and joints, and the use of lashing materials in place of nails.

As an organic material, bamboo is subject to decay and can also be destroyed by insects, particularly borers. However, research shows that bamboo treated with copper–chrome–arsenic and copper–chrome–acetic acid compositions are adequately preserved. Methods of treatment include the Boucherie process and dipping. It is estimated that the expected service life of treated bamboo is 15–20 years, compared with 2–3 years for untreated bamboo.

2.9 Wood

Wood is one of the oldest building materials used considerably in construction. There is a general botanical classification of trees into two groups: exogenous or outward-growing, and endogenous or inward-growing. The former includes all the commercial timber used for building purposes and the latter includes the plants and bamboos of the tropics. The exogenous trees are divided into two main groups: hardwoods and softwoods. This division rests on a purely botanical distinction and not upon the hardness or softness of the timber; many hardwoods are softer than certain softwoods.

Woods that are particularly important to the construction industry include teak, deodar (a member of the cedar family) and sal. If these woods are in short supply, there are more than 120 varieties of secondary species which could be used for construction if properly seasoned and given preservation treatment.

The process of drying timber under controlled conditions is called seasoning (see *Drying of Wood*). It is especially important for hard, heavy timbers located in countries with wet climates, such as India. Once seasoned, timber should be stored under cover, protected from direct sun and rain.

Timber deteriorates from a number of causes, among which are fungi, insects and marine borers (see *Wood: Degradation by Insects and Other Animals During Use*). It will also break down quickly if subjected to alternate dry and wet conditions or used in a dark, damp, unventilated area. However, timber kept in a dry, airy place or continuously submerged under fresh water will last indefinitely. The use of preservatives to protect against various organic pests further ensures the longevity of the wood.

Various types of wood preservatives are available: oils, organic-solvent-based compositions as well as water-soluble "leachable" or "fixed" preservatives (see *Protective Finishes and Coatings for Wood*). The timber can be treated by surface application, soaking treatment, the hot and cold process, the pressure process or the Boucherie process.

It is not only sawn timber which finds an application in construction. Branches and twigs are used in conjunction with mud to build low-cost houses, particularly in the construction of walls. A variety of agricultural wastes such as timber refuse, bagasse and sugar cane also find use in building.

See also: Bamboo; Building Materials: An Overview; Cement as a Building Material; Roofing Materials; Structural Clay Products; Wood as a Building Material

Bibliography

Central Building Research Institute 1983 *Gypsum as a Building Material*, Building Research Note 14. Central Building Research Institute, Uttar Predash, India

Central Building Research Institute 1983 *Precast Stone Block Masonry*, Building Research Note 7. Central Building Research Institute, Uttar Predash, India

Government of India 1976 *Status Report on Rural Building Materials*. Department of Science and Technology, Government of India, New Delhi

Mathur G C 1976 *Survey of Rural Housing and Related Community Facilities in Developing Countries of ESCAP Region*. United Nations Economic and Social Commission for Asia and the Pacific, New Delhi

Mathur G C 1983 *Development and Promotion of Appropriate Technologies in the Field of Construction and Building Materials Industries in India*. United Nations Industrial Development Organization, Vienna

United Nations 1976 *Use of Agricultural and Industrial Wastes in Low Cost Construction*. United Nations, New York

G. C. Mathur

Indium Production

Indium has not been found free in nature, but occurs in many minerals including sphalerite, cassiterite, chalcopyrite, coal, galena, hematite, pyrite and wolframite. Its crustal abundance is about $0.1\,g\,t^{-1}$. In the USA, indium can be found in minute quantities in various ores of copper, lead, molybdenum, vanadium and zinc. Indium has not been found in deposits with sufficient concentration to permit mining and processing of the indium alone. Thus indium is a byproduct, primarily from zinc or lead processing.

The indium present in zinc ores and concentrates appears in intermediate products as a constituent of flue and precipitator dusts, residues from zinc retorts, residues from cadmium recovery, lead blast-furnace slags, and crude zinc and lead bullion. In the distillation process for refining zinc, most of the indium remains with the crude zinc until the lead is removed by fractional distillation. Indium is recovered from the lead in an involved procedure of pyrometallurgy, leaching, purification and cementation.

Anode slime, containing about 25% indium, obtained from electrolytic zinc plants, is refined chemically to produce a crude indium metal (99% pure). This is further refined electrolytically to produce standard-grade indium or high-purity grades of indium. The metal is cast into ingots weighing 0.3–10 kg.

See also: Gallium, Germanium and Indium Resources; Metals Production: An Overview

Bibliography

Milner E F, White C E T 1981 Indium and indium compounds. In: Grayson M (ed.) 1981 *Kirk–Othmer Encyclopedia of Chemical Technology*, 3rd edn., Vol. 13. Wiley, New York, pp. 207–12

J. F. Carlin Jr.

Induction, Flame and Laser Hardening

A number of processes are available for heat treatment of the surfaces of steel parts to produce a hard martensite layer that increases resistance to wear and fatigue. Such heat-treatment techniques are widely used in the production of gears, shafts, crankshaft bearings and bearing races. Although flame and induction hardening are the most common processes used for surface hardening, laser hardening is also used for some special applications.

These surface-hardening methods are designed to supply energy to the surface of the part at a rate faster than it can diffuse into the interior. As a result, the surface temperature of the part is raised above the transformation temperature of the steel, and an austenite surface layer is formed. Subsequent quenching of this austenite layer from the high temperature transforms it into a hard martensite layer. The part may be tempered afterwards to increase the toughness of the martensite layer.

In these surface-hardening techniques, the heat treatment is performed in air so that the increased hardness of the surface layer arises solely from formation of a martensite surface layer with no alteration of the chemical composition of the steel. In other surface-hardening methods such as carburizing, nitriding or carbonitriding, the carbon and/or nitrogen content of the surface is increased by the use of selective atmospheres as part of the heat-treatment cycle (see *Carburizing and Carbonitriding*; *Nitriding*; *Steels: Surface Treatment*).

1. Induction Hardening

In induction hardening, the workpiece is placed inside a water-cooled induction coil through which a high-frequency current is passed. This current generates a highly concentrated alternating magnetic field within the coil. The magnetic field in turn generates a current in the conducting steel workpiece;

this current is concentrated near the surface because of the skin effect of high-frequency currents (Cable 1954). The surface of the piece is rapidly heated because of the Joule heating (I^2R) loss.

The pattern of heating obtained by induction is determined by (a) the shape of the induction coil, (b) the number of turns of the coil, (c) the operating frequency, and (d) the alternating-current power input (American Society for Metals 1981). One possible configuration of the induction coil is shown in Fig. 1. It consists of a multiturn coil for outside-diameter heating. Other examples of coil configurations include single-turn coils for outside-diameter heating and specially shaped coils for local surface heating.

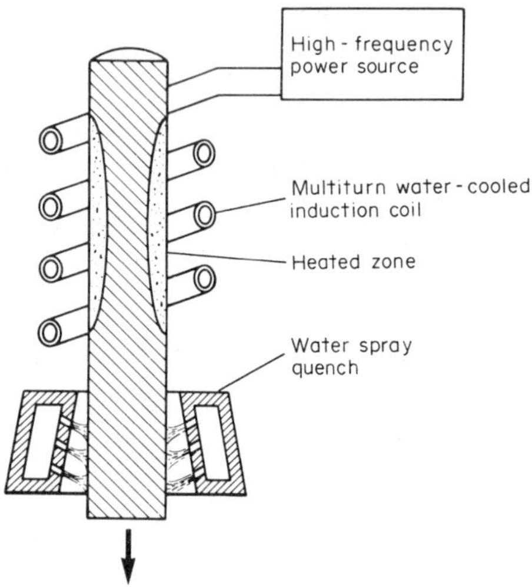

Figure 1
Multiturn induction coil and quenching unit for surface hardening of a steel shaft. Section through cylindrical shaft and heating coil

High-frequency power to the induction coil can be supplied from a motor–generator set, a spark-gap oscillator or a vacuum-tube oscillator. Since the energy transfer to the workpiece occurs only within the confines of the coil, surface hardening of a long shaft, for example, is accomplished by either moving the shaft through the coil or moving the coil along the shaft. The shaft is usually rotated at the same time to ensure more uniform heating.

Quenching of the part after the surface has been heated into the austenite temperature range (750–800 °C) may be accomplished by either removing the piece from the coil if it is small enough, or quenching it in place by a spray located at the exit side of the coil (Fig. 1). The short heating times of 3–4 s characteristic of induction heating and the continuous passage of a piece through the coil make induction heating ideally suited to automatic handling methods.

The rapid heating is also beneficial because oxidation and scaling of parts is reduced to a minimum. In addition, the rapid heating may be beneficial because of the grain-refining action caused by the rapid austenitizing treatment (Porter and Dabkowski 1970).

The case depth varies with the frequency of the power source. A shallow hardened case of 0.025–0.150 cm is achieved with 10 kHz to 2 MHz power sources; a deeper hardened case of 0.150–0.600 cm is achieved with 1–10 kHz power sources. Of course, use of lower frequencies and lower power densities may permit small steel parts to be through-hardened. Induction heating is widely used for this purpose.

Tempering of induction-hardened surfaces may be accomplished by use of a second induction coil at lower power densities, or can be done separately in a conventional furnace.

2. *Flame Hardening*

In flame hardening, the surface of the steel part is rapidly heated into the austenite temperature range by direct impingement of a high-temperature flame or of high-velocity combustion gases, and then cooled at a rate sufficient to produce a hard martensite layer.

Three configurations of the flame head are possible. In the spot (stationary) method, the heating head is stationary and only selected areas of the part are heated and quenched. In the progressive method, the heating head is moved over the surface area of the part, progressively heating a narrow band of area that is subsequently quenched. In the spinning method, circular or semicircular parts such as cams, wheels or shafts are rotated or spun in a fixed heating head. Combinations of these techniques are also possible in the progressive-spinning technique where a round part may be rotated and fed through a fixed head. The quenching fixture is usually attached directly after the heating head, as shown in Fig. 2.

Fuel gases that can be used in flame hardening are acetylene, propane and methane mixed in a suitable ratio with air or oxygen. The heating time will vary with the heating values of the specific fuel mixture and the velocity of burning, but is typically 10–60 s.

The thickness of the hardened case will vary from 1 to 6 mm. Shallow cases are obtained only with oxygen–fuel gas mixtures, which give higher flame temperatures. Air–fuel gas mixtures are used where deeper cases are desired.

3. *Laser Hardening*

In laser hardening, a high-power continuous-wave CO_2 laser is used in combination with a lens or mirror to produce a heated area on the material. The laser consists of an optical resonant cavity in which a mixture of CO_2, N_2 and helium gases flows. An electrical discharge excites the gas mixture so that a

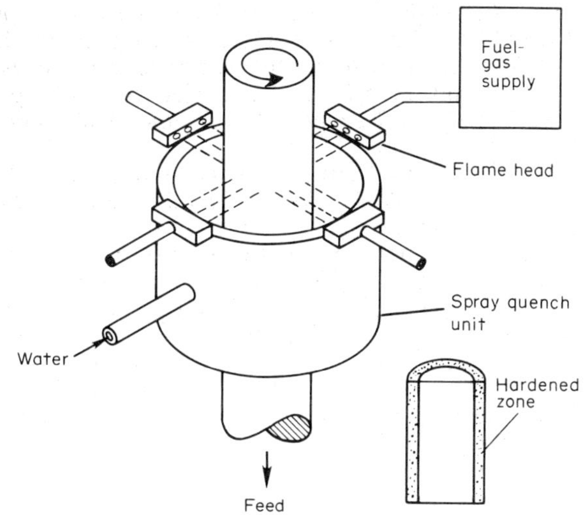

Figure 2
Flame hardening of a steel shaft by the progressive-spinning method

high-power-density beam with a wavelength in the infrared region (10.6 μm) is emitted from the end of the cavity. The beam can be focused with lenses or mirrors. The focused spot size is dependent on the focal length of the lens and the distance to the work-piece. Beam direction can be controlled by small mirrors so that the beam can be made to move over programmed paths of any complexity, as shown in Fig. 3 (Harth et al. 1976, Yessik and Schmatz 1975).

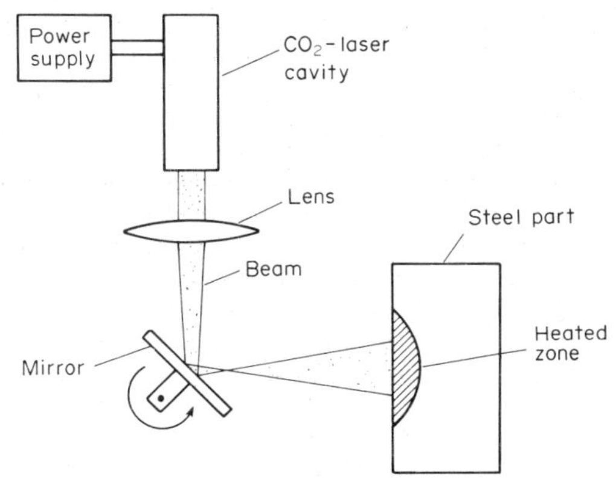

Figure 3
Laser surface hardening of a steel part

Since much of the energy in the laser beam is reflected, special precautions such as coating the surface with a thin layer of highly absorbing material or roughening the surface may be required to maximize the efficiency of this process. Because of the high power densities (up to 1.5 MW cm^{-2}), the case depths for laser hardening are much smaller than for flame or induction hardening, typically 0.020–0.080 cm. The depth of hardening is limited because surface melting will occur if the beam is focused to a smaller size to increase the power density.

In contrast to induction or flame hardening, laser hardening uses the mass of the underlying metal to self-quench the heated surface layer to form martensite, rather than an external quenching unit. This is possible because of the shorter heating times and thinner layers developed during laser hardening.

Laser hardening has two major advantages. The first is the ability to selectively harden very small areas. The second is the ability to harden areas such as small bores or complicated shapes which cannot be induction or flame hardened.

4. Metallurgical Considerations

Both steels and cast irons can be surface hardened by any of these three techniques. The most widely used steels for induction and flame hardening are carbon steels with carbon contents between 0.37 and 0.55%. These steels have sufficient hardenability for the thin surface layer to transform to martensite during water quenching. Also, the carbon content is sufficiently high for the martensite layer to have the necessary hardness to impart wear resistance. Alloy steels are usually justified only if water quenching causes too much distortion or if a high core strength must be obtained by heat treatment before surface hardening (*Metals Handbook* 1981).

The prior microstructure of steels may have an influence on the selection of the heating cycle during surface hardening because of the short heating times involved. Spheroidized steels with coarse carbides that dissolve slowly during austenitizing are less suitable than steels with a ferrite–pearlite microstructure that has been produced by normalizing, annealing or hot rolling. For the same reason, cast irons with a fine pearlite matrix are desired. Since graphite flakes or nodules remain undissolved in the austenite during these short heating times, cast irons in which most of the iron is present as graphite, such as nodular cast iron, are not suitable for surface hardening by these techniques.

See also: Quench and Strain Aging; Steels: Surface Treatment

Bibliography

Cable W J 1954 *Induction and Dielectric Heating*. Reinhold, New York

Harth G H, Leslie W C, Gregson V G, Sanders B A 1976 Laser heat treating of steels. *J. Met.* 28(4): 5–11

Metals Handbook, 9th edn., Vol. 4, 1981. Heat treating. American Society for Metals, Metals Park, Ohio, pp. 451–517

Oakley P J 1981 Laser heat treatment and surfacing techniques—A review. *Weld. Inst. Res. Bull.* 22(1): 4–10
Porter L F, Dabkowski D S 1970 Grain size control by thermal cycling. In: Burke J H, Weiss B (eds.) 1970 *Ultrafine Grain Metals*. Syracuse University Press, New York, pp. 133–61
Yessik M, Schmatz D J 1975 Laser processing at Ford. *Met. Prog.* 107(5): 61–66

G. R. Speich

Industrial Minerals: An Overview

Industrial minerals, including stone and rocks, may be defined as those naturally occurring materials used to build structures or supply products that are useful to an industrialized society. Since industrial minerals exclude the ores of metals, they have been called the "nonmetallics." But this is a term not now widely used and not properly inclusive, as fuels and water are also excluded from industrial minerals.

Industrial minerals are extensively used, although the average individual is unaware of their importance. Roads and buildings are conspicuous signboards of industrial minerals, but minerals used in processing or making products are not so apparent. For example, automobiles have metal parts molded in foundry sand, bodies polished by abrasives and covered with paint containing mineral fillers and pigments, and engines fuelled by petroleum products produced and refined with the help of clay.

Industrial minerals are an important part of commerce in the industrialized countries. For many years in the USA, the annual value of industrial-mineral production, which in 1983 was $15.5 billion, has been more than twice that of metal production. The tonnage of the construction materials, stone and sand and gravel, which together in 1983 was 1.5 Gt, surpassed the total tonnage of all metals produced.

1. Classification

The industrial minerals covered in this Encyclopedia include more than 50 rocks and minerals from igneous, metamorphic, and sedimentary sources. They have been classified as to principal use (see Table 1).

Table 1
Classification of industrial minerals according to use[a]

Abrasives
Nonsiliceous minerals
Precious and semiprecious minerals
Siliceous minerals
Binders
Industrial minerals
Clay minerals
Construction materials
Industrial minerals
Crushed stone
Sand and gravel
Dimension stone
Fill and soil
Granules
Electronic and optical minerals
Fillers and coatings
Industrial minerals
Siliceous minerals
Clay minerals
Carbonate minerals
Sulfate minerals
Filters, sorbents and ion exchangers
Siliceous minerals
Clay minerals
Foundry sands
Pigments
Industrial minerals
Iron oxide and related minerals
Soil additives
Industrial minerals
Phosphate, potash and sulfur
Well-drilling materials
Clay and non-clay minerals
Industrial minerals
Zeolite minerals

a Articles in the Encyclopedia on industrial minerals are grouped under the general headings shown in *italic type*

2. Characteristics

Most classifications draw sharp boundaries that are more apparent than real, and the classification for industrial minerals is no exception. Some broad generalizations can be made about the characteristics of industrial minerals (Table 2), but a classification based on use has inconsistencies because of the great number and diversity of the minerals.

Although industrial minerals can be thought of as a group of "nonmetallics," some ores of metals are included in the classification. For example, iron oxide, an ore of iron, is listed as a pigment, a usage that qualifies it as an industrial mineral, and so it is placed in the classification. Some minerals may have more than one use and are included under multiple entries. Discussions of clay minerals, because of their many applications, are found under binders; construction materials; fillers and coatings; filters, sorbents, and ion exchangers; and well-drilling materials. Clay is also an important abrasive and a principal component in most fill and soil.

Industrial minerals as used here exclude those that undergo major chemical conversion, a definition in itself that does not bear close analysis. Zeolites, for example, undergo ionic substitution, a chemical process, but because these minerals also have physical properties that make them useful as filters and sorbents, they are included as industrial minerals. Most potash and phosphate minerals used in commerce are chemically converted into fertilizer prod-

Table 2
Characteristics of natural industrial minerals (excluding those undergoing chemical conversion)

(*a*) *Mineral and rock types*
Igneous, metamorphic, and sedimentary rocks and minerals of many different types. Those of low unit value are generally rocks of widespread occurrence and associated with simple geologic conditions; those of high unit value are generally rocks and minerals of restricted occurrence and associated with complex geologic conditions

(*b*) *Exploration*
Those of low unit value generally cannot bear the cost of intensive geologic evaluation; exploration is simple or nonexistent, relying on use of regional studies or production trial and error. Those of high unit value, which require large expenditures for mine and plant development, can bear the cost of regional studies and site evaluations

(*c*) *Mining*
Either surface or underground mining may be used, but underground mining is generally more expensive and most often used for deposits that are not near the surface, for special structural conditions, for selective removal of particular beds, and for accommodation of weather or land-use considerations. In some places the extra cost of underground mining is compensated by the usable space that is created

(*d*) *Processing*
Those of low unit value require simple processing, but those of high unit value can bear the cost of complex processing

(*e*) *Transportation*
Those of low unit value are generally used near the point of production, require short haulage and use trucks extensively. Those of high unit value generally can absorb higher transportation costs, make use of rail and water haulage, and be traded in import and export markets

(*f*) *Marketing*
Those of low unit value are generally marketed near the deposit by the producing company; price is generally more important than specification to the consumer. Those of high unit value may be marketed far from the deposit by the producer, sales agents, or other companies; specification is generally more important than price to the consumer. For both those of high and low unit value, one mineral or synthetic product may be substituted for the same use

ucts, but some are not, and so they qualify for inclusion as additives to soil. In this classification, the quantity of material used is not so important as how the material is used. Gems and art objects are valuable for their intrinsic properties, but because they are not used in the sense of structures or products, they are not included. Industrial-grade diamonds and semiprecious minerals, however, are useful to industry because of their hardness and are included under abrasives.

Although chemical properties may affect physical properties of rocks and minerals, they are a secondary consideration for most uses of industrial minerals. Some industrial minerals are used just as they are taken from the ground; processing, compared with that of metals, is simple, and, if needed at all, usually consists of crushing, sizing, or washing. The physical properties inherent in the rock are the same properties that make it useful to industry, and processing seldom appreciably enhances these properties. Processing may be used to beneficiate the product. At times processing may be a sophisticated operation to remove unwanted material or produce particle sizes within close limits, but it does not alter or refine the material in the way that metals are altered or refined. Clay for filler and coating applications, for example, requires special processing, which may include dispersion, sizing, leaching, magnetic separation, drying, and pulverizing. But even after processing, the clay minerals are still similar in character to what they were before processing.

The wide range in values of industrial minerals indicates something of their diversity in occurrence and in properties. The cost of some construction materials is little more than the cost of the labor and materials to produce them; that is, the minerals have a value no greater than the surface value of the land. For example, in some parts of the USA, land can be purchased and stone can be quarried, crushed, sized, and loaded for shipment for no more than $4 per ton. Other industrial-mineral deposits (for example, industrial diamonds) may be much more valuable than the surface value of the land and on a weight basis sell for a price a million or more times that of crushed stone.

The industrial minerals of low unit value, such as the construction materials, are generally produced near the place where they are used, which is in or near populated areas. Reserves of these materials are widespread and abundant, and because they are heavy and bulky they cannot be shipped far before transportion costs exceed production costs. Operations that produce these types of materials generally have production capacities that far exceed product demand. Materials of high unit value generally have properties that are not so widespread; they can be shipped farther and can bear the cost of higher transportation in a competitive market. In general, as the demand grows, so does the distance a product can be shipped, but this increase is not without limits. More than one mineral may have the same use, and the one that is used will be the one that is more competitively priced.

Surface mining is more generally used than underground mining for industrial minerals because it is usually cheaper, but both methods have advantages and disadvantages. For the industrial minerals of low unit value, cost of production is an important factor in a competitive market, and so, if possible, surface mining will be the method of choice. But in some places where surface deposits are not available or where such factors as unfavorable geologic structures, thin mining seams, all-weather mining, environmental problems, governmental incentives, and post-mining use are considerations, underground mining may be preferred. For industrial minerals of high unit value, underground mining may be the more efficient method of extraction because their mineral occurrence is generally geologically more complex.

See also: Prices of Industrial Minerals: History; Building Materials: An Overview

Bibliography

Bates R L 1969 *Geology of the Industrial Rocks and Minerals*. Dover, New York

Berthoumieux G 1978 The industrial minerals of France. *Proc. 3rd Industrial Minerals Int. Congress*. Metal Bulletin, Worcester Park, pp. 15–24

Blunden J 1975 *The Mineral Resources of Britain: A Study in Exploitation and Planning*. Hutchinson, London

Borzunov V M 1969 *Nonmetallic Deposits, their Exploration and Economic Appraisal*. Nedra, Moscow

Bowie S H U, Kvalheim A, Haslam H W (eds.) 1978 *Mineral Deposits of Europe,* Vol. 1: *Northwest Europe*. The Institution of Mining and Metallurgy and the Mineralogical Society, London

Brazilian Ministry of Mines and Energy (1973 onwards) Analytical Profile Series. Brazilian National Department of Mineral Production, Brasilia. (Each bulletin deals with a specific industrial mineral commodity.)

Brobst D A, Pratt W P (eds.) 1973 *United States Mineral Resources*, US Geological Survey Professional Paper 820. US Government Printing Office, Washington, DC

Carta M, Ciccu R, Pretti S 1977 The industrial minerals of Sardinia: Present situation and future perspectives. *Proc. 2nd Industrial Minerals Int. Congress*. Metal Bulletin, Worcester Park, pp. 41–55

Committee on the Industrial Minerals Volume 1949 *Industrial Minerals and Rocks*, 2nd edn. American Institute of Mining, Metallurgical and Petroleum Engineers, New York

Council of Commonwealth Mining and Metallurgical Institutions 1957 Geology of Canadian industrial mineral deposits. *Proc. Commonwealth Mining and Metallurgical Congress*. Council of Commonwealth Mining and Metallurgical Institutions, London

Dickson T 1978 Industrial minerals of West Germany. *Ind. Miner. (London)* 131: 15–33

Dunn J R 1973 A matrix classification for industrial minerals and rocks. *Proc. 8th Forum on Geology of Industrial Minerals*. Iowa Geological Survey Public Information Circular No. 5. Iowa Geological Survey, Iowa City, Iowa, pp. 185–89

Fleming J M, McArthur J G 1977 Industrial minerals in Newfoundland and Labrador: Export potential and internal industrial growth. *Proc. 2nd Industrial Minerals Int. Congress*. Metal Bulletin, Worcester Park, pp. 17–40

Gillson J L (ed.) 1960 *Industrial Minerals and Rocks*, 3rd edn. American Institute of Mining, Metallurgical and Petroleum Engineers, New York

Harben P W, Bates R L 1984 *Geology of the Nonmetallics*. Metal Bulletin, Worcester Park

Jelić M, Joksimovic D, Milojković A 1977 Current activity and development prospects in the commercialisation of Yugoslavia's industrial mineral resources. *Proc. 2nd Industrial Minerals Int. Congress*. Metal Bulletin, Worcester Park, pp. 57–68

Knight C L (ed.) 1976 *Economic Geology of Australia and Papua New Guinea—4: Industrial Minerals and Rocks*, Australasian Institute of Mining and Metallurgy Monograph Series No. 8. Australasian Institute of Mining and Metallurgy, Parkville, Victoria

Kuzvart M 1984 *Industrial Minerals and Rocks*. Elsevier, Amsterdam

Lefond S J (ed.) 1975 *Industrial Materials and Rocks*, 4th edn. American Institute of Mining, Metallurgical and Petroleum Engineers, New York

Lefond S J (ed.) 1983 *Industrial Minerals and Rocks*, 5th edn. American Institute of Mining, Metallurgical, and Petroleum Engineers, New York

Lüttig G et al. 1977 The industrial minerals of the Federal Republic of Germany, and the role of the industry in the public scene. *Proc. 2nd Industrial Minerals Int. Congress*. Metal Bulletin, Worcester Park, pp. 3–16

Mineral Resources Consultative Committee (1971 onwards) Mineral Dossier Series. HMSO, London. (Each volume deals with a specific industrial mineral commodity.)

Raevskii V I, Fiveg M P (eds.) 1973 *Deposits of Potassium Salts in the USSR*. Nedra, Leningrad

US Bureau of Mines 1980 *Mineral Facts and Problems*, US Bureau of Mines Bulletin 671. US Government Printing Office, Washington, DC

Woodland A W, Highley D E 1975 Britain's industrial minerals: Their geological framework, their production today, and some aspects of their development. *Proc. 1st Industrial Minerals Int. Congress*. Metal Bulletin, Worcester Park, pp. 2–11

D. D. Carr

Ineffective Length (Transfer Length)

When an aligned fiber composite is axially stressed, the matrix, being more compliant than the fiber, tends to undergo a larger axial deformation. The longitudinal tensile stress in the fiber is then a maximum in the middle (neglecting the effect of adjacent fiber ends) and a minimum at the end. The region near the end of the fiber over which the longitudinal stress is smaller than the maximum is called the ineffective length or transfer length corresponding to some particular fraction (half, 0.9, etc.) of the maximum longitudinal stress in the fiber.

A. Kelly

Information Systems: Analysis and Design

A management information system (MIS) is a system that provides information and processing support for management decision making. The concept of such an information system preceded the advent of computers, but computers made the concept feasible.

The computer has added a new dimension to information systems design, so that computer-based information systems can be radically different from systems using manual processing.

Somewhere within the overall scheme of a manufacturing organization's MIS there is usually a vast collection of data on engineering materials. With over 100 000 engineering materials available, computerization is the only logical choice. The development of a materials database and its integration within the overall MIS can be crucial to the successful operation of a computer-aided materials selection system. Since information systems will play an increasing role in the metals industry it is essential that metallurgists and engineers understand the principles in the analysis and design of information systems.

1. Data and Information

Data, the raw material for information, are defined as nonrandom symbols which represent quantities, actions, attributes and such like. Information refers to data transformed into a form serving some useful purpose. Thus, information is data processed into a form that is meaningful to the recipient and is of real or perceived value in current or prospective decisions.

The relationship between data and information is analogous to that of raw material to finished product. The information processing system converts data in unusable form into usable data that are information to the intended recipient. The transformation of data into information may also involve separation of the significant data from the insignificant—discriminating that which is important and eliminating the mass of unrelated data, or noise.

The value of information is related to decisions. If there were no decisions, information would be unnecessary.

2. Value of Information

Since information is data transformed into a useful form, it follows that the value of information is not an inherent property of the information itself, but rather a consequence of its use. Hence, the value of information to an organization is the increment in performance that is realized due to having the information. If V_I is the value of information I, P_I is the performance with information I available and P_0 is the performance without information I, then

$$V_I = P_I - P_0 \tag{1}$$

One of the major problems in designing effective information systems is our inability to ascertain the value of information. Also, since information relates to a dynamic, real-world situation, it is constantly updated and, as shown in Fig. 1, the value of information changes with age. Hence,

$$V_I = p(A) - C \tag{2}$$

where p is the probability of use, A is the average economic benefit from use and C is the cost of obtaining and storing information.

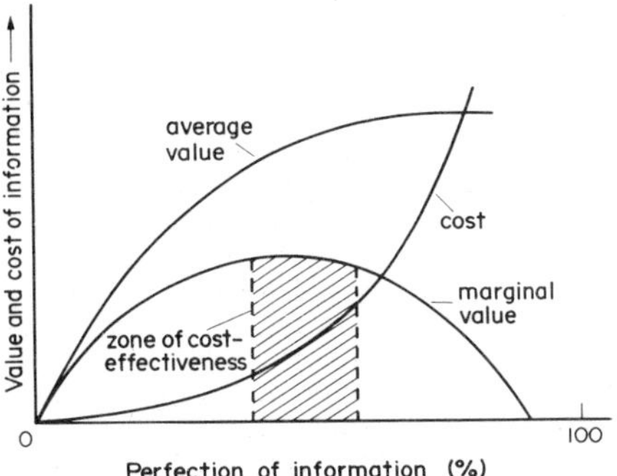

Figure 1
The cost of information

Figure 1 shows that the cost of continually trying to perfect information rises exponentially and the value of additional units of information starts to decline drastically at some point. The cost curve rises rather slowly at first, because the initial units of information are often obtained with relatively little effort. However, as more and more information is sought, it becomes increasingly difficult to obtain and the cost curve rises exponentially.

The average-value curve moves in a direction opposite to that of the cost curve. The early units of information increase in value as business alternatives resulting from their use begin to take shape at relatively low cost. However, as additional information is sought, the average-value curve declines, because the information is worth less than the cost of obtaining it. The average-value curve turns down rather slowly, because the cumulative value of the early units of information offsets the sharply rising cost of the later units. There comes a point when the cumulative cost outweighs the cumulative value; at this point the average-value curve begins to decline.

The marginal-value curve in Fig. 1 is an even more sensitive indicator of the cost of additional information. It reflects the value of the next unit of information rather than the average cumulative value of all units. The point of optimality is the most desirable point on the marginal-value curve, where the next unit of information will decline in value. It is that point where the value of the last unit is the highest obtainable and thus where the pursuit of additional information is not cost-effective. In practice, however, the determination of a precise point of optimality is difficult and therefore a zone of cost-effectiveness around the point of optimality, as shown in Fig. 1, is a much more useful concept. This zone allows the searcher to obtain enough information to formulate alternatives within reasonable time and cost constraints and at the same time prevents him from overlooking valuable data. However, it also stops him from proceeding with a search to the point where the marginal-cost curve becomes negative.

Two major components of the value of information are quantity and quality. The quantity of information is measured in terms of volume, completeness and accessibility. The volume of information refers to the capacity of the system and the amount of information that is available for use. There is a natural limit to the volume of information that a system can store and a user can reference. As shown in Fig. 1, this limit is reached when the cost of storing and maintaining this information exceeds its value. Most computer-based information systems utilize a storage hierarchy concept, wherein infrequently used information can be stored on relatively inexpensive storage media.

The capacity of an information system is also related to the efficiency of the system or the accessibility of information, since there exists a relationship between the volume of a storage medium and the speed of access. Completeness refers to the degree to which an information system satisfies the information needs of a user. Completeness can also be transformed into an economic variable since complete information is obviously of greater value than incomplete information but is more costly to maintain. The last attribute of the quantity of information is accessibility, which denotes the response time of the system and the facilities available for using it. To summarize, the quantity of information is not an absolute measure, but rather is a compromise between the value and cost of information.

The quality of information implictly relates to how information can be used and the degree of confidence that can be placed on it. The attributes of quality are timeliness, relevance, accuracy, reliability and flexibility. Timeliness refers to the process of collecting, storing and processing of data and the time factors involved. The relevance of information is the measure of how well the system meets the needs of the user. Accuracy refers to errors in collecting, storing and processing of data and may refer to explicit inaccuracies caused by bad data or implicit inaccuracies caused by information that is outdated. Reliability is an operational characteristic that measures the degree of confidence that the user can place in the system and the information it contains. Flexibility is the last attribute of information and of an information system and indicates the diversity of applications for which a given information set can be used.

3. Analysis of Information Systems

All information systems contain the following:

(a) computer hardware

(b) software

  (i) systems software

  (ii) applications software

(c) the database

(d) operating procedures

(e) operating personnel

The design and development of all information systems should begin with detailed systems analysis. The study of a system is not undertaken for its own sake but is aimed at determining effective solutions for planning and allocating resources to achieve desired goals. In systems analysis, the whole system under study is considered, rather than any part of it. Often the best solution for the whole system may not necessarily be the best for each of the subsystems. The overall cost-effectiveness for the given system is the essence of the systems approach (Optner 1973).

Systems analysis attempts to look at the entire problem, to investigate systematically the objectives of the system, to define the input, output and volume of data, and to evaluate the alternatives in terms of cost and effectiveness. Systematic and formal design procedures are now available which help in the analysis and design of effective operational systems (Lancaster 1968, Liston and Schoene 1971).

The major stages in a systems approach can be categorized as follows: (a) description of the organization or department; (b) definition of information requirements; and (c) specification of system requirements.

The starting point for the development of an information system must be an understanding of the organization department and how it functions. The organization must be viewed in terms of inputs, outputs and the processes that convert inputs to outputs. The organization must next be broken down into subsystems for further analysis. One approach to this problem is to study the information needs of each subsystem. Another approach is to organize the study around related groups of data.

The systems-analysis study next describes the basic characteristics of the future system, breaks it down into activities and specifies required inputs, outputs, operations and resources for each activity. In order to specify the information requirements of these activities, each activity must be defined precisely in terms of the output it must produce, the input it must accept, the resources it must use and the operations it must perform.

Specification of the system requirements via a

detailed report is the last activity of the systems-analysis phase. This report should include a section for each activity, preceded by a summary which integrates the activities and outlines the overall analysis of the entire system. The specification should also include documentation such as flow charts, decision tables and input–output requirements.

4. Information Systems Design

The two most common approaches in systems design are (a) the top-down approach and (b) the bottom-up approach (Davis 1974).

The top-down approach seeks to develop a model of information flow in the organization and to design the system to suit this information flow. The top-down approach begins by defining the objectives of the organization and the constraints under which it operates. From a functional analysis of the system the major information requirements are specified for the overall system and for the subsystems. The advantage of the top-down method is its very logical approach to the development of an overall plan. A disadvantage is the fact that it is difficult to make large-scale plans unless the information requirements of the various subsystems are known earlier in the development stage.

Since the basic elements of any system are the modules for processing transactions and updating files, the evolutionary or bottom-up approach states that the way to develop an overall design is to start with the operational modules for processing transactions and updating files and then to add planning, control, decision-making and other modules as demand develops. In other words, the system is assumed to grow in response to the needs expressed by management. The system design then reflects the orderly implementation and integration of the modules that have been requested. The advantages of the evolutionary approach are that the information system expands in response to real needs rather than assumed needs. It builds on the transaction-processing and file-updating capabilities that must be available for operational support. The disadvantage is the inability to integrate the subsystems as closely as might be desirable.

In both the approaches mentioned, defining the requirements of the various subsystems (e.g., production, sales, design) and their eventual integration is always the crucial activity. Two formal methods for determining the information requirements of a subsystem or an application are (a) decision analysis and (b) data analysis.

Decision analysis is a top-down approach which derives the information requirements for an application by analysis of the objectives and decisions to be made. The result of the decision-analysis method is a set of data required for the decision and some measure of the importance of each data element and the accuracy needed. The philosophy of this method is to collect and report only information that is useful in the decision-making process.

Data analysis is an evolutionary or bottom-up approach which seeks to derive the information requirements by analysis of the data currently or potentially used. The result of the data-analysis method is an identification of data currently collected and thought to be needed or not currently collected but thought to be of value. The philosophy of this method is to accept these representations of need unless they are found to be incorrect. The data-analysis method is very appropriate for applications involving mechanization of an existing administrative procedure.

Finally, any systems design activity will ultimately obey the zeroth law of reliability (Gilb and Weinberg 1977) which can be stated as follows:

> "If a system does not have to be reliable, it can meet any objective."

Figure 2 shows clearly the effect on development time of increasing the system reliability. The solution proposed by Gilb and Weinberg is to plan for evolutionary and early delivery of small subsystems. By early measurement of the cost of development against the quantity levels delivered, a realistic idea of how much the total system will cost and its development time can be deduced.

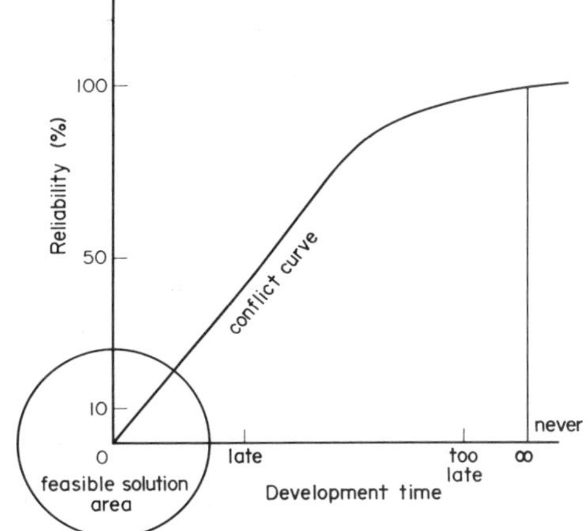

Figure 2
The zeroth law of system reliability

5. Information Systems Implementation

The life cycle of systems development consists of three main stages—analysis, design and implementation. The final stage in the development process, implementation, begins with testing. There are various methods of testing. The first is to test the system under simulated conditions; another is to test under actual conditions, operating in parallel with the existing system and procedures. It is never good practice to implement a complex system without a full system test.

It is also not considered good practice to build a system without proper documentation. Documentation includes anything written about the system in the process of analysis and design. Documentation begins with the forms used for data gathering and/or data entry. It includes procedural flow charts, input–output formats and file layouts. Documentation should also include test data that can be used to determine whether the system or programs are performing properly. Finally, documentation includes clear descriptions of the system written for the user who does not understand computer terminology and who only needs to know what he has to do to get the information he needs.

Documentation is important for several reasons. First, documentation is the major tool for coping with the tremendous complexity involved in information systems. Second, documentation provides the medium of communication from stage to stage in the analysis and design process. Documentation passes along information about the system that is accumulated as it develops from an idea into a working system. Also, documentation makes it possible to expand and modify the system to reflect the changing needs of the user.

The last phase of the implementation stage is a performance review or postaudit. In this stage, the objectives and cost–benefit representations made on behalf of the project are compared with actual achievements. The operational characteristics of the system are also determined to see whether they are satisfactory.

In addition to the problem of overcoming resistance to change, the systems analyst must take into account human factors during the design and implementation phases. Managers and scientists will tend not to use systems they do not understand. This does not mean a technical understanding of, say, the program structure or file design, but an understanding of the logic of the process they are asked to rely on. A system that is so complex that the user cannot understand why the inputs result in certain outputs will most probably never be used. This suggests the need for simple systems which grow as more complex features can be accepted by the users (Gall 1975).

Data processing has traditionally been a decentralized operation—a portion of the work of each department within an organization is concerned with data processing. When a computer system is implemented, a large part of the work may be removed from its previous organizational location and centralized for machine processing. One of the major advantages of a computerized information system is the integration of this work into a single system, thereby avoiding the duplication and wasted effort inherent in a manual system.

Another problem is that of attitudes towards "ownership" of information. With manual processing, each department or group produces most of its own information and prepares its own reports to higher management. Frequently, the information is considered to be the rightful property of the department that produces it. In a computer-based information system the processing is centralized and the information becomes the property of the organization. Lower levels of management and technical personnel may oppose the development of an integrated information system because they view this takeover of their information ownership as a threat to their position.

Perhaps the most serious organizational conflict associated with computerization of existing manual systems concerns the two quite dissimilar but closely interrelated activities—of systems development and current operations. Systems development and implementation are similar to research and development, that is, highly creative and rather ill-structured. Current operations, on the other hand, are concerned with efficient day-to-day procedures, accuracy and overcoming emergencies. Resolving this conflict is a prime requisite in the successful integration of any computer-based information system.

6. *Materials Information Systems*

One of the biggest problems in conducting a large-scale evaluation of materials and processes is the sheer volume of data that has to be processed to ensure that nothing is neglected. Taking into consideration approximately 2000 major materials and analyzing ten properties for each, there are therefore 20 000 units of data with which to cope. Further, data are given for representative materials only and do not reflect, for instance, all possible heat treatments, composition variations and size effects. Allowing a factor of 100 for these variations, there will be 2×10^6 units of engineering data for the major materials alone. Also, these data have to be assembled, analyzed and classified into a useful form. The use of a computer seems the only answer to these problems.

Another problem with materials information systems is the application of the traditional record-based structures. Record-based systems presume a horizontal and vertical homogeneity in data: horizontally each record of a given type contains the same fields, and vertically a given field contains the same kind of information in each record.

Information systems normally work best in areas which fit this pattern, but in many cases the pattern does not hold. Consider a materials information system. While ultimate tensile strength (UTS), yield strength and ductility are all attributes of materials, it would be impossible to define a conventional material record format. Consider: hardness, toughness, fracture resistance, yield point, UTS, wear resistance, corrosion resistance, fatigue strength, electrical conductivity, volume resistivity, rupture strength, and so on. How would one design a record format for engineering materials? A record format for high-temperature turbine materials would include: UTS, yield strength, elongation, rupture strength, oxidation resistance, thermal fatigue resistance and density; whereas a database for surgical implant materials would comprise: yield strength, tensile strength, elongation, modulus of elasticity, Rockwell hardness, endurance limit, corrosion resistance, tissue tolerance and density. Electrical insulating materials would have a totally differing database content with attributes such as dielectric strength, dissipation factor, thermal expansion coefficient, volume resistivity, maximum operating temperature and density.

To summarize, materials property databases exhibit a high degree of inhomogeneity and therefore cannot easily be handled by traditional record-based data structures such as network or hierarchical. What is required is a relational database model which will allow a single fact (field) to include both a type and value, where the syntax and structure of the value depended on the type (Date 1977, Codd 1979).

In the near future it should be possible to create a computerized materials-information center where, for a fee, an organization could obtain a complete screening and selection of materials to meet product requirements. Major engineering organizations are accumulating information about materials and processes for a computerized materials database (Wetzler 1977).

This is the age of information, and to meet this information explosion numerous information analysis centers have arisen, especially in the USA. Some of their characteristics are as follows.

(a) They provide technical information service to industry, universities and governmental organizations.

(b) They possess a genuine interest in the fields of materials engineering and information retrieval.

(c) They are service-oriented, to be of assistance to engineers, designers and management.

(d) They possess expertise in materials engineering and electronic data processing.

Recent development work in materials databases and computer-based materials selection systems has been covered by Westbrook and Rumble (1982). Other sources of materials information and the trend toward integration of materials and manufacturing data has been covered by White (1978) and by Appoo and Percival-Barker (1980).

The major benefits arising from using a computer-based materials selection system are as follows.

(a) It forces the designer to define in detail the important properties required in service.

(b) It allows the materials engineer to optimize the usage of materials, taking into consideration factors such as availability, energy cost and machinability.

(c) It allows the designer to consider the relative importance of the properties.

(d) With the aid of the system the user can carry out sensitivity analysis and consider compromises between properties. Judgement superimposed on this approach thereby permits selection of the best candidate materials, so that detailed design work can proceed with a fair chance that the cheapest product will be made.

As John Diebold expressed it:

> "The immutable fact is that the management team of any business or institutional organization will have new tools, new problems and new opportunities as a result of future developments in information technology. The task of the management team is to apply the new developments to its company effectively. The success with which this is done will be a significant factor determining the competitive position and growth of the company."

See also: Computerized Materials Databases; Computer-Aided Materials Selection

Bibliography

Appoo P M, Percival-Barker K 1980 Information and the forging industry. *Metallurgia* 47: 16–23

Codd E F 1979 Extending the database relational model to capture more meaning. *ACM Trans. Database Systems* 4(4)

Date C J 1977 *An Introduction to Database Systems.* Addison-Wesley, Reading, Massachusetts

Davis G B 1974 *Management Information Systems: Conceptual Foundations, Structure and Development.* McGraw-Hill, New York

Gall J 1975 *Systemantics—How Systems Work and Especially How They Fail.* Wildwood House, London

Gilb T, Weinberg G M 1977 *Humanised Input: Techniques for Reliable Keyed Input.* Studentlitteratur, Lund, Sweden

Lancaster F W 1968 *Information Retrieval Systems: Characteristics, Testing and Evaluation.* Wiley, New York

Liston D M, Schoene M L 1971 A systems approach to the design of information systems. *J. Am. Soc. Inf. Sci.* 22: 115–22

Optner S L (ed.) 1973 *Systems Analysis: Selected Readings.* Penguin, Harmondsworth

Westbrook J H, Rumble J 1982 *Selected Readings on Computerized Materials Data Systems.* CODATA, Paris

Wetzler F U 1977 Data banks for R and D. *Res. Dev.* 28(6): 54–64

White M S 1978 Information for the metals industry. *Metall. Mater. Technol.* 10: 257–61

P. M. Appoo

Infrared Laser Window Materials

Windows developed for high-power infrared lasers must have extremely small losses so that laser heating of the window will not optically distort the laser beam or thermally fracture the window. An ideal laser window should have a bulk absorption coefficient near 10^{-5} cm^{-1} with a surface and coating absorptance of less than 0.005% per surface. In addition, high-durability materials with strengths greater than 70 MN m^{-2} and hardness (Knoop) above 500 are preferable. Important considerations other than optical and mechanical properties are cost, scalability (sizes up to 1 m) and environmental stability. This latter consideration is important because windows are subjected to hostile environments like those found in HF–DF chemical lasers and in space radiation.

Many materials have been investigated for use as high-power infrared laser window materials, and five polycrystalline materials have emerged as the most successful window candidates: zinc selenide, potassium chloride, sodium chloride, calcium fluoride and strontium fluoride. These five materials have been developed to the point that they now have some of the ideal laser window properties. The alkali halides, for example, have very low absorption across the broad spectrum of high-power infrared laser wavelengths from 2.5 to 11 μm, but they suffer from low durability; their mechanical strength, however, has been greatly increased through the addition of certain cation dopants (such as Rb^+, Eu^{2+} and Sr^{2+}) and by press-forging to produce high-strength, small-grain-size polycrystalline material. Chemically vapor deposited (CVD) ZnSe, however, is intrinsically harder than the halide materials, but its absorption is approximately ten times greater. At the CO (5.2 μm), DF (3.8 μm) and HF (2.8 μm) laser wavelengths, alkaline-earth fluorides become particularly attractive window candidates. They combine good durability with low losses and minimal optical distortion to be the best materials at the shorter infrared laser wavelengths, but their scalability is limited, with diameters greater than 25 cm being difficult to produce routinely.

1. Optical and Mechanical Properties

The physical properties of the five polycrystalline materials, CaF_2, SrF_2, NaCl, KCl–Rb and ZnSe, chosen as being most viable for use as laser windows are listed in Table 1. In the design of a laser window, the usual compromises between physical parameters for the specific window application are encountered. For example, from Table 1 it can be seen that the high thermal conductivity and low thermal expansion of ZnSe make this material particularly appealing as a window candidate. The forged halides, however, are poor thermal conductors with rather large thermal-expansion parameters, although their refractive index is low enough to make antireflection coatings unnecessary for some applications. The alkaline-earth fluorides have properties intermediate between ZnSe and the forged halides, but their transmission restricts their use to 2–5 μm infrared lasers.

The mechanical properties of these five materials are summarized in Figs. 1 and 2. The average grain size of these materials varies widely. Traditionally, the smallest grain sizes (~10 μm) are produced in the forged halides, while cast CaF_2 and SrF_2 have grain sizes of the order of several centimeters. The dotted areas in Fig. 1 indicate that millimeter-grain-size forged CaF_2 and SrF_2 have also been produced but that these materials still have larger grain size than the forged halides or CVD ZnSe (average grain size

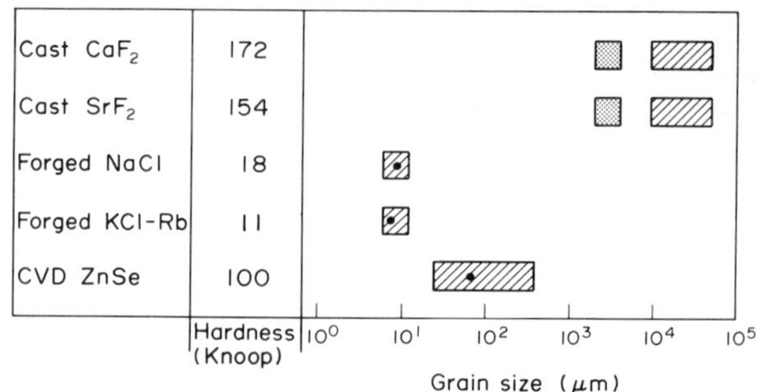

Figure 1
Mechanical properties of infrared laser window materials

Table 1
Physical properties of infrared laser window materials

Material	Thermal conductivity ($W cm^{-1} K^{-1}$)	Specific heat capacity ($J g^{-1} K^{-1}$)	Refractive index	Expansion coefficient ($10^{-6} K^{-1}$)
Cast CaF_2	0.080	0.812	1.4	21.2
Cast SrF_2	0.074	0.556	1.4	21.3
Forged NaCl	0.037	0.857	1.5	44.0
Forged KCl–Rb	0.033	0.647	1.46	40.0
CVD ZnSe	0.170	0.355	2.42	7.80

70 μm). The hardness (Knoop) and strength of these materials are greatest for the fluorides and, as expected, lowest for the chlorides. Further, the strength can vary greatly depending on grain size and surface conditions; cast CaF_2, for example, has exhibited fracture strengths up to 193 MN m^{-2} (well above the average value of 82 MN m^{-2}) in samples with very carefully controlled surface finishing. For practical purposes, however, the average strengths (indicated by black dots in Fig. 2) for fracture (CaF_2 and SrF_2) or yield (forged halides and ZnSe) are to be regarded as the most representative value for a particular material.

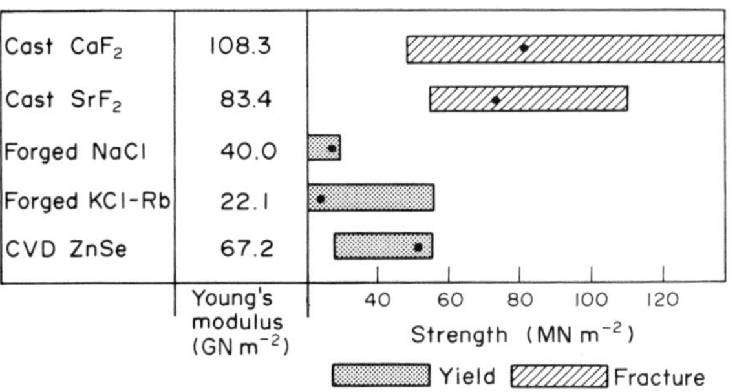

Figure 2
Strength of infrared laser window materials

The optical absorption coefficient β of these five window materials is summarized in Table 2. The total, bulk and surface absorption coefficients at the various infrared laser wavelengths in this table represent best values for single-crystal (SC) or CVD materials. Since these measurements were all made on state-of-the-art material under ideal laboratory conditions, these values would not necessarily be representative of actual window material. In fact, production window material can often be found to have absorptions two to ten times higher than the total absorptions listed in Table 2.

2. Alkali Halides

Alkali halide materials have had a long history of use as infrared-transparent optical components. During the past ten years, the optical and mechanical properties of these materials have been substantially improved as a result of two major accomplishments: the preparation of ultrapure, single-crystal material through reactive-atmosphere-process (RAP) chemistry (see *Reactive Atmosphere Processing*) and the successful forging (see *Forging of Optical Materials*) of fine-grain, large-size halides for superior mechanical properties without sacrifice of optical transmission. Today, not only can large-diameter (> 50 cm) polycrystalline halides be fabricated, but also finished optical components such as lenses and aspheric optics can be press-forged. RAP methods have reliably produced optically uniform material that when forged is particularly immune to grain regrowth.

Although RAP chemistry was very successful at reducing losses in KCl to near the intrinsic value of 8×10^{-5} cm^{-1} at 10.6 μm, problems developed in forging the pure KCl. It was found that some forged KCl would exhibit substantial grain growth. This recrystallization, which is caused by moisture, heat and impure material, is undesirable because it leads to a weakened structure. To inhibit grain growth and enhance mechanical strength, small amounts of cation dopants have been added to KCl. In particular, 1.75 mol% Rb^+, Sr^{2+} and Eu^{2+} added to KCl have increased the strength of KCl by a factor of ten over single-crystal KCl without seriously increasing the optical losses (β_T at 10.6 μm = 4×10^{-4} cm^{-1} for KCl–Rb^+). It has also been found that undoped RAP-forged material experiences very little grain growth.

Table 2
Lowest reported absorption coefficients for infrared laser window materials

Material	Absorption coefficient			
	10.6 (μm)	5.25 (μm)	3.8 (μm)	2.8 (μm)
SC CaF_2	3.60, T	6.7×10^{-4}, B	6.2×10^{-5}, B 1.0×10^{-4}, S	1.8×10^{-4}, B 1.4×10^{-4}, S
SC SrF_2	0.71, T	4.1×10^{-5}, B 3.9×10^{-5}, S	1.4×10^{-4}, B 9×10^{-5}, S	2.7×10^{-4}, B 6×10^{-4}, S
SC NaCl	1.0×10^{-3}, B 9.0×10^{-4}, S	3.4×10^{-5}, T	5.3×10^{-5}, T	5.8×10^{-5}, T
SC KCl	8.0×10^{-5}, B 3.5×10^{-4}, S	5×10^{-7}, B 4×10^{-6}, S	7.0×10^{-6}, B 4×10^{-5}, S	5.6×10^{-6}, B 5.5×10^{-5}, S
CVD ZnSe	4.0×10^{-4}, B 5.2×10^{-4}, S	9.5×10^{-4}, B 2×10^{-3}, S	4.8×10^{-4}, B 2.1×10^{-3}, S	7.3×10^{-4}, B 3.3×10^{-3}, S

B, bulk (cm^{-1}); S, surface; T, total (cm^{-1})

3. Alkaline-Earth Fluorides

Single-crystal fluorides of calcium, strontium and barium have long been used as transparent optics at infrared wavelengths up to 7–8 μm. However, since the single-crystal material cleaves easily, as in the alkali halides, polycrystalline forms of this material are more desirable for laser window applications. As in the development of alkali halides, fluoride material research has been along two fronts: purification and growth of single-crystal material and forging or casting for the fabrication of large, polycrystalline fluoride windows.

The purification of CaF_2, SrF_2 and BaF_2 has also been successfully achieved using RAP chemistry techniques (see *Reactive Atmosphere Processing*). Polycrystalline material has been prepared using conventional hot-forging techniques. Also, polycrystalline fluoride windows have been prepared using a fusion-casting technique. In this method, a two-zone vacuum furnace is used to first purify and then to grow, by unidirectional solidification, windows as large as 25 cm in diameter. The growth process must be carried out very slowly to prevent cracking since the thermal-expansion coefficient is large for these materials (see Table 1). Typical cooling rates are 40–60 °C h^{-1}. As might be expected, the grain size of these slow-cooled windows is very large—grain diameters are generally comparable to the ingot thickness (1–3 cm). The fusion-cast materials fabricated to date are free of voids and optically uniform and have total absorption coefficients in the low 10^{-4} cm^{-1} range at 3.8 μm.

4. Zinc Selenide

Zinc selenide holds great promise as a window material because of its many favorable mechanical properties (see Figs. 1 and 2) and its broad transmission range from 0.5 to beyond 18 μm. With the advent of a new, commercial CVD process, a microcrystalline form has been produced in sizes greater than 100 cm. In the CVD method, ZnSe is formed from the vapor-phase reaction

$$Zn + H_2Se \rightarrow ZnSe + H_2$$

This reaction, which takes place at temperatures above 600 °C and pressures below 100 torr, produces large areas (several square meters) of ZnSe deposited on hot mandrels at growth rates up to 100 μm h^{-1}. The optically finished material contains some particulate matter due to ZnH_2, which has a weak absorption band near 6 μm, and ZnO. As a result, there is some visible-light scattering (0.1 cm^{-1} typical scattering coefficient) although the optical uniformity is good. It should also be pointed out that the CVD process lends itself well to the fabrication of irregularly shaped optics even though uniformity of temperature and growth rates would not be trivial problems for large areas or convoluted shapes.

See also: Alkali Halide Crystals; Laser Materials

Bibliography

Bendow B 1978 Multiphonon infrared absorption in the highly transparent frequency regime of solids. *Solid State Phys.* 33: 249–316

Deutsch T F 1975 Laser window materials: An overview. *J. Electron. Mater.* 4: 663–719

Feldman A, Waxler R M 1979 Properties of crystalline materials for optics. *Proc. Soc. Photo-Opt. Instrum. Eng.* 204: 68–76

Harrington J A, Gregory D A, Otto W F 1976 Infrared absorption in chemical laser window materials. *Appl. Opt.* 15: 1953–59

Miles P 1976 High transparency infrared materials: A technology update. *Opt. Eng.* 15: 451–59

Sherman G H, Frazier G F 1978 Transmissive optics for high power CO_2 lasers: Practical considerations. *Opt. Eng.* 17: 225–31

J. A. Harrington

Infrared Spectroscopy

The principal uses of infrared (ir) spectroscopy in materials science are for identification of unknown substances, or their characterization as to chemical structure; identification and quantification of constituents in mixtures; and following chemical reactions either in the bulk or on the surface of materials. It is often used in problem solving, frequently in combination with other analytical techniques.

Most liquids and gases and many solids are amenable to ir analysis. The exceptions are monatomic and homonuclear gases (e.g., He, H_2, N_2 and O_2), metals and dilute aqueous solutions. The ir spectrum of a material is its "fingerprint," and an unknown material may be positively identified by matching its spectrum to that of an authentic standard. Even lacking standards, however, the presence or absence of certain groups can be correlated with the presence or absence of absorptions at specific wavelengths.

1. Basic Principles

Infrared spectra are usually presented as a plot of absorption (in percent transmission) as a function of wavelength (in μm) or wave number (in cm^{-1}). The units are related as follows:

$$10^4/\lambda_{vac} = \nu/n \qquad (1)$$

where λ is the wavelength, ν is the wave number and n is the refractive index of air.

Molecules in any state of matter can be viewed from a mechanical perspective as consisting of masses (atoms) connected by springs (chemical bonds). As in any mechanical system, energy can be absorbed with the result that certain frequencies are excited. The exact frequencies depend on the masses of the atoms, the force constants of the chemical bonds connecting them, and to a lesser extent on interactions between nonbonded atoms. In the simple-harmonic-oscillator approximation, the frequency of vibration f of a diatomic molecule with atoms of masses m and M is

$$f = (2\pi)^{-1}(k/\mu)^{1/2} \qquad (2)$$

where k is the force constant and μ is the reduced mass (i.e., $\mu = mM/(m + M)$).

The vibrational frequencies of atoms in molecules correspond to the frequencies found in the ir region of the electromagnetic spectrum. In order for interaction to occur, however, the vibration must involve an oscillation in electric dipole moment. Only under these circumstances is ir radiation of the same frequency absorbed and the vibration excited. The absorption intensity is

$$\int a_\nu \, d\nu = (N\pi/3c^2)(\partial\bar{\mu}/\partial Q)^2 \qquad (3)$$

where a_ν is the absorption coefficient at frequency ν, N is the number of molecules per cm^3, $\bar{\mu}$ is the bond moment and Q is the normal coordinate for the vibration.

Thus, metals and homonuclear gases do not show ir absorption patterns because the vibrating atoms do not give rise to dipole oscillations. Strongly polar substances, such as water, are very strong absorbers of ir radiation whereas nonpolar molecules, such as hydrocarbons, show rather weak absorption for many of their skeletal vibrations.

Any defined aggregation of atoms (i.e., a molecule) has a finite number of fundamental vibrational absorptions, equal to $3N - 6$ where N is the number of atoms in the molecule ($3N - 5$ for linear molecules). Overtone and combination bands can also appear, and the ir absorption pattern for each molecular species is unique. As might be expected, however, structurally identical groups have the same or similar frequencies in different molecules. Thus, the aliphatic CH group always absorbs near 2960 cm^{-1}. The carbonyl (C═O) group absorbs near 1710 cm^{-1} in organic acids; near 1720 cm^{-1} in ketones; and near 1740 cm^{-1} in esters. (Examples of other "group frequencies" are given in Table 1.) Deviations of the absorptions from their "normal" positions can be utilized to gain more sophisticated structural information.

2. The IR Spectrometer

Spectroscopy in the ir region presents several unique difficulties. The region is intrinsically of low energy, and thus the energy transmission of the spectrometer must be maximized. Many materials that are transparent to visible radiation do not transmit ir; optical glass, for example, is ir-opaque and so special materials such as single crystals of NaCl, KBr or AgCl

Table 1
Some typical group frequencies (cm^{-1}) in the infrared region[a]

The NH–OH region		
3460–3620 alcohol (solution)	3500–3300 secondary amine	3200–3180 primary amide
3620–3590 phenol (solution)	3520–3200 alcohol (H-bonded)	3460–3490 C═O overtone
3550–3330 primary amine (solution)	3350–3330 primary amide	3100–2500 carboxylic acid
3450–3250 primary amine (solution)		
The CH stretching region		
3100–3010 aromatic CH	2900–2800 aldehyde	2805–2780 $-NCH_3$
2970–2910 aliphatic CH	2900–2865 $-OCH_3$	
The "window" region		
2275–2263 isocyanates	2260–2100 acetylenes	2150–2050 isothiocyanates
2260–2220 nitriles	2175–2160 thiocyanates	2150–2110 isonitriles
2260–2090 SiH		
The double-bond region		
1870–1845 cyclic anhydride	1750–1735 aliphatic ester	1725–1705 ketone
1825–1780 aliphatic anhydride	1740–1715 aromatic ester	1690–1630 alkenes
1755–1745 aliphatic anhydride	1740–1720 aldehyde	1680–1630 secondary amide
1810–1760 acid halide	1735–1690 carbamates	1660–1620 primary amide (2 bands)
The fingerprint region		
1650–1590 primary amines	1415–1390 sulfate	995–985 vinyl
1600–1450 aromatic ring	1200–1187 sulfate	910–905 vinyl
1550–1510 secondary amide	1210–1100 tertiary alcohol	765–735 monosubstituted aromatic
1556–1545 nitro	1150–1075 secondary alcohol	710–685 monosubstituted aromatic
1390–1368 nitro	1075–1000 primary alcohol	

a For a more complete listing and discussion, see Colthup et al. (1975)

Table 2
Properties of some ir-transmitting materials

Window material	Transmission range[a] (μm)	Refractive index	Reflection loss[b] (%)	Remarks
Quartz (SiO_2)	0.2–3.6 (10)	1.43	6	
LiF	0.2–6.2 (12)	1.37	5	
Sapphire (Al_2O_3)	0.2–5 (8)	1.70	13	
Fluorite (CaF_2)	0.2–7.8 (10)	1.41	6	
Irtran I (MgF_2)	2–6.4 (12)	1.35	4	
Servofrax (As_2S_3)	1–11.1 (5)	2.59	35	
BaF_2	0.2–11.5 (9)	1.45	7	
Rock salt (NaCl)	0.2–17 (10)	1.52	8	
Irtran II (Zn_2S)	1–10.7 (12)	2.24	27	
Sylvite (KCl)	0.3–2.1 (10)	1.49	8	
Irtran III (CaF_2)	0.2–8.7 (12)	1.40	5	
Irtran IV (Zn_2Se)	1–15.2 (12)	2.4	31	brittle
KBr	0.2–32 (4)	1.53	9	hygroscopic
AgCl	0.6–24 (5)	2.0	21	reacts with metal
Ge	2–20	4.0	59	brittle
Si	1.5–150	3.4	51	brittle
KRS-5 (Tl_2BrI)	0.7–38 (5)	2.4	31	toxic
CsBr	0.3–40 (10)	1.66	12	soft
CsI	0.3–55 (5)	1.74	14	soft

a Thickness in mm in parenthesis b Two surfaces

must be used for window materials. Properties of some common ir-transmitting materials are given in Table 2. Water is strongly absorbing and thus unsuitable as a solvent. Finally, blackbody radiation originating from the spectrometer components is at its peak in the most commonly used region and is responsible for a high background of stray radiation.

Dispersive spectrometers are the most common type. A typical optical system is shown in Fig. 1. Infrared radiation from the source (a hot ceramic rod or Nichrome coil) is directed by front-surface optics through the sample compartment. A rotating sector mirror alternately allows the sample and reference beams to pass through the entrance slit into the monochromator, where the radiation is dispersed by a grating. (Older spectrometers may use a rock-salt prism instead of a grating.) The spectrum is scanned by rotating the grating. The dispersed radiation scans across an exit slit, through a band-pass filter (to eliminate overlapping orders and stray light), and is focused on a detector, usually a tiny thermocouple with a blackened target area. The thermocouple is evacuated for greater sensitivity.

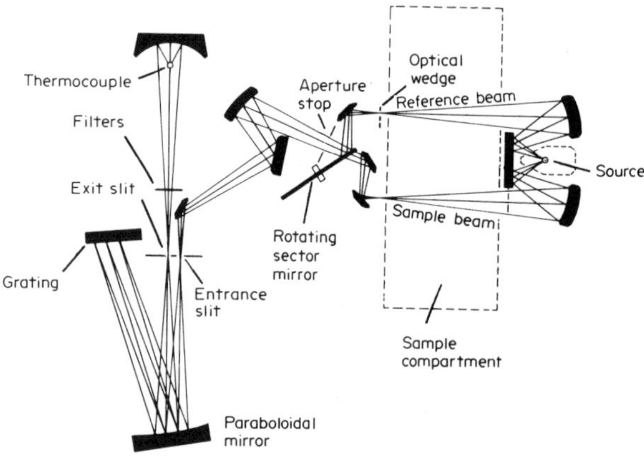

Figure 1
Optical system of an optical-null grating spectrometer (courtesy Perkin-Elmer Corporation)

The sample and reference beams fall alternately on the detector (the rotational frequency of the sector mirror is usually 10–15 Hz) and any difference in the intensities of the two beams (caused, for example, by sample absorption at that wavelength) results in an alternating signal at the detector. In the common optical-null spectrometer, this alternating signal activates a servosystem that moves a comb-shaped optical wedge in or out of the reference beam in such a way as to cancel the difference in the beams. The motion of the comb is recorded as a function of wavelength or wave number to produce the ir spectrum of the sample.

The operating parameters may be varied to suit the purpose of the spectrum determination. The operator can vary the slit width; the scanning speed of the spectrometer, which must be compatible with the response speed of the pen (the electrical signal is filtered to reduce the noise or random pen fluctuations); and the gain of the servoloop driving the pen. Thus, the three basic variables are resolution (slit width), scan speed (response time) and noise level (gain). Any two of these parameters may be arbitrarily chosen; the third is then fixed by that choice. The relationship between these variables is:

$$\text{resolution} = ct^{1/4}(S/N)^{-1/2} = 1/w \qquad (4)$$

where t is the response time of the pen, S/N is the signal-to-noise ratio, w is the slit width and c is a constant. This equation is the basis of "trading rules" which are useful in setting the spectrometer controls for special-purpose spectra. Note, for example, that the signal-to-noise ratio improves as the square of the slit width; for quantitative analysis, the noise can be reduced by a factor of four (and precision thus improved) by doubling the slit. To double the resolution by halving the slit, however, the scan time must be increased by a factor of sixteen. It is important that the gain be properly set to give a "live" pen that shows a small amount of noise (±0.2%) in the spectral record. Spectrometer response should be tested by scanning a material having a variety of sharp and broad bands (such as indene) at normal and then at quarter-normal scan speed—the two recordings should be identical. Wavelength accuracy should be periodically checked by scanning a standard material such as indene or polystyrene (see Table 3 and Fig. 2).

Ratio-recording ir spectrometers are more complex electronically but have some distinct advantages over opitical-null spectrometers. Instead of utilizing an

Table 3
Absorption maxima (cm^{-1}) for spectrometer calibration

Indene (undiluted)	
(0.2 mm cell)	(0.025 mm cell)
(1) 3927.2 ± 0.5	(6) 1361.1 ± 0.2
(2) 3139.5 ± 0.4	(7) 1205.1 ± 0.2
(3) 2770.9 ± 0.4	(8) 1018.5 ± 0.3
(4) 1915.3 ± 0.3	(9) 830.5 ± 0.3
(5) 1553.2 ± 0.2	(10) 590.8
	(11) 381.4
Polystyrene (solid film)	
(1) 3027.1 ± 0.3	(6) 1583.1 ± 0.3
(2) 2850.7 ± 0.3	(7) 1154.3 ± 0.3
(3) 1944.0 ± 1	(8) 1028.0 ± 0.3
(4) 1801.6 ± 0.3	(9) 906.7 ± 0.3
(5) 1601.4 ± 0.3	

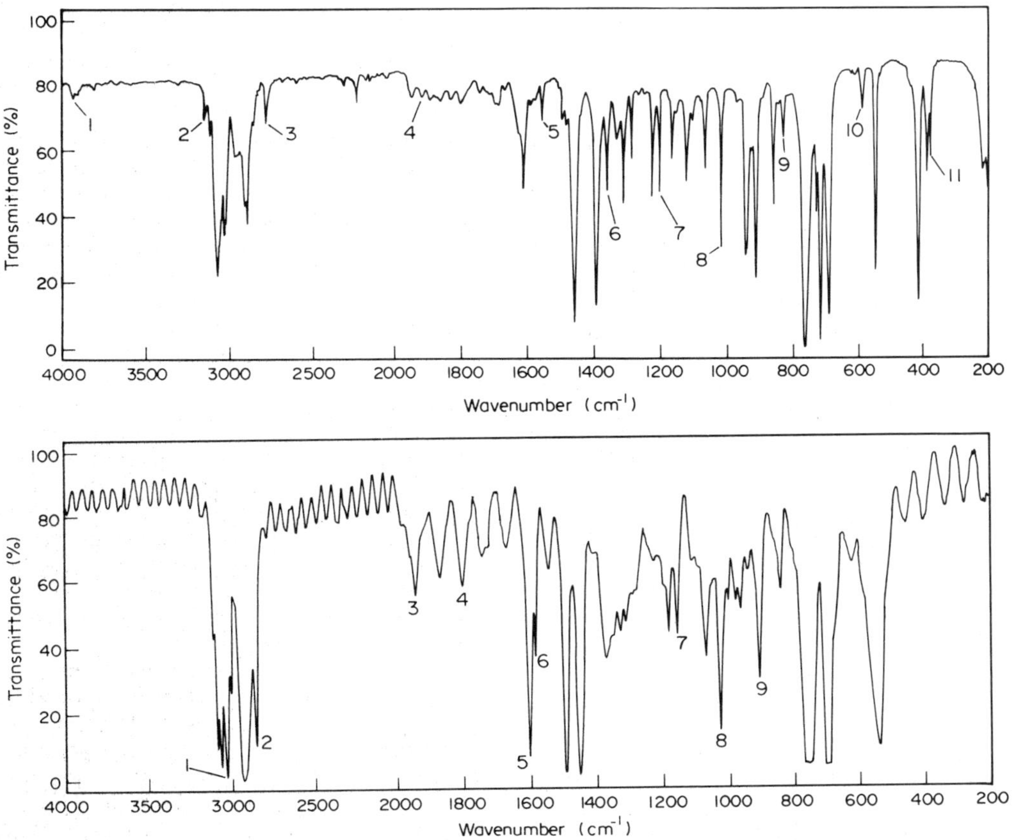

Figure 2
Spectra showing calibration bands for infrared spectrometers: top, indene; bottom, polystyrene film. Numbered absorption maxima correspond to those given in Table 3 (courtesy Perkin–Elmer Corporation)

optical attenuator, the instrument accurately measures the intensities of the sample and reference beams. The ratio of the two beams is compared electronically and recorded to give the spectrum.

Fourier-transform (FT) ir spectrometers utilize a different optical principle from dispersive spectrometers. Source radiation is directed through an interferometer as shown in Fig. 3, in which the position of the moving mirror is coded by means of a laser and a white-light reference. In effect, all frequencies are measured simultaneously during the scan time of the moving mirror. The resulting interferogram is transformed to the frequency domain by the Fourier method, using a computer. The resulting ir spectrum is, in principle, identical to that obtained from a dispersive spectrometer. Fourier-transform ir spectroscopy has significant energy advantages; since all frequencies are observed continuously during the scan instead of as a small bundle of adjacent frequencies as in dispersive spectrometers, much more energy per unit time is available (Felgett's advantage). Another energy gain results from the use of a circular entrance aperture rather than an entrance slit (Jacquinot's advantage). This energy advantage enables reduced scan time or increased resolution.

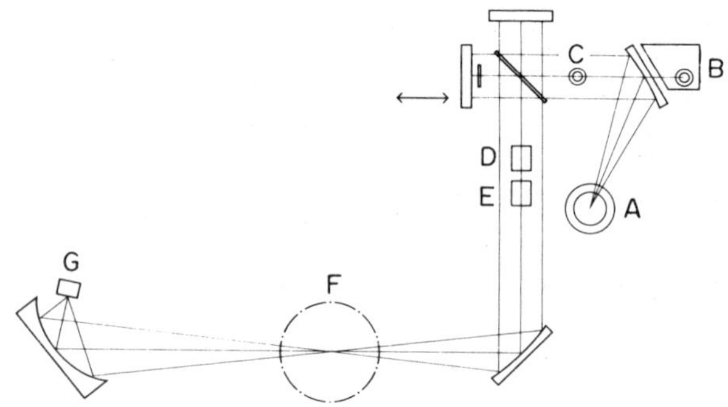

Figure 3
Optical schematic for an interferometer spectrometer: A, ir source; B, laser; C, white light; D, white-light detector; E, laser detector; F, sample chamber; and G, ir detector (reproduced with permission from IBM Instruments, Inc.)

Because of their need for fast response and wide dynamic range, Fourier-transform ir spectrometers do not use thermocouple detectors, but instead use pyroelectric (e.g., triglycine sulfate) or cooled semiconductor (Hg–Cd–Te) detectors. Other important detector parameters are sensitivity and wavelength range. A wide choice of detectors is available, but all have limitations, and some compromises are necessary in order to best match the requirements of the spectrometer and the user.

Special-purpose spectrometers that utilize tunable-laser radiation are also available. They offer much higher resolution (1–$10^{-6}\,cm^{-1}$ bandwidth) and greater intensity. Although they do not have the capability for continuous scanning over a broad wavelength range, they are useful for special problems, such as quantitative measurements at very high absorbances, and high-resolution scanning of limited spectral ranges, as for gas analysis.

The most common type of tunable ir lasers are semiconductor diode lasers, prepared from alloy crystals of Pb–Sn–Te, Pb–S–Se and Pb–Cd–S. The frequency at which they emit radiation depends on chemical composition and temperature. They are used at cryogenic temperatures, and may be tuned over a range of about $1\,cm^{-1}$ by varying the diode current.

3. Sampling

Rather thin layers of sample—approximately 0.01 mm for liquids and solids—are optimum for good quality ir spectra. Optimum sample thickness is attained when several bands have transmission of less than 30–40%, but no band has less than 5–10% transmission. The background in nonabsorbing regions should be relatively flat at about the 90–95% transmission level. The sample should be uniform over the area that intercepts the sample beam; that is, it should not be wedged or show holes. Gases are usually sampled in cells 2–10 cm in length, although multireflection cells of over 1 km in path length have been used for trace analysis. Many sampling methods are available, of which the most popular are: for liquids, solution in an inert solvent and liquid film; for solids, solution in an inert solvent, as a powder mulled with mineral oil (to reduce scattering of the radiation), powder suspended in a KBr matrix that has been pressed into a transparent pellet, and attenuated total reflectance. Each sampling method has advantages and drawbacks; the method should be matched to the problem.

Liquid films are prepared by placing a drop of liquid on the salt window and "wiping" the window against a flat surface. Alternatively, another window may be pressed on the sample to form a sandwich. Some trial and error will be necessary to achieve a spectrum having the proper intensities.

Soluble liquids and solids (except amines) may be dissolved in ir-transmitting solvents, such as CCl_4 or C_2Cl_4 (used in the 4000–1340 cm^{-1} region) and CS_2 (used in the 1340–600 cm^{-1} region). These solvents should be manipulated only in a chemical fume hood, however, as their vapors are toxic. A sample concentration of 10% (wt/vol) in a 0.1 mm cell usually gives optimum intensities. More polar solvents can be used in limited wavelength regions where their transmission exceeds 50%.

Grindable solids can be sampled as mineral oil or alkali halide mulls. The objective is to suspend the finely ground sample in a matrix of similar refractive index in order to reduce scattering of the radiation.

Mineral oil mulls are prepared by vigorously grinding 10–50 mg of sample to a finely divided state, using a large mortar and pestle. Mineral oil is added dropwise and light grinding is continued until the sample has the appearance and consistency of cold cream. It is then gathered with a rubber wiper and sandwiched between two ir windows. Absorptions from the mineral oil obscure the CH stretching and bending regions (2850–2950 cm^{-1}, 1450–1480 cm^{-1} and 1400–1420 cm^{-1}) but the remainder of the spectrum is relatively clear. A second mull that uses a perfluorinated or perchlorinated oil can be prepared if it is necessary to examine the CH regions.

Alkali halide mulls are usually made from KBr powder, into which about 1 mg of the finely ground sample is thoroughly mixed. The powder is pressed into a transparent pellet, using a commercial die (usually evacuated before pressing). This technique should not be used with samples that contain exchangeable halides (such as amine hydrochlorides) or those that display polymorphism, as it will be very difficult to obtain reproducible spectra.

Powdered solids can also be examined by diffuse reflectance, in which the sample beam is directed onto the powder. The diffusely scattered radiation is collected by a mirror system and routed back into the optical path of the spectrometer.

Attenuated total reflectance (ATR) is a technique in which the sample (usually a solid) is simply pressed against an internal-reflection element (Fig. 4). The radiation is internally reflected from the interior surface of the element, but also penetrates slightly (of the order of one wavelength) into the adjacent medium (in this case, the sample). Multireflection elements are used to increase absorption intensities.

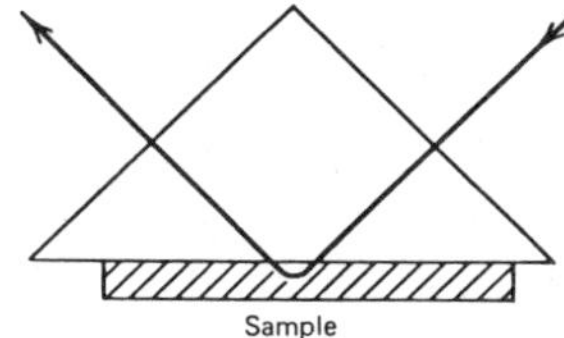

Figure 4
Schematic representation of the radiation path in a single-reflection internal-reflection element (after Smith 1979. © John Wiley and Sons, New York. Reproduced with permission)

Special techniques are used for microsamples. Absorption or reflection spectra can be obtained on particles as small as 20 μm in diameter using a microscope with reflecting optics. Thin films (monolayers or less) of organic materials on reflecting metal surfaces can be sampled by setting the ir beam at glancing incidence to the surface (reflection–absorption spectroscopy).

Pyrolysis is usually considered to be a technique of last resort. It is often applied to insoluble or heavily filled polymers, and to inorganic powders treated with unextractable organic material. The fragmented sample (0.1–0.5 g) is placed in a test tube and heated strongly over a flame with the tube horizontal. The volatile polymer fragments condense near the neck of the tube and can easily be transferred to a salt plate. Since the spectrum of the pyrolysate may be somewhat different from that of the parent compound, a reference library of pyrolysate spectra can be very helpful.

When combined with gas chromatography (GC), ir spectroscopy provides a powerful method of identifying the components of complex mixtures. The Fourier-transform ir spectrometer, because of its fast scan capability, is well suited to obtaining "on-the-fly" spectra of fractions as they are eluted. Specially designed gas cells, or "light pipes," and sensitive detectors permit the use of capillary GC columns (Griffiths et al. 1983). Detection limits of less than 40 ng have been reported. The GC/ir combination is often preferred to GC/mass spectrometry for polar samples, whereas the reverse is true for nonpolar materials.

A matrix isolation system, in which the chromatograph effluent is mixed with argon gas and frozen at 14 K on a reflecting substrate, is claimed to give even better sensitivity. Good spectra on fractions of 5 ng or less have been demonstrated.

Infrared spectroscopy has also been combined with liquid chromatography for characterization of nonvolatile samples such as polymers. Problems of solvent opacity often preclude direct scanning of the eluent solution, but ingenious solvent-removal systems have been devised that allow unrestricted sampling of the eluted fractions.

4. *Qualitative Analysis*

An ir spectrum contains an enormous amount of information, but much of it is not readily extractable. With some experience, it is possible to recognize certain well-known group frequencies (Table 1). Correlation charts that show band locations for the major chemical groups are often used to suggest structures for further study and corroboration. Existence of a chemical group is usually verified at several places in the spectrum. A priori structure determination from correlation charts is not feasible, however; comparison of the unknown spectrum with an authentic reference spectrum of the suspected substance is essential. The same sample preparation technique should be used for both unknown and reference, as details of the spectrum depend on the physical state of the sample. Reference spectra may be prepared by the user from carefully run pure samples, or purchased in a commercial collection.

Infrared spectroscopy is often teamed with other structural techniques such as NMR (see *Nuclear Magnetic Resonance Spectroscopy*) to obtain more complete structural information about a material.

Computerized search routines are often used to identify spectra of unknowns. In one system, reference spectra are coded by the positions of their strongest bands. A search based on band position will usually retrieve a spectrum that matches the unknown, if it is present in the library. Another computerized system utilizes a logical interpretation approach, based on group frequencies. In any case, identification is not complete until the unknown spectrum has been matched to an authentic reference spectrum.

5. *Quantitative Analysis*

Quantitative analysis is also possible using ir spectroscopy, and indeed was one of the earliest practical uses of the technique. Sampling is usually carried out using solutions so that the analyte concentration and the path length can be controlled. If I_0 is the intensity of the radiation incident on the sample cell, I is the intensity of the transmitted radiation and T is the transmittance, the Bouguer–Beer law states that the absorbance A is

$$A = \log 100/\%T = \log I_0/I = abc \qquad (5)$$

Here a is a proportionality constant related to the absorptive power of the analyte, b is the sample thickness and c is the analyte concentration. It should be noted that most ir spectrometers record

Table 4
Specific materials applications of ir spectroscopy

Area of application	Specific uses
Air pollution	Determination of ppm SO_2, SO_3 and H_2SO_4
	Studies of the chemistry of smog formation in a laboratory reactor
	Detection of hazardous vapors in the ppm and ppb range
Biomedical materials	Analysis of urinary stones or calculi
	Detection of allergenic substances on skin
	Analysis of anesthetic gases
Catalysts	Characterization of intermediates in the CO/H_2 reaction on SiO_2-supported Ru
	Studies of coking on zeolites
	Studies of chemisorption of H_2O and CO on Fe–Cr catalysts
Chemicals	Quality control of intermediates
	On-line process stream analysis
	Identification of impurities in products
Coatings, paints, varnishes and resins	Identification of polymers, pigments and solvents
	Quantitation of coating components
	Study of weathering effects on chemical structure
Drugs and pharmaceuticals	Quality control of products
	Analysis of competitive products
	Identification of antibiotics
Fibers and textiles	Analysis of cotton–polyester fiber blends
	Characterization of flame retardants, lubricants and softeners
	Determination of crystallinity, orientation and polymorphic form in fibers
Inorganic materials	Characterization of soils
	Identification of pigments
	Study of hydration reactions in portland cement
Metals	Determination of boron nitride in transformer steel
	Determination of oxygen in sodium metal
	Characterization of coatings on metals
Petroleum	Tracing crude oil spills as to origin
	Determination of additives in lubricants
	Analysis of crude oil fractions
Polymers and elastomers	Identification of polymers, blends and plasticizers
	Characterization of stereoregularity of isotactic and syndiotactic polypropylene
	Determination of hydroxyl groups in epoxy resins
Semiconductors	Analysis of silicon for ppb oxygen, boron, carbon and phosphorus
	Determination of epitaxial layer thickness
	Analysis of silicon halides used to produce silicon
Wood and paper	Analysis of paper for groundwood content
	Determination of lignin in birchwood pulp
	Determination of silicone release coating on paper

band intensities linearly in percent transmission, which is a logarithmic function of concentration. Thus, measurements in $\%T$ must be converted to absorbance before calculations of concentration can be done.

Analysis of mixtures is carried out by measuring a strong and preferably isolated band of each component. Either peak absorbance or integrated band area (on a linear absorbance ordinate) may be used, with the latter giving better precision. If mutual interference occurs (i.e., overlapping absorptions), corrections can be made for the interference as follows.

The absorbance A_ν at any frequency ν is the sum of the absorbances of the ith species at that frequency, or

$$A_\nu = a_1 b_1 c_1 + a_2 b_2 c_2 + \ldots + a_i b_i c_i \quad (6)$$

The absorption coefficients a_i are determined by measurements on the pure components, at frequencies chosen to give maximum absorbance for one component and minimum absorbance for the others. Mixtures are analyzed by making absorbance measurements at the selected frequencies and using the results in a set of simultaneous equations, which can be solved by matrix inversion to give the individual concentrations:

$$c_i = \sum_i \sum_\nu k_{i\nu} A_\nu \quad (7)$$

Multivariate least-squares calibration methods, which use mixtures instead of pure components for calibration, have some advantages (Haaland et al. 1985). Precision is sometimes improved by the overdetermination method; that is, measuring several wavelengths for each component.

For best quantitative accuracy, band intensities should fall in the range of 20–70$\%T$. Measurements should never be made below 10$\%T$ ($A = 1$) because of the large effect of small errors in establishing the $T = 0$ point and in measuring A.

Quantitative analysis for solids such as polymer blends can sometimes be accomplished by using band ratios. For binary mixtures, measurement of one band from each component is used to calculate an absorbance ratio that can be related graphically to the concentration ratio. Or, if one component of a mixture is constant in concentration or is the major constituent, the ratio of an analyte band to a band of the major constituent can be plotted against analyte concentration. It is good practice to derive statistical measures of accuracy and precision by repetitively analyzing independent standards over a period of several days.

6. *Applications*

Applications to numerous nonmetallic (and some metallic) materials have been documented. Reports of studies concerned with chemical composition, kinetics, structure and surfaces are numerous. Some specific materials applications are listed in Table 4. The listing is not comprehensive; it is intended only to illustrate the range of possible applications.

See also: Infrared Spectroscopy of Polymers; Investigation and Characterization of Materials: An Overview

Bibliography

American Society for Testing and Materials, Committee E-13. 1979 *Manual on Practices in Molecular Spectroscopy*. American Society for Testing and Materials, Philadelphia, Pennsylvania

Colthup N B, Daly L H, Wiberley S E 1975 *Introduction to Infrared and Raman Spectroscopy*. Academic Press, New York

Craver C D (ed.) 1977 *The Coblentz Society Deskbook of Infrared Spectra*. Coblentz Society, Kirkwood, Missouri

Griffiths P R 1975 *Chemical Infrared Fourier Transform Spectroscopy*. Wiley, New York

Griffiths P R, Dehaseth J A, Azzaraga L V 1983 Capillary GC/FT-IR. *Analyt. Chem.* 55A: 1361–87

Haaland D M, Easterling R G, Vopicka D A 1985 Multivariate least-squares methods applied to the quantitative spectral analysis of multicomponent samples. *Appl. Spectrosc.* 39: 73–84

Harrick N J 1967 *Internal Reflection Spectroscopy*. Interscience, New York

Haslam J, Willis H A, Squirrell D C M 1979 *Identification and Analysis of Plastics*, 2nd edn. Heyden, London

Infrared Spectroscopic Committee of the Chicago Society for Coatings Technology 1979 *An Infrared Spectroscopic Atlas for the Coatings Industry*. Federation of Societies for Coating Technology, Philadelphia, Pennsylvania

Koenig J L 1980 *Chemical Microstructure of Polymer Chains*. Wiley, New York

Pouchert C J 1975 *The Aldrich Library of Infrared Spectra*, 2nd edn. Aldrich Chemical Company, Milwaukee, Wisconsin

Smith A L 1979 *Applied Infrared Spectroscopy: Fundamentals, Techniques and Analytical Problem-Solving*. Wiley, New York

A. L. Smith

Infrared Spectroscopy of Polymers

The infrared spectrum of a polymer contains information on the vibrational frequencies of the long-chain molecules in the sample. The observed bands have frequencies and intensities that derive from the chemical and conformational structure, the intra- and intermolecular forces, and the electrical properties of the molecules. Vibrational spectroscopy (comprising infrared and Raman) is thus an important general technique for gaining an understanding of the above characteristics of polymer systems.

The key to this understanding lies in knowledge of the vibrational origins of the bands. Historically, this has been achieved at different levels, from the early compilations of chemical group frequencies to the present calculations of normal modes. The power of infrared spectroscopy is that useful applications to the study of polymers have been possible at all of these levels of understanding.

1. Experimental

Experimental techniques are available for readily obtaining infrared spectra of polymers (Siesler and Holland-Moritz 1980). Spectrometers can be either of the grating dispersive or interferometric Fourier-transform types. Both kinds of instruments can have data processing capabilities easily permitting a wide variety of spectral manipulation, such as background or solvent subtraction and multicomponent analysis. Each kind of spectrometer has its particular advantages, but the Fourier-transform instrument is significantly better for studies requiring fast recording of a spectrum. Accessories permit obtaining spectra as a function of many variables, such as temperature, pressure, stress, and polarization of incident light.

Polymer samples can be studied in many forms. They may be examined in solution or the melt state, using cells to contain the sample. In the more common situation samples are examined in the form of thin films, prepared by casting from a solvent, compression molding above the melting point, or microtoming. Powdered samples, which scatter highly, can be mixed with potassium bromide and pressed into pellets. Fibers can be handled in various ways, including microscopy. Samples from which it is difficult to get transmission spectra, or for which the surface is of particular interest, can be studied by attenuated total reflection.

2. Molecular Vibrations: General

The absorption of a photon in the infrared region corresponds to an increase in the energy state of one of the vibrations of the molecule (of which in general there are $3N-6$, N being the number of atoms). For a change between adjacent energy levels (usually between the ground and first excited states), the frequency ν of the photon is equal to the classical frequency of this particular harmonic (or normal) mode of vibration. Mathematical methods for calculating the normal frequencies and the atomic displacements in the normal modes (normal coordinates) are well developed (Wilson et al. 1955, Califano 1976). These require knowledge of the molecular geometry and the vibrational potential function (i.e., the force constants for bond stretch, angle bend, torsion and their interactions). Since the geometry (particularly the conformation) is often the main question being studied, the agreement between observed and calculated frequencies can be used as a measure of the validity of a hypothesized structure.

All normal modes of a molecule need not give rise to infrared absorption; this occurs only when there is a nonzero change in dipole moment during the vibration (i.e., when $\boldsymbol{M}_k(\equiv\partial\boldsymbol{p}/\partial Q_k)\neq 0$, where $\boldsymbol{p}$ is the dipole moment and Q_k the normal coordinate for the kth vibration). The observed absorptivity for such a vibrating dipole fixed in space is

$$a_k \propto (\boldsymbol{M}_k\cdot\boldsymbol{E})^2 \qquad (1)$$

where $\boldsymbol{E}$ is the electric vector of the incident radiation.

When a molecule has symmetry its normal modes fall into categories, called symmetry species, in which all modes in a given species behave the same with respect to the symmetry operations. Symmetry analysis (Wilson et al. 1955) can be very useful because it provides information on the numbers of normal modes in each species and on their infrared and/or Raman activity (so-called selection rules). This classification also defines the direction of $\boldsymbol{M}_k$ in the molecule, thus permitting the assignment of observed bands in polarized spectra to predicted normal modes.

3. Group Frequencies and Applications

In general, each normal mode involves significant displacements of all of the atoms in the molecule, but under certain circumstances the displacements are concentrated within a small (chemically distinct) group of atoms and are associated with a characteristic range of frequencies. Such group frequencies result when the coupling is small between the group and other parts of the molecule. This can occur for small mechanical coupling, as in the case of disparate masses in the group (e.g., CH, NH and OH) or small potential coupling, as in the case of disparate constants (e.g., C═O, C═C or C≡N compared to C—C) (Krimm 1960). This property has permitted an extensive empirical correlation of characteristic frequencies with chemical groups (Bellamy 1975, Colthup et al. 1964).

An important application of group frequencies is to the qualitative analysis of polymers (Hummel and Scholl 1969, Siesler and Holland-Moritz 1980). For example, the polymerization of butadiene yields a polymer that, depending on polymerization conditions, can have varying relative amounts of *trans*-1,4, *cis*-1,4, and 1,2 addition. Analysis for these different structures is possible because each type of double bond has a characteristic CH out-of-plane bend group frequency associated with it. This approach can be used to study the presence of additives, impurities, end groups, and oxidation and degradation products in polymers, and for many other general identification problems involved in polymer analysis.

It follows from the above considerations that if a band can be identified with a chemical group the quantitative amount of that group in a sample can be determined. According to the Beer–Lambert law, the absorbance A is given by

$$A \equiv \log(I_0/I) = abc \qquad (2)$$

where I_0 and I are, respectively, the incident and transmitted intensities through a sample of thickness b in which the concentration of the group in question is given by c. Since the spectrometer provides I_0 and I, and the absorptivity a and sample thickness are often known, c can be determined. When b can be kept constant (as in a solution cell) or an internal standard can be found, the use of Eqn. (2) can be significantly simplified.

Many external variables influence the local forces acting on chemical groups and, therefore, infrared spectroscopy provides a ready method for observing the presence of such perturbations. Solvent interactions can often be detected in this way, even though the frequency shifts may be small. Strong interactions are of course more easily seen; for example, hydrogen bonding causing frequency shifts of up to several hundred cm^{-1}. Stress on a polymer sample is also found to produce frequency shifts in some bands, probably due in part to changes in force constants.

If the direction of the oscillating dipole moment associated with a group frequency is known, then Eqn. (1) can be used to determine the orientation of the group in an oriented sample. This is achieved by using a polarized infrared beam and varying the direction between $\boldsymbol{E}$ and the orientation direction in the sample. The relative absorption with $\boldsymbol{E}$ parallel and perpendicular to this direction (the dichroic ratio) depends on $\boldsymbol{M}_k$ and its orientation distribution in the sample. The direction of $\boldsymbol{M}_k$ is not necessarily specified by the group frequency description of the band (e.g., for the C═O stretch of the amide group $\boldsymbol{M}$ makes an angle of about 20° with the C═O bond), so caution must be exercised in this respect. This is only one of the limitations of the group frequency concept. We now consider other extensions necessary for the study of polymers.

4. Molecular Vibrations: Chain Molecules

An ideally ordered polymer chain of infinite length has an important structural feature not shared by other molecules: it is a one-dimensional periodic chain of identical repeating units. Such a chain of coupled oscillators (generally helical) has special vibrational characteristics (Zbinden 1964, Zerbi 1969, Snyder 1980, Painter et al. 1982).

The solution of the vibrational problem for a periodic chain reveals the following features. (a) In a particular vibrational mode of the entire chain the same atomic displacements occur within each translational repeat unit (e.g., CH_2CH_2 in planar zigzag polyethylene), with the phase difference ϕ between motions in adjacent units being the same. (b) Despite the fact that the atomic displacements within each unit are of the same kind (e.g., CH_2 bend), the frequency of the chain vibration is a function of ϕ (which can vary from $-\pi$ to $+\pi$). (c) The curve that gives ν as a function of ϕ for a particular mode is called a dispersion curve, and the derivative $d\phi/d\nu$, which gives the number of vibrational modes per frequency interval, is a density of vibrational states. (d) Although the chain vibrates at frequencies associated with all phase differences, infrared absorption (and Raman scattering) can occur only for modes with $\phi = 0$; only when all unit cells move in phase is there a possibility that $\boldsymbol{M} \neq 0$ (no selection rules operate in inelastic neutron scattering, where modes with $\phi \neq 0$ also contribute). (e) In a helical structure with n monomers per repeat unit (e.g., three $CH_2CH(CH_3)$ in the 3_1 helix of isotactic polypropylene) a phase difference of $m(2\pi/n)$ ($m = 0, 1, 2\ldots$) between motions in adjacent monomers is still consistent with $\phi = 0$. Infrared activity is associated, however, only with $m = 0$ and 1, giving rise to modes with $\boldsymbol{M}$ parallel to the helix axis in the first case and perpendicular in the second.

The existence of a one-dimensional periodic structure permits the development of three-dimensional order, or crystallinity, which is indeed found in polymers. This lateral order introduces specific interchain interactions, which manifest themselves in three ways: shifting of frequencies due to the additional forces acting on a chain; splitting of single bands into multiplets whose number depends on the number of chains and their arrangement in the unit cell; and the presence of lattice modes associated with translatory and rotatory motions of the chains with respect to one another. Analysis of these effects can lead to an understanding of the nature of the crystalline structure and the interactions between adjacent chains.

Real polymer molecules differ from the above model systems in several ways. Firstly, even if regular, the chain is finite rather than infinite in length. Since exact translational symmetry no longer exists, the $\phi = 0$ condition for infrared activity no longer holds. If the chain is long enough, its vibrational modes will resemble those of the infinite chain for the discrete values of ϕ appropriate to the number of repeat units, and the infrared activity of these modes will be determined by the overall symmetry of the molecule. If the chain is short the situation can be more complex, with end effects becoming important. Secondly, even if the chain is long (and thus essentially periodic) but has a small concentration of "defects" (any structure that replaces a monomer by an oscillator of different frequency), then, depending on the relative frequencies of the defect and monomer oscillators, it is possible for the spectrum to exhibit frequencies other than the $\phi = 0$ modes of the regular chain. The effects can be complex (Fanconi 1980), and dependent on the nature of the density of vibrational states, but in interpreting a spectrum it is necessary to realize that they may be present. Thirdly, since even a chemically regular polymer chain can exist in a spatially nonregular (amorphous) structure, the spectral consequences of these irregularities must be recognized. In general, such structures are conformational isomers of the periodic chain and thus introduce new frequencies associated with the different three-dimensional arrangements of atoms. This conformational sensitivity of the frequency usually derives from the nonlocalized nature of the mode (Snyder 1980), and results in one of the most powerful uses of infrared spectroscopy, namely the study of changes in polymer chain conformation brought about by external perturbations.

5. *Polymer Chain Frequencies: Applications*

An understanding of the coupled oscillator nature of polymer chain frequencies leads to insights and applications of infrared spectroscopy beyond those possible only on the basis of the group frequency concept (Krimm 1960, Natta and Zerbi 1964, Zerbi 1969, Siesler and Holland-Moritz 1980, Snyder 1980, Painter et al. 1982).

The specific conformation and periodic structure of a regular polymer chain is a result of its stereoregular chemical structure. Therefore, bands characteristic of the regular conformation automatically identify the chain stereoregularity. Thus, the spectrum of the 3_1 helix of isotactic polypropylene is different from that of the fourfold helix conformation of syndiotactic polypropylene, and the latter differs from the spectrum of a planar zigzag syndiotactic chain obtained on cold-drawing. A departure from the regular conformation (such as occurs in solution, in the melt, in noncrystalline regions, or in an atactic polymer) will also result in spectral changes.

When a regular polymer chain crystallizes by lateral ordering we have seen that additional effects can usually be observed in the infrared spectrum. These can be used to define the presence of crystallinity in the sample, and in conjunction with other techniques (e.g., x-ray diffraction and density) will often provide useful indices of the degree of crystallinity. Thus, crystallinity in polyethylene results in the splitting of infrared bands into doublets, and an understanding of this effect has permitted a characterization of the crystalline–amorphous structure in this polymer.

The fact that chain vibrations have oscillating dipole moments only parallel or perpendicular to the chain axis makes possible a reliable analysis of orientation distributions in oriented samples.

6. *Molecular Vibrations: Normal-Mode Calculations*

While a knowledge of the special nature of the vibrations of regular chain molecules has expanded our ability to apply infrared spectroscopy to the study of polymers, it is only through a complete understanding of the spectrum, including calculations of the normal modes of vibration, that detailed insight into the structure and interactions of specific systems can be obtained (Krimm 1960).

The methods for calculating the normal vibrations of polymers are extensions of those for small molecules. The computational problem, using computers, is now relatively routine; the main uncertainty is the force field to be used in the calculation.

The process of developing a suitable force field for a polymer usually accompanies a normal-mode analysis. The underlying characteristic of this process is making appropriate assignments of observed bands (infrared and Raman) to the calculated modes, in terms of both frequency and nature of vibration. The procedure includes the following steps. (a) Selection of a general form to the potential function. A general valence force field is commonly used, involving the various force constants mentioned in Sect. 2. (b) The study of smaller model molecules and transfer of relevant force constants to the polymer calculation. Thus, a detailed analysis of *n*-paraffins not only led to a straightforward calculation of the normal modes of polyethylene but also provided the basis for force fields of other hydrocarbon polymers. (c) Symmetry analysis of the polymer structure being studied in order to define the expected band characteristics. (d) A variety of experimental studies whose goal is to identify the normal-mode origins of observed bands. These include numerous methods (e.g., temperature variation and polymerization conditions) to distinguish crystalline from noncrystalline bands; isotopic substitution, such as deuterium for hydrogen, to identify hydrogen modes; and infrared polarization studies on oriented samples to correlate bands with predicted symmetry species. As these components are only the basic ones, a critical examination of each element in the procedure is vital (Zerbi 1969).

7. *Polymer Vibrational Analyses*

Detailed normal-mode analyses have been performed for a number of polymer systems (Zerbi 1969, Snyder 1980). Polyethylene is the most studied polymer, both intra- and intermolecular force fields having been refined. As a result, it has been possible, for example, to characterize the nature of chain folding (i.e., adjacent vs random re-entry) in polyethylene crystals. The analysis of polypropylene, both isotactic and syndiotactic, has provided important insights into the relation between the spectrum and the tacticity and crystallinity of the polymer, including an understanding of how the spectrum can vary with the length of isotactic helical structure. In the case of poly(vinyl chloride), normal-mode analyses have permitted identification of specific conformational isomers of the chain, as well as providing an understanding of specific interchain interactions in the crystalline regions. Polyoxymethylene exists in two helical chain conformations, one of which crystallizes in a hexagonal and the other in an orthorhombic structure. Normal-mode analyses provide information on the different intra- and intermolecular interactions in these structures. Polypeptides can exist in many different conformations, and normal-mode analyses have permitted detailed correlations of spectra with structure.

Polymers that have been analyzed by normal-mode analysis include poly(vinylidene chloride), poly(vinylidene fluoride), polytetrafluoroethylene, polybutadienes and nylons.

See also: Elastomers: Spectroscopic Characterization; Infrared Spectroscopy; Raman Spectroscopy of Polymers

Bibliography

Bellamy L J 1975 *The Infra-Red Spectra of Complex Molecules*, 3rd edn. Chapman and Hall, London

Boerio F J, Koenig J L 1972 Vibrational spectroscopy of polymers. *J. Macromol. Sci., Rev. Macromol. Chem.* 7: 209–49

Califano S 1976 *Vibrational States.* Wiley-Interscience, New York

Colthup N B, Daly L H, Wiberley S E 1964 *Introduction to Infrared and Raman Spectroscopy*, 1st edn. Academic Press, New York

Fanconi B 1980 Molecular vibrations of polymers. *Annu. Rev. Phys. Chem.* 31: 265–91

Hummel D O, Scholl F 1969 *Infrared Analysis of Polymers, Resins, and Additives: An Atlas*, 1st edn. Wiley-Interscience, New York

Krimm S 1960 Infrared spectra of high polymers. *Adv. Polym. Sci.* 2: 51–172

Natta G, Zerbi G (eds.) 1964 Vibrational spectra of high polymers. *J. Polymer Sci., Polym. Symp.* 7: 3–224

Painter P C, Coleman M M, Kuenig J L 1982 *The Theory of Vibrational Spectroscopy and its Application to Polymeric Materials.* Wiley-Interscience, New York

Siesler H W, Holland-Moritz K 1980 *Practical Spectroscopy*, Vol. 4, *Infrared and Raman Spectroscopy of Polymers.* Dekker, New York

Snyder R G 1980 Infrared and Raman spectra of polymers. In: Marton L, Marton C (eds.) 1980 *Methods of Experimental Physics*, Vol. 16A. Academic Press, New York, pp. 73–148

Wilson E B Jr, Decius J C, Cross P C 1955 *Molecular Vibrations*, 1st edn. McGraw-Hill, New York

Zbinden R 1964 *Infrared Spectroscopy of High Polymers.* Academic Press, New York

Zerbi G 1969 Molecular vibrations of high polymers. *Appl. Spectrosc. Rev.* 2: 193–261

S. Krimm

Infrared-Thermal Nondestructive Evaluation

Infrared-thermal nondestructive evaluation (NDE) is normally performed by heating the test object and recording or observing the resulting surface temperature changes. Wide variations in the method of applying the excitation energy, as well as sensing and interpreting temperature, are possible. Energy application methods include direct heating of the test object's surface, heating the test object by the passage of an electrical current, and internally heating the test object by mechanical stress (see *Vibrothermography*). Temperature sensing methods include infrared scanners, infrared radiometers, infrared photography, temperature sensitive phosphors, liquid crystals, thermocouples, and sharp melting-point materials. The practical applications have included the detection of flaws in solar cells and microcircuits, as well as defects in nuclear fuel, helicopter rotors, aircraft honeycomb panels, solid fuelled rockets, stainless steel welds, electrical heating elements, automobile tires, carbon composite parts, wood glue joints, ceramic parts, and bonds of thin protective coatings.

1. Heating and Temperature Sensing

Radiant heating can be accomplished by the use of infrared heat lamps, projection lamps, quartz–halogen lamps or lasers. One disadvantage in the use of radiant heating lies in its dependence on surface absorption, which is equal to the emissivity (emittance). A region having lower surface emissivity not only causes a smaller infrared signal to be sensed by an infrared radiometer for a given surface temperature, but also causes lower heat absorption and, hence, lower temperatures in the region. For test articles with high uniform emissivity, however, radiant heating can be used successfully.

Hot-gas jet-heating is a useful method when testing for voids or delaminations between a substrate and a thin surface coating. It is also useful for samples with rough surfaces. Optimum spacing can be determined empirically by measuring heat input against distance and selecting the desired region. Heat input to the test article surface is much higher when a high conductivity gas such as helium is used. Commercial filament-type gas-jet heaters are available and, for very high heat inputs, plasma arc jets may be used (Green 1961).

Induction heating is a satisfactory method for electrically conductive test objects (Green 1961). This method also has its disadvantages. It is extremely dependent upon the spacing between the work coil and the surface of the test object. Thus, if there are any irregularities on the test object, those areas nearest the work coil will receive much more heat than the remainder of the test object. High-frequency induction heating deposits most of the heat near the surface; the heating intensity decreases exponentially with depth. The heating rate also depends upon the magnetic permeability of the test article. Hence, if the permeability varies from one location to another (as it can, for example, in cold-worked austenitic stainless steel), hot spots could arise in regions of higher permeability. Careful matching of the work coil to the output impedance of the induction heater is necessary, thus restricting the size and geometry of the coil. These difficulties have prevented induction heating from being more widely used in infrared and thermal NDE.

Electrical conduction heating has been used in cases where electrical contact with a conductive test article is permissible (McCullough and Green 1972, Green and Hassberger 1977). An experimental method has been developed for applying an electric current to stainless steel pipes without damaging their surface, demonstrating an electrothermal method for nondestructive weld testing in these pipes (Green 1979).

Thermocouples and thermistors are seldom used for direct sensing of temperatures in thermal NDE. The requirement for good thermal contact between the test object and thermocouple or thermistor makes it difficult to scan the surface temperature directly with these sensors. However, thermocouples and thermistors have been used for indirect NDE temperature measurements using noncontacting thermal transducers. Thermopiles have been used as infrared detectors for remote, noncontacting surface temperature measurements. However, thermistor–bolometer infrared detectors are more satisfactory in the lower temperature range usually found in NDE applications.

Infrared radiometers focus incoming infrared from a region in the object field onto a detector. Usually the infrared beam is chopped to provide an ac signal out of the detector. The amplified signal is subsequently rectified and filtered to produce a dc output. The two most common types of detectors are photovoltaic cells and thermistor–bolometers. Photovoltaic cells usually require cooling to liquid nitrogen temperature (77 K) and are often characterized by rapid response times (~1 μs). For this reason they are used in most infrared scanning cameras required to scan a two-dimensional field. Such scanners are often just a radiometer with deflection mirrors or prisms to sweep the instantaneous field of view of the camera from side-to-side and up or down in a raster fashion, thereby scanning a predetermined area in the object space. Output from the detector is displayed on a cathode-ray tube unit, producing images of the object. Another type of infrared scanning camera which has recently become commercially available is the pyroelectric vidicon. Here, the infrared scene is focused on a pyroelectric plate scanned by an electron beam. Special electronic features must be incorporated into the camera to prevent its output from being rate dependent (if the infrared image is not steady state). The output picture is also produced on a cathode-ray tube unit.

Signal-to-noise ratios in images from infrared scanners can be improved by averaging successive images together (ensemble averaging). However, the extent to which this can be applied is limited by the normal flaws and irregularities introduced by the image-producing device.

Infrared radiometers (microscopes) capable of resolving regions of 25 μm in diameter are available. However, the minimum detectable temperature difference with such a finely focused infrared radiometer is often 5–20 °C. Temperature differences as small as 0.05 °C or less can be detected with some types of infrared radiometers focused to resolve 1 mrad. Target resolution in this case is 0.001 times the distance from radiometer to target, down to a minimum diameter of about 1.5 mm. Scanning cameras with the same resolution are capable of detecting a temperature difference of about 0.1–0.2 °C on the test object. (Emissivity of the test object is assumed to be at least 0.9 in all cases; lower emissivity would mean correspondingly larger minimum temperature resolution.) In most commercial radiometers and scanning cameras the minimum resolvable target size is limited to about 1–3 mm in diameter, regardless of how close the instrument is focused (unless a special close-up lens or microscope attachment is used).

Another thermal imaging method makes use of temperature sensitive materials applied to the test object's surface. Thermographic phosphor, a finely divided powder of silver- and copper-doped zinc sulfide, when illuminated with ultraviolet light fluoresces to produce bright-yellow visible light. Heating the phosphor quenches the light, while upon cooling, the phosphor regains its brightness. The process is completely reversible and does not appear to suffer hysteresis. Temperature differences of about 0.1 °C can be detected repeatably using a phototube to detect changes in brightness of the phosphor. Normally the phosphor is suspended in a special lacquer and sprayed or brushed onto the test object. A range of temperatures can be detected, from slightly below room temperature to about 50 °C.

Liquid crystals are cholesteric compounds that selectively scatter different wavelengths of light. This causes white light to be scattered according to wavelength and produces the appearance of color when viewed against a dark background. These temperature sensitive compounds can be mixed to change color in any desired temperature range from about 1 to 100 °C, passing through the visible spectrum of color as the temperature passes through its active range. Ordinarily, matt-black background paint is applied to the test specimen and thoroughly dried. The liquid crystal compound is then applied over the background with a brush. Alternatively, the liquid crystal may be applied to thin black plastic film which is then pressed into contact with the test object to sense its surface temperature.

Materials possessing a sharp melting point have been successfully used for temperature mapping, for example, dry-ice cooling of test objects to cause formation of frost when breathed upon. Subsequent induction heating of the test objects causes melted spots to appear in the frost over the defects.

2. NDE Applications

One of the first noncontacting nondestructive thermal bond tests was applied to aluminum-clad uranium nuclear fuel elements. A coating of acenaphthene, which has a sharp melting point, was sprayed onto the fuel elements as a 50:7 mixture of carbon tetrachloride and acenaphthene. When the fuel elements were passed through a 30 kW, 10 kHz induction heater, unbonded regions in the fuel elements reached temperatures exceeding 95 °C, causing the acenaphthene to melt. As the coating had a frosty appearance, the method was called the Frost Test. Where the coating was melted it became permanently transparent, so the unbonded regions were visibly recorded on the surface of the fuel elements. The acenaphthene was removed in a hot carbon tetrachloride vapor tank.

An infrared radiometer has been used to map the surface temperature of aluminum cladding uranium core bonds in nuclear fuel elements (Green 1961). The fuel element was rotated and translated through a plasma arc jet. Emissivity differences on the 0.76 mm thick aluminum cladding limited the minimum size bond defect that could be detected to about 9.5 mm diameter.

An emissivity independent method makes use of an auxiliary infrared source to cancel the effect of emissivity variations on the surface of fuel elements. Radiant energy originating from the auxiliary source is reflected from the fuel element surface. This energy is intercepted along with energy radiated from the fuel element surface itself. Reflectivity is approximately $r = 1 - \varepsilon$, assuming the emissivity ε is not too dependent on wavelength. Therefore, it is possible to adjust the auxiliary source so emissivity differences do not cause variations in the infrared signal at the detector when a test object having a uniform constant surface temperature is scanned. Defects that cause temperature differences, on the other hand, result in infrared radiant emittance variations that signal the presence of a defect. The auxiliary source reflection method works well on test objects that have smooth symmetrical surfaces. However, it cannot be used on irregular surfaces where the relative angles between the radiometer, the auxiliary source and the test object surface are not constant.

Another emissivity independent method makes use of a dual-scan technique to nondestructively test nuclear fuel elements. A plasma arc jet applies heat to a 1 cm spot on the surface. The fuel element is rotated and the plasma jet, together with two identical infrared radiometers, is moved longitudinally along the fuel element during heating. One radiometer is mounted at 90° from the fuel element in the direction of rotation from the plasma jet, and the second at 180° about 1 cm further along the fuel element. An emissivity-independent output is obtained as

$$(E_A + C)/(E_D + C) = f(T_A)/f(T_D) = E_o$$

where E_A is the electrical signal from the first radiometer, E_D is that from the second, T_A is the fuel element's surface temperature under the first radiometer, T_D is the temperature under the second, C is a constant and E_o is the emissivity independent output. Even with a stripe of black paint applied down one side of a fuel element, true bond defects are easily seen and emissivity effects are eliminated (Green 1968).

Electrothermal NDE involves passing an electrical current directly through the test article while the surface temperature is mapped with an infrared scanner. Defects cause differences in electrical current density on the test object's surface, thus giving different local heating rates. Since the temperature differences are primarily due to electrical current density differences, this method is essentially instantaneous and does not require time for heat to diffuse to the defects. Stainless steel plates and stainless steel pipe welds have been nondestructively tested using this method. Slightly uneven surfaces normally found on stainless steel welds have little effect on the electrothermal NDE results provided there are no sharp angles of intersection. Microstructural differences that do not appreciably affect the average electrical resistivity in macroscopic regions do not cause problems in detecting defects with the electrothermal method.

Thermal image transducers (Green 1970) have been developed for nondestructively imaging bond defects and examining glue or braze material in metal and fiberglass honeycomb structures. They have also been used for detecting density differences and bond deficiencies in composites and ceramics. Basically, the transducers comprise a layer of metal foil coated with a temperature sensitive material (thermographic phosphor) backed with a transparent flexible backing of silicone rubber (Fig. 1). The electrically heated foil, pressed into contact with the test object, produces uniform heat generation. The heat flows into the test object's surface, causing temperature differences to appear over defects. If the test object is an electrical conductor, a thin dielectric layer must be placed between the foil and test object (Fig. 1). Thermal images produced with these devices are viewed at the end of a delay period after terminating the heat input to the foil. This allows the foil temperatures to reach equilibrium with the temperature of the adjacent test object surface. If viewed too early, the thermal image is dependent upon thermal contact resistances between the foil and test object;

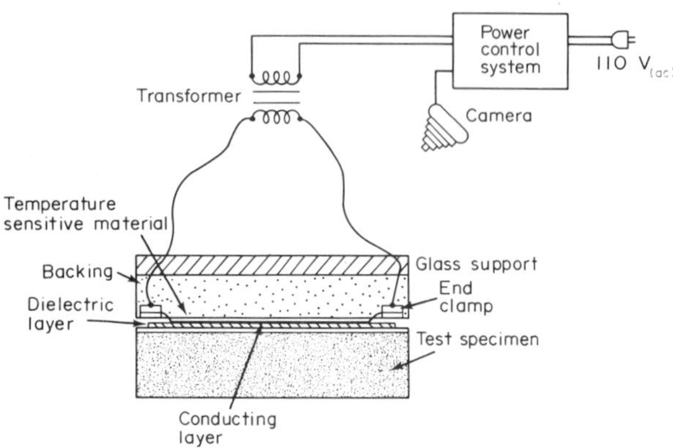

Figure 1
Thermal image transducer system

with the proper delay, the image is almost independent of contact resistance.

See also: Nondestructive Evaluation: An Overview; Temperature Sensor Materials

Bibliography

Adams J A, Rogers D F 1973 *Computer Aided Analysis in Heat Transfer.* McGraw-Hill, New York

Carslaw H S, Jaeger J C 1959 *Conduction of Heat in Solids,* 2nd edn. Clarendon Press, Oxford

Green D R 1961 An instrument for nondestructively testing fuel core to cladding heat transfer. *Nucl. Sci. Eng.* 12: 271–75

Green D R 1968 Principles and applications of emittance independent infrared nondestructive testing. *Appl. Opt.* 7: 1779–89

Green D R 1970 High speed image transducer for practical NDT applications. *Mater. Eval.* 28: 97–102

Green D R 1979 Experimental electro-thermal method for nondestructively testing welds in stainless steel pipes. *Mater. Eval.* 37: 54–60

Green D R, Day C K 1968 *Infrared Testing of Bonds Between Graphite and Protective Coatings,* ASTM Special Technical Publication No. 439. American Society for Testing and Materials, Philadelphia, Pennsylvania, pp. 4–17

Green D R, Hassberger J A 1977 Infrared electro-thermal examination of stainless steel. *Mater. Eval.* 35: 39–43

Kruse P W, McGlauchlin L D, McQuistan R B 1962 *Elements of Infrared Technology: Generation, Transmission and Detection.* Wiley, New York

McCullough L D, Green D R 1972 Electrothermal nondestructive testing of metal structures. *Mater. Eval.* 30: 87–91

Parker W J, Jenkins R J, Butler C P, Abbot G L 1961 Flash method of determining thermal diffusivity, heat capacity and thermal conductivity. *J. Appl. Phys.* 32: 1679–84

Vanzetti R 1972 *Practical Applications of Infrared Techniques.* Wiley, New York

D. R. Green

Ingot and Continuous Casting of Steel

Nearly all molten steel is cast into solid shapes by two different processes: ingot casting and continuous casting. With ingot casting (the older technique), steel is teemed from a ladle into the top or bottom of cast-iron molds (Fig. 1a). The molds are frequently equipped with "hot tops" consisting of insulating boards and exothermic compounds to minimize the depth of the shrinkage cavity that forms as the ingot solidifies. At a certain time the molds are stripped from the ingots, and the latter are charged into soaking pits, later to be rolled into finished shapes. In continuous casting, steel is first poured from a ladle into a refractory-lined tundish which then directs flow into the molds (Fig. 1b). Beneath the mold, which is constructed of copper, subjected to oscillating motion and water cooled, heat is extracted by water sprays and by radiation cooling. Before casting, a movable dummy bar is inserted into the bottom of the mold and is withdrawn after steel pouring commences. The strand of steel which follows is semisolid (a liquid core encased in a solidifying shell) over much of the length of the casting machine. After the point of complete solidification, the steel strand is cut into desired lengths. Most continuous-casting machines employ either curved molds with subsequent straightening of the strand into the horizontal position, or straight molds with bending and straightening several meters below the mold.

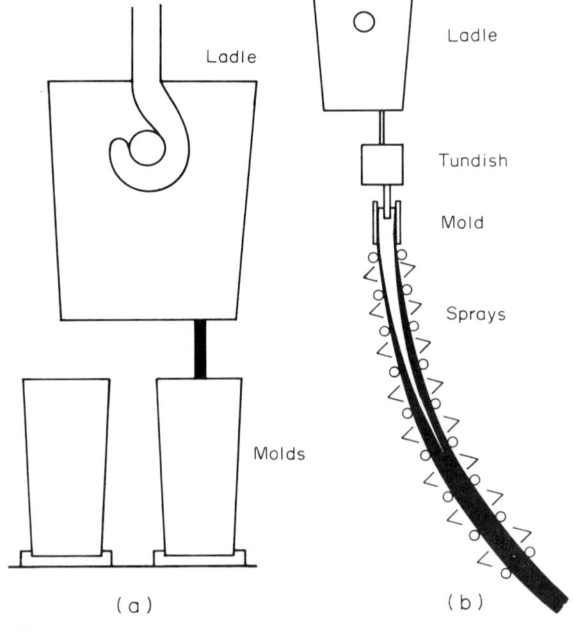

Figure 1
Schematic diagram of (a) ingot casting and (b) continuous casting

Continuous casting has many advantages because it produces a semifinished shape (billet, bloom or slab) in a near continuous manner, unlike ingot casting, which yields unfinished shapes of larger cross section. Thus continuous casting offers higher productivity, lower yield loss and lower energy consumption, and is the preferred casting technique for large-quantity production of steel.

Ingots are usually thicker in at least one transverse direction than continuously cast semifinished shapes. In continuous casting the depth of the liquid core, which may easily reach values of 20 m, is much greater than in a solidifying ingot, and therefore solidification of the center of a continuously cast section is far removed from the mold. Also, the continuously cast strand, unlike the ingot, normally changes its orientation from vertical to horizontal as it moves through the casting machine. Submerged pouring tubes and electromagnetic stirring are more commonly used in continuous casting than ingot casting. Most of these differences affect fluid flow conditions in the liquid core, which strongly influences structure and segregation.

1. Ingot Casting

Steel ingots are normally classified according to the amount of gas released during solidification. The gas is generated chiefly by the reaction between carbon and oxygen dissolved in the steel, which is favored thermodynamically at lower temperatures.

Killed steel ingots are fully deoxidized so that during solidification no gas is evolved. Ingots of this type are usually cast in "hot-topped, big-end-up molds" to minimize the depth of the shrinkage cavity. Killed steels are used where emphasis is placed on soundness and homogeneity of structure (e.g., alloy steels, forging steels and steels for carburizing).

Semikilled steels are deoxidized less than killed steels with the result that there is sufficient oxygen to combine with carbon and generate some CO during solidification. The gas is trapped in the steel as blowholes, primarily in the upper part of the ingot, and is sufficient in volume to compensate for solidification shrinkage. These steels are commonly employed in structural shapes, plates and merchant bar.

Rimmed ingots are characterized by only a small amount of deoxidation and a rapid evolution of gas during solidification. Depending on the rate of gas evolution, large blowholes may be virtually eliminated from the sides of the ingot by the washing action of the large number of ascending bubbles. If properly cast, rimmed ingots have a good surface and minimum shrinkage pipe but more segregation than killed ingots.

Capped ingots are produced by pouring steel with about the same (or slightly lower) oxygen content as rimmed steels into "big-end-down, bottle-top molds." The rimming action is then allowed to proceed normally, but after a short time it is terminated by sealing the mold with a cast-iron cap. Thus, compared with the rimmed ingot, the capped ingot has a thinner rim zone and less segregation in the core. Capped ingots are cast for the production of sheet, strip, skelp, tin plate, wire and bars.

1.1 Structure

The structure of an ingot, as revealed by macroetches, sulfur prints or radioactive-tracer additions, basically consists of three zones (see Fig. 2a). Within about 10 mm of the surface, a chill zone exists with a structure that is difficult to resolve. The crystals in this region are numerous and fine, owing to the high rate of heat extraction, which causes undercooling and enhanced nucleation. Other factors including the superheat of the liquid and convective flow also affect the number of crystals formed.

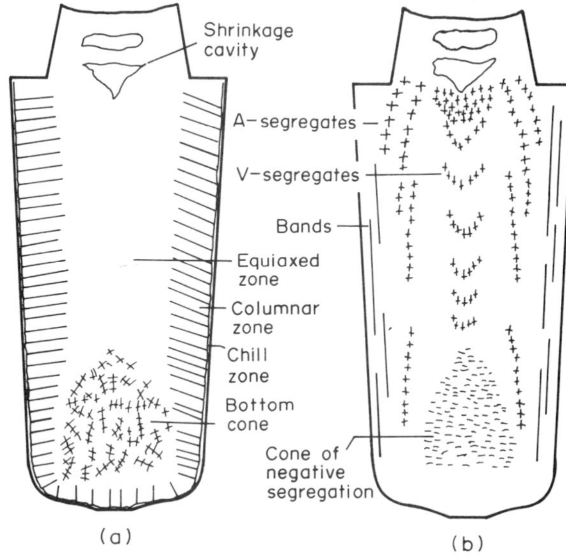

Figure 2
Schematic diagram of (a) structure and (b) segregation pattern in a killed, big-end-up ingot

Moving inward, the chill zone gives way to a columnar structure, as those crystals with a preferred orientation (⟨100⟩ for the face-centered-cubic structure) relative to the steepest thermal gradients, grow preferentially to form dendrites. Thus the dendrites are oriented roughly normal to the ingot surface along heat flow lines; in killed ingots the dendrites are usually inclined slightly upwards, which indicates that liquid steel was flowing down the advancing solidification front, at least in the early stages of solidification.

Beyond the columnar zone in the core of the ingot, an equiaxed structure consisting of randomly oriented dendrites is found. Although several theories have been advanced for the formation of the equiaxed zone, it most likely originates in the early stages of solidification. It has been suggested that the equiaxed crystals are first nucleated at the mold wall; then, due to thermal fluctuations in the liquid, the root of the crystals remelts leaving the crystals free to enter the liquid. Another source of free equiaxed crystals may be secondary dendrite arms that detach from the solidification front by a similar mechanism. The survival and growth of the equiaxed crystals depend strongly on the superheat of the liquid and convection. If the superheat is high the equiaxed crystals may remelt in the liquid, leaving the columnar dendrites to grow preferentially to form a large columnar zone. On the other hand, if the superheat is low and, owing to convection patterns, the crystals are allowed to grow by descending through regions of cold or constitutionally supercooled liquid (presumably near the solidification front), they may become attached to the tips of the advancing columnar dendrites and interfere with columnar growth. In this case, the size of the columnar zone is smaller and the structure is predominantly equiaxed. This mechanism is borne out in practice since the superheat has been found to influence the structure in killed steel ingots as postulated.

According to this picture of ingot solidification, convection in the liquid pool is an important phenomenon because it controls the extraction of superheat from the liquid, influences the formation of equiaxed crystals through temperature fluctuations, and affects the motion of the equiaxed crystals. In killed ingots, once the initial turbulence due to filling of the mold has damped out, natural convection takes hold with cold liquid near the walls descending under the influence of gravity and liquid at the center rising. In the case of semikilled or rimmed ingots, convection is generated by rapidly rising gas bubbles. Superimposed on this motion is that of the equiaxed crystals that are slightly denser than the liquid and which, depending on local fluid conditions, settle and accumulate at the bottom of the ingot. Thus a bottom cone of dendritic debris builds up with time, and as solidification continues the center of the ingot gradually becomes filled with equiaxed dendrites. At some stage in this process, free circulation of liquid is hampered by the network of equiaxed dendrites; and fluid flow, driven by solidification shrinkage and buoyancy arising from concentration gradients or bubbles, takes over in the interdendritic region. As will be seen, the movement of liquid at this stage and the settling of dendrites in the core are the cause of macrosegregation in ingots.

A structural feature of semikilled and rimmed ingots is the presence of blowholes, which may appear just inside the surface over much of the height and upper core of an ingot. The blowholes have a clean surface and reweld on rolling, if not exposed to air. Reoxidation in the air can be a problem if the blowholes are too close to the surface of an ingot and break through to the atmosphere during heating and rolling. The reoxidized blowholes will not reweld and will appear as defects in the product.

1.2 Segregation

Ideally an ingot with a uniform composition is preferred, but during the solidification of steel, solute elements such as carbon, manganese, sulfur and phosphorus become enriched in the liquid ahead of the advancing dendrites. The resulting segregation is manifested in ingots on both a microscale, as between individual dendrites, and on a macroscale over much greater distances. Macrosegregation is of greatest concern because it may carry through as a defect into the final product. In killed steel ingots three major forms of macrosegregation—negative segregation, A-segregates and V-segregates—are observed (Fig. 2b). Bands, probably arising from disturbances or fluid flow in the interdendritic region, may also be observed running parallel to the solidification front.

The negative segregation usually appears as a zone in the bottom third of the ingot, and is associated with the pile of dendritic debris that has settled in this region. These dendrites, having been formed early in the solidification process, are relatively solute poor. This region also contains a relatively high concentration of inclusions that are apparently caught by the descending equiaxed crystals and trapped in the core at the bottom of the ingot.

The A-segregates are pencil-like in shape and are oriented vertically but inclined slightly inward. They are the result of the buoyancy-driven flow of liquid, enriched in solute, through interdendritic channels that may actually be opened up through the remelting action of the liquid. Thus factors that influence the formation of A-segregates are the extent of solute enrichment, the dependence of steel density on solute concentration, and the morphology of the dendritic network.

The V-segregates appear in the center of the ingot and obviously form in the final stages of solidification. The shape of these segregates probably arises from the manner in which the equiaxed dendrites settle in the core of the ingot. As the dendrites settle and slump owing to liquid being drawn down to feed solidification shrinkage, fissures may be opened along shear planes oriented upward away from the ingot axis. Enriched liquid then fills these voids to form the V-segregates.

Other zones of positive segregation are also found near the centerline, particularly at the top of the ingot. Clearly all three types of segregation arise from the motion of liquid and solids in the unsolidified pool, and are affected by the structure of the ingot (e.g., the relative size of the columnar and equiaxed zones). It follows that any factors that affect structure and fluid flow must also influence the segregation in ingots.

Finally, as previously mentioned, more segregation is found in rimmed ingots than in killed ingots. The rim zone exhibits negative segregation, while positive segregation is measured at the center.

2. *Continuous Casting*

Continuously cast steels are normally fully killed with no evolution of gas in the mold. Continuous casting differs from ingot casting in that similar events occur in a moving section and thus time is measured with respect to a reference frame that moves with the strand (= axial distance from meniscus/casting speed), whereas in ingot casting, solidification takes place in a stationary space and time is viewed from the standpoint of a fixed observer. Hence events in the continuous-casting mold roughly correspond to the first minute or less of ingot casting, and moving down through the machine (i.e., with increasing time), events correspond more to the end of solidification in an ingot.

Viewed in this way, it is easy to see that many of the solidification phenomena discussed earlier with respect to the casting of killed steel ingots also take place in continuous casting. Fluid flow in the liquid pool is a good example. In the mold of a continuous-casting machine, fluid flow results from the momentum of the input stream(s), as is the case in ingot casting during and just after teeming. With increasing time (or distance down the machine in continuous casting), natural convection takes over until the latter stages of solidification (near the end of the pool in continuous casting), whereupon solidification shrinkage and buoyancy driven by concentration gradients become important. However, two important differences between ingot and continuous casting are that, in the latter, electromagnetic stirring is frequently induced at various locations, and bulging, particularly of slab shapes, can effect downward flow of liquid.

2.1 *Structure*

The broad structural features of continuously cast billets, blooms or slabs are similar to those of an ingot. Moving inward from the surface, a chill zone, columnar zone and central equiaxed zone are observed in most continuously cast sections. The origin of each of the zones is believed to be the same as in an ingot. The generation of equiaxed crystals that ultimately impede the growth of the columnar dendrites, for example, almost certainly begins in the mold by a crystal multiplication mechanism involving the remelting of secondary dendrite arms. The effect of superheat on the length of the columnar zone is the same as in ingot casting; increasing superheat increases the size of the columnar zone presumably because the small equiaxed crystals remelt. Application of electromagnetic stirring in the mold, however, can reduce the length of the columnar zone at a given level of superheat. This observation supports the postulate that the free equiaxed crystals are generated in the mold or just below it. The effect of superheat on structure is profoundly important to casting quality because segregation, central porosity and internal crack formation are all minimized by a large equiaxed zone. This is of particular importance to continuous casting, as compared with ingot casting, because continuously cast sections, being usually smaller in cross section, have a smaller equiaxed zone and receive less reduction in area during rolling to the final product.

The application of electromagnetic stirring adds a new dimension to the control of structure in continuous casting that is not feasible in ingot casting. Electromagnetic stirring makes it possible to cast steel with higher superheat, to take advantage of less frequent nozzle blockage and improved inclusion float-out, but at the same time to obtain a predominantly equiaxed structure.

The settling of equiaxed crystals into the lower part of the pool in a continuous-casting strand gives rise to a structure that is relatively uniform axially, unlike the bottom-core structure characteristic of ingot casting. The difference is due to one process being continuous in nature and the other being batch. However, the structure in continuously cast shapes, produced on a curved-mold machine, may be nonsymmetrical about the withdrawal axis. An example is shown in Fig. 3: an autoradiograph of a longitudinal section of a 133 mm square billet to which radioactive

Figure 3
Autoradiograph of a longitudinal section from a 133 mm square steel billet. Note that the columnar zone adjacent to the (upper) inside radius face is longer than the columnar zone next to the outside radius face (after Van Drunen et al. 1975. © The Metals Society, London. Reproduced with permission)

Au^{198} had been added during casting. The columnar zone adjacent to the inner-radius face is seen to be longer than that next to the outside-radius face. Moreover, a layer of dendritic debris that appears white in the autoadiograph, indicative of a low concentration of tracer, is seen sitting in a tracer-rich region adjacent to the outside-radius face. Clearly this reveals that equiaxed crystals generated in the upper part of the strand some time after the tracer was added have descended through the liquid pool and settled along the bottom of the curved strand into a tracer-rich region. Thus the crystals are an important link between conditions in the mold/upper sprays, as influenced by superheat, electromagnetic stirring and nozzle-induced stirring, and structure, which is determined many meters below.

Finally, the solidifying shell in a continuous casting strand does not always form uniformly in the mold. Corners are frequently thin relative to the midface, possibly owing to excessive superheat. Steel containing 0.1% carbon characteristically exhibits a wavy shell with locally thin regions that are not seen in higher-carbon (>0.4%) steels (Singh and Blazek 1974). This nonuniformity, which reduces heat transfer to the mold, is most likely related to the δ–γ phase transformation.

2.2 Segregation and Center Unsoundness

Although V-segregates, axial segregation and center unsoundness are less pronounced than in ingots, they may be seen in longitudinal sections of continuously cast steel. Depending on severity, the axial segregation and voids may pose serious quality problems. For example, a high concentration of manganese at the center of plate-grade slabs or billets for wire rod can lead to defects during subsequent processing operations. The origin of the segregation is different in billets than in slabs, but in both cases structure plays a significant role.

The axial segregation and center unsoundness in billets frequently appears periodically along the strand (Fig. 4a). The segregation is less severe under conditions that favor a predominantly equiaxed structure (i.e., low superheat and electromagnetic stirring in the upper part of the strand). Other factors, namely section size, aspect ratio and carbon content, also influence the severity of axial segregation and porosity. With increasing section size, the width of the central segregation zone increases but the fraction of total area occupied decreases. Rectangular sections exhibit less axial segregation than square sections. The carbon content affects both the width of the central segregation zone and the size of the axial voids: 0.3% < 0.1% < 0.6%C (Mori et al. 1972).

Segregation and porosity in billets arises because equiaxed crystals settling into the lower region of the liquid pool become caught periodically on the columnar dendrites growing in from the side of the billet, to form a succession of bridges. Owing to their high solid fraction, these bridges may solidify more quickly than regions immediately above and below, and effectively separate the billet into a series of mini-ingots. In each mini-ingot the liquid becomes progressively enriched and voids form as solidification proceeds in the absence of any liquid feeding. V-segregates form essentially in the same way as described for ingots; the mechanism involves the settling of dendrites, volume shrinkage and the flow of enriched liquid between dendrites.

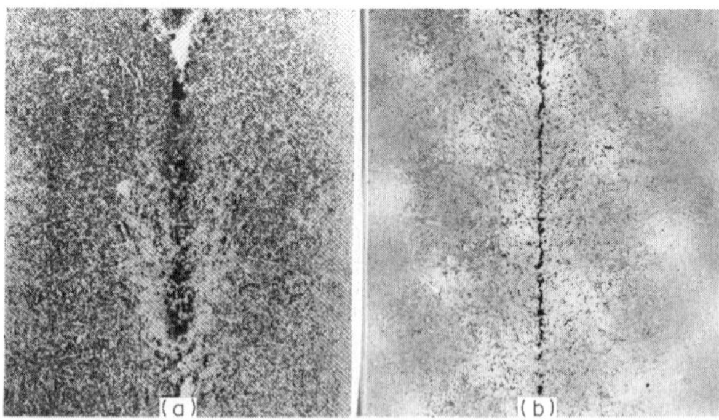

Figure 4
Sulfur prints showing axial segregation in (a) a billet and (b) a slab. Note that segregation in the billet is periodic (after Alberny and Birat 1976. © The Metals Society, London. Reproduced with permission)

The axial segregation typically found in slabs is shown in Fig. 4b. The major factors influencing the segregation are the stability of the strand and the superheat of the steel. The segregation results from bulging of the slab owing to inadequate roll containment close to the bottom of the pool. The bulging causes enriched liquid to be drawn down into the center of the slab where it quickly freezes. These effects are most pronounced in slabs with a predominantly columnar structure. Core cracks may also be generated by the same bulging mechanism. Methods of combatting these problems are to provide better roll containment, use electromagnetic stirring methods, decrease casting speed or increase spray water.

See also: Casting of Metals; Cast Structure of Alloys

Bibliography

Alberny R, Birat J P 1976 Electromagnetic stirring and product quality. *Continuous Casting of Steel.* Metals Society, London, pp. 116–24

Andrews K W, Gomer C R 1967 Use of radioactive isotopes in solidification studies of steel. *The Solidification of Metals.* Iron and Steel Institute, London, pp. 363–69

Davies G J 1973 *Solidification and Casting.* Applied Science, London

Flemings M C 1974 *Solidification Processing.* McGraw-Hill, New York

Lait J E, Brimacombe J K 1982 Solidification during continuous casting of steel. *Trans. ISS (AIME)* 1: 1–13
McGannon H E (ed.) 1971 *The Making, Shaping and Treating of Steel*, 9th edn. US Steel Corporation, Pittsburgh, Pennsylvania, pp. 546–52, 584–98, 706–26
Moore J J 1980 Review of axial segregation in continuously cast steel. *Iron Steelmaker* 7(10): 8–16
Mori H, Tanaka N, Sato N, Hirai M 1972 Macrostructure of and segregation in continuously cast carbon steel billets. *Trans. Iron Steel Inst. Jpn.* 12: 102–11
Myoshi S 1976 Influence of operating conditions and mechanical factors on center segregation of slabs. *Continuous Casting of Steel.* Metals Society, London, pp. 286–91
Ohno A 1976 *The Solidification of Metals.* Chijin Shokan, Tokyo
Pesch R, Etienne A 1977 Structure and defects in conventional ingots. *Casting and Solidification of Steel.* IPC Science and Technology, Guildford, pp. 170–211
Singh S N, Blazek K E 1974 Heat transfer and skin formation in a continuous-casting mould as a function of steel carbon content. *Open Hearth Proceedings*, Vol. 57. The Metallurgical Society (AIME), Warrendale, Pennsylvania, pp. 16–36
Van Drunen G, Brimacombe J K, Weinberg F 1975 Internal cracks in strand-cast billets. *Ironmaking Steelmaking* 2: 125–33
Weinberg F, Lait J, Pugh R 1979 Solidification of high-carbon steel ingots. *Solidification and Casting of Metals.* Metals Society, London, pp. 334–39

J. K. Brimacombe

Injection Molding of Advanced Ceramics

In the ceramics industry injection molding is used as a forming method for complicated shapes (Mangels and Trela 1983). In this process, granules consisting of powdered ceramics, a polymeric binder and appropriate additives are fed to the molding machine and heated until the binder melts. The melt is then forced into the mold. After cooling and solidification the mold is opened and the shaped product is removed.

Examples of injection-molded ceramic products include ferrite coils, Al_2O_3 thread guidance forms, oxide insulators and SiC and Si_3N_4 turbine blades (see *Injection Molding of Thermoplastics: Mold Filling*; *Injection Molding of Thermoplastics: Structural Foams*).

1. The Process

The process may involve either a thermoplast or a duroplast as the polymeric binder. With a thermoplast, solidification of the melt occurs on cooling; with a duroplast, a hardener is added to the feed mixture and solidification results from a binder–hardener reaction that occurs at elevated temperature. Figure 1 shows the viscosity–temperature relation for each type of binder. The reversibility of a thermoplast in terms of solidification makes recycling the reject a possibility but can lead to deformation of the compact during the subsequent burnout stage (see Sect. 3). For the duroplast process, solidification is irreversible and no deformation can occur during reheating, but the time for hardening is relatively long and the mold temperature is relatively high, and the reject cannot, of course, be recycled. The thermoplast process is the most widely used for ceramics and so this is discussed here.

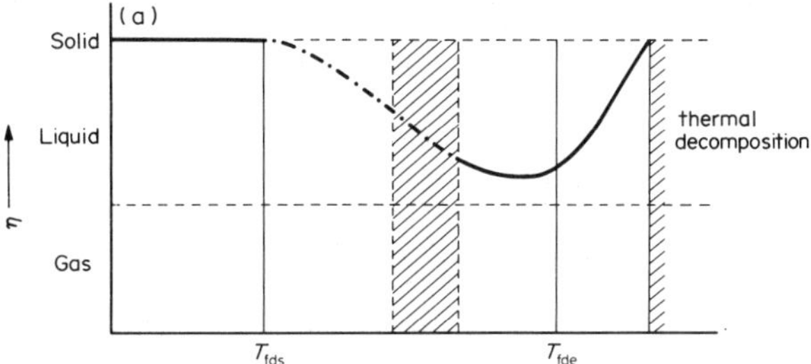

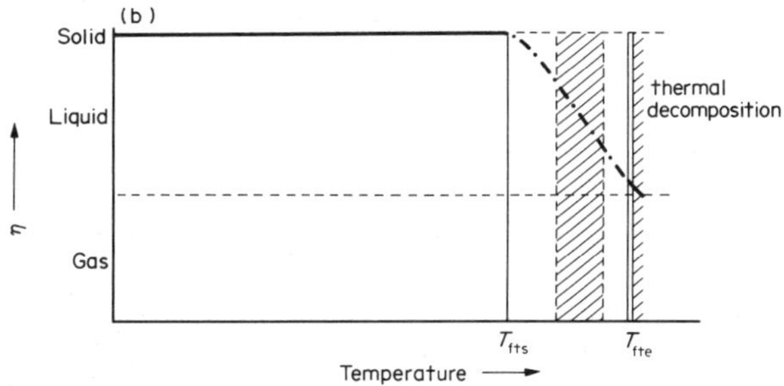

Figure 1
Fluidity of (a) a duroplast and (b) a thermoplast: T_{fds} (T_{fts}) and T_{fde} (T_{fte}) are the start and end temperatures, respectively, of fluidity of the duroplast (thermoplast); –·–, range of reversibility

The principle of the operation of the injection molding machine is illustrated in Fig. 2. Granules are fed to the machine and then transported by a screw or a plunger to the heating chamber, from which the fluid hot mixture is injected into the mold. In the mold the melt passes through a mean channel (sprue) to different side channels (runners) ending in cavities of the desired shapes. After filling the mold, the mass cools and solidifies; the mold is then opened and the products, runner and sprue are ejected. Nine major controls are available to the molding machine operation, and these govern (a) the amount of material introduced into the cylinder, (b) the pressure applied to the plunger, (c) the plunger speed, (d) the temperature of the heating cylinder (and nozzle if controlled), (e) the temperature of the mold, (f) the plunger forward time, (g) the mold closed time, (h) the clamping force, and (i) the mold open time.

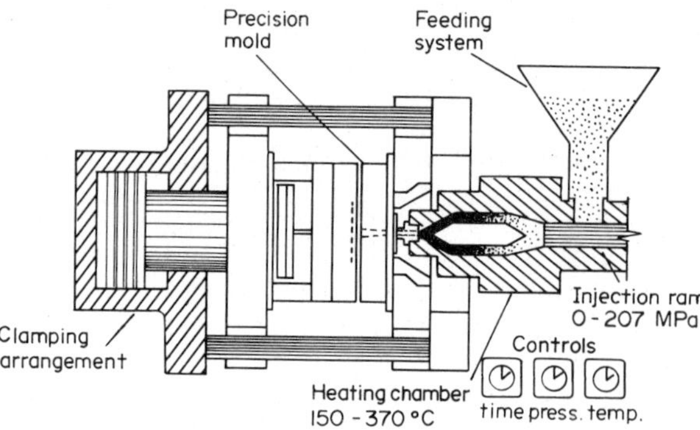

Figure 2
An injection molding machine (after Thayer et al. 1959)

Other controls may be provided, depending on the design and intended use of the machine.

Figure 3 shows the mold pressure cycle during one stroke. The subdivision of the molding cycle into the steps of Fig. 3 is based on consideration of the pressure along the material axis in the mold. The dead time (1) is the time before the material starts flowing into the mold. In the next period (2) the material fills the cavity. As soon as the cavity is filled, the pressure increases rapidly and packing (3) occurs. At this stage flow into the cavity is at a very slow rate. The compressibility of the material allows some flow during pressure buildup. Also, as the material cools in the mold it contracts, allowing more to enter. When the ram is returned, the gate is usually still relatively fluid in the case of a large gate and a thick part. Owing to pressure differences, reverse flow or discharge (4) occurs. Further cooling causes sealing (5) at the gate of the cavity, and no further flow in or out of the cavity is possible. With smaller gates, sealing occurs more rapidly, and common practice is to time the ram to return after sealing, thus preventing discharge. The remaining pressure in the cavity decays during the sealed cooling (6) until the mold is opened. Usually, there is pressure remaining in the cavity just before part ejection. Too high a residual mold pressure can cause sticking, scoring or cracking.

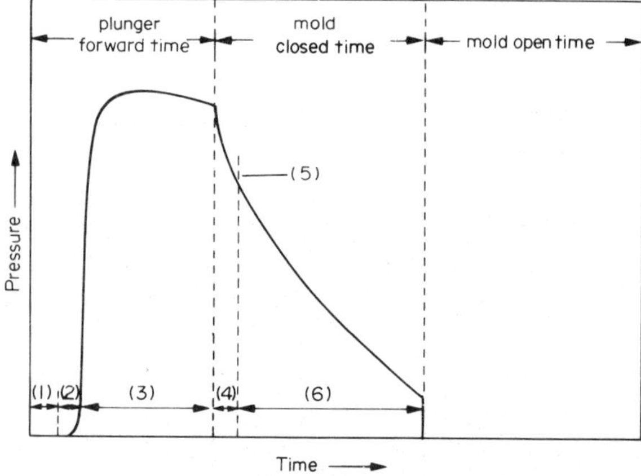

Figure 3
Mold pressure cycle during injection molding (after Thayer et al. 1959)

Compared with the dry pressing technique, injection molding can be used for making complicated forms owing to the high fluidity of the mass during the filling of the mold. In dry pressing, the "fluidity" of the granules disappears. Another advantage of the high fluidity of the injection molding mixture is the homogeneity of the compact. In contrast with dry pressing, the morphology of the feed granules disappears completely. Relative to dry pressing, the cycle time is very long and the cost of the mold is very high. Also, the binder content of injected compacts is much higher than that of dry-pressed ones.

2. *Feed Granules*

2.1 *The Powder*

The prepared powder has to be made suitable for the forming process; for example, for pressing, a granulated powder with good flow and filling properties is necessary. If the forming is done by injection molding the molding machine is fed by a homogeneous mixture of powder and a thermoplast, and the mixture must have good flow properties. The injection-molded compact consists of powder particles with organic materials completely filling the pore volume. For good injection molding properties (i.e., a relatively low viscosity at the injection temperature), the content of organic materials must exceed that necessary to fill the pore volume (Mostetzky and Hennicke 1975, Markhoff et al. 1983).

To be able to use a low binder content, the pore volume of the compact of the powder must be low. It is preferable, therefore, to start with a deagglomerated powder. This can be realized more easily with a relatively coarse powder.

2.2 *The Thermoplast*

Different systems of organic materials can be used as the thermoplast. In general, long-chain molecules and polymers result in strong compacts, but the fluidity at the injection temperature is reduced. On the other hand, short chains result in an easier injection process but lower strength. It is also necessary that the binder should disappear completely during the subsequent thermal decomposition (burnout).

The most important thermoplast systems are: (a) the paraffin types and waxes (Peltsman Corporation 1975), for which a fatty acid such as oleic acid can

Table 1
Firing of injection-molded compacts

Temperature range (°C)	Reactions	Possible problems
20–200	Softening, no decomposition	deformation
200–350	Exothermic decomposition of 80% of the organic materials	deformation, bubbles, cracks
350–450	Very strong exothermic decomposition of the rest of the binder	cracks, chipping off, breaks
450–700/1200	Some presintering	sticking

be used as a wetting agent; (b) resins from vinyl compounds such as polypropylene and polystyrene, in combination with lubricants and wetting agents such as microwaxes or phthalate esters (Litman et al. 1976).

2.3 The Feed Mixture

The powder and the plastic must be mixed well to allow the plastic polymer chains to attach themselves to the powder particles. An effective way of mixing is hot kneading the powder with a melt of the plastic under pressure. An advantage of this method is that the forces applied to the dough are effective in deagglomerating the powder. After kneading and cooling, the mixture has to be cut in a cutting mill into pieces of about 2–4 mm diameter. The granules produced must have good flow properties. Therefore the dust content must be low and the granules may not stick to each other, which happens if the plasticizer content is too high.

3. Firing of the Injection-Molded Compacts

The firing of the compact to form a ceramic product can be divided in two steps: thermal decomposition (burnout) of the binder and sintering. The former is the most critical and has to be done in an atmosphere containing sufficient oxygen for complete removal of the carbon and carbon compounds. Table 1 lists the reactions occurring at different temperatures, along with possible problems that may arise. To prevent these problems occuring it is important to heat the compact slowly to 450 °C (Johnsson 1983), the exact heating rate depending on the binder material and the thickness of the compacts.

In general, the sintering step is not critical. Owing to a higher and more uniform density before sintering compared with that of dry-pressed compacts, the sintering temperature can be lower and the microstructure after sintering more homogeneous.

See also: Ceramics Process Engineering: An Overview

Bibliography

Johnsson A, Carlström E, Hermansson L, Carlsson R 1983 Rate-controlled thermal extraction of organic binders from injection molded bodies. *Adv. Ceram.* 9: 241–45

Litman A M, Schott N R, Tozlowski S W 1976 Rheological properties of highly filled polyolefin ceramic systems suited for injection molding. *Soc. Plast. Eng. Tech. Pap.* 22: 549–51

Mangels J A, Trela W 1983 Ceramics by injection molding. *Adv. Ceram.* 9: 220–33

Markhoff J C, Mutsuddy B C, Lennon J W 1983 A method for determining critical ceramic powder volume concentration in the plastic forming of ceramic mixes. *Adv. Ceram.* 9: 246–50

Mostetzky H, Hennicke W 1975 Spritzgiessen von unbildsamen keramische Pulvern am Beispiel von Quarzsand. *Ber. Dtsch. Keram. Ges.* 2: 25–52

Peltsman Corporation 1978 *Preparation of Ceramic Slurry with Paraffin Type Binder.* Peltsman Corporation, Minneapolis, Minnesota

Thayer B S, Mighton J W, Dahl R B, Beyer C E 1959 Injection molding. In: Bernhardt E C (ed.) 1959 *Processing of Thermoplastic Materials.* Van Nostrand Reinhold, New York, pp. 308–14

L. J. Koppens

Injection Molding of Thermoplastics: Mold Filling

A variety of molding processes are used commercially, including injection, compression, transfer, blow and rotational molding. The particular process which is employed depends largely on the design, final application and requirements of the finished product. Of these molding processes, the most important commercially is injection molding. In this process, raw material is melted during the plastication stage and then injected under pressure into a mold. In order to achieve solidification of the product, the melt is injected into a cold mold cavity in the case of thermoplastics or into a heated mold cavity in the case of thermosets. Cooling or heating continues until the material has solidified sufficiently to permit ejection of the molded article from the mold without damage. The most commonly used injection molding systems are of the reciprocating screw type, although ram injection machines are not uncommon.

1. Injection Molding Sequence

The injection molding machine consists of two basic components: the injection unit and the clamping unit. The injection unit heats and melts the raw material and transfers it under pressure into a mold which is held by the clamping unit. Melting is achieved with the help of external heaters and, in most cases, by mechanical heating induced upon shearing and compressing the granular raw material. A typical operating sequence starts with feeding of the material by gravity from a hopper, the screw being in the forward position. The rotation of the screw plasticizes the material and conveys it forward in the screw barrel. Accumulation of the plasticized material at the front of the screw tip forces the screw backwards. A check valve prevents backflow over the screw flights.

When the injection unit is activated, the screw moves forward, as a piston, pushing the melt from the injection chamber through a nozzle into the mold. The hot melt flows into an empty cavity until the mold is filled. More polymer is then packed into the cavity to compensate for shrinkage due to solidification. The cycle continues until enough rigidity is achieved to permit ejection of the part without damage. The cycle is repeated to produce the next part.

2. Melt Delivery Channels: Sprues, Runners and Gates

A diagram of a typical injection mold arrangement is shown in Fig. 1. After leaving the nozzle, the melt generally flows through the sprue bushing and the runner system and finally enters the cavity through the gate. The opening in the sprue bushing is larger than the adjacent opening in the nozzle, and the sprue is tapered to facilitate removal of the plastic after solidification. Moreover, the outlet of the sprue should not be smaller than the runner diameter at the point where the two meet. The runner is usually circular or trapezoidal (if in one plate only). To minimize pressure loss, the runner cross section should be large, but not so large as to require a substantial rise in the cycle time due to cooling requirements.

When multiple cavities are used, it is important to ensure that all cavities fill at the same rate and that the material in the different cavities experiences comparable thermal histories. Sometimes this is achieved by a balance between the dimensions of the runner segments and the gates leading to the various cavities.

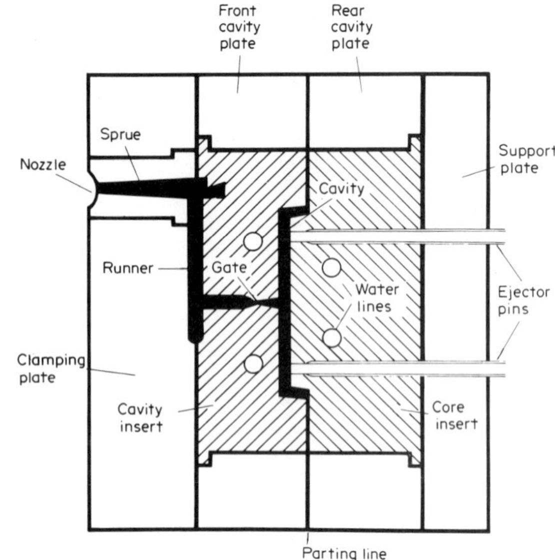

Figure 1
Schematic diagram of injection mold and melt delivery channels

In order to reduce regrinding and other materials handling requirements, hot runner molds have been designed to keep the material in the runner in the molten state.

The gate connects the runner to the mold cavity and controls the flow of the plastic melt into the cavity. It also provides a mechanism for separating the molded article from the runner system. Gates are either restricted (pinpoint gates) or large. Restricted gates are circular in cross section and generally do not exceed 1.5 mm in diameter, although they can be significantly larger for viscous melts. Large gates are square or rectangular, and typical side dimensions are ~6 mm. Gates may be also described by location: edge, tab, back, submarine, sprue, diaphragm, ring, fan and flash gates. Ideally the gate should be as short as possible, in order to minimize pressure loss and to avoid the development of flow lines. A long gate could also lead to sink marks and voids in the molding and causes premature freezing. On the other hand, some length is required to prolong the life of the cavity.

Although more accurate estimates may be made by using more complex methods, it is possible to obtain reasonable estimates of flow rates and pressure requirements in delivery channels by making the following two assumptions: first, that the flow is fully developed, isothermal and incompressible, and second, that the polymer melt behaves as a power law fluid, that is,

$$\tau = m\,\dot{\gamma}^{n} \qquad (1)$$

where τ is the shear stress, $\dot{\gamma}$ is the shear rate, and

m and n are appropriate power law constants. Thus, the following equations are applicable for the geometries indicated (Tadmor and Gogos 1979).

For a circular tube of radius r and length l,

$$Q = \left(\frac{\pi r^3}{s+3}\right)\left(\frac{r\Delta p}{2ml}\right)^s \qquad (2)$$

where $s = 1/n$, and Q is the volumetric flow rate.

For a converging conical channel of large radius r_1, small radius r_2 and length l,

$$Q = \left(\frac{n\pi r_1^3}{3n+1}\right)\left(\frac{r_1 \Delta p a_{13}}{2ml}\right)^s \qquad (3)$$

$$a_{13} = \frac{3n(r_1/r_2 - 1)}{r_1/r_2[(r_1/r_2)^{3n} - 1]} \qquad (4)$$

For a half-circle channel (a semicircle with a rectangular extension on the top) of radius r, depth d and length l,

$$Q = \frac{2M_0 r d^3 \Delta p}{12\mu l} \qquad (5)$$

where M_0 is a shape factor depending on the channel dimensions and μ is a Newtonian viscosity.

For a rectangular channel of length l, breadth b and depth h,

$$Q = \frac{bh^2}{M_1}\left[\frac{bh\Delta p}{2lm(b+h)}\right]^{1/n} \qquad (6)$$

where M_1 is a shape factor that depends on the channel dimensions.

3. The Injection Molding Cycle

After the material has traversed the delivery channels and the gate, the process of filling the cavity begins. From this point, the injection molding cycle may be considered to consist of three stages: filling, packing and cooling. During filling, the polymer melt is introduced into the empty cooled mold. After filling is complete, more material is packed into the cavity at high pressure to compensate for shrinkage as the material cools. Cooling continues after packing is completed. During the cooling state, the material solidifies and the pressure decreases. A typical pressure–time cycle during injection molding is illustrated in Fig. 2. In general, the cooling stage represents the longest segment of the cycle, usually accounting for >60% of the overall cycle.

Normally, commercial injection molding involves mold cavities of complex geometry, sometimes incorporating inserts and a variety of obstructions to direct the flow. Gates may be located at various positions and can have a multitude of shapes and dimensions, as mentioned earlier. In some cases, multiple gates are employed to fill a single cavity. All these considerations may have a profound influence on the dynamics of the molding process and on the ultimate characteristics of the molded articles. In particular, some molding arrangements lead to the formation of weld lines in regions where melt fronts intersect inside the cavity (e.g., after flowing around inserts and as a result of filling from multiple gates). Unless carefully manipulated and controlled, weld lines can represent very weak areas in the molding, leading to premature failure.

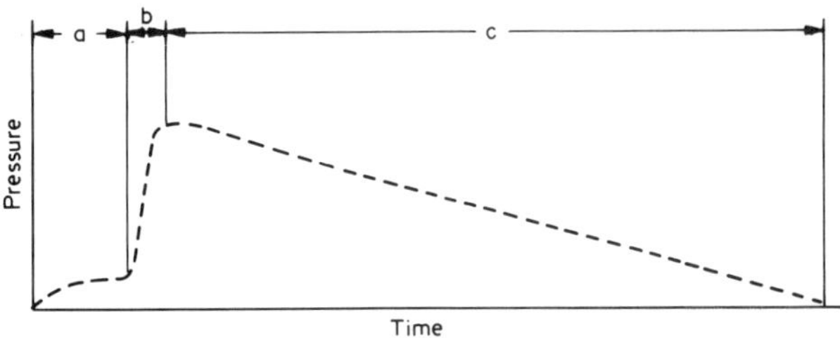

Figure 2
Typical pressure–time cycle for variation of pressure at the gate in injection molding: (a) filling; (b) packing; (c) cooling

Ideally, engineering analysis, mathematical modelling or computer simulation of the injection molding process should recognize the above factors and should thus permit the treatment of flow and heat transfer phenomena in complex injection molding cavities of the types described above. Modern finite element numerical techniques make this possible. Even with the finite element method, the polymers, molds and phenomena involved are so complex that further simplification is usually desirable.

One common approach involves the mold network method. According to this method, the complex mold cavity is divided into a number of units of reasonably simple geometry (e.g., rectangular, circular or cylindrical segments). The solutions for the melt behavior in each of the simpler units are treated separately and are sometimes readily available in graphical, numerical or analytical forms. To obtain the overall solution for the complete cavity, the solutions for the individual components of the network are coupled and dovetailed to ensure the systematic integration of the results from each segment (Hieber 1982).

4. The Filling Stage

The filling stage is concerned with the unsteady-state flow of a hot, compressible, viscoelastic melt into an empty cavity which is held at a temperature below the solidification temperature of the melt. In this sense, the process can be fully described in terms of the equations of conservation of mass, momentum and energy, coupled with appropriate thermodynamic relationships and a set of constitutive relations which describe the behavior of the material under the influence of stress and thermal fields. In general tensorial formulation, the conservation equations are as follows.

Continuity:

$$\frac{D\rho}{Dt} = -\rho(\nabla \cdot \boldsymbol{v}) \tag{7}$$

Momentum:

$$\rho\frac{D\boldsymbol{v}}{Dt} = -\nabla p - (\nabla \cdot \boldsymbol{\tau}) + \rho \boldsymbol{g} \tag{8}$$

Energy:

$$\rho c_V \frac{DT}{Dt} = -(\nabla \cdot \boldsymbol{q}) - T\left(\frac{\partial p}{\partial T}\right)_\rho (\nabla \cdot \boldsymbol{v}) - (\boldsymbol{\tau} : \nabla \boldsymbol{v}) \tag{9}$$

where ρ is the density, $\boldsymbol{v}$ is the velocity, p is pressure, $\boldsymbol{\tau}$ is the stress tensor, q is the heat flux, T is the temperature, and t is time.

To obtain a solution of the above equations, each stage of the process is associated with a set of initial and boundary conditions coupled with simplifying assumptions.

There are a variety of constitutive rheological equations that may be employed to describe the viscous or viscoelastic behavior of the polymer melt during the process. The most commonly employed rheological models in the injection molding process include the following: Newtonian liquid, power law model, second-order fluid (Pilate and Crochet 1977) and Leonov equation (Isayev and Hieber 1980). Whereas some of the models describe only the viscous behavior of the fluid, other models reflect some of the viscoelastic characteristics of the material.

In the filling stage, which starts with an empty cavity, a free surface exists at the advancing melt front. Thus, in a rectangular cavity, the part filled with the melt may be divided into three regions, as shown in Fig. 3. In region 1, near the gate, the flow is radial until the flow front reaches the wall. Away from the gate, region 2, the flow is characterized by a dominant axial velocity component. In region 3, called the front region, the melt, channelling through the center, tends to move towards the cavity wall in what is referred to as the "fountain flow." Although it is possible to analyze the details of the flow structure in the cavity by dealing with the appropriate three-dimensional flow equations, it is much simpler to treat the above regions separately and to solve the corresponding one- or two-dimensional flow equations.

For relatively thick mold cavities or for very short filling times, it is possible to obtain a reasonable analysis of the filling stage by assuming isothermal flow conditions (Berger and Gogos 1973). This is mainly due to the low thermal conductivity of the polymer melt. However, in general, temperature changes due to mold cooling and viscous dissipation must be taken into consideration. Since only moderate pressures are encountered during the filling stage, it is generally reasonable to assume incompressible flow.

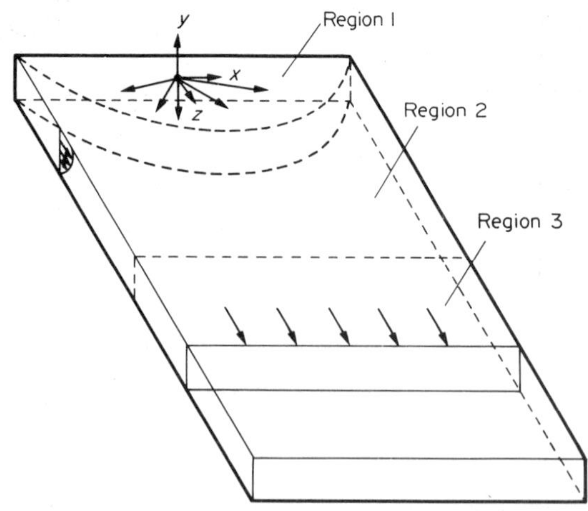

Figure 3
Flow regions during the filling of a rectangular injection molding cavity

Both finite element and finite difference numerical techniques may be employed to obtain computer solutions of the relevant equations. The former methods are more complex, but they are well suited for dealing with complex mold cavities and the free surface flow at the melt front. An interesting approach for dealing with the fountain flow near the free surface is based on the finite difference method in conjunction with the marker-and-cell technique (Kamal and Lafleur 1982).

The analysis of the filling stage yields information regarding the progression of the melt front and the distributions of velocity, pressure and temperature. Moreover, it supplies data regarding the development of the solid skin layer during the filling stage. Typical model predictions relating to the progression of the melt in the cavity and the pressure distribution during mold filling, for a high-density polyethylene resin, are shown in Figs. 4 and 5. By employing models incorporating appropriate rheological constitutive equations and expressions for crystallization kinetics, it is possible to estimate the distributions of crystallinity and frozen stresses in the system. In turn, this leads to reasonable estimates of the distributions of mechanical and optical properties in the molding.

5. *The Packing Stage*

The approach for the simulation of the packing stage is similar to that for the filling stage. For the packing stage, the deceleration motion of the flow reinforces the importance of the unsteady-state terms, and they

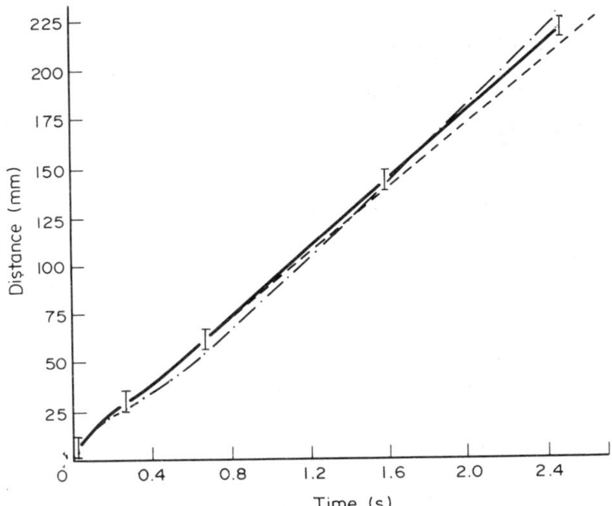

Figure 4
Comparison of model prediction with experimental data for the progression of the melt front during injection molding of high-density polyethylene: —— experimental transducer; --- experimental cinematography; —·— theoretical (after Doan 1974)

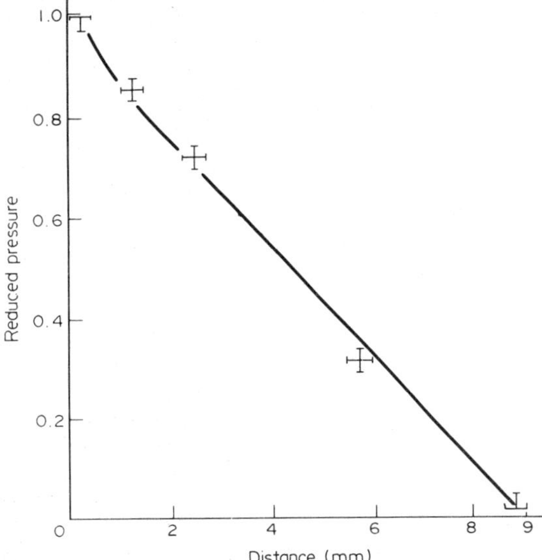

Figure 5
Comparison of model predictions with experimental pressure distribution at the end of filling a rectangular cavity with high-density polyethylene: ┼ experimental; —— theoretical (after Doan 1974)

should not be neglected. During packing, the contribution of the unsteady-state terms will tend to be larger than that of the convective terms in the equation.

The dominant feature in the packing stage is the compressibility of the polymer melt. This emphasizes the need for selecting an appropriate equation of state to describe the pressure–volume–temperature relationships for the melt. A number of empirical equations of state have been employed in conjunction with the processing of polymer melts (Tadmor and Gogos 1979). The following two equations have been found useful for injection molding analysis.

$$(p + \pi)(\bar{V} - w) = RT \tag{10}$$

where p is the pressure, T is the temperature, R is the gas constant and π and w are empirical constants (Spencer and Gilmore 1951).

$$\rho = \rho\infty + \left(\frac{\partial \rho}{\partial p}\right)_{p=0} T + (a + bT)p + \tfrac{1}{2}(c + dT)p^2 \tag{11}$$

where ρ is the density, $\rho\infty$ is the density at 0 K and zero pressure, and a, b, c and d are adjustable parameters (Kamal and Levan 1973).

6. *The Cooling Stage*

After the filling and packing of the cavity, the gate freezes and no additional material enters the cavity. Since fluid motion is negligible or absent during this stage, the problem is reduced to the transient heat transfer by conduction in the polymer melt trapped between the mold walls. For thin cavities, the one-dimensional conduction problem takes the form

$$\rho c_p \frac{\partial T}{\partial t} = \frac{\partial}{dx}\left(k \frac{\partial T}{\partial x}\right) + \frac{T}{V}\left(\frac{\partial V}{\partial T}\right)_p \frac{\partial p}{\partial t} \tag{12}$$

where c_p is the specific heat capacity, V is the specific volume and k is the thermal conductivity. All other terms are as defined earlier. The following analysis applies to partly crystalline polymers, which crystallize upon solidification. The treatment of amorphous, noncrystallizing polymers may be considered as a special case of this analysis.

In the cooling process of a crystallizing polymer, the above equation is applied to both the solid and melt phases, allowing the thermal properties to vary continuously with temperature and pressure. In order to allow for the effect of latent heat of crystallization, the latent heat is included in the specific heat as follows:

$$c_p = c_p' + \lambda \frac{dx}{dt} \tag{13}$$

where c_p is the effective specific heat capacity of the material, c_p' is the specific heat of the material at the same temperature in the absence of crystallization, λ is the latent heat of crystallization and x is the weight fraction crystallinity in the system. In fact, c_p

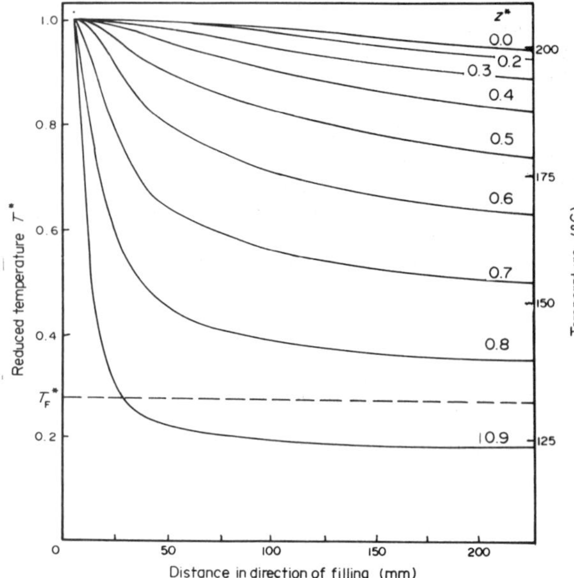

Figure 6
Calculated temperature distribution in a rectangular cavity at the end of filling with high-density polyethylene: z^*, dimensionless transverse distance (center = 0 and wall = 1.0); T_F^*, freezing temperature (after Doan 1974)

is the property normally measured during a crystallization experiment in the differential scanning calorimeter.

The rate of crystallization and the evolution of crystallinity during a nonisothermal crystallization experiment may be based on the extension of Avrami's isothermal treatment to nonisothermal systems (Nakamura et al. 1972):

$$x(t) = 1 - \exp\left[-\left(\int_0^t \kappa^- |T(t)|\, d\tau\right)^n\right] \qquad (14)$$

where n is the Avrami index determined from isothermal experiments and $\kappa^-(T)$ is connected with the crystallization rate constant of isothermal crystallization $\kappa(T)$ by the relation

$$\kappa^-(T) = [\kappa(T)]^{1/n} \qquad (15)$$

Figure 6 shows typical model predictions of temperature distributions in a rectangular cavity during the cooling stage for a high-density polyethylene resin.

See also: Injection Molding of Thermoplastics: Structural Foams; Injection Molding of Thermoplastics: Structure Development; Polymer Processing: Classification

Bibliography

Berger J L, Gogos C G 1973 A numerical simulation of the cavity filling process with PVC in injection molding. *Polym. Eng. Sci.* 13: 102–12

Doan P H 1974 Injection molding of thermoplastics in rectangular cavities. M.Eng. thesis, McGill University, Montreal

Dym J B 1979 *Injection Molds and Molding—A Practical Manual.* Van Nostrand Reinhold, New York

Gilmore G D, Spencer R S 1951 Photographic study of the polymer cycle in injection molding. *Mod. Plast.* 28(8): 117–85

Hieber C A 1982 Coupled-flow-path method for simulating the injection-molding filling Stage. *Soc. Plast. Eng., Tech. Pap.* 28: 356–58

Isayev A I, Hieber C A 1980 Toward a viscoelastic modelling of the injection molding of polymers. *Rheol. Acta* 19: 168–82

Kamal M R, Kenig S 1972 The injection molding of thermoplastics, Pt. I: Theoretical model. *Polym. Eng. Sci.* 12: 294–301

Kamal M R, Lafleur P G 1982 Computer simulation of injection molding. *Polym. Eng. Sci.* 22: 1067–75

Kamal M R, Levan N T 1973 Comparison between some empirical equations of state for polymers. *Polym. Eng. Sci.* 13: 131–38

Kamal M R, Moy F H 1981 Estimation and measurement of the distribution of tensile moduli in injection molded thermoplastics. *Chem. Eng. Commun.* 12: 253–65

Kuo Y, Kamal M R 1976 The fluid mechanics and heat transfer of injection mold filling of thermoplastic materials. *AIChE J.* 22: 661–69

Lafleur P G, Kamal M R 1981 Computer simulation of thermoplastic injection molding. *Adv. Plast. Technol.* 1(4): 8–13

Nakamura K, Watanabe T, Katayama K, Amano T 1972 Some aspects of nonisothermal crystallization of polymers, I: Relationship between crystallization temperature, crystallinity, and cooling conditions. *J. Appl. Polym. Sci.* 16: 1077–91

Pilate G, Crochet M J 1977 Plane flow of a second-order fluid past submerged boundaries. *J. Non-Newtonian Fluid Mech.* 2: 323–41

Rubin I I 1973 *Injection Molding: Theory and Practice.* Wiley, New York

Spencer R S, Gilmore G D 1951 Some flow phenomena in the injection molding of polystyrene. *J. Colloid Sci.* 6: 118–32

Tadmor Z, Gogos C G 1979 *Principles of Polymer Processing.* Wiley, New York

M. R. Kamal

Injection Molding of Thermoplastics: Structural Foams

The thermoplastic structural foam (TSF) process is a derivative plastics injection molding process for fabricating shapes having integral high-density surface layers or skins, nearly uniform cellular center sections or cores, a gradual transition in density from core to skin, and high stiffness–weight ratio compared with shapes conventionally molded of the same unfoamed material (Fig. 1). Typically, TSF shapes have bulk densities of ~50–90% of the bulk density of the unfoamed polymer. The ratio of foamed resin density to unfoamed resin density is known as the

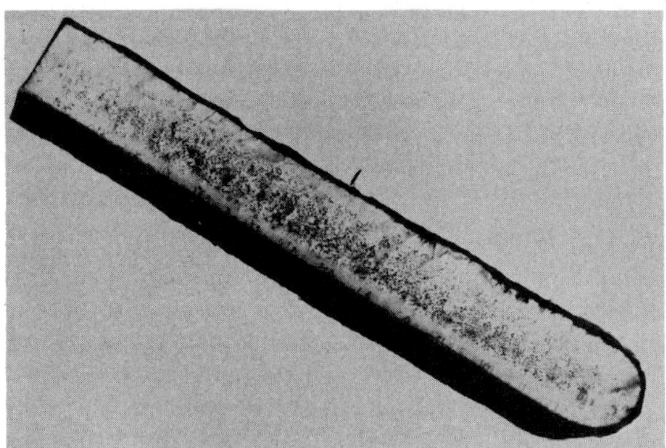

Figure 1
Cross section of a thermoplastic structural foam product

reduced bulk density, $\phi = \rho_f/\rho_0$. As a first approximation, the mechanical performance of TSF shapes can be related to the reduced bulk density.

In order to obtain optimum mechanical performance levels, TSF shapes are designed to have wall thicknesses no less than 3 mm and typically more than ~4 mm. The practical maximum thickness is governed by economics, affected for example by the time required to cool the foamed shape to a temperature where removal from the mold cavity will not cause postprocessing distortion (shrinkage, warpage and reheat blowing). It also depends upon the geometry of the article and competitive processing options, such as rotational molding, urethane RIM foam molding and blow molding. If the article is cylindrical or spherical, it can be as much as 100 mm in thickness.

Essentially, TSF processing requires dissolution of a carefully metered amount of a permanent or volatile gas, such as nitrogen, chlorofluorocarbon or carbon dioxide, in a softened or molten polymer under processing conditions approximating those for conventional injection molding. The foamable resin is then transferred from a high-pressure region (such as an accumulator of an injection molding machine) to a hydraulically clamped two-part injection mold cavity having the general dimensions of the shape to be fabricated but with a volume somewhat greater than the volume of the foamable resin being transferred. At reduced pressures and excess volume, the gas evolves into microbubbles, the foaming resin expands to fill the mold and the shape thus formed is cooled to a point where it retains the shape of the mold cavity.

1. History

Deliberate attempts to form a cellular thermoplastic composite structure by injection molding a gas-containing plastic apparently began in the mid-1950s with Dow Chemical Corporation's Frostwood process, where polystyrene was foamed with methylene chloride. In the early 1960s, Foster Grant, W. R. Grace, Union Carbide Corporation, and Bayer AG worked on various aspects of foam injection molding. The recognition of low mold cavity pressures during filling and expansion of the foaming resin led to the development of an isolatable melt accumulator in which the foamable molten resin could be stored prior to injection. The combination of a fixed plasticating extruder, the separate melt accumulator and a large surface area but relatively low-tonnage hydraulic mold clamp enabled rapid market development of very large fabricated TSF parts in the early 1970s. Computer housings of 45 kg or more have been fabricated, but there is no minimum practical size; 4.5 g paintbrush handles are an example of a small product.

2. Applications

Thermoplastic structural foam shapes have successfully competed with conventional materials in a number of ways: with wood when bulk at reduced weight is required, or when ornateness in carvings must be faithfully repeated; with cast metal when weight reduction, corrosion resistance or ease of assembly is needed; with sheet metal when stiffness is the objective; with sheet molding compound when reduced weight, reduced cost or many parts are needed; with assembled metal parts when labor costs require it; and with conventional injection-molded plastic parts when stiffness, low stress or an absence of sink marks is required. In 1979, about 20% of the TSF shapes fabricated in the USA went into utility conduits and closures and another 10% into each of the following markets: building and construction, furniture, materials handling, retail and janitorial, and business machines and computer housings. High-density polyethylene was used in 40% of these applications, with polystyrene and polypropylene in 20% each. Modified polyphenylene oxide was used in 10% and potentially important TSF materials such as ABS, polycarbonate and nylon in the rest.

3. Processing Equipment

Two TSF processing schools of thought developed in the early 1970s. The European school generated foaming gas primarily through decomposition of a thermally unstable chemical, such as azodicarbonamide, *p,p'*-oxybis(benzenesulfonylhydrazide) or 5-phenyltetrazole. These chemical foaming agents were dry-blended with powdered or granular polymer before addition to near-conventional injection molding machines. The machines were modified to include a positive shutoff nozzle between accumulator and mold cavity and a nonreturn ring on the reciprocating plasticating screw. These changes were necessary to prevent the gas generated by decomposition of the foaming agent from escaping from the polymer before injection.

The primary emphasis in the USA during this time was focused on the direct addition of nitrogen to the molten polymer through an orifice in the barrel wall at a low pressure zone along the screw. For obvious reasons, it was recognized both in Europe and in the USA that the heavy-walled TSF parts required considerably longer cooling times to achieve removal than did unfoamed injection-molded parts of the same material. This additional time could be used to advantage to plasticate and accumulate large quantities of molten foamable resin and this led to the development of an isolatable foamable-melt accumulator. The primary forces applied against the hydraulic system holding the mold closed are attributable to the gas pressure in the resin. These are typically 1–3 MPa, which are far below conventional packing pressures of 30–70 MPa for nonfoaming injection molding. As a result, the conventional hydraulic forces used to close the mold cavity against the melt pressure can be applied over a much greater mold surface area. This, coupled with the additional time available to accumulate foamable resin, allows for TSF part weights 10–50 times those for nonfoamed parts at the same hydraulic demands on the injection molding clamp.

In Europe, polymer was introduced to the mold cavity through a single nozzle. In order to spread the polymer throughout the mold cavity before foaming, very high pressures were required. As a result, the isolatable accumulator was fed through a nonreturn valve by a stationary plasticating extruder. Pressures in the region of 100 MPa are developed by applying nitrogen gas pressure of 20 MPa to hydraulic oil and then intensifying the resulting forces. Polymer transfer rates of up to 40 kg s^{-1} have been achieved with materials such as polystyrene. In the USA, the accumulator was the result of the requirement to introduce gas at a specific point along the extruder screw, in effect eliminating the traditional reciprocating motion of the screw. Further, multiple hydraulic nozzles were fed from the accumulator through a heated piping and manifold system. Accumulator pressures of 20–30 MPa were needed primarily to overcome the large pressure losses in the plumbing and this led to relatively slow injection speeds. In recent years, these limitations have for the most part been overcome through redesign of the manifold system.

The processes described above are considered to be low-pressure foaming processes. The lack of control of the foaming process once the accumulator is empty and the mold cavity sealed is considered to be the primary cause of uneven density distribution and varying skin–core thickness ratio across the fabricated shape, and the gas bubble surface splay or streaking so characteristic of low-pressure TSF parts. As will be discussed, a finite amount of pressure must be applied to the gas-containing resin to control foaming rate. There are many ways to achieve this. One is to essentially injection-mold the resin, then slowly expand the mold cavity. In another method, the mold cavity is pressurized with nitrogen or air just before injection, thus minimizing uncontrolled foaming at the entering wavefront of the injected polymer. In a third, the foamable resin is simultaneously injected with unfoamed resin. The latter forms a molten polymer balloon which is subsequently expanded by the foaming polymer in the center to fill the mold cavity. In a fourth, the mold cavity walls are heated to temperatures just below the polymer melt temperature, eliminating bubble freeze-off during injection and foam expansion. The walls are then rapidly cooled to remove the sensible heat of the polymer.

4. *Typical Resins and Foaming Agents*

Most conventional injection molding thermoplastic resins, either "neat" or containing additives, reinforcements and/or fillers, can be foam injection molded. The molecular weight of the resin should be relatively low to permit injection at relatively rapid rates, but high enough to minimize excessive bubble growth and agglomeration to large voids and abnormal in-cavity shrinkage during cooling. The resins should have sufficient melt strengths, as typified by film forming characteristics, thus minimizing bubble agglomeration and blowout during the expansion phase of the process. The resin should also be sufficiently thermally stable to minimize thermal degradation at the high temperatures and long times typical of the TSF process cycle. Materials should be stable for 50 000 K s or longer. These conditions also apply to additives (fire retardants, antioxidants, bubble stabilizers, colorants).

Typical physical foaming agents include nitrogen, carbon dioxide, chlorofluorocarbons (such as refrigerants 11 and 12), chlorinated hydrocarbons (such as methyl chloride) and the volatile hydrocarbons (e.g., propane or butanes). In order to achieve fine cellular products, admixtures such as inorganics (talc, silica) or sodium bicarbonate and citric acid are added as nucleants or cell formers.

Although chemical foaming agents can yield finer cell structures than physical gases, several stringent criteria must be met. Neither the chemical foaming agent itself nor any of its by-products should have an aggressive effect on the polymer, particularly at the long times and high temperatures typical of the TSF process. The agent should be relatively inexpensive at the characteristic dosages (normally <1 wt% and frequently <0.5 wt%) and must be in a form that allows for very fine dispersal in the molten or softened polymer. Normally this implies particle sizes of $<37\ \mu m$. The typical chemical foaming agent decomposes to produce nitrogen and inactive by-products. Regardless of the source, the foaming gas should be sufficiently soluble in the polymer at typical injection molding conditions to allow for dissolution under typical accumulator conditions. On the other hand, the gas should not be so soluble as to inhibit bubble formation and growth during the expansion phase in the mold cavity.

5. *Processing Characteristics*

To ensure capture of the foaming gas, the polymer must be molten or softened before decomposition of the chemical foaming agent or introduction of the physical foaming agent. The processing conditions everywhere ahead of the mold cavity must be such that essentially all the gas is dissolved in the polymer. According to Henry's Law, which has been shown to be valid up to relatively high pressures for many gases and commodity plastics:

$$S = S_0 p \exp(-E_s/RT)$$

where S is the solubility ($cm^3\,g^{-1}$ polymer at STP), p is the pressure (MPa), T is the temperature (K), R is the gas constant, S_0 is a preexponential constant and E_s is the activation energy for solubility. For a chemical foaming agent, the amount of gas generated per g of polymer is given as V_g(CBA), where V_g is the volume of gas generated per g of blowing agent and (CBA) is the actual dosage of blowing agent per unit weight of polymer. For physical foaming agents, V_g(CBA) is replaced by V_g', the amount of gas added directly to a fixed amount of polymer melt. The solubility equation can be solved for pressure:

$$P_{\text{equil}} = (V_g(\text{CBA})/S_0) \exp(E_s/RT)$$

Above this pressure, all the gas is soluble in the polymer melt. Similarly, below this pressure, gas begins to form into microbubbles, according to Rayleigh's Law.

For low-pressure foam molding, bubbles nucleate at accidental or deliberate sites in the foam and grow inertially until radii of about 100 μm are reached, at which point diffusion of dissolved gas to the growing bubble sites begins to control the growth rate. Initial bubble growth is significantly retarded by extremely small nucleating sites, very low diffusion coefficients and very low foaming agent concentration. Under certain conditions, bubble nucleation and growth can be several orders of magnitude slower than material transfer from accumulator to mold cavity. Typically, under optimal foaming conditions, the gas pressure causing the foaming melt to fill the mold cavity is in the region of 1–3 MPa, rather than the more typical injection pressure of 30–70 MPa found in conventional injection molding.

As the gas expands the polymer against the mold cavity walls, the surface layer begins to cool and the integral skin begins to form. It is thought that the final thickness of the skin depends upon the inhibition of bubble growth near the wall owing to the rapidly increasing viscosity, the partial collapse of formed bubbles owing to the temperature difference which allows for increasing solubility of the gas in the plastic near the mold cavity walls, and the internal, centerline gas pressure which allows the centerline bubbles to grow at the expense of the collapsing bubbles near the mold cavity walls. The internal gas pressure is seen to be directly proportional to the concentration of foaming agent in the plastic. Thus the dynamic balance between the internal gas pressure, which slowly falls with time as the polymer cools, and the building skins of solid or hardened polymer near the mold cavity walls determines the minimum time required to cool the shape so that the mold cavity can be opened without excessive shrinkage or postmold expansion. Note that this measure of cycle time differs considerably from the conventional concept of transient heat conduction in the thermoplastic which forms the basis for cycle time determination for unfoamed plastic. The rate of heat removal is dependent upon the thermal diffusivity of the material as well as the square of the part thickness. It has been shown that although the thermal conductivity of foam decreases with decreasing density, the thermal diffusivity actually increases with decreasing density, thus allowing foams to cool more quickly per unit thickness than unfoamed resins. On the other hand, TSF shapes are frequently substantially thicker than typical unfoamed shapes, and doubling the thickness increases the time to cool to a given temperature profile by a factor of four. In essence, then, structural foam parts require longer cycle times primarily because of internal gas pressure and only secondarily because of thicker cross sections.

In general, amorphous polymers exhibit thicker skins and more gradual demarcation between skin and foam core than do crystalline polymers. Also, amorphous materials exhibit more uniform cell structure in the foam core than do rapidly crystallizing materials, such as nylons. Polyolefin cell structure is intermediate. Highly filled materials are normally stiffer flowing than the equivalent neat materials. As a result, shapes of filled resins usually have somewhat higher reduced densities for the same foaming agent concentration and processing conditions than shapes of neat resins. However, the cell structure of filled resins is finer, probably due to the nucleating character of the filler. Since the process is characterized as one of high speed but low flow resistance, attrition of glass reinforcement is considerably less than in conventional injection molding. However, in TSF processing, the glass fibers tend to segregate in bands just below the surface in attitudes that are not necessarily parallel to the mold cavity walls. Therefore, the reinforcement of the structure does not occur at the maximum fiber stress in flexure, and the hoped-for enhancement of mechanical properties is not achieved.

6. Mechanical Performance

For typical low-pressure foamed shapes, it has been demonstrated repeatedly that mechanical design properties such as flexural strength and modulus, tensile strength and modulus and, for most materials, fatigue strength and creep resistance can be expressed in terms of the overall bulk density of the material and the property of the unfoamed material as:

$$X_{\mathrm{f}} = X_0\,(\rho_{\mathrm{f}}/\rho_0)^n = X_0\,\phi^n$$

where X_{f} is the foamed part property, X_0 is the unfoamed resin property and n is a coefficient with a value between 1 (law of mixtures rule) and 2 (experimentally determined to be a suitable design value for uniform-density foams). Significant deviations from this design equation can occur when the shapes have well defined thick skins or large voids in the cellular core, or for such design properties as impact strength (tensile, falling weight), deflection temperature and elongation to break. Some ductile polymers exhibit brittle loading behavior and some crystalline polymers exhibit atypical morphological characteristics when foamed. The surface roughness caused by gas bubble ebullition during foaming can lead to serious notch sensitivity in certain polymers and premature failure under impact loading.

See also: Injection Molding of Thermoplastics: Mold Filling; Injection Molding of Thermoplastics: Structure Development; Polymer Processing: Classification

Bibliography

Benning C J 1969 *Plastic Foams*, Vol. 2: *Structure, Properties and Applications*. Wiley–Interscience, New York, pp. 277–280

Berlin A A, Shutov F A 1980 *Strengthened Gas-Filled Plastics*. Khimiya, Moscow

Frados J (ed.) 1976 *Plastics Engineering Handbook*, 4th edn. Van Nostrand Reinhold, New York, pp. 580–99

Piechota H, Rohr H 1975 *Integralschaumstoffe*. Hanser, Stuttgart

Semerdjiev S 1982 *Introduction to Structural Foam*. Society of Plastics Engineers, Brookfield Center, Connecticut

Throne J L 1979 Principles of thermoplastic structural foam molding: A review. In: Suh N P, Sung N H (eds.) 1979 *Science and Technology of Polymer Processing*. MIT Press, Cambridge, Massachusetts, pp. 77–131

Throne J L 1982 Structural foam. In: Hilyard N C (ed.) 1982 *Mechanics of Cellular Plastics*. Macmillan, London, Chap. 7

Villamizar C A 1978 A study on thermoplastic structural foam. PhD thesis, Polytechnic Institute of New York

J. L. Throne

Injection Molding of Thermoplastics: Structure Development

There are many sources of structure in molded parts. Such structure may vary from the gross scale to the molecular level. If the melt enters the mold in the form of a jet rather than as a uniformly expanding front (Ballman et al. 1970, Oda et al. 1978, Sleeman and West 1974, Spencer and Gilmore 1951, Yamaguchi et al. 1968), the final molded part usually has both surface markings and internal structural discontinuities. This article, however, deals primarily with structure at a smaller scale. On the molecular level, molded parts exhibit polymer chain orientation (Ballman and Toor 1960, Fleissner 1973, Spencer and Gilmore 1950, Wales et al. 1972, White and Dietz 1979). This has been studied extensively using birefringence. Injection-molded parts produced from crystallizing polymer melts exhibit complex distributions of crystalline morphology (Clark 1967, 1973, 1974, Heckmann and Spilgies 1972, Heckmann and Johnsen 1974, Kantz et al. 1972, Matsumoto et al. 1979, Yamaguchi and Oyanagi 1964). Studies involving polarized light microscopy and both wide- and small-angle x-ray diffraction have been carried out.

1. Jetting

A basic study of jetting (Fig. 1) in injection mold filling was presented by Oda et al. (1978) (see also White and Dietz 1979). They found that jetting is associated with the swelling melt emerging from the gate and not striking the walls as it enters the mold. Thus, jetting is accentuated by melts exhibiting low extrudate swell and by gate designs which cause emerging melts not to strike the mold walls. Ballman et al. (1970) and Yamaguchi et al. (1968) have discussed characteristics of molded parts formed by jetting.

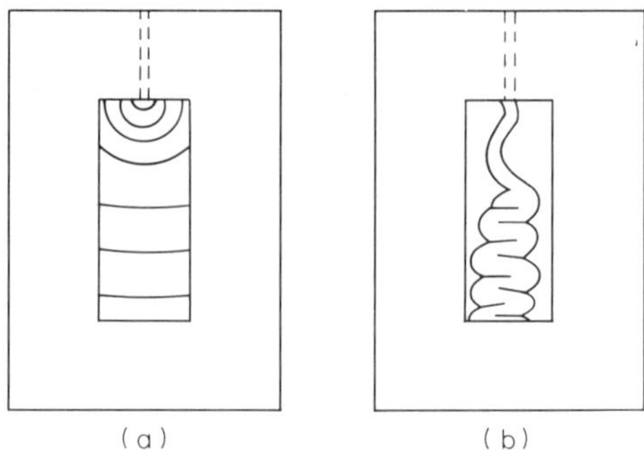

Figure 1
(a) Simple mold filling and (b) jetting

2. Orientation in Molded Glassy Plastics

The existence of orientation at the molecular level in injection-molded parts has been noted by various investigators. Ballman and Toor (1960) made the key basic experimental study of this, using end-gated injection-molded polystyrene bars. The bars were partly abraded away to various extents and birefringence was measured. With "1" taken as the direction of flow, "2" as the direction of the narrow thickness of the molded bar and "3" as the direction of the bar width, the birefringence Δn_{12} $(n_1 - n_2)$ exhibited a minimum at the center of the bar and a maximum at about 75% of the distance to the outer surface of the bar (Fig. 2). The magnitude of the birefringence maximum increased with ram pressure during the molding process and decreased with melt temperature. However, this result could be changed if mold designs with larger gates were used, where sealing occurred in the bar rather than in the gate. The birefringence was found to decrease with increasing mold thickness. Ballman and Toor suggested a mechanism for this behavior, the birefringence in the melt being determined by stresses developed during mold filling and relaxing with the stresses during the cooling process. The intermediate position of the birefringence maximum was due, they suggested, to a vitrified layer existing along the cold mold wall during the filling process, arising from the fountainlike fluid motion from the front (Fig. 3).

The hypothesis of Ballman and Toor was quantitatively developed in papers by Janeschitz-Kriegl (1977) and Dietz et al. (1978) (see also White and Dietz 1979). The rheooptical law

$$\boldsymbol{n} = \tfrac{1}{3}(\mathrm{tr}\,\boldsymbol{n})\boldsymbol{I} + C\boldsymbol{P} \qquad (1)$$

where $\boldsymbol{n}$ is the refractive index tensor, $\boldsymbol{I}$ the unit tensor, $\boldsymbol{P}$ the deviatoric stress tensor and C the stress optical constant, was applied together with a force balance of form

$$0 = -\partial p/\partial x_1 + \partial P_{12}/\partial x_2 \qquad (2)$$

where p is pressure, P_{12} shear stress and x_1 and x_2 distances in the "1" and "2" coordinate directions defined above. A heat balance was utilized to compute the cooling of the melt both during filling and in the filled mold. Birefringence profiles were predicted in molded parts and found to agree semiquantitatively with experiment (Fig. 2).

As the stresses P_{ij} are related through Eqn. (2) to the injection pressure, the pressure is predicted to be the primary factor influencing birefringence devel-

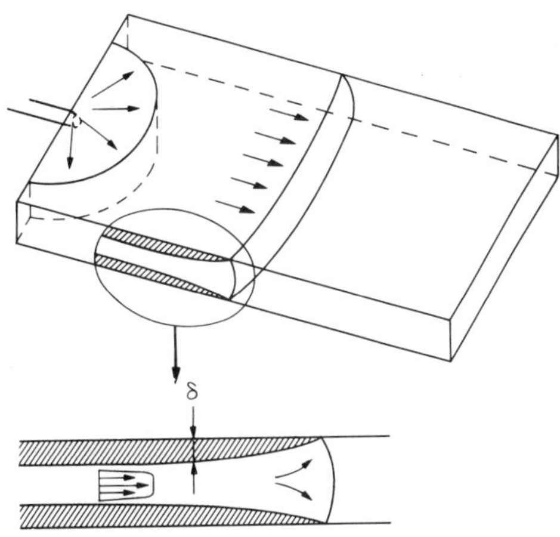

Figure 3
Filling of a mold by a melt front

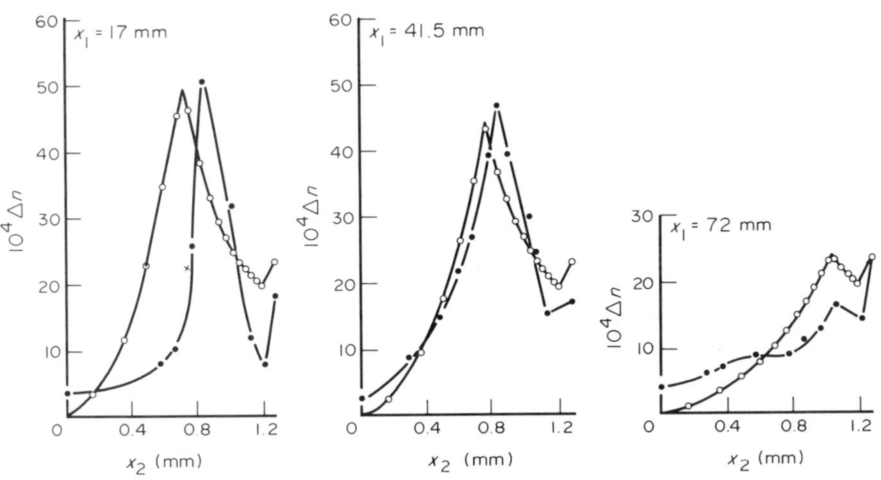

Figure 2
Birefringence distribution across a narrow section of an injection-molded part: ●, measured; ○, predicted

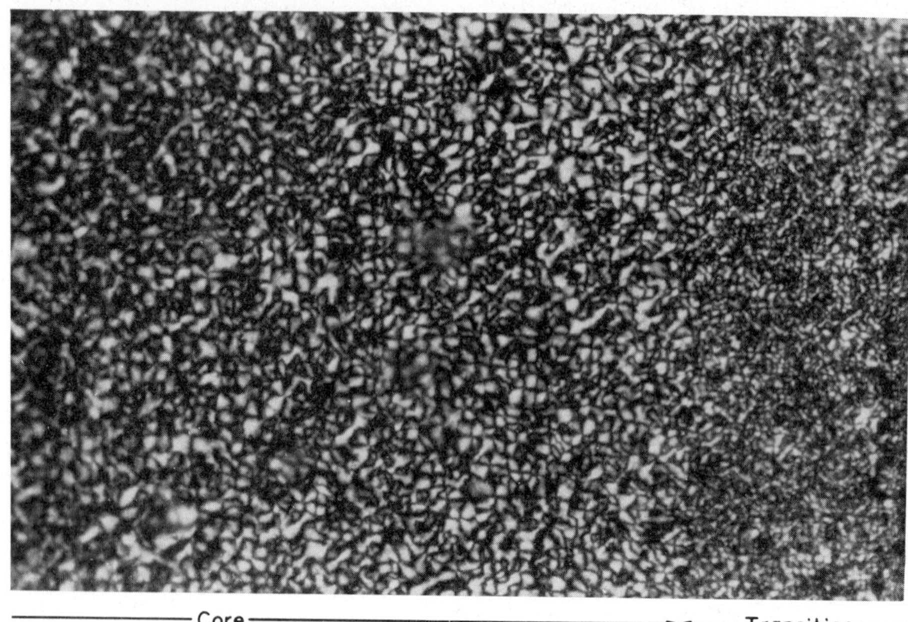

Figure 4
Cross section of molded polypropylene test bar

opment. Birefringence levels should also be increased by lower mold wall temperature.

3. Morphology in Molded Crystalline Plastics

Cross sections of injection-molded parts of crystalline polymers were first investigated by Yamaguchi and Oyanagi (1964). Using optical microscopy, they reported that polyethylene molded parts possessed an oriented skin layer and a spherulitic core. Between these two regions an intermediate morphology of ellipsoidal spherulites was observed. A typical cross-sectional morphology of an injection-molded polypropylene part showing similar characteristics is given in Fig. 4. These views were subsequently confirmed and extended in a series of papers by Clark (1967, 1973, 1974) dealing with polyoxymethylene molded products. The observations were reminiscent of those for glassy plastics. Decreasing mold wall temperature increased the extent of the oriented regions and decreased the amount of spherulites.

Clark also reported transmission electron microscopy (TEM) and small-angle x-ray scattering (SAXS) examinations of these injection-molded parts. These confirmed the oriented character of the outer layers. He used the results to develop a morphological model of the molded parts in terms of folded chain lamellae, randomly oriented in the center of the pot and oriented with chains in the flow direction in the outer layers.

Clark (1974) presented detailed optical microscopy observations of injection-molded springs, a cigarette lighter butane reservoir and a container made from acetal resin. In these studies, he noted the importance of factors such as the screw forward time; if the screw was brought back before the solidification was complete, morphology discontinuities were found in the molded parts.

The general observations described above were confirmed by the studies of Kantz et al. (1972) on polypropylene molded bars using optical microscopy and wide-angle x-ray scattering (WAXS) and of Heckmann and Spilgies (1972) and Heckmann and Johnson (1974) using WAXS and transmission electron microscopy. Heckmann and Spilgies interpreted their WAXS studies of the structure of injection-molded bars using pole figure analysis.

4. Molds with Moving Parts

Cleereman (1967, 1969) has described injection molding processes to produce plastic cups in which the core of the mold rotates during the mold filling process, inducing a helical orientation in the molded parts. This was noted as preferable to the usual uniaxial orientation resulting from end-gated molds, which suffer from inferior tensile characteristics transverse to the direction of melt flow.

See also: Injection Molding of Thermoplastics: Mold Filling; Injection Molding of Thermoplastics: Structural Foams; Polymer Processing: Structure Development

Bibliography

Ballman R L, Kruse R L, Taggart W P 1970 Surface fracture in injection molding of filled polymers. *Polym. Eng. Sci.* 10: 154–58

Ballman R L, Toor H L 1960 Orientation in injection molding. *Mod. Plastics* 38(2): 113–24, 205–07

Clark E S 1967 Molecular orientation in injection molding acetal homopolymer. *SPE J.* 23(7): 46–49

Clark E S 1973 Structure development in injection molding acetal homopolymer. *Appl. Polym. Symp.* 20: 325–32

Clark E S 1974 An in-depth look at molded parts. *Plastics Eng.* 30: 73–78

Cleereman K J 1967 Injection molding of shapes of rotational symmetry with multiaxial orientation. *SPE J.* 23(10): 43–47

Cleereman K J 1969 Injection molding of shapes of rotational symmetry with multiaxial orientation—Part II. *SPE J.* 25(1): 56–61

Dietz W, White J L 1978 Ein einfaches Modell zur Berechnung des Druckverlustes während des Werkzeugfullvorganges und der eingefrorenen Orientierung beim Spritzgiessen amorphol Kunststoffe. *Rheol. Acta* 17: 676–92

Dietz W, White J L, Clark E S 1978 Orientation development and relaxation in injection molding of amorphous polymers. *Polym. Eng. Sci.* 18: 273–81

Fleissner M 1973 Makroskopische und molekulare Orientierung in Hochpolymeren. *Koll. Z.-Z. Polym.* 251: 1006–14

Heckmann W, Johnsen U 1974 Scherinduzierte Kristallisation in Spritzgussteilen aus Niederdruck-Polyäthylen. *Coll. Polym. Sci.* 252: 826–35

Heckmann W, Spilgies G 1972 Röntgenographische Bestimmung unterschiedlichen Kristallitorientierungen in Polyäthylen-Spritzgussteilen. *Koll. Z.-Z. Polym.* 250: 1150–57

Janeschitz-Kriegl H 1977 Injection moulding of plastics: Some ideas about the relationship between mould filling and birefringence. *Rheol. Acta* 16: 327–39

Kantz M R, Newman H D Jr, Stigale F H 1972 The skin–core morphology and structure–property relationships in injection-molded polypropylene. *J. Appl. Polym. Sci.* 16: 1249–60

Matsumoto K, Miura I, Hayashida K 1979 Texture of injection-molded polypropylene. *Kobunshi Ronbushu* 36: 401–406

Oda K, White J L, Clarke E S 1978 Jetting phenomena in injection mold filling. *Polym. Eng. Sci.* 16: 585–92

Sleeman M J, West G H 1974 Rheology and product properties in injection molding. *Br. Polym. J.* 6: 109–18

Spencer R S, Gilmore G D 1950 Residual strains in injection molded polystyrene. *Mod. Plast.* 27: 97–155

Spencer R S, Gilmore G D 1951 Some flow phenomena in the injection molding of polystyrene. *J. Colloid Sci.* 6: 118–32

Wales J L S, van Leeuwen Ir J, van der Vijgh R 1972 Some aspects of orientation in injection molded objects. *Polym. Eng. Sci.* 12: 358–63

White J L, Dietz W 1979 Some relationships between injection molding conditions and the characteristics of vitrified molded parts. *Polym. Eng. Sci.* 19: 1081–91

Yamaguchi Y, Oyanagi K 1964 Study on structural non-uniformity in injection moldings of polyethylene. *Res. Rep. Kogakuin Univ.* 15: 1–9

Yamaguchi Y, Oyanagi K, Sugai Y, Kunitake M 1968 Molding pressure efficiency and gate sealing time in injection molding process of polyethylene. *Res. Rep. Kogakuin Univ.* 24: 71–82

J. L. White

Inorganic Adhesives

Before the development of synthetic polymeric adhesives, many bonding operations were accomplished using complex inorganic compositions. They were arrived at by trial and error, and little is known about their chemistry or their mechanical properties. A few of these inorganic adhesives continue to be used and are described in this article. Some, such as sodium silicate solution, are used in substantial quantities in the paper and lumber industry.

Inorganic adhesives function in various ways. A number of well-known mixtures of inorganic salts and metal oxides can be fused between substrates and thus act as adhesives. Other mixtures can form rigid continuous bodies by reaction with limited amounts of water to give a hydrated mass which then forms a coherent water-resistant matrix, reinforced in some cases by inert inclusions of other mineral substances or by sand or pebbles. Aqueous sodium silicate is an example of a third type of inorganic adhesive; it solidifies on substrates by the loss of water into porous adherends.

As a class, inorganic adhesives merge with materials such as enamels, which are not used as adhesives but must adhere to their metal substrates, and constructional cements, which may be used to fasten stone or brickwork but which are generally themselves formed into the desired building. Alone among these materials, sodium silicate is used in applications where it may be said to compete with dextrins, starches and gums on the one hand and modern synthetic adhesives on the other.

1. Sodium Silicate

Aqueous sodium silicate is widely used in the manufacture of corrugated paperboard for cartons and for similar applications in the packaging field. Although it is only used now with weak adherends such as paper, it can be a very strong adhesive, bonding hardwoods so that they can sustain loads of 10 GN m^{-2} and higher in direct tension, with failure eventually occurring in the wood.

Soluble silicates are prepared by melting a mixture of the required quantities of purified sand with sodium carbonate and leaching with alternate treatments of steam and water. The quantity of silica used with 1 mol of sodium carbonate may vary from 2 to 3.8 mol. The ratio is usually expressed in a form such as $Na_2O \cdot 3.3SiO_2$ without any implication that a chemical compound of this formula exists. In fact, when in aqueous solution, colloidal properties are evident and the colloid particles are probably those of silicic acid associated with both sodium metasilicate and disilicate.

For solutions with the same solids content the viscosity increases rapidly with increasing silica content. By varying the concentration in solution of sodium silicate of any $Na_2O:SiO_2$ ratio, both the viscosity and density are changed. In all cases the relation between viscosity and density for such solutions, when plotted as a curve, shows a marked change in direction at a density characteristic of the $Na_2O:SiO_2$ ratio of the sodium silicate. Thus, for a high-ratio material the curve shows a sudden change at a lower concentration and at a lower viscosity than shown by solutions of a lower-ratio material. The sudden rise indicates the onset of gelation. Hence, in order to allow for quite small losses of water, there is a limit to the concentration at which it is safe to store and handle a sodium silicate adhesive without danger of the solution setting. This leads to commercially available grades of sodium silicate solution differing in density, viscosity and total solids content, a situation which can be confusing. Typical figures are given in Table 1.

Table 1
Properties of commercial sodium silicate solutions (adapted from Macfarlane and Sewell 1954)

Density ($kg\,m^{-3}$)	Viscosity ($N\,s\,m^{-2}$)	Total solids (%)	Molar ratio $NaO_2:SiO_2$
1330	0.15–0.25	33.4	1:3.77
1395	0.2–0.5	37.8	1:3.4
1420	0.6–1.2	39.5	1:3.3
1500	1.5–2.0	44	1:2.9
1560	2.0–3.0	47	1:2.5
1700	60–160	54	1:2

Sodium silicate solutions lack adhesive tack and hence the adherends must be held in contact until setting has occurred. This is usually fairly rapid for, as already explained, the solution requires very little abstraction of water into the substrate before gelation takes place.

2. *Inorganic Hydration-Setting Cements*

The most important materials in this category, Portland cement and ciment fondu, are discussed elsewhere (see *Portland Cements, Blended Cements and Mortars*). Plaster of Paris is the hemihydrate of calcium sulfate (gypsum). When mixed with water it dissolves and then precipitates as the dihydrate, which is virtually insoluble. Its use as an adhesive rather than mixed with sand as a mortar is now rare, but it is often mixed with organic latex adhesives for fixing ceramic tiles.

3. *Ceramic or Refractory Cements*

A number of proprietary materials are available to withstand temperatures of several hundred degrees Celsius (Shields 1984). Some are glasses which devitrify after heat treatment to give an interlocking crystalline structure; others are based on alumina or zirconia.

4. *Miscellaneous Inorganic Cements*

4.1 Sorel Cement

Sorel cement is anhydrous magnesium chloride and magnesium oxide mixed in the weight ratio of 1:2, with water added to form a paste; it dries rather slowly to a hard white mass of magnesium oxychloride hydrate. This material is stronger in compression than Portland cement and very adherent to glass and ceramic materials. A tensile adhesion to shot-blasted glass of 700 $kN\,m^{-2}$ has been recorded (Shields 1984). Sorel cement is inert towards oil and grease but its weathering resistance is poor unless finely divided copper is added, probably acting as a biocide. The cement is antistatic and can therefore be used in hospital flooring and equipment. It is stable at elevated temperatures (to almost 200 °C), but on strong heating in air breaks down to give the oxide. The corresponding zinc oxychloride (1 part zinc chloride to 3 parts zinc oxide) is only used as a dental filling; it is not as strong as the magnesium compound but its use in the mouth suggests it is more water resistant.

4.2 Phosphate Cements

There are two types of phosphate cement: those made by mixing phosphoric acid with oxides in which zinc oxide predominates, and those which are more strictly phosphate–silicate cements, made by mixing phosphoric acid with calcium aluminosilicates or oxide mixtures containing silicates. The latter may have zinc phosphate present in the phosphoric acid, and tend to set more quickly than the former. However, both types set within 10 min of mixing at room temperature, but owing to the exothermic nature of the reaction may need cooling to remain at room temperature. They are used mainly in dental practice, being supplied as two-pack kits.

4.3 Sulfur Cements

Sulfur cements made by mixing sand into sulfur have been used for road works in California and for lining acid-containing tanks in chemical plants. Various additives have been tried in an attempt to improve thermal stability. Apart from carbon black, which acts as an inert filler, other additives such as dicyclopentadiene and styrene modify the polymer chain present in sulfur.

See also: Adhesion: Fundamental Principles; Adhesives in Civil Engineering

Bibliography

Macfarlane W S, Sewell J F 1954 Sodium silicate as an adhesive. *Adhesion and Adhesives Fundamentals and Practice.* Society of Chemical Industry, London

Shields J 1984 *Adhesives Handbook,* 3rd edn. Butterworth, London

Wills J H 1977 Inorganic adhesives and cements. In: Skeist I (ed.) *Handbook of Adhesives,* 2nd edn. Van Nostrand Reinhold, New York

W. Wake

Inorganic Fibers

This article covers synthetic inorganic fibers comprising aluminum oxide (alumina, Al_2O_3) with only a small (<1%) fraction of silicon dioxide (silica, SiO_2), or with a significant fraction of silica and having either a glassy or polycrystalline structure. The world manufacture of such fibers exceeds 20 kt per year. By far the largest part of this manufacture is a melt-spun glassy fiber containing roughly equal proportions by weight of alumina and silica, and used mainly in thermal insulation at temperatures up to 1400 °C. A much smaller amount of fiber is produced containing ≤5% silica; this material is made by the heat treatment of extruded slurries or solutions to produce polycrystalline fibers that are capable of withstanding higher temperatures (1600 °C), and have mechanical properties that make them candidates for composite manufacture. However, even fibers intended primarily for insulation are finding increasing application as a replacement for asbestos in friction materials, gaskets and packing.

It is convenient to divide Al_2O_3–SiO_2 fibers into glassy and polycrystalline varieties, since this also distinguishes the method of manufacture, range of properties and utility. Fibers containing less than about 65 wt% Al_2O_3 are spun from the melt and are usually called aluminosilicate fibers.

1. Aluminosilicate Fibers

The range of composition of melt-spun aluminosilicate fibers is about 45–60 wt% Al_2O_3, with SiO_2 as the other major component and with minor amounts of Fe_2O_3, TiO_2, CaO and other oxides. The fibers are usually manufactured from kaolin or a related clay mineral, or blends of alumina and silica (allowing closer control of composition than clay minerals). The raw material is melted and made into fibers in one of two ways. The melt may be poured into a stream of compressed gas (air or steam) which breaks the melt into fibers, or the melt may be fed to a rapidly rotating disk from which fibers are thrown by centrifugal force. The latter method produces longer and more silky fibers, but in both processes a wide range of fiber diameters (<1 to >10 μm) are formed, together with a fraction of nonfibrous material termed shot. Fiber length is from equidimensional fragments to several centimeters.

In the aluminosilicate fibers, the usual limit to temperature resistance is the devitrification of the glass with, for example, the nucleation and growth of mullite ($3Al_2O_3 . 2SiO_2$), which reduces mechanical strength drastically. Temperature resistance increases with increasing alumina concentration, fibers containing up to about 52% Al_2O_3 being suitable for use at temperatures up to 1250 °C, and those containing higher levels (up to about 65%) being capable of use at up to 1400 °C. Small additions of chromia (Cr_2O_3) improve temperature resistance. Aluminosilicate fibers are processed into a number of forms such as loose wool, blanket, felt, paper and board. These can be made by dewatering suspensions of the chopped fiber.

The major application of aluminosilicate fibers is in the insulation of furnaces in the metallurgical, ceramic and chemical industries, with a resulting considerable saving in weight and fuel usage. There may be up to a tenfold difference in the mass of a fiber-lined furnace and one constructed conventionally from refractory brick. This brings very considerable advantages in construction and maintenance; the low thermal mass allows rapid heating and cooling in cyclic operations, and a reduction in fuel usage of as much as 40% is not uncommon. Not surprisingly, the use of fibers in high-temperature technology has grown rapidly.

2. Manufacture of High-Alumina Fibers

The incentive to produce inorganic fibers higher in alumina content than can conveniently be manufactured by melt spinning is twofold. The success of melt-spun fibers in thermal insulation at temperatures up to 1400 °C led to a desire for fibers capable of use at higher temperatures, and the potential high stiffness of alumina-rich fibers was attractive for composite manufacture. The high modulus of alumina (α-alumina, 530 GPa), relatively low density (α-alumina, 4.0 g cm^{-3}) and nonreactivity make it an attractive candidate for metal reinforcement. The low viscosity of molten alumina and its high melting point (2070 °C) preclude melt spinning, so that processes had to be developed avoiding the melting step. Slurries of particulate alumina or an alumina hydrate may be extruded as fibers, and then dried and sintered to produce a polycrystalline filament. Similarly, viscous solutions of aluminum salts may be extruded or drawn into fibers, the gel being dried and fired. Other methods investigated include soaking an organic textile fiber (usually cellulosic) in an aluminum salt solution, then heating to burn out the organic material, and firing to sinter the alumina relic. Of the methods investigated, the slurry-extrusion and solution-spinning processes have been developed to commercial scale.

2.1 The Slurry Process

In the slurry process, an aqueous suspension of a particulate alumina is prepared. The aqueous phase may contain dissolved organic polymers to increase viscosity and stabilize the suspension, together with additives to control grain growth at the sintering stage. The aqueous phase may also contain a dissolved alumina precursor (e.g., an aluminum salt) to aid densification at the sintering stage. The slurry is then extruded into air, collected, dried, and fired at low temperature to allow shrinkage without distortion or cracking. Finally, it is flame-fired to complete conversion to the high temperature stable phase (α-alumina) and to eliminate porosity. The diameter of the continuous filaments produced by the slurry process is about 20 μm, and these may be in bundles of about 200. The α-alumina grain size is about 0.5 μm.

2.2 Solution Spinning

An alternative process to slurry spinning is the spinning of viscous, concentrated solutions of aluminum compounds, followed by drying and heat treatment of the gel fiber so produced. The starting material may be aluminum chloride. The hexahydrated aluminum ion, $Al(H_2O)_6^{3+}$, is stable only in strongly acidic solution, and as the solution becomes more basic, the hydrolysis of this ion occurs with the formation of polynuclear cations such as $[Al_6(OH)_{12}]^{6+}$ in which there are Al–O–Al bridges. Solutions containing such basic chlorides can be of high viscosity, and indeed, when highly concentrated, form aqueous glasses. For the preparation of alumina fibers, concentrated syrupy solutions of such basic chlorides are made. These may contain dissolved organic polymers to control rheology, together with phase stabilization and grain growth control additives. Of the latter, one of the most satisfactory is silica (at about 5 wt%), which may, for example, be added as a sol in which the silica particle size is about 200 Å. The syrupy solution is then formed into fibers by extrusion through a spinerette into dry air, or by ejection from a rapidly rotating disk (centrifugal spinning). On contact with dry air, the evaporation of water from the fiber causes a rapid increase in viscosity, so that the fiber becomes glassy. Thereafter, the filament is subjected to a progressive increase in temperature, first to remove chloride ion (as HCl) and to decompose and burn off organic matter, and finally to form and crystallize an anhydrous alumina phase. Fibers may be made in both continuous and staple form. The former will have diameters from 10–20 μm, whereas staple fibers may have a mean diameter of about 3 μm. In staple fiber production by solution spinning, the fiber diameter can be closely controlled, such that it is within the size range 1–7 μm. Such close control is important if fibers having no health hazard are desired, fibers of <1 μm diameter being toxicologically suspect.

In the heat-treatment sequence, the aqueous gel fiber first decomposes to form an amorphous or poorly crystalline alumina. On further heat treatment, this poorly defined alumina passes through a series of transition alumina-phases (η-, γ-, θ-, and δ-aluminas) before α-alumina appears. The process may be stopped at an intermediate stage, so that alumina fiber having a variety of properties may be prepared. The η-phase fiber, for example, has a high surface area of 200 $m^2 g^{-1}$ due to high internal porosity, which makes this form of fiber particularly useful as a catalyst support. To produce fibers for high-temperature use, the heat treatment is continued until the major phases present are δ- and α-alumina. The presence of a few percent of silica in the fiber is essential for the control of grain growth, since grain size must be a fraction of the fiber diameter if strength is to be preserved. The stiffness increases with the α-alumina content. Short staple fiber produced by solution spinning is converted to a variety of forms such as loose wool, blanket, paper and board.

3. Properties and Applications of High-Alumina Fiber

Generally, slurry-spun fibers are produced in continuous form, whereas solution-spun fibers are commonly made in short staple form. Continuous fibers are, of course, particularly adapted for use in composite manufacture. One product of this type is made as 20 μm diameter filaments in bundles of 200. It comprises >99% α-Al_2O_3. At a gauge length of some 10 mm, the tensile strength of typical material is about 1400 MPa. This strength is probably limited by surface imperfections, such as emergent grain boundaries, the grain size being 0.5 μm. The tensile strength is increased if such fibers are coated with silica, which heals surface flaws. The coated fibers have strengths up to 1900 MPa, although higher strengths (>2000 MPa) are said to be possible. When the major phase is α-alumina, the fibers have high stiffness, a typical modulus being 390 GPa at a density of 3.95 g cm^{-3}. Strength and modulus are retained at high temperature, an attribute valuable in the manufacture of metal–fiber composites. Typically, an α-alumina filament retains 100% of its room temperature strength after heating in air at 1000 °C for 300 h. Modulus is also retained at 1000 °C, and this resistance to temperature is a marked advantage over carbon, silicon carbide and metal fibers, all of which are prone to oxidative attack.

Whilst such fibers may be used as reinforcement in organic resins, their thermal stability is favorable for their incorporation into metals, and polycrystalline α-alumina fibers have been incorporated into aluminum, magnesium, lead, copper and zinc to form sheets, plates, rods, tubes and beams. Generally, a fiber preform is prepared using an organic binder such as poly(vinyl alcohol). The preform is packed into a casting mold—to give appropriate alignment and volume fraction—the organic binder is burned off, and the mold infiltrated with molten metal after evacuation. It is essential that the fiber is wetted by the metal, and that no brittle phase is formed at the metal–fiber interface. Whilst molten magnesium wets α-alumina well, no wetting occurs with molten aluminum even at 1000 °C. Alumina fiber coated with nickel or copper is wetted by molten aluminum, although brittle intermetallic phases then form at the interface. It has been mentioned previously that alumina fiber may be coated with silica to heal surface defects and so increase tensile strength. Fibers so coated are unsatisfactory for incorporation into molten aluminum, because the silica is reduced to elemental silicon ($3SiO_2 + 4Al \rightarrow 3Si + 2Al_2O_3$), and a poor fiber–metal bond is developed. The most satisfactory method of ensuring wetting and bonding has been found to be the addition of up to 3% lithium to the aluminum.

Unidirectional alumina fiber–aluminum composites have been prepared having a fiber volume fraction of 60%. The axial modulus was 262 GPa, the tensile strength 690 MPa, the flexural strength 1 GPa and the compressive strength 3.4 GPa. Composite density was 3.4 g cm^{-3}. The mechanical properties of the composite were retained at over 300 °C. Such materials are clearly of potential in aerospace engineering and in automobile construction.

Short staple, fine diameter fibers, of composition 95% Al_2O_3, 5% SiO_2 produced by solution spinning, are increasingly being used in the random reinforcement of aluminum alloys to preserve mechanical properties at elevated temperatures (300–350 °C). In diesel-engine pistons (British Patent 1567328, 1980), for example, it is important that the piston crown has an adequate tensile strength and wear resistance at engine temperature. A composite containing 30 vol% of randomly oriented staple alumina fiber in an aluminum alloy has been shown to have a tensile strength up to tenfold that of the unreinforced alloy at engine temperature. Fiber modulus depends on the alumina phase present, and increases with increasing α-alumina content. A fiber comprising predominantly η-alumina and of high porosity has a modulus of about 100 GPa, whereas δ-alumina fiber containing about 20% α-alumina has a modulus of about 320 GPa. Tensile strength rises as diameter is reduced, approaching 3000 MPa for 2 μm diameter fibers comprising the δ-phase with 6% α-alumina. An average value for tensile strength is 1500 MPa. Staple fiber comprising mainly the δ-phase has been found to be readily wetted by aluminum and its alloys, greatly facilitating the fabrication of composites by "squeeze casting." The composites, which show good retention of properties at engine operating temperatures, are fabricated by the pressure infiltration of molten alloy into a fiber preform.

The major application of short staple fiber is in thermal insulation at temperatures up to 1600 °C, in which use the high-alumina fibers show very marked advantages over aluminosilicate fibers. The fine and uniform diameter has led to other uses, such as hot-gas filtration (as in diesel exhaust cleaning), and since the fiber carries a positive surface charge in water (as does asbestos), it has been proposed as a filter for the removal of bacteria (usually negatively charged) from liquids. In several applications the fiber replaces asbestos (e.g., in friction materials), and here the freedom from toxic hazards—ensured by the absence of fibers below 1 μm in diameter—is valuable. The resilience of fine alumina fibers has prompted their use as a vibration-resistant packing to protect fragile ceramic components in automobile-exhaust catalytic cleaners.

See also: Aluminum Oxide; Inorganic Polymers; Composite Materials: An Overview

Bibliography

Bailey J E, Barker H A 1974 Trial of strength. *Chem. Br.* 10: 465–70

Birchell J D 1983 The preparation and properties of polycrystalline aluminum oxide fibre. *Trans. J. Br. Ceram. Soc.* 83: 143–45

Clyne T W et al. 1985 The use of a δ-alumina fibre for metal-matrix composites. *J. Mater. Sci.* 20: 85–96

Dhingra A K 1980a Alumina fiber FP. *Philos. Trans. R. Soc. London, Ser. A* 294: 411–17

Dhingra A K 1980b Metal matrix composites reinforced with FP fiber. *Philos. Trans. R. Soc. London, Ser. A* 294: 559–64

Dickson T 1979 Ceramic fibers: a growing refractory product. *Ind. Miner. (London)* 146 (November): 23–31

J. D. Birchall

Inorganic Polymers

Inorganic (nonhydrocarbon) polymers are very numerous, pervade the entire periodic table, and exhibit a wide variety of properties. Limiting consideration to entities which exhibit macromolecular properties in solution or under melt conditions still provides numerous giant inorganic species composed of linear chains, ladder polymers, two-dimensional sheets and three-dimensional lattices. The predominance of the latter causes inorganic polymers to be characterized as strong, hard and brittle.

1. Linear Chain Polymers

Inorganic chain polymers consisting of two bridging bonds per unit include homoatomic sulfur, selenium and tellurium; the alternating phosphorus–nitrogen phosphazenes; the alternating sulfur–nitrogen superconducting $(SN)_x$; the alternating phosphorus–oxygen polyphosphates; the alternating silicon–oxygen pyroxene silicates and the related silicones; the carboranes; and a variety of bridged metal chelates and organometallics.

Linear high-molecular-weight sulfur (S_n) can be prepared by the rapid quenching of hot (180 °C) molten sulfur. The polymeric material so obtained has elastomeric properties with a glass-transition temperature T_g of −30 °C. Removal of S_8 from the material raises its T_g to 75 °C. Unfortunately, polymeric sulfur is unstable below 160 °C relative to crystalline S_8, although the use of additives (phosphorus or arsenic) decreases the rate of conversion and the addition of 10% selenium produces stable polymeric species at 110 °C. Polymeric selenium (T_g = 31 °C) and tellurium are also known, but no practical uses appear to exist at present. The bonds in these "linear" polymers are actually at about 90°.

Phosphorus and nitrogen do not form linear polymers by themselves, but polymeric phosphazenes, $\leftarrow PX_2{=}N \rightarrow_n$, or phosphonitrilics are well known; in fact their organic derivatives are well enough established to be treated in a separate article (see *Polyphosphazene Elastomers*). When X = chloride or fluoride, the phosphazene chain is easily hydrolyzed, but organic substituents provide hydrolysis protection. A wide range of properties (including T_g values almost as low as the −90 °C value for the fluoride) and numerous uses (heart valves, O-rings, sutures, fuel pipes, flame resistant fibers, films) have been found for these polymers. More details can be found in the specific article on the phosphazenes.

Polymeric $(SN)_x$ has also recently achieved significance because of its superconductivity at very low temperatures (<0.3 K). Fiberlike crystals of $(SN)_x$ prepared indirectly from the tetramer S_4N_4 show essentially one-dimensional conductivity and metal-like properties (conductivity, malleability, softness). Unfortunately, the polymer decomposes slowly in the presence of water. Attempts to obtain higher temperature superconductors by other combinations of the Group V and VI elements have not yet proven successful.

Linear polyphosphate anions (**1**) approach the formulation of cyclic or metaphosphate $\leftarrow PO_3 \rightarrow_n$ as the chain lengthens, so that what appears to be a simple

$$-O \leftarrow \overset{\overset{\displaystyle O}{\|}}{\underset{\underset{\displaystyle O}{|}}{P}} - O -)_n^{(n+2)-}$$

(**1**)

polyphosphate salt, MPO_3, can be, and often is, a mixture of species both linear and cyclic. All polyphosphates have a tendency to hydrolyze slowly to the thermodynamically stable orthophosphate ion (PO_4^{3-}) upon standing in solution or in a moist environment. Polyphosphates, other than alkali-metal and ammonium salts, are typically insoluble in water. Divalent ion-precipitated polyphosphates, which retain appreciable water, are elastomeric or gummy, but the anhydrous salts are brittle or glassy materials. Alkylammonium polyphosphate salts possess greaselike properties, and are soluble in some organic solvents. Their T_g values range from less than room temperature for some alkylammonium polyphosphate salts, to 115 °C for the Ag^+ salt, 243–335 °C for K^+ to Li^+, through to 470–525 °C for Ba^{2+} to Mg^{2+}. Short polyphosphate chains also have biochemical significance; for example, they occur in adenosine triphosphate (ATP) and the diphosphate (ADP) analogue.

Chain silicates such as the lithium ore spodumene $[LiAl(SiO_3)_2]_n$ and diopside $[CaMg(SiO_3)_2]_n$ contain two connective pyroxenes, $\leftarrow SiO_3 \rightarrow_n^{2n-}$. A wide variety of silicones also fall into this category, where organic groups rather than oxygens are attached to silicon. These uncharged $+Si(R_2)O+_n$ polymers are treated in detail elsewhere (see *Silicones*; *Silicone Elastomers*). Suffice it to say that silicone rubber, silicone greases, silicone oils and waterproofing materials are such a part of our everyday life that it is often difficult to remember technology before silicones.

The extra thermal stability which the silicone oxygen linkages bring to macromolecules also extends to some of the polycarborane copolymers, in which $C(B_{10}H_{10})C$ icosohedral units are joined by short polysiloxane and dimethylsilyl bridges. One example of such copolymers is the dexsils, $-Si(CH_3)_2C(B_{10}H_{10})C[Si(CH_3)_2O]_n-$. When $n = 1$, T_g = 25 °C; longer chain dexsils have lower T_g values. Because of their stability, dexsils have found use as chromatography media and, after modification with vinyl groups, as useful elastomers which are stable in air to at least 260 °C. The usual icosahedral fragment in these polymers is the 1,7- or *m*-carborane. Dimethyl tin, carbon, polyphosphate and dimethyl germanium linkages between carborane units have also been investigated. Although extreme claims have been made for these polymers, air oxidation ensues at about 300 °C for practically all present formulations. In the absence of air some are stable to about 450 °C.

Metal-containing chain polymers include the water-soluble fungicides zineb and maneb, which are examples of bisdithiocarbomate polymers of the general formula (**2**) for tetrahedral zinc or manganese, respectively. These semiladder polymers are representative of a large number of metal coordination polymers. However, most metal coor-

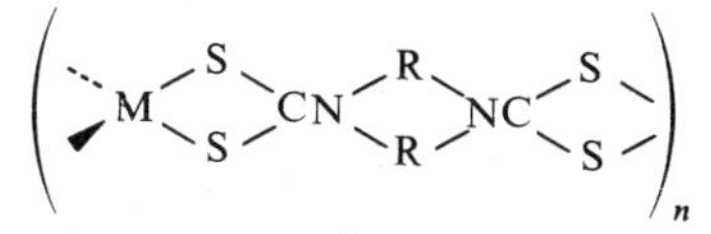

(2)

dination polymers have been formulated with square planar divalent ions (Ni^{2+}, Cu^{2+}, Pd^{2+}, Pr^{2+}), where stacking interactions tend to cause precipitation prior to the formation of large polymeric species.

Polymeric metal cyclopentadienyl sandwich compounds have also been prepared and offer promise for the future (e.g., polyferrocene (**3**, $n \geqslant 200$) has been isolated and appears to have good thermal stability. Coordination or organometallic coordination compounds possessing conjugated ligand systems offer exciting conducting polymer possibilities.

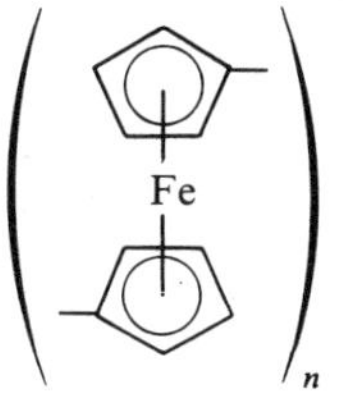

(3)

2. *Ladder Polymers*

Amphiboles are an important example of inorganic ladder polymers. They are double-chain silicates based on $(Si_4O^{6-})_n$ units as shown in formula (**4**).

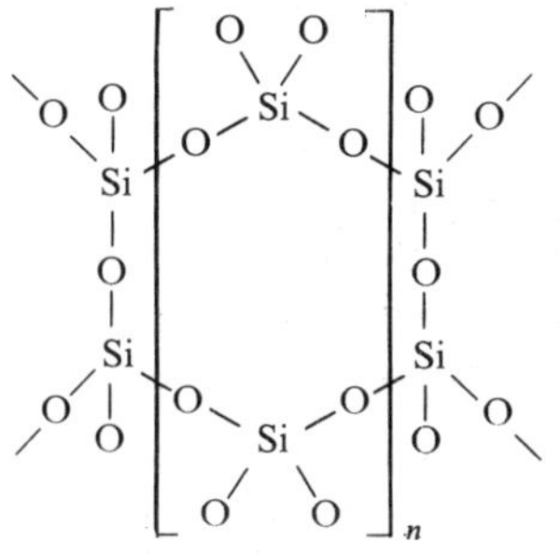

(4)

Fibrous, flame-resistant asbestos materials usually have an amphibole structure. Although such materials have been used extensively in the past, the linkage of increased incidence of lung cancer for smokers exposed to asbestos dust has caused a sharp reduction in asbestos use recently (see *Asbestos Hazards*). The properties of asbestos are consistent with the strength (Young's modulus = 140–175 GN m^{-2} for asbestos materials vs 5–15 for typical organic polymers) and brittleness (failure strain at 1% elongation vs 5–20% for typical organic polymers) considered general for inorganic macromolecules. Asbestos materials include anthophyllites, $M_7^{II}Si_8O_{22}(OH)_2$ (where M^{II} is typically magnesium(II) and a little iron(II)); crocidolite, $Na_2Fe_3^{II}Fe_2^{III}Si_8O_{22}(OH)_2$; and tremolite, $Ca_2Mg_5Si_8O_{22}(OH)_2$. At around 400 °C the hydroxy groups of these asbestos materials undergo dehydration and the fibers become even more brittle, although decomposition temperatures are typically above 1200 °C.

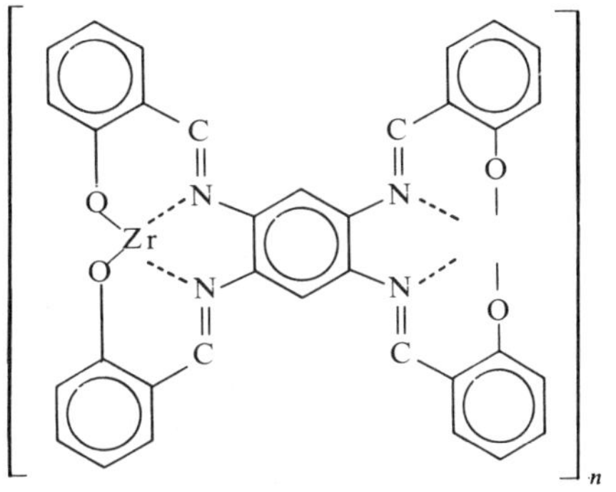

(5)

Metal coordination polymers with ladder type bridges include the zirconium tetrasalicyclidene-1,2,4,5-tetraaminobenzene polymer (**5**) and the bis(5,7-dichloroquinolinolato)quinoxaline-5,8-diolatotungsten(IV) polymer (**6**). Higher molecular-weight species occur for these eight-coordinate chelates because stacking is prevented.

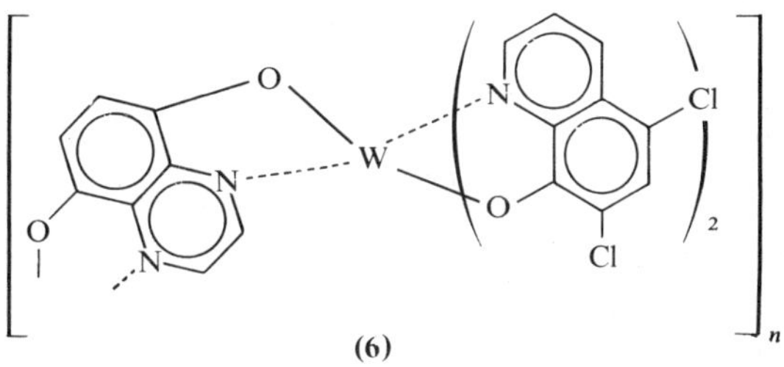

(6)

3. *Sheet Polymers*

Two-dimensional layer or sheet polymers are best exemplified by the graphite hexagonal layer structure, which is also found in purely inorganic species, such as molybdenum disulfide and the hexagonal form of boron nitride. All are soft, easily machined, and can serve as lubricants when finely divided. Boron nitride articles can be used in air up to about 800 °C, and in inert atmospheres to about 1600 °C.

It has a crystalline melting point of about 3000 °C under pressure (to prevent sublimation).

Layered structures are also observed in a variety of other crystalline inorganic species, such as mica, talc and related silicates, which possess $(Si_2O_5^{2-})_n$ infinite sheets with cations of counter charge between the anionic layers. The sheets possess the same twelve-membered ring system shown in formula (**4**), but the silicon atoms are all bridged via oxygen to three other silicon atoms, that is, each silicon has only one terminal oxygen. The typical inertness of these species is exemplified in the use of soapstone (primarily talc) for acid and solvent inertness and up to 700 °C. The vermiculites are related aluminosilicates with large cation-exchange capabilities. Other layered silicates include pyrophillite (used in electrical insulators), montmorillonite (used in absorbents), kaolinite and the asbestos fiber chrysotile.

4. Three-Dimensional Network Polymers

Three-dimensional or lattice polymers are best exemplified by silica, $(SiO_2)_n$. This exists as an array of silicon atoms tetrahedrally surrounded by four oxygen atoms, each of which is in turn bonded to a further silicon atom to form an infinite three-dimensional lattice, either symmetrical as in cristobalite or helical as in quartz. Replacement of some of the silicon atoms by aluminum results in the formation of ionic aluminosilicates. Aluminosilicates include the feldspar minerals (e.g., $KAlSi_3O_8$), the ultramarines (which possess S_n^- anions in addition to cations and aluminosilicate anions), and a wide variety of zeolites, which are open-structure aluminosilicates used as molecular sieves, catalysts, catalyst supports and ion exchangers. Naturally occurring zeolites include chabazite ($CaAl_2Si_4O_{12}.6H_2O$), mordenite ($NaAlSi_5O_{12}.3H_2O$), erionite ($Ca_9Al_{18}Si_{54}O_{144}.54H_2O$), fanjasite ($Ca_{59}Al_{108}Si_{266}O_{748}.470H_2O$) and clinoptilolite ($NaAlSi_5O.4H_2O$). In erionite and fanjasite, the calcium ions are typically replaced by Mg^{2+}, $2Na^+$ or $2K^+$. Hydrothermal syntheses can produce both naturally occurring types of zeolites plus a number of synthetic zeolites designated as A, X, Y, F, W, L, Ω, ZSM-5 (see *Zeolite Minerals*). The dehydrated zeolites are able to selectively absorb or reject molecules on the basis of size, shape and/or polarity. Catalytic properties are based on the use of acid sites, metals dispersed in the zeolite (e.g., catalytic hydrocracking with Pt), pore size and thermal stability. Zeolite powders are used as ion exchangers in heavy duty detergents designated as low phosphate or phosphate free (Na^+ replaces Ca^{2+} and Mg^{2+} in the water), in cesium and strontium radioisotope separation procedures, and for ammonium ion removal in waste treatment.

Whereas both boric oxide and phosphoric oxide are water soluble and low melting, boron phosphate, $(BPO_4)_n$, after annealing at 500 °C, is insoluble, chemically inert and sublimes above 1400 °C without melting. Structurally, boron phosphate has eight-membered BOPOBOPO rings multiply connected, with each boron and phosphorus tetrahedrally surrounded by oxygens, each of which is bonded to one boron and one phosphorus. Addition of alkali oxides or silica produces glasses, and boron phosphate gels in sulfuric acid are also known.

The borophosphate glasses provide insight into the relationships between crystalline and glassy inorganic networks. Whereas all of the boron atoms are four-coordinate in boron phosphate, in the borophosphate glasses some of the boron atoms are only three-coordinate with a vacant site which allows bond switching (Scheme 1). Phosphates can also expand their coordination number, so that bond interchange can occur even when the phosphate is tetrahedrally coordinated.

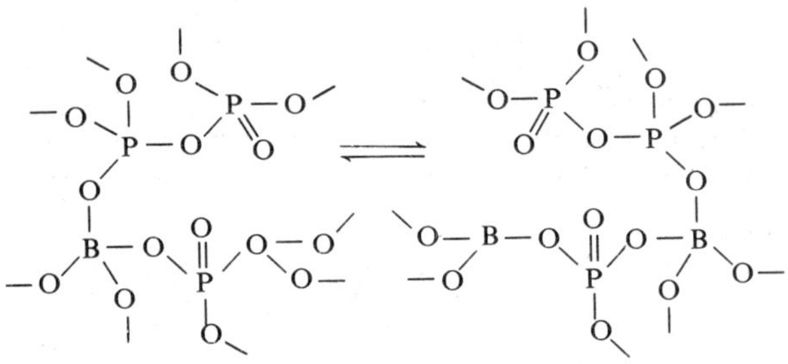

Scheme 1

Aluminum phosphate has the same network structure as boron phosphate with crystalline forms resembling silica. $AlPO_4$ melts at 1700 °C and forms an amorphous glass on cooling. The amorphous form is soluble in acids and forms films which are used as refractory cements and abrasive binders.

Phosphates with six-coordinate tetravalent titanium, zirconium, hafnium, cerium, thorium and tin all exist in both crystalline and amorphous modifications, and have good ion-exchange capabilities due to the formation of the corresponding $M(HPO_4)_2.xH_2O$ species. These species have only moderate thermal stability, with decomposition to the phosphate, MP_2O_7, ensuing around 400 °C. The analogous silicon and germanium acid phosphate compounds are known, but they decompose at even lower temperatures.

Polycrystalline alumina, Al_2O_3, a network polymer, can be fabricated as fibers with a melting point of over 2000 °C with long-term stability at up to 1400 °C. Analogous polycrystalline zirconia, ZrO_2, has a melting point of over 2700 °C and stability to 1600 °C. The ZrO_2 fibers are more resistant to both strong acids and alkalis than the Al_2O_3 fibers. Both are nonflammable, as are polycrystalline TiO_2

and $K_2Ti_6O_{13}$, but silica-based fiberglass materials predominate the inorganic fiber market.

The most widespread synthetic network inorganic polymer is hardened cement. Portland cement is primarily a mixture of anhydrous calcium silicates which set or harden after water is added.

Young's moduli for inorganic oxide glasses are upwards from 40 GN m^{-2}, appreciably above typical values for organic polymers (5–15 GN m^{-2}). Brittleness of inorganic oxide glasses is associated with fracture occurring at or below strains of only 1% of the modulus. The thermal stabilities of many of these inorganic polymers are extremely good; however, many have limited plasticity (high T_g values).

See also: Inorganic Fibers; Glasses: Structure; Advanced Ceramics: An Overview

Bibliography

Allcock H R 1985 Inorganic macromolecules. *Chem. Eng. News* March 18: 22–36

Archer R D, Batschelet W H, Illingsworth M L 1981 Linear conjugated coordination polymers containing 8-coordinate metal centers. *J. Macromol. Sci., A* 16: 261–71

Archer R D, Hardiman C J, Kim K S, Grandbois E R, Goldstein M 1985 Metal–chelate polymers: Structural/property relationships as a function of the metal ion. *Metal-Containing Polymeric Systems*. Plenum Press, New York, pp. 355–66

Breck D W, Anderson R A 1981 Molecular sieves. *Kirk–Othmer Encyclopedia of Chemical Technology,* 3rd edn., Vol. 15. Wiley–Interscience, New York, pp. 638–69

Peters E N 1981 Inorganic high polymers. *Kirk–Othmer Encyclopedia of Chemical Technology,* 3rd edn., Vol. 13. Wiley–Interscience, New York, pp. 398–413

Ray N H 1978 *Inorganic Polymers*. Academic Press, London

Sand L B, Mumpton F A 1978 *Natural Zeolites, Occurrence, Properties, Uses*. Pergamon, Oxford

Sheats J E 1981 Metal-containing polymers. *Kirk–Othmer Encyclopedia of Chemical Technology*, 3rd edn., Vol. 15. Wiley–Interscience, New York, pp. 184–220

Townsend R P (ed.) 1980 *The Properties and Applications of Zeolites*. Royal Society of Chemistry, London

R. D. Archer

Input–Output Analysis

Input–output analysis is a method of systematically quantifying the mutual interrelationships among the various sectors of a complex economic system. In practical terms, the economic system to which it is applied may be as large as a nation or even the entire world economy, or as small as the economy of a metropolitan area or even a single enterprise.

In all instances the approach is essentially the same. The structure of each sector's production process is represented by an appropriately defined vector of structural coefficients that describes in quantitative terms the relationship between the inputs it absorbs and the output it produces. The interdependence among the sectors of the given economy is described by a set of linear equations expressing the balances between the total input and the aggregate output of each commodity and service produced and used in the course of one or several periods of time.

The technical structure of the entire system can accordingly be represented concisely by the matrix of technical input–output coefficients of all its sectors. It constitutes at the same time the set of parameters on which the balance equations are based.

1. Input–Output Tables

An input–output table describes the flow of goods and services between all the individual sectors of a national economy over a stated period of time, say, a year. A simplified example of an input–output table depicting a three-sector economy is shown in Table 1. The three sectors are agriculture, whose total annual output amounts to 100 bushels of wheat; manufacture, which produced 50 yards of cloth; and households, which supplied 300 man-years of labor. The nine (3 × 3) entries inside the main body of the table show the intersectoral flows. Of the 100 bushels of farm products turned out by agriculture, 25 bushels were used up within the agricultural sector itself, 20 were delivered to and absorbed, as one of its inputs, by manufacture, and 55 were taken by the household sector. The second and third rows of the table describe in the same way the allocation of outputs of the two other sectors.

The figures entered in each column of the table thus describe the input structure of the corresponding sector. To produce the 100 bushels of its total output, agriculture absorbed 25 bushels of its own products, 14 yards of manufactured goods, and 80 man-years of labor received from the households; the manufacturing sector to be able to produce the 50 yards of its total output, had to receive and use up 20 bushels of agricultural—and 6 yards of its own (i.e., of manufactured)—products as well as 180 man-years of labor from households. In their turn, the households have spent the incomes—which they have received for supplying 300 man-years of labor—to pay for the consumption of 55 bushels of agricultural and 30 yards of manufactured commodities and 40 man-years of direct services of labor.

All entries in Table 1 are supposed to represent quantities, or at least physical indices of the quantities, of specific goods or services. A less aggregative, more detailed input–output table describing the same national economy in terms of not three but of 50, 100 or even 1000 different sectors would permit a more specific qualitative identification of all the individual entries. In a larger table, manufacturing would, for example, be represented not by one but

Table 1
Simplified input–output table for a three-sector economy

into / from	Sector 1: Agriculture	Sector 2: Manufacture	Sector 3: Households	Total output
Sector 1: Agriculture	25	20	55	100 bushels of wheat
Sector 2: Manufacture	14	6	30	50 yards of cloth
Sector 3: Households	80	180	40	300 man-years of labor

by many distinct industrial sectors; its output—and consequently also the inputs of the other sectors—would be described in terms of "yards of cotton cloth," "tons of paper products," or even "yards of percale," "yards of heavy cotton cloth," as well as "tons of newsprint" and "tons of writing paper."

1.1 Input–Output Tables and National Income Accounts

Although in principle the intersectoral flows, as represented in an input–output table, can be thought of as being measured in physical units, in practice most input–output tables are constructed in value terms. Table 2 represents a translation of Table 1 into value terms on the assumption that the price of agricultural products is $2 per bushel, the price of manufactured goods is $5 per yard and the price of services supplied by the household sector is $1 per man-year. Thus the values of total outputs of the agricultural, the manufacturing and the household sectors are shown in the new translated table as being equal to, respectively, $200 (=100 × 2), $250 (= 50 × 5) and $300 (=300 × 1). The last row shows the combined value of all outputs absorbed by each of the three sectors. Such column totals could not have been shown in Table 1 since the physical quantities of different inputs absorbed by each sector cannot be meaningfully added.

The input–output table expressed in value terms can be interpreted as a system of national accounts. The $300 showing the value of services rendered by the households over the period of the years obviously represent the annual national income. It equals the sum total of the income payments—shown in row 3—received by the households for services rendered to each sector; it also equals the combined value of goods and services—as shown in column 3—purchased by the households from themselves and from the other sectors. To the extent to which the column entries (showing the input structure of each productive sector) cover the current expenditures but not purchases made on capital account, the latter—being paid out of the net income—should be entered in the households column.

All figures in Table 2—except the column sums shown in the bottom row—can also be interpreted as representing physical quantities of the goods or services to which they refer. This only requires that the physical unit in which the entries in each row are measured be redefined as being equal to that amount of output of that particular sector which can be purchased for $1 at prices which prevailed during the interval of time for which the table was constructed.

National input–output tables are now constructed in some 80 countries; a large number of regional and metropolitan input–output tables have also been

Table 2
Simplified input–output table expressed in value terms

into / from	Sector 1: Agriculture ($)	Sector 2: Manufacture ($)	Sector 3: Households ($)	Total output ($)
Sector 1: Agriculture	50	40	110	200
Sector 2: Manufacture	70	30	150	250
Sector 3: Households	80	180	40	300
Total input ($)	200	250	300	

compiled. The number of sectors that describe the economic system has increased dramatically in recent years. Some of the more detailed tables describe a national economy in terms of five or six hundred separate sectors.

2. Technical Coefficients

Let the national economy be subdivided into $n + 1$ sectors: n industries, that is, producing sectors, and the $(n + 1)^{\text{th}}$ final demand sector, represented in input–output Tables 1 and 2 by the households. For the purposes of mathematical manipulation, the physical output of sector i is usually represented by x_1, while the symbol x_{ij} stands for the amount of the product of sector i absorbed—as its input—by sector j. The quantity of the product of sector i delivered to the final demand sector $x_{i,n+1}$ is usually identified in short as y_i.

The quantity of the output of sector i absorbed by sector j per unit of its total output j is described by the symbol a_{ij} and is called the input coefficient of the product of sector i into sector j:

$$a_{ij} = x_{ij}/x_j \tag{1}$$

A complete set of the input coefficients of all sectors of a given economy arranged in the form of a rectangular table—corresponding to the input–output table of the same economy—is called the structural matrix of that economy. Table 3 represents the structural matrix of the economy whose flow matrix is shown in Table 1. The flow matrix constitutes the usual—although not necessarily the only possible—source of empirical information on the input structure of the various sectors of an economy. The entries in Table 3 are computed according to Eqn. (1) from figures presented in Table 1. For example, $a_{11} = 25/100 = 0.25$, and $a_{12} = 20/50 = 0.40$.

Table 3
Simplified structural coefficient matrix of three-sector economy

into from	Sector 1: Agriculture	Sector 2: Manufacture	Sector 3: Households
Sector 1: Agriculture	0.25	0.40	0.133
Sector 2: Manufacture	0.14	0.12	0.100
Sector 3: Households	0.80	3.60	0.133

In practice the structural matrices are usually computed from input–output tables described in value terms, such as Table 2. In any case, the input coefficients (for analytical purposes described below) must be interpreted as ratios of two quantities measured in physical units. To emphasize this fact, the structural matrix (Table 3) in this example was derived from Table 1, not Table 2.

2.1 Static Input–Output System

The balance between the total output and the combined inputs of the product of each sector, as shown in the example in Tables 1 and 2, can be described by the following set of n equations:

$$\begin{array}{llll} (x_1 - x_{11}) & - x_{12} - \dots & & - x_{1n} = y_1 \\ - x_{21} + (x_2 - x_{22}) & - \dots & & - x_{2n} = y_2 \\ \dots & \dots \quad \dots & & \dots \quad \dots \\ -x_{n1} & -x_{n2} - \dots & + (x_n - x_{nn}) & = y_n \end{array} \tag{2}$$

Substitution of Eqn. (1) in Eqns. (2) yields n general equilibrium relationships between the total outputs, $x_1, x_2, \dots, x_n$, of all producing sectors and the final bill of goods, $y_1, y_2, \dots, y_n$, absorbed by households, government and other final users:

$$\begin{array}{llll} (1 - a_{11})\, x_1 & -a_{12}x_2 - \dots & & -a_{1n}x_n = y_1 \\ -a_{21}x_1 + (1 - a_{22})x_2 & - \dots & & -a_{2n}x_n = y_2 \\ \dots & \dots \quad \dots & & \dots \quad \dots \\ -a_{n1}x_1 & -a_{n2}x_2 - \dots & + (1 - a_{nn})x_n & = y_n \end{array} \tag{3}$$

If the final demand, $y_1, y_2, \dots, y_n$, that is, the quantities of all the different kinds of goods absorbed by households and all other sectors whose outputs are not represented by the variables appearing on the left-hand side of Eqns. (3), is assumed to be given, the system can be solved for the n total outputs, $x_1, x_2, \dots, x_n$.

The general solution of these equilibrium equations for the "unknown" x's in terms of the given y's can be presented in the following form:

$$\begin{aligned} x_1 &= A_{11}y_1 + A_{12}y_2 + \dots + A_{1n}y_n \\ x_2 &= A_{21}y_1 + A_{22}y_2 + \dots + A_{2n}y_n \\ &\dots \quad \dots \quad \dots \quad \dots \\ x_n &= A_{n1}y_1 + A_{n2}y_2 + \dots + A_{nn}y_n \end{aligned} \tag{4}$$

The constant A_{ij} indicates by how much the output x_i of the ith sector would increase if y_j, that is, the quantity of good j absorbed by households (or any other final users), had been increased by one unit. Such an increase would affect sector i directly (and also indirectly) if $i = j$, but when $i \neq j$ the output x_i is affected only indirectly, since sector i has to provide additional inputs to all other sectors which in their turn—directly or indirectly—must contribute to the increase in the delivery y_j made by sector j to the final users. From the computational point of view, that means that the magnitude of each coefficient A in the "solution" (4) in general depends on all the input coefficients a appearing on the left-hand side of the system of equilibrium equations (3).

In mathematical language, the matrix

$$\begin{bmatrix} A_{11} & A_{12} & \cdots & A_{1n} \\ A_{21} & A_{22} & \cdots & A_{2n} \\ \vdots & \vdots & & \vdots \\ A_{n1} & A_{n2} & \cdots & A_{nn} \end{bmatrix}$$

of constants appearing on the right-hand side of the "solution," Eqns. (4), is identified as the "inverse" of the matrix

$$\begin{bmatrix} (1-a_{11}) & -a_{12} & \cdots & -a_{1n} \\ -a_{21} & (1-a_{22}) & \cdots & -a_{2n} \\ \vdots & \vdots & & \vdots \\ -a_{n1} & -a_{n2} & \cdots & (1-a_{nn}) \end{bmatrix}$$

of constants appearing on the left-hand side of Eqns. (3). The computation involved in finding such a solution is called the inversion of the coefficient matrix of these original equations. The "inverse" of the matrix

$$\begin{bmatrix} (1-0.25) & -0.40 \\ -0.14 & (1-0.12) \end{bmatrix}$$

based on Table 3 is

$$\begin{bmatrix} 1.4570 & 0.6623 \\ 0.2318 & 1.2417 \end{bmatrix}$$

Inserted in Eqns. (4) this yields two equations:

$$\begin{aligned} x_1 &= 1.4570y_1 + 0.6623y_2 \\ x_2 &= 0.2318y_1 + 1.2417y_2 \end{aligned} \tag{5}$$

which permits us to determine what total outputs of agricultural and manufacturing sectors, x_1 and x_2, would correspond to any given combination of the deliveries of their respective products, y_1 and y_2, to the exogenous sector, households. To verify this result by comparing them with the corresponding entries in Table 1, set $y_1 = 55$ and $y_2 = 30$ on the right-hand sides of Eqns. (5) and they will yield $x_1 = 100$ and $x_2 = 50$ on the left.

Only if all elements A_{ij} of the inverted matrix are non-negative will there necessarily exist for any given set of final deliveries, $y_1, y_2, \ldots, y_n$, a combination of positive total outputs, $x_1, x_2, \ldots, x_n$, capable of satisfying it. A sufficient condition for this is that the determinant of the matrix,

$$\begin{vmatrix} (1-a_{11}) & -a_{12} & \cdots & -a_{1n} \\ -a_{21} & (1-a_{22}) & \cdots & -a_{2n} \\ \vdots & \vdots & & \vdots \\ -a_{n1} & -a_{n2} & \cdots & (1-a_{nn}) \end{vmatrix}$$

and of all its principal submatrices,

$$(1-a_{11}) > 0 \begin{bmatrix} (1-a_{11}) & -a_{12} \\ -a_{21} & (1-a_{22}) \end{bmatrix} > 0 \ldots$$

$$\begin{bmatrix} (1-a_{11}) & -a_{12} & \cdots & -a_{1n} \\ -a_{21} & (1-a_{22}) & \cdots & -a_{2n} \\ \vdots & \vdots & & \vdots \\ -a_{n1} & -a_{n2} & \cdots & (1-a_{nn}) \end{bmatrix} > 0$$

should be positive. If this so-called Hawkins–Simon condition is satisfied for one arbitrarily numbered sequence of sectors, it is necessarily satisfied for any other sequence too. The material interpretation of that condition is that if an economic system in which each sector functions by absorbing output of other sectors directly and indirectly is to be able not only to sustain itself but also to make positive delivery to final demand, each one of the smaller and smaller subsystems contained in it must necessarily be capable of doing so too. If even one of them cannot pass that test, it is bound to cause a leak that will destroy the sustainability of the entire system.

A simpler sufficient, but not necessary, condition of sustainability of an economy is that the sum of the coefficients of each column of its structural matrix should be $\leqslant 1$, with at least one of the column sums strictly <1.

In most cases in which the structural matrix of a national economy has been derived from a set of actually observed value flows such as are represented, for example, in Table 2, the condition stated above will be satisfied.

Since in an open input–output system the households are usually treated as a final, that is, an exogenous sector, its total output, x_{n+1}, that is, the total employment, usually does not appear as an unknown variable on the left-hand side of Eqns. (3) and on the right-hand side of its solution, Eqns. (4). After the outputs of the endogenous sectors, $x_1, x_2, \ldots, x_n$, have been determined, the total employment can be computed from the following equation:

$$x_{n+1} = a_{n+1,1}x_1 + a_{n+1,2}x_2 + \ldots + a_{n+1,n}x_n + y_{n+1} \tag{6}$$

The technical coefficients, $a_{n+1,1}, a_{n+1,2}, \ldots, a_{n+1,n}$, represent the inputs of labor absorbed by various industries (sectors) per unit of their respective output; y_{n+1} is the total amount of labor directly absorbed by households and other exogenous sectors. Such an employment equation constructed for the three-sector system with the structural matrix shown in Table 3 is:

$$x_3 = 0.80\,x_1 + 3.60\,x_2 + y_3 \tag{7}$$

Households must not necessarily be considered to be part of the exogenous sectors as they are in the example used above. In dealing with problems of income generation in its relation to employment, the quantities of consumer goods and services absorbed by households can be considered to be structurally dependent on the total level of employment in the same way as the quantities of coke and ore absorbed by blast furnaces are considered to be structurally related to the amount of pig iron produced by them. With households shifted to the left-hand side of Eqns. (2, 4), the exogenous final demand appearing on their right-hand side will comprise only such items as governmental purchases and exports and, in any case, additions to or reductions in stocks of goods, that is, real investment or disinvestment.

When all sectors and all purchases are considered to be endogenous, the input–output system is called closed. A static system cannot be truly closed, since endogenous explanation of investment or disinvestment requires consideration of structural relationships between inputs and outputs which occur in different periods of time (see Sect. 3).

2.2 Exports and Imports

In an input–output table of a country or a region which trades across its borders, exports can be entered as positive and imports as negative components of final demand. If the economy described in Table 1 ceased to be self-sufficient and started to import, say, 20 bushels of wheat and to export 8 yards of cloth—while letting the households consume the same amounts of both products as before—a new balance between all inputs and outputs would be established, as described in Table 4.

The input coefficients of the endogenous sectors, and consequently also the structural matrix of the system and its "inverse," remain the same as before. To form the new column of final demand, we have to add to the quantity of each good absorbed by the households the amount that was exported and subtract the amount that was imported (i.e., imports can be treated as negative exports):

$$y_1 = x_{1n} + e_1, \quad y_2 = x_{2n} + e_2, \ldots, y_n = x_{nn} + e_n \quad (8)$$

The corresponding sectoral outputs can then be derived (see above) from the general solution (4). For the present numerical example, we can use directly Eqns. (5). The total labor requirement of the economy—300 man–years—remains in this particular case unchanged after it enters foreign trade, because the total direct plus indirect labor content of the 20 bushels of imported wheat happens to be equal to the labor content of the 8 yards of exported cloth.

If the imports of good i—that is, the negative e_i—happens to exceed the final domestic consumption of that good x_{in}, the corresponding "net" final demand y_i will turn out to be negative. As y_i diminishes, the total output of all sectors and in particular the total output x_e must (ceteris paribus) diminish. At some point, that output will be reduced to zero, which means that the entire direct and indirect demand for that particular commodity will be covered by imports. The corresponding domestic industry will be automatically eliminated from the endogenous part of the input–output table. The imports of such goods are called noncompeting, particularly when—as in the case of coffee and certain minerals—even a large increase in demand does not call forth their domestic production. The magnitude of total domestic demand for noncompeting imports can be computed in the same way as the total demand for labor can be derived from Eqn. (6).

2.3 Prices in an Open Static Input–Output System

Prices are determined in an open input–output system from a set of equations which states that the price which each productive sector of the economy receives per unit of its output must equal the total outlays incurred in the course of its production. These outlays comprise not only payments for inputs purchased from the same and from the other industries, but also the "value added," which essentially represents payments made to the exogenous sectors:

$$\begin{array}{lllll} (1 - a_{11})p_1 & -a_{21}p_2 & - \ldots & -a_{n1}p_n & = v_1 \\ -a_{12}p_1 & + (1 - a_{22})p_2 & - \ldots & -a_{n2}p_n & = v_2 \\ \ldots & \ldots & \ldots & \ldots & \ldots \\ -a_{1n}p_1 & -a_{2n}p_2 & - \ldots & + (1 - a_{nn})p_n & = v_n \end{array} \quad (9)$$

Table 4
Simplified input–output table showing the effects of imports and exports on a three-sector economy

from \ into	Sector 1: Agriculture	Sector 2: Manufacture	Final demand: Sector 3: Households	Final demand: Exports (+) or Imports (−)	Final demand: Total final demand	Total output
Sector 1: Agriculture	19.04	22.12	(55)	(−20)	35	76.16 bushels
Sector 2: Manufacture	10.66	6.64	(30)	(+8)	38	55.30 yards
Sector 3: Households	60.93	199.07	(40)		40	300 man-years

Each equation describes the balance between the price received and the payments made by each endogenous sector per unit of its product; p_i represents the payments made by sector i—per unit of its product—to all exogenous (i.e., the final demand) sectors. These usually comprise wages, interest on capital and entrepreneurial revenues credited to households, taxes paid to the government and other final demand sectors.

In analogy to the solution (4) of output equations (3), the solution of the price equation (9) permits the determination of prices of all products from the given values added (per unit of output) in each sector:

$$\begin{aligned} p_1 &= A_{11}v_1 + A_{21}v_2 + \ldots + A_{n1}v_n \\ p_2 &= A_{12}v_1 + A_{22}v_2 + \ldots + A_{n2}v_n \\ \ldots &\quad \ldots \quad \ldots \quad \ldots \quad \ldots \\ p_n &= A_{1n}v_1 + A_{2n}v_2 + \ldots + A_{nn}v_n \end{aligned} \tag{10}$$

The constant A_{ij} measures the dependence of the price p_i of the product of sector j on the value added, v_i, earned per unit of its output in sector i.

Each row of the a_{ij} coefficients appearing in the output equations (3) makes up the corresponding column of coefficients appearing in price equations (9); the A_{ij} coefficients appearing in each row of the "output solution" equations (4) make up the corresponding coefficient column in the "price solution" equations (10).

Thus, inserting the inverse computed in the example used above in the Eqn. (10) solution of the price equation, we have:

$$\begin{aligned} p_1 &= 1.4570v_1 + 0.2318v_2 \\ p_2 &= 0.6623v_1 + 1.2417v_2 \end{aligned} \tag{11}$$

From Tables 2 and 3 we can see that in our example the values added paid out (i.e., the wages) by agriculture and in manufacture per unit at their respective outputs amounted to \$0.8 and \$3.6. According to the two equations above, this yields $p_1 = \$2$, $p_2 = \$5$, which are the prices of agricultural and manufactured products used in deriving the value figures presented in Table 2 from Table 1, which described the input–output flows only in physical units.

The internal consistency of the price and the quantity relationships within an open input–output system is confirmed by the following identity derived from Eqns. (4, 9):

$$x_1v_1 + x_2v_2 + \ldots + x_nv_n = y_1p_1 + y_2p_2 + \ldots + y_np_n \tag{12}$$

On the left-hand side stands the sum total of value added paid out by the endogenous to the exogenous sectors of the system; on the right-hand side are the combined values (quantities times prices) of their respective products delivered by all endogenous sectors to the final (exogenous) demand. This identity confirms, in other words, the accounting identity between the national income received and the national income spent, as shown in Table 2.

For the purposes of more detailed price analysis, the technical "cooking recipe" for producing, say, one ton of bread not only has to specify the requisite amounts of current inputs such as flour, milk and yeast but also has to list needed pots and pans and other kinds of capital goods required for that purpose. Thus the matrix A of technical flow coefficients has to be supplemented by a corresponding matrix of capital stock coefficients, B:

$$B = \begin{bmatrix} b_{11} & b_{12} & \ldots & b_{1n} \\ b_{21} & b_{22} & \ldots & b_{2n} \\ \vdots & \vdots & & \vdots \\ b_{n1} & b_{2n} & \ldots & b_{nn} \end{bmatrix}$$

A capital coefficient b_{ij} represents the technologically determined stock of the particular kind of goods—machine tools, equipment, industrial buildings, "working inventories"—produced by industry i that industry j has to employ per unit of its output. In other words, each column of matrix B describes the physical capital requirements (per unit of its total output) of a particular industry, in the same way that the corresponding column of matrix A describes its "current inputs" requirements.

The price analysis outlined above can be advanced one step further by splitting each of the "values added," appearing on the right-hand side of each Eqn. (9), into two parts: the returns on capital invested in buildings, machinery and other stocks of goods required for production of the output in question on the one hand, and wages on the other. The return on capital can be represented as the value (price times quantity) of all productive stocks (used per unit of its output) in each industry multiplied by the given rate of return.

The relationship between wage rates, the rate of return on capital (i.e., the "price" of capital) and the price of different goods and services takes on the following form:

$$p = (1 - A' - rB')^{-1}\boldsymbol{W} \tag{13}$$

where $\boldsymbol{W}$ is a column vector of wage costs paid by different industries per unit of their respective outputs.

Insertion in Eqn. (13) of the numerical values of the flow coefficient matrix A' as given in Table 3 and of capital coefficient matrix B as given above and "inverting" the bracketed expression on the right-hand side yields an explicit solution of that equation for various values of the rate of return on capital, r. For instance,

if $r = 10\%$	if $r = 20\%$
$p_1 = 1.55w_1 + 0.26w_2$	$p_1 = 1.65w_1 + 0.30w_2$
$p_2 = 0.76w_1 + 1.34w_2$	$p_2 = 0.87w_1 + 1.45w_2$

The magnitudes of all numerical parameters increase as the value of r is raised from 10 to 20%. That means that with given wage costs, w_1 and w_2, a rise in the rate of return, that is, in the costs of capital, must obviously result in higher prices p_1 and p_2. With higher prices, the purchasing power of money wages (i.e., the real wages) must necessarily fall.

3. *Dynamic Input–Output Analysis*

The following set of linear difference equations represents dynamic input–output relationships employed in description and analysis of the process of economic growth.

$$X(t) - AX(t) - B[X(t+1) - X(t)] = Y(t) \quad (14)$$

The column vectors $X(t)$ and $X(t+1)$ represent the output levels of different industries in time periods t and $t+1$, while the column vector $Y(t)$ represents the amounts of various goods and services delivered in year t by these producing sectors to households and other final users. A is the matrix of input coefficients referred to above, while B is the matrix of capital coefficients described above.

The balance relationship described by Eqn. (14) is based on the assumption that a good added to the capital stock in year t is put to use in the year $t+1$.

In a closed version of this dynamic system the "final demand" sectors are treated as if they were absorbing, like ordinary industries, inputs originating in other sectors and producing outputs—for instance, labor services—that they, in turn, deliver to other sectors. The flow and the capital coefficients reflecting the structure of households, government and other final demand sectors appear in a closed input–output model on the left-hand side of the dynamic balance equation side by side with all other industries, so that the column vector $Y(t)$ becomes zero on the right-hand side, its contents having been transferred to the left-hand side.

By setting the determinant of the characteristic matrix $|1 - A - \lambda B|$ of the resulting homogeneous system of linear difference equations equal to zero, we can determine the values of its n characteristic roots, $\lambda_{11}, \lambda_{21}, \ldots, \lambda_{n1}$. By the so-called Frobenius theorem the largest of these roots is necessarily simple and positive and so are all elements of its characteristic vector.

The reciprocal of that root, $1/\lambda_{max}$, represents the rate at which the closed economy described by dynamic equations will expand, while the relative magnitudes of the elements of the characteristic vector corresponding to that root represent the relative levels of sectoral outputs (including the output of labor produced by households) that have to expand evenly from year to year.

The set of difference equations stating the price relationships corresponding to the physical relationships, described by Eqn. (14), has to include, among other cost elements, interest payments on the stock of capital invested in each industry. The "real" rate of interest, that is, the money rate adjusted for the change in the general price level, turns out to be equal to $1/\lambda_{max}$, that is, to the growth rate of the economy.

The economy described in the following numerical example has the flow matrix shown in Table 2. The capital requirements of its three sectors are represented by the following matrix of capital coefficients.

$$B = \begin{bmatrix} 0.35 & 0.050 & 0.105 \\ 0.01 & 0.515 & 0.320 \\ 0 & 0 & 0 \end{bmatrix}$$

The right-hand column of coefficients describing the capital structure of a household refers to stocks of agricultural products normally held in family larders and textile products stored in linen closets.

Unlike agricultural and manufactured goods produced by the first two sectors, labor services supplied by the third cannot be stored and consequently cannot, according to the definitions used, be part of any capital structure. The bottom row of a B matrix therefore contains only zeros, which means that the matrix B is singular and cannot be inverted.

With only two industries contributing to capital formation, Eqn. (14) can be transformed (by expressing the magnitude of the third variable in terms of the other two through the use of the third equation) into a system containing only two linear difference equations of the same general form. Its two roots are those of the original system. In the present example they are found to be 0.39252 and 24.981.

The corresponding eigenvectors of the original three-equation system are

$$\begin{bmatrix} 0.26388 \\ 0.18210 \\ 1.0000_0 \end{bmatrix} \quad \text{and} \quad \begin{bmatrix} -0.041998 \\ 0.250070 \\ 1.00000_0 \end{bmatrix}$$

The reciprocal of the larger of the two roots is $1/24.981 = 0.04003$. That means that the economy as a whole, that is, all its three sectors, can expand at a rate of 4% per annum and that the relative level of their outputs will year after year be proportional to the relative magnitude of the three elements of a characteristic vector corresponding to that root.

The potential growth rate computed on the basis of the reciprocal of the much smaller second root would be much higher. However, since some of the elements of the corresponding characteristic vector have different signs, the output of some sectors would have to become negative with the passage of time, which of course is physically impossible.

Tests, based on empirically observed sets of flow and capital coefficients, have shown that in both the US and the Japanese economies the relative levels of outputs of different sectors do not deviate very much from those computed on the basis of the corresponding closed dynamic models. Nevertheless, for the purposes of most practical applications, the closed version of the dynamic model has proved to be too deterministic and too rigid; the input–output analysis is usually conducted in terms of the open version of the dynamic model described by Eqn. (14). The final bill of goods $\boldsymbol{Y}(t)$ of successive years is treated in this case as given, that is, prescribed or projected on the basis of some exogenous information or assumption. Then the vector $\boldsymbol{X}(0)$ describes the total output level of all producing sectors in the base year 0. The levels of output for subsequent years can be determined by a recursive computation based on rewriting Eqn. (14) in the following form:

$$X(1) = B^{-1}[(1 - A + B)X(0) - Y(0)] \qquad (15)$$

The following simple example of an open dynamic input–output model is based on information contained in the coefficient matrices A and B used in the example of a closed model above. Since in the present case the vector of final demand is considered as given, the structural relation, if it exists at all, between the output and the input of households is considered to be unknown; only the feedback relationships between the agricultural and the manufacturing sectors have to be taken into account. Accordingly, what might be called the dynamic core of the system to be solved is reduced to only two equations. After insertion of the appropriately reduced matrices A and B into Eqn. (15), starting with the given $\boldsymbol{X}(0)$ and using in the successive rounds of computation the externally determined vectors of final demand, $\boldsymbol{Y}(0)$, $\boldsymbol{Y}(1)$ and so on, one can compute, step-by-step, the levels of output (and investment) of both industries for all the years. The corresponding employment levels are determined by a separate subsidiary computation using labor input coefficients taken from Table 3. Some results are shown in Table 5.

4. Technological Change

Since the technological structure of each sector of the economy is represented by a column vector of input coefficients and the corresponding column vector of capital coefficients, technological change can be described concisely as a change in the magnitudes of the elements of these vectors. Introduction of new commodities or industries is represented through introduction of new and disappearance of old commodities (or industries) through elimination of old vectors from the structural matrix of the economy in question.

The choice between two (or more) alternative processes that might be available for production of a particular good or service must obviously be based on a comparison of the effects of a hypothetical shift from one technology to another. For instance, a shift from coal-generated to atomic energy would affect the cost of production, prices and the level of output and input of goods directly or indirectly. To determine these effects, several input–output computations have to be carried out, each based on the introduction into the flow (A) and into the capital (B) matrix of the economy in question of coefficient vectors characterizing the alternative technologies available for the industries in question. In the case where the choice of appropriate technology can be based on maximizing or minimizing an explicitly defined function of some variable—such as the aggregate input requirements for labor or specific natural resources, investment requirements or the cost of production of various goods (whose magnitudes can be determined by means of appropriate input–output computations)—it can be formalized and carried out with the help of an appropriate linear programming algorithm. Dr. George Danzig, the inventor of the well-known Simplex method of linear programming, actually developed it first as a means of automating input–output computations involving sequential substitution of alternative column vectors into square input–output coefficient matrices.

5. The Scenario Approach

Practical application of input–output analysis often takes the form of comparisons of the implications—described in terms of complete projected input–output tables—of several alternative scenarios, each based on a different set of assumptions concerning the level and the composition of the final demand,

Table 5
Simplified example of solution of an open dynamic input–output model

	Final demand Sector		Total output Sector		Investment Sector		Employment Sector			
Year	1	2	1	2	1	2	1	2	3	Total
0	55	30	115	60	7.3	6.7	92	216	11	319
1	57.7	31.5	134	73	13.6	13.7	107	261	12	380
2	60.6	33.3	169	99	26.8	29.9	135	355	12	502

changes in the magnitudes of input coefficients incorporated in various column vectors of the flow and capital coefficient matrices, or a combination of both.

Shortly before the end of World War II, President Roosevelt asked the US Labor Department to assess the probable effect on the American economy of the impending transition from the war to a peacetime footing. A static input–output model was constructed on the basis of a matrix of structural input coefficients derived from the 1939 input–output table of the US economy, the first such table compiled in the USA under government auspices. A comparison of the output and unemployment levels attained in all industries under war conditions with the hypothetical output and employment levels, computed on the assumption that a vector of final demand representing normal civilian consumption would be substituted for the vector of final delivery dominated by military goods, provided a detailed and internally consistent answer to the question raised. To the great surprise of experts who predicted a slump in steel—conventionally considered to be a "war industry"—these input–output computations led to the conclusion that substitution of a normal peacetime vector for the wartime vector of final demand would lead to a sharp rise in output and employment level in the steel sector. Subsequent development demonstrated that that conclusion was indeed correct.

Many, if not most, studies aimed at assessment of future energy demand and the effects of a shift from oil to coal or to atomic power have involved the use of supply and demand "elasticities" derived by means of simple or multiple correlation analyses applied to time series describing past changes in energy input, energy prices and prices of other goods. The contribution of an input–output approach to a consideration of the energy problem has consisted on the other hand in the construction of several alternative scenarios, each involving a different combination of input–output vectors describing the technical structure of various methods of producing and using energy.

Such computations have, for instance, shown that while alcohol distilled from grain does indeed improve the energy balance of Brazil, it would not do so in the USA. Taking into account the amounts of energy absorbed directly and indirectly in operating agricultural machinery and producing chemical fertilizer in the USA, it is found that more than one thermal unit would be used to supply one unit in the form of alcohol. A similar computation based on the Brazilian input–output table yields the opposite result.

In this connection, it has to be pointed out again that such computations necessarily take into account the entire input–output structure of the economy in question, including that of its foreign trade.

The first practical application of the input–output method to systematic study of materials flow was carried out at the end of World War II in the USA by the Western Electric company. Lead was, next to copper, one of the principal materials used at that time to manufacture electric cable. Expecting a rapid rise in demand for its own products, as well as the products of many other industries depending on the supply of lead, the management of that company carried out an input–output projection of production and consumption a few years ahead and came to the conclusion that critical shortages were bound to develop. On the basis of that finding, Western Electric initiated a crash research program aimed at replacing lead with a suitable polymeric material in cable manufacture.

A recent application of the input–output methodology to systematic study of materials has been reported by Leontief et al. (1983). It is based on a modified input–output model of the US economy imbedded into the previously constructed multiregional input–output model of the world economy. The core of the structural matrix consists of the official US input matrix with the five of its 106 sectors which depict production of nonferrous minerals expanded to 36 sectors, describing production, processing and consumption of 26 nonfuel metallic and nonmetallic minerals. Mine output supplemented by product output resulting from other mining operations and reprocessing of scrap is described in great detail in the corresponding matrix of technical coefficients.

The system of 321 equations (containing 328 variables) describing the balance between the total supply including imports and the total use including exports of various goods, among them all nonferrous metals in different forms, as well as the generation and elimination of major pollutants and allocation of labor, is presented in a schematic form below.

The system is described by the following set of equations:

$$p_1^{(f)} = (I - A_c) \,.\, q_1^{(f)} + I \,.\, q_3^{(f)} \tag{16}$$

$$p_2^{(f)} = -B_c \,.\, q_1^{(f)} + (I + C_c) \,.\, q_2^{(f)} + G_c \,.\, q_4^{(f)} \tag{17}$$

$$p_3^{(f)} = L_c \,.\, q_1^{(f)} - I \,.\, q_3^{(f)} \tag{18}$$

$$p_4^{(f)} = M_c \,.\, q_1^{(f)} + N_c \,.\, q_0^{(f)} + H_c \,.\, q_4^{(f)} \tag{19}$$

$$p_5^{(f)} = D_c \,.\, q_1^{(f)} - I \,.\, q_5^{(f)} \tag{20}$$

where the subscript c indicates a matrix of coefficients while superscript f refers to flows. Equation (16) states that the gross domestic output of each commodity plus imports minus intermediate consumption must satisfy final demand. Similarly, Eqn. (17) states that the domestic output of minerals (own industry plus by-product) plus competitive imports minus intermediate consumption must equal final demand for minerals. Equation (18) states that the level of imports for each commodity is equal to a specified fraction of domestic (own industry) output. Equation (19) states the same proposition for noncompetitive minerals (primary and scrap). Equation (20) states that the sum of each industry's value added, labor inputs and emissions "output" equals the respective total for the economy as a whole. As explained previously, noncompetitive imports are goods used to satisfy intermediate or final demand for which there is no corresponding domestic producing sector.

A solution vector has the following form:

$\boldsymbol{Q}^{\mathrm{f}} =$
- $q_1^{(\mathrm{f})}$ = a 106×1 vector of commodity output levels in time t
- $q_2^{(\mathrm{f})}$ = a 36×1 vector of mineral and scrap output levels in time t
- $q_3^{(\mathrm{f})}$ = a 106×1 vector of commodity imports levels in time t
- $q_4^{(\mathrm{f})}$ = a 39×1 vector of mineral and scrap import levels in time t
- $q_5^{(\mathrm{f})}$ = a 34×1 vector of value added, labor requirements, energy consumption, pollution emission and new scrap generation levels in time t

and:

$\boldsymbol{P}^{\mathrm{f}} =$
- $p_1^{(\mathrm{f})}$ = a 106×1 vector of final demand components minus imports for 106 commodities valued in dollars for time t
- $p_2^{(\mathrm{f})}$ = a 36×1 vector of final demand components minus imports for 36 mineral commodities in physical units for time t
- $p_3^{(\mathrm{f})}$ = a 106×1 null vector
- $p_4^{(\mathrm{f})}$ = a 39×1 vector with zeros everywhere except in rows 253, 254, 265 and 288, whose elements give final demand minus import levels for noncompetitive imports in time t
- $p_5^{(\mathrm{f})}$ = a 34×1 null vector

The symbols used in Eqns. (16–20) are defined as follows.

A_{c} a commodity by commodity matrix of input–output coefficients (106×106) whose elements $a_{ij}^{(\mathrm{c})}$ give dollar amounts of input i required to produce one dollar's worth of output j (valued in base year prices).

0 a null matrix or vector.

I an identity matrix.

$-B_{\mathrm{c}}$ an input–output coefficient matrix (36×106) whose elements $b_{ij}^{(\mathrm{c})}$ give the physical amount of mineral (primary or scrap) input i required to produce one dollar's worth of output j (only minerals which are produced in the USA are included in this submatrix).

C_{c} a diagonal by-product coefficients matrix (36×36) whose elements $c_{ij}^{(\mathrm{c})}$ give the physical amount of each mineral (primary or scrap) produced as a by-product per physical unit of its "own industry" output.

G_{c} a step-diagonal matrix (36×40) whose nonzero elements $g_{ij}^{(\mathrm{c})} = 1$.

L_{c} a diagonal import coefficient matrix (106×106) whose elements $l_{ij}^{(\mathrm{c})}$ give the dollar amount of imports per dollar's worth of j ($i = j$).

M_{c} a matrix (39×106) whose only nonzero elements appear in IEA–USMIN rows 253, 254, 265 and 288; the elements of these four rows, $m_{ij}^{(\mathrm{c})}$, give the physical or dollar amount of noncompetitive import i per dollar's worth of output j ($i = j$).

N_{c} a step-diagonal matrix (39×36) whose nonzero elements $n_{ij}^{(\mathrm{c})}$ give the physical amount of mineral (primary or scrap) i imported per physical unit of mineral j's "own industry" output ($i = j$).

H_{c} a diagonal matrix (39×39) whose nonzero elements $h_{ij}^{(\mathrm{c})} = -1$, except in rows 253, 254 and 288, where $h_{ij}^{(\mathrm{c})} = 1$.

D_{c} a matrix (34×106) whose elements d_{ij} give the amounts in dollars or physical units of value added, labor, energy, pollution emissions and new scrap associated with a dollar's worth of output of commodity j.

Such a system of equations was solved (to check its internal consistency) for the base year 1972 and for the years 1980, 1990, 2000 and 2030. Systematic projections of future changes in all sets of technological coefficients, particularly those reflecting efficient methods of extraction and refining and substitution between different materials, were the most demanding part of that task. The estimates of future changes in the exports and imports (that enter into the system as vectors of exogenously determined variables) were obtained by incorporating the system into the multiregional input–output analysis of the world economy constructed for the United Nations several years earlier.

Alternative projections were computed, based on eleven different scenarios. Each of these represented a different combination of specific assumptions concerning the dependence of the US economy on imports of nonferrous metals and future rates of technological change. Final conclusions were summarized in the form of separate observations on the present and expected future supply and demand for each nonferrous mineral on the domestic US and international markets.

One of the most ambitious applications of the input–output approach was the construction of a multiregional, multisectoral, dynamic input–output model of the entire world economy referred to above. That model was employed in the preparation of long-run projections based on alternative scenarios of prospective developments of the economic relationship between the developed and less developed regions. It also provided the basis for long-run projections of the economic growth (or decline as the case may be) of the various regions under alternative assumptions concerning population growth, technical change—particularly in the field of agriculture and energy production—and in the uncertain supply of various natural resources.

See also: Input–Output Analysis: Applications; Materials in the Industrial System

Bibliography

Carter A P 1970 *Structural Change in the American Economy.* Harvard University Press, Cambridge, Massachusetts

Davis C H, Lofting E M 1979 *Development of Interindustry Transactions Data on the Structure of United States Mining Industries for 1967 and a Comparison of Techniques for Updating Related Input–Output Coefficients.* US Department of the Interior, National Technical Information Service, Springfield, Virginia

Leontief W 1966 *Input–Output Economics.* Oxford University Press, New York (Expanded version, Oxford University Press, 1985)

Leontief W 1970 The dynamic inverse. In: Carter A P, Brody A (eds.) 1970 *Contributions to Input–Output Analysis.* North-Holland, Amsterdam

Leontief W 1977 *The Structure of the American Economy, 1919–39.* International Arts & Sciences Press, White Plains, New York

Leontief W, Carter A, Petri P 1977a *The Future of the World Economy.* Oxford University Press, Oxford

Leontief W, Chenery H B, Clark P G, Duesenberry J S, Ferguson A R, Carter A P, Grosse R N, Holzman M, Isard W, Kistin H 1977b *Studies in the Structure of the American Economy.* International Arts & Sciences Press, White Plains, New York

Leontief W, Koo J, Nasar S, Sohn I 1983 *The Future of Non-Fuel Minerals in the U.S. and World Economy.* Lexington Books, Lexington, Massachusetts

W. Leontief

Input–Output Analysis: Applications

Input–output models, also known as interindustry models and tools, can be applied to many areas of economic analysis. Major applications of interindustry analysis are structural analysis, impact analysis, projections of future events, and market analysis and specific industry studies.

1. Structural Analysis

A major question that is often asked is: How important are the materials industries to the economy? Input–output tables permit such a structural analysis, which may be defined as the analysis of the properties of an economic model that describes the structural interdependency of the economy. This type of analysis generally involves only two tools—forward linkage and direct backward linkage. The most common kind of structural analysis is the study of the role of various industries in the national economy. If SIC industry 3331 (primary copper metal) is taken as an example, the industry output distribution table of a US Department of Commerce interindustry structure study shows that about 60% of total output is sold to other industries, whereas SIC industry 121 (bituminous coal mining) indicates a forward linkage of 74%. Therefore, the bituminous coal mining industry has a higher forward linkage than that of copper metal. A strike in the bituminous coal mining industry will disrupt the national economy more than a strike in the primary copper metal industry. However, the copper metal industry shows a direct backward linkage of 92%, whereas bituminous coal mining shows only 41%. Thus, bituminous coal mining ranks lower than the copper metal industry in comparable structural analysis. In other words, the development of a new bituminous coal mine is likely to have less influence on the economy than developing a new copper primary metal plant with the same magnitude of new investment.

If the US input–output tables were available for every census year since 1958, the examination of the input–output relationships over an extended period of time by various intervals would show the trend in the changing economy as a whole, as well as in the various industrial sectors in the economy. Shifting patterns of input requirements and output distribution in any industry would indicate the impact of changing technology, innovation, consumers' choice and taste. It could also indicate shifts due to substitution of raw materials and other inputs, including capital equipment, based on price fluctuations. Furthermore, structural changes may be accounted for by variance in labor productivity and changes in relative wage levels, as well as relative price changes of inputs.

If input–output tables are available for two or more countries or regions within a country, a comparative analysis of the structure of the economies involved can be undertaken. Such analysis not only will show a simple comparison of the complexity of the economies involved, but also can be used by the policymakers in the respective countries or regions to assist in determining the type of economic development program that would be most advantageous in encouraging the desirable economic growth of the area in question. Currently, some 80 member nations of the United Nations have published input–output tables.

The statistical description of the interdependence of industry is in itself a valuable contribution to factual understanding of the structure of the economy. The systematic reconciliation of a vast amount of statistical data on output and cost of production will not only improve national statistics, such as disclosing data inconsistencies and gaps, but will also provide a sounder basis for estimating the national income and product accounts when they are worked out for the same period. The structural analysis of the national income and products accounts would be greatly facilitated by the by-products of interindustry analysis such as the deflation of gross national product by the industrial origin and also the analysis of production changes by sector, based on technical changes. Analyses of the national consumption of all types of materials in relation to economic growth and structure of countries can be conducted for the following: (a) intercountry comparisons of national requirements for materials, (b) input–output analysis of materials consumption in production sectors of national economies, (c) materials and energy requirements in relation to economic growth and structure, and (d) forecasting requirements for materials for a given final demand and technology.

2. *Impact Analysis*

The interindustry structure of the economy permits the computation of the quantitative impact of any specific changes in output of one industry on all other industries in the economy. The backward linkages and the variation of final demand schedule are the principal tools for impact analysis.

The most commonly used impact analysis is the determination of the impact of a change in the demand for an industry's output on all other industries in the economy. One can easily estimate such impacts by using the direct backward linkage and interindustry backward linkage tools. The impact of direct requirement in relation to the output of the mining industries may be rapidly ascertained by looking at a direct requirement coefficients table, whereas the impact of interindustry requirements of both the direct and indirect requirements in relation to the output of the mining industries can be quickly determined by looking at a total requirement coefficients table. It is easy to calculate the direct impact, but it is relatively difficult to ascertain the interindustry impact without the input–output tables.

Because of these features, the input–output tables can be readily used to assay the probable impact on all the industries in the economy of any foreseen or projected changes in the level of demand for the output of any specific industry that can be caused by changes in technology and innovation, consumers' choice and taste, price substitution and changes, foreign demand and supply, government policies, and/or business and industry decisions on investment expenditures. For example, the switching over from high explosives to the use of ammonium nitrate for mine blasting operations resulted in a decrease of the high explosives output and an increase of ammonium nitrate output. The total economic impact on other industries throughout the economy of these changes can be independently determined. Similarly, the interindustry impact of an increase in foreign demand for bituminous coal can be calculated. Another example would be if the government were to impose a tariff or quota on copper imports, which would result in an increase in output of the domestic copper industry. The interindustry impact on all industries in the economy could be estimated. More subtly, the total impact on the sand and gravel nonmetallic mineral industry from the government fiscal policy to maintain economic growth through a tax cut can be calculated.

So far, we have discussed only the measurement of the total impact upon output resulting from a given change in final demand, but business and government decisionmakers are also very interested in measuring the total impact upon income and employment. One of the more useful modern economic analytical techniques developed by Keynesian economics was that of the multiplier. Since Keynes dealt in broad aggregates, his income and employment multipliers were also highly aggregated. The concept of the aggregate multipliers is extremely useful in present-day public policy decisionmaking.

Aggregate multipliers are useful analytical tools, but they do not indicate the details of how multiplier effects are worked out throughout the economy by sectors, and at times both government and business decisionmakers are more interested in the sector details than in the overall impact. Some applications of interindustry analysis can be used to derive sector income and employment multipliers which can be used to estimate the probable total impact upon both income and employment by sectors.

Income effects trace the sectoral household income changes (wages and salaries generated in the row industry) resulting from a change in final demand; this is necessary because industry coefficients react differently to market influences than household coefficients with changes in demand. Interindustry or "Type I" income multipliers account for the direct and indirect effects on income with respect to the direct effect of a change in final demand in a particular industry. That is, as final demand changes in industry i, the income originating in that sector will be affected directly by this change. As all other industries adjust to a change in industry i, the income originating in these industries will be affected indirectly by the original change in industry i. The combination of these direct and indirect income effects, with respect to the direct income effect, are known as interindustry income multipliers.

The total or "Type II" income multipliers account for the direct and indirect effects on income, plus the induced income effects, which arise as consumers adjust their spending to changes in personal income, with respect to the direct income effect. Consequently, the total income multiplier will always be larger than interindustry income multipliers.

To generate income multipliers, the "opened" input–output model must be "closed" with respect to households. That is, the household sectors must be part of the endogenous mechanism of the model. In this respect, household consumers become much like machines, requiring inputs from the various sectors from which consumers purchase goods and services in order to produce an output called labor. The household coefficients are established when the household sector is made an endogenous sector by balancing personal consumption expenditures in the final demand vector with wages and salaries in the value-added row. These household coefficients serve as the basis from which income multipliers are calculated.

The income multipliers show different amounts of income which are generated by different industries as the level of output in each industry changes by the same amount. The differences in income changes are due to the interactions between industries and interrelations which are captured in the matrix of direct, indirect and induced income requirements. Information generated from income multipliers can indicate how changes in parts of the economy affect all other parts of the economy. Information of this nature can be valuable to decisionmakers in formulating policies and making rational choices among the various options available to them.

Type I and Type II multipliers have proved to be widely applicable for general magnitude estimating purposes. Specific impact analysis can be conducted for energy, materials and water requirements, pollutant output, and so on. For energy requirement impact analysis, the following procedure may be used: (a) divide the particular industry energy requirement by the industry gross output of the column to derive the average energy requirement coefficient for this industry; (b) develop the energy requirement coefficient industry-by-industry for the whole economy and the results will be a principal diagonal matrix of energy coefficients; (c) premultiply the diagonal matrix with the Leontief inverse matrix $(I-A)^{-1}$ of the input–output tables to derive the energy requirements interaction matrix; and (d) premultiply the energy interaction matrix with the final demand vector. The result will be the interindustry (direct and indirect) energy requirement by sector for the whole economy. The ratio of interindustry energy requirement divided by direct energy requirement yields the Type I energy requirement multiplier.

To derive the total energy requirement multiplier which embodies the direct, indirect and income-induced requirements, the energy requirements diagonal matrix is premultiplied with the Leontief inverse of the "closed" input–output model, with the endogenized household (personal consumption expenditure) column and wages and salaries value-added row. The total energy requirement interaction matrix can be premultiplied with the final demand vector to derive the total energy requirements by industry sectors. The total energy requirement is then divided by the original direct energy requirement to derive the Type II energy requirement multipliers. Similarly using a diagonal matrix of a specific material requirement coefficients, one can derive the interindustry specific material requirements and total specific material requirement, the Type I and Type II specific material multipliers, respectively.

Alternatively, using a diagonal matrix of average water requirement coefficients one can derive the interindustry water requirements and total water requirement, and Type I and Type II water multipliers, respectively.

Also, by dividing the industry-by-industry output of pollutants of solid or liquid waste with column totals of gross pollutant outputs one would get a principal diagonal matrix of average pollutant coefficients. Following a similar procedure, the interindustry and total pollutant output, as well as Type I and Type II pollutant output multipliers, can be calculated.

Similar procedures can be used to develop interindustry (direct plus indirect) labor requirements and total labor requirements that embody direct plus indirect and income-induced labor requirement, or Type I and Type II labor or employment multipliers, respectively. These can be derived using the developed principal diagonal matrix of average labor requirement coefficients, industry by industry.

By the use of supplementary data for capital investment requirements and a summary of available capacity pertaining to both construction and equipment cost including the material inputs consumed for the new plant and equipment construction and lead time, capital coefficients for every sector can be calculated and can be used to indicate the possible needs for and impact of additional plants and equipment. With the capital coefficients and the trend in production technology projections, the static input–output models can be transformed into dynamic models. Linear programming techniques can be incorporated into the model to find various constraints and ranges.

Industries are not located uniformly throughout the USA. For example, production of domestic copper ores is concentrated in the Rocky Mountain States. Production of oil and gas is concentrated in the southwest, the Rocky Mountain Region, and in California. With additional information on the geographic distribution of industry, regional input–output commodity flows (including trade and income accounts) can be used for developing models to obtain regional impacts. Thus, interindustry analysis can be used to shed light on the regional output, income and employment implications of many national and regional programs. This would make the input–output approach a powerful one for regional impact analysis as contrasted with industrial impact analysis; therefore, the impact analysis can be very useful in the formulation of programs of action. The various impacts of adding new industries to a region or the nation can be readily calculated by expanding the existing input–output model with relevant input data for new industries.

A new industry, such as a synthetic fuel industry, can be added to the input–output model with new industry input columns and new industry output distribution row coefficients. An old industry can be eliminated in similar fashion and input–output applications can be conducted on the revised models.

3. Projections of Future Events

It has been shown how the level and structure of production and its various requirements corresponding to a given final demand schedule (the autonomous sector) may be derived from an input–output matrix under the assumption of a constant input–output ratio. The basic tool used for this type of application is the variation of both the level and composition of the final demand schedule in which more than one industry's output is changed simultaneously.

The principal distinction between impact analysis and projections is that impact analysis generally involves only the change of one component of the final demand or the direct requirements placed on one or two industries.

A given or assumed change in the level and composition of the autonomous final demand schedule will project a new transactions table that would be consistent with the input–output matrix. This is also known as consistent forecasting. There is no assurance that consistent forecasting will turn out to be correct unless the structural input–output relations of the economy do not alter significantly over the projection period or unless proper allowance can be made for expected changes in the structural relations of the economy. The potentialities of input–output analysis for forecasting and programming or planning purposes are suggested by this production character of the input–output techniques.

These projections are used only to compute the various production requirements necessary to consistently sustain a certain level of final demand, but cannot be used in analyzing the size and character of final demand. Autonomous final demand must either be given as a goal or "plan" or else estimated by some techniques quite different from input–output analysis. The composition of final demand is determined independently, that is, outside of the input–output model. Despite some of the limitations, input–output analysis is probably one of the best available techniques for tracing the interindustry and total implications in terms of production levels and requirements that are consistent with any given input–output matrix for a specified bill of final demand, as stated in an economic plan or in an investment program for a country or region.

These derived data in economic programming may be of utmost importance by revealing in detail the new industrial structure and requirements implied by an economic plan or an investment program and by providing a foundation when resource availabilities are also taken into consideration for judging the reasonableness, for testing the feasibility, and for trying to locate potential bottlenecks. Analysis may indicate, for example, that a program is too ambitious and not feasible because the limited supply of some scarce resources such as skilled labor, foreign exchange, specific materials or specialized capital equipment is simply insufficient to support the given or projected bill of final demand. The fast growing or more desirable industries can also be identified by this analysis as logical candidates for future economic development programming.

A given investment program of a country or region, such as the construction of a multipurpose hydroelectric power dam project or of a new integrated iron and steel manufacturing complex, can be analyzed and evaluated by the use of input–output techniques based on the probable total requirements of domestic production, imports, and other scarce resources. In these cases, the direct requirements for investment in new or additional production capacity must be determined beforehand and must be added in with the additional final demand that will be generated from the higher level of investment activity to arrive at the new or desired bill of final demand. Using the new and given bill of final demand with the input–output tables, we can then determine the full impact on the economy with reference to those scarce resources of the program or project which must be estimated and evaluated simultaneously.

Since the 1960s, the Division of Economic Growth Bureau of Labor Statistics (BLS) of the US Department of Labor has prepared projections for 1970, 1980, 1985 and 1990 in an input–output framework, under the US Interagency Growth Project's US Economic Growth Model. The purpose of the Economic Growth Project is to develop a more comprehensive and integrated framework for analyzing the implications of long-term growth for a number of problem areas, particularly problems of manpower utilization. Similarly, materials problems can be analyzed.

The US Economic Growth Model is characterized by a number of distinct but interrelated steps. The first of these is the development of a potential growth rate for the economy, which is derived by projecting the labor force to the target year and projecting the change in average hours worked. The next step is to calculate and distribute this potential growth in real Gross National Product (GNP) among the major components of GNP. Then, the next stage is to calculate and distribute each of the major categories of expenditures into categories of demand including final demand matching the input–output producing industry classification used.

Demand in the national income accounting system covers only final demand, that is, only that of the ultimate consumers. Thus, in order to include those industries whose products are not sold to ultimate consumers, but are used up in the production process, input–output tables are utilized to translate final demand, such as personal consumption expenditure, government expenditure, investment, construction and exports, into the output required from all industries and services. These include primary ferrous and nonferrous metals, chemicals, textiles, iron ores, copper ores, coal, crude petroleum, electricity, basic agricultural products, lumber and other materials, as well as other sectors which do not usually sell a large fraction of their products to the ultimate consumers.

The basic input–output tables used in the Economic Growth Model are those produced by the US Department of Commerce and represent the technology and product mix for the year of the tables. A new set of projected input–output coefficients or sales–purchase ratios that would take into consideration changing technology and the anticipated changes in product mix due to different growth rates of product groups within the larger industry aggregates has been produced. These projected changes are related to the impact which they would have on an industry's purchases. Some changes are materials saving, others actually require larger inputs of materials; these factors are also introduced into the projections.

Given the demand projections and projected input–output tables, the necessary ingredients for estimating economic growth are present. The final step in the US Economic Growth Model is to develop employment estimates by industry. The Bureau of Labor Statistics accomplishes this by a set of industry projections (Miernyk 1981). Currently, the BLS is preparing similar projections for 1995 and later.

Using the BLS projected input–output tables, one can conduct impact analysis, such as adding an industry, eliminating an existing industry, incorporating new technology and/or changing product mix in an industry, measuring for water requirements, materials requirements, energy requirements, pollution output, and so on. Unfortunately, the BLS input–output industry classifications are usually not detailed enough to cover many of the specific materials that may be of interest in the analysis. Nevertheless, BLS projections can be utilized as the point of departure for further analysis.

Since 1977, Professor Wassily Leontief and his associates at the Institute for Economic Analysis, New York University (NYU), have been conducting consistent forecasting on production and consumption of basic minerals, and primary and secondary metal processing materials industries within the framework of multiregional input–output models of the US and world economies. The study was designed to address the growing national concern regarding the adequacy of nonfuel minerals in the US economy to the year 2030.

The study was intended to answer a host of questions which would affect the demand for mineral resourses. For example, what will be the effect on US mineral sectors' output levels, as well as on all other industrial sectors of the economy, if the reliance on foreign sources of supply is increased? How will the change in the rate of US population growth, and in the rest of the world's population, affect demand for primary materials resources? How will increased substitution of aluminum and/or precast concrete for steel in construction, or plastic pipes for copper pipes, or plastic tubes for lead tubes, affect the demands of these industries?

The NYU team projected the revised UN world input–output model to the year 2030 at 10-year intervals as well as its US input–output model. With this NYU long-term consistent forecasting model of the US and world economies, one can conduct many types of impact analyses on the demand and supply of minerals and other materials, with different scenarios and assumptions for analyzing and evaluating policy options.

4. Marketing Uses

The situations discussed so far involve evaluating government policies and operations, but input–output models also have many uses for corporate marketing analyses. Structural analysis has a number of implications for marketing research. Any single row in a base-year commodity flow or transactions table shows the detailed output distribution of the industry by its various immediate consumers. Such industry-wide marketing information is valuable for any firm in the industry to plan marketing strategy as well as location of future plants and warehouses. The forward linkage coefficients available in output distribution tables can be used for evaluating an individual firm's sales potential. For example, a businessman can compare his firm's present marketing distribution position with the industry-wide output distributing pattern and then recognize some of the probable areas of additional markets where his firm has yet to penetrate or exploit fully.

Furthermore, firms may want to rank and to determine the strength of indirect demand potentials, which may be very useful in planning sales efforts for developing new markets as well as intensifying efforts in existing markets. The structure of the table of commodity flows records the end results of demand patterns that give the final result.

Moreover, firms in industry frequently know their customers but seldom are aware of the industries which purchase the products of their customers. It would be beneficial to know their major customers' customers because an increase or decrease in the demand for their major customers' output would directly affect the future sales of the firm's own output. The input–output approach gives businessmen a better insight into such probable changes in the demand for its industry's output that may result from new activities of some seemingly unrelated industries.

The indirect linkage of iron ore and steel industries to the demand requirements for automobiles is apparent to nearly everyone; however, the indirect linkage of the steel industry to other end products may be quite complex and difficult to trace without the aid of the input–output tables. For instance, we may notice that there is a shift from single-family dwelling units to multiunit dwellings, such as large apartment/townhouse complexes. Using input–output analysis, not only can we derive the first-round impact on the construction industry, but we can also calculate the impact on the steel industry, iron ore industry, coal industry, limestone industry, and other materials industries, resulting from a shift in construction requirements. Furthermore, it is possible to determine the impact on iron and steel, coal and limestone industries of any changes in the final demands made by governments and households. All this information would be useful to a market research staff.

The production-requirement approach is used most often for backward analysis, which usually begins with the given final demand and then traces back through the processing industries and determines the production requirements upon the supplying materials industry, which are the interindustry requirement coefficients. It is also possible, given the final demand and input–output matrix, to trace the output of one industry forward to determine its ultimate dependence on any specific component of final demand.

Most of the mineral and mineral-processing materials industries have the common but distinguishing feature that their output distributions are principally shipped to other processing industries with very little shipped directly to final demand. This feature indicates that their respective output depends on two rather different factors: (a) the absolute production volume of the consuming industries, and (b) the relative importance of the supplying industry's output in meeting the direct requirement of the consuming industries. These two factors mean that the same changes in production volume of the different consuming industries may have very different effects on the demand of the supplying industry's output.

Any individual firm in a particular industry can not only use the input–output model to compare its own output and distribution pattern with that of the industry, but can also identify and measure the impact of its own operation within the economic system. The use of the model can complement other marketing research techniques used as a supplementary tool to provide longer range guidelines. For example, the National Planning Association and various economic consulting firms make regular and fairly reliable projections of GNP and its components, personal consumption (household) expenditures, private investments, net exports, and government purchases. Many firms subscribe to these sources, as an alternative to making their own projections. However, these projections are for end-product demands and not intermediate demand. Now with the interindustry model and the GNP projections, it is possible for the firm to project its market potential and goal for the next five to ten years in the future.

The firm may use the projections in conjunction with the interindustry model to determine the limits of its own and the industry's potential activities for the period of five to ten years hence. Given the range of projections of end products final demand, a range of intermediate demands may be established. Thus, the firm can zero in on the probable range of demand in the future, so that plans can be made for additions and sales, for additional plant and equipment investment, for shifting product markets, and for future sales promotion or curtailment, with reduction in risk in the decisionmaking process.

In the case of a small firm, whose volume of operation may be too low to integrate its activities meaningfully in the interindustry model, the market share of the industry's total output can be computed as given in the model. It can then make the projections for the industry and apply its market share, as well as apply its potential market expansion shares to the industry's outcome. It could also determine the impact on its operations of a gain or loss in existing market shares.

Thus, there are many possible marketing uses of interindustry analysis for firms in the private sector, all of which could enable businessmen to formulate better informed production and investment programs and to make wiser managerial decisions.

5. *Conclusion*

The input–output model and its applications have been tested and are workable for guiding today's decisionmakers on material policy and analysis. The model and its applications represent a definite progress and accomplishment in the state of the art in economic model building. The applications and uses are mostly static but useful. At the present time, advances in research design and applications are pushing the state of the art further into dynamic types of input–output analysis.

The principal thrust of input–output analysis, particularly in the business world, is centered in long-range contingency input–output forecasting, which complements the more traditional methods. This is probably because of several inherent advantages of interindustry forecasts over others.

The unique and advantageous characteristics of input–output or interindustry forecasting are: (a) it provides greater detail by specific industry and definite markets which the aggregate econometric models do not give, (b) it closes the gap between macro- and microforecasting, (c) it can take into full account changes in technology and in the production mix methods as well as altering the output production mix in the industry, (d) it explicitly takes into full account all the indirect and derived demands within the given time period, and (e) it provides an internally consistent framework of the economic structure that is fully integrated with the national income and products accounts for forecasting industry-by-industry sales.

As a word of caution, the input–output model in itself is not a self-sufficient forecasting model, since it depends on the future state of technology and on exogenous projections of the mix of final demand components. The consistent projections of each producing sector output can be accepted as forecasts only if the future state of technology and the detailed mix of final demand components are given exogenously. Accurate forecasts of both future technology and detailed final demand components are usually hard to come by. For this reason, the input–output model should not be used as a forecasting tool without fulfilling these necessary conditions. The model has not yet been used in making accurate forecasts. Only recently, Almon at the University of Maryland, and Leontief at New York University have begun to make consistent projections with their respective static interindustry models. It is hoped that in the future it will be possible to make some meaningful evaluation of these efforts.

Used in conjunction with modern computer technology, simulation for all major applications under various scenarios and assumptions, such as technology, economic growth rate, the level and components of government expenditures, tax and foreign trade policy, private consumption function, and private investment, can be performed readily with any interindustry model in conducting sound and consistent economic analysis. Input–output analysis can be a very powerful and versatile analytical tool, but one must use it carefully with full understanding of its limitations and problems.

See also: Input–Output Analysis; Materials in the Industrial System; Materials Economics, Policy and Management: An Overview

Bibliography

Almon C Jr 1974 *1985 Interindustry Forecasts of the American Economy.* Heath, Lexington, Massachusetts

Barna T (ed.) 1963 *Structural Interdependence and Economic Development.* St Martin Press, New York

Carter A P 1970 *Structural Changes in the American Economy.* Harvard University Press, Cambridge, Massachusetts

Carter A P, Brody A (eds.) 1970 *Applications of Input–Output.* North Holland, Amsterdam

Johnson R E 1965 *Input–Output Analysis: A Tool for Marketing Management: A User's Look.* National Planning Association, Washington, DC

Kutscher R E 1975 *The Structure of the U.S. Economy in 1980 and 1985*, Bureau of Labor Statistics Bulletin 1831. US Department of Labor, Washington, DC

Leontief W W, Carter A P, Petri P 1977 *The Future of the World Economy.* Oxford University Press, New York

Leontief W W, Koo J, Nasar S, Sohn I 1982 *Techniques for Consistent Forecasting of Future Demand for Major Minerals Using an Input–Output Framework.* Institute for Economic Analysis, New York University, New York

Lofting E M, Davis H C 1979 *The Input–Output Structure of US Mining Industries for 1958 and 1963: Transactions, Employment and Multipliers.* Dry Land Research Institute, University of California, Riverside, California

Miernyk W H 1981 Long range forecasting with regional input–output model. *West. Econ. J.* 6(3): 165–76

Radcliffe S V, Fischman L L, Schantz R Jr 1982 *Material Requirements and Economic Growth: A Comparison of Consumption Patterns in Industrialized Countries.* Resources for the Future, Washington, DC

Ritz P M 1965 *Input–Output Analysis Evaluation: Current Status and Uses.* National Planning Association, Washington, DC

Wang K L, Opyrchal A M 1981 *Economic Significance of the Florida Phosphate Industry: An Input–Output (I–O) Analysis*, Bureau of Mines Information Circular No. 8850. US Department of the Interior, Washington, DC

K. L. Wang

Intensity-of-Use of Materials

The intensity-of-use of raw materials (i.e., a ratio of the amount of a specific raw material used by a nation in a year to the final goods and services produced by the nation over that year) has been applied increasingly in analytical studies relevant to world minerals resources (Brooks and Andrews 1974). These were momentous years on the materials scene. World demand for industrial inputs of raw materials grew at record rates, notably during the years 1951–75. The world was introduced to a new materials

Malthusianism: unless there is a "major change in the present system . . . nonrenewable resource depletion . . . will certainly stop . . . population and industrial growth . . . within the next century at the latest . . ." (Meadows 1972). The world's less developed nations, striving to accelerate their limited economic modernization, formally launched a major drive for New International Economic Order (NEIO) at the 6th Special Session of the United Nations General Assembly in 1974, in order to reshape the world system so that it could better serve the economic progress of the 70% of the world's population in underdeveloped countries. Nonrenewable industrial raw materials have a key role in NEIO, as they now have in development policy and programs throughout the world.

This article presents theoretical and empirical aspects of the intensity-of-use concept in explanation of the role it has for materials analysis and policy over the decades ahead.

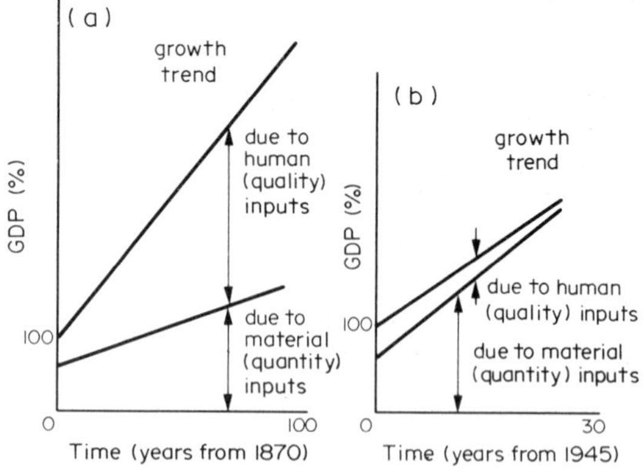

Figure 1
Human and material inputs in economic growth: (a) for rich countries and (b) for poor countries (after Malenbaum 1978)

1. Concept and Measurement

The intensity-of-use (IU) of a material *i* in area *j* over time *t* is defined as

$$_{ij}\mathrm{IU}_t = {_{ij}D_t}/{_j\mathrm{GDP}_t}$$

where $_iD$ is the tonnage of material *i* used and GDP is the total gross domestic product, the total real value of final goods and services produced in a specific period. Hence

$$_iD = \mathrm{GDP}\cdot{_i\mathrm{IU}} \equiv \mathrm{GDP}\cdot{_iD}/\mathrm{GDP}$$

The definition is a tautology: the usefulness of the decomposition of $_iD$ for the important purposes of comparisons over time and among regions depends upon independence of the determinants of changes in the two components GDP and IU for given *i*, *j* and *t*.

Modern economic-growth theory places primary emphasis on the role of human quality inputs; education, skills, attitudes and aspirations constitute the major forces for progress. More quantity inputs, increases in such traditional factors as labor and capital of constant productivity, are not primary contributors to dynamic economic and social change. Certainly the quantity $_iD$ cannot in itself be a determining force in GDP change.

Theoretical proposals for changes in IU have yet to be developed. The seeming input–output aspect of IU notwithstanding, technological explanations of the intensity ratio find limited support either in logic or measurement. Empirically, IU patterns appear to change with per capita product (GDP/population). Considered separately, neither size of GDP nor number of population is important in IU changes. Intensity increases with GDP/population as a poor region begins to develop; at a higher level of GDP/population, IU changes direction and declines with economic development (i.e., with expansion of GDP/population). The underlying patterns of change in the two components of $_iD$ are illustrated in Figs. 1 and 2.

The precise definition and the specific values of *D* and GDP must conform to what is available in accepted practice in the market places of the world and in international statistical agencies. Major applications of the concept involve comparisons over time and among regions of the world. For example, the amount of a metal *i* used in time *t* (i.e., $_iD_t$) refers to the weight of ore adjusted for metal concentration, or alternatively, the weight of the metal at a given stage of refining with or without scrap, as these alternatives are available in such regular international publications as the United Nations *Statistical Yearbook*, the United States Bureau of Mines *Minerals Yearbook*, the United Kingdom *Metal Bulletin*

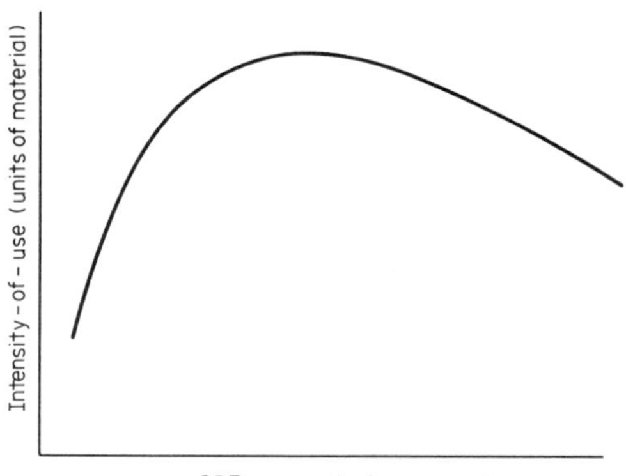

Figure 2
Typical intensity-of-use and GDP per capita (after Malenbaum 1978)

Handbook, the West German *Metal Statistics*, among others.

The metal or material "used" is apparent domestic consumption, which is domestic production plus imports minus exports, and is exclusive of materials traded in manufactured goods (e.g., in automobiles or their components) and exclusive of a nation's inventory changes of the material during the year. For the raw materials under consideration here, this definition is more straightforward for some materials (iron ore, refined copper, tin) than for others (manganese ore, chromium ore, platinum group metals). Problems with the numerator in the intensity-of-use ratio are well known in the minerals field: differences in the definition of the numerator are constantly being narrowed among regions. Thus, data for nonfuel natural resources that are relevant as inputs of modern industry are increasingly available. Data for their output, trade and use now tend to be delineated by the metal content of individual raw materials and are becoming standardized over most of the world.

For the denominator, the best measure is GDP, a quantity that national and international agencies report by nation, by year and at a constant price level over time. The growing importance in many applications of comparability among nations and time intervals assures persistent efforts at updating and improving national account data. In this regard, problems remain for standardizing comparisons of output among nations where factor endowments differ significantly: rich and poor nations have contrasting ratios of labor and fixed capital supplies. This complicates comparisons between rich and poor countries and their supplies and demands for industrial raw materials. The present IU measures rely primarily on United Nations data for real GDP, denominated for the world in US dollars and largely derived through converting national currencies by official exchange rates. Comprehensive alternative time series of international real GDP data are not yet available.

Thus assembled, measures of intensity-of-use, ${}_iD_t/\mathrm{GDP}_t$, have been computed for the years from about 1950 for most nations. This article draws on analyses of the following industrial raw materials: crude steel and iron ore; nickel, manganese ore, chromium ore, cobalt and tungsten; copper and aluminum; platinum metals, zinc and tin. At mid-1970s prices these 12 raw materials accounted for 80–90% of the value of world minerals production. Ten subdivisions of the world have been studied, each with a degree of homogeneity high enough to justify common assumptions about IU and GDP changes (Malenbaum 1978). These subdivisions are: (a) West Europe—Organization for Economic Cooperation and Development (OECD) countries in Europe; (b) Japan; (c) other developed lands (ODL)—Australia, Canada, Israel, New Zealand, South Africa; (d) the USSR; (e) East Europe—Soviet bloc, Albania and Yugoslavia; (f) Africa (excluding South Africa); (g) Asia (excluding Israel, Japan, China and related regions); (h) Latin America; (i) China (including Mongolia, North Korea, North Vietnam); and (j) the USA (including Puerto Rico and overseas islands). The poor nations, with over 70% of the world's population, consume about 10% of the annual world use of raw materials.

The measures give substance to the intensity concept: they differ significantly by region, time and material, yet the historical record suggests clear patterns of behavior of intensity-of-use.

2. *Intensity-of-Use Numbers*

Table 1 presents intensity-of-use data for refined copper. The copper account is illustrative of the intensities of other industrial raw materials. Thus, in the USA copper intensity-of-use fell almost 40% from 1951–55 to 1971–75; from 1934–38 the decline was over 60%. Other rich nations which were important consumers reflect the reconstruction activities during the years after World War II, notably Japan and West Germany. Nonetheless, individual nations in this group also tend to have lower ratios in recent years than prevailed in the 1934–38 period. For the rich countries, the decrease to the 1971–75 level from the 1956–60 level at the peak of the Marshall Plan development years was nearly 60%: the decrease was over 10% from 1951–56 when that program was only in its early phase. The underlying structure is clear: modern industrial countries expand national product with diminishing inputs of refined copper per unit of real GDP.

Table 1 also shows the very different pattern emerging in the developing countries. In each region (especially China) the intensity-of-use for copper has expanded. For the poor nations together, the data for 1971–75 relative to 1951–56 show an expansion in intensity of almost 110%. However, because of the comparatively lower intensity levels that still pertain in the poor nations relative to the rich nations and because of the poor nations' relatively small share of the total world gross product, the intensity changes in rich nations continue to dominate. Thus, refined copper per billion US dollars real GDP for the entire world in 1971–75 was 6% lower than in 1951–55. Hence, a smaller tonnage of copper was used in producing a given real volume of GDP in the world of the mid-1970s than was required in the early 1950s, despite major programs of traditionally copper-absorbing construction, electricity generation and transmission, and communications expansion throughout the world during 2–3 decades of unparalleled economic growth.

The record for copper is not unique: Malenbaum (1978) provides the same information for all 12 of the raw materials specified in Section 1. Only for primary aluminum and the platinum group metals are the historical intensity patterns significantly dif-

Table 1
Intensity-of-use: refined copper (metric tons per billion GDP, 1971 US prices)

Region	1934–38	1951–55	1956–60	1961–65	1966–70	1971–75
W. Europe	3307.0	2937.1	3315.1	3193.1	2675.9	2586.4
Japan	4116.0	2238.9	3067.3	3788.8	3998.1	3576.5
Other developed lands	1591.0	2314.0	2209.2	3003.7	2529.3	1943.8
USSR	1554.0	1839.4	1779.6	1906.0	1466.6	1776.8
E. Europe	NA	1697.2	1767.2	2008.3	2315.5	2378.7
Africa	NA	217.8	226.3	266.7	170.0	291.4
Asia	NA	452.2	644.1	812.4	416.6	550.7
L. America	NA	1175.5	974.1	1121.3	1028.5	1388.5
China	NA	122.6	795.9	1309.5	1489.8	2019.4
USA	2724.0	2330.1	1966.1	2174.4	2006.8	1680.9
Totals						
World	NA	2116.2	3007.0	2294.5	2060.1	2000.6
Non-US world	NA	2009.9	3502.9	2349.3	2082.7	2126.9
Poor nations	NA	546.1	735.0	959.1	828.7	1134.8
Rich nations	NA	2384.1	3417.6	2521.4	2272.5	2160.0

Source: Malenbaum 1978, p. 108

ferent. Aluminum has obverse patterns over past decades as circumstances permitted extensive substitutions for other raw materials inputs, particularly for copper and steel. Rates of growth of intensity-of-use of aluminum are decreasing; the prospect is the gradual emergence in aluminum of the declining intensity-of-use pattern shown by copper. The situation for the platinum group is that apparent consumption data contain a relatively large final demand for platinum as jewelry, especially in Japan. Available data for platinum metals do not permit intensity measures because their use primarily as consumption of jewelry is not relevant to the concept. However, preliminary analysis of incomplete statistics suggests that the intensity structure for the residual, intermediate demand for platinum metal might well conform to the concept and pattern illustrated for copper.

Table 2 offers further insight into materials use in the years 1951–75. For all materials and major regional components, comparison is made of change in materials use relative to GDP growth within this time interval, for the first decade 1951–60 and for the ten-year period 1966–75. The elasticity concept measures relative changes of materials use and of GDP as against the absolute change in the intensity ratio. Table 2 shows growth in materials use that declines relative to growth in the output flows of final

Table 2
Income elasticities of demand for raw materials, 1951–60 and 1966–75

	World[a,b]		USA[a,b]		Non-US[a,b]		Poor nations[a,b]		Rich nations[a,b]	
Raw material	(1)	(2)	(1)	(2)	(1)	(2)	(1)	(2)	(1)	(2)
Crude steel	1.6	1.0	0	0.1	2.2	1.3	5.2	2.4	1.3	0.8
Iron ore	1.9	0.7	1.0	0.6	2.3	0.7	4.8	1.0	1.8	0.6
Nickel[c]	2.1	0.9	1.1	0.3	2.7	1.1	6.4[d]	2.6[d]	1.6[d]	0.8[d]
Manganese ore	1.5	0.6	—	0.4	1.6	0.6	1.0	0.9	1.3	0.5
Chrome ore	1.2	0.8	1.2	−0.5	1.2	1.0	1.5	1.9	1.2	0.8
Cobalt	0.6	0.6	−0.5	0.8	1.1	0.6	1.2	1.0	0.5	0.6
Tungsten	1.0	0.4	0.6	−0.1	1.0	0.4	0.8	0.7	1.1	0.3
Refined copper	3.1	0.9	−0.2	0	4.3	1.1	2.4	2.5	3.3	0.7
Primary aluminum	2.0	1.6	1.8	1.5	2.5	1.8	7.0	1.9	2.0	1.6
Platinum group[c]	2.1	2.1	1.9	0.9	2.3	2.7	13.8	1.0	1.8	2.2
Zinc	0.9	0.8	0.2	0.1	1.3	1.0	2.4	2.9	0.9	0.7
Tin	0.7	0.3	0.4	0.5	1.0	0.5	1.6	0.2	0.6	0.3

Source: Malenbaum 1979 a Column 1: % change 1956–60 over 1951–55 b Column 2: % change 1971–75 over 1966–70 c % change 1961–65 over 1956–60 (column 1) d Market economies only

goods and services. The platinum group provides the only significant exception, which is attributable to the final demand phenomenon noted. The historical record is a strong one. As the world economy made rapid progress through the early 1970s, with output per person expanding in all regional components of the world, the relative use of raw materials in GDP was decreasing. For 10 of the 12 materials considered here, intensity-of-use actually declined for the world as a whole; even use of aluminum and platinum was declining relative to the expansion of real GDP.

What forces account for this pervasive pattern of declining materials use relative to world growth in general and in rich lands particularly?

3. *Determinants of Intensity-of-Use Patterns*

The use of intermediate inputs in the process of generating the flow of final goods output is appropriately identified with production functions and their theoretical framework. Crude-steel inputs contribute to the new steel plant and other capital creation or to an automobile and other goods for current use in a nation; this sequence pertains for the entire range of raw materials and the broad scope of final goods and services in GDP. The input–output relationship also reflects technological changes that in time alter the functional parameters. Research and development expenditures, whether in private or public entities, constantly press for such shifts. Innovations and inventions reflect the skills, knowledge and aspirations of man and society; there is a direct human contribution that serves to alter intensity-of-use through technological change.

The vast array of ratios for a single intermediate material (e.g., copper, Table 1) and the arrays cited for other materials which are so important in the elasticities of Table 2 reflect changes that can only in part be attributed to such shifts in production functions, even in a dynamic context. A very important explanation of the intensity changes measured is the changing structure of a nation over a period of time. Certainly the prototype process of development, with movement from agricultural and extractive enterprise to greater emphasis on industry and increasingly on services output, brings new materials intensity levels for the nation, even without any technological change in identified production processes. In addition to the technological and structural explanations for intensity change, adjustments are constantly and "normally" in process through market pressures toward the cost and efficiency goals of functioning economies. New national policies, changing ratios of capital and labor costs, as well as shifting prices for intermediate inputs and final products, serve to bring substitutions in the production of a given product, among, for example, copper, aluminum and steel inputs, or among the ferroalloys. Such changes are apart from those attributable to new technology.

Structural and efficiency forces for intensity change join the technological forces in determining patterns for intensity measures. Present documentation is only suggestive regarding the relative importance of these forces, partly because they often operate simultaneously. Thus, market costs induce technological change, and innovation or invention facilitates such structural shifts to increased manufacturing as against extractive enterprises. The literature on intensity measures suggests that structural forces have been predominant in the patterns since 1950, and that the influence of technological change may have been the least important of the three forces during those years.

Further research can yield more knowledge and thus enhance the value of the intensity-of-use concept for materials analysis and policy. Again note that all these determinants are functions of the quality of humans: skills and knowledge; aspirations and commitment. Human attributes more than material conditions impart the critical flexibility to the use of raw materials in any nation. As is true for modern growth theory, the forces for activating intensity changes of raw materials as for serving economic expansion are human quality forces more than material quantity forces (Fig. 1).

With respect to the impressionistic representation of the intensity-of-use relationship to GDP/population (Fig. 2), the documentation for world development is available only in part. The entire curve is hard to measure for specific regions: poor nations reflect only the upward movement as they are still poor, and rich nations reflect the declining movement as they were poor long ago. The data are revealing with respect to turning points so critical to the intensity concept. Thus, the record permits some approximation to a single empirical pattern for intensity changes for each material and each region as real GDP per capita grows.

Figure 3 represents intensity-of-use of iron ore; parallel data are available for all the materials. The historical pattern (solid lines) reveals the downward slope of the inverted "U" of Fig. 2 in the case of the more developed regions. Also revealed is the predominantly upward path for the lands where per capita GDP is still relatively low and where structural adjustments are very much in progress. The data also permit interpretation of the regional differences in intensity levels at a point of time (e.g., the high figures for Japan and those for the USSR, as compared to the USA in the case of iron ore), as well as rates of change. The emergent patterns provide powerful tools that warrant further research.

4. *Relevance of Intensity-of-Use Ratios*

Scientific analyses in recent decades have increased awareness of the importance of materials supply and

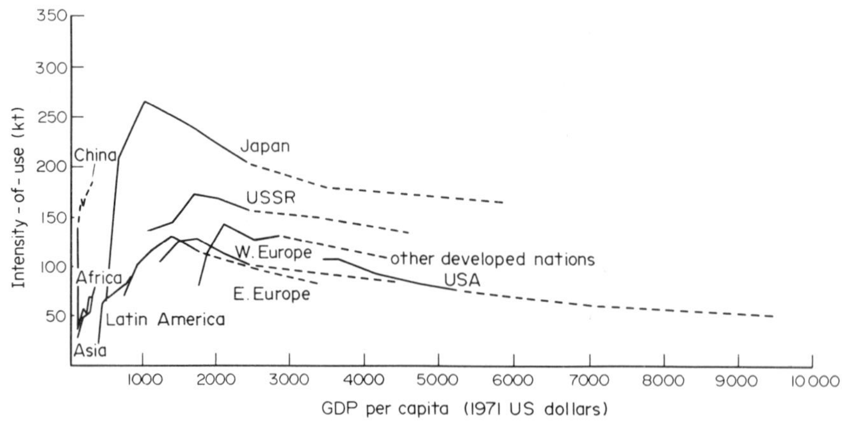

Figure 3
Iron ore intensity-of-use and GDP per capita: —— actual, – – – projected (after Malenbaum 1978)

demand to the future of man and society. On the one hand, all nations seek to expand their per capita GDP. Even conservative projections of world GDP imply use of industrial raw materials in tonnages 2–3 times the actual annual level of the early 1970s by the year 2000, and pressures for further such expansion exist. On the other hand, the possible depletion of the world's nonrenewable resources is also broadly accepted as a real threat to GDP and population growth, perhaps within the next century.

Research and analysis responsible for such propositions will continue to expand the world's capacity to live with alternatives in supply and demand of materials. In the early 1970s the projected tonnage for raw materials use in the year 2000 would have been projected at 4–5 times the actual 1970 tonnage levels. The idea of dynamic input–output relationships and the evidence of patterns of declining intensity-of-use of materials remain relatively unfamiliar. There is wide scope for additional research on other raw materials and for study directed towards a fuller understanding of the determining factors in intensity change. Perhaps these efforts will in time permit man and his organizations to decide the level of raw material inputs in a desired output and the required structure of gross domestic product together with the real costs of such goals.

The evidence from past studies associates these factors which determine demand for materials with forces now increasingly accepted as central in the entire process of the economic development of a nation. They offer a demand prospect that counters the growth-limiting significance attributable to marked inelasticity of raw materials supply. Generating the desired level of GDP in the year 2000 with 50–60% less tonnage of materials input is a contribution to world raw materials supply. Capacities of man are substitutes for quantities of material; intensity-of-use considerations are fundamental to supply analysis.

Concern about materials exhaustion often encompasses concern about US dependence on foreign sources of supply. The evidence that patterns of materials intensity-of-use have increasingly made the country a relatively smaller user of the world's industrial raw materials is fundamental. This derivative of intensity-of-use patterns reflects the relative strength of US economic potential: the greater capacity to meet the needs of the population with smaller recourse to scarcer resources.

In addition, the outlook for declining intensities in the materials demand of the rich nations and for expanding intensities in demand of the poor nations offers complementarities in the future materials interests in the world. There is greater scope for solving distribution conflicts critical in NEIO. Analysis of the imminence, significance and burden of world materials supply shortages is an integral part of analysis of intensity-of-use patterns of demand.

See also: Materials Markets: Modelling; Materials in the Industrial System; Prices of Materials: Theory; Substitution: Economics; Materials Economics, Policy and Management: An Overview

Bibliography

Brooks D B, Andrews P W 1974 Mineral resources, economic growth and world population. *Science* 185 (4145): 13–19

Congressional Research Service 1980 *Emerging Issues in Science and Technology*. US Government Printing Office, Washington, DC

Fischman L L 1980 *World Mineral Trends and US Supply Problems*. Resources for the Future, Washington, DC

International Iron and Steel Institute, Committee on Economic Studies 1972 *Projection 85: World Steel Demand*. IISI, Brussels

Landsberg H H, Fischman L L, Fisher J 1963 *Resources in America's Future: Patterns of Requirements and Avail-*

abilities, 1960–2000. Johns Hopkins University Press, Baltimore, Maryland
Leontief W W, Carter A P, Petri P A 1977 *The Future of the World Economy*. Oxford University Press, London
Leontief W W, Koo J C M, Nasar S, Sohn I 1983 *The Future of Non-Fuel Minerals in the US and World Economy: Input–Output Projections 1980–2030*. Lexington Books, Lexington, Massachusetts
Malenbaum W 1973 World resources for the year 2000. *Ann. Am. Acad. Polit. Soc. Sci.* 408: 30–46
Malenbaum W 1975 Laws of demand for minerals. *Proc. AIME*. American Institute of Mining, Metallurgical and Petroleum Engineers, New York, pp. 147–53
Malenbaum 1978 *World Demand for Raw Materials in 1985 and 2000*. McGraw Hill, New York
Malenbaum W 1979 Anatomy of world materials demand: Its role in supply availability. *Min. Congr. J.* 65(2): 59–64
Meadows D L, Meadows D H 1972 *The Limits to Growth*. Universe, New York

W. Malenbaum

Intercalation Compounds

Intercalation compounds are formed by the insertion of atomic or molecular layers of a guest chemical species between layers in a host material. The intercalation process occurs in highly anisotropic layered structures where the intraplanar binding forces are large in comparison with the interplanar binding forces. The guest species in an intercalation compound exhibits order, in contrast to doping where the guest species tends to occupy random locations. Examples of host materials for intercalation compounds are graphite, transition-metal dichalcogenides, some silicates, and metal chlorides. Intercalation provides the host material with a means for controlled variation of many physical properties over wide ranges. Of the various types of intercalation compounds, the graphite compounds are of particular physical interest because of their high degree of structural ordering.

1. Synthesis

Intercalation can be achieved starting from a solid, liquid or gaseous reagent, though preparation from the vapor or from solution is the most common. Electrochemical techniques are also used. A given intercalation compound can often be prepared by alternative growth techniques. Generally, simple molecules can be intercalated from the vapor, but large organic molecules are intercalated from solution. The intercalation rate and resulting intercalate concentration are strongly dependent on the intercalation conditions, such as pressure, the temperature difference between the host material and the intercalant, the physical sample dimensions, the degree of crystalline order and the defect density in the host material. When intercalating from the vapor, the initiation of the intercalation process requires the intercalate partial pressure to exceed a threshold value that depends on temperature, the host material and the intercalate species. Intercalation proceeds with the insertion of intercalate layers sequentially from the surface layers inward, but starting from the edges and advancing in the layer planes.

Of all the host materials, graphite is most commonly used. Techniques have been developed for the intercalation of large numbers ($\geqslant 10^2$) of chemical reagents into this host material, ranging from simple ionic species such as alkali metals, through diatomic molecules such as the halogens, to large organic molecules. The simpler binary and ternary compounds are usually prepared by direct synthesis, and the more complicated materials by a variety of stepwise intercalation procedures. A simple example of ternary intercalation is the synthesis of graphite compounds with the alloy Na_xK_{1-x} since sodium by itself does not readily intercalate into graphite but potassium does. Stepwise intercalation allows the introduction of interesting intercalants such as hydrogen and mercury into a graphite host into which potassium (or another alkali metal) has previously been intercalated. The hydrogen compounds allow packing densities of hydrogen atoms comparable to that in solid hydrogen, and the mercury compounds are of interest for their unusual superconducting properties.

Intercalation compounds are generally very reactive and must be stored in an inert atmosphere. A few compounds (e.g., graphite–$FeCl_3$ and graphite–$SbCl_5$) are quite stable in air. The reactivity of most compounds is greatly reduced by lowering the temperature to 77 K. Because of this reactivity, intercalation compounds are normally encapsulated after preparation (e.g., in glass or quartz ampoules) to prevent intercalate desorption.

However, if the intercalation compound is removed from the encapsulating ampoule and exposed to the normal room-temperature environment, desorption proceeds for a period of time, after which it is negligible. The resulting material, called a residue compound, is chemically stable and contains an intercalate concentration (usually small) dependent on the desorption temperature and on the intercalate concentration of the original lamellar compound.

2. Structure

Graphite intercalation compounds are of particular physical interest because of their high degree of ordering. The most striking type of ordering is the staging phenomenon, which is defined by a periodic arrangement of *n* graphite layers between sequential intercalate layers, where *n* is the stage index (see Fig. 1). It is possible to prepare single-staged materials with small (1–10%) admixtures of secondary-staged regions. X-ray diffractograms show that well-staged materials can be made up to stage ~10, indicating a c-axis repeat distance of $I_c \sim 40$ Å. A physical basis for the staging phenomenon is associated with the strong interatomic intercalant–intercalant binding relative to the intercalant–graphite binding, thereby favoring a close-packed in-plane arrangement, while the introduction of each intercalate layer adds a substantial strain energy, thereby favoring the minimum number of intercalate layers, consistent with a given average intercalate concentration.

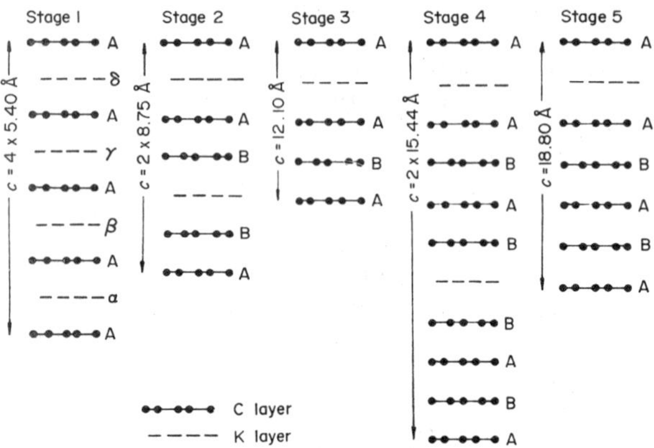

Figure 1
Schematic diagram illustrating the staging phenomenon in graphite–potassium intercalation compounds for stages $1 \leqslant n \leqslant 5$. The potassium layers are indicated by dashed lines and the graphite layers by solid lines connecting circles, indicating schematically a projection of the carbon atom positions. The ABAB graphite layer stacking for stages $n \geqslant 2$ is maintained between intercalate layers, although a rhombohedral stacking arrangement appears across intercalate layers. The stacking ordering is well-confirmed by x-ray diffraction (hkl) patterns

The intercalation process has a negligible effect on the in-plane ordering of the host layers and, for the case of graphite, also on the c-axis graphite interplanar spacing. In the transition-metal dichalcogenides staging does not generally occur but the c-axis expansion is proportional to the amount of intercalated species. The interlayer stacking of adjacent graphite planes remains ABAB as in pristine graphite, although departures from this stacking can occur across an intervening intercalate layer. The most stable stacking arrangement of the graphite layers is the one which minimizes the volume needed to accommodate the intercalate in the interlamellar space.

The intercalate layers also exhibit ordering which may be similar to that of the parent material and may be commensurate or incommensurate with respect to the adjacent graphite layers. An example of a commensurate superlattice is a first-stage alkali-metal compound with potassium or rubidium intercalated into the graphite host. The lattice constants of the intercalate layer are typically much larger than those of the graphite layers because the intercalate metallic radii or intramolecular bond lengths tend to be large compared with the carbon radius. In general, the metallic or molecular intercalate volume in the intercalation compound is similar to that in the parent material.

In some cases for first-stage compounds, site ordering on sequential intercalate layer planes occurs, resulting in interplanar intercalate stacking order, as, for example, the $\alpha\beta\gamma\delta$ stacking sequence of the intercalate layers in the first-stage C_8K and C_8Rb compounds (see Fig. 1). For the intercalate layers, intraplanar structural transitions, order–disorder transformations and interplanar stacking order transitions occur as the temperature is increased, with each type of transition characterized by its own transition temperature. Because the intercalate ions tend to be much larger than those of graphite, the intercalate ions are preferentially located in-plane over the centers of graphite hexagons, at positions of minimum potential energy. For the same reason, the intercalate ions are preferentially anticorrelated between planes. When the intercalate in-plane lattice spacing is incommensurate with respect to that for graphite, the preferential ion location over the hexagon centers often leads to preferred orientational alignment of the intercalant.

The principal technique used for the structural characterization of intercalation compounds is x-ray diffraction. For the case of graphite, this technique is primarily used for stage-index determination by taking ($00l$) scans. Because of the small size of an electron beam (1 μm diameter), the electron diffraction technique is particularly suitable for the determination of in-plane ordering, since the intercalant tends to form island domains ranging in size from less than 100 Å to greater than 1000 Å. Elastic neutron scattering has also been used to provide information on the stage index and on the interplanar intercalate stacking.

3. Electronic Structure

The observation that intercalation does not significantly change the in-plane ordering in the layers of the host material, nor the interlayer spacing between these layers, suggests that the electronic and lattice mode structures for intercalation compounds are closely related to those of the host material. In discussing the electronic and lattice properties of graphite intercalation compounds it is found that the perturbation introduced by the intercalate layers is largely confined to the two graphite layers adjacent to the intercalate layer (the graphite bounding layers), while the graphite layers not adjacent to an intercalate layer (the graphite interior layers) are largely unperturbed. A correlation with the parent intercalant material is also suggested by the similarities of the structural ordering in the intercalate layer and in the corresponding layer of the parent material. Thus, calculations of the electronic dispersion relations for intercalation compounds have been based on a hybridization of the electronic levels for the two parent materials. Furthermore, simplifications are also introduced by considering the symmetry for the intercalation compound. The superlattice symmetry arising from staging introduces zone-folding effects which have an important bearing on the form of the electronic energy band and lattice mode structures.

Detailed band calculations have been carried out for some intercalation compounds based on graphite and transition-metal dichalcogenide hosts, and the results have been useful for the interpretation of a large variety of experiments relevant to the electronic structure.

For the case of graphite intercalation compounds various experimental data suggest the applicability of a modified rigid-band approach to the calculation of the electronic band structure, the modifications being associated with interactions between the graphite bounding and intercalate layers, and the charge density distribution resulting from this interaction. A particularly sensitive experimental probe of the energy band structure is the magnetoreflection technique, which shows that the form of the graphite electronic energy band model is also applicable to the intercalation compounds within ± 0.2 eV of the band edges of the graphite valence and conduction bands. Approximately the same values of the band parameters which are appropriate to graphite also apply to the graphite layers in the intercalation compounds. The magnetoreflection experiment allows an accurate determination of the changes in these parameters as a function of intercalate concentration for each intercalant.

Because of the large changes in electron concentration, intercalation results in major changes in the transport properties. The in-plane electrical conductivity σ_a increases from that for graphite by an order of magnitude or more, depending on the intercalate species and its concentration. Of the various layers within the unit cell, the conductivity of the graphite bounding layers is dominant because of the high carrier density in these layers relative to the graphite interior layers and because of the much higher carrier mobility in the graphite bounding layers relative to the intercalate layer. Although smaller, the contribution of the graphite interior layers can in some cases be significant.

Hall-effect measurements show that the sign of the dominant carrier can be positive (for acceptors) or negative (for donors). Thus intercalation can cause the Fermi level to rise (donors) or to fall (acceptors). Although the c-axis conductivity σ_c also increases with increasing intercalate concentration for donor compounds, σ_c decreases in the case of acceptor compounds, resulting in materials with electrical conductivity anisotropy as large as $\sigma_a/\sigma_c \sim 10^6$. This wide range of behavior in the electrical conductivity occurs because intercalation increases the carrier density while decreasing the carrier mobility. The relative trade-off between these effects is dependent on intercalate species, concentration and polarization direction.

The alkali-metal intercalants donate essentially all their free carriers to the graphite layers, though predominantly to the graphite bounding layers which then screen the graphite interior layers from the positively charged intercalate layers. A rapid decrease in carrier density occurs with distance into the graphite interior layers. For the acceptor compounds, the intercalate layer becomes partially ionized by extracting electrons predominantly from the graphite bounding layers. Thus a high hole concentration is found in the graphite bounding layers, and the hole concentration decreases rapidly with distance into the graphite interior layers. The electronic charge distribution in the graphite and intercalate layers has been calculated in detail on the basis of an explicit band model in the case of the low-stage lithium compounds. The spatial dependence described above for the charge distribution in a general graphite intercalation compound has been calculated phenomenologically on the basis of a Fermi–Thomas screening model.

Quantum oscillatory effects have been observed in a number of graphite intercalation compounds and have yielded direct information on the shape of the Fermi surface and on the carrier density, independent of scattering processes and scattering times. The results show that intercalation increases the size of the Fermi surface and causes the formation of numerous carrier pockets while decreasing the carrier mobility by a smaller factor.

4. Optical Properties

The large increase in free-carrier concentration through intercalation gives rise to a sharp plasma edge which dominates the optical spectra of low-stage graphite intercalation compounds for light polarized in the layer planes electric field (***E*** parallel to the a axis). Just as for the dc conductivity, free-carrier effects in the reflectivity are highly anisotropic and only a weak plasma structure is found for the polarization ***E*** parallel to the c axis. Although interband transitions dominate the infrared and optical reflectivity spectra of graphite with free-carrier effects playing only a minor role, the opposite is the case for the infrared and optical spectra of the intercalation compounds. The reflectivity spectra for many acceptor and several donor compounds also show structure below the plasma edge; this structure has been identified with interband transitions between the bands arising from *c*-axis zone-folding effects associated with the staging periodicity. These optical transitions provide important information relevant to the modelling of the electronic structure of intercalation compounds.

Some information on the intercalate levels has been provided by electron energy loss spectra which are sensitive to electronic transitions, photoemission spectra which are sensitive to the electronic density of states, and by the resonant Raman effect which is sensitive to wave-vector-conserving electronic transitions. From these studies it has been concluded that the donor intercalate levels lie close to the Fermi level (and in some cases, such as stage 1 alkali-metal–graphite intercalation compounds, the alkali-metal *s* bands cross the Fermi level and are partly occupied). In contrast, the molecular acceptor levels tend to lie several electron volts away from the Fermi level. Because of the low concentration of the intercalate species relative to that for carbon atoms, relatively small contributions to the optical reflectivity spectra are expected from interband transitions between intercalate levels in the optical reflectivity spectra.

5. *Thermal Transport*

Thermal transport in intercalation compounds is highly anisotropic, as are the electrical transport and optical properties. Although the thermal conductivity in the host crystal is dominated by the lattice contribution, the great increase in carrier concentration results in the dominance of the electronic contribution at low temperatures ($T < 10$ K). Because of the increase in scattering due to increased defect concentration and electron–phonon interaction, the absolute magnitude of the room-temperature thermal conductivity is decreased by intercalation. The Wiedemann–Franz relation is found to relate satisfactorily the low-temperature electrical conductivity to the low-temperature electronic contribution to the thermal conductivity.

The anisotropy ratio in the room-temperature *c*-axis to *a*-axis thermal conductivity of graphite is essentially unaltered by intercalation. At low temperatures the electronic contribution to the *c*-axis thermal conductivity is insignificant because of the low *c*-axis electrical mobility. Thus the *c*-axis thermal conductivity is dominated by the phonon contribution throughout the entire temperature range, so that at low temperatures intercalated graphite exhibits thermal conduction by electrons in the layer planes and by phonons along the *c* axis.

Although the functional form of the temperature dependence of the thermopower for graphite intercalation compounds is not understood, the sign of the thermopower for acceptor compounds is positive and that for donor compounds is negative, as expected.

6. *Magnetic Properties*

Of all the intercalation systems, the graphite compounds have been studied most extensively with regard to magnetic properties, both in terms of modifications of the room-temperature magnetic susceptibility of the host material by intercalation and the formation of ordered magnetic phases at low temperatures.

Graphite is a highly diamagnetic material because of the large orbital contribution of the free carriers. At room temperature, the largest changes in the magnetic susceptibility χ are found in the alkali-metal donor compounds where the large Pauli free-carrier contribution results in a change in sign of χ from negative to positive for the in-plane susceptibility of the low-stage alkali-metal donor compounds. Intercalation also greatly reduces the magnetic anisotropy of graphite.

At low temperatures, intercalation can result in magnetically ordered phases when the intercalant itself is magnetic. The ability to vary the interplanar separation between magnetic layers by increasing the stage index permits study of the transition from three-dimensional magnetic interactions (stage 1 compounds) to two-dimensional magnetism (dilute compounds). Because of the smaller number of magnetic nearest neighbors, the magnetic ordering temperature of the intercalation compound tends to be lower than that in the parent pristine magnetic material. Ordered magnetic phases have been observed for both acceptor metal chloride compounds (e.g., $FeCl_3$, $FeCl_2$, $NiCl_2$, $CoCl_2$) and the donor compound C_6Eu. In general, the magnetic phase diagram appears to be significantly more complex as a function of temperature and magnetic field in the intercalation compound than in the parent pristine magnetic material. The detailed spin arrangements and magnetic phase diagrams for the magnetic intercalation compounds have not yet been firmly established.

7. *Superconductivity*

Intercalation compounds have particularly interesting superconducting properties. By increasing the stage index and the size of the unit cell I_c, it may be possible to decrease the *c*-axis coherence distance ξ_c so that it is less than I_c, and the possibility of two-dimensional superconductivity can be explored. Superconductivity has been observed in a number of intercalated transition-metal compounds and in intercalated graphite (see *Superconductivity in Alloys, Compounds, Mixtures and Organic*

Materials). In all cases, the critical magnetic fields are highly anisotropic and other unusual phenomena are observed.

For example, the first-stage intercalation compound C_8K is found to exhibit a sharp Meissner effect at ~150 mK even though neither of the constituents is superconducting by itself. Furthermore, C_8K exhibits type I superconductivity with the magnetic field $\boldsymbol{H}$ parallel to the *c* axis and type II superconductivity for $\boldsymbol{H}$ perpendicular to the *c* axis. The large anisotropy of the critical field H_{c_2} is well represented by the Ginsburg–Landau theory, yielding a large in-plane coherence distance (~2000 Å) relative to the *c*-axis coherence distance (~100 Å). The in-plane coherence distance is relatively insensitive to stage index, while the *c*-axis coherence distance decreases significantly with increasing stage. The superconducting transition temperature T_c tends to be higher (~1 K) for intercalants that are themselves superconducting (e.g., KHg), and it is expected that T_c decreases with increasing stage index, consistent with the decreases in the density of electronic states; notable exceptions are the stage 1 compounds C_4KHg and C_4RbHg, which have lower transition temperatures (0.77 K and 0.99 K, respectively) than the stage 2 compounds C_8KHg and C_8RbHg (1.90 K and 1.41 K, respectively). It is believed that the occurrence of superconductivity in intercalation compounds is connected with partial occupation of the intercalate band or incomplete charge transfer from the intercalant to the host.

Superconductivity has also been observed for organic intercalants such as pyridine in TaS_2, increasing T_c from 0.8 K in pristine TaS_2 to 3.5 K in TaS_2 $(\text{pyridine})_{1/2}$. This compound is a highly anisotropic superconductor with an anisotropy ratio of 10^4 in critical current and 30 in critical field H_{c_2}. Superconductivity has been observed in compounds (octadecylamine in TaS_2) with very large unit cells ($I_c \sim 57$ Å). For TaS_2 and MoS_2 hosts, intercalation tends to increase T_c, while in other hosts (e.g., $NbSe_2$) intercalation depresses T_c.

8. *Lattice Properties*

The high anisotropy of graphite is also reflected in the lattice properties. The lattice modes show large in-plane dispersion, but weak coupling and small dispersion along the *c* axis. Study of the lattice mode structure (for example, by the Raman effect) is of particular significance because the modes on the intercalate layer, the graphite bounding layers and the graphite interior layers occur at different frequencies and therefore can be investigated independently as a function of intercalate species and concentration. Evidence that the interior graphite layers are essentially unaffected by intercalation is provided by Raman spectra observed for a variety of donor and acceptor compounds, in which the principal in-plane Raman-active optical mode appears essentially unshifted from its frequency in pristine graphite, but with an intensity that decreases with increasing intercalate concentration, corresponding to the decrease in the relative number of interior graphite layers.

Modifications to the graphite lattice modes are, however, observed on the bounding graphite layers adjacent to the intercalate layer planes. The characteristics of the Raman spectra show that the perturbation due to the intercalate layer is largely confined to the graphite bounding layers, which effectively screen the graphite interior layers from the charged intercalate layer. The modes associated with the bounding graphite layers are upshifted in frequency relative to the interior layer modes by $\sim 20\ \text{cm}^{-1}$, almost independent of intercalate species and concentration. The intensity of the upshifted modes grows with increasing intercalate concentration as the relative number of bounding graphite layers increases. The lattice mode structure of the intercalate layer depends in detail on the intercalate species and is related to the corresponding modes in the parent crystalline solid.

Models for the lattice mode structure of graphite intercalation compounds confirm this interpretation of the Raman spectra and explain why no infrared modes are introduced into the Raman spectra and no Raman modes are introduced into the infrared spectra for the intercalation compounds.

Observation of resonant Raman enhancement effects also provides information on the electronic structure. This information is especially significant for an identification of electronic transitions within host or intercalate levels through the corresponding identification of the resonantly enhanced lattice modes.

Infrared spectroscopy has also been used to provide information on the lattice mode structure for several acceptor and alkali-metal donor compounds. The infrared spectra observed for graphite with $\boldsymbol{E}$ parallel to the *a* axis is associated with the infrared-active E_{1u} mode. In the intercalation compounds the infrared spectra show contributions from both the graphite interior and bounding layer modes. To determine the mode frequencies, linewidths and oscillator strengths, a lineshape analysis must be carried out. The oscillator strength provides information on the magnitude of the dipole moment and hence the effective charge associated with the dipole moment. There are no infrared-active modes for stage 1 compounds.

The lattice modes of the alkali-metal compounds in graphite have been studied extensively using Raman scattering, infrared spectroscopy and inelastic neutron scattering. Heat capacity measurements have also been carried out. The relatively small shifts in Debye temperature θ_D obtained from analysis of heat capacity data indicate that even for stage 1 compounds the identity of the in-plane modes with those in the parent compounds is maintained.

9. Summary

In recent years, improvements in materials preparation and characterization of intercalation compounds have made possible the quantitative measurement of their electronic, structural, lattice, magnetic and superconducting properties. Because of their unique types of ordering, intercalation compounds offer opportunities for the study of new types of phenomena. For example, novel phase transitions arise through the staging phenomenon which provides an ordered one-dimensional array of intercalate sheets that can independently undergo two-dimensional order–disorder and structural phase transitions. Staging provides a means for controlling the separation between the intercalate layers, thereby making possible the study of two-dimensional magnetism and two-dimensional superconductivity. With the current growing interest in submicrometer technology, graphite intercalation compounds provide natural submicrometer structures that may have potential for the exploration of submicrometer physical phenomena and new applications of submicrometer structures. Intercalation compounds are now being studied widely and intensively and it is expected that in the next few years great progress will be made in achieving a more detailed understanding of these materials and in their utilization.

Bibliography

Dresselhaus M S, Dresselhaus G 1981 Intercalation compounds of graphite. *Adv. Phys.* 30: 139–326

Gamble F R, Geballe T H 1976 Inclusion compounds. In: Hannay N B (ed.) 1976 *Treatise on Solid State Chemistry,* Vol. 3: *Crystalline and Noncrystalline Solids.* Plenum, New York, pp. 89–166

Nishina Y, Tanuma S, Myron H W (eds.) 1981 Proceedings of the Yamada conference IV on the physics and chemistry of layered materials, *Physica* 105B: 1–527

Pietronero K, Tosatti E (eds.) 1981 *Physics of Intercalation Compounds*, Springer Series in Solid-State Sciences Vol. 38. Springer, Berlin

Vogel F L (ed.) 1980, 1981 Proceedings of the second conference on intercalation compounds of graphite, *Synth. Met.* 2: 1–377; 3: 1–306

Vogel F L, Hérold A (eds.) 1977 Proceedings of the conference on intercalation compounds of graphite, La Napoule, France, 1977. *Mater. Sci. Eng.* 31: 1–355

M. S. Dresselhaus

Interface Effects on Mechanical Properties of Metals

In a polycrystalline material the boundaries of individual crystals or "grains" represent internal surfaces of crystallographic discontinuity. Since crystalline materials undergo both plastic deformation and brittle fracture on specific crystallographic planes, or sometimes along grain boundaries, the presence of grain boundaries and the grain-boundary area per unit volume can have profound effects on the mechanical behavior. This article reviews the various mechanisms by which grain boundaries exert their effects on plastic deformation and fracture both at ordinary temperatures, where diffusion is negligibly slow, and at high temperatures, where diffusive flow plays an important and sometimes dominant role.

1. Deformation at Ordinary Temperatures

Internal interfaces in polycrystalline materials are an important source of dislocations, the motion of which provides the dominant mechanism of plastic flow at ordinary temperatures. Dislocations can originate at steps in the interface or at particle sites, or may be generated by reactions of grain-boundary dislocations. In materials which deform by uniformly distributed dislocations ("wavy slip" materials, in which cross-slip is easy), rather than in sharply defined slip bands, the yield stress, or the flow stress at any given strain, is proportional to $d^{-1/2}$, where d is the average grain diameter. This experimentally discovered fact can be explained by coupling the well-known Taylor theory of strain hardening, which predicts that the flow stress is inversely proportional to the average dislocation spacing, with the assumption that dislocation sources are distributed uniformly over the entire grain-boundary area. Thus, if the dislocation density ρ is proportional to the grain-boundary area per unit volume, it must be inversely proportional to d. Since the average dislocation spacing is proportional to $\rho^{-1/2}$, it follows that the flow stress is proportional to $d^{-1/2}$.

On the other hand, it is found that some materials deform by means of shear along sharply defined slip bands. In such "planar-slip" materials the grain boundaries act as barriers to the propagation of slip from grain to grain, but it is still observed that the yield (or flow) stress is proportional to $d^{-1/2}$. This behavior can be accounted for as follows: If τ^* is the shear stress needed to activate slip in a neighboring grain then the stress concentration due to the blockage of a slip band at a grain boundary must reach this value in order to propagate slip into the next grain. This stress concentration is proportional to $\tau l^{1/2}$, where τ is the applied resolved shear stress on the slipping grain and l is the length of the blocked slip band in that grain. Thus, the condition for propagation of slip is

$$\alpha\tau l^{1/2} \geqslant \tau^* \tag{1}$$

and when $l \propto d$, it can be seen that $\tau \propto d^{-1/2}$. If allowance is made for a "friction stress" τ_i opposing dislocation motion, such that the effective stress $\tau_{eff} =$

$\tau - \tau_i$, then the Hall–Petch equation for the yield (or flow) stress τ_y as a function of grain size is obtained:

$$\tau_y = \tau_i + kd^{-1/2} \quad (2)$$

where $k \propto \tau^*$. In some materials τ^* can be quite large (e.g., in mild steel in which grain-boundary dislocation sources are pinned by segregated solutes such as carbon or nitrogen). In such materials the contribution of the grain boundaries to the yield strength can be quite large.

In certain materials, such as body-centered-cubic (bcc) metals at low temperatures or some hexagonal-close-packed (hcp) metals, plastic deformation can occur by twinning. This is a mechanism of homogeneous shear in which the crystallographic orientation of the twinned region is altered so as to take on a mirror image relationship with respect to the parent lattice. In general, twinning becomes easier as the grain size is increased because the propagation of twinning from grain to grain is analogous to the propagation of slip bands. However, the effect of grain size on the stress necessary to propagate twinning is generally not as great as it is for slip.

2. *Brittle Fracture at Ordinary Temperatures*

Grain size effects are important in both the nucleation and propagation of brittle fracture in polycrystalline metals. This can be illustrated by the behavior of mild steel, in which brittle fracture is nucleated by the piling up of dislocations at hard particles, usually lying in grain boundaries. The larger the grain size, the more dislocations can be found in a pile-up, and the greater the stress concentration on the particle. For a pile-up of n dislocations, the stress concentration is $n\tau$, where τ is the applied shear stress. Eshelby et al. (1951) have given the value of n as $n \simeq \pi\tau l/Gb$, where l is the length of the pile-up, b is the Burgers vector and G is the shear modulus. Thus, the concentrated stress is proportional to $\tau^2 l/Gb$. If the critical stress needed to fracture the particle (or to cause it to break away from the matrix) is σ^*, then putting $\tau \simeq \sigma/2$ and $l \simeq d$ and allowing for a friction stress σ_i leads to the condition for fracture nucleation:

$$(\sigma_f - \sigma_i)^2 d/Gb \geqslant \sigma^* \quad (3)$$

$$\sigma_f = \sigma_i + k_f d^{-1/2} \quad (4)$$

where $k_f \propto \sigma^*$. This analysis applies to situations in which fracture proceeds directly from (or is controlled by) nucleation of a microcrack.

The application of the above analysis is illustrated by Fig. 1, in which the line with the smaller slope is the yield stress of a low-carbon steel as a function of grain size at the test temperature 77 K; this is also the stress at which brittle fracture is nucleated due to dislocation pile-ups at carbide particles (McMahon and Cohen 1965). For grain sizes coarser than d', this stress coincides with the brittle fracture stress of the specimen. For finer grain sizes, however, cleavage microcracks are produced which do not propagate beyond one or two grain boundaries, as long as the fracture stress is not reached (i.e., the line with the greater slope). Thus, the grain boundaries can act as barriers to crack propagation, and because of this a decrease in grain size generally produces an increase in toughness. The propagation criterion represented by the fracture stress in Fig. 1 can be explained in terms of a Griffith-type analysis in which the fracture stress is given by

$$\sigma_f \geqslant (\alpha E\gamma/d)^{1/2} \quad (5)$$

which has the form of the original Griffith equation, and in which d is the length of the "Griffith crack," considered to be one grain diameter in length on the average; E is Young's modulus; and γ is the effective fracture energy which must be supplied to drive the microcrack past a grain boundary; α is of the order of unity. From Fig. 1 it can be seen that the barrier effect of grain boundaries for brittle crack propagation in fine-grained materials is even larger than the effect of grain boundaries in suppressing plastic flow.

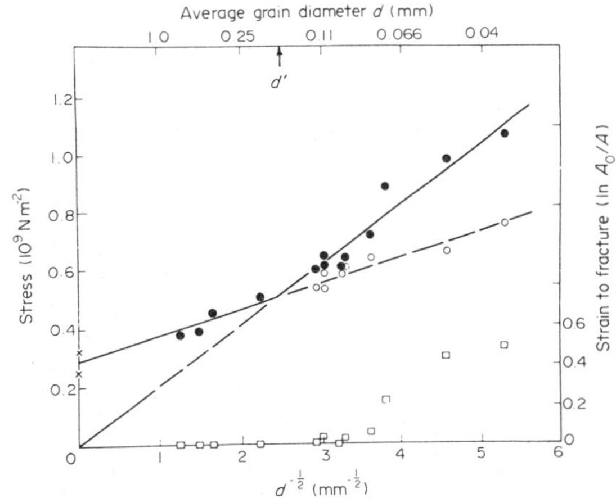

Figure 1
Yield and fracture stress, and strain to fracture, of low-carbon steel at 77 K as a function of grain size: ○, yield stress; ●, fracture stress; □, strain to fracture (after Low 1954)

Grain boundaries are also important in the brittle fracture of polycrystalline materials because of the segregation of certain solute elements to the boundaries; some of these segregating elements can reduce the cohesive strength of the base material. This phenomenon occurs in many metals and alloys, such as in nickel and iron owing to segregation of sulphur, and in many steels owing to segregation of various elements from groups IVb to VIb of the periodic table. Segregation occurs partly because the extra free volume in the grain boundary provides sites of relatively low energy for solutes which do not fit well into the crystal lattice, and also because the grain boundary can provide sites for the formation of pairs of elements between which there is a strong chemical attraction. The reduction in cohesion may occur because of the formation of strong chemical bonds in the plane of the grain boundary at the expense of bonding across the grain boundary, and also because of the local expansion due to the misfitting solutes. The occurrence of segregation is probably the main cause of "temper embrittlement" in alloy steels; as in other examples of brittle fracture, this embrittlement is exacerbated by an increase in grain size or in the strength of the matrix (see *Temper Embrittlement of Steel*). Another factor which can play a similar role in intergranular brittle fracture is the formation of hard or weakly bonded precipitates in the grain boundaries. For example, the formation of carbide particles in the grain boundaries of high-strength steels can cause a significant decrease in the fracture energy, even when the amount of segregated impurity is very low. This is the case in the phenomenon called "tempered martensite embrittlement."

Intergranular precipitation can produce a decrease in the fracture toughness of an alloy, even in the absence of impurity segregation on the atomic scale. In this case, the fracture mode is ductile, arising from the formation and plastic growth of cavities at the grain boundaries, the cavities being nucleated at the particles in the boundaries. If the intergranular particle density is high enough, the overall fracture strain can be quite low and the fracture appearance can be macroscopically brittle. An example of this kind of behavior is found in "overheated" steels in which MnS or CrS particles are dissolved owing to high-temperature austenitization and are reprecipitated during cooling. Other examples include AlN precipitates in steel forgings containing aluminum that are heated in air, carbide and sigma-phase formation in austenitic alloys during high-temperature service, and some precipitation-hardened aluminum alloys.

3. *Deformation at Elevated Temperatures*

Grain boundaries take on an added significance at elevated temperatures where mass transport by diffusion can occur. Since grain boundaries are regions of extra free volume compared with the crystal lattice, the interchange between atoms and lattice vacancies occurs more rapidly in the grain-boundary regions. Thus, when a grain boundary lies along a plane of high shear stress, a shear displacement can occur between adjacent crystals by diffusion-controlled motion of grain-boundary dislocations and by diffusion of atoms from regions of lattice compression to regions of lattice expansion. For example, Raj and Ashby (1971) have considered the strain rate that can be achieved by diffusion-accommodated grain-boundary shearing. For a two-dimensional array of equiaxed hexagonal grains, they showed that the strain rate $d\varepsilon/dt$ is given by

$$d\varepsilon/dt = A\sigma\Omega/kTd^2D_{\text{eff}} \quad (6)$$

where A is a constant, σ is the applied stress, Ω is the volume of a lattice vacancy, d is the grain size and D_{eff} is the effective diffusion coefficient (which depends on the proportion of the vacancy flux carried by grain boundaries and the crystal lattice, respectively).

An example of plastic deformation which occurs solely by diffusion along grain boundaries is Herring–Nabarro–Coble (H–N–C) creep. Here, grain-boundary diffusion of the most mobile atomic species occurs to those grain boundaries which lie perpendicular to the maximum tensile stress, from grain boundaries across which the tensile stress is minimum. Owing to this diffusive flow, the applied stress does work; that is, the material extends in the direction of maximum tension by the "pumping" of atoms from longitudinal to transverse grain boundaries. It can be shown that this process is one of Newtonian viscous flow and that the resulting strain rate is given by an expression having the same form as Eqn. (6). This mechanism is important in metals at low stresses, where dislocation motion is slow, and in polycrystalline ceramics, in which dislocation motion is difficult because of the nature of the bonding of the crystal.

In polycrystals with a stable, very fine grain size ($d \simeq 1$ μm) virtually all flow can occur by grain-boundary diffusion, and the material can then become "superplastic;" that is, the material can approach Newtonian viscous behavior at a certain value of $d\varepsilon/dt$. In this case, the material resists "necking" (i.e., local thinning) in tension, and very large strains can be imposed, sometimes at strain rates approaching $10^{-1}\,s^{-1}$, at relatively low applied stresses. Thus, this is the basis of a potentially attractive method of forming complex shapes, for example, from sheet materials. In this process the grains remain equiaxed (i.e., they do not elongate as in H–N–C creep), and they switch neighbors in a process described by Ashby and Verrall (1973).

4. *Fracture at Elevated Temperatures*

Fracture of metallic materials at elevated temperatures normally occurs along grain boundaries because of rapid diffusion in these regions. In fact, this failure usually occurs by rupture, that is, nucleation and growth of cavities, rather than "fracture" in the sense of brittle fracture. The growth of the cavities occurs by outward diffusion of atoms along the grain boundaries. This process is the opposite of sintering, which is the closure of internal cavities at high temperatures in an unstressed polycrystalline solid, usually formed from compacted powders, due to diffusion of atoms to the cavities. The driving force for sintering is the chemical potential gradient set up because of the surface tension of the cavities. In high-temperature rupture the opposite process occurs, because the component of tensile stress across the grain boundary does work as the solid extends in this direction owing to the "plating out" of atoms along the grain boundaries. The atoms come from the vicinity of the cavities and are exchanged for vacancies which have diffused down the boundaries. The resulting cavity growth continues to the point where rapid coalescence occurs by local plastic flow in the ligaments between the cavities. In this way, intergranular cracks can form; crack growth can then occur by the same process, and is more rapid at the crack tip because of the higher local stress in this region. The fracture is often macroscopically brittle as the overall plastic deformation of the specimen can be small. The process of cavity growth in a plastically deforming solid has been modelled recently (Beere and Speight 1978), and it has been shown that the rate of growth of cavities or cracks can be controlled either by grain-boundary diffusion or by the rate of plastic deformation in the uncavitated ligaments, depending on stress, temperature and microstructural variables.

Recently it has been found that true brittle fracture can also occur at elevated temperatures in grain boundaries if the material contains surface active impurity atoms which lower cohesion, since they can segregate to the region of high stress at the tip of any cracklike flaw from the free surface of the flaw. This can produce local brittle fracture in that region when the impurity concentration reaches a large-enough value. This mechanism, which was recently discovered in an alloy steel (Shin and McMahon 1984), produces a brittle fracture which apparently proceeds by diffusion-controlled decohesion. The rate of crack formation is therefore controlled by the local stress which drives the diffusion of the impurity, and by the diffusivity of the impurity in the grain boundary and on the free surface.

5. *Environmental Effects*

Grain boundaries can be sites of preferential chemical attack. This may occur because of a difference in chemical composition between the matrix and grain boundaries due to solute segregation or precipitation of some new phase. A now-classical example of the latter is "sensitization" of some austenitic stainless steels to intergranular anodic dissolution in aqueous environments due to intergranular precipitation of $Cr_{23}C_6$. This depletes the nearby regions of chromium, a certain level of which is required for adequate corrosion resistance (due to passivation, or formation of a protective film). This sensitization occurs when the steel is heated in a particular temperature range (especially 600–800 °C), and it is enhanced by plastic deformation, which produces a martensitic transformation in certain steels (Briant and Ritter 1980). The product of this transformation is ferrite, in which chromium diffuses relatively rapidly and sensitization is accordingly hastened. These conditions of heating and deformation are found in the heat-affected zones of welds.

As they offer paths of rapid diffusion, grain boundaries can be regions of contamination by deleterious species which diffuse into a material from the environment. Examples include preferential oxidation along grain boundaries in high-temperature alloys (Chaku and McMahon 1974) and probably some forms of liquid-metal embrittlement.

6. *Summary*

In summary, grain boundaries play a variety of roles in the deformation and fracture of metallic materials. In deformation they act both as a source of dislocations and as a barrier to dislocation motion; in fracture they are sites of brittle crack nucleation and barriers to crack propagation. In terms of "low-temperature" behavior, grain refinement is the only procedure by which one can simultaneously strengthen and toughen a material. Regarding high-temperature behavior, on the other hand, coarse-grain structures (even single crystals) are preferable, since diffusion along grain boundaries provides a mechanism for rapid flow and fracture, and preferential oxidation.

See also: Grain Boundary Structure; Dislocations; Intergranular Corrosion; Mechanics of Materials: An Overview; Surfaces and Interfaces: An Overview

Bibliography

Ashby M F, Verrall R A 1973 Diffusion–accommodated flow and superplasticity. *Acta Metall.* 21: 149–63

Beere W, Speight M V 1978 Diffusive crack growth in plastically deforming solid. *Met. Sci.* 12: 593–99

Briant C L, Ritter A M 1980 The effects of deformation induced martensite on the sensitization of austenitic stainless steel. *Metall. Trans., A* 11: 2009–17

Chaku P N, McMahon C J Jr 1974 The effect of an air environment on the creep and rupture behavior of a nickel–base high temperature alloy. *Metall. Trans.* 5: 441–50

Eshelby J D, Frank F C, Nabarro F R N 1951 The equilibrium of linear arrays of dislocations. *Philos. Mag.* 42: 351–64

Low J R Jr 1954 *Relation of Properties to Microstructure*. American Society for Metals, Metals Park, Ohio, p. 163

McMahon C J Jr, Cohen M 1965 Initiation of cleavage in polycrystalline iron. *Acta Metall.* 13: 591–604

Raj R, Ashby M F 1971 On grain boundary sliding and diffusional creep. *Metall. Trans.* 2: 1113–27

Shin J, McMahon C J J 1984 Mechanisms of stress relief cracking in a ferritic steel. *Acta Metall.* 32: 1535–52

C. J. McMahon Jr.

Interfaces in Diffusional Nucleation and Growth

The nucleus–matrix interfacial energy is the principal barrier to the nucleation of a new phase. This energy depends upon the crystallography of both the nucleus and matrix phases and has a minimum value for specific lattice orientation relationships. This orientational dependence of the interfacial energy is directly related to the interfacial structure observed during growth. Disordered-type interphase boundaries can migrate with long-range, mass-transport-controlled growth kinetics; partially or fully coherent interphase boundaries, however, probably migrate via the ledge mechanism when the crystal structures of the two phases are different.

1. *Diffusional Nucleation*

Figure 1 shows that interfacial energy is the major inhibitor of nucleation because it increases the free energy of the embryo. For small values of the radius r of the embryo, the interfacial energy is proportional to r^2, and therefore increases the free energy more rapidly than the free-energy change driving nucleation, proportional to r^3, can decrease it. The total free-energy change attending nucleation thus passes through a maximum ΔG^* at a critical radius r^*. Figure 1 also indicates that nucleation is a sequence of single-atom additions which increase the free energy of the embryo. These additions take place by statistical fluctuations. Decay of an embryo (with $r < r^*$) to single atoms is always far more likely than further development toward critical nucleus size. Hence, the smaller is ΔG^*, the greater is the probability that an embryo of a particular design will survive to become a critical nucleus.

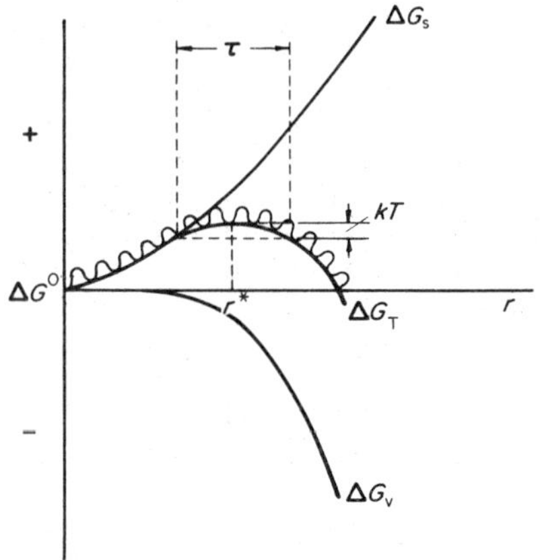

Figure 1
Interfacial free-energy change ΔG_s, volume free-energy change GRKPΔG_v, and their sum GRKPΔG_t as a function of radius r during embryo formation. The dashed substructure in the total free-energy curve incorporates activation free energy for the addition of single atoms to an embryo. τ is approximated as the incubation time for formation of critical nuclei

Nucleus–matrix boundaries are likely to be fully coherent at nearly all areas of the nucleus "surface." The composition change across coherent interphase boundaries is spread across several atomic layers, the number of which increases with temperature (Cahn and Hilliard 1958). When the radius of the nucleus exceeds the half-thickness of the interface, the composition of the nucleus is constant in its central region and the considerations of "classical" nucleation theory apply; this regime is applicable at small supersaturations. The classical nucleation rate J^*, expressed in number of critical nuclei per unit volume (or per unit area or unit length) in the matrix phase per unit time, is

$$J^* = Z\beta^* N \exp(-\Delta G^*/kT)\exp(-\tau/t) \quad (1)$$

where Z is the Zeldovich nonequilibrium factor, which corrects the "equilibrium" number to the actual number of nuclei present at steady state; β^* is the rate at which single atoms join the critical nucleus; N is the density of atomic nucleation sites; ΔG^* is the free energy of activation for critical nucleus formation; T is the temperature; τ is the incubation time (i.e., the time for the first nuclei to appear); and t is the isothermal reaction time. When the number of interfacial energies characterizing a critical nucleus is finite and each interface with a given energy is analytically describable, then

$$\begin{aligned} J^* &= [Dx_\beta \gamma^{1/2} K_Z K_{\beta^*} N v_\alpha/(kT)^{1/2} a^4] \\ &\quad \times \exp(-K_{\Delta G^*}\gamma^3/\Delta G_v^2 kT) \\ &\quad \times \exp[-K_\tau kT\gamma a^4/(Dx_\beta v_\alpha^2 \Delta G_v^2 t)] \end{aligned} \quad (2)$$

where D is the diffusivity in a β matrix; x_β is the atom fraction of solute in β; γ is the interfacial energy of a particular nucleus–matrix interface; v_α is the average volume per atom in an α nucleus; a is the average lattice parameter of α and β; ΔG_v is the volume free-energy change attending nucleation; and K_Z, K_{β^*}, $K_{\Delta G^*}$ and K_τ are constants applicable to their counterpart terms in Eqn. (1), incorporating the geometry and energies of the nucleus–matrix interfaces, and of the grain boundaries when nucleation occurs there.

D in Eqn. (2) is the volume interdiffusivity when nucleation occurs homogeneously, and can be the grain boundary, interphase boundary or dislocation pipe diffusivity when nucleation takes place at grain boundaries or dislocations. At sufficiently low temperatures, D becomes so small that $J^* \to 0$; otherwise, the influence of D upon the temperature dependence of J^* is comparatively modest. The dominant term in Eqns. (1) and (2) is ΔG^*. The ΔG_v component of $\Delta G^* \to 0$ as undercooling below the solvus decreases; hence $\Delta G^* \to \infty$ and $J^* \to 0$. At perceptible undercoolings, $K_{\Delta G^*}\gamma^3$ largely determines ΔG^*. Minimizing the product of these terms is vital if ΔG^* is to be confined to the level ($\lesssim 60kT$) required for a detectable J^*. Figure 2a illustrates a minimum $K_{\Delta G^*}\gamma^3$ shape for homogeneous nucleation; Fig. 2b exemplifies such a shape for nucleation at a planar, disordered grain boundary. The planar facets on these shapes correspond to coherent interphase boundaries and the spherically curved regions to disordered-type interfaces. Formation of such facets usually requires low-index lattice orientation relationships between nucleus and matrix phases; the marked reduction in $K_{\Delta G^*}$ when faceting is extensive makes these relationships ubiquitous in diffusional phase transformations.

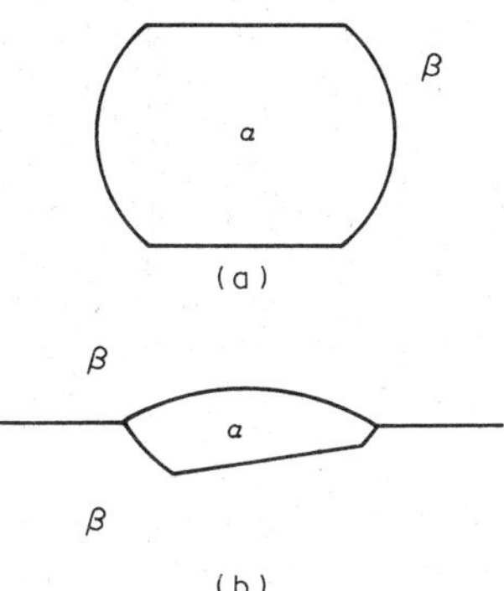

Figure 2
Polar cross sections through faceted critical nuclei formed (a) homogeneously and (b) at a planar, disordered grain boundary. The linearities are planes in three dimensions; the curved portions of the nucleus–matrix boundary are traces of spherical segments

J^* is markedly increased at grain boundaries, primarily through diminution of $K_{\Delta G^*}$. Physically this is achieved because the grain-boundary energy lost due to engulfment by the critical nucleus supplies some of the energy of newly created nucleus–matrix interfaces. Nucleation at grain edges and corners reduces $K_{\Delta G^*}$ more than at grain faces, but at appreciable cost to the density of nucleation sites N; the resulting balance favors corner nucleation at small undercoolings below the equilibrium temperature, then edge, face and finally homogeneous nucleation (if D is still sufficiently large) at successively lower temperatures (Cahn 1956).

When the volume strain energy W is small relative to ΔG_v, then W and ΔG_v may be algebraically combined in Eqn. (2). Nucleation at dislocations is favored primarily by the simultaneous reduction in W and in the dislocation strain energy (Cahn 1957) and is more pronounced at small undercoolings and in the formation of transition phases.

When undercooling is sufficiently large that compositional diffuseness of the interface reaches the center of the nucleus, nucleation becomes nonclassical and interfacial energy and ΔG_v can no longer be considered separately. When both matrix and precipitate have the same cubic-type crystal structure with identical orientations, then for homogeneous nucleation (Cahn and Hilliard 1959)

$$\Delta G^* = 4\pi \int_0^\infty [\Delta G_v + K(dx/dr)^2] r^2\, dr \qquad (3)$$

where K is the so-called gradient energy term and x is the solute concentration in the nucleus as a function of r, the radius of the spherical nucleus. The pre-exponential terms of Eqn. (2) apply approximately, though less accurately with increasing undercooling as the nuclei become increasingly spread out and finally merge smoothly with the product of spinodal decomposition. Application of nonclassical theory to transformations involving a change in crystal structure and/or heterogeneous nucleation has yet to be made. Effects of crystalline anisotropy upon ΔG^*, which Eqn. (3) does not include, have been studied in fcc→fcc transformations by the discrete lattice point approach (Cook et al. 1969) in a recent investigation (Le Goves et al. 1984a,b). Strain energy was also taken into account with this technique (Cook and De Fontaine 1969) and in an associated study (Le Goves et al. 1984a,b; Le Goves and Aaronson 1984).

2. *Diffusional Growth*

When an interphase boundary is sufficiently disordered so that displacement of the boundary is readily accomplished, and the boundary itself neither proves a "short circuit" for mass transport nor is in contact with a high density of dislocations in the matrix phase, the migration or growth kinetics of the boundary depend upon the volume diffusivity in the matrix, the driving force for mass transport of solute and the curvature of the boundary. The roles of the first two factors are readily perceived from the linearized concentration gradient approximation for the growth rate G of a planar, disordered interphase boundary of effectively infinite extent (Zener 1949):

$$G = \frac{K_1 D^{1/2}(x_\beta^{\beta\alpha} - x_\beta)}{2t^{1/2}(x_\beta^{\beta\alpha} - x_\alpha^{\alpha\beta})^{1/2}\,(x_\beta - x_\alpha^{\alpha\beta})^{1/2}} \equiv \frac{\xi D^{1/2}}{2t^{1/2}} \qquad (4)$$

where K_1 is a constant little greater than unity; ξ is the parabolic rate constant; t is the growth time; and the x's are the atom fractions of solute as indicated in the phase diagram of Fig. 3a. A linearized concentration–penetration plot normal to and through an α–β boundary is shown in Fig. 3b. As the solvus temperature is approached, $x_\alpha^{\beta\alpha} \rightarrow x_\beta$ and $\xi \rightarrow 0$. At low temperatures, D becomes very small and therefore so does G. Hence G passes through a maximum,

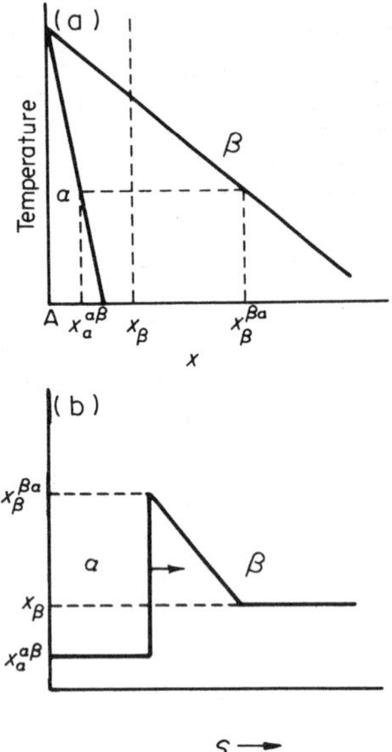

Figure 3
(a) Portion of an A-rich A–B phase diagram, showing the notation used to designate compositions; (b) linearized concentration–penetration (x–s) curve through an α–β boundary

and usually does so relatively close to the solvus because D is now the factor most sensitive to temperature.

Exact equations for ξ and for the concentration–penetration curve in the matrix are available for the infinite planar boundary, the sphere, the cylinder and the ellipsoid (Sekerka et al. 1975). The lengthening of a plate and of a needle with a disordered edge or tip can be modelled as the displacement of a parabolic cylinder and of a hyperboloid of revolution, respectively (Trivedi 1970a,b). Linear time laws prevail in both cases even though growth is volume diffusion controlled. Taking the influence of both capillarity and the "point effect of diffusion" upon G into account introduces the question of how the edge or tip radius, which controls the strength of both effects, is determined. Maximization of the lengthening rate is often assumed, but the "ragged edge of Mullins–Sekerka instability" (Mullins and Sekerka 1963) is increasingly preferred. It has been shown, however, that the two hypotheses become identical when side-branching is prohibited, for example, by the partially coherent broad faces of plates and needles (Trivedi 1983).

During growth, precipitate dimensions are usually attained at which the introduction of periodic arrays of misfit dislocations (Fig. 4) into originally fully coherent interphase boundaries becomes energetically advantageous (van der Merwe 1963). Acquisition of the equilibrium misfit dislocation structure,

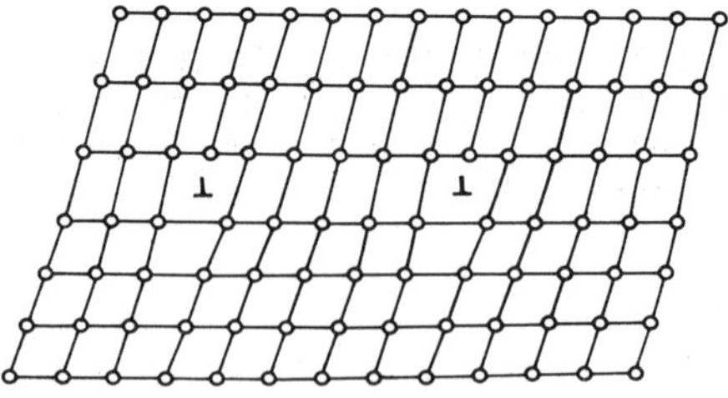

Figure 4
Misfit dislocations at an interphase boundary between two simple cubic crystals with identical orientations and differing only in lattice parameter

however, can be a tedious and idiosyncratic process (Aaronson 1974). When α and β have different crystal structures, both misfit dislocation (also termed partially coherent) and fully coherent interphase boundaries are immobile during diffusional growth because changing the stacking sequence in the coherent regions is kinetically unfeasible (Aaronson et al. 1970). Such interfaces can migrate only by means of the ledge mechanism, wherein the risers of the ledges either have a disordered structure or are themselves ledged (Fig. 5).

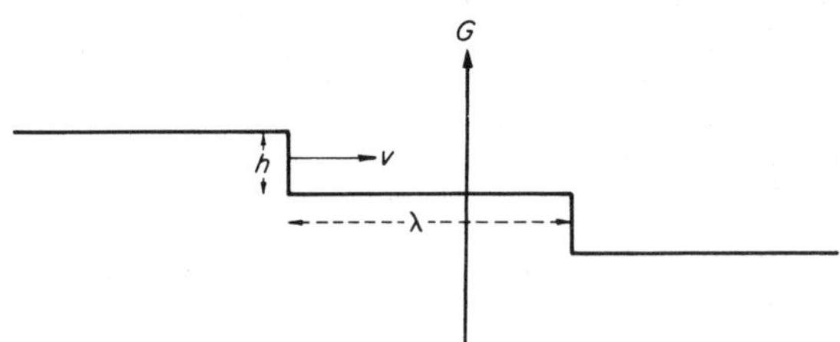

Figure 5
Growth of a partially coherent interphase boundary solely by the ledge mechanism

The growth rate G in the normal direction via the ledge mechanism is (Jones and Trivedi 1971)

$$G_{\text{ledge}} = D(x_\beta^{\beta\alpha} - x_\beta)/\varepsilon\lambda(x_\beta^{\beta\alpha} - x_\alpha^{\alpha\beta}) \tag{5}$$

where ε is a complex function of supersaturation and λ is the average spacing between growth ledges. Numerous proposals for the origins of ledges have been made; many have been experimentally confirmed but, at least on the broad faces of precipitate plates, which are almost entirely partially or fully coherent with the matrix, all appear to yield erratic and time-dependent λ's. Hence, plate thickening is difficult to describe analytically (Aaronson et al. 1970, Aaronson 1974, Sankaran and Laird 1974).

Theoretical prediction and experimental evaluation of the boundary orientation dependence of interfacial structure for various pairs of crystal lattices and the mechanism(s) and frequencies of ledge formation at partially coherent boundaries are the problems in diffusional growth most urgently in need of investigation at this time.

Acknowledgement

Appreciation is expressed for the support of nucleation studies by the National Science Foundation through Grant No. DMR 81-16905 from the Division of Materials Research and of growth studies by the Army Research Office on Contract DAAG-29-80-K0071.

See also: Phase Interfaces: Morphological Stability; Nucleation and Growth Transformations in the Solid State; Grain Boundary Structure; Surfaces and Interfaces: An Overview

Bibliography

Aaronson H I 1962 The proeutectoid ferrite and the proeutectoid cementite reactions. In: Zackay V F, Aaronson H I (eds.) 1962 *Decomposition of Austenite by Diffusional Processes.* Interscience, New York, pp. 387–546

Aaronson H I 1974 Observations on interphase boundary structure. *J. Microsc. (Oxford)* 102: 275–300

Aaronson H I, Laird C, Kinsman K R 1970 Mechanisms of diffusional growth of precipitate crystals. In: Aaronson H I (ed.) 1970 *Phase Transformations.* American Society for Metals, Metals Park, Ohio, pp. 313–96

Aaronson H I, Russell K C 1983 Nucleation—Mostly homogenous and in solids. In: Aaronson H I, Laughlin D E, Sekerka R F, Wayman C M (eds.) 1983 *Proc. Int. Conf. Solid-Solid Phase Transformations.* The Metallurgical Society (AIME), Warrendale, Pennsylvania, pp. 371–97

Cahn J W 1956 The kinetics of grain boundary nucleated reactions. *Acta Metall.* 4: 449–59

Cahn J W 1957 Nucleation on dislocations. *Acta Metall.* 5: 169–72

Cahn J W, Hilliard J E 1958 Free Energy of a nonuniform system, I: Interfacial free energy. *J. Chem. Phys.* 28: 258–67

Cahn J W, Hilliard J E 1959 Free energy of a nonuniform system, III: Nucleation in a two-component incompressible fluid. *J. Chem. Phys.* 31: 688–99

Cook H E, De Fontaine D, Hilliard J E 1969 A model for diffusion on cubic lattices and its application to the early stages of ordering. *Acta Metall.* 17: 765–73

Cook H E, De Fontaine D 1969 On the elastic free energy of solid solutions, I: Microscopic theory. *Acta Metall.* 17: 915–24

Jones G J, Trivedi R K 1971 Lateral growth in solid–solid phase transformations. *J. Appl. Phys.* 42: 4299–304

Le Goves F K, Aaronson H I 1984 Influence of crystallography upon critical nucleus shapes and kinetics of homogeneous fcc → fcc nucleation, IV: Comparisons between theory and experiment in Cu-Co alloys. *Acta Metall.* 32: 1855–64

Le Goves F K, Aaronson H I, Lee Y W 1984a Influence of crystallography upon critical nucleus shapes and kinetics of homogeneous fcc → fcc nucleation, III: The influence of elastic strain energy. *Acta Metall.* 32: 1845–53

Le Goves F K, Lee Y W, Aaronson H I 1984b Influence of crystallography upon critical nucleus shapes and kinetics of homogeneous fcc → fcc nucleation, II: The nonclassical regime. *Acta Metall.* 32: 1837–43

Mullins W W, Sekerka R F 1963 Morphological stability of a particle growing by diffusion or heat flow. *J. Appl. Phys.* 34: 323–29

Sankaran R, Laird C 1974 Kinetics of growth of platelike precipitates. *Acta Metall.* 22: 957–69

Sekerka R F, Jeanfils C L, Heckel R W 1975 The moving boundary problem. In: Aaronson H I (ed.) 1975 *Lectures on the Theory of Phase Transformations.* The Metallurgical Society (AIME), Warrendale, Pennsylvania, pp. 117–69

Trivedi R K 1970a The role of interfacial free energy and interface kinetics during the growth of precipitate plates and needles. *Metall. Trans.* 1: 921–27

Trivedi R K 1970b Growth of dendritic needles from a supercooled melt. *Acta Metall.* 18: 287–96

Trivedi R K 1983 Volume diffusion-controlled growth kinetics and mechanisms in binary alloys. In: Aaronson H I, Laughlin D E, Sekerka R F, Wayman C M (eds.) 1983 *Proc. Int. Conf. Solid-Solid Phase Transformations.* The Metallurgical Society (AIME), Warrendale, Pennsylvania, pp. 477–502

van der Merwe J H 1963 Crystal interfaces, Part 1: Semi-infinite crystals; Part 2: Finite overgrowths. *J. Appl. Phys.* 34: 117–27

Zener C 1949 Theory of growth of spherical precipitates from solid solution. *J. Appl. Phys.* 20: 950–53

H. I. Aaronson

Intergranular Corrosion

At the boundaries between the grains of metals, lattices of different orientations meet, and zones of less perfect structure than that of the interior of the grains are formed. Small amounts of impurities may segregate at the grain boundaries while remaining in solid solution during certain heat treatments. During exposure to a corrosive environment these impurities promote galvanic action by serving as cathodic sites (see *Galvanic Corrosion*). This increases the rate of anodic dissolution of the alloy matrix in the grain boundary zone.

Of greater practical importance is the fact that precipitates which may form during exposure of metals to elevated temperatures, e.g., during production, fabrication and welding, frequently nucleate and grow preferentially at grain boundaries. When such precipitates are rich in alloying elements essential for corrosion resistance, the matrix adjacent to the precipitates will be depleted in this alloying element. The metal is thus "sensitized" to intergranular attack. Such is the case with the formation of the $Cr_{23}C_6$ precipitate in austenitic stainless steels, e.g., Fe–18%Cr–8%Ni (AISI Type 304). The $Cr_{23}C_6$ precipitate forms because carbon is a residual impurity in these alloys.

The morphology of intergranular attack is a function of the ratio of the rate of penetration at the grain boundaries to the rate of corrosion at the grain faces, i.e., the unaffected matrix. When this ratio is one, there is either no intergranular attack or a mere outlining of the grain boundaries. High ratios indicate severe penetration without much attack on grain faces. In these cases, there is little attack on the walls of the deep grooves formed at susceptible grain boundaries. The grains tend to stay in place even though the entire cross section of the object (sheet, plate, rod) may be penetrated by attack along all the grain boundaries, but without any visible changes or significant weight loss. However, there is then a complete loss in strength, and the specimen can readily be broken by hand. The attack of hot sulfuric acid solution containing cupric sulfate on sensitized Type 304 is of this kind.

Intermediate ratios are indicative of significant attack on the undepleted matrix or grain faces. This means that there is significant attack on the walls of the narrow grooves surrounding the grains, and, as a result, the grains are dislodged and the material may be reduced to a pile of individual grains. An example of this is the corrosion of sensitized Type 304 steel in hot nitric acid solutions. Such attack also occurs in reducing acid solutions. The attack of austenitic stainless steels, in fact, is the most frequently encountered type of failure by intergranular corrosion in industry.

Another example of industrial importance is found in aluminum alloys when they contain Mg_5Al_8 and $CuAl_2$ intermetallic compounds at the grain boundaries. During exposure to chloride solutions the galvanic couples formed between these precipitates and the alloy matrix can lead to severe intergranular attack.

Intergranular attack can be prevented by (a) homogenizing the microstructure by means of heat treatments; (b) refining to lower the concentration of residual impurities; (c) adding stabilizing alloying elements (titanium or niobium) to combine with residual carbon in stainless alloys; and (d) adjusting the alloying elements to minimize the formation of intermetallic compounds, e.g., in 54%Ni–15%Cr–15%Mo alloys.

See also: Corrosion of Stainless Steels; Corrosion of Aluminum Alloys; Corrosion: Metallurgical Aspects

Bibliography

McLean D 1957 *Grain Boundaries in Metals.* Clarendon Press, Oxford

Streicher M A 1978 Theory and application of evaluation tests for detecting susceptibility to intergranular attack in stainless steels and related alloys—problems and opportunities. In: Steigerwald R F (ed.) 1978 *Intergranular Corrosion of Stainless Alloys,* Special Technical Publication 656. American Society for Testing and Materials, Philadelphia, Pennsylvania, pp. 1–84

M. A. Streicher

Internal Oxidation of Alloys

When a sufficiently dilute solid-solution alloy comprising a more noble solvent component A and a more reactive solute component B is exposed to an oxidizing gas at elevated temperature, particles of the stable oxide BO_ν may be precipitated within the A matrix by a reaction called internal oxidation. The dissolution and diffusion of the oxidant atoms in the A matrix require significant values for both the solubility and diffusivity of the oxidant in metal A. Although internal oxidation involving oxygen atoms is the most familiar, analogous internal precipitation reactions have been studied for the oxidants sulfur, phosphorus, nitrogen, hydrogen, carbon and fluorine.

1. Thermodynamics

The driving force for the precipitation of a compound BX_ν within an A matrix is quite clear: locally the solubility product of the compound BX_ν is exceeded, so that BX_ν precipitates are nucleated and grow as B is depleted from the surrounding matrix upon the arrival of X atoms by diffusion. In other words, the addition of the oxidant to the alloy system moves the solid solution into a two-phase field of matrix plus product compound. Consider the solubility product for CdO in the noble metal silver:

$$\mathrm{CdO} \rightarrow \mathrm{Cd\ (in\ Ag)} + \mathrm{O\ (in\ Ag)} \qquad (1)$$

At equilibrium,

$$a_{\mathrm{Cd}} a_{\mathrm{O}} = K_1 = \exp(-\Delta G_1^0/RT) \qquad (2)$$

where a_{Cd} and a_{O} are the activities of cadmium and oxygen, relative to Raoultian and Henrian standard states, respectively, and K_1 and ΔG_1^0 are the equilibrium constant and the standard Gibbs energy change for the reaction given by Eqn. (1). ΔG_1^0 can be evaluated from data for $\Delta G_{\mathrm{CdO}}^0$ (Gibbs formation energy for CdO) and the solubility of oxygen in silver, N_{O}, for any given partial pressure of oxygen p_{O_2}:

$$\Delta G_1^0 = \Delta G_{\mathrm{CdO}}^0 + RT \ln (p_{\mathrm{O}_2}^{\frac{1}{2}}/N_{\mathrm{O}})_{\mathrm{eq}} \qquad (3)$$

For the precipitation of CdO in solid silver at 1000 K, known values for the activity coefficient for dilutc cadmium in silver and for the solubility of oxygen in silver give the following result:

$$N_{\mathrm{Cd}} N_{\mathrm{O}} = 1.5 \times 10^{-9} \qquad (4)$$

where N_{Cd} and N_{O} are expressed as atom fractions. For the equilibrium of silver with air (p_{O_2} = 0.21 atm) at 1000 K, N_{O} equals 7.8×10^{-5}, so that N_{Cd} equals only 1.9×10^{-5} (Rapp 1980). Clearly, an oxide of even intermediate stability such as CdO exhibits only a very small solubility product in a given matrix metal.

2. Mechanism

Binary alloys containing dilute highly reactive components can be classified into two groups: (a) the less reactive solvent metal is noble in the high-temperature gaseous environment, or (b) the solvent metal is reactive and forms an external scale. The first group includes silver-base alloys exposed in air, or copper-, iron-, nickel- or cobalt-base alloys in CO/CO_2 or H_2/H_2O gases which are reducing to Cu_2O, FeO, NiO or CoO, respectively. For alloys with a noble solvent component, exposure to the hot gas which is oxidizing to the dilute solute B results in atoms of the oxidant dissolving into the alloy at its external surfaces. If the solute oxide, BO_ν, exhibits a low solubility product in the solvent, then nuclei of BO_ν are formed in a nearly pure A matrix just beneath the external surface. As each BO_ν nucleus grows upon the further dissolution of oxygen at the surface, B atoms arrive at the A–BO_ν interface by diffusion in the alloy from only a very small, localized volume around the nucleus. Finally, the solute B is severely depleted locally to approximately satisfy the solubility product, so that the reaction front (the site of nucleation and growth of BO_ν oxide particles) penetrates more deeply into the alloy.

Figure 1 illustrates the cross-sectional morphologies for the oxidation of silver–indium alloys in air at 550 °C. For the most dilute silver–indium alloys, tiny equiaxed precipitates are formed because the nucleation of new particles at the internal reaction front is favored over the growth of existing particles. For intermediate compositions, however, rods or platelets of BO_ν grow roughly parallel to the diffusion direction. Finally, a critical alloy composition is reached, $N_{In} = 0.15$ for the oxidation of silver–indium alloys in air at 550 °C, at which an external In_2O_3 scale is formed. For these concentrated alloys, internal oxidation is prevented by the presence of the adherent external scale. Most alloy systems exhibiting internal oxidation show analogous behavior with some differences in critical compositions for morphological changes, and perhaps some variation in particle morphology; for example, some precipitates form as imperfect internal oxide films.

Schematic concentration profiles for oxygen and component B are given in Fig. 2 for a noble matrix metal A. Figure 2a illustrates the usual case where the precipitate exhibits a negligibly small solubility product when the oxygen permeability product $N_O D_O$ is relatively large. In this case, the diffusion of B over an extended distance is not possible, resulting in a virtual step function in N_B. Then, at the reaction front, oxygen arrival via dissolution and diffusion controls the rate of oxidation. In Fig. 2b, if the supply of solute B is greater and the arrival of oxygen more limited, then the motion of the reaction front is controlled jointly by inward diffusion of oxygen and outward diffusion of B. In the zone of internal oxidation, B is enriched as BO_ν.

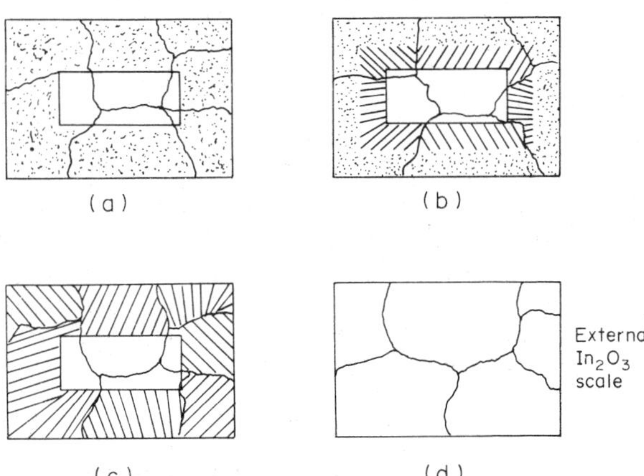

Figure 1
Schematic illustration of morphologies observed in the oxidation of silver–indium alloys at 550 °C in air: (a) $0 < N^0_{In} < 0.05$; (b) $0.05 < N^0_{In} < 0.07$; (c) $0.07 < N^0_{In} < 0.15$; (d) $N^0_{In} > 0.15$. The superscript 0 refers to the bulk (after Rapp 1965)

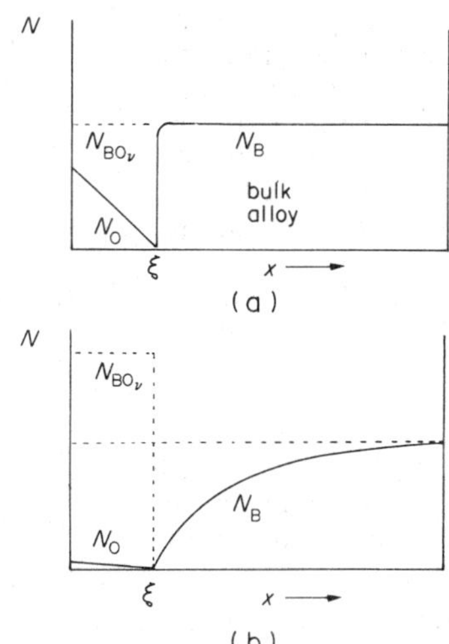

Figure 2
Concentration profiles for the exclusive internal oxidation of alloys: (a) $D_B/D_O \ll N^s_O/N^0_B \ll 1$; (b) $N^s_O/N^0_B \ll D_B/D_O \ll 1$. The superscript refers to the surface; the superscript 0 refers to the bulk (after Rapp 1965)

3. Simple Kinetics

For the realistic approximations of a zero solubility product and a linear oxygen profile, the kinetics for the internal oxidation described by Fig. 2a can be readily analyzed. The rate of oxygen arrival, j_O, at the precipitation front at $x = \xi$ is given by Fick's first law:

$$j_O(\text{mol s}^{-1}) = -D_O A\,[c_O(\text{front}) - c_O(\text{surface})]/\xi \tag{5}$$

where D_O is the oxygen diffusivity, A is the area and c_O is the oxygen concentration. For the migration of the precipitation front through an area A in a time dt, all of the B atoms in the volume $Ad\xi$ are oxidized to form BO_ν:

$$D_O A\,(N_O^s/V_A\xi)\,dt = \nu(N_B^0 A/V_A)\,d\xi \tag{6}$$

where V_A is the molar volume of A, N_O^s is the equilibrium oxygen mole fraction at the surface, and N_B^0 is the mole fraction of B in the bulk alloy. Upon separation of variables in Eqn. (6), and integration:

$$\xi^2 = 2N_O^s D_O t/\nu N_B^0 \tag{7}$$

Thus, the depth ξ of internal oxidation limited by oxygen diffusion to form insoluble BO_ν precipitates obeys a parabolic time dependence. Where the assumptions are obeyed, Eqn. (7) so accurately describes the isothermal internal oxidation kinetics that measurements of ξ for an alloy of known dilute N_B^0 have provided accurate values for the permeability product $N_O^s D_O$ in a noble metal matrix, particularly for silver- and copper-base alloys.

4. *Complex Kinetics*

When the assumptions leading to the simple kinetic description are not satisfied, a solution of Fick's second law for the oxidant and reactive-solute concentration profiles, using appropriate initial and boundary conditions, provides complex mathematical descriptions for the internal oxidation kinetics. These analyses include the following interesting cases:

(a) If, as shown in Fig. 2b, the reactive solute also provides significant diffusion, then the kinetic expression is still parabolic and includes the parameters N_O^s, D_O, N_B^0 and D_B. For a given time, the depth of internal oxidation is less than that calculated from Eqn. (7), and the product BO_ν enriches the solute in the reacted zone.

(b) If the BO_ν solubility product is relatively large and the solute B is relatively mobile, then some solute can reach the external surface to form a scale of BO_ν in addition to internal precipitates of BO_ν. An example of this, the oxidation of platinum–copper and palladium–copper alloys, was described by Wagner (1968).

(c) If internal oxidation occurs for cylindrical or spherical specimens, then the initial parabolic kinetics are speeded up as the reaction front approaches the center of the specimen. In comparing the times for "through oxidation" to the same depth, then $t_{\text{sphere}} = \frac{1}{3}t_{\text{slab}}$ and $t_{\text{cylinder}} = \frac{1}{2}t_{\text{slab}}$.

(d) If the reactive gas contains two oxidants (e.g., oxygen plus carbon, or oxygen plus sulfur), the solute may form two sequential product zones: BO_ν and BX_ν in noble matrix A. The product of B which is more stable in the gas phase occupies the outermost zone. Each precipitation zone grows parabolically, and a displacement reaction between the two zones inside the A matrix converts the less stable inner precipitate to the more stable outer precipitate.

(e) If the matrix metal A is not noble in the oxidizing gas but forms an AO_α scale, then oxygen can dissolve into the alloy to the oxygen solubility limit upon local dissociation of the scale at the AO_α–alloy interface. The internal oxidation of a dilute reactive solute B occurs in the same manner as for a noble matrix metal, except that the BO_ν precipitates become physically incorporated into the base of a cation-diffusing scale. The kinetic analysis takes into account the moving boundary at the scale–alloy interface to provide simultaneous parabolic growth of both the scale and the internal precipitation front. If an external scale is grown by inward diffusion (transport) of oxygen according to a linear rate law, then the width of the internal oxidation zone approaches a time-independent value, and the internal precipitates become distributed uniformly throughout the external scale.

5. *Transition to External Scale Formation*

As illustrated in Fig. 1, upon increasing the reactive-solute content in a given alloy system, a critical concentration is reached at which an external scale of BO_ν forms instead of internal precipitates. Because adherent scales of reactive components such as chromium, aluminum and silicon grow very slowly, the transition to external scaling generally represents a significant reduction in the rate of oxidation. Indeed, the external scale of BO_ν can prevent the formation of a much faster growing AO_α scale. Then the internal oxidation of a dilute alloy is an unfavorable occurrence leading to rapid alloy degradation. The rationalization for the cessation of internal oxidation with increasing alloy content concerns the blocking of the cross section of the A matrix to oxidant diffusion at the surface as the BO_ν particles occupy an increasing volume fraction with increasing B concentration. At a critical volume fraction of precipitated particles, the precipitates grow laterally to form a surface film and thereby prevent repeated nucleation in advance of the initial reaction sites. Pretreatments, surface preparations and alloy adjustments which assist external scale formation are used to avoid internal oxidation.

6. Mechanical Consequences

The tiny, perhaps coherent, internal precipitates formed in a dilute alloy serve as impediments to the glide of dislocations, and thereby introduce significant hardening of the matrix. Because the formation of internal precipitates introduces a large volume increase within the matrix, extensive plastic deformation must accompany internal oxidation. For some alloys requiring high-temperature strength, a part may be fabricated from the consolidation of alloy powders which have been intentionally internally oxidized to provide oxide-dispersion strengthening.

See also: Oxidation of Metals and Alloys

Bibliography

Meijering J L 1971 Internal oxidation in alloys. *Adv. Mater. Res.* 5: 1–81

Rapp R A 1965 Kinetics, microstructures and mechanism of internal oxidation: Its effect and prevention in high-temperature alloy oxidation. *Corrosion* 21: 382–401

Rapp R A 1980 *High Temperature Corrosion,* An audio-lecture course. American Chemical Society, Washington, DC, Manual pp. 97–98

Swisher J H 1971 Internal oxidation. In: Douglas D L (ed.) 1971 *Oxidation of Metals and Alloys.* American Society for Metals, Metals Park, Ohio, pp. 235–67

Wagner C 1968 Internal oxidation of copper–palladium and copper–platinum alloys. *Corros. Sci.* 8: 889–93

R. A. Rapp

International Trade in Materials

The international trade of materials has reached a new level of importance in the economic and political relations between nations. Factors which have long affected materials trade patterns have also received new prominence. Resource depletion in the industrialized countries has increased their materials imports from the rest of the world, particularly the developing countries. The unequal geographic distribution of these resources has now made geopolitics a critical factor in national materials decisionmaking. Multinational materials companies have also expanded their activities, as have their counterparts in the form of state trading organizations. Institutions also have arisen which would influence the competitive or free trade characteristic of materials flows. These range from producer organizations such as the Intergovernmental Council on Copper (CIPEC) to complex commodity arrangements such as those recently discussed within the Integrated Program for Commodities (IPC) of the United Nations Conference on Trade and Development.

1. Structure of Trade

The starting point for analyzing the structure of materials trade is the commodity composition of that trade, that is, the materials that are most important among world exports and how this importance has changed over time. Table 1 provides this information for the major commodity groups traded. Although the fuels are shown to dominate total trade, the main concern here is with the agricultural materials and minerals commodities. Among these, the agricultural raw materials and iron and steel are the most important. Exports of both of these materials groups have more than doubled since 1973, iron and steel alone increasing from \$28.46 billion in 1973 to \$76.50 billion in 1980. Next in importance are the nonferrous metals with an export total of \$52.00 billion in 1980. Ores and minerals exports also have increased sharply, reaching some \$39.00 billion in 1980, but represent a smaller fraction of minerals exports because of the high transport costs relative to those for refined metals. Nonmetals surprisingly are not very significant in world minerals trade. Since the magnitude of their transport costs is similar to that of ores, the low unit values of the nonmetals do not justify their transportation over long distances.

Table 1
Value of world export trade in major materials

Materials	Export trade (billion US\$)		
	1973	1978	1980
Agricultural raw materials	34.61	52.05	75.00
Ores and minerals	14.91	24.50	39.00
Nonferrous metals	17.28	27.80	52.00
Nonmetals (crude)	5.90	7.72	10.00
Iron and steel	28.46	57.15	76.50
Fuels	63.48	223.60	468.00
Total commodity exports	574.30	1303.00	1973.00

Source: *International Trade, 1980–81*. General Agreement on Tariffs and Trade (GATT), Geneva, 1981

Moving from trade composition to trading areas, it is obvious that these areas can be grouped only in a very general way. Different groupings are possible, but the one considered here is based on United Nations data for (a) industrial or developed market economy countries, (b) developing market economy countries, and (c) centrally planned economy countries. Table 2 provides the geographical patterns of materials trade over these areas. The materials data available for this purpose divide total materials trade into agricultural raw materials and into ores and metals. The percentage distributions reported show the destination of exports and sources of imports for each area.

The more important materials trade patterns both in terms of exports and imports are among the industrial economies. These countries, which basically

Table 2
Gross exports and imports of agricultural and minerals materials 1978

	Total value (billion US$)	Distribution (%)			
		World total	Industrial market countries	Developing countries	Centrally planned countries
Exports (fob)					
Industrial market countries					
Agricultural materials	31.39	100.0	80.6	13.6	5.8
Ores and metals	81.97	100.0	68.2	20.7	11.1
Total	871.99	100.0	70.8	23.8	5.4
Developing countries					
Agricultural materials	14.44	100.0	61.8	26.4	11.8
Ores and metals	16.24	100.0	73.3	20.9	5.8
Total	300.80	100.0	70.6	24.4	5.0
Centrally planned countries[a]					
Agricultural materials	4.83	100.0	55.9	10.0	34.1
Ores and metals	10.79	100.0	27.5	7.8	64.7
Total	112.43	100.0	26.1	15.2	58.7
Imports (cif)					
Industrial market countries					
Agricultural materials	37.95	100.0	67.6	23.4	9.0
Ores and metals	70.81	100.0	78.6	16.9	4.5
Total	863.54	100.0	71.5	24.6	3.9
Developing countries					
Agricultural materials	8.80	100.0	49.0	43.0	8.0
Ores and metals	21.24	100.0	79.2	16.0	4.8
Total	303.47	100.0	68.3	24.2	7.5
Centrally planned countries[a]					
Agricultural materials	4.31	100.0	31.4	26.6	42.0
Ores and metals	13.36	100.0	41.6	5.4	53.0
Total	110.45	100.0	31.2	10.3	58.5

Source: *Handbook of International Trade and Development Statistics*, Supplement, 1980. UNCTAD, United Nations, Geneva a Includes Eastern Europe only

comprise the Organization for Economic Cooperation and Development (OECD), dominate world trade. The developing countries export materials mostly to the industrial countries. They also import materials from them, most of which have been processed in some way. The centrally planned economies of Eastern Europe export and import mostly to and from each other, with the exception of agricultural raw materials exported to the industrial countries.

What Table 2 fails to enumerate in describing materials trade is the nature of primary material, ore and metals trade, compared with trade in materials that are processed or semiprocessed. It also fails to indicate trading patterns between individual countries. Although Table 2 cannot be disaggregated completely to show these more detailed trade flows, an attempt has been made in Table 3 to perform a disaggregation for the more important materials and countries. For each material the five most important countries have been selected and ranked according to percent market share.

Examining the agricultural raw materials first, one finds a similar trade pattern for cotton, jute, lumber and rubber. These materials are mostly exported from developing countries and are imported by the industrial countries. Some exceptions are the USA as a major lumber exporter and Pakistan as a major jute importer. A similar, though more mixed trade pattern emerges for the metals. Australia, for example, is a major exporter of bauxite, iron ore and lead, and Canada is a major exporter of lead and copper. All of the major importers of these metals are industrialized countries. While these trade patterns become less significant when viewed in terms of Table 2, which reflects the total mix of primary, semiprocessed and processed materials, they still suggest a flow of primary materials from the developing to the industrialized countries. This structure of trade has led to the so-called "North–South Debate" between materials producing and consuming countries. We will return to this important materials trade issue later.

Table 3
Geographical distribution of major materials exporters and importers (and percent market share in 1978)[a]

Materials	Major exporting nations	Major importing nations
Rubber	Malaysia (25.7), Indonesia (11.9), Singapore (18.1), Sri Lanka (1.2), Thailand (6.6)	USA (17.4), Singapore (13.3), Federal Republic of Germany (8.4), Japan (7.6), France (6.5)
Lumber	USA (27), Indonesia (20), Malaysia (16), Ivory Coast (5.2), Philippines (3.2)	Japan (54.6), Korea (10.1), Italy (6.3), Federal Republic of Germany (5.5), France (3.4)
Bauxite	Australia (38.3), Jamaica (19.1), Surinam (9.5), Guinea (4.4), USA (5.2)	USA (40.1), Canada (8.1), Japan (7.7), Norway (7.0), Federal Republic of Germany (7.0)
Copper	Canada (5.3), Chile (16.0), Zambia (8.8), Zaire (6.3), Peru (4.6)	Federal Republic of Germany (13.8), USA (13.3), Belgium (10.3), France (10.2), UK (8.8)
Iron ore	Australia (21.2), Brazil (20.9), Canada (13.4), India (7.6), Sweden (6.4)	Japan (38.9), Federal Republic of Germany (15.6), USA (15.8), UK (5.2), Belgium–Luxembourg (5.1)
Tin	Malaysia (39.6), Thailand (16.4), Bolivia (15.7), Indonesia (12.7), UK (4.9)	USA (31.2), Japan (18.6), Federal Republic of Germany (11.1), France (7.6), UK (5.3)
Cotton	USA (36.3), Egypt (7.3), Turkey (7.2), Sudan (5.6), Mexico (2.3)	Japan (22.1), Korea (9.3), Italy (7.9), France (6.7), Federal Republic of Germany (7.4)
Lead	Australia (23.6), UK (10.6), Federal Republic of Germany (12.0), Canada (9.1), Mexico (5.6)	UK (19.4), USA (17.8), Federal Republic of Germany (10.3), Italy (9.6), Netherlands (5.8)
Jute	Bangladesh (66.7), Thailand (12.3), Nepal (7.7), Burma (4.6), India (4.2)	UK (13.5), Pakistan (13.4), Egypt (7.0), Japan (5.9), France (5.7)
Tungsten	Australia (18.3), Thailand (16.5), Bolivia (13.0), Canada (11.7), Korea (7.5)	USA (23.7), Australia (16.6), UK (12.5), Sweden (9.0), Federal Republic of Germany (10.3)

Source: *Yearbook of International Trade Statistics*, Vol. 2, 1980: *Trade by Commodity*. United Nations, New York a A single country may be classified as both importer and exporter because stages of process have been aggregated

2. Determinants of Trade

A number of different determinants underlie the trade patterns described above. Geographic determinants relate to the geological distribution of resources; political determinants relate to particular strategic blocs such as the COMECON countries; and social determinants relate to the need to accelerate development in different parts of the world. However, economic determinants are the primary concern here.

These determinants may be understood by first abstracting trade flows from the above material by organizing them into a general trade table or matrix (Table 4). Its elements T_{ij} indicate the quantity of the material shipped from country i to country j during a period under consideration. Total exports of the ith country (T_i^X) can be calculated from this matrix simply by summing the ith row ($\Sigma_{j=1}^m T_{ij}$), and the total imports of the jth country (T_j^M) by summing the jth column ($\Sigma_{i=1}^n T_{ij}$). Total world trade in the material T is the sum of all elements ($\Sigma_{i=1}^n \Sigma_{j=1}^m T_{ij}$). Economic analysis of this trade is normally divided in three ways: (a) imports into a country T_j^m, (b) exports from a country T_i^X, and (c) flows between any pair of countries T_{ij}.

Materials import demands depend on a number of factors which range from a country's general economic level and its military goals to its relative prices and costs of producing that material domestically. Steel imports may thus depend on the size of the domestic automobile industry, the extent of the armaments industry, and the costs of producing steel domestically. Economic demand theory would refine this explanation by stating that the quantity of material imports purchased is based on real domestic income or product, the price level of that import relative to the price level of an equivalent domestic material, and other factors, such as foreign exchange reserves, credit availability, and so-called dummy variables which reflect unusual import-adjusting events or institutions. The substitutability of the material imports against other imported or domestic goods is also important (see *Substitution: Economics*).

Table 4
Generalized trade matrix

$$
\begin{array}{cccccc|c}
T_{11} & T_{12} & \cdots & T_{1j} & \cdots & T_{1m} & T_1^X \\
T_{21} & T_{22} & \cdots & T_{2j} & \cdots & T_{2m} & T_2^X \\
\vdots & \vdots & & \vdots & & \vdots & \vdots \\
T_{i1} & T_{i2} & \cdots & T_{ij} & \cdots & T_{im} & T_i^X \\
\vdots & \vdots & & \vdots & & \vdots & \vdots \\
T_{n1} & T_{n2} & \cdots & T_{nj} & \cdots & T_{nm} & T_n^X \\
\hline
T_1^M & T_2^M & & T_j^M & & T_m^M & T
\end{array}
$$

Materials exports depend on the demand for a material in other countries as well as the domestic capacity to produce that material. Economic trade theory normally considers exports in the demand context; they are primarily the demand for imports in other countries or areas of the world. A country's material exports are thus dependent on world income or product (or national income in the respective importing countries) and its export prices relative to that of the world or other countries.

Whether or not a country performs well in providing exports is often considered in terms of its ability to increase its market share. Since this depends on its relative export price level, export analysis must inevitably deal also with the conditions underlying the domestic supply of that material. These often include differential growth rates of available productive factors and the responsiveness of export supply to the domestic supply of these factors, differential rates of productivity increases, and the extent to which the country is concentrated in exports to very rapidly growing markets. Behind these factors is the quality of the materials resource base, whether it be the richness of mineral deposits or the high yields from agricultural production.

Instead of dealing just with total imports or exports, economic analysis also attempts to explain the individual trade flows between countries, as given in the above matrix. In this context, a country's choice of trading partners is seen as depending on a number of complex factors in addition to prices and incomes. The material trade flow between countries i and j is thus seen as depending on: (a) the potential of country i to export the material, (b) the potential of country j to import the material, and (c) the resistance to trade between country i and country j. The distance between the two countries is seen as a major factor determining resistance to trade, since transportation costs rise with distance. However, other factors can be equally if not more important. Membership in the same political bloc such as the European Economic Community could be expected to stimulate trade. Also in this category is the proximity of being a neighboring country. Similarly, when multinational companies based in the importing country own production facilities in the exporting country, ownership ties may promote trade. Production differentiation such as that arising when mineral concentrates contain particular impurities that only certain smelters can remove may also influence trade. Finally, the nature of purchase agreements such as long-term mineral contracts can also be significant.

An analysis of the relative importance of these determinants of trade flows in certain metals has recently been made (Demler and Tilton 1980). Materials studied included bauxite, alumina, aluminum, copper concentrate, blister copper, refined copper, iron ore and refined nickel. The results of the analysis generally support ownership ties and the multinational companies that create them as greatly influencing trade, at least at one stage of production. For aluminum and steel they are very important at early stages of processing but they are not so important for refined copper. Political blocs also are significant for trade in refined products but not for trade in ores and concentrates. Sharing a common border has also been found to stimulate trade in refined products such as aluminum, nickel and copper, but not in ores other than iron ore. Finally, long-term contracts have also been found to be an important determinant, particularly in more recent years.

An alternative view for explaining trade flows or why countries are net material importers or exporters concentrates on the concept of comparative advantage. This concept assumes international markets to be freely competitive and that trade improves in those countries which have lower production costs due to more efficient factors of production. If the factors of production can be extended to include resources as well as labor and capital, then countries with a higher quality (or lower cost) materials resource base will eventually become the major exporters. However, a number of imperfections in the world materials markets would complicate this simple concept. These include the stage of process of the material and the growth in demand for it, transport problems and costs, integrated and oligopolistic market conditions, and tariff and nontariff barriers levied on processed or semiprocessed materials as they move to other markets.

The latter are of particular importance here. In theory, they are seen as altering levels of economic welfare in particular countries. In practice, they are seen as influencing the benefit that can be derived from producing materials or having access to them. Thus tariffs on imports provide a form of domestic subsidy that would protect and foster growth of an infant materials industry. Domestic production of such materials might additionally be encouraged for security reasons. Tariffs could also preserve an inefficient yet active domestic mining industry to sustain exploration and to stimulate employment. Both arguments also seem to be influenced by industrialized countries having higher trade barriers for processed as compared to unprocessed materials. These tariffs are seen as being harmful for a number of reasons: materials cannot be obtained from lowest cost sources; rapid exhaustion of resources may occur, sometimes under premature technology; and trade patterns can develop that will lead to less than optimal growth in the long run.

3. *Important Trade Issues*

Several issues have tended to reappear frequently in recent discussions of international materials trade. In this section, they are briefly reviewed and their significance assessed.

3.1 *Commodity Arrangements*

Materials prices are known to fluctuate more frequently and widely than prices of semiprocessed or processed goods. The explanation normally given for this phenomenon is that materials markets experience variations in demand caused by changing industrial and economic activity in various nations. Given that materials producers cannot change their output rapidly to meet such demand variations in the short run, materials prices will then have to adjust or fluctuate considerably to move a particular market towards equilibrium. This instability has thus been seen as harmful to materials producers, particularly those located in developing countries where materials earnings dominate total export or foreign exchange earnings.

As a consequence, considerable international deliberations and negotiations arose during the 1970s. The focus of these negotiations has been the United Nations Conference on Trade and Development (UNCTAD). In particular, and at the request of the developing countries, it has launched an Integrated Program for Commodities (IPC). The set of policy initiatives contained therein relate to a set of corrective commodity arrangements as well as a so-called "Common Fund."

These commodity arrangements would attempt to maintain commodity prices at levels that in real terms provide producing countries with favorable terms of trade (i.e., maintain material export purchasing power in relation to manufactured goods import requirements). In addition they would also attempt to reduce price and volume fluctuations that cause fluctuations in export earnings and balance of payments. The issue of raising prices has tended to be met with disagreement because of the lack of international institutional structures that would accomplish this task efficiently.

Price stabilization would seem to be easier to accomplish and could benefit consumers as well as producers. However, the possibilities of a favorable outcome from enacting stabilization arrangements are difficult to assess and forecast (see *Prices of Materials: Stabilization*). The tin buffer stock stabilization mechanism associated with the International Tin Agreement is the only materials arrangement of this type which has come to fruition in the past. However, it has not been notably successful in its attempt to stabilize international tin prices.

The Common Fund is seen as a capital fund useful for financing buffer-stock stabilization as well as for assisting nonstocking efforts among developing countries such as diversification and marketing effects. While these countries can be viewed as benefiting from these activities, the financial viability of the Fund, particularly that of pooling buffer stock financing for as many as eight or ten commodities, has been questioned (Brown 1981). Several nations have agreed to make donations to the Fund, but no operational agreements have been set in motion.

3.2 Market Structure and Bargaining Power

A trade issue related to commodity arrangements is whether producers in individual materials markets can develop sufficient market power to control international materials supplies and prices. To a large extent, this issue has been raised because of the success of the Organization of Petroleum Exporting Countries (OPEC) in raising the levels of crude oil prices. The possibility of market control can be seen in the context of a recent study showing the relatively high degree of concentration in materials markets (Labys 1980). A small number of countries or multinational firms have the capacity to dominate the supplies of materials in a number of individual materials markets. This has given rise to two institutions in particular. First, materials producers have organized into producer associations such as CIPEC for copper, BAUPEC for bauxite, and PHOPEC for phosphate. Second, an increasing number of countries have concentrated their materials selling and buying power into state trading organizations (STOs).

There is considerable question, however, whether such organizations can attain price control power in the form of a cartel (see *Cartels and Producer Associations*). Such prospects are also further dimmed by the need of materials exporting developing countries for sizeable foreign investments in project development.

3.3 Import Dependence

In the above explanation of materials trade patterns, attention was drawn to the fact that many of the industrial or OECD countries import a considerable portion of their materials needs from other countries. Some of these other countries also come within the OECD bloc, but others are developing countries or fall within the COMECON bloc. This import dependence has raised problems concerning the availability of strategic and critical materials and the need for stockpiling them (see *Critical and Strategic Materials*; *Stockpiling: Economics*). Also relevant is the possibility of increased recycling and increased substitution.

The extent of materials import dependence varies from nation to nation. In the case of the USA, Japan and the UK this dependence is relatively high, particularly in the ferroalloying minerals such as manganese, chromium and nickel. At present, such dependence is not seen as particularly serious. New mineral investments are taking place in different parts of the world which tend to diversify sources of supply. In addition the maintenance of a competitive or free trade situation in raw materials can assure ready access to low-cost supplies, at least under normal conditions.

3.4 Tariffs and Processing

Materials trade between nations has been shown to be influenced by a variety of tariff and nontariff barriers. Among these are specific and ad-valorem tariffs, fixed and variable quotas, and price and export controls. Import tariffs or duties tend to be lower per unit of material contained in the commodity to be imported. For example, duties are lower per unit of metal contained in ores and concentrates than on smelter or refinery products. While such a differential allows in part for metallurgical losses, it also tends to encourage the importation of crude rather than refined materials. This would give a nation the benefit of the economic activity resulting from the processing of foreign crude materials for domestic consumption.

In recent years, developing countries have suggested that existing world tariff structures have not worked to their benefit. Nations are ready to purchase the materials they export, but impose relatively high tariffs on semiprocessed or processed materials they might purchase. This issue has received consideration in the context of the UNCTAD IPC, where the IPC would attempt to improve market access for processed materials sold to industrialized nations. It has also received attention in the international tariff negotiations, particularly in the context of the General Agreement on Tariffs and Trade (GATT). Some tariff concessions have taken place which would permit an expansion of materials processing in developing countries. However, the negotiation of tariff and nontariff barriers is an arduous and difficult process. In many cases the industrial nations have comparative advantage (greater efficiency) in processing these materials. In other cases, the industrial nations have large and long-established materials processing industries which cannot easily be displaced.

4. *Materials Trade and Development*

Primary commodity or materials exports have long been recognized as a base for importing goods and technology which can then serve to accelerate domestic economic growth. The argument for export-led growth stemming from a materials base applies principally to developing countries where materials are abundant and the manufacturing sector is not very strong (Bosson and Varon 1977). In this case materials exported are often those which reflect the natural resource base of a country and those in which it should be a reasonably efficient producer. By concentrating production in one or two commodities, the country should also be able to take advantage of productive economies of scale.

However, an even stronger rationale is that the export industry would feature certain linkages which can stimulate growth in the related national economy. Backward linkages exist because the materials export industry purchases other materials, energy and other inputs from the rest of the economy. Forward linkages exist because other export industries or domestic industries can process the material produced for export, thus increasing added value and employment. Final demand linkages exist because expenditures by the materials export industry in the form of salaries or wages create further demands for domestic goods. Fiscal linkages exist because the export industry through taxation provides a major source of government revenues, and technological linkages can exist in the form of knowledge, skills and other externalities passed on to other sectors of the economy.

As a consequence of this rationale, a number of developing countries have seized on their materials base as a way of accelerating their economic growth. For example, Bolivia has attempted to exploit its tin base, Chile its copper base, Indonesia its timber base, Jamaica its bauxite base, Liberia its iron ore base, Malayasia its rubber base, and Zaire and Zambia their copper bases. For a number of reasons, however, this opportunity base has not always led concomitantly to rapid growth. Firstly, materials prices which fluctuate considerably have often led to similar fluctuations in export earnings with a subsequent stop–go effect on development plans. This effect is particularly severe where one or only a few materials dominate the total export earnings of a country.

Secondly, dependence on one or two export materials is often coincident with a situation where the high costs of investment require that the country depends on foreign capital sources, such as that provided by a multinational company. The host country may then view the multinational company as a sole and major source of fiscal revenues. This unfortunately has led to an "obsolescing bargain" whereby the initial dependence lessens and the host country increases taxation and capitalization until nationalization or expropriation occurs. In this case, the host country may lose the required operational skills as well as possible sources of further investment.

Thirdly, the economic linkages obtained have not always been as strong as hoped for. For example, mining activity does not always produce strong backward or final demand linkages since its technology is generally highly capital and skill intensive, and hence it has input requirements that are highly divergent from domestic factor supplies. The fiscal linkage, though often stronger, has often not been combined with an ability of the host government to invest productively.

The developing countries with a strong materials base thus view themselves as being in a dilemma. They possess materials representing a potential source of wealth, yet they cannot develop their materials export industries sufficiently rapidly to capture this wealth. As a consequence, there has been to some extent a polarization of relations between the developing countries which export materials and the industrialized countries which import them. The results of this polarization have been discussed above as trade issues relating to commodity arrangements, market power, and tariffs and processing.

This dilemma, however, is not insoluble. Multinational materials companies have in many cases achieved a degree of cooperation with host governments and will continue to do so. The former can provide the stimulus for materials-based industrialization by helping to overcome the uncertainty of exploration and related investment decisions, by providing access to markets together with increased materials processing, and by organizing consortia to provide the large amounts of capital necessary for materials investment projects.

See also: Economic Theory of Mineral Resources and Markets; Materials Industry Structure; Prices of Materials: Theory

Bibliography

Bosson R, Varon B 1977 *The Mining Industry and the Developing Countries.* Oxford University Press, London

Brown C P 1980 *The Political and Social Economy of Commodity Control.* Macmillan, London

Demler F R, Tilton J E 1980 Modeling international trade flows in mineral markets. In: Labys W C, Nadiri I M, Núñez del Arco J (eds.) 1980 *Commodity Markets and Latin American Development: A Modelling Approach.* Ballinger, Cambridge, Massachusetts

Kindleberger C 1968 *International Economics*, 4th edn. Irwin, Homewood, Illinois

Labys W C 1980 *Market Structure, Bargaining Power and Resource Price Formation.* Heath Lexington, Lexington, Massachusetts

L'Huillier J 1968 *Les Organisations Internationales du Coopération Economique et le Commerce Extérieure des Pays en Voie de Développement.* Institut des Hautes Etudes Internationales, Geneva

Pöyhönen P 1963 A tentative model for the volume of trade between countries. *Weltwirtschaftliches Archiv* 90: 93–100

W. C. Labys

Interpenetrating Polymer Networks

An interpenetrating polymer network (IPN) is a combination of two polymers in network form, at least one of which was synthesized and/or cross-linked in the immediate presence of the other. Two broad classes of IPNs may be distinguished: sequential IPNs, where polymer I is cross-linked before the introduction of monomer II, and simultaneous interpenetrating networks (SINs), where monomer (or prepolymer) II is present before polymer I is cross-linked. In this notation, I and II represent the first and second component polymerized or present. The concept of an IPN can, of course, be broadened to include three or more polymers.

The IPNs are closely related to polymer blends, grafts and blocks. Each of these materials is a combination of two or more polymers, usually, but not always, phase-separated.

The term "interpenetrating polymer network" was coined and used before the full consequences of phase separation were realized. Molecular interpenetration occurs only in the case of total miscibility; however, most IPNs phase-separate to a greater or lesser extent. Thus, molecular interpenetration may be restricted or shared with supermolecular levels of interpenetration. In some cases, true molecular interpenetration apparently occurs only at the phase boundaries. In some IPN systems, dual phase continuity (interpenetrating phases) is thought to exist (see Fig. 1).

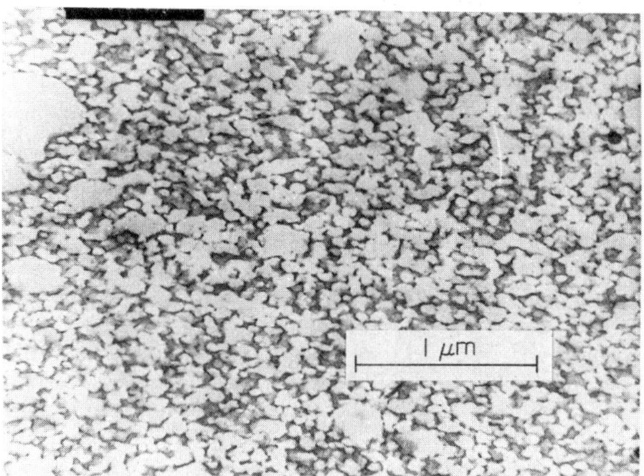

Figure 1
A styrene–butadiene rubber/polystyrene 20/80 sequential IPN. Rubber phase stained with osmium tetroxide

1. Theoretical Relationships

As in most new scientific fields, extensive experimentation preceded theoretical relationships. Gradually, an understanding of the factors that control phase domain size, shifts and broadenings in the glass transition temperature, stress–strain behavior and transport properties was achieved.

At first, equations were adopted from the polymer blend and composite theory. However, a small but growing number of analytical equations have been derived especially for IPNs as discussed, for example, by Lipatov and Sergeeva (1979) and Sperling (1981).

Experimentally, network I tends to form the continuous or more continuous phase in sequential IPNs. The domain diameter of network II, D_2, depends on the cross-link density of network I, ν_1, the interfacial tension between the two polymers, γ, and the weight fraction of network II, W_2

$$D_2 = \frac{2\gamma W_2}{RT\nu_1\left[\left(\frac{1}{1-W_2}\right)^{2/3} - \frac{1}{2}\right]} \tag{1}$$

Table 1
Recent patents on IPNs and related materials

Feature of combination	Application	Patent
Suspension particles in matrix	Low-profile plastics	US 4,048,257 (1977)
Anionic/cationic	Ion-exchange resin	US 4,152,496 (1979)
Water swellable/water swellable	Soft contact lenses	Ger. Offen. Pat. 2,518,904 (1975)
Plastic/rubber	Tough plastic	Ger. Offen. Pat. 2,153,987 (1972)
Plastic/plastic	Sheet molding compounds	US 4,062,826 (1977)
Rubber/crystalline plastic	Thermoplastic elastomers	US 3,806,558 (1974)
Block copolymer/crystalline plastic	High temperature elastomer	US 4,101,605 (1978)
Block copolymer/ionomer	Electrical wire insulation	US 4,468,499 (1984)

As usual, RT represents the gas constant times the absolute temperature.

Other equations describe the swelling behavior of a sequential IPN, and Young's modulus E of an SIN in the rubbery state (both polymers above the glass transition temperature) and of a sequential IPN, also in the rubbery state.

The latter two equations bear comparison:

$$E = 3(\nu_1 v_1^a + \nu_2 v_2)RT \qquad (2)$$

where v_i represents the volume fraction of the ith component. The exponent a in Eqn. (2) is 1 for the SIN form and $\frac{1}{3}$ for the sequential IPN form. For the sequential IPN, network II is assumed to swell and dilute network I, while network I only dilutes network II. For the SIN case, both networks merely dilute each other, the end-to-end distances of the chain approximate the bulk condition and the properties are additive.

Since v_1 is always a fraction less than unity, $v_1{}^{1/3}$ will always be larger than v_1. Hence, Young's modulus is predicted to be somewhat higher in a sequential IPN than in an SIN, a consequence of the stretched chains in network I.

2. *Applications*

Like most polymer blends, grafts and blocks, synergistic behavior is brought about because of phase separation, not in spite of it. The most important contribution of the IPN topology is the control over phase domain size (see Eqn. (1)). By definition, IPNs are thermoset, exhibiting a built-in resistance to flow at high temperatures.

For many applications, an IPN or SIN formed with an elastomer and a plastic yields outstanding properties. Rubber-toughened plastics, reinforced elastomers or leathery compositions can be obtained depending on the weight fraction of each polymer. The application may also be controlled by the miscibility of the two polymers. About 75 patents now describe applications, actual or proposed, for IPNs. A short list is illustrated in Table 1. Included are a few patents that describe physically cross-linked networks, cross-linked via ionomer formation, block copolymers or crystallinity. Such materials form thermoplastic IPNs, which flow at elevated temperatures, the subject of much recent research. As suggested by Linné et al. (1984), special functional triglyceride oils such as castor oil, vernonia oil or lesquerella oil (from popweeds in Arizona) can also be polymerized in IPN or SIN form to make tough plastics or reinforced elastomers.

See also: Elastomer Blends; Polymer Blends; Vulcanization (Cross-Linking)

Bibliography

Donatelli A A, Thomas D A, Sperling L H 1974 Poly(butadiene-to-styrene)/polystyrene IPNs, semi-IPNs and graft copolymers: straining behavior and morphology. In: Sperling L H (ed.) 1974 *Recent Advances in Polymer Blends, Grafts, and Blocks.* Plenum Press, New York, pp. 375–93

Frisch H L, Frisch K C, Klempner D 1977 Interpenetrating polymer networks. *Mod. Plast.* 54(4): 76–82; 54(5): 84–85

Klempner D 1978 Interpenetrating polymer networks. *Angew. Chem. (Int. Edn.)* 17: 97–106

Linné M A, Sperling L H, Fernandez A M, Qureshi S, Manson J A 1984 In: Riew C K, Gillham J K (eds.) 1984 *Rubber-Modified Thermoset Resins*, ACS Advances in Chemistry No. 208. American Chemical Society, Washington, DC

Lipatov Yu S, Sergeeva L M 1979 *Interpenetrating Polymeric Networks.* Naukova Dumka, Kiev

Manson J A, Sperling L H 1976 *Polymer Blends and Composites.* Plenum, New York

Sperling L H 1981 *Interpenetrating Polymer Networks and Related Materials.* Plenum, New York

L. H. Sperling

Interstitial Ordering in Alloys

Interstitial ordering is the phenomenon of ordering of the atomic distribution of solutes in interstitial solid solutions. Interstitial solid solutions are formed when solute atoms with small radii, such as H, C, N and O, are accommodated statistically in the interstices or interstitial sites of the lattice of a host metal.

As the interstitial ordering proceeds, the probability of finding the solute atoms in a specific set of interstitial sites increases. A phase transition between the ordered and disordered configurations of interstitial atoms is called an interstitial order–disorder transition. The ordering phenomena in interstitial alloys resemble those in intercalation compounds such as graphite intercalates; the intercalate species correspond to the interstitial solutes. The interstitial ordering is termed vacancy ordering in some cases, when the configuration of vacant interstices is mainly concerned.

1. *Interstitial Sites in Metal Structures*

In the bcc, fcc and hcp structures of metals, there are two main types of interstitial sites: octahedral and tetrahedral. The former is surrounded by six metal atoms at the corners of an octahedron and the latter by a tetrahedron of four metal atoms.

2. *Stress-Induced Ordering*

The interstitial ordering in alloys is sometimes termed stress-induced ordering or stress ordering, because the dominant contribution to the ordering energy is the stress-induced interaction caused by a modification of the elastic displacement field around interstitial atoms. The stress-induced interaction between atom pairs is long range, falling off in proportion to r^{-3}, where r is the interatomic distance.

If an oversize atom enters into an octahedral site of a bcc metal, for example, the metal lattice expands along the direction of the two nearest metal atoms, but contracts in the orthogonal directions of the four second-neighboring atoms. As the displacement field is anisotropic, the interstitial ordering is susceptible to internal as well as external stresses. An internal friction effect, called the Snoek effect (see *Diffusion in Crystalline Metals: Atomic Mechanisms*), appears when the stress-induced redistribution of the solute atoms occurs from one set of octahedral sites to the other.

3. *Interstitial Order–Disorder Transition*

In principle, there is no difference between the order–disorder transitions in interstitial and substitutional alloys, if an interstitial alloy is regarded as a substitutional one composed of interstitial atoms and vacant interstices. The degree of long-range order is defined in terms of the occupation probability for each of the sublattices of the interstitial sites; the sublattices gradually become equivalent at the order–disorder transition temperature. Above this temperature, there generally remains some degree of short-range order. The blocking model is useful in describing the short-range-ordered configuration; when an interstitial site is occupied by a solute atom, other atoms are prohibited from occupying some surrounding sites.

In the lattice-gas model, the interstitial order–disorder transition corresponds to a "solid–liquid" phase transition. Sometimes the polymorphic "solid–solid" transition takes place successively from one ordered configuration to another. Such multistep transitions are interpreted in terms of a branching scheme; the successive branching of energy levels for the sublattices occurs with falling temperature.

Typical examples of the interstitial order–disorder transitions are found in the alloys of transition metals with hydrogen, nitrogen and oxygen: for example, V–H, Pd–H, Fe–N and Ti–O. The interstitial order–disorder transition is also known in some interstitial compounds, the so-called Hägg compounds, such as Nb_2C, Fe_2N and V_6C_5.

Interstitial ordering appears in dilute interstitial alloys such as $Ta_{64}O$, $V_{16}N$ and Fe–C martensite (2.5–7.5 at.% C), but the occurrence of the order–disorder transition cannot be demonstrated experimentally, because these phases are metastable and decompose into stable ones at relatively low temperatures.

Appreciable changes in physical properties are observed at the interstitial order–disorder transition temperature, such as a sharp increase in the electrical resistivity and a λ-type peak of the specific heat. The distribution of interstitial atoms is investigated most suitably by neutron diffraction.

See also: Local Order in Crystalline Alloys; Long-Range Order in Alloys; Displacive and Order–Disorder Transformations

Bibliography

Fast J D 1965 *Interaction of Metals and Gases*, Vol. 1. Academic Press, New York

Fukai Y 1984 Site preference of interstitial hydrogen in metals. *J. Less-Common Met.* 101: 1–16

Hirabayashi M, Yamaguchi S, Asano H, Hiraga K 1974 Order–disorder transformations of interstitial solutes in transition metals of IV and V groups. In: Warlimont H (ed.) 1974 *Order–Disorder Transformations in Alloys*. Springer, Berlin, pp. 266–302

Khachaturyan A G 1978 Ordering in substitutional and interstitial solid solutions. In: Chalmers B, Christian J W, Massalski T B (eds.) 1978 *Progress in Materials Science*, Vol. 22. Pergamon, Oxford, pp. 1–150

Somenkov V A, Shil'stein S S 1980 Phase transitions of hydrogen in metals. In: Christian J W, Haasen P, Massalski T B (eds.) *Progress in Materials Science*, Vol. 24. Pergamon, Oxford, pp. 267–335

M. Hirabayashi

Intumescent Coatings

"Intumescent" is an adjective that describes anything that swells or expands, particularly when heated. Intumescent coatings can be formulated that are visually identical to conventional coatings but when exposed to high temperatures, they decompose and expand to form a thick, cellular, carbonaceous barrier. A properly formulated intumescent coating effectively insulates a substrate from heat beyond its reaction temperature.

When exposed to intense heat, an intumescent coating functions by three separate reaction mechanisms:

(a) the endothermic decomposition of the intumescent filler consumes heat energy;

(b) inert gases are evolved that expand the resulting carbonaceous char and drive back the hot, convective air currents; and

(c) the thick, expanded char layer which is permeated with entrapped, multicellular gas pockets serves as an effective heat insulator.

Figure 1 is a visual representation of the intumescent reaction in progress.

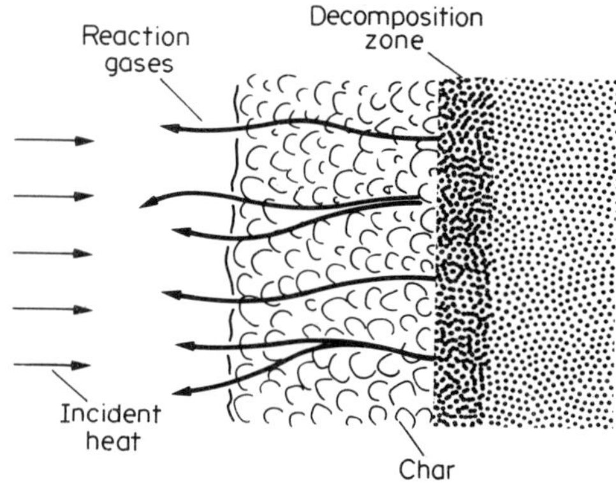

Figure 1
Schematic cross section of intumescent coating during reaction

Generally, the components of an intumescent system can be placed in one of five different categories:

(a) a polyhydric material rich in carbon (carbonific);

(b) an inorganic acid or acid salt;

(c) an organic amine or amide;

(d) a halogenated compound or other suitable material which can decompose and serve as a blowing agent (spumific); and

(e) a binder.

Most formulations contain all of these components. In some cases, two or more functions are included in a single component.

For intumescence to proceed, several distinct reactions must occur almost simultaneously, but in the proper sequence. First, the acid salt must decompose to yield the dehydrating acid which must esterify the carbonific. Decomposition of the salt must precede the thermal decomposition of the carbonific, so that dehydration can occur. Amines (or amides) catalyze this reaction, allowing it to occur at lower temperatures. Next, the ester must decompose by dehydration, resulting in the formation of a carbonaceous mass. Water vapor released from these reactions and gases evolving from the spumific cause the carbonaceous mass to foam and expand. These gases function as blowing agents only if they are released at the proper time (after melting, but before resolidification occurs). To ensure sufficient gas release at the proper point in the reaction, blowing agents with predetermined decomposition temperatures are employed. As the reaction nears completion, solidification occurs and a thick, multicellular char is formed with a volume that is 5–200 times the original volume of the coating. If any of these reactions does not proceed at the required time, intumescence will not take place. Thus, the temperature at which the specified reactions occur is extremely important.

1. Carbon Source (Carbonific)

The carbon char may be formed either by the action of a dehydrating acid on a polyhydric material or by the formation of a graphitic structure from a highly aromatic precursor. It may be chosen from the generic class of carbohydrates; starches and polyhydric alcohols are the materials most frequently used. A number of potentially suitable carbonifics are listed in Table 1. Their effectiveness is dependent on carbon content, which determines the size and stability of the char structure. The hydroxyl content determines the rate of dehydration and thus the rate of char formation. These two characteristics are, in general, inversely proportional. In other words, the higher the carbon content, the lower the number of reaction sites.

2. Acid Source

Since decomposition of the carbonific is not desirable before fire exposure, salts of the dehydrating component are used. These salts must be formed with a volatile or thermally degradable cation, so that the active component is released for dehydration during fire exposure. Evolution of the volatile component is thought to assist the fire retardation mechanism by diluting the flammable gases and the oxygen available for combustion. The pH of the salt must also be taken into consideration in the selection of a suitable material, since high levels of free acid can result in premature dehydration of the carbonific. The relative

Table 1
Carbonifics

	Formula	Carbon content (%)	Reactivity (OH^- sites per 100 g)
Sugars			
Glucose	$C_6H_{12}O_6$	40	2.8
Maltose	$C_{12}H_{22}O_{11}$	42	2.3
Arabinose	$C_5H_6O_4$	45	3.0
Starches	$(C_6H_{10}O_5)_n$	44	2.1
Polyhydric alcohols			
Erythritol	$C_4H_6(OH)_4$	39	3.3
Pentaerythritol	$C_5H_8(OH)_4$	44	2.9
Dipentaerythritol	$C_{10}H_{16}(OH)_6$	50	2.5
Tripentaerythritol	$C_{15}H_{24}(OH)_8$	53	2.4
Resorcinol	$C_6H_8(OH)_2$	63	1.8

effectiveness of these materials depends on the amount of acid character present, as represented by the percentage of the active element. A number of the more common acid sources are listed in Table 2.

Table 2
Acid salts

	Active element	Decomposition temperature (°C)
Melamine phosphate	14% P	300
Urea phosphate	20% P	130
Diammonium phosphate	24% P	87
Ammonium phosphate	27% P	147
Ammonium polyphosphate	32% P	215
Ammonium sulfate	24% S	
Ammonium borate	16% B	

3. *Nitrogen Source*

Nitrogenous compounds, such as amides, are used as catalysts to promote the esterification of the carbonific by a dehydration process. They are even more effective when used with orthophosphoric or sulfuric acid.

4. *Blowing Agent (Spumific)*

Blowing agents are needed to expand the melt formed during dehydration of the carbonific to a thick, multicellular char. They must decompose to release large quantities of nonflammable gases. It is often beneficial to use more than one blowing agent, so as to extend the temperature range over which the gases are evolved. Some of the more common blowing agents are listed in Table 3.

5. *Binder*

Binders serve two primary functions. First, they must provide the physical properties required for good serviceability. These properties include coating integrity, impact resistance, adherence and resistance to corrosion, weathering, chemicals and other agents. Secondly, the binder system must not impede and, indeed, preferably should contribute to the formation of a protective intumescent barrier in a fire. The major problem is to provide good physical properties with a binder system which may constitute <20 wt% of the total formulation. Urea– and melamine–formaldehyde resins are of particular interest when it is necessary to improve the cellular structure formed during intumescence. These resins melt at approxi-

Table 3
Blowing agents

	Gaseous by-products	Decomposition temperature (°C)
Urea	$NH_3 + CO_2 + H_2O$	130
Guanidine	$NH_3 + CO_2 + H_2O$	160
Dicyandiamide	$NH_3 + CO_2 + H_2O$	210
Glycine	$NH_3 + CO_2 + H_2O$	230
Melamine	$NH_3 + CO_2 + H_2O$	300
Chlorinated paraffin	$HCl + CO_2 + H_2O$	190

mately the same time as the onset of intumescence and then char into a thin membrane, which covers the foaming mass. This membrane sufficiently confines the blowing agents within the cellular structure to cause a homogeneous char to form.

6. *Novel Intumescent Salts: Nitroaromatic Amines*

The intumescent properties of *o*- and *p*-nitroanilines were first reported by NASA's Ames Research Center. Originally, it was believed that the reaction of nitroanilines could be caused only with concentrated sulfuric acid; however, it was later found that the use of concentrated phosphoric and polyphosphoric acids also promoted the reaction. The observation of the effect of heat and sulfuric acid on *p*-nitroaniline led to the development of the associated bisulfate salt as a dry compound. Upon heating the *p*-nitroaniline bisulfate to temperatures >220 °C, a yield of black char of ~50–54 wt% is obtained that is stable to temperatures >550 °C. Because the material is the product of a strong acid and a weak base, it is easily affected by moisture. This can degrade the vehicle in which it is combined and corrode metallic substrates upon humid exposure. The problem of hydrolytic stability is eliminated by the use of the ammonium salt of *p*-nitroaniline-*o*-sulfonic acid. The ring-substituted sulfonic acid acts as an in-situ source of sulfuric acid when this compound is heated beyond 300 °C. Thus, the reaction is similar to that of *p*-nitroaniline and sulfuric acid.

The key to the successful formulation of these coatings has been the selection of a binder that does not interfere with the intumescent process, is compatible with the intumescing salt, and does not contribute significantly to the flammability of the system. The binder should be either thermoplastic or easily degraded thermally in the same temperature range as that in which intumescence takes place. The use of the *p*-nitroaniline bisulfate salt also requires that the binder be relatively stable to strong acid.

7. *Hydrated Intumescent Salts*

Hydrated salts such as sodium tetraborate decahydrate (borax) function by a different reaction mechanism. First, they decompose endothermically. This process absorbs a large quantity of heat and effectively prevents the transfer of energy through the coating. The second process involves transpirational cooling by the water vapor produced from the decomposition as it leaves the coating surface. The sublimation of water molecules cools the upper layers of the coating by convection and reduces the thermal transport of energy through the boundary layer region immediately above the surface. The third process involves the formation of a viscous material. Because of the blowing action of the water, the molten material is transformed into an expanded char which solidifies and provides thermal insulation. The insulation efficiency of the char depends on cell structure. This structure depends on the timing of the blowing agent action relative to the formation of a critical viscosity in the molten layers formed by the active filler and binder system. If the blowing agent action occurs too early, the intumescent action is reduced because the molten material is too viscous. If the action occurs too late (when the viscosity is low), large cells are formed which are relatively ineffective as insulators due to their large convective heat-transfer coefficients. The problem in choosing a reactive filler such as borax is to balance the endothermicity of decomposition against other factors, such as the formation of a liquid mass in an acceptable viscosity range and the effective action of the residual decomposition products as blowing agents.

Aluminum hydroxide is another hydrated salt with particular advantages. Although the water of crystallization is lower than that of borax, the higher specific gravity allows for a 40% higher concentration by weight of aluminum hydroxide in formulations of equivalent filler volume. In addition, this compound is insoluble in water (resistant to high humidity) and has a much higher melting point. Table 4 summarizes the comparative properties of both salts.

8. *Applications*

Intumescent coatings have been used for such military applications as weapons insulation and shipboard fire containment. Specifications TT-P-26, TT-C-1883, MIL-C-46081, MIL-C-81945, MIL-C-81946

Table 4
Comparison of hydrated salts

	Sodium tetraborate decahydrate (borax)	Aluminum hydroxide
Molecular formula	$Na_2B_4O_7 . 10H_2O$	$Al(OH)_3$ or $Al_2O_3 . 3H_2O$
Water of crystallization (wt%)	47	34
Density (g cm^{-3})	1.73	2.42
Solubility in water (g l^{-1})	61.7	insoluble
Melting point (°C)	62	300

and MIL-C-81947 have been established by US government agencies for various compositions. In the private sector, intumescent coatings have been applied to steel superstructures and aircraft loading bridges. They have been proposed for use in chemical plants and tank farms and on engine canopies and electric motor casings. Many other possible applications exist that have not been fully explored. Intumescent coatings have demonstrated their potential to provide short-term thermal protection for vehicles and structures with minimum additional weight.

See also: Corrosion Protective Coatings for Metals

Bibliography

Marion F A, McSpadden H J 1977 Heat protective material and method of making the material. US Patent No. 4,001,126

Parker J A, Fohlen G M, Sawko P M, Griffin R N 1968 The use of a salt of *p*-Nitroaniline as a component for intumescent coatings. *SAMPE J.* 4(5)

Pulley D F 1979 Intumescent coatings. In: Leidheiser H (ed.) 1979 *Corrosion Control by Coatings.* Science Press, Princeton, New Jersey, pp. 411–19

Vandersall H L 1970 *Intumescent Coating Systems, Their Development and Chemistry*, University of Utah Polymer Conference Series. University of Utah, Salt Lake City, Utah

D. F. Pulley

Investigation and Characterization of Materials: An Overview

Present-day materials science depends heavily on understanding how the properties of a material relate to its composition and microstructure. However, this approach is relatively recent, and for centuries before Bohr's model of the atom, skilled craftsmen and builders produced many objects of remarkable utility and beauty without any knowledge of what is now considered basic chemistry or metallurgy. They started with materials found in their native state or those requiring minimal processing. Desired properties were achieved by trial and error experimentation or the utilization of experience passed down through generations.

For thousands of years certain metals were extracted from their ores by smelting, with variations in the resulting properties often being attributed to supernatural intervention rather than to subtle differences in the materials or processes used. As time passed and printing developed, practical experience was written down. For example, in the sixteenth century Agricola's *De re metallica* described many of the methods used for the extraction, measurement and classification of metals. At about the same time, empirical knowledge existed of how to use distillation to prepare a variety of simple compounds including various alcohols and oils.

Until the eighteenth century only a small number of elements were known, and physical attributes like density, color and hardness were sufficient to tell one from another. As chemistry and physics developed in the nineteenth century, the growth of more basic understanding led to the discovery of a number of elements and the extraction or synthesis of thousands of compounds. These new substances were applied to many areas including medicine, construction, transportation, weaponry and agriculture. By the mid-nineteenth century the amount of scientific information available had grown so large that various scientific disciplines began to split into specialties. In chemistry, general chemists were replaced by organic, inorganic, physical, physiological, agricultural and analytical chemists. In 1848 Fresenius founded the first analytical laboratory which, in addition to teaching, provided government agencies and industry with checks on material specifications and the analysis of water, food and physiological specimens. The idea caught on and today there are hundreds of independent analytical laboratories, as well as those associated with hospitals, industrial concerns, universities and government agencies. Following the specialization of analytical chemistry there has remained a strong interplay between the analysis and development of materials. Better analytical methods lead to both more systematic and complete classification of existing materials and insight required for the development of new ones.

1. Historical Background

The most significant event in laying the foundation for all phases of modern materials science took place at the very beginning of the nineteenth century when Dalton postulated that matter consisted of aggregates of "ultimate particles," or atoms. Atoms of any given element were indistinguishable from other atoms of the same element with respect to mass and all properties, while atoms of different substances could combine to form "atoms" (molecules) of a third substance. Dalton further recognized that elemental atoms combine only in certain ratios, so as to form products with unique stoichiometry. Knowing that he could not weigh individual atoms, he used this concept to determine the relative weights of atoms and molecules based on hydrogen. Thus, the formation of a given amount of water (assumed to be HO) with an atomic weight of 6.5 required the combination of oxygen with an atomic weight of 5.5 with hydrogen with an atomic weight of 1.0. Although Dalton did not recognize that two atoms of hydrogen combine with one atom of oxygen to form one molecule of water (a point which was later clarified by Avogadro) and his atomic weights were not particularly accurate, his ideas are of fundamental importance to modern chemistry.

The concept of definite proportions and atomic

weights combined with a fairly consistent symbolic nomenclature for the elements, introduced by Berzelius, led to the description of matter and chemical reactions in the form with which we are now familiar. This approach also led to the emergence of analysis techniques based on chemical behavior rather than physical property determinations. For example, the method known as gravimetric analysis was widely used in the nineteenth century. A chemical reaction between a solution of the material to be analyzed and a second reagent results in the formation of an insoluble compound containing the element of interest. The precipitate is isolated, dried and weighed, and the quantity of the element is determined from its stoichiometric fraction.

A second technique, known as volumetric analysis (titration), developed at about the same time. It involves measuring the volume of a reagent necessary to react completely with a fixed quantity of a sample in a predetermined chemical reaction. At first it was not as commonly used as gravimetric analysis, because of difficulty in measuring reaction end points, but as more indicators became available the method grew in popularity. A third technique, known as colorimetry, was also known in the nineteenth century. Selected reactions are performed to produce colored compounds containing the element of interest. Color intensity is then measured, relative to standard solutions, to estimate the amount of that element present. Although this approach was initially limited to a few elements and quantitative analysis was fairly subjective, its use increased as more colorimetric reagents were identified and instrumentation was developed to make measurements more accurate.

The nineteenth century also saw the initial development of electroanalytical chemistry and the use of polarized light to discriminate between materials, but the most significant instrumental method developed was the spectroscope. Its operation is based on the ability of a prism to disperse white light into its various constituents, a fact demonstrated by Newton in the seventeenth century. In 1814 Fraunhofer, a lens manufacturer, recognized that a number of both bright and dark lines could be observed in both sunlight and the flames produced by burning different fuels. The idea of using absorption spectra for more general chemical analysis was proposed by Brewster a few years later. In the 1850s Bunsen and Kirchhoff developed a spectroscope with an improved source and viewing optics which is the forerunner of instruments used today. With it they discovered both the elements cesium and rubidium.

At the beginning of the nineteenth century only about thirty elements were known; by the end this number had more than doubled. During this period it also became apparent that some type of systematic classification of the elements was needed. The concept of the periodic table was independently developed by Mendeleev in Russia and Meyer in Germany using schemes based on atomic weights and various other physical properties. Although ordering of the elements is now related to atomic number and the electronic structure of the elements, most of the changes made over the years have been to add new elements rather than to reorder known elements.

At the end of the nineteenth century and the beginning of the twentieth century a series of major events took place which revolutionized understanding of both the atom and electromagnetic radiation. Studies of the electrical conductivity of gases at reduced pressure led to the discovery of cathode rays by Hittorf in 1869, confirmed by Goldstein in 1876. Although initially the nature of these rays was unclear, work by Thomson, published in 1897, showed that they consisted of negatively charged particles, from their deflection in magnetic and electrostatic fields. He also demonstrated that these particles (later called electrons) had the same charge to mass ratio as those given off from hot filaments or released during the exposure of certain metals to electromagnetic radiation (the photoelectric effect, discovered by Hertz in 1887). Thus they appeared to be a key component of all matter, and when removed would leave atoms as positive ions.

The nature of this positive ion core was studied in the first decade of the twentieth century by Rutherford, who experimented on the scattering of α particles from thin metal foils. From this work came a model of the atom in which electrons revolve around a positively charged nucleus. Although this model was consistent with his observations, it was unacceptable from the point of view of classical physics, since the electrons would continuously radiate energy and should, therefore, spiral into the nucleus. However, the dilemma was soon reconciled by Bohr, who postulated that the electron orbits corresponded to discrete energy states and as long as an electron remained in its orbit no radiation would occur. The idea of discrete energy states was based on the principles of quantum mechanics introduced by Planck in 1900 to describe radiation emitted from a black body. Using these states, Bohr was able to describe correctly the line spectra of hydrogen observed in the 1880s by Balmer and mathematically formulated by Rydberg in 1890. Although the Bohr model was significantly modified by Sommerfeld and Schrödinger to describe multielectron atoms more accurately, the concept of electron transitions between discrete levels and the accompanying release or absorption of energy is of central importance to many of the different types of photon and electron spectrometries used today.

Thomson's use of magnetic and electric fields to determine charge to mass ratios was also applied to the characterization of positive ion beams (called Kanalstrahlen by Goldstein), which were also known to exist along with cathode rays, as shown in experiments on the electrical conductivity of gases. Thomson's apparatus, known as a parabolic mass spectrometer, produced a series of photographically recorded parabolas, one for each charge to mass ratio. The system lacked mass resolution, however, because it had no energy selectivity. This difficulty was overcome several years later (1919) when Aston developed a velocity focusing instrument which showed that the spectrum of neon actually contained two peaks, one at mass 20 and the other at mass 22. This demonstrated that an element of a given number can have more than one form (isotopes) and thus revealed the power of the mass spectrometer as an important analytical instrument. In the decades that followed, resolution and sensitivity continued to improve as a result of the efforts of Dempster, Mattauch and Herzog, and Nier and Johnson. Some commercially available instruments today have mass resolutions of 1 part in 100 000, compared with the 1 part in 12 of the first parabolic spectrometers.

Experiments on the electrical conductivity of gases also led to the accidental discovery of x rays by Roentgen in 1896. He recognized that their unusual penetrating power was influenced by the tube operating voltage as well as the thickness and type of absorbers used. Since x rays were not deflected by magnetic or electrostatic fields, it was believed that they were part of the electromagnetic spectrum, but that they must have a very short wavelength. Determining their wavelength distribution presented something of a problem, because although diffraction gratings had previously been developed for dispersing visible spectra, they were not fine enough to separate x rays. The solution was proposed by Laue in 1912, who believed that the interatomic spacing of crystals should be of the right order of magnitude to act as suitable diffraction gratings. He proved this hypothesis by passing a beam of x rays through a crystal of zinc sulfide and obtaining a pattern of spots on a photographic plate.

This work stimulated further studies by W H and W L Bragg (father and son), who built the first crystal diffraction spectrometer. Using a crystal of rock salt with a known interplanar spacing, they were able to characterize the spectral distribution of a platinum x-ray target and also develop the simple equation which relates interplanar spacing, x-ray wavelength and diffraction angle. The spectra they observed consisted of both a broad band of general radiation (the x-ray continuum) and discrete lines characteristic of the tube anode. Investigations by Moseley described in 1913 showed that there also existed a simple relationship between the wavelengths of the characteristic lines emitted from an element and its atomic number. This unique relationship forms the basis for elemental analysis by a variety of x-ray, electron and ion-induced x-ray spectroscopies. The work of the Braggs also provided the key to crystal structure determination by using a selected characteristic x-ray wavelength to determine a set of interplanar spacings for a single crystal.

The numerous scientific discoveries that took place at the turn of this century provided the physical foundation for the hundreds of instrumentally oriented analytical methods which are widely used today. It is astonishing that most of what we now consider standard laboratory instruments were developed within the past 50 years. The following techniques and key contributors are examples:

transmission electron microscopy, Ruska (1934)
scanning electron microscopy, von Ardenne (1938)
liquid chromatography, Martin and Synge (1941)
carbon-14 dating, Libby (1946)
nuclear magnetic resonance, Block and Purcell (1946)
electron microprobe analysis, Castaing (1951)
gas chromatography, Martin and James (1952)
secondary-ion mass spectrometry, Castaing and Slodzian (1960)

The names, events and dates mentioned so far tell only a small part of the chronological development of some of the methods described. Practical refinement of many of these instrumental methods often took many years and considerable ingenuity on the part of other key contributors. For example, the first commercially available scanning electron microscope did not appear until 27 years (1965) after von Ardenne's first instrument, and it has been only during the last few years that relatively easy-to-use computerized complete single-crystal structure determination equipment has been readily available. At present one of the most active areas of analysis is the study of surfaces by x-ray photoelectron spectroscopy but, as mentioned previously, the photoelectric effect was known before the turn of the century. Relatively recent developments in electronics, computers and high-vacuum systems have all had an enormous impact on analytical instrumentation in terms of ease of operation, sophistication of the analysis, quality of results and speed. Industrial laboratories now routinely undertake types of analysis that would formerly have required the effort for a PhD thesis.

With the great diversity of instrumentation available, the role of the traditional analytical laboratory has now been expanded to include determinations of morphology, microstructure, crystallography and a variety of physical properties. The term often associated with the combination of analytical chemistry and these other methods of analysis is materials characterization. The concept of a materials characterization group goes hand in hand with the evolution of interdisciplinary materials science and engineering centers which bring together experts in such areas as metallurgy, electronic materials, ceramics, solid state physics, polymers and surface chemistry. Not only have the variety of materials and the types of instrumentation available increased, but the reasons for looking at these materials have also multiplied. Materials characterization specialists must now be concerned with basic research, product and process development, failure analysis, regulatory compliance, patent issues, quality control, manufacturing productivity and the examination of competitive products.

It is important also to recognize the degree of analytical specialization which has been taking place. Individuals who are experts in nuclear magnetic resonance spectroscopy may not have expertise in single-crystal x-ray structure determination, even though information gained by the two techniques may be highly complementary and important to a polymer chemist. An expert in transmission electron microscopy may have only an incidental knowledge of secondary-ion mass spectrometry, even though both chemical and structural information may be the key to understanding the behavior of an electronic device. As the demand for materials with unique properties grows, so also does the complexity of the materials themselves and the methods needed to analyze them. Examples are the chemical identification of submicrometer second-phase particles which can significantly influence the properties of high-temperature alloys, the determination of pollutants at the parts per billion level and the characterization of a few atomic layers on the surface of a catalyst. Clearly, scientists and engineers have enough difficulty in keeping up with advances in their own fields without having to be materials characterization experts. However, it is essential to have enough basic understanding of currently used analytical methods to be able to interact effectively with such experts.

2. *Methods Available in the Modern Materials Characterization Laboratory*

Throughout this encyclopedia a series of articles describes various methods for the investigation and characterization of materials. While not every one is covered, most of the main ones in use today are represented. Each article includes a description of the underlying physical principles, the type of samples required and typical data output and some discussion of strengths and weaknesses. As described above, the structure and composition of matter can be studied by examining how it interacts with radiation, electrons and ions. A summary of some of these interactions and the related analytical techniques is given in Fig. 1. The approach that is best suited for a particular problem depends on a variety of factors, many of which are strongly interrelated.

Chemical or microstructural analysis in most laboratories of any size can be carried out by a centralized analytical facility which maintains core analytical instruments staffed by personnel who are experts in each area. The following methods are fairly essential in any type of multidisciplinary research or engineering laboratory:

classical chemical analysis techniques
thermal analysis techniques
optical microscopy and metallography
ultraviolet, visible and infrared spectroscopy
mass spectrometry
gas and liquid chromatography
scanning and transmission electron microscopy
nuclear magnetic resonance spectrometry
optical emission and absorption spectrometry
x-ray diffraction and fluorescence spectrometry
Auger and x-ray photoelectron spectrometry

Other techniques employing commercially available equipment which are somewhat less commonly used although often important include:

low-energy electron diffraction
secondary-ion mass spectrometry
Raman spectrometry
low-energy ion scattering spectrometry
field-ion microscopy and atom probe analysis
electron spin resonance spectrometry
Mössbauer spectrometry
laser microprobe analysis
acoustic microscopy
quantitative image analysis

What must be recognized also is that even if a well-equipped analytical laboratory has all the specialties listed in the first category, each of the methods can be substantially subdivided, requiring separate instruments and often separate skills. Taking x-ray diffraction as an example, the following subdivision can easily be made:

conventional diffractometry
high- and low-temperature diffractometry
back reflection and transmission Laue photography
residual stress measurment
single-crystal structure determination
pole figure determination
double-crystal diffractometry
microdiffraction
Debye–Scherrer and Guinier powder photography
Berg–Barrett and Lang topography measurements

Services not readily available in-house can often be purchased externally from general analytical laboratories or specialists in some of the above methods. This approach is particularly useful when the number of samples does not justify setting up an in-house facility or if in-house facilities are severely overloaded. No single laboratory can possess everything and the use of external analysts often provides an effective way of providing needed resources. In some cases, the use of external laboratories is the only approach when the instrumentation required is in the multimillion dollar price range and large support staffs are needed. Examples of such specialized facilities include:

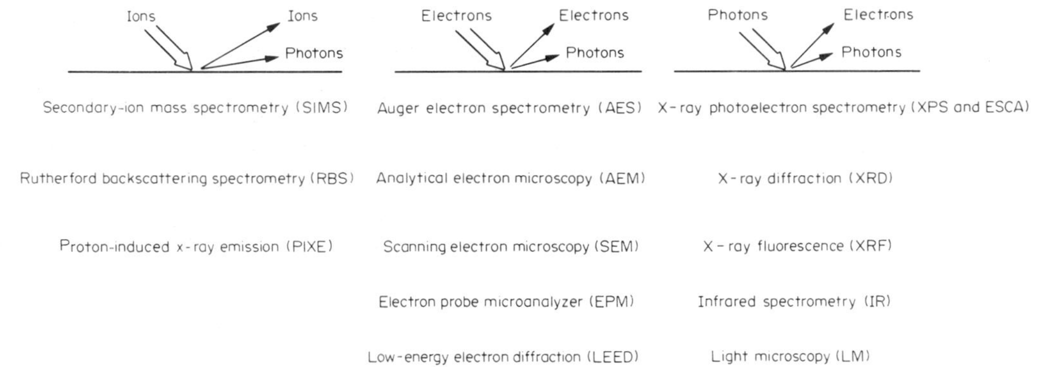

Figure 1
Analytical methods: probes and responses

extended x-ray absorption fine structure measurements (using synchrotron radiation)
neutron activation analysis
carbon-14 dating
picosecond laser fluorescence spectrometry
neutron activation spectrometry
high-voltage transmission electron microscopy
Rutherford backscattering and ion channelling spectrometry
proton-excited x-ray fluorescence

There are other factors in addition to cost and speed of analysis that should be considered in the decision to use in-house or external facilities. In an industrial environment, proprietary concerns often limit sample distribution. Another important factor is the nature of the interaction between the analyst and the research scientist or engineer. In those cases where the specific type of analysis needed is known and the right sample has been selected, it may be quite satisfactory to send a sample several thousand kilometers and receive the results in a few weeks. However, in other cases a much higher degree of interaction may lead more rapidly to meaningful results. This is particularly true if the technique or the sample selected has to be changed because the initial approach is unsuccessful. It can also be true if detailed data interpretation relative to the objectives of the overall investigation is required.

Sometimes the question of who should do the analysis becomes an internal one. Should scientists or engineers have their own characterization tools or should they use a centralized service? While there is no single answer to this question, it is clearly an issue of optimizing the utilization of resources. It is not uncommon for every chemist in a laboratory to have his own gas or liquid chromatograph and every metallurgist to have his own light optical microscope. In such cases an investment of a few thousand dollars could improve research productivity by providing quick answers expediting the next experiment. It is highly unlikely, however, that each chemist would have a private nuclear magnetic resonance spectrometer and each metallurgist a separate transmission electron microscope. These require very large initial capital as well as the expenditure of both money and time to acquire the needed expertise and to maintain a piece of equipment that may not even have long-term use. Obviously the overall impact could be a serious decrease in research productivity.

Occasions do arise, however, when it is important to have fairly sophisticated analytical instruments closely integrated with other types of experimental apparatus. For example, samples may be altered by being transferred to separate instrumentation, as in the characterization of thin films formed by molecular beam epitaxy by the use of integral low-energy electron diffraction, Auger spectrometry and residual gas analysis instrumentation. In a manufacturing environment a process may require rapid on-line monitoring to maintain quality control. Chemical reactors frequently have in-line gas chromatographs and infrared analyzers. Dedicated analytical equipment outside a centralized facility may also be required if it serves only the objectives of a single individual's research interest and both the degree of interaction needed and the duration of the project justify the associated cost. An example would be a laser Raman spectrometer uniquely set up to study combustion chemistry and dynamics.

3. *Interaction with the Analyst*

Given hundreds of analytical methods, it is obvious to ask how is an appropriate method selected for solving a given problem? Before the recent proliferation of so many new types of analytical instruments, a scientist or engineer would go to a designated analytical chemist in his organization and describe the information needed. The analytical chemist would either go to suitable references or know immediately from personal experience which technique to use. Today this dialogue could be considerably extended and would undoubtedly involve experts in several fields. The scientist or engineer initiating the study should be prepared to answer a number of questions which will aid in establishing the method or methods most suitable for solving the problem. The following is a list of representative questions frequently asked, separated into three general categories. In addition some comments are added which should help to understand the role of these categories in establishing an analytical strategy. If the scientist or engineer and the materials analyst are, in fact, the same person the discussion is replaced by introspection but the questions remain the same.

(a) Questions relating to the type and detail of information sought:
Is there a need to know what elements are present and/or how much of each?
Is there a need to know all the elements or only some of them?
Down to what level?
What degree of accuracy is required in the results?
Is crystal structure or microstructural information needed?

(b) Questions relating to what is already known about the sample:
What is the physical state of the sample?
Is the sample a metal, ceramic, semiconductor, chemical compound or polymer?
Does the sample consist of one or more than one phase?
Is the sample amorphous or crystalline?
Is special handling of the sample required, either because of toxicity or because the sample might transform to something different before or during analysis?
Was the sample (or its source) exposed to any unusual physical or chemical conditions which might have altered its composition or microstructure?
Does the sample consist of layers, and if it does, how many are there and how thick are they?
How much of each sample is there and how many samples are to be run?

(c) Questions relating to the working relationship between the analyst and the scientist or engineer:
How soon are results needed?
How much can be spent to obtain the results?
Is your presence needed during the analysis?
Is a detailed report on the results needed?
Are the samples proprietary?

These three categories of questions can be thought of as screens which help to identify one or more methods of approaching a given problem. They can easily be related to three general classes of analysis requests as follows:

(i) In the simplest case, the analysis is merely to repeat a particular type of measurement made earlier on a similar sample. Therefore, the method has been preselected and the only issues to be resolved are associated with the questions of category (c). The answers are determined by laboratory policy, backlogs, urgency of the work and availability of people and equipment. An example of this class of problem would be a request for mercury analysis in water by atomic absorption as part of periodic sampling of a body of water.

(ii) More frequently, a suitable method may be suggested based on earlier work, but the characteristics of the sample may be sufficiently different that the questions asked in category (b) also become important. The new sample may be of a different size or shape. It may also contain other elements or compounds that could complicate the analysis if the same analytical method is used. As an example, a researcher develops a new polymer containing a chlorinated flame retardant. He/she wants a chlorine analysis in the 50 ppm region but has only a 5 mg sample. Previously, successful results would be obtained by x-ray fluorescence.

(iii) Samples that can be classified as totally unknown occur rarely, since materials are usually not picked at random for analysis. The situation does arise, however, when one material behaves differently from another and the reasons for the differences are sought. Such samples may be the result of laboratory experiments or may be associated with a device or process that has changed or even failed for some unknown reason. The individual initiating the work may not even be aware of the right questions to ask in category (a) and must therefore work with the analyst to define a set of unique material characteristics which can be used both to classify the specimen and also to contribute to an explanation of its behavior. Although problems in this class are sometimes amenable to quick solutions, they tend more generally to be the most challenging and expensive ones to deal with, since they frequently involve the use of multiple techniques and often the samples are not optimized for those techniques. An example of this class of analysis would be a request to analyze a catalytic material obtained from a new vendor which, although nominally of the same composition as the previously used material, dramatically changes the rate of a chemical reaction. Also included in this class are problems in failure analysis such as determining why a turbine blade has failed prematurely. In this situation questions in category (b) relating to the sample environment can be extremely important.

4. Choosing a Sample and a Method

The quality of an analysis can be no better than that determined by the sample on which the analysis is performed. In trace level determination, a sample can easily be contaminated by handling or the use of unclean glassware or other laboratory supplies. Samples can also be affected by exposure to the atmosphere, as in the case of hygroscopic materials or hydrocarbon buildup on a sample to be studied by surface analysis techniques. Materials are also not always as homogeneous as expected. If a method is used with a precision of $\pm 1\%$ but the composition of a randomly picked sample varies by $\pm 10\%$, obviously the sampling error will dominate. This effect can take place at either the macro or the microscale. In other words, a 1 kg sample taken from a single location in a tank car can lead to the same magnitude of error in assessing the overall contents of the car as using a 1 ng microprobe measurement to determine the composition of a highly segregated alloy ingot.

Collecting samples from enough regions of a material to be analyzed to determine its homogeneity and/or average composition is an essential part of characterization measurements. If the method of analysis to be used in a particular problem is known in advance, as in the case of a research experiment, samples should be prepared so that they are truly representative of the problem and also of the proper size and geometry to be compatible with the method to be used. For most methods there is some correlation between sample size, sensitivity, accuracy and precision. Situations where the amount of sample is limited are frequently encountered, however, and in these cases the quantity of sample may strongly influence the analytical method selected and minimum detectability limits. Consider the following:

a geologist working in an inaccessible location gathering samples to map out its mineral content

a forensic scientist working with only a small piece of evidence

a metallurgical failure analysis expert working with a fragment of a broken device

a research chemist or metallurgist developing a new material in which it is important to explore hundreds of variations of composition and/or phase distribution

There are also times when an abundant amount of sample is available but either the concentration levels sought by a particular instrumental method are too low to detect because of maximum acceptable sample size limitations, or interferences from other constituents can introduce serious errors in the analysis. In such situations, determining whether the component of interest (phase or element) is dispersed homogeneously throughout the sample or is nonuniformly distributed is important not only from the sampling point of view but also as a means of selecting a technique to give enhanced localized analysis for inhomogeneous samples. Being able to detect 1 ng of material in a 1 g sample is a real challenge for any bulk analysis technique but is very simple by electron microprobe analysis if the relevant location is known. In the case of solids, this can often be done by microscopic examination of the sample either as it is, or specially prepared to enhance phase differences. When possible, conventional light microscopy should always be tried first. Low-magnification observation of an unmodified sample often reveals whether a sample is nonuniform. If the sample consists of particles observable under a low-power binocular microscope, they can often be separated for easier analysis by a variety of methods including sedimentation, magnetic separation or even hand selection with tweezers.

If magnifications above ~100× are needed and the sample consists of discrete particles, scanning electron microscopy is usually a better approach. However, since the attribute of color is not available, differences between particles tend to be based on morphology and in some cases on atomic-number-dependent electron backscattering. Furthermore, since particles smaller than a few tens of μm are difficult to handle, the analysis may have to be done in situ using energy-dispersive x-ray spectroscopy. This method primarily provides qualitative elemental analysis and can be applied to particulates using magnifications up to about 5000× if the particles are well separated. Light microscopy is used in the magnification range of ~100–2000× primarily to examine polished sections by reflection for opaque materials or transmission for transparent materials.

While some chemical information is obtainable by polarized light and refractive index measurements, the presence of multiple phases may indicate that the sample should be analyzed either by a high-spatial-resolution instrumental method, like electron microprobe analysis, or that the sample should be dissolved so as to isolate second-phase particles for analysis by other techniques. The transmission electron microscope extends the ability to perform combined microstructural and microchemical analysis to magnifications well above 5000×, but the complexities of specimen preparation and the relatively high cost and inaccessibility of this method in most laboratories, compared with light and scanning electron microscopy, tend to make it a method which is chosen only when the other methods are unsuccessful.

If by microscopic examination the sample appears to be fairly homogeneous, the advantages associated with the spatial selectivity of microanalysis cannot be realized and various methods of bulk analysis are probably more suitable. In addition to traditional wet chemical analysis, these methods include spark-

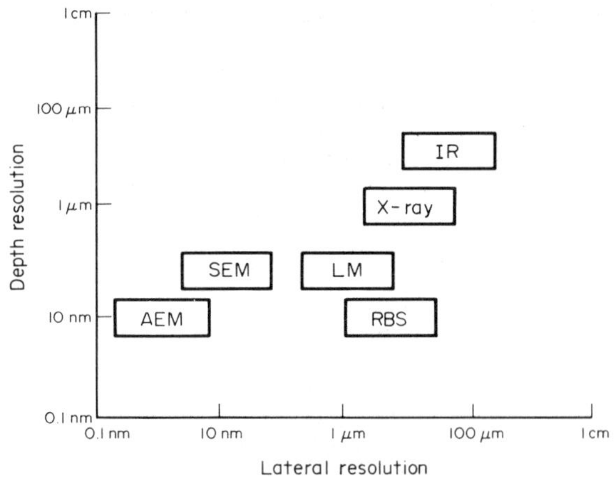

Figure 2
Lateral and depth resolution comparison for the methods indicated in Fig. 1 used for studying morphology

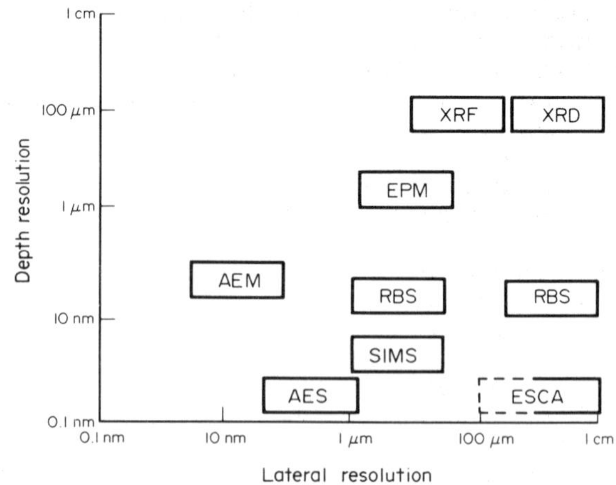

Figure 4
Lateral and depth resolution comparison for the methods indicated in Fig. 1 used for elemental analysis

source mass spectrometry, optical emission and absorption spectrometry and x-ray fluorescence spectroscopy. Each has a sensitivity extending down to the parts per million level and in some cases well below. In addition, the optical and x-ray techniques can both be made quantitative with relative ease. These techniques can often also be applied to inhomogeneous samples, provided that no breakdown by phases is required. Figures 2, 3 and 4 give an approximate summary of the performance of some of the methods described, with regard to their ability to determine elemental composition, crystal structure and phase morphology.

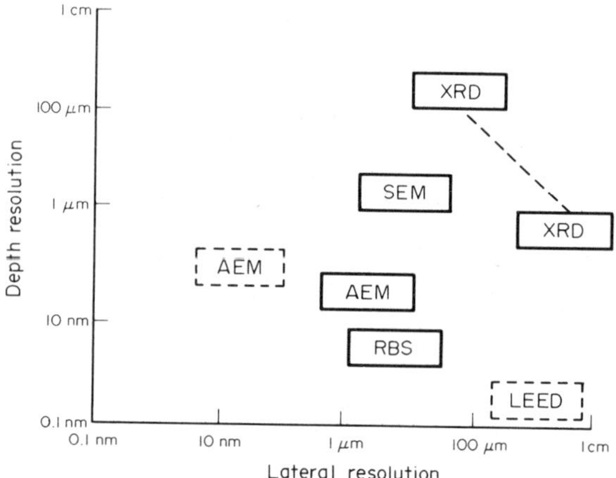

Figure 3
Lateral and depth resolution comparison for the methods indicated in Fig. 1 used for obtaining crystal structure information

The preceding comments about determining sample homogeneity refer principally to solid inorganic samples, although all the techniques mentioned can be used on polymers as well, within certain limitations associated with vacuum and electron beam compatibility requirements of the electron microscopic methods. Furthermore, x-ray measurements are primarily concerned with elemental analysis rather than compound identification. Since carbon, hydrogen, nitrogen and oxygen are ubiquitous in organic materials, qualitative elemental analysis by x-ray techniques (hydrogen is undetectable by these methods) does little to aid in the identification of the material other than to show that it is organic. Determining whether an organic sample consists of one or more phases or compounds is also an important issue, however, but the types of samples, the separation techniques used and the kinds of information required can differ considerably from those associated with the examination of inorganic samples. First of all, most samples are either submitted as solutions or are taken up into solution before analysis. Separations are usually achieved by solution chemistry, solvent extraction or chromatography (gas or liquid). Separated fractions are either analyzed by direct introduction into a physically coupled instrument like a mass or infrared spectrometer or are collected and transferred to another location for analysis. If the sample is a polymer of only one type, the information sought may be molecular weight distribution and chromatography (gel permeation) may be all that is required. In other cases the information needed is the structural formula. This could be determined from information obtained by a variety of methods including mass spectrometry, infrared spectroscopy and nuclear magnetic resonance, as well as by traditional carbon, hydrogen and oxygen analysis based on combustion and gas absorption. The sample size and concentration required again become an issue, because the characteristics of the various techniques differ broadly. While nuclear magnetic resonance is excellent for structure determination, it requires larger samples than the other methods mentioned and can rarely detect compounds or functional groups at a concentration level $<0.1\%$. Organic mass spectrometry, on the other hand, can work with much smaller samples and detect compounds even down to the parts per billion level in a well-separated sample. Structural information is harder to obtain, however, and compounds with high molecular weights (>2000) are difficult to analyze by conventional means. With sufficiently high resolution systems and soft ionization sources, it is possible to obtain molecular ions and determine stoichiometry based on direct packing fraction measurements. In some cases, structure can also be determined using conventional electron impact ionization and comparing fragmentation pattern spectra with those of known compounds. Infrared spectroscopy can also have relatively high sensitivity to specific functional groups and can be used to examine relatively small samples, although larger ones are preferred for higher sensitivity. It is used more effectively in relating the spectrum of an unknown sample to a reference spectrum rather than as a means of determining the structure of a complete unknown. Of particular value is the fact that it can be routinely used to examine solid samples such as cross-linked insoluble polymers.

5. *Cost and Complexity*

There can be an enormous range in the costs associated with materials characterization, depending on the complexity of the problem, the amount of information sought, the number of samples to be run, how quickly the analysis has to be done and the investment in equipment and manpower. Multielement analysis for large numbers of routine samples can be effected for less than US$25 per sample, while complex surface analysis can exceed $1000 per sample. Access to a particular type of instrumentation of the types described so far does not provide all the information needed about whether a given technique will have the sensitivity, resolution or sample-handling capability to do a given job. It is possible to purchase complete mass spectrometer systems for under $5000 or over $500 000. While the former may be perfectly adequate as a residual gas analyzer where unit mass resolution is sufficient, it would be totally unsatisfactory for the analysis of high-molecular-weight organic compounds. Even if the larger sum of money were available, a high-resolution mass spectrometer for isotopic analysis would be decidedly different from one used for combined gas chromatography–mass spectrometry.

The tradeoff between cost and performance features is characteristic of all major pieces of analytical instrumentation. To deal effectively with this issue when selecting an instrument to purchase or use for a given application requires the development of a specification for instrument performance relative to intended use. Again taking mass spectrometry as an example, if the need is for high-sensitivity, high-resolution analysis of gas chromatograph effluents, an instrument might be specified with a highly efficient electron impact source coupled to a fast-switching, double-focusing analyzer with high-sensitivity electrical detection. Furthermore, it should also be possible to specify a generally accepted performance standard. For example, the instrument should be able to detect 10^{-9} C of ion per μg of methyl stearate at a mass resolution of 1 part in 10 000 and a source voltage of 8 kV. Generally, when selecting an instrument of any type, cost rises with performance. It is therefore incumbent on the buyer to be realistic about how many special features and what level of performance are needed for a given range of applications. It makes little sense to pay the added cost for a scanning electron microscope capable of 100 000× magnification relative to a lower-resolution model if the details of interest are clearly observable at 5000× or if the nature of the sample is such that pictures at higher magnifications are not possible.

6. *Trends*

The major impact of computers and microelectronics on materials characterization techniques, already alluded to, will undoubtedly continue for a number of years. In addition to instrument control, data collection and the processing of results, computers will play a more important role in method selection and the management of laboratory information. It can be expected that more general spectral data bases and efficient means for searching them will lead both to faster and to more complete results. Since specimen preparation is often a rate-limiting step in many forms of analysis, the use of laboratory robots may also improve quality and productivity.

In very general terms, most analytical instruments can be thought of as consisting of a source, an analyzer and a detector. As was shown in Fig. 1, the source can be based on the interaction of the sample with probing radiation of any wavelength or particles of different types and energies. It may even be the specimen itself, as in the case of the analysis of radioactive samples. The analyzer serves as a means of separating the signals emitted or scattered by the sample, based on differences in their physical or chemical properties, and finally the detector converts the separated signals into a comprehensible form. A major trend which has emerged during the past decade is to try all manner of combinations of sources, analyzers and detectors and thereby create instruments which are optimized for particular applications. In organic mass spectrometry, for example, ion sources are based on electron impact, field ionization, field desorption, laser ionization and both ion and neutral particle beams. In gas chromatography, detectors include flame ionization, electron capture, thermal conductivity and infrared spectrometry.

If the sample is only partly consumed or not consumed at all by a given detection process, multiple detectors can be used in series to provide complementary information. This has been done in both gas and liquid chromatography. Sometimes the detector is itself a spectrometer and thus allows separated species to be characterized more fully. Examples include gas chromatography–mass spectrometry and both gas and liquid chromatography–infrared spectrometry. An important characteristic of spectrometric detectors used in combination methods is that they must either be very fast or function in a parallel detection mode. Thus, in the examples just mentioned, a mass spectrometer capable of scanning its complete mass range in a few seconds is used in one case and Fourier transform infrared spectrometry in the other. If separation techniques are to be used with atomic emission techniques, some of the newly developed types of parallel detection optical spectrometers will prove very valuable. In other analytical situations, several detectors capable of monitoring a variety of signals are used at the same time. For example, simultaneous x-ray and electron detection in scanning electron microscopes and analytical transmission electron microscopes makes it possible to relate chemical to microstructural information easily.

It is always difficult to predict where any field will go in the future, but clearly defined needs coupled with active technologies can provide some hints. All analytical methods share the common goals of higher speed, lower cost, greater sensitivity, more flexibility and increased accuracy. Over the past two decades, remarkable progress has been made in advancing localized elemental analysis capability to the limit of single-atom detection by atom microprobe analysis and ultrahigh-resolution scanning transmission electron microscopy, although only limited kinds of sample are suitable for these methods. Similar developments are needed in the high-spatial-resolution identification of compounds, particularly organics, since a knowledge of elemental constituents is usually of little value in distinguishing one material from another. Selected area diffraction, and more recently microdiffraction, are routinely used to identify inorganic compounds in transmission electron microscopes. The newer method has even given results from areas <5 nm in diameter. However, these approaches are of very limited value in the examination of organic materials, because of electron beam damage and the fact that many organics of interest are in a noncrystalline form.

The full characterization of thin organic films even when lateral spatial resolution is not an issue is complicated for other reasons. X-ray photoelectron spectroscopy has already proved itself to be a powerful tool, but the chemical shift information contained in a spectrum is often difficult to extract because of the inherent energy resolution limits of the technique. Perhaps one of the most interesting and challenging problems is the atomic level characterization of interfaces within materials, where any attempt to reach those interfaces by material removal, as is done in secondary-ion mass spectrometry and Auger electron analysis, so alters them as to make the information obtained useless. For these kinds of problems, nondestructive techniques like solid-state nuclear magnetic resonance and some forms of laser spectroscopy may provide valuable solutions.

See also: Nondestructive Evaluation: An Overview

Bibliography

Casper L A, Powell C J (eds.) 1982 *Industrial Applications of Surface Analysis.* American Chemical Society, Washington, DC

Czandema A W (ed.) 1975 *Methods of Surface Analysis.* Elsevier, Amsterdam

Ewing G W 1975 *Instrumental Methods of Chemical Analysis*, 4th edn. McGraw-Hill, New York

Friend J 1951 *Man and the Chemical Elements.* Griffin, London

Ihde A J 1964 *The Development of Modern Chemistry.* Harper and Row, New York

Kane P K, Larrabee G B (eds.) 1974 *Characterization of Solid Surfaces.* Plenum, New York

Mulvey T, Webster R K (eds.) 1974 *Modern Physical Techniques in Materials Science.* Oxford University Press, Oxford

Partington J R 1962 *A History of Chemistry*, Vols. 1–4. Macmillan, London

Peters P G et al. 1976 *A Brief Introduction to Modern Chemical Analysis.* Saunders, Philadelphia

Skoog D A, West D M 1982 *Fundamentals of Analytical Chemistry*, 4th edn. CBS College Publishing, New York

E. Lifshin

Investment Casting

Precision investment casting, or lost-wax casting, is a method in which a refractory mold is formed around an expendable pattern. The pattern is melted out, and the mold filled with metal. The fundamental process dates from prehistoric times: investment castings have been found which were made in China circa 4000 BC, and other castings have been discovered which date from early African civilizations and pre-

Columbian Central America. It was widely used during the Renaissance (most notably by Cellini) to cast statuary and ornamental objects. During the latter half of the nineteenth century the process was discovered by dentists, who used it to produce inlays. It was not, however, until World War II that investment casting assumed commercial importance because of its ability to produce exceptionally accurate parts having a fine surface finish. Requirements for intricately shaped parts of hard-to-machine alloys caused a rapid growth in the industry.

At present, investment casting is used worldwide to produce parts which require little or no machining for aircraft, aerospace and other applications. Sizes range from sewing machine parts, a few centimeters in length, to gas turbine engine casings, well over a meter in diameter. Dimensional accuracy is generally held to within 0.5% of a linear dimension, and special techniques may be used to improve this. Investment foundries are rapidly being automated, adding automated process control of critical manufacturing parameters to the already stringent controls exercised on raw materials.

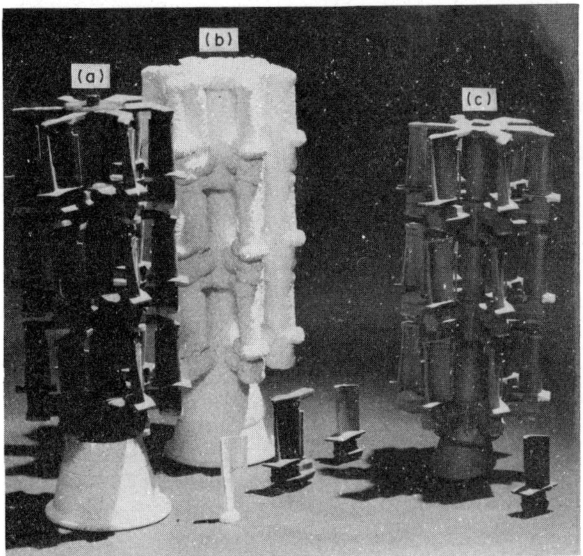

Figure 1
(a) Wax cluster, (b) investment mold and (c) cast tree (all inverted). Individual castings are shown at the bottom

1. Process Description

The investment casting process begins when pattern material, generally molten wax, is injected into a metal die to form a pattern. The patterns are attached to wax gates, runners and sprues, to form a tree or cluster. The cluster is then invested with mold material. Usually this consists of dipping the wax cluster in a refractory slurry, sprinkling refractory particles over the slurry, partially drying the cluster, and then repeating this step a number of times until a mold of sufficient thickness (normally 8–10 mm minimum) to withstand the pressure of molten metal is built up. The pattern and gating material is then melted or burned out of the mold, and the mold is fired to establish its ceramic bond. While still hot from firing (which aids in the filling of thin sections and intricate details frequently specified in investment castings), the mold is poured, and when the metal has solidified the mold material is knocked off. The castings are removed from the tree, then finished and inspected and shipped to the customer. Figure 1 shows a wax tree, mold and poured cluster, as well as the casting.

2. Pattern Materials

Pattern materials, usually waxes, have low softening or melting points (60–80 °C) so that they can be melted easily for injection and removal. This also makes possible the attachment of patterns to trees, by locally melting the attachment point (normally a gate) with a soldering iron or similar heated tool. Waxes are blended from a variety of hydrocarbon compounds selected to improve surface-finish dimensional stability, minimize the hardening time in the die, and increase strength. Nevertheless, pattern waxes generally suffer from low strength and large amounts of volumetric shrinkage when they cool from the molten state.

The shrinkage problem may be solved by using chills (solid pieces of wax placed in heavy sections of the die cavity prior to injection), which minimize the thickness of wax injected and hence its shrinkage. Some chills may be made of metal and incorporated as part of the dies. They are removed before the pattern is attached to the runner, leaving the patterns partially hollow.

The problem of low wax strength may be solved by the use of rigid plastics, such as polystyrene. These plastics have much higher melting temperatures and cannot be easily melted for attachment to sprues, or removal from the mold. Glue is usually used to attach them to the trees. They are widely used to make thin airfoil sections for gas turbine nozzles and rotors. It is common for patterns to be made of a combination of wax and plastic; for example, a turbine rotor pattern may have a wax hub and plastic blades. Figure 2 shows a variety of wax patterns and wax–plastic patterns.

Other substances, such as mercury, water-soluble urea compounds and mixtures of inorganic salts, have also been used as pattern materials. Waxes and plastics have, however, shown the best balance of properties and are by far the most widely used.

The dies used to produce the patterns are made of aluminum or steel, and may be simple or complex as

Figure 2
Patterns for investment casting may be made of wax (long blade in lower left and vane assembly in upper right) or wax and plastic assemblies (wheels and nozzles)

required by the part. For high production parts, it is common to inject an entire runner with its patterns attached integrally, thus saving the time required to attach the patterns to the runner. For larger parts, dies may be designed to inject only portions of the pattern (which are then assembled and wax welded together) or the entire part in one injection. As the die for an integral injection is usually quite expensive, its cost must be balanced against the cost of assembling individual pattern segments.

3. *Cores*

The interior cavities of investment castings are made by cores. If the internal cavity is large and accessible to air currents, it may be formed by the mold material itself. In this case, the cavity in the pattern is made either by portions of the metal die which are removed from the pattern before it is attached to the sprue, or by soluble cores. Soluble cores are waxes which dissolve in water or weak acids. They are leached from the pattern prior to its attachment to the runner system. Soluble cores generally lack the fine surface finish of metal or ceramic cores, and are not as dimensionally accurate as those materials. As a consequence, their use is limited.

Intricate internal passages in castings are formed by preformed ceramic cores. The cores, which are made of alumina–silica ceramic mixtures, are placed in the die prior to injection, and wax is injected around them. Care must be taken to leave a portion of the core (the core print) free of wax so that it may be supported securely by the mold during pouring. Cores must be refractory enough to withstand pouring temperatures without distorting, yet must also be able to be removed from the casting when it is solid. The core material usually contains at least 70% SiO_2, with the remainder alumina, zirconia and magnesia, alone or in combination with each other. These materials provide chemical stability, whereas the silica can be leached by hot caustic solutions.

Heavy cores may be made using Shaw process techniques, in which a refractory slurry containing a catalyst is poured into a die where it rapidly gels, and the volatile vapors from the slurry are burnt off. The core is then placed on a contoured refractory block and fired to develop strength. The core may be dipped in a slurry containing very fine particle-size ceramic material to give the core a smooth surface finish, and fired again. After final firing, the core will maintain its dimensions throughout the casting operation.

High production runs and smaller cores may also be formed by transfer or injection molding, or, if the core cross section is constant, by extrusion. Again, these cores are given multiple firings to develop adequate strength to the operations that precede pouring. Quartz rods are used for small cylindrical passages, and very complex passages may be made by cementing core segments and quartz rods together.

4. *Mold Materials*

Most of the mold systems in use in modern investment casting are bonded by silica. This silica is obtained either from ethyl silicate or colloidal silica.

Aqueous colloidal silica sols are prepared by hydrating sodium silicate. Silica particles range in size from 4 to 100 nm, and stable sols can be made which contain up to 50% silica. When the sol is mixed with a refractory aggregate and dried, the silica particles are forced together, and the hydroxyl groups covering their surface condense, splitting out water, forming siloxane (Si–O–Si) bonds giving the mold an unfired or green strength. On firing, the refractory particles and the silica form bonds which are strongest when the refractory used is fused silica, intermediate for refractory silicates such as mullite and zircon, and weakest with nonsilicate refractories such as alumina and zirconia. The slurries that are used for investment casting contain 12–21% binder (usually a 30% concentration of colloidal silica in water is used) and 70–80% refractory flour. Slurries are maintained with a basic pH (8.5–10.5) to prevent gelling.

Molds are built up by dipping clusters into these slurries, draining the excess, stuccoing the wetted surface with dry refractory particles either by sprinkling or by immersion in a fluidized bed, and then drying the resultant coating. The process is repeated until the required mold thickness is achieved. A very fine particle size is used in the first two dip coats, which provide the surface finish for the part and, to a large extent, determine its heat transfer properties. To give the mold sufficient strength to resist molten metal pressures, these first two coats are backed up with more coats, comprised of a slurry containing coarse particles of refractory materials. As many dips are added as are necessary to give the mold adequate strength. For small castings (up to 1 kg), seven coats are usually sufficient; more coats are added as the casting size is increased. The mold must be dried between each coat to remove excess water so that subsequent coats can be applied. Dipping and drying are carried out under tightly controlled conditions of temperature and humidity.

The refractory materials used in the slurries are usually silica, zircon or alumina sands. Back-up sands may contain zircon, alumina, mullite, fused silica, alumina–silicate sands, or mixtures of these. Slurries usually also contain wetting and deairing agents, and bacteriocides to improve shelf life.

Ethyl silicate, for investment castings, is made by a reaction between silicon tetrachloride and ethanol. Although the product contains 28% SiO_2, this can be concentrated to 40%. Before it can be used as a binder, this solution must be hydrolyzed by reacting it with water and a small amount of acid. The hydrolyzed ethyl silicate has a pH of 1–2, and can be stored indefinitely. When the pH is raised to 6, gellation will occur in 2–3 min. It is this feature which is taken advantage of in making an investment casting mold.

A slurry is formed by adding refractory fines (−325 mesh (−44 μm)) to the hydrolyzed ethyl silicate. The pattern is then dipped and stuccoed in the same manner as used for colloidal silica. However, drying can be shortened to 2–3 min by introducing ammonia into the drying atmosphere. This gels the coat by raising its pH, so that multilayered shells can be built up quite rapidly. Ammonia-hardened layers have about half the strength of dried layers because a network of very fine cracks is formed when the gelled dip coat is fired. To compensate for this condition, the number of dip coats is increased.

Ethyl silicate can also be used to make a monolithic mold. In this process, the pattern is sprayed with, or dipped into, the prime slurry, stuccoed with a very fine grain sand and then placed in a metal box or flask. The hydrolyzed binder–refractory particle slurry is poured over the pattern (investing), and the box vibrated to remove bubbles and compact the refractory. When gellation is complete the mold is placed in a low-temperature oven to evaporate the alcohol. This method of making molds has largely been replaced with the dip slurry method, because the latter uses less material.

5. *Dewaxing, Mold Firing and Cleaning*

When the mold has been dried, the wax must be removed. This is a critical operation, for when the wax inside a ceramic mold is heated it expands forty times as much as the ceramic. Therefore, if a mold containing wax is heated slowly the expansion of the wax will crack the mold. Wax, however, is a poor conductor of heat, so that rapid application of heat to the surface layer of wax at the interface between the mold and the pattern will cause it to melt and run out of the mold, producing space into which the remaining wax can expand when it heats up.

Steam autoclaves are widely used for dewaxing. When molds are exposed to steam at 148 °C, the steam penetrates the porous mold, condensing on the pattern surface and giving up its heat of vaporization, which rapidly melts the pattern. The cycle takes less than 10 min. Wax may be reclaimed from the autoclave operation by heating it to drive off the water picked up during autoclaving, and adding back those wax constituents which volatilize during the heating operations.

Flash-fire dewaxing is carried out by placing the mold in a gas or oil-filled furnace and removing the wax by a combination of burning and melting. The method can easily be made continuous and has the further advantage of combining mold firing with dewaxing, as temperatures in the furnace normally exceed 800 °C. This method can be used with both wax or plastic pattern materials. Since some of the wax is burnt, it cannot be recovered, so wax recovery is limited to 75%. Volatile additives must be replaced during reclamation of this wax.

Molds may be dewaxed by immersion in a bath of molten wax. The pressure of the expanding wax within the mold is balanced by the hydrostatic pressure of the wax in the tank. Bath temperature control is important: too low a temperature increases bath viscosity so that it is difficult to immerse the molds, whereas too high a temperature runs the risk of vaporizing volatile wax constituents. Mold green strength is also important, as the low thermal conductivity of the wax means that the heating rate is relatively slow and the risk of mold cracking is high. However, wax dewaxing has the advantage that wax recovery is complete, and it can be used with any type of wax.

Organic solvents (degreasing agents) may also be used to dissolve the wax. The mold is suspended upside down over a degreasing tank, and the vapors dissolve the wax. The method is used where the prime coat cannot be exposed to heat or steam, or where mold geometry is such that other methods of dewaxing cannot be carried out, or when delicate plastic patterns are used.

Microwaves may also be used to heat the mold and melt the wax. The method is continuous, may be used with all waxes, and requires only that the wax be filtered to be fully recovered. Since microwaves are not absorbed by metal, most of the energy generated is concentrated in the mold instead of being lost through furnace walls. Microwave dewaxing uses one fifth the energy required for autoclave dewaxing.

After dewaxing, molds must be fired at a temperature above 800 °C to set the silica bond. After the molds are fired, they are inspected and cleaned. Firing may be combined with dewaxing, or with preheating prior to pouring, or both.

Molds are cleaned to remove any particles of prime coat that may have flaked off, or refractory particles and dust that may have fallen into the mold during processing. They may be cleaned after they have cooled to room temperature, or cleaned hot, between the preheat and pour operations. In the cold cleaning operation, compressed air is used to blow the debris out of the mold, whereas in the hot cleaning operation air or nitrogen is blown through, or the mold is inverted for a few seconds over a vacuum cleaner hose. The mold is then poured immediately.

6. Preheating, Pouring and Finishing

Molds are preheated prior to pouring to facilitate the filling of thin sections, and to help in the control of as-cast grain size. Vacuum melting is frequently used to avoid metal contamination and to fill the thin sections.

After the mold has been removed from the furnace and allowed to cool, the ceramic is broken off and the castings removed from the tree. The castings are then sorted, cores removed, and sand-blasted before being inspected. Visual, dimensional, fluorescent penetrant and radiographic inspections are some of the methods which may be used to assure the quality of investment castings. Castings to be used in critical applications may be lightly etched to reveal their grain structure.

7. Advantages

Investment castings are generally more expensive than other castings, but are cost-effective because they can be cast accurately enough to avoid most subsequent machining steps. While the general investment casting tolerance is ±0.5% of linear dimensions, a number of techniques exist for improving this figure to ±0.1%. As a result investment castings are being substituted for complex fabrications in a variety of industries.

The second advantage of investment casting is that the thin molds make it relatively easy to control heat flow during solidification. It is for this reason that certain product lines (notably gas turbine engine parts) can be made with precisely defined metallurgical structures, ranging from multiple uniform equiaxed grains or crystals to single crystals.

See also: Casting of Metals; Cast Structure of Alloys; Ceramic Mold Casting

Bibliography

Bidwell H T 1969 *Investment Casting*. Machinery Publishing, London

T. S. Piwonka

Ion Backscattering Analysis

Ion backscattering provides a very simple direct and nondestructive method for microanalysis of materials. It utilizes the detection of particles backscattered from a given depth when an incident ion (MeV energy range) penetrates a target. It is best suited to the detection of heavy elements on light substrates, and has been applied to the determination of compositional variation or impurity distribution as a function of depth below the surface of a sample. As an ion beam has a finite range, this technique is most appropriate for the analysis of layered structures such as those encountered in thin films and thick solid surfaces. The technique has the ability to determine the concentration of atoms with high sensitivity as a function of depth and with a depth resolution of a few tens of nanometers over a depth of hundreds of nanometers without recourse to sample erosion or layer removal. Hence, the technique is an invaluable analytical tool for the investigation of solid-state phenomena and has been extensively utilized in areas of materials technology such as electronics and nuclear energy. In electronic materials, the applications range from routine ones such as quality control in the fabrication of semiconductor devices to specialized investigations of ion implantation (dopant profiles, thin-film deposition), impurity solubility, diffusion and reactions, contact formation, epitaxial layers, thermal and anodic oxidation, and metallization. In nuclear energy technology, the technique is used to investigate oxidation corrosion and cladding of materials. By combining channelling with backscattering measurements, crystallographic information such as lattice site location of impurities or lattice disorder can be determined. These topics have been covered extensively in the literature (Ziegler et al. 1975b, Chu et al. 1978, Mayer and Tu 1974, Foti et al. 1977).

1. Rutherford Backscattering

When a projectile penetrates a target, it loses energy throughout its trajectory to the electrons of the target atoms by ionization and by excitation, and it also loses energy by nuclear collisions. When this projectile encounters a hard collision, it can change its trajectory into an outward direction in a straight-line path—this is backscattering. During its outward path it also loses energy to the target atoms until the particle emerges from the target. The backscattered particles can thus be interpreted as single-collision events and the simplicity of two-body kinematics can be exploited. The energy loss suffered during the inward and outward paths provides the depth to which the projectile has penetrated. This analysis technique is known as Rutherford backscattering (RBS). A schematic diagram of a typical experimental setup is shown in Fig. 1. Ion beams of 0.1–

1 MeV hydrogen and 1–4 MeV helium ions are commonly used because they are readily available using small accelerators (see *Particle Accelerators in Materials Research*), they incur least damage among ions of comparable velocity and their stopping power (dE/dx) has been extensively investigated (Ziegler and Chu 1974, Andersen and Ziegler 1977, Ziegler 1977). Also, the Rutherford cross section is well known for these ions. In practice, a momentum-analyzed and well-collimated beam of 2 MeV $^4He^+$ from an accelerator is directed toward the target at angle θ_1 relative to the surface normal, and the elastically scattered yield emerging at angle θ_2 is recorded by a charged-particle detector. The 2 MeV helium ion is chosen as a projectile because it provides good mass resolution and is involved in no known nuclear reactions below 2300 keV. In special cases, ^{12}C and ^7Li ions have been used as projectiles. Although heavier ions provide better mass resolution for the high-mass elements, they have not been used owing to poor detector resolution and the large damage produced in the target by heavy ions. Increased depth resolution has been achieved using grazing incidence. Solid-state detectors are commonly employed because they are relatively inexpensive and can be used in an energy-dispersive mode with a reasonable energy resolution ($\delta E \approx 15$ keV). Distortions in the energy spectrum due to pile-up can be minimized by using either a lower counting rate or standard pile-up rejection circuits. Improved energy resolution can be achieved by using the bulkier electrostatic or magnetic spectrometers, but this must be weighed against a longer data accumulation time since each point in the energy spectrum must be recorded sequentially. Also, the higher accumulated beam dose contributes to increased radiation damage in the sample.

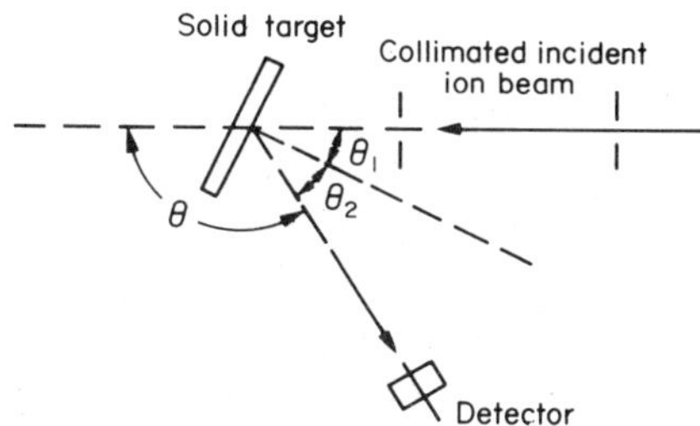

Figure 1
Schematic representation of the experimental geometry for the backscattering technique

There are three basic concepts in backscattering, and each provides an analytical capability: (a) the kinematic factor allows mass analysis; (b) the differential scattering cross section enables quantitative analysis; and (c) the energy loss makes possible depth analysis. In order to illustrate the principles and limitations of the technique, consider the backscattering analysis of a well-polished elemental target of atomic density N. Figure 2 gives the notation and schematic backscattering energy spectrum for a fixed bombarding energy and geometry. The edge occurring at an energy E' corresponds to the scattering from the atoms at the surface layer and is governed by the two-body kinematics of elastic scattering. For the energies of interest here, relativistic effects may be neglected. In terms of the incident-particle mass m and energy E_0, target mass M and laboratory scattering angle θ, kinematic factor K may be defined as

$$K = K_M = \frac{E'}{E_0} = \left[\frac{m\cos\theta + (M^2 - m^2\sin^2\theta)^{1/2}}{m + M}\right]^2 \qquad (1)$$

Here, the notation $K = K_M$ is applicable to a given backscattering experiment in which m and θ are predetermined. The yield at energies below the step (Fig. 2) therefore corresponds to the scattering from deeper layers because the scattered particle suffers energy loss in both the inward and outward paths. It can be seen that the presence of surface impurities shows up as a peak at an energy above the edge if their mass is greater and below if lower. This allows elements, and even isotopes up to about mass 40, to be distinguished. For elements of much greater mass, the separation becomes less and less, and close heavy elements cannot be separated. On the other hand, for light elements such as carbon, oxygen or nitrogen the problem is often that of interference due to a signal from the substrate. From the above equation it is seen that the best mass separation is obtained when $\theta = 180°$. In practice, $\theta = 160$–$179°$ can easily be achieved using annular solid-state detectors. The

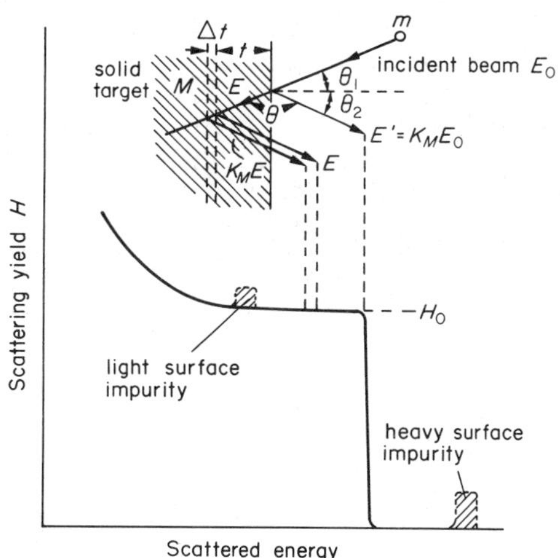

Figure 2
Schematic representation of the Rutherford backscattering yield from a thick target with surface impurities

energy difference ΔE represents the energy loss from the beam particles on inward and outward passage through the layer. There is a direct relation between the total thickness of the layer and the total width ΔE of the signal. In general, for small energy loss, where the rate of change of loss per unit path does not change appreciably, there is a linear relation between energy loss ΔE and depth t that can be expressed as

$$\Delta E = [S]t \tag{2}$$

where $[S]$ is called the backscattering energy-loss factor, which changes slowly as a function of E and t. This factor depends on K_M and dE/dx for the ingoing and outgoing path in the material, and can be calculated as

$$[S] = (K_M/\cos\theta_1)(dE/dx)_{\bar{E}_{in}} + (1/\cos\theta_2)(dE/dx)_{\bar{E}_{out}} \tag{3}$$

where $\bar{E}_{in} = E_0 - (\Delta E/4)$ and $\bar{E}_{out} = E_1 + (\Delta E/4)$ (Chu et al. 1978). Values of dE/dx are given in terms of the stopping cross section $\varepsilon = (1/N)(dE/dx)$, where N is the atomic density. For normal incidence, $\theta_1 = 0$, $\theta_2 = \pi - \theta$. Hence,

$$[S] = NK_M\varepsilon(\bar{E}_{in}) + (N/|\cos\theta|)\varepsilon(\bar{E}_{out}) \tag{4}$$

Values of stopping cross sections have been tabulated (Chu et al. 1978). Knowing the value of ΔE and S, sample thickness t can be determined. In the case of a thin heavy impurity layer on top of a light substrate, the heavy impurity corresponds to the peak above the edge. If the overall system resolution is small compared with the width of peak, this width gives a direct measure of the film thickness if the stopping power is known.

In backscattering analysis, the number of atoms Nt per unit area of the target is determined. The area A under each of the peaks in Fig. 2 is proportional to the total number of atoms per unit area of each type and to the differential scattering cross section σ of each atom (i.e., $A \propto \sigma Nt$). To determine an absolute value for Nt from the integrated number of counts requires careful determination of the scattering geometry, beam-current integration and measurement of the active area of the solid-state detector. For a total number of incident helium ions Q and a detector solid angle Ω, the total area A in counts under the peak is given by

$$A = Q\Omega\sigma Nt \tag{5}$$

The Rutherford differential scattering cross section for 1 MeV ^4He ions has been tabulated (Chu et al. 1978). The scattering cross section $\sigma \propto Z^2/E_0^2$, where Z is the atomic number; consequently, the total area under the peak is larger for heavier elements and for smaller projectile energy.

Backscattering analysis has been used to study compound or mixture targets. If the compound contains two elements A and B with a molecular composition A_mB_n, where m and n are fractions and $m + n = 1$, and if the ratio of the energy widths for the two elements is unity, then from the backscattered spectrum the ratio of the corresponding heights is $H_A/H_B \simeq (\sigma_A/\sigma_B)(m/n)$. The analysis for thick targets and the detailed formalism and a numerical method has been developed by Brice (1973). Computer analysis of complex backscattering spectra has been developed by Ziegler et al. (1975a).

There are limitations to backscattering spectrometry which must be borne in mind when considering the application of the technique to specific problems. The sample depth that can be analyzed with 2 MeV $^4He^+$ ions is about 1–2 μm. Also, the backscattering technique is not highly element specific, and for heavy-mass elements it is difficult to obtain good mass resolution. In this case, ion-induced x rays can be used for element identification. Another limitation is that it is not always possible to completely identify the chemical phases that are present in a compound. Complementary techniques like secondary-ion mass spectroscopy (SIMS) and x-ray diffraction in conjunction with backscattering studies provide information on phase formation and reaction kinetics. The sensitivity of the backscattering technique for detection of low-mass elements on a high-mass substrate is poor, and nuclear reactions or resonance techniques must be used to obtain quantitative information on trace amounts of these elements. These techniques have been reviewed by Wolicki (1975) and Chu et al. (1978).

2. Applications

As only a short time is required for data acquisition, backscattering spectrometry has been preferred for exploratory studies over the more lengthy and destructive analysis techniques involving sputtering for layer removal. The most straightforward application is the measurement of the thickness in atoms per cm^2 of deposited metal films. The concept of relating ΔE to layer thickness has been used in the determination of surface impurity concentration profiles (Ziegler and Baglin 1971). In general, systems are chosen where near-surface analysis is important; examples are thin films, layered structures, diffusion profiles, ion-implanted layers and dielectric layers. To take advantage of the capabilities of backscattering, there should be a change in the composition during the process treatment such as interdiffusion between thin-film elements or growth of oxide layers. This technique has been used to measure the composition of dielectric layers and the formation of silicides. Figure 3 shows a 2 MeV $^4He^+$ backscattering spectrum for a Pt–Si layer 140 nm thick on a silicon substrate. For comparison, a backscattering spectrum of pure platinum 100 nm film sputter deposited on a silicon substrate is also shown.

The silicon part of the spectrum is enlarged by a factor of five for clarity. The ratio of the shaded area of platinum to that of silicon is 32:1. The cross-section ratio σ_{Pt}/σ_{Si} is equal to 32:1. Therefore, the ratio of platinum to silicon atoms per unit volume, given by

$$(N_{Pt}/N_{Si}) = (A_{Pt}/A_{Si})(\sigma_{Pt}/\sigma_{Si}) \quad (6)$$

is equal to unity (to within 1%). Another advantage of these measurements is that the stoichiometry can generally be determined directly from the spectra.

The composition can also be determined from the heights of the platinum and silicon plateaus in the backscattering spectra as

$$N_{Pt}/N_{Si} = (\sigma_{Si}/\sigma_{Pt})(H_{Pt}/H_{Si})(\Delta E_{Pt}/\Delta E_{Si}) \quad (7)$$

In Fig. 3, the ratio $\Delta E_{Pt}/\Delta E_{Si}$ is 1.15 and the measured ratio H_{Pt}/H_{Si} is 27.5. Inserting values of σ, ΔE and H in the above equation gives $N_{Pt}/N_{Si} = 1$ (to within 2%). Hence, if the surface impurity is of high mass on a low-mass substrate, the precise values of Q and Ω need not be known.

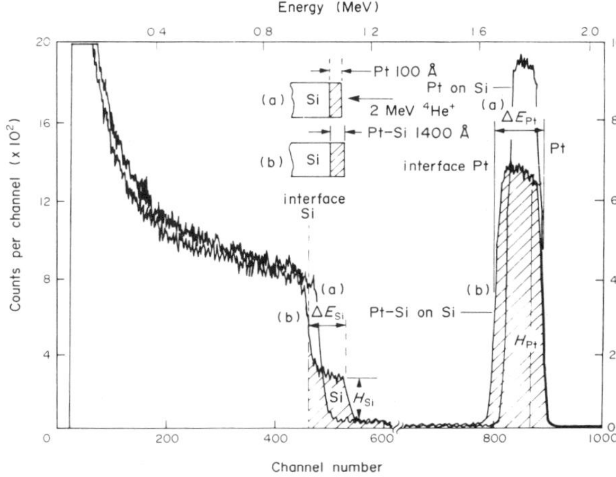

Figure 3
Backscattered energy spectra for 2 MeV $^4He^+$ ions from (a) 100 nm platinum sputter deposited on a silicon substrate and (b) sintered at 375 °C for 25 min in an argon atmosphere to form a Pt–Si layer 140 nm thick on a silicon substrate

Metallization plays a large role in the success or failure of integrated-circuit technology. In order to make electrical connections between the devices and the outside world, it is necessary to deposit thin metal layers for conducting paths. There are a host of problems associated with the metallization process, all of which stem from the fact that thin-film reactions occur at relatively low temperatures. To prevent compound formation an intermediate layer is deposited as a barrier layer. For example, a thin layer of SiO_2 has been deposited in between $NbSi_2$ and a silicon substrate (Chow et al. 1982). Figure 4a shows the 2 MeV $^4He^+$ spectrum for an $NbSi_2$ thin film (406.4 nm) on an SiO_2–Si substrate. The silicide films were sputtered from a hot-pressed alloy target of stoichiometric composition Si:Nb = 2 onto heated (350 °C) substrates. Oxidation was carried out in dry oxygen between 550 and 850 °C. ($NbSi_2$ films can be oxidized at lower temperatures than other refractory metal silicides, such as $MoSi_2$, WSi_2 and $TaSi_2$.) Figures 4b and 4c show the growth of oxide layers. From backscattering measurements, the oxide films exhibited the same Si:Nb ratio as the initial silicide layer. Although only Nb_2O_5 has been clearly identified, density considerations and experimental values of oxide thickness suggest that niobium and silicon oxides coexist. The backscattering technique not only provides the thickness and stoichiometry of new layers but also the composition of the remaining silicide layer.

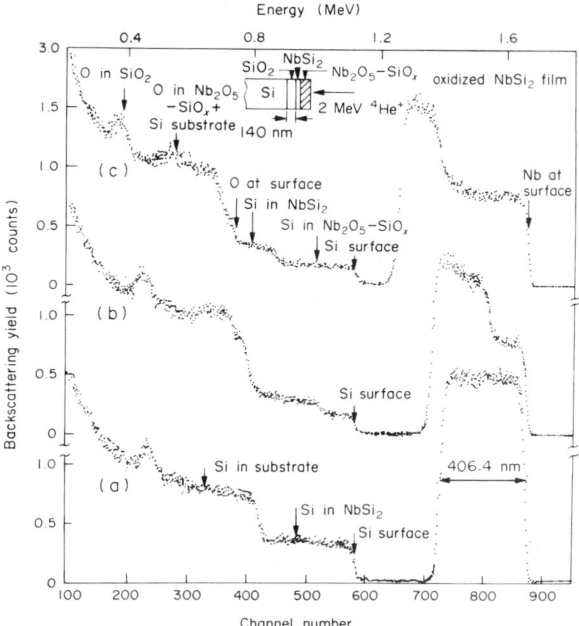

Figure 4
Backscattered energy spectra for 2 MeV $^4He^+$ ions from (a) 406.4 nm $NbSi_2$ polycrystalline film deposited on SiO_2–Si substrate; (b) thermal oxidation of $NbSi_2$ layer carried out in dry oxygen at 600 °C for 15 min; (c) thermal oxidation of $NbSi_2$ layer carried out in dry oxygen at 650 °C for 15 min. $NbSi_2$ layer is not completely oxidized (after Chow et al. 1982)

3. *Channelling*

Axial or planar channelling occurs when a collimated energetic beam of ions (e.g., 2 MeV helium ions) is incident on a single crystal along an axial or planar direction. The rows or planes of atoms act as potential barriers such that the ions undergo a series of correlated screened Coulombic collisions. These small-angle collisions result in the channelled ions being steered by the rows and planes towards the central channel region between the rows or planes. Hence, in terms of backscattering spectrometry, channelling effects produce strikingly large changes in the yield of backscattered particles as the orientation of the single crystal is changed with respect to the incident beam. Channelling of energetic ions is well described in the literature (Morgan 1973, Gemmell 1974, Dearnaley 1973, Picraux 1975, Mayer et al. 1970). By combining channelling with backscattering measurements, crystallographic information such as lattice site location of impurities or lattice disorder can be determined (Morgan 1973). Figure 5 shows a channelling backscattering spectrum from a BF_2^+-implanted silicon crystal. Figure 6 shows the effect of thermal and laser annealing on the damage created by a BF_2^+ implant. The amorphous portion of the as-implanted layer is annealed thermally or by laser.

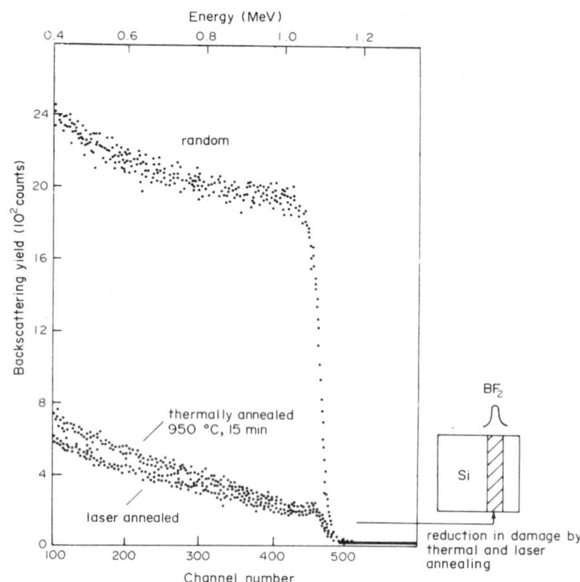

Figure 6
Random and (100)-aligned backscattering spectra for a silicon crystal implanted with BF_2^+

4. Microprobe Analysis

A recent application of the backscattering technique has been in the nondestructive analysis of microelectronic circuits. The utility of these ion-beam techniques has been greatly increased by the invention of lenses capable of focusing high-energy heavy-ion beams to micrometer dimensions. The combination of ion-beam analysis and micrometer-dimension ion beams yields a powerful, nondestructive method for materials analysis. Figure 7 shows a line-scan backscattering spectrum taken with a recently constructed microbeam system at the Sandia National Laboratories, New Mexico (Doyle et al. 1981). A 3 MeV $^4He^+$ beam of 2–5 μm was used to analyze an 80 nm thick (30 μm wide) gold grid on a silicon substrate.

Micrometer-size beam diameters are particularly important in experiments in which the sample is very small, as in the case of some biological studies, for studies of electronic circuits, or in metallurgy when

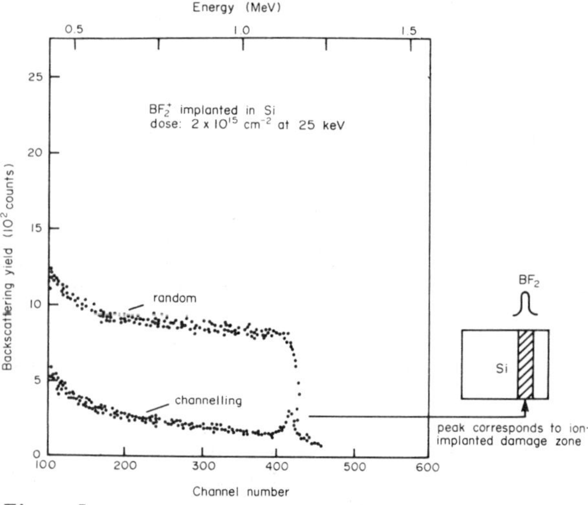

Figure 5
Random and (100)-aligned backscattering spectra for a silicon crystal implanted with (2×10^{15} cm^{-2} at 25 keV) BF_2^+ ions. The amorphous zone created by BF_2^+ implant is seen in the channelling spectrum

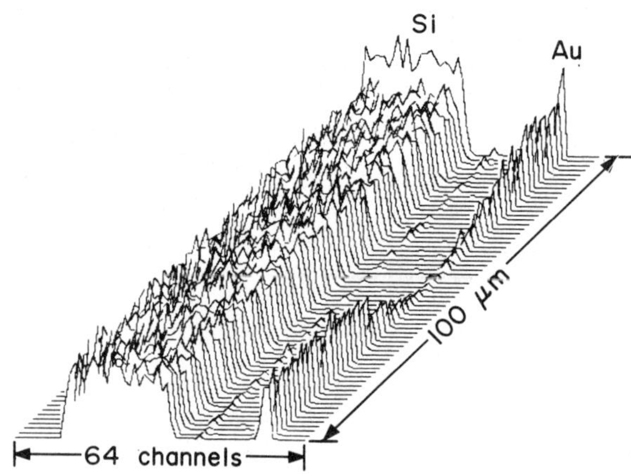

Figure 7
A line-scan backscattering spectra produced by analysis of a 3 MeV $^4He^+$ beam of 2–5 μm from 80 nm thick (30 μm wide) gold grid on a silicon substrate

individual grains are to be examined. The use of nuclear microprobes in microelectronics appears to be especially attractive.

See also: Investigation and Characterization of Materials: An Overview

Bibliography

Andersen H H, Ziegler J F 1977 *Hydrogen: Stopping Powers and Ranges in All Elements*. Pergamon, Oxford

Brice D K 1973 Theoretical analysis of the energy spectra of backscattered ions. *Thin Solid Films* 19: 121–35

Chow T P, Hamzeh K, Steckl A J 1982 Thermal oxidation of niobium silicide thin films. *J. Appl. Phys.* 54: 2716–19

Chu W-K, Mayer J W, Nicolet M-A 1978 *Backscattering Spectrometry*. Academic Press, New York

Dearnaley G 1973 Ion penetration and channeling. In: Dearnaley G, Freeman J H, Nelson R S, Stephen J (eds.) 1973 *Ion Implantation*. North-Holland, Amsterdam

Doyle B L, Wing N D, Peercy P S 1981 Nuclear microprobe analysis. In: Geiss R H (ed.) 1981 *Proc. Annu. Conf. Microbeam Analysis*. San Francisco Press, San Francisco, California, pp. 79–86

Foti G, Mayer J W, Rimini E 1977 Backscattering spectrometry. In: Mayer J W, Rimini E (eds.) 1977 *Ion Beam Handbook for Material Analysis*. Academic Press, New York, pp. 21–65

Gemmell D S 1974 Channeling and related effects in the motion of charged particles through crystals. *Rev. Mod. Phys.* 46: 129–227

Mayer J W, Eriksson L, Davies J A 1970 *Ion Implantation in Semiconductors, Silicon and Germanium*. Academic Press, New York

Mayer J W, Tu K N 1974 Analysis of thin-film structures with nuclear backscattering and x-ray diffraction. *J. Vac. Sci. Technol.* 11: 86–93

Morgan D V (ed.) 1973 *Channeling: Theory, Observation and Applications*. Wiley, New York

Picraux S T 1975 Lattice location of impurities in metals and semiconductors. In: Ziegler J F (ed.) 1975 *New Uses of Ion Accelerators*. Plenum, New York, pp. 229–81

Wolicki E A 1975 Material analysis by means of nuclear reactions. In: Ziegler J F (ed.) 1975 *New Uses of Ion Accelerators*. Plenum, New York, pp. 159–228

Ziegler J F 1977 *Helium: Stopping Powers and Ranges in All Elemental Matter*. Pergamon, Oxford

Ziegler J F, Baglin J E E 1971 Determination of surface impurity concentration profiles by nuclear backscattering. *J. Appl. Phys.* 42: 2031–40

Ziegler J F, Chu W K 1974 Stopping cross sections and backscattering factors for ^{4}He ions in matter. *At. Data Nucl. Data Tables* 13: 463–89

Ziegler J F, Lever R F, Hirvonen J K 1975a Computer analysis of nuclear backscattering. In: Meyer O, Linker G, Käppeler F (eds.) 1975 *Ion Beam Surface Layer Analysis*, Vol. 1. Plenum, New York, pp. 163–83

Ziegler J F, Mayer J W, Ullrich B M, Chu W K 1975b Material analysis by nuclear backscattering. In: Ziegler J F (ed.) 1975 *New Uses of Ion Accelerators*. Plenum, New York, pp. 75–158

H. Bakhru

Ion Bombardment of Materials

Bombardment of crystalline solids with ions in charged particle accelerators offers a rapid means of creating radiation damage similar to that caused by neutrons, and is a convenient way to study such damage. An energetic charged particle entering a solid target has a very much larger interaction cross section with lattice atoms than has a neutron of similar incident energy. Consequently the rate of atomic displacements is 10^3–10^5 times greater than with neutrons and this means that a level of atomic displacements (the dose, or number of displacements per atom, dpa) appropriate to a year in a nuclear reactor can be attained in a few hours with ion bombardment. Other advantages of ion bombardment are relatively low cost, wider availability, no residual radioactivity, and the capability for isolation and close control of individual bombardment parameters such as temperature, dose rate and effects of implanted impurities.

1. The Technique

The experimental details of ion bombardments, and the calculations of the atomic displacements and the depths at which they occur in the target, are described in detail in ASTM Recommended Practice ES21. Typical particle energies are in the range 0.01–20 MeV. The damage is produced quite close to the target surface, as shown in Fig. 1. At these energies, no transmutation products are created to simulate those in neutron irradiations. The important transmuted elements such as helium, which strongly influences formation of damage microstructure, must be artificially introduced. This should be done with another accelerator during the displacive bombardment, not before it, or the microstructural consequences may be strikingly different, as shown by the influence of helium and its method of implantation on cavity formation in Fig. 1.

2. Considerations

Since the damaged region is so shallow, heavy-ion bombardments are not suitable for assessing radiation-induced changes in bulk mechanical properties. However, highly energetic (>50 MeV) light ions such as protons and α particles, with penetrations of 50 μm or more but with low damage rates, are used for radiation creep and fatigue studies on appropriately thin specimens.

Ion bombardments are most useful for studies of microstructural changes. Interpretation of these microstructures with respect to neutron damage structures should include recognition of a number of contributing factors peculiar to ion bombardments. The most important of these is the need to perform the bombardments at temperatures 100–200 K higher

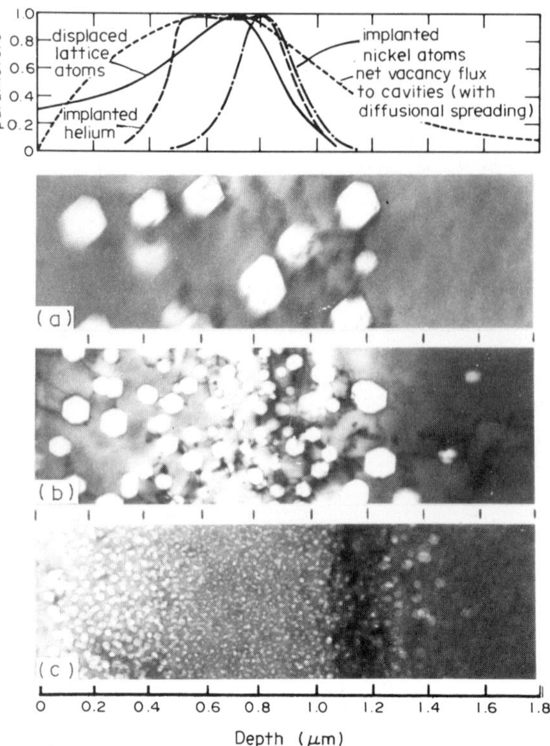

Figure 1
Electron micrographs of damage cavities lying along the ion range in a nickel target bombarded at 873 K with 4 MeV Ni^{2+} to a peak damage dose of 50 dpa. Helium ions are implanted at energies of 0.2–0.4 MeV to overlay the region where the maximum number of displaced atoms are created at a depth of 0.7 μm. The curves in the box represent the depth profiles for major damage parameters as they occur initially. Most displaced lattice atoms, vacant lattice sites and implanted atoms (point defects) are lost by recombination or by assimilation by the host lattice. Those that survive form clusters, seen as damage microstructure. At 873 K, some point defects migrate from their initial profiles, spreading out the damage structure, as illustrated by the curve for vacancies. These vacancies condense into cavities whose formation is assisted by gases, especially helium. (a) Without helium, the cavities are large and few. (b) Helium implantation simultaneously with nickel ion bombardment causes more cavities and reduces their size; this corresponds to the situation during neutron irradiation. (c) Room-temperature helium implantation before displacive bombardment considerably refines the cavities (adapted from Farrell et al. 1982)

than for comparable neutron irradiations in order to appropriately balance the point defect recombination rate with the high production rate. This alters the nucleation rates of defect clusters. Other factors are the nonuniform depth distribution of damage production, the nonsymmetry of profiles of the implanted "transmutation products" with the damage profile, diffusional spreading of point defects away from the damage region, and the residual effects of the damage-producing ions which become extra intersitities, sometimes alloying elements, in the system. Contributions from these factors can be estimated for phase-stable systems (Mansur and Yoo 1979).

3. Achievements

The proceedings of recent conferences on radiation damage in reactor materials (*Journal of Nuclear Materials* 1981, 1984) illustrate the utility of ion bombardments. They have proven valuable in at least five respects, namely:

(a) simulations of surface blistering of materials by formation of helium bubbles under conditions of plasma bombardment expected at the first wall of magnetic fusion reactors;

(b) comparative screening and selection of promising low-swelling alloys for prolonged neutron irradiations;

(c) rapid attainment of high-dose levels, and investigations of individual bombardment parameters and selected impurities;

(d) correlation with neutron damage structures;

(e) research into irradiation creep mechanisms.

It is now recognized that ion bombardments cannot simultaneously simulate all features of neutron damage microstructures because of nucleation and precipitation differences. However, ion bombardments have provided a great deal of information on damage microstructures. They have clearly demonstrated the vital role of gaseous impurities in microstructural evolution, and have allowed unequivocal demonstration of the phenomenon of radiation-enhanced solute segregation and precipitation. The good control and wide span of the bombardment parameters in ion bombardments is currently being used to explore irradiation features expected in future fusion reactors which cannot be reproduced in current fission reactors. These include high, and variable, ratios of implanted gas levels to dpa levels, pulsed irradiations and sudden temperature excursions. Radiation creep and fatigue experiments under light-ion bombardment are now being extended to studies of dynamic helium bubble growth under stress and strain in connection with the problem of helium embrittlement.

See also: Radiation Effects in Metals and Alloys; Swelling in Irradiated Metals and Alloys; Particle Accelerators in Materials Research

Acknowledgement

The authors acknowledge financial support from the Division of Materials Science, US Department of Energy, under Contract No. DE-AC05-840R21400 with Martin Marietta Energy Systems, Inc.

Bibliography

Farrell K, Packan N H, Houston J T 1982 Depth profiles of nickel ion damage in helium-implanted nickel. *Radiat. Eff.* 62: 39

Mansur L K, Yoo M H 1979 Advances in the theory of swelling in irradiated metals and alloys. *J. Nucl. Mater.* 85/86: 523

Journal of Nuclear Materials 103/104, 1981. Proceedings of Second Topical Meeting on Fusion Reactor Materials

Journal of Nuclear Materials 122/123, 1984. Proceedings of Third Topical Meeting on Fusion Reactor Materials

K. Farrell; M. B. Lewis

Ion Implantation Metallurgy

Ion implantation is the introduction into solid substrates of selected ionic species with accelerating energies in the keV–MeV range (typically 20–500 keV). The basic equipment requirements are an ion source, an ion accelerator, a mass separator (usually magnetic deflection) and a beam sweeper. All the elements of the periodic table can be implanted, and mass separation of the ionized beam permits implantation of individual isotopes, including radioactive species. The process has been used successfully in semiconductor technology, and an industry has grown up based on its application. More recently, considerable interest has emerged in the ion implantation of metals to synthesize alloys which are difficult or even impossible to obtain by traditional techniques. This versatility has made ion implantation a useful research tool for preparing novel metastable alloys, including superconductors and amorphous metal layers. For the same reason, ion implantation is being used as a materials modification technique to alter such surface-sensitive properties of metals as friction, wear, corrosion, oxidation, catalysis and fatigue. It also finds application as a means of simulating or accelerating the effects of neutron damage in nuclear materials. Considerable data exist regarding property changes in implanted metals and great effort is now being spent to correlate these effects with microstructural changes. There are currently only a few commercial applications of ion implantation metallurgy. As it continues to be used as a research tool, no doubt more practical aspects will be found.

1. General Features

Subjects of interest in ion implantation include the penetration of ions into a solid, the inevitable radiation damage which ensues, the recovery of such damage and the eventual fate of the implanted ion (Mayer et al. 1970).

The slowing down of ions in a solid is generally discussed in terms of stopping power ($-dE/dx$), the energy loss per unit distance travelled in the solid. There are two major processes of energy loss: (a) elastic Coulomb interactions between the screened nuclear charges of the ion and the target atom, and (b) inelastic interactions of the ion with bound or free electrons. At low energies (up to 200 keV), nuclear stopping dominates and electronic stopping may be ignored.

The penetration depth of ions into a material is a complex function of the energy, mass and size of the ion and the mass and crystalline structure of the substrate. However, it is possible to predict fairly accurately the depth and range distributions of the implanted ions, provided that the beam current density, ion energy and target material are known. Beams of light ions provide a convenient means for measuring the implanted distribution using Rutherford backscattering or nuclear reaction techniques. Focusing and rastering of the ion beam allow implantation over lateral dimensions of a few micrometers to several centimeters. The implanted layer is shallow, varying from tens of nanometers up to a few micrometers. The short mean free path of ions in matter requires that the implantation be done in a reasonable vacuum ($<10^{-2}$ Pa).

A typical range distribution is approximately Gaussian in shape and may therefore be characterized by a mean projected range R_p and a relative (straggling) range width $\Delta R_p/R_p$. To a first approximation, R_p depends on the ion mass and incident energy, whereas $\Delta R_p/R_p$ depends primarily on the ratio between ion mass and that of the substrate atoms. By varying the energy and dosage during implantation almost any type of dopant profile may in theory be achieved. Since the distribution is Gaussian, R_p corresponds to the position of the peak and ΔR_p is taken to be the full width at half maximum divided by $2(2\ln 2)^{1/2}$. An average concentration n (atoms per cm^3) near the maximum can therefore be estimated if R_p is known:

$$n = Nd/2(2\ln 2)^{1/2}R_p$$

where Nd is the number of implanted atoms per cm^2. There is a striking difference between the dopant profiles of amorphous materials and crystalline materials because channelling effects due to the periodic lattice enable ions to penetrate to much greater distances in crystalline solids; implantations near low-index crystallographic directions in the target produce distributions skewed toward greater depths.

Ion implantation inevitably produces radiation damage in the solid. The elastic collisions between ions and target atoms result in energy transfers ranging from electron volts to tens of keV. Provided that a recoiling lattice atom receives an energy in excess of the displacement energy (~25 eV), it can leave its lattice site and become permanently displaced in the solid. In general, a lattice atom receives sufficient kinetic energy from a primary collision to give rise to a succession of secondary collisions called a collision cascade. A collision cascade is completed in $\sim 10^{-13}$s, its excess energy being dissipated thermally to create a "thermal spike." Each cascade creates many vacancy–interstitial pairs and it is the accumulation of these cascades that gives rise to radiation damage. A vacancy and interstitial recombine spontaneously unless they are separated by a critical distance, and this sets an ultimate limit on the density of damage that can be done.

There is a difference in the nature of the radiation damage produced in solids with metallic bonding as opposed to solids with covalent bonding. Generally, metallic solids remain essentially crystalline even after bombardment. In covalent solids bombardment creates amorphous regions; as the ion dose is increased, these disordered zones increase in number until they eventually overlap.

The radiation damage in a solid can be repaired by annealing. The thermal activation of vacancies from a variety of sources results in vacancy–cluster shrinkage. Higher temperatures are required for interstitial cluster shrinkage. The temperatures required for annealing are usually well below those necessary for diffusion of dopants and hence no significant redistribution of implanted material occurs.

The eventual fate of the implanted ion is a key question. In a random-collision cascade the incident ion has a high probability of finally coming to rest as a consequence of a replacement collision and therefore remains in a substitutional site. However, the ion will probably be in a damaged region where lattice interstitials and vacancies are plentiful, and these may cause continued migration. If the ion dose is sufficiently high and the concentration of impurity exceeds the maximum solubility limit, a second phase may be expected to precipitate. However, this is not necessarily the case. The radiation damage and associated effects may sufficiently modify the conditions pertaining to thermodynamic equilibrium to allow metastable phase formation instead.

2. *Modification of Surface-Sensitive Properties of Metals*

There are several advantages to the treatment of surfaces by ion implantation: surface properties can be modified without affecting the bulk; any species can be incorporated with good control and at low temperatures; a finished product can be implanted with no change in shape; and the process is versatile, with a low power consumption, and conserves critical materials. However, on the negative side, disadvantages include the line-of-sight vacuum operation, the high capital cost and unfamiliarity of the process, and the shallowness of the treated layer.

Significant improvements in the hardness and wear of steels and carbide components have been observed by implanting with nitrogen (Dearnaley 1975, 1980). This is of particular interest for cutting and metal-forming tools and for bearings. For example, steel tools implanted with nitrogen have shown an increase in life of a factor of four. Wear rates are similarly improved, with the added advantage that as wear proceeds the implanted ion penetrates into the metal; thus the beneficial effects are still observed at depths a thousand times that of the implanted layer. Fatigue lifetimes have been increased as much as ten times in nitrogen-implanted steels. Corrosion and oxidation resistance of ferrous alloys and titanium have also been improved, presumably because it is possible to create supersaturated solid solutions at the surface which are beneficial to passivation.

Further details of surface modification by ion implantation are given in the article *Ion-Implanted Coatings*.

3. *Metastable Phase Formation*

Ion implantation is inherently a nonequilibrium process, which leads to the possibility of forming metastable phases. Characterization can be carried out with the help of such tools as ion channelling to determine the lattice site of the implanted ion, transmission electron microscopy to determine the microstructures, and x-ray diffraction to analyze crystal structure (Borders 1979).

Supersaturated solid solutions have been formed in some cases. An example is the silver–copper system, which has very limited solid solubility at room temperature. By implanting silver in copper, approximately 8 at.% Ag was found to be completely substitutional. This corresponds to the equilibrium solubility of silver at about 700 °C. The result is not surprising if ion implantation is considered to be effectively a rapid-quenching phenomenon brought about by the rapid cooling of material around the thermal spike generated during implantation. A more unusual example of metastable phase formation is that of tungsten in copper. These two elements are completely insoluble, even in the melt. After implantation at concentrations of 1–3 at.%, tungsten was found to be 90% substitutional.

Amorphous phases can also result from high-dose implantations. Examples are tantalum in copper, gold in platinum and bismuth and dysprosium in nickel. Metastable phases have been formed by implantation followed by epitaxial recrystallization of the implanted layer on a substrate of the desired crystal structure. This technique has been used for forming metastable A15 Nb_3Si (Clapp and Rose 1980).

High-dose implantations are often desirable in

metallurgy. However, there is a limit to the maximum concentration of implanted species. This is a result of sputtering, which is the ejection of atoms from the surface (see *Sputtering*). As more ions are implanted, more are removed from the surface, thus setting a concentration limit which depends on the type of incident ion and substrate. The maximum implanted concentration is approximately equal to the inverse of the number of atoms ejected per incident ion, which for typical implantations into metals varies between 2 and 20 atoms. Ion beam mixing is a method of overcoming this: a thin film is deposited on the substrate—for example, by sputtering or vapor deposition—and subsequent bombardment by energetic ions which penetrate through the layer–substrate interface causes intermixing of the layer and substrate materials because of the atomic displacements produced by the passage of ions through the solid (Mayer et al. 1979). An example is the gold–copper system: when gold was implanted directly into a single crystal of copper, concentrations of only 6 at.% Au were achieved; by ion beam mixing with xenon, concentrations of 30 at.% Au were obtained. Ion beam mixing also allows beams of easily obtainable gas ions to be used to produce alloyed surface layers at substantially lower doses than with direct implantation.

It is worth mentioning that ion implantation is being increasingly used as a tool in alloy development. The desired alloy composition is achieved by implantation and parameters such as diffusivities, solubilities and phase diagrams are then determined by varying subsequent heat treatments (Meyers 1978).

4. Ion Implantation in Superconductors

Ion implantation has been used to modify the three critical parameters of superconductivity: critical temperature T_c, critical current density J_c and critical magnetic field H_c (Meyer 1975, Picraux et al. 1973). Both the radiation damage and the doping can affect these properties.

Since ion implantation is inherently a nonequilibrium technique, there have been attempts to increase T_c by forming metastable phases. Buckel and Stritzker (Picraux et al. 1973 p. 3) have done interesting work with palladium alloys: palladium implanted with hydrogen and deuterium at liquid helium temperatures had T_c's of 9 and 11 K, respectively; palladium–copper implanted with nitrogen showed an increase in T_c from 8.6 to 16.6 K. In the case of nitrogen-implanted molybdenum films, T_c was increased from 1.2 to 9.2 K; x-ray diffraction revealed the presence of several phases, including an amorphous one which was responsible for the increase in T_c (Borders 1979).

Radiation damage can enhance both H_c and J_c. For example, the J_c of niobium and A15 compounds has been increased by ion bombardment. Neutron irradiation of niobium has resulted in more effective flux-pinning centers. If light ions and small doses are used T_c is only slightly affected. However, heavy ions or high doses decrease T_c significantly (see *Radiation Effects on Superconductors*).

5. Commercial Applications

Ion implantation metallurgy is likely to be most economical when applied to small expensive components that are costly to replace. For example, the cutting-tool industry involves hundreds of millions of dollars annually and improvement in tool life could result in considerable savings, not only in replacement costs, but also in plant productivity. Ion implantation may also have considerable benefits when applied to alloy systems involving elements that are in short supply or are dependent on foreign sources. For example, chromium is extensively used in electroplating and since ion-implanted layers are much thinner than electroplated layers the use of ion implantation could significantly reduce the consumption of chromium.

A considerable number of implanted tools and engineering components have already been tested under production conditions and have performed well. Some examples are press tools, mill rolls, metal-cutting tools, wiredrawing dies, tools used for the injection molding of plastics, ball bearings, fluid metering pumps and valve components (Dearnaley 1980).

Machines appropriate for commercial implantation of metals are now being designed. Some of the requirements are: very high beam currents, large target areas, and complex ion sources that direct the beams in a variety of ways to accommodate the different sizes and shapes of the metal components. The potential for metallurgical applications appears to be as great as it has been for semiconductor applications.

See also: Ion Bombardment of Materials

Bibliography

Borders J A 1979 Metastable phases in metals produced by ion implantation. *Annu. Rev. Mater. Sci.* 9: 313–39

Clapp M T, Rose R M 1980 On the synthesis of metastable A-15 "Nb_3Si" by ion implantation and on its superconducting transition temperature. *J. Appl. Phys.* 51: 540–44

Dearnaley G 1975 Ion implantation in metals. In: Ziegler J F (ed.) 1975 *New Uses of Ion Accelerators.* Plenum, New York, pp. 283–319

Dearnaley G 1980 Practical applications of ion implantation. In: Preece C M, Hirvonen J K (eds.) 1980 *Ion Implantation Metallurgy.* The Metallurgical Society (AIME), New York, pp. 1–20

Dearnaley G, Freeman J H, Nelson R S, Stephen J 1973 *Ion Implantation.* North-Holland, Amsterdam

Hirvonen J K (ed.) 1980 *Ion Implantation*. Academic Press, New York
Hubler G K, Holland O W, Clayton C R, White C W 1984 *Ion Implantation and Ion Beam Processing of Materials*, MRS Symposium Proceedings Vol. 27. North-Holland, Amsterdam
Mayer J W, Erikson L, Davies J A 1970 *Ion Implantation in Semiconductors*. Academic Press, New York
Mayer J W, Silvanus S L, Bor-Yeu T, Poate J M, Hirvonen J K 1979 High dose implantation and ion beam mixing. In: Preece C M, Hirvonen J K (eds.) 1980 *Ion Implantation Metallurgy*. The Metallurgical Society (AIME), New York, pp. 37–46
Meyer O 1975 Ion implantation in superconductors. In: Ziegler J F (ed.) 1975 *New Uses of Ion Accelerators*. Plenum, New York, pp. 323–53
Meyers S M 1978 Annealing behavior and selected applications of ion-implanted alloys. *J. Vac. Sci. Technol.* 15: 1650–55
Picraux S T, EerNisse E P, Vook F L (eds.) 1973 *Applications of Ion Beams to Metals*. Plenum, New York

M. T. Clapp; L. E. Rehn

Ion-Implanted Coatings

Ion implantation is a technique that involves the injection of high-velocity ions into the surface layers of a target solid. The ions are extracted from a suitable source, accelerated to high energies (typically 10–200 keV), mass analyzed to produce a pure beam and then directed at a target in a vacuum chamber. The ions penetrate the target, come to rest at depths typically 0.01–1 μm below the surface, and produce a parasurface alloy layer. The depth to which ions penetrate can be calculated from well-established theoretical considerations. The technique of ion implantation is extensively employed in the semiconductor industry, where it is used to inject dopant atoms into silicon during the manufacture of semiconductor electron devices. During the late 1970s, the technique has increasingly been used to modify and investigate the composition and structure-sensitive surface properties of metals.

The distance penetrated by an ion depends on its velocity and mass, and the mass of the target. The concentration profile of an implanted species is usually characterized by a Gaussian distribution centered about a mean projected range R_p, with a range straggling (or standard deviation) of ΔR_p. Values of R_p and ΔR_p are conveniently available for various ion/target combinations in tabulated or graphical form. An example of the latter is shown in Fig. 1a, together with the anticipated concentration profile (Fig. 1b). Departures from a Gaussian range distribution occur when high doses of ions are injected into a target. Energy transferred to target atoms can result in erosion or sputtering of the target surface, and this (together with other effects such as ion-beam induced atomic migration) often sets a limit on the ultimate alloy concentration attainable. An equilibrium is reached during high-dose implantation, such that sputtering removes as many of the implanted atoms as the ion beam injects. The resulting equilibrium profile is shown in Fig. 1c. It is relatively simple to obtain implanted concentrations of ~15 at.%, but a low sputtering coefficient is required to obtain concentrations higher than ~30 at.%. The processing time to achieve a required alloying concentration depends on the ion current density achievable at the workpiece surface, and can vary from minutes to hours.

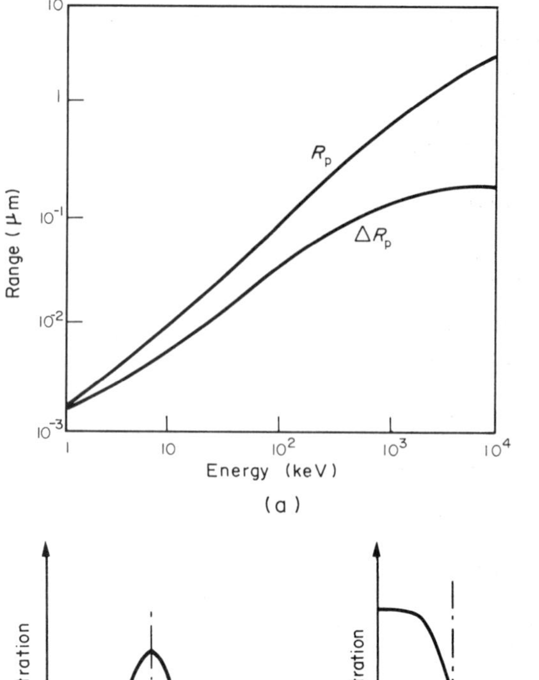

Figure 1
Ion-implanted nitrogen in iron: (a) the projected range R_p and range straggling ΔR_p as a function of ion energy; (b) sketch of the expected concentration of nitrogen following low-dose implantation; (c) sketch of the expected concentration of nitrogen following high-dose implantation. Sputtering sets a limit on the maximum concentration

Ion implantation as a technique for producing surface alloys has a number of distinct advantages. It can be used to coat a bulk material with a desirable surface alloy skin without sacrificing any of the bulk material properties. The technique is carried out at low temperatures and the resulting surface layer is continuous with the bulk; since there is no interface there are no adhesion problems. There are no significant dimensional changes, and the process, carried out in vacuum, is clean, highly reproducible and controllable. Ion implantation can be used to produce, relatively inexpensively, a wide range of conventional alloys of different compositions for screening tests in alloy development programs. Furthermore, the restrictions which derive from equilibrium phase diagrams do not apply to surface alloys produced by ion implantation, since the alloys are formed by virtue of the penetration of high-velocity ions into the solid surface. Ion implantation therefore has the considerable advantage of overcoming the limitations imposed by equilibrium phase diagrams and of producing novel, metastable, single-phase solid-solution alloys that are completely inaccessible by conventional metallurgical techniques. Against these advantages must be set the facts that ion implantation still requires relatively expensive and sophisticated equipment, that it is essentially a line-of-sight process and that treated layers are very shallow.

Implanted ions enter the target metal with kinetic energies typically five orders of magnitude greater than the binding energy of the solid. Atomic collisions between the ion and the target atoms cause considerable atomic movements, which then decay towards equilibrium values in $\sim 10^{-13}$ s. The essential long-range metal crystal structure is usually retained, although more localized defects such as dislocations are produced, and in certain circumstances highly noncrystalline (amorphous) alloys can be formed. The implanted species is randomly distributed and there is usually no evidence of second phases being formed. The movement of the system towards thermodynamic equilibrium then depends on the rate at which thermally activated processes proceed. This movement towards equilibrium usually only occurs during heat treatment after the implantation process has finished. During implantation, however, radiation-induced defects such as mobile point defects may produce solute gradients within the implanted layer, and these in turn can influence the local thermodynamic balance.

Work during the late 1970s has demonstrated that ion implantation can be used to alter certain surface properties of metals. The more significant results have been obtained in the following areas.

1. Aqueous Corrosion

The technique of alloying to produce corrosion-resistant alloys is well established, as is the use of corrosion-resistant coatings. A wide-ranging investigation has been initiated into the use of ion implantation to produce corrosion-resistant surface alloys on metals. The implantation process itself has been shown not to affect the corrosion resistance of a pure, film-free metal; any radiation damage is not of a sufficient level to affect corrosion mechanisms or kinetics. Conventional bulk alloys and corresponding ion-implanted surface alloys have been shown to exhibit similar corrosion-resistant properties. This similarity is illustrated in Fig. 2, which shows the anodic polarization behavior in deaerated acetic acid–sodium acetate buffer solution (pH 7.2) of (a) pure iron, (b) a conventional 7.3% Cr bulk alloy and (c) a Cr-implanted surface alloy on a pure Fe target. A general conclusion derived from a number of ion/target systems is that any species introduced by implantation will exhibit its normal chemical effect when the alloy is exposed to a corrosive environment. This conclusion has been confirmed, in the case of novel alloys, by implanting tantalum into iron (normally Ta is virtually insoluble in Fe) to produce a corrosion-resistant alloy, as might be expected from the known oxide-forming tendencies of tantalum.

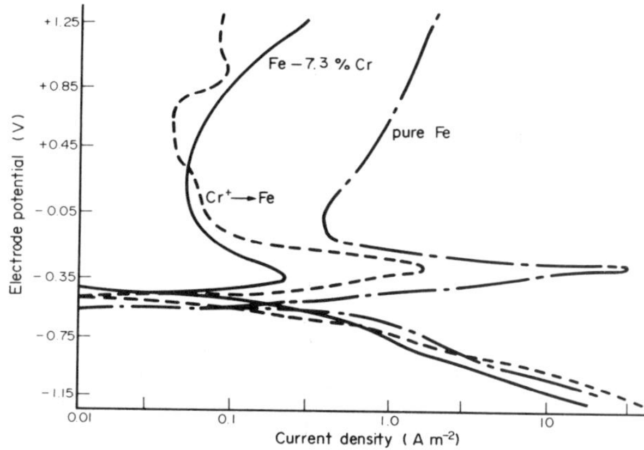

Figure 2
Potentiodynamic polarization curves for pure iron, a Fe–7.3% Cr conventional bulk alloy, and Cr-implanted iron (5×10^{20} ions m^{-2}) in sodium acetate–acetic acid buffer solution (pH 7.2)

This beneficial effect is not restricted to the control of uniform corrosion. Bulk molybdenum additions are often made to stainless steels to improve their resistance to pitting corrosion. There is evidence that Mo implantation can be used to improve the pitting resistance of steels which do not contain bulk molybdenum additions. More importantly, aluminum also suffers from pitting corrosion, but a single-phase Al–Mo alloy cannot be formed by conventional metallurgical techniques, since Mo is not soluble in Al. However, implantation of Mo into Al produces a pit-resistant alloy when tested in 0.01 M sodium sulphate solution containing 1000 ppm Cl^- (as sodium chloride).

2. *High-Temperature Oxidation*

The use of alloying techniques is also well established as a beneficial technique in controlling high-temperature oxidation. Yttrium, for example, is known to be an effective additive to Fe–Cr–Ni alloys exposed to oxidizing atmospheres at 700–1000 °C. It is an expensive material to alloy, and the presence of intermetallic precipitates often reduces the tensile strength and ductility of the steel. Ion-implanted yttrium has been shown to be as effective as conventional bulk additions in reducing oxidation (Fig. 3), even for oxide thicknesses much greater than the range of the implanted yttrium. Ion microprobe analysis has shown that the yttrium is transported inwards at the metal–oxide interface as the oxidation proceeds. The tarnishing of copper has also been shown to be substantially affected by ion implantation. For example, Al implantation into Cu reduces the oxidation at 300 °C in dry oxygen by approximately two orders of magnitude, whereas Cu implantation increases the oxidation rate. Extensive oxidation studies of ion-implanted metals have been reviewed (Hirvonen 1980 p.257).

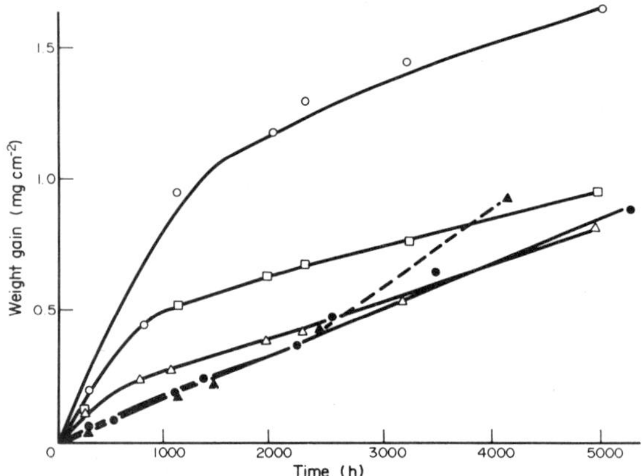

Figure 3
The effect of yttrium, either ion implanted or as an alloying addition, upon the oxidation of a 20%Ni–25% Cr, niobium-stabilized austenitic stainless steel, oxidized in CO_2 at 800 °C: ○ 20%Ni–25%Cr–0.7%Nb; □ 20%Ni–25%Cr–0.7%Nb–0.13%Y; △ 20%Ni–25% Cr–0.7%Nb–0.41%Y; ●, ▲ Y-implanted 20%Ni–25% Cr–0.7%Nb

3. *Wear, Friction, Hardness and Fatigue*

Ion implantation has been used in the important area of wear of metallic surfaces. Pin-on-disk tests on mild steel, En 40B nitriding steel and 440 C stainless steel implanted with N, B and Mo ions show reductions in wear rate by up to a factor of ten, depending on the applied load. Radiation damage effects alone cannot account for improvements in wear, since Ar and Ne implantations produce only short-lived effects, or are virtually ineffective. Many other wear tests on implanted metals have been reported. The coefficient of friction μ can be altered by ion implantation of metals. A wide range of ions implanted into various steels produce, in most cases, a reduction in μ, and occasionally an increase.

The hardness of a metal surface plays an important role in determining wear behavior. Measurements of the hardness of ion-implanted metal surfaces have been attempted. However, a fundamental difficulty in making these measurements is the shallowness of the implanted layer compared with the penetration depth of an indenter under conventional loads. Nevertheless, measurements that have been made suggest that ion implantation does affect the hardness of metal surfaces; boron implantation of beryllium, for example, has produced a microhardness value of four times that of the nonimplanted surface.

Fatigue is a further area in which ion implantation has been shown to be effective. The critical event in fatigue failure is the initiation of a surface crack that subsequently propagates into the bulk. The incorporation of selected ions by implantation into the surface layers of metals has been shown to be beneficial in inhibiting crack initiation. Implantation of B into copper gives a 60% increase in lifetime before fatigue failure.

4. *Catalysis and Electrocatalysis*

Heterogeneous catalysis and electrocatalysis are processes that depend on the structure and composition of the parasurface layers of metals and compounds. A small number of investigations suggest that ion implantation plays a role in both these fields. The hydrogen evolution reaction is of some importance for fuel-cell technology, and has been investigated on ion-implanted electrodes; Pt implantation of pyrocarbon and tungsten oxide targets, for example, results in an improved rate of hydrogen evolution. Implantation of Pt into titanium has been shown to produce an electrocatalyst for the oxidation of Cl^- to Cl_2 gas. A few papers show that heterogeneous catalysis can also be remarkably improved by implantation.

The above selection of published data indicates the wide variety of areas in which ion implantation is currently being used. Significant changes in a number of important surface-related properties have been reported, although the mechanisms underlying these effects are not well understood. The near future may therefore see ion implantation appear as a viable commercial coating technique, particularly for the tribological fields of wear, friction and hardness. Beneficial effects can be obtained in these fields using N implantation, and it is a relatively easy task to produce intense ion beams of nitrogen for the fast processing that is required commercially.

See also: Corrosion Protective Coatings for Metals; Ion Implantation Metallurgy

Bibliography

Antill J E, Bennett M J, Dearnaley G, Fern F H, Goode P D, Turner J F 1974 The effects of yttrium ion implantation upon the oxidation behavior of an austenitic stainless steel. In: Crowder B L (ed.) 1974 *Ion Implantation of Semiconductors and Other Materials.* Plenum Press, New York, pp. 415–22

Ashworth V, Grant W A, Procter R P M 1982 *Ion Implantation into Metals.* Pergamon, Oxford

Grant W A 1978 Amorphous metals and ion implantation. *J. Vac. Sci. Technol.* 15: 1644–49

Grant W A, Williams J S 1976 The modification of surface layers by ion implantation. *Sci. Prog. (Oxford)* 63: 27–64

Hirvonen J K (ed.) 1980 *Treatise on Materials Science and Technology,* Vol. 18, *Ion Implantation.* Academic Press, London

Preece C M, Kaufmann E N, Staudinger A, Buene L 1980 The structure and properties of boron implanted copper and nickel. In: Preece C M, Hirvonen J K (eds.) 1980 *Ion Implantation Metallurgy.* The Metallurgical Society of AIME, New York, pp. 77–91

W. A. Grant; R. P. M. Procter; V. Ashworth

Ionic Transport in Glass

Self-diffusion of network modifiers, especially those of the alkali ions, gives insight into mobility within the glassy network. Self-diffusion of network formers and of oxygen sheds light on the stability of the glass structure itself. Chemical reaction processes between glasses and salt melts, aqueous solutions or vapors, or between glasses of different compositions, demonstrate the possibility of altering the glass at the surface or in bulk. Chemical diffusion and self-diffusion are related.

1. Self-Diffusion

The measurement of self-diffusion processes is possible using radioactive or stable isotopes. Most alkalis, alkaline earths and many other elements important for glass technology have suitable radioactive isotopes with sufficiently long half-lives. Some other very important elements (e.g., silicon, boron, oxygen, lithium and magnesium) do not have suitable radioisotopes. Some of these elements have two or more stable isotopes, however, that can be used for the measurement.

In principle, a diffusion experiment with radioactive isotopes can proceed as follows: one drop of an aqueous solution of $^{22}NaCl$ (total amount of $^{22}Na \approx 10^{-12}$ kg) or of another isotope compound is put on the surface of a polished glass specimen. This specimen is covered with an identical disk to prevent loss of active material and heated in a furnace. Control of the atmosphere is recommended. The specimens are then carefully cleaned and sectioned by grinding or etching away thin layers parallel to the surface. The activity of a layer or the residual activity of the sample after each layer has been ground away is measured with a scintillation counter, using a discriminator for the 1.28 MeV ^{22}Na γ rays of the emission spectrum.

Figure 1 shows an example of such a diffusion profile. It can be evaluated according to the equation

$$I(x, t) = I_0 \operatorname{erfc}[x/2(D^*t)^{1/2}] \tag{1}$$

where I is the residual activity; I_0 the residual activity at the surface of the glass specimen; x the distance coordinate in the direction of diffusion; t the diffusion time; and D^* the self-diffusion coefficient.

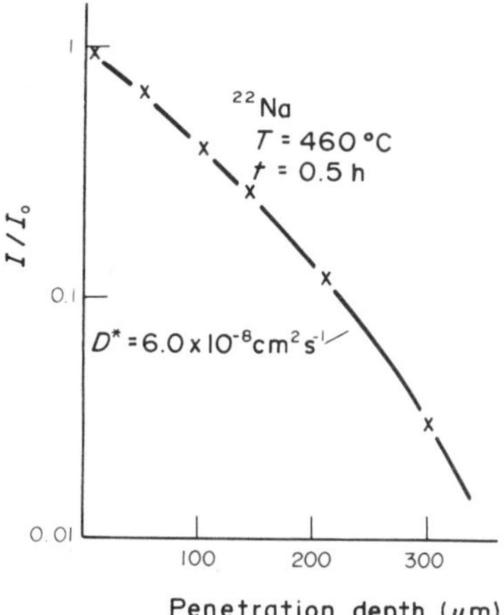

Figure 1
γ-ray residual activity following diffusion of ^{22}Na in a $Na_2O–Al_2O_3–SiO_2$ glass

The lowest D^* values obtainable in this way lie near 10^{-19} $m^2 s^{-1}$ for a diffusion anneal time of about 4 weeks. Values of the order of 10^{-22} to 10^{-23} $m^2 s^{-1}$ could be attainable if other sectioning methods were used, e.g., ion sputtering. For glasses and other insulating materials, a sputtering technique with neutralized atoms is advantageous.

Figure 2 shows some examples of sodium and calcium self-diffusion in different glasses, analyzed according to an Arrhenius-type equation such as

$$D^* = D_0 \exp(-Q/RT) \tag{2}$$

where D_0 is the frequency factor; Q the activation energy; R the gas constant; and T the absolute temperature. As can be seen, $\log D^* = f(1/T)$ displays a linear relationship at $T < T_g$ (glass-transition temperature), which may be an indication of a unique diffusion mechanism at these temperatures.

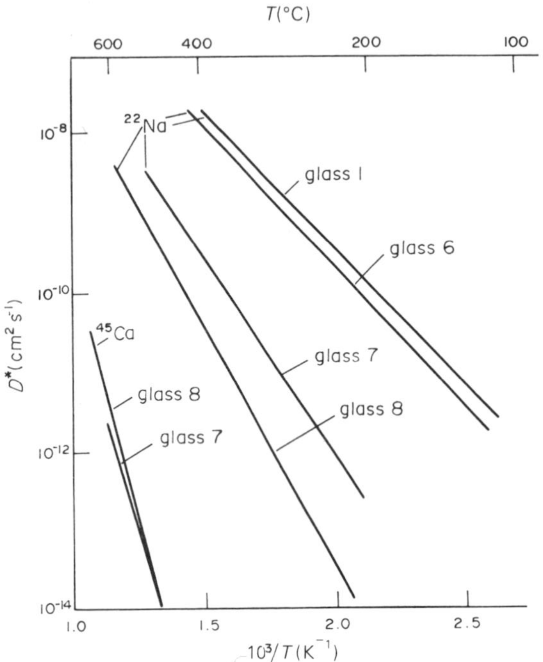

Figure 2
Arrhenius plots of ^{22}Na and ^{45}Ca diffusion in different glasses (composition in wt%). Glass 1: 34.6 Na_2O and 64.7 SiO_2, $D_0 = 2.46 \times 10^{-7}$ m^2 s^{-1}, $Q = 65.9$ kJ mol^{-1} (^{22}Na); glass 6: 29.9 Na_2O, 13.8 Al_2O_3 and 56.0 SiO_2, $D_0 = 2.75 \times 10^{-7}$ m^2 s^{-1}, $Q = 68.9$ kJ mol^{-1} (^{22}Na); glass 7: 15.8 Na_2O, 9.94 CaO and 73.9 SiO_2, $D_0 = 8.31 \times 10^{-7}$ m^2 s^{-1}, $Q = 95.8$ kJ mol^{-1} (^{22}Na), $D_0 = 5.97 \times 10^{-3}$ m^2 s^{-1}, $Q = 226$ kJ mol^{-1} (^{45}Ca); glass 8: 17.6 Na_2O, 31.7 CaO and 50.6 SiO_2, $D_0 = 3.76 \times 10^{-6}$ m^2 s^{-1}, $Q = 115$ kJ mol^{-1} (^{22}Na), $D_0 = 0.339$ m^2 s^{-1}, $Q = 236$ kJ mol^{-1} (^{45}Ca)

Sodium self-diffusion can be strongly dependent on glass composition. A relationship,

$$D_{Na}^* = a \exp(pc) \qquad (3)$$

where a and p are constants and c is the molar Na_2O concentration, has been proposed.

Substitution of SiO_2 by Al_2O_3 does not change D_{Na}^* considerably, as shown in the curves for glasses 1 and 6 in Fig. 2. Structurally, an $(SiO_{4/2})$ tetrahedron is substituted by an $(AlO_{4/2})$ tetrahedron that does not significantly change the glass structure and the sodium bonding.

The addition of divalent oxides (e.g., CaO) greatly changes D_{Na}^*, as shown in the D_{Na}^* curves for glasses 7 and 8 in Fig. 2. Structurally, this can be understood from the stabilizing action of divalent ions in glasses. By changing glass composition, it is thus possible to influence the self-diffusion coefficient of sodium and of other monovalent ions in a distinct way.

Diffusion of divalent network modifiers is much less than that of the monovalent cations, as shown by the curves for D_{Ca}^* in Fig. 2. From these data, it can be understood that calcium and other divalent cations are very tightly bonded into the glass structure. Self-diffusion of alkalis other than sodium is similar to that of sodium. However, quantitatively one must consider the different radii of these cations and also the fact that glasses with Li_2O as a network modifier are structurally slightly different from glasses with Na_2O. As far as data are available, one can state that $D_{Li}^* \gtrsim D_{Na}^* > D_K^* > D_{Rb}^* > D_{Cs}^*$. For binary glass systems, activation energies for the diffusion of all alkalis are similar; in multicomponent glasses, cesium diffusion has by far the highest activation energy, and Q values increase from about 85 for lithium to more than 200 kJ mol^{-1} for cesium. Nonsilicate oxide glasses behave in a way similar to silicate glasses.

The glasses discussed so far contain only one alkali oxide. By addition of a second alkali oxide (e.g., the glass system Na_2O–SiO_2), several properties, especially those which involve ionic transport, can change very drastically as a function of the alkali oxide ratio. Electrical conductivity, dielectric relaxation, internal friction, chemical attack and self-diffusion may be given as examples. In the glass system $(Na_2O + Rb_2O).3SiO_2$, D_{Na}^* decreases at 400 °C from about 10^{-12} to 4×10^{-15} m^2 s^{-1} with $\gamma_{Rb} = 0$–1 (where γ_{Rb} is the equivalent ratio of Rb_2O). D_{Rb}^* increases from 5×10^{-16} to 6×10^{-14} m^2 s^{-1} in the same way, and both curves intersect at $\gamma_{Rb} \simeq 0.7$. Many authors have tried to explain this "mixed-alkali" effect by a theoretical model. None of these models, however, has found general acceptance so far (Jain et al. 1984).

As already mentioned, Si and O as the most important constituents connected to the stability of the glass network itself do not have suitable radioisotopes for tracer diffusion measurements. However, both elements have two or more stable isotopes which have been used. Oxygen self-diffusion has been studied using the isotope ^{18}O and mass spectrometric analysis, silicon self-diffusion could be obtained with ^{30}Si and SIMS analysis. For oxygen diffusion in SiO_2 glass, D_0 is 4×10^{-15} m^2 s^{-1} and Q is 82 ± 17 kJ mol^{-1} (1150–1430 °C), and for silicon, D_0 is 2.9×10^{-2} m^2 s^{-1} and Q is 577 kJ mol^{-1} (1100–1400 °C) (Schaeffer 1979).

2. *Chemical Diffusion*

When an alkali-containing oxide glass is immersed into a salt bath, an ion exchange of the kind

$$(A^+)_{glass} + (B^+)_{melt} \rightleftharpoons (A^+)_{melt} + (B^+)_{glass} \qquad (4)$$

occurs. The exchange profiles in the glass can be measured by electron microprobe analysis. A similar quasi-binary exchange may also be obtained when two glasses with the cations A^+ and B^+, respectively, are contacted at their interfaces.

Figure 3 shows examples of such reactions between glasses of $Na_2O.3SiO_2$ and $Rb_2O.3SiO_2$ compositions. The plot of concentration versus $xt^{-1/2}$ demonstrates the consistence of the diffusion process. The diffusion coefficient can be evaluated by a modified Boltzmann–Matano equation

$$\tilde{D} = \frac{\left[(1-Y)\int_{-\infty}^{x(N_2)} Y/\bar{V}\,dx + Y\int_{x(N_2)}^{\infty} (1-Y)/\bar{V}\,dx\right]\bar{V}}{2t\,\dfrac{dY}{dx}} \qquad (5)$$

where Y is the normalized concentration, x the coordinate, N_2 the molar ratio of component 2, $\bar{V}$ the molar volume, t the reaction time and $\tilde{D}$ the binary interdiffusion coefficient.

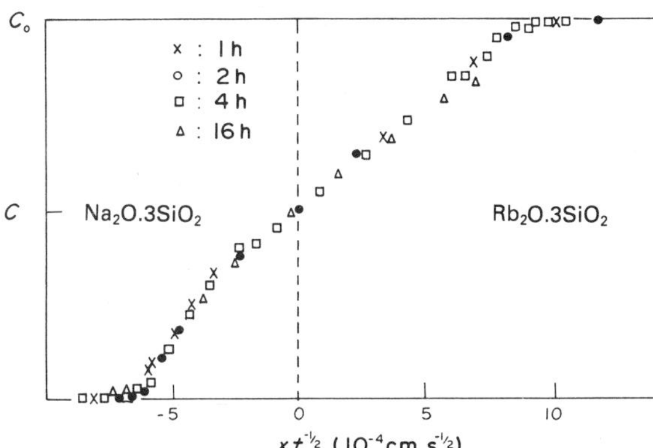

Figure 3
Concentration profiles of interdiffusion experiments at 650 °C for different times (electron-microprobe measurements)

Figure 4 displays the results of numerical evaluations of profiles at different temperatures. $\tilde{D}$ is strongly dependent on concentration and, due to the great difference in ionic radii ($r_{Na^+} = 0.098$ nm, $r_{Rb^+} = 0.148$ nm), the form of the curves demonstrates a pronounced mixed-alkali effect in the investigated glass system.

At least in part, similar results as in the case of an alkali–alkali ion exchange have been obtained with alkali–heavy metal ion exchanges. The Na^+–Ag^+ exchange ($r_{Na^+} = 0.098$ nm, $r_{Ag^+} = 0.113$ nm) also shows a kind of mixed-alkali effect. The Na^+–Cu^+ exchange ($r_{Na^+} = 0.098$ nm, $r_{Cu^+} = 0.096$ nm) does not show this effect, however, since a part of the Cu^+ is oxidized in the glass to much less mobile Cu^{2+} ions; these exchange profiles display anomalous behavior.

The ion-exchange reactions discussed so far could

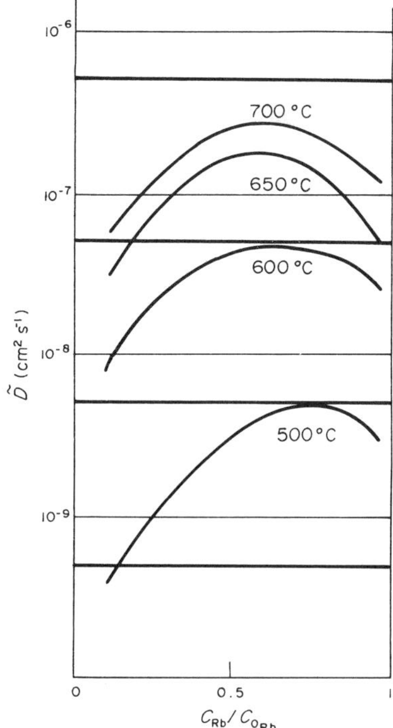

Figure 4
Dependence of the binary interdiffusion coefficient $\tilde{D}_{RbNa}$ on the concentration in the system $(Na_2O + Rb_2O).3SiO_2$

be described according to Eqn. (4) as quasi-binary processes. The interdiffusion of the dissimilar ions occurred within a rigid Si–O network. Except for a slight relaxation of this network with respect to the different sizes of the ion-exchanging cations, no changes could be found.

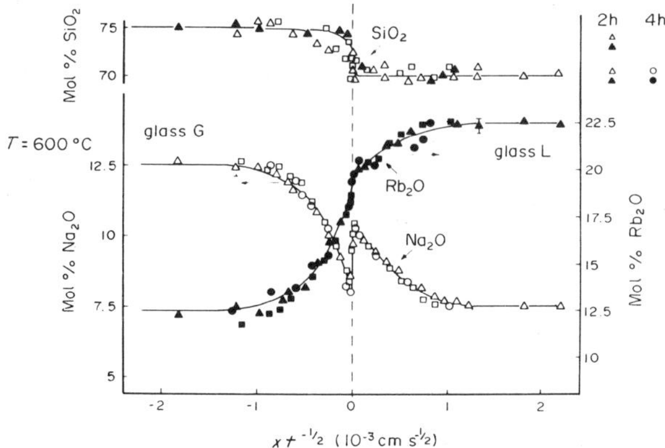

Figure 5
Reaction profiles between glass G (12.5 mol % Na_2O; 12.5% Rb_2O; 75% SiO_2) and glass L (7.5 mol % Na_2O; 22.5% Rb_2O; 70% SiO_2) at 600 °C, different times

Figure 5 shows an example of a more general reaction profile in the glass system $(Na_2O + Rb_2O).3SiO_2$. The reaction path is no longer linear but shows an s-shaped form. An evaluation of such profiles for a matrix of phenomenological interdiffusion coefficients (four in the case of this glass system) can be carried out. However, from a structural chemistry point of view such reactions may be discussed as follows. Far away from the original interface the reaction occurs as a quasi-binary interchange of Na^+ and Rb^+ ions according to Eqn. (4). Near the interface, SiO_2 also takes part in the reaction. The Na_2O profile shows up-hill diffusion. The reaction may be formulated as shown in Fig. 6.

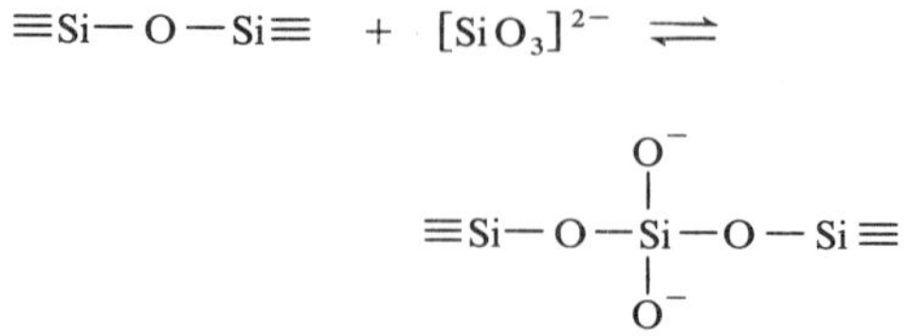

$$\equiv Si-O-Si\equiv \; + \; [SiO_3]^{2-} \rightleftharpoons \; \equiv Si-O-\underset{O^-}{\overset{O^-}{Si}}-O-Si\equiv$$

Figure 6
Ion-exchange reaction involving SiO_2 near the interface

This means the formation of nonbridging oxygens at the right-hand side which need positive charges for compensation. Na^+ ions, which are the most mobile species in this glass system, maintain this required electroneutrality.

3. Relation between Self-Diffusion and Chemical Diffusion

Relating chemical diffusion to self-diffusion by means of the Nernst–Planck model can answer the question as to what species controls the reaction process. In the case of a quasi-binary process, Si and O form a rigid network of anionic groups. Interdiffusion of the monovalent ions A^+ and B^+ occurs, then, in the holes of this network. The Nernst–Planck model for this simplest case reads

$$\tilde{D}_{AB} = \frac{D_A{}^* D_B{}^*}{(1-N_2)D_A{}^* + N_2 D_B{}^*}\, n \qquad (6)$$

where $D_i{}^*$ is the tracer diffusion coefficient of species i; N_2 the molar ratio of component 2; and n the thermodynamic factor. A test of this model according to Eqn. (6) was done by different authors for various glass systems. Values of n in the range 0.5–3.5 have been found so far, demonstrating a clear but not too strong deviation from the ideal solution case ($n = 1$). Below T_g, the influence of ion-exchange-induced stresses can also be important.

In the case of the more complex multicomponent reaction processes the Nernst–Planck model also holds, but the relation between the chemical diffusion coefficients and the self-diffusion coefficients is much less obvious.

4. Diffusion Mechanisms

As in the case of diffusion in crystalline materials, the concept of correlation effects has been applied to ionic migration in glasses. The correlation factor f accounts for the nonrandom distribution of successive jumps of a particle. This factor (or f_σ, the Haven ratio) can be calculated by the random walk theory of diffusion.

Possible diffusion mechanisms operating in solid materials are the vacancy mechanism, the direct interstitial mechanism and the indirect interstitial (interstitialcy) mechanism.

There are two possibilities of measuring this correlation factor f. One is based on the comparison of specific electrical conductivity and self-diffusion using the modified Nernst–Einstein equation, the other applies the isotope effect, if two suitable isotopes of the same element are present. Both methods were used successfully for solving problems connected with diffusion mechanisms in crystalline materials. Both methods were also applied to glasses and gave evidence that f_σ is about one for sodium diffusion in glasses with very low alkali contents, and less than one for sodium diffusion in glasses with higher alkali contents. The first case can be explained by uncorrelated jumps as for a direct interstitial mechanism, the second case is in line with a model given by Wegener and Frischat (1982) that transport may be understood as a superposition between collinear and noncollinear interstitialcy jumps. However, further experimental and theoretical evidence is necessary to solve this problem totally.

See also: Glass: An Overview

Bibliography

Doremus R H 1962 Diffusion in non-crystalline silicates. In: Mackenzie J D (ed.) 1962 *Modern Aspects of the Vitreous State*, Vol. 2. Butterworth, London, pp. 1–71

Doremus R H 1973 *Glass Science*. Wiley, New York

Frischat G H 1975 *Ionic Diffusion in Oxide Glasses*. Trans Tech Publications, Aedermannsdorf, Switzerland

Frischat G H 1979 Self diffusion in glasses. *Glastech. Ber.* 52: 143–54

Jain H, Downing H L, Peterson N L 1984 The mixed alkali effect in lithium-sodium borate glasses. *J. Non-Cryst. Solids* 64: 335–49

Schaeffer H A 1979 Application of stable isotopes for diffusion measurement in glasses. *Glastech. Ber.* 52: 155–63

Wegener W, Frischat G H 1972 Calculation of correlation factors for sodium self diffusion in glassy sodium disilicate. *J. Non-Cryst. Solids* 50: 253–61

G. H. Frischat

Ionizing Radiation: Acute Effects

Accidental exposures to ionizing radiation sufficient to cause detectable acute effects in humans are rare; since 1945 less than 200 such events involving fewer than 1000 persons have occurred worldwide. The most common cause of acute radiation-induced effects is exposure to external sources of radiation (x-ray machines, accelerators, large gamma-ray irradiators, industrial radiography sources). An individual thus exposed is not radioactive and poses no radiological threat to others. Acute effects are induced less frequently by radioactive materials deposited on skin surfaces or clothing (external contamination), and only rarely by the incorporation of radioactive materials (internal contamination). The severity of acute effects caused by such exposures is related directly to the radiation dose and sometimes to the chemical nature of the contaminant.

1. Radiation Sickness

External irradiation of the entire body sufficient to produce a dose of at least 1 Gy (100 rad) to the bone marrow causes no remarkable sensations or signs until about 2–12 h later, when nausea and vomiting can develop. A bone marrow dose of about 2 Gy is required for a 50% chance of symptoms developing, and a dose of more than 3 Gy is needed to cause radiation sickness within 6 h in everyone so exposed: the interval between exposure and the onset of symptoms tends to be shorter at the higher doses. These symptoms disappear spontaneously within 24–48 h irrespective of dose. Irradiation of the abdomen or head alone also may be followed by symptoms of radiation sickness, but exposures of extremities to large doses of radiation, sufficient to cause severe local burns, typically are not followed by symptoms of gastrointestinal distress. In addition to nausea and vomiting, there may be clinical signs and other symptoms such as loss of appetite, diarrhea, tiredness and apathy, collectively called the prodromal radiation syndrome, the extent of which roughly predicts the severity of subsequent systemic effects. The persistence of the prodromal syndrome portends death within 10–12 days. On the other hand, the disappearance of the symptoms does not necessarily mean that serious systemic damage has not occurred.

2. Bone-Marrow Damage

If the dose of radiation absorbed by the bone marrow is between 1.5 and 10 Gy, then a characteristic pattern of disease called the hematopoietic syndrome develops. Following the 1–2 day period during which symptoms of the prodromal syndrome may develop, there is a more-or-less symptom-free or latent period of 2–3 weeks before the disease, dominated by infection and/or hemorrhage as a result of radiation damage to blood-forming cells, develops. The treatment of severe radiation-induced bone-marrow damage includes isolation of the patient, antibiotics and transfusions of platelets and white blood cells; occasionally a bone-marrow transplantation can be performed. If death is to occur, it is to be expected between 30 and 60 days after the exposure. If, however, the patient can be kept alive for 5–6 weeks, bone-marrow regeneration will be established and the blood counts (except lymphocytes) will soon approach their preexposure levels. The hematopoietic syndrome is the one clinical entity of the acute radiation syndrome in which survival can be expected, especially with adequate treatment. Basically, whole-body irradiation—unless associated with, for example, severe trauma, burns or radioactive contamination—does not constitute a medical emergency. However, it is important to gather dosimetry information and make laboratory and clinical observations early on, for prognostic purposes and treatment planning.

3. Lethal Mechanisms of Radiation Syndrome

While many different mechanisms of radiation-induced death have been described in experimental animals, only three mechanisms have been described in human cases, following accidental whole-body irradiation; these mechanisms are bone-marrow damage, serum electrolyte loss, and shock from vascular damage that results in the leakage of fluids into the extravascular spaces.

The consequences of bone-marrow damage are insufficient white blood cells to combat infection and insufficient platelets (less than 20 000 ml^{-3}) to prevent hemorrhage. As a result a radiation "sore throat" or extremely painful "agranulocytic angina" (a hemorrhagic ulcerative, paratonsillar pharyngitis) develops.

Whole-body doses of 10 Gy or more cause variable degrees of vascular damage, which leads (first) to loss of sodium ions from the blood, and in more severe cases to loss of blood fluids into extravascular spaces. At high levels (50–100 Gy) the extravasation of fluid into brain tissue causes such severe cerebral edema that convulsions and coma develop rapidly.

Untreated high-dose radiation exposures, therefore, can result in death within 30–60 days from infection and/or hemorrhage as the result of bone-marrow damage; within 2 weeks from shock due to serum electrolyte imbalance and complications from infections that are incompatible with life; and within 48 h from shock due to blood-fluid loss that produces severe edema and anoxia of brain, lungs, kidneys and other tissues.

4. Acute Local Injuries

Exposure of small areas of the body and, more typically, of the extremities to radiation from external radiation sources is currently the most frequent type of radiation accident. In these accidents, the bone-marrow dose is usually too low to cause systemic symptoms of radiation sickness, although decreases in the lymphocyte and sperm counts and characteristic chromosome aberrations may be detectable. The onset of local symptoms (i.e., tingling, numbness and reddening of the skin) is delayed typically for days or even weeks, except at doses of more than several tens of grays; pain is absent immediately following exposures to all but very high doses (hundreds of grays). These characteristics are important in the differential diagnosis of burns, as they distinguish radiation-induced burns from thermal and chemical burns, which are immediately painful and develop to a maximum extent shortly after the insult.

The clinical course of local radiation injury is predictable according to dose, dose rate and the penetrating power of the radiation involved. Acute doses of less than 3 Gy locally produce little or no clinical effect. Loss of hair occurs during the second week after a dose of more than 3 Gy; the loss is temporary at doses of less than 40 Gy. Reddening of the skin occurs within minutes to weeks after an acute dose in excess of 6 Gy; the delay in the onset of symptoms is roughly related inversely to dose. Doses of less than 15 Gy are followed by erythema and edema (swelling) and bronzing of the skin, with subsequent peeling or scaling (dry desquamation); healing will occur spontaneously in most cases. The development of blisters indicates a dose of at least 15–20 Gy; with doses in excess of 30 Gy the blisters will break down (wet desquamation), forming ulcers; if healing of these injuries is to occur spontaneously, it will be quite slow; and skin grafting may be necessary to cover the more extensive lesions. High-dose damage (50–60 Gy) to the underlying blood vessels causes necrosis and dry gangrene within weeks or months; restorative surgery in areas so affected is difficult because of the compromised blood supply due to an obliterative endarteritis that may also extend for some distance beyond the necrotic area. Conservative treatment is recommended during the early clinical stages; exposed areas should be kept clean and protected from additional trauma; specifically, surgical intervention should be avoided during these early stages if the doses are estimated to exceed 20 Gy.

Repeated exposure of skin to radiation at doses of a few gray (but insufficient to cause acute injuries) in time may result in the development of chronic dermatitis, which heals only slowly and possibly incompletely. The affected skin is thin, scaly, has a tendency to crack, is easily injured and if untreated the condition is predisposed to the development of skin cancer after several years.

5. Internal Contamination

Although it is possible to have an acute accidental intake of a quantity of radioactive material, sufficient to give an internal high radiation dose resulting in the acute radiation syndrome, experience has shown this to be a rare occurrence. Typically, the quantities of radionuclides that are incorporated are small, and the resulting radiation doses delivered to tissues internally are proportionately small; consequently, the effects are delayed or more chronic in nature. However, some radioactive elements accumulate very rapidly in a specific site or target organ, and therefore can have a significant acute radiation effect on a particular organ. High doses of radioactive iodine may destroy the thyroid gland, radioactive phosphorus having bone as its target site might damage or destroy the radiosensitive cells of the bone marrow within the bone, or some ingested insoluble radioactive compounds may cause radiation damage to the intestine. In the USA, the only experience with large internal exposures is limited to a few cases of medical misadministrations of radiopharmaceuticals, and a small number of radiation accidents in industry and research. Decorporation (removal from the body) of internal contamination can be achieved, especially when certain treatments such as blocking or displacement therapy with stable isotopes, prevention of absorption, accelerated excretion, or chelation therapy are instituted soon after the uptake of radioactive material.

Depending on the chemical properties of a radioactive compound, the acute effects may be due to its chemical toxicity rather than the radiotoxicity. Examples are acute damage to the kidney from ingested uranium, or pulmonary edema from inhaling of radioactive halogens. The treatment of this type of early effect focuses on the chemical/toxic nature of the problem. Internal depositions are confirmed and quantified by bioassay for radioactivity in urine, feces and breath samples. The total amount of radioactivity taken into the body can be estimated and a total body or critical organ dose can then be calculated. A special type of bioassay, a whole-body count, is done by placing the person in a whole-body counter. Whole-body counters are radiation counters usually designed to determine only the amount of gamma-emitting radionuclides within the body.

Internally deposited radionuclides irradiate surrounding tissues continuously; although the doses to the tissues may be insufficient to cause acute effects, clinically detectable diseases can occur in target tissues at some time in the future. In the absence of acute injuries, the emergency treatment of persons contaminated with radionuclides should be directed towards the prevention or reduction of incorporation and internal deposition of the radionuclides in order to reduce the risk of late effects.

There are other acute biological effects of ionizing radiation that do not follow a set clinical pattern, but should be mentioned in this brief review. Examples of this type of acute somatic effect are temporary radiation-induced infertility (0.25 Gy whole-body dose to males, 0.5 Gy to females), reduction of thyroid function (threshold dose about 1 Gy penetrating radiation from an external source, and 0.2 Gy from internally deposited ^{131}I), and embryonic and fetal abnormalities after radiation exposure of the mother (effects have been reported for 0.05 Gy doses received *in utero*).

See also: Ionizing Radiation: Genetic Effects in Humans; Ionizing Radiation: Late Somatic Effects in Humans; Radioactive Materials: Safe Handling; Hazards and Materials: An Overview

Acknowledgement

This article is based on work performed under contract DE-ACO5-76OR00033 between the US Department of Energy and Oak Ridge Associated Universities.

Bibliography

Casarett G W 1980 *Radiation Histopathology*, Vols. I and II. CRC Press, Boca Raton, Florida

Hübner K F, Fry S A (eds.) 1980 *REAC/TS International Conference: The Medical Basis for Radiation Accident Preparedness, 1979.* Elsevier, New York

International Atomic Energy Agency 1978 *Manual on Early Medical Treatment of Possible Radiation Injury*, IAEA Safety Series No. 47. International Atomic Energy Agency, Vienna

National Council on Radiation Protection and Measurements 1980 *Management of Persons Accidentally Contaminated with Radionuclides*, NCRP Report No. 65. National Council on Radiation Protection and Measurements, Washington, DC

National Research Council, Committee on the Biological Effects of Ionizing Radiations 1980 *The Effects on Populations of Exposure to Low Levels of Ionizing Radiation: 1980*, BEIR Report III. National Academy Press, Washington, DC

United Nations Scientific Committee on the Effects of Atomic Radiation 1982 *Sources and Effects of Ionizing Radiation: 1982 Report.* United Nations, New York

C. C. Lushbaugh; K. F. Hübner; S. A. Fry; W. L. Beck

Ionizing Radiation: Genetic Effects in Humans

In humans, there is no unequivocal proof that ionizing radiation induces mutations that are transmitted to the progeny of exposed parents. Estimates of the genetic hazard of radiation are consequently based on experiments with animals, principally the laboratory mouse. Extensive information has been obtained with this animal on both of the major types of genetic damage induced by radiation: namely, gene mutations and chromosome aberrations. Studies have been performed which include the effect on mutation rate of a large array of physical and biological factors, including dose, dose rate, dose fractionation, radiation quality, sex, cell stage and age at exposure.

In humans, the germ-cell stages of primary concern are the long-lived stem-cell spermatogonia in the male and the immature arrested oocytes in the female. Maturing germ-cell stages are more sensitive to mutation induction by radiation, but the doses accumulated during their short lifespans will usually be much lower. In the event of a large single radiation exposure, the genetic risk can be reduced by postponing procreation for about six months for the male and one year for the female. This avoids transmitting any genetic damage induced in the more sensitive maturing germ-cell stages.

In mouse spermatogonia there is a marked effect of radiation dose rate over the range of approximately 100–1 R min^{-1}, a given dose at the lower dose rate being only about one-third as mutagenically effective as the same dose at the higher dose rate. This applies only to large doses. At small doses ($\leq$10 R) the response is like that from low-dose-rate irradiation, regardless of the dose rate delivered. At dose rates below 1 R min^{-1}, even down to 0.0007 R min^{-1}, there is no further reduction in mutation frequency. This indicates that there is no threshold dose rate for mutation induction in the male, and that below about 1 R min^{-1} mutational response will be linearly related to dose.

Arrested oocytes in the mouse are extremely resistant to mutation induction by radiation. There has been some question as to whether this finding can be extrapolated to women. Recent results increase the likelihood that it is valid to do so, and that genetic hazards of radiation in women will be much less than those in men. On a more conservative view, the risk in women would still be less than half that in men.

Fetal, newborn and young animals show no greater sensitivity to mutation induction than that in adults. For gene mutations and small deficiencies in spermatogonia, the relative biological effectiveness (RBE) for neutron irradiation relative to chronic gamma irradiation is approximately 20. For the two radionuclides incorporated into the body that have been investigated—namely, tritiated water and ^{239}Pu—the RBEs are 2 and 4, respectively. Arrested oocytes are as resistant to neutrons as they are to x rays and gamma rays.

A full assessment of genetic hazards requires not only information on mutation rate and the physical and biological factors that affect it, but also some knowledge of the actual anatomical and physiological damage caused by mutations. An estimate of this, now used by national and international committees, is based on the damage to one organ system (skeleton) of the mouse, observed in the offspring after parental exposure to radiation. Extrapolation to all body systems is made by using the proportion of all known human hereditary defects that affect the one system.

Taking all of the animal data for gene mutations and chromosome aberrations into consideration, the US National Academy of Sciences Advisory Committee on the Biological Effects of Ionizing Radiations concludes that, in humans, an average parental exposure of 1 rem before conception is expected to produce 5–75 additional disorders per million liveborn offspring in the first generation. To put this in perspective, the current incidence of human genetic disorders is approximately 107 000 cases per million liveborn. How well the animal data apply to humans is still an open question. The failure to detect genetic damage in the offspring of survivors of the Hiroshima and Nagasaki exposures provides some reassurance that the human risk is not underestimated by using the animal data.

See also: Ionizing Radiation: Acute Effects; Ionizing Radiation: Late Somatic Effects in Humans; Radioactive Materials: Safe Handling; Hazards and Materials: An Overview

Bibliography

National Research Council, Advisory Committee on the Biological Effects of Ionizing Radiations 1980 *The Effects on Populations of Exposure to Low Levels of Ionizing Radiation: 1980*, BEIR III Report. National Academy Press, Washington, DC

Neel J V, Kato H, Schull W J 1974 Mortality in the children of atomic bomb survivors and controls. *Genetics* 76: 311–26

Russell W L 1972 The genetic effects of radiation. In: International Atomic Energy Agency (ed.) 1972 *Peaceful Uses of Atomic Energy*, Vol. 13. International Atomic Energy Agency, Vienna, pp. 487–500

Selby P B 1979 Radiation-induced dominant skeletal mutations in mice: Mutation rate, characteristics, and usefulness in estimating genetic hazard to humans from radiation. In: Okada S, Imamura M, Terashima T, Yamaguchi H (eds.) 1979 *Radiation Research, Proc. 6th Int. Congr. Radiation Research.* Toppan Printing Company, Tokyo

W. L. Russell

Ionizing Radiation: Late Somatic Effects in Humans

Conditions or diseases that are late effects of radiation-induced damage in somatic cells of exposed individuals are clinically indistinguishable from the same conditions or diseases caused by other agents. The extent of the damage is influenced by physical factors including dose, dose rate, dose fractionation, half-life of the radionuclide and the penetrating power of the radiation. The damage can be lethal, causing cell death, or sublethal, which may result in production of abnormal cells. Subsequent expression of these cell changes can be described as non-neoplastic or neoplastic (tumorous). The latent period between initial exposure and appearance of late radiation effects varies from about one year for cataracts to several decades for some cancers.

1. Non-Neoplastic Effects

Non-neoplastic somatic effects include radiation-induced fibroatrophy and cataractogenesis; these effects are described also as threshold effects because their expression is predictable above a minimum dose. Fibroatrophy is an indirect effect of radiation-induced cell death that typically develops in lung, bone marrow, intestines, liver and kidneys following exposure to high doses of radiation (accidents, radiotherapy). Radiation-induced vascular damage and the consequently inadequate blood supply lead to death of functional tissue and its replacement by fibrous tissue. Doses in excess of 2 Gy (200 rad) of penetrating radiation to the optic lens induce the production of abnormal cells, resulting in cataract formation, typically in the posterior pole of the lens; the threshold for neutron-induced cataracts is considerably lower than 2 Gy.

2. Neoplastic Effects

Radiation-induced neoplastic somatic effects are considered to be nonthreshold; there is a probability that they can be expressed in response to any dose. In addition to physical factors, their development can be influenced by host factors including age, sex, genotype, the type of tissue irradiated and exposures to other hazardous agents. Much of the knowledge about the neoplastic effects of radiation has been derived from extensive animal studies. However, data continue to become available from studies of human populations exposed to relatively high doses of radiation as the result of (a) atomic bomb explosions (Japanese survivors, Marshallese Islanders); (b) irradiation for medical purposes (therapeutic irradiation of patients with ankylosing spondylitis, mastitis, metropathia hemorrhagica, enlarged thymus and tinea capititis, or fluoroscopic examinations of female patients with pulmonary tuberculosis); (c) occupational exposures to radiation from external sources (early radiologists) or from internally deposited radionuclides (early radium dial painters, uranium miners). Although exposure to external radiation may induce cellular changes in most types of exposed tissue, studies of exposed human populations indicate that bone marrow, breast and thyroid tissues are the most susceptible to radiation tumorigenesis. Statistically significant increases in the incidence of some types of solid tumors (breast carcinoma in females, cancers of the lung, stomach and colon, multiple myeloma, benign nodules and carcinoma of the thyroid) and leukemias (all types except chronic lymphocytic leukemia) have been observed. The ratio of solid cancers to leukemias may reach about 4:1 among populations exposed to external radiation. Leukemias may occur 2–20 years after irradiation, whereas solid tumors appear later and the risk may continue throughout the life span. Estimates of risks of mortality from radiation-induced leukemia, breast or thyroid cancer are 2×10^{-5}, 5×10^{-5} and 1×10^{-5}, respectively, per rad (1 rad = 0.01 Gy).

In the case of radiation from internally deposited radionuclides, tumor development is restricted to the sites of deposition. Statistically significant increases in the incidence of bone cancers have been noted in patients with ankylosing spondylitis treated with radium parenterally and in the early dial painters who ingested considerable quantities of radium in the course of "pointing" their brushes. Although internally deposited plutonium is highly carcinogenic in animals, there are no epidemiological data to suggest that it is equally carcinogenic in humans. Radioactive iodine deposited in the thyroid may give rise to benign nodules, or, less frequently, to cancer of the gland. An excess incidence of lung cancer due to exposure to alpha radiation from radon decay products in the air has been noted in uranium miners; a precise estimate of this risk is complicated by other carcinogens such as tobacco smoke. Studies are in progress of some populations exposed to low levels of radiation occupationally, for diagnostic purposes and as a consequence of nuclear weapons tests.

See also: Ionizing Radiation: Acute Effects; Ionizing Radiation: Genetic Effects in Humans; Radioactive Materials: Safe Handling

Acknowledgement

This article is based on work performed under contract DE-AC05-76OR00033 between the US Department of Energy and Oak Ridge Associated Universities, and research sponsored by the Office of Health and Environmental Research, US Department of Energy, under contract W-7405-eng-26 with the Union Carbide Corporation.

Bibliography

Boice J D, Fraumeni J F (eds.) 1984 *Radiation Carcinogenesis: Epidemiology and Biological Significance.* Raven Press, New York

Fullerton G D, Kopp D T, Waggener R G, Webster E W (eds.) 1980 *Biological Risks of Medical Irradiations.* American Institute of Physics, New York

Hall E J 1976 *Radiation and Life.* Pergamon, Oxford

Hübner K F, Fry S A (eds.) 1980 *REAC/TS International Conference: The Medical Basis for Radiation Accident Preparedness, 1979.* Elsevier, New York

International Atomic Energy Agency 1978 *Late Biological Effects of Ionizing Radiation,* Vols. 1 and 2. International Atomic Energy Agency, Vienna

National Council on Radiation Protection and Measurements 1977 *Medical Radiation Exposure of Pregnant and Potentially Pregnant Women,* NCRP Report No. 54. National Council on Radiation Protection and Measurements, Washington, DC

National Research Council, Advisory Committee on the Biological Effects of Ionizing Radiations 1980 *The Effects on Population of Exposure to Low Levels of Ionizing Radiation: 1980,* BEIR III Report. National Academy Press, Washington, DC

Woodward H Q 1980 *Radiation Carcinogenesis in Man: A Critical Review,* US Department of Energy Report EML-380. Memorial Sloan-Kettering Cancer Center, New York

United Nations Scientific Committee on the Effects of Atomic Radiation 1982 *Sources and Effects of Ionizing Radiation: 1982 Report.* United Nations Organization, New York

R. J. M. Fry; S. A. Fry

Ionomeric Elastomers

The incorporation into elastomeric materials of relatively low percentages (0–15 mol %) of groups bearing pendant and neutralized ionic units can modify the properties of the elastomers significantly. Specifically affected are their mechanical and thermal properties and their solvent resistance. The types of ionomeric (ion-containing) elastomers that have been investigated include the diene-, ethylene–propylene- and ethyl acrylate-based materials and ionene elastomers.

Structural studies performed on such polymers (in the class commonly known as ionomers) indicate that the ions in these materials exist in aggregates of some sort. This aggregation manifests itself in the viscoelastic properties of the ionomers as time- and temperature-dependent ionic cross-linking. The glass transition, density, melt viscosity, tensile strength, creep rate, modulus, degree of water absorption and resistance to nonpolar organic solvents are all influenced by the ionic forces. In general, the values of these properties increase with increasing ion content, the extent dependent on the counter-ion present. Elongation decreases with increasing ion content. Compared with equivalent covalently cross-linked materials, the lack of permanent vulcanization in metal-oxide-cured, carboxyl-containing rubbers greatly improves their tensile properties, tear strength, elongation and elongation at break. In addition, the ionic cross-linking allows the rubbers to be readily recovered. Residual creep, poor compression set and high stress relaxation, however, can be disadvantageous.

Stress-relaxation experiments and modulus-versus-temperature plots of ionomeric elastomers generally show a considerably broadened transition region or even a more- or less-developed rubber-like plateau between two transition regions. The viscoelastic behavior is, in fact, reminiscent of that of phase-separated block copolymers, having two different glass-transition temperatures. Most of these studies have been done on carboxylated elastomers vulcanized by metal oxide.

The unusual combination of properties—high initial modulus, high elongation at break and low permanent set—in oxide-vulcanized, carboxylated elastomers has been explained as follows. The low permanent set is attributed to the presence of very stable ionic cross-links. That those network elements, under strain, are able to relax by ionic bond interchange (rather than by rupturing, as in permanently cross-linked systems) can account for the high elongation at break. The large initial modulus may be due to larger phase-separated regions of high ionic content which act as reinforcing filler, but which fall apart under lower strain values than do the smaller ionic cross-links. Those phase-separated regions would also account for the modulus–temperature and stress-relaxation results observed.

Metal sulfonate groups appear to provide substantially stronger interactions than metal carboxylate groups. Sulfonated ethylene–propylene (EPDM) ionomers have extremely high melt viscosities. Those containing zinc cations have sufficiently reduced melt flow to make compression molding possible. Without plasticization, however, mixing, extrusion and injection molding operations are impossible. Furthermore, the compression-molded specimens have generally poor tensile properties because of their high melt viscosity and elasticity as well as their very low elongations, all of which are due to the very high level of ionic association in these materials.

The otherwise excellent physical properties can be retained, and the melt viscosity reduced, by an appropriate plasticizer of the ionic regions in these materials. Specifically, zinc-sulfonated EPDM plasticized with zinc stearate combines a marked improvement in both melt rheological and mechanical properties in a useful temperature range, to the point where injection-moldable rubbers are achieved. This result represents the most significant technological aspect of ionomeric elastomers—namely, that thermoplastic elastomers based on rubber ionomers are possible.

See also: Elastomers: An Overview

Bibliography

Eisenberg A, King M 1977 *Ion-Containing Polymers: Physical Properties and Structure*, Polymer Physics Series Vol. 2. Academic Press, New York, pp. 169–84

Eisenberg A (ed.) 1980 *Ions in Polymers*, Advances in Chemistry Series Vol. 187, Section on "Sulfonated Hydrocarbons". American Chemical Society, Washington, DC, pp. 3–66

Jenkins D K, Duck E W 1975 Carboxylated elastomers. In: Holliday L (ed.) 1975 *Ionic Polymers*. Applied Science Publishers, London, pp. 173–206

MacKnight W J, Earnest T R Jr 1981 The structure and properties of ionomers. *J. Polym. Sci., Macromol. Rev.* 16: 41–122

C. G. Bazuin; A. Eisenberg

Ionomers and Polyelectrolytes

Ionomers and polyelectrolytes can be considered as two very different classes of ion-containing polymers. The term ionomer generally refers to synthetic organic polymers of low ion content, perhaps up to 10–20 mol%. Although there is no sharp demarcation and the variation from one ionomer to another may be considerable, it may be useful to define the upper limit of ion content as that below which the material is insoluble in water. Ionomers of the lowest ion contents are soluble in the same organic solvents that dissolve the corresponding nonionic polymers. With increasing ion content, it becomes necessary to add a polar solvent to the solution, the amount required depending on the ionomer. The major component of an ionomer is nonionic units such as ethylene or styrene or any other monomer used to synthesize nonionic polymers. The minor component consists of ionic groups that may have been incorporated either through copolymerization with the major component or by chemical modification of a nonionic polymer. The ionic portion frequently is in the form of pendant carboxylic or sulfonic acid groups which are partially or completely neutralized by counterions such as sodium, zinc, or ammonium. Ionomers having positively charged components such as quaternized ammonium groups are often neutralized by halide or halide-containing ions.

The term polyelectrolyte usually refers to homopolymers of ionizable or ionic monomers, either organic or inorganic. The definition may be extended to include any ion-containing polymer of "sufficiently high" ion content. Again, partial or complete neutralization is effected through suitable counterions. Normally, because of the high ion content (and this would determine "sufficiently high"), polyelectrolytes are soluble in water. However, if they have been cross-linked to form a network system (as, for example, when used as ion-exchange resins), or if anionic homopolymer is mixed with cationic homopolymer in aqueous solution to yield stoichiometric polysalt complexes, the materials merely swell in water.

In the elucidation of the structure and properties of ionomers, the focus to date has been on the viscoelastic behavior in their bulk (solid) form. Solutions of ionomers have been receiving attention only recently. Investigation of the behavior of bulk polyelectrolytes is also scanty, whereas polyelectrolytes in aqueous solutions have been extensively studied. These solution studies have concentrated largely on chain dimensions, counterion distribution, thermodynamics, and transport properties.

1. Structure of Ionomers

The incorporation of ionic material into nonionic polymers can modify properties profoundly. In particular, there are major effects on the glass transition temperatures, transient and dynamic mechanical properties, and melt rheology. The structural explanations underlying these changes concern the development of microphase-separated regions composed largely of the ionic material. Two types of aggregates have been postulated. The first are called multiplets and contain only ionic material, specifically several ions or ion pairs. These are dispersed in the matrix and act as moderately strong, temporary, ionic crosslinks. The second type, the clusters, include nonionic material, have weaker interaction energies, and are considerably larger (of the order of 1–10 nm in radius). Due to their size, they act not only as crosslinks, but also as strongly reinforcing fillers. At sufficiently high ion contents, the presence of the clusters correlates with behavior resembling that of nonionic phase-separated block copolymers. The degree of aggregation, the amount of each type of aggregate, and the sizes and strengths of the aggregates depend particularly on the ion concentration, the dielectric constant of the nonionic component, the position and type of ionic group, the counterion type, and the degree of neutralization. For a given ion content, the largest effects are observed in ionomers containing small ions and whose nonionic portion is based on polymers of a low dielectric constant. Such a situation provides the greatest incompatibility between the ionic and nonionic regions of the system. Details of the structures of the aggregates have not yet been elucidated, although some models have been proposed and examined (MacKnight and Earnest 1981, Bazuin and Eisenberg 1981). The relative distributions of the two types of aggregates vary with ion content and with temperature.

2. Glass Transitions of Ionomers

In general, the glass transition temperature T_g of an ionomer initially rises linearly with ion content c, the value of dT_g/dc lying between 2 and 10 °C per mol%. The increase is greater when the corresponding nonionic polymer has a lower T_g and when the ions have larger q/a values (q being the charge of the counterion and a the distance between the centers of charge of the ion in the ionic group and its counterion). When T_g is plotted as a function of $c \cdot qa$ for ionomers where sufficient clustering develops, a sigmoidal rather than a linear plot is obtained. The region of the sigmoid is thought to correspond to a transition from multiplet-dominated to cluster-dominated rheological behavior.

3. Mechanical Properties of Ionomers

Simple, noncrystalline, nonionic polymers with a single relaxation mechanism (associated with the glass transition of the material) typically provide modulus against temperature and modulus against time plots having a glassy plateau that is followed first by a rapid drop in modulus values and then by a more or less developed rubbery plateau at low modulus values if the molecular weight of the polymer is sufficiently high. Time–temperature superposition is always possible for these polymers. The incorporation of ionic material broadens the transition region and causes development of a second rubber-like plateau whose modulus increases with increasing ion content. In fact, two distinct inflection points are frequently apparent, thus indicating at least two relaxation mechanisms. The additional relaxation has been attributed to the clusters. This interpretation is reinforced by the fact that time–temperature superposition characteristically fails for many ionomers beyond a certain ion concentration, the concentration depending on the system. The superposition breakdown occurs at the same ion concentration as does the curvature in the aforementioned plots of T_g against $c \cdot q/a$.

Investigations of anionic polystyrene-based ionomers indicate that, for these materials at least, the onset of thermorheological complexity is independent of the type of ionic group as well as its position relative to the backbone. Rather, the dielectric constant of the matrix is the significant factor in determining the degree of clustering and hence the failure of time–temperature superposition. Evidence indicates that the counterion type also exerts an influence.

On the other hand, the effectiveness of the ionic material in broadening and raising the modulus in the transition region of stress-relaxation curves does seem to depend on the position of the ionic group relative to the backbone, although not on the type of ionic group. This follows from the observation that a chain-carboxylated polystyrene ionomer yields higher modulus values above 6 mol% ion content than does a ring-carboxylated polystyrene ionomer at the same ion content whereas ring-sulfonated and ring-carboxylated polystyrenes behave similarly.

The significance of the type of ionic group lies in its effect on the strength of the ionic interactions. This manifests itself in particular in dynamic mechanical studies of ionomers. Simple polymers normally show one peak in the region of the glass transition when the loss tangent is plotted as a function of temperature. For ionomers, two such peaks appear. The lower temperature peak has been ascribed to the glass transition of the matrix. It moves to higher temperatures, but decreases in intensity, with increasing ion content. The higher temperature peak, associated with the "glass transition" of the clusters, also moves to higher temperatures, but increases in intensity, with increasing ion content. The type of ionic group has no effect on the low-temperature peak; however, a sulfonated ionomer will cause the high-temperature peak to appear at higher temperatures than will a carboxylated ionomer of the same backbone, counterion and ion content. Neither peak position is affected by the distance of the ionic group from the backbone. Thus, the sulfonated ionomers apparently allow for stronger interactions within the clusters than do carboxylated ionomers, irrespective of ionic group location.

4. *Other Properties of Ionomers*

Melt strengths and melt viscosities are much greater in ionomers than in the corresponding nonionic polymers. Again, the strengths of the ionic interactions are significant; sulfonated polystyrene ionomers yielding considerably higher melt viscosities than their carboxylated counterparts.

Ion introduction in crystalline polymers leads to more complicated behavior, particularly since the forces leading to ion aggregation and those causing crystallinity compete to some extent. In some cases, the presence of crystallinity dominates the properties and ion introduction serves to reduce that effect (for example, by decreasing the T_g and modulus values). In other cases, ion aggregation is dominant, and the presence of crystallinity reduces but does not eliminate the effect of ion aggregation.

Regarding solvent interactions, since an ionomer contains both largely ionic and largely nonionic regions, it becomes possible to plasticize preferentially either the backbone material or the ionic domains of the polymer. Solvents of low polarity can be used as backbone plasticizers; these lower viscosities and T_g values in the same manner as plasticizers for simple polymers. Sufficiently polar solvents, on the other hand, in predominantly affecting ionic interactions, are dramatically more effective in reducing ionomer viscosities, and at the same time have considerably less influence on the (matrix) T_g than do less polar solvents. Dual plasticization based on the above principles, using glycerol and dioctyl phthalate to plasticize sodium sulfonated polystyrene, has been shown to produce a material with properties similar to plasticized poly(vinyl chloride) (Eisenberg 1980).

From dilute solution studies, it has been found that sulfonated ionomers dissolved in a hydrocarbon solution with a small amount of polar cosolvent can yield curves where the viscosity actually increases with temperature or is constant over wide temperature ranges (Eisenberg 1980, Lundberg and Makowski 1980). This behavior has been interpreted, to a first approximation, as an equilibrium between a species where the sulfonate groups are solvated by the polar cosolvent (favored at low temperature) and an aggregated species with cosolvent (favored at high temperature). An investigation of sulfonated polystyrene ionomers in solvents spanning a range of polarities uncovered behavior that varies from ion-pair association in nonpolar solvents to classic polyelectrolyte behavior in sufficiently polar solvents (Lundberg and Phillips 1982).

5. *Bulk and Plasticized Polyelectrolytes*

The effect of the ionic forces on the glass transition and viscoelastic properties of solid polyelectrolytes parallels that observed for ionomers, the magnitude of the changes simply being greater. In fact, it was work with linear bulk polyphosphates that first led to the relationship $T_g \propto q/a$. Other completely ionized linear homopolymers, such as silicates and acrylates, as well as some partially neutralized ones show the same relationship. The q/a function can thus be viewed as a direct reflection of how the strength of the anion–cation interaction determines the T_g of the system by limiting segmental mobility. The viscoelastic response of the linear phosphate and silicate systems in the vicinity of T_g is very similar to that of rheologically simple polymers (that is, in the appearance of a single relaxation mechanism) and hence the success of time–temperature superposition.

Organic polyelectrolytes such as poly(acrylic acid) salts cannot be studied in the unplasticized state because they decompose near their T_g values, which tend to be very high. There is, moreover, a strong tendency for microphase separation due to the thermodynamically dissimilar components of the material. The extent of the aggregation does not vary significantly with the type and amount of plasticizer. However, as evidenced by the decrease in the glassy moduli and in the breadth of the transition region with increasing plasticizer content, the stability of the polymer superstructure does vary widely depending on the plasticizer. This variation indicates the importance of the plasticizer in influencing the lifetimes of the ionic clusters.

6. *Polyelectrolytes in Solution*

Typically, in dilute aqueous solutions of linear, flexible, fully ionized polyelectrolytes, the mutual repulsion of the fixed charges results in very extended chain conformations compared to neutral polymers in good solvents. The expansion increases with dilution, since the greater volume allows the counterions to distribute themselves at larger distances from the polyion and thus shield the fixed charges less, thereby leading to still greater mutual repulsion of the charges. This phenomenon is reflected in the often sharp increases observed in plots of reduced viscosities with dilution, a maximum appearing at very low concentration. The maximum must appear since there is a limit to chain expansion, and further dilution therefore shows the effects only of decreasing interchain interactions. In extremely dilute solutions, where intermolecular interactions cease to exist, the solution must be viewed as one consisting of small regions of high local charge density to which the counterions are strongly attracted and of a large, continuous region of low ion concentration.

Besides concentration, the pH of the solution or degree of ionization, the interaction with the counterions, and the presence of low-molecular-weight salts strongly affect the shape and dimensions of the polyelectrolyte chains. With increasing charge, the shape of the flexible chain changes from a more contracted to a fully extended random coil. The addition of low-molecular-weight salts causes chain contraction, again through shielding of the electrostatic repulsive interactions. Recent studies have given evidence of local order in dilute polyelectrolyte solutions; this has suggested that strong attractive forces between the macroions and counterions are at least as important as the repulsive effects between like charges in these solutions (Ise and Okubo 1980).

Unusual polyelectrolyte behavior has been observed when hydrophobic groups form part of the polyelectrolyte molecule, such as in the form of side chains. For certain of these materials, at low degrees of ionization or low pH, the polymer chains exhibit compact conformations and form hydrophobic aggregates or microdomains. Because such aggregation forms structures analogous to soap micelles, these materials are sometimes termed polysoaps. As the degree of ionization or pH is increased, some of these materials will undergo a reversible transition to random-coil conformation. This is evidenced, for instance, by a steep rise in the viscosity curve in that region, thus reflecting the dynamics of hydrophobic attractive interactions versus electrostatic repulsive forces. The phenomenon has been observed, in particular, with copolymers of maleic acid and *n*-alkyl vinyl ethers of varying alkyl lengths. As the solution pH is increased, the copolymers with small ($n < 4$) and large ($n > 8$) alkyl groups behave as random coils and compact structures, respectively, over the entire pH range. Copolymers with intermediate size alkyl groups, in contrast, show the transition from compact to extended conformations with increasing pH (Strauss and Schlesinger 1978). It is apparent, then, that polyelectrolyte structures and properties are extremely sensitive to both their chemical makeup and the microenvironment.

There has been a great deal of work in treating theoretically the behavior of polyelectrolytes in solution (Rice and Nagasawa 1961, Oosawa 1971). The greatest difficulties concern the complexities caused by the interdependence of polyion expansion and the distribution of the mobile ions in the vicinity of the charged chain. Regarding the latter aspect, counterions have been classified in three categories. The first are the unbound counterions moving freely outside the regions occupied by the macroions. The second are those that are bound by the "ionic atmosphere" of the macroions but still have mobility, albeit decreased. Finally, there are those ions which are specifically bound to individual charged groups of the macroion; this feature is often referred to as counterion condensation or site-binding. The nature and extent of counterion binding is dependent on the properties of both the fixed ionic group and the counterions, specifically on factors such as charge or charge density distribution, size, polarizability, availability of electrons and/or orbitals for covalent contributions, and, in some cases, the presence of hydrophobic groups. The interactions depend as well on the overall charge of the polyion and the possibility of cooperative binding sites and their location relative to the polymer contour.

See also: Ionomeric Elastomers; Polymers: An Overview of Structure, Properties and Structure–Property Relations

Bibliography

Bazuin C G, Eisenberg A 1981 Modification of polymer properties through ion incorporation. *Ind. Eng. Chem. Prod. Res. Dev.* 20: 271–86

Eisenberg A (ed.) 1980 *Ions in Polymers*, Advances in Chemistry Series, Vol. 187. American Chemical Society, Washington, DC

Eisenberg A, King M 1977 *Ion-containing Polymers, Physical Properties and Structure*, Polymer Physics Series, Vol. 2. Academic Press, New York

Holliday L (ed.) 1975 *Ionic Polymers*. Applied Science, London

Ise N, Okubo T 1980 "Ordered" distribution of electrically charged solutes in dilute solutions. *Acc. Chem. Res.* 13: 303–09

Lundberg R D, Makowski H S 1980 Solution behavior of ionomers. I. Metal sulfonate ionomers in mixed solvents. *J. Polym. Sci., Polym. Phys. Ed.* 18: 1821–36

Lundberg R D, Phillips R R 1982 Solution behavior of metal sulfonate ionomers. II. Effects of solvents. *J. Polym. Sci., Polym. Phys. Ed.* 20: 1143–54

MacKnight W J, Earnest T R Jr 1981 The structure and properties of ionomers. *J. Polym. Sci., Macromol. Rev.* 16: 41–122

Morawetz H 1975 *Macromolecules in Solution*, 2nd edn. Wiley, New York, pp. 344–96

Oosawa F 1971 *Polyelectrolytes*. Dekker, New York

Rice S A, Nagasawa M 1961 *Polyelectrolyte Solutions: A Theoretical Introduction*. Academic Press, New York

Sélégny E, Mandel M, Strauss U P (eds.) 1974 *Polyelectrolytes, Charged and Reactive Polymers*, Vol. 1. Reidel, Dordrecht, Holland

Strauss U P, Schlesinger M S 1978 Effects of alkyl group

size and counterion type on the behavior of copolymers of maleic anhydride and alkyl vinyl ethers. 2. Fluorescence of dansylated copolymers. *J. Phys. Chem.* 82: 1627–32

Wilson A D, Prosser H J (eds.) 1983 *Developments in Ionic Polymers*, Vol. 1. Applied Science, Barking, UK

C. G. Bazuin; A. Eisenberg

Iron and Low-Carbon Steels as Magnetic Materials

Iron and low-carbon steels are generally used as soft magnetic materials because of the ease with which they magnetize and demagnetize under the influence of a small magnetic field. These materials, low-carbon steels in particular, find extensive applications in small ac motors and other electrical machinery. The consumption rate for such steels is estimated to be about one million tons per year in the USA, which constitutes approximately 57% of the total electrical steel consumed. Rising energy costs have prompted the motor industry to produce energy-efficient motors by utilizing improved design and better-quality steels. Superior properties in steels are obtained with the proper selection of composition and processing. This has been made possible through advances in steelmaking technology and a better understanding of the relationship between the structure and properties of steels.

1. Basic Considerations

For dc applications, a soft magnetic material should possess as high a saturation magnetization as possible, and at the same time be able to reach the saturation value with a minimum magnetizing force. The performance of the material is usually described by quoting permeability, $\mu = B/H$, where H is the magnetizing force required to attain a particular induction level B. Iron is generally considered to be a suitable candidate for such purposes. However, in ac applications, low-carbon steels with high permeability and low core loss are required to reduce the total energy loss.

2. Permeability

Permeability improvements at inductions below the "knee" of the B/H (hysteresis) curve are obtained with large grain sizes and elimination of inclusion-forming impurities (such as carbon, nitrogen, oxygen and sulfur), because they interfere with the domain-wall movements. In the knee region, favorable crystallographic orientations are also required to achieve improved permeability because the magnetization process is controlled by both domain-wall motion and rotation. However, similar benefits at high inductions are only realized through a superior texture. Such a texture is generally characterized by the preponderance of the {100} and {110} components at the expense of {111} and {211} orientations. The simple reason for this is the presence of the easiest direction of magnetization ⟨100⟩. Based on this, an empirical texture-related parameter has been developed to determine the quantitative effects of textural changes on the permeability of low-carbon steels (Rastogi 1975). Improved textures in steels are developed with the selection of appropriate alloying elements and/or processing (Rastogi 1977).

3. Core Loss

According to the classical model, the total ac core loss P_T is composed of the hysteresis loss P_H, classical eddy current P_E, and anomalous or unaccounted loss P_A, and is described as

$$P_A = P_T - (P_H + P_E)$$

The calculation of P_E assumes complete flux penetration and constant permeability, which is not true for ferromagnetic materials because of the variation in permeability with magnetizing force. This apparently results in the underestimation of the eddy-current loss, and the observed discrepancy is represented by the anomalous loss. The ratio of the true eddy-current loss $(P_T - P_H)$ to the classical eddy-current loss P_E is called the anomaly factor, and is generally greater than one for low-carbon steels (Rastogi 1976, 1977), except when the domain size is extremely small. The reduction in P_E at a given test frequency and induction is readily accomplished by decreasing the sheet thickness and increasing the steel resistivity with the addition of alloying elements. Larger grain size, lower volume fraction of inclusions and favorable orientations are required to achieve improvements in the hysteresis loss P_H. Therefore, it is evident that at a particular frequency and induction P_H and P_E should be kept as low as possible to effect significant improvements in the core loss.

An empirical relationship (Rastogi and Shapiro 1973) has also been used to predict the relative importance of different parameters on the 1.5 T core loss, which is defined as

$$P_T = \frac{11.33\, t^{1.35}\, G^{0.065}}{(B_{30}/B_s)^{3.09}\, \rho^{0.40}}\ \mathrm{W\,kg^{-1}}$$

where t is thickness (m), G is the average number of grain-boundary intercepts per millimeter, ρ is resistivity (μΩ cm), and (B_{30}/B_S) is a texture-related parameter.

4. Effect of Processing and Composition

Magnetic properties of low-carbon steels (also known as lamination steels) are controlled by both processing and composition. Advances in steelmaking technology have led to the development of cleaner products with low levels of carbon and nitrogen. In addition, continuous casting is being heavily favored over ingot casting for producing more uniform products at lower cost. The processing scheme for most lamination steels normally involves hot rolling, pickling, cold rolling, annealing and temper rolling prior to decarburization and testing. Both batch-annealing and continuous-annealing processes are employed in the production of lamination steels. In many instances, continuous annealing is preferred to batch annealing to improve permeability in the final product through the formation of a more favorable texture. In the development of improved steels, the interactive effects of composition and processing are also considered.

Commercially pure iron is often used for electromagnets and other dc devices (see Table 1). Ideally, the highest saturation magnetization and permeability in iron is obtained with the removal of impurities (carbon, nitrogen, oxygen, sulphur). This can be achieved by annealing iron at high temperatures (1200–1300 °C) in a pure dry hydrogen atmosphere. The elimination of these impurities also favors grain growth, which is conducive to lower hysteresis loss. Since such a process is commercially difficult and very expensive, anneals for iron sheets are usually performed at ~800 °C to achieve a carbon level ≤0.005 wt%. The detrimental effects arising from nitrogen aging are minimized by controlling its level in steels to below 0.006 wt%. In comparison, the adverse effects resulting from the oxide and sulfide particles are less pronounced because of large particle sizes. Stabilized low-carbon steels are also used as inexpensive substitutes for pure iron in certain applications. Despite respectable permeability for nominally pure iron, it is not considered suitable for ac devices due to the high core loss associated with the low resistivity.

Table 1
Magnetic properties of iron (*Metals Handbook* 1981)

Grade	Maximum permeability	Hysteresis loss at 1.0 T ($W\,kg^{-1}$)
High-purity iron (99.99%)	100 000	~0.08
Ingot iron (99.6–99.8%)	5000	~2.12

5. *Low-Carbon Steels*

Relatively low prices and good magnetic properties have led to a greater use of low-carbon steels in motors and small transformers. These are used as sheet products in three different conditions: (a) full-hard, (b) fully processed, and (c) semiprocessed. The selection of a steel for a particular application is largely based on its magnetic characteristics. Cold-rolled steel can be supplied in one of three conditions: (a) full-hard (cold-rolled condition), (b) semiprocessed (cold-rolled, annealed and temper-rolled), and (c) fully processed (annealed). Both full-hard and semiprocessed steels are always decarburized to develop optimum magnetic properties through carbon removal and grain growth. This is accomplished by annealing steels at ~790 °C for 1.25 h in an atmosphere consisting of 9% H_2 and 91% N_2 with a dew point above 13 °C. A detailed description of this process employed by lamination-steel users is given by Shapiro (1978). Most steel is purchased in the semiprocessed condition, and the remainder finds applications as full-hard or fully processed products in less critical devices. The reason for this is the better magnetic properties of the semiprocessed products compared with the other two products.

Product quality is usually measured by permeability and core-loss values at 1.5 T. Of these properties, the core loss is used as a basis for selecting a product for a specific application. The following discussion of steels supplied in each of the three conditions is confined to 0.46 mm thick products. Most low-carbon steels for electrical applications generally contain 0.04–0.06 wt% C, 0.05–0.15 wt% P, 0.35–0.80 wt% Mn, 0.006–0.25 wt% Al, 0.05–0.25 wt% Si and 0.015–0.025 wt% S. Typical properties of these steels are listed in Table 2.

Table 2
Typical properties of 0.46 mm thick low-carbon steels at 1.5 T and 60 Hz

Steel	Condition	Core loss ($W\,kg^{-1}$)	Relative permeability
Standard product (0.05% C, 0.40% Mn, 0.12% P, 0.025% S)	full hard	8.50	2000
Con-Core[a]	fully processed	9.90	
Locore N[b] (silicon steel)	fully processed	9.90	
Vacuum-degassed extra-low-carbon steel (0.006% C)	fully processed	8.40	
Locore M-47[b] (silicon steel)	fully processed	8.40	
Standard product	semiprocessed	6.60	2200
Improved product (0.04% C, ≤0.60% Mn, 0.12% P, ~0.025% S)	semiprocessed	6.10	2300
New steel[c]	semiprocessed	5.50	2300

a US Steel trade name b Republic Steel trade names c Inland product

5.1 Full-Hard Steels

A full-hard steel is a cold-rolled steel which is decarburized in the form of punched laminations to develop optimum magnet properties. A typical decarburized full-hard steel (Table 2) exhibits a 1.5 T core loss of 8.50 W kg^{-1} with a relative permeability of 2000. Recent studies at the Inland Steel Company have shown that lowering both initial carbon and sulfur levels from 0.05 to 0.02 wt% and 0.025 to 0.015 wt%, respectively, results in a 0.75 W kg^{-1} decrease in the core loss along with a modest gain in permeability. Improved properties are primarily attributed to the lower volume fraction of nonmetallic inclusions and larger grain size. Additional improvement in the core loss can also be obtained with increased levels of manganese and silicon through higher resistivity. These observations indicate that the potential for producing low-loss full-hard steel has yet to be fully exploited.

5.2 Fully Processed Steels

Fully processed steels were developed as less costly alternatives to the low- and intermediate-level silicon steels. These materials are produced on a continuous-annealing line with the final carbon level as low as economically feasible, this being a function of the initial carbon content and the annealing practice. Annealing temperatures in the range 790–820 °C are usually employed to achieve efficient decarburization and larger grain size. Core-loss values in Table 2 indicate that these steels are comparable to the silicon steels. Fully processed products are also coated for the purpose of good interlaminar resistance and improved punching.

5.3 Semiprocessed Steels

Advances in steel metallurgy have led to significant improvements in the magnetic properties of decarburized semiprocessed steels. In recent years, changes in steel composition and processing have produced steady improvements in the 1.5 T core loss of a decarburized semiprocessed steel, as shown in Table 2. This approach has led to the development by the Inland Steel Company of a lamination product with a core loss of 5.50 W kg^{-1} and relative permeability over 2000. This core loss is about 1.10 W kg^{-1} lower than the typical value of 6.60 W kg^{-1} reported for a standard product. In the development of such products, silicon and aluminum levels are usually kept below 0.25 wt%, whereas the manganese content can be up to 0.80 wt%. The composition restrictions for silicon and aluminum are designed to achieve high permeability in these steels. Further improvement in these properties can be expected with the processing of such a steel containing $\leqslant$0.015 wt% C and $\leqslant$0.010 wt% S. Besides, a lower carbon level reduces the anneal time, thereby providing economic benefits to users. It has been recently demonstrated that further refinements in steel composition and processing lead to the development of a 4.40 W kg^{-1} lamination steel with high permeability. In addition to good magnetic properties, the surface roughness of the steel should be between 5 and 12 μm to achieve a high stacking factor for improved motor performance.

6. Applications

Low-carbon steels are largely used in small motors. It is recognized that steels with low core loss and high permeability are needed to produce energy-efficient motors. However, the relative significance of these properties in terms of improving motor performance is determined by motor-design parameters. It appears that proper design modifications may be necessary to optimize product performance in a particular ac device. The use of improved and new lamination steels in energy-efficient motors should result in substantial energy savings. Because of escalating energy costs, demands for better products are expected to grow rapidly in the future. The current drive for energy conservation has created a greater awareness among steel producers and users of the impact of both material properties and design parameters on motor performance.

See also: Magnet Steels; Magnetic Materials: An Overview; Magnetism in Materials: Basic Concepts

Bibliography

Metals Handbook, 9th edn., Vol. 3, 1980 Magnetically soft materials. American Society for Metals, Metals Park, Ohio, pp. 598–603

Rastogi P K 1975 Effect of chromium on the magnetic properties and texture of non-oriented steel. *AIP Conf. Proc.* 26: 556–57

Rastogi P K 1976 Magnetic properties and texture of unalloyed non-oriented electrical steel. *AIP Conf. Proc.* 34: 61–62

Rastogi P K 1977 Effect of manganese and sulfur on the texture and magnetic properties of non-oriented steel. *IEEE Trans. Magn.* 13: 1450–80

Rastogi P K, Shapiro J M 1973 A statistical model for the total AC core loss of lamination steels. *IEEE Trans Magn.* 9: 122–24

Shapiro J M 1978 Optimizing the decarburization of punched electrical steel laminations. *Mechanical Working and Steel Processing XVI.* Iron and Steel Society (AIME), New York, pp. 56–77

P. K. Rastogi

Iron-Based Biomedical Materials

As early as 1666, Fabricius described the use of iron wire loops for wound closure, and in 1886, Hansmann reported the application of nickel-plated sheet steel to fracture fixation. However, it was not until the development of the corrosion-resistant stainless steels, between 1900 and 1915, that iron-based alloys had the potential to be successfully employed as surgical implant materials.

Steels are iron-based alloys numbering in the thousands. Plain carbon steel is converted into alloy or special steel when sufficient concentrations of other elements have been introduced to alter the properties significantly. Stainless steel is one of the best known of the special steels, and is made by the addition of at least ~12% chromium. This minimum confers tarnish and corrosion resistance in gaseous and liquid media which would attack plain carbon steels. In 1947, the American College of Surgeons first recommended two types of stainless steel for implantation, and today this series of alloys is one of the primary materials for biomedical applications.

1. Composition, Atomic Structure and Microstructure

The American Iron and Steel Institute (AISI) uses a three-digit system to separate standard grades of wrought stainless steels into four general classes, based upon composition: series 200 (Cr, Ni, Mn), series 300 (Cr, Ni), series 400 (Cr) and series 500 (low Cr). The last two digits in each series indicate type, while a letter suffix is used to indicate a modification within a type (e.g., L: extra-low carbon). Precipitation-hardenable and duplex stainless steels are frequently designated by proprietary names or trademarks. Specifications generally used for cast alloys are those adopted by the Alloy Casting Institute (ACI). The compositions and structures of representative alloys are given in Table 1.

Addition to iron of body-centered-cubic (bcc) chromium, a ferrite former, produces an alloy system which is predominantly ferritic, restricting the γ (austenite) phase to a limited field or "loop." The resulting ferritic grade stainless steels range in chromium content from ~14.5 to 27%. While no desirable hardening heat treatment is possible, these alloys can be work-hardened. Their corrosion resistance is generally superior to that of the martensitic grade.

The martensitic grade ranges from ~11 to 18% Cr, with sufficient carbon present to enlarge the γ field. Chromium concentrations are minimized to avoid ferrite and to promote the γ–α' martensite (bcc) transformation, while nickel is restricted to avoid austenite retention. Strength and hardness are therefore maximized, but at the expense of corrosion resistance.

The high chromium (~16–19%) and nickel (~6–12%) concentrations of the austenitic type 300 series produce some of the most corrosion-resistant stainless steels, which also possess high ductility. Although these steels cannot be heat-treated, they can be significantly strengthened by cold working. For stable austenitic alloys containing more than ~11% Ni (e.g., type 316), however, the transformation to an acicular martensite structure is suppressed, thereby diminishing the strengthening effect of cold working.

Carbon is the most potent solid-solution-strengthening element in austenite, but high levels are undesirable, due to the precipitation of chromium carbides ($M_{23}C_6$) in the temperature range of 450–900 °C. Precipitation is favored along grain boundaries, thereby depleting the adjacent matrix in chromium.

Table 1
Representative stainless steel compositions and structure

Alloy designation	Composition (%)[a]									Structure
	C	Mn	P	S	Si	Cr	Ni	Mo	Others	
Wrought										
302	0.15	2.00	0.045	0.03	1.00	17.0–19.0	8.0–10.0			Austenitic
303	0.15	2.00	0.20	0.15 min	1.00	17.0–19.0	8.0–10.0	0.6[b]		Austenitic
304	0.08	2.00	0.045	0.03	1.00	18.0–20.0	8.0–10.5			Austenitic
305	0.12	2.00	0.45	0.30	1.00	17.0–19.0	10.5–13.0			Austenitic
316	0.08	2.00	0.045	0.03	1.00	16.0–18.0	10.0–14.0	2.0–3.0		Austenitic
316L	0.03	2.00	0.045	0.03	1.00	16.0–18.0	10.0–14.0	2.0–3.0		Austenitic
317	0.08	2.00	0.045	0.03	1.00	18.0–20.0	11.0–15.0	3.0–4.0		Austenitic
431	0.20	1.00	0.04	0.03	1.00	15.0–17.0	1.25–2.50			Martensitic
Precipitation hardenable										
17-7PH[c]	0.09	1.00	0.04	0.03	1.00	16.0–18.0	6.5–7.75		0.75–1.5Al	Austenitic/ martensitic
Cast										
CF-8M	0.08	1.50	0.04	0.04	2.00	18.0–21.0	9.0–12.0	2.0–3.0		Austenitic/ ferritic

a Single values are maximum values unless otherwise noted b Optional c Trademark of the Armco Steel Corporation

A marked depletion renders these areas susceptible to corrosion attack, as chromium is effective only when in solution. The alloy is then said to be sensitized. Precipitation is suppressed by quenching the alloy from the annealing temperature and is also further counteracted by use of extra-low-carbon alloys. Vacuum or electroslag remelting is now used to produce steels with extra-low carbon levels and exceptionally low nonmetallic inclusion contents. Vacuum-remelted stainless steels are frequently designated by manufacturers by a VM suffix, while the electroslag-remelted alloys are designated by the letters ESR.

Compositions of the precipitation-hardenable alloys are generally in the range ~11–18% Cr and 4–10% Ni, with the addition of such elements as aluminum, copper, titanium and molybdenum. The strengthening mechanism is believed to involve the precipitation of very fine intermetallics such as $Ni_3(Al,Ti)$ along slip planes or grain boundaries. These alloys offer the advantages of ease of fabrication in the annealed state and strengthening by simple heat treatments, while minimizing the reduction in ductility and corrosion resistance which can occur in producing steels of comparable strength levels.

2. *Fabrication and Finishing*

The methods of fabricating and finishing stainless steels have wide-ranging effects on mechanical properties and corrosion resistance, owing to such factors as recrystallization, phase transformations, cold working, carbide precipitation and modification of the surface state. Hot working, involving rolling or forging operations, refines the original cast structure (e.g., slab or billet), thereby enhancing mechanical properties. Hot-worked products can be further processed by cold finishing or cold working to raise mechanical properties or to adjust dimensions, or both. Stainless steels have excellent cold working behavior and are among the most readily fabricated and machined of the major implant alloys. Applicable forming operations include forging, drawing and extrusion. Most stainless steel implants are fabricated from cold-worked mill products and at these high strength levels they nevertheless retain considerable ductility and toughness compared with other implant alloys. This fact is utilized during surgical application in such operations as wire bending and twisting or bone plate contouring.

The production of a smooth surface enhances the service behavior of stainless steels by eliminating surface irregularities which can act as occluded corrosion sites or stress concentrators. A smooth surface is produced by either mechanical polishing and buffing or electrolytic polishing. A roughened or textured surface may also be employed in some instances in an attempt to provide improved mechanical interlocking for cemented implants. Exceptionally smooth finishes are developed with fine abrasives for the articulating surface of joint prostheses (roughness 0.025–0.05 μm). Minimum generation of defects on fine diameter wire during drawing and handling operations is especially important, as irregularities can constitute a large portion of the surface area.

Acid cleaning, to remove processing debris, is frequently referred to as a passivation treatment, since a highly oxidizing bath is used. This treatment develops a passive film of greater stability than the air-formed film. Nitric acid is normally the solution of choice, and considerable latitude exists in the selection of concentration, bath temperature and immersion time. The avoidance of disturbing the surface film during subsequent handling operations, including surgical implantation, remains a controversial subject. Although *in vitro* tests demonstrate a temporary reduction in corrosion resistance as a result of scratching or abrasion of the passivated surface, the clinical significance of this reduction is not known.

3. *Mechanical Properties*

Cold-worked, austenitic stainless steels compare favorably with other implant alloys for high-stress applications, as they possess high yield strength and good fatigue resistance. However, for more extended implantation periods, maximum corrosion resistance becomes increasingly important and therefore influences metallurgical processing. For example, forged type 316L stainless steel hip-joint prostheses are usually annealed to remove the effects of plastic deformation and variations in grain size. Larger cross-sectional areas are therefore required, to compensate for the reduction in yield strength and fatigue resistance caused by the annealing process.

Fatigue is generally believed to be the most common cause of fracture in orthopedic implants, while pure overload fracture rarely occurs. For stainless steels, it is frequently observed in engineering applications that increasing tensile strength and pitting resistance, and decreasing grain size, improve fatigue resistance. Retrieval analyses of implants reveal that fatigue failure of all alloys is primarily associated with material defects, implant design, unstable internal fixation, delayed or absence of union and premature weight-bearing. These categories are not mutually exclusive. Considerable disagreement continues as to whether these types of failure may be environmentally promoted by corrosion.

For joint prostheses, metal-on-metal stainless steel counterfaces are not used, because of high wear rates and galling. The total hip-joint prosthesis, for example, utilizes a type 316L stainless steel femoral component which articulates on an ultrahigh-molecular-weight polyethylene (UHMWPE) socket. *In vitro* and *in vivo* wear rates and coefficients of friction are low, and compare favorably with other alloy–UHMWPE pairs.

In terms of mechanical biocompatibility, attention has been drawn to complications arising from the high rigidity of metallic implants when used in stress-bearing applications involving bone plates, intramedullary rods or joint prostheses. This rigidity is exemplified by Young's modulus, which for stainless steel is approximately ten times the value for compact cortical bone from the femur. In the case of bone plate fixation, for example, it appears that at later stages a nonphysiological condition results, in which the healing site is overprotected from stress. Consequently, during long-term maintenance, the biomechanical feedback required for the remodelling phase may be suppressed as a significant portion of functional loading is borne by the more rigid plate, to the detriment of the healing site. Efforts to overcome this problem include a hollowed plate design which retains the high bending and torsional stiffness of the traditional solid plate, while possessing a low axial stiffness.

4. Chemical Properties

The corrosion resistance of stainless steel is conferred by a very thin, passive film (1–5 nm) of low ionic conductivity. The film is generally considered to be composed of hydrous oxides, enriched primarily with chromium relative to the bulk composition. Its protective capacity is reduced at microstructural inhomogeneities such as inclusions, chromium carbides, second phases and grain boundary precipitates, and by the effects of cold working.

General corrosion or uniform dissolution of the entire alloy surface is adequately prevented by the passive film. In contrast, localized film breakdown constitutes a problem. This is related to the close proximity of the corrosion or rest potential E_r to the critical breakdown potential of the film E_c in the presence of chloride ions. Small changes in solution composition, especially at film defects, cause E_c to be exceeded. The mechanism of chloride ion attack is unknown, however. If sufficient oxygen does not reach the rupture site, the area remains anodic and localized corrosion of the underlying metal ensues. The surrounding, unattacked area is cathodic and the resulting large cathodic–anodic area ratio contributes to acceleration of the process. Types of localized attack of stainless steel include crevice, pitting, intergranular and galvanic corrosion.

The incidence of crevice corrosion is very high on multicomponent devices, particularly orthopedic bone plates and screws, between the screw and countersunk hole site. It is frequently found accompanied by fretting corrosion (Fig. 1). Although crevice corrosion is most often cited as the predominant mechanism, Brown and Simpson (1981) recently proposed, on the basis of *in vitro* and *in vivo* studies, that crevice corrosion of bone plates and screws is initiated by fretting rather than by electrochemical film breakdown. Additionally, fretting corrosion is believed to be the most important cause of corrosion products associated with joint prostheses. The other localized corrosion types are much less frequently reported. Whatever their respective frequencies, however, corrosion reactions alone are seldom reported to cause implant failure, and this includes crevice corrosion.

Figure 1
Micrograph of a retrieved tibia plate after 8 months implantation, showing burnished perimeter and corroded area (lower left), suggestive of fretting and crevice corrosion, respectively (after Brown and Simpson 1981. © Wiley, New York. Reproduced with permission)

The combination of dissimilar metals in solution (i.e., mixing metals) has generally been avoided because of the concern for promoting galvanic corrosion. However, Mears (1979) has proposed that this practice is not always appropriate, considering the good corrosion resistance of present implant alloys. On the basis of electrochemical theory, Mears views galvanic coupling primarily as an increase in the less passive metal surface area which experiences passive dissolution and cathodic reduction. Consequently, an increased incidence or rate of corrosion should not be expected. The use of dissimilar metals is desirable as a means to optimize selection for mechanical properties, such as employing higher-strength stainless steel bone screws with less rigid Ti–6%Al–4%V alloy bone plates. While supportive evidence has recently been reported for the safe mixed use of current implant alloys, some conflicting observations require that careful selection and scrutiny be continued.

The possible occurrence of environmentally assisted mechanical failures in the form of corrosion fatigue or stress-corrosion cracking remains controversial. While some retrieval analyses of load-bearing implants have purported to show that corrosion fatigue was the major mechanism operating to cause fracture, other reports have claimed that pure fatigue was the predominant fracture mode and that corrosion, even when present, did not contribute to the fracture process.

Stress-corrosion cracking of austenitic stainless steels is believed to be unlikely at 70 °C, even in chloride environments which are more aggressive than *in vivo* conditions. Nevertheless, observations of this failure mode are increasingly reported and it has been suggested that some failures attributed to corrosion fatigue may have originated as stress-corrosion cracking. Metallographic examination of what may be a mixed failure mode is difficult, and the problem is compounded by the frequent fracture surface damage resulting from rubbing, pounding and additional corrosive attack after implant breakage. Resolution of this controversy therefore requires much additional research.

Conventional steam sterilization, electropolishing and passivation in nitric acid have each been shown to increase corrosion resistance *in vitro*. For passivation, enhancement of crevice corrosion resistance has also been demonstrated. Although the clinical significance of these treatments has not been established, their continued use is nevertheless warranted by the recognized susceptibility of stainless steel to localized attack.

5. *Biocompatibility*

All the alloying elements of austenitic stainless steels present in concentrations of at least 2% (iron, chromium, nickel, molybdenum and manganese) are essential in the human diet but are toxic in excess. These transition elements are strongly electropositive, exhibit variable oxidation states and can form strong complexes with inorganic and organic ligands. Of these five metals, the only specific detoxication mechanism known is for iron, involving the accumulation of iron compounds in lysosomes. For chromium, a possible mechanism is the increased formation of ribonucleoproteins in the liver. Otherwise, normal tissue levels must be maintained by homeostasis, whereby stability of the internal environment is achieved by control mechanisms which are activated by negative feedback.

Extensive clinical experience continues to show that dissolution products of austenitic stainless steels are generally well tolerated in the absence of a severe and prolonged corrosion process. Implantation frequently results in the short-term appearance of an encapsulating, fibrotic membrane, ~1 mm thick. The capsule represents a nonspecific response to the presence of a solid, nonporous foreign body, and increasing membrane thickness develops when corrosion rates are elevated above passive current densities. Histological response (e.g., severe inflammation, tissue necrosis) to gross corrosion, however, can necessitate premature implant removal. When applicable, wear particles must also be considered in terms of their size and shape, in addition to their enhanced corrosion rate due to a relatively high surface area. The following clinical and laboratory observations illustrate these points.

Winter (1974) analyzed tissue adjacent to orthopedic stainless steel devices which had been implanted in humans. The devices were plates, screws, nails, wires and joint prostheses, which are susceptible at times to severe crevice, fretting and pitting corrosion and the production of wear particles. The following observations are from cases which usually required a second operation, and therefore represent a negative selection.

Three types of deposits were optically observed. Type I deposits were small, opaque, irregularly shaped particles, 0.3 mm–0.1 μm, and were believed to be alloy wear particles from joint prostheses. Smaller alloy particles were found primarily in macrophages, while the larger particles were surrounded by multinucleate giant cells. There was no granulomatous reaction and the impregnated tissue was remarkably healthy, considering the quantity of foreign material present. Type II deposits were large platelets (sometimes green), 5.0–0.5 mm, and highly variable in size and morphology. They were identified as alloy corrosion products containing iron, chromium, phosphorus and sulfur, and were observed in 11 of 44 specimens. Type II deposits provoked a vigorous foreign body giant cell reaction, with fibrosis, and impregnated tissue frequently consisted of relatively acellular collagen or was necrotic. Winter proposed that type II deposits were cytotoxic, producing phagocytosis, cell death, fibrosis and lastly necrosis in the presence of large accumulations.

Type III deposits were somewhat spherical, yellow–brown, hemosiderin-like granules, 3–0.1 μm. Tissue analysis revealed mixtures of two or more of the iron oxides α-Fe_2O_3 and γ-Fe_2O_3 and the hydrated oxides α-FeOOH (goethite) and γ-FeOOH (lepidocrocite). No recognizable disturbance was observed when few particles were present. However, with heavy contamination, abnormalities included cytotoxicity and fibrosis or necrosis. It was proposed that type III deposits might have been the result of an iron detoxication mechanism, while in some cases, pathological changes indicated a local tissue iron overload or iron toxicity. Types II and III deposits were frequently found together, but the former were considered to be more cytotoxic.

Considering another aspect of biocompatibility, it is recognized that nickel and chromium are capable of producing delayed metal sensitivity reactions. However, the percentage of metal-sensitive individuals who actually respond is small. The proposed mechanism views the released metal ion as a hapten, which, when complexed with protein, stimulates the immune response. Because of the deep tissue situation of implant dissolution, a metal sensitivity reaction is not detectable until it has progressed to cause pain, fluid accumulation or a dermatological disorder. Reports in the literature conclude that it is primarily reactions to nickel that are clinically significant. While it appears that several months' implantation is necessary to cause sufficient dissolution, visible corrosion of the implant may not necessarily occur. Removal of the implant relieves the symptoms.

Implantation of a foreign body is believed to increase the risk of late infection. For metallic implants, a facilitated disease process may result from a mechanical barrier effect or from inhibition of host defense systems and enhancement of bacterial virulence by corrosion products. Animal and bacteriological studies have provided evidence for the potential role of alloys and individual elements. Implantation of type 316L stainless steel microspheres in rabbits has been shown to produce a significantly higher incidence of clinical infection, compared with unoperated or sham-operated controls. In another study, nickel was shown to inhibit the phagocytosis of *Staphylococcus aureus* and *Streptococcus faecalis* by macrophages *in vitro*. While Fe(II) and Fe(III) have been shown to alter the efficacy of some bactericidal mechanisms, it is recognized that decreasing iron content in the blood is a response to bacterial invaders, which deprives the organisms of this element essential to growth. Iron release from stainless steel implants may play a clinically significant role in counteracting this mechanism. For example, iron has been shown to increase the virulence of *Escherichia coli* and *Listeria monocytogenes* in guinea pigs and mice, respectively.

Chronic exposure to metal implant corrosion products introduces concern about chemical carcinogenesis, particularly since carcinogenic effects have been observed with nickel, chromium and iron. The literature records four cases of malignant neoplasms associated with metallic surgical implants in humans (Tayton 1980) and at least nine cases in canines (Harrison et al. 1976, Sinibaldi et al. 1976). In humans, two of the cases involved stainless steels (Table 2). It was reported in the first case (1924) that dissimilar metals were used and that corrosion was significant. A measured potential difference of 80 mV between the retrieved plate and screws was considered to be a source of irritation for 30 years. It should be noted also that the screws contained only 12% chromium, the minimum required for corrosion resistance. In the second case (1944), dissimilar metals were also used, but the observed corrosion was not considered significant. All nine cases in dogs

Table 2
Malignant neoplasms reported associated with metallic orthopedic implants in humans

Year; patient's age at implantation (years)	Lesion type and site	Implantation period (years)	Implantation to detection of neoplasm (years)	Histology	Implant type and alloy
1924; 12	Fracture of humerus	30	30	Ewing's sarcoma[a]	Fixation plate: Fe–18%Cr–8% Ni; screws: Fe–12%Cr
1944; 58	Fracture of tibia and fibula	26	26	Hemangioendothelomia (tibia)	Fixation plate: type 316 stainless steel; screws (8): 6 type 316, 2 type 304 stainless steel
1953; 37	Fracture of tibia	3	3	Unclassified sarcoma	Fixation plate and screws (2): unidentified metal
1969; 4	Congenitally dislocated right hip and a subluxating left hip	1	7.5	Ewing's sarcoma (femur)	Fixation plate and screws (6): cast Co–Cr–Mo

a Most probable diagnosis

reported in 1975 and 1976 involved stainless steel, and in most, significant corrosion was observed. The neoplasms comprised seven osteosarcomas and two undifferentiated sarcomas. The period between implantation and neoplasm detection was 1–11 years (average 4). These observations were of particular concern because the osteosarcomas were atypical for the canine breeds and sites, thus suggesting causality. This may be a species-specific effect.

For humans, the recorded incidence of malignant neoplasm development associated with metallic surgical implants is well below statistical expectations, and no causal relationship has been identified. Nevertheless, neoplasm induction is a long term process, lasting perhaps 7 or more years. Realization that stainless steels are highly susceptible to corrosion in multicomponent devices clearly indicates that increased research is particularly needed in this area of biocompatibility also.

6. *Applications*

Iron-based alloys currently form one of the predominant groups of metallic materials for biomedical applications. Of this group, only the corrosion-resistant stainless steels, primarily the austenitic or semiaustenitic precipitation-hardenable types, are used in situations requiring fluid contact. For implantation, and particularly in those circumstances which include multicomponent, weight-bearing or articulating devices, type 316L stainless steel is now most frequently selected. This is because of its superior corrosion resistance in addition to good mechanical properties. Introduction of vacuum and electroslag remelting has considerably improved material performance. For less rigorous environments and functions, such as some dental applications, austenitic stainless steels of lower corrosion resistance are also used, thereby permitting a wider selection for other material properties. In the absence of fluid contact, especially for limb prosthetics and orthotics, both steel and stainless steel components are utilized. A representative list of material applications is presented in Table 3. All the surgical and nonsurgical iron-based alloys have good fabricability and are therefore generally employed in the wrought and machined forms.

In the vast majority of implant applications, stainless steels perform their intended function without adverse physiological effects and have consequently achieved a large degree of success. Although they are clearly not without shortcomings in their mechanical properties for some applications, their major problem rests with crevice and fretting corrosion. For example, a great number of total hip prostheses have been inserted to date, including an estimated 80 000 during 1976 in the USA alone. A large percentage of these implants utilized stainless steel for the femoral stem. Fracture of this component is reported to account for ~0.1–0.2% of failures for all applications and is primarily due to fatigue. In contrast, despite the very high clinical success rate of millions of stainless steel fracture fixation plates, retrieval analyses record a crevice and fretting corrosion incidence in excess of 90%, and in one report 100%. In another application area, stainless steel stimulating electrodes experience marked pitting corrosion when

Table 3
Representative biomedical applications of iron-based alloys

Application	Description[a]
Bone screws and pins	Internal fixation of diaphyseal fractures of cortical bone, and metaphyseal and epiphyseal fractures of cancellous bone: screw comprised of hexagonal or Phillips recess driving head, threaded shaft and self-tapping or non-self-tapping tip; type 316L stainless steel
Onlay bone plates	Internal fixation of shaft and mandibular fractures: thin, narrow plate with slots or holes for retaining screws; type 316L stainless steel
Blade and nail bone plates	Internal fixation of fracture near the ends of weight-bearing bones: plate and nail, either single unit or multicomponent; type 316L stainless steel
Intramedullar bone nails	Internal fixation of long bones: tube or solid nail; type 316L stainless steel
Percutaneous pin bone fixation	External clamp fixation for fusion of joints and open fractures or infected non-unions: external frame supporting transfixing pins; stainless steel
Total joint prostheses	Replacement of total joints with metal and plastic components (shoulder, hip, knee, ankle, great toe): humeral, femoral (hip and knee), talus and metatarsal components; type 316L stainless steel
Wires	Internal tension band wiring of bone fragments or circumferential cerclage for comminuted or unstable shaft fractures; type 316L stainless steel
Harrington spine instrumentation	Treatment of scoliosis by application of correction forces and stabilization of treated segments: rod and hooks; type 316L stainless steel
Mandibular wire mesh prostheses	Primary reconstruction of partially resected mandible; types 316 and 316L stainless steel

Table 3—continued
Representative biomedical applications of iron-based alloys

Application	Description[a]
Fixed orthodontic appliances	Correction of malocclusion by movement of teeth: components include bands, brackets, archwires and springs; types 302, 303, 304 and 305 stainless steel
Preformed dental crowns	Restoration for extensive loss of tooth structure in primary and young permanent teeth: preformed shell; type 304 stainless steel
Preformed endodontic post and core	Restoration of endodontically treated teeth: post fixed within root canal preparation, with exposed core providing a crown foundation; types 304 and 316 stainless steel
Retention pins for dental amalgam	Retention of large dental amalgam restorations: cemented, friction lock and self-threading pins, placed ~2 mm within dentin with ~2 mm exposed; types 304 and 316 stainless steel
Wire mesh	Inguinal hernia repair, cranioplasty (with acrylic), orthopedic bone cement restrictor; types 316 and 316L stainless steel
Sutures	Wound closure, repair of cleft lip and palate, securing of wire mesh in cranioplasty, mandibular and hernia repair and realignment, tendon and nerve repair; types 304, 316 and 316L stainless steel
Stapedial prostheses	Replacement of nonfunctioning stapes: various types comprised of wire and piston or wire and cup piston (Teflon–stainless steel piston, platinum and stainless steel cup piston, and all stainless steel prostheses); types 316 and 316L stainless steel
Neurosurgical aneurysm and microvascular clips	Temporary or permanent occlusion of intracranial blood vessels; tension clips of various configurations, ~2 cm or less in length and constructed of one piece or jaw, pivot and spring components (similar and dissimilar compositions); 17-7PH, 17-7PH(Cb), PH 15-7Mo and types 301, 304, 316, 316L, 420, 431 stainless steel
Hydrocephalus drainage valve	Control of intercranial pressure: one-way valve; type 316 stainless steel
Trachea tube	Breathing tube following tracheotomy and laryngectomy: tube-within-a-tube construction; type 304 stainless steel
Electronic laryngeal prosthesis system	Electromagnetic voicing source following total laryngectomy: implanted unit comprised of subdermal transformer, rectifier pack and transducer encased in type 316 stainless steel, with spring steel diaphragm
Electrodes and lead wires	Anodic, cathodic and sensing electrodes and lead wires: intramuscular stimulation, bone growth stimulation, cardiac pacemaker (cathode), EMG, EEG and lead wires in a large number of devices; types 304, 316 and 316L stainless steel
Arzbaecher pill electrode	Atrial electrocardiograms: swallowed sensing electrode of short metal tubing segments forced over plastic tubing; stainless steel
Cardiac pacemaker housing	Hermetic packaging of electronics and power source: welded capsule; type 316L stainless steel
Variable capacitance transducer	Measurement of pressure on sound: metal diaphragm, mounted in tension; stainless steel
Variable resistance transducer	Measurement of respiratory flow: metal arms supporting wire strain gauge; stainless steel
Intrauterine device (IUD)	Contraception: stainless steel (Majzlin spring, M-316, M device), stainless steel and silicone rubber (Comet, M-213, Ypsilon device), stainless steel and natural rubber (K S Wing IUD), stainless steel and polyether urethane (Web device)
Intrauterine pressure sensor case	Protective shroud for transducer: stainless steel
Osmotic minipump	Continuous delivery of biologically active agents: implanted unit comprised of elastomeric reservoir, osmotic agent, rate-controlling membrane and stainless steel flow moderator and filling tube
Radiographic marker	Facilitation of postoperative angiography of bypass graft: open circle configuration of 25 gauge suture wires; stainless steel
Butterfly cannula	Intravenous infusion: stainless steel
Cannula	Coronary perfusion: silicone rubber reinforced with an internal wire spiral; stainless steel
Acupuncture needle	Acupuncture: 0.26 mm diameter × 5–10 cm length needles; stainless steel
Limb prostheses, orthoses and adaptive devices	Substitution, correction, support or aided function of movable parts of the body, and technical aids not worn by the patient: components such as braces, struts, joints and bearings of many items; steel and stainless steel

a Stainless steel types other than those listed for each application may also be used

larger current densities are required (e.g., 50 μA cm^{-2}).

While corrosion-mediated implant failure remains very infrequent, increasing attention is rightly being focused on the biological burden of dissolution products. Additionally, a more cautious approach is necessitated as younger patients become implant recipients and the average life span is extended. Increased quality control, informed material selection and implant usage and correct surgical application will all undoubtedly help to minimize the incidence and severity of corrosion and therefore enhance the continuing benefits derived from the present generation of stainless steels.

See also: Biocompatibility of Biomedical Materials; Biomedical Materials: An Overview

Bibliography

American Society for Testing and Materials 1984 ASTM standards for medical and surgical materials and devices. *1984 Annual Book of ASTM Standards*. American Society for Testing and Materials, Philadelphia

Brown S A, Simpson J P 1981 Crevice and fretting corrosion of stainless-steel plates and screws. *J. Biomed. Mater. Res.* 15: 867–78

Harrison J W, McLain D L, Hohn R B, Wilson G P, Chalman J A, MacGowan K N 1976 Osteosarcoma associated with metallic implants. *Clin. Orthop. Relat. Res.* 116: 253–57

Hawkins D T, Hultgren R 1973 Constitution of binary alloys. In: Lyman T (ed.) 1973 *Metals Handbook. Metallography, Structures and Phase Diagrams*, 8th edn., Vol. 8. American Society for Metals, Metals Park, Ohio, p. 291

Luckey T D, Venugopal B 1977 *Metal Toxicity in Mammals—1*. Plenum, New York

Mears D C 1979 *Materials and Orthopaedic Surgery*. Williams and Wilkins, Baltimore

Mears D C 1983 *External Skeletal Fixation*. Williams and Wilkins, Baltimore

Peckner D, Bernstein I M 1977 *Handbook of Stainless Steels*. McGraw–Hill, New York

Sinibaldi K, Rosen H, Liu S, De Angelis M 1976 Tumors associated with metallic implants in animals. *Clin. Orthop. Relat. Res.* 118: 257–66

Sutow E J, Pollack S R 1981 The biocompatibility of certain stainless steels. In: Williams D F (ed.) 1981 *Biocompatibility of Clinical Implant Materials*. CRC Press, Boca Raton, Florida, pp. 45–98

Tayton K J J 1980 Ewing's sarcoma at the site of a metal plate. *Cancer* 45: 413–15

Venugopal B, Luckey T D 1978 *Metal Toxicity in Mammals—2*. Plenum, New York

Waldron H A (ed.) 1980 *Metals in the Environment*. Academic Press, London

Williams D F, Roaf R 1973 *Implants in Surgery*. Saunders, London

Winter G D 1974 Tissue reactions to metallic wear and corrosion products in human patients. *J. Biomed. Mater. Res. Symp.* 5(1): 11–26

E. J. Sutow

Iron Production

Iron is usually produced by smelting or reduction of iron ore, or by melting of ferrous scrap. A number of industrial processes have been developed for iron production, of which the blast furnace accounts for the majority of worldwide iron production. The metallic iron produced by primary extraction (hot metal or pig iron) contains substantial quantities of impurities, and must be refined to yield cast iron or steel. Small amounts of high-purity iron powders and electrolytically deposited iron are produced from high-purity iron oxide powders and aqueous solutions by more specialized processes involving hydrogen reduction and carbonyl reactions. This article covers the primary extraction of metallic iron from ores containing iron oxide.

1. Raw Materials

The principal raw materials for iron production in the blast furnace include coke derived from coal, iron oxide concentrates, pellets and agglomerates derived from iron ores, and limestone and other fluxes (Fig. 1). In addition, the blast-furnace charge may include small amounts of iron and steel scrap, steelmaking slag, and consolidated waste materials such as dusts and sludges collected in gas-cleaning equipment. A small but significant portion of the fuel and reductant requirements in the blast furnace is supplied by inclusion of hydrocarbons, heavy oils and tars, and, to a lesser extent, natural gas and pulverized coal added with the preheated air blast entering the blast furnace through the tuyeres. Figures 2 and 3 provide some representative information on the design and relative consumption of raw materials in modern

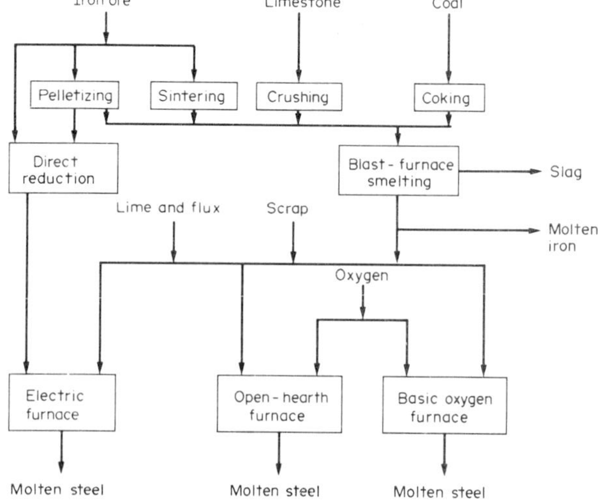

Figure 1
Flowchart showing the major steps in steelmaking

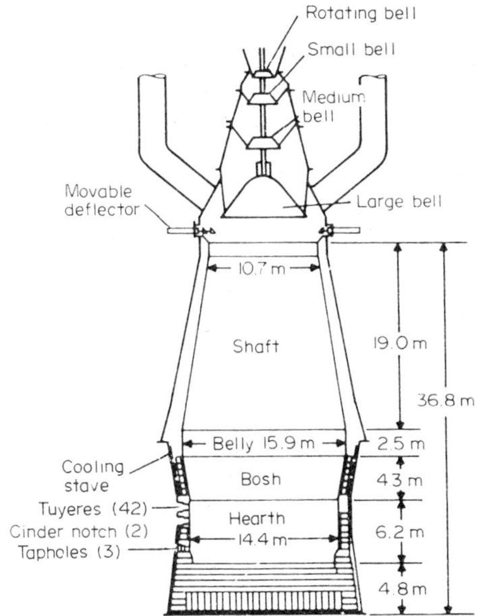

Figure 2
Features and approximate dimensions of a modern blast furnace (after Sugarawa et al. 1976)

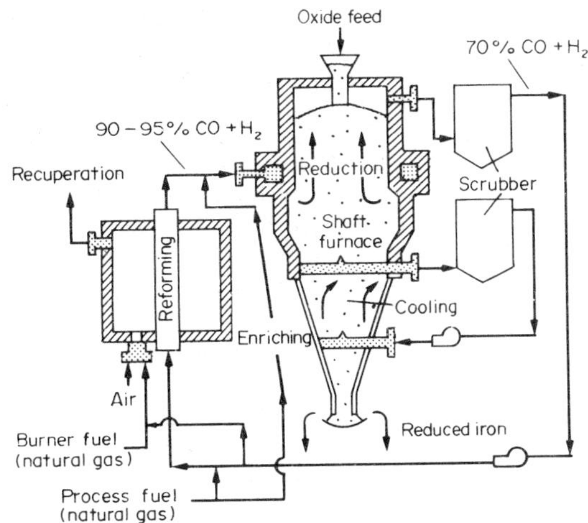

Figure 4
Simplified flowsheet of the standard Midrex direct reduction process (after Stephenson and Smailer 1980)

blast furnaces. In Fig. 3, the quantities of materials shown are based on the production of one tonne of iron of the indicated composition. Specific raw material consumption depends very strongly on the furnace design, composition and physical characteristics of available raw materials, and operational practices.

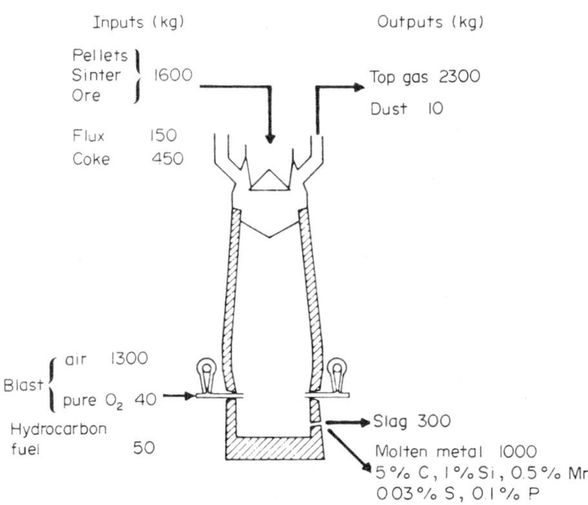

Figure 3
Representative materials balance for a modern blast furnace based on the production of 1 t of iron (after Peacey and Davenport 1979)

The principal raw materials for production of iron by direct reduction processes include iron oxide concentrates and pellets, and natural gas. A few of the direct reduction processes, particularly those that employ rotary kilns, use solid carbonaceous materials such as lignite and reactive coals. Figure 4 is a schematic diagram of the Midrex system, which is based on the use of natural gas as a reducing agent, and Fig. 5 is a schematic diagram of the SL/RN process, which is based, in part, on the use of solid carbonaceous materials as fuel and reductant. These are only two of many different direct reduction processes that exist. Continued interest in the direct reduction processes based on natural gas is expected, particularly in those countries where natural gas is abundant. The commercial development of coal-based, direct reduction processes has been slow, and significant technological progress is required before a

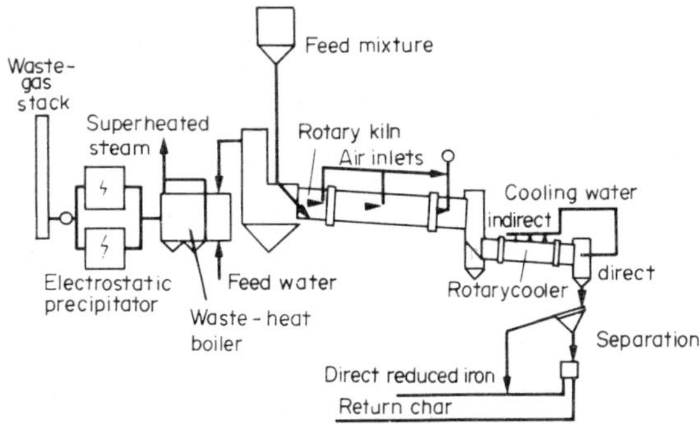

Figure 5
Schematic arrangement of SL/RN direct reduction process (after Stephenson and Smailer 1980)

significant competitive position in iron production can be attained. However, in specialized situations, the coal-based direct reduction processes may offer advantages.

In the production of coke, coals having both high and low volatile matter content are ground and blended to produce a suitable mixture for the destructive distillation process. Here, it is essential that the carbon formed from the plastic mass develops suitable strength, wear resistance and chemical reactivity for use in blast furnaces. The details of cokemaking and the characteristics of the resulting coke, coke-oven gas and by-products may be found in several sources cited in the bibliography. In general, coke is produced adjacent to, and as an integral part of, blast furnace iron production. In this way, the fuel value of the coke-oven gas may be used directly.

Available forms of iron oxide include high-grade lump ores, beneficiated iron ore fines, iron ore pellets, sintered iron ore, mill scale, and agglomerated and sintered iron oxide dusts. Iron oxide pellets are produced by reconsolidating iron oxide particles that are primarily finer than 325 mesh (0.044 mm). The iron oxide particles are obtained by crushing low-grade ores, typically of the taconite class, and separating the iron oxide from the siliceous gangue. The fine particles are reconsolidated into moist pellets about 1 cm in diameter which are then indurated by heating to temperatures sufficient to bring about recrystallization of the oxides. Because of the high gas velocities and abrasive conditions in most ironmaking processes, fine particles are not suitable as charge materials. They tend to be carried out with the gas streams, from which they must be separated and recirculated. The fluidized-bed direct reduction processes are an exception. Information on the types of ores and charge oxide materials that are acceptable as charge materials for ironmaking processes can be found in works cited in the bibliography.

2. *Principles, Operations and Products*

In the simplest instance, iron production consists of removing oxygen from iron oxide with a suitable reducing agent such as carbon monoxide or hydrogen:

$$Fe_2O_3 + 3CO = 2Fe + 3CO_2 \tag{1}$$

or

$$Fe_2O_3 + 3H_2 = 2Fe + 3H_2O \tag{2}$$

All of the ironmaking processes depend, in part, on these equations. The manner in which hematite (Fe_2O_3) is converted to metallic iron through the intermediate oxides magnetite ($\sim Fe_3O_4$) and wustite ($\sim FeO$) is very complex. At temperatures greater than 565 °C, the overall reaction represented in Eqns. (1) and (2) requires three major structural changes. (At about 565 °C, wustite decomposes by a eutectoid reaction to magnetite and metallic iron.) The rates at which these gas–solid reactions occur and the morphological features of the products, in particular the metallic iron, are highly variable. If the starting oxide materials are very dense, the reduction usually proceeds topochemically in accord with the shrinking core model. That is, the outside of each reacting particle or pellet is covered with a layer of metallic iron beneath which there are well-defined layers of wustite and magnetite and an unreacted core of hematite. If the oxide particles are very porous, the reducing gases, H_2 and CO, penetrate easily and reduction to metallic iron proceeds throughout. Within a single particle, there may be many grains of hematite undergoing topochemical reduction and there may be many interconnected pores along which the reduction steps are proceeding. The term sponge iron is used to describe particles of oxide that have been reduced to metallic iron under conditions which lead to a highly porous structure. Under certain conditions iron ore pellets may expand dramatically during reduction (catastrophic swelling). This phenomenon is attributed to the formation of filamentary iron growing from wustite at a limited number of sites.

Each reduction step is limited by the usual mass-action principles (e.g., at 1000 °C, reduction of Fe_3O_4 to FeO and FeO to Fe can occur only when p_{CO}/p_{CO_2} is greater than about 0.40 and 2.50, respectively, so that CO and H_2 cannot be fully utilized in iron oxide reduction alone). The exhaust gas always contains considerable quantities of the useful reactants. A variety of schemes are employed for using the values of the exhaust gas. Most of these depend upon blending with richer fuel gases to produce a satisfactory fuel for auxiliary plant operations. In a few instances, a portion of the exhaust gas is chemically regenerated and recycled (e.g., the Wiberg process).

The production of sponge iron (directly reduced iron, DRI) in most of the direct reduction processes involves little more than the removal of oxygen from iron oxide as just described, although many technical problems, such as prevention of interparticle sticking, deactivation of the iron by light carburization to suppress pyrophoric qualities and subsequent transportation and handling, have to be solved. However, these do not include major chemical or physical changes in the material. DRI is charged directly to electric-arc steelmaking furnaces in the same manner as steel scrap.

On the other hand, production of iron in the blast furnace does involve a great deal more than removal of oxygen from oxide particles. As the iron oxide particles descend in the shaft, they yield oxygen (by conversion of CO to CO_2 and H_2 to H_2O) to the ascending gas stream. The CO and H_2 content of the ascending gas is regenerated by "solution-loss" reactions:

$$CO_2 + C = 2CO \quad (3)$$

$$H_2O + C = CO + H_2 \quad (4)$$

Optimization of blast-furnace operations depends, in part, on proper control of the extent to which the reactions shown in Eqns. (3) and (4) occur. Coke consumed by Eqns. (3) and (4) is unavailable for combustion at the tuyeres. Furthermore, solution loss at a position too high in the furnace may lead only to some enrichment of the exhaust gas, a highly unprofitable way to use coke. The temperature of the descending solids in the blast furnace rises as shown in Fig. 6. The materials move through the upper preparation zone into the middle elaboration zone where the rate of temperature change is much less, and then into the fusion and smelting zone where the temperature rises rapidly. The reactions occurring in the smelting zone are very complex. In the vicinity of the tuyeres, temperatures as high as 2100 °C are developed in the raceways where the incoming preheated air and additives react with coke. In the bosh, around and above the raceways, fluxes, coke ash, ore gangue and unreduced iron oxide combine to form a fluid slag within the incandescent mass of coke particles. At the same time molten iron saturated with carbon is formed. Important reactions occur between the molten iron, molten slag and gas phases in this region. Sulfur is captured entirely by the slag and iron phases. Although the mechanism is not fully understood, the gaseous species SiS is an important part of the reaction sequence. As the slag and iron settle and separate in the hearth, almost all of the phosphorus and about three quarters of the manganese charged to the furnace are transferred to the iron phase. At the same time, the iron absorbs some silicon. Although it is possible to produce iron of high silicon content (silvery pig iron), blast-furnace iron usually contains <1 wt% silicon. Typical compositions of blast-furnace iron and DRI are shown in Table 1. Conditions in the hearth zone are highly favorable for the capture of sulfur and almost all is found either in the iron or slag phases. Representative information on the partitioning of sulfur between the two molten phases is given in Fig. 7. In recent years, advances in blast-furnace productivity and fuel consumption rates have tended to create conditions less favorable for transfer of sulfur from iron to slag. Coupled with higher sulfur content in

Table 1
Representative compositions of blast-furnace iron and directly reduced iron

Component	Composition[c] (%)	
	Blast-furnace iron	Directly reduced iron
Total Fe	~94	>92
Metallic Fe	~94	>85
Carbon	4.0–5.0	0.8–1.7
Silicon	0.2–1.2	1.3–2.0[a]
Manganese	0.3–2.0	0.1–0.2[b]
Phosphorus	0.1–1.0	<0.05
Sulfur	~0.03	~0.005

a SiO_2 b MnO c Balance includes small amounts of Al_2O_3, CaO, MgO and TiO_2

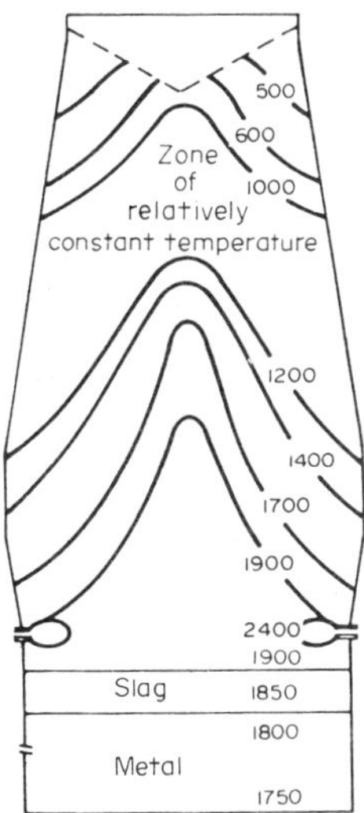

Figure 6
Representative temperature profile of a modern blast furnace, based on a quenched Nakamura furnace. Temperatures are in K (after Peacey and Davenport 1979)

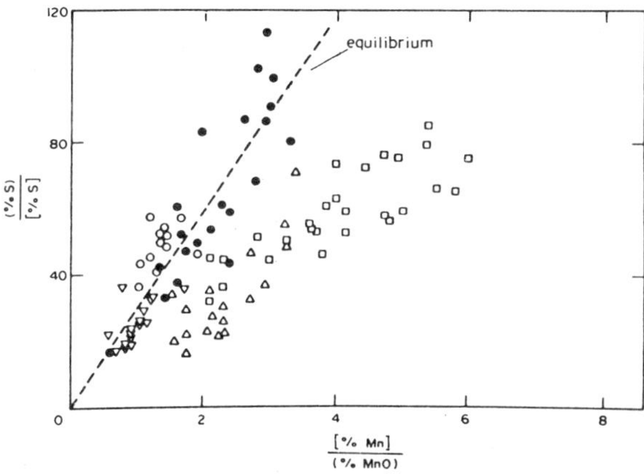

Figure 7
Representative operating data for sulfur and manganese partitioning between slag and iron at 1500 °C (after Turkdogan 1978)

coke and the demand for steels of much lower sulfur content, it has become desirable to adopt methods for desulfurizing blast-furnace hot metal in a separate processing unit between the blast furnace and the oxygen steelmaking processes where all hot metal is refined. Final desulfurization of steel to less than 50 ppm by ladle metallurgy injection techniques following electric or oxygen steelmaking is routine.

Typical operating data for blast furnaces are shown in Table 2. The slag from blast furnaces is used primarily as fill or road-bedding material. When properly prepared, it is suitable as a starting material for certain glass, brick, aggregate and insulation products.

Table 2
Representative operating data for modern blast furnaces

Iron production rate	2–10 kt d^{-1}
Number of tuyeres	20–40
Blast temperature	1050–1280 °C
Slag composition	40% CaO, 10% MgO, 35% SiO_2, 10% Al_2O_3
Top-gas analysis	22% CO, 20% CO_2, 4% H_2, 4% H_2O, 50% N_2
Top pressure	0.3–2.2 atm (gauge)
Coke consumption	0.400–0.450 kg per kg iron
Oil injection	0.040–0.120 kg per kg iron

The consumption of energy in the production of iron is relatively high but it must be remembered that the iron product is itself a fuel (the oxidation of 1 g of Fe to Fe_2O_3 releases 7.37 kJ, while the oxidation of 1 g of C to CO_2 releases 32.79 kJ). Heat balances on modern blast furnaces show remarkably high recovery of thermal values. Direct heat losses are usually <10% of total energy input; however, ~40% of total energy input is transferred as calorific value in the blast-furnace gas and there is substantial heat loss in its subsequent cleaning and reuse. About 3% of the input energy is carried off with the slag and unrecovered. Table 3 shows a simplified energy balance for a blast furnace. Care must be taken in evaluating such data because some energy is expended in preparing the charge materials (e.g., production of coke and beneficiation and induration of iron ore pellets). For comparison, the production of equal masses of copper and aluminum consume about three and twelve times as much energy, respectively, as the production of iron.

The consumption of energy in direct reduction processes is highly variable. Direct comparison with blast furnaces is not valid because the products are so dissimilar. Proper comparison is made by evaluating the blast furnace–oxygen steelmaking route and the direct reduction–electric furnace steelmaking route separately. The former method uses less energy.

Table 3
Simplified representative heat balance for a blast furnace

Component	Energy (MJ per kg iron)
Input	
calorific value of coke	15.0
calorific value of injected fuel	1.0
sensible heat of air blast	2.0
Total	18.0
Output	
calorific value of top gas	7.5
calorific value of iron	7.4
sensible heat of iron	1.3
sensible heat of slag	0.5
miscellaneous losses	1.3
Total	18.0

3. Further Developments

Recent gains in productivity and fuel consumption in blast furnaces have been achieved primarily through improvements in both the chemical and physical character of the solid charges and the uniform charging, distribution and descent of the solid materials. In addition, the use of increased top pressure and improved control of tuyere conditions (e.g., adiabatic flame temperature) have been highly beneficial. Proper care in the control of size distrbutions in solid feeds and proper blending and distribution of the solids is receiving attention.

The direct reduction processes offer an economical route to iron and raw steel where natural gas is cheap. However, the current coal-based direct reduction processes have not established a generally competitive position and their commercial adoption will occur only in special and limited circumstances until significant technological improvements are made. Therefore, the annual demand for iron of about 500 Mt throughout the world will be met primarily by blast-furnace production through the year 2000. Gas-based direct reduction processes may assume a 10% share over that time span.

See also: Iron Resources; Steel Production: Primary Processing; Metals Production: An Overview; Environmental Control in Metals Production

Bibliography

American Iron and Steel Institute 1981 *Annual Statistical Report 1981.* American Iron and Steel Institute, Washington, DC

Desy D H 1980 Iron and steel. *Minerals Yearbook.* US Government Printing Office, Washington, DC

Fine H A, Geiger G H 1979 *Handbook on Material and Energy Balance Calculations in Metallurgical Processes.* The Metallurgical Society of AIME, Warrendale, Pennsylvania

International Iron and Steel Institute 1982 *World Steel in Figures 1982.* International Iron and Steel Institute, Brussels

McGannon H E (ed.) 1971 *The Making, Shaping and Treating of Steel.* US Steel Corporation, Pittsburgh, Pennsylvania

Peacey J G, Davenport W G 1979 *The Iron Blast Furnace: Theory and Practice.* Pergamon, Oxford

Poveromo J J, Hlinka J W 1982 Development and application of Bethlehem Steel's blast furnace charging model. *Iron Steelmaker* 9(10): 19–25

Stephenson R L, Smailer R M (eds.) 1980 *Direct Reduced Iron: Technology and Economics of Production and Use.* Iron and Steel Society of AIME, Warrendale, Pennsylvania

Sugarawa T, Ikeda M, Shimotsuma T, Higuchi M, Izuka M, Kuroda K 1976 Construction and operation of no. 5 blast furnace, Fukuyama Works, Nippon Kokan K.K. *Ironmaking Steelmaking* 3(5): 242

Szekely J, Evans J W, Sohn H Y 1976 *Gas–Solid Reactions.* Academic Press, New York

Turkdogan E T 1978 Blast furnace reactions. *Metall. Trans. B* 9: 163–79

G. R. St Pierre

Iron Resources

Iron is the fourth most abundant element in the earth's crust, representing 5.06% of the lithosphere, and is an important constituent element in several hundred minerals. Iron readily combines with carbon dioxide, oxygen, sulfur or silica to form carbonates, oxides, sulfides or silicates, and only rarely occurs as the native element in meteorites and in basalts as at Ovifak, Disko Island, Greenland, where it is found as small grains through to masses up to 20 t in weight. Iron is an important element in sedimentary, igneous and metamorphic rocks and is particularly stable at and near the earth's surface under oxidizing conditions as ferric iron oxide or hydrated ferric iron oxide, which gives many surface materials a red, red–brown, brown or yellow color.

About 98% of all iron ore produced is used in the manufacture of iron and steel. Lesser amounts are used in the manufacture of Portland cement, as an aggregate material to add weight to concrete, as a heavy media material in mineral or coal processing, as a source of iron supplement in animal feed and in some fertilizers, as a paint pigment, and in the production of ferroalloys, ferrites and abrasive rouge.

1. Iron Ores

Iron-bearing materials in deposits that are of sufficient size and iron content to be either used or potentially used as commercial sources of iron are termed iron ores. The vast majority of iron ore bodies are composed of iron oxides in the minerals goethite ($Fe_2O_3.H_2O$), hematite (Fe_2O_3) or magnetite (Fe_3O_4) with lesser amounts of other iron minerals. In a few regions, as in the Michipicoten district, Ontario, Canada, siderite ($FeCO_3$) deposits are mined as iron ore. Commercial iron ores may be grouped into two major types: (a) "merchantable ores" or "natural ores" or "direct ores" are ores that can be shipped to the steel works, as mined, after minor processing that is chiefly crushing and sizing, and (b) concentrating-grade ores that require considerable processing which may include fine grinding to yield a concentrate. Current iron ore markets require that most iron ore be beneficiated in some way before shipment as there are considerable cost and energy savings at the blast furnace stage that result from the use of high-grade, low-impurity, sized and/or agglomerated iron ores.

Natural, merchantable iron ores commonly contain 55–68% Fe on a dry-analysis basis with low phosphorus and other impurity levels. They are produced on a large scale in Australia, Brazil, India, Liberia, Mauritania, South Africa, USSR and Venezuela. A merchantable iron ore with an iron content of 32–38% is mined in France and Luxembourg. These ores are economic because of their location near the steel works.

Concentrating-grade iron ores include a wide range of iron ore types that are processed using a variety of procedures. Process technologies have been developed that can concentrate a wide range of crude ores containing goethite, hematite or magnetite that have favorable grades and textures. These technologies have transformed many iron-rich deposits formerly of no economic value into iron ores or potential iron ores. Most iron ores currently mined in the USA and Canada and major deposits in China, Norway, Sweden and the USSR are classed as concentrating-grade ores.

1.1 Iron Ore Production

Iron ores are mainly produced at open-pit mines that use large materials-handling equipment. Important amounts of ore are mined by underground methods in France, Luxembourg and Sweden, with relatively few underground mines elsewhere.

Processing of off-grade natural, merchantable ores is commonly by crushing, sizing, washing and gravity methods that include heavy media and spirals. The sizing yields shipping-grade coarse ore material at about minus 4 in. (10.2 cm) and plus $\frac{1}{4}$ to $\frac{3}{8}$ in. (6.37–9.55 mm). Fine ore is commonly sintered or pelletized.

Concentrating-grade ores are processed in complex plants using gravity, magnetic, flotation or selective flocculation–flotation systems. The process used is designed to yield a uniform, high-quality iron ore pellet or sinter with a maximum recovery of the contained iron.

1.2 Iron Ore Market

Iron ores are a large-volume, relatively low-value commodity and are commonly sold in an oxide form. The ore markets are located at the various iron and steel plants throughout the world. The iron ores are prepared for market by a variety of sizing and concentration procedures but they still contain varying quantities of impurities that affect iron and steel production and metal quality. Silica, alumina, lime and magnesia determine the volume of slag produced whereas impurities such as Cr, Mn, P, S and Ti influence metal quality. Phosphorus is of particular importance and may have a considerable effect on ore value and acceptability. The commonly quoted iron ore price is at Lake Erie ports for a base-grade ore of 51.5% Fe content as Bessemer grade if less than 0.045% phosphorus (dry) and non-Bessemer if the phosphorus content is 0.045–0.18% (dry). The world trade in iron ore exceeds 335 million t per year and is growing as high-grade iron ores and pellets replace lower-grade, local ore sources. Most world trade ore contains in excess of 63% Fe (dry) with low impurities.

There is a growing market for prereduced pellets which have been treated by a direct reduction process to yield a pellet containing 87–92% Fe. Prereduced pellets can be used in an electric furnace to make steel.

2. *Geology of Iron Ore Deposits*

Iron ores occur in a wide range of geological environments in deposits that formed over a period that exceeds 3 billion years. Important deposits were formed under various sedimentary, igneous, metamorphic and surficial conditions that can be grouped as shown in Table 1.

2.1 Banded Cherty Iron-Formations

A sedimentary, bedded rock termed iron-formation occurs in extensive formations in early and middle Precambrian terrains throughout the world. Iron-formations are less commonly found in areas underlain by rocks younger than about 1.9 billion years, while a few minor occurrences may be of Cambrian age. Iron-formations are banded, layered sedimentary rocks of possible chemical or biochemical origin that contain ⩾15% iron and typically consist of iron-rich layers alternating with silica-rich layers. The iron-rich layers may be composed of siderite, iron oxides or iron silicates, or mixtures of these materials with admixed silica as chert or quartz. The silica-rich layers are composed of silica as chert or quartz with varying amounts of iron minerals. In some iron-formations, as in the Cuyuna District, Minnesota, manganese may reach several percent. The iron content varies considerably but averages about 30%.

Table 1
A simplified classification of iron deposits

Bedded sedimentary and metasedimentary deposits
banded cherty iron-formations
ironstones
river terrace
black sands
Igneous deposits
magmatic segregations
titaniferous magnetites
Hydrothermal deposits
contact zone replacements
replacements in iron-formations
Residual deposits
Lake Superior type ores
brown ores
laterites

Iron-formations in early Precambrian terrains (+3.2–2.5 billion years) are associated with volcanic rocks and volcanogenic sediments. In middle Precambrian terrains (2.5–1.7 billion years), iron-formations are most commonly associated with metasedimentary rocks and may occur as extensive stratigraphic units. Major iron-formations of this age occur in the Lake Superior region in the USA, Labrador–Quebec region, Canada, Minas Gerais and Carajas regions in Brazil, the Hamersley region, Australia, Krivoi Rog and Kursk regions in Russia, and in China, India, Africa and Venezuela. These iron-formations are often from 100 to several hundred meters thick and extend for tens or hundreds of kilometers. In the Hamersley region, Australia, iron-formations have an aggregate thickness of about 1100 m and extend along strike for several hundred kilometers. The tonnages of iron-formations are enormous.

In the Lake Superior and Labrador–Quebec regions of North America the following types of iron-formation ores are being mined:

(a) Magnetite taconite is mined extensively in Minnesota along the Mesabi Range with a 1981 combined productive capacity in eight plants of about 65 million gross tons of pellets. Taconite-type ores are concentrated by magnetic methods after fine grinding. This type of ore is also being produced in Michigan, Wisconsin and Wyoming in the USA, in Ontario, Canada, in the Kursk region, Russia, in Norway, and in Tasmania, Australia.

(b) Jaspilite ore is mined in Michigan on the Marquette Range. This ore is a layered metasediment with alternating layers of jasper and finely crystalline hematite. This ore type is concentrated by flotation.

(c) Itabirite type ores are being processed extensively in the Labrador–Quebec region of Canada. These ores represent metamorphosed iron-formations that consist of alternating layers of quartz and crystalline hematite with more or less magnetite and iron silicate minerals. These ores are concentrated by gravity and magnetic methods. Extensive deposits of the itabirite type occur in Brazil, Liberia and Venezuela.

(d) Oxidized banded iron-formation ores at the Tilden Mine, Marquette Range, Michigan, are being processed after very fine grinding, using a selective flocculation–flotation method. This type of iron-formation has been oxidized and partially leached by weathering processes, with most or all magnetite, siderite and iron silicates converted to hematite with more or less associated goethite. Extensive deposits of oxidized iron-formation are known in Michigan, Minnesota and Wisconsin, in the Laborador–Quebec region, Canada, in South Africa and elsewhere in the world, but their possible concentratability is not known.

(e) Siderite ores are mined in the Michipicoten District, Canada, from siderite bodies which are associated with volcanogenic sediments, tuffs and volcanics of Archean age and have an overlying low-iron, banded silica formation. These commercial deposits, unique to this district, are sintered to yield a commercial iron ore.

2.2 Ironstone Ores

Ironstones are a group of sedimentary rocks containing 20–40% Fe that typically consist of oolitic or pelletal grains of goethite, hematite or chamosite $[(Fe,Mg,Al)_6(Si,Al)_4O_{14}.(OH)_8]$ in a matrix that contains varying amounts of iron oxide, chamosite, siderite, calcite or dolomite, with common, variable amounts of clastic quartz. Ironstones are associated with sandstones, shales and limestones, are often lenticular, and commonly range in thickness from about 30 cm to a maximum of about 10 m. These iron deposits range in age from late Precambrian to Tertiary. Ironstone ores are relatively high in phosphorus, which, coupled with a common 30–38% Fe content, has resulted in a worldwide decline in use. France and Luxembourg are the major producers of ironstone ores.

2.3 River Terrace Ores

River-terrace iron ore deposits with reserves of hundreds of millions of tons occur in some ancient river valleys in the Hamersley region in western Australia. The iron ore beds consist of oolitic or pisolitic grains, locally with some interbedded clastic grains and pebbles of goethite or hematite, in a goethitic matrix. These deposits are from about 1 m to about 70 m in thickness and form mesas and dissected river terraces that are often several miles long. The ores contain from 50 to 60% Fe with a low phosphorus content. They are being mined on a large scale in the Robe River area. These iron ores appear to have formed in a geological environment of solution and redeposition of iron similar to conditions that result in bog iron deposits.

2.4 Black Sands

Accumulations of magnetite and ilmenite in beach sands are common. Many such deposits have been studied as potential sources of iron ore, but few are of commercial value. Sands containing magnetite are mined and concentrated on the west coast of North Island, New Zealand. About 3 million t of iron ore were exported to Japan in 1979.

2.5 Igneous Deposits

Two major classes of iron ores are igneous in origin: (a) magmatic segregations that occur as intrusive bodies, veins or extrusive deposits; and (b) titaniferous magnetite segregations in basic igneous rocks.

Iron ore deposits in the Kiruna region in northern Sweden and deposits in Missouri in the USA occur as large, tabular, rather massive bodies of magnetite with lesser hematite in rhyolite, quartz porphyry or syenite porphyry. These deposits commonly contain from 1 to 3% phosphorus. A unique iron deposit that appears to be of similar origin occurs in the high Andes Mountains in Chile at El Laco where shallow intrusives and lava flows are composed of almost pure magnetite–hematite with associated apatite. The iron deposits in northern Sweden produce 20 million or more tons of iron ore per year, which are largely exported.

Titaniferous magnetite bodies occur as segregations in gabbro, pyroxenite and anorthosite, but have limited potential as iron ore sources because of the relatively high titanium content.

2.6 Hydrothermal Deposits

Iron ores that were deposited from hydrothermal solutions may be grouped into two classes: (a) contact zone replacements, sometimes termed "contact metamorphic" or "pyrometasomatic" deposits, where iron oxides replace the intruded rock sequence generally near the contact with an intrusive igneous body; and (b) replacements in Precambrian iron-formations. The geological processes active in the two classes appear to be similar.

Contact zone replacement deposits are chiefly composed of magnetite with lesser hematite and minor amounts of carbonates, pyrite, chalcopyrite and pyrrhotite. The most common host rock is limestone. Skarn minerals such as garnets, pyroxenes and amphiboles are often associated with these ores. The deposits vary in shape from tabular bodies to irregular to vein-like. In size, they range up to several hundred million of tons. Major deposits were mined in the Cornwall area, Pennsylvania, and are mined in the Iron Springs District, Utah, at Eagle Mountain, California, at Mount Magnitnaya, USSR, and at Marcona, Peru.

Replacement iron deposits in Precambrian iron-formations are known in many parts of the world where hydrothermal solutions have replaced silica with iron oxide, commonly hematite with lesser magnetite. These deposits are often high grade, some containing about 70% Fe. The ore bodies are often tabular in shape and retain the layered appearance of the original cherty iron-formation. Deposits occur in Brazil, India, Liberia, Mauritania and Venezuela.

2.7 Residual Deposits

Weathering of iron-bearing and iron-rich rocks has been an important iron ore forming process as iron oxides are relatively insoluble compared to silica and most other materials in the surficial geological environment. Oxidation converts iron to an insoluble, ferric oxide and leaching removes the more soluble elements by the water system. This oxidation–leaching system may be most active in warm and moist climates with wet and dry seasons. The zone most affected is from the surface to depths of about 100 m, but locally under very favorable conditions, the oxidation and leaching can extend to depths of 1000 m or more. As an example, soft, Lake Superior type iron ores were mined to depths of about 1000 m at the Mather Mine, Marquette Range, Michigan, with ore known to occur to depths in excess of 1500 m. The residual ores are commonly grouped as: (a) Lake Superior type or soft ores, (b) brown iron ores, and (c) laterites.

(*a*) *Lake Superior type.* These iron ores were formed by the oxidation of the iron minerals and the leaching of silica from a Precambrian, banded cherty iron-formation, leaving a residual accumulation of hematite and goethite with associated varying amounts of silica and clay. These ores are one of the most important iron ore types in the world. The time of ore formation is often unknown, as some ore bodies could have formed in Precambrian time and other bodies during later epochs that may extend to quaternary time. The ore bodies have a wide range in size and shape that often are influenced by the structural setting and stratigraphy of the iron-formation. Iron deposits of this association have yielded over 3.5 billion t of shipping-grade ore in the Lake Superior region of the USA but are now largely depleted. Major production from Lake Superior type residual ore deposits is continuing from deposits in Australia, Brazil, India, Liberia, Mauritania, South Africa, the USSR and Venezuela.

(*b*) *Brown ores.* Weathering of Paleozoic limestones and dolomites in the southern Appalachian region of the USA may result in the development of irregular bodies of clays that contain lumps and nodules of iron oxide that can be shipped as iron ore after separation from the clay. In northeastern Texas the weathering of sedimentary beds containing siderite and glauconite has formed a series of iron ore deposits. These surficial deposits and others that are largely nodules and lumps of goethite and/or hematite with clay and sands that can be beneficiated by simple methods are collectively termed brown ores. Deposits are usually small.

(*c*) *Laterites.* Weathering of ultramafic rocks such as serpentinized peridotites, pyroxenite and dunites under tropical conditions often yields a surficial accumulation of iron oxide as goethite and hematite that is termed lateritic iron ore. This residual mantle is locally extensive and often ranges in thickness from about 1 m to 15 m or more. These deposits commonly contain from 40 to 55% Fe with varying amounts of alumina, chromium, nickel and cobalt, which makes them generally unsatisfactory as iron ore.

Table 2
World iron ore resources (million t) (after Reno and Brantley 1973)

Region	Reserves	Potential ore	Total resources
Africa	6693	24113	30806
Middle East, Asia and the Far East	17027	53344	70371
Australia, New Zealand and New Caledonia	16535	vast	>16535
Canada and West Indies	35727	87988	123715
Europe	20964	12598	33561
South America	33561	57478	91039
USSR	108755	190739	299494
USA, Mexico and Central America	60000	44620	104620
	299262	470880	>770141

3. Iron Ore Reserves and Resources

Iron ore reserves as commercial tonnages that include all ore types and grades are very large. Iron ore resources in known potential iron ore areas are enormous. A summation of iron ore reserves and resources, updated for the USA, is shown on Table 2. Studies made for the US Bureau of Mines in 1978 estimated concentrating-grade iron ore reserves of 47.52 billion t in Minnesota and 4.17 billion t in Wisconsin. These reserve tonnages have been included on Table 2 without changing the total resource tonnage for the USA, Mexico and Central America.

See also: Iron Production; Steel Production: Primary Processing; Mineral Resources: An Overview

Bibliography

Klemic H, James H L, Eberlein G D 1973 Iron. In: Brobst D A, Pratt W P (eds.) 1973 *United States Mineral Resources*, US Geological Survey Professional Paper 820. US Government Printing Office, Washington, DC, pp. 291–306

McGannon H E (ed.) *The Making, Shaping, and Treating of Steel*, 9th edn. United States Steel Corporation, Pittsburgh, Pennysylvania, pp. 1–35, 178–239

Peterson E C 1980 Iron ore. *Mineral Facts and Problems*, US Bureau of Mines Bulletin 671. US Department of Interior, Washington, DC, pp. 433–53

Pounds N J G 1971 *The Geography of Iron and Steel*, 5th edn. Hutchinson University Library, London, Chap. 2

Reno H T, Brantley F E 1973 *Iron—A Materials Survey*, US Bureau of Mines Information Circular 8574. US Department of Interior, Washington, DC

United Nations 1970 *Survey of World Iron Ore Resources*. United Nations, New York, 5th edn.

R. W. Marsden

Irreversible Thermodynamics

Irreversible thermodynamics has in earlier times been referred to as thermodynamics of the steady state to distinguish it from equilibrium thermodynamics. Specialists now distinguish between the near-equilibrium linear regime, wherein linear flux–force relations and the steady-state principle of minimum entropy production are used, and the far from equilibrium nonlinear regime where various instabilities and branching in kinetic phase space occur.

The terms thermokinetics and thermostatics are used to identify the disciplines pertaining to irreversible and reversible processes, respectively, the term thermodynamics being inclusive of both the kinetic and static domains. Thermodynamics is a discipline which seeks to identify macroscopic invariants in nature (e.g., mass, energy and charge) and to enter these or their derivatives into a differential manifold. Elementary thermokinetics attempts to establish an extended macroscopic state space which is an analytic continuation (e.g., a Taylor expansion consistent with the entropy optimum) about the thermostatic states described by the Gibbs equation. The discipline focuses on both open (discontinuous and continuous) systems and closed (isolated) systems, the latter case being the simplest to treat since the second law of thermodynamics takes the compact form

$$dS = 0; \qquad d^2S \geqslant 0 \tag{1}$$

Following Onsager (1931a,b), the entropy change in a Taylor series about the isolated optimum can be expanded to

$$\Delta S = -\tfrac{1}{2}\sum_i \sum_k g_{ik}\alpha_k\alpha_i \tag{2}$$

where α_k and α_i measure independent deviations of internal state variables from their optimal values and the matrix g_{ik} is positive definite for stable systems. ΔS can then be differentiated with respect to time to obtain the rate of entropy production (or dissipation):

$$\Delta\dot{S} = \sum_i \dot{\alpha}_i \sum_k (-g_{ik}\alpha_k) \tag{3}$$

which may be written in the bilinear form as

$$\Delta\dot{S} = \sum_i J_i X_i \tag{4}$$

where the fluxes or rates are identified as $J_i = \dot{\alpha}_i$ and the forces as

$$X_i = -\sum_k g_{ik}\alpha_k \tag{5}$$

Thus for each of the thermokinetic paths or order parameters α_i a term can be identified in the dissipation analogous to the Joule heat in a resistive electric circuit. The second law of thermodynamics requires that $\Delta\dot{S}$ be positive semidefinite; that is,

$$\Delta\dot{S} \geqslant 0 \tag{6}$$

the equality corresponding to equilibrium and to all $J_i = X_i = 0$. Then to assure the validity of Eqn. (6) it is logical to propose a linear flux–force relation

$$J_i = \sum_k L_{ik}X_k \tag{7}$$

(L_{ik} is constant) since it converts Eqn. (4) to the quadratic form:

$$\Delta\dot{S} = \sum_i \sum_k L_{ik}X_kX_i = \sum_i \sum_k [(L_{ik} + L_{ki})/2]\, X_kX_i \tag{8}$$

the transformed matrix of coefficients on the right being symmetric. The relation of Eqn. (6) is then assured by the standard condition that the symmetric matrix has a nonnegative determinant and minors; that is,

$$L_{ii},\ L_{kk},\ L_{ii}L_{kk} - [(L_{ik} + L_{ki})/2]^2 > 0 \tag{9}$$

Note particularly that there is no *a priori* requirement here that L_{ik} be a symmetric matrix. Nonzero off-diagonal values of L_{ik} define cross effects often designated as Onsager coupling.

Onsager (1931) summarized this description via an optimal fluctuation–dissipation theorem which is inclusive of the reciprocal relations in the J_i ($L_{ik} = L_{ki}$) and the principle of minimum entropy production (de Groot 1952, Prigogine 1955). Note that the rate equations in terms of α variables have the diagonalized form

$$\dot{\alpha} = -\mathscr{L}\alpha; \qquad \mathscr{L} > 0 \tag{10}$$

so the temporal solutions will be exponentially decaying, for thermodynamically stable systems.

The foregoing treatment applies strictly only to scalar processes, such as order–disorder transformation. An elegant statistical mechanical analogue has been developed by Kikuchi (1962) for simultaneous relaxation of short- and long-range order parameters, leading to the expected linear relations with a symmetric 2×2 matrix.

For most open systems the mathematical structure must be more elaborate. Discontinuous open systems have the property that the thermokinetic force can be defined simply as a difference in potential between the source and sink, and the fluxes (together with the entropy production rates) as flows in and out of the terminals. Examples, such as thermoelectric effects, have played a major part in testing the Onsager symmetry principle (Miller 1960).

The treatment of open, continuous systems starts with the Gibbs equation (de Groot 1952) in a time-differentiated and local form, and each differential of an extensive variable is then eliminated via the corresponding local differential conservation relation. The result is a nonconservative local entropy balance of the form

$$\rho\, ds/dt = -\operatorname{div} J_s + \sigma \tag{11}$$

where ρ is density and s is entropy per unit mass. The unit volume quantity σ takes the bilinear form

$$\sigma = \sum J_i X_i \tag{12}$$

producing a term for each independent conservation relation. J_s is then interpreted as the entropy flux, and by analogy with Eqn. (4), σ is interpreted as a positive semidefinite entropy source term or entropy production rate per unit volume. For example, in certain systems involving heat conduction and diffusion, Eqn. (11) takes the form for an n-component system

$$\rho\, ds/dt = -\operatorname{div}\left[\left(J_q - \sum_k^n \mu_k J_k\right)\Big/ T\right] + \left(J_q X_q + \sum_k^{n-1} J_k X_k\right)\Big/ T \tag{13}$$

where J_q is the energy flux, the J_k terms are mass fluxes, the thermal force is

$$X_q = -\nabla(1/T) \tag{14}$$

and the independent diffusive forces in terms of chemical potentials are

$$X_k = -\nabla(\mu_k - \mu_n)/T \tag{15}$$

The usual $n-1$ linear and symmetric flux–force relations of the form in Eqn. (7) are then conjectured to complete the structure. Elementary statistical or kinetic theories generally succeed in generating the same structure.

It is to be noted that the potential number of terms on the right-hand side of Eqn. (13) is myriad and may contain scalars (as in homogeneous chemical reaction), vectors (as shown) or tensors (for viscous dissipation or diffusion in anisotropic crystals). The most complete accounting of variants is to be found in de Groot and Mazur (1962). Clearly the potential for Onsager cross effects is also myriad. There is, however, an important simplifying restriction residing in a group-theoretical principle due to Curie. This states in effect that Onsager coupling cannot arise between processes of different tensorial character. For example, although for a system undergoing simultaneous diffusion and order–disorder transformation the processes will be directly coupled through shared concentration variables, all off-diagonal terms in the L matrix involving the two processes must vanish.

The foregoing may be characterized as the linear first-order theory of irreversible processes. Materials scientists have found it to be a very useful structure for the investigation of the interactions of charge, solute, defect and heat flows, of anisotropic diffusion, of anisotropic heat flow, and particularly of multicomponent diffusion in neutral and ionic solids and liquids.

Following Onsager's definitive paper (1945–46), an elegant structure has been devised and widely applied to isothermal multicomponent systems (Kirkaldy 1970). The key problem is to relate Fick-type concentration equations

$$J_i = -\sum D_{ik} \nabla C_k \tag{16}$$

with a matrix of coefficients D to the fundamental equations involving the forces X_k (Eqn. (15)) with their positive definite matrix of coefficients, L. In the simplest case D can be expressed as being proportional to a matrix product $D \sim \mu L$, where in terms of mole fractions N_k and for the ternary case

$$\mu = \partial(\mu_i - \mu_3)/\partial N_k \tag{17}$$

Although μ is symmetric, and for thermodynamically stable solutions is positive definite, its most interesting behavior lies near critical points and within miscibility gaps.

It is to be noted that D is not positive definite,

even though both μ and L may be so. Rather, in this case the trace and determinant are positive; for example, in the ternary 2×2 case,

$$D_{11} + D_{22} > 0 \quad (18)$$

$$D_{11}D_{22} - D_{12}D_{21} > 0 \quad (19)$$

or alternatively, the eigenvalues of the D matrix are both positive. For dilute stable solutions it can be demonstrated further that

$$D_{11}, D_{22}, D_{12}D_{21} > 0 \quad (20)$$

In the stable region near a critical point, one of the diagonal D's may change sign subject to the constraint of Eqn. (18). On crossing the critical line and the chemical spinodal surface, the determinant changes sign from positive to negative.

The combination of Eqn. (16) with the mass balances

$$(\partial C_i/\partial t) + \operatorname{div} J_i = 0 \quad (21)$$

leads to analogues of Fick's second equation with constant D's which in one dimension may be written in diagonalized form as

$$\partial C_i/\partial t = \lambda_i \, \partial^2 C_i/\partial x^2 \quad (22)$$

where the λ_i terms are the eigenvalues of the D matrix. In stable solutions the λ_i are all positive, and so all solutions decay. Referring to the 2×2 case, when the determinant changes sign at the spinodal surface, one of the λ's changes sign and growing (spinodal) solutions become chemically possible along the appropriate eigenvector in composition space.

There exists a linear, second-order formulation of irreversible thermodynamics which recognizes a combination of mechanically or electrically reversible processes and irreversible processes (Machlup and Onsager 1953); for example, systems of interest here are LCR networks. Rheological deformation models have much in common with this structure.

The nonlinear regime of thermokinetics is characterized by intensive research, some controversy and a predilection for jargon (Nicolis and Prigogine 1977). The continuous nonlinear regime, generally defined by large differences in boundary values, presents no serious problems because it can usually be dealt with by the introduction of variable phenomenological coefficients. The difficult problems arise when differences in boundary conditions introduce Gibbs phase rule constraints or the potential of the free-energy source is sufficiently high relative to that of the heat bath that localized energy states are created which substantially exceed in magnitude the value of kT for the heat bath. Surface states attending phase transformation (e.g., eutectic and cellular crystal growth), kinetic energy states attending rotational fluid flow (e.g., Benard cells) and photosynthetic states of inanimate and living systems (e.g., lasing states) fall into these symmetry-breaking categories.

See also: Entropy Principles in Irreversible Thermodynamics; Onsager Reciprocal Relations; Rate Theory; Spinodal Decomposition

Bibliography

de Groot S R 1952 *Thermodynamics of Irreversible Processes.* North-Holland, Amsterdam

de Groot S R, Mazur P 1962 *Non-equilibrium Thermodynamics.* North-Holland, Amsterdam

Kikuchi R 1960 Irreversible cooperative phenomena. *Ann. Phys. (N.Y.)* 10: 127–51

Kirkaldy J S 1970 Isothermal diffusion in multicomponent systems. *Adv. Mater. Res.* 4: 55–100

Machlup S, Onsager L 1953 Fluctuations and irreversible processes II: Systems with kinetic energy. *Phys. Rev.* 19: 1512–15

Miller D C 1960 Thermodynamics of irreversible processes: The experimental verification of the Onsager reciprocal relations. *Chem. Rev.* 60: 15–37

Nicolis G, Prigogine I 1977 *Self-Organization in Non-equilibrium Systems.* Wiley, New York

Onsager L 1931a Reciprocal relations in irreversible processes I. *Phys. Rev.* 37: 405–26

Onsager L 1931b Reciprocal relations in irreversible processes II. *Phys. Rev.* 38: 2265–79

Onsager L 1945–46 Theories and problems of liquid diffusion. *Ann. N.Y. Acad. Sci.* 46: 241–65

Prigogine I 1955 *Introduction to Thermodynamics of Irreversible Processes.* Charles C. Thomas, Springfield, Illinois

J. S. Kirkaldy

Isothermal Forging

Isothermal forging is one of the more important metalworking processes to have evolved in recent years. In this type of forging the dies are maintained at the same temperature as the workpiece during deformation. The chief advantages resulting from the use of this process are that workability is improved and forging loads are dramatically reduced. On the negative side, die-set costs are substantially higher and production rates much lower. Although studies concerned with the feasibility of isothermal forging of aluminum and steel date back over thirty years, it is only in the last decade that the technique has been extended, with good results, to strain-rate-sensitive metals such as titanium alloys and nickel-base superalloys. These materials are difficult or in some cases impossible to forge by conventional methods because of the effects of die chilling, strain hardening, and the sensitivity of strain rate to stress at high strain rates. In filling a complex die, local stresses will develop which can vary considerably depending on the shape of the preform and the geometry and temperature of the die. At high loading rates, characteristic of conventional forging, these high-stress regions will deform faster, while metal flow in the low-stress regions will be stagnant. Flaws will develop at the interface between the moving metal and the dead metal. Chilling of the preform by the relatively cold (~450 °C) die surfaces, combined with the effect of strain hardening, increases press and die loads sharply. In isothermal forging, die chilling is avoided and, as a result of high-temperature restoration processes, strain hardening is insignificant. The problem of the sensitivity of metal flow rates to local stress levels can be largely mitigated if the material to be forged is in a superplastic state (see *Superplastic Forming*).

1. Material Behavior

Titanium alloys and nickel-base superalloys are particularly good candidates for isothermal forging because both can be produced in a superplastic condition. Superplasticity is characterized by a low flow stress over a strain-rate regime (10^{-5}–10^{-2} s^{-1}) intermediate between conventional forging and creep, combined with only a modest increase in rate with increasing stress. The basic requirement for superplasticity in a metal is that it possesses an equiaxed microstructure with a grain size less than about 8 μm. Additionally, the grain size must remain stable for extended periods of time at high temperatures, a condition usually satisfied by a two-phase or microduplex structure. Superplasticity can also be achieved in large grain size materials with anomalously high diffusivities (e.g. β-titanium alloys and β-brass) if they also possess or quickly develop a very fine subgrain structure.

The customary thermomechanical processing used for α- and β-titanium-alloy billets or bar stock normally yields a superplastic microstructure. In nickel-base superalloys the consolidation of atomized prealloyed powders into preform shapes is necessary. One of several possible routings to produce such preforms is first to seal the powder in a metal can and then to hot isostatically press it into a dense billet. The billet can then be extruded at a temperature just below the γ'-$Ni_3(Al,Ti)$ solvus so that adiabatic heating causes momentary solutioning of the γ' phase followed by recrystallization of the matrix and rapid reprecipitation of the γ' phase. The resultant fine-grained microstructure will be superplastic under isothermal forging conditions. In many cases the intermediate extrusion step is unnecessary. The atomized powder inherently has a very fine grain size and by suitable choice of processing parameters a superplastic microstructure can be obtained directly.

An example of a good candidate material for isothermal forging is INCO 713LC, a high-strength casting-type nickel-base superalloy (Immarigeon and Floyd 1981). This alloy contains approximately 50 vol% of the γ' intermetallic phase. After hot isostatic pressing of argon-atomized powder at 1175 °C, a fine-grained (4–7 μm) two-phase microstructure was obtained. When isothermally forged at 1050 °C the microstructure was refined further; the greater the strain rate, the finer the grain size produced.

2. Practical Aspects

Die sets for isothermal forging are both complex and expensive. The preferred die material for forging titanium alloys is IN 100, a highly alloyed superalloy. It has an upper use temperature of about 950 °C and can be used in air. For forging superalloys, TZM molybdenum-alloy dies are used. This material has excellent high-temperature strength but requires a protective gas atmosphere or vacuum to prevent oxidation. It is also sometimes used for titanium alloys because its low thermal expansion coefficient reportedly makes ejection of the finished forging from the die easier.

IN 100 dies are customarily heated with cartridge-type resistance elements that may pass directly through holes machined in the die block. Gas flames have also been used. Induction heating is necessary for the TZM molybdenum alloy. Structural ceramics and water-cooled panels are used to protect the bed and superstructure of the press from the heat.

Isothermal forging is performed in hydraulic presses. Because of the low flow stress of the material under superplastic conditions, press loads of only 10% of those necessary in a mechanical press-forging operation are required. In theory, the press should be strain-rate controlled; however, such presses are not common. Instead, the load is normally ramped up to full pressure at an empirically determined rate. Often, the press plattens are separated by stop blocks so that the final increment of deformation is in the creep-forging regime.

3. Advantages and Disadvantages

One of the early promises of isothermal forging was that preform development would be simplified and the final forging would be more tolerant of variations in preform geometry. In general, however, this does not appear to have been realized. Like other forging methods, isothermal forging requires carefully developed, consistent preform shapes.

Since the workpiece remains fine grained and essentially strain-free throughout the forging operation, very high reductions can, in theory, be taken. In fact, a severe limitation on the permissible reduction is generally established by the thinning and consequent breakdown of the forging lubricant. At isothermal-forging temperatures, neither graphite nor boron nitride are lubricants; however, both are still useful as parting compounds. Molten-glass lubricants are used exclusively. These are normally applied in a slurry form to the workpiece and fuse during the heat-up to the forging temperature. For titanium alloys, the glass also serves to prevent oxygen absorption by the metal. Unfortunately, the molten glass sticks to the workpiece and the die surfaces, and it has a tendency to accumulate in deep sections of the die cavity. This can quickly affect the dimensional accuracy of the forging while hampering ejection of the part from the die. Removal of excess glass lubricant normally requires cooling the dies to room temperature. The overall economics of the process sometimes depends on how often this proves to be necessary. The most successful lubricants combine the lubricating properties of glass with the parting characteristics of graphite and boron nitride. The use of graded coatings, where the glass content of the lubricant is greatest nearest to the workpiece, has been moderately successful.

Forgings produced by this process are generally isotropic in crystallography and mechanical properties. They exhibit excellent reproducibility in their response to heat treatment. In many cases, mechanical properties have been found to be slightly to much improved over those of hardware machined from oversized forgings or plate stock. This is particularly true of the time-dependent properties: creep, stress rupture, and fatigue. It is believed this results from the extra mechanical working imparted to the workpiece in forging to a near-net-shape configuration. In the case of superalloys, the thinner cross sections allow faster quenching with resulting finer γ' precipitates.

The economics of the process depend on the balance between the ability of isothermal forging to form complex shapes to extremely close tolerances and the low production rates and high die costs. High capital costs associated with heating and insulating the dies are partially offset by the use of less expensive hydraulic presses of lower tonnage ratings. Improving raw-material utilization by forging to near-net shapes can represent substantial cost savings as well as conserving strategic metals. Savings become most pronounced when alloys that are difficult to machine are fabricated into intricate configurations. In theory, the larger the component, the greater the incentive to forge it isothermally.

See also: Metals Processing and Fabrication: An Overview

Bibliography

Chen C C, Coyne J E 1980 Recent developments in hot-die forging of titanium alloys. In: Kimura H, Izumi O (eds.) 1980 *Titanium '80*, Vol. 4. Metallurgical Society of AIME, New York, p. 2513

Coyne J E 1980 Technology is key to production of superalloy disc forgings. *Met. Prog.* 118(6): 35–37

Griffiths P, Hammond C 1973 Superplasticity in large grained beta titanium alloys. In: Jaffee R I, Burte H M (eds.) 1973 *Titanium Science and Technology*, Vol. 2. Plenum, New York, p. 1155

Immarigeon J-P A, Floyd P H 1981 Microstructural instabilities during superplastic forging of a nickel-base superalloy compact. *Metall. Trans. A* 12: 1177–86

J. Snow